CW01212639

# Plant Physiology, Development and Metabolism

Satish C Bhatla • Manju A. Lal

# Plant Physiology, Development and Metabolism

Springer

Bibliotheca hospes

MÄNGELEXEMPLAR

Satish C Bhatla
Department of Botany
University of Delhi
New Delhi, Delhi, India

Manju A. Lal
Department of Botany
Kirori Mal College, University of Delhi
New Delhi, Delhi, India

ISBN 978-981-13-2022-4      ISBN 978-981-13-2023-1   (eBook)
https://doi.org/10.1007/978-981-13-2023-1

Library of Congress Control Number: 2018961393

© Springer Nature Singapore Pte Ltd. 2018
This work is subject to copyright. All rights are reserved by the Publisher, whether the whole or part of the material is concerned, specifically the rights of translation, reprinting, reuse of illustrations, recitation, broadcasting, reproduction on microfilms or in any other physical way, and transmission or information storage and retrieval, electronic adaptation, computer software, or by similar or dissimilar methodology now known or hereafter developed.
The use of general descriptive names, registered names, trademarks, service marks, etc. in this publication does not imply, even in the absence of a specific statement, that such names are exempt from the relevant protective laws and regulations and therefore free for general use.
The publisher, the authors and the editors are safe to assume that the advice and information in this book are believed to be true and accurate at the date of publication. Neither the publisher nor the authors or the editors give a warranty, express or implied, with respect to the material contained herein or for any errors or omissions that may have been made. The publisher remains neutral with regard to jurisdictional claims in published maps and institutional affiliations.

This Springer imprint is published by the registered company Springer Nature Singapore Pte Ltd.
The registered company address is: 152 Beach Road, #21-01/04 Gateway East, Singapore 189721, Singapore

# Preface

Plants serve as a source for sustainable food and biofuel and also play crucial roles in maintaining human health and ecosystem. Thus, it becomes imperative to understand the mechanisms of plant growth and development. Plant physiology is that significant branch of plant science which deals with understanding the process of functioning of plants at cell, molecular, and whole plant levels and their interaction with the surrounding environment. In spite of being static in nature, plants can withstand adverse growth conditions due to a variety of adaptive mechanisms. Intracellular compartmentalization of biochemical pathways, expression of membrane-associated transporter proteins specific for various ions and metabolites, production of secondary metabolites with multiplicity of protective functions, and a wide variety of photoreceptors biochemically synchronized with various environmental and developmental conditions are some of the noteworthy adaptive features of plants enabling them to survive in almost all possible situations. The plethora of information available today has been made possible through interaction of cell and molecular biology, biochemistry, and genetics to understand plant processes.

Plant physiology is an experimental science. Plant water relation is the first area of research in plant physiology which caught attention of scientists. Stephen Hales, also called as the Father of Plant Physiology, published the book *Vegetable Staticks* in 1727, highlighting various experimental studies on transpiration and root pressure. In the beginning of twentieth century, the development of physicochemical and biochemical techniques further facilitated the understanding of the plant processes. These techniques include spectral analysis, mass spectrometry, differential centrifugation, chromatography, electrophoresis, and the use of radioisotopes, besides many others. In the last two decades, plant physiologists made an extensive use of the molecular tools and *Arabidopsis* as a model organism to facilitate learning about the role of genes and the crosstalk among various biomolecules affecting plant functions and development. Lately, chemical biology has also contributed significantly through the use of small molecules to identify intracellular targets, thereby facilitating development of new herbicides and plant growth regulators. They are also used to identify novel signaling pathways. Small molecules are used to alter protein structure and explore the biological roles of target proteins (an area termed as chemical genetics). Low-molecular mass molecules are used as probes to modify biological processes. Major areas in plant physiology which have gained a lot of new information include growth and development (both vegetative and reproductive), physiology of nutrition, metabolism, and plant responses to the environment.

Compilation of this volume was very enlightening as it demonstrated the extent to which information and concepts in plant physiology have changed over the years. The writing of this book began in July 2015 and took almost 3 years of persistent reading, assimilating, and consolidating of relevant information from various sources into 34 chapters. While presenting the current concepts in an understandable manner, due emphasis has also been laid on historical aspects, highlighting how the concepts evolved. All contributors are associated with Delhi University and have firsthand experience of the problems being faced by undergraduate students of plant science discipline in assimilating meaningful information from the vast literature available in plant physiology. So, the need for an easy-to-understand, systematic, and up-to-date account of plant physiology has led to writing this book. The book is well illustrated, and all illustrations have been either drawn in original by an expert or designed from experiments in the laboratory or field. The volume has been brought into its present form through strong technical support from the very supportive bright members of the research group of Professor Bhatla.

Dr. Manju A. Lal would like to thank her father, late Shri V. P. Gupta, who was instrumental in her taking up teaching science as a career choice. Dr. G. S. Sirohi, former head of the Division of Plant Physiology, Indian Agricultural Research Institute, initiated her into research and guided her Ph.D. work. Thanks are due to him. Last but not the least, Dr. Manju A. Lal would like to acknowledge the unstinted support of her husband, Dr. Anandi Lal, and son- Nitin A. Lal, during the long and arduous task of writing this book.

Professor Bhatla takes this opportunity to dedicate this work to his teachers, Professor R. C. Pant (former Head and Dean, College of Basic Sciences at G. B. Pant University of Agriculture and Technology, Pantnagar, India) and Professor Martin Bopp (former Director, Botanical Institute, University of Heidelberg, Germany). Professor Bhatla remains highly appreciative of the strong support and encouragement from his wife, Dr. Rita Bhatla, and children- Rajat, Vrinda, and Sahil. They were fully aware of the intensity with which this work was being pursued and also exhibited lot of patience with a smile. Thank you all for your understanding.

New Delhi, India  
Satish C Bhatla  
Manju A. Lal

# Contents

**Part I  Transport of Water and Nutrients**

**1  Plant Water Relations**........................................... 3
Renu Kathpalia and Satish C Bhatla
  1.1  Water Potential and Its Components................... 4
    1.1.1  Solute Potential............................ 4
    1.1.2  Pressure Potential.......................... 6
    1.1.3  Gravitational Potential..................... 7
    1.1.4  Matric Potential............................ 7
  1.2  Intercellular Water Transport........................ 8
    1.2.1  Diffusion................................... 9
    1.2.2  Mass Flow................................... 10
    1.2.3  Osmosis..................................... 11
  1.3  Short-Distance Transport............................. 14
    1.3.1  Water Absorption by Roots................... 15
  1.4  Long-Distance Transport.............................. 18
    1.4.1  Water Transport Through Xylem............... 18
    1.4.2  Mechanism of Transport Across Xylem......... 19
  1.5  Water Movement from Leaves to the Atmosphere......... 23
    1.5.1  Transpiration............................... 25
    1.5.2  Stomatal Movement........................... 30
  1.6  Guttation............................................ 33
  Multiple-Choice Questions................................. 35
  Suggested Further Readings................................ 36

**2  Plant Mineral Nutrition**....................................... 37
Renu Kathpalia and Satish C Bhatla
  2.1  Plant Nutrition...................................... 42
  2.2  Essential Elements................................... 43
    2.2.1  The Criteria of Essentiality................ 44
    2.2.2  Roles of Essential Elements................. 44
  2.3  Macroelements and Microelements...................... 46
    2.3.1  Macroelements or Macronutrients............. 46
    2.3.2  Microelements or Micronutrients............. 46

|     | 2.4   | Beneficial or Functional Elements | 46 |
|-----|-------|-----------------------------------|-----|
|     | 2.5   | Micronutrient Toxicity | 47 |
|     | 2.6   | Deficiency Symptoms of Mineral Elements in Plants | 50 |
|     | 2.6.1 | Mineral Deficiencies in Older Tissues | 58 |
|     | 2.6.2 | Mineral Deficiencies in Younger Tissues | 58 |
|     | 2.7   | Role, Deficiency Symptoms, and Acquisition of Macronutrients and Micronutrients | 59 |
|     | 2.7.1 | Macronutrients | 59 |
|     | 2.7.2 | Micronutrients | 69 |
|     | Multiple-Choice Questions | | 79 |
|     | Suggested Further Readings | | 81 |

## 3 Water and Solute Transport ................................ 83
Satish C Bhatla

| 3.1 | Water and Ion Uptake from Soil into Roots | 84 |
|-----|-------------------------------------------|-----|
| 3.2 | Symplastic Transport Across Plasmodesmata | 86 |
| 3.3 | Diffusion vs Bulk Transport of Water and Solutes | 89 |
| 3.4 | Structural Features of Xylem Elements Which Facilitate Water and Solute Transport | 90 |
| 3.5 | Membrane Transport System | 92 |
| 3.6 | Uniporters and Cotransporters | 94 |
| 3.7 | Ion Channels | 97 |
| 3.7.1 | Potassium Channels | 99 |
| 3.7.2 | Calcium Channels | 101 |
| 3.7.3 | Anion Channels | 102 |
| 3.7.4 | Aquaporins | 103 |
| 3.8 | Pumps | 106 |
| 3.8.1 | P-Type ATPases | 106 |
| 3.8.2 | Endomembrane-Associated $Ca^{2+}$ Pump | 108 |
| 3.8.3 | F-Type ATPases | 108 |
| 3.8.4 | V-Type ATPases | 109 |
| 3.8.5 | $H^+$-Pyrophosphatase (PPase) | 110 |
| 3.8.6 | ABC-Type Pumps | 110 |
| Multiple-Choice Questions | | 113 |
| Suggested Further Readings | | 115 |

## Part II Metabolism

## 4 Concepts in Metabolism ................................ 119
Manju A. Lal

| 4.1 | Basic Energetic Principles that Govern Metabolism | 122 |
|-----|---------------------------------------------------|-----|
| 4.2 | Energy Coupled Reactions | 126 |
| 4.2.1 | Structure of ATP | 126 |
| 4.2.2 | ATP Is the High-Energy Molecule | 128 |
| 4.2.3 | ATP Is the Energy Currency of the Cell | 130 |

|       |       |                                                                 |     |
|-------|-------|-----------------------------------------------------------------|-----|
| 4.3   |       | Reduction-Oxidation Coupled Reactions                           | 131 |
| 4.4   |       | Enzymes                                                         | 135 |
|       | 4.4.1 | Nomenclature and Classification of Enzymes                      | 136 |
|       | 4.4.2 | General Characteristics of Enzyme-Catalyzed Reactions           | 139 |
|       | 4.4.3 | Enzyme Kinetics                                                 | 142 |
|       | 4.4.4 | Factors Affecting Enzyme-Catalyzed Reactions                    | 145 |
|       | 4.4.5 | Role of Inhibitors                                              | 147 |
|       | 4.4.6 | Regulatory Enzymes                                              | 149 |
| Multiple-Choice Questions |  |                                                | 156 |
| Suggested Further Readings |  |                                               | 158 |

## 5 Photosynthesis ... 159
Manju A. Lal

|       |        |                                                                      |     |
|-------|--------|----------------------------------------------------------------------|-----|
| 5.1   |        | General Concepts                                                     | 160 |
|       | 5.1.1  | Properties of Light                                                  | 161 |
|       | 5.1.2  | Mechanism of Light Absorption and Emission                           | 162 |
|       | 5.1.3  | Photosynthetic Pigments                                              | 164 |
|       | 5.1.4  | Action Spectrum Relates to Absorption Spectra                        | 168 |
| 5.2   |        | Phases of Photosynthesis                                             | 172 |
| 5.3   |        | Light Reactions in Photosynthesis                                    | 174 |
|       | 5.3.1  | Organization of Photosynthetic Apparatus into Photosystems           | 176 |
|       | 5.3.2  | Organization of Chlorophylls and Other Pigments in LHCII and LHCI    | 179 |
|       | 5.3.3  | Photochemical Reaction Centers                                       | 179 |
|       | 5.3.4  | Cytochrome $b_6f$ (Plastoquinol-Plastocyanin Oxidoreductase)         | 181 |
|       | 5.3.5  | Two Mobile Electron Carriers                                         | 183 |
|       | 5.3.6  | Electron Transport Pathway During Light Reaction of Photosynthesis   | 183 |
|       | 5.3.7  | Photosystem II (Splitting of Water)                                  | 183 |
|       | 5.3.8  | Q-Cycle Results in Pumping of Protons                                | 185 |
|       | 5.3.9  | Photosystem I (Production of NADPH)                                  | 187 |
|       | 5.3.10 | Non-cyclic and Cyclic Electron Transport                             | 188 |
|       | 5.3.11 | ATP Generation During Electron Transport in Light Reaction           | 189 |
|       | 5.3.12 | Balancing Distribution of the Light Energy in Between the Two Photosystems | 190 |
|       | 5.3.13 | Elimination of Excess Light Energy as Heat                           | 191 |
| 5.4   |        | Photosynthetic Carbon Dioxide Assimilation                           | 192 |
|       | 5.4.1  | Calvin-Benson Cycle                                                  | 193 |
|       | 5.4.2  | Carboxylation Phase                                                  | 195 |
|       | 5.4.3  | Reduction Phase                                                      | 197 |
|       | 5.4.4  | RuBP Regeneration Phase                                              | 197 |

|  |  | 5.4.5 | ATP and NADPH (Energy Sources in $CO_2$ Fixation) | 199 |
|---|---|---|---|---|
|  |  | 5.4.6 | Autocatalytic Regulation of Regeneration of RuBP for Continuous $CO_2$ Assimilation | 200 |
|  |  | 5.4.7 | Regulation of Calvin-Benson Cycle | 200 |
|  | 5.5 | Photorespiration | | 204 |
|  |  | 5.5.1 | Significance of Photorespiration | 207 |
|  | 5.6 | C4 Pathway and Types of C4 Plants | | 211 |
|  |  | 5.6.1 | Regulation of C4 Pathway | 215 |
|  |  | 5.6.2 | Energy Requirement for $CO_2$ Fixation by C4 Pathway | 216 |
|  |  | 5.6.3 | Evolutionary Significance of C4 Pathway | 217 |
|  | 5.7 | Crassulacean Acid Metabolism (CAM): $CO_2$ Fixation in Dark | | 218 |
|  |  | 5.7.1 | Ecological Significance of CAM Plants | 221 |
|  | 5.8 | Summary | | 223 |
|  | Multiple-Choice Questions | | | 225 |
|  | Suggested Further Readings | | | 226 |
| **6** | **Photoassimilate Translocation** | | | **227** |
|  | Rashmi Shakya and Manju A. Lal | | | |
|  | 6.1 | Source-Sink Relationship | | 228 |
|  | 6.2 | Transition of Leaf from Sink to Source | | 230 |
|  | 6.3 | Pathway of Photoassimilate Translocation | | 231 |
|  |  | 6.3.1 | Experimental Evidence | 231 |
|  | 6.4 | Features of Phloem Cells with Reference to Photoassimilate Translocation | | 232 |
|  |  | 6.4.1 | Phloem Sealing Mechanism | 233 |
|  |  | 6.4.2 | Sieve Tube-Companion Cells Interaction | 234 |
|  |  | 6.4.3 | Composition of the Phloem Sap | 235 |
|  |  | 6.4.4 | Photoassimilate Translocation: Unique Features | 238 |
|  | 6.5 | Mechanism of Photoassimilate Translocation | | 239 |
|  |  | 6.5.1 | Photoassimilate Loading | 239 |
|  |  | 6.5.2 | Photoassimilate Unloading | 244 |
|  | 6.6 | Photoassimilate Allocation and Partitioning | | 248 |
|  | Multiple-Choice Questions | | | 250 |
|  | Suggested Further Readings | | | 251 |
| **7** | **Respiration** | | | **253** |
|  | Manju A. Lal | | | |
|  | 7.1 | Glycolysis | | 255 |
|  |  | 7.1.1 | Preparatory Steps | 258 |
|  |  | 7.1.2 | Entry of Molecules in Glycolysis Other than Glucose | 263 |
|  |  | 7.1.3 | *Payoff* Phase | 263 |

|  |  | 7.1.4 | Stoichiometry of Glycolysis................... | 265 |
|---|---|---|---|---|
|  |  | 7.1.5 | Significance of Phosphorylated Intermediates..... | 265 |
|  |  | 7.1.6 | Regulation of Glycolysis in Plants.............. | 267 |
|  | 7.2 | Oxidative Pentose Phosphate Pathway (OPPP)............ | | 269 |
|  |  | 7.2.1 | Oxidative Phase............................ | 270 |
|  |  | 7.2.2 | Non-oxidative Phase........................ | 270 |
|  |  | 7.2.3 | Significance of OPPP....................... | 271 |
|  |  | 7.2.4 | Regulation of OPPP........................ | 272 |
|  | 7.3 | PEP Metabolism in Cytosol........................ | | 273 |
|  | 7.4 | Pyruvate Metabolism.............................. | | 273 |
|  |  | 7.4.1 | Fermentation.............................. | 273 |
|  |  | 7.4.2 | Pyruvate Metabolism in Mitochondria........... | 275 |
|  | 7.5 | TCA Cycle.................................... | | 279 |
|  |  | 7.5.1 | General Features of the Cycle................. | 279 |
|  |  | 7.5.2 | Acetyl-CoA Enters the TCA Cycle.............. | 282 |
|  |  | 7.5.3 | Stoichiometry of TCA Cycle.................. | 286 |
|  |  | 7.5.4 | Amphibolic Role of TCA..................... | 287 |
|  |  | 7.5.5 | Anaplerotic Reactions....................... | 289 |
|  |  | 7.5.6 | Role of TCA in Plants Under Stress Conditions.... | 291 |
|  |  | 7.5.7 | Regulation of TCA......................... | 291 |
|  |  | 7.5.8 | TCA Cycle and GABA Shunt................. | 293 |
|  | 7.6 | Oxidation of the Reduced Coenzymes Produced During | | |
|  |  | TCA Cycle................................... | | 293 |
|  |  | 7.6.1 | Electron Transport Chain..................... | 297 |
|  |  | 7.6.2 | Electron Carriers Are Arranged in a Sequence..... | 297 |
|  |  | 7.6.3 | Components of the Electron Transport Chain Are Present as Multienzyme Complexes.......... | 298 |
|  |  | 7.6.4 | Proton Translocation Creates Proton Motive Force (PMF)............................... | 305 |
|  | 7.7 | NADH Shuttles................................. | | 305 |
|  | 7.8 | Alternate Mechanisms of NADH Oxidation in Plants........ | | 308 |
|  | 7.9 | Cyanide-Resistant Respiration...................... | | 309 |
|  | Multiple-Choice Questions................................. | | | 312 |
|  | Suggested Further Readings................................ | | | 314 |
| 8 | **ATP Synthesis**................................... | | | **315** |
|  | Manju A. Lal | | | |
|  | 8.1 | Proton Gradient Coupled ATP Synthesis................. | | 316 |
|  | 8.2 | ATP Synthase.................................. | | 324 |
|  | 8.3 | Mechanism of ATP Synthesis....................... | | 325 |
|  |  | 8.3.1 | Rotatory Model (Binding Change Mechanism).... | 326 |
|  |  | 8.3.2 | Rotatory Movement of c-Ring of Fo............ | 327 |
|  | 8.4 | Stoichiometry of $O_2$ Consumption and ATP Synthesis (P/O Ratio).................................... | | 329 |
|  | 8.5 | Substrate-Level Phosphorylation...................... | | 330 |

|   |   |   |   |
|---|---|---|---|
| | 8.6 | Electrochemical Gradient-Driven Transport of Various Metabolites Across Inner Mitochondrial Membrane......... | 331 |
| | 8.7 | Oxidative Phosphorylation and Photophosphorylation: A Comparative Account.............................. | 333 |
| | Multiple-Choice Questions................................. | | 335 |
| | Suggested Further Readings................................ | | 337 |

**9 Metabolism of Storage Carbohydrates**...................... 339
Manju A. Lal
- 9.1 Metabolite Pool and Exchange of Metabolites............ 339
- 9.2 Sucrose Metabolism................................. 342
  - 9.2.1 Sucrose Biosynthesis......................... 342
  - 9.2.2 Sucrose Catabolism.......................... 349
- 9.3 Starch Metabolism.................................. 353
  - 9.3.1 Starch Biosynthesis.......................... 357
  - 9.3.2 Transitory Starch............................ 362
  - 9.3.3 Starch Catabolism........................... 362
- 9.4 Fructans.......................................... 370
  - 9.4.1 Structure of Fructans......................... 371
  - 9.4.2 Metabolism of Fructans....................... 372
- Multiple-Choice Questions................................. 375
- Suggested Further Readings................................ 377

**10 Lipid Metabolism**....................................... 379
Manju A. Lal
- 10.1 Role of Lipids...................................... 380
- 10.2 Diversity in Lipid Structure........................... 382
  - 10.2.1 Fatty Acids................................. 382
  - 10.2.2 Storage Lipids (Neutral Fats, Waxes)............ 386
  - 10.2.3 Membrane Lipids............................ 387
- 10.3 Fatty Acid Biosynthesis.............................. 391
  - 10.3.1 Synthesis of Acetyl-CoA...................... 392
  - 10.3.2 Synthesis of Malonyl-CoA.................... 393
  - 10.3.3 Transfer of Malonyl Moiety from Coenzyme-A to Acyl Carrier Protein (ACP)................. 395
  - 10.3.4 Condensation Reaction....................... 399
  - 10.3.5 Reduction of 3-Ketobutyryl-ACP to Butyryl-ACP.............................. 399
  - 10.3.6 Extension of Butyryl-ACP..................... 400
  - 10.3.7 Termination of Fatty Acid Synthesis in Plastids.... 400
  - 10.3.8 Hydrocarbon Chain Elongation in Endoplasmic Reticulum.................................. 401
  - 10.3.9 Synthesis of Unsaturated Fatty Acids............ 402
- 10.4 Biosynthesis of Membrane Lipids...................... 404

|       | 10.5   | Biosynthesis of Storage Lipids in Seeds | 405 |
|-------|--------|------------------------------------------|-----|
|       | 10.6   | Lipid Catabolism | 409 |
|       |        | 10.6.1 β-Oxidation of Fatty Acyl-CoA | 409 |
|       |        | 10.6.2 Glyoxylate Cycle | 413 |
|       |        | 10.6.3 Gluconeogenesis | 414 |
|       |        | 10.6.4 α-Oxidation of Fatty Acids | 416 |
|       |        | 10.6.5 ω-Oxidation of Fatty Acids | 416 |
|       |        | 10.6.6 Catabolism of Unsaturated Fatty Acids | 419 |
|       |        | 10.6.7 Catabolism of Fatty Acids with Odd Number of Carbon Atoms | 420 |
|       | Multiple-Choice Questions | | 422 |
|       | Suggested Further Readings | | 424 |
| **11** | **Nitrogen Metabolism** | | **425** |
|       | Manju A. Lal | | |
|       | 11.1   | Biogeochemical Cycle of Nitrogen | 426 |
|       |        | 11.1.1 Ammonification | 428 |
|       |        | 11.1.2 Nitrification | 428 |
|       |        | 11.1.3 Denitrification | 429 |
|       |        | 11.1.4 Nitrogen Fixation | 429 |
|       | 11.2   | Nitrogen Nutrition for the Plants | 431 |
|       |        | 11.2.1 Ammonium Ions | 431 |
|       |        | 11.2.2 Nitrate Uptake | 432 |
|       |        | 11.2.3 Nitrate Assimilation | 435 |
|       |        | 11.2.4 Fixation of Molecular Nitrogen | 442 |
|       | 11.3   | Ammonia Assimilation | 462 |
|       |        | 11.3.1 Ammonia Assimilation by GS/GOGAT | 462 |
|       |        | 11.3.2 Ammonia Assimilation by Reductive Amination | 465 |
|       | 11.4   | Nitrogenous Compounds for Storage and Transport | 467 |
|       | 11.5   | Amino Acid Biosynthesis | 470 |
|       |        | 11.5.1 Aminotransferase Reaction (Tansamination) | 471 |
|       | Multiple Choice Questions | | 478 |
|       | Suggested Further Readings | | 479 |
| **12** | **Sulfur, Phosphorus, and Iron Metabolism in Plants** | | **481** |
|       | Manju A. Lal | | |
|       | 12.1   | Sulfur Metabolism | 481 |
|       |        | 12.1.1 Biogeochemical Cycle of Sulfur | 482 |
|       |        | 12.1.2 Uptake of Sulfur | 484 |
|       |        | 12.1.3 Sulfate Metabolism | 485 |
|       |        | 12.1.4 Cysteine Metabolism | 490 |
|       |        | 12.1.5 Sulfated Compounds | 494 |
|       | 12.2   | Phosphorus Metabolism | 497 |
|       |        | 12.2.1 Biogeochemical Cycle of Phosphorus | 498 |
|       |        | 12.2.2 Phosphate Transporters | 498 |
|       |        | 12.2.3 Role of Phosphorus in Cell Metabolism | 500 |
|       |        | 12.2.4 Mobilization of Phosphorus | 502 |

| | 12.3 | Iron Metabolism | 503 |
|---|---|---|---|
| | | 12.3.1 Biogeochemical Cycle of Iron and Iron Uptake by the Plant | 506 |
| | | 12.3.2 Transport of Iron Within Plant | 508 |
| | | 12.3.3 Redistribution of Iron at the Subcellular Level | 510 |
| | Multiple-Choice Questions | | 513 |
| | Suggested Further Readings | | 515 |

## Part III Development

**13 Light Perception and Transduction** .................................. 519
Satish C Bhatla

| | 13.1 | Light Absorption by Pigment Molecules | 521 |
|---|---|---|---|
| | | 13.1.1 Quantitative Requirement for Pigment Excitation | 522 |
| | 13.2 | Nature of Light | 524 |
| | 13.3 | Absorption and Action Spectra | 526 |
| | 13.4 | Light Parameters Which Influence Plant Responses | 527 |
| | 13.5 | Light Absorption Depends on Leaf Anatomy and Canopy Structure | 528 |
| | 13.6 | Photoreceptors | 530 |
| | 13.7 | Protochlorophyllide | 531 |
| | 13.8 | Phycobilins | 533 |
| | 13.9 | Phytochromes | 534 |
| | | 13.9.1 Photoreversibility of Phytochrome-Modulated Responses and Its Significance | 535 |
| | | 13.9.2 Chemical Nature of Phytochrome Chromophore | 536 |
| | | 13.9.3 The Multidomain Structure of Phytochrome Protein | 537 |
| | | 13.9.4 Forms of Biologically Active Phytochrome | 538 |
| | | 13.9.5 Phytochrome-Mediated Responses | 538 |
| | | 13.9.6 Phytochrome Action Involves Its Partitioning Between Cytosol and Nucleus | 539 |
| | | 13.9.7 Phytochrome Signaling Mechanisms | 541 |
| | 13.10 | Blue Light-Mediated Responses and Photoreceptors | 542 |
| | 13.11 | Cryptochromes | 543 |
| | 13.12 | Phototropins: Molecular Nature and Associated Phototropic Bending Response | 547 |
| | 13.13 | Phototropin-Modulated Chloroplast Movement | 548 |
| | 13.14 | Phototropin Signaling-Dependent Light-Induced Stomatal Opening | 550 |
| | 13.15 | UVR 8: A Photoreceptor for UV-B-Mediated Photomorphogenic Responses | 551 |

| | 13.16 | Other LOV Domain-Containing Photoreceptors in Plants | 552 |
|---|---|---|---|
| | 13.17 | Rhodopsin-Like Photoreceptors | 553 |
| | 13.18 | Summary | 553 |
| | | Multiple-Choice Questions | 556 |
| | | Suggested Further Reading | 558 |

**14 Plant Growth Regulators: An Overview** ................. 559
Satish C Bhatla
- 14.1 Plant Growth Regulators (PGRs) .................. 560
- 14.2 Estimation and Imaging of Hormones in Plant Tissue ....... 562
- 14.3 Experimental Approaches to Understand Perception and Transmission of Hormone Action ................ 564
- Multiple-Choice Questions ............................ 567
- Suggested Further Readings ........................... 568

**15 Auxins** ................................................ 569
Satish C Bhatla
- 15.1 Discovery of Auxin ............................. 569
- 15.2 Synthetic Auxins ............................... 572
- 15.3 Auxin Distribution and Biosynthesis ................ 574
  - 15.3.1 Tryptophan-Dependent Pathways ............. 575
  - 15.3.2 Tryptophan-Independent Pathways ........... 576
- 15.4 Conjugation and Degradation of Auxins .............. 578
- 15.5 Auxin Transport ................................ 579
  - 15.5.1 Auxin Influx ............................ 580
  - 15.5.2 Auxin Efflux ............................ 581
  - 15.5.3 Chemiosmotic Model for Auxin Transport ..... 582
- 15.6 Physiological Effects of Auxins .................... 584
  - 15.6.1 Cell Expansion (Acid Growth Hypothesis) ..... 584
  - 15.6.2 Apical Dominance ........................ 585
  - 15.6.3 Floral Bud Development .................. 589
  - 15.6.4 Vascular Differentiation ................. 590
  - 15.6.5 Origin of Lateral and Adventitious Roots ..... 591
- 15.7 Signaling Mechanisms Associated with Auxin Action ..... 592
  - 15.7.1 Changes in Gene Expression ............... 594
  - 15.7.2 The Process of AUX/IAA Degradation ......... 594
- Multiple-Choice Questions ............................ 598
- Suggested Further Readings ........................... 601

**16 Cytokinins** .......................................... 603
Geetika Kalra and Satish C Bhatla
- 16.1 Bioassay ...................................... 605
- 16.2 Biosynthesis ................................... 605

|  |  |  |  |
|---|---|---|---|
| 16.3 | Transport | | 607 |
| 16.4 | Metabolism | | 607 |
| 16.5 | Physiological Role of Cytokinins | | 607 |
| | 16.5.1 | Cell Division | 607 |
| | 16.5.2 | Regulation of Cell Cycle | 609 |
| | 16.5.3 | Morphogenesis | 609 |
| | 16.5.4 | Lateral Bud Formation | 609 |
| | 16.5.5 | Bud Formation in Mosses | 610 |
| | 16.5.6 | Delay of Leaf Senescence | 610 |
| | 16.5.7 | Movement of Nutrients | 611 |
| | 16.5.8 | Chloroplast Development | 611 |
| | 16.5.9 | Mechanical Extensibility of Cell Wall | 611 |
| 16.6 | Mode of Cytokinin Action | | 612 |
| Multiple-Choice Questions | | | 614 |
| Suggested Further Readings | | | 615 |

**17 Gibberellins** .................................................. 617
Geetika Kalra and Satish C Bhatla

|  |  |  |  |
|---|---|---|---|
| 17.1 | Biosynthesis | | 617 |
| 17.2 | Modulation of Gibberellin Biosynthesis | | 619 |
| 17.3 | Enzymes Involved in Gibberellin Metabolism | | 620 |
| 17.4 | Gibberellin Metabolism | | 620 |
| 17.5 | Physiological Roles of Gibberellins | | 621 |
| | 17.5.1 | Internode Elongation | 621 |
| | 17.5.2 | Floral Initiation and Sex Determination | 622 |
| | 17.5.3 | Seed Germination | 622 |
| | 17.5.4 | Fruit Production | 622 |
| | 17.5.5 | Stimulation of Cell Elongation and Cell Division | 622 |
| | 17.5.6 | Regulation of Transcription of Cell Cycle Kinases | 623 |
| 17.6 | Mode of Action | | 623 |
| Multiple-Choice Questions | | | 627 |
| Suggested Further Reading | | | 628 |

**18 Abscisic Acid** .................................................. 629
Geetika Kalra and Satish C Bhatla

|  |  |  |  |
|---|---|---|---|
| 18.1 | Bioassay | | 630 |
| 18.2 | Biosynthesis, Catabolism, and Homeostasis | | 631 |
| 18.3 | Translocation of Abscisic Acid | | 632 |
| 18.4 | Developmental and Physiological Effects of ABA | | 633 |
| | 18.4.1 | ABA Levels Increase in Response to Environmental Stress | 633 |
| | 18.4.2 | Seed Development | 633 |
| | 18.4.3 | Desiccation Tolerance | 634 |

|  |  | 18.4.4 | Inhibition of Precocious Germination and Vivipary........................... | 634 |
|---|---|---|---|---|
|  |  | 18.4.5 | Counteraction of GA Action................. | 634 |
|  |  | 18.4.6 | Seed Dormancy........................... | 635 |
|  |  | 18.4.7 | Stomatal Closure During Water Stress.......... | 635 |
|  |  | 18.4.8 | Promotion of Root Growth and Inhibition of Shoot Growth at Low Water Potential......... | 635 |
|  | 18.5 | Mode of Action..................................... | | 636 |
|  | 18.6 | Mechanism of ABA Signaling......................... | | 637 |
|  | Multiple-Choice Questions.................................. | | | 640 |
|  | Suggested Further Readings................................. | | | 641 |
| 19 | **Ethylene**................................................... | | | 643 |
|  | Satish C Bhatla | | | |
|  | 19.1 | Ethylene vs Non-gaseous Plant Hormones: Some Interesting Facts................................... | | 645 |
|  | 19.2 | Biosynthesis of Ethylene............................ | | 645 |
|  | 19.3 | Regulation of Ethylene Biosynthesis................... | | 649 |
|  | 19.4 | Bioassay and Mutant Analysis........................ | | 650 |
|  | 19.5 | Physiological and Developmental Effects................ | | 652 |
|  |  | 19.5.1 | Fruit Ripening............................ | 652 |
|  |  | 19.5.2 | Leaf Epinasty............................. | 653 |
|  |  | 19.5.3 | Induction of Lateral Cell Expansion............ | 654 |
|  |  | 19.5.4 | Breaking Bud and Seed Dormancy............. | 654 |
|  |  | 19.5.5 | Elongation of Submerged Aquatic Species....... | 654 |
|  |  | 19.5.6 | Formation of Adventitious Roots and Root Hairs... | 655 |
|  |  | 19.5.7 | Leaf Senescence and Abscission............... | 655 |
|  |  | 19.5.8 | Ethylene in Defense Response................ | 655 |
|  |  | 19.5.9 | Abscission............................... | 655 |
|  | 19.6 | Ethylene Receptors and Signal Transduction.............. | | 656 |
|  | Multiple-Choice Questions.................................. | | | 660 |
|  | Suggested Further Readings................................. | | | 661 |
| 20 | **Brassinosteroids**............................................ | | | 663 |
|  | Satish C Bhatla | | | |
|  | 20.1 | Biosynthesis and Homeostasis........................ | | 663 |
|  | 20.2 | Functions of Brassinosteroids........................ | | 665 |
|  |  | 20.2.1 | Regulation of Photomorphogenesis............. | 665 |
|  |  | 20.2.2 | Unrolling and Bending of Grass Leaves.......... | 665 |
|  |  | 20.2.3 | Other Effects............................. | 665 |
|  | 20.3 | Brassinosteroid Mutants............................. | | 665 |
|  |  | 20.3.1 | Mutants with Impaired Brassinolide Synthesis (Brassinolide-Sensitive Mutants).............. | 665 |
|  |  | 20.3.2 | Brassinosteroid-Insensitive Mutants (bri)......... | 667 |

|      |       | 20.4    | Brassinosteroid Signaling Mechanism .................... | 667 |
|------|-------|---------|------------------------------------------------------------|-----|
|      |       |         | Multiple-Choice Questions ................................. | 669 |
|      |       |         | Suggested Further Readings ............................... | 670 |
| 21   | **Jasmonic Acid** ........................................................ | | | 671 |
|      | Satish C Bhatla | | | |
|      |       | 21.1    | Biosynthesis ................................................ | 671 |
|      |       | 21.2    | Metabolism and Homeostasis ............................. | 672 |
|      |       | 21.3    | Physiological and Developmental Roles ................ | 673 |
|      |       |         | 21.3.1  Trichome Formation ............................. | 673 |
|      |       |         | 21.3.2  Reproductive Functions ........................ | 673 |
|      |       |         | 21.3.3  Induction of Secondary Metabolites Production | 674 |
|      |       |         | 21.3.4  Role as a Senescing Promoting Factor .......... | 675 |
|      |       |         | 21.3.5  JA and Photomodulation of Plant Development | 675 |
|      |       |         | 21.3.6  Response to Herbivores ........................ | 676 |
|      |       |         | 21.3.7  Mycorrhizal Interactions and Modulation ...... | 676 |
|      |       | 21.4    | JA-Induced Gene Expression ............................. | 676 |
|      |       | 21.5    | JA-Mediated Signaling ................................... | 676 |
|      |       |         | Multiple-Choice Questions ................................. | 678 |
|      |       |         | Suggested Further Readings ............................... | 679 |
| 22   | **Recently Discovered Plant Growth Regulators** ................ | | | 681 |
|      | Satish C Bhatla | | | |
|      |       | 22.1    | Salicylic Acid ............................................. | 681 |
|      |       |         | 22.1.1  Biosynthesis ................................... | 681 |
|      |       |         | 22.1.2  Physiological Functions ....................... | 683 |
|      |       | 22.2    | Nitric Oxide ............................................... | 686 |
|      |       |         | 22.2.1  Physicochemical Properties of NO ............. | 687 |
|      |       |         | 22.2.2  NO Biosynthesis in Plants ..................... | 688 |
|      |       |         | 22.2.3  NO as a Signaling Molecule and Its Effect on Gene Expression | 689 |
|      |       |         | 22.2.4  Physiological Functions of NO in Plants ...... | 690 |
|      |       |         | 22.2.5  NO Metabolism ................................. | 695 |
|      |       |         | 22.2.6  NO Transport ................................... | 695 |
|      |       | 22.3    | Indoleamines (Serotonin and Melatonin) ................ | 695 |
|      |       |         | 22.3.1  Biosynthesis of Serotonin and Melatonin ...... | 697 |
|      |       |         | 22.3.2  Melatonin: Structure and Activity Relationship | 698 |
|      |       |         | 22.3.3  Roles of Serotonin and Melatonin .............. | 699 |
|      |       |         | 22.3.4  NO-Melatonin CrossTalk ........................ | 701 |
|      |       | 22.4    | Strigolactones ............................................. | 702 |
|      |       |         | 22.4.1  Physiological Roles of Strigolactones ......... | 703 |
|      |       |         | 22.4.2  SL Crosstalk with Auxin, Ethylene, and Cytokinins | 705 |
|      |       |         | 22.4.3  Signaling Mechanism for Strigolactone Action . | 706 |

|   | 22.5 | Polyamines | | 708 |
|---|------|-----------|---|-----|
|   |   | 22.5.1 | Most Common Polyamines in Plants and Their Distribution | 708 |
|   |   | 22.5.2 | Polyamine Homeostasis | 709 |
|   |   | 22.5.3 | Biosynthesis of Polyamines | 710 |
|   |   | 22.5.4 | Catabolism of Polyamines | 711 |
|   |   | 22.5.5 | Functions of Polyamines | 712 |
|   |   | 22.5.6 | Ionic Interactions | 712 |
|   |   | 22.5.7 | Crosstalk with Hormones | 713 |
|   |   | 22.5.8 | Polyamines as Signaling Molecules and in Modulating Stress | 714 |
|   |   | 22.5.9 | Role in Plant-Microbe Interactions | 714 |
|   | 22.6 | Peptide Signaling Molecules | | 715 |
|   |   | 22.6.1 | Systemins | 715 |
|   |   | 22.6.2 | Types of Signaling Peptides | 717 |
|   |   | 22.6.3 | Perception of Signal Peptides by the Cells | 717 |
|   |   | 22.6.4 | Signal Peptides and Their Potential Benefits in Agriculture | 717 |
|   | 22.7 | Karrikins: A New Class of Plant Growth Regulators in Smoke | | 718 |
|   |   | 22.7.1 | Chemical Nature | 720 |
|   |   | 22.7.2 | Karrikin-Sensitive Plants | 720 |
|   |   | 22.7.3 | Mode of Karrikin Action in Plants | 721 |
|   | Multiple-Choice Questions | | | 725 |
|   | Suggested Further Readings | | | 728 |
| 23 | **Signal Perception and Transduction** | | | **729** |
|   | Satish C Bhatla | | | |
|   | 23.1 | Routes of Signal Perception, Transduction, and Response in Plants | | 730 |
|   | 23.2 | Spatial and Temporal Aspects of Signal Transduction (Table 23.1) | | 732 |
|   | 23.3 | Signal Perception | | 735 |
|   |   | 23.3.1 | Membrane Potential as a Receptor | 738 |
|   |   | 23.3.2 | Characteristic Features of Membrane Receptors | 743 |
|   |   | 23.3.3 | Tissue Sensitivity for Receptor-Mediated Signaling Responses | 743 |
|   | 23.4 | Signal Perception at the Plasma Membrane | | 744 |
|   |   | 23.4.1 | Receptor Kinases | 744 |
|   |   | 23.4.2 | G-Protein-Coupled Receptors (GPCRs) | 746 |
|   |   | 23.4.3 | Ion Channel-Linked Receptors | 747 |

|  |  |  |  | |
|---|---|---|---|---|
| | 23.5 | Signal Transduction and Amplification via Second Messengers................................. | 748 | |
| | | 23.5.1 Significance of Second Messengers in Signal Transduction............................. | 749 | |
| | | 23.5.2 Ca$^{+2}$: The Most Ubiquitous Second Messenger..... | 749 | |
| | | 23.5.3 Modulation of Cytosolic or Cell Wall pH as a Second Messenger........................ | 750 | |
| | | 23.5.4 Reactive Oxygen Species and Reactive Nitrogen Species as Second Messengers in Environmental and Developmental Signals................... | 751 | |
| | | 23.5.5 Lipid-Signaling Molecules.................... | 753 | |
| | | 23.5.6 Mitogen-Activated Protein (MAP) Kinase Cascade................................. | 756 | |
| | | 23.5.7 Cyclic Nucleotides......................... | 758 | |
| | 23.6 | Adaptive Mechanisms of Plant Signaling and Their Termination........................................ | 759 | |
| | Multiple-Choice Questions................................. | | 764 | |
| | Suggested Further Readings............................... | | 765 | |
| **24** | **Embryogenesis, Vegetative Growth, and Organogenesis**......... | | **767** | |
| | Rama Sisodia and Satish C Bhatla | | | |
| | 24.1 | Embryogenesis...................................... | 768 | |
| | | 24.1.1 Acquisition of Polarity....................... | 768 | |
| | | 24.1.2 Stages of Embryo Development................ | 770 | |
| | | 24.1.3 Developmental Patterns...................... | 771 | |
| | 24.2 | Genetic Control of Patterning During Embryogenesis........ | 775 | |
| | | 24.2.1 Mutants Defining Genes Involved in Axial Patterning................................. | 776 | |
| | | 24.2.2 Genetic Control of Radial Patterning............ | 779 | |
| | 24.3 | Role of Auxin in Establishing Polarity During Embryogenesis and Vegetative Development......................... | 781 | |
| | 24.4 | Plant Development and Meristems..................... | 782 | |
| | | 24.4.1 Shoot Apical Meristem...................... | 783 | |
| | | 24.4.2 SAM Development: Role of Auxin and Transcription Factors........................ | 784 | |
| | | 24.4.3 Root Apical Meristem....................... | 785 | |
| | | 24.4.4 Role of Auxin and Cytokinin in Establishment of RAM and Root Development................ | 787 | |
| | 24.5 | Growth and Differentiation of Lateral Roots............. | 789 | |
| | 24.6 | Cell Growth and Differentiation....................... | 791 | |
| | | 24.6.1 Cell Growth: Role of Wall Extensibility.......... | 792 | |
| | | 24.6.2 A Role for Cell Wall Components............. | 793 | |
| | Multiple-Choice Questions................................. | | 794 | |
| | Suggested Further Readings............................... | | 796 | |

## 25 Physiology of Flowering ... 797
Geetika Kalra and Manju A. Lal
- 25.1 Juvenile Phase ... 798
- 25.2 Flower Induction ... 800
- 25.3 Photoperiodism: The Light-Dependent Pathway ... 801
  - 25.3.1 Critical Day Length ... 802
  - 25.3.2 Critical Role of Dark Period ... 803
- 25.4 Photoinductive Cycle ... 803
  - 25.4.1 Perception of Photoperiodic Signal and Florigen ... 805
- 25.5 Circadian Rhythm ... 806
- 25.6 Photoreceptors ... 808
- 25.7 Vernalization ... 809
- 25.8 Role of Gibberellins ... 811
- 25.9 Flower Development ... 813
- Multiple-Choice Questions ... 818
- Suggested Further Readings ... 819

## 26 Pollination, Fertilization and Seed Development ... 821
Rashmi Shakya and Satish C Bhatla
- 26.1 Development of Male Gametophyte ... 821
- 26.2 Development of Female Gametophyte ... 827
- 26.3 Pollination and Double Fertilization ... 828
  - 26.3.1 Pollen Adhesion and Hydration ... 829
  - 26.3.2 $Ca^{2+}$ Triggered Polarization of Pollen Grain Before Emergence of Pollen Tube ... 830
  - 26.3.3 Apical Growth of Pollen Tube Tip and Its Regulation ... 831
  - 26.3.4 Signaling Events at the Tip of Growing Pollen Tube ... 833
  - 26.3.5 Directional Growth of Pollen Tube in the Pistil ... 834
  - 26.3.6 Double Fertilization ... 837
- 26.4 Pre-zygotic Barriers to Self-Fertilization ... 838
  - 26.4.1 Genetic Basis of Self-Incompatibility (SI) ... 839
  - 26.4.2 Receptor-Ligand Interactions Mediated Rejection of Self-Pollen at the Stigma Surface ... 840
  - 26.4.3 Programmed Cell Death of the Pollen Tubes After Penetration into Stigma Surface ... 842
  - 26.4.4 Inhibition by Cytotoxic Stylar RNAses and Degradation of Pollen Tube RNA ... 843
- 26.5 Seed Development ... 844
  - 26.5.1 Embryogenesis ... 844
  - 26.5.2 Endosperm Development ... 845
  - 26.5.3 Cellularization of Endosperm ... 845

|    |       | 26.5.4 | Hormonal and Genetic Regulation of Aleurone Development in Cereal Grains | 848 |
|----|-------|--------|--------------------------------------------------------------------------|-----|
|    |       | 26.5.5 | Development of Seed Coat                                                 | 849 |
|    |       | 26.5.6 | Seed Maturation and Desiccation Tolerance                                | 851 |
|    |       | 26.5.7 | Molecular Basis of Desiccation Tolerance                                 | 851 |
|    | Multiple-Choice Questions |  |                                                    | 854 |
|    | Suggested Further Readings |  |                                                   | 856 |
| 27 | **Fruit Development and Ripening** | | | 857 |
|    | Rashmi Shakya and Manju A. Lal | | | |
|    | 27.1  | Stages of Fruit Development and Ripening Stages | | 858 |
|    | 27.2  | Physiological Changes During Fruit Ripening | | 862 |
|    |       | 27.2.1 | Fruit Texture and Softening | 862 |
|    |       | 27.2.2 | Changes in Pigmentation | 865 |
|    |       | 27.2.3 | Flavor and Fragrance | 868 |
|    |       | 27.2.4 | Antioxidants and Bioactive Compounds | 868 |
|    | 27.3  | Climacteric and Non-climacteric Fruit Ripening | | 869 |
|    | 27.4  | Fruit Ripening: An Oxidative Phenomenon | | 872 |
|    |       | 27.4.1 | Role of Alternative Oxidase (AOX) | 873 |
|    | 27.5  | Role of Phytohormones | | 874 |
|    | 27.6  | Nitric Oxide and Ethylene CrossTalk | | 876 |
|    | 27.7  | Transcriptional Regulation | | 878 |
|    | 27.8  | Epigenetic Regulation of Gene Expression | | 880 |
|    | Multiple-Choice Questions | | | 882 |
|    | Suggested Further Readings | | | 883 |
| 28 | **Seed Dormancy and Germination** | | | 885 |
|    | Renu Kathpalia and Satish C Bhatla | | | |
|    | 28.1  | Seed Morphology and Structure | | 886 |
|    | 28.2  | Food Reserves in Seeds | | 888 |
|    | 28.3  | Seed Dormancy | | 888 |
|    | 28.4  | Hormonal Regulation of Seed Dormancy | | 891 |
|    | 28.5  | Mechanisms to Overcome Seed Dormancy | | 892 |
|    | 28.6  | Release from Seed Dormancy | | 892 |
|    |       | 28.6.1 | Afterripening | 893 |
|    |       | 28.6.2 | Impaction | 893 |
|    |       | 28.6.3 | Scarification | 893 |
|    |       | 28.6.4 | Temperature | 894 |
|    |       | 28.6.5 | Light | 894 |
|    |       | 28.6.6 | Light-Temperature Interaction | 895 |
|    |       | 28.6.7 | Leaching of Inhibitors by Rainfall | 895 |
|    |       | 28.6.8 | Leaching of Chemicals | 895 |
|    | 28.7  | Sequence of Events Breaking Seed Dormancy | | 895 |

|       |        |                                                              |     |
|-------|--------|--------------------------------------------------------------|-----|
| 28.8  |        | Significance of Seed Dormancy                                | 896 |
|       | 28.8.1 | Seasonal Synchrony                                           | 896 |
|       | 28.8.2 | Geographical Distribution                                    | 897 |
|       | 28.8.3 | Spreading Germination Time                                   | 898 |
|       | 28.8.4 | Prevention of Pre-harvest Sprouting                          | 898 |
| 28.9  |        | Seed Germination                                             | 898 |
|       | 28.9.1 | Imbibition                                                   | 900 |
|       | 28.9.2 | Respiration                                                  | 900 |
|       | 28.9.3 | Light Requirement                                            | 901 |
|       | 28.9.4 | Mobilization of Reserves and Growth Regulators               | 901 |
|       | 28.9.5 | Development of Embryonic Axis into Seedling                  | 903 |
| Multiple-Choice Questions                                                    || 905 |
| Suggested Further Readings                                                   || 906 |

## 29 Plant Movements ............................................................. 907
Rama Sisodia and Satish C Bhatla

|       |        |                                                              |     |
|-------|--------|--------------------------------------------------------------|-----|
| 29.1  |        | Tropic Movements                                             | 908 |
|       | 29.1.1 | Phototropism                                                 | 908 |
|       | 29.1.2 | Gravitropism                                                 | 913 |
|       | 29.1.3 | Chemotropism                                                 | 916 |
| 29.2  |        | Nastic Movements                                             | 916 |
|       | 29.2.1 | Epinasty and Hyponasty                                       | 917 |
|       | 29.2.2 | Nyctinasty                                                   | 919 |
|       | 29.2.3 | Thermonasty                                                  | 919 |
|       | 29.2.4 | Thigmonasty                                                  | 919 |
| 29.3  |        | Autonomous Movements                                         | 920 |
|       | 29.3.1 | Diurnal Movements and Circadian Rhythms                      | 921 |
|       | 29.3.2 | Photoperiodism                                               | 923 |
| 29.4  |        | Mechanisms of Movement                                       | 924 |
|       | 29.4.1 | Turgor-Mediated Movement                                     | 924 |
|       | 29.4.2 | Growth-Mediated Movement                                     | 925 |
|       | 29.4.3 | Movement by Change in Conformation                           | 926 |
|       | 29.4.4 | Movement by Contraction                                      | 926 |
|       | 29.4.5 | Twining Plants                                               | 927 |
| 29.5  |        | Prey-Driven Movements                                        | 927 |
|       | 29.5.1 | Parasitic Plants                                             | 927 |
|       | 29.5.2 | Carnivorous Plants                                           | 928 |
| 29.6  |        | Movements for Dispersal                                      | 931 |
|       | 29.6.1 | Cohesion-Mediated Seed Propulsion                            | 931 |
|       | 29.6.2 | Turgor-Mediated Seed Propulsion                              | 932 |
| Multiple-Choice Questions                                                    || 934 |
| Suggested Further Readings                                                   || 935 |

## 30   Senescence and Programmed Cell Death . . . . . . . . . . . . . . . . . . . .   937
Geetika Kalra and Satish C Bhatla
- 30.1   Patterns of Senescence . . . . . . . . . . . . . . . . . . . . . . . . . . .   938
  - 30.1.1   Cellular Senescence . . . . . . . . . . . . . . . . . . . . . . . . .   939
  - 30.1.2   Tissue Senescence . . . . . . . . . . . . . . . . . . . . . . . . .   940
  - 30.1.3   Organ Senescence . . . . . . . . . . . . . . . . . . . . . . . . .   942
- 30.2   Types of Cell Death . . . . . . . . . . . . . . . . . . . . . . . . . . . . . . .   942
  - 30.2.1   Vacuolar-Type PCD . . . . . . . . . . . . . . . . . . . . . . . .   943
  - 30.2.2   Hypersensitive Response-Type PCD . . . . . . . . . . . . .   943
- 30.3   Autophagy . . . . . . . . . . . . . . . . . . . . . . . . . . . . . . . . . . . . . .   944
- 30.4   PCD During Seed Development . . . . . . . . . . . . . . . . . . . . . .   948
- 30.5   PCD During Tracheary Element Differentiation . . . . . . . . . . .   948
- 30.6   PCD During Gametogenesis . . . . . . . . . . . . . . . . . . . . . . . . .   950
- 30.7   Leaf Senescence . . . . . . . . . . . . . . . . . . . . . . . . . . . . . . . . .   950
- 30.8   Hormonal Regulation of Senescence . . . . . . . . . . . . . . . . . . .   953
  - 30.8.1   Cytokinins . . . . . . . . . . . . . . . . . . . . . . . . . . . . . . . .   953
  - 30.8.2   Auxin . . . . . . . . . . . . . . . . . . . . . . . . . . . . . . . . . . . .   955
  - 30.8.3   Gibberellins . . . . . . . . . . . . . . . . . . . . . . . . . . . . . . .   955
  - 30.8.4   Jasmonic Acid . . . . . . . . . . . . . . . . . . . . . . . . . . . . .   956
  - 30.8.5   Abscisic Acid . . . . . . . . . . . . . . . . . . . . . . . . . . . . . .   956
  - 30.8.6   Ethylene . . . . . . . . . . . . . . . . . . . . . . . . . . . . . . . . . .   957
  - 30.8.7   Salicylic Acid . . . . . . . . . . . . . . . . . . . . . . . . . . . . . .   957
- 30.9   Developmental Regulation of Senescence . . . . . . . . . . . . . . . .   958
- 30.10   Role of ROS in Leaf Senescence . . . . . . . . . . . . . . . . . . . . . .   959
- 30.11   Role of Sugar Accumulation in Leaf Senescence . . . . . . . . . .   959
- 30.12   Role of Pigment Composition in Senescence . . . . . . . . . . . . .   959
- 30.13   Leaf Abscission . . . . . . . . . . . . . . . . . . . . . . . . . . . . . . . . . . .   962
- 30.14   Mechanism of Abscission . . . . . . . . . . . . . . . . . . . . . . . . . . .   963
  - 30.14.1   Leaf Maintenance Phase . . . . . . . . . . . . . . . . . . . . .   964
  - 30.14.2   Abscission Induction Phase . . . . . . . . . . . . . . . . . . .   964
  - 30.14.3   Abscission Phase . . . . . . . . . . . . . . . . . . . . . . . . . .   964
- 30.15   Whole-Plant Senescence . . . . . . . . . . . . . . . . . . . . . . . . . . . .   964
Multiple-Choice Questions . . . . . . . . . . . . . . . . . . . . . . . . . . . . . . . . .   966
Suggested Further Readings . . . . . . . . . . . . . . . . . . . . . . . . . . . . . . . .   966

## Part IV   Stress Physiology

## 31   Abiotic Stress . . . . . . . . . . . . . . . . . . . . . . . . . . . . . . . . . . . . . . . . . .   969
Satish C Bhatla
- 31.1   Plant Responses to Abiotic Stress . . . . . . . . . . . . . . . . . . . . . .   970
  - 31.1.1   Decline in Crop Yield . . . . . . . . . . . . . . . . . . . . . . .   970
  - 31.1.2   Physiological Adjustment . . . . . . . . . . . . . . . . . . . .   970
  - 31.1.3   Resistance Mechanisms . . . . . . . . . . . . . . . . . . . . . .   970
  - 31.1.4   Alteration in Gene Expression Patterns . . . . . . . . . .   972

| | | | |
|---|---|---|---|
| 31.2 | Oxidative Stress | | 974 |
| | 31.2.1 | Production of Reactive Oxygen Species (ROS) | 974 |
| | 31.2.2 | Dual Role of ROS | 975 |
| | 31.2.3 | Cellular Antioxidative Defense System | 976 |
| | 31.2.4 | Ozone Exposure Leads to Oxidative Stress | 977 |
| 31.3 | Salt Stress | | 979 |
| | 31.3.1 | Disruption of Ionic Homeostasis due to Salt Stress | 980 |
| | 31.3.2 | Sodium Entry Through Symplastic and Apoplastic Pathways | 982 |
| | 31.3.3 | Salt Stress-Induced Signal Transduction Events | 984 |
| | 31.3.4 | Salt Stress Tolerance Mechanisms | 985 |
| | 31.3.5 | Accumulation and Sequestration of $Na^+$ and $Cl^-$ | 985 |
| | 31.3.6 | Expulsion of Salt Ions | 986 |
| | 31.3.7 | Accumulation of Organic Solutes | 987 |
| | 31.3.8 | Morphological Adaptations to Salt Stress | 987 |
| | 31.3.9 | Physiological Adaptations to Salt Stress | 989 |
| 31.4 | Water Deficit | | 990 |
| | 31.4.1 | Avoidance of Water Stress | 990 |
| | 31.4.2 | Developmental Adaptations | 990 |
| | 31.4.3 | Specialized Xeromorphic Features | 991 |
| | 31.4.4 | Implications of Water-Deficit Stress | 994 |
| | 31.4.5 | Water-Deficit Tolerance | 994 |
| 31.5 | High Temperature Stress | | 997 |
| | 31.5.1 | Thermal Injury | 1000 |
| | 31.5.2 | Developmental Adaptations to Heat Stress | 1000 |
| 31.6 | Low Temperature Stress | | 1002 |
| | 31.6.1 | Freezing Versus Chilling Stress | 1002 |
| | 31.6.2 | Low Temperature and Water Deficit | 1003 |
| | 31.6.3 | Plant Species of Warm Climates | 1003 |
| | 31.6.4 | Low Temperature Sensing Systems in Plants | 1003 |
| | 31.6.5 | Membrane Destabilization as a Response to Chilling and Freezing | 1004 |
| | 31.6.6 | Adaptive Mechanisms for Low Temperature Stress | 1006 |
| 31.7 | Flooding (Anaerobic) Stress | | 1009 |
| | 31.7.1 | Flooding Sensitivity of Plants | 1010 |
| | 31.7.2 | Methane Emissions from Wetlands | 1012 |
| | 31.7.3 | Adaptations of Wetland Plants | 1013 |
| | 31.7.4 | Ethylene Action in Anaerobic Stress | 1017 |

| | | | |
|---|---|---|---|
| 31.8 | | Signaling Pathways in Response to Abiotic Stress Conditions | 1018 |
| | 31.8.1 | Rapid Signaling Stress Sensors | 1019 |
| | 31.8.2 | Involvement of Transcriptional Regulatory Networks (Regulons) During Stress Acclimation | 1020 |
| | 31.8.3 | Acquisition of Systemic Acquired Acclimation (SAA) | 1020 |
| | 31.8.4 | Role of Epigenetic Mechanisms and Small RNAs in Stress-Response (Long-Term Stress Adaptive Mechanisms) | 1020 |
| | 31.8.5 | Regulation of Abiotic Stress Responses by Hormonal Interactions | 1021 |
| 31.9 | | Summary | 1022 |
| Multiple-Choice Questions | | | 1026 |
| Suggested Further Readings | | | 1028 |

## 32 Biotic Stress ........ 1029
Manju A. Lal, Renu Kathpalia, Rama Sisodia, and Rashmi Shakya

| | | | |
|---|---|---|---|
| 32.1 | | Interactions with Pathogens | 1031 |
| 32.2 | | Susceptibility and Resistance | 1037 |
| | 32.2.1 | Entry of Pathogen | 1038 |
| | 32.2.2 | Hypersensitive Response | 1039 |
| 32.3 | | Plant Defense Mechanisms | 1042 |
| | 32.3.1 | Pathogen- or Microbial-Associated Molecular-Pattern (PAMP/MAMP)-Triggered Immunity (PTI) | 1045 |
| | 32.3.2 | Effector-Triggered Responses | 1050 |
| | 32.3.3 | Signal Transduction | 1057 |
| | 32.3.4 | Systemic Acquired Resistance (SAR) | 1058 |
| | 32.3.5 | Phytohormones in Plant Defense Response to Pathogens | 1059 |
| | 32.3.6 | Pathogenesis-Related Proteins | 1063 |
| 32.4 | | Viruses as Plant Pathogens | 1063 |
| 32.5 | | Plant Responses to Herbivory | 1064 |
| | 32.5.1 | Defense Mechanisms in Plants Against Herbivory | 1068 |
| | 32.5.2 | Signal Transduction Triggering Plant Responses | 1071 |
| | 32.5.3 | Herbivore Response to Plant Defense | 1076 |
| | 32.5.4 | Nematodes | 1077 |
| 32.6 | | Plant-Plant Interactions | 1079 |
| | 32.6.1 | Parasitic Plants | 1079 |
| | 32.6.2 | Establishment of Contact Between Parasitic and Host Plants | 1080 |
| | 32.6.3 | Physiological Interactions Between the Parasite and Host | 1083 |
| | 32.6.4 | Competition for Resources | 1084 |

|  |  32.7 | Allelopathy | 1085 |
| --- | --- | --- | --- |
|  |  | 32.7.1 Molecular Mechanisms of Allelopathy | 1086 |
|  |  | 32.7.2 Plant Physiological and Biochemical Processes Affected by Allelochemicals | 1088 |
|  | Multiple-Choice Questions |  | 1093 |
|  | Suggested Further Readings |  | 1095 |

## Part V  Applied Plant Physiology

**33  Secondary Metabolites** .................................................. 1099
Satish C Bhatla

|  | 33.1 | Terpenes | 1101 |
| --- | --- | --- | --- |
|  |  | 33.1.1 Structure | 1106 |
|  |  | 33.1.2 Classification | 1107 |
|  |  | 33.1.3 Biosynthesis | 1107 |
|  |  | 33.1.4 Functions | 1110 |
|  |  | 33.1.5 Industrial Applications | 1115 |
|  |  | 33.1.6 Metabolic Engineering in Terpenes | 1116 |
|  | 33.2 | Phenolic Compounds | 1117 |
|  |  | 33.2.1 Functions | 1119 |
|  |  | 33.2.2 Biosynthesis | 1120 |
|  |  | 33.2.3 Classification | 1121 |
|  | 33.3 | Nitrogen-Containing Compounds | 1138 |
|  |  | 33.3.1 Alkaloids | 1139 |
|  |  | 33.3.2 Glycosides | 1148 |
|  |  | 33.3.3 Glucosinolates | 1160 |
|  |  | 33.3.4 Non-protein Amino Acids | 1161 |
|  | Multiple-Choice Questions |  | 1163 |
|  | Suggested Further Reading |  | 1166 |

**34  Plant Physiology in Agriculture and Biotechnology** ............ 1167
Satish C Bhatla

|  | 34.1 | Optimizing Nutrient and Water Uptake | 1167 |
| --- | --- | --- | --- |
|  |  | 34.1.1 Plant and Soil Elemental Analysis Facilitates Correct Application of Fertilizers | 1168 |
|  |  | 34.1.2 Managing Soil Acidity Through Liming for Better Nutrient Absorption by Plants | 1168 |
|  |  | 34.1.3 Green Manure as an Alternative to Chemical Fertilizers | 1169 |
|  |  | 34.1.4 Antitranspirants for Enhancing Water Stress Tolerance in Ornamental Plants | 1171 |
|  | 34.2 | Eradicating Weeds | 1172 |
|  |  | 34.2.1 Parasitic Weeds in Crop Fields | 1172 |
|  |  | 34.2.2 Weed Control by Introducing Glyphosate-Resistant Crops | 1172 |
|  |  | 34.2.3 Blockers of Photosynthetic Electron Transport as Potential Herbicides | 1173 |

| | 34.3 | Making Plants More Energy Efficient | 1174 |
|---|---|---|---|
| | 34.4 | Plant Growth Regulators (PGRs) in Agriculture and Horticulture | 1174 |
| | | 34.4.1 Some Specific Applications of PGRs in Agriculture and Horticulture | 1178 |
| | 34.5 | Biotechnological Approaches | 1180 |
| | | 34.5.1 Transgenic Fruits | 1180 |
| | | 34.5.2 Genetic Engineering and Conventional Breeding Approaches to Reduce Caffeine Content in Coffee and Tea | 1181 |
| | | 34.5.3 Bulk Production of Secondary Metabolites | 1182 |
| Multiple-Choice Questions | | | 1186 |
| Suggested Further Readings | | | 1188 |

**Glossary** ................................................................. 1189

# About the Authors, Co-contributors, and Technical Support

## Authors

**Professor Satish C Bhatla** is serving in the Department of Botany, University of Delhi, since 1985. After obtaining a specialized Master's degree in Plant Physiology in 1976, and Ph.D. in 1980, Professor Bhatla undertook post-doctoral training as a Fellow of Alexander von Humboldt Foundation (Germany), in the University of Heidelberg, Freie University Berlin, and University of Freiburg in Germany. He has been teaching and undertaking research in plant physiology for the past 38 years. He has supervised 22 Ph.Ds. so far and has international collaborations with the University in Bonn (Germany), Minsk (Belarus), and Tel Aviv (Israel). Professor Bhatla is a member of Steering Committee of International Society of Plant Signaling and Behavior and is an Associate Editor of its journal. He has more than 90 international research publications to his credit. His recent past and current areas of research include nitric oxide signaling, oil body mobilization mechanisms, salt stress tolerance mechanisms in plants, seed germination, and the regulation of adventitious and lateral roots. He is also a Fellow of Indian National Science Academy (India) and has been Dean of Faculty of Science and Head of the Department of Botany at the University of Delhi. Contact: bhatlasc@gmail.com

**Dr. Manju A. Lal** obtained her Ph.D. in plant physiology from Indian Agricultural Research Institute at New Delhi in 1976. She specialized in abiotic stress in relation with nitrogen metabolism in crop plants. Dr. Manju A. Lal has taught plant physiology to undergraduate students for more than four decades. She superannuated in December 2017 as Associate Professor from the Department of Botany, Kirori Mal College, University of Delhi. Beyond classroom teaching, Dr. Manju A. Lal has also contributed significantly in disseminating information on key topics in plant physiology to undergraduate students through e-learning projects as well. Contact: lalmanjua@gmail.com

## Co-contributors

**Dr. Renu Kathpalia** obtained her Ph.D. from the University of Delhi in 1990. She is an associate professor and has been teaching plant physiology to undergraduate students at Kirori Mal College, University of Delhi, for the last 28 years. Contact: renukathpalia@gmail.com. (Chapters 1, 2, 28, 32)

**Dr. Geetika Kalra** obtained her Ph.D. from the University of Delhi in 1995. She specializes in physiology of adventitious root formation. She is an associate professor and has been teaching undergraduate students at Acharya Narendra Dev College, University of Delhi, since 2005. Contact: geetskalra18@gmail.com. (Chapters 16, 17, 18, 25, 30)

**Dr. Rama Sisodia** obtained her Ph.D. from the University of Delhi in 2003. She specializes in hormonal regulation in flowering. Subsequently she worked as a research associate in the University of Delhi. She has been teaching undergraduate students at Maitreyi College, University of Delhi since 2008. Additionally, Dr. Rama coordinated the development of e-learning material for undergraduate students. Contact: ramasisodia@gmail.com. (Chapters 24, 29, 32)

**Dr. Rashmi Shakya** obtained her Ph.D. from the University of Delhi in 2008. She specializes in physiology of reproduction in plants. She undertook advanced training in bioinformatics and proteomics from the University of Edinburgh (UK). Dr. Rashmi has been teaching undergraduate students at Miranda House College, University of Delhi, as assistant professor since 2006. Contact: rashmishakya@gmail.com. (Chapters 6, 26, 27, 32)

About the Authors, Co-contributors, and Technical Support xxxiii

## Technical Support

Dr. Prachi Jain (Research Associate, Indo-Israel Research Project, Department of Botany, Delhi University)

Dr. Harmeet Kaur (Assistant Professor, SGTB Khalsa College, Delhi University)

Dr. Sunita Yadav (Assistant Professor, Sri Venkateshwara College, University of Delhi)

Dr. Neha Singh (Research Associate, Indo-Israel Research Project, Department of Botany, Delhi University)

Dr. Dhara Arora (Post-Doctoral Fellow, Department of Botany. Delhi University)

Monika Keisham (Members of research group, c/o Prof. Satish C Bhatla, University of Delhi)

Aditi Tailor (Members of research group, c/o Prof. Satish C Bhatla, University of Delhi)

Archana Kumari (Members of research group, c/o Prof. Satish C Bhatla, University of Delhi)

Mansi Gogna (Members of research group, c/o Prof. Satish C Bhatla, University of Delhi)

Mr. Harsh Sharma (Illustrator)

# Part I

# Transport of Water and Nutrients

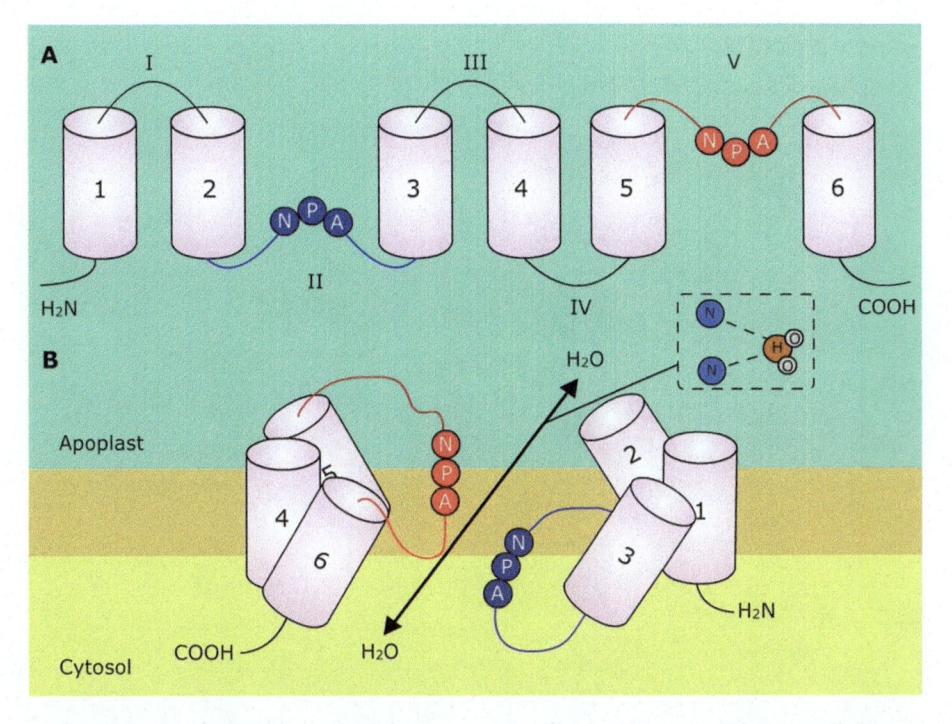

Mechanism of water transport across plasma membrane through aquaporins. More details are provided in Chap. 3, Sect. 3.7.4, Fig. 3.10

Chapter 1 Plant Water Relations
Chapter 2 Mineral Nutrition
Chapter 3 Water and Solute Transport

# Plant Water Relations

Renu Kathpalia and Satish C Bhatla

Water is one of the most important constituents of life. Chemically, water is the hydride of oxygen. Oxygen, being more electronegative, exerts a strong attractive pull on its electrons. This unequal attraction results in small positive charge on two hydrogen molecules and a small negative charge on the oxygen molecule. The two lone pairs of electrons of the oxygen molecule result in bending of water molecule. The partial charges on oxygen and hydrogen molecules result in high electric dipole moment and polarity of water molecule. The distinct physical and chemical properties of water, namely, cohesion, surface tension, high specific heat, high heat of vaporization, lower density of ice, and solubility, are due to hydrogen bonding between water molecules (Fig. 1.1). Water forms solution with large number of compounds. It is thus usually referred as a universal solvent. The solvent action of water is of tremendous importance for the cells. All cells require water, dissolved ions, and sugars to survive. Cells undergo oxidation-reduction reactions, and water serves as the medium in which all these reactions are carried out. Plants, being immobile and autotrophic, have to depend on the supply of water and minerals from the soil and carbon dioxide and light from the atmosphere. Transport of water and minerals in the vascular strands is based on the differences in pressure and concentration gradients of both solutes and the solvent (water). The transport of minerals and water from the soil to xylem and from xylem to substomatal cavity is referred as **short-distance transport**. Once water enters the xylem elements, it is transported up to 100 m or more by the transpirational pull created in the leaves. Therefore, there is need to have an essential **long-distance transport** by two different transport systems involving transport in opposite directions. This chapter shall focus on various mechanisms of short- and long-distance transport in plants.

© Springer Nature Singapore Pte Ltd. 2018
S. C Bhatla, M. A. Lal, *Plant Physiology, Development and Metabolism*,
https://doi.org/10.1007/978-981-13-2023-1_1

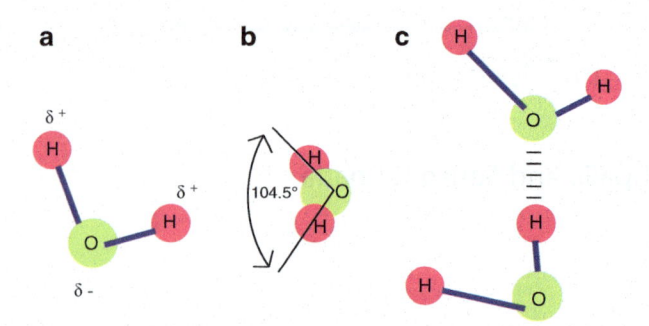

**Fig. 1.1** Structure and unique features of water. (**a**) The two hydrogen atoms in water molecule carry partial positive charges, whereas oxygen atom carries a partial negative charge. (**b**) Water molecule has a bent geometry, thereby causing asymmetric distribution of charge. So, water is polar. (**c**) Neighboring water molecules are linked through hydrogen bonding between oxygen atom of one with hydrogen of the other

## 1.1 Water Potential and Its Components

The **free energy** of any molecule determines its capacity to perform work. The chemical potential is the free energy per mole of any substance in a chemical system. Chemical potential is a relative term. It is expressed as the difference between the potential of a substance in a given state and the potential of the same substance in a standard state. The chemical potential of water (plant physiologists use the term **water potential**) is the free energy of water. It is the chemical potential of water divided by partial molar volume. In other words, water potential is a measure of free energy of water per unit volume ($Jm^{-3}$). In terms of pressure units, water potential is expressed as MPa (megapascal). The lower the water potential of the plant, the greater is its ability to absorb water and vice versa. It also helps to measure the water deficit and stress in plants. Water potential is not an absolute value and is symbolized by the Greek letter $\Psi_w$ (psi). Water potential of pure water is maximum and its value is zero at the atmospheric pressure. In a living cell, water potential refers to the sum of the following components:

$$\Psi_w = \Psi_s + \Psi_p + \Psi_m + \Psi_g$$

where $\Psi_w$ is the water potential, $\Psi_s$ solute/osmotic potential, $\Psi_p$ pressure potential, $\Psi_g$ gravitational potential, and $\Psi_m$ matric potential.

### 1.1.1 Solute Potential

**Osmotic potential ($\Psi_s$)** or **solute potential** refers to the effect of solutes on water potential. Solutes reduce the free energy of water. The addition of solutes also changes the colligative properties of solutions. Macromolecules, like proteins,

## 1.1 Water Potential and Its Components

nucleic acids, and polysaccharides, have far less effects on the solute potential as compared to their respective monomers. The cell stores fuel as macromolecules (starch in plant cells or glycogen in animal cells), rather than glucose or other simple sugars to avoid drastic changes in osmotic potential.

For dilute solutions, the osmotic potential can be calculated by using Van't Hoff equation

$$\Psi_s = -iCRT$$

where $i$ is the ionization constant (for sucrose the value is 1, but for ionic solutes, it is multiplied by the number of dissociated particles), $C$ solute concentration in moles, $R$ gas constant (8.314 J.mol$^{-1}$), and $T$ temperature in Kelvin (273 + $°$C of solution). The −ve sign indicates that dissolved solutes reduce the water potential of the solution. This equation is valid for dilute solutions only. Using this equation, water potential and solute potential of a cell or tissue can be measured (Box 1.1).

---

**Box 1.1: Methods to Measure Osmotic Potential**

**Plasmolytic method**: A series of sucrose solutions with different osmotic potentials are prepared. Epidermal peels of Rhoeo, red onion containing anthocyanin, are placed in different solutions, and after about half an hour, the peels are observed under the microscope. Some of the peels from different solutions will show the cells to be fully turgid, or they become flaccid (plasmolyzed). The number of plasmolyzed and unplasmolyzed cells is counted and their percentage calculated. In some epidermal peels, 50% of the cells are just beginning to show the signs of plasmolysis (*incipient plasmolysis*). During incipient plasmolysis, the turgor pressure of the cell is zero, and the osmotic potential of the cell contents is equal to the water potential and osmotic potential of the external solution.

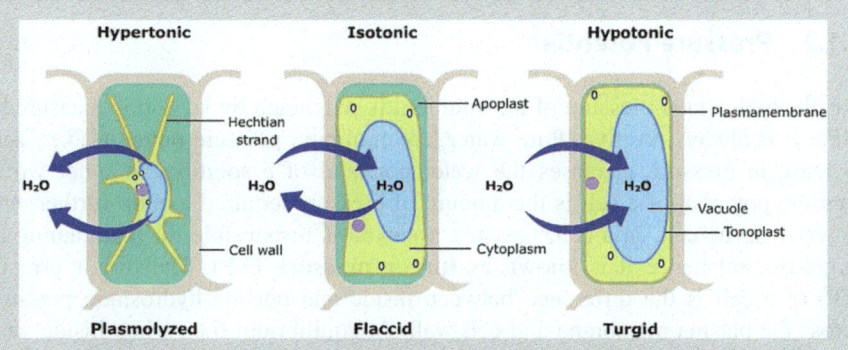

Cell placed in hypertonic solution (left) undergoes plasmolysis due to loss of water from vacuole. In an isotonic solution (center), cell remains flaccid, but in a hypotonic solution, the cell is fully turgid (right). Plasma membrane is

(continued)

**Box 1.1** (continued)

connected to cell wall. These connections are retained in the form of thin strands at the time of plasmolysis. These are known as *Hechtian strands*.

**Methods to Measure Water Potential**

**Gravimetric method**: This method is the same as volumetric method. It involves the placement of pre-weighed plant tissue (potato tissue cylinders) into graded series of solutions of sucrose or other **osmoticum** (osmotically active solutes) at a defined osmotic potential. After 1 h of incubation, Tissue weight gain or loss is plotted against the water potential ($\Psi_w = \Psi_s$) of each solution. The water potential of the solution in which there is no net gain is equal to that of tissue.

**Charadakov's or Falling drop method**: Two series of sucrose solutions (ranging from 0.15 to 0.5 molal, in increments of 0.5 molality) are placed in two sets of test tubes. An equal amount of tissue is placed in the two sets of test tubes. In another set, equal and minute amount of methylene blue is added. After half an hour, incubated tissue is removed; a small drop of respective control solution is added in test solution. If the drop rises in the test solution, this indicates that the test solution is concentrated, i.e., water from the solution has entered into the tissue, making it more concentrated. If the drops fall, this indicates that water has come out from the tissue making the tissue lighter than the control solution. On the other hand, if drops float, it means there is no net movement of water. At this point, water potential of the tissue is equal to that of the sucrose solution of that particular concentration.

### 1.1.2 Pressure Potential

It is the hydrostatic pressure of the solution. It is denoted by $\psi_p$ and is measured in MPa. It is always positive. Pure water has minimum pressure potential, i.e., zero. Increase in pressure increases the water potential of a solution. In other words, pressure potential of a cell is the amount of pressure required to stop further entry of water in the cell. In a cell, pressure potential is responsible for maintaining the turgidity, and hence it is known as **turgor pressure (TP)**. The turgor pressure (TP) of a cell is the difference between inside and outside hydrostatic pressures across the plasma membrane and cell wall. At equilibrium (i.e., when inside water potential is equal to outside water potential), TP will be equal to the difference in internal solute potential and that of external solute potential.

$$TP = \Psi_{s(inside)} - \Psi_{s(outside)}$$

## 1.1 Water Potential and Its Components

In plants a fully turgid cell experiences an equal and opposite pressure, known as **wall pressure**. The presence of cell wall in plant cells allows it to withstand a wide range of osmotic variations. In contrast, an animal cell can only survive in an **isotonic** solution. The plant cell placed in pure water swells but does not burst. The negative osmotic potential in vacuolated solution (cell sap) causes the movement of water only in the cell. Due to entry of water, plasma membrane is pressed against the cell wall.

### 1.1.3 Gravitational Potential

Gravitational pull has the same effect on water potential as the effect of pressure, and it is expressed as **gravitational potential** ($\Psi_g$). It is the sum of three forces: gravitational acceleration, height of water column, and density of water. When dealing with water transport at the cell level, the gravitational component is usually omitted because it is negligible as compared to solute potential and pressure potential. However, gravitational pull definitely affects water movement in tall trees.

### 1.1.4 Matric Potential

It is expressed as the adsorption affinity of water to colloidal substances and surfaces in plant cells. **Matric potential** is negligible in a hydrated cell but is of considerable importance in dehydrated cells and tissues such as seeds and desert plants. Under these conditions, water exists as a very thin layer bound to solid surfaces by electrostatic interactions. These interactions are not easily separated into their effects on solute and pressure potential and thus sometimes combined into matric potential. The adsorption of water by hydrophilic surfaces is known as hydration or **imbibition** (Box 1.2). Matric potential is measured in the same unit as water potential. A dry or hydrophilic substance has extremely negative potential (as low as −300 MPa). Figure 1.2 illustrates the changes in water potential, pressure potential, and osmotic potential as the cell volume changes. The relationship is expressed considering that there is no movement of solute into or out of the cell, and the cell volume changes because of movement of water only. It describes the status of a mature cell when placed in solutions of different osmotic potentials. There is gain or loss of water from the cells, but the total quantity of solute within the cell remains constant. Water potential and osmotic potential are always negative in a cell. In a fully turgid cell, the water potential becomes zero even in the presence of solute. As in plants, cell wall exerts pressure, and with the dilution of the cell sap, the water potential at that pressure becomes zero. The plant cell volume changes, but small change in it causes large changes in turgor pressure. A 10% increase in cell volume due to uptake of water gives rise to a fully turgid cell with a small change in solute potential. But

**Box 1.2: Imbibition**
It is a special type of diffusion involving adsorbents. During **imbibition**, the molecules of a liquid or gas diffuse into a solid substance, causing it to swell. In biological systems, it is the liquid water or water vapor diffusing into the colloidal matrix, causing it to swell. There is a need to have strong binding force between the molecules of the imbibant and the substance being imbibed. Such an intermolecular force exists between cellulose and water. If dry plant material is placed in water, it swells and results in considerable increase in volume. For imbibition to occur, a water potential gradient must exist between the surfaces of the adsorbent and liquid being imbibed, and an affinity is also essential between the components of the adsorbent and the imbibed substance. In some dry seeds, water potential of up to −900 bars has been observed. Therefore, dry seeds immediately imbibe water, and this is the first step in seed germination. This is followed by swelling of seeds and mobilization of reserve food materials and, finally, the emergence of root and shoot. The great reduction in the potential of the imbibed water is a result of the greatly decreased kinetic energy. The lost energy appears as heat. Imbibition invariably accompanies release of heat. Unlike osmosis imbibition does not require semipermeable membrane. Because of the high matric force of the cellulose, cell wall still retains water after complete desiccation of the cell. Only complete desiccation of a plant tissue, as by drying in an oven at relatively high temperature, will remove all the water imbibed in the cellulosic cell walls. Cell wall of dry seeds has minimum water, but it germinates only when it acquires water by imbibition. Imbibition also takes place in plants and animal products such as wood, cork, and sponge. In ancient times, this concept was used for stone-cutting technique. People used to place dry wooden logs into holes in rocks. The swelling of these logs would result in breaking the rocks. In nature, plants survive in the crevices of rocks because swollen seeds create space in the rocks and allow roots to grow.

growing cells do not show this relationship because, in these cells, cell wall expands and cell wall pressure decreases.

## 1.2 Intercellular Water Transport

All living cells contain approximately 60–95% of water, and water is required for their growth and reproduction. Even the dormant cells and tissues also have 10–20% of water. Water flow is a passive process, and it moves in response to physical forces toward regions of low water potential or low free energy. The intercellular or short-distance water transport takes place through diffusion, mass flow, or osmosis.

## 1.2 Intercellular Water Transport

**Fig. 1.2** The relationship between change in cell volume and change in water potential, solute potential, and pressure potential. In a fully turgid cell, the water potential, pressure potential, and solute potential are maximum. The pressure potential and solute potential are equal and opposite at zero water potential. $V_i$, initial volume; $V_{eq}$, volume at equilibrium

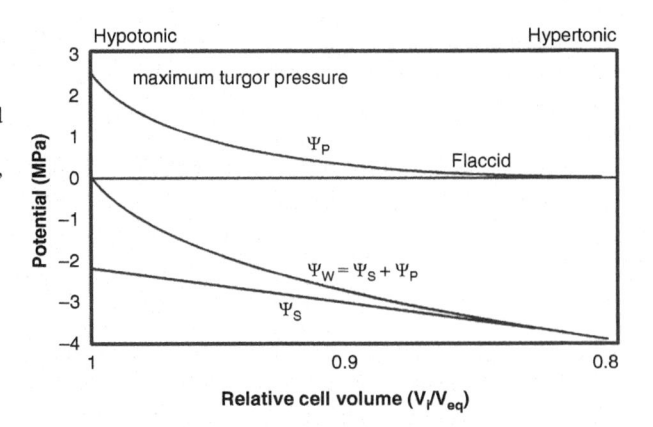

### 1.2.1 Diffusion

It is a physical process where movement of any substance or molecule takes place from the region of its higher concentration to the region of its lower concentration due to kinetic energy. **Diffusion** can take place in all three phases of matter. The rate of diffusion is affected by temperature, molecular density, diffusion medium, and chemical potential gradient. In plants, diffusion occurs in stomata to facilitate the exchange of carbon dioxide, oxygen, and water vapors between leaf cells and external atmosphere. Diffusion also plays an important role in gas exchange in lenticels present in the stem. The apoplastic and symplastic pathway of intercellular transport also involve diffusion. Imbibition during seed germination is also a special type of diffusion.

The rate of diffusion is directly proportional to the concentration gradient ($\Delta C_S/\Delta X$), which can be expressed as (Fick's equation)

$$Js \propto \left(\frac{\Delta C_s}{\Delta X}\right)$$

$$Js = -Ds\left(\frac{\Delta C_s}{\Delta X}\right)$$

where $Js$ is the flux density, $\Delta C_S$ is the difference in the concentration of substance in two media, and $\Delta X$ is the distance travelled in cm by substance, i.e., substance (s) crossing a unit area per unit time. It is expressed in $mol.m^{-2}.s^{-1}$. $Ds$ is the **diffusion coefficient** that measures the movement of a particular substance through a medium. The diffusion coefficient is characteristic of a substance and depends on the medium in which diffusion is taking place. The larger molecules have smaller diffusion coefficient, and diffusion in air is much faster than diffusion in a liquid medium. The negative sign in the equation is due to movement taking place down the gradient. In other words, **Fick's first law of diffusion** states that a substance diffuses faster if there is more difference in its concentration gradient or the diffusion coefficient is higher. Diffusion is rapid if the distance is small but it is extremely slow over long distances.

The time taken ($t_c$) for diffusion to reach one half of the concentration from the starting point is given by the formula

$$t_c = \frac{1}{2} = \frac{(\text{Distance})^2}{Ds} \times K$$

where $K$ is constant which depends on the shape of the system. For example, in a cell of 50 μm diameter with diffusion coefficient of glucose being $10^{-9}$ m$^2$.s$^{-1}$, the diffusion of glucose molecule from one point to another to reach half the concentration will take 2.5 s.

$$t_c = \frac{1}{2} = \frac{(50 \times 10^{-6} \text{m})^2}{10^{-9} \text{m}^2.\text{s}^{-1}} = 2.5 \text{ s}$$

But if we consider the diffusion of glucose molecule to 1 meter distance, it will take up to 32 years.

$$t_c = \frac{1}{2} = \frac{(1 \text{ m})^2}{10^{-9} \text{m}^2.\text{s}^{-1}} = 32 \text{ years}$$

At intercellular level, the movement of water can be due to diffusion, but it is not a viable option for long-distance travel.

### 1.2.2 Mass Flow

**Mass flow** or **bulk flow** is pressure-driven movement of molecules. In contrast to diffusion and osmosis, mass flow is independent of solute concentration. The protoplasm streaming in plant cells during active growing season is an example of mass flow. The rate of flow of protoplasm is approximately 25 mm.h.$^{-1}$ at room temperature. The upward longitudinal transport of water and dissolved substances in the xylem elements and in phloem transport takes place through mass flow. The xylem and phloem strands undergo many development changes to increase their conductance for bulk flow. The rate of flow in xylem vessels can be calculated by **Hagen-Poiseuille's equation**:

## 1.2 Intercellular Water Transport

$$\text{Volume flow rate} = \frac{dV}{dt} = \frac{\pi.\Delta P.R^4}{8\eta L}$$

where $\Delta P$ is the pressure difference, $R$ is the radius of xylem vessel, $\eta$ is the viscosity of xylem sap, and $L$ is the length of xylem vessel. The bulk flow rate is very sensitive to the radius of the vessel. If radius is doubled, then the rate of flow of water in xylem vessels will increase by 16-fold. Mass flow has a significant role in long-distance transport of water and minerals. During intercellular transport, mass flow faces many barriers, viz., cell wall, plasma membrane, tonoplast, and plasmodesmata. Vascular cells have adopted many changes for bulk movement of water, e.g., removal of end walls in xylem vessels, enlargement of plasmodesmata, and depletion of protoplasm.

### 1.2.3 Osmosis

It is a special type of diffusion which takes place in solvents through semipermeable membrane. **Osmosis** is a biological process where the solvent molecules move from their higher concentration (lower solute concentration) to lower concentration (higher solute concentration) through a semipermeable membrane (Box 1.3). It is controlled both by concentration and pressure gradient. The rate of osmosis can be increased by the addition of osmotically active substances and can be measured by **osmometer.** Osmosis is the process by which water is transported into and out of the cell. The growth, development, and turgidity of the cells are maintained by the process of osmoregulation (Box 1.4). The process of osmosis can be stopped or reversed by applying pressure. This is called **reverse osmosis**. This process is now commercially used for purifying water (Box 1.5).

> **Box 1.3: Demonstration of the Phenomenon of Osmosis**
> To understand the process of osmosis, pure water is taken in a beaker, and a thistle funnel containing sucrose solution is inserted in it, with the dipped end of the funnel being sealed with a semipermeable membrane. Keep this apparatus undisturbed for some time. Depending on the concentration of sucrose, water will move from beaker into the funnel by osmosis. Since pure water has maximum water potential, it will move from its higher potential to its lower potential, i.e., water will move in the funnel. In other words, water moves from **hypotonic** to **hypertonic** solution or from the direction of lower solute potential to higher solute potential.

(continued)

**Box 1.3** (continued)

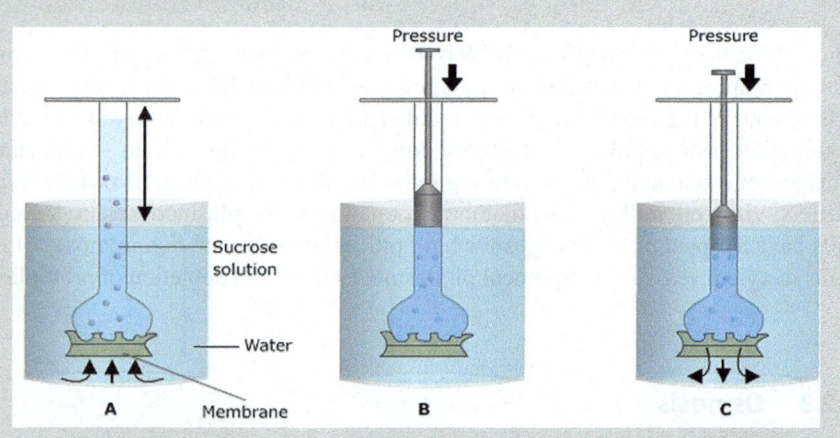

Water will keep on moving upward as long as the water potential difference exits between the two solutions. However, osmosis can be stopped by putting pressure. If pressure is continuously increased, water moves out from thistle funnel against the water potential difference.

**Box 1.4: Significance of Osmosis**
All living cells are like bags carrying solutions in semipermeable membranes. The survival of a cell depends on the ability to maintain intracellular solute concentration. Living organisms develop many strategies to maintain a steady influx of water by osmosis. Some single-celled organisms, such as *Paramecium*, extrude water through special organelles called *contractile vacuoles*. Marine organisms adjust their internal concentration of solutes to match that of sea water. Many terrestrial animals have a circulating fluid which is isotonic and cells are bathed in it. The blood flowing in our body contains a high concentration of the protein albumin which elevates solute concentration of the blood to match with that of cell cytoplasm. The lack of cell wall in animal cells makes it more susceptible to osmotic lysis. Storing of blood, injecting of blood and drugs intravenously, and maintaining the pH of blood are very challenging and require maintaining osmotic concentration. The RBCs immediately burst if there is a slight change in their osmotic concentration. Most of the plant cells are hypertonic to their immediate environment. They contain a

(continued)

## 1.2 Intercellular Water Transport

**Box 1.4** (continued)
central vacuole which has high solute concentration, resulting in high hydrostatic pressure inside the cell. This hydrostatic pressure, known as turgor pressure, keeps the cell turgid and maintains its shape. During osmotic adjustment, vacuole losses solutes and thus protects the cells from turgor loss. The presence of rigid cell wall in the plant cell makes it sturdier, and it can withstand adverse environmental conditions. The plasmolyzed cells remain connected to the cell wall with the help of strands called **Hechtian strands.**

The transport of water and minerals from roots to leaves and carbohydrates from leaves to roots is dependent on osmosis. The opening and closing of stomata, touch response in plants, movement to catch insect by insectivorous plants are all regulated by osmosis. **Dialysis**, one of the major vital processes, where blood is purified by artificial kidney, is also based on the osmosis. Osmosis is used for preserving fruits, jams, and pickles. Preservation requires dehydration, which is achieved by the use of salt or sugar solution. The use of hypertonic solution causes the movement of water from the food tissue and protects it from microbes. During food preservation, more than 50% of the tissue water is removed by osmotic dehydration.

**Box 1.5: Reverse Osmosis**
Water contains significant amount of salts even after purification with high-grade filters, which poses health hazards. **Reverse osmosis** (RO) technology is used to demineralize or deionize water for making it potable. This process is reverse of osmosis. In reverse osmosis, water moves from its lower potential to higher potential under pressure through semipermeable membrane. Since it involves semipermeable membrane and the direction of flow is reverse, it is known as reverse osmosis. The RO system consists of membrane filters (usually 5) which do cross filtration. The solution passes through filters with two outlets. The clean or deionized water having few salts goes one way, and the rest of it comes out from the waste pipe. Cross filtration helps in cleaning the filters so that there are no depositions on the filter. The cleaned water is free from dissolved salts, organic matter, colloids, bacteria, and viruses. In addition to solvent, solute of smaller size also moves through semipermeable membrane.

(continued)

**Box 1.5** (continued)

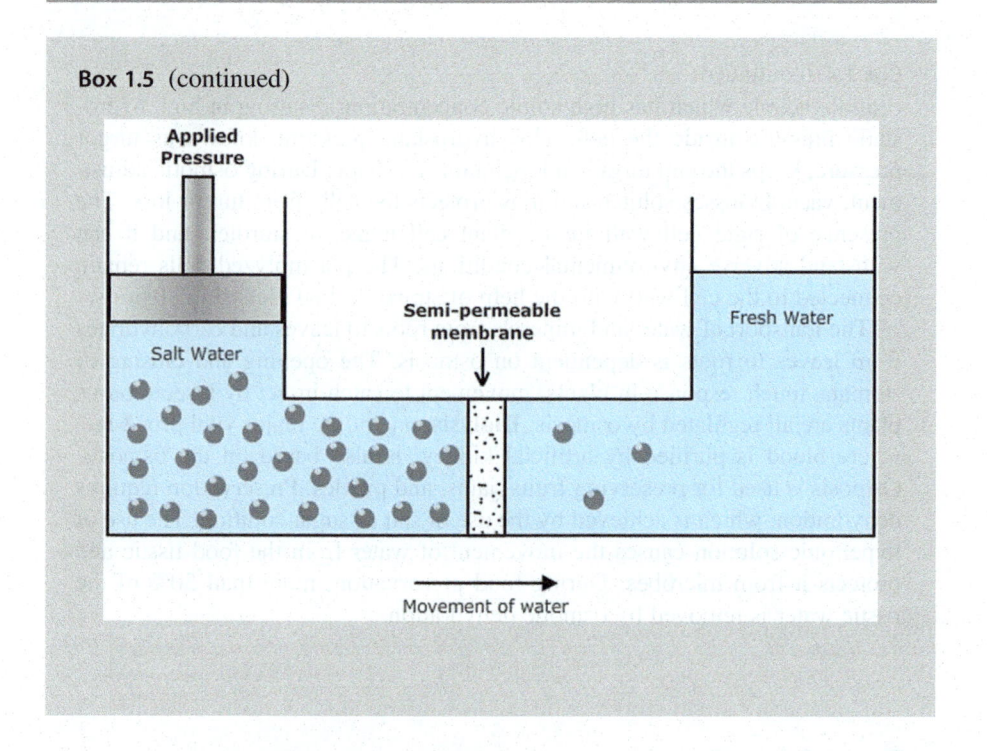

*Figure: Diagram showing applied pressure on salt water side, separated from fresh water by a semi-permeable membrane, with movement of water indicated from fresh water to salt water side.*

## 1.3 Short-Distance Transport

The movement of water from root to upper foliar parts of the plant requires integration of many levels of transport. The long-distance transport of water in xylem is supplemented by short-distance transport of water from the root to cortex and ultimately to the xylem. Basically, there are three levels of water transport in plants. The first level of water transport refers to the entry of water from the soil into the root cells, through the plasma membrane. At this level, the selective permeability of the plasma membrane regulates the movement of solutes and the solvent between the cell and extracellular solutions. The molecules tend to move down their concentration gradient. This has been already discussed in the above sections. In addition to active and passive transport, proteins present in the membrane speed up the movement across the membrane. The second level of transport is from root epidermal cells to the innermost layer of the cortex. This is referred as short-distance transport, and it includes **apoplast**, **symplast**, and **transcellular pathways** including transporters, channels, and plasmodesmata. The short-distance transport also includes water transport from xylem in the leaf veins to substomatal cavity. The third level of transport is the long-distance transport in xylary elements.

## 1.3.1 Water Absorption by Roots

The area covered by the plant roots is generally very high. It is, therefore, appropriate to call root as *the hidden half* of the plant body. Growth and development of roots can be analyzed by special subterranean camera, known as **rhizotrons.** The region of maximum water uptake is the region of active root hair development. The root hair density is an important factor in water uptake, and its extent of proliferation depends on the ecological conditions in which plants grow. The influx of water into roots is only through the apical zone just behind the zone of elongation of primary root where the root hair density is maximum (Fig. 1.3). Increased suberization in the older parts does not allow entry of the water. Water passes through the lipid layer (despite nonpolar in nature of lipids) due to water potential gradient between root cells and soil. As long as the water potential of root cells is more negative than that of soil, water is continuously being taken into the plant. Depending on the requirement by the plant, rate of water absorption keeps changing. The water absorption can be active (ATP is involved) or passive (through osmosis), or it can be facilitated by special membrane channels for water transport, mainly through aquaporins.

Once water and minerals are absorbed by the roots, they follow three pathways to reach vascular tissue of the root. These are apoplastic, symplastic, and transcellular pathways. Water may follow one or more possible pathways across the root depending on the requirement of water by the plant.

The diffusion of solute or solvent across the lipid membrane or any porous structure depends on the **permeability coefficient** ($P_s$). It is constant and depends on the type of membrane and the size of molecule to be diffused ($m.s^{-1}$). Thus, diffusion flux ($J_s$; $mol.m^{-2}.s^{-1}$), driven by a concentration gradient ($\Delta C_s$) of uncharged molecules across a membrane, is given by the equation

$$J_s = P_s \times \Delta C_s$$

Thus, if molecules of equal size have to cross a membrane, the one with greater solubility in lipids will pass more quickly into the cell. On the other hand, if

**Fig. 1.3** The graph showing that maximum uptake of water in plants is from the small portion of root which is in between the zone of root growth and the zone of maturation (suberin gets deposited in the walls of root cells), i.e., zone of elongation

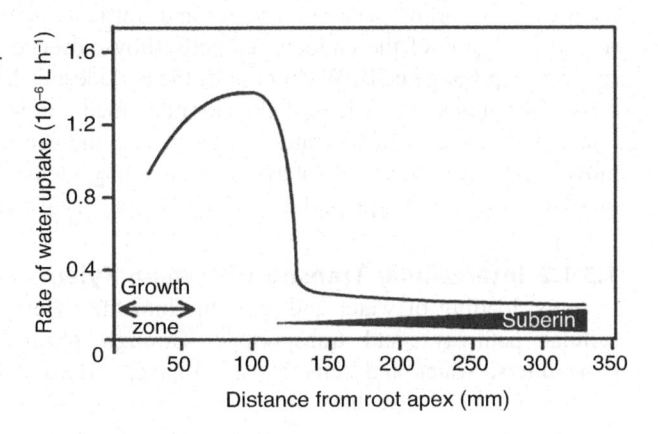

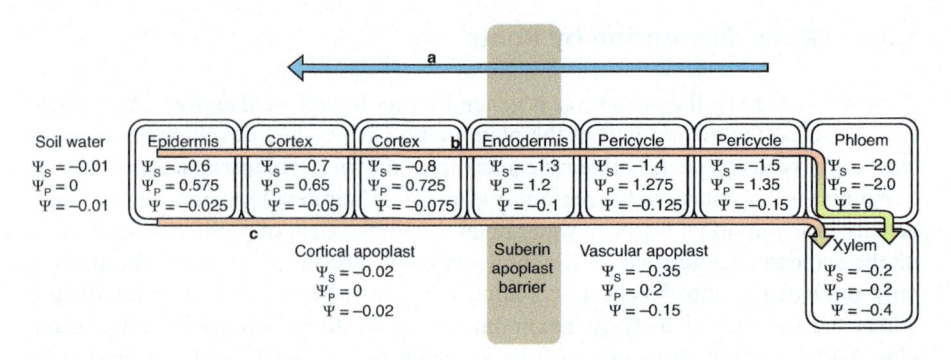

| | Epidermis | Cortex | Cortex | Endodermis | Pericycle | Pericycle | Phloem |
|---|---|---|---|---|---|---|---|
| Soil water $\Psi_s = -0.01$ $\Psi_p = 0$ $\Psi = -0.01$ | $\Psi_s = -0.6$ $\Psi_p = 0.575$ $\Psi = -0.025$ | $\Psi_s = -0.7$ $\Psi_p = 0.65$ $\Psi = -0.05$ | $\Psi_s = -0.8$ $\Psi_p = 0.725$ $\Psi = -0.075$ | $\Psi_s = -1.3$ $\Psi_p = 1.2$ $\Psi = -0.1$ | $\Psi_s = -1.4$ $\Psi_p = 1.275$ $\Psi = -0.125$ | $\Psi_s = -1.5$ $\Psi_p = 1.35$ $\Psi = -0.15$ | $\Psi_s = -2.0$ $\Psi_p = -2.0$ $\Psi = 0$ |

Cortical apoplast
$\Psi_s = -0.02$
$\Psi_p = 0$
$\Psi = -0.02$

Suberin apoplast barrier

Vascular apoplast
$\Psi_s = -0.35$
$\Psi_p = 0.2$
$\Psi = -0.15$

Xylem
$\Psi_s = -0.2$
$\Psi_p = -0.2$
$\Psi = -0.4$

**Fig. 1.4** The entry and pathway of water movement from and across the root. None of the root cells and apoplast fluid are in equilibrium. Water and solute potential gradient shown are responsible for the flow of water. Arrow "a" shows sucrose movement which maintains the solute potential across the root. The symplastic pathway is shown by arrow "b" and apoplastic by arrow "c"

molecules of equal solubility have to cross the membrane, the smaller ones penetrate faster. If diffusion flux is high, then permeability of the membrane will be more. The permeability of plasmodesmata is 10,000 times more than that of plasma membrane when the size of the molecule is less than 700 Dalton. The plasmodesmatal connections present in different regions shows different *limits of size exclusion* (ranging from molecule of 1 to 65 kDa). In roots, the size of molecules passing the plasmodesmatal connection is maximum, i.e., 65 kDa. It increases the flow of water and solutes between the adjacent cells. None of the root cells and apoplast fluids are in equilibrium (Fig. 1.4). Macromolecules, like RNA, proteins, and viruses, can also move through plasmodesmata.

### 1.3.1.1 Movement of Water Across Endodermis

Once water reaches the endodermis, further passage through the cell wall (apoplastic) is blocked by Casparian strips. The cells of endodermis have connecting walls embedded with the waterproof material—suberin. This layer blocks the apoplastic pathway and controls water and nutrient flow to the xylem. The Casparian band does not allow passage of water and nutrients across it into the deeper cells (Fig. 1.5). Some of the endodermal cells show absence of thickenings. These are referred as **passage cells**. Water crosses the endodermis layer through these passage cells or through symplasm. Radial water movement through cortical cells is fast, and transport rate increases as water moves toward the central stele. The rate of water movement from xylem parenchyma to tracheary elements (tracheids and vessels) increases because of aquaporins, which plays a very pivotal role in radial transport.

### 1.3.1.2 Intercellular Transport Between Xylem and Nonvascular Cells

In roots, loading of water and nutrients into the xylem strands involves multiple cellular pathways and transporters. Through plasmodesmata and membrane transporters, water and minerals are transported to xylem parenchyma. Xylem parenchyma cells also have highly active transporters. Parenchyma surrounding

## 1.3 Short-Distance Transport

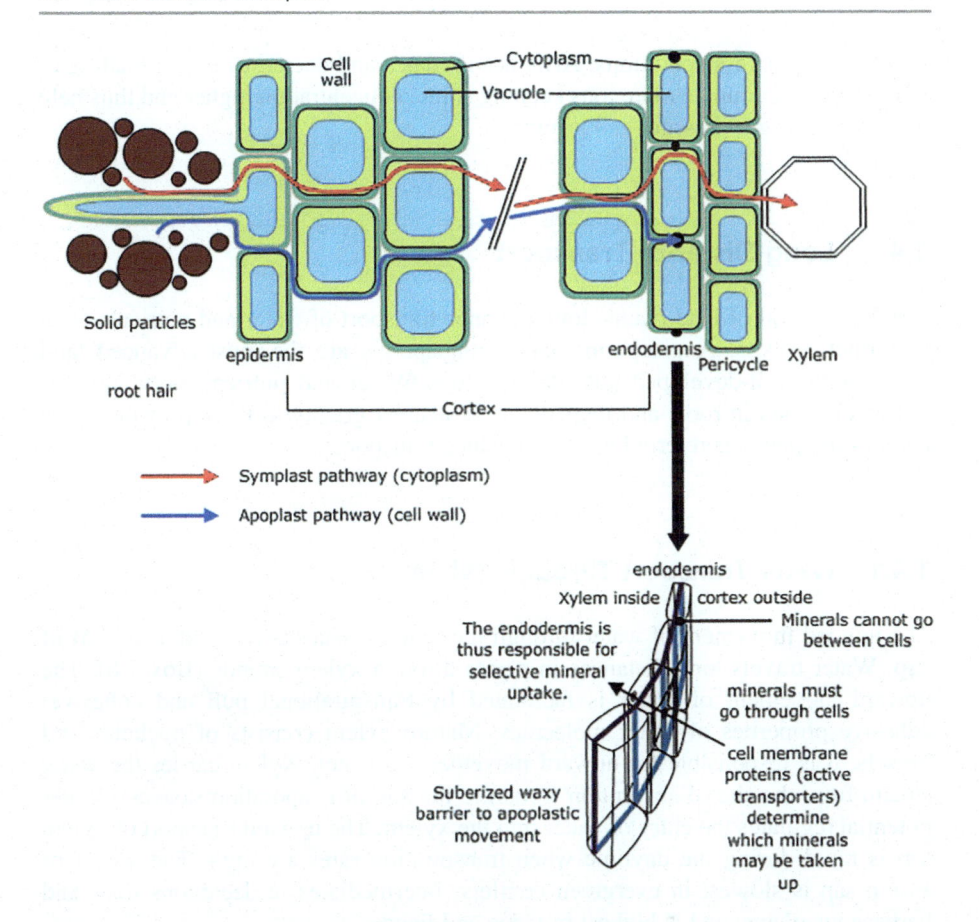

**Fig. 1.5** The pattern of radial transport or short-distance transport across root. The entry of water from the soil to root takes place due to water potential difference by osmosis or through water channels. The intercellular transport through the cortex follows apoplast or symplast pathway. Once it reaches the endodermis, the apoplastic entry of water is stopped by the Casparian bands in the endodermis. Water crosses this layer through symplast pathway or passage cells and reaches the xylary elements

the xylary elements invaginate the cell wall under the pit membrane to increase the surface area for transport. In addition to water, some of the nutrients are also transported through xylem. At the root surface, some nutrients and salts enter into the xylem. One such example is $K^+$ which is delivered to xylem at the root surface and transported to the shoot through xylem. Anions, such as $NO^-_3$ and $Cl^-$, are also released in xylem sap by anion channels. The concentration of nutrients in xylem varies with water flow rate, soil composition, and nutritional status of the plant. Similarly, unloading of water and nutrients in the leaves also involves multiple cellular pathways and specific transporters. These transporters are responsible for silicon uptake and its loading and unloading in xylem. Loading and unloading of element in xylem are under the control of abscisic acid (ABA) and signals from the shoot. During stress conditions, ABA accumulates and decreases xylem loading of ions. Roots accumulate more ions keeping solute concentration higher and thus help to maintain turgor pressure and growth.

## 1.4 Long-Distance Transport

For the survival of land plants, long-distance transport of water and nutrients is an important process. Angiosperms and gymnosperms are the most advanced land plants with well-developed vascular systems. Water and nutrient entry into the xylary elements in roots and their transport against gravity to the top of the aerial parts of the plants is referred as long-distance transport.

### 1.4.1 Water Transport Through Xylem

The upward movement of water through the xylem tissues is referred as **ascent of sap**. Water travels long distances in plants through xylem strands (Box 1.6). The upward movement of water is facilitated by transpirational pull and cohesive-adhesive properties of water molecules. Mature xylem consists of tracheids and vessels. It is responsible for upward movement of water. Xylem carries the water stream from the site of absorption (roots) to the site of evaporation (leaves). Water potential regulates the entry of water into the xylem. The upward transport of xylem sap is rapid during the daytime when transpiration rates are high. The ascent of xylem sap is slowest in evergreen conifers, intermediates in deciduous trees and herbaceous plants, and is highest in vines and lianas.

> **Box 1.6: Experimental Evidence for Xylem Transport**
> In a potted plant, phloem is separated from xylem (approximately at 25 cm height above soil) with the help of wax paper. The separation of xylem and phloem stops lateral transport of water in this region. The segment above the stripped zone is labeled SA and segment below it as SB. There are six segments in the stripped area numbered as $S_1$–$S_6$. The roots are exposed to a soluble dye containing $^{42}K$, and the different sections are analyzed for radioactive potassium in both xylem and phloem. The section where xylem is separated from phloem, the content of potassium is highly reduced in phloem while it is high in xylem. The sections below and above the stripped region have the equal content of potassium in both phloem and xylem. This indicates that water moves through xylem and is laterally transported in phloem.

(continued)

## 1.4 Long-Distance Transport

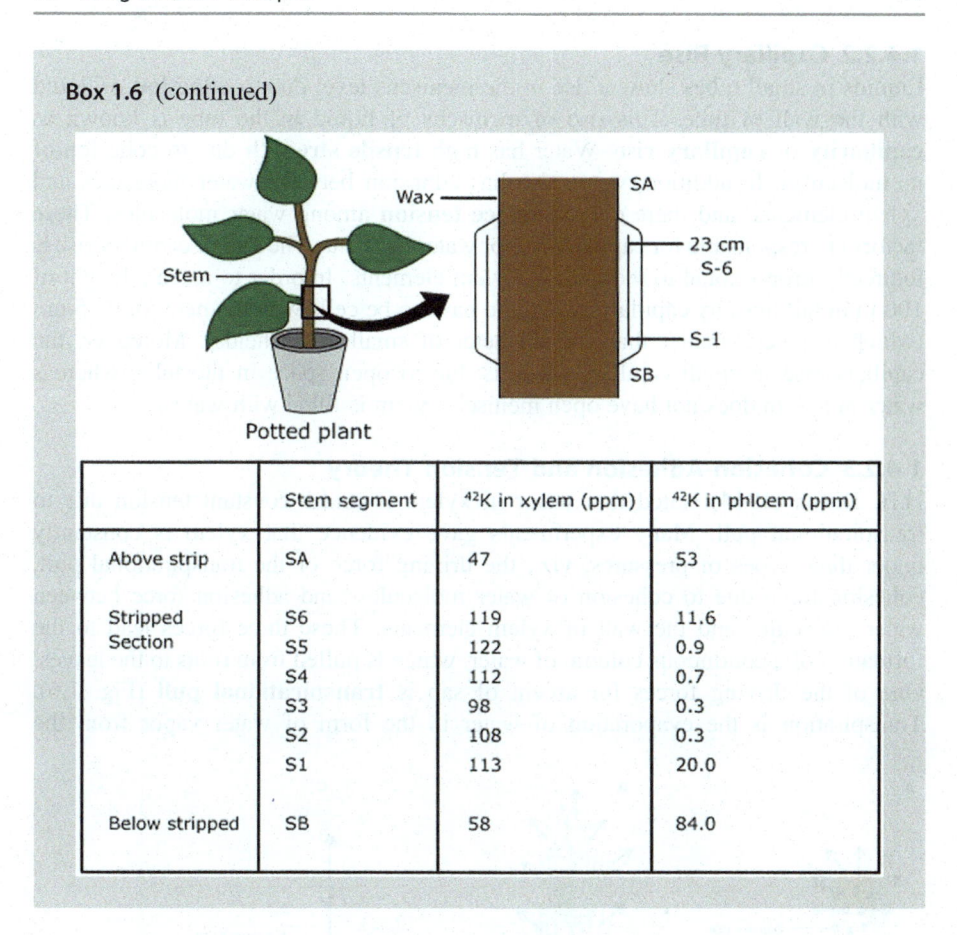

**Box 1.6** (continued)

| | Stem segment | $^{42}$K in xylem (ppm) | $^{42}$K in phloem (ppm) |
|---|---|---|---|
| Above strip | SA | 47 | 53 |
| Stripped Section | S6 | 119 | 11.6 |
| | S5 | 122 | 0.9 |
| | S4 | 112 | 0.7 |
| | S3 | 98 | 0.3 |
| | S2 | 108 | 0.3 |
| | S1 | 113 | 20.0 |
| Below stripped | SB | 58 | 84.0 |

### 1.4.2 Mechanism of Transport Across Xylem

#### 1.4.2.1 Root Pressure

It is a positive hydrostatic pressure (1–2 bars) developed due to the difference in solute potential between the soil solution and xylem sap. It often develops at night when the rate of transpiration is low or absent and humidity is high. In roots, the development of pressure takes place due to high salt concentration and presence of Casparian bands in endodermis. Ion accumulation decreases the solute potential of roots, and despite the absence of transpiration, water enters into the root and into the xylem. Root pressure is insignificant in tall trees but has a significant role in the young plants (Sect. 1.6; Guttation). During seed germination and bud growth, prior to the development of leaf and transpiration stream, water uptake is due to root pressure.

### 1.4.2.2 Capillary Rise

Liquids in small tubes show a rise in the meniscus level due to **adhesion** of liquid with the wall of tube. This rise of meniscus of liquid in the tube is known as **capillarity** or **capillary rise**. Water has high **tensile strength** due to cohesion of its molecules. In addition, water also has adhesion between water molecules and xylem elements, and there exists **surface tension** among water molecules. These factors are responsible for capillary rise of water in plants. The rate of capillary rise is indirectly proportional to the radii of xylem elements. In order to reach a height of 100 m in tall trees by capillary rise, there have to be cells with diameter of 0.15 μm (which is much smaller than the diameter of smallest tracheids). Moreover, the capillary rise in small capillary tubes is due to open space in the tube, whereas water in xylem does not have open menisci (xylem is filled with water).

### 1.4.2.3 Cohesion-Adhesion and Tension Theory

H.H. Dixon (1914) stated that water in xylem is under constant tension due to transpirational pull. Many experiments gave evidence that xylem is constantly under three types of pressures, viz., the driving force or the transpirational pull, cohesion force due to cohesion of water molecules, and adhesion force between water molecules and the wall of xylem elements. These three forces lead to the formation of a continuous column of water, which is pulled from roots to the leaves. One of the driving forces for ascent of sap is **transpirational pull** (Fig. 1.6). Transpiration is the evaporation of water in the form of water vapor from the

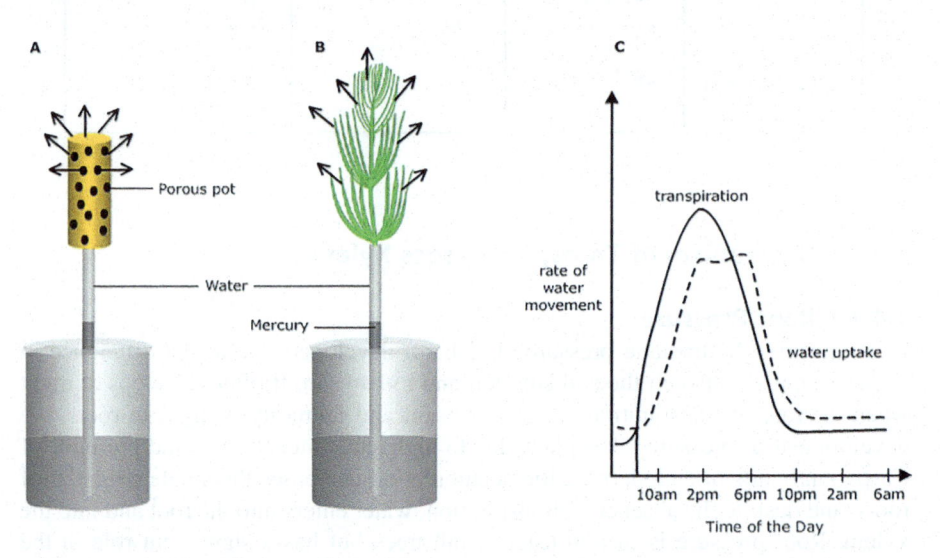

**Fig. 1.6** Demonstration of transpiration pull. (**a**) Water molecules evaporated from porous pot create suction pressure leading to a rise in the mercury level. (**b**) The pot is replaced by a healthy twig. Due to transpiration pull, suction is created in xylem vessels and mercury level rises. The suction pressure due to transpiration is very high due to which it can pull heavy metal like mercury. (**c**) Pattern of uptake of water in plant begins after a rise in the rate of transpiration. The rate of transpiration is maximum at noon, and water uptake is also maximum during that period

aboveground parts of the plant. During rapid transpiration period, water moves in xylem at an average rate of 4 mm. s$^{-1}$, causing development of pressure of more than $-5$ MPa. Tension in xylem can be measured using different methods. Water can withstand tension of $-21$ MPa which is enough to overcome the gravitational force in large trees. These tensions and water potential difference in xylem are transferred to root and finally to the soil, leading to soil-plant-atmosphere continuum (Table. 1.1). This continuous column can only be maintained due to cohesion of water molecules by hydrogen bonding. Sometimes the gaps in certain tracheary elements cannot withstand large tensions, and continuation of column is not maintained, leading to disruption in transport.

### 1.4.2.4 Cavitation and Embolism

The continuity of water in the xylem strands is broken down despite filtration provided by the roots. The reason for breakage of water column in xylem is the high tension in xylem cells. This tension in xylem causes water to change into vapor state. The breakage of column can be due to air getting trapped in or water getting converted to vapors in the xylem. This sudden reversal of liquid water into water vapors, leading to breakage of water column, is known as **cavitation**. The presence of large number of pores in xylem strands breaks surface tension of water, and it draws minute bubbles in the xylem column. Cavitation due to trapping of air is called "air seeding." Air seeding is a very common phenomenon in plants. In this process, the air enters tracheids through pit membranes. Tension in tracheids causes damage to pit membrane and increases its pore size, and air gets filled into it. The process of filling of the vessel or tracheids with air is called **embolism** (Fig. 1.7). The main causes of cavitation are water stress during high transpiration, freezing of xylem sap in winter, and pathogen attack (Fig. 1.8). Whenever there is a breakage of water column in the tracheids, it creates click-like sound which can be recorded using acoustic methods. As the plant experiences water stress due to high rate of transpiration during the daytime, the frequency of cavitation increases. The loss of xylem

**Table 1.1** Transpiration flow in the leaf is maintained by decreasing water potential from xylem sap to external atmosphere immediately around the leaf

|  |  | Solute potential | Pressure potential | Gravitational potential | Water potential |
|---|---|---|---|---|---|
| Leaf |  |  |  |  |  |
|  | Xylem sap | −0.2 | −2.5 | 1.5 | −1.2 |
|  | Bundle sheath | −3.3 | 0.1 | 1.5 | −1.7 |
|  | Apoplast | −3.5 | −0.7 | 1.5 | −2.7 |
|  | Substomatal cavity |  |  |  | −8 |
| External atmosphere |  |  |  |  | −100 |

The solute potential and pressure potential show variations, but gravitational pull is the same at different locations in the leaf cell. All the values are in MPa

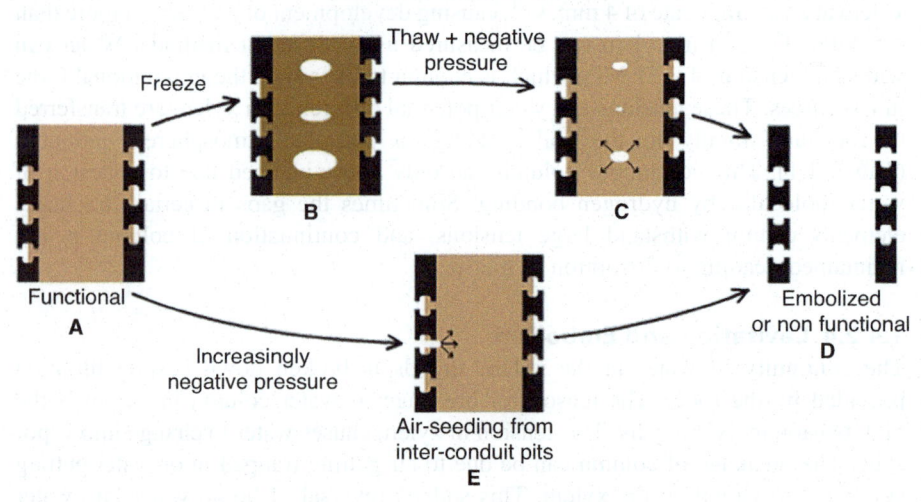

**Fig. 1.7** The process of cavitation in xylem. (**a**) A functional xylary element. (**b**) and (**c**) During the process of thawing, after freezing temperature, water molecules experience very high negative pressure resulting in breaking of water column. Air gets trapped and spreads resulting in air seeding from inter-conduit pits (**d**) followed by complete embolization of the cell. (**e**) The air is pulled through pit membrane from adjacent embolized cell or under water stress conditions. The process of air getting trapped is called air seeding which results in embolized cell

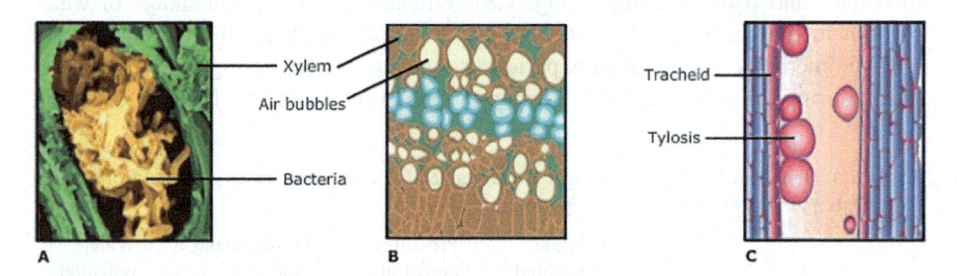

**Fig. 1.8** Causes of cavitation in xylem. (**a**) Xylem-filled with bacterial pathogen. (**b**) Xylem tracheids showing inward bursting of air bubbles. (**c**) Blockage of tracheid due to tyloses in gymnosperms

function due to cavitation varies in different plants. Bordered pits present in gymnospermous tracheids act like valves. The torus present in it closes the pit and avoids spreading of cavitation to adjacent functional tracheids. Vessels are more susceptible to air seeding as compared to tracheids. The absence of xylem vessels in gymnosperms growing in the cold temperate zones is one of the adaptations to avoid air seeding during freezing temperatures. The embolized xylem cells are repaired at night when transpiration stops or slows down. Water moves from the adjacent tracheids and maintains water uptake. At the root level, the solutes are pumped into xylem, lowering its water potential as compared to the soil. Water moves from the soil and develops a positive root pressure. This pressure forces water to move up to 10 m and fills the embolized xylem cells. In herbaceous plants, the refilling of tracheary elements occurs during night due to root pressure. Aquaporins in xylem-associated cells are also involved in refilling of the embolized tracheids.

### 1.4.2.5 Unloading of Water and Nutrients from Xylem in Leaves

In the leaves, a network of major and minor veins helps in the distribution of water and nutrients throughout. The xylary elements in the leaves are present in minor leaf veins. The leaf veins are enclosed by a bundle sheath. Water and nutrients in the minor veins are unloaded in the spongy parenchyma. Water transport from the spongy parenchyma to the site of evaporation (stomata) may follow symplastic or transcellular pathways, as in roots. However, apoplastic pathway is blocked in the leaves. As in roots, unloading of water and nutrients again involves transporters and specific intercellular pathways. Some of these transporters block the entry of nutrients in the leaves, and such nutrients are again taken back by the phloem to the root. $Na^+$ is one such example. Plants protect leaves from salt stress by recirculating $Na^+$. Nutrients are targeted to different cells of the leaves according to the plant type, and water moves through spongy parenchyma to the substomatal cavity.

## 1.5 Water Movement from Leaves to the Atmosphere

The driving force for water movement from soil to leaves and from the leaves to atmosphere is transpiration. The upward translocation of water in the xylem from roots to the leaves is coupled with the pressure of transpiration and is referred as **transpirational stream** (Fig. 1.9). During the process of rapid evapotranspiration, the rate of water transport in xylem is about 4 $mm.s^{-1}$. Water lost by transpiration must be replenished by the absorption of an equivalent amount of water from the soil. One sunflower plant "imbibes" and "perspires" 17 times more water than a human every 24 h. A single maize plant transpires approximately 150 l of water in its average life span (Box 1.7).

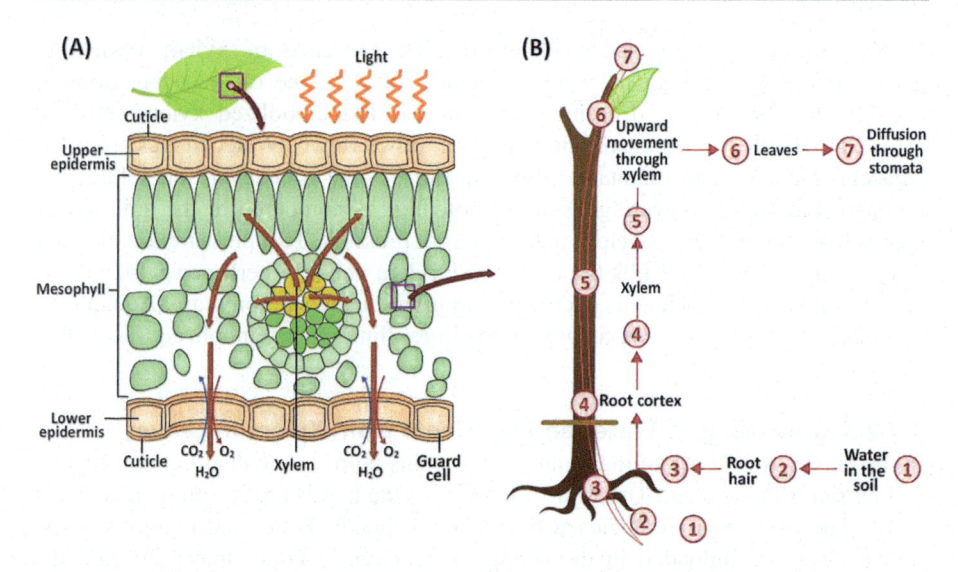

**Fig. 1.9** (a) Evaporation of water from leaf surface. Water from xylem enters into air spaces of spongy parenchyma and diffuses out through stomata present on the lower epidermis. Gas exchange takes place when stomata open up. Carbon dioxide is taken in and oxygen is released. (b) Transpiration from leaf creates a continuous water stream up to the soil

> **Box 1.7: Transpiration Ratio**
> It is used to determine the efficiency of plants to fix $CO_2$, relative to the amount of water loss by transpiration. The opening of stomata during daytime is a compromise between photosynthesis and transpiration. Plants balance water loss with photosynthesis. Carbon dioxide concentration in the substomatal cavity plays a key role in balancing the two processes. This shows that plant regulates stomatal movement in response to the extent of photosynthesis. Plants adapt many strategies to reduce transpiration. The relative amounts of photosynthesis ($CO_2$ in) and transpiration ($H_2O$ out) occurring at any time depend on the amount of $CO_2$ and $H_2O$ in the atmosphere inside and outside the leaf, respectively. For every unit of $CO_2$ used in photosynthesis, plants lose about 600 units of $H_2O$. The amount of water loss is quite high. Thus, transpiration as a necessary evil is justified. This ratio of $CO_2$ taken in and water lost by the plant is known as **transpiration ratio** or **water use efficiency**. $C_4$ and CAM plants are more efficient because they are well adapted to reduce transpiration. Transpiration ratio usually varies from 100 to 1000 among different plants, depending on the environmental conditions. There is more than twofold difference in the transpiration ratio between $C_3$ and $C_4$ plants. This difference is due to leaf anatomy and pathway of $CO_2$ fixation.

## 1.5.1 Transpiration

The uptake of $CO_2$ for photosynthesis requires moist surface, but when water is exposed, it gets evaporated. Plants face this challenge of taking more carbon dioxide and at the same time loss of water. Plants lose 400–600 molecules of water while gaining 1 molecule of carbon dioxide. Thus, both photosynthesis and loss of water by transpiration are inseparable processes in the life of green plants. Water has high **latent heat of vaporization**. At 30 °C, 1000 gm (1 kg) of water absorbs 580 Kcal of heat from its environment. Large amount of water is being evaporated from plants, and the heat required for vaporization is being drawn from leaf. It helps plants to maintain the temperature and tolerate harsh environmental pressures. The main advantage of transpiration is creation of suction pressure for uptake of water and minerals from the soil. Once plants build up enough transpirational pull during early hours of the day, water uptake by the plants begins. Transpiration is definitely a necessary evil. There are three modes of transpiration in plants, viz., cuticular transpiration, stomatal transpiration, and lenticular transpiration. The cuticular transpiration is the loss of water from the surface of the plant. This type of transpiration takes place when cuticle is very thin and there is no water scarcity. It accounts for only 2% of water loss. The stomatal transpiration takes place through stomata, and more than 95% of water loss takes place through stomatal openings present on the leaf epidermis. The third mode of transpiration is through lenticels present on the bark and in the peel of some fruits (e.g., apple). Water loss through lenticels is minimal. The driving forces for transpiration are **vapor pressure** and **vapor density** in the substomatal area. The concentration of water molecules in vapor phase is expressed as vapor mass per unit volume (g.m$^{-3}$) and is referred as vapor density. Vapors cause pressure on the walls of the stomatal chamber.

The substomatal air space is saturated with water vapors, while the immediate air around leaves is unsaturated. This gradient around the substomatal area and air around the leaves lead to vapor loss through transpiration (Table 1.1). The rate of transpiration ($T$) is governed by vapor pressure gradient between the leaf ($e_{leaf}$) and the surrounding ($e_{air}$):

$$T \propto e_{leaf} - e_{air}$$

Transpiration also undergoes considerable resistance by stomata and vapors present in the atmosphere. There are two main boundaries that create high resistance to water movement. The first one is at the stomatal pore, called the **leaf stomatal resistance** ($r_{leaf}$), and the other is the air immediately around the leaf, called **boundary layer resistance** ($r_{air}$).

This can be expressed as

$$T = \frac{e_{\text{leaf}} - e_{\text{air}}}{r_{\text{leaf}} + r_{\text{air}}}$$

In other words, the rate of transpiration is directly proportional to the difference in vapor pressure between the leaves and atmosphere divided by the sum of resistance encountered by air and the leaves. The number and size of stomatal pores contribute to leaf stomatal resistance ($r_{\text{leaf}}$) and the degree of difference in vapor pressure between leaf and air to boundary layer resistance ($r_{\text{air}}$) (Box 1.8).

> **Box 1.8: Measurement of Transpiration Rate**
> **Lysimeter or gravimetric method:** The amount of water used by plants for their growth is just 1% of the total water absorbed from the soil. In this method, it is presumed that loss of weight in plants is due to loss of water, i.e., transpiration. A potted plant with soil covered with plastic bag is taken. Plant is weighed at different intervals. The higher the rate of transpiration, the more will be the loss of weight. On the same principle, a simple method to find out the rate of transpiration in small twigs is using a device called Ganong's **potometer**. The rate of transpiration is measured by tracking the distance travelled by the air bubble in graduated tube per unit time.
>
> **Gas exchange or cuvette methods**: In this method, the twig is kept in a sealed container with two openings, and relative humidity and temperature of air going inside and air coming out are measured at the beginning of the experiment. After an interval, again the relative humidity and temperature are measured. Absolute humidity (vapor density and vapor pressure) is measured by **hygrometer**. The rate of transpiration will be equal to the change of vapor density in the air going in and air coming out of the container. It can be used for large number of plants. The relative humidity, however, changes with time and that affects the rate of transpiration.
>
> **Stemflow or thermocouple method:** The amount of water flowing through stem helps to get a reliable calculation of the rate of transpiration. Thermocouple method involves giving a heat pulse to the stem at a given point. The temperature is then measured above at some point. The time required to reach the same temperature at the second position indicates the velocity of xylem sap flow.

(continued)

### Box 1.8 (continued)

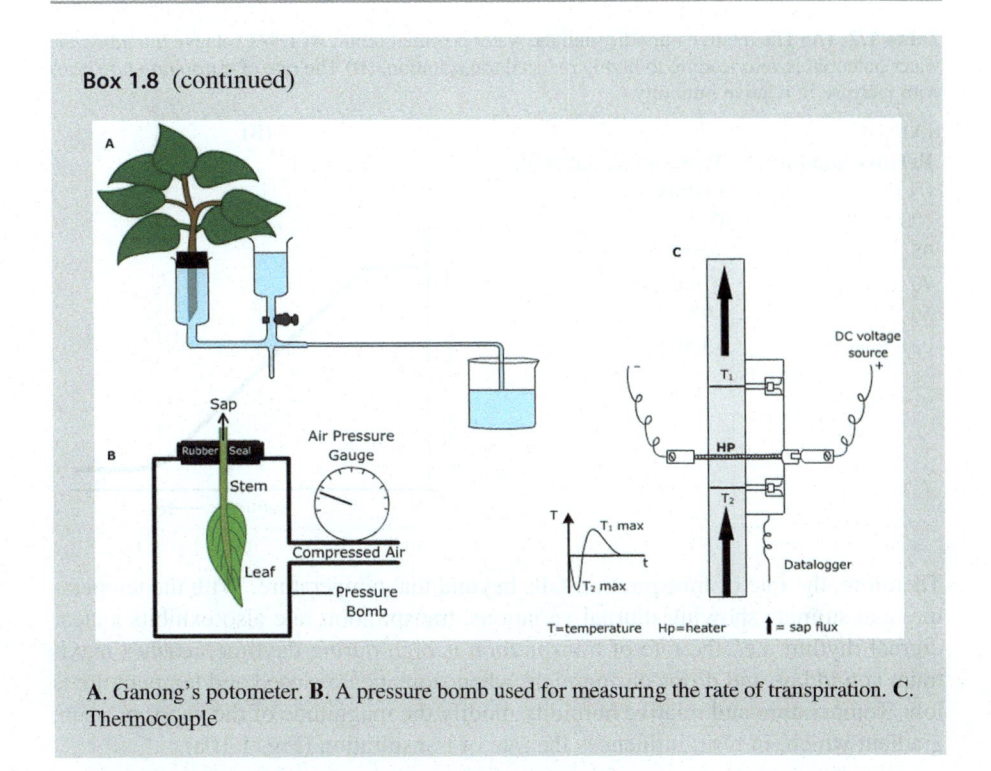

A. Ganong's potometer. B. A pressure bomb used for measuring the rate of transpiration. C. Thermocouple

#### 1.5.1.1 Factors Affecting Rate of Transpiration

The rate of transpiration is affected by the plant water status, its anatomy, as well the environmental factors. The most important environmental factors are humidity, light, temperature, wind velocity, and availability of soil water. These factors are discussed here individually, but in natural habitat, they are influencing each other and affect the rate of transpiration.

**Humidity** Atmospheric humidity changes with change in temperature or vapor pressure. A rise or fall in temperature with no change in vapor pressure will drop or raise relative humidity, respectively. Similarly, with no change in temperature, rise or fall of vapor pressure will cause rise and fall of relative humidity, respectively. Whenever there is a rise in relative humidity, the rate of transpiration decreases (Table 1.2). A vapor pressure gradient in the vicinity of the leaves and in substomatal chamber determines the rate of transpiration. Whenever this gradient is steep, the rate of transpiration rises.

**Temperature** The rise in temperature, with all other factors nearly constant, increases the rate of transpiration. The increase in temperature also increases the difference in vapor pressure of the leaf and outside the atmosphere. Hence, the rate of transpiration increases. However, stomata close at temperature higher than 35 °C.

**Table 1.2** (**A**) The relative humidity and the water potential of air. At 100% relative humidity, the water potential is zero leading to highly reduced transpiration. (**B**) The rate of transpiration reduces with increase in relative humidity

(A)

| Relative humidity (%) | Water potential of air (MPa) |
|---|---|
| 100 | 0 |
| 95 | −2.50 |
| 90 | −14.2 |
| 50 | −93.6 |
| Less than 10 | −300 |

(B)

Therefore, the rate of transpiration falls beyond that temperature. With the temperature and stomata showing diurnal variations, transpiration rate also exhibits a clear diurnal rhythm, i.e., the rate of transpiration is high during daytime, reaches maximum at midday, and drops during night when stomata are closed and temperature is low. Temperature and relative humidity modify the magnitude of the vapor pressure gradient which, in turn, influences the rate of transpiration (Fig. 1.10a).

**Wind Velocity** An increase in the rate of transpiration is not directly proportional to wind velocity. During high wind velocity, stomata close. Therefore, the rate of transpiration also drops. But wind at low velocity increases the rate of transpiration. As wind disperses the air around the leaf and reduces the vapor pressure in the immediate vicinity of stomata, it increases the rate of transpiration (Fig. 1.10b).

**Internal Factors** One of the most important factors affecting transpiration rate is leaf-shoot ratio. The magnitude of transpiration will be more in leaves with greater leaf area. However, the rate of transpiration has no correlation with leaf size when per unit leaf area is considered. The leaf structure is very important in regulating the rate of transpiration. Leaf adapts many strategies to reduce transpiration (Sect. 1.5.1.3; Antitranspirants). Stomatal frequency (number of stomata present in per unit area of leaf), pore size, and distribution show tremendous variation in plants growing in different habitats (Table 1.3).

### 1.5.1.2 Ecological Adaptations to Reduce Transpiration

Plants develop many adaptations to avoid water loss due to transpiration. The presence of thick cuticle, sunken stomata, and stomata only on the lower side of the leaves are some of the adaptive mechanisms developed by plants. Plants growing in desert experience more water crisis. They need to restrict the rate of transpiration.

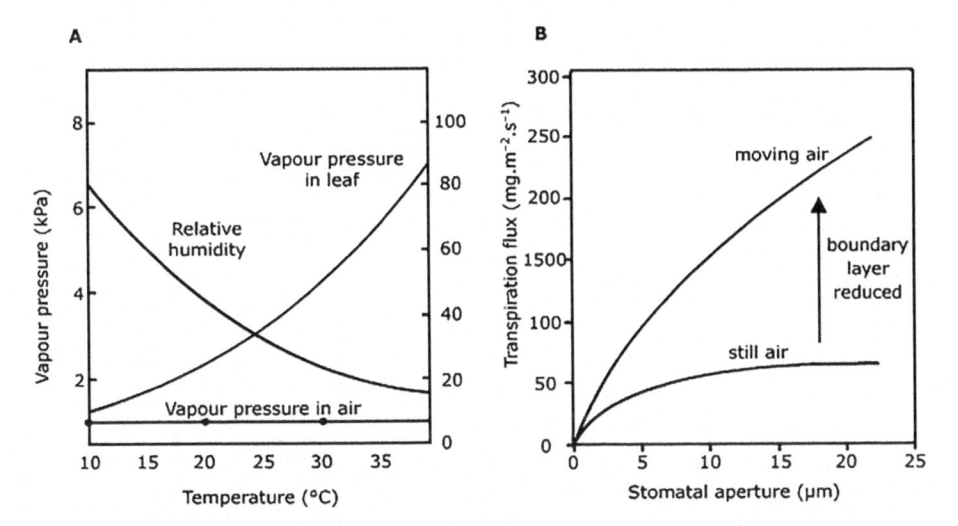

## 1.5 Water Movement from Leaves to the Atmosphere

**Fig. 1.10** (a) With the increase in temperature, the relative humidity decreases resulting in high vapor pressure gradient in leaf and outside air, leading to stomatal opening and increase in the rate of transpiration. (b) Effect of wind velocity on the rate of transpiration. As wind velocity increases, it reduces the boundary layer resistance around stomata leading to opening of stomata and hence more transpiration. In still air, the boundary layer resistance is high. Therefore, the rate of transpiration is same

The ecological adaptations of desert plants are reduction of leaf area, presence of thick cuticle, sunken stomata, and stomata opening during night, special water storage capacity, modified root structure, and $C_4$ photosynthesis.

### 1.5.1.3 Antitranspirants

These are the chemicals which decrease water loss from plants due to transpiration. They are also useful for avoiding transplantation shock to the nursery plants and plants raised in tissue culture. An antitranspirant should be nontoxic and affect only stomata. It should not cause permanent damage to stomatal mechanism, and the effect should persist only for short duration. There are mainly four types of antitranspirants: **stomatal-closing types** – some fungicides, like phenyl mercuric acetate (PMA), and herbicides, like atrazine, act as antitranspirants by closing the stomata. **Film-forming type**: These chemicals form coating on the surface of leaves. Hence, stomatal pores are blocked resulting in reduction in the rate of transpiration. The material used is either plastic or waxy in nature. **Reflectance type:** These are white materials which form a coating on the leaves and increase the leaf reflectance. Materials like kaolin, hydrated lime, calcium carbonate, magnesium carbonate, and zinc sulfate are used to reduce transpiration in this category. **Growth retardant**: ABA can bring stomatal closure. Chemicals like cycocel reduce shoot growth and increase root growth, thus enabling the plants to resist drought. It is useful for improving water status of the plant.

**Table 1.3** (A) Distribution and frequency of the stomata in horizontally and vertically oriented leaves. (B) Stomatal frequency in monocot leaves is nearly equal in both upper and lower surfaces, while in dicot leaves, stomatal frequency is more on lower surface

(A)

| Orientation of the leaf | Upper epidermis | Lower epidermis |
|---|---|---|
| **Horizontal** | | |
| *Malus* | 0 | 38,660 |
| Bean | 4051 | 24,906 |
| Oak | 0 | 58,040 |
| Pumpkin | 2791 | 26,932 |
| **Vertical** | | |
| Corn | 9800 | 10,900 |
| Pine | 12,000 | 11,000 |
| Onion | 16,900 | 16,900 |

(B)

| Type | Name of the plant | Upper epidermis | Lower epidermis |
|---|---|---|---|
| Monocot | Wheat | 50 | 40 |
| | Onion | 160 | 160 |
| | Barley | 75 | 80 |
| Dicot | Sunflower | 115 | 165 |
| | Alfalfa | 169 | 182 |
| | Geranium | 29 | 175 |

## 1.5.2 Stomatal Movement

The stomata make up 15–40% of the total leaf volume. However, the stomatal pore accounts for only 1% of the total surface area, and it is responsible for more than 90% of water loss due to transpiration. The guard cells are bean-shaped in dicots and dumbbell-shaped in monocots (Fig. 1.11a). In addition to structural differences from epidermal cells, guard cells lack plasmodesmatal connections and have chloroplasts. The wall of the lining of pore is thicker than the outer wall of the guard cells, which are thin. A fully open stomatal pore measures 5–15 μm wide and is about 20 μm long. The ratio of $CO_2$ diffusion through a perforated membrane varies in proportion to the diameter of pore and not the area, i.e., as the pore size decreases, the efficiency of $CO_2$ diffusion per unit area increases severalfolds. This is due to spillover effect for diffusion of $CO_2$ through a stomatal aperture (Fig. 1.11b).

### 1.5.2.1 Mechanism of Guard Cell Movement

The stomatal opening and closing take place due to the movement of guard cells. Microfibrils are located around the circumference of the elongated guard cell. This arrangement of radiating microfibrils is called as **radial micellation** (Fig. 1.11). When water enters into the guard cells, it cannot increase much in its diameter due to the presence of radial microfibrils. However, it increases in length, especially along its outside thin wall. With the increase in size, guard cells exert pressure on the microfibrils, and, in turn, microfibrils pull the inner (thicker) wall of the stomata, leading to the opening of pore. In intact leaf, the guard cells obtain $K^+$ from adjacent cells. There is evidence to show that the level of $K^+$ controls stomatal movement. $K^+$

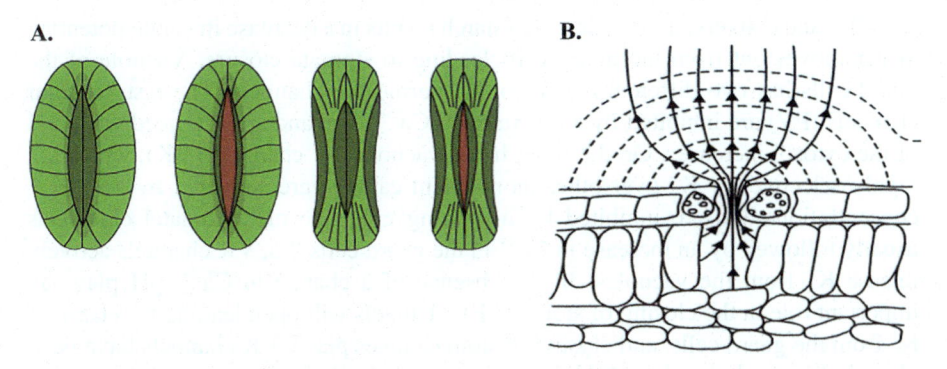

## 1.5 Water Movement from Leaves to the Atmosphere

**Fig. 1.11** (a) Bean-shaped stomata present in dicot leaves (left) and dumbbell-shaped as observed in monocot leaves. Note the presence of radial micelle in both types of stomata and the deposition of thickening in boundary around stomatal pore. (b) The spillover effect for diffusion of $CO_2$ through a stomatal pore. The dashed lines are isobar representing region of equivalent $CO_2$ partial pressure

**Table 1.4** The amount of $K^+$(mM) in open and close stomata

| Species | $K^+$(mM) in stomata Open | Close |
|---|---|---|
| Vicia faba | 552 | 112 |
| Commelina communis | 448 | 95 |

levels are high when stomata are open, and it is low when stomata are closed (Table 1.4). The amount of $K^+$ present in the vacuole of guard cells is enough to cause a decrease in the osmotic potential by 2.0 MPa. The fungal toxin fusicoccin is known to cause an active proton extrusion, and its treatment stimulates stomatal opening. Proton extrusion is balanced with $K^+$ movement in guard cells resulting in opening of stomata and wilting of plant. In *Vicia* leaves, it has been observed that when $K^+$ moves out of guard cells, an equal amount of $H^+$ are also released. These $H^+$ are, during the formation of organic acids, mainly malic acid. The rise in $H^+$ leads to a drop in the pH of guard cells. The excessive $H^+$, if not extruded out, will further drop the pH of the guard cells. Thus, $H^+$ is extruded out and is exchanged with $K^+$ influx into the guard cells. If the proton pumps are inhibited, stomata remain closed even in blue light. Application of inhibitors of photosynthesis prevents organic acid formation, and stomatal opening is blocked.

$$\text{Starch} \rightarrow \text{Phosphoglyceric acid} \xrightarrow{\text{carboxylase}} \text{Phosphoenol pyruvic acid} + CO_2$$
$$\downarrow$$
$$\text{Malic acid} \leftarrow \text{Oxaloacetic acid}$$

The increase in malic acid content and entry of $K^+$ further drop the osmotic potential. $K^+$ uptake is balanced by uptake of chloride ion ($Cl^-$) into the guard cells and transport of $H^+$ released from organic acid (malic acid). Thus, all these factors are responsible for the opening of stomata. The stomata closure is due to efflux of $K^+$

and Cl⁻, and conversion of malate into starch results in a decrease in solute potential. Water moves out from the guard cells leading to stomata closure. Vacuole of the guard cells has two different classes of K⁺-permeable channels. The fast-vacuolar channels (FV) are inhibited by an increase in $Ca^{2+}$ level and are activated when pH of the cytosol increases. On the other hand, vacuolar K⁺ channels (VK), which are highly selective for K⁺ over other monovalent cations, are activated by low $Ca^{2+}$ concentrations and are inhibited by increasing cytosolic pH. Stomatal closure is usually followed by an increase in $Ca^{2+}$ in the guard cells. So, VK channels actively release K⁺ from the vacuoles. In the absence of a change in $Ca^{2+}$, pH plays an important role in the closing of stomata. FV channels will open leading to release of K⁺ from the guard cells, and closing of stomata takes place. VK channels have been identified in *Arabidopsis* and these channels have $Ca^{2+}$-binding domain.

### 1.5.2.2 Factors Affecting Stomatal Movement

Environment immediately around the plants keeps changing and so is the stomatal aperture. Stomata keep adjusting to these changes. Environment signals are perceived by the guard cells. These signals are transduced into ion fluxes leading to change in the turgor pressure of the guard cells. Some of the important environmental factors affecting stomatal movement are as follows.

**Light** Stomata of most plants open during daytime and close during nighttime. However, in succulent plants stomata open during nighttime and close in the presence of light. This adaption of CAM plants leads to carbon dioxide absorption during nighttime when transpiration stress is low. The effect of light on stomatal movement is independent of photosynthesis. Stomata respond to light even in leaves where photosynthesis has been reduced to zero by the application of photosynthesis inhibitor (e.g., cyanazine). Blue light has direct effect on the opening of stomata, independent of $CO_2$ concentration. Thus, in *Allium cepa*, guard cells swell in the presence of K⁺ when blue light is given. Blue light enhances the H⁺ transport leading to proton gradient which, in turn, causes the opening of K⁺ channels. Red light also stimulates stomatal opening, but blue light is ten times more effective than red light.

**Carbon dioxide** It is mainly the internal level of $CO_2$ which controls the opening and closing of stomata. In most plants, low $CO_2$ concentration in intercellular spaces causes the stomatal opening during nighttime as well. High concentrations of $CO_2$ in intercellular spaces cause partial closure of stomata during daytime. Once stomata closes, external $CO_2$ concentration does not affect the stomatal movement. High level of intracellular $CO_2$ induces closure of stomata in the same way as ABA-induced closure.

**Relative Humidity** Stomata close when the vapor content of the air and that of intercellular space exceeds a critical level. The water potential of the leaf also affects stomatal movement. As water potential decreases (water stress), the stomata are closed.

**pH** Generally high pH favors closing while low pH favors opening of stomata. pH regulates $K^+$ movement in the guard cells. Mostly alkaline pH enhances the $K^+$ outflow and results in stomatal closure.

**Temperature** The stomatal pore closes as the temperature rises above 30–35 °C. To ensure gas exchange at high temperature, stomata opens during nighttime. Temperature has indirect effect on stomatal movement. It affects the balance of respiration and photosynthesis. Whenever there is increase in temperature, it leads to rise in respiration rate. Due to high rate of respiration, the level of $CO_2$ increases and it causes closure of stomata. In some plants, high temperature induces stomatal opening to increase the transpiration rate (cooling effect) so that temperature of leaf is maintained.

**Water Stress** Sometimes, due to water stress, guard cell loses more water as compared to its intake. It loses its turgidity and stomatal closure takes place. This is called **hydropassive closure** of stomata. On the other hand, stomata also experience ABA-mediated **hydroactive closure**. ABA is synthesized in leaves, but during water stress, it is synthesized in roots and quickly transported to leaves. ABA induces stomatal closure by opening $Cl^-$ channels and allows the $Cl^-$ efflux from the guard cells. ABA also inhibit $H^+$ pump. This loss of ions further opens up $K^+$ channels and $K^+$ also diffuses out of the guard cells. The solute concentration decreases followed by loss of water and closure of stomata. ABA is also referred as an antitranspirant. It regulates stomatal closure and thus reduces transpiration rate. When stomatal closure is induced by ABA, large numbers of ion channels are regulated by $Ca^{2+}$ signals.

## 1.6 Guttation

Guttation is the process of extrusion of liquid droplets from the leaves through special structures called water stomata or **hydathodes**. During early morning in summer, when relative humidity is very high, small droplets appear on the vein endings of grasses or serrate margins of certain leaves (Fig. 1.12a). Guttation takes place only when relative humidity is high and the rate of transpiration is extremely low because only in these conditions significant root pressure develops. Thus, this phenomenon can only be seen in small plants where root pressure has significance. Although water coming out due to guttation looks like dew drops, it contains minerals, organic acid, sugars, and even enzymes. On evaporation, the solutes that remain on the leaves cause salt burning, which is called **guttation burn**. Hydathodes are restricted to the apex or the serrated edges of the margins of leaves. It consists of a simple pore in the epidermal layer found at the tip and is surrounded by a special parenchymatous tissue, called **epithem**. The cells of epithem are isodiametric in shape, are loosely arranged, and enclose lot of intercellular spaces. The xylem elements of leaf vein terminate in epithem (Fig. 1.12b). Cells of epithem contain a large number of finger-like projections (a characteristic of transfer cells) to increase

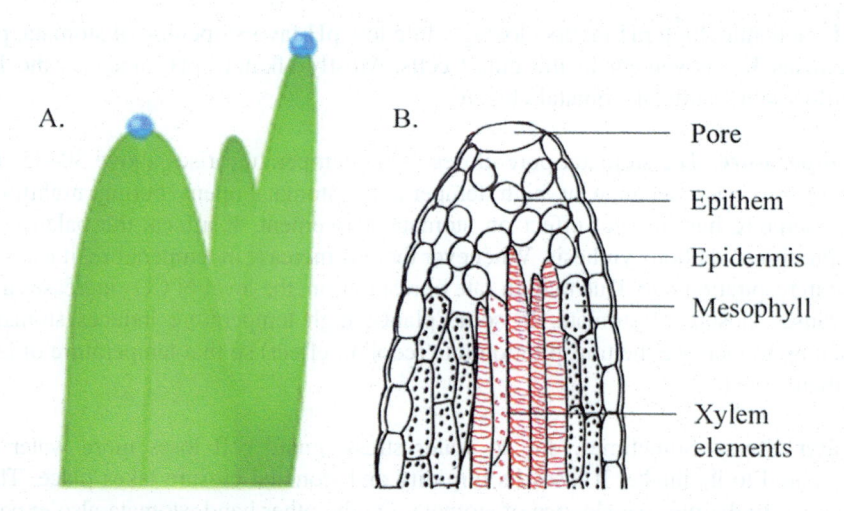

**Fig. 1.12** (**a**) Grass leaves showing guttation through hydathodes. (**b**) Sectional view of a hydathode

the surface area for transport of water. Plants which loose water by guttation are restricted to few taxons in the plant kingdom. There are about 150–250 plants showing this phenomenon, e.g., tomato, grasses, *Colocasia*, cucurbita members, balsam, etc.

**Summary**

- The chemical potential of water is the free energy of water. Water moves from its higher potential to lower potential. The gradient of water potential from soil to leaf, which creates soil-root-air continuum, is the key to the transport of water. The additive effects of solute potential and pressure potential ultimately affect water potential.
- Mode of water transport changes at three different levels. The first level of water transport refers to entry of water from the soil into the root cells, through the plasma membrane. The absorption and release of water and solutes by individual cells occur through diffusion, osmosis, or mass flow. A key component of cellular transport is crossing the plasma membrane. The second level of transport is from root epidermal cells to the innermost layer of cortex. This is referred as short-distance transport, and it includes apoplast, symplast, and transcellular pathways, including transporters, channels, and plasmodesmata. The third level of transport is the long-distance transport in xylary elements, also known as ascent of sap. Water moves against gravity by three forces, viz., transpirational force, cohesion, and tension. Loading of water into xylem again involves transporters, pumps, and channels.
- The driving force for water movement from soil to leaves and from leaves to atmosphere is transpiration. Transpiration is the evaporation of water from the

aerial part of the plants. The most important environmental factors are humidity, light, temperature, wind velocity, and availability of soil water. Water vapor leaves the air spaces of the plant via the stomata. Stomata open up during the day for gaseous exchange. At night when transpirational pull is almost negligible, roots actively absorb solutes. Water passively enters into roots and creates pressure. Due to this pressure, water is forced out of the leaf tips through structures called as hydathodes by a process called guttation.

## Multiple-Choice Questions

1. Which of the following phenomenon is involved in shrinkage of grapes when placed in hypertonic sugar solution?
   (a) Deplasmolysis
   (b) Imbibition
   (c) Exosmosis
   (d) Osmosis
2. Guard cells differ from epidermal cells in having:
   (a) Mitochondria
   (b) Chloroplast
   (c) Nucleus
   (d) Golgi body
3. When water enters a cell, it builds up positive:
   (a) Osmotic pressure
   (b) Turgor pressure
   (c) Vapor pressure
   (d) Atmospheric pressure
4. The surface of leaves remains cool due to:
   (a) Transpiration
   (b) Guttation
   (c) Transport of water
   (d) Evaporation
5. During daytime, if carbon dioxide concentration around the leaves increases:
   (a) Stomata will open gradually.
   (b) Stomata will open suddenly.
   (c) No change in transpiration.
   (d) Decrease in transpiration due to closure of stomata.
6. The movement of water in xylem is due to:
   (a) Cohesive force among water molecules
   (b) Adhesive force between water and wall of xylem element
   (c) Transpirational pull
   (d) Cohesion and transpirational pull

7. How would you treat an epidermal peel if stomatal pore is to be opened?
   (a) Float the peel in abscisic acid.
   (b) Take buffer of pH 7 and add KCl and immerse the peel in it.
   (c) Use sucrose solution.
   (d) Use fusicocin.
8. A potato tuber weighing 0.5 gm and water potential of 1 MPa is immersed in coconut for 1 h. The tuber is removed and again weighed. What do you conclude about the water potential of coconut water if potato tuber weight after the treatment is reduced to 0.35 gm?
   (a) Less than 1 MPa.
   (b) More than 1 MPa.
   (c) 0 MPa.
   (d) It is not possible to find water potential of coconut water.
9. Four solutions contain 1 g/L of the following molecules. Which one would have lowest water potential?
   (a) Sucrose
   (b) Glucose
   (c) DNA
   (d) Starch
10. Guttation is shown by the plants when:
    (a) Root pressure is equal to transpiration.
    (b) Temperature and relative humidity are low.
    (c) Temperature is high and relative humidity is low.
    (d) Root pressure is high due to high temperature and relative humidity.

**Answers**

1. c   2. b   3. b   4. a   5. d   6. d   7. b
8. a   9. d   10. d

## Suggested Further Readings

Jones RL, Ougham H, Thomas H, Waaland S (2013) The molecular life of plants. Wiley-Blackwell, Chichester, pp 504–533

Patrik JW, Tyerman SD, van Bel AJE (2015) Long distance transport. In: Buchanan BB, Gruissem W, Jones RL (eds) Biochemistry and molecular biology of plants. Wiley-Blackwell, Chichester, pp 658–701

Ridge I (ed) (2002) Plants. Oxford University Press, New York, pp 105–165

Taiz L, Zeiger E (2010) Plant physiology, 5th edn. Sinauer Associates Inc, Sunderland, pp 85–105

# Plant Mineral Nutrition

Renu Kathpalia and Satish C Bhatla

Elements mainly derived from soil in inorganic form are known as **mineral elements**. The three main sources of nutrients for plants are air, water, and soil. Elements obtained from air are known as **non-mineral elements**, such as carbon, oxygen, and hydrogen. Irrigation water is also a source of mineral elements from dissolved salts, mainly NaCl, $Na_2SO_4$, $NaHCO_3$, $MgSO_4$, $CaSO_4$, $CaCl_2$, KCl, and $K_2SO_4$. The third environmental source of nutrition for autotrophic plants is soil. Figure 2.1 shows the relative distribution of various elements in the earth's crust, most of which are required for plant growth. Of these, oxygen constitutes about 46.5%, copper being 0.01%, and zinc, nickel, and selenium are present in traces. The relative percentage of some of the well-known macronutrients required for plant growth (calcium, potassium, magnesium, and phosphorus) is in the range of 3.6–0.12%. Phosphorus is the least abundant among the four elements. Among macroelements, nitrogen is the most abundant (44%) followed by potassium, calcium, magnesium, phosphorus, and sulfur. Iron is the most abundant (51%) microelement followed by manganese, boron, zinc, and copper. Mineral elements are essentially used in the synthesis of a variety of important organic compounds. They also play a variety of roles as ions or as components of inorganic compounds (Table 2.1). Mineral nutrients continually cycle through all living organisms. However, they enter biosphere primarily through the roots. Elemental composition of plants reflects the composition of soil. In soil, more than 60 elements are present, but not all are absorbed by the plants. At the same time, many of the ions absorbed by the plants remain in ionic state inside the cells for an indefinite period. Many of these ions are incorporated either into the structure of more complex molecules, such as storage proteins, calcium oxalate, glycosides, etc., or into the protoplasm or cell walls. Some mineral elements are utilized in one organ of a plant and get subsequently released by disintegration of cellular constituents. These are further translocated to other organs of the plant for reutilization. Redistribution of minerals which have accumulated in cells but have not been actually utilized is also common in plants. Plants have the ability to accumulate **essential elements** and exclude

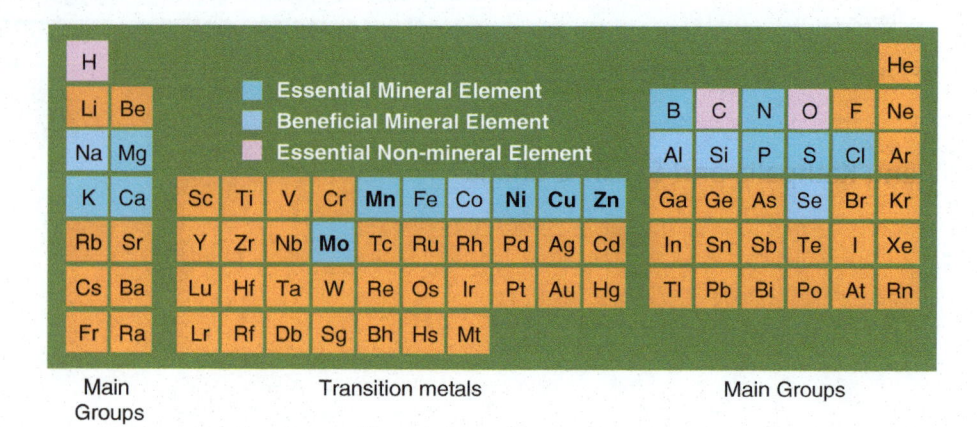

**Fig. 2.1** Distribution of essential and beneficial mineral elements. All, except molybdenum, are among the 30 lightest elements. Elements in bold letters are hyperaccumulators in plants

**Table 2.1** The roles of various mineral nutrients in plant cells

| Mineral nutrients | Role in metabolism |
| --- | --- |
| **As a constituent of organic compounds** | |
| Nitrogen | Amino acids, proteins, purines and pyrimidines, chlorophyll, and many coenzymes |
| Phosphorus | Nucleic acids, phospholipids, ADP, ATP, NAD, NADP, and phosphate esters of sugars |
| Sulfur | Amino acids like cysteine and methionine, vitamin (thiamine and biotin) |
| **As enzyme activators** | |
| Iron | Cytochromes, peroxidases, catalases, and metalloflavoproteins |
| Calcium | Hydrolysis of ATP and phospholipids |
| Magnesium | Enzymes involved in carbohydrate metabolism and synthesis of DNA, RNA |
| Manganese | Nitrite reductase and hydroxylamine reductase |
| Molybdenum | Nitrate reductase, nitrogenase (in symbiotic association) |
| Zinc | Alcohol dehydrogenase, glutamic dehydrogenase, lactic dehydrogenase, and alkaline phosphatase |
| Copper | Cytochrome oxidase, ascorbic acid oxidase, polyphenol oxidase, and plastocyanin |
| **Other roles** | |
| $K^+$ and $Na^+$ | Increase membrane permeability |
| $Ca^{2+}$ and $Mg^{2+}$ | Reduce permeability |
| Calcium | As calcium pectate in cell wall |
| Magnesium | Component of chlorophyll molecule and pectate in the middle lamellae |

nonessential elements. Some minerals are required in minute quantities, e.g., plant requires 60 million times less molybdenum than hydrogen (Box 2.1). Such minute quantities could not be detected by the crude methods which were used earlier. In

**Box 2.1: Mineral Nutrition: Historical Background**
**Aristotle** (384–322 BC) and his student **Theophrastus** (371–285 BC) observed that plants require soil for their growth as it provides nutrition to the plants. Aristotle considered soil as vast stomach for the plants that prepares and supplies food to the plants. This observation was referred to as "humus theory of plant nutrition." According to this theory, humus provides carbon to the plants. However, according to **Thales of Miletus** (460–547 BC), "If water can change into ice and air then perhaps under some circumstances it changes into a tree or a rock." In 1648, **Van Helmont** noticed the requirement of inorganic nutrition in plants. In an experiment carried with willow plants, he observed that after 5 years, weight of dry soil decreased by 2 oz., while weight of the plant increased by 3 oz. He further observed that 61 pounds of gases and 1 pound of ash were produced when he burned 62 pounds of oak charcoal. He concluded that weight of plant increased by water only. In 1656, **Glauber** observed promotion of plant growth when potassium nitrate was added to the soil. He concluded that potassium nitrate is an "essential principle of vegetation." Later in 1699, **Woodward** observed that the plants grow better in muddy water as compared to rainwater. Thus, besides water, something else was believed to be contributing to the growth of plants. Water was only acting as a vehicle to transport nutrients from soil to the plants. In 1804, **Nicolas Théodore De Saussure** provided foundation for present knowledge of mineral nutrition. He rejected the theory of humus and demonstrated that plants obtain minerals from the soil through their root system and gases from air. **Sprengel** and **Boussingault** examined the relationship between fertilizer application and crop yield. Boussingault is also credited for providing the first evidence that legumes have the unique capacity to assimilate atmospheric nitrogen. **Sachs** and **Knops** started liquid culture of plants to study the nutrient requirements of the plants. They identified ten elements which are essential for the plant growth, viz., C, H, O, N, P, K, Ca, S, Mg, and Fe. **Justus von Liebig** was pioneer of agricultural chemistry and proposed the **law of minima** which states that productivity depends on the amount of deficient minerals. He stressed upon the importance of replenishing the mineral nutrients by fertilizer application. He analyzed ash composition of many plants and recorded 109 mineral elements to be present. Out of these, 20 elements were present in all plants, and these elements were called as **essential elements. J.B. Lawes** and **J.H. Gilbert** successfully converted insoluble rock phosphate into soluble phosphate (called superphosphate) by treating with sulfuric acid. **Arnon** and **Stout** in 1939 gave three criteria for an element to be considered as essential element.

addition to seven **ash elements** (the incombustible fraction of plant tissue), all elements which are required in large quantity (except for iron) are now known as **macronutrients**. There is another group of minerals, the **micronutrients**, needed in only minute quantities. It includes molybdenum, copper, zinc, manganese, boron, chlorine, and nickel.

In addition to essential elements, there is a small group of **beneficial elements** (sodium, silicon, aluminum, selenium, and cobalt) required by some plants. Sodium is essential for animals but is also required by $C_4$ plants. Although aluminum is found in both plants and animals, it is not essential for either of them. It is toxic to plants and may even be detrimental to animals. In some plants, aluminum acts as a functional element. For example, it is beneficial for the growth of tea plants and combats heavy metal toxicity. Accumulation of some micronutrients leads to toxicity in plants. Toxicity of elements generates reactive oxygen species which causes cellular damage. Some highly toxic elements, like lead and cadmium, are not distinguishable by the roots, and they enter the food web via nutrient uptake. Plants utilize various strategies for mobilization and uptake of nutrients. Soil properties like humidity, soil pH, and aeration are responsible for the availability of nutrients to the plants (Box 2.2). Most plants grow in nutrient-limited soils by changing root structure and increasing overall surface area to increase nutrient acquisition. They may exhibit elongation of the root system to access new nutrient sources. The most common change is inhibition of primary root growth (often associated with phosphorus deficiency). Other changes include increase in lateral root growth and density (during nitrogen, phosphorus, iron, and sulfur deficiency) and increase in root hair growth and density (during iron and phosphorus deficiency). An element present at a low level in the soil may cause deficiency symptoms in plants, while the same element at a higher level may cause toxicity. Sometimes, symptoms of deficiency of one element are similar to symptoms of toxicity of another element. Furthermore, abundance of one nutrient may cause deficiency of another nutrient. For example, lower availability of $SO_4^{2-}$ can affect the uptake of $NO_3^-$, and $K^+$ uptake can be influenced by the amount of available $NH_4^+$. In certain plants, abnormal growth can be attributed to mineral deficiency, e.g., **"heartrot"** of sugar beet (boron), **"dieback"** in citrus tree (copper), **"whiptail"** of cauliflower (molybdenum), and **"little leaf"** of apple (zinc). All these are visible symptoms and can also be treated. This chapter emphasizes on inorganic plant nutrition requirement and their roles in plant growth and development and also on their deficiency symptoms.

**Box 2.2: Soil-Nutrients Relationship**
The concentration of free ions in the soil solution is generally low. The major portion of the cations is absorbed on the negatively charged sites of clay micelle and organic material in the soil. Anions, such as nitrate and sulfate, usually occur as free ions in the soil solution, whereas phosphate is firmly bound to the clay particles and is found at low concentrations in the soil solution. Anions are absorbed from the soil solution, but cations are exchanged directly between roots and soil particle by cation exchange. The soil cations essential for plant growth include ammonium, calcium, magnesium, and potassium ions. There are three additional cations which are not essential plant elements but affect soil pH. These include sodium, aluminum, and hydrogen. Soil cations are further divided into two categories. Ammonium, calcium, magnesium, potassium, and sodium are known as *base cations*, whereas aluminum and hydrogen are known as *acid cations*. When soil particles have a negative charge, they attract and retain cations. Most of the soils are negatively charged and attract cations. The ability or capacity of the soil colloid to hold cations is called **cation exchange capacity** (CEC). Cations in the soil compete with one another for a spot on the basis of CEC. However, some cations are attracted and held more strongly than other cations. Holding strength order of cations as held by soil particles is $Al^{+3} > H^+ > Ca^{+2} > K^+ > Na^+$. This is directly dependent on the amount of charge on the soil colloid. The number of cations a soil can hold is dependent upon the quality and composition of soil, pH of soil, and the presence of hydrous oxides of iron and aluminum. CEC is expressed as milliequivalents per 100 g of soil. CEC is indicative of nutrient-holding capacity of a soil. Addition of lime and how often it should be added are determined by CEC. In tropics, many highly weathered soils retain anions, rather than cations. The anions that are held by soil particles include phosphate, sulfate, nitrate, and chlorine (in order of decreasing strength). In comparison with soils with CEC, soils with an anion capacity have net positive charge. Soils that have an **anion exchange capacity** (AEC) typically contain weathered kaolin (hydrated aluminum silicate) minerals, iron and aluminum oxides, and amorphous materials. AEC is also dependent upon the pH of the soil and increases as the pH of the soil decreases. Base saturation is a measurement that indicates the relative amounts of base cations in the soil. It is the percentage of calcium, magnesium, potassium, and sodium cations that make up the total CEC. For example, a base saturation of 25% means that 25% of the CEC is occupied by the base cations. If soil does not exhibit AEC, then remaining 75% of the CEC will be occupied by acid cations. Generally, the base saturation is relatively high in moderately weathered soil that is formed from the basic igneous rocks. The pH of soil increases as base saturation increases. In contrast, highly weathered and acidic soils tend to have low base saturation. Whenever there is depletion of any free ions from the soil solution, respective ions are released from clay particles into the soil solution to maintain the equilibrium.

(continued)

> **Box 2.2** (continued)
> 
> This is achieved by a process called ion exchange process. This may be due to contact ion exchange mechanism or by carbonic acid ion exchange mechanism. Roots respire and liberate significant quantities of $CO_2$, which when dissolved in soil water produces carbonic acids.
> 
> **Contact ion exchange mechanism**: Plant roots are in contact with soil clay particles which have colloidal dimensions. Root cells, which are living, secrete hydrogen ions. Hydrogen ions displace cations like $K^+$, $Na^+$ ions that are bound to clay particles. Order of the strength of adsorption of cations by the soil is $Al^{+3} > H^+ > Ca^{+2} = Mg^{+2} > K^+ > NH^{4+} > Na^+$. This is called the *lyotropic series* or *Hofmeister series*. It is a classification of ions in the order of their ability to salt out or salt in water. The sequence of ions is determined by their charge, size, and hydration. It lists the common soil cations in order of their strength of bonding to the cation exchange surface.
> 
> The charged soil particle with adsorbed cations and anions at different pH

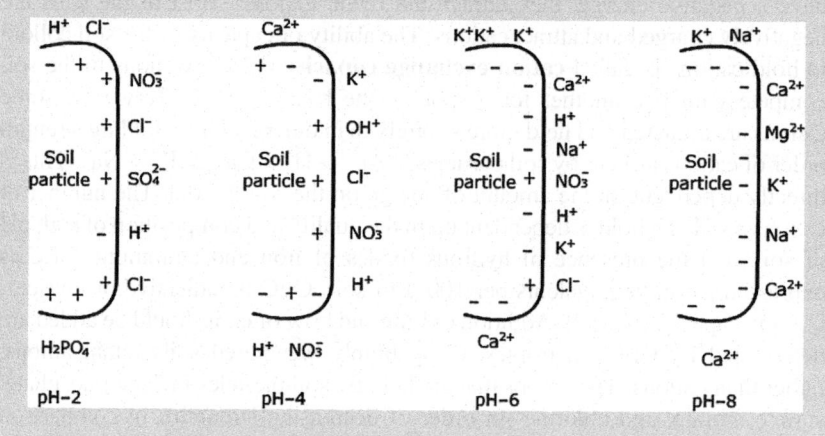

## 2.1 Plant Nutrition

On the basis of modes of nutrition, living organisms have been classified into **autotrophs** (Greek: auto, self; trophe, nourishing) and **heterotrophs** (Greek: hetero, different; trophe, nourishing). Green plants obtain energy from sunlight and synthesize their own food using raw material by the process known as photosynthesis and are hence called as autotrophs. Since green plants use light to synthesize their food, they are called **photoautotrophs**. Heterotrophs are those living organisms which cannot synthesize their own food. The heterotrophic plants are further categorized into three main groups, viz., **saprophytes** (Greek: sapros, rotten; phyton, plants), **parasites**, and **insectivorous/carnivorous plants,** which

## 2.2 Essential Elements

> **Box 2.3: Mineral Nutrition of Carnivorous Plants**
>
> Plant carnivory is an adaptation strategy to unfavorable conditions, mostly low-nutrient availability in wet and acidic soils. There are about 600 terrestrial and 50 aquatic or amphibious species of carnivorous plants (CPs). Some of the common CPs are *Nepenthes* sp. (pitcher plant), *Drosera* sp. (sundew), *Dionea* sp. (Venus flytrap), and *Utricularia* sp. (bladderwort). All CPs are green and are able to fix $CO_2$ (autotrophy) but are partly dependent on organic carbon uptake from prey (*facultative heterotrophy*). Nearly all plants with glandular hairs are potentially carnivorous. Foliar nutrient uptake from prey is "ecologically significant" for CPs. The glandular hairs contain phosphatase, phosphodiesterase, and protease activities. They are generally present in nitrogen-deficient soil and use insects as a source of nitrogen. CPs are classified into three main ecophysiological groups. The first group of plants is "nutrient-requiring species" which increase their growth due to both soil and leaf nutrient supply. CPs in the group of "root-leaf nutrient competitors" grow better and accumulate more nutrients because of both root and leaf nutrient uptake, resulting in a competition between the root and leaf nutrient uptake. CPs in the third group of "nutrient modest species" have roots with very low nutrient uptake capacity. Therefore, they depend on leaf for their nutrient supply. However, terrestrial CPs have considerably lower content of macroelements per dry weight as compared to aquatic CPs. Nutrient uptake from prey is advantageous because animal prey is relatively rich in mineral nutrients. The total nutrient contents found in insects ($g.kg^{-1}$ DW) are N, 99–121; P, 6–14.7; K, 1.5–31.8; Ca, 22.5; and Mg, 0.94.

eat insects. Saprophytic plants grow on dead decaying matter of plant and animal origin. They release extracellular enzymes which break down complex organic compounds into simpler forms. Parasitic plants derive food from the host by penetrating their **haustoria** into the phloem of the host plants, for example, *Cuscuta*. There are certain autotrophic plants which derive nutrition from insects to supplement the deficiency of a specific mineral in the soil. Such plants are called insectivorous or carnivorous plants (Box 2.3).

## 2.2 Essential Elements

At present the list of essential elements includes 13 elements which are essential for all angiosperms and gymnosperms. Adding carbon, hydrogen, and oxygen makes it 16, and after the addition of nickel (a trace element), it makes a total of 17 essential elements. The known essential elements have interesting distribution in periodic table (Fig. 2.1). All, except molybdenum, are among the 30 lightest elements and are clustered in such a way that every essential element (except hydrogen) lies adjacent to another horizontally or vertically placed element. Only molybdenum is present diagonally to manganese.

### 2.2.1 The Criteria of Essentiality

The **dry matter** residue from any plant tissue obtained after desiccation can be separated into **combustible** and **incombustible** fractions. The combustible fraction represents organic matter, while incombustible element is called **ash**. Ash roughly corresponds to the mineral salts absorbed by the plant from the soil. Nitrogen is not included in ash because it is released during combustion process along with carbon, hydrogen, and oxygen. The mineral elements occur as oxides in the ash. Hence, the ash content of a plant tissue gives a crude estimation of mineral content of the tissue. The presence of a particular element in plant ash does not necessarily mean that it is an important element for plant growth and development. Roots absorb around 60 elements from the soil, but not all are required for plant growth. The nutrients or elements necessary for growth or completing life cycle of a plant are considered as essential elements. Only 16 elements are essential for most of the plants. Criteria to find out the essentiality of microelements are difficult. The criteria of essentiality of elements are as follows:

- Deficiency of essential elements prevents completion of life cycle and produces deficiency symptoms in the plant.
- They cannot be replaced by another element with similar properties.
- They are directly involved in plant metabolism.
- In the absence of essential elements, plants are unable to produce viable seeds.
- Essential elements should be constituents of some essential plant metabolites, e.g., $Mg^{2+}$ is a constituent of chlorophyll molecule.

### 2.2.2 Roles of Essential Elements

Essential elements play two major roles in plants: (i) structural role and (ii) activators of enzymes. There is no sharp distinction between their roles because many elements form structural components of essential enzymes and help to catalyze the chemical reactions in which enzymes participate. Carbon, hydrogen, oxygen, nitrogen, sulfur, and magnesium perform both the functions. Magnesium is a structural constituent of chlorophyll molecules and it also activates many enzymes. A number of elements also regulate osmotic potential. Monovalent ions, potassium and chloride, control osmotic potential as well as act as activators of certain enzymes (Table 2.1). Depending upon the role played by essential elements, they can be broadly classified as:

**Framework/Structural Elements** These elements are components of biomolecules present in protoplasm, cell wall, and storage products in plants, e.g., sulfur in proteins, phosphorus in nucleoproteins and lecithins, magnesium in chlorophyll, and calcium in calcium pectate. Carbon (C), hydrogen (H), and oxygen (O) are elements which make the structural framework of plants.

**Colloidal Elements** Cations and anions such as calcium, magnesium, and chloride ions influence the degree of hydration of the colloidal micelles in the protoplasm and affect the permeability of the membrane. In general, potassium and sodium enhance cell permeability, whereas divalent ions, such as calcium and magnesium, reduce membrane permeability.

**Elements Associated with pH Modulation and Buffering Action** The mineral salts absorbed from the soil influence the pH of cell sap. The metabolic activity of the cell depends on its pH. Two important buffer systems found in the plants are the phosphate and carbonate systems.

**Balancing Elements** They minimize the toxic effects of heavy elements, e.g., $Ca^{2+}$, $Mg^{2+}$, and $K^+$.

**Activator, Cofactor, or Constituent of Enzyme** Elements like iron, copper, and molybdenum are constituents of many enzymes. Some minerals are either the components of enzymes or are constituents of cofactors or they act as activators of the enzymes, e.g., $Ca^{2+}$, $Mg^{2+}$, $K^+$, Mn, Cu, Ni, Zn, etc. Calcium is required as a cofactor or enzyme activator for hydrolytic enzymes involved in hydrolysis of ATP and phospholipids hydrolysis. Magnesium is required by a large number of enzymes involved in phosphate transfer. Manganese is involved in the activity of some dehydrogenases, decarboxylases, kinases, and peroxidases.

**Osmotic Pressure and Turgor Movements** The osmotic pressure of the cell sap is due to the presence of dissolved mineral salts. The difference in osmotic pressure in different cells is responsible for intercellular transport. For example, expansion and contraction of guard cells involve rapid $K^+$ fluxes across the cell.

## 2.3 Macroelements and Microelements

### 2.3.1 Macroelements or Macronutrients

These are mineral nutrients consumed in larger quantities and are present in plant tissue in quantities ranging from 0.2% to 4.0% on dry matter weight basis. The **primary macronutrients** are nitrogen, phosphorus, and potassium, while **secondary macronutrients** include calcium, sulfur, and magnesium. Usually, potassium, calcium, and magnesium are grouped together as they are present as cations ($K^+$, $Ca^{2+}$, $Mg^{2+}$). Similarly, nitrogen, phosphorus, and sulfur are grouped together because they are present as anions ($NO_3^-$, $SO_4^-$, and $H_2PO_4^-$).

### 2.3.2 Microelements or Micronutrients

These are the mineral nutrients present in plant tissue in quantities measured in parts per million (ppm), ranging from 5 to 200 ppm or less than 0.02% of dry weight. Micronutrients or trace minerals are boron, chlorine, manganese, iron, zinc, copper, molybdenum, and nickel. Microelements are present as inorganic ions, oxyanions (anion with one or more oxygen atom), or undissociated molecules of boron or as organic compound complexes (chelates).

## 2.4 Beneficial or Functional Elements

Mineral elements which either stimulate growth, but are not essential, or which are essential only for certain plant species under given conditions are referred to as beneficial or functional elements. In addition to 17 essential elements, some plants require certain other elements for their growth. In some cases, a particular element may substitute an essential element when that element is deficient and thus produces its beneficial effects. Alternatively, it may stimulate absorption and transport of an essential element which is in limited supply or inhibit uptake and distribution of an element that is present in excess. Sodium is required for regeneration of phosphoenolpyruvate from pyruvate in chloroplast in CAM and $C_4$ plants. It was first observed in bladder saltbush (*Atriplex vesicaria*) in which the absence of sodium resulted in reduced growth, **chlorosis**, and **necrosis** of the leaves. Silicon is the second most abundant element in the earth's crust. Higher plants differ characteristically in their capacity for silicon uptake. Depending on their silicon content, they can be divided into three major groups: (1) wetland Gramineae, which includes wetland rice and horsetail (10–15%); (2) dryland Gramineae, which includes sugarcane, most of the cereal species and few dicotyledon species (1–3%); and (3) most of dicotyledons including legumes (<0.5%).

The long distance transport of silicon in plants is confined to xylem. Therefore, its distribution within the shoot is determined by transpiration rate. Silicon can stimulate growth and yield of plants due to several indirect actions. These include decrease in mutual shading by improving leaf erectness and decrease in susceptibility to lodging and preventing manganese and iron toxicity. It strengthens cell wall and thus improves drought and frost resistance and boosts the immune system of the plants. It is responsible for improvement of root mass and density and hence results in an increase in biomass as well as yield. It is an essential element for plant growth and development (except for specific plant species, such as sugarcane, and members of the horsetail family). Silicon is considered a beneficial element in many countries due to its numerous benefits to various plant species. Silicon is currently under consideration by the Association of American Plant Food Control Officials (AAPFCO) for its elevation to the status of a "plant beneficial substance." Cobalt is essential for some plants, such as legumes, where it is required by nitrogen-fixing bacteria rather than the host plant. When legumes are provided with nitrates, cobalt is not required. In nonleguminous plants, it is beneficial.

## 2.5 Micronutrient Toxicity

Micronutrient toxicity is the result of accumulation of high levels of trace elements (primarily iron and manganese) in plant tissues. The critical concentrations (above which plants exhibit toxic effects) of copper and zinc are 20 and 200 $\mu g.g^{-1}$ dry weight, respectively; critical toxicity level of manganese is 200 $\mu g.g^{-1}$ dry weight in corn, while it is 5300 $\mu g.g^{-1}$ dry weight for sunflower. Root growth is the first one to be affected by micronutrient toxicity. In vineyards and orchards, excess use of copper-containing fungicides and subsequent soil pollution lead to copper toxicity. Acidic soils increase zinc toxicity. The most common symptoms of zinc toxicity are interveinal chlorosis, necrotic speckling, and leaf deformity. It usually occurs in older or lower leaves and in mature tissues. The toxicity symptoms are difficult to predict as an excess of one nutrient may induce deficiency of another nutrient, e.g., the toxicity symptoms of manganese (due to $MnO_2$) are brown spots surrounded by chlorotic veins. Excess of manganese also induces deficiency of iron, magnesium, and calcium. Manganese competes with iron and magnesium for their uptake. It also competes with magnesium for binding sites in some enzymes. It inhibits calcium translocation into the shoot apex and induces deficiency commonly known as "crinkle leaf" (leaf cups upward and is smaller in size and has chlorotic edges). The symptoms resemble those of calcium deficiency. Now it has been shown that these symptoms are due to manganese-induced calcium deficiency. Therefore, the symptoms of manganese toxicity overlap with the deficiency symptoms of iron, magnesium, or calcium. Moist, organic soils under acidic conditions are especially susceptible to manganese toxicity. Manganese toxicity results in the "sudden crash" syndrome of watermelon which results in wilting and death of plant. When the soil pH drops below 5.2, manganese minerals become highly soluble and perhaps toxic. Manganese oxide gets solidified and becomes unavailable to plants, while manganese ions are soluble and are readily available for plant uptake.

$$MnO_2 + 4H^+ + 2e^- \rightarrow Mn^{2+} + 2H_2O$$

Farmers may reduce manganese toxicity by liming and aerating the soil. Plants fertilized with ammonium-nitrogen ($NH_4^+$-N) salts may exhibit $NH_4^+$ toxicity symptoms, accompanied with carbohydrate depletion and reduced plant growth. Excess phosphorus, manganese, or zinc can cause iron deficiency (chlorosis in young leaves) and symptoms of nutrient toxicity (old leaves). High nickel concentrations can also cause iron displacement. Tolerance to aluminum varies among plant species. Certain crops, such as sugarcane, pineapple, and corn, can tolerate relatively high levels of aluminum. Aluminum toxicity inhibits root development and limits crop growth. It occurs readily under acidic conditions, especially when pH values are equal to or less than 5.4. In acidic soils of the tropics, aluminum toxicity may become a serious problem and limit crop yield. Management of soil pH is a key factor in avoiding aluminum toxicity. Tea plants exhibit a high degree of tolerance for aluminum toxicity and their growth is stimulated by its application. The possible reason is the prevention of copper, manganese, or phosphorus toxicity effects. There have been reports that aluminum may serve as a fungicide against certain types of root rot. Plants have homeostatic mechanisms to cope with changing concentrations of toxic inorganic ions (Box 2.4). A variety of naturally produced organic acids released in soil also play a crucial role in mineral nutrient acquisition by plants.

---

**Box 2.4: Metal Homeostasis**

In plant cells, the major sinks of micronutrient metal cations are chloroplasts and mitochondria (50% of copper and 80% of iron are present in the chloroplasts of leaf), whereas the major repository for toxic ions in roots as well as shoots is vacuoles. Therefore, it is very important to have efficient mechanism of transport of these toxin ions through cytoplasm to organelles. The transport of metal ions through cytoplasm is facilitated by proteins that act as **metallochaperones.**

**Metallothioneins** act as metallochaperones and control toxic ion concentrations in plant cells. These are also present in mycorrhizal fungi. There are molecules that routinely detoxify or chelate reactive metals. In general, glutathione and amino/carboxylic acids perform these functions in plant cells.

(continued)

## 2.5 Micronutrient Toxicity

**Box 2.4** (continued)

Mechanism of metal homeostasis in plants

## 2.6 Deficiency Symptoms of Mineral Elements in Plants

Plants respond to inadequate supply of essential mineral elements by exhibiting deficiency symptoms (Table 2.2). The characteristic deficiency symptoms can be correlated with specific deficient mineral element using "hydroponics," but under natural environment conditions, this correlation becomes difficult to observe (Box 2.5). Inadequate supply of an element can be due to its low concentration in soil, the presence of an element in the form that the plant cannot access, or the influence of other factors, such as soil pH, aeration, water status, or high concentration of antagonistic elements. Under natural conditions, these symptoms are the indicators of mineral deficiencies in the soil. Most of the symptoms appear on the shoot system and are easily observed (Fig. 2.2). Deficiency symptoms are of many types (Table 2.3):

**Table 2.2** Deficiency symptoms of different nutrient elements in plants

| Elements | Deficiency symptoms | Symptoms due to excess supply |
| --- | --- | --- |
| Nitrogen | Lateral bud dormancy, wrinkled cereal grains | Dark bluish-green leaves; high shoot/root ratio; new growth will be succulent and susceptible to disease; insect infestation and drought stress; flower abort and lack of fruit set |
| Phosphorus | Premature leaf fall and flower bud; delay in seed germination; chlorosis; and necrosis first in older leaves | No direct effect on the plant, but will show deficiency symptoms of Zn, Fe, Mn, or Ca; maturity often delayed; poor vascular tissues; shoot growth is less and root growth is more |
| Potassium | Loss of apical dominance; interveinal chlorosis first in older leaves; scorched leaf tips; short internodes; and dieback | Excess of potassium causes Mg and Ca deficiency |
| Calcium | Stunted growth; degeneration of meristems, especially root meristem; growing tips of roots and leaves will turn brown and die; fruit quality will be affected; decay of the conductive tissue in lower region of stem wilt easily | Deficiency symptoms of magnesium can be seen, and if concentration further increases, potassium deficiency may also occur |
| Magnesium | Interveinal chlorosis and anthocyanin pigmentation appear; older leaves affected first; premature leaf abscission | Occurs rarely; results in cation imbalance; plant shows calcium and/or potassium deficiency |
| Sulfur | Chlorosis first in young leaves; reduced nodulation in legumes, stunted and delayed growth; anthocyanin accumulation; defoliation in tea | Premature senescence of leaves |
| Iron | Interveinal chlorosis appearing first in young leaves, slow growth of the plant | Occurs rarely; results in bronze-colored leaves with brown spots; |

(continued)

## 2.6 Deficiency Symptoms of Mineral Elements in Plants

**Table 2.2** (continued)

| Elements | Deficiency symptoms | Symptoms due to excess supply |
|---|---|---|
| | | symptoms frequently seen in rice leaves |
| Manganese | Interveinal chlorosis with gray spots on young leaves; malformed leaves, white streaks on the leaves of some plants; plant growth is slow | Older leaves show brown spots surrounded by a chlorotic zone; tree fruits, referred to as measles |
| Zinc | Upper leaves show interveinal chlorosis; stunted growth; dieback; internodes will be short and plants will be stunted | Fe deficiency develops; toxicity is severe; plants severely stunted and eventually die |
| Copper | Leaf tip necrosis; blackening of potato tubers; bark becomes rough and splits; loss of apical dominance | Fe deficiency may be induced with very slow growth; root tips may die |
| Boron | Death of root and shoot tips; abscission of flowers; reduced nodulation in legumes; interveinal chlorosis with marginal necrosis and in folding; pollination is reduced; no internode elongation giving a compressed appearance | Leaf tips and margins turn brown and die; stunted growth toxic to many plants |
| Molybdenum | Similar to nitrogen; slight retardation of growth; chlorosis and necrosis of old and middle leaves; sometimes leaf margins get rolled, new growth is malformed, and flower formation is restricted | Not of common occurrence |
| Chlorine | Bronze color in leaves; wilting of leaves; swollen root tips; stunted root growth | Premature yellowing of the lower leaves with burning of leaf margins and tips; wilting and leaf abscission in woody plants |

1. *Chlorosis*—it refers to yellowing of leaf tissue due to lack of chlorophyll. Poor drainage in soil, damaged roots, compact roots, high alkalinity, and nutrient deficiencies are some of the reasons for yellowing of leaves. The affected plants or leaves are unable to manufacture carbohydrate and they ultimately die.
2. *interveinal chlorosis*—yellowing in between leaf veins though veins remain green.
3. *Necrosis*—the affected plant tissue usually turns brown to black in color. It is caused due to the death of plant cells. Necrotic symptoms could appear in any part of the plant, such as in storage organs, green tissues, or woody tissues.
4. *Cessation of growth* or terminal growth resulting in rosette appearance.
5. *Pigmentation*—when sugars are not metabolized in the plant cells, it results in the accumulation of anthocyanin. The leaves may become purple because of anthocyanin (glycosylated form of anthocyanidins) accumulation. Reduction in

nitrogen or phosphorus level favors anthocyanin accumulation in various plant parts. This symptom can be particularly difficult to diagnose because low temperatures, disease, drought, and even maturation of some plants can also cause anthocyanin to accumulate.
6. *Stunting* or reduced growth.
7. *Premature fall of leaves and buds.*
8. *Delayed flowering.*

> **Box 2.5: Hydroponics**
> In 1880, two German botanists Julius von Sachs and Knop working independently demonstrated that plants can absorb minerals from solution as well. The plants were grown in Knop's nutrient solution which consisted of potassium nitrate, calcium nitrate, potassium dihydrogen phosphate, magnesium sulfate, and an iron salt. This technique of growing plants in nutrient solution without soil is known as **hydroponics**. Inert materials like vermiculite, sand, coir, etc. can be used as supporting material. Vigorous air bubbling is done to provide oxygen to the root system.
> 
> The most common solution used in hydroponics is **Hoagland's solution**. It consists of all the essential elements in the correct proportion necessary for growth of almost all plants. The concentration of minerals in most nutrient solutions is many times greater than in soils in order to maintain continuous supply of nutrients. The most commonly encountered problem with hydroponics is providing air to roots. The absence of oxygen in solution culture can result in **anoxia** or **hypoxia** in plant roots. Iron absorption poses another challenge as it needs to be provided in chelated form for ready absorption by the roots. Hydroponics offers the following advantages: (1) soil-less raising of plants, (2) the water stays in the system and can be reused, (3) it is possible to control the nutrition levels in the system, and it reduces nutrients loss, (4) no nutritional pollutants are released into the environment because of the controlled system, (5) plants show uniform growth and high yields, (6) pests and diseases are easily eradicated as compared to plants growing in soil, (7) ease of harvesting, and (8) no pesticide used and consequently no damage to the plants. However, some of the hurdles experienced in hydroponics are as follows: (1) any failure in the system that leads to rapid plant death, (2) the need to replace the solution after every few days in order to achieve good growth (this is because the composition of nutrient solution changes as certain ions are absorbed more rapidly as compared to others), and (3) selective uptake of ions that also changes the pH of the medium. For example, when nitrogen is given as nitrate, it is rapidly absorbed by the plants along with $H^+$ resulting in a rapid rise of pH. At high pH, iron and other elements precipitate as hydroxides

(continued)

**Box 2.5** (continued)

and hence are not available to the plant. It can be minimized by adding ammonium salt. Pathogen attack is associated with hydroponics, and overwatering of soil-based plants can lead to wilting, e.g., damp-off due to *Verticillium* wilt caused by the high moisture levels. The mineral requirement of each plant is different. Therefore, it is important to find optimal medium required for growth of different plants.

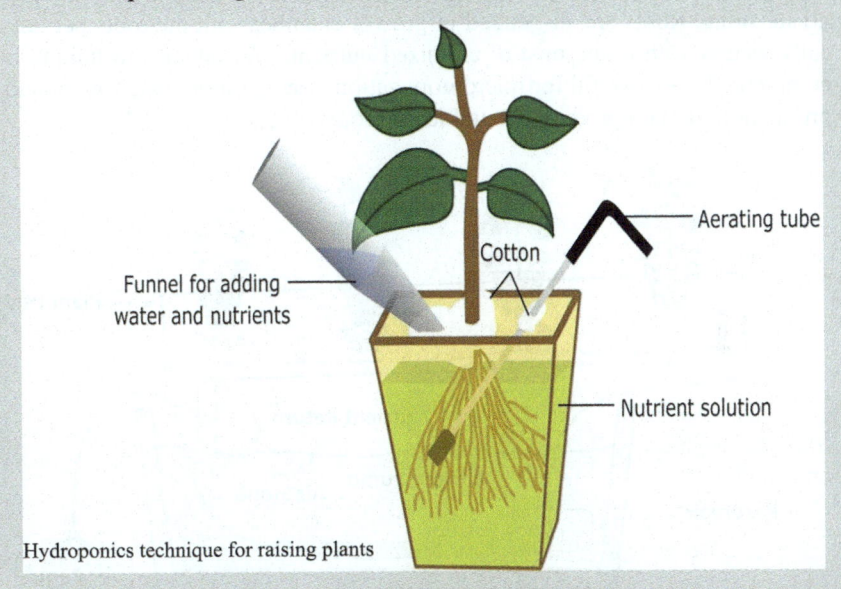

Hydroponics technique for raising plants

Two main types of hydroponics are **solution culture** and **solid culture**. There are three types of solution cultures, viz., **static solution culture**, **continuous-flow solution culture**, and **aeroponics**. **(i) Static solution culture**: Plants are grown in containers with nutrient solution. If it is unaerated, the solution level is kept low enough so that enough roots are above the solution and can get adequate oxygen. Aeration in the medium can also be provided by a small pump. The nutrient solution is changed either on a schedule, such as once per week, or when the concentration of nutrients drops below a certain level, which can be monitored with an electrical conductivity meter. **(ii) Continuous-flow solution culture**: In this method, the nutrient solution constantly flows past the roots. It is much easier to automate than the static solution culture because sampling and adjustments for the temperature and nutrient concentrations can be made in a large storage tank that has potential to serve thousands of plants. **Nutrient film technique (NFT)** is a popular

(continued)

**Box 2.5** (continued)

variation of continuous-flow culture, in which a very shallow stream of water containing all the dissolved nutrients required for plant growth is recirculated around a thick root mat. Subsequent to this, an abundant supply of oxygen is provided to the roots of the plants. The main advantage of the NFT system over other forms of hydroponics is that the plant roots are exposed to adequate supply of water, oxygen, and nutrients. **(iii) Aeroponics:** This method requires no substrate. Roots are suspended in growth chamber with the roots periodically wetted with a fine mist of atomized nutrients. Aeroponics technique is commercially successful for micropropagation, seed germination, seed potato production, tomato production, and leafy crops.

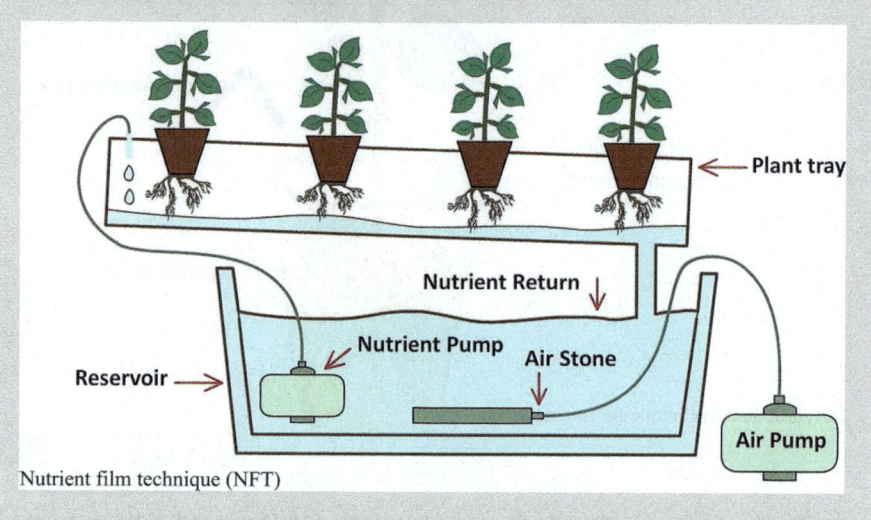

Nutrient film technique (NFT)

The limitation of conventional hydroponics is aeration of roots. One kilogram of water can only hold 8 milligrams of air, no matter whether aerators are utilized or not, but excellent aeration is achieved only by aeroponics. Almost all species of plants can be grown in a true aeroponics system because the microenvironment around the plants can be finely controlled. However, in conventional hydroponics, only certain species of plants can survive due to waterlogged conditions. Plants receive adequate oxygen for the roots which accelerates biomass growth and reduces rooting time. NASA research has shown that plants grown using aeroponics have an 80% increase in dry weight biomass (essential minerals) as compared to plants grown by using conventional hydroponics. Aeroponic-grown plants do not suffer from transplant shock when transplanted to soil and are less prone to diseases.

(continued)

**Box 2.5** (continued)

**Aeroponics**

Advantage of aeroponics over conventional hydroponics

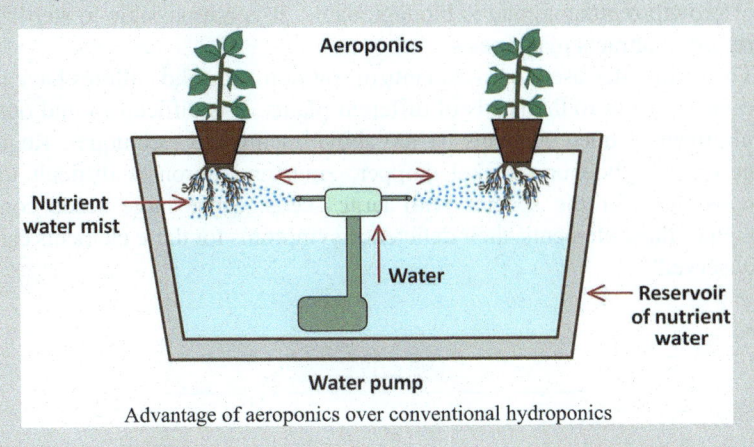

**Solid Medium Culture**

This method uses solid media for the roots and is named on the basis of the type of medium used, as discussed below:

**Vermiculite:** Vermiculite is a hydrated magnesium, aluminum, and silicate mineral which resembles mica in appearance. Vermiculite has a natural "wicking" property to draw water and nutrients in a passive hydroponic system. Vermiculite improves aeration, slightly raises pH, enhances drainage, and does not interfere nutrient availability to the plants.

**Sand:** Sand is easy to recharge with nutrients and can be washed easily. However, it is heavy and does not hold water very well. It is recommended for growing succulents, sand-loving trees, drought-tolerant plants, and *Euphorbia* species.

**Gravel:** Gravel, as used in aquarium, can be used after washing. Plants are grown in a typical traditional gravel filter bed, with water circulated using electrically powered head pumps. Gravel drains well and does not get waterlogged. However, it is heavy, and, if not kept wet, the plant roots may dry out.

**Polystyrene packing peanuts:** These are standard packing peanuts used in shipping industry. Polystyrene packing peanuts have excellent drainage for growing plants. Biodegradable packing peanuts, however, decompose into sludge, and plants may absorb styrene and may cause health risk to their consumers.

**Wood fiber:** It is a very efficient organic substrate for hydroponics. For organic farming the wood fiber is the best medium to supply nutrients. It maintains its structure for a very long time. However, it reduces the effect of

(continued)

**Box 2.5** (continued)

plant growth regulators and is biodegradable. It is not possible to sterilize it; therefore, it attracts many pests.

The techniques used to raise plants in solution and solid cultures have been modified to cater to the needs of different plants. It is difficult to find out the requirement of trace elements by the above mentioned techniques. Requirement for molybdenum, nickel, copper, zinc, and boron is difficult to be demonstrated for the species with large seeds. Since large seeds contain enough of these elements, their deficiency symptoms for these elements cannot be observed.

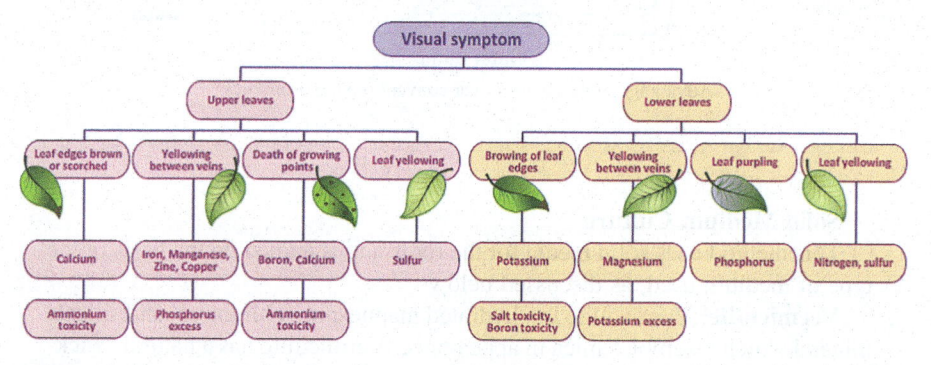

**Fig. 2.2** Major diagnostic features to identify the nutritional status of different mineral nutrients in plants

**Table 2.3** Different types of deficiency symptoms and the elements responsible for those symptoms

| Deficiency symptoms | Elements showing the symptom |
|---|---|
| Chlorosis | K, Mg, N, S, Fe, Mn, Zn, Mo |
| Necrosis | P, K, B, Cu |
| Lack of new growth or terminal growth resulting in rosette | N, K, S, Mo |
| Anthocyanin formation | N, P, S, Mg |
| Stunted/retarded plant growth | N, P, K, Zn, Ca |
| Premature fall of leaves and buds | K, P |
| Delayed flowering | N, S, Mo |

It is, however, not easy to diagnose elemental deficiency symptoms. For example, chlorosis of leaves can be caused by the deficiency of iron or nitrogen or due to low light intensity or insect or fungal infestations, albinism, and senescence. Visual symptoms of elemental deficiencies have limited use in field crops because they appear only when deficiency is severe. Appearance of these symptoms depends upon the role played by these elements or upon their mobility (Table 2.4). Depending on the mobility of the element, leaf symptoms can occur in the upper, middle, or lower

## 2.6 Deficiency Symptoms of Mineral Elements in Plants

**Table 2.4** Mobility patterns of elements essential for plant growth

| Essential element | Mobility pattern |
|---|---|
| Boron (Bo) | Mobile in phloem, degree of mobility varies in plants |
| Calcium (Ca) | Mobile through xylem (and not phloem) |
| Copper (Cu) | Poorly mobile element |
| Iron (Fe) | Mobile as ferrous ions ($Fe^{3+}$) |
| Magnesium (Mg) | Good mobility in plants, transported through phloem |
| Manganese (Mn) | Less mobile |
| Molybdenum (Mo) | Relatively less mobile and least abundant microelement |
| Nitrogen (N) | Mobile in the form of nitrates across ammonium transporter |
| Phosphorus (P) | Least mobile, taken up in inorganic form ($P_i$) |
| Potassium (K) | Highly mobile element due to potassium channel |
| Sodium (Na) | Mobile |
| Sulfur (S) | Mobile |
| Zinc (Zn) | Immobile |

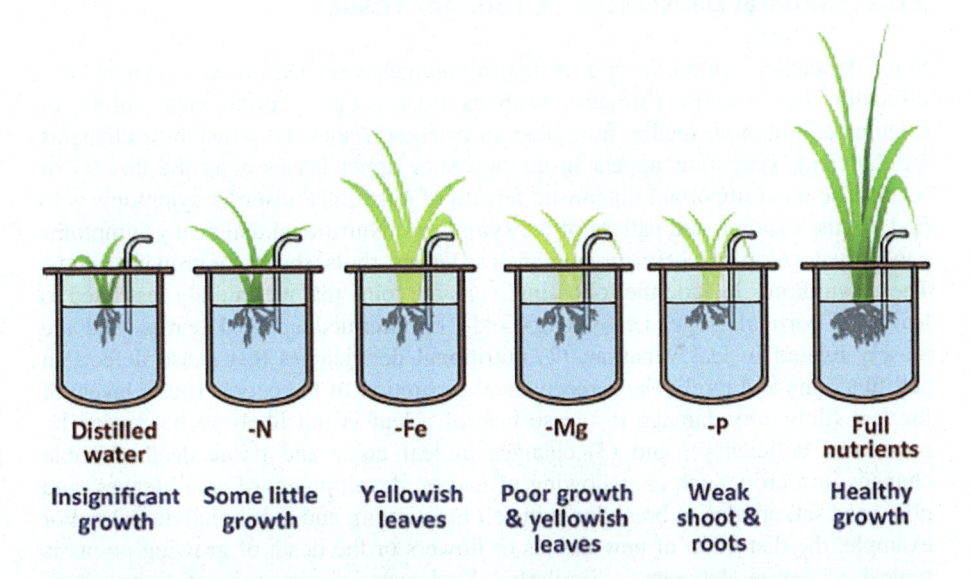

**Fig. 2.3** The experiment to demonstrate the role of inorganic nutrients in plant growth. Plants are grown in solution with all nutrients (minus one) and compared with plants grown in full nutrient medium and distilled water

sections of a plant. This can be demonstrated by adding all the nutrients in solution culture and then transferring plants to solution culture lacking one element each time (Fig. 2.3). If older leaves remain normal, while the newer leaves develop the deficiency symptoms, then the deficient element is immobile. On the other hand, if the deficiency is there in old leaves as well, then the element is mobile. In other words, those elements which are translocated fast are called **mobile elements** and those which are not translocated are called as **immobile elements**. Mobile nutrients include nitrogen, phosphorus, potassium, magnesium, and molybdenum. Immobile nutrients include calcium, sulfur, boron, copper, iron, manganese, and zinc.

### 2.6.1 Mineral Deficiencies in Older Tissues

Magnesium is highly mobile within the plant. It is a component of chlorophyll molecule and its deficiency results in less chlorophyll formation. During magnesium deficiency, older leaves first turn yellow, and gradually yellowing occurs in younger leaves as well. Mobile elements like nitrogen, magnesium, phosphorus, chlorine, molybdenum, cobalt, or potassium satisfy local shortages, particularly in new shoots or developing seeds. When one of these mobile elements is deficient, older leaves are first to be depleted and show symptoms.

### 2.6.2 Mineral Deficiencies in Younger Tissues

Some deficiency symptoms appear first in younger parts and move elsewhere with difficulty. Less mobile elements, such as iron, copper, boron, zinc, sulfur, or calcium, do not move readily from older to younger tissues. So, when these elements are deficient, symptoms appear in the newer or upper leaves or in the flowers or seeds. The most important diagnostic feature of nutritional disorder symptoms is to find out the location and pattern of the symptoms. Nutritional deficiency symptoms generally develop in specific organs, such as leaves, roots, shoots, or growing points. These symptoms include the following: (1) symptoms that are initially restricted to leaves of particular age, i.e., young-, old-, or intermediate-aged leaves, and are closely related to leaf venation; (2) nutritional deficiencies that cause defects in cell functions and rarely cause mechanical disruption of the cuticle (outer layer) of the leaf (thus, any damage to the surface of a leaf is not likely to be caused by nutritional deficiency); and (3) changes in leaf color and tissue death. Visible changes in a crop, such as yellowing of leaves, development of small leaves, and poor seed set, are due to breakdown in cell functioning and nutritional disorder. For example, the distortion of new tissues or flowers or the death of growing points is typical of boron deficiency. Similarly, the leaves of nitrogen- or magnesium-deficient plants are pale because nitrogen and magnesium are components of chlorophyll. The nature of symptoms is a useful guide to identify the nutritional disorder. Diagnostic features help in identifying deficiencies, but sometimes it is difficult to identify them because of the following reasons:

- Sometimes a disorder is quite advanced before clear visual symptoms appear and this results in loss of yield or quality.
- The absence of symptoms in a plant does not mean that nutrition is adequate. "Hidden hunger" is the condition in which yield is poor due to inadequate nutrition, but no symptoms can be observed.

- Visual symptoms can be unreliable when more than one element leads to same deficiency symptoms.
- Some symptoms may appear due to modification in some environmental stress.

## 2.7 Role, Deficiency Symptoms, and Acquisition of Macronutrients and Micronutrients

### 2.7.1 Macronutrients

#### 2.7.1.1 Carbon
Carbon and oxygen account for almost 90% of the dry weight of higher plants. Carbon forms bonds in the shape of a tetrahedron and is the backbone of many biomolecules, like starch and cellulose. It is fixed as photoassimilate through photosynthesis using carbon dioxide drawn from the atmosphere. Nearly, all the complex molecules of living organisms have carbon.

#### 2.7.1.2 Hydrogen
It plays a central role in plant metabolism and is necessary for the synthesis of sugars. In its oxidized form, it is responsible for creating proton gradient which in turn regulates electron transport chain in photosynthesis and respiration. Protons are ubiquitous and are important for maintaining ionic balance.

#### 2.7.1.3 Oxygen
Like carbon, oxygen is present in all organic compounds of living organisms. Free oxygen is primarily involved as an electron acceptor in respiration. It functions as a substrate in reactions involving oxidases. Some hydroxylation reactions use free oxygen instead of hydroxyl ions. Plants produce oxygen during photosynthesis and form glucose, while they undergo aerobic cellular respiration by the breakdown of glucose to generate ATP. The role of oxygen as an inhibitor of metabolism (photorespiration) has more problem than its deficiency.

#### 2.7.1.4 Nitrogen
Nitrogen is the most abundant gas (80%) in the atmosphere, but only certain bacteria and cyanobacteria can utilize gaseous nitrogen directly. It exists in a number of oxidized and reduced forms and cycles in the atmosphere between organic and inorganic pools. A number of plant species can fix nitrogen by having symbiotic association with diazotrophic microorganisms. Nitrate and ammonium ions are two sources of nitrogen in nonleguminous plants. Most of the nitrogen taken up by the plants is derived from the soil in the form of nitrate ($NO^{3-}$), which is then converted into nitrite by nitrate reductase in the cytosol. Nitrite is transported to plastids and is then converted into $NH_4^+$, by the action of nitrite reductase. Soluble forms of nitrogen are transported as amines and amides. Nitrogen is a constituent of all proteins, enzymes, and various metabolic processes involved in the synthesis and

**Table 2.5** Roles of macronutrients in plants

| Nutrients | Functions in cell metabolism | Functions at whole-plant level |
|---|---|---|
| Nitrogen | Constituent of amino acids, nitrogenous bases, cofactors, alkaloids, coenzymes, and chlorophyll, including some hormones (IAA) | Enhanced seed and fruit production; improved leaf and forage crop production |
| Phosphorus | Required as phosphate in sugar; as ester in DNA, RNA; as phospholipids in membrane and is constituent of ATP | Rapid growth; encourages blooming and root growth |
| Potassium | As an activator of enzymes; essential ion for protein synthesis; manufacture of sugar and starches | Regulates opening and closing of stomata |
| Sulfur | Constituent of amino acids (cysteine and methionine); vitamins (thiamine and biotin); coenzyme A (required in respiration); formation of sulpholipids | Improves root growth and seed production; helps with vigorous plant growth and resistance to cold |
| Calcium | Cell wall formation; maintenance of membrane structure and permeability; cell signaling | Regulates fruit quality, protects against heat stress and disease |
| Magnesium | Constituent of chlorophyll molecules and required as an enzyme activator; essential for binding of ribosome subunit | Nutrient uptake control, root formation |

transfer of energy. It is also a constituent of many other important biomolecules, such as hormones (indole-3-acetic acid and cytokinins) and chlorophyll. Some plants, such as corn (*Zea mays*), require very high dosage of nitrogen as compared to other plants. It facilitates rapid plant growth and helps increase seed and fruit production (Table 2.5).

Nitrogen is a mobile element. Therefore, older leaves exhibit chlorosis and necrosis earlier than younger leaves during nitrogen deficiency. During severe nitrogen deficiency, leaves become completely yellow and fall off. Nitrogen deficiency causes stunted and slow growth because cell division is inhibited and lateral buds become dormant. Chlorosis and purple appearance on the stems as well as petiole and underside of leaves are also caused by nitrogen deficiency. Some plants, such as tomato and maize, also accumulate anthocyanins which accompany nitrogen deficiency. Plants grown in the presence of excess nitrogen produce dark green leaves and vigorous foliage as root system is highly reduced resulting in high shoot/root ratio. In nitrogen deficiency, reverse situation is evident, i.e., low shoot/root ratio. Potato plants grown in the presence of abundant nitrogen exhibit more foliage and small tubers. Flower and seed formation are highly reduced due to high nitrogen in the soil. Excess nitrogen also results in the splitting of tomato fruits as they ripen. The crops become susceptible to disease, insect infestation, and drought stress, leading to lodging, when nitrogen content is high. Soils are usually deficient in nitrogen as compared to other elements. In addition to atmosphere, the primary source of nitrogen is often provided to cultivated plants from fertilizer application.

### 2.7.1.5 Phosphorus

It is required in young meristematic cells as it is utilized in the formation of nucleoproteins and other phosphorus-containing compounds in the growing tissues. Requirement of phosphorus in the annual plants is more during the first few weeks of germination and again near the end of their life cycle (fruit and seed development). Addition of phosphates in the soil also promotes root development. Like nitrogen, phosphorus is essential for the process of photosynthesis. It is needed as the structural component of ATP which is synthesized during light reaction of photosynthesis. It is an essential component of many sugars involved in photosynthesis, respiration, and other metabolic processes. Organic phosphates play an important role in metabolism. For example, in the metabolism of sugars (which have hydroxyl groups, -OH), phosphate esters are often formed as intermediate compounds. In plants, phosphorus is mainly present as phosphate esters which include sugar phosphates. Phosphate esters are needed for the synthesis of DNA, RNA, and phospholipids present in the membrane. Phosphorus plays an important role in membrane biochemistry in the form of phospholipids. Phosphorus is also a constituent of ATP, ADP, AMP, and pyrophosphate ($PP_i$), which are important components of energy metabolism (Table 2.5). It participates in signal transduction pathway by phosphorylation and dephosphorylation of receptors, secondary messengers, and target enzymes. Modification of activity of various enzymes requires phosphorylation and is also used for cell signaling as inositol triphosphate ($IP_3$). $IP_3$ is a secondary messenger involved in signal transduction and lipid signaling. In many species, the amount of phosphorus and nitrogen regulates plant maturation process. Excess nitrogen delays maturation, while abundant phosphorus speeds up the process of maturation. Phosphate is withdrawn from older senescing leaves and is redistributed in different plant organs. As a result, first symptoms of phosphorus deficiency appear in older leaves. It results in stunted growth, poor vascular tissue formation, dark green coloration in leaves, and necrosis of leaves. Plant may show premature fall of leaves and flower buds (Fig. 2.4).

Phosphorus is present in various forms in the earth's crust. It is present in mineral deposits, as inorganic and organic phosphorus in soil and water and in different organisms. Although it can react with other elements to make hydrides, halides, sulfides, and metal phosphides, it is present in neutral state in combination with oxygen as phosphates. Unlike nitrates and sulfates, phosphates are not reduced in plants during assimilation. It remains in its oxidized state forming phosphate esters in a wide range of organic compounds. It combines with hydrogen and oxygen to form phosphoric acid. Phosphoric acid is tribasic (having three replaceable hydrogen atoms) and can form monophosphate, diphosphate, and triphosphate salts in which one, two, or three of the hydrogens of the acid are replaced, respectively. Because replaceable hydrogen remains in monophosphates and diphosphates, they are called acid phosphates. The most important inorganic phosphate is calcium phosphate [$Ca_3(PO_4)_2$]. It makes up the larger part of phosphate rock, a mineral that is abundantly distributed throughout the world. Since calcium phosphate is only slightly soluble in water, it is not very suitable as a source of phosphorus. However,

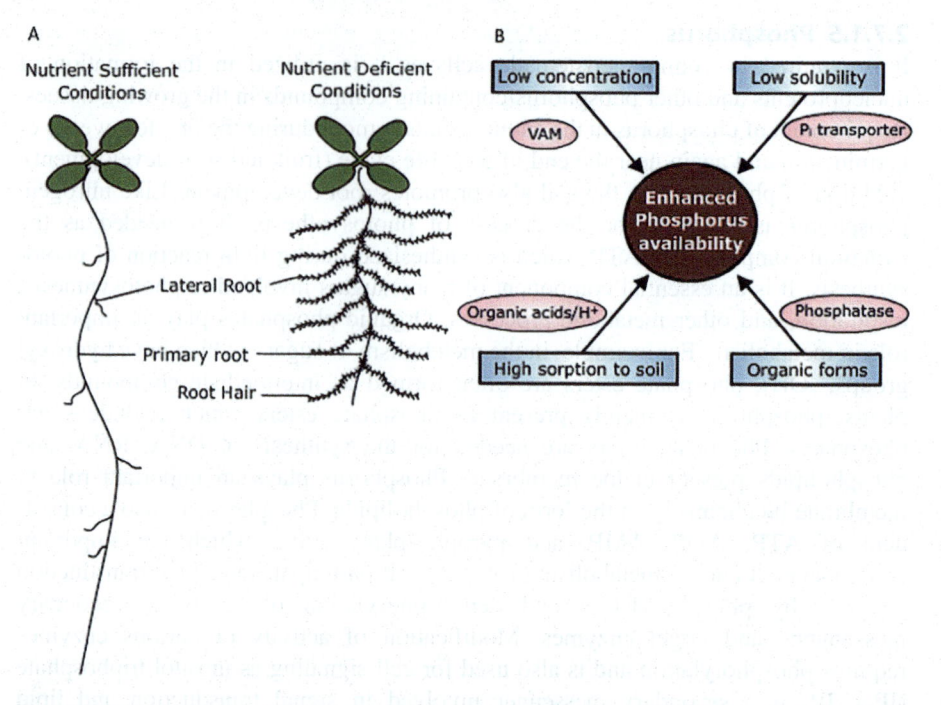

**Fig. 2.4** (**a**) Plant roots exhibiting change in response to phosphorous deficiency. The deficiency induces inhibition of primary root elongation and increase in growth and density of lateral roots and root hairs. (**b**) Plants can increase phosphorous availability by secreting phosphatase, organic acids, protons, or via involving Pi transporters or gets associated with vesicular arbuscular mycorrhiza (VAM) to

by treating it with sulfuric acid, the soluble calcium acid phosphate known as superphosphate [$Ca(H_2PO_4)_2$] is formed. Other important inorganic phosphates include ammonium phosphate, which is an important fertilizer.

Phosphorus is readily absorbed in the form of monovalent anion $H_2PO_4^-$ and less rapidly in the form of divalent anion, $HPO_4^{2-}$. Below pH 7, phosphorus is absorbed as monovalent anion $H_2PO_4^-$ and above pH 7 as divalent anion $HPO_4^{2-}$. Most cations (except $Na^+$, $K^+$, $NH_4^+$, and $Li^+$) form insoluble salts with phosphorus leading to nonavailability of phosphorus for plant growth. Thus, in soils rich in iron and aluminum, most of the phosphates are not available to the plants. Addition of chelating agent releases inorganic phosphate ($P_i$) from aluminum and iron. The concentration of phosphorus in root cells is in the millimolar range, whereas in soil the concentration is 1 μM or less.

In addition to high-affinity transporters, two groups of low-affinity transporter have been identified for intracellular transfer of inorganic phosphate across membranes. During sufficient supply of phosphorus, more than 85% of the inorganic phosphate is stored in vacuoles. However, when there is deficiency of phosphorus, vacuolar $P_i$ is mobilized to maintain cytosolic $P_i$ homeostasis. The import of $P_i$ is inhibited until vacuolar reservoir is over. Availability and acquisition of phosphorus

## 2.7 Role, Deficiency Symptoms, and Acquisition of Macronutrients and Micronutrients

from soil are very low. It is because of low solubility, low concentrations in soil (>1 µM), presence of $Al^{3+}$ and/or $Fe^{2+}$, and conversion of soil $P_i$ to organic forms. Phosphorus starvation response is observed at whole-plant level. As phosphate starvation occurs, there is an increased expression of high-affinity transporters. Phosphorus deficiency also activates several purple acid phosphatases (PAPase) and ribonucleases (RNase) which accelerate movement of phosphorus from old tissues to new tissues. Under $P_i$-deficient conditions, root diameter decreases, and the number of root hairs as well as their length increases which enhances $P_i$ uptake by increasing the root-soil contact area. Plants release phosphatases, organic acids, and protons to solubilize $P_i$. The other strategy adopted by the plants is mycorrhizal-mediated phosphorus acquisition with the help of **arbuscular mycorrhizal fungi** (Box 2.6).

---

**Box 2.6: Plants and Mycorrhizae**

Plant growth is highly dependent on bacteria and saprophytic and mycorrhizal fungi which facilitate the cycling and mobilization of nutrients. In addition to bacteria, more than 80% of plants have symbiotic relationship with fungi (mycorrhizae). This association is mainly of two types and these are *endomycorrhizae* and *ectomycorrhizae*. Endomycorrhizae are those fungi which develop the association by penetrating into the cortical cells of the roots of the host plants. On the other hand, ectomycorrhizae develop the association by developing a vast hyphae network between cortical cells without penetrating the host plants. A variety of plant species develop the endomycorrhizal association with *arbuscular mycorrhizal fungi* (AMF) also known as *vesicular-arbuscular mycorrhiza* (VAM). Plants release chemicals which induce the germination of mycorrhizal spores present in the soil. The germinating spores form a network of hyphae which penetrate the cortical cells of the roots of host plants forming a highly branched structure called arbuscule. This symbiotic association facilitates phosphorus uptake from the soil by increasing root absorptive area of the host plant. The ectomycorrhizal fungi form symbiotic association with many tree species. The fungi form extensive hyphal growth around compatible root cap forming a *hartig net* surrounding the cells within the root cortex and provide phosphorus and nitrogen to the host plants. The ectomycorrhizae produce enzymes that digest organic material present in the litter and mobilize the nutrients to the hartig net, making it available to the plant. This interaction is essential for the trees, as the nutrients present in the litter are otherwise not available to long deep roots of the trees. Absorption of nutrients around the roots depletes nutrients and results in nutrient depletion zone in the region of the soil near the plant roots. Root association with mycorrhizal fungi helps plants to overcome this problem by moving the nutrients from high concentration zone to the depletion zone. The benefit to the fungal partners in this relationship involves transfer of

(continued)

**Box 2.6** (continued)

carbohydrates from the plant to the microorganism. The symbiotic fungi obtain water and nutrients, primarily S, $P_i$, and N, from the soil and translocate them to the host plants helping their growth and development. The presence of AMF enhances sulfur uptake in maize, clover, and tomato. At molecular level AMF can influence the expression of plant sulfate transporters which improves the sulfur status of the host plant. Additionally, inoculation with AMF has been shown to increase both root colonization and the magnitude of the sulfonate mobilizing bacterial community in the rhizosphere. Whenever there is lack of readily available sulfate in soil, it leads to reduction in plant exudates, and as a consequence, soil microbial activity decreases due to reduced availability of photosynthate as a source of carbon. Inoculation practices, therefore, have huge potential to sustainably increase crop yield in areas where sulfur is becoming a limiting factor for plant growth.

Under $P_i$ deficiency, plants need to minimize the production of new shoot branches and direct limited $P_i$ resources to already existing shoots while maximizing $P_i$ acquisition from the soil. AMF spores treated with root exudates from plants grown under $P_i$ starvation have more hyphal branching activity than those treated with exudates from $P_i$-sufficient plants. Moreover, increased soil $P_i$ levels resulted in a decreased AMF colonization of the roots. Subsequently, a stimulant of hyphal branching and root colonization of symbiotic AMF was isolated. This stimulant was found to be a terpenoid lactone derived from carotenoids and was named *strigolactones* (SLs). It was originally derived from plant root exudates and recognized as germination stimulant for root parasites such as *Striga*, *Orobanche*, and *Phelipanche*. SLs play a dual role in the modulation of $P_i$ acquisition and utilization under $P_i$-deficient conditions. $P_i$ deficiency stimulates SL biosynthesis in roots and exudation to soil. Elevated SLs (acting as endogenous hormones) act locally by modifying root system to increase root coverage that provides more surface area to explore more soil volumes and allow higher $P_i$ uptakes. SLs are also transported through the xylem to suppress shoot branching (a means to reduce $P_i$ utilization). SL exudation into the soil serves as a rhizosphere signal for symbiotic interaction between some host plants and arbuscular mycorrhizal fungi (AMF), a means to increase $P_i$ acquisition. More recently, it has been demonstrated that SLs act as long-distance signaling molecules that can be transported from roots to shoots for their specific functional control on shoot branching. The transport of SLs from roots to shoots is partly mediated by ATP-binding cassette (ABC) transporters. Nutrient uptake soon depletes nutrients near the roots and forms nutrient depletion zone in the region of the soil near the plant roots. Root associations with mycorrhizal fungi help the plant to overcome this problem.

### 2.7.1.6 Potassium

The name potassium is derived from "pot ash." Plant ash mainly contains potassium, which constitutes nearly 50% of its total weight. It is absorbed by plants in larger amounts than any other mineral element except nitrogen. It occurs in plants mainly as soluble inorganic salts. Cytoplasmic potassium concentration varies from 80 mM to 200 mM. It varies considerably in subcellular compartments. This fluctuation of potassium levels is regulated by its accumulation in the vacuoles of plant cells. Vacuolar potassium can be exchanged against sodium to maintain potassium concentration in cytosol. The young and active regions of the plants, especially buds, young leaves, and root tips, are rich in potassium. Older tissues, such as wood, contain much less potassium. It is supplied to plants from soil minerals, organic materials, and fertilizers. Potassium is not used in the building up of any cell constituents. However, it mainly has catalytic and regulatory roles. It is an activator of enzymes used in photosynthesis and respiration. It is used to build cellulose and aids in photosynthesis by the formation of a chlorophyll precursor. $K^+$ is highly mobile and helps in balancing the anion charges within the plant. $K^+$-$Na^+$ pumps facilitate active transport. $K^+$ regulates the opening and closing of stomata by regulating the activity of potassium ion pumps. It also reduces water loss from the leaves, increases drought tolerance, and maintains turgidity of the cell. Potassium helps in the buildup of proteins and fruit quality and disease resistance. Starch and protein syntheses are also affected by potassium ions (Table 2.5).

Like nitrogen and phosphorous, $K^+$ is also easily redistributed from mature to younger organs, so its deficiency symptoms first appear in older leaves. $K^+$ deficiency causes necrosis or **interveinal chlorosis** and leads to the formation of scorched leaf tips and short internodes. Potassium-deficient plants also exhibit loss of apical dominance and cambial activity because of its high solubility in water, and $K^+$ leaches out of rocky or sandy soils resulting in potassium deficiency. It may result in higher risk of **pathogen** attack, **wilting**, chlorosis, brown spotting, and chances of damage from frost and heat. In most of the monocots, cells at the tips and margins of leaves become necrotic first, and the symptom moves basipetally along the margins toward younger as well as lower parts of leaf bases. Maize and other cereals develop weak stalk during deficiency. Their roots are easily infected, and plants show root-rotting, leading to lodging of the plants by wind and rain. After nitrogen and phosphorus, potassium is one of the deficient elements in soil. Potassium is generally provided to agricultural crops as potash (potassium carbonate, $K_2CO_3$). In soil, potassium exists in four different forms, viz., exchangeable, fixed, solution, and hidden, in molecular lattice of clay particles. Potassium is highly mobile in soil as well as in plants. More recent work shows that plants contain different transport systems to acquire potassium from the soil and distribute it within the plants. Plants utilize both high- and low-affinity transport systems (**HATS** and **LATS**, respectively) to acquire potassium from the soil. Low-affinity transport systems generally function when potassium levels in the soil are adequate. This process is mediated by ion channels in the plasma membrane of root cells, allowing passive transport of $K^+$ from external areas of relatively high concentration. Under low concentration of potassium, plants usually induce high-affinity $K^+$ transport systems. In both cases, plasma membrane proton pumps are activated to restore membrane potential and generate a proton gradient. High-affinity potassium transport is an active process. This pattern of K⁺ uptake was described as dual isotherm and reflects the activity of two families of transporters (Fig. 2.5). In addition to K⁺ uptake from root surface, potassium channels are also involved in its loading and unloading in both xylem and phloem. Voltage-gated potassium channels are involved in the regulation of stomatal movement.

### 2.7.1.7 Sulfur

Sulfur, an essential macroelement, may be supplied to the soil from rainwater. Use of gypsum also increases soil sulfur levels. The global sulfur cycle involves microbial conversion between its oxidized and reduced forms. There are many microorganisms capable of oxidizing sulfides or decomposing organic sulfur compounds. Heavy consumption of fossil fuel and natural phenomena such as hot sulfur springs, volcano, and geysers release large amount of sulfur oxides into the atmosphere. Sulfur dioxide, an environmental pollutant, can be absorbed by stomata in the leaves. It is then converted to bisulfate ($HSO_3^-$) upon reaction with water in the cells which inhibits photosynthesis and causes chlorophyll destruction. Bisulfate present in air is oxidized to $H_2SO_4$ which is responsible for acid rain. Sulfur is a structural component of some amino acids (cysteine and methionine), vitamins (thiamine and biotin), and coenzyme A (Table 2.5). It is essential for the biogenesis of chloroplasts. It is important for the structure of certain proteins where disulfide bonds (-S-S-) between neighboring cysteine and methionine residues result in folding of polypeptide chain, producing a tertiary structure. Sulfur improves root growth and seed production and facilitates vigorous plant growth and resistance to cold. It is also present in the form of iron-sulfur protein complexes, such as ferredoxin, in the electron transport chain in photosynthesis. It enhances root development and nodule formation in legumes. Sulfur-containing thiocyanates and isothiocyanates (also known as mustard oils) are responsible for the pungent flavor of mustard, cabbage, turnip, horseradish, and

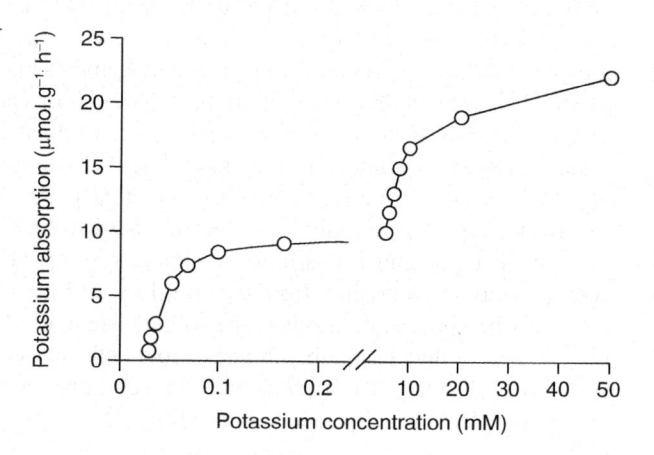

**Fig. 2.5** Dual isotherm of K⁺ influx reflects the activity of two families of transporters

other members of Brassicaceae. The presence of sulfur in Brassicaceae members makes it fatal for livestock and forms a line of defense against insects and herbivores.

Sulfur deficiency is not very common and it is an immobile element. Therefore, sulfur deficiency symptoms appear first on the younger tissues. Yellowing of leaves, stunted growth, and anthocyanin accumulation are also evident due to its deficiency. In tea plants, sulfur deficiency causes defoliation, and in legumes it leads to reduced nodulation. Approximately 95% of sulfur present in soil is bound in the form of sulfate esters or sulfonates. It is absorbed as divalent sulfate anions ($SO_4^{2-}$) through roots. Plants and microorganisms assimilate sulfur by reducing sulfate and synthesizing the sulfur-containing amino acid, cysteine, and other organic sulfur compounds. Some of the sulfur is reduced and assimilated in root plastids, but most of the sulfur in plants is transported to shoots. Chloroplasts are sites for light-driven assimilation of sulfate into cysteine, glutathione, and other metabolites. Some sulfate is transported across tonoplast and stored in the vacuoles. Sulfate is taken against electrochemical gradient at the plasma membrane by the proton gradient generated by the plasma membrane $H^+$-ATPase. It is transported into the cytosol by an electrogenic symport that moves three $H^+$ per sulfate ions transported. Sulfite, selenate, molybdate, and chromate compete with sulfate for binding to sulfate transporter proteins. The electrochemical gradient favors the diffusion of sulfate into the vacuole. The transfer across the tonoplast occurs through sulfate-specific channels. Multiple transporters with variable affinities for sulfates are present in the roots. In *Arabidopsis*, 14 genes encoding sulfate transporters have been identified. Two of these transporters are high-affinity $SO_4^{2-}$ transporters, SULTR1.1 and SULTR1.2, which are present in the root epidermis and cortex. Four transporters are low-affinity transporters: SULTR1.3, SULTR2.1, SULTR3.5, and SULTR2.2. They are present in the vascular system. SULTR4.1 and SULTR4.2 are tonoplast transporters facilitating the efflux of vacuolar sulfate.

### 2.7.1.8 Calcium

Calcium is the second most abundant element in plant ash ($K^+$ being the most abundant). A large proportion of calcium is located in the leaves. In plant cells, calcium is present in central vacuoles, ER, and mitochondria, and it is also bound in cell walls as calcium pectate. High concentration of calcium inhibits cytoplasmic streaming. All organisms maintain low concentration of free $Ca^{2+}$ in the cytosol (~100–200 nM) to prevent formation of insoluble calcium salts of phosphates. In the vacuoles of some plants, calcium gets precipitated as insoluble crystals of calcium oxalate and in some species as insoluble phosphate, sulfate, or carbonate. It is an essential constituent of plant cell wall, and as calcium pectate, it helps in joining the cells together. It plays a key role in transport and retention of other elements and affects the permeability of the cytoplasmic membrane and hydration of colloids in the protoplasm. It is required for normal functioning of the membrane, where it binds phospholipids and membrane proteins. Calcium is supposed to counteract the effect of alkali salts and organic acids within plants. In meristematic tissues, calcium is required for cell division (spindle formation) and cell enlargement. It activates the enzymes required for the growth of root and shoot tip. Calcium regulates the transport of other nutrients into the plants and activates various other enzymes (Table 2.5). Various studies have shown the effect of calcium on diverse developmental processes, such as embryogenesis in sandalwood, cotton fiber elongation, tuberization in potato, and pollen development.

As calcium is not loaded into phloem and is not highly mobile, deficiency symptoms first appear in younger leaves. Meristematic tissues of roots, stems, and leaves are quickly affected by its deficiency as it is required to form middle lamellae in the dividing cells. Deficiency of calcium causes formation of twisted and deformed tissues or stunted growth, leading to rapid death of meristematic tissue especially root meristems. In tomatoes, deficiency of calcium causes degeneration of young fruits. Natural sources of calcium are dolomite, lime, gypsum, and superphosphate. Most soils contain enough calcium, but acidic soils with high rainfall are often supplemented with lime fertilizer (a mixture of CaO and $CaCO_3$) to raise soil pH. Calcium is taken up as divalent cations ($Ca^{2+}$) due to which it is unable to diffuse through the lipid bilayers without channels or pumps. Calcium entry into root xylem takes place at the apical region of root tip where endodermis has not differentiated. Casparian strips in endodermis block the diffusion of calcium. It is distributed within the plant in free form or complexed with organic acid. Pectin and lignin (both are negatively charged) in the xylem wall do not allow mass flow of calcium. Significant amount of calcium is lost in the form of calcium oxalate. $Ca^{2+}$ serves as a universal second messenger whose cytosolic concentration is tightly regulated by $Ca^{2+}$ transporters. Cells use energy to pump it out across the plasma membrane or into the storage organelles, such as vesicles, vacuole, or the space between the inner and outer nuclear membranes (nuclear membrane lumen). In these organelles, calcium-binding proteins sequester calcium ions to minimize their harmful effects. The low level of calcium in cytosol is sensed by calmodulin (CaM). After binding, it interacts with target proteins, such as protein phosphatases and protein kinases. Vacuolar calcium is released through voltage- or ligand-gated calcium permeable channels in the tonoplast.

### 2.7.1.9 Magnesium

Plant ash is rich in magnesium. Oilseed crops are richer in magnesium than non-oily seeds. Magnesium is an important constituent of chlorophyll (mainly in the porphyrin moiety) and also acts as an enzyme cofactor for the production of ATP. It is also essential for binding of ribosome subunits. It is also an activator of ribulose bisphosphate carboxylase (Rubisco) and phosphoenolpyruvate carboxylase (PEP carboxylase), two important enzymes involved in dark reaction in photosynthesis. It is also required for the activity of many enzymes of respiration and nucleic acid biosynthesis (Table 2.5). Magnesium is a mobile element. The absence of magnesium results in interveinal chlorosis. It also results in accumulation of anthocyanin pigment in older leaves. Magnesium deficiency results in premature leaf abscission. Magnesium is absorbed as divalent cation ($Mg^{2+}$) and is transported through the xylem vessels in free or chelated form. It is never limited in soil, but because of soil pH, it is not available to the plants growing in acidic and sandy soils. The concentration of magnesium in soil solutions and cytosol is approximately 0.1–8.5 mM and

0.4 mM, respectively. Entry of magnesium in roots takes place through plasma membrane magnesium channels of the MSR2 family. Movement across tonoplast takes place through MHX $Mg^{2+}/H^+$ antiporter and TPC1 $Mg^{2+}$-permeable cation channel.

### 2.7.2 Micronutrients

#### 2.7.2.1 Iron

Iron is the fourth most abundant element in the earth's crust. It is an essential micronutrient with numerous cellular functions, and its deficiency represents one of the most serious problems in human nutrition worldwide. Plants face two major problems with iron as a free ion, i.e., its insolubility and its toxicity. To ensure iron acquisition from soil and to avoid iron excess in the cells, uptake and homeostasis are tightly controlled. Iron is stored in chloroplasts as iron-protein complexes known as **phytoferritin**. The chemical properties of iron are also responsible for its limited accumulation in plants. Ferrous ($Fe^{2+}$) and ferric ($Fe^{3+}$) ions catalyze the reduction of molecular oxygen to damaging ROS (reactive oxygen species). In the symplasm, iron is maintained in a soluble and transportable form. Iron is necessary for the synthesis of many proteins (ferredoxin and cytochromes) that carry electrons during photosynthesis and respiration. Iron is also present as an enzyme cofactor in plants and activates catalase and peroxidase. Iron is not a structural part of chlorophyll but is required for its synthesis (Table 2.6). Iron deficiency in plants is caused largely due to its insolubility in soil rather than its absence. The concentrations of soluble iron optimal for most plants are in the range of $10^{-4}$–$10^{-8}$ M (optimal soils are usually slightly acidic). However, $10^{-9}$ M or lower concentrations of soluble Fe (calcareous or alkaline soils with low bioavailable Fe) are insufficient for plant growth, and plants may develop iron deficiency triggered leaf chlorosis. Iron is one of the most immobile elements in the plants. Its deficiency causes interveinal chlorosis and necrosis, like magnesium, but unlike magnesium, symptoms are first evident in younger leaves. Interveinal chlorosis is followed by chlorosis of the veins, turning the whole leaf yellow. During severe deficiency, young leaves turn white with necrotic lesions. Iron deficiency is most common among the members of Rosaceae, maize, sorghum, and fruit trees. Source of iron for plants is soil, which is available in the form of iron sulfate and iron chelates. Iron undergoes oxidation and reduction, forming $Fe^{2+}$ and $Fe^{3+}$, alternatively. Both these forms have limited solubility, and in well-aerated soil, their concentration is less than $10^{-15}$ M. This means, it is not readily available to the plants at physiological pH. Soils with alkaline pH and bicarbonates are deficient in iron. Iron can be solubilized in soil by its detachment from mineral soil particles. Mobilization of iron is a prerequisite for its uptake into the roots. Higher plants can mobilize iron via two distinct strategies (Fig. 2.6).

According to strategy I, Fe acquisition is regulated by the availability of iron in the soil and by the developmentally controlled need of the plant. When plants require additional iron, they are able to enhance the activities of the necessary transporters

**Table 2.6** Role of micronutrients

| Micronutrients | Function in cell metabolism | Function at plant level |
|---|---|---|
| Iron | Chlorophyll formation and synthesis of ferredoxin and cytochromes, activates catalase and peroxidase and many enzymes with iron-based cofactor | Provides resistance against plant pathogens |
| Molybdenum | Nitrogen metabolism and nitrogen fixation | Optimizes plant growth; aids in nodule formation in leguminous crops |
| Boron | Cell wall formation along with calcium, necessary for sugar translocation | Promotes maturity; essential for pollen grain formation and pollen tube elongation; fruit yield and quality of temperate fruits |
| Copper | Activates enzymes necessary for photosynthesis and respiration, constituent of cytochrome oxidase and polyphenol oxidase, present in the receptor of ethylene signal | Provides resistance against plant pathogens |
| Manganese | Acts as a catalyst in growth process; constituent of oxygen evolving complex | Accelerates seed germination and maturity; increases the availability of phosphate and calcium |
| Zinc | Chlorophyll formation; involved in respiration and nitrogen metabolism; regulates functioning of DNA/RNA polymerase | Provides resistance against plant pathogens |
| Chlorine | Formation of cytochromes; light reaction of photosynthesis | Regulation of stomatal movement; delayed senescence |
| Nickel | Constituent of urease | Increases crop yield |

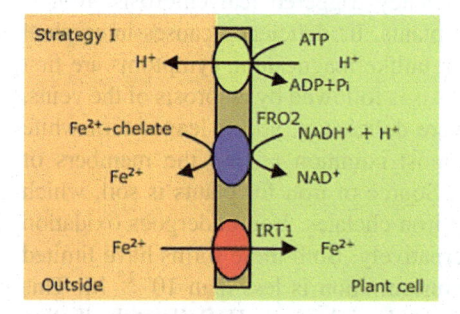

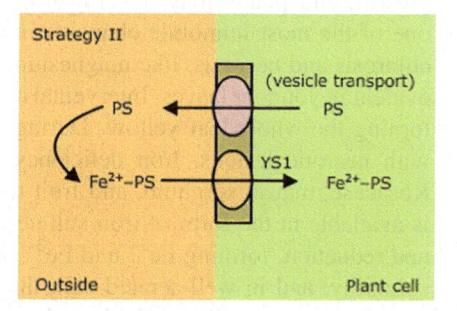

**Fig. 2.6** The acquisition of iron by two different strategies in higher plants. Strategy I involves transporters and enzymes, while strategy II involves phytosiderophores

and enzymes performing this strategy. This strategy is found in most monocotyledonous and dicotyledonous plants, except in grasses, and involves the solubilization of iron by soil acidification.

## 2.7 Role, Deficiency Symptoms, and Acquisition of Macronutrients and Micronutrients

$$Fe(OH)_3 + 3H^+ \leftrightarrow Fe^{3+} + 3H_2O$$

The lowering of pH from 8 to 4 increases the concentration of ferric iron in the soil from $10^{-20}$ M to $10^{-9}$ M. This lowering of pH and increase in concentration are carried out by a root plasma membrane bound $H^+$-ATPase. In *Arabidopsis*, this enzyme is encoded by AHA gene family. Before the uptake of iron by roots, it is essential that $Fe^{3+}$ is reduced to $Fe^{2+}$. This reduction is catalyzed by an inducible plasma membrane bound ferric reductase oxidase (FRO). This enzyme is encoded by the gene FRO2, which is expressed in iron-deficient epidermal root cells. The enzyme has eight membrane-spanning domains, two histidine-coordinated heme groups, and site for FAD and NADPH. There are other FRO gene families of metal reductases in *Arabidopsis* which are expressed in other cells of the root, vascular tissues, and shoot. $Fe^{2+}$ is then transported into the root cells by IRT1 (iron-regulated transporter1), a member of ZIP metal transporter family (Zn transporter). IRT1 is expressed in the plasma membrane of epidermal cells in iron-deficient roots. Studies with *Arabidopsis* mutants and yeast cells transformed with IRT1 gene have shown that this transporter protein can also transport other divalent metal ions such as zinc, manganese, and cadmium. Strategy II, found in grasses, is based on solubilization of $Fe^{3+}$ by chelation with **phytosiderophores** (Box 2.7). Strategy II is restricted to members of Poaceae family (barley, rice, and maize). According to this strategy, plants release low-molecular-weight phytosiderophores (PSs) of the mugineic acid family. PSs are chelators that bind ferric iron in the rhizosphere and make it available to the roots. This chelation-based response makes it easier for grasses to grow on iron-deficient soils more efficiently as compared to eudicots and other monocots. Mugineic acid and other related PSs are produced from S-adenosyl methionine (SAM). Nicotianamine synthase (NAS) acts upon SAM resulting in the formation of nicotianamine which is a precursor of PS. Nicotianamine is a common chelator for various metal ions found in all plants. The enzyme nicotianamine aminotransferase (NAAT) is specific to grasses and catalyzes the step toward PS production along with a number of other enzymes.

$$3\ SAM \xrightarrow{\text{nicotianamine synthase}} nicotianamine \xrightarrow{\text{nicotianamine transferase}} 2'deoxymugineic\ acid \xrightarrow{\text{dioxygenase}} mugineic\ acid$$

The $Fe^{3+}$-PS complex is transported into the epidermal cells of iron-deficient roots by high-affinity transporters. These transporters are encoded by YS1 (yellow stripe 1) and have 12 transmembrane domains. YS1 is expressed in a number of tissues. This protein is a proton-coupled symporter for PS and mugineic acid metal chelates. PSs are extruded by the roots and can be reimported as $Fe^{3+}$-PS complexes. Iron homeostasis is required for building heme and Fe-S prosthetic groups and assembling them correctly to apoproteins. Storage and buffering of iron at subcellular level are very essential to ensure protection against iron toxicity and deficiency. Plastids store iron in the form of **phytoferritin**, a protein molecule that encloses

**Box 2.7: Chelation and Mineral Nutrition**
The word chelate is derived from the Greek word "chel," meaning a crab's claw. Chelation is a natural process that prevents absorbed nutrients from precipitation. It allows the nutrients to move freely in soil and increases its availability to plants. In plants, proteins, peptides, porphyrins, carboxylic acids, and amino acids act as natural chelating agent. Other naturally occurring organic acids, such as malonic acid and gluconic acid, also play an important role in plant mineral nutrition. Organic acids and amino acids, such as citric acid and glycine, are also naturally occurring chelating agents. Chemically, a chelate is complex of cations with organic compounds resulting in a ring structure. Chelates of glycine with cations, such as iron, zinc, and copper, have been well investigated. They usually contain two moles of ligand (glycine) and one mole of metal. There are also synthetic chelating agents with high stability with divalent and trivalent ions. The proteoid root (clustered roots) in phosphorus-starved plants releases organic acids, mainly citrate and malate. Release of these acids binds aluminum and iron in the soil leading to availability of phosphorus to the plants. A strong chelating agent may bind the mineral too strongly and make it unavailable to plants. On the other hand, a weak chelating agent may not be able to protect the chelated minerals from chemical reactions with other compounds and thereby reduce their availability to plants. A combination of chelating agents can improve product stability and broaden product effectiveness. Organic substances in the soil either applied or produced by plants or microorganisms are the natural chelating agents. The most important compounds exhibiting this nature are hydroxamate siderophores, organic acids, and amino acids. Hydroxamate siderophores are naturally produced by soil microorganisms and are essential in natural ecosystems to solubilize and transport nutrients, especially iron to plant roots. Under iron-deficient conditions, microorganisms produce siderophores to overcome iron starvation. In plant tissue culture, usually Fe-EDTA is added to the medium to improve the availability of the element. Although low concentrations of EDTA stimulate the growth of whole plants in tissue as well as hydroponic cultures, at high concentration tissue may be damaged. For some plant species, EDTA is inhibitory. In a cell, high concentrations of chelating acid are phytotoxic, as they competitively withdraw essential elements from enzymes.

The significance of chelation process are:

- It increases the availability of nutrients, e.g., chelating agent binds relatively insoluble iron in high-pH soil and makes it available to the plants.

(continued)

**Box 2.7** (continued)
- Chelation prevents mineral nutrients to form insoluble precipitates. At high pH, iron reacts with hydroxyl group and forms ferric hydroxide which is not available to the plants.
- It reduces toxicity of some metal ions to plants.
- It prevents nutrients from leaching.
- It increases mobility of nutrients.
- It suppresses the growth of plant pathogens.

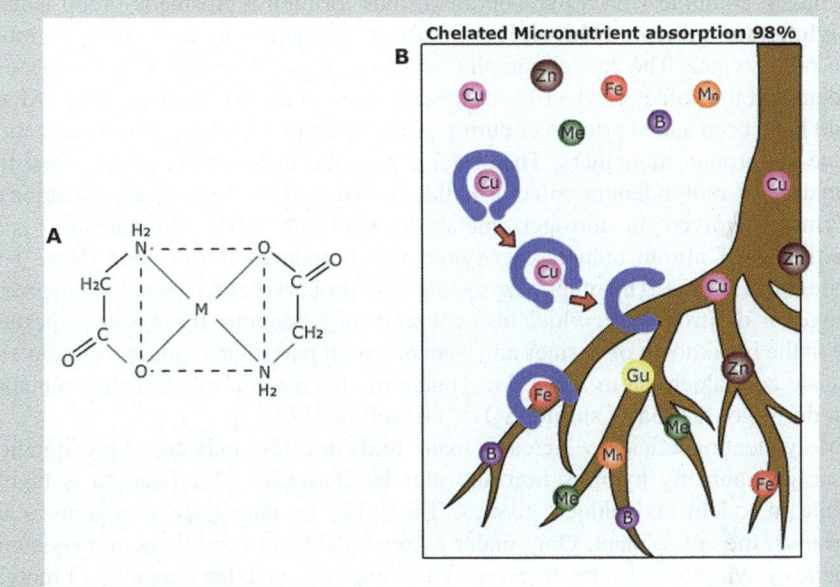

**C**

**EDTA:** Ethylenediaminetetraacetic acid

**EGTA:** Ethyleneglycol-bis(2-aminoethylether) tetraacetic acid

**EDDHA:** Ethyleanediamine-di(o-hydroxyphenyl)acetic acid

**DTPA:** Diethylenetriaminepentacetic acid

**DHPTA:** 1,3-diamino-2-hydroxypropane-tetraacetic acid

**A.** A chelator with two ligand molecules (glycine) around metal ion (M) in the center forms a ring-like structure. **B.** The chelator binds to the nutrients present in the soil and prevents them from precipitating and leaching, thereby increasing their mobility and making them available to root. **C.** Common synthetic chelating agents

4500 $Fe^{3+}$. *Arabidopsis* encodes four ferritins, viz., FER1, FER2, FER3, and FER4. Root and seeds have FER1 and FER2, respectively, whereas FER3 and FER4 are expressed in shoot tissues. The sensing mechanism of iron deficiency is not yet clearly understood. However, the expression of a transcription factor F1 T1 (Fe-induced deficiency transcription factor) is upregulated as a result of iron deficiency.

### 2.7.2.2 Molybdenum

Molybdenum is essential to plants but is toxic to animals. Among all the nutrients, molybdenum is required in least concentration. In order to gain biological activity, Mo has to combine with a pyranoprotein, thus forming a prosthetic group named **molybdenum cofactor (Moco)**. It is involved in maintaining the activity of more than 60 enzymes. The crystallographic analysis of molybdenum enzymes made it evident that the cofactor (Moco) is deep seated within the holoenzyme. Thus, Moco could have been added prior to or during the completion of folding and dimerization of the apoprotein monomers. Therefore, during the biosynthesis of enzymes, the formation of molybdenum cofactor is the first step. It is required as a cofactor of enzymes involved in nitrogen metabolism (Table 2.6). Molybdenum is a constituent of nitrate reductase enzyme, which reduces nitrate ions ($NO_3^-$) to nitrite ions ($NO_2^-$). The other enzyme used by prokaryotes to reduce atmospheric nitrogen is dinitrogenase, which also contains molybdenum. It plays an important role in the breakdown of purines and is an essential part of an oxidase that converts abscisic acid aldehyde to ABA. Molybdenum also plays a role in sulfur metabolism during oxidation of sulfite ($SO_3^{2-}$) to sulfate ($SO_4^{2-}$).

Molybdenum deficiency increases many folds in acidic soils due to precipitation of molybdenum by hydrous iron and aluminum oxides. Molybdenum is highly mobile in xylem and phloem tissues. Therefore, its deficiency symptoms often appear on the entire plant. Only under extreme deficient conditions, molybdenum deficiency symptoms can be observed. The complication in the diagnosis of molybdenum deficiency symptoms is due to its manifestation as nitrogen deficiency symptoms, which are clearly visible in legumes. These symptoms are related to the function of molybdenum in nitrogen metabolism. In legumes, the requirement of molybdenum is higher as compared to other crops. Its deficiency symptoms in legumes are chlorosis, stunted growth, and small root nodules. In other dicotyledonous species, its deficiency leads to drastic reduction in leaf size and yellowing of the leaves. The absence of leaf tissue at the edges of the leaf results in the formation of narrow, distorted leaves that are usually slightly thickened, causing leaf edges to curl upward, a symptom commonly referred as "whiptail." The whiptail disorder is observed in crucifers with cauliflower being the most sensitive to molybdenum deficiency. Twisting of young leaves, which eventually die, can be seen in whiptail of cauliflower and broccoli. Marginal and interveinal necrosis is associated with elevated nitrate concentration indicating a lack of nitrate reductase activity under molybdenum deficiency.

In soils, molybdenum can occur in four different fractions, viz., as dissolved molybdenum in soil solution, with oxides, as a constituent in minerals, and associated with organic matter. The availability of molybdenum for plant growth is highly dependent on soil pH, concentration of adsorbing oxides, soil drainage, and interaction with organic compounds present in the soil colloids. It exists in soil as molybdate ($MoO_4^{2-}$) and as sulfide ($MoS_2$). It is absorbed by roots as $MoO_4^{2-}$ under neutral or slightly alkaline conditions. It can be stored as $MoO_4^{2-}$ in vacuoles. However, in acidic soils, the availability of molybdenum is limited due to fixation of $MoO_4^{2-}$ by iron, aluminum, and manganese oxides. Molybdenum uptake increases with liming of soil (addition of lime increases the soil pH). The use of phosphate helps in the release of adsorbed $MoO_4^{2-}$ from iron oxides (phosphate has high affinity for iron oxides) and increases water-soluble $MoO_4^{2-}$ concentration in the soil. There are many chemical similarities in $SO_4^{2-}$ and $MoO_4^{2-}$ acquisition by plants due to which $SO_4^{2-}$ inhibits uptake of $MoO_4^{2-}$ as both compete with each other during root absorption. Fertilizers containing phosphorus and sulfur facilitate higher $MoO_4^{2-}$ uptake. It is highly mobile, and its long-distance transport occurs through xylem and phloem. The high-affinity ABC transporters encoded by the *modA*, *modB*, and *modC* genes are responsible for molybdenum uptake in bacteria. Specific transporters of molybdenum in plants have still not been identified. However, the $MoO_4^{2-}$ behaves similar to $SO_4^{2-}$, and their uptake is decreased in the presence of high concentrations of $SO_4^{2-}$. So, it is possible that both these anions use the same transporters.

### 2.7.2.3 Boron

At biochemical and physiological level, the role of boron is not clearly understood. It is only when boron is supplemented to the plant-growth medium/soil that its role becomes evident. It is present in the cell wall and is an important part of pectins. Boron is required for maintaining structural stability of cell wall since primary walls of boron-deficient cells show deformities. Boron plays an important role in the elongation of pollen tubes (Table 2.6). In diatoms, it forms a part of silicon-rich cell wall. It is also involved in sugar translocation and is an essential element for seed and fruit development. It also helps in the use of nutrients and regulates other nutrients. Other secondary roles of boron may be in sugar transport, cell division, and synthesizing certain enzymes. Sources of boron are organic matter and borax. It is absorbed from soil as undissociated boric acid ($H_3BO_3$) at pH < 8. Its deficiency is not very common, but several disorders related to disintegration of internal tissues result. These include "heartrot" of beets, "stem crack" of celery, "water core" of turnip, and "drought spot" of apples. It causes necrosis in young leaves and also stunting of growth. Boron is involved in nucleic acid synthesis during cell division in apical meristems, resulting in the loss of apical dominance, death of root and shoot tips, abscission of flowers, shortened internodes, and reduced nodulation in legumes.

### 2.7.2.4 Copper

It is an important component of the electron transport chain in photosynthesis and is also involved in manufacture of lignin. It is a component of oxidase enzymes (cytochrome oxidase) and plastocyanin (chloroplast protein). It also aids in root metabolism and helps in the utilization of proteins. It is absorbed as a divalent cupric ion ($Cu^{2+}$) in aerated soils or as monovalent ion ($Cu^+$) in wet soils. The browning of freshly cut apple and potato is due to the activity of copper-containing polyphenol oxidases which leads to the production of red or brown-colored polyphenols. Superoxide dismutase (SOD) is another copper-containing enzyme, which is an antioxidant and protects the cell from oxidation from reactive oxygen species. Copper deficiency symptoms are chlorosis, leaf tip necrosis, and the bark becomes rough and splits. Leaves turn dark green and develop necrosis. Citrus orchards show dying young leaves which is commonly referred to as "dieback" disease. In potato, it causes blackening of tubers. The Cu transporter (COPT) mediates $Cu^+$ uptake in plants.

### 2.7.2.5 Manganese

Manganese exists in three oxidation states, viz., $Mn^{2+}$, $Mn^{3+}$, and $Mn^{4+}$, as insoluble oxides as well as in chelated form in soils. It is mostly absorbed as divalent manganese cation ($Mn^{2+}$) after its release from chelates or reduction in higher valence oxides. Manganese ions ($Mn^{2+}$) are readily taken up by roots and transported to shoots. The movement of manganese ions from roots to shoots is very rapid due to which it is less toxic to roots as compared with other metals present in the soil. $Mn^{2+}$ is required for chloroplast development and also for activation of many enzymes of photosynthesis, respiration, and nitrogen metabolism. It acts as electron donor for chlorophyll b and is involved in decarboxylation reaction during respiration (Table 2.6). $Mn^{2+}$ deficiency causes interveinal chlorosis and gray spots on leaves. It also results in coloration abnormalities, such as discolored spots on the foliage. Various disorders, such as "gray speck" of oats, "marsh spot" of peas, and "speckled yellows" of sugar beets, are due to $Mn^{2+}$ deficiency. The absence of manganese ions also causes disorganization of thylakoid membranes.

### 2.7.2.6 Zinc

It is distributed within the cytoplasm (50%), nucleus (30–40%), and cell membrane (10%). It can bind tightly to metalloproteins as a structural component or to metalloenzymes as a cofactor. Zinc binds metallothioneins (MTs) with low affinity, which constitutes about 5–15% of the total cellular zinc pool. It can be compartmentalized into intracellular organelles and vesicles for storage, which serve as a supply for zinc-dependent proteins. Cytosolic free zinc is maintained at very low concentrations. MTs and two zinc transporter families, Zrt- and Irt-like proteins (ZIP) and Zn transporters (ZnT), play crucial roles to maintain this cellular zinc homeostasis. Zinc plays a pivotal role as a structural, catalytic, and signaling component that functions in numerous physiological processes. It participates in chlorophyll formation and prevents its destruction. It is a component of the enzyme carbonic anhydrase (CA). It also regulates the transformation of carbohydrates and consumption of sugars. Zinc has a role in the formation of tryptophan synthase, an enzyme responsible for the synthesis of tryptophan. Tryptophan is a precursor of indole acetic acid (IAA). Thus, zinc has indirect role in IAA synthesis and activates a large number of enzymes, e.g., dehydrogenases (e.g., alcohol dehydrogenase, ADH, and carboxylases). It is also associated with important enzymes, such as SOD. It plays an essential role in maintaining the structure and function of DNA transcription factors, including Zn finger, Zn cluster, and RING finger domains (Table 2.6). Zinc deficiency results in malformed or stunted leaves, commonly known as "little leaf" and "rosette" of apples and peaches. It is caused by oxidative degradation of auxin, the growth hormone. Auxin level in zinc-deficient plants is very low. Leaf margins are often distorted and puckered. Other zinc deficiency symptoms include interveinal chlorosis and stunted growth in the leaves of maize, sorghum, beans, and fruit trees. Sources of zinc are soil, zinc oxide, zinc sulfate, and zinc chelates. It is absorbed as divalent cations ($Zn^{2+}$) from zinc chelates. A group of genes that encode $Zn^{2+}$ micronutrient transporters (ZIPs) have been isolated. ZIP transporters are ubiquitous having been identified in bacterial fungi, mammals, and plants. Most ZIP proteins have eight transmembrane helices, and in many cases, a loop region is present between transmembrane domains 3 and 4 containing a histidine-rich sequence which binds to metal and regulates zinc transport. ZIP1, ZIP3, and ZIP4 are high-affinity zinc transporters which bind to other divalent cations like $Cd^{2+}$ and $Cu^{2+}$. Both ZIP1 and ZIP3 are expressed in response to zinc deficiency in roots, whereas ZIP4 is expressed in both root and shoot (Box 2.8).

> **Box 2.8: Hyperaccumulators**
> Some plants take up high concentration of metal elements from the soil and store them in their aerial tissues. These plants are known as *hyperaccumulators* or *metallophytes*. The elemental concentration in the above ground part ranges from 100- to 1000-fold higher than the observed concentration in non-hyperaccumulators species. They are unique as they can be utilized in biogeochemical and phytoremediation studies. There are around 450 hyperaccumulators plants and nickel is the most accumulated metal. In addition to nickel, arsenic, cobalt, manganese, lead, cadmium, zinc, selenium, and copper too are also being accumulated by the plants. For example, *Brassica* can accumulate up to 30,000 $\mu g.g^{-1}$ zinc and 1300 $\mu g.g^{-1}$ cadmium. The three main characteristic features of hyperaccumulators are:
>
> (i) *Greater capability of heavy metal uptake*: The metal uptake by roots is more because of constituted overexpression of the genes responsible for normal uptake of nutrients. In some species zinc uptake is more because of the overexpression of ZIP (zinc- and iron-regulated protein transporter) family gene coding for plasma membrane cation transporters.
> (ii) *Higher root to shoot translocation of metals*: In non-accumulator plants, the metal ions are detoxified by chelation and stored in vacuoles of root cells. On the contrary, the hyperaccumulators rapidly translocate the

(continued)

**Box 2.8** (continued)

metals to shoot via xylem. Overexpression of HMA (heavy metal-transporting ATPase) proteins is responsible for rapid loading of metals into xylem.

(iii) *Detoxification and sequestration of metals*: It mainly consists of ligation organic components with metal ions and their removal from metabolically active cytoplasm to non-active compartments of the cell, mainly vacuoles and cell wall. Comparative genomics have shown that overexpression of CDF (cation diffusion facilitator) genes removes divalent metal cations from the cytoplasm to vacuole.

Why should some plant accumulate metals at such high concentration? Most probably hyperaccumulated metals provide defense against herbivores and pathogens. In *Nicotiana caerulescens*, there is significant inhibition of the bacterial pathogen *P. syringae* by zinc accumulation. They also have significant roles in phytoremediation, an eco-friendly method of removal of heavy metals from the polluted soils. Hyperaccumulators also have potential significance in phytomining, recovering, or phytoextraction of metals from plants.

### 2.7.2.7 Chlorine

Chlorine is universally present in plants in the form of inorganic chlorides. Plants growing in salt marshes and saline soils can tolerate high concentrations of chlorides. Chlorine is essential for photolysis of water leading to oxygen evolution during photosynthesis. It is also essential for roots and for cell division in leaves and maintains ionic balance in cells. It is one of the osmotically active elements in vacuoles. It is involved in transporting cations, such as potassium, calcium, and magnesium, using antiporters. It is required to chemically balance potassium ion concentration that increases during the opening and closing of stomata (Table 2.6). Chloride ions are rarely deficient because of their high solubility and availability in soils as well as in dust or in tiny moisture droplets. Chlorine deficiency causes reduced growth, wilting and bronze coloration of leaves, and swelling of root tips. Leaf mottling and chlorine deficiency in cabbage are marked by the absence of cabbage odor from the plant. Chlorine is absorbed as chloride ions ($Cl^-$) and remains in the same form in approximately 130 organic compounds, but still it is present in trace amounts in plants. Most species absorb 10–100 times more $Cl^-$ than required. Asparagus requires sodium chloride for its profuse growth.

### 2.7.2.8 Nickel

It is an abundant metallic element in soil, absorbed in the form of $Ni^{+2}$ ion. It is present in plant tissues in the range of 0.05–5.0 mg.kg$^{-1}$ dry weight. It is essential for activation of urease, an enzyme involved in nitrogen metabolism. In legumes, removal of nickel from nutrient solutions leads to accumulation of large amount of urea in leaves resulting in necrotic spots. Beneficial effects of nickel on the growth of oats, wheat, and tomato have been reported.

## Summary

- Plant nutrition is the study of the nutrients necessary for plant growth and development. Roots absorb around 60 elements from the soil, but not all are required by the plant growth. The nutrients or elements necessary for growth or completing life cycle of a plant are considered as essential elements. There are 17 essential plant nutrients. They mainly serve structural roles, act as enzyme activators, and act as osmotic regulators in plants. The elements which stimulate growth but are not essential, or which are essential only for certain plant species, are referred to as beneficial or functional elements.
- Carnivorous, insectivorous, and parasitic plants are different in acquiring mineral nutrients. Macronutrients are consumed in larger quantities and constitute 0.2–4.0% on a dry matter weight basis. Micronutrients are present from 5 to 200 ppm, or less than 0.02% dry weight, in plant tissues. Most of the micronutrients have very narrow adequate range and very minute change in their concentration leads to symptoms.
- Mobility of nutrients within the soil is related to chemical properties of the soil, such as cation exchange capacity and anion exchange capacity, as well as the soil conditions, such as moisture, pH, etc. The movement of nutrients from soil to root takes place when root comes in physical contact with nutrients. Root hair, along with the rest of the root surface, is the major site for water and nutrient uptake by the plants. The selective permeability features of plasma membrane make it impermeable to certain ions and allow entry of other ions.
- The absence or deficiency of any nutrient causes the development of specific symptoms. Appearance of these symptoms depends on the role played by these elements or upon their mobility in the plants. Depending on the mobility of the element, leaf symptoms can occur in the upper, middle, or lower regions of a plant. When nutrients are mobile, deficiency symptoms are apparent first on the older leaves, e.g., in case of nitrogen, phosphorus, and potassium deficiency. When nonmobile nutrients are deficient, the younger leaves show deficiency symptoms, as the nutrients are utilized in the older leaves and do not move up to the young leaves. This phenomenon is significant in determining which nutrients a plant may be lacking.

## Multiple-Choice Questions

1. Which element plays an important role in pollen germination?
    (a) Potassium
    (b) Magnesium
    (c) Zinc
    (d) Boron

2. Which of the following elements is immobile in plants relative to all others listed below?
   (a) Magnesium
   (b) Potassium
   (c) Calcium
   (d) Nitrogen
3. Which of the following cannot be considered as a criterion for essentiality of an element for plants?
   (a) Element must be an essential for normal growth and reproduction.
   (b) Element must be easily absorbed by plant roots.
   (c) Specificity for element's role in plant growth and development.
   (d) Element's direct involvement in plant metabolism.
4. Senescing leaves export much of their mineral content to the younger, healthy leaves. Element most mobilized is:
   (a) Calcium
   (b) Sodium
   (c) Sulfur
   (d) Magnesium
5. The macronutrients potassium, calcium, and magnesium are the examples of:
   (a) Metallic essential elements
   (b) Nonmetallic essential elements
   (c) Nonmetallic nonessential elements
   (d) Metallic nonessential elements
6. Which of the following element is required in least quantity?
   (a) Zn
   (b) Mn
   (c) Mo
   (d) Co
7. Which of the following element is a constituent of biotin and coenzyme A?
   (a) Copper
   (b) Molybdenum
   (c) Sulfur
   (d) Iron
8. An element is considered essential if it:
   (a) Is found in plant ash
   (b) Induces flowering
   (c) Is present in the soil where the plant is growing
   (d) Is not replaceable and is indispensable for the growth of the plant
9. Which of the following mineral nutrients is not involved in redox reactions in plant cells?
   (a) Iron
   (b) Zinc
   (c) Copper
   (d) Sodium

10. The photosynthetic and mitochondrial transports are affected by which of the following three elements?
    (a) Cu, Mn, and Fe
    (b) Co, Mn, and Fe
    (c) Cu, Mg, and Cl
    (d) Zn, Cu, and Fe

## Answers

1. d   2. c   3. b   4. c   5. b   6. c   7. c
8. d   9. d   10. a

## Suggested Further Readings

Delhaize E, Schachtman D, Kochian L, Ryan PR (2015) Mineral nutrient acquisition, transport and utilization. In: Buchanan BB, Gruissem W, Jones RL (eds) Biochemistry and molecular biology of plants. Wiley Blackwell, Chichester, pp 1101–1131

Jones RL, Ougham H, Thomas H, Waaland S (2013) The molecular life of plants. Wiley-Blackwell, Chichester, pp 455–503

Taiz L, Zeiger E (2010) Plant physiology, 5th edn. Sinauer Associates Inc, Sunderland, pp 107–126

# Water and Solute Transport

## Satish C Bhatla

Movement of water and solutes from soil solution to the seed tissue is one of the first processes occurring during seed germination in soil. Mature seeds contain less than 10% water and imbibition leads to hydration of its cells and tissues. With the exception of oxygen and carbon, which are readily available to plants from air, terrestrial plants generally take up water and dissolved nutrient elements from the soil through the root system. Molecular and ionic movements from one site to another are known as **transport**. Long-distance transport of solutes from one tissue system to another is referred to as **translocation**. Intracellular and intercellular distribution of water, ions, and organic molecules is crucial for plant growth, cell signaling, nutrition, and cellular homeostasis. To fulfill these essential functions, plants have evolved various transport mechanisms through apoplast and symplast. Membranes act as barriers which separate cells from the environment. The hydrophobic nature of the lipid bilayer of cell membranes ensures that hydrophilic compounds, including most metabolites, are sequestered in one or the other organelles or in the cytosol. Development of endomembrane system in the cells has facilitated homeostatic functions of membranes through compartmentalization of solutes. The major advantage of the compartmentalization of solutes and macromolecules within the membrane-bound organelles is that it concentrates the reactants and catalysts. It also segregates incompatible processes taking place in a cell. Recent advancements in our understanding of the membrane transport process have benefitted significantly from the isolation and characterization of a variety of mutants. Electrophysiological analysis, using techniques like patch clamp, has provided useful information on the modulation of the activity of a number of membrane transport proteins. In this chapter, we will discuss the physical and chemical principles which govern movement of water and ions into and across the plant cells. Attention is further being paid to understand the molecular mechanisms of various transport processes taking place across cells, which are mediated by the large variety of transport proteins, and also about the intracellular distribution of proteins required for maintaining the required ionic balance.

© Springer Nature Singapore Pte Ltd. 2018
S. C Bhatla, M. A. Lal, *Plant Physiology, Development and Metabolism*,
https://doi.org/10.1007/978-981-13-2023-1_3

## 3.1 Water and Ion Uptake from Soil into Roots

Water in soil is adsorbed on the surface of soil particles (sand, clay, silt, and organic material). It enters plant roots most readily across the cells near the root tip. Root hairs further increase the surface area of roots for absorption of water and mineral ions. Once water reaches inside the epidermis, it can be transported up to the endodermis through one or more of the three pathways—apoplast, symplast, and transcellular pathways (Fig. 3.1). **Apoplast** represents a continuous system of cell walls in which water moves without crossing any membrane as it travels across the root cortex. The **symplast** refers to continuation of the cytosol of neighboring cells via cytoplasmic canals in the plasmodesmata. In the transcellular pathway, water enters a cell from one side and exits the cell from the other side across the plasma membrane, again entering the next cell in series and so on. In this pathway water crosses the membrane of each cell twice, once to enter and second time to leave the cell. Solutes must be dissolved in soil solution before they can be taken up by roots. Ions vary in their solubility, which is also affected by the soil pH. With the absorption of water by the plant, soil solution recedes into small pockets, channels, and crevices having concave menisci (curved interface between air and water). As a result, soil solution develops negative pressure due to surface tension. Clay particles,

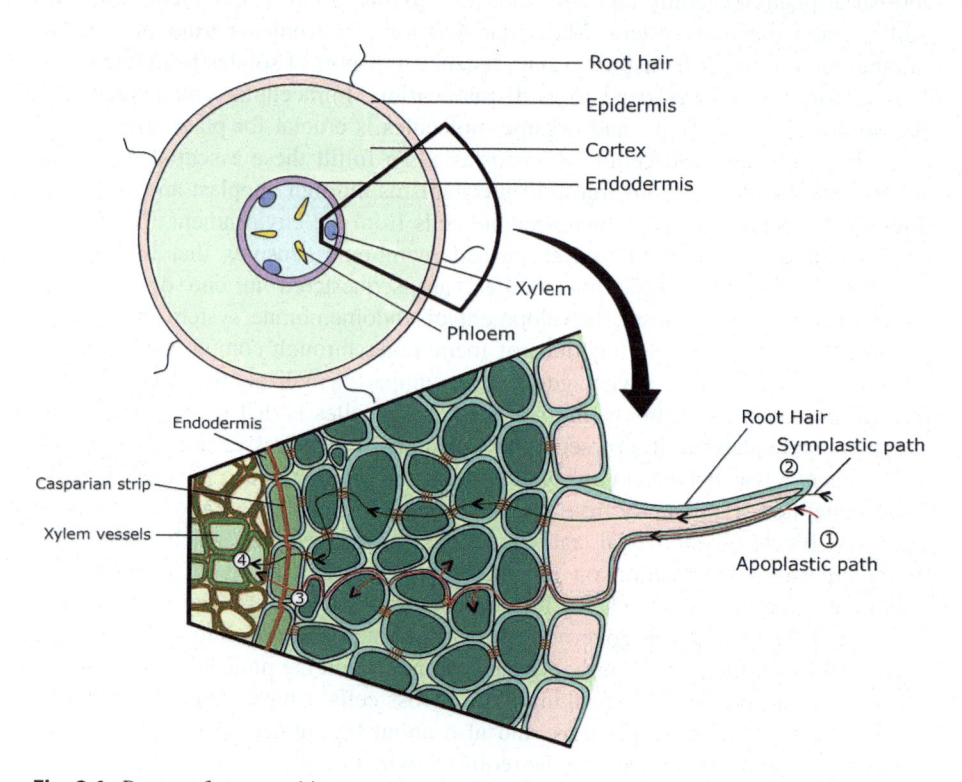

**Fig. 3.1** Routes of water and ion transport across various regions of plant root

## 3.1 Water and Ion Uptake from Soil into Roots

being negatively charged, attract positively charged cations. Protons can displace these cations by cation exchange, releasing them into the soil solution. Plant roots may facilitate cation exchange to acquire bound cations either by releasing $H^+$ directly from their surface or because of carbonic acid formed in soil solution through $CO_2$ emission from roots during respiration (Fig. 3.2). During the passage of a solute from soil solution into the cortical cells in a root, it has to traverse several regions exhibiting varying degrees of resistance in different regions. First, solutes encounter an unstirred film of water which adheres to the exterior of the cell wall of root hairs and other water-absorbing regions of the growing root. Generally, ions penetrate this layer quickly by diffusion. Next, solutes have to penetrate and pass through cell wall. Three major polysaccharide constituents are recognized in the primary cell wall. (i) *Cellulose*: It consists of linear chains of 2000–20,000 $(1 \rightarrow 4)$-β-D-linked glucose molecules. These chains are packed together in regular, partially crystalline arrays (microfibrils) embedded in an amorphous matrix of noncellulosic polysaccharides. (ii) Cross-linking glycans (earlier called *hemicelluloses*): These are mainly composed of xyloglucan polymers and glucuronoarabinoxylans. (iii) *Pectic substances*: They are a group of polysaccharides rich in polygalacturonic acid. They have weakly acidic carboxyl groups (-COOH) which ionize and give rise to fixed negative charges (-COO$^-$) and loosely held $H^+$ charges. Positively charged cations ($K^+$, $Mg^{2+}$, $Ca^{2+}$) passing through cell wall displace hydrogen ions on the carboxyl groups of polygalacturonic acid molecules and are held there by relatively weak interionic forces on the fixed negative charges (COO$^-$). These fixed negative charges in plant cell wall (carboxyl groups of polygalacturonic acid) are called **cation adsorption sites** or **cation exchange sites**. Different cations dissolved in soil solution exhibit different affinities for cation exchange sites, depending on their charge. A cation with a relatively high

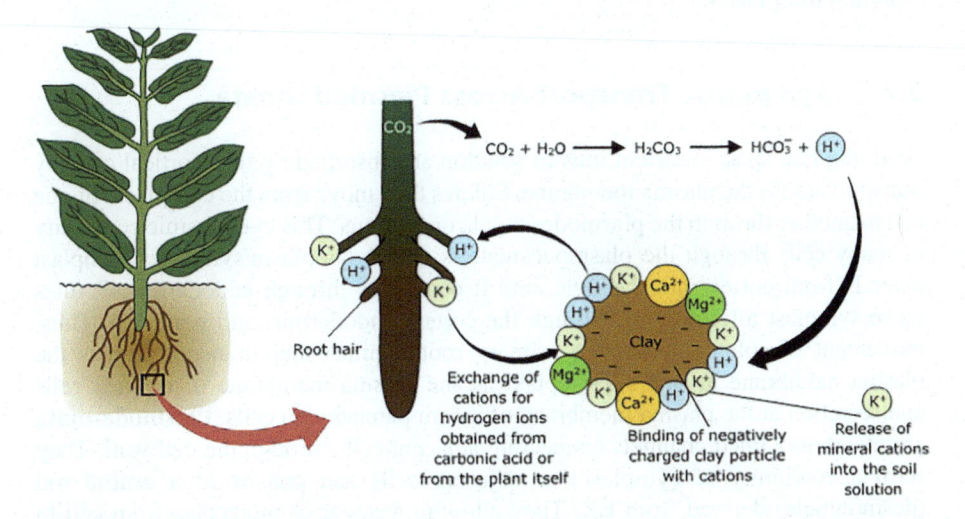

**Fig. 3.2** The process of cation exchange between soil particles and roots through displacement of protons

adsorption affinity (e.g., $Ca^{2+}$) will displace another one with lesser adsorption affinity (e.g., $K^+$). The adsorption affinity of a cation will determine its ability to diffuse across the wall in a "leap frog" manner by migrating from one negatively charged adsorptive site to another by displacing other cations. Water forms a large part of the cell wall. It adheres to both cellulosic and noncellulosic cell wall components. Primary cell wall has an "open" structure with large passageways filled with water for migration of ions. Except for the relatively weak cation exchange sites, primary cell wall does not offer any resistance to solute movement across it. Cell wall also contains a variety of noncellulosic constituents, such as extensins and lignin, for providing rigidity to cell wall.

Generally, water, nutrient molecules, and ions dissolved in water readily diffuse across primary cell wall. The fraction of the plant tissue (apoplast) readily accessible for diffusion of an externally applied solute dissolved in water is referred to as **free space**. Free space includes primary wall since it offers relatively less hindrance to diffusion of dissolved solutes. The plasma membrane forms the boundary of free space because most solutes do not diffuse readily across it without active transport. Limit of free space (in a primary root) is up to endodermis. Casparian strips in the endodermal cells are suberized in the transverse and radial walls. Suberization of cell walls blocks the movement of water and dissolved ions. But in young primary roots undergoing gradual maturation, the limit of free space may be even further because Casparian strips in the endodermal cells are either poorly developed or they exhibit discontinuity. Free space is a functional concept and its dimensions can be measured by physiological experiments. Apoplast, on the other hand, is an anatomical term and includes all interconnected cell walls of a plant tissue. Inward movement of ions dissolved in water is most rapid in the root region where root hairs have attained their maximum length. In this region, vessels and tracheids are fully mature (dead and are without protoplasts).

## 3.2 Symplastic Transport Across Plasmodesmata

From the free space, nutrient ions in solution are absorbed by root cortical cells by transport across the plasma membrane. Solutes then move from the cytoplasm of one cell to another through the plasmodesmatal connections. This cytoplasmic continuity of many cells through the plasmodesmata is called symplasm/symplast. Symplast extends from cortex into the stele, and it penetrates through endodermis. Solutes leave symplast after passing through the cortex, endodermis, and pericycle. Thus, movement of solutes across the primary root requires their transport across the plasma membrane at two sites: uptake at the plasma membrane of cortical cells and secretion at the plasma membrane of xylem parenchyma cells. **Plasmodesmata** are membrane-lined channels connecting adjacent cells through the cell wall. They form a continuity of cytoplasm of adjacent cells and consist of a central rod (desmotubule) derived from ER. They allow movement of molecules from cell to cell through symplasm. Plasmodesmata may be formed during cell division (*primary plasmodesmata*) or later in preexisting cell walls as well (*secondary*

## 3.2 Symplastic Transport Across Plasmodesmata

*plasmodesmata*). Secondary plasmodesmata are generally branched, and their formation is most evident during the development of host-parasite connections, in graft unions and in organs post-genitally fused (e.g., some carpels). In vascular plants, the basic plasmodesmatal structure consists of a cell-to-cell tubule of the plasma membrane which surrounds a cylindrical strand of appressed endoplasmic reticulum (the desmotubule). A cytoplasmic sleeve lies between the desmotubule and the plasma membrane (Fig. 3.3). A central rod occupies the center of the desmotubule which contains lipid polar groups and some proteins. Outer surface of the desmotubule and inner surface of the plasma membrane are studded with protein subunits. The gaps between the protein particles constitute the physical basis of molecular sieving during transport across plasmodesmata. Actin filaments have been reported to spiral along the length of plasmodesmatal channels. They regulate channel diameter by an actin-myosin-based mechanism of contraction or expansion. Furthermore, actin can also serve as a track to facilitate movement of solutes along the length of plasmodesmata. Calcium-regulated centrin filaments are localized along the neck region of plasmodesmata. They may be involved in the closing of the neck region of the plasmodesmata by pulling the endoplasmic reticulum and plasma membrane

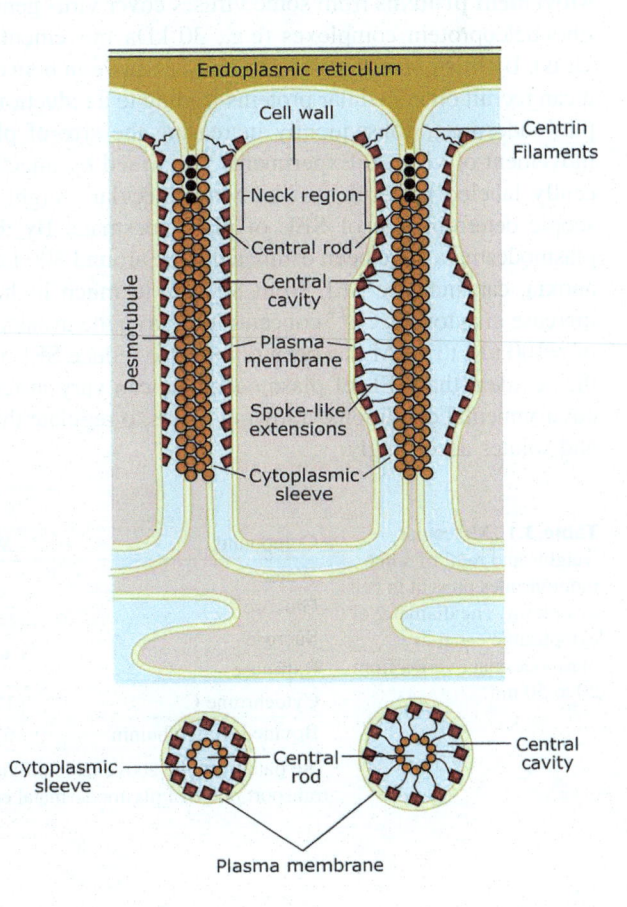

**Fig. 3.3** Structure of plasmodesmata along its length and in cross section

closer to each other. The size of solute particles is the principal factor governing their symplastic mobility across plasmodesmata. It may also depend on charge of the molecules. Since plasmodesmatal channels are aqueous pathways lined with polar-charged groups and hydrogen-bonding groups, these structural components can be expected to interact with solutes being transported, especially those with size close to **size exclusion limit** (SEL). The size exclusion limit of a symplastic pathway through plasmodesmata is typically referred to as the molecular mass of the smallest solute excluded from movement across the plasmodesmatal channels. Molecular weights and radii of some common cytoplasmic constituents presented in Table 3.1 indicate the range of biomolecules which can pass through the cytoplasmic canal of the plasmodesmata (generally 20–50 nm in diameter). SEL of plasmodesmatal channel depends on the type of tissue, its developmental state, and physiological conditions. The plasmodesmatal connections between meristematic cells in roots permit the passage of macromolecules up to 65 kDa. A variety of proteins, such as actin and myosin, which are involved in macromolecular trafficking, have also been detected in plasmodesmatal channels. At times, plant viruses use plasmodesmata for their spread from cell to cell using "movement proteins" encoded by virus genome. **Movement proteins** from some viruses cover virus genome (mainly RNA) forming ribonucleoprotein complexes (e.g., 30 kDa movement protein of tobacco mosaic virus). By this process, virus genome can move in between the cells in leaves where it can recruit other cellular proteins leading to a reduction in callose deposition in the plasmodesmata, consequently increasing the size of plasmodesmatal pore for the movement of viruses. Experiments performed by injecting plant cells with fluorescently labeled molecules of known molecular weight have facilitated the microscopic determination of SEL of plasmodesmata. By this approach, SEL of most plasmodesmata have been estimated to be around 800 Da. Stress conditions, such as anoxia, can increase SEL from 800 Da to much higher values. On the contrary, increase in cytosolic $Ca^{2+}$ concentration from the usually resting state concentrations of $\approx$100 nM to 1 µM has been reported to reduce SEL in the stamen hair cells. It is thus evident that SEL of plasmodesmata can vary in response to a wide variety of environmental conditions, allowing plants to regulate the intercellular flow of water and solutes accordingly.

**Table 3.1** Molecular weights and radii of some biomolecules present in cell cytoplasm. The diameter of cytoplasmic canal in plasmodesmata varies from 20 to 50 nm

| Compound | MW (Da) | Radius (nm) |
|---|---|---|
| Water | 18 | 0.15 |
| Glucose | 180 | 0.35 |
| Sucrose | 342 | 0.47 |
| Raffinose | 504 | 0.57 |
| Cytochrome C | 12,400 | 1.65 |
| Bovine serum albumin | 67,000 | 3.55 |

The data from the above table provides information about the solute transport range of plasmodesmatal connections in terms of their SEL

## 3.3 Diffusion vs Bulk Transport of Water and Solutes

Diffusion is a spontaneous process which does not involve energy. It occurs in biological systems within the cells or from cell to cell across the plasma membrane. The rate of diffusion is related to the size of the molecule, its concentration gradient, viscosity of the medium, and temperature (Boxes 3.1 and 3.2). Diffusion across a cell membrane begins with the solute partitioning from solution into the membrane, then diffusing across it, and then partitioning back into the solution on the other side. Apoplastic or symplastic diffusion is the simplest way for intercellular transport of water and solutes in plant cells. According to diffusion theory, sucrose molecules require 4.8 s to reach 37% of their equilibrium concentration over a distance of 100 µm in a typical cell. The time required for diffusion of sucrose across ten such cells arranged end-to-end and interconnected by plasmodesmata is about 8 min. Thus, ion movement due to diffusion is not a viable option to meet metabolic demands for transcellular migration of metabolites beyond 1 mm. *Cytoplasmic streaming* does, however, partially enhance diffusion rate within the cells. Cytoplasmic streaming is facilitated by the action of cytoskeletal motor proteins and is powered by ATP hydrolysis. This mode of intracellular transport (cytoplasmic streaming) is critical not only for ordinary parenchyma cells in vascular plants but

---

**Box 3.1: Diffusion is a Spontaneous Process Which Obeys Fick's Law**

The rate of diffusion of a molecule, from one point in a cell to another or across a membrane, is related to its size, its concentration gradient, viscosity of the medium, and temperature. This relationship is referred to as **Fick's law** and is represented in an equation as follows:

$$J_s = -D_s(\Delta C_s / \Delta x) \quad (3.1)$$

where $J_s$ represents rate of diffusion of the molecular species which is determined by its diffusion coefficient ($D_s$) and concentration gradient ($\Delta C_s / \Delta x$), i.e., the concentration difference ($\Delta C_s$) between two points ($\Delta x$). $J_s$ is expressed as moles per unit area per unit time. The negative sign in the equation shows that substances move by diffusion down a concentration gradient. The time taken by a substance to a move distance $L$ is expressed as $L^2/D$. Thus, the time taken by molecules to diffuse increases with the square of the distance. The value of $D_s$ for ions is $10^{-9}$ $m^2\,s^{-1}$ and for larger molecules $D_s$ is around $10^{-11}$ to $10^{-10}$ $m^2\,s^{-1}$. A molecule with $D_s$ of $10^{-9}$ $m^2\,s^{-1}$ will take 2.5 s to move a distance of 50 µm in a plant cell but will take 32 years (!) to move a distance of 1 m. Thus, the process of diffusion is effective for migration of molecules in cellular distance but is not effective for long-distance movement of ions.

> **Box 3.2: Nernst Equation Predicts Internal and External Ion Concentrations at a Given Membrane Potential**
>
> According to Nernst equation:
>
> $$\Delta E_s = 59 \text{ mV } \log\left(C_s^o/C_s^i\right) \qquad (3.2)$$
>
> Solutes moving into and out of cells by diffusion are expected to reach equilibrium. $\Delta E_s$ represents the electric potential difference between the inside and the outside of the cell. $\Delta E$ for a specific ion is known as **Nernst potential**. Nernst equation can be used to predict whether or not ions will accumulate against their chemical gradient. If the solute concentrations inside $(C_s^i)$ and outside $(C_s^o)$ a cell are known, and if the membrane potential of cell can be determined, then it is possible to know whether the solute moves down the electrochemical gradient.

also for solute transport in giant algal cells (e.g., *Nitella* sp.) which can be several millimeters in diameter and many centimeters in length. In view of these limitations of diffusive movement of water and solutes, plants have evolved pressure-driven **bulk flow** for long-distance transport of water and dissolved solutes in xylem and also through the cell wall in plant tissues. In contrast to diffusion across membranes, pressure-driven bulk flow is independent of solute concentration gradients. Pressure potential gradients generated through different means are responsible for bulk flow of water and dissolved solutes in opposite directions through xylem (tension) and phloem (hydrostatic pressure). In this context, it may be noted that vessel elements up to 500 μm in diameter occur in the stem of climber plants. Such large vessels permit vines to transport large volume of water in spite of the slenderness of the stem.

## 3.4 Structural Features of Xylem Elements Which Facilitate Water and Solute Transport

Water-conducting cells of xylem (vessels and tracheids) are collectively known as tracheary elements. Mature vessels and tracheids are dead and hollow, lack plasma membrane, and have a rigid cell wall impregnated with lignin (Fig. 3.4). Water movement between adjacent tracheids occurs through pits in the cell wall. Pits lack secondary wall but primary wall is retained. Among vessels, water movement is unobstructed from one vessel element to another, across scalariform perforation plates in the end walls. Physical forces, like wind, can sometimes cause trapping of air bubbles in the tracheary elements leading to disturbance in the process of water

## 3.4 Structural Features of Xylem Elements Which Facilitate Water and Solute...

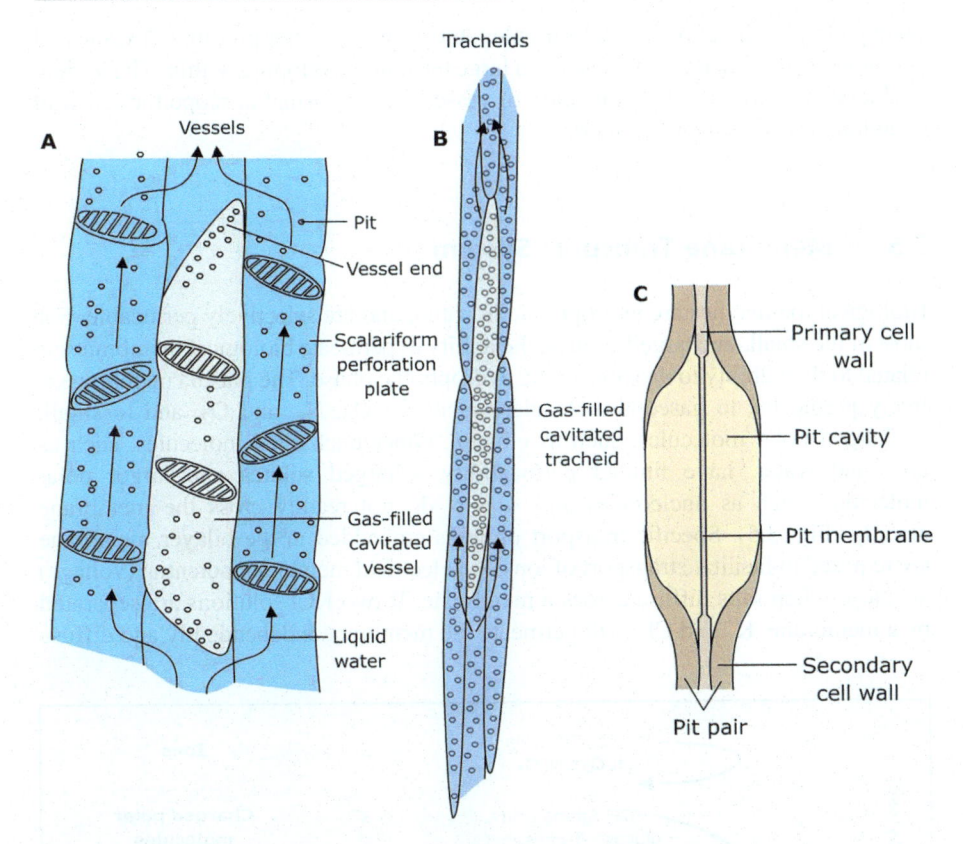

**Fig. 3.4** Structure of (**a**) vessels, (**b**) tracheids, and (**c**) pits in higher plants

and solute transport. Absence of lipid membranes, significant length and diameter of the trachedial elements, and hydrogel nature of pectic membrane of pits and lignification of cell walls are the major structural features of xylem elements which regulate water and solute transport through them. The absence of lipid membranes in vessels and tracheids enhances hydraulic conductance severalfold as compared to cells with intact plasma membrane. Since end walls account for 60–80% of hydraulic resistance accompanying water transport, length of vessels is an important determinant of their water transport efficiency. It may be noted that tracheids generally have smaller diameter (10–15 μm) and are longer than individual vessel elements (50–100 μm long). Thus, the smaller diameter and greater number of end walls per unit length are also responsible for much lower water transport efficiency of tracheids than vessels. Pit membrane consists of microfibrils and a pectin/hemicellulose matrix with a number of pores (5–20 nm in diameter). The pectin component of the pit membrane can respond like a hydrogel to alterations in ion concentrations in the xylem sap which can alter pore size in the pit membrane. Thus, high $K^+$ concentrations increase the hydraulic conductance of pit membrane, thereby acting as an important regulatory mechanism for diverting the flow of water to different

## 3.5 Membrane Transport System

Biological membranes are hydrophobic in nature and are selectively permeable. For most of the small, uncharged solutes, the ability to permeate biological membranes is related to their ability to dissolve in the hydrophobic phase. The plasma membrane is freely permeable to gaseous molecules, such as $CO_2$, $N_2$, and $O_2$, and to small, uncharged polar molecules, such as ethanol. Other uncharged molecules, such as urea and water, have limited permeability. Charged solutes and larger polar molecules, such as nucleotides and sugars, do not readily cross the membrane directly (Fig. 3.5). Specific transport proteins embedded in the bilayer membrane are required to facilitate transport of ions. An electrical membrane potential (voltage) develops when salts diffuse across a membrane. If two KCl solutions are separated by a membrane, $K^+$ and $Cl^-$ will permeate the membrane independently and diffuse

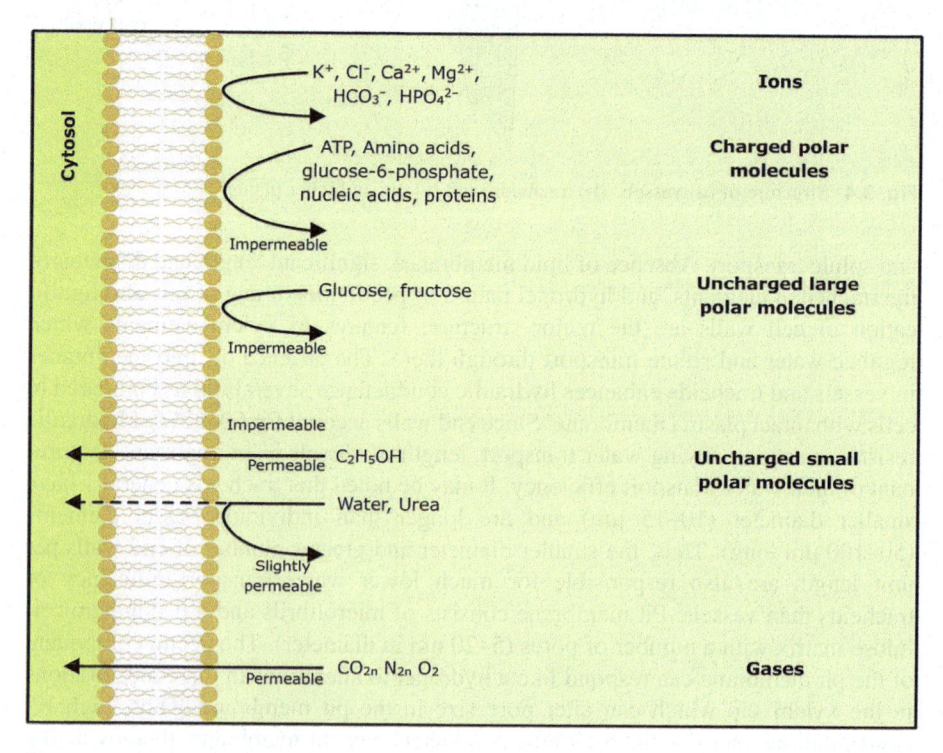

**Fig. 3.5** Differential permeability of phospholipid bilayer membrane

## 3.5 Membrane Transport System

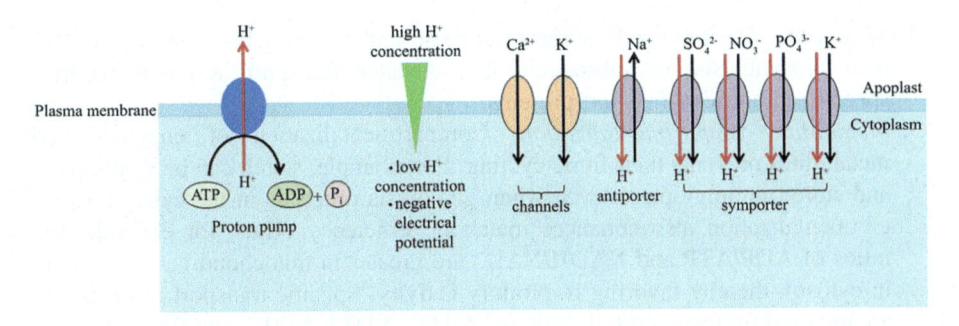

**Fig. 3.6** Various ways of solute transport across the phospholipid bilayer membrane

according to their respective gradients of **electrochemical potential**. This creates an electrical potential across the membrane (Fig. 3.6). Biological membranes are usually more permeable to $K^+$ than to $Cl^-$, leading to faster migration of $K^+$ and causing the cell to develop an inside negative charge with respect to extracellular medium. A potential that develops as a result of diffusion is known as **diffusion potential**. Upon attainment of equilibrium, both concentration gradient and diffusion potential across the membrane collapse. All living cell membranes exhibit membrane potential due to asymmetric distributions of ions inside and outside the cell. Membrane potential can be determined by inserting a microelectrode into the cell and measuring the voltage difference between the inside of the cell and the extracellular medium. When the rate of influx and efflux of a given solute are equal, the cell is said to have attained *steady state*. During steady state, the occurrence of active transport across the membrane prevents many diffusive fluxes from attaining a state of equilibrium. A number of ions permeate cell membrane simultaneously, but $K^+$ has the highest concentration in plant cells and also highest permeability across membrane. Since transport proteins mostly exhibit specificity for the solutes they transport, cells require a great diversity of transport proteins. *Haemophilus influenzae*, a simple prokaryote and the first organism for which complete genome was sequenced, has only 1743 genes. Out of this, more than 200 genes encode various proteins involved in membrane transport. In *Saccharomyces cerevisiae*, nuclear genome, about 2000 of the nearly 6000 genes encode for membrane-associated proteins, of which a large proportion are transport proteins. In *Arabidopsis*, out of a predicted 33,000 protein-coding genes, as many as 1300 may encode proteins with transport functions.

1. Membrane transport in plants is crucial for a wide range of essential processes. These include: (1) *Nutrient acquisition*: Uptake of a number of inorganic nutrients (nitrogen as $NH_4^+$ or $NO_3^-$, sulfur as $SO_4^{2-}$, and phosphorus as $PO_4^{3-}$, $K^+$, and $Ca^{2+}$ and a number of trace elements) is mediated by specialized nutrient transport proteins. Their uptake is vital for plants since, unlike animals, plants synthesize organic biomolecules from inorganic nutrients, most of which must be absorbed by roots from soil.

2. *Metabolite distribution*: Loading of sucrose and amino acids from the sites of their biosynthesis for subsequent long-distance transport in plants requires specialized nutrient transport proteins.
3. *Metabolite compartmentalization*: Compartmentalization of enzymes and metabolites prevents their futile cycling. For example, starch can be synthesized and stored in amyloplasts even when glycolysis proceeds in the cytosol. Compartmentalization also enhances metabolic efficiency. Thus, for example, the ratios of ADP/ATP and NADH/NAD$^+$ are greater in mitochondrial matrix than in cytosol, thereby favoring respiratory activity. Specific transport mechanisms are required for the export of ATP and NAD$^+$. NDT1, NDT2, and PXN are some NAD$^+$ transporters found in plants.
4. *Energy transduction*: Membrane transport is crucial for conversion of free energy into biologically useful forms. Thus, light energy stimulates photosynthetic electron transport chain to pump H$^+$ into the thylakoid lumen.
5. *Turgor generation*: This is accomplished in plant cells by accumulating salts via specific membrane transporters. In most mature cells of plants, potassium ions accumulate in the cytosol and vacuole which are balanced with anion (mostly Cl$^-$) uptake to maintain electroneutrality. This leads to water accumulation in cells.
6. *Waste product excretion*: Proton pumps play critical role in the removal of H$^+$ from the cytosol. Likewise, other energy-driven transporters in the tonoplast membrane can play significant roles in removing metabolic wastes from the cytosol and accumulating them in vacuoles.
7. *Signal transduction*: Modulation of intracellular Ca$^{2+}$ concentration in the cells is achieved through a regulated activity of membrane localized Ca$^{2+}$-ATPases and Ca$^{2+}$ channels, thereby affecting signaling mechanisms operating through calcium.

## 3.6 Uniporters and Cotransporters

Although particular transporter proteins are usually specific for the solute to be transported across membrane, their specificity is at times not absolute. Likewise, proteins involved in the transport of neutral amino acids, such as glycine, alanine, and valine, may not accept acidic or basic amino acids as aspartic acid or lysine, respectively. Solutes may get freely transported across the lipid bilayer of membranes, moving down their electrochemical potential gradient by simple diffusion (Fig. 3.6). Movement of specific solutes using membrane transporter proteins to diffuse across the membrane according to their concentration gradient is referred to as **facilitated diffusion**. Transporter proteins associated with this process include

## 3.6 Uniporters and Cotransporters

*uniporters* (also called carriers) and **ion channels** (Fig. 3.6). On the contrary when transporter proteins facilitate solute movement against their electrochemical gradient, the process is called *active transport*. Active transport can be of two types: **primary active transport** and **secondary active transport**. Primary active transport accompanies ATP or pyrophosphate hydrolysis to provide energy for ion transport against concentration gradient. Such transporters which perform primary active transport are called **pumps**. Secondary active transporters (also called cotransporters), however, utilize ion gradients established by primary active transport to move other solutes against their electrochemical gradients. Thus, secondary active transporters facilitate movement of two solutes simultaneously. One solute moves down its electrochemical gradient (which is created due to the activity of pumps), whereas the second one moves up along its electrochemical gradient. Secondary active transporters (or cotransporters) can be further divided into two groups: symporters and antiporters (Fig. 3.6). Symporters move both the solutes across the membrane in the same direction. For example, proton flux according to concentration gradient across membrane provides energy to move K+ and NO3- into the cell along with proton flux. Antiporters move the two solutes in opposite directions across the membrane. In plants, secondary active transport normally exploits $H^+$ gradients established by proton pumps. Thus, sodium ions move out of cells against the flux of protons using specific antiporters. Potassium level is maintained high in plant cells as compared to sodium. This is just opposite of what is seen in animal cells (high sodium and low potassium levels).

Proton-coupled symporters are generally required for substrate uptake into the cytosol, while antiporters function to export solute out of the cytosol. In both types of secondary transport, the ions or solutes being transported move against their concentration gradient. The energy driving secondary transport is provided by **proton motive force** (PMF) rather than directly by ATP hydrolysis. Figure 3.7 shows an overview of the subcellular location of $H^+$ pumps in plant cells. In chloroplasts and mitochondria, energy in $H^+$ gradients is utilized to synthesize ATP. Proton gradients are also established across the plasma membrane and tonoplast by pumps which utilize ATP or $PP_i$. The electrochemical potential thus generated is used by plant cells to transport various ions and metabolites across the plasma membrane and tonoplast using various integral membrane channels and cotransporters. In the transport mediated by uniporters and cotransporters, the solute to be transported binds to the transporter and causes confirmation change in the transporter protein, leading to solute movement across the membrane. The solute-transporter protein interaction is generally selective to the extent that these proteins can even distinguish between stereoisomers of sugars and amino acids.

In addition to their association with plasma membrane and tonoplast, uniporters and cotransporters are found in the **endomembrane system** as well as in the envelopes of chloroplasts and mitochondria. They play roles in the uptake of

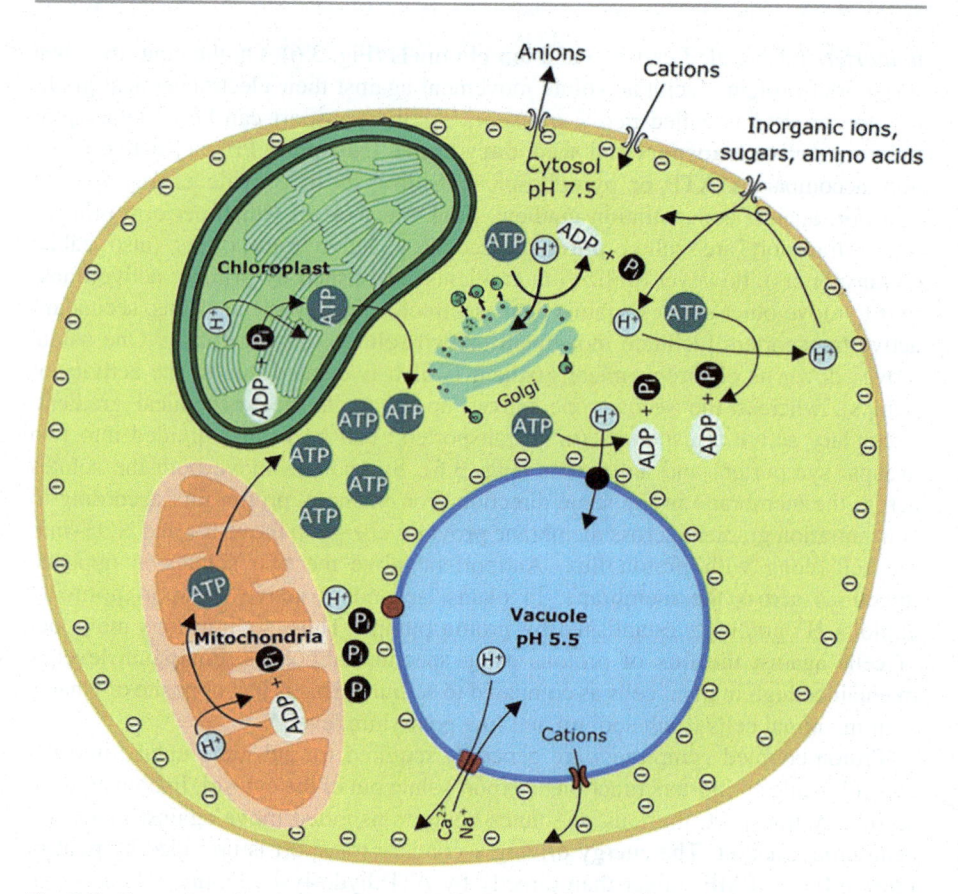

**Fig. 3.7** Generation of proton gradients across the plasma membrane and cell organelles leading to solute transport by various transporters

inorganic nutrients, such as $NH_4^+$, $NO_3^-$, $SO_4^{2-}$, and $H_2PO_4^-$. They are also important for loading of sugars into phloem for long-distance transport. Proton-coupled symporters include $H^+$/sucrose symporter (involved in loading of sucrose into phloem), several $H^+$/anion symporters, and a number of $H^+$/amino acid symporters. The most abundant protein in the chloroplast membrane is a phosphate translocator that exchanges inorganic phosphate for triose phosphate. Proton-coupled antiporters include $H^+/Ca^{2+}$ antiporter in the tonoplast and $Na^+/H^+$ antiporter in the plasma membrane. $Na^+/H^+$ antiporters on the plasma membrane of glycophytes enhance their sensitivity to salt. These antiporters, identified as "salt overly sensitive" or SOS1 in roots, extrude $Na^+$ from the cell, thereby lowering their internal concentration. Vacuolar $Na^+$ sequestration also occurs by the activity of a subset of cation/H$^+$ antiporter (CPAs)—a Na$^+$/H$^+$ antiporter, which couples downhill movement of H$^+$ into the cytosol across the tonoplast with Na$^+$ uptake into the vacuole. The *Arabidopsis* AtNHX1 Na$^+$/H$^+$ (NHX) antiporter gene overexpression confers increased salt tolerance in various crop species, such as wheat, maize, and tomato, as also in *Arabidopsis*. In contrast with SOS1 and NHX antiporters, which reduce cytosolic Na$^+$ concentration, HKT1 transporters transport Na$^+$ from the apoplast into the cytosol.

## 3.7 Ion Channels

Hodgkins and Huxley, while working on squid exons in 1950, described ion channels as:

> *elements in the plasma membrane that respond to electrical stimuli by opening and facilitating fluxes of selective ions during action potentials.*

Action potentials are generated when the bilayer membrane is depolarized to a voltage more positive than the threshold voltage. A few terrestrial plants are excitable; e.g., in *Mimosa pudica*, leaf stroking evokes an action potential that causes the pulvinus (at the leaf base) to lose turgor, causing collapse of the leaf. Some insectivorous plants (e.g., *Dionaea muscipula*, *Drosera* sp.) also use action potential to couple the sensing of prey to subsequent leaf movement. First reports on the existence of ion channels in non-excitable plant cells came in 1980s in guard cells. The guard cell ion channels were identified to play key role in mediating solute fluxes accompanying stomatal opening and closing. Ion channels are now known to be present in all plant cells on the plasma membrane and tonoplast and are studied with electrophysiological techniques—such as **patch clamp** (Box 3.3). It utilizes pushing a blunt-tipped glass micropipette (microelectrodes) against biological membrane to measure pico ampere currents. In essence, the patch clamp technique allows detection of tiny electrical currents that ions carry as they flow through channels. The outstanding capability of patch clamp is that it can resolve the activity of single protein molecules (channels) as they catalyze ion translocation. Using the whole-cell patch clamp technique, the activity of several ion channels in a single cell can be quantified. Ion fluxes through channels are driven solely by electrical potential difference. Ion flow through channels is passive. Thus, in contrast to ATPases or carriers, the direction of flow through ion channels is dictated simply by electrochemical potential gradient for that ion. In addition to their activation by electropotential difference, channels exhibit two additional properties essential to their function, i.e., *ionic selectivity* and *gating*.

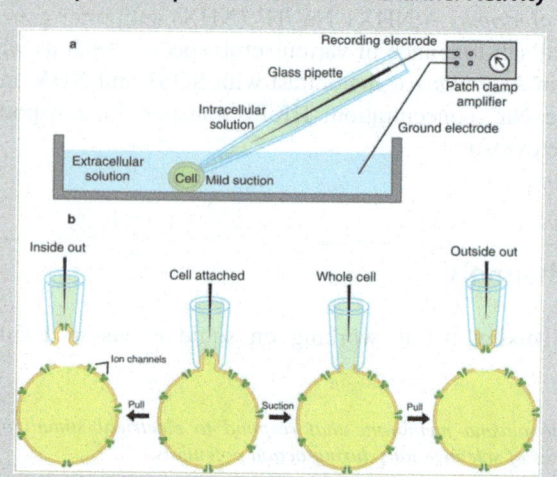

### Box 3.3: Patch Clamp Technique to Measure Ion Channel Activity

The patch clamp technique involves pushing a blunt-tipped, glass micropipette against a biological membrane and simultaneously applying suction to the inside of the micropipette to form an electrically tight seal (**a**). Initial seal formation attains cell-attached mode which can record the activity of individual ion channels without any control over the cytosolic ionic composition. Pulling the pipette away from the membrane generates an inside-out patch in which cytosolic face is exposed to the bathing solution. In this situation, the solution composition on both sides of the membrane is defined, and the electrical activity across single channels can be assessed under controlled conditions. In the third alternative, the membrane patch that covers the pipette tip can be disrupted with a high-voltage pulse or suction, leading to electrical access to the inside of the cell. Current flow can thus be monitored across the entire membrane in this whole-cell mode of recording. The large volume of medium in the pipette exchanges with cell contents, thereby defining the intracellular solution composition. Finally, if the pipette is pulled away from the cell after attaining the whole-cell mode, a membrane bleb is also pulled away, and it reseals itself across the pipette tip as an outside-out patch. This recording mode is useful for testing the effects of putative cytosolic regulators on ion channel activity. In this way patch clamp technique can be used to record electrical properties of pumps and carrier proteins on plant protoplasts and endomembranes.

Major cation channels in plants include $K^+$-selective (inward or outward) and $Ca^{2+}$-selective channels. Anion channels, in general, allow permeation of a wide range of anions, including $Cl^-$, $NO_3^-$, and organic acids. Specific anion channels exist for organic acids, such as malate. Channels are tightly controlled by conformational shifts between permeable (or open) and non-permeable (or closed) states. This alteration between open and closed states of channels is known as "gating." Gating is controlled either by voltage or by a ligand (chemicals that bind to channel proteins), such as hormones, $Ca^{2+}$, G-proteins, and pH. Some channels are stretch-sensitive and are gated by changes in turgor pressure of a cell. Guard cells have at least four types of $Ca^{2+}$ channels, one regulated by voltage, one by $Ca^{2+}$, and the other two by ligands. This kind of multiple channels for $Ca^{2+}$, gated by different signals, allows for dynamic changes in cytosolic $Ca^{2+}$ in response to a variety of stimuli. Of the six types of plant cation channels, the *shaker channels* are well characterized. They are named so after a *Drosophila* $K^+$ channel whose mutation causes flies to shake or tremble. Plant Shaker channels are highly $K^+$ selective and are responsible for $K^+$ flux across guard cell plasma membrane. They also help in $K^+$ uptake from the soil, participate in $K^+$ release from living stelar cells to xylem vessels, and also play a role in $K^+$ uptake in pollen.

## 3.7.1 Potassium Channels

The ability of an ion channel to permit transport of ions only in one direction is called **rectification**. Inward and outward currents are carried by separate classes of ion channels. The channels carrying these currents are said to rectify. **Rectifying channels** carry current in one direction only. $K^+$ inward rectifiers and $K^+$ outward rectifiers have been identified and characterized in guard cells and are now known to be present in a wide variety of plant cells. Inward-rectifying $K^+$ channels ($K^+$ influx channels) are activated by hyperpolarization of the membrane. First eukaryotic $K^+$ influx channels were cloned from plants and characterized by expression in *Xenopus* oocytes. $K^+$ influx channel subunits in plants are products of a multigene family and exhibit tissue-specific expression. Thus, one member, KAT1, is expressed in guard cells, and another, AKT1, is expressed in roots and hydathodes. The *akt1* mutant of *Arabidopsis* exhibits reduced $K^+$ uptake.

$K^+$ channel AKT1 has six transmembrane spans, S1 through S6, with a membrane-intrusive loop between S5 and S6. This loop makes up the pore domain of the channel (P-domain). The fourth transmembrane helix, known as S4 domain, exhibits a regular pattern of positively charged residues (lysine or arginine) every third residue, so that the charged residues tend to project from one side of the helix (Fig. 3.8a). This region of the protein forms the *voltage sensor*, which is involved in the opening and closing of the channel in response to permissive voltage. The functional channel works as a tetramer, with the P-domains (between S5 and S6) of four monomers interacting to form a narrow constriction that contains the $K^+$ binding and recognition site. P-domain is responsible for ion selectivity due to a conserved sequence of amino acids (Gly-Tyr-Gly). TPK1 from *Arabidopsis* has been identified at the molecular level as an outward $K^+$ rectifier. It has just four transmembrane spans but two P-domains (Fig. 3.8b). Toward the C-terminus are two high-affinity $Ca^{2+}$-binding motifs known as *EF hands*. These channels are activated

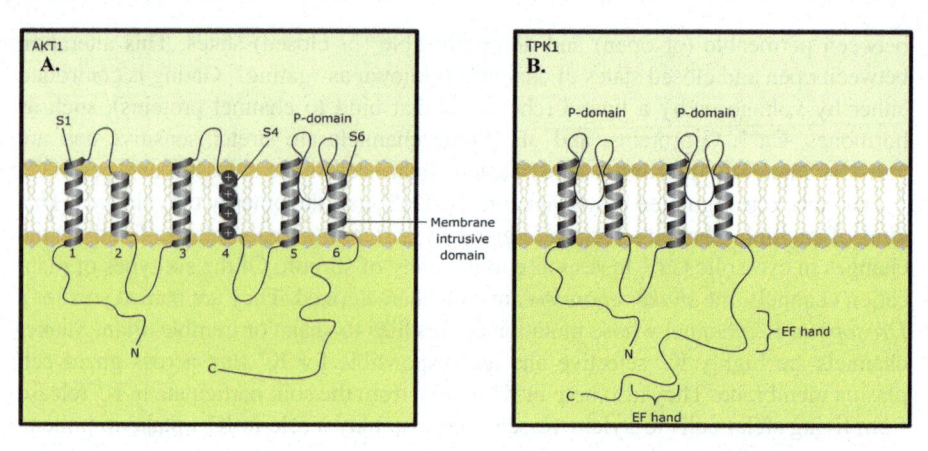

**Fig. 3.8** Structure of (**a**) an inward-rectifying K⁺ channel (AKT1), (**b**) an outward-rectifying K⁺ channel (TPK1)

upon elevation of cytosolic free calcium concentration in plant cells. Thus, K⁺ efflux channels are activated as a result of $Ca^{2+}$ at the EF hand on the C-terminal of the channel protein. Outward K⁺ rectifiers in yeast and humans do not possess EF hands.

Four P-domains form the ion selectivity filter for K⁺ outward-rectifying channel. The structure of P-domains is highly homologous in all known K⁺ channels and different from that in other ion channels. Sodium ions are smaller than K⁺. Still they cannot pass through K⁺ channels. Pore segment contains conserved Gly-Tyr-Gly residues. As K⁺ enters the selectivity filter, it loses its water of hydration and gets bound instead in the same geometry to eight backbone carbonyl oxygens of Gly-Tyr-Gly sequence (carbonyl oxygens). Thus, little energy is required to strip off the eight waters of hydration of a K⁺, and a relatively low activation energy is required for passage of K⁺ ions through the channel. A dehydrated Na⁺, however, is too small to bind to all eight carbonyl oxygens which line up at the selectivity filter. As a result, Na⁺ prefers to remain in water rather than enter the selectivity filter. Thus, the activation energy for Na⁺ passage across the potassium channel is high. This difference in activation energy favors K⁺ by a factor of 1000 over Na⁺ to pass through K⁺ channels. Like Na⁺, $Ca^{2+}$ is also smaller than K⁺ and cannot interact properly with the oxygen atoms in the selectivity filter. Also, more energy is required to strip the water of hydration from $Ca^{2+}$ than from K⁺. The inward and outward K⁺ channels in plant cells are also regulated by factors other than membrane voltage and $Ca^{2+}$, respectively. Increase in cytosolic pH in the guard cells due to increase in ATPase mediated ATP H⁺ pumps elicited by abscisic acid also activates outward K⁺ channels. Inward channels have also been reported to be regulated by phosphorylation. In guard cells, two different K⁺ channels are responsible for K⁺ transport across vacuole. Both are voltage sensitive for gating. One of them (FV, fast vacuolar) is inhibited by high cytosolic $Ca^{2+}$ concentration (>1 μM). It is activated with an increase in cytosolic pH and does not exhibit much selectivity among monovalent cations. The other K⁺-permeable channel (VK, vacuolar K⁺) is, however, highly

## 3.7 Ion Channels

**Table 3.2** Salient features of vacuolar potassium channels in guard cells

| Fast vacuolar channels (FV) | Vacuolar K⁺ channels (VK) |
|---|---|
| Nonselective for monovalent cation efflux into cytosol | Selective for K⁺ efflux into cytosol |
| Inhibited by high $[Ca^{2+}]_{cyt}$ (> 1 µM) | Activated by $[Ca^{2+}]_{cyt}$ in nanomolar range |
| Activated by increase in cytosolic pH | Inhibited by increase in cytosolic pH |
| Molecular identity not yet known | A member of "two-pore" H⁺ channel |
| Both channels are voltage sensitive for gating | |

selective for K⁺ over other monovalent cations. It is activated by cytosolic $Ca^{2+}$ in nanomolar to micromolar range and gets inhibited by increasing cytosolic pH (Table 3.2). Closure of stomata generally precedes an increase in $[Ca^{2+}]_{cyt}$, thereby triggering the opening of VK channels leading to K⁺ release from vacuole. Closure of stomata in the absence of a change in $[Ca^{2+}]_{cyt}$ coincides with an increase in the sensitivity of signaling proteins to $[Ca^{2+}]_{cyt}$. Increase in guard cell cytosolic pH is also likely to play a role in stomata closure and FV channels open in this situation. A coordinated action of FV and VK channels thus demonstrates how parallel mechanisms often mediate biological processes. The molecular identity of FV channels is not yet known. A VK channel, TPK1, has been identified in *Arabidopsis*. It is a member of "two-pore" K⁺ channel family, has only four transmembrane spans in each subunit, and has two Ca2+ motifs at the C-terminus.

### 3.7.2 Calcium Channels

Calcium uptake into the cytosol largely takes place by ion channels rather than proton-coupled transporters presumably because cytoplasmic $[Ca^{2+}]$ is maintained at very low levels (0.0001 mM or $10^{-7}$ M) in the eukaryotic cells by the activity of $Ca^{2+}$-ATPase. In contrast, $[Ca^{2+}]$ in soil, apoplast, and vacuole can be in the range of 0.1–1 mM. Because of the existence of such a steep $[Ca^{2+}]$ gradient (soil cell or vacuole cytosol), the plasma membrane- and tonoplast-associated calcium channels can easily pull $Ca^{2+}$ into the cytosol (Table 3.3). Plant $Ca^{2+}$ channels are usually not specific for $Ca^{2+}$ transport but are also permeable to other cations. But because of the steep $[Ca^{2+}]$ gradients across the plasma membrane and tonoplast, opening of even nonselective $Ca^{2+}$-permeable channels will cause a rapid rise in $[Ca^{2+}]_{cyt}$. One of the major calcium channels in the guard cell plasma membrane is activated by hyperpolarizing potentials, ABA and reactive oxygen species (ROS). The molecular identity of these $Ca^{2+}$ channels is still not known. Other well-characterized plant $Ca^{2+}$ channels are cyclic nucleotide-gated (CNGC), calcium-permeable channels in *Arabidopsis* which function in transducing a $Ca^{2+}$ signal in response to pathogen-specific molecules, such as lipopolysaccharides. Another category of calcium channels called glutamate (Glu) receptors has been implicated in microtubule depolymerization and growth inhibition in *Arabidopsis* roots. Induction of transient

**Table 3.3** Major calcium channels in plants

| Channel | Location | Function | Activated by |
|---|---|---|---|
| Two-pore channel 1 (TPC1) | Tonoplast | Turgor regulation, cation homeostasis | ROS, ABA, hyperpolarization |
| Cyclic nucleotide-gated channels (CNGC) | Not known | Ion homeostasis | Lipopolysaccharides |
| Glutamate receptors (GLR) | Ubiquitous | Microtubule polymerization, $NH_4^+$ uptake | Not known |

depolarization and $Ca^{2+}$ influx upon exposure of roots to wounding, glutamate (Glu), and few other amino acids is correlated with the expression of GLR3.3 gene. The biological function of Glu receptors in plants still needs to be investigated.

$Ca^{2+}$-permeable channels in plant cell endomembranes are activated both by voltage and ligands. Several different classes of $Ca^{2+}$-permeable channels may be present in the intracellular membranes of plants. The slow vacuolar (SV) $Ca^{2+}$-permeable cation channels located on the plant vacuolar membrane are activated slowly in response to membrane depolarization and are strongly activated by $Ca^{2+}$-calmodulin. SV channels are permeable to several cations including $Ca^{2+}$, $K^+$, $Na^+$, and $Mg^{2+}$. Their role in conditional release of $Ca^{2+}$ from vacuoles has been established in plant roots. SV channel activity is regulated by multiple factors. Enhanced $[Ca^{2+}]_{cyt}$ levels induce channel opening through the presence of calmodulin and magnesium. Vacuolar $[Ca^{2+}]$ has opposite effect and promotes channel closure. SV channel activity is also regulated by phosphorylation and acidification on either side of tonoplast. Positive membrane potentials promote SV opening. A SV channel, TPC1, has been cloned from *Arabidopsis*. It contains two homologous domains, each having six transmembrane helices and one pore domain in each set. The channel is regulated by calcium through two $Ca^{2+}$-binding EF hands (EF1 and EF2) in the cytosolic loop linking transmembrane helix S6 with S7 (Fig. 3.9).

### 3.7.3 Anion Channels

Some of the major inorganic anions in plant cells are $NO_3^-$, $Cl^-$, $SO_4^{2-}$, and $H_2PO_4^-$, and malate is a major organic anion. The free energy gradients of anion channels are in the direction of passive efflux. Among plants, anion channels were first characterized in guard cells as $Ca^{2+}$-activated, rapidly activating R-type and slowly activating S-type. These channels are also gated by voltage, and their opening leads to massive efflux of $Cl^-$ ions from the cells and membrane depolarization. The resulting depolarization activates outward rectifying $K^+$ channels during salt loss. Anion channels, thus, serve as pacemakers of plant turgor reduction. S-type anion channels are gated by ABA, elevated $CO_2$, pathogenic elicitors, and ozone, thereby triggering stomatal closure. Both $Ca^{2+}$-dependent and $Ca^{2+}$-independent protein kinases regulate anion channel activation in guard cells indicating their activity modulation through phosphorylation. Malate is a major organic anionic vacuolar

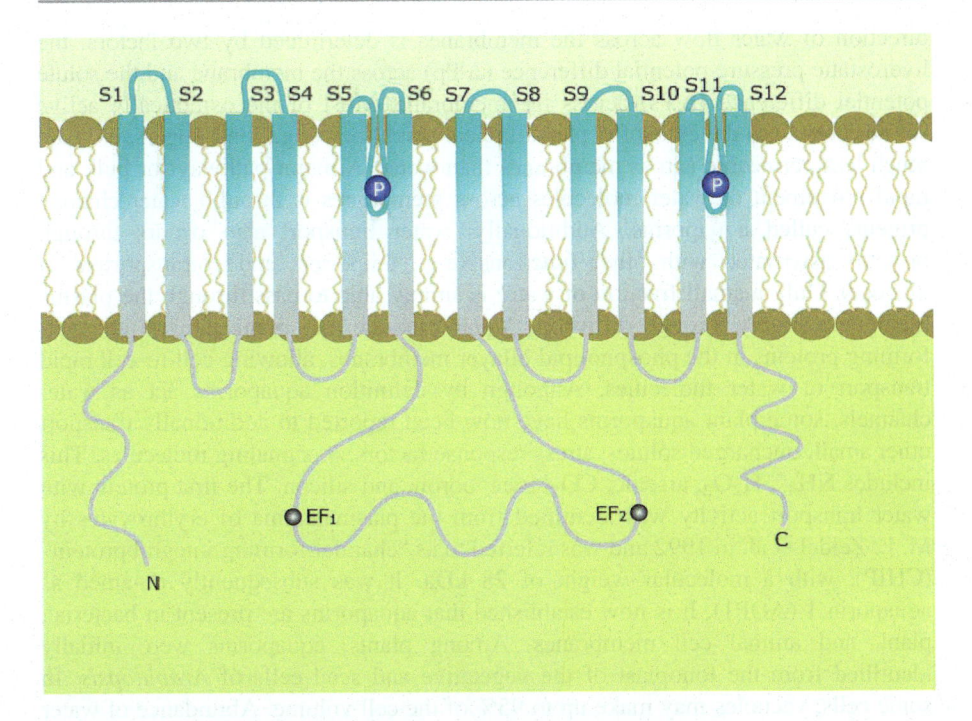

**Fig. 3.9** Predicted structure of calcium channel (SV) from *Arabidopsis*

constituent in many glycophytic plants. CAM plants diurnally take up or release malate from vacuoles. Malate uptake channels are activated by $Ca^{2+}$-dependent protein kinases. Due to acidic pH inside the vacuole, malate$^{2-}$ rapidly gets protonated upon influx into the vacuole as H•malate$^-$ and H$_2$• malate forms. The pH difference of the tonoplast allows maintenance of malate concentration difference to facilitate malate$^{2-}$ entry into the vacuole. Reverse flow (to cytosol) is energetically very unfavorable. Therefore, malate migration out of vacuoles in CAM plants is likely to take place through an independent route involving cotransporters. Analysis of *Arabidopsis* mutants has shown at least two kinds of vacuolar malate channels, namely, ALMT (aluminum-activated malate transporter) and AttDTa (tonoplast dicarboxylic acid transporter). Both genes are preferentially expressed in mesophyll cells.

### 3.7.4 Aquaporins

Most biological membranes exhibit high degree of permeability toward water in spite of the fundamentally hydrophobic characteristics of the fatty acyl chains in the phospholipid bilayer. The rapid rate of plasmolysis of plant tissues under hyperosmotic conditions, making tissue flaccid, provides clear evidence for rapid water permeability across the plasma membrane. In terms of water potential, the direction of water flow across the membranes is determined by two factors: the hydrostatic pressure potential difference ($\Delta\Psi p$) across the membrane and the solute potential difference ($\Delta\Psi s$) across the membrane. Most of the osmotically active solutes present in the cytosol in physiological conditions (e.g., ions, sugars, etc.) are much less permeable across membranes than water. A major pathway for bulk and rapid movement of water molecules across membranes is through water channel proteins called aquaporins. Additionally, water transport also occurs through proteins associated with other functions (e.g., uniporters and cotransporters of glucose). Only a small fraction of water is simply able to pass through the plasma membrane simply through diffusion. Aquaporins are membrane integral channel-forming proteins in the phospholipid bilayer membranes, allowing cell-to-cell rapid transport of water molecules. Although by definition aquaporins act as water channels, some plant aquaporins have now been reported to additionally transport other small, uncharged solutes, stress response factors, or signaling molecules. This includes $NH_4^+$, $H_2O_2$, arsenic, $CO_2$, urea, boron, and silicon. The first protein with water transport activity was identified from the plasmalemma of erythrocytes by M. L. Zeidel et al. in 1992 and was referred to as "channel-forming integral protein" (CHIP), with a molecular weight of 28 kDa. It was subsequently renamed as aquaporin 1 (AQP1). It is now established that aquaporins are present in bacterial, plant, and animal cell membranes. Among plants, aquaporins were initially identified from the tonoplast of the vegetative and seed cells of *Arabidopsis*. In some cells, vacuoles may make up to 95% of the cell volume. Abundance of water channels (aquaporins) on the tonoplast allows rapid and regulated movement of water across the tonoplast in response to changing extracellular and cytosolic solute concentrations. Aquaporins belong to large gene families. Thus, *Arabidopsis* has 35 aquaporin genes, maize has 36, and rice has 33. Aquaporins in plant cells can be grouped into four categories: plasma membrane intrinsic proteins (PIPs), tonoplast intrinsic proteins (TIPs), nodulin intrinsic proteins (NIPs) found on the peribacteroid membranes of symbiotic nitrogen-fixing nodules, and small basic intrinsic proteins (SIPs) found in the endoplasmic reticulum. The four categories of aquaporins have been reported from practically all groups of land plants, ranging from mosses to angiosperms.

Structurally, aquaporins are very hydrophobic proteins (25–30 kDa), with six transmembrane spans (Fig. 3.10). The C- and N-terminals of aquaporins are localized in the cytosol. Four aquaporin monomers make a functional complex, but each subunit of the tetramer can form a water channel. The connecting loops I, III, and V face the apoplast. Highly conserved asparagine-proline-alanine (NPA) residues are located on the connecting loops II and V. The pore (water channel) is formed by the two loops (II and V) containing NPA domains. In a functional monomer, these hydrophilic loops reconfigure themselves so that two NPA motifs face each other across the bilayer membrane to form a water-selective channel. The hydrogen-bonding properties of the amino side groups of the two Asn residues facing the lumen of the channel are thought to regulate water transport across the

## 3.7 Ion Channels

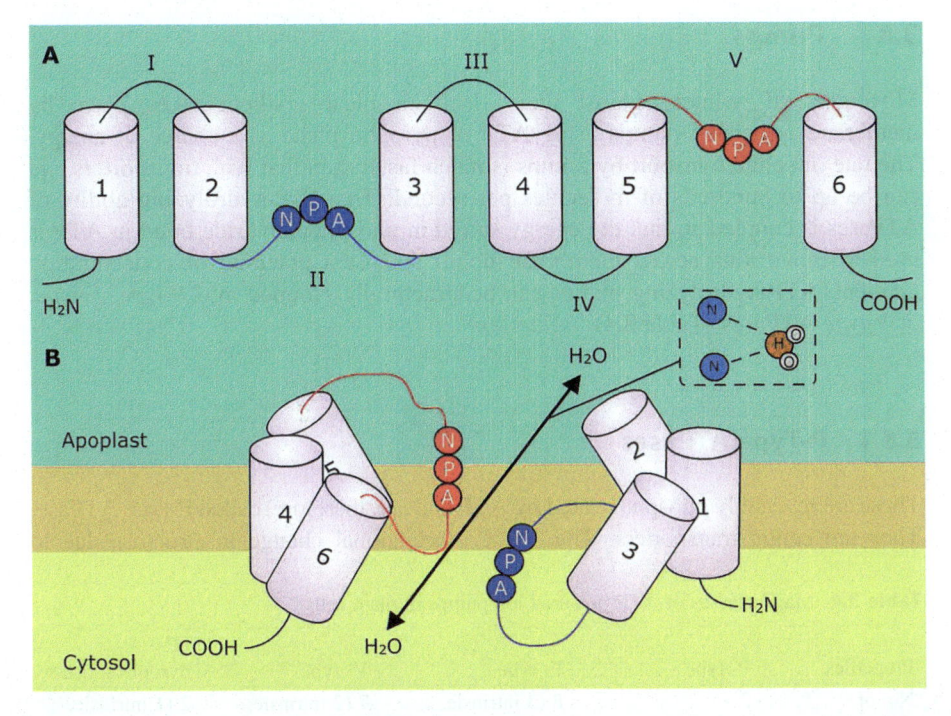

**Fig. 3.10** (a) Structure of an aquaporin monomer, (b) mechanism of water transport across the aquaporin channel

channel. The channel has a passageway of about 3 Å which is only slightly larger than a typical water molecule (2.8 Å). As the water molecule approaches the channel, its oxygen atom orients toward the two Asn residues creating a positive electrostatic field. As a result, water molecules break their hydrogen bonds with each other and instead form hydrogen bonds with the amino groups on the Asn residues. By this mechanism a string of hydrogen bonds is generated, enabling rapid transport of water across the aquaporin channel at a rate of about $10^9$ water molecules per second.

The expression of aquaporin genes in plants is regulated by a variety of environmental factors, such as water and nutrient availability, salt stress, drought, anoxia and light quality, and intensity. Changes in cytosolic pH also alter water movement across aquaporins. Thus, in the roots of *Arabidopsis* plants subjected to anoxic conditions, cytosolic pH becomes highly acidic and leads to dramatic reduction in water transport. Aquaporins are extensively post-translationally modified by phosphorylation and methylation at the C- and N-terminals, respectively. Phosphorylation of aquaporins at the conserved serine residues also leads to channel closing. Since this phosphorylation is catalyzed by $Ca^{2+}$-dependent protein kinases, a link between $Ca^{2+}$ signaling and the regulation of water movement is evident.

## 3.8 Pumps

Movement of solutes against their electrochemical gradient takes place by membrane-associated pumps using ATP or pyrophosphate as a source of energy. The rate of solute transport by pumps is much faster than that with transporters and can be up to hundreds of molecules per second. The ATP-hydrolyzing ability of ATPases is coupled so that the energy stored in phosphoanhydride bond in ATP is used to move ions across membrane uphill against a potential or concentration gradient. ATP-hydrolyzing pumps can be structurally grouped into *P-type*, *F-type*, or *V-type* ATPases (Table 3.4).

### 3.8.1 P-Type ATPases

These are reversibly phosphorylated by ATP and are, therefore, called P-type ATPase. They are cation transporters. There is conformational change in structure due to

**Table 3.4** Major classes of ATP-powered ion pumps in plant cells

| Properties | P- type | F- type | V- type | $H^+$-pyrophosphatase |
|---|---|---|---|---|
| No. of subunits | 2 | 8 (3 intrinsic, 5 cytosolic) | 7 (2 intrinsic, 5 cytosolic) | 2 (1 intrinsic, 1 cytosolic) |
| Ions transported | $H^+$, $Na^+$, $K^+$, $Ca^{2+}$ | $H^+$ | $H^+$ | $H^+$ |
| Location | Plasma membrane, chloroplast | Mitochondrial membrane, thylakoids of chloroplast | Tonoplast | Tonoplast, Golgi bodies, plasma membrane |
| Characteristics | Activity of H+ pump modulated by phosphorylation at C-terminal | Function to synthesize ATP, powered by the movement of $H^+$ down its electrochemical gradient | Causes acidification of vacuolar contents, no phosphorylated intermediates formed | 80 kDa protein |
| | 100 kDa protein ($H^+$-ATPase) | | | 17 subunits, functions as a homodimer |
| | 110 kDa ($Ca^{2+}$-ATPase) | | | |
| | 10 membrane-spanning domains | | 750 kDa protein | |
| | Establish PMF | 320 kDa (cytosolic component) | 13 subunits | Activated by cytosolic $K^+$ or $Ca^{2+}$ |
| | $Ca^{2+}$-ATPase activity modulated by CaM binding at N-terminal | | | |

phosphorylation and dephosphorylation. Plant and fungal plasma membrane-associated $H^+$-ATPase and $Ca^{2+}$-ATPase are members of P-type ATPases. Members of ABC (**A**TP-**b**inding **c**assette) superfamily of transporters are also of P-type ATPases. $H^+$-ATPase expression is high in cells involved in nutrient movement. Plasma membrane $H^+$-ATPase is regulated by ATP concentration in the cell, pH, and temperature. Their activity is modulated by the phosphorylation/dephosphorylation of the autoinhibitory domain at the C-terminal. $H^+$-ATPases are encoded by 11 genes in *Arabidopsis*. Some $H^+$-ATPases exhibit cell-specific expression. Several $H^+$-ATPases are expressed in guard cells to drive solute uptake during stomatal opening. Plasma membrane-associated $H^+$-ATPase is a $Mg^{2+}$-dependent, P-type ATPase. Its single subunit is a 100 kDa protein with ten membrane-spanning domains. For each ATP hydrolyzed, one $H^+$ is pumped out of the cytosol. This pump establishes the proton motive force in the cell which enables the activity of various symporters and antiporters.

Movement of protons by $H^+$ pumps is generally not balanced by the movement of anions and establishes a charge gradient leading to a change in membrane potential (Box 3.4). Such pumps which create a charge gradient are called **electrogenic**

---

**Box 3.4: Electrophysiological Processes in the Plant Cells**

Most cytosolic proteins bear a net negative charge at the physiological pH of 7.2–7.5. This excess of negative charge on proteins is countered by cytosolic $K^+$ accumulation. Membranes are impermeable to proteins, but they show finite $K^+$ permeability through $K^+$ channels. Cytosolic $K^+$ concentration is generally maintained high (about 100 mM) compared to outside the cell. Thus, the tendency of $K^+$ to leak outside generates an inside negative charge. This combination of fixed (immobile) negative charge of proteins and mobile positive charge of $K^+$ is known as **Donnan potential**. **Membrane potential** refers to the difference in electrical potential between two aqueous media separated by a membrane. It is presented as $Vm$. Normally, $Vm$ across the plasma membrane of plant cells is about $-150$ mV. Other membranes are less polarized. A value of $-20$ mV is commonly cited for tonoplast. A membrane potential is created by an imbalance in the number of cations and anions in the aqueous media separated by the membrane. Electrogenic transport of $H^+$ (by $H^+$-ATPases) out of the cytosol tends to drive $Vm$ to more negative values. This results in **hyperpolarization** of the membrane. In some conditions, a high flow of anions out of the cells (or of $Ca^{2+}$ into the cell) leads to transient positive swing in $Vm$. In other words, membrane **depolarization** leads to **action potential** and opening of ion channels. The electrochemical potential difference for an ion is the sum of chemical and electrical potential difference. When chemical potential is equal and opposite to electric potential, the sum is zero, and there is no overall driving force.

**pumps**. Activity of H$^+$ pumps at the plasma membrane makes it more positively charged on the outer surface, leading to **hyperpolarization**. Reverse happens (**depolarization**) when H$^+$-ATPase activity decreases. Cell wall acidification by plasma membrane H$^+$-ATPase influences growth and development. Auxins are known to activate H$^+$ pumps causing cell wall acidification followed by cell wall extension in a turgid cell. Plasma membrane H$^+$-ATPase is also responsible for the maintenance of cytosolic pH in the range of 7.3–7.5 in spite of the fact that many reactions of the intermediary metabolism generate excess of H$^+$. Since the pH optimum for plasma membrane H$^+$-ATPase is 6.6, an accumulation of H$^+$ in the cytosol activates its activity. Plant cells differ from animal cells in exploiting H$^+$ concentration across the plasma membrane to drive movement of solutes through generation of *proton motive force (pmf)*. Animal cells, on the other hand, maintain a gradient of Na$^+$ and K$^+$ through the activity of Na$^+$-/K$^+$-ATPase. Like pmf, Na$^+$ accumulation in animal cells is used to drive uptake of other solutes.

### 3.8.2 Endomembrane-Associated Ca$^{2+}$ Pump

Like H$^+$-ATPase, Ca$^{2+}$ pump belongs to P-type of ATPases. It is a single polypeptide of 110 kDa, and ATP hydrolysis by this enzyme brings about transport of two Ca$^{2+}$ across the membrane. Ca$^{2+}$ pumps are found on almost all membranes, including those of chloroplasts and mitochondria. Since calcium can form insoluble salts with phosphate, its (Ca$^{2+}$) concentration is maintained low in the cytosol. Ca$^{2+}$ pumps serve to maintain cytosolic [Ca$^{2+}$] in the range of 50–200 nM by pumping Ca$^{2+}$ out of the cell or its sequestration into the vacuoles or in the lumen of ER. The calmodulin-binding Ca$^{2+}$-ATPases possess an N-terminal autoinhibitory domain which binds with calmodulin in a Ca$^{2+}$-dependent manner, leading to inhibition of pump activity and consequent increase in [Ca$^{2+}$]$_{cyt}$. Thus, a negative feedback loop maintains cytosolic Ca$^{2+}$ homeostasis.

### 3.8.3 F-Type ATPases

These pumps are identified as energy coupling factors and are, therefore, known as F-type ATPases. The cytosolic component of F-type ATPases has a molecular mass of about 320 kDa. They use energy of ATP hydrolysis to drive protons across the membrane and also are responsible for the synthesis of ATP in reverse direction utilizing the proton gradient. They are located on mitochondrial and thylakoid membranes.

## 3.8 Pumps

### 3.8.4 V-Type ATPases

V is for vacuolar, since these pumps are proton-transporting ATPases and are responsible for maintaining acidic pH of vacuoles (between 3 and 6) in the cells of fungi and higher plants. These pumps are responsible for acidification of lysosomes, endosomes, and Golgi complex. The size of plant cells is regulated by water uptake into the vacuole. Cytosolic entry of water under these conditions is possible if the osmotic pressure of the vacuole is kept high. The pH of most plant vacuoles is mildly acidic (about 5.5). In lemon fruits, however, pH of the vacuole is much lower—a phenomenon termed as **hyperacidification** (Fig. 3.11). Vacuolar acidification is the cause of sour taste of lemon fruits (pH around 1.7). This is due to a combination of

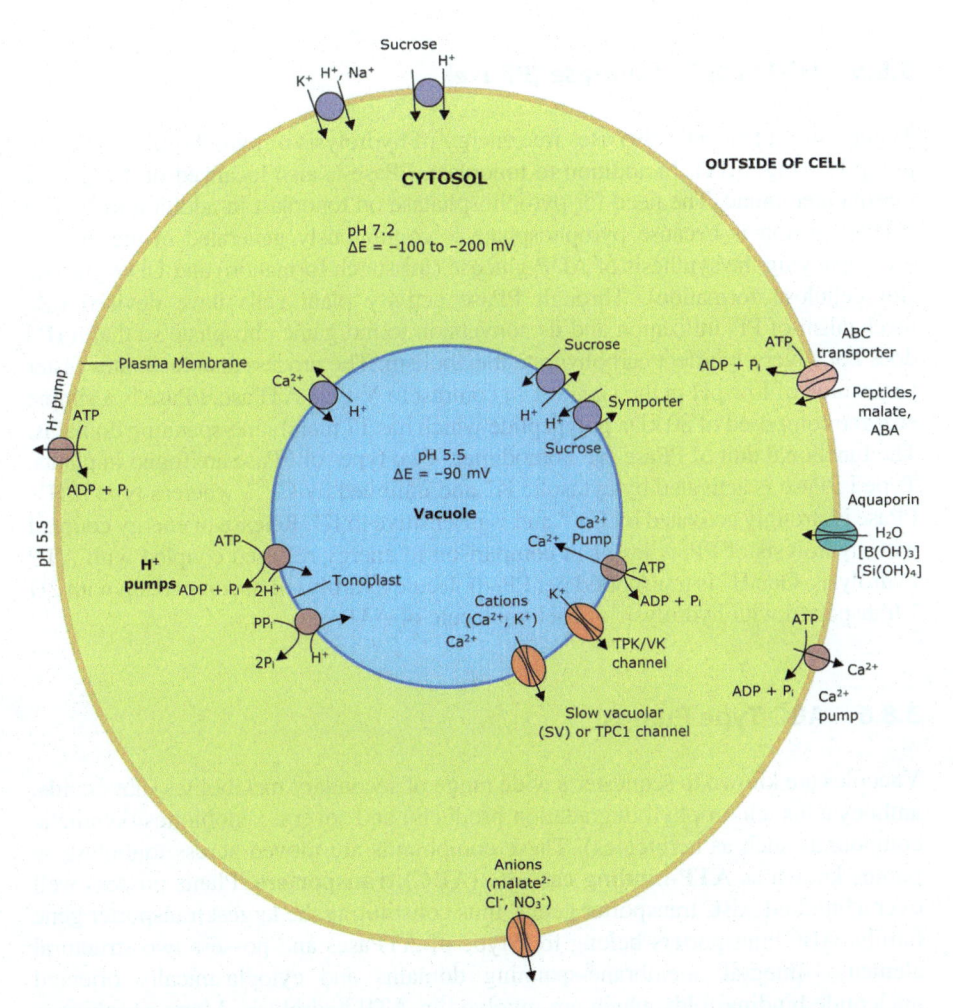

**Fig. 3.11** An overview of various transport proteins on the plasma membrane and tonoplast of plant cells

factors: (i) low vacuolar membrane permeability to protons, leading to buildup of steep pH gradient; (ii) more efficient vacuolar ATPase; (iii) accumulation of organic acids, such as citric, malic, and oxalic acids, in the vacuole to help maintain low pH; and (iv) $H^+$-pyrophosphatase activity on the tonoplast. In *Arabidopsis*, V-type ATPases are 750 kDa enzymes made up of 13 subunits. They pump three $H^+$ per ATP hydrolyzed. They are electrogenic pumps and contribute to generation of proton motive force (pmf) and membrane potential of the tonoplast. Unlike plasma membrane $H^+$-ATPases, V-ATPases do not form phosphorylated intermediates during ATP hydrolysis. Because of their similarity and common origin with F-type ATPase (located in chloroplast and mitochondrial membranes), V-ATPase are assumed to operate like tiny motors.

### 3.8.5 H⁺-Pyrophosphatase (PPase)

Tonoplast of plant cells also uses free energy of hydrolysis of pyrophosphate ($PP_i$) to pump $H^+$ (Fig. 3.11). In addition to tonoplast, PPase is also localized on Golgi and plasma membrane. The need for pyrophosphatase on tonoplast in addition to V-type ATPase action is because pyrophosphate is continuously generated in the cytosol accompanying the synthesis of ADP-glucose (for starch formation) and UDP-glucose (for cellulose formation). Through PPase activity plant cells have devised safe mechanism of $PP_i$ utilization and its conversion to inorganic phosphate so that it ($P_i$) does not adversely affect carbohydrate metabolism. The process also facilitates faster acquisition of low pH in the vacuole. In contrast to V-type ATPase, PPase is a simple enzyme composed of 80 kDa polypeptide which has 17 membrane-spanning domains. The functional unit of PPase is a homodimer. Two types of PPase are found in plants. Type I PPase is activated by cytosolic $K^+$ and inhibited by $Ca^{2+}$, whereas type II $H^+$-PPase is strongly activated by $Ca^{2+}$ and is insensitive to $K^+$. Release of energy coupled with hydrolysis of $PP_i$ is lesser in comparison to energy released coupled with ATP hydrolysis. One $H^+$ is transported per $PP_i$ molecule hydrolyzed in comparison with $2H^+$ transported with hydrolysis of each molecule of ATP.

### 3.8.6 ABC-Type Pumps

Vacuoles are known to sequester a wide range of secondary metabolites (flavonoids, anthocyanins, chlorophyll degradation products) and several xenobiotics (synthetic compounds such as herbicides). These compounds are moved across tonoplast by pumps known as **ATP-binding cassette (ABC) transporters**. Plants possess well over a hundred ABC transporter genes, thus constituting the largest transporter gene family. ABC transporters belong to P-type of ATPases and possess two structural elements—integral, membrane-spanning domains and cytoplasmically oriented nucleotide-binding folds which are involved in ATP hydrolysis. Many xenobiotics are sequestered in plant vacuoles after glycosylation, and ABC pumps are possibly

## 3.8 Pumps

involved in their sequestration. Flavonoids are known to be transported by ABC transporters as glutathione conjugates (GS-conjugates) facilitated by glutathione S-transferase reaction. Chlorophyll degradation products are, however, transported without prior conjugation. ABC transporters have also been implicated in the transport of waxes to the surface of leaf cells.

**Summary**

- Water can be transported up to endodermis by one or more of the three pathways—apoplast, symplast, and transcellular pathways. Cations ($K^+$, $Mg^{2+}$, $Ca^{2+}$) passing through the cell wall displace hydrogen ions on the carboxyl groups of polygalacturonic acid molecules and are held there by relatively weak interionic forces on the fixed negative charges ($COO^-$) (carboxyl groups of polygalacturonic acid) called "cation absorption sites" or "cation exchange sites." Size of solute particles is the principal factor governing their symplastic mobility across plasmodesmata. The "size exclusion limit" of a symplastic pathway through plasmodesmata is typically referred to as the molecular mass of the smallest solute excluded from movement across the plasmodesmatal channels. SEL of most plasmodesmata has been estimated to be around 800 Daltons. Stress conditions, such as anoxia, can increase SEL from 800 to much higher values. Pressure potential gradients generated through different means are responsible for bulk flow of water and dissolved solutes in opposite directions through xylem (tension) and phloem (hydrostatic pressure). Absence of lipid membranes, significant length and diameter of the trachedial elements, hydrogel nature of pectic membrane of pits, and lignification of cell walls are the major structural features of xylem elements which regulate water and solute transport through them. The absence of lipid membranes in vessels and tracheids enhances hydraulic conductance severalfold as compared to cells with intact plasma membrane.
- Solutes may get freely transported across the lipid bilayer of membranes, moving down their electrochemical potential gradient by simple diffusion. Movement of specific solutes using membrane transporter proteins to diffuse across the membrane according to their concentration gradient is referred to as "facilitated diffusion." Primary active transport accompanies ATP or pyrophosphate hydrolysis to provide energy for establishing ion gradients. Such transporters which perform primary active transport are called pumps. Secondary active transporters (also called cotransporters) utilize ion gradients established by primary active transport to move another solute against its electrochemical gradient. Secondary active transporters (or cotransporters) can be further divided into two groups: symporters and antiporters. Symporters move both the solutes across the membrane in the same direction, while antiporters move the two solutes in opposite directions across the membrane. Proton-coupled symporters are generally required for substrate uptake into the cytosol, while antiporters function to export solute out of the cytosol. The energy driving secondary transport is provided by

proton motive force rather than directly by ATP hydrolysis. The solute-transporter protein interaction is very selective to the extent that these proteins can distinguish between stereoisomers of sugars and amino acids.
- Ion fluxes through channels are driven solely by electrical potential difference. Ion flow through channels is passive. Channels also exhibit two additional properties essential to their function, i.e., ionic selectivity and gating. Anion channels in general allow permeation of a wide range of anions, including $Cl^-$, $NO_3^-$, and organic acids. Channels are tightly controlled by conformational shifts between permeable (or open) and non-permeable (or closed) states. This alteration between open and closed state of channels is known as "gating." Gating is controlled either by voltage or a ligand (chemicals that bind to channel proteins), such as hormones, $Ca^{2+}$, G-proteins, and pH. Some channels are stretch-sensitive and are gated by changes in turgor pressure of a cell. The functional $K^+$ inward channel works as a tetramer, with the P-domains of each subunit interacting to form a narrow constriction that contains the $K^+$ binding and recognition site. P-domain is responsible for the ion selectivity due to a conserved sequence of amino acids (Gly-Tyr-Gly). Outward $K^+$ rectifiers are activated with elevated cytosolic free calcium in plant cells. The difference in the activation energy favors $K^+$ by a factor of 1000 over $Na^+$ to pass through $K^+$ channels. Inward channels are downregulated by increased cytosolic $[Ca^{2+}]$ and have also been reported to be regulated by phosphorylation. Plant $Ca^{2+}$ channels are usually not specific for $Ca^{2+}$ transport but are also permeable to other cations. Anion channels, thus, serve as pacemakers of plant turgor reduction. S-type anion channels are gated by ABA, elevated $CO_2$, pathogenic elicitors, and ozone, thereby triggering stomatal closure. Malate uptake channels are activated by $Ca^{2+}$-dependent protein kinases. Due to acidic pH inside the vacuole, $malate^{2-}$ rapidly gets protonated upon influx into the vacuole as H•$malate^-$ and $H_2$•malate forms. The pH difference of the tonoplast allows maintenance of malate concentration difference to facilitate $malate^{2-}$ entry into the vacuole.
- Aquaporins are membrane integral channel-forming proteins in the phospholipid bilayer membranes, allowing cell-to-cell rapid transport of water molecules. Aquaporins in plant cells can be grouped into four categories: plasma membrane intrinsic proteins (PIPs), tonoplast intrinsic proteins (TIPs), nodulin intrinsic proteins (NIPs) found on the peribacteroid membranes of symbiotic nitrogen-fixing nodules, and small basic intrinsic proteins (SIPs) found in the endoplasmic reticulum. Structurally, aquaporins are very hydrophobic proteins (25–30 kDa), with six transmembrane spans. Four aquaporin monomers make a functional complex, but each subunit of the tetramer can form a water channel. The expression of aquaporin genes in plants is regulated by a variety of environmental factors, such as water and nutrient availability, salt stress, drought, anoxia, and light quality and intensity.
- Movement of solutes against their electrochemical gradient takes place by membrane-associated pumps using ATP or pyrophosphate as a source of energy. ATP-hydrolyzing pumps can be structurally grouped into P-type, F-type, or

V-type ATPases. Members of ABC superfamily of transporters are also of P-type ATPases. Movement of protons by $H^+$ pumps is generally not balanced by the movement of anions and establishes a charge gradient leading to a change in membrane potential. Such pumps which create a charge gradient are called electrogenic pumps. $Ca^{2+}$ pumps serve to maintain cytosolic [$Ca^{2+}$] in the range of 50–200 nM by pumping $Ca^{2+}$ out of the cell or its sequestration into the vacuole or in the lumen of ER. V-type ATPases are electrogenic pumps and contribute to PMF and membrane potential on the tonoplast. Through pyrophosphatase (PPase) activity plant cells have devised safe mechanism of $PP_i$ utilization and its conversion to inorganic phosphate so that it ($PP_i$) does not adversely affect carbohydrate metabolism. Vacuoles are known to sequester a wide range of secondary metabolites (flavonoids, anthocyanins, chlorophyll degradation products) and several xenobiotics (synthetic compound such as herbicides). These compounds are moved across tonoplast by pumps known as ATP-binding cassette (ABC) transporters. ABC transporters also belong to P-type class of ATPases.

## Multiple-Choice Questions

1. The energy source utilized for antiport:
   (a) ATP hydrolysis.
   (b) The movement of one of the transported substances up its concentration gradient.
   (c) The movement of one of the transported substances down its concentration gradient.
   (d) It requires no energy.
2. Which of the following is true for $H^+$-ATPase?
   (i) It uses energy of hydrolysis of ATP.
   (ii) It maintains a high $H^+$ concentration inside the cell.
   (iii) It is also responsible for the maintenance of cytosolic pH in the range of 7.3–7.5.
   (iv) It results in the generation of proton motive force (pmf).
      (a) Only i and ii
      (b) Only i and iii
      (c) Only i, iii, and iv
      (d) All of the above
3. Long-distance transport of solutes from one tissue system to another is known as:
   (a) Transport
   (b) Translocation
   (c) Transduction
   (d) None of the above

4. Which of the following mechanisms of transport involves the movement of water through the cytoplasm of cells?
   (a) Apoplastic transport
   (b) Symplastic transport
   (c) Transcellular transport
   (d) Both (b) and (c)
5. Which of the following is true for the movement of water and nutrients through "free space"?
   (i) It can occur only up to the endodermis.
   (ii) It is restricted by the presence of suberized Casparian strip.
   (iii) It involves the movement across the plasma membrane.
   (a) Only i
   (b) Only ii
   (c) Only i and ii
   (d) All of the above
6. Which of the following is not a property of SOS1—a sodium ion transporter?
   (a) It is a $Na^+/H^+$ antiporter.
   (b) It helps in lowering of the internal $Na^+$ concentration of the cell.
   (c) It is present on the plasma membrane.
   (d) It is involved in the cotransport of $H^+$ and $Na^+$ outside the cell.
7. Rectification is:
   (a) The ability of a channel to transport ions only in one direction
   (b) The ability of an ATPase pump to hydrolyze ATP
   (c) The ability of a channel to transport ions in both direction
   (d) The property of a channel to cotransport two species of molecules or ions
8. The transport of sugars and amino acids is facilitated by:
   (a) Diffusion
   (b) Osmosis
   (c) Symport with $Na^+$ or $H^+$
   (d) Antiport with $Na^+$ and $H^+$
9. The cylindrical strand of endoplasmic reticulum running through the cytoplasmic connections is known as:
   (a) Plasmodesmata
   (b) Desmotubule
   (c) Plasmalemma
   (d) None of the above
10. Which of the following is not true for electrogenic pumps?
    (i) They create a charge gradient.
    (ii) Their activity results in generation of proton motive force (pmf).
    (iii) They cause hyperpolarization of the membrane.
    (a) Only i and ii.
    (b) Only i and iii.
    (c) All of the above.
    (d) None of the above.

## Answers

1. c   2. c   3. b   4. d   5. c   6. d   7. a
8. c   9. b   10. c

## Suggested Further Readings

Patrik JW, Tyerman SD, van Bel AJE (2015) Long distance transport. In: Buchanan BB, Gruissem W, Jones RL (eds) Biochemistry and molecular biology of plants. Wiley-Blackwell, Chichester, pp 658–701

Taiz L, Zeiger E, Møller IM, Murphy A (2015) Plant physiology and development, 6th edn. Sinauer Associates Inc, Sunderland, pp 143–168

# Part II

# Metabolism

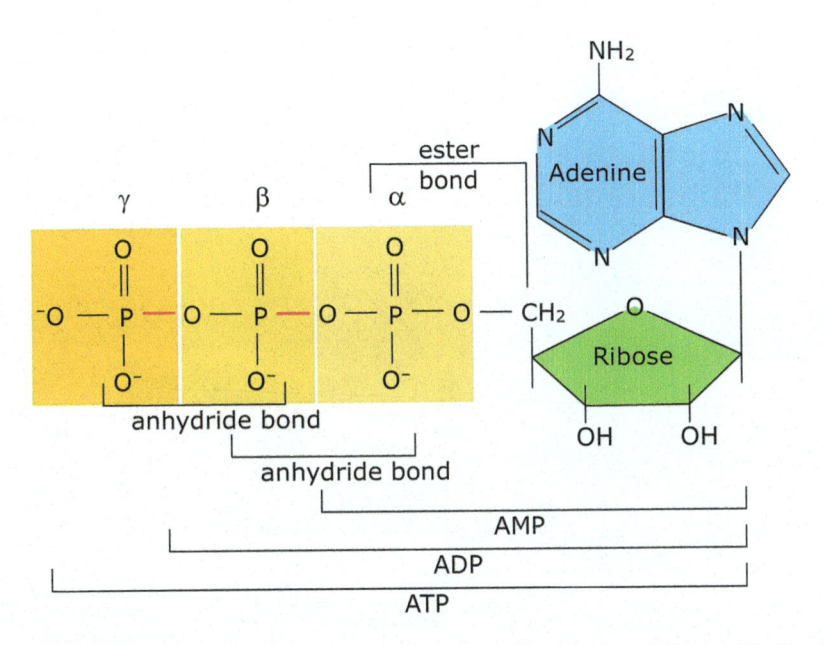

Structure of ATP and its constituents. More details are provided in Chap. 4, Sect. 4.2.1, Fig. 4.6

Chapter 4 Concepts in Metabolism
Chapter 5 Photosynthesis
Chapter 6 Photosynthate Translocation
Chapter 7 Respiration
Chapter 8 ATP Synthesis
Chapter 9 Metabolism of Storage Carbohydrates
Chapter 10 Lipid Metabolism
Chapter 11 Nitrogen Metabolism
Chapter 12 Sulphur, Phosphorus and Iron Metabolism

# Concepts in Metabolism

## Manju A. Lal

Though the vitalist group initially used the term "organic" for compounds produced only by organisms, it was later on used for carbon compounds. Wohler (1928) discovered that urea, which otherwise was thought to be produced only in the living beings, could also be produced in the laboratory from ammonia and bicarbonate. In 1897, German chemists Eduard Buchner and Hans Buchner demonstrated that fermentation could be carried out by the **cell-free extract** of yeast. These observations lead to the development of the science of biochemistry. In the early twentieth century, due to discovery of various metabolic pathways, biochemistry was dominated by organic chemistry, followed by enzymology and **bioenergetics**. Some of the analytical techniques which made study of biochemistry possible included isolation of organelles, high-performance liquid chromatography, electrophoresis, use of radioactive tracers, plant transformation techniques using *Agrobacterium tumefaciens,* gene silencing, forward genetics, reverse genetics, mass spectrometry, and DNA microarray, among others. With computational technology, it is now possible to have complete understanding of the interconnectivity of metabolic pathway.

Autotrophs are able to synthesize organic material from $CO_2$ and $H_2O$, deriving energy either from the chemical reactions (chemoautotrophs) or by utilizing light energy (photoautotrophs). Heterotrophs, including mammals, have to depend on the autotrophs for the availability of complex carbon containing organic substances. Sum total of all chemical reactions occurring in a living being are called as **metabolism**. These occur through enzyme-catalyzed reactions that constitute metabolic pathways. Metabolic pathways include precursors, which are converted to products. Various intermediates are called metabolites. Combined activity of all metabolic pathways involved in interconversion of precursors, metabolites, and products is called **intermediary metabolism**. **Primary metabolites** are the intermediates or the products of a pathway, which are used for growth, development,

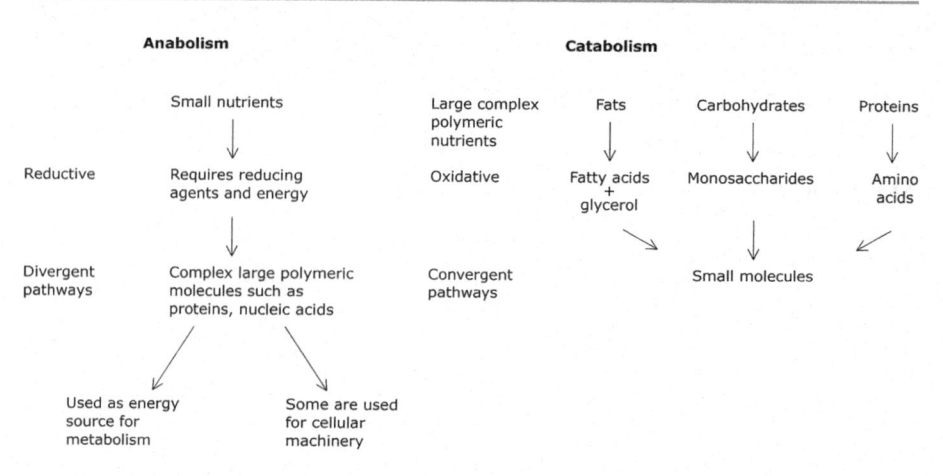

**Fig. 4.1** Comparison of anabolism and catabolism

and reproduction of the organism. **Secondary metabolites** are bioactive specialized compounds produced in a metabolic pathway which are used to protect plants against herbivory and microbial pathogen infection or to attract pollinators or seed dispersal animals. Metabolism includes both anabolic and catabolic reactions. The terms anabolism and catabolism were first used by the physiologist Gasket in 1886. Anabolism includes all the reactions involved in conversion of simpler molecules to complex ones. This requires input of energy and the pathway involved is a divergent pathway (Fig. 4.1). On the contrary, catabolism involves conversion of complex substances into simpler molecules, which is coupled with release of energy. The catabolic pathways are convergent pathways since many of the metabolic reactions converge to join the pathway involved with release of simpler molecules. The energy transitions in these pathways are mediated through two high-energy molecules which are reduced form of nicotinamide dinucleotide (NADH), and adenosine triphosphate (ATP). ATP is a high-energy phosphate compound, which mediates energy transfer, while NADH is the donor for high-energy electron transfer (Fig. 4.2).

All living organisms have the unique ability to adjust to the changing environment through alteration in their metabolism even though maintaining their internal cellular environment. Unlike animals, plants are sessile and are exposed to harsher conditions. They have more robust metabolism, which is evident from flexibility in their metabolism and **metabolic redundancy.** Not only that, **metabolic flux** (rate of movement of metabolites in a pathway) should also be regulated according to the need of the cell, tissue, or the organism. The metabolic flux is achieved through regulation of metabolism by pacemaker enzymes which are responsible for

## 4 Concepts in Metabolism

**Fig. 4.2** Role of mobile electron carriers $NAD^+$/$NADP^+$ and ATP (the energy currency of the cell)

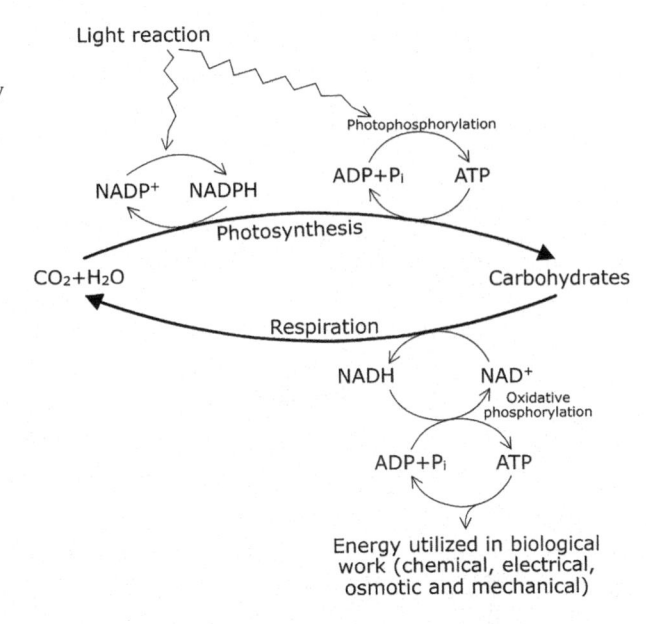

catalyzing rate-determining steps of metabolic pathways. Understanding the regulation of such enzymes at the expression of gene level or at the level of protein degradation would help in producing plants with altered metabolism (**metabolic engineering**). Many enzymes involved in a pathway have been proposed to exist as multi-enzyme complexes (**metabolons**) as means of **metabolic channeling**. Metabolic channeling allows direct transfer of biosynthetic intermediates from one enzyme to another in a pathway, minimizing their loss due to diffusion. Each cell compartment provides optimal conditions for specific metabolic pathways to occur at the optimal level (Fig. 4.3). Exchange of metabolites between the compartments is regulated by the transporters localized in the membranes. Plants are unique in having plastids which house the enzymes for photosynthesis. Besides photosynthesis, the enzymes for lipid and terpenoid biosynthesis, for biosynthesis of chlorophyll and related pigments, and for starch biosynthesis as well as many enzymes of nitrogen metabolism are also present. Both plastids and cytosol contain enzymes of glycolysis as well as for oxidative pentose phosphate pathway. Besides occurrence of various metabolic activities among cell organelles, various metabolic processes are also compartmentalized between soluble phase and the membranes. Thus, enzymes required for $CO_2$ reduction are present in stroma of the chloroplasts, while those involved in harvesting the solar energy and electron transport process are localized in the thylakoids.

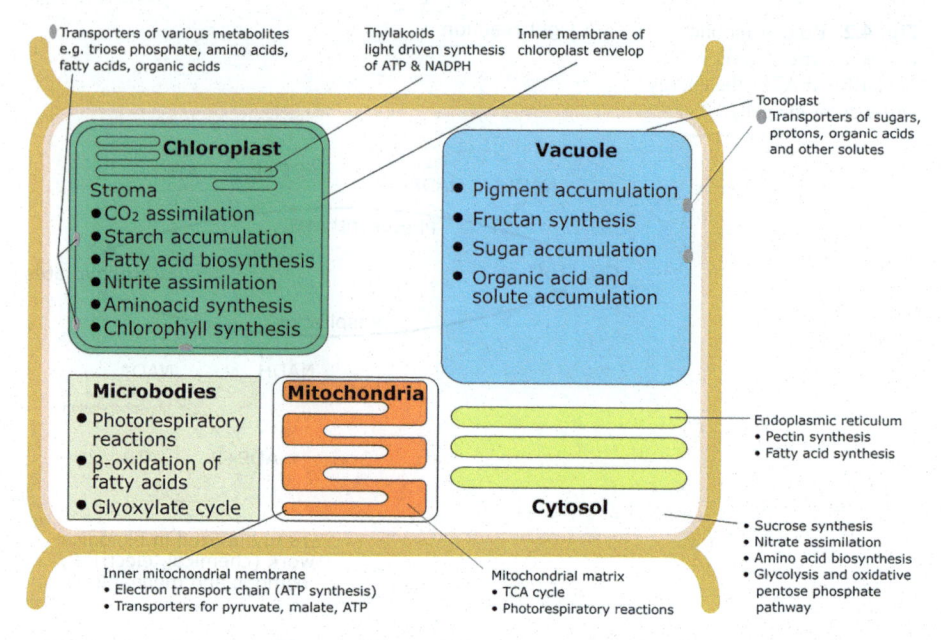

**Fig. 4.3** Compartmentalization of metabolic pathways in a plant cell

## 4.1 Basic Energetic Principles that Govern Metabolism

Sun is the ultimate form of energy for most carbon-based life forms. Thermonuclear fusion reactions in sun convert four protons ($4H^+$) to one helium (He). During this conversion, there is 0.72% loss in total mass of $H^+$ (atomic weight of $H^+$ is 1.0079, while helium has an atomic weight of 4.0026). The missing mass is converted into energy in the form of electromagnetic radiations. Flow of energy is central to maintenance of life (Fig. 4.4 and 4.5). A living cell is a **system** in which all the reactants and products of a reaction are present along with the solvent. It is neither an **isolated** nor a **closed** system since there a continuous exchange of energy and matter with the surroundings, making it an **open** system. The science which deals with energy transduction within a living system is called **bioenergetics**. An understanding of integration of bioenergetics with biochemical reaction is central for understanding cell physiology. The energy released during reactions is utilized by the cell to perform work, e.g., for creation of proton gradient across the membrane. To understand the energy transduction within a cell, we need to revise the laws of thermodynamics (Box 4.1).

## 4.1 Basic Energetic Principles that Govern Metabolism

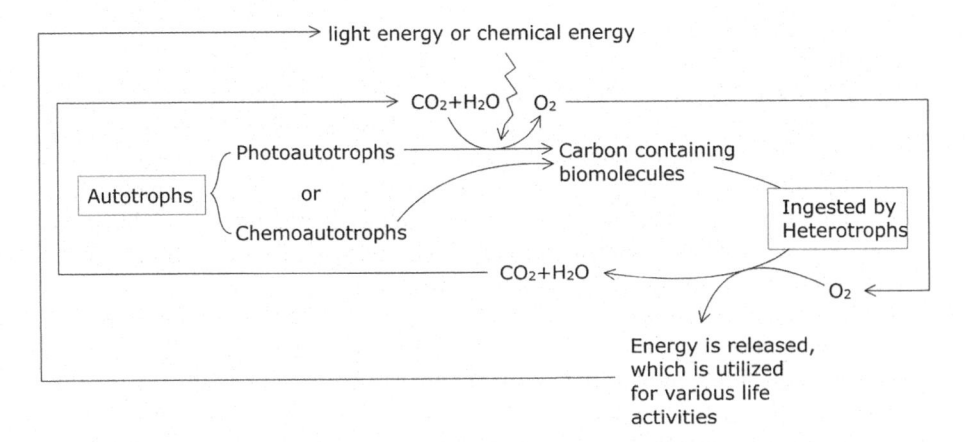

**Fig. 4.4** Global energy cycle

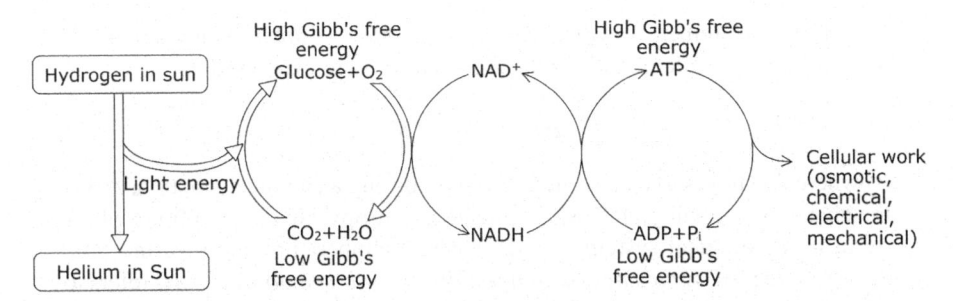

**Fig. 4.5** Sun is the ultimate source of energy. Autotrophs are able to convert solar energy into organic compounds. These compounds are oxidized and are the source of ATP, which primarily is used for cellular work

> **Box 4.1: Laws of Thermodynamics**
> First law of thermodynamics states that the total energy of the universe is constant and it can neither be created nor destroyed. Different forms of energy such as light energy, chemical energy, thermal energy, mechanical energy, etc. are interconvertible. A living cell is an open system which can exchange energy and the matter with the surroundings. According to the second law of thermodynamics, there is increase in disorder during any spontaneous reaction which is coupled with release in energy. Energy is required to put the system in order. Extent of disorder of any system is measured by **entropy**, a term which was coined by the German physicist Rudolf Clausius. It is a thermodynamic quantity which represents the amount of energy that is no longer available for doing mechanical work. Higher is value of entropy, high is the disorder of the

(continued)

**Box 4.1** (continued)

system and lesser energy will be available to do the work. **Enthalpy** (H) refers to the total potential energy of a molecule which is determined by its chemical structure. It includes the number and the type of chemical bonds which make up the molecule. During any spontaneous reaction, a complex molecule (having more ordered structure) is converted to the simpler molecules (having less ordered structure), which is coupled with release of energy. In a living cell, which is an isothermal system, released energy may be utilized for doing work. Thus, in an isothermal system out of total energy of a system (H), only some amount of energy is available for doing work, which is called as **Gibbs free energy**, in the, honor of J. Willard Gibbs, who developed the theory of energy exchanges during chemical reactions. Relationship between these thermodynamic quantities at the absolute temperature (T) can be expressed as,

$$G = H - TS$$

Since measurement of absolute values is not possible, changes in these three thermodynamic quantities during a reaction are expressed as

$$\Delta G = \Delta H - T\Delta S$$

where $\Delta G$ refers to change in free energy during a chemical reaction, $\Delta H$ is the change in enthalpy, and $\Delta S$ is change in entropy. Since conditions which occur in a biological system include constant temperature and pressure, *free energy is defined as the energy isothermally available to do work*. $\Delta G$ refers to the difference in free energy of the products and the free energy of the reactants during a reaction. In a spontaneous reaction, $\Delta G$ value is negative, i.e., energy is released during reaction (exergonic). On the contrary, a positive value of $\Delta G$ indicates the reaction to be endergonic and would require input of energy. In case value of $\Delta G$ is 0, the reaction will be at equilibrium and will occur in either forward or backward direction depending upon concentrations of reactants and the products. $\Delta G$ values are expressed in terms of calories (cal) or joules (J) per mole (1 cal = 4.184 J). Joule is the official term used now. The magnitude of free energy change is also determined by the conditions in which the reaction is taking place, which include the molar concentrations of the reactants, pH, and temperature of the medium. **Standard free energy** change refers to free energy change during a reaction that occurs at physiological pH (7.0), at 25 °C and under conditions when both reactants and products are at unit concentrations (1 M) and is expressed as $\Delta G^0$. Relationship between $\Delta G$ and $\Delta G^0$ in a reaction, which is not at equilibrium, is expressed as

(continued)

**Box 4.1** (continued)

$$\Delta G = \Delta G^{0\prime} + RT \ln K_{eq}$$

where R is the universal gas constant, T is the absolute temperature, and $K_{eq}$ is the equilibrium constant of a reaction. The standard free energy changes are directly related to the equilibrium constant. For a reaction at equilibrium, value of $\Delta G$ will be 0. The relationship between $\Delta G^0$ and $K_{eq}$ is expressed as

$$\Delta G^{0\prime} = -RT \ln K_{eq}$$

The standard free energy change ($\Delta G^{0\prime}$) is a constant value which tells us a characteristic unchanging value for a given reaction, while actual free energy change ($\Delta G$) is a function of reactants and products concentrations in the cell and the temperature prevailing at the time of occurrence of chemical reactions. Value of $\Delta G$ changes with the reaction proceeding spontaneously toward equilibrium and becomes 0 at the point of equilibrium. Thus, criterion for spontaneity of the reaction is $\Delta G$ and not $\Delta G^{0\prime}$. Equilibrium of the reaction plays a very important role. Value of $\Delta G^{0\prime}$ of a reaction at equilibrium is zero. It is important to maintain a reaction far from equilibrium since amount of work done depends on how far a reaction is maintained away from equilibrium. Thus, maintaining a disequilibrium is key to all the life processes.

The standard free energy changes occurring during sequential reactions in a metabolic pathway are additive. The overall change in standard free energy ($\Delta G^{0\prime}_{total}$) in two sequential reactions having standard free energy change values of $\Delta G^{0\prime}_1$ and $\Delta G^{0\prime}_2$, respectively, sharing common intermediate will be $= \Delta G^{0\prime}_1 + \Delta G^{0\prime}_2$. This explains how a thermodynamically unfavorable reaction (endergonic) is driven forward by its coupling with the thermodynamically favorable reaction (exergonic) through a common intermediate.

$$\text{Glucose} + P_i \rightarrow \text{Glucose 6-phosphate} + H_2O \quad \Delta G^{0\prime}_1 = 13.8 \text{ kJ/mol}$$

$$ATP + H_2O \rightarrow ADP + P_i \quad \Delta G^{0\prime}_2 = -30.5 \text{ kJ/mol}$$

These two reactions share the common intermediates $P_i$ and $H_2O$. Overall reaction is the sum of these reactions, which can be written as,

$$\text{Glucose} + ATP \rightarrow ADP + \text{Glucose 6-phosphate}$$

Overall standard free energy change ($\Delta G^{0\prime}_{total}$) is obtained by adding the values of $\Delta G^{0\prime}_1$ and $\Delta G^{0\prime}_2$.

(continued)

> **Box 4.1** (continued)
>
> $$\Delta G^{0\prime}_{total} = \Delta G^{0\prime}_1 + \Delta G^{0\prime}_2$$
> $$= 13.8 \text{ kJ/mol} + (-30.5 \text{ kJ/mol}) = -16.7 \text{ kJ/mol}$$
>
> Overall reaction is exergonic. Exergonic ATP hydrolysis is coupled to the endergonic reaction involving synthesis of glucose 6-phosphate. The common intermediate-strategy is used by all of living cells.

There are many cellular reactions, which cannot occur spontaneously without input of required energy. These reactions are coupled to the energy releasing reactions. This is possible as long as the net $\Delta G$ (free energy change) value of the combined reactions is negative. Such reactions are known as coupled reactions. Coupled reactions occur simultaneously since one reaction is necessary for the other one to occur. These are reactions, which share common intermediates, and the product of one reaction becomes the reactant for another. For example, the product of ATP hydrolysis is the reactant for phosphorylation of glucose. Coupled reactions can be either energy-coupled reactions or oxidation-reduction reactions.

## 4.2 Energy Coupled Reactions

Adenosine triphosphate (ATP) is synthesized from ADP and $P_i$ during exergonic reactions and is hydrolyzed to provide energy for reactions requiring energy. It was first isolated from muscles in 1929 by Cyrus H. Fiske in the USA and Yellapragada Subbarao and Karl Lohman in Germany independently. Fritz Lipmann along with Herman Kalckar proposed in 1941 the possible involvement of ATP in bioenergetic processes in cells. Lipmann was awarded Nobel Prize in 1953 for his work. He introduced the "squiggle" notation (~) for the energy-rich bonds of biomolecules such as ATP and ADP. A high-energy bond generally refers to unstable or labile bond. For the sake of simplicity, high-energy bond refers to ATP or any other phosphate compound with large, negative, standard free energy. The P-O bond itself does not contain energy. Free energy that is released from hydrolysis of P-O bond does not come from breaking of specific bond. It results from the products of the reaction that have lower free energy content than the reactants. ATP, ADP, AMP are charged molecules which are not able to diffuse through cell membrane. Since cells cannot get them from outside, each cell synthesizes the entire molecules by itself. Intracellular ATP/ADP exchange occurs between different compartments of the cell.

### 4.2.1 Structure of ATP

The phosphate groups attached to $5'$ hydroxyl group of a nucleoside result in formation of trinucleoside phosphates, which include UTP, GTP, CTP, and ATP. ATP is the nucleoside triphosphate most widely used as a high-energy phosphate

## 4.2 Energy Coupled Reactions

compound. The three phosphates are labeled as α, β, and γ (Fig. 4.6). The bond between ribose and α-phosphate is an ester bond, while α-β and β-γ linkages are **phosphoanhydrides**. Hydrolysis of ester bond yields about 14 kJ/mole under standard conditions, while phosphoanhydride bonds yield 30.5 kJ/mole. However, actual free energy change during the hydrolysis of phosphoanhydride bonds in cellular conditions also known as **phosphorylation potential** ($\Delta G_p$) is very different since concentrations of ATP, ADP, and $P_i$ are not identical and is very much lower than 1.0 M. Secondly since $Mg^{2+}$ forms complex with ATP, it is the Mg-ATP which is the true substrate for enzyme-catalyzed reactions (Fig. 4.6). ATP is not complexed with $Mg^{2+}$ when being transported across membranes within the cell.

**Fig. 4.6** (A) Structure of ATP; the two phosphoanhydride bonds γ-β and β-α are high-energy bonds. (B) $Mg^{2+}$ is complexed with ATP and ADP. $Mg^{2+}$ partially shields the negative charge and influences the conformation of phosphate groups in molecules, such as ATP and ADP

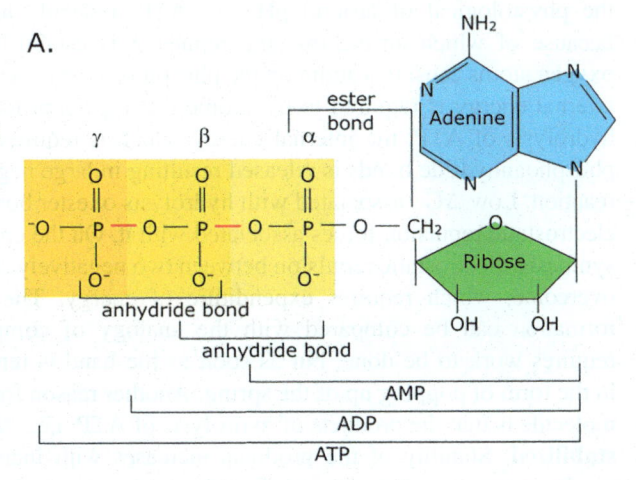

## 4.2.2 ATP Is the High-Energy Molecule

Standard free energy change ($\Delta G^{o\prime}$) is determined by unstability or stability of the reactants/products of the chemical reaction and also by subsequent fate of the products. Large negative value of standard free energy ($\Delta G^{o\prime}$) of ATP hydrolysis is associated with unstability of the reactant (ATP) and stability of its products (ADP + $P_i$) of hydrolysis. The electronegative oxygen in the P=O bond of ATP molecule attracts electrons creating a partial negative ($\delta^-$) charge on oxygen atom and a partial positive ($\delta^+$) charge on phosphorus atom. As a result, two strongly electron withdrawing groups must compete for the lone pair of electrons of its bridging oxygen making the molecule less stable than its hydrolysis products. At the physiological of around pH 7.0, ATP molecule has four negative charges because of which an electrostatic repulsion is established between the adjacent oxygen atoms causing a strain on the phosphoanhydride bond (Fig. 4.6). A sufficient internal energy is required to overcome this repulsion like charges. At the time of hydrolysis of ATP, the internal energy which is required to maintain the strained phosphoanhydride bonds is released resulting in large negative value of $\Delta G^{o\prime}$ of the reaction. Low $\Delta G^{o\prime}$ associated with hydrolysis of ester bond of AMP is due to fewer electrostatic repulsion forces associated with it. On the contrary, at the time of ATP synthesis, electrostatic repulsion between two negatively charged groups needs to be overcome, which requires expenditure of energy. The phosphoanhydride bond formation can be compared with the analogy of compressing a spring, which requires work to be done, but as soon as the hand is removed, energy is released in the form of popping up of the spring. Another reason for ATP to be a high-energy molecule is that the products of hydrolysis of ATP, i.e., ADP and $P_i$, are **resonance stabilized**. Stability of the products increases with increase in resonance, which results increase in entropy and decrease in the energy level of the products of the reaction. As a result, there is release of energy coupled with ATP hydrolysis (Fig. 4.7). Probability of reverse reaction decreases due to stability of the products. In aqueous environment of the cell, hydration of the reactants and products also plays a significant role in $\Delta G$ of the reaction. Thus, large negative value of $\Delta G^{o\prime}$ of ATP hydrolysis is due to electrostatic repulsion in the molecule and resonance and hydration of the products.

It is not only the intrinsic property of ATP molecule that determines high value of $\Delta G$ of its hydrolysis but also the cellular reactions, which are responsible for holding high cellular ATP concentrations far above than required to maintain the equilibrium of the hydrolysis reactions. $\Delta G$ of a reaction is also determined by how far the rate constant of the reaction is from the **equilibrium constant** at a given time. Potency of the hydrolysis of ATP is lost at the equilibrium of the reaction; thus it is necessary that effective intracellular concentration of ATP should be maintained high so as to keep the rate of the reaction higher than the equilibrium rate constant. When the ATP level is dropped, not only the amount of fuel decreases, but there is also a decline in phosphorylation potential of the molecule. Thus, living cells have developed effective mechanisms to maintain high intracellular ATP concentrations. Even though

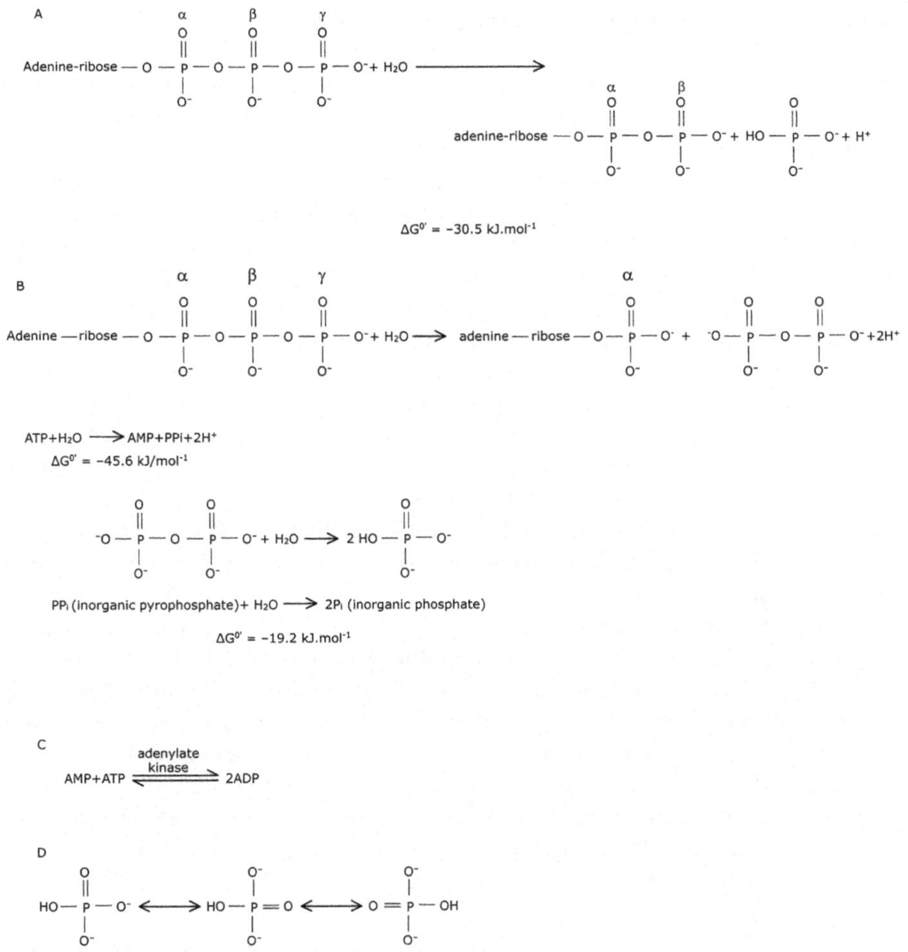

## 4.2 Energy Coupled Reactions

**A**

Adenine-ribose—O—P(α)(=O)(O⁻)—O—P(β)(=O)(O⁻)—O—P(γ)(=O)(O⁻)—O⁻ + H₂O ⟶ adenine-ribose—O—P(α)(=O)(O⁻)—O—P(β)(=O)(O⁻)—O⁻ + HO—P(=O)(O⁻)—O⁻ + H⁺

$\Delta G^{0'} = -30.5 \text{ kJ.mol}^{-1}$

**B**

Adenine—ribose—O—P(α)(=O)(O⁻)—O—P(β)(=O)(O⁻)—O—P(γ)(=O)(O⁻)—O⁻ + H₂O ⟶ adenine—ribose—O—P(α)(=O)(O⁻)—O⁻ + ⁻O—P(=O)(O⁻)—O—P(=O)(O⁻)—O⁻ + 2H⁺

ATP + H₂O ⟶ AMP + PPi + 2H⁺
$\Delta G^{0'} = -45.6 \text{ kJ/mol}^{-1}$

⁻O—P(=O)(O⁻)—O—P(=O)(O⁻)—O⁻ + H₂O ⟶ 2 HO—P(=O)(O⁻)—O⁻

PPi (inorganic pyrophosphate) + H₂O ⟶ 2Pi (inorganic phosphate)

$\Delta G^{0'} = -19.2 \text{ kJ.mol}^{-1}$

**C**

AMP + ATP $\xrightleftharpoons{\text{adenylate kinase}}$ 2ADP

**D**

HO—P(=O)(O⁻)—O⁻ ⟷ HO—P(O⁻)(=O)—O⁻ ⟷ ⁻O—P(O⁻)(=O)—OH

**Fig. 4.7** (**a**) ATP hydrolysis leading to formation of ADP and P$_i$; (**b**) ATP hydrolysis leading to formation of AMP and P$_i$; (**c**) conversion of AMP to ADP in a reaction catalyzed by adenylate kinase. (**d**) Products of hydrolysis of ATP (P$_i$) are resonance stabilized. This increases entropy and therefore decreases the energy level of the products so that on breaking of the bond there is larger yield of energy. In an inorganic phosphate ion, all the P-O bonds are partially double-bonded in character rather than proton being associated with any one oxygen, resulting in increase in entropy and a lowering of their energy level. ↔ symbol indicates the structure which exists in an intermediate one in which all oxygen have partial negative charge and proton is not associated with any form

ATP hydrolysis is a highly exergonic reaction, it is kinetically stable because high amount of activation energy is required for uncatalyzed hydrolysis (200–400 kJ/mol) of the phosphoanhydride bonds of the molecule. Enzymes lower the requirement of activation energy, and phosphoryl group transfer occurs either to water or to any other acceptor of the group.

## 4.2.3 ATP Is the Energy Currency of the Cell

Besides ATP, there are other phosphoryl group carrying compounds which can be divided into two categories. One category of compounds includes those which have $\Delta G^{o\prime}$ of hydrolysis larger than $-25$ kJ/mol. These compounds are high-energy phosphate compounds, while the other category of compounds are low-energy phosphate compounds whose hydrolysis is associated with negative $\Delta G^{o\prime}$ value of about 9–20 kJ/mol. ATP serves as the energy currency in the cell since it has $\Delta G^{o\prime}$ value—30.5 kJ/mol. It occupies intermediate position in phosphoryl group transfer potential. It can carry energy from high-energy phosphate compounds, which are produced during catabolism (such as phosphoenolpyruvate) to compounds such as glucose, converting them into more reactive compounds, such as glucose 6-phosphate (Fig. 4.8). Transfer of phosphoryl group results in adding more free energy to a molecule, which is given up during subsequent metabolic reactions. This activation is called "priming" of the molecule. ATP itself can be synthesized by coupling with exergonic reactions of hydrolysis of compounds which have higher value of $\Delta G^{o\prime}$ of hydrolysis than ATP itself. ATP synthesis by this means is referred to as **substrate-level phosphorylation** and has been dealt in the chapter dealing with ATP synthesis.

During group transfer all three phosphate groups of ATP molecule can take part. Depending on the site of action of **nucleophilic** group (e.g., –OH group of attacking molecule) on ATP molecule, it can be phosphoryl (attack at γ-phosphate), pyrophosphoryl (attack at β-phosphate), or adenylyl moiety (attack at α-phosphate) transfer. It is the phosphoryl group ($-PO_3^{2-}$) of ATP which is transferred and not the phosphate ($PO_4^{2-}$) since oxygen (-O-) that bridges the group with the attacking molecule does not come from ATP; rather it comes from the attacking molecule. Free energy release during pyrophosphoryl ($PP_i$) transfer (hydrolysis of α-β

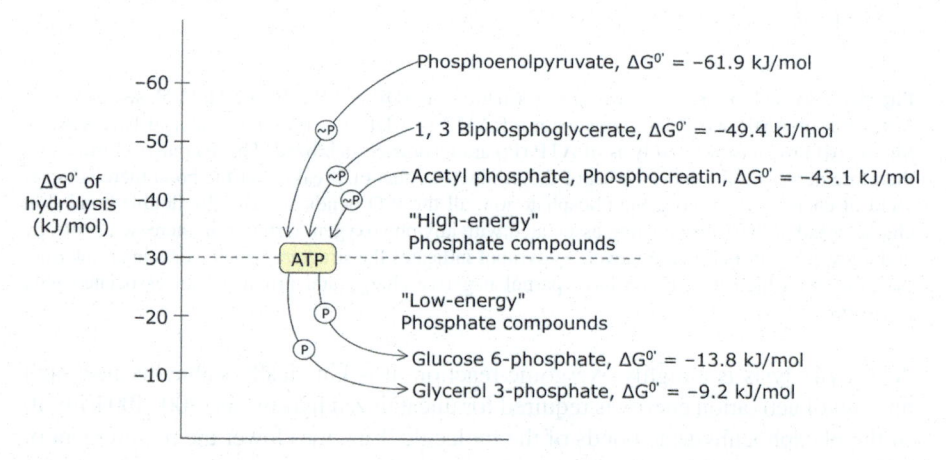

**Fig. 4.8** ATP is the "energy currency" of the cell since free energy of hydrolysis ($\Delta G^{o\prime}$) of ATP is between the "high-energy" and "low-energy" phosphate compounds. Transfer of phosphoryl group from "high-energy" compounds to "low-energy" acceptor compounds occurs via ATP-ADP system

phosphoanhydride bond) is much more (~46 kJ/mol) than the hydrolysis of β-γ bond (~31 kJ/mol), making the reaction irreversible. $PP_i$ formed is hydrolyzed to two $P_i$ by the ubiquitous enzyme inorganic **pyrophosphatase**, resulting in further release of energy (19 kJ/mol). The adenylation reaction is particularly useful to drive thermodynamically unfavorable reactions, e.g., fatty acid activation. During fatty acid activation, first adenylyl (AMP) is transferred from ATP to carboxylic group of the fatty acids resulting in the formation of fatty acid adenylate and $PP_i$. The adenyl group is replaced by thiol group of Coenzyme A, resulting in the formation of thioester. Net free energy change in these two reactions is negative and energetically equivalent to hydrolysis of ATP to AMP and $PP_i$ (−45.6 kJ/mol). AMP has to be converted to ADP since it the ADP, which is required for conversion to ATP (Fig. 4.7).

$$AMP + ATP \rightarrow 2ADP$$

The reaction is catalyzed by **adenylate kinase** (or AMP kinase). Kinase is the term used for the enzymes which transfer the phosphoryl group from ATP to other molecules. Such reactions, in which hydrolysis of ATP is not involved, occur frequently during the metabolic pathway.

## 4.3 Reduction-Oxidation Coupled Reactions

Photoautotrophs utilize radiant energy from the sun to remove electrons from water (oxidation of water) and transfer them to $CO_2$ resulting in its reduction. As a result, solar energy is trapped in the form of reduced organic molecules:

$$6CO_2 + 12\ H_2O \rightarrow C_6H_{12}O_6 + 6O_2 + 6H_2O$$

Chemoautotrophs derive energy from the oxidation of the chemical compounds. For doing work, organisms obtain energy by removing electrons from organic molecules (oxidation) and recycle them to $O_2$ (reduction) resulting in synthesis of water.

$$C_6H_{12}O_6 + 6O_2 \rightarrow 6CO_2 + 6H_2O$$

Thus, there is global recycling of $O_2$ and $CO_2$ accompanied with the **oxidation-reduction** reactions which are responsible for energy release and energy conservation. In a cell, many reactions involved in energy transductions require electron flow from one molecule to another. These are called **reduction-oxidation reactions or redox reactions** (Box 4.2). In cells oxidation-reduction reactions are part of metabolic pathways. Electrons removed during hundreds of oxidative reactions are channeled through only just a few types of universal electron carriers such as $NAD^+$, $NADP^+$, FMN, and FAD (Figs. 4.9, 4.10 and 4.11). These undergo reversible oxidation and reduction in many of the redox reactions of cellular metabolism.

**Box 4.2: Redox Potential and Redox Couples**
In an oxidation-reduction reaction,

$$AH_2 + B \rightarrow A + BH_2$$

$AH_2$ is oxidized to A and B is reduced to $BH_2$. The reaction can be understood as two half reactions; oxidation of $AH_2$ (removal of electrons coupled with removal of protons) and reduction of B (acceptance of electrons and protons) separately,

$$AH_2 \rightarrow A + 2e^- + 2H^+$$

$$B + 2e^- + 2H^+ \rightarrow BH_2$$

$AH_2$ is the reductant (electron donor), while B is the oxidant (electron acceptor) in this redox reaction. In any redox reaction electron transfer may or may not be coupled with proton transfer. The following redox reaction involves transfer of electrons only,

$$Fe^{3+} + Cu^+ \rightarrow Fe^{2+} + Cu^{2+}$$

$Cu^+$ is the reductant since it is the electron donor while $Fe^{3+}$ as an oxidant as it is the electron acceptor.
The two half reactions can be written as,

$$Cu^+ \rightarrow Cu^{2+} + e^- \quad \text{(oxidation)}$$

$$Fe^{3+} + e^- \rightarrow Fe^{3+} \quad \text{(reduction)}$$

Both reactions occur simultaneously as oxidation of $Cu^+$ to $Cu^{2+}$ is coupled with reduction of $Fe^{3+}$ to $Fe^{2+}$. $Cu^+$ and $Cu^{2+}$ are called conjugate **redox couple** since $Cu^+$ serves as electron donor while $Cu^{2+}$ will serve as conjugate electron acceptor. In the similar way $Fe^{3+}/Fe^{2+}$ will be another redox couple of the reaction.

There are four ways in reduction/oxidation, i.e., involving only electron transfer, as hydrogen atom (hydrogen atom consists of a proton and an electron), as **hydride** ion ($H^-$) (hydride ion has two electrons; net charge on hydrogen atom will be negative), e.g., hydride ion transfer in the reduction of $NAD^+/NADP^+$, involvement of oxygen in the redox reaction which is covalently incorporated in the product. **Reducing equivalent** term is used to express transfer of single electron, which participates in the redox reaction either as electron or hydrogen atom or hydride ion. Direction of electron flow is determined by affinity of the compounds for electrons. Electron will flow

(continued)

### Box 4.2 (continued)

from the compounds having lower affinity to those which have higher affinity for the electrons. Relative affinity of compounds for the electrons is expressed as their **redox potential. Standard reduction potential ($E^0$)** is a measure of this affinity, which is expressed in volts and standard of reference is half-cell where hydrogen ion in aqueous solution is in equilibrium with hydrogen gas. The standard reduction potential ($E^0$) of the conjugate redox pair represents the potential difference at 1M concentration, 25 °C and pH 7.0 with reference to the standard (pH 0) hydrogen electrode.

The oxidized form of a redox couple with a large positive standard reduction potential has a high affinity for electrons and is a strong oxidizing agent, while its conjugate reductant is a weak electron donor. Electron flow occurs from the redox couple with less positive reduction potential to the redox couple with more positive values. Thus, direction of the electron flow in between electron donor of a redox couple to electron acceptor of another redox couple is determined by difference in their standard reduction potential ($\Delta E^0$). It is measured in volts (V).

$$\Delta E^0 = E^0_{(e^- \text{ acceptor})} - E^0_{(e^- \text{ donor})}$$

Standard reduction potential is used to calculate free energy change during the electron transfer which can be calculated by the following formula,

$$\Delta G = -nF\Delta E \quad \text{or} \quad \Delta G^{0'} = -nF\Delta E^0$$

where $n$ is the number of electrons transferred in the reaction and $F$ is Faraday's constant (a proportionality constant that converts volts to joules (F = 96,480 J/V.mol)).

---

NAD$^+$ and NADP$^+$ are loosely bound with enzyme protein. So, they can move from one enzyme protein to another enzyme protein, while FMN and FAD are tightly bound to the enzyme protein and form prosthetic group of the enzyme. Besides these iron-sulfur proteins, cytochromes also have tightly bound prosthetic groups that undergo reversible reduction and oxidation on accepting and removal of electron. Quinones, such as ubiquinone and plastoquinone, serve as the mobile carriers of electrons as they become lipid soluble on being reduced. NADH produced in mitochondria during catabolic oxidative reactions is oxidized, and electrons which are removed are finally accepted by $O_2$. Since $O_2$ accepts one electron at a time, electrons removed in pairs (from a metabolite to NADH) are transferred further to other intermediate carriers that can undergo both two-electron and one-electron redox reactions. $O_2$ accepts electrons from cytochromes which facilitate one-electron transfer. Electrons move through various electron carriers in order of their increasing positive $\Delta E^{0'}$ values.

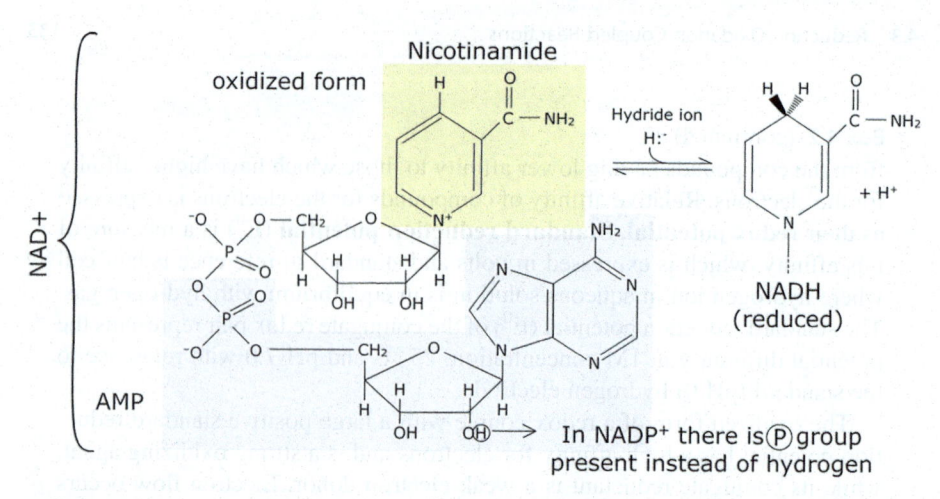

**Fig. 4.9** Reduction of NAD⁺ to NADH requires two-electron transfer. Only nicotinamide ring is affected. Hydride ion (a proton with two electrons) transfer results in reduction of NAD⁺ to NADH. NADP⁺ differs from NAD⁺ only in presence of phosphoryl group on the 2′-hydroxyl group of ribose sugar. Plus (+) sign on NAD⁺ and NADP⁺ indicates that nicotinamide ring is in oxidized form and there is positive charge on the nitrogen atom of the ring. In many cells ratio of NAD⁺ to NADH is high which favors hydride transfer to NAD⁺ to form NADH. Contrary to this ratio of NADPH to NADP⁺ is high which favors hydride transfer from NADPH to substrates. In most of the cells, total concentration of NAD⁺ + NADH is about $10^{-5}$ M, while that of NADPH + NADP⁺ is $10^{-6}$ M

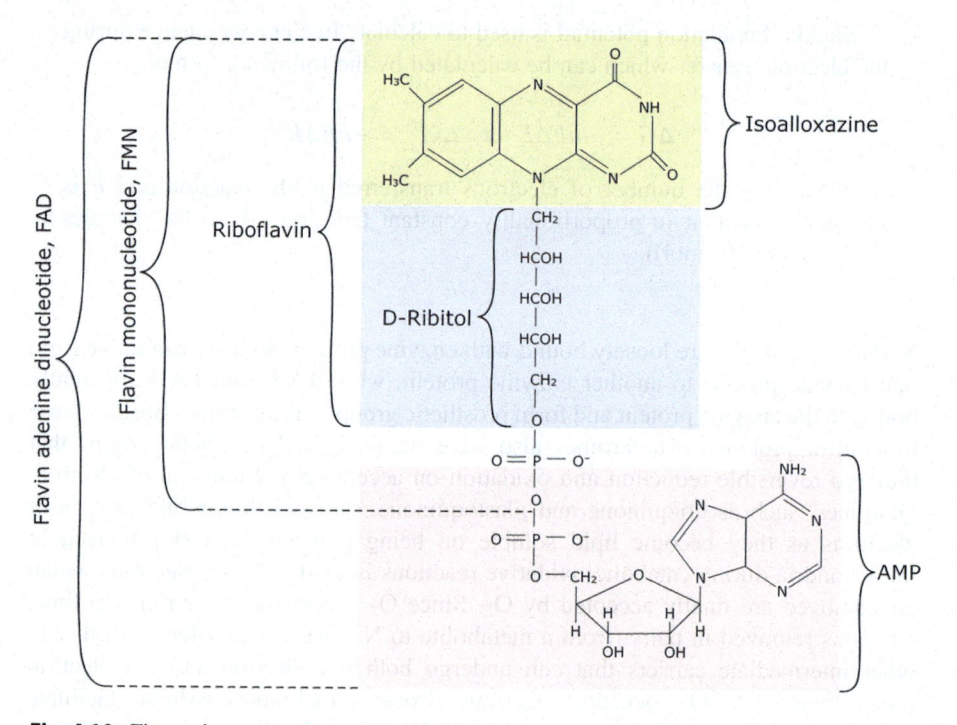

**Fig. 4.10** Figure shows structures of riboflavin, flavin mononucleotide (FMN), and flavin adenine dinucleotide (FAD). Flavin coenzymes are stronger oxidizing agents than NAD⁺ and NADP⁺. These can be reduced both by single electron and two electron pathways. These are reoxidized by molecular oxygen

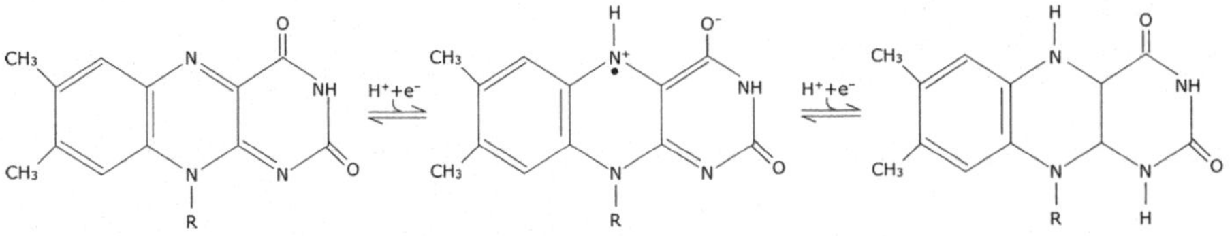

**Fig. 4.11** Isoalloxazine ring of flavin nucleotide (FMN and FAD) undergoes reversible reduction. Unlike reduction of NAD$^+$ and NADP$^+$, reduction of these nucleotides occurs on accepting either one or two electrons in the form of one or two hydrogen atoms (each atom is in the form of an electron and one proton). On accepting one hydrogen atom, semiquinone form of isoalloxazine ring is formed. These are abbreviated as FADH (FMNH) which on accepting one more hydrogen atom is fully reduced to FADH$_2$ (FMNH$_2$). Since these nucleotides can participate in either one or two electron transfer reactions, flavoproteins are involved in a greater diversity of reactions

## 4.4 Enzymes

Enzymes are central to metabolism since these are the biocatalysts catalyzing almost all cellular reactions. These include slow but thermodynamically feasible reactions at ambient cellular conditions, which include biological pH, temperature, as well as the molar concentrations of the reactants and the products. The metabolic pathways need to be modified in response to the cellular needs. This occurs through regulation of the activity of various enzymes. Discovered in yeast for the first time, the biocatalysts were called "enzyme" by Wilhelm Kuhne in 1878 which in Greek means "in yeast" (*en*, in; *zyme*, yeast). Earlier Louis Pasteur had called these vital factors present in intact yeast cells as "ferments," since these were thought to be responsible for carrying out fermentation. Chemical nature of enzymes was not established till the time Sumner crystallized urease from jack beans and established their proteinaceous nature in 1926. Earlier enzymes were thought to be small biologically active molecules analogous to hormones. Sumner postulated that all enzymes were proteins. It was after John Northrop and Moses Kunitz had crystallized trypsin and pepsin in 1930 that Sumner's conclusions were widely accepted. Sumner was awarded the Nobel Prize in 1946. At the same time in a treatise entitled "enzymes", J.B.S. Haldane postulated that the weak bonding between enzyme and substrate might be responsible for the reactions catalyzed by them. Since then thousands of enzymes have been isolated and characterized. A new science called "enzymology" developed, which dealt with study of enzymes. Barring ribozymes (catalytic RNA molecules), all enzymes are proteins. While some of the enzymes consist of proteins only (simple proteins), in others a non-protein part is also part of their structures

**Fig. 4.12** Enzymes consist of non-protein part in addition to the proteinaceous structure

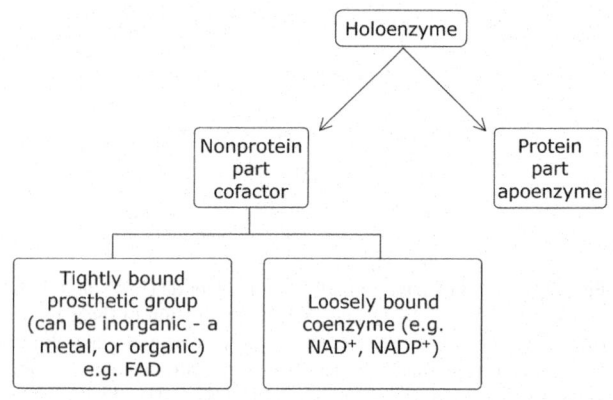

```
                    Holoenzyme
                   /          \
          Nonprotein          Protein
             part              part
           cofactor          apoenzyme
          /        \
Tightly bound      Loosely bound
prosthetic group   coenzyme (e.g.
(can be inorganic - a   NAD+, NADP+)
metal, or organic)
e.g. FAD
```

(conjugated proteins). The non-protein part of these enzyme-conjugated proteins is called **cofactor**. In case where cofactor is inorganic, such as metals ($Mg^{2+}$, $Zn^{2+}$, $Fe^{2+}$), the enzymes are called as metalloenzymes. The organic cofactors are called **coenzymes**. Cofactors may be loosely bound with the enzyme proteins or may be tightly associated through a covalent bond. Cofactors, which are tightly associated with the protein part of the enzymes, are called **prosthetic group**, which may be inorganic or organic in nature (Fig. 4.12). Sometimes both metal and the organic molecules are required as the cofactors for enzyme activity. In case of cytochromes, the prosthetic group **heme**, along with a metal ion ($Fe^{3+}$), is bound to enzyme protein through hydrogen bonding, hydrophobic interactions, and the covalent bonding to a specific site of the enzyme protein. The functionally active enzyme, in case of conjugated proteins, is called **holoenzyme,** and the protein portion of the enzyme is called **apoenzyme**. Loosely bound coenzymes, such as $NAD^+$ or $NADP^+$, are transiently associated with enzyme proteins. These function as **co-substrates**, which need to be regenerated maybe through independent reactions. Contrary to this in case of prosthetic groups, regeneration of the group occurs as a part of enzyme-catalyzed reaction. Catalytic activity of enzymes depends on the integrity of constituent protein conformation, which is determined by the primary, secondary, and tertiary protein structures. Besides, in case of enzyme molecule requiring two or more than two subunits, the intact quaternary structure is also important for the catalytic activity. Any factor, which is responsible for destroying the conformation, would lead to loss in their activity.

### 4.4.1 Nomenclature and Classification of Enzymes

After thousands of enzymes had been discovered, different strategies were adopted, namely:

(i) By adding suffix "-ase" to the name of the substrate: Substrate is the substance on which the enzyme acts upon. For example, enzymes acting upon proteins were called proteinases, those acting upon lipids were called as "lipases," and the ones acting on nucleic acid were named as "nucleases." Specific names were also given to enzymes acting on specific substrates, such as "urease," "lecithinase," or "maltase," for the enzymes acting on urea, lecithine, or maltose, respectively.

(ii) Another strategy adopted to name the enzymes was to add suffix "-ase" to the kind of reaction catalyzed by the enzymes, e.g., isomerases (which catalyze isomerization), hydrolases (catalyzing hydrolysis reactions), transaminases (catalyzing transamination), etc.

(iii) Both of the above systems of naming the enzymes appeared inadequate since the naming was either based on the type of molecules on which the enzyme acted upon or the type of reaction catalyzed by them. Another system was adopted in which some of the enzymes were named both on the basis of substrate utilized and the reaction catalyzed by them. For example, succinic acid dehydrogenase signifies both the name of the substrate succinic acid as well as the reaction catalyzed by them dehydrogenation.

To maintain uniformity in naming enzymes the International Union of Biochemistry and Molecular Biology (IUBMB) set up an Enzyme Commission (EC) on enzyme nomenclature, which gave its first recommendations in 1961. Some of the recommendations given by EC are as follows:

1. Each enzyme can have a **trivial name**, which is short and is easy to use. The **systematic name** of the enzyme however should be formed according to the definite rules showing the action of the enzyme as much as possible. It should have two parts: the first name denotes the substrate, the second one with the suffix "-ase" which specifies the reaction catalyzed by them. Additional information, if there is any, is given in parenthesis. For example, malate dehydrogenase which catalyzes the following reaction:

$$L - Malate + NAD^+ \rightarrow Pyruvate + CO_2 + NADH + H^+$$

The enzyme can be called L-malate: NADH oxidoreductase (decarboxylating).

2. All enzymes were classified into six classes depending upon the type of reaction catalyzed by them (Table 4.1). Each enzyme is given a **classification number**. The classification number is known as Enzyme Commission (EC) number assigned by the nomenclature committee of IUBMB. The classification number has four digits, e.g., if the classification number of an enzyme is a, b, c, and d, a stands for the number of the class given in the classification number, b is the

**Table 4.1** Major classes of enzymes

| Class No. | Name of the class | Nature of the reaction catalyzed |
|---|---|---|
| 1. | Oxidoreductases | Catalyze transfer of hydrogen or oxygen atoms or electrons from one substrate to another, also called oxidases, dehydrogenases, or reductases. Substrate that is oxidized is electron donor. Systematic name is based on *donor: acceptor oxidoreductase*. Common name will be dehydrogenase except where electron acceptor is oxygen, then called oxidases |
| 2. | Transferases | Catalyze group transfer reactions. Systematic names are formed according to the scheme *donor: acceptor group transferase*. Common name according to acceptor group transferase or donor group transferase |
| 3. | Hydrolases | Catalyze hydrolytic cleavage of C-C, C-O, and C-N bonds and some other bonds including phosphoanhydride bonds. Common name in many cases formed by the name of the substrate with suffix "-ase" |
| 4. | Lyases | Catalyze cleavage of C-C, C-O, C-N, or other bonds by elimination, leaving double bonds or rings, or catalyze addition of groups to double bonds. Systematic name is formed according to the pattern substrate group-lyase. Hyphen is important part of the name |
| 5. | Isomerases | Catalyze transfer of groups within molecule to yield isomeric form. According to type of isomerism, these can be accordingly called isomerases, epimerases, mutases, etc. Subclass is formed according to type of isomerism and sub-subclass according to type of substrate |
| 6. | Ligases | Catalyze joining together two molecules forming C-C, C-O, C-S, and C-N bonds by condensation reactions coupled with hydrolysis of ATP or similar triphosphate |

number of subclass, c is the number of sub-subclass, while d represents the number of sub-sub-subclass which specifies the actual substrate of the enzyme which distinguishes it from other enzymes catalyzing similar reactions. In the following enzyme-catalyzed reaction:

$$ATP + D - glucose \rightarrow ADP + glucose\ 6 - phosphate$$

The trivial name of the enzyme is hexokinase/glucokinase, which is commonly used. The systematic name of the enzyme catalyzing the reaction is ATP: glucose phosphotransferase which indicates that the enzyme catalyzes transfer of phosphoryl group from ATP to glucose. The classification number (Enzyme Commission number) is E.C 2.7.1.1. The first number 2 signifies class number (transferase); the second number 7 is about the phosphate group transferred; the third number 1 is about the number of the sub-subclass which signifies a phosphotransferase with a hydroxyl group as an acceptor, while the last digit 1 is the number of sub-sub-subclass which includes D-glucose as the phosphoryl group acceptor.

### 4.4.2 General Characteristics of Enzyme-Catalyzed Reactions

Most of the chemical reactions require presence of a catalyst and generally occur in extreme conditions, such as high temperatures or a low or high pH, or may require organic solvents. However, enzymes enable the chemical reactions to occur in ambient cellular conditions, i.e., at temperature 37 °C, biological pH of 6.5–7.5, and in aqueous medium. Enzymes are very efficient in catalyzing the reactions $10^6$–$10^{14}$ times faster than those not catalyzed by enzymes. One of the most catalytically potent enzymes, carbonic anhydrase has a **turnover number** of 600,000 per second. Even in a spontaneous thermodynamically feasible reaction where free energy of the product is less than that of reactants, i.e. $\Delta G^{o'}$ of the reaction is negative, the reaction does not start by itself. A substrate needs to be converted into an intermediate state before being converted to products. The intermediate state known as **transition state** refers to energy requiring molecular arrangement in a substrate molecule which makes it easy for the substrate to get converted to product. Free energy of transition state is higher than either substrate or the product. The starting point for either in forward or reverse direction is known as **ground state**. Difference in free energy of ground state and transition state is known as **energy of activation** of that reaction. During interconversion of substrate (S) and the product (P), change in free energy is plotted against progress of a reaction in a reaction coordinate diagram (Fig. 4.13). Substrate exists for a very short period in transition state, i.e., $10^{-14}$ to $10^{-13}$ of a second, after which it is converted to product. Energy of activation refers to the energy required to initiate a reaction. It constitutes the barrier to any chemical reaction. Higher activation energy of a reaction corresponds with slower reaction rate. Enzymes do not alter equilibrium constant but enhance reaction rates by lowering activation energies. Though activation energy is lowered, there is no alteration of $\Delta G^{o'}$ of enzyme-catalyzed reaction. Interconversion of two sequential reaction intermediates constitutes a reaction step. In case there are several reaction steps in a pathway, the one which requires highest activation energy is the rate-limiting step.

Transition state is achieved when enzymes bind with the substrates to form enzyme-substrate (ES) complex. The region of an enzyme molecule by which it binds with substrate is called the **active site**. Active site of an enzyme molecule is a three-dimensional structure formed due to the folding of constituent polypeptides leading to specific conformation of the molecule. Though active site occupies only a very small fraction of the large structure of an enzyme, it is needed to keep the interacting groups properly positioned so as to prevent active site from collapsing. Residues which constitute active site are responsible both for binding with substrate and holding it in specific orientation (binding residues), and also for carrying out

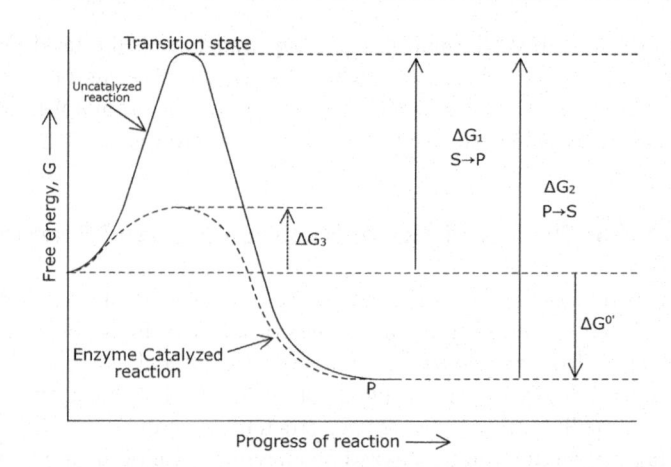

**Fig. 4.13** Reaction coordinate diagram showing changes in free energy during uncatalyzed and enzyme-catalyzed reaction. Changes in free energy during reaction is plotted against progress of reaction. $\Delta G_1$ is the energy of activation for uncatalyzed conversion of substrate to product ($S \rightarrow P$) which is required for bond breakage and formation. $\Delta G_2$ is energy of activation for reversible reaction $P \rightarrow S$. $\Delta G^{o'}$ is the overall change in standard free energy in a spontaneous exergonic reaction during $S \rightarrow P$. $\Delta G_3$ is the energy of activation for enzyme-catalyzed reaction

catalysis (catalytic residues). In some enzymes, binding and catalytic residues may be same. Any change in protein conformation will result in alteration of the structure of active site, and enzyme will not be able to carry out catalysis. Enzymes differ from other catalysts in being highly specific for a particular substrate. Specificity is derived from formation of many weak interactions between the active site of the enzyme and the substrate. Specific groups of R-side chains of both binding and catalytic residues interact with specific substrate, which provides specificity to the enzyme-catalyzed reactions. For example, if a hydroxyl group of a substrate interacts with a specific residue of the active site, any compound lacking a hydroxyl group will be a poor substrate for that enzyme. Many enzymes act only on one biological substrate (absolute substrate specificity); others act on broader range of substrates, which are structurally similar (relative group specificity). Glucose 6-phosphatase catalyzes hydrolysis of only glucose 6-phosphate, while acid and alkaline phosphatase can act on various phosphorylated substrates, thus displaying absolute substrate specificity and relative group specificity, respectively. Hexokinase adds a phosphate group to D-glucose and not to its optical isomer (L-glucose) displaying stereospecificity. Stereospecificity makes them unique and highly useful in pharmacology industry. This property of enzyme is due to their inherent chirality (proteins consist of only L-amino acids), which leads to formation of asymmetric active site.

Binding of substrate with active site of the enzyme involves non-covalent bonds such as ionic, hydrogen, and hydrophobic bonds and van der Waals interactions. It is the **binding energy** which is responsible for lowering of activation energy. Two models have been proposed to describe the binding process. According to **lock and key** model proposed by Emil Fischer in 1894, there is a structural similarity between

## 4.4 Enzymes

**Lock and key model**

Active site of enzyme molecule fits with substrate

Substrate

Active site

Enzyme (large molecule)

**Induced-fit model**

Substrate

Active site of enzyme molecule has different structure so does not fit with substrate molecule

Active site

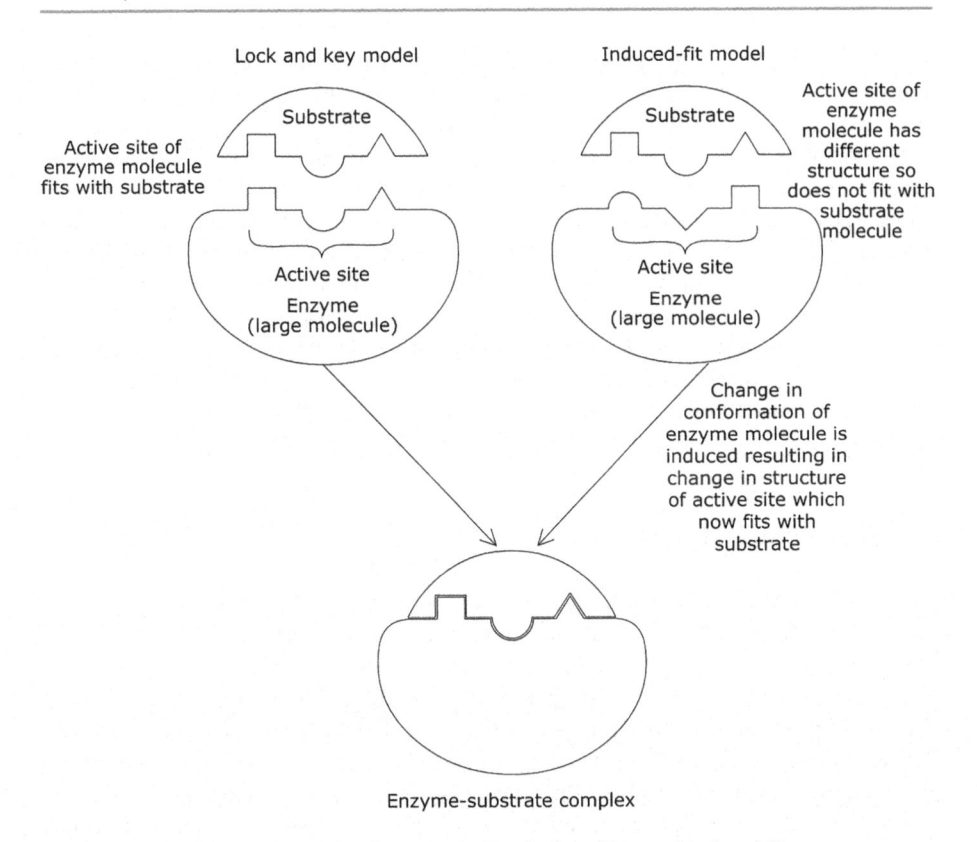

Enzyme (large molecule)

Change in conformation of enzyme molecule is induced resulting in change in structure of active site which now fits with substrate

Enzyme-substrate complex

**Fig. 4.14** Two proposed models of enzyme action: lock and key and induced fit

active site of the enzyme and the substrate. As there is a specific key, which fits into the grooves of a lock, a compound having a unique structure, which fits into the active site, will be the substrate of the enzyme (Fig. 4.14). Compounds having structural similarity to the substrate have been found to inhibit the enzyme activity (**competitive inhibition**). A competitive inhibitor binds with the active site of the enzyme forming enzyme-inhibitor complex, thus preventing the binding of the substrate molecule, which does not get dissociated to form products. However, the model suggests structural rigidity for the enzyme, which is otherwise a dynamic structure. Change in protein conformation is possible due to formation and breaking of non-covalent bonds giving flexibility to the enzyme molecule. X-ray studies have indicated that the active sites undergo conformational changes on binding with substrate. In 1959, Daniel E. Koshland proposed **induced-fit** model, according to which the active site of an enzyme is flexible. Presence of a substrate induces a conformational change resulting in alteration of the active site, which can now bind with the substrate. "Induced-fit" model is more attractive since it provides flexibility and dynamicity to enzyme molecule. The model was first established for the enzyme **hexokinase**. Hexokinase catalyzes transfer of phosphoryl group from ATP to glucose. Active site of enzymes may be complimentary not to the substrate rather to transition state of the substrate. Substrate in transition state binds more tightly with the active site, resulting in lowering of activation energy requirement in enzyme-catalyzed reactions. **Transition state analogs** (molecules which have structure similar to transition state) have been found to bind with active site of the enzyme more tightly than either the substrate or the product. The fact that active site is less perfect fit to the substrate than to the transition state, binding of substrate will cause strain on the substrate molecule to fit properly into the active site which favor formation of its transition state resulting in lowering of activation energy. The products are released since these bind less tightly with the active site resulting in enhancement of rate. A very small reduction in activation energy can increase rate of the reaction many times. Decrease in activation energy by 84 kJ/mole by urease can result in increase in reaction rate by a factor of $10^{14}$.

### 4.4.3 Enzyme Kinetics

Enzyme kinetics is the study of enzymes by determining their reaction rates. Laboratory measurement of rate of enzyme-catalyzed reaction is called enzyme assay. Enzyme assays are developed either to measure the amount of substrate used up during the reaction or amount of products formed in a unit time. Measuring amount of products formed is preferred since it is a direct method. In case the product formed is colored or produces colored compounds on reacting with some chemical, colorimetric assay is possible. Enzyme assay by spectrophotometric method is possible for the enzymes which utilize $NAD^+$ or NADH during a reaction. Since NADH (not $NAD^+$) has absorption peak at 340 nm, change in absorption of that wavelength will indicate appearance or disappearance of NADH. Rate of an enzyme-catalyzed reaction decreases with time because of either depletion of substrate or accumulation of products while enzyme is kept constant. The decrease may also be due to denaturation of protein in case of a sensitive enzyme. The most rapid reaction rate is observed at the start of reaction and is known as **initial velocity** ($v_0$). $v_0$ is used in enzyme kinetic studies. Adrian Brown had started enzyme kinetic studies in 1902. Substrate concentration is one of key factors affecting velocity of enzyme-catalyzed reaction. In case initial velocity of a reaction ($v_0$) is plotted against substrate concentration [S], the graph obtained is hyperbolic (Fig. 4.15). At low substrate concentration, $v_0$ is considered as function of [S]. Increase in $v_0$ becomes smaller with increasing [S] until a plateau-like region for $v_0$ is achieved which is close to maximum velocity ($V_{max}$), beyond which substrate concentrations do not increase the reaction rate substantially. $V_{max}$ is the function of the amount of enzyme present in a given experiment. Briggs and Haldane introduced the concept of steady state in 1925 when P is produced at the same rate at which S is consumed. Lower region of the graph displays first-order kinetics since increase in $v_0$ is proportional to increase in substrate concentration. At lower substrate concentrations, active sites of the enzyme molecules are not saturated and are free to bind with the substrate molecules. It is this transient pre-steady phase when the concentration of ES builds

## 4.4 Enzymes

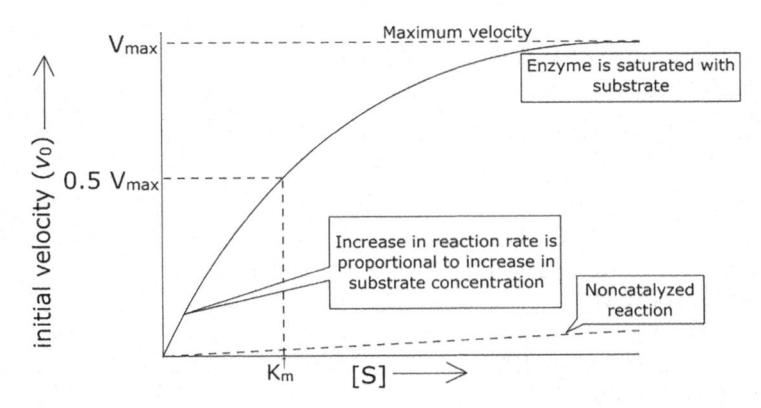

**Fig. 4.15** Effect of substrate concentration on the reaction rate catalyzed by an enzyme following Michaelis-Menten kinetics. $K_m$ (Michaelis constant) is a constant, which is the substrate concentration at which velocity of the reaction is half of the maximum. The dashed line which is below represents the reaction rate of a non-catalyzed reaction which has been given for comparison

up for a very short duration and generally lasts microseconds. Steady state is achieved when all of the enzyme molecules are saturated and the reaction becomes independent of further increase in substrate concentrations. Since pre-steady state lasts only for a very short time, $v_0$ generally reflects steady state, and analysis of these reaction rates refers to steady-state kinetics. The reaction displays mixed order in the intermediate portion of the curve when there is increase in reaction rate with increasing substrate concentration, but increase is not proportional with substrate concentration. Victor Henri in 1903 had proposed the idea of formation of enzyme-substrate complex as an explanation for the kinetic pattern of the enzyme-catalyzed reactions. This was further expanded in 1913 by Leonor Michaelis and Maud L. Menten into a general theory of enzyme action. An enzyme displaying this kinetics is referred as **Michaelis-Menten enzyme** and the kinetics as hyperbolic kinetics or **Michaelis-Menten kinetics**. In 1913, Michaelis and Menten proposed a general theory of enzyme action and enzyme kinetics. Hyperbolic kinetics of enzymes can be expressed algebraically by Michelis-Menten equation.

Derivation of the Michaelis-Menten equation has been given in the Box 4.3. Enzymes initially interact with the substrate in a relatively faster reaction forming enzyme-substrate complex [ES]. This (ES) then breaks down in a slower second step to yield free enzyme and the product. Both reactions are considered reversible reactions. The substrate concentration at which the velocity of an enzyme-catalyzed reaction is half of its maximum is defined as **Michaelis constant** ($K_m$). $K_m$ is expressed in mM. For many enzymes, $K_m$ is in the range of $10^{-4}$ to $10^{-6}$ M. Measured values of $K_m$ can provide an estimate of intracellular concentration of the substrate. Generally, most of the enzymes function at subsaturating levels in the cell. In case enzymes can use different substrates, their $K_m$ values can be used for differences in their relative affinity for the substrates. Greater $K_m$ value indicates lesser affinity of enzyme for the substrate and vice versa. An enzyme's biological

**Box 4.3: Derivation of Michaelis-Menten Equation**

Michaelis-Menten theory explains the course of enzyme catalyzed reaction as follows:

$$E + S \underset{k_{-1}}{\overset{k_1}{\rightleftharpoons}} Es \overset{k_2}{\rightarrow} E + P$$

Assuming that reverse reaction P→S is negligible, $v_0$ can be determined by breakdown of [ES]:

$v_0 = k_2[ES]$    (i)

Since neither $k_2$ nor [ES] can be measured in a reaction, an alternative expression was found.

- Rate of formation of [ES]

$$\frac{d[ES]}{dt} = k_1([Et] - [ES])[S]$$

- Rate of breakdown of [ES]

$$-\frac{d[ES]}{dt} = k_{-1}[ES] + k_2[ES]$$

- Since initial rate ($v_0$) represents steady state, i.e., in which [ES] is constant—i.e., rate of formation of ES is equal to rate of its breakdown,

$$k_1([Et] - [ES])[S] = k_{-1}[ES] + k_2[ES]$$

- Equation is simplified to find value of [ES]

$$[ES] = \frac{k_1[E_t][S]}{k_1[S] + k_{-1} + k_2} = \frac{[E_t][S]}{[S] + \frac{(k_{-1} + k_2)}{k_1}} = \frac{[E_t][S]}{[S] + Km}$$

Since rate constants can be combined into one expression $Km$, which is defined as Michaelis constant.

By substituting value of [ES] in equation (i),

$v_0 = \frac{k_2[E_t][S]}{[S] + Km}$    (ii)

Maximum velocity ($V_{max}$) occurs when the enzyme is saturated, i.e.,

$[ES] = [E_t]$

$V_{max} = k_2[E_t]$

Substituting the value in equation (ii),

(continued)

> **Box 4.3** (continued)
> $$v_0 = \frac{V_{max}[S]}{Km+[S]}$$
> This is rate equation for one-substrate enzyme catalyzed reaction (Michaelis-Menten equation)
> In case $v_0$ is exactly one-half of $V_{max}$.
> $$\frac{V_{max}}{2} = \frac{V_{max}[S]}{Km+[S]}$$
> On dividing by $V_{max}$, the equation will be,
> $$\frac{1}{2} = \frac{[S]}{Km+[S]}$$
> i.e., $Km = [S]$, when $v_0$ is $1/2\ V_{max}$.
>
> Thus Km (Michaelis constant) can be defined as the substrate concentration at which velocity is half of the maximum. The term is sometimes used as an indicator of the affinity of the enzyme for its substrate.

function can be estimated from the $K_m$ value for its substrate. For example, $K_m$ values of glutamate dehydrogenase and glutamine synthetase with reference to utilization to $NH_4^+$ were found to be 30 mM and 0.015 mM, respectively, during an experiment conducted with *Lemna*. Tissue concentration of $NH_4^+$ was estimated to be 1/30 of those required for glutamate dehydrogenase but was saturating for glutamine synthetase, indicating that the enzyme glutamine synthetase may be having a primary role in $NH_4^+$ assimilation. Function of glutamate dehydrogenase predominantly is to release $NH_4^+$ from glutamate by the reverse of assimilatory reaction. $V_{max}$ is related to turnover number, which refers to the number of moles of substrate that react to form product per mole of enzyme per unit time. This assumes that the enzyme is fully saturated with substrate and the reaction is proceeding at maximum rate. It is also expressed as $K_{cat}$. A straight line obtained, when reciprocal of $v_0$ is plotted against reciprocal of [S], is known as **Lineweaver-Burk double reciprocal plot** which is used to study enzyme kinetics (Box 4.4).

### 4.4.4 Factors Affecting Enzyme-Catalyzed Reactions

**Effect of pH** Protein conformation is influenced by the state of its ionizable groups, thereby affecting the structure and function of active site of the enzyme. pH of the medium influences ionic status of the R-side chains of the amino acid residues of proteins affecting the non-covalent bonds responsible for holding the protein molecule in correct conformation. Furthermore, residues present at the active site need to be in appropriate ionic status required for binding with the substrate as well as for catalyzing the reaction. pH also affects the ionic status of the substrate. Generally, a typical bell-shaped plot is obtained on studying the effect of pH on enzyme activity.

**Box 4.4: Lineweaver-Burk Double Reciprocal Plot**
It is quite difficult to estimate $V_{max}$ in the hyperbolic curve which describes the rate of non-allosteric enzymatic reaction. $V_{max}$ value is never reached with any finite substrate concentration that could be used in lab and it becomes difficult to determine Km of the enzyme. Michaelis-Menten equation can be algebraically transformed into equation by which a straight line is obtained instead of hyberbolic curve which becomes more useful. Michaelis-Menten equation,

$$v_0 = \frac{V_{max}[S]}{Km+[S]}$$

By taking reciprocals on both sides of equation:

$$\frac{1}{v_0} = \frac{Km+[S]}{V_{max}[S]}$$

The reaction is simplified to give:

$$\frac{1}{v_0} = \frac{Km}{V_{max}} \frac{1}{[S]} + \frac{1}{V_{max}}$$

This equation now has a form of a straight line equation, y = mx+b, where $\frac{1}{v_0}$ is plotted on y-axis and $\frac{1}{[S]}$ on x-axis; it gives straight line, where $\frac{Km}{V_{max}}$ is represented by slope of reaction. Intercept of straight line on y-axis represents $\frac{1}{V_{max}}$, while on the horizontal axis it gives value of $-\frac{1}{Km}$.

This form of the Michaelis-Menten equation is called Lineweaver-Burk equation and the graphic representation of same is called Lineweaver-Burk double reciprocal plot.

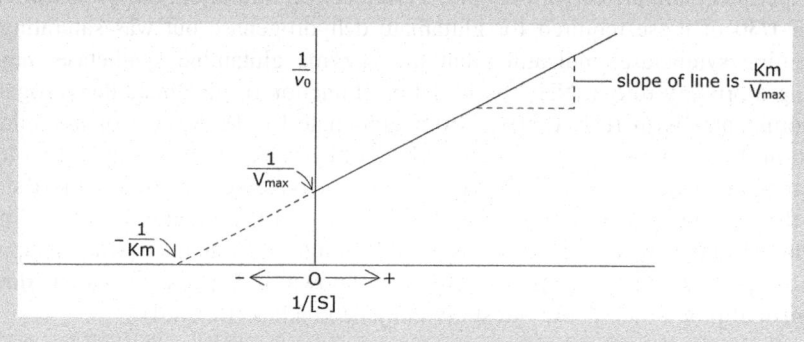

**Effect of Temperature** Increase in temperature is responsible for increase in kinetic energy of the reactant molecules as it increases their chances of collision and therefore increasing the rate of any chemical reaction. In an enzyme-catalyzed reaction, temperature adversely affects enzyme structure because of the thermolabile nature of the non-covalent bonds maintaining the protein structure. An optimum temperature for the enzyme-catalyzed reaction is the balance of the two. Optimum temperature is also determined by the time of exposure of enzyme to that temperature. Temperatures above 50 °C are generally destructive for the enzyme protein. However, some enzymes are stable even at high temperatures.

## 4.4.5 Role of Inhibitors

Various compounds inhibit or alter the activity on binding with enzyme molecules. Inhibitors act through a variety of mechanisms. An inhibitor which binds reversibly to the enzyme molecule thereby lowering its activity causes reversible inhibition. On the contrary, the inhibitor which results in permanent damage to enzyme molecule results in irreversible inhibition. Removal of such inhibitor does not result in resumption of the enzyme activity. **Reversible inhibitors** generally bind with the enzyme molecules through non-covalent bonds, altering their conformation temporarily or they bind with the active sites due to similarity of their structure with the substrate molecules. Reversible inhibition can be competitive, uncompetitive, or noncompetitive. In presence of competitive inhibitors, availability of free enzyme to bind with the substrate is reduced since, unlike ES, EI complex does not break down to form product. This type of inhibition can be reversed by increasing the substrate concentration since it increases the possibility of the substrate binding to the active sites of enzyme molecules rather than with the inhibitor. $V_{max}$ of the reaction is not altered; however, $K_m$ increases (Fig. 4.16). This demonstrates a decrease in sensitivity of the enzyme for the substrate in presence of competitive inhibitor. Inhibition of succinic acid dehydrogenase by malonate is an example of competitive inhibition. Succinic acid dehydrogenase catalyzes conversion of succinate to fumarate. Malonate competes for binding with the active site of the enzyme since it resembles succinate in its structure. Inhibition of Rubisco by $CO_2$ and $O_2$ is also an example of competitive inhibition in plants. These two gases compete with each other for binding with the active site of the enzyme. Oxygenase activity of the enzyme can be reduced by increasing concentration of $CO_2$. Transition state analogues are specifically effective competitive inhibitors since active site of the enzyme specifically catalyzes the reaction on binding with transition state of the substrate. Noncompetitive inhibition is reversible inhibition. It is also known as "mixed type" of inhibition as the inhibitor binds either with free enzyme or with the enzyme-substrate complex. Inhibitor binds to a site of the enzyme distinct from the active site. As a result, active site of the enzyme is not blocked for binding with the substrate, but subsequent reaction is inhibited, resulting in decrease in $V_{max}$ of the reaction. Since affinity of the enzyme for substrate is not reduced, $K_m$ does not change. This kind of inhibition is not reversed by an increase in substrate concentration because inhibitor and substrate are not competing for the same active site. In the **uncompetitive inhibition**, inhibitor binds only with enzyme-substrate complex and not with free enzyme. As a result of this, inhibition increases with increase in substrate concentration. Both $V_{max}$ and $K_m$ are affected. $V_{max}$ is reduced while there is increase in $K_m$ (Fig. 4.16).

During **irreversible inhibition,** an inhibitor binds covalently to the enzyme protein forming a complex which does not dissociate to release free enzyme from the product. Organophosphorus compound—diisopropylfluorophosphate (DIFP)—is an irreversible inhibitor of the enzyme acetylcholine esterase which catalyzes hydrolysis of ester bond in acetylcholine, producing inactive molecules, acetate and choline. DIFP covalently binds to the seryl residue of active site of the enzyme,

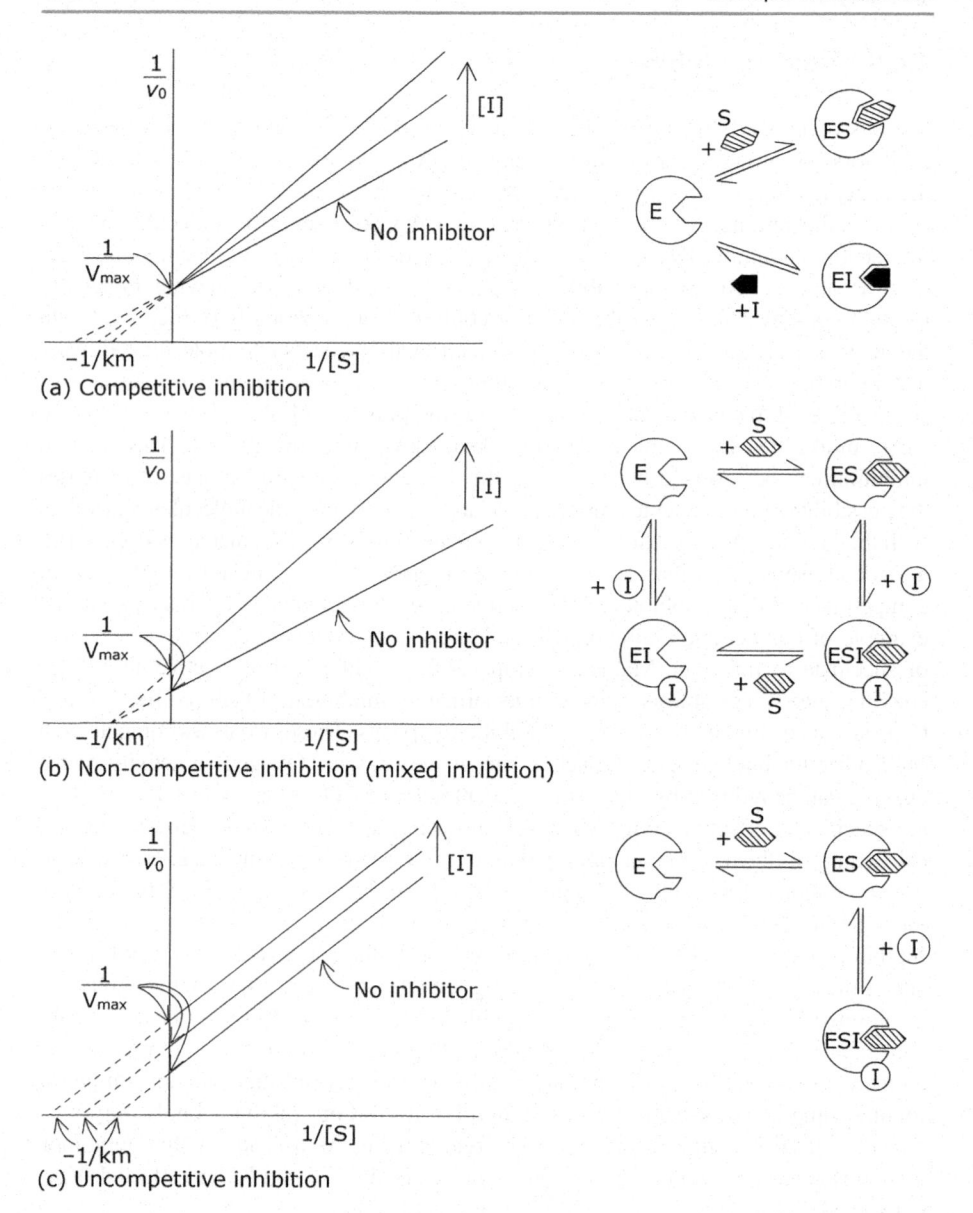

**Fig. 4.16** Double reciprocal plot showing three types of reversible inhibition of enzyme activity

thereby inhibiting its activity (Fig. 4.17). Irreversible inhibitor may not bind covalently in some cases, but binding is strong enough so that the inhibitor does not get dissociated easily from the enzyme, e.g., transition state analogues. Though bonding is non-covalent, these compounds bind with active site of the enzyme so tightly that the two rarely get dissociated, thus inhibiting the enzyme activity. Transition state

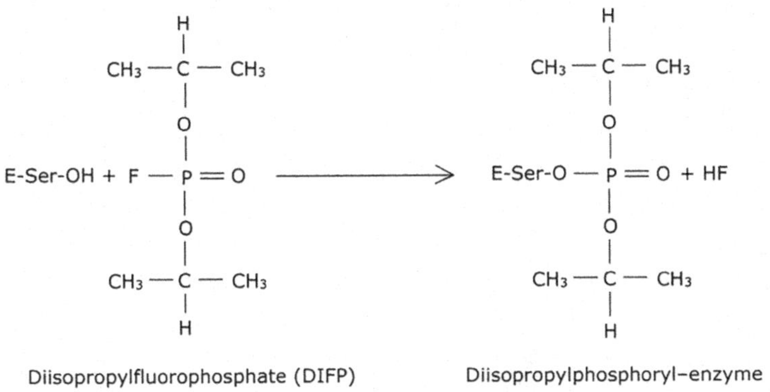

## 4.4 Enzymes

$$\text{E-Ser-OH} + \text{F}-\overset{\underset{|}{O}-CH(CH_3)_2}{\underset{|}{\overset{|}{P}}}=O \longrightarrow \text{E-Ser-O}-\overset{\underset{|}{O-CH(CH_3)_2}}{\underset{|}{\overset{|}{P}}}=O + \text{HF}$$

Diisopropylfluorophosphate (DIFP)     Diisopropylphosphoryl–enzyme

**Fig. 4.17** Irreversible inactivation of enzyme acetylcholine esterase. Active serine residue of the enzyme binds covalently with DIFP, and the resulting complex of the enzyme is not reactive toward its own substrate

analogues cannot perfectly mimic the transition state. Even then they bind to the target enzyme $10^2$–$10^8$ times more tightly. This concept of enzyme inhibition by transition state analogs is important in pharmaceutical industry for designing new drugs. Another class of irreversible enzyme inhibitors includes **suicide inactivators.** After allowing first few reactions to happen normally, instead of being converted to product, these compounds are converted to highly reactive molecules which combine irreversibly with the enzyme and inhibit its activity.

### 4.4.6 Regulatory Enzymes

Metabolic pathways in a cell are regulated according to the requirement of a cell. One of the mechanisms for regulation of metabolism is determined by the amount and availability of a particular enzyme, indicating control being at the transcription or translational level. This is a slower process and regulation through this mechanism will require a longer time period. A quicker regulation of the metabolic pathway occurs through regulation of activity of enzymes. Regulatory enzymes catalyze the slowest reactions of a pathway and the occurrence and pace of that pathway is determined by activity of these enzymes. Generally, it is the first reaction which is regulatory besides other steps in a pathway. Metabolic pathways may be linear or branched. In addition to the first step, the reactions at the branching point of the pathway are also regulated since conversion of common metabolite to the products may depend upon the need of the cell for a particular end product (Fig. 4.18). In case the end product of a particular branched pathway is not required and its production in the cell is stopped and the enzyme at the branching of the pathway is inactivated, activity of the regulatory enzymes may increase or decrease in case of positive or negative regulation, respectively, which in turn influences the metabolic reactions

**Fig. 4.18** Feedback regulation in (**a**) a linear metabolic and (**b**) branched pathway. Linear pathway is regulated by the end product of the pathway if accumulated. In a branched metabolic pathway, regulation may occur at the branching or at the starting reaction of the pathway in case D accumulates

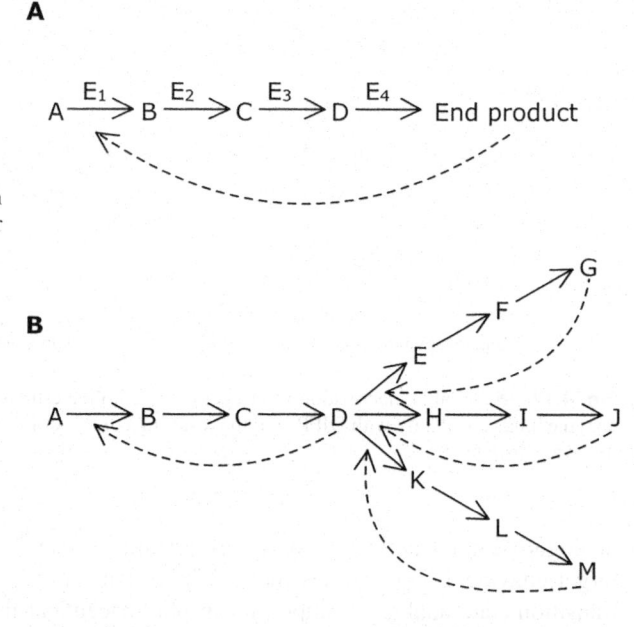

accordingly. There are different mechanisms in the cell by which the activity of an enzyme is regulated.

### 4.4.6.1 Allosteric Regulation

Allosteric regulation of enzyme activity is important in control of metabolism. The term allosteric is derived from Greek word *allos* which means "other" and *stereos* meaning "three-dimensional." Activity of allosterically regulated enzymes is determined by the metabolites known as **allosteric modulators**. These modulators act by inducing change in the conformation of the enzyme upon binding to a site other than the active site by non-covalent bonds. Enzyme activity may be either inhibited or activated on binding with the allosteric modulator which are called allosteric inhibitor or allosteric activator, respectively. **Ligand** refers to the end product of a reaction (in case of feedback regulation) or a metabolite. Feedback inhibition is instantaneous and can be reversed quickly. Enzyme activity is inhibited in case the product of a reaction accumulates and will start functioning when the concentration of the product falls down. NADH/$NAD^+$ and ADP/ATP are some of the important allosteric modulators of enzyme activity. For example, ADP may act as positive modulator for several enzymes which are involved in the oxidation of sugars, thereby stimulating the conversion of more ADP to ATP. Allosteric enzymes are much larger and complex in structure. They generally consist of more than two subunits. The catalytic sites present on subunits are different from the modulator sites. Different subunits communicate with each other through change in conformation. Binding of a modulator to the modulator site of the enzyme subunit induces change

## 4.4 Enzymes

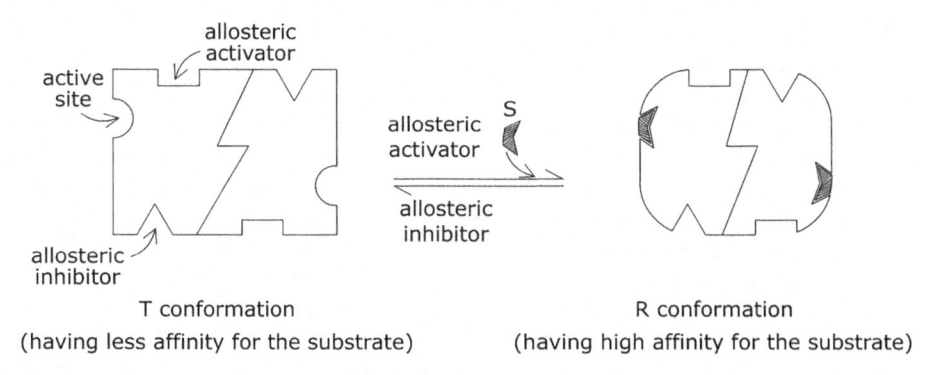

**Fig. 4.19** Subunits of allosteric enzymes can exist in two conformations, T (tense) conformation, which has less affinity for substrate, or R (relaxed) conformation, which has more affinity for binding with substrate. Substrate may act as allosteric activator (homotropic allosteric enzymes) or other metabolite may act as allosteric activator or allosteric inhibitor (heterotropic allosteric enzymes)

in the conformation, resulting in increased or decreased affinity of the catalytic site for the substrate in case of positive or negative regulation, respectively. Generally, modulators have shapes different from that of the substrates. Besides enzymes, there are non-enzymatic proteins which alter their conformation on binding with ligand. The protein subunits can exist in two conformations, one which has lesser affinity for the substrate, i.e. T (tense) conformation, and the one having more affinity for the substrate R (relaxed) conformation (Fig. 4.19). Two principal models have been proposed to explain the behavior of the enzyme subunits upon binding with the modulators. These are the "concerted model" and "sequential model." Both of these models are used as the basis for interpreting experimental results. The "concerted model" was proposed by Jacques Monod, Jeffries Wyman, and Jean-Pierre Changeux in 1965. According to this model, conformation of all the subunits of an enzyme changes from T state to R state simultaneously on binding with the positive modulator, and vice versa on binding with negative modulator. Binding of positive modulator to the subunit is cooperative and stabilizes all the subunits in R conformation resulting in shifting of equilibrium, while reverse happens when negative modulator is bound. This results in T conformation of all the subunits simultaneously (Fig. 4.20). Positive cooperativity is also seen on binding with substrate in case of homotropic allosteric enzymes. The "sequential model" was proposed by Daniel Koshland in 1966 based on "induced-fit theory" of substrate binding. According to this model, binding of substrate induces conformation change in the subunit of allosteric enzyme from T to R which makes conformational change in other subunits of the enzyme easier. Similarly, binding of activators or inhibitors also takes place by induced-fit mechanism. Conformation change in one subunit influences the conformation change in other subunits also. In presence of an inhibitor, substrate is less likely to bind to the active site in T conformation, thus affecting the sensitivity of the subunits. Similarly, in presence of an activator, sensitivity of the

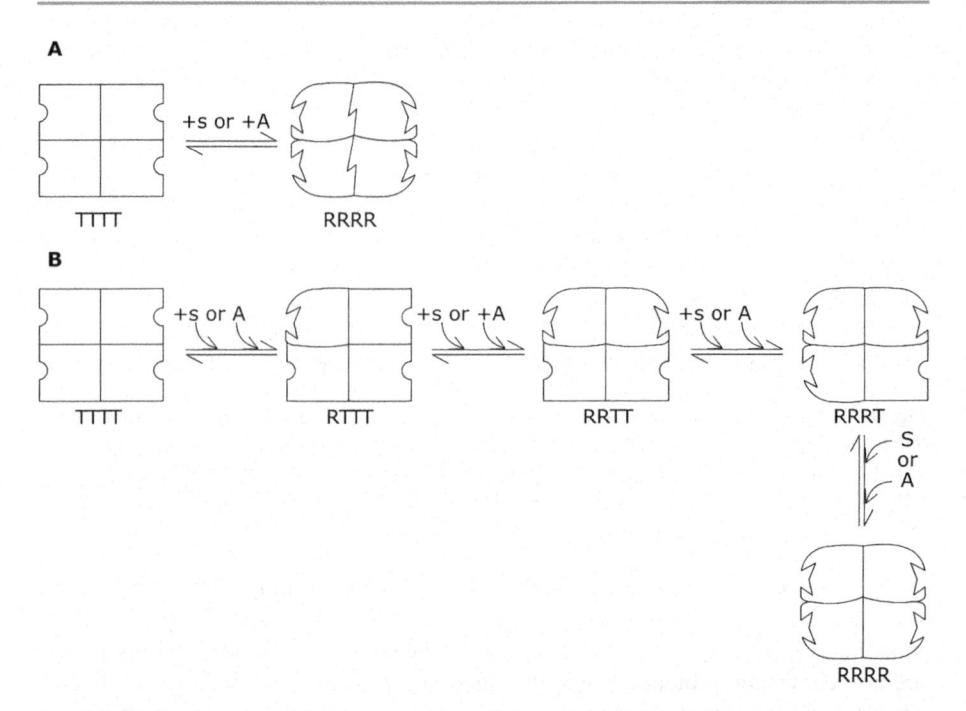

**Fig. 4.20** Hypothetical models given for allosterically regulated enzymes. (**a**) Concerted model: on binding with substrate S (homotropic) or activator (A) other than substrate (heterotropic) affinity of all the subunits is increased for binding with substrate. (**b**) Sequential model: on binding with the substrate (homotropic) or activator other than substrate (heterotropic) affinity of other subunits to bind with substrate changes one by one as a result of which all intermediate species are present

subunits to bind with the substrate increases substantially. Thus, conformational change is passed to all other subunits making them more or less likely to bind with substrate in presence of activator or inhibitor, respectively. Sequential model has also been able to incorporate negative cooperativity, which did not find any provision in concerted model (Fig. 4.20).

**Kinetics of Allosteric Enzymes** When effect of [S] on $v_0$ of enzyme-catalyzed reaction is studied, allosteric enzymes do not display Michelis-Menten kinetics. Instead of hyperbolic graph, the allosteric enzymes usually produce sigmoid saturation curve (Fig. 4.21). Sigmoid kinetics usually displays cooperative interactions between multiple subunits. Change in conformation in one subunit triggers change in all other subunits mediated by non-covalent interactions at the interface between subunits. Sigmoid kinetic behavior of allosteric enzymes is explained by subunit interactions in both concerted and sequential models. Value of $v_0$ at half of [S] is not referred as $K_m$ because enzymes do not follow hyperbolic kinetics, rather it is expressed as $K_{0.5}$. One characteristic of sigmoid curve is that a small change in the concentration of modulator can bring about large changes in the velocity of the

## 4.4 Enzymes

**Fig. 4.21** An allosterically regulated enzyme displays a sigmoid curve when effect of [S] is studied on initial velocity ($v_0$) of the reaction, contrary to hyperbolic curve obtained in case of a Michaelis-Menten enzyme

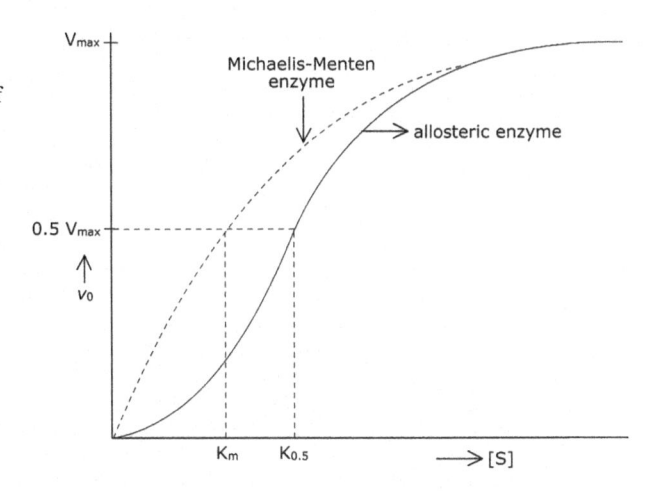

reaction. For heterotropic allosteric enzymes, a positive modulator may change the sigmoid curve to more of a hyperbolic curve with a decrease in $K_{0.5}$. On the contrary, presence of a negative modulator may increase sigmoid nature of the curve with an increase in $K_{0.5}$. There are some heterotropic modulators which increase or decrease $V_{max}$ with little changes in $K_{0.5}$ (Fig. 4.22).

### 4.4.6.2 Covalently Modulated Enzymes

Another way by which activity of enzymes is regulated is by reversible covalent attachment of a group such as phosphoryl, adenylyl, adenosine, ribosyl, etc. to specific amino acid residues of the enzyme protein. Covalent attachment of a protein such as ubiquitin may also alter the activity of an enzyme. The most significant group which alters the enzyme activity on being covalently attached is phosphoryl group. Phosphoryl group is generally attached reversibly to a specific serine, threonine, or tyrosine residue of the enzyme protein, resulting in alteration of structural and functional properties of the molecule. Since phosphoryl group carries two negative charges, it will attract positively charged amino acids of the molecule while repelling amino acids with negatively charged side chains. As a result, conformation of the enzyme protein is altered. Phosphoryl group bound to the enzyme protein may also influence interaction with substrate molecule. Removal of phosphoryl group reverses the effect of phosphorylation. Phosphorylation and dephosphorylation of enzyme protein are catalyzed by protein kinases and phosphoprotein phosphatases, respectively. Phosphorylation may result in activation or inhibition of enzyme activity which will depend on a particular enzyme. However, reverse will be true on dephosphorylation of that enzyme (Fig. 4.23). One example is the regulation of pyruvate dehydrogenase activity, which is a component of pyruvate dehydrogenase complex and catalyzes conversion of pyruvate to acetyl-CoA. Phosphorylation of pyruvate dehydrogenase makes it inactive, while removal of

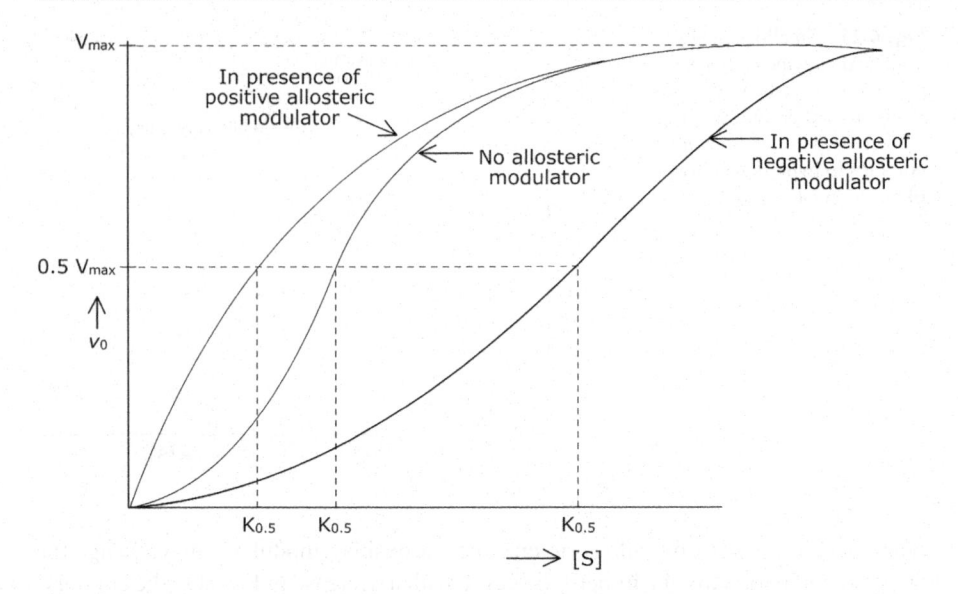

**Fig. 4.22** Effect of a modulator on the kinetics of an allosterically regulated enzyme. Curves are drawn arbitrarily

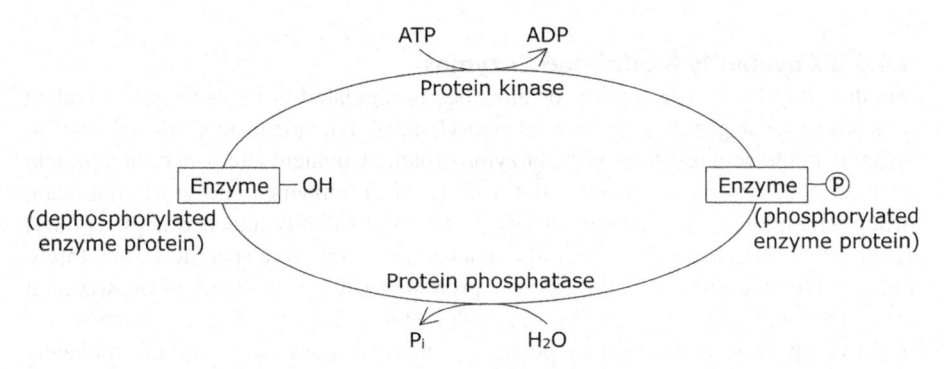

**Fig. 4.23** Interconversion of phosphorylated and dephosphorylated forms of an enzyme through the action of protein kinase and protein phosphatase. Some enzymes are active when they are phosphorylated while others are active when they are dephosphorylated

the phosphate group restores the active enzyme. Phosphorylation and dephosphorylation of the enzyme protein are carried out by ATP-dependent pyruvate dehydrogenase kinase and phosphopyruvate dehydrogenase phosphatase, respectively. Additionally, covalently modulated enzymes include enzymes in which protein conformation changes in response to reversible reduction and oxidation of sulfur-containing groups of cysteine residues, which occurs in response to the redox status of the cell. Interconversion of sulfur-containing groups of the cysteine residues

## 4.4 Enzymes

between dithiol (-SH SH-) and disulfide (–S-S-) is mediated by thioredoxin. Activities of four of the Calvin cycle enzymes are regulated through this type of covalent modification. Alternate oxidase activity is also regulated through reversible oxidation and reduction of sulfur-containing groups of cysteine. Covalent addition of a hydrophobic group may also affect conformation of some of the enzymes altering their activity. Enzyme regulation also occurs through proteolytic cleavage or through calcium-mediated calmodulins.

**Summary**

- A sum total of all the chemical reaction occurring in a cell is called as metabolism and study of metabolites of a cell is called metabolomics. Since plants are sessile organisms and are exposed to harsher environmental conditions, they have more flexible and diverse metabolic pathways. Metabolic pathways are compartmentalized in a cell and regulated movements of metabolites occur through the cell membranes.
- Metabolism is classified as anabolism and catabolism. Anabolism includes biosynthetic reactions which involve synthesis of complex molecules from simple ones requiring input of energy. On the contrary complex molecules are broken down into simpler forms during catabolism releasing energy in the process.
- Spontaneous reactions are exergonic reactions with a negative free energy change during the reaction, while free energy change during endergonic reactions is positive. Exchange of energy occurs through ATP, which is the energy currency of the cell. ATP is synthesized during exergonic reactions, while endergonic reactions occur at the expense of ATP. Another kind of reactions involves removal and acceptance of electrons which are mediated by electron carriers such as $NAD^+/NADP^+$ and FAD besides others.
- All enzymes, except ribozymes, are proteins. Additionally, some enzymes consist of non-protein part also, which is called cofactor. A conjugated enzyme protein is called holoenzyme, while the protein part is called apoenzyme. A covalently bound cofactor is called prosthetic group, while a loosely bound cofactor is called coenzyme. All enzymes are classified into six major classes and named according to guidelines given by International Union of Biochemists.
- Enzymes are very efficient catalysts. Active site of the enzyme refers to the part of protein molecule with which substrate binds. It is a three-dimensional structure and consists of grooves and crevices. Substrate binds with the active site by non-covalent bonds. Enzymes act by lowering the activation energy since they bind with the transition state of the substrates.
- Michaelis and Menten proposed the equation known as Michaelis-Menten equation, which explains the relationship between the substrate concentration and velocity of the reaction. A new kinetic constant, Michaelis constant, was derived which refers to the substrate concentration at which velocity of the enzyme-catalyzed reaction is half of the maximum. Michaelis-Menten equation is

rearranged to get a straight-line equation which is used to get Lineweaver-Burk double reciprocal plot. Enzyme activity is altered by various factors, including pH and temperature of the medium. Enzyme activity is also inhibited by the compounds known as inhibitors. Inhibition can be reversible or irreversible.
- There are regulatory steps in a metabolic pathway, which are catalyzed by regulatory enzymes. Enzymes can either be allosterically regulated or covalently modulated besides other means. Kinetics of allosterically regulated enzymes is different from that of a typical enzyme following Michaelis-Menten kinetics. One of the most important groups which is responsible for regulating the enzyme activity of covalently modulated enzymes is the phosphoryl group.

## Multiple-Choice Questions

1. Bioenergetics refers to:
   (a) Energy exchange in between the cell and the surroundings
   (b) Science which deals with energy transductions within the cell
   (c) Energy release during a chemical reaction
   (d) None of the above
2. A living cell is an open system because:
   (a) It neither exchanges energy nor matter with the surroundings.
   (b) It can exchange both energy and matter with the surroundings.
   (c) It can exchange energy with the surroundings but not the matter.
   (d) It can exchange matter but not energy with the surroundings.
3. According to second law of thermodynamics, spontaneous reaction will occur:
   (a) When less complex molecules will be converted to more complex ones.
   (b) When there is absorption of energy from the surroundings.
   (c) Molecules having higher entropy are converted to molecules having lesser entropy.
   (d) Molecules having lesser entropy are converted to molecules having high entropy.
4. $\Delta G$ of a cellular reaction will be negative if:
   (a) Products of the reaction have lesser entropy than the reactants.
   (b) Products of the reaction have more entropy than the reactants.
   (c) The reaction is non-spontaneous.
   (d) There is requirement of input of energy for the reaction to occur.
5. Which of the following statement is true?
   (a) In a living cell, equilibrium constant is maintained at 0.
   (b) Free energy is the total energy present in a molecule.
   (c) Free energy is the energy isothermally available to do work.
   (d) $\Delta G^{o'}$ is defined as the change in free energy during a reaction which is not at equilibrium.

6. High-energy bond (~) of ATP indicates:
   (a) Formation of this bond requires energy.
   (b) Hydrolysis of this bond releases energy.
   (c) Products of hydrolysis have lesser energy than the molecule itself.
   (d) Products of hydrolysis have more energy than the molecule itself.
7. ATP is the high-energy molecule because:
   (a) It is nucleoside triphosphate.
   (b) ATP is more resonance stabilized than the products of its hydrolysis.
   (c) ATP is present as Mg-ATP complex in the cell.
   (d) The products of its hydrolysis are stabilized by resonance.
8. In a redox reaction, electrons move:
   (a) From the compounds having more positive redox potential to compounds having lesser positive redox potential
   (b) From compounds having lesser positive redox potential to more positive redox potential
   (c) From compounds having lesser negative redox potential to compounds have more negative redox potential
   (d) None of the above
9. In an enzyme molecule consisting of a protein and non-protein structure, the protein part is known as:
   (a) Cofactor
   (b) Holoenzyme
   (c) Apoenzyme
   (d) Coenzyme
10. A non-protein structure covalently attached to the protein part of an enzyme molecule is called:
    (a) Coenzyme
    (b) Cofactor
    (c) Apoenzyme
    (d) Prosthetic group
11. Michaelis constant ($K_m$) of an enzyme is:
    (a) The substrate concentration at which enzyme is fully saturated
    (b) The substrate concentration at which $V_{max}$ is half of the maximum
    (c) The enzyme concentration at $V_{max}$
    (d) The enzyme concentration at which $V_{max}$ is half of the maximum

**Answers**

1. b   2. b   3. d   4. b   5. c   6. c
7. d   8. b   9. c   10. d   11. b

## Suggested Further Readings

Jones RL, Ougham H, Thomas H, Waaland S (2013) The molecular life of plants. Wiley-Blackwell, Chichester, pp 42–70

Nelson DL, Cox MM (2017) Lehninger principles of biochemistry, 7th edn. W.H. Freeman, New York, pp 495–525

Voet DJ, Voet JG, Charlotte WP (2008) Principles of biochemistry, 3rd edn. Wiley, Hoboken, pp 448–484

# Photosynthesis

## Manju A. Lal

All forms of life in this universe require energy for growth and maintenance. Plants and some forms of bacteria capture light energy directly from solar radiation and utilize it for synthesis of food materials besides producing basic raw materials from which other cellular biomolecules are produced. The term **photosynthesis** describes the process by which green plants synthesize organic compounds from inorganic raw materials using light. Photosynthesis is the source of all biological energy, viz., food, biological fuels, and biomass, and is also most important for availability of free oxygen. Whatever free oxygen is there in the atmosphere is the result of photosynthesis. Since heterotrophic organisms including animals cannot use sunlight as direct source of energy, they consume plants as the source of energy. Photosynthesis is the means for solar energy to enter into the global ecosystem, and it alone is the essential biological process by which solar energy is transformed into metabolic form of energy for all forms of life on earth. An understanding of the fundamental and applied aspects of the process comes from a wide range of studies including agriculture, forestry, plant biochemistry, plant molecular biology, tissue culture, and metabolic engineering.

Primitive life is believed to have existed in anaerobic conditions which utilized energy stored in chemical compounds for the biosynthesis of the biomolecules required for their growth. However, hundreds of millions of years ago, with the depletion of these compounds, photoautotrophs might have originated which utilized solar energy to produce reduced organic compounds, which either oxidized water and released oxygen (**oxygenic photosynthesis**), ferrous ions ($Fe^{2+}$) to ferric ions ($Fe^{3+}$) (e.g., purple photosynthetic bacteria), or used $H_2S$ as source of electrons (e.g., green sulfur bacteria). In the latter case, sulfur got deposited by the organisms (**anoxygenic** photosynthesis). However, in due course of time almost $3.5 \times 10^9$ years ago, because of free availability, water is believed to have been utilized by cyanobacteria in place of $H_2S$. It was almost $2.7 \times 10^9$ years ago when oxygen is presumed to have been released as a waste product which started to accumulate on the earth surface resulting

© Springer Nature Singapore Pte Ltd. 2018
S. C Bhatla, M. A. Lal, *Plant Physiology, Development and Metabolism*,
https://doi.org/10.1007/978-981-13-2023-1_5

in an oxygenic environment. Accumulation of oxygen shielded living organisms from damaging effects of UV irradiations.

The site of photosynthesis in eukaryotes (algae and higher plants) are the cells that contain few to numerous (about 1–1000) **chloroplasts** which vary in size and shape. Chloroplasts are unique double-membrane-bound organelles that originated through an endosymbiotic association between free-living oxygen-evolving photosynthetic bacteria which might have been incorporated into the growing eukaryotic cells as chloroplast. Outer chloroplast membrane is relatively freely permeable, while the inner membrane exhibits more selective permeability. The sites of light reactions in the chloroplast are the saclike structures, known as **chloroplast lamellae** or **thylakoids**. The space within the chloroplasts is divided into two compartments, viz., one enclosed within the thylakoids called **lumen** and the other outside the thylakoids, which is called **stroma**. Stroma, the matrix around the thylakoid, is the site where $CO_2$ is assimilated, leading to the synthesis of sugars. Thylakoids exist either as stacks called **grana** or are unstacked and are interconnected to form **stroma lamellae**. Each chloroplast contains 10–100 grana. Light is captured by various pigments which includes chlorophyll molecules as the photoreceptors for photosynthesis. These exist as the chlorophyll-protein complexes which are involved in harvesting light energy and transporting electrons, resulting in generation of reductant and synthesis of ATP. In cyanobacteria, photosynthetic machinery required for light reactions exists in plasma membrane which forms invaginations or folded structures resembling grana of chloroplasts in eukaryotic cells.

**Photosynthesis** is an oxidation-reduction process in which oxidation of water (electrons being removed from water) is coupled with the release of oxygen and reduction of carbon dioxide leads to synthesis of carbohydrates. It is a two-stage process. During stage I, known as light reaction, photolysis of water takes place:

$$2H_2O \xrightarrow{Light} O_2 + 4[H^+] + 4e^- \qquad \text{(Stage I)}$$

Electrons removed from water are used to reduce $CO_2$ in the subsequent stage II, known as $CO_2$ assimilation:

$$4e^- + 4[H^+] + CO_2 \rightarrow (CH_2O) + H_2O \qquad \text{(Stage II)}$$

Thus, light energy is converted into chemical energy and is conserved in the form of carbohydrates. Stage I is photochemical, while stage II is a purely chemical reaction. At present the molecular mechanism involved in photosynthesis is fairly understood.

## 5.1 General Concepts

Learning basic concepts of photosynthesis is necessary before understanding its mechanism. This includes properties of light, photosynthetic pigments, mechanism of light absorption, and emission.

## 5.1 General Concepts

### 5.1.1 Properties of Light

Human eye is sensitive to a narrow range of light spectrum from i.e. 400 to 700 nm, which is called as visible light (Fig. 5.1). The wavelengths shorter than 400 nm (UV light) have very high energy, and they are hazardous for biomolecules, while wavelengths longer than 700 nm (infrared) have much less energy. It is the light with wavelengths ranging from 400 to 700 nm that is significant for most of the photobiological processes. Light and all other electromagnetic radiations are transmitted in form of waves, while absorption and emission of light occur in the form of particles. Parameters such as wavelength ($\lambda$) or frequency ($v$) characterize the wave aspect of the light. When light having a particular wavelength ($\lambda$) passes across an observer at a velocity ($c$), the number of waves passing per second is the frequency ($v$), and the relationship between these parameters is represented as

$$v = c/\lambda$$

where $c$ is the velocity of light, i.e., $2.99 \times 10^8$ m.s$^{-1}$.

Light beam can be imagined as a stream of particles or photons. Associated unit of energy of each photon is called a **quantum**. The energy value of a quantum ($E$) is related to the frequency ($v$) of the light which is represented by the equation known as **Planck's equation**,

$$E = hv$$

where $h$ is Planck's constant ($6.62 \times 10^{-34}$ J.s). By replacing $v$ by $c/\lambda$ from the previous equation, it can be understood that the energy of a photon is inversely related to the wavelength of the light. But at the time of absorption and emission of light, single photon is rarely dealt with, and in biochemical processes, conversion of light energy to chemical energy is usually expressed on molar basis, i.e., energy of a mole ($6.02 \times 10^{23}$) of photons (which is an Avogadro's number, $N_A$). A mole of photons is called one **Einstein**. Energy of a mole of photon amounts to

$$E = hv.N_A$$

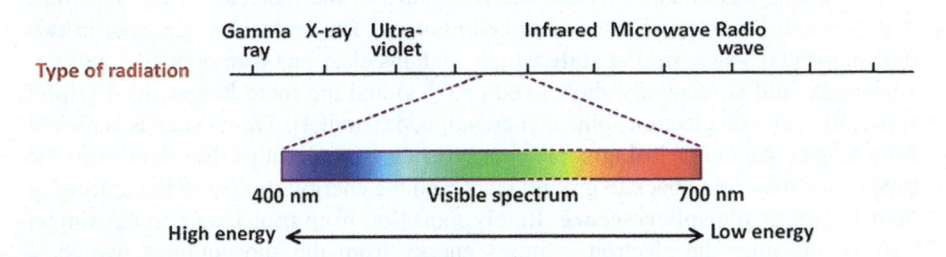

**Fig. 5.1** Electromagnetic spectrum of sunlight

Not all colors of light have equal energy. The energy content of light is inversely proportional to its wavelength, e.g., one mole of photon (an Einstein) of 490 nm blue light will have energy of 240 kJ, whereas 700 nm red light will have only 170 kJ. For any photochemical reaction energy of a photon should be at least as much as that required for the reaction ($\Delta G$).

$$\Delta G = E = h\upsilon . N_A$$

It is more useful to state electric potential ($\Delta E$) of irradiance instead of energy ($\Delta G$), when photosynthetic reactions are compared with redox reactions. This can be calculated by the following equation:

$$\Delta E = -\Delta G/F$$

F is Faraday's constant which refers to number of charges per mole (96,480 Amp.s.mol$^{-1}$).

The solar energy, which is radiating toward earth, is nearly $13 \times 10^{23}$ calories per year. Out of this 30% is reflected back straight away into outer space, 20% is absorbed by the atmosphere, and the remaining almost 50% is absorbed by earth which is converted to heat. Plants convert, utilize, and store less than 1% of the solar energy which is responsible for all chemical, mechanical, and electrical energy driving all organisms on earth. Oxygenic photosynthetic organisms use visible light with wavelength of 400–700 nm, whereas anoxygenic photosynthetic organisms can harness less energetic wavelengths in the infrared region at wavelengths greater than 700 nm.

### 5.1.2 Mechanism of Light Absorption and Emission

For light energy to be used by plants, it has to be absorbed. The absorption of photons by a pigment molecule results in bringing the pigment from its lowest ground energy state ($E_g$) to an excited state ($E_e$), which causes difference in distribution of electrons in the excited molecule. According to the law of quantum mechanics, a given molecule can absorb photons of only certain wavelengths, so that that the energy difference between the two states of the molecule ($E_e - E_g$) must match exactly the energy of the absorbed photons. The molecules can exist in two type of excited states, **singlet state** which contains electrons with opposite (antiparallel) spins and is relatively short lived ($10^{-9}$ s) and the more longer-lived **triplet state** ($10^{-3}$ s) with electron spins that are aligned (parallel). Triplet state is achieved from singlet state after losing some energy to the surroundings. Electron from the triplet state may come back to ground level, and the energy is released in the form of light known as **phosphorescence**. Rarely transition from triplet state to the singlet state occurs after the electron acquires energy from the surroundings, which is followed by release of energy in the form of light wavelength known as delayed **fluorescence** (Fig. 5.2). In the singlet state, the excited chlorophyll molecules have

## 5.1 General Concepts

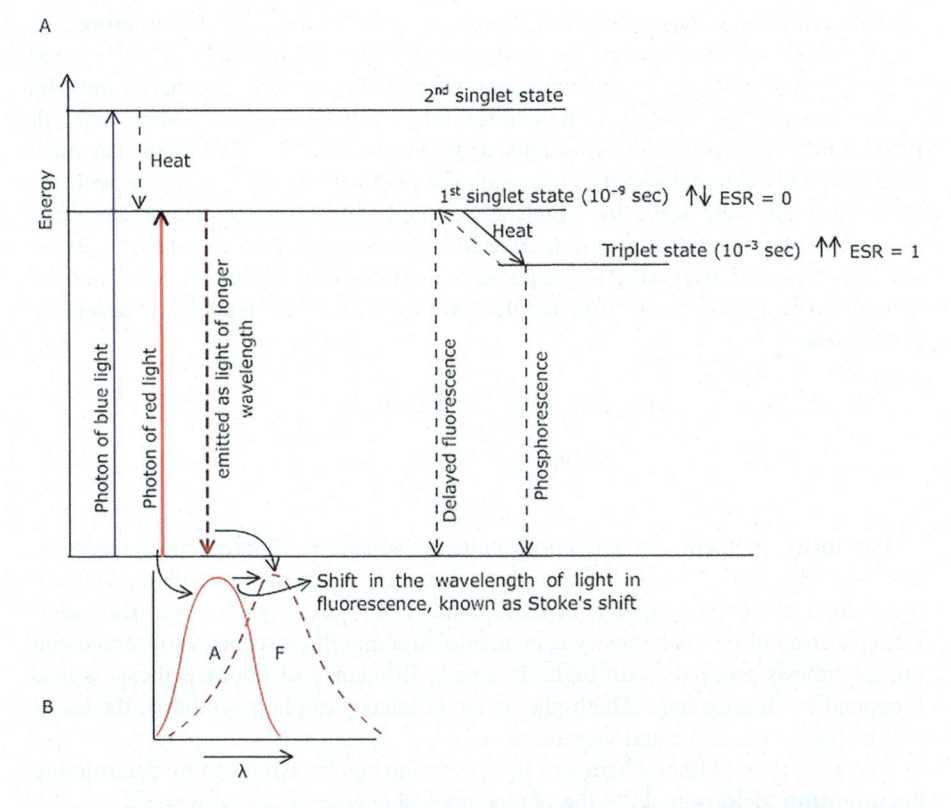

**Fig. 5.2** (**a**) Mechanism of light absorption and emission. Electron is excited to higher energy level (second singlet) by absorption of blue-light photon than when photon of red light is absorbed since blue light has higher energy. In this excited state, chlorophyll molecule is extremely unstable, and some energy is lost as heat resulting in electron attaining new lower energy level, first singlet state which has natural lifetime of $10^{-9}$ s after which the electron may come to the ground level after losing the energy as light. (**b**) Some of the absorbed excitation energy is lost due to vibrations and rotations; as a result energy of the emitted light is lesser than that of absorbed light and wavelengths of emitted light are longer, a phenomenon known as fluorescence. *ESR* electron spin resonance, *F* fluorescence

alternative pathways for dissipation of their available energy. Energy may be lost by internal conversions, which refers to non-radiative decay and is common mode of energy loss from the excited molecule in singlet state. The electronic energy is converted to the kinetic energy of molecular motion, i.e., in the form of heat. Chlorophyll molecules usually relax and attain ground state by this common mode of energy release as heat without any emission of photon. Alternatively, an excited pigment molecule may decay to its ground state by emitting a photon (fluorescence). A fluorescently emitted photon generally has a longer wavelength (lower energy) than the initially absorbed wavelength since some of the excitation energy has been lost as heat. Fluorescence accounts for dissipation of only 3–6% of the light energy absorbed by living plants. The excited molecules may directly transfer their excitation energy to nearby unexcited molecules with similar electronic properties by resonance energy transfer. Because of the inherent instability of the excited singlet state of chlorophyll molecule, any process that captures its energy must be extremely rapid. The process with the faster rate will be favored over others and will predominate. Photochemical reactions of photosynthesis are the fastest chemical reactions. This extreme speed is necessary for photochemistry to compete with the three other alternate ways by which the excited state loses its energy described above. In photochemical reaction excited pigment may lose an electron to an acceptor molecule triggering a charge separation event in which excited molecule is oxidized on losing the electron and the acceptor molecule is reduced on accepting the electron.

$$\text{Pigment} + hv \rightarrow \text{Pigment}^+ + e^-$$

$$\text{Acceptor} + e^- \rightarrow \text{Acceptor}^-$$

For many pigments in the photosynthetic apparatus, fluorescence occurs in nanoseconds ($10^{-9}$ s), whereas photochemical reactions (in photosynthetic organisms) occur more rapidly, in picoseconds ($10^{-12}$ s). When a faster (thousand-fold) pathway of photochemistry is available, minimal fluorescence is observed and photosynthesis proceeds with high efficiency. Efficiency of photosynthesis is also measured by fluorescence. The higher is the efficiency of photosynthesis, the lesser will be the fluorescence and vice versa.

The efficiency of photochemistry in any system can be estimated by determining the **quantum yield** (phi, ϕ) of the photochemical event.

$$\phi = \frac{\text{Number of products formed photochemically}}{\text{Number of photon absorbed}}$$

According to this equation, a quantum yield of 1 would indicate that every absorbed photon is converted into a chemical product. Quantum yield less than 1 would indicate that other decays are responsible for decreasing the efficiency of the photochemical reactions. Under optimum conditions, the measured quantum yield of photochemical reaction in a photosynthetic system is approximately 1 which is indicative of highly efficient photosynthetic process. This indicates that almost all absorbed photons are utilized for photochemical charge separation and other significant decay routes do not occur. Quantum requirement is reverse of $\phi$ and refers to number of quanta required for the production of one photochemical product.

### 5.1.3 Photosynthetic Pigments

In order for light to be absorbed, plants must possess light-absorbing molecules, which occur as complexes bound with proteins. These complexes are called

## 5.1 General Concepts

**pigments**. The pigments consist of **chromophores** (Greek, carrier of color)—the light-absorbing component, and the associated proteins. Absorption of light by the chromophore-protein complex differs from that of free chromophores. On the basis of structure of the chromophore, photosynthetic pigments are classified as follows:

**Chlorophylls** The principal photoreceptor in photosynthesis is chlorophyll which has cyclic tetrapyrrole ring structure termed as porphyrins or chlorin. Structural formula of the green leaf pigment chlorophyll was given by Richard Willstatter and his collaborators as a result of work conducted during 1905–1913 in Zurich and Berlin. He was awarded the Nobel Prize in Chemistry for the same work in 1915. The structure shows similarity with porphyrin ring of hemoglobin. However, chlorophyll molecules have $Mg^{2+}$ instead of $Fe^{2+}$, which occupies the central position. The hydrophobic alcohol, phytol, is derived from four isoprenoid units. The presence of phytol tail facilitates its location along with membrane proteins in thylakoids due to hydrophobic interactions, and it makes the chlorophyll molecules soluble in organic solvents. The five-ring heterocyclic structure surrounding $Mg^{2+}$ has an extended polyene structure with alternating single and double bonds which is responsible for absorption in the visible region of light spectrum. The major forms of chlorophylls in higher plants and green algae are chlorophyll a (Chl a) and chlorophyll b (Chl b) which generally are present in a ratio of 3:1. Chl a is universally present in all organisms which carry out oxygenic photosynthesis. The major forms of chlorophylls in purple photosynthetic bacteria are bacteriochlorophyll a (BChl a) and bacteriochlorophyll b (BChl b). The difference between Chl a and Chl b lies in the substitution of methyl group by an aldehyde group in ring b of the porphyrin ring of Chl b (Fig. 5.3). The difference between Chl a and BChl a is the presence of double bond in ring b of the porphyrin ring of Chl a, while it is saturated in case of BChl a. This lack of alternating double and single bond in the ring b of porphyrin ring of BChl a causes a significant difference in the absorption of light by BChl a as compared to Chl a and b. Phytol tail is identical in all forms of chlorophylls. In angiosperms chlorophyll does not accumulate in dark. However, in gymnosperms and some algae, it can be synthesized in dark. Chl b is synthesized from Chl a through the action of an oxygenase enzyme that converts methyl group present in ring b to formyl side group. These small changes in chemical structure of the chlorophyll molecules and non-covalent interactions with membrane proteins in thylakoids significantly alter the absorption properties of different chlorophyll species. Pigments are also named after the wavelength of their absorption maximum; e.g., chlorophyll $a_{700}$ refers to chlorophyll a molecules which has absorption maximum at 700 nm. Chl c is a member of chlorophyll family largely associated with marine photosynthetic organisms, especially golden-brown eukaryotic algae. They are associated with Chl a and carotenoids to harvest light energy to perform photosynthesis in these algae. Chl c differs from other chlorophylls by having **Mg-phytoporphyrins** (ring d is unsaturated) instead of Mg-phytochlorins. Chl c also possesses trans-acrylic (propionic) acid at C-17 region instead of propionic acid side chains present in chlorophylls a and b. However, chlorophyll c also differs from other forms of chlorophylls in not having phytol tail. Chl c absorbs moderately in red

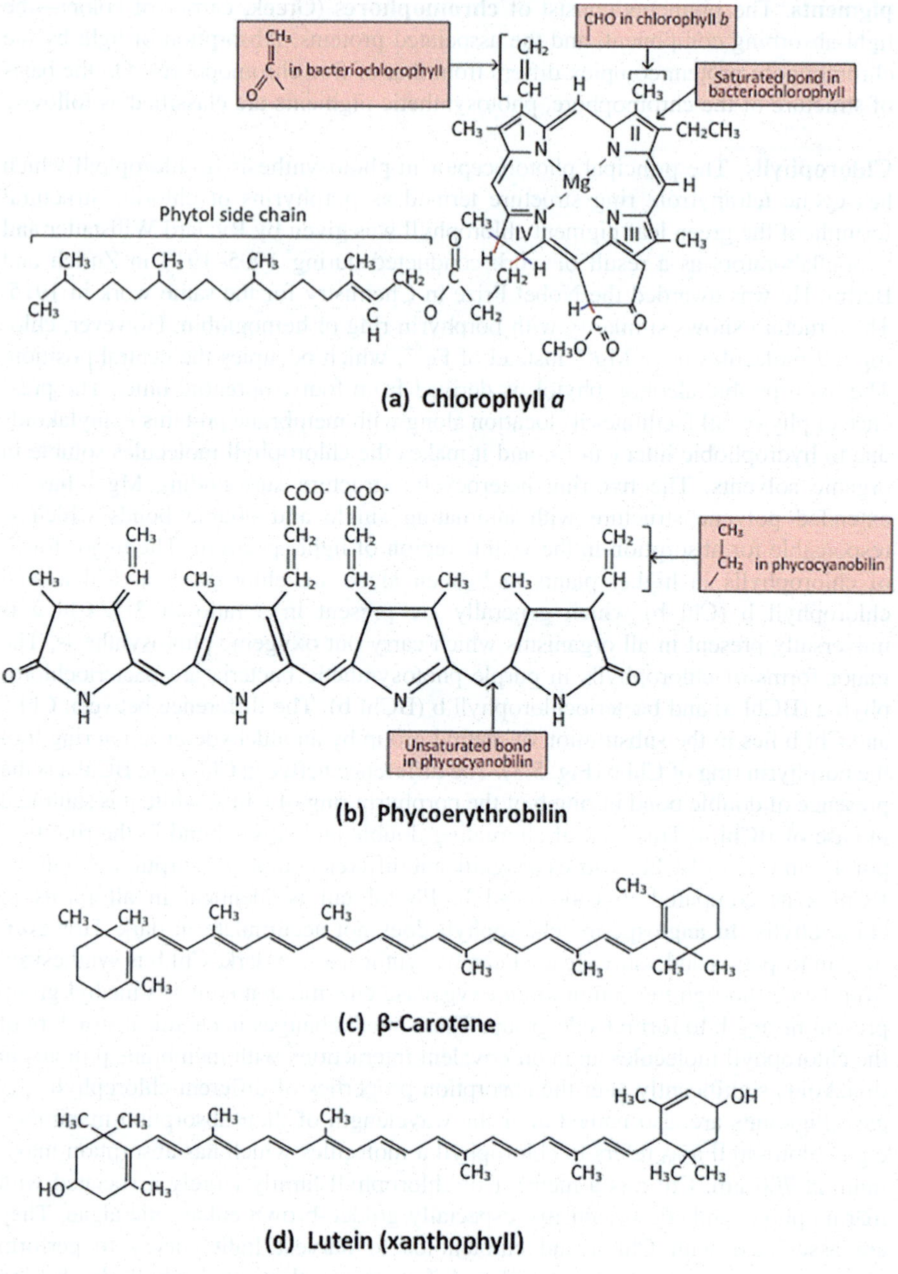

**Fig. 5.3** Structures of chlorophylls, phycobilins, and carotenoid

region, i.e., around 620 nm, but approximately tenfold more strongly in 400–450 nm. Chl d has formyl group present in place of vinyl group found in Chl a. It shows absorption peak in infrared region of light spectrum (700 nm instead of 665 nm). It is present in cyanobacteria which grow as epiphytes under the fronds of red algae.

**Carotenoids** Another group of photosynthetic pigments present along with chlorophylls are carotenoids. Carotenoids are linear polyenes which can be yellow, purple, or red. These absorb blue and green light and serve both as antenna pigments as well as photoprotective agents. Carotenoids include **carotenes** (it includes β-carotene and lutein) and **xanthophylls**. These are tetraterpenes consisting of eight isoprene units ($C_{40}H_{64}$). They comprise conjugated double bond between carbon and hydrogen, while xanthophylls additionally contain one oxygen atom in each of their terminal rings. These are primarily responsible for the orange-yellow coloration of senescent leaves of the plants. Carotenoids serve as accessary pigments and have a secondary role in photosynthesis. They act as antenna pigments which absorb light between 400 and 500 nm and transfer it to chlorophyll molecules. Fucoxanthin, a type of xanthophyll, is particularly efficient in harvesting blue and green light in brown seaweeds. These have an additional function to give structural stability to the assembly of light-harvesting complexes (LHCs). Xanthophylls also protect the photosynthetic apparatus from photooxidative damage (**Xanthophyll cycle,** Fig. 5.4). In the presence of light and oxygen, mutants of carotenoid biosynthesis result in generation of reactive oxygen species (ROS), which are responsible for causing damage to the photosynthetic apparatus.

**Phycobilins** Phycobilins are the chromophores consisting of linear tetrapyrroles derived from the chlorophyll biosynthetic pathway. These are group of accessory pigments which are found in algae. These are water soluble since the phytol tail is absent and structurally resemble bile pigments. These are present as complex protein-containing structures called **phycobilisomes**. Phycobilins are covalently linked via thioester bond to specific proteins which are known as phycobiliproteins.

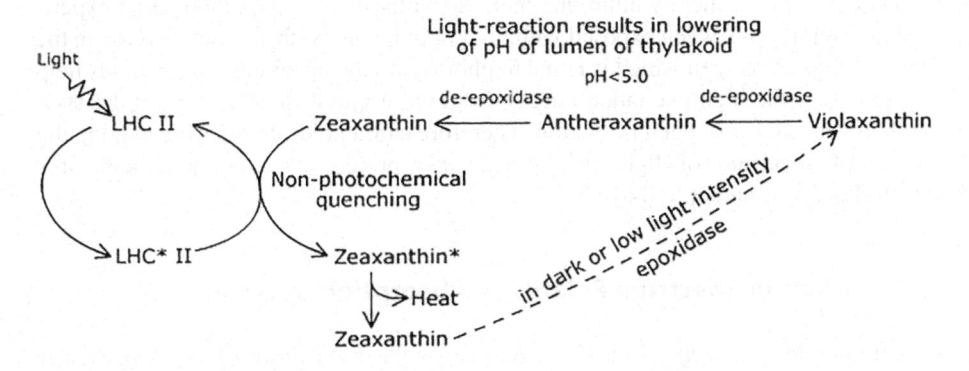

**Fig. 5.4** Xanthophyll cycle

The cysteine residue in the protein and vinyl side chain of the phycobilins forms this bond. Basic structure of phycobiliproteins is a heterodimer (α- and β-proteins). Each of the protein subunit contains one to four phycobilins as the chromophores. The prefix, *phyco*, designates algal origin. Three of the phycobilins, i.e., phycoerythrin, phycocyanin, and allophycocyanin, serve as photosynthetic pigments in algae, while the fourth phytochromobilin is an important photoreceptor (phytochrome) in plant growth and development. The chromophores of phycoerythrin and phycocyanin are known as phycoerythrobilin and phycocyanobilin. The presence of phycobilins enables red algae and cyanobacteria to absorb light and carry out photosynthesis even in dim light. Cyanobacteria and red algae are able to survive in deep water because of the presence of phycobilisomes which absorb green light. Light reaching to the bottom of the ocean is rich in green light because it is not absorbed by the chlorophylls of green algae **(green window)** which are growing in the upper regions. Algae growing in deeper water are able to mainly get dim green light. Red algae appear black in daylight because of absorption of almost all the wavelengths of visible light spectrum. All oxygen-evolving photosynthetic organisms contain Chl a. These organisms also contain other forms of chlorophyll, like chlorophyll b (higher plants), chlorophyll c (diatoms), or chlorophyll d (red algae). However, most prokaryotic cyanobacteria usually contain only Chl a. In addition, photosynthetic organisms contain carotenoids. Some of them contain phycobilins as well (Table 5.1). Therefore, the range of wavelengths that can be absorbed by the organisms is broadened by the presence of accessory pigments. This results in more effective utilization of visible light energy than could be achieved with only one pigment. In organisms which are present in submerged aquatic niches where penetration of red light is limited, a variety of light-absorbing pigments makes it possible for maximum absorption of light.

High light intensities allow plants to absorb more light energy than that which can actually be used for photosynthesis. This leads to excessive excitation of chlorophyll molecules resulting in generation of more of "triplet state" of chlorophyll and singlet state of oxygen. High levels of singlet oxygen can decrease the efficiency of photosynthesis by a process called **photoinhibition**. In plants which have lower carotenoid level (either by inhibiting their biosynthesis or by mutation under experimental conditions), when exposed with high light intensity, there is an increase in the level of singlet oxygen which is lethal to photosynthetic apparatus. Carotenoids help in accepting the high excitation energy from chlorophyll molecules in triplet state and prevent singlet oxygen formation. Therefore carotenoids, besides broadening the range of spectrum of light absorption, also protect photosynthetic apparatus photodamage (xanthophyll cycle).

### 5.1.4 Action Spectrum Relates to Absorption Spectra

A photobiological response is the outcome of the light absorbed by a particular photoreceptor. In order to find out which photoreceptor is responsible for carrying

## 5.1 General Concepts

**Table 5.1** Photosynthetic pigments

| Name of the pigment | Unique features of the structure | Solubility | Absorption maxima in their respective solvents (nm) | Distribution in oxygenic photosynthetic organisms, examples | Role in photosynthesis |
|---|---|---|---|---|---|
| Chlorophylls | | | | | |
| Chl a | Porphyrin structure with long hydrocarbon phytol tail | Organic solvents, petroleum ether | 435, 670–680 | All organisms carrying out oxygenic photosynthesis, higher plants, green algae, diatoms, brown algae, cyanobacteria | Primary photosynthetic pigment, responsible for carrying out photochemical reaction |
| Chl b | Porphyrin structure with long hydrocarbon phytol tail, presence of -CHO group in place of $CH_3$ in ring B | Methyl alcohol | 480, 650 | Higher plants green algae | Accessory pigment helps in harvesting and funneling light energy to reaction center |
| Chl c | Phytol tail is absent | Soluble in ether, acetone, methanol; insoluble in water and petroleum ether | 445, 625 | Diatoms, brown algae | Accessory pigments of PSII |
| | Presence of formyl group in place of vinyl group found in porphyrin ring of Chl a | Soluble in ether and acetone, very slightly soluble in petroleum ether | 450, 456, 696 | Red algae, in some cyanobacterium discovered lately, *Acaryochloris marina* | Major chlorophyll present in cyanobacterium *Acaryochloris marina* in place of Chl a |
| Carotenoids | Linear molecules with multiple conjugated double bonds, oxygen-free hydrocarbons | Organic solvents | 420–525 | In all photosynthesizing organisms | Accessory pigment helps in harvesting and funneling light energy to reaction center |

(continued)

**Table 5.1** (continued)

| Name of the pigment | Unique features of the structure | Solubility | Absorption maxima in their respective solvents (nm) | Distribution in oxygenic photosynthetic organisms, examples | Role in photosynthesis |
|---|---|---|---|---|---|
| Carotenes | Oxygenated derivatives of carotenes | Organic solvents | 500–650 | In all photosynthesizing organisms | Accessory pigment helps in harvesting and funneling light energy to reaction, also protects photosynthetic apparatus from photooxidation |
| Xanthophyll | Open tetrapyrrole ring chromophore covalently bound to protein | Water soluble | Phycoerythrin $\lambda_{max} = 565$ Phycocyanin, $\lambda_{max} = 620-638$ | Present in red algae and cyanobacteria | Accessory pigment helps in harvesting and funneling light energy to reaction center |

## 5.1 General Concepts

out a particular photobiological response, absorption spectrum of the photoreceptor should match with the action spectrum of that process. **Absorption spectrum** of a pigment is obtained when relative absorbance is plotted as a function of wavelengths. Height of absorption of any wavelength in the given absorption spectrum reflects the probability by which light of that energy level will be absorbed. The absorption spectrum of a pigment is unique and is used for identification of that molecule (Fig. 5.5). An **action spectrum** shows the scale of response of a biological system as a function of wavelength. The action spectrum of photosynthesis is obtained by plotting the rate of oxygen evolution by a plant under different wavelengths of light. Studying action spectrum has been crucial in understanding the photosynthetic light reaction besides establishing the relationship of chlorophyll pigments with photosynthetic process. Theodore W. Englemann gave the first action

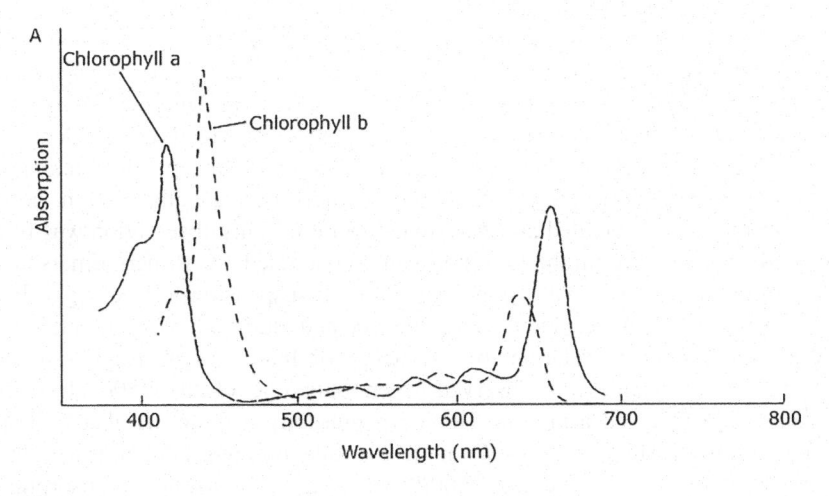

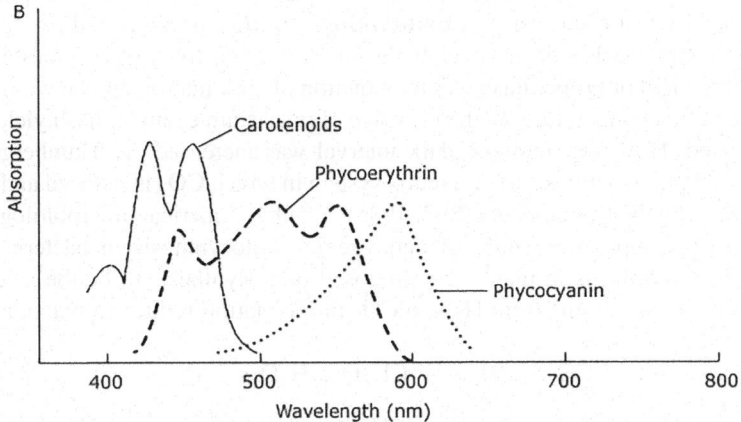

**Fig. 5.5** Absorption spectrum (**a**) of chlorophylls a and b and (**b**) carotenoids and phycobiliproteins

spectrum of photosynthesis in *Cladophora*. He had taken aerophilic bacteria along with the alga, which was exposed to a spectrum of visible light. Maximum accumulation of bacteria occurred in the regions of alga exposed to blue and red light which indicated maximum evolution of oxygen occurred in blue and red region of the light spectrum. This indicated rate of photosynthesis being maximum in blue and red light. Chlorophylls absorb maximally in the blue and red region of light spectrum which indicates the role of chlorophylls in photosynthesis.

## 5.2 Phases of Photosynthesis

In 1905, F.F. Blackman, a British plant physiologist, interpreted the light curve of photosynthesis as an evidence of it being a two-step process. According to Blackman, initial part of the light curve, which shows increase in photosynthesis with increasing light intensity, corresponds to light-limited phase of photosynthesis. He proposed that the intermediates produced during light reaction need to be supplied for further conversion during dark. Horizontal bending of the curve observed due to light saturation of photosynthesis provided the evidence that chemical apparatus of the plant is being overtaxed and is incapable of taking care of the intermediates as quickly as these are produced during light reaction. Initially dark reaction was also called as "Blackman reaction." This observation was further validated by using flashlight in experiments conducted by Robert Emerson and William Arnold in 1932. They exposed *Chlorella* suspension to brief light flashes lasting $10^{-5}$ s. They measured $O_2$ evolution and duration of dark intervals in between the flashes in relation to the energy of light flashes. Light saturation, measured as $O_2$ yield, was observed when only one of the 2500 chlorophyll molecules had received a photon. However, presently it is known that 8 photons are required for the release of one $O_2$ molecule (quantum requirement is 8). Thus, it is the 300 chlorophyll molecules (2500/8) which are responsible for absorption of one photon. They are termed as **photosynthetic unit**. Duration of dark interval following the light flash is determined by the rate-limiting activity of enzymes. It was observed that yield of $O_2$ per flash was the function of dark interval which was found to increase at low temperature. In the presence of potassium cyanide, flash yield was not influenced. However, required dark interval was increased. T. Thunberg proposed in 1923 photosynthesis to be a redox system in which $CO_2$ is reduced and $H_2O$ is oxidized. Cornelis B. van Niel (1897–1985), a Dutch-American microbiologist in 1931, made a comparative study of anoxygenic photosynthesis in bacteria with oxygenic photosynthesis in plants. He proposed photosynthesis to be the result of transfer of hydrogen atoms from $H_2A$ to $CO_2$ (an oxidation-reduction reaction):

$$CO_2 + 2H_2A \rightarrow CH_2O + H_2O + 2A$$

$$CO_2 + 2H_2O \rightarrow CH_2O + H_2O + O_2$$

## 5.2 Phases of Photosynthesis

In green plants $H_2A$ is water, while in purple sulfur bacteria (anoxygenic photosynthesis) it is $H_2S$. In case of bacteria, accumulation of sulfur reinforced the reduction-oxidation concept of photosynthesis and photolysis of water. This observation also gave evidence in support of $H_2O$ being the source of released $O_2$ rather than $CO_2$. Using water ($H_2^{18}O$) labeled with heavy isotope of oxygen ($^{18}O$), Sam Ruben also demonstrated in 1941 that oxygen released during photosynthesis originated from water. Role of photochemical light reaction has been developed from many experiments. Pioneer among them was a study in 1939 by Robert Hill at the University of Cambridge. He made the crucial observation that isolated chloroplasts (when illuminated) could promote reduction of artificial electron acceptors. Ferric salts ($Fe^{3+}$), used in the experiment, were reduced to ferrous form ($Fe^{2+}$). The experiment was conducted in the absence of $CO_2$.

$$4Fe^{3+} + 2H_2O \xrightarrow[\text{Light energy}]{\text{Isolated chloroplasts;}} 4Fe^{2+} + 4H^+ + O_2$$

Robert Hill observed that a variety of compounds can act as electron acceptors. This reduction of artificial electron acceptors and release of oxygen by isolated chloroplasts in presence of light and absence of $CO_2$ is called **Hill's reaction**. The reaction has been demonstrated in the laboratory using different oxidants such as the dye, DCPIP (2,6-dichlorophenolindophenol), which acts as artificial electron acceptors. The dye changes from its blue color to a colorless state on being reduced.

$$H_2O \rightarrow 2H^+ + \tfrac{1}{2}O_2 + 2e^-$$

$$\underset{\text{Blue}}{\text{DCPIP (oxidized)} + 2e^- + 2H^+} \longrightarrow \underset{\text{Colorless}}{\text{DCPIPH}_2 \text{ (reduced)}}$$

This demonstrates that the photosynthetic systems can oxidize water to $O_2$ without any involvement of $CO_2$. Dye reduction could be measured spectrophotometrically at 620 nm. Hill's landmark experiments lead to the following findings: (i) evolution of oxygen in photosynthesis is independent of reduction of carbon dioxide, (ii) released oxygen originates from water and not from carbon dioxide since no $CO_2$ was used in Hill's experiment, and (iii) fragments of isolated chloroplasts are capable of performing partial reactions of photosynthesis and some intermediates are reduced prior to reduction of $CO_2$. The above observations gave clear indications that light and dark reactions are separate processes in photosynthesis. It was observed that spinach grana had the ability to reduce NADP under the influence of light. In 1951, three scientists, Wolf Vishniac, S. Ochoa, and Dan Arnon, independently reported the role of pyridine nucleotide ($NADP^+$) then called TPN as the natural electron acceptor in vivo for the light reactions in photosynthesis yielding NADPH. An overview of photosynthesis is given in Fig. 5.6.

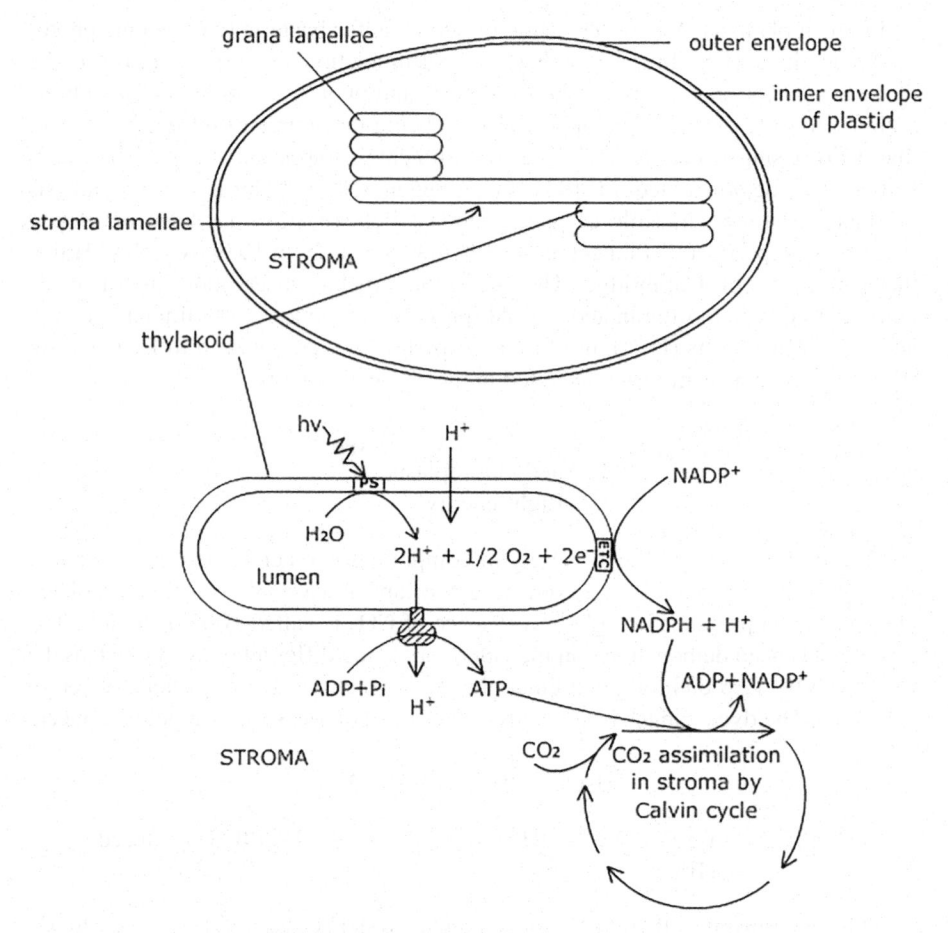

**Fig. 5.6** Overview of two steps of photosynthesis, light reaction which occurs in thylakoids and $CO_2$ assimilation which occurs in the stroma of chloroplasts

## 5.3 Light Reactions in Photosynthesis

In 1943, Robert Emerson and Charlton Lewis explained the action spectrum for photosynthesis in the visible region of the light spectrum, while performing experiment with the green alga *Chlorella pyrenoidosa*. It was visualized that if light is absorbed by chlorophyll molecules, its energy should be utilized for $O_2$ evolution. Since absorption of 8 photons is required for release of one oxygen, they assumed that quantum yield of 0.11 should be fairly constant for light absorbed by the chlorophyll molecule. (Theoretically quantum requirement was 8, but practically the value was calculated to be 10). They reported that value of 0.1 (quantum yield) was remarkably constant over most of the spectrum. This indicates that any photon absorbed by chlorophyll is more or less equally effective in driving photosynthesis.

## 5.3 Light Reactions in Photosynthesis

**Fig. 5.7** Graph showing effect of visible light on quantum yield of photosynthesis—"red drop phenomenon." There is sudden decline in quantum yield of oxygen evolution in plants irradiated with light of longer wavelength greater than 680 nm, which is still absorbed by chlorophylls, as is depicted in their absorption spectrum

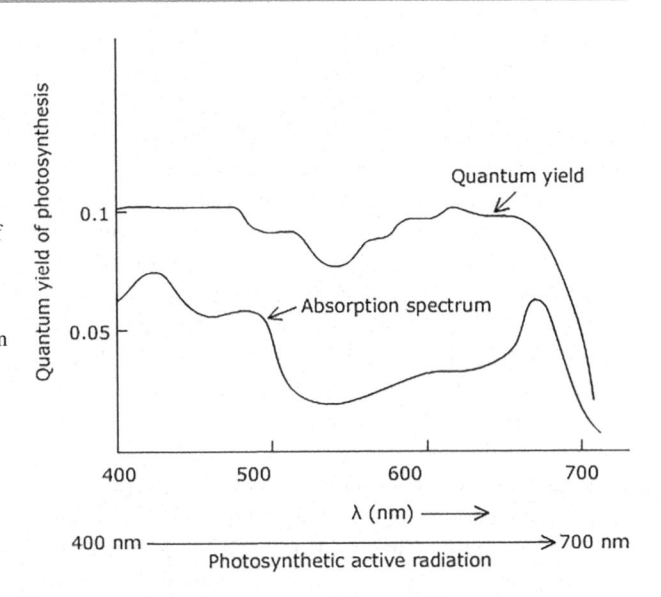

However, a sudden drop in the quantum yield at wavelengths greater than 680 nm was observed, even though chlorophyll still absorbed in that range. This puzzling drop in quantum yield in the red region of the spectrum is called the **red drop phenomenon** (Fig. 5.7). This red drop effect was a strange observation since chlorophylls show appreciable absorbance at 700 nm when quantum yield declines. It showed that the energy was just not being used as efficiently above 680 nm. Interestingly, it was observed by Emerson in 1960 that simultaneous illumination with lower wavelengths of red light (650–680 nm) along with far-red (700–720 nm) light produced a marked two to three times increase in rate of photosynthesis as compared to when both the wavelengths were given separately and value of rate of photosynthesis was added (rate of photosynthesis was measured as the rate of $O_2$ evolution). Therefore, it was predicted that action of both wavelengths (680 nm and 700 nm) must be required simultaneously for photosynthesis to proceed with maximum efficiency. This phenomenon of enhancement of photosynthetic efficiency under simultaneous irradiation was termed as the **Emerson enhancement effect** (E), which can be expressed as the ratio of rate of oxygen evolution ($\Delta O_2$) in presence of far-red light as the supplementary beam and the same in absence of it.

$$E = \frac{\Delta O_2 \text{ (in combined beam)} - \Delta O_2 \text{ (short beam alone)}}{\Delta O_2 \text{ (long wave beam alone)}}$$

The enhancement effect suggested that photosynthesis must involve two photochemical events or systems, one driven by short wavelengths of light ($\leq$680 nm) and another driven by long wavelengths of light (>680 nm). For optimal photosynthesis both wavelength ranges must be utilized by the photosynthetic apparatus simultaneously or in rapid succession. Initially, observations were made by Louis Duysens in *Rhodospirillum rubrum* in 1952, and later on Bessel Kok made similar observations while working with chloroplasts that bleaching of chlorophyll by longer wavelengths or shorter wavelengths occurred which was reversed when these were placed in dark. Bleaching could be attributed to primary reaction of photosynthesis, and the concept of existence of two **pigment systems I and II** was held responsible for catalyzing two light reactions. Light reaction I refers to reduction of $NADP^+$, while light reaction II refers to photolysis of water. Pigment system I was found to be responsible for light reaction I and pigment system II was for light reaction II. These were later on called as photosystem I and II (PSI and PSII), respectively. Photosystem I and Photosystem II were so called according to the order of their discovery. In 1960, Robin Hill and Fay Bendall proposed role of cytochrome b and f as the intermediates in the electron transfer. They demonstrated coupling of the two light reactions involving two photosystems in a linear electron transfer chain which formed the basis of Z-scheme. The excited PSI was oxidized when exposed to long wavelengths (700 nm), because an electron was lost to an electron acceptor. The oxidized PSI replenished its electron obtaining it from cytochrome resulting in their oxidation. PSII gets oxidized on receiving shorter wavelength (680 nm) and is responsible for reducing cytochromes. In this way cytochrome mediates electron transport and the two photosystems operated in a linear manner. This observation was the basis of Z-scheme. The components of electron transport chain were identified because of changes in their redox potentials. When these components were placed according their redox potential, the arrangement appeared in the shape of Z. Consequently, this scheme of electron transport in the light reaction of photosynthesis was called **Z-scheme** (Fig. 5.8).

## 5.3.1 Organization of Photosynthetic Apparatus into Photosystems

The structure and composition of photosynthetic apparatus are responsible for its functional characteristics. In eukaryotic photoautotrophs, photosynthesis takes place in the subcellular organelles known as chloroplasts. Its extensive system of internal membranes, known as thylakoids, contains all the chlorophyll molecules where the light reactions of photosynthesis take place. A variety of proteins essential to photosynthesis are embedded in the thylakoid membranes. These integral membrane proteins contain a large portion of hydrophobic amino acids which are, therefore, much more stable in a nonaqueous hydrophobic region of the membrane. For maximum optimization of energy transfer in antenna complexes and for electron transfer in the reaction centers, chlorophylls and light-harvesting pigments in the thylakoid membranes are always associated with proteins through non-covalent but highly specific bindings, forming pigment-protein complexes within the thylakoid

## 5.3 Light Reactions in Photosynthesis

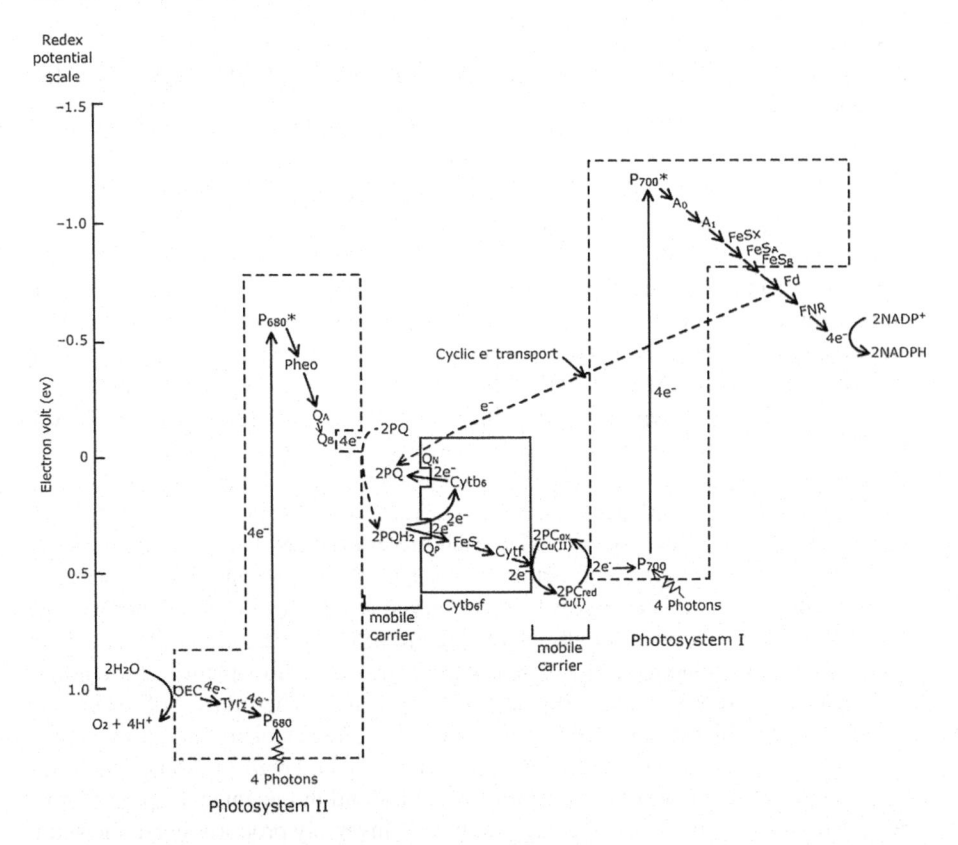

**Fig. 5.8** Detailed Z-scheme consisting of photosystem II, photosystem I, cytochrome $b_6f$, and two mobile carriers plastoquinone and plastocyanin for $O_2$ evolution. Out of the four electrons removed from two molecules of water, two electrons are passed to plastocyanin (PC). Electrons are replenished in oxidized $P_{700}$ from reduced PC. In the absence of oxidized $NADP^+$, electrons are passed from reduced ferredoxin to PQ which are cycled back through $P_{700}$ (cyclic electron transport). *OEC* oxygen evolving complex

membranes. In 1960, Govindjee and Rabinowitch suggested two spectroscopically different forms of Chl a in vivo, which had different photochemical functions, one absorbing shorter wavelength of light, i.e., 680 nm, while the other form of Chl a absorbs longer wavelength of light, i.e., 700 nm. These two forms are primarily responsible for carrying out the two light-driven photochemical reactions. These two pigment systems are now known as photosystems II and photosystem I, respectively.

Both photosystems contain a reaction center where the photosynthetic electron transfer begins with the removal of an electron. As a result, the electron donor pigment is oxidized resulting in its positive charge (Chl $a^+$). The primary electron acceptor (A) is reduced on receiving the electron thus carries a negative charge ($A^-$).

This process is the charge separation event and is known as photochemical reaction. The rest of the components of the reaction center are involved in stabilizing this charge separation.

$$\text{Chl a} \xrightarrow{hv} \text{Chl a}^+ + e^-$$

$$A + e^- \rightarrow A^-$$

Besides primary pigment, which is a pair of chlorophyll a molecules, other components of the electron transport are associated with specific proteins in the photosystems. The reaction center of plants shares a conserved core structure with that of purple photosynthetic bacteria. The second component of photosystems is an array of antenna pigments forming **light-harvesting complex (LHC)**. LHCs associated with photosystem II and photosystem I are called LHCII and LHCI, respectively. Their proteins are encoded by nuclear genes belonging to LHCA and LHCB gene families. The electrophoretic analysis shows a spectrum of individual chlorophyll-protein complexes in which each protein retains its natural chlorophyll array. These antenna pigments function to absorb light energy and transfer it to the reaction center where this energy is then conserved as chemical form of energy. The antenna pigments involved in light harvesting are of two types, peripheral and core antenna pigments, which are associated with proteins. In most plants, 250 chlorophyll molecules are associated with each reaction center. The transfer of excitation energy from one pigment molecule to another (by a mechanism known as Foster energy transfer or resonance) does not require emission and reabsorption of photons. The proximity of the donor and acceptor molecules within the antennae pigments is critical because the efficiency of energy transfer is inversely proportional to the sixth power of the distance separating the two molecules. For two pigments separated by approximately 10 Å, an energy transfer time of less than 1 picosecond has been estimated. The relative orientation of the pigments in LHC is also significant, and the absorbance spectra of a pigment must overlap with the fluorescence spectrum of another one for efficient energy transfer. The sequence of pigments within the antenna molecules which funnel the absorbed energy toward the reaction center has an absorption maxima that is progressively shifted toward red wavelength. This red shift in the absorption maxima means that the energy of the excited state is somewhat lower nearer the reaction center than in the more peripheral portions of the antenna system. Because of this arrangement, when excitation energy is transferred from one chlorophyll molecule to another, the difference in energy between the two excited chlorophylls is lost to the environment as heat. Chlorophyll a is found in all reaction center complexes as well as in the antenna, whereas chlorophyll b is found only in the antenna complexes and thus have role in light harvesting. Approximately 15 different chlorophyll-binding proteins (CP) have been identified. Some are associated with PSI and others with PSII. They are all encoded in nucleus and, therefore, must be imported into the chloroplasts before binding with chlorophylls and associating with their proper photosystems. In addition to the chlorophyll pigments, carotenoids are also commonly found in the antenna complexes. The antenna complexes in most plants typically have a ratio of 0.5 for carotenoid/total chloroplast pigments. Association with specific proteins causes a shift in peak absorption wavelength by chloroplasts toward the red (lower energy) wavelength because the reaction center complex absorbs longer wavelength than the antenna. The reaction center chlorophyll (Chl a) acts as an energy trap, promoting transfer of energy from the antenna toward the reaction center complex. Photosystems have their own complement of chlorophyll-binding antenna proteins. Properties of these pigments protein complexes are optimal for the particular reaction center chlorophyll in the photosystem. Therefore, the antenna chlorophylls of PSI with a reaction center Chl a absorbing maximally at 700 nm would be expected to absorb wavelengths longer than the antenna of PSII, which has a reaction center Chl a that absorbs maximally at 680 nm ($P_{680}$).

### 5.3.2 Organization of Chlorophylls and Other Pigments in LHCII and LHCI

The major pigment-binding protein of LHCII represents about half of the total protein in the thylakoid membranes. It is the second most abundant protein after Rubisco. It is a trimeric complex. The three polypeptides are encoded by nuclear family of genes Lhcb1, Lhcb2, and Lhcb3. Each monomeric unit consists of 230–250 amino acid residues, has a molecular weight of 24–29 kDa, and consists of three transmembrane helices. It is bound non-covalently to eight Chl a, six Chl b, and four xanthophyll molecules. Two lutein molecules serve as scaffolding, while a neoxanthin and violaxanthin are bound at the periphery. Besides these proteins of LHCII, other polypeptides Lhcb4, Lhcb5, and Lhcb6 are also present which facilitate forming bridges between PSII and major LHCII components. Core proteins of LHCII include CP43 and CP47 where CP refers to chlorophyll proteins and number refers to their molecular weight (Fig. 5.9). Small proteins such as cytochrome$_{b559}$ encoded by plastid gene psbE and psbF are required for correct assembly of PSII. The detergent fractionation of thylakoid membranes yields intact PSI and PSII complexes with full array of pigment molecules. Proteins associated with PSI are called LHCI proteins. The structure of the LHCI proteins is generally similar to that of the LHCII proteins. All these proteins have a significant sequence similarity and, therefore, reflect common ancestral proteins. The LHCI consists of four peripheral polypeptide subunits Lhca1–4, which belong to a family of 25 kDa proteins. These are arranged in two dimers. Each Lhca protein is associated with 13 chlorophylls and 2–3 carotenoid molecules. The amount of Lhca proteins is variable depending on light intensity and nutrient availability.

### 5.3.3 Photochemical Reaction Centers

There are two reaction centers identified on the basis of the primary electron donor, Chl a molecule. These reaction centers are the components of the two photosystems

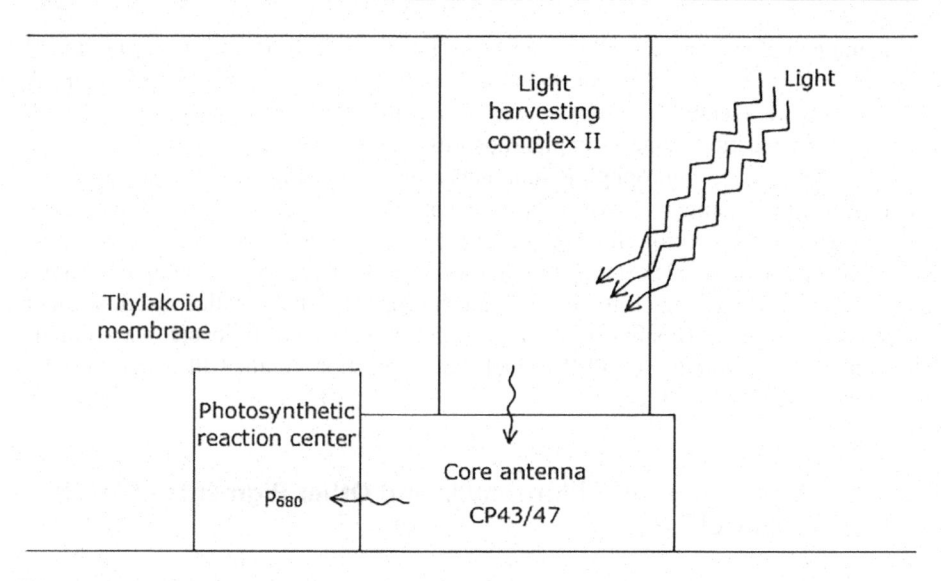

**Fig. 5.9** Basic scheme of light-harvesting complex

in plants. On the basis of electron acceptor, there are two types of reaction centers. Reaction center which reduces a quinone is a type II reaction center, while type I ultimately reduces Fe-S cluster. The organisms which carry out oxygenic photosynthesis contain both type II and type I reaction centers. Unlike plants there is only one reaction center in the bacteria which carry anoxygenic photosynthesis. Green sulfur bacteria contain type I reaction center, while purple bacteria have only type II. Electron carriers of type II reaction center in plants resemble the type present in purple bacteria. There is similarity of electron carriers of type I in the reaction center of oxygenic organisms with those present in green sulfur bacteria. PSII is present in all the organisms which carry out oxygenic photosynthesis. It is also called as **water-plastoquinone oxidoreductase.** Reaction center of PSII is associated with the oxygen-evolving complex (OEC) which is present toward the lumen side of thylakoids. The two proteins associated with reaction center of PSII are D1 and D2 (Fig. 5.10). D1 is a hydrophobic 32 kDa protein which is encoded by the plastid gene. Another protein which is 34 kDa protein is also encoded in the plastid genome. Since D1 is exposed to extremely oxidizing environment created by excited $P_{680}$, it leads to high light-dependent turnover of the protein. On the contrary D2 represents inactive branch of the reaction center. There is a similarity of these proteins with L and M proteins of reaction center complex present in *Rhodopseudomonas viridis*. Hartmut Michel, Johann Deisenhofer, and Robert Huber carried out the X-ray crystallography studies of the reaction center in these bacteria and elucidated for the first time the three-dimensional structure of a membrane protein. These three scientists were awarded Noble Prize in chemistry in 1988 for their work. Chl a dimer is associated with the heterodimeric proteins in plants (D1 and D2) and is called $P_{680}$ because of its characteristic absorption maxima. Besides dimeric $P_{680}$, the other

## 5.3 Light Reactions in Photosynthesis

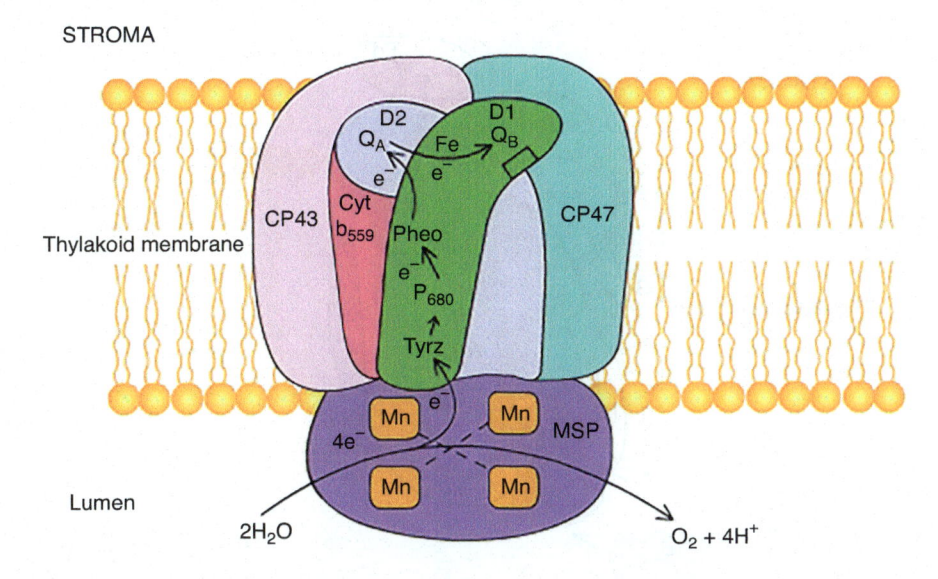

**Fig. 5.10** Molecular structure of photosystem II, MSP (manganese-stabilizing protein), CP43 and CP47 are part of inner core of light-harvesting complex

components of PSII which are involved in charge separation and electron transport are two pheophytin molecules (chlorophyll lacking $Mg^{2+}$ ions), a non-heme iron, and two plastoquinones designated as $Q_A$ and $Q_B$ which act as the terminal electron acceptor. The site for binding $Q_A$ is present in D2 and for $Q_B$ is present in D1. $P_{680}$, pheophytin, and plastoquinone are the prosthetic groups of the reaction center proteins. The reaction center in **PSI** (also known as **plastocyanin-ferredoxin oxidoreductase**) contains dimeric $P_{700}$ Chl a molecule which is associated with the proteins PsaA and PsaB (Fig. 5.11). The reaction center chlorophyll $P_{700}$ absorbs a photon and subsequently transfers an electron through a series of acceptors which includes an acceptor Chl a molecule $A_0$, a phylloquinone $A_1$, and the proteins having iron-sulfur (Fe-S) center. Comparison between PSII and PSI is given in Table 5.2. The reaction center subunits of PSI are encoded in plastid genome referred to as *psa*, while the genes encoding the components of PSII are referred to as *psb*. D1, a hydrophobic 32 kDa protein, is encoded by *psb*A in plastid genome, while D2, a 34 kDa protein, is encoded by the gene *psb*B in plastid genome.

### 5.3.4 Cytochrome b₆f (Plastoquinol-Plastocyanin Oxidoreductase)

Another component which mediates transfer of electrons during light reaction between PSII and PSI is cytochrome b₆f which functions as plastoquinol-plastocyanin oxidoreductase. It is similar to mitochondrial cytochrome bc₁. Both complexes are dimers with a molecular mass of 220 kDa, each monomeric unit

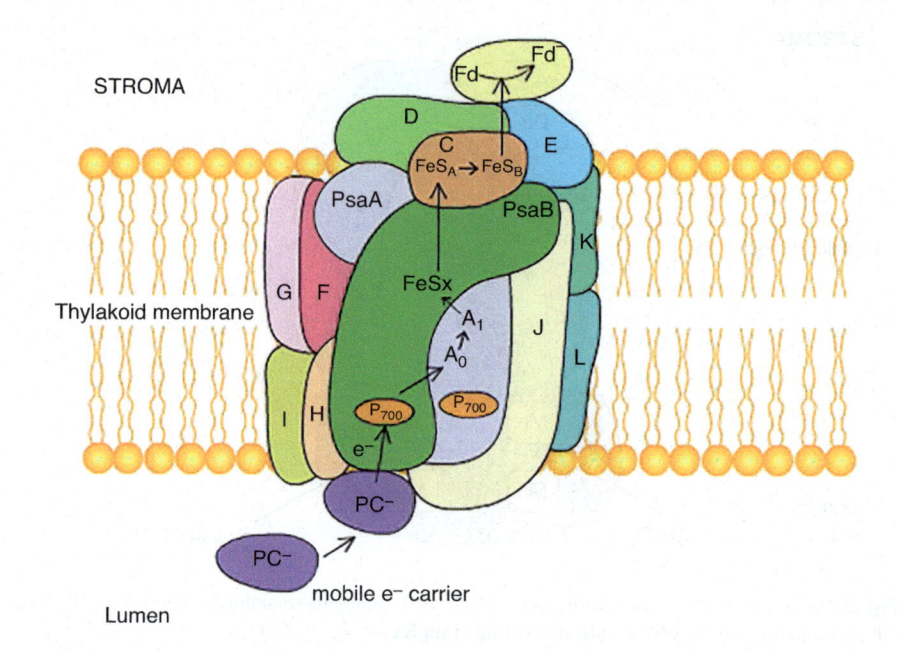

**Fig. 5.11** Molecular structure of photosystem I, psa A and psa B, are major proteins associated with PSI. Proteins labeled from C to L are minor proteins. $A_0$ is a chlorophyll (a molecule which is a primary electron acceptor from $P_{700}$); $A_1$ is a phylloquinone; Fe-$S_x$, Fe-$S_A$, and Fe-$S_B$ are Fe-S centers; Fd is a soluble iron-sulfur protein ferredoxin

**Table 5.2** Comparison of PSII and PSI

| Characteristics | PSII | PSI |
| --- | --- | --- |
| Chlorophyll a present in reaction center | $P_{680}$ | $P_{700}$ |
| Location in thylakoids | Appressed regions of thylakoids | Non-appressed region of thylakoid |
| Primary electron acceptor ($A_o$) | Pheophytin-a | Chlorophyll a |
| $A_1$ | Plastoquinone ($Q_A$) | Phylloquinone |
| $A_2$ | Plastoquinone ($Q_B$) | $F_X$ (Fe-S center) |
| Primary function | Photolysis of water | NADP$^+$ reduction |

containing a conserved core of four electron carriers which contains two **cytochromes** (the proteins which have heme as the prosthetic group) belonging to b type cytochromes (cyt $b_6$ in chloroplasts or cyt b in mitochondria) and c type of cytochromes (cyt f in chloroplasts or cyt $c_1$ in mitochondria). These are linked by a protein having **Rieske Fe-S center** which constitutes the third redox component of the complex. Fourth component of the complex is a small 17 kDa peripheral protein which has got two binding sites for quinone, $Q_p$, site for binding with quinol (QH$_2$), which is present toward the lumen of the thylakoid, while another site $Q_n$ is present toward the stroma side; significance of these two sites will be discussed later on. $Q_p$ and $Q_n$ refer to quinone-binding sites toward the positive and negative sides of the thylakoids, respectively.

### 5.3.5 Two Mobile Electron Carriers

There are two mobile electron carriers—plastoquinone and plastocyanin—which facilitate electron transfer between PSII and cytochrome $b_6f$ and cytochrome $b_6f$ and PSI, respectively. Plastoquinone facilitates two-electron transfer from PSII and is highly lipophilic in reduced plastoquinone form, which enables its lateral diffusion within the lipid bilayer of thylakoid. Plastocyanin is a small copper-containing protein (11 kDa) which is located in the aqueous phase of thylakoid lumen. Almost 50% of the copper is utilized for plastocyanin synthesis in a photosynthetic cell. Unlike plastoquinone, plastocyanin facilitates the transfer of single electron at a time. The genes for plastocyanin are encoded in nuclear genome.

### 5.3.6 Electron Transport Pathway During Light Reaction of Photosynthesis

The primary step in light reaction involves transfer of electrons (excited by light) from the reaction center $P_{680}$ or $P_{700}$ (in PSII or PSI, respectively) into an electron transport chain. The ultimate source of the electrons is the water molecule, which is photolyzed to release electrons, protons, and $O_2$, while the final destination is $NADP^+$, which is thereby reduced to NADPH. Energy derived during electron transport is coupled to transfer of protons from stroma into the lumen, resulting in pH gradient which drives ATP synthesis similar to ATP synthesis in mitochondria. Products of the light reactions, ATP and NADPH, are utilized subsequently during $CO_2$ assimilation.

### 5.3.7 Photosystem II (Splitting of Water)

As a result of electron transport during light reaction, an organized series of oxidation-reduction reactions occur. On losing an electron, a molecule is oxidized and attains positive charge (+). On the contrary, molecules on gaining electrons are reduced and become negatively charged (−). In a series of electron transport chain, electrons move from molecules having less positive redox potential (act as reductant) to molecules having more positive redox potential (act as oxidant) spontaneously. LHCII acts like an antenna, which on receiving the energy of photons, transfers it to chlorophylls bound to CP43/47—the core proteins of LHCII. Similar to cyanobacteria, X-ray crystallographic studies of PSII in plants have shown it to be a heterodimeric structure. However, its molecular structure in plants has not been determined. Reaction center of PSII consists of dimeric $P_{680}$, one on each polypeptide D1 and D2. There are other proteins bound with PSII toward the lumen of the thylakoid which are involved in water oxidation (oxygen-evolving complex, OEC). Various prosthetic groups which are bound with proteins in PSII participate in electron transport. Wavelengths of light absorbed by LHCII are slightly shorter (650–670 nm). These are funneled to the Chl a ($P_{680}$) at the reaction center by fluorescence. The absorbed photons raise the molecule, i.e., $P_{680}$, from the ground state to an excited state. The excited $P_{680}^*$ molecule becomes an excellent reducing agent which is now capable of transferring an electron quickly to the primary electron acceptor, pheophytin "a" (Pheo), which carries lower redox potential than excited $P_{680}^*$. Pheophytin is identical to chlorophylls except that two protons substitute for the centrally bound magnesium ion. The electron is then transferred quickly to plastoquinone. There are quinones which are bound to the two binding sites: $Q_A$ is bound tightly to D2, while $Q_B$ is loosely bound to D1. Pheo$^-$ immediately transfers the electron to $Q_A$ located in D2 to form semiquinone ($PQ^-$), which transfers electron to $Q_B$ which is located in D1. $Q_A^-$ is returned to $Q_A$ form and is ready to receive another electron from Pheo$^-$. Transferring of electron to $Q_A$, which is located in D2 at a site different from Pheo, must be to stop short circuiting since $Q_B$ is located at a different site, i.e., in D1. Unlike $Q_A$, $Q_B$ requires transfer of two electrons from $Q_A$ which occurs in two steps and is fully reduced to $PQ^{2-}$ on receiving two electrons. This transfer of electron from $Q_A$ to $Q_B$ is believed to involve non-heme iron and $HCO_3^-$. After receiving two electrons at $Q_B$ site, $PQ^{2-}$ picks up two protons from stroma and is converted to plastoquinol ($PQH_2$), which is lipid soluble in the lipid bilayer of the thylakoid membrane. After losing electrons $P_{680}$ becomes a strong oxidant $P_{680}^+$ (redox potential $E'^0 = -1.0$ V). It gains electron from the closely placed tyrosine residue (Tyr161) located on the luminal side of D1. On becoming deficient, tyrosine residue collects electron from water. However, first of all electrons are removed from oxygen-evolving complex (OEC), a protein, which is present toward the luminal side of thylakoid. OEC contains a cluster of four oxygen-bridged manganese atoms. Electrons are removed from OEC one at a time. As a result, there is accumulation of positive charge. OEC is always positive (+) in dark and will require three photochemical events for accumulation of four positive charges. This results into the strongest biological oxidant having redox potential +1.2 eV. It was first observed by P. Joliot and coworkers in 1969 and was interpreted by B. Kok and coworkers in 1970 as the S-cycle. The metal cluster in OEC can exist in a series of oxidation states ($S_0$–$S_4$) depending upon number of electrons removed. On removal of four electrons, four positive charges are accumulated on OEC ($S_4$) (Fig. 5.12). It is now capable of oxidizing water by removing electrons, resulting in photolysis of water with simultaneous release of $O_2$ and $4H^+$ in the thylakoid lumen. Electrons are removed one by one, resulting in accumulation of charges on OEC. If electrons are removed directly from water one by one, this would result in formation of free radicals. OEC consists of three extrinsic proteins bound to D1 and D2 proteins of PSII projecting toward the luminal side of thylakoid. Core of OEC consists of four manganese atoms, a calcium atom, and a chlorine atom. It is the manganese atoms of OEC which provide electrons to Tyr$_z$, while calcium and

## 5.3 Light Reactions in Photosynthesis

**Fig. 5.12** The S-state cycle of oxygen-evolving complex (OEC). $P_{680}$ is oxidized by a photon of 680 nm wavelength of light. It derives the electron from OEC, which accumulates positive charge and becomes an oxidant. After four electrons have been removed, OEC becomes a strong oxidant because of the accumulation of four positive charges and is able to remove electrons from water

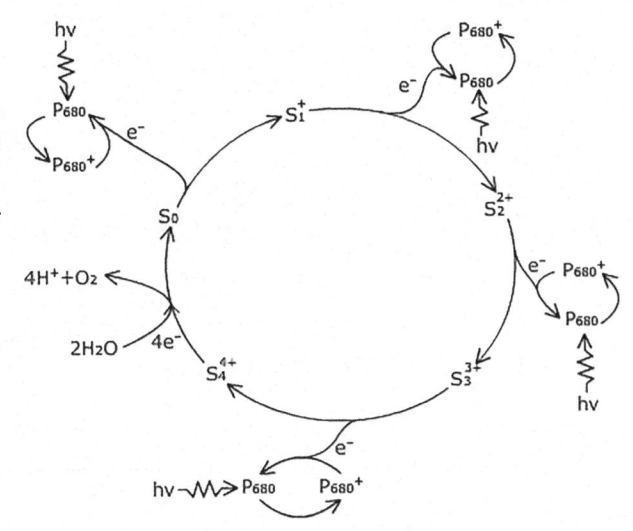

chlorine are thought to facilitate combining of OEC with oxygen atoms of water molecule. Released oxygen diffuses out of the chloroplasts. The four protons produced from photolysis of $2H_2O$ add to pH gradient between the lumen and stroma of the chloroplast. Photolysis of water by thylakoids is summarized as:

$$2H_2O \rightarrow 4H^+ + 4e^- + O_2$$

### 5.3.8 Q-Cycle Results in Pumping of Protons

Plastoquinol ($PQH_2$) is lipophilic and is released in the membrane lipid bilayer from $PQ_B$ which is present toward stroma on D1 of PSII. It interacts with thylakoid-bound cytochrome $b_6f$ complex. The major components of this complex are two cytochromes ($b_6$ and f) and an iron-sulfur protein (ISF). Cytochrome $b_6f$ complex is a dimer having two plastoquinone binding sites facing each other. The sites are called $Q_P$ and $Q_N$ sites toward lumen and stroma, respectively. These are called as $Q_P$ and $Q_N$ since the lumen side of thylakoids is positive because of accumulation of protons, while the side toward the stroma is negative because of proton movement from the stroma to lumen during light reaction (Fig. 5.13). Cytochrome $b_6f$ complex carries out three functions; it regenerates PQ from $PQH_2$, catalyzes transfer of electron to PSI via plastocyanin, and transports protons from the stroma to lumen of the thylakoid. Energy for transport of two electrons received from $PQH_2$ is utilized for transport of 4 $H^+$, which occurs through Q-cycle, thus contributing to proton gradient. Q-cycle occurs in two steps. In the first step, one of the electrons

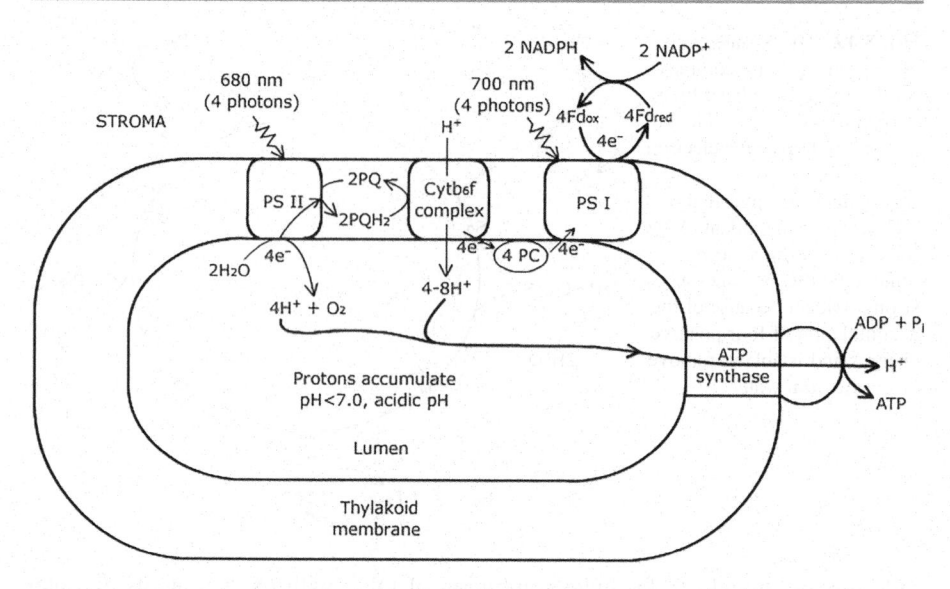

**Fig. 5.13** Figure showing positioning of the photosynthetic electron transport complexes in thylakoids and proton accumulation in lumen

received from PQH$_2$ is transferred via Fe-S protein and Cyt f to a small copper-containing protein, plastocyanin (PC), which is present toward the luminal side. Cu (II) of plastocyanin is reduced to Cu (I) which further is oxidized by PSI. When PQH$_2$ is oxidized, protons (2H$^+$) are released in the lumen of thylakoid. Complex b$_6$f serves two purposes. First, one of the two electrons of PQH$_2$ is transferred to PS I through plastocyanin. At the same time, it pumps two protons from stroma into thylakoid lumen. The order of transfer of one of the electron is mentioned below.

$$\text{Plastoquinol} \rightarrow \text{ISF} \rightarrow \text{Cytochrome f} \rightarrow \text{Plastocyanin}$$

The other electron is cycled back via cyt b$_6$ to plastoquinone bound to Q$_N$ site of the complex. This transfer of electron is facilitated by two hemes of cyto b$_6$ (b$_l$ and b$_h$), which refers to low-potential and high-potential hemes, respectively. PQ, on receiving the electron at Q$_N$ site of the complex, is reduced to semiquinone (PQ$^-$), which, on receiving another electron in second cycle, is further reduced to PQ$^{2-}$. PQ$^{2-}$ picks up two protons from stroma and is reduced to PQH$_2$ followed by similar cycle (Fig. 5.14). Net result is translocation of 4H$^+$ across thylakoid coupled with transport of two electrons to plastocyanin. The cytochrome b$_6$f complex plays a role analogous to that of the cytochrome oxidoreductase (cytbc$_1$) complex in mitochondria. In this process, copper in plastocyanin is first reduced to Cu(I) on receiving electron and then reoxidized to Cu (II) when electron is given to PSI. Reactions of the cyt b$_6$f are rate limiting in the electron transport since quinol oxidation requires 10–20 ms, while PQH$_2$ is formed in 100 μs.

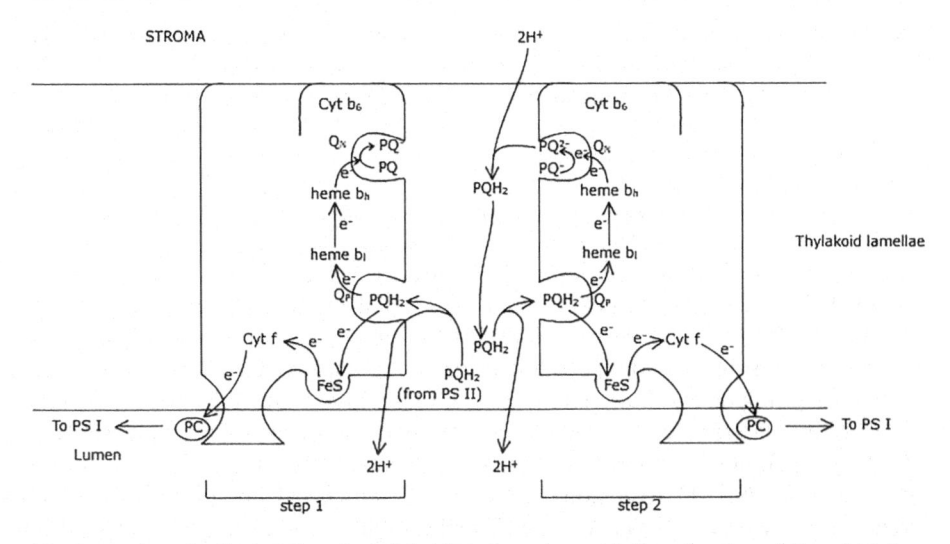

**Fig. 5.14** Q-cycle. Cyt b$_6$f dimer in thylakoid having quinone-binding sites Q$_P$ and Q$_N$, which face each other. PQ$^-$ is a semiquinone, PQ$^{2-}$ is a fully reduced plastoquinone, PQH$^{2-}$ is a plastquinol, heme b$_l$ and b$_h$ are cytochrome b with heme moiety having low and high midpoint potential, respectively

### 5.3.9 Photosystem I (Production of NADPH)

Through X-ray diffraction techniques, the structure of photosystem I complex in plants has been shown to be a supercomplex with monomeric reaction center unlike in cyanobacterium where PSI is trimer of reaction centers. PS I is a multiprotein complex containing at least 11 polypeptide chains. It contains many antenna chlorophylls (LHCI) and reaction center chlorophyll P$_{700}$ which can absorb light of 700 nm. The number of Chl a molecules is double of Chl b. In plants central heterodimer of the reaction center is bound with major electron carriers, P$_{700}$ and accessory pigments, chlorophyll A$_0$ acceptor molecule, phylloquinone (A$_1$), and the bound Fe-S centers F$_A$ and F$_B$. Excitation by a photon, absorbed and transferred by antenna chlorophylls, raises electrons in P$_{700}$ from ground state to an excited state ($-1.5$ eV). Electrons are released from the excited reaction center (P$_{700}$). As a result, P$_{700}$ is oxidized to P$_{700}^+$ (redox potential +0.5 eV). Excited electron then passes through an electron transport chain. It is first accepted by Chl a-like molecule A$_0$, which is reduced to A$_0^-$ on receiving the electron. Electron is transferred from A$_0^-$ to phylloquinone and then is passed through a series of proteins (F$_x$, F$_B$, and F$_A$) which contain iron-sulfur clusters. Finally, the electron is accepted by 2Fe-2S soluble protein ferredoxin (Fd), which is present in the stroma of chloroplast. Electron from ferredoxin is not directly transferred to NADP$^+$. The transfer occurs through the activity of an intermediate enzyme ferredoxin-NADP reductase (FNR). There is complex electrostatic interaction between ferredoxin and FNR. FNR is an FAD-containing enzyme, which is reduced by two single-electron steps. The enzyme FNR catalyzes transfer of electrons to NADP$^+$ reducing it to NADPH after ferredoxin has been reduced by photosystem I. Reduction of NADP$^+$ to NADPH requires transfer of two electrons. However, only one proton is covalently bound, while the other H$^+$ remains free in the medium.

$$2Fd^- \rightarrow 2Fd^+ + 2e^-$$

$$NADP^+ + 2e^- + 2H^+ \rightarrow NADPH + H^+$$

It is ferredoxin, rather than NADP$^+$, which can be considered as the ultimate recipient of electrons during light reaction. Much of the reduced ferredoxin is used to reduce NADP$^+$, but some of it is also used for other reductive reactions. Reduced ferredoxin is a source of low-potential electrons for many reductive processes such as $NO_2^-$ assimilation, sulfur assimilation, and lipid biosynthesis. Daniel Arnon had observed in 1951 for the first time that isolated chloroplasts could reduce NADP$^+$ to NADPH and found out in 1954 that it is coupled with ATP synthesis. In 1962, Arnon discovered that NADP$^+$ reduction requires two electrons and a proton. The NADPH produced by ferredoxin oxidation is released into the stroma where it is utilized in $CO_2$ reduction.

## 5.3.10 Non-cyclic and Cyclic Electron Transport

When both the photosystems PSII and PSI are working, electrons for NADP$^+$ reduction are ultimately replenished from $H_2O$. This movement of electrons, which requires participation of both photosystems, occurs through open chain and is known as non-cyclic electron transport, since it is a one-way transport of electrons. There are inhibitors which interfere electron transport on binding with specific sites of the complexes in the electron transport chain. This includes atrazine and DCMU (3–3-4-dichlorophenyl)-1,1-dimethyl urea) which bind with $Q_B$ site of PSII, blocking electron flow from $Q_A$ to $Q_B$. As a result, non-cyclic electron transport is blocked, and there are no photolysis of water and no NADP$^+$ reduction to NADPH. DCMU is commercially known as diuron. DBMIB (dibromothymoquinone) blocks electron flow to cyt $b_6f$ at the $Q_P$ site of the complex. The herbicide, methyl viologen (commercially known as paraquat), accepts electrons instead of NADP$^+$ resulting in generation of oxygen radicals which is responsible for destroying the photosynthetic apparatus. In the absence of $CO_2$ reduction or if it is slow, NADPH will not be oxidized. In the absence of availability of oxidized NADP$^+$, electrons are transferred to PQ bound to $Q_N$, resulting in cycling back of the electrons through cyt $b_6f$, PC, and PSI. This closed cycle of electron transport is known as cyclic electron transport. Cyclic electron transport is operational when only long wavelength of light (700 nm) is available. There is no photolysis of water, resulting in zero quantum yield in the absence of functional PSII. As a result, red drop phenomenon is observed (Fig. 5.7).

## 5.3.11 ATP Generation During Electron Transport in Light Reaction

ATP synthesis during light reaction of photosynthesis is known as **photophosphorylation** (see Chap. 8). Daniel Arnon observed in 1954 that along with reduction of NADP$^+$ to NADPH, ATP is produced by isolated chloroplasts in light. When exposed to shorter wavelengths, P$_{680}$ is oxidized to P$_{680}^+$ making it a very strong oxidant with redox potential of +1.2 eV, and a weak relatively stable reductant Q$_A^-$, plastosemiquinone, is generated. Q$_A^-$ donates electrons to PSI via Q$_B$, cyt b$_6$f, and PC. When excited by the long wavelengths (700 nm), a strong reductant (reduced Fe-S center) having redox potential of $-0.73$ eV and a weak oxidant (i.e., P$_{700}^+$) with the redox potential of +0.49 eV are produced. Energy released during the flow of electrons downhill from the weak reductant (Q$_A^-$) to the weak oxidant (P$_{700}^+$) is conserved as a gradient of protons across thylakoid membranes with H$^+$ getting accumulated in the lumen. Protons flow back in response to the gradient through ATP synthase resulting in ATP synthesis (Fig. 5.13). Proton gradient between the lumen of thylakoid and stroma is the result of (1) photolysis of water in the lumen, (2) transport of H$^+$ from stroma to the lumen during Q-cycle, (3) and utilization of H$^+$ for reduction of NADP$^+$ to NADPH at the stroma side.

Six protons accumulate in the lumen coupled with transport of two electrons through non-cyclic electron transport. 2H$^+$ are released during photolysis of a water molecule and 4H$^+$ are transported through Q-cycle. Release of an O$_2$ molecule requires non-cyclic electron transport of four electrons. Accumulation of protons makes the lumen more acidic than the stroma. This light-triggered proton gradient results into pH difference of 3–4 units. Unlike oxidative phosphorylation in mitochondria, a pH gradient is the main contributing factor for generation of proton motive force (PMF) in chloroplasts. In chloroplasts, H$^+$ accumulates in the lumen of thylakoid, which is an isolated compartment. In lumen lowering of pH to as low as 4 will not affect any enzyme activity as most of them are present in stroma. On the contrary, in case of mitochondria PMF is mainly contributed by the membrane potential ($\Delta$E). H$^+$ accumulation in the intermembrane space will adversely affect cytosolic enzymes by lowering the pH of the cytosol. In chloroplasts ATP synthase complexes are called **CF$_0$-CF$_1$** which are similar to **F$_0$-F$_1$** complexes of mitochondria. In chloroplasts most of the evidence indicates the stoichiometry of number of protons translocated for synthesis of one ATP to be 4.67 (H$^+$/ATP = 4.67). The pH gradients across the membrane correspond to a $\Delta$G of about 20 kJ.mol$^{-1}$ for the passage of one proton. Photosystems I and II, the cytochrome b$_6$f complex, and ATP synthase (CF$_0$–CF$_1$) are all separate entities embedded in the thylakoid membrane but are not contiguous. The mobile components that link the photosystems and the b$_6$f complex are plastoquinone in the lipid phase of the membrane and plastocyanin in the thylakoid lumen. Therefore, it seems that electrons can be transported over long distances in this system. Such a range of transport is necessary because two photosystems are separated from each other. The entire process of electrons being displaced from PSI due to excitation by light, getting replaced from photosystem II, and PSII obtaining the electrons then from water is called **non-cyclic electron flow**. ATP production during this process is called **non-cyclic photophosphorylation.**

Cyt $b_6f$ complex and $NADP^+$ are the competitors for the electrons from reduced ferredoxin. In the limited availability of $NADP^+$, electrons pass from reduced ferredoxin to the cytochrome $b_6f$ complex via PQ instead of $NADP^+$ and are cycled back to $P_{700}$ via plastocyanin. This results in pumping of protons across thylakoid membrane, creating $H^+$ gradient which leads to ATP generation. This process is called **cyclic photophosphorylation**. No $O_2$ is released since there is no photolysis of water and no $NADP^+$ is reduced. Cyclic electron flow generally serves to generate ATP when availability of oxidized $NADP^+$ is limiting or if plants are irradiated only with long wavelengths (700 nm). Requirement for ATP in the $CO_2$ reduction is large, and only non-cyclic electron flow may not be able to generate sufficient ATP to meet the requirement. Since assimilation of each $CO_2$ molecule requires 2NADPH and 3ATP molecules, requirement of ATP cannot be met only through non-cyclic photophosphorylation. **Cyclic electron transport produces ATP and no NADPH, which helps to maintain the necessary balance between ATP and NADPH production.** In bundle sheath chloroplasts of C4 plants cyclic electron transport meets ATP requirement for $CO_2$ assimilation.

### 5.3.12 Balancing Distribution of the Light Energy in Between the Two Photosystems

Ideally there should be even distribution of energy between the two photosystems for maximum utilization of light energy for photosynthetic process. However, it does not happen since photons are preferentially transferred to pigments requiring the least energy for excitation. Since PSI requires lesser energy, there is a chance of excitation of PSI more in comparison with PSII. One of the ways by which this problem is handled by plants is by spatial separation of the two photosystems. PSI and PSII are not distributed randomly throughout the thylakoid membrane. PSI, along with ATP synthase, is primarily present in the unstacked stroma membranes, whereas PSII is mostly present in the stacked grana membranes. PSI and ATP synthase are present in the outer membrane of the stacked lamellae also and are thus exposed to stroma. This phenomenon is called **lateral heterogeneity**, indicating that the photosystems which participate in transferring electrons from $H_2O$ to $NADP^+$ in the photosynthetic electron transport chain are spatially separated from each other. One to one stoichiometry of the two photosystems is also not required. Most commonly the ratio of PSII to PSI is 1.5:1 which can change depending upon the light conditions in which plants are grown. ATP synthase, which is involved in ATP synthesis during photophosphorylation, is located almost entirely in the exposed membranes in stroma (Fig. 5.15). Not all membrane-binding proteins, which participate in photosynthetic electron transport, are distributed unequally in the thylakoid. Cytochrome $b_6f$ complex, which facilitates electron transfer between the two photosystems, is distributed evenly in both types of lamellae. Excitation of PSII

## 5.3 Light Reactions in Photosynthesis

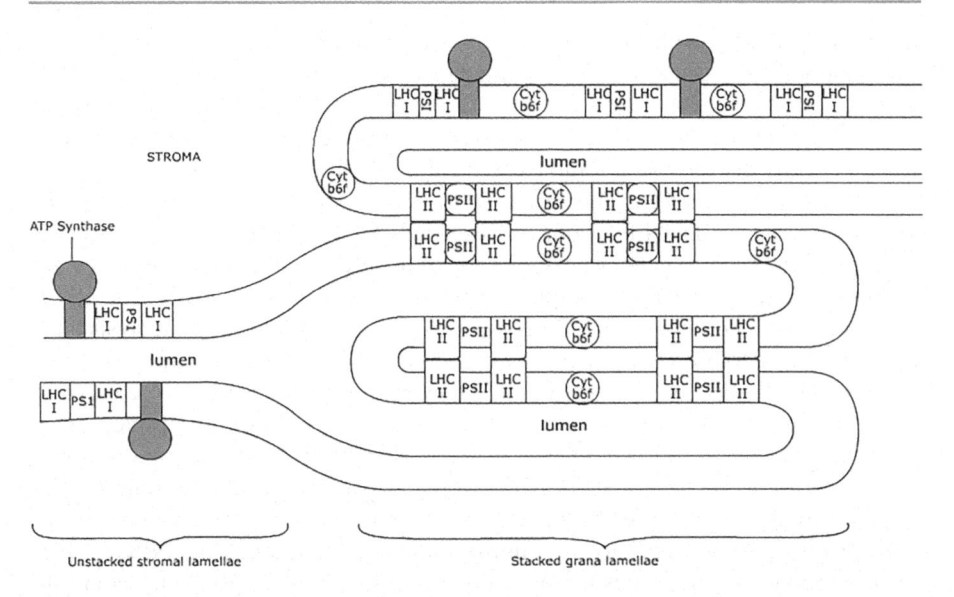

**Fig. 5.15** Lateral heterogeneity. LHCII has hydrophobic region protruding out from thylakoid, which is responsible for stacking of thylakoids resulting in grana lamellae. In case PQ is present as PQH$_2$, a protein kinase phosphorylates a specific threonine of LHCII, resulting in its detachment from PSII, and it gets associated with PSI. As a result, there is a balance in distribution of energy in between the two photosystems

reduces the common intermediate pool of lipid soluble mobile carrier, plastoquinone, which is oxidized by the intermediates of electron transport. Oxidized PSI (P$_{700}^+$) obtains electrons from the common pool of reduced intermediates. Absence of direct energy transfer from PSII to PSI prevents over-excitation of PSI. Thus, spatial separation of photosystems prevents light energy from being preferably transferred to PSI, and PSII also functions effectively. Another mechanism involved in the regulation of energy distribution between the two photosystems is the participation of LHCII. Less functional PSI in comparison of PSII will lead to accumulation of PQH$_2$ because of which a protein kinase is activated which phosphorylates hydroxyl residue of threonine residue of peripheral LHCII. As a result, LHCII get dissociated from PSII and bind with PSI because of the changed conformation, resulting in PSI getting more light energy. Reverse happens when PQ is oxidized (PSI being more active than PSII) due to activation of a protein phosphatase. Thus accumulation of reduced PQ regulates energy distribution in between the two photosystems so that a balance is maintained.

### 5.3.13 Elimination of Excess Light Energy as Heat

At times NADPH and ATP accumulate because of much faster photochemical reaction that can be consumed by CO$_2$ assimilation, especially when high light intensity is coupled with low temperature, which lowers the metabolic rate of $CO_2$ assimilation. This results in ATP accumulation and shortage of oxidized $NADP^+$, which causes reduction in the availability of oxidized pheophytin (Pheo). In the absence of oxidized Pheo, there will be no acceptor of excited electron of $P_{680}$* at the reaction center of PSII. As a result excited chlorophyll molecules close to $P_{680}$* are not able to lose the energy. This results in triplet state of the chlorophyll molecules which has much longer life (from microseconds to milliseconds). Triplet state of chlorophyll molecules generates **singlet $O_2$ molecule** which is highly reactive and causes damage of the adjacent photosynthetic machinery, especially to the D1 protein of PSII. Following this, DI is targeted for degradation by proteolysis since it becomes vulnerable to breakdown by the action of proteases due to alteration in its conformation by the singlet $O_2$ or by the superoxide anion radicals. It is replaced by new D1 which is synthesized on the chloroplast ribosomes and is inserted in the thylakoid membrane followed by the reassembly of PSII. D1 has a high turnover. Excess rate of D1 damage than its synthesis reduces the rate of photosynthesis. This is called **photoinhibition**. Plants have devised mechanism to control the energy loss of singlet $O_2$ by a process known as **non-photochemical quenching (NPQ)**, which refers to molecular mechanisms for removing this trapped energy (quenching the excited state). Carotenoids present adjacent to $P_{680}$ in the core of the reaction center of PSII play an important role in protecting PSII apparatus. Because of movement of $H^+$ from the stroma to lumen of thylakoids in high light intensity conditions, there is a decrease in pH in lumen which increases the quenching process. Decrease in pH results in the activation of violaxanthin de-epoxidase which converts violaxanthin to antheraxanthin and then to zeaxanthin. Zeaxanthin accepts energy from excited chlorophyll molecules and, on returning to ground state, loses energy as heat. Antiquenching becomes necessary at low light levels or in dark. It occurs due to activation of the enzyme epoxidase with rise in luminal pH under low light conditions. The process is reversed and zeaxanthin is converted back to antheraxanthin and violaxanthin. There is a possible involvement of a very hydrophobic transmembrane protein which undergoes protonation in low pH of lumen and facilitates quenching of excited PSII by zeaxanthin (Fig. 5.4).

## 5.4 Photosynthetic Carbon Dioxide Assimilation

Photosynthesis consists of reactions involving two redox phases. The first phase includes generation of NADPH and ATP in thylakoids during light reaction, while the second set of redox reactions occurs in the stroma. The second set of redox phase includes $CO_2$ assimilation which uses chemical energy, NADPH and ATP generated during the first phase. $CO_2$ reduction was earlier known as dark reactions. However, it is a misnomer since activity of many enzymes of $CO_2$ assimilation is light-regulated. Instead, $CO_2$ assimilation or photosynthetic carbon reduction are the terms which are now used. It is because of these reactions that carbon is incorporated into living beings. In green plants, chloroplasts contain unique enzymatic machinery which catalyzes $CO_2$ assimilation. Plants convert simple products of photosynthesis into more complex biomolecules, including various sugars, polysaccharides, and other metabolites, which are derived from them through various associated metabolic pathways.

## 5.4.1 Calvin-Benson Cycle

$CO_2$ assimilation makes up a cyclic pathway in which key intermediates are constantly regenerated. This pathway was elucidated by Melvin Calvin, James Bassham, and Andrew Benson (1946–1953) for which Melvin Calvin was awarded the Nobel Prize for chemistry in 1961. The cycle is often called the *Calvin-Benson cycle* **or** *reductive pentose phosphate pathway*. In 1940, Samuel Ruben and Martin Kamen discovered a new radioisotope of carbon, $^{14}C$, which has a very long half-life (>5000 years) in comparison with $^{11}C$ (half-life of 20 min). They conducted a series of experiments to trace the metabolic fate of the labeled $^{14}CO_2$ during $CO_2$ fixation. Liquid cultures of *Chlorella* were exposed to $^{14}CO_2$ for varying time periods, and plants were exposed to different illumination conditions. Cells were then dropped into the boiling alcohol so as to stop the metabolic pathway by killing them while preserving the labeling pattern. The radioactive products were subsequently identified using two-dimensional paper chromatography coupled with autoradiography. They observed that a three-carbon compound, 3-phosphoglycerate, was first to get the labeled carbon, followed by a series of other compounds. By following the pattern of getting labeled compounds and relating it to the time of exposure of plants to $^{14}CO_2$ as well as by studying the distribution pattern of radioactivity within the compound, path of $^{14}CO_2$ assimilation was traced. Interconversion of various intermediates during assimilation of $^{14}CO_2$ could also be understood. Intramolecular labeling studies were carried out chiefly by Andrew Benson. Since the first compound identified was a three-carbon compound, it was envisaged that the primary acceptor for $^{14}CO_2$ could be either a two-carbon compound or a five-carbon compound. A five-carbon compound, ribulose 1,5-bisphosphate, was identified as the acceptor molecule. Convincing evidence for RuBP as the acceptor molecule for $^{14}CO_2$ came from studies involving transferring algae from light to dark, which were conducted by James Bassham when increase in PGA pool was coupled with decrease in RuBP (Fig. 5.16). Since RuBP also gets the labeling, $CO_2$ assimilation was considered as a cyclic process, and it was called as Calvin-Benson cycle.

Calvin-Benson cycle can be studied in three stages—the first stage involves condensation of $CO_2$ with ribulose 1,5-bisphosphate to generate 3-phosphoglycerate. In the second stage, 3-phosphoglycerate is reduced to yield glyceraldehyde 3-phosphate (a triose phosphate) at the expense of ATP and NADPH produced during light reaction. In the third stage, known as the regeneration phase, 5/6 of glyceraldehyde 3-phosphate is used for regeneration of ribulose 1,5-bisphosphate (Fig. 5.17), while 1/6 of glyceraldehyde 3-phosphate is either transported out of the chloroplasts through the transporters located in the inner envelope of chloroplasts or is stored as transient starch during the daytime. The transported glyceraldehyde 3-phosphate is used for sucrose synthesis in cytosol

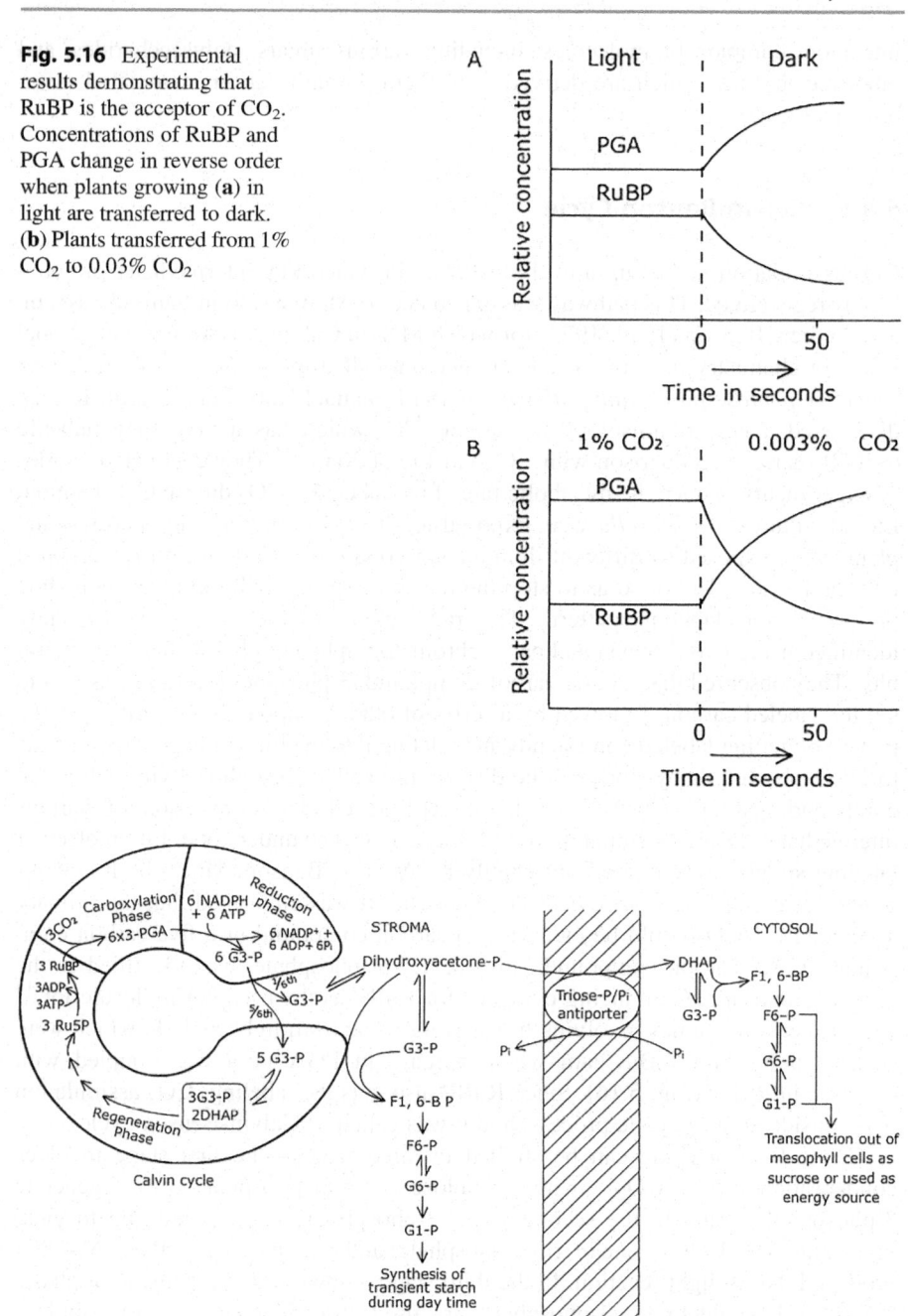

**Fig. 5.16** Experimental results demonstrating that RuBP is the acceptor of $CO_2$. Concentrations of RuBP and PGA change in reverse order when plants growing (**a**) in light are transferred to dark. (**b**) Plants transferred from 1% $CO_2$ to 0.03% $CO_2$

**Fig. 5.17** Triose P/P$_i$ antiporter localized on the inner membrane of plastid facilitates the exchange of triose-P between stroma of the plastid and cytosol

### 5.4 Photosynthetic Carbon Dioxide Assimilation

which is then transported out to other parts of the plant. Thus, the cyclic process allows continuous conversion of $CO_2$ into triose phosphates, hexose phosphates, and other intermediates besides generating acceptor molecules of $CO_2$. Calvin-Benson cycle involves 13 enzyme-catalyzed reactions which occur in stroma of chloroplasts.

#### 5.4.2 Carboxylation Phase

Melvin Calvin and his associates (Box 5.1) observed that in the algae when exposed to $^{14}CO_2$ for a few seconds only, first stable radioactively labeled compound was

> **Box 5.1: Melvin Ellis Calvin (1911–1997)**
> M.E. Calvin was born to Jewish immigrant parents in 1911. His parents owned a grocery store in Detroit where Calvin had his school education. He did his graduation in chemistry major and obtained his Ph.D. with George C. Glockler in 1935 from University of Minnesota. He joined the chemistry faculty at the University of California in 1937 as an instructor where he remained for the rest of his life. Calvin in 1947 begun his Nobel Prize winning work on photosynthesis using alga *Chlorella pyrenoidosa*. After illuminated algal cells were exposed to $^{14}CO_2$, its growth was stopped at different stages starting from 5 s. He used paper chromatography to isolate and identify the minute quantities of radioactive intermediates. Calvin supported interdisciplinary research. He has written 7 books and almost 600 research articles. He was the sole recipient of Noble Prize in Chemistry in 1961 for his work on Calvin cycle, which is also known as Calvin-Benson cycle. In 1963 he was given an additional title of Professor in Molecular Biology. The search of NASA for extraterrestrial life was greatly influenced by Professor Calvin. Later on his research area of interest included artificial photosynthesis and plants, producing hydrocarbons as substitute for fuels.

identified to be 3-phosphoglycerate (3-PGA). Initially 3-PGA was labeled on its carboxyl (-COOH) group. This suggested 3-PGA to be an early intermediate generated during $CO_2$ fixation. Experiments were conducted in plants which were either shifted from light to dark or from high to low $CO_2$ concentration. Increase or decrease in 3-PGA was found to be coupled with decrease or increase in five-carbon compound, ribulose-I,5-biphosphate (RuBP) (Fig. 5.16). This indicated formation of 3-PGA occurred at the expanse of RuBP. Carboxylation of RuBP is catalyzed by the enzyme *ribulose-1,5-bisphosphate carboxylase/oxygenase*, for which the acronym Rubisco was proposed by David Eisenberg in 1979. Carboxylation of RuBP is followed by cleavage of the unstable six-carbon intermediate leading to formation of two molecules of 3-PG, one of which bears the new carbon introduced as $CO_2$ in its carboxyl group. Driving force for the highly exergonic reaction ($\Delta G^{\circ\prime} = -35.1$ kJ.mol$^{-1}$) is provided by the cleavage of the β-keto acid intermediate to yield an additional resonance-stabilized carboxylate group. Rubisco is located in the chloroplast stroma. It is a very slow enzyme catalyzing the fixation of 1–12 $CO_2$ molecules on each active site per second, unlike many other enzymes which catalyze at the reaction rate to the order of $10^3$ to $10^5$ s$^{-1}$. Slow activity of Rubisco is compensated by the large amount of the enzyme being present, so much so that it can constitute up to 50% of the leaf protein. It is the most abundant protein in the biosphere. Enormous amount of nitrogen is invested for synthesis of Rubisco. Another shortcoming of the enzyme is failing to discriminate between $CO_2$ and $O_2$. It wastes one carbon every time enzyme reacts with $O_2$. During oxygenation reaction, a molecule of 3-PGA and a two-carbon compound 2-phosphoglycolate are produced (Fig. 5.22).

Besides being the slowest, Rubisco is also one of the largest enzymes. Four distinct forms of Rubisco have been identified in the organisms which fix $CO_2$. In all the eukaryotic photoautotrophs, form I Rubisco is present which consists of 16 subunits of two types, i.e., L and S (a hexadecamer, $L_8S_8$, molecular mass 560 kDa). Each larger subunit (L) has a molecular weight of 55 kDa, while molecular weight of each small subunit (S) is 15 kDa. Large subunits are present as four dimers $(L_2)_4$. There are two of tetrameric small subunits which are present on top and bottom of aggregate of large subunits. Rubisco holoenzyme is expressed as $(L_2)_4(S_4)_2$. Larger subunit has got the catalytic side and is encoded in plastid genome (rbcL), whereas a family of nuclear rbcS genes encodes nearly identical smaller subunits in all photoautotrophic land plants and green algae, which are synthesized by cytosolic ribosomes and are transported into the plastids. Following post-translational processing, molecular chaperones help in the assembly of subunits in plastids. Form I Rubisco is the only type of Rubisco which consists of small subunits also along with the large subunits. Homodimers of catalytic large subunits are common in all forms. Form II consists of dimer of only large subunits and is present in dinoflagellates which is encoded in nuclear genome (unlike in algae and land plants). In archaebacterial, Rubisco is neither form I nor II but a decamer of five large subunits dimers.

### 5.4.3 Reduction Phase

Reduction of 3-PGA to glyceraldehyde 3-phosphate (GAP) occurs in two steps. In the first step transfer of phosphate from ATP to 3-PGA occurs, yielding 1,3-bisphosphoglycerate (BPG). Reaction is catalyzed by phosphoglycerate kinase. In the second step, NADPH-specific glyceraldehyde 3-phosphate dehydrogenase catalyzes the reduction of BPG to GAP. Reduction of BPG is the major energy-requiring step of the cycle. Triose phosphate isomerase catalyzes the conversion of GAP to dihydroxyacetone phosphate (DHAP). Equilibrium of this reversible reaction favors synthesis of ketone. Both GAP and DHAP are together called triose phosphates. In addition to being used for RuBP generation, GAP has several other fates in the metabolism. It may be metabolized via glycolysis in the plastid itself or is used for starch biosynthesis (in chloroplasts) or is transported out of chloroplast to cytosol for biosynthesis of sucrose. Export of triose phosphates occurs in exchange of cytosolic $P_i$. This is facilitated by the antiporters located in the inner membrane of plastid (Fig. 5.17).

### 5.4.4 RuBP Regeneration Phase

As RuBP is consumed during initial carboxylation phase, it must be constantly regenerated. This phase is significant for continuation of $CO_2$ assimilation. Out of every six molecules of triose phosphates, three molecules of RuBP (15 C-atoms) are regenerated by reshuffling of carbon atoms. Reactions involved are the following:

**Aldol Condensation Reaction** Condensation of two triose phosphates, i.e., dihydroxyacetone-3-phosphate (DHAP) with GAP, is catalyzed by aldolase yielding fructose-1,6-bisphosphate (FBP).

**Dephosphorylation Reaction** Fructose 1,6-bisphosphatase catalyzes the irreversible dephosphorylation of FBP to yield fructose 6-phosphate and inorganic phosphate. This is followed by transfer of two-carbon unit (C-1 and C-2) of fructose 6-phosphate to the third molecule of GAP to form xylulose 5-phosphate. The reaction is catalyzed by **transketolase**. The remaining carbons (C-3 to C-6) of fructose-6-phosphate form erythrose 4-phosphate.

**Second Aldol Condensation Reaction** Another condensation reaction, catalyzed by aldolase, involves erythrose 4-phosphate and DHAP (fourth molecule of triose phosphate), which yields sedoheptulose 1,7-bisphosphate (seven-carbon sugar). This is followed by the **dephosphorylation** of sedoheptulose 1,7-bisphosphate to yield sedoheptulose 7-phosphate. This is another irreversible reaction of the pathway which is catalyzed by a specific phosphatase. Further reaction involves transfer of two-carbon unit (C-1 and C-2) of sedoheptulose 7-phosphate to a fifth molecule of GAP to form xylulose 5-phosphate, while remaining carbon (C-3 to C-7) of sedoheptulose 7-phosphate generate ribose 5-phosphate. Reaction is catalyzed by

**transketolase**. Phosphopentose epimerase catalyzes **epimerization** of the two molecules of xylulose 5-phosphate to produce two molecules of ribulose 5-phosphate, while ribose 5-phosphate isomerase catalyze isomerization of ribose 5-phosphate into ribulose 5-phosphate. The last step in the Calvin cycle involves phosphorylation of three molecules of ribulose 5-phosphate consuming three ATP molecules to generate ribulose 1,5-bisphosphate (RuBP). The reaction is catalyzed by ribulose 5-phosphate kinase also known as Phosphoribulo kinase (PRK). With this regeneration stage is completed (Fig. 5.18). In this way three RuBP molecules, consumed during carboxylation using 3$CO_2$, are regenerated. Reactions occurring during interconversion of various intermediates during regeneration phase of Calvin

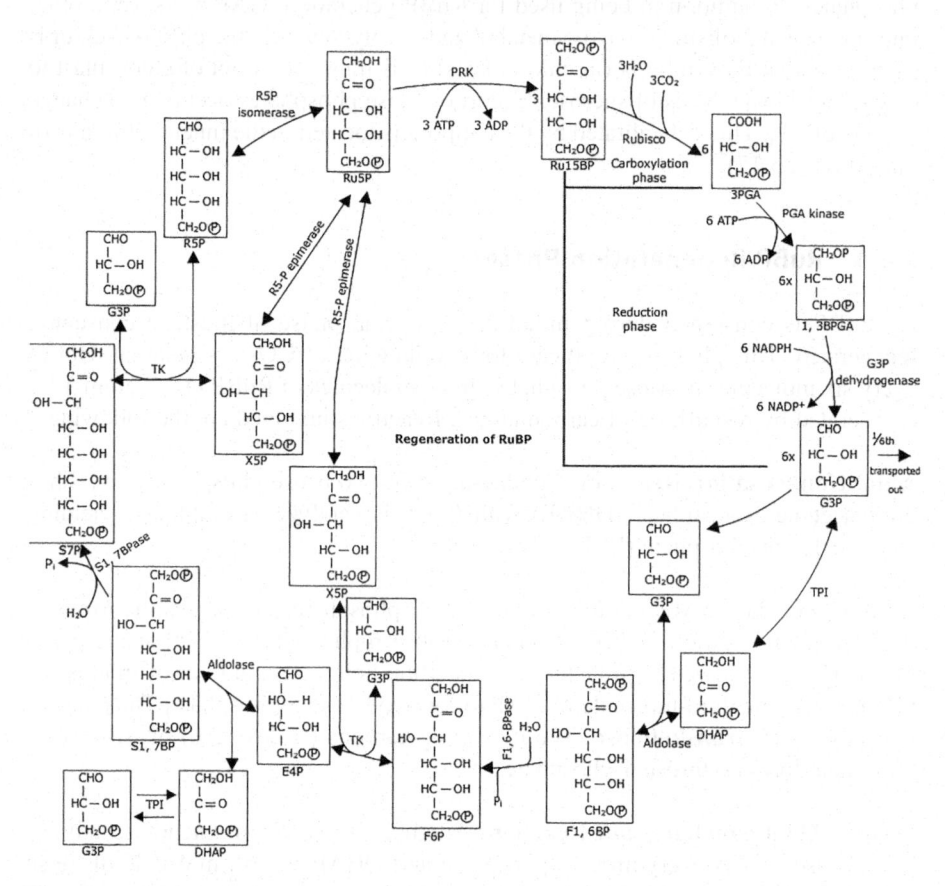

**Fig. 5.18** Calvin cycle includes carboxylation phase, reduction phase, and phase involving regeneration of RuBP. 1/6 of the trioses produced are transported out of plastids, while 5/6 are utilized for regeneration of RuBP. Reversible and irreversible reactions have been demonstrated by using appropriate arrows. *PRK* phosphoribulokinase, *PGA* phosphoglyceric acid, *BPGA* biphosphoglyceric acid, *G3P* glyceraldehyde 3-Phosphate, *TPI* triose phosphate isomerase, *TK* transketolase, *E4P* erythrose 4-Phosphate, *S1,7BP* sedoheptulose 1,7-biphosphate, *DHAP* Dihydroxyacetone 3-phosphate, *S1,7BPase* sedoheptulose 1,7 biphosphatase, *F1,6-BPase* fructose 1,6 biphosphatase

cycle (also known as reductive pentose phosphate pathway) are quite similar to those which occur during oxidative pentose phosphate pathway. However, the two unique reactions in photosynthesis are dephosphorylation of sedoheptulose 1,7-biphosphate to sedoheptulose 7-phosphate and ribulose 5-phosphate to ribulose 1, 5-bisphosphate which are catalyzed by a phosphatase and phosphoribulose kinase, respectively. Reason for production of so many intermediates during the regeneration of RuBP is not obvious, but possibly besides being utilized for RuBP generation, these are to be used for other metabolic processes also. Erythrose 4-phosphate is utilized in shikimic pathway for generation of aromatic amino acids in plastids, while ribose 5-phosphates are used as precursor for nucleotide biosynthesis.

### 5.4.5 ATP and NADPH (Energy Sources in $CO_2$ Fixation)

The carboxylation reaction is exothermic ($\Delta G^{o'} = -35$ kJ.mol$^{-1}$) and is, thus, energetically favorable. But formation of GAP from 3-PGA and regeneration of RuBP require energy. This is provided by ATP and NADPH which are generated during light reaction. Three molecules of RuBP condense with three $CO_2$ molecules to form six molecules of 3-PGA. These are reduced to six molecules of GAP, using six ATP and six NADPH molecules. One out of every six molecules of GAP is either used for synthesis of temporary starch (transient starch) in plastid or is transported out of the plastids in exchange of $P_i$ through antiporters located in the inner envelope of chloroplast which is utilized for synthesis of sucrose to be transported out of mesophyll cells. Carbon atoms of the remaining five GAP molecules ($3 \times 5$) = 15 carbon atoms) are rearranged to regenerate 3 R5P molecules ($5 \times 3$) = 15 carbon atoms) which are further required for $CO_2$ fixation and continuation of the cycle. Phosphorylation of 3 R5P molecules requires three ATPs to produce three molecules of RuBP. Thus, for every molecule of triose phosphate transported out, six NADPH and nine ATP are required.

$$3CO_2 + 3RuBP + 3H_2O + 6NADPH + 6H^+ + 6ATP$$
$$\rightarrow 6\text{Triose phosphate} + 6NADP^+ + 6ADP + 6P_i$$

Out of six triose phosphates, five are used for regeneration of RuBP (Fig. 5.22).

$$5\text{Triose phosphate} + 3ATP + 2H_2O \rightarrow 3\ RuBP + 3ADP + 3P_i$$

Since it is only one triose phosphate out of six that is used for storage or further metabolism, net reaction can be written as:

$$3CO_2 + 5H_2O + 6NADPH + 9ATP \rightarrow GAP + 6NADP^+ + 9ADP + 9P_i$$

Efficiency for energy use in photosynthesis can be calculated easily. For each mole $CO_2$ reduction $\Delta G^{o'}$ is +478 kJ. Energy of light of 600 nm wavelength is

1593 kJ/mole. Energy of 680 nm and 700 nm wavelength will be lesser than this. Thus efficiency of photosynthesis is at least 30% (478/1593 × 100).

## 5.4.6 Autocatalytic Regulation of Regeneration of RuBP for Continuous $CO_2$ Assimilation

Autocatalytic regulation of Calvin cycle is responsible for maintaining balance in between RuBP regenerated and RuBP consumed for $CO_2$ fixation. If all ten molecules of triose phosphates, produced for every five molecules of $CO_2$ fixed in Calvin cycle, are utilized for regeneration of RuBP, there will be net gain of one RuBP, since six molecules of RuBP will be produced while five are consumed. Increase in RuBP will result in increased enzyme activity. In that case none of the triose phosphates will be diverted for further utilization. Shutting down the photosynthetic process during night results in reduction of intermediates of the Calvin cycle, because these are consumed in various metabolic reactions. Consequently, at the start of photosynthesis the following day, $CO_2$ fixation is severely affected because of availability of RuBP being low. As a result, all of ten triose phosphates generated during fixation of five $CO_2$ molecules are consumed for regeneration of RuBP, and none of the triose phosphates is transported out of chloroplast. This will result in net gain of one RuBP, resulting in lag phase for triose phosphate being transported out. This lag phase is known as **photosynthetic induction period**. Once photosynthesis reaches steady state of $CO_2$ assimilation, one-sixth of triose phosphates generated during Calvin cycle start getting diverted for carbohydrate biosynthesis maintain the balance between RuBP produced and RuBP being consumed. As a result, sucrose is synthesized in cytosol.

$$6CO_2 + 12H_2O + 18ATP + 12NADPH + 12H^+$$
$$\rightarrow C_6H_{12}O_6 + 18ADP + 18P_i + 12NADP^+ + 6O_2 + 6H_2O$$

## 5.4.7 Regulation of Calvin-Benson Cycle

During day, green plants, being autotrophs, carry out photosynthesis to fulfill their energy requirements by using light as an energy source, whereas at night, like other heterotrophic organisms, they use their nutritional reserves to generate NADH and NADPH through glycolysis and oxidative pentose phosphate pathway, respectively. $CO_2$ assimilation occurs during Calvin cycle at the expense of NADPH and ATP. On the contrary, release of $CO_2$ occurs coupled with reduction of $NADP^+$ to NADPH during oxidative-reductive pentose phosphate (ORPP). ORPP also occurs in stroma of chloroplasts. If both cycles occur simultaneously, it will result in **futile cycle** since NADPH generated in ORPP will be consumed during Calvin cycle. This is checked by the regulation of enzymes of both pathways. Plants have a light-sensitive control mechanism to prevent Calvin cycle operating in the dark (Fig. 5.19). The activity of Rubisco is light-dependent. Rubisco has very low catalytic activity and must be

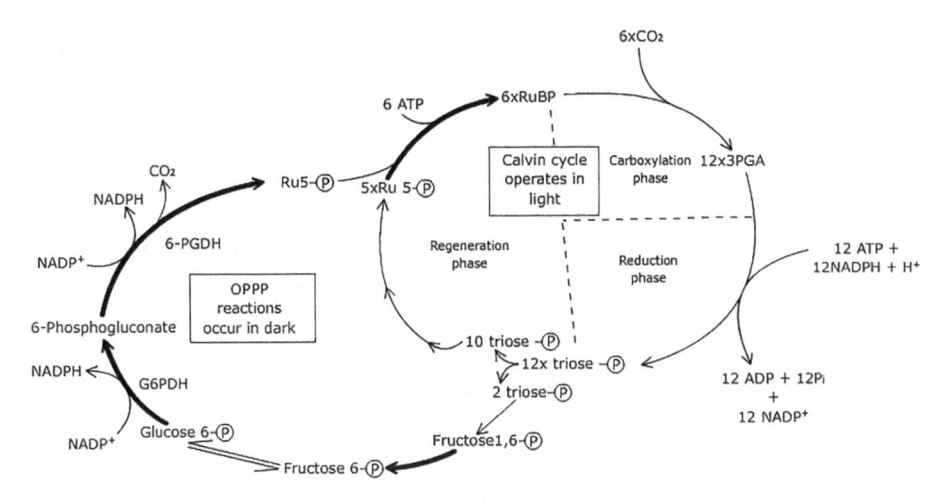

## 5.4 Photosynthetic Carbon Dioxide Assimilation

**Fig. 5.19** Potential for futile cycle in chloroplast is avoided by difference in timings of the two cycles, i.e., OPPP (oxidative pentose phosphate pathway) and Calvin cycle. OPPP occurs during the night, while Calvin cycle (reductive pentose phosphate pathway) occurs during the day. Bold lines indicate irreversible regulatory steps of Calvin cycle and OPPP

activated before acting as a catalyst. Since Rubisco is the first enzyme in $CO_2$ fixation by Calvin cycle, it is the prime target for regulation. The enzyme is inactive until carbamylated, which occurs due to its complex interaction with $CO_2$, high pH of stroma, and $Mg^{2+}$. pH of stroma increases from 7 (in dark) to 8, since during daytime protons move into the thylakoid lumen in the light reaction. Consequently, $Mg^{2+}$ moves out of the lumen to compensate for the proton flux in the opposite direction, increasing $Mg^{2+}$ concentration from 3 to 6 mM in the stroma. The allosteric site is present in larger subunit (LSU) of the enzyme, which is separate and distinct from the substrate binding site of the enzyme. The model proposed for in vitro activation of Rubisco takes into account all three factors: $CO_2$, $Mg^{2+}$, and pH. According to the proposed model, the $CO_2$ molecules first binds with ε-amino group of lysine residue present in position 201 of the 470 amino acid long protein of large subunit (LSU) at the allosteric site to form carbamate. This is different from the one in which $CO_2$ is required as the substrate by the enzyme. Carbamate formation requires release of two protons from the ε-amino group of lysine residue, which is favored by increase in pH of stroma. $Mg^{2+}$ which has moved out of lumen to stroma coordinates with the carbamate to form carbamate-$Mg^{2+}$ complex, which is the active form of Rubisco (Fig. 5.20). However, binding of sugar phosphates to the enzyme, such as ribulose 1,5-bisphosphate, prevents carbamylation of Rubisco. In vitro experiments have indicated that the $Mg^{2+}$, $CO_2$, and pH differences alone are not sufficient to account for more than half of the expected activation level of Rubisco. An *Arabidopsis* mutant, *rca*, has been isolated, in which Rubisco fails to be activated in light, even though the enzyme which had been isolated from the mutant is apparently identical to the one isolated from the wild type. Electrophoretic analysis of chloroplast proteins has revealed that the mutant lacked a soluble chloroplast protein, and full activation of Rubisco could be

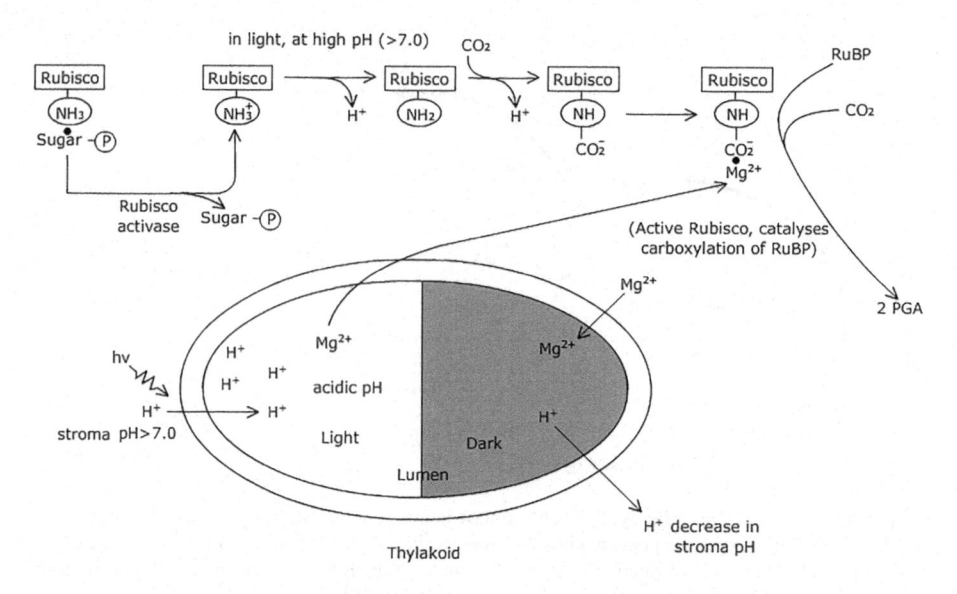

**Fig. 5.20** In dark activity of Rubisco is inhibited because of binding of sugar phosphates such as RuBP or CA1P (2-carboxyarabinitol 1-phosphate) to the active site of the enzyme. CA1P and RuBP bind to the uncarbamylated and carbamylated Rubisco, respectively. Inhibition is relieved in light by removal of sugar phosphates in an ATP-dependent catalysis by Rubisco activase. Increase in stromal pH, because of protons being transferred from stroma to lumen in light, results in loss of H$^+$ from ε-amino group of lys$^{210}$ of the enzyme protein followed by carbamate formation by $CO_2$. The carbamylated form of the enzyme forms a complex with $Mg^{2+}$ resulting in the activation of enzyme

restored in vitro simply by adding that missing protein to a reaction mixture containing Rubisco, RuBP, and physiological levels of $CO_2$. This protein has been named **Rubisco activase**, which signifies its role in regulation of Rubisco activity. Rubisco activase brings about a conformational change in the Rubisco protein, causing release of bound sugar phosphates and allowing carbamylation to occur. This is an ATP-dependent process. In dark, a proton gradient created during light reaction is dissipated which causes $Mg^{2+}$ to flow back to the lumen from the stroma. The decrease in $Mg^{2+}$ concentration as well as decrease in pH in stroma results in the inactivation of Rubisco. 2-carboxyarabinitol-1-phosphate (CA1P) also acts as an inhibitor of Rubisco. CA1P is the structural analog of six-carbon intermediate of the carboxylation reaction, which is generally synthesized in high concentrations in dark in the leaves of legumes. It alters the activity of Rubisco by binding to its activation site and thus keeping a control on Calvin cycle. But in light, specific phosphatase releases the phosphate group from CA1P, thereby making it incompetent to bind to the activation site of the enzyme. The activity of Rubisco activase is also regulated by light making it active in the light and inactive in the dark. Interconversion of active and inactive forms of Rubisco activase is regulated by ferredoxin-thioredoxin system discussed below.

Ferredoxin-thioredoxin system regulates Calvin-Benson enzymes. Calvin cycle enzymes, which includes fructose-1,6-bisphosphatase, sedoheptulose-1,7-

**Fig. 5.21** Regulation of Calvin cycle enzymes by thioredoxin system

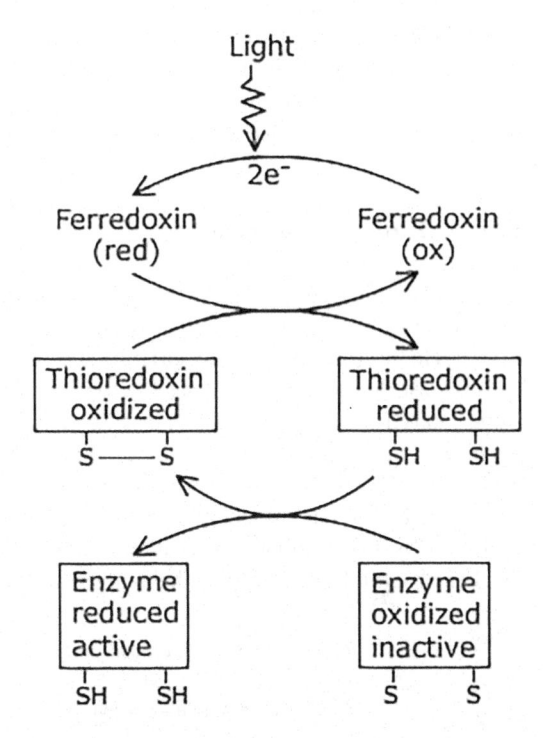

bisphosphatase, phosphoribulokinase, and NADP-glyceraldehyde 3-phosphate dehydrogenase, are regulated due to interconversion of thiol (reduced) and sulfhydryl (oxidized) forms of cysteine residues of the enzyme proteins. This is mediated by ferredoxin-thioredoxin system, which requires proteins ferredoxin (Fdx), ferredoxin-thioredoxin reductase (FTR), and thioredoxin (Trx), a small protein-disulfide reductase. The activation of photosystem I (P700) results in the reduction of ferredoxin in the light. Reduced ferredoxin transforms the disulfide bond (-S-S-) of the ubiquitous regulatory protein **thioredoxin** to its reduced state (-SH-HS-). The reaction is catalyzed by the enzyme ferredoxin-thioredoxin reductase. Reduced thioredoxin subsequently reduces the appropriate disulfide bonds of the target Calvin cycle enzymes, resulting in their activation (Fig. 5.21). This process is reversed in the dark, and inactivation of the target enzymes is observed due to oxidation of cysteine residues involved. Oxidation converts thioredoxin and target enzyme from their reduced active state (-SH HS-) to the oxidized inactive state (-S-S-) leading to their inactivation.

Another means for regulation of activity of Calvin cycle enzymes is due to their non-covalent interactions with proteins to form supramolecular complex. This mechanism is important especially when plants are exposed to quick-changing light conditions either in cloudy conditions or due to change in plant canopy. This includes enzymes viz., phosphoribulose kinase and glyceraldehyde 3-phosphate dehydrogenase whose activity is also regulated by NADPH directly. In the dark, these enzymes associate with a nuclear-encoded small chloroplast protein CP12

(~8.5 kDa protein) to form a large complex, which results in their inactivation. NADPH, generated in light reactions, on binding to this complex, leads to the release of the enzymes. Thus, activity of these enzymes depends both on reduction by thioredoxin and NADPH-mediated release from CP12.

## 5.5 Photorespiration

$CO_2$ is also released in a pathway which is distinct from mitochondrial respiration. The pathway is known as **photorespiration** or **C2 oxidative photosynthetic carbon cycle** or **photosynthetic carbon oxidation cycle (PCO)**. Rubisco possess carboxylase as well as oxygenase activities, since both $CO_2$ and $O_2$ compete for same catalytic site of the enzyme. Rubisco reacts with its second substrate, RuBP, to generate an unstable intermediate that splits into 2-phosphoglycolate and 3-phosphoglycerate in the presence of light and $O_2$ (Fig. 5.22). In 1920, Otto Warburg, a German biochemist, observed in *Chlorella* that photosynthesis was inhibited by $O_2$. Similar observations were made in many other plants. In 1955, Decker observed that photosynthesizing plants when transferred from light to dark,

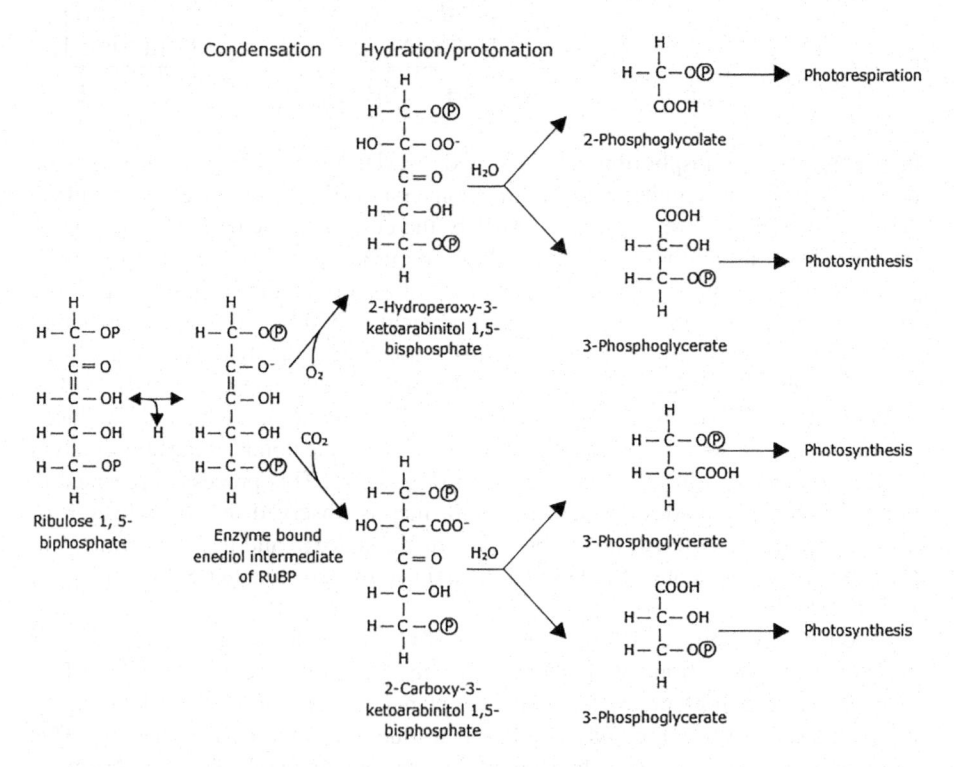

**Fig. 5.22** Carboxylation and oxygenation of RuBP catalyzed by Rubisco. Binding of RuBP with Rubisco results in formation of an enzyme bound enediol intermediate which can react either with $CO_2$ or $O_2$ depending upon their availability

## 5.5 Photorespiration

$CO_2$ production during initial 1–4 min was much more than $CO_2$ produced later on, a phenomenon which he called as "post-illumination burst of $CO_2$." Factors which influenced rate of photosynthesis during the light period also affected post-illumination burst of $CO_2$. This light-stimulated $CO_2$ production was called photorespiration. Several scientists including Zelitsch, Tolbert, and their associates related glycolate metabolism to photorespiration. Though glycolate pathway was different from mitochondrial respiration (also called dark respiration), this was called as photorespiration, since, similar to respiration, $O_2$ was consumed and $CO_2$ was released. However, unlike mitochondrial respiration there is no ATP production, but rather ATP is consumed, making it an energy-wasteful process. Since Rubisco possess both carboxylase and oxygenase activity, it is named RuBP carboxylase/oxygenase, which is abbreviated as Rubisco. Affinity of the enzyme for oxygen is much lower than for carbon dioxide. $K_m$ value of Rubisco for oxygen is 535 $\mu mol.L^{-1}$, while for $CO_2$ it is 5 $\mu mol.L^{-1}$. Both gases, $O_2$ and $CO_2$, are present in low concentrations in chloroplasts. $O_2$ concentrations in chloroplasts are 250 $\mu mol.L^{-1}$, while concentrations of $CO_2$ are about 8 $\mu mol.L^{-1}$. Though affinity of the enzyme for $O_2$ is much lower, oxygenase activity of the enzyme is quite significant, because of comparatively high oxygen concentration in chloroplast. Oxygenase activity of the enzyme usually proceeds at the rate which is 25% of the carboxylation rate at normal $O_2$ and $CO_2$ concentration. It means that one oxygenation reaction occurs for every three carboxylation reactions. At high levels of $O_2$, oxygenation reaction of the enzyme results in formation of 3-phosphoglycerate and 2-phosphoglycolate. Photorespiratory glycolate pathway was discovered in 1972 by American scientist Edward Tolbert. 2-phosphoglycolate, produced during oxygenation of RuBP, is recycled to generate RuBP through this pathway. Phosphate group of 2-phosphoglycolate is hydrolyzed to form glycolate by chloroplast-specific enzyme, glycolate phosphatase. Subsequent metabolism of the glycolate involves the participation of other two organelles: peroxisomes and mitochondria (Fig. 5.23). Glycolate is exported out of chloroplasts through specific transporters present in the inner envelope of the plastid. It diffuses into peroxisomes through porins present in their membranes. In peroxisomes, oxidation of glycolate to glyoxylate is catalyzed by glycolate oxidase which contains flavin mononucleotide (FMN) as the cofactor. $H_2O_2$, which is produced during the reaction, is highly toxic and is degraded by peroxisome-localized catalase to yield $O_2$ and water.

$$H_2O_2 \rightarrow H_2O + \tfrac{1}{2}O_2$$

Glycolate oxidation does not occur in chloroplasts, since both products of the reaction are toxic for chloroplast-localized enzymes. Glyoxylate is toxic for Rubisco activity, and $H_2O_2$ being highly oxidizing agent may cause damage to thylakoids, besides inactivating Calvin cycle enzymes. Glyoxylate undergoes transamination to form glycine. Transamination of glycine occurs by two reactions which occur in 1:1 ratio. One of them is catalyzed by glutamate:glyoxylate aminotransferase, requiring glutamate as the amino group donor. The enzyme may also use alanine as the amino

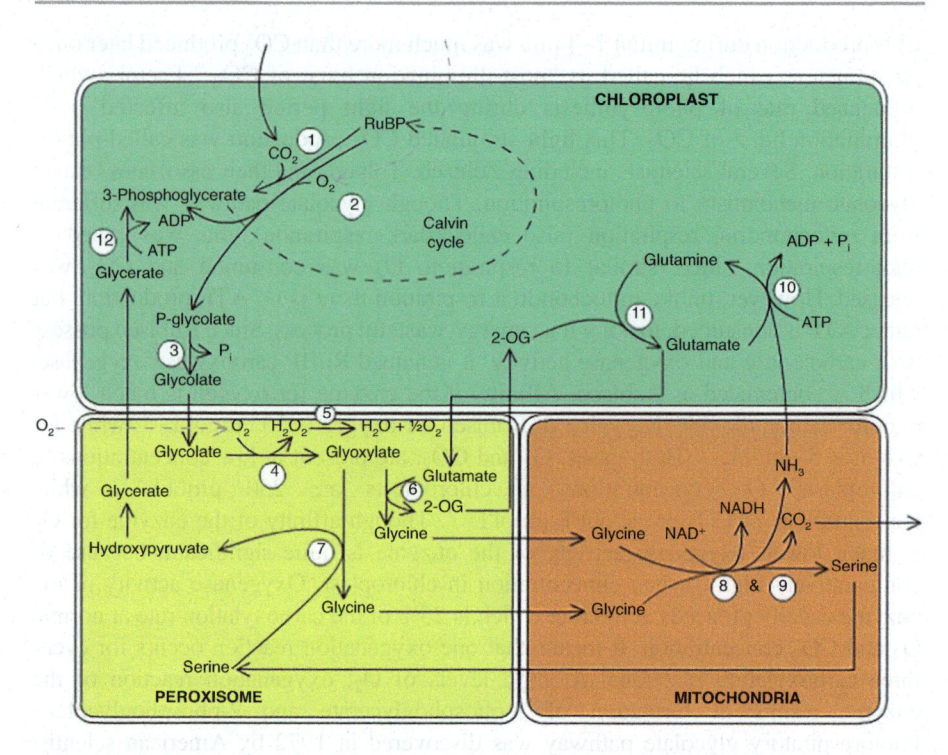

**Fig. 5.23** Photorespiratory cycle involving three organelles chloroplast, peroxisomes, and mitochondria. Enzymes catalyzing reactions include (1) RuBP carboxylase, (2) RuBP oxygenase, (3) P-glycolate phosphatase, (4) glycolate oxidase, (5) catalase, (6) glutamate/glyoxylate aminotransferase, (7) serine/glyoxylate aminotransferase, (8) glycine decarboxylase complex, (9) serine hydroxymethyltransferase, (10) glutamine synthetase, (11) GOGAT (12) glycerate kinase

group donor. Another transaminase catalyzing the reaction is serine:glyoxylate aminotransferase, which use serine as the amino group donor. Glycine is released from peroxisomes via porins and is transported to mitochondria through specific transporters located in the inner mitochondrial membrane. In mitochondria, glycine decarboxylase, a multienzyme complex, catalyzes the condensation of two molecules of glycine (4 carbon), which are oxidized to yield one molecule of three-carbon amino acid serine, $CO_2$ and $NH_3$. This oxidative reaction is coupled with reduction of $NAD^+$ to NADH and is catalyzed by the enzyme glycine decarboxylase-serine hydroxymethyl transferase complex, which is present in large concentrations in the matrix of plant mitochondria. It consists of four proteins, namely, H-protein (a lipoamide-containing polypeptide), P-protein (a 200 kDa, homodimer, pyridoxal phosphate-containing protein), T-protein (containing tetrahydrofoliate as the prosthetic group), and L-protein (a dihydrolipoate dehydrogenase). Glycine decarboxylase constitutes as much as 30–50% of the total soluble protein in mitochondria. The enzyme serine hydroxymethyltransferase is in close proximity which facilitates transfer of the formyl group to another molecule of glycine resulting in synthesis of serine. Ammonia, which is released during the reaction in mitochondria, diffuses rapidly to chloroplasts, where it is fixed through GS/GOGAT pathway involving Fd-GOGAT (Chap. 11). One of the glutamates, produced during the reaction, is transported to peroxisomes. When photorespiration rate is high, the rate of $NH_3$ release in mitochondria is much higher. It has been found to be ten times higher than produced during nitrate reduction. NADH is either oxidized through electron transport chain localized in mitochondria resulting in ATP generation or is transported out of mitochondria to cytosol through malate-oxaloacetate NADH shuttle system to be used for reduction reactions such as nitrate reduction. Serine is transported from mitochondria to peroxisomes through the transporters localized in the inner mitochondrial membrane, where it is converted to hydroxypyruvate by transamination. In peroxisomes hydroxypyruvate is reduced to glycerate utilizing NADH. As there is no endogenous source for NADH in peroxisomes, it is produced during malate dehydrogenase catalyzed oxidation of malate to OAA. Malate is imported from cytosol through malate-oxaloacetate shuttle. Glycerate is transported to chloroplast via glycolate-glycerate transporter located in inner membrane of the plastid where it is phosphorylated to 3-phosphoglycerate by glycerate kinase and is metabolized through the Calvin cycle (Table 5.3). C2 oxidative photosynthetic carbon cycle is dependent on Calvin cycle for RuBP formation, and the balance between the two cycles is determined by three factors: Rubisco kinetics, substrate concentration ($CO_2$ and $O_2$), and temperature. With increase in temperature, affinity of oxygenase activity of Rubisco for $O_2$ increases. On a hot bright day, when photosynthesis has depleted $CO_2$ in the chloroplasts and raised the level of $O_2$, the rate of photorespiration may approach that of photosynthesis (Fig. 5.24). This phenomenon is, in fact, a major limiting factor in growth of many plants. Indeed, plants possessing a Rubisco with significantly less oxygenase activity would not only have increased photosynthetic efficiency but would need less water because they could spend less time with their stomata (the pores leading to their internal leaf spaces) open acquiring $CO_2$ and would have a reduced need for fertilizers because they would require less Rubisco. The control of photorespiration is, therefore, an important unsolved agricultural problem that is presently being attacked through genetic engineering research.

## 5.5.1 Significance of Photorespiration

Overall stoichiometry of the PCO cycle is:

$$2RuBP + 3O_2 + 2Fd_{red} + 2ATP \rightarrow 3,3\,PGA + CO_2 + 2Fd_{ox} + 2ADP + 2P_i$$

It indicates that one in ten carbon atoms from carbon acceptor molecule (RuBP) is lost during photorespiration besides NADPH and ATP which also are consumed during assimilation of $NH_3$. Thus photorespiration increases the energy cost of $CO_2$ assimilation. A number of photons which are required for $CO_2$ assimilation increase from 8 to nearly 14. Thus photorespiration reduces the efficiency with which plants fix $CO_2$, which is usually measured as the quantum efficiency (i.e., moles of $CO_2$

**Table 5.3** Reactions of oxidative photosynthetic carbon cycle

| S. No. | Reaction | Enzyme | Site |
|---|---|---|---|
| 1. | $2RuBP \rightarrow 2P - Glycolate + 23 - PGA$ | RuBP oxygenase | Chloroplast |
| 2. | $2P - Glycolate + 2H_2O \rightarrow 2Glycolate + 2P_i$ | Phosphoglycolate phosphatase | Chloroplast |
| 3. | $2Glycolate + 2O_2 \rightarrow 2Glyoxylate + 2H_2O_2$ | Glycolate oxidase | Glyoxysome |
| 4. | $2H_2O_2 \rightarrow 2H_2O + O_2$ | Catalase | Glyoxysome |
| 5. | $2Glyoxylate + 2Glutamate \rightarrow 2Glycine + 2x2 - oxoglutarate$ | Glutamate: glyoxylate aminotransferase | Glyoxysome |
| 6. | $2Glycine + NAD^+ + H_2O \rightarrow Serine + NADH + NH_4^+ + CO_2$ | (i) Glycine decarboxylase complex | Mitochondria |
|  |  | (ii) Serine hydroxymethyltransferase |  |
| 7. | $Serine + 2 - oxoglutarate \rightarrow Hydroxypyruvate + Glutamate$ | Serine:2-oxoglutarate aminotransferase | Glyoxysome |
| 8. | $Hydroxypyruvate + NADH + H^+ \rightarrow Glycerate + NAD^+$ | Hydroxypyruvate reductase | Glyoxysome |
| 9. | $Glycerate + ATP \rightarrow 3 - PGA + ADP$ | Glycerate kinase | Chloroplast |
| 10. | $Glutamate + NH_3 + ATP \rightarrow Glutamine + ADP + P_i$ | Glutamine synthetase | Chloroplast |
| 11. | $2 - oxoglutarate + Glutamine + 2Fd_{red} + 2H^+ \rightarrow 2Glutamate + 2Fd_{ox}$ | GOGAT | Chloroplast |

## 5.5 Photorespiration

**Fig. 5.24** Graph showing effect of light intensity on respiratory $CO_2$ loss and photosynthetic $CO_2$ assimilation. Light intensity, at which respiratory $CO_2$ loss equals to $CO_2$ assimilated during photosynthesis resulting in no apparent change in $CO_2$ concentration, is known as light compensation point

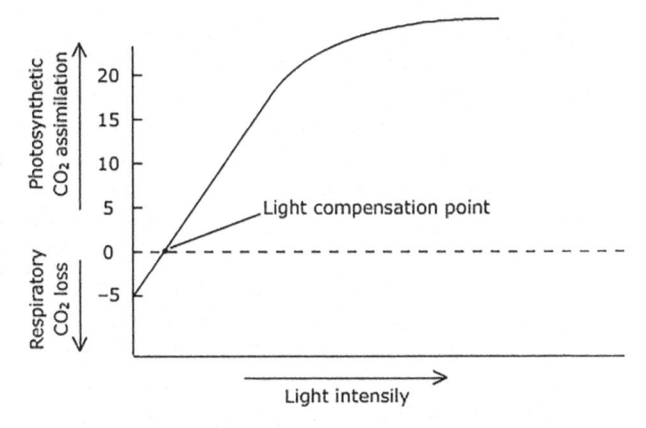

fixed per quanta absorbed). With increase in temperature, quantum efficiency decreases further, because photorespiration increases more in comparison with photosynthesis. Relative rates of carboxylase and oxygenase activity depends upon the affinity of Rubisco for $CO_2$ and $O_2$ and also on the ratio of $CO_2$ and $O_2$ present in the cell solution. With rise in temperature not only solubility of $CO_2$ decreases more in comparison with that of $O_2$, but affinity of enzyme for $CO_2$ also decreases in comparison with $O_2$. As a result, glycolate is produced more in comparison with 3PGA especially in C3 plants. Though being a wasteful process, photorespiration is significant for plants.

- One of the significance associated with photorespiration is believed to its scavenging role. Since glycolate production could not be avoided, because of inherent oxygenase activity of Rubisco, photorespiratory cycle must have evolved to remove glycolate causing minimum loss of carbon. Seventy-five percent of carbon lost from Calvin cycle as 2-phosphoglycolate is recovered in the C2 oxidative photosynthetic carbon cycle. Out of ten carbons of RuBP (two molecules), only one carbon is lost as $CO_2$, while nine carbons are recycled as 3-PGA.
- Involvement of three organelles for removal of glycolate possibly is because glyoxylate and $H_2O_2$ are toxic for Rubisco and Calvin cycle enzymes. Glycolate is transported out of chloroplasts and is oxidized to produce glyoxylate in peroxisomes and not in chloroplast. Otherwise photosynthesis would have been inhibited in chloroplasts. Peroxisomal matrix has got a very efficient system for removal of $H_2O_2$ as well for conversion of glyoxylate to glycine. Glyoxylate, $H_2O_2$, and hydroxypyruvate, the intermediates of glycolate metabolism, are not released because of substrate channeling in a multienzyme complex. It is observed that even if peroxisomal membrane is broken, the enzyme complex remains intact, indicating its protective function. However, in case of any leakage, there are enzymes present in cytosol required for conversion of glyoxylate to glycolate.
- Another significance associated with photorespiration is believed to be its role in preventing **photoinhibition.** In high light intensity and high temperature, stomata

of the plants close in order to prevent water loss. Because of restricted supply and constant utilization in photosynthesis, there is a decrease in concentration of $CO_2$, in comparison with that of $O_2$. As a result of this, $CO_2$ concentration decreases to much below the **compensation point**. Because of the decrease in availability of $CO_2$, intermediates reduced during light reaction accumulate. In absence of $CO_2$, ATP is not consumed leading to reduction in proton gradient created across thylakoids. As a result of this photosynthetic electron transport is also not dissipated leading to ROS production, which causes damage to photosynthetic apparatus. Consumption of NADPH and ATP through photorespiration may be a safety measure strategy adopted by plants, which can utilize NADPH as well as ATP thus protecting the photosynthetic apparatus and thylakoids from damaging effects of light. ATP is also required in photorespiration for conversion of glycerate to phosphoglycerate and for assimilation of $NH_3$ in chloroplasts. Net result of this complex photorespiration cycle is useless dissipation of some of the ATP and NADPH generated by the light reaction. Thus, under insufficient $CO_2$ concentration, Rubisco may undergo oxygenase activity to protect the photosynthetic apparatus from photooxidative damage but resulting in decrease in photosynthetic efficiency also.

- NADH, generated during conversion of glycine to serine, is oxidized through ETC in mitochondria.
- In addition to role of photorespiration in carbon economy, it plays significant role in nitrogen economy. When two molecules of glycine are converted to serine in mitochondria, one nitrogen atom, which is lost as $NH_3$, is re-assimilated in chloroplasts by GS/GOGAT reactions, thus conserving nitrogen. During $NH_3$ assimilation, reducing equivalents are utilized as reduced ferredoxin and NADPH.
- Photorespiratory cycle also plays an important role in synthesis of two essential amino acids, glycine and serine. Alternate pathways for synthesis of these amino acids function when photorespiration is suppressed.

Photorespiration is generally believed to be primarily an evolutionary remnant. It is speculated that Rubisco in bacteria which might have existed in anoxic conditions almost 3.5 billion ($3.5 \times 10^9$) years ago has the sequence similarity with that of plants. Later around 1.5 billion years ago when concentration of free $O_2$ in air increased due to oxygenic photosynthesis, RuBP oxygenase activity increased possibly because of inability of the enzyme to discriminate in between $CO_2$ and $O_2$. Ratio of oxygenase to carboxylase activity decreased possibly due to selection of modification of the active site of the enzyme. Rubisco in anaerobic bacteria has higher oxygenase activity than carboxylase activity, when exposed to aerobic conditions which suggest possible alteration in the enzyme activity having lower oxygenase activity in plants. One of the possible reasons for retaining oxygenase activity of Rubisco activity might have been because oxygenase activity could not be modified without affecting the carboxylase activity.

## 5.6 C4 Pathway and Types of C4 Plants

Plants have developed strategies to reduce carbon loss by photorespiration. This is achieved by increasing $CO_2$ concentration at the Rubisco site, especially by plants growing at high light intensities, high temperatures, and under arid conditions. In these plants, the first stable product of $CO_2$ fixation is a four-carbon compound, viz., oxaloacetic acid, instead of 3-PGA. Earliest observations in this context were made in 1950s by Kortschak and coworkers at the Hawaiian Sugar Planters Association Experimental Station. In order to identify photosynthetic intermediates, leaves of sugarcane were exposed to $^{14}CO_2$, and surprising results were obtained. Unlike in Calvin cycle, the first stable compounds formed as a result of $^{14}CO_2$ fixation in sugarcane were four-carbon compounds—malate and aspartate. Since labeled carbon appeared first in four-carbon dicarboxylic acid and not in 3-PGA, it was known as C4 dicarboxylic acid pathway. The pathway is also called as **Hatch and Slack pathway** after the name of the scientists who elucidated the pathway. This pathway is operative in many tropical grasses, such as sugarcane, maize, sorghum, and *Amaranthus*. In contrast to C3 plants, C4 plants exhibit a different type of leaf anatomy. Vascular bundles in leaves of these plants are surrounded by large parenchyma cells which form a sheath. Chloroplasts present in bundle sheath cells are larger, lack grana and contain starch grains, and are morphologically and functionally distinct from the chloroplasts of mesophyll cells which are smaller and contain grana. This peculiar anatomy in leaves of C4 plants is called **Kranz anatomy**. The bundle sheath cells are bigger and look like a ring or wreath. Kranz in German means wreath, and hence this anatomical feature of the leaves of C4 plants is also referred to as Kranz anatomy. Mesophyll and bundle sheath cells have plasmodesmata connections, which provide symplasmic connections in between them. Unlike mesophyll cells in many C4 plants, there is deposition of suberin in cell walls of the bundle sheath cells. However, in some aquatic plants dimorphic chloroplasts occur within the same cell. Initial fixation of $CO_2$ occurs in cytosol of mesophyll cells which are exposed to both $O_2$ and $CO_2$ in stomatal chamber. Phosphoenolpyruvate, a three-carbon compound, is the acceptor of $CO_2$, and the enzyme which catalyzes the carboxylation reaction is PEP carboxylase (PEPC) (Fig. 5.25). Unlike Rubisco substrate for PEPC is $HCO_3^-$, and oxaloacetate is formed as a result of carboxylation reaction. Carboxydismutase catalyzes the formation of $H_2CO_3$, which gets ionized to produce $HCO_3^-$.

$$H_2O + CO_2 \rightarrow H_2CO_3$$

$$H_2CO_3 \rightarrow H^+ + HCO_3^-$$

$$PEP + HCO_3^- \rightarrow OAA + P_i$$

In C4 plants, both C4 and C3 pathways are spatially separated; C4 pathway occurs in mesophyll cells, while C3 in bundle sheath cells. Basic scheme for C4 pathway involves four steps which include: (i) carboxylation of PEP in cytosol of mesophyll cells is catalyzed by PEPC, resulting in synthesis of C4-dicarboxylic acid OAA; (ii) conversion of OAA to malate occurs in cytosol of mesophyll cells, which is transported to bundle sheath cells through plasmodesmatal connections; (iii) in bundle sheath cells malate is decarboxylated to generate $CO_2$, and a three-carbon compound, pyruvate, is produced. The released $CO_2$ is fixed by Rubisco through Calvin cycle in the chloroplasts of bundle sheath cells; (iv) the three-carbon compound, pyruvate, is transported back to the chloroplasts of mesophyll cells, where it is phosphorylated to produce 2-phosphoenolpyruvate in an enzyme-catalyzed reaction, which then is transported back to cytosol and is ready for another carboxylation reaction (Fig. 5.25). The last step is catalyzed by the enzyme pyruvate orthophosphate dikinase (PPDK) consuming two ATP equivalents.

$$\text{Pyruvate} + \text{ATP} + P_i \xrightarrow{\text{PPDK}} \text{PEP} + \text{AMP} + PP_i$$

The characteristics of this pathway are (1) high $CO_2$ concentration at the site of Rubisco (bundle sheath cells), which is produced as a result of decarboxylation of four-carbon compound. As a result, oxygenase activity of Rubisco is reduced to great extent because of high $CO_2/O_2$ ratio since the enzyme is localized deep inside the tissue and is not exposed to $O_2$ directly. (2) In some C4 plants deposition of suberin in cell wall of bundle sheath cell prevents any escape of $CO_2$. Besides having different pathways operating in bundle sheath and mesophyll cells, the chloroplasts of the two type of cells differ in their structure and function. (3) Bundle sheath cell chloroplasts lack grana and PSII. PSI operates in the bundle sheath chloroplasts which is responsible for cyclic photophosphorylation and ATP generation.

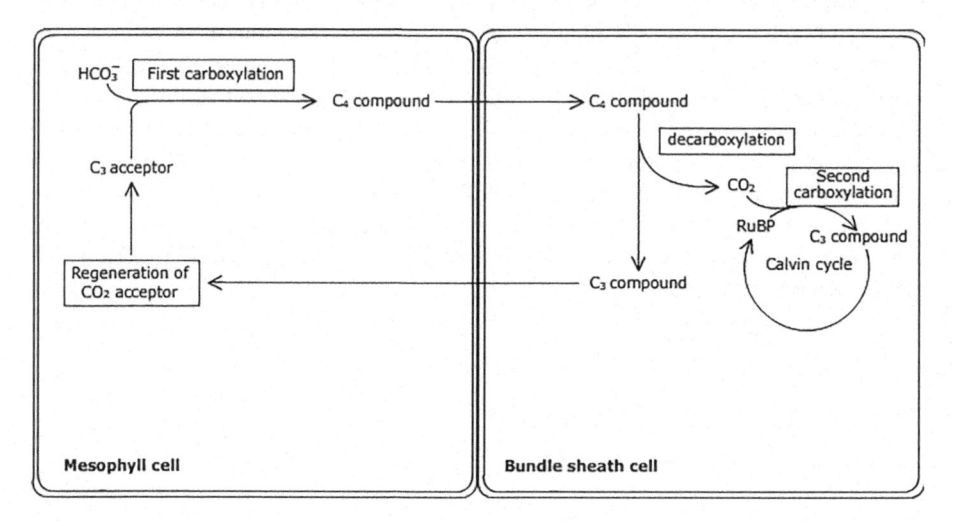

**Fig. 5.25** Overview of C4 pathway. Mesophyll cells are interconnected with bundle sheath cells through plasmodesmata

## 5.6 C4 Pathway and Types of C4 Plants

3-phosphoglycerate, the product of RuBP carboxylation in bundle sheath cells, is transported to mesophyll cell chloroplasts where it is reduced to triose phosphate using NADPH and ATP produced during photosynthetic non-cyclic electron transport since chloroplasts of mesophyll cells have grana and PSII. Triose phosphates are transported back to bundle sheath chloroplasts. (4) Unlike in maize and sorghum, there are some aquatic and land plants which do not have Kranz anatomy. There are single-cell C4 plants in which dimorphic chloroplasts may be present in the same cell; e.g., in *Bienertia sinuspersici* chloroplasts of peripheral and central domain of the cell differ in their structure and function. Peripheral chloroplasts are similar to mesophyll cell chloroplasts, while chloroplasts of the central domain are similar to those of bundle sheath cells. Carboxylation reaction catalyzed by PEPC occur toward outside, while Rubisco has been pushed inside to reduce oxygenase activity (not exposed to $O_2$) so that photorespiratory loss of carbon skeleton is reduced.

**Variations in the Basic Pathway** However, variation of the scheme of C4 pathway has been observed. On the basis of the decarboxylation of four-carbon compound, C4 plants belong to three categories.

**NADP-ME Types (NADP-Malate Enzyme Types)** These include *Sorghum bicolor* and *Zea mays*. In these types of C4 plants, four-carbon exported out of mesophyll chloroplasts to bundle sheath cell is malate, while three-carbon which comes back to mesophyll cells is pyruvate. Oxidative decarboxylation of malate, in chloroplasts of bundle sheath cells, results in release of $CO_2$. The decarboxylation reaction is coupled with reduction of $NADP^+$ to NADPH. NADPH produced during oxidative decarboxylation of malate is used for reduction phase in Calvin cycle as no PSII is present in the bundle sheath chloroplasts, since they lack grana. However, ATP requirement is met through cyclic photophosphorylation which occurs in the agranal thylakoids, as PSI is present. Around 50% of the requirement of NADPH is met through oxidative decarboxylation of malate. 3-PGA is transported back to mesophyll cell chloroplasts where PSII operates, and NADPH generated due to non-cyclic photosynthetic electron transport is used for reduction of 3-PGA to G3P, which is the transported back to chloroplasts of bundle sheath cells (Fig. 5.26).

**NAD-ME Types (NAD-Malate Enzyme Types)** *Panicum miliaceum* and *P. virgatum* are included in this category. In these type of C4 plants exchange of carbon compounds between mesophyll cells and bundle sheath cells occurs as amino acids. Four-carbon compound exported out of mesophyll cells is aspartate, and the three-carbon compound which is returned is alanine. Transamination is carried out by cytosolic aspartate aminotransferase and alanine aminotransferase which are localized in cytosol of mesophyll and bundle sheath cells, respectively. Once in bundle sheath cells, aspartate is first converted back to OAA by aminotransferase in mitochondria, followed by its reduction to malate. The reduction is catalyzed by

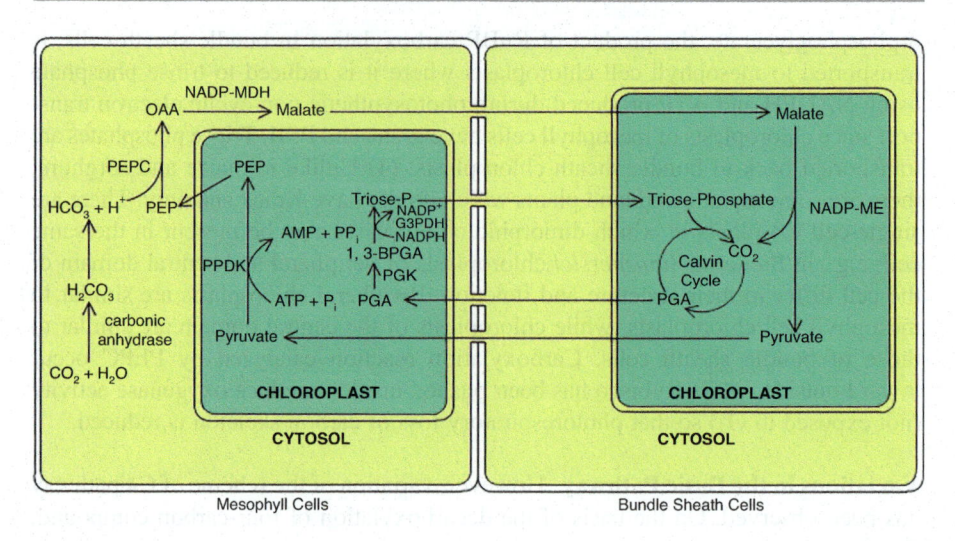

**Fig. 5.26** NADP-ME type of C4 plants, e.g., *Sorghum bicolor* and *Zea mays*. *PEPC* phosphoenolpyruvate carboxylase, *G3PDH* glyceraldehyde 3-phosphate dehydrogenase, *NADP-ME* NADP-malate enzyme

mitochondrial NADH-dependent malate dehydrogenase. Decarboxylation of malate occurs in mitochondria, which is catalyzed by the NAD$^+$-dependent malate enzyme. $CO_2$ produced during decarboxylation diffuses into the chloroplasts and is fixed through Calvin cycle. Pyruvate is transported out of the mitochondria to cytosol where it's conversion to alanine is catalyzed by the aminotransferase. Alanine returns to mesophyll cells through plasmodesmatal connections and is converted to pyruvate by another cytosolic alanine aminotransferase of the cell. In mesophyll cells pyruvate is transported to chloroplasts, where it is converted to PEP catalyzed by the enzyme pyruvate dikinase. PEP is transported out of the chloroplasts to cytosol, where carboxylation reaction catalyzed by PEPC takes place (Fig. 5.27).

**PEPCK (PEP Carboxykinase)** These types of C4 plants include *Megathyrsus maximus*. There are two types of pathway operate in these category of plants. In one type of pathway, 4C compound and 3C compound which are exchanged in between mesophyll cells and bundle sheath cells are aspartate and PEP. Aspartate is converted back to OAA by the cytosolic aminotransferase in bundle sheath cells, which is decarboxylated, resulting in the formation of PEP and release of $CO_2$. Reaction, catalyzed by cytosolic PEPCK, is ATP requiring. Released $CO_2$ is fixed through Calvin cycle in chloroplasts, and PEP is returned to mesophyll cells. The other pathway includes exchange of C4 and C3 compound as malate and alanine, respectively. In mesophyll cells some OAA is reduced to malate in the chloroplasts and is transported to mitochondria of bundle sheath cells. Malate is decarboxylated and oxidized by NAD$^+$-dependent malate enzyme. The released $CO_2$ is fixed by Calvin cycle in the chloroplasts of bundle sheath cells. Pyruvate is converted to alanine by aminotransferases in cytosol of bundle sheath cells. Alanine returns to the

## 5.6 C4 Pathway and Types of C4 Plants

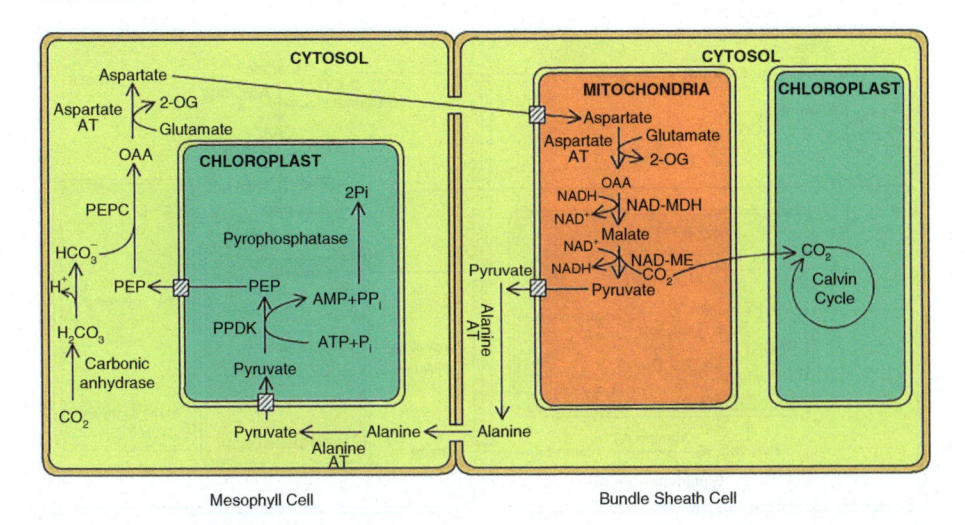

**Fig. 5.27** NAD-ME type of C4 plants, e.g., *Panicum miliaceum* and *P. virgatum*. *PEPC* phosphoenolpyruvate carboxylase, *2-OG* 2-oxoglutarate, *NAD-MDH* NAD-malate dehydrogenase, *NAD-ME* NAD-malate enzyme, *PPDK* pyruvate-phosphate dikinase

mesophyll cells, where it is converted to pyruvate by aminotransferases in the cytosol. Pyruvate is converted to PEP after being transported to the chloroplasts of mesophyll cells (Fig. 5.28).

### 5.6.1 Regulation of C4 Pathway

Similar to regulation of Calvin cycle in C3 plants, C4 pathway is also regulated by light, so that there is coordination in between the two pathways. Three enzymes of C4 pathway which are activated in light, include PEPC, PPDK, and NADP-ME. NADP-ME is regulated by ferredoxin-thioredoxin system. The regulation of PEPC (PEP carboxylase) and PPDK (pyruvate-phosphate dikinase) occurs through phosphorylation and dephosphorylation of the enzyme proteins. Phosphorylated PEPC is active while dephosphorylated form is inactive. In the presence of light, activation of PEPC occurs due to phosphorylation of the enzyme protein at a specific serine residue, which is catalyzed by PEPC kinase. PEPC kinase is active in light, and inactive in dark possibly due to ferredoxin-thioredoxin system. PEPC has more affinity for PEP in light, and less affinity for malate, which acts as inhibitor in dark. Unlike PEPC, phosphorylated PPDK is inactive and dephosphorylated form is active (Fig. 5.29). Phosphorylation of PPDK at threonine residue occurs in dark. Dephosphorylation of the enzyme protein in light restores its activity. Phosphorylation and dephosphorylation of PPDK is carried on by a regulatory protein (RP). ADP is the phosphate group donor in case of PPDK whose availability is influenced by light. In dark ADP concentration is high and ATP concentration is low, reverse is true in light.

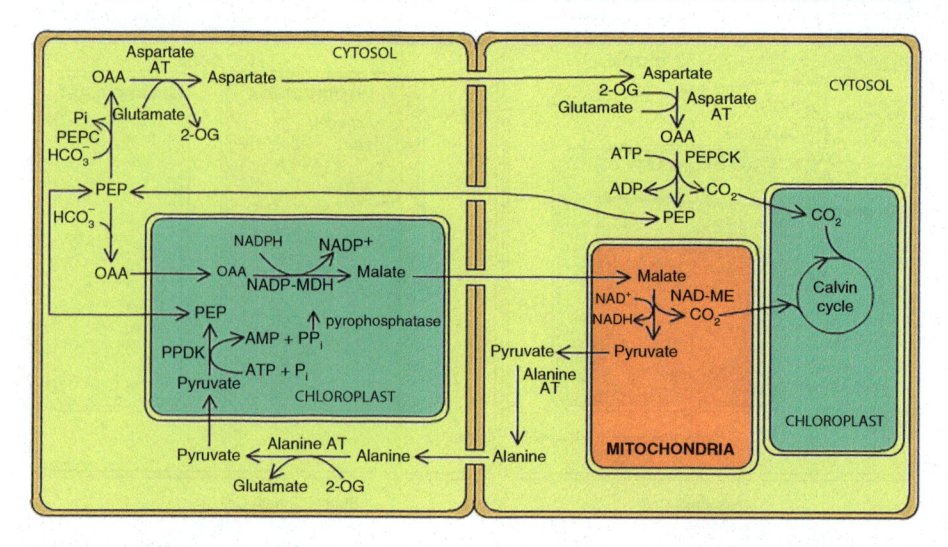

**Fig. 5.28** PEPCK type of C4 plants, e.g., *Megathyrsus maximus*. *PEPCK* PEP carboxykinase, *AT* aminotransferase, *PPDK* pyruvate-phosphate dikinase, *NADP-MDH* NADP-malate dehydrogenase, *NAD-ME* NAD-malate enzyme

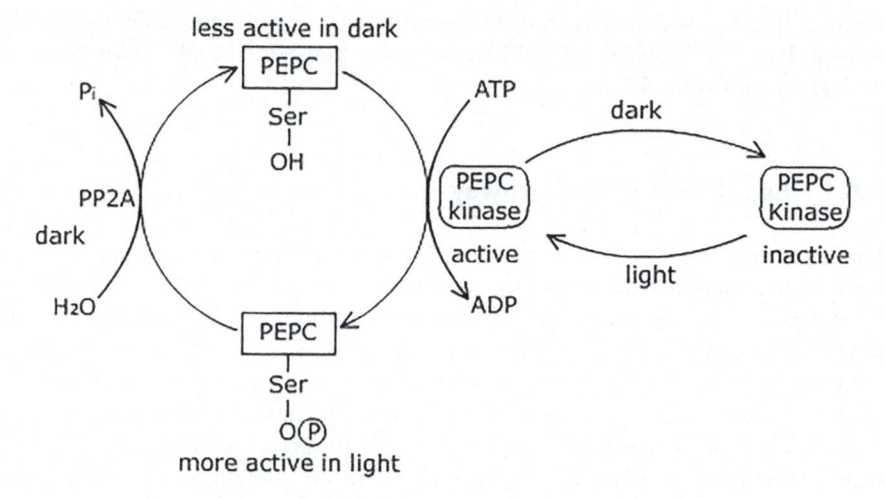

**Fig. 5.29** Regulation of PEP carboxylase (PEPC) by light in C4 plants. *PP2A* type 2 protein phosphatase, *PEPC* PEP carboxylase

### 5.6.2 Energy Requirement for CO₂ Fixation by C4 Pathway

Since in all C4 plants, both C4 and C3 pathways operate, it indicates that additional energy must be required. Two ATPs and three NADPH are required for fixation of one $CO_2$ through C3 pathway, while two additional ATPs are required for regeneration of PEP from pyruvate in C4 pathway. One additional ATP is used by the enzyme PPDK and another by adenylate kinase. The reaction catalyzed by phosphoenolpyruvate dikinase (PPDK) is:

$$\text{Pyruvate} + \text{ATP} + P_i \xrightarrow{\text{PPDK}} \text{PEP} + \text{AMP} + PP_i$$

$$PP_i \xrightarrow{\text{Pyrophosphatase}} 2P_i$$

One molecule of ATP is used for generation of ADP from AMP, in a reaction catalyzed by adenylate kinase.

$$\text{AMP} + \text{ATP} \rightarrow 2\text{ADP}$$

AMP is produced in the reaction catalyzed by PPDK. Thus, for assimilation of one molecule of $CO_2$, at least five ATPs are consumed. Overall equation for synthesis of glucose by C4 plants is as follows:

$$6CO_2 + 24H_2O + 12NADPH + 12H^+ + 30ATP$$
$$\rightarrow C_6H_{12}O_6 + 12NADP^+ + 30ADP + 30P_i$$

### 5.6.3 Evolutionary Significance of C4 Pathway

Though C4 plants consume additional energy for $CO_2$ fixation, they are photosynthetically more efficient than C3 plants. Maize, sorghum, and sugarcane have been found to be world's most productive crops. Most of the C4 plants are tropical plants. When exposed to higher temperature and reduced water supply, stomata of a plant are closed so as to reduce water loss. As a result $CO_2$ uptake is reduced causing decrease in the ratio of $CO_2/O_2$ in stomatal chamber. At high temperature, affinity of Rubisco for $O_2$ also increases due to change in kinetic property of the enzyme. Temperature also influences solubility of the gases. At high temperature $O_2$ is less affected, while there is decrease in solubility of $CO_2$. As a result Rubisco is exposed to relatively higher $O_2$ concentration resulting in higher oxygenase activity of the enzyme. There will be increase in carbon loss because of photorespiration. In C4 plants, Rubisco is localized deep inside the tissue, i.e., in bundle sheath cells where it is not exposed to atmospheric $O_2$. There is increased $CO_2$ concentration at the Rubisco site because of $CO_2$ concentration mechanism. Initial fixation of $CO_2$ by PEPC occurs at mesophyll cells resulting in synthesis of four-carbon compound which is transferred to the site of Rubisco where it gets decarboxylated, thus acting like a $CO_2$ pump. $CO_2$ is concentrated as much as 20-folds in mesophyll cells which is the site of functioning of Rubisco, thus reducing oxygenase activity of the enzyme. Another structural feature of C4 plants is the presence of suberin in the cell walls of bundle sheath cells, which prevents any leakage of $CO_2$ adding to increased $CO_2$

concentration at the site of Rubisco action. Unlike Rubisco substrate for PEPC is $HCO_3^-$. This has two advantages, (i) it is distinct from $CO_2$ in its structure thus there is no competition from $O_2$ for binding with active site of the enzyme, and (ii) $HCO_3^-$ concentration in the medium is five times higher than $CO_2$. Some scientists believe that it is not hot and dry climate rather it is the decrease in $CO_2$ concentration which might have led to development of C4 pathway. Until the end of crustacean period almost 100 million years ago, $CO_2$ would have been higher which fell dramatically during past 100 million years to around 200 ppm. Decrease in $CO_2$ concentration might have led to development of C4 pathway. Increase in $CO_2$ probably would result in reduced photorespiratory loss and increase in photosynthetic efficiency, a phenomenon known as "$CO_2$ fertilization effect."

## 5.7 Crassulacean Acid Metabolism (CAM): $CO_2$ Fixation in Dark

CAM plants are usually succulents which are adapted to grow under extremely adverse xeric conditions. Mesophyll cells of the plants have larger number of chloroplasts, and, unlike $C_4$ plants, vascular bundles are not surrounded by the well-defined bundle sheath cells. In these plants, stomata remain open during night when temperature is low and there is a humid condition but are closed during daytime so as to reduce water loss. They are not as photosynthetically efficient as C4 plants but are better suited to conditions of extreme desiccation. Only 5% of the vascular plants use CAM as their photosynthetic pathway, while some of these may exhibit CAM activity when needed. Earliest observations were made by Romans that leaves of some of the plants tasted bitter early in the morning than the leaves which were harvested later in the day. Benjamin Heyne had made similar observations in 1815 with the leaves of *Bryophyllum calycinum*, a member of family Crassulaceae. In 1804 *Theodore de Saussure* had published his observations that gas exchange in plants such as cactus differed from the thin-leaved plants. Ranson and Thomas possibly had given the term Crassulacean acid metabolism in 1940, but they were not able to discover the cycle. The discovery of CAM can be attributed to the cyclical acidification and deacidification. It was after the discovery of C4 pathway, almost 150 years after the observation were made by Benjamin Heyne, that an understanding about the CAM pathway was made. $C_4$ and CAM pathways are not alternate pathway to C3 cycle, rather they occur in addition to Calvin cycle ($C_3$ pathway). In $C_4$ plants there is spatial separation, while in CAM, there is temporal separation of $C_3$ and $C_4$ pathways (Fig. 5.30).

*Acidification* Diurnal fluctuations of acidity in leaves of succulents were observed due to the accumulation of malic acid during night and decrease in malic acid during daytime. The amount of stored carbohydrates (starch) also fluctuated. Starch accumulated during daytime, while it disappeared during night. Stomata of succulents are open in night so as to reduce water loss, when temperature is low

## 5.7 Crassulacean Acid Metabolism (CAM): CO$_2$ Fixation in Dark

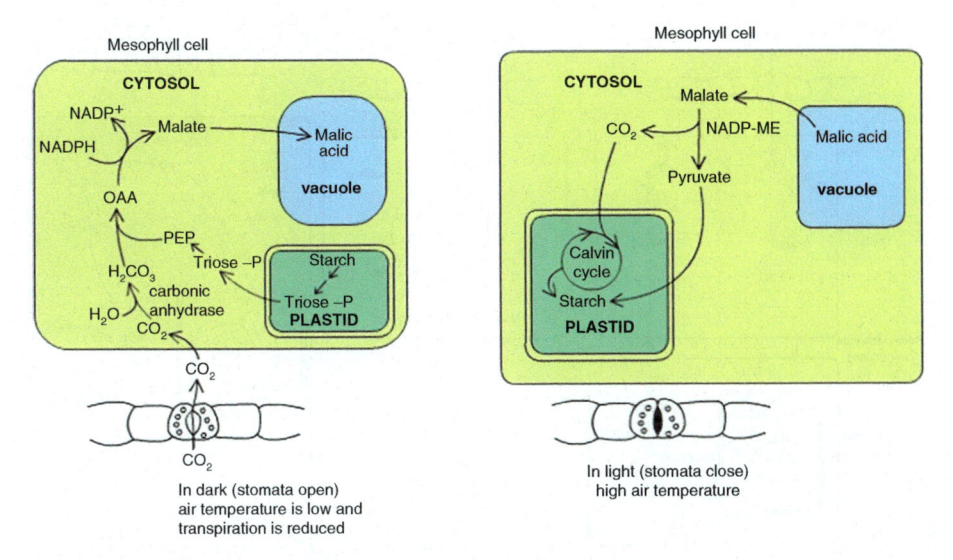

**Fig. 5.30** Comparison of C4, C3, and CAM plants

and air is humid and are open during the daytime. During night CO$_2$ diffuses through the open stomata and is converted to HCO$_3^-$ in cytosol of mesophyll cells. PEPC catalyzes carboxylation of phosphoenolpyruvate (PEP) using HCO$_3^-$ as the substrate.

$$\text{Phosphoenol pyruvate} + \text{HCO}_3^- \xrightarrow{\text{PEP Carboxylase}} \text{Oxaloacetate} + P_i$$

Oxaloacetate is reduced to malate by the enzyme malic acid dehydrogenase. The reaction requires NADPH.

$$\text{Oxaloacetate} + \text{NADPH} + \text{H}^+ \xrightarrow{\text{Malic acid dehydrogenase}} \text{Malate} + \text{NADP}^+$$

Malate which is produced is transported to vacuole through a translocator located in tonoplast and is stored there as malic acid which is responsible for the increase in acidity of the tissues during night.

*Deacidification* Following morning, when the stomata are closed, malic acid moves out of the vacuole and is decarboxylated in cytosol to produce CO$_2$ and pyruvate. The reaction is catalyzed by NADP$^+$-specific malate enzyme. NADP$^+$ is reduced during the reaction. Since stomata are closed CO$_2$ can't escape and is refixed in chloroplasts through Calvin cycle. Calvin cycle enzyme becomes active in light. pH increases due to decrease in malic acid. The process is called deacidification.

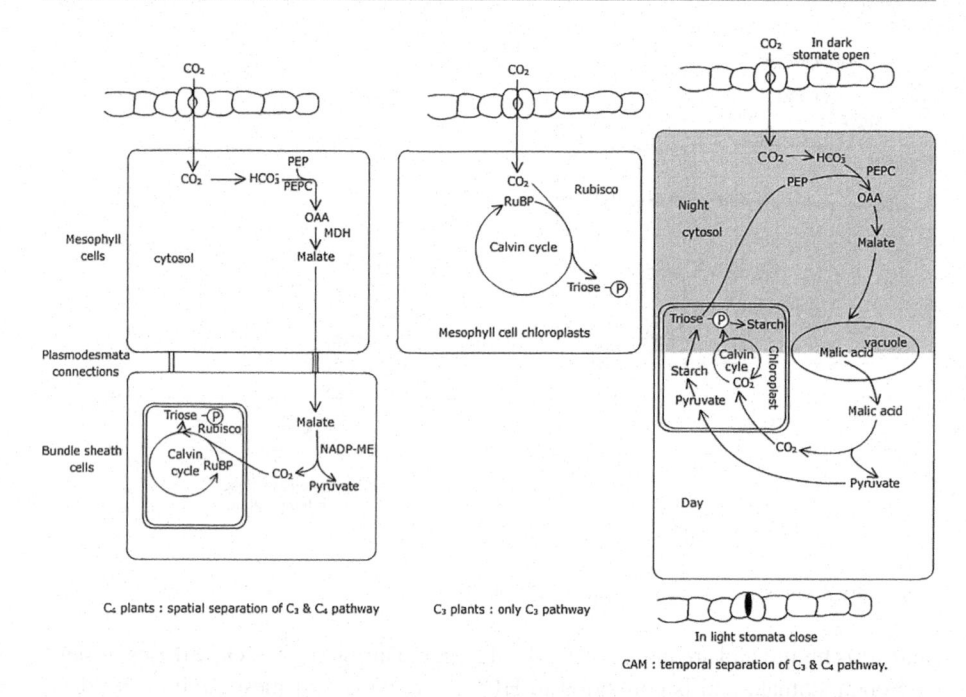

**Fig. 5.31** Crassulacean acid metabolism (CAM). In these plants the occurrence of two pathways of $CO_2$ fixation, i.e., C4 and C3 pathways, is separated by the time. C4 occurs during night when stomata are open, while C3 occurs during daytime when stomata are closed

$$\text{Malate} + NADP^+ \rightarrow \text{Pyruvate} + CO_2 + NADPH + H^+$$

Pyruvate is converted to triose phosphate during Calvin cycle which accumulates in plastids as **transitory starch**. Transitory starch is the source of PEP during dark, which is required as an acceptor of $CO_2$ by PEPC (Fig. 5.31).

Carboxylation of PEP by PEPC during nighttime and decarboxylation of malate during daytime are called phase I and phase III. However, phase II and phase IV are also included in CAM. Phase II refers to $CO_2$ fixation in the morning both by PEPC and Rubisco when stomata open briefly. When plants are subjected to extreme xeric conditions, both light signal and low internal $CO_2$ concentration act as signal for opening of stomata. As malate is decarboxylated $CO_2$ concentration increases and stomata close and phase III begins. At times during later part of the day when malate is depleted and $CO_2$ is exhausted stomata open and $CO_2$ diffuse into the stomatal chamber which is fixed both by PEPC and Rubisco activity resulting in phase IV of CAM. Phase IV occurs in response to environmental signal especially when the plants are exposed to adequate water supply. However, under extreme xeric conditions, stomata remain closed both during day and night, and $CO_2$ is produced because of respiration that is fixed through PEPC activity. There are some CAM

## 5.7 Crassulacean Acid Metabolism (CAM): $CO_2$ Fixation in Dark

which are facultative CAM, e.g., *Mesembryanthemum crystallum*, which change from C3 to C4 when there is decrease in rainfall. In *Kalanchoe blossfeldiana* also CAM is expressed in warmer long days and reduced water supply. Phytochrome might be involved in perception of signal. Unlike C4 plants in which PEPC is active in light, in CAM it is active in dark and is inactive in light. Similar to C4 plants phosphorylated PEPC is active. PEPC kinase is responsible for phosphorylation of PEPC. In CAM, the source of ATP during dark is respiration, while in C4 plants, it is photophosphorylation. Another regulatory factor for the activity of PEPC is malate. When malate accumulates at the end of dark period, PEPC activity is inhibited, while decrease in malate at the end of day PEPC is activated. A comparison of C3, C4, and CAM plants is given in Table 5.4.

### 5.7.1 Ecological Significance of CAM Plants

CAM plants open their stomata during night when exposed to xeric conditions so that water loss due to transpiration is reduced. These have high water-use efficiency (WUE) which can be calculated by the following formula:

$$\text{Water use efficiency} = \frac{\text{Moles of } CO_2 \text{ assimilated}}{\text{Moles of } H_2O \text{ transpired}}$$

WUE of CAM plants is high (4–20 mmol mol$^{-1}$) when compared to C4 and C3 plants which have value of WUE 4–12 mmol mol$^{-1}$ and 2–5 mmol mol$^{-1}$, respectively. Though CAM plants are superior to other types in WUE but their photosynthetic rates and growth rates are much lower. Another way to relate transpiration rates to uptake of $CO_2$ is measurement of transpiration ratio (TR), which is reciprocal of WUE.

$$\text{Transpiration ratio} = \frac{\text{Moles of } H_2O \text{ transpired}}{\text{Moles of } CO_2 \text{ assimilated}}$$

CAM plants have low transpiration ratio (TR) in the range of 50–100, which is substantially lower than that of C3 and C4 plants, which are typically in the range of 200–350 and 500–1000, respectively. Low TR of CAM and C4 plants indicate their capacity to maintain high rates of photosynthesis while effectively conserving water.

**Table 5.4** Comparison among C3, C4 and CAM plants

| Characteristics | C3 plants | C4 plants | CAM plants |
|---|---|---|---|
| Optimal conditions for their growth | Temperate | Tropical | Arid |
| Examples | Wheat, barley, rice | Maize, sugarcane, millets (*Panicum* sp.) | Cacti, *Agave*, pineapple, succulents |
| Pathways for $CO_2$ fixation | C3 | C4 and C3 | C4 and C3 |
| Anatomy of leaves | Mesophyll cells, bundle sheath cells are absent | Kranz anatomy, presence of large chloroplast containing bundle sheath cells present around vascular bundle which are distinct from other mesophyll cells | Mesophyll cells, bundle sheath cells are absent |
| Chloroplasts | Only one type of chloroplasts is present in all mesophyll cells | Chloroplasts are morphologically and functionally dimorphic | Only one type of chloroplasts is present |
| Location of the pathway | Only C3 in all mesophyll cells | C4 in bundle sheath chloroplasts while C3 in mesophyll cell chloroplasts (spatial separation of the two pathways) | In mesophyll cells only, C4 in night and C3 during daytime (temporal separation of the two pathways) |
| First fixation product of $CO_2$ site of its production | 3-PGA in all mesophyll cells | Oxaloacetate in mesophyll cells in daytime | Oxaloacetate in mesophyll cells in night |
| Carboxylating enzyme carrying out initial fixation of $CO_2$ | Rubisco | PEPC | PEPC |
| Energy requirement for fixation one $CO_2$ | 3 ATP + 2 NADPH | 5 ATP + 2 NADPH | 6.5 ATP + 2 NADPH |
| Photorespiration | High | If present very less | If present very less |
| Net assimilation rate, i.e., g dry matter produced per unit leaf area (m$^2$) per day | 10–25 | 40–80 | 6–10 |
| Transpiration ratio | 450–950 g g$^{-1}$ | 250–350 g g$^{-1}$ | 18–125 g g$^{-1}$ |
| Optimum temperature for photosynthesis (°C) | 15–25 | 30–47 | 35 |
| Water use efficiency (moles of $CO_2$ fixed mole$^{-1}$ of water loss) | 1.05–2.22 | 2.85–4.00 | 8.0–55.0 |

## 5.8 Summary

- Green plants and certain bacteria are photoautotrophs since these are able to convert light energy into chemical energy by photosynthesis and store it in the form of organic compounds. Green plants carry out oxygenic photosynthesis thus releasing free $O_2$ as the by the product of the process.
- Light is transmitted in the form of waves, while it is absorbed and emitted in the form of particles known as photons. Energy of a photon is expressed as quanta. Energy of light is generally expressed as mole of quanta, which is known as Einstein. Electromagnetic spectrum of light includes those frequencies which are utilized by the biological systems. Less than 1% of the solar light reaching earth is utilized for photosynthesis.
- Photosynthetic pigments absorb light to be used for photosynthesis. These include chlorophylls, carotenoids, and phycobiliproteins. In nature multiple chlorophylls are present, Chl a, Chl b, Chl c, and Chl d. All chlorophylls consist of porphyrin ring which is esterified to phytol (except in Chl c in which phytol tail is not present). Chl a is present in all photosynthesizing organisms. Role of carotenoids is in harvesting the sunlight and also in protecting photosynthetic apparatus from photooxidative damage. Phycobiliproteins help in harvesting light energy in red and blue-green algae. It is the characteristic for each pigment to absorb specific wavelengths of light which when plotted that gives absorption spectrum. Similarly, each photobiological process requires specific light frequency, which when plotted give action spectrum. By comparing absorption spectrum of a pigment and action spectrum of a photobiological process, the role of a pigment can be understood.
- A pigment is excited after absorbing specific photon since its electron is elevated to higher energy level (singlet state). The excited electron can lose its energy through various pathways including through fluorescence. The electron may be lost to another molecule. As a result the electron donor is oxidized, while the molecule on accepting the electron is reduced. Photochemical reaction is very fast; thus the excited state of the molecule having the lowest lifetime will be responsible for carrying out photochemical reactions.
- It was the work of Blackman, Emerson, and Arnold which led to understanding that photosynthesis includes one light-dependent phase and a light-independent enzymatic phase. Experiment of Robert Hill is also a landmark in photosynthetic research since it led to understanding that light reaction is responsible for photolysis of water and generation of reduced intermediates before $CO_2$ is reduced. These intermediates were later identified to be $NADP^+$.
- While studying action spectrum of photosynthesis in *Chlorella*, Emerson et al. observed sudden decline in quantum yield of photosynthesis in red region of spectrum, while light was still absorbed by chlorophylls. This led to the discovery of two light reactions in photosynthesis, one requiring shorter wavelengths while another requiring longer wavelengths. Duysens conceptualized existence of two pigment systems, which are called photosystem I and photosystem II.

- Each photosystem consists of light-harvesting complex, LHC, and a reaction center. LHC is responsible for absorbing light and funneling it to the reaction center where photochemical reaction occurs. Photochemical reaction includes charge separation event which occurs because of losing of an electron of the primary chlorophyll molecule ($P_{680}$ and $P_{700}$ in PSII and PSI, respectively) and acceptance of the electron by another variants of chlorophyll a. Various electron carriers are involved in transfer of electrons from water to $NADP^+$ including $cyt_{b6f}$ complex.
- Mobile electron carriers, plastoquinone and plastcyanine, mediate electron transport in between the three complexes, PSII, $Cyt_{b6f}$, and PSI, which are localized in thylakoid membrane. Energy of electron transport downhill is coupled with proton transport across thylakoids resulting in accumulation of $H^+$ in thylakoid lumen, as a result of which proton motive force is created, and ATP synthesis coupled with protons movement is known as photophosphorylation. Thus the net outcome of light reaction is photolysis of water resulting in release of $O_2$, NADP reduction to NADPH, and ATP generation.
- NADPH and ATP generated during light reaction are utilized in $CO_2$ assimilation, which occurs in stroma. Path of $CO_2$ assimilation has been traced by Melvin Calvin using radioactive carbon, $^{14}C$, and autoradiography. First compound was identified to be a three-carbon compound, 3-phosphoglycerate, and the $CO_2$ acceptor was identified to be ribulose 1, 5 bisphosphate, which is regenerated during $CO_2$ fixation. The pathway was called as Calvin cycle. Calvin cycle includes three phase, carboxylation phase, reduction phase, and regeneration phase. Rubisco is the enzyme catalyzing the carboxylation phase. During reduction phase, reducing power NADPH and ATP, which is generated during light reaction, is utilized. 1/5 of the triose phosphate produced is diverted out of the chloroplast to cytosol for product biosynthesis, while the remaining 5/6 are used for regeneration of RuBP. Calvin cycle occurs only during light because many of Calvin cycle enzymes are active in light. As a result of this occurrence of futile cycle is avoided.
- In some tropical grasses, initial $CO_2$ fixation product is not a three-carbon compound but is a four-carbon compound. The pathway was elucidated by Hatch and Slack. The plants in which this pathway occurs are called C4 plants. C4 plants have characteristic leaf anatomy known as Kranz anatomy. C4 pathway is not alternative to C3 but rather occurs in addition to C3 pathway. In C4 plants both C4 and C3 pathways are spatially separated. Initial fixation of $CO_2$ in C4 plants involves activity of PEP carboxylase in mesophyll cell cytosol. C4 compound is transported to bundle sheath chloroplasts and is decarboxylated. Released $CO_2$ is refixed through Calvin cycle. Depending upon decarboxylation reaction of C4 compound, three types of C4 plants have been identified.
- Plants belonging to family Crassulaceae have developed a mechanism to conserve water while still carrying out photosynthesis. $CO_2$ is fixed in dark when stomata are open, and the 4C compound is stored in vacuoles in the form of malic acid, while during daytime when stomata are closed, malic acid comes out of vacuole and is decarboxylated. Released $CO_2$ is refixed through C3 pathway.

Thus both C4 and C3 pathways operate in C4 and CAM plants, but these are spatially separated in C4 plants, while there is temporal separation of these two pathways in CAM plants.
- Because of RuBP oxygenase activity in C3 plants, glycolate is also produced in addition to 3PGA which is the substrate for a process known as photorespiration. The process results in loss of almost 25% of the carbon fixed through photosynthesis. Plants have developed a metabolism to get rid of glycolate, which is known as glycolate metabolism. This involves many reactions and various organelles which include chloroplast, glyoxysome, and mitochondria. Photorespiration may act as energy spill-over mechanism.

## Multiple-Choice Questions

1. Excited state of chlorophyll molecule responsible for the photochemical reaction of photosynthesis is:
   (a) First singlet state
   (b) Second singlet state
   (c) Triplet state
   (d) None of the above
2. Photochemical reaction requires:
   (a) $10^{-3}$ seconds
   (b) $10^{-12}$ seconds
   (c) $10^{-9}$ seconds
   (d) $10^{-2}$ seconds
3. Phenomenon of loss of energy of excited pigment molecule as light wavelength of longer wavelength than the wavelength of absorbed light is known as:
   (a) Homogenous energy transfer
   (b) Resonance
   (c) Fluorescence
   (d) Phosphorescence
4. Primary electron acceptor from excited $P_{680}$ in PSII is:
   (a) $Q_A$
   (b) Phylloquinone
   (c) $Tyr_z$
   (d) Pheophytin
5. Primary electron donor to $P_{680}^+$ in PSII is:
   (a) $Tyr_z$
   (b) $H_2O$
   (c) Oxygen-evolving complex
   (d) $Q_B$

6. Quantum yield in photosynthesis is defined as:
   (a) Number of quanta required for release of one $O_2$
   (b) Number of $O_2$ molecules produced per quanta absorbed
   (c) Number of chlorophyll molecules required to absorb one quanta
   (d) Number of chlorophyll molecules responsible for release of one $O_2$
7. (Tick which is not right) Activation of Rubisco by light is due to:
   (a) Decrease in pH of lumen
   (b) Increase in pH of stroma
   (c) Due to release of Rubisco from thylakoids into the stroma
   (d) Due to $Mg^{+2}$ moving out of thylakoid lumen to stroma
8. Which of the following reactions in Calvin cycle is reversible?
   (a) Conversion of sedoheptulose 1,7-bisphosphate to sedoheptulose 7-phosphate
   (b) Synthesis of ribulose 1,5-bisphosphate from ribose 5-phosphate
   (c) Synthesis of 3-phosphoglyceraldehyde from 3PGA
   (d) Conversion of xylulose 5-phosphate to ribulose 5-phosphate
9. The three types of C4 plants differ from each other on the basis of:
   (a) Chemical nature of C4 compound transported out of mesophyll cells
   (b) Decarboxylation reaction of C4 compound in bundle sheath cells
   (c) Chemical nature of C3 compound which is returned to mesophyll cells
   (d) The enzyme which is responsible for initial carboxylation of $CO_2$ acceptor
10. Tick incorrect statement:
    (a) Oxidation of glycolate to glyoxylate occurs in peroxisomes
    (b) P-glycolate is dephosphorylated to glycolate in peroxisomes
    (c) $NH_3$ is produced in mitochondria
    (d) α-hydroxypyruvate is reduced to glycerate in glyoxysomes

## Answers

1. a   2. b   3. c   4. d   5. a   6. b   7. c
8. d   9. b   10. b

## Suggested Further Readings

Jones R, Ougham H, Thomas H, Waaland S (2013) The molecular life of plants. Wiley-Blackwell, Chichester, pp 284–326

Nelson DL, Cox MM (2017) Lehninger principles of biochemistry, 7th edn. W.H. Freeman, New York, pp 755–798

Niyogi KK, Wolosiuk RA, Malkin R (2015) Photosynthesis. In: Buchanan BB, Gruissem W, Jones RL (eds) Biochemistry and molecular biology of plants. Wiley-Blackwell, Chichester, pp 508–565

Taiz L, Ziegler E, Moller IM, Murphy A (2015) Plant physiology and development, 6th edn. Sinauer Associates Inc, Sunderland, pp 171–239

# Photoassimilate Translocation

Rashmi Shakya and Manju A. Lal

During evolution, early land plants were challenged by serious environmental pressures for their survival. The main challenge was the absorption and retention of water. This pressure for survival led to the differentiation of roots for absorption of water and inorganic nutrients and leaves for light absorption, photosynthesis, and gaseous exchange. Development of leaves rendered plants capable of carrying out photosynthesis. Xylem strands are responsible for transport of water and minerals from roots to the aerial parts of the plants, while translocation of photosynthetic products (**photoassimilates**) is facilitated by phloem elements. It is estimated that as much as 80% of the photosynthetically fixed carbon can be exported out of mature leaves. Storage or photosynthesizing organs, which have surplus sugars, can either metabolize or export them. These are known as **source**. On the contrary, actively metabolizing organs or the ones which store carbohydrates need to import them. These plant parts are known as **sinks**. Life cycle of a plant is characterized by source-sink transitions due to changes in sink strength. A plant consists of series of sources and sinks, with many sinks competing for sugars exported by the source organs. Phloem plays a major role in connecting source and sink. In the early developmental stage of a plant, roots and shoots majorly compete for receiving photoassimilates, and later on many other organs become effective sinks. These include reproductive structures, buds and flowers, and developing grains or the underground storage organs, such as tubers. **Sink strength** or **sink dominance** refers to the capacity of sink organs to acquire sugars from the transporting vascular strands. Distribution of sugars in the sink is the key factor in determining the **harvest index (HI)**, which refers to the ratio of dry weight of harvestable part (economically important) of the plant to the total dry weight of the plant. The higher the ratio (high value of HI), the higher the plant productivity. Thus, transport of photoassimilates is targeted as the key factor determining plant productivity. Various abiotic and biotic factors adversely affect translocation of sugars. Accumulation of sugars in the cytosol of mesophyll cells at the source is the key factor for inhibiting photosynthesis. Studying phloem structure is important since it would facilitate in learning about the molecular mechanism involved in sugar translocation. Mineral deficiency may result in increase in the ratio of roots to shoots. Various sucrose transporter proteins (SUTs), facilitating intracellular sugar transport between various subcellular compartments as well as from cell to cell, have potential roles in controlling sucrose movement to the desired sinks. The chapter deals with the source and sink concept, pathways involved in photoassimilate transport, and unique features of sugar translocation. Later part of the chapter shall explore the mechanisms of phloem translocation, including phloem loading and unloading of photoassimilates and distribution of photoassimilates that encompasses allocation and partitioning.

## 6.1  Source-Sink Relationship

Translocation of photoassimilates occur in phloem which can be functionally characterized into three different zones along the source-to-sink pathway (Fig. 6.1). At the sources, these are often referred as **collection phloem**, while at the sinks, these are termed as **release phloem**. The connecting pathways of the two

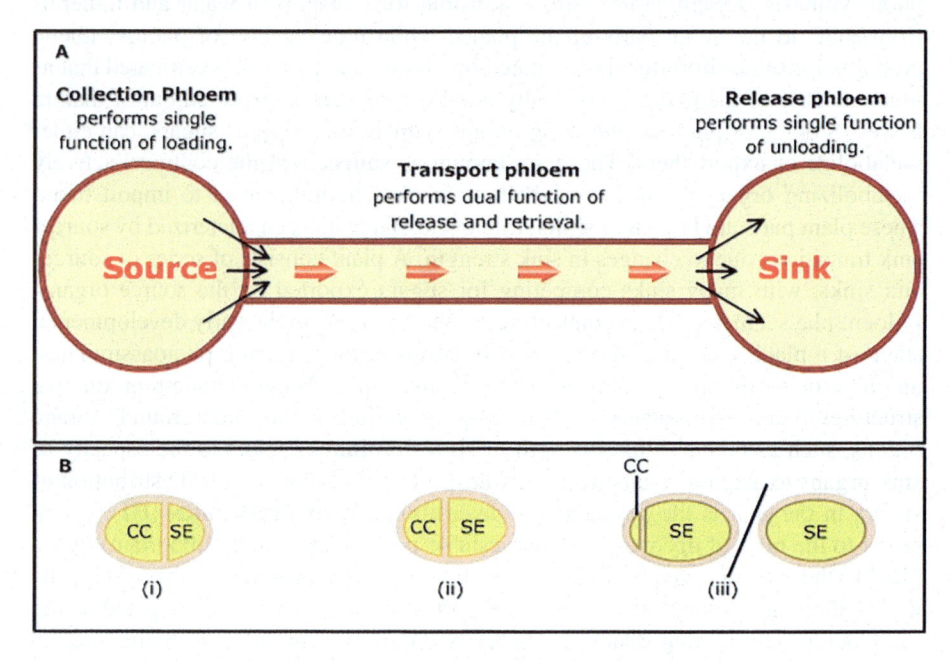

**Fig. 6.1** (a) Three functional zones identified in phloem. (b) Three different phloem zones can be characterized by the size ratios between sieve element and companion cell. *CC* companion cell, *SE* sieve element

are known as **transport phloem**. Supply of photoassimilates from all sources does not reach all sinks. Instead, specific sinks are preferred over others by certain sources. Factors affecting the movement of photoassimilates from source to sink are as follows:

1. *Proximity of source to sink*—Mature leaves (sources) located on the upper region of the aerial parts of plants usually provide photoassimilates to the developing immature leaves (sinks) on the same orthostichy (a vertical row of leaves arranged one directly above another). Leaves present on the lower portion of stem predominantly supply underground parts of plants, whereas leaves present in the middle portion of the stem supply in both upward and downward directions.
2. *Developmental stage*—Root and shoot apices are usually the major sinks during vegetative growth, whereas developing fruits become the major sinks during reproductive phase. At the time of senescence, mature leaves serve as sink. Thus, there is change in the source and sink status of the growing organs during plant development.
3. *Vascular connections*—Sinks which have direct vascular connections with source are preferred.
4. *Modification in translocation pathways*—Wounding interferes with the translocation pathway and leads to the alteration of translocation patterns in relation to proximity and vascular connections. In fact, vascular interconnections, known as anastomoses (reconnection of leaf veins that were previously branched out), act as alternative pathway for translocation in the absence of direct vascular connections between source and sink.
5. *Sink strength*—Ability of a sink to store or to metabolize sugar imports determines its capacity to compete for sugars exported by various source tissues.

Removal of a sink results in increased translocation of sugars to other available and competing sinks. Young leaves act as stronger sinks in comparison to roots when supply from a source is compromised. Rapid utilization of sugars by sink cells results in lowering the concentration of photoassimilates in sieve elements of the young leaves resulting in lowering of the hydrostatic pressure. As a result, an increase in pressure gradient between source and sink is evident, and it leads to change in the translocation of photoassimilates. This ability of young leaves to mobilize sugars toward themselves is due to their relatively high **sink strength**. Sink strength is dependent on the size (total weight of sink tissue) and activity of sink (rate of uptake of transport sugars per unit weight of sink tissue). Translocation patterns may be modulated because of alteration in either sink size or activity.

$$Sink\ strength = sink\ size \times sink\ activity$$

## 6.2 Transition of Leaf from Sink to Source

In eudicots, leaves act as sinks during the initial stages of their development. The transition of leaves from sink to source takes place gradually. This transition is initiated when the leaf attains 25% of its mature size and transition is usually completed in 40–50% expanded leaves. Transition during leaf development is accompanied by several anatomical and physiological changes responsible for the reversal of function from sinks to source. Sugar export from the leaf is initiated at the apex of the leaf blade and gradually moves toward the base until the entire leaf functions as a photoassimilate exporter. During this transition, leaf tip exports sugars, whereas the base imports it from other source leaves. Initiation of export and cessation of import are two independent processes. In tobacco leaves, it has been shown that loading and unloading of sugars occurs through entirely different veins (Fig. 6.2a). When leaf is young, it is in its sink phase, and it imports photosynthates from mature leaves, which gets distributed throughout lamina via major veins (Fig. 6.2b, thick lines marked with arrows). The photosynthate import unloads from the major veins into the mesophyll. The minor veins (marked with IV) which are enclosed by major veins of third-order (III) do not function in import and unloading as they are immature. The minor veins do not attain maturity until the import ceases from the source leaves. The cessation of import involves blockage of unloading from the major veins for sometime during the leaf development. The closing of plasmodesmata and a decrease in plasmodesmatal frequency are instrumental in stopping the import of sugars. Leaves begin to export sugars after the closure of sugar import route and accumulation of sufficient photoassimilates in the sieve elements necessary to initiate translocation (acts as source). In a source leaf, photoassimilates get loaded into the minor veins, and the major veins function for export only, i.e., they can no longer unload the photosynthate (Fig. 6.2c).

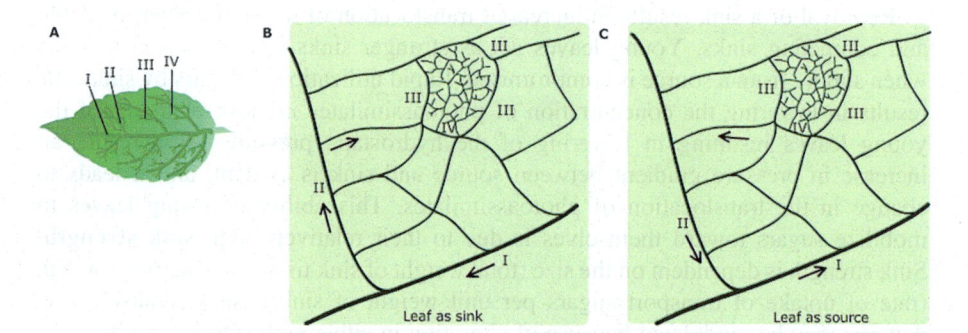

**Fig. 6.2** (a) Veins in the leaves show division of labor. (b) When leaf is immature, it acts as a sink. (c) When the leaf is mature, it starts functioning as source. (I) Larger major vein of first order (midrib), (II) major veins of second order, (III) major veins of third order, (IV) minor veins. Arrows indicate the direction of flow of photoassimilate

## 6.3 Pathway of Photoassimilate Translocation

### 6.3.1 Experimental Evidence

#### 6.3.1.1 Girdling

In 1686, Marcello Malpighi (an Italian anatomist) performed a classical experiment on the translocation of organic solutes. In this experiment, bark of a tree was removed in the form of a ring around the trunk. In 1727, Stephan Hales (an English clergyman) repeated the girdling experiment. In this experiment, a strip of bark around a tree trunk was removed which effectively eliminated the phloem elements. Subsequently, swelling in the region of the bark just above the girdle was observed. Contrary to this, bark region which was present immediately below the girdle shrank (Fig. 6.3). The experiment demonstrated that due to girdling, transport of sugars from photosynthesizing leaves to the roots was obstructed. However, transport of water through xylem remained unaffected. The experiment demonstrated that transport of sugars from leaves to roots occurs through phloem. The plant died after some time which demonstrated that the photoassimilates are essential for growth of those plant parts which cannot perform photosynthesis.

#### 6.3.1.2 Autoradiography

Availability of radioactive compounds after World War II provided scientists an opportunity to use them in various experiments. **Reverse-flap** technique was used

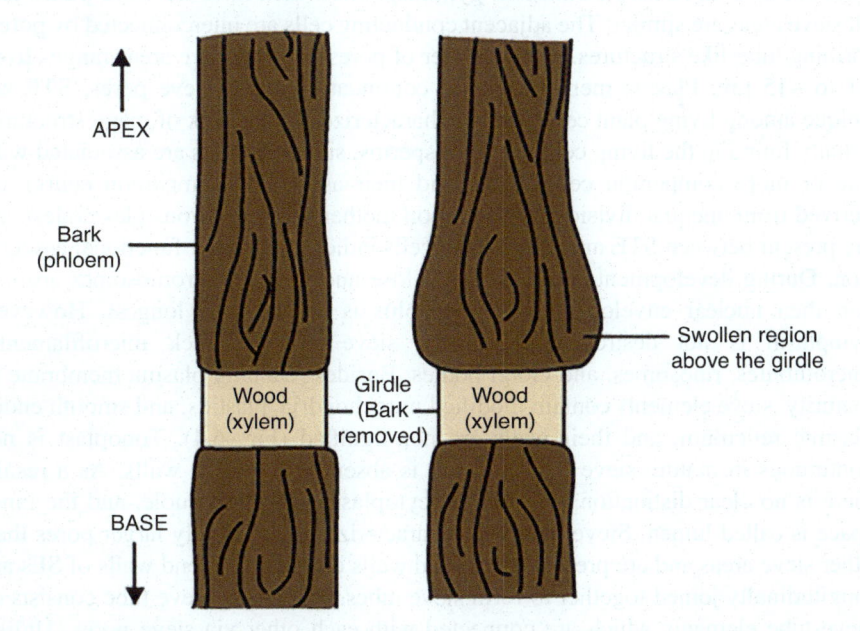

**Fig. 6.3** Diagrammatic representation of girdling experiment in stem of an intact plant, showing swelling of the region above the girdle due to accumulation of sugars

for application of radioactive tracers on the leaf. In another approach, the leaf cuticle was removed by abrasion, and then radioactive compounds were applied directly to the leaf. Alternatively, leaf was exposed to labeled carbon dioxide ($^{14}CO_2$) in a closed chamber. Subsequently, $^{14}CO_2$ incorporated into the photoassimilates was analyzed which then got exported via translocation stream. The labeled $^{14}C$ first gets incorporated into sucrose, but later it is incorporated in many other organic compounds. Localization of these radioactive compounds can be done with the help of autoradiography technique.

## 6.4 Features of Phloem Cells with Reference to Photoassimilate Translocation

Phloem consists of sieve elements (SEs), companion cells, and phloem parenchyma. In addition, phloem tissue may include fibers and sclereids also which primarily provide strength and protection. Phloem parenchyma stores and releases food molecules. Mature sieve elements (SE) are phloem cells which are specialized for translocation. Sieve elements were discovered by Theodor Hartig in 1837, and their role in long-distance transport was demonstrated by him in 1860. A sieve element is a composite term used for sieve-tube elements (in angiosperms, STE) and sieve cells (in gymnosperms). Sieve areas are present on the cell walls of sieve elements. STEs in angiosperms are characterized by the presence of sieve plates which are differentiated sieve areas, unlike in gymnosperms which have no sieve plates and all sieve areas are similar. The adjacent conducting cells are interconnected by pores, forming tube-like structures. The diameter of pores in other sieve areas ranges from <1 to ~15 µm. Plasma membrane is in continuation at the sieve pores. STE are unique among living plant cells and are characterized by the lack of many structures usually found in the living cells. In angiosperms, sieve elements are associated with one or more companion cells. STEs and their associated companion cell(s) are derived from unequal division of a common mother cell. Numerous plasmodesmata are present between STE and companion cells indicating of their functional association. During development, sieve elements lose nuclei. First, chromosomes are lost and then nuclear envelop is lost. Nucleolus is retained the longest. However, cytoplasm is not destroyed. Generally, sieve elements lack microfilaments, microtubules, ribosomes, and Golgi bodies. Besides retaining plasma membrane at maturity, sieve elements contain modified mitochondria, plastids, and smooth endoplasmic reticulum, and their walls are non-lignified (Fig. 6.4). Tonoplast is not continuous in mature sieve elements and is absent at the cross walls. As a result, there is no clear distinction between the cytoplasm and the vacuole, and the inner space is called lumen. Sieve plates are characterized by relatively larger pores than other sieve areas and are present on the end walls of STEs. The end walls of SEs are longitudinally joined together to form sieve tubes. Thus, long sieve tube consists of sieve-tube elements which are connected with each other via sieve pores. Unlike

## 6.4 Features of Phloem Cells with Reference to Photoassimilate Translocation

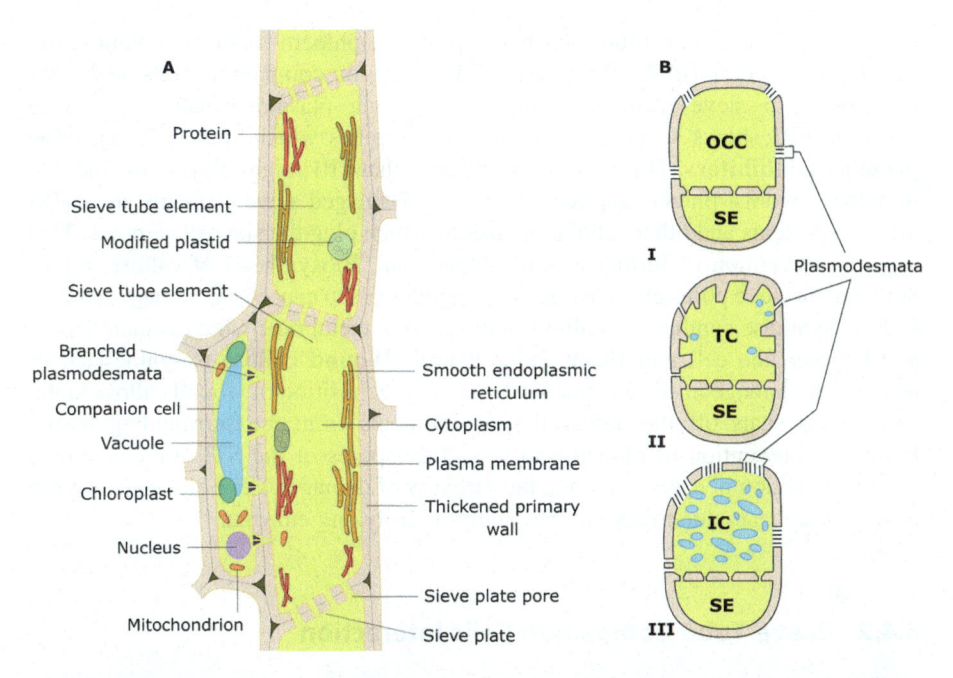

**Fig. 6.4** (a) Diagrammatic representation of the ultrastructural features of a mature sieve element. (b) Diagrammatic representation of three different types of companion cells (arrows pointing plasmodesmata). *SE* sieve elements, *OCC* ordinary companion cell (I), *TC* transfer cell (II), *IC* intermediary cell (III)

xylem vessels, these are living cells. These cells can be plasmolyzed since plasma membrane of these cells is maintained because of which sieve-tube elements retain the capacity to generate turgor pressure.

### 6.4.1 Phloem Sealing Mechanism

During translocation process, sieve tubes are under very high internal turgor pressure. Whenever there is damage to a sieve tube (e.g., a cut), the release of pressure results in oozing of its contents from the cut end which causes loss of **phloem sap** containing photoassimilates. Plants employ sealing mechanism to avoid loss of photoassimilates. Short-term plugging involves P-proteins, while callose is involved in the long-term damage control mechanism. **P-proteins** (earlier known as slime) are found in STE of most of the angiosperms (except few monocots) but are absent in gymnosperms. P-proteins may exist in tubular, granular, fibrillar, or crystalline forms depending upon the species and maturity of the cell. In immature cells P-proteins appear as discrete bodies and are referred as **P-protein bodies**. P-protein bodies may be spindle-shaped, spheroidal, coiled, or twisted. In *Cucurbita* *maxima*, P-proteins constitute two major proteins, phloem filament protein (PP1) and phloem lectin (PP2). They are synthesized in companion cells and later transported to sieve elements symplastically (via plasmodesmata). The gene encoding PP1 shows sequence similarity to the genes which encode for **cysteine proteinase inhibitors**. This homology indicates that PP1 might play a possible role in defense against phloem sap feeding insects. Damaged sieve elements are sealed off by P-proteins and other cellular inclusions which plug the sieve plate pores. This leads to prevention of further loss of phloem sap. Biosynthesis of callose, a β-1, 3-glucan, in sieve pores also prevents loss of phloem sap caused by damage to STEs. Callose synthase synthesizes callose in the plasma membrane which is then deposited between the cell membrane and cell wall. **Wound callose** is synthesized in response to damage in sieve elements. Gradually, deposition of wound callose in the sieve pores seals off the damaged sieve elements from the surrounding tissue, leading to prevention of phloem sap loss. Sieve pores of the STE are constricted due to deposition of callose. During the recovery of damaged sieve elements, wound callose disappears due to activity of callose-hydrolyzing enzyme.

### 6.4.2 Sieve Tube-Companion Cells Interaction

Companion cells are developmentally and functionally closely associated with the STE. Each STE is associated with one or more companion cells. These are connected to STE through numerous plasmodesmatal connections. Since companion cells are not longitudinally continuous, they do not constitute pathway for photoassimilate transport. Transport of photoassimilates to STEs in minor veins of the leaves is facilitated by companion cells. Venation pattern in leaves is designed to optimize photoassimilate transport from mesophyll cells to SE/CC complex. Minor veins in the leaves are the catchment area which consists of one or few SE/CC complexes. Structurally, minor veins have few sieve elements and much larger parenchyma cells. Minor veins constitute collection phloem. CCs are believed to provide energy to STEs in the form of ATP (by virtue of having several mitochondria) and proteins because the cytoplasm of STEs lacks mitochondria, which are crucial for maintenance of cell structure. Companion cells have nucleus with large degree of endoploidy. The presence of dense cytoplasm with large amount of RNA and many mitochondria in CCs suggest that they are metabolically active. High ATPase activity has been demonstrated in companion cells. Species have been categorized depending on apoplasmic and symplasmic continuity of the minor veins with the phloem parenchyma cells which surround them (discussed later). Type 1 species include the ones in which there are symplasmic connections in between the STE/CC complex and the surrounding parenchyma cells because of the presence of many plasmodesmatal connections. The companion cells in type 1 species are known as intermediary cells (Fig. 6.4b). Intermediary cells possess several small vacuoles but have chloroplasts with poorly developed thylakoids. They also lack starch grains in the chloroplasts. Several plasmodesmata are present at the interface of these types of companion cells with the surrounding cells, especially with the bundle sheath cells. Type 2 species include those in which there are moderate-to-low number of plasmodesmata between STE/CC and the surrounding parenchymatous cells. In type 2A category, ordinary companion cells are characterized by relatively fewer plasmodesmatal connections with their surrounding phloem parenchyma cells. They possess chloroplasts with well-developed thylakoids, and the inner surface of the cell wall is smooth. Apparently, it appears that these might be involved in phloem loading mainly through apoplastic way or at times symplastic way since they either have none or variable number of plasmodesmatal connections with the surrounding cells. The second category of type 2, i.e., type 2B, possess companion cells which have finger-like wall ingrowths on the cell walls away from the sieve elements. These are known as **transfer cells**. The wall ingrowths increase the surface area of plasma membrane, thus increasing the transfer of solutes across the membrane. Similar to ordinary companion cells, transfer cells also possess very few plasmodesmatal connections with the surrounding cells and appear to be symplastically isolated. It is believed that these cells are specialized in solute uptake via apoplastic pathway. However, the role of few plasmodesmatal connections is presently not known. Transfer cells may also possess chloroplasts with well-developed thylakoids. Ordinary companion cells and transfer cells are features of those plants in which photosynthates are transferred apoplastically from mesophyll cells to sieve elements. On the other hand, intermediary cells function in symplastic transport of photosynthates from mesophyll cells to sieve elements. Table 6.1 depicts the relationship between the structure of various types of companion cells in minor veins and their role in phloem loading.

### 6.4.3 Composition of the Phloem Sap

It is difficult to obtain pure sap from STE of the phloem for analysis because of the possibility of contamination from the other injured tissues at the time of sample collection. Another difficulty in collecting phloem sap is due to the sealing

**Table 6.1** Relationship between the structure of companion cells in minor veins and mode of phloem loading

| Type of companion cells | Ordinary companion cell | Intermediary companion cell | Transfer cells |
|---|---|---|---|
| No. of plasmodesmata at the interface of companion cell and its neighboring cells | 10–0.1 per $\mu m^2$ | More than 10 per $\mu m^2$ | Less than 0.1 per $\mu m^2$ |
| Predominant transport sugar | Sucrose, sugar alcohols, and RFOs[a] | Sucrose, galactosyl oligosaccharides (RFOs) | Mainly sucrose |

[a]*RFO* Raffinose family oligosaccharides

mechanism which occurs in the STEs at the time of mechanical injury. One of the methods used to reduce sealing due to injury is by a chelating agent such as EDTA, responsible for chelating $Ca^{2+}$ ions required for callose synthesis. Another mechanism involved is collecting the sap from the severed stylet of an aphid after it has been anesthetized (Box 6.1). Possible presence of some compounds in the stylet does not allow the sealing in STE during suction by aphids. Since sap in the STE is under pressure, it keeps on oozing out of the stylet and can be collected. Chemical

---

**Box 6.1: Technique for Collection of Phloem Sap**

High turgor pressure in the sieve-tube elements and wound reactions make the collection of the phloem sap a challenging task. Sometimes during the process of severing, phloem damage is caused, and this leads to the contamination of the phloem sap. Sudden pressure released due to damage causes disruption of cellular organelles and proteins. Dilution of the phloem sample occurs due to influx of water from the xylem. Thus, analysis of phloem sap collected after severing the phloem is usually not a preferable approach.

A reliable approach for collection of phloem sap exploits aphid stylet as a "natural syringe." Aphids are small sap-sucking insects belonging to the superfamily Aphidoidea. Their mouth part consists of four tubular stylets which are inserted into shared cell wall between epidermal cells and steer them through cortical parenchyma cells (CPCs) and phloem parenchyma cells (PPCs) and finally to sieve tubes (STs). During this navigation, aphids secrete gel saliva which hardens into tubule and facilitate the forward movement of stylet. Aphids are anesthetized with $CO_2$ and their stylets are cut using laser. The phloem sap oozes out from the cut end due to high turgor pressure in the sieve elements which is collected for further analysis. The amount of phloem sap collected is small. This method of phloem sap collection is technically difficult but is believed to yield relatively pure phloem sap and, therefore, gives fairly accurate details about the composition of phloem sap.

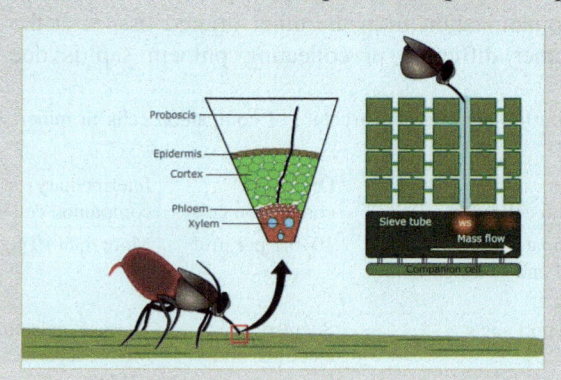

analysis of the phloem sap, collected either from an aphid stylet or from wounds that have severed sieve elements, has revealed that it contains water, sugars, RNA, proteins, amino acids, organic acids, hormones, and mineral ions. Table 6.2 shows different types of materials translocated in the phloem. Ninety percent of the organic molecules present in phloem sap consists of carbohydrates which is mainly sucrose. Sucrose (0.3–0.9 M) has emerged as the most preferred sugar for long-distance translocation of photosynthates. This could probably be attributed to its nonreducing nature. The nonreducing sugars are less reactive in comparison to the reducing sugars. This is due to the fact that their ketone or aldehyde groups are reduced to an alcohol or they combine with a similar group on another sugar. The acetal link between the subunits is stable and nonreactive in alkaline conditions. On the contrary, reducing sugars such as glucose and fructose have a free aldehyde or ketone functional group which is capable of reducing mild oxidizing agents. Contrary to starch, sucrose is water soluble. In the members of families Rosaceae and Cucurbitaceae, oligosaccharides, such as raffinose (a trisaccharide), stachyose (a tetrasaccharide), and verbascose (a pentasaccharide), are also translocated in the phloem. In the stem internodes of *Cucurbita maxima*, stachyose (sucrose + two molecules of galactose) accounts for about 46% of the total translocated sugars, while in Oleaceae and Rosaceae families, polyols, such as mannitol and sorbitol, are also translocated. In *Sorbus aucuparia* (family Rosaceae), sorbitol is the main carbohydrate component of phloem sap next to sucrose. There are specific polyol transporters. pH of the phloem sap is high (7.4–8.7). Reduced glutathione, a translocated form of sulfur, is also present in phloem sap. Cations, such as $K^+$ and $Mg^{2+}$, are also found in the phloem sap besides the presence of trance amounts of $Ca^{2+}$.

In phloem, nitrogen is translocated in the form of amino acids, mainly glutamic acid and aspartic acid, and their respective amides, glutamine and asparagine. However, concentration of amino acids and organic acids is variable in the same species and is usually low as compared to sugars. Due to the presence of high amounts of carbohydrates, C/N ratio of phloem sap is much higher than that of xylem sap and ranges between 15 and 200. Levels of nitrogenous compounds being transported are quite high during leaf senescence since these get mobilized to the

**Table 6.2** Composition of phloem exudates from the stem of castor bean (*Ricinus communis*)

| Constituents | Concentration (mg.mL$^{-1}$) |
|---|---|
| Sugars | 80–106 |
| Amino acids | 5 |
| Potassium | 2–4 |
| Organic acids | 2–3 |
| Proteins | 1–2 |
| Chloride | 0.4–0.7 |
| Phosphate | 0.4–0.6 |
| Magnesium | ~0.1 |

woody tissues for storage. Plants having nitrogen-fixing nodules transport nitrogen in the form of ureides, such as allantoic acid, allantoin, and citrulline. Proteins and RNA are also transported in low concentrations. P-proteins are the predominant form of proteins which occur in the phloem sap. Several enzymes, such as protein kinases, thioredoxin, ubiquitin, and chaperones, have also been reported. RNAs, which include mRNA, pathogenic RNAs, and small regulatory RNA, also occur in the phloem sap. RNA molecules usually form complexes with proteins as ribonucleoproteins (RNPs). RNAs and proteins probably act as signal molecules. Reportedly, phloem sap contains plant hormones, such as auxin, gibberellin, cytokinin, and abscisic acid. It has been suggested that, to some extent, long-distance transport of auxin occurs in the sieve elements. Plant defense hormone, jasmonic acid, has also been reported in the phloem sap. Recently, translocation of **xenobiotic** agrochemicals has also been reported in the phloem. Xenobiotics are biologically active molecules which are alien to an organism's body. The efficacy of xenobiotic agrochemicals as herbicides and insecticides is often determined by their rate of absorption and subsequent translocation. Leaf application of a broad-spectrum herbicide, N-(phosphonomethyl) glycine, also known as glyphosate, results in its rapid translocation to the meristematic regions, including underground roots, through phloem.

### 6.4.4 Photoassimilate Translocation: Unique Features

The rate of movement of various substances in the sieve elements can be expressed either in terms of velocity or mass transfer rate. Velocity is the linear distance traveled per unit time. Mass transfer rate refers to the quantity of material passing through a given cross section of sieve elements per unit time (**specific mass transfer**, i.e., SMT). The measurement of velocity can be achieved by employing simple conventional radioactive labeling technique. It involves exposing the source leaf to $^{14}CO_2$ for brief period. The arrival of $^{14}C$-labeled sugars at the sink tissue is monitored with the help of a detector. The measurement of velocities through these techniques has shown that velocity of phloem transport ranges from 0.08 to 0.42 mm.sec$^{-1}$ (average ~ 0.28 mm/sec). However, more sophisticated techniques, such as nuclear magnetic resonance (NMR) spectroscopy and magnetic resonance imaging (MRI), are now being used for more accuracy. In castor bean, the velocity of photoassimilate transport, measured through NMR and MRI, has been found to be 0.25 mm.sec$^{-1}$. Presently, rates are measured as the mass (in kg) gained per unit time. Studies concerning the measurement of mass transfer rates have indicated that it ranges from 2.8 to 41.7 μg.sec$^{-1}$.mm$^{-2}$. The rate of sugar transport is higher than can be explained simply by diffusion.

## 6.5 Mechanism of Photoassimilate Translocation

Movement of photosynthates from source tissues to sieve elements is known as **phloem loading**. Energy is required for the movement of photoassimilates from sieve elements to the sink tissues. Movement of photoassimilates from sieve elements to the sink tissues is known as **phloem unloading**. Movement of sucrose along the transport route of phloem does not require any energy. Factors, such as structure of sieve elements, rates of translocation, and simultaneous translocation in different directions, should be taken into consideration, while any hypothesis for mechanism of sugar translocation is proposed. Thus, mechanism of photoassimilate translocation would be dealt under three headings, i.e., phloem loading, phloem unloading, and sugar movement in the conduits after the sugar has been uploaded at the source and sugars have been unloaded at the sink.

### 6.5.1 Photoassimilate Loading

Triose phosphate synthesized during daytime as a result of photosynthesis is transported from chloroplasts to the cytosol where these are utilized for synthesis of sucrose. Transitory starch stored in chloroplasts during daytime is converted to sucrose during nighttime. In some species, other forms of transport sugars, such as raffinose family of oligosaccharides (RFOs), are later synthesized from sucrose. Sugar alcohols, like mannitol, are also synthesized from hexose phosphates. During phloem loading sugars are transported into SE-CC complex. Phloem loading can occur via the apoplastic or symplastic pathways. In apoplastic transport, sugars have to cross the plasma membrane at least once, while sugars move from cell to cell through plasmodesmatal connections without crossing the plasma membrane during symplastic transport. In photosynthesizing leaves the initial transport of photosynthates (mainly sugars) from the mesophyll cells to parenchymatous cells adjacent to SE-CC complex is symplastic in nature. However, the subsequent movement of sugars to the companion cells can occur either through symplast via plasmodesmata, or sugars might enter the apoplast before phloem loading. In some species, one of the two transport pathways is predominant, while many species may exhibit more than one mechanism for loading of sugars.

#### 6.5.1.1 Apoplastic Loading
In the apoplastic phloem-loading pathway, sugars are transported symplastically from mesophyll cells to phloem parenchyma cells located adjacent to SE-CC complex and are released in the apoplast (Fig. 6.5a), or alternatively sugars may be released in the apoplast near mesophyll cells and may diffuse in the apoplast region up to SE-CC and are uploaded in SE-CC complex. Sugars usually become more concentrated in the SE-CC complex as compared to the mesophyll cells which is confirmed by measurement of osmotic potential ($\Psi_s$) of these cells. In sugar beet, the osmotic potential of SE-CC complex has been observed to be about $-3.0$ MPa as compared to the osmotic potential of the mesophyll cells, which is $-1.3$ MPa. Since

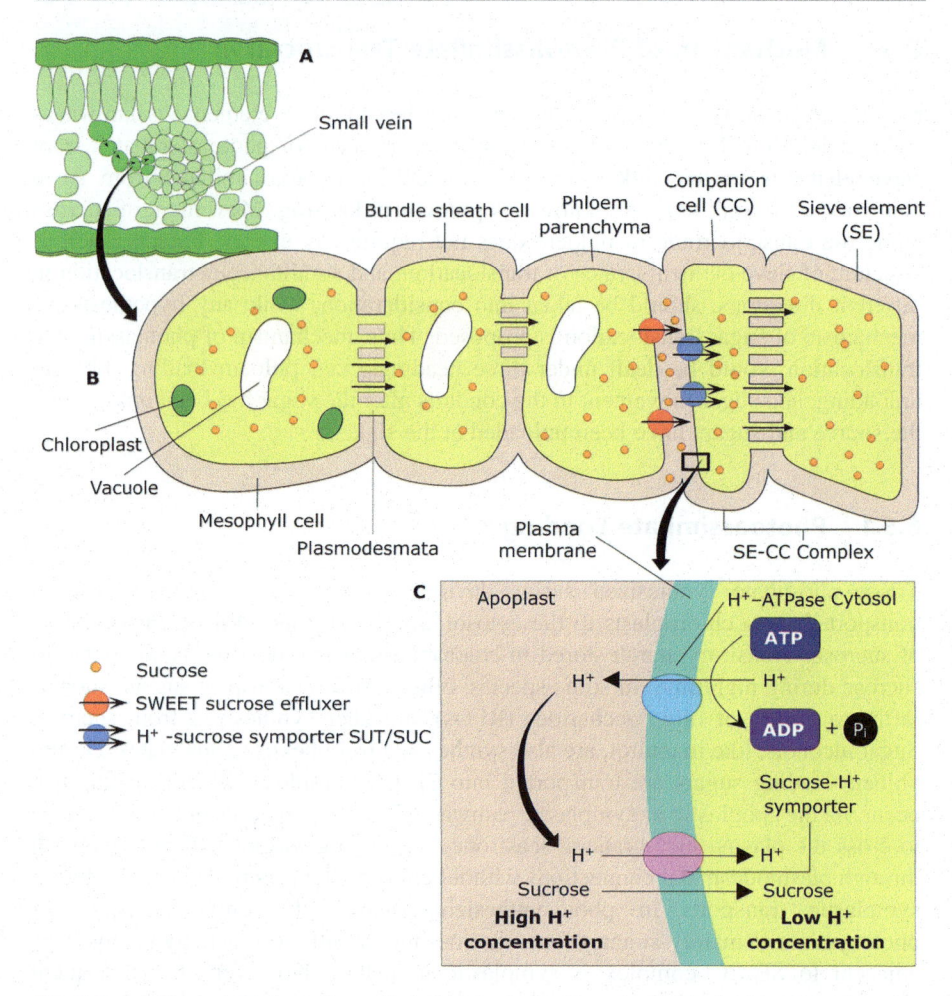

**Fig. 6.5** (**a**) Vertical section of a typical dicot leaf. (**b**) A portion of leaf enlarged to show the apoplastic phloem loading. Transport of sugars from mesophyll to the phloem parenchyma cell is symplastic. Later, sugars are released in apoplast, or they may be released in apoplast near mesophyll cells and diffuse in the apoplast region near SE/CC for apoplastic loading. (**c**) ATPase located in plasma membrane of companion cell implicated in apoplastic loading of phloem

sucrose accumulation in SE-CC complex is against concentration gradient, the uptake of sucrose in the apoplastic pathway requires metabolic energy. The active loading of sucrose has been experimentally confirmed by demonstrating inhibition of loading of exogenously supplied sugars in the presence of respiratory inhibitors. There are sucrose efflux carriers located in the plasma membrane of phloem parenchyma cells or in the plasma membrane of mesophyll cells which are known as SWEET transporters (sugars will eventually be exported transporters). These are responsible for the release of sucrose in apoplastic region, and they play a significant role in apoplastic mode of phloem loading. Pathogens induce expression of SWEET genes in order to access plant's sucrose pool. Transport of sugars from apoplast to the SE-CC complex again requires crossing of plasma membrane and is an energy-requiring process. It is facilitated by sucrose-$H^+$ symporters. These are specific sucrose transporters located in the plasma membrane (Fig. 6.5b). These sucrose transporters belong to SUT/SUC family. It is a secondary transport process in which energy generated by proton pump is used. ATPase located in plasma membrane pumps protons out of SE-CC complex into the apoplast (Fig. 6.5c). This leads to the establishment of higher proton concentration in the apoplast, and a membrane potential of about $-120$ mV is established. Energy in the proton gradient is subsequently exploited to drive cotransport of sucrose into the symplast of the SE-CC complex, which is coupled with $H^+$ transport. Several sucrose-$H^+$ symporters have been localized in phloem. To name a few, sucrose transport proteins SUT1 and SUC2 have been identified in *Solanum tuberosum* and *Arabidopsis thaliana*, respectively. On the contrary tonoplast sucrose carriers function as sucrose-$H^+$ antiporters. In many species polyol transporters have also been identified for transport.

### 6.5.1.2 Active Symplastic Loading

Several species are characterized by the presence of intermediary cells which have numerous plasmodesmata with the surrounding cells. Species such as coleus (*Coleus blumei*), pumpkin (*Cucurbita pepo*), and melon (*Cucurbita melo*) have intermediary cells in the minor veins. All the cells right from mesophyll cells to SE-CC complex are connected to each other through plasmodesmata (Fig. 6.6). How does SE-CC complex contain higher solute concentration as compared to mesophyll cells? How does accumulation of specific sugars take place against a concentration gradient? Following two possible mechanisms have been proposed to answer these questions.

### 6.5.1.3 Polymer Trapping

It is essential that sucrose concentration should be high in the mesophyll cells as compared to the intermediary cells for successful diffusion of sucrose into the intermediary cells. According to the model proposed by Turgeon (1996), larger sugar molecules belonging to RFOs (raffinose family of oligosaccharides), such as raffinose a trisaccharide (fructose-glucose-galactose), stachyose-a tetrasaccharide (Fru-Glu-Gal-Gal), and verbascose, a pentasaccharide (Glu-Fru-gal-gal-Gal) (Fig. 6.7a), are synthesized inside the intermediary cells from the transported sucrose and galactinol (a metabolite of galactose). Enzymes essential for the synthesis of RFOs are preferentially located in the intermediary cells. The utilization of sucrose for synthesis of RFOs in intermediary cells and its synthesis in mesophyll cells maintains the concentration gradient required for the diffusion of sucrose into the intermediary cells. These polymers synthesized in the intermediary cells cannot diffuse back to mesophyll cells. Possibly the size exclusion limit (SEL) of plasmodesmata at the interface of phloem parenchyma and the intermediary cells is small so that molecules larger than sucrose are excluded by them. The SEL of plasmodesmata present at the interface of the intermediary cells and the sieve

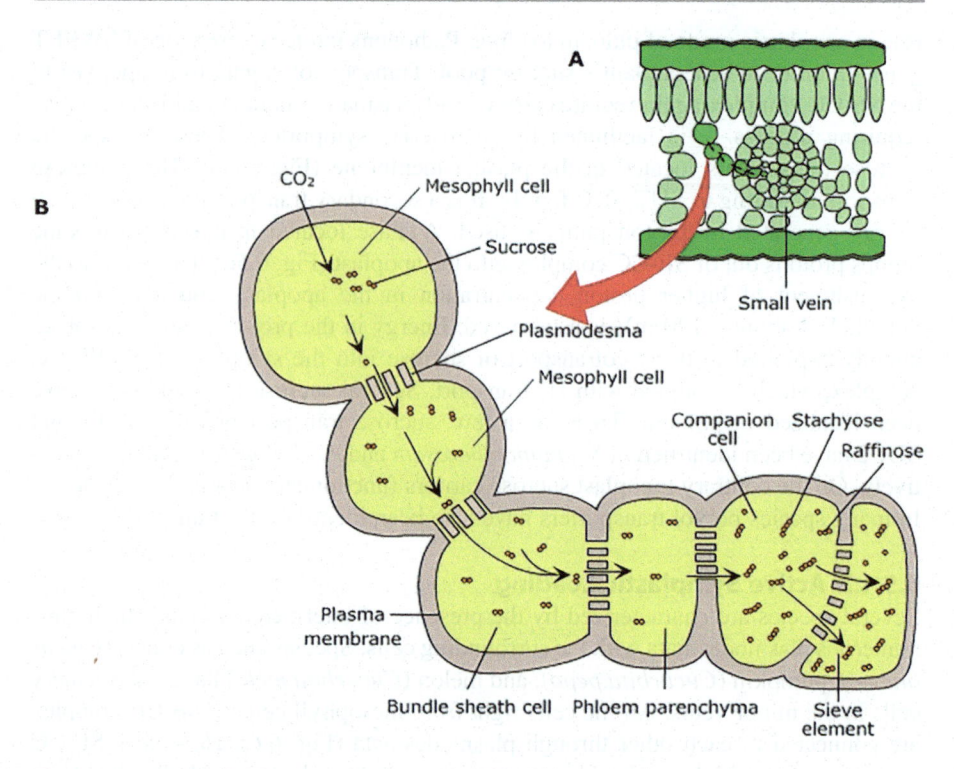

**Fig. 6.6** (a) Vertical section of a typical dicot leaf. (b) A portion of leaf enlarged to show the symplastic pathway of phloem loading

elements should be large enough to allow the passage of the polymerized sugars such as raffinose, stachyose, and verbascose molecules. Since according to this model sugars in polymeric form (raffinose, stachyose, and verbascose) are trapped because of their large size and cannot flow back, this model is known as polymer-trapping model (Fig. 6.7b). Sucrose gradient maintained in between SE-CC complex and mesophyll cell facilitates diffusion of sucrose. Though sucrose moves passively from mesophyll cells to intermediary cells in response to concentration gradient, polymer-trapping model is an active mechanism. Unlike apoplastic mode of sugar translocation, energy is not consumed for the transport of sucrose across the plasma membrane; rather it is consumed for linking of sucrose to one, two, or three molecules of galactinol which is ATP-requiring process.

### 6.5.1.4 Passive Symplastic Loading

Symplastic loading of photoassimilates in some plants is passive and is driven by simple diffusion. Passive loading is more widespread in woody species especially in trees. In passive symplast phloem-loading transport of sucrose occurs in response to solute concentration in between that of mesophyll cells and SE-CC complex. Comparison of solute concentrations and osmotic potentials of whole leaves has shown that sucrose and sugar alcohols are more concentrated in the cytosol as

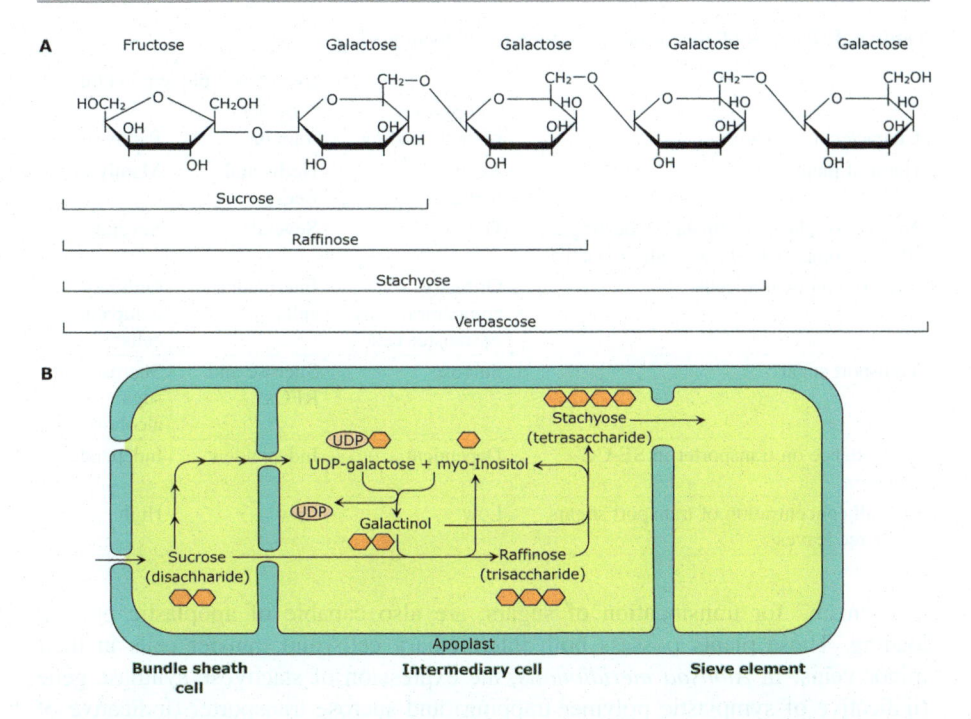

**Fig. 6.7** (a) Different sugar polymers depicting their constituent units. (b) Diagrammatic representation of polymer-trapping model

compared to vacuoles of mesophyll cells. This results in increasing the driving force for passive loading in species that employ this strategy.

### 6.5.1.5 Patterns of Apoplastic and Symplastic Phloem Loading

In tropical and subtropical regions, trees and shrubs possess abundant plasmodesmatal connections between phloem and surrounding cells. Contrary to this, herbaceous plants found in temperate and arid regions have less plasmodesmata between phloem and the surrounding cells. This indicates that apoplasmic pathway of phloem loading is preferred in plants growing in temperate climate, while symplasmic phloem loading may be preferred in plants growing in tropical climate. This may be due to decrease in solubility of raffinose at lower temperature which can increase the viscosity of phloem sap resulting in decrease in photosynthesis. That may be the reason for apoplastic mode of photoassimilate translocation in temperate plants. Reduced capacity of photoassimilate transport in plants with symplasmic phloem loading may further be associated with limited capacity of intermediary cells to convert sucrose to RFOs. Table 6.3 summarizes the comparison in patterns observed in apoplastic and symplastic phloem-loading mechanisms. There are plants which are loaded entirely by one mechanism, such as tobacco (*Nicotiana tabacum*), an apoplastic loader, and mullein (*Verbascum phoeniceum*), a symplastic loader. Some plants, such as oyster plant (*Acanthus mollis*) using "polymer-trapping

**Table 6.3** Patterns of apoplastic and symplastic phloem loading

| Characteristic features | Apoplastic phloem loading | Symplastic phloem loading | |
| --- | --- | --- | --- |
| | | Polymer trapping | Passive |
| Habit of plant | Mainly herbaceous | Herbs and trees | Mainly trees |
| Number of plasmodesmata connecting the SE-CC complex to the surrounding cells | Few | Several | Several |
| Type of companion cells | Ordinary companion cells or transfer cells | Intermediary cells | Ordinary companion cells |
| Transport sugar | Sucrose | Sucrose and RFOs | Sucrose and sugar alcohols |
| Dependence on transporter in SE-CC complex | Dependent | Independent | Independent |
| Overall concentration of transport sugars in source leaves | Low | Low | High |

mechanism" for translocation of sugars, are also capable of apoplastic phloem loading. These plants possess both intermediary cells and transfer cells in their minor veins. In *Alonsoa meridionalis*, the expression of stachyose synthase gene (indicative of symplastic polymer trapping) and sucrose transporter (indicative of apoplastic phloem loading) has been found to be specific to the intermediary cells and ordinary companion cells, respectively.

### 6.5.2 Photoassimilate Unloading

Phloem unloading involves import of sugars from sieve elements to the sink tissues by an energy-demanding short-distance transport pathway. It is also called post-sieve element transport. Rapid phloem unloading occurs when an aphid feeds on phloem sap or also because of other pathogens. Aphids feed on nitrogenous compounds and carbohydrates are secreted as honeydew. Transported sugars are either stored or metabolized in sinks.

#### 6.5.2.1 Phloem Unloading Occurs via Apoplast or Symplast

Due to variation in the structure and function of sinks, more than one possible pathway exists for phloem unloading and short-distance transport. The phloem-unloading pathway often depends on the developmental stage of the sinks. Like phloem loading, the process of phloem unloading can occur entirely through symplast via plasmodesmata. In few dicots, such as sugar beet and tobacco, phloem unloading is completely symplastic in young leaves. Root tips, the meristematic and elongating regions of roots, also exhibit symplastic pathway for phloem unloading. Mostly symplastic pathway predominates in sink tissues having involved in cell division and metabolic activities, but sinks involved in storage activities, such as in

## 6.5 Mechanism of Photoassimilate Translocation

**Fig. 6.8** (a) Apoplastic phloem unloading and short-distance transport. **1.** Apoplastic step is located at the site of unloading itself, i.e., during movement from SE-CC complex to phloem parenchyma. **2A.** Apoplastic step is located away from SE-CC complex, i.e., during movement from one parenchyma cell to another parenchyma cell or bundle sheath cell or **2B** from bundle sheath cell to sink cell. (**b**) Symplastic phloem unloading and short-distance transport

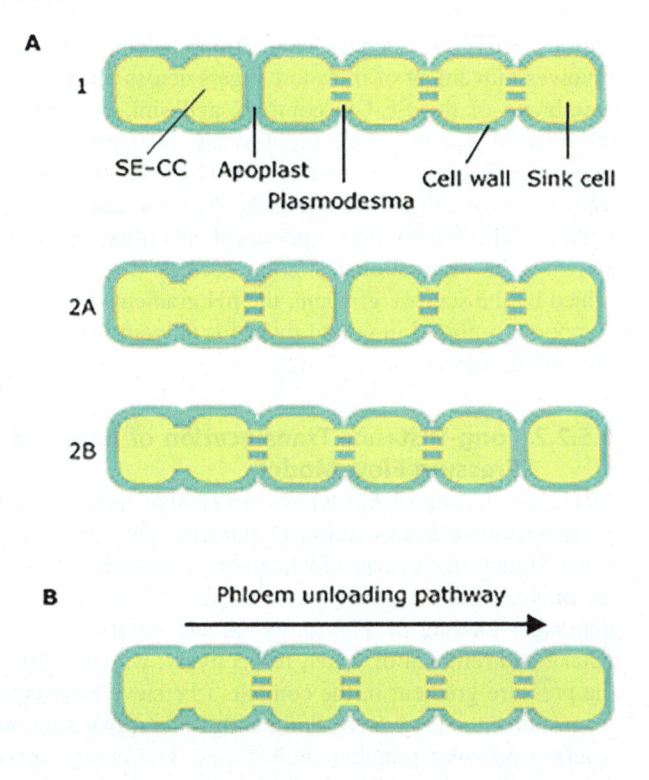

fruits and seeds, the short-distance pathway is partly apoplastic at some stage of development. In developing seeds, there is a requirement of an apoplastic pathway for uploading sugars. This is attributed to the absence of plasmodesmata between maternal and the embryonic tissues (Fig. 6.8a). There is switching of apoplastic or symplastic pathway in these sinks, with apoplastic pathway becoming functional when concentration of sugars in sinks is high. Apoplastic step could be located either at the site of unloading itself (Fig. 6.8a1) or away from SE-CC complex (Fig. 6.8a, 2A-2B). In the apoplastic pathway, the transport sugar can be partly metabolized in the apoplast or may cross the apoplast unchanged. For instance, sucrose can be hydrolyzed into glucose and fructose in the apoplast by cell wall invertase. Subsequently, glucose and/or fructose will cross plasma membrane to reach the sink cells. Cell wall invertase activity adds up to sink strength. Sucrose will be translocated in STE as long as gradient is maintained between source and sink. In the parasitic plant—*Orobanche* sp.—mannitol accumulates following haustorial connection which lowers the osmotic potential of the plant in comparison to host plant. This indicates that the enzyme responsible for mannitol synthesis in the *Orobanche sp.* can be targeted to ward off the parasitic plant. During symplastic unloading energy is not required for transport of sugar from sieve elements into sink tissues. Sugar unloading via plasmodesmata takes place passively (Fig. 6.8b). Movement of transport sugars from sieve elements to sink tissues takes place in response to the concentration gradient. However, metabolic energy is required in sink tissues for various activities, such as respiration. In contrast, apoplastic phloem unloading involves movement of transport sugars across plasma membrane twice, i.e., plasma membrane of the SE-CC complex and sink cell. Furthermore, tonoplast is also traversed in case translocation of transport sugars takes place in vacuole for storage purposes. Experimental studies on soybean have shown that energy-dependent transporters facilitate sucrose unloading into apoplast and sucrose uptake into the embryo. The transporters implicated in efflux and uptake of sucrose have been shown to be bidirectional in some studies. The direction of the transport is determined by the sucrose gradient, the pH gradient, and the membrane potential. SUT1 symporter reported in potato tuber for phloem loading has also been reported from the sink tissues.

### 6.5.2.2 Long-Distance Translocation of Photoassimilates in STE-Pressure Flow Model

Mass flow is one of the widely acceptable models used to explain long-distance photoassimilate translocation in phloem. This was initially proposed in 1930 by Ernst Münch (a German biochemist). According to this model, the mechanism for the photoassimilate translocation in phloem is passive in nature (Fig. 6.9). It states that sugar loading in STE at source and unloading at sink are followed by water uptake and removal of water, respectively, which is then responsible for setting up the pressure gradient in the conduits (the sieve elements). Mass transfer of solutes from source to sink occurs along with water movement, which moves in response to a turgor pressure gradient (bulk flow). The energy-driven phloem loading in the source tissues causes accumulation of sugars in the sieve elements which generates a negative solute potential ($\Delta\Psi_s$) and leads to a drop in the water potential ($\Delta\Psi_W$). This results in development of water potential gradient, and water tends to enter sieve elements from xylem, causing an increase in turgor pressure ($\Psi_P$). Reverse happens in the sink tissues where unloading of sugars causes a decrease in sugar concentration in the sieve elements, leading to a decrease in solute potential and increased water potential as compared to that of xylem resulting in movement of water from sieve elements to xylem and subsequently decreased turgor pressure in the sieve elements of the sink. Movement of water in phloem takes place from source to sink in response to pressure potential gradient. Such water movement does not take place against the law of thermodynamics. The movement of water occurs by bulk flow rather than by osmosis during its movement from one sieve tube to another. Furthermore, the movement of solutes and water takes place at the same rate. Under these conditions, solute potential cannot contribute to drive water movement. Thus, in translocation pathway water movement is driven by pressure gradient rather than by water potential gradient. Sieve plates causes increase in resistance along the translocation pathway in sieve elements. Due to this substantial pressure is generated and maintained in the sieve elements between source and sink.

Certain objections have been proposed against the pressure flow hypothesis. These include the presence of sieve plates with sieve pores between STEs. Pores are likely to create obstruction to the solute transport. Studies carried out with

## 6.5 Mechanism of Photoassimilate Translocation

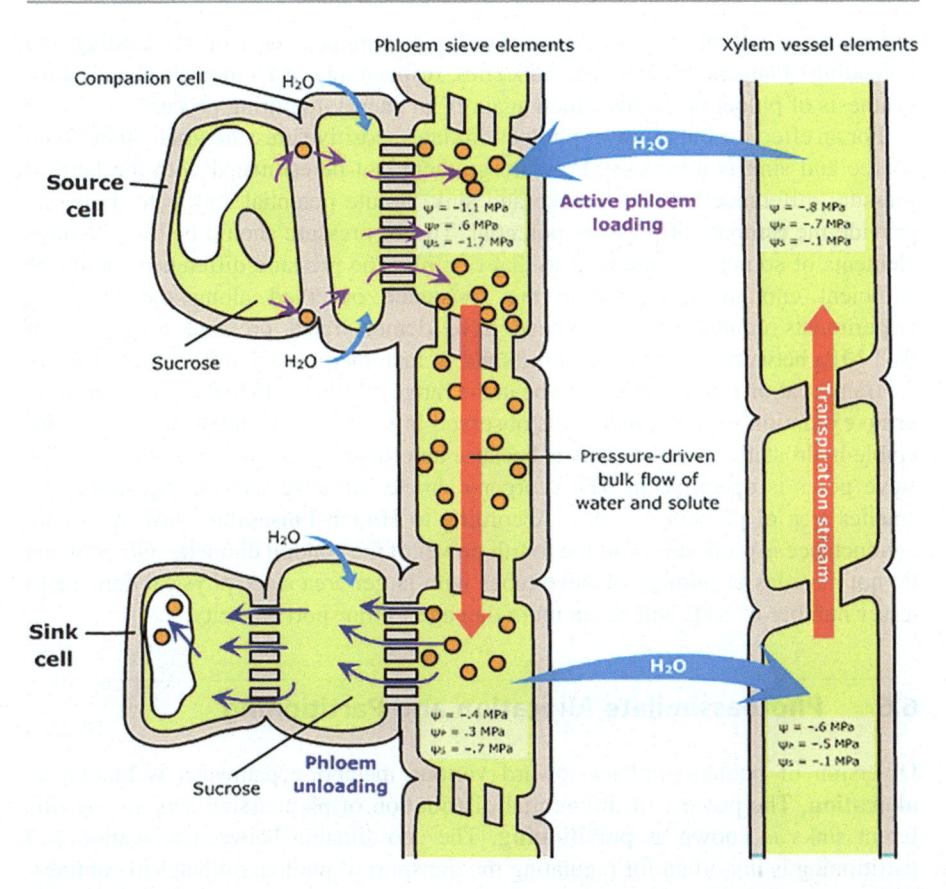

**Fig. 6.9** Diagrammatic representation of pressure flow model for photoassimilate transport in phloem

confocal laser microscopy have facilitated analysis of intact. The proponents of mass flow hypothesis regard pores to be free from the obstruction in the natural state and dismiss the densely filled pores as the artifacts. Secondly, bidirectional flow of solutes has been observed for solute transport which can be explained by solute transport occurring in a single STE to be independent of another STE. Single sieve elements never exhibit bidirectional transport of photosynthates. Bidirectional transport occurs in sieve elements from different vascular bundles. Adjacent sieve elements of the same vascular bundle in petioles of leaves, which are undergoing transition from sink to source, also show bidirectional transport.

According to mass flow hypothesis, long-distance translocation of sugars is a passive process. Experiments involving treatment with low temperature or giving chilling stress or the use of inhibitors, which adversely affect the supply of ATP to the path tissues, do not influence translocation of photoassimilates. However, normal translocation is slowly resumed after a brief period of inhibition. However, ATP turnover rate is high in phloem sap. Energy requirement may be for loading and unloading of sugars at source and sink sites, respectively. ATP may also be used for synthesis of phloem proteins which again is an energy-requiring process.

For an effective bulk flow of photoassimilates, positive pressure gradient between source and sink is necessary. Pressure gradient can be estimated with the help of pressure difference between source and sink. Solute potential and water potential provide the estimate of pressure potential. Turgor pressure should be high in sieve elements of source as compared to that of sink. The pressure difference should be sufficient enough to overcome the resistance observed along the pathway. Experiments conducted on soybean have demonstrated pressure difference of 0.41 MPa between the source and the sink. In fact, the pressure difference sufficient for translocation has been found to be in the range of 0.12–0.46 MPa. However, little or no exudation from cut phloem is observed in spite of sieve tubes having considerable hydrostatic pressure. This is because an efficient mechanism of plugging the sieve pores is operative in STE. Larger diameter of sieve tubes compensates for smaller area of phloem in vines. According to **Hagen-Poiseuille's law**, hydraulic conductance is proportional to the fourth power of the conduit diameter. Other plants do not have lesser number of sieve tubes with larger area since physical damage to lesser number of STE will cause more damage to transport capacity.

## 6.6 Photoassimilate Allocation and Partitioning

Diversion of photoassimilates toward various metabolic pathways is known as **allocation**. The process of differential distribution of photoassimilates among different sinks is known as **partitioning**. The coordination between allocation and partitioning is important for regulating the transport of photoassimilates to commercially important organs of plants. Research on photoassimilate allocation and partitioning is primarily focused on increasing the **harvest index** (HI) which is determined by the competition among various sinks for the photoassimilates exported by the source tissues. The signals transmitted between source and sink tissues could be physical or chemical in nature. The change in turgor pressure, which gets transmitted rapidly via sieve elements, accounts for the physical signal. Chemical signals include phytohormones found in the phloem sap, proteins, mRNAs, small RNAs, and sometimes translocated sugars. Coordination between source and sink activities in part is regulated by turgor pressure. It has been observed that if the utilization of photoassimilates is rapid in sink tissues at the time of phloem unloading, then it leads to the reduction of turgor pressure in the sieve elements of sink tissues which acts as a signal to the source tissues. Consequently, loading of photoassimilates would increase in response to this signal from the sink tissues. Conversely, if the unloading of photoassimilates is slow in sink tissues, then high turgor pressure signal present at the sink is transmitted to the source tissues along the STEs and accounts for slower loading of photoassimilates in source tissues. Phytohormones play significant roles in the regulation of source-sink relationships. Plant growth regulators, such as auxins, produced in shoot are rapidly transported to the roots via the phloem. In some source tissues, loading is stimulated by exogenous auxin, whereas ABA inhibits it. However, a reverse response is observed in sink tissues, with ABA enhancing the uptake of sucrose in some sink tissues and auxin inhibiting it. Hormonal regulation of apoplastic loading and unloading is due to their influence on the active transporters present in plasma membrane and tonoplast (unloading only). Plant defense hormone, such as jasmonic acid, affects the photoassimilate allocation and partitioning partially due to responses against the pathogens and herbivores. Accumulation of sugars favors expression of genes involved in storage and utilization of sugars. Thus, transport sugars are not only translocated in the phloem but also act as signals in regulating the activities at sources and sinks. It has been proposed that reduced sink demand leads to elevated sucrose levels in phloem which leads to downregulation of the sucrose-$H^+$ symporter in the source, ultimately resulting in the increased concentration of sucrose in the source.

**Summary**

- Translocation in the phloem is responsible for the movement of photoassimilates from sources (mature leaves) to the areas of growth and storage (sinks). Depending on development stage, leaves can serve as sink or source for photoassimilates. The pattern of phloem translocation is independent of gravity. Factors, such as proximity of source to sink, development stage of plant, vascular connections, and modification of translocation pathway, sometimes play an important role in determining the translocation pathway.
- Transition of leaves from sink to source is a gradual process involving two separate events, first cessation of import followed by initiation of export. This transition necessitates several conditions, such as expression of the sucrose-$H^+$ symporter. Translocation of sugars takes place in the sieve elements of phloem. Mature sieve elements are unique structures involved in the translocation of sugars. They possess a variety of structural adaptations making them suitable for translocation process. Companion cells help in the functioning of the sieve elements. They transport photoassimilates from the mature leaves (mesophyll cells) to the sieve elements and also provide energy and proteins to the sieve elements for their maintenance as sieve elements lack mitochondria and machinery for protein synthesis.
- Sucrose is the most commonly translocated sugar in plants. However, other nonreducing sugars, such as raffinose, stachyose, and verbascose, are also transported. The rate at which various substances move in the phloem during translocation process is more than can be justified by diffusion. Pressure flow model explains the mechanism of translocation in the phloem. According to this model, the bulk flow of photoassimilates and other materials present in the phloem takes place in response to an osmotically generated pressure gradient developed as a result of phloem loading at the source and phloem unloading at the sink. Phloem loading involves short-distance transport of sugars into sieve elements via companion cells. The pathway of phloem loading can be apoplastic

or symplastic. In the apoplastic pathway, sucrose is actively transported in the SE-CC complex. The polymer-trapping model suggests that the synthesis of polymeric sugars (raffinose, stachyose, and verbascose) takes place in the intermediary cells which can easily diffuse into the sieve elements due to the presence of plasmodesmata with large SEL in between companion cell and sieve element.
- Phloem unloading involves unloading of photoassimilate into sink cells, short-distance transport, storage, and metabolism. Phloem unloading can also occur through apoplastic and symplastic pathways. The transport of sugars in sinks is an energy-dependent process.
- Allocation directs the regulation of the quantities of fixed carbon that are channeled into various metabolic pathways. It determines the quantities of photoassimilate that will be used either for storage or utilized for biosynthesis of transport compounds. Partitioning refers to differential distribution of the quantities of photoassimilate delivered to various sinks.

## Multiple-Choice Questions

1. The ability of sink to mobilize photoassimilates toward it is known as:
   (a) Sink activity
   (b) Sink strength
   (c) Sink size
   (d) Sink power
2. Decrease in sink-source ratio causes rate of photosynthesis to:
   (a) Decrease
   (b) Increase
   (c) Remain unaffected
   (d) First increase followed by decline
3. Which of the following subcellular constituents is not present in mature sieve elements?
   (a) Nucleus, microfilaments, and plastids
   (b) Nucleus, Golgi bodies, and ribosomes
   (c) SER, Golgi bodies, and plastids
   (d) Microtubules, mitochondria, and plastids
4. Which type of companion cells is present in minor veins of leaf to facilitate symplastic transport of photoassimilate from mesophyll cells to sieve elements?
   (a) Ordinary companion cells
   (b) Transfer cells
   (c) Intermediary cells
   (d) Normal companion cells

5. Translocation of sugars from mesophyll cells to sieve-tube elements in leaves is known as:
   (a) Phloem unloading
   (b) Phloem loading
   (c) Sieve element unloading
   (d) Photoassimilate loading
6. Who proposed the pressure flow model to explain long-distance photoassimilate translocation in phloem?
   (a) Marcello Malpighi
   (b) Turgeon
   (c) Ernst Münch
   (d) Stephan Hales
7. During partitioning process, the distribution of photoassimilates among different sinks is:
   (a) Uniform
   (b) Differential
   (c) Normal
   (d) Similar

**Answers**

1. b   2. a   3. b   4. c   5. b   6. c   7. b

## Suggested Further Readings

Chen LQ (2014) SWEET sugar transporters for phloem transport and pathogen nutrition. New Phytol 201(4):1150–1155

Fritz E (1973) Microautoradiographic investigations on bidirectional translocation in the phloem of *Vicia faba*. Planta 112(2):169–179

Lemoine R, La Camera S, Atanassova R et al (2013) Source-to-sink transport of sugar and regulation by environmental factors. Front Plant Sci 4:272. https://doi.org/10.3398/fpls.2013.00272

Turgeon R (1996) Phloem loading and plasmodesmata. Trends Plant Sci 1(12):418–423

# Respiration 7

Manju A. Lal

Energy requirement for growth and functions in all living beings is met through ATP generated during respiration. Additionally in plants, light energy is conserved as ATP and NADPH, which are subsequently utilized for $CO_2$ assimilation. Besides, carbohydrates thus synthesized are consumed by animals as the primary source of energy. Though in animals lipids may also be consumed, in plants carbohydrates remain the main energy source. Plants store carbohydrates mainly as starch since it is osmotically inactive. All carbon accumulation in plants is the result of photosynthesis, which remains the source of energy as well as for biosynthesis of various other biomolecules. In cells hexose monophosphate pool constitutes an important component of carbohydrate metabolism. **Hexose monophosphate pool** maintained in cells consists of glucose 6-phosphate, glucose 1-phosphate, and fructose 6-phosphate. At the sinks, sucrose translocated from source is consumed during respiration. Respiration is an exergonic redox reaction with $\Delta G^{0'}$ value $-5760$ kJ.mol$^{-1}$ of sucrose. Sucrose is oxidized to $CO_2$ coupled with the reduction of $O_2$ to water as follows:

$$C_{12}H_{22}O_{11} + 13H_2O \rightarrow 12C + 48H^+ + 48e^-$$

$$12O_2 + 48H^+ + 48e^- \rightarrow 24H_2O$$

Overall reaction can be written as:

$$C_{12}H_{22}O_{11} + 12O_2 \rightarrow 12CO_2 + 11H_2O$$

The reaction is apparently reverse of photosynthesis which also is a redox reaction. In photosynthesis, oxidation of $H_2O$, resulting in release of $O_2$, is coupled with reduction of $CO_2$ to carbohydrates. It is an endergonic reaction for which light energy is used. Since glucose is primarily utilized as energy source, starch is converted to simple sugars in storage organs of plants. The translocated form of carbohydrates, i.e., sucrose, is also hydrolyzed to produce monosaccharides. As a result, it is mainly glucose which is catabolized. Respiration may take place either in presence or in absence of oxygen depending upon the availability of oxygen. However, generally oxygen is not limiting for plants except when they grow under certain conditions including waterlogged situations and roots do not get any $O_2$ or enough $O_2$. Under aerobic conditions, complete oxidation of glucose results in free energy change of $-2840$ kJ.mol$^{-1}$. Respiration is a multistep process in which energy released during oxidation of the substrate is conserved as ATP. **Combustion** is also an energy-releasing process, but it differs from respiration, the latter being a multistep process catalyzed by enzymes, while combustion is burning of fuel with release of energy in a single step. In this chapter mainly the catabolism of glucose has been taken up since it is the major molecule which is produced either by hydrolysis of sucrose or starch. However, carbohydrates in other forms, such as triose phosphates or hexose phosphates, may also enter at relevant steps in the pathway. The major steps in respiration include catabolism of glucose in the cytosol, which involves glycolysis (synthesis of pyruvate from glucose), metabolic fate of pyruvate in the absence of oxygen (fermentation), and oxidative pentose phosphate pathway. Studying pyruvate metabolism in the presence of $O_2$, which occurs in mitochondrial matrix, will involve conversion of pyruvate to acetyl-CoA and tricarboxylic acid (TCA) cycle. Reduction of NAD$^+$ to NADH and FAD to FADH$_2$ occurs during TCA cycle. These reduced cofactors are oxidized through electron transport chain which is located in the inner mitochondrial electron transport. Proton gradient is created across the inner membrane of mitochondria coupled with electron transport which is responsible for ATP synthesis (Fig. 7.1). Mechanism of ATP synthesis is dealt in Chap. 8.

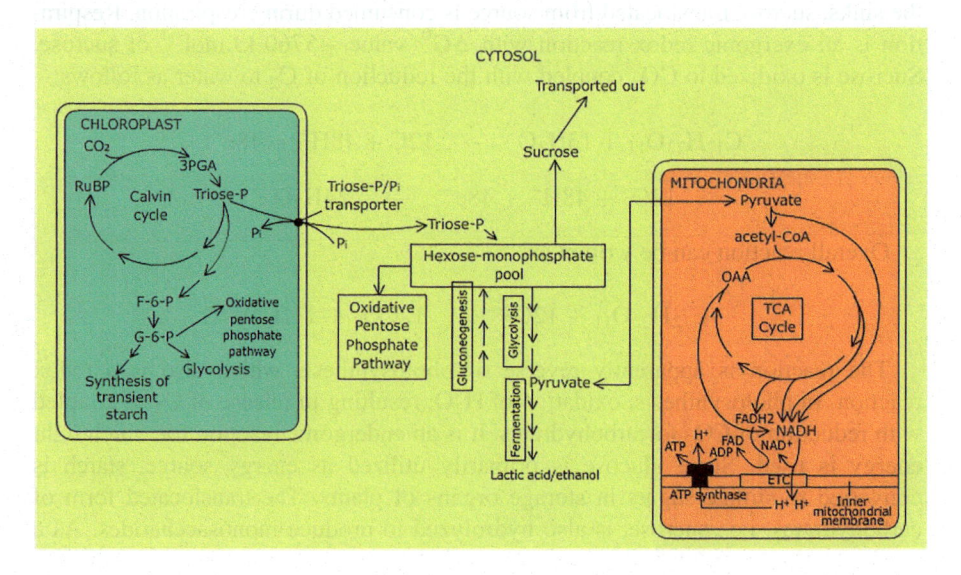

**Fig. 7.1** Overview of respiration in a plant cell

## 7.1 Glycolysis

Glycolysis (Gr. glykos, sweet, and lysis, splitting) is a ten-step catabolic pathway during which a glucose molecule is converted to two molecules of three-carbon compound pyruvate. Two molecules of ATP are produced, and 2 NAD$^+$ molecules are reduced to NADH simultaneously. Many scientists, working with various strategies, have helped in elucidating the path of glucose to pyruvate, a pathway occurring in all organisms. Glycolytic pathway was understood mainly while working on fermentation process, the oldest biochemical process known. In 1860 Louis Pasteur had postulated that fermentation required intact living yeast cells, i.e., some kind of vital force was required for the fermentation to occur. However, this observation was found to be incorrect when in 1897 Hans Buchner and Eduard Buchner, German biologist brothers, demonstrated that fermentation could be carried out by cell-free extract of yeast as well. This leads to conclusion that since no vital force is required, fermentation could have been a chemical process. When the fermentation declined after sometime, it could be restored on adding inorganic phosphate to the medium. They demonstrated the presence of fructose 1,6-bisphosphate in the medium. After observing glucose 6-phosphate in the medium along with fructose 1,6-bisphosphate in a ratio of 3:1, Robinson concluded that phosphorylated intermediates are produced during fermentation. In one experiment with fermentation, Harden and Young used cell-free extract of yeast. The extract lost its fermentation capacity when heated or dialyzed, the catalytic activity could be restored after mixing both the fractions, i.e., dialyzable components in the heated fraction and functional proteins in dialyzed fractions. Loss in the catalytic activity of the heated fraction was due to denaturation of proteins, while the dialyzable components, which included smaller molecules such as NAD$^+$, ATP, etc., were lost due to dialysis of the second fraction. The two components were named as zymase (heat-labile) and co-zymase (dialyzable). Another approach used by Embden for elucidating the pathway was adding inhibitors of certain enzymes to the medium. He used iodoacetate, an inhibitor of enzymes containing –SH groups, which lead to the accumulation of fructose 1,6-bisphosphate. Otto Warburg identified the enzyme aldolase and the reaction catalyzed by it. Embden also observed that adding fluoride (another inhibitor which inhibits activity of enolase) to the fermentation medium leads to the accumulation of 3-phosphoglycerate and 2-phosphoglycerate, thus demonstrating the role of both aldolase and enolase. Complete pathway of fermentation was worked out. In RBCs or in some bacteria, glycolytic pathway is the sole source of energy. Even in plants growing under waterlogged conditions, since their roots are exposed to hypoxia or anoxia, this pathway is the only source of energy. Besides providing energy, various intermediates of the glycolytic pathway serve as precursors for many biosynthetic reactions. Glycolytic pathway is universally present in all organisms and is similar in fermentation as well as aerobic respiration. Unlike animal cells, glycolysis occurs both in cytosol and plastids of the plant cells, and the pathway is almost complete in both the locations, except that one or two enzymes may be missing in the plastidial pathway. These pathways do not occur in isolation, and exchange of various intermediates between the cytosol and plastids is facilitated by the transporters localized on the inner membrane of plastids. Universality of this pathway indicates its significance. The ten reactions of the glycolysis can be considered into two phases: the *preparatory* phase is an energy investment phase in which two ATP molecules are consumed for each glucose molecule, which is converted to two molecules of triose phosphates. This phase is also known as mobilization phase since it correlates with mobilization of carbohydrates other than glucose, which include storage forms or the translocating form (sucrose) of carbohydrates. The second phase of glycolysis represents *payoff* phase which is an energy generation phase. During this phase triose phosphates are converted to pyruvate coupled with synthesis of four molecules of ATP and two molecules of NADH. This is also known as Embden-Meyerhof-Parnas (EMP) pathway after the name of the scientists who facilitated the elucidation of this pathway. Figure 7.2 represents the abbreviated scheme of the conversion of glucose to pyruvate. In case glucose enters glycolysis, first five reactions of the *investment phase* involve two phosphorylation reactions, two isomerization reactions, and one hydrolytic reaction, catalyzed by kinases, isomerases, and aldolase, respectively, thereby yielding two molecules of triose phosphate, i.e., glyceraldehyde 3-phosphate and dihydroxyacetone phosphate. Both triose phosphates are interconvertible and maintain equilibrium, but glyceraldehyde 3-phosphate is on the main pathway. During the latter *payoff* phase of glycolysis, glyceraldehyde 3-phosphate undergoes further activation and oxidation to yield compounds containing energy-rich phosphate bonds, i.e., 1,3-bisphosphoglycerate and phosphoenolpyruvate which are considered as super high-energy compounds since these have higher energy of hydrolysis than required for ATP synthesis. ATP is synthesized by the transfer of high-energy phosphate group of these compounds to ADP. ATP synthesis by this process without involvement of any electron transport is called **substrate-level phosphorylation**, i.e., transfer of phosphoryl group from a high-energy compound to ADP, yielding ATP. Substrate-level phosphorylation is different both from oxidative phosphorylation and photophosphorylation, since ATP synthesis is not coupled to any electron transport. Net reaction of glycolysis can be summarized as follows:

$$C_6H_{12}O_6 + 2ADP + 2P_i + 2NAD^+ \rightarrow 2CH_3COCOOH + 2ATP + 2NADH + 2H^+ + 2H_2O$$

Initially produced phosphorylated sugars in glycolysis are low-energy compounds. In enzyme-catalyzed reactions, these are converted into compounds with high phosphoryl group transfer potential which subsequently generate ATP. Glycolytic equation can be resolved into two processes. The first one is exergonic reaction involving synthesis of pyruvate from glucose.

$$\text{Glucose} + 2NAD^+ \rightarrow 2\text{Pyruvate} + 2NADH + 2H^+$$

## 7.1 Glycolysis

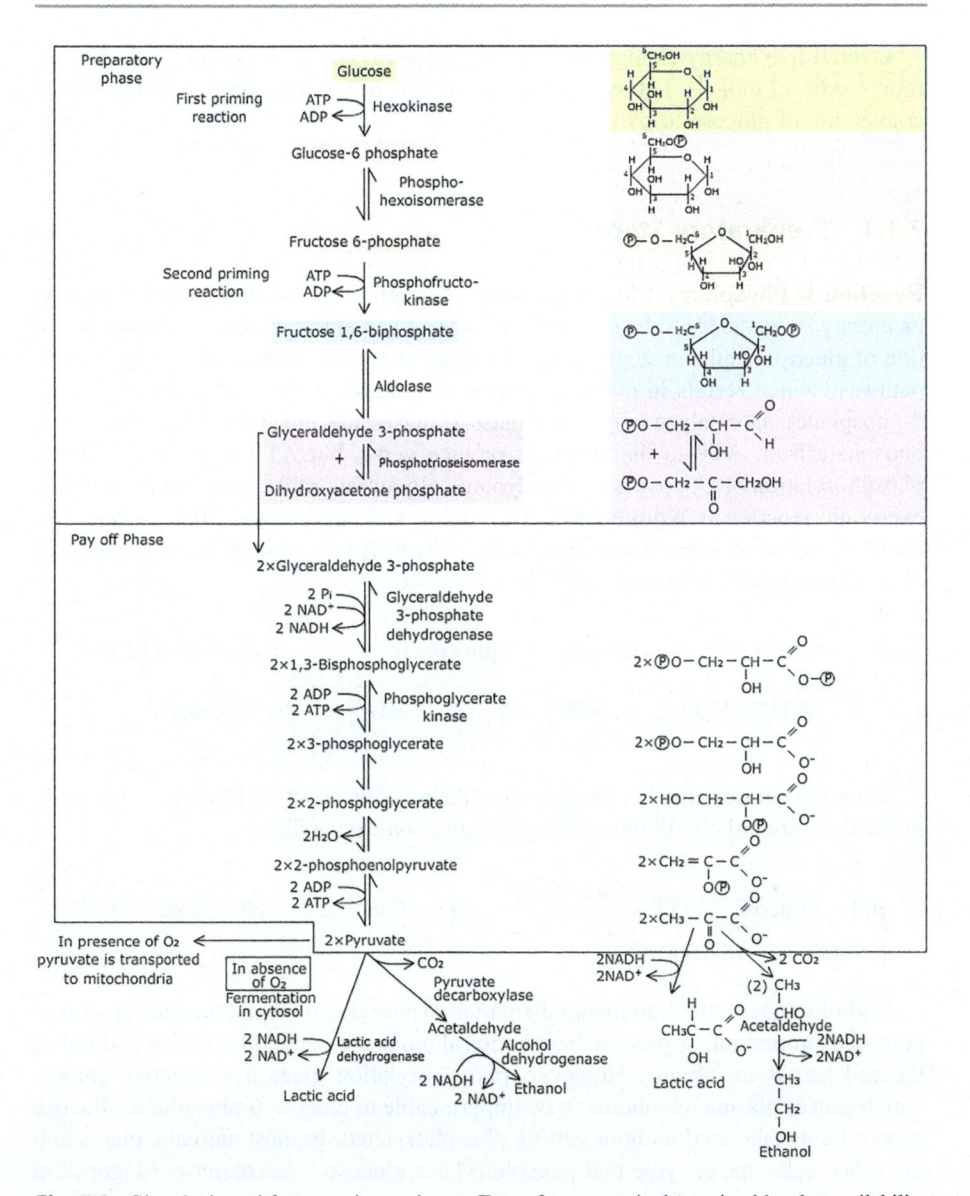

**Fig. 7.2** Glycolysis and fermentation pathway. Fate of pyruvate is determined by the availability of $O_2$

$\Delta G^{0'}$ of the reaction is $-146$ kJ.mol$^{-1}$. The second process involves synthesis of ATP from ADP and is endergonic.

$$2ADP + 2P_i \rightarrow 2ATP$$

$\Delta G^{0'}$ of the reaction is 61 kJ.mol$^{-1}$ (2 × 30.5 kJ.mol$^{-1}$).

Overall free energy change ($\Delta G^{0'}$) during glycolysis is $-85$ kJ.mol$^{-1}$ ($-146$ kJ.mol$^{-1}$ + 61 kJ.mol$^{-1}$). Let us consider the details of ten reactions occurring during conversion of glucose to pyruvate.

### 7.1.1 Preparatory Steps

**Reaction 1: Phosphorylation of glucose:** Though glucose is the source of energy, its energy status needs to be stepped up before it can enter glycolysis. Phosphorylation of glucose results in stepping up its energy state. The first reaction of glycolytic pathway, which results in phosphorylation of glucose at the sixth carbon (glucose 6-phosphate), is catalyzed by hexokinase (kinases are enzymes which transfer γ phosphate from ATP to the substrate or vice versa, but ATP is involved). Since phosphorylation of glucose is thermodynamically unfavorable, coupling this with an exergonic reaction is required which involves ATP hydrolysis. 30.5 kJ.mol$^{-1}$ of energy is released when a molecule ATP is hydrolyzed, while phosphorylation of glucose requires 13.8 kJ.mol$^{-1}$ (Table 7.1).

$$\alpha D - \text{Glucose} + P_i \rightarrow \text{Glucose 6} - \text{phosphate} \qquad \Delta G^{0'} = 13.8 \text{ kJ.mol}^{-1}$$

$$\text{ATP} + H_2O \rightarrow \text{ADP} + P_i \qquad \Delta G^{0'} = -30.5 \text{ kJ.mol}^{-1}$$

Thus, the net free energy change during the reaction is $-16.7$ kJ.mol$^{-1}$ of glucose under the standard conditions making the reaction irreversible.

$$\alpha D - \text{Glucose} + \text{ATP} \xrightarrow{\text{Hexokinase;Mg}^{2+}} \alpha D - \text{Glucose 6} - \text{phosphate} + \text{ADP}$$
$$\Delta G^{0'} = -16.7 \text{ kJ.mol}^{-1}$$

Carbohydrates enter into metabolism in phosphorylated form. Phosphorylation of glucose is important, as glucose being neutral molecule could diffuse into and out of the cell across membrane. However, phosphorylation gives it a negative change which causes plasma membrane to be impermeable to glucose 6-phosphate. Glucose cannot be metabolized without getting phosphorylated. In most animals, plants and microbial cells, the enzyme that phosphorylates glucose is hexokinase. Magnesium ion (Mg$^{2+}$) is required for this reaction since substrate for the hexokinase reaction is Mg. ATP$^{2-}$ not ATP$^{4-}$. Mg$^{2+}$ shields two negative charges of ATP making the terminal phosphorus atom an easier target for nucleophilic attack by an –OH group of glucose. In plants, glucose cannot be phosphorylated by any other enzyme, except hexokinases. The other way by which phosphorylated form of glucose can be produced is due to phosphorylatic breakdown of starch. No hexose phosphate phosphatase has been identified in plants. Hexokinases are generally characterized by their broad specificity for sugars. Thus they catalyze phosphorylation of various hexose sugars, which include fructose and mannose besides glucose, leading to their

## 7.1 Glycolysis

**Table 7.1** Reactions of glucose catabolism during glycolysis

| | Reactions | Enzyme | $\Delta G^{0'}$ of the reaction in kJ mol$^{-1}$ |
|---|---|---|---|
| 1. | Glucose + ATP → Glucose 6 – phosphate + ADP | Hexokinase | −16.7 |
| 2. | Glucose 6 – phosphate ↔ Fructose – 6 phosphate | Glucose-6-phosphate isomerase | 1.7 |
| 3. | Fructose 6 – phosphate + ATP → Fructose 1, 6 – biphosphate + ADP | Phosphofructokinase | −14.2 |
| | (In plants) Fructose 6 – phosphate + PP$_i$ ↔ Fructose 1, 6 bisphosphate + P$_i$ | PP$_i$-dependent phosphofructokinase | −2.9 |
| 4. | Fructose 1, 6 – bisphosphate ↔ 3 – Phosphoglyceraldehyde + dihydroxyacetone phosphate | Aldolase | 23.8 |
| 5. | 3 – Phosphoglyceraldehyde ↔ Dihydroxyacetone phosphate | Triose phosphate isomerase | 7.5 |
| 6. | 3 – Phosphoglyceraldehyde + P$_i$ + NAD$^+$ ↔ 1, 3 – bisphosphoglycerate + NADH + H$^+$ | G3P dehydrogenase | 6.3 |
| 7. | 1, 3 – bisphosphoglycerate + ADP ↔ 3 – Phosphoglycerate + ATP | Phosphoglycerate kinase | −18.5 |
| 8. | 3 – Phosphoglycerate ↔ 2 – Phosphoglycerate | Phosphoglyceromutase | 4.4 |
| 9. | 2 – Phosphoglycerate + ADP ↔ Phosphoenolpyruvate + ATP | Enolase | 7.5 |
| 10. | Phosphoenolpyruvate + ADP → Pyruvate + ATP | Pyruvate kinase | −31.4 |

utilization via glycolysis. Hexokinases are inhibited by the product of the reaction glucose 6-phosphate, which regulates the influx of hexose sugars into the glycolytic pathway. Glucose 6-phosphate can also be metabolized through oxidative pentose phosphate pathway. Structure of hexokinase provides striking evidence for induced fit model of enzyme catalysis. Hexokinase occurs in single form and consists of two lobed 50 kDa subunits. The enzyme is characterized by the presence of ATP-binding site. Daniel Koshland had proposed "induced fit" model and predicted many years back that hexokinase undergoes conformational change on binding with substrates, $Mg.ATP^{2-}$ and glucose. On binding with $Mg.ATP^{2-}$, there is a conformational change in the enzyme protein which brings it closer to the bound glucose. As a result, access to the water molecules is stopped, and phosphate group of the ATP is transferred to glucose molecule rather than being attacked by water. These predictions were later on confirmed in the enzyme obtained from yeast, both in presence or in absence of glucose. Unlike animals, hexokinase in plants is the only enzyme that phosphorylates glucose. However, fructose can be phosphorylated either by hexokinase or fructokinase. Almost three to ten genes have been identified which encode hexokinase. While different isoforms have been identified from the cytoplasm, only one hexokinase isoform is present in the plastids. Four isoforms of the enzyme have been identified in wheat germ. One isoform, called glucokinase, is found in the human liver, which is responsible for lowering blood sugar level.

**Reaction 2: Isomerization of glucose 6-phosphate to fructose 6-phosphate:** The second step in glycolysis is the isomerization of glucose 6-phosphate. The carbonyl oxygen of glucose 6-phosphate is shifted from C-1 to C-2. This results in isomerization of an aldose, i.e., glucose 6-phosphate, to a ketose—fructose 6-phosphate. Enzyme which catalyzes this reaction is called either phosphohexoisomerase or phosphoglucoisomerase or glucose phosphate isomerase. The enzyme facilitates conversion of pyranose ring of glucose to furanose ring of fructose. Isomerization has two relevances: First in the subsequent step of glycolysis when phosphorylation occurs at C1, the hemiacetal group –OH of glucose 6-phosphate would be more difficult to phosphorylate than a simple primary hydroxyl –OH. Second, for C-C bond cleavage, which occurs in the next step, isomerization to fructose with carbonyl group coming to C-2 activates C-3, making cleavage of the –C-C bond easy. In humans, this enzyme requires $Mg^{2+}$ for the activity, and it is highly specific for glucose 6-phosphate. The $\Delta G^{0'}$ is 1.67 kJ.mol$^{-1}$. This small value means that the reaction operates near equilibrium in the cell and is easily reversible.

**Reaction 3: ATP-driven second priming reaction: investment of another ATP:** In the previous reaction, carbonyl group of glucose 6-phosphate moves from C-1 to C-2 portion, and the hydroxyl group at C-1 becomes available. Phosphorylation of this group occurs which is catalyzed by the enzyme **phosphofructokinase (PFK)**. Similar to reaction 1, phosphoryl group is provided by ATP and requires $Mg^{2+}$. Phosphorylation of fructose 6-phosphate is endergonic and is another priming reaction.

Fructose 6 − phosphate + $P_i$ → Fructose 1, 6 − bisphosphate

$$\Delta G^{0'} = 16.4 \text{ kJ.mol}^{-1}$$

When coupled with the hydrolysis of ATP, the overall reaction becomes exergonic.

Fructose 6 − phosphate + ATP → Fructose 1, 6 − bisphosphate

$$\Delta G^{0'} = -14.2 \text{ kJ.mol}^{-1}$$

Reaction catalyzed by ATP-dependent PFK is essentially irreversible. At pH 7 and 37 °C, reaction catalyzed by phosphofructokinase lies mostly toward right, similar to the reaction catalyzed by the enzyme hexokinase, and is irreversible. Conversion of fructose 6-phosphate to fructose 1,6-bisphosphate, catalyzed by phosphofructokinase, is a committed step for glucose to be used for energy-releasing process rather than converting it to another sugar or just storing it. Thus, reaction catalyzed by PFK is an important regulatory step in glycolysis. ATP acts both as a substrate as well as allosteric inhibitor of the enzyme. Therefore, phosphofructokinase has two distinct binding sites for ATP, a high-affinity substrate site and a low-affinity regulatory site. In the presence of high ATP concentration, phosphofructokinase subunits behave cooperatively, i.e., the activity of one subunit influences the activity of all other subunits. When enzyme activity is plotted against fructose 6-phosphate, a sigmoidal curve is obtained which is typical of an allosterically regulated enzyme. Inhibition is reversed by AMP. Rise in the levels of AMP within the cell is coupled with decrease in ATP, and it is the ratio of ATP/AMP which regulates the activity of phosphofructokinase (PFK). Thus, the activity of phosphofructokinase depends on the levels of both ATP and AMP and is a function of cellular energy status. Glycolytic rate is decreased when ATP is in excess and increases when more ATP is required by the cell. In plants an additional cytosolic enzyme $PP_i$-dependent phosphofructokinase (pyrophosphate: fructose 6-phosphate 1-phosphotransferase) is present which phosphorylates fructose 6-phosphate using pyrophosphate not ATP in a reversible reaction.

Fructose 6 − phosphate + $PP_i$ → fructose 1, 6 − biphosphate + $P_i$

$$\Delta G^{0'} = -2.9 \text{ kJ.mol}^{-1}$$

The $PP_i$-dependent PFK (PP-PFK-1) is present in quite an appreciable amount in the cytosol of plant cells, but ATP-dependent PFK can take over the function of PP-PFK-1 in case of its absence or is present with reduced activity. This shows that quite a flexible system present in plants to provide more alternative reactions. Other molecules which are potent allosteric activators of PFK are fructose 2,6-bisphosphate and ribulose 5-phosphate. Besides, in plants the activity of PFK is also inhibited by PEP. Thus, there are multple layers of regulation for the activity of PFK, which indicates reaction catalyzed by this enzyme to be the primary regulatory step of glycolysis.

**Reaction 4: Hydrolytic cleavage of fructose 1,6-bisphosphate to two molecules of triose phosphate:** Fructose 1,6-bisphosphate aldolase cleaves fructose 1,6-bisphosphate between the C-3 and C-4 carbons to yield two molecules of triose phosphates. The products are dihydroxyacetone phosphate (DHAP) and glyceraldehyde 3-phosphate (G3P). The reaction is reversible but is in favor of fructose 1,6-bisphosphate. The reaction has $\Delta G^{0'}$ of +23.9 kJ.mol$^{-1}$. Equilibrium of the reaction involving the formation of two molecules of triose phosphates from one molecule fructose 1,6-bisphosphate depends on the concentration of all the three (the substrate and the products). The enzyme is called aldolase as the reaction is similar to the reverse of an aldol condensation reaction and is strongly endergonic. Aldolase is a tetrameric protein and an example of covalent catalysis. The enzyme activates the reaction by condensing the keto carbon at position two with ε-amino group of lysine residue present at the active site of the enzyme.

**Reaction 5: Interconversion of Triose Phosphates:** The two triose phosphates, 3-phosphoglyceraldehyde (G3P) and dihydroxyacetone phosphate (DHAP), produced from the previous reaction are interconvertible with the help of enzyme triose phosphate isomerase (TPI). The two triose phosphates occur in equilibrium favoring DHAP over G3P in a ratio of 22:1. Since G3P is on the main pathway of glycolysis and is substrate in the next glycolytic reaction, more of dihydroxyacetone phosphate (DHAP) is converted to G3P to maintain the equilibrium. The reaction is endergonic under standard conditions, but low intracellular concentration of glyceraldehyde 3-phosphate makes the reaction shift to right. Thus both triose sugars produced from glucose molecules are permitted to proceed further in glycolytic pathway. C-1, C-2, and C-3 of the initial molecules become equivalent to C-6, C-5, and C-4, respectively. Triose phosphate isomerization reaction completes the first phase of glycolysis. Each glucose molecule that passes through glycolysis is converted to two molecules of glyceraldehyde 3-phosphate. Till this point, two ATP molecules are consumed. In plants sucrose is the main carbohydrate to be metabolized. In case of sucrose catabolism, the number of ATP consumed, during this phase, will depend upon the mechanism of sucrose hydrolysis. If sucrose is hydrolyzed by invertase, 4 ATP will be consumed per molecule of sucrose, since two molecules of hexoses, i.e., glucose and fructose, will be produced as a result of hydrolysis of sucrose. In case hydrolysis of sucrose occurs by sucrose synthase activity, 2 ATP per molecule of sucrose will be consumed. Hydrolysis by this enzyme will result in the formation of one molecule of fructose, and another hexose will be phosphorylated, i.e., glucose 6-phosphate. As a result of this, 2 ATP will be saved. G3P is metabolized further to generate high-energy compounds which are responsible for synthesis of ATP. The next phase of glycolysis involves the *payoff* phase of glycolysis.

## 7.1.2 Entry of Molecules in Glycolysis Other than Glucose

Since in plants generally sucrose is translocated, this is the principal form of carbohydrates, which enters into the metabolism. Plant metabolism is characterized by **metabolic redundancy** because of the presence of alternate pathways. Starch is hydrolyzed either by hydrolytic or phosphorolytic enzymes in chloroplasts or in plastids of storage tissue. In chloroplasts, **transistory starch** is hydrolyzed during night, while storage starch hydrolysis occurs in the storage organs either in daytime or nighttime. Since glucose 1-phosphate is the product of starch hydrolysis, phosphorolytic enzymes have the advantage over hydrolytic enzymes as one ATP, required for the first priming reaction, is saved. Triose phosphates produced in light during $CO_2$ assimilation may enter glycolysis in chloroplasts directly during daytime. Glycerol is an important simple substance which can also be metabolized through glycolytic pathway. This metabolite, which is produced in substantial amount in cytosol as a result of hydrolysis of triglycerides especially in germinating oil seeds, is converted to glycerol 3-phosphate by the action of glycerol kinase. Glycerol 3-phosphate is oxidized to dihydroxyacetone phosphate by cytosolic glycerol phosphate dehydrogenase, with $NAD^+$ as coenzyme. The dihydroxyacetone phosphate thus produced enters the glycolytic pathway as a substrate of triose phosphate isomerase.

$$\text{Glycerol} + \text{ATP} \xrightarrow[\text{Glycerol kinase;Mg}^{2+}]{} \text{Glycerol phosphate} + \text{ADP}$$

$$\text{Glycerol phosphate} + NAD^+ \longrightarrow \text{Dihydroxyacetone phosphate} + NADH + H^+$$

## 7.1.3 *Payoff* Phase

The second half of the glycolytic pathway involves partial oxidation of triose phosphates as well as the reactions that convert terminal phosphate groups of the phosphorylated sugars into ATP.

**Reaction 6: Generation of first energy-rich compound in glycolysis:** It is the first oxidation reaction of glycolysis in which glyceraldehyde 3-phosphate is oxidized to 1,3-bisphosphoglycerate by the action of glyceraldehyde 3-phosphate dehydrogenase. Oxidation of glyceraldehyde to the acid is coupled to reduction of $NAD^+$ to NADH. Energy released during oxidation is conserved as carboxylic acid anhydride bond with phosphate group, known as acyl phosphate. The compound formed is 1,3-bisphosphoglycerate (BPG) which has very high $\Delta G^{0'}$ of hydrolysis, i.e., $-49.4$ kJ.mol$^{-1}$. Hydrolysis of 1,3-bisphosphoglycerate drives synthesis of ATP from ADP, as acyl phosphate group of BPG is much more energy rich than the phosphate anhydride of ATP. Glyceraldehyde 3-phosphate dehydrogenase activity is inhibited by iodoacetate, which blocks cysteine -SH group present at the active site of the enzyme, necessary for carrying out reaction.

**Reaction 7: First ATP generation reaction in glycolysis:** Because of its high-energy transfer potential, 1,3-bisphosphoglycerate has a strong tendency to transfer its acyl phosphate group to ADP, resulting in the formation of ATP and 3-phosphoglycerate. This **substrate-level phosphorylation** reaction is catalyzed by phosphoglycerate kinase. Since each molecule of glucose produces two triose phosphates, two ATPs will be produced as a result of this reaction. The phosphoglycerate kinase reaction pays off the debt created by the priming reactions. This phosphoryl transfer is facilitated by $Mg^{2+}$ since true nucleotide substrate for the reaction is $Mg.ADP^{2-}$. It is appropriate to consider the sixth and seventh reactions of glycolysis as coupled reactions since 1,3-bisphosphoglycerate is as an intermediate produced during the sixth reaction. The sum of both reactions can be written as:

$$\text{Glyceraldehyde 3 - phosphate} + NAD^+ + ADP + P_i$$
$$\rightarrow 3 - \text{phosphoglycerate} + NADH + H^+ + ATP$$

The reaction is exergonic and net change in standard free energy ($\Delta G^{0'}$) is $-12.2$ kJ.mol$^{-1}$. Under cellular conditions, the reaction is reversible.

**Reaction 8: Phosphoryl Transfer Reaction:** Phosphoglycerate mutase catalyzes the transfer of phosphate group from the third to second position in 3-phosphoglycerate to produce 2-phosphoglycerate. $Mg^{2+}$ is required for this reversible endergonic reaction. $\Delta G^{0'}$ value of this reaction is 4.4 kJ.mol$^{-1}$. The intracellular level of 3-phosphoglycerate is high relative to that of 2-phosphoglycerate so that in vivo the reaction proceeds uninterrupted to the right. The enzyme phosphoglycerate mutase requires 2,3-bisphosphoglycerate as an activator initially to phosphorylate histidine, which is present at the active site of the enzyme, to phosphohistidine. During the reaction, firstly the phosphate group is transferred from phosphohistidine to the substrate 3-phosphoglycerate, resulting in the formation of an intermediate 2,3-bisphosphoglycerate. This is followed by the breakdown of the enzyme-bound intermediate regenerating the phosphorylated histidine of the enzyme. The product 2-phosphoglycerate is released. $Mg^{2+}$ is required for the reaction. Enzyme for this particular reaction has been found to be absent in the chloroplasts, as 3-phosphoglycerate is also produced during initial reactions of Calvin cycle. The enzyme, phosphoglycerate mutase, if present would draw carbon from Calvin cycle to glycolysis.

**Reaction 9: Synthesis of High-Energy Compound Phosphoenolpyruvate (PEP):** Enolase catalyzes conversion of 3-phosphoglycerate, having very low potential as phosphoryl group donor, to phosphoenolpyruvate, which has a very high potential for phosphoryl group transfer. The $\Delta G^{0'}$ for this reaction is very low, i.e., 7.5 kJ.mol$^{-1}$. It may be difficult to understand how enolase reaction transforms a substrate with a relatively low free energy of hydrolysis into a product (PEP) with a very high free energy of hydrolysis. 2-Phosphoglycerate and PEP contain about the same amount of potential metabolic energy with respect to their respective decomposition to $P_i$, $CO_2$, and $H_2O$. Enolase is responsible for rearrangement in the structure of the 2-phosphoglycerate by removing a water molecule into a form which is capable of releasing more of the potential energy upon hydrolysis. Enzyme is inhibited by fluoride ions in the presence of phosphate. Yeast enolase is a dimer of identical subunits.

**Reaction 10: Generation of Second Molecule of ATP** PEP produced in the previous reaction has a very high phosphoryl group transfer potential and $\Delta G^{0'}$ for hydrolysis of PEP is $-61.9$ kJ.mol$^{-1}$. This reaction is catalyzed by the enzyme pyruvate kinase. It is another example of substrate-level phosphorylation in which phosphoryl group of phosphoenolpyruvate is transferred to ADP. It is $Mg^{2+}$ requiring reaction which is stimulated by $K^+$. Significant change in the free energy for the conversion of PEP to pyruvate is mainly because of highly favorable and spontaneous conversion of enol **tautomer** of pyruvate to the more stable keto form after the phosphoryl transfer step. About half of the energy, which is released on hydrolysis of PEP, is conserved as ATP phosphoanhydride bond ($\Delta G^{0'} = -30.5$ kJ.mol$^{-1}$), while the remaining half ($-31.4$ kJ.mol$^{-1}$) is responsible for pushing the reaction forward.

### 7.1.4 Stoichiometry of Glycolysis

For each molecule of glucose metabolized through glycolysis, 2 ATP molecules are consumed during initial five reactions which constitute the preparatory phase of glycolysis. During *payoff* phase, there is generation of 4 ATPs and 2 NADH (since one glucose molecule produces two triose phosphates, and each triose phosphate produces 2 ATP and 1 NADH). As a result, there is net gain of 2 ATP and 2 NADH. However, the number of ATP molecules may vary depending upon the kind of carbohydrates metabolized and also the process of hydrolysis involved in sucrose or starch.

Glycolysis is the sole source of energy under anaerobic conditions which plants rarely experience except in certain situations such as in germinating seeds or in roots exposed to hypoxia or anoxia under waterlogged conditions. However, since the amount of energy produced is much less because of partial oxidation under anaerobic conditions, more amount of glucose is consumed (**Pasteur effect**). Since oxidized form NAD$^+$ is required for continuation of glycolysis, under anaerobic conditions NADH is oxidized during fermentation. Many glycolytic intermediates serve as precursors (Fig. 7.3).

### 7.1.5 Significance of Phosphorylated Intermediates

Nine out of ten intermediates of the glycolytic pathway are phosphorylated. This is significant because of the following reasons:

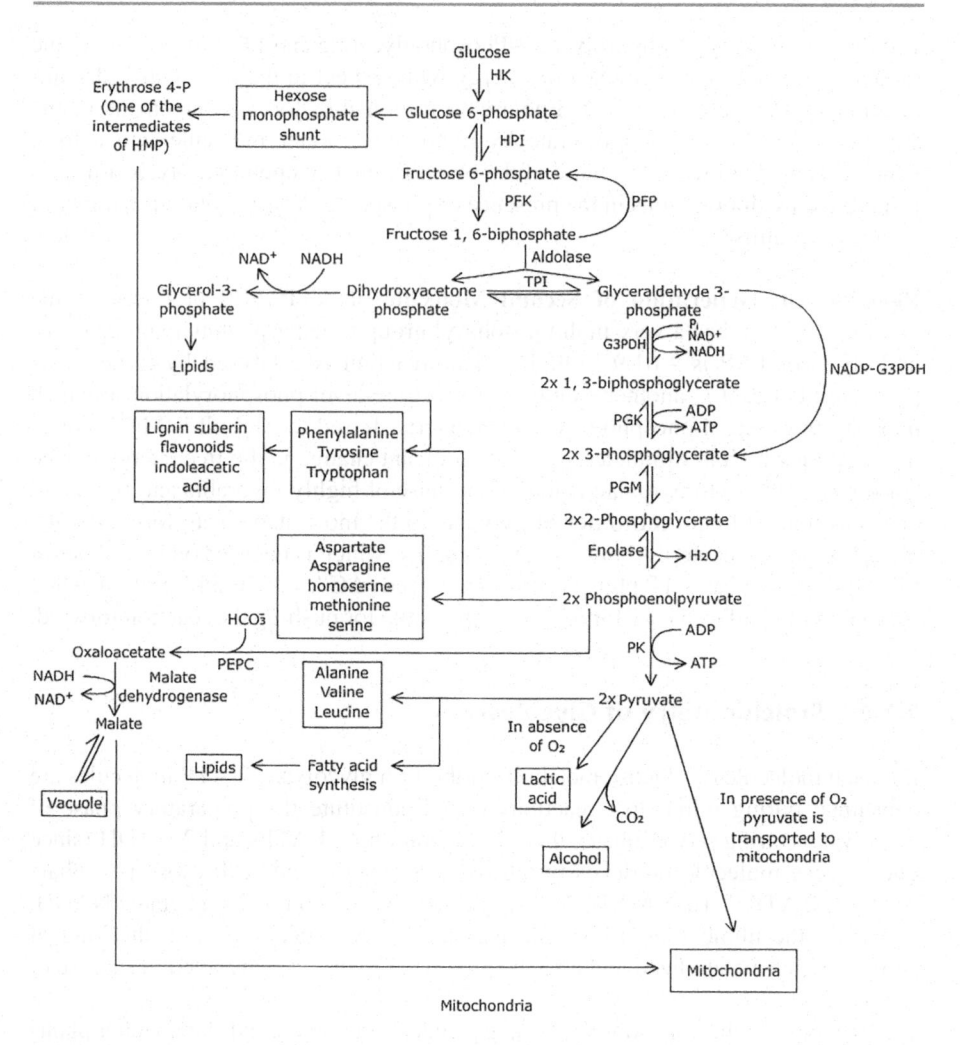

**Fig. 7.3** Cytosolic glycolytic pathway and its role in metabolism of plants. HK, hexokinase; HPI, hexose phosphate isomerase; PFK, phosphofructokinase; G3PDH, glyceraldehyde 3-phosphate dehydrogenase; PGK, phosphoglycerate kinase; PGM, phosphoglyceromutase; PK, pyruvate kinase; PEPC, PEP carboxylase; TPI, triose phosphate isomerase; HMP, hexose monophosphate pathway

1. Plasma membrane is impermeable to phosphorylated sugars because of the absence of their transporters. As a result, the phosphorylated intermediates of glycolysis are retained inside the cell without any expenditure of energy involved.
2. There is a decrease in activation energy when phosphorylated substrates bind with the active sites of the enzymes. Active sites of many of the enzymes are specific for binding with $Mg^{2+}$ complexes of phosphorylated sugars.

3. Energy consumed during priming of sugars at the expense of ATP is conserved as phosphorylated sugars, such as glucose 6-phosphate. High-energy compounds, formed due to rearrangement in the molecules in enzyme-catalyzed reactions, subsequently donate the phosphoryl group to ADP, resulting in the synthesis of ATP.

### 7.1.6 Regulation of Glycolysis in Plants

Regulation of any pathway is mainly determined by the requirement of the products. Irreversible reactions are the sites for regulation in glycolysis. Standard free energy changes ($\Delta G^{0'}$) vary from positive to negative during the ten reactions of glycolysis. Under cellular conditions, $\Delta G^{0'}$ of the reactions can be grouped into two distinct classes: (1) for reactions 2 and 4 through 9, free energy change is very close to zero, meaning these reactions operate essentially at equilibrium. Small changes in the concentration of reactants and products can push the reaction either forward or backward to maintain $K_{eq}$. (2) However, reactions catalyzed by hexokinase, phosphofructokinase, and pyruvate kinase are exothermic, exhibiting large change in free energy under cellular conditions, and are not reversible. These reactions are the sites of glycolytic regulations. When these enzymes are active, gylcolysis proceeds in forward direction, and glucose is readily metabolized to pyruvate. Inhibition of these enzymes brings glycolysis to halt. The main point of regulation of glycolysis is the reaction which involves conversion of fructose 6-phosphate to fructose 1,6-bisphosphate. Both in plants and animals, there are two major enzymes which regulate this irreversible reaction, ATP-dependent PFK and fructose 1,6-biphosphate phosphatase, which regulate flux of carbon through glycolysis and **gluconeogenesis**, respectively. In plants, additional enzyme—$PP_i$-dependent phosphofructokinase—is present. The reaction catalyzed by this enzyme is readily reversible. The activity of the three enzymes (discussed above) is regulated to keep a balance between respiration and synthesis of sucrose or polysaccharides. A prominent difference between animal and plant glycolysis is the primary control point of regulation of glycolysis. In plants, primary control point is at the reaction catalyzed by pyruvate kinase. The enzyme is allosterically inhibited by pyruvate and ATP, resulting in the accumulation of PEP. PEP is an allosteric inhibitor of PFK. Reaction catalyzed by PFK is secondary regulatory point in which $P_i$ acts as an activator of the enzyme. Thus, ratio of PEP and $P_i$ regulates the activity of PFK, with PEP inhibiting and $P_i$ promoting enzyme activity. This type of regulation is known as "bottom-up" regulation. It is significant in plants, since PEP is also metabolized by pathways other than glycolysis. This includes carboxylation of PEP by PEP carboxylase, which results in synthesis of OAA and malate. Malate may be transported to mitochondria. In plants instead of ATP and AMP, accumulation of PEP is the primary control point of glycolysis. PEP is the allosteric inhibitor of PFK, which is the secondary control point (Fig. 7.4). Activity of pyruvate kinase is also inhibited by intermediates of

```
                F6P                        F6P
  Secondary      |                          |
  control point →| PFK                      ↓ PFK  ←――― Primary
                 ↓                          ↓            control point
               F 1,6-BP                   F 1,6-BP
                 ↓                          ↓
                 ↓                          ↓
                 ↓                          ↓
                 ↓                          ↓
                 ↓                          ↓
               PEP                        PEP
  Primary      ↓ PK                       ↓↓ PK  ←――― Secondary
  control point                                        control point
               Pyruvate                   Pyruvate

        In plants 'bottom         In animals 'feed
        up' regulation            forward' regulation
```

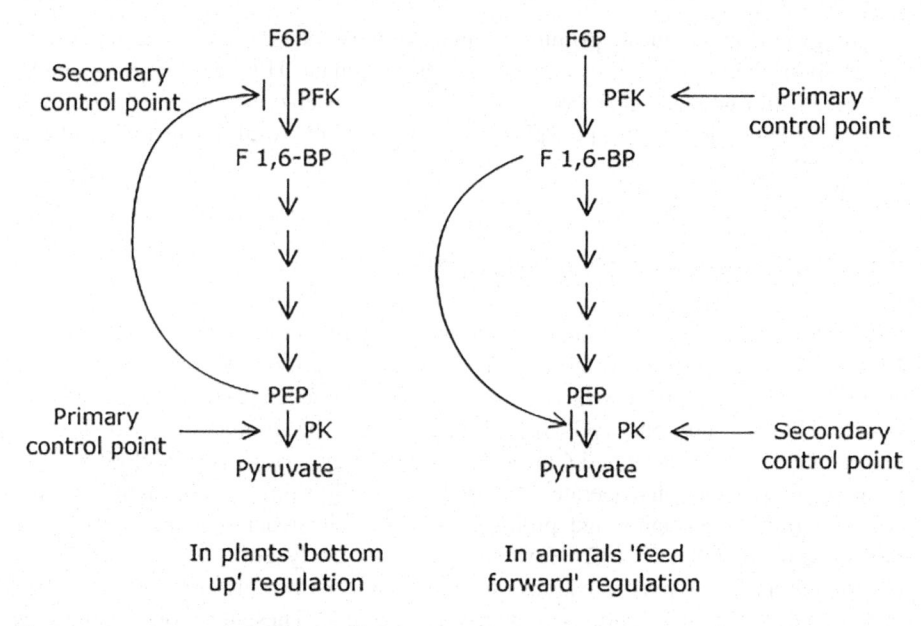

**Fig. 7.4** Regulation of glycolysis in plants and animals

TCA cycle. Contrary to this, regulation of glycolysis in animals is "feed forward" in which the primary control point is the reaction catalyzed by PFK, which leads to accumulation of F 6-P. F 6-P inhibits the activity of pyruvate kinase. Thus, reaction catalyzed by pyruvate kinase is the secondary control point for glycolysis in animals. PEP carboxylase (or other enzymes that metabolize PEP) releases the inhibition of phosphofructokinase by PEP, hence allowing glycolysis to proceed. Reduced levels of cytosolic phosphoenolpyruvate result in elevated levels of fructose 2,6-bisphosphate, which affects the reaction catalyzed by PFK in both directions. Plant pyruvate kinase is not activated by fructose 1,6-bishosphate (which is seen in the case of animals). The advantage of "bottom-up" regulation of glycolysis in plants is that it allows them to control net glycolytic flux of pyruvate independent of related metabolic processes, such as the Calvin cycle and sucrose-triose phosphate-starch interconversion. Activity of hexokinase is allosterically inhibited by ATP or fructose 6-phosphate. However, no glucose 6-phosphatase has been identified in plants. Besides soluble cytosolic pool, a substantial pool of glycolytic enzymes bound to the surface of mitochondria exists under high respiratory rates. These are subjected to regulation by TCA cycle intermediates. Because of **substrate channeling** (Box 7.1), positioning of glycolytic intermediates on mitochondrial surface increases the efficiency of the process.

## Box 7.1: Multienzyme Complex (Significance)

Enzymatic reaction rates are limited by the frequency at which enzymes collide with their substrates. If a series of reactions occurs within a multienzyme complex, the distance that substrates must diffuse between active sites is minimized; thereby achieving a rate enhancement. Multienzyme complex formation provides the means for substrate **channelling**, i.e., passing of metabolic intermediates between successive enzymes in a metabolic pathway, thereby minimizing diffusion of intermediates and avoiding side reactions. Enzymes in pyruvate dehydrogenase pathway are coordinated in multienzyme complex.

## 7.2 Oxidative Pentose Phosphate Pathway (OPPP)

In the 1930s Otto Warburg made an observation that in addition to glycolysis, an alternate metabolic pathway exists in living cells for glucose metabolism. The enzyme was isolated from yeast and erythrocytes and was named as *zwischenferment*. Instead of diphosphopyridine, DPN (now known as $NAD^+$), the enzyme used another coenzyme triphosphopyridine nucleotide TPN (now known as $NADP^+$) for oxidation of glucose to 6-phosphogluconate. However, details of the pathway, result of the work done by Horeckert et al., were presented in 1955. Enzymes, which catalyze the reactions, have been isolated, and regulatory mechanism involved has also been worked out. In plants, besides cytosol, the pathway also operates in plastids. It is not an alternate pathway but occurs in addition to glycolysis. Almost 10–15% of glucose is metabolized through this pathway. It consists of an oxidative phase and a non-oxidative phase. It is known as oxidative pentose phosphate pathway (OPPP), since first two irreversible reactions are oxidative, which are coupled with reduction of $NADP^+$ to NADPH. The subsequent reversible reactions are similar to those involved in regeneration of RuBP during Calvin-Benson cycle. Calvin-Benson cycle is also known as reductive pentose phosphate pathway since that is an assimilatory reductive pathway. OPPP is also called as **hexose monophosphate shunt (HMP)**, as hexose phosphates are regenerated during the cycle, or **phosphogluconate pathway** as first compound formed is phosphogluconate. The pathway provides cells with NADPH, which are required for biosynthetic anabolic reactions, and is especially significant in the plastids of the dark-grown plants or in animal cells (such as RBC) which do not have any other source of NADPH. NADPH is required for glutathione production which is essential for maintaining membrane integrity. Non-operation of this pathway, because of mutant forms of the enzyme glucose 6-phosphate dehydrogenase, causes hemolysis in animal cells.

OPPP includes an oxidative phase, which involves two irreversible reactions, while non-oxidative phase consists of reversible reactions.

### 7.2.1 Oxidative Phase

Oxidation of glucose 6-phosphate by the enzyme glucose 6-phosphate dehydrogenase results in the formation of 6-phosphoglucono-δ-lactone, an unstable intermediate which is spontaneously decomposed to form 6-phosphogluconate. This oxidative reaction is coupled with reduction of $NADP^+$ to $NADPH$ since the enzyme is $NADP^+$-specific. It is highly exergonic reaction and is irreversible. This is the regulatory step of the pathway, which determines metabolic fate of glucose 6-phosphate, either through OPPP or glycolysis. The second reaction, the oxidative decarboxylation of 6-phosphogluconate, results in the formation of ribulose 5-phosphate. It is catalyzed by $NADP^+$-specific 6-phosphogluconate dehydrogenase which again is coupled with reduction of $NADP^+$ to $NADPH$, and one carbon is lost as $CO_2$ during the reaction (Fig. 7.5).

### 7.2.2 Non-oxidative Phase

Carbon atoms of ribulose 5-phosphate are reconstituted to regenerate glucose 6-phosphate during non-oxidative phase of the pathway which involves reversible reactions. Ribulose 5-phosphate is interconverted to ribose 5-phosphate and xylulose 5-phosphate through the action of ribose phosphate isomerase and ribulose phosphate epimerase, respectively. The two enzymes **transketolase** (TK) and **transaldolase** (TA) catalyze the interconversion of the intermediate sugars. The formation of sedoheptulose 7-phosphate and glyceraldehyde 3-phosphate from ribose 5-phosphate and xylulose 5-phosphate is catalyzed by transketolase. Transaldolase is the characteristic enzyme of OPPP, which catalyzes the transfer of non-phosphorylated 3-carbon fragment from sedoheptulose 7-phosphate to glyceraldehyde 3-phosphate after hydrolyzing –C-C bond. As a result of this reaction, fructose 6-phosphate and erythrose 4-phosphate are synthesized. The next step also

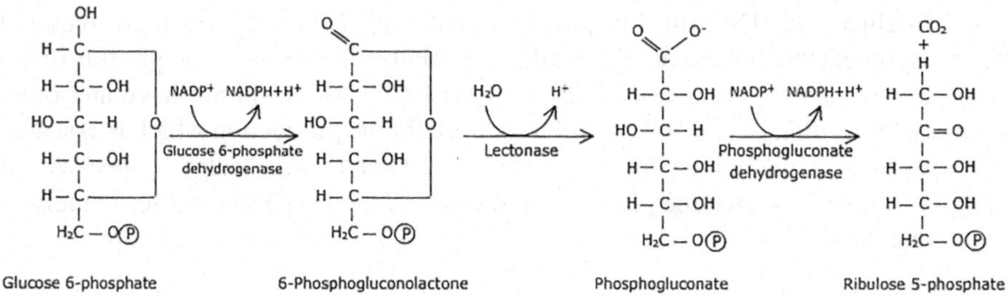

**Fig. 7.5** Reactions of oxidative phase of oxidative pentose phosphate pathway

## 7.2 Oxidative Pentose Phosphate Pathway (OPPP)

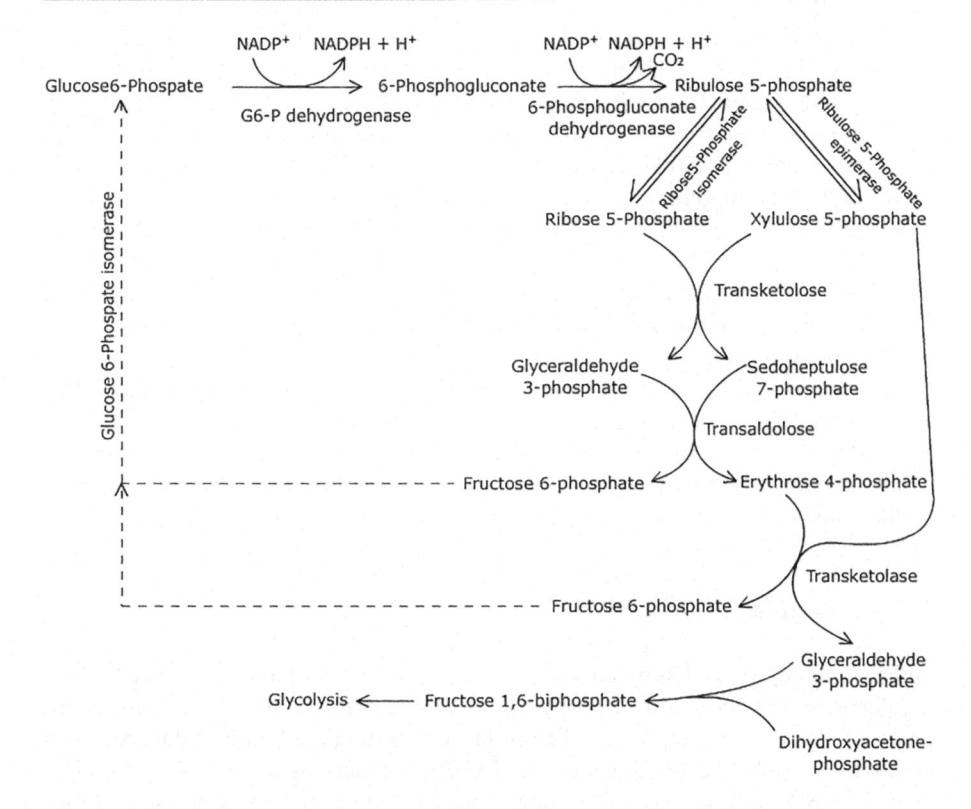

**Fig. 7.6** Oxidative pentose phosphate pathway (OPPP)

involves transketolase, which results in the formation glyceraldehyde 3-phosphate and fructose 6-phosphate from erythrose 4-phosphate and xylulose 5-phosphate because the enzyme catalyzes transfer of 2-carbon fragment from xylulose 5-phosphate to erythrose 4-phosphate. The two triose phosphates condense together to form hexose phosphates (Fig. 7.6).

If removal of any of the intermediates is bypassed, overall reaction of the pathway can be written as:

$$6\text{Glucose } 6-P + 12\,NADP^+ \rightarrow 5\text{Glucose } 6-P + P_i + 6CO_2 + 12NADPH + 12\,H^+$$

### 7.2.3 Significance of OPPP

1. In plants growing in dark or during greening of leaves when photosynthetic apparatus has not yet developed fully and the light generated NADPH is not available, OPPP is the sole source of NADPH, which is required for various

biosynthetic reactions. In plastids, it is required for fatty acid biosynthesis and nitrogen metabolism. Unlike plants, in animal cells, cytosolic OPPP is the only source for NADPH. NADPH is also required under oxidative stress as it is needed for the reactions which are involved in removing reactive oxygen species (ROS).
2. In plant cells, NADPH is also the source of ATP generation, since external NADPH dehydrogenase present on the cytosolic side of the inner mitochondrial membrane can oxidize NADPH, and the electrons move downhill through the electron transport chain.
3. NADPH produced during oxidation of glucose 6-phospate may also signal the sugar status in amyloplasts and may regulate enzymes of starch synthesis through thioredoxin system.
4. The pathway also provides ribose 5-phosphate which is the precursor for nucleotide biosynthesis as well as for ATP, NAD$^+$, and NADP$^+$.
5. Erythrose 4-phosphate along with phosphoenolpyruvate (a glycolytic intermediate) is required for the biosynthesis of plant phenolic compounds, such as lignin, flavonoids, phytoalexins, and aromatic amino acids (Fig. 7.3).

### 7.2.4 Regulation of OPPP

OPPP is mainly regulated by the redox status of the cell. Ratio of NADPH/NADP$^+$ regulates the activity of glucose 6-phosphate dehydrogenase, the first enzyme of the pathway. The enzyme is active if the ratio is low. Strong inhibition of the activity of glucose 6-phosphate dehydrogenase by NADPH indicates important role of OPPP in providing NADPH. In case not required, the nucleotide precursor ribose 5-P may enter glycolysis after it is converted back to glucose 6-P by the reversible non-oxidative reactions. If NADPH is not needed, ribose 5-P may be generated by alternate reactions than by OPPP. In plants regulation of OPPP is significant especially in plastids, where reductive pentose phosphate pathway (Calvin cycle) occurs. If both pathways operate simultaneously at the same place, fructose 6-phospate, an intermediate of the Calvin cycle, is oxidized through OPPP producing 2 NADPH and ribulose 5-phosphate. Ribulose 5-phosphate will enter Calvin cycle requiring 2 NADPH and 3 ATP for $CO_2$ assimilation. As a result, there is a net loss of 3 ATP molecules. NADPH produced during OPPP is also consumed. Such cycles are called **futile cycles**. Loss of ATP molecules is avoided since these two cycles operate at different times. Calvin cycle operates during daytime since the enzymes of the cycle are activated in light by the thioredoxin-ferredoxin system. On the contrary OPPP operates during dark since, unlike Calvin cycle enzymes, the first two enzymes of the cycle are inactive when reduced, which again is regulated by the same thioredoxin-ferredoxin system. Redox status of specific cysteine residues of OPPP enzymes plays an important role in their regulation. OPPP also shares some of the reactions with glycolysis. Operation of glycolysis and OPPP is interdependent since the two share some common intermediates and are operating at the same place at the same time. In both the cycles, phosphotrioseisomerase is one of the most active enzymes which catalyzes reversible interconversion of glyceraldehyde 3-phosphate and dihydroxyacetone phosphate. In OPPP, the formation of dihydroxyacetone phosphate is favored which is transported out of the plastids by triose phosphate/$P_i$ antiporters. On the contrary, glyceraldehyde 3-phosphate is favored in glycolysis and on the main pathway.

## 7.3 PEP Metabolism in Cytosol

The fate of PEP produced during glycolysis is determined by the cytosolic enzymes. Conversion of PEP to pyruvate is catalyzed either by pyruvate kinase or by PEP phosphatase. In the former case, ATP is synthesized, while in the reaction catalyzed by PEP phosphatase, there is no ATP synthesis (Fig. 7.7). Alternatively, PEP is metabolized by PEP carboxylase which catalyzes its carboxylation, resulting in the synthesis of oxaloacetate, which may either be transported to mitochondria as such or may be reduced to malate in cytosol using NADH. Malate may be transported to mitochondria. Thus reducing power (NADH) equivalents are also shuttled to mitochondria as malate. Once inside mitochondria, malate is oxidized to OAA coupled with reduction of $NAD^+$ to NADH. The reaction is catalyzed by malate dehydrogenase. In mitochondria, alternatively malate is converted to pyruvate by $NAD^+$ malic enzyme. Pyruvate enters TCA cycle through link reaction.

## 7.4 Pyruvate Metabolism

Pyruvate, the end product of glycolysis, is metabolized differently under aerobic and anaerobic conditions. Under aerobic conditions pyruvate is oxidized to acetyl-CoA in mitochondria, which enters the citric acid cycle, and the NADH formed in glycolysis is oxidized to $NAD^+$ through NADH shuttle system. NADH is not transported across inner mitochondrial membrane directly because of its transporters being absent. Under anaerobic conditions pyruvate is metabolized through fermentation.

### 7.4.1 Fermentation

Plants are rarely exposed to anaerobic conditions. Anaerobic conditions prevail when plants are either completely submerged or $O_2$ is completely exhausted by respiration resulting in anoxia or in germinating seeds. $O_2$ is scarce or even absent in waterlogged soils, and, therefore, respiration is inhibited in roots. Because of mitochondrial electron transport not functioning in the absence of $O_2$, which is the terminal acceptor of electrons, NADH will not be oxidized. Failure to regenerate $NAD^+$ will leave the cell with no electron acceptor for the oxidation of glyceraldehyde 3-phosphate, resulting in stopping of energy yielding reactions of glycolysis. Continuation of glycolysis will require an alternate path for regeneration of $NAD^+$. In the absence of $O_2$, the end product of glycolysis, pyruvate, is metabolized either

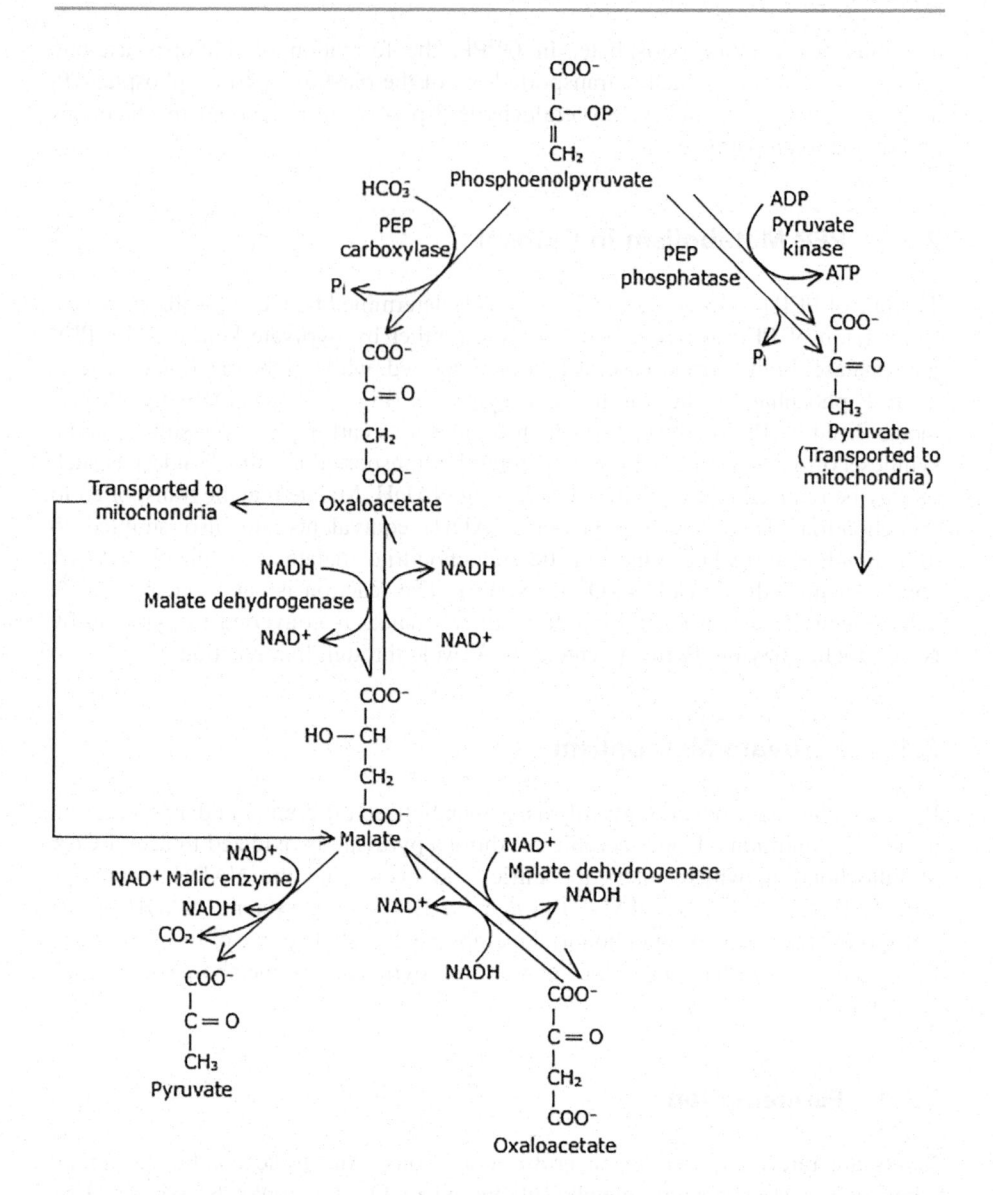

**Fig. 7.7** During respiration, phosphoenolpyruvate is converted to either pyruvate or oxaloacetate in the cytosol which can also be transported to mitochondria. OAA is transported either as OAA or malate after it is reduced by the enzyme malate dehydrogenase. Inside mitochondria, malate is metabolized either by malate dehydrogenase or NAD$^+$ malic enzyme

through lactic acid or ethanol fermentation, resulting in oxidation of NADH to NAD$^+$. Breakdown of carbohydrates to ethanol and $CO_2$ is the main fermentative pathway in plants. Some lactate is also formed due to the activity of lactic acid dehydrogenase, but in contrast to lactic acid fermentation, ethanol fermentation can continue for days in anoxic tissues since ethanol is nontoxic and can leach out of the plasma membrane. Initially oxygen deficiency induces synthesis of lactic acid dehydrogenase. Lactic acid formation lowers the cytosolic pH, resulting in inactivation of lactic acid dehydrogenase and upregulation of alcohol dehydrogenase. Oxidation of NADH to NAD$^+$ is coupled with reduction of pyruvate to either lactate or ethanol. There is no net change in oxidation state of carbon in glucose ($C_6H_{12}O_6$) when converted to lactate, and the H/C ratio also remains the same. Most of the energy is still retained in the molecule of lactate. During fermentation there is net production of 2 ATP for every molecule of glucose being converted to either lactate or alcohol.

In alcoholic fermentation, decarboxylation of pyruvate is catalyzed in an irreversible reaction by the action pyruvate decarboxylase. The enzyme requires $Mg^{2+}$ and a tightly bound cofactor—thiamine pyrophosphate. Alcohol dehydrogenase catalyzes the reduction of acetaldehyde to ethanol using NADH, which is oxidized to NAD$^+$. End products of ethanol fermentation are ethanol and $CO_2$. $CO_2$ produced by pyruvate decarboxylase action in brewer's yeast is responsible for the characteristic carbonation of champagne. Modern-day brewing involves several enzymatic processes. Alcohol dehydrogenase occurs in organisms that metabolize ethanol which also includes human beings. In the liver the enzyme catalyzes oxidation of ethanol, followed by reduction of NAD$^+$ to NADH. This reaction is the reverse of production of ethanol during fermentation.

## 7.4.2 Pyruvate Metabolism in Mitochondria

Glycolysis is not the only energy-yielding catabolic pathway. Cells that rely exclusively on glycolysis to meet their energy requirement actually waste most of the potential energy of the carbohydrates. During fermentation glucose is converted to either lactate or ethanol which are relatively lesser reduced product and still retain most of the energy. The end product of glycolysis, pyruvate, is instead further oxidized, and the cell can recover considerably more energy. In the presence of oxygen, pyruvate is transported to mitochondria and is oxidized. Oxidation of an organic compound requires an electron acceptor such as $NO_3^-$, $SO_4^{2-}$, $Fe^{3+}$, or $O_2$, all of which are exploited as oxidants in different organisms. In aerobic organisms the electrons which are removed during oxidative metabolism are ultimately transferred to $O_2$. Studying pyruvate oxidation will involve reactions of conversion of pyruvate to acetyl-CoA, citric acid cycle, and oxidation of reduced intermediate compounds produced during TCA cycle through electron transport chain. Pyruvate derived from glucose is split into $CO_2$ and a two-carbon fragment in mitochondria that enters citric acid cycle for oxidation as acetyl-CoA. It is sometimes convenient to think of the citric acid cycle as an addition to glycolysis. However, it is really misleading to think of the citric acid cycle as merely a continuation of carbohydrate catabolism. The citric acid cycle is a central pathway for recovering energy from several metabolic fuels, including carbohydrates, fatty acid, and amino acids, especially in animals. TCA cycle also serves to provide precursors for a variety of biosynthetic pathways. First of all we explore how acetyl-CoA, its starting compound, is formed from pyruvate followed by its metabolism through TCA cycle.

*Oxidative Decarboxylation of Pyruvate: The Link Reaction* Pyruvate produced in glycolysis is transported to mitochondria for further metabolism. The outer membrane allows passage of **metabolites** having molecular mass up to 1 kDa. However, free movement of proteins does not occur. Since the inner membrane does not allow free passage, the molecules need to be transported through various transporters. The inner mitochondrial membrane is infolded to form cristae and contains almost 50% of the mitochondrial proteins. The two membranes of mitochondria enclose a compartment known as matrix, and almost 50% of the molecules of mitochondria by weight are present in the matrix. Mobility is restricted since there is little water being present, and possibly the matrix proteins are organized in multiprotein complexes (**metabolons**) to facilitate substrate channeling. Pyruvate which enters mitochondria is converted to acetyl-CoA by oxidative decarboxylation. One carbon of the molecule is lost as $CO_2$. Oxidation is coupled to reduction of $NAD^+$ to NADH. The reaction is catalyzed by pyruvate dehydrogenase complex (PDC). In plants, PDC is present in plastids also, but there it plays a role in lipid biosynthesis. This reaction is also called link reaction since it joins glycolysis and TCA cycle.

$$\text{Pyruvate} + \text{CoA} - \text{SH} + NAD^+ \xrightarrow{\text{Pyruvate dehydrogenase complex}} \text{acetyl} - \text{CoA} + \text{NADH} + CO_2 + H^+$$

Pyruvate dehydrogenase complex (PDC) is a multienzyme complex consisting of three catalytic units. These are pyruvate dehydrogenase (E1), dihydrolipoyl transacetylase (E2), and dihydrolipoyl dehydrogenase (E3). Conversion of pyruvate to acetyl-CoA is a multistep process which requires five cofactors along with the three enzymes. E1 is covalently linked to TPP, lipoic acid is covalently linked to E2, and FAD is the prosthetic group of E3. The other cofactors which are present as free molecules are $NAD^+$ and coenzyme A. Besides, the enzymes which regulate the activity of pyruvate dehydrogenase, are also present. In plants, the core of the PDC in mitochondria consists of 60 copies of E2 in 12 sets of 5, each having lipoic acid-binding region (studies from the purified form of PDC of mitochondria in maize) and binding site for E1. E1 is a tetraheteromer consisting of $\alpha$ and $\beta$ subunits and is covalently linked with TPP. Phosphorylation of two conserved serine residues (at 300 and 306 position) in $\alpha$ subunit of the enzyme makes E1 inactive, while dephosphorylation of these residues converts the enzyme to active form. E1 subunits are covalently linked to the E2 core. E3, which is present as dimer, has got binding sites for FAD and $NAD^+$. E3 is also linked to E2 core toward the inner side of the complex. Besides this, the regulatory enzymes pyruvate dehydrogenase kinase (PDK), which catalyzes ATP-dependent phosphorylation of E1, and phosphopyruvate dehydrogenase phosphatase (PDP), which catalyzes dephosphorylation of the enzyme protein of E1, are also present as components of the complex. PDK and PDP are responsible for inactivation and activation of the enzyme, respectively.

## 7.4 Pyruvate Metabolism

Oxidative decarboxylation of pyruvate occurs in the following five steps (Fig. 7.8):

1. Pyruvate dehydrogenase (E$_1$-TPP), a TPP-requiring enzyme, decarboxylates pyruvate, with the formation of an intermediate hydroxyethyl-TPP. This reaction is identical with the one catalyzed by yeast pyruvate decarboxylase.

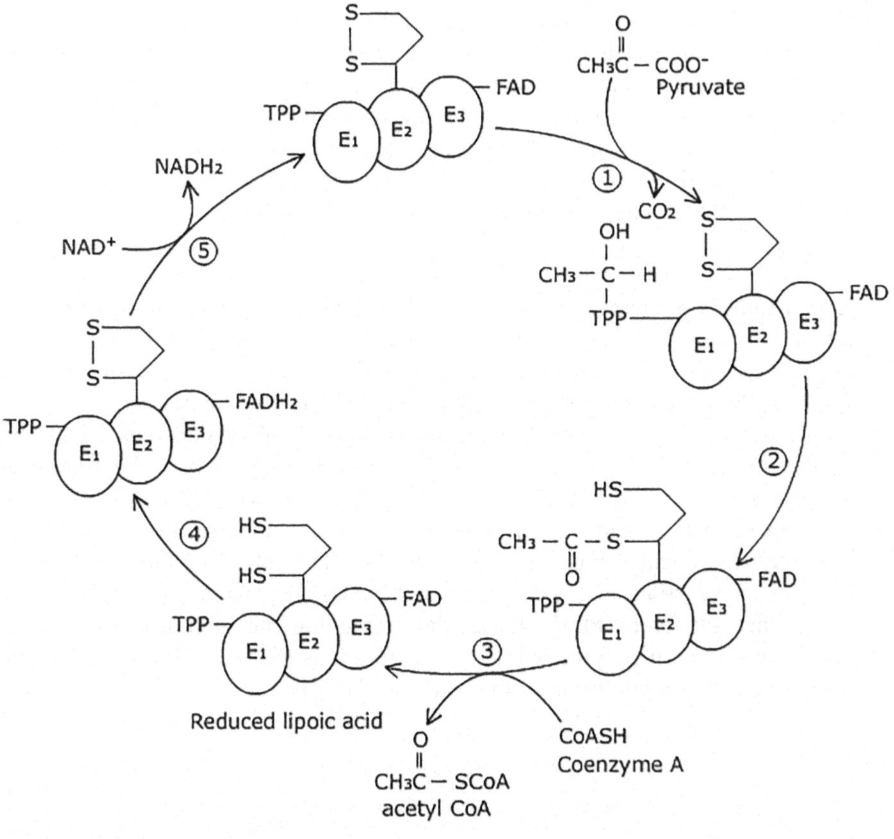

**Fig. 7.8** Figure showing the reaction steps, which occur during conversion of pyruvate to acetyl-CoA catalyzed by pyruvate dehydrogenase complex (PDC). PDC is a multienzyme complex which consists of three enzymes, E1 (pyruvate dehydrogenase), E2 (dihydrolipoyl transacetylase), and E3 (dihydrolipoyl dehydrogenase), which require cofactors TPP, lipoic acid, and FAD, covalently linked to them, respectively. Besides these the other cofactors required are NAD$^+$ and CoA-SH. The reaction occurs in five steps: Step 1. E1 catalyzes decarboxylation of pyruvate generating acetaldehyde to produce hydroxyethyl-TPP complex. 2. Hydroxyethyl group forms acetyl-dihydrolipoamide, a thioester of lipoic acid of E2. 3. E2 catalyzes transfer of acetyl group from acetyl-dihydroipoamide to coenzyme A, resulting in the formation of acetyl-CoA and fully reduced dihydrolipoamide group of E2. 4. E3 catalyzes reoxidation of reduced dihydrlipoamide resulting in reduction of FAD to FADH$_2$. 5. FADH$_2$ is oxidized to FAD and is coupled with reduction of NAD$^+$ to NADH, and PDC is regenerated in original form

2. Unlike pyruvate decarboxylase, pyruvate dehydrogenase does not convert hydroxyethyl-TPP into acetaldehyde and TPP. Instead, the hydroxyethyl group is transferred to lipopic acid of the next enzyme dihydrolipoyl transacetylase (E2) of the complex, forming thioester bond. The reaction is coupled with the elimination of TPP and formation of acetyl dihydrolipoamide with simultaneous release of active E1. The hydroxyethyl is thereby oxidized to an acetyl group by the concomitant reduction of the lipoamide disulfide bond.
3. E2 catalyzes the transfer of the acetyl group from acetyl dihydrolipoamide to SH-CoA, yielding acetyl-CoA and dihydrolipoamide-E2.
4. Dihydrolipoyl dehydrogenase (E3, also called dihydrolipoamide dehydrogenase) reoxidizes dihydrolipoamide completing the catalytic cycle of E2. Oxidation of dihydrolipoamide is coupled with reduction of the covalently bonded cofactor of E3, i.e., FAD to $FADH_2$, and formation of the lipoamide disulfide bond (-S-S-).
5. Reduced $FADH_2$ is reoxidized by the simultaneous reduction of $NAD^+$ to NADH. E3 has got the binding site both for FAD and $NAD^+$.

As a result of sequence of these five reactions, the net result of oxidative decarboxylation of pyruvate is formation of acetyl-CoA and NADH and loss of one carbon as $CO_2$. This reaction is very significant and is regulated by various metabolites. Pyruvate dehydrogenase (PDH), the first enzyme catalyzing these set of reactions, is subjected to regulation. Ratio of ATP/ADP, which determines the metabolic status of the cell, has an important role to play in its regulation. The enzyme E1 is inactivated by ATP-dependent phosphorylation catalyzed by pyruvate dehydrogenase kinase, while a phosphatase removes the phosphoryl group making the enzyme active again (Fig. 7.9). Activity of PDH is also inhibited by the end products of the reaction, i.e., acetyl-CoA and NADH. Pyruvate activates PDH by inhibiting the activity of PDH kinase, thus removing the inhibition because of phosphorylation of the enzyme, while ammonium ions ($NH_4^+$) activate the activity of kinase, resulting in inhibition of the activity of PDH.

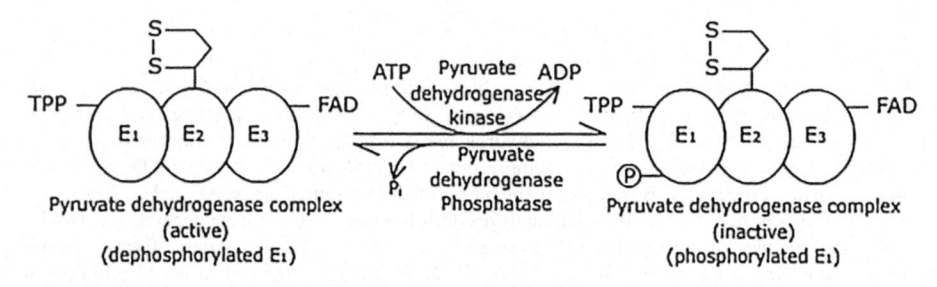

**Fig. 7.9** Phosphorylation and dephosphorylation of E1 of the pyruvate dehydrogenase complex regulate the activity of pyruvate dehydrogenase complex

## 7.5 TCA Cycle

Acetyl group of acetyl-CoA is completely oxidized during citric acid cycle releasing the remaining energy. Released energy is conserved as reduced pyridine nucleotides, and the two carbons of the molecule are released as $CO_2$. The cycle is called as Krebs cycle after the name of the scientist who was responsible for putting together the various reactions which had been studied by different scientists. In fact this cycle was the first of series of reactions to have been put together to give the concept of cyclic reactions in which substrate of the first enzyme is regenerated. Hans Krebs demonstrated formation of citrate, a tricarboxylic acid, when OAA condenses with acetyl-CoA. The cycle was also named as citric acid cycle after the product of its first reaction, citrate. One complete round of cycle starting with acetyl-CoA yields two molecules of $CO_2$, three NADH, one $FADH_2$, and one high-energy compound (ATP/GTP in plants or animals, respectively). Citric acid cycle first came to light in the 1930s, when Hans Krebs proposed a circular reaction scheme for the interconversion of certain compounds containing two or three carboxylic acid groups. At that time, many of the citric acid cycle intermediates were already well-known plant products: citrate from citrus fruit, aconitate from monkshood (*Aconitum*), succinate from amber (*Succinum*), fumarate from the herb *Fumaria*, and malate from apple (*Malus*). Two other intermediates α-ketoglutarate and oxaloacetate are known by their chemical names because they were synthesized before they were identified in living organisms. Discovery of the citric acid cycle ranks as one of the most important achievements of metabolic chemistry, and Hans Krebs was awarded Noble Prize in 1953 for that.

### 7.5.1 General Features of the Cycle

1. It is a circular pathway during which acetyl groups from many sources, not just pyruvate, are oxidized. The citric acid cycle is often considered the hub of cellular metabolism.
2. The net reaction of citric acid cycle is

$$3NAD^+ + FAD^+ + ADP + P_i + acetyl - CoA$$
$$\rightarrow 3NADH + FADH_2 + ATP + CoA + 2CO_2$$

Oxaloacetate that is consumed in the first step of the citric acid cycle is regenerated in the last step of the cycle. The citric acid cycle acts as multistep catalyst that can oxidize an unlimited number of acetyl groups. In eukaryotes, all the enzymes of the citric acid cycle are located in the mitochondria. All the products of the citric acid cycle must be consumed in the mitochondria or transported to the cytosol. Thus there is lot of exchange of metabolites between cytosol and mitochondria through various transporters located in the inner membrane (Fig. 7.10, Table 7.2).

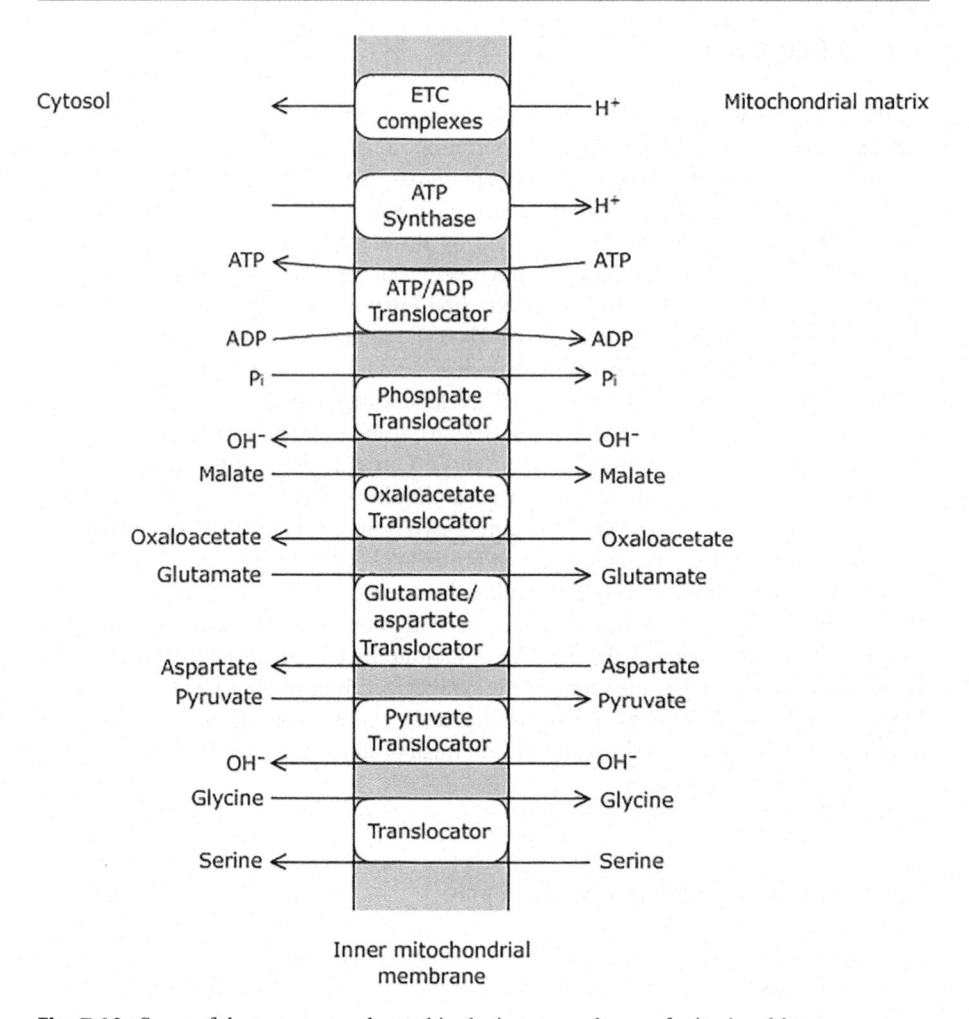

**Fig. 7.10** Some of the transporters located in the inner membrane of mitochondria

3. The two carbon atoms of the $CO_2$ produced in one round of TCA cycle are not the same two carbons of the acetyl-CoA that began the round. These acetyl carbon atoms are lost in subsequent rounds of the cycle. However, the net result of each round of the cycle is the oxidation of one acetyl group to $2CO_2$.
4. Citric acid cycle plays **amphibolic role** since, besides playing a catabolic role, intermediates of the cycle serve as precursors for various biomolecules.
5. Oxidation of an acetyl group to $2CO_2$ requires the transfer of four pairs of electrons. Reduction of $3NAD^+$ to 3NADH accounts for three pairs of electrons; reduction of FAD to $FADH_2$ accounts for the fourth pair. Much of the free energy of oxidation of the acetyl group is converted in these reduced coenzymes. Energy is also recovered as ATP (or GTP in animals).

## 7.5 TCA Cycle

**Table 7.2** Reactions of TCA cycle

| | Reactions | Enzyme catalyzing the reaction | $\Delta G^{0'}$ of the reaction in kJ.mol$^{-1}$ |
|---|---|---|---|
| 1. | Acetyl – CoA + OAA $\rightarrow$ Citrate + CoASH + H$^+$ | Citrate synthase | –32.2 |
| 2. | Citrate $\leftrightarrow$ Isocitrate | Aconitase | 13.3 |
| 3. | Isocitrate + NAD$^+$ $\rightarrow$ 2 – oxoglutarate + NADH + CO$_2$ + H$^+$ | Isocitrate dehydrogenase | –17.5 |
| 4. | 2 – oxoglutarate + NAD$^+$ $\rightarrow$ Succinyl – CoA + NADH + CO$_2$ | 2-Oxoglutarate dehydrogenase complex | –33.5 |
| 5. | Succinyl – CoA + ADP + P$_i$ $\leftrightarrow$ Succinate + ATP + CoASH | Succinyl-CoA synthetase | –2.9 |
| 6. | Succinate + FAD $\leftrightarrow$ L – Fumarate + FADH$_2$ | Succinate dehydrogenase | 0 |
| 7. | L – Fumarate + H$_2$O $\leftrightarrow$ L – malate | Fumarase | –3.8 |
| 8. | L – Malate + NAD$^+$ $\leftrightarrow$ OAA + NADH + H$^+$ | NAD-malate dehydrogenase | –29.7 |

## 7.5.2 Acetyl-CoA Enters the TCA Cycle

Only a fraction of the energy stored in glucose is harvested in the form of ATP during glycolysis. Aerobic metabolism of pyruvate is the source for the majority of the ATP generated in respiration, when there is complete degradation and oxidation of glucose, resulting in release of both the carbons as two molecules of carbon dioxide. Oxidation during citric acid cycle includes a series of reactions (Fig. 7.11).

*Condensation Reaction: The First Reaction of the Cycle* The first reaction of the TCA cycle involves condensation of acetyl-CoA with oxaloacetate to produce citrate catalyzed by the enzyme citrate synthase (CS). There is an aldol condensation between methyl group of acetyl-CoA and carbonyl group of oxaloacetate, resulting in synthesis of citroyl-CoA. This reaction has an equilibrium constant near 1, but the overall reaction is driven to completion by subsequent hydrolysis of the high-energy thioester to citrate and free CoA. The overall $\Delta G^{0'}$ is $-31.4$ kJ.mol$^{-1}$, and under standard conditions the reaction is essentially irreversible. Although mitochondrial

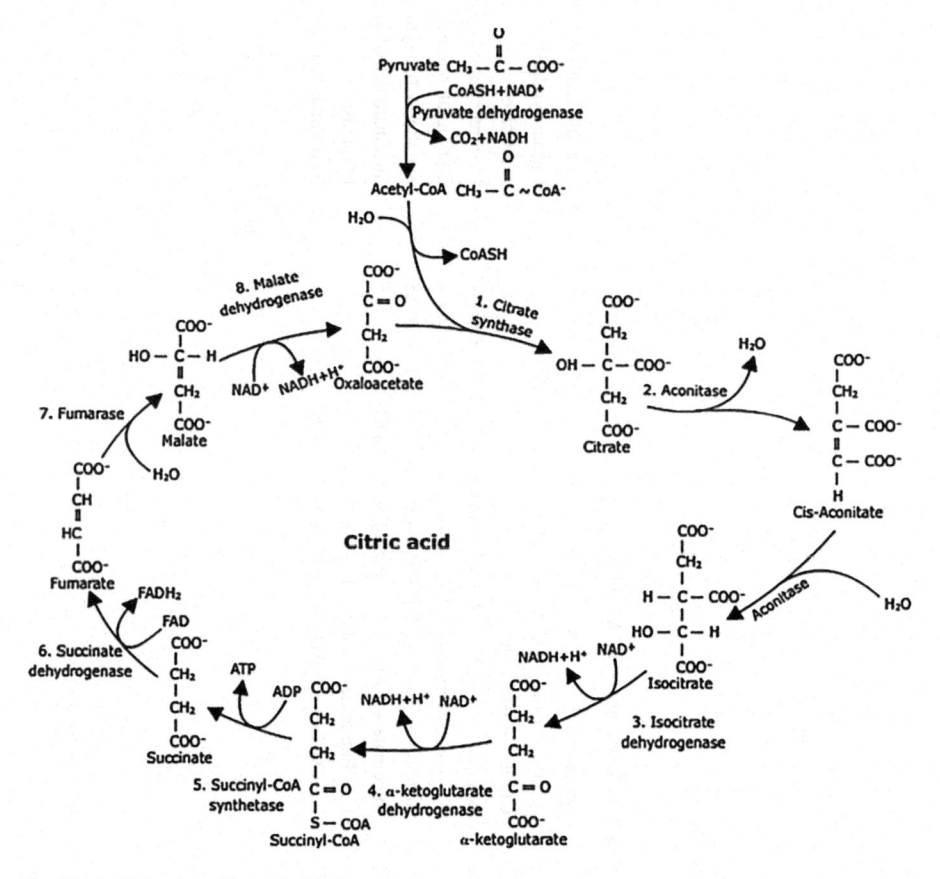

**Fig. 7.11** Tricarboxylic acid (TCA) cycle

concentration of oxaloacetate is very low (much less than 1 M), the strong, negative $\Delta G^{0'}$ drives the reaction forward. Citrate synthase is a unique enzyme of mitochondria, but in plants this enzyme is also present in glyoxysomes which catalyzes reaction during glyoxylate cycle. The mitochondrial enzyme is homodimeric with each subunit having two domains—one large and rigid, while the other one is small and flexible. Active site of the enzyme is present in between the two domains. There is conformational change in the enzyme protein (induced fit) on binding of the first substrate, oxaloacetate, with small and flexible unit of the enzyme. As a result, the site for binding with second substrate acetyl-CoA is created, resulting the formation of an intermediate citroyl-CoA. This is followed by another conformational change resulting in the hydrolysis of the thioester bond. As a result acetyl-CoA and citrate are released. CS is inhibited by ATP, but this inhibition is not very significant since ATP concentration in mitochondria is very low. Reaction catalyzed by CS is the first step in this metabolic pathway having a large negative $\Delta G^{0'}$. As might be expected, it is a highly regulated enzyme. NADH, a product of the TCA cycle, is an allosteric inhibitor of citrate synthase, as is succinyl-CoA, the product of the fifth step in the cycle (which is also an analog of acetyl-CoA).

*Citrate Is Isomerized by Aconitase to Form Isocitrate* Citrate is isomerized to isocitrate by the action of aconitase in a two-step reversible reaction, involving cis-aconitate as an intermediate. The net effect is the conversion of a symmetrical tertiary alcohol, i.e., citrate to an asymmetric secondary alcohol, i.e., isocitrate. The active site of aconitase has iron-sulfur cluster. Cysteine residues of the enzyme protein form coordinate bonds with the three iron atoms, and the fourth iron is bound to one carboxylic group and non-covalently to hydroxyl group of citrate. In this reaction, water is first removed from citrate to yield cis-aconitate, which is then rehydrated resulting in interchange in positions of groups H- and HO- to produce isocitrate. The $\Delta G^{0'}$ of the reaction is 6.7 kJ.mol$^{-1}$. At equilibrium though only 10% isocitrate is present, its utilization in subsequent reaction drives the reaction toward right. Aconitase is present both in mitochondria and cytosol of the plant cell. In mitochondria it participates in TCA cycle, while in cytosol especially in germinating fatty seeds, it participates in glyoxylate cycle. Along with NADP-dependent isocitrate dehydrogenase in cytosol, it also plays an important role in providing 2-oxoglutarate for amino acid biosynthesis. Aconitase is responsible for catalyzing the first step in citric acid catabolism during fruit ripening. Fluoroacetate is an extremely poisonous agent that blocks the TCA cycle in vivo, although it has no apparent effect on any of the isolated enzymes. The action of fluoroacetate has been traced to aconitase in vivo.

*Oxidative Decarboxylation of Isocitrate* The next step of the TCA cycle involves oxidative decarboxylation of isocitrate, resulting in the formation of α-ketoglutarate (2-oxoglutarate). The reaction is coupled with reduction of NAD$^+$ to NADH and loss of one carbon as $CO_2$. It is catalyzed by the enzyme NAD-isocitrate dehydrogenase (IDH). Net $\Delta G^{0'}$ of the reaction is $-8.4$ kJ.mol$^{-1}$, which is sufficiently exergonic to pull the previous reaction catalyzed by aconitase forward. This two-step reaction

**Fig. 7.12** NADPH generation in mitochondria

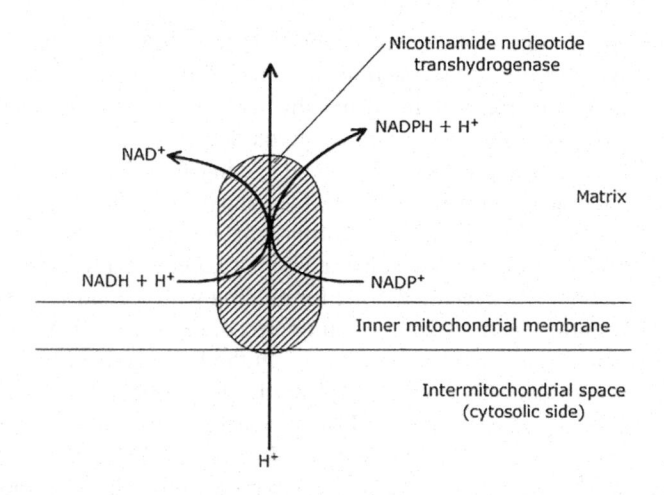

involves synthesis of an intermediate oxalosuccinate. The first step of the reaction results in the oxidation of the –OH group present at C-2 of isocitrate to form oxalosuccinate, which is followed by a decarboxylation of central carboxyl group, resulting in the formation of α-ketoglutarate. Decarboxylation of oxalosuccinate makes the reaction irreversible. Besides NAD-IDH, NADP-IDH is also present in mitochondria. In mitochondria NADPH may be generated either due to the activity of nicotinamide nucleotide transhydrogenase (Fig. 7.12) or because of the activity of $NADP^+$-specific isocitrate dehydrogenase ($NADP^+$-IDH). In plants it is the latter activity which is more significant. $NADP^+$ isoforms of isocitrate dehydrogenase are present in cytosol and plastids, while $NAD^+$-IDH isoform is present only in mitochondria. It is the NAD-IDH form which is important for TCA cycle. Citrate exported from mitochondria is metabolized by the cytosolic aconitase and cytosolic NADP-IDH, resulting in synthesis of α-ketoglutarate for amino acid metabolism. The main function of NADP-IDH enzyme both in mitochondria and cytosol may be to provide NADPH to be used in biosynthetic anabolic reactions. NADH produced by NAD-IDH is oxidized by the electron transport chain in mitochondria, which is the first reaction of the TCA cycle to be connected with oxidative phosphorylation. Oxidative decarboxylation of isocitrate is an important regulatory step of TCA cycle. Both NADH and ATP are allosteric inhibitors of NAD-IDH. Enzyme activity is also inhibited by NADPH. The product of the reaction, α-ketoglutarate, can enter nitrogen metabolism through biosynthesis of glutamic acid.

*Oxidative Decarboxylation of α-Ketoglutarate* The next reaction of TCA cycle involves another oxidative decarboxylation during which α-ketoglutarate is converted to succinyl-CoA. The reaction is catalyzed by *α*-ketoglutarate dehydrogenase enzyme complex which also is a multienzyme complex. Similar to pyruvate dehydrogenase complex, it also consists of three enzymes, i.e., α-ketoglutarate dehydrogenase, dihydrolipoyl transsuccinylase, and dihydrolipoyl dehydrogenase. Reaction occurs in five steps which require five different coenzymes, i.e., TPP, lipoic acid, CoASH, FAD, and NAD$^+$. Mechanism is analogous to the reaction catalyzed by pyruvate dehydrogenase complex. E1 of the α-ketodehydrogenase complex binds with α-ketoglutarate and differs from E1 of PDH complex in some of the amino acids; however, dihydrolipoyl dehydrogenase of both complexes is quite similar. Both the complexes seem to have the same ancestral origin. However, unlike PDH, E1 of this complex is not regulated by phosphorylation; rather AMP is the activator of the enzyme. Similar to pyruvate dehydrogenase reaction, NADH is also produced here along with a thioester product—in this case, succinyl-CoA. One carbon is lost as $CO_2$. The products of the reaction, succinyl-CoA and NADH, are energy-rich compounds which are important sources of metabolic energy. NADH provides ATP when oxidized through electron transport chain, while succinyl-CoA is metabolized through the following reaction.

*Succinyl-CoA Is Converted to Succinic Acid* Succinyl-CoA is a high-energy intermediate having negative free energy of hydrolysis. $\Delta G^{0\prime}$ of the reaction is −36 kJ. mol$^{-1}$, which is utilized to drive the phosphorylation ADP to ATP (in plants and bacteria) by means of substrate-level phosphorylation. In animals similar reaction results in phosphorylation of GDP resulting in synthesis of GTP. The reaction is catalyzed by succinate thiokinase also called succinyl-CoA synthetase (synthetases are the enzymes which catalyze condensation reactions involving nucleoside triphosphates, unlike synthase for which no nucleoside triphosphates are required). The free energies of hydrolysis of succinyl-CoA and GTP or ATP are similar. The net $\Delta G^{0\prime}$ of the reaction is $-3.3$ kJ.mol$^{-1}$. In addition to some of the reactions of glycolysis, this reaction is only such reaction in the TCA cycle, which is the source of substrate-level phosphorylation, in which a substrate (here succinyl-CoA), rather than an electron transport chain or proton gradient, provides the energy for phosphorylation.

*Oxidation of Succinate to Fumarate* Oxidation of succinate to fumarate is carried out by a flavoprotein-linked succinate dehydrogenase. This is the only enzyme of TCA cycle which is bound to inner mitochondrial membrane and is a component of the electron transport chain. Succinate-binding site of the enzyme is present toward the matrix. All other enzymes of TCA cycle are present as soluble proteins in mitochondrial matrix. Succinate dehydrogenase is also known as succinate ubiquinone oxidoreductase, a component of the electron transport chain. Succinate oxidation involves removal of electrons from succinate and passing them to UQ, a component of ETC resulting in the formation of trans-unsaturated fumarate. Succinate dehydrogenase is a dimeric protein, with subunits having molecular masses 70 and 27 kDa. The larger subunit of the enzyme protein is covalently linked to the cofactor FAD. The enzyme also contains non-heme iron, and the electrons removed from succinate pass through FAD and Fe-S centers before being passed to ubiquinone of the ETC. Activity of the enzyme has been demonstrated to be competitively inhibited by malonate in vitro. The enzyme in vivo is strongly inhibited by OAA but is activated by UQH$_2$, ATP, and ADP. Effect of ATP or ADP may be the indirect result of possibly activating the enzyme by releasing OAA. Activity of plant SDH may be regulated by energy status of the mitochondrial membrane, physiological relevance of which is unclear, since there are reciprocal changes in the concentrations of ATP and ADP in the cell.

*Trans-hydration of Fumarate to Form L-Malate* The next reaction of TCA cycle involves trans-hydration of fumarate which requires addition of –H and –OH groups to the C=C double bond of fumarate resulting in synthesis of L-malate. Fumarase catalyzes the reversible reaction which has $\Delta G^{0'}$ value $-3.8$ kJ.mol$^{-1}$. The reaction is stereospecific, as in the reversible reaction, it is L-malate which is the substrate for the enzyme and not D-malate. Aconitase carries a similar reaction, i.e., trans-addition of -H and -OH occurs across the double bond of cis-aconitate.

*Oxidation of L-Malate to Oxaloacetate* This is the last reaction of TCA cycle in which malate is oxidized to regenerate OAA, thus completing the cycle. Oxidation of malate is coupled to reduction of NAD$^+$ molecule, the third one in the cycle. This energy-requiring reaction has $\Delta G^{0'}$ value 29.7 kJ.mol$^{-1}$ which favors synthesis of malate rather than that of OAA. However, removal of OAA in subsequent reaction and oxidation of NADH to NAD$^+$ is responsible for the direction of the reaction toward right. Malate dehydrogenase is a ubiquitous enzyme, and its isoforms are found in mitochondria, cytosol, glyoxysomes, and peroxisomes. NADP$^+$-specific MDH is found in chloroplasts of C4 plants, where it is involved in production of malate from OAA during C4 pathway of $CO_2$ assimilation. It is NAD-MDH which participate in TCA cycle. Its isoform is also involved in exchange of reducing equivalents between cell compartments across the membranes (NADH shuttle).

## 7.5.3 Stoichiometry of TCA Cycle

Net reaction for complete oxidation of one molecule of acetyl-CoA through TCA cycle can be written as

$$Acetyl - CoA + 3NAD^+ + FAD + ADP + P_i + 2H_2O$$
$$\rightarrow 2CO_2 + Co - ASH + 3NADH + 3H^+ + FADH_2 + ATP$$

The cycle is exergonic with a net $\Delta G^{0'}$ for one round of the cycle $= -40$ kJ.mol$^{-1}$

Glucose metabolized via glycolysis produces two molecules of pyruvate, which in the presence of $O_2$ are converted to acetyl-CoA in mitochondria. 2NADH are produced during this conversion of 2 pyruvate to 2 acetyl-CoA. These two molecules of acetyl-CoA thus produced enter the TCA cycle. Combining glycolysis, link reaction, and the TCA cycle, the following net reaction is obtained:

$$\text{Glucose} + 2H_2O + 10NAD^+ + 2FAD + 4ADP + 4P_i$$
$$\rightarrow 6CO_2 + 10\ NADH + 10H^+ + 2\ FADH_2 + 4\ ATP$$

All six carbons of glucose are released as $CO_2$, and a total of four molecules of ATP are produced as a result of substrate-level phosphorylation. This includes net production of 2 ATP during glycolysis and 2 ATP during two TCA cycles (for two molecules of acetyl-CoA). There are 12 reduced coenzymes produced which include 2 NADH produced during glycolysis, 2 NADH during link reaction, while 6 NADH and 2 $FADH_2$ during two TCA cycles. During oxidative phosphorylation, oxidation of NADH generates proton motive force sufficient enough for synthesis of 2.5 ATP per NADH and 1.5 ATP per $FADH_2$. Thus during complete oxidation of a glucose molecule, 30–32 ATP molecules are produced (Table 7.3). This will be equivalent to 976 $kJ.mol^{-1}$ (32 × 30.5 $kJ.mol^{-1}$ for hydrolysis of ATP) which constitutes 34% of the potential energy of a glucose molecule (2840 $kJ.mol^{-1}$).

### 7.5.4 Amphibolic Role of TCA

TCA cycle occupies the central role in metabolism. Apparently TCA cycle is the terminal point at least in terms of carbon turnover. The cycle is important not only in catabolism but has **amphibolic** role since it has a significant role in anabolism as well (Fig. 7.13). Many intermediates of TCA cycle serve as precursors for various cellular compounds. Almost 50% of the carbon is retained by plant tissues through incorporation of intermediates of glycolysis, OPPP, and TCA for biosynthesis of the compounds required by the cell, instead of being released as $CO_2$. One of the important roles of the TCA cycle is to provide carbon skeleton for glutamate biosynthesis in the form of 2-oxoglutarate. There are transporters present in the inner mitochondrial membrane which facilitate exchange of 2-oxoglutarate with cytosolic malate. There are three sources of 2-oxoglutarate, which is required for glutamate biosynthesis. First is from TCA cycle in mitochondria, second is cytosolic NADPH-IDH, and third is in peroxisomes where it is generated during photorespiration by glutamate:glyoxylate aminotransferase. 2-Oxoglutarate is transported to chloroplasts where it is used for glutamate biosynthesis by GS-GOGAT enzymes. In the inner envelope of chloroplasts, there are transporters present for both 2-oxoglutarate and glutamate, which facilitate import of 2-oxoglutarate and export of glutamate. Another intermediate of TCA, i.e., citrate, also is transported out of mitochondria by tricarboxylic acid transporters present in the inner membrane of the organelle, and once in the cytosol, it is converted into isocitrate by the cytosolic aconitase. Isocitrate is the source of cytosolic 2-oxoglutarate, produced due to the activity of $NADP^+$-IDH. Export of tricarboxylate or dicarboxylate from mitochondria is accompanied with uptake of malate. Glutamate serves as the precursor of various other nitrogenous compounds, which include many other amino acids. Another intermediate of TCA, succinyl-CoA, is the source of carbons for various

**Table 7.3** Stoichiometry of aerobic respiration of one molecule of glucose

| | Reactions | ATP used or produced/reduced | ATP equivalent |
|---|---|---|---|
| | **Glycolysis** | | |
| 1. | Glucose → Glucose 6 – phosphate | ATP consumed | −ATP |
| 2. | Fructose – 6 – phosphate → Fructose 1, 6 – bisphosphate | ATP consumed | −ATP |
| 3. | 2X3 – Phosphoglyceraldehyde → 2X1, 3 – Phosphoglycerate | 2NADH produced | $2 \times 1.5$ or $2.5^a$ = +3 or 5 ATP |
| 4. | 2X1, 3 – Bisphosphoglycerate → 2X3 – Phosphoglycerate | 2ATP produced | + 2 ATP |
| 5. | 2Phosphoenolpyruvate → 2Pyruvate | 2ATP produced | + 2 ATP |
| | **Link reaction in mitochondria** | | |
| 6. | 2 Pyruvate → 2 Acetyl – CoA | 2NADH and 2ATP produced | $2 \times 2.5 + 2$ = +7 ATP |
| | **TCA cycle** | | |
| 7. | 2Isocitrate → 2X2 – oxoglutarate | 2NADH produced | $2 \times 2.5$ = +5 ATP |
| 8. | 2X2 – oxoglutarate → 2Succinyl – CoA | 2NADH produced | $2 \times 2.5$ = +5 ATP |
| 9. | 2Succinyl – CoA → 2Succinate | 2ATP produced | + 2 ATP |
| 10. | 2Succinate → 2Fumarate | 2FADH$_2$ produced | $2 \times 1.5$ = +3 ATP |
| 11. | 2Malate → 2Oxaloacetate | 2NADH produced | $2 \times 2.5$ = +5 ATP |
| | | **Net production of ATP** | **30–32 ATP molecules** |

[a]The number of ATP produced depends on the nature of NADH shuttle, since NADH is transported to mitochondria for ATP generation. Those reactions have not been included in the table during which there is no ATP turnover

## 7.5 TCA Cycle

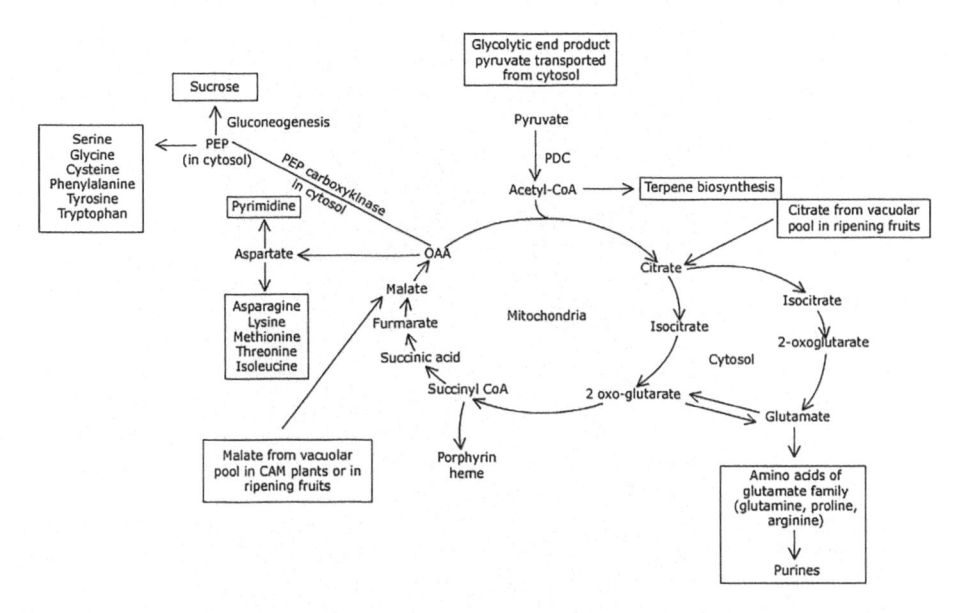

**Fig. 7.13** Amphibolic role of TCA

porphyrin-containing molecules such a chlorophylls, cytochromes, etc. Oxaloacetate is transaminated to produce aspartate which itself is the precursor for asparagine, methionine, threonine, lysine, isoleucine, and pyrimidine nucleotides. On decarboxylation oxaloacetate produces phosphoenolpyruvate, which is the precursor of various aromatic amino acids including phenylalanine, tyrosine, tryptophan, and many other aromatic compounds. Besides PEP can be converted to 3-phosphoglycerate, which is required as the key precursor for biosynthesis of amino acids serine, glycine, and cysteine. In animals, citrate, exported from mitochondria, is responsible for synthesis of acetyl-CoA produced from oxaloacetate and acetyl-CoA by cytosolic ATP-citrate lyase. Oxaloacetate is transported to mitochondria as malate. In an animal cell, cytosolic acetyl-CoA is used as precursor for biosynthesis of fatty acids.

### 7.5.5 Anaplerotic Reactions

If there is no way of replenishing, removal of intermediates of TCA (which are to be used as precursor for anabolic pathway) will result in stopping the cycle. Various reactions, which act as "filling-up" pathways, are called as anaplerotic reactions (derived from Greek word *anaplerotikos* which means filling up) (Fig. 7.14). In plant cells one of the ways of replenishing the intermediates is by cytosolic oxaloacetate. Plant mitochondria are characterized by specific oxaloacetate transporters present in the inner membrane. Carboxylation of phosphoenolpyruvate generates OAA in the cytosol catalyzed by PEP carboxylase:

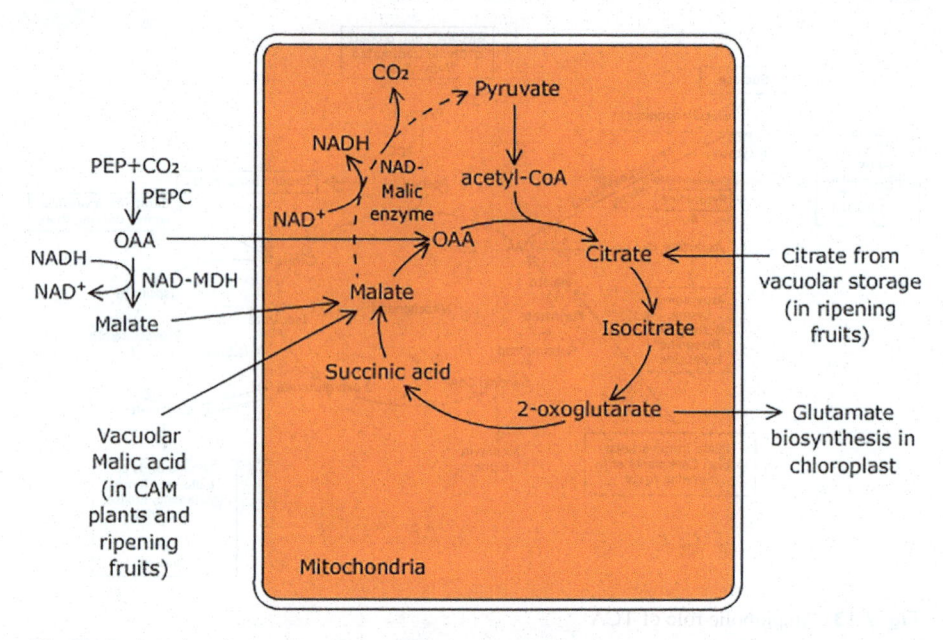

**Fig. 7.14** Anaplerotic reactions. Some of the intermediates of TCA serve as precursors of some of the biomolecules. The intermediates of TCA are replenished by anaplerotic reactions. *PEPC* PEP carboxylase, *NAD-MDH* NAD-malate dehydrogenase

$$PEP + HCO_3^- \rightarrow OAA + P_i$$

There is also a gain of one carbon in this reaction. Reduction of oxaloacetate by cytosolic malate dehydrogenase results in production of malate, which enters into the mitochondria. Malate may also be released and transported into the mitochondria from storage sites of the cell such as vacuoles through the translocators present in tonoplast and inner mitochondrial envelope. In mitochondria, malate is oxidized to OAA by NADH-malate dehydrogenase. Alternatively, OAA is decarboxylated to produce pyruvate by the catalytic action of **NAD-dependent malic enzyme.**

$$Malate + NAD^+ \rightarrow Pyruvate + NADH + CO_2$$

In this case requirement of glycolysis is substituted due to the formation of pyruvate by alternate reaction. In this case there is loss of carbon, first in the reaction catalyzed by NAD-dependent malic enzyme, secondly during conversion of pyruvate is acetyl-CoA. Though carbon is lost, the combined reactions of NAD-malate enzyme and PDC provide flexibility to the metabolism, since malate can be the source both for OAA and for acetyl-CoA. The resultant citrate can be oxidized to produce 2-oxoglutarate, which easily can be removed to provide carbon skeleton for glutamate biosynthesis without any requirement of pyruvate from glycolysis.

Besides sucrose, glutamate is also present in high concentration in the plant cells, which can be oxidized to 2-oxoglutarate by NADH-glutamate dehydrogenase located in mitochondrial matrix, which may join TCA. Photorespiratory glycine is also oxidized in mitochondria.

### 7.5.6 Role of TCA in Plants Under Stress Conditions

Under normal conditions, it is generally the sucrose which is translocated and metabolized in plants. However, reduced $CO_2$ assimilation under stress results in carbon starvation, especially when plants are exposed to prolonged darkness, low light conditions, low temperature, or drought. As a result, carbohydrate supply is reduced, and plants may need to metabolize other biomolecules, such as fatty acids or proteins. In plants fatty acids are oxidized by β-oxidation in peroxisomes, or in some of the case by α-oxidation, resulting in producing acetyl-CoA, which is metabolized through TCA cycle. Amino acids may be deaminated to produce α-keto acids, which are the intermediates of TCA cycle, e.g., glutamic acid may be oxidized and deaminated to produce 2-oxoglutarate by the NAD-dependent glutamate dehydrogenase. The non-protein amino acid such as proline accumulates under water stress conditions which is also oxidized to 2-oxoglutarate. Under carbohydrate starvation, oxidation of lysine, catalyzed by an enzyme lysine-2-oxoglutarate reductase, produces 2-oxoglutarate which enters TCA cycle. Various other amino acids including valine, isoleucine, and leucine are also thought to be degraded in mitochondria. In C3 plants, glycine produced during photorespiration is also metabolized in mitochondria. Glycine decarboxylase is responsible for glycine metabolism, and NADH generated during the reaction is either oxidized through ETC in mitochondria or is oxidized through OAA/malate shuttle system.

### 7.5.7 Regulation of TCA

Most important regulators of TCA are NADH and ATP. Ratio of NADH/NAD$^+$ and ATP/ADP is responsible for regulation of the cycle. These act as allosteric modulators of the TCA cycle enzymes. Enzymes are activated when the ratio falls below the threshold value. Under environmental stress, demand for ATP decreases in plants, resulting in increase in ATP levels. As a result, activity of some of the enzymes of TCA, such as citrate synthase, is inhibited. In plants there is a high ATP/ADP ratio in light since ATP is also generated during light reaction in chloroplasts. In that case ATP production in mitochondria is reduced because of which NADH is not oxidized. Accumulated NADH will ultimately either shut off TCA cycle or reduce it. The enzymes which are sensitive to inhibition by NADH include NAD-IDH and NAD-MDH. Since alternative mechanisms for oxidation of NADH are present in plants, the mechanisms for regulation of TCA cycle may be different and less rigidly controlled by NADH/NAD$^+$ ratio in comparison to that in animals. One of the important regulatory points is the entry of pyruvate into TCA through link reaction. This reaction provides a balance between glycolysis and TCA. If TCA cycle is not operating, accumulated acetyl-CoA will inhibit the activity of PDC. On the contrary, supply of pyruvate from glycolysis will alleviate the inhibition. Pyruvate dehydrogenase complex is also regulated through phosphorylation and dephosphorylation of PDH. Major regulatory steps of TCA include reactions catalyzed by citrate synthase, isocitrate dehydrogenase, 2-oxoglutarate dehydrogenase, succinic acid dehydrogenase, and fumarase. In plants another mechanism involved is regulation of enzyme activities by modulating their thiol redox status by cysteine-cysteine redox interconversion (-SH + HS$^-$ ⇌ S - S-). It is well established for chloroplast enzymes. It is of considerable significance even with reference to the mitochondrial enzymes. Plant mitochondria are known to have thioredoxin system, which might be playing important role in regulating activity of TCA cycle enzymes. Activity of AOX has also been demonstrated to be regulated through thioredoxin system. Studies with Arabidopsis mutants have demonstrated that succinic acid dehydrogenase and fumarase are deactivated when reduced by thioredoxin, whereas cytosolic ATP-citrate lyase is activated. Activity of citrate synthase is also inhibited when oxidized. Similar to the mammalian system, isocitrate dehydrogenase and 2-oxoglutarate dehydrogenase activities are inhibited by NADH. However, knowledge about the regulation of TCA in plants is limited especially under in vivo conditions since most of the studies have been carried out with isolated enzymes (Fig. 7.15).

In addition to mitochondria, plants have another source for ATP, i.e., the light reaction in chloroplasts. Excess reducing power produced in the form of NADPH during light reaction needs to be transported out of the chloroplasts so as to protect thylakoids from excessive reduced conditions. OAA/malate shuttle is responsible for transporting reducing equivalents out of the chloroplasts. Once in cytosol malate may be transported to mitochondria. NADH/NADPH do not accumulate in mitochondria in plants, because of presence of alternate mechanisms for their

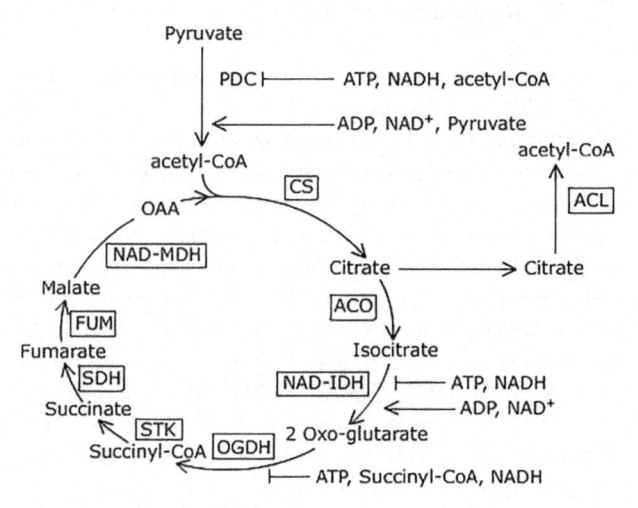

**Fig. 7.15** Regulation of TCA cycle. Activities of citrate synthase (CS), succinic acid dehydrogenase (SDH), and fumarase (FUM) are inhibited when reduced by the thioredoxin system, while activity of ATP citrate lyase (ACL) is activated. *PDC* pyruvate dehydrogenase complex, *ACO* aconitase, *NAD-IDH* NAD-specific isocitrate dehydrogenase, *OGDH* 2-oxoglutarate dehydrogenase, *STK* succinic acid thiokinase, *NAD-MDH* NAD-malate enzyme

oxidation. Many studies have indicated reduction in TCA cycle enzyme activities, or TCA cycle does not operate in green parts of the plants under illuminated conditions. In mitochondria NAD-IDH is particularly sensitive to inhibition by light. NADH is also produced during oxidation of glycine by glycine decarboxylase complex during photorespiration. It is observed that glycine oxidation is coupled with decreased TCA cycle activity, may be because of inhibition by increase in levels of NADH. Ammonia released during oxidation of glycine inhibit PDC, thus inhibiting TCA at the entry point.

### 7.5.8 TCA Cycle and GABA Shunt

γ-Aminobutyric acid (GABA) is a non-proteogenic amino acid. Since GABA has been found to accumulate in plants in response to abiotic stresses, such as salt stress or low light intensities, its significance in metabolism has been studied. In animals role of GABA has been found to be that of neurotransmitter. Synthesis of GABA occurs as a result of bypass reactions of TCA which is also known as GABA shunt. GABA shunt is also known to provide interlinking of carbon and nitrogen metabolism. The intermediate of TCA, 2-oxoglutarate, is transaminated to glutamate, which is transported out of mitochondria. In the cytosol, glutamate is converted to GABA as a result of decarboxylation reaction catalyzed by glutamate decarboxylase. After GABA is transported back to mitochondria, it is converted to succinic acid semialdehyde (SSA) due to another transamination reaction which is catalyzed by GABA transaminase. The reaction is coupled with conversion of pyruvate to alanine. Succinic acid semialdehyde is oxidized to succinic acid with simultaneous reduction of $NAD^+$ to NADH. Reaction is catalyzed by succinic semialdehyde dehydrogenase (Fig. 7.16).

## 7.6 Oxidation of the Reduced Coenzymes Produced During TCA Cycle

Though discovered in the beginning of the century, it was in 1948 when Eugine Kennedy and Albert Lehninger discovered that mitochondria are the sites of oxidative reactions. 1100 proteins have been identified in mitochondria out of which functions of almost 300 proteins are not known. Reduced coenzymes, NADH and $FADH_2$, which are produced during glycolysis or TCA, need to be oxidized back to $NAD^+$ and FAD, respectively, so that glycolysis and TCA keep on functioning. In absence of the oxidized coenzymes, dehydrogenases in the oxidative pathways are inhibited. Energy released during oxidation of the coenzymes is conserved in the form of proton gradient across inner mitochondrial membrane which is coupled with ATP synthesis. ATP synthesis, coupled with oxidation of substrates, is termed as oxidative phosphorylation. It will be dealt in the chapter dealing with ATP synthesis. Electrons removed from NADH are passed through a number of electron carriers which are organized in the form of complexes. These are present in the inner

**Fig. 7.16** GABA shunt—an alternative pathway in TCA cycle which gives flexibility against environmental changes. GABA is now considered a metabolite to play a significant role in C-metabolism in response to stress. *GABA* gamma-aminobutyric acid, *GAD* glutamate decarboxylase, *GABAT* GABA transaminase, *SSA* succinic semialdehyde, *SSADH* succinate-semialdehyde dehydrogenase, *OAA* oxaloacetate

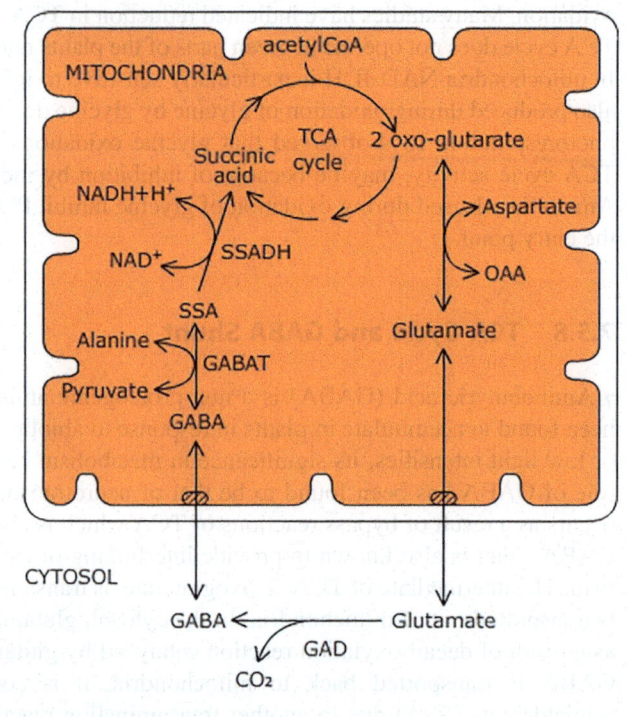

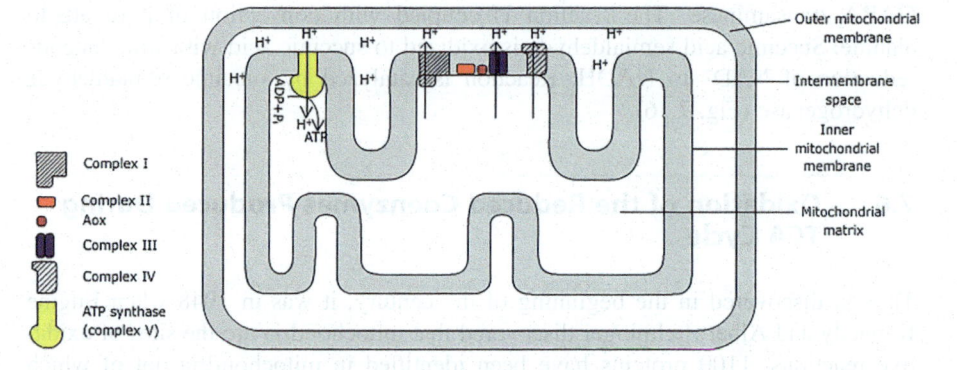

**Fig. 7.17** Role of four complexes present in the inner mitochondrial membrane in generation of proton gradient. Protons accumulate in the intermembrane space and flow back through ATP synthase synthesizing ATP

mitochondrial membrane which divides the mitochondrial compartment in two spaces (Fig. 7.17). The space in between the outer and inner mitochondrial membrane is known as perimitochondrial space which directly is in contact with cytosol since **porins** present in the outer membrane allow free movement of the molecules having molecular weight up to 10 kDa, and most of the biomolecules can pass through these channels except the proteins. Unlike the outer membrane, the inner

## 7.6 Oxidation of the Reduced Coenzymes Produced During TCA Cycle

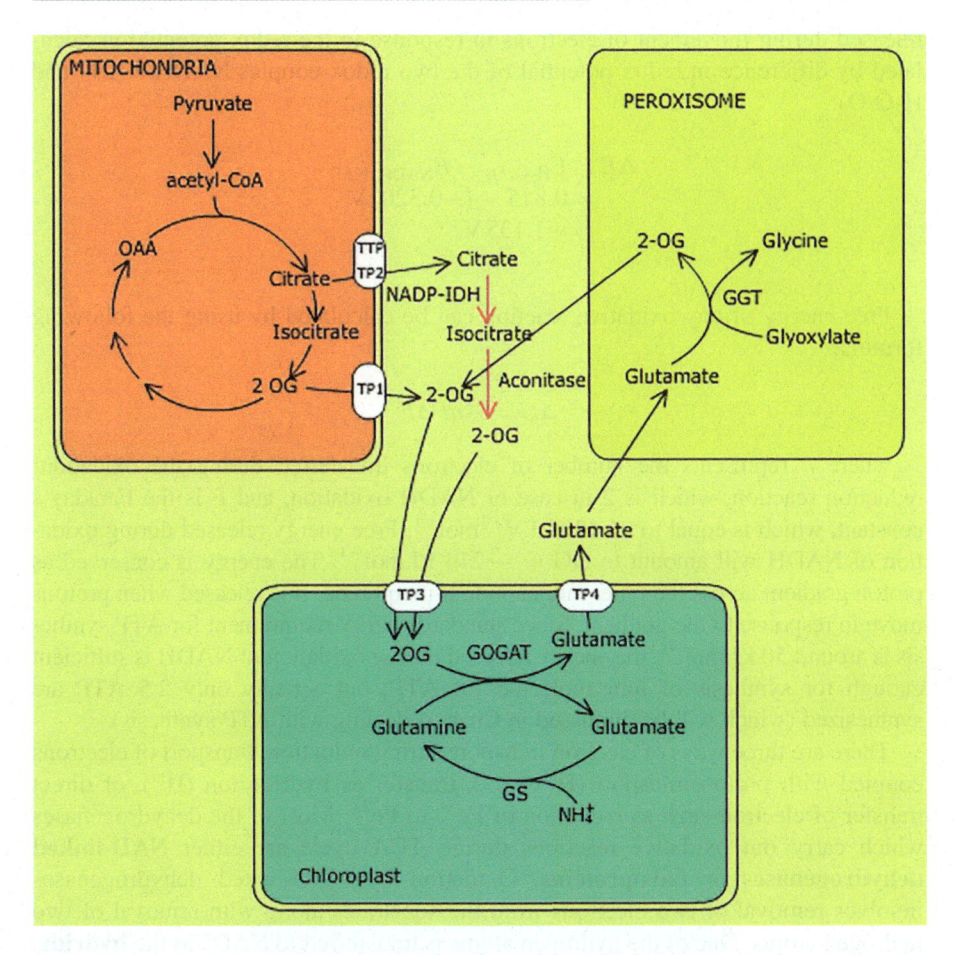

**Fig. 7.18** Source of carbon skeleton of glutamate: *TP1* mitochondrial oxoglutarate transporter, *TTP* mitochondrial tricarboxylate transporter, *TP2* mitochondrial dicarboxylate transporter, *TP3* chloroplast 2-oxoglutarate transporter, *TP4* chloroplast glutamate transporter, *NADP-IDH* NADP-isocitrate dehydrogenase, *GS* glutamine synthetase, *GGT* gamma-glutamyltransferase

membrane is impermeable to most of the biomolecules, and movement of substance across the membrane is facilitated by transporters present in it which are responsible for exchange of various biomolecules, such as pyruvate, dicarboxylic and tricarboxylic acids (Fig. 7.18). ATP/ADP antiporters present in the membrane facilitate exchange of ATP with cytosolic ADP. The matrix side of the membrane is referred to as the N (negative) side, while the side facing the cytosolic side is known as P (positive) side according to the potential of the membrane which is created as a result of proton movement. Energy released during oxidation of NADH to NAD$^+$ can be calculated by their redox potentials. During oxidation final acceptor of electrons is $O_2$ which is reduced to $H_2O$. The two redox couples NADH/NAD$^+$ and $O_2/H_2O$ have standard reduction potential ($E^{0'}$) $-0.320$ V and $0.815$ V, respectively. Energy released during movement of electrons in response to the redox potential is calculated by difference in redox potential of the two redox couples NADH/NAD$^+$ and H$_2$O/O$_2$:

$$\Delta E = E_{H_2O/O_2} - E_{NADH/NAD^+}$$
$$= 0.815 - (-0.320)\,V$$
$$= +1.135\,V$$

Free energy of any oxidative reaction can be calculated by using the following formula:

$$\Delta G = -nF\Delta E$$

where $n$ represents the number of electrons transferred during the oxidation-reduction reaction, which is 2 in case of NADH oxidation, and $F$ is the Faraday's constant, which is equal to 96.480 kJ.V$^{-1}$mol$^{-1}$. Free energy released during oxidation of NADH will amount to $\Delta G = -214$ kJ.mol$^{-1}$. The energy is conserved as proton gradient across the inner mitochondrial membrane. It is released when protons move in response to the gradient. Since standard energy requirement for ATP synthesis is around 50 kJ.mol$^{-1}$, the energy released during oxidation of NADH is sufficient enough for synthesis of four molecules of ATP, but actually only 2.5 ATP are synthesized (which will be discussed in Chap. 4 dealing with ATP synthesis).

There are three ways of electron transport during reduction, transport of electrons coupled with proton transport (H$^+$ + e$^-$), transfer as hydride ion (H$^-$), or direct transfer of electron such as reduction of Fe$^{3+}$ to Fe$^{2+}$. Most of the dehydrogenases which carry out oxidative reactions during TCA cycle are either **NAD-linked dehydrogenases** or **flavoproteins**. Oxidation by NAD-linked dehydrogenases involves removal of two electrons from the substrates along with removal of two hydrogen atoms. One of the hydrogen atoms is transferred to NAD$^+$ as the **hydride**, while the other one is released as H$^+$ in the medium. Though NADPH is used as reducing power for the anabolic reactions, plants have NADPH dehydrogenases which are capable of removing electrons from NADPH and channelizing them to the electron transport chain. Since NADH and NADPH are not able to cross the membranes, each cell compartment maintains its own pool, and there is exchange of reducing power equivalents. Additionally flavoprotein, such as succinic dehydrogenase, is also a component of the electron transport chain which has covalently bound flavin nucleotide. Unlike NADPH, standard reduction potential of flavin nucleotides is that of the flavoprotein and not of isolated FAD or FMN since it depends upon the proteins to which these are bound. Oxidized flavin nucleotide accepts either one or two electrons, yielding semiquinone form or FADH$_2$, respectively. Since flavoproteins can participate in either single- or two-electron transport, these mediate the electron transport in between the compounds which are oxidized on removal of two electrons and the compounds which are reduced on accepting single electron.

## 7.6.1 Electron Transport Chain

Besides $NAD^+$ dehydrogenases and flavoproteins, the other components of electron transport chain involve a quinone, heme proteins, and non-heme proteins. Quinone is a small hydrophobic molecule which is also known as **coenzyme Q (ubiquinone)** in ETC of respiration. It is an abundant mobile electron transfer component, which consists of a $C_{45}$ to $C_{50}$ prenyl side chain having a substituted $p$-benzoquinone head group. On accepting two electrons, it is reduced to quinol. Electron transport to quinone occurs in two steps. On accepting one electron, it is converted to semiquinone ($UQ^{e-}$), while on accepting another electron also, it is reduced completely ($UQ^{2e-}$) to quinol ($UQH_2$) on combining with two protons ($H^+$) from the medium. Both in fully oxidized and fully reduced forms, it is highly mobile. It acts at the junction of two-electron and single-electron transfer of ETC. Since its reduction involves both electrons and protons, it plays an important role in coupling electron flow with the proton movement.

Other components of the electron transport chain are **cytochromes**. These are heme proteins, and because of iron of the heme, which has variable valency, it can exist in oxidized forms, as $Fe^{3+}$ which gets reduced to $Fe^{2+}$ on receiving electron or vice versa. These participate in transfer of single electrons. Three categories of cytochromes are designated as a, b, and c which are characterized by their wavelength absorbing spectra. Cytochrome a absorbs the longest wavelength, i.e., near 600 nm, type b absorbs near 560 nm, while cytochrome c absorbs near 550 nm. Sometimes cytochromes are characterized by specifying exact absorption maxima such as cytochrome $b_{562}$. Reduction potential of the cytochromes is determined by the proteins to which the cytochromes are covalently attached through the cysteine residues of the apoenzyme. All cytochromes in mitochondria are integral parts of the inner membrane except cytochrome c which is present as soluble protein in the matrix and is associated with the inner membrane through electrostatic attractions. Non-heme proteins are also integral part of the electron transport chain. These are **iron-sulfur proteins** in which non-heme iron is present in association with inorganic sulfur or sulfur of the cysteine residues of the protein apoenzyme. Rieske iron-sulfur protein is a type of iron-sulfur protein in which, instead of binding with two cysteines of the protein, Fe is covalently bonded to two histidine residues of the proteins. These proteins participate in the transport of single electron at a time since only one Fe is oxidized or reduced at a time. Reduction potential of these proteins varies from $-0.65$ V to $+0.45$ V which depends upon the microenvironment of the protein in which it is working. Standard reduction potential of various components of electron transport chain is given in Table 7.4.

## 7.6.2 Electron Carriers Are Arranged in a Sequence

Electron carriers involved in transport of electrons from NADH to molecular $O_2$ were arranged in ordered sequence by using the following three approaches:

**Table 7.4** Redox potential of some of the redox couples involved in respiratory chain complexes

| Redox couple | $E^{0'}$ (volts) |
|---|---|
| NAD$^+$/NADH | −0.320 |
| FMN/FMNH$_2$ | −0.30 |
| UQ/UQH$_2$ | 0.045 |
| Cytochrome b (Fe$^{3+}$/Fe$^{2+}$) | 0.077 |
| Cytochrome c$_1$ (Fe$^{3+}$/Fe$^{2+}$) | 0.22 |
| Cytochrome c (Fe$^{3+}$/Fe$^{2+}$) | 0.254 |
| Cytochrome a (Fe$^{3+}$/Fe$^{2+}$) | 0.29 |
| Cytochrome a$_3$ (Fe$^{3+}$/Fe$^{2+}$) | 0.36 |
| ½ O$_2$/H$_2$O | 0.815 |

1. By withholding the supply of O$_2$ which is the terminal electron acceptor in the electron transport chain, all components of the chain remain in reduced state. When O$_2$ is supplied rate and oxidation of the electron carriers occurs according to the sequence in which these are present. Components adjacent to O$_2$ are oxidized first, and the components last to be oxidized are the farthest away from O$_2$. Oxidized and reduced state of the compounds can be monitored spectroscopically since these two states of the molecules have different absorption maxima.
2. The second approach includes use of inhibitors. In presence of rotenone, NADH accumulates since it is not oxidized. However, addition of reduced coenzyme Q results in resumption of electron transport to O$_2$. Similarly, in presence of antimycin A, cytochrome b is not oxidized; addition of reduced cytochrome c1 results in resumption of electron transport. Cyanide inhibits the terminal step of electron transport which involves transfer of electrons to molecular O$_2$. In the presence of cyanide, all components of ETC accumulate in reduced state (Fig. 7.19).
3. The sequence of components of the electron transport chain, when arranged according to the above experimental approaches, is similar to when arranged according to their redox potentials. NADH/NAD$^+$ has the redox potential −0.320 V, while O$_2$/H$_2$O has the redox potential +0.815 V, and redox potential of the other components of ETC is between these is given in the Fig. 7.20.

### 7.6.3 Components of the Electron Transport Chain Are Present as Multienzyme Complexes

Oxidation of NADH releases 220 kJ.mol$^{-1}$ of energy. This will be wasted if all of the energy is released in a single step. However, multistep release of the energy is used for translocation of protons across membrane and is conserved as proton gradient. ATP synthesis occurs in response to the proton movement through ATP synthase. This process of ATP synthesis is known as oxidative phosphorylation. In 1962, Youssef Hatefi successfully isolated four different complexes (Complexes I–

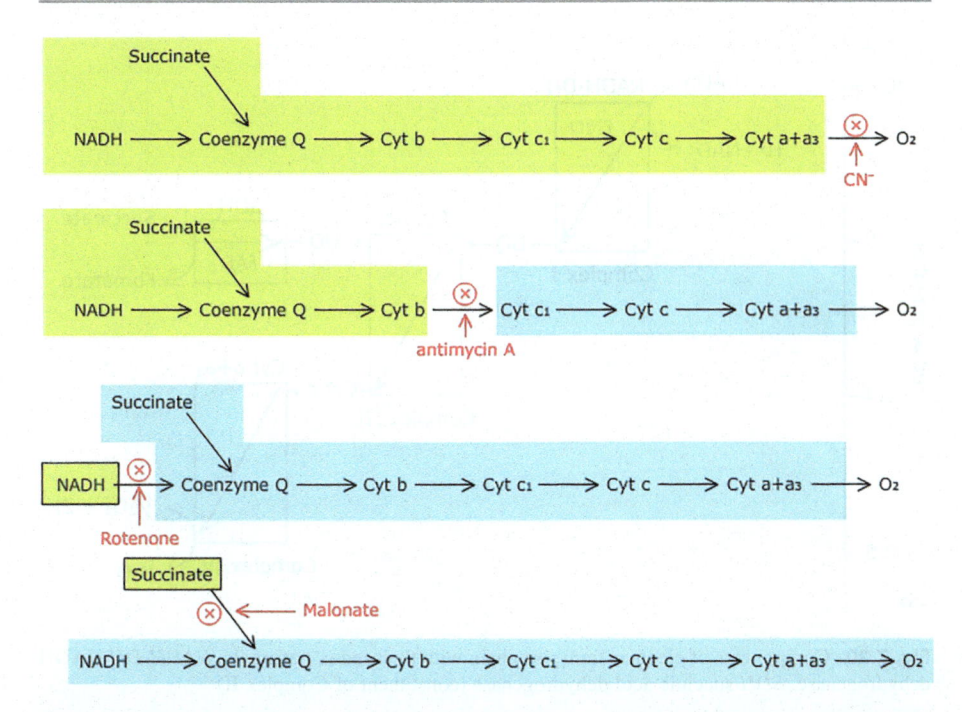

## 7.6 Oxidation of the Reduced Coenzymes Produced During TCA Cycle

**Fig. 7.19** The use of inhibitors in elucidating the sequence of components of electron transport chain. Yellow color indicates the components in reduced state, while blue color indicates the components in oxidized state

IV) while working with beef heart mitochondria, indicating that the components of the electron transport chain are associated into multienzyme complexes which are conserved in the mitochondria of all the organisms. Proteomic studies, using **peptide mass spectrometry,** have revealed that almost 85% of the proteins of these complexes are common in all eukaryotic systems. However, 15% of the proteins are plant specific or similar to those present in algae but are absent from fungi and animals. The role of these proteins has not been specified yet. Some of the proteins of these complexes are encoded in mitochondrial genome, while others are encoded in the nuclear genome. There are studies which indicate mitochondrial e-transport complexes to be associated to form supercomplexes which are called **respirosomes**. Complexes I, III, and IV have been found to be associated with different degrees and configurations, which suggest the possibility of channelling electrons from NADH to $O_2$ with higher transfer rate. The formation of supercomplexes may help in increasing the stability of the individual complexes as well as providing a regulatory mechanism that controls the passage of electrons through ETC (Fig. 7.21).

*Complex I: NADH Dehydrogenase (NADH-Ubiquinone Oxidoreductase)* This is the first supercomplex in the electron transport chain associated with NADH oxidation. It is a large multi-subunit complex consisting of about 45 polypeptides. It is similar to mammalian and fungal Complex I; however, there are a few proteins

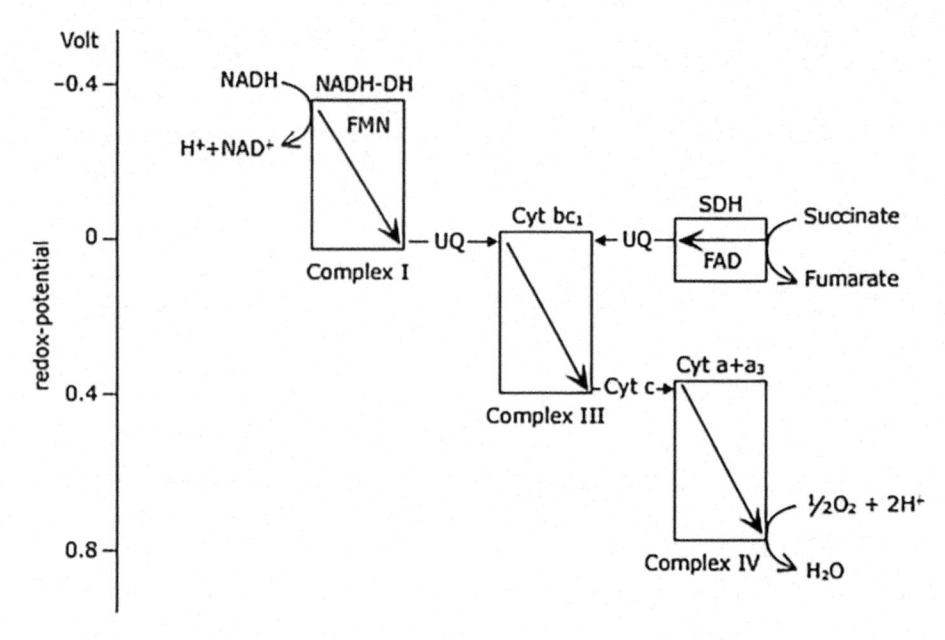

**Fig. 7.20** Components of electron transport chain with their redox potentials. *NADH-DH* NADH dehydrogenase, *SDH* succinic acid dehydrogenase (constituent of Complex II)

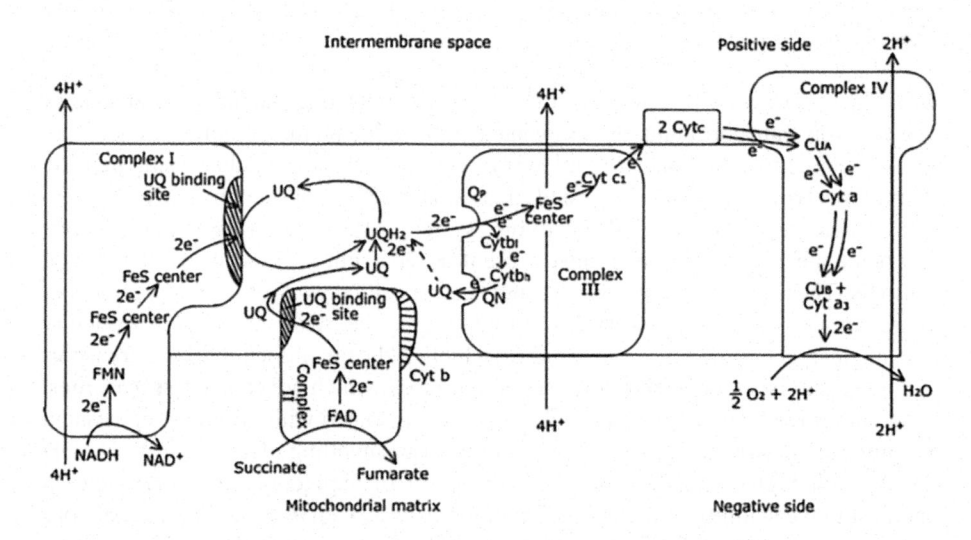

**Fig. 7.21** Pathway of electron transport through four supramolecular complexes present in the inner mitochondrial membrane. Transport of two electrons removed from NADH is coupled with translocation of 10 H$^+$ across the membrane

## 7.6 Oxidation of the Reduced Coenzymes Produced During TCA Cycle

which are plant specific. Plant mitochondrial genome may encode as many as nine components of Complex I. It has been found to be L-shaped in studies carried out with *Arabidopsis* mutants. The complex oxidizes NADH which is produced in mitochondrial matrix during TCA. The electrons removed from NADH are transferred to ubiquinone. While electrons are transferred to UQ, conformational changes occur in the antiporter proteins present in the membrane arm of the complex as a result of which four protons are translocated across the membrane. Besides presence of various proteins which make up this complex, the cofactors associated with the complex include FMN-containing protein and at least six Fe-S centers. Reaction catalyzed by the complex includes transfer of one hydride ion from NADH and one proton from the matrix to ubiquinone. It can be written as,

$$NADH + H^+ + Q \rightarrow NAD^+ + QH_2$$

The two electrons and protons removed from NADH are transferred to FMN and then through a series of Fe-S proteins are transferred to ubiquinone which is bound to the ubiquinone-binding site of the complex. Transport of two electrons to UQ by the complex is coupled with transport of 4 $H^+$ across the membrane. Thus Complex I serves as the proton pump. The reaction can be written as

$$NADH + 5 H^+{}_N + Q \rightarrow NAD^+ + QH_2 + 4H^+{}_P$$

The electron transfer can be inhibited by rotenone (a plant-derived flavonoid), amytal (a barbiturate drug), and piericidine (an antibiotic) which inhibit electron transfer from the Fe-S centers of the complex to UQ (Fig. 7.22). Plants are characterized by the additional presence of both external and internal NAD(P)H dehydrogenases which are insensitive to rotenone and are capable to oxidize both externally and internally generated NADH and NADPH (NADex and NADin). This again indicates flexible systems present in plants so that they are capable of adjusting according to their needs (Fig. 7.23).

*Complex II: Succinate Dehydrogenase (Succinate-Ubiquinone Oxidoreductase)*
This is the only membrane-bound enzyme of TCA cycle. It catalyzes oxidation of succinate to fumarate by removing two electrons and transferring them to ubiquinone which is reduced to $UQH_2$. Unlike Complex I, electron transport is not coupled with proton translocation across membrane, and this complex does not contribute to the proton gradient. This is the smallest complex of the electron transport chain, consisting of only four different protein subunits, which range in sizes from 13.5 to 70 kDa. In the majority of plants including angiosperms and gymnosperms, none of these proteins are encoded by the mitochondrial genome. Two of the smaller hydrophobic protein subunits are present as integral membrane proteins, while the other two large hydrophilic protein subunits extend in the matrix. Succinate-binding site, FAD, and three Fe-S centers are associated with the peripheral proteins which are extending toward the matrix of mitochondria. The two binding sites for ubiquinone are present in the hydrophobic integral proteins, which are also associated with

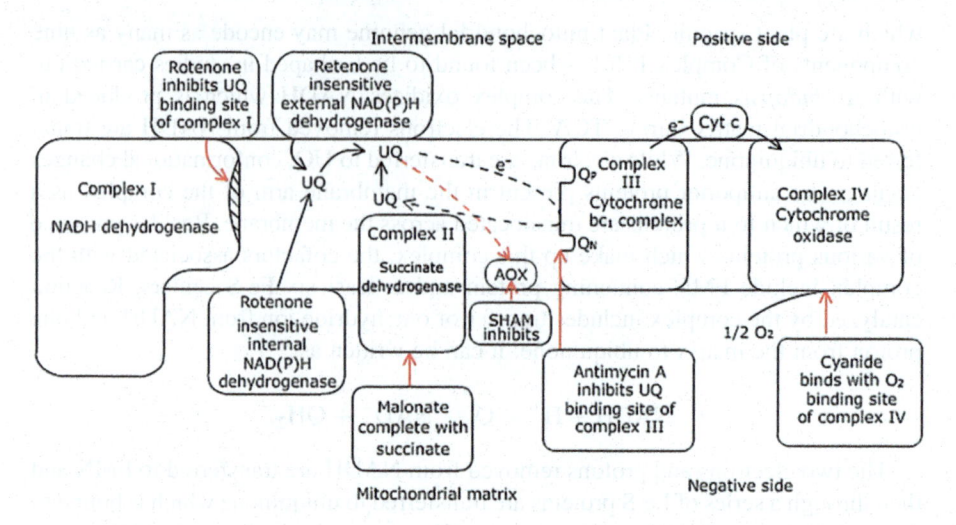

**Fig. 7.22** Various inhibitors and their site of action on the transport of electrons

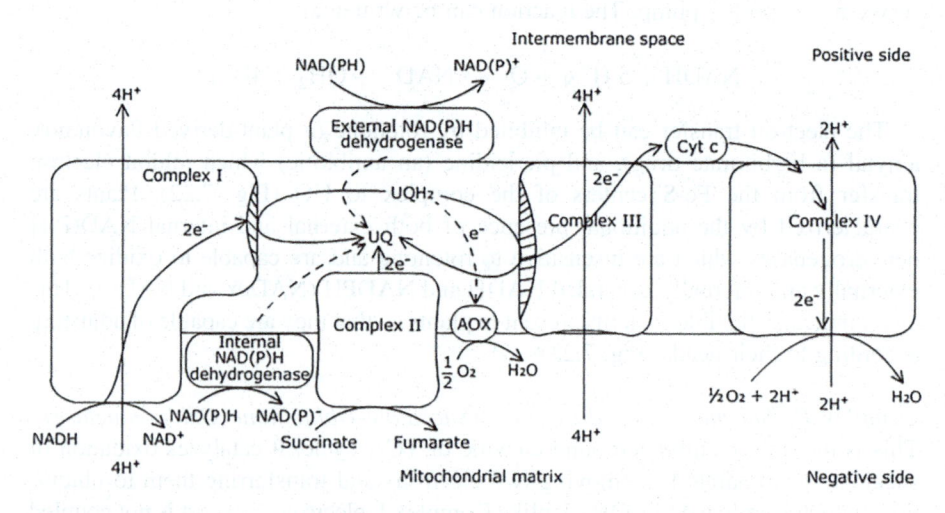

**Fig. 7.23** Components of electron transport chain are present as four complexes: Complex I (NADH dehydrogenase), Complex II (succinate dehydrogenase), Complex III and Complex IV (cytochrome oxidase). In plants, there are additional NAD(P)H dehydrogenases; both external and internal are present which oxidized NAD(P)H by transferring electrons directly to UQ, thus missing the first step of $H^+$ translocation. $UQH_2$ is oxidized by Complex III which is coupled with translocation of $4H^+$ across the membrane. In plants, there is also an alternate oxidase (AOX) present which can oxidize $UQH_2$ directly without transferring electrons to Complex III and Complex IV. From Complex III electrons are carried to cytochrome c and are delivered to Complex IV which pass on the electrons to $O_2$ which is reduced to water after receiving electrons combined with $H^+$ from mitochondrial matrix

cyt b. Electrons removed from succinate are transferred to UQ through FAD and three Fe-S centers. Electron transport by Complex II is inhibited by malonate which is an analogue of the substrate. Cytochrome b is not in the main pathway for electron transport; however, its role is believed to be in reducing the leakage of electrons to $O_2$ and thus in preventing the formation of reactive oxygen species.

*Complex III: Cytochrome bc₁ Complex (Ubiquino-Cytochrome c Oxidoreductase)*
Complex III is responsible for removing electrons from $UQH_2$ and transferring them to cytochrome c, which is a small protein present toward the N side of the membrane (inter-mitochondrial space). It is responsible for shuttling of electrons from Complex III to Complex IV. This is a huge complex whose complete structure was determined between 1995 and 1998 using X-ray crystallography. The functional unit of *cytochrome bc₁* has been found to be consisting of 11 polypeptide subunits, out of which only one is encoded in mitochondrial genome. This complex has the structure similar to cytochrome $b_6f$ which is present in chloroplasts and is part of photosynthetic electron transport. It consists of cytochromes having heme of two b types, $b_{566}$ ($b_l$) and $b_{560}$ ($b_h$), and a cytochrome $c_1$, a Rieske-type Fe-S protein, and the other polypeptides. The complex has largest polypeptide in the center which is known as the core protein having a molecular weight ranging from 51 to 55 kDa. There are two binding sites present in the Complex III for binding with ubiquinone which are known as $Q_P$ and $Q_N$ sites, which indicate the Q-binding site present toward P (positive) side of the membrane and toward N (Negative) side of the membrane, respectively. The terminology is based on the fact that the intermembrane space of the mitochondria is the positive side because of accumulation of $H^+$, while the matrix side is negative because of lesser $H^+$ concentration. The complex is present at the interface of two-electron transfer and one-electron transfer. Two electrons are removed from $UQH_2$ at the P side of the membrane. One electron is passed on to cytochrome $c_1$ via Fe-S protein and then to cytochrome c which contains single heme. It is present toward the intermembrane space, held to the membrane by electrostatic attraction. It can move and deliver the electrons to Complex IV. The second electron removed from $QH_2$ is cycled through two hemes to UQ at the $Q_N$ site reducing it partially to semiquinone state ($UQ^{e-}$). The cycle is repeated twice; as a result $UQ^{e-}$ will receive another electron in similar manner and will further be reduced to $UQ^{2e-}$, and then it picks up two $H^+$ from the matrix and will be reduced to $UQH_2$. In this form, it is freely mobile within membrane and is oxidized at the $Q_P$ site. The two protons will be released in the intermembrane space. This cycle is known as Q-cycle which is significant for transporting protons across the membrane, thus adding to proton gradient. Q-cycle is similar to the one occurring during photosynthetic electron transport. For transport of two electrons in the chain, four protons are transported across the membrane (Fig. 7.24).

The net redox reaction of Q-cycle will be

$$QH_2 + 2 \text{ cyt } c_1(\text{oxidized}) + 2 H^+_N \rightarrow Q + 2 \text{ cyt } c_1 (\text{reduced}) + 4 H^+_P$$

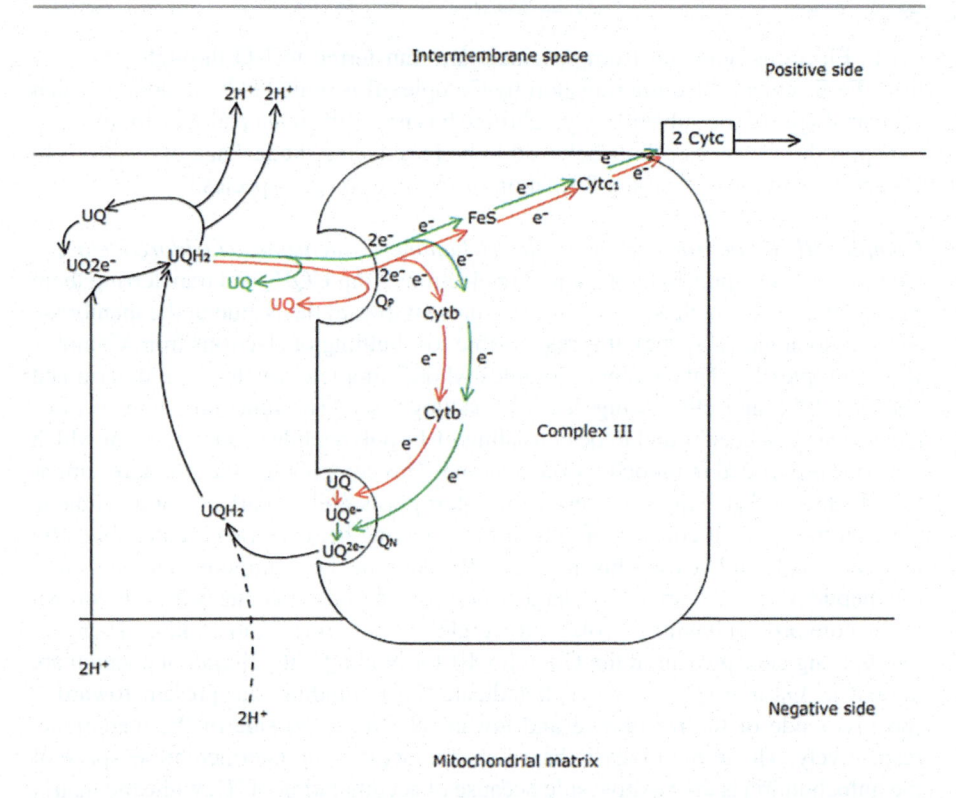

**Fig. 7.24** Q-cycle. In the first cycle, out of the two electrons removed from UQH$_2$, one electron is transferred to cyt c, and one is cycle back to UQ via cyt b (demonstrated in red). Similar cycle is repeated from another molecule of UQH$_2$ (demonstrated in green). This results in reduction of UQ $^{e-}$ to UQ$^{2e-}$

The inhibitor antimycin A binds with the Q$_N$ site of the complex, thus inhibiting reduction of UQ (Fig. 7.23).

*Complex IV: Cytochrome Oxidase (Cytochrome c: O$_2$ Oxidoreductase)* Reduced cytochrome c is positively charged, so it can diffuse on the negatively charged inner membrane of mitochondria toward the intermembrane space and deliver the electrons to cytochrome oxidase. Structure of the cytochrome oxidase has been worked out using X-ray crystallography. It consists of 13 protein subunits, out of which at least 3 are encoded in the mitochondrial genome. Redox centers of the complex consist of two "a" type cytochromes, i.e., cytochrome a and a$_3$, and two copper centers Cu$_A$ and Cu$_B$. Redox centers are located in the mitochondrial encoded protein subunits. The complex consists of a large hydrophilic area, which protrudes in the intermembrane space and has got the binding site for oxidation of cytochrome c. The electrons removed from cytochrome c are transferred to two Cu$_A$ atoms of the proteins which are held by their interactions with S of the cysteine of the protein. Electrons are then passed through heme of cytochrome "a" to another redox center consisting of $Cu_B$, bound to histidine and cytochrome $a_3$. This second redox center probably acts as binuclear center, $Cu_B$ and heme of $cyta_3$, which can accumulate two electrons. These electrons are then transferred to $O_2$, which on receiving the two electrons can bind with 2 $H^+$ from the matrix side and are reduced to water, completing the electron transport process.

$$[Fe^{+++}.Cu^{++}] + 2\,e^- \rightarrow [Fe^{++}.Cu^+]$$

$$\tfrac{1}{2}O_2 + 2\,e^- + 2\,H^+ \rightarrow H_2O$$

Both CO and $CN^-$ can bind to the $O_2$-binding site of cytochrome $a_3$, thus compete with $O_2$, and act as inhibitor of the Complex IV. Thus, in the presence of $CN^-$, electron transport to $O_2$ is inhibited, and the whole process of electron transport is halted. However, in plants, respiration occurs even in presence of cyanide, which is known as cyanide-resistant respiration (Fig. 7.23).

### 7.6.4 Proton Translocation Creates Proton Motive Force (PMF)

There are three sites in the electron transport chain during NADH oxidation, which are coupled with transport of $H^+$ across the membrane:

1. Complex I, which is responsible for transport of 4 $H^+$ across the membrane coupled with the removal of two electrons from NADH
2. Complex III, which transports 4 $H^+$ coupled with transport of two electrons
3. Complex IV, which transports 2 $H^+$ coupled with transport of two electrons to $O_2$

Thus, energy released during movement of two electrons downhill from NADH to $O_2$ results in creating a gradient of 10 $H^+$ across the inner mitochondrial membrane, creating membrane potential also. As a result, there will be a positive side of the membrane toward the intermembrane space due to proton accumulation and negative side toward the matrix side of the membrane. These proton gradient and voltage gradient together generate *proton motive force* (PMF) which is responsible for backflow movement of protons from the cytosolic side toward matrix. Since the membrane is impermeable to protons, protons move through specific complex, Complex V or also known as $F_o$-$F_1$ or ATP synthase since these are associated with ATP synthesis (Fig. 7.23).

## 7.7 NADH Shuttles

"Redox status" of the cell is important for the cellular metabolism. Each compartment of the cell has got its own pool of pyridine molecules ($NAD(P)^+$) which may either be present as $NAD(P)^+$ or $NAD(P)H$. Many of the cellular reactions will require $NAD(P)^+$ in the oxidized state. In case $NAD(P)^+$ pool of the cell compartments is exhausted, the cell metabolism will stop. Direct transport of NAD(P)H does not occur since transporters for NAD(P)/NAD(P)H are absent in the cell membranes. Transport occurs indirectly as exchange of reduced compound with the oxidized form. This exchange of reducing power indirectly is called NADH shuttle. In order to avoid **oxidative stress**, it is very important for the cells to modulate ratio of NAD(P)/NAD(P$^+$)H and is significant for maintaining redox homeostasis of the cell. ATP synthesis coupled with NADH oxidation occurs only in mitochondria because of the presence of electron transport chain. Various NADH shuttles operate in both plants and animals, which are responsible for reducing power to be exchanged in between different cell compartments. These include malate/oxaloacetate NADH shuttle, malate aspartate shuttle, and glycerol phosphate/dihydroxyacetone phosphate shuttle mechanism.

*Glycerol phosphate/dihydroxyacetone phosphate shuttle*: NADH produced during glycolysis is reoxidized to NAD$^+$ coupled with reduction of dihydroxyacetone phosphate to glycerol phosphate. Reaction is catalyzed by the cytoplasmic glycerol phosphate dehydrogenase. Glycerol phosphate diffuses through outer mitochondrial membrane and transfers electrons to FAD of the Complex II reducing it to FADH$_2$. The reaction is catalyzed by the mitochondrial glycerol phosphate dehydrogenase. Electrons from FADH$_2$ enter the electron transport chain at the coenzyme Q level (Fig. 7.25).

*Malate/oxaloacetate shuttle*: This shuttle is the most extensively studied redox exchange mechanism and is one of the most important NADH shuttles in plants. It is responsible for transfer of reducing equivalents in between chloroplasts, cytosol, mitochondria, and also peroxisomes. Oxaloacetate transporters are present in membranes which transport oxaloacetate in exchange of malate or vice versa. Excess NADPH produced during light reaction is transported out of chloroplasts in the form of malate, which is produced because of reduction of oxaloacetate. Malate is

**Fig. 7.25** Glycerol-3-phosphate shuttle

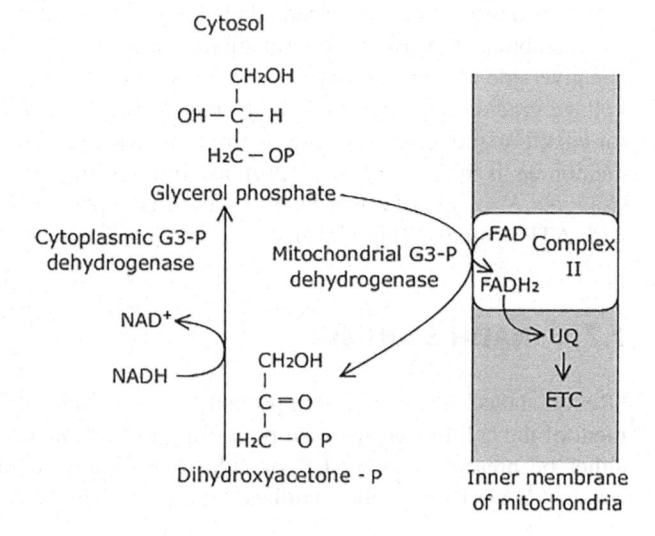

## 7.7 NADH Shuttles

exchanged with cytosolic oxaloacetate, which is then transported to mitochondria in exchange of oxaloacetate through dicarboxylate transporters. Once inside mitochondria, malate is oxidized to oxaloacetate by NAD-malate dehydrogenase with simultaneous reduction of NAD$^+$ to NADH, which is oxidized through electron transport chain. In this way reducing power, NADPH, is transferred from chloroplasts to mitochondria as NADH. Direction of the shuttle depends upon the relative concentrations of the participating molecules. An alternative use of malate released from chloroplasts may be in generation of cytosolic NADH, which may be utilized for nitrate reduction. NADH, produced in mitochondria during photorespiration, may also be exported out to peroxisomes where it is consumed for reduction of α-hydroxypyruvate (Fig. 7.26).

*Malate/aspartate NADH shuttle*: This shuttle has been demonstrated in isolated mitochondria and is responsible for transporting cytosolic reducing equivalents in the form of malate. There are 2-oxoglutarate transporters and glutamate/aspartate transporters present in the mitochondrial membrane which facilitate the movement as malate and glutamate/aspartate, respectively. Oxidation of malate in mitochondria generates NADH which may be used for ATP generation through ETC. More ATP is generated through this shuttle than the glycerol phosphate/dihydroxyacetone phosphate shuttle. Exchange of glutamate with aspartate is electrogenic in nature through glutamate/aspartate carriers present in the mitochondrial membrane. This shuttle is significant especially in actively respiring mitochondria, when aspartate is driven out of mitochondrial matrix because of $\Delta\Psi$, mitochondrial matrix side being negative (Fig. 7.27).

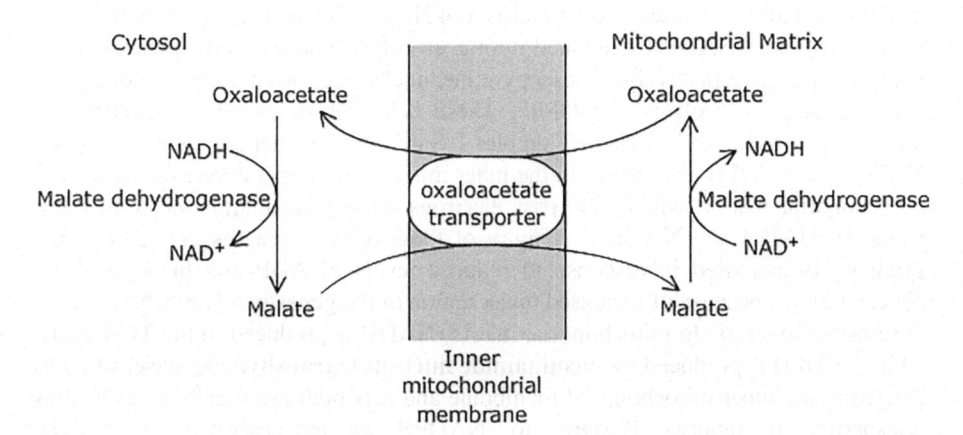

**Fig. 7.26** Malate/oxaloacetate shuttle transports reducing equivalents from cytosol to mitochondrial matrix. Exchange occurs in the form of malate from cytosol to mitochondrial matrix and OAA from mitochondrial matrix to cytosol mediated by OAA transporters present in the inner mitochondrial membrane. Isoforms of malate dehydrogenase catalyze conversion of malate to OAA in cytosol and from OAA to malate in mitochondrial matrix. NADH produced in mitochondrial matrix can now be oxidized through ETC

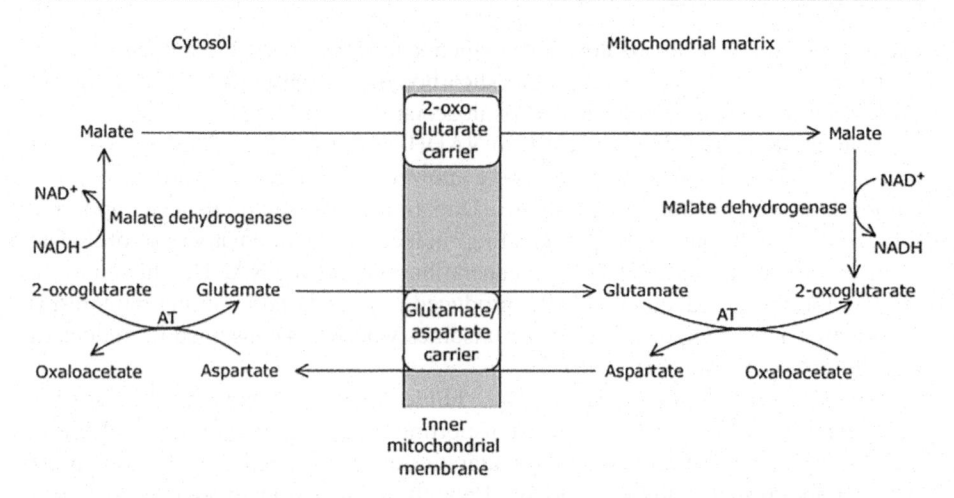

**Fig. 7.27** Malate/aspartate NADH shuttle. Reducing equivalents are transported across inner mitochondrial membrane in the form of malate and aspartate. Isoforms of malate dehydrogenase and aspartate aminotransferase (AT) are found on both sides of inner mitochondrial membrane

## 7.8 Alternate Mechanisms of NADH Oxidation in Plants

Certain features of mitochondrial electron transport in plants are different from those in animals. In plants there are four more dehydrogenases present in the inner mitochondrial membrane, which are not inhibited by rotenone. Two of these rotenone-insensitive dehydrogenase, NADH and NADPH, are responsible for oxidizing cytosolic NADH and NADPH generated during glycolysis and oxidative pentose phosphate pathway, respectively. These pyridine nucleotides need to be oxidized for continuation of glycolysis and OPPP. These dehydrogenases donate electrons to ubiquinone directly bypassing the Complex I. Similarly the other two dehydrogenases NADH and NADPH are present in the inner mitochondrial membrane facing toward mitochondrial matrix, which also pass electrons directly to ubiquinone from mitochondrial NADH and NADPH. Activity of these dehydrogenases as well as their synthesis is increased in response to reduced levels of ADP and increased $Ca^{2+}$ concentrations, because of increased transcription of the genes involved which occurs in response to stress. In mitochondrial matrix, NADH is produced in the TCA cycle, while NADPH is produced by **nicotinamide nucleotide transhydrogenase**, which is located in the inner mitochondrial membrane and acts both as an enzyme as well as transporter. It reduces $NADP^+$ to NADPH at the expense of NADH. Transhydrogenase involves only mitochondrial $NAD^+$ and $NADP^+$, and cytosolic pools remain unaffected. The enzyme reaction is coupled to translocation of proton.

$$NADH + NADP^+ + H^+_{(out)} \rightarrow NAD^+ + NADPH + H^+_{(in)}$$

## 7.9 Cyanide-Resistant Respiration

**Box 7.2: Mitochondrial Diseases Are Being Discovered**
As can be understood, mitochondria are central to the metabolism in all the eukaryotes, since it is the hub where majority of ATP is produced. In plants however chloroplasts provide additional ATP especially in presence of light. Any dysfunctional mitochondrial disorder due to any of its defective enzymes, whether they are encoded in nucleus or mitochondria, results in various diseases in humans. The first mitochondrial disease to be understood was Leber hereditary optic neuropathy (LHON), a form of blindness that strikes in midlife as a result of mutations to the NADH-Q oxidoreductase component of Complex I. Some of these mutations impair NADH utilization, whereas others block electron transfer to Q. A knowledge about few of the diseases in human is there. However, more studies are required regarding diseases in plants due to defective mitochondrial functions. Respiratory chain complexes are composed of around 90 proteins, out of which almost 13 are encoded in the mitochondrial genome. Besides providing ATP, mitochondria are also the sites for ROS production especially when ubiquinone reduction exceeds its oxidation, and there is possibility of leakage of electrons to $O_2$ resulting in ROS generation. Little information is there regarding the role of mitochondrial ROS in plants. Studies carried out with Complex II mutants of *Arabidopsis*, SDH-1, have demonstrated that these exhibited reduced response to stress.

Since the electrons are directly donated to ubiquinone and Complex I is bypassed, less proton gradient is created. Thus, the main role of these dehydrogenases is in oxidation of extra NADH and NADPH, so that the components of electron transport chain remain in oxidative condition rather than in reduced form and ROS formation is avoided under **oxidative stress**. At times NADH is not required, but TCA occurs to provide precursors for various anabolic pathways. These dehydrogenases play an important role in oxidation of NADH for continuation of TCA. It is evident that mitochondria are the hub of metabolic activities. Impaired mitochondrial metabolism leads to various diseases both in animals and plants (Box 7.2).

## 7.9 Cyanide-Resistant Respiration

In animals respiration is completely inhibited by cyanide, which blocks the electron transport between $O_2$ and the Complex IV. However, in plants respiration is inhibited 10–100% by cyanide, depending upon the species variations. There is an **alternate oxidase (AOX)** present in the inner mitochondrial membrane, which facilitates transfer of the electrons from reduced ubiquinone (semiquinone or quinol) directly to $O_2$ bypassing the Complex III and Complex IV (Figs. 7.22 and 7.23).

Respiration which occurs in presence of cyanide is known as cyanide-resistant respiration. Unlike alternative dehydrogenases in which branch point of electron transport is before reducing ubiquinone, AOX provides the branch point after ubiquinone has been reduced, and electrons instead of being transported through cytochromes oxidases are directly transferred to $O_2$. As a result, proton gradient is not created across inner mitochondrial membrane, and ATP is not produced. Energy of the electron transport is dissipated as heat, and ATP yield decreases by up to 60%. The pathway is also believed to be associated with **thermogenic** events of the plants. An increase in as much as 25 °C has been observed in voodoo lily (*Sauromatum guttatum*) at the time of inflorescence (spadix) formation. Increase in temperature is associated with volatilization of various compounds, such as amines, which give putrid smell and serve to attract pollinators. Cyanide-resistant respiration may also be responsible for generating heat in plants growing in cold climate. Besides having thermogenic significance, cyanide-resistant respiration may have a role in metabolism of plants subjected to oxidative stress. Inhibition of electron transport under stress conditions results in increased ratio of $NADH/NAD^+$, which will result in slowing of TCA cycle. Alternate oxidase catalyzes oxidation of NADH. The formation of salicylic acid is induced under stress, which leads to synthesis of AOX. CN-resistant respiration is considered as "spill-over" mechanism or can be compared to a short circuit. It is considered significant in plants growing under high light intensities or other kinds of stress. ROS accumulate in AOX mutants of *Arabidopsis* when exposed to adverse conditions. Two forms of AOX have been reported in dicots (AOX1 and AOXII), while in monocots there is only one form, i.e., AOXI. AOX is a homodimer with each subunit having the molecular mass of 36 kDa, which are tightly bound to the matrix side of the inner mitochondrial membrane. Cysteine residues, located at the active sites of AOX, are exposed to matrix. When oxidized they are bonded to each other because of the disulfide bonds (-S-S-), making the enzyme inactive. When reduced (-SH) these are not bonded to each other, and the enzyme remains active. This interconversion of redox state of AOX is regulated by the mitochondrial NADP-dependent thioredoxin system, which also regulates activity of many of TCA cycle enzymes. Regulation of redox status of Calvin cycle enzymes is by ferredoxin-dependent thioredoxin system.

$$NADPH + Thioredoxin_{ox} \rightarrow NADP^+ + Thioredoxin_{red}$$

Thioredoxin$_{red}$ is responsible for reduction of the cysteine residues, which makes the enzyme susceptible to activation by pyruvate, and also it has more affinity with reduced $UQH_2$. Increased ratio of $NADH/NAD^+$ will decrease the TCA cycle. As a result of this, pyruvate accumulates which in turn increase activity of AOX and NADH is oxidized. **Salicylhydroxamic acid (SHAM)** interferes with the electron transport to AOX; that is why the pathway is also known as **SHAM pathway**.

## Summary

- Respiration is a multistep catabolic process catalyzed by enzymes. Energy released in some of the steps is sufficient enough to be used for ATP synthesis. In all living beings, carbohydrates are the main source of energy. In plants sucrose or starch is the main source of glucose which is primarily metabolized during respiration.
- Initial steps of glucose catabolism, known as glycolysis, are similar both in the presence or absence of $O_2$. Many scientists have contributed in elucidating the path of glycolysis. Unlike in animals, glycolysis occurs both in cytosol and plastids in plants, and there are various isoforms of the glycolytic enzymes. Regulation of glycolysis in plants is different from that of animals. In plants, there is "bottom up," while in animals it is "feed-forward" regulation of glycolysis.
- In the absence of $O_2$, pyruvate is metabolized through fermentation in cytosol. It is converted either to lactic acid or ethanol. For continued production of 2 ATP per glucose molecule, in absence of $O_2$, fermentation is necessary so that NADH is oxidized to $NAD^+$ to be available for glycolysis. Besides energy, glycolytic pathway also provides precursors to be used for biosynthesis of various biomolecules.
- Additionally, glucose is metabolized through oxidative pentose phosphate pathway (OPPP). OPPP consists of an irreversible oxidative phase, during which two molecules of NADPH are produced per molecule of glucose 6-phosphate and one carbon is lost as $CO_2$ and non-oxidative reversible phase which results conversion of ribulose 5-phosphate back to glucose 6-phosphate. Both in plants and in animal cells, OPPP operates in cytosol. However, in plants the pathway occurs in plastids also, where OPPP operates in dark, while reductive PPP operates in light. The enzymes of both pathways are regulated by light through the thioredoxin-ferredoxin system. OPPP plays a significant role in providing NADPH especially in fungi, in animal cells, and in plants during dark. Various intermediates of the pathway also serve as precursors for other pathways, such as ribose 5-phosphate is used for nucleotides biosynthesis, and erythrose 4-phosphate along with glycolytic intermediate PEP is a source of phenolic compounds.
- In the presence of $O_2$, pyruvate is converted to acetyl-CoA in mitochondria by oxidative decarboxylation and catalyzed by pyruvate dehydrogenase complex (PDC), which is a complex reaction and involving five steps. PDC is a multienzyme complex of three enzymes and five cofactors.
- Acetyl-CoA condenses with oxaloacetate to form citrate, which is the first reaction in citric acid cycle. Hans Krebs had put together the works of many scientists and proposed the reactions to be cyclic in nature and proposed citric acid cycle. TCA cycle is primarily regulated by $NADH/NAD^+$ and ATP/ADP ratio. Besides providing NADH, which is oxidized by electron transport chain to generate ATP,

TCA cycle also plays an anabolic role. Various intermediates of the cycle serve as precursors for many biomolecules. As some of the intermediates of TCA are consumed, there are anaplerotic reactions, which replenish the cycle intermediates.
- Reduced coenzymes, NADH and $FADH_2$, are oxidized through electron transport chain. Energy released during downhill movement of electron is conserved as the proton gradient across inner mitochondrial membrane, which is coupled with ATP synthesis. Components of the electron transport chain are present in four complexes, Complexes I, II, III, and IV. Movement of electrons between these complexes is facilitated by the mobile carriers, ubiquinone and cytochrome c. $O_2$-binding site is present in Complex IV.
- Plants have unique features like the presence of external and internal NAD(P)H dehydrogenases in addition to the electron transport chain, which are rotenone insensitive. There is an alternate oxidase (AOX) present in plants, which accepts electrons from ubiquinone directly bypassing Complexes III and IV. This is called as cyanide-resistant respiration, since the site of action of cyanide, which is present in Complex IV, is bypassed during the electron transport. Cyanide-resistant respiration does not result in any ATP generation. It is associated with dissipation of extra energy and might be associated with coping mechanism in plants when subjected to stress.

## Multiple-Choice Questions

1. While studying glycolysis, adding iodoacetate to the medium leads to accumulation of:
   (a) Fructose 6-phosphate
   (b) 3-Phosphoglyceraldehyde
   (c) 1,3-Bisphosphoglycerate
   (d) Fructose 1,6-bisphosphate
2. Net gain of energy during conversion of one glucose molecule to two molecules of pyruvate in glycolysis is:
   (a) 2 ATP + 2 NADH
   (b) 4 ATP
   (c) 2 NADH
   (d) 2 ATP
3. Which of the following reaction in glycolysis is not reversible?
   (a) Glucose 6-phosphate to fructose 6-phosphate
   (b) Glyceraldehyde 3-phosphate to 1,3-bisphosphoglycerate
   (c) 2-Phosphoglycerate to 2-phosphoenolpyruvate
   (d) Fructose 6-phosphate to fructose 1,6-bisphosphate

4. Under anaerobic conditions in the cell, fermentation is necessary because:
   (a) Lactate is produced.
   (b) Ethanol produced leaches out of the cell.
   (c) NADH is oxidized to $NAD^+$.
   (d) ATP is produced.
5. Pyruvate dehydrogenase is active when the enzyme is:
   (a) Dephosphorylated
   (b) Phosphorylated
   (c) Reduced
   (d) Oxidized
6. TCA has got amphibolic role in cell metabolism because:
   (a) Both ATP and NADH are produced in the cycle.
   (b) It is the main pathway for generation of metabolic form of energy.
   (c) It is responsible for oxidative as well as reductive reactions.
   (d) Precursors of various pathways are also produced during the cycle besides their oxidation.
7. Glycolysis in plants differs from that of in animals in the presence of:
   (a) Fermentation in animals
   (b) $PP_i$-dependent phosphofructokinase
   (c) Transport of glycolytic NADH to mitochondria
   (d) Synthesis of pyruvate from PEP in animal cells
8. Antimycin A inhibits electron transport at the:
   (a) UQ-binding site of Complex I
   (b) UQ-binding site of Complex III
   (c) $O_2$-binding site of Complex IV
   (d) With Complex II
9. The inhibitor for cyanide respiration is:
   (a) Rotenone
   (b) Antimycin A
   (c) Salicylhydroxamin
   (d) Cyanide
10. The number of ATP equivalents produced during complete oxidation of one glucose molecule is:
    (a) 30–32
    (b) 32–34
    (c) 34–36
    (d) None of the above

**Answers**

1. d  2. a  3. d  4. c  5. a  6. d  7. b
8. b  9. c  10. a

## Suggested Further Readings

Browsher C, Steer M (2008) In: Tobin A (ed) Plant biochemistry. Garland Science, Taylor & Francis Group, New York, pp 143–192

Campbell MK, Farrell SO (2012) Biochemistry, 7th edn. Cengage Learning, Stanford, pp 481–506

Nelson DL, Cox MM (2017) Lehninger principles of biochemistry, 7th edn. WH Freeman, MacMillan Learning, New York, pp 533–570

Taiz L, Zieger E, Moller IM, Murphy A (2015) Plant Physiology and Development, 6th edn. Sinauer Associates, Inc. Publishers, Sunderland, pp 317–339

# ATP Synthesis

## 8

Manju A. Lal

Adenosine triphosphate (ATP) is the energy currency of the cell and is required in cell metabolism of all living beings including bacteria, plants, and animals. It is utilized for the endergonic reactions and various energy-requiring processes in the cell. Various energy-requiring processes of the cell include transport of ions and biomolecules across plasma membrane against concentration gradient, movement of substances within the cell, cell growth including cell division, anabolic reactions of the cell, and various other cellular processes. ATP synthase catalyzes ATP synthesis during oxidative phosphorylation and photophosphorylation. Structure of ATP synthase and the mechanism of ATP synthesis have remained conserved through evolution. ATP was discovered in 1929 by German chemist Karl Lohmann. In 1941, it was demonstrated by Fritz Albert Lipmann that ATP was the universal energy carrier of the cells. Later on, Lohmann shared Nobel Prize with Hans Krebs in 1953 for his work on citric acid cycle. Lipmann was the first to use the squiggle (~) symbol and the term "high-energy bond" for the compounds having high phosphate group transfer potential. In 1948, Alexander Todd undertook chemical synthesis of ATP and deciphered its structure, for which he was awarded Nobel Prize in 1957. The standard free energy of ATP hydrolysis is 30.5 kJ.mol$^{-1}$. However, under cellular conditions, the free energy value is around $-50$ kJ.mol$^{-1}$. Thus, synthesis of 1 mole of ATP requires around 50 kJ of energy. There are two ways by which ATP can be synthesized in a cell. One of the ways is known as substrate-level phosphorylation, in which high-energy bond of the substrate is hydrolyzed and the energy released is utilized for ATP synthesis. Another mechanism involves coupling of energy of proton gradient with the synthesis of ATP. Proton gradient is created by electron transport chain either during oxidation of NADH/FADH$_2$ or during the light reaction of photosynthesis in mitochondria and chloroplasts, respectively.

## 8.1 Proton Gradient Coupled ATP Synthesis

ATP synthesis in mitochondria and chloroplasts is coupled with dissipation of proton gradient which is the result of electron transport during oxidation of substrates or during light reaction of photosynthesis. Energy required for movement of electrons is derived from oxidation of substrates in mitochondria, while in chloroplasts light energy is responsible for movement of electrons. ATP synthesis during electron transport in mitochondria is known as **oxidative phosphorylation**, while in chloroplasts it is known as **photophosphorylation.** In 1961, Peter Mitchell proposed **chemiosmotic model** for ATP synthesis according to which an electrochemical gradient is established across mitochondrial membrane when electrons move downhill during oxidation of the substrate through electron transport chain. Energy so released is utilized in the movement of protons across the inner mitochondrial membrane from matrix to intermembrane space, resulting in accumulation of protons. Thus, an electrochemical potential is developed which is the result of chemical potential energy (due to gradient in pH ($\Delta$pH) created as a result of accumulation of $H^+$ in the intermembrane space) and electrical potential energy which contributes to membrane potential ($\Delta\Psi$) (intermembrane space being more positive in comparison to the matrix because of accumulation of $H^+$) (Fig. 8.1).

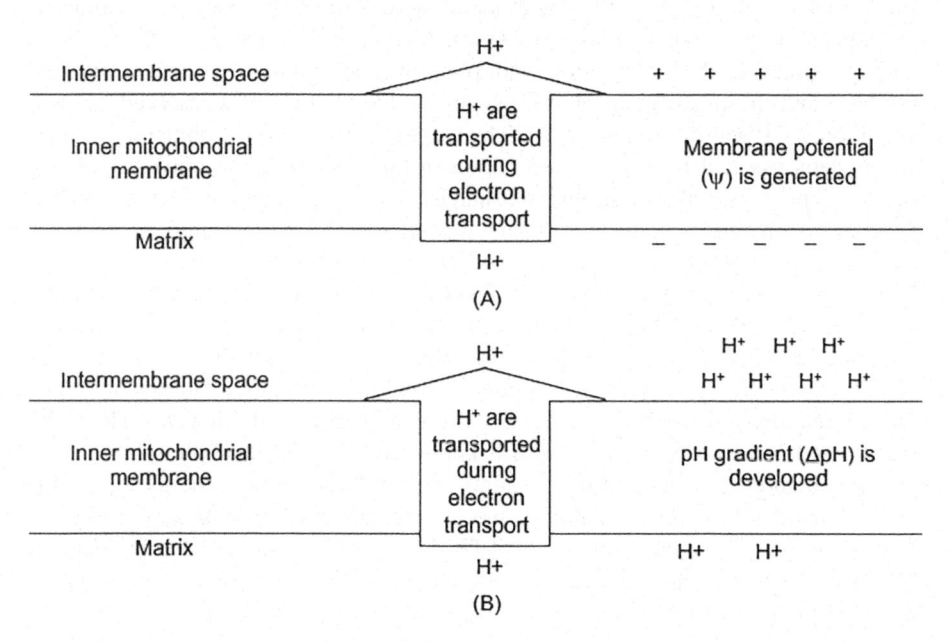

**Fig. 8.1** Two components of PMF. (**a**) Membrane potential due to charge difference. (**b**) pH gradient

## 8.1 Proton Gradient Coupled ATP Synthesis

NADH oxidation in mitochondria through electron transport chain involves removal and transport of two electrons from NADH to $O_2$. This results in conservation of around 200 kJ energy in the form of 10 $H^+$. Mitchell called this energy due to electrochemical gradient as **proton motive force (PMF)**, which is measured in volts (V). This energy is the result of transmembrane potential ($\Psi$) of about ~160 mV and a $\Delta pH$ of about 1 unit, which is equivalent to ~60 mV.

$$\text{PMF}\ (\Delta p) = \Delta \Psi - 59\ \Delta pH\ (pH_i - pH_o)$$

where $\Delta\Psi$ refers to potential gradient across the inner mitochondrial membrane while $\Delta pH$ is the difference in pH across the membrane between matrix ($pH_i$) and intermembrane space ($pH_o$). This value will be negative, in case $pH_o$ is lesser because of accumulation of $H^+$ in comparison to $pH_i$. 59 mV per pH unit is the constant of proportionality, which demonstrates that a transmembrane pH difference of 1 unit will contribute to a transmembrane potential of 59 mV at 25 °C. Thus, PMF refers to the energy responsible for the passive movement of protons from outside the mitochondria (intermembrane space) to the inside (matrix). Since the inner membrane of mitochondria is impermeable to protons, proton movement in response PMF occurs only through **ATP synthase** localized in the membranes, which provides the channel for proton movement. ATP synthase, which is also known as Complex V, is responsible for utilizing energy of proton movement for ATP synthesis.

$$ADP + P_i + nH_p^+ \rightarrow ATP + H_2O + nH_N^+$$

$nH_p^+$ and $nH_N^+$ refer to $H^+$ concentration on the positive (intermembrane space) and negative sides (matrix of mitochondria). Mitchell proposed **chemiosmotic** model in 1961. However, its significance was realized later, and Mitchell was awarded Nobel Prize in Chemistry in 1978 for his contribution. "Chemiosmotic" term used by Mitchell reflects a link between the chemical bond-forming reactions that generate ATP ("chemi") and membrane transport processes ("osmotic"). Sometimes it is referred as chemiosmotic coupling, where coupling refers to coupling of ATP synthesis and flow of electrons through electron transport chain in the inner mitochondrial membrane. Initially a mechanism involving formation of high-energy bond, such as in substrate-level phosphorylation, was favored. However, no intermediates with high-energy bonds could be isolated. Various experimental evidences are proposed in support of chemiosmotic model (Boxes 8.1 and 8.2). In chloroplasts, protons are pumped from stroma to the lumen of thylakoids. As a result, lumen becomes more acidic in comparison to stroma. Since thylakoid membrane is permeable to anions such as $Cl^-$, $\Delta pH$ plays more significant role in comparison to $\Delta\Psi$ for generating PMF in chloroplasts. In case of mitochondria, it is the $\Delta\Psi$ which contributes more to PMF (Figs. 8.2 and 8.3).

**Box 8.1: Evidence in Support of Chemiosmotic Hypothesis**
Coupling of ATP synthesis with electron flow from a substrate (succinate) to $O_2$ can be demonstrated in an experiment by measuring change in $O_2$ concentration in the medium containing ADP and $P_i$.

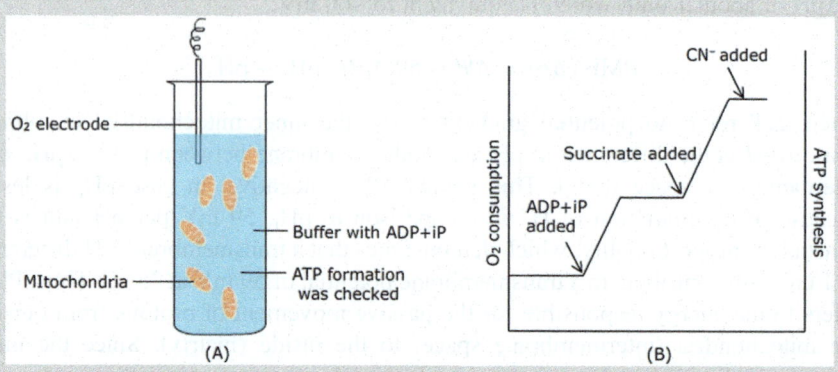

(A) Isolated mitochondria are suspended in buffer. Change in $O_2$ concentration is monitored using $O_2$ electrode. Reduction of $O_2$ is indicative of electron transport since on accepting electrons, $O_2$ will be converted to water. Addition of ADP and $P_i$ alone does not result increase in $O_2$ consumption. However, an increase in $O_2$ consumption was observed on addition of ADP + $P_i$ along with succinate, which stops after some time indicating that ADP and $P_i$ have been consumed. This is validated by checking ATP synthesis. Addition of $CN^-$ blocks electron transfer. (B) Graphical representation of the experiment.

**Use of uncoupling agents**: Uncouplers are the amphiphilic compounds which are soluble both in water and lipids. As a result, they can permeate through the membranes by diffusion, transferring $H^+$ or $OH^-$ across, and resulting in dissipation of proton gradient without affecting electron flow, but ATP synthesis does not occur. Uncouplers provide pathway for proton movement, dissipating proton gradient bypassing ATP synthase activity. During the experiment addition of uncouplers results in continuation of electron transport and proton pumping, without generation of any proton gradient. ATP synthesis does not occur without affecting uptake of oxygen. In the absence of proton gradient, however, protons are transported in reverse direction through ATP synthase at the expense of ATP. Uncouplers which transfer protons across the membrane are known as protonophores. These include FCCP (carbonylcyanide-*p*-trifluoromethoxyphenylhydrazone) and 2,4-dinitrophenol (DNP). Protonated DNP (a weak acid) diffuses from high proton concentration side of the membrane to low proton concentration side where it gets dissociated to generate protons resulting in dissipation of proton gradient. Membrane is permeable to both protonated and anionic forms of these

(continued)

**Box 8.1** (continued)

uncouplers. The anionic form diffuses back and combines with more protons and carries them to the other side of the membrane creating short circuiting for protons. As a result, no ATP is synthesized. However, electron transport is not affected, and oxygen will be reduced to water in the experimental conditions. Another class of uncouplers, known as ionophores, transfer alkali cations (e.g., $K^+$ ions) along with them across the membrane. As a result, membrane potential is dissipated. These include valinomycin.

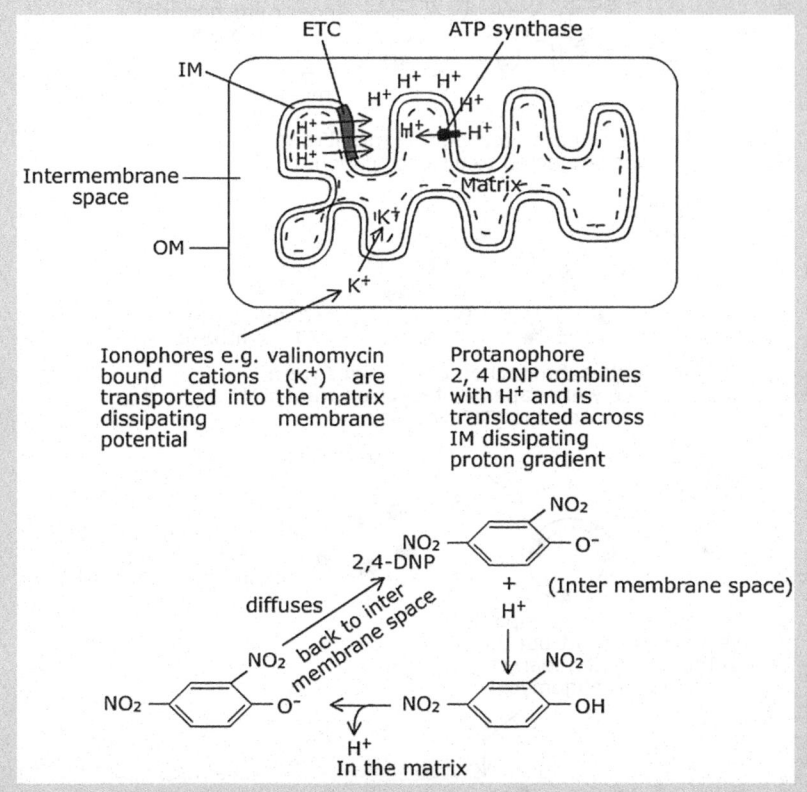

Ionophores e.g. valinomycin bound cations ($K^+$) are transported into the matrix dissipating membrane potential

Protanophore 2, 4 DNP combines with $H^+$ and is translocated across IM dissipating proton gradient

**Studies with reconstituted membrane vesicles**: Coupling of proton gradient with ATP synthesis was also demonstrated by Efraim Racker and colleagues (Cornell University in the mid-1960s) in an experiment with isolated mitochondria. Besides supporting coupling of proton gradient with ATP synthesis, the experiment also demonstrated that $H^+$ / ATP synthase of mitochondria consists of two parts, an oligomycin-sensitive membrane-bound factor, $F_o$, which is required for proton translocation, and a soluble fraction $F_1$ which is required for ATP synthesis. Racker called these particles $F_1$ ATPases. Mordechai Avron from Israel demonstrated the existence of similar particles in

(continued)

**Box 8.1** (continued)

chloroplast membranes which also had ATPase activity. Though H⁺ ATP synthase of chloroplasts is not inhibited by oligomycin, membrane-bound part of these is designated as $F_o$. Similar particles are also reported in bacteria, and these collectively are called F-ATP synthases or F-ATPases or $F_o$-$F_1$ ATP synthases or $F_o$-$F_1$ ATPases.

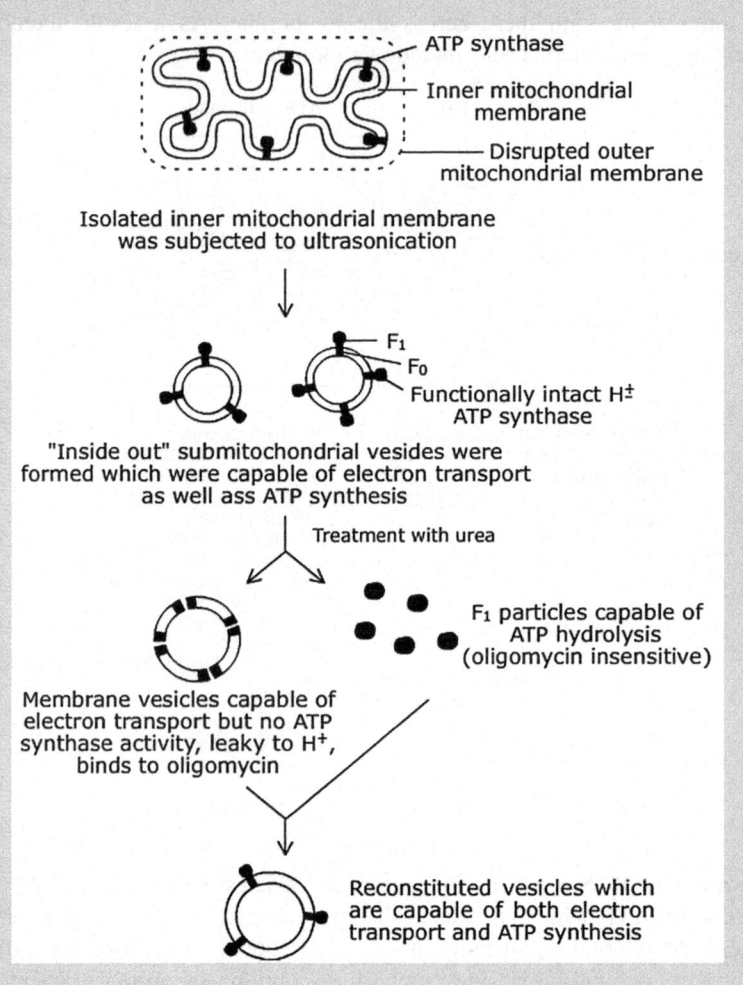

- ATP synthase
- Inner mitochondrial membrane
- Disrupted outer mitochondrial membrane

Isolated inner mitochondrial membrane was subjected to ultrasonication

- $F_1$
- $F_o$
- Functionally intact H⁺ ATP synthase

"Inside out" submitochondrial vesides were formed which were capable of electron transport as well ass ATP synthesis

Treatment with urea

Membrane vesicles capable of electron transport but no ATP synthase activity, leaky to H⁺, binds to oligomycin

$F_1$ particles capable of ATP hydrolysis (oligomycin insensitive)

Reconstituted vesicles which are capable of both electron transport and ATP synthesis

(continued)

**Box 8.1** (continued)

Experiment was performed with artificial vesicles prepared from purified phospholipid (liposomes) to demonstrate coupling of light-driven proton gradient with ATP synthesis. Mitochondrial $F_o$-$F_1$ particles and bacteriorhodopsin were incorporated in the membranes of liposomes. In the presence of light, bacteriorhodopsin pumped $H^+$ into the lumen of the vesicles, which resulted in synthesis of ATP from ADP and $P_i$. Photosynthetic archaebacteria utilizes bacteriorhodopsin to generate proton motive force.

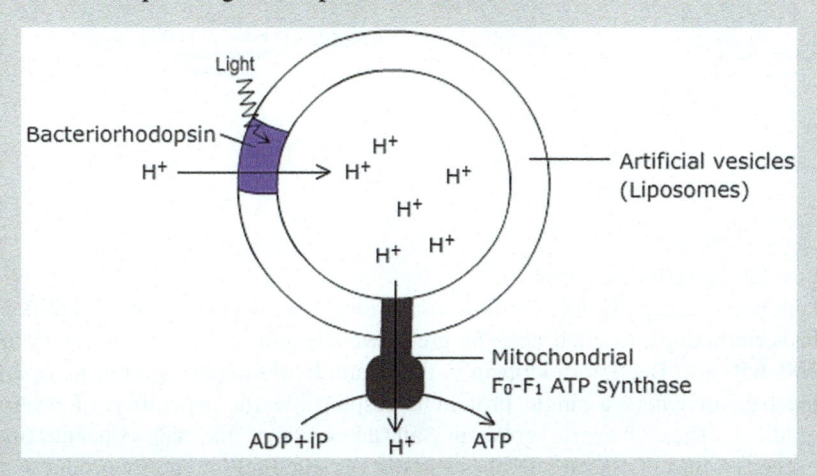

**Jagendorf's experiment**: Daniel Arnon (Berkley) had observed in 1954 that ATP was synthesized from ADP and $P_i$ when suspended thylakoid membranes are illuminated. This process of ATP synthesis in the presence of light was called photophosphorylation. In earlier observations ATP synthesis had been found to be coupled with oxidation of NADH in mitochondria. Photophosphorylation, however, was coupled with NADPH generation during light reaction. American scientist Andre Jagendorf carried out an experiment in which he experimentally demonstrated that by subjecting thylakoids to a proton gradient ($\Delta pH = 4$), ATP could be synthesized from ADP and $P_i$, even in the absence of light. The experiment demonstrates that during light reaction of photosynthesis, the role of light is to create proton gradient. ATP synthesis in thylakoid could occur even in dark, if pH gradient could be established across thylakoids. This observation further supported chemiosmotic hypothesis.

(continued)

**Box 8.1** (continued)

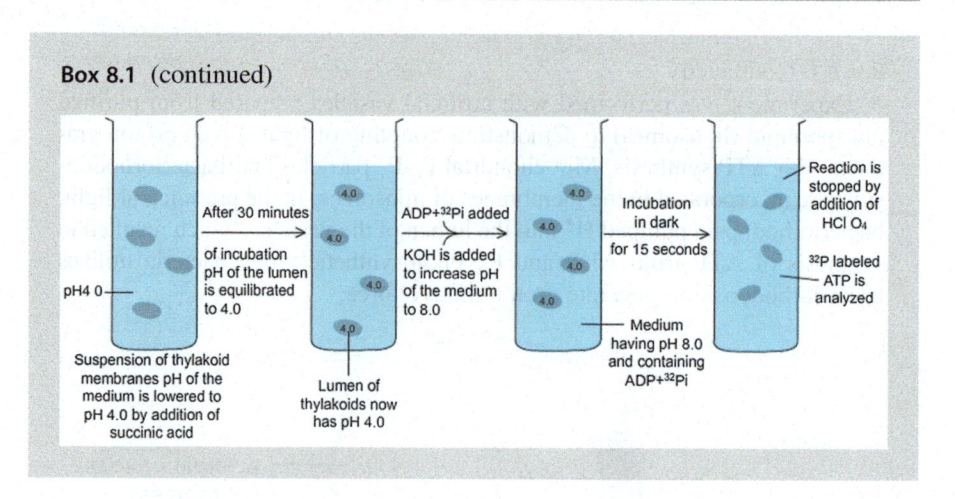

[Figure: Experimental setup showing thylakoid membrane suspension at pH 4.0 by addition of succinic acid → After 30 minutes of incubation pH of the lumen is equilibrated to 4.0 (Lumen of thylakoids now has pH 4.0) → ADP+$^{32}$Pi added, KOH is added to increase pH of the medium to 8.0 (Medium having pH 8.0 and containing ADP+$^{32}$Pi) → Incubation in dark for 15 seconds → Reaction is stopped by addition of HClO$_4$; $^{32}$P labeled ATP is analyzed]

**Box 8.2: Bacteriorhodopsin**

Halobacteria, e.g., *Halobacterium salinarum*, contain a purple-colored protein bacteriorhodopsin which absorbs green wavelength of light in the range of 500–650 nm. Bacteriorhodopsin is the example of simplest proton pumping mechanism where a single protein is responsible for generation of proton gradient. These bacteria grow in salt lakes where the salt concentration exceeds 5 mM. These are aerobes and use organic fuel for energy production. However, at times availability of $O_2$ becomes limiting because of which the oxidative mechanism of energy production needs to be supplemented by capturing light energy. On absorbing photons, bacteriorhodopsin can transport protons from inside of the cell to extracellular side resulting in generation of proton gradient across the plasma membrane. Protons flow back in response to the proton gradient through ATP synthase localized in the plasma membrane resulting in ATP synthesis. Bacteriorhodopsin consists of a light-sensitive cofactor retinal which is an aldehyde derivative of vitamin A and a protein bacteriopsin. The protein has seven transmembrane helices. On absorbing proton, retinal undergoes photoisomerization. This results in conformational change of the surrounding protein bacteriopsin followed by proton pumping action toward extracellular side of the membrane. ATP generation by this means in these bacteria is not associated with any release of $O_2$ or any $CO_2$ assimilation.

## 8.1 Proton Gradient Coupled ATP Synthesis

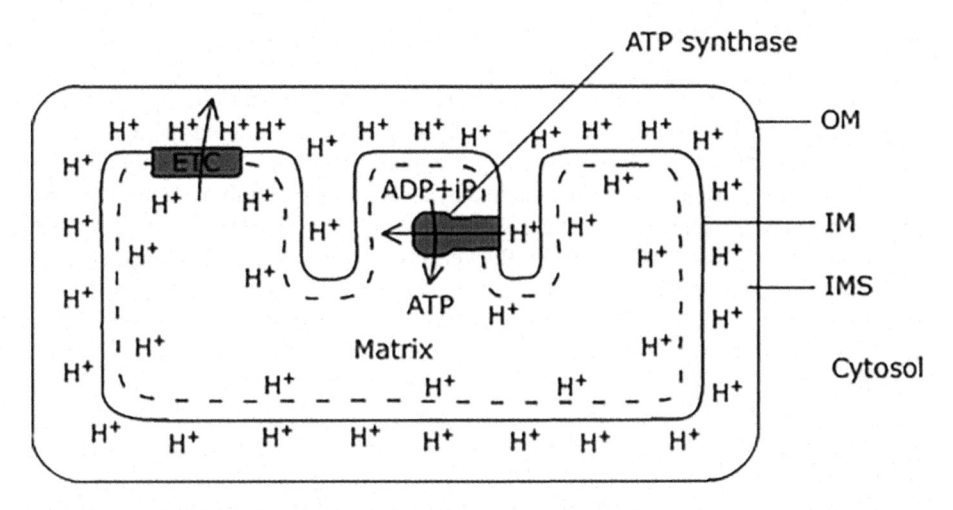

**Fig. 8.2** Oxidative phosphorylation in mitochondria. Membrane potential ($\Delta\Psi$) is the major component of proton motive force in mitochondria. OM, outer mitochondrial membrane; IM, inner mitochondrial membrane; IMS, intermembrane space

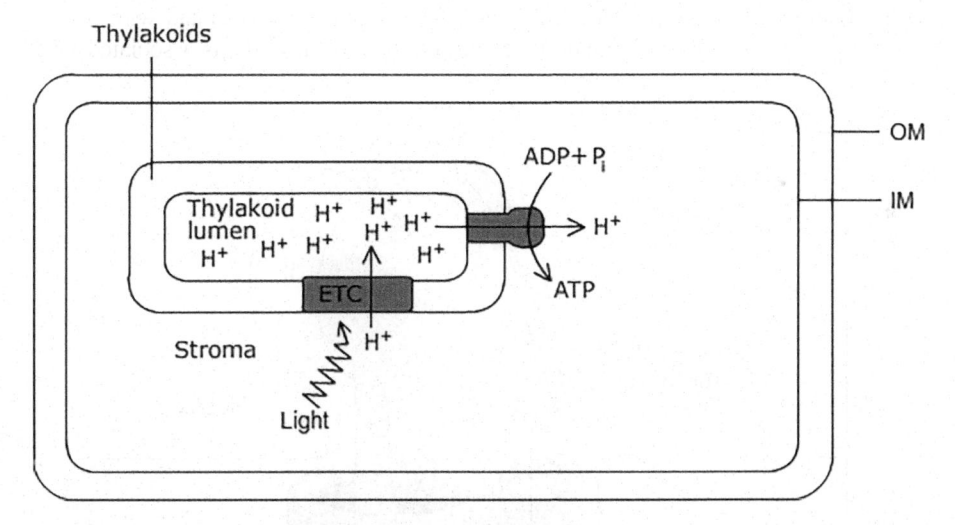

**Fig. 8.3** Photophosphorylation in chloroplasts. $\Delta$ pH in between stroma and lumen of thylakoid is the major component of PMF. OM, outer membrane of chloroplasts; IM, inner membrane of chloroplasts; ETC, electron transport chain localized in thylakoid membrane. Light energy is used for pumping $H^+$ from stroma to thylakoid lumen

## 8.2 ATP Synthase

Parsons and Fernández-Moran first demonstrated elementary particles in 1963–1964 in their preparations of animal mitochondria using negative staining method. These "elementary particles" were present along the inner membrane of mitochondria and consisted of a spherical headpiece attached by a stalk to the membrane inner surface. However, a relationship among these subunits with oxidative phosphorylation was subsequently demonstrated by Efraim Racker. It was demonstrated in the experiment that during oxidative phosphorylation, the headpiece of these subunits acts as a coupling factor ($F_1$) for ATP synthesis. $F_1$ was so called since it was the first factor found essential for ATP synthesis. These structures did not have any electron transport capacity. These are of ancient origin, and similar structures are also found in chloroplasts and bacteria. These are F-type ATPases since these were demonstrated to catalyze ATP hydrolysis when isolated and these were called **$F_1$ ATPases**. Alan Senior and Harvey Penefsky purified bovine $F_1$ complex. Penefsky demonstrated it to be an assembly of five different kinds of polypeptides, which he called α, β, γ, δ, and ε. These were found to be having a mass of more than 500,000 Daltons. In 1994, John Walker and his colleagues at Cambridge carried out crystallographic determination and gave the three-dimensional structure of $F_1$ part of ATP synthase isolated from beef heart mitochondria (Fig. 8.4). Though amino acid composition of the three β-subunits is identical, these three subunits differ in their conformations (β-ATP, β-ADP, and β-empty). γ-subunit is associated with one of them having β-empty conformation. Membrane-bound factor $F_o$ is associated with

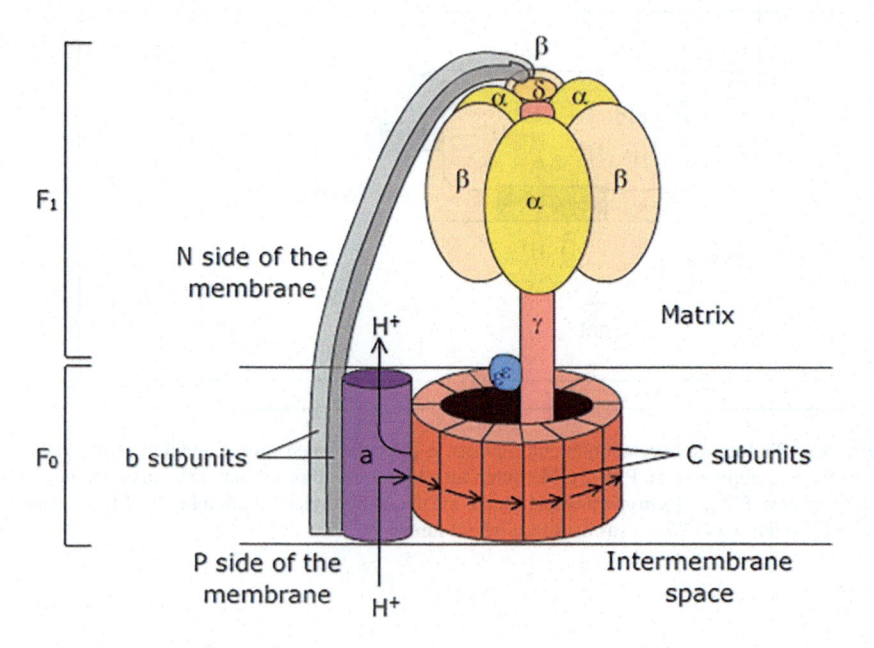

**Fig. 8.4** A model depicting ATP synthase structure

proton translocation, which is inhibited by adding oligomycin; thus, it is called as $F_o$. "c" is a small hydrophobic protein which has two transmembrane-spanning helical regions with a loop extending in the matrix of mitochondria. The two transmembrane helices of c-subunits run perpendicular to the plane of membrane and are arranged in two concentric rings. The inner ring consists of N-terminus of the polypeptide, while the outer ring consists of C-terminus. The outer circle has a diameter of 55 Å. The c-ring makes a rotatory movement associated with proton movement. Hence $F_o$ is also called "rotor." The two b-subunits anchor $F_o$-$F_1$ to the membrane through binding with a-subunit of $F_o$ and δ-subunit on the other side, holding α- and β-subunits in place. Thus, two b-subunits along with a-subunit and F1 headpiece (consisting of $α_3β_3$) form "stator." The ε-subunit along with γ-subunit forms leg and foot that projects from the bottom of the $F_1$. Subunit "a" of $F_o$ consists of several hydrophobic transmembrane helices and is in close association with one of the c-subunits of the c-ring (Fig. 8.4). These provide a transmembrane channel for protons. Structures of ATP synthase in mitochondria and chloroplast differ significantly in their $F_o$ domain though there is similarity in the structure of $F_1$. Orientation of the enzyme is similar in both the cases. $F_1$ is present toward the alkaline side (N) of the membranes. ATP synthase in chloroplasts is known as $CF_o$-$CF_1$. It is a 400 kDa complex consisting of 9 different subunits, some of which are encoded in chloroplast genome while others in nuclear genome. Similar to $F_1$, $CF_1$ also consists of three large subunits of α and β (three copies of each) and three small subunits, i.e., γ, δ, and ε (one copy each). δ-subunit links $CF_o$ to $CF_1$, γ-subunit appears to control proton gating through the enzyme, while ε-subunit blocks catalysis in the dark as a result of which breakdown of ATP is prevented. γ-subunit may also participate in the regulatory mechanism by the Fd/thioredoxin system. As a result, ATP synthase may be activated in light at low PMF, and enzyme may be deactivated in dark. $CF_o$ consists of four types of subunits, i.e., I, II, III, and IV. Except III all other subunits are present in single copy. There may be 14 copies of subunit III per complex in plant chloroplasts, which form the pathway for translocation of protons from luminal space to stroma.

## 8.3 Mechanism of ATP Synthesis

Paul Boyer proposed **rotational catalysis** as a key to **binding change mechanism** for ATP synthesis for which he was awarded Nobel Prize in 1997 together with John E. Walker. According to this model, the major energy-requiring step is not ATP synthesis; rather energy is required for change in the affinity of β-subunits of the enzyme complex for the substrates (ADP + $P_i$) and release of the product (ATP), which occurs due to change in conformation of the protein subunits. The active sites of $F_1$ are responsible for catalyzing ATP synthesis. It is also proposed that these affinity changes are coupled with rotation of a shaft (consisting of γ- and ε-subunits) associated with rotation of c-ring which itself is associated with proton translocation.

Mechanism of ATP synthesis, both in chloroplasts and mitochondria, is believed to be identical. ATP synthesis by rotational catalysis will be dealt as (i) ATP synthesis due to rotation of γ- and ε-shaft of $F_o$-$F_1$ complex (ii) rotation of γ- and ε-shaft due to proton movement in response to electrochemical gradient.

### 8.3.1 Rotatory Model (Binding Change Mechanism)

$F_1$ headpiece is a hexamer consisting $α_3β_3$. The subunits constituting headpiece do not move since it is bound to two "b" polypeptides through δ-subunit, and the two "b" polypeptides are bound to static a-subunit of $F_o$ constituting "stator." The active sites for catalysis of ATP synthesis are localized in the β-subunits of $F_1$ at α and β interfaces. There are three possible conformations for each of β-subunits, i.e., β-ADP (loose), β-ATP (tense), and β-empty (open). In β-ADP (loose) conformation, the subunit can loosely bind with ADP and $P_i$, followed by change in its conformation to β-ATP (tense) resulting in ATP synthesis. Change in conformation of β-tense to β-empty results in release of ATP. Each of the β-subunit changes from β-ADP (loose) to β-ATP (tense) to β-empty (open). The adjacent β-subunits are not in similar conformations. The β-subunit prior to β-ADP (loose) will be in β-empty (open) conformation, while the β-subunit prior to β-empty will be in β-ATP (tense) conformation. Thus, on one side of β-empty subunit, it is β-ATP and on the other side β-ADP. Change in conformation of β-subunits occurs due to the rotation of γ-polypeptide. The γ-subunit brings about the conformational change in β-subunit so that ATP is released. Thus γ-subunit is linked with β-empty conformation. Rotation of γ-subunit occurs in anti-clock direction (Fig. 8.5). With each anti-clock rotation of 120$°$, γ-subunit gets attached to different β-subunits, which is forced to change its conformation from β-ATP to β-empty. One complete rotation of γ-subunit (360$°$) will cause each of the three β-subunits to acquire all the three possible conformations. The rotation is not smooth rather occurs in three discreet steps of 120$°$ each. It is not the binding or synthesis of ATP rather it is the release of ATP which is energy-requiring process. Direction of the rotation of γ-subunit determines the enzymatic nature of the complex. In the absence of proton gradient, the isolated enzyme complex catalyzes ATP hydrolysis. This model was confirmed in an experiment conducted by Masasuke Yoshida and Kazuhiko Kinoshita in Japan. They attached a fluorescent molecule to the upper end of γ-subunit contained in an $F_o$ particle present in the membrane. Using a special video microscopy documentation, they were able to demonstrate that during ATP hydrolysis, γ-subunit rotates in the opposite direction, i.e., in a clockwise manner. However, ATP synthesis due to movement of γ-subunits in anti-clockwise manner could not be demonstrated in the absence of proton gradient. Movement of shaft consisting of "γ" and "ε" occurs due to rotatory movement of c-ring of $F_o$.

## 8.3 Mechanism of ATP Synthesis

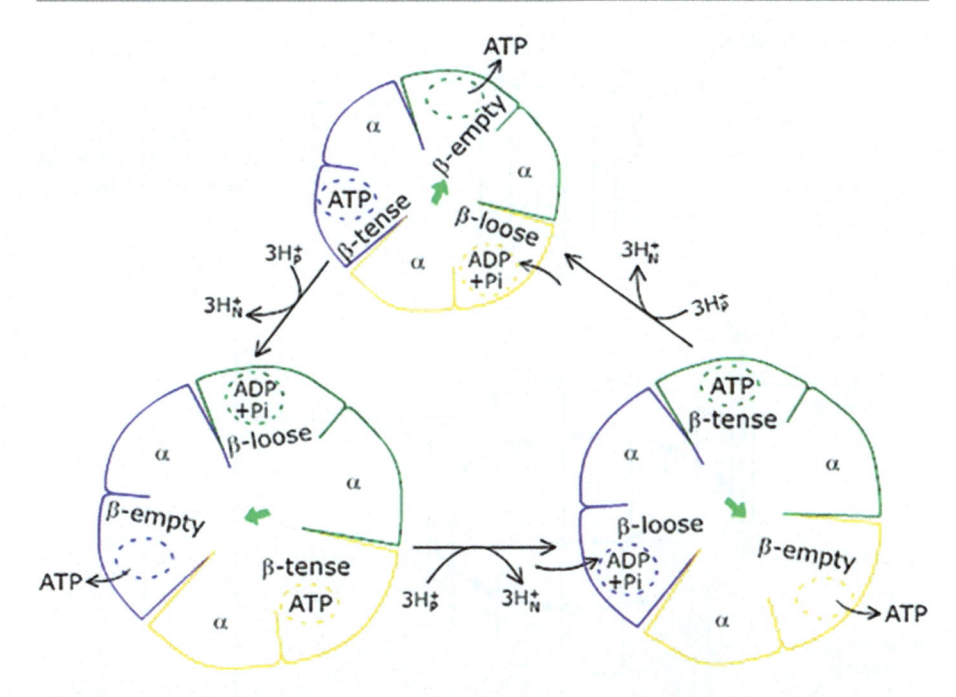

**Fig. 8.5** Rotational model for binding change mechanism proposed by Paul Boyer for ATP synthesis

### 8.3.2 Rotatory Movement of c-Ring of Fo

How is the movement of shaft, consisting of γ- and ε-subunits, coupled to proton translocation through proton channel? Proton movement occurs in response to electrochemical gradient created by proton transport. $F_o$ consists of two types of subunits, single a-subunit and 8–15 subunits of "c." The number is variable depending upon the source of ATP synthase. In animal mitochondria, the number of c-subunits is 8, in yeast mitochondria it is 10, in chloroplasts the number is 12–14, while it can be as high as 15 as in cyanobacterium *Spirulina platensis*. $F_o$ functions like a nanomotor. In chloroplasts, its velocity of rotation has been estimated to be 160 revolutions per second. It is the c-ring which rotates causing the movement of shaft consisting of γ- and ε-subunits, which are arranged in leg and foot formation, while the a-subunit does not move and is part of "stator" along with two b polypeptides, a δ-subunit and $F_1$ headpiece. The hydrophobic helices of c-subunits are arranged perpendicular to the membrane around the axis forming c-ring. The two transmembrane helices of c-subunits which run perpendicular to the plane of membrane are arranged in two concentric rings. The outer circle has a diameter of 55 Å and consists of C-terminus of the polypeptide, while the inner ring consists of N-terminus of the polypeptide. Wolfgang Junge from Germany developed a model for explaining how proton gradient drives the nanomotor. Two hydrophilic half channels for protons are localized in the a-subunit, one is leading from "P" side of

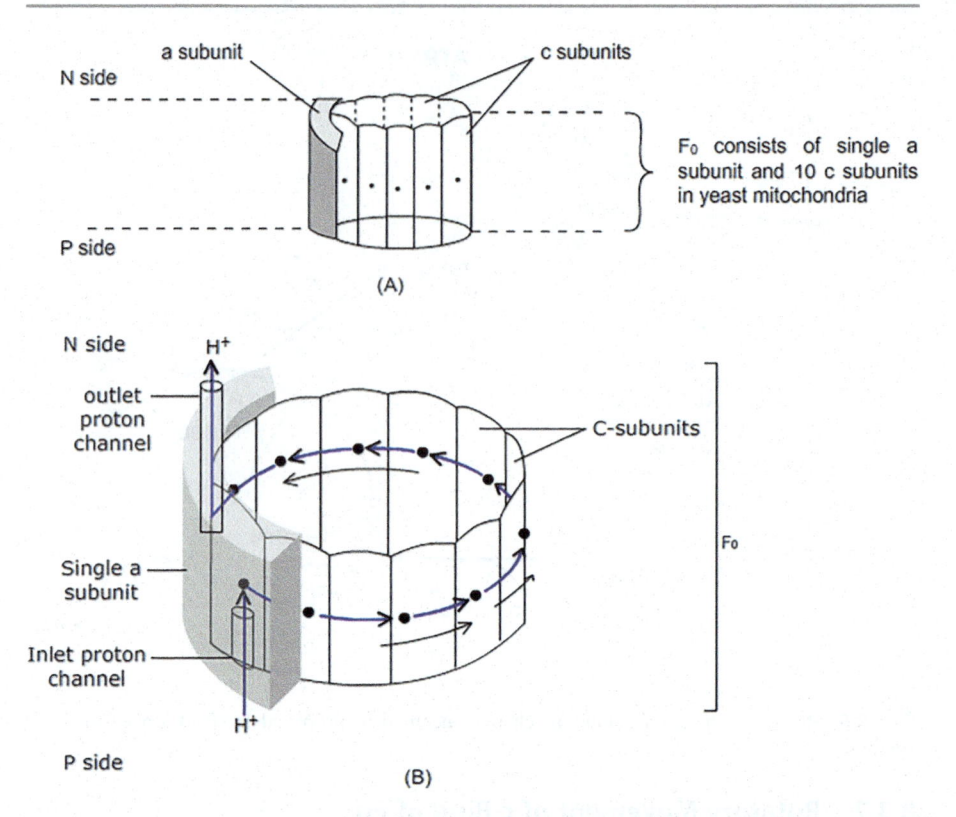

**Fig. 8.6** (a) $F_o$ of ATP synthase. (b) Figure depicting a model demonstrating rotation of c-ring. Rotation of c-ring is driven by proton translocation from acidic to alkaline side of the membrane in response to electrochemical gradient

the membrane to the middle of the membrane, while the other is leading from middle of the membrane to "N" side (Fig. 8.6). The a-subunit is believed to have five hydrophobic membrane-spanning regions and has positively charged arginine (Arg$^{210}$) which plays active role in channeling protons across. Subunit "b" is anchored in the membrane by a single N-terminal α-helix. Each of c-subunits has a critical aspartate in the middle (Asp$^{61}$ in *E. coli*) or glutamate buried in the membrane. The side-chain carboxyl group of the residue plays a critical role in proton translocation. This carboxyl group is conserved throughout in all known subunits of "c" and is required for transmembrane proton transport. The c-subunits form a ring by interactions through their C-terminal α-helices with the N-terminal α-helices outside the ring. The annular structure of c-ring has been visualized by **atomic force microscopy**. The proton (H$^+$) travels from "P" side to the middle of the membrane through the half channel where it protonates carboxylate group (COO$^-$) of Asp$^{61}$ of c-subunit. As a result of this, electrostatic attraction between the protonated carboxyl (COOH) group of Asp$^{61}$ of c-subunit and positively charged Arg$^{210}$ of a-subunit is weakened, and the positively charged Arg$^{210}$ side arm swings and displaces the proton of carboxyl group of its Asp$^{61}$ of c-subunit present on the other side. This results in rotation of c-ring because of Brownian movement. The c-ring subunit (having deprotonated carboxylate group) now faces the other half channel of the stationary a-subunit. The displaced proton enters the half channel and is released toward "N" side of the membrane (Fig. 8.6). The half channel of a-subunit facing "P" side is ready to receive another proton from "P" side, and the cycle is repeated again. Proton translocation occurs as long as the proton gradient is there. In the absence of proton gradient, ATP synthase will act as ATPase and will generate proton gradient at the expense of ATP. Why does Brownian movement occur only in one direction? This may be due to the presence of two half channels, and possibly the repulsion between the positively charged Arg$^{210}$ residue of a-subunit and the proton loaded c-subunit prevents the backward movement of the rotor. ATP synthase may form a complex, known as **ATP synthasome**, with adenine nucleotide translocase (ADP/ATP symporter) and phosphate carrier (P$_i$/H$^+$ symporter). A complete rotation of c-ring with 10 subunits will involve translocation of 10 H$^+$. This will result in rotation of γ- and ε-shaft by 360°. As a result of this, all the three β-subunits will undergo changes in their three conformations, resulting in synthesis of three ATPs. This indicates that translocation of 10 H$^+$ across the membrane will result in synthesis of three ATPs. Transport of a pair of electron from NADH to ½ O$_2$ results in the release of 214 kJ.mol$^{-1}$ free energy, which is conserved as gradient of 10 H$^+$ across the membrane. Free energy (ΔG) required for ATP synthesis is 50 kJ.mol$^{-1}$. Energy conservation equivalent to 214 kJ.mol$^{-1}$ justifies that. The rest of the energy may be released as heat.

## 8.4 Stoichiometry of O$_2$ Consumption and ATP Synthesis (P/O Ratio)

P/O ratio or P/2e$^-$ refers to the number of moles of ATP synthesized per ½ O$_2$ reduced (i.e., per pair of electron transferred to O$_2$) during oxidative phosphorylation. Generally, the experimental value varies from 2.5 in case of NADH oxidation to 1.5 in FADH$_2$ oxidation. The value also depends upon the number of c-subunits which constitute F$_o$ of the F$_1$-F$_o$ complex of mitochondria. One full rotation of the c-ring results in translocation of 8 H$^+$ and synthesis of 3 ATPs, in case there are 8 subunits. In addition, three P$_i$ would need to be translocated for ATP synthesis. Thus, that a total of 11 H$^+$ would need to be translocated for synthesis of 3 ATP molecules, i.e., 3.7 protons (11 H$^+$ and 3 ATP) are required for generation of one ATP. Oxidation of one NADH is coupled with generation of a gradient of 10 H$^+$, which indicates that the number of ATPs produced coupled with oxidation of one NADH will be 10/3.7, i.e., 2.7 ATPs. In yeast mitochondria, a complete rotation of c-ring will result in translocation of 10 H$^+$ (since there are 10 subunits) and in the synthesis of 3 ATP molecules. In addition to this, three H$^+$ are translocated along with three P$_i$. It makes a total of 13 H$^+$ translocated for synthesis of 3 ATP molecules, i.e., 13/3 = 4.3 H$^+$ are translocated for the synthesis of one ATP. Thus P/O ratio of NADH during oxidative phosphorylation in yeast will be 10/4.3 = 2.3 ATPs. Since a gradient of 6 H⁺ is created in case of oxidation of FADH$_2$, P/O ratio will be 6/3.7 = 1.6 or 6/4.3 = 1.4, respectively, if c-ring has 8 or 10 subunits. The experimental values of P/O ratio for NADH and FADH$_2$ oxidation generally used are 2.5 and 1.5, respectively. In case of chloroplasts (photophosphorylation), 14 copies of subunit III (c) appear to be present. Hence, translocation of 14 H⁺ will be required for complete rotation of c-ring, resulting in the synthesis of three ATP molecules. Thus H⁺/ATP ratio will be 14/3 ATP, i.e., 4.67.

## 8.5 Substrate-Level Phosphorylation

ATP can also be synthesized directly by coupling bond energy of high-energy compounds with ATP synthesis. For example, there are two steps of glycolysis and one step of TCA cycle during which ATP is synthesized directly without involvement of any electron transport. Energy released during the hydrolysis of high-energy compound during these reactions is more than that required for ATP biosynthesis. In one of the glycolytic reactions, phosphate group is transferred from 1, 3-bisphosphoglycerate to ADP, resulting in ATP synthesis. The reaction is catalyzed by the enzyme **glycerate kinase**.

$$1,3 - \text{Bisphosphoglycerate} + \text{ADP} \rightarrow 3 - \text{Phosphoglycerate} + \text{ATP}$$

The standard free energy change during hydrolysis of 1,3-bisphosphoglycerate is $-49.3$ kJ.mol$^{-1}$, and for ATP synthesis, it is $+30.5$ kJ.mol$^{-1}$. Reaction is favored by net $\Delta G^{o\prime}$ of the reaction being $-18.8$ kJ.mol$^{-1}$. Another reaction of glycolysis involves synthesis of pyruvate from 3-phosphoenolpyruvate, which is also coupled with synthesis of ATP. The reaction is catalyzed by **pyruvate kinase**. Phosphoenolpyruvate is a high-energy compound with high group transfer potential.

$$\text{Phosphoenolpyruvate} + \text{ADP} \rightarrow \text{Pyruvate} + \text{ATP}$$

Free energy of hydrolysis PEP ($-61.9$ kJ.mol$^{-1}$) is more negative than that required for ATP biosynthesis ($+30.5$ kJ.mol$^{-1}$). The two reactions are coupled, and the net reaction that occurs is the sum of hydrolysis of PEP and ATP synthesis with $\Delta G^{o\prime} = -31.4$ kJ.mol$^{-1}$. Substrate phosphorylation also occurs in TCA during synthesis of succinate from succinyl-CoA. Free energy of hydrolysis of high-energy **thioester** bond of succinyl-CoA is $-33.4$ kJ.mol$^{-1}$, while standard free energy requirement for ATP synthesis is 30.5 kJ.mol$^{-1}$. Overall reaction will be slightly exergonic with a net $\Delta G^{o\prime}$ is $-2.9$ kJ.mol$^{-1}$.

$$\text{Succinyl} - \text{CoA} + \text{ADP} + \text{P}_i \rightarrow \text{Succinic acid} + \text{CoA} + \text{ATP}$$

The reaction is slightly exergonic and is catalyzed by the enzyme **succinyl-CoA synthetase** (SCS), also known as **succinic thiokinase**. The mammalian SCS consists of two subunits, α and β, which have molecular weight 32,000 Daltons and 42,000 Daltons, respectively. α-subunit has got the binding site for CoA and a phosphate-binding site, i.e., histidine (His$^{246}$). The β-subunit confers specificity to either GDP or ADP. Active site of the enzyme is present at the interface of the two subunits. Reaction occurs in three steps: (i) In the first step of the reaction catalyzed by SCS, the –CoA group of succinyl-CoA is replaced by a phosphate group, resulting in formation of high-energy acyl-phosphate. (ii) In the second step, high-energy phosphate group of acyl phosphate is transferred to His$^{246}$ of the α-subunit, resulting in phosphorylation of the residue, and succinate is released free. (iii) The third step involves transfer of this high-energy phosphate group to ADP/GDP, resulting in the synthesis of ATP/GTP, respectively, and the enzyme is released free (Fig. 8.7). This reaction is the only reaction of TCA cycle which is the source of ATP generation in the absence of functional electron transport chain, especially when $O_2$ is absent. Equilibrium constant of the reaction is nearly 1, so that the ratio of nucleotide present (ADP/ATP in plants or GDP/GTP in animals) will determine the direction of the reaction. Net result of the reaction is conservation of energy as ATP, without involvement of any electron transport.

## 8.6 Electrochemical Gradient-Driven Transport of Various Metabolites Across Inner Mitochondrial Membrane

Inner mitochondrial membrane is impermeable to most of the biomolecules carrying charge. However, the metabolic pathways which operate in mitochondria require exchange of various biomolecules across the membrane. Majority of the cell's requirements are met by ATP generated in mitochondria which needs to be exported, while ADP and $P_i$ need to be transported inside. Since the membranes are impermeable for ATP and ADP, their transport is facilitated by ADP/ATP transporters. ATP$^{4-}$ carries four negative charges, while ADP$^{3-}$ carries three negative charges. Movement of ATP$^{4-}$ is driven by membrane potential generated due to proton transport across the membrane. Net charge in the intermembrane space is positive due to accumulation of H$^+$, while toward the matrix side, it is negative. The transport is facilitated by **adenine nucleotide translocases**, which are antiporters. The same protein moves ATP$^{4-}$ out and ADP$^{3-}$ in. **Phosphate translocases** are symporters responsible for transporting one $P_i$ into the matrix, coupled with transport of one H$^+$. This transport is also favored by the transmembrane proton gradient (ΔpH). A complex of ATP synthase along with the adenine nucleotide and phosphate translocases is known as **ATP synthasome**, which has been isolated from mitochondria by gentle treatment with detergents. Pyruvate is also cotransported along with H$^+$ into the matrix of mitochondria in response to proton gradient.

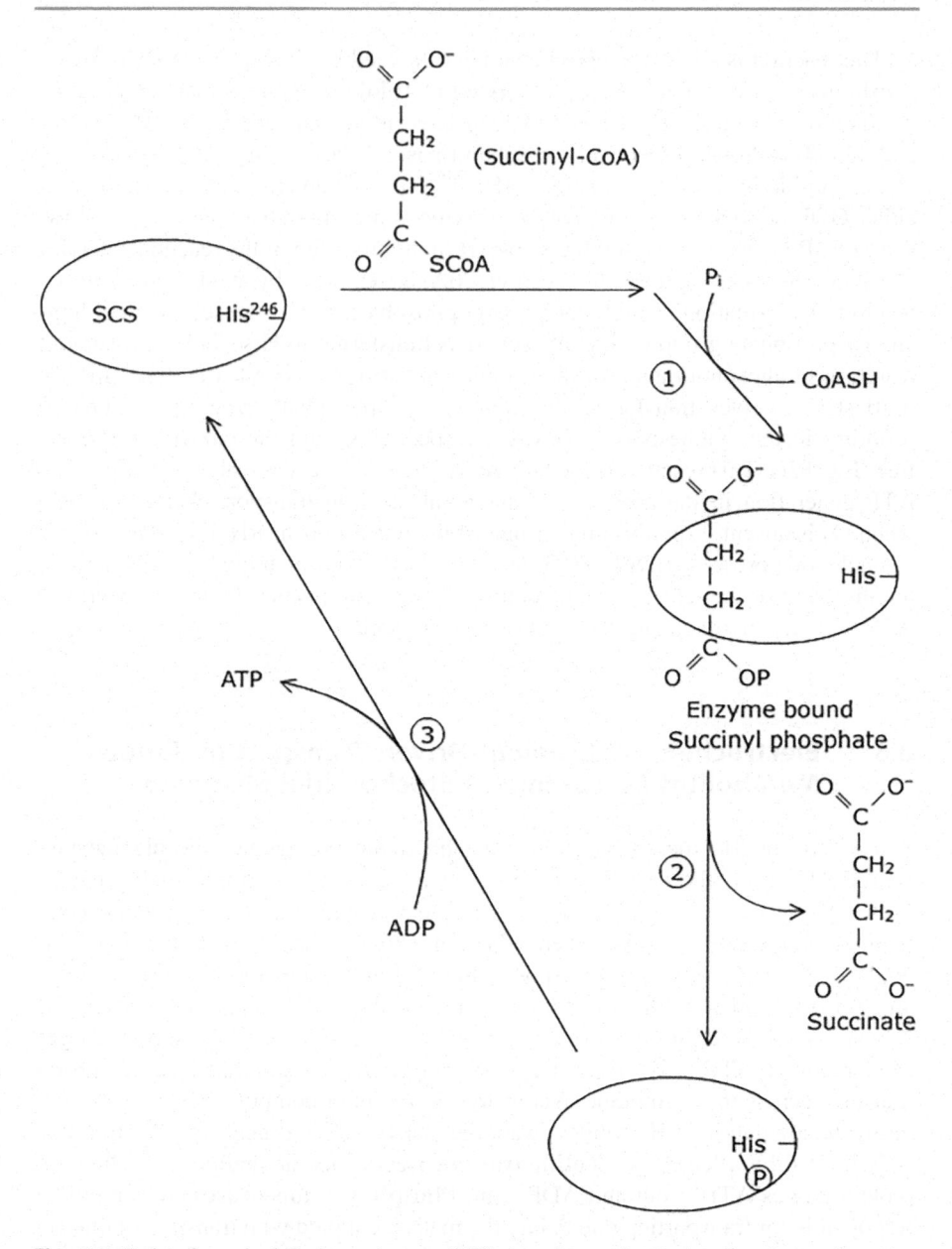

**Fig. 8.7** Role of succinyl-CoA synthetase (SCS) in substrate-level phosphorylation. Reaction occurs in three steps. Step 1, SCS catalyzes replacement of CoA of succinyl-CoA with $P_i$ resulting in enzyme-bound succinyl phosphate; Step 2, $P_i$-group of succinyl-P is transferred to His[246] of the enzyme protein resulting in formation of N-phosphohistidine and succinate is released free; Step 3, finally, phosphoryl group is transferred from phosphorylated His[246] to ADP resulting in synthesis of ATP

## 8.7 Oxidative Phosphorylation and Photophosphorylation: A Comparative Account

Unlike in animals, ATP synthesis in plants occurs both in chloroplasts and mitochondria during photophosphorylation and oxidative phosphorylation, respectively. The two phosphorylation processes differ from each other in relative contributions by the two components of PMF. In mitochondria, accumulation of $H^+$ in the intermembrane space will lower cytosolic pH adversely, affecting activity of many cytosolic enzymes since intermembrane space is in continuation with cytosolic space. Thus, membrane potential ($\Delta\psi$) plays more significant role in building up the electrochemical gradient instead of $\Delta pH$ in mitochondria. On the contrary, in chloroplasts, $H^+$ are accumulated in the enclosed space of lumen. As a result, the activity of the enzymes present in stroma will not be affected. Membrane potential of thylakoids is negated due to the movement of many anions such as $Cl^-$ to the lumen. Thus, in chloroplasts, it is the pH gradient which is the major component contributing to the electrochemical gradient. A comparison of oxidative phosphorylation, photophosphorylation, and substrate-level phosphorylation is given in Table 8.1.

**Summary**

- ATP is the universal energy currency of the cell in all living beings. Both its structure and mechanism of synthesis have been conserved during course of evolution. Synthesis of ATP occurs by oxidative phosphorylation, photophosphorylation, and substrate-level phosphorylation. Peter Mitchell had proposed chemiosmotic model for ATP synthesis which states that the energy released during electron transport is conserved as electrochemical gradient across inner membrane of mitochondria which contributes to proton motive force. There are two components of proton motive force, i.e., pH gradient ($\Delta pH$) and membrane potential ($\Delta\Psi$). Since the membrane is impermeable, the protons flow back in response to proton motive force through ATP synthase, the enzyme localized in the inner mitochondrial membrane. Evidence which support the hypothesis include experiments conducted using unsouplers of proton gradient, with reconstituted membrane vesicles and isolated thylakoids.
- ATP synthases are F-type ATPases, known as $F_o/F_1$ in mitochondria or $CF_o/CF_1$ in chloroplasts. These consist of a knob-like structure – $F_1$, which is projecting out, and $F_o$ embedded in the membrane. The two are joined together through a stalk. $F_1$ consists of eight subunits of five different types of polypeptides, i.e., $\alpha_3\beta_3\gamma\delta\varepsilon$. The headpiece consists of three $\alpha$- and three $\beta$-subunits which alternate with each other and are arranged like pieces of an orange. $\delta$-subunit holds $\alpha$- and $\beta$-subunits in place and is connected with two b-subunits. $F_1$, along with two b- and one a-subunits, forms "stator," while c-subunits of $F_o$ along with $\gamma$-subunit is the "rotor" of ATP synthase. $\gamma$-subunit is connected to c-ring through $\varepsilon$-subunit in leg and foot manner.
- Paul Boyer had proposed binding change mechanism for ATP synthesis. According to this model, catalytic site for ATP synthesis is present in $\beta$-subunits

**Table 8.1** Comparison of oxidative phosphorylation, photophosphorylation, and substrate-level phosphorylation

| Characteristics | Oxidative phosphorylation | Photophosphorylation | Substrate-level phosphorylation |
|---|---|---|---|
| Site of occurrence | Mitochondria | Chloroplasts | Cytosol and matrix of mitochondria |
| Electron transport, resulting in proton gradient across membrane | Proton gradient is created across inner mitochondrial membrane | Proton gradient is created across thylakoid membrane | Energy of high-energy bond of the substrate is coupled with synthesis of ATP. Electron transport is not required rather |
| Metabolic pathway associated with ATP generation | During respiration. Reduced cofactors NADH, FADH$_2$ are produced | During light reaction of photosynthesis | During two reactions of glycolysis and one reaction of TCA cycle |
| Electron transport chain | | | Not involved |
|   Location | Inner membrane of mitochondria | Thylakoid membrane | |
|   Components | Immobile components consist of Complex I, Complex II, Complex III, and Complex IV; mobile carriers are UQ and Cyt c | Immobile components are PSII, cytb$_6$f, and PSI; mobile carriers are PQ and PC | |
| Source of energy for ATP synthesis | Oxidation of NAD$^+$, FADH$_2$ | Light energy | Energy released from hydrolysis of high-energy bonds |
| Electron donor | NADH/NADPH/FADH$_2$ | H$_2$O | |
| Electron acceptor | O$_2$ | NADP$^+$ | Not involved |
| Main contributor to PMF | Voltage gradient across inner mitochondrial membrane and intermembrane space. pH in intermembrane space is only 0.2–0.5 units lower than that of matrix | pH gradient across thylakoid membrane. Lumen being acidic and stroma alkaline | Not involved |
| ATP synthase | F$_o$–F1 in inner mitochondrial membrane, | CF$_o$-CF$_1$ in thylakoid membrane | Not involved |
| | F$_1$ is located toward alkaline side (N) of the membrane, i.e., the matrix of mitochondria, | CF$_1$ is located toward alkaline side of the membrane, i.e., stroma of chloroplasts | |
| | F$_1$ consists of $\alpha_3\beta_3\gamma\delta\varepsilon$ and OSCP[a] | CF$_1$ consists of $\alpha_3\beta_3\gamma\delta\varepsilon$ | |
| | F$_o$ consists of a, b, and c, i.e., ab$_2$c$_n$ where n = 8–15. | CF$_0$ consists of a, b and b' (I and II), and c (III), 14 subunits of c (III) are present | |

(continued)

**Table 8.1** (continued)

| Characteristics | Oxidative phosphorylation | Photophosphorylation | Substrate-level phosphorylation |
|---|---|---|---|
| Ratio of number of ATP molecules formed to electrons transferred (P/O ratio) | High since number of c-subunits present in $F_o$ are less; high P/O ratio is predicted | Low since 14 c-subunits are present in the $F_o$ complex; low P/O ratio predicted | |
| Inhibition of proton translocation | Proton translocation is inhibited by oligomycin, since it binds with OSCP which limits $O_2$ uptake | Proton translocation is not inhibited by oligomycin | |

[a]OSCP Oligomycin-sensitivity conferring protein

which can exist in three possible conformations, i.e., loose, tight, and empty. These conformations of β-subunits differ in their affinity for ATP. In loose conformation, β-subunits are loosely bound with ADP and $P_i$, while in tight conformation, ATP synthesis occurs, and in empty conformation, ATP is released. There is a change in conformation of these subunits due to rotation of γ-subunit which occurs due to rotation of the "rotor" in response to proton movement from positive side of the membrane to the negative side. The energy of proton movement is coupled with the movement of rotor.
- Thermal movement of "rotor" occurs due to interaction of positively charged group of a-subunit with the c-subunits because of the periodic protonation and deprotonation of carboxyl group present in residues of c-subunit polypeptides. Protonation and deprotonation of the carboxyl group occur due to movement of protons through two half channels present in a-subunit, the inlet channel and an outlet channel.
- ATP synthesis also occurs by substrate-level phosphorylation during which ATP is synthesized directly due to transfer of high-energy phosphate bond of the substrate without involvement of any electron transport. The reactions are mediated by kinase enzyme.

## Multiple-Choice Questions

1. Proton motive force across the membrane is created due to:
   (a) pH gradient
   (b) Membrane potential
   (c) Both of them
   (d) None of them

2. Uncouplers are chemical agents which:
   (a) Interfere with electron transport across ETC.
   (b) Interfere proton movement across membrane.
   (c) Electron transport occurs but interfere with proton movement.
   (d) Electron transport occurs but proton gradient is dissipated.
3. Protonophores are the uncouplers which dissipate proton gradient due to the following:
   (a) Their acidic nature combines with protons on the acidic side and diffuses to the other side of the membrane.
   (b) They make the membrane permeable to protons.
   (c) They interfere with electron transport as a result proton gradient is not formed.
   (d) They allow the movement of other cations.
4. Energy released during proton movement in mitochondrial membrane in response to PMF is required by ATP synthase:
   (a) For binding of ADP and $P_i$ to β-subunits of $F_1$
   (b) For ATP synthesis
   (c) For release of ATP
   (d) For transport of electrons
5. The "stator" component of $F_1$–$F_o$ particles consists of:
   (a) c-Ring
   (b) γ- and ε-shaft of the particle
   (c) 3 α- and 3 β-subunits of $F_1$
   (d) 3 α- and 3 β-subunits, δ-subunit, b-subunit of $F_1$ attached to "a" of $F_o$
6. PMF in mitochondria is because of:
   (a) Gradient of pH and membrane potential
   (b) Mainly pH gradient
   (c) Mainly membrane potential gradient
   (d) None of the above
7. Jagendorf's experiment demonstrated that:
   (a) Light is required for ATP synthesis.
   (b) ATP synthesis can occur even if light is not present.
   (c) ATP synthesis can occur when thylakoids are suspended in buffer of pH 4.0.
   (d) ATP synthesis in thylakoid can be promoted by addition of KOH.
8. Chemiosmotic hypothesis for ATP synthesis was proposed by:
   (a) Paul Boyer
   (b) John Walker
   (c) Peter Mitchell
   (d) Charles Darwin

**Answers**

1. c   2. d   3. a   4. c   5. d   6. c   7. b   8. c

## Suggested Further Readings

Boyer PD (1997) The ATP synthase-A splendid molecular machine. Ann Rev Biochem 66:717–749
Junge W, Nalson N (2015) ATP synthase. Ann Rev Biochem 84:631–657
Prebble JN (2012) Contrasting approaches to a biological problem: Paul Boyer, Peter Mitchell and the mechanism of the ATP synthase, 1961–1985. J Hist Biol 46(4):699–737
https://www.nobleprize.org/nobel_prizes/chemistry/laureates/1997/walker-lecture.html

# Metabolism of Storage Carbohydrates

## 9

Manju A. Lal

Besides other roles carbohydrates are the major source of energy for all living beings. Almost 30% of the carbohydrates in plants are utilized for cell wall biosynthesis by each cell. Carbon skeleton also needs to be diverted for synthesis of defense chemicals (secondary metabolites) in order to deter herbivory. This requires a continuous flow of carbohydrates from source to sink. In autotrophic plants $CO_2$ is fixed in green parts resulting in production of simple sugars (monosaccharides). During night, surplus photosynthates stored as **transitory starch** in chloroplasts are transported to other parts of the plant after being converted to soluble form (Fig. 9.1). Sugars are transported in the least reactive soluble form, primarily as sucrose from source (site of its synthesis) to sink (site of its utilization) though carbohydrates other than sucrose are also translocated, which include raffinose, verbascose, and stachyose. Starch is the primary storage form of carbohydrates. However, there are instances of sucrose being stored also, e.g., in sugarcane and beetroot. In some plants fructans are the storage form of carbohydrates. In members of grass family, starch is stored in the grains. Though carbohydrate metabolism is similar in other organisms, there are certain unique features which make plants distinct. These include their autotrophic nature and the presence of specialized class of organelles, i.e., plastids. Besides plastids, cytosol and vacuoles are also involved in the metabolism of carbohydrates (Fig. 9.2). Because of their inability to move to a safer place under unfavorable environmental conditions, plants have a flexible metabolism and many a times exhibit alternate metabolic pathways. In this chapter, the main focus will be the metabolism of storage form of carbohydrates in plants.

## 9.1 Metabolite Pool and Exchange of Metabolites

Many metabolic pathways are interconnected and share intermediates. These intermediates are maintained at equilibrium with each other by reversible enzymatic reactions. These are considered as "pool" from which the compounds can be

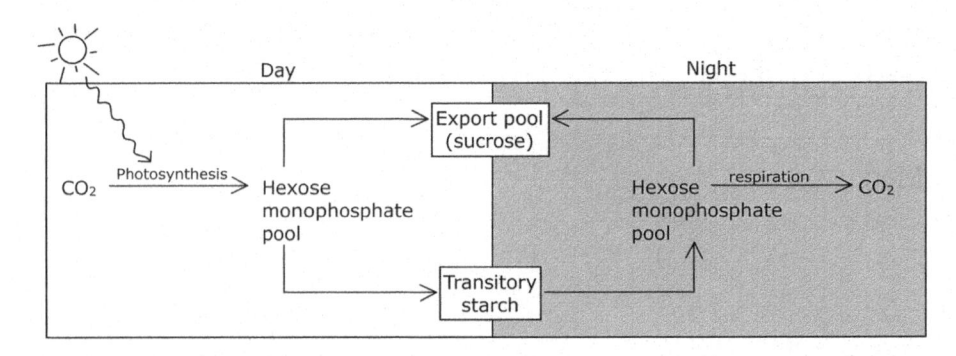

**Fig. 9.1** During daytime, occurrence of photosynthesis in chloroplasts adds to hexose monophosphate pool which is used for synthesis of transitory starch. During nighttime, carbon skeleton stored as starch is mobilized and translocated out of chloroplasts to other parts of plants, either to be used as the substrate for respiration or for storage in non-green plastids of storage organs or is translocated to be used as the source for respiration. Intercellular transport occurs in the form of sucrose

withdrawn as and when required and added when produced, maintaining almost constant availability of these compounds. Direction of interconvertibility of the intermediates in a "pool" is determined by the removal or addition of particular compound because of reversibility of reactions in order to maintain equilibrium. "Metabolite pool" occurs in the subcellular compartments, such as plastids and cytosol. Interchange of the metabolites occurs through cell membranes which separate the subcellular compartments. There are **hexose monophosphate pool**, **triose phosphate pool**, and a metabolite pool consisting of intermediates of **pentose phosphate pathway** (Figs. 9.3 and 9.4). Hexose monophosphate pool is composed of hexose monophosphates, glucose 1-phosphate, glucose 6-phosphate, and fructose 6-phosphate. All of these forms are interconvertible, and the reversible reactions are catalyzed by the enzymes phosphoglucomutase and glucose-6-phosphate isomerase. Triose phosphate pool consists of interconvertible forms of trioses, glyceraldehyde 3-phosphate, and dihydroxyacetone phosphate which are catalyzed by triose phosphate isomerase. Carbon compounds are added in the pool which are either produced during photosynthesis or other metabolic pathways (Fig. 9.5). In germinating oil seeds, there is an additional pathway, i.e., gluconeogenesis, which is also responsible for addition to the metabolite pools. Some metabolites of a pool are transported across the membranes which separates the subcellular compartments of a cell. An exchange of triose phosphate with inorganic phosphate between plastid and cytosol is mediated by an antiporter triose phosphate translocator (TPT). TPT is located in the inner plastid membrane. Triose phosphate is transported out of the plastid in exchange of inorganic phosphate obtained from cytosol. Exchange of triose phosphate will stop in case of non-availability of inorganic phosphate. In chloroplasts, triose phosphate is produced during daytime as a result of $CO_2$ fixation by Calvin cycle, while during night hydrolysis of transient starch is the source of triose phosphate. Non-phosphorylated hexose sugars such as glucose and maltose which

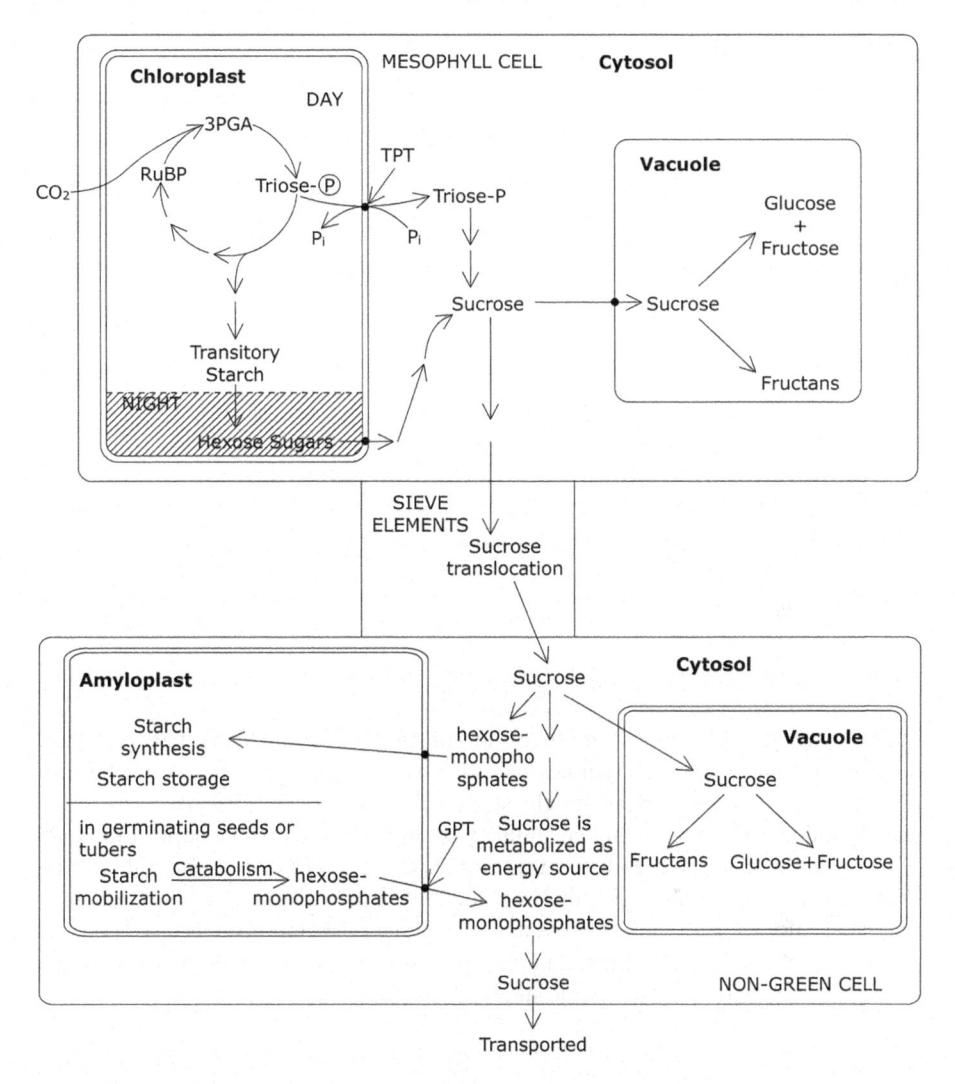

## 9.1 Metabolite Pool and Exchange of Metabolites

**Fig. 9.2** Carbohydrate metabolism: An overview of the major pathways and organelles involved. *TPT* triose phosphate transporter, *GPT* glucose phosphate transporter

are derived from starch hydrolysis can also be transported across inner envelop of the green plastids; however, transporters for phosphorylated hexose sugars are not present in the inner envelop of green plastids. However, in amyloplasts, glucose 6-phosphate/phosphate translocator (GPT) is present and is localized in the inner envelop of the plastid. GPT is responsible for transport of glucose 6-phosphate in exchange of inorganic phosphate. Besides glucose 6-phosphate, it is also responsible for exchange of triose phosphate and xylulose 5-phosphate with cytosolic inorganic phosphate. Thus, there is a direct correlation in between the metabolite pools of

**A.**

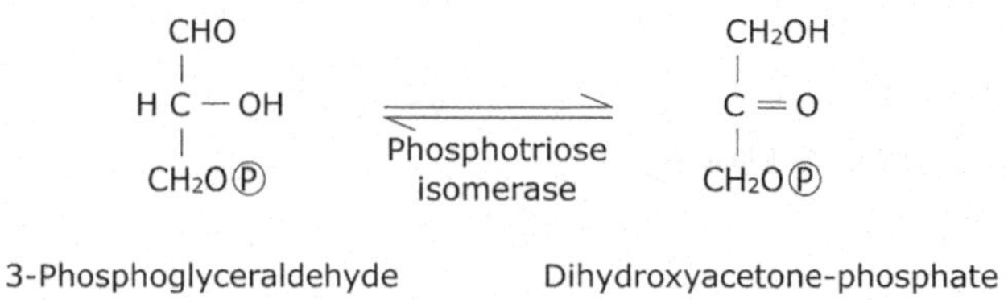

CHO                                  CH$_2$OH
|                                     |
H C — OH    ⇌    C = O
|        Phosphotriose   |
CH$_2$O ⓟ    isomerase      CH$_2$O ⓟ

3-Phosphoglyceraldehyde      Dihydroxyacetone-phosphate

**B.**

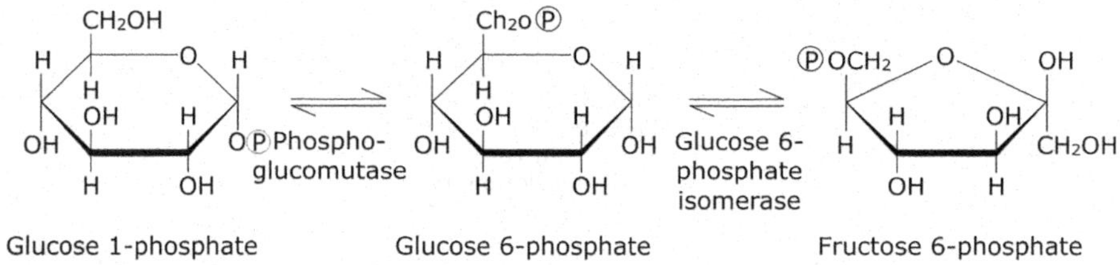

Glucose 1-phosphate ⇌ (Phosphoglucomutase) Glucose 6-phosphate ⇌ (Glucose 6-phosphate isomerase) Fructose 6-phosphate

**Fig. 9.3** (a) Triose phosphate pool. (b) Hexose monophosphate pool

cytosol and plastids, which involves lot of exchange of sugars between these compartments. Since carbohydrates in plants are primarily transported in the form of sucrose and are stored as starch, study involving metabolism of sucrose and starch, both their synthesis and catabolism, is significant and is the subject for study in this chapter. Since some plants store carbohydrates as fructans, a brief discussion about fructan metabolism is also included. A brief discussion about enzyme-regulated allocation of carbon for sucrose and starch metabolism is also included. Activity of enzymes are regulated in response to internal factors of the plant (stages of growth), as well as by the environmental factors.

## 9.2 Sucrose Metabolism

### 9.2.1 Sucrose Biosynthesis

Sucrose is the major form of carbohydrates which is translocated from source to sink in sieve elements of plants. It is the most ubiquitous and abundant disaccharide (α-D-glucopyranosyl-β-D-fructofuranoside) in plant tissue which is synthesized from two monosaccharides (α-D-glucopyranose and β-D-fructofuranose) by joining C-1 of α-D-glucose to C-2 of β-D-fructose by 1α→2β glycosidic bond. It is a nonreducing sugar since glycosidic bond joins the carbonyl carbons of glucose and fructose and protects potentially reactive groups from oxidation and from non-specific enzyme attack making it structurally stable. Since it is very stable, highly soluble, and

## 9.2 Sucrose Metabolism

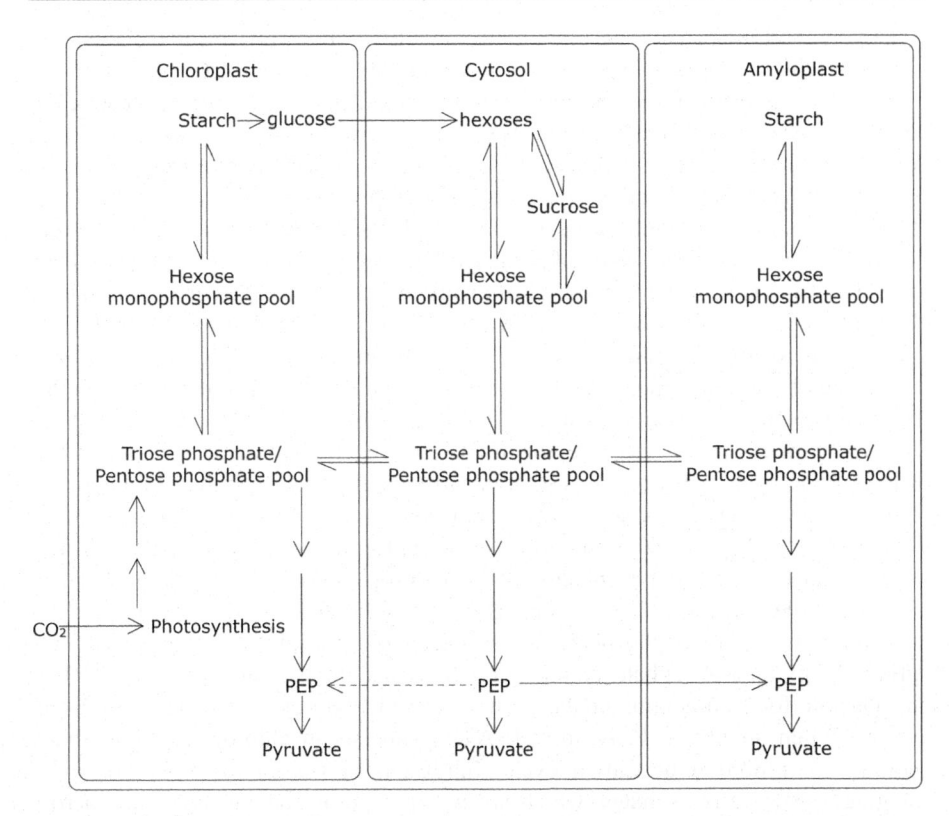

**Fig. 9.4** Metabolite pools involved in carbohydrate metabolism. Duplicate pathways are present in cytosol, chloroplast, and non-green plastid

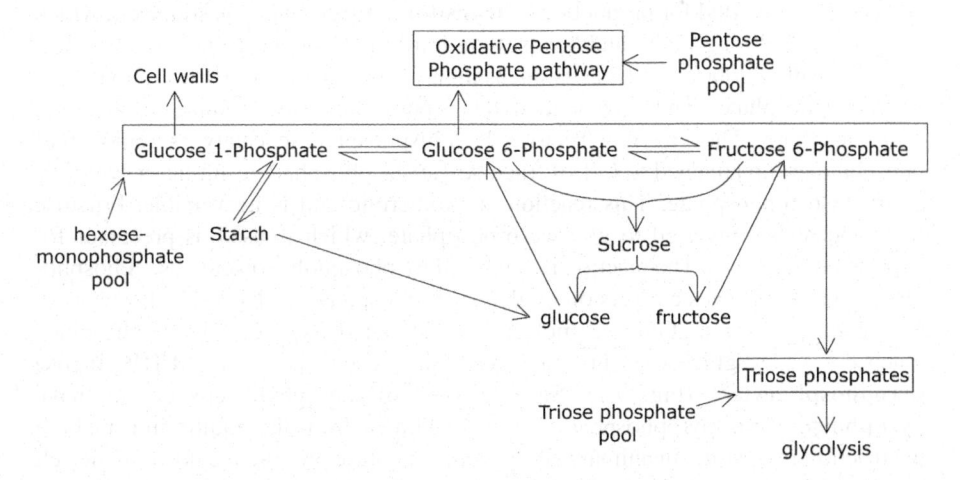

**Fig. 9.5** Hexose phosphate, triose phosphate, and pentose phosphate pools and exchange of intermediates in between these pools (both in cytosol and in plastid)

relatively inert molecule, it becomes quite suitable for transport. In some plants, e.g., beetroot (*Beta vulgaris*), sugarcane (*Saccharum* sp.), and carrot (*Daucus carota*), sucrose can also be stored. Besides osmotic effect, it does not affect most of the biological processes. Though sucrose is the major form for translocation, other forms of carbohydrates such as raffinose, verbascose, stachyose (in members of family Cucurbitaceae) and sorbitol (in many plants of family Rosaceae) are also translocated. Sucrose synthesis occurs only in plants and bacteria. In plants, sucrose is synthesized mainly in cytosol of photosynthetic tissues or in germinating seeds. In mesophyll cells, source for triose phosphates, which are the precursor for sucrose biosynthesis, is the Calvin cycle which occurs in chloroplasts. Almost 1/6 of triose phosphates synthesized during the Calvin cycle are transported out to cytosol in exchange of inorganic phosphate ($P_i$), to be used for sucrose biosynthesis, while 5/6 of the triose phosphates are consumed for regeneration of ribulose 1, 5-bisphosphate (RuBP). Since RuBP is the acceptor of $CO_2$, its regeneration is mandatory; otherwise, Calvin cycle will stop. In case of occurrence of photorespiration (in C3 plants in light), 1/8 of the triose phosphate are transported out of chloroplast to be used for sucrose biosynthesis. TPT is an antiporter present in the inner envelope of plastids, which transports triose phosphates out of chloroplasts to cytosol in the form of dihydroxyacetone phosphate in exchange of cytosolic inorganic phosphate. In cytosol, dihydroxyacetone phosphate is predominant from of triose phosphate than glyceraldehyde 3-phosphate and is present almost in the ratio of 22:1. TPT is responsible for maintaining a balance in between Calvin cycle and sucrose biosynthesis. When supply of $P_i$ is limited, ATP synthesis in chloroplasts is inhibited leading to inhibition of Calvin cycle. Inhibition of Calvin cycle will result in shortage of triose phosphates, which are not transported out. As a result, sucrose biosynthesis in cytosol is inhibited. During daytime, when Calvin cycle results in production of triose phosphates in excess than can be utilized for sucrose biosynthesis, these are diverted for production of **transitory starch** in the plastid itself, which is hydrolyzed during night, and hexoses are transported out (Fig. 9.6).

In cytosol, aldolase catalyzes condensation of the triose phosphates, dihydroxyacetone phosphate, and glyceraldehyde 3-phosphate, to produce fructose 1, 6-bisphosphate. The enzyme fructose 1, 6-bisphosphatase catalyzes removal of phosphate group attached to C-1 of fructose 1, 6-bisphosphate resulting in synthesis of fructose 6-phosphate. This reaction is exothermic and is irreversible. Fructose 6-phosphate is converted to glucose 6-phosphate, which is used as precursor for sucrose biosynthesis. The reaction is catalyzed by phosphohexoisomerase. Phosphoglucomutase catalyzes conversion of glucose 6-phosphate to glucose 1-phosphate in a reversible reaction. Glucose molecule is activated as nucleoside diphosphate glucose (UDP-glucose) in a reaction catalyzed by **UDP-glucose pyrophosphorylase** (Fig. 9.7). Since cytosol of mesophyll cells do not have **pyrophosphatase,** pyrophosphate is not withdrawn from the equilibrium making the reaction reversible. In animal cells, pyrophosphatase present in cytosol of the cell rapidly hydrolyzes $PP_i$ produced during the reaction. In plant cells, metabolism of cytosolic $PP_i$ is coupled with various other metabolic processes. This includes activity of **$PP_i$-dependent phosphofructokinase** and tonoplast-localized

## 9.2 Sucrose Metabolism

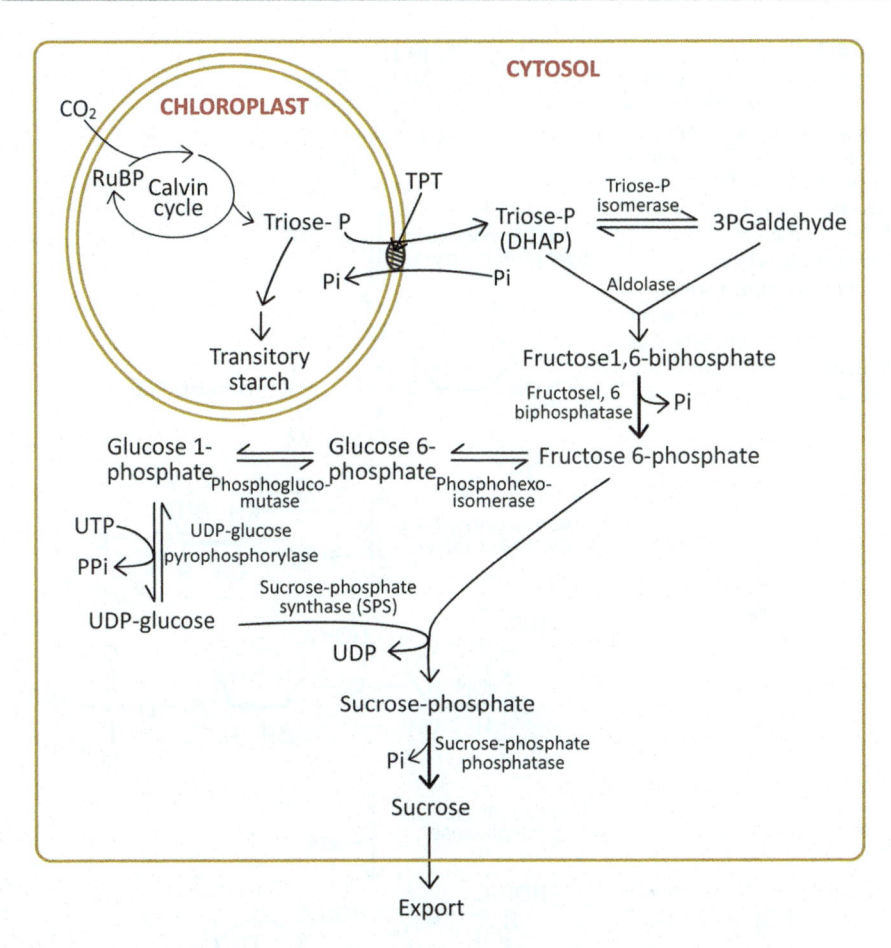

**Fig. 9.6** Synthesis of sucrose in a mesophyll cell. Glucose is activated by UDP to form UDP-glucose. Activated glucose is transferred to fructose 6-phosphate resulting in synthesis of sucrose phosphate. The two reactions catalyzed by fructose 1,6-biphosphatase and sucrose-phosphate phosphatase are irreversible. These two reactions make sucrose synthesis irreversible

pyrophosphatase, which utilize energy of the hydrolysis of pyrophosphate for driving protons across tonoplast resulting in acidification of vacuole. The electrochemical gradient so generated is used to power transmembrane transport. Activated glucose is transferred from UDP-glucose to fructose 6-phosphate resulting in synthesis of sucrose 6-phosphate (phosphate is linked to 6-C of fructose) in a reversible reaction catalyzed by the enzyme sucrose-phosphate synthase. Sucrose-phosphate phosphatase catalyzes hydrolysis of sucrose-phosphate resulting in formation of sucrose and release in $P_i$. This is an exothermic reaction coupled with release of energy ($\Delta G^{o'} = -18.4$ kJ.mol$^{-1}$). Release of $P_i$ is necessary for sucrose biosynthesis because it makes the reactions irreversible. Additionally, recycling of $P_i$ to chloroplast is essential. The enzymes sucrose-phosphate synthase and sucrose-phosphate

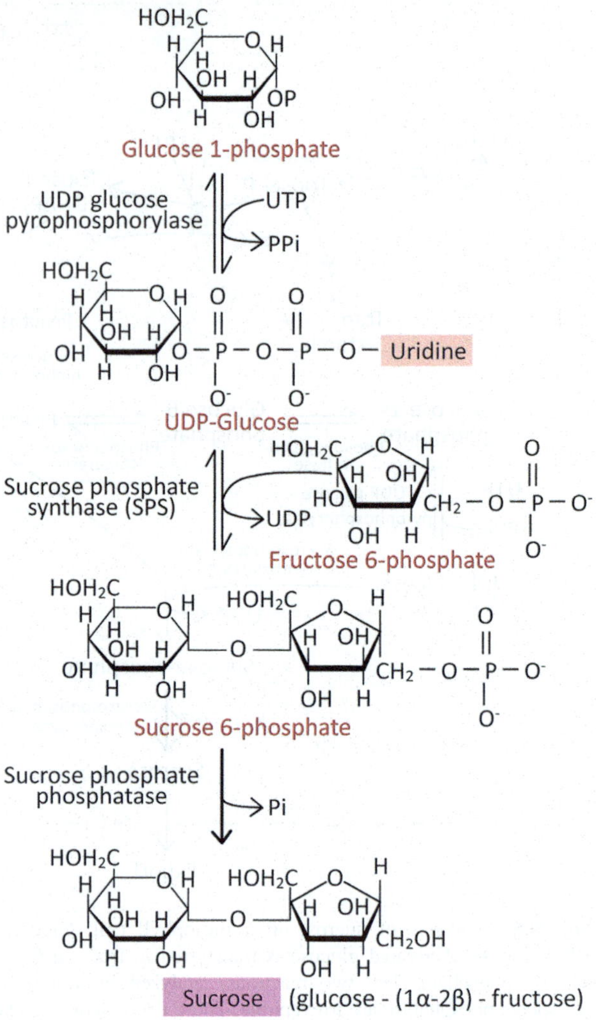

**Fig. 9.7** Synthesis of sucrose: Activation of glucose occurs by transferring glucose molecules from glucose 1-phosphate to UTP resulting in synthesis of UDP-glucose. Reaction is catalyzed by UDP-glucose pyrophosphorylase. The reaction is coupled with release of PP$_i$ (pyrophosphate) and becomes irreversible due to hydrolysis of sucrose-phosphate to sucrose

phosphatase are associated into macromolecular complexes, which facilitate direct transfer of sucrose 6$^F$-phosphate to sucrose phosphatase without mixing with other metabolites. The two irreversible reactions of the pathway, i.e., hydrolysis of fructose 1,6-bisphosphate to fructose 6-phosphate and sucrose phosphate to sucrose, catalyzed by fructose 1,6-bisphosphatase and sucrose-phosphate phosphatase, respectively, make sucrose biosynthesis irreversible. Both reactions are coupled with simultaneous release of P$_i$, which is cycled back to chloroplast, thus are responsible for continuation of Calvin cycle.

## 9.2 Sucrose Metabolism

$$\text{Glucose 1 - phosphate} + \text{UTP} \Leftrightarrow \text{UDP - glucose} + \text{PP}_i \qquad (9.1)$$

$$\text{UDP - glucose} + \text{fructose 6 - phosphate} \\ \Leftrightarrow \text{sucrose 6 - phosphate} + \text{UDP} \qquad ((9.2)$$

$$\text{Sucrose 6 - phosphate} \Rightarrow \text{sucrose} + P_i \qquad (9.3)$$

Reaction (9.2), catalyzed by sucrose-phosphate synthase (SPS), is reversible, while reaction (9.3) catalyzed by sucrose-phosphate phosphatase is irreversible. Direction of the pathway is determined by the irreversible reactions. Activity of the enzymes, catalyzing the reactions, is regulated depending upon the need of the plant. Overall $\Delta G^{o\prime}$ of the reaction for sucrose synthesis is $-25$ kJ/mole by this route, which primarily is associated with large negative energy change associated with dephosphorylation of sucrose 6-phosphate ($\Delta G^{o\prime} = -16.5$ kJ.mol$^{-1}$).

### 9.2.1.1 Regulation of Sucrose Synthesis

In chloroplasts, photosynthesis occurs actively during daytime leading to accumulation of triose phosphates, while inorganic phosphate ($P_i$) becomes short in supply because of active ATP synthesis. Short supply of $P_i$ in chloroplast leads to inhibition of photosynthesis. Triose phosphates in chloroplasts are exchanged with cytosolic $P_i$ through TPT, an antiporter, which is responsible for this exchange on one-to-one basis keeping a balance in triose phosphates and $P_i$ in chloroplasts (Fig. 9.8). Sucrose synthesis occurs in cytosol and is regulated through interaction of hexose monophosphate pool status, as well as the activity of sucrose-phosphate synthase. There is daily variation in sucrose concentrations in the plants, which primarily is due to changes in the activity of sucrose-phosphate synthase. Both allosteric modulation and covalent modulation of the enzyme protein due to phosphorylation seem to contribute to regulation of SPS activity. SPS is allosterically modulated by glucose 6-phosphate and $P_i$. Glucose 6-phosphate activates the enzyme, while its activity is inhibited by $P_i$. An increase in cytosolic hexose monophosphates is coupled with reduction in $P_i$, thus activating SPS. Activity of SPS is also regulated by covalent modulation of the enzyme protein. Activity of the enzyme is inhibited on phosphorylation of Ser$^{158}$ of the enzyme protein, which is catalyzed by SPS kinase, while dephosphorylation of Ser$^{158}$, catalyzed by SPS phosphatase, activates SPS. Activity of SPS kinase is inhibited by glucose 6-phosphate, and activity of SPS phosphatase is inhibited by high concentrations of $P_i$. Thus glucose 6-phosphate promotes the activity of SPS resulting in increased sucrose synthesis, while reverse is true $P_i$ (Fig. 9.9). SPS activity is also inhibited by phosphorylation at another site, i.e., Ser$^{229}$ of the enzyme protein, because of interaction of the phosphorylated enzyme with a 14-3-3 regulatory protein. On the contrary, phosphorylation at Ser$^{424}$ results in activation of the enzyme. Thus, there are multiple sites in enzyme protein which when phosphorylated are responsible for regulation of the enzyme activity differently. Various isoforms of the enzyme SPS and sucrose-phosphate phosphatase have been reported from different sources.

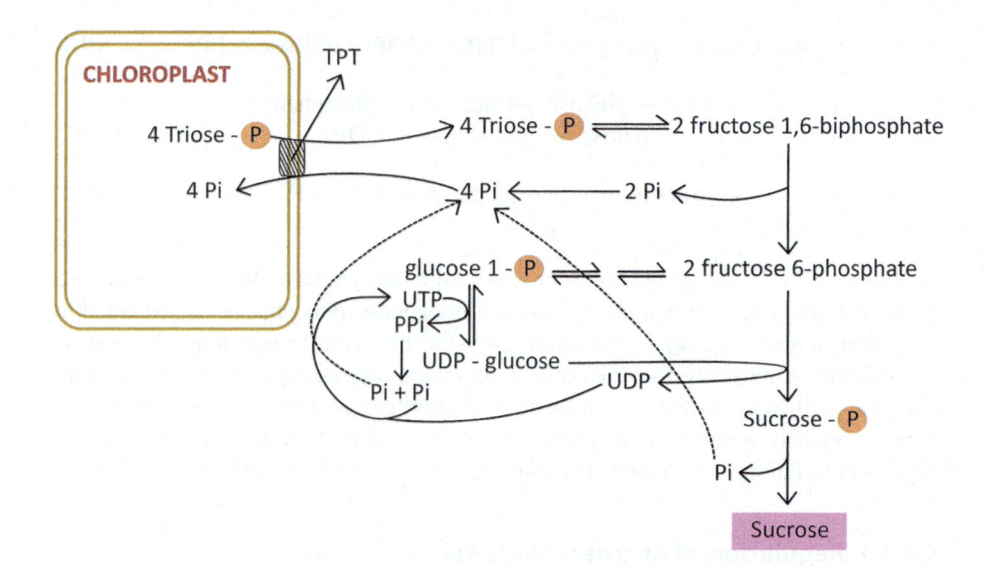

**Fig. 9.8** Figure showing balance of triose-phosphate transported out of chloroplasts in exchange of $P_i$. TPT (triose-P/phosphate transporter) is an antiporter localized in inner envelope of chloroplast which is responsible for exporting Calvin cycle intermediate triose-P in exchange of $P_i$ from the cytosol on one-to-one basis. Three of $P_i$ are released in the reactions of sucrose biosynthesis, while fourth $P_i$ is released from hydrolysis of pyrophosphate by the action of pyrophosphatase

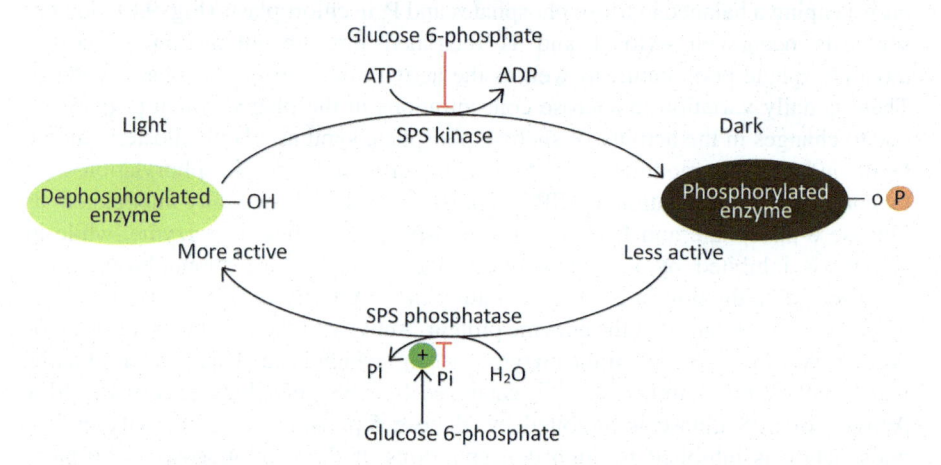

**Fig. 9.9** Regulation of activity of sucrose-phosphate synthase. The enzyme activity is regulated both by allosteric and covalent modulations. Phosphorylation at Ser[158] downregulates the activity of the enzyme, while its dephosphorylation activates the enzyme. Phosphorylation and dephosphorylation of enzyme protein are catalyzed by SPS kinase and SPS phosphatase, respectively. Glucose 6-phosphate inhibits activity of SPS kinase and activates SPS phosphatase resulting in inhibition of phosphorylation and promoting dephosphorylation of the enzyme protein

## 9.2 Sucrose Metabolism

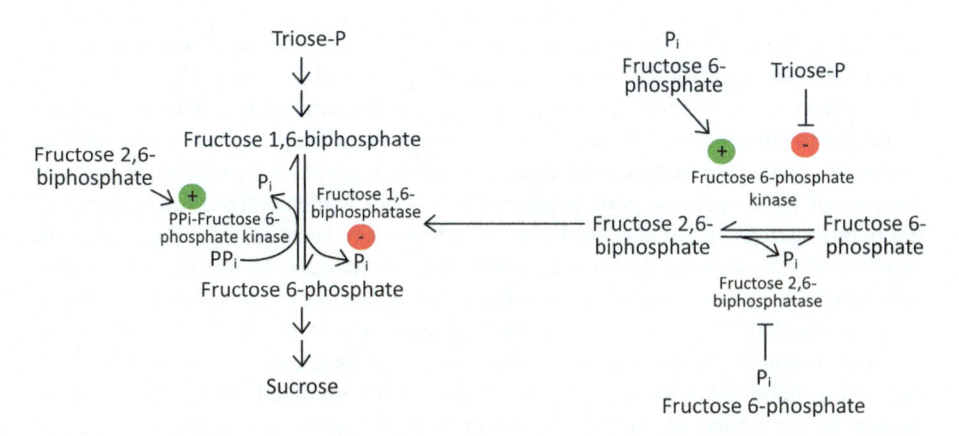

**Fig. 9.10** Fructose 2,6-biphosphate is the regulator for one of the regulatory step-in sucrose biosyntheses. It can inhibit sucrose biosynthesis at this regulatory step by either inhibiting fructose 1,6-biphosphatase or activating PP$_i$-fructose 6-phosphate kinase. Fructose 2,6-biphosphate is regulated by triose phosphate which inhibits its synthesis (i.e., promoting sucrose synthesis), while P$_i$ and fructose 6-phosphate inhibit its breakdown. Green circles with (+) sign represent activation, while red circles with (−) sign represent inhibition

**Fructose 2, 6-bisphosphate** (F2,6-BP) plays an important regulatory role in sucrose metabolism by regulating activity of fructose 1, 6-bisphosphatase. F2,6-BP is different from fructose 1,6-bisphosphate only in positioning of one phosphate group. Its role is found to be significant in regulation of glycolysis and gluconeogenesis in animals, fungi, and plants. It inhibits sucrose biosynthesis by inhibiting fructose 1, 6-bisphosphatase. On the contrary, pyrophosphate-dependent fructose 6-phosphate kinase (the enzyme which is responsible for conversion of fructose 6-phosphate to F1, 6-BP in a plant cell) is activated by F2,6-BP. Its cytosolic concentration in a plant cell is extremely low. Its synthesis is catalyzed by ATP-dependent fructose 6-phosphate 2 kinase (F6P2K), whose activity is inhibited by triose phosphate, while fructose 6-phosphate and P$_i$ act as activators of the enzyme. As a result, triose phosphates accumulate lower F2,6-BP concentration, while P$_i$ and fructose 6-phosphate are responsible for synthesis of the molecule, in turn increasing and decreasing sucrose biosynthesis, respectively. Sucrose biosynthesis is regulated both by **feed-forward** mechanism, i.e., by triose phosphate which promotes sucrose biosynthesis by decreasing fructose 2,6-bisphosphate, and **feedback** mechanism, i.e., which is inhibited by fructose 6-bisphosphate and P$_i$ (Fig. 9.10).

### 9.2.2 Sucrose Catabolism

Sucrose is the major form of carbohydrate that is transported in plants from source (e.g., photosynthetic tissues) to sinks (non-photosynthetic tissues), where it is metabolized to give energy or it is stored to be used later on as and when required. In a cell, sucrose is generally stored in vacuoles. Larger size of vacuoles is advantageous for storage. In response to increase in metabolism, such as when plants are subjected to environmental stresses, rate of sucrose consumption is much faster than rate of its entry in the metabolism. This increase in requirement of sucrose is buffered because of large vacuolar pool, since cytosolic pool is metabolized faster. Some of the plants such as sugarcane and beetroot, which store sucrose in the stem or roots, respectively, have been domesticated for obvious reasons. However, their wilder relatives have lesser amount of stored sucrose, since these are subjected to more unfavorable conditions. As a result, more carbon may be diverted for their growth and development. Sucrose may be transported in apoplastic or symplastic way, depending upon the type of the plant or the tissues involved. Sugar translocation occurs in sieve-tube elements of phloem in response to the gradient, which is maintained due to loading at the source and unloading at the sink. In developing grains or in germinating seeds, apoplastic mode of sugar transport occurs, since there are no plasmodesmata connections in between the embryo and the maternal tissues (conduits of translocation), and sucrose needs to be released in the apoplast before being actively transported across cell membrane (Fig. 9.11). Since sucrose is the major form of translocation in plants, its catabolism in sinks is of significant importance. Enzymes involved in sugar catabolism are invertase and sucrose synthase (SS). Unlike invertase, sucrose synthase catalyzes reversible reaction and thus is responsible for synthesis of sucrose also. However, in comparison to sucrose-phosphate synthase (SPS), its role has not found to be significant in the tissues involved in sucrose synthesis. Instead, SS is more significant during catabolism of sucrose, either when it is used as an energy source or is metabolized for starch biosynthesis. Invertase is hydrolytic enzyme, which hydrolyzes sucrose to glucose and fructose. Several isoforms of invertase occur in a plant cell. Apoplastic and vacuolar invertases have acidic pH optima of 5.0, while cytosolic invertase has neutral or alkaline pH optima of 7.5. Vacuolar invertases are soluble acid invertases, while cell wall invertases are insoluble acid invertases. Each of the isoforms has a distinct role in plant metabolism, and corresponding genes are temporarily or spatially expressed during plant development. High level of vacuolar invertases has been found in tissues accumulating hexoses, such as developing fruits, or in rapidly growing tissues, such as elongation zones of roots and hypocotyls. Vacuolar invertases are involved in cell expansion, sugar storage, and in regulation of cold-induced sweetening. At the time of requirement, sucrose is hydrolyzed by vacuolar invertase into glucose and fructose, which are then transported out of vacuole. Apoplastic invertase has significant role when sucrose transport is apoplastic. Sucrose is released in the apoplast, where it is hydrolyzed to glucose and fructose by apoplastic invertase before being transported across cell membrane or it is transported as such through the transporters present in the membrane against osmotic concentration (Fig. 9.12). Once in cytosol, sucrose is either consumed as an energy source after being hydrolyzed by cytosolic invertase or is transported across tonoplast to be stored in the vacuole. Cell wall invertases play a key role in sucrose partitioning, plant development, and cell differentiation. Cytosolic invertases are

## 9.2 Sucrose Metabolism

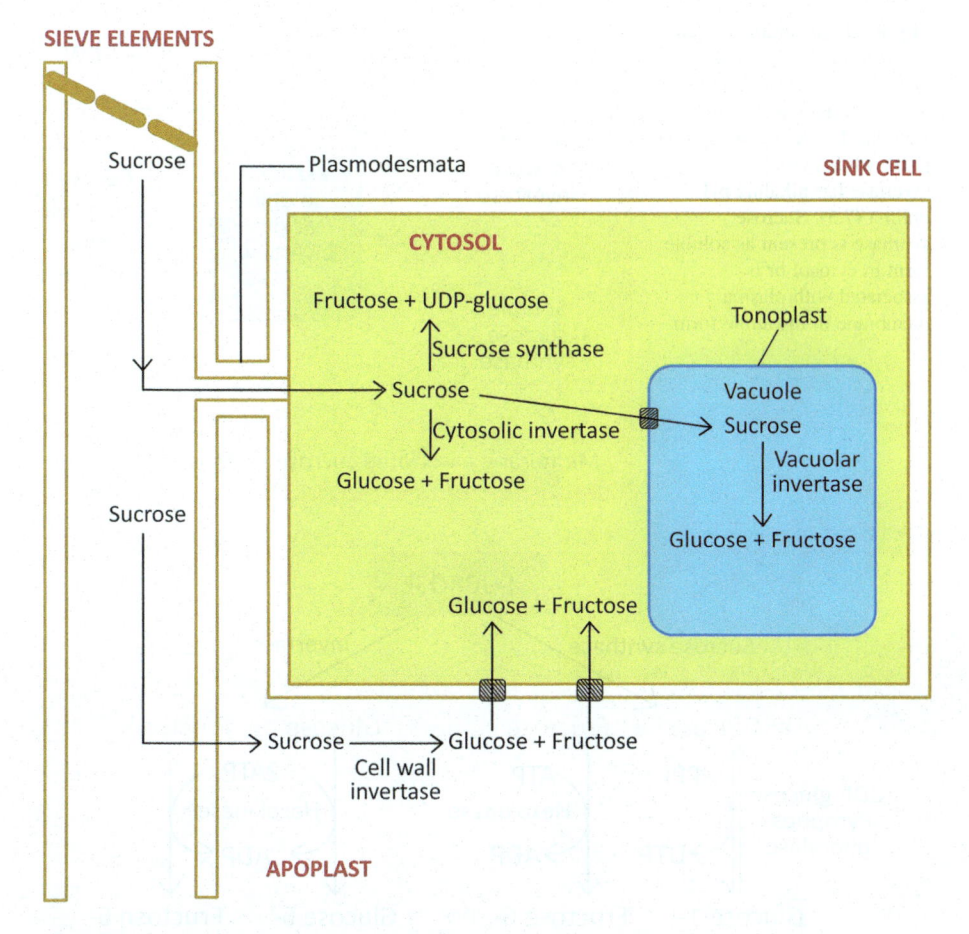

**Fig. 9.11** Diagram showing sucrose catabolism by either invertase or sucrose synthase. Sucrose is translocated in sieve elements. Sucrose may reach sink cells either symplastically (through plasmodesmata) or may be released first in the apoplast from where it is hydrolyzed by apoplastic invertase to glucose and fructose, which are actively transported through hexose transporters localized in plasma membrane. Once inside cytosol, sucrose is metabolized either by sucrose synthase or cytosolic invertase, or it is transported to vacuoles where it is hydrolyzed by vacuolar invertase. Products of sucrose hydrolysis are metabolized or are used for synthesis of polymeric molecules such as starch or cellulose

important for plant growth, especially roots. There are proteinaceous inhibitors of invertase, which interact with acid invertases and repress their activity; however, much is not known about them.

Another reaction involved in sucrose breakdown includes reaction catalyzed by sucrose synthase (SS). Hydrolysis of sucrose by invertase yields glucose and fructose, which must subsequently be phosphorylated at the expense of ATP, in order to be utilized further. Contrary to invertase, SS utilizes UDP for catalyzing sucrose breakdown, releasing UDP-glucose and fructose. The reaction is different

**Fig. 9.12** Several isoforms of invertase and sucrose synthase: Invertase found in apoplast (cell wall) and vacuoles has acidic pH optima (5.0), while cytosolic invertase has alkaline pH optima (7.5). Sucrose synthase is present as soluble form in cytosol or is associated with plasma membrane in insoluble form

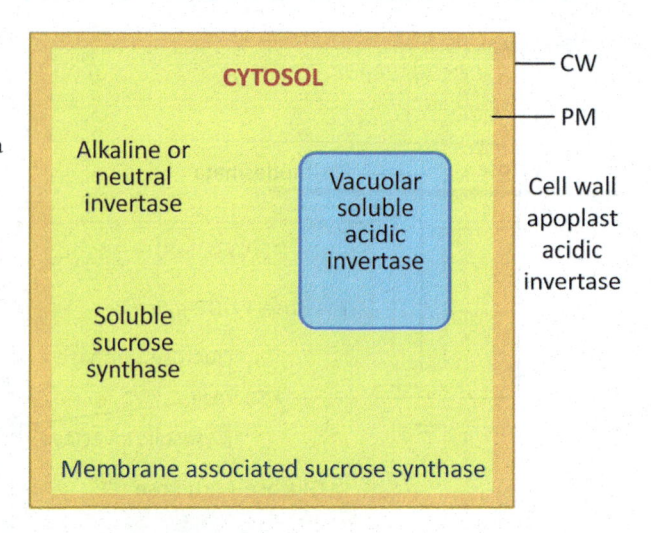

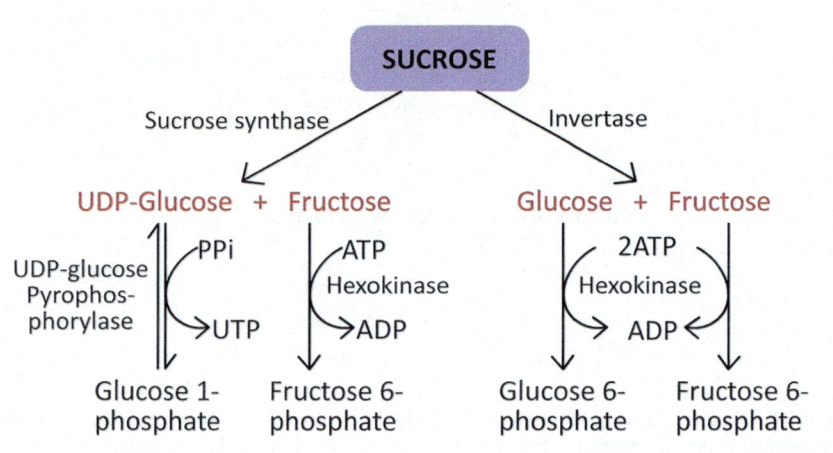

**Fig. 9.13** Sucrose is catabolized by either invertase or sucrose synthase (SS). Sucrose synthase along with UDP-glucose pyrophosphorylase catalyzes synthesis of glucose 1-phosphate by an ATP-independent pathway. The products of hydrolysis catalyzed by invertase require ATP for hexokinase-dependent generation of phosphorylated hexose phosphates

from the one catalyzed by invertase, in the energy status of the products. UDP-glucose is converted to glucose 1-phosphate by the enzyme UDP-glucose pyrophosphorylase. During the reaction enzyme use $PP_i$ and UTP is synthesized. Product of the reaction glucose is in phosphorylated form (glucose 1-phosphate), in the reaction catalyzed by SS. Thus, SS-catalyzed hydrolysis of sucrose is different from the one catalyzed by invertase (Fig. 9.13). Both SS and UDP-glucose pyrophosphorylase are responsible for generating phosphorylated hexoses in an ATP-independent pathway. These are thought to be important in tissues such as in amyloplasts of developing grains, or in potato tuber, which are involved in starch biosynthesis for storage purposes from the translocated sucrose. Invertase activity was found to be almost nil in potato tuber, while activity of SS was significantly higher. Poor development of potato tubers is observed in the mutants for SS enzyme. Similarly, mutants of this enzyme were associated with poor grain development. SS has also been found to have significant role in synthesis of cellulose and callose. Models have been proposed which suggest a direct correlation of SS with cellulose synthase complex. A direct channeling of UDP-glucose, produced by SS, to the active site of cellulose synthase complex, has been suggested. Phosphorylation status of SS determines its cell wall-bound nature or its soluble cytosolic form. Phosphorylation of the enzyme protein makes it soluble, while removal of the phosphate groups exposes the hydrophobic amino acids which make it insoluble. As a result, it is bound to the membrane.

## 9.3 Starch Metabolism

Starch is the dominant form of carbohydrates which is used by almost half of world's population as the energy source. There are widespread uses of starch for commercial purposes other than to be used as a food source. The commercial use of starch includes production of biodegradable plastics and packaging material, in building material as adhesives, in paper and textiles as coating material, in pharmaceuticals as the inert carrying agent, etc. Almost 30% of the carbon, fixed in leaves, is incorporated in the form of starch. Approximately 65–80% of the dry weight of cereals, and potato tuber, consist of starch. Thus, studying starch metabolism in plants is of prime importance. Starch is a polysaccharide, consisting of α-D-glucose molecules, which are linked through α-1,4-glycosidic linkages. These linkages protect aldehyde groups of glucose molecules against oxidation, except that of terminal glucose molecule, which is unprotected and is called reducing end of the molecule. Additionally, α-1,6-glycosidic linkages are also present, which add to branching of the molecule. Depending upon the number of glucose molecules and number and frequency of α-1,6-glycosidic linkages present in the molecule, starch is classified into two categories, i.e., amylose and amylopectin (Table 9.1). Amylose is almost a linear polymer consisting of about 600–3000 glucosyl residues, held together by α-1,4-glycosidic linkages. Occasionally, 1,6-glycosidic branch point linkages are present after almost every 1000 residues (Fig. 9.14). Amylose polymer forms a single helical structure with six glucose residues being present in each helix. Contrary to this, amylopectin is much larger molecule, 200–400 nm long and 15 nm wide, consisting of 6000–60,000 glucosyl residues linked by α-1,4-glycosidic linkages with approximately 5% of the glucose residues linked through α-1,6-glycosidic branch points after every 20–26 residues, which make the molecule as highly branched. There are many hydroxyl groups present which form hydrogen bonds. Adjacent chains of amylopectin form double helical structures with themselves or with amylose polymer (Fig. 9.15). At the beginning of the chain, a terminal glucose molecule with free reducing group is present, which is called reducing end, while reducing aldehyde groups of all other glucosyl are blocked due to formation of

**Table 9.1** Comparison of amylose, amylopectin, and phytoglycogen

| Characteristics | Amylose | Amylopectin | Phytoglycogen |
|---|---|---|---|
| Number of glucose residues | $\approx 10^3$ | $\approx 10^4$–$10^5$ | $\approx 10^5$ |
| Nature of glycosidic linkages | Straight-chain polymer, more rigid structure | Branched polymer, less rigid in structure because of branching | Extensively branched as in glycogen in animals |
| Structure | α-D-glucose are linked by α-1,4-glycosidic linkages; single helical structure due to formation of intra-molecular hydrogen bonds; six residues in each loop | α-D-glucose are linked by α-1,4 and α-1,6-glycosidic linkages; adjacent branches form double helical structure. Six residues in each loop | α-D-glucose are linked by α-1,4 and α-1,6-glycosidic linkages |
| Branching | Very less, may be after 1000 glucose residues | Branching after every 20–26 glucose residues, branching accounts for about 5% of the residues | Extensive branching after every 10–15 glucose residues, branching accounts for about 10% of the bonds |
| Solubility | Less soluble in cold water, soluble in hot water without swelling, does not form gels and paste | Soluble in water, crystallization occurs, forms gel or paste | Soluble in water |
| Crystallization | Is not responsible for crystallization of starch | Crystallization occurs | Cannot crystallize |
| Hydrolyzing enzyme | α-amylase and β-amylases in addition to α-glucosidase | α-amylase and β-amylases in addition to α-glucosidase, in addition debranching enzyme | α-amylase and β-amylases in addition to α-glucosidase, in addition debranching enzyme |
| Color with iodine, absorption of the glucan-iodine complex | Blue, 660 nm | Reddish brown, 530–550 nm | Violet, 430–450 nm |
| % in plant starch | 20–30 | 70–80 | In some plants up to 20 |

glycosidic bonds with the adjacent glucosyl molecule. New glucosyl residue is added from the nonreducing end at the time of starch biosynthesis. Both amylose and amylopectin assemble together to make a complex structure, known as semi-crystalline granule. Ratio of amylose to amylopectin by weight is approximately 1:3, which is variable depending upon the source. Ratio of amylose to amylopectin as well as degree of branching in amylopectin determines the quality and type of starch which is to be used for commercial purposes (Box 9.1). Food produced from

## 9.3 Starch Metabolism

**Fig. 9.14** Amylose is a homopolysaccharide consisting of α-D-glucose which are joined by α-1-4-glycosidic linkages to make it a linear polymer. Occasional branching after almost 1000 glucosyl residues occurs due to the presence of α-1-6-glycosidic linkages. Due to the presence of extensive –OH groups, there are hydrogen bonds which makes the molecule to form helical ring. Ring of the molecule is planar; however, binding of C-O linkages of glycosidic bonds results in single helical formation with six glycosyl residues in each helix

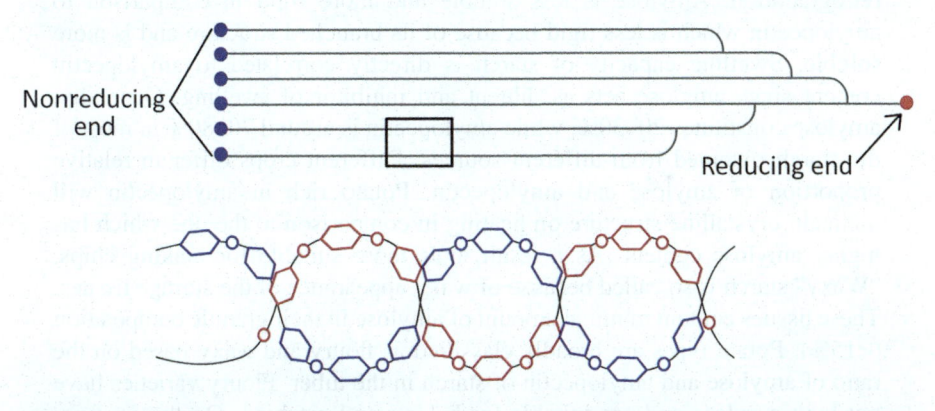

**Fig. 9.15** In amylopectin, α-D glucosyl residues are joined by α-(1-4)-glycosidic bonds similar to those in amylose. However, unlike amylase, amylopectin is a highly branched polymer because of the presence of α-(1-6)-glycosidic linkages after every 20–26 glucosyl residues. Hydrogen bonds are formed in between hydroxyl groups of adjacent chains as a result of which double helical structure is formed in amylopectin either in between adjacent branches of the molecule itself or with amylose polymer

amylopectin rich starch, e.g., from waxy mutants of maize, can be frozen or thawed with minimal loss of texture since these do not contain any amylose. Amylose molecules associate more closely forming dense crystals on being frozen—a process known as **retrogradation.** This results in watery, unappetizing mess, which to some

**Box 9.1: Starch Composition Determines Its Uses**
Relative proportion of amylose and amylopectin determines physicochemical properties of starch. This is of great commercial significance since starch has multipurpose uses, for example, used as a source of food and in various other industries including as an adhesive, as packaging material, as coating agents, in pharmaceuticals and chemical industries, etc. Amylose chain consists of linear molecules, which has more extensive hydrogen bonding; thus, more energy is required to break these bonds. On the contrary, branched amylopectin molecules cannot align easily and thus form weak hydrogen bonds. Upon heating, semicrystalline structure of starch granules is broken, and water molecules associate to hydroxyl groups exposed on amylose and amylopectin by hydrogen bonding. These associations cause irreversible swelling of starch and its crystalline structure collapses—a phenomenon known as gelatinization. Hydration of amorphous phase causes loss of crystallinity. The presence of amylose also tends to reduce crystallinity of amylopectin. It influences penetration of water into the starch granule because of its hydrophilic nature. The extent of this interaction is influenced by the proportion of amylose and amylopectin in starch since amylose molecules tend to aggregate on cooling, while amylopectin molecules generate more desirable gels. Upon cooling, starch granules reassociate in a complex recrystallization process known as retrogradation. Amylose is less soluble and more rigid in comparison to amylopectin which is less rigid because of its branched structure and is more soluble. Swelling capacity of starch is directly correlated to amylopectin content since amylose acts as diluent and inhibitor of swelling. Generally, amylose constitutes 20–30%, while amylopectin is around 70–80% in most of the starch obtained from different sources. Different crops differ in relative proportion of amylose and amylopectin. Potato rich in amylopectin will maintain crystalline structure on heating in comparison to the one which has higher amylose content. As a result, it is more suitable for making chips. "Waxy" starch is so called because of waxy appearance of the storage tissues. These tissues contain minimal amount of amylose in their granule composition (<15%). Potato types are broadly classified as floury and waxy based on the ratio of amylose and amylopectin of starch in the tuber. Floury varieties have got high amylose and are suitable for baking and mashing. On the contrary waxy potatoes retain their structure when boiled or cooked otherwise, since they are low in amylose content. High-amylose starch is used in adhesive products and also in production of paper because of their high gelling properties. Besides, it has film-forming ability which is useful as a coating on fried products to keep them crispy while reducing fat uptake while cooking. Mutants deficient in GBSS activity will accumulate starch low in amylose. Downregulation of both SBEI and SBEII expression in potato tuber using antisense techniques resulted in starch with more than 60% amylose. It is

(continued)

**Box 9.1** (continued)
possible to produce starches with desired levels of amylose and amylopectin. What makes rice fluffy? Ratio of amylose and amylopectin will determine texture of cooked rice whether it will be fluffy or sticky. As the rice cooks, both liquid and heat penetrate the grain, and the starch molecule inside the grain breaks down. Since amylose does not gelatinize during cooking, grains with high amount of amylose will be fluffy and separated once cooked. Long grain rice usually has high amount of amylose (about 20%) and least amount of amylopectin. Amylopectin gelatinizes and is responsible for making rice sticky after it is cooked. Rice with high amount of amylopectin will be very sticky once cooked. Medium rice contains about 15–17% amylose and a good amount of amylopectin. Waxy rice has almost no amylose.

extent can be rescued by reheating. Starch containing only amylopectin can be produced from the mutants which lack granule-bound starch synthase (GBSS). Transgenic crops are produced to meet the required demand; thus, an understanding of the structure and biosynthesis of starch is important. Starch consists of semicrystalline structure which is stable, insoluble, and osmotically inert. There is extensive hydrogen bond formation in between the hydroxyl groups of adjacent polymers resulting in helical formation. Amylose is linear and slightly branched. It forms a single helical structure which is entangled with amylopectin. On the contrary amylopectin has a nonrandom branching pattern. The "outer" unbranched chains of the molecule are known as A chains, the inner branched chains are B chains, while there is only one C chain containing single reducing group. These A and B chains form double helical structure with six glucose molecules per turn between the linear unbranched segments. When analyzed, the X-ray structure shows starch granules to be consisting of concentric layers of alternating amylose and amylopectin. Amylopectin molecules are arranged in radial manner with the reducing glucose residue directed toward inside while other glucose molecules radiating toward outside. Double helices of amylopectin are arranged into ordered crystalline lamellae, which alternate with amorphous lamellae containing the branched points. These are often referred as growth ring because of similarity with growth rings of a tree. However, these two are different, since in case of starch granules the rings are the structural features rather than the features of periodic growth in case of trees (Fig. 9.16).

## 9.3.1 Starch Biosynthesis

Plants synthesize starch in plastids. In mesophyll cells, carbon is fixed through Calvin cycle during the day. During a normal day, $CO_2$ fixed in mesophyll cells is able to meet the requirement of green and non-green parts of plants because of its

**Fig. 9.16** (a) Semicrystalline structure of a starch granule: It consists of alternating layers of semicrystalline and amorphous zones, (b) and (c) clusters of α1→4 chains are packed in double helical structures to make crystalline lamellae. Alternating amorphous lamellae is due to the branch points (α-1-6-glycosidic linkages) which form the crystalline structures

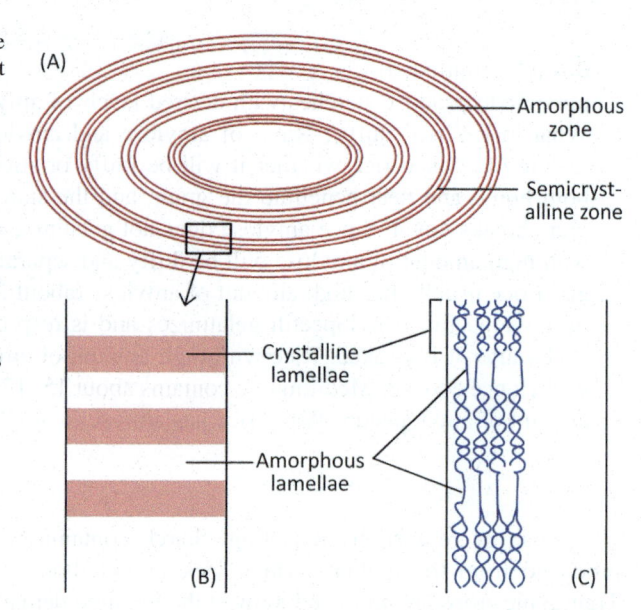

translocation as sucrose. In addition to chloroplasts (synthesis of transient starch), starch is also synthesized and stored in amyloplasts of storage organs of the plant, including in grains and tubers. Basic mechanism of starch synthesis, in chloroplasts and amyloplasts, is similar. There is a hexose monophosphate pool also in plastids. Fructose 6-phosphate is converted to glucose 6-phosphate which is further converted to glucose 1-phosphate catalyzed by phosphohexoisomerase and phosphoglucomutase, respectively. Mutant for any of these enzymes will result in impaired starch synthesis. The first committed step-in starch biosynthesis is catalyzed by **ADP-glucose pyrophosphorylase (AGPase)**, which includes activation of glucose to ADP-glucose. Pyrophosphate released during the reaction is immediately hydrolyzed to produce two molecules of $P_i$, which ensures irreversibility of the reaction (Fig. 9.17).

$$\text{Glucose 1} - \text{phosphate} + \text{ATP} \rightarrow \text{ADP} - \text{glucose} + PP_i$$

$$PP_i + H_2O \rightarrow 2P_i$$

ADP-glucose pyrophosphorylase is a heterotetrameric ($\alpha_2\beta_2$) protein, consisting of small and large subunits. These are encoded by two different nuclear genes. Catalytic activity is present in small subunits, while large subunits are believed to have regulatory role. Activity of AGPase is allosterically regulated by the key metabolites of carbon assimilation. PGA is an activator of the enzyme, while $P_i$ is the inhibitor. In studies with AGPase purified from wheat, spinach, and barley leaves, a tenfold activation by PGA and tenfold decrease in enzyme activity with

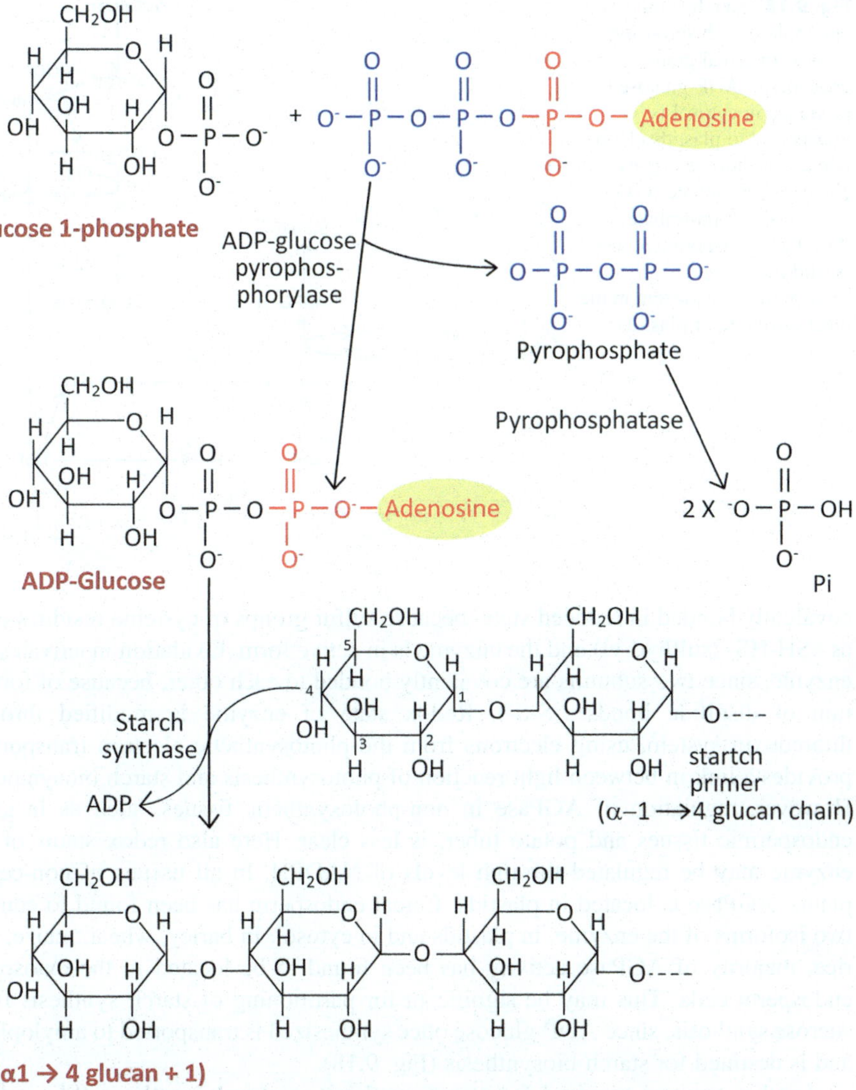

## 9.3 Starch Metabolism

**Fig. 9.17** Starch synthesis catalyzed by starch synthase. Glucose from ADP-glucose is added to C-4 (nonreducing end of starch primer) catalyzed by starch synthase

$P_i$ were observed. Factors, which influence PGA/$P_i$ ratio, will also affect starch synthesis in chloroplasts. In light, $P_i$ is kept low in chloroplasts because of ongoing photophosphorylation, which results increase in starch synthesis, while reverse happens in dark. Another mechanism of regulation is through reversible redox activation of the enzyme, AGPase. Cysteine residues present on small subunits of the enzyme can exist either in reduced or in oxidized states. The two subunits are not

**Fig. 9.18** Starch synthesis in amyloplasts of heterotrophic tissues. In cereal grain endosperm, ADP-glucose produced in cytosol is transported to plastids (1). In non-cereal storage organs, glucose 1-phosphate (G1P) and glucose 6-phosphate (G6P) are transported across plastid membrane. Transporters are present in the inner envelope of plastids

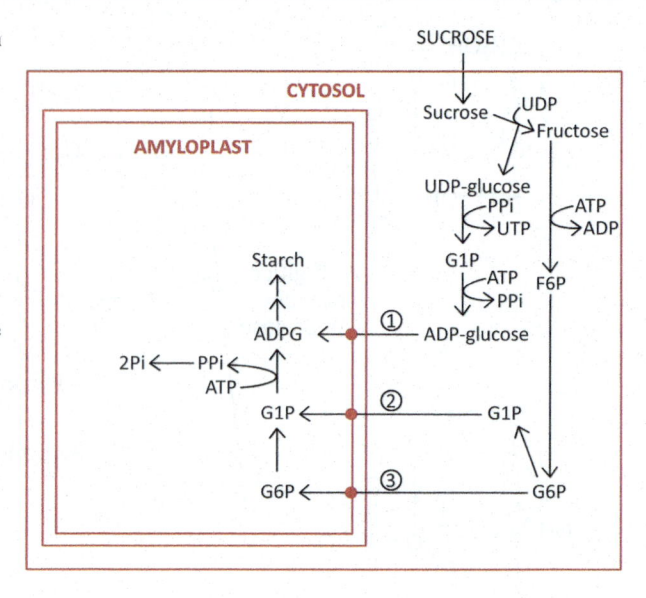

covalently bonded in reduced state, because sulfur groups of cysteine residues exist as –SH-HS- (sulfhydryl) and the enzyme is in active form. Oxidation inactivates the enzyme, since two subunits are covalently bonded to each other, because of formation of disulfide bonds (-S-S-). Redox state of enzyme is modified through thioredoxin system, using electrons from the photosynthetic electron transport. It provides a link in between light reaction of photosynthesis and starch biosynthesis. However, regulation of AGPase in non-photosynthetic tissues, such as in grain endospermic tissues and potato tuber, is less clear. Here also redox status of the enzyme may be regulated through levels of NADPH. In all tissues of non-cereal plants, AGPase is located in plastids. Cereal endosperm has been found to contain two isoforms of the enzyme, in plastids and in cytosol. In barley, wheat, maize, and rice, majority of AGPase activity has been found to be located in the cytosol of endosperm cells. This may be significant for partitioning of starch synthesis from sucrose synthesis, since ADP-glucose once synthesized is transported to amyloplasts and is destined for starch biosynthesis (Fig. 9.18).

Another enzyme required for starch synthesis is **starch synthase (SS)**, which catalyzes transfer of glucose from ADP-glucose to preexisting starch primer as amylose or amylopectin. Glucose is added to the C-4 of the glucosyl residue at the nonreducing end of starch.

$$\text{ADP} - \text{glucose} + \alpha - \text{glucan}_n \rightarrow \alpha - \text{glucan}_{n+1} + \text{ADP}$$

Starch synthase remains bound with the growing starch granule. Five isoforms of starch synthase encoded by five genes have been identified, i.e., GBSS (granule-bound starch synthase), SSI, SSII, SSIII, and SSIV. GBSS is tightly associated with starch granule and is responsible for synthesis for amylose (Figs. 9.19 and 9.20). On

## 9.3 Starch Metabolism

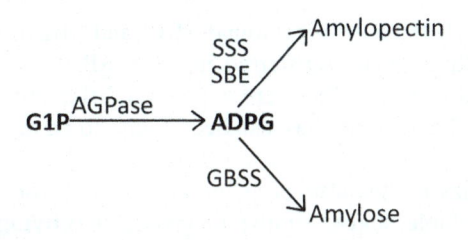

**Fig. 9.19** Pathway for starch biosynthesis. *AGPase* ADP-glucose pyrophosphorylase, *GBSS* granule-bound starch synthase, *SSS* soluble starch synthase, *SBE* starch branching enzyme

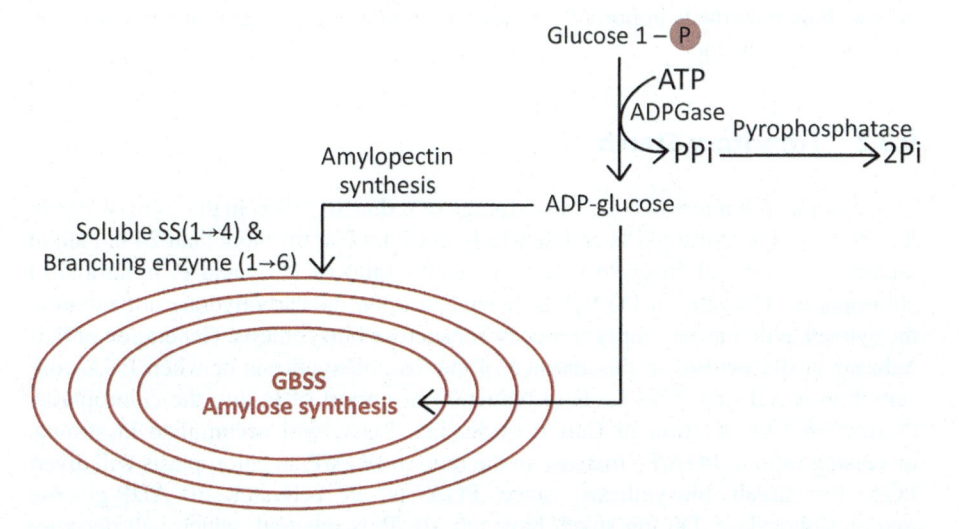

**Fig. 9.20** Starch synthesis in stroma of plastids. *GBSS* granule-bound starch synthase, *SS* starch synthase, *ADPGase* ADP-glucose pyrophosphorylase

the contrary, SS isoforms are the soluble forms, which are partly bound with starch granules and are located in the plastid stroma. These work with starch branching enzyme and synthesize amylopectin. Mutants for any of these enzymes will produce starch with altered properties. Besides GBSS, starch granule is also associated with other enzymes of starch catabolism, which have the potential for restructuring the growing starch granule or dismantling it. The biochemical versatility of starch has commercial significance, since its various uses are dependent on its structure (Box 9.1). Different species differ in the presence and activity of various isoforms of SS. The waxy mutants of maize are deficient in GBSS; thus, only amylopectin is produced with no amylose.

**Starch branching enzyme (SBE)** is responsible for introducing the branches in the molecule. This cleaves the α-1,4-glycosidic linkages almost 20 glucosyl residues down the chain and rejoins them through formation of α-1,6-glycosidic linkages, thus adding branching to the molecule during synthesis of amylopectin. These chains are further elongated until a new branching appears. Two isoforms of starch branching enzymes (SBE) have been found SBEI and SBEII, which differ in their specificity for length of chain transferred in vitro. SBEII transfers longer glucan chain in comparison to SBEI. These are expressed differently during the development of the plants. Their activity has also been found to be regulated due to protein phosphorylation.

**Debranching** enzyme has also been found to play a role in starch synthesis. Highly branched soluble starch, similar to glycogen (**phytoglycogen**), has been found to be produced in the mutants for specific class of this enzyme. **Isoamylase-1** type debranching enzyme is responsible for removing unwanted branches of the molecule, which is required for highly ordered crystalline starch granules. Though debranching enzyme is important for starch degradation, its role in starch biosynthesis is also significant.

## 9.3.2 Transitory Starch

Transitory starch refers to temporary storage of reduced carbon in the form of starch. It is produced in chloroplasts and generally does not last for more than 24 h. Fate of intermediates of Calvin cycle is determined by ratio of PGA and $P_i$ in stroma of chloroplasts. The ratio of PGA/$P_i$ is dependent upon the carbohydrate utilization in the cytosol, either as an energy source or for sucrose biosynthesis. Generation of $P_i$ is reduced in the cytosol in the absence of sucrose biosynthesis or when PGA consumption is reduced. This leads to reduction in import of $P_i$ into the chloroplasts. Product of $CO_2$ fixation in Calvin cycle, i.e., PGA, will accumulate in stroma, increasing ratio of PGA/$P_i$. Increase in the ratio of PGA/$P_i$ in chloroplasts will divert PGA for starch biosynthesis, since PGA is an activator of ADP-glucose pyrophosphorylase. During starch biosynthesis, $P_i$ is released, which will decrease the PGA/$P_i$ ratio. Availability of $P_i$ will restore Calvin cycle. This temporary starch accumulated in leaves during daytime is called **transitory starch**. Glucose cannot be stored since it adds to osmotic concentration, and too much glucose accumulation will lead to lowering of water potential of plastids resulting in water uptake leading to their bursting. Starch is osmotically inactive and can be accumulated. Transient starch biosynthesis acts as buffer. During night, the absence of $CO_2$ fixation results in lowering of PGA/$P_i$ ratio. As a result, starch synthesis will stop. Starch is hydrolyzed to produce hexose sugars, which are exported out of the plastids through the hexose translocators located in the inner envelope of the plastid. A hexokinase also located in the inner envelope catalyzes phosphorylation of glucose generating glucose 6-phosphate, which adds to the cytosolic hexose monophosphate pool (Fig. 9.21).

## 9.3.3 Starch Catabolism

Starch, whether stored temporarily in chloroplasts (transient starch) or stored in amyloplasts of cereal grain or tubers (storages starch), needs to be utilized either as a

## 9.3 Starch Metabolism

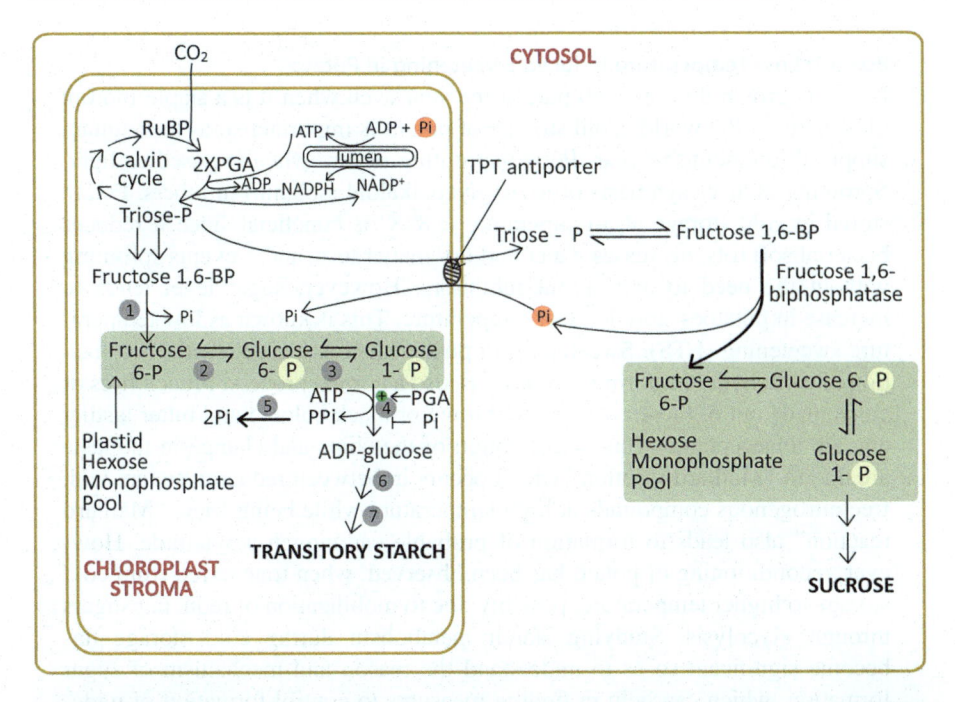

**Fig. 9.21** Synthesis of transitory starch in chloroplast. Two irreversible reactions are responsible for synthesis of starch, which are catalyzed by fructose 1,6-biphosphatase and by ADP-glucose pyrophosphorylase. Pyrophosphate (PP$_i$) hydrolysis by pyrophosphatase makes the reaction irreversible. ADP-glucose pyrophosphorylase (AGPase) catalyzes the synthesis of ADP-glucose, which is a precursor for starch synthesis. AGPase is sensitive to the ratio of PGA/P$_i$ with PGP promoting the activity of enzyme and P$_i$ inhibiting the activity. High rate of ATP synthesis during light reaction results in shortage of P$_i$ in plastids, thereby increasing PGA/P$_i$ ratio. During daytime starch synthesis is favored in plastids which is stored as transient starch. This results in decreased export of triose-P. Decrease in synthesis of sucrose in cytosol also results in decreased release of P$_i$, and it is not imported into plastid which results in decreasing the ratio of PGA/P$_i$. This promotes synthesis of starch in the plastids catalyzed by fructose 1,6-biphosphatase (*1*), phosphohexoisomerase (*2*), phosphoglucomutase (*3*), ADP-glucose pyrophosphorylase (*4*), pyrophosphatase (*5*), starch synthase (*6*), and starch branching enzyme (*7*)

source of energy or as a source for carbon skeleton. Degradation pathways are different in different organs, and there are distinct pathways which operate within the same organ. Pathway for starch breakdown is fairly understood, both in chloroplasts and in amyloplasts of germinating cereal grains. However, the steps do not seem to be very clear in other storage organs of plants, such as tuber and roots and in other non-cereal crops (Box 9.2). Range of enzymes with many isoforms are involved in degradation of starch along with the related glucans which vary from tissue to tissue. The large polymeric molecule is cleaved into small monomeric molecules, glucose by either of the two classes of enzymes, **phosphorolytic enzymes** or **hydrolytic enzymes**.

### Box 9.2: Low Temperature-Induced Sweetening in Potato

Potato is grown during small part of the year even when it is a staple food in most parts of the world. Cold storage of potato is important so as to maintain supply throughout the year. Potato sprouting occurs when exposed to light. Sprouting induces synthesis of toxic glycoalkaloid (solanin) in tubers. Potato stored in cold storage at a temperature $< 8^0$ C is beneficial since it reduces bacterial soft rots, decreases water and dry matter loss, and prevents sprouting without the need to add sprout inhibitors. However, sugar level tends to increase in potatoes stored at low temperature. This is known as low temperature sweetening (LTS). Sweetening in potato tuber associated with mobilization of carbohydrates during sprouting is also a normal process. French fries or chips made out of these potatoes are brown or black colored and bitter tasting and are unacceptable. This discoloration of the chips and change in the taste are due to "Maillard reaction" which occurs in between reducing sugars with free nitrogenous compounds at high temperature while being fried. "Maillard reaction" also leads to formation of probable carcinogen acrylamide. However, reconditioning of potato has been observed, when transferred from cold storage to higher temperature, possibly due to mobilization of reducing sugars through glycolysis. Studying starch metabolism during cold storage has become significant so as to understand the reason and mechanism of sugar formation, which can help in finding measures to control formation of undesirable reducing sugars during cold storage. It was observed that there was increase in extractable activity of both endo- and exoamylases and starch phosphorylase in potato upon cold storage, which might contribute to increase in sugars due to hydrolysis of starch. Change in kinetic property of sucrose-phosphate synthase (SPS) due to cold storage also might be responsible for sucrose biosynthesis. Another reason possibly was increase in sugars due to restriction in entry of hexose phosphates into glycolysis. Conversion of fructose 6-phosphate to fructose 1,6-bisphosphate is the first dedicated step of glycolysis. The reaction is catalyzed by the enzyme $PP_i$-fructose 6-phosphate phosphotransferase, which is more sensitive to temperature than other enzymes. Using RNAi technique by silencing vacuolar invertase (VI) gene, a direct correlation between lightness of potato chip color and decreased VI transcripts was observed. In some studies, potatoes containing high amylose were found to be more resistant to low temperature sweetening, possibly because amylose was less susceptible to hydrolysis by α-amylase and is more stable due to hydrogen bonds within and in between molecules.

### 9.3.3.1 Phosphorolytic Starch Degradation

The enzymes, which use $P_i$ to break the glycosidic bonds, are called phosphorolytic enzymes. At least three categories of enzymes are involved in phosphorolytic degradation of starch. This includes **starch phosphorylase, debranching enzyme**, and **glucosyltransferase**. Starch phosphorylase cleaves one glucose molecule from

## 9.3 Starch Metabolism

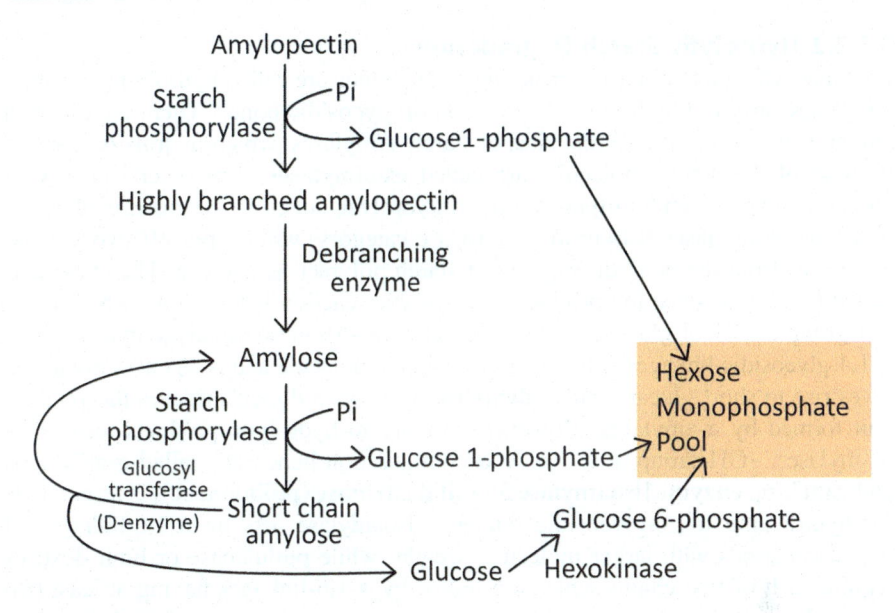

**Fig. 9.22** Phosphorolytic starch breakdown

the nonreducing end of the polysaccharide as glucose 1-phosphate. Unlike the hydrolytic enzymes, amylases, during phosphorolytic degradation of starch, energy of glycosidic bond is conserved as the phosphate esters. In further mobilization, one ATP is consumed less, thus saving one ATP per molecule of glucose. Reaction catalyzed by starch phosphorylase is as follows:

$$\alpha-(\text{Glucan})_n + P_i \rightarrow \alpha(\text{Glucan})_{n-1} + \text{Glucose 1}-\text{phosphate}$$

The enzyme acts at glycosidic bonds, which are at least four residues away from the branch points. Debranching enzyme, which is also known as **pullulanase** or **R-enzyme**, acts on branch points, i.e., on α-(1-6)-glycosidic linkages, releasing linear glucan polymer, which are acted upon by starch phosphorylases at the nonreducing end, resulting in removing terminal glucose as glucose 1-phosphate, one at a time. Condensation of shorter glucan chains, to produce substrate for starch phosphorylase, is catalyzed by the **D-enzyme** also known as **glucosyltransferase** (Fig. 9.22). Role of glycogen phosphorylase in animals is similar to that of starch phosphorylase in plants; however, regulation of glycogen phosphorylase is better understood, which is believed to be regulated through phosphorylation and dephosphorylation of the enzyme protein. Starch phosphorylase is also regulated by $P_i$, whose availability fluctuates with the growth conditions, such as plants exposed to various kinds of stresses. However, an understanding regarding regulation of the enzyme activity is required.

## 9.3.3.2 Hydrolytic Starch Degradation

Enzymes which use water to break glycosidic bonds are called hydrolytic enzymes. Amylases carry out hydrolytic degradation of glycosidic bonds. There are different amylases which act at different sites of starch. Amylases, which hydrolyze starch at the end of the starch molecule, are called **exoamylases.** The other category of amylases includes **endoamylases,** which hydrolyze starch at the interior of starch molecule. β-amylases belong to exoamylase category, which split off two glucose molecules from the nonreducing end of starch polymer as maltose. The enzyme is called β-amylase since the product of hydrolytic reaction is β-maltose which has – OH group at C-1 of glucose in β configuration. α-Amylases catalyze hydrolysis of α-1,4-glycosidic linkages at the interior of starch molecule and is an endoamylase. It gives rise to short glucans called **dextrins**. Maltose, a disaccharide, is the smallest unit formed by α-amylases. However, contrary to hydrolytic products formed by β-amylases, –OH group at C-1 is in α configuration hence are called α-amylases. **Debranching enzyme [isoamylase 3/limit dextrinase (pullulanase)]** is responsible for hydrolyzing α-1,6-glycosidic linkages. Isoamylase acts on α-1,6-linkages of oligosaccharides with larger molecular weight, while pullulanase or limit dextrins require each of two chains which are linked by α-1,6-linkages having at least two glucosyl residues linked by α-1,4 linkages (a tetrasaccharide, maltosyl maltose). Pullulanase name was originally used for the debranching enzyme from microbial origin, while limit dextrinase or R-enzyme is used for the enzyme of plant origin. Degradation products of α-amylases, which include dextrins and maltoses, is further degraded to glucose molecules by the enzyme **α-glucosidase** (Figs. 9.23 and 9.24). Thus, complete degradation of amylopectin would require debranching enzyme, both the amylases and glucosidase. Hexoses, which are formed as a result of starch hydrolysis, are transported out of amyloplasts through the translocators located in the inner envelope of the plastid. The process of starch degradation in photosynthetic tissues or non-photosynthetic tissues, where starch is stored temporarily (transient starch), is different from other starch-storing organs. In photosynthetic tissues, the tissue integrity is maintained, while starch degradation in germinating cereal grains coincides with destruction of endosperm.

## 9.3.3.3 Catabolism of Transitory Starch

During dark, transitory starch is degraded and is transported out of chloroplast to be translocated as sucrose. Mechanism of starch degradation is well studied in plastids. Transitory starch is present in semicrystalline form which results from packing of double helical structures of the molecule. This needs to be solubilized before being degraded, since the enzymes responsible for degradation will not be able to get an access to starch, which is present in helical form. In addition to enzymes required for hydrolysis of α-1,4- and α-1,6-glycosidic linkages, some starch-phosphorylating enzymes are also required for starch mobilization. Some glucosyl residues of amylopectin are phosphorylated. Hydrolysis of semicrystalline structure is facilitated by two phosphorylating enzymes, i.e., **glucan, water dikinase (GWD)** and **phosphoglucan, water dikinase (PWD)**. GWD phosphorylates at C-6 of some of the glucosyl residues at the surface of starch granules by transferring β-phosphate

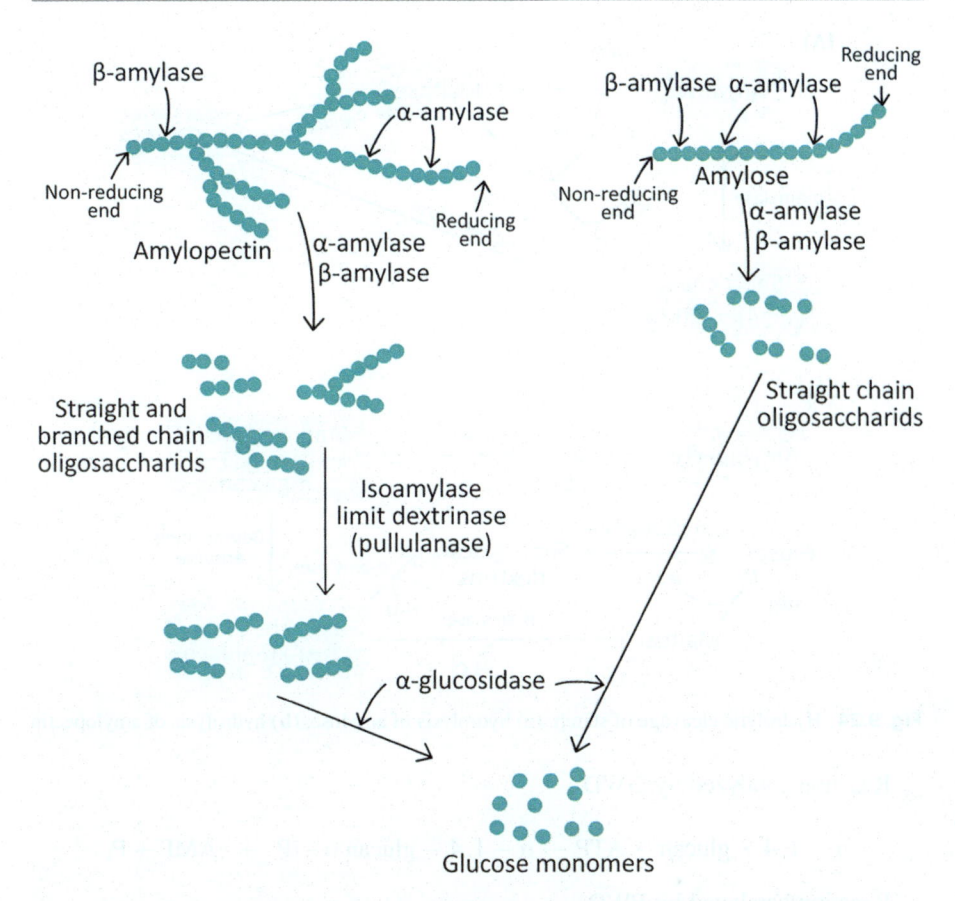

**Fig. 9.23** Starch polymers amylose/amylopectin is hydrolyzed by α-amylase and β-amylase enzymes to straight-chain and branched-chain oligosaccharides. Limit dextrins (pullulanase) act on the chains (each with two glucosyl residues) linked by α-1,6-glycosidic linkages, while isoamylase also acts on α-1,6-glycosidic linkages of branched oligosaccharides with large molecular weight

group of ATP. Another phosphorylating enzyme, PWD, catalyzes, less frequently, phosphorylation of glucosyl residue at C-3 position of another glucosyl residue of phosphorylated glucans. Activity of PWD is preceded by that of GWD. Addition of phosphate group to glucan influences the hydrophilicity and structure of starch, leading to disruption and destabilization of the packing of double helices. As a result, glucosyl residues are exposed, thus facilitating accessibility to attack by hydrolytic enzymes on rather hydrophobic starch. GWD may also increase activity of β-amylase. Catalytic mechanism of both dikinases is regulated by autophosphorylation of the enzyme protein. Mutants, with reduced activity of GWD and PWD, display increased level of starch in chloroplasts of their leaves. Activity of these enzymes is regulated by redox modification.

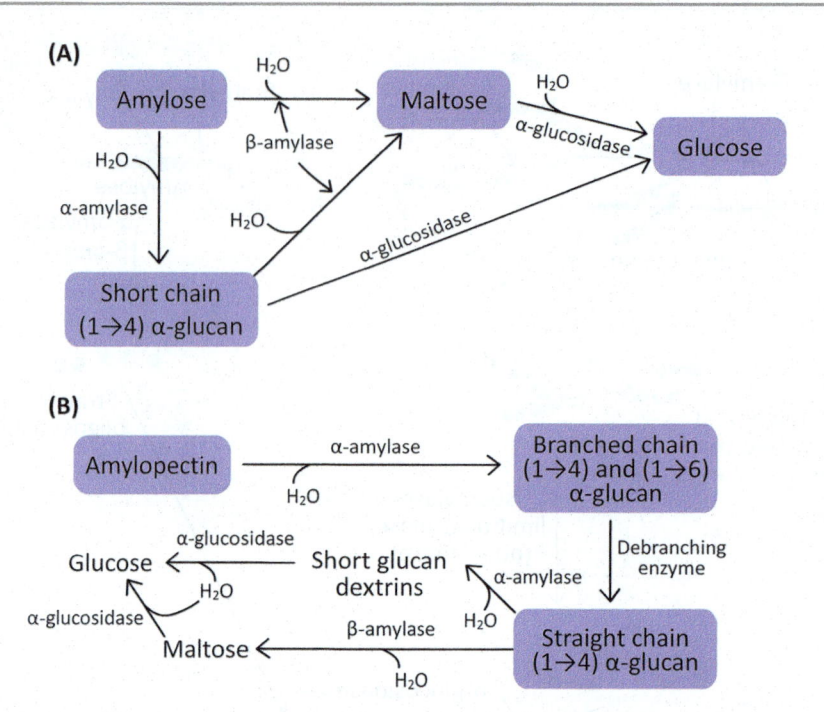

**Fig. 9.24** Hydrolytic cleavage of starch. (**a**) hydrolysis of amylose; (**b**) hydrolysis of amylopectin

Reaction catalyzed by GWD

$$\alpha - 1,4 - \text{glucan} + \text{ATP} \rightarrow \alpha - 1,4 - \text{glucan } 6 - P + \text{AMP} + P_i$$

Reaction catalyzed by PWD

$$\alpha - 1,4 - P - \text{glucan} + \text{ATP} \rightarrow \alpha - 1,4 - \text{glucan} - 3 - P + \text{AMP} + P_i$$

Further hydrolysis of starch is done by the hydrolytic enzymes, α-amylases and β-amylase. When action of β-amylase is blocked by a phosphate group, its removal is catalyzed by the enzyme **phosphoglucan phosphatase**.

$$\alpha - 1,4 - \text{glucan} - 3 - P + H_2O \rightarrow \alpha - 1,4 - P - \text{glucan} + P_i$$

Chloroplasts also contain α-amylase, which results in the release of spectrum of linear and branched malto-oligosaccharides from the starch granule surface. Branched malto-oligosaccharides can be debranched by the debranching enzyme in the stroma of chloroplast. Combined action of β-amylase, α-amylase, and debranching enzyme results in the release of malto-oligosaccharides, no longer than three glucosyl residues (maltotriose) from the starch granule surface. Since maltotriose are too small for the action of β-amylases, α-1,4-P-glucanotransferase catalyzes transfer of a part of maltotriose to the acceptor molecule resulting in synthesis of longer chain molecule with simultaneous release of a glucose molecule. Longer chain glucan is further attacked by β-amylase resulting in synthesis of

## 9.3 Starch Metabolism

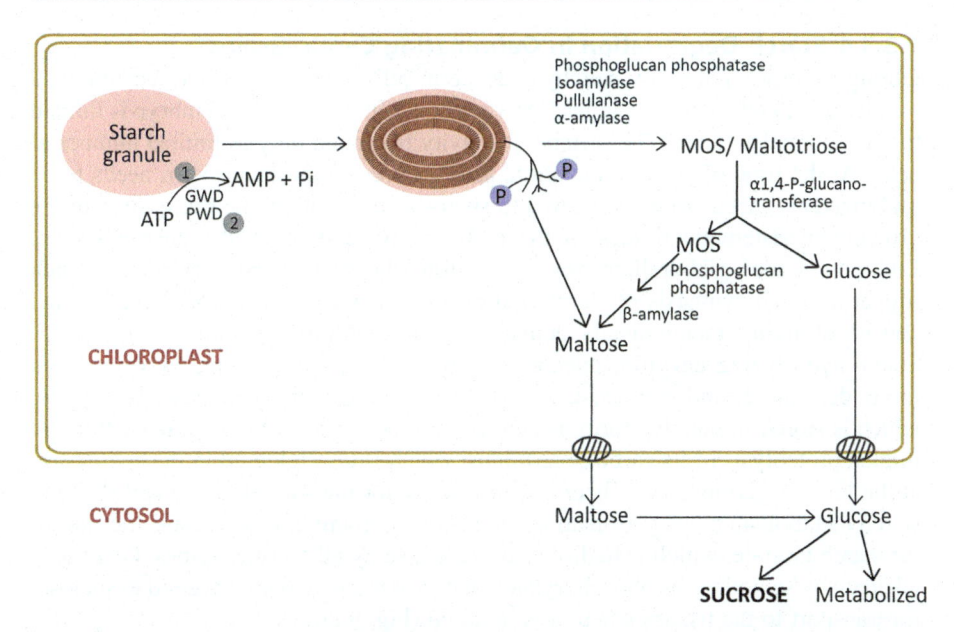

**Fig. 9.25** Catabolism of transitory starch in chloroplast. C-6 of some glucosyl residues of the starch polymer of semicrystalline starch granule are phosphorylated by glucan, water dikinase (GWD) (1), followed by phosphorylation of some other glucosyl residues at C-3 by phosphoglucan water dikinase (PWD). Phosphorylated branches of amylopectin become accessible to the enzymes α-amylose, debranching enzyme, and β-amylases resulting in formation of malto-oligosaccharides not bigger than maltotrioses. α 1,4-glucan transferase transfers part of maltotriose to another molecule of maltotriose resulting in the synthesis of malto-oligosaccharides (MOS) and glucose. Phosphate groups are removed by phosphoglucan phosphatase which is hydrolyzed to generate maltose. Both maltose and glucose are transported out of chloroplast

maltose. Maltose and glucose are transported out of the chloroplasts since distinct transporters for both are present in the inner envelope of the plastids (Fig. 9.25).

$$2\text{Maltotriose} \rightarrow \text{Maltopentose} + \text{Glucose}$$

Chloroplasts also possess α-glucan phosphorylase (starch phosphorylase), which catalyzes release of glucose from nonreducing end of starch by phosphorolysis, generating glucose 1-phosphate, which supports plastid metabolism through hexose monophosphate shunt or glycolysis, instead of being transported out of chloroplast. Unlike in amyloplasts, transporters for phosphorylated hexose have not been reported in the chloroplasts. Post-translational modifications, which include phosphorylation-dephosphorylation of protein and redox modulation, might be responsible for regulation of starch degradation. Starch degradation is regulated by requirement of the cell, since it is observed that starch degradation will not begin in dark till the time sucrose level in leaf has fallen.

### 9.3.3.4 Starch Degradation in Germinating Cereal Grains

During maturation of seeds, starchy endosperm cells die because of desiccation. In a starchy seed, endosperm occupies larger portion of the seed, while embryo is limited to only a smaller space. The metabolic activity is limited only to limited number of cells. At the time of seed germination, starch stored in the endosperm needs to be mobilized and needs to be converted to sucrose, so that it can be transported to the growing meristematic tissues of the embryo, to be used as the energy source. Embryo releases gibberellins from **scutellum** (the specialized cotyledon), which trigger *de novo* synthesis of starch-hydrolyzing enzymes in aleurone layer, which consist of living metabolically active cells surrounding the starchy endosperm. Starch-hydrolyzing enzymes released are: $\alpha$-amylase, debranching enzyme which limits dextrinase, and $\alpha$-glucosidase which is also known as maltase. $\beta$-amylase, which is stored in inactive form, is converted to active form by a **protease** through limited proteolysis of its carboxy-terminus, which is also released in the endosperm at the time of germination. There is no evidence for the role of GWD and PWD in starch metabolism in the germinating cereal grains. $\alpha$-amylase acts at specific sites of the starch granule, which is hydrolyzed to glucose by all these enzymes. Glucose is taken up by scutellum, is phosphorylated by hexokinase, and is converted to sucrose, for transport to the tissues where it is required (Fig. 9.26).

## 9.4 Fructans

There are many plants in which primary form of storage carbohydrates is neither sucrose nor starch. In almost 15% of the angiosperms (around 40,000 species), carbohydrates are stored as fructans, which unlike starch is water soluble and is synthesized and stored in plant vacuoles. Fructans were first found in tubers of the ornamental plant dahlias. Later on, these have been reported to be present in a number of economically important plants. Though content of fructans is very low in mature grains of wheat, barley, and other temperate grasses, large quantities are found in foliage, leaf bases, and stem of the plant. Large quantities of fructans are reported in bulbs of *Allium* species, tap roots of *Cichorium* (chicory), and tubers of *Helianthus tuberosus* (Jerusalem artichoke). Fructans are osmotically active and have stabilizing effect on cell membranes, thus having antistress role, besides being a storage carbohydrate. Unlike starch, which is insoluble polyglucose molecules formed in plastids, these are soluble polyfructose molecule, synthesized and stored in vacuoles (Fig. 9.27). Small fructans are sweet in taste but cannot be digested by human beings. Food industry has suggested their use as a sweetener, because of their low calorific value. Large fructans have fat-like texture, form emulsions, and have a neutral taste. Their use as low calorie fat has also been suggested by food industry.

**Fig. 9.26** Starch hydrolysis in endosperm of germinating cereal grains. (*1*) Starch is hydrolyzed to dextrins by α-amylase which is synthesized de novo in aleurone layer of the seed. (*2*) Dextrins are hydrolyzed by β-amylases which have been converted to active form from inactive form by the action of proteases synthesized de novo. (*3*) Resulting maltose fragments are hydrolyzed to glucose by glucosidase. Glucose is absorbed by the embryo and is converted to sucrose and is translocated out to the growing parts of the seedling

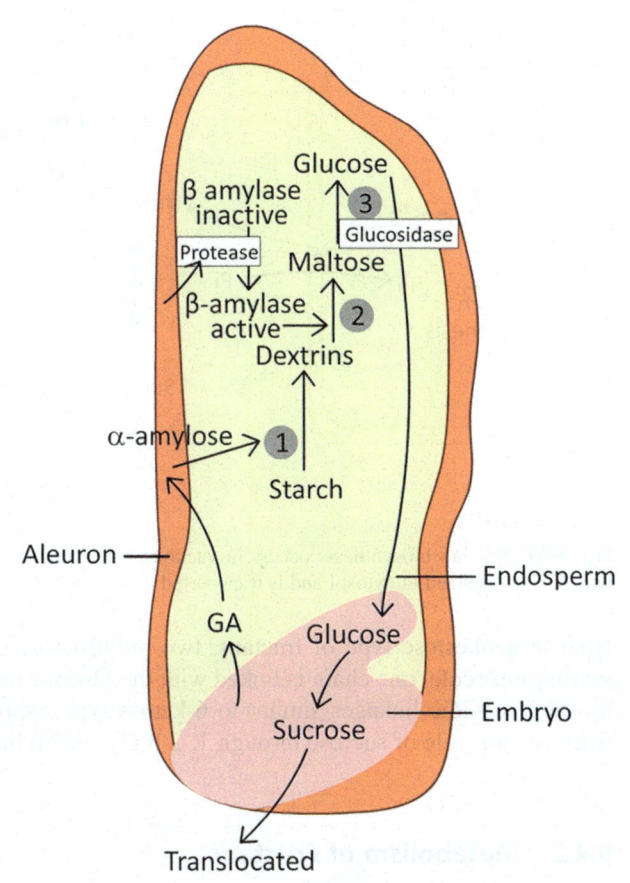

## 9.4.1 Structure of Fructans

Fructans are polymers based on sucrose, to which chains of fructose are linked by glycosidic linkages. A trisaccharide, **kestose,** is the parent molecule for fructans, which consists of fructose residue linked to a sucrose molecule. Depending upon the kind of linkage between sucrose and the other fructose molecule, there are three types of kestose, **1-kestose, 6-kestose,** and **neokestose,** from which three types of fructans are derived (Fig. 9.28). 1-Kestose types, also called inulin type of fructans, are found in dahlia tubers. These consist of a sucrose and up to 50 fructose molecules. In this type, fructose of sucrose molecule is linked to fructose, and to each other, by means of 1,2-β-glycosidic linkages. 6-Kestose types are levan type of fructans which are often found in grasses. In these type of fructans, C-6 of fructose residue of sucrose is linked to C-2 of another fructose molecule at β position by means of 6,2-β-glycosidic linkages and also to other fructose molecules by similar linkages. The number of fructose residues in the chain varies from 10 to 200. Third category of fructans is found in wheat and barley, which are known as neokestose

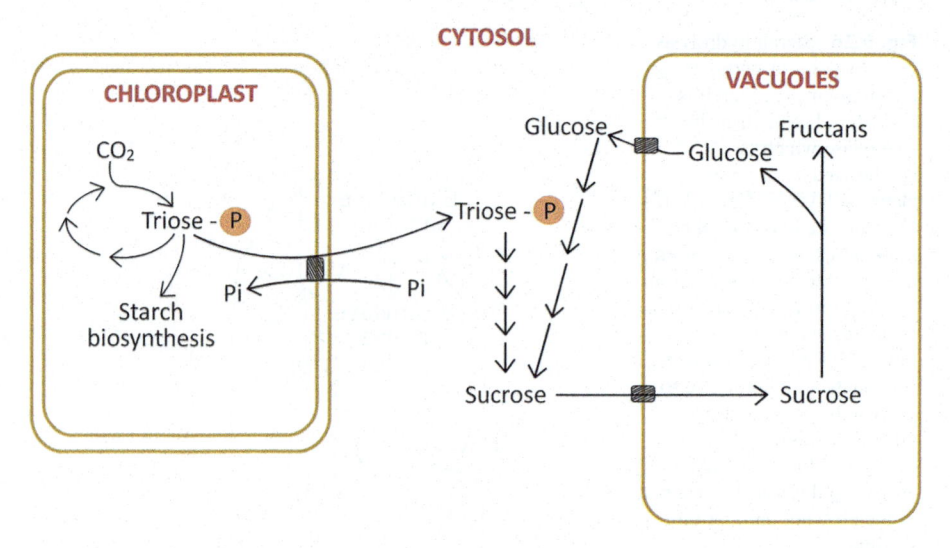

**Fig. 9.27** Fructan biosynthesis occurs in vacuole. Precursor for fructan biosynthesis is sucrose which is synthesized in cytosol and is transported to vacuoles

type. In neokestose type of fructans, two polyfructose chains are linked with the sucrose molecule, one chain is linked with the glucose molecule of sucrose through 6, 2-β glycosidic linkages similar to 6-ketose type, while other chain is linked to fructose molecule of sucrose through 1, 2-β glycosidic linkages as in 1-ketose type.

### 9.4.2 Metabolism of Fructans

Synthesis of fructans occurs in vacuoles of the cell of storage tissues. Large size of leaf vacuoles, often comprising of about 80% of the total leaf volume, offers a very storage advantage to the plant. Fructans are stored in addition to sucrose and starch. These can comprise up to 30% of the dry matter. Fructans are preferred as a reserve over other polysaccharides sometimes, especially preceding flowering or rapid seed growth or when plants are subjected to unfavorable conditions, such as cold or water stress. It is also suggested that fructan biosynthesis may occur to control sucrose synthesis in cytosol. Sucrose accumulation in cytosol may result in sugar-induced inhibition of photosynthesis. On the contrary, fructans may accumulate up to 70%, without inhibiting photosynthesis. Unlike synthesis of sucrose and starch, where UDP-glucose and ADP-glucose are used for their synthesis, respectively, fructans are synthesized without activation of their precursors. Thermodynamic requirement is entirely met through breakage of glucose-fructose bond ($\Delta G^{o'} = -29.3$ kJ.mol$^{-1}$). Precursor required for synthesis is sucrose molecule. If only sucrose is available as a substrate, the enzyme sucrose-fructan 6-fructosyl transferase (6-SFT) produces 6-ketose, which becomes substrate for further transfers, resulting in building of polymer with a series of chain length to produce levans. Elongation of chain is

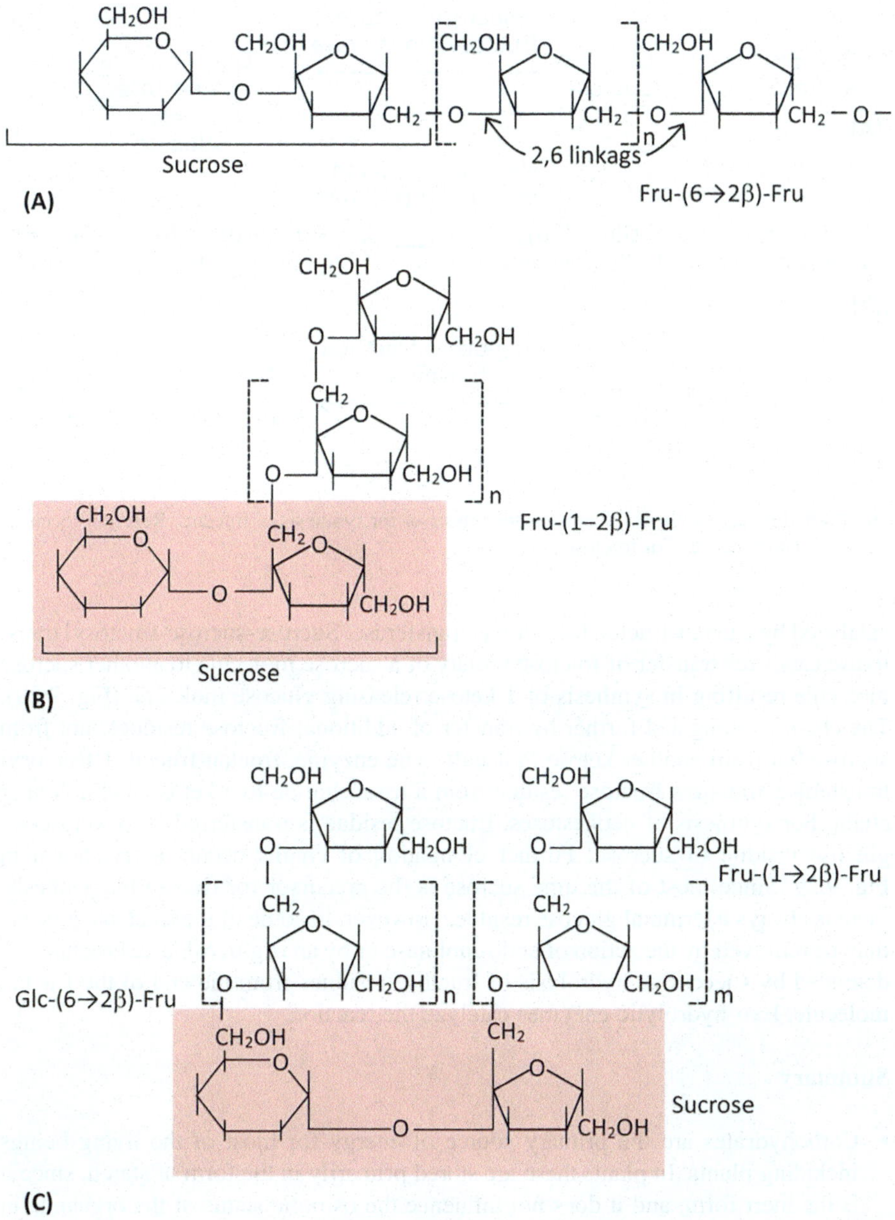

## 9.4 Fructans

**Fig. 9.28** (a) 6-Kestose consists of (2,6)-linked β-D fructosyl units to sucrose, and it is levan type of fructans, where n = 10–200; (b) 1-kestose consists of (2–1) linked β-D fructosyl unit to sucrose and is the shortest inulin molecule n > 50. (c) Neokestose is the β-D fructosyl unit to the C-6 of glucose moiety of sucrose n ≤ 10

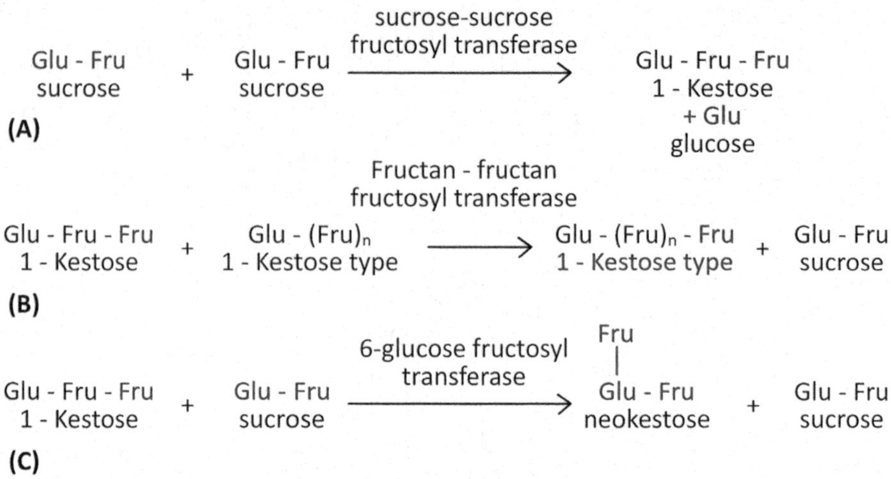

**(A)**

Glu - Fru sucrose + Glu - Fru sucrose $\xrightarrow{\text{sucrose-sucrose fructosyl transferase}}$ Glu - Fru - Fru 1 - Kestose + Glu glucose

**(B)**

Glu - Fru - Fru 1 - Kestose + Glu - (Fru)$_n$ 1 - Kestose type $\xrightarrow{\text{Fructan - fructan fructosyl transferase}}$ Glu - (Fru)$_n$ - Fru 1 - Kestose type + Glu - Fru sucrose

**(C)**

Glu - Fru - Fru 1 - Kestose + Glu - Fru sucrose $\xrightarrow{\text{6-glucose fructosyl transferase}}$ Glu - Fru neokestose (with Fru attached) + Glu - Fru sucrose

**Fig. 9.29** Fructan synthesis: Sucrose is the precursor for synthesis of fructans. Reactions occur in vacuoles. *Glu* glucose, *Fru* fructose

catalyzed by fructan-fructan 6-fructosyl transferase. Sucrose-sucrose-fructosyl transferase catalyzes transfer of fructose moiety of a sucrose molecule to another sucrose molecule resulting in synthesis of 1-ketose releasing glucose molecule (Fig. 9.29). The chain is elongated further by transfer of additional fructose residues, not from sucrose but from another ketose molecule. The enzyme, fructan-fructan 1-fructosyl transferase, transfers fructose residue from a trisaccharide to a ketose with a longer chain. For synthesis of neokestoses, fructose residue is transferred for 1-ketoses to glucose residue of sucrose. Further elongation of chains occurs as is shown in Fig. 9.29. Since most of the time sucrose is the precursor for fructan biosynthesis, there is always a terminal glucose residue. However, in some of the fructans, glucose may be removed by the action of endo-inulinase or by an α-glucosidase. Fructans are degraded by successive hydrolysis of fructose residues from the end of the fructan molecule. Exo-hydrolytic enzymes catalyze the reaction.

**Summary**

- Carbohydrates are the primary source of energy for most of the living beings including plants. In plants these are stored primarily in the form of starch, since it is the inert form, and it does not influence the osmotic status of the organelle in storage tissue. Starch is stored in plastids. In chloroplast, starch is temporarily stored during daytime (transitory starch) which is mobilized during night. Starch is also stored in amyloplasts of grains, as well as of tubers and roots of some plants. Both plastidial and cytosolic hexose monophosphate pools, consisting of glucose 1-phosphate, glucose 6-phosphate, and fructose 6-phosphate, contribute to carbohydrate metabolism.

- Sucrose is synthesized in cytosol of the cell, using triose phosphates. Export of triose phosphates from plastids is facilitated by the antiporters, localized in the inner envelope of the organelle. Triose phosphates from plastids are exchanged with cytosolic inorganic phosphate ($P_i$). Sucrose synthesis requires activation of glucose as UDP-glucose, which involves UDP-glucose pyrophosphorylase. Sucrose-phosphate synthase results in synthesis of sucrose phosphate, which is hydrolyzed by sucrose phosphatase to sucrose, with simultaneous release of $P_i$. This step-in sucrose synthesis makes the reaction irreversible. Sucrose is mainly hydrolyzed by sucrose synthase and invertase. Three isoforms of invertase are significant, apoplastic and vacuolar invertase, which function at acidic pH, while pH optima of cytosolic invertase are neutral or alkaline.
- Starch is synthesized in the plastids of storage organs, as well in chloroplasts (during day). Starch, which includes both amylose (linear polysaccharide) and amylopectin (branched polysaccharide), generally is present in semicrystalline granule form. In most cases ratio of amylose and amylopectin is 1:3; however, the ratio may vary in starches from different sources.
- Starch synthesis requires activated form of glucose as ADP-glucose. Synthesis of ADP-glucose is catalyzed by ADP-glucose pyrophosphorylase. This step is the dedicated step-in starch biosynthesis. Glucose as ADP-glucose is added to the nonreducing end of a starch primer. Reaction is catalyzed by starch synthase. Addition of branching to the polymer is catalyzed by branching enzyme. Excess branching of the molecule is removed by disproportion enzyme also known as D-enzyme.
- Starch is degraded both by phosphorolytic and hydrolytic enzymes. Phosphorolytic cleavage requires starch phosphorylases and debranching enzyme. α-amylase, β-amylases, debranching enzyme, and glucosidase are hydrolytic enzymes. Hydrolysis of transitory starch during night, and in cereal grains at the time of germination, is well understood. Degradation of transitory starch is facilitated by action of GWD and PWD, the enzymes which phosphorylate starch in the granules having semicrystalline structure, to increase hydrophilicity of the molecule, resulting in accessibility of hydrolytic enzymes.
- Some plants store carbohydrates in the form of fructans, which are water-soluble polymers, stored in the cell vacuoles. These are polymers of fructose having a terminal glucose molecule since these are synthesized from addition of fructose to sucrose molecule.

## Multiple-Choice Questions

1. Inner envelope of chloroplasts have got the:
   (a) Antiporters for glucose 6-phosphate/$P_i$ transport
   (b) Symporters for triose phosphates and $H^+$
   (c) Antiporters for triose phosphates/$P_i$
   (d) Symporters for xylulose 5-phosphate/$P_i$

2. Temporary starch stored in chloroplasts is known as:
   (a) Soluble plastid starch
   (b) Transitory starch
   (c) Insoluble plastid starch
   (d) Phytoglycogen
3. Precursor for sucrose biosynthesis in cytosol is:
   (a) ADP-glucose
   (b) Glucose 6-phosphate
   (c) Fructose 6-phosphate
   (d) UDP-glucose
4. Enzyme which catalyzes synthesis of precursor for sucrose biosynthesis is:
   (a) UDP-glucose pyrophosphorylase
   (b) PPi-dependent phosphofructokinase
   (c) Pyrophosphatase
   (d) Sucrose-phosphate synthase
5. Sucrose-phosphate synthase is:
   (a) Allosterically inhibited by glucose 6-phosphate
   (b) Allosterically activated by Pi
   (c) Inhibited by phosphorylation of Ser158 residue of the enzyme protein
   (d) Activated by phosphorylation of Ser158 residue of the enzyme protein
6. Activity of ADP-glucose pyrophosphorylase is promoted by:
   (a) High PGA/$P_i$ ratio
   (b) Low PGA/Pi ratio
   (c) ATP
   (d) When the subunits of the enzymes are oxidized
7. GBSS is responsible for synthesis of:
   (a) Semicrystalline starch
   (b) Phytoglycogen
   (c) Amylopectin
   (d) Amylose
8. Phosphorolytic enzymes are so called because they use:
   (a) $H_2O$ to break glycosidic bonds
   (b) ATP to break glycosidic bonds
   (c) Pi to break glycosidic bonds
   (d) Phosphorylate starch before cleavage
9. β-Amylases are so called because:
   (a) They act in addition to another enzyme α-amylase.
   (b) They produce β-maltose on hydrolysis of starch.
   (c) They are endoamylases.
   (d) Maltose is produced on hydrolysis of starch.
10. Phosphorylation of some glucosyl residues of starch by GWD/PWD at the starch grains is required because:
    (a) It increases accessibility by hydrolytic enzymes.
    (b) It increases hydrophobic nature of the glucosyl residues.
    (c) Glucose will be released as glucose 1-phosphate.
    (d) Phosphorylated starch is the substrate for the hydrolytic enzymes.

**Answers**

1. c  2. b  3. d  4. a  5. c  6. a  7. d
8. c  9. b  10. a

## Suggested Further Readings

Browsher C, Steer M, Tobin A (eds) (2008) Plant biochemistry. Garland Science, Tailor & Francis Group, New York, pp 195–235
Heldt HW (2005) Plant biochemistry, 3rd edn. Elsevier Academic Press, Amsterdam/Boston/Heidelberg, pp 243–269
Jones R, Ougham H, Thomas H, Waaland S (2013) The molecular life of plants. Wiley-Blackwell, Chichester, pp 211–213
Zeeman SC (2015) Carbohydrate metabolism. In: Buchanan BB, Gruissem W, Jones RL (eds) Biochemistry and molecular biology of plants. Wiley-Blackwell, Chichester, pp 567–609

# Lipid Metabolism 10

Manju A. Lal

Lipids are water-insoluble heterogeneous group of compounds, which are soluble in organic solvents. Besides being an important constituent of the diet, lipids have many other uses. A great diversity of lipids occurs in nature, and it is very important to get the right kind of lipid in a diet. Lipids are among the biomolecules required for structure and functions of the cell. Since they are mostly stored in seeds or fruits of some plants, these are the main sources for obtaining dietary lipids. Studying the diversity and the metabolism of lipids is important. A brief introduction to diversity in structure and functions of lipids is taken in this chapter besides learning about the classification and nomenclature of fatty acids. The chapter includes the pathways, the role of enzymes, and the organelles involved in the lipid metabolism. Studying the catabolism of lipids is significant in germinating fatty seeds, since they have primarily oils as the storage form. In fatty seeds the stored lipids are not the source of energy directly. Lipids are metabolized to produce soluble form of sugars in the storage organs, which are translocated to the meristematic regions in the growing seedlings. A study of pathways, enzymes, and organelles involved in lipid metabolism in germinating seeds is also included in the chapter. Though there is similarity in lipid metabolism of plants and animals, a great degree of disparity also exists. In plants lipid metabolism is more complex and involves many cell organelles. Fatty acid synthesis and fatty acid oxidation occur in different subcellular locations and involve different metabolic pathways and different sets of enzymes. There is successive removal of two-carbon units as acetyl-CoA during oxidation of fatty acids. Elongation of hydrocarbon chain also requires, at the time of fatty acid synthesis, addition of two carbons at a time. However, precursors for fatty acid synthesis are three-carbon compounds, malonyl-CoA, except one acetyl-CoA, which is required as a primer. The condensation reaction is coupled with simultaneous release of one carbon as $CO_2$ (Fig. 10.1).



S. C Bhatla, M. A. Lal, *Plant Physiology, Development and Metabolism*,
https://doi.org/10.1007/978-981-13-2023-1_10

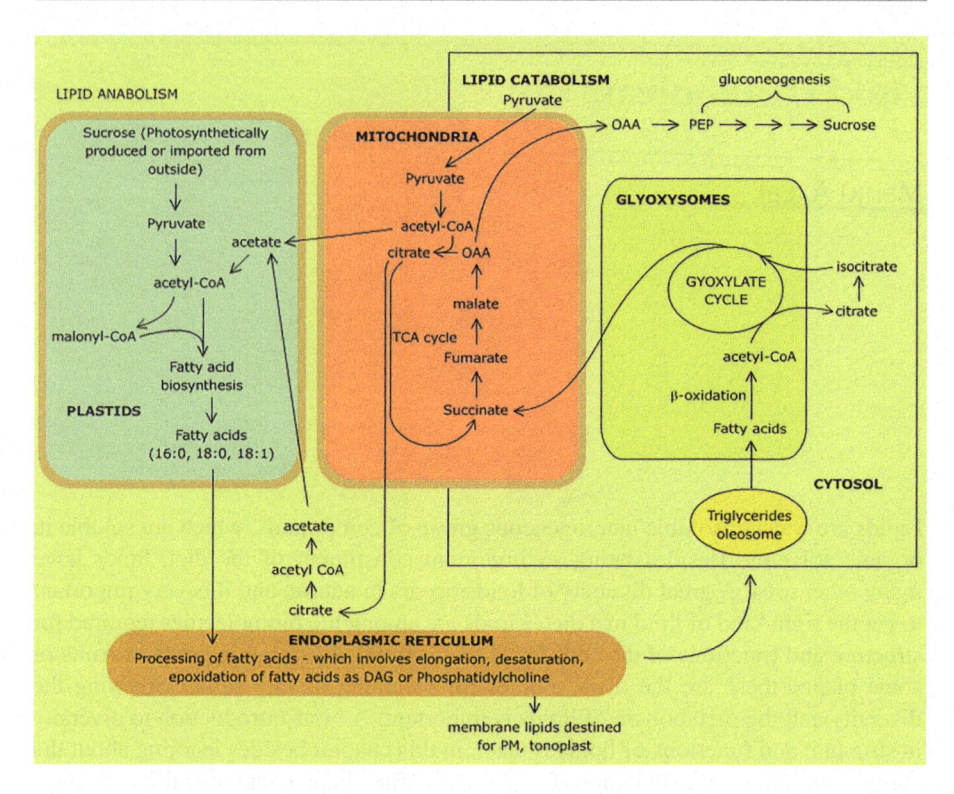

**Fig. 10.1** Overview of lipid metabolism in plants

## 10.1 Role of Lipids

Lipids are stored as neutral fats in seeds of many plants or are stored in fruits or other vegetative parts of some of the plants. Lipids are integral constituents of all cell membranes. Membrane lipids are **amphipathic** in nature, i.e., their one end is hydrophilic, and the other one is hydrophobic. This facilitates them to organize in the form of a bilayer with their hydrophilic ends facing toward the aqueous side of the cell (cytosolic side and outside aqueous environment of the cell), and hydrophobic ends face each other. Both hydrophobic interactions between the hydrophobic ends of the lipids and hydrophilic interactions of the polar groups with aqueous environment are important factors for organization of the membranes. Bimolecular layer of lipids along with proteins keeps the membrane structure intact and is key for existence of life. The membrane is responsible for maintaining the optimal chemical environment in the cell and various subcellular compartments for various metabolic

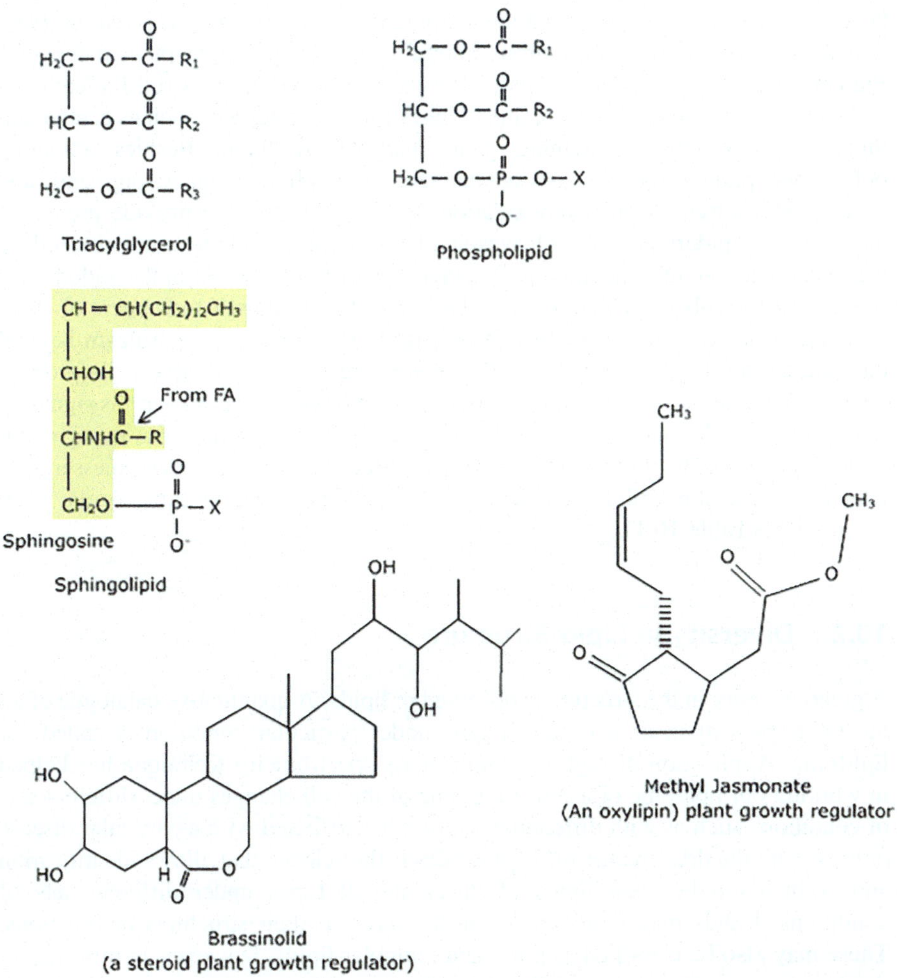

## 10.1 Role of Lipids

**Fig. 10.2** Some of the lipid biomolecules in plants. *FA* fatty acids

reactions to occur by providing limited permeability to selected molecules into and from the cell. Unique structures of thylakoids and those of mitochondria are responsible for carrying out light reactions of photosynthesis and oxidative reactions during respiration, respectively. Cofactors of many enzymes are lipids. Lipoic acid, ubiquinone, and plastoquinone are few of them. Vitamins such as vitamin D, A, E, and K are lipid soluble and are derived from five-carbon isoprene units. Many of the hormones are derived from sterols both in animal and plants. In plants, brassinosteroids are the growth regulators which are steroid in nature, while jasmonates are derived from linolenic acid (Fig. 10.2). Jasmonates are plant hormones which regulates plant development besides being involved in plant defense against insects, herbivorous animals, and many fungal plant pathogens. Pigments such as carotenoids impart colors to plants and animals (bird's feathers). Colors of feathers of the birds are due to ingestion of carotenoid (such as canthaxanthin and zeaxanthin) containing plant material by them. Besides attracting pollinators, plant pigments also act as secondary photosynthetic pigments and protect chlorophylls from photooxidation. Lipids such as cutin provide protective barrier to the epidermal cells of leaves and being impermeable to water also reduce water loss because of transpiration. Another type of lipid, present in the endodermis of root cells, is suberin which provides barrier to the **apoplastic** movement of water from the cortex to vascular tissues. Water has to travel through **symplasm** beyond the endodermis layer in the root cell. Suberin present in epidermal cells protects some of the tubers from infections. In plasma membranes lipids also act as signaling molecules and play an important role in cell signaling. In addition to being membrane lipid, phosphatidylinositol serves as a precursor for secondary messengers, which regulate intracellular concentration of calcium, and in turn regulate cell metabolism (Table 10.1).

## 10.2 Diversity in Lipid Structure

A great diversity in the structure is observed in lipids. A quantitative catalogue of all the lipids present in a specific cell type under particular condition is called the **lipidom**e. Application of high-resolution mass spectrometry technique has helped in working out lipidome of a cell. Lipidome of the cell changes under different sets of conditions such as with differentiation or when afflicted with a particular disease or even with the drug treatment. **Lipidomics** is the science that allows identification and comparison between lipids of different cell types under different sets of conditions. Lipids may be classified on the basis of their structures or functions. These may also be classified as polar and nonpolar lipids. However, in this chapter these are organized under the following categories.

### 10.2.1 Fatty Acids

Fatty acids are the structural components of lipids. These are derivatives of hydrocarbons. These consist of hydrocarbon chain and a carboxylic group. Generally fatty acids contain an even number of carbon atoms varying from 4 to 36. However, most common fatty acids contain 12–24 carbon atoms. Hydrocarbon chain may be branched or unbranched. The most common fatty acids have unbranched hydrocarbons. Hydrocarbon chain of fatty acids may not have any

## 10.2 Diversity in Lipid Structure

**Table 10.1** Diverse roles of lipids in a plant

| Lipids | Functions | Chemical nature |
|---|---|---|
| Triacylglycerol | Storage | Esters of glycerol, all three hydroxyl groups of glycerol are esterified with fatty acids |
| Wax[a] | Storage (only in jojoba plant), protection of plant surface from dehydration and predators | Esters of long-chain alcohol and fatty acids having long hydrocarbon chain |
| Phospholipids | Membrane lipids | Amphipathic molecules, the polar phosphate group esterified with another polar group, which is choline (phosphatidylcholine), serine (phosphatidylserine), ethanolamine (phosphatidylethanolamine), or inositol (phosphatidylinositol) |
| Cardiolipins | Inner membrane of the mitochondria, involved in energy metabolism and also in mitochondrial apoptosis | Diphosphatidyl glycerol (two phosphatidic acid moieties connect with the glycerol backbone in the center) |
| Galactolipids | Lipids in thylakoids | Membrane lipids without any phosphate group. Third hydroxyl group of diacylglycerol esterified with galactose sugars |
| Sulfolipids | Lipids present in thylakoids of chloroplasts | Esterification of two fatty acids to glycerol and a sulfoquinovosyl group |
| Sphingolipids | Membrane lipids, may be involved in programmed cell death | Amphipathic lipid, instead of glycerol an amino alcohol sphingosine |
| Sterols | Membrane lipids, growth regulator brassinosteroid | Fused ring structure, cholesterol is absent in plants. Phytosterols are present |
| Cutin[a] | Prevents free diffusion of hydrophilic substances so protects from dehydration, protects aerial organs of plants from predators | A tough sheet like complex structure made of long carbon chain oxygenated fatty acids esterified with other fatty acid containing hydroxyl groups in their hydrocarbon chains |
| Suberin | Prevents flow of water and hydrophilic substance, generally present in the plant parts not exposed to air | Similar to cutin, except the presence of dicarboxylic fatty acids and phenolic compounds |
| Oxylipins | Defense against herbivorous animals | Jasmonates derived from linolenic acid |
| Inositol phosphates | Cell signaling | Derived from phosphatidylinositol phosphate |

[a]In aerial organs bulk of fatty acids are devoted to production of wax and cutin for protection

double bond and be saturated, while others may be unsaturated because of the presence of double bonds in them. The number and position of double bonds are characteristics of unsaturated fatty acids. If hydrocarbon chain of a fatty acid contains double bonds, two consecutive double bonds are separated by $-CH_2$ (methylene) group. A simplified system for abbreviations has been adopted for fatty acid structures by giving two digits which are separated by colon. The first digit indicates the number of carbon atoms in the hydrocarbon chain, while the number after colon indicates the number of double bonds in the hydrocarbon chain. For example, palmitic acid is abbreviated as 16:0. The first digit 16 indicates that hydrocarbon chain consists of 16 carbons, and the second digit (0) after colon indicates that the number of double bond is zero. Stearic acid is abbreviated as 18:1, which indicates that the hydrocarbon chain of fatty acid is 18 carbons long and it is an unsaturated fatty acid with one double bond being present. The position of double bond may be specified by delta nomenclature according to which structure of stearic acid can be abbreviated as $18:1^{\Delta 9}$ which indicates the presence of double bond in between the ninth and tenth carbon atom, when the carbon of carboxylic group is numbered as one. If two double bonds are present, the number following the colon will be two. The structure of linoleic acid will be written as $18:2^{\Delta 9,12}$. Table 10.2 provides the structure and abbreviations for some of the fatty acids. There is another way of nomenclature for unsaturated fatty acids. Instead of indicating position of double bond from the carboxylic group (as in delta system of nomenclature), the position of double bond is shown with reference to methylene group ($-CH_3$) of fatty acid as omega ($\omega$), which refers to the carbon most distant from carboxylic group and is numbered as carbon 1. This type of nomenclature is significant for polyunsaturated fatty acids (PUFA) in which double-bond near-omega ($\omega$) carbon is of physiological significance. PUFAs with double bond present in between the third and fourth carbon from the methylene group ($\omega$) are called **omega-3 ($\omega$-3) fatty acids**, while PUFAs with double bonds in between the sixth and seventh carbon atoms are called **omega-6 ($\omega$-6) fatty acids**. Fatty acids may remain in gel or fluid state which is determined by the surrounding temperature and also by the features of their hydrocarbon chain. Saturated fatty acids with 12–24-carbon-long hydrocarbon chain have waxy consistency at 25 °C, while unsaturated fatty acids remain in oil form. This property of fatty acids is due to the nature of packing of the extended and flexible hydrocarbon chain. The presence of double bonds in *cis* configuration causes a kink in the hydrocarbon chain. As a result, their compact packing in membranes is not possible, and they remain in fluid consistency. Almost all the naturally occurring unsaturated fatty acids contain double bonds in *cis* configuration (Box 10.1).

## 10.2 Diversity in Lipid Structure

**Table 10.2** Some of the fatty acids present in plant

| Name of the fatty acids | Abbreviation | Systematic name | Structure |
|---|---|---|---|
| **Saturated fatty acids** | | | |
| Lauric acid | 12:0 | n-Dodecanoic acid | $CH_3(CH_2)_{10}COOH$ |
| Palmitic acid | 16:0 | n-Hexadecanoic acid | $CH_3(CH_2)_{12}CH_2CH_2COOH$ |
| Stearic acid | 18:0 | n-Octadecanoic acid | $CH_3(CH_2)_{12}CH_2CH_2CH_2COOH$ |
| Arachidic acid | 20:0 | n-Eicosanoic acid | $CH_3(CH_2)_{12}CH_2CH_2CH_2CH_2COOH$ |
| **Monounsaturated fatty acids (one double bond)** | | | |
| Palmitoleic acid | 16:1$^{\Delta 9}$ | CIS-9-Hexadecenoic acid | $CH_3(CH_2)_5C\overset{H}{=}\overset{H}{C}(CH_2)_7COOH$ |
| Oleic acid | 18:1$^{\Delta 9}$ | CIS-9-Octadecenoic acid | $CH_3(CH_2)_7C\overset{H}{=}\overset{H}{C}(CH_2)_7COOH$ |
| Petroselenic acid | 18:1$^{\Delta 6}$ | CIS-6-Octadecenoic acid | $CH_3(CH_2)_{10}C\overset{H}{=}\overset{H}{C}(CH_2)_4COOH$ |
| Erucic acid | 22:1$^{\Delta 13}$ | CIS-13-Docosenoic acid | $CH_3(CH_2)_7C\overset{H}{=}\overset{H}{C}(CH_2)_{11}COOH$ |
| **Polyunsaturated fatty acids (>1 double bonds)** | | | |
| Linoleic acid | 18:2$^{\Delta 9,12}$ | CIS, CIS-9, 12 Octodecadienoic acid | $CH_3(CH_2)_4-\overset{H}{C}=\overset{H}{C}-CH_2-\overset{H}{C}=\overset{H}{C}-(CH_2)_7COOH$ |
| α-Linolenic acid | 18:3$^{\Delta 9,12,15}$ | all CIS 9, 12, 15 Octadecatrienoic acid | $CH_3(CH_2)_2-\overset{H}{C}=\overset{H}{C}-CH_2-\overset{H}{C}=\overset{H}{C}-CH_2-\overset{H}{C}=\overset{H}{C}-(CH_2)_7COOH$ |
| **Some unusual fatty acids** | | | |
| Ricinoleic acid | 12-OH-18:1$^{49}$ | 12-Hydroxyoctadec-9-enoic add | $CH_3(CH_2)_5-\overset{OH}{\underset{H}{C}}-CH_2-\overset{H}{C}=\overset{H}{C}(CH_2)_7COOH$ |

**Box 10.1: Adaptation in Lipid Composition of Plant Membranes in Response to the Change in Environment Temperature**

At room temperature, fatty acids occur either in waxy consistency or as oils. This property of fatty acids is determined by length, saturation status, and configuration of double bonds in their hydrocarbon chain. The presence of more double bonds will result in reduced melting temperature of fatty acids. Free rotation around carbon atom in fatty acids with saturated hydrocarbons allows their compact packing unlike unsaturated fatty acids where the presence of double bond in the hydrocarbon chain restricts the free rotation. Unsaturated fatty acids can either be *trans* or *cis* depending upon the arrangement of hydrogen atoms around double bonds. Compact packing of lipids occurs in membranes when double bonds of hydrocarbon chains of unsaturated lipids are in *trans* configuration because of the chain being straight as in the case of saturated fatty acids. Compact packing of lipids will restrict their movement resulting in their waxy consistency and affecting permeability of the membrane. Contrary to this, *cis*- configuration causes a 30° kink in the hydrocarbon chain because of which close packing of lipid is not possible and the membrane will be more fluid. Plants adjust to changing temperature by altering the ratio of lipids with unsaturated fatty acids to lipids with saturated hydrocarbon chains. One way by which plants adapt to freezing temperature (freezing tolerance) is by increasing lipids with unsaturated fatty acids.

Melting temperature of some of the fatty acids:

| Name of fatty acid | Abbreviations for the fatty acids | Melting temperature °C |
|---|---|---|
| Lauric acid | 12:0 | 40 |
| Palmitic acid | 16:0 | 63 |
| Stearic acid | 18:0 | 70 |
| Oleic acid | $18:1^{\Delta 9}$ | 13 |
| Linoleic acid | $18:2^{\Delta 9,12}$ | −5 |
| Linolenic acid | $18:3^{\Delta 9,12,15}$ | −11 |
| Arachidonic acid | $20:4^{\Delta 5,8,11,14}$ | −49.5 |

## 10.2.2 Storage Lipids (Neutral Fats, Waxes)

Most common form of storage lipids are triglycerides (neutral fats). For some plants, neutral fats are preferred form of storage in seeds, mainly because fats contain more reduced carbon so it will be the source for more energy in comparison to carbohydrates. Secondly, neutral lipids are anhydrous, so they are stored without water unlike carbohydrates which retain hydration of water. Thus, additional weight due to retention of water is not added in case of fats being stored, and seeds remain lighter. This makes them suitable for their dispersal. Triacylglycerol are esters of

## 10.2 Diversity in Lipid Structure

glycerol in which all three hydroxyl groups of glycerol are esterified with fatty acids. The three hydroxyl groups of a glycerol may be esterified by similar fatty acids or may be esterified by different fatty acids. In case all three hydroxyl groups are esterified by similar fatty acids, these are named according to the fatty acids, e.g., tripalmitin, tristearin, or triolein if fatty acids are palmitic acid, stearic acid, or oleic acid, respectively. In case hydroxyl groups of glycerol are esterified by two or three different fatty acids (mixed), the name and position of each fatty acid are specified. Carbon atoms of glycerol are given stereospecific numbers ($sn$) with $sn1$ being the top carbon of the glycerol, $sn2$ is the middle one, and $sn3$ is the bottom carbon atom. In plants, synthesis of neutral fats occurs within the endoplasmic reticulum membranes, and these are stored in oil bodies of endosperm cells of the seed. In some cases, triglycerides may be stored in fruits, stems, or roots. *Tetraena mongolica* Maxim is an example of plant species where triacylglycerols are stored in stem tissues. This plant is also called as "oil firewood." Almost 9% of the dry weight of plant is triacylglycerol. In *Manihot esculenta*, triacylglycerol accumulation occurs in roots in which about 25% of the dry weight is oil and 30% is starch, which are accumulated in tubers. Waxes are complex molecules consisting of mixture of long-chain hydrocarbons, alcohols, aldehydes, acids, and esters of long-chain alcohols (C14–C36) with fatty acids having long hydrocarbon chains (C16–C30). These are diverse group of molecules whose composition varies from plant to plant. They have higher melting point (60–100 °C) than triacylglycerols. These are highly hydrophobic and have water-repellent property. Waxes are present within the cutin network as an amorphous substance. They are spread on the cutin surface in leaves of many plants such as those of rhododendrons. It makes cutin highly impermeable to water. As a result, plants are protected from excessive loss of water and also from parasites. Waxes are produced through action of **wax synthase** in the ER membrane and are exported to the cell wall. Their transport across plasma membrane involves **ABC transporters**. The mechanism of transfer of wax monomers from epidermal cells to the plant surface is poorly understood. It possibly involves **lipid transfer proteins**. In aerial organs of plants, a bulk of fatty acids is directed toward synthesis of wax and cutin. Some plant waxes are commercially important, such as carnauba wax, which is found on the surface of Brazilian palm tree. The only angiosperm in which wax is used for storage purpose is jojoba desert plant (*Simmondsia chinensis*). Liquid wax obtained from the fruits of the plant is used in cosmetic industry.

### 10.2.3 Membrane Lipids

Amphipathic nature of membrane lipids makes them different from storage lipids. These are classified as below:

*Glycerophospholipids* These lipids have glycerol as the backbone of their structure. Carbons 1 and 2 are esterified with fatty acids, while carbon 3 is linked with phosphate group by phosphoester linkage. This molecule is known as phosphatidic acid (PA). The presence of nonpolar fatty acids at one end of the molecule makes it hydrophobic, while a polar group (an alcohol) linked via phosphodiester linkage at the other end makes it hydrophilic. Depending upon the type of polar group, which are choline, ethanolamine, serine, inositol, or glycerol, these are called as phosphatidylcholine (PC), phosphatidylethanolamine (PE), phosphatidylserine (PS), phosphatidylinositol (PI), or diphosphatidylglycerol, respectively (Fig. 10.3). Phospholipids are present in membranes of bacteria, animals, and plants. Tables 10.3 and 10.4 show distribution of membrane lipids in different cell membranes of a plant cell.

*Glycolipids* Plant membranes contain glycolipids and sulfolipids in addition to phospholipids. Glycolipids and sulfolipids constitute major membrane lipids in thylakoids and in cyanobacteria. They have one or two galactose residues attached to carbon 3 of glycerol in place of phosphate. Depending upon the number of galactose residues attached, these are named as monogalactosyldiacylglycerol (MGD) and digalactosyldiacylglycerol (DGD). Sulfolipids (SL) are glycerolipids in which polar group at carbon 3 of diacylglycerol consists of a glucose molecule which is attached to sulfonic acid group at the sixth carbon atom. These are also known as sulfoquinovosyldiacylglycerol (SOD). Galactolipids are the chief constituents of thylakoids and chloroplast envelop. In green plants, 70–80% of the membrane lipids constitute thylakoids. Galactolipids are the most abundant lipids on this earth. This indicates the adaptation of the plants since plants get limited supply of phosphorus and major membrane lipids of plants (both galactolipids and sulfolipids) do not require phosphorus.

*Sphingolipids* Sphingolipids are also the membrane lipids which are amphipathic. Instead of glycerol they consist of a long-chain amino alcohol **sphingosine**, which is derived from serine and fatty acyl-CoA. In place of carbons from glycerol in glycerolipids, the three carbons of sphingolipids are derived from serine. Hydrophobic end of sphingolipids consists of two hydrophobic acyl chains. One of the acyl group linked to C 1 of the molecules is long-chain base (most common chain being 18 carbon). Another acyl group having 14–26 carbon atoms is a N-linked fatty acid which is linked to C-2 of the molecule via an amide linkage. A molecule, with two acyl groups attached to it, is known as **ceramide**. Hydrophilic end of the molecule consists of a large array of polar head groups which are bound to hydroxyl group at C 3 of ceramide. In plants two major classes of complex sphingolipids have been identified. They are glucosylceramides or glucocerebrosides and glycosylinositol phosphoceramides. Sphingolipids are major components of endomembranes, and $\geq$40% of plasma membrane also has sphingolipids. They are also present in tonoplast. These are synthesized in the endoplasmic reticulum and Golgi apparatus. In plants sphingolipids are involved in **programmed cell death (PCD)** associated with defense.

*Sterols* Sterols are membrane lipids consisting of four fused rings. Three of the rings consist of six carbons, while the fourth one has five carbons. The presence of hydroxyl group at 3 C position in ring A makes this end of the molecule polar,

## 10.2 Diversity in Lipid Structure

| Name of the lipid | Structure | |
|---|---|---|
| Phosphatidic acid | $H_2C-O-\overset{O}{\underset{\|}{C}}-R$ <br> $HC-O-\overset{O}{\underset{\|}{C}}-R$ <br> $H_2C-O-\overset{O}{\underset{\|}{P}}-O^-$ <br> $\quad\quad\quad O^-$ <br> diacylglycerol | Minor membrane component, a signal substance. |
| Phosphatidylcholine | Diacylglycerol$-O-\overset{O}{\underset{\|}{P}}-O-CH_2-CH_2-\overset{CH_3}{\underset{CH_3}{N^+}}-CH_3$ <br> $\quad\quad\quad\quad\quad\quad\quad\quad O^-$ <br> Choline | ER membrane and Plasma membrane |
| Phosphatidylserine | Diacylglycerol$-O-\overset{O}{\underset{\|}{P}}-O-CH_2-\overset{NH_3^+}{\underset{COO^-}{C}}-H$ <br> $\quad\quad\quad\quad\quad\quad O^-$ <br> Serine | Plasma membrane and ER membrane |
| Phosphatidyl-ethanolamine | Diacylglycerol$-O-\overset{O}{\underset{\|}{P}}-O-CH_2-CH_2-NH_3^+$ <br> $\quad\quad\quad\quad\quad\quad O^-$ <br> Ethanolamine | ER membrane and Plasma membrane |
| Phosphatidylinositol | Diacylglycerol$-O-\overset{O}{\underset{\|}{P}}-O$ — Inositol ring (with OH groups) | ER membrane and Plasma membrane |
| Phosphatidylglycerol (cardiolipin) (CL) | Diacylglycerol$-O-\overset{O^-}{\underset{\|}{P}}-O-CH_2$ <br> $\quad\quad\quad\quad O \quad HC-OH$ <br> $\quad\quad\quad\quad O^- \quad \|$ <br> Diacylglycerol$-O-\overset{\|}{\underset{O}{P}}-O-CH_2$ | Inner mitochondrial membrane |
| Monogalactosyl diacylglycerol (MGD) | Galactose ring with CH$_2$OH, O-Diacylglycerol, OH groups | Thylakoid membrane |
| Digalactosyldiacyl glycerol (DGD) | Two galactose rings with O-CH$_2$ linker, O-Diacylglycerol | Thylakoid membrane |
| Sulfolipids (SL) sulfoquinovosyl glycerol (SL) | $O^--\overset{O}{\underset{O}{S}}-CH_2$ — sugar ring — O-Diacylglycerol | Thylakoid membrane |

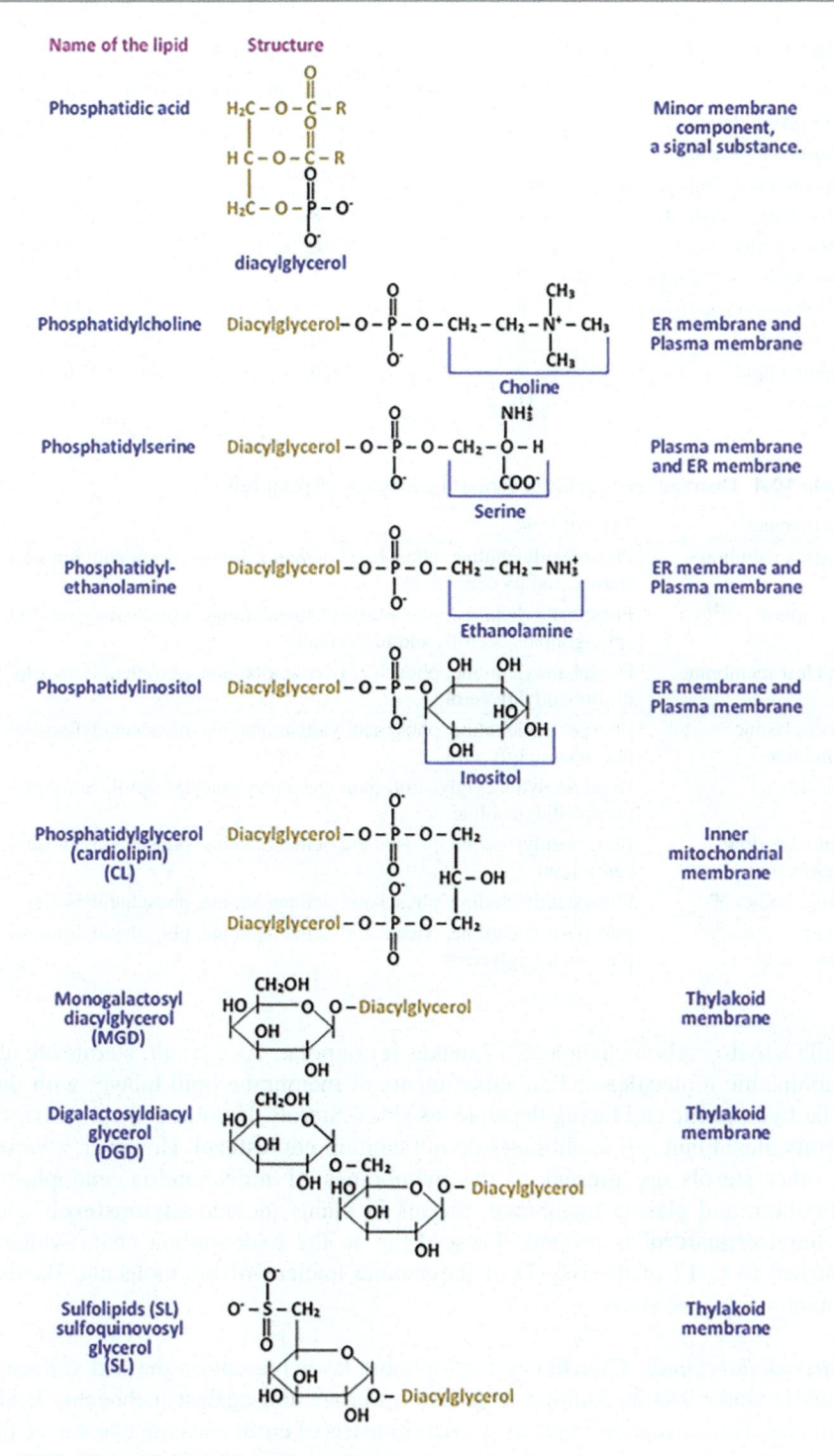

**Fig. 10.3** Glycerolipids of the cell membranes

**Table 10.3** Lipid composition of some of the membranes of plant cell

| Lipid | Plasma membrane | Endoplasmic reticulum | Chloroplast |
|---|---|---|---|
| Phosphatidylcholine | 19 | 45 | 5 |
| Phosphatidylserine | 3 | 1 | 0 |
| Phosphatidylethanolamine | 17 | 15 | 1 |
| Phosphatidylinositol | 2 | 8 | 0 |
| Phosphatidylglycerol | 12 | 6 | 11 |
| Monogalactosyldiacylglycerol | 2 | 1 | 42 |
| Digalactosyldiacylglycerol | 3 | 2 | 33 |
| Sulfolipid | 0 | 0 | 5 |
| Sphingolipid | 7 | 10 | 0 |
| Sterols | 31 | 5 | 0 |

**Table 10.4** Distribution of lipids in various membranes of plant cell

| Membranes | Type of lipid |
|---|---|
| Plasma membrane | Phosphatidylcholine, phosphatidylethanolamine, phosphatidylinositol, sterols, and its derivatives |
| Tonoplast | Phosphatidylcholine, phosphatidylethanolamine, phosphatidylinositol, sphingolipids, sterols, and its derivatives |
| Nuclear membrane | Phosphatidylcholine, phosphatidylethanolamine, phosphatidylinositol, phosphatidylglycerol |
| Endoplasmic reticulum | Phosphatidylcholine, phosphatidylethanolamine, phosphatidylinositol, phosphatidylglycerol |
| Plastids | Digalactosyldiacylglycerol, monogalactosyldiacylglycerol, sulfolipids, phosphatidylcholine |
| Mitochondrial membrane | Phosphatidylcholine, phosphatidylethanolamine, phosphatidylinositol, cardiolipin |
| Golgi bodies | Phosphatidylcholine, phosphatidylethanolamine, phosphatidylserine |
| Glyoxysomes, peroxisomes | Phosphatidylcholine, phosphatidylethanolamine, phosphatidylinositol, phosphatidylglycerol |

while a hydrocarbon chain at C 17 makes it nonpolar. As a result, sterols are also amphipathic molecules and are constituents of membrane lipid bilayer with their polar hydrophilic end facing the aqueous sides. Sterols determine the properties of membranes. Plant cell membranes do not contain **cholesterol**. However, a variety of other sterols are present in the membranes of mitochondria, endoplasmic reticulum, and plasma membrane. Sterols in plants include **stigmasterol**, while in fungi **ergasterol** is present. These differ in the hydrocarbon chain which is attached to C 17 of the ring D of the sterane nucleus of the molecule. Bacteria cannot synthesize sterols.

*Extracellular Lipids* Cuticle is a hydrophobic layer present on the leaf surface. It restricts water loss in addition to providing protection against pathogens. It also provides skeletal support to plants. Cuticle consists of **cutin** and cuticular wax. Cutin is characteristics of the plant cells. It is present exterior to leaf epidermal cells and is cross-linked with the cell wall. It is produced by the non-photosynthetic epidermal cells of leaves. Cutin is a polymeric molecule consisting of oxygenated fatty acids with a carbon chain length of 16 or 18. Monomers of cutin are monohydroxy, polyhydroxy, or epoxy fatty acids. Constituent fatty acids of cutin form ester linkages either with glycerol or with primary or secondary hydroxyl groups present in other fatty acids. This results in the formation of inelastic hydrophobic framework. Because of the space in between, water can be lost from the leaf surface. These pores are plugged with cuticular wax. Pathogens need to produce cutinase in order to facilitate their entry into the plant cells. Another extracellular lipid significant for plants is **suberin**. Suberin is also a polymeric molecule but differs from the cuticle in its composition of fatty acids. In suberin the fatty acids have longer carbon chains of 18–22 carbon than cutin. Unlike in cutin, fatty acids in suberin are not oxygenated. They do not have secondary alcohol or epoxy groups. A high amount of dicarboxylic acids is also present in suberin. Additionally, phenylpropanoids are also present which are partly linked with each other as in lignin. Suberin differs from cutin in getting deposited toward the inner side of the primary cell wall, while cutin is deposited toward the outside of the epidermal cell wall. Deposition of cutin is polar since it is deposited only on one side of the epidermal cells. Cutin is present at the interface of plants and the environment, while suberin is generally deposited all around the plasma membrane of the cells. **Casparian strips** of endodermis consist of suberized cells which make the cell wall impermeable to water. As a result, water and minerals are not transported to the vasculature of the plant through the apoplast; rather they have to move through the symplast beyond the endodermis in roots. Therefore, water and minerals do not diffuse back from the plant vasculature, thereby contributing to root pressure. In C4 plants, bundle sheath cells have suberin depositions in the cell wall, which makes it impermeable to $CO_2$. Suberin is present in the cork tissue which consists of dead cells surrounded by alternating layers of suberin and wax. Suberin is deposited in response to abiotic stress and mechanical injury. Waxy layer present on pollen grains plays a significant role in pollen-pistil interactions. It is known as exine in which other lipids are also embedded. Exine consists of complex polymer—sporopollenin—which is similar to cutin in its structure. Studies carried out with *Arabidopsis* mutants have demonstrated that altered lipid composition of exine induces male sterility, indicating significance of exine in pollen-pistil interaction.

## 10.3 Fatty Acid Biosynthesis

Unlike carbohydrates, lipid biosynthesis occurs in every cell at the site of their utilization, since the mechanism for the transport of lipids in plants is not known. Unlike in animals where the site for fatty acid (FA) biosynthesis is cytosol, in plants fatty acids having hydrocarbon chain up to 16 or 18 long are synthesized in the chloroplasts of green cells and leucoplasts or chromoplasts in nongreen cells. However, both in plants and animals, further elongation of the hydrocarbon chain occurs in the endoplasmic reticulum. Biosynthesis of fatty acids in plants is quite similar to that in bacteria. Precursors required for FA biosynthesis are acetyl-CoA (required as primer) and malonyl-CoA. Enzymes required are acetyl-CoA carboxylase (ACCase) and fatty acid synthase (FAS).

## 10.3.1 Synthesis of Acetyl-CoA

Since synthesis of fatty acids involves condensation of two-carbon fragments, most fatty acids consist of an even number of carbon atoms. Acetyl-CoA is required as the precursor. Since acetyl-CoA is not able to cross the membranes, it is produced in plastids. In chloroplasts, acetyl-CoA is derived from an intermediate of Calvin cycle, i.e., 3-phosphoglycerate, which is converted to pyruvate during glycolysis (plastids also have glycolytic enzymes). Pyruvate dehydrogenase complex (PDC) catalyzes oxidation and decarboxylation of pyruvate to acetyl-CoA. A substantial activity of pyruvate dehydrogenase (PDH), a component of PDC, has been demonstrated in chloroplasts. In nongreen plastids pyruvate is obtained from cytosol of the cell. Pyruvate is produced by the conversion of sucrose to pyruvate in the cytosol through glycolysis. Sucrose is imported from the outside. After being transported to plastids, pyruvate is converted to acetyl-CoA by PDC. It is also produced in plastids from acetate, which is imported either from mitochondria or cytosol. In the cytosol, **ATP citrate lyase** is responsible for the synthesis of acetyl-CoA from citrate, which has been exported from mitochondria. Acetate is activated to acetyl-CoA in the stroma of plastids by acetyl-CoA synthase (ACS) (Fig. 10.4). In the cytosol, acetyl-CoA is also used as the precursor of malonyl-CoA, which is required for elongation of the fatty acids in the endoplasmic reticulum.

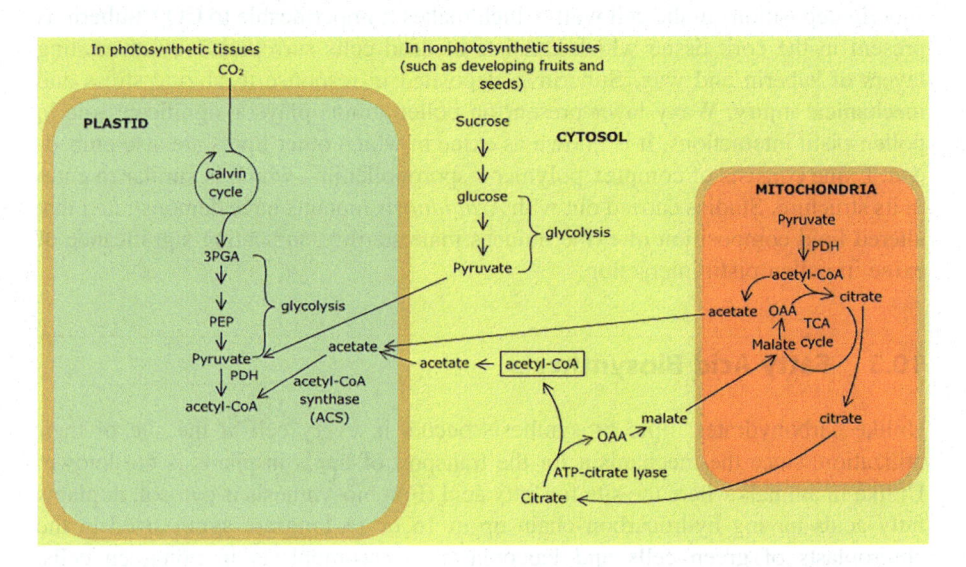

**Fig. 10.4** Different sources for acetyl-CoA, a precursor required in lipid biosynthesis

## 10.3.2 Synthesis of Malonyl-CoA

Carboxylation of acetyl-CoA produces malonyl-CoA, which is the activated form of acetyl-CoA. Reaction is catalyzed by **acetyl-CoA carboxylase (ACCase)**. This is the first step in biosynthesis of fatty acids and is a rate-limiting step. Cofactor for the enzyme ACCase is biotin, which is covalently linked to ε-amino group of a specific lysine in the enzyme protein. There are two types of ACCase in plant cells. Plastidial ACCase is similar to that of bacteria, while cytosolic ACCase is eukaryotic in origin. Plastidial ACCase catalyzes production of majority of malonyl-CoA required for fatty acid biosynthesis in plants, while malonyl-CoA, produced outside plastids by cytosolic ACCase, is used for elongation of fatty acid chains in the ER, for flavonoid biosynthesis, and for synthesis of aminocyclopropane carboxylic acid, the precursor of ethylene (Fig. 10.5). Plastidial ACCase is a heteromeric protein (HET-ACCase) with a molecular mass 650 kDa. It consists of four subunits: biotin carboxyl carrier protein (BCCP, the subunit to which biotin is covalently linked), biotin carboxylase (BC), and two subunits (α and β) of carboxyl transferase (CT). Three out of the four subunits of plastidial ACCase are encoded in nuclear genome, while β subunit of carboxyl transferase is encoded in plastid genome itself. On the contrary, cytosolic ACCase in plants is similar to ACCase present in animals and fungi. It is a

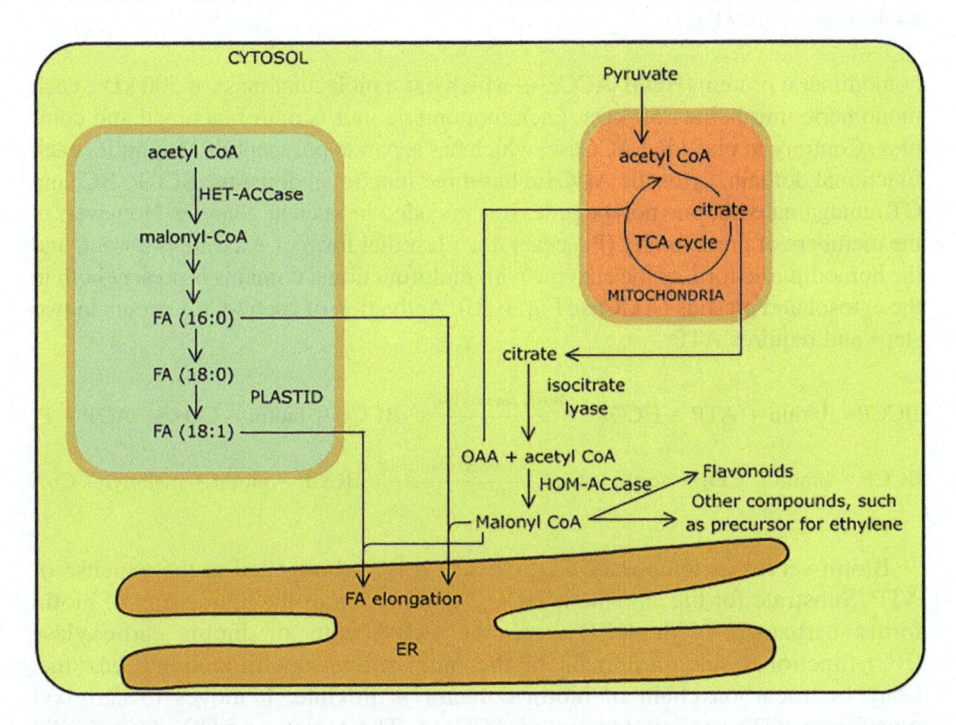

**Fig. 10.5** Fate of malonyl-CoA in a plant cell; HOM-ACCase multifunctional homomeric acetyl-CoA carboxylase in the cytosol of plant cells, HET-ACCase multi-subunit heteromeric ACCase in plastids of plant cells

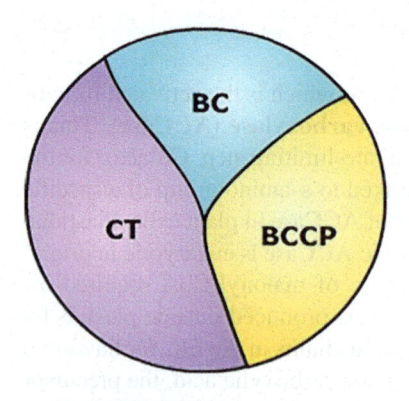

Multifunctional ACCase
HOM-ACCase (located in cytosol of all plants and in plastids and cytosol of members of grass family)

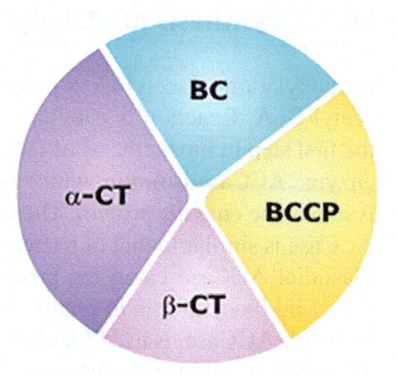

Multisubunit ACCase
HET-ACCase (located in plastids of all plants - except in members of grass family)

**Fig. 10.6** Composition of ACCase. *BC* biotin carboxylase (50 kDa), *BCCP* biotin carboxylase carrier protein (21 kDa), *CT* carboxylase transferase, *α-CT* α-subunit of CT (91 kDa), *β-CT* β-subunit of CT (67 kDa)

homodimeric protein (HOM-ACCase) which has a molecular mass of 500 kDa, each monomeric unit being 250 kDa. Each monomeric unit is multifunctional and complex. Contrary to plastidial ACCase, which has separate polypeptide subunit for each functional domain, cytosolic ACCase has three functional domains, BCCP, BC, and CT amalgamated in one polypeptide. It is encoded in nuclear genome. However, in the members of grass family (Poaceae), the plastidial form of ACCase is absent, and the homodimeric form of the enzyme with multifunctional domains is present both in the cytosol and plastids (**ACCase** Fig. 10.6). Activation of acetyl-CoA occurs in two steps and requires ATP:

$$BCCP - biotin + ATP + HCO_3^- \xrightarrow{\text{Biotin carboxylase}} BCCP - biotin - CO_2 + ADP + P_i$$

$$BCCP - biotin - CO_2 + acetyl - CoA \xrightarrow{\text{Carboxyl transferase}} BCCP - biotin + malonyl - CoA$$

Biotin serves as temporary $CO_2$ carrier. It is carboxylated at the expense of ATP. Substrate for the enzyme is $HCO_3^-$. NH group of the ureide ring of biotin forms carbamate with $HCO_3^-$ at the active site of biotin carboxylase (BC) functional subunit/domain of the multi-subunit/multifunctional enzyme. Long hydrocarbon chain of biotin cofactor is flexible. It moves to carboxyl transferase (CT) subunit/domain of ACCase. The transfer of $CO_2$ from biotin to acetyl-CoA is facilitated by carboxyl transferase, resulting in the formation of

## 10.3 Fatty Acid Biosynthesis

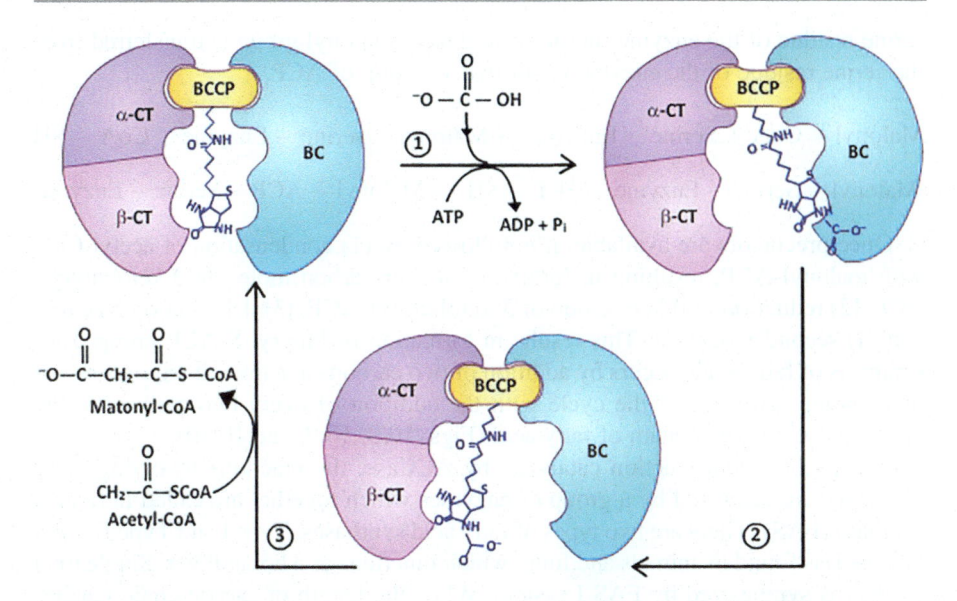

**Fig. 10.7** Acetyl-CoA carboxylase (ACCase) reaction. *1*. Biotin (attached with biotin carboxyl carrier protein, BCCP) activates $CO_2$ by attaching it to the nitrogen ring of biotin. *2*. Flexible arm of biotin transfers activated $CO_2$ to carboxyl transferase (CT) *3*. CT transfers $CO_2$ to acetyl-CoA converting it to malonyl-CoA. *BC* biotin carboxylase

malonyl-CoA and releasing biotin which is ready to receive another molecule of $HCO_3^-$. The reaction requires $Mg^{2+}$ (Fig. 10.7).

### 10.3.3 Transfer of Malonyl Moiety from Coenzyme-A to Acyl Carrier Protein (ACP)

After the synthesis of precursors for fatty acid biosynthesis, i.e., acetyl-CoA and malonyl-CoA, the first step involves the transfer of malonyl moiety from malonyl-CoA to **ACP** resulting in the formation of malonyl-ACP. Reaction is catalyzed by the enzyme **malonyl-CoA:ACP transacylase** (a serine enzyme). The following reactions include a condensation, two reductions, and a hydration, while the extending acyl group remains attached to the ACP. ACP is a small protein of about 80 amino acids long. The cofactor phosphopantetheine is covalently linked to a serine residue present in almost the center of the protein. It acts like a flexible arm carrying the acyl group from enzyme to enzyme. Similar to ACP, phosphopantetheine group is also present in coenzyme A, but the two differ in the protein being absent in coenzyme A. There is terminal –SH group present in both coenzyme A and ACP. Synthesis of malonyl-ACP from malonyl-CoA occurs in two steps. The first step involves the transfer of malonyl group from coenzyme A to serine residue of the enzyme. In the second step, malonyl group is transferred from the serine residue of the enzyme to sulfhydryl group of ACP.

Malonyl − CoA + Serine − Enzyme → Malonyl − serine − Enzyme + CoA − SH

Malonyl − Serine − Enzyme + ACP − SH → Malonyl − ACP + Serine − Enzyme

Once precursors are available, it is followed by (1) condensation of acetyl-CoA with malonyl-ACP, resulting in formation of four-carbon molecule 3-ketobutyryl-ACP, (2) reduction of β-keto group of 3-ketobutyryl-ACP, (3) dehydration reaction, and (4) second reduction. This results in formation of butyryl-S-ACP group. Biosynthesis of fatty acids occurs by addition of two carbons at a time. It requires repeat of four-step sequence of the cycle with the addition of every two carbons to the growing hydrocarbon chain of fatty acid (Figs. 10.8, 10.9, and 10.10).

Except the initial reaction catalyzed by ACCase, the reactions involving these four steps are catalyzed by a group of enzymes which together are called fatty acid synthase (FAS). There are two types of fatty acid synthases, Type I and Type II. FAS I (Type I) is found in animals and fungi which functions as a homodimer. Single final product is synthesized by FAS I system. When the length of carbon chain reaches 16, the final product (palmate) is released, with no intermediates. Contrary to FAS I, FAS II in plants and bacteria consists of many polypeptides. Each step in the series of reactions is catalyzed independently by individual polypeptides. Intermediates of the reaction can diffuse out and are diverted to other pathways. Fatty acids of

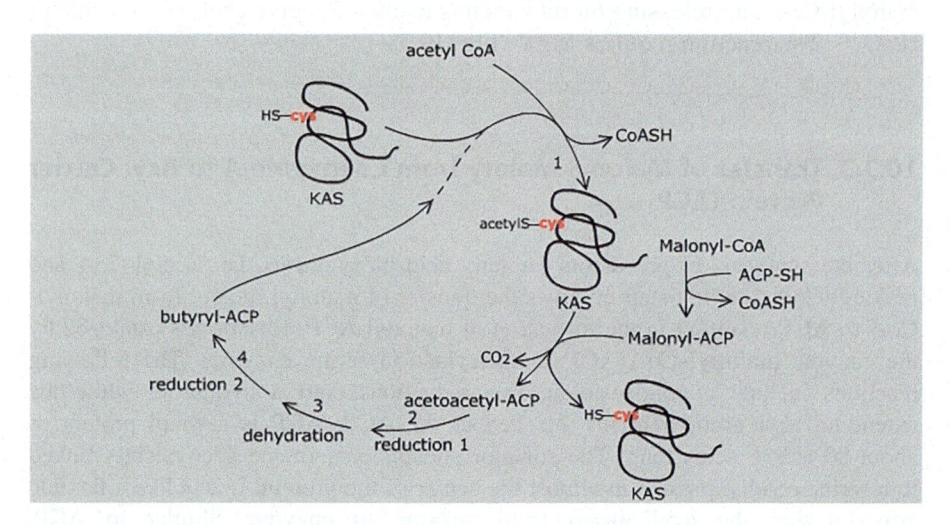

**Fig. 10.8** Fatty acid biosynthesis. Condensation of acetyl-CoA and malonyl-ACP is facilitated by the enzyme ketoacyl synthase (KAS). One carbon of malonyl-ACP is released as $CO_2$ and KAS are released. Condensed acetoacetyl-ACP group enters reduction 1, dehydration, and reduction 2 resulting in synthesis of butyryl-ACP. Butyryl-ACP enters into another cycle of chain elongation in a similar way after butyryl group is transferred to KAS

## 10.3 Fatty Acid Biosynthesis

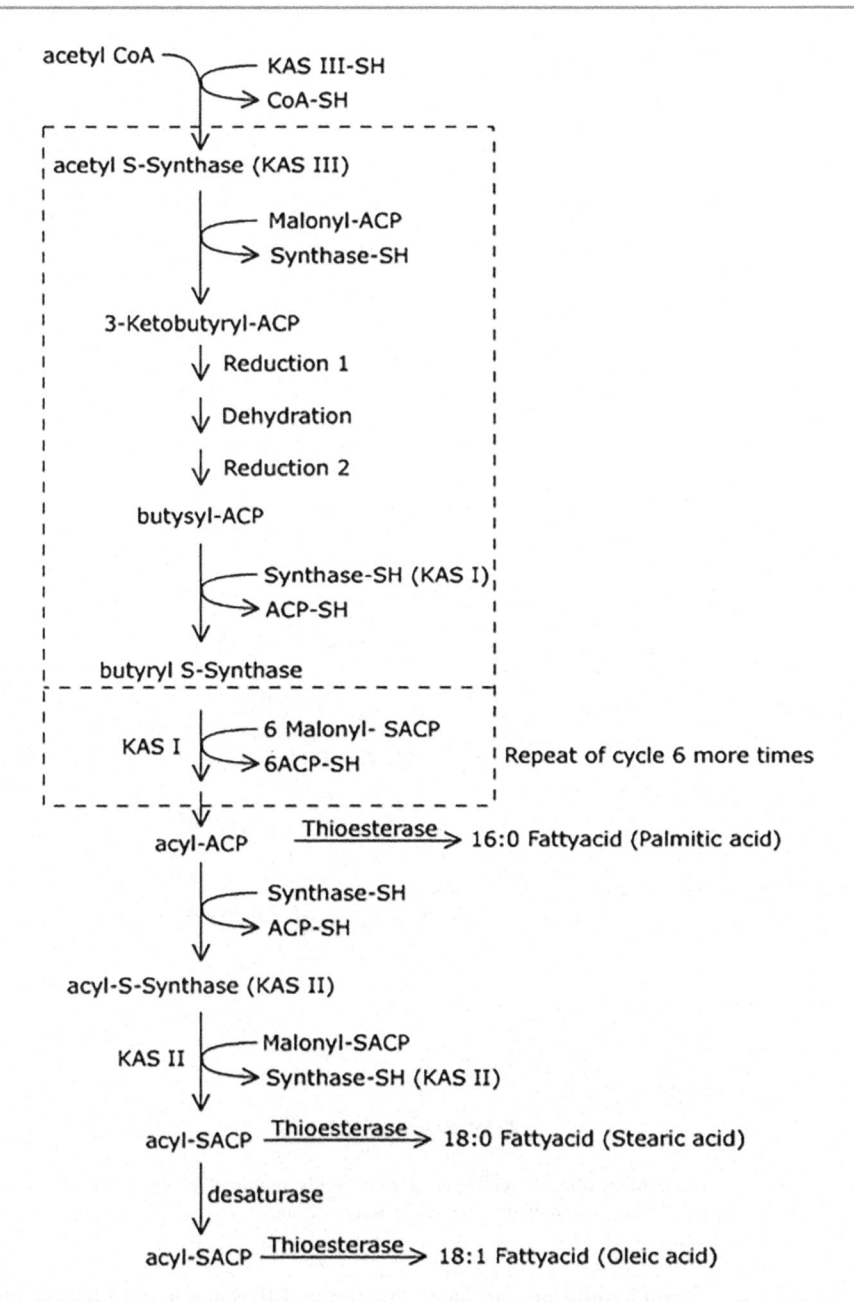

```
acetyl CoA ─┐  ┌── KAS III-SH
            ├──┤
            │  └─→ CoA-SH
            ▼
acetyl S-Synthase (KAS III)
            │  ┌── Malonyl-ACP
            ├──┤
            │  └─→ Synthase-SH
            ▼
  3-Ketobutyryl-ACP
            ↓ Reduction 1
            ↓ Dehydration
            ↓ Reduction 2
  butysyl-ACP
            │  ┌── Synthase-SH (KAS I)
            ├──┤
            │  └─→ ACP-SH
            ▼
  butyryl S-Synthase

  KAS I  ┌── 6 Malonyl- SACP        Repeat of cycle 6 more times
         └─→ 6ACP-SH
            ▼
  acyl-ACP  ──Thioesterase──→ 16:0 Fattyacid (Palmitic acid)
            │  ┌── Synthase-SH
            ├──┤
            │  └─→ ACP-SH
            ▼
  acyl-S-Synthase (KAS II)

  KAS II ┌── Malonyl-SACP
         └─→ Synthase-SH (KAS II)
            ▼
  acyl-SACP ──Thioesterase──→ 18:0 Fattyacid (Stearic acid)
            │ desaturase
            ▼
  acyl-SACP ──Thioesterase──→ 18:1 Fattyacid (Oleic acid)
```

**Fig. 10.9** Role of isoforms of 3-ketoacyl synthase (KAS), i.e., KAS III, KAS I, and KAS II in fatty acid synthesis in plastids. KAS III catalyzes condensation of acetyl-CoA with malonyl-ACP, and KAS I catalyzes condensation of acyl groups that contain carbon chain length from 4 to 16, while KAS II catalyzes extension of carbon chain length of fatty acids from 16 to 18 carbons

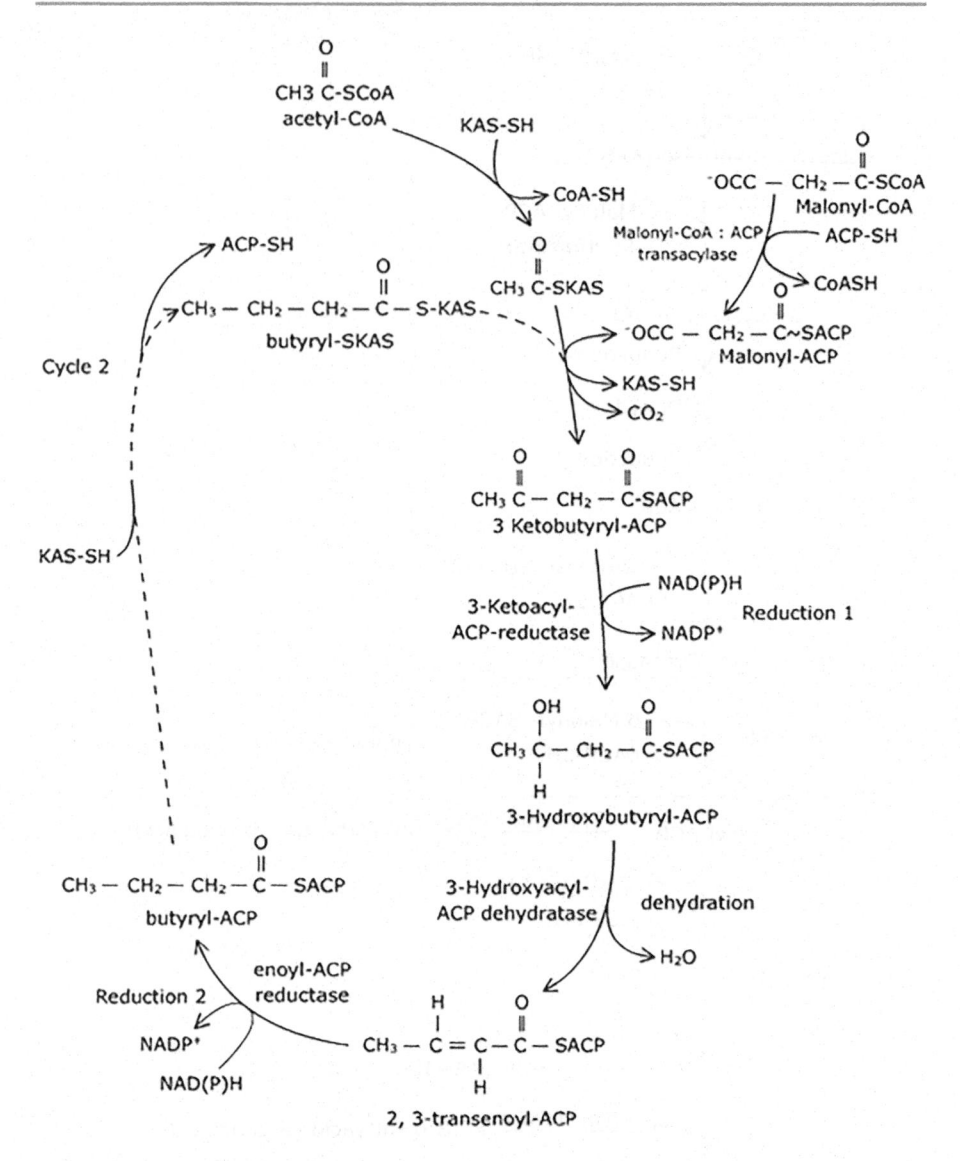

**Fig. 10.10** Fatty acid biosynthesis: synthesis of fatty acids requires repeat sequence. In one sequence of reactions, hydrocarbon chain extends by two carbons

variable carbon chain lengths are produced due to the difference in termination sites for extending hydrocarbon chain. The independent catalytic units possibly are associated in **supramolecular organization**. In such a kind of structural association, efficiency of metabolic pathway is enhanced due to **substrate channeling**.

## 10.3.4 Condensation Reaction

Condensation of acetyl-CoA (primer) with malonyl-ACP results in formation of carbon-carbon bond, and a four-carbon compound, 3-ketobutyryl-ACP, is synthesized. Reaction is catalyzed by **ketoacyl-ACP synthase (KAS)**. KAS has SH groups present at the active site (KAS-SH). Reaction occurs as follows:

$$Acetyl - CoA + SH - KAS \rightarrow Acetyl - S - KAS + CoA - SH$$

$$Acetyl - S - KAS + Malonyl - ACP \rightarrow 3 - Ketobutyryl - ACP + CO_2 + KAS - SH$$

Reaction involves condensation of two carbons, donated from acetyl unit of acetyl-CoA, and two carbons from malonyl-ACP. Carboxyl group of malonyl moiety of malonyl-ACP is released as $CO_2$. Release of $CO_2$ makes the reaction irreversible. 3-Ketoacyl-ACP synthase (KAS) is commonly called condensing enzyme. There are three isoforms of the KAS: KAS I, KAS II, and KAS III. They differ in their specificity for the substrate, acyl-ACP, which have hydrocarbon chains with variable carbon numbers. KAS III catalyzes condensation of acetyl-CoA with malonyl-ACP. KAS II prefers a substrate with a longer carbon-chain acyl-ACP (with C10-14/16 long hydrocarbon chain), while KAS I prefers condensation reaction of the acyl-ACP having C4-C14 long hydrocarbon chain. Condensation reaction is initiated by KAS III, hydrocarbon chain is extended by KAS I, and it is completed by KAS II, resulting in fatty acids with C16/C18 long hydrocarbons (Fig. 10.9).

## 10.3.5 Reduction of 3-Ketobutyryl-ACP to Butyryl-ACP

Conversion of 3-ketobutyryl-ACP to butyryl-ACP involves two reduction reactions and a dehydration reaction. This set of three reactions is similar to reactions of Krebs cycle in reverse order during which succinate is converted to malate. Krebs cycle involves two oxidations and one hydration reaction, while during fatty acid synthesis, two reduction and one dehydration reactions occur. Reduction requires reducing power mostly in the form of NADPH, which is provided in chloroplasts by the light reaction. In nongreen plastids, the source for NADPH is oxidative pentose phosphate pathway. During reduction 1,3-ketobutyryl-ACP is reduced to 3-hydroxybutyryl-ACP (β-D-hydroxyacyl-ACP). Reaction is catalyzed by **3-ketoacyl-ACP reductase.**

$$3 - ketoacetyl - ACP + NADPH \rightarrow 3 - (D) - hydroxybutyryl - ACP + NADP^+$$

Dehydration of 3-hydroxybutyryl-ACP results in formation of 2,3-trans-enoyl-ACP. Reaction is catalyzed by **3-hydroxyacyl-ACP dehydrase.**

$$3 - hydroxybutyryl - ACP \rightarrow 2, 3 - trans - enoyl - ACP + H_2O$$

Second reduction involves reduction of 2, 3-trans-enoyl-ACP to butyryl-ACP, catalyzed by **enoyl-ACP reductase** which requires reducing power as NAD(P)H/NADH depending upon the isoform of the enzyme which exists in two isoforms.

$$2, 3 - \text{trans} - \text{enoyl} - \text{ACP} + \text{NAD(P)H} \rightarrow \text{butyryl} - \text{ACP} + \text{NAD(P}^+)$$

Flexible arm of ACP shuttles the four-carbon moiety from enzyme to enzyme during the reactions steps.

### 10.3.6 Extension of Butyryl-ACP

Extension of hydrocarbon chain in butyryl-ACP (acyl-ACP) occurs through a repeat of similar set of reactions from condensation to reduction 2 in a cyclic manner. In the beginning of the cycle, a growing acyl chain is transferred from ACP to the active site (-SH) of condensing enzyme (KAS-SH). At the time of condensation, the acyl chain is transferred from KAS to ACP, followed by a similar set of reactions as with acyl-ACP (Fig. 10.8). During each cycle of reactions, hydrocarbon chain of acyl group is extended by two carbons ($-CH_2-CH_2$) till hydrocarbon chain is extended up to 16- or 18-carbon chain length depending upon the plant (Figs. 10.9 and 10.10).

### 10.3.7 Termination of Fatty Acid Synthesis in Plastids

Release of acyl groups from fatty acyl-ACP is catalyzed by ACP thioesterase which is localized in the inner membrane of plastids. After the extension of hydrocarbon chain up to 18-carbon length in plastids, there are three possible fates of the fatty acids:

1. Acyl group is directly transferred from ACP to glycerol in plastids, resulting in the synthesis of glycerolipids which are the precursors for triglycerides to be used for storage or for biogenesis of membrane lipids.
2. A double bond is introduced in the hydrocarbon chain of acyl group by desaturase present in the stroma of plastids, resulting in the formation of oleoyl-ACP (18:1).
3. Release of fatty acids from fatty acyl-ACP is catalyzed by the soluble fatty **acyl-ACP thioesterase** (Fat). Thioesterase hydrolyzes sulfhydryl bond of acyl group with ACP, releasing ACP and the acyl group (generally fatty acids having hydrocarbons with carbon chain length of 16 or 18). Fatty acids are converted to fatty acyl-CoA at the outer membrane of plastids. Plastidial ACP thioesterase primarily hydrolyzes palmitoyl-ACP (16:0) or oleate acyl-ACP (18:1). Stearate-ACP (18:0) is hydrolyzed to a lesser extent. ACP-thioesterase is localized in the inner envelop of the plastids releasing fatty acyl groups in the stroma. The fatty acyl groups are exported out of plastid and are reactivated to fatty acyl-CoA at the

## 10.3 Fatty Acid Biosynthesis

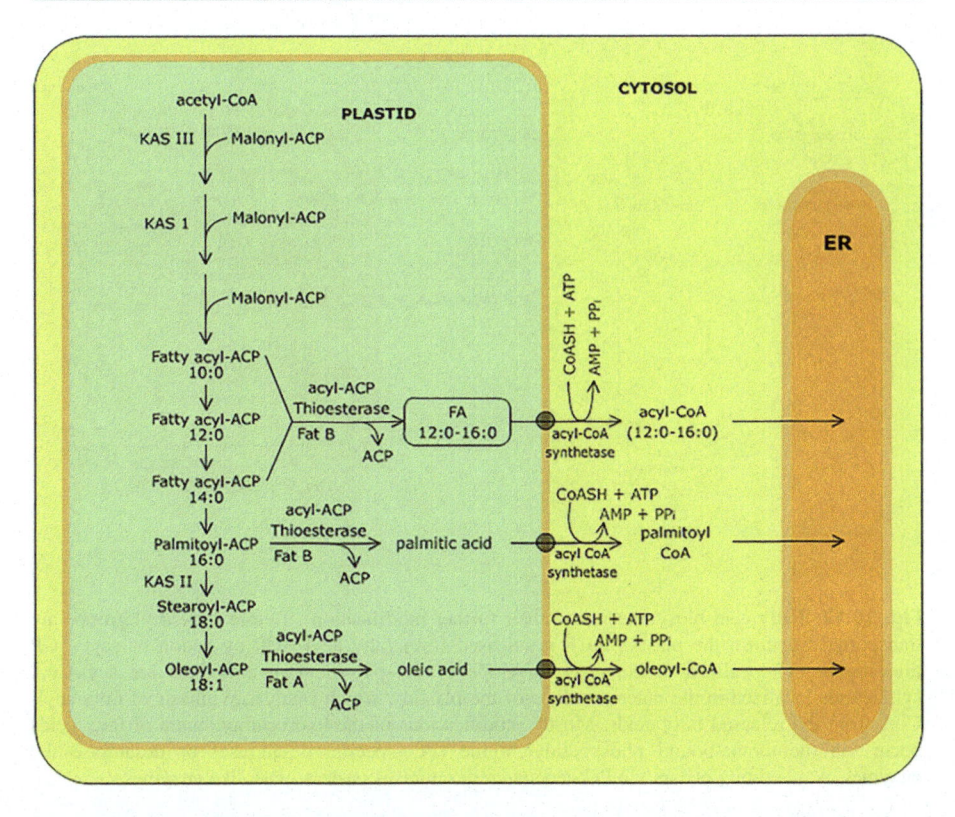

**Fig. 10.11** Role of thioesterase. Acyl-ACP thioesterase catalyzes hydrolysis of acyl moiety from ACP stopping further extension of hydrocarbon chain and yielding fatty acid and ACP. Two types of acyl-ACP thioesterases, fat A and fat B, occur in plants. Fat B is active with shorter chains FA, while fat A is active with $18:1\Delta^9$-ACP. Some of the fatty acids with shorter hydrocarbon chain have commercial significance. This characteristic has been exploited to incorporate ACP-thioesterases in plants to be able to produce FAs with shorter hydrocarbon chains

outer envelope by **acyl-CoA synthetase** (Fig. 10.11). Further elongation of hydrocarbon chain or introduction of double bond in acyl groups occurs as acyl-CoA and not as fatty acyl-ACP in the endoplasmic reticulum.

### 10.3.8 Hydrocarbon Chain Elongation in Endoplasmic Reticulum

In plants, fatty acids with C20–C24 carbon atoms are required either for the synthesis of wax (which are esters of long-chain fatty acids having C20–C24 carbon atoms and long-chain acyl alcohols of C26–C32 length) or for the synthesis of sphingolipids, which contain C22–C24 long fatty acids. In some plants, triacylglycerols, containing fatty acids with longer hydrocarbon chain (C20 and C22), are also present. Palmitate (16:0) or oleate (18:1) synthesized in plastids is

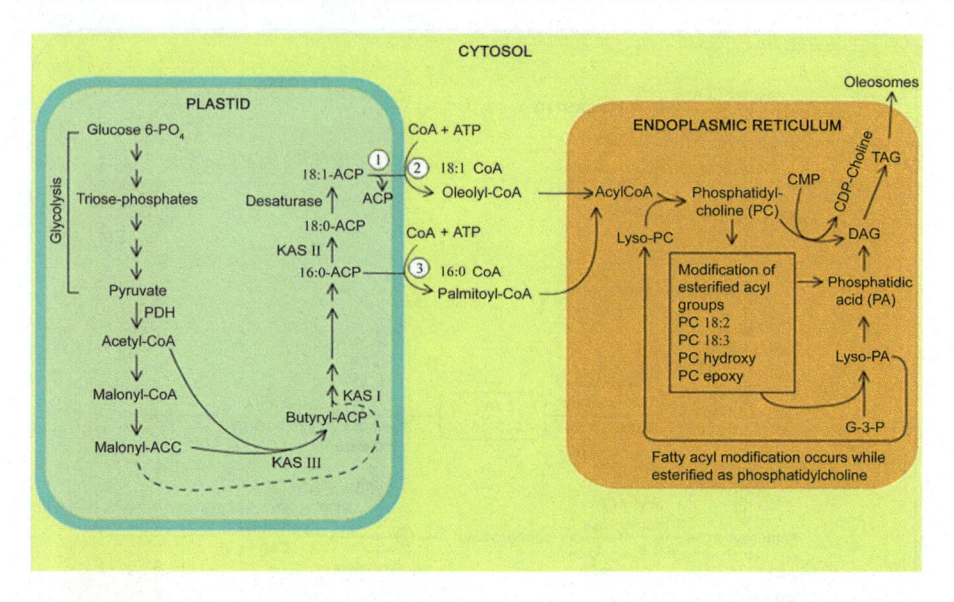

**Fig. 10.12** Fatty acid biosynthesis and their further modifications. Before fatty acyl groups are transported out form the plastid, ACP is released from fatty acyl-ACP by action of acyl ACP thioesterase (1) localized in the inner envelope of the plastid. Two and three are acyl-CoA synthetases localized in the outer envelope of the plastids, which catalyze synthesis of fatty acyl-CoA from the released fatty acids. Mostly modifications of the hydrocarbon chain of fatty acids occur via membrane-bound phosphatidylcholine (PC)-specific desaturases or desaturase-like enzymes. Acyl editing occurs via PC-dependent desaturases or desaturase-like enzymes

exported to the endoplasmic reticulum for further extension (Fig. 10.12). They require membrane-bound fatty acid elongase system. It involves condensation with two-carbon unit at a time, followed by reduction, dehydration, and another reduction of the intermediates in sequential order similar to FAS reactions. NADPH is used as the reductant. However, reactions of fatty acid elongation differ from FAS reactions in the following: (1) FAS reactions occur in plastids, while the reaction involving fatty acid elongation occurs in ER membranes. (2) ACP is not involved, and all the intermediates are activated as Coenzyme A esters. (3) The enzyme is elongase 3-ketoacyl-CoA synthase (elongase KCS), which catalyzes condensation with malonyl- CoA. (4) After condensation of two carbon groups from malonyl-CoA with elongating acyl groups of fatty acid, fatty acids undergo another cycle involving two reductions and a dehydration. Fatty acids are also utilized in other pathways of lipid metabolism (Fig. 10.13).

## 10.3.9 Synthesis of Unsaturated Fatty Acids

First double bond in fatty acid molecules is introduced in plastids itself. Stearoyl-ACP (18:0) is desaturated to oleoyl-ACP (18:1), catalyzed by a soluble, plastid localized **stearoyl-ACP Δ9-desaturase**. However, a family of this enzyme is

## 10.3 Fatty Acid Biosynthesis

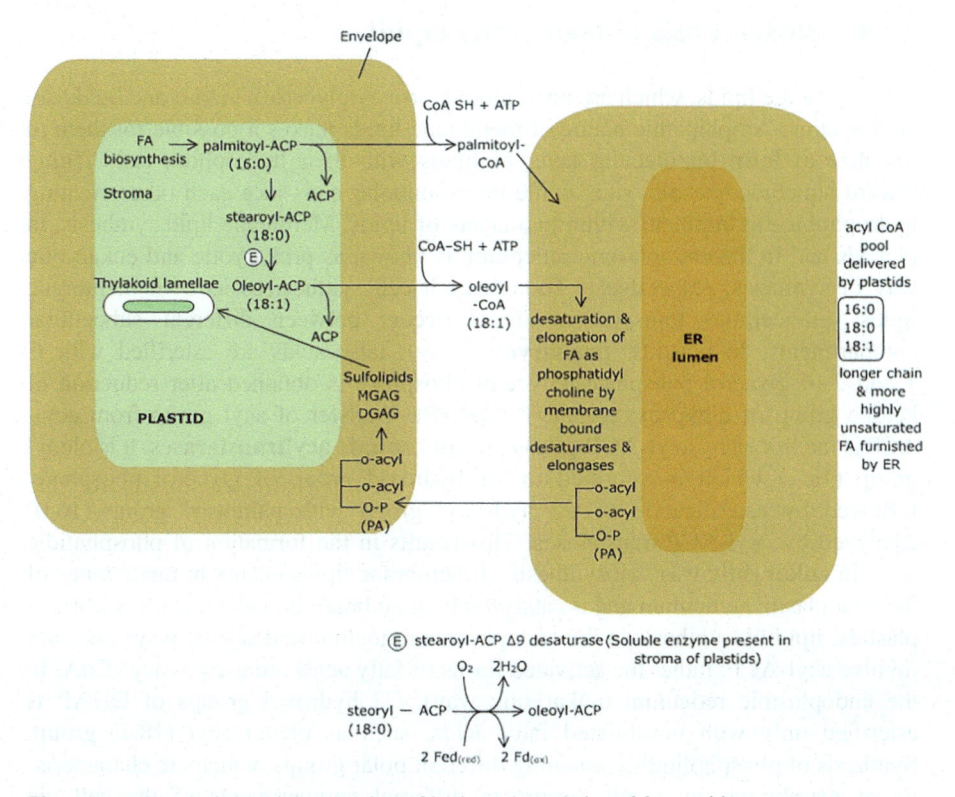

**Fig. 10.13** Elongation and desaturation of hydrocarbon chains of fatty acids

responsible for introduction of double bonds at various locations of acyl chain, including the one which catalyzes desaturation of palmitic acid at $\Delta 4$ and $\Delta 6$ locations. Studies with mutants of *Arabidopsis* have shown the presence of eight genes which encode fatty acid desaturases (FAD). Out of these three have been found to encode for the soluble chloroplast desaturase enzyme and three for the membrane-bound enzyme in the endoplasmic reticulum. Chloroplast enzymes use soluble ferredoxin as the immediate electron donor, and the desaturated fatty acids are esterified as galactolipids, sulfolipids, and phosphatidylglycerol. ER membrane-localized plant desaturases utilize cytochrome b5 instead of ferredoxin and act on fatty acids esterified to phosphatidylcholine and possibly other phospholipids. Ferredoxin and cytochrome b5 are used as intermediate electron donors for the reaction. Except stearoyl-ACP $\Delta 9$-desaturase, which is soluble and is present in the stroma of plastids, other desaturases are membrane bound. FAD in ER membrane is responsible for desaturation of acyl groups as phosphatidylcholine. Possible expression of genes encoding desaturases is regulated by temperature in which plants grow or some post-translational modification of the fatty acids is required, resulting in alteration of fatty acid composition of membranes (Box 10.1).

## 10.4 Biosynthesis of Membrane Lipids

Unlike storage lipids, which are neutral lipids (triacylglycerol), membrane lipids are amphipathic. Amphipathic nature of membrane lipids makes it possible for them to organize to form bimolecular layer of lipids with their hydrophilic ends facing toward aqueous cytosolic side, while their nonpolar ends face each other creating hydrophobic environment within two layers of lipids. Membrane lipid synthesis, in plastids and in the endoplasmic reticulum, is known as prokaryotic and eukaryotic way of synthesis, respectively. Though each cell synthesizes its own membrane lipids, intracellular transport of lipids occurs between different subcellular compartments. In plastids (**prokaryotic way**), fatty acids are esterified with C 1 and 2 of glycerol phosphate. Glycerol phosphate is obtained after reduction of ketone group of dihydroxyacetone-P (DHAP). Transfer of acyl group from acyl-ACP is facilitated by **acyl-ACP-glycerol-3-phosphate acyltransferases**. It is oleoyl group (18:1) which is esterified to C1 hydroxyl group of glycerol phosphate, followed by esterification of C2 hydroxyl group with palmitoyl group (16:0) catalyzed by acyl-ACP transferases. This results in the formation of phosphatidic acid. In **eukaryotic way**, biosynthesis of membrane lipids occurs in membranes of the endoplasmic reticulum and is catalyzed by membrane-bound enzymes. Unlike in plastids, lipid biosynthesis in the endoplasmic reticulum (eukaryotic way) does not involve acyl-ACP; rather the activated forms of fatty acids are used as acyl-CoA. In the endoplasmic reticulum (eukaryotic way), C2 hydroxyl groups of DHAP is esterified only with unsaturated fatty acids, such as oleoyl acyl (18:1) group. Synthesis of phospholipids containing different polar groups, which are characteristic of membranes in a cell, occurs in different compartments of the cell via CDP-DAG (cytidine diphosphate-diacylglycerol) and DAG (diacylglycerol) pathway (Fig. 10.14). Contribution of each pathway varies among organisms.

At least some of the lipids which are synthesized and modified in the endoplasmic reticulum are exported to plastids to be incorporated in thylakoids. After initial desaturation of stearic group to oleoyl in plastids, further desaturation occurs as acyl groups of phosphatidylcholine in ER membranes. After being transported back to plastids as phosphatidylcholine, fatty acids are transferred to the glycerol backbone to form diacylglycerols. One or two galactose groups are added to diacylglycerol to form galactolipids in the chloroplast envelope which are then inserted in thylakoids. Acyl groups can be further desaturated by FADs present in plastid envelope as well. Galactolipids contain 18:2 and 18:3 acyl groups of the fatty acids. Thylakoids make 80% of the acyl pool of leaf cells, either as monogalactolipids or digalactolipids. In pea and barley leaves, a majority of galactolipids of thylakoids are derived from lipids synthesized in the endoplasmic reticulum by eukaryotic pathway. Phosphatidylglycerol (PG) is the only phospholipid present in thylakoids, which is derived by prokaryotic pathway. However, this (PG) forms only a minor component in thylakoids. In some plants, such as in spinach, a majority of plastid lipids are derived from prokaryotic pathway (Figs. 10.14 and 10.15).

## 10.5  Biosynthesis of Storage Lipids in Seeds

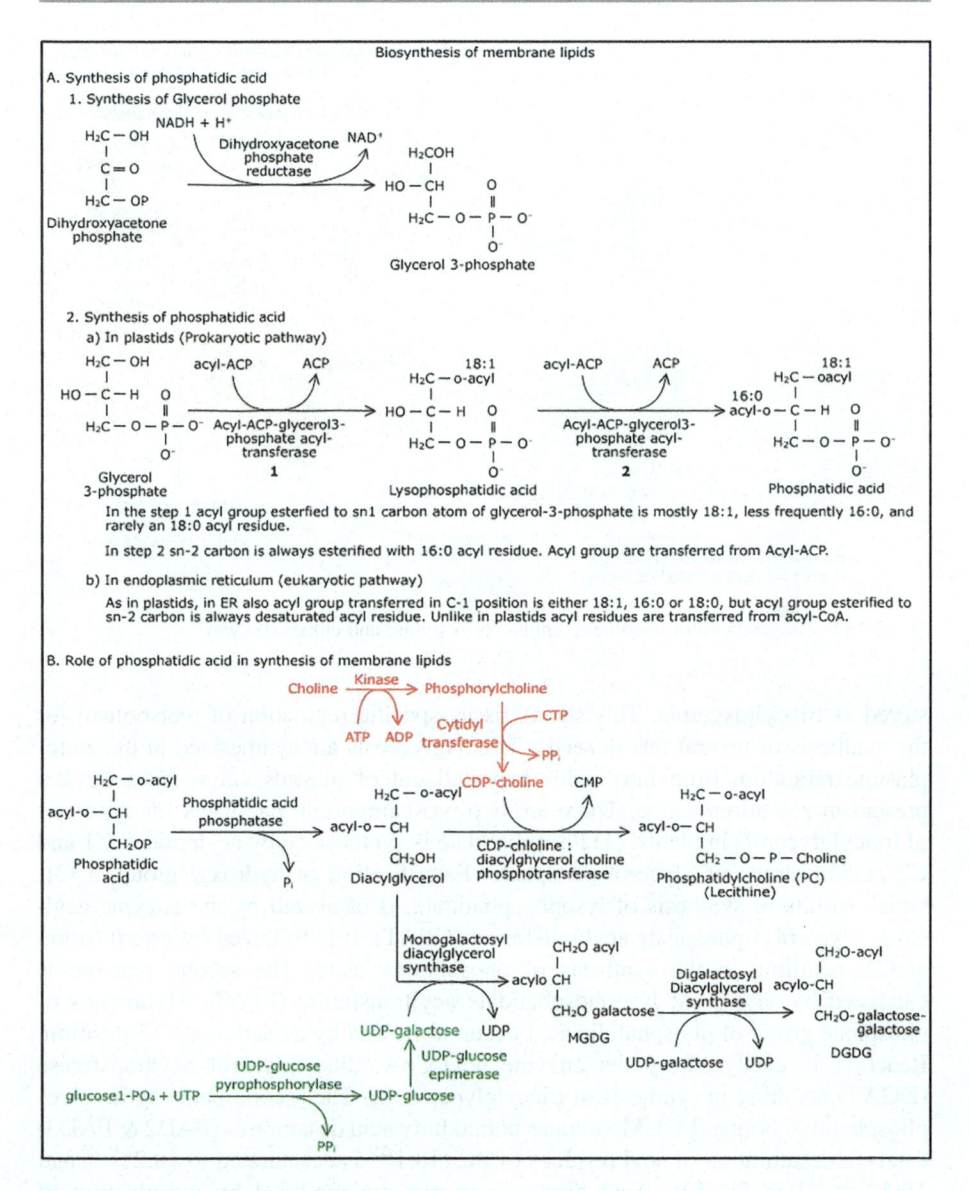

**Fig. 10.14** Biosynthesis of membrane lipids

## 10.5  Biosynthesis of Storage Lipids in Seeds

Lipids constitute less than 5% of the total dry weight of most of the vegetative tissues in plants, and only 1% are present as triacylglycerols. Contrary to this, lipids may constitute as much 60% of the dry weight in oil seeds, and 95% of these lipids are

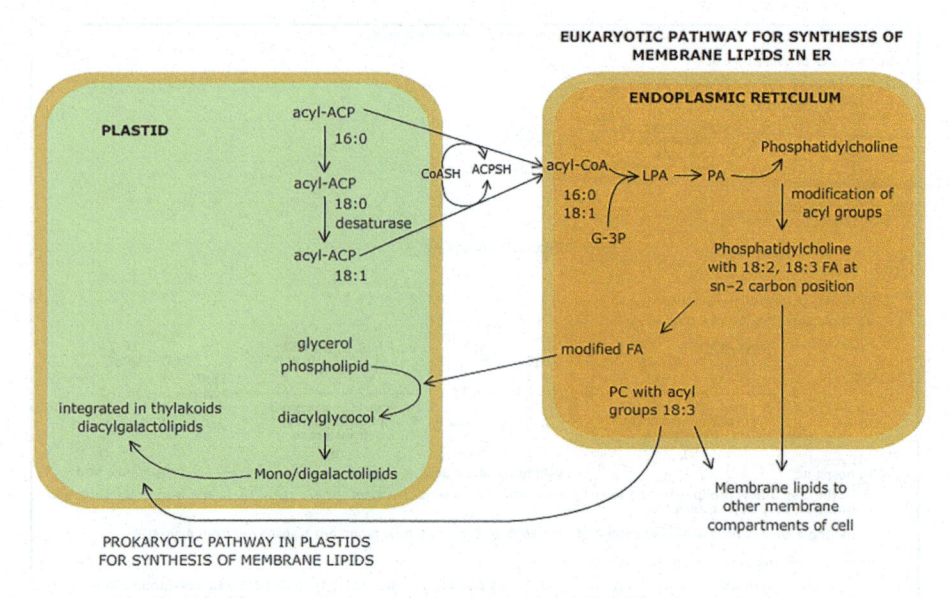

**Fig. 10.15** Biosynthesis of membrane lipids—prokaryotic and eukaryotic way

stored as triacylglycerols. This shows tissue-specific regulation of metabolism for the synthesis of neutral fats in seeds. Triacylglycerols are synthesized in the endoplasmic reticulum from fatty acids exported out of plastids via specific carriers present in plastid envelope. There are two most important pathways for synthesis of triacylglycerols in plants: (1) Phosphatidate is synthesized by acylation at C1 and C2 carbon atoms of glycerol phosphate. Esterification of hydroxyl group at C1, which results in synthesis of lysophosphatidate, is catalyzed by the enzyme acyl-CoA: glycerol-3-phosphate acyltransferase (GPAT). It is followed by esterification at C2, resulting in the synthesis of phosphatidic acid. The second reaction is catalyzed by acyl-CoA: lysophosphatidate acyltransferase (LPAT). Hydrolysis of phosphate group of phosphatidic acid occurs followed by acylation at C-3 position. Reaction is catalyzed by the enzyme acyl-CoA: diacylglycerol acyltransferase (DGAT) resulting in synthesis of triacylglycerol. (2) The second pathway involves phosphatidylcholine (PC). Membrane-bound fatty acid desaturases (FAD2 & FAD3) catalyze desaturation of acyl residues of PC ($18:1^{\Delta 9}$ is desaturated to $18:2^{\Delta 9,12}$ and $18:3^{\Delta 9,12,15}$) in the ER. Acyl groups may also be modified by introduction of hydroxy or epoxy groups by desaturase like enzymes. This is followed by hydrolysis of choline phosphate, resulting in synthesis of diacylglycerol. Reaction is catalyzed by the enzyme phosphatidylcholine: diacylglycerol cholinephosphotransferase (PDCT). Esterification of diacylglycerol at C-3 position, to produce triacylglycerol, is catalyzed by acyl-CoA: diacylglycerol acyltransferase (DGAT) (Fig. 10.16). Relative contribution of these two pathways for in vivo biosynthesis of triacylglycerol in seeds varies in different species. **Oleosomes** are specialized subcellular structures, which are specifically adapted for storage of neutral fats in

## 10.5 Biosynthesis of Storage Lipids in Seeds

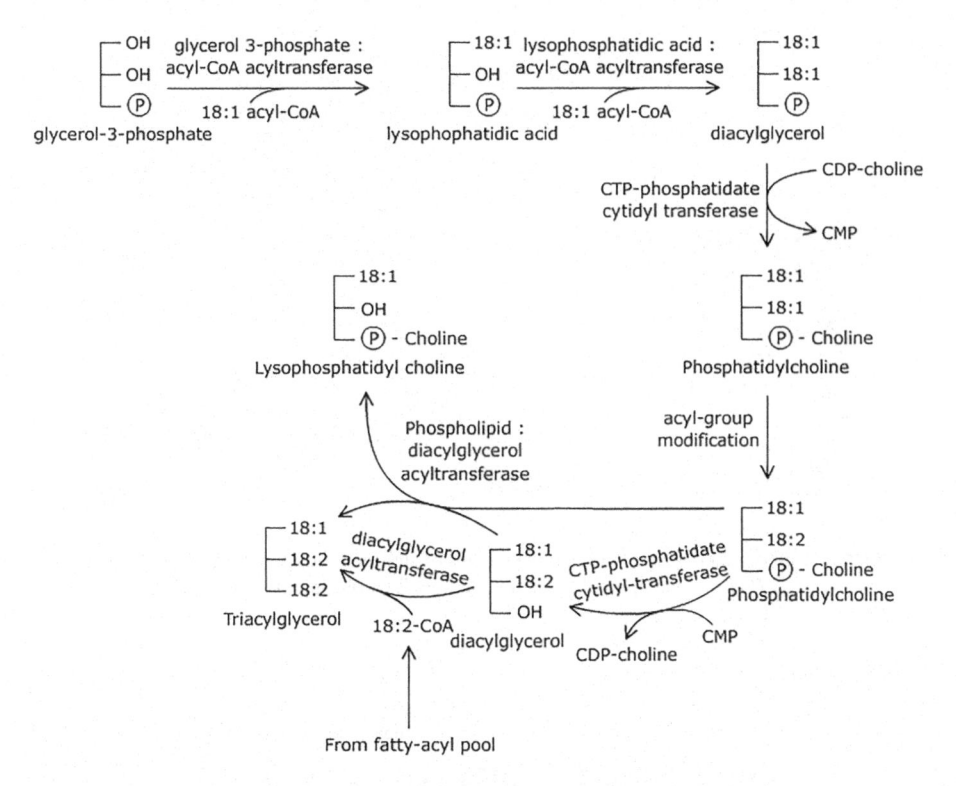

**Fig. 10.16** Acyl group diversity in triacylglycerols

oilseeds and pollen grains. Their membrane consists of a monolayer of lipids with hydrophilic ends facing toward cytosolic side and hydrophobic ends facing inside toward the stored triacylglycerols. **Oleosin** are specific types of proteins present in lipid membranes of oleosomes. These are low-molecular-mass (15–25 kDa) proteins having a central domain consisting of about 70–80 hydrophobic amino acids, which protrudes toward the triacylglycerol core of the oil body. Oleosomes originate on the endoplasmic reticulum membrane. Within the ER membrane bilayer, triacylglycerols are synthesized, and the monolayer of membrane along with triacylglycerols is pinched off resulting in the formation of oil bodies. Diameter of oil bodies varies from 0.5 to 2 μm; however, it may vary from 0.5 to 30 μm. Seeds are extremely dehydrated structures, and oleosins play an important role in stability of the oil body. Oleosins probably also have an important role in maintaining the viability of seeds during dehydration. At the time of seed germination, oleosins play a critical role during extreme hydration associated with imbibition of water. In addition to oleosins, oil bodies also contain caleosins and steroleosins which play a signaling role during oil body mobilization. In fruits such as avocado and olive, oil bodies contain triacylglycerols. Triacylglycerols in seed oil may include unusual acyl groups exhibiting a great diversity. This includes fatty acyl groups with more unsaturated, branched, and longer hydrocarbon chains (Figs. 10.17 and 10.18).

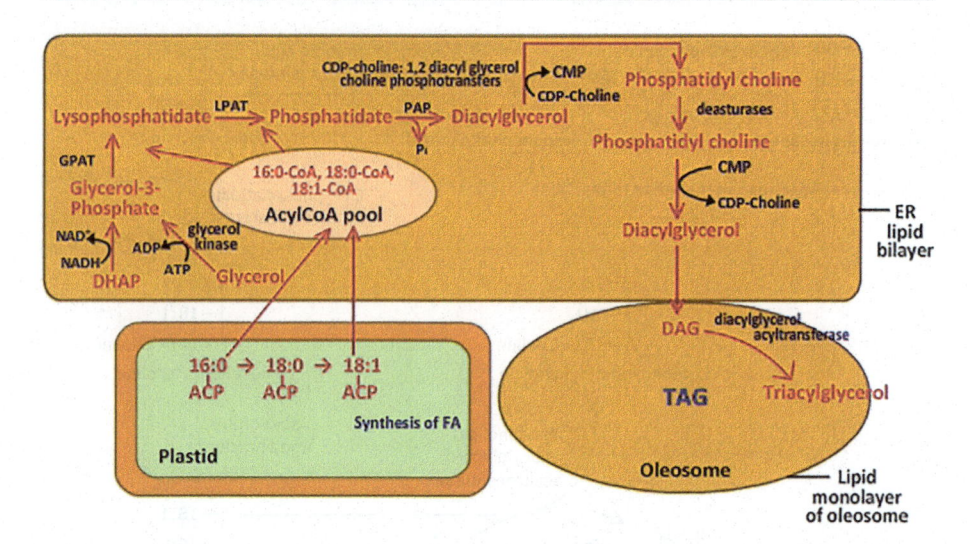

**Fig. 10.17** Cell organelles involved in biosynthesis of storage lipids. *GPAT* glycerol-3-phosphate acyltransferase, *LPAT* lysophosphatidic acyltransferase, *PAP* phosphatidic acid phosphatase

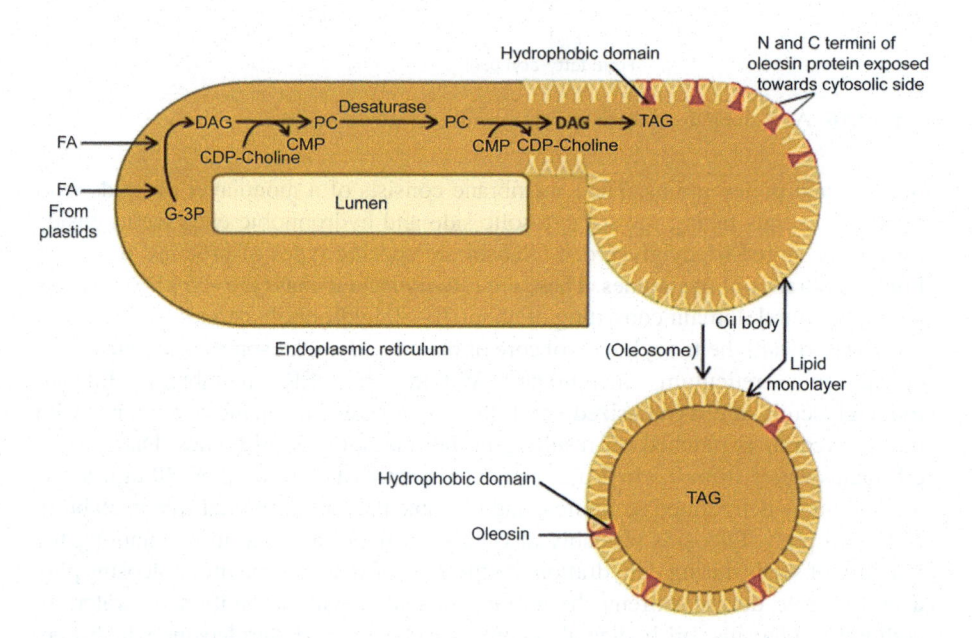

**Fig. 10.18** Formation of oleosomes

## 10.6 Lipid Catabolism

In oil seeds and pollen grains, lipids are stored as neutral lipids which constitute as much as ~60% of dry weight of the seeds. Triglycerides are stored either in the endosperm or in cotyledons of the oil seeds. At the time of germination of oil seeds, triacylglycerides (TAG) are hydrolyzed by lipases resulting in the formation of glycerol and fatty acids. Lipases bind to oleosins of the oil bodies and hydrolyze TAG at the oil/water interface. In castor bean, there are two types of lipases, acid lipases which act on mono-, di-, or triglycerides. These are active in dry seeds also and are localized in oil bodies. On the contrary, alkaline lipases are present in membranes of **peroxisomes**, which are active only after water is imbibed by the seeds. Alkaline lipases act on monoglycerides.

$$\text{Triglycerides} + \text{water} \xrightarrow{\text{lipase}} \text{glycerol} + 3 \text{ Fatty acids}$$

Glycerol is metabolized in the cytosol of the cell, while fatty acids are transported to peroxisomes (specialized peroxisomes present in nongreen parts are called glyoxysomes). Glycerol is converted to dihydroxyacetone phosphate (DHAP) by the action of glycerol kinase (GLK) and FAD-linked glycerol-3-phosphate dehydrogenase (GDH). DHAP is metabolized through glycolysis. Fatty acids are transported to peroxisomes. A transporter, identified in *Arabidopsis*, is an **ATP-binding cassette (ABC) transporter**, which is localized in the membrane of peroxisomes. The peroxisomal **ABC transporter comatose (CTS)** is required for transporting acyl groups to peroxisomes. Once fatty acyl groups have been transported to peroxisomes, these are activated to form fatty acyl-CoA. This reaction is catalyzed by **long-chain fatty acyl-CoA synthetase (LACS)**, LACS6, and LACS7. ATP and coenzyme A are required for activation of fatty acyl groups to fatty acyl-CoA. Since fats are not transported, these need to be converted to sugars at the time of germination in seeds. During senescence, β-oxidation of degradation products of the membrane lipids occurs in leaves, though at much lower rates. The fatty acids, which are no longer required, are recycled, and the carbohydrates so synthesized are transported to other plant parts. Conversion of fatty acids to sugars involves (1) β-oxidation of fatty acids, (2) glyoxylate cycle, and (3) gluconeogenesis. The organelles involved are peroxisomes (glyoxysomes), mitochondria, and cytosol. Because of the absence of glyoxylate cycle in animals, fats are not converted to sugars (Fig. 10.19).

### 10.6.1 β-Oxidation of Fatty Acyl-CoA

β-Oxidation involves breaking down of fatty acyl-CoA with the removal of two-carbon fragment at a time, as acetyl-CoA. Through repeated cycles of β-oxidation, there is complete degradation of fatty acyl-CoA to acetyl-CoA. The number of acetyl-CoA molecules produced depends upon the length of hydrocarbon chain in fatty acids. Unlike in animal cells, where β-oxidation occurs in the

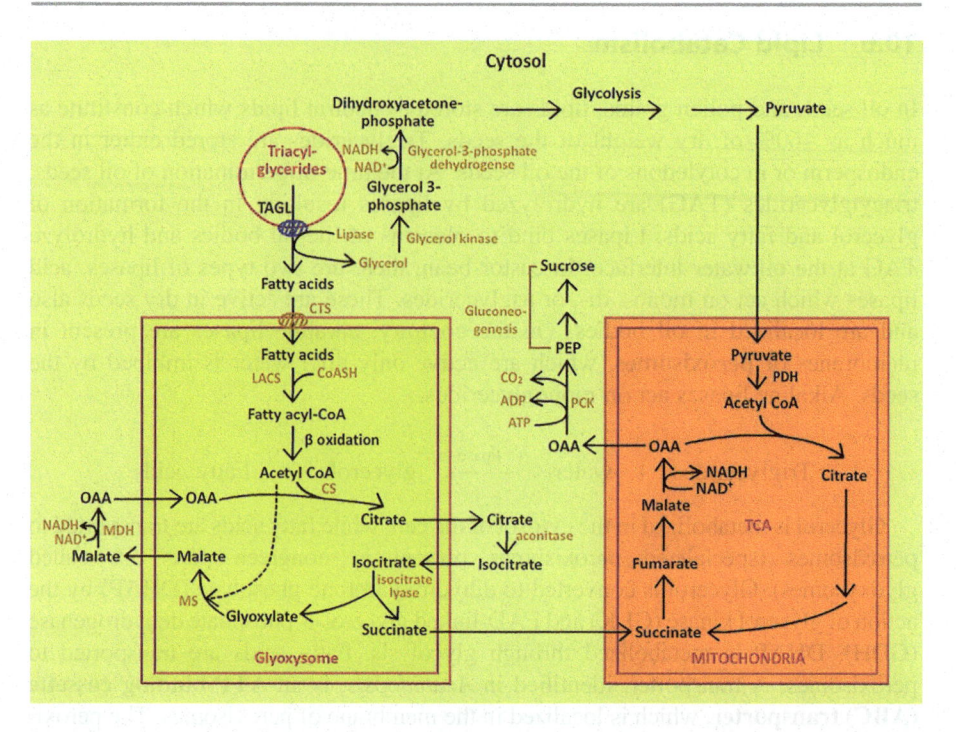

**Fig. 10.19** Organelles involved in lipid catabolism. *TAGL* triacylglycerol lipase, *LACS* long-chain acyl synthase, *CTS* comatose, *ATP*-binding cassette transporter, *PCK* PEP carboxykinase, *CS* citrate synthase, *MS* malate synthase

mitochondria, in plants it occurs in **glyoxysomes**. Glyoxysomes are specialized peroxisomes present in endosperm cells of the germinating oil seeds. Glyoxysomes are so called because glyoxylate cycle occurs there. These are found adjacent to oil bodies. β-Oxidation of fatty acids involves the following steps which are similar in plants and animals: (1) Dehydrogenation of fatty acyl-CoA at two- and three-carbon atoms resulting in production of a Δ-2 trans-unsaturated fatty acyl-CoA molecule (trans-Δ2-enoyl-CoA) is catalyzed by **acyl-CoA oxidase**; (2) hydration of trans-Δ-2-enoyl, which results in formation of ʟ-hydroxyacyl-CoA, is catalyzed by **enoyl-CoA hydratase**; (3) oxidation of ʟ-hydroxyacyl-CoA is catalyzed by **β-hydroxyacyl-CoA dehydrogenase** resulting in synthesis of a ketone, β-ketoacyl-CoA; (4) thiolytic cleavage of β-ketoacyl-CoA by coenzyme A resulting in the formation of acetyl-CoA and fatty acyl-CoA, which is two carbons shorter. Reaction is catalyzed by **acyl-CoA acetyltransferase (thiolase)**. Similar sets of four steps of β-oxidation of fatty acyl-CoA are repeated in a cyclic manner. During each cycle, a two-carbon fragment would be released as acetyl-CoA, and a two-carbon shorter fatty acyl-CoA is produced. Thus, 16-carbon long palmitoyl-CoA will require 7 cycles of β-oxidation to be completely converted to 8 molecules of acetyl-CoA (Fig. 10.20).

## 10.6 Lipid Catabolism

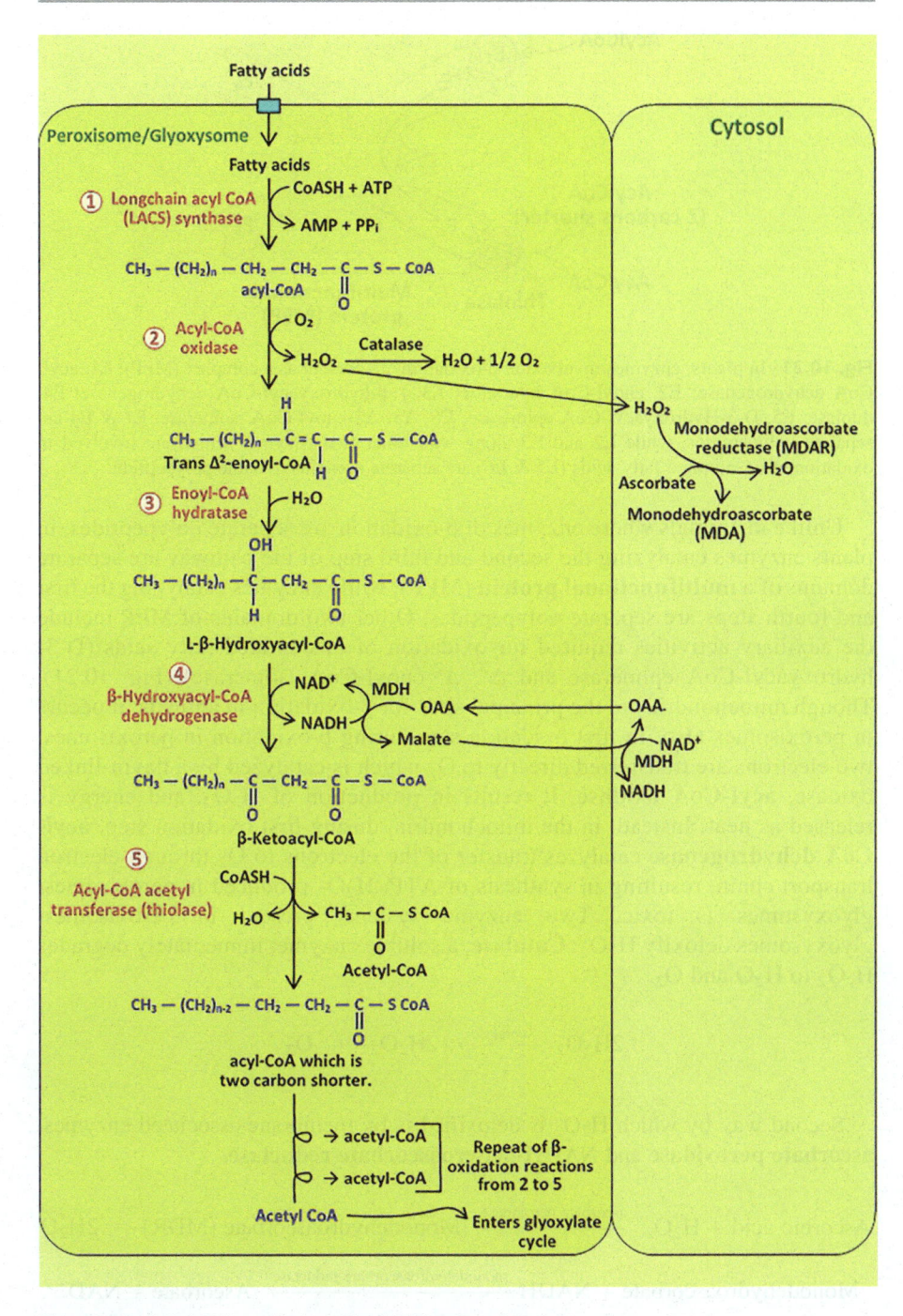

**Fig. 10.20** β-Oxidation of fatty acids. *MDH* malate dehydrogenase

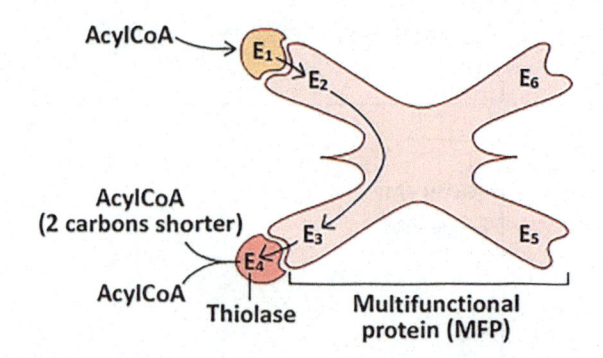

**Fig. 10.21** In plants, enzymes involved in β-oxidation are present as a complex (MFP); E1, acyl-CoA dehydrogenase; E2, enoyl-CoA hydratase; E3, L-β-hydroxyacyl-CoA dehydrogenase; E4, thiolase; E5, D-3-Hydroxyacyl-CoA epimerase; E6, Δ3, Δ2-enoyl-CoA isomerase; E1 & E4 are separate polypeptides, while E2 and E3 along with other two enzymes which are involved in oxidation of unsaturated fatty acids (E5 & E6) are separate domains of same polypeptide

Unlike in animals where enzymes of β-oxidation are separate polypeptides, in plants enzymes catalyzing the second and third step of the pathway are separate domains of a **multifunctional protein (MFP)**, while enzymes catalyzing the first and fourth steps are separate polypeptides. Other two domains of MFP include the auxiliary activities required for oxidation of unsaturated fatty acids (D-3-hydroxyacyl-CoA epimerase and $\Delta^3$, $\Delta^2$-enoyl-CoA isomerase) (Fig. 10.21). Though mitochondria are the principal sites for β-oxidation in animals, it occurs in peroxisomes also. In first oxidation step during β-oxidation in peroxisomes, two electrons are transferred directly to $O_2$ which is catalyzed by a flavin-linked oxidase, **acyl-CoA oxidase**. It results in production of $H_2O_2$, and energy is released as heat. Instead, in the mitochondria, during first oxidation step, **acyl-CoA dehydrogenase** catalyzes transfer of the electrons to $O_2$ through electron transport chain, resulting in synthesis of ATP. $H_2O_2$ produced in peroxisomes/glyoxysomes is toxic. Two enzyme systems present in peroxisomes/glyoxysomes detoxify $H_2O_2$. **Catalase**, a soluble enzyme, immediately degrades $H_2O_2$ to $H_2O$ and $O_2$.

$$2H_2O_2 \xrightarrow{\text{catalase}} 2H_2O_2 + O_2$$

Second way by which $H_2O_2$ is detoxified is by membrane-associated enzymes, **ascorbate peroxidase** and **NADH-dehyroascorbate reductase**.

$$\text{Ascorbic acid} + H_2O_2 \xrightarrow{\text{ascorbate peroxidase}} \text{Monodehydroascorbate (MDR)} + 2H_2O$$

$$\text{Monodehydroascorbate} + NADH \xrightarrow{\text{monodehydroascorbate reductase}} \text{Ascorbate} + NAD^+$$

## 10.6 Lipid Catabolism

**Table 10.5** Comparison of biosynthesis and β-oxidation of fatty acids in plants

| Characteristics | Fatty acid biosynthesis | β-Oxidation |
| --- | --- | --- |
| Cellular location | Plastids | Glyoxysomes |
| Activation | As thioesters of acyl carrier protein | As thioesters of CoA |
| Fatty acid sizes | Biosynthesis of fatty acids up to 18 carbon long occurs in plastids, further elongation in the endoplasmic reticulum | All sizes are degraded in glyoxysomes |
| Enzymes | Fatty acid synthase, which includes six enzymes besides ACCase | β-Oxidation requires complex of 4 enzymes |
| Redox reaction cofactors | NADPH | Flavoprotein dehydrogenase which directly pass electrons to $O_2$ and NADH |
| Process | Two-carbon elongation using malonyl-ACP | Two-carbon fragments are removed as acetyl-CoA |
| β-Hydroxyacyl intermediate | D-Isomer | L-Isomer |

Second oxidation step in β-oxidation involves dehydrogenation of β-L-hydroxyacylacyl-CoA by the enzyme β-hydroxyacyl-CoA dehydrogenase. This oxidation reaction is coupled with reduction of $NAD^+$ to NADH. NADH needs to be oxidized for continuation of β-oxidation of fatty acids. Since, unlike in the mitochondria, electron transport chain is not present in peroxisomes/glyoxysomes, NADH is oxidized, coupled with reduction of monodehydroascorbate (MDA). The reaction is catalyzed by the enzyme MDAR (MDA reductase). NADH oxidation also occurs by malate dehydrogenase (MDH), which catalyzes reduction of oxaloacetate to malate in peroxisomes. Malate, after being transported to cytosol, is oxidized to oxaloacetate, which is coupled with reduction of $NAD^+$ to NADH. Apparently, reactions of β-oxidation of fatty acids appear to be reversible of the reactions which occur during synthesis of fatty acids. However, a comparison between the two is listed in Table 10.5.

### 10.6.2 Glyoxylate Cycle

Since in animal cells, β-oxidation occurs mainly in the mitochondria, acetyl-CoA is metabolized through TCA cycle. In germinating oil seeds, however, β-oxidation occurs in glyoxysomes. Acetyl-CoA, produced during β-oxidation in glyoxysomes, is the source for many of the precursors required at the time of germination of seeds and growth of the seedlings. Acetyl-CoA is also the precursor for sugar biosynthesis. Conversion of acetyl-CoA to sugars involves glyoxylate cycle, TCA cycle, and gluconeogenesis which occur in glyoxysomes, mitochondria, and cytosol, respectively. Two enzymes of glyoxylate pathway, i.e., **isocitrate lyase** and **malate synthase**, are specific for glyoxysomes. Isocitrate lyase catalyzes hydrolysis of isocitrate to a four-carbon dicarboxylic acid succinate and a two-carbon compound, glyoxylate. Succinate is exported out of glyoxysomes and is transported to mitochondria where it is converted to OAA through TCA cycle. Oxaloacetate is exported out of mitochondria to the cytosol and is converted into phosphoenolpyruvate (PEP) through the action of **phosphoenolpyruvate carboxykinase (PCK)**. Synthesis of sugars from PEP occurs through gluconeogenesis. Other enzyme, which is specific for glyoxysomes, is malate synthase, which catalyzes synthesis of malate from glyoxylate and acetyl-CoA. Enzymes for two of the reactions of glyoxylate pathway are present in the cytosol. These include **aconitase** and **malate dehydrogenase (MDH)**, which catalyze synthesis of isocitrate from citrate and dehydrogenation of malate, respectively. In the cytosol MDH catalyzes dehydrogenation of malate resulting in the production of OAA and reduction of $NAD^+$ to NADH. Oxaloacetate is reduced to malate in glyoxysomes after it is transported back which is associated with oxidation of NADH. Oxidation of NADH is required for continuity of oxidation of fatty acids. Entry of β-oxidation product, acetyl-CoA, directly into TCA will result in saving of one carbon since the link reaction is bypassed.

## 10.6.3 Gluconeogenesis

Oxaloacetate exported from the mitochondria is converted to sugars in the cytosol. Phosphoenolpyruvate carboxykinase catalyzes the conversion of oxaloacetate to phosphoenolpyruvate (PEP). The reaction is irreversible and requires ATP, and one carbon is lost as $CO_2$.

$$\text{Oxaloacetate} + \text{ATP} \xrightarrow{\text{PCK}} \text{PEP} + CO_2 + \text{ADP}$$

PEP is converted to sugars by reverse reactions of glycolytic pathway, which are known as gluconeogenesis. Since gluconeogenesis occur in the cytosol along with glycolysis, an understanding of regulation of both pathways is required. Only three carbons of OAA are used for conversion to sugars, since there is a loss of one carbon as $CO_2$ in the reaction catalyzed by PCK. Thus, only 75% of the carbon is utilized for the production of sugars, and there is a loss of 25% of the carbon. Seven out of ten reactions of glycolysis are reversible, while three reactions are irreversible. These include conversion of PEP to pyruvate, fructose-6-phosphate to fructose-1,6-bisphosphate, and glucose to glucose 6-phosphate, catalyzed by PEP kinase, phosphofructokinase, and hexokinase, respectively. During gluconeogenesis, reaction involving conversion of pyruvate to PEP is bypassed, since PEP produced from OAA is available. Conversion of fructose-6-phosphate to fructose-1,6-biphosphate is the main regulatory reaction, which determines whether sugars would be catabolized through glycolysis or sugars will be produced by gluconeogenesis (Fig. 10.22). During glycolysis, the reaction is catalyzed by phosphofructokinase, while another enzyme **fructose-1,6-bisphosphatase (FBPase)** catalyzes the

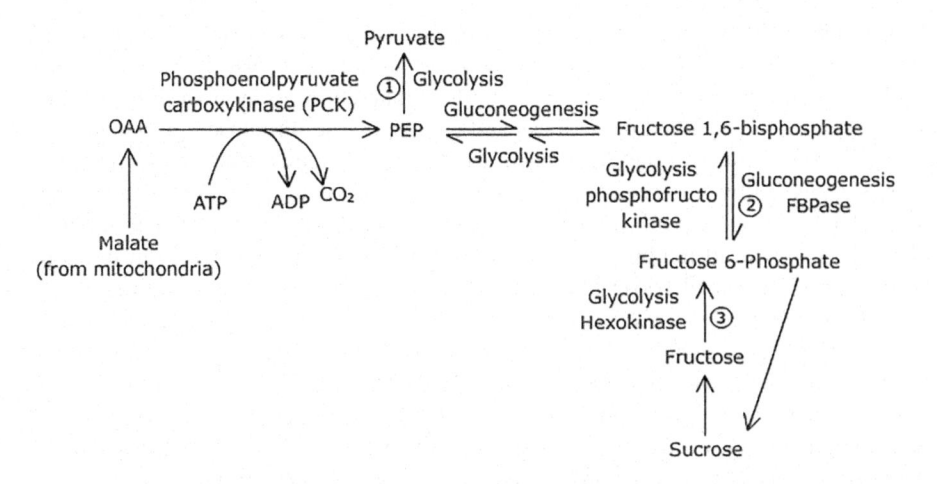

## 10.6 Lipid Catabolism

```
                    Pyruvate
                       ↑
  Phosphoenolpyruvate  ①  Glycolysis
  carboxykinase (PCK)  |   Gluconeogenesis
OAA ──────────────────→ PEP ⇌ ⇌ Fructose 1,6-bisphosphate
 ↑      ╱    ↓  ↓         Glycolysis
 |     ATP  ADP CO₂       Glycolysis  ↑  Gluconeogenesis
 |                        phosphofructo ║ ② FBPase
Malate                    kinase       ║↓
(from mitochondria)       Fructose 6-Phosphate
                          Glycolysis ↑
                          Hexokinase │ ③
                          Fructose
                             ↑
                             |  ↓
                          Sucrose
```

1, 2 & 3 reactions of glycolysis are irreversible. 1 & 3 are bypassed in gluconeogenesis, Enzyme catalyzing reaction 2 is different in gluconeogenesis. It is F1, 6-bisphosphatase whose activity is regulated by Fructose 2, 6-bisphosphate

```
                    Fructose 6-Phosphate
                          ↑  ╱── ATP
                    Pi ←  ║ ╱
                          ║╱ Kinase (activity promoted
         Phosphatase      ║  by high F, 6-P & high Pi)
        (activity is promoted ║ ─→ ADP
         by high triose-P)  ║↓
                    Fructose 2, 6-biphosphate
```

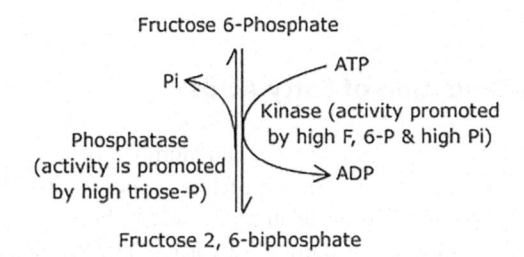

Low level of fructose 2, 6-biphosphate promotes FBPase activity promoting gluconeogenesis, white low levels of fructose 2, 6-biphosphate inhibit FBPase activity

**Fig. 10.22** Gluconeogenesis and glycolysis occur in the cytosol of cells. Gluconeogenesis involves synthesis of sugars by the reversal of glycolysis except reactions 1, 2, and 3

hydrolysis of phosphate group linked to C-1 of fructose-1,6-bisphosphate, in gluconeogenesis.

$$\text{Fructose} - 1,6 - \text{bisphosphate} + H_2O \xrightarrow{\text{FBPase}} \text{fructose} - 6 - \text{phosphate} + P_i$$

Thus, the enzymes catalyzing the reactions are different in gluconeogenesis and glycolysis. Regulation of FBPase and PFK determines the direction of the pathways, whether sugars will be synthesized by gluconeogenesis or will be catabolized through glycolysis. Both pathways occur in the cytosol. Metabolites which regulate activity of FBPase include fructose-2,6-bisphosphate. Low levels of fructose-2,6-

bisphosphate (FBP) promote activity of FBPase. As a result, sucrose synthesis will be favored, whereas high concentration of fructose-2,6-bisphosphate inhibits FBPase, inhibiting sucrose synthesis. Intracellular concentration of fructose-2,6-bisphosphate is controlled by a kinase and a phosphatase. Fructose-2,6-bisphospate kinase is responsible for the synthesis of fructose-2,6-bisphosphate, and its degradation is catalyzed by fructose2,6-bisphosphatase. Activity of FBP kinase is promoted by high concentrations of F-6,P and $P_i$, thereby increasing F2,6-BP concentrations in the cell. This would inhibit gluconeogenesis and promote glycolysis. On the contrary, accumulation of triose phosphates and glycerate 3-phosphate promotes the activity of FBPase, in turn decreasing F2,6-BP concentrations, leading to promotion of gluconeogenesis (Fig. 10.22). Third irreversible step of glycolysis includes conversion of glucose to glucose 6-phosphate. This reaction is also bypassed in gluconeogenesis. Glucose 6-phosphate is converted to glucose 1-phosphate by phosphoglucomutase. Glucose 1-phosphate is the precursor for sucrose biosynthesis. Sucrose thus synthesized is translocated to the growing parts of the seedlings from the storage regions of the germinating seeds, i.e., either from the endosperm (castor seeds) or from cotyledons of the seeds (sunflower). A comparison of lipid metabolism in plants and animals is given in Table 10.6.

### 10.6.4 α-Oxidation of Fatty Acids

Generally fatty acids contain an even number of carbon atoms since these are synthesized as a result of addition of two carbon units. However, fatty acids with odd-number carbons also occur in plants and in brain cells of animals. α-Oxidation of fatty acids with an even number of carbon atoms results in the synthesis of fatty acids with odd-number carbon atoms. Generally, fatty acids with C13-C18 are oxidized by α-oxidation which involves sequential removal of α-carbon as $CO_2$ from carboxylic end of the fatty acids. Unlike β-oxidation, there is no energy release and no involvement of CoA-SH intermediates during α-oxidation. α-Oxidation of fatty acids is possibly involved in the degradation of long-chain fatty acids. In humans also, it occurs in peroxisomes and is involved in the oxidation of dietary **phytanic acid** (a branched-chain fatty acid obtained by humans from dairy products), which cannot undergo β-oxidation because of its methyl branches.

### 10.6.5 ω-Oxidation of Fatty Acids

Some fatty acids, which are constituents of cutin and suberin, have hydroxyl group or carboxylic group at the ω-methyl terminal. These are produced as a result of ω-oxidation. Fatty acids having carboxylic groups at both of their ends are dicarboxylic acids which are characteristic of suberin. During ω-oxidation, molecular oxygen is used, and it requires involvement of **cytochrome P450**. ω-Oxidation occurs in the endoplasmic reticulum.

**Table 10.6** Comparison of lipid metabolism in plants and animals

| | Plants (plastids) | Plants (cytosol/ ER) | Animals (cytosol/ER) |
|---|---|---|---|
| *Fatty acid biosynthesis* | | | |
| Site of biosynthesis | Plastids | Further elongation and modifications in FA | Only cytosolic ACCase present, which generates malonyl-CoA used both for FA elongation and further elongation in ER |
| ACCase | Molecular mass 650 kDa, heteromeric, HET-ACCase consists of four polypeptide subunits, BC, BCCP, α- and β-CT | Molecular mass 500 kDa. Homodimeric protein, HOM-ACCase, three functional domains, BC, BCCP, and CT amalgamated in one polypeptide | Molecular mass 500 kDa, Hom-ACCase, Homodimeric protein, each monomeric protein is complex multifunctional unit |
| | Encoded in nuclear genome and synthesized by cytosolic ribosome's except β-subunit which is encoded in plastid genome | Encoded in nuclear genome | Encoded in nuclear genome |
| | ACCase regulated by stromal pH and $Mg^{2+}$ concentration | – | Regulated by citrate and phosphorylation and dephosphorylation |
| FAS | Type II present in plastids only | Not present in the cytosol | Type I present in the cytosol only |
| | Multi-subunits, each polypeptide subunit catalyzes a separate reaction | – | Multifunctional polypeptide, homodimer, molecular mass 480 kDa, activities of FAS are localized in separate domains of polypeptide |
| | Intermediates of the reaction may diffuse and be diverted for other pathways | – | No intermediates are released |
| | Fatty acids of variable length (<16) may also be generated in certain plants | – | Fatty acids with hydrocarbon chains <16 are not generated |
| Introduction of double bond/bonds in fatty acids | Soluble stearoyl-CoA desaturase is present in the stroma of plastids | In the ER, desaturases are membrane bound | Desaturation of FAs only in ER by membrane-bound desaturases |
| | Membrane-localized desaturases in plastids use soluble Fd as electron donor; act on galactolipids | Desaturation of FA occurs as glycerolipids. ER membrane-bound desaturases act on FAs esterified to PC and utilize cyt b5 as an | Some desaturases act on acyl-CoA |

(continued)

**Table 10.6** (continued)

| | Plants (plastids) | Plants (cytosol/ ER) | Animals (cytosol/ER) |
|---|---|---|---|
| | | intermediate electron donor | |
| Elongation of FAs | Fatty acids having up to 18-carbon length synthesized | Membrane-bound elongases localized in the ER use malonyl-CoA and not malonyl-ACP | Fatty acid elongase systems are present in the smooth ER and in the mitochondria, use malonyl-CoA and not malonyl-ACP |
| *Lipid catabolism* | | | |
| Site | Peroxisomes (in leaf cells) and glyoxysomes (in nongreen tissues) | – | Mitochondrial matrix |
| | Generally, triglycerides are not used as the source of energy | – | Triglycerides stored in adipose tissues are used as the source of energy |
| | Stored triglycerides as seed oil are converted to sugars at the time of seed germination | | Fats cannot be converted to sugars because of the absence of glyoxylate cycle |
| β-Oxidation | Two of the enzymes, enoyl-CoA hydratase and ι-β-hydroxyacyl-CoA dehydrogenase, are present as the two domains of a single polypeptide, while other enzymes of β-oxidation are separate polypeptides | | Catalytic sites of enzymes are present on separate polypeptides |
| | β-Oxidation occurs in peroxisomes/ glyoxysomes | | β-Oxidation occurs in mitochondrial matrix |
| | Acetyl-CoA generated in β-oxidation enters glyoxylate cycle | | Acetyl-CoA produced during β-oxidation enter TCA cycle in the mitochondria |
| | In the first oxidation reaction during β-oxidation, $H_2O_2$ is produced, which is detoxified without any ATP formation | | $FADH_2$ is produced which is oxidized by the electron transport chain resulting in ATP generation |
| | NADH produced during second oxidation is oxidized by malate dehydrogenase located outside of the glyoxysomes since the enzyme is not present in glyoxysomes | | NADH is oxidized in the mitochondria through electron transport chain |

## 10.6.6 Catabolism of Unsaturated Fatty Acids

Besides saturated fatty acids, catabolism of unsaturated fatty acids is also significant both among animals and plants since most of the fatty acids of triacylglycerol and phospholipids are unsaturated. Catabolism of unsaturated fatty acids involves additional enzyme-catalyzed reactions. Most of the unsaturated fatty acids have double bonds present in *cis* configuration, the first double bond being present between nineth and tenth carbon atom from the carboxylic end. These double bonds are not acted upon by the enzymes of β-oxidation, which recognize only *trans* configuration. Thus, catabolism of unsaturated fatty acids requires a change in configuration of double bonds from *cis* to *trans*. One of the abundant unsaturated fatty acids, i.e., oleic acid, has the structure $18:1^{\Delta 9}$. Three cycles of β-oxidation will remove 6 carbons as three acetyl-CoA, with a 12-carbon fatty acid which is produced with double bond present at the third carbon atom having a structure $12:1^{\Delta 3cis}$. This intermediate compound is not the substrate for enoyl-CoA hydratase which recognizes the *trans* double bonds only. An isomerase, $\Delta^3, \Delta^2$-enoyl CoA isomerase will shift the double bond from third to second position and *cis* form of double bond to *trans* configuration, resulting in conversion of 12:1, *cis*-$\Delta^3$ enoyl-CoA to 12:1 *trans*-$\Delta^2$ enoyl-CoA. This is followed by the action of enoyl-CoA hydratase which results in a hydroxy intermediate, L-β-hydroxy-CoA. This is the substrate for another dehydrogenase (similar to β-oxidation), followed by subsequent β-oxidation reactions (Fig. 10.23).

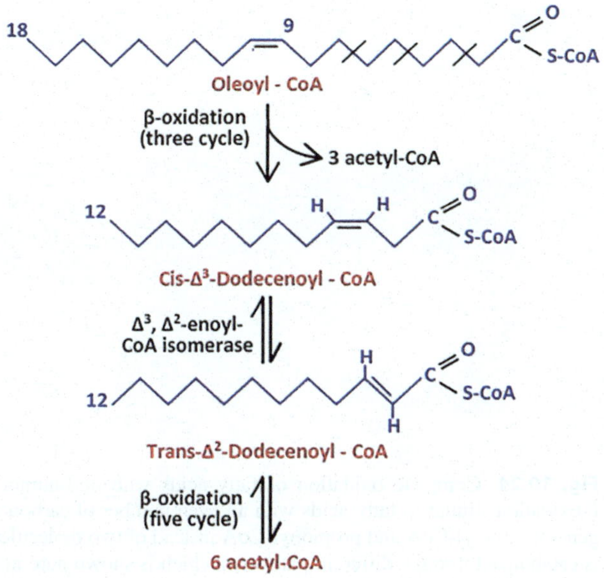

**Fig. 10.23** Oxidation of monounsaturated fatty acids (oleic acid, 18:1). Δenoyl-CoA hydratase catalyzes addition of water to trans-Δ 2 enoyl CoA. Thus, cis-isomer is converted to trans-isomer (which is a normal intermediate of β-oxidation) by enzyme Δ3, Δ2 enoyl-CoA isomerase

## 10.6.7 Catabolism of Fatty Acids with Odd Number of Carbon Atoms

Plants have fatty acids with odd numbers of carbons as constituents of lipids. These are oxidized and degraded till propionyl-CoA and acetyl-CoA are produced as a result of the last cycle of β-oxidation, unlike the formation of two molecules of acetyl-CoA. Propionyl-CoA is catabolized due to additional enzymes, i.e., propionyl-CoA carboxylase, methyl malonyl-CoA epimerase, and methyl malonyl-CoA mutase. The reaction catalyzed by propionyl-CoA carboxylase is an ATP-requiring reaction, which requires biotin as the cofactor, and it results in production of D-methyl malonyl-CoA. D-methyl malonyl-CoA is converted to its isomer L-methyl malonyl-CoA by the enzyme methyl malonyl-CoA epimerase. Intramolecular rearrangement of the L-methyl malonyl-CoA molecule is catalyzed by the enzyme methyl malonyl-CoA mutase resulting in the production of succinyl-CoA which is metabolized by the cell (Fig. 10.24).

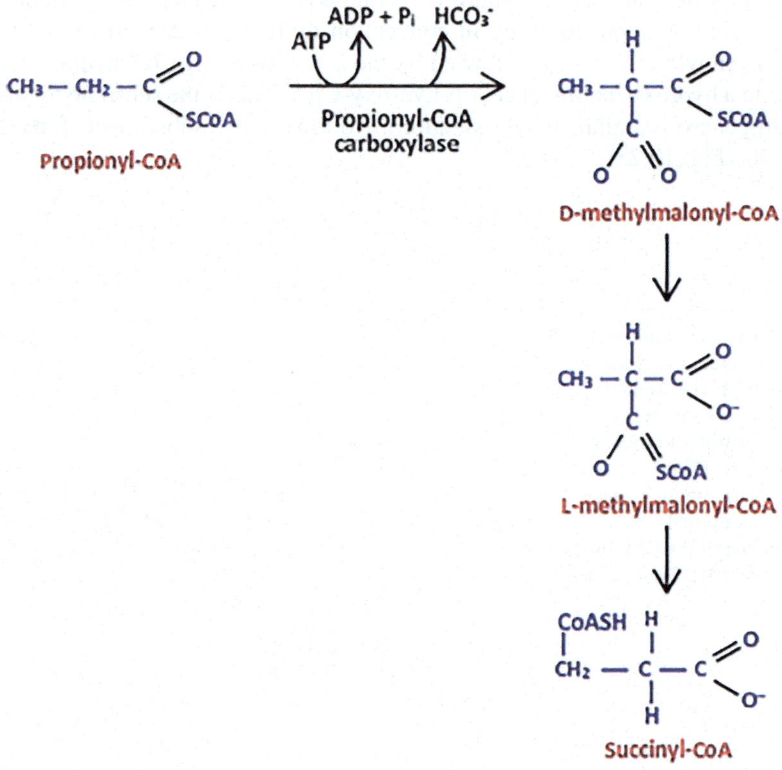

**Fig. 10.24** Complete oxidation of fatty acids with odd-number carbon atoms occurs through β-oxidation similar to fatty acids with an even number of carbon atoms. Last cycle of β-oxidation generates acetyl-CoA and propionyl-CoA instead of two molecules of acetyl-CoA. Propionyl-CoA is catabolized through different pathways which is shown here in this figure

## Summary

- Lipids are important biomolecules having diverse functions. These are stored as neutral fats and waxes. Membrane lipids include glycerophospholipids, glycolipids, sphingolipids, and sterols. Extracellular lipids, such as cutin and suberin, serve to protect plants from water loss by evaporation. Lipids also function as signaling molecules and as cofactors of various enzymes.
- In plants, fatty acids with carbon atom numbers up to 18 are synthesized in plastids, unlike in animals where fatty acid biosynthesis occurs in the cytosol. However, both in animals and plants, further elongation of the carbon chain and desaturation of fatty acids occur in the endoplasmic reticulum.
- Precursors for fatty acids biosynthesis are acetyl-CoA and malonyl-CoA. Acetyl Co-A is synthesized from pyruvate which is generated during glycolysis of sugars from either Calvin cycle in plastids or has been imported from the outside in nongreen plastids. Malonyl-CoA is produced as a result of carboxylation of acetyl-CoA catalyzed by acetyl-CoA carboxylase (ACCase), which has biotin as the cofactor. In plants, ACCase consists of four polypeptide subunits, while in animals, the enzyme consists of single polypeptide with separate catalytic domains (multifunctional protein).
- Biosynthesis of fatty acids is catalyzed by the enzyme fatty acid synthase (FAS). Except ACCase, FAS refers to all other catalytic activities during fatty acid biosynthesis. These include malonyl/acetyl-CoA ACP transferase (MAT), β-ketoacyl-ACP synthase (KAS), β-hydroxyacyl-ACP dehydratase, enoyl-ACP reductase, and β-ketoacyl-ACP reductase. In plants FAS II consists of separate polypeptides for different catalytic activities, while in animals, FAS I is a multifunctional protein having separate catalytic domains for each catalytic activity. Acyl moiety is hydrolyzed from ACP by thioesterase.
- Fatty acids with a chain length up to 16 carbons or 18 carbons are synthesized in plastids depending on the plant type. Further elongation or desaturation occurs in the endoplasmic reticulum after these are transported out as CoA derivatives. Elongated of fatty acids occurs as CoA derivatives and not as derivatives of ACP. The mechanism of elongation of fatty acid hydrocarbon chains and the enzymes required are similar to those of fatty acid synthesis. Unlike in animals, a soluble desaturase present in the stroma of plastids catalyzes formation of oleoyl-ACP. Other desaturases are present in bound form within membranes of the endoplasmic reticulum.
- In plants, membrane lipids are synthesized both in plastids (prokaryotic way) and the endoplasmic reticulum (eukaryotic way). At least some of the membrane lipids in plastids are imported from the endoplasmic reticulum.
- Neutral fats are synthesized in the endoplasmic reticulum by two ways. In one of the ways, glycerol 3-phosphate is acylated first at C1 and C2 carbons followed by hydrolysis of phosphate group and its acylation. Second way includes

modification of the acyl groups as phosphatidylcholine followed by hydrolysis of choline phosphate and acylation. Triacylglycerol synthesis occurs within the lipid bilayer of the endoplasmic reticulum. The outer lipid layer is pinched off as oleosomes.
- Oleosomes are surrounded by a monolayer of lipids in which specific types of proteins including oleosins are present. Oleosins have their hydrophobic domains exposed toward the inside of the oleosomes, while the hydrophilic N- and C-terminal ends are exposed toward the cytosolic side.
- Lipid catabolism is significant in germinating oil seeds. It involves hydrolysis of fats, β-oxidation of fatty acids, glyoxylate cycle, and gluconeogenesis, which occur in the organelles oleosomes, glyoxysomes, mitochondria, and cytosol, respectively. Sugars thus synthesized are translocated to the meristematic regions of the seedlings.
- Catalytic domains for β-oxidation of fatty acids in plants are present in separate polypeptides, unlike in animals, where these are present as multifunctional proteins.

## Multiple-Choice Questions

1. The lipids which form the major component of thylakoids are:
   (a) Phospholipids
   (b) Sphingolipids
   (c) Galactolipids
   (d) Neutral lipids
2. In plants fatty acid biosynthesis occurs in:
   (a) Plastids
   (b) Peroxisomes
   (c) Mitochondria
   (d) Cytosol
3. First step in fatty acid biosynthesis is catalyzed by:
   (a) Acetyl-CoA synthase
   (b) Carboxyl transferase
   (c) Malonyl-CoA:ACP transacylase
   (d) Acetyl-CoA carboxylase
4. At the time of fatty acid synthesis, the cofactor to which acyl group is attached during reactions involving two reductions and dehydration is:
   (a) Coenzyme A
   (b) Fatty acid synthase
   (c) ACP
   (d) Ketoacyl-ACP synthase

5. Initial condensation of acetyl CoA with malonyl-ACP resulting in production of acetoacetyl-ACP is catalyzed by:
   (a) KAS I
   (b) KAS II
   (c) KAS III
   (d) 3-Ketoacyl-ACP reductase
6. In the endoplasmic reticulum, modification of hydrocarbon chain of fatty acids requires:
   (a) Fatty acyl-ACP
   (b) Fatty acyl-CoA
   (c) Thioesterase
   (d) FAS
7. In a prokaryotic way, synthesis of membrane lipids involves:
   (a) Esterification of C1 hydroxyl group of glycerol phosphate with 16:0 FA and C2 hydroxyl group with 18:0 FA
   (b) Esterification of C1 hydroxyl group of glycerol phosphate with 18:1 FA and C2 hydroxyl group with 16:0 FA
   (c) Esterification of both C1 and C2 of glycerol phosphate with 18:1 FA
   (d) Esterification of both C1 and C2 of glycerol phosphate with 16:0 FA
8. Conversion of stearoyl-ACP to oleoyl-ACP (tick the correct answer):
   (a) Is catalyzed by a soluble desaturase present in the stroma of plastids
   (b) Is catalyzed by desaturase present in the envelope of the plastid
   (c) Is catalyzed by desaturase present in the ER membrane
   (d) Is catalyzed by a soluble desaturase present in the cytosol
9. In plants β-oxidation of fatty acids occurs in the:
   (a) Mitochondria
   (b) Cytosol
   (c) Oleosomes
   (d) Glyoxysomes
10. Which of the following statements is correct?
    (a) In plants acyl-CoA oxidase catalyzes transfer of electrons $O_2$ during first oxidation step in β-oxidation of fatty acids.
    (b) Acyl-ACP is transported out of plastids to be used for further modifications in the hydrocarbon chain of fatty acids.
    (c) Oleosin is the membrane lipid of oleosomes.
    (d) Glyoxylate cycle occurs both in plants and animals.

**Answers**

1. c  2. a  3. d  4. c  5. c  6. b  7. b
8. a  9. d  10. a

## Suggested Further Readings

Browsher C, Steer M, Tobin A (eds) (2008) Plant biochemistry. Garland Science, Tailor & Francis Group, New York, pp 303–329

Heldt HW (2005) Plant biochemistry, 3rd edn. Elsevier Academic Press, Burlington, pp 363–396

Nelson DL, Cox MM (2017) Lehninger principles of biochemistry, 7th edn. WH Freeman, MacMillan Learning, New York, pp 649–668

Ohlrogge J, Browse J, Jaworski J, Chris S (2013) Lipids. In: Buchanan BB, Gruissem W, Jones RL (eds) Biochemistry and molecular biology of plants. Wiley-Blackwell, Chichester, pp 337–396

# Nitrogen Metabolism

Manju A. Lal

Nitrogen is one of the most important macronutrients required by the plant. All proteins consist of nitrogen-containing amino acids. Heme component of the chlorophylls, nitrogenous bases of DNA and RNA, and **phenylpropanoids** (such as **flavonoids**) are all nitrogen-containing biomolecules (Fig. 11.1). Plants absorb carbon in the form of $CO_2$ fixed by photosynthetic process; hydrogen and oxygen are taken up in the form of $H_2O$. Oxygen is also absorbed from the air. In contrast, plants are not able to utilize molecular nitrogen directly. In spite of the fact that nitrogen constitutes almost 80% of the air, it needs to be provided to plants in the form of fertilizers. There are some prokaryotes, however, which have the ability to fix molecular nitrogen by means of biological fixation leading to enhancement of soil nitrogen as well. Thus, it becomes necessary to study the forms of nitrogen which plants can absorb, mechanisms of their uptake, biological nitrogen fixation, and also the mechanisms involved in their further metabolism. Plants absorb nitrogen through their roots either in the form of ammonium ions or nitrates which are available in the soil. Nitrate uptake can take place through leaves, only in case of epiphytes or only through foliar spray of fertilizers. Molecular nitrogen can be fixed directly by the nitrogen-fixing prokaryotes (*diazotrophs*), which are either free living or growing in symbiotic associations with plants. When uptake of nitrogen by plants is in the form of nitrate, it needs to be reduced to ammonium before further assimilation for amino acid biosynthesis, followed by biosynthesis of purines and pyrimidines. Within plants nitrogen is transported in the form of nitrates, amino acids, amines, and ureides. This chapter shall focus on nitrogen uptake in the form of ammonium and nitrate ions, mechanisms and the enzymes involved in their reduction, physiology and biochemistry of molecular nitrogen fixation, ammonium assimilation, transport of the assimilated nitrogen, and finally amino acid biosynthesis (Fig. 11.2).

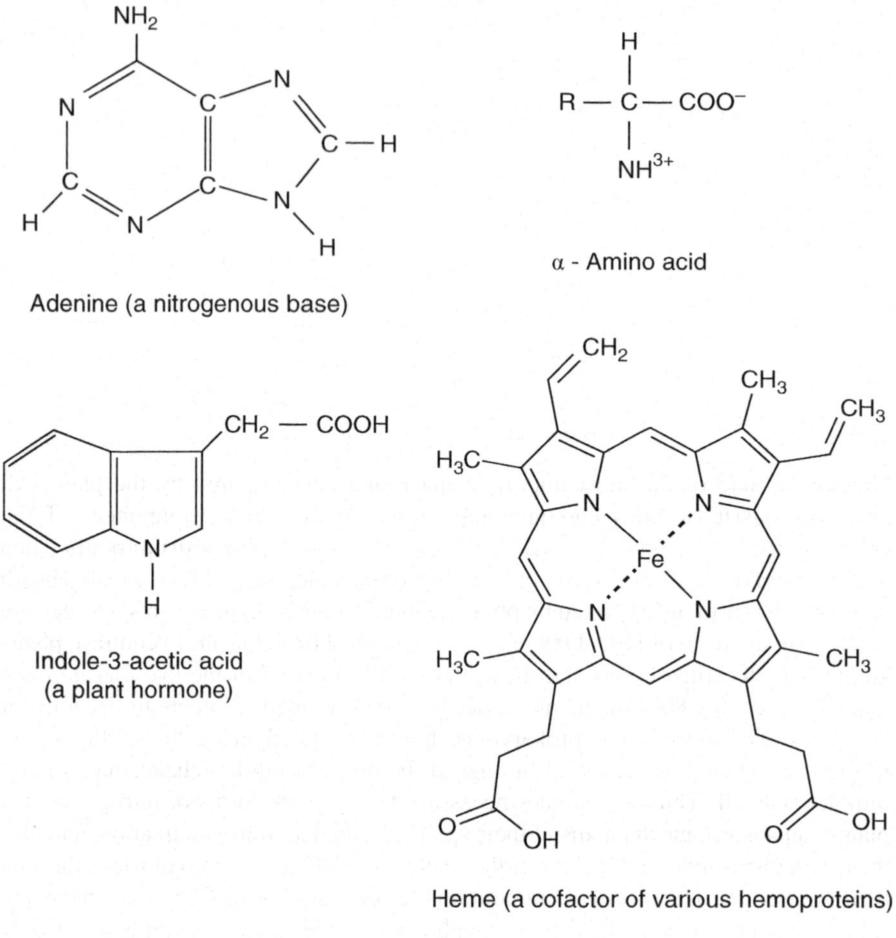

Fig. 11.1 Some of the nitrogenous biomolecules

## 11.1 Biogeochemical Cycle of Nitrogen

Nitrogen occurs in different forms in nature. It is present as molecular nitrogen in the air, and among living organisms, it is present in various organic forms. In soil, inorganic nitrogen is present as nitrates and ammonium ions, which are mostly derived from supplementation of fertilizers to the soil. Organic nitrogen in soil originates either from dead animals or plants or is derived from the excreta of the animals. Its source can also be the fertilizers. Some prokaryotic organisms also contribute to the nitrogen pool of the soil. These organisms are either free living or are growing in symbiotic associations with certain plants. These organic and inorganic forms of nitrogen are interconvertible which is facilitated by a number of processes responsible for the geochemical and biochemical cycle of nitrogen (Box 11.1).

## 11.1 Biogeochemical Cycle of Nitrogen

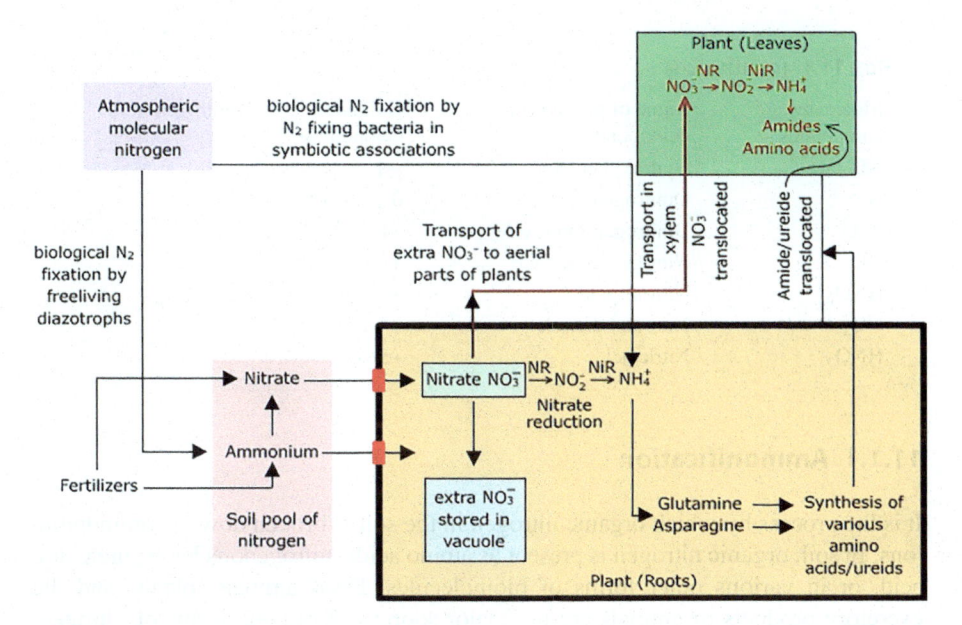

**Fig. 11.2** An overview of nitrogen metabolism

---

**Box 11.1: Oxidation States of Nitrogen**

Oxidation state of an element (often called oxidation number) refers to the state of the element in a compound when it has either gained or lost electrons. Since, in free state, the number of protons is equal to number of electrons, oxidation state of the element will be zero. In bound forms, oxidation state of the element can be calculated by subtracting the number of electrons from the number of protons of the element. The number refers to number of lost or gained electrons, in which loss of an electron raises the oxidation state of the element by one. In similar way, addition of an electron lowers the oxidation state by number one and is known as reduction. In compound form, nitrogen can have the oxidation levels ranging from $-3$ (in ammonia) to $+5$ (in nitric oxide). In the following table, some of the nitrogenous compounds are listed along with the oxidation levels of nitrogen present in them.

| Molecule/ compounds | Name of the molecule/ compounds | Oxidation level of nitrogen in the compounds |
|---|---|---|
| $NH_3$ | Ammonia | $-3$ |
| $NH_2NH_2$ | Hydrazine | $-2$ |

(continued)

**Box 11.1** (continued)

| Molecule/compounds | Name of the molecule/compounds | Oxidation level of nitrogen in the compounds |
|---|---|---|
| $NH_2OH$ | Hydroxylamine | −1 |
| $N_2$ | Dinitrogen | 0 |
| $N_2O$ | Dinitrogen monoxide | +1 |
| NO | Nitrogen monoxide | +2 |
| $HNO_2$ | Nitrous acid | +3 |
| $NO_2$ | Nitrogen dioxide | +4 |
| $HNO_3$ | Nitric acid | +5 |

### 11.1.1 Ammonification

It is the process by which organic nitrogen in the soil is broken down to ammonium ions. In soil, organic nitrogen is present as amino acids, nitrogenous bases, urea, uric acid, or in various other forms of biomolecules. Dead animals, plants, and the excretory products of animals are the major sources of organic forms of nitrogen. Proteins are hydrolyzed by **proteases** released from microorganisms in the soil, resulting in the release of amino acids by the process known as **proteolysis**. At the time of ammonification, amino acids undergo **oxidative deamination**, and organic nitrogen is released as ammonia. Microorganisms such as *Clostridium* sp., *Proteus* sp., *Micrococcus* sp. etc. are responsible for deamination. The process occurs under aerobic conditions. Ammonia produced in this process gets dissolved in water resulting in the formation of ammonium ions. In alkaline soil, $H^+$ are not available, so $NH_3$ formed, which leaks out of the soil resulting in loss of soil nitrogen.

$$C_3H_7NO_2 + \tfrac{1}{2}O_2 \longrightarrow C_3H_4O_3 + NH_3$$
$$\text{Alanine} \qquad\qquad \text{Pyruvic acid} \quad \text{Ammonia}$$

### 11.1.2 Nitrification

In acidic soils, ammonia exists as ammonium ions ($NH_4^+$). Ammonium ions do not leach out of the soil because these are positively charged and are held by the negatively charged clay particles. Certain plants are capable of absorbing $NH_4^+$. However, in most cases ammonium ions are converted to nitrates by a process known as nitrification. The nitrifying bacteria are **chemoautotrophs** which oxidize ammonium ions. Oxidation of ammonium ions is an exergonic reaction and the released energy is utilized by chemoautotrophs for biosynthetic reactions. Oxidation of ammonium ions occurs in two steps:

## 11.1 Biogeochemical Cycle of Nitrogen

$$NH_4^+ + 1\tfrac{1}{2} O_2 \rightarrow NO_2^- + H_2O + 2H^+ \quad \left(\Delta G^{0'} = -272 \text{ kJ.mol}^{-1} \text{ NH}_4^+\right)$$

$$NO_2^- + \tfrac{1}{2} O_2 \rightarrow NO_3^- \quad \left(\Delta G^{0'} = -76 \text{ kJ.mol}^{-1} \text{ NO}_2^-\right)$$

These two reactions are carried out by autotrophic bacteria which include *Nitrosomonas* sp. and *Nitrobacter*, respectively, which exist in well-aerated soil. The two reactions occur simultaneously so that there is no accumulation of nitrite in the soil. These reactions are the source of energy for bacteria to form ATP, which is used by them for fixation of $CO_2$. The reactions are complex and are catalyzed by various enzymes such as **ammonia monooxygenase, hydroxylamine oxidoreductase,** and **nitrite oxidoreductase** present in these bacteria. The first two enzymes are present in *Nitrosomonas* and are required for oxidation of ammonium ions, while nitrite oxidoreductase required for the oxidation of nitrite to nitrate is present on the inner side of the cell membrane of *Nitrobacter*. In some microorganisms $NO_3^-$ serves as the electron acceptor in place of $O_2$ during oxidation of the substrate which gets reduced to $NH_4^+$ in a process known as anaerobic nitrate ammonification or **nitrate respiration.**

$$C_6H_{12}O_6 + 3NO_3^- + 6H^+ \rightarrow 6CO_2 + 3NH_4^+ + 3H_2O \quad \left(\Delta G^{0'} = +1766 \text{ kJ.mol}^{-1}\right)$$

Many bacteria do not reduce nitrogen for their own sake. Rather, there is a need to eliminate reducing equivalents produced during respiration. So, in place of $O_2$, nitrate is used. This process is different from assimilatory nitrate reduction.

### 11.1.3 Denitrification

In some microorganisms, nitrate is degraded to molecular nitrogen under anaerobic conditions. $NO_3^-$ serves as the electron acceptor in place of oxygen during oxidation of the substrates. This process is known as denitrification, and it occurs in the soil especially under anaerobic conditions. Nitrogen so formed escapes in the air.

$$2NO_3^- + 10e^- + 12H^+ \rightarrow N_2 + 6H_2O$$

In waterlogged soils, high nitrate concentration causes a great loss of soil nitrogen. Some of the denitrifying bacteria are *Thiobacillus denitrificans, Micrococcus denitrificans,* etc.

### 11.1.4 Nitrogen Fixation

Nitrogen pool of the soil in enriched from different sources. Atmospheric molecular nitrogen can be fixed by physical phenomenon, such as electrical discharge during lightning. Fixed nitrogen reaches soil in the form of nitrates. Molecular nitrogen is also fixed by some organisms by a process known as biological nitrogen fixation. Biological nitrogen fixation is of considerable importance and is carried out by various prokaryotes, which are either free living or are growing in symbiotic association with the plants (Fig. 11.3).

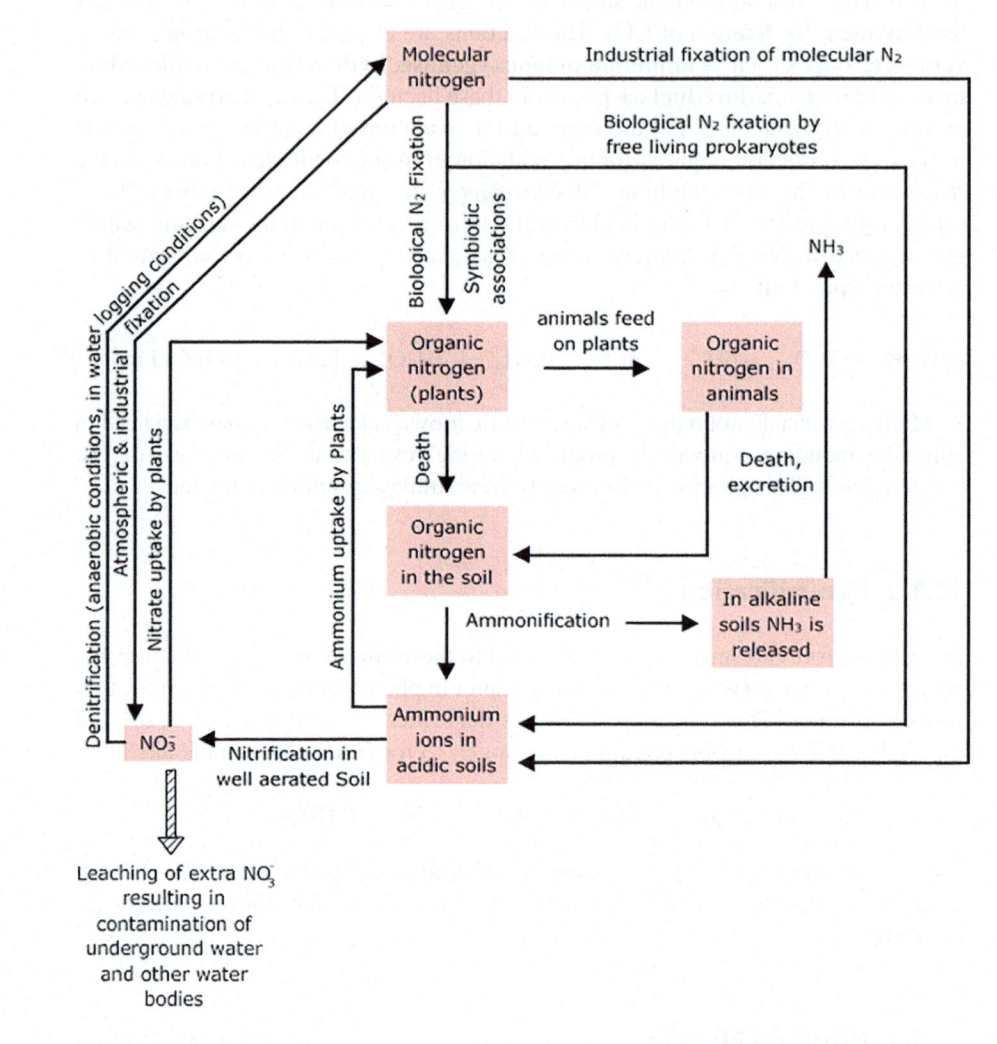

**Fig. 11.3** Nitrogen cycle

## 11.2 Nitrogen Nutrition for the Plants

As stated in the previous section, animals obtain their nitrogen supply through food, whereas plants have to depend solely upon the supply of fixed nitrogen present in the **rhizosphere** as ammonium or nitrate ions. Mostly these ions are absorbed through roots except in case of nitrogen fertilizer which are applied as the foliar spray. It is only in prokaryotic organisms growing in associations (symbiotic or asymbiotic) with higher plants that free nitrogen can be fixed and made available to plants. In the following section, various forms of nitrogen available to plants and their uptake mechanisms are being discussed.

### 11.2.1 Ammonium Ions

Under waterlogged conditions and in acidic soils, nitrogen is available in the form of ammonium ions because of low rates of nitrification. Unlike nitrates, ammonium ions do not easily leach out from the soil. Since they are positively charged ($NH_4^+$), they bind to the soil particles. Plant roots can absorb ammonium form of nitrogen into the symplasm via ammonium transporters located on the plasma membrane. High ammonium concentrations suppress the expression of the genes encoding for nitrate transporters. There are two classes of ammonium transporters (AMT) in the plants—low-affinity transporters, which are not saturated, and high-affinity saturable transporters. In the low-affinity non-saturable transporters, there is a possibility of the involvement of the aquaporins or nonselective ion channels. Proteins of low-affinity ammonium transporters, however, have not been identified so far. High-affinity transporters include transporters encoded by the AMT/Rh gene family. $K_m$ values of ammonium transporters range from 10 to 70 micromoles. Plant AMTs function as $NH_4^+$ uniporters. These transporters have multiple membrane-spanning domains. Five members of the gene family encoding AMT1 have been identified in *Arabidopsis*. All of them are expressed in roots, while two are expressed in shoots. $NH_4^+$ diffuses from soil to the symplasm of the root cells down the electrochemical gradient through the AMT1 (Fig. 11.4). Once inside the cytosol, some of $NH_4^+$ may be converted to $NH_3$ due to alkaline cytosolic pH, and it may enter the vacuoles since the membranes are permeable to ammonia. Ammonia may also enter the vacuoles through aquaporins called as **tonoplast intrinsic proteins (TIPs)**, located in the tonoplast membrane. Since $NH_3$ can bind with $H^+$ and is converted to $NH_4^+$ ions, it is trapped inside the vacuoles in ammonium form. $NH_3$ in the gaseous form can diffuse through the cell membranes and dissipate the proton gradient across the membranes, which is required for the transport of substances across the membrane. It may also be responsible for the dissipation of proton gradient created across the thylakoids by the photosynthetic electron transport or the proton gradient across inner mitochondrial membrane created during oxidative electron transport. The proton gradient across membranes is essential for ATP synthesis and transport of ions and other biomolecules. Since excess of ammonia/ammonium ions is toxic for the cells, animals have developed an aversion for ammonia (Box 11.2).

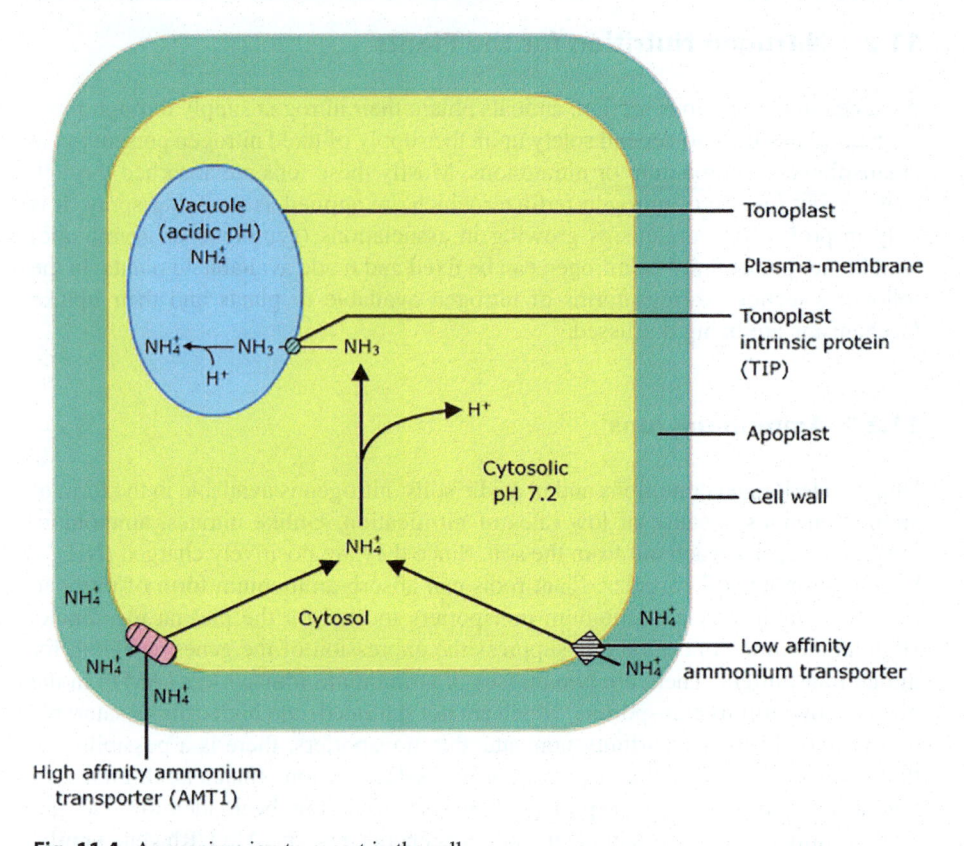

**Fig. 11.4** Ammonium ion transport in the cell

---

**Box 11.2: Toxicity of Ammonia**
Ammonia is a polar molecule. It gets readily dissolved in water to form ammonium hydroxide, which is dissociated to form ammonium ($NH_4^+$) ions. It is the ammonia which is toxic and not the ammonium ions because the cell membrane is permeable to ammonia and it can combine with $H^+$ ion to form $NH_4^+$ and disturb the proton gradient across the membranes. Among animals, ammonia gets converted to lesser toxic forms such as urea and uric acid before it is excreted out.

---

### 11.2.2 Nitrate Uptake

Ammonium ions are short-lived in the soil as nitrifying bacteria convert them into nitrate ions. Nitrate is the most common form in which nitrogen is absorbed by plants from the soil. Generally, epidermal and cortical cells of the root absorb nitrate

> **Box 11.3: Toxicity Due to Excess Nitrates in Plants**
> Excess nitrate ions accumulated in plants may be dangerous to the humans or other animals feeding upon them. Nitrate may be reduced to nitrite, nitric oxide, or nitrosamines by the liver. Nitrosamines are potent carcinogens. Nitric oxide is a biosignaling molecule and is involved in many physiological processes, such as widening of blood vessels. Nitrite ions ($NO_2^-$), upon combining with hemoglobin, render it incapable of binding with oxygen, resulting in a disease known as methemoglobinemia. High nitrate concentration in plants is deterrent to herbivorous animals.

from the soil solution, although it can also be taken up from leaves during foliar application of fertilizers. Nitrate concentration in the soil may vary from micromolar to millimolar range. Nitrate may be derived in the soil from the nitrification of the organic forms or from fertilizer application. In spite of energy consumed for reduction of nitrate to ammonium in the plant, nitrate is still the preferred form of nitrogen because, unlike ammonium, excess nitrate can either be safely stored in the vacuoles up to a concentration $\geq$ 20–70 mM or it can be transported to the leaves through xylem, where again it may be stored or metabolized (Box 11.3). If plants are supplied with ammonium nitrate as fertilizer, after plants have been subjected to nitrogen deficiency, accumulation of glutamine in cells will downregulate the transporter for ammonium ions, AMT1. Once inside the epidermal, cortical, or endodermal cells, further transport of nitrate into the root cells is mediated by two classes of symporters, located in plasma membrane and tonoplast, and an antiporter localized only in the tonoplast. In order to deal with varying concentrations of nitrates in soil and in tissues, plant cells possess nitrate transporters having different affinities for nitrates. On the basis of their affinity with nitrate, nitrate transporters are classified as high-affinity transporter systems (HATS) and low-affinity transporter systems (LATS). These two classes of transporters function at micromolar to millimolar nitrate concentrations. HATS function at 0.2–0.5 mM nitrate concentration, while LATS function above 0.5 mM and do not display any saturating effect for nitrate. HATS can either be inducible (iHATS) or constitutive (cHATS). Constitutive HATS are expressed even in the absence of nitrate, while inducible HATS are expressed only in the presence of nitrates. These transporters are expressed differently in different tissues and at various stages of development. Nitrate transporters consist of 12 transmembrane domains which are distributed in two sets of six helices in each, which are separated by a hydrophilic domain extending into the cytosol.

The gene families encoding the symporters that have been discovered so far are NRT1 and NRT2. NRT2 encodes for the inducible type of HATS nitrate transporters and is conserved in algae, fungi, and plants. NRT2 is downregulated by several reduced forms of nitrogen such as ammonium and glutamine. This helps the plant to regulate gene expression for transporters in response to changing needs of the plant. NRT2 is expressed mainly in roots. There are seven members in NRT2 gene family in *Arabidopsis* genome. NRT1 gene family encodes both constitutive and inducible nitrate transporters and is more complex. Fifty-three members of this gene family have been identified in *Arabidopsis* genome so far. These are mainly expressed in roots and encode either low-affinity or high-affinity nitrate transporters. Nitrate uptake, mediated by symporters present in plasma membrane, is driven by H$^+$ gradient. It is a 2H$^+$/NO$_3^-$ symport. H$^+$ gradient is created across the plasma membrane by the ATPase-mediated H$^+$ pumps. Flow of H$^+$ in response to their gradient is coupled with nitrate uptake. NO$_3^-$ concentration in the cytosol is maintained by translocating excess nitrate through xylem or by storing it in vacuoles. One class of nitrate transporters present in tonoplast belongs to **chloride channel family (CLC)** (Fig. 11.5). CLC can be either H$^+$ gated anion channels or 2NO$_3^-$/H$^+$ antiporter. It functions to transport nitrate across tonoplast. CLC can have dual-affinity system. On being phosphorylated at threonine (at position 101), it functions

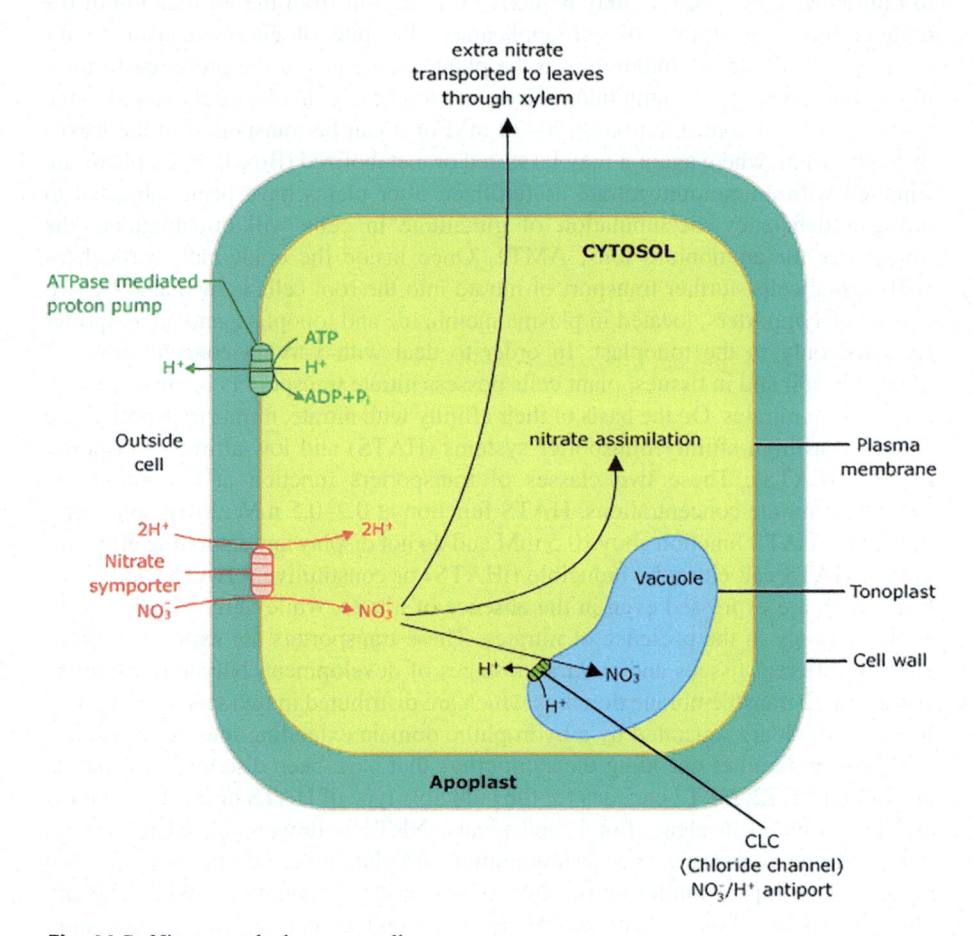

**Fig. 11.5** Nitrate uptake by a root cell

## 11.2 Nitrogen Nutrition for the Plants

as high-affinity system, while its dephosphorylation makes it a low-affinity transporter. Homeostasis of nitrate in cytosol is determined by the balance between nitrate uptake, nitrate efflux, reduction of nitrate, or storage of nitrate in the vacuoles.

### 11.2.3 Nitrate Assimilation

Nitrate reduction occurs throughout in all parts of the plant. However, some species, such as white clover (*Trifolium repens*) or cranberry (*Vaccinium macrocarpon*), exhibit nitrate reduction in roots only. In cocklebur (*Xanthium* sp.), it occurs in leaves only. Distribution of NR activity within plant is also governed by nitrate concentrations of available nitrate. At low concentrations (up to 1 mM), nitrate is directly assimilated in the roots of temperate legumes. If $NO_3^-$ concentration is higher than 1 mM, its transportation to shoot becomes significant. $NO_3^-$ transport occurs via xylem sap. In tropical and subtropical cereals and grain legumes, nitrate assimilation occurs in shoot irrespective of the external concentration of nitrate. In maize, nitrate reduction occurs in mesophyll cells (and not in bundle sheath cell) since mesophyll cells have greater capacity for generating reductant [NADPH], which is required for nitrate reduction.

Nitrate assimilation occurs in two steps (Fig. 11.6). First step requires reduction of nitrate to nitrite. It occurs in the cytosol of the cells and is catalyzed by the enzyme nitrate reductase (NR). However, nitrite reduction to ammonium occurs in plastids since the enzyme nitrite reductase (NiR) is located in the plastids.

$$NO_3^- \xrightarrow{NR} NO_2^- \xrightarrow{NiR} NH_4^+$$

**Nitrate Reductase (NR)** Reduction of nitrate to nitrite requires transfer of two electrons from the electron donor to $NO_3^-$ and consumption of one proton. Reaction is coupled with oxidation of NADH or NADPH, which serves as electron donor for the reaction in most of the plants. NADH specific reduction of $NO_3^-$ occurs in land plants and algae, while NADPH is used as the electron donor in fungi. However, some NR forms in land plants and algae are able to use either NADH or NADPH as the reductant in the reaction. The reaction occurs in cytosol of the cell. Active form of NR is a **homodimer**. It has the molecular weight of about 100–114 kDa. Each monomeric unit consists of about 1000 amino acids. It has three domains associated with cofactors. The three cofactors are molybdenum cofactor (MoCo), heme Fe, and FAD. N-terminal region of each monomeric unit is associated with MoCo, while the C-terminal region of each monomeric unit is associated with the cofactor FAD. Central region of each polypeptide unit is complexed with heme Fe. MoCo center is the molybdenum complexed with organic molecule **pterin** (Box 11.4). Heme Fe contains cytochrome $b_5$. FAD-binding domain consists of 260–265 amino acids;

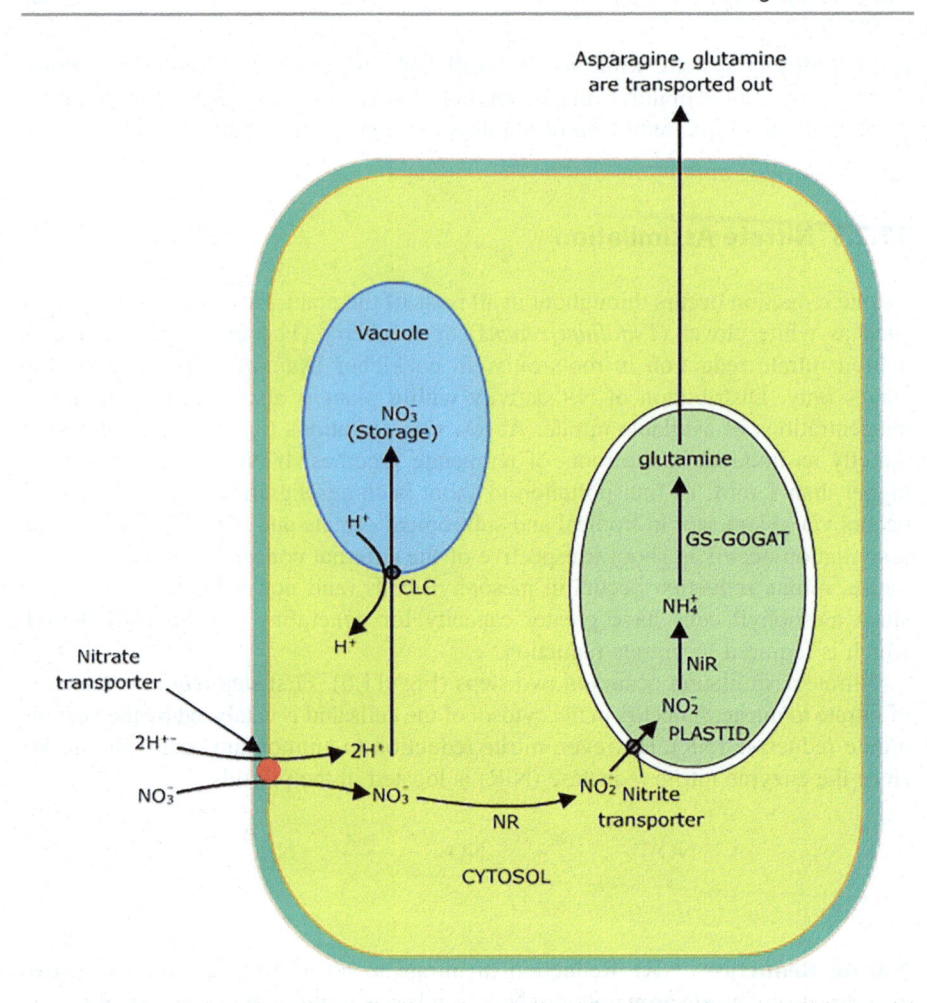

**Fig. 11.6** Fate of nitrate in a plant cell. Nitrate is assimilated by nitrate reductase (NR) in cytosol and nitrite reductase (NiR) in plastids. Excess nitrate is stored in vacuoles. Ammonium ($NH_4^+$) ions are assimilated in plastids

central **heme** domain consists of 75–80 residues, while MoCo-binding domain of enzyme protein is large and consists of 360–370 amino acids. Each domain of the enzyme is an independent unit and is believed to belong to a distinct protein family. These three domains of each monomeric unit of the homodimeric enzyme are associated with hinge (hI and hII) regions, hI being present between MoCo center and heme domain of the polyptide, while hII is present between heme domain and the domain complexed with FAD. hI region exhibits a regulatory role in maintaining enzyme activity (Fig. 11.7).

## 11.2 Nitrogen Nutrition for the Plants

**Box 11.4: Molybdopterin**

Molybdenum is a transition element. It is catalytically active when complexed with the cofactor pterin (molybdopterin). Molybdenum cofactor is known as MoCo. Pterin-bound Mo is the cofactor for many plant enzymes involved in redox reactions, e.g., xanthine dehydrogenase, sulfite oxidase, nitrate reductase, etc. Sulfite oxidase has sequence quite similar to NR. It has a cytochrome b$_5$, heme domain attached to it, but contrary to NR (where heme terminal is C-terminal to MoCo domain), heme domain of sulfite oxidase is N-terminal to MoCo domain.

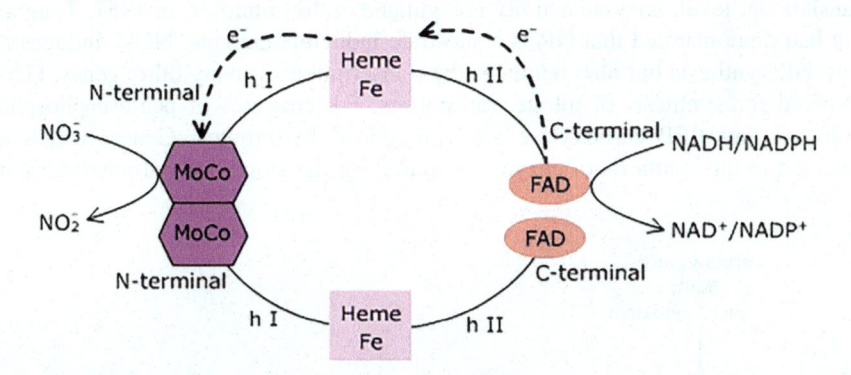

**Fig. 11.7** Nitrate reductase (NR) is a homodimer. Each monomeric unit has three domains associated with FAD at the C-terminal, MoCo (molybdenum cofactor) with N-terminal, and central region complexed with Heme Fe. The three domains of each monomeric unit is associated with hinge regions hI and hII. Electron transfer from NADH/NADPH at the C-terminus is facilitated by the thiol group of cysteine of the polypeptide. MoCo facilitates transfer of electron to NO$_3^-$ at the active site present at the N-terminal of the protein

Reduction of $NO_3^-$ by NR requires transfer of two electrons and one proton from NADH/NADPH; another proton is obtained from the medium.

$$NO_3^- + NADH + H^+ \rightarrow NO_2^- + H_2O + NAD^+$$

Midpoint potentials of the three centers are $-272$ mV for FAD, $-160$ mV for heme, and $-10$ mV for MoCo center. Thiol group of cysteine present at the C-terminus of the enzyme has been found to facilitate transfer of electrons between NADH/NADPH and FAD (the cofactor complexed with the enzyme protein at the C-terminus). However, the presence of cysteine at the C-terminus of the enzyme has not been found to be mandatory. In case cysteine is substituted with serine, even then the enzyme is able to catalyze electron transfer but with lesser affinity. FAD-binding domain is believed to belong to the **ferredoxin-NADP reductase (FNR)** family of oxidoreductases. Reduction of FAD cofactor is followed by reduction of heme, the central domain of enzyme belonging to the **cytochrome $b_5$ family** of proteins, which further passes electrons to MoCo center. The domain containing MoCo factor belongs to the family of enzymes which includes xanthine oxidase and sulfite oxidase. Reduced molybdenum passes the electrons to $NO_3^-$ at the active site of the enzyme. NR functions as mini electron transport chain oxidizing NADH/NADPH and reducing the terminal electron acceptor $NO_3^-$ (Fig. 11.8).

**Regulation of NR Activity** NR activity is highly regulated. Regulation may be at the transcription level of gene NIA, a nuclear gene which encodes NR. Regulation at the transcription level is time requiring (may be hours or days), while at the post-translational level, enzyme activity is regulated within minutes. In 1957, Tang and Wu had demonstrated that NR is a substrate-inducible enzyme. $NO_3^-$ induces not only NR synthesis but also regulates the transcription of many other genes. Genes involved for synthesis of nitrate transporters, Fd, enzymes of pentose phosphate pathway, and PEP carboxylase are upregulated by nitrates. Genes, which are involved in the synthesis of enzymes required for the storage of carbohydrates, are

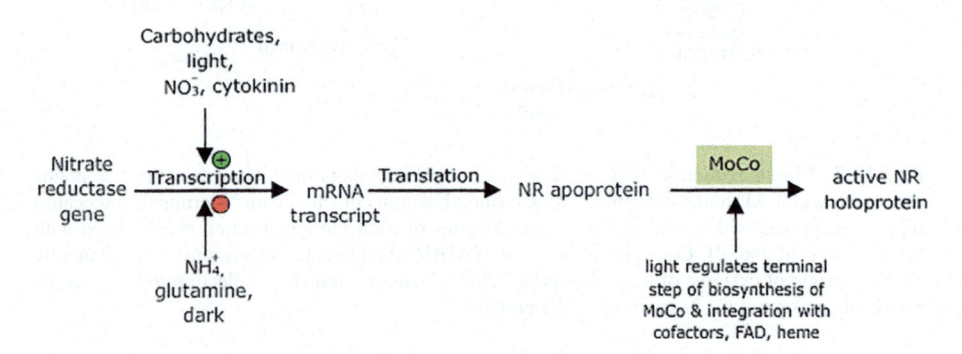

**Fig. 11.8** Regulation of biosynthesis of nitrate reductase (NR). Red circle represents inhibition and green circle represents promoting effect on gene transcription

## 11.2 Nitrogen Nutrition for the Plants

downregulated, while genes involved in the synthesis of amino acids and organic acids are upregulated. Thus, NR is responsible for redirecting the carbon from storage as carbohydrates to synthesis of amino acids. In *Arabidopsis*, $NO_3^-$ induces the synthesis of MoCo factor as well. $NO_3^-$ also regulates the synthesis for NiR, so that $NO_2^-$ generated in nitrate reduction does not accumulate. Since nitrite is toxic for the cell it must be assimilated fast. NR activity is highly regulated. Regulation may be at the transcription level of gene NIA, a nuclear gene which encodes NR. Regulation at the transcription level is time requiring (may be hours or days), while at the post-translational level, enzyme activity is regulated within minutes. In 1957, Tang and Wu had demonstrated that NR is a substrate-inducible enzyme. $NO_3^-$ induces not only NR synthesis, but also regulates the transcription of many other genes. Genes involved for synthesis of nitrate transporters, Fd, enzymes of pentose phosphate pathway and PEP carboxylase are upregulated by nitrates. Genes, which are involved in synthesis of enzymes required for the storage of carbohydrates are downregulated while genes involved in synthesis of amino acids and organic acids are upregulated. Thus, NR is responsible for redirecting the carbon from storage as carbohydrates to synthesis of amino acids. In *Arabidopsis*, $NO_3^-$ induces the synthesis of MoCo factor as well. $NO_3^-$ also regulates the synthesis for NiR, so that $NO_2^-$ generated in nitrate reduction does not accumulate. Since nitrite is toxic for the cell, it must be assimilated fast. Light stimulates transcription of NIA. In some of the cases, light requirement can be replaced by providing sucrose in the dark. Besides generation of reductants and photosynthates, light may also be acting through signaling pathways of photomorphogenesis. Once transcription has been initiated, mRNA transcript exhibits diurnal fluctuations which shows possible involvement of phytochrome. At the end of the dark period, there is decrease in mRNA transcript of NR, which may be because of inhibition of transcription of NR gene due to accumulation of glutamine. Post-translational regulation of NR is very quick and occurs within minutes. It occurs through phosphorylation and dephosphrylation of the enzyme protein. The phosphorylated enzyme protein is inactive, while dephosphorylated form of the enzyme protein is active. ATP-dependent phosphorylation occurs at one specific conserved serine residue present in hI region (in spinach it is serine $_{543}$) and is catalyzed by $Ca^{2+}$-dependent NR kinase. Activity of NR kinase is inhibited by light, triose phosphates, and hexose phosphates. Post-translational regulation of the enzyme activity is mediated by a highly conserved 14-3-3 class of regulatory proteins. These proteins (14-3-3) are inactivator proteins, and the name 14-3-3 comes from the separation pattern of the proteins during purification (Box 11.5). Binding of 14-3-3 proteins with phosphorylated serine present in hI region of the enzyme interferes with the electron transfer in between the heme and MoCo cofactor of the enzyme. NR/14-3-3 complex formation is promoted in dark due to accumulation of $Mg^{2+}$, while it is broken down in light due to low $Mg^{2+}$ concentrations. This NR/14-3-3 complex is sensitive to attack by cytosolic proteases. Dephosphorylated NR protein prevents 14-3-3 regulatory proteins from binding with the enzyme protein. Dephosphorylation of NR is catalyzed by a protein phosphatase (PP2A). Dephosphorylation of the

> **Box 11.5: 14-3-3 Proteins**
> 14-3-3 proteins belong to a class of conserved multifunctional regulatory proteins. These proteins are present in animals and plants. These specifically bind to proteins, which have got a specific binding site of six amino acids with phosphorylated serine being present at position 6. The name 14-3-3 refers to particular elution fraction number (14) and band position (3.3) while purifying the protein from bovine brain homogenate on DEAE cellulose chromatography. A large number of gene families of 14-3-3 proteins have been identified. These are involved in diverse functions in plants, including regulation of plasma $H^+$-ATP pumps, transmembrane receptors, sucrose synthetase, and nitrate reductase. They have a potential role in regulating plant development in response to abiotic and biotic stress conditions as these regulatory proteins bind to various protein kinases. One of the examples for the role played by 14-3-3 class of proteins has been studied with reference to attack on plant by the fungus *Fusicoccum* spp. the substance produced by the fungus is fusicoccin which specifically binds to 14-3-3 binding sites of various proteins, thereby interfering with the regulatory functions of 14-3-3 proteins. This results in the disruption of the metabolism of the plant to such an extent that the plant does not survive.

enzyme protein is also promoted by AMP and iP. This reversible post-translational regulation of enzyme activity occurs within minutes of exposing plants to light or darkness or stress conditions. (Fig. 11.9).

**Nitrite Reductase** Nitrite produced as a result of nitrate reduction is immediately transported from the cytosol into the plastids, chloroplasts in green tissues, and plastids in nongreen tissues where it is further reduced. Efforts are underway for the identification of nitrite transporters in higher plants. Initially diffusion of nitrite was proposed. However, very low cytosolic nitrite concentration does not justify its diffusion across inner chloroplast membrane. Furthermore, saturation kinetics of nitrite transport suggests the presence of a nitrite transporter in the thylakoid membranes of the chloroplasts. In cucumber, the existence of transporter—Nitr1—belonging to **proton-dependent oligopeptide transporters (POT)**, has been suggested. However, it has been suggested that in *Arabidopsis* a member of **CLC family** (chloride channel) is involved.

Reduction of nitrite occurs in a single step, catalyzed by the enzyme nitrite reductase (NiR). Nitrite reduction requires transfer of six electrons. Reduced ferredoxin is used as the donor for electrons, and eight protons are obtained from the medium.

## 11.2 Nitrogen Nutrition for the Plants

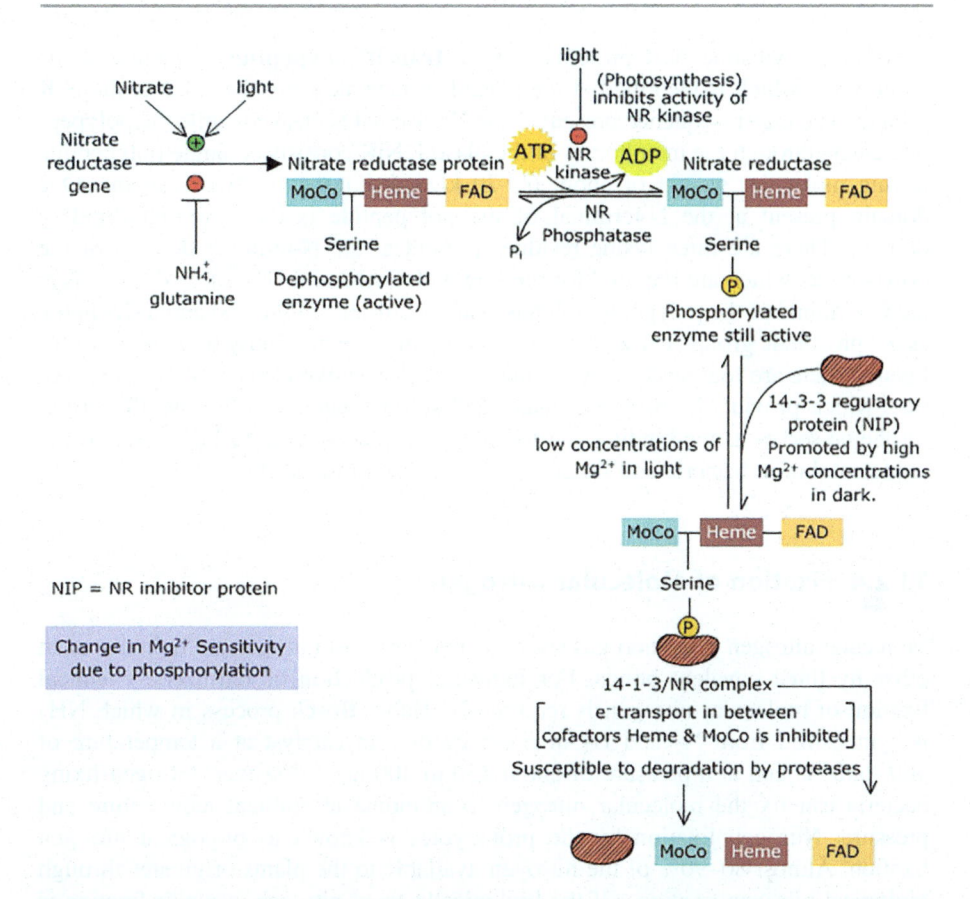

**Fig. 11.9** Post-translational regulation of NR. Phosphorylation of NR enables binding of 14-3-3 regulatory protein which makes the NR protein susceptible to degradation by proteases

$$6\ Fed_{red} \rightarrow 6\ Fed_{ox} + 6\ e^-$$

$$NO_2^- + 6\ e^- + 8\ H^+ \rightarrow NH_4^+ + 2H_2O$$

The reduced ferredoxin used in $NO_2^-$ reduction is generated in chloroplasts during light reaction of photosynthesis. In nongreen tissues, such as roots, reduced ferredoxin is produced from NADPH, which is generated during oxidative pentose phosphate pathway. Ferredoxin in non-photosynthetic tissues has been found to be more electropositive than ferredoxin of the chloroplasts. Reduction of ferredoxin by NADPH in nongreen tissues is facilitated by the enzyme **ferredoxin-NADP reductase**. Since nitrite is toxic, it should not accumulate in the cells. That is why activity of NiR (expressed as per fresh gram weight basis) is several times higher than the activity of NR. This ensures no accumulation of nitrite in the cells. NiR is a constitutive enzyme. In some plants NiR synthesis is also induced by nitrate. NiR is encoded in nuclear genome, and the transcribed mRNA is translated by cytosolic ribosomes. Cytosolic NiR proteins carry a **transit polypeptide** sequence at the N-terminal, which is identified at the plastid surface allowing the entry to the NiR protein. The transit sequence present at the N-terminal is cleaved while the polypeptide enters through the membranes of the plastid. NiR consists of single polypeptide having molecular weight of about 60–70 kDa. It consists of two domains. One domain present at the N-terminal of the polypeptide is the ferredoxin-binding domain. There are three lysine residues present at the N-terminal domain of the polypeptide, which are required for binding with ferredoxin. The other domain is at the C-terminal of the polypeptide. It has a single cluster of [4Fe-4S] and a **siroheme** as the prosthetic group. These two cofactors are in close proximity bridged by sulfur ligand. There are four cysteine molecules associated in two clusters on the polypeptide, which provide the bridging ligand and sulfur ligands for the 4Fe-4S cluster. Siroheme serves as the binding site for $NO_2^-$. In plants, reduced Fd is used as the reductant, but in bacteria and fungi, NADPH is used instead (Fig. 11.10).

### 11.2.4 Fixation of Molecular Nitrogen

Molecular nitrogen is an inert gas since the two atoms of nitrogen are linked to each other by three covalent bonds. For industrial production of fertilizers, chemical fixation of molecular nitrogen is required by Haber Bosch process in which $NH_3$ is synthesized from $N_2$ and $H_2$ in presence of iron catalyst at a temperature of 400 -650 °C and at a pressure of about 150 to 400 atm. However, nitrogen-fixing bacteria can fix the molecular nitrogen to ammonia at ambient temperature and pressure. Nitrogen fixation by the prokaryotes is known as biological nitrogen fixation. Almost 80–90% of the nitrogen available to the plants originates through biological nitrogen fixation. Of the biologically fixed nitrogen, a major fraction is fixed through symbiotic associations. Process of nitrogen fixation is limited to prokaryotes only. The organisms which fix molecular nitrogen are known as

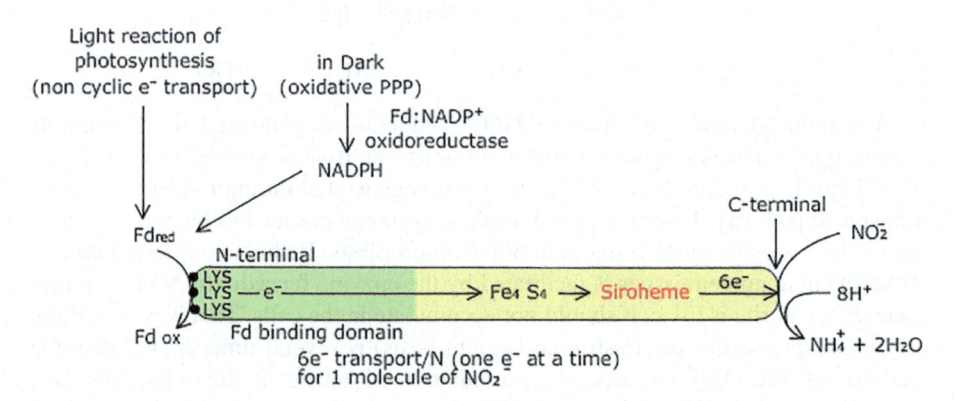

**Fig. 11.10** Nitrite reductase (NiR) structure. NiR (Fd nitrite oxidoreductase) has two functional domains, Fd-binding domain at N-terminal and nitrite-binding domain at C-terminal

## 11.2 Nitrogen Nutrition for the Plants

*diazotrophs* (diazo, molecular nitrogen; trophs, eater). Free-living diazotrophs include prokaryotes in the soil without any type of associations with plants. Aerobic nitrogen-fixing bacteria, such as *Azotobacter*, maintain low partial pressure of oxygen at the site of nitrogen fixation by maintaining high respiration rate and in the process also provide ATP as the energy source. *Gloeothece* evolve photosynthetic oxygen during day while fix nitrogen during night. Cyanobacteria such as *Anabaena* and *Nostoc* grow in flooded rice fields and can fix nitrogen in anaerobic conditions. Anaerobic conditions can also be created in the heterocystous forms of cyanobacteria. Heterocysts are the specialized cells of these organisms since photosystem II is not present in them and there is no photosynthetic oxygen evolution in these cells. Thus, anaerobic conditions prevail which are required for nitrogen fixation. Diazotrophs also include facultative anaerobes (e.g., *Klebsiella*) which can fix nitrogen only under anaerobic conditions. However, there are obligate anaerobes, which grow and fix nitrogen only under anaerobic conditions. These anaerobes can be photosynthetic (e.g., *Rhodospirillum* sp.) or non-photosynthetic (e.g., *Clostridium* sp.).

Diazotrophs also fix nitrogen in symbiotic associations with plants and in turn are benefitted from the host plant in getting the nutrients, carbon skeletons, and the energy source required for their growth and nitrogen fixation. These are classified under three categories. The first category includes symbiotic association between cyanobacteria *(e.g., Anabaena)* and diverse group of plants. This includes liverworts (such as *Anthoceros*), coralloid roots of cycads, *Azolla* (a water fern in which *Anabaena* grow in the leaf cavities), and one angiosperm genus, i.e., *Gunnera* in which *Anabaena* is present in the glands found at the base of leaf petioles. *Azolla* meets the nitrogen requirement of the rice plants when grown along with them. Second category includes the actinorhizal symbiosis. It is the symbiotic association of *Frankia* species with various dicots. *Frankia* are the gram-positive actinomycetes, which grow in symbiotic associations with as many as 20 genera of dicots belonging to 8 or more families. These include *Alnus, Casuarina, Ceanothus*, and many other species. These plants are known as actinorhizal plants. This symbiotic association plays an important role in maintaining the nitrogen content of soil in forests. The third category includes the symbiotic association of rhizobia with plants. Unlike *Frankia* spp. rhizobia are gram-negative bacteria, which in most cases form symbiotic associations with legumes (Fabaceae). One non-leguminous plant (*Parasponia spp.* family Ulmaceae) also forms symbiotic associations with rhizobia. In both actinorhizal and rhizobial symbiotic associations, root nodules are formed due to division of cortical and hypodermal cells. In grasses as well, symbiotic associations develop but without any nodule development. In this case nitrogen-fixing organism either anchors on the root surface near the elongation zone or it sticks on the root hairs. These may also colonize the plant tissues as **endophytes**. This category of nitrogen fixers includes *Acetobacter diazotrophicus* and *Herbaspirillum* spp. Nitrogen-fixing bacteria colonize in the apoplastic regions in the stem tissues of sugarcane. These diazotrophs in symbiotic association meet 30% of the nitrogen requirement of the sugarcane plant (Table 11.1).

**Table 11.1** Diversity of diazotrophs

| Diazotrophs | Types of diazotrophs | Examples |
|---|---|---|
| *Free-living diazotrophs* | | |
| | Cyanobacteria | |
| | Heterocystous forms | *Anabaena* sp., *Nostoc* sp. |
| | Non-heterocystous forms | *Gloeothece* sp. |
| | Other bacteria | |
| | Obligate aerobic | *Azotobacter vinelandii* |
| | Facultative aerobic | *Klebsiella pneumoniae* |
| | Anaerobic | *Clostridium pasteurianum* |
| | Photosynthetic bacteria | *Rhodospirillum* sp. |
| *Diazotrophs in symbiotic associations* | | |
| | Endosymbiotic associations (nodule formation) | |
| | Legumes | *Rhizobium, Azorhizobium, Bradyrhizobium, Mesorhizobium* |
| | Non-legumes symbiosis | |
| | *Parasponia* sp. | *Bradyrhizobium* sp. |
| | Actinorhizal (*Alnus* sp.) | *Frankia* sp. |
| | Endosymbiotic (no nodules formed) | |
| | *Gunnera* sp. (Angiosperm) | *Nostoc* sp. |
| | *Azolla* (Pteridophyte) | *Anabaena* sp. |
| | Associative symbiosis | |
| | Sugarcane | *Acetobacter* sp. |

**Legume Rhizobia Symbiosis** This is the most common type of symbiosis, which occurs between the rhizobia (includes species of *Azorhizobium, Bradyrhizobium, Mesorhizobium, Rhizobium,* and *Sinorhizobium*) and leguminous plants. Symbiosis is not obligatory as either member of the symbiotic association can survive independently. However, both partners are benefitted from each other. Rhizobia are free-living bacteria in the soil and can multiply without the host plant, but they fix nitrogen only after forming symbiotic association with the host plant. Legumes develop nodules to encase the rhizobia, which develop well only when they are grown in nitrogen-poor soils. This symbiotic association is of commercial significance since it contributes nitrogen in the form of proteins to the seeds, such as soybean, lentil, peanut, and also to the forage crops such as alfalfa (*Medicago sativa*) and clover (*Trifolium* sp.) Nodulation process begins with identification of the specific host plant by a particular rhizobial species. The symbiotic association is very specific since one particular species of rhizobia (also called biovar) infects either only specific host plant or its range of the host plants is very small. This specificity is because of interaction of certain chemical signals between the bacteria and the host plant, which causes changes in gene expression in both the symbiotic partners resulting in effective nodulation.

## 11.2 Nitrogen Nutrition for the Plants

**Table 11.2** Some of the *Rhizobium* strains and their specific hosts releasing chemotactic compound

| Rhizobial strain | Host plants | Nod factor inducer released by the plant |
|---|---|---|
| *Sinorhizobium meliloti* | *Medicago truncatula* | Luteolin |
| *Bradyrhizobium japonicum* | *Glycine* (soybean), *Vigna* (cowpea) | Daidzein |
| *Mesorhizobium loti* | *Lotus japonicus*, *Lupinus* | Aldonic acid |
| *Rhizobium leguminosarum* bv. *viciae* | *Vicia*, *Pisum* | Flavanone |

### Steps of Nodulation

1. The first step of identification begins with the release of specific signals by the host plant which generally are flavonoids and secondary metabolites derived from **phenylpropanoid pathway**. Specific flavonoids, released by the prospective host plant, are recognized by compatible rhizobia (Table 11.2).
2. These flavonoids are recognized by the specific rhizobial spp. followed by synthesis of nodulation factors (Nod) by the bacteria. Synthesis of Nod factors is under the regulation of nodulation (*nod*) genes of the bacteria (Box 11.6).
3. Nod factors produced by rhizobia are recognized by the receptors present in the membrane of root hairs of the host plant. After recognition of Nod factors, various

changes in plant-specific proteins (nodulins) are produced within the host plant which trigger physiological responses associated with nodulation process. First step is the curling of the root hair causing trapping of the bacteria inside the curl.
4. Bacteria enter the root hair through infection thread which is formed as a result of invagination of the plasma membrane of the root hair cell. Formation of the infection thread is followed by triggering of cell division in the cortical or cambial cells of the root, causing nodule development.
5. Bacteria reach these dividing cells, which later on develop into nodules, through the growth of infection thread and are released while still enclosed by the cell membrane of the host plant. These bacteria are differentiated into bacteroids inside **symbiosome**. Bacteroids do not divide further and are capable to fix nitrogen.
6. Nodule-specific proteins (nodulins) are synthesized as a result of expression of the nodulin genes of the host plant (Fig. 11.11).

---

**Box 11.6: Nod Factors**
The core structure of Nod factor consists of a backbone of three to four N-acetyl-D-glucosamine which are linked by β-1,4 linkages. It is linked to a fatty acid on the terminal sugar. Synthesis of core Nod factor is under the control of the products of three common *nod* genes, i.e., *nodA, nodB, and nodC* (NodA, NodB, and NodC, respectively). NodC catalyzes polymerization of N-acetyl-D-glucosamine. NodB removes the acetyl group from the end glucosamine residue followed by addition of the fatty acyl group to the free amino group of the oligomer by NodA. Further alterations are done in the core structure of this polysaccharide by the products of the host-specific *nod* genes of the bacteria. The alterations include changes in the length and saturation of the fatty acyl group or addition of the groups which determine the host specificity. The host-specific genes *nodE* and *nodF* encode the enzymes required for the synthesis of factors NodE and NodF. These enzymes catalyze the synthesis of an 18:4 and 16:2 fatty acyl groups, respectively. The enzyme product of *nodL* catalyzes the synthesis of factor NodL through addition of specific groups at the reducing and nonreducing N-acetylglucosamine residues of the Nod factors. This includes C6 modification on the reducing and nonreducing ends. This may include N-methylation and O-substitution of the inner N-acetyl-D-glucosamine residues or other modifications. These modifications help the bacteria to determine its host specificity. The rhizobia infecting broad range of the hosts produce multiple Nod factors.

(continued)

## 11.2 Nitrogen Nutrition for the Plants

**Box 11.6** (continued)

1. Polymerization of N-acetylglucosome by enzyme NodC

2. acetyl group removed by enzyme Nod B

3. a fattyacid is added in the position of acetyl group by enzyme NoeI A

There is variation of fatty acid and R group in nodulation factor produced by various species of rhizobia.

Common basic structure of Nod factor consists of 4-5 units of N-acetylglucosmine

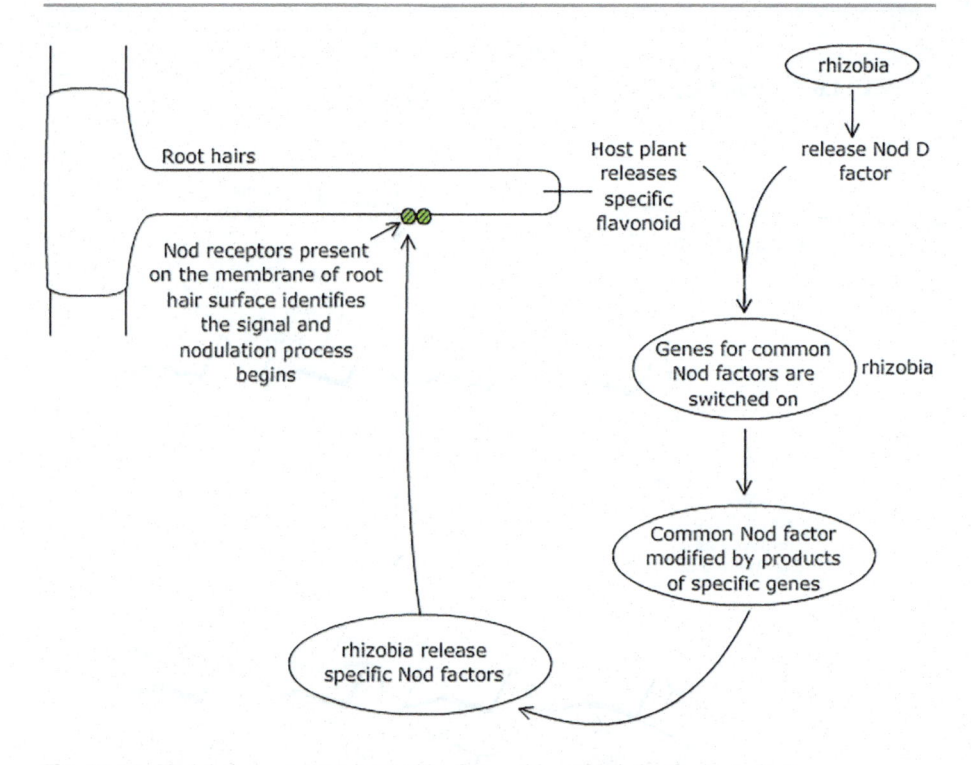

**Fig. 11.11** Nodulation process triggered by flavonoid synthesis by the host plant

Only about 1–5% of root hairs become infected, and about 20% of these infections result in nodules. Plants produce chitinase that breaks down Nod factors of incompatible rhizobia sp. Chitinase produced later during the infection process also acts against rhizobia to control further nodulation. The biochemical signal molecules released by the prospective host plant induce gene expression in rhizobia, which results in the synthesis of Nod factors. Nod factors are **lipochitooligosaccharides (LCO)**. Rhizobial genes responsible for the synthesis of Nod factors are called *nod* genes. Expression of *nod* genes in bacteria depends on the interaction of a transcription inducer NodD in bacteria and chemical signals released by the prospective host plant. The gene responsible for the expression of the regulatory factor NodD is *nodD*, which is expressed constitutively. It regulates the transcription of other *nod* genes, which are of two types, the common *nod* genes and host-specific *nod* genes. Common *nod* genes include *nodA, nodB, and nodC*. These common nod genes are present in all rhizobial strains and encode enzymes required for the synthesis of basic structure of the Nod factor. The host-specific nod genes include *nodP, nodQ, and nodH* or *nodF, nodE,* and *nodL*. These host-specific genes differ in rhizobial sp. These are involved in the modification of the basic structure of

Nod factors. These modified Nod factors are recognized by the legume receptors present in plasma membrane of root hairs of specific host. Besides these, the invading bacteria may also require some extracellular polysaccharides (EPS) for their recognition by the host plant for nodule formation. For example, **succinoglycan** (EPS-I) produced by *Sinorhizobium meliloti* is required for the invasion of its host plant *Medicago* sp.

**Nodulins** Nodulins are the proteins synthesized by the host plant in response to bacterial infection. These are required for effective nodulation of host plants and for establishing symbiotic associations with the invading bacteria. The host specificity is determined by the interaction of specific Nod factors with plant receptors present on the plasma membrane of root hairs. The very first response of the host to specific rhizobial sp. is expressed as root hair curling. The proteins, which are essential for the initial identification of the Nod factor, have been identified in *Lotus* mutants which fail to nodulate in the presence of *Mesorhizobium loti*. These are a type of **pattern recognition receptors (PRRs)**. These receptors are transmembrane kinases, which have extracellular region-bearing LysM domains, a transmembrane domain, and a cytosolic domain. These include "entry" receptors and "signaling" receptors, which are encoded by specific genes. These receptors resemble other LysM receptors which are involved in identifying the pathogenic bacteria (Box 11.7). Mutations in any of the genes encoding these plant membrane receptors can cause severe nodulation defects, and plant is unable to nodulate. Extracellular domain of the receptors may be interacting with the Nod factors. If extracellular domains of the receptor proteins from the two different host plants are swapped, there will be change in their specificity to infection by the compatible species of rhizobia. The cytosolic

---

**Box 11.7: LysM Motif and LRR Domains**

It is a peptidoglycan-binding module. LysM motif of the proteins consists of 44–65 amino acid residues. It binds to peptidoglycan and chitin non-covalently. It serves as a module which mediates recognition of different N-acetylglucosamine-containing ligands. The glucans (e.g., Nod factor oligosaccharides consisting of N-acetyl-D-glucosamine residues) probably might have served as the immunogenic patterns, and this class of proteins might have evolved as the sensor for these kinds of ligands. So, LysM protein receptor-mediated immune system might have been developed to stop microbial infection. It has been identified in plant proteins that are also involved in detection of fungus. LRR domain of the protein refers to leucin-rich repeat domain. There are 2–45 motifs of 20–30 amino acids in each, which are rich in hydrophobic amino acid leucine. These motifs are folded in the form of a horseshoe-like structure. The LRR domain has been found in the proteins of diverse functions. These comprise the largest subfamily of transmembrane receptor-like kinases (LRR-RKs) in plants. These LRR domains are particularly suitable for protein-protein interactions.

domain of "entry receptor" exhibits typical kinase function. It has a phosphorylation loop, a catalytic domain, and an activation loop. Upon phosphorylation, cytosolic domain changes its conformation and is responsible for initiating a signaling cascade by phosphorylating the target proteins. Target proteins might be the proteins involved in $Ca^{2+}$ channel regulation. On the contrary, P-loop and activation loop are missing in the cytosolic domain of the "signaling receptor." Signal receptor might be involved in transmitting the signal to the proteins via a kinase activity. The "entry receptor" is required for infection, while the "signaling receptor" is required for the host response to the nodulation factor. Besides these receptors, there is another class of receptors present in the membrane of the host plant, i.e., **symbiosis receptor kinase (SymRK)**, which is involved in protein recognition. This receptor is required both for symbiosis with bacteria as well as for establishing symbiosis with mycorrhizal fungi which help plants in acquisition of minerals. This receptor has three extracellular LRR (leucine-rich repeats) domains in a large N-terminal segment of the receptor (Fig. 11.12). SYMRK gene is responsible for the expression of extracellular LRR domain, a membrane-spanning domain, and a cytoplasmic intracellular kinase. The kinase domain is capable of **autophosphorylation** and can act as a kinase for the target proteins in vitro.

**Nodulation** Once Nod factors are identified by the receptors present on the root hair surface, and the signal is transduced, calcium ion oscillations are activated in the nuclear and perinuclear regions of the root hair cell. This causes periodic **calcium spiking**, i.e., a transient increase or decrease in $Ca^{2+}$ levels. The reservoir for the release of $Ca^{2+}$ is believed to be the lumen of nuclear membrane. $Ca^{2+}$ spiking requires a channel for the release of $Ca^{2+}$ from the reservoir and an energy-requiring pump for uptake of $Ca^{2+}$ into the reservoir. Oscillations in the calcium ion concentrations are recognized by the **calcium- and calmodulin-dependent protein kinase (CCaMK)**. CCaMK is required for subsequent nodule development. Kinase activity at the C-terminal of protein is inhibited when N-terminal of the protein is autophosphorylated in response to $Ca^{2+}$. When autoinhibitory N-terminal domain of CCaMK is deleted or mutated, nodule development occurs in the absence of bacteria. This indicates that CCaMK acts as a central coordinating switch in early nodule development. Another protein required by the CCaMK for its action is a nuclear protein, i.e., **CYCLOPS**. CYCLOPS is a DNA-binding transcriptional activator and is a direct phosphorylation substrate of CCaMK. It induces nodule organogenesis. CCaMK and CYCLOPS are the key regulators in interpreting $Ca^{2+}$ spiking in the nucleus and nodule development (Fig. 11.13). Proteins, which are expressed in plants early during the infection, are called **early nodulins** and are encoded by **ENOD** genes. By using Nod factors in absence of bacteria, synthesis of these proteins can be triggered. However, further development of the nodules will require infection with live bacteria. Proteins synthesized by the plants during the later phase of infection process are called late nodulins. After the recognition of Nod factors by the receptors present on root hair and epidermal cells of the host plant, there is signaling for the development of the nodules. It brings about curling of the

## 11.2 Nitrogen Nutrition for the Plants

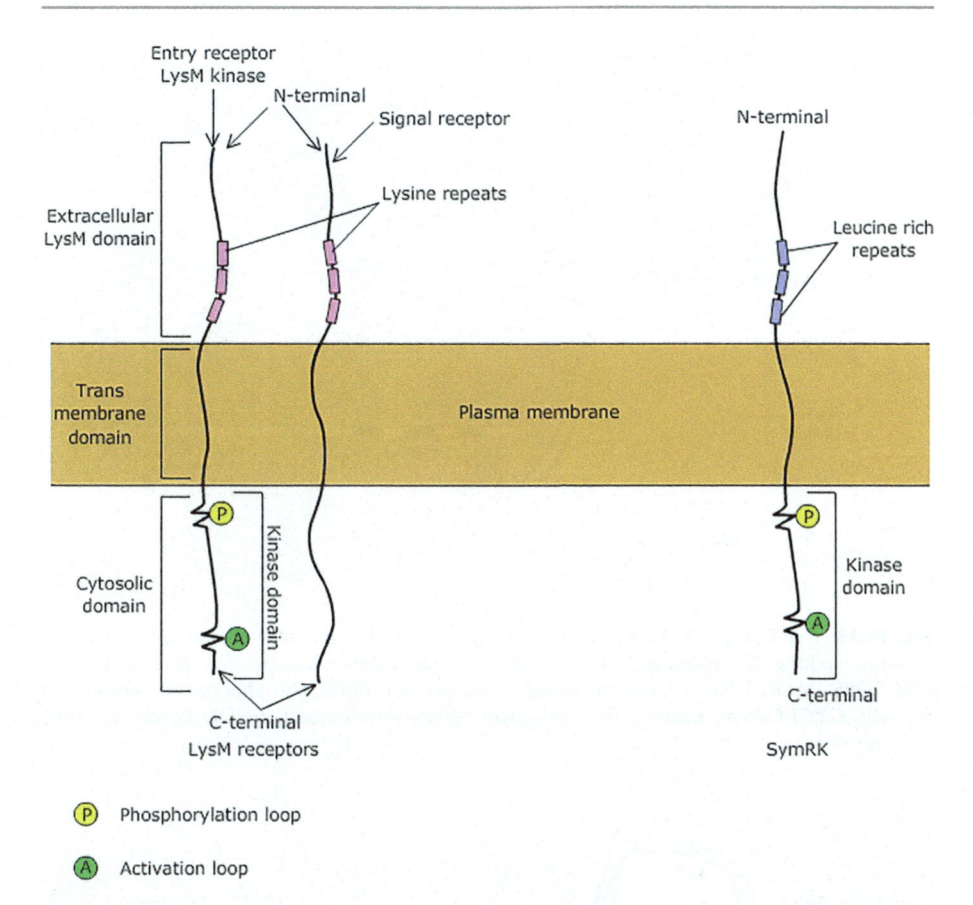

**Fig. 11.12** Nodulation factor receptors present on root hair surface of the host plant. Each of these proteins has membrane-spanning domain, an extracellular domain, and cytosolic domain. Proteins, bearing LysM domains in extracellular region, are Nod factor receptors, which are "Entry receptors" or "Signal receptors." "Entry receptor" is required for infection, and "signaling factor" is required for host response to nodulation factors. Cytosolic domain of "Entry receptor" has a phosphorylation loop (P) and an activation loop (A). It exhibits a typical kinase function. P-loop and A-loop are missing in cytosolic domain of "signaling receptor." SymRK (symbiosis receptor kinase) has three extracellular leucine-rich repeats (LRR) and is required both for symbiosis with rhizobia and establishing symbiosis with broad host range mycorrhizal fungi that enhance nutrient uptake

tip of the root hair in which bacteria are trapped. Cell wall of the root hair adjacent to bacteria partially degrades followed by invagination of plasma membrane, forming a tunnel-like structure. Bacteria enter into this, and it develops into an infection thread. Infection thread grows through the cell by coalescing with Golgi vesicles at the tip (Fig. 11.14). Matrix of the infection thread is rich in products of bacterial origin and the one synthesized by the host plant. These include various extracellular products, such as protein of plant origin, **arabinogalactan** protein (of the **extensin** family). On reaching the cortical cells, bacteria are released in the apoplastic region followed by

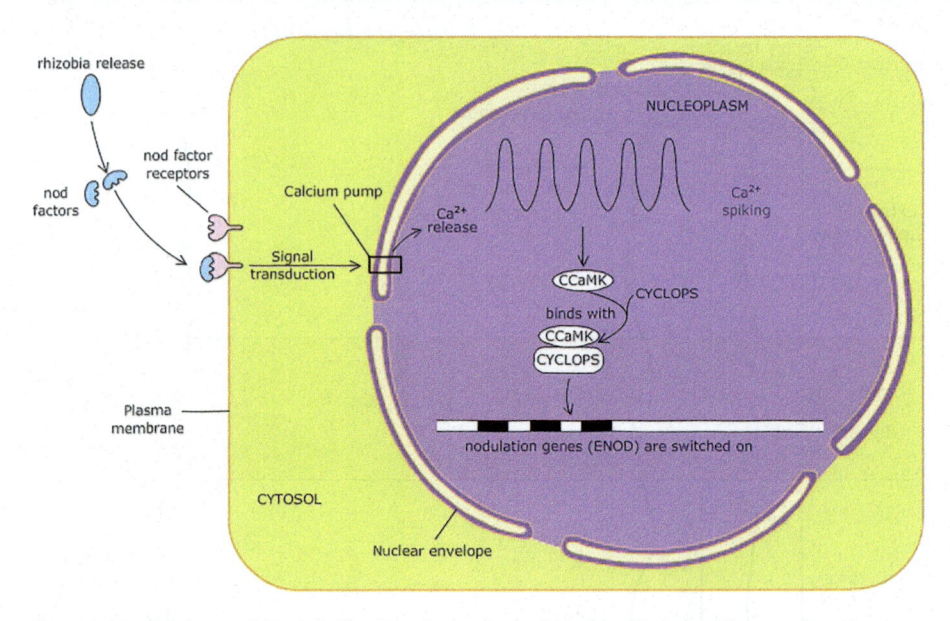

**Fig. 11.13** Calcium spiking is induced in the nucleus of the rhizobia-infected root cells of legume. Calcium spiking is recognized by CCaMK (calcium-calmodulin-activated protein kinase). CYCLOPS (a DNA-binding transcriptional activator) is a direct phosphorylation substrate of CCaMP. CYCLOPS is required for subsequent nodule development. ENOD (genes encoding early nodulins)

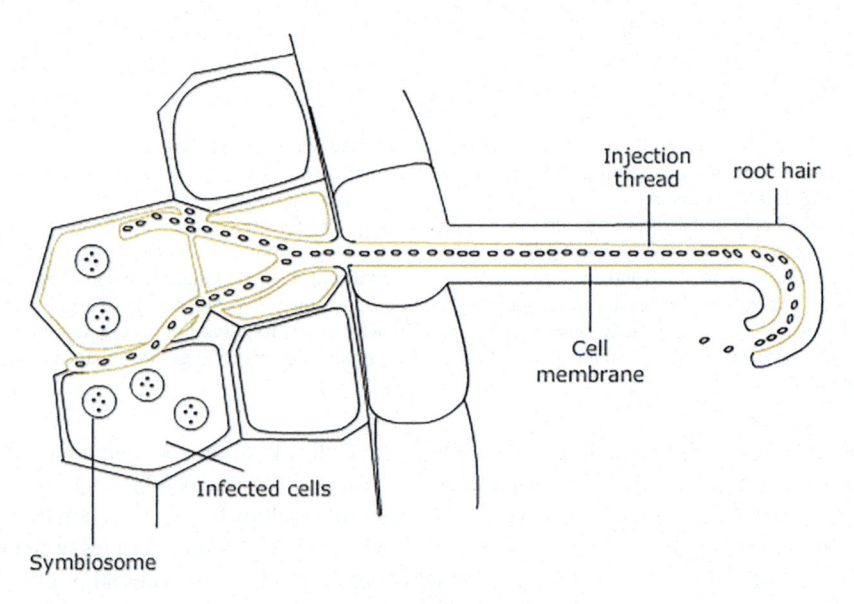

**Fig. 11.14** Formation of infection thread and nodule development

## 11.2 Nitrogen Nutrition for the Plants

**Table 11.3** Characteristic features of determinate and indeterminate type of nodules

| Characteristics | Determinate | Indeterminate |
|---|---|---|
| Nodule initiation | In outer cortex | In inner cortex |
| Meristem | Meristem does not persist in mature nodules; meristem is active for days | Meristem persists throughout nodule growth; meristem is active for months |
| Shape | Spherical | Elongate, cylindrical |
| Infected cells | Non-vacuolate | Vacuolate |
| Nod genes inducer | Isoflavones | Flavones, isoflavones |
| Bacteroids in symbiosomes | Each symbiosome contains several smaller rod-shaped bacteroids; number of bacteroids per symbiosome may be up to 20 | Each symbiosome contains single enlarged pleomorphic bacteroid |
| Nitrogen transported from the nodules | In the form of ureides, presence of ureides in the xylem sap is indicative of nitrogen being fixed by the plants | In the form of amides (glutamine and asparagine), which can be found in xylem even when nitrogen is not being fixed by the plants. Presence of amides in xylem sap is not marker for nitrogen fixation by the plants |
| Geographical origin | Tropical | Temperate |
| Examples | *Vigna unguiculata* (cow pea), *Glycine max* (soybean) | *Trifolium* sp., *Pisum* sp. |

further invagination of the plasma membrane of cortical cells. As the infection thread is established, there is dedifferentiation of the cortical cells near the xylem, which enter cell cycle, triggering cell division and forming a distinct area of dividing cells, called **nodule meristem**. This results in the development of protuberances on the roots. These protuberances, which are formed due to cell division, develop into nodules. Depending upon the type of meristem, nodules can be **determinate** (spherical nodules with the limited growth potential, e.g., soybean, *Lotus japonicus*) or **indeterminate** (continually growing which are cylindrical, e.g., in *Medicago* sp.) type. Determinate nodule cells mature synchronously so that no meristematic cells are present in mature nodules. In the indeterminate type of nodules, newly formed cells are constantly formed (Table 11.3). The tip of the infection thread is pinched off, and bacteria are released in the cells in the form of vesicles, surrounded by plasma membrane of the host plant. Bacteria stop dividing thereafter and enlarge. These are differentiated into nitrogen-fixing forms called **bacteroids**. Bacteroids have diminished ability to grow in culture medium. They may be greatly enlarged and possess various shapes. Vesicles are surrounded by the plasma membrane of the host plant, known as **peribacteroid membrane (PBM)**. One of the reasons for this effective symbiotic association of the host plant with rhizobia is that the bacteria do not trigger plant defense mechanism. The polysaccharides, which are present within the vesicles around bacteria, may be the possible reason for this. Some of the

*Bradyrhizobium* sp., which lack Nod factors, may adopt an alternate mechanism of infection. These may enter through the cracks in the epidermis and invade through the short infection thread. Possibly these invade the host cells by endocytosis-like mechanism and establish symbiosis.

**Role of Phytohormones in the Nodule Development** During early stage of the nodule formation, an increase in the cytokinin activity is reported, while auxin transport in the nodules is decreased. Nodule formation can be inhibited in mutants of cytokinin receptors, which indicates that cytokinin has an important role in the formation of nodule primordia. Cytokinins act downstream of the Nod factor signaling pathway. Auxin transport is inhibited in developing nodules. Use of an inhibitor of the auxin transport has, in fact, been found to induce nodule development. However, auxin may be required in the later part of nodule development. Ethylene diminishes the sensitivity of the host plant to the Nod factors released by the bacteria. With ethylene available, host plants require higher amount of Nod factors for the nodulation responses (calcium spiking or synthesis of early nodulin transcription factors). Ethylene mutants exhibit higher degree of nodulation.

**Symbiosomes** Symbiosomes are the vesicular structures formed in the infected cortical cells of nodules of the host plant. When infection thread reaches the cells, which later on develop into nodules, there is an endocytosis-like process. By pinching of infection thread, rhizobia are released in the form of vesicles still surrounded by the host cell membrane. These vesicles are called symbiosomes, which refer to the new organelle-like structures formed as a result of symbiosis. They contain one or more bacteria (Fig. 11.15). Membrane of symbiosome is derived from the host plant and is known as peribacteroid membrane. It is thought to mediate the flow of the nutrients between bacteria and host plant. Various intrinsic proteins are present in the peribacteroid membrane including **nodulin 26-like intrinsic protein (NIP)**, which transports water, glycerol, and ammonia. The space between the peribacteroid membrane and bacteroid membrane is known as peribacteroid space. Some of the plant proteins, which are uniquely expressed only in the infected cells, are synthesized in the cytosol of the infected cells and are transported across symbiosome membrane. These proteins carry a distinct targeting peptide which facilitates their entry into the symbiosome. It indicates symbiosome to be a unique cellular compartment. Symbiosome provides the microaerobic nodule environment, which is suitable for $N_2$ fixation by the bacteroids. Inside the symbiosomes, bacteria develop nitrogen-fixing enzyme dinitrogenase and are differentiated into bacteroids. These bacteroids appear to be terminally differentiated. There may be single enlarged pleomorphic bacteroid in a vesicle, or there may be several (up to 20) smaller, rod-shaped bacteroids. Nitrogen fixation is an energy (ATP)-requiring process. Requirement of ATP is met through aerobic respiration by the bacteria. At the same time, nitrogenase requires reducing environment since it is highly sensitive to oxygen. Thus, it becomes challenging for the plant to provide suitable conditions for the efficient functioning of the enzyme and

## 11.2 Nitrogen Nutrition for the Plants

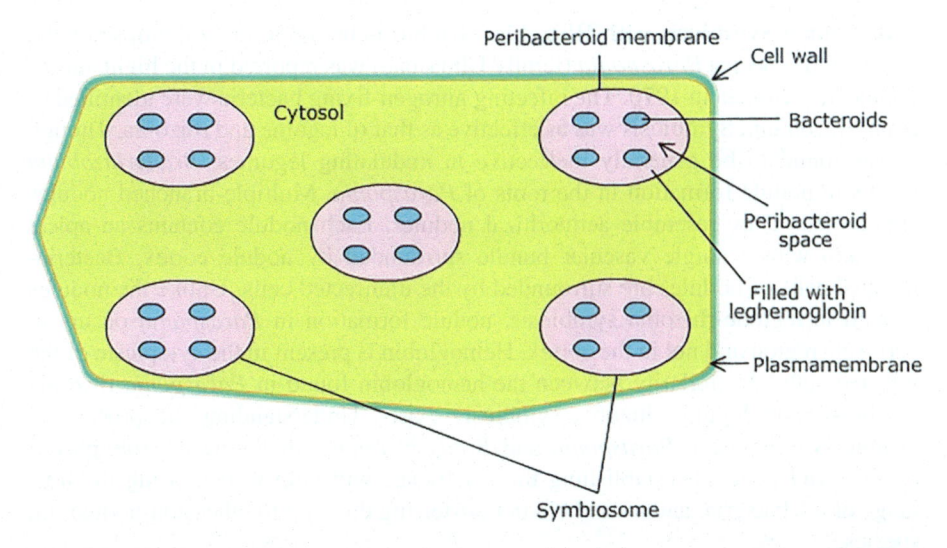

**Fig. 11.15** Infected cell of a nodule. Vesicles are formed as a result of pinching of infection thread containing rhizobia. Rhizobia divide inside vesicles and get transformed into nitrogen-fixing bacteroids. Bacteroids are surrounded by membrane derived from the host plant plasma membrane (peribacteroid membrane) and a polysaccharide matrix. The structure so developed is known as symbiosome

simultaneously to be able to meet the high ATP demand for the nitrogen fixation. This requirement is met through by the following factors:

1. High respiration rate in bacteria serves as sink for the oxygen. $K_m$ value of the cytochrome oxidase of the bacteroids is much lower than that of free-living bacteria, which indicates the efficiency of the bacteria to utilize oxygen more efficiently. $K_m$ value of cytochrome oxidase of the bacteroids is 8 nM, while that of free-living bacteria is 50 nM. $K_m$ for plant mitochondrial **cytochrome oxidase** is near 100 nM.
2. There is some control on the permeability for the entry of oxygen in the nodule parenchyma, which can change depending upon the external $O_2$ availability.
3. Presence of **leghemoglobin** (Lb) in the space of symbiosome surrounding the bacteroids. Although it is well established that apoprotein parts of the leghemoglobin are synthesized by the host plant, the source for heme moiety is uncertain (plant or bacteria!). In nodules, it is present in millimolar concentration. Nodules need to be kept under low oxygen concentration in order to prevent nitrogenase enzyme from getting inactivated by oxygen. Lb has an active role in delivering $O_2$ to the actively respiring nodules and in regulating the amount of $O_2$ near the site of nitrogen fixation. Lb acts like a buffer to regulate changes in concentration of oxygen which might arise due to changes in the rate of respiration in the bacteria. Leghemoglobin reduces the availability of free oxygen but increases the flux of oxygen to the respiring bacteroids.

**Non-legume Symbiosis with Rhizobia** Symbiosis between a rhizobium sp. and a non-legume species *Parasponia* (family Ulmaceae) was reported in the highlands of Papua New Guinea in 1970. The infecting nitrogen-fixing bacteria were identified as *Bradyrhizobium*. Symbiosis was as effective as that of legume and rhizobia. Though it was found to be generally ineffective in nodulating legumes, *Bradyrhizobium* results in nodule formation in the roots of *Parasponia*. Multiple-branched nodules are formed. They resemble actinorhizal nodules. Each nodule contains an apical meristem with a single vascular bundle surrounded by nodule cortex. Bacteria-infected cells of nodules are surrounded by the uninfected cells. Unlike the nodules formed in legume-rhizobia symbiosis, nodule formation in *Parasponia* occurs in pericycle region and not in the cortex. Hemoglobin is present in the cytoplasm of the infected cells. A similarity between the hemoglobin found in *Parasponia*-rhizobia symbiosis and legume-rhizobia symbiosis exists. Understanding the process of symbiosis in between *Parasponia* and *Bradyrhizobium*, including the role played by the plant genes in establishing the symbiosis, will help in increasing the host range of the bacteria, and it can help in transferring this trait to other non-nodulating species.

*Frankia*-**Actinorhizal Symbiosis** In these symbiotic associations also, nodules are formed in the roots of host plant for nitrogen fixation. The contribution of this type of nitrogen fixation to the soil nitrogen is as important as by rhizobia and legume symbiosis, especially in forests and other natural ecosystems. *Frankia* sp., a gram-positive bacterium belonging to family Frankiaceae (Actinomycetales), are involved in nitrogen fixation in symbiotic associations with some of the woody plants which include trees or woody shrubs predominantly of temperate zones, for example, *Alnus, Casuarina, Ceanothus, and Myrica* sp. *Casuarina* is more prevalent in tropical climates. The term actinorhizal is given to the nodules formed by *Frankia*. Actinorhizal plants are dicotyledon angiosperms characterized by their symbiotic associations with nitrogen-fixing actinobacteria of *Frankia* sp. *Frankia alni* is the only named species in the genus *Frankia*. Unlike rhizobia, *Frankia* are filamentous gram-positive bacteria. Their **nitrogenase** enzyme is also sensitive to oxygen similar to nitrogenase in legume-rhizobial symbiosis. *Nif* genes in both cases share sequence homology. *Frankia* also infects host plants through root hairs. In case of *Frankia*-actinorhizal symbiosis, prenodules are formed which consist of both infected and uninfected cells. Hemoglobin is synthesized in most of the actinorhizal plants. Hemoglobin present in actinorhizal symbiosis shares sequence homology with the one present in *Parasponia*-rhizobia symbiosis but is quite different from that of soybean. Unlike rhizobia, common *nod* genes, *nodA, nodB, nodC, and nod*, are located on the resident chromosome of *Frankia* sp., while in rhizobia these are located on megaplasmid. Mutations in any of these genes ($nod^-$) will result in a phenotype, which is not able to nodulate. The actinorhizal spp. are the pioneer species which are able to grow in nitrogen-poor soils and are able to enrich the soil with nitrogen because of their ability to form symbiotic associations with $N_2$-fixing *Frankia* sp.

## 11.2 Nitrogen Nutrition for the Plants

**Biochemistry of Nitrogen Fixation** Reduction of nitrogen to ammonia is an exergonic reaction:

$$N_2 + 3H_2 \rightarrow 2NH_3 \quad \Delta G^{0'} = -27 \text{ kJ.mol}^{-1}$$

However, molecular nitrogen in the atmosphere is an inert gas since the two atoms of nitrogen are triple bonded to each other. Bond energy of the nitrogen molecule is 930 kJ.mol$^{-1}$. Breaking of these bonds requires quite a high amount of activation energy. However, biological nitrogen fixation can occur at 0.8 atm and at normal temperature of the cell. Reduction of molecular nitrogen to two molecules of ammonia requires transfer of 8 electrons and 16 ATPs. The reaction that is catalyzed by the enzyme nitrogenase is

$$N_2 + 10\,H^+ + 8e^- + 16ATP \rightarrow 2NH_4^+ + 16ADP + 16P_i + H_2$$

Reduction of nitrogen by nitrogenase is coupled with the reduction of two protons resulting in the release of hydrogen. Production of H$_2$ during reduction of nitrogen by nitrogenase is obligatory. The biological role of the same has not been understood.

**Nitrogenase enzyme** is a highly complex and conserved protein. It consists of two components, component I and component II. These components are also called as dinitrogenase and dinitrogenase reductase, respectively. Active site for binding with molecular nitrogen is present in the dinitrogenase, while reduction of dinitrogenase is catalyzed by dinitrogenase reductase. Dinitrogenase reductase (component II) is an iron protein with a molecular weight of 62–64 kDa. It has two identical subunits of 30–32 kDa each. The protein is associated with single metal center [4Fe-4S]. Each of the two subunits has a site for binding with Mg.ATP. MoFe protein (component I), a heterotetramer. It is also known as dinitrogenase. It has a molecular weight of 220 kDa and consists of four polypeptide subunits of two types α and β (α2β2). Each α subunit has the molecular weight of approx 56 kDa, while molecular weight of each of β subunit is about 60 kDa. Binding site for nitrogen reduction is present in this component. Component I consists of two units. Each unit is a dimer consisting of one α and one β polypeptides. Each dimer (αβ) is associated with two metal centers called P cluster and M cluster. P cluster has a pair of [4Fe-4S] centers. Two [4Fe-4S] centers share one sulfur atom, so the stoichiometry will be 8Fe-7S. P cluster is bound to αβ dimer through covalent bond with cysteine residue. P cluster is also linked to the dimer through O-side chain of serine and amide from the peptide backbone. P cluster accepts electrons from Fe protein. Each dimer is also associated with another cofactor FeMo cofactor (FeMoCo). It is also known as M cluster, which is composed of Mo, Fe, S, carbide, and an organic molecule homocitrate, with an overall stoichiometry [1Mo-7Fe-9S-1C-homocitrate]. Organic homocitrate is bound to Mo of the cofactor on one side and to the protein on the other side through covalent bond with cysteine and histidine (Fig. 11.16).

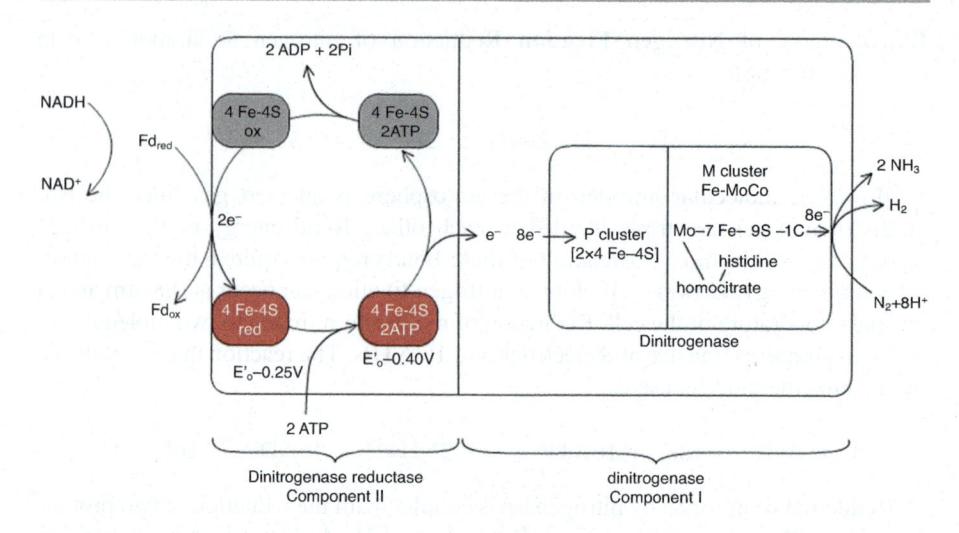

**Fig. 11.16** Nitrogenase complex consists of dinitrogenase and dinitrogenase reductase. Dinitrogenase reductase is a homodimer which is associated with single metal center (4Fe-4S). Each of the two subunit has binding site with Mg.ATP. Dintrogenase is a heterotetramer which consists of two units. Each unit of dinitrogenase is dimer consisting of two types of polypeptide subunits (α,β). Each dimer is associated with two metal centers, called P cluster and M cluster. P cluster has a pair of [4Fe-4S] centers. M cluster is composed of Mo, Fe, S, carbide (C), and an organic molecule homocitrate with overall stoichiometry 1Mo-7Fe-9S-1C-homocitrate. Homocitrate is bound to Mo on one side and histidine of the protein on the other side. Reduction of $N_2$ requires eight electron transfers from dinitrogenase reductase to dinitrogenase, one at a time. MoFe protein must accumulate eight electrons for $N_2$ reduction. Both ATP binding and ATP hydrolysis brings about conformational changes in dinitrogenase reductase which help in overcoming high activation energy of nitrogen reduction

**$N_2$ reduction to $NH_3$ by nitrogenase** In most of the cases, it is the reduced ferredoxin, which reduces dinitogenase reductase (component II). Reduction is accompanied with transfer of single electron from reduced Fd to dinitrogenase reductase. This is accompanied with binding of two ATP molecules at their binding sites bringing about the conformational change. As a result, reduction potential of the enzyme shifts from −300 mV to −420 mV, enhancing the reduction potential of the enzyme. 4Fe-4S center of the reductase comes closer to the P center of dinitrogenase which facilitates transfer of electron between dinitrogenase reductase and dinitrogenase accompanied with ATP hydrolysis. In oxidized state, Fe protein (component II) gets dissociated, and the other reduced Fe protein will take its place. Fe protein gets dissociated at least eight times during the transfer of eight electrons. MoFe protein must accumulate a minimum of eight electrons to complete the nitrogen reduction cycle. In the initial stage, possibly $2H^+$ might bind with the active site, resulting in the release of $H_2$. Release of $H_2$ is inevitable, causing wastage of reducing energy. Almost 30–60% of the energy supplied to nitrogenase gets wasted due to release of hydrogen. Dinitrogenase can possibly exist in different oxidation states depending upon number of electrons received by it. It is only after the

## 11.2 Nitrogen Nutrition for the Plants

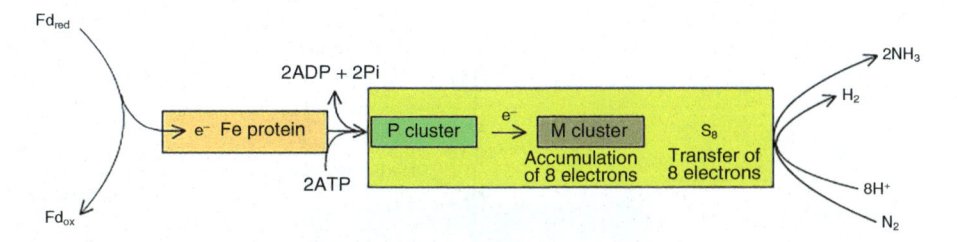

**Fig. 11.17** Molecular nitrogen reduction by dinitrogenase. $N_2$ binding site is present in M cluster which should accumulate eight electrons ($S_8$) before $N_2$ is reduced to $NH_3$. In the initial stage, possibly two $H^+$ might bind with the active site resulting in release of $H_2$

accumulation of eight electrons that the enzyme can possibly reduce molecular nitrogen to two molecules of ammonium ions (Fig. 11.17). Besides dinitrogenase, which has molybdenum as a part of the structure, other nitrogenases have also been reported, which are called **alternate nitrogenases**. These contain vanadium or Fe in place of Mo, but not much information is available on this aspect. Reduction carried out by nitrogenase is a slow process, which is compensated by large amount of nitrogenase synthesized by the bacteria. Nitrogenase can amount up to 20% of the total soluble protein in the bacteria.

**Endosymbiont-Host Plant Partnership** Host plant provides ecological niche to bacteria by providing suitable conditions for nitrogen fixation, such as low partial oxygen pressure, besides providing the source for ATP. Since nitrogen fixation is an energy-requiring process, it consumes up to 20% of the total photosynthates generated in plants. According to one of the calculations, for every mole of nitrogen fixed, plant will require 9.3 moles of $CO_2$. Carbon skeleton provided by the host plant is mainly in the form of dicarboxylic acid—malate. Source of malate is sucrose produced in the leaves during photosynthesis. It is metabolized through glycolysis, resulting in the formation of phosphoenolpyruvate, which is carboxylated to produce oxaloacetate. Oxaloacetate is reduced to malate, which is taken up by the bacteria. Malate is metabolized in bacteria through TCA cycle to provide reducing equivalents for generation of ATP, which is required for nitrogen fixation. Peribacteroid membrane allows transport of malate. Nitrogen fixation by nitrogenase requires anaerobic conditions. Most of the available oxygen is only allowed to pass through the nodule apex. As a result, a longitudinal gradient of oxygen concentration is created. Oxygen concentration is less than 50 nM in the center of nodule, where most of the nitrogen fixation takes place. Leghemoglobin present in peribacteroid space of the symbiosome regulates the supply of oxygen to rhizobia. Oxygen is required for active respiration in bacteria to meet the demand for ATP during nitrogen fixation. Bacteroids have developed a cytochrome oxidase, which has got higher affinity for oxygen, and it operates at maximal rate at lower oxygen concentration than the free-living rhizobia (Fig. 11.18). Rhizobia provides nitrogen to the host plant in the form ammonia, which is converted to ammonium ions after combining with protons once it diffuses into the peribacteroid space. Ammonium

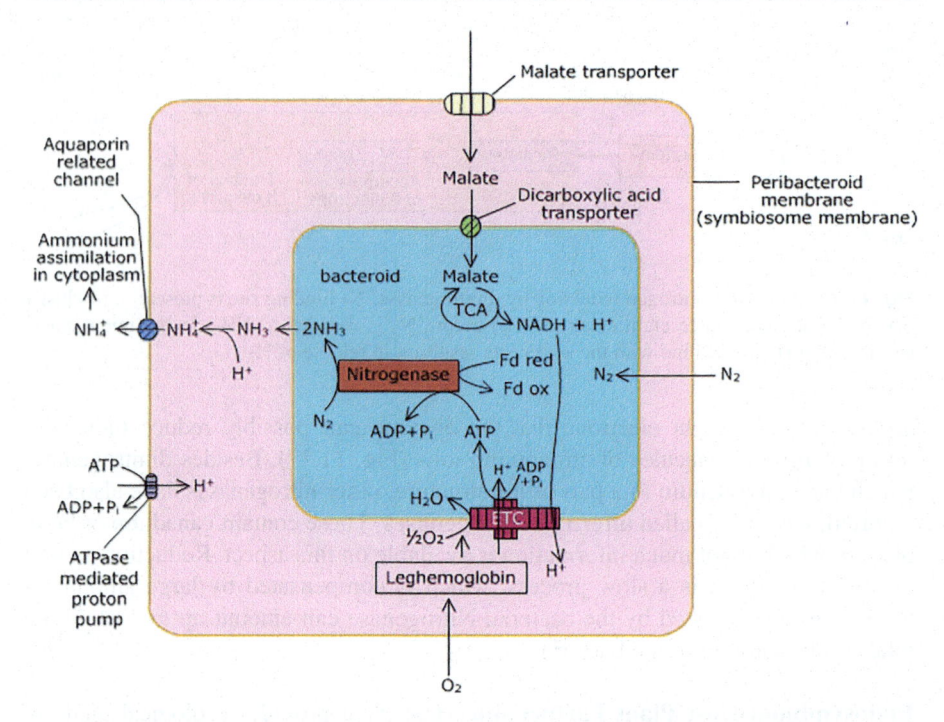

**Fig. 11.18** Host-bacteria partnership: Bacteroid metabolism in symbiosome. Bacteroids receive malate as energy source from the host which is then metabolized to generate ATP. NH$_3$ traverse through symbiosome membrane as NH$_4^+$ by facilitated diffusion, mediated by the NOD-26 aquaporin-related channel

ions are released into the cytosol of the infected cell of the host from symbiosomes through the channels present in the peribacteroid membrane. The host plant immediately assimilates ammonium ions in the form of **amides** (glutamine or asparagine) or **ureides,** either in the cytosol or in the organelles of the infected host cell. Host plant provides dicarboxylic acids, ATP, and reducing equivalents for assimilation of ammonium ions. Amides/ureides are then transported through xylem to rest of the plant.

**Genetics of Nitrogen Fixation** Genes involved in the effective symbiotic relationship in between the bacteria and the host plant can be considered under two categories. One set of genes is required to develop effective symbiotic relationship. This includes the *nod* genes present in the plasmid of bacteria (Sym). Mutant strains of bacteria, which are not able to form the symbiotic relationship with the host are called *nod*$^-$ bacteria. Second set of genes is required for the synthesis of nitrogen fixation apparatus and bacteroid metabolism. In nitrogen-fixing bacteria, these can be grouped into two categories: 1. The *nif* genes are required for nitrogen fixation by the symbiotic bacteria and by free-living nitrogen fixers, while 2. *fix* genes are essential for nitrogen fixation in symbiotic nitrogen fixers but do not have

## 11.2 Nitrogen Nutrition for the Plants

counterparts in free-living forms. *fix* genes are involved in the development and metabolism of bacteroids. *nifD* and *nifK* encode for alpha and beta subunits of nitrogenase, while the monomeric Fe protein is encoded by *nifH*. Ferredoxin is encoded by *nifF*. In *Rhizobium meliloti, nifHDK* genes are organized in an operon along with *nifE*. Over 20 *nif* genes are known. Synthesis of FeMo cofactor requires the products of *nifE, nifN,* and *nifB*. However, exact biochemical functions of these proteins are not known. Host plant can contribute to the formation of nitrogenase by providing homocitrate, a component of nitrogenase, to the bacteria. Expression of the *nif* genes is regulated by the availability of nitrogen and oxygen. Some of the *nif* genes are involved in regulatory functions. Transcription of *nif* genes is activated by the product of *nifA,* which is expressed in the absence of oxygen. FixK is a protein in symbiotic bacteria, as in rhizobia, which regulates the expression of *fix* group of nitrogen-fixing genes. Expression of *fixK* and *nifA* is regulated by FixL, which is an oxygen-sensitive heme-containing protein kinase. This protein is anchored to the bacterial plasma membrane. In the absence of $O_2$, this protein is phosphorylated and, in turn, regulates the activity of FixJ by phosphorylating it, which regulates expression of *nifA* and *fixL* genes. These genes, in turn, will control the expression of other *nif* and *fix* genes. *fix* genes are present only in those bacteria which are in symbiotic relationship. One of the *fix* genes is involved in the synthesis of specific ferredoxins, and others may be involved in the synthesis of high-affinity cytochromes. Host plant genes required for the nodulation process encode nodulins. Plant proteins, which are formed early before any infection by bacteria has taken place, are called "early nodulins." They are encoded by *ENOD* genes. Late nodulins are the proteins, which are synthesized later in the symbiotic process. Leghemoglobin is an example of late nodulin. Other late nodulins include proteins required for metabolizing carbon and nitrogen and in facilitating transport of compounds across symbiosome membrane. Some of the nodulins may be involved in transducing signal for cell differentiation in plants and bacteria to accomplish nitrogen fixation.

**Transport of Ammonium from Nodules** Ammonia formed in the bacteroids is released in the peribacteroid space, where it combines with proton ions and is converted to ammonium ions. Protons are pumped into the peribacteroid space by ATPase-mediated proton pumps localized in the symbiosome membrane and also by the electron transport chain present in the bacteroid membrane. Ammonia is transported as ammonium ions to the cytosol of the infected cell of the nodules via monovalent cation channels. Glutamine synthetase (GS) and GOGAT (glutamine oxoglutarate aminotransferase) enzymes are expressed in the cytosol of the infected nodule cells, which assimilate ammonium ions firstly to glutamine and then to glutamate. Fixed nitrogen is transported out of the nodule either in the form of amides or ureides via xylem. In the legumes of temperate origin, nitrogen is transported in the form of amides, i.e., glutamine or asparagine. These products are generally synthesized in the nodules with indeterminate apical meristem. Examples of the amide transporter legumes are pea (*Pisum sp.*), broad beam (*Vicia sp.*), clover (*Trifolium sp.*), etc. In the legumes of tropical origin, e.g., soybean (*Glycine max*), common bean (*Phaseolus vulgaris*), peanut (*Arachis hypogea*), etc., fixed nitrogen is transported out of the nodules in the form of ureides. Possibly ureides are sparingly soluble in temperate conditions, so it is the tropical legumes in which fixed nitrogen travels in the form of ureides. Major ureides transported are allantoin, allantoic acid, and citrulline. Allantoin is synthesized from uric acid in the peroxisomes of the infected cells of the nodules, while allantoic acid is synthesized from allantoin in the endoplasmic reticulum of the cell. The site for synthesis of citrulline is not known. It is the determinate type of nodules which serve as ureide exporter. However, all the three ureides are released in xylem and are transported to the shoot. In shoot, these compounds (ureides) are converted to ammonium ions, which are further assimilated by the GS/GOGAT system (Fig. 11.22).

## 11.3 Ammonia Assimilation

Ammonia assimilation in plants is a complex process since it is produced in plants from different sources. These sources can be classified as the primary or secondary. Primary source refers to the ammonia which is produced from the inorganic nitrogen, such as ammonium ions absorbed by the plants and ammonia generated due to nitrite reduction or due to nitrogen fixation by the nitrogen-fixing bacteria in symbiotic association with plants. Secondary sources refer to the ammonia generated from the organic compounds during metabolism. These include ammonia generated due to (i) oxidation of glycine to serine during photorespiration; (ii) degradation of nitrogenous compounds, such as asparagine, arginine, and ureides; and (iii) protein degradation and deamination of amino acids. Since ammonia is produced at different sites in plants, the enzymes catalyzing ammonia assimilation function in different cellular conditions. As a result, various isozymes of the enzymes are involved in assimilation of ammonia.

### 11.3.1 Ammonia Assimilation by GS/GOGAT

Ammonia produced in nodules by the bacteroids is released through the bacteroid membrane. Since ammonia is nonpolar, it can easily diffuse out of the membrane. Once in the peribacteroid space, ammonia accepts up proton and is converted to $NH_4^+$. Ammonium ions are transported to the cytosol of the infected cell of the nodule through the channels present in the peribacteroid membrane of the symbiosome by facilitated diffusion. $NH_4^+$ transport is mediated by an aquaporin-related channel (NOD−26) across the peribacteroid membrane in soybean nodules. $NH_4^+$ is assimilated in the cytosol of the infected cells of the host plant. Two enzymes responsible for the initial assimilation of $NH_4^+$ are **glutamine synthetase** (GS) and **glutamate synthase** (which is also known as **GOGAT, glutamine 2-oxoglutarate aminotransferase**) (Fig. 11.19).

## 11.3 Ammonia Assimilation

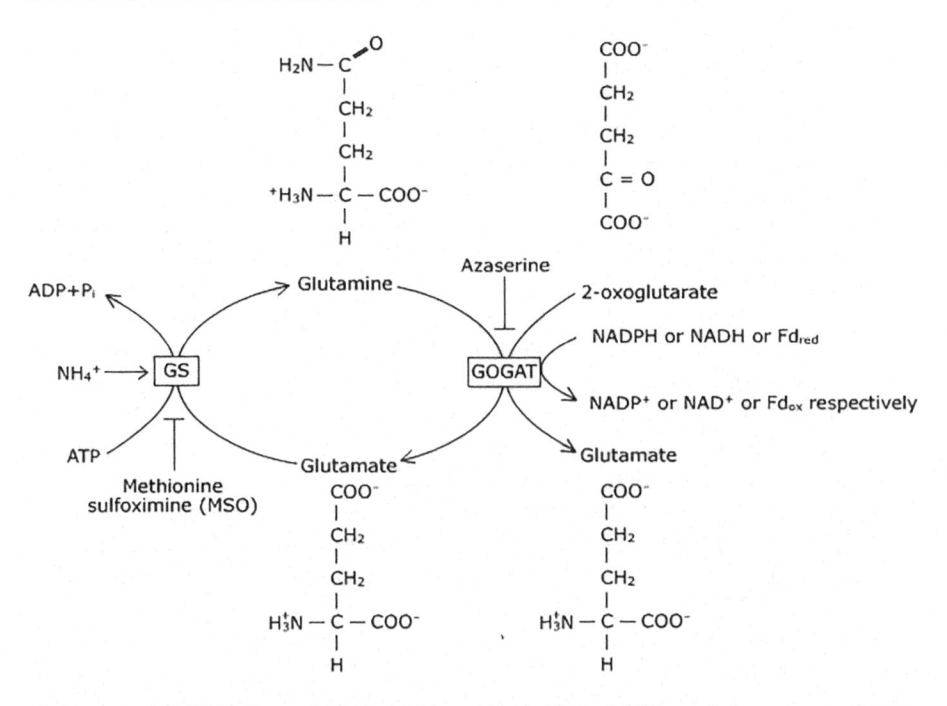

**Fig. 11.19** GS/GOGAT reaction. Methionine sulfoximine is inhibitor of glutamine synthetase. Azaserine is substrate analogue of glutamate and is inhibitor of GOGAT (also known as glutamate synthase). GS, glutamine synthetase; GOGAT, glutamine 2-oxoglutarate aminotransferase

Glutamine synthetase (GS) catalyzes the synthesis of glutamine from $NH_4^+$ and glutamate. (Box 11.8). This ATP-requiring reaction is

$$\text{Glutamate} + NH_4^+ + ATP \rightarrow \text{Glutamine} + ADP + P_i$$

A divalent cation such as $Mg^{2+}$ or $Mn^{2+}$ or $Co^{2+}$ is also required as the cofactor for the enzyme. There are two classes of glutamine synthetase (GS) in plants: cytosolic (GS1) and plastidial (GS2). GS1 is also present in leaves in low concentrations, while it is present in higher concentrations in roots. Role of GS1 is important in primary assimilation of $NH_4^+$. GS1 is expressed in germinating seeds or in vascular bundles of roots or shoots and generates glutamine for intercellular transport. GS2 is responsible for assimilating $NH_4^+$ produced during photorespiration. GS2 is the primary enzyme of glutamine synthesis in the leaves. Gene encoding GS2 is expressed in mesophyll cells, while gene for GS1 is expressed in phloem, indicating its primary role in producing glutamine for long-distance transport. Mutants of GS2 do not survive in conditions which favor photorespiration, while they survive in conditions which suppress photorespiration. Glutamate synthase transfers amino group from glutamine to 2-oxoglutarate, resulting in the synthesis of two molecules of glutamate. One of the glutamate molecules serves as the substrate for GS, while the other one is available for further metabolism. The combined reaction for ammonium metabolism is known as **GS/GOGAT** reaction.

> **Box 11.8: Glutamine Synthetase (A Molecular Computer)**
> Glutamine synthetase (GS) is an important enzyme of nitrogen metabolism since it is the entry point for assimilation of ammonia. The products of the ammonia assimilation are responsible for regulation of the GS activity. Thus, GS monitors nitrogen requirement of the plant. The enzyme exists as multi-subunit complex consisting of similar subunits. The number of subunits varies from 8 to 12 depending upon the source. Bacterial GS consists of 12 subunits, while in eukaryotes, including plants, it has 10 subunits. Structure of the enzyme is stabilized due to non-covalent interactions between the subunits. Regulation of GS activity is both due to allosteric and covalent modulation. Covalent binding of Tyr$^{397}$ residue (present near the active site of the enzyme subunits) with AMP increases the sensitivity of subunits to allosteric inhibition by the end products of glutamine metabolism, such as alanine, glycine, and histidine. Binding of subunits with one of the inhibitors partially decreases GS activity, and the effect of multiple inhibitors is additive, thereby completely shutting down its activity. Thus, GS monitors the nitrogenous status of the plant and is active accordingly. That is why the enzyme may also be called "molecular computer."

There are two types of GOGAT in plants. One of them accepts electrons from NADH (NADH-GOGAT), while the other type accepts electrons from reduced ferredoxin (Fd-GOGAT).

$$\text{Glutamine} + 2-\text{oxogutarate} + \text{NADH} \rightarrow 2\,\text{Glutamate} + \text{NAD}^+$$

$$\text{Glutamine} + 2-\text{oxoglutarate} + \text{Fd}_{red} \rightarrow 2\,\text{Glutamate} + \text{Fd}_{ox}$$

Unlike GS, which has both cytosolic and plastidic isoforms, GOGAT is present only in the plastids. NADH-GOGAT is present in non-photosynthetic tissues such as roots or in vascular bundles of developing leaves. NADH-GOGAT, present in roots, is involved in glutamine assimilation which is produced due to $NH_4^+$ assimilation in the roots, either absorbed from the rhizosphere or has been fixed by the bacteria in the symbiotic association (Fig. 11.20). The NADH-GOGAT present in vascular bundles is responsible for assimilation of glutamine, which is destined to be translocated either from roots or senescing leaves. The other type, Fd-GOGAT, is present in chloroplasts. It catalyzes metabolism of glutamine, produced from assimilation of $NH_4^+$ generated during photorespiration. Glutamine utilization by the Fd-GOGAT accounts for 95–97% of the total GOGAT activity in leaves. This shows its predominant role in leaves in metabolism of glutamine, which is produced due to ammonia assimilation generated both from primary sources (produced as a result of $NO_3^-$ reduction) or from secondary sources (produced due to photorespiration) (Fig. 11.21). Most plants have been found to contain a single gene for

## 11.3 Ammonia Assimilation

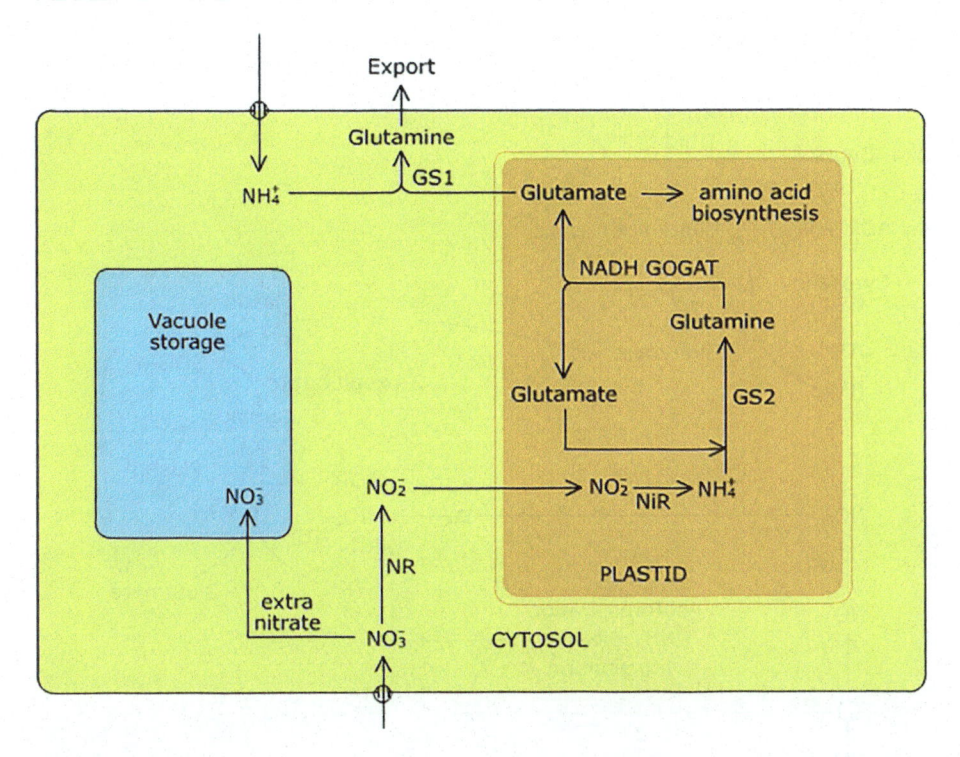

**Fig. 11.20** Ammonium assimilation in roots. Isozymes of glutamine synthetase, GS 1 and GS 2, are present in cytosol and plastids, respectively

Fd-GOGAT. However, in *Arabidopsis* Fd-GOGAT is encoded by two genes (GLU1 and GLU2), while NADH-GOGAT is encoded by a single gene (GLT). Both Fd-GOGAT and NADH-GOGAT exist as monomeric proteins (Fig. 11.22).

### 11.3.2 Ammonia Assimilation by Reductive Amination

Reductive amination was thought earlier to play primary role in ammonia assimilation. The enzyme **glutamate dehydrogenase (GDH)** catalyzes the reaction in which ammonia, in the form of ammonium ions, is assimilated and α-ketoglutarate is aminated to form glutamate. Two types of GDH are present in plants. One class of GDH, NADH-dependent GDH, functions in mitochondria and requires NADH as the reductant for the reaction. The other is NADPH-dependent GDH found in chloroplasts. The reaction is reversible as GDH catalyzes both reductive amination, in which 2-oxoglutarate is converted to glutamate, and **oxidative deamination** in which glutamate is deaminated to generate 2-oxoglutarate. In reductive amination, reaction is coupled with oxidation of NADH/NADPH to $NAD^+/NADP^+$,

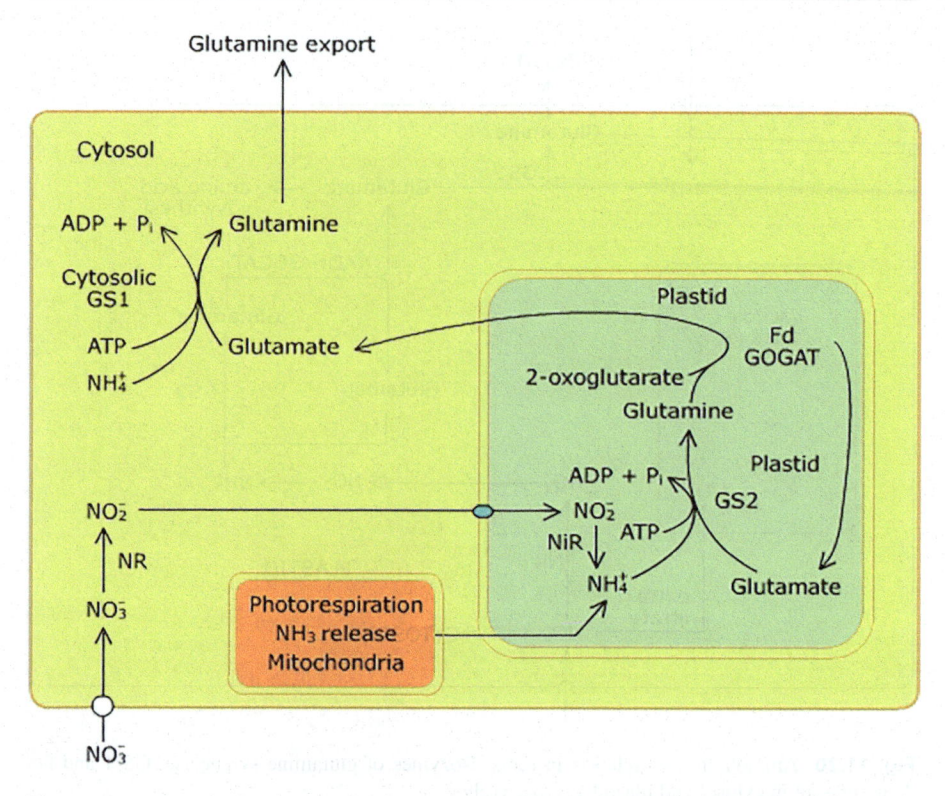

**Fig. 11.21** Ammonium assimilation in leaves

respectively. On the contrary, reducing power is generated during oxidative deamination (Fig. 11.23). Ammonium assimilation by GDH was earlier considered to be of primary importance. But currently its role has been found to be more significant in the catabolism of glutamate. $K_m$ value of GDH for ammonium ions is significantly higher (10–80 mM), while ammonium concentration in the cell has been found to be very low (0.2–1.0 mM). On the contrary, GS has high affinity for ammonium ions ($K_m$ 3–5 micromoles) and can operate effectively at cellular concentrations. Studies carried out with $^{15}NH_4^+$ have provided evidence that labeled nitrogen first appears in the amide group of glutamine and then in the glutamate. Use of inhibitors of GS (**MSO, methionine sulfoximine**) leads to inhibition of $NH_4^+$ incorporation into glutamate even in presence of GDH and higher concentrations of ammonium. GDH might possibly be involved in assimilation of higher amount of ammonium ions (generated from photorespiration) in mitochondria. GDH activity increases in presence of high ammonium concentrations, indicating a possible role of GDH in detoxification. GDH may be important in glutamate catabolism generating 2-oxoglutarate which might be used as the TCA intermediate for energy generation in dark, in germinating seeds, or in senescing leaves. The reaction catalyzed by

## 11.4 Nitrogenous Compounds for Storage and Transport

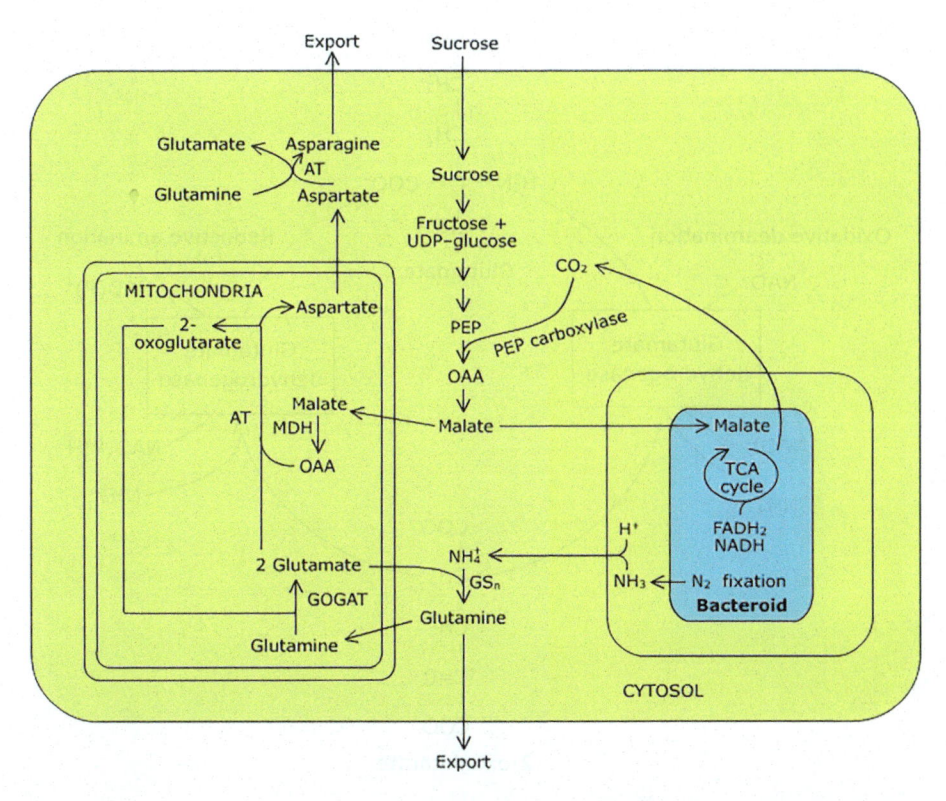

**Fig. 11.22** Nitrogen metabolism in the infected cells of root nodules. $GS_n$ nodule-specific cytosolic GS isoenzyme, *AT* aminotransferase, *MDH* malate dehydrogenase

GS/GOGAT requires ATP and is irreversible. On the contrary, there is no ATP requirement by GDH for the formation of glutamate. Reaction catalyzed by GDH is reversible, and the equilibrium is unfavorable for glutamate formation. In plants GDH is encoded by two genes, *GDH1* and *GDH2*. GDH has a **hexameric** structure, and seven isoforms have been reported in plants depending upon the products of these two genes.

## 11.4 Nitrogenous Compounds for Storage and Transport

The amino group of the glutamate is transferred to oxaloacetate, producing aspartate and 2-oxoglutarate. Regeneration of 2-oxoglutarate is required for the continuation of ammonium metabolism by GS/GOGAT. Amino group transfer is catalyzed by an **aminotransferase**, i.e., **aspartate aminotransferase (AspAT)**, which is also known as **glutamate-oxaloacetate aminotransferase (GOT)**. It

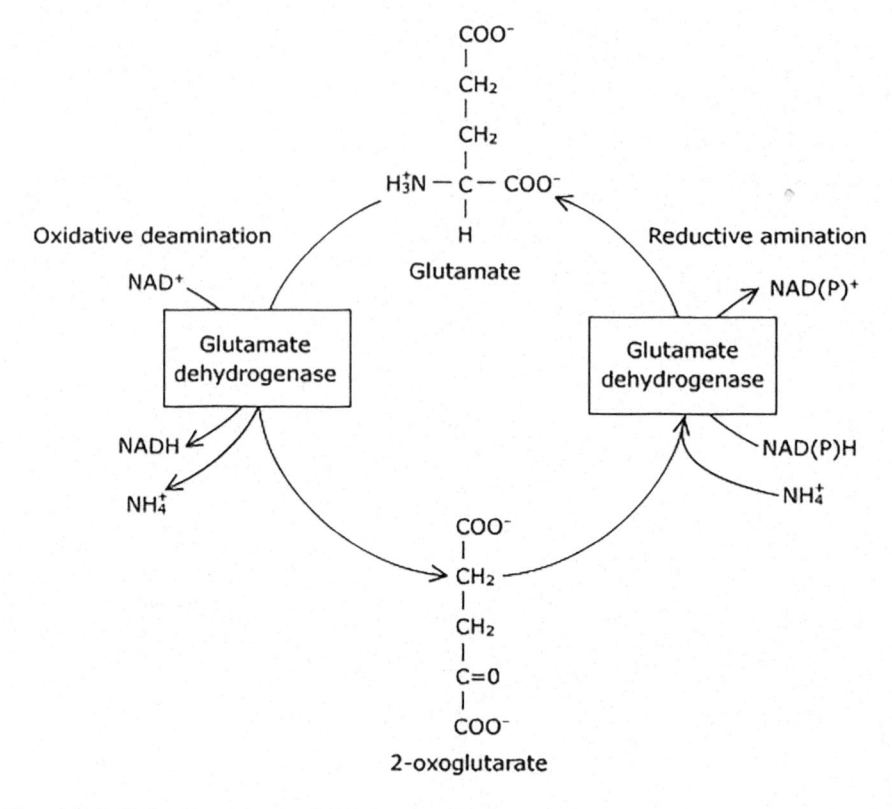

**Fig. 11.23** Reductive amination/oxidative deamination reactions

requires pyridoxal phosphate as a cofactor like all other aminotransferases (Box 11.9). The reaction is known as transamination reaction.

Glutamate + Oxaloacetate → 2 − oxoglutarate + aspartate

In addition to other compounds, **asparagine** is also used for storage and transport of nitrogen. High stability of asparagine and high N/C ratio of the molecule (it is 2:4 in case of asparagine, while in case of glutamine, it is 2:5) make it suitable for storage and transport of nitrogen. [*Asparagine was the first amino acid to be isolated almost 200 years ago from the extracts of Asparagus. It is responsible for the unique flavor of Asparagus. When plants are grown in dark, asparagine content in Asparagus increases, while there is decrease in woody tissues; that is why etiolated plants are both tastier and softer.*] Asparagine is synthesized from

## 11.4 Nitrogenous Compounds for Storage and Transport

> **Box 11.9: Pyridoxal 5′-Phosphate (PLP)**
>
> PLP serves as the coenzyme for variety of enzymes. Its primary role is in the metabolism of molecules with amino groups since it is also the cofactor of aminotransferases. All aminotransferases catalyze similar reactions. Pyridoxal phosphate (PLP) functions as an intermediate carrier of the amino group. It is derivative of vitamin $B_6$. Vitamin $B_6$ in diet consists of three different forms of PLP, pyridoxamine, pyridoxine, and pyridoxal. In the cell, all of these three forms of PLP can be converted to pyridoxal phosphate.
>
> Pyridoxamine phosphate         Pyridoxal phosphate

glutamine and aspartate, catalyzed by the enzyme **asparagine synthetase (AS)** (Fig. 11.24).

$$\text{Glutamine} + \text{Aspartate} + \text{ATP} \rightarrow \text{Glutamate} + \text{Asparagine} + \text{AMP} + \text{PP}_i$$

AS is present in the cytosol of leaf and root cells and in nodules. Similar to GS, activity of AS is also regulated by the adenylation of the enzyme. Two types of AS have been reported in plants. One type of AS utilizes glutamine, while the other type is ammonium-dependent asparagine synthetase. However, $K_m$ value of the ammonium-dependent AS is higher, so possible direct amidation of aspartate to asparagine is of no significance. Transport of nitrogen in the form of glutamine is induced by light or availability of sucrose, while transport of asparagine is promoted by dark and in the conditions when sucrose supply is limited.

In some legumes, such as alfalfa and pea, nitrogen is transported as amides, i.e., glutamine and asparagine, while in tropical legumes, such as soybean (*Glycine max*) and cowpea (*Vigna sp.*), nitrogen is transported in the form of **ureides,** such as allantoin and allantoic acid (Fig. 11.25). Biosynthesis of these compounds involves various compartments of the cells, both infected as well uninfected. Purine biosynthesis and their degradation to uric acid occur in the infected cells of nodules. Uric acid is transported to uninfected cells and is degraded to ureides. Peroxisomes in the uninfected cells are also involved in formation of ureides. The ureides, allantoin and allantoic acid, are transported through xylem to other parts of the plants. Ureides have an advantage over amides since they have a very high N/C ratio which is 1:1 in comparison to glutamine and asparagine which have N/C ratio of 2:5 and 2:4, respectively (Fig. 11.25).

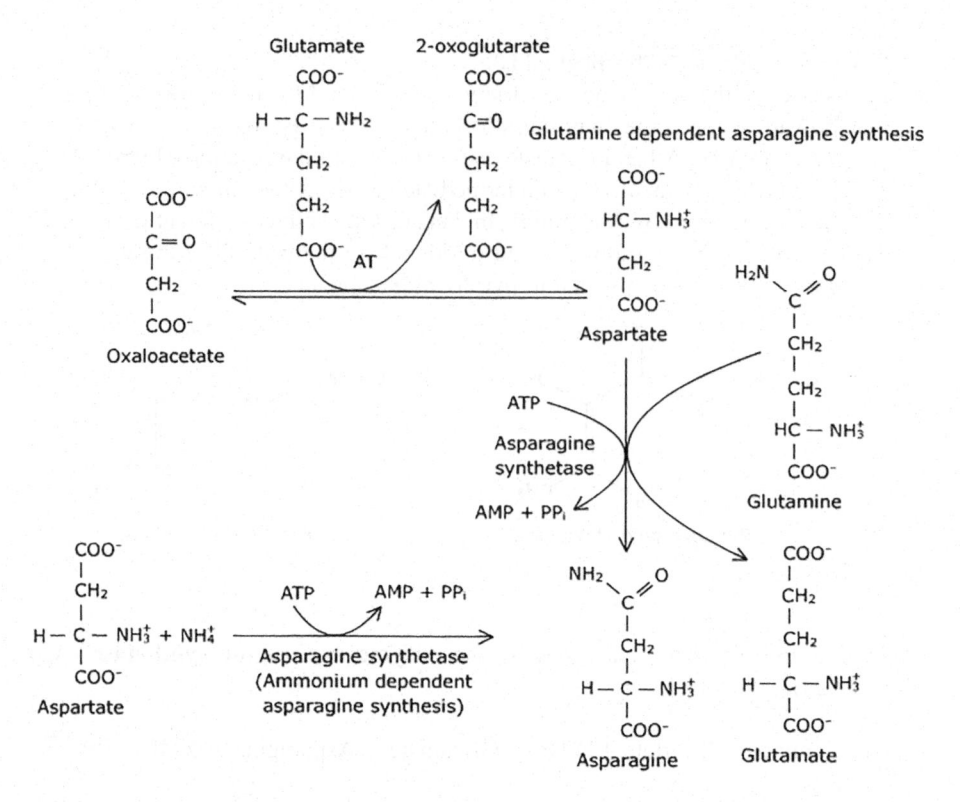

**Fig. 11.24** Aspartate and asparagine biosynthesis

## 11.5 Amino Acid Biosynthesis

All 20 amino acids, which are integrated in the proteins during protein synthesis, can be synthesized in plants and bacteria (Fig. 11.26). On the contrary, only ten of these amino acids are synthesized in the human body. Rest of the amino acids need to be provided in the diet, which are called essential amino acids (Table 11.4). There is no mechanism by which surplus amino acids can be stored in the human body. So, unutilized amino acids are oxidized and stored in the form of glycogen or fats, and nitrogen released is converted to urea, which is excreted out of the body. On the contrary, nitrogen is precious for the plants since they have to depend upon limited nitrogen supply, so they cannot afford to lose it, and hence whatever nitrogen released in the form of ammonia is recycled and reassimilated by the plant. Since plants are the major source of essential amino acids for the animals, it is significant to study their biosynthesis. Carbon skeleton for the amino acid biosynthesis is derived from compounds which include 3-phosphoglycerate, pyruvate, phosphoenolpyruvate, oxaloacetate, and 2-oxoglutarate. These are the intermediates in the glycolytic

## 11.5 Amino Acid Biosynthesis

**A.**

Allantoin

Allantoic acid

**B.**

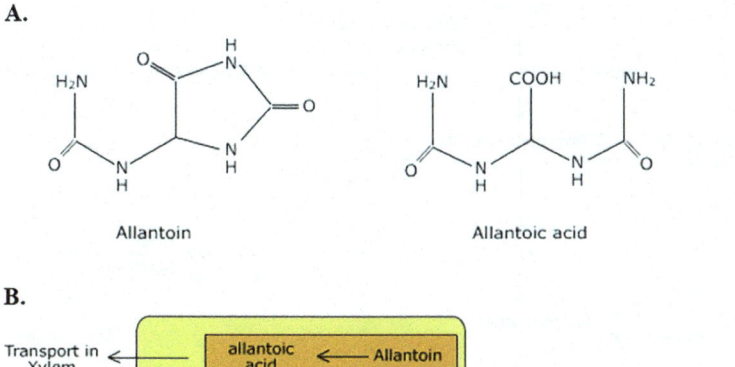

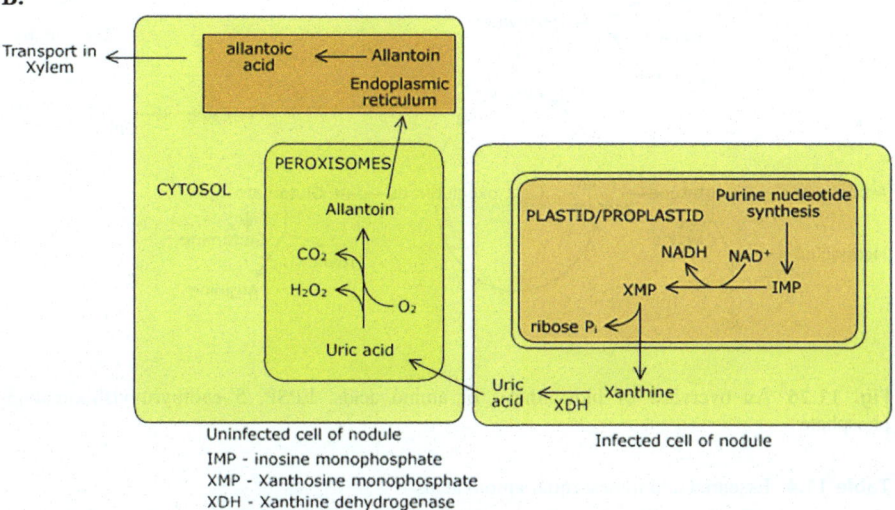

Uninfected cell of nodule
Infected cell of nodule
IMP - inosine monophosphate
XMP - Xanthosine monophosphate
XDH - Xanthine dehydrogenase

**Fig. 11.25** (a) Structure of ureides (compounds which are related to urea). (b) Cellular and subcellular compartmentation in nodule cells for ureide biosynthesis

pathway and citric acid cycle, besides other sugars such as ribose 5-phosphate and erythrose 4-phosphate, which are produced either in reductive or oxidative pentose phosphate pathway.

### 11.5.1 Aminotransferase Reaction (Tansamination)

Key reaction during amino acid biosynthesis is transamination, catalyzed by transaminases (also known as aminotransferases). The reaction involves transfer of an amino group from one α-amino acid to α-carbon of an α-keto acid resulting in the formation of a new α-amino acid and a new α-keto acid. The reaction is catalyzed by the enzymes which are known as aminotransferases.

$$\alpha - \text{amino acid}_1 + \alpha - \text{keto acid}_2 \rightarrow \alpha - \text{keto acid}_1 + \alpha - \text{amino acid}_2$$

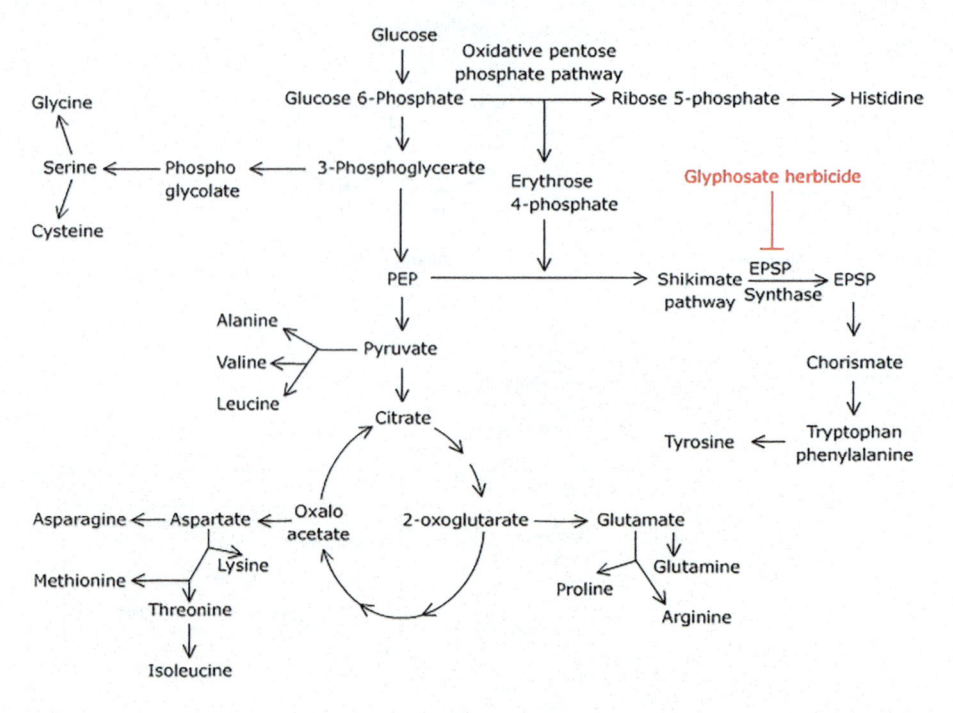

**Fig. 11.26** An overview of biosynthesis of amino acids. EPSP, 5′-enolpyruvylshikimate-3-phosphate

**Table 11.4** Essential and nonessential amino acids

| Essential amino acids | Nonessential amino acids | Conditional essential amino acids[a] |
|---|---|---|
| Histidine | Asparagine | Arginine |
| Leucine | Aspartate | Cysteine |
| Isoleucine | Alanine | Tyrosine |
| Lysine | Glutamate | |
| Methionine | Glycine | |
| Phenylalanine | Proline | |
| Valine | Serine | |
| Tryptophan | Glutamine | |
| Threonine | | |

[a]Conditional essential amino acids are those which may be required in specific condition, e.g., arginine is required at the growing stage; cysteine can be synthesized provided methionine is provided in diet. Tyrosine can be produced only from phenylalanine

In most of the cases, either glutamine (an amine) or glutamate are the amino group donors, while an α-keto acid is the amino group acceptor. Some of the aminotransferases prefer aspartate or asparagine as the group donor, which

## 11.5 Amino Acid Biosynthesis

themselves are produced at the expense of glutamate/glutamine. There are various aminotransferases. Some of them may be specific for 2-oxoglutarate as the amino group acceptor but may differ for the amino group donor α-amino acids. Aminotransferases are named on the basis of the amino acids which serve as the amino group donors, e.g., alanine aminotransferase and aspartate aminotransferase, the enzymes which catalyze amino group transfer from alanine and aspartate, respectively.

All aminotransferases require **pyridoxal 5′-phosphate (PLP)** as the cofactor. PLP is covalently bound to the protein (E-PLP) at the active site of the enzyme through an **aldimine (Schiff base)** linkage with the ε-amino group of a lysine residue. PLP can undergo reversible transformations. The incoming amino acid donates its amino group to the aldehyde form of the cofactor (PLP) (Box 11.9). The amino group donor, amino acid, is converted to the respective keto acid and leaves the active site of the enzyme, while PLP, on receiving the amino group, is converted to its amino form, i.e., **pyridoxamine phosphate (PMP).** The second substrate (another α-keto acid) binds to the active site of the enzyme and receives the amino group from PMP in a typical "Ping-Pong" reaction and gets converted to respective amino acid, and PLP is regenerated.

$$E - PLP + \text{amino acid}_1 \Leftrightarrow E - PMP + \text{keto acid}_1$$

$$E - PMP + \text{keto acid}_2 \Leftrightarrow E - PLP + \text{amino acid}_2$$

Reactions catalyzed by the aminotransferases are reversible reactions with a $\Delta G^{0\prime} \approx 0$.

Understanding of the amino acid biosynthesis pathways in plants has mainly been inferred from the pathways worked out in bacteria and fungi. However, the pathways occurring in plants are much more complex than those in bacteria, since bacterial genome is small and organelles are absent in bacteria. There are multiple genes present in plants, which regulate each step of the pathway. Most of the pathways involved in amino acid biosynthesis are quite complex and involve a number of enzymes. Various amino acid transporters have been identified for inter- and intracellular transport of the amino acids in plants; however the mechanism of amino acid transport is still poorly understood. Enzymatic pathways for the biosynthesis of many of amino acids have been worked out, which are commercially exploited for regulating synthesis of especially the essential amino acids, as these are not synthesized in human body and need to be provided in their diet. Additionally, since enzymes for the synthesis of essential amino acids are absent in humans and present in plants and bacteria, using inhibitors of these enzymes can be exploited for producing potential antibiotics against animal pathogens and also for developing **herbicides** (Box 11.10). Amino acids are classified under the following categories on the basis of their precursors and the amino group donors required for their synthesis.

> **Box 11.10: Blockers of Synthesis of Essential Amino Acid in Bacteria Can Be Potential Antibiotics**
>
> Unlike plants, fungi, and bacteria, animals are not able to synthesize essential amino acids. Some of the enzymes required for biosynthesis of these amino acids are missing in animals, so they have to depend on the supply of these amino acids in their diet. Learning about the biosynthetic pathways for these amino acids can be exploited in developing new antibiotics and antifungal medicines. This becomes significant with the growing antibiotic resistance in bacteria. One of the potential targets is the enzyme aspartate-β-semialdehyde dehydrogenase (ASADH) which catalyzes the second reaction of aspartate biosynthetic pathway. Almost a quarter of the amino acids are synthesized through this pathway. Besides essential amino acids, this pathway provides many other compounds required for various critical functions of the organisms, and any disruption of the pathway will be fatal. Blocking lysine biosynthesis in *Mycobacterium tuberculosis* is fatal for the organism which requires de novo synthesis of the amino acid for its survival. Another significance is in developing the herbicides, which can work by blocking the biosynthesis of essential amino acids in weeds. One of the examples is the use of glyphosate as the potential herbicide. Glyphosate is the inhibitor of enzyme 5-enolpyruvylshikimate-3-phosphate (EPSP) synthase, which is the sixth enzyme of shikimic acid pathway. The herbicide-tolerant crop plant can be produced by genetic engineering.

**Glutamate-Derived Amino Acids** Amino acids included in this category are proline, arginine, and a non-protein amino acid ornithine. Initial reactions require ATP and NADPH leading to formation of glutamate semialdehyde. This is followed by the formation of a ring because of condensation of the carbonyl group of glutamate semialdehyde with α-amino group of the molecule, resulting in the formation of proline. The reaction is coupled with oxidation of yet another molecule of NADPH. Proline accumulates in plant tissues in response to drought and salt stress. It serves as a compatible solute, so it can accumulate at higher concentration without disturbing normal cellular activities and can protect the cell against osmotic stress. Proline also serves in scavenging ROS. Thus, synthesis of proline by the plants is significant in developing transgenics, which are more suitable to grow under drought. Arginine, another amino acid, has a high N/C ratio (4:6). It serves as the nitrogenous compound for storage in the seeds, both in protein bound form and in free form. Arginine also serves as a precursor of many alkaloids and signaling molecule—NO (nitric oxide). Synthesis of arginine requires acetyl CoA along with glutamate. Acetylation of α-amino group of glutamate protects it from cyclization. Three more amino groups are added from glutamate, **carbamyl phosphate,** and aspartate, resulting in synthesis of arginine. Prior to arginine, ornithine is produced, which is a non-protein amino acid. In animals, arginine is the immediate precursor of urea. So, it is important in urea cycle.

**Aspartate-Derived Amino Acids** Amino acids included in this category are lysine, threonine, isoleucine, and methionine. Since these amino acids are not synthesized in animals, these are required in their diet. Some of the foods, such as corn, are poor in lysine, methionine, and tryptophan, while soybean is rich in lysine. Corn can be supplemented with soybean even for the grain-fed animals for their better growth. Initial steps in the biosynthesis require ATP and NADPH. There is a bifurcation of the pathway after intermediate aspartate semialdehyde is synthesized. For biosynthesis of lysine, pyruvate is required. For synthesis of threonine and methionine, NADH and ATP are consumed. Synthesis of methionine is sulfur requiring and is a complicated multistep process. All steps of aspartate-derived amino acids, including methionine, are known to occur in plastids. Methionine occupies central position in sulfur metabolism. It is entirely obtained by the mammals from their diet. Carbon skeleton of the methionine is provided by aspartate, methyl group comes from β carbon of serine, and sulfur comes from cysteine. Methionine has two functions. It is the structural component of proteins and is required for generation of S-adenosylmethionine (SAM). It is of considerable interest for the plant scientists to produce plants which are able to synthesize more lysine. Generally, lysine accumulation leads to inhibition of its biosynthesis because of feedback regulation of the enzyme dihydrodipicolinate synthase. Plants with mutant dihydrodipicolinate synthase have been produced in which enzyme is insensitive for lysine accumulation. Another strategy adopted to produce transgenic plants is by incorporating bacterial dihydrodipicolinate synthase in plants, since the bacterial enzyme is less sensitive to lysine accumulation. However, only a small amount of lysine is increased, which is incorporated in the proteins because free lysine gets degraded. So, protecting lysine from degradation can be a better strategy, which has been tested in *Arabidopsis*.

**Aromatic Amino Acids** Synthesis of aromatic amino acids occurs in plastids. This category of amino acids includes tryptophan, phenylalanine, and tyrosine. Precursors for biosynthesis of aromatic amino acids are phosphoenolpyruvate and erythrose 4-phosphate. Shikimate is formed as an intermediate, so the pathway is known as shikimate pathway. Pathway is common till the formation of an intermediate chorismate. Subsequent to chorismate formation, the pathway branches off. Enzymes of this pathway are located in plastids and have not been found in animals, so inhibitors for these enzymes would be potential antibiotics against animal pathogens. Besides synthesis of aromatic amino acids, numerous other aromatic compounds (plant hormones, e.g., auxins, salicylic acid; pigments, e.g., anthocyanins; signal molecules, e.g., isoflavonoids and many others) are also synthesized through this pathway. Products of the shikimate pathway can be as high as up to 50% of the total dry matter of the plant; therefore it is regarded as one of the major pathways.

**Branched-Chain Amino Acids** Isoleucine, leucine, and valine are included in this category. These amino acids represent three of the ten essential amino acids and are nutritionally important. Two molecules of pyruvate (precursor) are converted to α-acetolactate. The reaction is catalyzed by the enzyme acetolactate synthase, which requires thiamine pyrophosphate (TPP) as the cofactor for this reaction. α-Acetolactate is reduced to valine using NADPH, a dehydration reaction and transamination by glutamate. Branching of pathway results in the synthesis of leucine. Isoleucine is synthesized from threonine. Biosynthetic pathway of these amino acids exists in chloroplasts.

**Histidine** The pathway for histidine biosynthesis has been completely worked out in plants. Eight enzymes are involved in the biosynthesis of histidine. Complete pathway for histidine biosynthesis occurs in the plastids. Precursors for biosynthesis of histidine are ribose-5-phosphate and ATP.

**Alanine** Alanine is derived from pyruvate, an intermediate of glycolytic pathway, by a single transamination step.

**Glycine and Serine** The major route for the biosynthesis of these amino acids is during photorespiration. Serine may also be synthesized by an alternate route as well, i.e., 3-phosphoglycerate-dependent synthesis in chloroplasts.

**Cysteine** Mammals are not able to reduce sulfates, so they are dependent on the plants for the supply of sulfur-containing amino acids: methionine and cysteine. Cysteine is required for maintaining protein structure. Sulfur-containing R groups of cysteine can be easily converted from thiol (-SH) to the disulfide bond formed in between two cysteine amino acids (-S-S-) on being oxidized and vice versa. So, it plays an important role in maintaining protein structures. In plants, enzymes for cysteine biosynthesis are present in the cytosol, plastids, and mitochondria. This might be due to the possibility of cysteine being unable to cross the cell membranes. Cysteine is synthesized from serine, which first gets acetylated followed by addition of the thiol (-SH) group, which is coupled with the removal of acetate group. Methionine is synthesized from cysteine.

**Non-protein Amino Acids** Various amino acids are present in free form and are not bound to form the proteins. In legume seeds, many of these types are present, which function in chemical defense against predation and disease. Toxic action of many of the non-protein amino acids may be through a range of their actions, e.g., through inhibiting catalytic activity of the key enzymes. One of the examples of non-protein amino acids is γ-aminobutyric acid (GABA) which is produced in response to abiotic and biotic stress.

Nucleotide biosynthesis steps are similar in all the organisms. This includes purine and pyrimidine biosynthesis. Purine biosynthesis requires 5-phosphoribosyl 1-pyrophosphate (PRPP). Nitrogen in the purines is furnished by the glutamine, glycine, and aspartate. Pyrimidine biosynthesis requires carbamoyl phosphate and aspartate. Ribose-5-phosphate is attached afterward yielding pyrimidine

## 11.5 Amino Acid Biosynthesis

ribonucleotides. These are phosphorylated to form triphosphates. Ribonucleotides are converted to deoxyribonucleotide by the action of an enzyme known as ribonucleotide reductase.

**Summary**

- Nitrogen is an essential element for the plants. Many biomolecules in plants such as proteins, nucleic acids, plant growth regulators, and plant pigments contain nitrogen as their constituent.
- In nature, nitrogen is present either in free form as molecular nitrogen in the air or is present in the soil in bound form, such as nitrates and ammonium or as organic forms derived from dead plants or animals and animal excretion. Nitrogen is cycled in nature since all these forms of nitrogen are interconvertible due to microbial action.
- Plants are able to absorb nitrogen from the soil as nitrates or ammonium. Absorption of nitrate and ammonium ions is facilitated by the receptors present on plasma membrane of the root cells. Excess nitrates can be stored in the vacuoles, while excess ammonium, if not metabolized, becomes toxic for the plants. Nitrate can also be translocated to leaves and is metabolized there.
- Nitrate needs to be reduced in the plant cells before it is further metabolized. There are two enzymes, i.e., nitrate reductase and nitrite reductase, which catalyze reduction of nitrate to ammonium in cytosol and plastids of the plant cell, respectively. Nitrate reductase is an inducible enzyme, while nitrite reductase is a constitutive enzyme. The latter is present in much more amount so that there is no accumulation of nitrite in the cell which is otherwise toxic.
- Molecular nitrogen present in the air can be fixed by some of the prokaryotes, which are called *diazotrophs*. *Diazotrophs* can either be free living or grow in symbiotic associations with plants. Two most important nitrogen-fixing symbiotic associations are *Rhizobium*, legume, and *Frankia*, actinorhizal.
- A great degree of specificity exists in between the specific strain of rhizobia and the leguminous plant, which is due to the interaction of biochemical signals produced by the plants in the form of flavonoids, and the Nod factors of rhizobia. Specific Nod factors produced by the rhizobia species are identified by the Nod receptors present on the plasma membrane of the root hair. This is followed by the formation of an infection thread and subsequently nodules. Effective symbiosis is established in the form of symbiosomes due to the involvement of both rhizobial and host genes.
- Bacteroids are the nitrogen-fixing forms of rhizobia because of nitrogenase being synthesized by them, which is responsible for the reduction of molecular nitrogen to ammonium ions. Nitrogen-fixing genes of the bacteria, i.e., *nif* genes, are involved in the synthesis of nitrogenase.
- Ammonium ions, either absorbed by the plants or produced as a result of nitrate reduction or nitrogen fixation, need to be assimilated which is carried out by glutamine synthetase and 2-oxoglutarate aminotransferase, which are collectively abbreviated as GS/GOGAT.

- Nitrogen is transported either in the form of glutamine or asparagine. There are however some legumes in which nitrogen is translocated in the form of ureides. It is the ratio of nitrogen to carbon in the molecules which is significant for the nitrogen to be translocated. Ureides have the highest N/C ratio.
- Plants and bacteria can synthesize all the 20 amino acids, while animals are able to synthesize only about half of them since some of the enzymes required for the biosynthesis of these amino acids do not exist in animal systems. Since, plants, fungi, and bacteria are able to synthesize all the 20 amino acids, there is potential for developing new herbicides, antibiotics, and antifungal medicines by selectively blocking the pathway for the synthesis of these essential amino acids in weeds, bacteria, and fungi, respectively.

## Multiple Choice Questions

1. The process in which $NO_3^-$ serves as electron acceptor in place of $O_2$, is known as:
   (a) Ammonification
   (b) Nitrate reduction
   (c) Nitrate respiration
   (d) Denitrification
2. Nitrate uptake into roots of the plants from soil is facilitated by:
   (a) Uniporters
   (b) $2H^+/NO_3^-$ symporters
   (c) $2H^+/NO_3^-$ antiporters
   (d) Nitrate transporters belonging to chloride channel family
3. During reduction by nitrate reductase, electrons are passed to $NO_3^-$:
   (a) From heme cofactor of the enzyme
   (b) From FAD cofactor at C-terminal of the enzyme
   (c) From reduced molybdenum of MoCo at the C-terminal of the enzyme
   (d) From reduced molybdenum of MoCo at the N-terminal of the enzyme
4. Electrons for $NO_2^-$ reduction are obtained from:
   (a) NADH
   (b) Reduced ferredoxin
   (c) NADPH
   (d) Reduced glutathione
5. An example of symbiotic association of non-leguminous plants with rhizobium is:
   (a) *Gunnera*
   (b) *Anthoceros*
   (c) *Casuarina*
   (d) *Parasponia*

6. Chemical nature of Nod factors is:
   (a) Lipoproteins
   (b) Oligosaccharides
   (c) Lipochitooligosaccharides
   (d) Oligopeptides
7. The term symbiosome is used for depicting:
   (a) Vesicular structure formed in the infected cells of nodules
   (b) The symbiotic relationship in between the endosymbiont and host plant
   (c) The infection thread formed at the time of nodulation
   (d) Infected roots cells of nodules in legumes
8. Site for $N_2$ reduction in nitrogen-fixing enzyme is present in:
   (a) Dinitrogenase reductase
   (b) P cluster of dinitrogenase
   (c) M cluster of dinitrogenase
   (d) O-side chain of serine from the peptide backbone of P cluster
9. $N_2$ fixed by bacteroids is released in the cytosol of the infected cell as:
   (a) $NH_3$
   (b) $NH_4^+$
   (c) Glutamine
   (d) Ureides
10. Which of the following statements is true?
    (a) *nifD* and *nifK* encode for nodulins.
    (b) Fe protein is encoded by *nifF*.
    (c) The *nif* genes are required for nitrogen fixation only by the symbiotic bacteria and not by free-living nitrogen fixers.
    (d) *fix* genes are essential for nitrogen fixation in symbiotic nitrogen fixers but do not have counterparts in free-living forms.

## Answers

1. c   2. b   3. d   4. b   5. d   6. c
7. a   8. c   9. b   10. d

## Suggested Further Readings

Browsher C, Steer M, Tobin A (eds) (2008) Plant biochemistry. Garland Science, Tailor & Francis Group, New York, pp 237–300

Heldt HW (2005) Plant biochemistry, 3rd edn. Elsevier Academic Press, Cambridge, pp 275–302

Long SR, Kahn M, Seefeldt L, Tsay Y, Kopriva S (2015) Nitrogen and Sulfur. In: Buchanan BB, Gruissem W, Jones RL (eds) Biochemistry and molecular biology of plants. Wiley-Blackwell, Chichester, pp 711–745

Jones R, Ougham H, Thomas H, Waaland S (2013) The molecular life of plants. Wiley-Blackwell, Chichester, pp 463–475

Taiz L, Zieger E, Moller IM, Murphy A (2015) Plant physiology and development, 6th edn. Sinauer Associates, Inc. Publishers, Sunderland, pp 354–367

# Sulfur, Phosphorus, and Iron Metabolism in Plants

Manju A. Lal

## 12.1 Sulfur Metabolism

Sulfur is an essential element for plant growth. Though it constitutes only 0.1% of the dry weight of a plant, its requirement for the plant is crucial. It is a constituent of sulfur-containing amino acids, cysteine, and methionine which are integral to the protein structure. Cysteine residues are responsible for holding proteins in proper conformation because of disulfide linkages (-S-S-) between two –SH containing amino acids. Iron-sulfur (4Fe-4S) clusters present in various proteins are engaged in electron transport reactions. Sulfur is also a constituent of a number of molecules such as lipoic acid, thiamin, biotin, ACP, and coenzyme A, which are required as cofactors by various enzymes. Sulfur-containing lipids, sulfoquinovosyldiacylglycerol, are structural constituents of thylakoids. Various secondary metabolites produced from cysteine and methionine have diverse roles in plants. Many molecules synthesized by the plants in response to abiotic and biotic stress contain sulfur. These include **phytoalexins,** thioredoxin, **alliins, glucosinolates**, etc. Alliins are found in onion and garlic, while glucosinolates are found in members of family Brassicaceae and are responsible for their flavor and smell. In some plants, elemental sulfur is deposited which functions as a potent fungicide (Fig. 12.1).

Sulfur is available to plants in oxidized form as sulfate ions from soil and/or as a pollutant from air in the form of $SO_2$. Plants absorb sulfate through their roots as well as through the stomatal chambers in the leaves. Once absorbed by roots, sulfate is transported to plastids where it is metabolized or translocated to leaves where sulfur metabolism is significant and occurs in chloroplasts. Extra sulfate ions may be stored in vacuoles. Metabolism of sulfur includes reduction of sulfate ions to sulfide before its incorporation into cysteine, which can further be converted to methionine. Various other sulfur-containing compounds are synthesized from cysteine (Fig. 12.2).

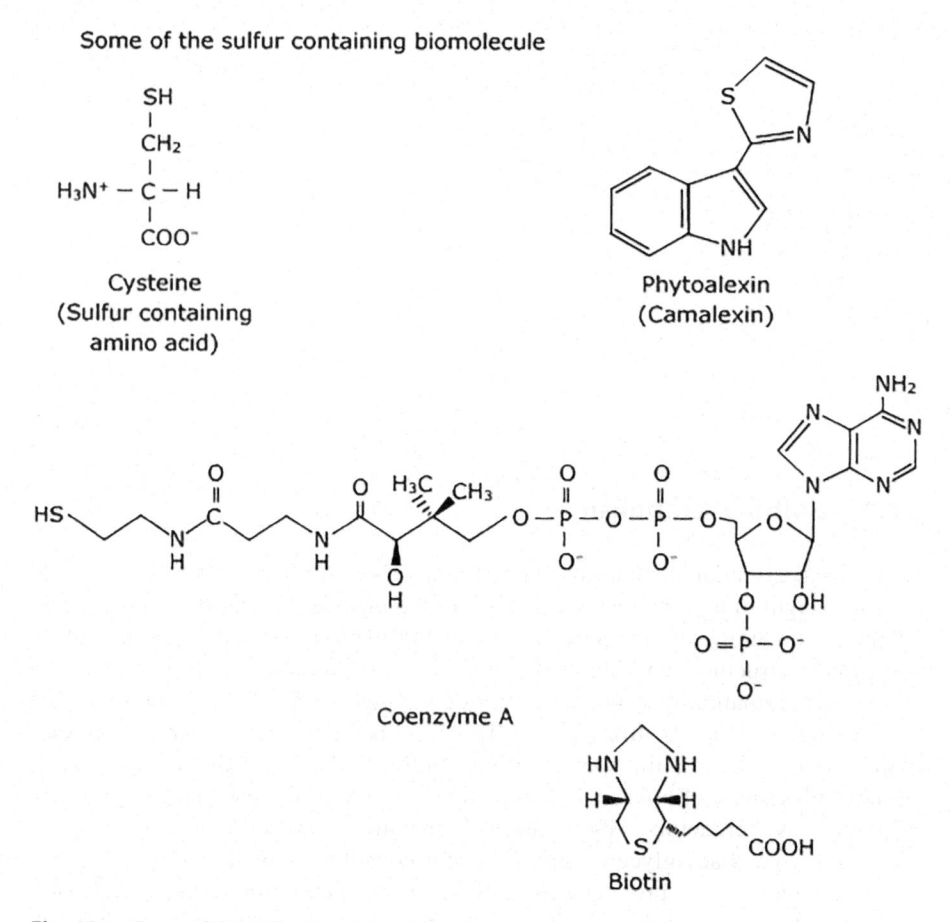

**Fig. 12.1** Some of the sulfur-containing biomolecules

## 12.1.1 Biogeochemical Cycle of Sulfur

In nature, sulfur is present both in oxidized and reduced forms. Interconversion of these forms constitutes the **biogeochemical cycle**. Oxidized forms of sulfur include sulfates present in the soil, which are absorbed by the plants. Once sulfates gain entry into the plants, they are reduced to sulfides which are metabolized further or lead to cysteine biosynthesis. Sulfides also act as precursor for the biosynthesis of other biomolecules. Some bacteria also reduce $SO_4^{2-}$. Reduced form of sulfur is oxidized during animal metabolism or by some of the chemoautotrophs, which use reduced sulfur either as $H_2S$, sulfur, or $SO_3^-$ as the terminal electron acceptors in place of oxygen. Since animals, including human, are not able to reduce sulfur, it needs to be provided in their diet. Since plants are the major source for providing reduced sulfur to animals, studying sulfur metabolism in plants is significant. Plants also absorb

## 12.1 Sulfur Metabolism

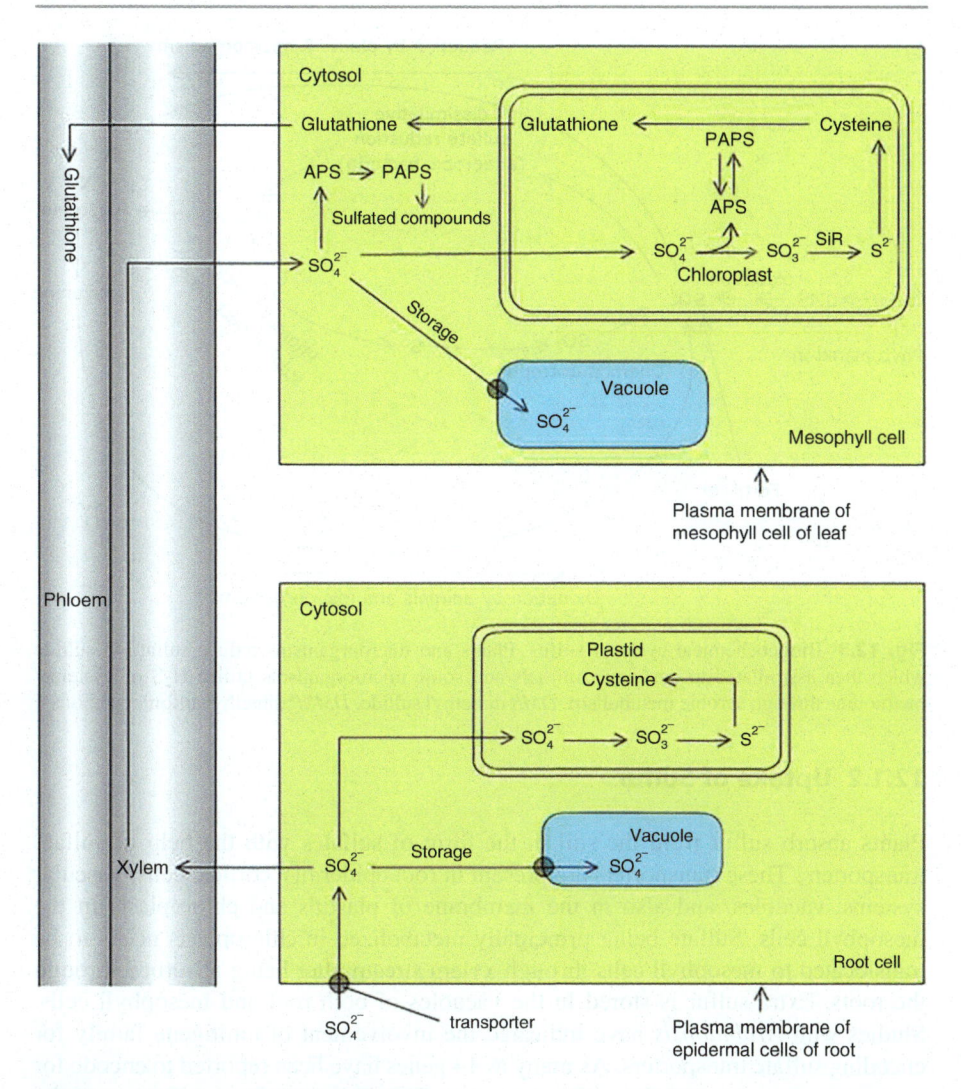

**Fig. 12.2** Overview of sulfur metabolism

sulfur in the form of $SO_2$ from the air where it accumulates because of the volcanic activity or due to burning of fossil fuels. The major source of sulfur in plants is through soil in the form of sulfate fertilizers. Phytoplanktons also produce **dimethylsulfoniopropionate (DMSP)** which functions as an osmoprotectant and a cryoprotectant. After DMSP is released from algae, it is converted to **dimethyl sufide (DMS)**, which is a volatile sulfur-containing compound released into the atmosphere. DMS is oxidized to **DMSO (dimethyl sulfoxide)**, sulfite, and sulfate (Fig. 12.3).

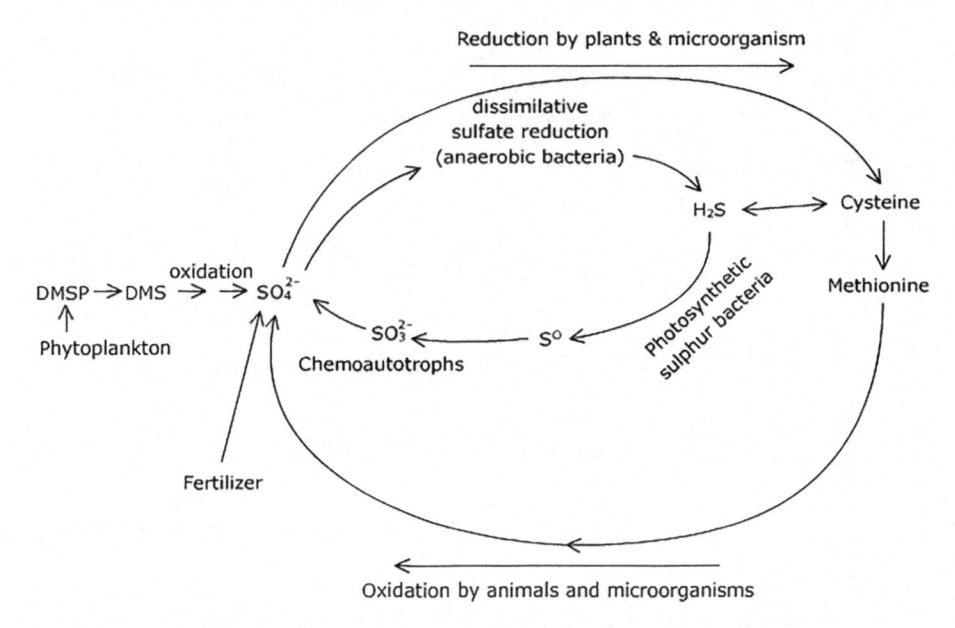

**Fig. 12.3** Biogeochemical cycle of sulfur. Plants and microorganisms reduce sulfate to sulfide which then assimilated to cysteine. Animals and some microorganisms utilize H₂S or cysteine/methionine through aerobic metabolism. *DMS* dimethyl sulfide, *DMSP* dimethylsulfoniopropionate

### 12.1.2 Uptake of Sulfur

Plants absorb sulfur from the soil in the form of sulfates with the help of sulfate transporters. These transporters are present in root epidermis, cortical cells, vascular systems, vacuoles, and also in the membrane of plastids and chloroplasts in the mesophyll cells. Sulfate being principally metabolized in chloroplasts needs to be translocated to mesophyll cells through xylem stream after being absorbed through the roots. Extra sulfur is stored in the vacuoles of both root and mesophyll cells. Studies with *Arabidopsis* have indicated the involvement of multigene family for encoding sulfate transporters. As many as 14 genes have been reported to encode for the transporters required for sulfate transport. This indicates the significance of the sulfate uptake. Different transporters are expressed in various compartments of the plant cell involved in accumulation of sulfates from micromolar to millimolar concentration. These transporters belong to different groups which include high-affinity sulfate transporters (operate in the range of 1–7 μM range) and low-affinity transporters (operate in the range of 100 μM to 1.2 mM sulfate). In plasma membrane of root epidermal cells, high-affinity $SO_4^{2-}$ cotransporters are also present which are $H^+/SO_4^{2-}$ symporters. These transporters accumulate sulfate against electrochemical gradient and are powered by ATPase-mediated proton pumps in the plasma membrane. Movement of $3H^+$ occurs in response to proton motive force (PMF), which is coupled with the transport of one $SO_4^{2-}$. SULTR-2 has been shown to be the major sulfate transporter in *Arabidopsis* and, it is located in plasma

## 12.1 Sulfur Metabolism

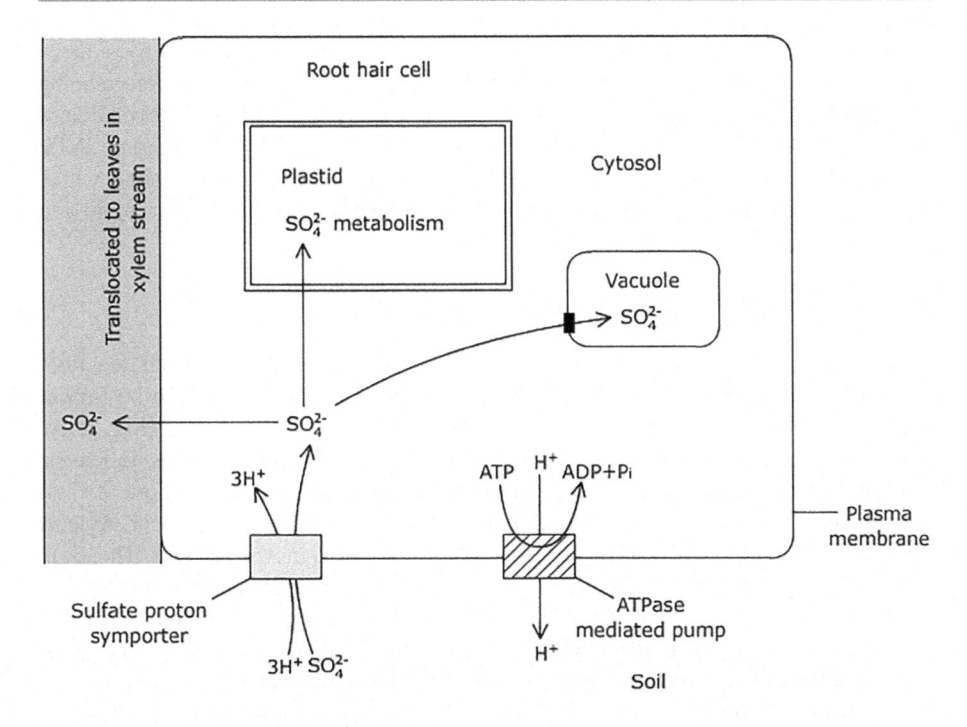

**Fig. 12.4** Sulfate uptake by root cells

membrane of root epidermis. It consists of 500–700 amino acids having 10–12 predicted transmembrane helices. Low-affinity transporters are found to function in the vascular system of plants. These are involved in translocation of sulfate within the plant. Unlike plasma membrane-localized sulfate transporters, electrochemical gradient is thought to drive diffusion of sulfate into vacuoles through sulfate-specific channels present on the tonoplast membranes. However, in vacuoles, there seems to be an involvement of uniporters. Involvement of transporters similar to those of plasma membrane might be possible in plastids (Fig. 12.4).

### 12.1.3 Sulfate Metabolism

Almost 80–85% sulfate accumulated in roots is reduced and incorporated into cysteine, mainly in chloroplasts. Translocation of sulfate to leaves occurs through xylem. Similar to nitrate reduction, sulfate is also reduced in two steps. First step involves reduction of sulfate ($SO_4^{2-}$) to sulfite ($SO_3^{2-}$), while sulfite is reduced to sulfide ($S^{2-}$) before being incorporated into cysteine which is metabolized further.

$$SO_4^{2-} \rightarrow SO_3^{2-} \rightarrow S^{2-} \rightarrow \text{Cysteine}$$

Reduction of sulfate to sulfite requires two electrons transfer provided by reduced glutathione. Being relatively inert molecule, sulfate requires activation before being reduced to sulfite. Sulfate is activated by reacting with ATP, resulting in formation of **adenosine 5′-phosphosulfur** (APS). Reaction is catalyzed by the enzyme **ATP sulfurylase**. APS contains a phosphoric acid-sulfuric acid anhydride bond which makes $SO_4^{2-}$ reduction possible. Free energy of the reaction ($\Delta G^{0\prime}$) is estimated to be +41.8 kJ.mol$^{-1}$.

$$SO_4^{2-} + Mg.ATP \rightarrow Mg.PP_i + APS$$

Forward reaction is possible with removal of the products of the reaction. Pyrophosphatase hydrolyzes $PP_i$, while APS is at branch point and can be further metabolized by either of the two enzymes APS kinase or APS reductase (Fig. 12.5). APS sulfurylase is a homotetramer consisting of 52–54 kDa polypeptides. In leaves, it exists in two isoforms. Enzyme found in the leaf plastids represent about 90% of the total enzyme, while the one found in the cytosol is the minor component. At least three genes encoding APS sulfurylase have been identified in *Arabidopsis*. These are APS1, APS3, and APS4, while the cytosolic isoform is encoded by the fourth gene-APS2. The enzyme binds to $Mg.ATP^{2-}$ and $SO_4^{2-}$ sequentially. Demand of the end products might regulate activity of the enzyme since expression of the genes encoding the enzyme is downregulated by the addition of reduced glutathione.

Adenosine 5′-phosphosulfur (APS) is a high-energy compound due to the presence of high-energy phosphoric acid-sulfuric acid anhydride bond. It is because of this high-energy bond that it facilitates further metabolism of $SO_4^{2-}$ to $SO_3^{2-}$. There are two alternative pathways for APS. Sulfur present in APS in the form of $SO_4^{2-}$ is either reduced to $SO_3^{2-}$ by the enzyme **APS reductase**, or APS is converted to 3′-phosphoadenosine 5′-phosphosulfate (PAPS) through the activity of **APS kinase**. Although PAPS acts as a sulfur donor in plants, however in some organisms such as cyanobacteria, PAPS can be reduced to sulfite or may also provide precursor for cysteine biosynthesis. Reaction catalyzed by APS kinase is as follows:

$$APS + ATP \rightarrow PAPS + ADP$$

Three reactions are involved in $SO_4^{2-}$ reduction. First APS is reduced to sulfite, catalyzed by APS reductase. The reaction occurs as follows:

$$APS + 2glutathione_{red} \rightarrow SO_3^{2-} + 2glutathione_{ox} + AMP + 2H^+$$

$SO_3^{2-}$ is reduced to sulfide ($S^{2-}$) which is followed by formation of cysteine. Reduction of APS to sulfite by APS reductase is unique in plants and occurs in plastids. **Glutathione** functions as a reductant for this reaction. Multiple isoforms of APS reductase are expressed in plants in response to environmental signals such as the availability of reduced sulfur. Three genes in *Arabidopsis* encode for APS

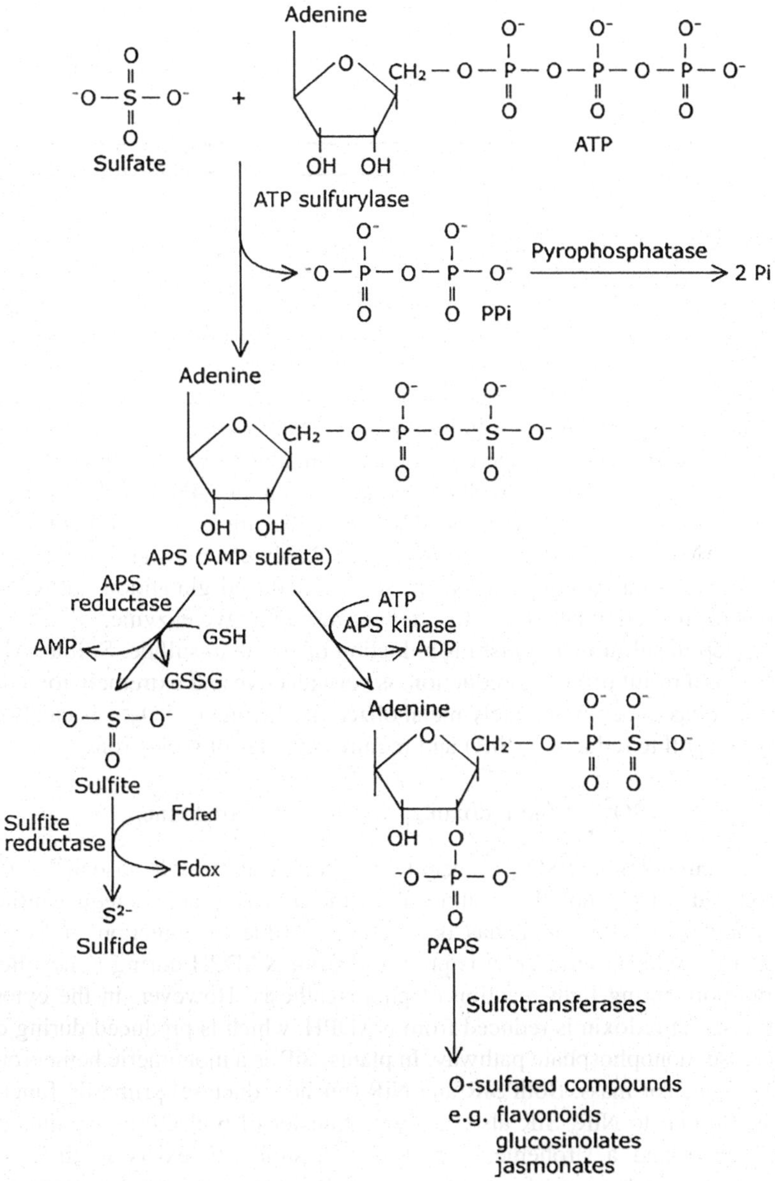

## 12.1 Sulfur Metabolism

**Fig. 12.5** Reduction of sulfate to sulfide and synthesis of some of the sulfur-containing compounds. *APS* adenosine-5'-phosphosulfates, *PAPS* 3'-phosphoadenosine-5'-phosphosulfate

reductase. In the presence of reduced sulfur, expression of the gene encoding this enzyme is downregulated, resulting in the accumulation of $SO_4^{2-}$. However, overexpression of the gene also results in accumulation of reduced sulfur which may cause cellular damage. The enzyme is synthesized in cytosol by cytosolic

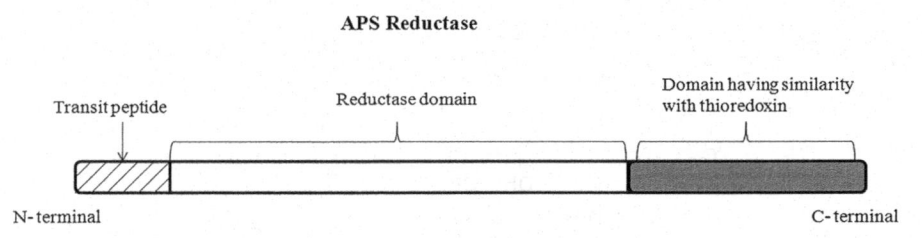

**Fig. 12.6** Structure of APS reductase

ribosomes, and the presence of transit peptide at the N-terminus of the polypeptide facilitates its movement into the plastid. The mature enzyme is a homodimer consisting of 43 kDa monomeric subunits. It is characterized by the presence of a C-terminal redox domain and a N-terminal reductase domain (Fig. 12.6). Reaction between the redox region of the enzyme and glutathione is the source of electrons used for reduction of $SO_4^{2-}$ to $SO_3^{2-}$. Enzyme contains 4Fe-4S cluster at the N-terminal-localized reductase region. $SO_4^{2-}$ binds to this region with simultaneous release of AMP and its reduction to $SO_3^{2-}$. Probably electrons are provided by the reductase with simultaneous release of $SO_3^{2-}$. Reduced glutathione provides two electrons for reduction necessary for the recovery of active enzyme.

Next step in sulfur metabolism is reduction of sulfite to sulfide. Sulfite, which is produced as a result of $SO_4^{2-}$ reduction, is very reactive and extremely toxic for the plants, so it has to be immediately metabolized. Reduction of $SO_3^{2-}$ is catalyzed by the enzyme sulfite reductase (SiR) and requires transfer of 6 electrons.

$$SO_3^{2-} + 6 \text{ferredoxin}_{red} \rightarrow S^{2-} + \text{ferredoxin}_{ox}$$

Since accumulation of $SO_3^{2-}$ is very toxic, level of sulfite reductase in the tissue is half saturated at $10^{-6}$ mole/L of sulfite. Thus, the enzyme is able to reduce efficiently any amount of sulfite which has been produced due to reduction of $SO_4^{2-}$. In chloroplasts, reduced ferredoxin is produced from NADPH during noncyclic electron transport during light reaction of photosynthesis. However, in the cytosol or chloroplasts, ferredoxin is reduced from NADPH, which is produced during oxidative pentose monophosphate pathway. In plants, SiR is a monomeric hemeprotein of 65-kDa molecular mass. Both SiR and NiR (nitrite reductase) primarily function in plastids. Similar to NiR, SiR also catalyzes transfer of 6 electrons facilitated by a $Fe_4$-$S_4$ center and a siroheme (Fig. 12.7). A similarity exists in their tertiary structures and mechanism of action as well. SiR isolated and purified from spinach is capable of reducing nitrite, though with lower efficiency. Both enzymes show about 20% similarity in their amino acids composition, and probably a primitive bifunctional reductase might have been a common ancestor for both the enzymes.

Detoxification of sulfite also occurs due to its oxidation by **sulfite oxidase** which protects plants against surplus $SO_3^{2-}$. Though this enzyme has been reported earlier

## 12.1 Sulfur Metabolism

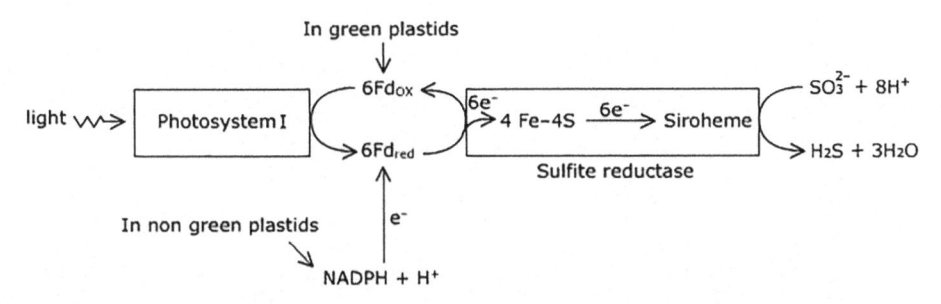

**Fig. 12.7** Sulfite reductase: sulfite is reduced to sulfide in chloroplasts. Reduced ferredoxin produced during light reactions is used as the reductant

both in animals and microorganisms, its role in plants is not very clear. Reports about sulfite oxidase in plants are comparatively recent. Reaction catalyzed by this enzyme is as follows:

$$SO_3^{2-} + H_2O + O_2 \rightarrow SO_4^{2-} + H_2O_2$$

Conversion of $SO_3^{2-}$ back to $SO_4^{2-}$ is **futile cycle** and is prevented by compartmentalization. Both pathways including reduction of $SO_4^{2-}$ to $SO_3^{2-}$ and oxidation of $SO_3^{2-}$ to $SO_4^{2-}$ occur in different compartments of the cell. Sulfite reduction occurs in plastids, while oxidation of sulfite occurs in peroxisomes where $H_2O_2$ produced can be degraded. Sulfite oxidase belongs to family of molybdenum-containing enzymes.

Final step in reductive sulfate assimilation is biosynthesis of cysteine from sulfide ($S^{2-}$). Precursors required are O-acetylserine (OAS) and sulfide ($S^{2-}$). Sulfide exists as bisulfide (HS) at physiological pH. O-acetylserine is produced from serine and acetyl-CoA. Acetylation of serine is catalyzed by **serine-acetyl-CoA transferase (SAT)**.

$$\text{Serine} + \text{acetyl} - \text{CoA} \rightarrow \text{O} - \text{acetylserine} + \text{CoA}$$

Formation of acety-CoA from acetate and CoA requires ATP, and the reaction is catalyzed by the enzyme **acetyl-CoA synthetase** present in chloroplasts. Pyrophosphate released in the process is hydrolyzed by pyrophosphatase. Activation of serine to O-acetylserine costs two ATPs in chloroplasts. SAT is a dimer of two trimers. Each trimer consists of 29-kDa monomers. There are six active sites in the hexamer, and a histidine residue is present in each active site which can bind to hydroxyl group of serine and activate it to bind with the acetyl group from acetyl-CoA. Three isoforms of SAT have been isolated which differ in regulation of their activity, especially in their sensitivity to inhibition by cysteine accumulation. The enzyme is indispensable for plant growth, and the mutants for this enzyme results in growth

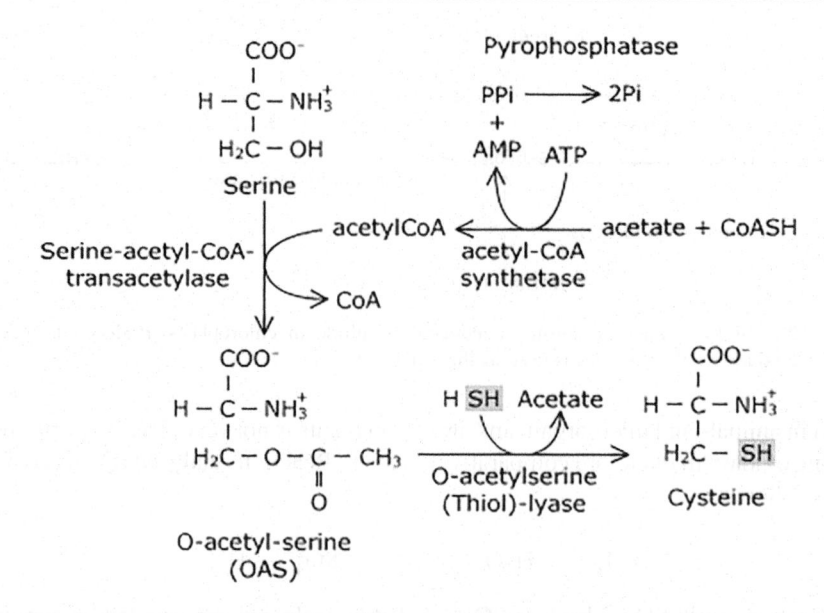

**Fig. 12.8** Cysteine biosynthesis from serine using the hydrogen sulfide produced as a result of sulfite reduction

retardation. Subsequent fixation of $H_2S$ in the form of cysteine from O-acetylserine requires the enzyme **O-acetylserine (thiol) lyase**. The enzyme requires pyridoxal phosphate as the prosthetic group. It has high affinity both for $H_2S$ and O-acetylserine. Ester linkage of O-acetylserine is cleaved, and SH group from $H_2S$ is incorporated with simultaneous release of acetyl unit and cysteine biosynthesis (Fig. 12.8). Cysteine biosynthesis primarily occurs in cytosol. O-acetylserine is synthesized in mitochondria, and sulfide production occurs in the plastids. OAS and OAS (thiol) lyase form a complex called as **cysteine synthase**. Active complex consists of a hexamer of OAS and two dimers of OAS (thiol) lyase. Protein-protein interaction is responsible for regulation of activity of the enzyme complex in response to concentration of OAS and sulfide. OAS may act as the signal for fine-tuning of the pathway involved.

### 12.1.4 Cysteine Metabolism

Cysteine is an important sulfur-containing amino acid since it readily undergoes dithiol (-SH-HS-) and disulfide (-S-S-) interconversion similar to glutathione molecule. It plays an important role in protein structure if the polypeptide contains two cysteine residues. Protein folding is promoted if two cysteine residues are covalently

## 12.1 Sulfur Metabolism

linked to each other by disulfide (-S-S-) bonds under oxidative cellular conditions, while reducing conditions will promote unfolding due to reduction of disulfide bonds to sulfhydryl (-SH) groups. Cysteine is also a precursor of various biomolecules, such as methionine (another sulfur-containing amino acid), glutathione, glucosinolates, etc. Besides being structural component of proteins, methionine has many other important roles in plant metabolism. It is the methyl group donor for lignin biosynthesis. And its derivative, S-adenosyl methionine, is the precursor for ethylene biosynthesis. Besides crop plants, many legumes also have low methionine content. So, enriching the plant dietary sources with these sulfur-containing amino acids is one of the goals of the biotechnologists. Methionine is synthesized from cysteine. Precursor for methionine biosynthesis is homoserine which is produced as a result of reduction of aspartic acid. Phosphorylation of homoserine results in the production of O-phosphohomoserine. The phosphate group is displaced by sulfhydryl group of cysteine, resulting in the formation of cystathionine, followed by cleavage of alanine group to yield homocysteine. Cofactor methyl-tetrahydrofolate of the enzyme methyl transferase is responsible for methylation of sulfhydryl group of homocysteine resulting in the formation of methionine (Fig. 12.9).

**Glutathione (GSH)** is a tripeptide which plays an important role as cellular redox buffer. It acts as an antioxidant since it provides protection to cellular constituents against reactive oxygen species (ROS) together with ascorbate. It also provides protection against toxicity by heavy metals since it is required for formation of **phytochelatins** (Box 12.1), herbicides, and **xenobiotic** by forming complexes with them. It consists of amino acid residues glutamate, cysteine, and glycine in which γ-carboxylic group of glutamate is linked to amino group of cysteine which forms peptide bond with glycine. When in oxidized state, thiol groups of cysteine amino acids present in two glutathione molecules are linked together to form disulfide (-S-S-) bond, while under reduced conditions, these exist as thiol groups (-SH). Conversion of oxidized glutathione to reduced form is catalyzed by a flavoprotein, **glutathione reductase** which requires NADPH (Fig. 12.10). Oxidized form of glutathione is present in micromolar concentrations, predominantly in endoplasmic reticulum and Golgi apparatus, while reduced form is present in millimolar concentrations. Glutathione biosynthesis requires two enzymes and three amino acid precursors. **γ-glutamyl-cysteine synthetase** (the enzyme found in plastids) catalyzes an ATP requiring reaction by forming an amide linkage between γ-carboxylic group of glutamic acid and amino group of cysteine. A second peptide bond is formed between carboxylic group of cysteine and amino group of glycine. This bond formation requires another molecule of ATP and is catalyzed by the enzyme **glutathione synthetase** (Fig. 12.11). Isoforms of the enzyme are found both in plastids and cytoplasm. Glutathione also acts as the reservoir for organic sulfur, and cysteine is released free when required by the plant.

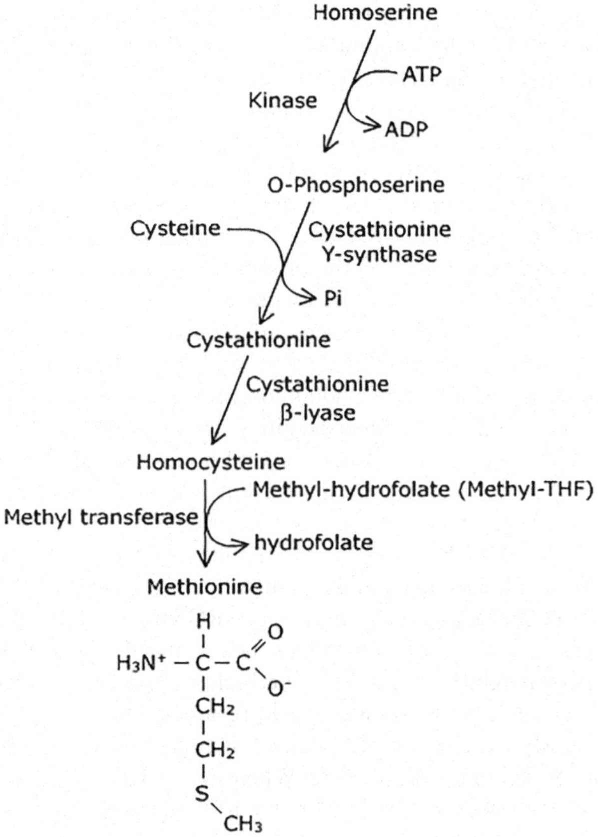

**Fig. 12.9** Biosynthesis of methionine from cysteine

```
                          Homoserine
                              |
                   Kinase    / ← ATP
                            ↓ → ADP

                       O-Phosphoserine
                            /
          Cysteine ──\    / Cystathionine
                      \  /  γ-synthase
                       ↓ → Pi
                       ↓
                     Cystathionine
                        /
                       / Cystathionine
                      /  β-lyase
                     ↓
                  Homocysteine
                       | ← Methyl-hydrofolate (Methyl-THF)
   Methyl transferase  |
                       ↓ → hydrofolate
                    Methionine

                        H    O
                        |   ‖
          H₃N⁺ — C — C
                   |     \
                   CH₂    O⁻
                   |
                   CH₂
                   |
                   S
                    \
                     CH₃
```

---

**Box 12.1: Glutathione as a Precursor of Phytochelatins**

Phytochelatins are oligomers of glutathione produced by the enzyme phytochelatin synthase, a transpeptidase which transfers free amino group of glutamate component of glutathione to carboxyl group of cysteine of another glutathione molecule. Reaction is accompanied by release of glycine and is repeated resulting in the formation of a chain of up to 11 repeats of dipeptide (Glu-Cys). Synthesis of enzyme phytochelatin synthase is triggered in

(continued)

## 12.1 Sulfur Metabolism

**Box 12.1** (continued)

response to heavy metals, such as $Zn^+$, $Cd^+$, $Pb^+$, $Hg^+$, etc. which is then responsible for de novo synthesis of phytochelatins. Heavy metals form tight complexes with the thiol group of cysteines of phytochelatin, which are then transported across tonoplast to the vacuoles. Transport of the phytochelatin-heavy metal complex requires ATP. Once inside the vacuoles, the complex breaks down releasing heavy metal which gets accumulated there. This property of phytochelatins has been exploited by certain plants effectively for phytoremediation. Since glutathione is the precursor for the biosynthesis of phytochelatins, glutathione level in plant falls down in response to heavy metals.

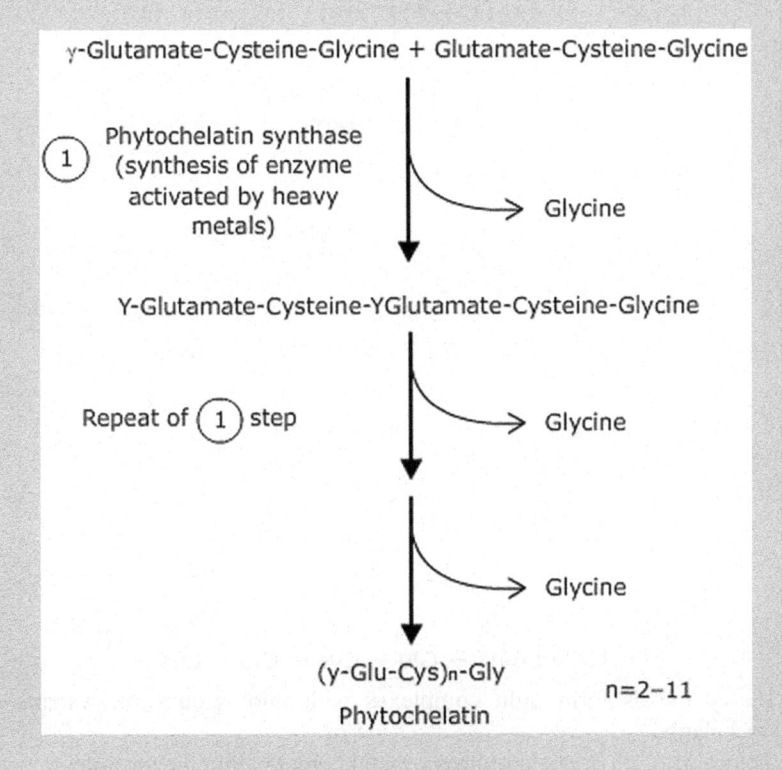

γ-Glutamate-Cysteine-Glycine + Glutamate-Cysteine-Glycine

(1) Phytochelatin synthase (synthesis of enzyme activated by heavy metals) → Glycine

Y-Glutamate-Cysteine-YGlutamate-Cysteine-Glycine

Repeat of (1) step → Glycine

→ Glycine

(γ-Glu-Cys)n-Gly    n=2–11
Phytochelatin

(continued)

**Box 12.1** (continued)

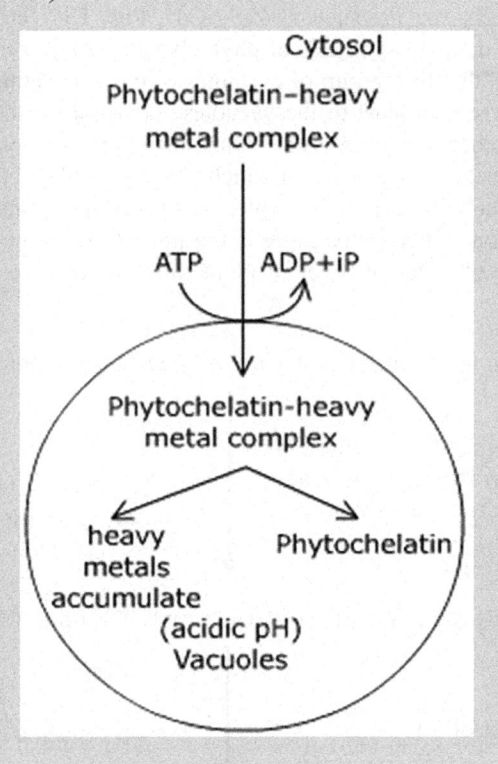

```
                        Cytosol
              Phytochelatin–heavy
                metal complex
                        │
              ATP    ADP+iP
                ╲      ╱
                 ╲    ╱
                  ▼
           Phytochelatin–heavy
              metal complex
              ↙         ↘
          heavy        Phytochelatin
          metals
        accumulate
          (acidic pH)
           Vacuoles
```

```
Phytochelatins — Glu — Cys — Glu — Cys —
                       │                │
                      HS ·.            S
                          ·.          ╱
                            ·. -Cd- ·
                           ╱          ·.
                          S             SH
                          │             │
Phytochelatins — Glu — Cys — Glu — Cys —
```

Heavy metals form tight complexes with thiol groups of cysteine of phytochelatins

Transport of phytochelatin-heavy metal complex into the vacuoles

### 12.1.5 Sulfated Compounds

Besides being used for sulfate reduction to sulfite, APS is also metabolized by alternate pathway. APS is phosphorylated by APS kinase resulting in production of *PAPS* (**adenosine-3′-phosphate-5′-phosphosulfate**) which is sulfuryl group

## 12.1 Sulfur Metabolism

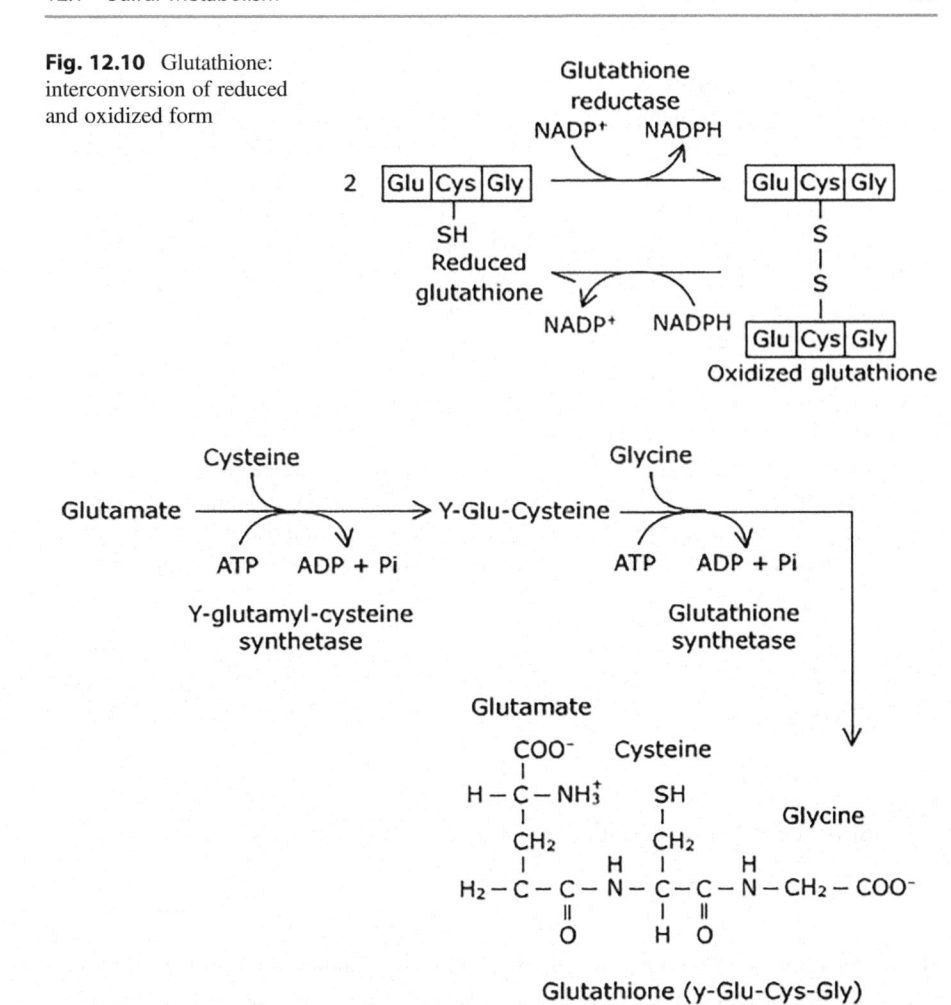

**Fig. 12.10** Glutathione: interconversion of reduced and oxidized form

$$2 \;\boxed{\text{Glu}|\text{Cys}|\text{Gly}} \xrightarrow[\text{NADP}^+ \;\; \text{NADPH}]{\text{Glutathione reductase}} \;\; \boxed{\begin{array}{c}\text{Glu}|\text{Cys}|\text{Gly}\\|\\S\\|\\S\\|\\\text{Glu}|\text{Cys}|\text{Gly}\end{array}}$$

Reduced glutathione (left); Oxidized glutathione (right). Reverse reaction uses NADP⁺ ← NADPH.

**Fig. 12.11** Glutathione biosynthesis

Glutamate + Cysteine → γ-Glu-Cysteine (ATP → ADP + Pi); enzyme: γ-glutamyl-cysteine synthetase

γ-Glu-Cysteine + Glycine → Glutathione (ATP → ADP + Pi); enzyme: Glutathione synthetase

Glutathione (γ-Glu-Cys-Gly) structure:

Glutamate — Cysteine — Glycine

$$\begin{array}{c}\text{COO}^-\\|\\\text{H}-\text{C}-\text{NH}_3^+\\|\\\text{CH}_2\\|\\\text{H}_2-\text{C}-\text{C}-\text{N}-\text{C}-\text{C}-\text{N}-\text{CH}_2-\text{COO}^-\\\phantom{xx}\|\phantom{xxx}|\phantom{x}\|\\\phantom{xx}\text{O}\phantom{xxx}\text{H}\phantom{x}\text{O}\end{array}$$

with SH on Cys CH₂ side chain and H on the N atoms.

donor for synthesis of various sulfated compounds required by plants for defense against abiotic and biotic stress, such as **glucosinolates** (Box 12.2) and sulfated derivatives of jasmonic acid, etc. (Fig. 12.12). Besides, another large group includes sulfated compounds which are medicinally important, such as alliin (Box 12.3). Some of the sulfated compounds are also involved in seismonastic movements in *Mimosa pudica*. Small sulfated peptides such as **phytosulfokines** are used as regulators of plant growth. APS kinase is present both in plastids and cytosol. Since many of the sulfated compounds produced from PAPS are required for stress responses, synthesis of APS kinase is believed to be regulated in response to stress.

> **Box 12.2: Glucosinolates**
> Glucosinolates are natural compounds present in the members of family Brassicaceae such as in mustard, cabbage, or horseradish. These are derived either from cysteine or from compounds of cysteine biosynthetic pathway. Pungency of these plants is due to the presence of glucosinolates. These are responsible for providing protection to plants against herbivores. On being hydrolyzed by the enzyme myrosinase, glucosinolates produce isothiocyanates which are toxic to insect herbivores since these bind with the digestive enzymes of the insects. Some of the derivatives of isothiocyanates are volatile, because of which insects avoid these plants by detecting their odor. These volatile compounds are responsible for the characteristic odor of *Brassica* sp. Glucosinolates and myrosinase are stored in separate compartments which get mixed when the tissue is damaged. Isothiocyanates are detoxified by glutathione in human liver. Glucoraphanin is the major type of glucosinolate found in broccoli. Sulforaphane is the hydrolysis product of glucoraphanin which may protect humans against certain forms of cancers.
>
> Structure of glucoraphanin

Reactions requiring PAPS as the sulfate group donor and compound with hydroxyl group as the acceptor molecule are catalyzed by the enzymes sulfotransferases (SOT). Multiple sulfotransferases have been isolated which are required for the structural diversity of the biological molecules with hydroxyl groups, which act as the acceptor of sulfate groups from PAPS. Cysteine is the key amino acid whose biosynthesis is influenced by carbon, nitrogen, and sulfate metabolism. In plants exposed to nitrogen starvation, sulfate transporters as well as APS reductase are downregulated. Sulfate starvation leads to reduced nitrate uptake as well as reduction in nitrate reductase activity. In a similar way, under reduced $CO_2$ assimilation, both sulfur and nitrogen metabolisms are reduced. Under stress situations, growth regulators such as salicylate and ethylene upregulate APS reductase, which is the key enzyme in sulfur metabolism. Under biotic stress, genes involved in sulfate assimilation and glucosinolate biosynthesis are also upregulated by the jasmonates.

## 12.2 Phosphorus Metabolism

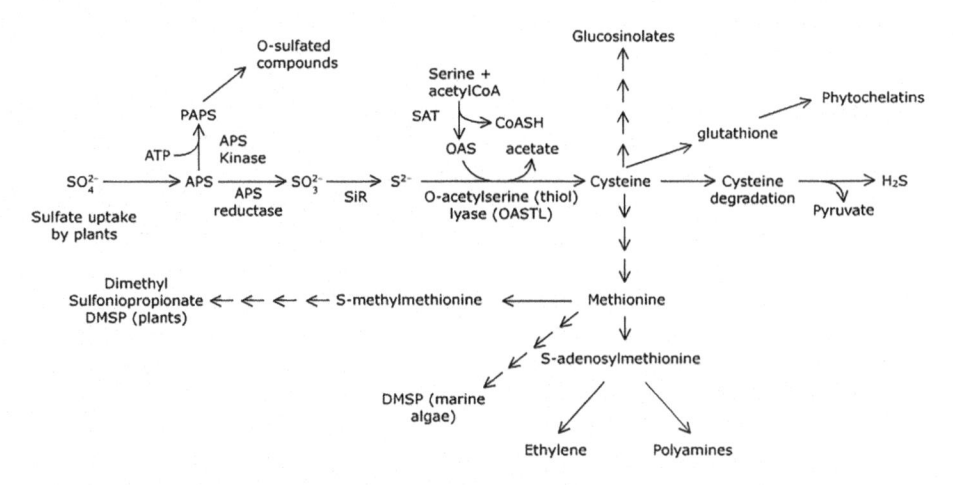

**Fig. 12.12** Cysteine metabolism

---

**Box 12.3: Alliin**

Alliin is a sulfoxide which is a derivative of amino acid cysteine. It is generally found in *Allium* species which includes onion and garlic. Plants store alliin in the inactive form as conjugated with sugar molecules. When plant cells are injured, an enzyme **alliinase** is released from vacuoles which results in the removal of the sugar moiety. An odoriferous compound is released which has antibacterial properties. In onion, alliin is further decomposed to produce propane thiol-S oxide which triggers tears in the eyes. Alliins have an important role in protecting plants against herbivory. In cell culture, alliins have been found to be effective in inhibiting cell division and are effective in inducing apoptosis in human cells. So, they may be effective in protection against cancer.

Structure of alliin

---

## 12.2 Phosphorus Metabolism

Phosphorus is essential macronutrient which is central to plant metabolism. It is the constituent of nucleic acids as **phosphodiester** linkages join two subsequent nucleotides. Many of the intermediates of glycolysis and Calvin cycle are phosphorylated sugars. Phosphate-containing nucleotides play an important role in almost all the energy transformation reactions. ATP occupies the central role in energy metabolism. Almost 70% of the proteins in a cell need phosphorylation at some point of time. Phosphorylation of proteins is key metabolic requirement as part of post-translational regulation of enzyme activity. These include regulation of nitrate reductase, phosphoenolpyruvate dehydrogenase, and the enzyme involved in starch biosynthesis besides many other enzymes. Protein kinases and phosphatases are responsible for catalyzing the post-translational modifications of the proteins which is also required for cell signaling. Phosphorus is an essential structural component of membranes as phospholipids. Inositol phosphate (produced as a result of hydrolysis of a membrane lipid phosphatidylinositol by phospholipase) is required as the signaling molecule. Thus, phosphorus is constituent of the biomolecules which are indispensable for life.

In this section, we will briefly discuss about the forms in which phosphorus is available to plants, transport of inorganic phosphates from soil to root cells and at the subcellular level, phosphorus metabolism. This will include role of inorganic phosphate as the potential phosphate group donor, role of various phosphorylated compounds in energy transfer reactions, role of ATP as the phosphate group donor and reactions involving transphosphorylation between nucleotides, and role of polyphosphates as the storage molecule as well as its significance in cell signaling and remobilization of phosphorus during phosphorus deficiency and leaf senescence.

## 12.2.1 Biogeochemical Cycle of Phosphorus

Phosphorus is a macronutrient found in lowest concentration in the soil. Despite being an essential element, its availability in the soil is limited. Unlike nitrogen ($N_2$), phosphorus is not biologically fixed. In contrast to sulfur which can be taken up by plant in form of $SO_2$, phosphorus is also not available in gaseous form. However, phosphorus is available in both inorganic and organic forms in the soil. Phosphorus is found in soil solutions in the form of $H_2PO_4^-$, $HPO_4^{2-}$, and $PO_4^{3-}$, depending upon the pH of the soil solution. In the pH range of 3–7, it is the $H_2PO_4^-$ which dominates. This form of phosphorus ($H_2PO_4^-$) which is denoted by inorganic phosphate ($P_i$) is absorbed by the plant. Organic form of phosphorus is present in organic molecules as ester derivatives. It is made available by the action of certain enzymes produced by the plants. Bones and many other structures of dead animals are rich source of phosphorus. Phosphorus derived from organic wastes of animals and plants is recycled by the plants, and through plants, by the animals. Unlike $NO_3^-$ and $SO_4^-$, phosphates are not reduced, and it is the oxidized form which is incorporated in the biomolecules (Fig. 12.13).

## 12.2.2 Phosphate Transporters

Phosphorus concentration in soil is very low (0.1 µM–1.0 µM) which is about three times lower than present inside plants. Phosphorus is present in the soil as $H_3PO_4^-$,

## 12.2 Phosphorus Metabolism

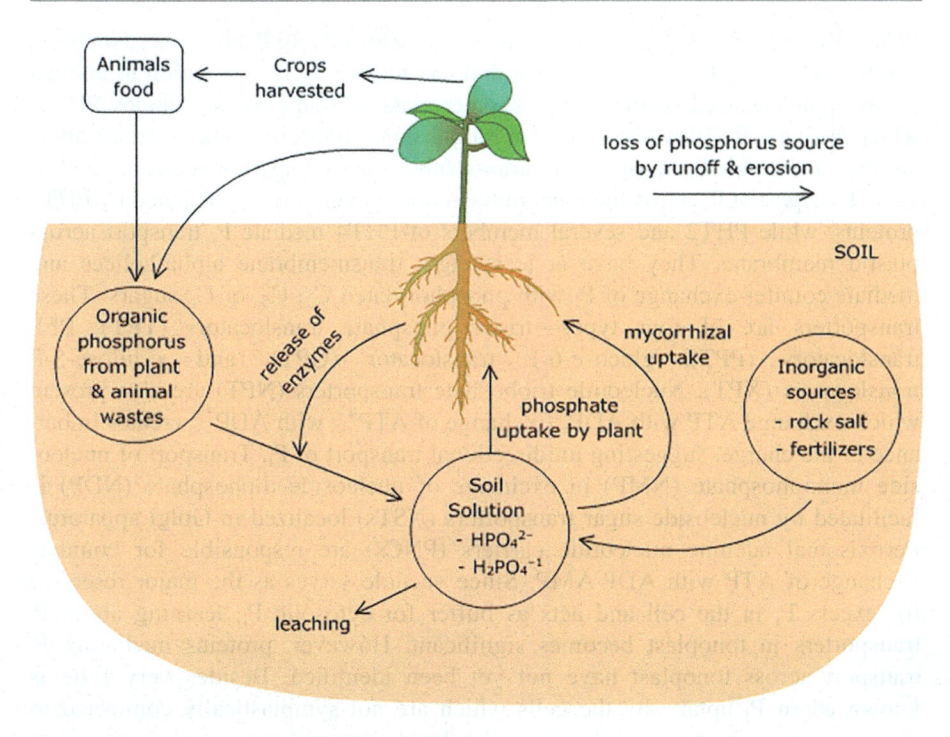

**Fig. 12.13** Biogeochemical cycle of phosphorus

$H_2PO_4^-$, $HPO_4^-$, and $PO_4^-$. $PO_4^-$ is not available since it forms insoluble salts with different mineral elements, making them unavailable for the plant. $H_2PO_4^-$ is the predominant form of phosphorus available in soil at pH 5–6. Plants absorb phosphorus in the form of $H_2PO_4^-$. There are high-affinity and low-affinity transporters which work in the range of 2.5–12 μM and 50–100 μM, respectively. Low-affinity transporters are constitutive, while high-affinity transporters are expressed in response to phosphorus deficiency. Phosphorus starvation increases its uptake capacity by increasing the high-affinity transporters. Phosphate transporters belong to major facilitator family of transporters known as **MFS transporters.** These transporters consist of single polypeptide chain mediating transport of small molecules in response to electrochemical gradient. These transporters are formed by 12 transmembrane domains and one long hydrophilic domain in the middle. These differ from **ABC transporters** since these do not have any binding site for ATP. However, energy demand is met through ATPase-mediated $H^+$ pumps. Phosphate transporters are $H^+/P_i$ symporters with 2 or 4 $H^+$ being transported for each $P_i$ transported. Various categories of $P_i$ transporters are involved in transport of $P_i$ from soil solution to the plant through root systems, while others are involved in subcellular redistribution and long-distance transport. PHT1 transporters are involved in transport of $P_i$ from apoplast/soil solution to the root cell through the plasma membrane. It belongs to multigene family. In *Arabidopsis*, 9 genes encode for

PHT1, while 13, 15, and 8 genes have been found to code for PHT1 in rice, soybean, and tomato, respectively. $P_i$ is mobile and can migrate from one region to another based upon the need of the plant. Another class of transporters includes PHO1, which mediates $P_i$ efflux into xylem for long-distance transport. Other $P_i$ transporters are responsible for $P_i$ transport in various subcellular compartments of the cell as well. Transport of $P_i$ across the inner mitochondrial membrane is mediated by PHT3 proteins, while PHT2 and several members of PHT4 mediate $P_i$ transport across plastid membrane. They have at least eight transmembrane alpha-helices and mediate counter-exchange of $P_i$ with phosphorylated $C_3$, $C_5$, or $C_6$ sugars. These transporters are of four types—triose phosphate translocators (TPT), PEP translocators (PPT), glucose-6-P translocator (GPT), and xylulose-5-P translocators (XPT). Nucleotide triphoshate transporters (NPT) are also present which exchange ATP with ADP. Exchange of $ATP^{4-}$ with $ADP^{3-}$ creates imbalance of the charge, suggesting unidirectional transport of $P_i$. Transport of nucleoside monophosphate (NMP) in exchange of nucleoside diphosphate (NDP) is facilitated by nucleoside sugar transporters (NSTs) localized in Golgi apparatus. Peroxisomal adenine nucleotide carriers (PNCs) are responsible for counter-exchange of ATP with ADP/AMP. Since vacuole serves as the major reservoir for excess $P_i$ in the cell and acts as buffer for cytosolic $P_i$, learning about $P_i$ transporters in tonoplast becomes significant. However, proteins mediating $P_i$ transport across tonoplast have not yet been identified. Besides very little is known about $P_i$ uptake by the cells which are not symplastically connected to their neighbors, such as guard cells or the developing embryo.

### 12.2.3 Role of Phosphorus in Cell Metabolism

Many biomolecules participating in metabolic reactions are charged species since it keeps them water soluble and also prevents them from diffusing across plasma membrane. One of the most common biologically ionic groups is the phosphate group. Almost all Calvin-Benson cycle intermediates are phosphorylated with $P_i$ being released at the time of synthesis of the end products such as sugars and starch. Except starch, most of the end products of photosynthesis are synthesized in cytosol. Triose phosphates need to be transported out of the plastids and $P_i$, which is released in the cytosol and is cycled back to plastids. This transport is facilitated by antiporters located on the inner envelope of the plastids (Fig. 12.14). Photosynthesis is regulated by the availability of $P_i$ and is inhibited when $P_i$ supply becomes limiting.

Phosphorus is capable of forming five covalent bonds. In the inorganic phosphoryl form, phosphorus is covalently linked to oxygen by three P-O and one P=O bond. Since oxygen is more electronegative than phosphorus, sharing of bond between phosphorus and oxygen is unequal. At physiological pH, the electronegative oxygen in P=O bond of ATP attracts electrons away from phosphorus, resulting in slight electropositive charge accumulation ($\delta^+$) on phosphorus than oxygen, which carries partial negative charge ($\delta^-$). This provides an electrophilic property to phosphorus. As a result, phosphoryl group is transferred from ATP to a

## 12.2 Phosphorus Metabolism

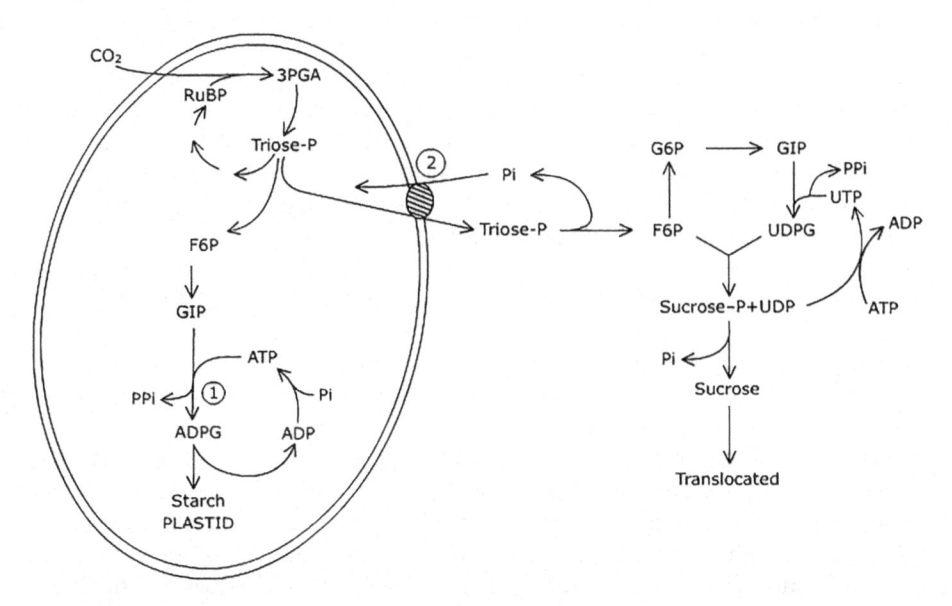

**Fig. 12.14** Role of phosphates in carbohydrate metabolism and its regulation. (1) ADP-glucose pyrophosphorylase; (2) triose-P antiporter

nucleophilic group, such as a carboxylic group, forming an ester. Phosphoryl group transfer reactions are important group transfer reactions required for activation of intermediates of the pathway. A good leaving group is central to metabolic reactions. "Leaving group" is a metabolic fragment which departs with unpaired electrons due to **heterolytic bond cleavage**. Generally, a good "leaving group" is attached to activate an intermediate of the metabolic reactions so that it can participate in the reaction. Though intermediates with good leaving groups are fairly stable, their low activation energy requirement in the presence of enzymes makes them suitable for faster reactions. Inorganic orthophosphate groups which include $H_2PO_4^-$ and $HPO_4^-$ (which are commonly referred as inorganic phosphates) are good "leaving groups" which participate in nucleophilic substitution reactions. Nucleophilic substitutions, which include $-PO_3^-$ as the leaving group, occur in many metabolic reactions. Phosphorus plays vital role in plant processes that involve energy transfer. It is a constituent of various biomolecules involved in energy transfer reactions, such as ATP, NTPs, and NADPH.

One important role of phosphorus is in post-translational regulation of activities of various enzymes by phosphorylation and dephosphorylation, thus regulating metabolism. Phosphoryl group is transferred to –OH group of either serine, threonine, or tryptophan residues of the protein. $P_i$ acts as potent inhibitor of starch synthesis by acting as negative modulator of the enzyme ADP-glucose pyrophosphorylase. Another example which is significant in plants is post-translational modification of PEP dehydrogenase. About 70% of all eukaryotic proteins are possibly phosphorylated at some point of time. Protein kinases and "super family of proteins" possibly amount to 4% of the entire genome. O-Phosphorylation of proteins at serine, threonine, or tyrosine residues occurs in the ratio of 75:25:5.

Phosphorylation of proteins as a regulatory mechanism is significant during development as well as in response to stress conditions. Thus, understanding phosphorylation of the proteins is significant. Use of mass-spectrometry has helped in identification of the phosphoproteins. Phosphorus homeostasis includes regulating the cytosolic concentration of $P_i$. If there is deficiency of phosphorus, it is mobilized from vacuolar storage pools, or genes involved in acquisition or transport of phosphorus are upregulated. When phosphorus is present in excess amounts, it can be stored in vacuoles so as to avoid toxicity. In case of deficiency, mobilization of phosphorus from senescing leaves occurs, which is known as **retrotranslocation**. Many of the senescence-induced genes (because of $P_i$ deficiency) are involved in lipid degradation. This includes phospholipase A, phospholipase D, etc. This results in the production of $IP_3$ which acts as the signal molecule. There is overlapping of upregulation and downregulation of genes involved in $P_i$ deficiency and senescence which signifies the importance of phosphorus recycling. Major pool of organic phosphorus is present in nucleic acids, phosphorylated proteins, various phosphorylated metabolites, and phospholipids. During senescence, phosphatases, nucleases, and phosphoesterases play important roles in the release of $P_i$ from the organic forms. Transporters are required for uploading $P_i$ into phloem so that it can be supplied to younger parts of the plant. In thylakoids, phospholipids might be replaced by sulfolipids or galactolipids under $P_i$ deficiency. However, phosphorus is irreplaceable in many phosphorus-containing biomolecules.

### 12.2.4 Mobilization of Phosphorus

In plants phosphorus is mostly present in organic forms as organic esters. In seeds, it is present as phytic acid (inositol hexakisphosphate) (Box 12.4). Phytic acid can bind to various anions resulting in the formation of phytate. At the time of germination, phytase is responsible for release of phosphates from phytates. Numerous enzymes including phosphatases and diesterases present in the plant are responsible for breaking down of the organic forms (prevalent as phosphomonoesters and diesters) thus making phosphorus available to plants. After reserves are exhausted, activity of these enzymes declines. Mobile nature of phosphorus is responsible for maintaining the homeostasis at the time of its deficiency or during leaf senescence. There are phosphatases which differ in their optimal activity at different pH. Many plant phosphatases which have acid optima absorb light at 560 nm wavelength in pure solution. These are known as purple acid phosphatases (PAPs). PAPs play very important role in plant nutrition. $PO_4^-$ is an inhibitor of PAPs. A large pool of phosphorus is in the form rRNA. In developing leaves, rRNA pool increases, while it decreases in the senescing leaves where phosphorus concentration decreases up to 78%. During $P_i$ deficiency and senescence of leaves, genes required for synthesis of PAPs are upregulated. However, mobilization of $P_i$ from senescing leaves remains the major source for $P_i$ which is required by various sinks in the plants, such as young leaves, reproductive structures, and storage tissues. Under $P_i$ deficiency, many genes required for

## Box 12.4: Phytase Enzyme

A number of plants store phosphorus in seeds in the form of phytate. Phytic acid is chemically inositol hexakisphosphate which on dissociation in aqueous solutions bind to cations such as Ca, Mg, Fe, or Zn in anionic form (phytate). This can be hydrolyzed by phytase enzyme which is present in microorganisms and plants. However, the presence of this enzyme has not been reported in animals. Bacteria present in ruminant stomachs of Artiodactyla not only release cellulase required for the digestion of cellulose but also phytase. Plant-derived feed for poultry and animals is often supplemented with phytase, which results in making inorganic phosphorus available for growth.

Structure of phytic acid (Inositol hexakisphosphate)

hydrolysis of phospholipids and for biosynthesis of galactolipids and sulfolipids are upregulated. Understanding the recycling and mobilization of phosphorus will help in improving $P_i$ use efficiency of crop plants which will further minimize the usage of $P_i$ fertilizers.

## 12.3 Iron Metabolism

Iron is an essential element in all living beings. It is required as a component of various proteins and as an activator in various enzyme-catalyzed reactions. It is important in redox reactions of the cell because of its variable valency. Since it is a constituent of heme, it has structural role in all heme proteins, including cytochromes. The cofactor of cytochromes consists of porphyrin ring with iron in the center (Fig. 12.15). Various cytochromes, which absorb different light

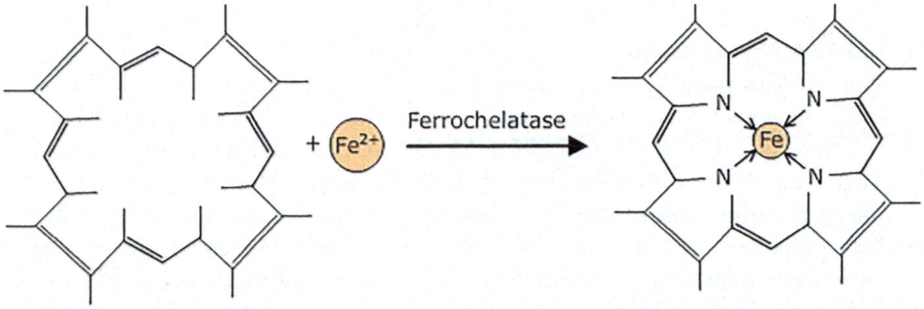

**Fig. 12.15** Ferrochelatase catalyzes the terminal step in heme biosynthesis. It catalyzes chelation of ferrous ion into the protoporphyrin IX ring to form protoheme. In plants, heme is synthesized both in mitochondria and chloroplasts

wavelengths, have been identified. These include cytochromes a, b, and c, which differ in the side groups attached to porphyrin ring (Fig. 12.16). The longest wavelength absorbed is near 600 nm in type "a" cytochromes, near 560 nm in type "b," while it is near 550 nm in case of cytochrome "c." Cytochromes are constituents of electron transport chain both in mitochondria and chloroplasts, which differ in their redox potential. Additionally, non-heme Fe also participates in various subcellular redox reactions as **iron-sulfur proteins (ISP)**, in which Fe is present as a complex with sulfur. A coordination between the metabolism of these two nutrients, i.e., Fe and S, is required since most of the metabolically important Fe is covalently bound to S as Fe-S complex. Chelated iron and reduced sulfur are required for the biosynthesis of Fe-S clusters. Fe-S clusters (Fig. 12.17) are component of various proteins involved in electron transport. There are multiple types of Fe-S clusters, the simplest one having one Fe coordinated to four cysteine -SH groups. The complex may also be present as $Fe_2$-$S_2$, $Fe_3S_3$, and $Fe_4$-$S_4$. **Rieske iron-sulfur proteins** are $Fe_2S_2$ type in which one Fe is coordinated with two histidine residues of the apoprotein rather than with -SH of cysteine, while one Fe is coordinated with two -SH groups of cysteine. These are called **Rieske protein** after the name of the scientist John S. Rieske and coworkers who discovered and isolated them in 1964. In each cluster, iron is present either in reduced state (Fe II) or in oxidized state (Fe III). Each center is able to carry only one electron at a time. Rieske proteins are part of the electron transport system with electron reduction potential ranging from −150 to +400 mV. These are components of cytochromes $bc_1$ complexes in respiratory electron transport chain (ETC) and cytochrome $b_6f$ in ETC of photosynthesis. Ferredoxin type [$Fe_2S_2$] iron clusters are coordinated with four cysteine residues in a protein. Nitrogenase also has Fe-S containing cofactor, which mediates electron transfer between reduced ferredoxin and nitrogen. Catalase and peroxidase are also iron requiring heme proteins. Iron is also required for chlorophyll biosynthesis. But unlike heme, it is not a structural component of chlorophyll. The rate-

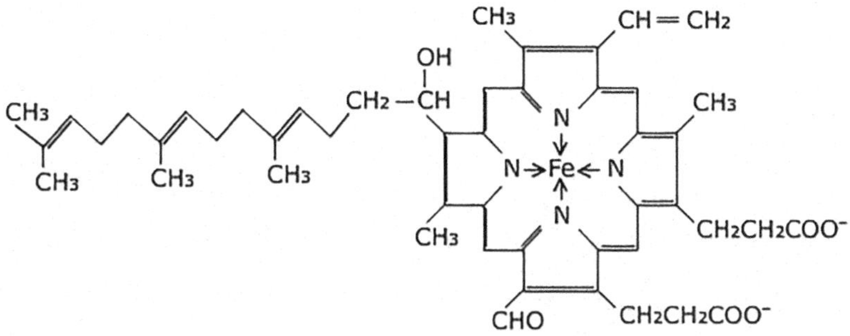


Heme a
(in a-type cytochromes)

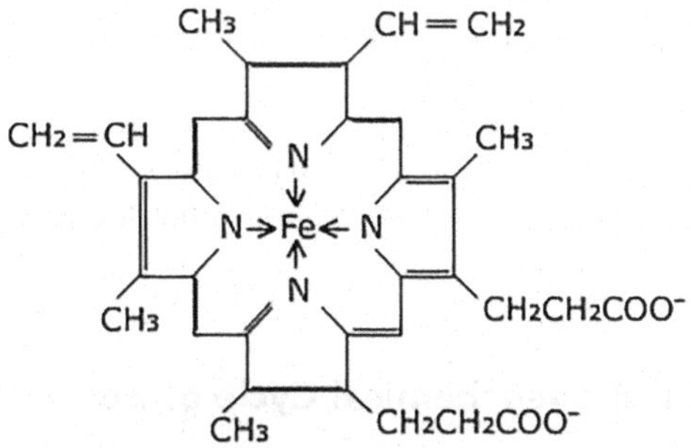


Iron protoporphyrin IX
(in b-type cytochromes)

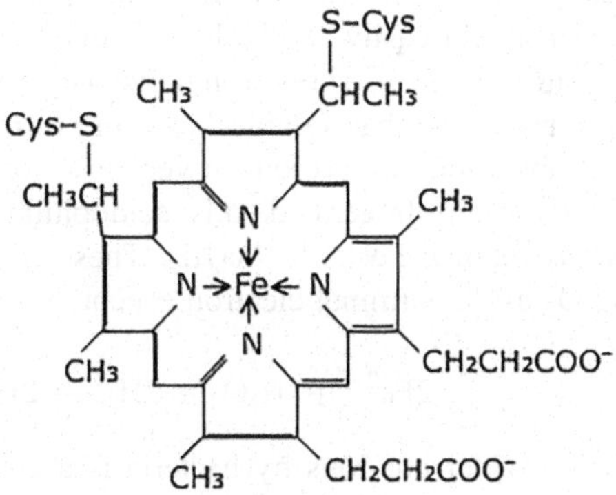


Herne c
(in c-type cytochromes)

**Fig. 12.16** Prosthetic group of cytochromes. Heme is a tetrapyrrole that binds ferrous iron ($Fe^{2+}$) at four coordinated nitrogens in the protophyrin ring system and is incorporated into various apoproteins as a prosthetic group. Protophyrin IX is the immediate precursor for heme biosynthesis. Protophyrin IX complexed with iron to form iron protophyrin IX, i.e., heme of cyt b type of proteins. Heme c is bound covalently through thioester bonds in cyt c type of proteins, while heme a exhibits long isoprenoid tail attached to one of the five-membered rings, which is present in a-type cytochromes

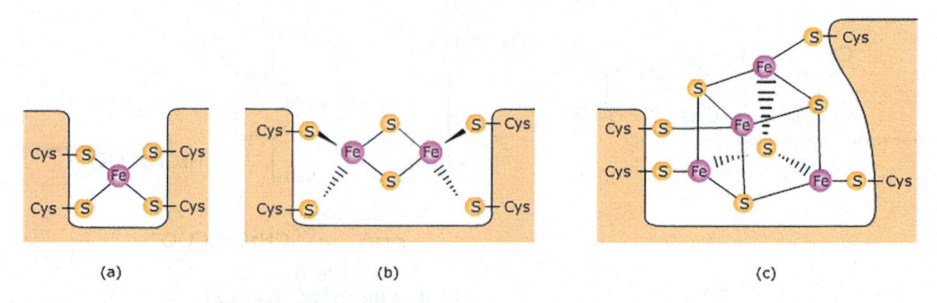

**Fig. 12.17** Iron-sulfur clusters (Fe-S) of Fe-S proteins. (**a**) Single Fe is coordinated to four sulfur atoms of cysteine residues of the protein. (**b**) In 2Fe-2S cluster 2 Fe and, in (**c**) 4 Fe are coordinated with 4 Cys-SH groups of the protein. All iron-sulfur proteins are involved in single electron transfer, since single Fe takes part in electron transfer

limiting step in chlorophyll biosynthesis is catalyzed by iron-dependent enzyme glutamyl-tRNA reductase. Iron also has a critical role in DNA replication.

## 12.3.1 Biogeochemical Cycle of Iron and Iron Uptake by the Plant

In all terrestrial forms of life, including human beings, iron is obtained from the plants, which sequester it from the soil. Thus, iron enters the food chain through the plants. Iron constitutes approximately 3% of the earth's crust. It ranks fourth among all the elements which are available in the earth crust. Though present in high amount, iron is not easily available, and plants exhibit deficiency symptoms. The reason most of the plants show iron deficiency even when iron is present in soils in significant amount is that soluble inorganic forms of iron ($Fe^{2+}$) which plants are capable of absorbing are present in very low concentrations in the soil, i.e., in the range of 0.01–1 nm. In aerated soils, acidophilic bacteria such as *Acidithiobacillus ferrooxidans* oxidize Fe(II) to Fe(III). These bacteria grow autotrophically on Fe(II) using $O_2$ as the terminal electron acceptor.

$$2Fe^{2+} + 1/2 O_2 + 2H^+ \rightarrow 2Fe^{3+} + H_2O$$

Oxidation of ferrous ions by bacteria is used both for generating ATP and for reduction of $CO_2$ to carbohydrates.

$$4Fe^{2+} + HCO_3^- + 10H_2O + h\nu \rightarrow 4Fe(OH)_3 + (CH_2O) + 7H^-$$

Ferric hydroxide is insoluble, and its formation is favored in pH 5–8. However, if pH of the soil is more than 8, iron is present as Fe$(OH)_4$, which is sparingly used thus not available to plants. Aerobic soils have moderate to basic pH, and therefore, iron is unavailable to plants. Iron is present in oxidized state or is complexed with

## 12.3 Iron Metabolism

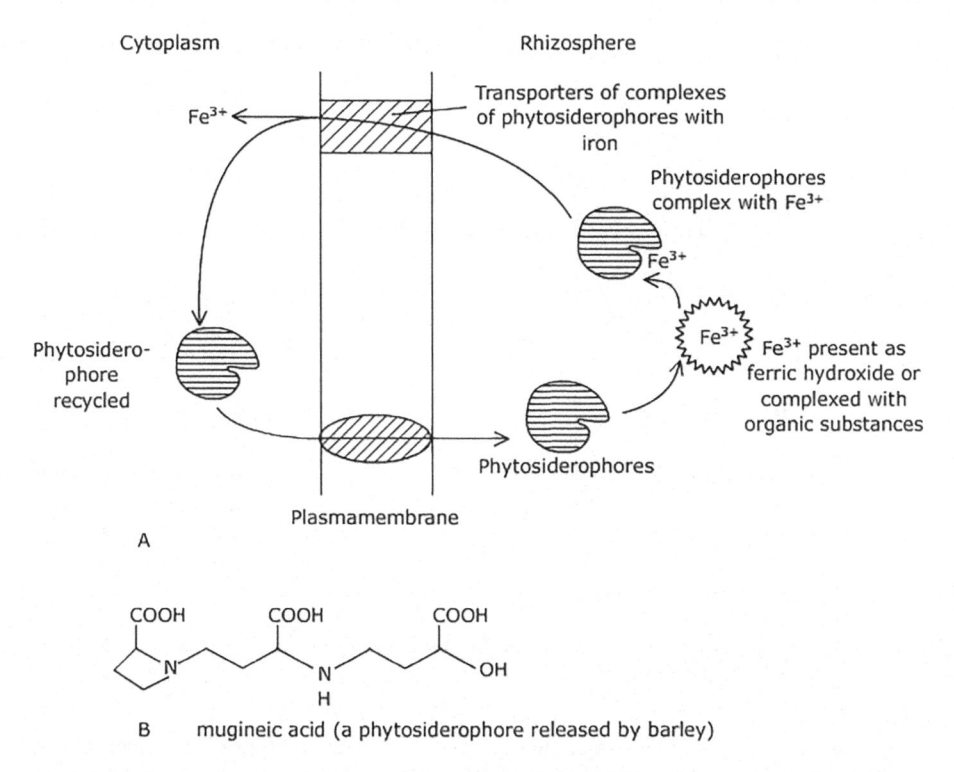

**Fig. 12.18** (a) Iron uptake as complex with phytosiderophore released by the plant. (b) Mugineic acid, an amino acid secreted by some members of the grass family under iron deficiency

organic material in soils having neutral and basic pH. However, in waterlogged soils, iron is available as $Fe^{2+}$. Plants can absorb reduced form of iron ($Fe^{2+}$) since transporters for these are present in plasma membrane of the root epidermis. Two strategies are adopted by plants for the uptake of iron. In grasses, plants release organic compounds known as **phytosiderophores (PS)** that efficiently chelate and solubilize ferric ions. The Fe (III)-PS complex is the soluble form of the iron complexes which can be absorbed by high-affinity transporters present in the plasma membrane of root epidermis (Fig. 12.18). Transporters of Fe (III)-PS complexes are integral membrane protein with 12 putative transmembrane domains encoded by the gene **yellow stripe 1 (YS1)**. The protein is responsible for proton-coupled symport of Fe (III)-PS complexes. In oat and barley, phytosiderophores have been characterized which consist of non-protein amino acids avenic acid and mugineic acid. These amino acids are synthesized by the plants in response to iron deficiency from methionine. Function of phytosiderophores is less affected by pH. ATPase-mediated $H^+$ pumps in root epidermis pump out protons altering pH of rhizosphere adjacent to roots to acidic. In acidic pH, $Fe^{3+}$ is released from organic complexes. In

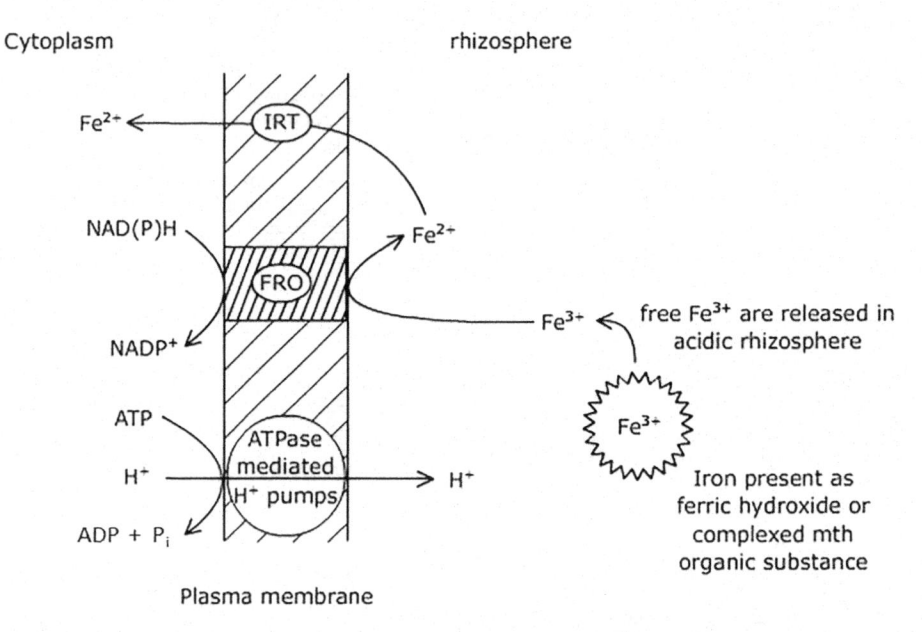

**Fig. 12.19** Mechanism of iron uptake in roots of the plant. FRO, ferredoxin reductase/oxidase; IRT, iron-regulated transporter

many species of monocots other than grasses, enzymes of **ferredoxin reductase/oxidase (FRO)** family are synthesized (Fig. 12.19). These are **flavocytochromes** with intramembrane heme moieties and large cytoplasmic loops having binding sites for FAD and NAD(P)H. FROs facilitate electron transfer from NAD(P)H to $Fe^{3+}$ mediated by heme as a result of which $Fe^{3+}$ is reduced to $Fe^{2+}$. The presence of **iron-regulated transporter (IRT1)** in root epidermis is responsible for the absorption of $Fe^{2+}$ ions. Coordinated role of ATPase-mediated $H^+$ pumps, FROs and IRT, is required for iron uptake by the monocots and eudicots.

### 12.3.2 Transport of Iron Within Plant

After uptake from the soil through different strategies adopted in different plants, iron is made available in the cytosol of the root epidermal cells complexed with some organic compounds which are released in xylem. The pH of the xylem sap is acidic (5.5). Iron released in the xylem form complex with citrate or some other compounds (in grasses this may be a phytosiderophore) which facilitates iron transport to shoots along with the transpiration stream. In case iron is required to be transported to plant parts with lower transpiration rate, it takes place in phloem (pH 7.2). The iron-chelating compound identified in *Ricinus communis* is a 17-kD protein, which is called **iron transport protein (ITP)**. It binds with Fe (III) form of iron rather than with Fe (II). Another compound **nicotianamine (NA)** forms a stable complex with Fe

(II) rather than Fe (III) at neutral pH (Box 12.5). According to a proposed model, iron must be transported as Fe (III) complexed possibly with IRT. However, at the time of unloading, iron forms a complex with NA, and the Fe:NA complex crosses plasma membrane through yellow stripe-like (YSL) transporters. A reductase may reduce Fe (III) to Fe (II) prior to binding with NA.

**Box 12.5: Nicotianamine (NA)**

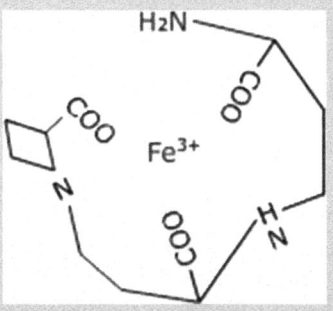

Nicotianamine (NA) is a metal chelator which plays an important role in long-distance transport and homeostasis of Fe in plants. It is an intermediate in the biosynthesis of **phytosiderophores**. NA is synthesized from methionine as a result of single-step condensation of three molecules of S-adenyl-methionine catalyzed by the enzyme NA synthase. NA has three amine and three carboxylic groups in an **azetidine** ring. At cytosolic pH (around 7.2), NA can form chelates with both forms of Fe ($Fe^{2+}$ and $Fe^{3+}$). NA aminotransferase (NAAT) catalyzes amino group transfer from NA during the biosynthesis of phytosiderophores by graminaceous plants. Activity of NAAT affects both iron homeostasis and its transport. Introduction and overexpression of the NAAT gene in tobacco result in reduction of endogenous NA, which significantly affects iron transport and iron homeostasis, resulting in the production of abnormal phenotype of the plant. It also affects the genes involved in iron uptake machinery of the plants. NA-Fe complexes are transported across membranes, and this process seems to be very significant for intracellular transport of Fe in the plants. NA-Fe transporters are members of oligopeptide transporters (OPT) groups which are found in archaea, fungi, bacteria, and plants.

Nicotianamine (NA)
Chelation of $Fe^{3+}$ with NA

## 12.3.3 Redistribution of Iron at the Subcellular Level

Long-distance transport of iron as well as intracellular transport requires specialized proteins and small molecular weight chelators because the presence of $O_2$ iron is insoluble and toxic. For long-distance transport, iron may get chelated with nicotianamine or organic acids. Ferric reductase and transporters may possibly be involved in the transport of iron into the mesophyll cells. However, exact mechanism is not known. Iron metabolism in chloroplasts and mitochondria is particularly important since these organelles are the major sinks for iron. Both organelles may have a prokaryotic type of iron transport system. Iron may be able to cross the outer membrane of both the organelles freely through porins. Mitochondria are the major sinks for iron because it is the major site for biogenesis of Fe-S cluster and for heme biosynthesis. There is a possible role of metalloreductase in iron transport across inner mitochondrial membrane. An iron transporter known as **mitochondrial iron transporter (MIT)** has been identified in rice. Iron-sulfur cluster biogenesis machinery has a very important role in maintaining iron homeostasis. Two proteins play a significant role in iron metabolism of mitochondria. **Frataxin (FH)**, a conserved mitochondrial protein, is a molecular chaperon which regulates biogenesis of Fe-S cluster and heme biosynthesis. Maintaining iron concentration in mitochondria is a challenge since pH of the matrix is alkaline. Role of frataxin and a storage protein **ferretin** is significant in maintaining iron homeostasis. Ferretin is the iron-storage protein which stores excess iron and protects the cellular machinery from ROS production. Role of ferretin as the iron-storage protein complex is more significant in chloroplasts, while frataxin has more significance in mitochondria (Fig. 12.20). Iron in the form of free ions is toxic in mitochondria and chloroplasts since these are the major sites for ROS generation in the cell. The conserved proteins in ferretin, which are oligomerized to form a hollow

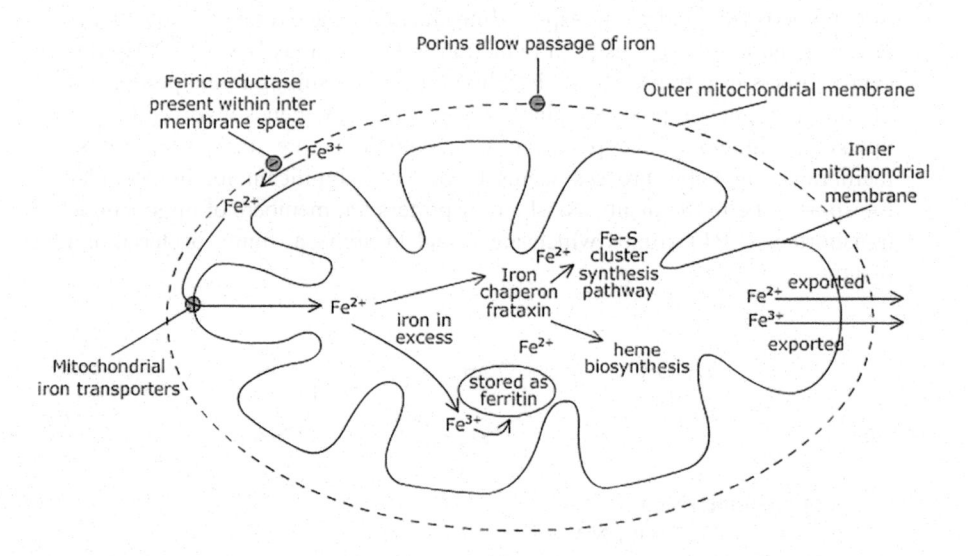

**Fig. 12.20** Iron metabolism in mitochondria

## 12.3 Iron Metabolism

sphere, exhibit ferroxidase activity (Fig. 12.21). These oxidize $Fe^{2+}$ to $Fe^{3+}$ which is stored within the core of the protein complex in the form of hydrous ferric oxide along with phosphates. There may be as many as 2000–4000 $Fe^{3+}$ atoms per molecule of ferritin. Mechanism of the release of $Fe^{3+}$ from ferretin is not yet understood. Transport of iron into chloroplasts requires ferric chelate reductase (FRO7). Another cell organelle involved in iron storage is vacuole (Fig. 12.22). Transport of iron into the vacuoles is mediated by **vacuolar iron transporter (VIT 1)**. Once inside the vacuole, iron is chelated with nicotianamine (NA). Release of Fe from chloroplasts and vacuoles is possibly mediated by efflux transporters YSL (yellow stripe-like) and by **NRAMP** family of transporters, respectively. Since iron can catalyze formation of deleterious ROS, it may be used by the host as a defense mechanism against invading pathogens (iron homeostasis). A cell needs to have a sensing and signaling mechanism

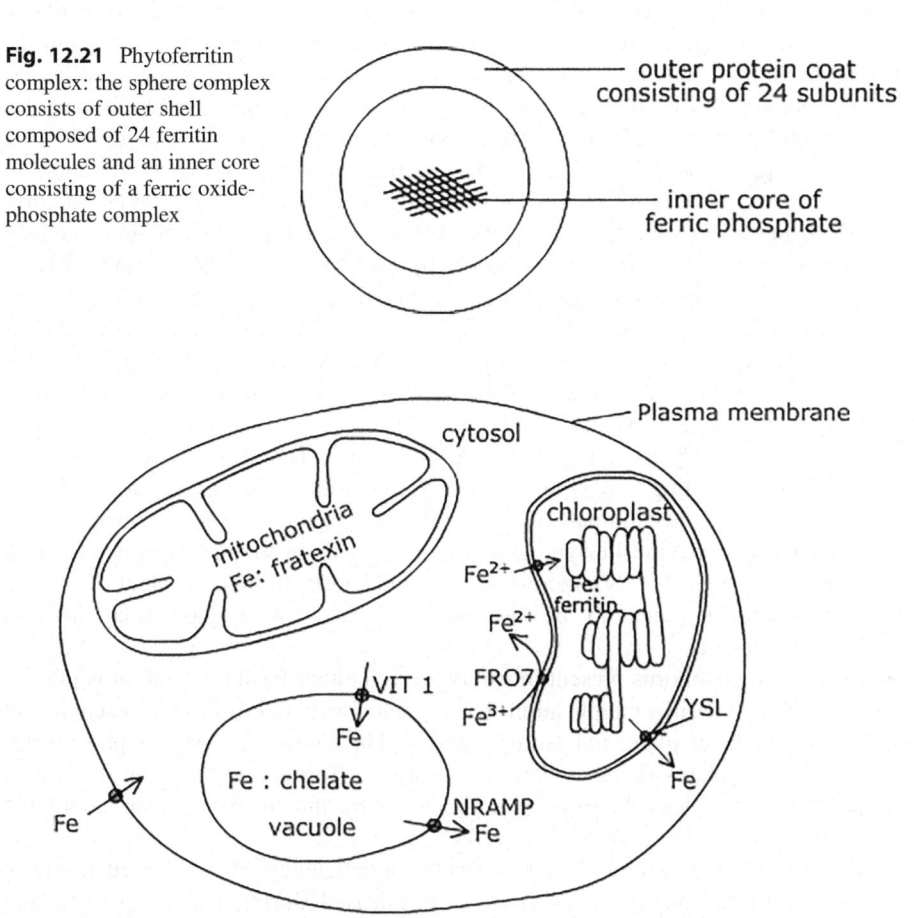

**Fig. 12.21** Phytoferritin complex: the sphere complex consists of outer shell composed of 24 ferritin molecules and an inner core consisting of a ferric oxide-phosphate complex

**Fig. 12.22** Cell organelles involved in iron metabolism. *VIT* vacuolar iron transporter, *YSL* yellow stripe-like, an efflux transporter, *FRO7* ferric chelate reductase, *NRAMP* family of natural resistance-associated macrophage protein (metal ion transporters, which play a major role in metal iron homeostasis)

by which iron homeostasis can be maintained because both deficiency and excess iron can affect various physiological processes in the cell. Iron homeostasis is required for heme biosynthesis, for assembly of Fe-S proteins as well as incorporation of iron with apoproteins. Under iron deficiency, expression of FIT1 is upregulated, which is an iron-induced transcription factor and might be involved in sensing iron deficiency. There may also be possible involvement of hormonal signaling in sensing iron deficiency.

**Summary**

- Sulfur is an essential macronutrient which is a structural component of various biomolecules, including amino acids cysteine and methionine, cofactors, coenzymes, lipoic acids, sulfolipids, and many others. In soil, sulfur is available to plants as sulfate ions and as $SO_2$ in the air present as a pollutant. Sulfate transporters in the root hair epidermis facilitate uptake of sulfate. Sulfate is metabolized in root plastids, and the excess amount is either stored in vacuoles or is transported to leaves via xylem where it is metabolized in chloroplasts.
- Primary route for assimilation of $SO_4^-$ is similar to that of nitrate assimilation. In chloroplasts, ATP sulfurylase catalyzes reaction of $SO_4^-$ with ATP to form adenosine 5′-phosphate sulfate (APS). While attached to APS, sulfate is reduced to $SO_3^-$ using the electrons provided by glutathione and $SO_3^-$, and AMP is released free. Sulfite reductase catalyzes reduction of $SO_3^-$ to $S^{2-}$ (sulfide) using six electrons from reduced ferredoxin. Sulfide is added to O-acetyl serine (OAS) to form cysteine with simultaneous release of acetate. Cysteine is the precursor for various other sulfur-containing biomolecules like methionine, glutathione, phytochelatins, glucosinolates, etc.
- APS kinase catalyzes conversion of APS to PAPS. This is followed by sulfation reactions resulting in biosynthesis of various other sulfur-containing biomolecules. Since animals, including human, do not have the enzyme for $SO_4^-$ reduction, cysteine biosynthesis does not occur in them, and it needs to be provided in their diet. So, increasing cysteine and methionine content of various crop plants remains to be an important and challenging goal for the biotechnologists.
- Inorganic phosphorus present in soil is derived either from rock salt or is present in soil due to fertilizer application. The organic form of phosphorus present in soil is derived from plant and animal wastes. The inorganic form of phosphorus available to plants is released from the organic forms due to action of various phosphatases released by the plant roots or by the microorganism. Inorganic phosphorus refers to $HPO_4^-$ and $H_2PO_4^-$.
- Phosphorus is a mobile element. During its deficiency, $P_i$ is released from the organic forms in the older leaves or is released during leaf senescence and translocated to younger leaves as well as the reproductive parts of the plant. Phosphorus homeostasis is maintained by increasing $P_i$ supply from vacuoles under $P_i$ deficiency and promoting storage in vacuoles if phosphorus is in excess so as to avoid phosphorus toxicity in cytosol of the cell.

- Phosphorus is required in the cell as the structural component of biomolecules which include phospholipids, nucleic acids, ATP, etc. Various intermediates of Calvin cycle and respiration are phosphorylated. Phosphorylation and dephosphorylation of proteins are essential for regulation of activities of various enzymes. Many of the intermediates of various metabolic pathways including photosynthesis and respiration are phosphorylated because phosphorylation lowers the activation energy requirement during enzyme-catalyzed reactions.
- Various phosphate transporters are present in plasma membrane and in other cellular membranes, which regulate uptake of phosphates from the soil and its distribution within cell and within the plant. At the root hair epidermis, a symporter is responsible for exchange of 2 or 4 $H^+$ with inorganic phosphate.
- In spite of iron being present as the fourth major element in earth crust, its availability to plants is limited which is mainly influenced by pH of the soil. Plants adopt different strategies to absorb iron from the soil which include release of chelating organic compounds, phytosiderophores, or release of $H^+$ by the roots, which facilitate release of iron from the organic complex forms, and reducing $Fe^{3+}$ to $Fe^{2+}$ form. This is followed by its absorption through transporters present on the root epidermis.
- Iron is an essential element which is required by all living beings as the structural component of heme proteins as well as Fe-S protein complexes which participate in electron transport reactions besides its requirement as an activator of some of the enzymes.
- Once inside the cytosol, iron is either transported to the shoots or there is intracellular redistribution of Fe within the root cell/mesophyll cell mediated by various transporters which include mitochondrial iron transporters and vacuolar iron transporters or is transported to chloroplasts. Mitochondria and chloroplasts are the major sink for iron since these two organelles house the electron transport chain.
- In case of nonavailability of iron or when excess free iron is present which is toxic, plants show deficiency symptoms. To maintain iron homoeostasis, it is either transported out to the subcellular organelles or is stored as frataxin in mitochondria or ferritin in chloroplasts. Iron is also transported as the complex of nicotianamine.

## Multiple-Choice Questions

1. Phytoplanktons produce:
   (a) Dimethyl sulfide
   (b) Sulfur dioxide
   (c) Dimethylsulfoniopropionate
   (d) Dimethyl sulfoxide

2. $SO_4^-$ uptake in root hair cell is facilitated by the plasma membrane localized:
   (a) 3 $H^+/SO_4^-$ symporter
   (b) $H^+/SO_4^-$ symporter
   (c) 3 $H^+/SO_4^-$ antiporter
   (d) $H^+/SO_4^-$ antiporter
3. The enzyme ATP sulfurylase catalyzes synthesis of:
   (a) Cysteine
   (b) Disulfide
   (c) Glutathione
   (d) Adenosine 5'-phosphosulfur
4. Which of the following statements is not correct?
   (a) Reduction of $SO_3^-$ to $S^{2-}$ involves transfer of 6 electrons.
   (b) Sulfite reductase draws electrons from reduced ferredoxin for reduction of $SO_3^-$.
   (c) Sulfite reductase draws electrons from NADH for reduction of $SO_3^-$.
   (d) Both nitrite reductase and sulfite reductase are similar since these consist of $Fe_4$-$S_4$ and siroheme.
5. Which of the following statements is correct?
   (a) Methionine is the precursor of cysteine.
   (b) Methionine is synthesized from cysteine.
   (c) Cysteine biosynthesis occurs in the plastids.
   (d) Glutathione acts as buffer for the changes in cytosolic pH.
6. Rieske iron-sulfur proteins are:
   (a) $Fe_4S_4$ types in which the iron atoms are coordinated with four cysteine residues of the apoprotein.
   (b) $Fe_2S_2$ types in which the iron atoms are coordinated with the cysteine residues of the apoprotein.
   (c) $Fe_3S_3$ types in which all iron are coordinated with histidine of the apoprotein.
   (d) $Fe_2S_2$ types in which one iron is coordinated with two histidine residues while another Fe is ordinated with two cysteine residues of the apoprotein.
7. The enzyme, ferredoxin reductase/oxidase (FRO), which is localized in plasma membrane of root hair epidermis facilitates Fe absorption because:
   (a) It is transporter of the iron phytosiderophore complex.
   (b) It facilitates reduction of $Fe^{3+}$ to $Fe^{2+}$ which can pass through the transporters of the membrane.
   (c) It pumps protons out of the cell to make the soil pH acidic which results in release of Fe from the complex with organic compounds.
   (d) It oxidizes $Fe^{2+}$ to $Fe^{3+}$ which makes the iron soluble and it can be absorbed.
8. The iron-storage protein is:
   (a) Ferretin in chloroplasts
   (b) Ferretin in mitochondria
   (c) Frataxin in mitochondria
   (d) Frataxin in chloroplasts

## Answers

1. c  2. a  3. d  4. c  5. b  6. d  7. b
8. a

## Suggested Further Readings

Jones RL, Helen O, Howard T, Susan W (2013) The molecular life of plants. Wiley-Blackwell, Chichester, pp 477–491

Smith AM, Coupland G, Liam D, Harberd N, Jones J, Martin C, Sablowski R, Amey A (2010) Plant biology. Garland Science/Taylor & Francis Group, New York, pp 284–293

Taiz L, Zeiger E, Moller IM, Murphy A (2015) Plant physiology and development, 6th edn. Sinauer Associates, Inc, Sunderland, pp 367–370

Long SR, Kahn M, Seefeldt L, Tsay Y, Kopriva S (2015) Nitrogen and sulfur. In: Buchanan BB, Gruissem W, Jones RL (eds) Biochemistry and molecular biology of plants. Wiley-Blackwell, Chichester, pp 746–767

# Part III
# Development

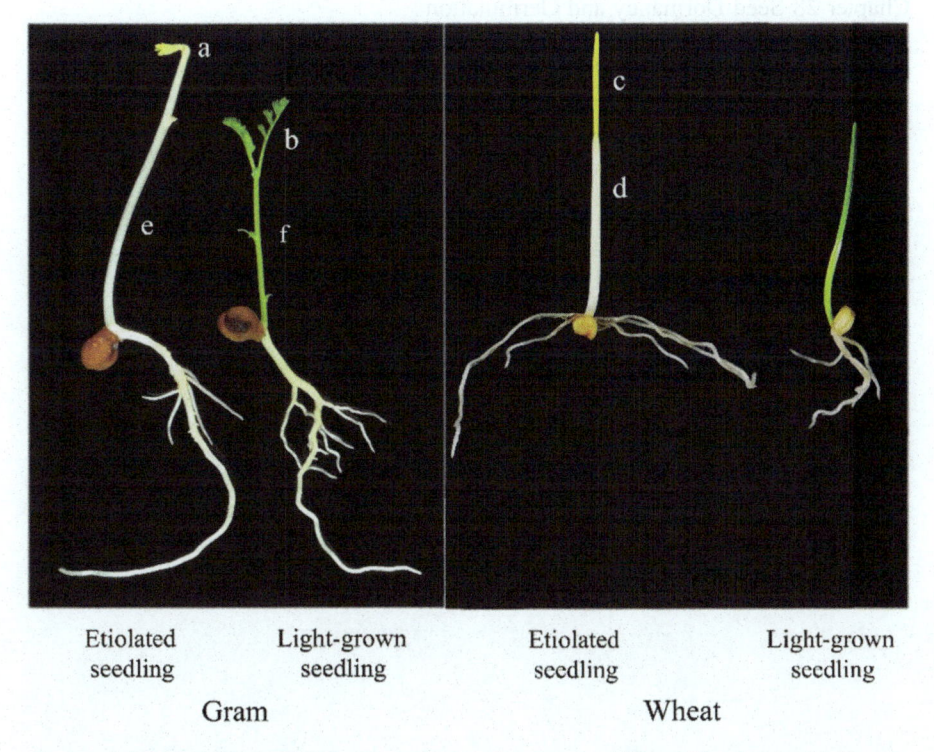

Characteristics of etiolated and light-grown dicot and monocot seedlings. (**a**) apical hook; (**b**) open, expanded green leaves; (**c**) shoot; (**d**) coleoptile; (**e**) elongated hypocotyl; (**f**) short hypocotyl. More details are provided in Chap. 13, Fig. 13.1

Chapter 13 Light Perception and Transduction
Chapter 14 Plant Growth Regulators – An Overview
Chapter 15 Auxins

Chapter 16 Cytokinins
Chapter 17 Gibberellins
Chapter 18 Abscisic Acid
Chapter 19 Ethylene
Chapter 20 Brassinosteroids
Chapter 21 Jasmonic Acid
Chapter 22 Recently Discovered Plant Growth Regulators
Chapter 23 Signal Perception and Transduction
Chapter 24 Embryogenesis, Vegetative Growth and Organogenesis
Chapter 25 Physiology of Flowering
Chapter 26 Pollination, Fertilization and Seed Development
Chapter 27 Fruit Development and Ripening
Chapter 28 Seed Dormancy and Germination
Chapter 29 Plant Movements
Chapter 30 Senescence and Programmed Cell Death

# Light Perception and Transduction

## 13

Satish C Bhatla

Plants utilize light not only for photosynthesis but also as an environmental signal for many developmental processes. Plants are very sensitive to seasonal, daily, and moment-to-moment changes in solar radiation. They are thus capable of perceiving wavelength, intensity, direction, duration, and other attributes of light to bring about appropriate physiological and developmental changes. These light-triggered growth and developmental responses are known as photomorphogenic responses (Box 13.1). Thus, **photomorphogenesis** may be defined as the developmental response of an organism to the information in light, which may be its quantity, quality (i.e., the wavelength), and direction or relative length of day and night (photoperiod). The importance of light in plant development can be most dramatically illustrated in case of early seedling growth. A dark-grown seedling is said to be **etiolated**. In etiolation, embryonic stem (hypocotyl in dicots and epicotyl in monocots) of seedlings exhibits very rapid and extensive elongation of internodes. There is no cotyledon/leaf expansion, and seedlings appear pale as there is no chloroplast development. So, etiolated plants are pale yellow, and the hypocotyl remains "hooked" at the apex. Curving of the hypocotyl is thought to protect the apical meristem from damage during seedling growth through the soil. On exposure to light, hypocotyl elongation slows down; cotyledons and leaves expand and become green. The apical hook straightens. In monocots, the etiolated coleoptile exhibits extended growth which gets decelerated (slows down) in light and developing leaves pierce its tip (Fig. 13.1). In contrast with **scotomorphogenesis**, seedlings grown in light exhibit photomorphogenesis and exhibit relatively shorter embryonic stem, lose apical hook, and develop expanded green cotyledons, and there is rapid initiation of leaf development at the shoot meristem. Upon sensing light, a seedling emerging from the soil switches from scotomorphogenesis to photomorphogenesis. This process is known as **de-etiolation**, and it involves the coaction of red/far-red light-absorbing phytochrome and blue light-sensitive cryptochrome.

**Box 13.1 History of Photobiology**
- Joseph Priestley (1772) discovered that green plants utilize light as their source of energy for production of organic substances.
- Julius Sachs (1864) demonstrated that blue region of visible light results in phototropic bending of plants.
- Charles Darwin and his son Francis (1881) examined light signal transduction of phototropism and demonstrated that photoreceptive site (shoot tip) is different from the region showing bending response (subapical region) in monocot seedlings.
- Julien Tournois (1914) discovered that night length rather than day length determines flowering time.
- Wightman Garner and Harry Allard (1920) discovered that most plants could be classified as "short-day" or "long-day" plants and established the concept of "photoperiodism."
- Karl Hamner and James Bonner (1938) made a decisive contribution to photoperiodism by finding that a brief exposure of light in midnight, given under normally inductive conditions for flowering, caused cocklebur, a short-day plant, to remain completely vegetative.
- Harry Borthwick and his colleagues (1952) discovered the red (R) and far-red (FR) photoreversible effect on seed germination in lettuce and night-break of photoperiodic floral induction in cocklebur.
- Warren Butler, Karl Noris, Bill Siegelman, and Sterling Hendricks (1959) showed photoreversible absorption changes at 660 nm and 730 nm upon alternately given R and FR actinic light in etiolated maize tissues and a crude extract of the relevant proteinaceous pigment.
- The term "phytochrome" was half-jokingly used by Butler in his laboratory and then published by Borthwick and Hendricks (1960).
- From the 1960s to 1980s, only phytochrome was the known photoreceptor for photomorphogenesis.
- Hans Mohr and Schäfer (1983) in Freiburg (Germany) extensively investigated the effect of blue and far-red light on photomorphogenesis in terms of sensor pigments, signal amplification, and gene expression and established the concept of "high-energy reaction."
- Margaret Ahmad and Tony Cashmore (1993) isolated an *Arabidopsis* mutant, *hy4*, which was defective in blue light-dependent photomorphogenesis. The protein encoded by Hy4 was a member of the photolyase family and was named cryptochrome (cry).
- Chentao Lin and colleagues (1996) cloned and characterized a second member of the cry family containing a distinct C-terminal sequence, which was named cry 2, and Hy-4 encoded cry was renamed as cry 1.
- Winslow Briggs and colleagues cloned and characterized genes of *nph* (non-phototropic hypocotyl) mutants and showed that the gene product of NPH1 was a blue light receptor.

Huala et al. (1997) cloned and characterized phototropin from *Arabidopsis thaliana*.

## 13.1 Light Absorption by Pigment Molecules

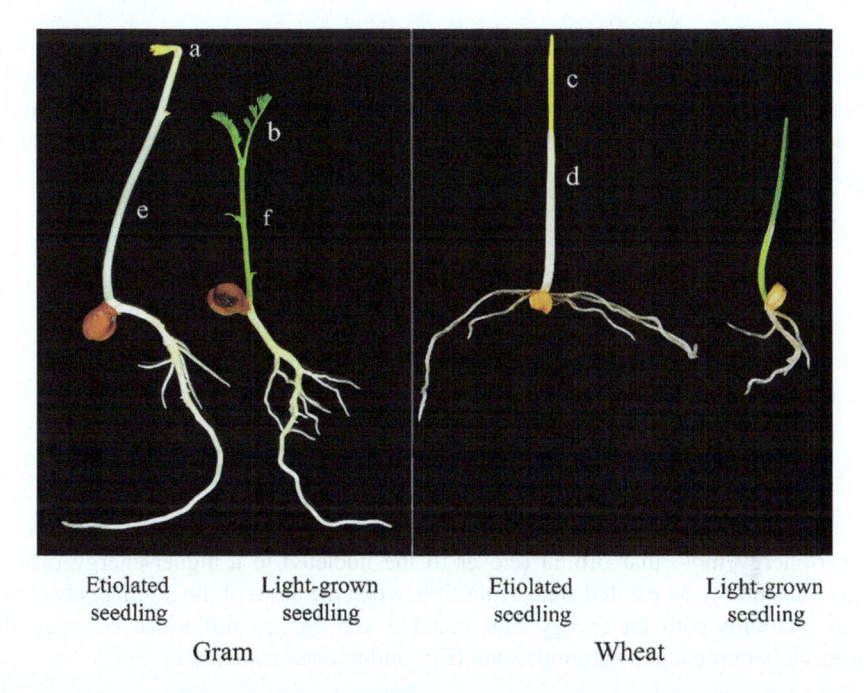

**Fig. 13.1** Characteristics of etiolated and light-grown dicot and monocot seedlings. (**a**) Apical hook; (**b**) open, expanded green leaves; (**c**) shoot; (**d**) coleoptile; (**e**) elongated hypocotyl; (**f**) short hypocotyl

Many seeds require light for germination, a process called **photoblasty** [e.g., lettuce (*Lactuca sativa*)]. In some plants, leaves fold up at night (**nyctinasty**) [e.g., lotus (*Lotus japonica*)] and open at dawn (**photonasty**) [e.g., common evening-primrose (*Oenothera biennis*)]. In contrast with directional bending of shoots in response to unilateral or unequally distributed light (**phototropism**), photonastic movements take place in response to nondirectional light. Many plants flower at specific times of the year in response to changing day length, a phenomenon called as **photoperiodism**. In order to understand and appreciate the importance of light to plants, it is necessary to understand the physical nature of light.

## 13.1 Light Absorption by Pigment Molecules

For light to be effective in inducing a response in any living being, it must first be absorbed. A molecule that can interact with photons from the visible portion of the electromagnetic spectrum is called a **pigment**. Any photobiological process requires a light-absorbing molecule or pigment. Plants contain a variety of pigments. Since pigments selectively absorb certain parts of the visible wavelength range, they appear colored to the human eye. The absorption of photons by the pigment molecule shifts the pigment molecule from its lowest-energy (ground) state to an excited state.

$$\text{Pigment} \xrightarrow{h\nu} \text{Pigment*}$$
$$\text{(ground state)} \qquad \text{(excited state)}$$

$h$ = Planck's constant; $\nu$ = photon's frequency

This is caused by the shifting of one of the electrons of the pigment from its lower-energy molecular orbital (closer to the nucleus) to a higher-energy orbital. This transition to an excited state is possible when the atom of the pigment absorbs a light quantum with an energy that matches the energy difference between the molecule's non-excited (ground) state (Eg) and excited state (Ee).

$$E_e - E_g = hc/\lambda$$

$h$ = Planck's constant, $c$ = speed of light, $\lambda$ = wavelength (nm)

### 13.1.1 Quantitative Requirement for Pigment Excitation

Two types of excited states can exist in light-sensitive molecules. The *singlet state* is relatively short-lived than the *triplet state*, in which electrons take longer to de-excite. An excited pigment molecule has a very short life (around $10^{-9}$ s), and it must get rid of excess energy and return to ground state. This dissipation of excess energy by the **exciton** (energized electron) is accomplished in several ways (Fig. 13.2):

1. *Thermal deactivation*: It involves loss of energy as heat into the environment and electron comes back to singlet state from ground state.
2. *Fluorescence*: It involves emission of light photons during a relatively slow process of return of exciton from first excited singlet state to ground state. The emitted photon has lower energy than its excited form, and so emission is in longer wavelength range. Thus, for example, chlorophyll molecules emit red fluorescence in solution upon excitation by blue light absorption.

## 13.1 Light Absorption by Pigment Molecules

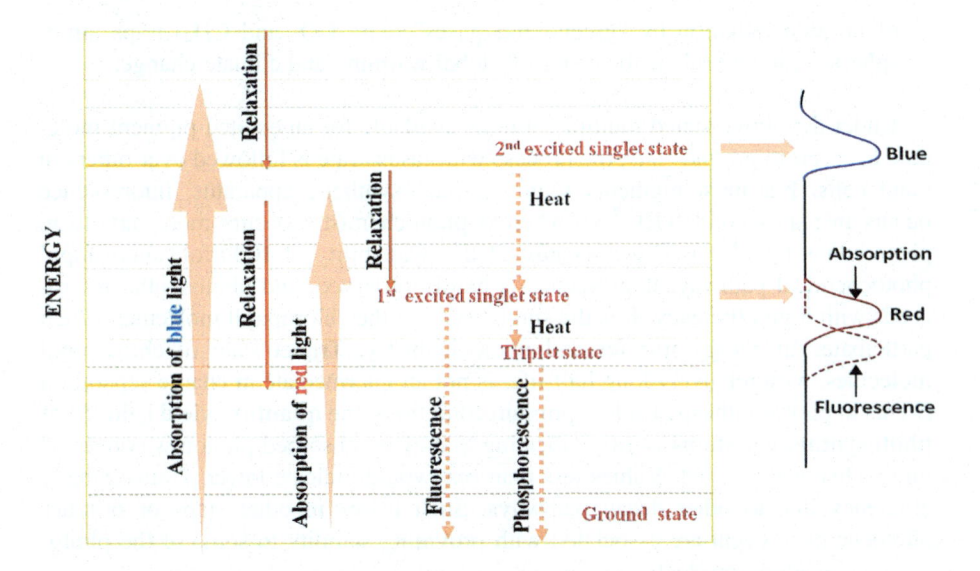

**Fig. 13.2** Different energy levels attained by chlorophyll molecules upon absorption of blue or red light

3. *Energy transfer (inductive resonance)*: This involves transfer of energy to another molecule, generally involving an array of pigment molecules. This process accounts for much of the transfer of energy between pigment molecules in chloroplasts.
4. *Charge separation (photochemistry)*: An excited molecule may transfer energy (**hv**) to another molecule, leaving the donor pigment molecule positively charged and the acceptor molecule negatively charged. This phenomenon is central to the process of photosynthesis.

$$\text{Pigment} + \text{acceptor} \xrightarrow{hv} \text{pigment}^* + \text{acceptor} \rightarrow \text{pigment}^+ + \text{acceptor}^-$$

5. *Shift to triplet state*: A molecule in singlet state may revert to another excited state, called the **triplet state**. Triplet state is more important than singlet state because longer lifetime of the metastable triplet state (around $10^{-3}$ s) allows photooxidation of the pigment (donor), and the acceptor molecule is reduced.
6. *Vibrational energy*: Although the energy levels of infrared and other longer wavelengths are very low to facilitate electron jump, their absorption by molecules can create vibrational energy in bonding systems. Thus, absorption

of infrared radiation by "greenhouse gases" (e.g., $CO_2$ and $CH_4$) traps atmospheric heat, and this is the cause of global warming and climate change.

Under the above-stated multiple options available for an excited pigment molecule to revert to ground state, the process with fastest rate is followed over others in plant cells. For most pigments found in photosynthetic apparatus, fluorescence occurs in nanoseconds ($10^{-9}$ s), whereas photochemistry occurs more rapidly in picoseconds ($10^{-12}$ s). Thus, because of the availability of 1000-fold more rapid photochemical pathway, fluorescence is hardly observed, and photosynthesis proceeds with high efficiency. It is the singlet state of the chlorophyll molecules which participates in energy transfer and photochemistry. Triplet state of chlorophyll molecules, with relatively long half-life, is not an intermediate in charge separation events in photosynthesis. Under optimum conditions, the **quantum yield (phi; $\phi$) of photosynthetic systems** is 1.0, indicating that every absorbed photon is converted into a chemical product. Values less than one would indicate lower photosynthetic efficiency due to other decay pathways. Such losses in other steps of primary photochemical event are associated with providing stability to some of the photochemical reaction products.

$$\text{Quantum yield }(\phi) = \frac{\text{Number of photochemically formed products}}{\text{Number of quanta involved}}$$

$$\text{Maximum value} = 1.0$$

## 13.2 Nature of Light

By passing white light through a prism, English physicist Sir Isaac Newton (1642–1727) separated light into a spectrum of visible colors, thereby demonstrating that white light consists of different colors, ranging from violet at one end and red at the other end of the spectrum. Separation of light of different colors is possible because they are refracted at different angles while passing through a prism. Subsequently, another British physicist, James Clerk Maxwell (1831–1879), demonstrated that white light is a small component of the vast spectrum of the **electromagnetic radiation**. The radiations in the electromagnetic spectrum travel in waves. The **wavelength** refers to the distance from the crest of one wave to the crest of the next (Fig. 13.3). The wavelengths in the electromagnetic spectrum range from cosmic rays (measured in nanometer; 1 nm = $10^{-14}$ meter) to those of low-frequency radio waves (measured in kilometers; 1 km = $10^3$ meter = 0.6 mile) (Fig. 13.4). Shorter wavelengths have greater energy associated with them. Conversely, longer wavelengths have lower energy. In the visible spectrum of light,

## 13.2 Nature of Light

**Fig. 13.3** The wave nature of light

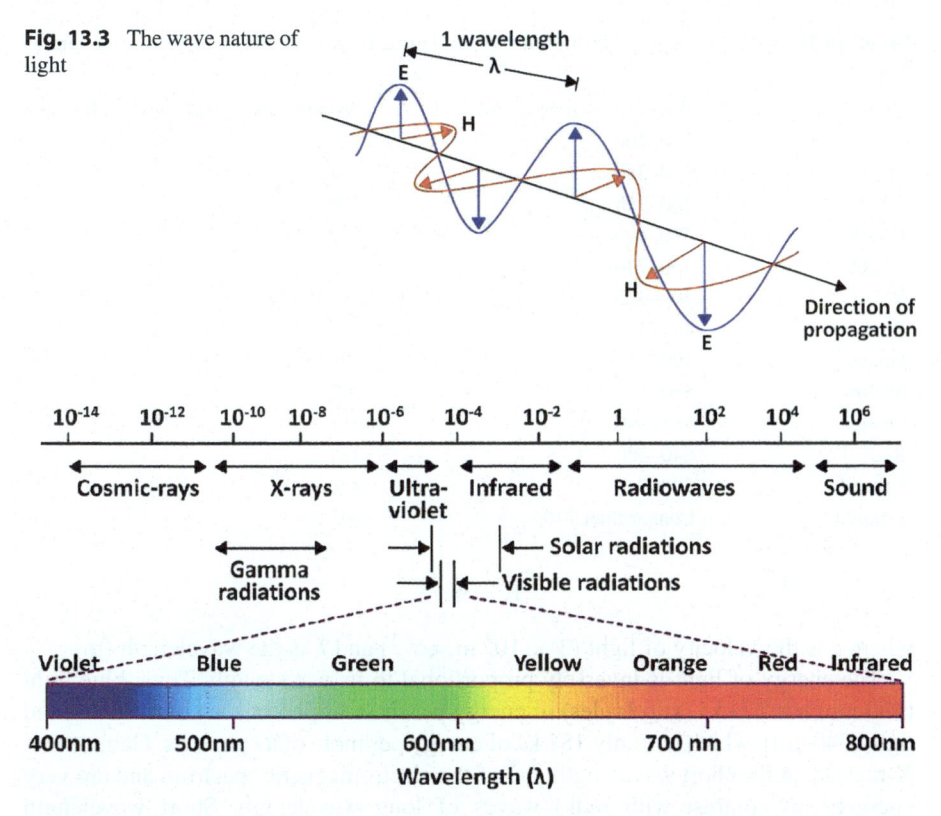

**Fig. 13.4** The electromagnetic spectrum. Visible light is only a small part of the electromagnetic spectrum

violet light has the shortest wavelength and has almost twice the energy as compared to of the longest rays of far red (Table 13.1).

According to the **particle model of light** proposed by Albert Einstein in 1905, light is composed of particles of energy, called **photons** or **quantas**. The energy of a photon or quanta of light is inversely proportional to its wavelength, i.e., the longer the wavelength, the lower is the energy associated with it. Both the above-stated models of light (wave model and particle model) are complementary to each other and are required for a full understanding of the properties of light.

Each quantum or photon of light contains energy equal to **Planck's constant**, h [$6.626 \times 10^{-34}$ Joule ($s^{-1}$) multiplied by the frequency of the radiation, $\nu$, in cycles per second ($s^{-1}$)]

$$E = h\nu$$

Since different wavelengths of light have different energy levels, the energy of photons of a particular wavelength can be described as

**Table 13.1** Wavelengths capable of causing biological responses and their respective energy levels

| Color | Wavelength range (nm) | Average energy (kJ. mol$^{-1}$ photons) |
|---|---|---|
| Ultraviolet | 100–400 | |
| UV-C | 100–280 | 471 |
| UV-B | 280–320 | 399 |
| UV-A | 320–400 | 332 |
| Visible | 400–740 | |
| Violet | 400–425 | 290 |
| Blue | 425–490 | 274 |
| Green | 490–550 | 230 |
| Yellow | 550–585 | 212 |
| Orange | 585–640 | 196 |
| Red | 640–700 | 181 |
| Far-red | 700–740 | 166 |
| Infrared | Longer than 740 | 85 |

$$E = hc/\lambda$$

where c is the velocity of light ($3 \times 10^8$ m.sec$^{-1}$) and $\lambda$ is the wavelength (m).

The energy of light is inversely proportional to its wavelength. Thus, blue light (490 nm) has 274 kJ (kilo Joules) of energy per mole of photons in contrast with red light (740 nm), which has only 181 kJ of energy per mole of its photons. Gamma and X-rays are at the short wavelength end of the electromagnetic spectrum and are very energetic in contrast with radio waves of long wavelength. Short wavelength photons are energetic enough to provide sufficient kinetic energy to break electrons free from atoms. Therefore, these radiations are referred as **ionizing radiations**. These radiations (wavelength < 295 nm) are filtered by the atmosphere and do not reach earth, thereby avoiding their hazardous effect on living beings. Ultraviolet (UV) radiation is generally subdivided into UV-A (400–320 nm), UV-B (320–280 nm), and UV-C (280–100 nm). It may be noted that oxygenic photosynthetic organisms use visible light (400–700 nm) for photosynthesis, growing in anoxygenic, whereas many photosynthetic organisms can photosynthesize in near-infrared wavelengths greater than 700 nm.

## 13.3 Absorption and Action Spectra

Absorption of light by a pigment (e.g., chlorophyll) plotted as a function of wavelength is known as **absorption spectrum**. Such a spectrum provides information about the extent to which principal bands of light spectrum are absorbed by the pigment. The height and/or width of the absorption curve indicates the extent by which light of that wavelength is absorbed. Each light-absorbing molecule has a unique absorption spectrum. Thus, an absorption spectrum is like a fingerprint of the pigment molecule which serves as a key to its identification. Varying forms of

**Fig. 13.5** (A) Action and (B) absorption spectra of chlorophyll pigments. The peaks in red and blue regions of the action spectrum correspond with principal absorption peaks

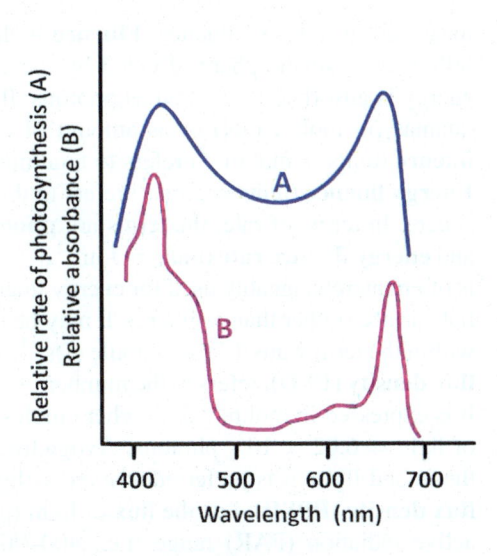

chlorophyll molecules absorb optimally at specific wavelengths, indicated as peaks of absorbance in the absorption spectrum wavelengths of light. An **action spectrum** is a graph which depicts light action/response. One of the earliest action spectra was generated by T.W. Engelmann in the late 1800s by using a prism to disperse sunlight into a spectrum. These separated wavelengths of visible light were used to illuminate *Spirogyra* filaments which lead to preferential accumulation of oxygen-seeking bacteria in the filament's regions exposed to blue and red light. These regions of light are strongly absorbed by chlorophyll molecules leading to enhanced photosynthesis (and oxygen evolution). Thus, analysis of action spectra provides critical information for the discovery of photosynthesis in oxygen evolving photosynthetic organisms. In modern laboratories, action spectra are also generated by using **spectrographs** fitted with monochromatic sources of light. A comparison of the absorption and action spectra of a pigment can provide useful information on the identity of the pigment responsible for a photosynthetic process. Figure 13.5 shows typical absorption and action spectra of chlorophyll from leaves. It may be noted that the action spectrum has pronounced peaks in red and blue regions of the spectrum which correspond with absorption maxima for chlorophyll, thereby indicating the role of chlorophyll in photosynthesis.

## 13.4 Light Parameters Which Influence Plant Responses

Light can influence the physiology and development of plants in many ways. Each light source may have different effects on plant development and behavior due to variations in spectral distribution. Three parameters of light play significant role in modulating plant development. These are 1. light intensity, 2. light composition/(quality), and 3. light duration (photoperiod). Light intensity is most commonly measured in terms of fluence. **Fluence** is defined as the amount of radiant energy falling on a small sphere, divided by the cross section of the sphere. Since light energy is absorbed or emitted as photons, fluence is expressed either as photons or quantas (in moles, mol) or as amount of energy (in Joules, J). The term **photon fluence** (units = $mol.m^{-2}$) refers to total number of photons incident on the sphere. **Energy fluence** (units = $J.m^{-2}$) refers to the total amount of energy incident on the sphere. In terms of rate, the terms are **photon fluence rate** (units = $mol.m^{-2}.s^{-1}$) and **energy fluence rate** (units = $J.m^{-2}.s^{-1}$ or $W.m^{-2}$). **Irradiance** ($W.m^{-2}$), which is often interchangeably used for energy fluence rate, refers to the flux of energy on a flat surface (rather than a sphere). It may be noted that time (seconds, s) is contained within the term watts $1 W = 1$ Joule $(J)s^{-1}$. Another term, **quantum flux** or **photon flux density (PED)**, refers to the number of incident quantas of light striking the leaf. It is expressed in $mol.m^{-2}.s^{-1}$, where moles refers to the number of photons (1 mol of light = $6.02 \times 10^{23}$ photons; Avogadro number). When considering photosynthesis and light, it is preferred to express light intensity as **photosynthetic photon flux density (PPFD)**, i.e., the flux of light ($\mu mol.m^{-2}.s^{-1}$) in the photosynthetically active radiation (PAR) range (i.e., 400–700 nm). On a clear sunny day, plants usually experience a PPFD of about 200 $\mu mol.m^{-2}.s^{-1}$ at the top of the forest canopy. It may be as little as 10 $\mu mol.m^{-2}.s^{-1}$ at the bottom of the canopy.

## 13.5 Light Absorption Depends on Leaf Anatomy and Canopy Structure

On an average, each square meter of earth surface receives about 340 W of solar energy every day. But a major fraction of solar radiation (50%) is either of too short range or too long range to be absorbed by the photosynthetic pigments. Furthermore, only a small percentage of photosynthetically active radiation (PAR, 400–700 nm) incident on a leaf is surface transmitted through the leaf, and the rest (15%) is reflected or transmitted from the leaf surface (Fig. 13.6). Since chlorophylls absorb blue and red wavelengths of light most strongly, green wavelengths are maximally transmitted and/or reflected, giving green color to the vegetation. A fraction of PAR absorbed by the leaf is also lost through metabolism (20%) and as heat (10%). Thus, at the end, about 5% of the sun's total radiant energy falling on leaf surface is utilized in photosynthesis to produce carbohydrates (Fig. 13.7).

Leaf anatomy is well adapted for maximum utilization of light. The epidermal cells are generally convex so that they focus light for enhanced capture by the chloroplasts. Epidermal focusing of light is common among herbaceous plants and in tropical plants growing on the forest floor where light levels are very low. The palisade cells below the epidermis are shaped like pillars arranged in parallel columns which provide sieve effect and channeling of light. The **sieve effect** is caused by nonuniform distribution of chlorophyll molecules within the chloroplasts which results in shaded areas in chloroplasts which do not receive light. Consequently, much less light is absorbed in palisade cells than would be expected if chlorophyll molecules are uniformly distributed. Presence of large vacuoles and air

## 13.5 Light Absorption Depends on Leaf Anatomy and Canopy Structure

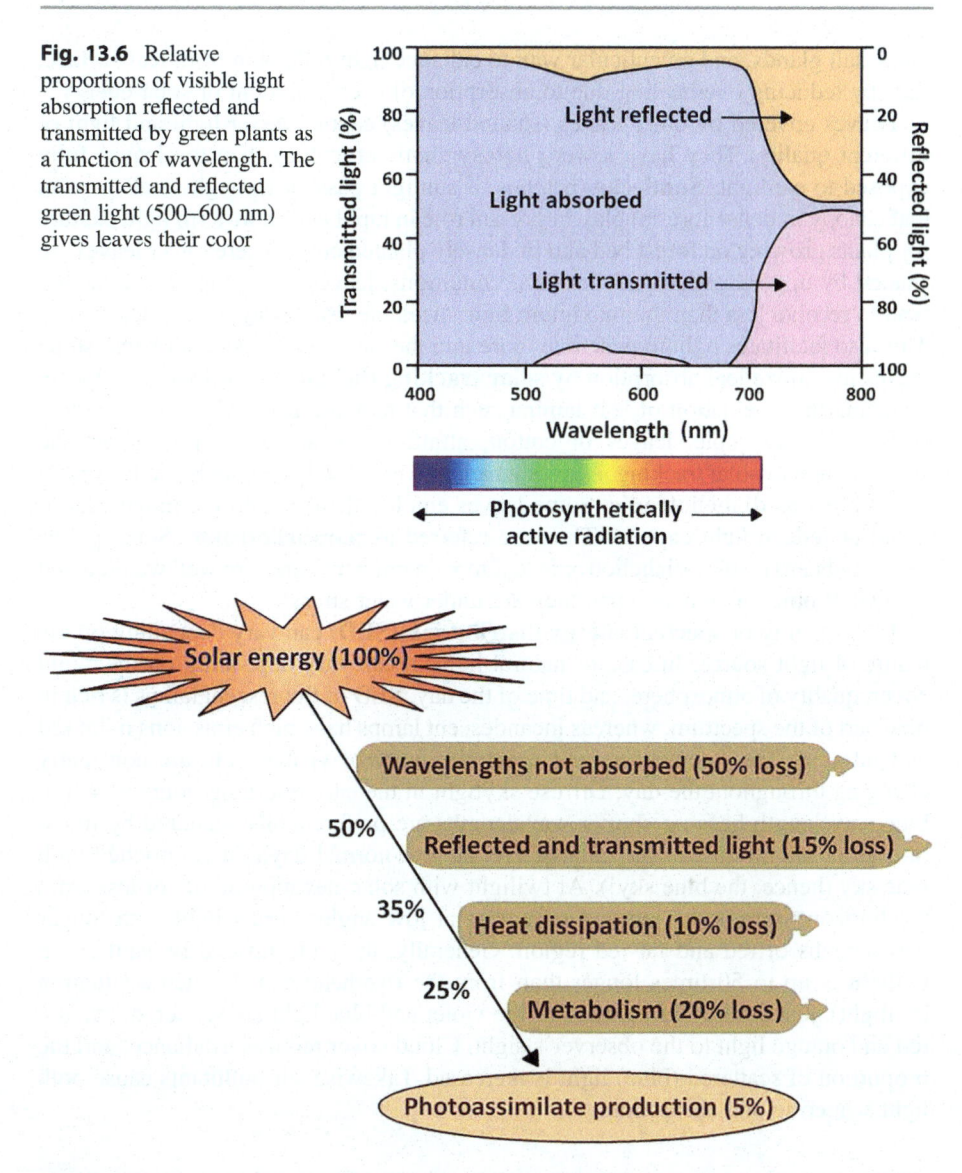

**Fig. 13.6** Relative proportions of visible light absorption reflected and transmitted by green plants as a function of wavelength. The transmitted and reflected green light (500–600 nm) gives leaves their color

**Fig. 13.7** Utilization of solar energy in metabolism and photoassimilate (carbohydrate) production in leaves

spaces in the zone of palisade cells causes **light channeling** into deeper regions of the leaf. The large air spaces in the cluster of irregular-shaped spongy mesophyll cells reflect and refract light, thereby causing light scattering. Thus, sieving effect of palisade cells and scattering effect of spongy cells together contribute in an efficient absorption of light throughout the leaf. Desert plants exposed to excess light develop hairs, salt glands, and epicuticular wax to enhance light reflection from leaf surface, thereby reducing overheating due to absorption of excess amount of solar energy.

Leaves covered by other leaves (shaded leaves) capture lesser light and light of different quality. They have lower photosynthetic efficiency than the leaves fully exposed to sunlight. **Sunflecks** (patches of sunlight passing through small gaps in leaf canopy in dense forests) play important role in rapid and short-term light capture for plants growing on forest bed and in densely planted crops where lower leaves are shaded by upper leaves. Under natural conditions, leaves at the top of the canopy tend to receive less than the maximum light available, by having a steep leaf angle. This also facilitates light to penetrate more into the canopy. On the other hand, some leaves maximize leaf absorption by **solar tracking (heliotropism)** by continuously adjusting the orientation of leaf lamina such that they are perpendicular to the leaf surface. For example, leaves of cotton, alfalfa, soybean, and lupine show the phenomenon of solar tracking. Leaves which maximize light capture by heliotropism are referred as **diaheliotropic**. Some leaves employ their heliotropic movement to avoid or reduce light capture. They are referred as **paraheliotropic**. Some plants (e.g., soybean) display diaheliotropic leaf movement when they are well watered and paraheliotropic movement when they are under water stress.

Light quality or **spectral energy distribution (SED)** can vary depending on the nature of light source. In case of natural light, it can vary depending on the cloud cover, quality of atmosphere, and time of the day. SED of fluorescent lamps is rich in blue part of the spectrum, whereas incandescent lamps have high emissions in far red and infrared. The fluence rate and spectral quality of visible light are constantly changing throughout the day. Diffuse skylight in the early morning hours is rich in blue wavelength because shorter wavelengths are preferentially scattered by moisture droplets and dust of the atmosphere. Hence, normal daylight is enriched with blue sky (hence, the blue sky!). At twilight with solar elevation of $10°$ or less from the horizon, light scattering and refraction at low angles enrich light with longer wavelengths of red and far-red region. Generally, the path traveled by sunlight at twilight is up to 50 times longer than from the overhead sun. In such a situation (twilight), just before sunset much of the violet and blue light are scattered, leaving red and orange light to the observer's sight. Cloud cover reduces irradiance, and the proportion of scattered (blue) light is increased. Likewise, air pollutants cause both light scattering and absorption.

## 13.6 Photoreceptors

Plant pigments, such as chlorophyll and accessory pigments of photosynthesis, absorb visible light at specific wavelengths and reflect or transmit non-absorbed wavelengths, which are perceived as colors. **Photoreceptors**, unlike plant pigments (such as chlorophyll), are chromophore-containing biomolecules which absorb photons of a given wavelength and use the energy derived from it as a signal to initiate a photoresponse. All photoreceptors contain a protein (**chromoprotein**) attached to a light-absorbing, non-protein, prosthetic group called as **chromophore**.

The protein part of the chromoprotein is called **apoprotein**. The complete molecule, or **holochrome**, consists of chromophore plus the protein. Other common features of all photoreceptors are their sensitivity to light quantity (number of photons), quality (wavelength), and duration of exposure. All photoreceptors perceive light and initiate cellular signals leading to a response. In the following pages, a detailed account of plant photoreceptors is being presented.

## 13.7 Protochlorophyllide

Etioplasts are plastids whose development from proplastids to chloroplasts has been arrested due to the absence of light. They lack chlorophyll but produce **protochlorophyllide—a precursor of chlorophyll a**. Etioplasts contain a prominent structure called **prolamellar body** which contains membrane lipids, protochlorophyllide, and the light-requiring enzyme—**protochlorophyllide oxidoreductase (POR)**. Prolamellar bodies are formed when membrane lipid synthesis continues in the absence of corresponding amounts of thylakoid protein synthesis, which requires light. The high concentration of lipids in the prolamellar bodies (75%) results in the formation of lipid tubes which branch in three dimensions to form a semicrystalline lattice. When etioplasts are illuminated, they begin to develop into chloroplasts. Light triggers chlorophyll biosynthesis from protochlorophyllide. It also results in the assembly of stable chlorophyll-protein complexes, resulting in the outgrowth of thylakoid membranes from the prolamellar body. Protochlorophyllide is structurally similar to chlorophyll a, except that it has a double bond between carbon C17 and C18 in D ring (Fig. 13.8). Protochlorophyllide is not green and cannot absorb light for photosynthesis. It is a photoreceptor with an absorption maximum at 650 nm. Light absorption by protochlorophyllide takes place in the presence of NADPH-protochlorophyllide oxidoreductase (POR), resulting in the addition of protons to the double bond between C17 and 18 in ring D, and protochlorophyllide gets converted into chlorophyll a. It is now established that aerobic photosynthetic bacteria, liverworts, and gymnosperms contain a light-independent form of POR. At some point during evolution, angiosperms lost this enzyme, and they have only light-dependent form of POR. Thus, angiosperms have an absolute requirement of light for chlorophyll a biosynthesis in chloroplasts. Light-

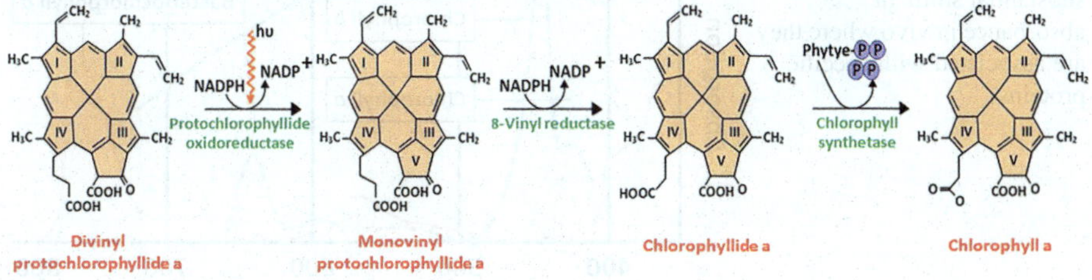

**Fig. 13.8** Conversion of protochlorophyllide a into chlorophyll a in response to light

dependent POR is a monomeric enzyme of 35–38 kDa. Barley has two POR genes, while *Arabidopsis* has three. In *Arabidopsis*, POR A is specifically expressed in dark and is downregulated by phytochrome. POR B is constitutively expressed, and POR C is induced by light. The lattice-like structure of prolamellar bodies in etioplasts is largely composed of a complex of protochlorophyllide, NADP and POR. Upon illumination, POR rapidly converts protochlorophyllide into chlorophyllide, and the internal structure of etioplasts is reorganized into a typical chloroplast. Subsequently, POR is released and degraded by proteolysis. Thus, newly developed chloroplasts do not contain POR protein.

$$\text{Protochlorophyllide} + \text{NADPH} + \text{H}^+ \xrightarrow[\text{Light}]{\text{POR}} \text{Chlorophyllide a} + \text{NADP}$$

The photoreceptor properties of POR are comparable with those of phytochrome in the following ways:

1. POR possesses a tetrapyrrole chromophore which is responsible for triggering photomorphogenic changes (chloroplast development).
2. The form and abundance of POR are regulated by light.

Most of the chlorophyll molecules in chloroplasts are conjugated with thylakoid proteins. The absorption spectra of pigment-protein complexes are markedly different from that of free pigment in solution (Fig. 13.9). This chlorophyll-protein conjugation helps to maintain the pigment molecules in the precise relationship for efficient light absorption and energy transfer. It also provides a unique environment to each pigment molecule, resulting in their specific absorption maxima. These slight differences in absorbance maxima facilitate an orderly transfer of energy through various pigments to the reaction center for photochemical reactions. Lastly, chlorophyll-protein interaction also reduces photosensitization of plants which

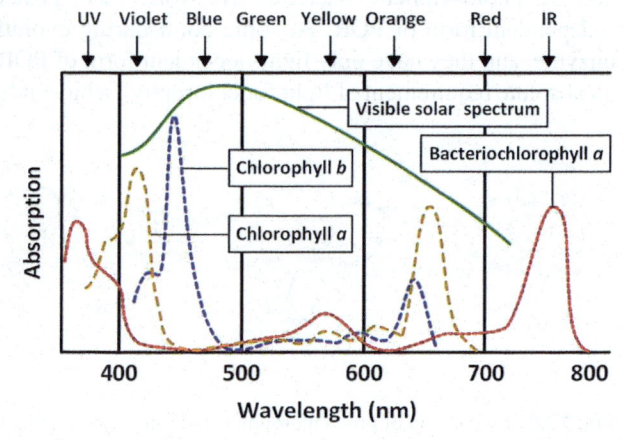

**Fig. 13.9** Absorption spectra of chlorophyll a, b and bacteriochlorophyll in solution. Note that the spectra of these pigments exhibit substantial shifts in absorbance in vivo where they are associated with specific proteins

would lead to destruction of chloroplasts. Free form of chlorophyll is less efficient in photosynthetic energy transfer, but it reacts more readily with oxygen to produce reactive oxygen species (ROS) such as oxygen free radicals, hydroxyl radicals, and singlet oxygen, which can disrupt chloroplasts.

## 13.8 Phycobilins

Phycobilins are open-chain tetrapyrrole pigment molecules present in red algae and cyanobacteria (Fig. 13.10). Of the four phycobilins, three (phycoerythrin, phycocyanin, and allophycocyanin) are involved in photosynthesis, and the fourth one (phytochromobilin) is an important photoreceptor (phytochrome) that regulates various aspects of growth and development. In addition to their open-chain tetrapyrrole structure, the phycobilin pigments differ from chlorophylls by the fact that the tetrapyrrole group is covalently linked with a protein, and these phycobilins are organized into macromolecular complexes called **phycobilisomes**. Phycocyanin, phycoerythrin, and allophycocyanin are exclusively found in cyanobacteria and red algae where they function for light harvesting during photosynthesis. In addition to other wavelengths, these pigments (particularly phycoerythrin) also absorb energy in green region of the visible spectrum where chlorophylls do not absorb. The red algae appear almost black because chlorophyll and phycoerythrin together absorb almost all wavelengths of the visible spectrum (Fig. 13.11). The fourth phycobilin

**Fig. 13.10** Structure of phycocyanin chromophore—an open-chain tetrapyrrole molecule

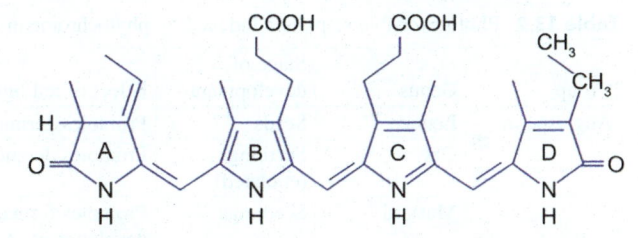

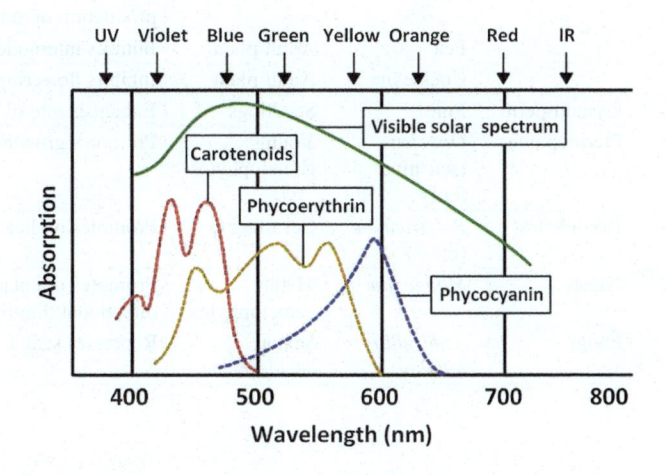

**Fig. 13.11** Absorption spectra of accessory photosynthetic pigments

(phytochromobilin) is **phytochrome** which is a receptor in higher plants and plays important roles in many photomorphogenesis phenomena.

## 13.9 Phytochromes

Phytochromes were first identified among flowering plants as photoreceptors for photomorphogenesis in response to red light and far-red light. They strongly absorb red (620–700 nm) and far-red (700–800 nm) wavelengths but can also absorb blue light (350–500 nm) and UV-A radiation (320–400 nm). Phytochromes were discovered through experiments on **red/far-red reversibility** of seed germination in lettuce. It is now known that phytochromes are members of a gene family present in all land plants, streptophyte algae, cyanobacteria, other bacteria, fungi, and diatoms. Since red or far-red light cannot penetrate deep water, phytochromes in deep water aquatic organisms, mainly algae, have been reported to sense orange, green, or even blue light indicating the potential of phytochromes to be separately tuned to absorb different wavelengths during natural selection. Table 13.2 highlights some of the established photoreversible roles of phytochrome in a variety of organisms.

**Table 13.2** Photoreversible responses induced by phytochrome in a variety of organisms

| Group | Genus | Stage of development | Effect of red light |
|---|---|---|---|
| Angiosperms | Lettuce | Seeds | Promotes germination |
|  | Oat | Seedlings (etiolated) | Promotes de-etiolation (e.g., leaf unrolling) |
|  | Mustard | Seedlings | Promotes formation of leaf primordial, development of primary leaves, and production of anthocyanin |
|  | Pea | Adult plant | Inhibits internode elongation |
|  | Cocklebur | Adult plant | Inhibits flowering (photoperiodic response) |
| Gymnosperm | Pine | Seedlings | Enhances rate of chlorophyll accumulation |
| Pteridophytes | *Onoclea* (sensitive fern) | Young gametophytes | Promotes growth |
| Bryophytes | *Polytrichum* (moss) | Germling | Promotes replication of plastids |
| Algae | *Mougeotia* | Mature gametophytes | Promotes orientation of chloroplasts to directional dim light |
| Fungi | *Aspergillus nidulans* | Sexual development | Represses sexual development |

## 13.9.1 Photoreversibility of Phytochrome-Modulated Responses and Its Significance

**Photoreversibility** (**photoconversion**; also referred as **photochromism**) is a defining feature of phytochromes. Phytochromes are synthesized as an inactive red light-absorbing isomer, called $P_r$. Red light absorption by $P_r$ form converts it into an active, far-red-absorbing isomer called $P_{fr}$. Absorption of far-red light by $P_{fr}$ form converts it back to $P_r$ form. Thus, red light promotes phytochrome responses by converting inactive $P_r$ into active $P_{fr}$ form. Far-red light inhibits the red light-induced response by converting $P_{fr}$ back to $P_r$ spontaneously as well. Such a spontaneous reversion of $P_{fr}$ to $P_r$ occurs both in light and dark. Since it is measured by a decrease in $P_{fr}$ concentration, it is referred as **dark reversion**. Phytochrome pool is never fully converted to $P_{fr}$ or $P_r$ form because the absorption spectra of $P_{fr}$ and $P_r$ forms overlap (Fig. 13.12). When $P_r$ molecules are exposed to red light, most of it is converted to $P_{fr}$, but some of $P_{fr}$ is spontaneously converted back to $P_r$. The proportion of phytochrome in $P_{fr}$ form after providing saturating red light condition is about 88%. Likewise, a very small amount of far-red light absorbed by $P_{fr}$ form leads to incomplete conversion of $P_{fr}$ to $P_r$. Thus, an equilibrium of 98% $P_r$ and 2% $P_{fr}$ is established. This state of equilibrium is termed as **photostationary state**. Photoconversion of phytochrome provides assessment of the available red-to-far-red light ratio in the plant's environment. Low R:FR ratio has a variety of development effects on various species. It inhibits seed germination in *Arabidopsis* and induces shade escape responses during plant growth. Plants growing under a canopy use phytochrome to sense R:FR ratio so as to regulate shade avoidance, seed germination, etc. In natural conditions, plants are exposed to a broad spectrum of light. R:FR ratio is strongly affected by plant canopy because chlorophyll absorbs red light but not far-red light, thereby affecting R:FR ratio for plants growing below the canopy.

**Fig. 13.12** Absorption spectra of two forms of phytochrome ($P_r$ and $P_{fr}$)

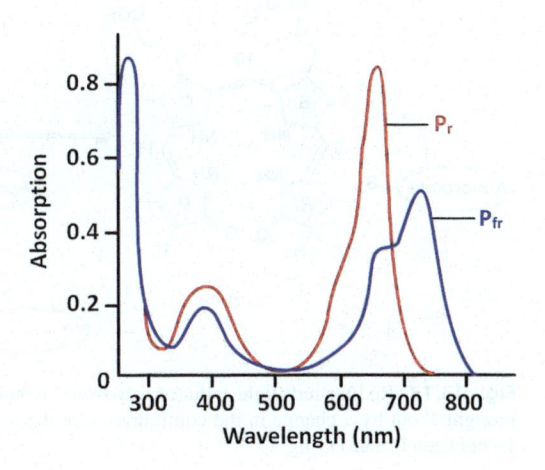

## 13.9.2 Chemical Nature of Phytochrome Chromophore

Phytochrome is a biliprotein consisting of 120 kDa apoprotein with a photosensitive prosthetic group (**chromophore**). The chromophore is a straight chain tetrapyrrole, a bilin called as **(3E)-phytochromobilin** or PϕB, where 3E refers to the isomeric form of the photoactive molecule. Phytochromobilin is synthesized inside the plastids and shares its biosynthetic origin with cyclic tetrapyrrole molecules, such as heme and chlorophyll. Phytochromobilin is exported from plastids into the cytosol where it gets attached to the apoprotein through a thioether linkage (ethers in which oxygen is replaced by sulfur in a cysteine residue of the protein). The four pyrrole rings are referred as A, B, C, and D. Phytochromobilin exists in two molecular forms: PϕB$_r$ and PϕB$_{fr}$. The red wavelength-absorbing form of phytochromobilin (PϕB$_r$) has a peak absorbance at 660 nm and gets converted into PϕB$_{fr}$, which has a peak absorbance at 730 nm in the far-red region. PϕB$_{fr}$ gets converted into PϕB$_r$ form upon exposure to far-red light (Fig. 13.13). The phytochrome biliprotein, in association with PϕB$_r$ chromophore, is the **red-absorbing form** of the phytochrome and is referred as P$_r$. The **far-red-absorbing form** is P$_{fr}$ with chromophore in PϕB$_{fr}$ configuration. Red light exposure brings about a cis to trans change in configuration in the methane bridge between C and D pyrrole rings. *Far-red light drives it back into cis form*. PϕB$_{fr}$ also gets converted to PϕB$_r$ in dark through a slow process (dark reversion). *It is the PϕB$_{fr}$ form which triggers photomorphogenesis*. The biological output of phytochrome cycle is estimated by the ratio of PϕB$_r$ and PϕB$_{fr}$ forms of the chromophore. This, in turn, gives an indication of the relative levels of red and far-red light in the plant's environment.

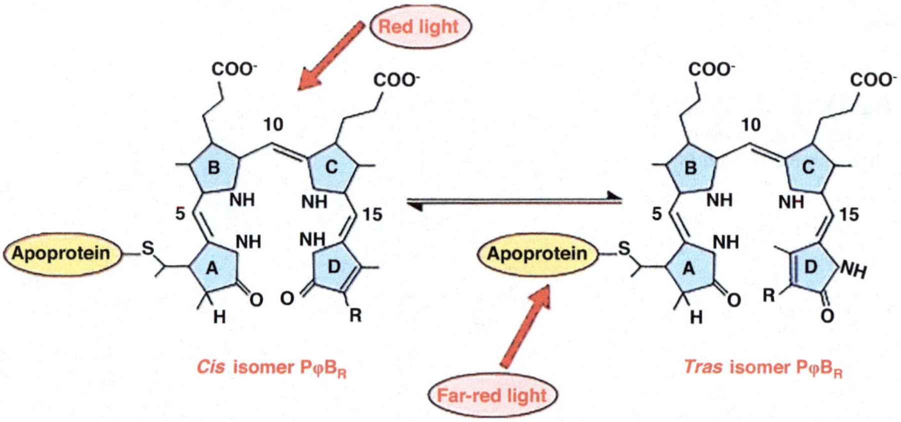

**Fig. 13.13** Red/far-red light-induced *cis-trans* isomerization of phytochrome chromophore brought about by a change in the configuration at the double bond between carbon atoms 15 and 16 between C and D ring

## 13.9.3 The Multidomain Structure of Phytochrome Protein

Phytochromes are soluble proteins and exist as dimers. Each subunit of the dimer is bound to a phytochromobilin molecule. Light absorption by phytochromobilin leads to its isomerization which results in changes in its absorption spectrum and conformation of the phytochrome protein. The phytochrome protein consists of an N-terminal **photosensory region** and a C-terminal **regulatory domain**. The two are attached with a **hinge** between them (Fig. 13.14). The photosensory part has four domains (P1 to P4). P1 and P4 function in the inhibition of dark reversion of $P_{fr}$ to $P_r$. $P_2$ (a PAS type of domain) is involved in sensing of signals, and $P_3$ is the binding site for the phytochromobilin molecule. The regulatory region of phytochrome protein consists of two PAS-type domains. PAS *(Per-ARNT-Sim) refers to some (α/β) conserved structural domains in a wide range of prokaryotic and eukaryotic proteins involved in sensing of signals*. C-terminal domain of the regulatory region exhibits kinase activity. It may be noted that higher plant phytochrome lacks a functional histidine kinase (HK) domain which is present in bacteria. They (higher plants) contain histidine kinase-related domain (HKRD) in place of functional histidine kinase of prokaryotes. HKRDs are 13–17% identical to HKs of the bacterial two component system.

In the red-absorbing form of the holoprotein ($P_r$), both subunits of the phytochrome dimer are folded at the hinge so that N-terminal photosensory and C-terminal regulatory regions are brought into contact with each other. Red light exposure opens up the structure in the far-red-absorbing form of phytochrome (Pfr), making its surface accessible to **phytochrome-interacting factors (PIFs)** in the signal transduction pathway. To sum up, phytochrome is activated in the cytoplasm by the absorption of light and enters the nucleus as $P_{fr}$, and $P_{fr}$ interact with PIFs to alter gene transcription. The opened-up structure of phytochrome protein allows **light-dependent phosphorylation** of exposed serine and threonine residues, possibly

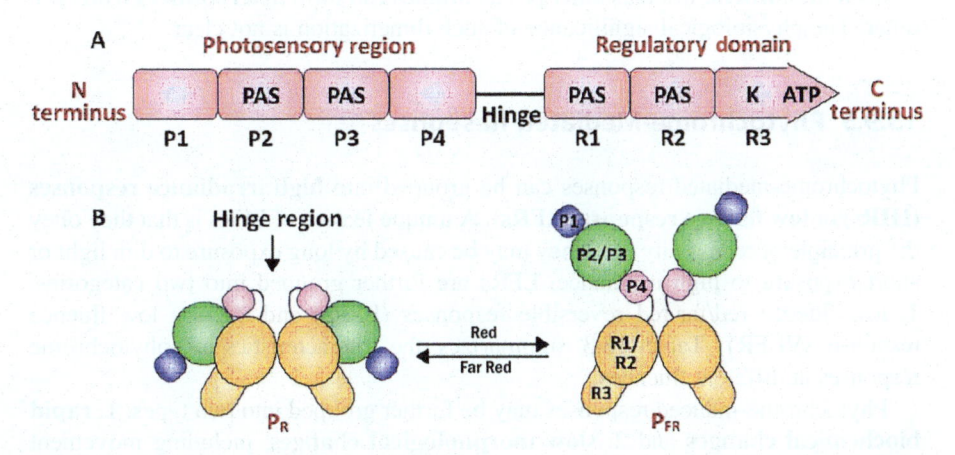

**Fig. 13.14** Structure of phytochrome apoprotein and modulation of the folding pattern of its dimer by red/far-red light

triggered by autophosphorylation by R3 kinase at the C-terminal. A $P_{fr}$-specific **phosphatase** removes phosphate group from the amino acids in the hinge region, thereby enhancing the affinity of phytochrome protein for interaction with PIFs. Phytochrome probably also acts as a signal in cytoplasm in addition to its action through nuclear gene transcription. Such cytoplasmic signaling might occur via phosphorylation of substrate proteins by phytochrome. Conformational differences between $P_r$ and $P_{fr}$ are also responsible for differences in their relative stability. *Thus, the half-life of $P_r$ is about 1 week, whereas that of $P_{fr}$ is 2 h.* This also reflects the difference in the compactness of the protein. Thus, phytochrome action is under a fine control of red/far-red modulated phosphorylation/dephosphorylation events.

### 13.9.4 Forms of Biologically Active Phytochrome

Different forms of phytochrome are produced by plants. Five genes encode for phytochrome in *Arabidopsis* (PHY A–E). Rice has three genes (PHY A–C). Most conifers have four genes, cycads have three, ferns have two, and lycopods have one gene encoding phytochrome. Phytochromes produced by these genes have different molecular properties. Five different phytochromes (phy A–E) produced in *Arabidopsis* possess identical chromophores, showing the same absorption spectra, but have different apoproteins displaying distinct functions. Only phy A is light labile, and all others (phy B-E) are light stable. The light-labile form of phytochrome (phy A) is predominant in etiolated seedlings which strongly express PHY A gene, resulting in high accumulation of $P_rA$. Prolonged light exposure causes phy A concentration to drop by almost 100-fold. Light also inhibits PHY A transcription. Though less abundant, phy A still functions in light-grown plants. In contrast, phy B–E (light-stable phytochromes) have the similar abundance in both light- and dark-grown plants and are majorly found in mature plants. In *Arabidopsis*, only phy A forms **homodimers**, whereas other phytochromes can form **heterodimers** with each other. The physiological significance of such dimerization is not clear.

### 13.9.5 Phytochrome-Mediated Responses

Phytochrome-mediated responses can be grouped into **high irradiance responses (HIRs)** or **low fluence response (LFRs)**. A unique feature of LFRs is that they obey the principle of reciprocity; i.e., they may be caused by long exposure to dim light or short exposure to high irradiance. LFRs are further grouped into two categories: 1. low fluence red/far-red reversible responses (LFRs) and 2. very low fluence response (VLFR). Table 13.3 summarizes the characteristics of phytochrome responses at different fluences.

Phytochrome-induced responses may be further grouped into two types: 1. **rapid biochemical changes** and 2. **slow morphological changes**, including movement and growth. Rapid biochemical changes induced by phytochrome often affect a number of developmental responses taking place later through a series of signal

## 13.9 Phytochromes

**Table 13.3** Characteristics of phytochrome responses at different fluences

| Very low fluence response (VLFR) | Low fluence response (LFR) | High irradiance response (HIR) |
|---|---|---|
| Fluence range | | |
| 0.0001–0.1 µmol m$^{-2}$ | 1–1000 µmol m$^{-2}$ | >1000 µmol m$^{-2}$ |
| R-FR reversibility | | |
| Not FR reversible | FR reversible | Not FR reversible |
| Fluence-time reciprocity | | |
| Dependent on product of fluence rate and time of irradiation | Dependent on product of fluence rate and time of irradiation | Requires long irradiation times, dependent on fluence rate (e.g., I1 > I2 > I3) but not on rate X time |
| Examples | | |
| Red light stimulation of coleoptile and inhibition of mesocotyl growth in etiolated seedlings | Promotion of lettuce seed germination, regulation of leaf movements | Anthocyanin biosynthesis, ethylene production, plumular hook opening in lettuce |

*FR* far red, *R* red

transduction events. The morphological responses due to photoactivation of phytochrome generally exhibit a **lag period**, the time between phytochrome activation and the biological response. The lag period may vary from a few minutes to several weeks, depending on the response. Noteworthy among the **rapid responses** are phytochrome-induced reversible movement of organelles and cell volume changes (swelling or shrinking). Even some growth responses are very fast. Thus, for example, inhibition of stem elongation in *Arabidopsis* and *Chenopodium album* occurs within minutes after red light exposure. In contrast, lag period of several weeks is required for phytochrome-mediated floral induction in *Arabidopsis* and some other species. A number of phytochrome responses also exhibit the phenomenon of **escape from photoreversibility**. In these events, red light-induced response is reversed by far-red light in a limited time after red light exposure, after which the response is said to have "escaped" from reversible control by far-red light. This is due to the multistep sequence of linked biochemical reactions leading to the response. Early stages in the sequence of biochemical reactions may be reversible by removing P$_{fr}$, and beyond some point, the reactions proceed irreversibly in the forward directions, leading to response. The escape period for different phytochrome-mediated responses ranges from less than a minute to few hours.

### 13.9.6 Phytochrome Action Involves Its Partitioning Between Cytosol and Nucleus

Seedlings raised in dark synthesize P$_r$ (inactive) form of phytochrome. Under these conditions, phytochrome is exclusively cytosolic. Conversion of P$_r$ to P$_{fr}$ upon exposure to red light causes its movement from the cytosol to the nucleus. Since phytochromes are too large to exhibit passive diffusion across nuclear pores, they are transported by active means across the nuclear membrane. The process differs between phy A and phy B. Nuclear import of $P_{fr}$ A is extremely rapid and occurs within minutes of red light exposure. Since dark-grown seedlings contain high concentration of phy A, sufficient $P_{fr}$ A gets transported into the nucleus. Red light exposure to phy B exposes the nuclear localization sequence associated with it during conformational changes from $P_r$ B to $P_{fr}$ B. This facilitates binding of nuclear import proteins leading to transport of phy B across the nuclear membrane. Compared to nuclear migration of phy A, transport of phy B by the above-stated process is slow and occurs in a few hours after red light exposure (Fig. 13.15).

Both phy A and phy B accumulate in the nucleus as a cluster of small particles, called "**speckles**." Together these clusters of speckles are also referred as "**nuclear bodies**" *(NBs)*. The number and size of these speckles are correlated with light responsiveness. The nuclear bodies associated with $P_{fr}$ B disappear upon exposure to far-red light, accompanying the conversion of $P_{fr}$ to $P_r$. Accumulation of nuclear bodies also exhibits circadian pattern. Thus, NBs disappear at night and reappear shortly before dawn, indicating an integration between light and clock signaling. Nuclear $P_r$ formed as a result of far-red light exposure is exported from the nucleus to the cytosol. Keeping in view the fact that the rate of phytochrome photoconversion ($P_r$-$P_{fr}$) is more rapid than the rate of phytochrome migration (cytosol to nucleus and reverse), both forms of phytochrome are present in both cytosol and nucleus in light-grown plants.

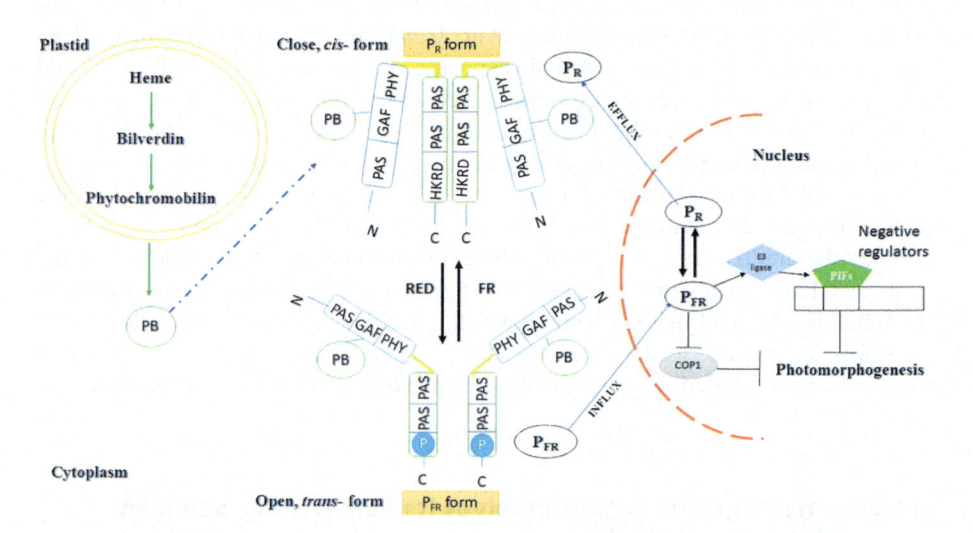

**Fig. 13.15** Photoconversion of phytochrome and its intracellular migration between the cytoplasm and nucleus. *HRKD* histidine kinase-related domain, *PIF₃* phytochrome integrating factor 3, *PAS* Per-ARNT-Sim domain, *PB* phytochromobilin, *GAF* cGMP-specific phosphodiesterase, adenyl cyclase, and FhlA domain

## 13.9.7 Phytochrome Signaling Mechanisms

Light sensing by phytochrome chromophore modulates the interaction of phytochrome protein with other cellular components which ultimately leads to changes in growth, development, or position of the organ. Phytochrome action either involves **changes in ion fluxes** (leading to rapid turgor responses) or brings about alteration of **gene expression** (resulting in slower, long-term responses).

1. *Regulation of membrane potential and ion fluxes*: Rapid changes in membrane properties can be brought about by phytochrome on sensing light. Thus, membrane potential of roots and oat coleoptiles changes within few seconds of red light exposure. This leads to ion flux changes across the plasma membrane suggesting that some of the phytochrome-induced rapid responses are initiated at or near the plasma membrane.
2. *Regulation of gene expression*: Long-term alterations in metabolism brought about by changes in gene expression are responsible for **de-etiolation** of seedlings. As a result, etiolated seedlings with elongated stem, folded cotyledons, and lack of chlorophyll exhibit slowing down of stem growth, opening of cotyledons, and chlorophyll formation in light. The stimulation and repression of transcription can be very rapid, and lag time can be as short as 5 min. Nuclear import of phytochrome proteins (phy A and phy B) as a response to light quality represents a major control point in phytochrome signaling. A number of transcription factors are upregulated following a shift of seedlings from dark to light, and they subsequently activate the expression of several genes. Such genes which encode rapidly upregulating proteins are called **primary response genes**. Expression of primary response genes is dependent on signal transduction pathways and is independent of protein synthesis. The expression of later or **secondary response genes** requires new protein synthesis. Phytochrome signaling brings about changes in gene expression largely by affecting the stability of transcription factors. It promotes the degradation of transcription factors that act as negative regulators of light response, whereas it stabilizes the transcription factors which act as positive regulators.
3. *Phytochrome-interacting factors (PIFs)*: PIFs are proteins which primarily act as regulators of photomorphogenic responses. They are basic helix-loop-helix transcription factors and consist of 15 proteins in *Arabidopsis*. PIF$_3$ was the first PIF identified. PIFs specifically bind with P$_{fr}$. They regulate phytochrome-mediated seed germination, shade avoidance, and hypocotyl elongation. PIFs promote **skotomorphogenesis** (etiolated development in dark) by acting as transcriptional activators of dark-induced genes and also by repressing some light-induced genes. Upon light exposure, PIF$_3$ co-localizes with phytochrome in nuclear bodies and is subsequently degraded, resulting in the inhibition of etiolation and promotion of photomorphogenesis.

Degradation of PIF3 correlates with its phosphorylation by the phytochrome and is further dependent on its ubiquitination followed by targeting to 26S proteasome.

In contrast with promotion of ubiquitination of PIF3, phytochrome signaling inhibits ubiquitination of transcription factors HY5, LAF1, and HFR1, which promote expression of light-induced genes. These transcription factors are ubiquitinated by the E3 enzyme-COP1 in dark, leading to degradation by 26S proteasome. COP1-dependent ubiquitination also leads to phy A degradation in light. Inhibition of COP1 activity is caused by phytochrome binding to COP1, causing exclusion of COP1 from the nucleus. Consequently HY5, LAF1, and HFR accumulate and induce light-responsive genes. Lastly, phytochrome activity is also modulated by its autophosphorylation, more so in $P_{fr}$ than in $P_r$ form. Phosphorylation of $P_{fr}$ affects its stability and affinity for interacting partners, such as $PIF_3$. Phosphorylated phytochrome can again be dephosphorylated by a phosphatase (type 5 protein phosphatase; PAPP5). In this way, the amplitude of phytochrome signal can be modulated through a balance of its phosphorylation and dephosphorylation status.

## 13.10 Blue Light-Mediated Responses and Photoreceptors

Blue light (320–500 nm) affects a wide variety of growth responses in higher plants, algae, ferns, fungi, and prokaryotes. The major responses involving a role of blue light include:

- Activation of gene expression (e.g., for chalcone synthase, chlorophyll biosynthesis, smaller subunit of Rubisco, chlorophyll binding protein, D-II subunit of PS-II reaction center)
- Membrane depolarization due to opening of anion channels
- Chloroplast movement
- Inhibition of hypocotyl elongation
- Stomatal opening
- Phototropism
- Phototaxis—movement of motile unicellular organisms toward or away from light

All blue light-mediated responses exhibit three unique features:

- A typical "three-finger" action spectrum with $\lambda_{max}$ at around 420, 450, and 475 nm (Fig. 13.16)
- A significant lag period between the sensing of light signal and the response
- Persistence of the response after light has been switched off

Whereas photosynthesis response is fully activated immediately after sensing light and ceases as soon as light is switched off, blue light-mediated responses exhibit a lag period of variable duration after sensing light stimulus. The persistence of blue light response is explained on the basis of photochemical cycle in which a blue light receptor activated by blue light is slowly converted to its inactive form after

**Fig. 13.16** Typical "three-finger" action spectrum of blue light-mediated responses in plants

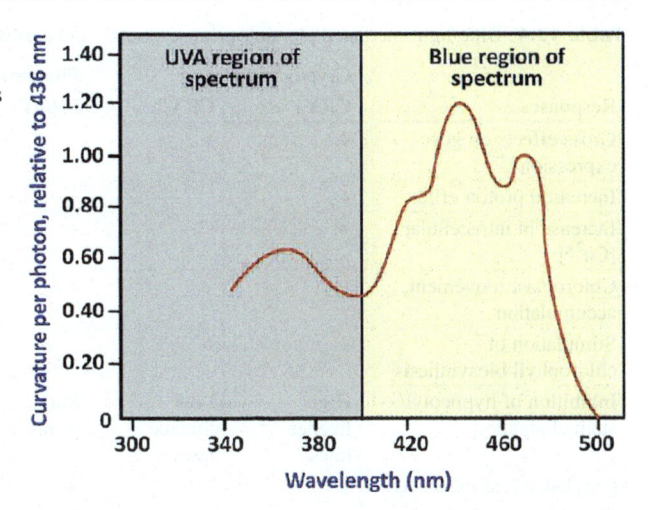

switching off blue light. This reversal of blue light-mediated responses appears to involve four main processes: receptor dephosphorylation by protein phosphatases, breaking of covalent carbon-sulfur bond, dissociation of the receptor from its target molecules, and dark reversion of blue light-driven conformational changes in the receptor. Thus, the rate of response reversal depends on the time taken by the photoreceptor to convert back to its inactive form from the active state. Plants contain three types of UV-A and blue light-sensitive photoreceptors. These are cryptochromes (CRY 1 and 2), phototropins (PHOT 1 and 2), and zeitlupe (ZTL). All of them contain flavins as chromophores. Table 13.4 summarizes the various physiological responses mediated by these photoreceptors.

## 13.11 Cryptochromes

Cryptochromes are blue light-sensitive photoreceptors which are responsible for cotyledon expansion, suppression of hypocotyl elongation, membrane depolarization, anthocyanin production, and circadian clock function (Box 13.2). Cryptochromes are known to exist in cyanobacteria, ferns, and algae. Three types of cryptochromes have been reported so far from the analysis of *Arabidopsis* mutants. These are CRY1, CRY2, and CRY3. CRY1 was initially isolated by detecting *Arabidopsis* seedlings with abnormally long hypocotyls when grown in blue or UV-A (320–400 nm) light but exhibiting normal hypocotyl length in R or FR light. Analysis of this mutant showed that long hypocotyl phenotype of the said mutant (*hy4*) was specific to blue light inhibition of hypocotyl elongation. It is now established that cryptochromes are responsible for long-term blue light-induced inhibition of hypocotyl elongation, whereas phototropins trigger rapid inhibitory responses (Fig. 13.17).

**Table 13.4** Blue light-mediated physiological responses and corresponding photoreceptors

| Responses | Cryptochromes CRY1 | CRY2 | Phototropins PHOT1 | PHOT2 | Zeitlupe (ZLT/ADO) |
|---|---|---|---|---|---|
| Gross effects on gene expression | + | + | | | |
| Increased proton efflux | + | | + | + | |
| Increase in intracellular [Ca$^{2+}$] | + | | + | + | |
| Chloroplast movement, accumulation | | | + | + | |
| Stimulation of chlorophyll biosynthesis | + | + | | | |
| Inhibition of hypocotyl/stem elongation | High fluence rates | Low fluence rates | Rapid response | | + |
| Cotyledon/leaf expansion | + | | + | + | |
| Phototropism | | | Low intensity | High intensity | |
| Circadian clock entrainment | + | + | | | |
| Flowering time | + | + | | | + |

---

**Box 13.2 Effects of Green Light on Plant Development**

Cryptochromes and phytochromes are known to absorb green light (500–550 nm) to initiate photomorphogenic responses. It is now evident that green light exerts specific and frequently antagonistic functions in directing light responses in two ways:

1. Those that antagonize normal light-mediated responses
2. Those that function to forward normal developmental processes

With reference to the first category of responses, Fritz Went (1957) described that green light retards growth of tomato seedlings by opposing the effects of red and blue light. Later, Klein (1964) observed that green light inhibition of crown gall callus was correlated with fluence rate. Green light was also found to inhibit root gravitropism which could be reversed with an orange/red treatment. Green light induces phototropic curvature in *Arabidopsis* and lettuce seedlings with characteristics distinct from blue light-mediated phototropism. These observations suggest a separate green light-sensing pigment.

There is substantial evidence that green light also modulates stomatal aperture. A blue light pulse leads to an increase in stomatal aperture. If the blue pulse is immediately followed by a green light pulse or if blue and green

(continued)

## 13.11 Cryptochromes

> **Box 13.2** (continued)
> pulses are delivered simultaneously, stomatal opening does not occur. The action spectrum for reversal of stomatal opening has a peak at 540 nm. The absence of response in **npq1** mutants (containing a lesion in zeaxanthin de-epoxidase activity and therefore no zeaxanthin production) suggests that blue and green light-induced changes in stomatal aperture may be regulated through trans-cis isomerization of bulk zeaxanthin in chloroplasts. Narrow-bandwidth green light also causes hypocotyls of dark-grown seedlings to be slightly longer than dark-grown seedlings, in a phytochrome-independent manner.

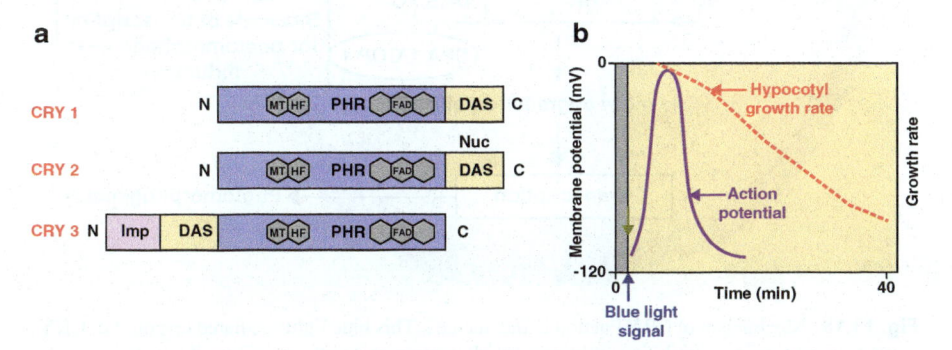

**Fig. 13.17** (a) Domain structures of three cryptochromes (CRY1, CRY2, and CRY3). (b) Blue light-triggered generation of action potential followed by reduction in hypocotyl elongation rate

The term "cryptochrome" was initially coined for blue light photoreceptors as a reflection of the "cryptic" (mysterious) nature of these chromoproteins. Subsequently, CRY genes were found to encode 70–80 kDa proteins which form dimers in active state. CRY1 and CRY2 exhibit 58% homology in their amino acid sequence at the amino-terminal region but only 14% homology at the C-terminal region. The sequence significantly matches with microbial **photolyase**, which is a blue light-activated enzyme responsible for the repair of pyrimidine dimers in DNA damaged by exposure to UV radiation. CRY1 and 2 proteins, however, do not exhibit any photolyase activity. The function of CRY3 is not yet fully known, but it does not exhibit photolyase activity for single-stranded DNA lesions. Cryptochromes contain **two cofactor chromophores—flavin adenine nucleotide (FAD) and pterin (methenyltetrahydrofolate, MTHF)**. The cofactor-binding region of the protein is referred as PHR (photolyase-related) domain. This domain of the protein is structurally similar to photolyase, and, in addition to serving as the binding site for the chromophore cofactors, it also mediates dimerization of photoprotein (Fig. 13.18). All three cryptochrome (CRY1, 2, and 3) proteins contain a highly conserved motif called DAS (DQXVP-acidic-STAES). In CRY1 and

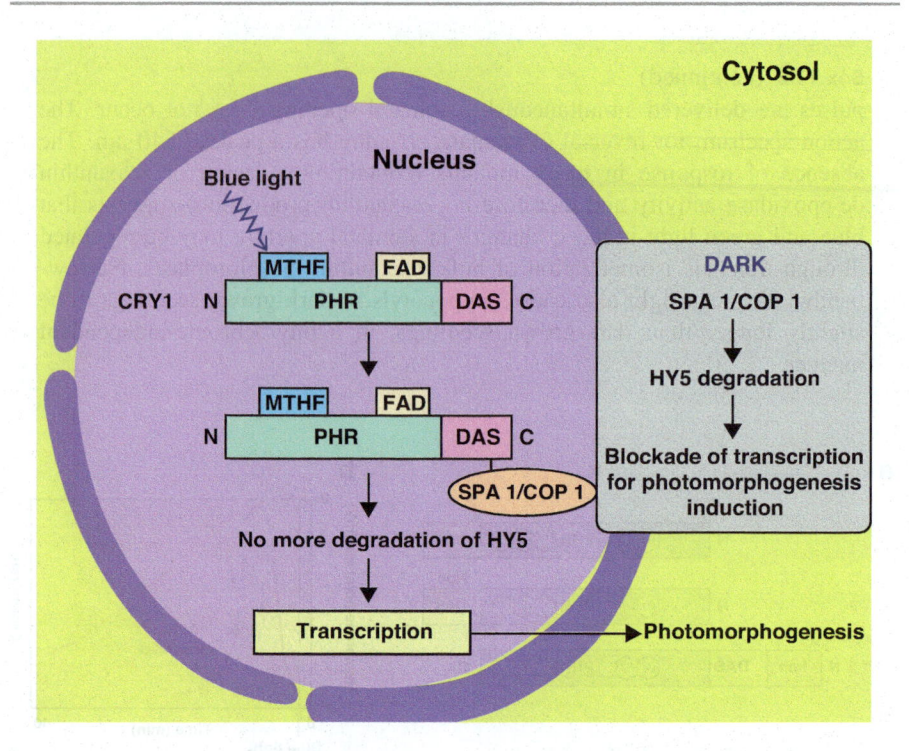

**Fig. 13.18** Mechanism of CRY1 action in the nucleus. This blue light-mediated response of CRY1 plays a primary role in inhibition of hypocotyl elongation and anthocyanin production

CRY2, DAS is localized at the C-terminal. It also contains a nuclear localization signal (*Nuc*), whereas CRY3 contains an N-terminal extension required for the import of protein into chloroplasts and mitochondria (**Imp**). Another major difference between CRY1 and 2 is that CRY1 protein is much more stable in light than CRY2 which gets preferentially degraded under blue light. Overexpression of CRY1 protein in transgenic *Arabidopsis* or tobacco plants brings about stronger blue light-induced inhibition of hypocotyl elongation and anthocyanin production, whereas similar overexpression of CRY2 protein brings about only a small enhancement of the inhibition of hypocotyl elongation in wild plants. These overexpression studies indicate that CRY1 plays a primary role in the inhibition of hypocotyl elongation and anthocyanin production. Both CRY1 and CRY2 are present in the nucleus and cytoplasm, whereas CRY3 is localized in chloroplasts and mitochondria. Absorption of blue light by the chromophore FAD brings about conformational changes at the C-terminus, leading to its (cry protein) physical interactions with other signaling factors. Dimerization of photoprotein is essential for it to undertake subsequent signaling activities. It has been further observed that nuclear and cytoplasmic pools of CRY1 have distinct biological functions. One of the fastest blue light-mediated responses, i.e., changes in membrane depolarization, is regulated by nuclear (rather than cytoplasmic) localized CRY1. There is no evidence that cryptochrome moves from the cytosol to the nucleus in response to light. COP1 is one of the signaling factors known to mediate blue light-induced morphogenic/biochemical changes through cryptochromes. COP1, along with another signaling factor, SPA, is known to degrade transcription factors such as HY5 in dark, which induces gene expression required for photomorphogenesis. Blue light triggers binding of C-terminus of CRY1 in the nucleus with SPA1 and COP1, thereby preventing the degradation of HY5, and other transcription factors, leading to promotion of blue light-mediated photomorphogenesis. Blue light-induced phosphorylation could also be contributing to maintain C-terminus of CRY1 in an active form (Fig. 13.18).

## 13.12 Phototropins: Molecular Nature and Associated Phototropic Bending Response

Isolation of several non-phototropic hypocotyl (*nph*) mutants of *Arabidopsis* leads to identification of the phototropin genes (*phot 1* and *phot 2*) with partially overlapping functions. In addition to their effect on phototropic bending response to blue light, phototropin protein receptors also modulate stomatal opening, ion transport, chloroplast movement, and cotyledon and hypocotyl growth (Table 13.5). The term "**phototropin**" should not be confused with "**phytotropin**" which refers to noncompetitive inhibitors of polar transport of auxins in plants. Some commonly used phytotropins are triiodobenzoic acid (TIBA), morphactin, and naphthylphthalamic acid (NPA). Unlike cryptochromes, phototropin receptors are associated with plasma membrane and function as light-activated serine-threonine kinases. The N-terminal of phototropin protein has the photosensory domain, and it is connected by a hinge with C-terminal serine-threonine kinase domain. The photosensory region has two similar motifs of 110 amino acids each. They are called LOV 1 and LOV 2 because of their sensitivity to light, oxygen, and voltage and their ability to bind the light-sensing cofactor-FMN (flavin mononucleotide). In dark, these two LOV domains are appressed and kinase at the C-terminal is inactive. Blue light absorption by the LOV domains leads to covalent binding of LOV domains with FMN via a conserved cysteine residue. LOV 2 interacts with a conserved helical region called Jα-helix located on the C-terminal side of the photosensory domain. The N-terminal photosensory region controls the activity of the C-terminal of the phototropin, which contains a serine-threonine kinase domain. In dark, the N-terminal region and the

Table 13.5 Unique and overlapping functions of PHOT 1 and PHOT 2

| Functions | PHOT 1 | PHOT 2 |
|---|---|---|
| Phototropism | ✓ (Low and high fluence) | ✓ (High fluence) |
| Cotyledon growth | ✓ | |
| Hypocotyl growth | ✓ | |
| Chloroplast movement | ✓ | ✓ |
| Stomatal opening | ✓ | ✓ |
| Ion transport | ✓ | ✓ |

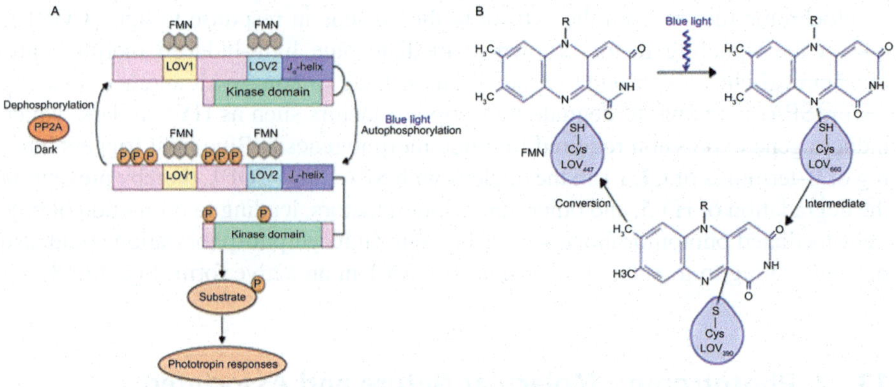

**Fig. 13.19** Model for blue light-mediated autophosphorylation of phototropin. PHOT often leads to generation of auxin concentration gradient leading to bending response toward light

LOV domains "cage" the C-terminal and inhibit kinase activity. Blue light absorption by LOV domains leads to the uncaging of kinase domain and its activation by the unfolding of Jα-helix. The activated C-terminal kinase domain brings about autophosphorylation of the phototropin molecule on serine residues. Phototropin autophosphorylation seems to be critical for all phototropin-mediated morphogenic responses. A protein phosphatase action on the autophosphorylated phototropin results in its inactivation in dark. With reference to phototropic bending response of grass seedlings, it is now established that unilateral blue light induces a gradient of PHOT 1 autophosphorylation, which leads to generation of auxin concentration gradient (Fig. 13.19). Higher auxin concentration on the darker side of the shoot results in greater growth on that side causing its bending response toward light.

## 13.13 Phototropin-Modulated Chloroplast Movement

Light absorption and photodamage in green plants are effectively modulated by changes in the intracellular distribution of chloroplasts in leaves in response to changing light conditions. Chloroplasts gather near the upper and lower walls of palisade cells under weak illumination, thereby maximizing light absorption. On the other hand, strong illumination leads to chloroplast movement to the lateral walls so as to minimize light absorption and avoid photodamage. At night, chloroplasts move to the bottom of the cell. The physiological significance of such a migration is, however, not yet clear (Fig. 13.20). Analysis of chloroplast movement pattern in low and bright light in *phot 1*, *phot 2*, and *phot 1/phot 2* double mutants has demonstrated the specific and common roles of PHOT 1 and PHOT 2 in the avoidance and accumulation responses. Thus, *phot 1* mutants have a normal avoidance response and a poor accumulation response. *phot 2* mutants lack the avoidance response but retain normal accumulation response. Cells from the double mutant (*phot 1/phot 2*) lack both avoidance and accumulation responses. *phot 2* mutant plants are unable to survive under bright sunlight due to photooxidative damage. Based on the molecular

## 13.13 Phototropin-Modulated Chloroplast Movement

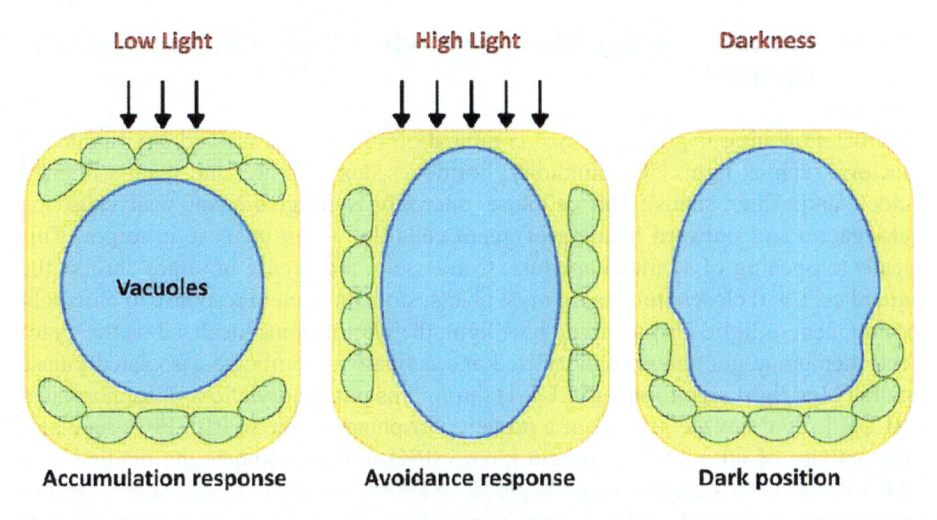

**Fig. 13.20** Changes in chloroplast distribution pattern in leaf cells in response to varying light intensity and in dark

**Fig. 13.21** Mechanism of phototropin-mediated chloroplast movement. Chloroplast movement in cytoplasm in response to light is aided by cytoskeletal elements majorly actin (G-actin and F-actin), profilin, and thrumin. F-actin binding protein CHUP1 (chloroplast unusual positioning 1) brings about change in movement

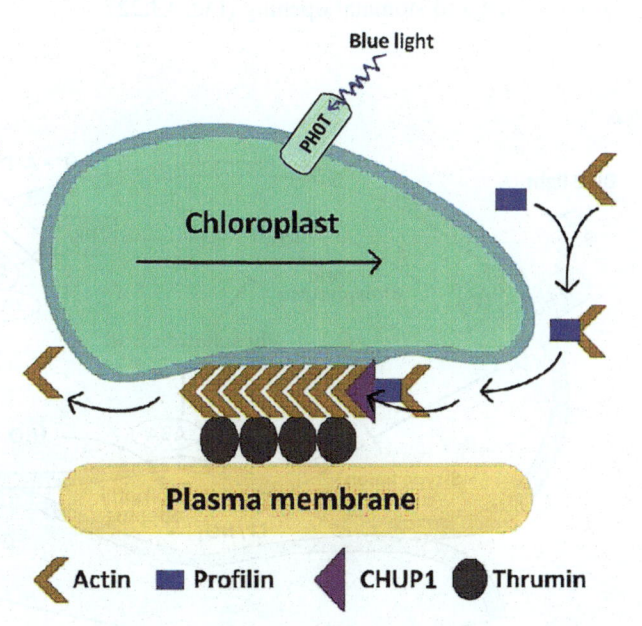

analysis of these mutants, it is now evident that both PHOT 1 and PHOT 2 are localized on the plasma membrane, whereas PHOT 2 is also localized on the chloroplast membrane. Chloroplast movement is brought about by changes in the cytoskeleton. These changes are brought about by a novel, F-actin binding protein called as **chloroplast unusual positioning 1 (CHUP 1)**. In bright light, CHUP 1 localizes itself on the plasma membrane via protein interactions and binds to the chloroplast envelop. CHUP 1 recruits G-actin and profilin (actin-polymerizing protein) to extend an existing F-actin filament which pushes the chloroplast from its earlier position (Fig. 13.21).

## 13.14 Phototropin Signaling-Dependent Light-Induced Stomatal Opening

Stomatal opening is governed by a combination of endogenous and environmental factors, such as light, $CO_2$, humidity, hormones, temperature, and plant's internal clock and water status. The cellulose microfibrils in guard cell wall cause an elongation and outward bending of guard cells due to an increase in turgor. This leads to opening of stomatal aperture. Conversely, a decrease in turgor shrinks the guard cells and closes stomata. In most plants, stomata open in response to blue light component of light. On absorbing blue light, the phototropins localized on the guard cell membrane get autophosphorylated and activate a membrane-associated kinase called **blue light signaling** *1* (BLUS 1) through its phosphorylation. Phosphorylated BLUS 1 regulates the activity of a protein phosphatase called PP1c. PP1c regulates the activity of an unknown protein kinase (PK) which promotes the binding of a 14-3-3 protein to the plasma membrane $H^+$-ATPase, leading to its activation. Proton-pumping activity of $H^+$-ATPase leads to membrane hyperpolarization. This causes $K^+$ uptake through $K^+$ channels, and the resulting decrease in water potential drives water uptake and stomatal opening (Fig. 13.22).

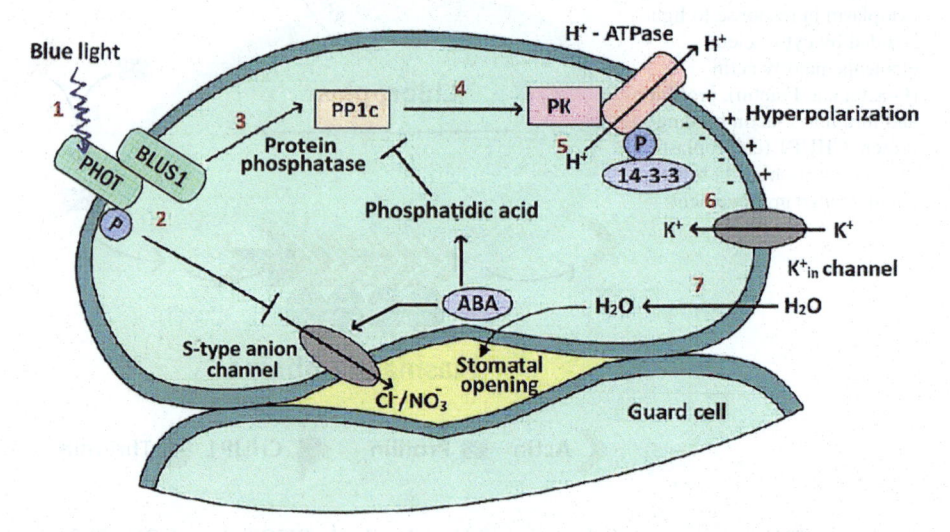

**Fig. 13.22** Phototropin-mediated stomatal opening mechanism. The opening and closing are turgor mediated. On blue light absorption, phototropin on guard cell membrane undergoes autophosphorylation and activates BLUS. BLUS further regulates protein phosphatase PP1c, which, in turn, regulates protein kinase, leading to activation of $H^+$-ATPase. This causes hyperpolarization of the membrane. Subsequent $K^+$ uptake results in decreased water potential and eventually stomatal opening

## 13.15 UVR 8: A Photoreceptor for UV-B-Mediated Photomorphogenic Responses

UV-B (280–315 nm) radiation is known to affect gene regulation, flavonoid biosynthesis, hypocotyl growth suppression, epidermal cell expansion, stomatal density, entrainment of circadian clock, and increase in photosynthetic efficiency. The photoreceptor for UV-B-mediated photomorphogenic responses is a seven-bladed β-propeller protein called as **UVR 8**. UVR8 lacks a prosthetic group and exists as a homodimer in its inactive form. The two subunits are linked by a network of salt bridges formed between tryptophan residues, which serve as sensors of UV-B radiation. UV-B absorption by tryptophan residues leads to breaking of salt bridges leading to separation of active monomers of UVR8. An interaction of UVR8 monomers with COP1-SPA complex brings about gene expression. Contrary to the role of COP1-SPA as a negative regulator for transcription factors for degradation during phytochrome and cryptochrome responses, it acts as a positive regulator during UV-B signaling by interacting with UVR8 in the nucleus. UVR8-COP1-SPA complex activates the transcription of the transcription factor HY5, thereby regulating the expression of genes induced by UV-B (Fig. 13.23). RUP

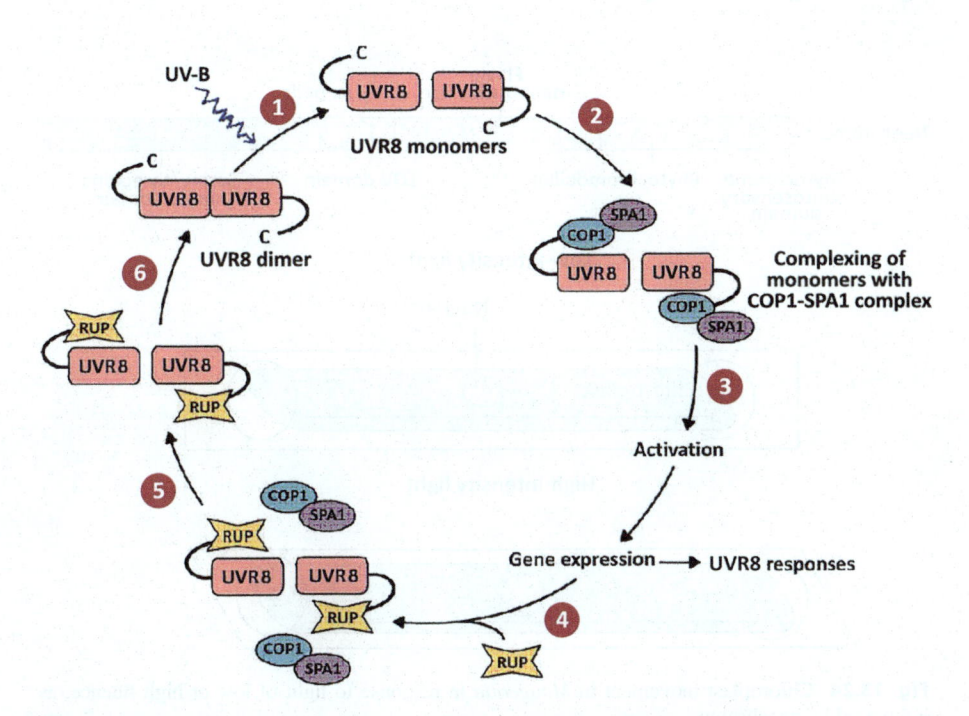

**Fig. 13.23** UVR8-mediated signaling mechanism via action of RUP protein

(REPRESSOR OF UV-B PHOTOMORPHOGENESIS 1 and 2) is a negative regulator of UVR8 receptor. It facilitates the dimerization of the UVR8 receptor.

## 13.16 Other LOV Domain-Containing Photoreceptors in Plants

*Zeitlupe (ZTL/ADO)* [German word for "slow motion"] photoreceptors reported from *Arabidopsis*, rice, poplar, maize, and pine are involved in the targeted proteolysis of signaling components which control flowering time and circadian clock. Like phototropins, ZTL/ADO photoreceptors resemble PHOTs but have a photoactive FMN-binding LOV region along with a F-box motif of about 50 amino acids for protein-protein interaction. **Neochrome** is another plant photoreceptor with LOV domain, found in ferns and algae. Neochrome has a phototropin-like protein sequence fused with a phytochrome chromophore-binding domain. Therefore, it acts as dual red/blue light photoreceptor. *Mougeotia*, a filamentous green alga, has a large single, ribbon-like chloroplast in each cell. It displays a striking light-avoidance response through neochrome and phototropin-mediated movement of its chloroplast. In dim light, the flat side of the chloroplast faces light, and in bright light, the edges face the light (Fig. 13.24).

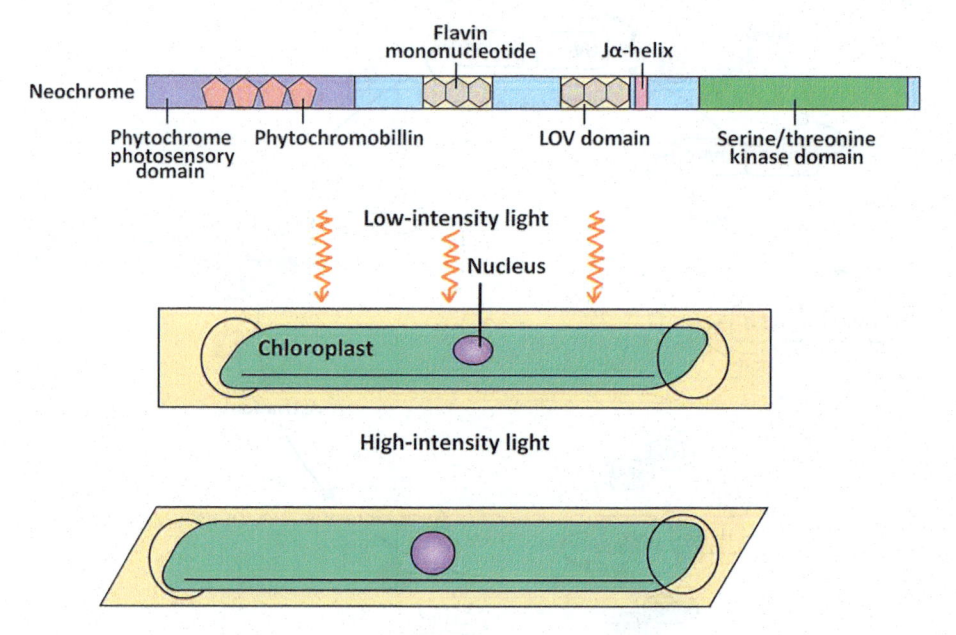

**Fig. 13.24** Chloroplast movement in *Mougeotia* in response to light of low or high fluence, as influenced by neochrome

## 13.17 Rhodopsin-Like Photoreceptors

Movement of *Chlamydomonas reinhardtii*—a unicellular alga—toward or away from the directional light (**phototaxis**) is controlled by two rhodopsin-like molecules called as channel rhodopsins (ChR1 and ChR2) located in the eye spot. These photoreceptors are light-gated proton channels which regulate photoaxis response to blue-green light ($\lambda_{max}$ 500 nm).

## 13.18 Summary

- Photomorphogenesis may be defined as the developmental response of an organism to the information in light, which may be its quantity, quality (i.e., the wavelength), and direction or relative length of day and night (photoperiod). A dark-grown seedling is said to be etiolated. In etiolation, embryonic stem (hypocotyl in dicots and epicotyl in monocots) of seedlings exhibits very rapid and extensive elongation of internodes. There is no cotyledon/leaf expansion, and seedlings appear pale as there is no chloroplast development. So, etiolated plants are pale yellow, and the hypocotyl remains "hooked" at the apex. Upon sensing light, a seedling emerging from the soil switches from skotomorphogenesis to photomorphogenesis. This process is known as **de-etiolation**, and it involves the coaction of red/far-red light-absorbing phytochrome and blue light-sensitive cryptochrome.
- Any photobiological process requires a light-absorbing molecule or pigment. Plants contain a variety of pigments. Absorption of light by a pigment (e.g., chlorophyll) plotted as a function of wavelength is known as absorption spectrum which depicts light. In modern laboratories, action spectra are also generated by using **spectrographs** fitted with monochromatic sources of light. Three parameters of light play significant role in modulating plant development. These are 1. light intensity, 2. light composition/(quality), and 3. light duration (photoperiod).
- Only a small percentage of photosynthetically active radiation (PAR, 400–700 nm) incident on a leaf is surface transmitted through the leaf, and the rest (15%) is reflected or transmitted from the leaf surface. Since chlorophylls absorb blue and red wavelengths of light most strongly, green wavelengths are maximally transmitted and/or reflected, giving green color to the vegetation. The palisade cells below the epidermis are shaped like pillars arranged in parallel columns which provide sieve effect and channeling of light. The **sieve effect** is caused by nonuniform distribution of chlorophyll molecules within the chloroplasts which results in shaded areas in chloroplasts which do not receive light. Desert plants exposed to excess light develop hairs, salt glands, and epicuticular wax to enhance light reflection from leaf surface, thereby reducing overheating due to absorption of excess amount of solar energy. **Sunflecks** (patches of sunlight passing through small gaps in leaf canopy in dense forests) play important role in rapid and short-term light capture for plants growing on

forest bed and in densely planted crops where lower leaves are shaded by upper leaves. Some leaves maximize leaf absorption by **solar tracking (heliotropism)** by continuously adjusting the orientation of leaf lamina such that they are perpendicular to the leaf surface.
- *Photoreceptors,* unlike plant pigments (such as chlorophyll), are chromophore-containing biomolecules which absorb photons of a given wavelength and use the energy derived from it as a signal to initiate a photo response. All photoreceptors contain a protein (**chromoprotein**) attached to a light-absorbing, non-protein, prosthetic group called as **chromophore**.
- Etioplasts contain a prominent structure called **prolamellar body** which contains membrane lipids, protochlorophyllide and the light-requiring enzyme—**protochlorophyllide oxidoreductase (POR)**. Prolamellar bodies are formed when membrane lipid synthesis continues in the absence of corresponding amounts of thylakoid protein synthesis, which requires light. Light triggers chlorophyll biosynthesis from protochlorophyllide. Protochlorophyllide is not green and cannot absorb light for photosynthesis. It is a photoreceptor with an absorption maximum at 650 nm. Light absorption by protochlorophyllide takes place in the presence of NADPH-protochlorophyllide oxidoreductase (POR). Upon illumination, POR rapidly converts protochlorophyllide into chlorophyllide, and the internal structure of etioplasts is reorganized into a typical chloroplast. Subsequently, POR is released and degraded by proteolysis.
- Photoreversibility of phytochrome-modulated responses and its significance: Photoreversibility (photoconversion; also referred as photochromism) is a defining feature of phytochromes. Phytochromes are synthesized as an inactive red light-absorbing isomer, called $P_r$. Red light absorption by $P_r$ form converts it into an active, far-red-absorbing isomer called $P_{fr}$. Absorption of far-red light by $P_{fr}$ form converts it back to $P_r$ form. Phytochromes were discovered through experiments on red/far-red reversibility of seed germination in lettuce. Phytochromes are soluble proteins and exist as dimers. The phytochrome protein consists of an N-terminal photosensory region and a C-terminal regulatory domain. The two are attached with a hinge between them. Red light exposure opens up the structure in the far-red-absorbing form of phytochrome (Pfr), making its surface accessible to phytochrome-interacting factors (PIFs) in the signal transduction pathway. The opened-up structure of phytochrome protein allows light-dependent phosphorylation of exposed serine and threonine residues, possibly triggered by autophosphorylation by R3 kinase at the C-terminal. The light-labile form of phytochrome (phy A) is predominant in etiolated seedlings which strongly express PHY A gene, resulting in high accumulation of $P_rA$. Prolonged light exposure causes phy A concentration to drop by almost 100-fold. Light also inhibits PHY A transcription. Though less abundant, phy A still functions in light-grown plants. Phytochrome-mediated responses can be grouped into high irradiance responses (HIRs) or low fluence response (LFRs). Phytochrome-induced responses may be further grouped into two types: 1. rapid

## 13.18 Summary

biochemical changes and 2. slow morphological changes, including movement and growth. Conversion of $P_r$ to $P_{fr}$ upon exposure to red light causes its movement from the cytosol to the nucleus. Compared to nuclear migration of phy A, transport of phy B by the above-stated process is slow and occurs in a few hours after red light exposure. They are transported by active means across the nuclear membrane. Phytochrome action either involves changes in ion fluxes (leading to rapid turgor responses) or brings about alteration of gene expression (resulting in slower, long-term responses).

- Both CRY1 and CRY2 are present in the nucleus and cytoplasm, whereas CRY3 is localized in chloroplasts and mitochondria. Absorption of blue light by the chromophore FAD brings about conformational changes at the C-terminus, leading to its (cry protein) physical interactions with other signaling factors. One of the fastest blue light-mediated responses, i.e., changes in membrane depolarization, is regulated by nuclear (rather than cytoplasmic) localized CRY1. Unlike cryptochromes, phototropin receptors are associated with plasma membrane and function as light-activated serine-threonine kinases. The N-terminal of phototropin protein has the photosensory domain, and it is connected by a hinge with C-terminal serine-threonine kinase domain. The photosensory region has two similar motifs of 110 amino acids each. They are called LOV 1 and LOV 2 because of their sensitivity to light, oxygen, and voltage and their ability to bind the light-sensing cofactor-FMN (flavin mononucleotide).
- On absorbing blue light, the phototropins localized on the guard cell membrane get autophosphorylated and activate a membrane-associated kinase called **blue light signaling 1** (BLUS 1) through its phosphorylation. Phosphorylated BLUS 1 regulates the activity of a protein phosphatase called PP1c. PP1c regulates the activity of an unknown protein kinase (PK) which promotes the binding of a 14-3-3 protein to the plasma membrane $H^+$-ATPase, leading to its activation. Proton-pumping activity of $H^+$-ATPase leads to membrane hyperpolarization. This causes $K^+$ uptake through $K^+$ channels and the resulting decrease in water potential drives water uptake and stomatal opening.
- The photoreceptor for UV-B-mediated photomorphogenic responses is a seven-bladed β-propeller protein called as **UVR 8**. UVR8 lacks a prosthetic group and exists as a homodimer in its inactive form. The two subunits are linked by a network of salt bridges formed between tryptophan residues, which serve as sensors of UV-B radiation. UV-B absorption by tryptophan residues leads to breaking of salt bridges leading to separation of active monomers of UVR8. UVR8-COP1-SPA complex activates the transcription of the transcription factor HY5, thereby regulating the expression of genes induced by UV-B.
- Neochrome has a phototropin-like protein sequence fused with a phytochrome chromophore-binding domain. Therefore, it acts as dual red/blue light photoreceptor. *Mougeotia*, a filamentous green alga, has a large single, ribbon-like chloroplast in each cell. It displays a striking light-avoidance response through neochrome and phototropin-mediated movement of its chloroplast.

## Multiple-Choice Questions

1. Match the correct option

   1. Requirement of light for germination   (i). Nyctinasty
   2. Folding of leaves at night   (ii). Photonasty
   3. Opening of leaves at dawn   (iii). Phototropism
   4. Directional bending of shoots to light   (iv). Photoblasty

   Chose the correct option

   (a) 1, (iv); 2, (iii); 3, (ii); 4, (i)
   (b) 1, (i); 2, (iv); 3, (iii); 4, (ii)
   (c) 1, (iv); 2, (i); 3, (ii); 4, (iii)
   (d) 1, (ii); 2, (iii); 3, (ii); 4, (iv)

2. The process of transfer of energy from one molecule to another molecule is known as:
   (a) Fluorescence
   (b) Inductive resonance
   (c) Vibrational energy
   (d) Quantum yield

3. "Spectrographs" fitted with monochromatic source of light are used for the generation of:
   (a) Action spectra
   (b) Absorption spectra
   (c) Both (a) and (b)
   (d) None of these

4. "Epidermal focusing" for enhanced capture of light by chloroplasts is seen in:
   (a) Grasses
   (b) Deciduous plants
   (c) Conifers
   (d) Tropical plants

5. Patches of sunlight passing through small gaps in leaf canopy is called:
   (a) Para-heliotropism
   (b) Sieve effect
   (c) Sunflecks
   (d) Light channeling

6. Protochlorophyllide and prolamellar bodies are found in:
   (a) Chloroplast
   (b) Amyloplast
   (c) Etioplast
   (d) Leucoplast

7. Protochlorophyllide:
   (i) Is synthesized using chlorophyll b
   (ii) Has an unsaturated double bond between carbon C17 and C18
   (iii) Has a saturated bond between carbon C17 and C18
   (iv) Is synthesized using chlorophyll a
   (v) Absorbs light in the presence of NADH-dependent POR
   (vi) Absorbs light in the presence of NADPH-dependent POR
   Choose the correct combination:
   (a) i, iii, vi
   (b) ii, iv, vi
   (c) i, ii, v
   (d) iii, iv, vi
8. Out of the four phycobilins listed below, which one is known to regulate various aspects of growth and development?
   (a) Phycoerythrin
   (b) Phycocyanin
   (c) Phytochromobilin
   (d) Allophycocyanin
9. Presence of which pigment(s) is responsible for absorption of all wavelengths along the visible spectrum in red algae, giving it almost black appearance:
   (a) Chlorophyll a and d
   (b) Phycocyanin, allophycocyanin
   (c) Phycocyanin, phycoerythrin
   (d) Phycoerythrin, chlorophyll
10. Phytochromes are synthesized as _____, _____ light-absorbing isomer called Pr. Light absorption converts it into an _____ isomer, Pfr. Absorption of _____ light of wavelength _____nm converts Pr back to Pfr.
    (a) Inactive, far-red, active, red, 650 nm
    (b) Active, red, inactive, far-red, 700 nm
    (c) Inactive, red, active, far-red, 700 nm
    (d) Inactive, red, active, far-red, 650 nm
11. Which of the phytochrome-mediated responses follow the "principle of reciprocity"?
    (a) High irradiance responses (HIRs)
    (b) Low fluence responses (LFRs)
    (c) Very low fluence responses (VLFRs)
    (d) Both (a) and (b)
12. "Speckles" or "nuclear bodies" are clusters or aggregates of small particles which accumulate:
    (a) Phycocyanin and allophycocyanin
    (b) Sodium and potassium ions
    (c) phy A and phy B
    (d) Cryptochromes

13. Phytochrome-interacting factors regulate:
    (a) Photomorphogenesis, positively
    (b) Skotomorphogenesis, positively
    (c) Photomorphogenesis, negatively
    (d) Both (b) and (c)
14. Which of these is NOT a unique feature of blue light-mediated response?
    (a) Three-finger action spectra
    (b) Significant lag between sensing the light signal and the response
    (c) Persistence of response after light has been switched OFF
    (d) Appearance of nuclear bodies or speckles
15. Phytotropin is involved in:
    (a) Modulation of stomatal opening
    (b) Phototropic bending in response to blue light
    (c) Noncompetitive inhibition of polar transport of auxin
    (d) Chloroplast movement and ion transport
16. Determine the correct order of phototropin-dependent light-induced stomatal signaling.
    (i) K+ uptake and decrease in water potential
    (ii) Autophosphorylation of phototropin localized on the guard cell membrane and activation of BLUS through phosphorylation
    (iii) Membrane hyperpolarization
    (iv) Binding of 14-3-3 protein to plasma membrane H+ ATPase
    (v) Regulation of protein phosphatase PP1c
        (a) ii, v, iv, iii, i
        (b) i, iii, iv, iii, v
        (c) v, iv, ii, iii, i
        (d) iv, i, iii, ii, i
17. Neochrome, a dual red/blue light photoreceptor, is found in:
    (a) Filamentous green algae, *Mougeotia*
    (b) *Arabidopsis* rhizogenes
    (c) *Arabidopsis thaliana*
    (d) Unicellular algae, *Chlamydomonas reinhardtii*

## Answers

1. c   2. b   3. a   4. d   5. c   6. a   7. b
8. c   9. d   10. c   11. b   12. c   13. d
14. d   15. c   16. a   17. a

## Suggested Further Reading

Ulm R, Jenkins GI (2016) Q&A: how do plants sense and respond to UV-B radiation? BMC Biol 13:45

# Plant Growth Regulators: An Overview

## 14

Satish C Bhatla

Plants respond to external stimuli, sense changes in their environment, and their biological clock works in a precise manner. Many deciduous plants shed their leaves in a particular season or develop flowers and fruits during a specific time of the year. All these and many more rather intriguing responses of plants indicate that they have unique mechanisms to sense and communicate with the environment they live in. In order to achieve this, plants synthesize several biomolecules of varying chemical nature which facilitate cell-to-cell communication and affect varied aspects of growth and development. In higher plants, regulation and coordination of metabolism, growth, and morphogenesis depend on the transport of various chemical signals from one part to another. In the nineteenth century, a German botanist Julius von Sachs hypothesized that certain chemical messengers are responsible for the initiation and growth of different plant organs. Since then it has become evident that these messengers or signaling molecules translate environmental signals into growth and developmental responses and regulate metabolism by their redistribution. A number of signaling molecules have been identified in the animal systems and prokaryotes and are referred as "**hormones**." The term "hormone" was initially used about 100 years ago in medicine for a stimulatory factor. It is a Greek word meaning "to stimulate" or "to set in motion." Hormones have since long been defined as biochemicals produced in specific tissues which are transported to some distant cells to trigger response at low concentrations. The classic definition of a plant hormone according to Went and Thimann (1937) states that it is a substance which is produced in any part of the organism and is transferred to another part to influence a specific physiological process. Another term, "**morphogen**," is used for those biomolecules which possess the ability to modify cell pattern formation in a concentration-dependent manner, and at different concentrations, they lead to different outputs. Cells have the ability to assess their respective positions in response to the gradient of a particular morphogen, and cells use this information to establish their identity.

© Springer Nature Singapore Pte Ltd. 2018
S. C Bhatla, M. A. Lal, *Plant Physiology, Development and Metabolism*,
https://doi.org/10.1007/978-981-13-2023-1_14

Currently, **plant hormones** are defined as a group of endogenous organic substances which influence various physiological and developmental processes in plants at low concentrations. A number of biological processes in plants are affected by internal and external stimuli through the involvement of plant hormones. Biosynthesis of plant hormones can occur at different morphological and cytological locations, and their actions may be observed at their sites of production, or they may get translocated to some distant target tissues to evoke a response. Generally, plant hormones are produced in meristems, leaves, and developing fruits. Different plant hormones can cause opposite effects on a particular developmental process at varying concentrations. Antagonistic and synergistic interactions among various plant hormones further add to the complexity of hormonal actions in higher plants. Different hormones are able to influence the biosynthesis of some other hormones or may interfere with their signaling mechanisms. A hormone can evoke different responses in different tissues or at different times of development in the same tissue. Different tissues require varying amounts of hormones of specific kinds for their development. Such differences are referred to as differences in *sensitivity*.

## 14.1 Plant Growth Regulators (PGRs)

PGRs are a broader category of compounds which include plant hormones, their synthetic analogs, inhibitors of hormone biosynthesis, and blockers of hormone receptors. Precisely, plant growth regulators (PGRs) may be defined as naturally occurring or synthetic compounds which affect metabolic and developmental processes in higher plants, mostly at low concentrations. PGRs do not possess any nutritive or phytotoxic value. *"True"* PGRs are compounds which directly and specifically interfere with the actions of naturally occurring hormonal systems in higher plants. *"Atypical"* PGRs, however, act by bringing about a local or transient phytotoxic effect. A set of PGRs are known to reduce stem elongation and are often named as "growth retardants."

Principal PGRs modulating plant development can be grouped into four categories (Fig. 14.1):

**Category I** PGRs derived from **isoprenoid pathway,** starting from isopentenyl diphosphate. This includes cytokinins, gibberellins, brassinosteroids, abscisic acid, and strigolactones.

**Category II** PGRs derived from **amino acids,** viz., indole-3-acetic acid (IAA); serotonin and melatonin derived from tryptophan; ethylene derived from methionine, and polyamines derived from arginine or ornithine; nitric oxide derived from arginine.

**Category III** PGRs derived from lipids, e.g., jasmonic acid derived from α-linolenic acid.

## 14.1 Plant Growth Regulators (PGRs)

**Plant growth regulators derived from amino acids**

Fructose 6- Phosphate
↓
Glycerate 3- Phosphate
↓
Phosphoenol Pyruvate
↓
Erythrose 4- Phosphate
↓
Phosphoserine
↓
Serine → Oxaloacetate
↓
Cysteine
↓
Glutamate
↓
Methionine
↓
Ornithine → **Polyamine**
↓
Shikimate
↓
**Ethylene**
↓
Citrulline
↓
Arginine
↓
**NO**
↓
Tryptophan
↓
**IAA**
↓
**Serotonin**
↓
**Melatonin**

**Plant growth regulators derived from isoprenoid pathway**

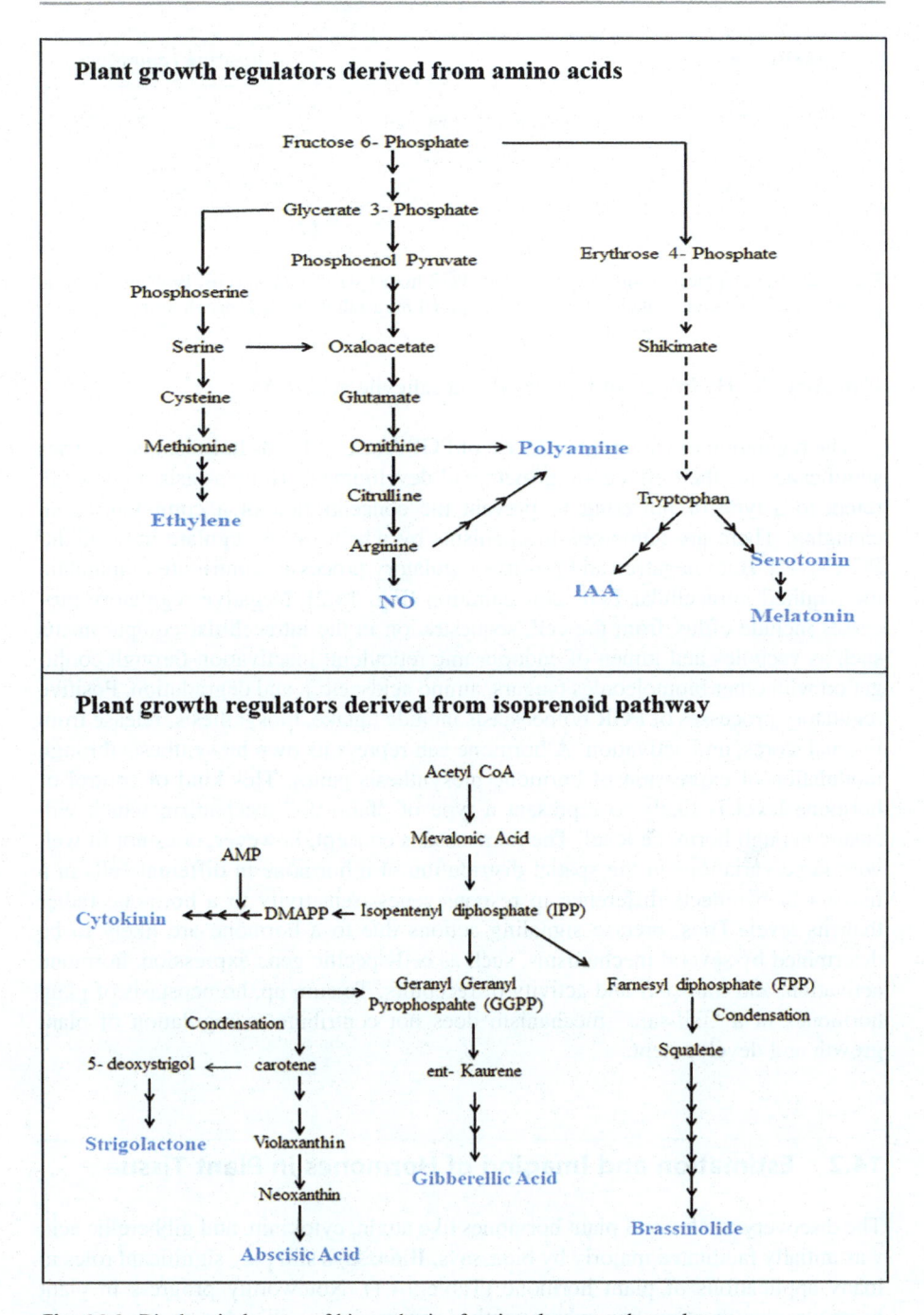

Acetyl CoA
↓
Mevalonic Acid
↓
AMP
↓
**Cytokinin** ←←← DMAPP ← Isopentenyl diphosphate (IPP)
↓
Geranyl Geranyl Pyrophospahte (GGPP)
↓
Farnesyl diphosphate (FPP)
↓ Condensation
Squalene
↓
Condensation
5- deoxystrigol ← carotene
↓
**Strigolactone**
↓
Violaxanthin
↓
Neoxanthin
↓
**Abscisic Acid**
↓
ent- Kaurene
↓
**Gibberellic Acid**
↓
**Brassinolide**

**Fig. 14.1** Biochemical routes of biosynthesis of major plant growth regulators

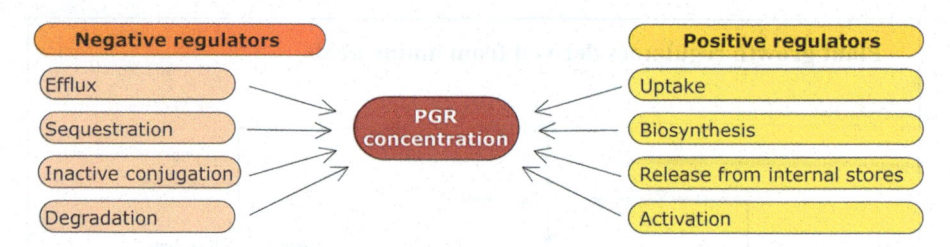

**Fig. 14.2** Various mechanisms responsible for PGR homeostasis in plant cells. Both negative and positive regulators contribute to maintain the required intracellular PGR concentration

**Category IV** Hydrogen sulfide ($H_2S$) and salicylic acid (SA).

The regulation of endogenous levels of PGRs, referred to as **homeostasis**, carries significance for their effects on growth and development. Homeostasis, in general, refers to a type of buffering to prevent the concentration of a compound from changing. There are numerous mechanisms by which plants regulate intracellular PGR levels. Both negative and positive regulatory processes contribute to maintain the required intracellular PGR concentration (Fig. 14.2). Negative regulatory processes include efflux from the cell, sequestration in the intracellular compartments such as vacuoles and lumen of endoplasmic reticulum, inactivation through conjugation with other biomolecules (sugars, amino acids, etc.), and degradation. Positive regulatory processes of PGR homeostasis include uptake, biosynthesis, release from internal stores, and activation. A hormone can repress its own biosynthesis through modulation of expression of hormone biosynthesis genes. This kind of control of hormone level is likely to represent a type of "fail-safe" mechanism which will ensure enough hormone level. The homeostasis concept, however, does not fit well with large variations in the spatial distribution of a hormone in different cells in a tissue which reflects difference in responsiveness/sensitivity to a hormone rather than its level. Thus, precise signaling actions due to a hormone are likely to be determined by several mechanisms, such as cell-specific gene expression, hormone activation, and transport and activity of receptors. To sum up, homeostasis of plant hormones in a "fail-safe" mechanism does not contribute to regulation of plant growth and development.

## 14.2 Estimation and Imaging of Hormones in Plant Tissue

The discovery of classical plant hormones like auxin, cytokinin, and gibberellic acid was initially facilitated majorly by bioassays. **Bioassays** still play significant roles in many applications of plant hormone (Table 14.1). Noteworthy progress in plant hormone identification has subsequently been made possible due to advances in chemical analysis, chromatographic techniques, and mass spectrometry. As compared to major metabolites, such as sucrose, hormone concentrations in plant tissues

## 14.2 Estimation and Imaging of Hormones in Plant Tissue

**Table 14.1** Bioassays for plant hormone analysis: advantages and limitations

| Advantages | Limitations |
| --- | --- |
| Sensitivity of detection: 1–1000 ng. $g^{-1}$ fw | Concentration difference of the hormone applied due to possible uptake problems |
| Monitor biological activity from a crude homogenate | Probable metabolism of hormone applied |
| Determine the concentration as well as potency of the test compound | Provide large statistical variations in estimated hormone concentration |
| Compare the activity of new compounds with reference to natural hormones | Lower sensitivity as compared to other high-end techniques like HPLC, GC-MS |
| Simple and easy to perform | A greater likelihood of false-positive results |
| Generally inexpensive | Precision of the bioassay is hampered due to large biological variability as well as temporal changes in the sensitivity of the tissues |

are generally very low, up to more than ten orders of magnitude. Moreover, the complexity of plant extracts also poses problems during the separation and analysis of plant hormones. Not only do the wide range of metabolites in the tissue extracts exhibit a wide range of chemical properties, many hormones (e.g., gibberellic acid and brassinosteroids) are quite hydrophobic, while hormones like auxins and cytokinins are more hydrophilic. Keeping the diversity of solubility and chemical nature in view, tissue homogenates are subjected to "clean up" or enrichment methods by partitioning between aqueous and organic solvents prior to separation by liquid chromatography (LC) and gas chromatography (GC). Mass spectrometry (MS) further offers fast and sensitive analysis of plant hormones. "Triple quadrupole" instrument allows analysis of multiple target ions, whereas chosen ions can be captured and fragmented for structural analysis using "ion trap" instruments. In the current era, direct coupling of various techniques with LC or GC instruments offers efficient separation, identification, and quantification of hormones. Latest technological advancements now permit simultaneous analysis of several hormones in a single extract of biological sample. Characterization of plant hormone action in cells and tissues is greatly facilitated by the use of reporter genes like β-glucuronidase (GUS) and green fluorescent protein (GFP) linked to a promoter sequence from a gene for plant hormone response or biosynthesis. GUS assays have the disadvantage that they are destructive in nature. Moreover, the diffusive nature of blue product of GUS reaction reduces resolution. In contrast, fluorescent proteins, such as GFP, have high resolution and sensitivity, and are nondestructive in nature, thus allowing live cell imaging. Signaling action of some hormones results in the degradation of certain target proteins. This feature of hormone action has also been exploited in visualizing plant hormones by developing a "degron" system which encodes hormone binding domain of the corepressor protein fused with a fluorescent protein and expressed using a promoter. Such degron reporters allow screening of hormone distribution in transgenic plants. "Degron" approach is now used to monitor distribution of auxins, gibberellic acid (GA), and jasmonic acid (JA). A gene encoding DELLA (GA-mediated transcription repressor protein)-GFP translational fusion protein, expressed from DELLA promoter, is used to monitor cellular changes in GA action since GA perception brings about degradation of DELLA-GFP fusion. Degron systems thus allow high resolution of detection and spatial and temporal information about hormone signaling and distribution in plant cells. New *biosensor* methods are now coming up to detect hormones and their biomolecular interactions. These include fluorescence resonance energy transfer (FRET), surface plasmon resonance (SPR), and luminescence techniques.

## 14.3 Experimental Approaches to Understand Perception and Transmission of Hormone Action

Four major approaches are used by plant physiologists to decipher the mechanisms of perception, transmission, and action of various plant hormone signals. These approaches generally bring about varying degrees of success:

1. Isolation of **plant hormone receptor** protein is the most direct approach and can be achieved using a labeled hormone or its close analog. Hormone-binding proteins are subsequently purified to undertake further analysis. In the early 1970s through mid-1980s, this has been the approach to understand the action of five major plant hormones (auxins, cytokinins, gibberellins, abscisic acid, and ethylene). Subsequently, this approach proved more difficult than anticipated earlier and has not been found to be very useful for detection and characterization of plant hormones.
2. Isolation and **cloning of genes** regulated by specific hormones at transcription level. Since gene transcription is usually a terminal event in a signal transduction pathway, it is expected that the information from the common regulatory sequence and transcription factors of a number of such genes can facilitate analysis of backward signaling events toward the receptor.
3. **Inhibitors of hormone action** are also useful in deciphering hormone action.
4. Screening and analysis of **mutants**, with altered hormone synthesis or sensitivity to specific hormone for one or more responses, are currently the most preferred method of choice for analysis of hormone signaling mechanisms. Response mutants do not respond to a hormone due to a defective protein constituent in the signaling chain for the said hormone. Such mutants may be specific to a particular signal transduction pathway. Isolation of mutant allele facilitates cloning of wild-type genes leading to a clue to the function of encoded protein. In the past decade, several mutants have been isolated and screened from *Arabidopsis* which is now an established model plant system for such analyses (Box 14.1).

Information obtained from all these approaches put together is preferred to get a clear picture of hormone action as no single approach offers enough crucial leads. Traditionally, five categories of plant hormones referred as "classical hormones" have received much attention. These are auxins, cytokinins, gibberellins, ethylene, and abscisic acid. More recently additional biochemical constituents of plant tissues

> **Box 14.1: *Arabidopsis* as a Model Organism for Decoding the Biochemical Mechanisms**
>
> Genetic potential of *Arabidopsis* has for long been utilized to characterize several plant-specific growth and developmental processes owing to its small genome size, simple life cycle, and self-pollination ability, but it was the development of mutant screens that established *Arabidopsis* as a model organism worldwide for the study of biochemical and biophysical phenomenon in plants. Twenty-five years ago, German botanist **Friedrich Laibach** proposed *Arabidopsis thaliana* as the model organism of choice for research in plant biology. He became the **founder of experimental Arabidopsis research** by describing its correct chromosome number in 1970. Many significant discoveries, like the universally accepted ABCE model of flowering, genes involved in embryogenesis, organogenesis, and transport mechanisms and mechanism of hormone action, have been made using *Arabidopsis* as the model system. *Arabidopsis*, a Brassicaceae member, is a winter annual which has a short and rapid life cycle of about 6 weeks, starting from seed germination to maturity. Its small genome size of 140 Mb makes it an excellent choice for genetic, recessive mutant and transformation studies. The seed production is profuse, and its cultivation can be easily done in limited spaces.

have been shown to act as hormones. These include brassinosteroids, salicylic acid, jasmonic acid, nitric oxide, indoleamines, and some more. Plant cells respond to hormones only through receptors. Some hormone receptors are localized on the plasma membrane, while others may be intracellular—cytosolic or localized on intracellular membranes such as those of endoplasmic reticulum. The hormone-receptor complex may cause immediate metabolic change, or the complex may migrate to another site of the cell, such as the nucleus. For many hormone-induced responses, some genes are activated or repressed. Cells are likely to have receptors for several hormones. In the following text, the terms PGRs and hormones would often be used interchangeably for the specific biomolecules under discussion.

## Summary

- Plants synthesize several biomolecules of varying chemical nature which facilitate cell-to-cell communication and affect varied aspects of growth and development. A number of signaling molecules have been identified in the animal systems and prokaryotes and are referred as "hormones."
- A number of biological processes in plants are affected by internal and external stimuli through involvement of plant hormones. Biosynthesis of plant hormones can occur at different morphological and cytological locations, and their actions may be observed at their sites of production, or they may get translocated to some distant target tissues to evoke a response.
- Plant growth regulators (PGRs) are a broader category of compounds which includes plant hormones, their synthetic analogs, inhibitors of hormone biosynthesis, and blockers of hormone receptors.
- There are numerous mechanisms by which plants regulate intracellular PGR levels. Both negative and positive regulators contribute to maintain the required intracellular PGR concentration. Negative regulators include efflux from the cell, sequestration in the intracellular compartments such as vacuoles and lumen of endoplasmic reticulum, inactivation through conjugation with other biomolecules (sugars, amino acids, etc.), and degradation. Positive regulators of PGR homeostasis include uptake, biosynthesis, release from internal stores, and activation.
- Noteworthy progress in plant hormone identification has been made possible due to advances in chemical analysis, chromatographic techniques, and mass spectrometry. As compared to major metabolites such as sucrose, hormone concentrations in plant tissues are generally very low in up to more than ten orders of magnitude.
- Keeping the diversity of solubility and chemical nature in view, tissue homogenates are subjected to "clean up" or enrichment methods by partitioning between aqueous and organic solvents prior to separation of plant hormones by liquid chromatography (LC) and gas chromatography (GC). Mass spectrometry (MS) further offers fast and sensitive analysis of plant hormones.
- Characterization of plant hormone action in cells and tissues is greatly facilitated by the use of reporter genes like β-glucuronidase (GUS) and green fluorescent protein (GFP) linked to a promoter sequence from a gene for plant hormone response or biosynthesis.
- In the past decade, several mutants have been isolated and screened from *Arabidopsis* which is now an established model plant system for analysis of plant hormone action.

# Multiple-Choice Questions

1. Biomolecules possessing the ability to modify cell pattern formation in a cell-dependent manner are known as:
   (a) Hormones
   (b) Morphogens
   (c) Plant growth regulators
   (d) None of the above
2. Polyamines are derived from:
   (a) Methionine
   (b) Arginine
   (c) Ornithine
   (d) Both (b) and (c)
3. Hydrophilic hormones include:
   (a) Auxin
   (b) Gibberellin
   (c) Brassinosteroids
   (d) None of the above
4. "Triple quadrupole" instrument carries out:
   (a) Analysis of specific chosen ions
   (b) Efficient separation of hormones
   (c) Analysis of multiple target ions
   (d) Fluorescent visualization of plant hormones
5. The measurement of the concentration of a physiologically active compound by a bioassay is dependent on:
   (a) Exact measurement of concentration in the biological tissue
   (b) Measurement of the biological activity of the compound in the biological tissue
   (c) Assessment of the interaction of the compound with other molecules
   (d) None of the above
6. Bioassays can be performed to provide:
   (a) Quantitative assessment of biological activity of a compound
   (b) Qualitative assessment of biological activity of a compound
   (c) Both (a) and (b)
   (d) None of the above
7. Which of the following techniques allows fast and sensitive analysis of plant hormones in living tissue?
   (a) Liquid chromatography (LC)
   (b) β-Glucuronidase (GUS)
   (c) Mass spectrometry (MS)
   (d) Green fluorescent protein (GFP)

8. Which of the following approaches can be employed to assess the hormone action?
   (a) Isolation and cloning of genes regulated by the hormone
   (b) Using inhibitors of the hormone action
   (c) Screening and analysis of mutants with altered synthesis and sensitivity with respect to the hormone
   (d) All of the above

**Answers**

1. b   2. d   3. a   4. c   5. b   6. c   7. d   8. d

## Suggested Further Readings

Cleland RE (1996) Growth substances. In: Salisbury FB (ed) Units, symbols and terminology for plant physiology. Oxford University Press, New York, pp 126–128

Ludwig-Müller J, Lüthen H (2015) From facts and false routes: how plant hormone research developed. J Plant Growth Regul 34:697–701

Smith SM, Li C, Li J (2017) Hormone functions in plants. In: Li J, Li C, Smith SM (eds) Hormone metabolism in Plants. Academic Press, San Diego, pp 1–38

# Auxins

15

Satish C Bhatla

The term "auxin" is derived from the Greek word *auxein* which means "to increase." Auxins constitute an important group of naturally occurring hormones which have been detected practically in all land plants and in several soil- or plant-associated microbes. Naturally occurring auxins include indole-3-acetic acid (IAA), indole-3-butyric acid (IBA), phenyl acetic acid, and 4-chloroindole-3-acetic acid (4-Cl-IAA). Of all these, IAA is the most extensively studied auxin (Fig. 15.1). Chemically, a common feature of all these molecules exhibiting auxin activity is the presence of acidic side chain on the aromatic ring. All the above natural auxins, except for phenyl acetic acid, are also indole derivatives.

## 15.1 Discovery of Auxin

Auxin was the first hormone discovered in plants with history dating back to the days of De Candolle (1832) who made the first observations on phototropism. He observed that indoor plants turn themselves toward the windows since they seek light (and not air). Ciesielski (1872) observed loss of gravitropism upon removal of root tips and its restoration by the replacement of severed tip. From this he concluded that some "influence" is transmitted from the root tip to the bending part of the root before it is amputated. Subsequent detailed investigations by Charles Darwin and his son Francis, published in their book entitled *The Power of Movement in Plants* in 1881, showed that when coleoptiles of the dark-grown canary grass, *Phalaris canariensis*, are exposed to unilateral light, some influence is transmitted from the upper part of the coleoptile downward causing the lower part to bend. When the seedling coleoptile is decapitated or an opaque cap is placed on the tip, the bending response to unilateral light is prevented. It was obvious that the tip of the coleoptile was the site of light perception. Paal (1914–1919) showed that removal of the coleoptile tip in dark stops growth and replacing it on one side causes curvature

© Springer Nature Singapore Pte Ltd. 2018
S. C Bhatla, M. A. Lal, *Plant Physiology, Development and Metabolism*,
https://doi.org/10.1007/978-981-13-2023-1_15

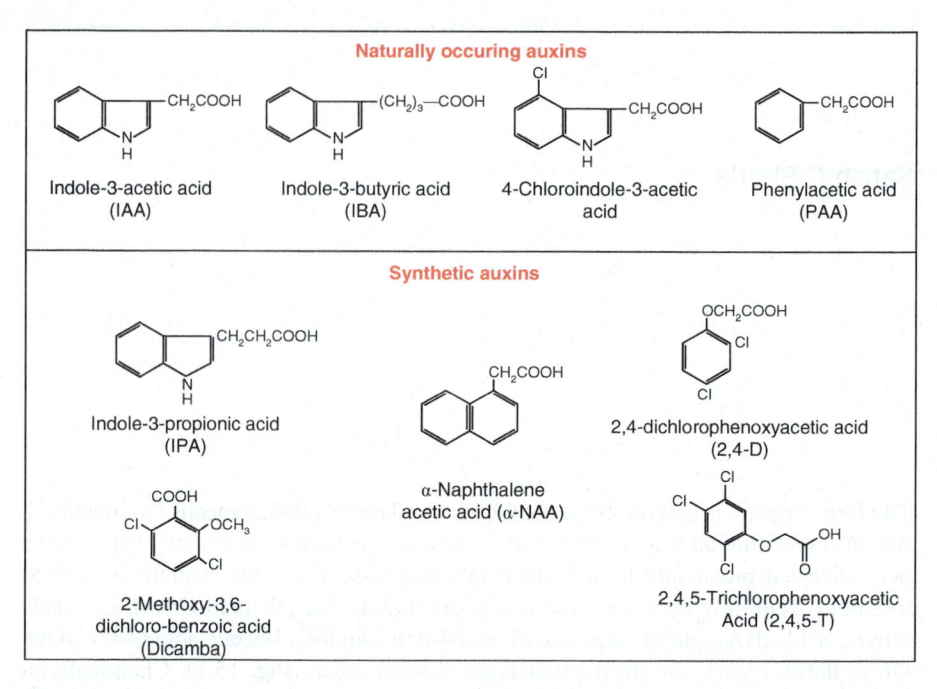

**Fig. 15.1** Structures of some naturally occurring and synthetic auxins

(Box 15.1). Boysen-Jensen (1910–1913) demonstrated that phototropism would fail when the downward flow of material on the shaded side is interrupted by a piece of mica, but not when mica is placed on the lighted side, indicating the presence of a growth **promoter**, rather than an **inhibitor**. Later workers proposed the presence of a growth-inducing substance (*Wuchsstoff*) which was proposed to exhibit polar movement from the tip to the zone of elongation, thereby causing asymmetric growth. Söding (1923) and Cholodny (1924) were the first to call the "Wuchsstoff" a hormone. In 1928, F. W. Went reported a quantitative assay for the diffusible substance responsible for curvature response in *Avena sativa* (oat) coleoptiles. The assay was referred as *Avena* curvature assay, and it established that the extent of curvature is proportional to the amount of growth substance diffusing from the tip into the lower region (Box 15.2). This concept is still called the **Cholodny-Went theory**. Went's experimental approach was to place coleoptile tips on agar blocks to let the "Wuchsstoff" accumulate in the agar block. These "Wuchsstoff"-containing agar blocks, when placed asymmetrically on the cut surface of decapitated coleoptiles, induced curvature. This was, in fact, the turning point in auxin research. Went's work is significant because (1) it confirmed the existence of a regulatory substance in the coleoptile apex, and (2) a means for isolation and quantitative analysis of the active substance was established. Subsequently, Kögl and Haagen-Smit (1934) proposed the term "**auxin**" for the "Wuchsstoff" crystalized from human urine. They reported isolation of auxin a and auxin b and heteroauxin from human urine. Almost during the same period, Thimann group (1935) reported auxin from *Rhizopus* culture medium. Investigations in

## 15.1 Discovery of Auxin

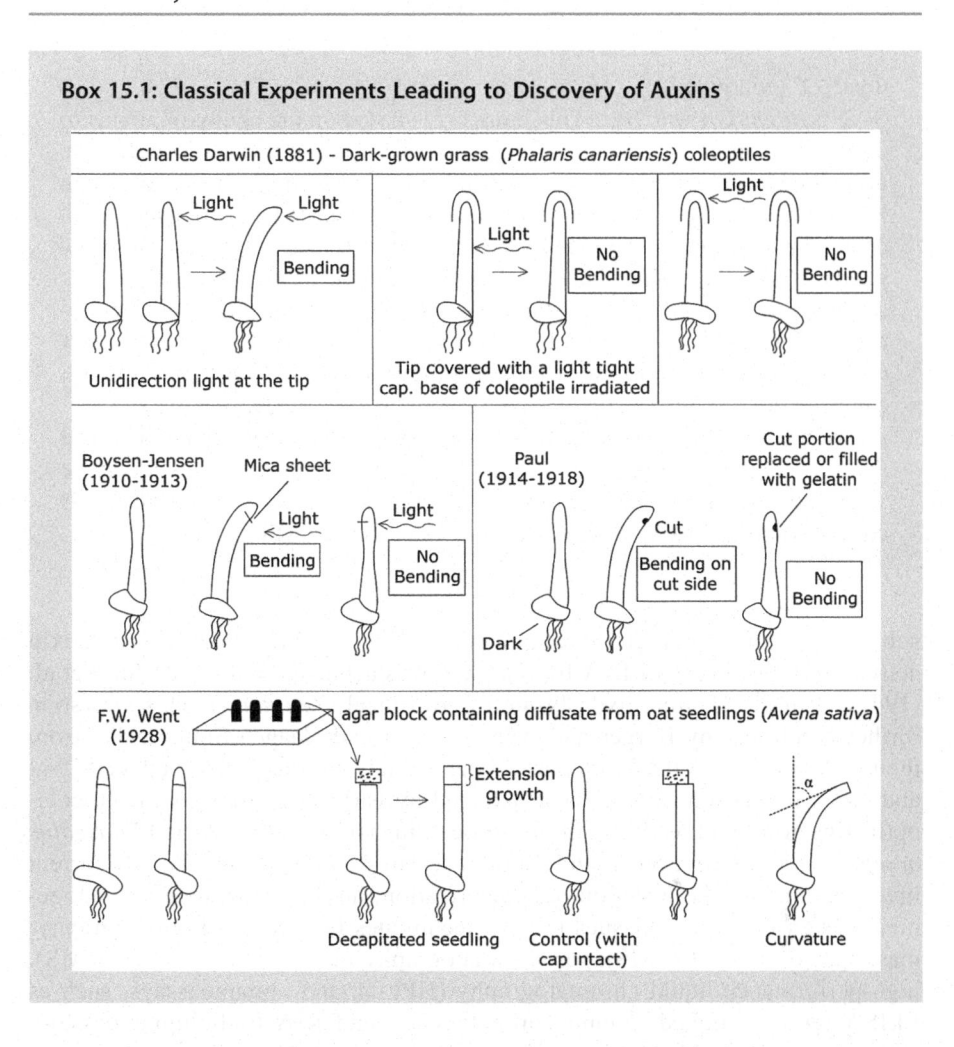

### Box 15.1: Classical Experiments Leading to Discovery of Auxins

**Charles Darwin (1881)** - Dark-grown grass (*Phalaris canariensis*) coleoptiles

- Unidirection light at the tip → Bending
- Tip covered with a light tight cap, base of coleoptile irradiated → No Bending
- Light → No Bending

**Boysen-Jensen (1910-1913)** — Mica sheet
- Light → Bending
- Light → No Bending

**Paul (1914-1918)**
- Cut, Dark → Bending on cut side
- Cut portion replaced or filled with gelatin → No Bending

**F.W. Went (1928)** — agar block containing diffusate from oat seedlings (*Avena sativa*)
- Decapitated seedling → Extension growth
- Control (with cap intact)
- Curvature

### Box 15.2: Bioassay for Auxins

*Bioassay*: An assay using biological samples to estimate the concentration or activity of a substance.

*1. Avena Curvature Test:* In this bioassay, an auxin-containing agar block is placed on one side of a decapitated oat coleoptile. As auxin from the agar block diffuses down through the coleoptile, it causes rapid growth of cells in the region just below the agar block (auxin source). This results in bending of the coleoptile to the opposite side. Degree of bending can be measured by placing the coleoptiles on a graph paper. Up to 200 µg. litre$^{-1}$, the degree of coleoptile bending shows linearity with increasing IAA concentration. Using this bioassay, unknown auxins removed from a plant can be estimated.

(continued)

> **Box 15.2** (continued)
> 
> 2. *Straight Growth Test:* This bioassay is based on the ability of auxins to cause elongation of coleoptile or stem cells. Dark-grown coleoptiles are decapitated to remove endogenous auxin source. An equal number of 5 mm long sections of the remaining coleoptile stump are placed in petri dishes, each containing solution of known IAA concentration. The length of coleoptile increases with increasing concentration of IAA in the incubation medium over a defined period (24–36 h). A standard curve thus generated (IAA concentration vs increase in coleoptile length in mm) can be used to estimate IAA concentration in a plant tissue exudate homogenate (unknown sample), and it can be expressed as units of IAA equivalents. Alternatively, sections of dark-grown pea seedling stem can also be used for this bioassay. It may be noted that stem segments obtained from light-grown pea seedlings are less effective since they exhibit lesser extension growth in response to added auxin. A control solution (-IAA) will not result in extension growth.

subsequent years, however, revealed that auxin a and b did not contain any auxin activity. First isolation of IAA from a plant was achieved by Haagen-Smit et al. (1942) from the alkaline hydrolysate of corn meal (*Zea mays*). This was soon further confirmed by Berger and Avery (1944) and Haagen-Smit (1946) from their isolation of IAA from immature maize kernels. Subsequently, between 1944 and 1974, indole-3-acetic acid was isolated by various investigators from coleoptile tips, roots and seedlings of maize, seedlings of *Avena sativa* and *Phaseolus mungo*, *Pinus radiata* buds, and phloem sap of *Ricinus communis*. All current investigations on the isolation, characterization, and quantification of endogenous auxins from plant tissues rely on techniques like liquid chromatography-mass spectrometry (LC-MS), gas chromatography-mass spectrometry (GC-MS), high-performance liquid chromatography (HPLC), and immunoassays, such as ELISA (enzyme-linked immunosorbent assay) and RIA (radioimmunoassay), which possess variable sensitivity for auxin detection.

## 15.2 Synthetic Auxins

After the discovery of indole-3-acetic acid and establishment of its role in cell elongation, attempts were made to examine the biological activity of a number of compounds with substituted indoles, such as indole-3-propionic acid (IPA). Like natural auxins, IPA is biologically active and is commonly used as a rooting hormone in horticultural practices. Like IAA, it has an indole ring and a terminal carboxyl group but differs in its side chains (Box 15.3). Compounds such as α-naphthalene acetic acid (NAA) are also biologically active and are used as rooting hormone for certain plants. NAA lacks indole ring but retains acetic acid side chain present in IAA. Another biologically active synthetic auxin is 2,4-dichlorophenoxy acetic acid (2,4-D) (Box 15.4). Like NAA, it also lacks indole ring.

### Box 15.3: Synthetic Auxins

Synthetic auxins include 2,4-dichlorophenoxyacetic acid (a herbicide), naphthalene acetic acid (a component of commercially available rooting chemical), 2-methoxy-3,6-dichlorobenzoic acid (dicamba, a herbicide), and indole butyric acid (IBA). Propagation of plants by vegetative means is commonly practiced in horticulture. Auxin applications are generally beneficial in bringing about rooting of cuttings for vegetative propagation. Another property of auxins is their ability to initiate development of fruits without pollination (**parthenocarpy**). Because of the difficulty of satisfactory pollination in some plants as well as to improve the quality of the fruits, auxin treatment in the form of spray or aerosol is commonly practiced. Fruits so induced are seedless. So, besides increasing yield, auxin treatment also makes parthenocarpy feasible.

In pineapple, flower induction at the appropriate time can be a problem. It can, however, be made to flower at any time of the year by the application of synthetic auxins. In the USA and Russia, growers of citrus, apple, and pear extensively use auxin spray for the prevention of premature fall of fruits. Premature fruit fall can lead to extensive damage to crop. Thus, growers must either harvest fruits before the best quality is obtained or else risk a heavy loss. Auxin sprays have proved highly successful in preventing premature fruit fall, and now growers can obtain reasonable assurance against loss of apple and pear crops by using auxin sprays.

### Box 15.4: 2,4-D: The First Selective Herbicide

During Vietnam War (1961–1971), the US Air Force sprayed "Agent Orange" over Vietnamese agricultural land with the intention of destroying crops. Agent Orange or Herbicide Orange (HO) is one of the herbicides and defoliants used by the US military as a part of its herbicidal warfare program. HO is a mixture of equal parts of two herbicides, 2,4,5-trichlorophenoxy acetic acid (2,4,5-T) and 2,4-dichlorophenoxy acetic acid (2,4-D). The mission to destroy crop, however, failed. 2,4-D is a selective herbicide which kills dicots while leaving monocots alone. In other words, it can be sprayed on grasses (like wheat, rice, corn, and other cereal crops) without any major harm while killing broad leaf weeds (dicots). It gives dicots a form of symbolic cancer, and the plants grow uncontrollably (the plant grows itself to death). It affects only dicots because the two classes of plants (monocots and dicots) transport these herbicides using different mechanisms. Broad-leaved weeds have their growing point at the tips of the leaves, while grasses grow from their base.

(continued)

**Box 15.4** (continued)

2,4-D works by interfering with growth, either by blocking photosynthesis and protein synthesis or by destroying or inhibiting root formation. 2,4-D gets absorbed through stomata and is transported to the meristems of the plant. This causes uncontrolled and unsustainable growth, and the plants wilt and die. The underlying molecular mechanism regarding selectivity of this synthetic auxin is because of either limited translocation or rapid degradation, altered vascular anatomy, or altered perception of auxin in monocots. Also, auxin transport is influenced by plant vascular systems. The differences in vascular tissue structure between dicots and monocots may contribute to the selectivity of auxinic herbicide. In monocots stems, the vascular tissues are scattered in bundles and lack vascular cambium. In dicot stems, the vascular tissues are formed in rings and possess a cambium. To conclude, in monocots, general mode of action of auxin is not affected but an accessory pathway clears up auxin in levels excess of any normal amounts that may be present.

Differences/similarities in response of dicots and monocots to synthetic auxin application (spray or through soil)

| Dicots (e.g., weeds) | Monocots (crops) |
|---|---|
| Accumulates in free form; so remains active (e.g., 2,4-D) | Conjugates with other biomolecules, so becomes inactive soon after absorption |
| Slow transport (due to lack of compatible transporter proteins), thus acting longer | Slow transport |
| Stays longer in metabolically active form, hence causes toxic effects due to accumulation | Low affinity with auxin-binding proteins, hence less tissue sensitivity |

## 15.3 Auxin Distribution and Biosynthesis

The amount of IAA present in the tissue will depend on the type of tissue and its age. In vegetative tissues, IAA concentration varies between 1 and 100 μg (5.7–570 nanomoles) kg$^{-1}$ fresh weight. Seeds, however, generally have an abundance of IAA. Thus, the endosperm of a maize seed may contain up to 308 picomoles of IAA. On the contrary, maize shoot contains much less IAA (in the range of 27 picomoles). High level of auxin has also been reported in legume seeds. Cereal grains mainly contain esterified IAA, while legume seeds largely contain peptidyl IAA. Although most plant tissues are able to biosynthesize low levels of IAA, it is largely produced in shoot apical meristem, developing fruits and seeds and in young leaves. High amount of IAA present in root tips is due to its transport from other sites of production rather than its biosynthesis in the root tip alone. Auxin biosynthesis has also been demonstrated in the tips of dicot leaves (Fig. 15.2). With the passage of

## 15.3 Auxin Distribution and Biosynthesis

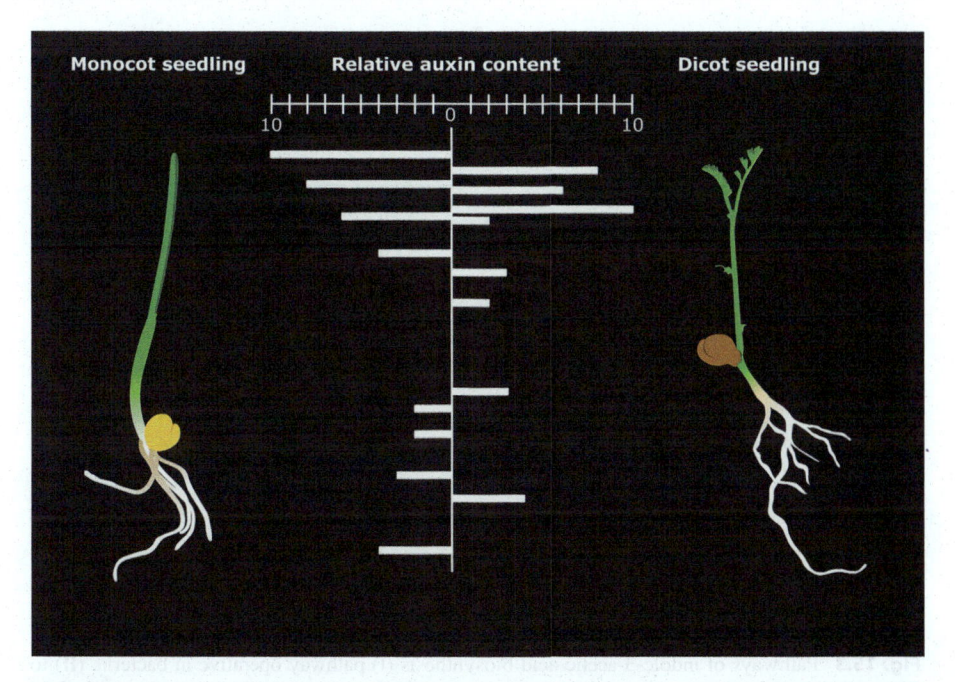

**Fig. 15.2** Relative auxin distribution in a monocot and a dicot seedling. Relative units of auxin: 10=high; 1=low

development of leaves, the site of IAA biosynthesis gradually shifts toward the base of the leaves and then to the central region representing the vascular system. Intense auxin biosynthesis is also evident in the cells that differentiate into hydathodes (gland-like structures through which water is released as a result of positive pressure). A decreasing auxin gradient from the tip of the hydathodes to the developing vascular strand down below indicates the role of auxin in vascular tissue differentiation.

Biosynthesis of tryptophan (auxin precursor) principally takes place in chloroplasts. Subsequent formation of IAA may occur both in cytosol and chloroplasts. About one-third of IAA in a cell is found in chloroplasts and rest in the cytosol. IAA biosynthesis can take place both via tryptophan-dependent and tryptophan-independent pathways.

### 15.3.1 Tryptophan-Dependent Pathways

Tryptophan (an aromatic amino acid) is the precursor of IAA. Plants convert tryptophan to IAA through several pathways (Fig. 15.3).

1. *Bacterial pathway*: In some pathogenic bacteria, such as *Agrobacterium tumefaciens* and *Pseudomonas savastanoi*, tryptophan is converted into indole-3-acetamide (IAM) in the presence of enzyme tryptophan monooxygenase. IAM is then hydrolyzed to IAA in the presence of enzyme IAM hydrolase.

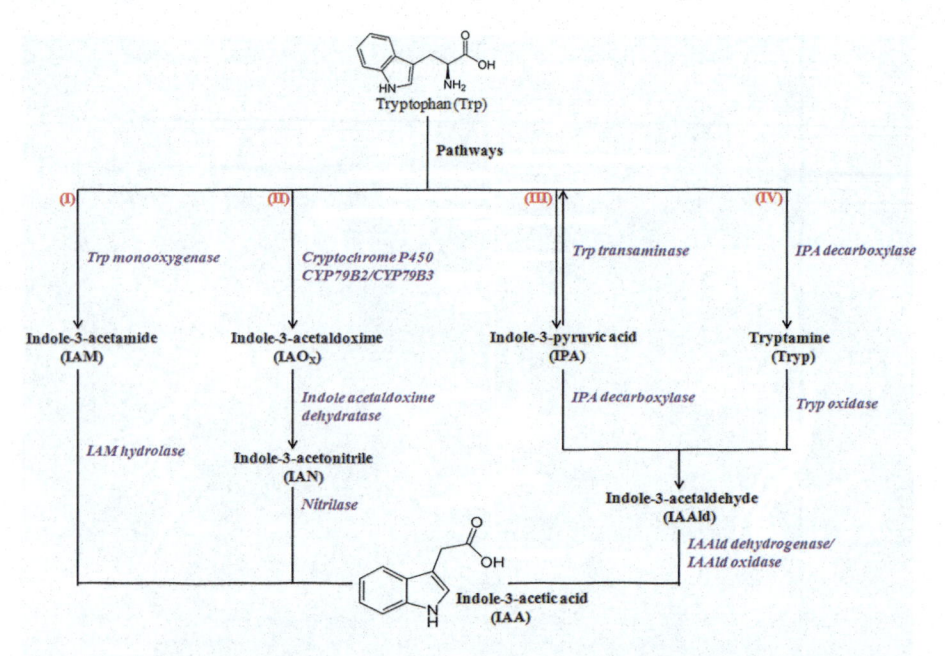

**Fig. 15.3** Pathways of indole-3-acetic acid biosynthesis (I) pathway operative in bacteria. (II) to (IV) pathways operative in plants—(II) Indole-3-acetonitrile pathway. (III) Indole-3-pyruvic acid pathway. (IV) Tryptamine pathway

2. *IAN (indole-3-acetonitrile) pathway*: It is prevalent in members of Brassicaceae, Poaceae, and Musaceae. Tryptophan is converted to IAA in the presence of the enzyme nitrilase. Indole-3-acetaldoxime and indole-3-acetonitrile are the intermediates.
3. *IPA (indole-3-pyruvic acid) pathway*: Tryptophan →IPA →Indole-3-acetaldehyde →IAA.
4. Tryptophan is deaminated to form indole-3-pyruvic acid followed by decarboxylation, resulting in the formation of indole-3-acetaldehyde. The enzymes involved are transaminase and IPA decarboxylase, respectively.
5. *TAM (tryptamine) pathway*: Tryptophan is decarboxylated to form tryptamine (TAM), followed by deamination to form indole-3-acetaldehyde (IAld). The enzymes involved are tryptophan decarboxylase and tryptamine oxidase, respectively. IAld is readily oxidized to form IAA by the enzyme IAld dehydrogenase.

### 15.3.2 Tryptophan-Independent Pathways

In this pathway IAA is also synthesized from indole or indole-3-glycerol phosphate. Evidence for the same was obtained from orange pericarp mutant of maize. In this mutant, both subunits of the enzyme tryptophan synthase are inactive. It requires

## 15.3 Auxin Distribution and Biosynthesis

exogenous tryptophan to survive, and it is unable to convert tryptophan to IAA. Despite this, the mutants contain high level of auxin. Thus, it was concluded that IAA could be synthesized from indole or indole-3-glycerol phosphate (Fig. 15.4).

Chorismate → Phenylalanine Tyrosine

*Anthranilate synthase* — Glutamine ↓ Glutamate + pyruvate

Anthranilate

*Phosphoribosylanthranilate transferase* — 5-phosphoribosyl-1-pyrophosphate ↓ ℗℗ᵢ

Phosphoribosylanthranilate

*Phosphoribosylanthranilate isomerase* ↓

1-(o-carboxyphenylamino)-1-deoxyribulose phosphate

Tryptophan-independent IAA biosynthesis

*Indole-3-glycerol-phosphate synthase* ↓

→ Indole-3-glycerol phosphate

*Tryptophan synthase α subunit* ↘ Glyceraldehyde 3-phosphate

→ Indole

*Tryptophan synthase β subunit* — Serine ↓ H₂O

Indole-3-acetic acid

Tryptophan-dependent IAA biosynthesis

Tryptophan

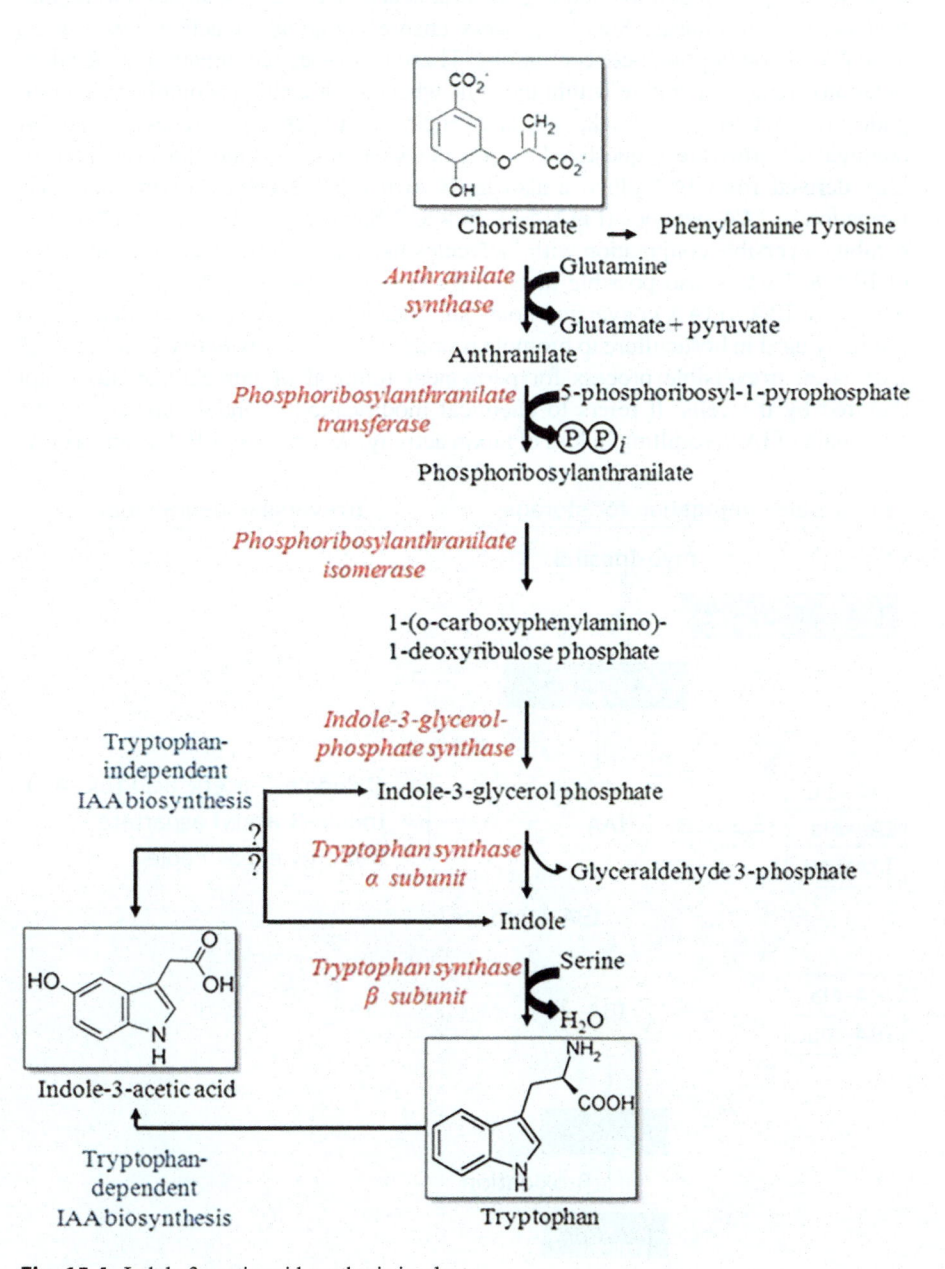

**Fig. 15.4** Indole-3-acetic acid synthesis in plants

## 15.4 Conjugation and Degradation of Auxins

At high concentrations, intracellular auxin can be toxic. Therefore, its homeostatic control through conjugation and degradation is also necessary in addition to regulation of its biosynthesis. Such a process ensures removal of active auxin when available at supraoptimal concentrations. These processes (conjugation and/or degradation) are also operative within the cells when auxin action is completed. Conjugation of IAA to glucose, alanine, and leucine is a reversible process, and most conjugated forms are sequestered within the vacuoles. Myoinositol derivative of IAA derived from IAA-glucose also leads to reversible release of free IAA via formation of IAA peptides (Fig. 15.5). Indole-3-butyric acid (IBA), like IAA, also exhibits reversible conjugation with molecules like alanine and glucose. Conversion of IBA to IAA is also possible through β-oxidation of IBA-CoA derivative. This process of IBA to IAA conversion holds significance in view of the fact that IBA is routinely used in horticulture to promote rooting of cuttings. Oxidative catabolism of IAA is an irreversible process for permanent removal of intracellular auxin not required by the cells. It refers to chemical modification of indole nucleus or the side chain of IAA, resulting in loss of auxin activity. As early as in 1940s, an enzyme

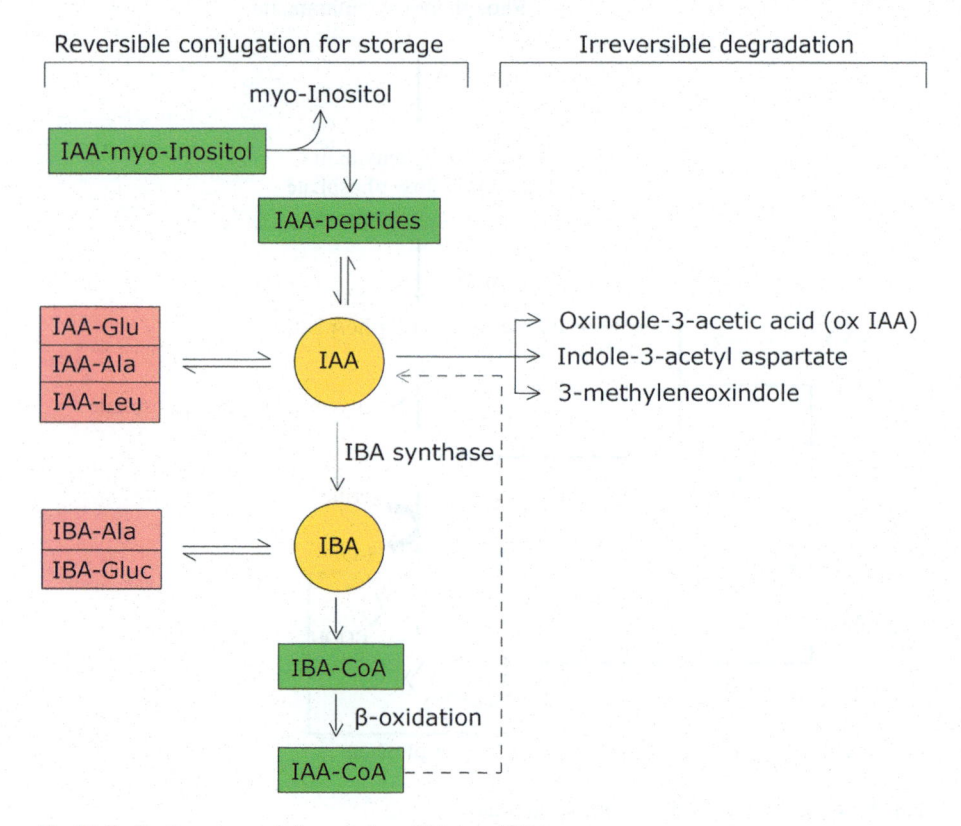

**Fig 15.5** Conjugation and degradation of IAA and IBA

responsible for inactivation of IAA was isolated from plant extracts and was called **IAA oxidase**. IAA catabolism is now known to be mediated by specific isoforms of **peroxidase**, resulting in the formation of 3-methyleneoxindole derivative. IAA can also be irreversibly oxidized to oxindole-3-acetic acid (ox-IAA) or indole-3-acetyl aspartate (Fig. 15.5).

## 15.5 Auxin Transport

Auxin is the only plant hormone which exhibits polar transport in almost all plants, including bryophytes and ferns. Polar transport of auxin is always basipetal, i.e., toward the base, both in shoots and roots. Thus, it is transported from shoot tip and leaves downward in the stem and from the root tip toward the base of the root to the root-shoot junction (Fig. 15.6). The rate of polar auxin transport varies from 5 to 20 cm/h. Auxin biosynthesized in the leaves has been reported to be transported via the phloem cells to the other parts of the plant. Most of the auxin transport within the plant occurs through vascular parenchyma (specially xylem parenchyma) and also through sieve-tube elements. Polar auxin transport proceeds in a cell-to-cell fashion. It exits the cell through the plasma membrane, diffuses across the middle lamella, and crosses the next plasma membrane of the next cell (*apoplastic movement*). Auxin entry is referred as influx and its exit is efflux.

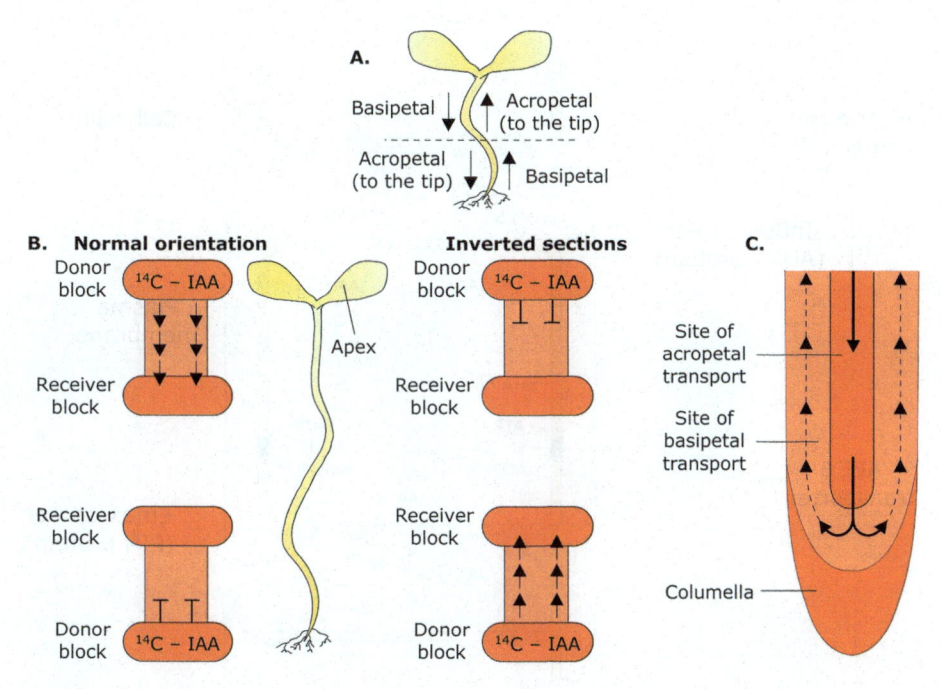

**Fig 15.6** (a) Nomenclature used to define the direction of auxin transport. (b) Demonstration of polar auxin transport in the hypocotyl segments of a dicot seedling. (c) Direction of auxin transport in root

## 15.5.1 Auxin Influx

Two mechanisms are operative for auxin influx.

1. Passive diffusion of protonated form (IAAH, a lipophilic molecule) from any direction across the phospholipid bilayer
2. Secondary active transport of dissociated form (IAA$^-$) using symporter protein, AUX1

The plasma membrane H$^+$-ATPase activity normally maintains the cell wall pH around 5.5. As a result, about 25% of IAA in the apoplast remains in protonated form (IAAH) which diffuses passively across the plasma membrane according to concentration gradient. The secondary active transport of auxin allows greater auxin accumulation than simple diffusion does because it is driven across the membrane by proton motive force (pmf). An auxin uptake carrier protein (AUX1) functions in leaf vascular tissue and root apices. AUX1 belongs to a family of proteins similar to permeases in prokaryotes. [*Permeases are a group of membrane-bound carriers (enzymes) that affect solute transport across the semipermeable membrane.*] AUX1 exhibits polar localization at the ends of protophloem cells in root apex where it is thought to function in the process of unloading phloem-derived auxin, moving acropetally toward the tip. It also exhibits a polar localization in root cap for basipetal transport (Fig. 15.7).

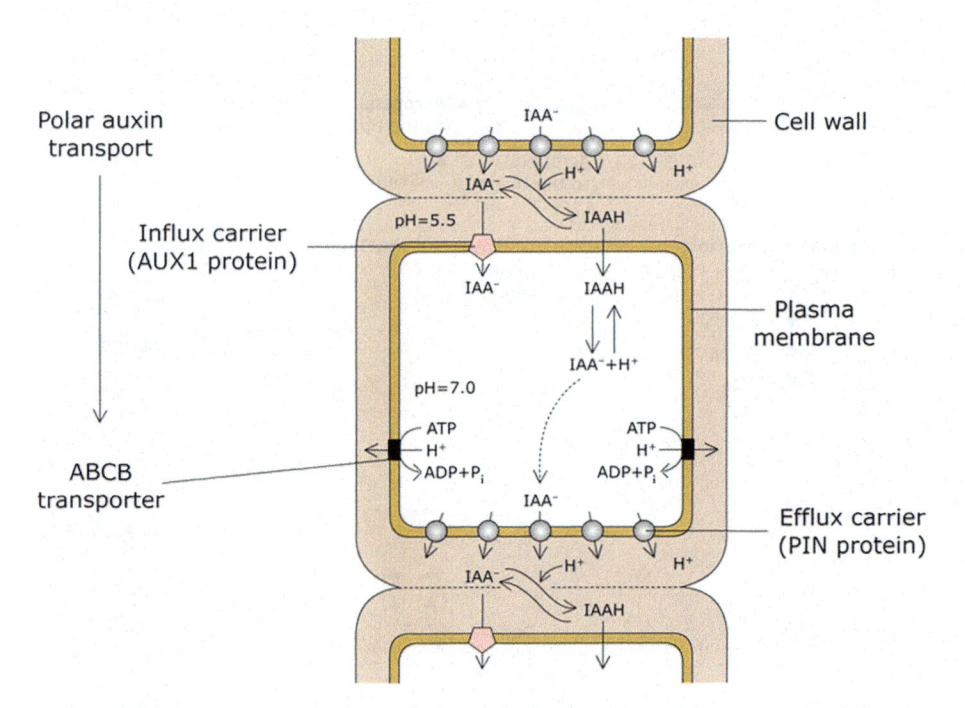

**Fig. 15.7** Schematic chemiosmotic model for polar transport of auxin

## 15.5.2 Auxin Efflux

Is largely brought about by a set of PIN proteins which are mainly located in the basal region of the cells. PIN derives its name from the *Arabidopsis pin 1* mutant because of the needle-like shape of inflorescence and the stem which lacks leaves, buds, and flowers (Fig. 15.8). At least eight PIN genes have been isolated from *Arabidopsis*. Polar auxin transport is significantly reduced in *pin* mutants, and this characteristic feature of these mutants can be mimicked by blocking polar auxin transport using certain compounds categorized as **phytotropins**. Phytotropins are noncompetitive inhibitors of polar auxin transport. These include TIBA (2,3,5-triiodobenzoic acid), morphactin (9-hydroxyfluorine-9-carboxylic acid), and NPA (N-1-naphthylphthalamic acid). PIN1 is mainly located in xylem parenchyma both in shoot and root and is crucial for polar auxin transport (Fig. 15.8). PIN2 is localized in the root, cortex, and epidermis and regulates basipetal movement of auxin in roots. PIN3 is localized in endodermis in the shoot and in columella in the root. In both the tissue systems (shoot and root), PIN3 is involved in lateral distribution of auxin. PIN4 is located in the quiescent center and appears to function in the establishment of auxin sink below the quiescent center in the root apical meristem. PIN7 has been reported to play a role in forming and maintaining apical-basal auxin gradient in the

**Fig. 15.8** Phenotypes of wild-type *Arabidopsis* (**a**) and PIN1 mutant (**b**) plants. Note the lack of inflorescence in the mutant (**b**) giving it a pin-like appearance. AUX1 (**c**) and PIN1 (**d**) proteins localized in the vascular parenchyma cells in the roots of *Arabidopsis*

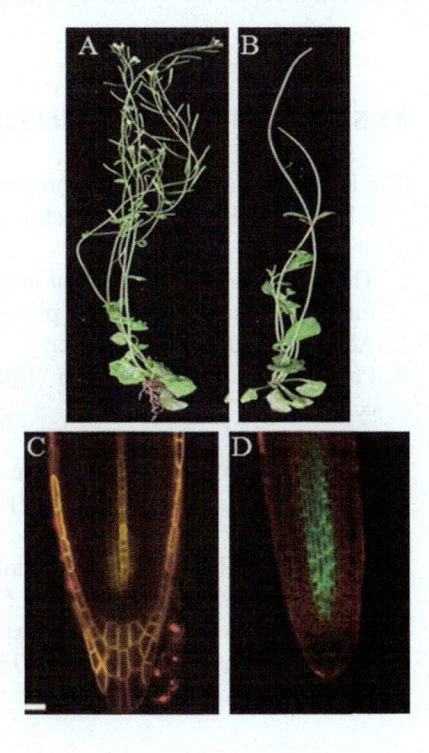

**Table 15.1** Various types of PIN proteins found in plants

| Auxin efflux protein | Location | Role |
|---|---|---|
| PIN1 | Xylem parenchyma in shoots and roots | Basipetal IAA transport in shoots and acropetal transport in roots |
| PIN2 | Roots | Basipetal redistribution of auxin through cortical and epidermal cells |
| PIN3 | Cells of shoot epidermis; columella of roots and pericycle | Lateral redistribution of auxin |
| PIN4 | Quiescent center | Establishment of auxin sink below the quiescent center in root apical meristem |
| PIN7 | Embryo | In forming and maintaining apical-basal gradients for embryonic polarity |

embryos (Table 15.1). Polar auxin transport is largely controlled through the recycling of PIN proteins between the plasma membrane and various endomembrane compartments by an actin-dependent mechanism. This involves secretion of vesicles carrying PIN proteins from their site of biosynthesis and their migration to specific sites on the plasma membrane of auxin-conducting cells. This secretion is regulated by endocytotic cycling of vesicles along the actin track.

### 15.5.3 Chemiosmotic Model for Auxin Transport

The model for auxin transport proposed by Rubery, Sheldrake, and Raven in the mid-1970s has three principle features:

1. The existence of a pH gradient or proton motive force (pmf) as the driving force for IAA transports across the plasma membrane.
2. Activity of auxin influx carrier.
3. Preferential location of auxin efflux carrier proteins in the basal region of the cells.

On the basis of these three features, chemiosmotic model of auxin transport can be summarized as follows (Fig. 15.7):

1. IAA is a lipophilic weakly acid molecule. The apoplast/cell wall space is moderately acidic with a pH of about 5.5. At this pH, about 25% of IAA remains protonated (IAAH), and the rest exists in anionic form (IAA$^-$). Thus, the cell wall region contains both protonated (IAAH) and anionic (IAA$^-$) forms of indole-3-acetic acid (Box 15.5).

## 15.5 Auxin Transport

**Box 15.5: Dissociation Pattern of IAA in the Cells**

Auxins, gibberellins, abscisic acid (ABA), and jasmonic acid (JA) are weak acids, which in solution dissociate into anions and protons (*weak acids and bases are those acids which are not completely dissociated in solution. They exist in solution as an equilibrium mixture of undissociated and dissociated species*). pKa value of IAA, ABA, and JA at 20 °C is about 4.7–4.8 (pKa of $GA_1$ is lower, 3.85). Accordingly, at neutral pH in free solution, these hormones exist predominantly in dissociated form, but at pH closer to pKa, the proportion of undissociated form increases. The equilibrium constant (K) for the ionization of an acid (Ka) is

$$K = \frac{[H^+][A^-]}{[HA]}.$$

The pKa of IAA is defined as: $pKa = -\log K = \log 1/K$

At pH 4.7 $\frac{[IAA^-][H^+]}{[IAA]} = 1$

In undissociated form, the carboxyl group of IAA is protonated and is lipophilic. It readily diffuses across the membrane. Dissociated form is negatively charged ($IAA^-$) and, therefore, does not cross plasma membrane unaided.

At pH 5.5, the two forms maintain a balance as:

| IAA⁻ | ⟷ | IAAH⁺ |
|---|---|---|
| (75%) | | (25%) |

Driving force for auxin uptake is proton motive force (pmf) across the plasma membrane (*proton motive force refers to the gradient of electrochemical potential for $H^+$ across the plasma membrane. It is the sum of proton chemical potential and transmembrane electric potential*). Driving force for auxin efflux is membrane potential. Auxin efflux is driven by the inside negative membrane potential.

2. Because of its lipid solubility, IAAH slowly diffuses across the plasma membrane from the cell wall region into the cytosol.
3. Most of the IAA, however, enters the cell across the plasma membrane as IAA- using AUX1 symporter proteins which are uniformly distributed along the plasma membrane.
4. Inside the cells, at the cytoplasmic pH which is generally close to 7, IAAH dissociates into IAA− and H+. Thus, auxin gets trapped inside the cell because IAA- cannot spontaneously diffuse across the plasma membrane.

5. The critical component of the chemiosmotic model is, however, the basal location of PIN proteins which is responsible for efflux of IAA- from the cells. It is this unique basal location of PIN proteins which is responsible for establishing polarity in auxin transport.
6. The major driving force causing efflux of IAA− from the cell is the negative plasma membrane potential, which varies between −200 and −300 mV.

Another set of auxin efflux proteins recently reported from some "loss of function" mutants of *Arabidopsis* is referred as **ABCB transporters** (Fig. 15.7). They have been reported to facilitate IAA⁻ efflux or influx across the plasma membrane (and also tonoplast), but unlike PIN proteins, they do not show preferential basal localization in the cells. So, although these transporter proteins are involved in auxin efflux or influx, they do not play any significant role in maintaining polar transport of auxin. Furthermore, they require energy in the form of ATP hydrolysis to transport IAA⁻.

## 15.6 Physiological Effects of Auxins

### 15.6.1 Cell Expansion (Acid Growth Hypothesis)

A common feature of the cell wall in growing cells is that they extend much faster at acidic pH than at neutral pH. This phenomenon is called **acid growth**. Cleland and Rayle proposed in 1970 a theory to explain auxin-stimulated cell wall extension. According to them, auxin causes acidification of cell wall environment through the release of protons from the cells. Lower (acidic) apoplastic pH conditions so created then activate the process of cell wall loosening through enzymatic action. During the same period in 1970s, another scientist named Hager from Germany further proposed the role of plasma membrane-bound H$^+$-ATPases in auxin-stimulated proton release. These two proposals earlier referred as Cleland-Hager proposal are now known as **acid growth hypothesis** for plant cell enlargement. This hypothesis proposes that:

1. Auxin activates plasma membrane-localized H$^+$-ATPase resulting in cell wall acidification.
2. Lower/acidic cell wall pH enhances the activity of **expansins** which results in breakage of hydrogen bonds between the cellulose microfibrils.
3. This loosening of cellulose microfibrils in the cell wall allows turgor-induced cell expansion.

Auxin brings about activation of H$^+$-ATPases through its binding with a membrane-associated protein called **ABP1** (auxin-binding protein 1). ABP1 is thus an auxin receptor which has been localized in the ER and plasma membrane.

## 15.6 Physiological Effects of Auxins

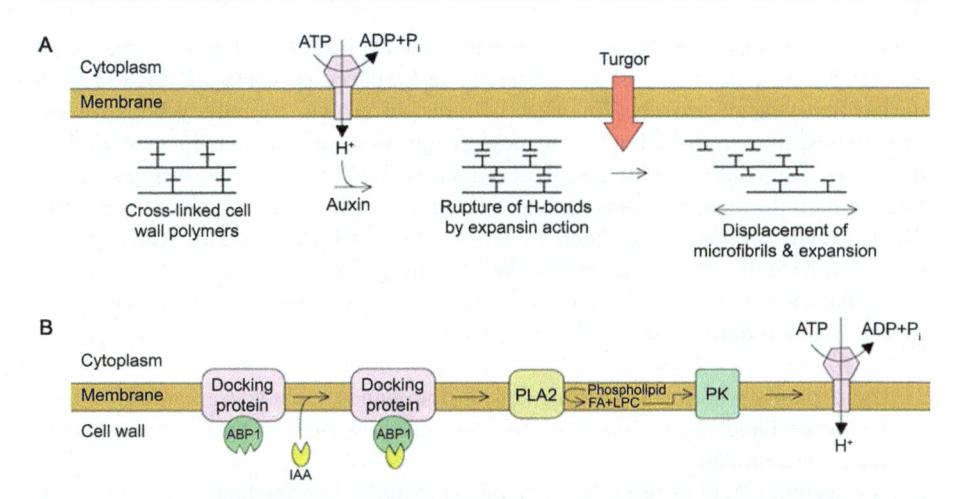

**Fig. 15.9** Mechanism of cell enlargement. (**a**) Cell wall polymers (cellulose microfibrils) are extensively cross-linked which limits their capacity to expand. An auxin-activated ATPase proton pump acidifies the cell wall space by pumping protons from the cytoplasm. The lower pH activates expansins which loosen the load-bearing bonds. Turgor pressure causes the polymers to displace and cell enlarges. (**b**) Proposed signal transduction chain linking auxin with ATPase proton pump activity. *ABP1* auxin-binding protein 1, *PLA2* phospholipase A2, *FA* fatty acids, *LPC* lysophosphotidylcholine, *PK* protein kinase

It is a 43 kDa glycoprotein dimer of two subunits of 22 kDa each. It is proposed that ABP1 forms a complex with a transmembrane docking protein which provides lipid solubility to anchor ABP1 to the membrane. This complex (ABP1-docking protein complex) then migrates from ER to the plasma membrane where ABP1 gets lodged facing outside (apoplastic side). In this position, ABP1 attaches itself to an auxin molecule. This auxin-ABP1-docking protein complex initiates a signal transduction pathway leading ultimately to activation of ATPase proton pumps. There is evidence to implicate phospholipase A2 (PLA2) in this auxin-induced signal transduction chain. It is suggested that PLA2 follows ABP1 in this chain of events followed by formation of lysophospholipids and fatty acids through the cleavage of phospholipids by PLA2 action. The lysophospholipids have been reported to activate $H^+$-ATPase through the involvement of protein kinase cascade (Fig. 15.9).

### 15.6.2 Apical Dominance

Removal of the shoot apex is a common horticultural practice to produce bushy plants by stimulating the growth of axillary buds. Apical dominance refers to the inhibition of axillary bud growth by the continued meristematic activity in the shoot apex. This process can be easily demonstrated by the decapitation of the shoot apices. Apical dominance is best observed in herbaceous plants and also in trees during first year of growth. This state of arrest of growth of axillary buds (quiescence) is different from bud dormancy, in physiological terms. In case of bud dormancy (in contrast with apical dominance), both terminal as well as lateral buds remain dormant in plants native to temperate and colder regions of the earth. Their growth activity is regulated by temperature and photoperiod. Apical dominance, on the other hand, determines the branching patterns of plants.

Herbaceous plants can be grouped into three categories with reference to the extent of apical dominance exhibited by them.

1. Those plants which show strong apical dominance include *Helianthus annuus* (sunflower) and *Tradescantia* sp. such plants show little or no lateral branching unless decapitated.
2. The second category of herbaceous plants exhibits intermediate or partial apical dominance. Thus, they exhibit some branching. For example, *Phaseolus vulgaris* (common bean), *Pisum sativum* (pea), *Ipomoea nil* (Japanese morning glory), and *Vicia faba* (broad bean).
3. The third category of plants shows very weak or no apical dominance. These plants continue to exhibit substantial branching even when intact. For example, *Coleus* and *Arabidopsis*.

Plants with strong to moderate apical dominance exhibit greater inhibition of axillary buds close to the apex, and the impact of apical dominance is gradually released on the axillary buds at a greater distance from the apex where branching becomes more common. A change in the reproductive phase of the plant also brings about significant changes in their branching patterns. Thus, for example, in the case of oats, growth inhibition of lateral buds is nullified with the onset of flowering. In contrast, in plants such as *Phaseolus*, fruiting and fruit set can reimpose inhibition of growth of lateral buds.

Attempts to understand the physiology of apical dominance began with the observations of Thimann and Skoog in 1933 whereby they showed that application of lanolin paste containing IAA to the cut stumps of *Vicia faba* plants inhibits the growth of lateral buds below. Subsequently, various investigators demonstrated similar response using both IAA and the synthetic auxin-NAA, thereby highlighting that auxins can substitute for shoot apex and result in inhibition of axillary bud growth (Fig. 15.10). Application of other hormones, such as cytokinins, gibberellins, or abscisic acid to the cut stumps, does not bring about similar inhibition of lateral bud growth. Inhibition of lateral bud growth is released upon application of compounds, such as TIBA or NPA (auxin transport inhibitors), in the lanolin paste placed over the decapitated stump (Fig. 15.10). All these observations lead to conclusions that auxin produced in the shoot apex inhibits lateral bud growth

## 15.6 Physiological Effects of Auxins

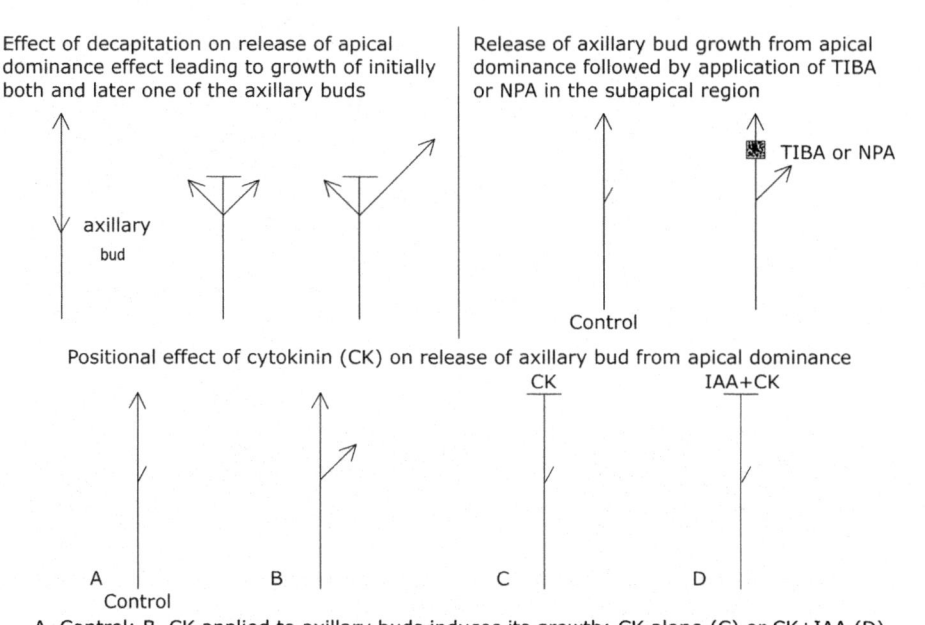

**Fig. 15.10** Experiments demonstrating the effect of decapitation, auxin transport inhibitor, and cytokinin in the release of axillary bud growth from apical dominance effect in pea plants

through its transport downward. The most recent model to understand the mechanism of apical dominance proposes as follows (Figs. 15.11 and 15.12):

1. The initial signal for axillary bud growth has been found to be an increase in sucrose availability to the bud. The use of radioactively labeled sucrose has demonstrated that the concentration of sucrose derived from leaves decreases in the stem region adjacent to the axillary bud as early as within 2 h of decapitation. This decline in sucrose in the stem is due to its uptake by the axillary bud. It is thus evident that following decapitation, sucrose depletion in the stem adjacent to the bud takes place prior to auxin depletion leading to bud outgrowth. Apical dominance is thus regulated by limiting sugar availability to the axillary buds in intact plants, and sustained growth of the axillary bud requires depletion of auxin in the stem adjacent to the bud as well.
2. Auxin itself does not accumulate in the cells of axillary buds whose growth is suppressed. It, in fact, acts via intermediary signals.
3. Activation of axillary bud growth following decapitation of the apical bud depends on the supply of cytokinins and on the ability of the axillary bud to export auxin.

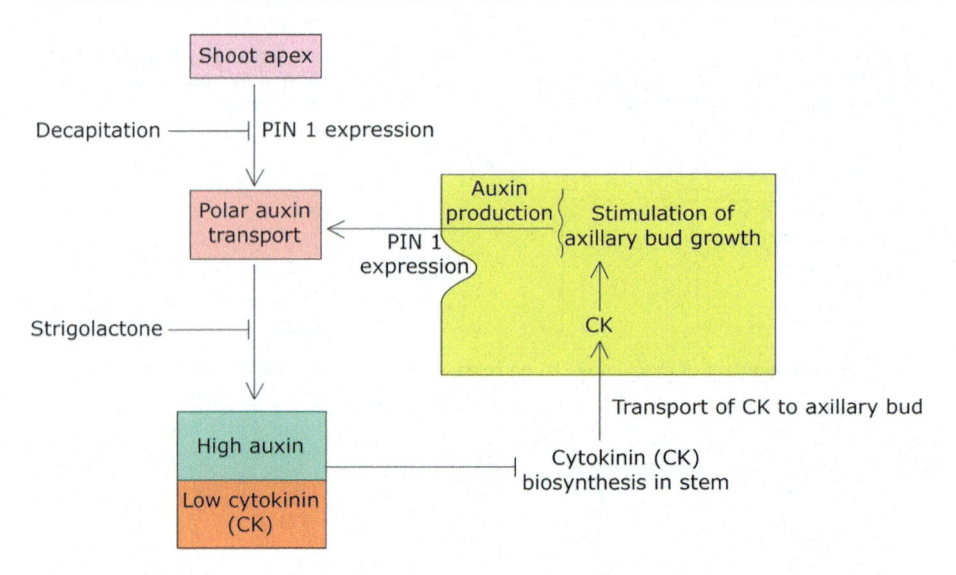

**Fig. 15.11** Model depicting interaction among auxin, strigolactone, and cytokinin in controlling apical dominance

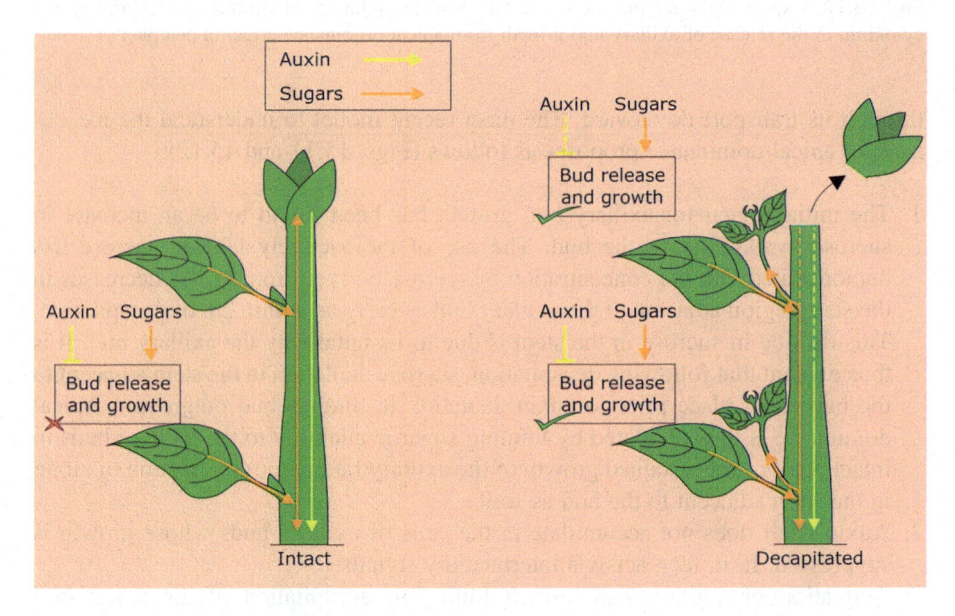

**Fig. 15.12** Regulation of apical dominance by sugar availability

4. Strigolactones (a family of terpenoid derivatives) negatively regulate basipetal auxin transport from the shoot apex. In this way, strigolactones reduce the sink strength of the stem for auxin.
5. Auxin is a positive regulator of strigolactone synthesis, and it also inhibits cytokinin synthesis in the stem. As a result of decapitation of the shoot apex, auxin source is lost. Consequently, suppression of cytokinin biosynthesis genes is reversed, and they become active.
6. Enhanced cytokinin biosynthesis in the suppressed axillary buds initiates meristematic activity in them resulting in axillary bud growth. Thus, a new cycle of apical dominance/axillary bud growth is established.

Strigolactones are believed act in coordination with auxin during apical dominance. It has been observed that *Arabidopsis* mutants deficient in strigolactone biosynthesis (*max1* [*more axillary growth 1*], *max3* and *max4*) or strigolactone signaling mutant (*max2*) exhibit increased shoot branching without decapitation. If cytokinin is applied directly to axillary buds in intact plants, it stimulates their growth, indicating the involvement of cytokinin in breaking apical dominance. It has been shown that the cytokinins involved in breaking apical dominance are synthesized locally in the axillary buds.

## 15.6.3 Floral Bud Development

Polar auxin transport is required for floral development in the inflorescence meristem. *pin1* mutant of *Arabidopsis* lacks auxin efflux carrier and exhibits abnormal flowers. This shows that in the absence of auxin efflux carriers, the meristem is starved of auxin, and normal floral development is disrupted. PIN1 encodes a transmembrane protein that has been shown to localize with polarity in cells and to participate in auxin efflux. Interestingly, disruption of auxin polar transport does not appear to prevent the transition from vegetative growth to reproductive growth as *pin1* still forms an inflorescence. However, converting an inflorescence meristem to a floral meristem apparently requires normal polar auxin transport. Furthermore, treatment of plants with polar auxin transport inhibitor, naphthylphthalamic acid (NPA), also leads to the formation of pin-shaped inflorescence. It is possible that inhibition of auxin transport leads to accumulation of auxin in the meristem to levels that are inhibitory for the formation of new primordia. Alternatively, blocking of auxin transport may also lead to depletion of auxin in the meristem depending on where auxin is synthesized. Furthermore, disruption of auxin transport may not change the overall auxin levels. Rather, the phenotypes may be the result of a change in the local auxin gradients, which are presumably essential for organogenesis. Two other *Arabidopsis* mutants, *pinoid*, and *monopteris* (mp) also fail to initiate floral buds. PINOID encodes a serine/threonine protein kinase and is suggested to participate in auxin signaling and polar auxin transport, supporting the theory that auxin plays an essential role in the formation of floral buds. The role of Yucca (YUC) family of flavin monooxygenases in auxin biosynthesis provides further evidence regarding the role of auxin in flower formation. Overexpression of YUC1 gene leads to auxin overproduction, and mutations in some members of this gene lead to development of inflorescence without flower primordia.

## 15.6.4 Vascular Differentiation

Both localized and transported auxins play crucial role in the differentiation of the vascular tissues. Since differentiation of phloem tissue is usually difficult to follow, therefore, most reports on the effect of auxins on vascular tissue differentiation are confined to differentiation of xylem elements, a process called as **xylogenesis**. In a growing shoot or root, the procambial strands invariably differentiate in an acropetal order toward the shoot and root apices, respectively. Within the procambial strands as well, the primary xylem and phloem elements also differentiate acropetally. The effect of auxin on xylogenesis is twofold: (1) it acts as a vascular inducing signal and (2) "canalization" of newly differentiated vascular strand leading to polar transport of auxin.

1. Localized effect of auxin in the induction of xylogenesis is evident from the application of auxin to explants derived from cortex or pith or in callus tissues. Under such situations, addition of suitable concentrations of IAA or NAA to the culture medium containing the explants of lettuce (*Lactuca sativa*), *Coleus blumei*, tobacco (*Nicotiana tabacum*), or pea (*Pisum sativum*) leads to differentiation of tracheary cells. Likewise, differentiation of tracheary elements is also released in the mesophyll cells derived from *Zinnia* leaves in the presence of suitable concentrations of NAA. From this study, it has been observed that the first step of vascular differentiation is dedifferentiation of mature cells which requires no hormone. Both auxin and cytokinin regulate the second phase, i.e., the phase of induction. In the examples discussed so far, auxin has been found to be necessary for xylogenesis without involving its long-distance polar transport.
2. Xylogenesis in wounded tissue in an intact plant best depicts the impact of polarly transported auxin on xylogenesis. Auxin gradient created by its basipetal transport regulates vascular tissue differentiation in the shoot. New vascular elements usually develop toward the preexisting vascular strands and ultimately unite with them (Fig. 15.13). It is thus evident that wounded site acts as auxin source and the existing vasculature in the stem acts as auxin sink. This source-sink model is also referred as **canalization model** for auxin flow leading to reestablishment of continuous strands of vascular elements.

Xylogenesis proceeds in well-defined stages. These include radial expansion of cells, deposition of secondary wall, lignification, and, finally, sequential degradation of nucleus and cytoplasmic contents. Whereas the involvement of auxin in radial expansion of cells is well established, it is also linked to the process of lignification.

**Fig. 15.13** Effect of wounding on auxin-induced xylogenesis

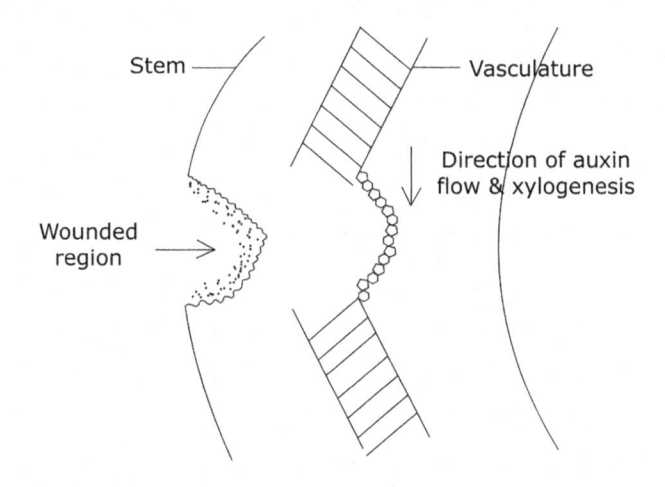

It is further expected that auxin may also play a role in the terminal stage of xylogenesis (i.e., loss of contents/autolysis or cell death) through induction of ethylene biosynthesis.

### 15.6.5 Origin of Lateral and Adventitious Roots

Lateral root (LR) primordia are initiated through localized cell divisions in the pericycle which results in a protrusion in the otherwise differentiated pericycle zone. A lateral root primordium thus formed penetrates through the cells of the cortex by developing its own root apex and root cap and emerges out of the root epidermis as a lateral root. Likewise, adventitious roots are also initiated by cell divisions in the parenchyma cells close to the vascular tissues in the stem (note: stem and leaves of seed plants generally do not have pericycle). A root primordium so organized subsequently proliferates as a new adventitious root.

Creation of local auxin gradient through equilibrium between auxin biosynthesis and transport has been observed to be critical for LR formation (Fig. 15.14). Auxin seems to act through regulation of cell cycle during LR initiation. In addition to the promotive role of auxin in LR initiation, certain other hormones also interplay in its regulation. Thus, increased ethylene levels have been reported to inhibit LR initiation. Auxin effect is also antagonized by abscisic acid by blocking cell cycle during G1 to S transition. On the other hand, LR initiation is inhibited by cytokinins by blocking G2 to M transition. It has further been observed that expression of cytokinin biosynthesis genes in the pericycle cells reduces LR formation and also disturbs cell division patterns. On the other hand, cytokinin degradation in the pericycle cells enhances LR initiation. Expression of PIN is negatively influenced by cytokinins,

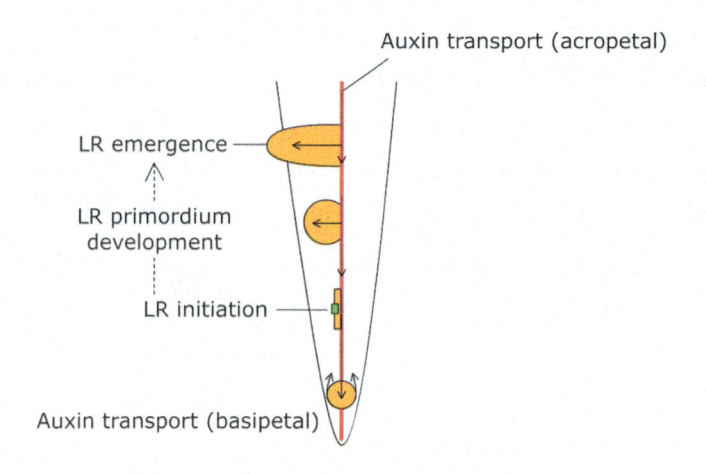

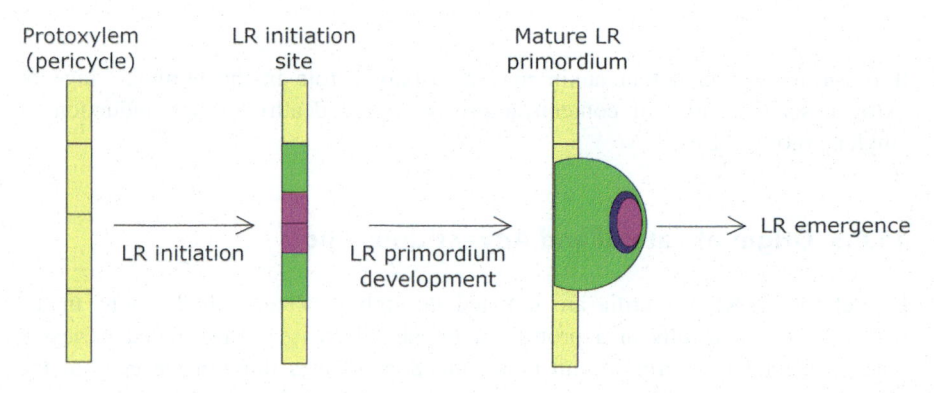

**Fig. 15.14** Process of lateral root (LR) formation

which also induces degradation of PIN1 proteins resulting in lowering of auxin levels in the meristematic cells.

## 15.7 Signaling Mechanisms Associated with Auxin Action

Auxin action in plant cells is brought about in two major ways: (1) fast, non-transcriptional responses and (2) changes in gene expression. The non-transcriptional responses include activation of plasma membrane proton pump and ion channels, thereby elevating the apoplastic hydrogen ion concentration which will ultimately bring about loosening of cell wall through the activation of expansins. The non-transcriptional auxin response also brings about reorientation of microtubules.

Two major categories of **auxin receptors** are known. These include auxin-binding proteins (ABPs) and F-box protein, which is a component of E3 ligase involved in ubiquitin-mediated proteolysis. During auxin action through non-transcriptional means, auxin (IAA molecule) binds ABPs located either on the plasma membrane or on the ER. This binding (IAA-ABP complex) triggers enhanced trafficking of vesicles carrying newly synthesized $H^+$-ATPase from the ER through the vesicles, and they are transported to the plasma membrane to elicit enhanced $H^+$-ATPase activity (Fig. 15.15). Thus, IAA-ABP complex not only facilitates faster migration of $H^+$-ATPase to the requires site on the plasma membrane, it also further enhances its activity on the membrane so long as $H^+$-ATPase has the proximity of IAA-ABP complex on the membrane. Enhanced $H^+$-ATPase activity results in greater release of $H^+$ ions into the apoplast leading to cell wall loosening process.

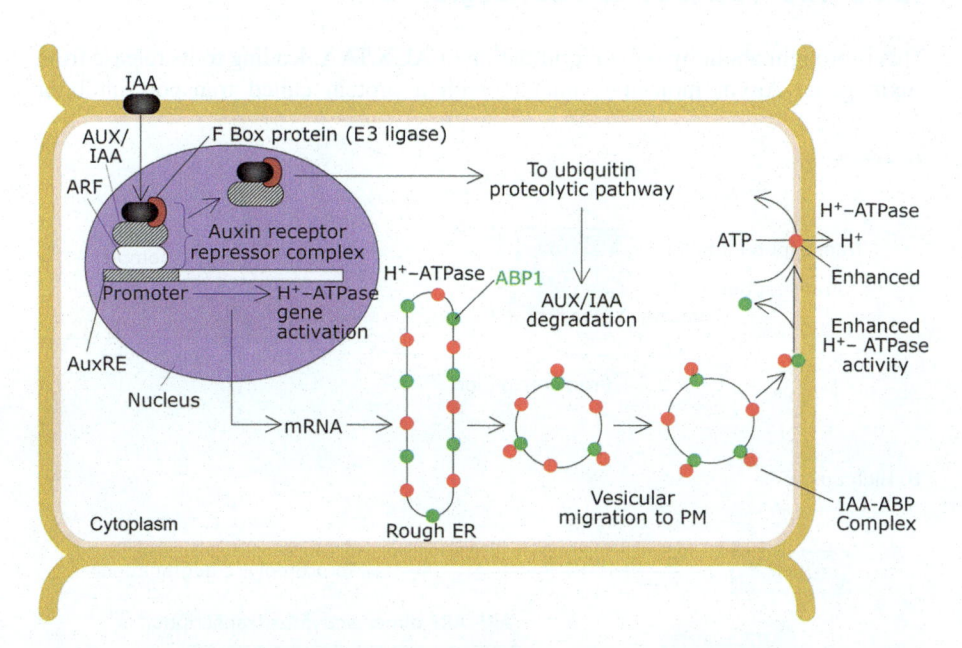

**Fig. 15.15** Auxin-modulated gene expression leading to activation and enhanced activity of PM-associated $H^+$- ATPase. *IAA* indole-3-acetic acid, *AUX/IAA* transcription repressor, *ARF* auxin response factor, *AuxRE* auxin sensitive promoter region, *ABP1* auxin-binding protein 1

## 15.7.1 Changes in Gene Expression

A typical auxin responsive gene has auxin response element (AuxRE) which is the binding site located in the promoter region. Certain transcription factors known as auxin response factors (ARFs) bind to these AuxRE motifs to stimulate or repress transcription. The third component of an auxin-inducible gene is a proteinaceous repressor of the transcription factor. These repressors, referred as AUX/IAA, remain associated with ARFs when auxin concentrations are low, thereby repressing transcription activity (Fig. 15.16). In the presence of auxin, AUX/IAA repressors get separated from the auxin-inducible gene and get tagged with ubiquitin for degradation through 26S proteasome pathway. [*Ubiquitin is a small (76 kDa), single peptide protein that is found in almost all living cells and plays an important role in degradation of defective and superfluous proteins.*] **Ubiquitination**, i.e., addition of ubiquitin to the target protein, marks AUX/IAA for degradation via the proteasome pathway. As a result of this dissociation of AUX/IAA repressors from the ARFs (transcription factors), the ARFs are able to dimerize or even oligomerize leading to gene activation.

## 15.7.2 The Process of AUX/IAA Degradation

This is brought about by polyubiquitination of AUX/IAA, leading to its release from auxin gene. Auxin molecules interact with a protein called transport inhibitor

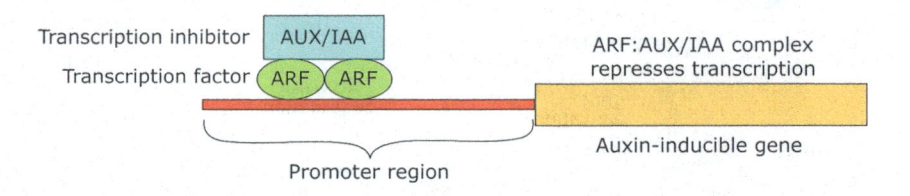

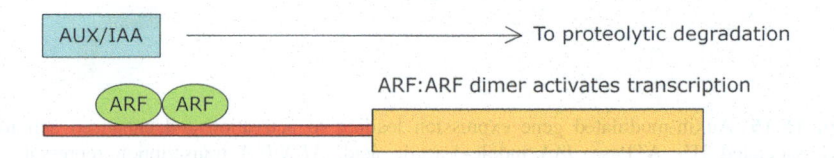

**Fig. 15.16** Activation of auxin responsive gene by auxin

## 15.7 Signaling Mechanisms Associated with Auxin Action

response protein 1 (TIR1) in the SCF complex, which is now known to be the auxin receptor associated with auxin-induced gene action. TIR1, therefore, functions as an auxin receptor with auxin acting as a molecular glue. TIR1 is a part of a protein complex called SCF. SCF refers to the three polypeptides found in the complex: *Skp*, *Cullin*, and *F-box*. SCF catalyzes ATP-dependent covalent addition of ubiquitin molecules to proteins targeted for degradation. Auxin binds to TIR1 at the conserved leucine-rich repeats. Four steps lead to auxin perception through auxin-induced gene expression: (1) Binding of TIR1 to IAA; (2) attachment of AUX/IAA proteins with the activated TIR1, i.e., IAA/TIR1 complex leading to its ubiquitination; (3) degradation of AUX/IAA proteins by proteasome pathway; and (4) activation of transcription factor ARF due to non-availability of transcription inhibitor (AUX/IAA) which has been degraded facilitating transcription of auxin response gene (Figs. 15.16 and 15.17).

Auxin-induced gene expression can lead to two types of responses:

1. *Early or primary response genes*: Their expression is insensitive to protein synthesis inhibitors, and their expression time is short-lived. Their expression has three main functions. These encode transcription factors for the transcription of late or secondary response genes. The expression products of these genes are

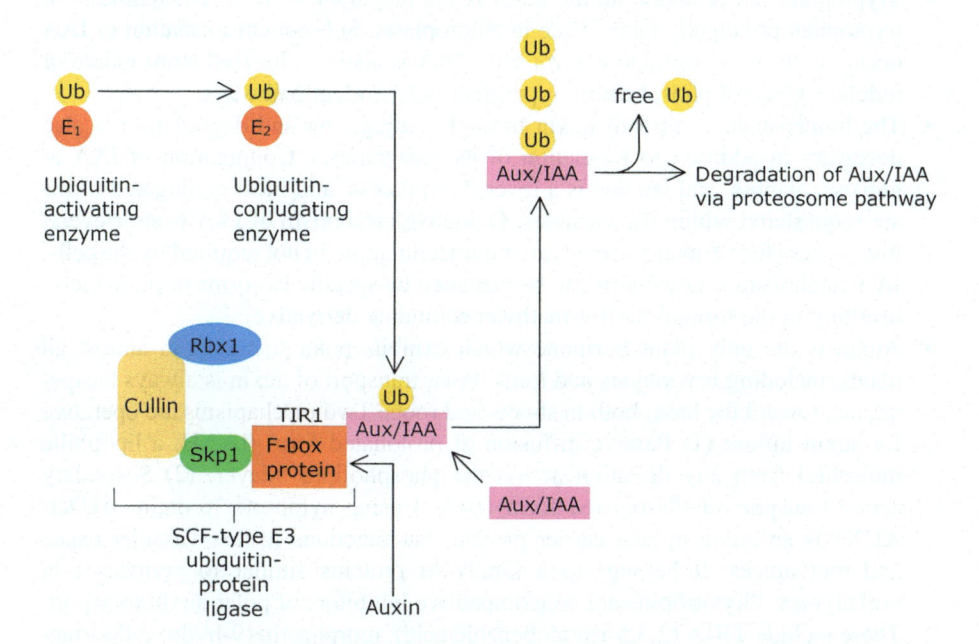

**Fig. 15.17** Targeted protein degradation mediates auxin signal transduction. Auxin binding to the F-box protein TIR1 enhances interaction between TIR1 and AUX/IAA proteins. This allows polyubiquitination of Aux/IAA proteins through the sequential actions of E1, E2, and E3 enzymes and so targets Aux/IAAs for degradation
*Ub* Ubiquitin

involved in signaling of intracellular communication or cell-to-cell signaling. They also include genes encoding proteins involved in adaptation to stress.
2. *Late or secondary response genes*: They require de novo protein synthesis, and they trigger the expression of enzymes such as glutathione S-transferase and ACC synthase.

The highly diversified response due to auxin-induced gene action is brought about by the existence of multiple forms of the three components of auxin signaling. Thus, *Arabidopsis* is known to have 6 TIRs, 29 AUX/IAA, and 23 ARF proteins. The auxin response signaling route discussed above primarily derives information from investigations on *Arabidopsis*, maize, and rice as model systems. Similar auxin response pathway has also recently been found to operate in the earliest diverging land plants, liverworts, and mosses. Thus, auxin response in plant cells has an ancient history.

## Summary

- Auxins constitute an important group of naturally occurring hormones which have been detected practically in all land plants and in several soil- or plant-associated microbes.
- Tryptophan (an aromatic amino acid) is the precursor of IAA. Biosynthesis of tryptophan principally takes place in chloroplasts. Subsequent formation of IAA occurs both in cytosol and chloroplasts. IAA is also synthesized from indole or indole-3-glycerol phosphate in tryptophan-independent pathways.
- The homeostatic control of auxin through conjugation and degradation is also necessary in addition to regulation of its biosynthesis. Conjugation of IAA to glucose, alanine, and leucine is a reversible process, and most conjugated forms are sequestered within the vacuoles. Oxidative catabolism of IAA is an irreversible process for permanent removal of intracellular auxin not required by the cells. IAA catabolism is now known to be mediated by specific isoforms of peroxidase, resulting in the formation of 3-methyleneoxindole derivative.
- Auxin is the only plant hormone which exhibits polar transport in almost all plants, including bryophytes and ferns. Polar transport of auxin is always basipetal, i.e., toward the base, both in shoots and roots. Two mechanisms are operative for auxin influx: (1) Passive diffusion of protonated form (IAAH, a lipophilic molecule) from any direction across the phospholipid bilayer. (2) Secondary active transport of dissociated form (IAA$^-$) using symporter protein, AUX1. AUX1 is an auxin uptake carrier protein that functions in leaf vascular tissue and root apices. It belongs to a family of proteins similar to permeases in prokaryotes. Phytotropins are noncompetitive inhibitors of polar auxin transport. These include TIBA (2,3,5-triiodobenzoic acid), morphactin (9-hydroxyfluorine-9-carboxylic acid), and NPA (N-1-naphthylphthalamic acid).

## 15.7 Signaling Mechanisms Associated with Auxin Action

- The critical component of the chemiosmotic model is, however, the basal location of PIN proteins which is responsible for efflux of IAA$^-$ from the cells. It is this unique basal location of PIN proteins which is responsible for establishing polarity in auxin transport.
- The major driving force causing efflux of IAA$^-$ from the cell is the negative plasma membrane potential, which varies between $-200$ and $-300$ mV. ABCB transporters facilitate IAA$^-$ efflux or influx across the plasma membrane (and also tonoplast), but unlike PIN proteins, they do not show preferential basal localization in the cells.
- Acid growth hypothesis proposes that auxin brings about activation of H$^+$-ATPases through its binding with a membrane-associated protein called ABP1 (auxin-binding protein 1). ABP1 is an auxin receptor which has been localized in the ER and plasma membrane. It is a 43 kDa glycoprotein dimer of two subunits of 22 kDa each. It is proposed that ABP1 forms a complex with a transmembrane docking protein which provides lipid solubility to anchor ABP1 to the membrane. This auxin-ABP1-docking protein complex initiates a signal transduction pathway leading ultimately to activation of ATPase proton pumps.
- Apical dominance refers to the inhibition of axillary bud growth by the continued meristematic activity in the shoot apex. It is best observed in herbaceous plants and also in trees during first year of growth.
- Auxin produced in the shoot apex inhibits lateral bud growth through its transport downward. Apical dominance is regulated by limiting sugar availability to the axillary buds in intact plants, and sustained growth of the axillary bud requires depletion of auxin in the stem adjacent to the bud as well.
- Polar auxin transport is required for floral development in the inflorescence meristem.
- Both localized and transported auxins play crucial role in the differentiation of the vascular tissues. The effect of auxin on xylogenesis is twofold: (1) it acts as a vascular inducing signal and (2) "canalization" of newly differentiated vascular strand leading to polar transport of auxin. Creation of local auxin gradient through equilibrium between auxin biosynthesis and transport has been observed to be critical for lateral root (LR) formation. Auxin seems to act through regulation of cell cycle during LR initiation.
- Auxin action in plant cells is brought about in two major ways: (1) fast, non-transcriptional responses and (2) changes in gene expression. Two major categories of auxin receptors are known. These include auxin-binding proteins (ABPs), and F-box protein, which is a component of E3 ligase involved in ubiquitin-mediated proteolysis. IAA-ABP complex not only facilitates faster migration of H$^+$-ATPase to the required site on the plasma membrane, it also further enhances its activity on the membrane so long as H$^+$-ATPase has the proximity of IAA-ABP complex on the membrane. In the presence of auxin, AUX/IAA repressors get separated from the auxin-inducible gene and get tagged

with ubiquitin for degradation through 26S proteasome pathway. Auxin-induced gene expression can lead to two types of responses. (1) *Early or primary response genes*: These encode transcription factors for the transcription of late or secondary response genes. They also include genes encoding proteins involved in adaptation to stress. (2) *Late or secondary response genes*: They require de novo protein synthesis and they trigger the expression of enzymes such as glutathione S-transferase and ACC synthase.

## Multiple-Choice Questions

1. Which of the following is a natural auxin?
   (a) NAA
   (b) 2,4-D
   (c) 2,4,5,-T
   (d) IAA
2. Phytotropins refer to:
   (a) Conjugated forms of auxin
   (b) Noncompetitive inhibitors of polar auxin transport
   (c) Proteins responsible for efflux of auxin
   (d) Transporters of auxin
3. Which of the following is true regarding the auxin-induced expression of early or primary response genes?
   i. Expression of these genes is insensitive to protein synthesis inhibitors.
   ii. These encode transcription factors for the transcription of late or secondary response genes.
   iii. These encode proteins involved in adaptation to stress.
   iv. They are expressed for a rather longer period of time.
   (a) Only ii and iv
   (b) Only i, ii, and iv
   (c) Only i, ii, and iii
   (d) Only i and iv
4. Which of the following *does not* inhibit auxin transport?
   (a) TIBA
   (b) NAA
   (c) NPA
   (d) Morphactin
5. Canalization model of auxin flow states that:
   (a) Auxin flow majorly occurs through xylem.
   (b) Auxin flow majorly occurs through phloem.
   (c) Auxin is transported from the source to sink in an acropetal manner.
   (d) Auxin transport results in the reestablishment of continuous strands of vascular elements in a self-organizing pattern.

6. Which of the following is/are auxin receptor(s)?
    i. ABP
    ii. AUX/IAA
    iii. F-box proteins
    iv. ARF
    (a) Only i
    (b) Only ii
    (c) Only i and iii
    (d) Only i, ii, and iii
7. Which of the following is true regarding the role of ABP?
    (a) ABP forms complex with IAA and enhances trafficking of $H^+$-ATPase carrying vesicles to plasma membrane.
    (b) ABP binds with IAA and brings about degradation of AUX/IAA and subsequent transcription of auxin responsive genes.
    (c) ABP facilitates the influx of auxin into cells.
    (d) ABP is responsible for establishing polar transport of auxin.
8. Which of the following does not come under the category of tryptophan-dependent pathway of IAA production?
    (a) Tryptamine pathway
    (b) Indole-3-acetonitrile pathway
    (c) Indole-3-glycerol pathway
    (d) Indole-3-pyruvic pathway
9. Auxin transport is:
    (a) Always basipetal both in roots and shoots
    (b) Always acropetal both in roots and shoots
    (c) Acropetal in roots and basipetal in shoots
    (d) Basipetal in roots and acropetal in shoots
10. Influx of auxin:
    i. Takes place through passive diffusion of the dissociated form of IAA, i.e., IAA$^-$
    ii. Takes place through passive diffusion of the protonated form of IAA, i.e., IAAH
    iii. Takes place through active transport of dissociated form of IAA, i.e., IAA-
    iv. Takes place through active transport of protonated form of IAA, i.e., IAAH
    (a) Only i and ii
    (b) Only i and iv
    (c) Only ii and iii
    (d) None of the above

11. Which of the following is true for the processes of conjugation and oxidation both of which regulate auxin homeostasis?
    i. Both work to remove the active auxin when present at supraoptimal concentrations.
    ii. Both are irreversible processes, causing permanent removal of auxin.
    iii. Conjugation is reversible, while oxidation of IAA is irreversible and brings about loss of auxin activity.
    iv. Both conjugation and oxidation are reversible and allow for reversible release of free IAA when required.
    (a) Both i and ii
    (b) Both i and iii
    (c) Both i and iv
    (d) None of the above
12. PIN1 is:
    (a) A transmembrane protein that is responsible for establishing polar basipetal auxin transport.
    (b) An inhibitor of polar auxin transport.
    (c) A protein for auxin uptake.
    (d) A positive regulator of auxin biosynthesis.
13. Which of the following is responsible for establishing polarity of auxin transport?
    (a) ABP1
    (b) AUX1
    (c) TIBA
    (d) PIN1
14. Which of the following is true for apical dominance?
    i. It is the phenomenon of inhibition of lateral bud growth by continuous auxin production in shoot apex.
    ii. It is regulated by limitation of sugar availability to axillary buds.
    iii. It is due to the accumulation of cytokinins within cells of axillary buds.
    iv. It is controlled by both strigolactones and auxin.
    (a) Only i, ii, and iv
    (b) All of the above
15. Decapitation of apical bud results in overcoming apical dominance due to:
    (a) Reversion of auxin-induced cytokinin biosynthesis inhibition and initiation of meristematic activity of axillary buds
    (b) The efflux of cytokinins from the axillary buds that promotes their growth
    (c) The rapid accumulation of auxin in axillary buds that facilitates their growth
    (d) Increased meristematic activity of axillary bud resulting from efflux of growth inhibiting substances from them

**Answers**

1. d  2. b  3. c  4. b  5. d  6. c
7. a  8. c  9. a  9. a  11. b  12. a
13. d  14. c  15. a

## Suggested Further Readings

Bishop G, Sakakibara H, Seo M, Yamaguchi S (2015) Biosynthesis of hormones. In: Buchanan BB, Gruissem W, Jones RL (eds) Biochemistry and molecular biology of plants. Wiley Blackwell, Chichester, pp 769–833

Jiang Z, Li J, Qu LJ (2017) Auxins. In: Li J, Li C, Smith SM (eds) Hormone metabolism and signaling in plants. Academic Press, San Diego, pp 39–76

Leyser O, Day S (2015) Signal transduction. In: Buchanan BB, Gruissem W, Jones RL (eds) Biochemistry and molecular biology of plants. Wiley Blackwell, Chichester, pp 834–871

Lüthen H (2015) What we can learn from old auxinology. J Plant Growth Regul 34:697–701

Paque S, Weijers D (2016) Q&A: Auxin: the plant molecule that influences almost anything. BMC Biol 14:67

Tivendale ND, Cohen JD (2015) Analytical history of auxin. J Plant Growth Regul 34:708–722

# Cytokinins                                                                      16

Geetika Kalra and Satish C Bhatla

Cytokinins (CK) are a class of plant growth substances which promote cell division. The first cytokinin was discovered from Herring (an oily fish from genus *Clupea*) sperm DNA by Miller et al. in 1955. In the 1940s and 1950s, Skoog and his coinvestigators tested many substances for their ability to initiate and sustain proliferation of cultured tobacco pith tissue. They observed stimulation of cell division when cultured pith tissue was treated with autoclaved Herring sperm DNA. This indicated that DNA degradation product caused stimulation of cell division in tobacco pith culture. This compound was identified as **kinetin** since it caused cytokinesis (Fig. 16.1). It is now characterized as 6-furfurylaminopurine. Although kinetin is a natural compound, it is not synthesized in plants. It is, therefore, considered a "synthetic cytokinin" with reference to plants. Subsequently, immature endosperm from corn (*Zea mays*) was found to contain a substance with biological activity similar to kinetin. This substance stimulates mature plant cells to divide when added to a culture medium along with auxin. The active ingredient was later identified as **zeatin** [trans-6-(4-hydroxy-3-methyl-2-butenylamino) purine]. Zeatin was also the first natural cytokinin reported from unripe maize kernels by Miller and Letham in 1963. Zeatin can exist in *cis* or *trans* configuration. These forms can be interconverted by an enzyme known as *zeatin isomerase*. The *trans* form is biologically more active, although *cis* form has been found in high levels in a number of plant species. Cytokinins can be present in plants as ribosides (in which ribose sugar is attached to the 9 nitrogen of the purine ring), ribotides (in which the ribose sugar moiety contains a phosphate group), or a glycosides (in which a sugar molecule is attached to 3, 7, or 9 nitrogen of the purine ring).

Many synthetic compounds have been synthesized and tested for cytokinin activity. Some of these are benzylaminopurine (BAP); N,N′-diphenylurea; thidiazuron (TDZ); and benzyladenine. Also, a range of natural cytokinins have now been isolated like isopentenyladenine (iPA) and dihydrozeatin in addition to zeatin. With the exception of diphenylurea, all native and synthetic CKs are derivatives of the purine base adenine. Cytokinins occur both in free and bound forms. They are detectable across all plant

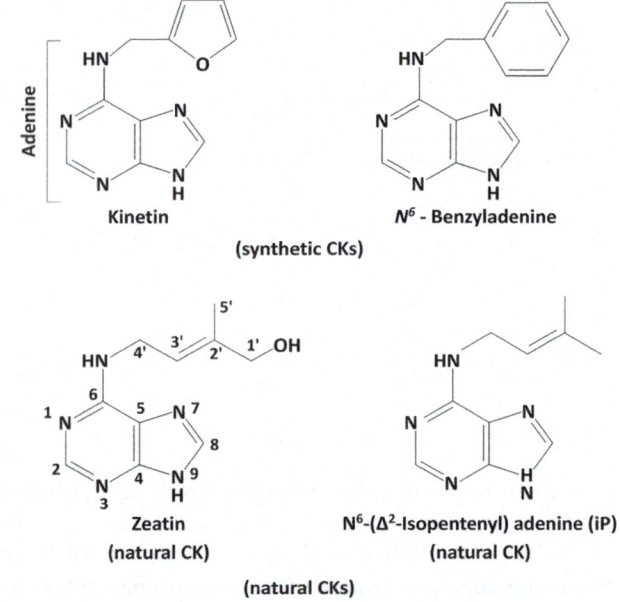

**Fig. 16.1** Structures of some natural and synthetic cytokinins (CKs) in plants

**Fig. 16.2** Overexpression of cytokinins causes (**a**) tumor formation and (**b**) abnormally located meristems

groups. Some plant pathogenic bacteria, insects, and nematodes also secrete free cytokinins. For example, *Corynebacterium fascians* is a major cause of the growth abnormality known as "witches broom" in which lateral buds are stimulated to grow by the bacterial cytokinin (Fig. 16.2). Cytokinins are also known to occur abundantly in coconut milk. The most abundant cytokinin in coconut water is *trans*-zeatin riboside.

## 16.1 Bioassay

Cytokinin activity in stimulating cell division forms the basis of various cytokinin bioassays.

1. *Tobacco pith culture*: Tobacco pith cultures supplemented with cytokinin can be used as a bioassay for this hormone. Increase in fresh weight of the tissue over the control is a measure of cell divisions and thereby cytokinin activity.
2. *Retardation of leaf senescence*: Cytokinins are known to delay leaf senescence by inhibiting chlorophyll degradation. Leaf discs kept in cytokinin containing solution show delayed senescence against control after 48–72 h. The test is sensitive at CK concentration as low as 1 pg.litre$^{-1}$. All these bioassays, however, have certain limitations as they utilize relatively large volumes of solutions and/or relatively long time for completion of response.
3. *Excised cotyledon expansion*: Cotyledons when placed in cytokinin-containing solution exhibit enlargement against control. This is taken as an indication of cytokinin activity. This bioassay is considered more efficient since cytokinin response can be obtained using very small solution volume and it is sensitive at concentration as low as $10^{-8}$ M.

## 16.2 Biosynthesis

Understanding the biosynthetic route for cytokinins has been relatively more difficult as compared to other hormones since it is practically not possible to isolate cytokinin-impaired mutants. Cytokinin, being principally responsible for cell divisions, proves lethal for plants in attempts to isolate mutants lacking it. Two pathways have been proposed for CK biosynthesis. The *direct pathway* involves formation of N$^6$-isopentyladenosine monophosphate (iPMP) from 5'-AMP and dimethylallyl pyrophosphate (DMAPP), followed by hydroxylation of the side chain to form zeatin-type compounds (Fig. 16.3). Conversion of DMAPP and AMP to iPMP occurs in plastids because DMAPP is synthesized in plastids. This reaction is the rate limiting step in cytokinin biosynthesis. The next step is the hydroxylation of the isopentenyl side chain by a member of the cytochrome P450 monooxygenase (CYP) family of enzymes. Hydroxylation reaction occurs largely on ER membranes and is regulated by several hormones. iPMP is then converted to zeatin by unidentified hydroxylases. Various phosphorylated forms can be interconverted, and free zeatin can be formed from the riboside by the action of enzymes of general purine metabolism. Also, reduction of the double bond in isopentenyl side chain of zeatin gives rise to **dihydrozeatin (diHZ).**

The *indirect pathway* of cytokinin biosynthesis involves turnover of tRNA containing *cis*-zeatin. tRNAs with anticodons that start with a uridine and carrying an already-prenylated adenosine adjacent to the anticodon release the adenosine as a

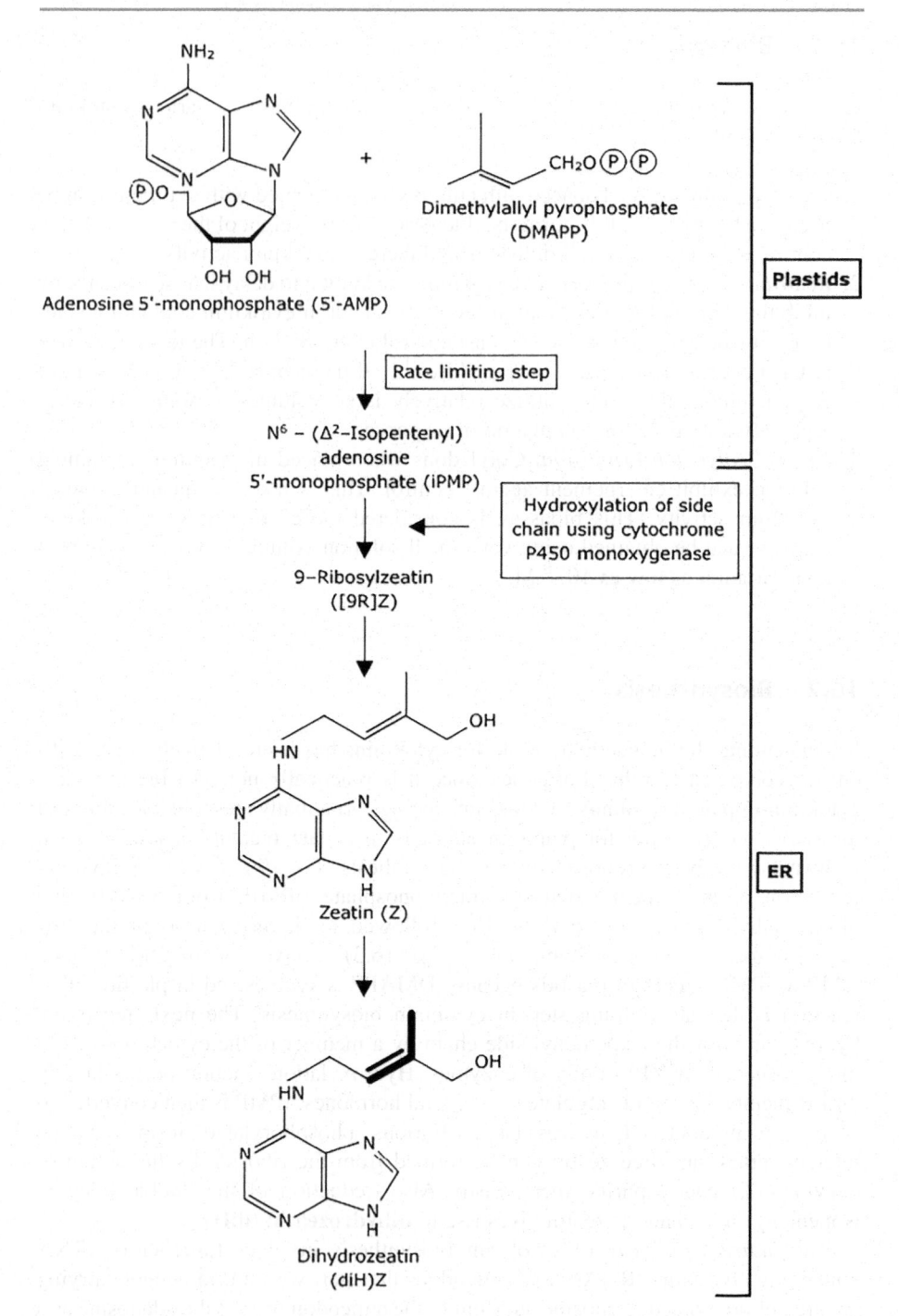

**Fig. 16.3** Direct pathways of cytokinin biosynthesis from 5'-AMP and dimethylallyl pyrophosphate (DMAPP)

cytokinin upon degradation. Prenylation of these adenines is carried out by tRNA isopentenyltransferase. (*Prenylation refers to addition of hydrophobic molecule to a chemical compound.*)

## 16.3 Transport

Root apical meristem is a major site of cytokinin (CK) biosynthesis in plants. CKs move through the xylem into the shoot along with water and minerals. When shoot is separated from a rooted plant, xylem sap continues to flow from the root stock. Analysis of this root exudate shows the presence of cytokinins. Even if the flow of xylem exudate continues, the cytokinin content does not diminish. This shows that cytokinins are synthesized in the roots. Environmental factors like water stress reduce the cytokinin content of the xylem exudate.

## 16.4 Metabolism

Free cytokinins are readily converted to their respective nucleosides and nucleotides. Many plant tissues contain the enzyme cytokinin oxidase which cleaves the side chain from isopentenyladenine. Cytokinin oxidase irreversibly inactivates cytokinin and is thus important in regulating cytokinin effects. Cytokinin oxidase activity is induced by high cytokinin concentration. Cytokinin levels can also be regulated by conjugation of the hormone at various positions. Conjugated forms of cytokinin are inactive in bioassay. Their conjugation at the side chain can be removed by glucosidase enzyme to yield free cytokinin. Dormant seeds generally contain high levels of cytokinin glucosides but very low level of free cytokinins. As seed germination is initiated, levels of free cytokinin increase rapidly with simultaneous decrease in cytokinin glucoside (Fig. 16.4). Cytokinins can also be inactivated by linkage with sugars or amino acids. Glycosylation can occur either at N-7 or N-9 positions of the adenine moiety or on the OH group of the isopentenyl side chain. N-Glycosylation is irreversible and results in inactivation of CK, but O-glycosylation can be reversed. Alanine can be added to the N-9 position. This reaction is also reversible. Reversible modification of CKs by the addition of sugars or amino acids may be an important mechanism for storage since conjugated CKs are located in the vacuole.

## 16.5 Physiological Role of Cytokinins

### 16.5.1 Cell Division

When plants are cultured on auxin-containing medium without cytokinin, they increase in size but cells do not divide. Addition of cytokinin to the medium leads to cell division and differentiation. Presence of both auxin and cytokinin in equal concentration leads to formation of undifferentiated callus. Increased cytokinin

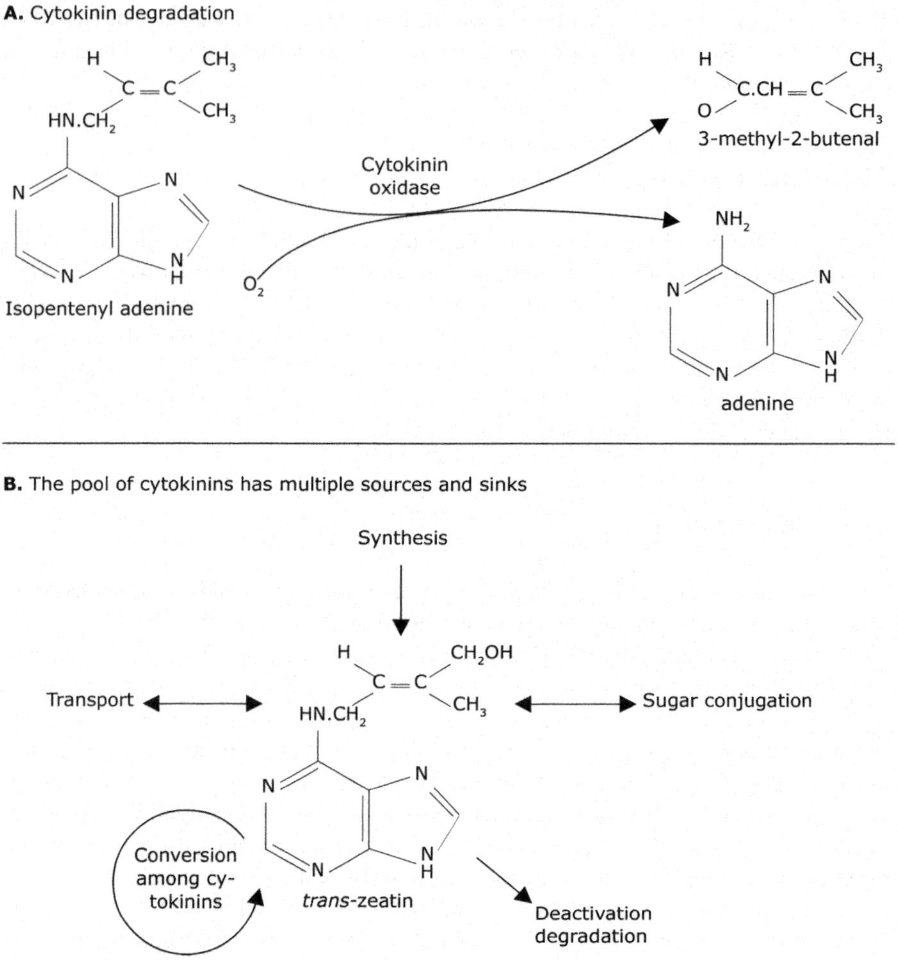

**A. Cytokinin degradation**

Isopentenyl adenine →(Cytokinin oxidase, $O_2$)→ 3-methyl-2-butenal + adenine

**B. The pool of cytokinins has multiple sources and sinks**

trans-zeatin: Synthesis ↓, Transport ↔, Sugar conjugation ↔, Conversion among cytokinins ↻, Deactivation degradation →

**Fig. 16.4** Routes of cytokinin homeostasis via degradation, conjugation, synthesis, and transport

concentration induces growth of shoot buds, whereas increased auxin induces root formation. Evidence indicates that localized expression of *ipt* gene (responsible for overexpression of cytokinin) from *Agrobacterium* in the somatic cells of tobacco leaves causes the formation of ectopic (abnormally located) meristems. This indicates that elevated levels of cytokinins are sufficient to initiate cell divisions in these leaves. Overexpression of several Arabidopsis cytokinin oxidase genes in tobacco results in a reduction of endogenous cytokinin levels. This results in retardation of shoot development due to reduction in the rate of cell proliferation and shoot apical meristem. These findings strongly support the observation that endogenous cytokinins regulate cell division.

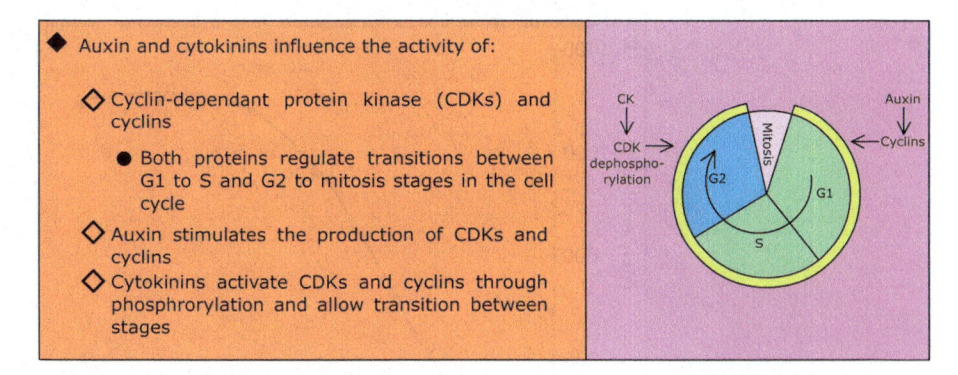

- Auxin and cytokinins influence the activity of:
  - ◇ Cyclin-dependant protein kinase (CDKs) and cyclins
    - Both proteins regulate transitions between G1 to S and G2 to mitosis stages in the cell cycle
  - ◇ Auxin stimulates the production of CDKs and cyclins
  - ◇ Cytokinins activate CDKs and cyclins through phosphrorylation and allow transition between stages

**Fig. 16.5** Auxin-cytokinin interaction in regulating cell cycle through modulation of the activity of cyclin-dependent protein kinase and cyclins

### 16.5.2 Regulation of Cell Cycle

Cytokinins regulate cell proliferation, affecting mitosis and endoreplication. They control cell cycle both at $G_1/S$ and $G_2/M$ transition. There is evidence to suggest that both auxins and cytokinins participate in the regulation of cell cycle. This is accomplished by controlling the activity of cyclin-dependent kinases (CDKs). The expression of gene which encodes major CDKs is regulated by auxins. However, CDKs induced by auxins are enzymatically inactive. Cytokinins induce the removal of an inhibitory phosphate group from CDKs. This action of cytokinin provides a link between cytokinins and auxins in regulating the cell cycle (Fig. 16.5).

### 16.5.3 Morphogenesis

High auxin: cytokinin ratio stimulates the formation of roots whereas low ratio leads to the formation of shoots.

### 16.5.4 Lateral Bud Formation

Cytokinins are known to regulate axillary bud growth and apical dominance. It is postulated that auxin from apical buds travels through shoots to inhibit axillary bud growth. This promotes shoot growth and restricts lateral branching. On the other hand, cytokinins move from root to the shoot and stimulate lateral bud growth. Direct application of cytokinin to the axillary buds stimulates cell divisions and growth of buds. Removal of apical bud leads to uninhibited growth of axillary buds. Cytokinins also modify apical dominance. Cytokinin-overproducing plants generally exhibit a bushy appearance.

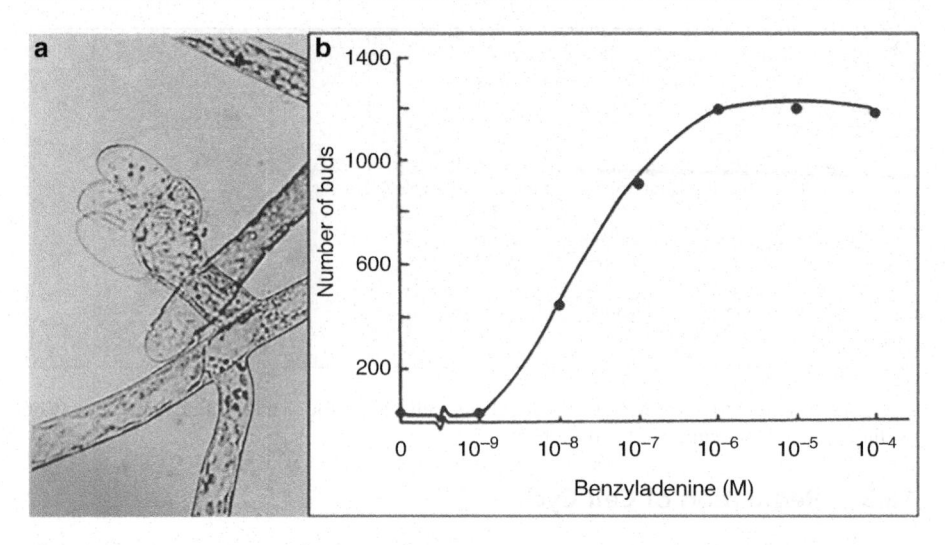

**Fig. 16.6** (a) Bud induction in response to cytokinin on caulonema filaments in moss protonema. (b) Dose response curve of the effect of benzyladenine on bud formation in moss protonema

### 16.5.5 Bud Formation in Mosses

Life cycle of a moss includes the gametophytic protonemal phase followed by formation of leafy shoots from buds developing on the protonema. Cytokinins are known to stimulate bud development and also increase the total number of buds formed in a concentration-dependent manner (Fig. 16.6).

### 16.5.6 Delay of Leaf Senescence

Cytokinins have been shown to slow aging of plant organs by preventing protein breakdown, activating protein synthesis, and assembling nutrients from nearby tissues. Isolated leaves when treated with cytokinins exhibit delayed senescence. If the leaf of a plant is treated with cytokinin, it remains green for a long period, while untreated leaves of the same developmental age develop yellow color and eventually drop off. In soybean leaves, senescence is initiated by seed maturation. This phenomenon is known as *monocarpic senescence* and can be delayed by seed removal. Seed pods control the onset of senescence by controlling the transport of cytokinins derived from roots to the leaves. Role of cytokinins in senescence can be tested by transforming tobacco plants with chimeric gene in which senescence-specific promoter is used to drive the expression of *ipt* gene. At the time of senescence, elevated cytokinin levels block senescence and also limit further expression of *ipt* gene, hence preventing cytokinin overproduction (Box 16.1).

## 16.5 Physiological Role of Cytokinins

> **Box 16.1 Cytokinin Overproduction in Plants and Its Significance**
> Cytokinin overproduction is highly beneficial for agriculture. Since CKs promote plant cell division and growth, they are used by farmers to increase crop productivity. *Ipt* gene from *Agrobacterium* Ti plasmid is introduced into many plant species resulting in overproduction of cytokinin. These plants exhibit all characteristics which point to the role of cytokinins in plant growth and development, viz., (i) more leaf production from shoot apical meristems, (ii) higher chlorophyll levels in leaves, (iii) adventitious shoots development from unwounded leaf veins and petioles, (iv) retardation of leaf senescence, (v) reduced apical dominance, and (vi) reduced rooting on stem cutting.
>
> Application of cytokinin to cotton seedlings has been reported to increase the yield by 5–10% under drought conditions. If leaf senescence could be delayed in plants, it is also possible to extend their photosynthetic productivity. Cytokinin production is also linked to the damage caused by predators, and it plays role in plant pathogenesis. For example, cytokinins have been known to induce resistance against *Pseudomonas syringe* in *Arabidopsis thaliana* and *Nicotiana tabacum*. Tobacco plants transformed with *ipt* gene under the control of promoter from a wound-inducible protease inhibitor II gene are more resistant to insect damage. Also in context of biological control of plant diseases, cytokinins seem to have potential functions.

### 16.5.7 Movement of Nutrients

Cytokinins promote movement of nutrients by creating a new source-sink relationship. Cytokinin-treated tissue acts as strong sink and thereby nutrients are channelized toward it. If a plant requires excessive nutrients, cytokinins accumulate in the root zone to stimulate growth.

### 16.5.8 Chloroplast Development

Etiolated leaves, when treated with cytokinin, develop chloroplasts with more extensive grana, and photosynthetic enzymes are synthesized at a greater rate upon illumination. This suggests that cytokinins along with other factors, like light and nutrition, regulate the biosynthesis of photosynthetic pigments and proteins.

### 16.5.9 Mechanical Extensibility of Cell Wall

Cotyledons of dicots, like mustard and sunflower, expand on treatment with cytokinin.

## 16.6 Mode of Cytokinin Action

Cytokinin receptor is related to bacterial two-component receptors. First clue to the nature of cytokinin receptor came from the discovery of CKI1gene. Plants require cytokinins in order to divide in culture. However, a cell line of *Arabidopsis* which overexpresses CKI1 gene is capable of growing in culture in the absence of added cytokinin. Phenotype resulting from CKI1 overexpression has suggested that histidine kinases are cytokinin receptors. Support for this model has come from the identification of the CRE1 gene. *cre1*, a loss-of-function mutant, exhibits the absence of shoot development from the undifferentiated tissue culture cells in response to cytokinin. CKI1 encodes a protein similar in sequence to bacterial two-component sensor histidine kinase. CK receptor is composed of two functional elements, a histidine kinase to which CK binds and a downstream response regulator whose activity is regulated via phosphorylation by histidine kinase. Histidine kinase is a membrane-bound protein that contains two distinct domains called input and transmitter domains. Detection of signal (CK) by the input domain alters the activity of the histidine kinase domain. Active sensor kinases are dimers which transphosphorylate a conserved histidine residue. The phosphate group is then transferred to a conserved aspartate residue in the receiver domain of a response regulator, and this phosphorylation alters the activity of the kinases. Most response regulators also contain an output domain which acts as a transcription factor.

Cytokinins also cause a rapid increase in the expression of response regulator genes. The first set of genes reported to be upregulated in response to cytokinin are the ARR (*Arabidopsis* response regulator) gene. Response regulators in *Arabidopsis* are encoded by a multigene family. These genes are grouped into two basic classes, namely, type-A ARR genes and type-B ARR genes. Type A gene is solely made up of a receiver domain, whereas type B contains a transcription factor domain in addition to the receiver domain. Rate of transcription of type A gene is enhanced within 10 min in response to applied cytokinin. This rapid induction is specific for cytokinin and does not require new protein synthesis. The rapid induction of type A gene suggests that these elements act downstream of the CRE 1 cytokinin receptor to mediate primary cytokinin response. Also, type A gene, ARR5, is expressed primarily in the apical meristem of shoot. This is consistent with its role in regulating cell proliferation.

The complete model for cytokinin signal transduction cascade is as follows (Fig. 16.7):

Cytokinin binds to an extracellular portion of CRE 1 (a dimer) called the CHASE domain of the CK receptor. Two other hybrid sensor kinases (AHK2 and AHK3) containing a CHASE domain are also likely to act as cytokinin receptors in *Arabidopsis*. Cytokinin binding to these receptors activates their histidine kinase activity. The phosphate group is transferred to an aspartate residue (D) on the fused receiver domain. The phosphate is then transferred to a conserved histidine present in an AHP protein. Phosphorylation causes the AHP protein to move into the nucleus where it transfers the phosphate group to an aspartate residue located within the receiver domain of a type-B ARR. The phosphorylation of the type-B ARR activates

## 16.6 Mode of Cytokinin Action

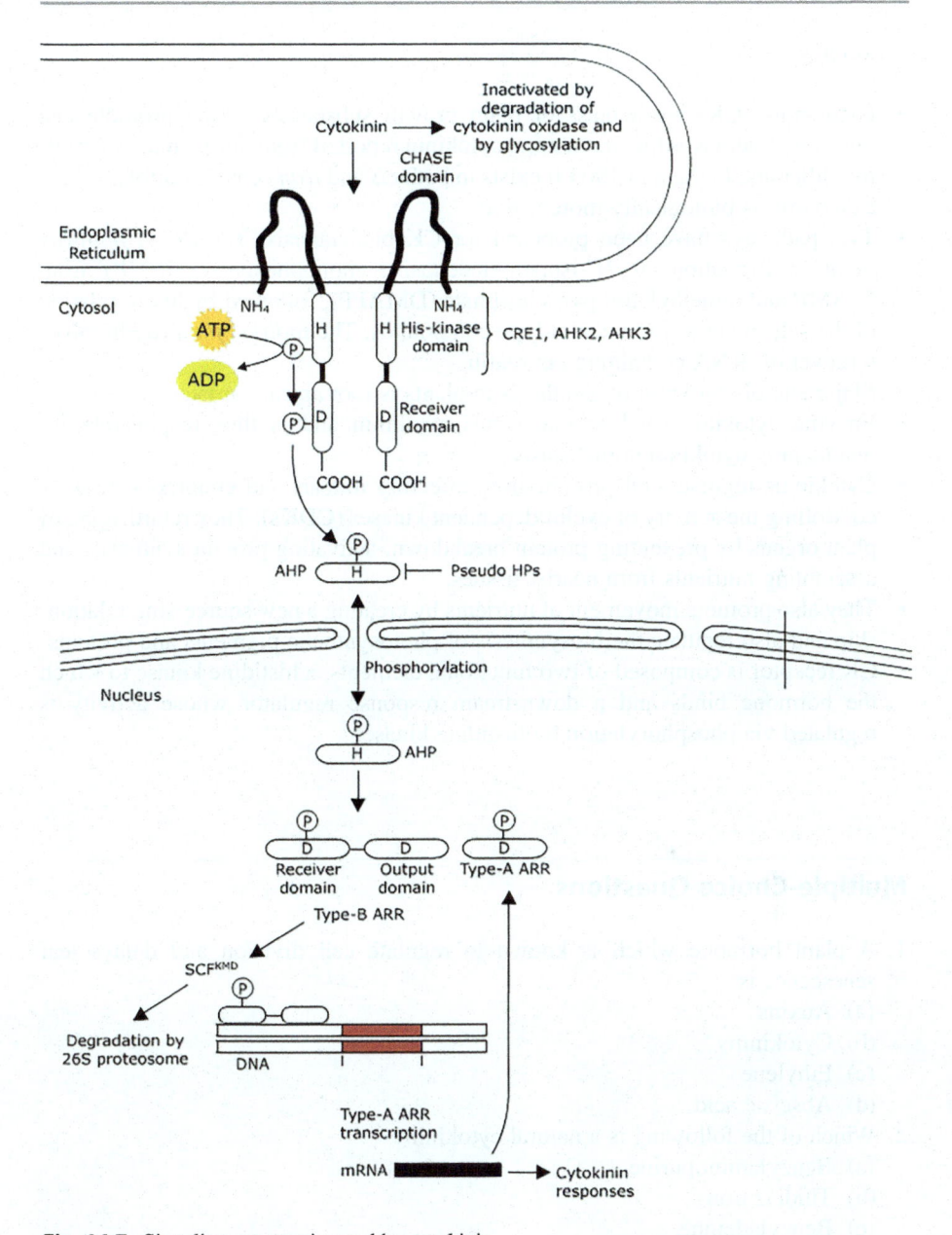

**Fig. 16.7** Signaling events triggered by cytokinin

the output domain to induce transcription of genes encoding type-A ARRs. These are, in turn, phosphorylated by the AHP proteins. The phosphorylated type A ARRs interact with various effectors to bring about cytokinin response.

## Summary

- Cytokinins (CK) are a class of plant growth substances which promote cell division. Zeatin was the first natural cytokinin reported from unripe maize kernels by Miller and Letham in 1963. It exists in both *cis* and *trans* configurations, where *trans* form is biologically more active.
- Two pathways have been proposed for CK biosynthesis. The *direct pathway* involves formation of $N^6$-isopentyladenosine monophosphate (iPMP) from 5′-AMP and dimethylallyl pyrophosphate (DMAPP), followed by hydroxylation of the side chain to form zeatin-type compounds. The *indirect pathway* involves turnover of tRNA containing *cis*-zeatin.
- Major site of cytokinin biosynthesis in plants is root apical meristem.
- Enzyme cytokinin oxidase inactivates cytokinin and is thus responsible for maintaining cytokinin homeostasis.
- Cytokinins regulate cell proliferation, affecting mitosis and endoreplication by controlling the activity of cyclin-dependent kinases (CDKs). They retard aging of plant organs by preventing protein breakdown, activating protein synthesis, and assembling nutrients from nearby tissues.
- They also promote movement of nutrients by creating a new source-sink relationship and also regulate the biosynthesis of photosynthetic pigments and proteins.
- CK receptor is composed of two functional elements, a histidine kinase to which the hormone binds and a downstream response regulator whose activity is regulated via phosphorylation by histidine kinase.

## Multiple-Choice Questions

1. A plant hormone which is known to regulate cell division and delays leaf senescence is:
   (a) Auxins
   (b) Cytokinins
   (c) Ethylene
   (d) Abscisic acid
2. Which of the following is a natural cytokinin?
   (a) Benzylaminopurine
   (b) Thidiazuron
   (c) Benzyladenine
   (d) Kinetin
3. _____ is the major site of cytokinin biosynthesis.
   (a) Shoot apical meristem
   (b) Root apical meristem
   (c) Floral tips
   (d) Leaf tips

4. Which of the following hormones resemble nucleic acids with reference to their structures?
    (a) Auxins
    (b) Gibberellins
    (c) Cytokinins
    (d) Ethylene
 5. Which hormone is responsible for a disease called "witch's broom" where lateral buds are stimulated to grow abnormally?
    (a) Cytokinins
    (b) ABA
    (c) Gibberellins
    (d) Ethylene

**Answers**

1. b   2. d   3. b   4. c   5. a

## Suggested Further Readings

Feng J, Shi Y, Yang S, Zuo J (2017) Cytokinins. In: Li J, Li C, Smith SM (eds) Hormone metabolism and signaling in plants. Academic Press, San Diego, pp 77–106

Kamínek M (2015) Tracking the story of cytokinin research. J Plant Growth Regul 34:723–739

Osugi A, Sakakibara H (2015) Q&A: how do plants respond to cytokinins and what is their importance? BMC Biol 13:102

# Gibberellins 17

Geetika Kalra and Satish C Bhatla

Gibberellins are growth hormones known to stimulate cell elongation and influence various developmental processes like stem elongation, seed germination, dormancy, flowering, sex expression, enzyme induction, and leaf and fruit senescence. Japanese scientists observed a common disease leading to excessive growth of rice plants. Eiichi Kurosawa (1926) investigated this *bakanae* (foolish seedling) disease in rice and found that tallness of diseased rice plants was induced by a chemical secreted by the fungus that had infected the plants. This chemical was isolated from the filtrates of the cultured fungus and was called **gibberellin**, after *Gibberella fujikuroi* (now renamed as *Fusarium fujikuroi*), the said fungus infecting rice plants. Kurosawa also noted that this active factor could promote the growth of maize, sesame, millet, and oat seedlings. In 1935, Yabuta and Hayashi successfully crystallized the fungal growth-inducing factor called gibberellin from the fungus *Gibberella fujikuroi*. All gibberellins are technically diterpene acids. They are either 19 or 20 carbon structures. A number of gibberellins are found in plants, of which only few are biologically active as hormones. The 19-carbon forms are, in general, biologically active gibberellins. Three most common biologically active gibberellins are $GA_1$, $GA_3$, and $GA_4$. All other GAs serve either as active GAs or their degradation products (Fig. 17.1). In view of their acidic nature, gibberellins are also referred as gibberellic acids (GAs). GAs are named $GA_1$ through $GA_n$ in order of discovery, and $GA_3$ was the first GA to be structurally characterized. So far, 126 GAs have been identified in plants, fungi, and bacteria.

## 17.1 Biosynthesis

Gibberellins are tetracyclic diterpenoid acids and are synthesized from terpenoid pathway in plastids and subsequently modified in endoplasmic reticulum and cytosol into biologically active forms. Gibberellin biosynthetic pathway can be considered in three stages, each operating in different cellular compartments:

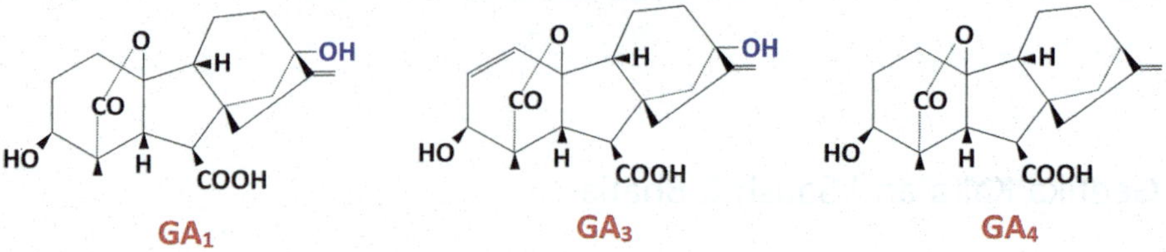

**Fig. 17.1** Chemical structures of selected active gibberellins

*STAGE 1: Production of terpenoid precursors and ent-kaurene in plastids*

Gibberellins are generally synthesized in plastids from glyceraldehyde-3-phosphate and pyruvate. In the endosperm of pumpkin seeds, isopentenyl diphosphate (IPP) is formed in the cytosol from mevalonic acid, which is itself derived from acetyl CoA. IPP isoprene units are added successively to produce intermediates of 10 carbons (geranyl diphosphate), 15 carbons (farnesyl diphosphate), or 20 carbons (geranylgeranyl diphosphate, GGPP). GGPP is the precursor for a number of terpenoid compounds, like carotenoids and many essential oils. After GGPP formation the pathway becomes specific for gibberellin biosynthesis. The cyclization reactions which convert GGPP to ent-kaurene represent the first step that is specific for gibberellin biosynthesis. The two enzymes which catalyze these reactions are localized in the proplastids of shoot meristem, and they are not localized in mature chloroplasts. Thus, green leaves lose their ability to synthesize gibberellins from IPP once their chloroplasts mature. Compounds such as AMO-1618, cycocel, and phosphon D are specific inhibitors of the first stage of gibberellin biosynthesis, and they are used to reduce plant height in agriculture.

*STAGE 2: Oxidation reactions on the ER leading to $GA_{12}$ and $GA_{53}$ biosynthesis*

Kaurene is transported from the plastid to the ER and is oxidized to kaurenoic acid by kaurene oxidase, which is associated with the plastid envelope. Further conversions to $GA_{12}$ take place on the ER. Paclobutrazol and other inhibitors of P450 monooxygenases specifically inhibit this stage of gibberellin biosynthesis before $GA_{12}$-aldehyde formation, and they act as growth retardants. A methyl group on ent-kaurene is oxidized to a carboxylic acid. This is followed by contraction from six to five carbon rings to produce $GA_{12}$-aldehyde, which is then oxidized to $GA_{12}$. $GA_{12}$ is the first gibberellin produced in this pathway. Many gibberellins in plants are also hydroxylated on carbon 13, which forms $GA_{53}$ from $GA_{12}$. All enzymes involved are monooxygenases which utilize cytochrome P450 in their reactions. These P450 utilizing monooxygenases are localized on the ER.

## 17.2 Modulation of Gibberellin Biosynthesis

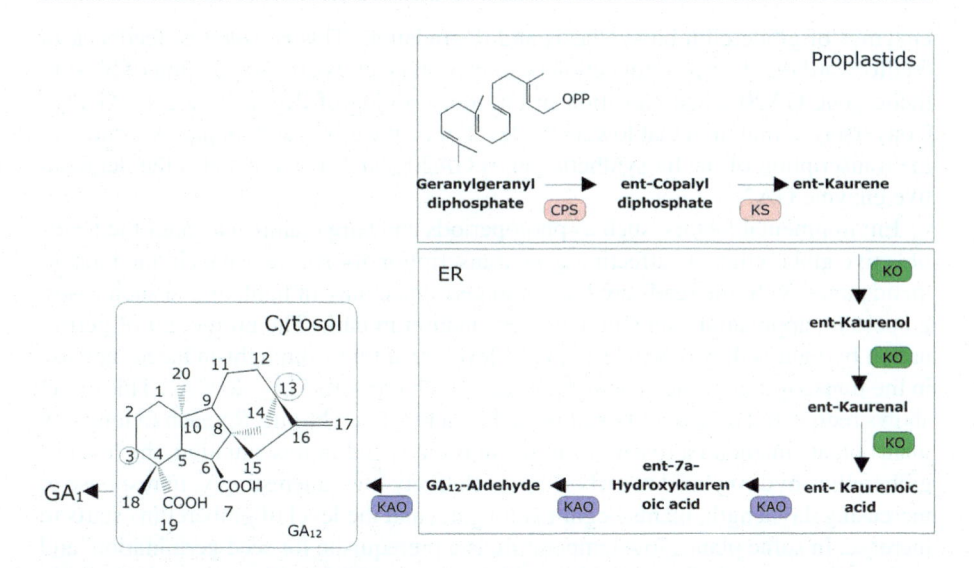

**Fig. 17.2** Gibberellin biosynthesis. *CPS ent*-copalyl diphosphate synthase, *KS ent*-kaurene synthase, *KO ent*-kaurene oxidase, *KAO ent*-kaurenoic acid oxidase; encircled carbon atoms on $GA_{12}$ are oxidized for the synthesis of $GA_1$

*STAGE 3: All other gibberellins are formed from $GA_{12}$ or $GA_{53}$ in the cytosol*

All subsequent steps in the pathway of GA biosynthesis are carried out by a group of soluble dioxygenases in the cytosol. These enzymes require 2-oxoglutarate and molecular oxygen as co-substrates and use $Fe^{2+}$ and ascorbate as cofactors. Two basic chemical changes occur in most plants: (i) hydroxylation at carbon 13 (on ER) or carbon 3 or both and (ii) a successive oxidation at carbon 20. Final step of this oxidation process is the loss of carbon 20 as $CO_2$. When these reactions involve gibberellins initially hydroxylated at C-13, the resulting gibberellin is $GA_{20}$. This is then converted to the biologically active form, $GA_1$, by hydroxylation of carbon 3. Inhibitors of the third stage of the gibberellin biosynthetic pathway interfere with enzymes that utilize 2-oxyglutarate as co-substrates (Fig. 17.2).

## 17.2 Modulation of Gibberellin Biosynthesis

Highest levels of gibberellins are found in immature seeds and developing fruits. Gibberellin biosynthetic enzymes are localized in young, actively growing buds, leaves, and upper internodes. In *Arabidopsis*, $GA20_{ox}$ is expressed primarily in the apical buds and young leaves which thus appear to be the principal sites of gibberellin biosynthesis. The gibberellins produced in shoots can be transported to the rest of the plant via the phloem stream. Gibberellins regulate their own metabolism by either switching on or inhibiting the transcription of the genes which encode enzymes of gibberellin biosynthesis and degradation. This is called as feedback or feedforward regulation. Application of gibberellin causes downregulation of biosynthetic gene $GA20_{ox}$ and elevation in the transcription of degradative gene $GA2_{ox}$. Conversely, a mutation that lowers the level of active gibberellin in plants stimulates the transcription of the biosynthetic genes $GA20_{ox}$ and downregulates the degradative enzyme $GA2_{ox}$.

Environmental factors, such as photoperiods and temperature, can alter the levels of active gibberellins by affecting gene transcription for specific steps in the biosynthetic pathway. Some seeds are known to germinate only in light, and in such cases, gibberellin application can stimulate germination in dark. The promotion of germination by light is due to increase in $GA_1$ level, resulting from light-induced increase in the transcription of the gene for $GA3_{ox}$, which converts $GA_{20}$ to $GA_1$. This effect shows red/far-red reversibility and is mediated by phytochrome. Spinach (a long day plant) plants maintain a rosette form in short days, and in this situation, the level of gibberellins hydroxylated at carbon 13 is relatively low. Interestingly, in response to increasing day length, plants begin to elongate, and the level of gibberellins starts to increase. In some plants, low temperature is a prerequisite for seed germination, and gibberellins can substitute for the cold treatment. Gibberellins can induce auxin biosynthesis and vice versa. In decapitated pea plants, reduced stem elongation is accompanied with reduced auxin and gibberellin content. Replacing the supply of auxin restores gibberellin levels.

## 17.3 Enzymes Involved in Gibberellin Metabolism

GA 20-oxidase: It catalyzes all the reactions involving the successive oxidation steps of C-20 between $GA_{53}$ and $GA_{20}$, including the removal of C-20 as $CO_2$.

GA 3-oxidase: It functions as a 3β-hydroxylase and adds a hydroxyl group at C-3 position to form the active gibberellin, $GA_1$.

GA 2-oxidase: It inactivates $GA_1$ by catalyzing the addition of a hydroxyl group to C-20.

## 17.4 Gibberellin Metabolism

Although active gibberellins exist in free forms, a variety of gibberellin glycosides are formed by a covalent linkage between gibberellin molecule and a sugar. These gibberellin conjugates are particularly prevalent in some seeds. The conjugating sugar is usually glucose and may be attached to the gibberellin molecule via a carboxyl group forming a gibberellin glycoside or via a hydroxyl group forming a gibberellin glycosyl ether. Glycosylation of gibberellins represents a mode of their inactivation.

## 17.5 Physiological Roles of Gibberellins

### 17.5.1 Internode Elongation

Evidence for the role of gibberellins in stem elongation comes from the study of rosette plants. A rosette is an extreme case of dwarfism in which the absence of internode elongation results in closely spaced leaves. The failure of internode elongation may result from genetic mutation or it may be environmentally induced. Exogenous application of gibberellin is known to promote internodal elongation in a wide range of species, a phenomenon known as **bolting** (Fig. 17.3). This leads to decrease in stem thickness and leaf size and pale green coloration of leaves. Some plants assume a rosette form in short days and undergo shoot elongation in long days. $GA_3$ results in bolting in plants kept in short days, and normal bolting is regulated by endogenous gibberellin. Thus, the target of gibberellin action is intercalary meristem (meristem near the base of the internode). Deepwater rice shows pronounced effect of gibberellin on internode elongation so that its foliage may remain above water in the field. Striking growth rates of as much as 25 cm per day have been observed in rice plants under flooding conditions. Partial submergence of rice plants is believed to reduce partial pressure of $O_2$ which triggers biosynthesis of ethylene precursor ACC. ACC transported to the aerial parts get converted to ethylene. Ethylene, in turn, reduces the level of ABA which acts as an antagonist to GA. Submerged rice tissue thus becomes more responsive (enhanced sensitivity) to endogenous GA, resulting in marked elongation of internodes. Gibberellins also regulate the transition from juvenile to adult phase of plant growth. Numerous perennials do not flower till they reach a certain stage of maturity. Till then they

**Fig. 17.3** Effect of gibberellic acid on bolting in *Arabidopsis thaliana*

are said to be juvenile. In English ivy (*Hedera helix*), gibberellin application can cause reversion from mature to juvenile state, and juvenile conifers can be induced to enter the reproductive phase by the application of nonpolar gibberellins, like $GA_4 + GA_7$. Thus, gibberellins can regulate juvenility in both directions depending on species (i.e., mature to juvenile or vice versa).

### 17.5.2 Floral Initiation and Sex Determination

Gibberellins can substitute for long day or cold requirement for flowering in many plants. Thus, it is a component of flowering stimulus. In plants where flowers are unisexual, sex determination is regulated genetically as well as by environment and nutritional status. For example, in maize plants, staminate flowers are observed in tassel. Exposure to short days and cool nights increases endogenous gibberellin levels and causes feminization of tassel flowers. On the other hand, in dicots, like cucumber and spinach, gibberellins have an opposite effect, i.e., they promote formation of staminate flowers.

### 17.5.3 Seed Germination

Light requirement of seeds for germination can be overcome by treatment with gibberellins. This might be either through weakening of growth-restricting endosperm layer surrounding the embryo or due to mobilization of food reserves stored in the endosperm.

### 17.5.4 Fruit Production

Gibberellins are sprayed to increase the stalk length of seedless grapes. Because of shortness of individual stalks, bunches are too compact and the growth of the berries is restricted. $GA_3$ spray promotes elongation of fruit and also increases its sweetness. Gibberellins also delay senescence in citrus fruits. This allows fruits to be left on the tree for a longer period to extend their shelf life.

### 17.5.5 Stimulation of Cell Elongation and Cell Division

Gibberellins increase both cell elongation and cell division. Mitotic activity increases markedly in the subapical region of the meristem of the rosette of long-day plants after treatment with gibberellins. Unlike auxins, gibberellins enhance cell wall extensibility without acidification. The elongation rate is influenced by both cell wall extensibility and the osmotically driven rate of water uptake. Gibberellins cause an increase in both mechanical extensibility of the cell walls and stress relaxation of the walls of living cells. There is evidence that the enzyme xyloglucan endotransglycosylase (XET) is involved in gibberellin-promoted cell wall extension. The function of XET may be to facilitate the penetration of expansins into the cellulose microfibrils of the cell wall.

### 17.5.6 Regulation of Transcription of Cell Cycle Kinases

Growth rate of the internodes of deepwater rice dramatically increases in response to submergence, and this is due to enhanced rate of cell divisions in the intercalary meristem. In these plants, gibberellins activate the cell division cycle first during the transition phase from $G_1$ to S phase, leading to an increase in mitotic activity. This is followed by regulation of transition from $G_2$ to M phase. To achieve this, gibberellins induce the expression of genes for several cyclin-dependent protein kinases (CDKs). Three types of gibberellin response mutants are useful in the identification of genes involved in the gibberellin signaling pathway involved in stem growth: (i) gibberellin-insensitive dwarfs (e.g., *gai*-1), (ii) gibberellin deficiency reversion mutants (e.g., *rga*), and (iii) constitutive gibberellin responders (slender mutants, e.g., *spy*). GAI and RGA are nuclear transcription factors responsible for the repression of growth. In the presence of gibberellins, these transcription factors are degraded. Mutant *gai*-1 and the related wheat dwarfing gene mutant *rht* have lost the ability to respond to gibberellins. SPY encodes a glycosyl transferase that is a member of a signal transduction chain prior to GAI/RGA. Whenever a mutation interferes with the repressor function of any of these, the plants grow tall.

## 17.6 Mode of Action

Gibberellin signal transduction with reference to stimulation of α-amylase synthesis in cereal aleurone layers has been extensively worked out to understand mode of GA action (Fig. 17.4). Cereal grains are composed of three main parts, viz., diploid embryo, triploid endosperm, and fused testa called pericarp (morphologically fused seed coat and fruit wall). Embryo has a specialized absorptive organ, the scutellum, which functions in absorbing the solubilized food reserves from the endosperm and transmits them to the growing embryo. Endosperm is composed of a central starchy component surrounded by aleurone layer. Cells of aleurone layer contain protein storage vacuoles (PSVs), also called protein bodies. During seed germination, food stored in the endosperm is broken down and transported to the growing embryo by the activity of α and β amylases. Main function of aleurone layer is the synthesis and release of hydrolytic enzymes. Cells of aleurone layer undergo programmed cell death thereafter.

When de-embryonated half seeds of barley are kept close to excised embryo, they trigger the release of α-amylase, and starch is digested. $GA_3$ is known to substitute for embryo in stimulating starch degradation. It exerts its effect by enhancing the

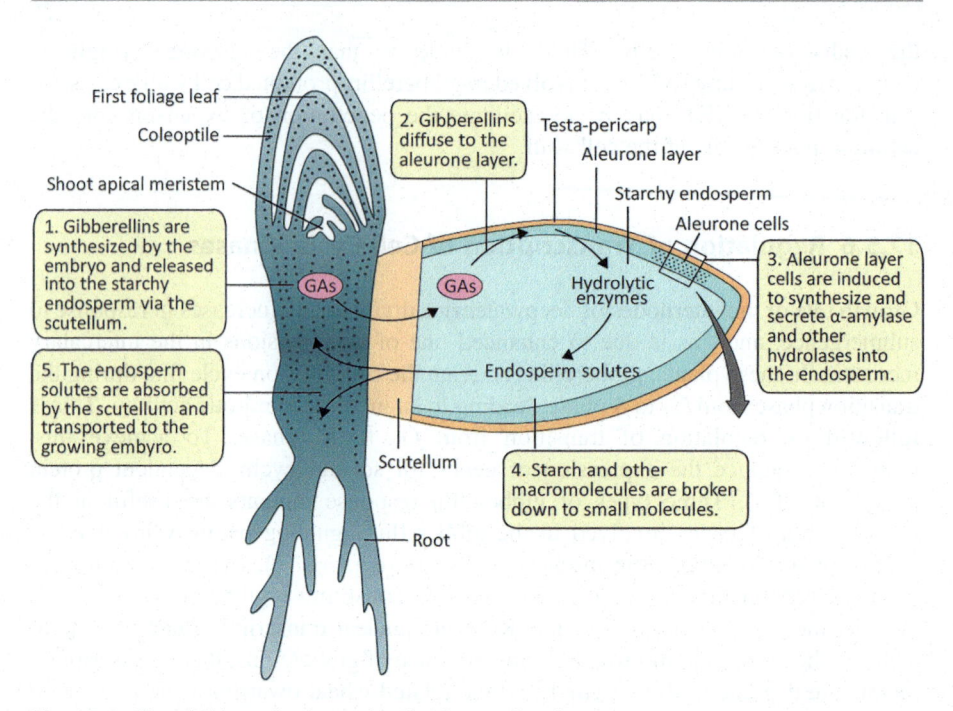

**Fig. 17.4** Site of GA production and action in cereal grains

transcription of α-amylase mRNA. This is evident from the observation that GA$_3$-stimulated α-amylase production is blocked by the inhibitors of transcription and translation such as actinomycin D and cycloheximide. Radioactive isotope-labeling studies have demonstrated that the stimulation of α-amylase activity by gibberellin involves de novo synthesis of the enzyme from amino acids rather than activation of preexisting enzymes. In another experiment, the α-amylase gene tagged to the promoter sequence of GUS reporter gene gives blue color in the presence of an artificial substrate, upon gene expression. When such a chimeric gene is introduced into aleurone protoplasts, the blue color intensity is stimulated by gibberellins. In addition to α-amylase, proteolytic enzymes (proteases), β-amylase, and other starch-degrading enzymes are also involved in mobilizing the endosperm reserves. The stimulation of α-amylase gene expression by gibberellins is mediated by a special transcription factor which binds to the promoter region of α-amylase gene. Certain regulatory DNA sequences (gibberellin response elements) have been identified in the promoters that are involved in binding the protein. The sequence of gibberellin response element in the α-amylase gene promoter has been identified to be similar to that of the binding site for MγB transcription factors which regulate phytochrome responses. Synthesis of GA-MγB mRNA in the aleurone cells increases within 3 h of GA$_3$ application. This happens to be several hours before the increase in α-amylase mRNA level. Protein synthesis inhibitors have no effect on MγB mRNA production. Thus, GA-MγB is a **primary response gene** or **early gene**. α- amylase gene is a

## 17.6 Mode of Action

**A. Gibberellin (GA) activates the production of α-Amylase**

a. The embryo absorbs water and the germination process begins.
b. Gibberellins diffuse from the embryo to the aleurone layer.
c. Cells in the aleurone layer respond by releasing digestive enzymes such as α-amylase.
d. The enzymes digest starch, releasing sugars and other molecules for the growing plant.

**B. Hypothesis for mechanism of activation of α-amylase by GA**

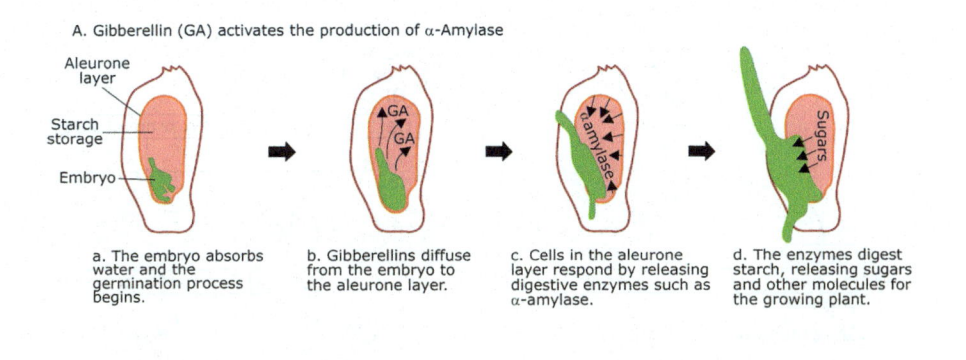

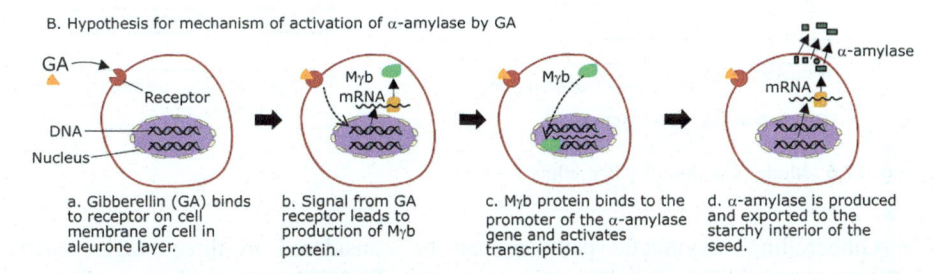

a. Gibberellin (GA) binds to receptor on cell membrane of cell in aleurone layer.
b. Signal from GA receptor leads to production of Mγb protein.
c. Mγb protein binds to the promoter of the α-amylase gene and activates transcription.
d. α-amylase is produced and exported to the starchy interior of the seed.

**Fig. 17.5** Sequence of events leading to α-amylase biosynthesis in aleurone layer of cereal grains in response to GA

**secondary response gene** or **late gene**, as its transcription is blocked by protein synthesis inhibitor, like cycloheximide (CHI) (Fig. 17.5).

GA signaling is mediated through the DELLA protein (nuclear protein) which functions as negative regulators of growth. GA is perceived by a soluble receptor protein GID1 (GA-insensitive dwarf 1). Binding of GA with GID1 proteins enhances its interaction with DELLA which in turn enhances its interaction with F-box protein (SLY1). This leads to ubiquitination of DELLA protein followed by its degradation by 26S proteasome pathway. Removal of DELLA proteins facilitates the transcription of GA-regulated genes (Fig. 17.6).

## Summary

- Gibberellins are growth hormones known to stimulate cell elongation and influence various developmental processes like stem elongation, seed germination, dormancy, flowering, sex expression, enzyme induction, and leaf and fruit senescence.
- Gibberellins are tetracyclic diterpenoids made up of four isoprenoid units. They are either 19 or 20 carbon structures. The 19-carbon forms are, in general, biologically active gibberellins. Three most common biologically active gibberellins are $GA_1$, $GA_3$, and $GA_4$.

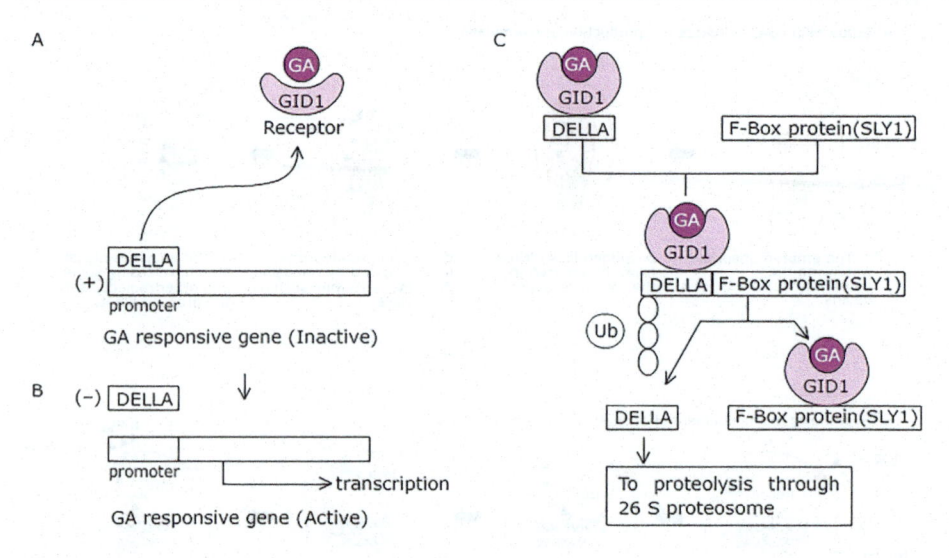

**Fig. 17.6** Mode of action of gibberellins

- Gibberellin biosynthetic pathway can be considered in three stages, each operating in different cellular compartments. The first step involves production of terpenoid precursors and ent-kaurene in plastids followed by oxidation reactions on the ER leading to $GA_{12}$ and $GA_{53}$ biosynthesis. All other gibberellins are formed from $GA_{12}$ or $GA_{53}$ in the cytosol.
- Compounds such as AMO-1618, cycocel, and phosphon D are specific inhibitors of the first stage of gibberellin biosynthesis, and they are used to reduce plant height in agriculture.
- A variety of gibberellin glycosides are formed by a covalent linkage between gibberellin molecule and a sugar (usually glucose). Glycosylation of gibberellins represents a mode of their inactivation.
- $GA_3$ spray promotes elongation of fruit and also increases its sweetness. Gibberellins also delay senescence in citrus fruits.
- Gibberellins induce the expression of genes for several cyclin-dependent protein kinases (CDKs) thereby activating the cell division cycle ultimately leading to an increase in growth rate of internodes of deepwater rice. Gibberellins also enhance cell wall extensibility without acidification, unlike auxin.
- $GA_3$ is known to substitute for embryo in stimulating starch degradation. It exerts its effect by enhancing the transcription of α-amylase mRNA.
- The stimulation of α-amylase gene expression by gibberellins is mediated by a special transcription factor which binds to the promoter region of α-amylase gene. Synthesis of GA-MγB mRNA in the aleurone cells increases within 3 h of $GA_3$ application.
- GA signaling is mediated through the DELLA protein (nuclear protein) which functions as negative regulators of growth. Binding of GA with GID1 receptor proteins enhances its interaction with DELLA which in turn promotes its

interaction with F-box protein (SLY1). This leads to ubiquitination of DELLA protein followed by its degradation by 26S proteasome pathway. Removal of DELLA proteins facilitates the transcription of GA-regulated genes.

## Multiple-Choice Questions

1. The specialized absorptive organ in cereals which functions in absorbing the solubilized food reserves from the endosperm is known as:
   (a) Scutellum
   (b) Pericarp
   (c) Plumule
   (d) Coleoptile
2. Gibberellin was first isolated from which of the following fungus:
   (a) *Gibberella fujikuroi*
   (b) *Gibberella acuminate*
   (c) *Gibberella africana*
   (d) *Gibberella gaditjirrii*
3. Gibberellins are chemical derivatives of:
   (a) Phenols
   (b) Terpenoids
   (c) Alkaloids
   (d) None of the above
4. Which of the following is not an inhibitor of gibberellin biosynthetic pathway?
   (a) Phosphon D
   (b) Cycocel
   (c) Maleic hydrazide
   (d) AMO-1618
5. Functions of gibberellic acid involve:
   (a) Internode elongation
   (b) Seed germination
   (c) Stimulation of cell elongation and division
   (d) All of the above
6. Which of the following was the first GA to be structurally characterized?
   (a) $GA_1$
   (b) $GA_3$
   (c) $GA_{12}$
   (d) $GA_4$

**Answers**

1. a   2. a   3. b   4. c   5. d   6. b

## Suggested Further Reading

Gao X, Zhang Y, He Z, Fu X (2017) Gibberellins. In: Li J, Li C, Smith SM (eds) Hormone metabolism and signaling in plants. Academic Press, San Diego, pp 39–76

# Abscisic Acid

18

Geetika Kalra and Satish C Bhatla

Some molecules produced by plants exert their effects as negative regulators of various plant responses. **Abscisic acid** is an inhibitory hormone that helps plants adapt to stress. It also maintains water balance, prevents seed embryos from germinating, and induces seed and bud dormancy. Initial attempts to identify abscisic acid were made by Fredrick T. Addicott and his coworkers in cotton fruits in 1963. It was earlier suspected that seed and bud dormancy are caused by some inhibitory compounds and attempts were made to extract these compounds from various plant tissues. Acidic compounds separated by paper chromatography from these tissues were tested for their ability to promote growth in oat coleoptile, but instead these compounds inhibited coleoptile elongation. This compound was referred as "β-inhibitor complex." Subsequently, high β-inhibitor levels were correlated with suppression of sprouting in *Solanum tuberosum* tubers, abortion of *Lupinus arboreus* pods, and bud dormancy in trees like *Betula pubescens*. This compound was subsequently named abscisin II, since it was identical to a substance that promotes abscission in cotton fruits (commercially important for mechanization of cotton picking). Another substance isolated a substance from *Betula pubescens*, a deciduous plant, inhibits growth and induces bud dormancy. It was named "**dormin**." "Dormin" was subsequently found to be structurally similar to "abscisin II." The compound was subsequently renamed as abscisic acid (ABA)—a compound which inhibits growth and stomatal opening when plants are under environmental stress.

ABA is ubiquitous in vascular plants, and in several genera of fungi, it is produced as a secondary metabolite. In higher plants, it is found in all living tissues from root cap to the apical bud. It is synthesized in all cells which contain chloroplasts or amyloplasts. ABA is a 15-carbon sesquiterpene characterized by a cyclohexane ring with a keto and hydroxyl group and a side chain with a terminal carboxyl group in its structure. The orientation of carboxyl group at C2 position determines the *cis* and *trans* isomers of ABA. ABA has an asymmetric carbon atom at position 1 in the ring resulting in S- and R-enantiomers. S-enantiomer is the active

© Springer Nature Singapore Pte Ltd. 2018
S. C Bhatla, M. A. Lal, *Plant Physiology, Development and Metabolism*,
https://doi.org/10.1007/978-981-13-2023-1_18

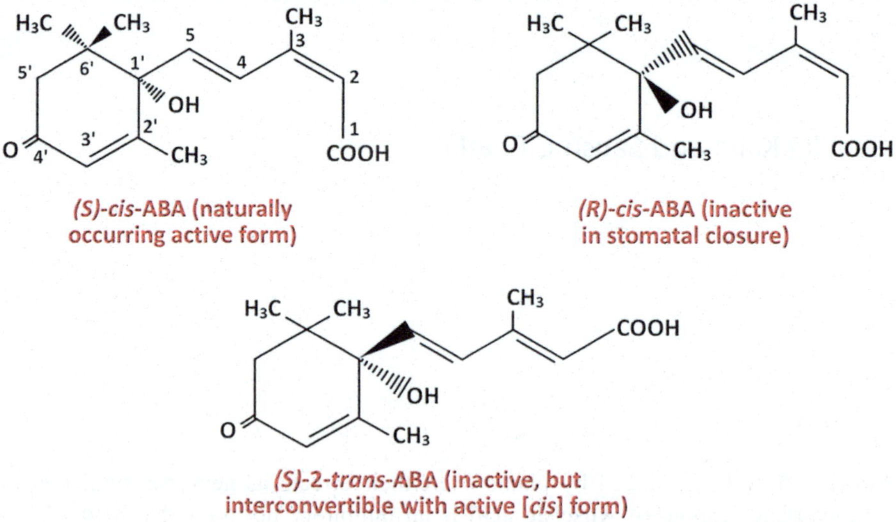

**Fig. 18.1** Structures of "S" and "R" forms of abscisic acid (ABA)

form leading to fast responses like stomatal closure. In long-term responses, both the enantiomers are active. (Fig. 18.1). Contrary to its name, ABA does not induce abscission. Accumulation of ABA during seed development has been associated with maturation of seeds, desiccation tolerance, and suppression of vivipary. ABA, when produced in terminal buds, slows plant growth and directs leaf primordia to develop scales to protect the dormant buds from low temperature during cold season. It also inhibits cell division in vascular cambium and acclimatizes plants to cold conditions by suspending primary and secondary growth. ABA is synthesized in green fruits at the beginning of the winter or in maturing seeds while establishing dormancy. It is rapidly translocated through xylem vessels from the roots to the leaves along the transpiration stream. ABA is also produced in response to environmental stresses, such as heat, water, and salt stress.

## 18.1 Bioassay

A number of bioassays have been developed to detect the presence of ABA in plant systems. These include stimulation of stomatal closure, inhibition of germination of isolated wheat embryos, inhibition of growth of duckweed-*Lemna minor*, acceleration of the abscission of cotyledons in cotton seedlings, inhibition of IAA-induced straight growth of *Avena* (oat) coleoptile, and inhibition of GA-induced α-amylase in the cells of aleurone layer in germinating barley seeds. These bioassays have now been replaced with more accurate and reliable methods for ABA detection and estimation. These include high-performance liquid chromatography (HPLC) followed by mass spectrometry (MS). These modern methods are useful for

## 18.2 Biosynthesis, Catabolism, and Homeostasis

determination of the molecular structure. Highly sensitive and specific immunoassays are also applied for quantification and spatial distribution of intracellular ABA.

## 18.2 Biosynthesis, Catabolism, and Homeostasis

Plant responses due to ABA depend on its concentration within the tissue and on the sensitivity of the tissue to the hormone. ABA biosynthesis takes place in chloroplasts or amyloplasts. Complete ABA biosynthetic pathway has been elucidated with the help of ABA-deficient mutants which exhibit blockade of ABA biosynthesis at specific steps. Two pathways have been proposed for ABA biosynthesis:

1. *Isoprenoid pathway:*

$$Mevalonate \rightarrow Isopentanyl\ pyrophosphate \rightarrow Geranyl\ pyrophosphate \rightarrow Farnesyl\ pyrophosphate$$

ABA

2. *Carotenoid pathway* (Fig. 18.2):

Carotenoid pathway begins with isopentenyl diphosphate (IPP), the biological isoprene unit which leads to the synthesis of $C_{40}$ xanthophyll called violaxanthin. The reaction is catalyzed by zeaxanthin epoxidase (ZEP). Violaxanthin leads to the formation of $C_{40}$ compound, 9-cis-neoxanthin, which is further cleaved to form the $C_{15}$ compound called xanthoxin, by the action of NCED (9-*cis*-epoxycarotenoid dioxygenase). Xanthoxin is a natural growth inhibitor that has physiological properties similar to ABA. Synthesis of NCED is rapidly induced by water stress and constitutes the key regulatory step for ABA synthesis. Lastly, xanthoxin leads to the formation of ABA via oxidative steps involving the intermediates ABA-aldehyde or xanthoxic acid in the cytoplasm.

ABA catabolism is brought about by hydroxylation and conjugation steps. Its hydroxylation at C-8′ position is the predominant ABA catabolic route. The 8-′-hydroxy ABA, however, contains substantial biological activity like ABA, but its spontaneous cyclization results in the formation of **phaseic acid (PA)** which does not exhibit any ABA-like activity. ABA-binding proteins from barley aleurone layer are also unable to bind to phaseic acid suggesting that PA is an inactive catabolite of ABA for some physiological processes. Very recently, however, PA has been observed to activate a subset of ABA receptors. Additionally, ABA homeostasis is maintained through glucosylation of ABA's hydroxylated catabolites.

ABA concentration in tissues is highly variable during development or in response to changing environmental conditions. ABA biosynthesis is precisely

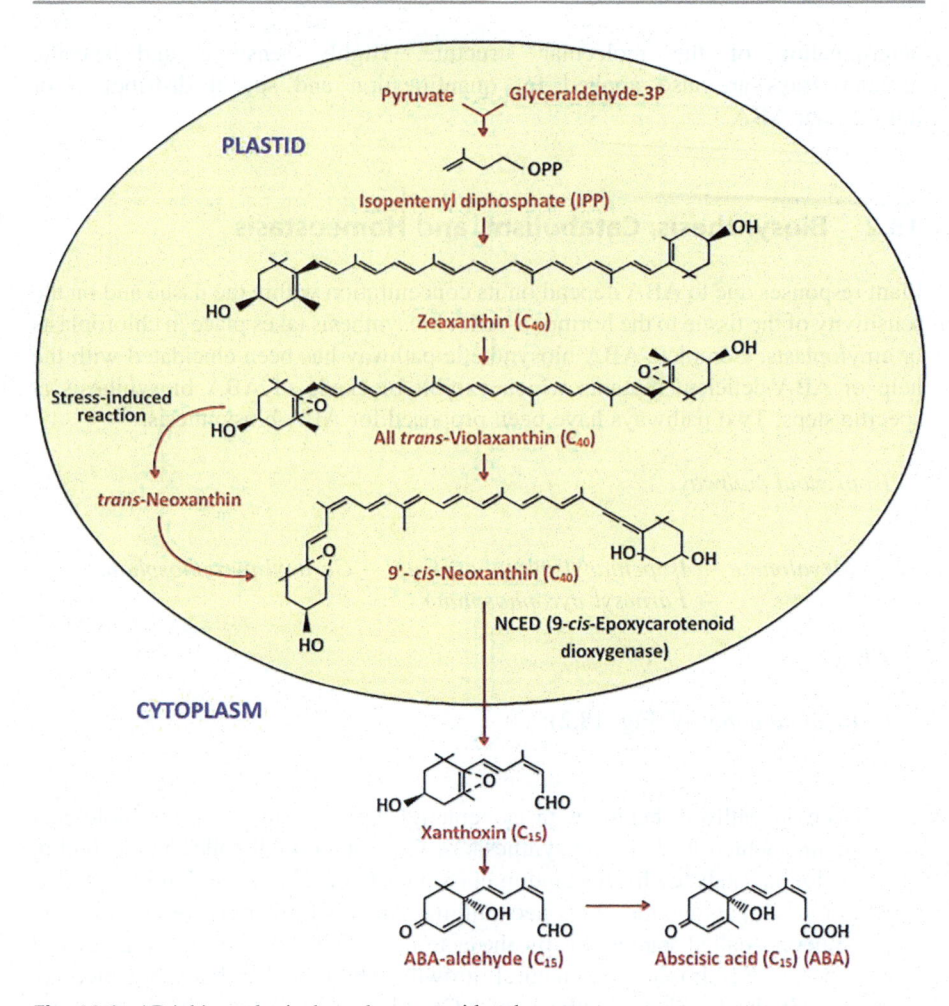

**Fig. 18.2** ABA biosynthesis through carotenoid pathway

regulated with reference to its timing and location. Its availability is enhanced during desiccation of vegetative tissues. It is synthesized in green tissues at the beginning of winter period. ABA establishes dormancy when synthesized in maturing seeds. Under conditions of water stress, ABA concentration in the leaves increases up to 50-fold within 4–8 h. Later, its content declines upon rewatering.

## 18.3 Translocation of Abscisic Acid

ABA is transported through xylem and phloem, but it is much more abundant in the phloem sap. With the onset of water stress, ABA is synthesized in roots, and it gets transported to aerial parts before the low water potential of soil causes any measurable change in the water status of the leaves; ABA is believed to be a root signal which helps reduce transpiration rate by closing stomatal pores in the leaves. During early stages of water stress, pH of xylem sap becomes more alkaline. The dissociated anionic form of the weak acid accumulates in alkaline compartments and may be redistributed according to the steepness of the pH gradient across membrane. Also, specific uptake carriers contribute in maintaining a low apoplastic ABA concentration in unstressed plants. Stress-induced alkalinization of apoplast favors dissociation of ABA, which does not cross membranes. Hence, less ABA enters the mesophyll cells and more reaches the guard cells via transpiration stream.

## 18.4 Developmental and Physiological Effects of ABA

### 18.4.1 ABA Levels Increase in Response to Environmental Stress

ABA is an essential mediator in triggering plant responses to adverse environmental stimuli. Increase in the ABA content is observed during hardening and cold acclimation. The response also depends upon variety of the plant. For example, a freeze-resistant variety of wheat had a higher ABA level than a less resistant variety. Salt stress also results in intercellular accumulation of ABA during growth of tobacco cells and also increased ABA following infection by TMV. These observations indicate that ABA production is positively related to the degree of resistance to a given stress factor. Moreover, exogenous application of ABA increases plant adaptive response to various environmental conditions in variety of plants (Table 18.1).

### 18.4.2 Seed Development

Seed development in flowering plants proceeds in three discrete stages, i.e., early, mid-, and late embryogenesis. During late embryogenesis, seeds lose water and desiccate. They develop hard and resistant seed coat and have negligible metabolism and respiration rate. Four hormones play major role in seed development. IAA and cytokinin levels are high during early phase of embryogenesis when cell divisions occur at high frequency. IAA also plays a major role in patterning and polarity

**Table 18.1** ABA-induced proteins in response to stress

| Crop | Name of the protein | Nature of stress |
| --- | --- | --- |
| Rice | RAB 16A, 16B | Salt and desiccation |
| Wheat | EM, RAB 15 | Drought |
| Barley | Dehydrins B8, B9, B17, B18 | Desiccation |
| Maize | pMAH9 | Drought and wounding |
| Tobacco | Osmotin | Salt |
| Tomato | TAS 14 | Salt and drought |
| Potato | Pt II | Wounding |

establishment in the young embryos. Gibberellic acid is involved in the growth of embryo and endosperm, thus creating a sink for photoassimilates. ABA content rises when the levels of all other hormones decline. It plays a central role during mid to late embryogenesis transition. Once development is complete, ABA content falls.

### 18.4.3 Desiccation Tolerance

ABA induces desiccation tolerance in seeds. Desiccation can damage membranes and other cellular components. During mid- to late stages of seed development, ABA content is high, and numerous metabolic processes are activated that contribute to the acquisition of desiccation tolerance. One of them is synthesis of **late embryogenesis abundant (LEA) proteins**. These are hydrophilic and thermostable proteins which perform protective function against desiccation. These proteins have the ability to form hydrogen bonds with sucrose, which is known to accumulate during seed maturation. This interaction turns it (sucrose) into highly viscous liquid with slow molecular diffusion rate due to which it participates in limited chemical reactions, thereby rendering desiccation tolerance to the seeds. ABA affects the accumulation and composition of storage proteins. ABA-deficient mutants have reduced storage protein accumulation. It also maintains embryos in dormant state until the environmental conditions are optimal for growth. ABA induces changes in cellular metabolism by activating a network of transcription factors. For example, ABI3 induces the synthesis of storage proteins and LEA proteins through interactions with bZIP transcription factors, such as ABI5.

### 18.4.4 Inhibition of Precocious Germination and Vivipary

When immature embryos are removed from their seeds and placed in culture, midway through development before the onset of dormancy, they germinate precociously. ABA added to the culture medium inhibits precocious germination. This suggests that ABA is the natural constraint that maintains developing embryos in their embryogenic state. ABA levels are high during mid to late seed development phases. Further evidence was provided by ABA-deficient mutants of maize, which germinate directly on the cob. **Vivipary** in these mutants is prevented by treatment with exogenous ABA.

### 18.4.5 Counteraction of GA Action

ABA counters many effects of GA, such as induction of hydrolases and α-amylases in barley seedlings. ABA inhibits the synthesis of hydrolytic enzymes which are essential for the breakdown of storage reserves in seeds. It inhibits transcription of α-amylase mRNA via two mechanisms: (i) UP1, a protein which is an activator of ABA-induced gene expression, acts as a transcriptional repressor of some

GA-regulated genes; and (ii) ABA represses GA-induced expression of GA-MγB, a transcription factor that mediates the induction of α-amylase expression by GA.

### 18.4.6 Seed Dormancy

It has also been observed that ABA can be replaced by solution of high osmolarity. Many stress-related genes are induced by both ABA and high osmoticum. ABA-deficient *(aba)* mutants in *Arabidopsis* have been instrumental in demonstrating the role of ABA in seed dormancy. *Aba* mutants are nondormant at maturity. The reciprocal cross between *aba* and wild-type plants exhibits dormancy only when embryo itself produces ABA. The overall effect of ABA on seed dormancy is also influenced by other hormones. Interestingly, rise in ABA production in seeds coincides with decline in IAA and GA level. An experiment to demonstrate the effect of ABA/GA ratio was conducted on *Arabidopsis* seeds that could not germinate in the absence of exogenous GA. These seeds were mutagenized and grown in green house. The seeds produced by these mutagenized plants were screened for **revertants** (seeds that regain the ability to germinate). These revertants were found to be mutants of ABA synthesis, wherein dormancy could not be induced and hence subsequent synthesis of GA was not required to overcome it. This leads to a general conclusion that the balance of plant hormones is more critical than their absolute concentrations in regulating development.

### 18.4.7 Stomatal Closure During Water Stress

Accumulation of ABA in stressed leaves plays an important role in the reduction of water loss by transpiration under water-stressed conditions. Mutants that lack the ability to produce ABA exhibit permanent wilting due to their inability to close stomata. Application of exogenous ABA to such mutants causes stomatal closure and a restoration of turgor pressure. It has been suggested that the permeability of chloroplast membranes in mesophyll cells increases in response to ABA. As a result, ABA diffuses to cytoplasm and moves to guard cells through plasmodesmata. When water potential is restored, movement of ABA into the guard cells stops. ABA causes stomatal closure by inhibiting the ATP-mediated $H^+/ K^+$ exchange pumps in the guard cells. Figure 18.3 summarizes major biochemical routes of ABA action during regulation of stomatal opening and closing and in release of seed dormancy.

### 18.4.8 Promotion of Root Growth and Inhibition of Shoot Growth at Low Water Potential

Under dehydrating conditions, when ABA levels are high, the endogenous hormone exerts a strong positive effect on root growth by suppressing ethylene production and negative effect on shoot growth. This increases the root:shoot ratio at low water

```
                      Abscisic acid
                            ↓
            Transient membrane depolarization
                            |
          ┌─────────────────┴─────────────────┐
          ↓                                   ↓
Alkalinization of cytosolic pH (upto 7.9)   Increase in Ca²⁺ influx and release
          |                                   from intracellular stores
          ↓                                   ↓
Activation of K⁺ efflux channels       Opening of slow (S- type) anion channels
          |                                   ↓
          |                             Efflux of Cl⁻ and malate
          |                                   ↓
          |                             Membrane depolarization
          |                                   ↓
          └───────────────┬───────────────────┘
                          ↓
                      K⁺ Efflux
```

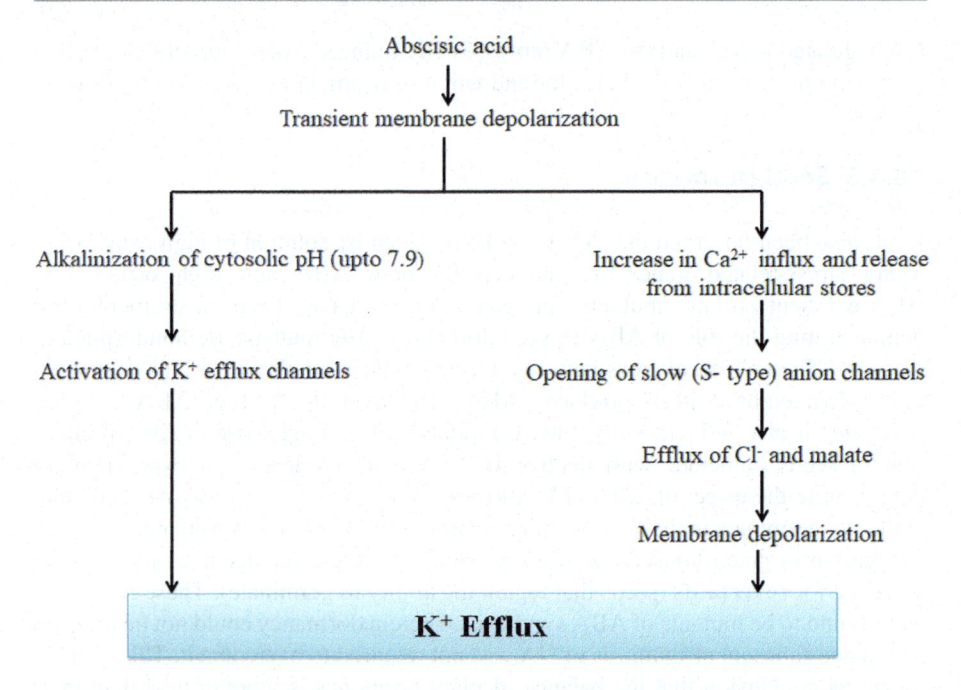

**Fig. 18.3** ABA induces depolarization of plasma membrane and increase in cytosolic calcium which are required for activation of S-type anion channels leading to opening of K⁺ efflux channels on the plasma membrane of guard cells

potential which, along with the effect of ABA on stomatal closure, helps the plant to cope up with water stress.

## 18.5 Mode of Action

ABA influences both short-term physiological effects, like stomatal closure, and long-term developmental processes, like seed maturation, and combats drought stress. Rapid responses involve alterations in the fluxes of ions across membranes and may also involve gene regulation. Long-term responses involve major changes in the pattern of gene expression. ABA receptor is a protein, apparently located on the outer surface of the membrane. Stomatal closure is driven by reduction in guard cell turgor pressure caused by efflux of K⁺ ions. Extracellular application of ABA is effective in inhibiting stomatal opening at pH 6. At this pH, ABA is protonated and is readily taken up by guard cells. The first change detected, after exposure of guard cells to ABA, is transient membrane depolarization caused by the net influx of hydrogen ions and increase in cytosolic $Ca^{2+}$ concentration. ABA stimulates this influx of calcium through the release of calcium from plasma membrane channels and also from internal compartments like vacuoles. Calcium release can also be induced by second messengers like inositol 1,4,5-trisphosphate (IP3) and cyclic

ADP-ribose (cADPR). ABA also causes alkalinization of cytosol from pH 7.6 to 7.9. This activates K⁺ efflux channels on the plasma membrane by increasing the number of channels available for activation. The rapid, transient depolarization induced by ABA is insufficient to open the K⁺ efflux channels. They require long-term membrane depolarization to open K⁺ channels. ABA activates slow ion channels leading to long-term membrane depolarization. Long-term depolarization is triggered by two factors: (1) ABA-induced transient depolarization of plasma membrane and (2) increase in cytosolic calcium. Both of these conditions are prerequisites for opening of calcium-activated slow (S-type) anion channels on the plasma membrane. Prolonged opening of these slow anion channels permits large quantities of chloride and malate ions to escape from the cell, moving down their electrochemical gradients. The outward flow of these negatively charged ions depolarizes the membrane and triggers the opening of voltage-gated K⁺ efflux channels (Fig. 18.3).

In addition to causing stomatal closure, ABA prevents light-induced stomatal opening. In this case, ABA acts by inhibiting the inward K⁺ channels which open when the membrane is hyperpolarized by the proton pump. Inhibition of the inward K⁺ channels is mediated by ABA-induced increase in cytosolic calcium concentration. Thus, calcium and pH affect guard cell plasma membrane channels in two ways: (1) They prevent stomatal opening by inhibiting inward K⁺ channels and plasma membrane proton pumps; (2) They promote stomatal closing by activating outward anion channels, thus leading to activation of K⁺ efflux channels. It is thus evident that ABA elicits two types of responses in plants—rapid responses like stomatal closure and more gradual responses like effect on seed dormancy and reactions to abiotic stresses. It is also likely that there are multiple ABA receptors influencing these responses.

## 18.6 Mechanism of ABA Signaling

Three classes of ABA receptors have been identified till date. These include (i) the *plasma membrane-localized G-proteins* GTG1 and GTG2, (ii) a *plastid-localized enzyme* that coordinates nucleus to plastid signaling, and (iii) *cytosolic ligand-binding proteins* of START (steroidogenic acute regulatory protein-related lipid transfer) domain superfamily. Soluble START domain proteins are the principal ABA receptors which function in short-term as well as long-term responses. A well-worked out pathway for ABA signaling has been designated as PYR/PYL/RCAR pathway. It constitutes members of protein superfamily with a predicted hydrophobic ligand-binding pocket. Their nomenclature reflects their discoveries as PYRABACTIN RESISTANCE 1 (PYR 1) shows resistance to the synthetic sulphonamide compound pyrabactin which mimics ABA action and PYR 1-like (PYL) and regulatory components of ABA receptors (RCARs). This protein subfamily is conserved among plants. They interact with protein phosphatase PP2C in an ABA-dependent manner to regulate the downstream activity of serine/threonine protein kinases of the sucrose non-fermenting related kinase 2 (SnRK2) family. In the absence of ABA, the protein phosphatase (PP2C) keeps the protein kinase

**Fig. 18.4** Mechanism of ABA signaling

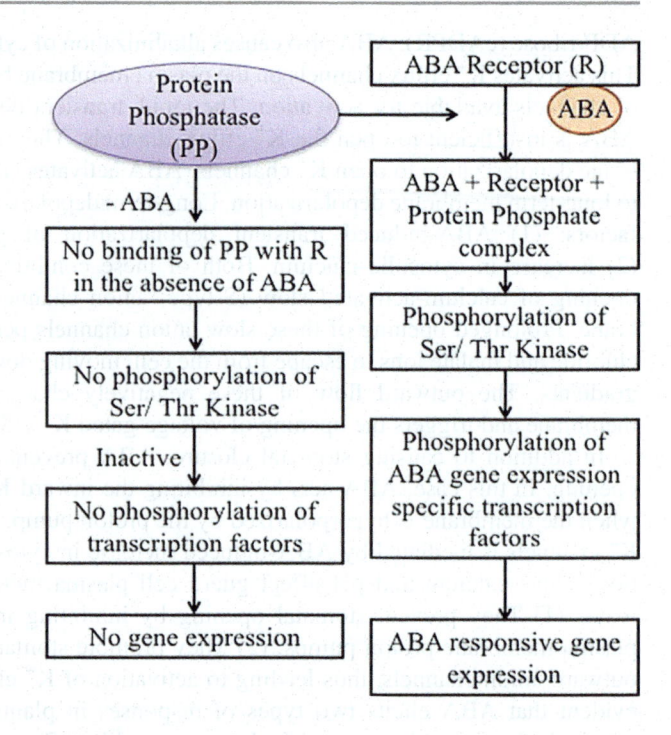

SnRK2 dephosphorylated and thereby inactivated. For inactivation, PP2Cs bind to the C-termini of SnRK2s and block its activity by removing phosphate group from a region within the kinase domain, termed the activation loop. ABA binding changes the conformation of PYR/PYL/RCAR receptors to permit or enhance interaction with PP2C and thereby repress PP2C phosphatase activity. This releases SnRK2 kinases from inhibition. SnRK2 proteins now activated phosphorylate ABA-responsive transcription factors, thereby inducing ABA responses (Fig. 18.4).

To sum up, ABA action on gene expression operates through a cascade of phosphorylation of transcription factors for ABA-sensitive genes. This phosphorylation remains blocked by keeping a serine/threonine protein kinase in non-phosphorylated state by a protein phosphatase. But with the availability of ABA receptor activated by ABA, this phosphatase activity is reduced, thereby allowing phosphorylation of serine/threonine protein kinase, which then phosphorylates the transcription factors for ABA-responsive genes, resulting in its activation.

## Summary

- ABA is ubiquitous in vascular plants and is found in all living tissues. It is a 15-carbon sesquiterpene characterized by a cyclohexane ring with a keto and hydroxyl group and a side chain with a terminal carboxyl group in its structure.

## 18.6 Mechanism of ABA Signaling

- The asymmetric carbon atom at position 1 in the ring structure of ABA results in formation of S- and R-enantiomers. S-enantiomer is the active form leading to fast responses like stomatal closure. In long-term responses, both the enantiomers are active.
- Various bioassays have been developed to detect and estimate ABA in plant systems. Modern methods such as high-performance liquid chromatography (HPLC) followed by mass spectrometry (MS) are useful for determination of the molecular structure. Highly sensitive and specific immunoassays are also applied for quantification and spatial distribution of intracellular ABA.
- ABA biosynthesis takes place in chloroplasts or amyloplasts via two pathways, namely, isoprenoid pathway and carotenoid pathway.
- The catabolism of ABA is brought about by its hydroxylation at C-8' position resulting in the formation of 8'-hydroxy ABA, which contains substantial biological activity like ABA. However, phaseic acid (PA) formed as a result of spontaneous cyclization of 8'-hydroxy ABA does not exhibit any ABA-like activity. Very recently, PA has been observed to activate a subset of ABA receptors. ABA is transported through xylem and phloem, but it is much more abundant in the phloem sap.
- ABA plays a central role in triggering plant responses to adverse environmental stimuli and during transition from mid to late embryogenesis. It induces desiccation tolerance in seeds through the synthesis of late embryogenesis abundant (LEA) proteins which are hydrophilic and thermostable proteins that perform protective function against desiccation.
- ABA counters many effects of GA, such as inhibiting the synthesis of hydrolases and α-amylases which are essential for the breakdown of storage reserves in barley seedlings. It is also responsible for maintaining the embryos in their embryogenic state, thereby preventing precocious germination and vivipary. The overall effect of ABA on seed dormancy is influenced by other hormones. Interestingly, rise in ABA production in seeds coincides with decline in IAA and GA level.
- Accumulation of ABA in stressed leaves plays an important role in the reduction of water loss by transpiration under water-stressed conditions. The endogenous hormone, under such stressed conditions, exerts a strong positive effect on root growth by suppressing ethylene production and negative effect on shoot growth. This increases the root:shoot ratio at low water potential which, along with the effect of ABA on stomatal closure, helps the plant to cope up with water stress.
- ABA influences both short-term physiological effects, like stomatal closure, and long-term developmental processes, like seed maturation, and combats drought stress. Rapid responses involve alterations in the fluxes of ions across membranes and may also involve gene regulation. Long-term responses involve major changes in the pattern of gene expression.
- Three classes of ABA receptors have been identified till date. These include (i) the *plasma membrane-localized G-proteins* GTG1 and GTG2, (ii) a *plastid-localized enzyme* that coordinates nucleus to plastid signaling, and (iii) *cytosolic ligand-*

*binding proteins* of START (steroidogenic acute regulatory protein-related lipid transfer) domain superfamily.
- ABA action on gene expression operates through a cascade of phosphorylation of transcription factors for ABA-sensitive genes.

## Multiple-Choice Questions

1. "Dormin" was isolated from which of the following plant species?
   (a) *Solanum tuberosum*
   (b) *Lupinus arboreus*
   (c) *Betula pubescens*
   (d) *Lemna minor*
2. The cyclohexane ring of ABA has an asymmetric carbon at which position?
   (a) 1
   (b) 2
   (c) 3
   (d) 4
3. Which of the following is not an ABA-mediated response in plants?
   (a) Stomatal closure
   (b) Desiccation tolerance
   (c) Seed dormancy
   (d) Promotion of seed germination
4. ABA is synthesized in which of the following cellular organelle?
   (a) Mitochondria
   (b) Endoplasmic reticulum
   (c) Chloroplast
   (d) Golgi bodies
5. Which of the following is a natural growth inhibitor which has physiological properties similar to ABA?
   (a) Xanthoxin
   (b) Violaxanthin
   (c) Phaseic acid
   (d) Isopentenyl diphosphate
6. During ABA-mediated stomatal closure, membrane depolarization is caused by net influx of which of the following ions?
   (a) Chloride
   (b) Potassium
   (c) Hydrogen
   (d) Calcium

## Answers

1. c    2. a    3. d    4. c    5. a    6. c

## Suggested Further Readings

Dörffling K (2015) The discovery of abscisic acid: a retrospect. J Plant Growth Regul 34:795–808

Nambara E, Marion-Poll A (2005) Abscisic acid biosynthesis and catabolism. Annu Rev Plant Biol 56:165–185

Zhang DP (2014) Abscisic acid: metabolism, transport and signalling. Springer, Dordrecht ISBN (Hardcover) 978-94-017-9423-7

## Suggested Further Readings

# Ethylene

## 19

Satish C Bhatla

**Ethylene (C₂H₄)** is a gaseous hydrocarbon molecule which is not only of great significance in various industries (such as rubber, plastics, paints, toys, and detergents), it is also the first reported gaseous plant hormone which can serve as a signal molecule. Ethylene is both biosynthesized and sensed by plants. Since it can diffuse to and affect surrounding plants, it also has the characteristic of pheromone. Responses to ethylene in plants can be either harmful or desirable depending on its concentration, stage of plant/tissue development, and the plant species. It also mediates adaptive response to a variety of stress factors such as salinity, pathogen attack, flooding, and drought. Ethylene is, however, best known for its essential role in the ripening of climacteric fruits such as tomato, apple, pear, and banana. A commercial source of ethylene, i.e., ethephon, is also used as a spray to induce flowering in pineapple plants and also to prevent lodging (bending over) in wheat plants (Box 19.1). It may be noted that ethylene is a by-product of partial combustion of organic fuels and is, therefore, present in the atmosphere as a result of forest fires, car exhaust, and volcanic eruptions. The greatest risk of working with pure ethylene is that it can explode because of its flammable nature.

The initial observations on the discovery of ethylene came from a report by George Fahnestock in 1858 that leakage of illuminating gas used during that period for lighting in homes and streetlights caused damage to plants. In 1896, Dimitry Neljubow from the University of St. Petersburg (Russia) identified ethylene as the active component in the illuminating gas which affects plant growth. In 1901, Neljubow further showed that ethylene alters the growth of etiolated pea seedlings and leads to shorter, thicker hypocotyls which display diageotropism. This was the first biological effect of ethylene reported in plants and became subsequently known

© Springer Nature Singapore Pte Ltd. 2018
S. C Bhatla, M. A. Lal, *Plant Physiology, Development and Metabolism*,
https://doi.org/10.1007/978-981-13-2023-1_19

> **Box 19.1: Ethylene, Acetylene, and Compounds Releasing or Inducing Ethylene**
>
> Ethylene accelerates ripening in apples and other climacteric fruits. In the fruit storage rooms, ethylene gas can be provided from high-pressure cylinders or from generators which catalyze the conversion of ethanol into ethylene. Field-grown plants can also be supplied acetylene in solution which is generated from calcium carbide in water.
>
> Cl–CH$_2$–CH$_2$–PO$_3$H$_2$ + H$_2$O $\xrightarrow{pH > 3.5}$ CH$_2$=CH$_2$ + H$_3$PO$_4$ + Cl$^-$
>
> 2–chloroethylphosphoric acid    H$_2$O      Ethylene    Phosphoric acid
>
> (Ethephon)
>
> Release of ethylene from ethephon in aqueous solution
>
> A nongaseous source of ethylene is ethephon (also called ethrel or florel by different manufacturing firms) which can be applied much more effectively. Once absorbed by plant, ethrel decomposes into ethylene, phosphoric acid, and hydrochloric acid in acidic conditions. Ethrel has been found to have many commercial uses. Thus, in cereals, its spary reduces stem elongation thereby reducing lodging. In cotton, its spary acts as a ball opener and defoliant prior to harvesting. It also intensifies flow of latex in rubber tress. In pineapples, ethrel is used to induce flowering. Bayer Crop Science (Germany) is a key producer of ethephon for various applications in agriculture.

as the **triple response** of etiolated seedlings in response to ethylene. Subsequently, Crocker et al. (1913) established triple response as a sensitive bioassay for ethylene. Cosins (1910) was the first to indicate that plants biosynthesize ethylene. He reported that fungus-damaged oranges produced a gas which accelerated fruit ripening in banana. Richard Gane (1934) estimated ethylene released from apples and demonstrated that plants biosynthesize ethylene. Finally, in 1935, Crocker et al. proposed that ethylene is a plant hormone. Our current understanding of ethylene receptors has been obtained from the characterization of ethylene insensitive mutants.

*Illuminating gas used in the 1700s–1800s for lighting in homes, businesses, and street lights used to be initially manufactured by the destructive distillation of bituminous coal. Thus, it was also sometimes referred as coal gas. Later, carbureted water gas was also used for lighting purpose.*

## 19.1 Ethylene vs Non-gaseous Plant Hormones: Some Interesting Facts

Ethylene is biologically active at very low concentrations which may vary from 0.1 to 1 ppm. However, lower or higher sensitivity is also observed depending upon the plant species and the response. In fact, some climacteric fruits, such as apple and tomato, can generate up to tens of ppm of ethylene. Ethylene is different from nongaseous plant hormones in the following ways:

1. Ethylene is more soluble in lipid bilayers than in aqueous phase. It can freely pass through the plasma membrane. It is thought to be biosynthesized at or near its site of action as is also the case with nitric oxide. Its precursor, 1-aminocyclopropane-1-carboxylic acid (ACC), is water soluble and is thought to get transported in plants via xylem.
2. Ethylene does not require any transporter proteins for its delivery to the target cells. In fact, no such transporters have been identified.
3. Unlike other plant hormones, ethylene does not conjugate or get broken down for intracellular storage or deactivation. When not required, it simply diffuses out of the plant.

## 19.2 Biosynthesis of Ethylene

Ethylene biosynthesis in plants takes place in a three-step pathway starting from methionine (Fig. 19.1). (1) Methionine gets converted to S-adenosyl-L-methionine (SAM) by the action of SAM synthetase. (2) SAM gets converted into 1-aminocyclopropane-1-carboxylic acid (ACC) by the action of ACC synthase (ACS). (3) ACC is converted to ethylene by the action of ACC oxidase (ACO). ACC synthase (ACS) is induced by various factors like stress, fruit ripening, etc. Ethylene biosynthesis is also controlled by ACC synthase turnover which is regulated by phosphorylation. This enzyme was first characterized in a semi-purified preparation from *Solanum* pericarp. It was subsequently isolated from a number of plant tissues following induction by factors such as exogenous IAA, wounding,

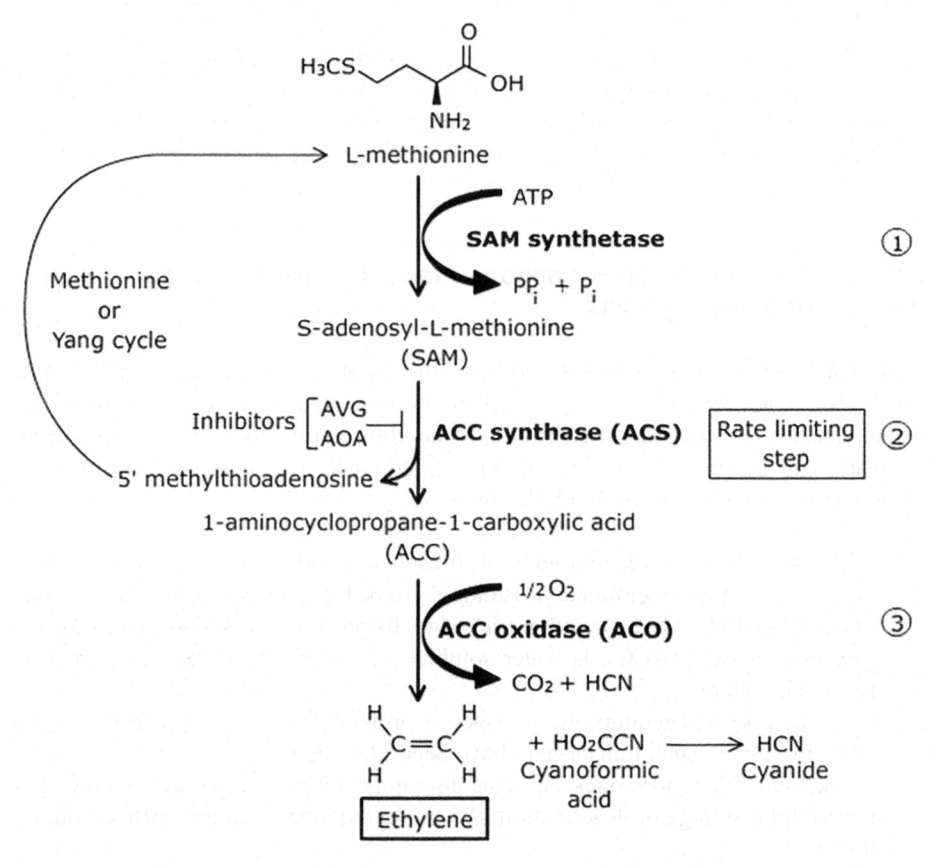

**Fig. 19.1** Biochemical pathway for ethylene biosynthesis in plants. In the first step, SAM synthetase converts methionine to SAM. In the second step, ACC synthase converts SAM into ACC. This is the rate-limiting step. In the third step, ACC oxidase converts ACC into ethylene. SAM synthetase and ACC synthase are part of the yang or methionine cycle. *AVG* aminoethoxyvinyl glycine, *AOA* aminooxyacetic acid

lithium chloride stress, and climacteric fruit ripening. ACC synthase is an unstable enzyme and is encoded by a multigene family which is differentially regulated by various inducers of ethylene biosynthesis (Table 19.1). Ethylene synthesis is inhibited by aminoethoxyvinyl glycine (AVG) and aminooxyacetic acid (AOA). Both these compounds affect conversion of SAM to ACC. In plants, methionine is present in limited amount and hence is regenerated through a cycle called **Yang's cycle**. Without this recycling, the amount of reduced sulfur present would limit the available methionine and also the synthesis of ethylene. Thus, during ethylene production, methionine depletion is prevented due to recycling of methionine. ACS catalyzes the rate-limiting step in ethylene biosynthesis, and ACS enzyme levels are controlled by transcription and through protein stability. Enhanced ethylene production associated with seed germination, fruit ripening, flooding,

## 19.2 Biosynthesis of Ethylene

and chilling stresses are invariably accompanied by increased ACC production due to induction or activation of ACS.

ACC oxidase (ACO) catalyzes the last step in ethylene biosynthesis. In tissues that show high rate of ethylene production, ACO activity can be a rate-limiting step during ethylene biosynthesis. This enzyme requires $Fe^{2+}$ and ascorbate for its activity (Table 19.1). Cyanoformic acid produced as a by-product of this ACO-catalyzed reaction is a source of cyanide in plants. Plants have devised unique biochemical routes for cyanide detoxification (Box 19.2). Many fruits

**Table 19.1** Characteristics of ACC synthase and ACC oxidase

| Enzyme | Mol. Wt. | Cofactors | Inhibitors | Activators |
|---|---|---|---|---|
| ACC synthase | 50–60 kDa | Pyridoxal phosphate | AVG, AOA | Climacteric fruit ripening, wounding, chilling, seed germination |
| ACC oxidase | 35–36 kDa | $Fe^{2+}$ | $Co^{2+}$ | – |
|  |  | Ascorbate | $Fe^+$ chelators (e.g. EDTA) |  |

---

**Box 19.2: How Do Plants Protect Themselves from Endogenous Cyanide Poisoning?**

Cyanide is produced stoichiometrically as a co-product during ethylene biosynthesis in plants. The first step is the conversion of S-adenosylmethionine (S-AdoMet) to 1-aminocyclopropane-1-carBOXylic acid (ACC) by ACC synthase (S-adenosyl-l-methionine methylthioadenosine-lyase). ACC is oxidized by ACC oxidase (ACO) to form ethylene and cyanoformic acid, which spontaneously degrades to carbon dioxide and cyanide. Thus, cyanide is produced whenever ethylene synthesis is induced. Ethylene production is known to increase during plant developmental events, such as seed germination, senescence, abscission flowering, and fruit ripening. Abiotic and biotic stress factors in plants also elicit ethylene synthesis which leads to "stress cyanide" production. Healthy plant cells have inbuilt protection from the risk of cyanide poisoning. Cyanide is also produced from cyanohydrin glycosides stored in plant vacuoles, as a defense response against herbivores. A grazing herbivore will ingest plant material, the vacuolar and cytosolic contents will be mixed, and poisonous cyanide will be liberated by the action of cytoplasmically localized glycosidase and hydroxynitrile lyase on cyanohydrin glycoside and cyanohydrin, respectively. Endogenous and exogenous cyanide can be detoxified by the cyanoalanine synthase (CAS) pathway. CAS is localized in the mitochondria and is responsible for the formation 3-cyano-L-alanine from a reaction between cysteine and cyanide. The 3-cyano-L-alanine formed is further metabolized into asparagine (nontoxic) by cyanoalanine hydratase.

(continued)

**Box 19.2** (continued)

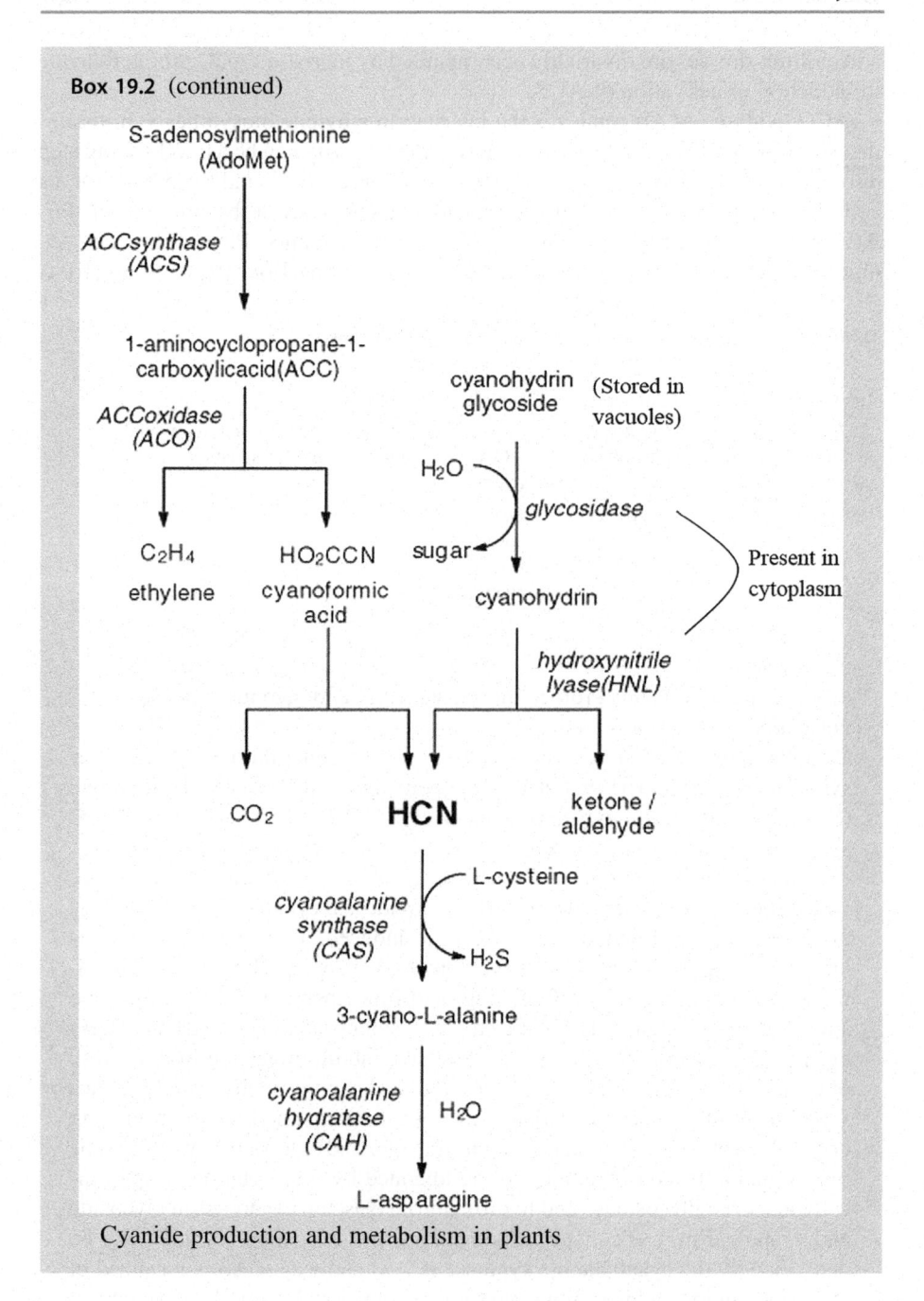

S-adenosylmethionine (AdoMet)

*ACCsynthase (ACS)*

↓

1-aminocyclopropane-1-carboxylicacid(ACC)

*ACCoxidase (ACO)*

$C_2H_4$ ethylene    $HO_2CCN$ cyanoformic acid

cyanohydrin glycoside (Stored in vacuoles)

$H_2O$ → *glycosidase* → cyanohydrin (Present in cytoplasm)

*hydroxynitrile lyase(HNL)*

↓ CO$_2$    ↓ **HCN**    ↓ ketone / aldehyde

*cyanoalanine synthase (CAS)* — L-cysteine → H$_2$S

↓

3-cyano-L-alanine

*cyanoalanine hydratase (CAH)* | H$_2$O

↓

L-asparagine

Cyanide production and metabolism in plants

**Fig. 19.2** Four ways to reduce ethylene production by plants: (*1*) silencing of ACS gene expression, (*2*) silencing of ACO gene expression, (*3*) deamination of ACC, and (*4*) hydrolysis of S-adenosylmethionine

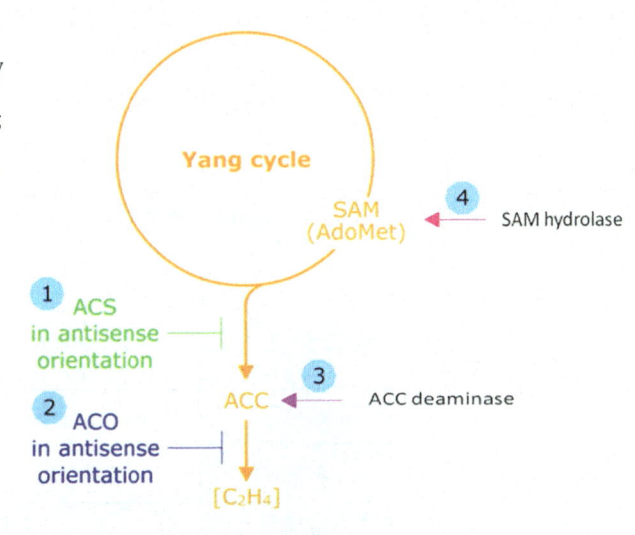

produce ethylene, and it is instrumental in their ripening. $CO_2$ and potassium permanganate can reduce the concentration of ethylene in apple storage areas from 250 to 10 µl liter$^{-1}$, thus extending the storage life of the fruit. An understanding of the steps in ethylene biosynthesis and cloning of genes of the enzymes ACC synthase and ACC oxidase has allowed the production of transgenic plants with altered levels of ethylene production. Transgenic tomatoes have been produced that have low levels of ethylene and thus have fruits that mature but fail to ripen and have extended shelf life by 6 months. Introduction of ACS or ACO cDNA in an antisense orientation silences the expression of the corresponding genes and results in plants that produce very low levels of ethylene (Fig. 19.2). Ethylene levels in plants can also be reduced by deamination of ACC or hydrolysis of S-adenosylmethionine. Ethylene biosynthesis has also been blocked in *Petunia* using biotechnological techniques to increase longevity of cut flowers for several weeks.

## 19.3 Regulation of Ethylene Biosynthesis

Under normal growth conditions, ethylene is produced at low levels in practically all regions of the plant body. Its tissue levels are, however, known to increase in response to both developmental cues and environmental signals. Thus, accumulation of ethylene in many fruits during ripening is the result of dramatic increase in its biosynthesis. Because of the significant accumulation of ethylene in response to a wide variety of biotic and abiotic stress factors, ethylene is often called the "stress" hormone. Regulation of ACC synthase activity is known to be the critical regulatory factor for ethylene biosynthesis (Fig. 19.3). *eto1* is a constitutive ethylene response mutant of *Arabidopsis* which along with two other mutants of similar traits (*eto2* and *eto3*) possesses higher ACC synthase activity than the wild plants. Out of the three

```
                    Environmental
                       stress
                          ⇓
                 Activation of MAPKs
                  (stress activated)
                          ⇓
           ┌──────────────┬──────────────┐
           │ ACC synthase │  Associated  │
           │   isoforms   │    motifs    │
           ├──────────────┼──────────────┤
           │    ACS 1     │ MAPK & CDPK  │
           ├──────────────┼──────────────┤
           │    ACS 2     │    CDPK      │
           ├──────────────┼──────────────┤
           │    ACS 3     │      –       │
           └──────────────┴──────────────┘
Dephosphorylation          ⇓                 Lowering of
    of ACS        ⇐ Phosphorylation of ⇐  environmental stress
      ⇓            specific ACS isoforms
 Reduced ethylene          ⇓
    production      ┌──────────────────┐
                    │ Enhanced ethylene│
                    │   biosynthesis   │
                    └──────────────────┘
```

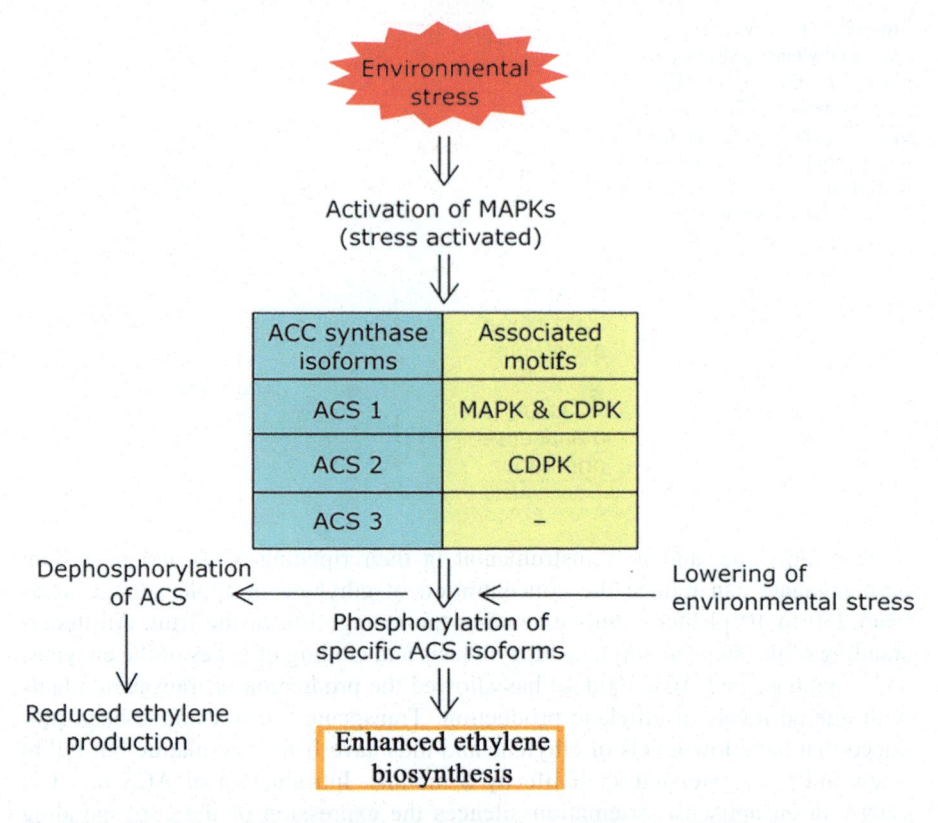

**Fig. 19.3** Regulation of ethylene biosynthesis by environmental stress. *MAPK* mitogen activated protein kinases, *CDPK* cyclin-dependent protein kinases, *ACS* ACC synthase

isoforms of ACC synthase so far characterized (ACS1, ACS2, and ACS3), ACS1 contains target motifs for mitogen-activated protein kinases (MAPKs) as well as calcium-dependent protein kinases (CDPKs), ACS2 contains only target motifs for CDPKs, and ACS3 does not contain either of the motifs. Increased phosphorylation of specific ACS isoforms is triggered by stress-activated MAPKs and/or CDPKs, leading to protein stabilization and increased ethylene production (Fig. 19.3).

## 19.4 Bioassay and Mutant Analysis

Analysis of dark-grown dicot seedlings and mutants exposed to ethylene shows unique concentration-dependent morphological changes referred as **triple response**. The triple response is a highly specific and sensitive bioassay for the detection of

## 19.4 Bioassay and Mutant Analysis

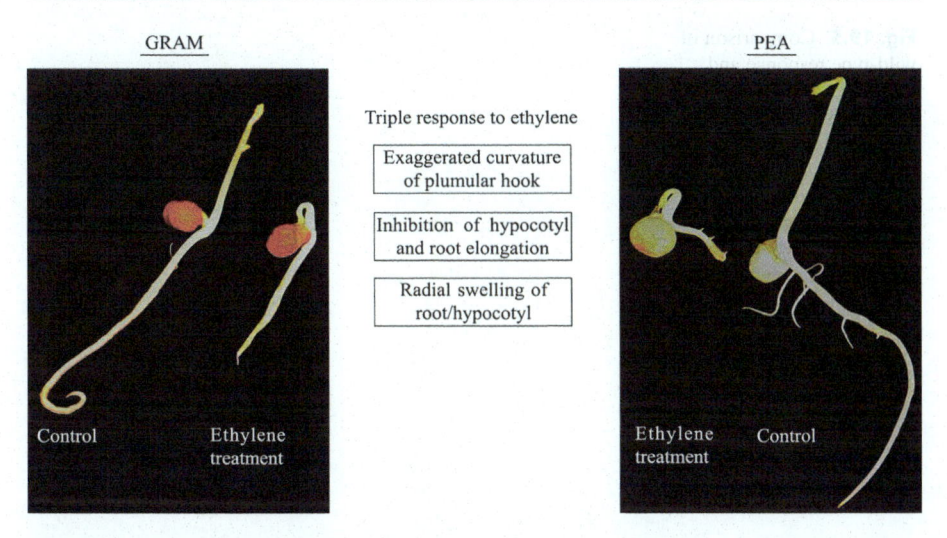

**Fig. 19.4** Ethylene-mediated triple response in gram and pea. Dark-grown (6-day-old) seedlings were raised in half-strength Hoagland medium in the absence or presence of 0.26 M of ethrel as ethylene source

ethylene. Triple response due to ethylene can be demonstrated in various dark-grown dicot seedlings (Fig. 19.4). This response is characterized by the inhibition of hypocotyl and root elongation, a pronounced radial swelling of the hypocotyl, and exaggerated curvature of the plumular hook. The response is rapid and allows large populations of seedlings to be screened for ethylene mutations. Absence of triple response in the presence of exogenous ethylene has been used to identify ethylene-resistant mutants and other ethylene response defects. These mutants generally fall into three categories: (1) constitutive triple response (*ctr1*) mutants that exhibit triple response in the absence of ethylene, (2) ethylene-insensitive response (*etr1*) mutants, and (3) mutants in which ethylene insensitivity is limited to specific tissues, such as plumular hook or root elongation. As a result of these experiments, several ethylene receptors and downstream elements in ethylene signal transduction chain have been identified. Ethylene insensitive mutants (*etr1*) have long rather than short hypocotyls (Fig. 19.5). These mutants are deficient in perception or transduction of ethylene signal. The other category of mutants (*ctr1*) exhibits a triple response in the absence of ethylene exposure. They have a constitutively activated ethylene signaling pathway. This can be either due to more production of ethylene in the mutant than the wild type or because the signaling pathway is activated in the absence of ethylene.

**Fig. 19.5** Comparison of wild-type response and ethylene triple response mutants of *Arabidopsis*. (**a**) The wild-type (WT) plants exhibit a classic triple response (short root, short hypocotyls, exaggerated apical hook) when grown in ethylene, whereas the *etr1* mutant shows growth like that of a wild-type plant in the absence of ethylene. (**b**) In the absence of ethylene, the *ctr1* mutant exhibits a constitutive triple response

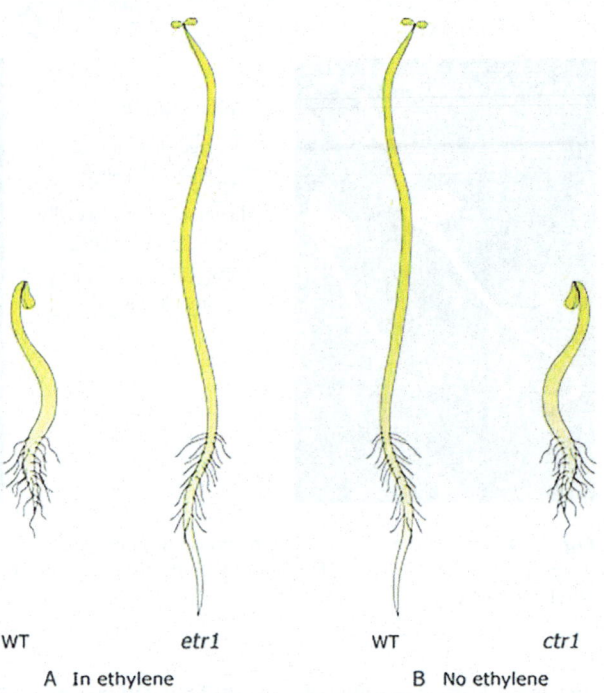

## 19.5 Physiological and Developmental Effects

### 19.5.1 Fruit Ripening

Climacteric fruits, such as avocado, banana, mango, tomato, and apple, ripen slowly as they mature. But in final stages, various developmental changes occur rapidly. These include conversion of starch to soluble sugars, breakdown and softening of cell wall, development of aroma, and color change. Ethylene stimulates these changes. One effect of ethylene in climacteric fruits is the production of more ethylene which constitutes a **positive feedback system** (Table 19.2). Ethylene concentration increases exponentially and rapidly during final phase of maturation of climacteric fruits. This sudden burst of ethylene and rapid completion of maturation of fruits are known as **climacteric**. During this period, ethylene production can be as high as 320 nL/g/hr. In non-climacteric fruits, such as orange, lemon, grapes, watermelon, and cherries, ethylene does not stimulate its own production. Therefore, in non-climacteric fruits, ethylene levels remain stable, and no sudden change occurs just before maturity, but their maturation can be accelerated when they are placed along with maturing climacteric fruits (Fig. 19.6). Exogenous ethylene treatment not only stimulates production of additional ethylene in climacteric fruits, it is also followed by marked increase in rate of respiration. In non-climacteric fruits, however, externally applied ethylene brings about minor increase in the endogenous ethylene production and no increase in respiration.

## 19.5 Physiological and Developmental Effects

**Table 19.2** Estimated rate of ethylene production in some maturing fruits, leaves, and fungi

| Tissue | Rate of ethylene production (upper limit reported)[a] (nL.g$^{-1}$.hr$^{-1}$) |
| --- | --- |
| *Climacteric fruits* | |
| Avocado | 130 |
| Banana | 20 |
| Tomato | 20 |
| *Non-climacteric fruits* | |
| Lemon | 0.1 |
| Orange | 0.1 |
| Tobacco leaves | 2 |
| Orchid flowers | 3400 |
| *Fungi* | |
| Penicillium digitatum | 6000 |

[a] Rate of ethylene production in the listed fruits may vary depending on variety and environmental conditions

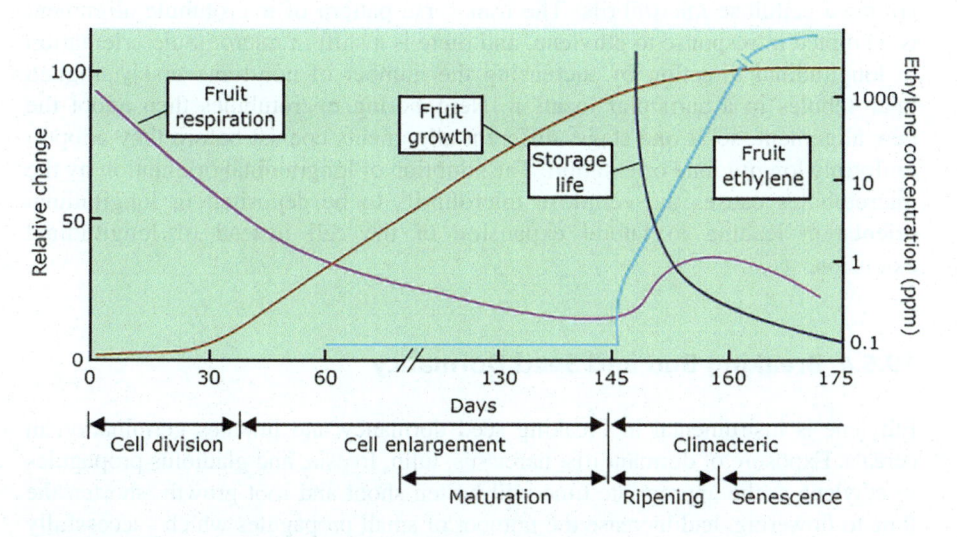

**Fig. 19.6** Development associated changes in climacteric fruits

### 19.5.2 Leaf Epinasty

The downward curvature of leaves that occurs when the upper side of the petiole grows faster than the lower side is termed as epinasty. It is induced by ethylene and high concentration of auxin. In tomato plants, anaerobic conditions around the roots enhance the ethylene accumulation in shoots, leading to the epinastic response. These environmental stresses are sensed by the roots, and the response is displayed by the shoots. Signal from roots is possibly transported to shoots in the form of ACC, the immediate precursor of ethylene, which accumulates in and around the roots. It is then transported to shoots along the transpiration stream where it is converted to ethylene and induces epinasty. The epinastic strategy during flooding may be significant because the downwardly directed leaves act as sails in the wind, and the wind action on the leaf acts like a lever to water pump. This adaptation helps to pump water out of spaces surrounding the roots as fast as possible with ordinary transpiration.

### 19.5.3 Induction of Lateral Cell Expansion

The orientation of cellulose microfibrils in the cell wall determines the direction of plant cell expansion. Transverse microfibrils reinforce the cell wall in the lateral direction, so that turgor pressure is channeled into cell elongation. The orientation of microfibrils, in turn, is determined by the orientation of the cortical array of microtubules in the cortical cytoplasm. In typical elongating plant cells, the cortical microtubules are arranged transversely, giving rise to transversely arranged cellulose microfibrils. The transverse pattern of microtubule alignment is disrupted in response to ethylene, and there is a shift in microtubule orientation in longitudinal direction by increasing the number of non-transversely aligned microtubules in a particular location. Neighboring microtubules then adopt the new alignment, so at one stage, different alignments coexist before they adopt a uniformly longitudinal orientation. The adoption of longitudinal orientation by the microtubules causes the cellulose microfibrils to be deposited in longitudinal orientation leading to lateral expansion of the cell instead of longitudinal extension.

### 19.5.4 Breaking Bud and Seed Dormancy

Ethylene is instrumental in breaking seed dormancy and initiates germination in cereals. Exposure of dormant iris, narcissus, tulip, freesia, and gladiolus propagules to ethylene at the appropriate time will hasten shoot and root growth, shorten the time to flowering, and increase the number of small propagules which successfully flower. Also, the pronounced effect of ethylene on sprouting of potatoes is associated with increased respiration and mobilization of carbohydrates.

### 19.5.5 Elongation of Submerged Aquatic Species

In deep water rice, submergence induces rapid internode or petiole elongation which allows the leaves or upper parts of the shoot to remain above water. Treatment with ethylene mimics the effect of submergence. Ethylene stimulates internode elongation by increasing the sensitivity of the tissue to gibberellins. This increased sensitivity to gibberellins in turn is brought about by decrease in level of abscisic acid, indicating a crosstalk between various hormones to bring about a particular response.

### 19.5.6 Formation of Adventitious Roots and Root Hairs

Ethylene induces adventitious root formation on leaves, stem, flower, etc. and also acts as a positive regulator of root hair formation. In *Arabidopsis*, it is seen that root hairs are located on the epidermal cells which overlie a junction between the underlying cortical cells. In ethylene-treated roots, additional hairs are formed in abnormal locations. Treatment with ethylene inhibitors reduces root hair formation.

### 19.5.7 Leaf Senescence and Abscission

Exogenous application of ethylene accelerates leaf senescence. Enhanced ethylene production is associated with chlorophyll loss. Inhibitors of ethylene synthesis retard leaf senescence. 1-methylcyclopropane (MCP), an inhibitor of ethylene action, acts as a competitive and irreversible inhibitor of ethylene binding. It effectively eliminates the effect of ethylene on cut flowers to increase their shelf life. The role of ethylene-mediated reduction of auxin transport in natural and ethylene-induced leaf abscission is well established. In plants where cotyledons are allowed to senesce naturally, there is a decline in auxin-transport capacity of petioles and increase in ethylene synthesis in cotyledons.

### 19.5.8 Ethylene in Defense Response

Pathogen infection and disease occur only if the interactions between host and pathogens are genetically compatible. However, increase in ethylene production has been found in response to pathogen attack in both compatible and non-compatible interactions. In compatible interactions, elimination of ethylene responsiveness prevents the development of disease symptoms even though the growth of pathogen appears to be unaffected.

### 19.5.9 Abscission

Ethylene is the primary regulator of abscission process, with auxin as a suppressor of ethylene effect. During the early phase of leaf maintenance, auxin from the leaves prevents abscission by maintaining the cells in ethylene insensitive state. During leaf shedding phase, there is decrease in auxin content in the tissue, while ethylene level rises. Ethylene decreases the activity of auxin both by reducing its synthesis and its transport and also by promoting auxin destruction. Reduction in the concentration of free auxin increases the response of specific target cells to ethylene. This phase is characterized by the induction of genes encoding specific hydrolytic enzymes for degradation of cell wall polysaccharides and proteins.

## 19.6 Ethylene Receptors and Signal Transduction

Ethylene receptors are similar to "two component histidine kinase receptors" found in bacteria, which usually involve a sensor molecule carrying a histidine kinase domain and a response regulator (Fig. 19.7). Ethylene receptors have an N-terminal ethylene binding domain and a histidine kinase domain. Five ethylene receptors are known from *Arabidopsis*. They are ETR1 and ETR2 (ethylene receptors 1 and 2), ERS1 and ERS2 (ethylene response sensors 1 and 2), and EIN4 (ethylene insensitive 4) (Fig. 19.8). Since the loss of any receptor has negligible effect on the plant phenotype, it is obvious that their functions overlap and one receptor can take the place of another if full complement is not present. The individual receptors might have slightly different roles in plant's life cycle. Expression of some receptor genes is inducible, for example, accompanying fruit ripening or flooding. *ETR1* gene was the first to be identified. *etr1* mutations abolish triple response to endogenously produced ethylene, and these mutations are genetically dominant. They render the mutant receptor insensitive to ethylene. Thus, the mutant receptor continues to signal in presence of ethylene, unlike wild type. Ethylene receptors exist as dimers across the lipid bilayer of membranes. The N-terminal contains the ethylene binding site on the apoplastic side (Fig. 19.9). This binding site requires copper ions probably supplied by RAN1, a product of *responsive antagonist 1 (RAN1)* gene. The histidine-kinase domain is cytoplasmic. The receiver domain is also cytoplasmic and is located toward the C-terminal. The ethylene binding site is a hydrophobic site embedded in the PM. Ethylene binding to receptor requires copper ions. In wild

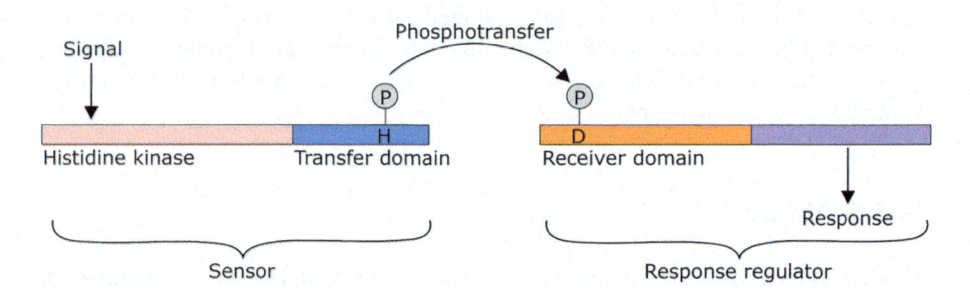

**Fig. 19.7** Bacterial two-component regulation. A signal causes autophosphorylation of the sensor; transfer of a phosphate group (phosphotransfer) from a histidine residue (H) in the sensor to conserved aspartate residue (D) in the response regulator causes phosphorylation of the response regulator, ultimately resulting in a response based on the signal received

## 19.6 Ethylene Receptors and Signal Transduction

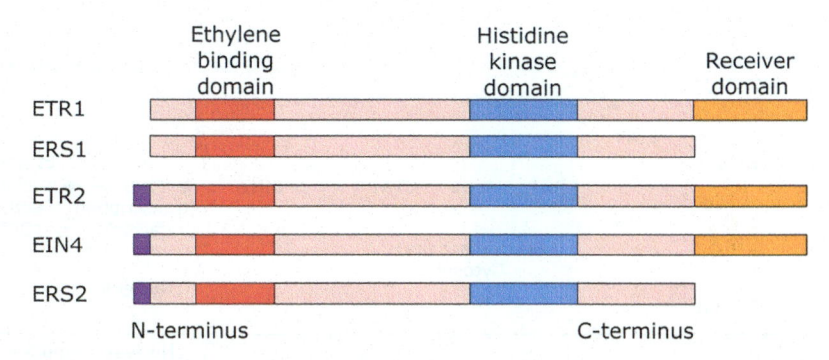

**Fig. 19.8** The *Arabidopsis* family of ethylene receptors. ETR1, ETR2, and EIN4 consist a sensor and receiver domain, whereas ERS1 and ERS2 lack receiver domain. ERS1 and ERS2 might recruit ETR1, ETR2, or EIN4 receiver domains or might use other response regulators

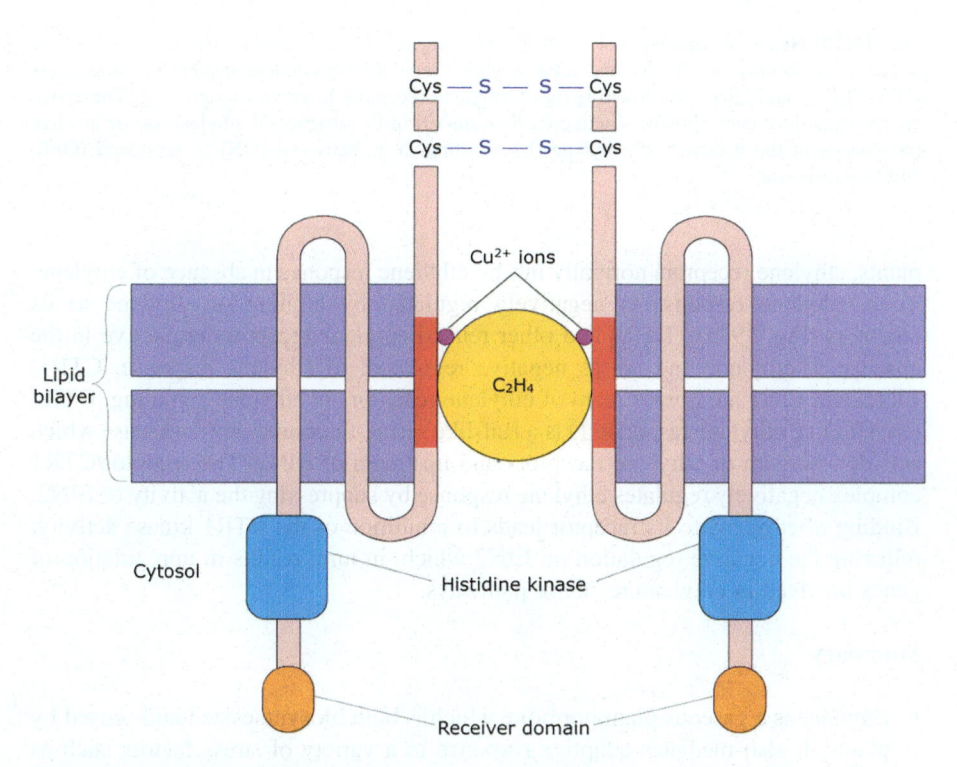

**Fig. 19.9** The domains of ETR1. ETR1 spans the lipid bilayer of the plasma membrane. Ethylene ($C_2H_4$) binding, which occurs in a hydrophobic pocket in the plasma membrane, requires the binding of copper ($Cu^{2+}$ ions)

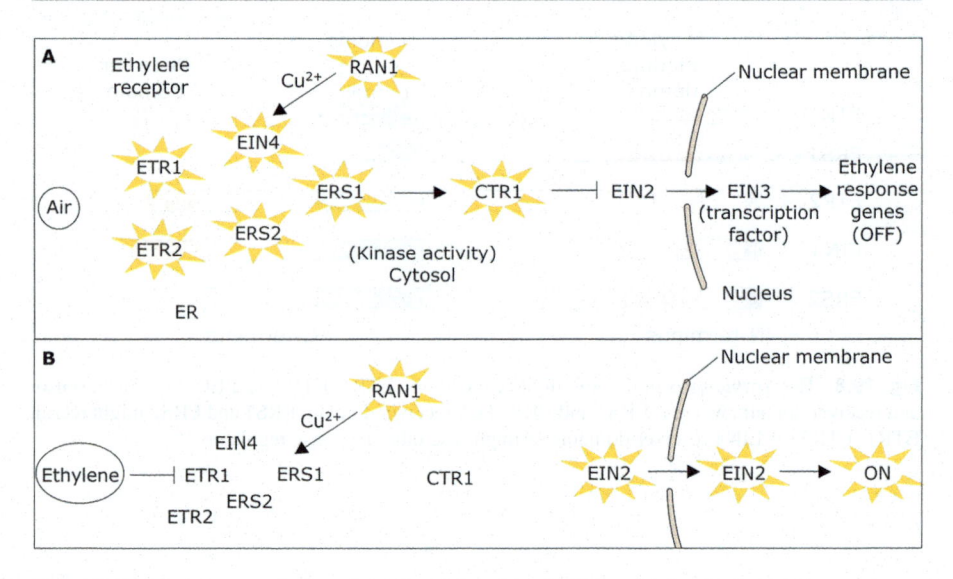

**Fig. 19.10** Negative regulation of ethylene signaling. ETR1 and related ethylene receptors are active in the absence of ethylene and act as negative regulators of ethylene response. Components CTR1, EIN2, and EIN3 are downstream of ethylene receptors in ethylene signaling. These two components have been shown as active (*yellow spikes*) in the absence of ethylene (**a**) or inactive (no spikes) in the presence of ethylene (**b**) resulting in an activated (ON) or repressed (OFF) ethylene response

plants, ethylene receptors normally inhibit ethylene response in absence of ethylene. Thus, ethylene response is negatively regulated by binding of ethylene to its receptors (Fig. 19.10). ETR1 and other related ethylene receptors are active in the absence of ethylene and act as negative regulators of ethylene response. CTR1, EIN2, and EIN3 are downstream of ethylene receptors in ethylene signaling. CTR1 (constitutive ethylene response 1) is a Raf-like serine-threonine protein kinase which acts downstream of ethylene receptors and upstream of EIN2. The receptor/CTR1 complex negatively regulates ethylene response by suppressing the activity of EIN2. Binding of ethylene to its receptor leads to inhibition of the CTR1 kinase activity, relieving the negative regulation on EIN2 which, in turn, results in upregulation of genes involved in ethylene response pathways.

**Summary**

- Ethylene is a gaseous plant hormone which is both biosynthesized and sensed by plants. It also mediates adaptive response to a variety of stress factors such as salinity, pathogen attack, flooding, and drought. It is biologically active at very low concentrations which may vary from 0.1 to 1 ppm. Unlike other plant

## 19.6 Ethylene Receptors and Signal Transduction

hormones, ethylene does not conjugate or get broken down for intracellular storage or deactivation. When not required, it simply diffuses out of the plant. Ethylene is often called the "stress" hormone because of involvement in a wide variety of biotic and abiotic stresses.
- Ethylene biosynthesis in plants takes place in a three-step pathway starting from methionine involving the enzymes SAM synthetase, ACC synthase, and ACC oxidase. In plants, during ethylene production, methionine depletion is prevented by its recycling through a cycle called Yang's cycle. Transgenic plants with lower levels of ethylene have been produced which have extended shelf life by several months.
- Dark-grown seedlings exposed to ethylene show concentration-dependent morphological changes referred as "triple response," characterized by the inhibition of hypocotyl and root elongation, a pronounced radial swelling of the hypocotyl, and exaggerated curvature of the plumular hook.
- Climacteric fruits ripen slowly as they mature and show a rapid exponential increase in ethylene concentration during final phase of maturation. Non-climacteric fruits, on the other hand, show no such sudden change in ethylene levels. However, their maturation is enhanced when they are placed with maturing climacteric fruits.
- Ethylene and high concentration of auxin induce epinasty which is the downward curvature of leaves that occurs when the upper side of the petiole grows faster than the lower side.
- Ethylene causes a shift in the microtubule alignment from transverse to longitudinal plane leading to deposition of cellulose microfibrils in the longitudinal orientation resulting in lateral expansion instead of longitudinal extension.
- Ethylene is instrumental in breaking seed dormancy and initiating germination in cereal. It stimulates internode elongation by increasing the sensitivity of the tissue to gibberellins. It also induces adventitious root formation on leaves, stem, flower, etc. and also acts as a positive regulator of root hair formation. Exogenous application of ethylene accelerates leaf senescence. An increase in ethylene production has been found in response to pathogen attack in both compatible and non-compatible interactions. It is the primary regulator of abscission process, with auxin as a suppressor of ethylene effect.
- Ethylene receptors are similar to "two-component histidine kinase receptors" found in bacteria, which usually involve a sensor molecule carrying a histidine kinase domain and a response regulator. Five ethylene receptors are known from *Arabidopsis*. They are ETR1 and ETR2 (ethylene receptors 1 and 2), ERS1 and ERS2 (ethylene response sensors 1 and 2), and EIN4 (ethylene insensitive 4). Ethylene receptors exist as dimers across the lipid bilayer of membranes. Ethylene binding to receptor requires copper ions. Ethylene response is negatively regulated by binding of ethylene to its receptors.

## Multiple-Choice Questions

1. Which of the following is not a characteristic feature of the seedlings exhibiting "triple response"?
   (a) Exaggerated apical/plumular hook
   (b) Shortened hypocotyl
   (c) Elongated roots
   (d) Radially swollen hypocotyl
2. Which of the following is an ethylene action inhibitor?
   (a) Aminooxyacetic acid
   (b) Aminoethoxyvinyl glycine
   (c) 1-Methylcyclopropane
   (d) Ammonium sulfate
3. Precursor for ethylene biosynthesis is:
   (a) Tryptophan
   (b) Methionine
   (c) Arginine
   (d) Ornithine
4. Which of the following *is not* a role of ethylene?
   (a) Lateral cell expansion
   (b) Fruit ripening
   (c) Promotion of senescence
   (d) Apical dominance
5. Which of the following mutants *does not* exhibit insensitivity to ethylene?
   (a) *etr1*
   (b) *ein2*
   (c) *ctr1*
   (d) *ers1*
6. Climacteric fruits include all of the following except:
   (a) Banana
   (b) Grapes
   (c) Mango
   (d) Tomato
7. Why is ethylene response said to be negatively regulated by its receptors?
   (a) Ethylene receptors cause inhibition of further ethylene biosynthesis in presence of ethylene
   (b) Ethylene receptors are active in absence of ethylene and negatively regulate ethylene response genes
   (c) Ethylene receptors cause inhibition of expression of ethylene response genes as soon as they bind ethylene
   (d) None of the above
8. Ethylene receptor that lacks a histidine kinase domain is:
   (a) ETR1
   (b) ETR2
   (c) ERS1
   (d) EIN4

**Answers**

1. c   2. c   3. b   4. d   5. d   6. b
7. b   8. c

## Suggested Further Readings

Bakshi A, Shemansky JM, Chang C, Binder BM (2015) History of research on the plant hormone ethylene. J Plant Growth Regul 34:809–827

Wen CK (2015) Ethylene in plants. Springer Science and Business Media, Dordrecht

# Brassinosteroids

## 20

Satish C Bhatla

The discovery of brassinosteroids can be traced back to 1941 from Mitchell and Whitehead of the United States Department of Agriculture (USDA, Beltsville, MD) who reported that pollen extracts often have growth-promoting properties on other plant tissues. In 1970, Mitchell et al. reported that the crude extract of pollen from *Brassica napus* (rape) contained "**brassins**" which promoted rapid elongation of internodes in *Phaseolus vulgaris*. This elongation response was distinct from GA-mediated stem elongation. Subsequently, their work leads to the isolation and identification of brassinolide (BL) as the first steroidal plant growth regulator. Another steroid hormone called castasterone (SC) was isolated from insect galls of chestnut (*Castanea* sp.). Since then, a number of related steroidal compounds have been isolated which are collectively called as **brassinosteroids (BRs)**. Structurally, BRs are C-27, C-28, and C-29 steroids with different functional groups on A and B rings and on the side chain. BRs occur in algae, ferns, gymnosperms, and angiosperms. They have not been detected in microorganisms. Brassinolide is a C-28 brassinosteroid and exhibits highest activity among all BRs so far known.

## 20.1 Biosynthesis and Homeostasis

BRs are synthesized from campesterol (a plant sterol) through formation of many intermediates. Castasterone is the immediate precursor for the formation of brassinolide (BL), which is the most active BR (Fig. 20.1). Bioactive BR levels (homeostasis) are regulated by a variety of catabolic reactions, including hydroxylation, oxidation, sulfonation, epimerization, and conjugation to lipids or glucose. Only a few enzymes responsible for BR catabolism are so far reported. Level of active BR in plant tissue is also regulated by negative feedback mechanism, whereby BR concentration in the tissue above a certain level results in reduction in BR biosynthesis by the downregulation of BR biosynthesis genes and upregulation of genes for BR catabolism.

**Fig. 20.1** Biosynthetic pathway of brassinosteroids (BRs). Campesterol and campestanol are the inactive precursors of the biologically active compounds castasterone and brassinolide. Brassinolide is the most active naturally occurring BR and is metabolized to inactive 26-hydroxybrassinolide by a cytochrome P450 enzyme encoded by the gene *BAS1*. The biological activity of BRs is dependent on the presence of hydroxyl groups at carbons 2 and 3 of ring A and positions 22 and 23 of the side chain and a seven-membered lactone B-ring

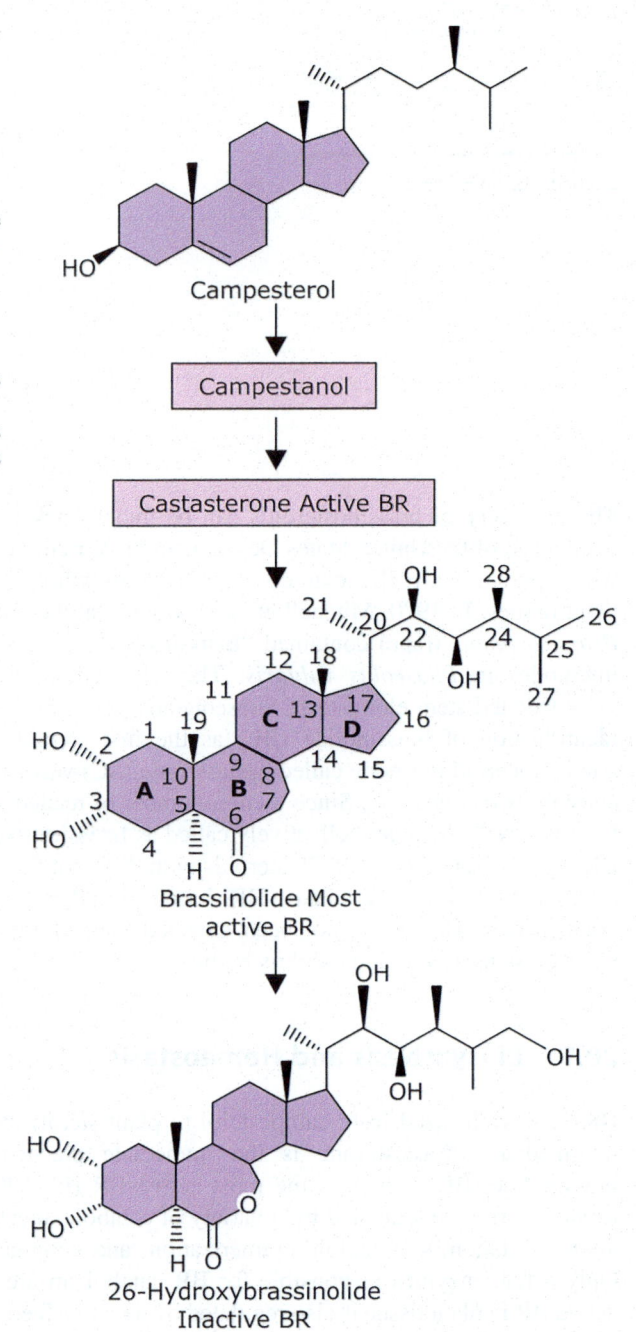

## 20.2 Functions of Brassinosteroids

### 20.2.1 Regulation of Photomorphogenesis

BRs prevent photomorphogenesis in etiolated seedlings. This has been extensively worked out using BR mutants with impaired BR synthesis. The role of BRs in repressing photomorphogenesis has further been observed to be operative through different mechanisms as compared to other regulatory mechanisms (e.g., those mediated by COP). Like GA, BRs also suppress photomorphogenesis in dark.

### 20.2.2 Unrolling and Bending of Grass Leaves

Brassinolide (BL) and other BRs induce a dose-dependent swelling of the adaxial cells of the joint between the rice seedling's leaf blade and sheathing base. This bioassay is very sensitive and is used to determine structure-activity relationship of BR by measuring the angle of lamina bending in response to a defined concentration of BR applied over a period of time (Fig. 20.2).

### 20.2.3 Other Effects

BRs also induce stem elongation, pollen tube growth, proton pump activation, reorientation of cellulose microfibrils, and xylogenesis. They are required for promotion of GA-mediated cell elongation. GA-induced degradation of DELLA repressor protein also enhances BR response. The signal transduction pathways of BRs and GA also interact with phytochrome pathway by regulating the activity of PIFs (phytochrome-interacting factors).

## 20.3 Brassinosteroid Mutants

### 20.3.1 Mutants with Impaired Brassinolide Synthesis (Brassinolide-Sensitive Mutants)

Genetic analysis of de-etiolated mutants grown in dark has led to identification of DE-ETIOLATED2 (DET2) gene which encodes a brassinosteroid biosynthetic gene. *det2* are "loss of function" mutants with low levels of BR and de-etiolated appearance of seedlings even in dark growth conditions. DET2 encodes an enzyme associated with the biosynthesis of brassinolide. The endogenous brassinolide levels in *det2* mutants are less than 10% of that found in wild plants of *Arabidopsis*. The phenotype of dark-grown det2 mutants can be corrected by exogenous application of brassinolide for subsequent photomorphogenic response. Similar to *det2* mutant, some more mutants have also been isolated whose phenotype can be corrected for normal growth by exogenous brassinolide application. The corresponding genes in

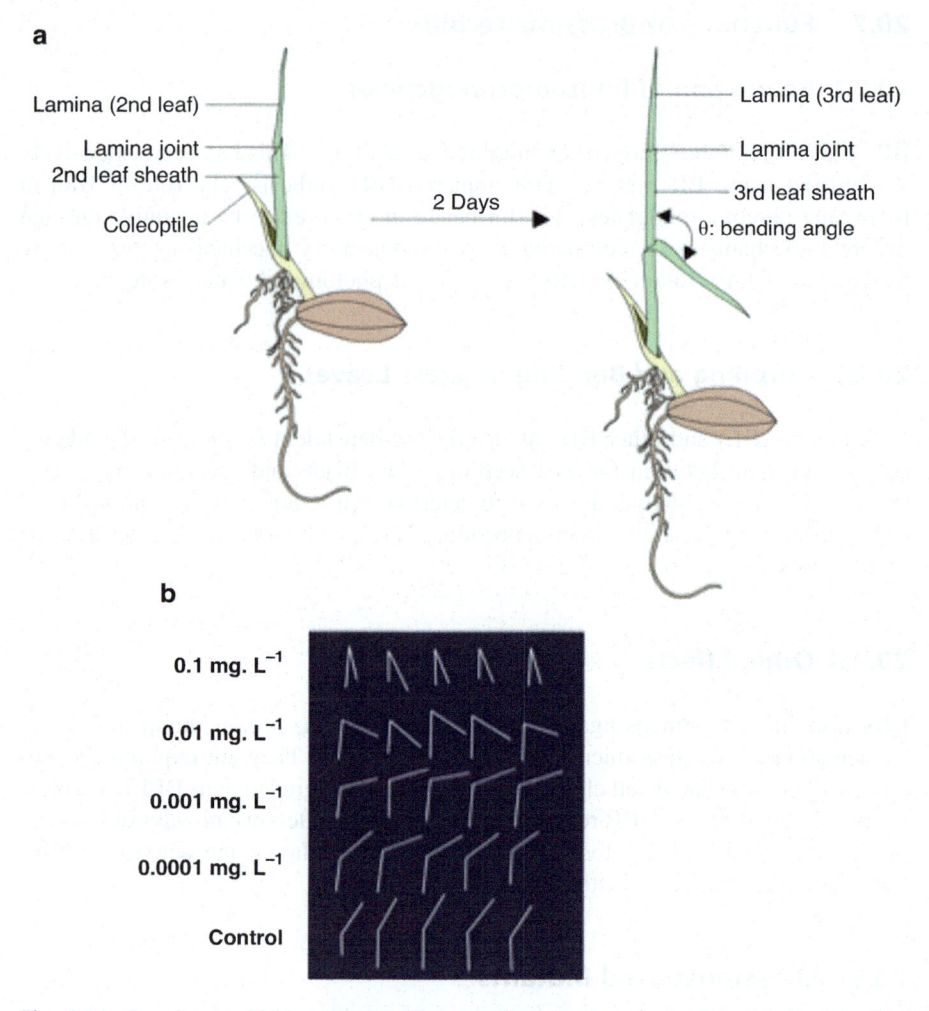

**Fig. 20.2** Brassinosteroid bioassay; (**a**) Rice seedling showing the lamina and sheathing leaf sheath. Brassinosteroids to be assayed are applied at the leaf sheath-lamina joint, and the angle of lamina is measured after 2 days of growth. (**b**) Effect of increasing BR concentration on the angle of lamina bending

all these mutants are involved in brassinosteroid biosynthesis, and these mutations have noteworthy effects on the phenotype of light-grown plants and de-etiolation of dark-grown seedlings. The light-grown mutant plants are dwarfed, have short internodes, possess dark green curved leaves, and exhibit reduced apical dominance and fertility. All these observations indicate the role of BR in the inhibition of photomorphogenesis in dark and plant development in light. Dwarf mutants with altered brassinolide synthesis have also been reported from tomato and pea. Like in *Arabidopsis*, the mutants reported from tomato and pea also have aberrations in brassinolide biosynthesis.

## 20.3.2 Brassinosteroid-Insensitive Mutants (bri)

Another class of mutants were identified in *Arabidopsis* on the basis of their insensitivity to exogenously applied brassinolide. These mutants called as *bri* mutants have a phenotype similar to *det2* mutants, but their growth cannot be normalized by exogenous brassinolide application. The analysis of these mutants enabled identification of proteins associated with BR signal transduction. In these mutants (*bri1*), brassinolide accumulation is higher than in wild-type plants, and the expression of CPD (an enzyme involved in brassinolide biosynthesis) is also increased. This indicates that brassinolide represses its own biosynthesis by feedback mechanism, thereby modulating BR homeostasis. Another class of *Arabidopsis* mutants—*bin2* (brassinosteroid insensitive 2)—are impaired in the expression of a serine/threonine kinase that is a negative regulator of BR signaling.

## 20.4 Brassinosteroid Signaling Mechanism

BR receptor (BRASSINOSTEROID INSENSITIVE 1; BRI1) on the plasma membrane binds with BR at its extracellular domain. The receptor contains leucine-rich repeats (LRRs) consisting of a series of 20–29-amino acid residues rich in leucines in the extracellular domain. Regions of LRRs are thought to function in protein-protein interactions. It is, however, a 70-amino acid "island domain" interrupting the LRRs on BRI1 receptor, which most likely acts as the site for BR binding on the plasma membrane-associated receptor. BR signaling follows the following steps (Fig. 20.3):

1. Inactive state (homodimer): Inactive BR receptor BRI1 exists as a homodimer on the PM, and in this situation, it coexists with BRI1 KINASE INHIBITOR (BKI1).
2. Phosphorylation: Binding of BR to the receptor (BRI1) induces phosphorylation of the inhibitor (BAK1) at its cytosolic loop and its own phosphorylation (BRI1-P).
3. Loss of inhibitor (BAK1) activity: Phosphorylation of inhibitor makes it inactive, and it is lost from its normal association with the receptor on the PM.
4. Phosphorylation of other cytosolic kinases (BSKs): In the absence of inhibitor activity (BKI1), the phosphorylated receptor (BRI1-P) phosphorylates a small family of cytosolic kinases (BSKs 1, 2, and 3).
5. Cytosolic kinases (BSKs) modulate the activity of phosphatase (BSU1) and another kinase (BIN 2).
6. Phosphatase (BSU1) and kinase (BIN2) regulate the phosphorylation status of transcription factors—BZR1 and BES1 in the cytosol and nucleus.
7. Phosphorylation inhibits the activity of transcription factors BZR1 and BES1 and modulates gene expression by enhancing the rate of their degradation and reduction in their DNA-binding ability.

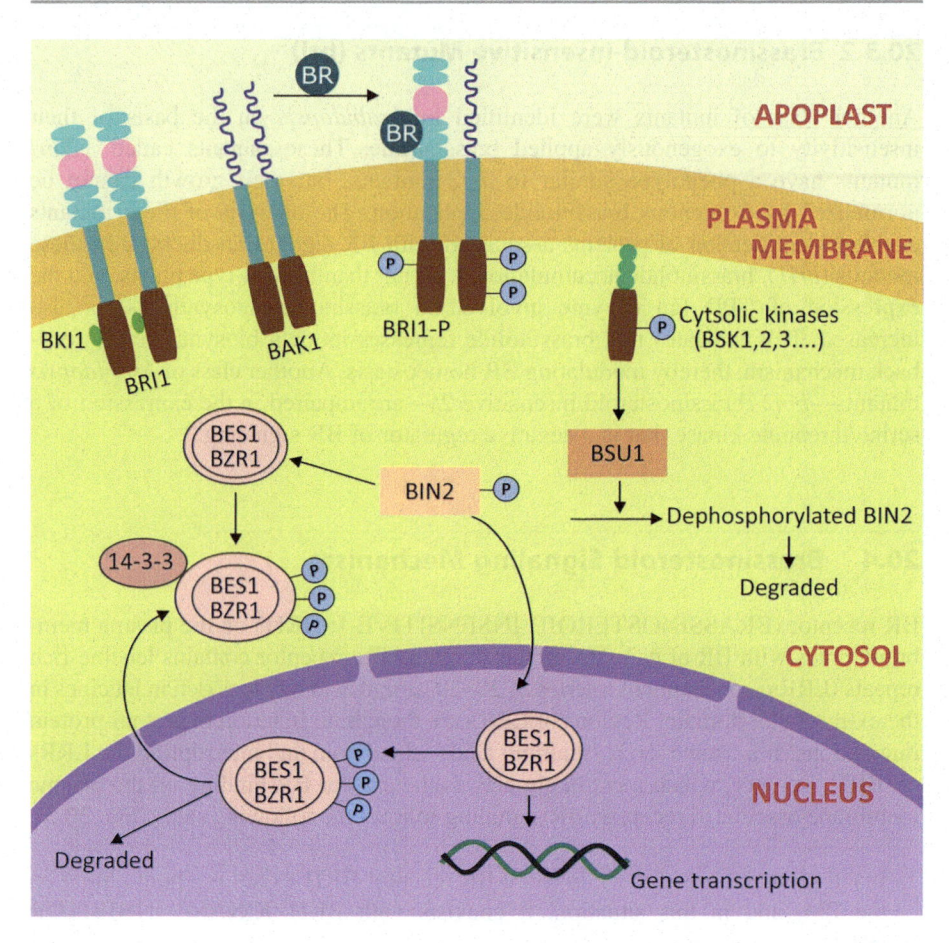

**Fig. 20.3** Signal transduction accompanying the action of brassinosteroids

8. BR perception leads to inhibition of BIN2 kinase and stimulation of BES1 (transcription factors), leading to accumulation of unphosphorylated BZR1 and BES1 in the nucleus and causing the regulation of BR-responsive genes.

**Noteworthy Features of Brassinosteroid Receptor BRI1 Are**
1. In tomato, BR receptor has the ability to bind two different signals—brassinosteroids and **systemin**. Systemin is an 18-amino acid peptide produced by tomato plants and other Solanaceae members in response to insect attack, which is capable of inducing defense-related genes.
2. BRI1 can probably signal from both plasma membrane and **endosomes**. Endosomes are intracellular vesicles created by endocytosis. The significance of using endosomes in addition to PM is that it increases the membrane area

available for signaling interaction. Migration of endosomes by cytoplasmic streaming also enables their proximity to the nucleus to evoke a response.
3. Brassinosteroid receptor inhibitors (BAK1) is a member of kinases called SOMATIC EMBRYOGENESIS RECEPTOR-LIKE KINASEs; SERKs). Four SERKs have so far been reported which play significant roles in microspore development (SERK1, SERK2), defense signaling (SERK3), brassinosteroid signaling (SERK1, SERK3), and regulation of cell death (SERK3, SERK4). Functions of another kinase (SERK5) are not yet known.

## Summary

- Brassinosteroids (BRs) are C-27, C-28, and C-29 steroids with different functional groups on A and B rings and on the side chain. BRs occur in algae, ferns, gymnosperms, and angiosperms. Brassinolide is a C-28 brassinosteroid and exhibits highest activity among all BRs so far known.
- Bioactive BR levels (homeostasis) are regulated by a variety of catabolic reactions, including hydroxylation, oxidation, sulphonation, epimerization, and conjugation to lipids or glucose. Level of active BR in plant tissue is also regulated by negative feedback mechanism, whereby BR concentration in the tissue above a certain level results in reduction in BR biosynthesis by the downregulation of BR biosynthesis genes and upregulation of genes for BR catabolism.
- BRs prevent photomorphogenesis in etiolated seedlings. Brassinolide (BL) and other BRs induce a dose-dependent swelling of the adaxial cells of the joint between the rice seedling's leaf blade and sheathing base.
- BR receptor (BRASSINOSTEROID INSENSITIVE 1; BRI1) on the plasma membrane binds with BR at its extracellular domain. The receptor contains leucine-rich repeats (LRRs) consisting of a series of 20–29-amino acid residues rich in leucines in the extracellular domain. Regions of LRRs are thought to function in protein-protein interactions.
- BR perception leads to inhibition of BIN2 kinase and stimulation of BES1 (transcription factors), leading to accumulation of unphosphorylated BZR1 and BES1 in the nucleus and causing the regulation of BR-responsive genes.
- In tomato, BR receptor has the ability to bind two different signals—brassinosteroids and systemin. BRI1 can probably signal from both plasma membrane and endosomes.

## Multiple-Choice Questions

1. Brassinosteroids were first reported in:
    (a) *Arabidopsis thaliana*
    (b) *Brassica napus*
    (c) *Brassica juncea*
    (d) *Phaseolus vulgaris*

2. Which of the following is true for brassinosteroids?
   (a) Stimulate photomorphogenesis in dark
   (b) Stimulate skotomorphogenesis in light
   (c) Prevent photomorphogenesis in dark
   (d) Both a and b
3. Brassinosteroid mutants—bri1 and det2—have similar phenotypes, but exogenous application of brassinolide:
   (a) Can normalize their growth
   (b) Can normalize the growth of only det2
   (c) Cannot normalize the growth
   (d) Can normalize the growth of only bri1
4. CPD enzyme is involved in:
   (a) Brassinosteroid degradation
   (b) Brassinosteroid phosphorylation
   (c) Brassinosteroid biosynthesis
   (d) None of the above
5. bin2 (brassinosteroid insensitive 2) mutants exhibit:
   (a) Overexpression of BR signaling
   (b) Impaired expression of phosphatase
   (c) Impaired expression of ser/thr kinase
   (d) Both b and c

**Answers**

1. b    2. c    3. b    4. c    5. c

## Suggested Further Readings

Clouse SD (2015) A history of brassinosteroid research from 1970 through 2005: thirty-five years of phytochemistry, physiology, genes and mutants. J Plant Growth Regul 34:828–844

Fariduddin Q, Yusuf M, Ahmad I, Ahmad A (2014) Brassinosteroids and their role in response of plants to abiotic stress. Biol Plant 58:9–17

# Jasmonic Acid

## 21

Satish C Bhatla

Jasmonic acid (JA) and its various derivatives are lipid-derived signaling molecules which participate in the regulation of a number of plant processes, including growth, reproductive development, photosynthesis, and responses to biotic and abiotic stress factors. As compared to auxins, ABA, cytokinins, GAs, and ethylene, discovery of JA and elucidation of its roles in plants have been made in the relatively recent past. JA was first isolated from the fungus *Lasiodiplodia theobromae* as a plant growth inhibitor. Among green plants, the free acid (−JA) was first detected and identified as a growth inhibitor from the pericarp of *Vicia faba* (broad bean). The immature fruits of *Vicia faba* contain a mixture of (−) JA and its stereoisomer (+)-7-iso-jasmonic acid (+7-iso-JA) in 65–35% ratio (Fig. 21.1). Hydroxy-jasmon ester of chrysanthemic acid in pyrethrins (a class of insecticides), isolated from *Chrysanthemum cinerariifolium* (*Pyrethrum*), was the first JA metabolite discovered from plants. Active jasmonates, most importantly the methyl esters of jasmonates (JAMe), were initially detected as odorants from the flowers of *Jasminum grandiflorum*.

## 21.1 Biosynthesis

The biochemical steps leading to JA biosynthesis in plants are partitioned into two organelles—the plastids and peroxisomes. The substrate for JA biosynthesis—α-linolenic acid (a $C_{18}$ polyunsaturated acid)—is released from the plastid membrane lipids by lipase action. Subsequent sequential action of three enzymes, namely, 1,3-lipoxygenase (LOX), allene oxide synthase (AOS), and allene oxide cyclase (AOC) on α-linolenic acid within the plastids, leads to the formation of oxo-phytodienoic acid (OPDA). OPDA transport across the plastids into the peroxisomes is facilitated by an ABC transporter. Within the peroxisomes, OPDA is converted to (+)-7-iso-jasmonic acid [(+)-7-iso-JA)] through reduction and β-oxidation steps. In the cytosol, (+)-7-iso-JA gets conjugated with isoleucine to form jasmonyl isoleucine (JA-Ile) which is considered the physiologically active form of JA (Fig. 21.1).

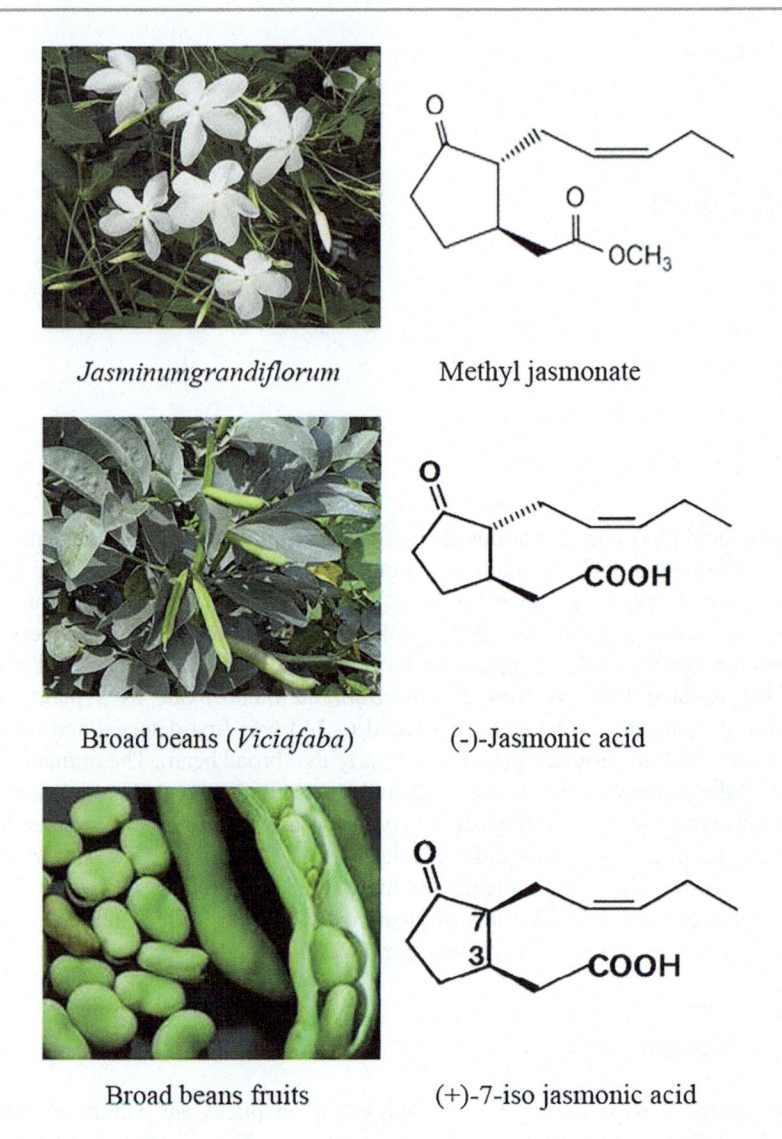

**Fig. 21.1** Naturally occurring jasmonic acid, its derivatives, and the plants from which they were first isolated

## 21.2 Metabolism and Homeostasis

JA modulates various aspects of plant development through a variety of its conjugated forms, namely, cis-jasmone, methyl jasmonate, jasmonyl-1-β-glucose, cucurbic acid, and some more. Conjugation of jasmonic acid with isoleucine and other amino acids is catalyzed by an enzyme encoded by JAR1 (JASMONATE

RESISTANT 1) locus in *Arabidopsis*. *jar1* mutants are resistant to exogenous JA in a root growth inhibition assay. In addition to the biosynthesis of JA isoleucine (JA-Ile) from (+)-7-iso-JA, JA is also the precursor for a volatile derivative, namely, methyl jasmonate, the product of a SAM-dependent carboxymethyl transferase, (−)-JA, and methyl (−) JA are the major naturally occurring jasmonates in plant tissues (Fig. 21.1). Their respective stereoisomers, (+)-7-iso-JA and methyl (+)-7-iso-JA, also exhibit significant biological activity, though not in all systems. Cis-jasmone is another volatile compound synthesized by the decarboxylation of JA. Jasmonyl-1-β-glucose has been reported from the cell cultures of tobacco and tomato where it is formed by the conjugation of the carboxyl group of JA with glucose. 12-hydroxy-(+)-7-iso-jasmonic acid, also called as tuberonic acid, has been reported from potato, where it exhibits tuber-inducing activity (Figs. 21.2 and 21.3).

JA levels in plant tissues are determined by: (1) substrate availability, (2) a feedback loop which affects the expression of JA biosynthesis genes, (3) cell- and tissue-specific distribution of enzymes associated with JA biosynthesis, and (4) post-translational regulation of JA.

*Evidence* includes the following: (1) Observation of low JA levels in transgenic lines overexpressing JA biosynthesis enzymes and their induction upon wounding suggests a requirement of release of α-linolenic acid as the substrate for JA biosynthesis. (2) Induction of genes involved in JA biosynthesis following JA application suggests a positive feedback loop. (3) Differential distribution of various JA biosynthesis enzymes in different tissues has also been observed. (4) A rapid accumulation (burst) of JA within few seconds of wounding suggests post-translational regulation of JA biosynthesis.

## 21.3 Physiological and Developmental Roles

### 21.3.1 Trichome Formation

Biosynthesis of alkaloids, flavones, terpenoids, and defense proteins in trichomes correlates with their active role in plant defense mechanisms. Genetic analysis has shown a strong involvement of JA in trichome formation as well as monoterpene synthesis within them. On similar lines, another phenomenon of agricultural importance is JA-mediated formation of cotton fibers, which represent single-cell trichomes from seed surface.

### 21.3.2 Reproductive Functions

JA has been detected as an odorant in the flowers of some plants, and thus it plays a role in attracting pollinating insects. Characterization of some *Arabidopsis* mutants has demonstrated a role of JA in flower development with reference to the male sterility trait. Sex determination in maize also requires JA through modulation of 13-LOX activity.

**Fig. 21.2** Pathway for jasmonic acid biosynthesis in plants

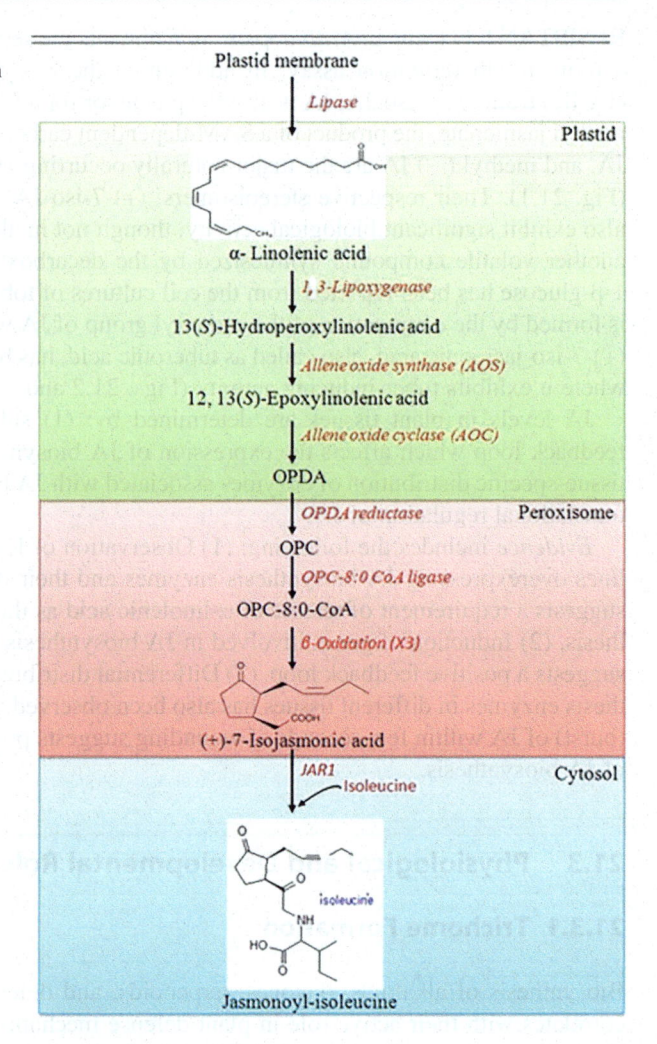

### 21.3.3 Induction of Secondary Metabolites Production

Subjecting plant cell cultures to fungal elicitors enhances JA content which triggers the formation of a number of alkaloids in several plant species. It includes SA-induced biosynthesis of nicotine, vinblastine, artemisinin, glucosinolates, anthocyanins, and benzylisoquinoline alkaloids. JA is also responsible for the rapid induction of some transcription factors responsible for the expression of several genes for the production of secondary compounds such as benzylisoquinoline, nicotine, and morphine.

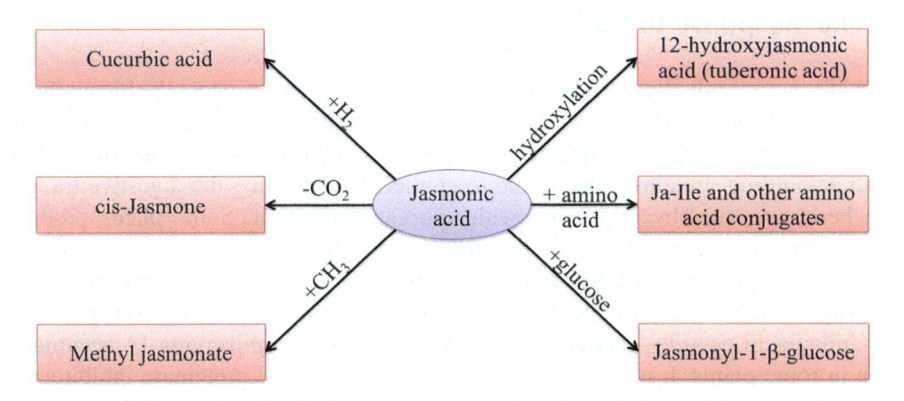

**Fig. 21.3** Jasmonic acid metabolism in plants through modification of various groups associated with this molecule

## 21.3.4 Role as a Senescing Promoting Factor

JA-specific proteins, called JIPs (JA-inducible proteins) of varying molecular masses, accumulate in senescing barley leaf segments subjected to JA treatment. This accumulation is accompanied by enhanced degradation of several housekeeping proteins, such as Rubisco (small and large subunits). A thionin was one of the first JIPs identified in senescing barley leaf segments which is induced in response to JA treatment. Its overexpression in tobacco leads to repression of leaf proteins. The function of JIPs is, however, still not fully resolved. 13-LOX, a 90–95 kDa JIP localized in chloroplasts, provided first evidence for a link between JA-induced gene product (JIP) and activation of JA biosynthesis. Other JA-inducible proteins associated with senescence include chlorophyllase (which catalyzes chlorophyll breakdown) and proteins involved in chlorophyll stability. Senescence-promoting activity of JA also leads to repression of mRNA and protein levels of Rubisco activase in a COI1-dependent manner.

## 21.3.5 JA and Photomodulation of Plant Development

Seed development, photomorphogenesis, shade avoidance, and skotomorphogenesis are some of the light-mediated developmental responses in plants and are governed by the actions of GA, ABA, auxin, and CK. It is now known that light regulates JA biosynthesis. JA-related genes mediate hypocotyl growth in far red light. Furthermore, light dependence of JA biosynthesis and JA-dependent processes is now evident during some plant developmental processes. This includes extrafloral nectar formation and responses to necrotrophic pathogens and herbivores.

## 21.3.6 Response to Herbivores

In addition to production of secondary metabolites, JA also triggers the biosynthesis of defense proteins which interfere with herbivore digestive system. This includes the synthesis of **α-amylase inhibitors** in legumes and production of **lectins** (carbohydrate binding proteins) which bind to epithelial cells lining the digestive tract of the herbivore, thereby interfering with nutrient absorption. Some plants produce **cysteine proteases** which disrupt the membrane of the gut epithelium of insects. A number of **proteinase inhibitors** produced in some legumes and tomato block the activity of herbivore proteolytic enzymes such as trypsin and chymotrypsin. An 18-amino acid peptide—**systemin**—is biosynthesized accompanying JA accumulation in some plants. It is believed to induce the synthesis of proteinase inhibitors.

## 21.3.7 Mycorrhizal Interactions and Modulation

An increase in JA level in wounded tissue correlates with increased mycorrhizal association in roots. JA also appears to play a role in plant resistance to nonpathogenic plant growth-promoting rhizobacteria and fungi. This is referred as **induced systemic resistance**.

## 21.4 JA-Induced Gene Expression

In the late 1980s to the early 1990s, various researchers provided evidence for JA-induced changes in gene expression. The JA-induced proteins (JIPs) of varying molecular masses have been reported to accumulate in plants following long-term JA treatment. Evidence has also been provided for induction of vegetative storage proteins (VSPs) in response to wounding and JA treatment. This was followed by the observation of alkaloid formation upon elicitation of cell cultures with yeast elicitor which follows a preceding rise in JA levels in the tissue. These observations paved way for numerous reports on JA-responsive pathways and genes and also gave directions for research in JA-mediated plant-herbivore interactions and plant-to-plant communication via volatiles, such as sesquiterpenes, leaf aldehydes, and leaf alcohols.

## 21.5 JA-Mediated Signaling

Similar to auxin, JA acts through a conserved ubiquitin ligase-based signaling mechanism (Fig. 21.4). JA receptors are F-box proteins, COI1, a component of SCF-COL E3 ubiquitin ligase which targets **JAZ (JASMONATE ZIM DOMAIN)** family of transcriptional repressors for degradation, thereby activating the expression of JA-sensitive genes. Although unconjugated JA is active as a hormone, most JA responses require its conjugation with isoleucine (JA-Ile) by the action of the

## 21.5 JA-Mediated Signaling

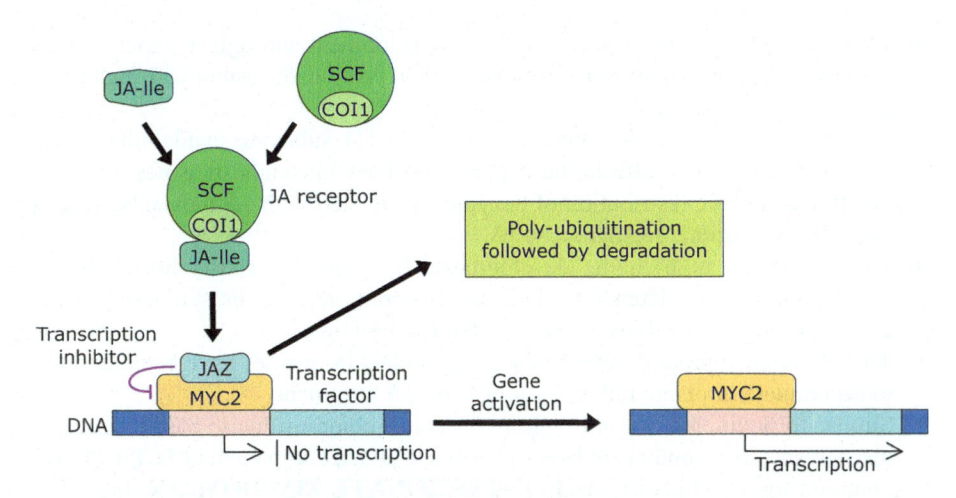

**Fig. 21.4** Jasmonic acid signaling. Jasmonic acid needs to be first conjugated to an amino acid to bind to COI1 (JA receptor) as part of a SCF$^{COI1}$ protein complex. This complex targets JAZ (transcription repressor) leading to the degradation of this protein via polyubiquitination followed by degradation in a proteasome. MYC2 (transcription factor) initiates transcription of JA-dependent genes, including those for defense

enzyme called as **jasmonic acid resistance (JAR) protein**. When endogenous levels of JA are low, the **JAZ** protein family repress the expression of JA-responsive genes. JAZ repressors function by binding to the **transcription factor MYC2** which is a major switch in the activation of JA-inducible genes. The sequential action during JA perception is as follows: (1) JA-Ile complexes with COI1 (as SCF$^{COI1}$); (2) COI1-JA-Ile complex binds with JAZ (repressor protein) and releases it from the MYC2 (transcription factor). JAZ then undergoes polyubiquitination and subsequent degradation; (3) MYC2 (transcription factor), which does not any more have binding with the transcription factor (JAZ), is activated, and it transcribes JA-inducible genes.

## Summary

- Jasmonic acid (JA) and its various derivatives are lipid-derived signaling molecules which participate in the regulation of a number of plant processes, including growth, reproductive development, photosynthesis, and responses to biotic and abiotic stress factors.
- The biochemical steps leading to JA biosynthesis in plants are partitioned in two organelles—the plastids and peroxisomes. The substrate for JA biosynthesis—α-linolenic acid (a $C_{18}$ polyunsaturated acid)—is released from the plastid membrane lipids by lipase action. In the cytosol, (+)-7-iso-JA gets conjugated with isoleucine to form jasmonyl isoleucine (JA-Ile) which is considered the physiologically active form of JA.

- JA modulates various aspects of plant development through a variety of its conjugated forms, namely, cis-jasmone, methyl jasmonate, jasmonyl-1–βglucose, cucurbic acid, and some more.
- JA levels in plant tissues are determined by (1) substrate availability, (2) a feedback loop which affects the expression of JA biosynthesis genes, (3) cell- and tissue-specific distribution of enzymes associated with JA biosynthesis, and (4) post-translational regulation of JA.
- JA is involved in trichome development, formation of cotton fibers, flower development with reference to male sterility trait, alkaloid biosynthesis, senescence promotion, and biosynthesis of defense proteins.
- The JA-induced proteins (JIPs) of varying molecular masses have been reported to accumulate in plants following long-term JA treatment.
- Similar to auxin, JA acts through a conserved ubiquitin ligase-based signaling mechanism. JA receptors are F-box proteins, COI1, a component of SCF-COL E3 ubiquitin ligase which targets **JAZ (JASMONATE ZIM DOMAIN)** family of transcriptional repressors for degradation, thereby activating the expression of JA-sensitive genes.

## Multiple-Choice Questions

1. Among green plants, the first jasmonic acid was isolated from:
    (a) *Chrysanthemum cinerariifolium*
    (b) *Jasminum grandiflorum*
    (c) *Vicia faba*
    (d) *Arabidopsis thaliana*
2. Organelle where JA biosynthesis occurs is plants are:
    (a) Plastids and mitochondria
    (b) Plastids and peroxisomes
    (c) Mitochondria and chloroplasts
    (d) Peroxisomes and mitochondria
3. JA performs which of the following roles in plants:
    (a) Senescing promoting factor
    (b) Photomodulation of plant development
    (c) Response to herbivores
    (d) All of the above
4. Which of the following is the receptor for JA-mediated signaling?
    (a) COI1 (coronatine insensitive 1)
    (b) JAZ (jasmonate zim domain)
    (c) JAR (jasmonate acid resistance)
    (d) MYC2

**Answers**

1. a    2. c    3. d    4. a

## Suggested Further Readings

Miersch O, Meyer A, Vorkefeld S, Sembdner G (1986) Occurrence of (+)-7-iso-jasmonic acid in *Vicia faba* L. and its biological activity. J Plant Growth Regul 5:91–100

Turner JG, Ellis C, Devoto A (2002) The jasmonate signal pathway. Plant Cell 14:S153–S164. https://doi.org/10.1105/tpc000679

Wasternack C (2015) How jasmonates earned their laurels: past and present. J Plant Growth Regul 34:761–794

# Recently Discovered Plant Growth Regulators

## 22

Satish C Bhatla

## 22.1 Salicylic Acid

Salicylic acid is a phenolic plant growth regulator known to regulate various aspects of plant growth and development. It also functions in various ways in modulating biotic and abiotic stress responses. Plants, such as willow tree (*Salix* sp.) and poplar (*Populus* sp.), have been used since the fourth century BC to relieve pain in the human body. But it was only in the nineteenth century that salicylic acid (SA) and related compounds such as methyl salicylate, saligenin, and their glycosides were isolated from the bark of willow tree and were found to be analgesic. Oil of wintergreen, extracted from the American plant *Gaultheria procumbens*, which was widely used as analgesic during the mid-nineteenth century, is also rich in methyl salicylate. SA was chemically synthesized in 1858 in Germany, and it replaced wintergreen oil as an analgesic. The sharp bitter taste and gastric irritation caused by SA, however, did not make it popular for its application as an analgesic. Subsequently, Bayer and Co. (a German pharmaceutical company) produced acetyl derivative of SA, i.e., acetyl salicylic acid, with the trade name Aspirin which became popular as an analgesic since then (Fig. 22.1). In the recent past, action of this phenolic compound, i.e., SA (chemical name: 2-hydroxybenzenecarboxylic acid), has also been discovered in various aspects of plant growth and development and acquisition of disease/wound resistance.

### 22.1.1 Biosynthesis

Biosynthesis of salicylic acid in plants occurs via two routes (Fig. 22.2):

1. **The *trans*-cinnamic acid** (tCA) pathway whereby tCA is synthesized from phenylalanine in a reaction catalyzed by phenylalanine ammonium lyase (PAL). tCA is then converted to benzoic acid which gets converted to salicylic

**Fig. 22.1** Chemical structures of salicylic acid and acetyl salicylic acid (aspirin) and picture of a weeping willow (*Salix babylonica*) tree whose bark is known to contain salicylic acid and its derivatives

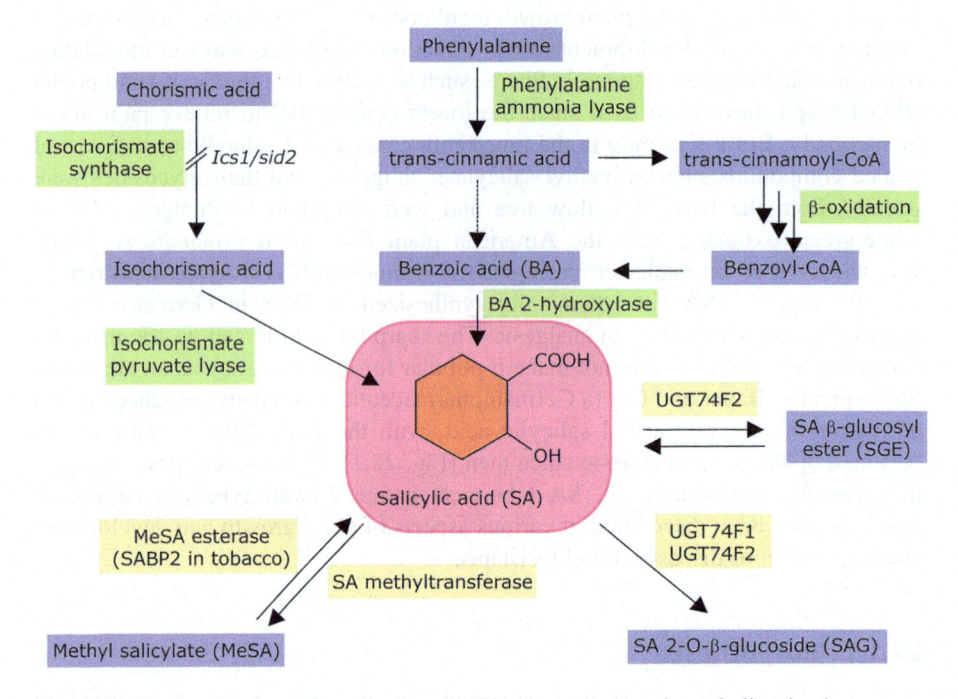

**Fig. 22.2** Proposed pathway for salicylic acid (SA) biosynthesis and metabolism in plants

acid through hydroxylation catalyzed by benzoic acid 2-hydroxylase. In support of this pathway of SA synthesis, it has been observed that when tobacco plants are inoculated with tobacco mosaic virus (TMV), PAL activity is suppressed, and this coincides with low levels of SA.

2. **The chorismate pathway** of SA biosynthesis occurs in two steps. The first step is catalyzed by isochorismate synthase (encoded by ICS1/SID2) leading to conversion of chorismic acid into isochorismic acid which subsequently gets converted to salicylic acid by the activity of isochorismate pyruvate lyase. The ICS1/SID2 mutant of *Arabidopsis* fails to accumulate SA under otherwise inductive conditions suggesting the operation of chorismate pathway of SA biosynthesis in this species. The information available so far suggests the operation of both the pathways of SA biosynthesis in plants though species specificity and treatment specificity may be evident for the operation of one or the other of the two pathways. SA has been reported to get converted to methyl salicylate (MeSA). Methylation of SA is catalyzed by a carboxyl methyl transferase. It can also be converted into a glucoside derivative called SA-2-*O*-β glucoside, SAG, or an ester (SA β-glucosyl ester, SGE).

## 22.1.2 Physiological Functions

### 22.1.2.1 Senescence

Addition of aspirin (acetyl salicylic acid) to a flower vase containing cut flowers is known to retard senescence of petals and thus enhance their longevity. This is due to the conversion of acetyl salicylic acid into SA in water and consequent lowering of the rate of ethylene biosynthesis in cut flowers. SA positively regulates leaf senescence. In *Arabidopsis* leaves SA content increases as the chlorophyll concentration begins to decline with the onset of senescence. Senescing leaves also exhibit an upregulation of many genes involved in SA biosynthesis. These observations suggest that SA plays a significant role not only in the onset of senescence but also in its progression.

### 22.1.2.2 Regulation of Thermogenesis

During thermogenesis, most of the electron flow in mitochondria gets diverted from cytochrome respiration pathway to cyanide-insensitive non-phosphorylating electron transport pathway which is unique to plant mitochondria. The energy released by electron flow through this alternative respiratory pathway is not conserved as chemical energy but is released as heat. Some plants are known to generate large amount of heat as a result of the above-stated alternative respiratory pathway. Thus, in voodoo lily (*Sauromatum* sp.), the part of the inflorescence becomes much warmer than the surrounding air causing the release of amines and other chemicals as vapors which serve as chemical attractants for pollinators. The cyanide-resistant oxygen uptake is catalyzed by the **alternative oxidase**. When electrons pass to the alternative pathway, there is no energy conservation site between ubiquinone and oxygen. Thus, the free energy which is normally conserved as ATP is lost as heat when electrons are shunted through this pathway. The functional usefulness of the activity of alternative oxidase in thermogenic flowers is evident with a rise in temperature of the upper appendix in voodoo lily to as much as 25 degrees over and above the ambient temperature. It has been observed that salicylic acid acts as a signal which initiates this thermogenic event in voodoo lily. The thermogenic-inducing principle was initially called **calorigen** and was identified as SA in the 1980s. SA application to the arum spadix can bring about thermogenecity accompanying an increase in endogenous SA levels as well (an example of positive feedback) (Box 22.1). Through this mechanism the inflorescence in voodoo lily can

---

**Box 22.1: Thermogenic Plants**

Thermogenicity (heat production) in plants, first described by Lamarck in 1778 for the genus *Arum*, is now known to occur in the male reproductive structures of cycads and in the flowers or inflorescence of some angiosperm species belonging to families *Annonaceae, Araceae, Aristolochiaceae, Cyclanthaceae, Nymphaeaceae* and *Palmae*. The heating is believed to be associated with a large increase in the cyanide-insensitive non-phosphorylating electron transport pathway unique to plant mitochondria. The increase in this so called alternative respiratory pathway is so dramatic that oxygen consumption in the inflorescence of *Arum* lilies at the peak of heat production is as high as that of a humming bird in flight. In addition to the activation of the alternative oxidase, thermogenicity involves activation of the glycolytic and Krebs cycle enzymes which provide substrates for this remarkable metabolic explosion. In one of the *Arum* lilies, *Sauromatum guttatum* (voodoo lily), the inflorescence develops from a large corm, and can reach 80 cm in height. Early on the day of anthesis, a large bract (spathe) which surrounds the central column of the inflorescence (spadix) unfolds to expose the upper part of the spadix known as the appendix. Soon thereafter, the appendix starts to generate heat, which facilitates volatilization of foul-smelling amines and indoles attractive to the insect pollinators. By early afternoon the temperature of the appendix can increase by 14 °C above ambient, but it returns to ambient in the evening. The second thermogenic episode in the lower spadix starts late at night and ends the following morning after the maximum temperature increases more than 10 °C. In 1937, Van Herk suggested that the burst of metabolic activity in the appendix of the voodoo lily is triggered by "calorigen", a water-soluble substance produced in then male flower primordia located just below the appendix. Van Herk believed that calorigen begins to enter the appendix on the day preceding the day of anthesis. At that time attempts to isolate and characterize calorigen were not successful and hence his ideas met the scepticism. Mass spectroscopic analysis of the purified calorigen have indicated the presence of salicylic acid.

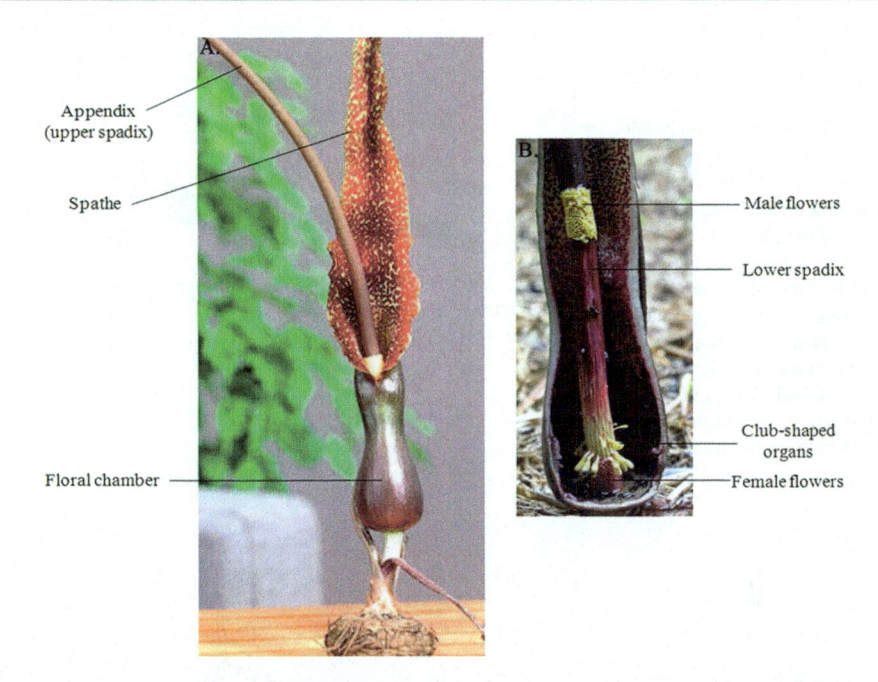

**Fig. 22.3** (a) *Sauromatum* sp. (Voodoo lily) inflorescence. (b) A cut-open inflorescence

mimic the scent of rotting flesh and is used to deceive insects that lay their eggs (Fig. 22.3). Likewise, skunk cabbage (*Symplocarpus foetidus*) often begins its floral development while still covered with snow. But as a result of generation of large quantities of heat, it melts the snow cover and exposes its flowers. Thus, in this case too, thermogenic respiration appears to be a metabolic event taking place rarely to help plants survive stress conditions.

### 22.1.2.3 Systemic Disease Resistance

SA is an important participant in systemic acquired resistance (SAR) to pathogenic infections. Initial experiments demonstrated that treatment of tobacco leaves (susceptible to TMV) with SA induces the synthesis of pathogenesis-related (PR) proteins and resistance to TMV infection. It is now well established that SA induces SAR in a variety of pathogen-infected plants. Furthermore, an increase in the endogenous SA concentration in TMV-resistant plants accompanies an enhanced expression of genes responsible for encoding PR proteins both in inoculated and un-inoculated leaves. Tobacco plants susceptible to TMV do not show an increase in SA levels or PR protein gene expression when infected with the virus, indicating that SA plays a key role in signal transduction events leading to the development of SAR (Fig. 22.4).

**Fig. 22.4** In systemic acquired resistance (SAR), a hypersensitive response to invasion by a pathogen results in the generation of a signaling molecule that moves from the infected leaf to other parts of the plant via the phloem and triggers the expression of defense-related resistance in the remote parts of the plant. SAR requires production of the signaling molecule salicylic acid, although this is not the signal that moves in the phloem

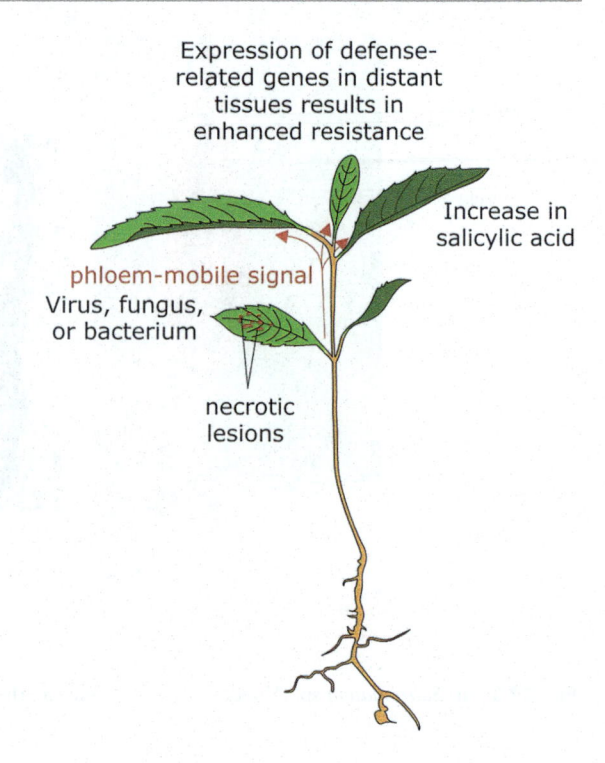

**System Acquired Resistance (SAR)**

### 22.1.2.4 Induction of Flowering

SA has been observed to induce flowering in duckweeds—*Lemna gibba*, *Spirodela polyrhiza*, and *Wolffia microscopica*—when these long-day plants are grown under noninductive short-day photoperiods.

## 22.2 Nitric Oxide

Nitric oxide (NO) is an established bioactive signaling molecule which was first described in mammals and was shown to be involved in several physiological processes, such as relaxation of smooth muscles, apoptosis, immune regulation, and neural communication. Although NO is a noxious chemical in the atmosphere, in the human body, it brings about great benefits when available in small and controlled doses. The great significance of the physiological roles of NO in mammals was recognized in 1998 with the award of Nobel Prize for Physiology and Medicine to three US scientists for their pioneering work on its role as a signaling molecule in the relaxation of smooth muscles (Box 22.2). In plants, NO has emerged as an important signaling molecule with diverse physiological functions in growth and development, starting from seed germination to flowering,

## Box 22.2: 1998 Nobel Prize for Physiology and Medicine to Three US Scientists

**Prof. Murad (Univ. of Texas)**
Identification of NO as a signaling molecule in cells

**Prof. Ignarro (UCLA School of Medicine, California)**
Role of nitrates in vasodilation through release of NO

**Prof. Furchgott (State Univ. of New York)**
Identified nitric oxide as a biological agent for relaxation of vascular smooth muscles

fruit ripening, and also during senescence of organs. Under environmental stress conditions caused by various biotic and abiotic stress factors, different plant species and organs exhibit enhanced NO generation. In plant cells, NO can induce both harmful and beneficial effects depending on its local concentration, translocation, rate of biosynthesis, and ability to get removed as various reactive nitrogen species (RNS).

### 22.2.1 Physicochemical Properties of NO

NO is a highly diffusible, gaseous free radical which is soluble both in hydrophilic and hydrophobic phases. Thus, it easily migrates both in the hydrophilic regions of the cell, such as in the cytoplasm, and can also freely diffuse across the lipid phase of the membranes. In biological systems, NO has a half-life of about 3–5 s. It is a very reactive species in the presence of atmospheric oxygen and readily reacts with superoxide anions ($O_2^{-\cdot}$) resulting in the formation of peroxynitrite ions ($ONOO^-$). Peroxynitrite ions are unstable at the physiological pH and consequently react with thiol groups of proteins and polyunsaturated radicals of fatty acids, thereby damaging the cell structure. NO radical ($NO^\cdot$) also readily reacts with transition metals,

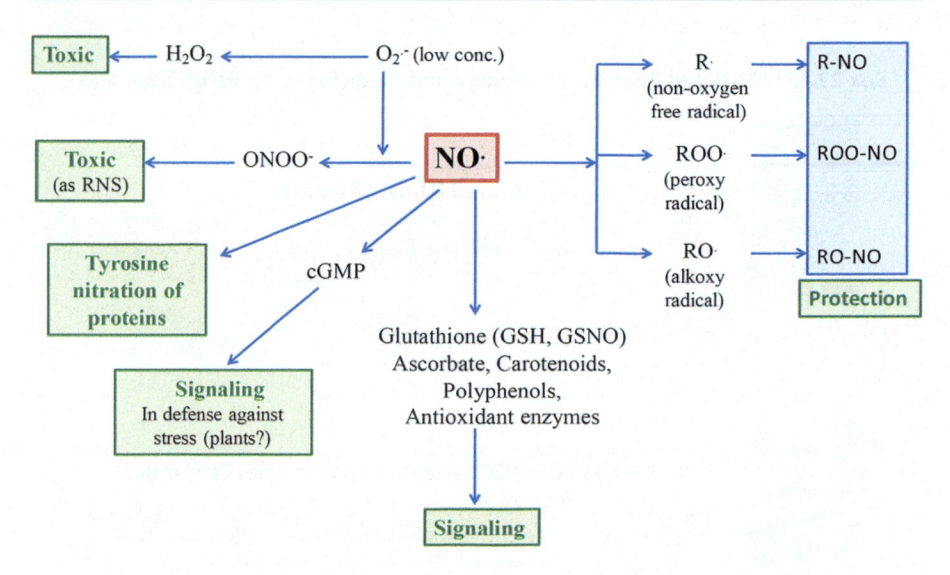

**Fig. 22.5** Is nitric oxide toxic or protective for cells?

such as heme iron and iron-sulfur centers of proteins. It also leads to reversible nitrosylation of sulfhydryl groups in proteins. Thus, S-nitrosylation/denitrosylation of proteins is responsible for modulating the biological activities of these biomolecules. NO also readily complexes with $Fe^{2+}$ and low molecular weight thiols. It is known to interact with plant hemoglobins under aerobic conditions resulting in the formation of $NO_3$. Thus, depending on its endogenous concentration in the cells and its ability to react with various biomolecules, NO can be toxic or protective for the cells (Fig. 22.5).

### 22.2.2 NO Biosynthesis in Plants

There are three probable routes of NO biosynthesis in plants: (1) L-arginine-dependent pathway, (2) nitrite-dependent pathway, and (3) nonenzymatic NO production (Fig. 22.6). Differential activity of these three probable NO biosynthesis routes is collectively responsible for NO generation in plant cells. The L-arginine-dependent pathway is prevalent in plastids and mitochondria. In terms of substrate and product formation, this pathway of NO production in plants is presumed to be catalyzed by the action of putative nitric oxide synthase (NOS) on L-arginine, resulting in the formation of L-citrulline and NO. There is, however, enough controversy about the nature of NOS activity in plants. The putative NOS in plants, though functionally similar in terms of its substrate requirement (arginine), has been found to be a structurally different protein as compared to the well-characterized NOS in animals. The AtNOS1 (putative NOS in plants) is a mitochondrial protein which bears a

## 22.2 Nitric Oxide

**L-arginine dependent pathway**

L-arginine + NAD(P)H + O$_2$

↓ Putative nitric oxide synthase (AtNOS) characterized as GTPase

L-citrulline + NAD(P)$^+$ + H$_2$O + NO

Plastids, Mitochondria

**Nitrite dependent pathway**

NO$_2^-$ + e$^-$ + 2H$^+$

↓ Nitrate reductase /Xanthine oxidase dehydrogenase

NO + H$_2$O

Chloroplasts

**Non-enzymatic NO production**

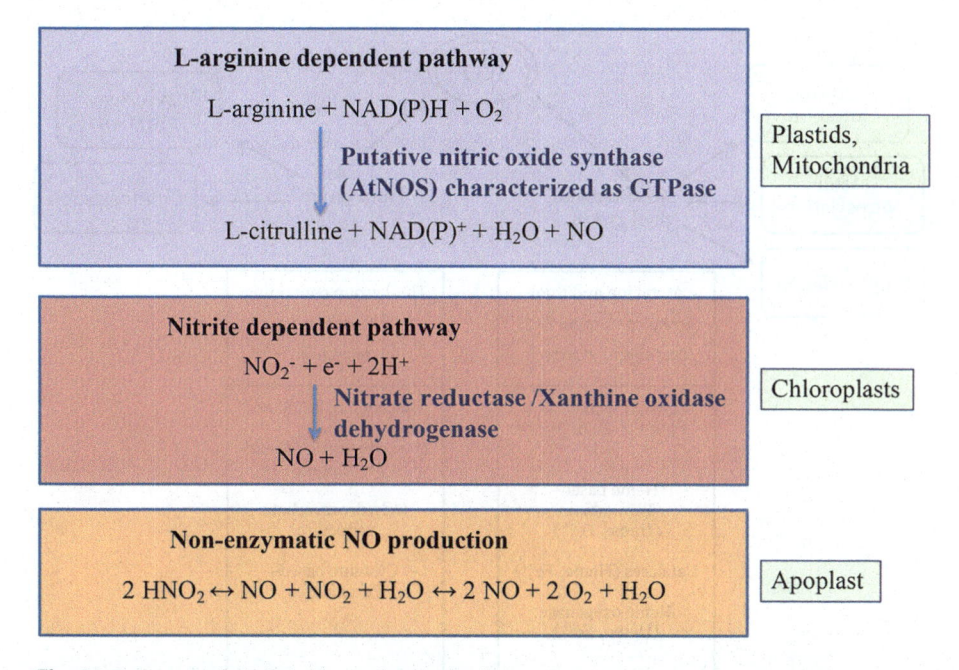

2 HNO$_2$ ↔ NO + NO$_2$ + H$_2$O ↔ 2 NO + 2 O$_2$ + H$_2$O

Apoplast

**Fig. 22.6** Routes of nitric oxide production in plants

centrally positioned GTP-binding domain. NO production through nitrite-dependent pathway is evident in chloroplasts, whereas nonenzymatic NO production is evident in the apoplast. Additionally, NO is also produced in the peroxisomes and cytoplasm in plant cells (Fig. 22.6).

### 22.2.3 NO as a Signaling Molecule and Its Effect on Gene Expression

Like in most animals, NO exerts its signaling function through cGMP-mediated pathways leading to various post-translational modifications of proteins, such as S-nitrosylation/denitrosylation. It also brings about its effect through calcium mobilization by modulating the activity of intracellular calcium channels leading to elevation of free cytosolic calcium levels in plants. NO treatment activates protein kinases. Thus, it brings about its cascade of signaling roles through more than one biochemical routes. In plant cells exposed to varied biotic or abiotic stress conditions, NO modulates the activity of various ROS-scavenging enzymes through their tyrosine nitration, metal nitrosylation, or S-nitrosylation thereby facilitating homeostatic regulation of enhanced ROS levels generated in plant cells under stress conditions (Fig. 22.7). There is enough evidence that NO affects gene expression in plants. Additionally, protein products of S-nitrosylation or products of NO activity, for example, histidine and cysteine units of various proteins, may also directly alter

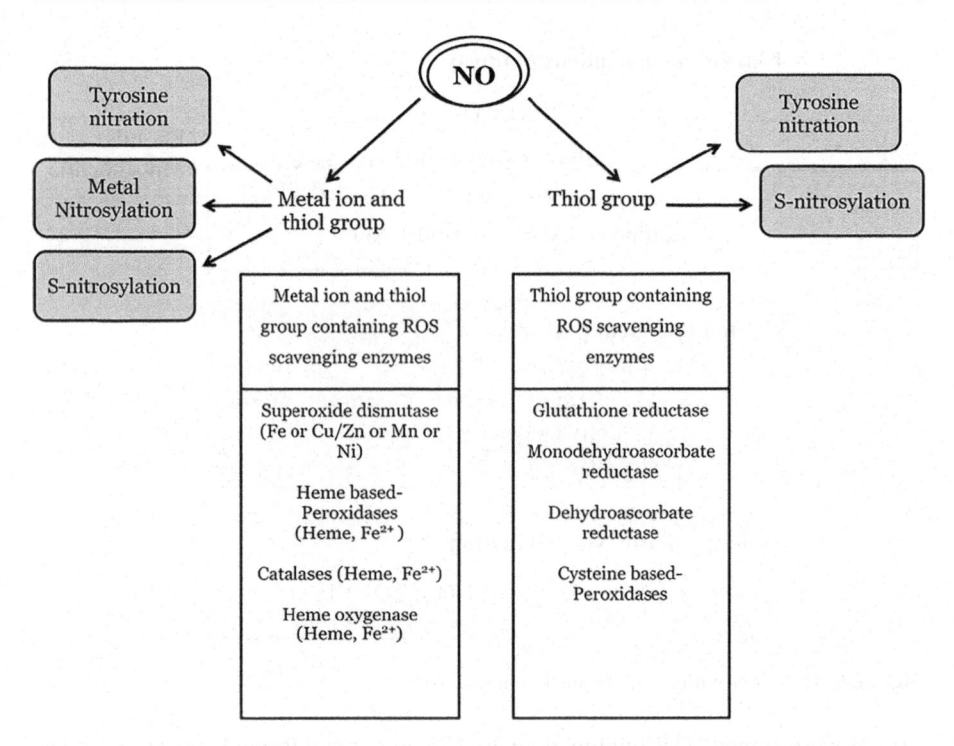

**Fig. 22.7** NO modulated post-translational modifications of ROS-scavenging enzymes

their transcript profile. They may diffuse to the nucleus and modify transcription factors. Indirectly, NO may affect transcription process through the activation of signaling cascade, such as synthesis of cyclic GMP and calcium channel proteins.

### 22.2.4 Physiological Functions of NO in Plants

#### 22.2.4.1 NO and Plant Growth and Development

Under normal growth conditions, plants are known to emit NO. In tomato plants, high levels of NO (40–80 pphm) have been reported to inhibit growth, whereas in lettuce low levels of NO (0–20 pphm) enhance growth. Excessive NO from the atmosphere generated through industrial pollution can have detrimental effects on photosynthesis through reversible suppression of electron transport and ATP synthesis in chloroplasts. Application of NO through NO donors inhibits hypocotyl growth, stimulates de-etiolation process, and increases chlorophyll content in potato, lettuce, and *Arabidopsis*. The positive effect of NO on chlorophyll accumulation reflects its effect on iron availability. In wild maize plants, NO treatment has in fact been shown to inhibit chlorosis induced by iron deficiency. Iron-deficient *yellow stripe* mutants of maize exhibit chlorophyll accumulation by NO application (Fig. 22.8). In pea leaves, NO donor application inhibits ethylene biosynthesis

## 22.2 Nitric Oxide

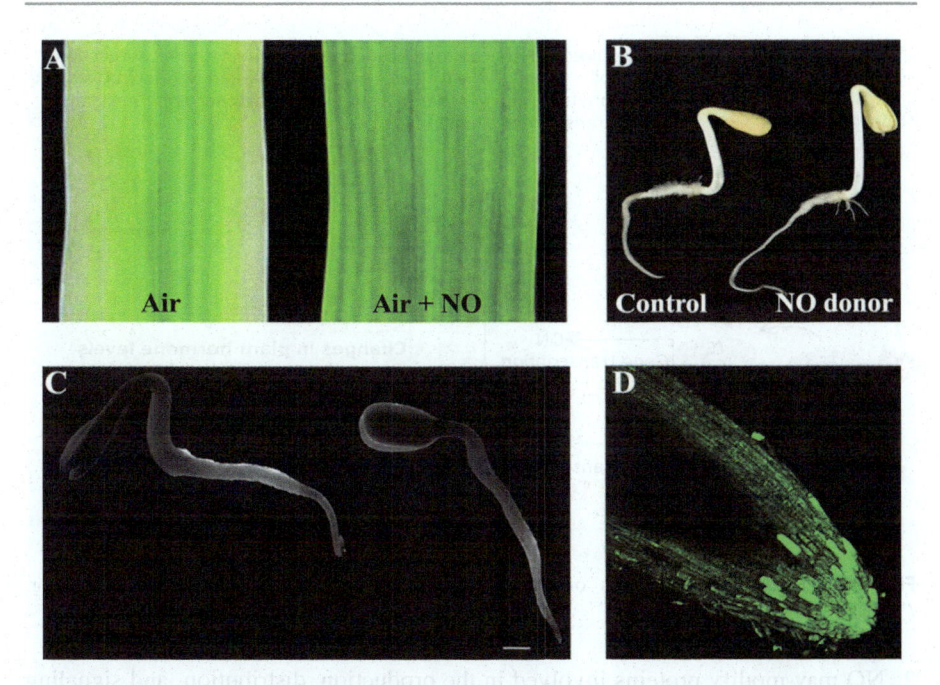

**Fig. 22.8** Role of NO in plant growth and development. (**a**) Leaves of *yellow stripe* mutant of maize seedlings raised hydroponically in a nutrient solution containing 50 μM Fe(III)-EDTA and transferred to a translucent chamber with air (control) or air supplemented with 100 μL.L$^{-1}$ gaseous NO. Note enhanced chlorophyll accumulation in the leaf in response to NO treatment. (**b**) Two-day old sunflower seedlings grown in Hoagland solution (control) or Hoagland solution containing 250 μM sodium nitroprusside (NO donor). Note enhanced extension growth of hypocotyl in response to NO application. (**c**) Endogenous NO in sunflower seedlings visualized (fluorescing sites) using MnIP-Cu (a fluorochrome to detect NO). (**d**) Root tip treated with DAF showing NO accumulation (green fluorescence) predominantly in the apical region

thereby delaying senescence. Several varieties of cut flowers exhibit enhanced longevity upon application of NO donors in solution. Thus, NO delays flower senescence as well. Application of NO also extends postharvest life of several vegetables and fruits. It can break seed dormancy and stimulate seed germination in some plants (e.g., lettuce, sunflower, tomato). Endogenous nitric oxide accumulation is also observed in the zone of elongation of primary roots indicating the role of endogenous NO in the extension growth of the roots (Fig. 22.8). Many more new roles of NO in plant growth and development are likely to be reported in the near future.

### 22.2.4.2 NO-Phytohormone Interaction
NO can modify phytohormone action in plants in three different ways:

1. It can chemically modify the transcription factors (TFs) and other proteins (P) which are involved in phytohormone metabolism, transport, or signaling.

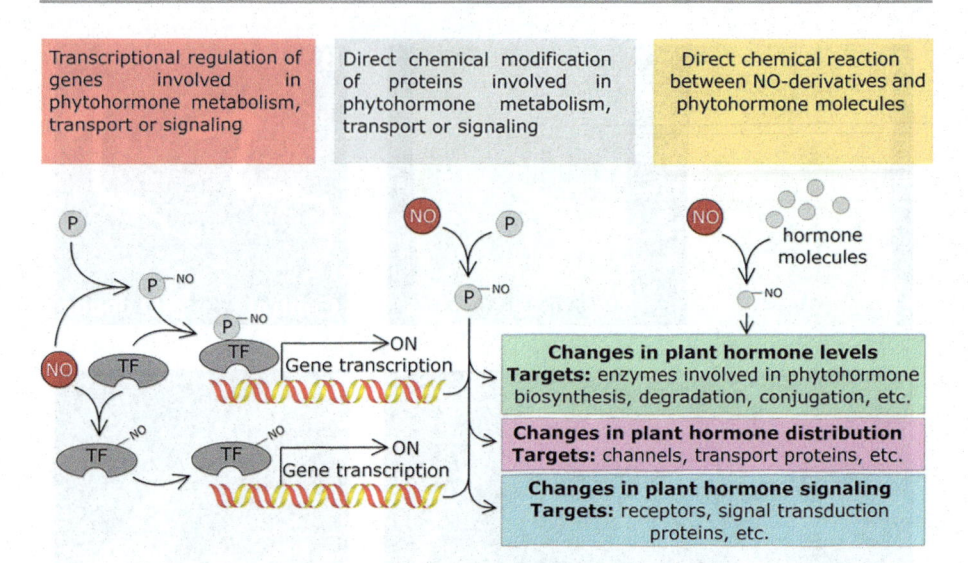

**Fig. 22.9** Potential mechanisms of NO-phytohormone interaction. *TF* transcription factor, *P* protein

2. NO may modify proteins involved in the production, distribution, and signaling of plant hormones at post-translational level.
3. NO or RNS may also directly react with certain plant hormones thereby altering their biological activity (Fig. 22.9).

Synergistic action of auxin and NO has been observed during the regulation of various plant responses, such as root differentiation, gravitropic response, nodule formation, and embryogenesis. A direct influence of NO on auxin perception and signal transduction has also been suggested through S-nitrosylation of auxin receptor protein (TIR1) which is a part of E3-ubiquitin ligase complex. Both synergistic and antagonistic actions between NO and cytokinins have been reported depending on the physiological response, plant species, and experimental approach. One of the initial reports of NO-cytokinin interaction pertains to accumulation of red pigment betalain in *Amaranthus cordatus* seedlings which is positively affected both by cytokinins and NO donors. In *Arabidopsis* seedlings, zeatin treatment triggers a rapid increase in NO production in the tissue. Synergistic interaction between cytokinins and NO has also been reported to regulate leaf senescence, programmed cell death (PCD), and plant's adaptability to drought stress in terms of photosynthesis. NO and abscisic acid (ABA) have been reported to interact during water deficit and UV-B radiation stress. Thus, under such stress conditions, their interaction regulates stomatal closure and antioxidant defense responses. NO may act as a downstream element in ABA-signaling pathway since removal of NO from the tissue usually decreases or eliminates ABA-triggered responses, while the inhibition of ABA production does not affect the induction of these responses by exogenous

NO application. NO has also been reported to directly modulate calcium-independent outward rectifying potassium channels through post-translational modifications of these channel proteins or their regulatory moieties. NO affects ABA catabolism by regulating the activity of ABA 8′ hydroxylase, a key enzyme in ABA catabolism. It may also affect the sensitivity of plants to ABA. NO has been reported to influence several GA-modulated developmental events in plants. These include seed germination, hypocotyl elongation, primary root growth, pollen tube growth, and acquisition of photomorphogenic traits. In quite a few instances, NO and GA antagonize each other's effects on the physiological processes. It is further evident that DELLA proteins represent a key crosstalk component in GA and NO signaling interaction. DELLA proteins are a relatively small family of transcriptional regulators which play important role in diverse hormonal signals, such as those of GA, ethylene, jasmonate, and ABA. NO enhances the cellular concentration of DELLA proteins thereby bringing about its negative impact on GA signal transduction. Thus, NO-driven DELLA accumulation alters the tissue sensitivity to GA. Additionally, NO and GA may also control each other's endogenous levels. In wheat roots, exogenously supplied NO induces apical growth which is also accompanied with increase in $GA_3$ levels. NO may stimulate germination not only by breaking seed dormancy but also be alleviating the influence of inhibitory environmental factors on the process of germination. NO and ethylene exhibit an antagonistic relationship with each other in a number of ethylene-mediated developmental processes. Exogenously supplied NO has been reported to delay senescence of both vegetative and reproductive organs by negatively regulating a number of elements involved in ethylene production. In many climacteric fruits, NO can modulate both the transcription and activity of ACS and ACO, the key enzymes in ethylene biosynthesis. Besides interacting with "classical phytohormones," NO has also been reported to crosstalk with other plant hormones such as jasmonic acid, salicylic acid, polyamines, and brassinosteroids. During the induction of plant defense responses against biotic challenges, NO positively affects the production of both SA and JA. A rapid NO production has also been observed in plant tissues exposed to millimolar concentrations of polyamines.

### 22.2.4.3 NO and Abiotic Stress

Abiotic stresses, such as low and high temperature, salinity, and drought, generally lead to enhanced generation of ROS which oxidatively destroy/disturb several metabolic and signaling pathways. NO interacts with ROS in various ways, and it might serve an antioxidant function during various stresses (Fig. 22.10). Modulation of superoxide formation and inhibition of lipid peroxidation by NO demonstrate some of its potential antioxidant roles. NO can also result in nitrosative stress; thus a favorable balance of ROS/NO is important.

### 22.2.4.4 NO and Biotic Interactions

NO plays a key signaling role during the induction of hypersensitive response (HR). HR is a defense process activated in plants in response to pathogen attack. HR is also associated with enhanced ROS generation, PCD, and induction of signaling

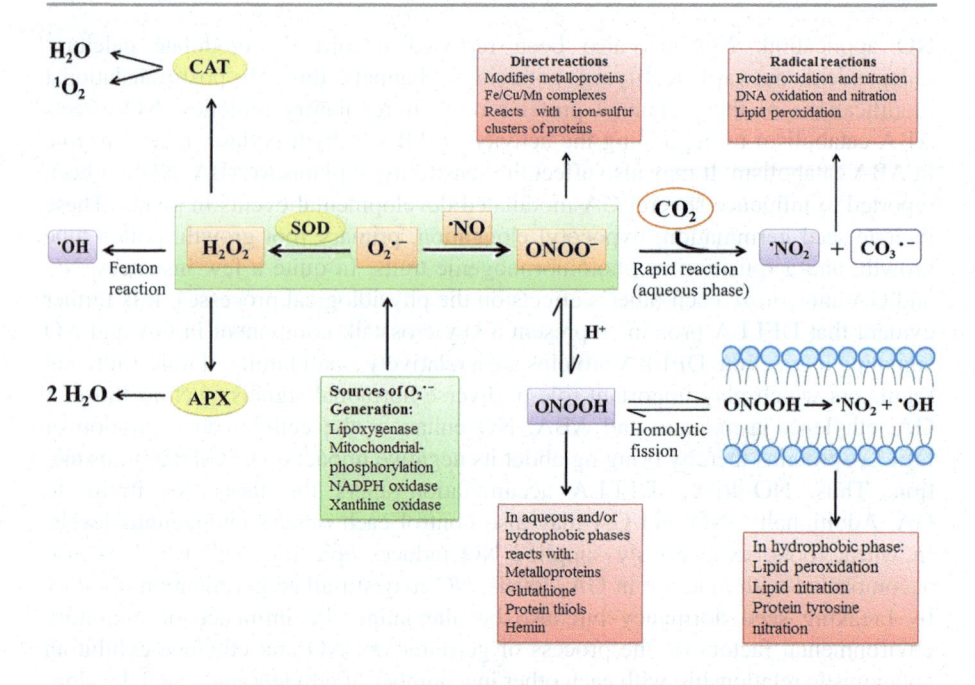

**Fig. 22.10** Interactions between NO and ROS and their possible roles

pathways leading to expression of various defense-related genes. NO donors have been reported to induce cell death indicating an interaction between NO and ROS accumulation. NO treatment of soybean cotyledons triggers biosynthesis of phytoalexins which are normally produced as a response to pathogen attack. NO has also been implicated during bacteria-legume interaction accompanying nodule formation. There are reports that NO acts as a negative regulator of nitrogen fixation due to its interaction with leghemoglobin.

### 22.2.4.5 NO and Programmed Cell Death (PCD)

PCD is a genetically determined and metabolically directed cellular process during which cells die because of activation of intrinsic signaling processes rather than due to necrosis. In soybean cell culture NO, superoxide ratio determines the extent of PCD. If superoxide levels are greater than NO, then it results in peroxynitrite formation which slows down PCD, but if NO levels are higher, then it reacts with $H_2O_2$ to induce cell death. In barley aleurone layer, GA-induced PCD is delayed in presence of NO due to a loss of activity of antioxidant enzymes, namely, catalase and superoxide dismutase. PCD is correlated with altered mitochondrial function, and it seems that NO has an emerging role in this process.

### 22.2.5 NO Metabolism

Nitric oxide can be endogenously removed through the activity of an enzyme called GSNO reductase (GSNOR) which degrades S-nitrosoglutathione (GSNO). GSNOR thus prevents excessive formation of S-nitrosylated proteins to damaging levels by transnitrosation reactions with GSNO. Other NO-metabolizing mechanisms have also been characterized during plant-pathogen interactions. Bacterial flavohemoglobin serves a protective role in bacteria such as *Salmonella typhimurium* and *Escherichia coli* against nitrosative stress imposed within the host environment. These proteins possess NO dioxygenase activity thereby converting NO to nitrite and nitrate.

### 22.2.6 NO Transport

Although NO is a diffusible gas, it is possible that NO precursors or NO adducts could serve as NO reservoirs or transport forms operative over short or long distances in plant system. In animals, S-nitrosohemoglobin circulates in the blood as a source of NO for varying distances. Since nitrite is a precursor of NO in plant cells, it is possible that nitrite can serve as a mobile source of NO across the xylem stream. Additionally, GSNO can serve as another mobile source of NO in plants through its migration across phloem or xylem elements. Plants also contain hemoglobin genes, thereby indicating the possibility of hemoglobin-NO complexes as transport molecules for NO.

## 22.3 Indoleamines (Serotonin and Melatonin)

Two major indoleamines, serotonin (5-hydroxytryptamine) and melatonin (*N*-acetyl-5-methoxytryptamine), known to function as neurotransmitters in animals, have also been reported to play major roles in the regulation of growth and development in plants. These compounds perform varied biological functions in plants which include flowering, shoot-root morphogenesis, apoptosis, and defense mechanisms associated with stress induction. They have been detected in various plant parts (leaf, root, fruit, seed, etc.) at different concentrations, ranging from picogram to microgram. **Melatonin** (N-acetyl-5-methoxytryptamine) is a pleiotropic molecule with numerous physiological and cellular actions. It was first discovered in bovine pineal gland in 1958. It is a hormone which is secreted into the cerebrospinal fluid and blood stream by the pineal gland. It regulates various physiological processes, including sleep, body temperature, circadian rhythm, appetite, retina function, immunological system, sexual behavior, and mood. It is involved as an antioxidant in various cellular actions and possesses strong antioxidative properties. Plant tissues, in general, contain higher melatonin levels as compared to that in vertebrate blood or tissues. **Serotonin** (5-hydroxytrytamine) is produced from the same biochemical route prior to melatonin biosynthesis. Both these compounds have great antioxidant potential. **Serotonin** was first identified in mammalian systems in the 1930s and was initially called as "enteramine" since it was reported from enterochromaffin cells of the gut and was found to cause contraction of smooth muscles. Serotonin has also been established as an essential neurotransmitter in the central nervous system and is known to play significant roles in various diseases, particularly in neurological disorders, such as Parkinson's, Alzheimer's, and depression. It has been detected in a number of plant species and was first reported from the fruits of cowhage plant (*Mucuna pruriens*).

Since the 1950s, serotonin has been identified in over 90 species from 37 families. Its endogenous concentration varies significantly depending on plant species, cultivar, tissue types, and stage of maturity (Table 22.1). Like melatonin, serotonin is also a multifunctional indoleamine. Edible tissues of cranberry, corn, banana, walnut, coffee bean, and ginger are also known to have moderately high levels of serotonin. Immunolocalization analysis has shown the abundance of serotonin in the vascular parenchyma cells (xylem and companion cells) of rice plants. It helps in the maintenance of the cellular integrity during senescence which facilitates efficient

**Table 22.1** Plants known to contain high concentration of serotonin and melatonin

| Plant | Common name | Serotonin Tissue | Concentration (ng.g$^{-1}$) | Melatonin Tissue | Concentration (ng.g$^{-1}$) |
|---|---|---|---|---|---|
| *Brassica rapa* L. | Chinese cabbage | Edible tissue | 110,900 | Edible tissue | 0.113 FW |
| *Cucumis sativus* L. | Cucumber | Edible tissue | 23,700 | Seeds | 11 FW |
| *Helianthus annuus* L. | Sunflower | Vegetative tissue | 9100–27,800 | Seeds | 29 DW |
| *Hordeum vulgare* L. | Barley | Edible tissue | 44,900 | Seeds | 0.58 FW |
| *Juglans regia* L. | Walnut | Edible tissue | 87,000 | Edible tissue (nuts) | 3.5 DW |
| *Oryza sativa* L. | Rice | Edible tissue | 77,300 | Edible tissue | 1.006 FW |
| *Solanum lycopersicum* L. | Tomato | Edible tissue | 221,900 | Fruit | 4.1–114.5 FW |
| *Vitis vinifera* L. | Grapevine | Fruit | 9000–10,000 | Fruit | 3–18 FW |
| *Zea mays* L. | Sweet corn | Edible tissue | 108,200 | Edible tissue | 1.37 FW |

*DW* Dry weight, *FW* Fresh weight

## 22.3 Indoleamines (Serotonin and Melatonin)

recycling of nutrients from senescing to sink tissues. In banana, it has been detected in the vascular bundles of fruit wall. Protein bodies of cotyledons also show serotonin accumulation in the developing embryos of *Juglans regia*.

### 22.3.1 Biosynthesis of Serotonin and Melatonin

Tryptophan is the common precursor for the biosynthesis of IAA and serotonin and melatonin (Fig. 22.11). Biosynthesis of serotonin is a two-step enzymatic process. Tryptophan decarboxylase (TDC; EC 4.1.1.28) catalyzes the conversion of tryptophan to tryptamine which is followed by serotonin formation by the action of tryptamine hydroxylase (T5H). *N*-acetylserotonin, the precursor for melatonin,

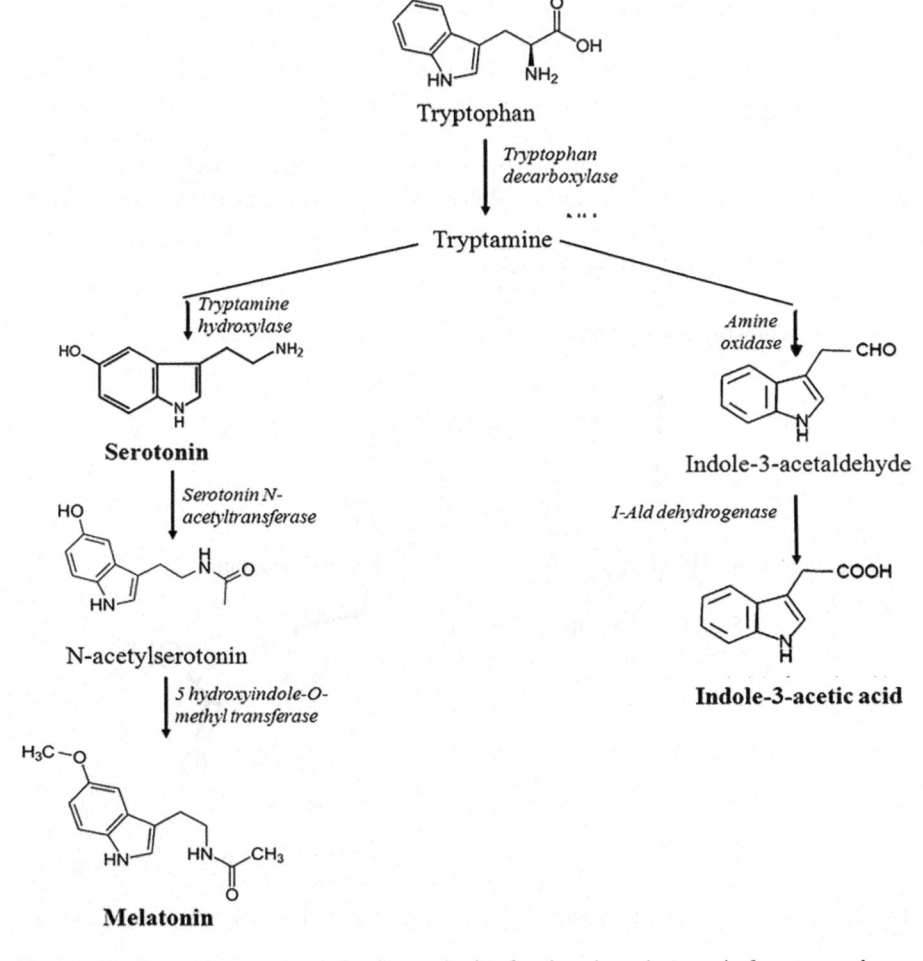

**Fig. 22.11** Biosynthetic pathway for the synthesis of melatonin and serotonin from tryptophan

leads to the formation of melatonin by the conversion of serotonin further downstream in this pathway. Tryptophan decarboxylase (TDC), with a high $K_m$ (690 μM) for tryptophan (its substrate), is the rate-limiting key regulatory enzyme of this pathway. The major sites for serotonin and auxin biosynthesis, i.e., root tips and stele, exhibit tissue-specific differential expression of tryptophan biosynthesis genes in rice.

## 22.3.2 Melatonin: Structure and Activity Relationship

An electron-rich indole moiety together with two side chains, a 5-methoxy group and a 3-amide group, constitutes the melatonin molecule (Fig. 22.12). The high resonance stability, electroreactivity, and low activation energy barriers make melatonin a potent free radical scavenger. The side chains also have a significant contribution in the antioxidative properties of the molecule. Carbonyl moiety present in the C3 amide side chain (in the functional group (N–C=O)) plays a major role in scavenging a number of reactive oxygen species. One molecule of melatonin can scavenge up to four or more ROS. Various mechanisms are involved in the interaction of melatonin with free radicals, such as addition reaction, nitrosation, substitution, hydrogen donation from nitrogen atom, and melatoninyl cation radical formation due to electron donation. Melatonin can also repair oxidized molecules

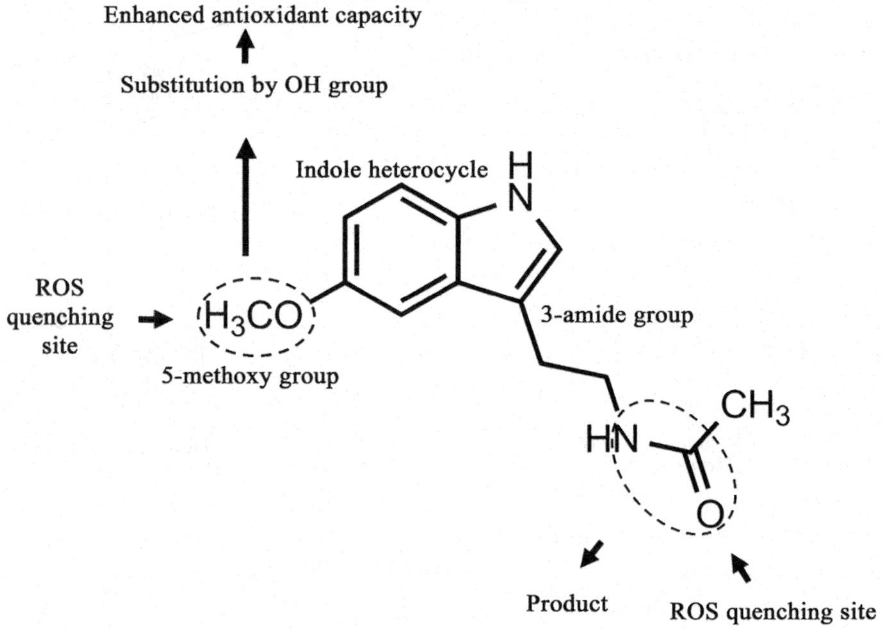

**Fig. 22.12** Structure of melatonin showing the reactive groups on the indole heterocycle and two side chains

and thus shows damage repairing abilities. Melatonin exhibits several functional similarities to serotonin and auxin. It is a potential antioxidant and acts as a photoregulatory molecule in plants. Melatonin possesses high affinity for calcium-dependent protein kinase and calmodulin and thus plays a role in the regulation of microtubule polymerization and auxin activity in plants.

### 22.3.3 Roles of Serotonin and Melatonin

#### 22.3.3.1 Morphogenesis

Serotonin plays a crucial role in root development and shoot morphogenesis. It modulates auxin-induced primary root growth and lateral root branching. It is involved in stimulation of pollen germination, growth regulation, plant morphogenesis, flowering, and ion permeability, and it has been detected in the exudates of root xylem sap. It also stimulates germination of radish seeds and acts as a growth regulator. In *Arabidopsis thaliana*, exogenously applied serotonin (10–160 µM) promotes lateral root development. The endogenous levels of serotonin and melatonin may also play a crucial role in in vitro plant morphogenesis. Indoleamines also regulate secondary metabolite levels in plants. Exogenous application of both these indoleamines also augments caffeine and polyamine levels in in vitro callus cultures of *Coffea canephora*. The influence of serotonin and melatonin has also been reported on somatic embryogenesis in *C. canephora*. SER stimulates the turnover of phosphoinositide (PI) which mimics the effect of red light and enhances the transcript levels of nitrate reductase and releases second messengers in maize.

#### 22.3.3.2 Plant Defense Responses

Serotonin is known to accumulate in the sting nettle (*Urtica dioica*) and in trichomes in the pods of *Mucuna pruriens*. Pathogenic infection also leads to synthesis of serotonin in rice leaves which then accumulates in the cell walls, thereby strengthening it. Rice plants accumulate tryptamine, serotonin and their phenolic acid coupled amides in response to fungal pathogen infection. All these compounds prevent pathogen invasion by forming a physical barrier. Serotonin plays an important role as a potential antioxidant by quenching ROS, and it exhibits in vitro antioxidative properties (Fig. 22.13). Serotonin has the highest antioxidant activity among tryptamine, tryptophan, and related derivatives. It helps in maintaining the reducing potential of cells due to its antioxidative properties by detoxifying the cells (relieves the accumulated toxic tryptamine). It delays senescence by efficient scavenging of ROS and protects young reproductive tissues from different environmental stresses in *Datura metel*. Increased level of serotonin has been reported in flowers of *Datura* undergoing cold stress. Some foods, such as pineapple, nuts, banana, milk, and plum, are rich in this indoleamine, and tryptophan may elevate mood by increasing serotonin levels in the brain.

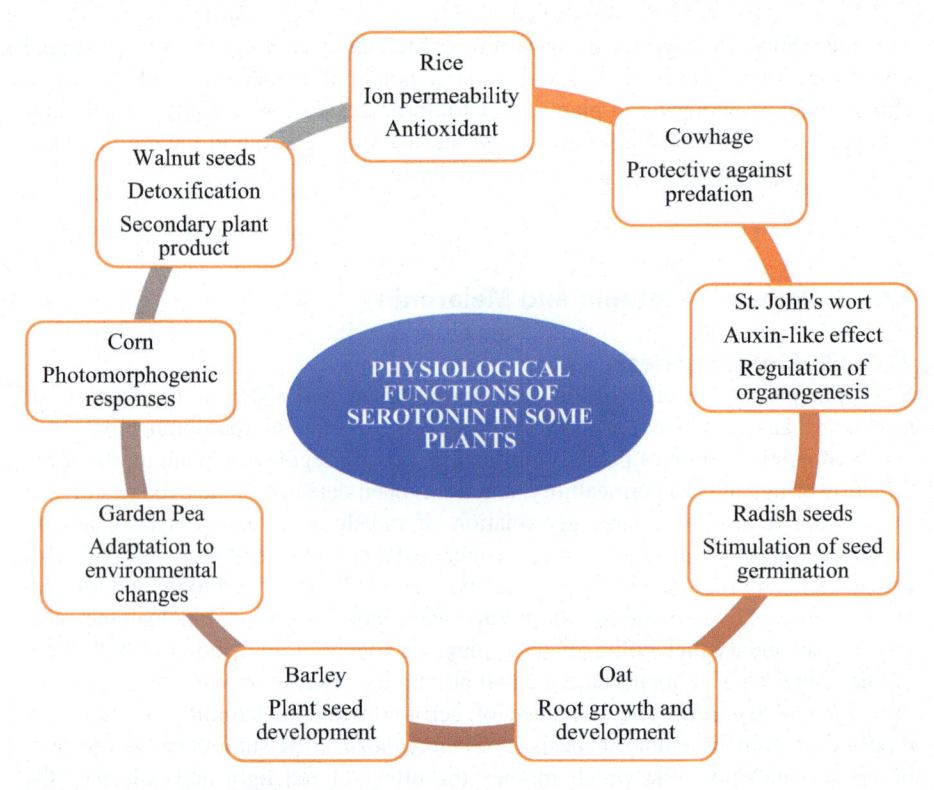

**Fig. 22.13** Physiological functions of serotonin in some plants

### 22.3.3.3 Gene Expression of Auxin Responsive Genes

Serotonin regulates gene expression associated with auxin response pathways. In plant tissues, polarity, growth, and gravitropism are regulated by auxin transport and its spatiotemporal distribution. Serotonin accumulation is likely to increase due to the abiotic stress-induced inhibition of auxin biosynthesis in plant tissues. NaCl stress leads to enhanced distribution of serotonin in the vascular cells of primary root and sunflower seedlings. At the sites of primary roots, adventitious roots, and lateral root primordia, serotonin has been reported to elicit tissue-specific inhibitory effects on various auxin responsive genes, and, therefore, it modulates root growth partially independent of auxin activity. Inhibition of lateral root primordia development and primary root length has been reported under salt stress due to inhibition of auxin accumulation in the emerging lateral root primordial in *Arabidopsis*. Under normal conditions, plant tissues have lower serotonin levels, but stress signals can lead to enhanced serotonin accumulation and, therefore, inhibit synthesis of auxin and impair its functions. These two indoleamines have been proposed to show auxin-like action and thus affect shoot and root growth in in vitro conditions. Exogenously applied melatonin and serotonin also alter root development in seedlings. Enhanced primary root elongation and lateral root branching have also been reported in sunflower seedlings due to supplementation of melatonin. In *Brassica juncea* and *Lupinus albus*, role of exogenously provided melatonin in the regeneration of lateral root has been suggested.

#### 22.3.3.4 Modulation of Responses Due to Abiotic and Biotic Stress

Serotonin and melatonin prevent oxidative damage by regulating cellular ROS levels and serve as line of defense against stress. Involvement of serotonin has been reported in the senescence of leaves in rice due to overexpression of enzyme tryptophan decarboxylase (TDC). Significantly high melatonin content has been reported in mature seeds which act as a free radical scavenger during seedling growth and seed germination in sunflower and in various other edible seeds. Variable ozone sensitivity, depending on the melatonin content, has been reported in the leaves of *Nicotiana tabaccum* and *Lycopersicon esculentum*. Enhanced melatonin formation in the dark is associated with the protection of photosynthetic pigments from ROS (by quenching) produced during photosynthesis. Melatonin has a protective role against leaf senescence. It helps in increasing the photosynthetic efficiency in plants. Under light and dark conditions, melatonin increases the efficiency of photosystem II in apple trees by mitigating the inhibitory effect of drought stress on photosynthesis and, therefore, allows plant leaves to maintain a higher capacity for stomatal conductance and $CO_2$ assimilation. Exogenously applied melatonin increases photosynthetic rate and activities of various ROS-scavenging enzymes and reduces chlorophyll degradation. Melatonin reduces oxidative damage by decreasing the ROS burst. It leads to increase in the activity of antioxidants such as SOD, peroxidase, catalase, and glutathione reductase under different stresses (Fig. 22.14).

### 22.3.4 NO-Melatonin CrossTalk

It has been established through investigations on animal systems that under nitrosative stress melatonin acts as a mediator of RNS and ROS crosstalk. It interacts with nitrogen-centered radicals to form nitrosated products, such as N-nitrosomelatonin (formed due to the reaction between NO and melatonin). It also maintains NO levels by the inhibition of activity of nitric oxide synthase. Melatonin can scavenge peroxynitrite anions ($ONOO^-$) which are formed due to interaction of $O_2^{\cdot-}$ and NO or peroxynitrous acid by interacting with the indole moiety present in melatonin. This, however, needs to be verified in plant cells.

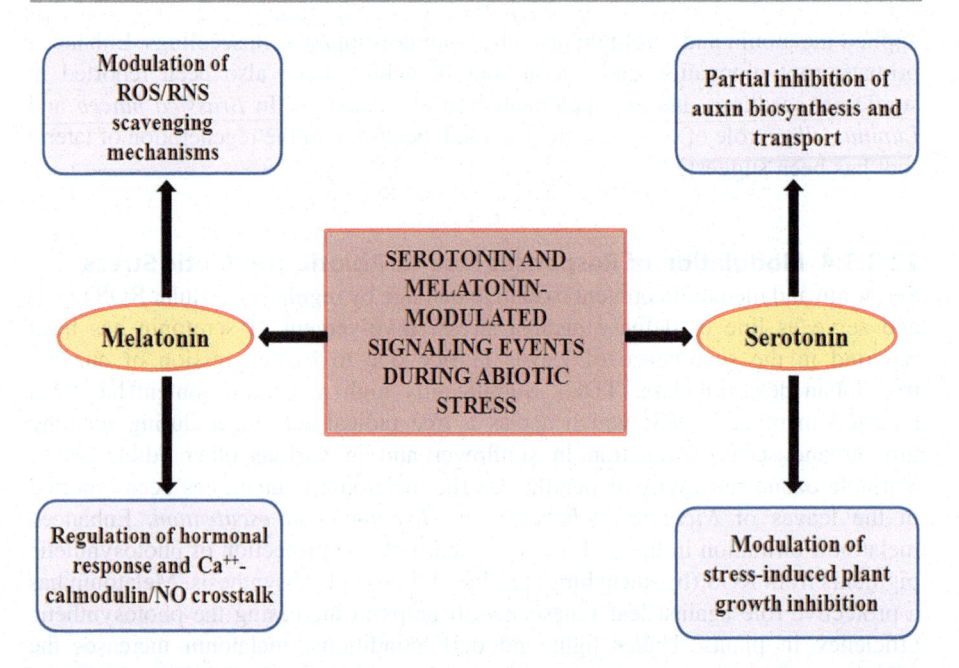

**Fig. 22.14** Scheme depicting the signaling events during serotonin and melatonin-modulated abiotic stress

## 22.4 Strigolactones

**Strigolactones** (SLs) are carotenoid derivatives which were first identified as germination stimulants for the seeds of root parasitic weeds such as *Striga*, *Orobanche*, and *Phelipanche* spp. A variety of SLs differing in the side group associations on the four-ringed basic structure have been identified from a number of plants.

To date, more than 15 SLs have been characterized from various plant spp. (Fig. 22.15). Some of the common naturally occurring SLs are strigol, orobanchol, and sorgomol. GR24 is an extensively used synthetic SL. SL biosynthesis occurs in plastids, starting from β-carotene and using three sequentially acting enzymes, D27, β-carotene isomerase, and CCD7 and CCD8, which are carotenoid cleavage dioxygenases (Fig. 22.16). The reaction product is an apocarotenoid (carlactone) which can move between cells and undergo two oxygenation steps to produce **5-deoxystrigol**. These oxygenation steps are catalyzed by a cytosolic enzyme—P450. The common precursor for the major natural SLs is 5-deoxystrigol. Through the process of hydroxylation, decarboxylation, oxidation and dehydration, various SLs are finally synthesized in the cytosol. The ABC ring of 5-deoxystrigol (basic moiety) is derived from a C15-carotenoid cleavage product. The enzyme D27 isomerizes trans-β-carotene into cis-β-carotene (C40). Then CCD7 (carotenoid cleavage dioxygenase 7) cleaves all cis-β-carotene into cis-β-apo-10-carotenal

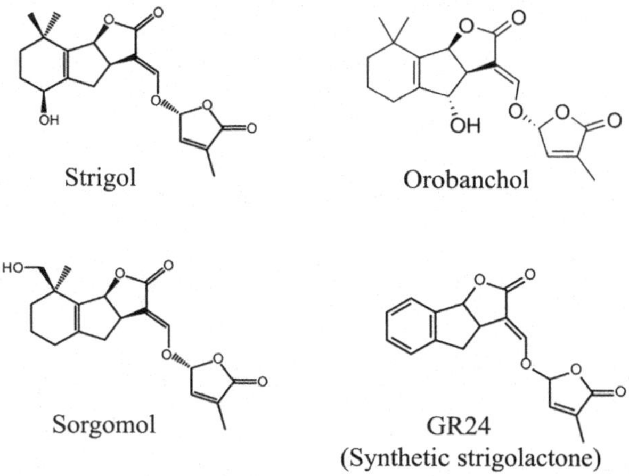

**Fig. 22.15** Commonly known natural and synthetic strigolactones

Strigol

Orobanchol

Sorgomol

GR24
(Synthetic strigolactone)

(C27) and β-ionone (C13). Subsequently, CCD8 cleaves C27 molecule to the C18 ketone (carlactone), which moves into the cytosol and cyclizes to form various strigolactones.

## 22.4.1 Physiological Roles of Strigolactones

### 22.4.1.1 Shoot Architecture
SLs appear to act in coordination with auxin to modulate apical dominance. Mutants of *Arabidopsis*, which are either defective in SL biosynthesis [*max1* (*more axillary growth1*), *max3*, or *max4*] or signaling (*max2*), exhibit enhanced branching without decapitation. Grafting the shoots of SL biosynthetic mutants (*max1*, *max3*, and *max4*) to wild-type roots restores apical dominance, indicating the movement of SLs from root to the shoot. Bud growth repressing SL normally comes from the shoot itself. SLs promote the development of interfascicular cambium. Under low phosphate conditions, SLs repress shoot branching. They, however, inhibit shoot branching in the presence of auxin source. It is now evident that SLs promote or inhibit shoot branching, depending on the auxin status of plants. SL signaling is believed to trigger PIN1 depletion from the xylem parenchyma cells. Accordingly, a reduction in auxin level enables the growth of lateral buds.

### 22.4.1.2 Root Growth
The expression of SLs biosynthetic genes is primarily evident in the vascular parenchyma cells of the root. In general, SL concentrations are higher in roots compared to other plant parts. The activity of CCD8 is upregulated in the primary root and cortical tissue of root apex upon treatment with 1-naphthaleneacetic acid (NPA) an auxin efflux blocker. SL-deficient or SL-insensitive mutant of *Arabidopsis* exhibits shorter **primary root** than those of wild-type plant. The root length is normalized in SL-deficient mutants by exogenous GR24 treatment but not in

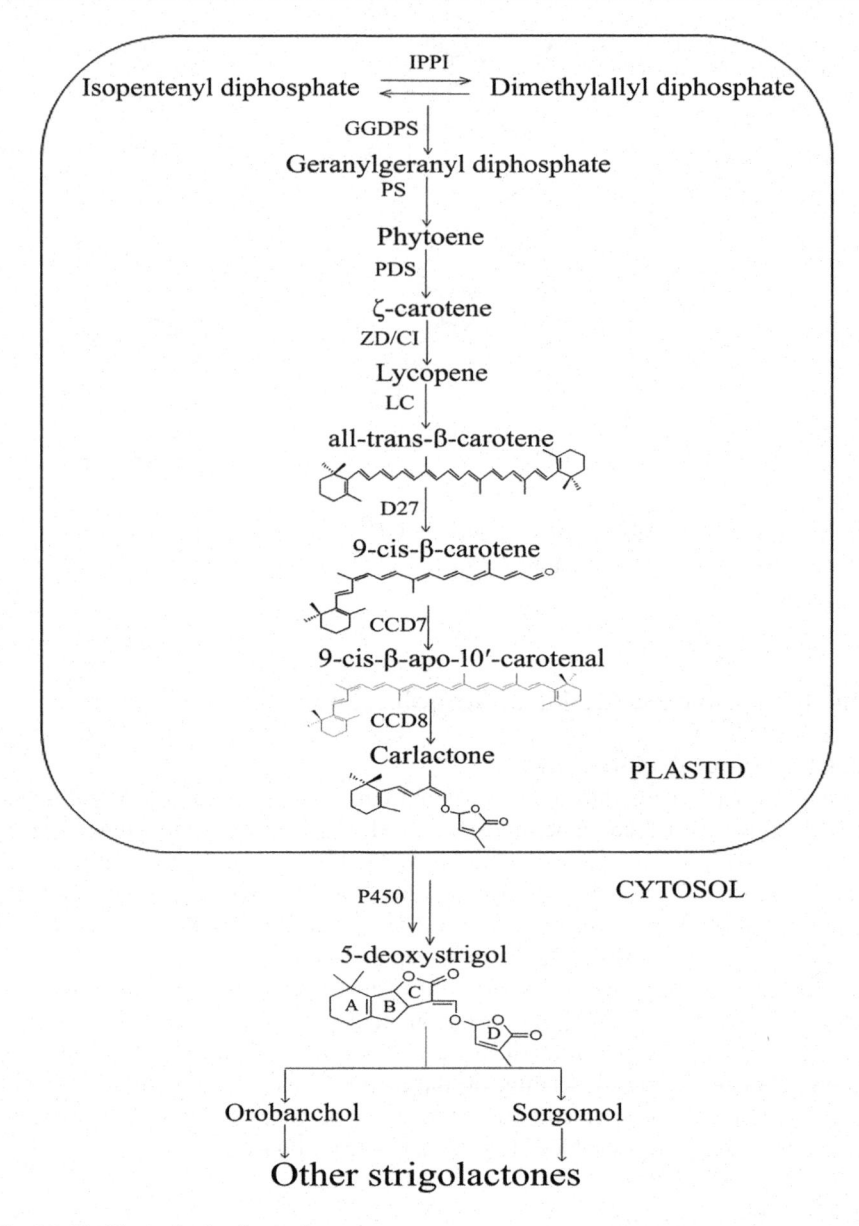

**Fig. 22.16** Biosynthesis of strigolactone

SL-insensitive mutant. Furthermore, auxin-induced inhibition of primary root elongation in tomato plants can be reversed by GR24 application. Availability of phosphate in soil modulates the development of lateral roots (LRs) in response to SLs. Thus, high phosphate conditions negatively control **lateral root formation**.

Under low phosphate availability, a treatment with GR24 enhances LR formation. Thus, it has been further observed that GR24 affects LR formation but not its elongation. SLs have also been reported to suppress **adventitious root** (AR) formation in *Arabidopsis* and pea. Thus, SL-deficient and response mutants of both these plants exhibit enhanced adventitious rooting response. It has been demonstrated that SLs negatively regulate auxin accumulation in the pericycle, thereby reducing AR initiation. Treatment with GR24 has been observed to increase root hair length in both wild-type and SL-deficient mutant in *Arabidopsis*. The impact of SL on this response varies according to phosphate availability in the external environment, thereby, indicating the influence of SLs on root development.

## 22.4.2 SL Crosstalk with Auxin, Ethylene, and Cytokinins

Strigolactones play a significant role in shaping root architecture, whereby auxin-SL crosstalk has been observed in SL-mediated responses of primary root elongation, lateral root formation, and adventitious root (AR) initiation. Whereas GR24 (a synthetic strigolactone) inhibits LR and AR formation, the effect of SL biosynthesis inhibitor (fluridone) is just the opposite (root proliferation). Naphthylphthalamic acid (NPA) leads to LR proliferation but completely inhibits AR development. The diffusive distribution of PIN1 in the provascular cells in the differentiating zone of the roots in response to GR24, fluridone, or NPA treatments further indicates the involvement of localized auxin accumulation in LR development responses. Inhibition of LR formation by GR24 treatment coincides with inhibition of ACC synthase activity. Differences in the spatial distribution of NO in the primary and lateral roots further highlight the involvement of NO in SL-modulated root morphogenesis in sunflower seedlings. Thus, a negative modulation of SL biosynthesis through modulation of CCD activity by endogenous nitric oxide during SL-modulated LR development is evident (Fig. 22.17). SLs are synthesized primarily in the root, and they can be transported acropetally to the shoot in the xylem. SLs have been reported to get transported from the sites of their synthesis using plasma membrane-associated ABC transporters, and they move in the xylem from the root to shoot. SLs and auxins primarily interact in the shoot. In the root tissue system, auxins have been shown to induce SL synthesis. It is, thus, apparent that SLs are modulators of auxin flux and reduce auxin import to the root leading to inhibition of LR formation. This is further evident from the distorted expression of PIN-auxin efflux carrier in tomato and *Arabidopsis* upon GR24 treatment.

SLs can induce ethylene biosynthesis in the seeds of parasitic plant—*Striga*—leading to seed germination. Ethylene biosynthesis inhibitor 2-aminoethoxyvinylglycine (AVG) abolishes the effect of SL on root hair elongation. GR24 elevates the transcription of *At-ACS2*, which encodes one of the rate-limiting enzymes in ethylene biosynthesis. It is evident from the investigations undertaken so far that SLs induce ethylene biosynthesis, both SL and ethylene regulate root hair elongation through the same pathway, and ethylene may be epistatic to SLs in this hormonal pathway. According to a simplified model, antagonistic interaction exists between cytokinins and SLs.

**Fig. 22.17** (a) Seed germination in *Orobanche* induced by application of a commercial strigolactone (GR24). (b) Strigolactone (GR24) inhibits lateral root extension, fluoridone (Fl), an inhibitor of SL biosynthesis does the opposite, i.e., LR proliferation, and NPA (auxin efflux blocker) completely suppresses LR induction in sunflower seedlings

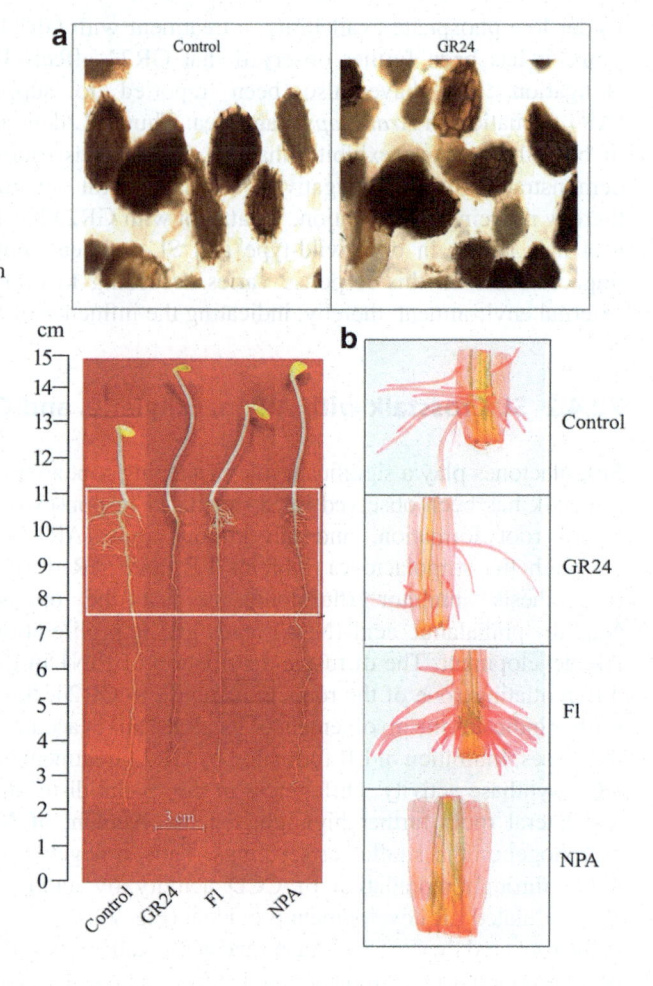

Auxins maintain apical dominance by stimulating SL synthesis via the *MAX4* gene. SLs then activate *BRC1*, a transcription factor known to suppress axillary bud growth. Additionally, SLs also inhibit cytokinin biosynthesis by negatively regulating the expression of *IPT* genes which otherwise would prevent BRC1 production (Fig. 22.18).

## 22.4.3 Signaling Mechanism for Strigolactone Action

The probable signaling mechanism for SL actions involves targeting of proteins by degradation through ubiquitination (Fig. 22.19). SLs are perceived by a protein complex which contains an α-/β-hydrolase protein and F-box protein D14 and MAX2, respectively. D14 binds and reacts with SL, thereby changing its (D14) confirmation to the active form—D14*. D14* interacts with the F-box protein

22.4 Strigolactones

**Fig. 22.18** Probable crosstalk among auxin, cytokinin, and strigolactone and their influence on axillary bud growth

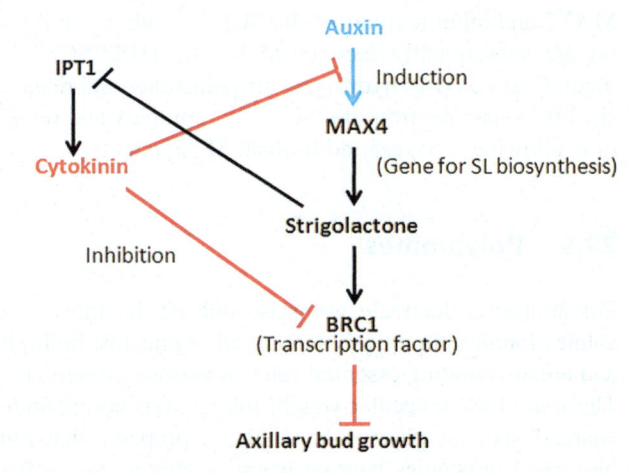

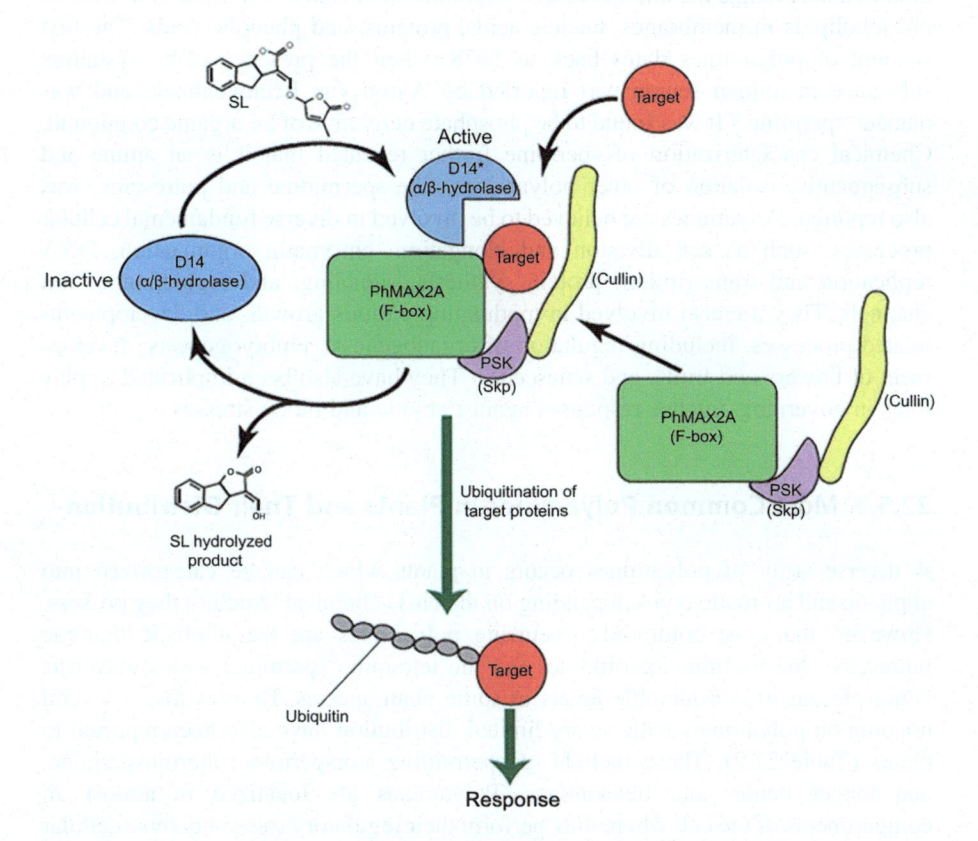

**Fig. 22.19** A model for the reception and the signal transduction of strigolactone

MAX2 and other partners of the SCF$^{MAX2}$ ubiquitin ligase complex. Target protein (s) are subsequently recognized by the D14*-SCF$^{MAX2}$ complex and are then ubiquitinated. D14* hydrolyzes strigolactones and releases the products of hydrolysis. D14 separates from the SCF$^{MAX2}$ complex and returns to its original confirmation, allowing it to respond to fresh SL signal.

## 22.5  Polyamines

Polyamines collectively comprise one of the most efficient class of compatible solutes found ubiquitously among all organisms, both prokaryotes and eukaryotes, and are also finding essential roles in various growth and developmental processes. These are low molecular weight nitrogenous compounds which exist in positively charged state at physiological pH, a property that allows them to bind to all biological molecules bearing negative charge. So, polyamines have the ability to bind to and stabilize the anionic macro- and micromolecules within the cell, such as phospholipids in membranes, nucleic acids, proteins, and phenolic acids. The first account of polyamines dates back to 1678, when the presence of a crystalline substance in human semen was reported by Anton van Leeuwenhoek, and was named "spermine." It was found to be phosphate derivative of an organic compound. Chemical characterization of spermine further revealed that it is an amine and subsequently, isolation of other polyamines, like spermidine and putrescine, was also reported. Polyamines are believed to be involved in diverse fundamental cellular processes, such as cell division and elongation, chromatin organization, DNA replication and transcription, protein synthesis, signaling, and regulation of ion channels. They are also involved in modulating various growth- and development-related processes, including regulation of organogenesis, embryogenesis, development of flowers and fruits, and senescence. They have also been implicated to play roles in governing defense responses against abiotic and biotic stresses.

### 22.5.1  Most Common Polyamines in Plants and Their Distribution

A diverse array of polyamines occurs in plants which can be categorized into aliphatic and aromatic types depending on the basic chemical structure they possess. However, the most commonly occurring polyamines are the aliphatic diamine putrescine, the triamine spermidine, and the tetramine spermine, with cadaverine being present in considerable levels in some plant species. Besides these, several uncommon polyamines with a very limited distribution have also been reported in plants (Table 22.2). These include norspermidine, norspermine, thermospermine, and longer penta- and hexaamines. Polyamines are localized in almost all compartments of the cell where they perform their regulatory roles in diverse cellular events. The relative abundance of polyamines may vary between micro- to millimolar range, majorly depending upon the species and the stage of development.

## 22.5 Polyamines

**Table 22.2** Some common and uncommon polyamines existing in plants

| Types of polyamines | Structure |
| --- | --- |
| *Common polyamines* | |
| Putrescine | $NH_2(CH_2)_4NH_2$ |
| Cadaverine | $NH_2(CH_2)_5NH_2$ |
| Spermidine | $NH_2(CH_2)_3NH(CH_2)_4NH_2$ |
| Spermine | $NH_2(CH_2)_3NH(CH_2)_4NH(CH_2)_3NH_2$ |
| *Uncommon polyamines* | |
| 1,3-Diaminopropane | $NH_2(CH_2)_3NH_2$ |
| Homospermidine | $NH_2(CH_2)_4NH(CH_2)_4NH_2$ |
| Norspermidine | $NH_2(CH_2)_3NH(CH_2)_3NH_2$ |
| Homospermine | $NH_2(CH_2)_3NH(CH_2)_4NH(CH_2)_4NH_2$ |
| Norspermine | $NH_2(CH_2)_3NH(CH_2)_3NH(CH_2)_3NH_2$ |
| Thermospermine | $NH_2(CH_2)_3NH(CH_2)_3NH(CH_2)_4NH_2$ |
| Caldopentamine | $NH_2(CH_2)_3NH(CH_2)_3NH(CH_2)_3NH(CH_2)_3NH_2$ |
| Caldohexamine | $NH_2(CH_2)_3NH(CH_2)_3NH(CH_2)_3NH(CH_2)_3NH(CH_2)_3NH_2$ |

However, they are known to exert their biological effects in millimolar range. Additionally, they can occur in soluble fractions in free (cytoplasmic) and/or conjugated (attached to organic acids like hydroxycinnamic acids) forms as well as insoluble fractions bound to the macromolecules, like phospholipids in the plasma membrane, nucleic acids, proteins, and cell wall polysaccharides.

Putrescine and cadaverine are both diamines and chemically very similar, having a primary amine group on either end of a short hydrocarbon chain. The difference in their structure is due to their different precursors—putrescine is produced from arginine by removal of carboxylic acid group, whereas cadaverine is derived from lysine. Both of these diamines are foul-smelling and are responsible for the foul odor associated with the process of putrefaction (decaying of dead tissue). They also contribute to the distinctive odor of semen and urine. Spermidine and spermine belong to higher amines which are synthesized from putrescine by addition of extra chains of three carbons, each ending with another amine group, to one and both ends of putrescine, respectively.

### 22.5.2 Polyamine Homeostasis

The endogenous levels of polyamines are very closely regulated to maintain homeostasis and achieve desired requirements and functions during various stages of growth. This homeostasis is extremely crucial since both the depletion and excess of polyamines may be deleterious to cells. Several processes together help in maintenance of appropriate levels of polyamines including their synthesis, uptake, degradation, compartmentalization, conjugation, and transport.

## 22.5.3 Biosynthesis of Polyamines

Putrescine is synthesized from arginine via two different biosynthetic pathways (Fig. 22.20). In one pathway, arginine is converted in three subsequent reactions into first agmatine, followed by N-carbamoylputrescine, and finally into putrescine via the action of arginine decarboxylase (ADC), agmatine iminohydrolase (AIH), and N-carbamoylputrescine aminohydrolase (CPA), respectively. In the alternate pathway, arginine is first converted into ornithine, which in turn is converted into putrescine in a reaction catalyzed by ornithine decarboxylase (ODC). Both the pathways of putrescine synthesis exist in plants, while in animals, putrescine is synthesized exclusively via the ODC pathway. Spermidine and spermine synthesis requires putrescine as well as decarboxylated S-adenosylmethionine as precursor molecules. Decarboxylated S-adenosylmethionine serves as the aminopropyl group donor for higher polyamine synthesis and is synthesized by decarboxylation of S-adenosylmethionine by the action of S-adenosylmethionine decarboxylase. Spermidine and spermine are synthesized from putrescine in reactions catalyzed by two different aminopropyltransferases, i.e., spermidine synthase (SPDS) and spermine synthase (SPMS). Spermidine synthase uses putrescine as the basic moiety and transfers the aminopropyl group from a decarboxylated S-adenosylmethionine molecule to it to produce spermidine. Spermidine is subsequently converted to spermine by spermine synthase by utilizing another decarboxylated S-adenosylmethionine molecule.

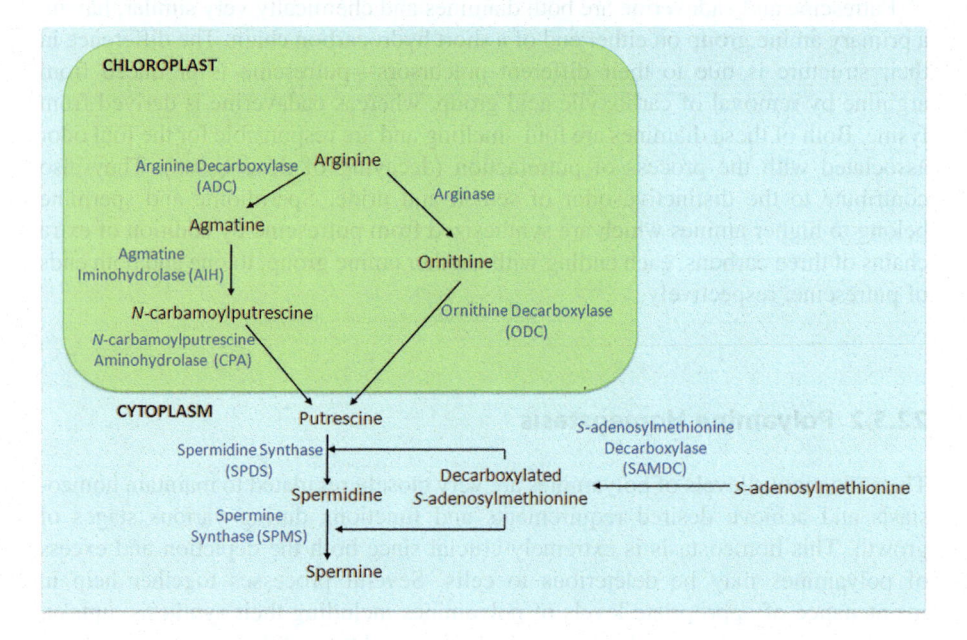

**Fig. 22.20** Biosynthetic pathways for polyamines

## 22.5.4 Catabolism of Polyamines

Degradation of polyamines is mediated by diamine and polyamine oxidases (Fig. 22.21). The diamine oxidases preferentially work on diamines, such as putrescine and cadaverine, but can also act on spermidine and spermine, though with low affinity. Putrescine is oxidized to produce pyrroline, along with $H_2O_2$ and ammonia as side products. Polyamine oxidases carry out the oxidation of higher amines, such as spermidine and spermine, producing pyrroline, 1,3-diaminopropane and $H_2O_2$ from spermidine and 1-(3-aminopropyl)-pyrroline, 1,3-diaminopropane, and $H_2O_2$ from spermine. Pyrroline and 1,3-diaminopropane are further converted to succinate and β-alanine, respectively. Apart from being involved in oxidation of polyamines, polyamine oxidases are also involved in back-conversion of spermine to spermidine and spermidine to putrescine, with the release of pyrroline and $H_2O_2$. In such back-conversions, spermidine and spermine are first transformed by spermidine/spermine *N*-acetyl transferase (SSAT), by utilizing acetyl-CoA, into their acetyl derivatives, *N*-acetylspermidine and *N*-acetylspermine, respectively, before being acted upon by polyamine oxidases. Another enzyme, spermine oxidase (SMO), also exists which is responsible for conversion of spermine to its acetyl derivative.

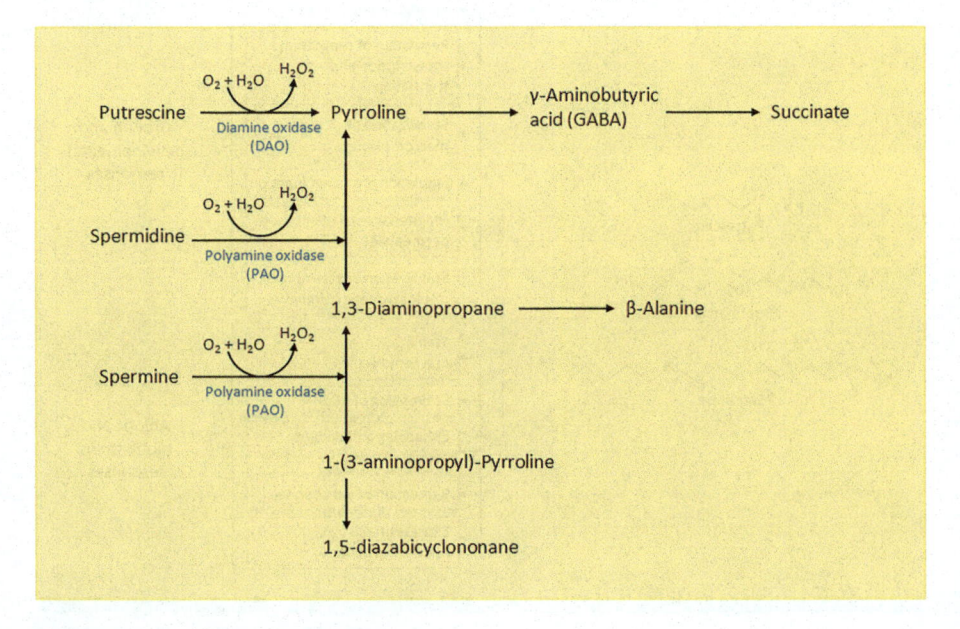

**Fig. 22.21** Catabolism of polyamines

## 22.5.5 Functions of Polyamines

Polyamines play diverse roles throughout the life span of plants regulating various processes such as germination organogenesis, embryogenesis, flower initiation, fruit development and ripening, and senescence (Fig. 22.22). They have also been shown to play crucial roles in acclimation to both abiotic and biotic stresses. Accumulation of polyamines during various stresses has been reported to serve as an important adaptive measure to improve resistance.

## 22.5.6 Ionic Interactions

The growth promoting activities of polyamines are majorly accounted for by their roles in regulation of basic cellular events, like nucleic acid synthesis and protein synthesis. They bind to and modulate conformation of DNA and RNA, thereby facilitating the processes of replication, transcription, and RNA processing. They stabilize the negative charge associated with phosphate groups in nucleic acids and help in condensation of chromatin. They also interact with proteins via ionic interactions or through covalent conjugation at specific glutamyl residues (via the activity of enzyme transglutaminase). Binding of polyamines regulates the 3D

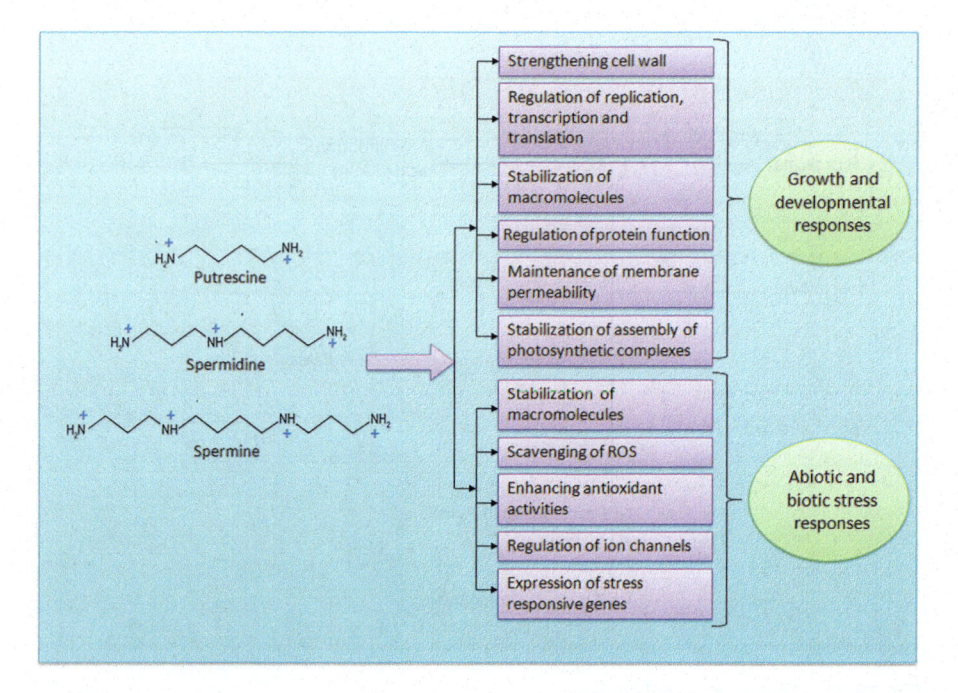

**Fig. 22.22** Some regulatory roles of polyamines in plant growth and development and under abiotic and biotic stresses

conformation of proteins and also their functions. They interact with negatively charged membrane lipids and strengthen their binding with other lipids and membrane-associated proteins and, therefore, increase the stability of the membrane. In the cell wall, they bind to polysaccharides, pectin, and phenolic acids in lignin to enhance the strength of primary cell wall as well as deposition of secondary cell wall and lignification. Thus, they play important role in cell wall extension and development of vasculature. All of these roles played by polyamines help in regulating cell growth and wall extension as well as cell division and therefore help in overall growth of plant. Other important roles played by polyamines in plant life cycle include induction of root initiation and growth, pollen development and pollen-pistil interaction, and promoting flower and fruit development.

### 22.5.7 Crosstalk with Hormones

Because of their involvement in numerous growth- and development-related processes, polyamines have since long been argued to be "hormones" in their action. One contradictory argument to this belief is that polyamines are active at millimolar concentrations, while the hormones elicit their effects in much smaller amounts. However, they have been well-documented to interact with different phytohormones in synergistic or antagonistic manner to regulate many processes, including the stress responses. For instance, polyamines act antagonistically to ethylene to control leaf senescence and fruit ripening. They are thus considered "anti-senescent" molecules as they prevent loss of chlorophyll and retard fruit ripening. On the other hand, they have been reported to have a synergistic interaction with abscisic acid in reducing stress-induced damage to plants. A remarkable example of such an interaction is in facilitating stomatal closure under water-deficit conditions. Polyamines are also known to interact with hormones associated with plant defense against biotic stress, like jasmonic acid and salicylic acid. Both of these hormones have been demonstrated to induce polyamine biosynthesis and enhance accumulation of conjugated form of polyamines in response to pathogen attack, thereby increasing the tolerance to infection. Accumulation of conjugated polyamines, or polyamines in general, contributes to reduction of the symptoms associated with diseases. Jasmonates are also known to induce polyamine oxidation which plays essential role during pathogen attack. Polyamine oxidation generates $H_2O_2$ which is associated with defense responses, like induction of the hypersensitive (HR) response to elicit host cell death and improve tolerance against pathogens. The other reaction products of polyamine oxidation give rise to intermediaries which have substantial role in acclimation during pathogen invasion. For example, pyrroline and 1,3-diaminopropane are metabolized to γ-aminobutyric acid (GABA) and β-alanine, respectively, both of which are involved in improving plant resistance.

## 22.5.8 Polyamines as Signaling Molecules and in Modulating Stress

Recently polyamines have also been implicated to serve as signaling molecules in many cellular signaling routes, particularly via their interaction with other such signaling molecules, like nitric oxide, $Ca^{2+}$, and $H_2O_2$. They have been suggested to induce nitric oxide biosynthesis, a molecule which has been known to mediate several stress responses in plants, especially under stressful conditions. One of the best examples of the interaction of these signaling molecules is seen under salinity and drought stress, where polyamines regulate cytoplasmic $Ca^{2+}$ levels and induce nitric oxide, which work in concert with each other to trigger ABA-induced stomatal closure. Polyamine oxidation product $H_2O_2$ has also been reported to play a crucial role in stress-induced stomatal closure. Other responses to stresses include enhanced production and accumulation of compatible solutes, modification of cell wall, and maintenance of cellular redox and ionic homeostasis. Polyamines have a role in all of the abovementioned responses which collectively help in overcoming detrimental effects of stress. The major roles of polyamines during stress can be summarized as follows:

1. Preserving cell membrane integrity by reducing lipid peroxidation and stabilizing lipids
2. Reinforcement of cell wall during pathogen attack and wound healing
3. Preserving thylakoid membrane, assembly of photosystems, and increasing abundance of photoprotective molecules like zeaxanthin and carotenoids, therefore, increasing photosynthetic efficiency and improving dissipation of excess energy
4. Modulating the activities of ion channels and maintaining ionic balance
5. Scavenging of reactive oxygen species (ROS) and increasing antioxidant activities, therefore maintaining redox homeostasis
6. Regulating stress responses, such as HR, accumulation of GABA, and other compatible solutes, and inducing expression of pathogen-related (PR) proteins

## 22.5.9 Role in Plant-Microbe Interactions

Besides providing tolerance to pathogens, polyamine metabolism also influences the establishment and development of beneficial plant-microbe associations, such as those with nitrogen-fixing bacteria (rhizobia), mycorrhizal fungi, and growth-promoting rhizobacteria. Both accumulation of polyamines and their oxidation seem to be important for initiation of nodulation and maintenance of nodules. $H_2O_2$ formed as a result of polyamine oxidation promotes cross-linking of plant matrix glycoproteins in the lumen of infection threads during initial stages of rhizobia colonization. Polyamine accumulation enhances nodule tolerance to osmotic stress caused by establishment of the symbiotic association with rhizobia in plant roots. Similarly, they also facilitate establishment of mutualistic associations with mycorrhizal fungi by stimulating the colonization of roots by the fungus and hence contribute significantly to increased absorption of minerals by the host. Therefore, polyamines not only provide important measure against pathogens (and abiotic stresses) to control plant diseases, but also have important roles during establishment of beneficial plant-microbe interactions.

## 22.6 Peptide Signaling Molecules

A number of plants are routinely wounded by treading and grazing. The cells so damaged are prone to easy colonization by microbes than the healthy cells of unwounded regions of the plants. Wounded and exposed cells thus provide a point of microbe entry into plant body since it is no longer protected by any thick epidermal cell wall and cuticle. Wounding of plant tissue leads to enhanced ROS production which, in turn, induces certain defense responses including gene expression for the lignin and polyphenol biosynthesis. Lignin and similar other hydrophobic polymers seal the wound against pathogen invasion, whereas polyphenols provide further protection to the wounded tissue which is exposed to an oxidizing environment. Polyphenols are also the cause of browning of the cut surface (such as in potatoes and apples) due to accumulation of cross-linked phenolic compounds synthesized by the action of polyphenol oxidases. In addition to these localized defenses, host plants also exhibit activation of several genes in the unwounded cells near the wounding tissue, leading to synthesis of defense-related proteins. Such an activation of genes in the intact zones of a wounded plant is referred as a **systemic response** which is triggered in the uninfected parts of the plant by a broad spectrum of pathogens, and it leads to plant immunization, thereby limiting pathogen growth. Plants achieve this systemic resistance through long distance transport of certain peptides known as **signal peptides** (Fig. 22.23).

### 22.6.1 Systemins

Small polypeptide signaling molecules play important roles in short-range intercellular communications. First plant signaling peptide was discovered by Clarence E. Ryan in 1970 from tomato and was named **systemin** (TomSys). When healthy young plants were irrigated with water containing wounded tomato leaves, it induced the production of proteinase inhibitor (TomSys) in healthy plants. TomSys or systemin was characterized as an 18-amino-acid signaling peptide and was shown to be involved in the production of jasmonic acid, the main wound response signaling molecule. Systemin is capable of imitating defense responses at very low concentrations in the range of few femtomoles per plant. Systemin is synthesized at the wounding site by cleavage from the C-terminus of a 200 residue precursor protein called **prosystemin**. It then moves rapidly (within 1–2 h) through phloem to the unwounded parts of the plant (Fig. 22.24). Transgenic tomato plants which

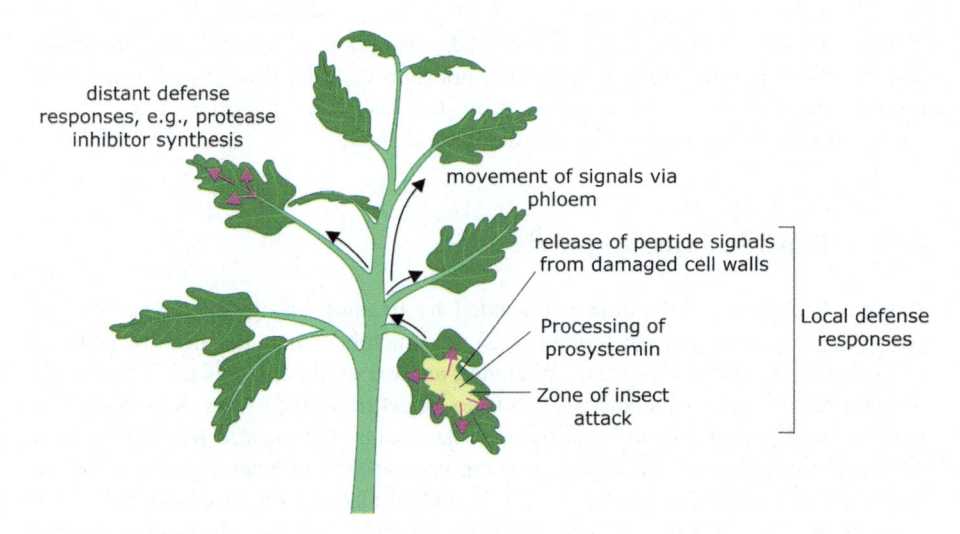

**Fig. 22.23** The process of induction of defense response in unwounded plant parts by peptide signal transport from the zone of damage by wounding/insect attack

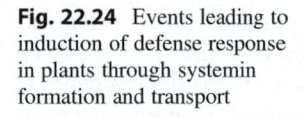

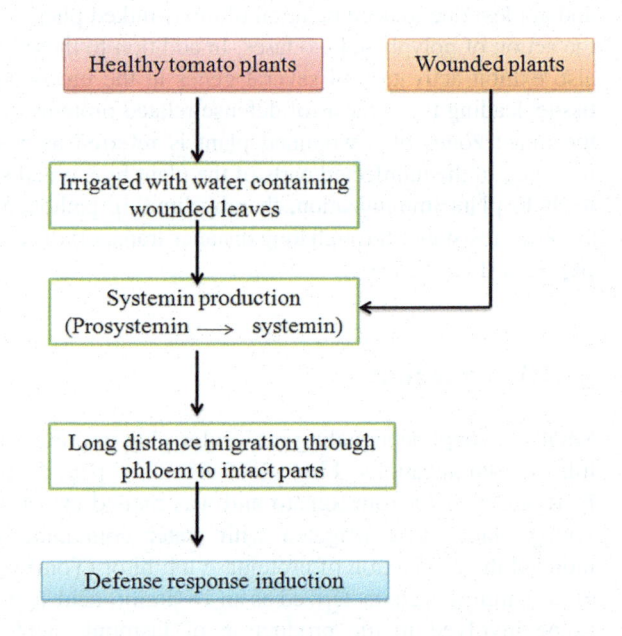

**Fig. 22.24** Events leading to induction of defense response in plants through systemin formation and transport

produce lower than normal levels of systemin exhibit severely reduced systemic induction of defense mechanisms and have reduced resistance to herbivory by the larvae of tobacco hornworm (*Manduca sexta*). Transgenic plants producing high levels of systemin production, constitutively express defense mechanisms even if they are not wounded.

## 22.6.2 Types of Signaling Peptides

About 13 families of signaling peptides have been identified so far. They can be grouped into two categories:

1. **Small post-translationally modified peptides:** They are composed of 5–20 amino acids, e.g., CLE, IDA, and RGF (ROOT GROWTH FACTOR) peptides.
2. **Cysteine-rich polypeptides:** They have a length of approx. 50 amino acids and are synthesized as precursor proteins, e.g.: RALF (RAPID ALKALINIZATION FACTOR) and PDF (PLANT DEFENSIN) peptides.

Most of the signal peptides are derived from the proteolysis of precursor proteins. Some of them are produced by non-ribosomal synthesis, e.g., glutathione and phytochelatins. Processing of the precursor molecules can take place both in the cytosol and apoplast since proteases are a part of the plant secretome as well.

## 22.6.3 Perception of Signal Peptides by the Cells

The perception of signal peptides occurs via plasma membrane localized receptor-like kinases (RLKs). RLKs exhibit an extracellular domain, a transmembrane domain, and intracellular kinase domain. RLK gets activated by the binding of signal peptide to the extracellular domain, leading subsequently to induction of different pathways monitoring cell growth, differentiation, or defense response (Fig. 22.25). Thus, some peptides regulate meristematic activity in shoot or root meristem (e.g., CLV3 and CLV40) or promotion of abscission (IDA). STOMAGEN is a cysteine-rich signal peptide and is involved in stomata development. Overexpression of STOMAGEN enhances the number of stomata on the leaves (Fig. 22.26). Signal peptides are also important in stress responses and in symbiotic interactions. Thus, FLS 2 (FAGELLIN SENSING 2) is a signal peptide that senses pathogens by binding to bacterial flagellin. This binding triggers intracellular calcium signaling and further activation of downstream responses.

## 22.6.4 Signal Peptides and Their Potential Benefits in Agriculture

Our understanding of the roles of signal peptides in plant development can benefit agriculture. A knowledge about novel signal peptides which influence the size of meristem, is likely to help in finding mutants which can result in enhanced yield. Identification and characterization of mutations carrying genes for signal peptides or their receptors hold promise for benefits. in agriculture. For example, mutations in the maize orthologs of CLV1 thick tassel dwarf 1 (td1) influence male and female inflorescence ear and have been found to modulate seed production. The male inflorescence shows increased density of spikelets and female inflorescence exhibits more kernels. So far, information is available for only very few signal peptide-

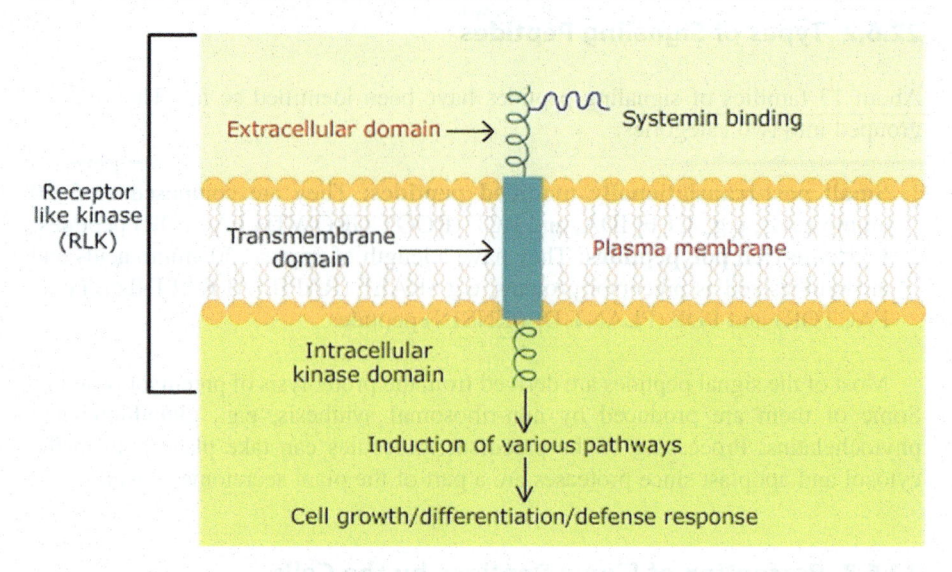

**Fig. 22.25** Perception of signal peptides by the cells

triggered pathways in plants. A deeper understanding of more of such pathways is likely to provide us information about the biochemical/metabolic routes to modify plant architecture and development for crop improvement.

## 22.7 Karrikins: A New Class of Plant Growth Regulators in Smoke

Karrikins are a class of butanolide compounds derived from burnt plant material which can stimulate seed germination in a number of plants. The term "karrikin" is derived from the aboriginal word "karrik" for smoke used by Western Australian Noongar people. It is a common observation that some plants grow immediately after wildfires or bushfires and their seeds remain dormant in soil until a fire generates karrikins which get bound to soil particles as a result of subsequent rainfall following fire. Such plants which remain dormant until a fire generates fresh karrikins in soil to stimulate their germination are referred as "fire followers" or "fire ephemerals." Fires release plant-bound nutrients and create an open habitat where seedlings establish themselves before other competing plants colonize the land. Such seeds can remain viable in soil for decades between fires and undergo many cycles of wetting (imbibition) and drying (dehydration). Smoke water is sometimes used to promote germination of horticultural seeds. The smoke from cigarettes also stimulates seed germination, probably due to the presence of karrikins.

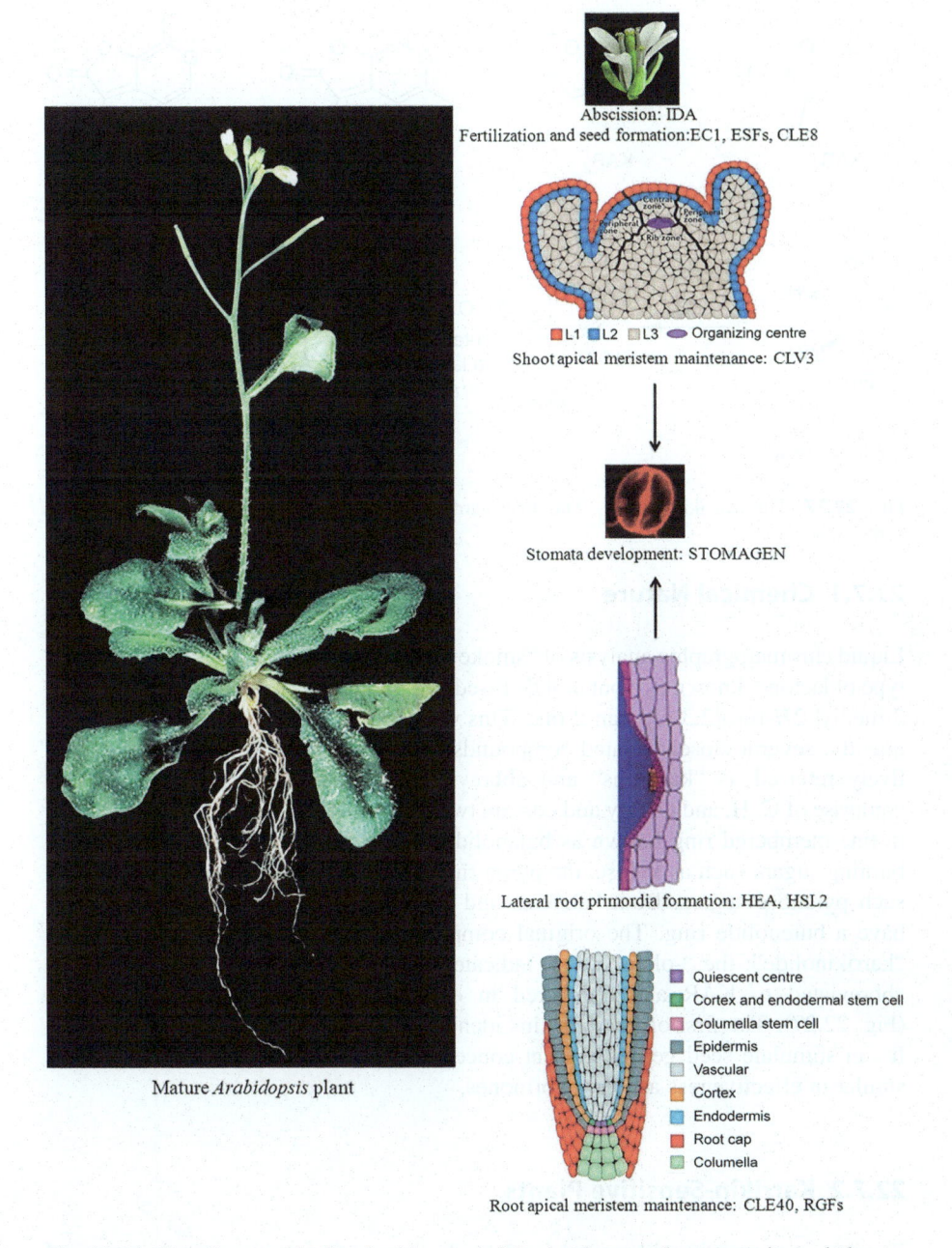

**Fig. 22.26** Signaling peptides involved in some of the developmental processes in *Arabidopsis*

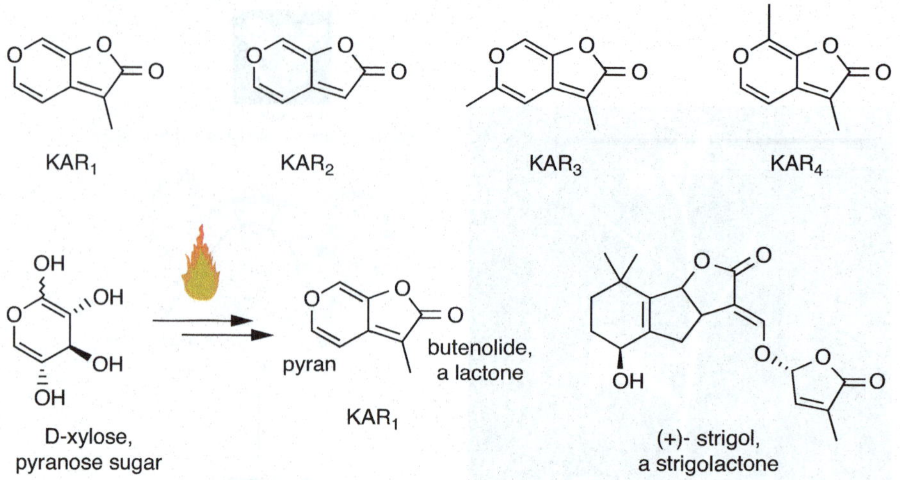

**Fig. 22.27** The karrikin family. The first karrikin discovered was KAR1, also known as karrikinolide

## 22.7.1 Chemical Nature

Liquid chromatographic analysis of "smoke water" lead to identification of a specific type of lactone, known as a butanolide fused to a pyran ring, with the systemic name: 3-methyl-2$H$-furo[2,3-c]pyran-2-one. This was the first karrikin discovered. Subsequently, several closely related compounds were discovered in smoke and collectively referred as "karrikins" and abbreviated as KAR (Fig. 22.27). Karrikins comprise of C, H, and O only and contain two ring structures—a pyran and a lactone (a fine-membered ring known as butanolide). Since karrikins can be produced by burning sugars such as xylose, the pyran ring of karrikins is probably derived from such pyranose sugars. Both karrikins and strigolactone hormones such as strigol have a butenolide ring. The original compound identified is often referred to as "karrikinolide": the "-olide" suffix indicates that it is a lactone. The karrikins are abbreviated to KAR and numbered in order of their identification in smoke (Fig. 22.27). The first of the karrikins identified from smoke is referred as KAR1. It can stimulate seed germination at concentrations as low as $10^{-10}$ M, which is similar in effectiveness as many hormones.

## 22.7.2 Karrikin-Sensitive Plants

Karrikin response is widespread, and seeds from different flowering plants and conifers respond to karrikins by exhibiting seed germination. Plants with smoke-responsive seeds are found in both fire-prone and non-fire-prone environment. Many weeds from agricultural land and seeds of horticultural plants, such as lettuce and

**Table 22.3** Some plants known to exhibit enhanced seed germination response due to karrikin

| |
|---|
| *Aristolochia debilis* (birthwort) |
| *Arabidopsis thaliana* (Arabidopsis) |
| *Brassica tournefortii* (Asian mustard) |
| *Lactuca sativa* (lettuce) |
| *Apium graveolens* (celery) |

tomato, also exhibit improved germination in response to karrikins. Karrikins have also been reported to bring about vigorous growth of maize and *Arabidopsis* seedlings (Table 22.3).

### 22.7.3 Mode of Karrikin Action in Plants

The genes for karrikin response are conserved in all seed plants. Two genes are essential for karrikin action. They are MAX2 (MORE AXILLARY GROWTH) and KAI2 (KARRIKIN INSENSITIVE 2). MAX2 is already known for its role in response to strigolactone hormone. Karrikins and strigolactones are perceived by plant cells separately and plants respond differently to the two classes of compounds. But the two classes of compounds are structurally closely related. All plants apparently contain KAI2 gene, but there is no evidence so far that plants produce karrikins. KAI2 gene encodes karrikin receptor—KARRIKIN INSENSITIVE 2 (KAI 2). KAI 2 is an $\alpha/\beta$ serine hydrolase involved in seed germination and seedling development. This protein has a key function in plant development, and it also responds to an endogenous signaling compound similar to karrikins (Fig. 22.28). This unidentified signaling compound is also likely to be similar to strigolactones since KAI2 is very similar to strigolactone receptor DWARF14. It is believed that plants take up karrikins and convert them to some active compounds which interact with KAI2 protein to cause subsequent seed germination response.

**Summary**

- Biosynthesis of salicylic acid in plants is occurs via two routes the *trans*-cinnamic acid (tCA) pathway and the chorismate pathway. The information available so far suggests the operation of both the pathways of SA biosynthesis in plants though species specificity and treatment specificity may be evident for the operation of one or the other of the two pathways. SA plays a significant role not only in the onset of senescence but also in its progression. During thermogenesis, most of the electron flow in mitochondria gets diverted from cytochrome respiration pathway to cyanide-insensitive non-phosphorylating electron transport pathway which is unique to plant mitochondria. The thermogenic inducing principle was initially called **calorigen** and was identified as SA in the 1980s. Thus, thermogenic respiration appears to be a metabolic event taking place rarely to help plants survive stress conditions. SA is an important participant in systemic acquired

**Fig. 22.28** Scheme of events leading to karrikin-induced seed germination

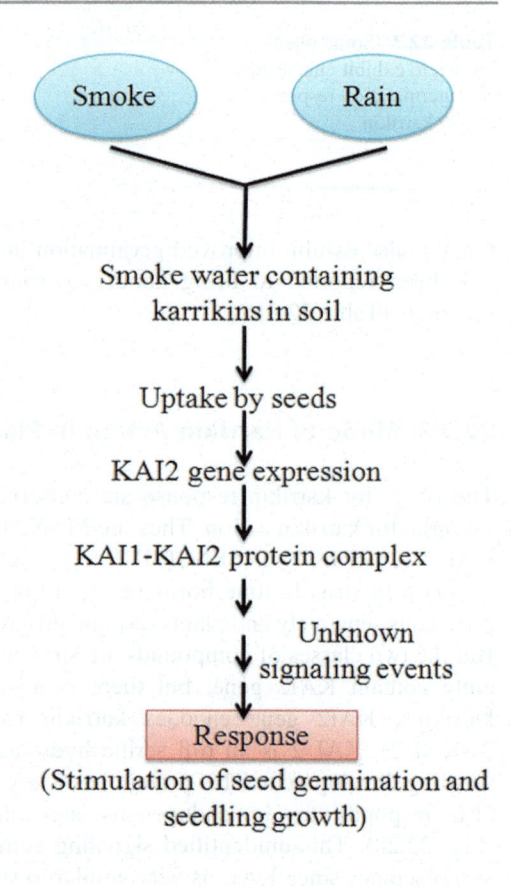

Smoke → / ← Rain
↓
Smoke water containing karrikins in soil
↓
Uptake by seeds
↓
KAI2 gene expression
↓
KAI1-KAI2 protein complex
↓
Unknown signaling events
↓
Response
(Stimulation of seed germination and seedling growth)

resistance (SAR) to pathogenic infections. SA induces SAR in a variety of pathogen-infected plants. SA has been observed to induce flowering in duckweeds, when these long-day plants are grown under noninductive short-day photoperiods.

- NO has emerged as an important signaling molecule with diverse physiological functions in growth and development, starting from seed germination, flowering, fruit ripening, and also during senescence of organs. Under environmental stress conditions caused by various biotic and abiotic stress factors, different plant species and organs exhibit enhanced NO generation. In plant cells, NO can induce both harmful and beneficial effects depending on its local concentration, translocation, rate of biosynthesis, and ability to get removed as various reactive nitrogen species (RNS). There are three probable routes of NO biosynthesis in plants: (1) L-arginine-dependent pathway, (2) nitrite-dependent pathway, and (3) nonenzymatic NO production. There is, however, enough controversy about the nature of NOS activity in plants. The AtNOS1 (putative NOS in plants) is a mitochondrial protein which bears a centrally positioned GTP-binding domain. In plant cells exposed to varied biotic or abiotic stress conditions, NO modulates the activity of various ROS-scavenging enzymes through their tyrosine nitration, metal nitrosylation, or S-nitrosylation thereby facilitating homeostatic regulation of enhanced ROS levels generated in plant cells under stress conditions. NO can modify phytohormone action in plants in three different ways. It can chemically modify the transcription factors (TFs) and other proteins (P) which are involved in phytohormone metabolism, transport, or signaling, 2. NO may modify proteins involved in the production, distribution, and signaling of plant hormones at post-translational level, 3. NO or RNS may also directly react with certain plant hormones thereby altering their biological activity.

- NO plays a key signaling role during the induction of hypersensitive response (HR). NO has also been implicated during bacteria-legume interaction accompanying nodule formation. There are reports that NO acts as a negative regulator of nitrogen fixation due to its interaction with leghemoglobin. Nitric oxide can be endogenously removed through the activity of an enzyme called GSNO reductase (GSNOR) which degrades S-nitrosoglutathione (GSNO). Since nitrite is a precursor of NO in plant cells, it is possible that nitrite can serve as a mobile source of NO across the xylem stream. Plants also contain hemoglobin genes, thereby indicating the possibility of hemoglobin-NO complexes as transport molecules for NO.
- Melatonin (N-acetyl-5-methoxytryptamine) is a pleiotropic molecule with numerous physiological and cellular actions. Like melatonin, serotonin is also a multifunctional indoleamine. Tryptophan is the common precursor for the biosynthesis of IAA and serotonin and melatonin. Tryptophan decarboxylase (TDC), with a high $K_m$ (690 µM) for tryptophan (its substrate), is the rate-limiting key regulatory enzyme of this pathway. One molecule of melatonin can scavenge up to four or more ROS. Serotonin plays a crucial role in root development and shoot morphogenesis. Serotonin has the highest antioxidant activity among tryptamine, tryptophan, and related derivatives. It helps in maintaining the reducing potential of cells due to its antioxidative properties by detoxifying the cells (relieves the accumulated toxic tryptamine). It delays senescence by efficient scavenging of ROS and protects young reproductive tissues from different environmental stresses. Serotonin regulates gene expression associated with auxin response pathways. Serotonin accumulation is likely to increase due to the abiotic stress-induced inhibition of auxin biosynthesis in plant tissues. Serotonin and melatonin prevent oxidative damage by regulating cellular ROS levels and serve as line of defense against stress. Melatonin has a protective role against leaf senescence.
- Strigolactones (SLs) are carotenoid-derivatives which were first identified as germination stimulants for the seeds of root parasitic weeds such as *Striga*, *Orobanche*, *Phelipanche* spp. Some of the common naturally occurring SLs are strigol, orobanchol, and sorgomol. GR24 is an extensively used synthetic SL. The common precursor for the major natural SLs is 5-deoxystrigol. SLs appear to act in coordination with auxin to modulate apical dominance. SLs promote the development of interfascicular cambium. SL signaling is believed to trigger PIN1 depletion from the xylem parenchyma cells. Accordingly, a reduction in auxin level enables the growth of lateral buds. SLs have also been reported to suppress **adventitious root** (AR) formation in *Arabidopsis* and pea. SLs

negatively regulate auxin accumulation in the pericycle, thereby reducing AR initiation. SLs are synthesized primarily in the root and they can be transported acropetally to the shoot in the xylem. SLs have been reported to get transported from the sites of their synthesis using plasma membrane-associated ABC transporters, and they move in the xylem from the root to shoot. SLs can induce ethylene biosynthesis in the seeds of parasitic plant—*Striga*—leading to seed germination. Auxins maintain apical dominance by stimulating SL synthesis via the *MAX4* gene. SLs then activate *BRC1*, a transcription factor known to suppress axillary bud growth. Additionally, SLs also inhibit cytokinin biosynthesis by negatively regulating the expression of *IPT* genes which otherwise would prevent BRC1 production. SLs are perceived by a protein complex which contains an α-/β-hydrolase protein and F-box protein D14 and MAX2, respectively.

- Polyamines are low molecular weight nitrogenous compounds which exist in positively charged state at physiological pH, a property that allows them to bind to all biological molecules bearing negative charge. So, polyamines have the ability to bind to and stabilize the anionic macro- and micromolecules within the cell, such as phospholipids in membranes, nucleic acids, proteins, and phenolic acids. The most commonly occurring polyamines are the aliphatic diamine putrescine, the triamine spermidine, and the tetramine spermine, with cadaverine being present in considerable levels in some plant species. Polyamines are localized in almost all compartments of the cell where they perform their regulatory roles in diverse cellular events. Putrescine is synthesized from arginine via two different biosynthetic pathways. Both the pathways of putrescine synthesis exist in plants. Spermidine and spermine synthesis requires putrescine as well as decarboxylated *S*-adenosylmethionine as precursor molecules. The growth promoting activities of polyamines are majorly accounted for by their roles in regulation of basic cellular events, like nucleic acid synthesis and protein synthesis. They bind to and modulate conformation of DNA and RNA, thereby facilitating the processes of replication, transcription, and RNA processing. Binding of polyamines regulates the 3D conformation of proteins and also their functions. In the cell wall, they bind to polysaccharides, pectins, and phenolic acids in lignin to enhance the strength of primary cell wall as well as deposition of secondary cell wall and lignification. Polyamines act antagonistically to ethylene to control leaf senescence and fruit ripening. They are thus considered "antisenescent" molecules as they prevent loss of chlorophyll and retard fruit ripening. Besides providing tolerance to pathogens, polyamine metabolism also influences the establishment and development of beneficial plant-microbe associations, such as those with nitrogen-fixing bacteria (rhizobia), mycorrhizal fungi, and growth-promoting rhizobacteria.

- An activation of genes in the intact zones of a wounded plant is referred as a **systemic response** which is triggered in the uninfected parts of the plant by a broad spectrum of pathogens, and it leads to plant immunization, thereby limiting pathogen growth. Plants achieve this systemic resistance through long distance transport of certain peptides known as **signal peptides.** Systemins are small polypeptide signaling molecules that play important roles in short-range intercellular communications. The first plant signaling peptide was discovered by Clarence E. Ryan in 1970 from tomato and was named systemin (TomSys). TomSys or systemin was characterized as an 18-amino-acid signaling peptide and was shown to be involved in the production of jasmonic acid, the main wound response signaling molecule. Systemin is capable of imitating defense responses at very low concentrations in the range of few femtomoles per plant. Systemin is synthesized at the wounding site by cleavage from the C-terminus of a 200 residue precursor protein called **prosystemin**. It then moves rapidly (within 1–2 h) through phloem to the unwounded parts of the plant. About 13 families of signaling peptides have been identified so far.

- Karrikins are a class of butanolide compounds derived from burnt plant material which can stimulate seed germination in a number of plants. Karrikins comprise of C, H, and O only and contain two ring structures—a pyran and a lactone (a fine-membered ring known as butanolide). Two genes are essential for karrikin action. They are MAX2 (MORE AXILLARY GROWTH) and KAI2 (KARRIKIN INSENSITIVE. All plants apparently contain KAI2 gene, but there is no evidence so far that plants produce karrikins. KAI2 gene encodes karrikin receptor—KARRIKIN INSENSITIVE 2 (KAI 2). KAI 2 is an α/β serine hydrolase involved in seed germination and seedling development. This protein has a key function in plant development, and it also responds to an endogenous signaling compound similar to karrikins. It is believed that plants take up karrikins and convert them to some active compounds which interact with KAI2 protein to cause subsequent seed germination response.

## Multiple-Choice Questions

1. Perception of signal peptides by cells occur via:
   (a) RLKs
   (b) MAPKs
   (c) CDPKs
   (d) PKC
2. First plant signaling peptide was named as:
   (a) Systemin
   (b) Prosystemin
   (c) Defensin
   (d) None of the above
3. Signaling peptide involved in the development of stomata in *Arabidopsis* is:
   (a) IDA
   (b) STOMAGEN
   (c) CLV3
   (d) HEA

4. SA has the following physiological functions:
   (a) To retard senescence
   (b) To induce flowering
   (c) To regulate thermogenesis
   (d) All of the above
5. Plants like *Sauromatum* sp. are known to generate large amount of heat as a result of:
   (a) Cytochrome respiration pathway
   (b) Alternative respiratory pathway
   (c) Release of amines
   (d) Photophosphorylation
6. Sequentially acting enzymes for synthesizing strigolactones in plastids are:
   (a) P-450, D27, CCD8
   (b) CCD7, D27, P-450
   (c) D27, β-carotene isomers, CCD7, CCD8
   (d) CCD7, CCD8, D27
7. Which other hormones functions in coordination with auxin to modulate apical dominance?
   (a) Jasmonic acid
   (b) Brassinosteroids
   (c) Salicylic acid
   (d) Strigolactones
8. Two indoleamines known to function as neurotransmitters in animals and growth and development in plants are:
   (a) Brassinosteroids and jasmonic acid
   (b) Serotonin and melatonin
   (c) Tryptophan and brassinosteroids
   (d) Salicylic acid and tryptophan
9. Two enzymes leading to the formation of serotonin are:
   (a) Tryptophan decarboxylase (TDC) and tryptamine hydroxylase
   (b) Tryptophan decarboxylase and tryptophan synthase
   (c) Tryptophan decarboxylase and anthranilate synthase
   (d) Tryptophan synthase and anthranilate synthase
10. Rate-limiting step in the synthesis of serotonin, melatonin and IAA is the one catalyzed by:
    (a) TDC
    (b) Tryptamine hydroxylase
    (c) Amine oxidase
    (d) 5-hydroxyindole-o-methyl transferase
11. The first polyamine reported by Anton von Leuwenhoek was:
    (a) Spermidine
    (b) Putrescine
    (c) Spermine
    (d) Cadaverine

12. Compounds responsible for distinctive odor of sperm and urine are:
    (a) Spermidine
    (b) Putrescine
    (c) Non-spermines
    (d) Thermospermine
13. Compounds helping in condensation of chromatin by stabilizing negative charges associated with phosphate groups in NAs:
    (a) Polyamines
    (b) Arginine
    (c) Methionine
    (d) S-adenosylmethionine
14. Anti-senescent molecules to control leaf senescence and fruit ripening are:
    (a) Ethylene
    (b) Polyamines
    (c) Abscisic acid
    (d) Jasmonic acid
15. A compound important for initiation of modulation and maintenance of nodules is:
    (a) NO
    (b) Polyamines
    (c) Salicylic acid
    (d) Strigolactones
16. NO is synthesized in plants through:
    (a) L-arginine-dependent pathway
    (b) Nitrite-dependent pathway
    (c) Nonenzymatic NO production
    (d) All of the above
17. NO can be endogenously removed through activity of enzyme:
    (a) Nitrate reductase
    (b) Nitrite reductase
    (c) S-nitrosoglutathione reductase (GSNO)
    (d) Oxidoreductase xanthine
18. Putative NOS (At NOS) is known to be present in:
    (a) Plastids and mitochondria
    (b) Apoplast
    (c) Chloroplasts
    (d) Cytosol
19. Compounds derived from burnt plant material, responsible for stimulating seed germination in plants are:
    (a) Strigolactones
    (b) Polyamines
    (c) Jasmonic acid
    (d) Karrikins

## Answers

| | | | | | | |
|---|---|---|---|---|---|---|
| 1. a | 2. a | 3. b | 4. d | 5. b | 6. c | 7. d |
| 8. b | 9. a | 10. a | 11. c | 12. b | 13. a | 14. b |
| 15. b | 16. d | 17. c | 18. a | 19. d | | |

## Suggested Further Readings

Flematti GR, Waters MT, Scaffidi A, Merritt DJ, Ghisalberti EL, Dixon KW, Smith SM (2013) Karrikin and cyanohydrin smoke signal provide clues to new endogenous plant signaling compounds. Mol Plant 6:29–37

Kaur H, Mukherjee S, Bhatla SC (2015) Regulatory roles of serotonin and melatonin in abiotic stress tolerance in plants. Plant Signal Behav 10:1–8

Liu JH, Wang W, Wu H, Gong X, Moriguchi T (2015) Polyamines function in stress tolerance: from synthesis to regulation. Front Plant Sci 6:1–10

Smith SM (2014) Q&A: what are strigolactones and why are they important to plants and soil microbes? BMC Biol 12:1–7

Vicente MR, Plasencia J (2011) Salicylic acid beyond defence: its role in plant growth and developmental. J Exp Bot 62:3321–3338

# Signal Perception and Transduction

**23**

Satish C Bhatla

While investigating the effect of unilateral light on the bending response of canary grass (*Phalaris canariensis*) seedlings, Charles Darwin (1881) observed that although light signal is perceived at the shoot tip, the bending of coleoptile due to differential growth occurs in the subapical region. This classic case of environmental signal perception and transduction resulting in growth response leads to the concept of signal transduction. It is now known that plant growth and development are modulated by a variety of environmental (external) and physiological (internal) signals. Some of the major signals (stimuli) to which plant cells are sensitive include light, mineral nutrients, organic metabolites, gravity, water status, soil quality, turgor, mechanical tensions, heat, cold, wind, freezing, growth hormones, pH, gases ($CO_2$, $O_2$, NO, $C_2H_4$), volatile compounds (e.g., jasmonates), electrical fluxes, wounding, and disease (Fig. 23.1). These signals can vary in quality and quantity over a period. Some signals penetrate across the plasma membrane, while others are carried over long transcellular distances through vessel elements (xylem) and sieve tubes (phloem). Plasmodesmata also facilitate symplastic migration of a number of signaling biomolecules. With the advent of molecular genetic studies on *Arabidopsis thaliana* in the current era of plant biology research, there has been a flood of information on signal perception and transduction mechanisms in plants. A variety of receptors for various plant hormones have been identified and characterized. Mutant analysis has facilitated the identification of many new signal transduction components which act downstream of receptors for various environmental and internal signals perceived by plants. Thus, an entirely new level of understanding of the complexities of signaling mechanisms in plants has been unfolded. New signaling molecules continue to be discovered, and sophisticated signaling mechanisms are being explained through the development of models which explain interactions between signaling pathways and modulation of various signaling networks (Fig. 23.2). The application of knowledge thus acquired on signaling

© Springer Nature Singapore Pte Ltd. 2018
S. C Bhatla, M. A. Lal, *Plant Physiology, Development and Metabolism*,
https://doi.org/10.1007/978-981-13-2023-1_23

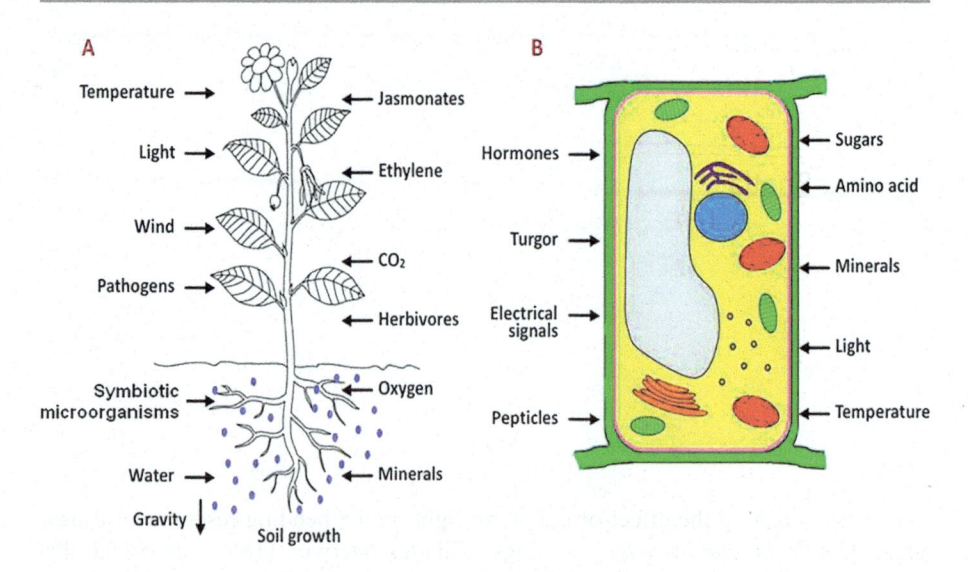

**Fig. 23.1** External (**a**) and internal (**b**) signals which modify cell metabolism, defense, and development

mechanisms using model plant systems such as *Arabidopsis thaliana*, to agriculturally significant species, is likely to provide practical benefits in understanding plant responses to varied environmental stresses.

## 23.1 Routes of Signal Perception, Transduction, and Response in Plants

Any environmental or intracellular input, which initiates one or more responses in the cell/plant, is referred as a **signal**. Plant cells are sensitized to the signals (**perception**) by employing specialized sensor proteins, termed **receptors**. Receptors may be located on the plasma membrane, cytoplasm, endomembrane system, or in the nucleus. In quite a few cases, receptors can move from one cell compartment to another. Sensing of signal by the receptor is followed by the transfer of the information from the signal from one biochemical form to another (**transduction**) so as to amplify the impact of signal leading to a cellular response. Signal transduction is achieved by the receptor-mediated modifications of other proteins. The response to a signal is regulated by the nature of plant part experiencing the signal (root, leaf, meristematic tissue, etc.), the stage of tissue development (physiologically active cells or fully differentiated cells), the previous environmental sensing, and the circadian clocks. The response due to a signal can be biochemical, physiological, morphological, or developmental. Also, the responses due to more

## 23.1 Routes of Signal Perception, Transduction, and Response in Plants

**Rapid signaling**

Dim light | Strong light | Tentacular movements

Blue light sensitive chloroplast movement | Insect trapping by Drosera

Wind-mediated bending | Reaction wood formation

**Long term signaling**

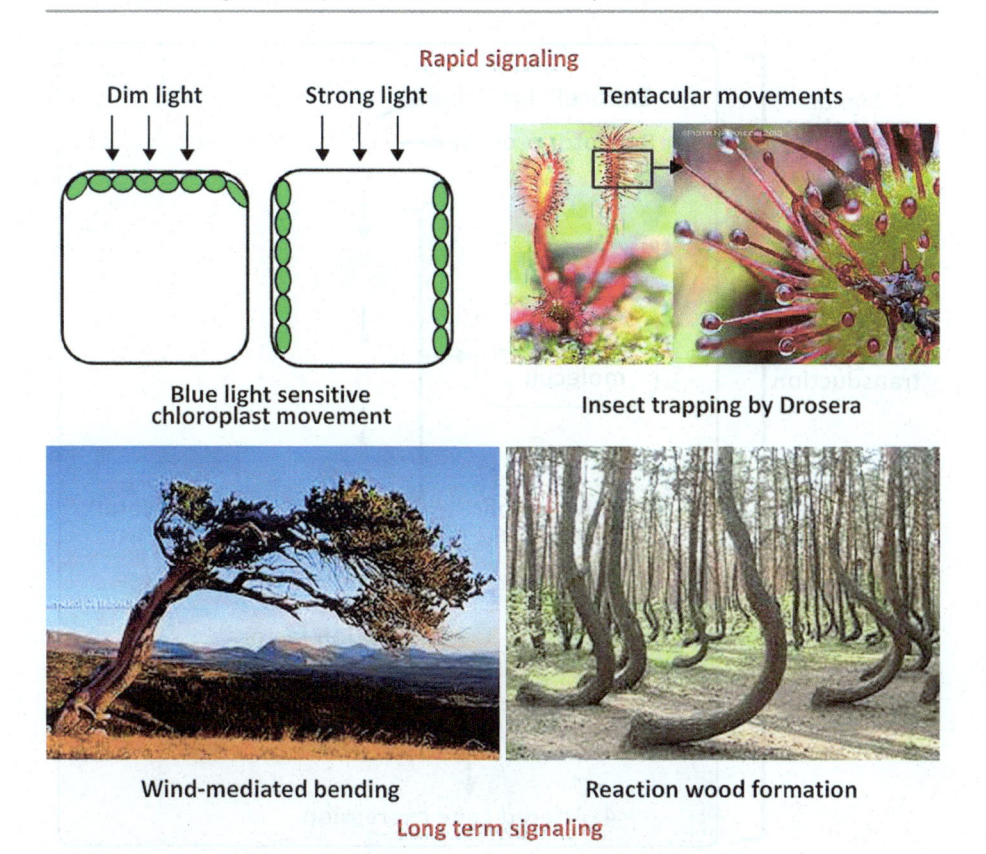

**Fig. 23.2** Some examples of rapid and long-term signaling events in plants

than one signal can often integrate to determine the final response and multiple responses from any signal (Fig. 23.3).

Plants cells contain **genetic and epigenetic information systems**. Although phenotypic characteristics of plants are governed by the genetic system (DNA → RNA → Protein → Phenotype), quite a few phenotypic characteristics are strongly modified by the environment as well. This includes biomass production, duration of growth, branching, partitioning of photosynthate between reproductive and vegetative structures, and responses to stress. Such phenotypic characters are modulated by **epistatic genes**. The products of epistatic gene expression regulate the expression of other genes inherited independently. They (epistatic gene expression products) also regulate the expression of **pleiotropic genes** (single genes which influence multiple traits). Thus, epigenetic characters (heritable phenotypes resulting from changes in chromosomes without alterations in the DNA sequence) are resulting from a web of interacting gene products derived through different signal transduction networks regulated by plant environment.

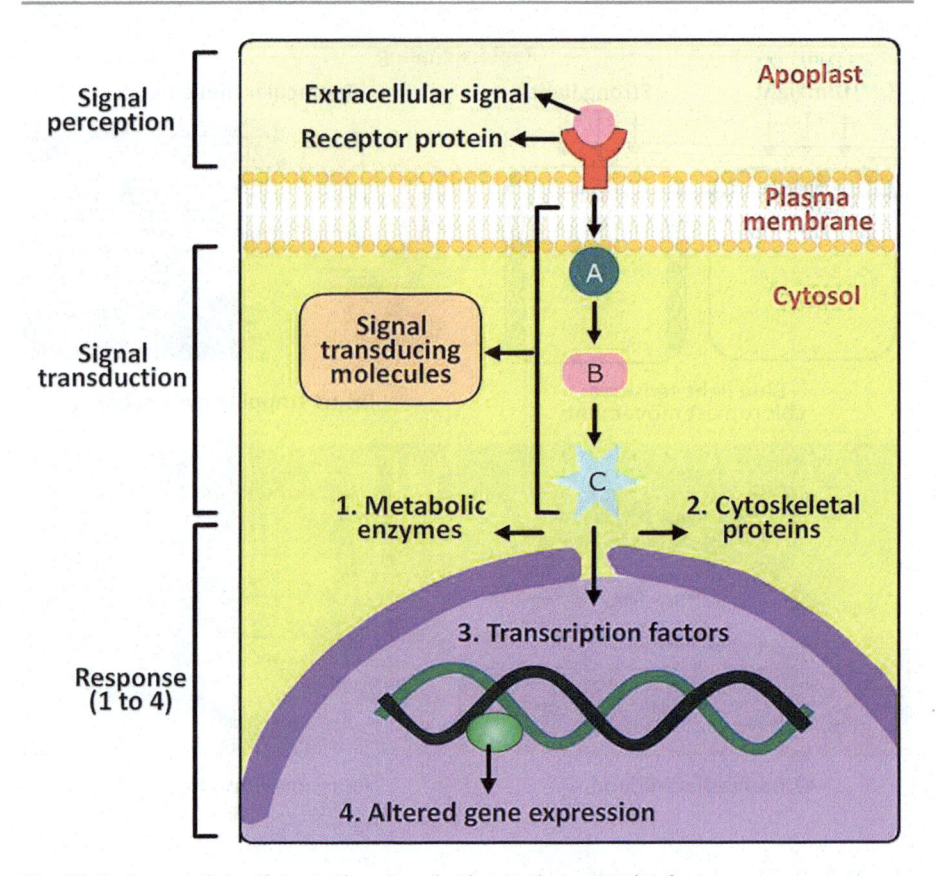

**Fig. 23.3** Routes of signal perception, transduction, and response in plants

## 23.2 Spatial and Temporal Aspects of Signal Transduction (Table 23.1)

A number of signaling events in plants involve signal reception and response in the same cell. For example, opening of stomatal aperture involves activation of membrane-associated ion transporters in response to blue light, leading to swelling of guard cells via phototropin (blue light receptors). Such plant responses to environmental signals are called **cell autonomous responses**. Many other responses involve signal perception in one cell and response in distal cells, tissues, or organs. Such signal responses are called **non-cell autonomous responses**. Some examples of non-cell autonomous responses include:

1. Short-distance signaling response of radial patterning of *Arabidopsis* primary root is regulated by the **transcription regulator SHORT-ROOT (SHR)**. SHR gene is transcribed and translated in the cells of stele, and SHR protein moves to

## 23.2 Spatial and Temporal Aspects of Signal Transduction (Table 23.1)

**Table 23.1** Spatial and temporal modulation of signal transduction events

| Spatial modulation | Temporal modulation |
|---|---|
| I. Cell autonomous (intracellular) responses | I. Involving change in protein/enzyme activity (most rapid responses) |
| 1. Stomatal opening and closing | 1. Closing of leaf traps of Venus flytrap upon contact with insect |
| 2. Chloroplast movement | 2. Chloroplast movement in response to light |
| II. Non-cell autonomous (transcellular) responses | II. Rapid response involving electrochemical changes |
| 1. Radial patterning in roots mediated by SHR transcription regulator | 1. Stomatal opening |
| 2. Gravitropic bending of roots modulated by differential auxin flow | |
| 3. Floral induction through transmission of FT from leaves to shoot apex | |
| | III. Moderately fast responses involving gene expression modulation |
| | 1. Insect attack on plants resulting in release of volatile compounds to attract insect predators |
| | IV. Long-term responses (involving chromatin remodeling) |
| | 1. Vernalization |
| | 2. Seed dormancy |
| | 3. Modulation of root branching in response to nutrition availability |
| | 4. Lateral bud outgrowth |

endodermis via plasmodesmata where it activates the expression of cell plate regulator, such as SCARECROW (SCR). Thus, the transcription factor, SHR, can be considered as a short-distance signal transducer operating between adjacent cells and tissue (Fig. 23.4a).

2. **Auxins** can be considered as a short- to long-distance signaling molecules operating between adjacent cells, tissue, and organs. Thus, changes in the position of statoliths accompanying gravitropism result in rapid redistribution of PIN3 (auxin efflux protein) to create an auxin gradient at the root apex. Laterally located root cap cells facilitate rapid transport of auxin from columella cells to the zone of elongation by expressing uniform distribution of AUX1 (an auxin influx carrier) and asymmetric localization of PIN2 (auxin efflux carrier) (Fig. 23.4b).

3. At times, **transcription regulators** are able to act over long distances to control developmental programs. For example, during the transition from vegetative to reproductive phase (floral induction), a long-day inductive signal induces the expression of FLOWERING TIME (FT) gene in the companion cells of *Arabidopsis* leaves. The resulting FT protein is then transported to the shoot

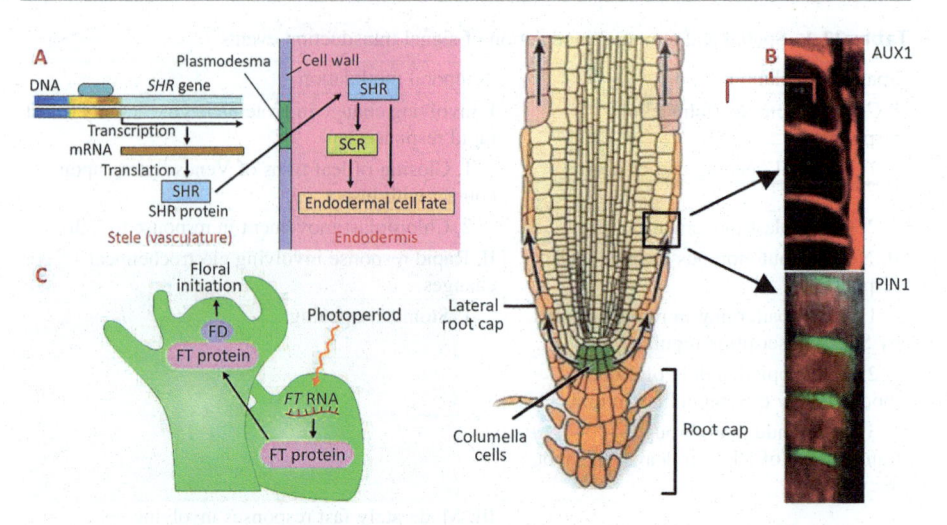

**Fig. 23.4** (**a**) Short-distance signaling of SHR protein between stele and endodermal tissue regulates radial patterning in roots; (**b**) regulation of auxin transport by the modulation of AUX1 and PIN1 proteins in the gravity sensing cells of root; (**c**) long-distance transport of FT protein from leaf to shoot tip leading to floral initiation

apex to interact with another transcription factor, FLOWERING LOCUS D (FD). FT-FD protein interaction coactivates other targets genes, triggering flowering (Fig. 23.4c).

Signal transduction processes in plants operate over time scales ranging from few seconds to several days. Those involving just a **change in protein/enzyme activity** and **ion fluxes** are most rapid and take place in few seconds. For example, Venus flytrap (*Dionaea muscipula*) closes its modified leaf traps within seconds of touch by an insect. The same holds true for the folding of leaflets of *Mimosa pudica* upon being touched. Likewise, reorientation of chloroplasts is regulated by PHOTOTROPIN (PHOT) class of blue light photoreceptors located on the plasma membrane. Light-induced autophosphorylation of PHOT induces rapid changes in the cell's cytoskeleton, resulting in chloroplast reorientation within seconds. Quite a few rapid signaling responses involve **electrochemical changes** (e.g., stomatal opening). Signaling responses involving gene expression are moderately fast since plant cells take up to 30 min to transcribe a gene, process its mRNA, and then export mRNA to the cytoplasm. Additional time is further required for protein synthesis and intracellular trafficking. For example, plants attacked by insects are known to emit volatiles to attract insect predators within a few hours. Processes taking place in such a time scale often involve new transcription and translation activity. Long-term responses involving multiyear signaling mechanisms also affect plant development. They include vernalization in biennial species, flowering in trees, seed dormancy, modulation of root branching in response to nutrition availability, growth of sun or shade leaves, and activation of lateral bud outgrowth. **Chromatin remodeling** is often associated with such long-term signaling responses.

## 23.3 Signal Perception

Although some events of signal transduction in plants can be mediated by physical forces generated by tissue growth and also by electrical events at the membrane, most of the signaling events are mediated by specific molecules which result in a change in receptor protein activity (Box 23.1). Interactions between various signal-transducing molecules often result in covalent modifications and/or allosteric changes, ultimately leading to transfer of signal information from the site of perception to the site of response within the cell. Neighboring cells communicate with each other both through **apoplast** (formed by the interconnection of cell walls) and **symplast** (the cell-to-cell cytoplasmic continuum via plasmodesmata). Plasma membrane is the site for apoplastic sensing of environmental signals. Receptors located on the plasma membrane can perceive both physical signals (e.g., mechanical forces, such as touch and blue light) and chemical signals (such as hormones and signal peptides). Signaling via symplast permits regulated movement of RNAs and transcription factors, which induce response in the signal receiving cells. In addition to apoplastic and symplastic routes of signal perception, a third category of small lipophilic signaling molecules (e.g., ethylene, auxin) can cross the **plasma membrane** and is perceived in the cytoplasm or nucleus to evoke a response. Thus, ethylene is diffusible across the plasma membrane of plant cells and binds with its receptors located on the endoplasmic reticulum membrane. Likewise, auxins and gibberellins are perceived by the cytosolic (soluble) receptors which interact with intracellular components of protein degradation pathway to evoke a response.

> **Box 23.1: Contributions of Jagadis Chandra Bose (1858–1937) in Plant Electrophysiology**
>
> (continued)

**Box 23.1** (continued)

*All creative scientists know that the
true laboratory is the mind, where behind
illusions they uncover the laws of truth.* J.C.BOSE

Jagadis Chandra Bose, who is generally acknowledged as the father of modern scientific research in India, began biophysical experiments in plants in 1897. The main focus of his investigations in plant physiology was to establish that all characteristics of responses to external factors exhibited by animal tissues are equally exhibited by plant tissues. He theorized that regular wave-like "pulsations" in cell electric potential and turgor pressure were an endogenous form of cell signaling. Bose selected plants which gave quick responses to external stimuli and displayed intrinsic rhythms. These include *Mimosa pudica*, which folds leaflets or dips the entire leaf upon stimulation. Likewise, Bose worked on the telegraph plant, *Desmodium*, which exhibits remarkable "spontaneous" gyration of the lateral leaflets as a result of rhythmic up and down movements. Bose devised an instrument called **crescograph** to measure mechanical responses, such as drooping of leaves at narrow time intervals (<1 to 2 s), in response to touch, heat, and chilling. He also experimented with the effects of changes in turgor pressure, sudden changes in temperature, and inhibitory poisons, such as KCN or anesthetics (e.g., chloroform and ether). Bose concluded that, like in animals, a stimulus was transmitted electrically to the motor organ, the pulvinus, both in *Desmodium* and *Mimosa*. The major conduction pathway (established with electric probe) was established to be phloem. Transmission of the response was strongly temperature-dependent and also influenced by light. He further observed that turgor decrease and cell contraction, accompanied by an electrical response, lead to leaf drooping in *Mimosa* and "downstroke" of *Desmodium* leaflet. Bose also used electric probes to measure periodic electrical "pulsations" in the inner cortical cells abutting endodermis of the herbaceous plant, *Impatiens*, *Chrysanthemum*, *Canna*, tomato, and potato. He reasoned that if all the inner cortical cells pulsed or contracted at the same moment, there would be no flow of water through xylem. Thus, there has to be a phase difference. J.C. Bose died in 1937, and in the current era of research in plant electrophysiology, his theories on electrical signaling and electromechanical oscillations in plant cells are being highly appreciated.

(continued)

**Box 23.1** (continued)

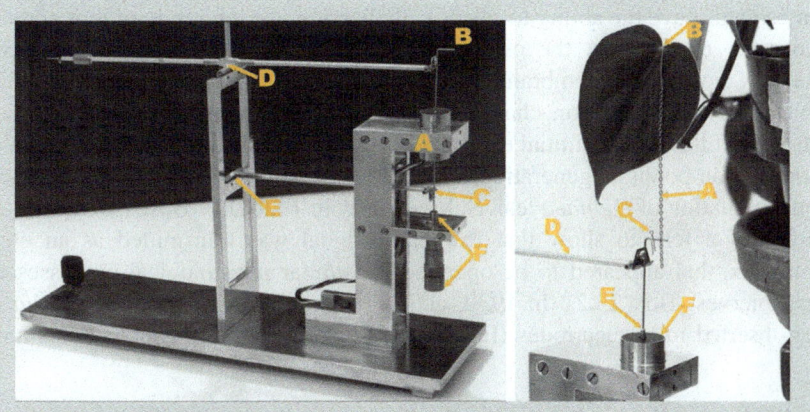

**Modern crescograph:** The component which actually measures growth movement is a differential transformer (A). Its movable core is hinged between points (B) and (C). Two lever arms hinged at points (D), (B), (E), and (C) form a parallelogram that holds the core centered within the opening of the differential transformer as it moves up and down. A micrometer (F) is used to adjust and calibrate the system. Here a *Philodendron oxycardium* leaf is attached by a hook (b) and a gold chain to the plant movement detector. The lever arm (d) and transformer core (e) are held taut against the chain (a) by a 5 mg weight (c). A movement of the core (e) within the opening of the differential transformer (f) produces a change in the output voltage of the transformer which is proportional to the movement.

Selected publications of J.C. Bose:

1. Bose J.C. (1902) Response in Living and Non-Living. Longmans Green, London, UK.
2. Bose J.C. (1907) Comparative Electrophysiology. Longmans Green, London, UK.
3. Bose J.C. (1913) Researches on Irritability of Plants. Longmans Green, London, UK.
4. Bose J.C. (1918) Life Movements in Plants. B.R. Publishing Corp., Delhi, India.
5. Bose J.C. (1926) The Nervous Mechanism of Plants. Longmans Green & Co., London, UK.
6. Bose J.C. (1928) The Motor Mechanism of Plants. Longmans Green, London, UK.

## 23.3.1 Membrane Potential as a Receptor

Many signals have the ability to selectively discriminate migration of ions across the membrane, resulting in the establishment of a potential difference of −80 to −200 mV across the membrane. The plasma membrane potential, thus modified, activates voltage-gated ion channels which further activate a signal transduction sequence. Following the initial experiments conducted by Charles Darwin on Venus flytrap, action potential generation was first recorded in 1873 by Burdon Sanderson upon stimulation of *Dionaea* leaf. Subsequently in 1926, J.C. Bose isolated vascular bundles of a fern to show that excitation signal was transmitted as an electric disturbance that appeared to be operated by similar physiological processes as in animal nerves (Box 23.2). In 1930, action potential was recorded in *Nitella* cells using inserted microelectrodes (Fig. 23.5). Finally, in 1950s, Sibaoka was able to

---

**Box 23.2: Sensitive Plants**

A number of plants with compound leaves (e.g., *Mimosa pudica* and *Cassia fasciculata* from Leguminosae and *Biophytum sensitivum* from Oxalidaceae) are called as sensitive plants because they exhibit rapid folding of leaflets upon mechanical disturbance by touch or shaking. In nature, this behavior may protect leaves from being eaten by animals and also protect them from damage due to rain or wind. Such movements in the double pinnate leaves are regulated by motor cells in the pulvinules of individual leaflets (pinnules), in the pulvini at the juncture between pinnae and the rachis, and even in the pulvinus at the base of the petiole (the juncture with the stem). The motor cells in the lower (abaxial) sector of the pulvinus undergo immediate loss of water and turgor, and folding of leaves is completed within 1–2 s. Original configuration is recovered by the uptake of osmotically active solutes ($K^+$ and $Cl^-$) into the vacuole, followed by water uptake. The process of recovery is slow and is completed within 15 min. Recovery of pinnules and pinnae (but not of the base of petiole) takes place only in light, which is required to activate plasma membrane $H^+$-ATPase to enable $K^+$ and $Cl^-$ influx into the cell. Sensors of mechanical perturbation (**motor cells**) are located in pulvinus, and their excitation spreads rapidly from one to the neighboring motor cell, probably through electric signals, leading to leaf folding. Such responses are called **seismonastic responses**. Insect-trapping movements of insectivorous plants are also driven by structural deformations caused by mechanical perturbation of specialized sensory organs, such as bristles (in *Dionaea muscipula*) or tentacles (in *Drosera tentaculata*).

## 23.3 Signal Perception

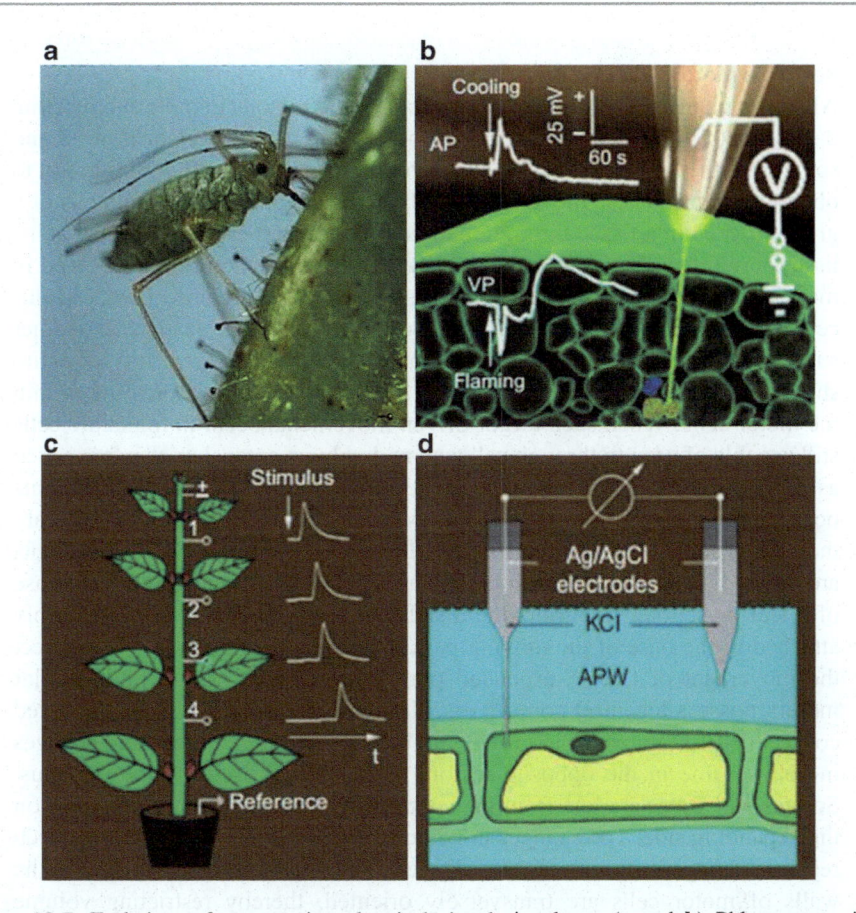

**Fig. 23.5** Techniques for measuring electrical signals in plants. (**a** and **b**) Phloem potential measurement. (**a**) An aphid with its stylet inserted into a stump to obtain phloem exudate. (**b**) A tip of the microelectrode is attached to the stylet after separating it from aphid. (**c**) Extracellular recording of electrical signals with four channels and a reference electrode inserted in the soil ± electrical stimulation. An action potential generated by electrical stimulation appeared successively at electrodes 1–4. (**d**) Intracellular measurement of membrane potential with a microelectrode inserted into the cytoplasm of a cell while the reference electrode is in contact with bathing solution

demonstrate propagation of electrical signals in *Mimosa pudica*. Electrical signals not only trigger leaf movements in "sensitive" plants, such as *Mimosa pudica* or *Dionaea muscipula*, they also stimulate physiological processes in other plants through generation of action potential (Box 23.3). It is likely that plants have developed pathways for electrical signals leading to rapid response to external stimuli, such as environmental stress. In contrast to chemical signals, such as hormones, electrical signals are able to rapidly transmit information over long distances (Table 23.2). Touch-sensitized drooping of leaflets of *Mimosa pudica* is a classic example where membrane potential serves as a receptor (Fig. 23.6). Each leaf has a pulvinus at the base of its petiole. Many secondary pulvini are located at the bases of individual leaflets. When turgid, pulvini hold leaflets away from the stem and remain expanded. Loss of turgor in pulvini results in the folding of leaflets.

> **Box 23.3: Motor Cells and Tissues**
> Although plants are rooted in place, their specific regions in roots and in aerial parts are constantly in motion. Thus, some move their leaves or flowers and some twist their whole body. Likewise, roots grow deeper and penetrate soil to obtain nutrients. These movements are brought about by individual cells or a group of cells in a tissue through their contraction and expansion. Such cells/tissues are referred as **motor cells/tissues**. Single-cell motors are involved in movements of stomatal guard cells, root hairs, and pollen tube growth. Multicell motors are responsible for moving entire leaves or leaflets and other subterranean plant parts, apical buds, flowers or inflorescence, fruits, or entire shoots. Most movements are driven by turgor or hydrostatic pressure within the motor cells. When such turgor-driven movements occur in immature cells still capable of growth through cell wall and volume extension, they are called as **growth-mediated movements**. When these turgor-mediated movements occur in mature cells, which undergo reversible, elastic extension of cell wall, they are designated as **turgor-mediated movements**. Most plant movements are driven by multicellular motors such as pulvinus—a tissue found at the base of a leaf or petiole. As a rule, the distal part of multicellular motor is firmly attached to the base of the moving part, thereby allowing the motor to direct the movement of freely suspended portion. Most, but not all multicellular motors possess a central core of vascular tissue surrounded by multilayered cortical cells. Differences in growth are brought about by differential changes in cell volume in the opposite sectors of the motor tissue of the pulvinus. Solutes and water are lost from the contracting side and taken up by cells on the expanding side. The changes in volume of the pulvinar motor tissue, which result in change in curvature, are fully reversible. Cellulose microfibrils in the walls of motor cells are transversely oriented, thereby restricting volume changes in motor cells along the pulvinar axis. Furthermore, epidermis restricts radial expansion of the pulvinus. Veins in the pulvinus coalesce thereby allowing any change in the radius of curvature without changing length. Motor cells generate enough metabolic energy to execute these processes, as is evident from large nuclei, abundance of mitochondria, chloroplasts, and ER. Root hairs represent unicellular, growth-mediated motors, as are pollen tubes. On the other hand, stomata represent unicellular, turgor-mediated motors. Such motors bring about change in conformation (spatial architecture) of the cell.

Leaf stroking (touch) generates an action potential, i.e., membrane is depolarized to a voltage more positive than the threshold voltage. This results in massive movement of $K^+$ and $Cl^-$ in the motor cells which causes loss of turgor in the pulvini and consequent folding of leaflets. It may be noted that the upper and lower halves of pulvinus exhibit differential changes in action potential, which facilitates leaflets folding and unfolding (Box 23.4).

## 23.3 Signal Perception

**Table 23.2** Some physiological effects of electrical signals in plants

| Plant | Stimulus | Physiological effects | Signal |
|---|---|---|---|
| *Mimosa pudica* (Touch Me Not Plant) | Cold shock, mechanical (touch) | Leaf movement | Action potential |
| *Drosera tentaculata* (Sundews) | Mechanical | Movement of tentacles around insect | Action potential |
| *Dionaea muscipula* (Venus flytrap) | Mechanical | Release of digestive enzymes; trap closure | Action potential |
| *Hibiscus sp.* (Shoeblack plant) | Pollination | Increase in respiration | Action potential |
| *Pisum sativum* (Garden Pea) | Wounding | Inhibition of protein synthesis | Variation potential |
| *Luffa sp.* (Sponge gourd) | Electrical, cooling | Decreased elongation and growth of stem | Action potential |

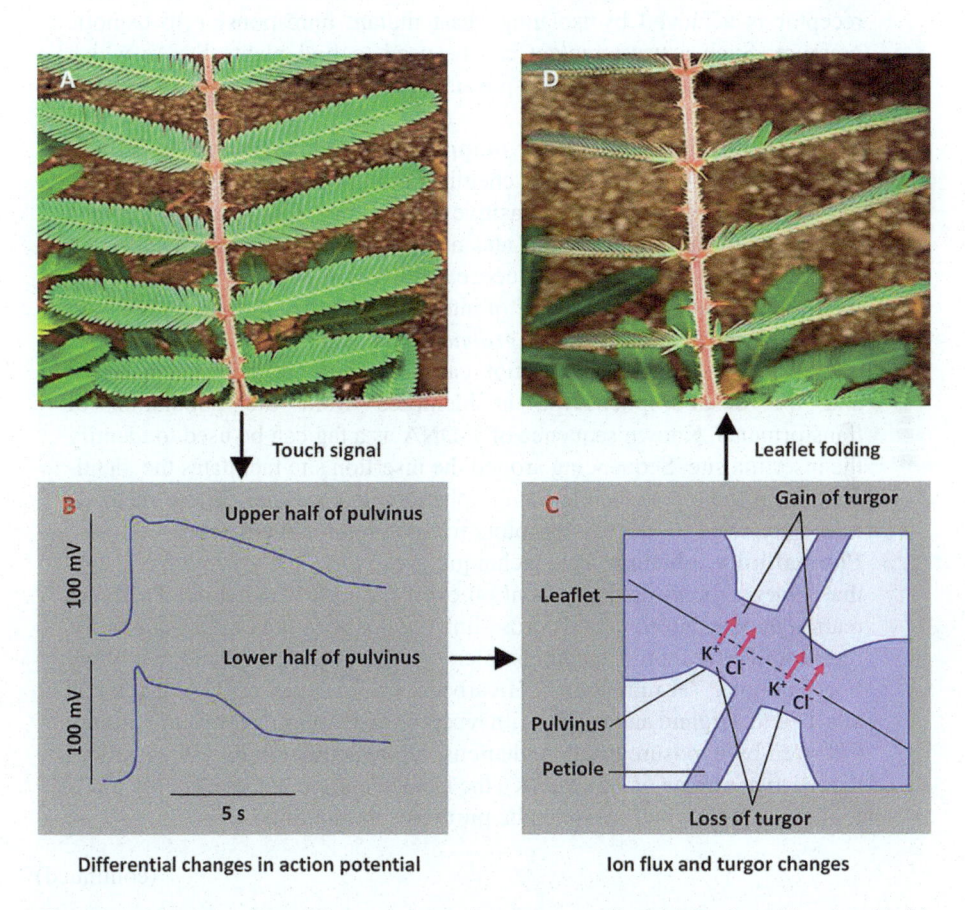

**Fig. 23.6** Membrane potential as a receptor in *Mimosa pudica*. (a) Leaf stroking (touch) (b) generates an action potential (c) resulting in massive movement of $K^+$ and $Cl^-$ in the motor cells in the pulvinus. (d) This causes loss of turgidity and consequent folding of leaflets

**Box 23.4: Detection and Identification of Plant Receptor Proteins**
As compared to enzymes and other proteins, the cellular concentrations of receptor proteins are much less. Therefore, their detection requires special techniques. Some of the approaches being used to detect plant receptor proteins are as follows:

1. **Functional complementation for gene identification:** In this approach, a plant gene restores a wild-type phenotype in a mutant strain of yeast. Complementation requires isolation of a yeast mutant which is deficient in the plant character under investigation. The yeast is transformed with a plant cDNA library, followed by isolation of a complemented yeast clone and sequencing of plant transgene. For example, osmotic pressure responses are shared by plants and yeast. Identification of osmosensing receptor is achieved by isolating yeast mutant unresponsive to osmotic pressure. Such a yeast mutant can be used to find plant cDNAs which might encode plant receptors responsible for signaling changes in osmotic pressure.
2. **Mutant analysis by molecular mapping and "chromosome walking":** Mutant populations obtained by chemical mutagenesis are used to identify desired phenotypes that are insensitive to the signal. Chromosomal regions are identified through selection and mapping. Chromosomal "walks" are used to identify the mutant sequence, which is then used to identify the wild-type sequence and the gene of interest. Genes can also be tagged with T-DNA from the bacterium (*Agrobacterium tumefaciens*) which causes crown gall disease. Transformation can be carried out in plants or seeds, and the desired phenotype is identified from the population of transformants. Known sequence of T-DNA as a tag can be used to identify the insertion site. Sequencing around the insertion site facilitates the identification of gene responsible for the phenotypic character. These methods have been used to identify receptors for blue light and ethylene.
3. **Photoaffinity labeling:** This technique uses isotopically labeled reagents that undergo bond rearrangement on exposure to UV radiation. The bond rearrangement can covalently cross-link the ligand to the binding site of the receptor. The receptor-ligand complex can then be purified and identified. For example, $^3$H-nitrene or $^3$H-carbene (radioactive affinity labels) is attached to a ligand and mixed with receptor preparation to prevent binding, followed by exposure to UV radiations. Upon activation by UV radiation, the reactive nitrene or carbene (R), the labeled compound irreversibly binds to the receptor, which can then be purified and identified.

(continued)

**Box 23.4** (continued)

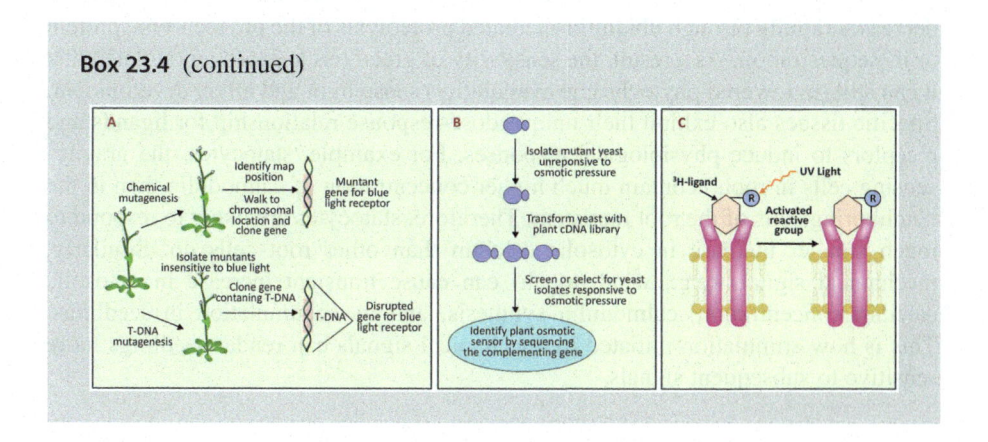

## 23.3.2 Characteristic Features of Membrane Receptors

A number of plant cell membrane-localized receptors identified so far are proteins and possess following features:

1. High binding affinity for ligands.
2. Reversibility of ligand-receptor binding. This allows the system to respond to changes in ligand concentration since receptors are generally present in low abundance.
3. Saturation of ligand-receptor binding at certain concentration.
4. Selectivity of receptors for biologically active ligands.
5. Ligand-receptor binding mimics physiological activity.
6. Affinity constant ($K_d$) for receptor-ligand binding correlates with ligand concentration active in vivo.

## 23.3.3 Tissue Sensitivity for Receptor-Mediated Signaling Responses

Various tissues or cell types respond differently to the variety of signals they are exposed to. Thus, for example, fruit tissues become sensitive to ethylene after achieving a certain stage of ripening. In contrast, guard cells are totally insensitive to high concentrations of ethylene. Same signal can lead to different responses in different tissues through its ability to bind to different types of receptors. Thus, auxin can induce pericycle cells to form adventitious or lateral roots, but it promotes cell elongation in coleoptile cells. Both, different receptors and divergent downstream elements of signal transduction pathway, may be responsible for such variations in auxin (or any other hormone)-modulated responses in different cell types. It is also likely that different tissues may adapt or desensitize themselves to continuous signals and receptor concentrations during development. Thus, for example, when etiolated seedlings are exposed to red or white light, phytochrome concentration decreases rapidly through ubiquitin-mediated proteolysis of the phytochrome protein or its sequestration. As a result, the sensitivity of green tissue gets so modulated that it can still (in lowered phytochrome availability) sense light and affect development. Specific tissues also exhibit their unique dose-response relationship for ligands and receptors to induce physiological responses. For example, statocytes (the gravity-sensing cells in roots) contain much higher concentration of calmodulin than in the neighboring cells of the root meristem. Therefore, statocytes are likely to respond to much smaller increase in cytosolic calcium than other root cells do. Similarly, mechanical signals (e.g., wind, touch) can cause transient increase in cytosolic calcium concentration, calmodulin synthesis, and its accumulation in seedlings. This is how stimulation initiated by mechanical signals can render seedlings more sensitive to subsequent signals.

## 23.4 Signal Perception at the Plasma Membrane

Large and hydrophobic molecules, for which import channels do not exist on the plasma membrane, can only be perceived by plasma membrane receptors. Additionally, membranes also perceive physical signals, such as mechanical forces and blue light (via phototropins). Small lipophilic molecules can, at times, cross the plasma membrane and are perceived in the cytoplasm or nucleus. Thus, ethylene can cross plasma membrane and is perceived by receptors on the endoplasmic reticulum membrane. Likewise, auxins and gibberellins are perceived by soluble receptors. Phytochrome and cryptochrome receptors also have cytoplasmic and nuclear locations. Three major groups of plasma membrane receptors can be distinguished on the basis of how they interact with signaling components. These are receptor kinases, G-protein-coupled receptors, and ion channel receptors.

### 23.4.1 Receptor Kinases

They constitute the largest group of membrane receptors in plants and are responsible for transducing extracellular signals through phosphorylation of intracellular targets. One of the most common post-translational modifications used by cells to alter protein stability is through phosphorylation catalyzed by kinases. Kinases have the ability to alter protein stability, subcellular location, binding properties, enzyme activity, and their susceptibility to further modifications. It is also possible that a protein may be phosphorylated at more than one amino acid residue, involving more than one kind of kinases, thereby eliciting different effects. Thus, kinases and phosphatases (which dephosphorylate proteins) play crucial roles in signal transduction mechanisms. **Receptor-like kinases (RLKs)** constitute a family of over 600 proteins in *Arabidopsis* and over 1100 proteins in rice, which are known to phosphorylate target proteins in plant cells at serine and threonine residues. RLKs possess ligand-binding sites on an extracellular domain and on an intracellular kinase domain. RLKs are involved in a wide range of plant signaling processes,

## 23.4 Signal Perception at the Plasma Membrane

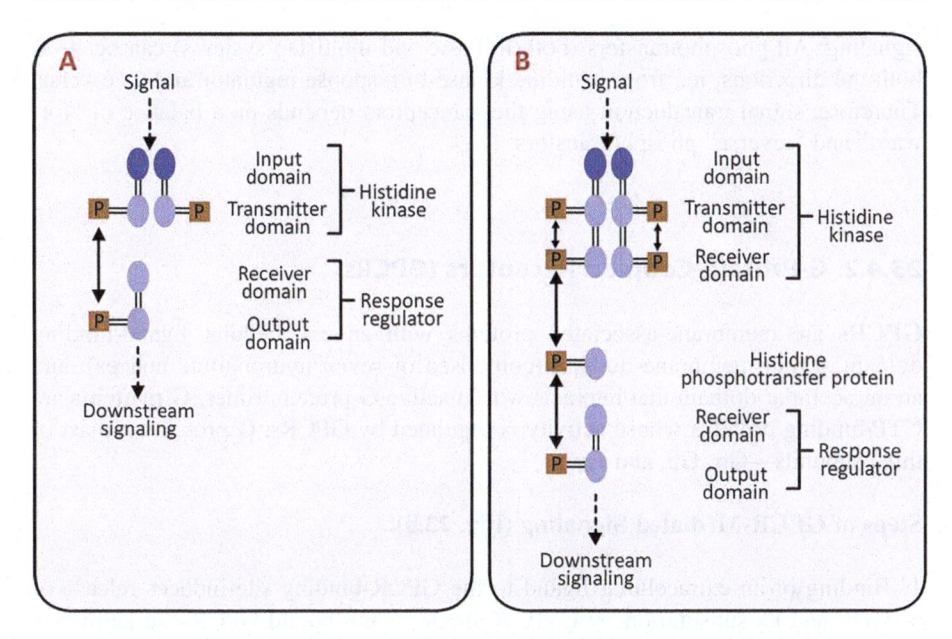

**Fig. 23.7** Histidine kinase receptors. (a) The two-component system and (b) its multistep phosphorelay derivative

including hormone perception, defense, development, symbiosis, and pollen tube germination and guidance. In addition to RLKs, plants also possess **histidine kinase receptors** derived from bacterial two-component system. In addition to RLKs, plants possess **histidine kinase receptors** for ethylene and cytokinin signaling. These kinases are multistep derivatives of bacterial **two-component systems** which consist of two proteins—a **histidine kinase** and a **response regulator** (Fig. 23.7a). The histidine kinase may be membrane spanning or soluble and exists as a dimer. Each kinase monomer has an input domain and a transmitter domain. The input domains of both monomers perceive signal and reciprocally autophosphorylate a histidine residue in each other's transmitter domain. The response regulator also contains a receiver domain and an output domain. Phosphoryl groups from the autophosphorylated histidine kinases's transmitter domain are transferred to an aspartate residue in the receiver domain of the response regulator, thereby activating the output domain. This triggers onward signaling events including gene transcription by activated response regulators which act as transcription factors. Prokaryotes, plants, and fungi (but not animals), in contrast with bacterial two-component system, **multistep phosphorelay** with three components: histidine kinase, histidine phosphotransfer protein, and response regulator (Fig. 23.7b). The histidine kinase often contains its own receiver domain into which phosphoryl group is transferred after autophosphorylation of its transmitter domain following signal perception. The phosphoryl group from the receiver domain histidine is transferred to an intermediate histidine phosphotransfer protein (HPt) and then to response regulator for further signaling. All phosphotransfers (both in basic and multistep systems) can occur in both the directions, ie., from histidine kinase to response regulator and vice versa. Therefore, signal transduction using these receptors depends on a balance of "forward" and "reverse" phosphotransfers.

### 23.4.2 G-Protein-Coupled Receptors (GPCRs)

**GPCRs** are membrane-associated proteins with an extracellular ligand-binding domain, a transmembrane domain (composed of seven hydrophobic helices), and an intracellular domain that interacts with inactive G-protein trimer. **G proteins** are GTP-binding proteins whose activity is regulated by GPCRs. G proteins consist of three subunits—Gα, Gβ, and Gγ.

**Steps of GPCR-Mediated Signaling (Fig. 23.8):**

1. Binding of an extracellular ligand to the GPCR-binding site induces release of GDP and its substitution by GTP. A single ligand bound GPCR can amplify a signal by activating multiple G proteins. Thus, GTP binding to G protein at the Gα subunit marks the first stage of signal transduction and activates G protein.
2. Activated G protein uncouples itself from the receptor (GPCR) and dissociates into Gα subunit and a Gβ/Gγ dimer.
3. Gα and Gγ possess short covalently attached lipid tails which anchor them to the inside of PM and activate downstream signaling proteins in the cytoplasm.
4. Active Gα (with GTP bound to it) is inactivated by the hydrolysis of bound GTP to GDP, which is stimulated by GTPase-activating protein (GAP).
5. Signaling is terminated by the hydrolysis of GTP to GDP by GTPase-activating proteins (GAPs). GDP-bound Gα rejoins with Gβ/Gγ dimer to form an inactive trimer which again binds to GPCR to complete the cycle. In contrast with the role played by GAPs in activating GTPase, guanine nucleotide exchange factors (GEFs) are known to activate GTPases by replacing GDP with GTP.

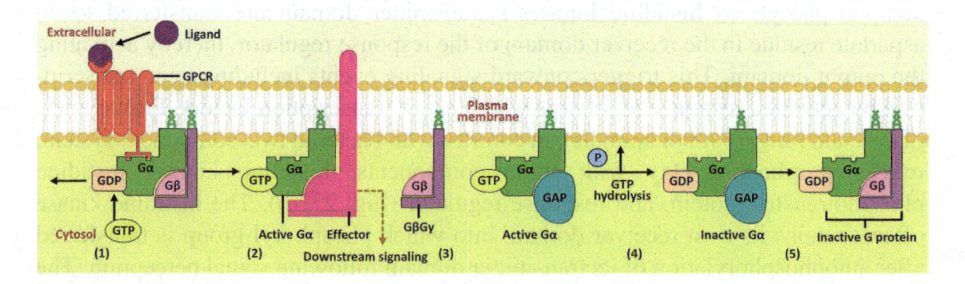

**Fig. 23.8** Steps of G-protein-coupled receptor (GPCR)-mediated signaling. Gα, Gβ, and Gγ are three subunits of GTP; *GAP* GTPase-activating protein, *GDP* guanine diphosphate, *GEF* guanine nucleotide exchange factor, *GPCR* G-protein-coupled receptor, *GTP* guanine triphosphate

## 23.4 Signal Perception at the Plasma Membrane

6. Monomeric G proteins are activated by guanine nucleotide exchange factors (GEFs) which induce conversion of GDP to GTP. The hydrolysis of GTP to GDP is stimulated by GTPase-activating proteins (GAPs). Lastly, GDP dissociation inhibitors (GDIs) prevent the spontaneous release of GDP by monomeric G proteins.

Tip growth and polar extension in plants are regulated by a family of small GTPases called ROPs (for Rho-like GTPase). ROP1 GTPases are localized on the PM at the tip of the growing pollen tube (Fig. 23.9). As with other GTPases, ROP1 activity can be switched on and off by GEFs and GAPs, respectively. PM-localized RLK is believed to activate GEF, which activates ROP1. ROP1 stimulates NADPH oxidase activity leading to ROS production. ROS, in turn, promotes $Ca^{2+}$ influx from the extracellular space, which enhances tip growth.

### 23.4.3 Ion Channel-Linked Receptors

Ion channels allow inorganic ions to migrate across membranes according to their electrochemical gradients. Selectivity of ion channels is determined by the pore diameter and charges on the amino acid residues which line the channel passage way. Opening of ion channels can be regulated by their association with a ligand, change in voltage, or by mechanical force. So far, no plasma membrane-associated ion channels have been reported to function as receptor in response to ligand binding. However, PM-associated mechano-sensing calcium channels do play a role in plant responses to physical force. Plants exhibit developmental responses to touch, gravity, wind, and soil pressure, mediated by rapid and transient increase in

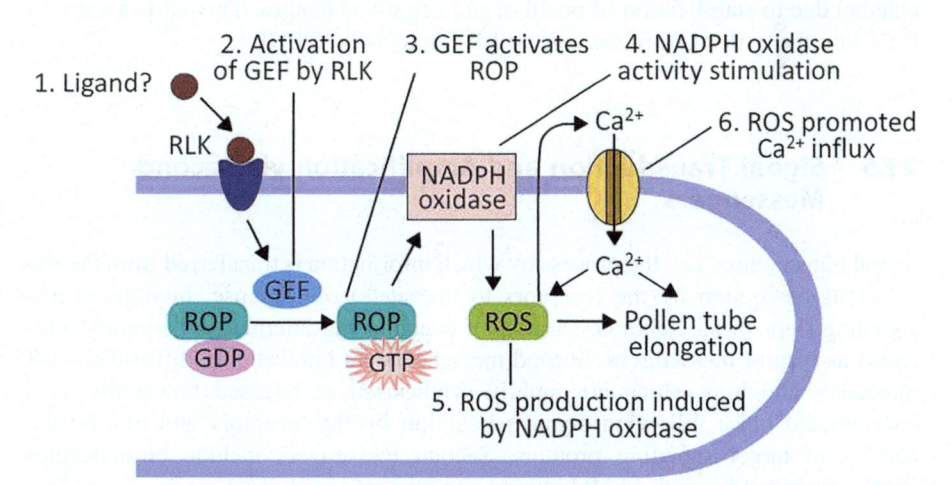

**Fig. 23.9** Crosstalk between receptor-like kinases and ROP GTPases in the regulation of pollen tube growth. *GEF* guanine nucleotide exchange factor, *GTP* guanine triphosphate, *RLK* receptor-like kinase, *ROP* small GTPase, *ROS* reactive oxygen species

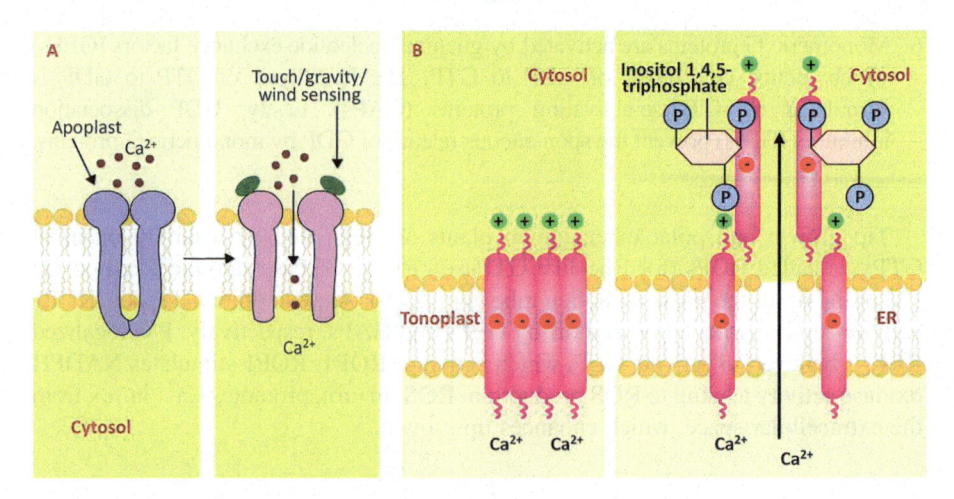

**Fig. 23.10** Ion channel-linked receptors. (**a**) PM-associated $Ca^{2+}$ channels open up as a result of mechanosensing (due to touch, wind, gravity, soil pressure, etc.). (**b**) $IP_3$ binding to the four-domain ion channel on the tonoplast leads to opening of $Ca^{2+}$ channels due to conformation changes in the receptor, which are caused by stabilization of positive and negative charges

cytosolic calcium ion concentration indicating the existence of mechano-sensing ion channels. MID 1 COMPLIMENTING ACTIVITY1 (MCA1) is believed to be a PM-associated calcium channel in *Arabidopsis* which responds to mechanical stimulation in roots. Some ion channel receptors are also located on the internal membranes. The receptor for inositol 1,4,5-triphosphate ($IP_3$) is located on the tonoplast and the ER membranes. These receptors are composed of four subunits. $IP_3$ binding to the receptor leads to conformational changes resulting in opening of channel due to stabilization of positive and negative changes. This allows entry of $Ca^{+2}$ into the cytosol from vacuole or ER lumen (Fig. 23.10).

## 23.5 Signal Transduction and Amplification via Second Messengers

Signal transduction, i.e., the process by which information is transferred from the site (s) of its perception by the receptors to the site(s) of response, involves a few signaling steps or an elaborate cascade of events using intermediary biomolecules called as **second messengers**. Second messengers are intracellular, diffusible small molecules and ions which are rapidly synthesized or released transiently as a concentrated pulse following signal perception by the receptors and modify the activity of target signaling proteins. Second messengers include biomolecules (ROS, phosphatidic acid, MAP kinases) and mineral ions ($Ca^{+2}$).

## 23.5.1 Significance of Second Messengers in Signal Transduction

The length of signal transduction pathway is determined to some extent by the location of receptor where signal is perceived. Perception of signals by the PM often involves subsequent signal transduction via many intermediates. It is well known that in the event of low abundance of receptors, generation of second messengers helps in further amplifying an otherwise low signal before it reaches nucleus to modulate gene expression or to any other site of response. In the absence of second messengers, the signal is likely to get diluted due to diffusion and deactivation (e.g., by dephosphorylation, degradation, or sequestration). In order to further elevate weak initial signals, cells employ amplification mechanisms such as phosphorylation cascade and second messengers. Major categories of second messengers operating in plant cells are:

1. $Ca^+$
2. Cytosolic or cell wall pH
3. ROS
4. Cyclic nucleotides
5. MAP kinases
6. Lipid-signaling molecules

## 23.5.2 $Ca^{+2}$: The Most Ubiquitous Second Messenger

$Ca^{+2}$, a divalent cation, is involved in a vast variety of signaling pathways, including pathogen infection, modulation, touch, wind, changes in gravity vector, anaerobic conditions, drought and salt stress, and responses to GA, ABA, and red and blue light. Plant cells maintain a very low resting cytosolic $Ca^{+2}$ concentration (100–350 nM) but maintain large $Ca^{+2}$ reserves sequestered in vacuoles and endoplasmic reticulum (1–2 mM). Plant cell wall also contains relatively high calcium concentration (0.5–1 mM). Any of the above-stated biological factors can result in spikes (rapid rise and fall) of $Ca^{+2}$ in the cytosol (up to 1 μM), which triggers subsequent signaling action. Cytosolic $Ca^{+2}$ increase is tightly regulated by the opening and closing of $Ca^{+2}$ channels localized on PM, ER, and tonoplast. Plasma membrane is known to possess $Ca^{+2}$ channels **called glutamate-like receptors (GLRs) and cyclic nucleotide-gated channels (CNGCs)**. Calcium channel opening is also regulated by changes in transmembrane electric potential, membrane tension, or binding of ligands. Intracellular spikes of cytosolic $Ca^{+2}$ brought about by the activation of calcium channels through various activation routes result in the regulation of downstream signaling through **$Ca^{+2}$-sensor proteins** as the intermediates. Four major categories of $Ca^{+2}$-sensor proteins are calmodulin (CaM) and calmodulin-like proteins, calcium-dependent protein kinases (CDPKs), $Ca^{+2}$/calmodulin-dependent protein kinases (CCaMKs), and calcineurin B-like proteins (CBLs). These target proteins, upon activation by $Ca^{+2}$, modulate the activity of target proteins. **Target proteins** include transcription factors, various

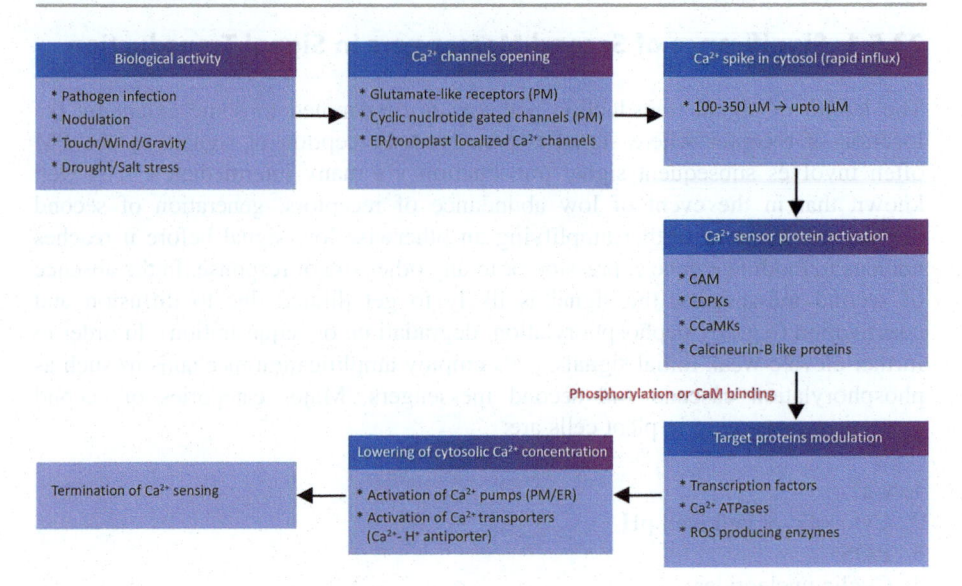

**Fig. 23.11** Sequence of signal transduction events employing $Ca^{2+}$ as a second messenger

protein kinases, $Ca^{+2}$-ATPases, ROS-producing enzymes, and ion channels. Soon after eliciting the response, excess $Ca^{+2}$ is removed from the cytosol by $Ca^{+2}$ pumps and $Ca^{+2}$ exchangers on PM or ER/tonoplast to terminate $Ca^{+2}$ signaling.

**How does $Ca^{+2}$ signaling link each stimulus to its appropriate response?** A number of stimuli lead to $[Ca^{+2}]_{cyt}$ spikes, and a large number of $Ca^{+2}$ and a large number of $Ca^{+2}$- and/or CaM-regulated kinases activate downstream pathways (Fig. 23.11). Plant cells achieve linking of these $Ca^{+2}$ spikes with specific response through one of the following means:

1. Gradation of intracellular $Ca^{+2}$ gradients, e.g., in growing pollen tubes and root hairs. Spatial segregation of $[Ca^{+2}]_{cyt}$ changes leads to varied subsequent biochemical events, leading to tip growth.
2. Temporal changes in $[Ca^{+2}]_{cyt.}$, e.g., oscillation of $[Ca^{+2}]_{cyt}$ in guard cells during stomatal opening.
3. Localization of different components of $Ca^{+2}$ signal transduction machinery in different cells at a given time.

### 23.5.3 Modulation of Cytosolic or Cell Wall pH as a Second Messenger

One of the major roles of electrochemical proton gradient (i.e., proton motive force) across the cell membrane is to drive ATP synthesis and accelerate secondary active transport. Cytosolic pH is generally maintained at 7.5 in contrast to cell wall pH of

## 23.5 Signal Transduction and Amplification via Second Messengers

**Table 23.3** Possible routes of pH modulation as a second messenger in plant cell signaling event

| Cell wall | Cytosol |
|---|---|
| H⁺-ATPase phosphorylation leading to extrusion of H⁺ in the apoplast | Presence of acidic and basic amino acids on proteins |
| 1. Expansion activation leading to cell wall loosening | 1. pH-sensitized activity modulation |
| 2. Alteration of protonation status of weakly acidic hormones (e.g., indole-3-acetic acid) leading to their altered permeability across PM | 2. Effect on protein interaction with other proteins, substrates, or ligands |
| 3. pH-sensitive gating of K⁺ channels and aquaporins | |

about 5.5 or less. Protons exert their signaling role both in the cell wall and cytosol. Activation of H⁺-ATPase activity through its phosphorylation in the PM triggers extrusion of hydrogen ions into the apoplast where it triggers expansions, leading to breaking of hydrogen bonds between closely placed cellulose microfibrils and causing subsequent cell expansion due to turgor pressure. This is commonly observed in hypocotyls. Altered pH of the cell wall is also likely to affect protonation status of weakly acidic plant hormones, thereby affecting their ability to enter cells by diffusion. Gating of potassium channels and aquaporins has also been observed to be affected by pH. Intracellularly, presence of acidic and basic amino acids makes all proteins sensitive to pH changes in the cytosol, thereby altering their activity/function. The pKa value (dissociation constant) of these amino acids and the impact of their protonation are likely to affect the ability of the concerned proteins to interact with other proteins, substrates, or ligands (Table 23.3).

### 23.5.4 Reactive Oxygen Species and Reactive Nitrogen Species as Second Messengers in Environmental and Developmental Signals

Reactive oxygen species (ROS) production is an unavoidable consequence of aerobic life. ROS include both oxygen radicals, like superoxide ($O_2^{·-}$), hydroxyl ($^·OH$), peroxyl (ROO$^·$), etc. and non-radicals, such as hydrogen peroxide ($H_2O_2$), singlet oxygen ($^1O_2$), ozone ($O_3$), etc. Reactive nitrogen species (RNS) include radicals, like nitric oxide (NO·) and nitric dioxide ($NO_2^·$), and non-radicals, such as nitrous acid ($HNO_2$) and dinitrogen tetroxide ($N_2O_4$). Apart from their roles in plant development, ROS and RNS, particularly $H_2O_2$ and NO, respectively, act as second messengers during plant responses to several abiotic and biotic stress factors, like salinity, drought, temperature, heavy metals, and pathogen attack. ROS signals originate from different organelles and induce transcriptional changes and cellular reprogramming, leading to either protection of the plant cell or induction of programmed cell death (PCD). The sensitive equilibrium between ROS production and scavenging at the specific intracellular location and time determines whether

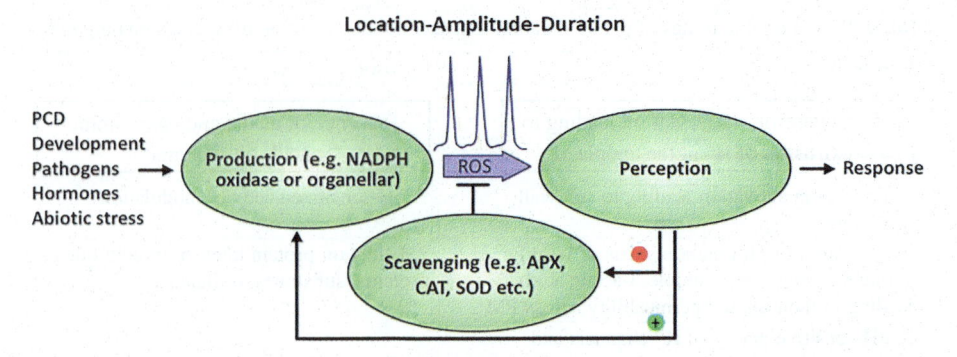

**Fig. 23.12** ROS signaling in plant cells. An equilibrium between ROS production and scavenging at specific intracellular locations and timing determine the nature of its role as a damaging, protective, or signaling factor. *APX* ascorbate peroxidase, *CAT* catalase, *PCD* programmed cell death, *ROS* reactive oxygen species, *SOD* superoxide dismutase

ROS will act as damaging, protective, or signaling factor (Fig. 23.12). Cells have evolved strategies to use ROS as biological signals which activate and control various genetic stress/response programs as ROS influences the expression of a number of genes and operation of signal transduction pathways. Plants are able to regulate developmental processes and respond to environmental cues with the involvement of hormones, like auxins, cytokinins, ethylene, ABA, jasmonic (JA), and salicylic (SA) acids, in signaling together with ROS signaling. Increased levels of ROS can cause SA accumulation which, in turn, is involved in SA-induced stomatal closure. Gibberellic acid (GA) signaling is linked with ROS by stimulating the destruction of DELLA proteins which regulate transcript levels of antioxidant enzymes.

Oxidation of thiol groups by $H_2O_2$ may lead to inactivation of enzymes. $H_2O_2$ plays a concentration-dependent dual role in plants. It can act as a signal molecule in acclimatory signaling, triggering tolerance to different biotic and abiotic stress factors at low concentration, whereas it leads to PCD at high concentration (50 mM). $H_2O_2$ acts as a key regulator in a wide range of physiological processes, like photorespiration and photosynthesis, stomatal movement, cell cycle, and growth and development. It acts as a second messenger for signals generated by ROS due to its relatively long life and high permeability across membranes. $H_2O_2$ plays the role of messenger in numerous signaling pathways for regulation of many of the stress genes. In response to environmental (abiotic) stress in plants, RNS also play a significant role as signaling molecules. During defense response to pathogen attacks in plants, NO acts as a key mediator together with ROS. In general, a rapid overproduction of ROS and RNS takes place when plants are subjected to biotic and environmental stresses. PCD is an important mechanism which regulates varied aspects of growth and development and also eliminates damaged or infected cells produced in response to environmental stress and pathogen attack. Both NO and ROS play key functions during PCD. In the presence of $O_2$, NO can react with reduced glutathione (GSH) by an *S*-nitrosylation reaction to form *S*-

## 23.5 Signal Transduction and Amplification via Second Messengers

**Fig. 23.13** Post-translational modifications mediated by nitric oxide in plant cells. The presence of GSNO, NO, and ONOO⁻ leads to covalent post-translational modifications (PTMs) like S-nitrosylation and nitration of proteins and fatty acids. *GSH* reduced glutathione, *GSNO* S-nitrosoglutathione, *NOS* nitric oxide synthase, *NR* nitrate reductase, $ONOO^-$, peroxynitrite anion, $O_2^{\cdot-}$ superoxide anion

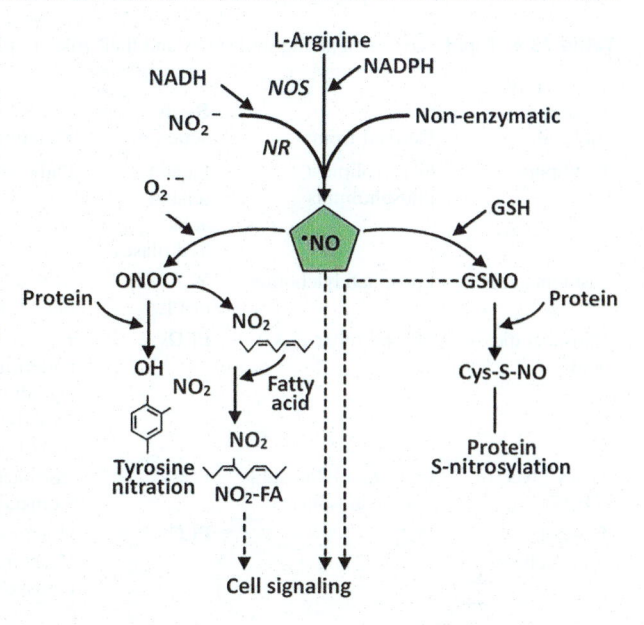

nitrosoglutathione (GSNO), an important mobile reservoir of bioactive NO (Fig. 23.13). The presence of GSNO has been demonstrated in different plant species. On the other hand, peroxynitrite (ONOO⁻), a powerful oxidant/nitrating RNS species, is formed by the rapid reaction between $O_2^{\cdot-}$ and NO, and its occurrence has been reported in plant organelles, such as peroxisomes. As a result of the presence of NO and GSNO in plant tissue and generation of ONOO⁻, important covalent post-translational modifications (PTMs), like S-nitrosylation and the nitration of proteins, can take place under natural and stress conditions. In peroxisomes, S-nitrosylation inhibits the activities of catalase and glycolate oxidase, and this could regulate the cellular level of key signaling molecules like $H_2O_2$. Generation of ONOO⁻ can cause tyrosine nitration of plant proteins leading to nitrosative damage in plant cells, although a basal endogenous nitration could also have a regulatory function. $H_2O_2$ and NO have also been shown to play an important role in avoiding pathogen advancement as they activate transcriptional factors of pathogenesis-related (PR) proteins during the induction of resistance.

### 23.5.5 Lipid-Signaling Molecules

The physical properties of cell membranes such as changes on their surface and fluidic nature are modulated by their primary lipidic constituents, i.e., phosphoglycerolipids and sphingolipids. A typical phospholipid molecule consists of two fatty acid chains esterified to a glycerol backbone, a phosphate group, and a variable head group. Common head groups include choline, ethanolamine, glycerol, serine, inositol, inositol monophosphate, inositol bisphosphate, and OH group. The resulting lipids from these head groups are phosphatidylcholine (PC),

**Table 23.4** Lipid-derived signaling molecules and their roles in plants

| Lipid-derived signaling molecule | Derived from | By the action of | Signaling roles |
|---|---|---|---|
| Oxylipins | Glycerolipids/ phospholipids | Patatin-related acyl hydrolases | Pathogen defense (?)/abiotic stress (?) |
| Jasmonic acid | Phosphatidylcholine | $PLA_1$ (?) or $PLA_2$ | Senescence, flower opening, anther dehiscence, pollen maturation |
| Phosphatidic acid | PC and other PLs | PLDs | Freezing tolerance, root hair patterning, cell differentiation, ABA signaling for stomatal closure, modulation of actin and microtubule dynamics |
| Diacylglycerol (DAG) | Phosphatidylinositol 4,5-bisphosphate | PLC | PKC activation; pollen tube growth; stomatal opening |
| Inositol triphosphate ($IP_3$) | $PIP_2$ | PLC | Opening of ER/tonoplast-localized $Ca^{2+}$ channels; interaction with cytoskeletal proteins |

phosphatidylethanolamine (PE), phosphatidylglycerol (PG), phosphatidylserine (PS), phosphatidylinositol (PI), phosphatidylinositolphosphate (PIP), phosphatidylinositolbisphosphate, and phosphatidic acid (PA), respectively (Table 23.4). Phospholipases ($PLA_1$, $PLA_2$, PLC, PLD) act at specific positions on these phospholipid molecules, resulting in the formation of a variety of lipid-signaling molecules which modulate a number of cellular and physiological processes. Members of **phospholipase A (PLA)** enzyme are responsible for cleaving one of the acyl ester bonds on the phospholipids, thereby releasing a fatty acid molecule and a lysophospholipid. **Lysophospholipids** are small lipid molecules consisting of a single carbon chain and a polar head group. They are more hydrophilic than their corresponding phospholipids, and they are known to regulate a number of plant processes. PLAs cleave either of the acyl bonds on the phospholipid molecule, releasing a fatty acid and a lysophospholipid. Two forms of PLAs, $PLA_1$ and $PLA_2$, are known, but only $PLA_2$ can be regulated by G proteins, protein kinases, and $Ca^{2+}$ in animal cells. Phospholipase C **(PLC)** hydrolyzes the glycerophosphate bond to yield diacylglycerol (DAG) and a phosphorylated head group, such as $IP_3$ (inositol 1,4,5-triphosphate). Both these molecules (DAG and $IP_3$) are involved in the regulation of $Ca^{2+}$ fluxes essential for a number of physiological processes. **Phospholipase D (PLD)** action on phosphatidylcholine (PC) leads to the formation of **phosphatidic acid (PA)**, which is a crucial signaling molecule to counter environmental stresses. Different PLD isoforms are expressed in different cellular location at different stages of development.

PM-associated PLC has been purified from several plants and is involved in signaling events associated with inositol phospholipid synthesis and breakdown, resulting in an increase in cytosolic calcium (Fig. 23.14). PLC is activated by its

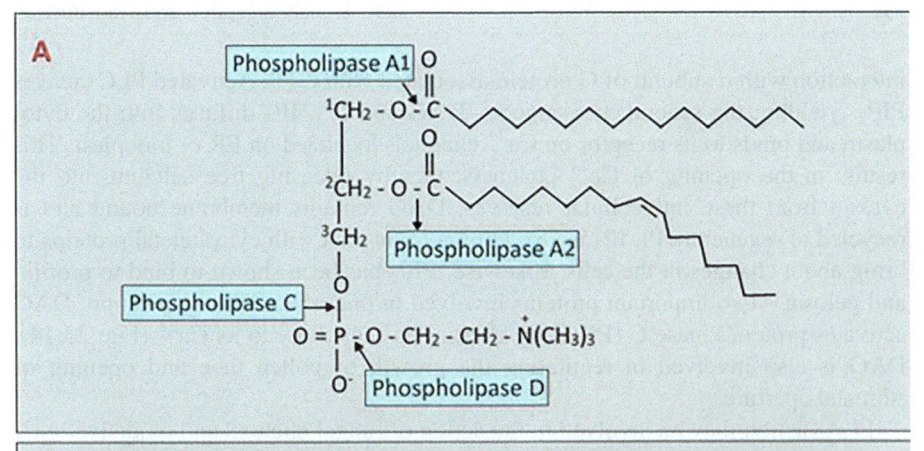

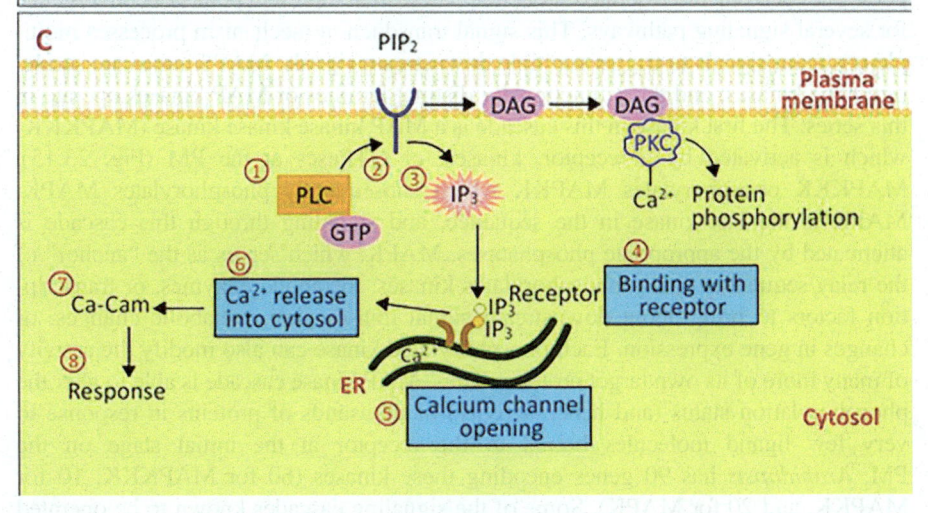

**Fig. 23.14** (a) Cleavage sites of specific phospholipases. (b) Products of action of phospholipases on phospholipids. (c) Major signal transduction events involving lipid-signaling molecules: (1) activation of PLC2, (2) activation of activated PLC on $PIP_2$ (DAG and $IP_3$ formation), (3, 4) diffusion and binding of $IP_3$ with the receptor on ER, (5) opening of $Ca^{2+}$ channels in ER, (6) release of free $Ca^{2+}$ from ER stores, (7) $Ca^{2+}$-CaM complex formation, and (8) response

interaction with α subunit of G protein associated with GTP. Activated PLC cleaves PIP$_2$, yielding the second messengers—DAG and IP$_3$. IP$_3$ diffuses into the cytoplasm and binds to its receptor on Ca$^{2+}$ channels localized on ER or tonoplast. This results in the opening of Ca$^{2+}$ channels, thereby releasing free calcium into the cytosol from these intracellular reserves. DAG remains membrane bound and is recycled to regenerate PI. IP$_3$ is also known to interact with cytoskeletal proteins to bring about changes in the cells. Likewise, PIP$_2$ has been shown to bind to profilin and gelosin—two important proteins involved in microfilament organization. DAG activates protein kinase C (PKC) by altering its sensitivity to its Ca$^{2+}$ (Fig. 23.14). DAG is also involved in regulating the growth of pollen tube and opening of stomatal aperture.

PLA$_2$ action may be involved in the action of fungal elicitors and oxidative burst responses associated with them (1, 3). β-Glucanase is regulated by PLA$_2$ activation. Linolenic acid, a product of PLA$_2$ action, is a precursor for jasmonic acid biosynthesis via octadecanoid pathway. Phosphatidic acid (PA) generated through PLD action has been implicated in senescence, wounding, stress, and fruit ripening. PA can also act as a calcium ionophore, thereby permitting free movement of calcium ions across PM. PA regulates several protein kinases, GTP-binding proteins, vesicle trafficking, and oxidative burst response due to pathogens.

### 23.5.6 Mitogen-Activated Protein (MAP) Kinase Cascade

MAP kinase cascade derives its name from a series of protein kinases which phosphorylate each other in a defined sequence. They often represent points of convergence for several signaling pathways. This signal transduction mechanism processes multiple upstream signals to trigger multiple downstream signals. MAP kinases are serine/threonine kinases, and three functionally distinct families of MAP kinases operate in this series. The first kinase in this cascade is a MAP kinase kinase kinase (MAPKKK) which is activated by a receptor, kinases, or GTPases at the PM (Fig. 23.15). MAPKKK phosphorylates MAPKK which subsequently phosphorylates MAPK. MAPK is the last kinase in the sequence, and signaling through this cascade is attenuated by the appropriate phosphatases. MAPK, which serves as the "anchor" of the relay sequence, finally phosphorylates kinases, metabolic enzymes, or transcription factors to bring about downstream signal transduction, metabolic changes, or changes in gene expression. Each phosphorylated kinase can also modify the activity of many more of its own target proteins. Thus, MAP kinase cascade is able to alter the phosphorylation status (and hence activity) of thousands of proteins in response to very few ligand molecules bound to the receptor at the initial stage on the PM. *Arabidopsis* has 90 genes encoding these kinases (60 for MAPKKK, 10 for MAPKK, and 20 for MAPK). Some of the signaling cascades known to be operated through MAP kinase route include jasmonic acid and ABA signal transduction,

## 23.5 Signal Transduction and Amplification via Second Messengers

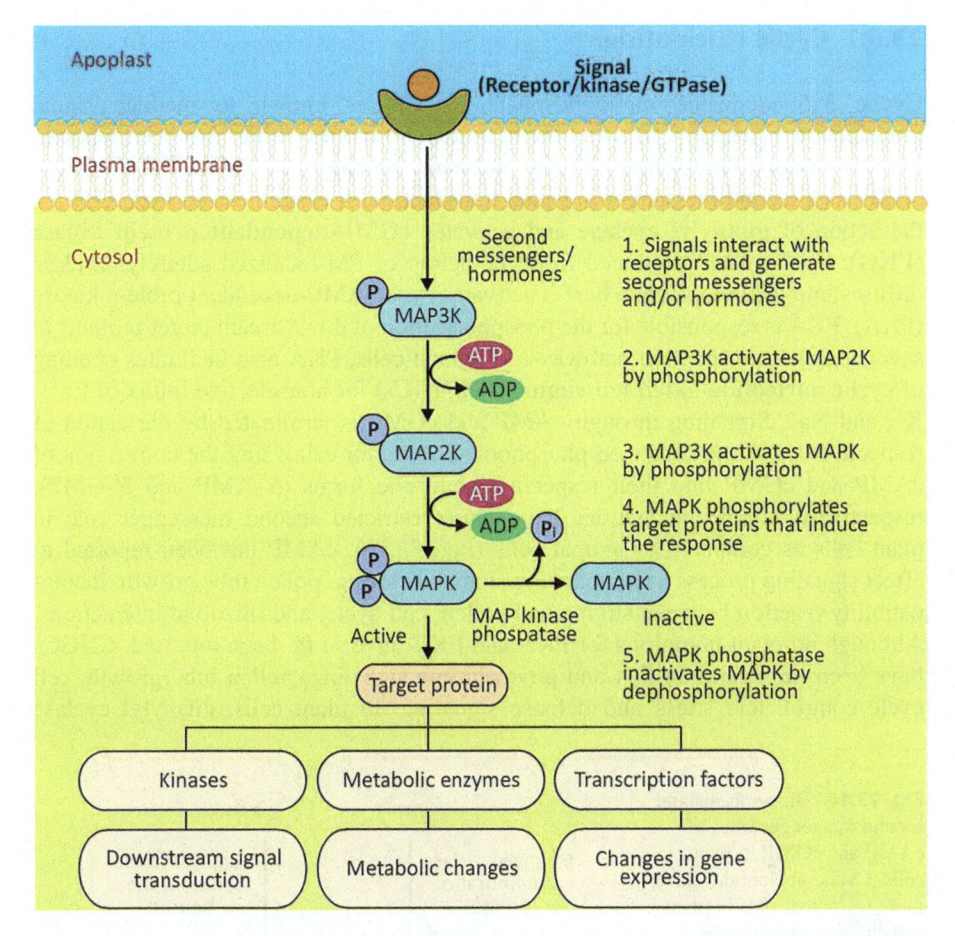

**Fig. 23.15** Mitogen-activated protein (MAP) kinase-mediated signaling through a series of phosphorylation of MAP3K, MAP2K, and MAPK. This signaling route brings about multiple downstream signals

regulation of asymmetric formation of division plate during cell division, stomatal patterning, and cellular responses to abiotic stresses and pathogens.

In view of the multiple inputs and outputs, how does MAP kinase cascade maintain signal specificity? This is made possible by one of the two ways: (1) selective co-localization of kinases into a cascade forming combination to restrict kinase interactions and (2) selective repression of some cascades by phosphatases to permit only certain transduction pathways to operate.

## 23.5.7 Cyclic Nucleotides

Cyclic 3',5'-adenosine monophosphate (cAMP) is known to mediate signal transduction in response to a number of animal hormones. It has also been reported to act as a second messenger in fungi and many prokaryotes. Animal cells also use cGMP as a second messenger in signal transduction. It is synthesized from GTP by the action of **guanylyl cyclase** and activates **cGMP-dependent protein kinase (PKG)**. Cyclic AMP produced from the action of PM-localized adenylyl cyclase diffuses into the cytoplasm where it activates cyclic AMP-dependent protein kinase (PKA). PKA is responsible for the phosphorylation of downstream target proteins in several signal transduction pathways. In animal cells, PKA also facilitates opening of **cyclic nucleotide-gated ion channels (CNGCs)** for nonselective influx of $Ca^{2+}$, $K^+$, and $Na^+$. Signaling through cAMP and cGMP is terminated by the action of respective cytoplasm localized phosphodiesterases for catalyzing the conversion of cAMP and cGMP into their respective noncyclic forms (5'-AMP and 5'-GMP), respectively. Cyclic nucleotides have more restricted second messenger role in plant cells as compared to animal cells (Fig. 23.16). cAMP has been reported to affect signaling process accompanying stomatal closing, pollen tube growth, incompatibility reaction between stigma and pollen, cell cycle, and rhizobial interactions. Although no plant homologs for PKA and PKG have so far been reported, CNGCs have been implicated in GA and phytochrome signaling, pollen tube growth, cell cycle control, and stress and defense signaling. In plant cells, guanylyl cyclase

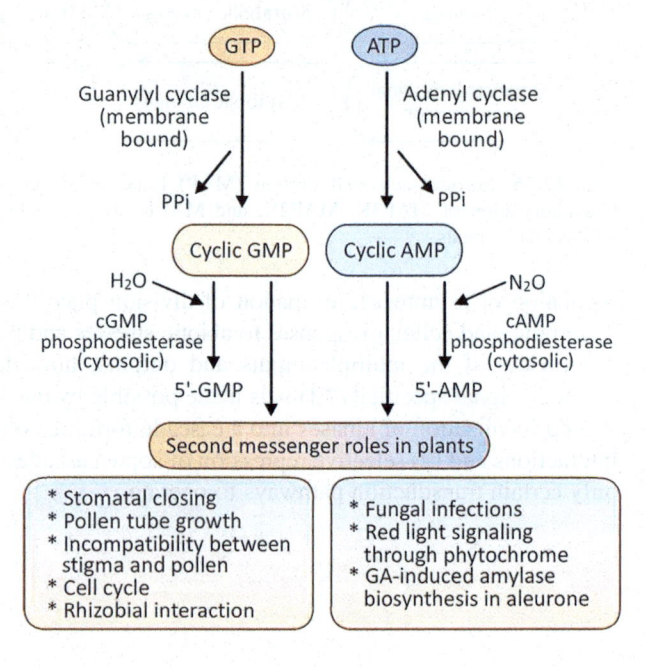

**Fig. 23.16** Biosynthesis and second messenger roles of cAMP and cGMP in plant cells. Cyclic nucleotides have more restricted roles in plant cells than in animal cells

activity may be more important than adenyl cyclase in signal transduction of various processes. cGMP may participate in transducing signals of fungal infection, red light signals through phytochrome, and synthesis of amylase in aleurone layer in response to GA action.

## 23.6 Adaptive Mechanisms of Plant Signaling and Their Termination

1. A majority of plant hormone signal transduction pathways induce a response by **negative regulation** through inactivation of repressor proteins. This is in contrast with animal signal transduction pathways which induce responses mostly through activation of a cascade of positive regulators. Thus, for example, brassinosteroid binding to the receptor kinase BRI1 leads to inactivation of repressor protein BIN2. This results in the inactivation of transcription factors—BES1 and BZR1. Why have plant cells evolved signaling pathways based on negative regulation rather than positive regulation (based on animal cells)? Mathematical modeling has suggested that negative regulators result in faster induction of downstream response genes. Since the speed of response to environmental stress conditions is crucial for survival of sessile plants, the adoption of negative regulatory signaling is likely to confer selective advantage during evolution. Some of the major molecular mechanisms adopted by plant cells to inactivate repressor proteins include dephosphorylation, retargeting of repressors to other subcellular compartments and degradation.
2. **Plants have evolved mechanisms for switching off or attenuating signal transduction responses**: Plant hormones can be degraded or inactivated by oxidation or conjugation to sugars or amino acids. Receptors and signaling intermediates can also be inactivated by dephosphorylation. Signal amplification can further be switched off by the lowering of elevated concentrations of secondary messengers by ion transporters and cellular scavengers. Proteolytic degradation of the components of signal transduction pathways serves as an important mechanism to regulate signaling events. Feedback regulation is another mechanism for attenuation of a signal response. For example, degradation of AUX/IAA proteins thereby terminates the response. Likewise, gibberellin negatively regulates its own intracellular concentration through DELLA-mediated feedback loops. In the absence of GA, DELLA transcriptional regulators promote GA biosynthesis through enhanced expression of $GA_{20\ ox}$ and $GA_{30\ ox}$ enzymes. DELLA also inhibits genes encoding GA catabolism ($GA_{2\ ox}$), thereby causing GA degradation. As a result of these two effects of DELLA, GA concentration in the cell increases. Soon thereafter, as soon as DELLA proteins are degraded by proteosomal pathway, GA biosynthesis decreases and GA catabolism increases. Thus, GA negatively regulates its own concentration in the affected cells.

3. **Tissue specificity of signal responses**: Not all responses evoked by a signal are evident in any cell type. Any tissue exhibits some of the responses due to a signal. Thus, auxin promotes cell expression in growing aerial tissues but inhibits lateral root formation in the cells of pericycle, while leaf primordia are initiated at the shoot apical meristem. These differential auxin responses in different cells are due to auxin-dependent interaction of TIR/AFB receptors and AUX/IAA repressor proteins and variation in the response by the tissue in plants where these components are expressed, the abundance of their expression, the degree of their binding affinity and cellular auxin level.
4. **Agonistic (additive or positive) and antagonistic (inhibitory or negative) crosstalk among different signaling mechanisms**: Signal transduction pathways mostly operate as part of a complex web of signaling interactions rather than in isolation, e.g., antagonistic interaction between gibberellins and ABA during seed germination. Such interactions among various signaling routes are referred as **cross regulation**. It can be of one of the following three types:
   (a) **Primary cross regulation** where distinct signaling pathways regulate a shared transduction pathway. Both pathways have the same effect on the response leading to positive interaction. When one of the pathways inhibits the effect of the other, it leads to negative primary interaction. For example primary cross regulation is evident when both cytokinin and ABA induce transcription of ABI4 to further regulate transcription of a number of biosynthetic and response genes (Fig. 23.17).
   (b) **Secondary cross regulation** where output of one signaling route regulates the abundance or perception of another signal. It is positive interaction when one pathway enhances the input levels or perception of the other pathway and

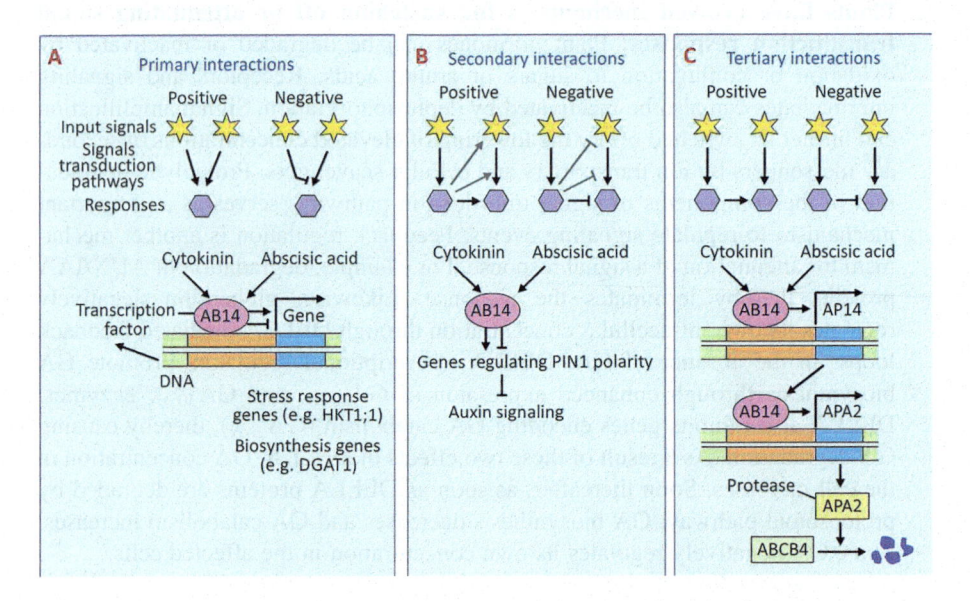

**Fig. 23.17** Agonistic and antagonistic crosstalk among different signaling mechanisms

is negative when the second pathway is repressed by the first. For example, induction of transcription factor ABI4 acts on auxin signaling by lowering auxin flow and polar localization and abundance of PIN 1 in root vasculature, which is responsible for lateral root initiation.

(c) **Tertiary cross regulation** where outputs of two pathways influence each other positively or negatively. ABI4 positively regulates $APA_2$ protease expression which induces degradation of ABCB4 auxin transporter that regulates auxin flow in root epidermal cells during lateral root elongation. ABCB4 abundance on PM is reduced after ABA or CK treatment in wild plants (Fig. 23.17).

An understanding of the complexities of these cross regulations is possible through mathematical and computational models to simulate these biological networks, an approach often referred as **system biology**.

## Summary

- Plant growth and development are modulated by a variety of environmental (external) and physiological (internal) signals. A number of signaling events in plants involve signal reception and response in the same cell (**cell autonomous responses**). Many other responses involve signal perception in one cell and response in distal cells, tissues, or organs (**non-cell autonomous responses**). Same signal can lead to different responses in different tissues through its ability to bind to different families of receptors. Different tissues may adapt or desensitize themselves to continuous signals and receptor concentrations during development. Signal transduction pathways mostly operate as part of a complex web of signaling interactions rather than in isolation.
- Any environmental or intracellular input, which initiates one or more responses in the cell/plant, is referred as a **signal**. Plant cells are sensitized to the signals (**perception**) by employing specialized sensor proteins, termed **receptors**. Receptors may be located on the plasma membrane, cytoplasm, endomembrane system, or in the nucleus. Receptors located on the plasma membrane can perceive both physical and chemical signals. Sensing of signal by the receptor is followed by the transfer of the information from the signal from one biochemical form to another (**transduction**) so as to amplify the impact of signal leading to a cellular response.
- Plants have also developed pathways for electrical signaling to rapidly respond to external stimuli. In contrast to chemical signals, electrical signals rapidly transmit information over long distances. Touch-sensitized drooping of leaflets of *Mimosa pudica* is a classic example where membrane potential serves as a receptor.

- Three major groups of plasma membrane receptors can be distinguished on the basis of how they interact with signaling components. These are receptor kinases, G-protein-coupled receptors (GPCRs), and ion channel receptors. **Receptor kinases** are responsible for transducing extracellular signals through phosphorylation of intracellular targets. Plants also possess **histidine kinase receptors** derived from bacterial two-component system. Signal transduction using these receptors depends on a balance of "forward" and "reverse" phosphotransfers.
- **G-protein-coupled receptors (GPCRs)** are membrane-associated proteins with an extracellular ligand-binding domain, a transmembrane domain (composed of seven hydrophobic helices), and an intracellular domain that interact with inactive G-protein trimer. **G proteins** are GTP-binding proteins whose activity is regulated by GPCRs. G proteins consist of three subunits—Gα, Gβ, and Gγ. Tip growth and polar extension in plants are regulated by a family of small GTPases called ROPs (for Rho-like GTPase). ROP1 GTPases are localized on the PM at the tip of the growing pollen tube.
- So far, no plasma membrane-associated ion channels have been reported to function as receptor in response to ligand binding. However, PM-associated mechanosensing calcium channels do play a role in plant responses to physical force.
- **Second messengers** are intracellular, diffusible small molecules and ions which are rapidly synthesized or released transiently as a concentrated pulse following signal perception by the receptors and modifying the activity of target signaling proteins. $Ca^{+2}$ is involved in a vast variety of signaling pathways. Cytosolic $Ca^{+2}$ increase is tightly regulated by the opening and closing of $Ca^{+2}$ channels localized on PM, ER, and tonoplast. Plasma membrane is known to possess $Ca^{+2}$ channels **called glutamate-like receptors (GLRs) and cyclic nucleotide-gated channels (CNGCs)**. Intracellular spikes of cytosolic $Ca^{+2}$ brought about by the activation of calcium channels through various activation routes result in the regulation of downstream signaling through $Ca^{+2}$**-sensor proteins** as the intermediates. Four major categories of $Ca^{+2}$-sensor proteins are calmodulin (CaM) and calmodulin-like proteins, calcium-dependent protein kinases (CDPKs), $Ca^{+2}$/calmodulin-dependent protein kinases (CCaMKs), and calcineurin B-like proteins (CBLs). These target proteins, upon activation by $Ca^{+2}$, modulate the activity of target proteins. **Target proteins** include transcription factors, various protein kinases, $Ca^{+2}$-ATPases, ROS-producing enzymes, and ion channels.
- Protons also exert their signaling role both in the cell wall and cytosol. Altered pH of the cell wall is also likely to affect protonation status of weakly acidic plant hormones, thereby affecting their ability to enter cells by diffusion. Gating of potassium channels and aquaporins has also been observed to be affected by pH. ROS signals originate in different organelles and induce transcriptional changes and cellular reprogramming, leading to either protection of the plant cell or induction of programmed cell death (PCD). $H_2O_2$ plays the role of messenger in numerous signaling pathways for regulation of many of the stress genes. RNS also play a significant role as signaling molecules. As a result of the presence of NO and GSNO in plant tissue and generation of $ONOO^-$, important covalent post-translational modifications (PTMs), like *S*-nitrosylation and the

nitration of proteins, can take place in plants under natural and stress conditions. In peroxisomes, S-nitrosylation inhibits the activities of catalase and glycolate oxidase, and this could regulate the cellular level of key signaling molecules, like $H_2O_2$. $H_2O_2$ and NO have also been shown to play an important role in avoiding pathogen advancement as they activate transcriptional factors of pathogenesis-related (PR) proteins during the induction of resistance.
- PM-associated PLC is involved in signaling events associated with inositol phospholipid synthesis and breakdown, resulting in an increase in cytosolic calcium. PLC is activated by its interaction with α subunit of G protein associated with GTP. Activated PLC cleaves $PIP_2$, yielding the second messengers—DAG and $IP_3$. $IP_3$ diffuses into the cytoplasm and binds to its receptor on $Ca^{2+}$ channels localized on ER or tonoplast. This results in the opening of $Ca^{2+}$ channels, thereby releasing free calcium into the cytosol from these intracellular reserves. $IP_3$ is also known to interact with cytoskeletal proteins to bring about changes in the cells. $PLA_2$ action may be involved in the action of fungal elicitors and oxidative burst responses associated with them. Phosphatidic acid (PA) generated through PLD action. PA can also act as a calcium ionophore, thereby permitting free movement of calcium ions across PM. PA regulates several protein kinases, GTP-binding proteins, vesicle trafficking, and oxidative burst response due to pathogens.
- MAP kinases are serine/threonine kinases, and three functionally distinct families of MAP kinases operate in this series. The first kinase in this cascade is a MAP kinase kinase kinase (MAPKKK) which is activated by a receptor, kinases, or GTPases at the PM. MAPKKK phosphorylates MAPKK which subsequently phosphorylates MAPK. MAPK is the last kinase in the sequence, and signaling through this cascade is attenuated by the appropriate phosphatases. MAPK, which serves as the "anchor" of the relay sequence, finally phosphorylates kinases, metabolic enzymes, or transcription factors to bring about downstream signal transduction, metabolic changes, or changes in gene expression.
- Cyclic nucleotides have more restricted second messenger role in plant cells as compared to animal cells. In plant cells, guanylyl cyclase activity may be more important than adenyl cyclase in signal transduction of various processes. cGMP may participate in transducing signals of fungal infection, red light signals through phytochrome, and synthesis of amylase in aleurone layer in response to GA action.
- A majority of plant hormone signal transduction pathways induce a response by negative regulation through inactivation of repressor proteins. This is in contrast with animal signal transduction pathways which induces responses mostly through activation of a cascade of positive regulators. Thus, for example, brassinosteroid binding to the receptor kinase BRI1 leads to inactivation of repressor protein BIN2. Since the speed of response to environmental stress conditions is crucial for survival of sessile plants, the adoption of negative regulatory signaling is likely to confer selective advantage during evolution. Proteolytic degradation of the components of signal transduction pathways serves as an important mechanism to regulate signaling events. Feedback regulation is another mechanism for attenuation of a signal response.

## Multiple-Choice Questions

1. Which of the following is NOT a non-cell autonomous response?
   (a) Radial patterning of *Arabidopsis* primary root regulated by transcription regulator-SHR
   (b) Auxin gradient in the root by rapid redistribution of PIN3 (auxin efflux protein)
   (c) Opening of stomata in response to blue light
   (d) Transition from vegetative to reproductive phase (floral induction)
2. Signal perception via symplast regulates the movement of which of the molecules listed below?
   (a) Lipophilic molecules
   (b) Chemicals like signal peptides
   (c) Mechanical forces like touch and blue light
   (d) RNAs and transcription factors
3. In GPCRs, the Gα subunit of G protein is active when bound with _____. Inactivation of Gα subunit is mediated via the action of _____.
   (a) GTP, GTPase-activating protein (GAP)
   (b) GDP, Guanine nucleotide exchange factors (GEFs)
   (c) GDP, GTPase-activating protein (GAP)
   (d) GTP, Guanine nucleotide exchange factors (GEFs)
4. Tip growth and polar extension in plants is regulated by:
   (a) GSNO
   (b) Phospholipase A (PLA)
   (c) Rho-like GTPase (ROP1)
   (d) Phospholipase D (PLD)
5. In peroxisomes, S-nitrosylation has the ability to inhibit the activity of:
   (a) Superoxide dismutase (SOD)
   (b) Catalase
   (c) Glutathione peroxidase
   (d) b and c
6. Small lipid molecules consisting of a single carbon chain and a polar head group are known as:
   (a) Sphingolipids
   (b) Phosphoglycerolipids
   (c) Lysophospholipids
   (d) None of these
7. Calcium channels—glutamate-like receptors (GLRs) and cyclic nucleotide-gated channels (CNGCs)—are located in:
   (a) Tonoplast
   (b) Plasma membrane
   (c) Endoplasmic reticulum
   (d) All of these

8. Activated _____ cleaves PIP 2 to yield secondary messengers- _____ and _____.
   (a) PLD, IP 5, and DAG
   (b) PLC, IP 3, and DAG
   (c) PLC, IP 2, and DAHA
   (d) PLD, IP 3, and DAG

**Answers**

1. c  2. d  3. a  4. c  5. b  6. c  7. b  8. b

## Suggested Further Readings

Leyser O, Ray S (2015) Signal transduction. In: Buchanan BB, Gruissem W, Jones RL (eds) Biochemistry and molecular biology of plants. Wiley-Blackwell, Chichester, pp 834–871

Taiz L, Zeiger E (2010) Plant physiology, 5th edn. Sinauer Associates Inc, Massachusetts, pp 407–445

# Embryogenesis, Vegetative Growth, and Organogenesis

## Rama Sisodia and Satish C Bhatla

The life cycle of all seed plants can be divided into three main stages: embryogenesis, vegetative growth, and reproductive development. During embryogenesis, the single-celled zygote follows a defined pattern of cell divisions and differentiation to form the mature embryo and eventually the seedling, which contains all tissues and organs that develop into the mature plant body. In angiosperms, early phase of the embryogenesis is marked by laying down of the basic body plan of embryo, including establishment of apical—basal polarity and formation of embryonic organs (root and shoot primordial, cotyledons, etc.) and embryonic tissue layers. The embryonic layers – **protoderm, ground meristem**, and **procambium**, are the progenitors of the future epidermal, cortical, and vascular tissues, respectively. At the end of embryogenesis phase, several physiological changes occur within the embryos enabling them to sustain the period of dormancy and adverse environmental conditions. Upon the availability of factors like ambient moisture, light, temperature, etc., the seeds germinate, and with accompanying mobilization of stored reserves, the vegetative development commences, wherein tissue primordia in the seedlings grow and cells differentiate. During this period, the meristematic zones corresponding to shoot and root become active. Seedlings display a simple body plan which consists of shoot meristem, cotyledons, hypocotyl, root, and root meristem along the apical-basal axis and a concentric arrangement of tissues along the radial axis: epidermis at the periphery and subepidermal ground tissue and conductive tissue in the center. Vegetative growth is typically **indeterminate**, i.e., with no definite end and characterized by development of lateral organs—the leaves. As the plant undergoes the transition to reproductive development, the shoot apical meristem (SAM) changes shape, becoming an inflorescence meristem, and gives rise to flowers or flowering shoots, thus acquiring a **determinate** growth pattern. Plants exhibit variations in size ranging from a few mm to several hundred meters acquiring the architecture that is suited to the local environment. They grow in size throughout their lives and produce new organs like leaves, flowers, fruits, etc. The growth of plants is defined in terms of increase in number of cells as well as an increase in size

© Springer Nature Singapore Pte Ltd. 2018
S. C Bhatla, M. A. Lal, *Plant Physiology, Development and Metabolism*,
https://doi.org/10.1007/978-981-13-2023-1_24

of cells. It is important to note that this growth is not in terms of increase in dry mass as up to 95% of plant cells is made up of water. Plant growth is accompanied with differentiation of cells, tissues (**histogenesis**), and organs (**organogenesis**) that are specialized structures with distinct functions. Thus, growth and differentiation result in morphogenesis with the plant acquiring an overall form.

## 24.1 Embryogenesis

Details of embryo development have been studied in several plants using molecular techniques, including the use of mutants, cell-specific markers, etc. Embryogenesis begins with double fertilization in which one sperm cell fuses with the egg cell and another fuses with the central cell to form the zygote and endosperm mother cell, respectively. Embryo development occurs within the highly polarized environment of the ovule. Both, the ovule and the embryo sac, have a distinct polarized axis, and also the egg cell is intrinsically polar as judged from the localization of its organelles on the basal (e.g., vacuole) or apical end (e.g., nucleus).

### 24.1.1 Acquisition of Polarity

As mentioned earlier, basic features of the body plan, which includes the development of polarity of the embryo and the establishment of the apical-basal axis, are laid down during the early phase of embryogenesis. **Polarity** is the term used to describe the fact that differences can arise along an axis such that one end of the axis is different from the other. Polarity is evident in the embryo sac, egg cell, zygote, and embryo-suspensor complex. The zygote undergoes an asymmetric transverse division to form two daughter cells that are unequal in size and follow distinct developmental pathways. In *Arabidopsis*, the larger basal cell develops from the vacuolar region of the zygote, while the smaller upper cell develops from the cytoplasm-rich region. The upper cell then divides to form the embryo proper, while the basal cell forms a single file of typically six to nine cells, the suspensor. This asymmetric division of the zygote is an important factor in the establishment of polarity in morphogenesis. A number of different mechanisms are known to account for asymmetry in the products of eukaryotic cell division, though the precise factors that are responsible for the asymmetry in plant embryogenesis are yet to be identified (Box 24.1). An important factor in the mechanism of cell divisions, which lead to nonequivalent daughter cells, involves the reorganization of actin and associated motor proteins leading to orientation of mitotic spindle. Yet another hypothesis includes the localization of regulatory factors in a specific region of the cell prior to cell division that results in two daughter cells.

**Box 24.1: Polarity Induction in *Fucus* Zygote**
There are hardly any cells or multicellular systems that are not polarized. One of the few exceptions is the zygote of the brown alga of the genus *Fucus*. The alga is commonly found growing in the upper intertidal shores of temperate oceans. The large-sized (up to 1 mm) egg cells of *Fucus* are released into the seawater, and fertilization occurs in the seawater independent of the maternal influences. The study of ultrastructural details of the egg and early zygote shows these are symmetrical and apolar. As the zygotes develop, the cellular components redistribute in an asymmetrical manner with the zygote now acquiring a distinct polarity. Experimental data has revealed that several factors can affect polarity. Temperature, light, pH value, different chemical gradients, and neighboring *Fucus* zygotes all play a role. The trigger for polarization of the fucoid zygote includes a range of stimuli, including unidirectional light and fertilization. Associated with axis formation, there is an observed localization or redistribution of plasma membrane components, including ion channels, a redistribution of calcium to the basal shaded end, a localization of F-actin at the rhizodermis, and a polarized secretion of Golgi-derived cell wall components toward the "basal" region from which the rhizoid cell will develop.

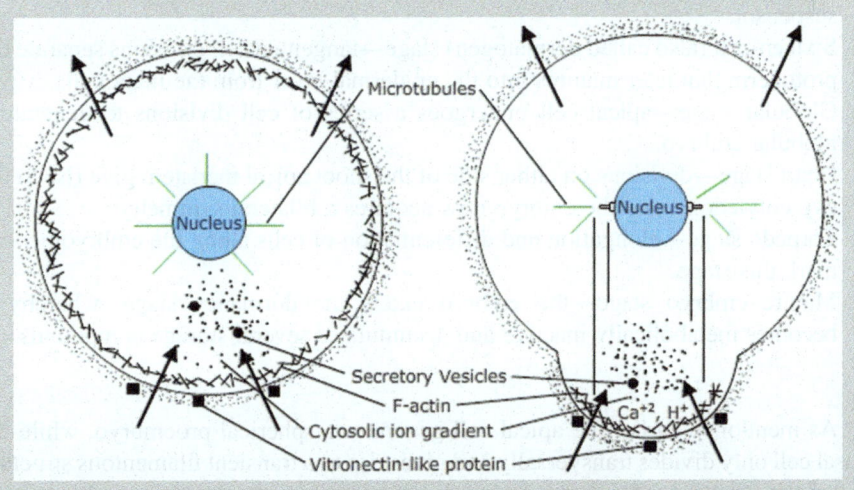

Figure shows axis formation in zygote of *Fucus*. Zygote lacks polarity with more or less uniform distribution of cellular components. As the zygote germinates, these are unevenly distributed along the polar axis. Arrows indicate polar distribution of intracellular constituents.

## 24.1.2 Stages of Embryo Development

Fertilization induces stretching of the zygote, which is followed by an asymmetric division that gives rise to two daughter cells with different composition, shape, and developmental fate. The apical cell is smaller and densely cytoplasmic, whereas the basal cell is larger and has a big vacuole. The fates of these two cells are dramatically different. Divisions in the apical cell give rise to embryo proper, which goes through several, but continuous, developmental phases in continuity: an eight-celled proembryo is followed by globular, heart-shaped, and torpedo stages. Details of different stages of embryo development have been well described in the model organism *Arabidopsis* (Fig. 24.1):

1. Zygotic stage—haploid egg fuses with sperm to form diploid single-celled zygote.
2. Two-cell stage—asymmetric transverse division of the zygotic cell gives rise to a small apical cell and a larger basal cell.
3. Four-cell stage—apical cell first divides by two longitudinal divisions at right angles to each other.
4. Eight-cell stage—apical cell is further subdivided to form an upper tier and a lower tier of four cells each. The basal cell forms the hypophysis cell and the suspensor.
5. Sixteen-cell (also called dermatogen) stage—tangential cell divisions separate the protoderm that later matures into the epidermal layer from the inner cells.
6. Globular stage—apical cell undergoes a series of cell divisions to generate a globular embryo.
7. Heart stage—divisions on either side of the shoot apical meristem give rise to the two cotyledons, and the embryo now acquires a bilateral symmetry.
8. Torpedo stage—elongation and differentiation of cells along the embryonic axis mark this stage.
9. Mature embryo stage—the embryo enters the dormancy stage, whereby it becomes metabolically inactive and accumulates several storage compounds.

As mentioned earlier the apical cell generates a spherical proembryo, while the basal cell only divides transversally and gives rise to a transient filamentous structure called the suspensor. This extra-embryonic suspensor connects the proembryo to the maternal tissue and pushes it into the lumen of the ovule. Later during embryogenesis, only the uppermost suspensor cell, the hypophysis, becomes incorporated in the embryonic root meristem, as the precursor of the quiescent center (QC) and central root cap cells. The development of embryo in dicots and monocots is identical up to the globular stage. However, later in dicots while the development of the shoot apex continues along the axial axis and the cotyledons develop laterally, in monocots the shoot apex acquires a lateral position, and the single cotyledon is terminal (Table 24.1).

## 24.1 Embryogenesis

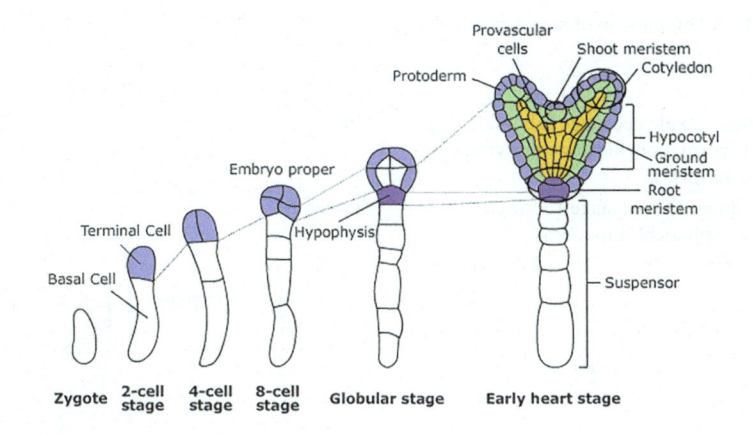

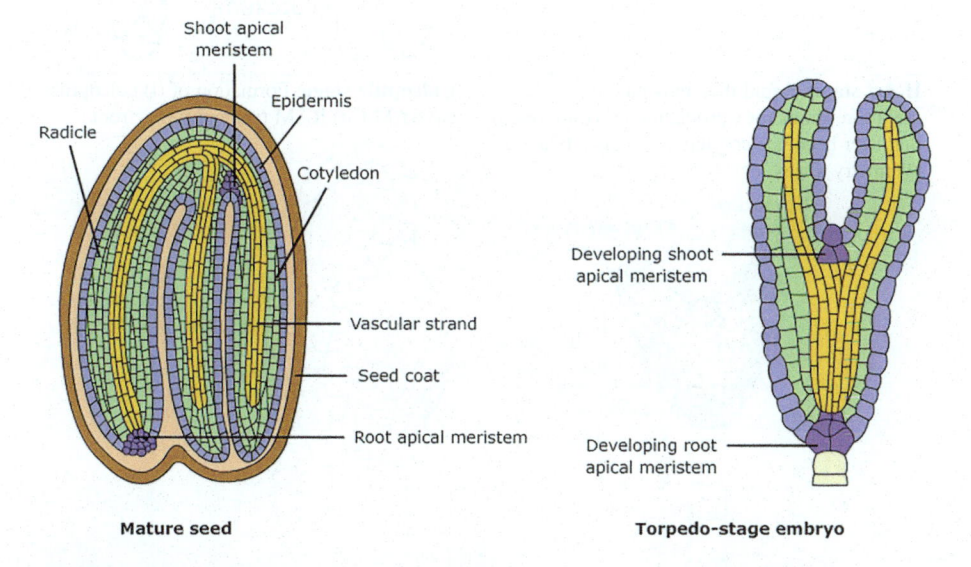

**Fig. 24.1** An orderly sequence of events establishes the body plan in *Arabidopsis*

### 24.1.3 Developmental Patterns

Fate of individual cells within the embryos in animals becomes fixed to produce a defined form. This is defined as lineage-dependent mechanism of embryo development. However, in plants position-dependent signaling mechanisms operate, and the behavior of cells depends on the position of these cells within the developing embryos. Such mechanisms ensure the formation of equivalent forms in spite of having different patterns of cell division.

**Table 24.1** A comparison of stages of embryogenesis as observed in *Arabidopsis* (dicot) and rice (monocot)

| *Arabidopsis* | Rice |
|---|---|
| **Zygote stage:** Single cell stage following fusion of egg and sperm; concludes with asymmetric division (transverse). | Similar |
| **Globular stage:** Octant stage followed by formation of spherical protoderm. | Similar |
| **Heart stage:** Rapid divisions on either side of the future shoot apex producing two outgrowths that later becomes cotyledons. Thus a bilateral symmetry arises. | **Coleoptile stage**: Formation of (i) coleoptile (ii) SAM (iii) RAM (iv) embryonic root |

(continued)

**Table 24.1** (continued)

| Arabidopsis | Rice |
|---|---|
| **Torpedo stage:** Results from cell elongation throughout the embryo axis and further development of cotyledons. 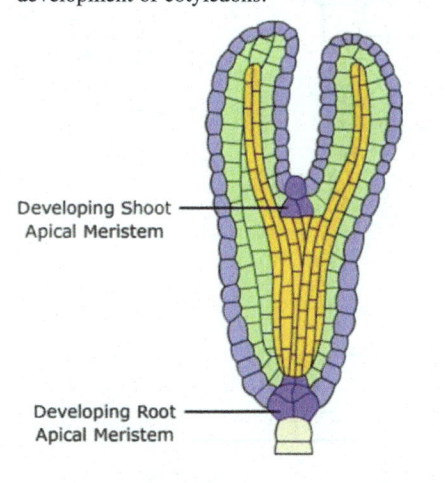 | **Juvenile vegetative stage**: SAM initiates several vegetative leaves. Development of scutellum. Scutellum represents the modified cotyledon specialized to absorb sugars from endosperm during germination. 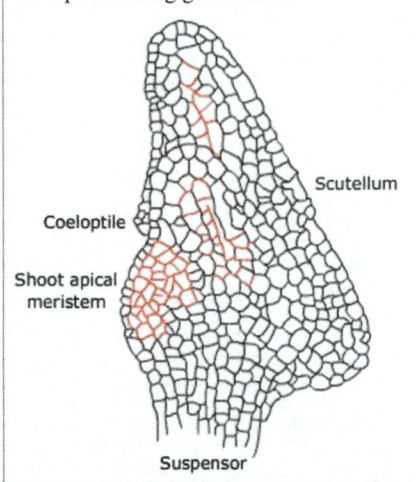 |
| **Mature stage:** The dicotyledonous embryo is oriented along an apical–basal axis and consists of the shoot apical meristem (SAM) located between two cotyledons, a hypocotyl, the primary root and the root meristem. 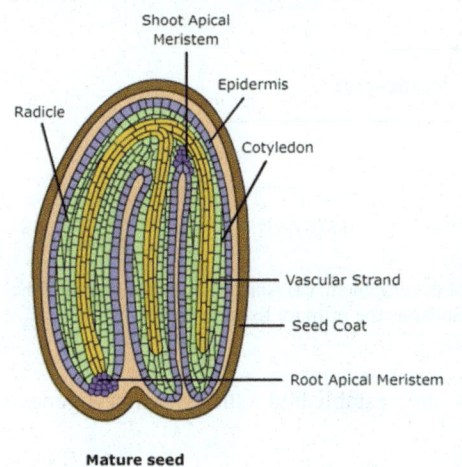 | **Mature stage:** The embryonic axis is displaced laterally relative to the scutellum. The SAM is established at a lateral position at the adaxial side of the transition stage embryo opposite to the scutellum. The shoot and root apical meristems at distal ends of the embryonic axis are protected by the coleoptile and coleorhiza, two organs exclusively found in monocots. 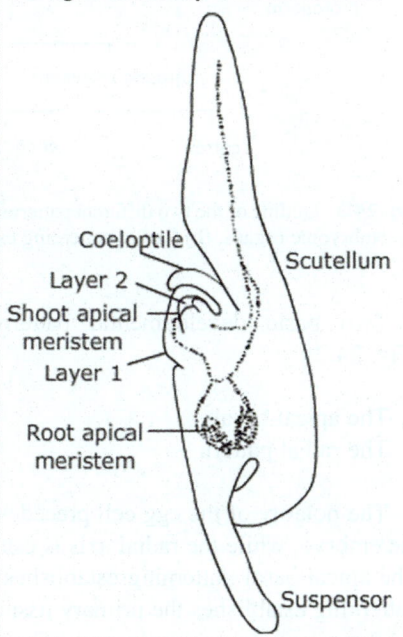 |

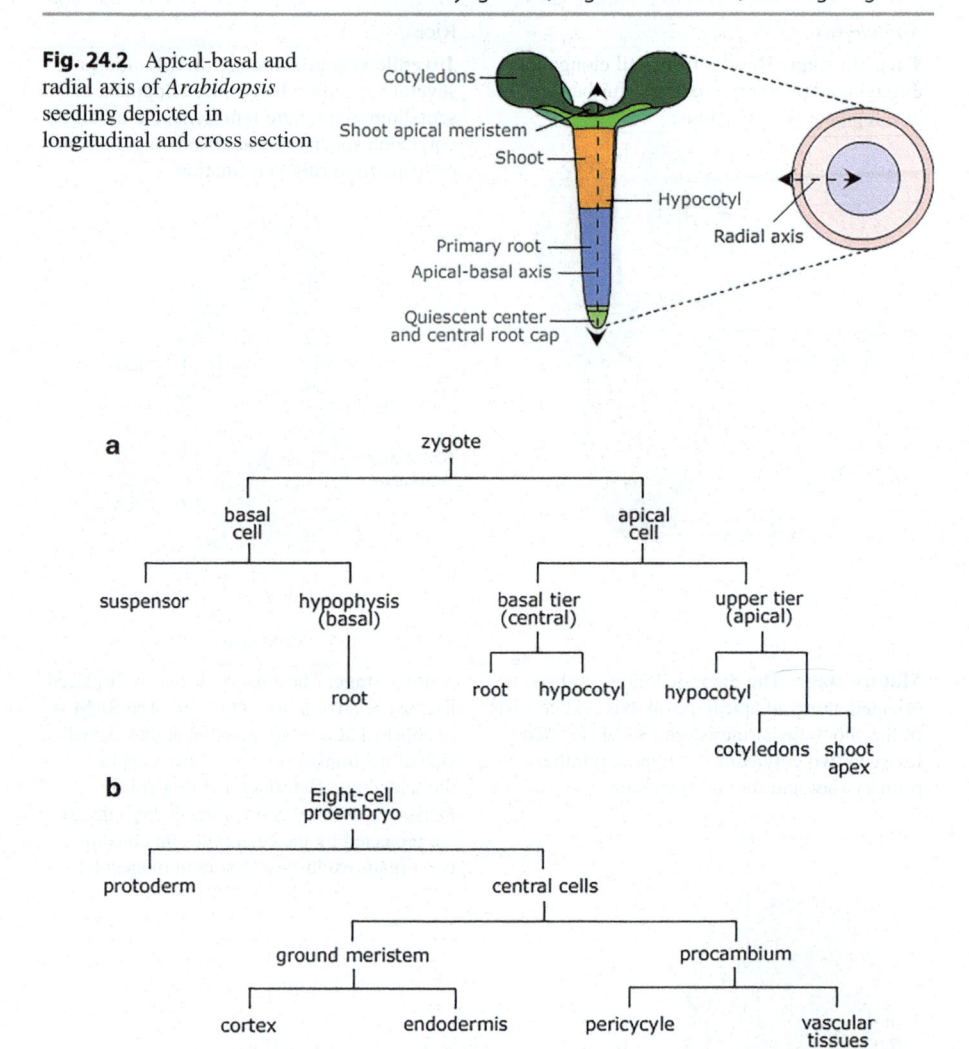

**Fig. 24.2** Apical-basal and radial axis of *Arabidopsis* seedling depicted in longitudinal and cross section

**Fig. 24.3** Outline of the two different patterns of development. (**a**) Apical-basal pattern establishes the embryonic organs. (**b**) Radial patterning establishes the primary tissue layers

Two basic developmental patterns are established during embryogenesis (Fig. 24.2):

1. The apical-basal
2. The radial pattern

The polarity of the egg cell precedes the development of the apical-basal axis of the embryo, while the radial axis is established after several cycles of cell division. The apical-basal patterning establishes the main embryonic organs, and the radial patterning establishes the primary tissues (Fig. 24.3).

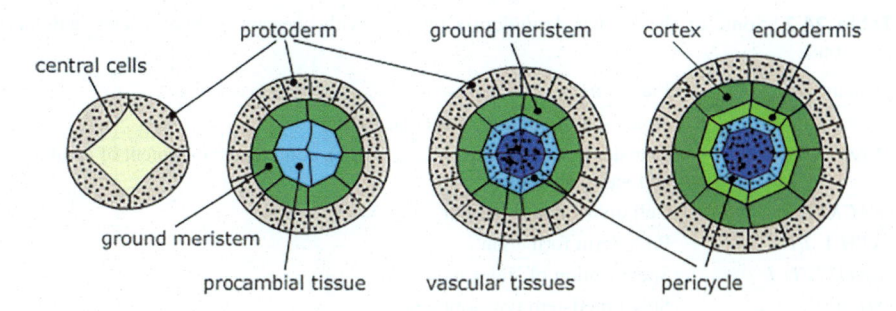

**Fig. 24.4** Demarcation of radial pattern in *Arabidopsis*. Periclinal divisions in the octant stage of the embryo separate the central cells from the outer protoderm. Central cells divide to form the ground meristem and the procambial tissues. The ground meristem further gives rise to cortex and endodermis, and the procambial tissues give rise to vascular tissues and pericycle

#### 24.1.3.1 Development of the Axial Axis

The tissues and organs in a typical plant body are arranged along a linear polarized axis with the shoot apical meristem at one end of the axis and the root apical meristem at the other. In the seedlings, this linear array of shoot apical meristem is followed by cotyledons and hypocotyl, and the root with the root apical meristem and the root cap are established during embryogenesis. Different segments of the root possess distinct structural and physiological features. For example, the adventitious roots develop from the lower end of the stem, while shoot buds develop from the apical end even when inverted.

#### 24.1.3.2 Development of the Radial Axis

Radial patterning of the tissues and organs of the plant body is also established during embryogenesis. All cells in the octant stage of the proembryo divide along a tangential plane, aligned along the apical-basal axis. This divides the proembryo in two different regions with different identities: an outer layer of eight cells, the protoderm, which is the precursor of the epidermis, and eight cells in the center of the proembryo, the inner cells, which are the precursors of ground and vascular tissues. A cross section of a root or a stem reveals the presence of three concentric rings of tissues arranged along the radial axis. These include the outermost epidermal layer of cells constituting the epidermis followed by the inner cortical layer (cortex) which encircles the vascular tissues comprising of the endodermis, pericycle, phloem, and xylem (Fig. 24.4).

## 24.2 Genetic Control of Patterning During Embryogenesis

Genetic and molecular studies using the plant model *Arabidopsis* have opened new arenas of embryo research. The major advances have been based on the analysis of specific mutant phenotypes. These have helped gain an insight into the processes involved in establishing the basic polarity of the embryo. Seedling defective mutants

**Table 24.2** Some of the known *Arabidopsis* mutants with aberrant embryogenesis and their functions

| Mutant/gene | Functions |
|---|---|
| GNOM | Apical-basal polarity, vesicular traffic |
| MONOPTEROS | Encodes auxin-response factor (ARF), controls development of axial axis and vascular tissues |
| FACKEL | Cell division and growth |
| ATM 1, PDF 2 | Protoderm formation |
| KNOTTED 1 | Specification of shoot meristem |
| SHOOT MERISTEMLESS | Shoot meristem development |
| WUCHSEL | Maintenance of shoot meristem |
| CLAVATA 1 | Regulation of size of shoot meristem |
| AINTEGUMANTA | Regulation of size of lateral organs |
| L1 | Specification of L1 layer of tunica |
| WOL | Vascular tissue specification |

are screened to isolate mutations that affect the processes of embryonic patterning. The mutants that affect the essential metabolic activity are avoided. The mutants generated are capable of developing into mature seeds that germinate but display abnormal organization, for example, those with defective apical-basal morphologies. These loss-of-function mutants are indicative of the corresponding genes that are required for normal apical-basal pattern. More than 80 genes with functions in embryogenesis have been examined for tissue expression patterns and interactions. Many genes with functions in pattern formation have been identified by studies of embryogenesis mutants. The genetic analysis of these mutants has allowed development to be modeled as a regulatory network of interaction among various transcription factors. A large number of mutants exhibiting aberrations in embryogenesis are known in *Arabidopsis* (Table 24.2) and a few in some other species such as maize and rice.

### 24.2.1 Mutants Defining Genes Involved in Axial Patterning

Among the different classes of mutations are the ones with defects in the apical-basal organization of the seedling (Fig. 24.5). These have been given names that indicate the observed mutant phenotypes. *MONOPTEROS* (*MP*) mutant lacks both the central and the basal regions of the seedling, consisting of most of the root and root apex, and so consists essentially of an apical piece of axis to subtend the cotyledons and shoot meristem. The apical structures are not structurally normal and the cotyledons are disorganized. The lower part of the globular embryo that gives rise to hypocotyl and root fails to develop the procambial tissue. Later during development some vascular tissue does develop in the cotyledons, but the vascular strands are not properly connected. The *mp* mutants lack the primary root, but later when they germinate to form seedlings, adventitious roots develop which, however,

## 24.2 Genetic Control of Patterning During Embryogenesis

**Fig. 24.5** Mutants from *Arabidopsis* with defects in apical-basal organization of the seedling. The shaded area refers to the deleted region

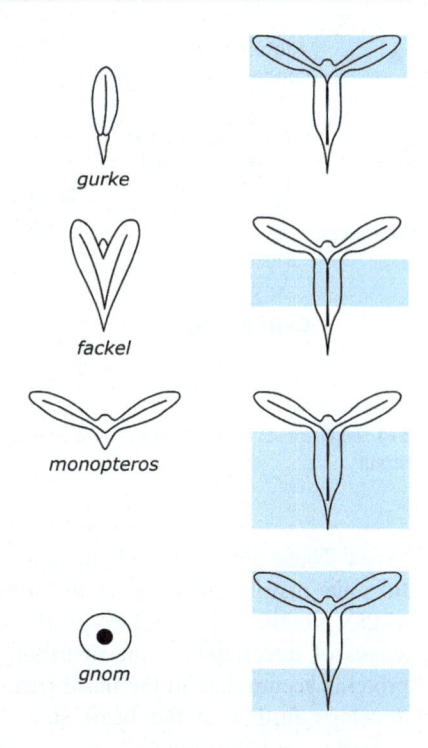

*gurke*

*fackel*

*monopteros*

*gnom*

lack well-developed vascular tissues. The *MP* gene is thus essential for the formation of vascular tissue during the postembryonic development. It encodes an auxin response transcription factor (ARF) suggesting an important role in auxin-dependent mechanisms. The activity of ARF is negatively regulated due to their association with repressors like IAA/AUX proteins. Auxin stimulates the activity of their target genes by targeted degradation of the ARF-associated repressors. MP-regulated vascular tissue development therefore results from the auxin-dependent regulation of ARF activities. Additional evidence for such an effect is provided by studies on *BDL (Bodenlos)* mutants that exhibit a similar phenotype as *MP* mutants. *BDL* encodes an IAA/AUX repressor that associates with MP to repress its activity, and this effect can be overcome by auxin-induced degradation of BDL. *GNOM (GN)* mutants lack both the apical and basal regions including the cotyledons and the root and so consist of a cellular mass with no obvious signs of apical-basal polarity. *GNOM* gene expression is required for the establishment of axial patterning. The gene encodes a guanine nucleotide exchange factor (GEF) that establishes a polar distribution of an auxin efflux carriers (PIN) that are involved in directional transport of auxin. Disruption of GEF activity in *GNOM* mutants disrupts the polar distribution of PIN proteins. The resulting change in the auxin distribution results in developmental defects resulting from the disruption of PIN protein-encoding

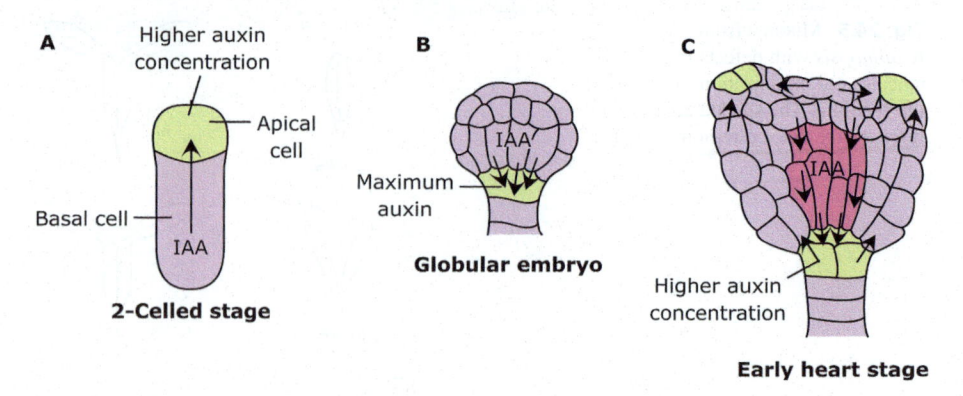

**Fig. 24.6** (a–c) Early stages of embryogenesis are marked by the PIN1-dependent movement of auxin

genes. These defects include the apical-basal patterning of the embryo that relies on the polar distribution of auxin in addition to other factors. Preferential accumulation of PIN proteins occurs in the apical cell in the two-celled stage. However, in the later stages of development, the distribution of PIN proteins is reversed, and the PIN proteins accumulate in the basal part of the embryo (Fig. 24.6a, b). As the embryo develops further to the heart stage, the distribution of these proteins becomes complex, and the upward flow of auxin through the superficial layer of cells is balanced by the downward flow in the internal layers (Fig. 24.6c). Both MP and GNOM therefore form a part of the mechanism that guides the auxin-dependent axial patterning of the embryonal axis (Box 24.2). *GURKE (GK)* mutants lack the apical region consisting of the shoot apex, the cotyledons, and the upper hypocotyl. The *GK* gene encodes an enzyme—acetyl-CoA carboxylase—that is required for the synthesis of long chain fatty acids and for the formation of pattern in the apical portion of the embryo. *FACKEL (FK)* mutants lack the hypocotyl and so have the cotyledons attached directly to the root. The gene is therefore interpreted to be required for hypocotyl formation and encodes a sterol C-14 reductase. Other genes implicated in embryo development include the *SHOOT MERISTEMLESS (STM)* and *WUSCHEL (WUS)*, which are critical for the initiation of the shoot apical meristem, and *ROOTLESS (RTL)*, which is critical for the initiation of the root apical meristem in the embryo. The expression of *HOBBIT (HBT)* gene is essential for the development of root apical meristem.

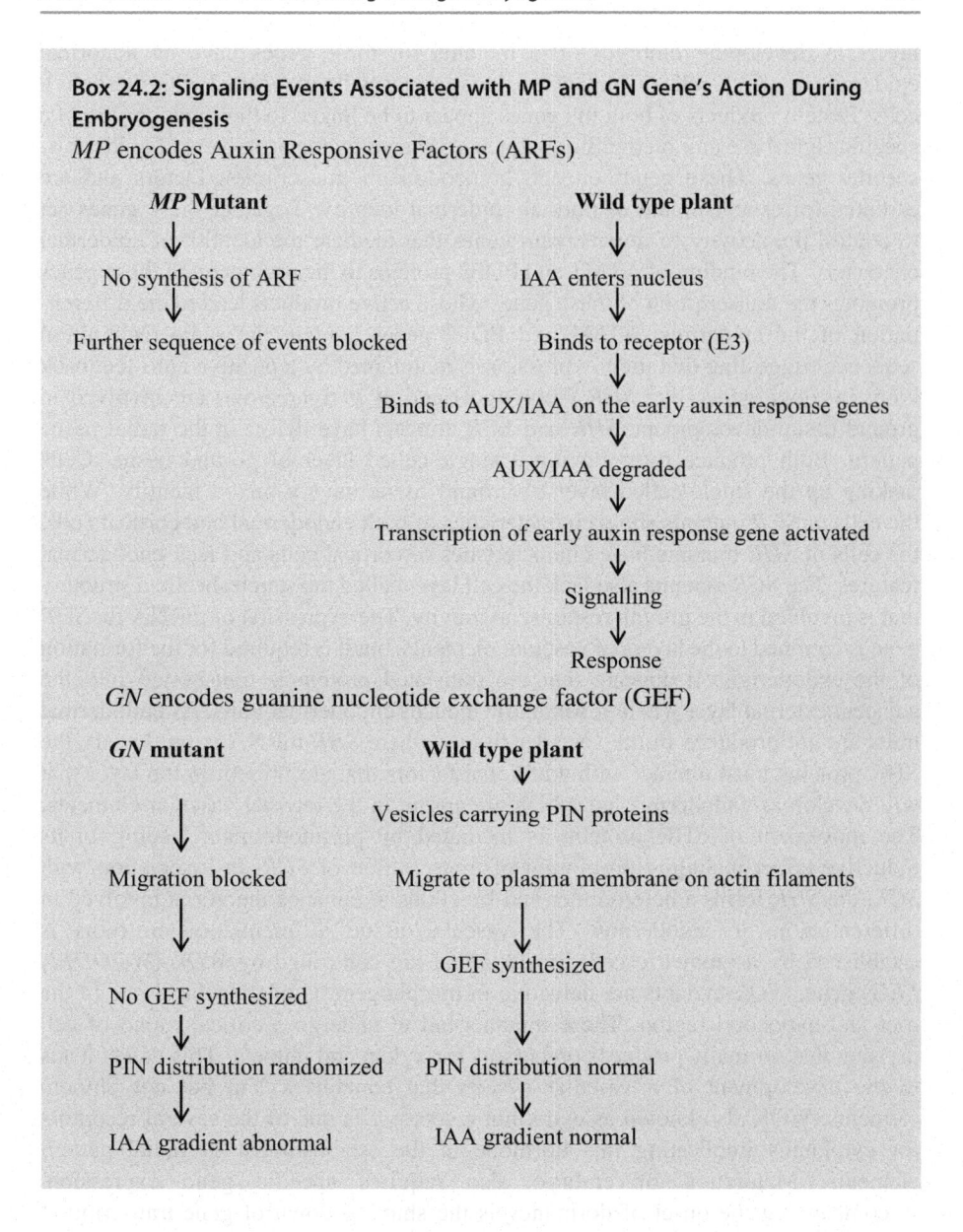

> **Box 24.2: Signaling Events Associated with MP and GN Gene's Action During Embryogenesis**
>
> *MP* encodes Auxin Responsive Factors (ARFs)
>
> **_MP_ Mutant**  
> ↓  
> No synthesis of ARF  
> ↓  
> Further sequence of events blocked
>
> **Wild type plant**  
> ↓  
> IAA enters nucleus  
> ↓  
> Binds to receptor (E3)  
> ↓  
> Binds to AUX/IAA on the early auxin response genes  
> ↓  
> AUX/IAA degraded  
> ↓  
> Transcription of early auxin response gene activated  
> ↓  
> Signalling  
> ↓  
> Response
>
> *GN* encodes guanine nucleotide exchange factor (GEF)
>
> **_GN_ mutant**  
> ↓  
> Migration blocked  
> ↓  
> No GEF synthesized  
> ↓  
> PIN distribution randomized  
> ↓  
> IAA gradient abnormal
>
> **Wild type plant**  
> ↓  
> Vesicles carrying PIN proteins  
> ↓  
> Migrate to plasma membrane on actin filaments  
> ↓  
> GEF synthesized  
> ↓  
> PIN distribution normal  
> ↓  
> IAA gradient normal

### 24.2.2 Genetic Control of Radial Patterning

Genetic studies have pointed toward several genes involved in radial tissue patterning. Genes like *ATML1* (*ARABIDOPSIS THALIANA MERISTEM LAYER 1*) and *PDF2* (*PROTODERMAL FACTOR 2*) are involved in shoot epidermal cell differentiation and are expressed from early stages of embryogenesis in the protodermal layers in developing embryos. The mutants for these genes have an abnormal epidermis in which cells display characteristics normally associated with mesophyll cells. Protein products of both the genes appear to be linked to their recognition of a specific eight-base pair recognition sequences shared by the promoters of epidermis-specific genes. These genes encode homeodomain transcription factors and are essential for establishment of normal epidermal identity. Together these genes act to control the activity of downstream genes that mediate the identity of epidermal characters. The binding of ATM1 and PDF2 proteins to the promoters of these genes promotes the transcription of these genes whose active products lead to the differentiation of the epidermis. ATM1 and PDF2 genes contain the same recognition sequence suggesting that their expression is maintained by a positive auto-feedback loop. In roots, genes like *SHR* (short root) and *SCR* (scarecrow) are involved in ground tissue development. *SHR* and *SCR* mutants have defect in the radial tissue pattern. Both produce roots having a single-celled layer of ground tissue. Cells making up the single-celled layer of ground tissue have a mixed identity. While the cells of *SCR* mutants show characteristics of both endodermal and cortical cells, the cells of *SHR* mutants have characteristics of cortical cells and lack endodermal features. The *SCR* mutants also lack the cell layer called the starch sheath, a structure that is involved in the growth response to gravity. The expression of mRNA for *SHR* gene is confined to the layers of vascular elements, but it is required for the formation of the endodermis. It appears that the translated protein is transported into the adjacent external layer where it somehow induces endodermal traits. As endodermal traits are not produced in the vascular tissue (where *SHR* mRNA is produced), the SHR proteins must interact with additional factors that are present in the layer that will develop as endodermis but which are absent in the internal vascular elements. The movement of SHR proteins is mediated by plasmodesmata leading to its inductive effect including promotion of transcription of *SCR*. In association with *SCR*, the *SHR* forms a heterodimer and functions to enhance the genes involved in differentiation of endodermis. The vascular tissue differentiation in roots is established by asymmetric cell divisions that are controlled by *WOL* (*WOODEN LEG*) gene. *WOL* mutants are defective in morphogenesis of vascular tissue in the root and hypocotyl region. These mutants fail to undergo a critical round of cell division that normally produces precursors for xylem and phloem. This defect leads to the development of a vascular system that contains xylem but not phloem elements. WOL also known as cytokinin receptor 1 is one of the several receptors for cytokinins implicating this hormone in the establishment of radial pattern elements. Maturation of embryo also requires specific gene expression. Accompanying the onset of dormancy is the shutting down of gene transcription and protein synthesis. Additionally, expression of specific genes like *ABI3* (*ABSCISIC ACID INSENSITIVE 3I*) and *FUSCA3* genes are essential for initiation of dormancy. These genes are sensitive to the hormone abscisic acid which is known to be the molecule involved in signaling that initiates dormancy. The *LEC1* (*LEAFY COTYLEDON1*) gene which is active during late embryogenesis is another gene involved in embryo maturation. It acts as a repressor of vegetative development, and its expression is required throughout embryogenesis.

## 24.3 Role of Auxin in Establishing Polarity During Embryogenesis and Vegetative Development

Auxin plays a critical part in embryogenesis, providing positional information for the coordination of correct cellular patterning from globular stage onward. The term **morphogens** is used for auxins and its synthetic analogs since these can induce the formation of embryos from somatic cells and can induce responses in target tissues in a concentration-dependent manner. The main sites of synthesis of auxin are the shoot meristems and leaf primordia. The polar auxin transport is important for modulation of several key responses. The polar auxin transport is thus created by a combination of localized synthesis and intercellular processes. Auxin controls much of postembryonic development, including plant architecture through the modulation of meristem activity, organogenesis, vascular tissue differentiation, tropic growth, etc., in response to environmental factors. The movement of auxin toward the root apex and the shoot apex is referred to as rootward and shootward transport. This polar transport of auxin occurs in cell-to-cell manner across the cell wall and plasma membrane of adjacent cells, and the overall process requires expenditure of energy. The polar auxin gradients are initially established by auxin synthesized locally, but later these gradients are extended with the directed auxin transport by specific transporter proteins present on the plasma membrane.

**The chemiosmotic theory** proposes that auxin requires an **influx** and **efflux** carrier proteins in order to move through cells and tissues. In *Arabidopsis* roots the AUX1 proteins have been identified as the influx carrier, while the PIN proteins (named after the pin-shaped inflorescences formed by the *pin1* mutant of *Arabidopsis*) constitute the putative transport protein of the efflux carrier complex. A crucial feature of the polar transport model is that the auxin efflux carriers are localized at the basal ends of the conducting cells. The auxin flow is therefore determined by the differential cellular positioning of the PIN proteins. Several *PIN* genes have been identified, and more than ten different *PIN* homologues have been found in *Arabidopsis*. *PIN* genes have also been identified in maize, rice, and poplar, and the high degree of conservation between the monocot and dicot species indicates that the PIN proteins have a conserved function throughout the plant kingdom. In *Arabidopsis* members of this family of transporters have different expression patterns within time and space and so offer the plant a means by which auxin can be transported precisely (Fig. 24.7). *PIN1* mediates organogenesis and development of vascular tissue and is located at the basal end of cells within the vascular stele. Gravitropic growth of roots is mediated by *PIN2*; differential growth of shoots is controlled by *PIN3*. *PIN4* regulates the activity of the root meristem, and the early embryo development is mediated by *PIN7*. Auxins provide the positional information to a developing and patterning tissue. Studies on the *POLARIS* gene of *Arabidopsis* provide further information on the role of auxin in defining position and cell activities during embryonic and seedling root development. It encodes a very short transcript that appears to regulate root sensitivity to ethylene and to modulate root growth. The *POLARIS* gene promoter is upregulated by auxin very rapidly, within minutes, and its spatial expression pattern represents a useful marker

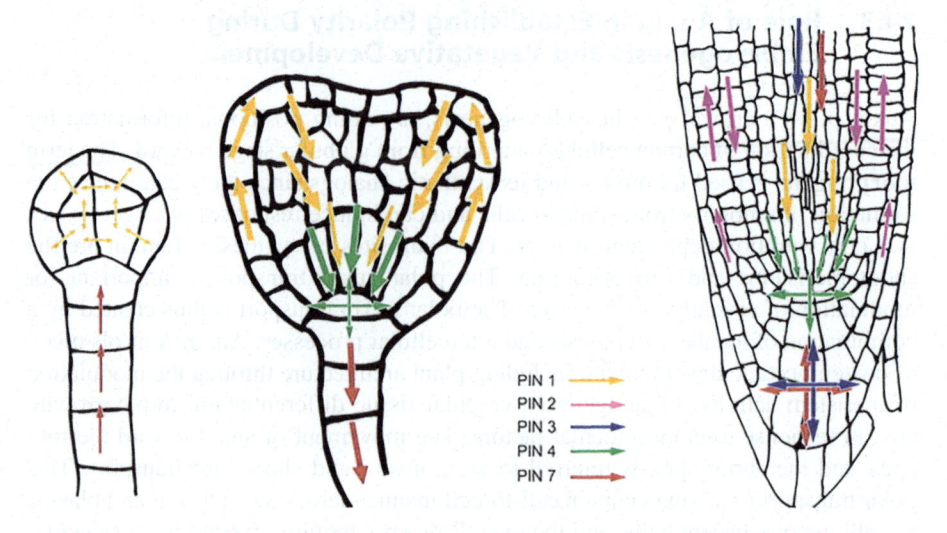

**Fig. 24.7** *PIN* gene expression and polar localization in preglobular, early heart embryos and in seedling root. Arrows indicate presumed directions of auxin flow based on subcellular PIN polarity

of auxin localization in the root. Interestingly, correct spatial patterning of *POLARIS* expression is disrupted significantly only in the most severe, ball-shaped *gnom* seedlings, suggesting that these individuals, but not the more conical-shaped *gnom* seedlings, are defective in polar auxin transport. This is consistent with the observed defective PIN1 localization in *gnom* embryos and suggests that auxin provides a chemical framework for the patterning of apical-basal gene expression and cellular activity in both embryo and seedling.

## 24.4 Plant Development and Meristems

Cell division in plants is localized in meristems that act as source of cells for the formation of new tissues and organs. Meristems can be defined as a group of cells that retain the capacity to proliferate and whose ultimate fate remains undetermined. Based on the position, several types of meristems are distinguished: (1) *shoot apical meristem (SAM)* and (2) *root apical meristem (RAM)*. Shoot and root apical meristems are formed in embryogenesis and are called **primary meristems.** The primary tissues and organs that comprise the primary plant body arise from these primary meristems. Meristems that arise later during the postembryonic phase of development are called as **secondary meristems**. These include the axillary, inflorescence, floral, intercalary, and lateral meristems (vascular cambium and cork cambium). (1) Axillary meristems—these are derived from the shoot apical meristems and are formed in the axil of leaves. (2) Intercalary meristem—it represents meristematic region within the developing organs, and these enable further localized growth (e.g., at the base of elongating grass leaves). (3) Meristemoids—they arise in differentiated region such as leaves to give rise to specific structures such as stomata. (4) Vascular cambium—these are secondary meristems that develop from procambium within the vascular cylinder. Two types of cells are distinguished in the vascular cambium—the **fusiform** and **ray initials**. While the former gives rise to the conducting cells of secondary xylem and phloem, the latter are divided to form the parenchyma cells arranged in radial files (rays). (5) Cork cambium—these are secondary meristems that develop in the cortex and the secondary phloem and give rise to outer protective periderm or bark of the secondary plant body.

Apical meristems localized in the tips of stems (SAM) and roots (RAM) contribute toward growth of the apical-basal axis of the plant. These meristems contribute toward the formation of primary meristems like the **periderm**, the **procambium**, and the **ground meristem** that are the source of epidermis, vascular tissue, and ground tissues, respectively. The apical meristems are organized into distinct regions. Although the meristems—SAM and RAM—differ in their anatomical characteristics and the organs or cells that are produced, the basic organization is similar. In both, an organizing center (OC in the shoot meristem, QC in the root meristem) is surrounded by stem cells (also called as initial cells in plants) for various tissues. These stem cells remain undifferentiated and retain the capacity for cell divisions indefinitely and are not committed to a differentiation pathway.

### 24.4.1 Shoot Apical Meristem

The shoot apical meristems (SAM) comprise of the central region (CZ), the peripheral zone (PZ), and the rib zone (RZ). The central zone (CZ) is located at the summit of the SAM and contains the slowly dividing stem cell population. This region does not consist of permanent initial cells but rather acts as a reservoir through which individual stem cells pass. Stem cell division replenishes the CZ and displaces the daughter cells outward into the surrounding PZ, where cells are rapidly dividing and new organs are initiated, as well as downward into the RZ that provides cells for the internal tissues of the stem and lateral organs. At the top of the RZ is positioned a small group of cells called the organizing center (OC), which acts as a niche that sustains the overlying stem cell reservoir. Although clear boundaries between the three domains have not been observed, expression studies reveal that the cells in each zone have different gene expression patterns and thus distinct molecular as well as functional characteristics. Classical studies indicate that the angiosperm SAM is organized into overlapping cell layers and domains (Fig. 24.8). The *Arabidopsis* SAM consists of three cell layers, L1–L3, which remain clonally distinct from one another due to their specific cell division patterns. The single-layered L1 and L2 cells comprise the **tunica** and divide anticlinally, perpendicular to the plane of the meristem. These cell layers generate the epidermis and subepidermis, respectively, of the shoots, leaves, and flowers. The multitiered L3 cells comprise the **corpus** and

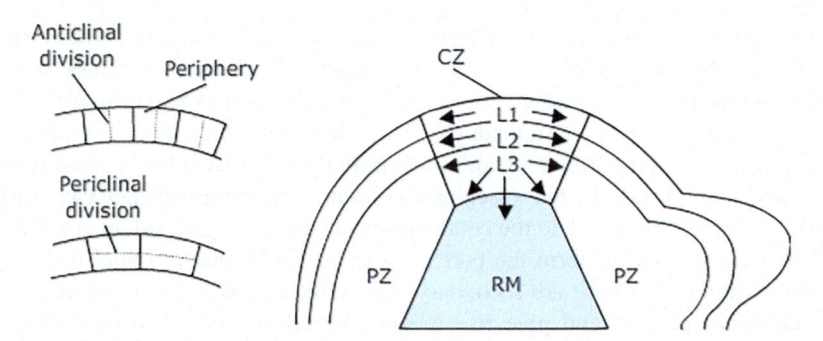

**Fig. 24.8** Shoot apical meristem is distinguished into distinct zones. L1, L2, and L3 represent successive layers of the CZ. *CZ* central zone; *PZ* peripheral zone; *RM* rib meristem

divide in all planes, forming the internal tissues such as the stem vasculature and pith. SAM produces leaf **primordia** that develop into new leaves and cells that will produce lateral (axillary) bud primordia, which have the potential to form new branches, in the axils of leaves. The SAM is initiated within the apical half of the proembryo and as such does not involve cell-cell signaling across the apicobasal boundary. Many factors are known to control aspects of shoot meristem function, but the vast majority of these act in the homeostatic control of meristem size rather than the initiation of the meristem. The OC cells and the overlying stem cells mutually control the size of the other population, leading to a stable meristem size. The installation of the OC and stem cell area occurs during the mid-globular stage, and both are maintained throughout plant life. It is important to note that the identities and the characteristic patterns of division of different initial cells are determined by their position rather than their genetic programming.

### 24.4.2 SAM Development: Role of Auxin and Transcription Factors

SAM develops at a position where auxin is low. Auxin activity in embryo development leads to the formation of terminal domains maintained in pluripotent state, i.e., capable of differentiating into many cell types. It leads to the formation of stem cells, which are cells that are undifferentiated and which can both perpetuate themselves and give rise to daughter cells capable of differentiating into specialized cells. SAM development is related to auxin-related gene activities. PIN-mediated auxin flow directs auxin to regions that flank SAM, while the central zone becomes relatively auxin-deficient. The early embryogenesis is marked with the accumulation of PIN1 protein in the apical regions. This is however reversed to the basal region in the early heart stage. The change in the concentration is mediated by the phosphorylation on PIN proteins by kinases like PINOID and phosphatase PP2. The effect is also affected by the alteration in the transcription of several genes that are involved in the distribution of PIN proteins. Four genes (auxin related) are responsible for SAM development:

1. *MP (MONOPTEROS)*: leads to differences in auxin activity in central and flanking regions. *MP* expression is relatively weak in central region of the apex but stronger in flanking regions, which will later develop into cotyledons.
2. *WUS* (WUSCHEL) genes: these express in the subapical region in the early 16-cell stage of embryo development. These play an important in specifying and maintaining the identities of initials in the SAM.
3. *CUC* (Cup-shaped cotyledon): expressed in central region between the cotyledons, and it helps in the development of cotyledons. *CUC* mutants have no cotyledon development.
4. *STM* (Shoot meristemless) gene: the expression of *STM* gene coincides with *CUC* expression and is confined more around the central region. Both *CUC* and *STM* mutants have similar phenotypes.

Both *CUC* and *STM* help to maintain the initial cells in a proliferative state. The expression of these two genes in the central region of the SAM is related to the low auxin concentration in this region as compared to the flanking regions. This auxin-dependent patterning in the shoot apex is thus related to the low levels of auxin in the SAM and the intercotyledonary regions and the related high expression of CUC in these regions (Fig. 24.9). An opposite pattern is seen in the cotyledon primordia. Mutations in *MP* and *NPH4 (NON PHOTOTROPIC HYPOCOTYL 4)* block the auxin signaling in the cotyledonary region which leads to ectopic expression of CUC genes in this area. This further initiates the localized expression of *STM* gene and later the *CLAVATA3* (*CLV3*). The latter plays an important role in limiting the number of apical initial cells in the SAM. The movement of auxin away from the central region of SAM along with the upward flowing stream of auxin results in auxin accumulation in the tips of the growing cotyledons and it further flows down into the hypocotyl. This directional transport of auxin results in the activation of ARF proteins MP and NPH4, which are involved in the differentiation of vascular tissues.

### 24.4.3 Root Apical Meristem

The root apical meristem has a simple structure with distinct zones (Fig. 24.10):

1. Root cap (RC): At the distal tip of the root is the root cap, consisting of the central columella (COL) and lateral root cap (LRC).
2. Meristematic zone (MZ): Located just beneath the root cap, it consists of a group of initials that divide and differentiate to form the mature tissues of the root. The MZ can be divided in two zones: the apical MZ, consisting of the most actively dividing cells, and a basal MZ, in which cells become larger, divide less frequently, and begin to acquire distinct fates. The RC can thus be distinguished into RC meristem (RCM) and, the main, proximal meristem (PM).
3. Zone of elongation: As cells exit mitosis, they leave the MZ and undergo a period of elongation, thereby defining the elongation zone (EZ). The rate of cell division progressively decreases as the distance from the apical meristem increases.

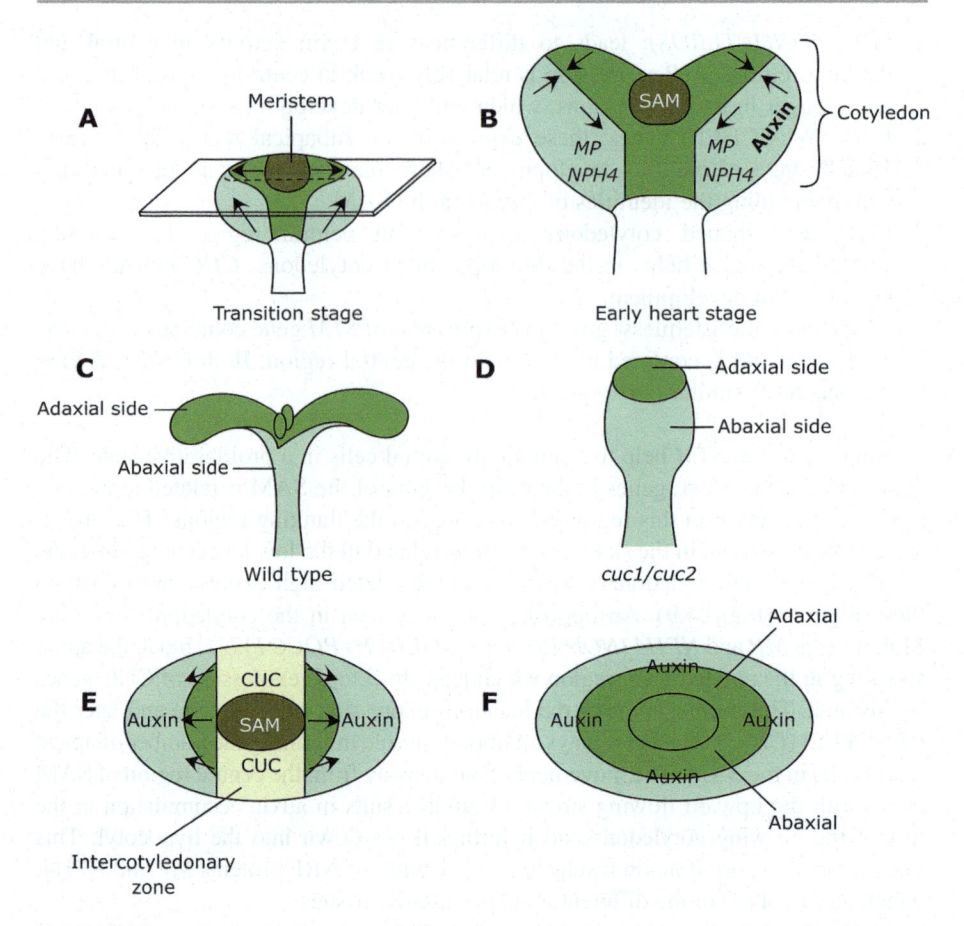

**Fig. 24.9** Patterning in the shoot apex is dependent on relative auxin concentrations in the central region versus in the flanking region of the developing shoot apex. (**a, b**) Arrows depict the flow of auxin in the transition and the early heart stage. The auxin-dependent genes such as *MP* and *NPH4* are thereby differentially regulated in these regions to promote patterning in these regions. (**c**) and (**d**) represent the morphological features of the wild-type and *cuc* mutants. (**e**) and (**f**) represent a cross-sectional view of the wild-type and *cuc* mutants, respectively

4. Zone of maturation: Finally, cells cease elongation and enter the differentiation zone (DZ), which is typically defined by the appearance of root hairs and vascular cells with visible secondary cell wall formation.

Located just above the RC lies the **quiescent center** which is so named because of the relatively low rates of cell divisions that occur in this region in comparison to surrounding tissues. The close association between the cells of the QC and the initial cells in the MZ suggests a functional interdependence between the two. Experiments involving surgical removal of QC leads to disruption of normal cell division and

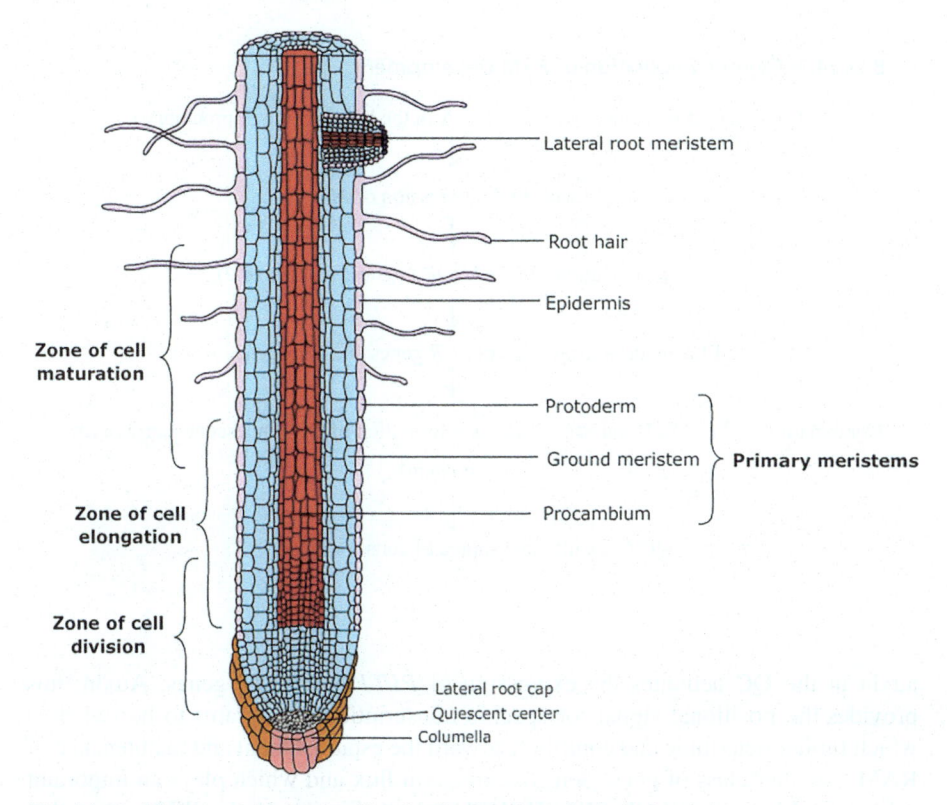

**Fig. 24.10** Diagrammatic sketch of longitudinal section of a typical root

precocious differentiation in the adjacent initial cells. This suggests that as proposed for the SAM, position-dependent mechanisms also play an important role in specifying the different cell types in RM.

### 24.4.4 Role of Auxin and Cytokinin in Establishment of RAM and Root Development

Auxin plays an important role in the determining the complex structure and function of RAM. A higher concentration of auxin is found in the QC. A shift in the position of auxin maxima on using chemical treatments leads to a corresponding change in the position of QC. A gradient of auxin occurs across the root and is responsible for a number of responses including localized areas of cell division and cell differentiation (Box 24.3). Auxin triggers the breakdown of transcription repressor of *MP (MONOPTEROS)* the IAA/AUX. MP encodes ARF that has an auxin-dependent role in maintenance of root structure in vegetative growth. Several transcription factors act downstream to the MP and coordinate the specific growth of roots. These include the PLETHORA 1 (PLT1) and PLETHORA 2 (PLT2) that belong to the AP2/ETHYLENE RESPONSIVE FACTOR class of proteins. The accumulation of

> **Box 24.3: Genetic Regulation of RAM Development**
>
> Occurs due to high accumulation of IAA in the basal region of proembryo
> ↓
> Leads to de-repression of *MP*
> ↓
> Leads to activation of *PLETHORA* genes (*PLT 1* and *2*)
> ↓
> PLT induces expression of *SCR* genes and *SHR* genes
> ↓
> Combination of *PLT, SCR* and *SHR* directs the formation of QC (Quiescent centre in root meristem)
> ↓
> QC is the ultimate source of cells for RAM

auxin in the QC activates the expression of *PLT1* and *PLT2* genes. Auxin thus provides the positional signal for specific transcriptional programs to be switched which further determine and contribute toward the establishment and maintenance of RAM. Another class of genes sensitive to auxin flux and which plays an important role in RAM are the *WOX* (*WUSHCHEL*) genes. Similar to the *WUS* genes that operate in the SAM, the *WOX* helps to maintain the population of undifferentiated initials in the QC and prevents their premature differentiation.

Another class of hormone cytokinin is also known to play an important role in normal root development. While auxin is synthesized in the shoot and transported toward the roots, the cytokinins are synthesized in the root and then moves toward the shoot. Cytokinin signaling begins during early phase of root development in the hypophysis of the globular embryo. As the hypophysis divides in two, the basal cell loses the cytokinin expression, while the apical cell retains it and in fact divides to form the QC. Auxin on the other hand has a reverse expression. This supports the premise that the two hormones have an opposing effect in shoot and root development. The loss of cytokinin expression in the basal cell is due to activity of two genes *ARR7* and *ARR15* that have auxin response elements (AuxRE) in their promoters. Antagonistic activity of cytokinin and auxin in the basal cell is essential for normal development of RAM and enables the particular pattern of cell divisions that occur in the QC.

## 24.5 Growth and Differentiation of Lateral Roots

The architecture of a plant's root system is important for carrying out its functions. The **radicle** of the seedling together with the lateral (branch) roots it produces constitutes the primary root system of the plant. The root systems of most vascular plants are formed by branching of **lateral roots** (LRs) from a **primary root** (PR) that first develops during embryogenesis. Root hairs, each of which is an outgrowth of a single epidermal cell, provide most of the surface through which roots absorb water and dissolved minerals. Stems and leaves may have the ability to produce **adventitious roots**. In this case, adventitious root primordia originate from parenchyma cells that are adjacent to vascular bundles. The roots of monocots and dicots are similar in some features and also differ in certain characteristics. While they have roughly similar features like the embryonically derived primary root, lateral roots, and adventitious root, they differ significantly in the type of root systems they possess. The monocot root system consists of a primary root that develops from the radicle and adventitious or **seminal roots** that branch from the scutellar nodes. Also present are the postembryonically developed crown roots or **prop roots** that develop from the lower most nodes of the stem. These prop roots unlike the primary and seminal roots develop and branch throughout the vegetative growth of the plant and make up the majority of the root system in monocots. In dicots, the root system consists of the main primary or tap root and its branches. Additionally, adventitious roots arise from the subterranean stems or from the hypocotyl. Root hairs perform the important function of water and nutrient uptake in plants. Additionally, they also serve to anchor plants to soil. The initiation and growth of lateral roots has been studied in detail in several flowering plants. Three patterns of root hair differentiation have been seen in plants. Type I pattern is one in which all the cells of the root epidermis have the potential to form root hairs. Majority of the plants exhibit Type I root hair development pattern. The other types include those in which some of the cells have the potential to form root hairs (trichoblasts) and some cells are incapable in forming root hairs (atrichoblasts). These are further divided into Type II, in which the cells of the root meristems divide by an asymmetric division and the trichoblast is formed from the smaller cell. Examples of plants exhibiting Type II pattern of root hair development include the primitive vascular plants *Lycopodium*, *Selaginella*, etc. Type III is exhibited by members of the Brassicaceae family in which trichoblasts and atrichoblasts occur in alternate files. LRs initiate from a specialized cell layer in the pericycle overlying the developing xylem tissue (Fig. 24.11). LR development involves stimulation and dedifferentiation of pericycle founder cells, which increase in size, reenter the cell cycle, and divide asymmetrically to give rise to a lateral root primordium (LRP), which then emerges through the outer layers of the PR. The endodermis, the cell layer immediately overlying the pericycle, has recently been identified as a key regulator of LR developmental progression. Feedback from the endodermis to the pericycle is required for LR initiation. LR development and changes in root architecture are brought about through a combination of hormone signaling, environmental cues, and hormone-independent protein activity. The initiation and growth of lateral roots are controlled by the concentration and transport of

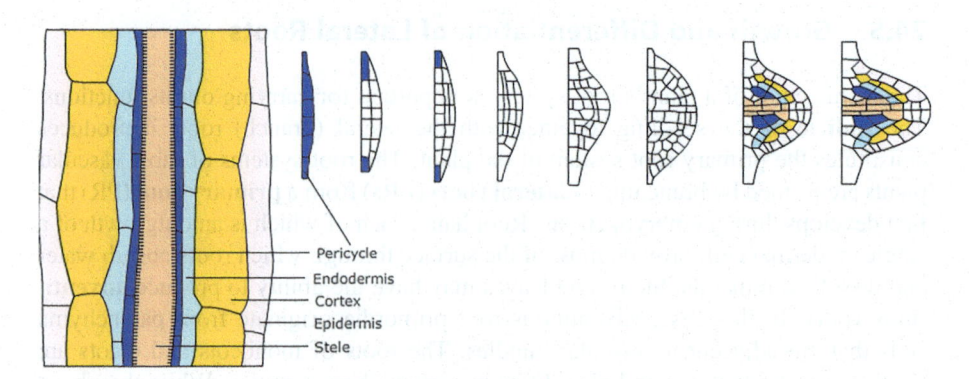

**Fig. 24.11** Stages of development of lateral roots which originate from pericycle of the primary root

plant hormones, particularly auxin. Some of the key genes regulating lateral root development are auxin-independent, however. Molecular genetic studies using *Arabidopsis* mutants have revealed that the auxin transport system with a balance of influx and efflux is important for LR initiation and subsequent LR primordium development. In addition, normal auxin signaling is mediated by two families of transcriptional regulators, Aux/IAAs and ARFs, which are necessary for LR formation. Gain-of-function Aux/IAA mutants have altered auxin sensitivity and pleiotropic defects in growth and development that include differences in lateral root formation. Auxin synthesized in the aboveground part of the plant is transported to the roots; more recent work has revealed that some auxin is also synthesized in the roots themselves. Auxin is transported directionally within roots in a process known as polar auxin transport. Auxin moves toward the root tip in cells associated with the vascular cylinder (stele) and moves away from the tip in cells of the epidermis. Localized auxin concentrations are regulated both by diffusion across membranes and by the action of several auxin transport proteins; these include AUX1, which facilitates influx of auxin into cells, and PIN, which controls auxin efflux. Taken together, the actions of these proteins establish local auxin maxima as well as concentration gradients, and it is in the areas of these maxima that lateral roots are initiated. Any mutation that prevents the establishment of normal auxin gradients disrupts the patterning of root development. The promotion of lateral root formation by auxin is inhibited by cytokinins (CKs), which act directly on the lateral root founder cells in the primary root to bring about this inhibition. There is evidence that cytokinin may interfere with PIN gene expression.

The role of auxin in initiating and maintaining the quiescent center of the maize root meristem is hypothesized to be mediated through the variation in the enzyme ascorbic acid oxidase (AAO) which is involved in the breakdown of ascorbic acid. Ascorbic acid is a compound which is necessary for the transition from $G_1$ to S phase in the cell cycle, and which is broken down by AAO. Formation of AAO is increased in response to auxin in the quiescent center than surrounding cells. Auxin influences

AAO levels within the root meristem and that this ensures the continued stem cell ability of the quiescent center.

Proteins like ALF4, ARABIDILLO-1, and ARABIDILLO-2 are known to promote lateral root initiation in an auxin-independent manner. The aberrant lateral root formation *ALF4* mutation in *Arabidopsis* blocks the initiation of lateral roots, thus greatly altering root system architecture. The *ALF4* mutant completely lacks lateral roots, and it appears that the cell cycle in pericycle cells is blocked in the mutant plants. ALF4 protein functions in maintaining the pericycle in the mitotically competent state needed for lateral root formation.

## 24.6 Cell Growth and Differentiation

An essential feature of plant development is the ability of plant cells to grow and differentiate. **Growth** of an organism is defined as an irreversible increase in mass. Because mass is related to cell volume and cell number, growth refers to an irreversible increase in cell size (enlargement) or to an increase in cell size as well as cell number (cell division). Cell division, by itself, is not sufficient to result in growth. **Differentiation**, in contrast, refers to the cells acquiring qualitative differences among other cells of common origin, i.e., those derived from a cell or group of cells. It is by differentiation that cells in an organ or tissue become different from each other or specialized for different functions, e.g., the epidermis, or mesophyll, or xylem or phloem cells in a leaf. **Morphogenesis** is the acquisition of form, how a plant or organ acquires its distinctive shape or form. The control of these two processes is central to a study of plant morphogenesis.

The growth and development in higher plants involves cell division, expansion, and differentiation along the apical-basal axis and the radial axis. The radial axis especially in dicotyledonous species grows in size as the concentric rings of new cell layers are added following cell divisions in the vascular cambium of seedling stem, hypocotyl, and roots. On the other hand, development of the apical-basal axis involves meristems like the shoot apical meristem and the root apical meristem and includes patterning of symmetry and formation of functionally distinct structures. This meristematic activity of plant cells is based on the essential property of every plant cell—**totipotency**—i.e., the ability of the plant cells to divide, grow, and differentiate into whole organisms. **Differentiation** refers to differences, other than size, that arise among cells, tissues, and organs. Differentiation occurs when cells assume different anatomical characteristics and functions or form patterns. Differentiation begins in the earliest stages of development, such as when division of the zygote gives rise to cells that are destined to become either root or shoot. Later, unspecialized parenchyma cells may differentiate into more specialized cells such as xylem vessels or phloem sieve tubes, each with a distinct morphology and unique function. The plant cells exhibit extreme plasticity with their ability to differentiate, dedifferentiate, and redifferentiate, i.e., even though some plant cells may appear to be highly differentiated or specialized, they may often be stimulated to revert to a more embryonic form. For example, mature somatic cells isolated from any body part of the plant, cultured on an artificial medium, may be stimulated to reinitiate cell division, to grow as undifferentiated **callus** tissue, and eventually to give rise to a new plant. It is as though the cells have been genetically reprogrammed, allowing them to reverse the differentiation process and to differentiate along a new and different path. This ability of differentiated cells to revert to the embryonic state and form new patterns without an intervening reproductive stage is called totipotency. Plants can therefore be propagated indefinitely in the vegetative state.

### 24.6.1 Cell Growth: Role of Wall Extensibility

Growth of cells derived from meristems is accompanied by cell enlargement mediated by water uptake. An essential requirement for such increase in cell volume requires wall extension, which is related to turgor pressure. Wall extension is related to turgor pressure and is a quantitative measure of the capacity of the wall to irreversibly increase its surface area. Turgor pressure develops because the cell wall resists the force of the expanding protoplast pushing against the wall, which thus generates stress within the wall. However, in order to prevent excessive water uptake and to avoid rupturing the plasma membrane due to high turgor pressure, plant cells are surrounded by a very strong and relatively rigid wall. The strength and rigidity of the cell wall impose critical restrictions on the capacity of plant cells to grow. The ability of cell wall to extend is essential for plant cell growth and morphogenesis. Modification of cell wall strength and rigidity is essential in order to change the water potential of the cell, permit water uptake, and, consequently, allow the cell to enlarge. Plant cell wall is a composite polymeric structure comprising of cellulose microfibrils coated with heteroglycans (hemicelluloses such as xyloglucan) and embedded in a dense, hydrated matrix of various neutral and acidic polysaccharides and structural proteins. Plant cell walls have the ability to extend at acidic pH. This pH-dependent extension is known as acid growth. Cell wall proteins such as **expansins** that are activated by acidification at pH between 4.5 and 6 disrupt noncovalent binding between cellulose and hemicelluloses. Other wall proteins such as xyloglucan endotransglycosylase/hydrolases (XTHs) which hydrolyze glycosidic bonds also contribute to wall loosening by modifying wall properties. Acid growth is mediated by an auxin-stimulated proton pump in the plasma membrane, which decreases the pH of the cell wall solution, promoting the activity of wall-loosening proteins (Fig. 24.12). Within minutes of treatment with auxin, there is induction of rapid cell elongation in stem, coleoptile, and hypocotyl segments. This rapid effect is believed to result from the activation of a proton pump ATPase at the plasma membrane, inducing extrusion of $H^+$, extracellular acidification, activation of expansins, and subsequent wall loosening. Activation of the plasma membrane $H^+$ ATPase causes hyperpolarization of the membrane potential and activation of voltage-dependent $K^+$ inward channels. Uptake of $K^+$ is likely to contribute to the water uptake necessary to sustain expansion. In addition to the stimulation of their activity, auxin also induces expression of both $H^+$ ATPase and $K^+$ channels. Two types of modes of cell expansion are there in plants: diffuse and tip growth. The

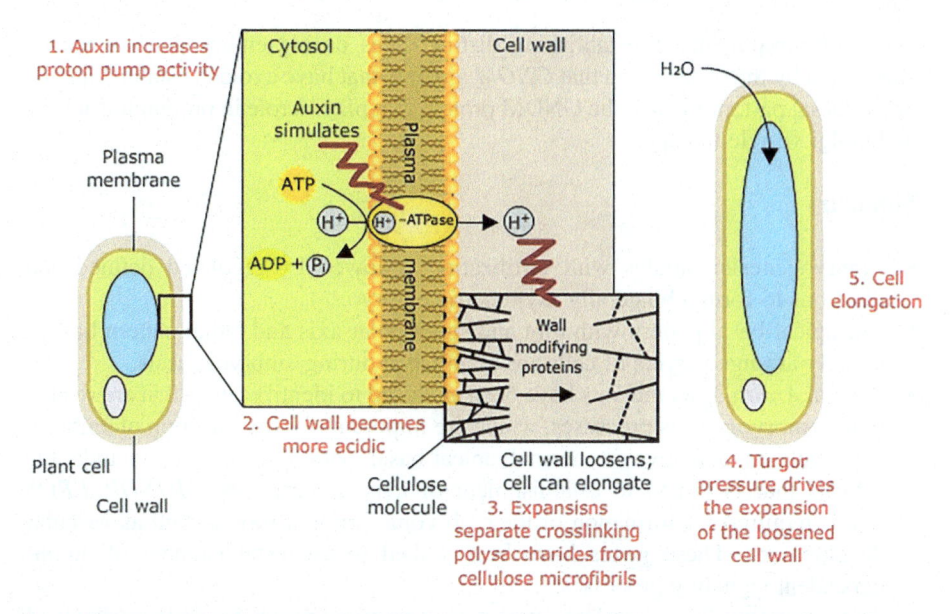

**Fig. 24.12** Cell expansion is mediated by auxin-stimulated proton pump, which reduces the pH and promotes the activity of wall-loosening enzymes leading to loosening of cross-linking interactions of glycans with cellulose microfibrils

diffuse growth is affected by changes in the orientation of cellulose microfibrils in the cell wall in response to developmental and environmental cues. The tip growth mode of cell expansion is controlled by the dynamic cytoskeletal element F-actin that can change from single filaments and associate to from a dense meshwork forming a cytoskeletal array in the growing tips.

### 24.6.2 A Role for Cell Wall Components

The nature of cell wall varies between different cells and tissue types and is also a prerequisite for asymmetric divisions. Asymmetric division during pollen mitosis I is essential to the consequent differential expression of genes in the two daughter cells and is important for establishing the structurally and functionally different generative and vegetative cells. Similarly, in *Fucus* zygote asymmetric division leads to the formation of basal rhizoid and upper larger thallus cell with distinct fates. Evidences indicate the role of cell wall components such as sulfonated polysaccharides in determining the identities of the thallus and rhizoid cells by providing positional information. Recognition and localization of specific cell surface polysaccharide epitopes such as **arabinogalactans** (AGP) are possible using specific monoclonal antibodies. AGPs are differentially expressed during embryogenesis in *Brassica* highlighting cell differences between embryo proper and suspensor. The role of cell wall components in establishing distinct fates during embryogenesis can be speculated by the observation that *GNOM* mutants that have a role in establishing the apical-basal patterning lack the GNOM protein that plays a role in protein trafficking and Golgi vesicle transport.

**Summary**

- Embryogenesis initiates with fertilization followed by regulated defined cell divisions to form a bilaterally asymmetric embryo.
- The apical-basal pattern with root and shoot polar axis and radial pattern having radially arranged layers of cells are established during embryogenesis.
- Several *Arabidopsis* mutants have been analyzed to identify genes that are related to the embryogenesis processes, which are important for organization of embryo. The mutants defective in normal apical-basal pattern formation include the *GNOM* that is linked to establishment of axial polarity and *MONOPTEROS* that is required for formation of the embryonic primary root as well as vascular development. These genes have been linked to the establishment of auxin-dependent signaling processes.
- Auxin functions as a chemical signal during embryogenesis. Discrete gradients of auxin are created during embryonic development through localized auxin synthesis and polar auxin transport.
- Polar auxin transport over long distances from the site of synthesis in apical tissues to the root tip regulates stem elongation, apical dominance, and lateral branching.
- Plant growth is limited to discrete regions called the meristems. Two such regions are the apical meristems located at the tips of roots and stems. These regions of active cell division are responsible for primary growth or the increase in the length of roots and stems.
- Plant cell growth is an irreversible change in cell size involving stretching of the cell wall driven by the internal turgor pressure of the cell walls.
- Cell growth appears to be initiated when these stresses are relaxed the activity of endogenous wall-loosening proteins, called expansins that characteristically weaken cross-links between cellulose molecules, increase wall extensibility, and allow for turgor-induced cell expansion.

**Multiple-Choice Questions**

1. Which of the following statements is **false** about the polarity of an embryo?
    (a) It is a term used to describe the differences along an axis such that one end of the axis is different from the other.
    (b) It is established only after the zygote divides.
    (c) It is important for development of structural axis of the body.
    (d) It is key to biological pattern formation.

# Multiple-Choice Questions

2. The body plan of plant is distinguished by the _____ pattern.
   (a) Apical, basal
   (b) Radial, apical
   (c) Apical-basal, radial
   (d) Apical-radial, basal
3. Growth and development of a plant body are the result of:
   (a) Differentiation
   (b) Cell division and enlargement
   (c) Morphogenesis
   (d) All of the above
4. Which of the following developmental mutants of *Arabidopsis* will lack both the apical and basal regions including the cotyledons and the root and so consist of a cellular mass with no obvious signs of apical-basal polarity?
   (a) *GNOM*
   (b) *GURKE*
   (c) *MONOPTEROS.*
   (d) *FACKEL*
5. The hormone that provides positional information for the coordination of correct cellular patterning from the globular stage onward:
   (a) Cytokinin
   (b) Gibberellin
   (c) Auxin
   (d) Ethylene
6. *SHR* (short root) and *SCR* (scarecrow) are involved in:
   (a) Ground tissue development
   (b) Apical-basal pattern
   (c) RAM development
   (d) Establishment of polarity
7. PIN-mediated auxin flow in the SAM directs auxin to:
   (a) Regions that flank SAM
   (b) Central zone
   (c) Rib zone
   (d) Leaf primordia
8. Antagonistic activity of the following two hormones in the basal cell is essential for normal development of RAM and enables the particular pattern of cell divisions that occur in the QC.
   (a) Cytokinin and gibberellin
   (b) Auxin and ethylene
   (c) Gibberellin and auxin
   (d) Cytokinin and auxin

9. Characteristic feature that distinguishes the development of a dicot embryo from a monocot embryo:
   (a) Asymmetric division of the zygote
   (b) Octant stage
   (c) Establishment of bilateral asymmetry
   (d) SAM is established at a lateral position opposite to the scutellum
10. The cell wall proteins that are activated by acidification and disrupt noncovalent binding between cellulose and hemicelluloses:
    (a) Cellulases
    (b) Hemicellulases
    (c) Expansins
    (d) Pectinases

**Answers**

1. b    2. c    3. d    4. a    5. c    6. a    7. a    8. d    9. d    10. c

## Suggested Further Readings

Fosket DE (1994) Plant growth and development- a molecular approach. Academic, San Diego

Howell SH (1998) Molecular genetics of plant development. Cambridge University Press, Cambridge

Raghavan V (1997) Molecular embryology of flowering plants. Cambridge University Press, Cambridge

# Physiology of Flowering

## 25

Geetika Kalra and Manju A. Lal

About 90% of the ≈ 350,000 known plant species are the flowering plants. Flowering is the most enigmatic phase in the life of a plant. It provides a mechanism to plants for genetic outcrossing which provides a means of securing a greater variety of genetic recombination. Flowers are specialized structures which differ extensively from the vegetative plant body in form and cell types. Numerous physiological and biochemical changes take place within the shoot apex when it prepares itself for transition into floral bud. The precise time of flowering is important for reproductive success of the plant. Plants need to sense when to produce flowers so that fruit and seed development can be attained which will ensure its survival in the next season. Synchronous flowering is significant in outcrossing plants. Since long, people have wondered how plants are able to flower in a particular season. Plants possess the ability to anticipate and sense change of seasons. It has always been a fundamental question as to how environmental signals influence flowering and how these signals are perceived.

Transition from vegetative to reproductive development is generally marked by an increase in the frequency of cell divisions within the central zone of SAM. The process by which the shoot apical meristem becomes committed to forming flowers is termed **floral evocation**. SAM has an undifferentiated dome of cells at the center which, after the signal is perceived, triggers **quiescent** cells to enter into the phase of cell division and leads to transition from vegetative meristem to floral meristem. To some extent, timing of flowering can be influenced by external factors. The competence resides in the leaf, the shoot apex, or both. The regulation of competence to flower is complex and is variable across variable taxa. Floral induction requires both the perceptive organ and the shoot apex to acquire competence during plant maturation. Apical incompetence is a characteristic feature of the shoots of young trees, which are usually unable to transit into flowering even when grafted on to mature plants. This chapter deals with factors responsible for transition from vegetative to reproductive mode of differentiation and the associated physiological and biochemical changes, including present understanding about the role of genes in this process.

© Springer Nature Singapore Pte Ltd. 2018
S. C Bhatla, M. A. Lal, *Plant Physiology, Development and Metabolism*,
https://doi.org/10.1007/978-981-13-2023-1_25

Onset of flowering can be considered under three steps: first, the plant must acquire competence to respond to inductive signals; second, perception of the signal, which is then sent to shoot apex, resulting its transition from vegetative to reproductive phase; and third, the differentiation of shoot apex into floral organs.

## 25.1 Juvenile Phase

The internal developmental changes allow plants to obtain competence to respond to both external and internal signals that trigger flower formation. Such a transition is called as **heteroblasty** or **phase change**. Plants pass through a series of developmental stages starting from seed germination, root and shoot development, flower evocation and, finally, seed formation (Fig. 25.1). The transition between different phases is tightly regulated developmentally according to the integrated information perceived from the environment. Broadly, plant development is categorized into four phases: embryonic phase, postembryonic juvenile phase, adult vegetative phase, and reproductive phase. The main difference between the juvenile and the adult vegetative phase is that the latter is sensitive to various factors leading to development of reproductive structures. Distinct phases of juvenility and maturity can be identified in many species. Each phase is associated with a characteristic package of

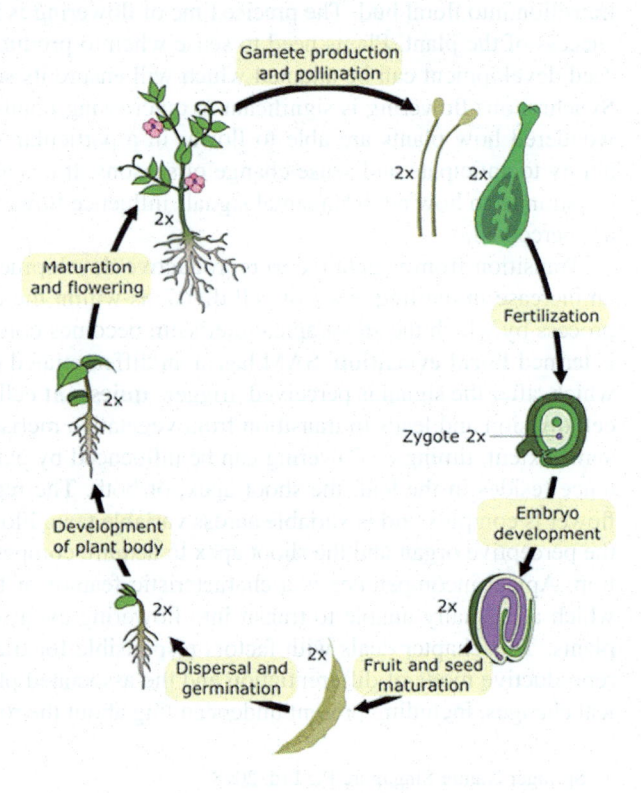

**Fig. 25.1** Life cycle of an angiosperm showing various phases of development. Transition to each phase involves remarkable biochemical and physiological changes

## 25.1 Juvenile Phase

**Fig. 25.2** Characteristic features of ivy (*Hedera helix*) in its juvenile (nonflowering) and adult (flowering) phases of growth

**Table 25.1** Distinguishing characteristics of juvenile to adult ivy

| Character | Juvenile plants | Adult plants |
| --- | --- | --- |
| Leaves | 3–5 lobed, palmate | Entire, ovate |
| Phyllotaxy | Alternate | Spiral |
| Shoot apex | Relatively narrow with large cells | Wide apex with small cells |
| Rate of internode growth | High | Low |
| Stem | Hairy | Smooth |
| Habit | Climbing and spreading | Upright or horizontal |
| Shoots | Unlimited growth and lack terminal bud | Slow limited growth terminated by buds with scales |
| Flowering | Absent | Present |
| Adventitious roots | Present | Absent |
| Rooting ability of cutting | Good | Poor |

morphological and physiological features (Fig. 25.2). For example, ivy (*Hedera helix*) exhibits characteristic juvenile and adult phases of life cycle (Table 25.1). In this plant numerous morphological changes take place with maturity, including leaf form and phyllotaxy, growth habit, and shoot and root development. During juvenile phase the plant will not flower even if exposed to suitable environmental conditions. Probably juvenile phase of the plant is required so that there are sufficient number of photosynthetically efficient leaves which are able to support flower development and seed setting later on. The transition of shoot apex from juvenile to the adult phase is affected by transmissible factors from the rest of the plant. Carbohydrate supply may play a major role in the transition from juvenility to maturity. Gibberellins are another important factor responsible for phase change. Treatments, like removal of roots, water stress, and nitrogen starvation, may result in accumulation of gibberellins in the plant. In *Arabidopsis*, carbohydrate is transmitted as a small signaling molecule—trehalose-6-phosphate—a disaccharide. This molecule activates flowering pathway in the shoot apex. Factors such as exposure to low light conditions prolong juvenility or may cause reversal to juvenility. Transition from juvenile to adult phase (vegetative phase change) is influenced by environmental factors such as day length, light intensity, ambient temperature, and gibberellic acid. The signals that trigger phase change are perceived by leaf primordia. Transition from juvenile to adult phase shares some regulatory steps with reproductive phase transition. Two evolutionary conserved microRNAs (miRNAs), i.e., miR156 and miR172, and their targets have been identified as key components of genetic control mechanisms that underlie phase changes. miR156 targets the transcripts of the gene which have been found to promote transition from juvenile to adult phase. Level of miR156 declines with increasing age of the plant. miR172 targets the mRNA that encodes proteins which promote transition to flowering and floral development. Studies in *Arabidopsis* indicate miR156 promoting juvenile phase and delaying adult phase. Expression of miR172 appears to be under photoperiodic control.

After the plant has acquired the competence to flower, the perceived inductive signal is sent to the vegetative shoot apex which triggers transition of shoot bud to the reproductive form. Recently, genes have been identified that play crucial role in the formation of flowers, indicating that reproductive development in plants is genetically controlled. SAM becomes an inflorescence meristem when it produces structures like bracts and floral meristem instead of leaf primordia or stem. A cascade of gene expression leads to transition from SAM to floral meristem, which has been well established in the model plant, *Arabidopsis thaliana*. The *embryonic flower* (*EMF*) gene of *Arabidopsis* prevents early flowering. Mutant plants that lack EMF protein flower as soon as they germinate, indicating that wild-type allele is responsible for suppression of flowering. These observations suggest that flowering is a default state, and plants have evolved mechanisms to delay flowering. This delay allows plants to store more energy which may be allocated for reproduction at a later stage. Another example of inducing juvenile to adult transition comes from overexpressing a gene for flowering, namely, LEAFY (LFY). This gene was cloned in *Arabidopsis*, and its promoter was replaced with a viral promoter that results in high levels of LFY transcription. LFY with its viral promoter was then introduced to cultured aspen cells that were used to regenerate plants. When LFY is overexpressed in aspen, flowering occurs in weeks instead of years. Phase change thus requires both a strong promotive signal and the ability to perceive the signal. The final outcome depends on the increase in promotive signal or decrease in inhibitory signal in conjunction with production of adequate receptors on the shoots to perceive the signal.

## 25.2 Flower Induction

Transition to flowering involves major changes in the pattern of morphogenesis and cell differentiation at the shoot apical meristem (SAM), which is accompanied with reallocation of biomolecules and associated energy to the shoot tip to initiate floral differentiation. After the plant has acquired competence to flower, it is able to perceive environmental or endogenous stimulus. The environmental factors which have been specifically monitored are day length and temperature. The endogenous factors include carbohydrates (nutrients), hormones, and circadian rhythm. The interaction of external and internal factors enables plants to synchronize their reproductive development with the environment. Some plants exhibit an absolute requirement for specific environmental conditions in order to flower. Such floral induction responses are referred as **obligate** or **qualitative** responses. If flowering is promoted by certain environmental conditions but eventually occurs in the subsequent absence of such conditions as well, the flowering response is said to be **facultative** or **quantitative**. There are plants which flower strictly in response to internal developmental factors and do not depend on any particular environmental condition. These are said to exhibit **autonomous regulation**, such as garden pea in which retuning of vegetative apex to form flowers is genetically determined. There can be early cultivars or late-flowering cultivars. Four genetically regulated pathways leading to flowering have been identified: (i) light-dependent pathway, (ii) temperature-dependent pathway, (iii) gibberellin-dependent pathway, and (iv) the autonomous pathway. Plants can rely on one of these pathways, but all the four pathways may be essential for floral induction.

## 25.3 Photoperiodism: The Light-Dependent Pathway

Flowering is so predictable in plants that it is used as a floral calendar. As we know that roses bloom in summer and chrysanthemums in winter. It is generally the length of day that gives the most reliable indication of advancing season. An organism's capacity to measure day length is known as **photoperiodism.** Initial experiments on photoperiodism were conducted by a French scientist J. Tournois in 1912. He observed that *Cannabis* plants flower vigorously when planted early in the spring but remain vegetative if planted in late spring or summer. He concluded that shortening of day length was not as important for early flowering as lengthening of night. At about the same time, George Klebs from Germany demonstrated that *Sempervivum funkii* could be induced to flower in winter in greenhouse when exposed to artificial light although normal time is June. First clear-cut hypothesis of photoperiodism was given by W.W. Garner and H.A. Allard from the US Department of Agriculture (Beltsville, Maryland) in 1920. They observed that *Biloxi* soybean flowers around same time in September/October even if it is germinated over a 3-month period from May to July, i.e., irrespective of how long they have been growing, they flower around same time. Garner and Allard hypothesized a seasonal timing mechanism in soybean. They also observed flowering response of tobacco (Maryland strain) which normally flowers in summer. A mutant of the plant called Maryland Mammoth was observed to grow up to the height of 3–5 m in summer without any flowering. The plants growing in green house under relatively short photoperiods flowered profusely in mid-December when the relative length of day was shorter than the length of the dark period. The mutants could be made to flower when exposed to short-day length next year in summer by placing the plants in darkness after placing the plants in light equivalent to that of winters. These observations lead to the discovery of the phenomenon as well as for the coining of the term "photoperiodism" by Garner and Allard. These observations also lead to the fact that plants vary considerably in their response to day length.

### 25.3.1 Critical Day Length

On the basis of photoperiodic requirement for floral induction, plants have been classified under different categories. **Short-day plants (SDPs)** will flower if the day length is shorter than a critical photoperiod. Hillman (1959) showed that SDPs are capable of flowering even if kept continuously in dark provided with adequate sucrose. This shows that the SDPs require light only for carrying on photosynthesis. Examples of SDPs are soybean, poinsettia, potato, sugarcane, cosmos, chrysanthemum, etc. **Long-day plants (LDPs)** require a photoperiod of more than a critical length which varies from 14 to 18 h. The best flowering usually occurs in continuous light. A flash of light during a long dark period can induce flowering even under short- day conditions. Since dark phase has inhibitory effect on flowering, these plants can also be called as *short-night plants*. Examples of this category are spinach, lettuce, radish, alfalfa, sugar beet, larkspur, etc. The critical value of the photoperiod requirement is not absolute rather varies according to species (Fig. 25.3). In **day-neutral plants**, flowering is not affected by day length. For example, tomato, cucumber, cotton, pea, and sunflower. Within this category, there are obligate or facultative requirements for a particular photoperiod. Plants having absolute requirement for a particular photoperiod for flowering are called **qualitative photoperiod** types. For example, *Xanthium strumarium* does not flower unless it receives an appropriate short photoperiod. It is a qualitative SDP. In quantitative SDPs, flowering is accelerated by short days, e.g., *Cannabis sativa* (hemp) and *Helianthus annuus* (sunflower). Spring cereals, like *Triticum aestivum* (spring wheat) and

**Fig. 25.3** Photoperiodic control of flowering. Decrease in critical dark period leads to vegetative stage. Also a flash of light during dark period inhibits flowering

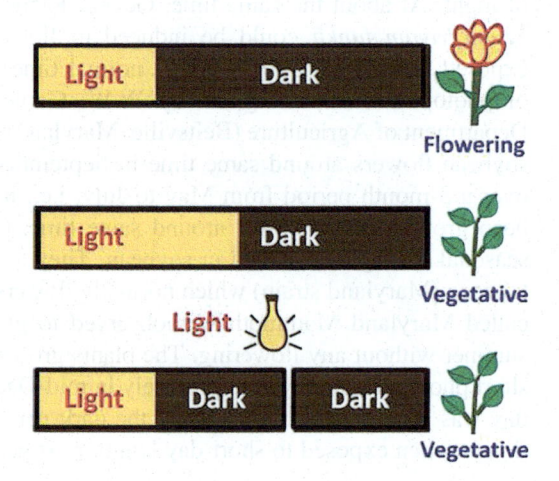

*Secale cereale* (winter rye), are quantitative LDPs. They do flower under short days, but flowering is accelerated under long days. Qualitative LDPs include *Hyoscyamus niger* (black henbane) and *Arabidopsis thaliana*. Photoperiod requirement is often modified by external conditions like temperature. There are also other response types in which plants respond to long and short days in some combination. Thus, *Bryophyllum* is a **long-short-day plant**. It flowers when a certain number of short days are preceded by a specific number of long days. *Trifolium repens* exhibits a reverse condition of **short-long-day plant**. Some plants, like winter cereals, require a low temperature treatment before they become responsive to photoperiod, while others may have a qualitative photoperiodic requirement at one temperature but a quantitative requirement at another temperature. Some plants are **intermediate-day length** plants. They flower in response to day length of intermediate range but remain vegetative when the day is too long or too short. Interestingly, flowering is delayed in *Madia elegans* under intermediate-day length (12–14 h) but occurs under day length of 8 or 18 h. It may be noted here that this classification is based on whether a particular plant will flower when subjected to photoperiod that exceeds or is less than a critical length.

## 25.3.2 Critical Role of Dark Period

Plants neither measure relative length of day and night nor the length of photoperiod. They measure the length of dark period. This was demonstrated by K.C. Hammer and J. Bonner (1938) in experiments conducted with *Xanthium*. In a 24 h cycle of light and dark periods, *Xanthium* flowers only when dark period exceeds 8.5 h but remains vegetative when provided with 16 h of light followed by 8 h of dark (Fig. 25.4). Similarly, long-day plants require a dark period shorter than some critical maximum. In LDPs, a flash of light in the middle of an otherwise noninductive long dark period will shorten the dark period requirement to less than the maximum and permit flowering to occur. Measuring the time of dark period is central to photoperiodic time keeping.

## 25.4 Photoinductive Cycle

In nature, plants are exposed to photoperiodic cycles which consist of alternate periods of light and dark diurnally (Fig. 25.5). Any photoperiodic cycle which induces flowering in a plant is called **photoinductive cycle**. On the contrary the photoperiodic cycle which does not induce flowering in a plant is non-photoinductive cycle. A photoperiodic cycle consisting of 16 h light and 8 h dark period generally induces flowering in LDPs, while a cycle consisting of 8 h light and 16 h dark period induces flowering in SDPs (Fig. 25.6). The number of cycles required to induce flowering in a plant varies. One SD photoinductive cycle is sufficient to induce flowering in *Xanthium strumarium* and *Pharbitis nil*, while *Salvia occidentalis*, a SDP, may require at least 17 cycles. *Plantago lanceolata*, a

|  | Long-day plants | Short-day plants |
|---|---|---|
| Early summer | **Clover:** short length of dark required for bloom | **Cocklebur:** Long length of dark required for bloom |

**Fig. 25.4** Critical role of dark period in long-day and short-day plants

**Fig. 25.5** Flowering response in LDP and SDP in 24 h cycle

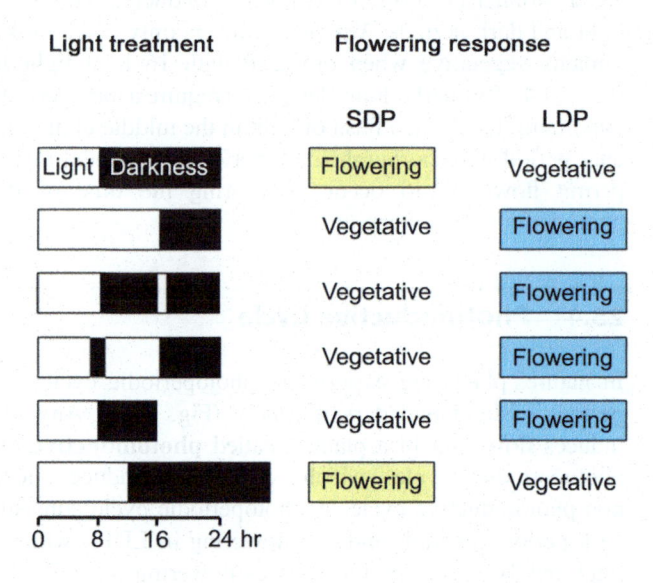

## 25.4 Photoinductive Cycle

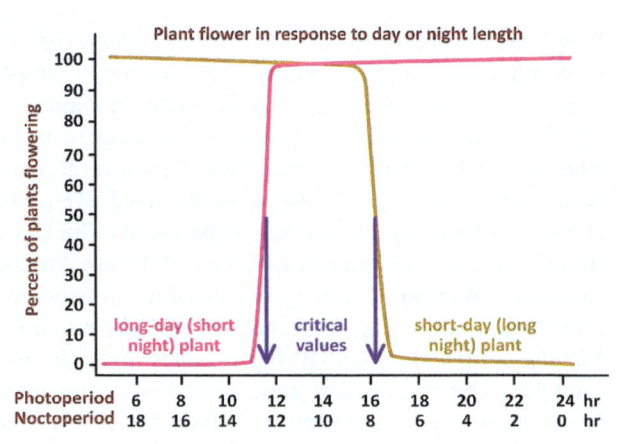

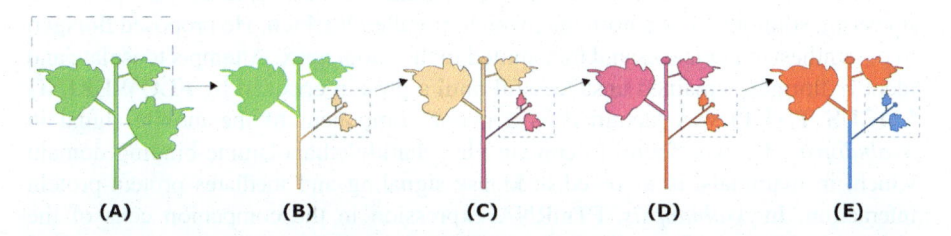

**Fig. 25.6** Categorization of plants according to their response toward length of day and night

**Fig. 25.7** Transmissible nature of florigen as depicted through grafting experiments in *Xanthium*. (a) Grown under short-day condition, (b–e) grown under long-day condition

LDP, requires 25 photoinductive cycles for maximum floral response. If the plant is returned to non-photoinductive cycle after ten cycles, it will not flower. However, if returned to photoinductive cycle, only 15 cycles are required. This indicates that some factor responsible for flowering response gets accumulated during inductive cycle.

### 25.4.1 Perception of Photoperiodic Signal and Florigen

The photoperiodic signal for floral induction is perceived by the leaves and not by SAM. This was demonstrated in experiments conducted by the Russian plant physiologist M. Chailakhyan in 1937. He reported flowering in *Chrysanthemum morifolium*, a SDP, when a leafy portion of the plant was subjected to short days and the apical meristem and defoliated portion of the shoot were subjected to long days. However, the plants remained vegetative when conditions were reversed, i.e., the upper defoliated portion kept in short days and the leafy portion in long days (Fig. 25.7). In another set of experiments, SDPs *Perilla* and *Xanthium* could be induced to flower even when all the leaves had been removed except one leaf which was kept in SD conditions. When leaves taken from the induced plants were grafted to non-induced ones, it resulted in induction of flowering in the non-induced ones.

Even the excised leaves of *Perilla frutescens* (SDP), when exposed to photoinductive cycle and grafted back to non-induced plants, induced flowering even when plants were maintained under noninductive long-day conditions. A rapidly expanding leaf is most sensitive to perceive the photoperiodic stimulus when it is half of its final size. Even when several *Xanthium* plants were joined to each other through grafts, all plants flowered even when only the first plant was exposed to short days. From these experiments Chailakhyan suggested that the floral stimulus might be a hormone which could diffuse through graft union. He called this stimulus as **florigen**. Grafting was also done in between the plants belonging to the same family but having different photoperiodic requirements, such as between SDP *Nicotiana tabacum* and LDP *Hyoscyamus niger*. *Hyoscyamus niger* flowers under short days if tobacco is kept under short days. Conversely, the grafted tobacco plants flower if *Hyoscyamus* is kept under long days. This experiment indicates that floral stimulus might be same in all photoperiodic classes. Chailakhyan proposed flowering stimulus to be a hormone, which he called **florigen**. He proposed **florigen** to be synthesized in leaves and transmitted to the shoot apex. Attempts to isolate and identify florigen remained unsuccessful until a protein encoded by **FLOWERING LOCUS T (FT)** was identified as a major component of the mobile signal in *Arabidopsis*. FT was found to contain phosphatidylethanolamine-binding domain which, in mammals, is involved in kinase signaling and mediates protein-protein interaction. In *Arabidopsis*, FTmRNA expression in the companion cells of the phloem in leaves triggers flowering when FT protein is transported to the apical meristem through phloem sieve elements where it interacts with bzip transcription factor encoded by **FLOWERING LOCUS D (FD)**, and it is responsible for the regulation of genes involved in the change of vegetative meristem to produce flowers. The florigen model was replaced by **nutrient diversion hypothesis**. According to this hypothesis, an inductive treatment stimulates the flow of nutrients into the apical meristem. A high level of nutrients has been found to stimulate flowering. This hypothesis is based on the observation that induction of flowering in white mustard (*Sinapis alba*), a LD plant, gives rise to a rapid and transient increase in the export of sucrose from leaves to the shoot apex. The third hypothesis, the **multifactorial hypothesis**, proposes that flowering occurs when a number of factors, including promoters, hormones, and nutrients, are present in the apex at an appropriate time and in appropriate concentrations. This hypothesis points at multiple genes that control flowering, out of which some genes respond to photoperiod and temperature, while others act independent of environment.

## 25.5 Circadian Rhythm

In addition to photoperiodism, plants also display other time measuring systems. Endogenous rhythms persist even when plants are placed in constant environmental conditions. They are based on a cycle of approx. 24 h and are known as circadian rhythms (circa = about, diem = day). Circadian rhythms are synchronized with the daily day-night cycle, which is known as **entrainment**. Erwin Bunning (1936) proposed that daily rhythms consist of two phases, i.e., **photophil phase** (light-loving phase) and **skotophil phase** (dark-loving phase). Photophil phase is characterized by intensive photosynthesis and weak respiration (anabolic processes predominate). On the contrary, skotophil phase is characterized by intensive respiration. In this phase, hydrolytic activity increases, and decomposition of starch into sugars takes place (predominance of catabolic processes). According to *Bunning hypothesis*, the two phases alternate about every 12 h. Under constant environmental conditions, photophil phase would probably correspond to subjective day, while skotophil phase is equivalent to subjective night. The ability of light to promote or inhibit flowering depends on the phase in which light is given. When light signal is applied during light-sensitive phase of the rhythm, the effect is either to promote flowering in LDPs or to prevent flowering in SDPs. In an experiment, *Chenopodium rubrum* plants (a SDP in which exposure to single photoinductive cycle is sufficient to induce flowering) were shifted to 72 h. dark period after being exposed to a photoperiod. Two minutes of night break was given at different time intervals in the dark period before transferring the plant to continuous light. Inhibition of flowering was most effective if night break was given at 6, 33 or 60 h after the start of dark period. This is the time when the plant might have been in darkness in a normal 24 h cycle, i.e., skotophilous phase. However, night breaks do not result in inhibition of flowering if the night breaks are given near 18 and 46 h after the start of dark period. This is the time when plant would have been in light in a 24 h cycle, i.e., photophil phase. This indicates interaction of photoinduction with endogenous rhythm of the plant. Flowering in both LDPs and SDPs is induced when light exposure is coincident with the appropriate phase of the rhythm. Some kind of regulating mechanism is present which is called circadian regulator. Bunning's hypothesis has evolved into **coincidence model**. According to this model, a key regulator accumulates in LDPs and reaches a maximum concentration during LDs. The regulator also requires light for its activation, i.e., the presence of light coincides with the accumulated regulator, followed by cascade of events leading to flowering. In *Arabidopsis* (a quantitative LDP), the genes which have been identified and characterized as key regulators of flowering include GIGANTEA (GI), CONSTANS (CO), and FLOWERING LOCUS (FT). Isolation of a mutant (*co*) of *Arabidopsis*, in which flowering was delayed under LD but without affecting the response under SD, leads to identification and isolation of CO gene. The gene has been found to be a key regulator in photoperiodic control of flowering. In *Arabidopsis*, mRNA for CO (which encodes a nuclear zinc-finger transcription factor) starts accumulating and reaches a peak in LD and is translated in light. CO protein is stabilized by exposure to blue and FR light which is absorbed via the pigments CRY2 (cryptochrome) and PHYA (phytochrome), respectively. CO expression and activation of FT gene occur in the companion cells. As a result, FT protein is transported to the shoot apex. Thus, flowering in *Arabidopsis* occurs only when transcription and translation of CO gene coincide with exposure to light, which occurs under LD. There is an overlapping (coincidence) between CO mRNA synthesis and day light so that light can permit active CO protein to accumulate to a level that promotes flowering. Thus, rhythmicity of accumulation of COSTANS mRNA in photoperiod and its light-dependent translation to CO protein provide the molecular basis for external coincidence model. Interestingly, FT is a target gene downstream of CO. FT is expressed in the companion cells. Thus, CO activity is mediated by the expression of FT. Movement of FT from the companion cells to the sieve elements requires ER-localized protein called FT INTERACTING PROTEIN (FTIP1). Once in floral meristem, FT protein enters the nucleus and forms a complex with bzip transcription regulator FD, which is encoded by the gene FLOWERING LOCUS D (FD). FT-FD activates expression of **floral meristem-identity** genes, the MADS box transcription factors, such as SUPPRESSOR OF OVEREXPRESSION OF CONSTANS-1 (SOC1) and AP1 (Fig. 25.9). These genes specify that the vegetative shoot meristem of the plant gets differentiated into floral meristem. Investigations have been undertaken on the flowering behavior of rice (SDP) plants. The major genes, i.e., CO and FT, which have regulatory function in *Arabidopsis*, are conserved in rice (SDP). However, their specific regulation has been altered by evolution to promote flowering under short days. The genes **Heading-date1 (Hd1)** and **Heading-date3a (Hd3a)** are homologous to *Arabidopsis* CO and FT, respectively. Similar to FT in *Arabidopsis*, overexpression of Hd3a in rice results in rapid flowering irrespective of photoperiod. Besides the expression of FT in *Arabidopsis* and that of Hd3a gene in rice, flowering is elevated during the inductive photoperiods, i.e., LD and SD, respectively. However, unlike in *Arabidopsis* (LDP), where coincidence of CO with light period promotes flowering in rice, coincidence of Hd1 expression with the light period suppresses flowering since Hd1 acts as the suppressor of Hd3a. The lack of coincidence between Hd1mRNA expression and day light prevents accumulation of Hd1 protein, which acts as a repressor of the gene encoding the transmissible floral stimulus, Hd3a, in rice. In the absence of the Hd1 protein repressor, Hd3a mRNA is expressed, and the protein it encodes is translocated to the apical meristem where it causes flowering. Under long days (sensed by phytochrome), the peak of Hd1 mRNA expression overlaps with the day, allowing the accumulation of the Hd1 repressor protein. As a result, HD3a mRNA is not expressed, and the plant remains vegetative.

## 25.6 Photoreceptors

Phytochrome and cryptochrome play important roles in photomorphogenesis of plants. One of the best studied SDPs in terms of effect of light on flowering is *Pharbitis nil*. It is a qualitative SDP in which 4–5-day-old cotyledonary photoresponsive tissue can receive the stimulus for floral induction when given a single photoinductive cycle. In experiments with this plant, night breaks given during photoinductive dark period, which prevent attainment of critical dark period, inhibit flowering. Night breaks were found to be most effective if red light was used. However, the effect was reversed if red light treatment was immediately followed by exposure to far-red light. The photoreversible effect of R/FR light suggested the role of phytochrome (Fig. 25.8). Phytochrome comprises of nuclear encoded proteins. The *Arabidopsis* genome encodes five phytochromes (PHYA to PHYE) that are

**Fig. 25.8** Influence of phytochrome on flowering

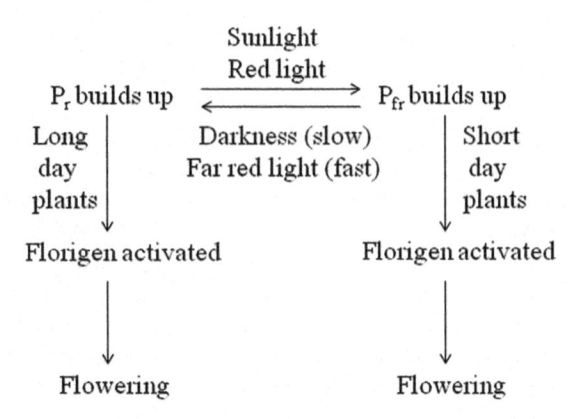

```
                        Sunlight
                        Red light
        P_r builds up  ⇌              P_fr builds up
        Long          Darkness (slow)    Short
        day           Far red light (fast) day
        plants                           plants
          ↓                                ↓
        Florigen activated            Florigen activated
          ↓                                ↓
        Flowering                      Flowering
```

involved in floral induction. Late-flowering mutants (phyA) are defective in genes that promote flowering, while early-flowering mutants (phyB) are defective in genes that ordinarily repress flowering.

Blue light promotes effect on flowering in LDPs, especially in members of family Cruciferae. Two members of cryptochrome gene family (CRY1 and CRY2) are present in *Arabidopsis*. Cryptochromes are flavoproteins that act as the blue light receptors. Both CRY1 and CRY2 function in stabilizing CO protein along with PHYA toward the end of light period, whereas in other plants this role is taken up by PHYA alone. CRY2 mutants of *Arabidopsis* flower later than the wild type under inductive long days. Under continuous white light exposure, phy1 mutant (any type of phytochrome cannot be synthesized in them because of defective enzyme, which is required for synthesis of chromophore of the pigment) plants flower similar to the wild types. This indicates that in continuous white light exposure, no phytochrome is required, and the blue light receptor is involved. Mutation in one of the cryptochrome genes (CRY2) causes a delay in flowering.

According to the coincidence model, CO gene is expressed during light period. The effect of light on CO stability further depends on the photoreceptor involved. In morning hours (after dark), phyB signaling enhances CO degradation, whereas in the evening (when CO protein accumulates after long day), cryptochromes and phyA antagonize this degradation and allow CO protein to build up. CO, a transcriptional regulator, promotes flowering by stimulating the expression of a key floral signal, FLOWERING LOCUS T (FT).

## 25.7 Vernalization

In many long-day plants, exposure to low temperature is critical for the acquisition of competence to respond to photoinductive conditions for flowering. This cold temperature requirement is called **vernalization**, which acts as a time computing mechanism that measures the passage of winter and ensures that flowering does not begin until the favorable conditions of spring arrive. The concept was introduced by T. D. Lysenko (1920) who observed the ability of cold treatment to make the winter cereal behave as spring cereal. This could be of practical utility like: (1) crops can be harvested much earlier, (2) crops can be grown in regions where they are not naturally productive, and (3) plant breeding experiments can be accelerated. Generally, it is the stem apex which perceives the cold temperature signal. The dividing cells in plants perceive vernalization stimulus. Period of chilling can vary from few days to weeks and from plant to plant, but longer exposure to low temperature will be more effective for early flowering. Response due to vernalization decreases if it is interrupted by heat treatment. In contrast to photoperiodic effect, which leads to flower initiation, vernalization prepares plants for flowering. G. Melchers and A. Lang (1948) demonstrated that the biennial LDP *Hyoscyamus niger* (which requires a low temperature season before flowering unlike the annual type which flower in one season) should be at least 10 days old before becoming responsive to the low temperature treatment. However, Gregory and Purvis in 1930s suggested that hydrated seeds of Petkus winter rye (*Secale cereale*) may be vernalized making them sensitive to LD photoperiod. The cold treatment of the seeds reduces the number of photoinductive period required for flowering since the Petkus winter rye does not have obligate requirement for vernalization. That vernalization is an energy-dependent process was demonstrated in an experiment in which excised embryos were supplemented with carbohydrates and oxygen. Melchers had demonstrated that vernalization stimulus could be transmitted through graft union. He was the first to coin the term vernalin for the hypothetical active factor required for vernalization. It was observed that once a plant has been vernalized, it remembers the cold treatment throughout its life. The memory is maintained in cell derived from the induced cell through mitotic division but not the one which are derived through meiotic division. Lang stated a direct connection between vernalin and florigen.

Low Temperature → Vernalin → Florigen

One of the pathways for flowering is through vernalization, where low temperature treatment leads to accumulation of vernalin which in turn stimulates the flowering stimulus florigen. Vernalization affects competence of a plant to flower by bringing about stable changes in the pattern of gene expression in the meristem after cold treatment. Such changes are termed as epigenetic changes. Requirement of vernalization is conferred by two genes, FRIGIDA (FRI) and FLOWERING LOCUS C (FLC). FRI acts in upregulation of FLC. FLC encodes MADS- domain DNA-binding protein that functions as a repressor of flowering. Levels of FLC are the primary determinant of vernalization requirement in *Arabidopsis*. It is highly expressed in the shoot apical meristem of non-vernalized plants. It represses flowering by repressing the expression of **floral integrators**, such as FT, FD, and SOC1. Floral integrators are the genes that are involved in regulation of meristem-identity genes. These are so named because these integrate the floral stimulus which is due to some environmental cues and trigger the vegetative to reproductive transition. Binding of FLC with the promoters of SOC1, FD, and FT decreases the ability of the photoperiods to activate these integrators. During vernalization FLC is epigenetically switched off for the rest of plant's life cycle. These are stable changes in gene expression that do not involve alterations in DNA sequence and which can be passed on to descendent cells through mitosis or meiosis. This is achieved by repressive changes in FLC which includes **chromatin remodeling**. This includes histone methylation of lysine-27 and lysine-9 residues which are characteristics of heterochromatin, and acetyl groups are removed from lysine-9 and lysine-14 of H3 which otherwise are characteristics of euchromatin. Thus, low temperature induces conversion of FLC from active to inactive form. The importance of histone modification was further clarified after mutants of *Arabidopsis* have been identified which do not respond to vernalization. These mutants included *vernalization insensitive* (*vin*) and *vernalization* (*vrn*) mutants. These mutants prevent vernalization and alter histone modifications. Thus, photoperiod pathway, vernalization pathway, and autonomous pathway form a regulatory network which converges to modulate the activities of a set of genes that integrate the floral stimulus and trigger the transition from vegetative to reproductive phase (Fig. 25.9).

## 25.8 Role of Gibberellins

Gibberellins play important role/s during transition of vegetative to reproductive meristem. This includes their role in competence, promotion of bolting, and flowering in *Arabidopsis* and many other long-day plants. Flowering in perennial species tends to be insensitive to gibberellins. In an interesting observation, when extract from photoinduced leaves of *Xanthium* is applied, it induces flowering in *Lemna* kept under noninductive conditions. However, the extracts need to be supplemented with gibberellin. The leaf extract alone or gibberellin alone has no effect. Extract prepared from spinach leaves grown under short days suggests that a critical step in GA biosynthesis is inhibited. Plants remain vegetative and rosetted under short days. This shows that gibberellin is partially responsible for flowering. There is a possibility that GA is a mobile signal that transmits the photoperiodic floral stimulus and its action is independent of FT, the phloem mobile protein that relays the floral induction signal from leaf to shoot apex. Expression of both SOC1 and LFY in *Arabidopsis* is promoted by GA via DELLA-mediated signaling mechanism. SOC1 is thus regulated in a multifactorial manner and integrates the autonomous, vernalization, and GA pathways (Fig. 25.10). Chailakhyan stated that vernalin hormone may be a precursor of gibberellin. Under long-day conditions, it is converted to gibberellin. Another hormone called **anthesin** is present in long-day plants which, along with vernalin, causes flowering in long-day plants. In short-day conditions, vernalin is not converted to gibberellin. Hence, flowering does not occur. Gibberellin treatment to long-day non-vernalized plants kept under long day leads to flowering as these plants possibly contain anthesin. Gibberellin is ineffective in flower induction in short-day plants as they lack anthesin. Auxin application induces flowering in pineapple and litchi. In pineapple, the effect of auxin may be due to stimulation of ethylene production.

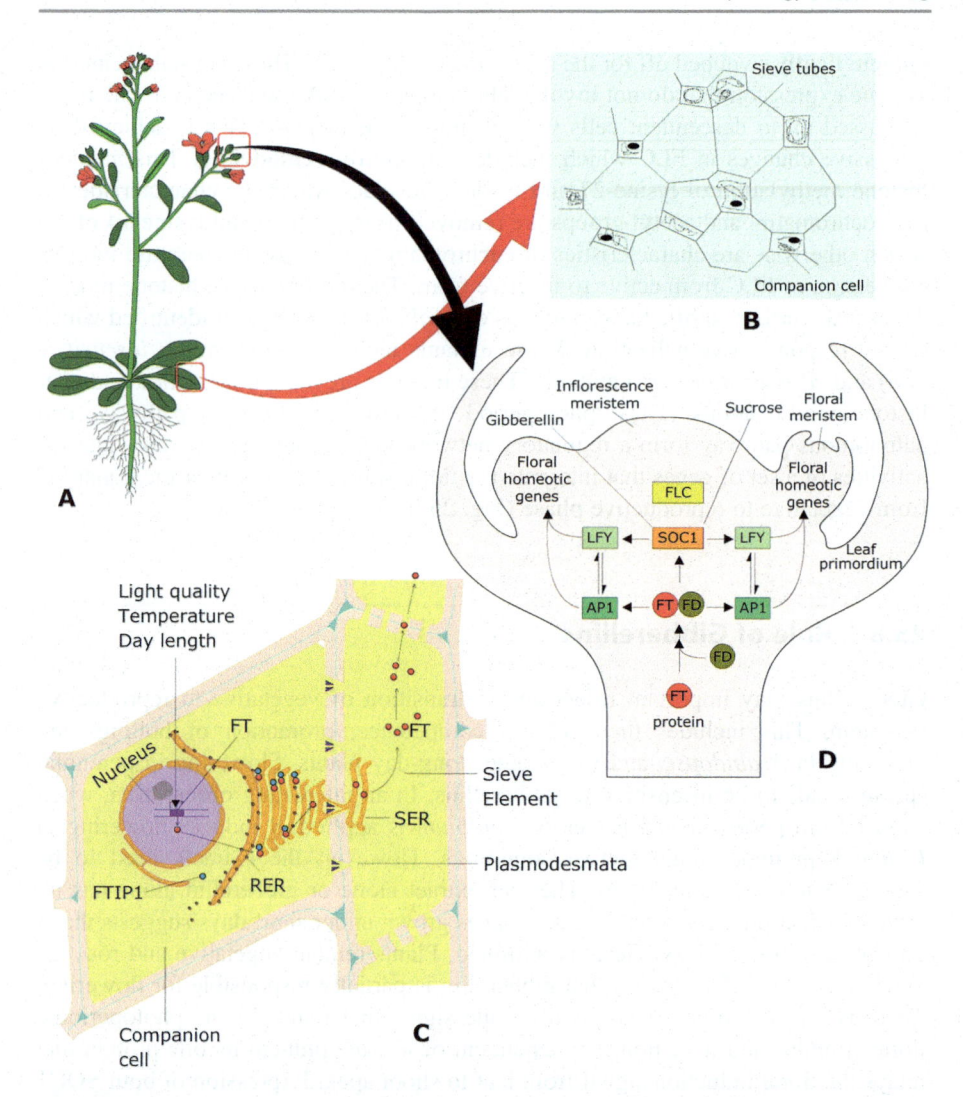

**Fig. 25.9** Flowering is regulated by multiple factors in *Arabidopsis* (**a**); (**b**) FT mRNA is expressed in companion cells of leaf vein in response to multiple signals, including day length, light quality, and temperature; and (**c**) FTIP1 mediates through a continuous ER network between the companion cells and the sieve tube elements. FT moves in the phloem from the leaves to the apical meristem. (**d**) FT is unloaded from the phloem in the meristem and interacts with FD. Then FT-FD complex activates SOC1 in the inflorescence meristem and AP1 in the floral meristem, which triggers LFY gene expression. LFY and AP1 trigger expression of the floral homeotic genes. The autonomous and vernalization pathways negatively regulate FLC, which acts as a negative regulator of SOC1 in the meristem and as a negative regulator of FT in the leaves. *FD* FLOWERING LOCUS D, *FT* FLOWERING LOCUS T, *FTIP1* FT-interacting protein 1, *SOC1* suppressor of constans1, *AP1* apetella1, and *LFY* leafy

**Fig. 25.10** ABC model whereby floral organ identity is controlled by three homeotic genes, namely, A, B, and C

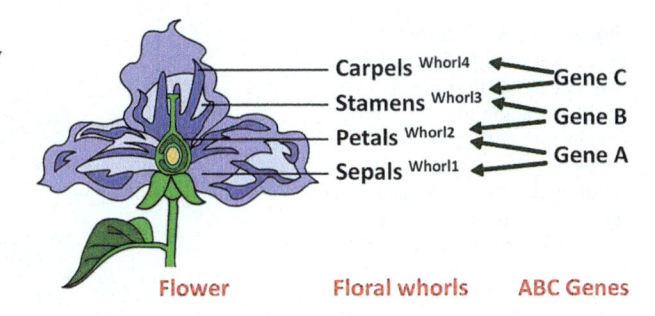

## 25.9 Flower Development

Two categories of genes are responsible for flower development ,viz., floral meristem identity genes and floral organ identity genes. The floral meristem identity genes are responsible for the transition of vegetative meristem to floral meristem. In *Arabidopsis*, these genes include LEAFY (LFY), FLOWERING LOCUS D (FD), SOC1, and APETALA1 (AP1). LFY, FD, and SOC1 play a critical role in integrating the signals—both environmental and internal. These genes act as master regulators for the initiation of floral development. Floral meristems can be distinguished from vegetative meristem by its larger size. The transition from vegetative to reproductive phase is marked by an increase in the frequency of cell division within the central zone of shoot apical meristem. Four different types of floral organs are initiated in separate whorls, namely, sepals, petals, stamens, and carpels. They develop in concentric rings called **whorls**, numbered 1, 2, 3, and 4, respectively. Molecular basis of floral morphogenesis has been studied extensively in *Arabidopsis*. Floral organ identity genes were discovered in **homeotic gene** mutants. Homeotic genes encode transcription factors that determine the location where specific structures develop. Five key genes have been identified in *Arabidopsis* which specify floral organ identity, namely, APETALA1 (AP1), APETALA2 (AP2), APETALA3 (AP3), PISTILLATA (P1), and AGAMOUS (AG). Influence of organ identity genes on floral development in *Arabidopsis* can be understood by loss-of-function mutants of these genes. Mutations in these genes change the floral organ identity without affecting the initiation of flowers. The genes that determine the four basic whorls in flower have been grouped into three classes, A, B, and C. Each group does not necessarily represent a single gene. This view is expressed as **ABC model**. This model postulates that organ identity in each whorl is determined by a unique combination of the activities of three organ identity genes (Fig. 25.10). Type A gene alone specifies sepals, while A and B together are required for petal formation. Genes of B and C category are required for stamen differentiation, while type C genes are responsible for carpel formation (Fig. 25.11). According to ABC model, Class A and C genes are mutually repressive to each other. Loss of type A activity (encoded by AP1 and AP2) results in the formation of carpels instead of sepals in the first whorl and stamens instead of petals in the second whorl. Loss of

**Fig. 25.11** ABC model that postulates when C-function is lost, A extends into whorl 3 and 4, leading to development of sepals and petals in whorl 3 and 4. Similarly, loss of A gene leads to extension of expression of C gene and formation of carpels and stamens in whorls 1 and 2

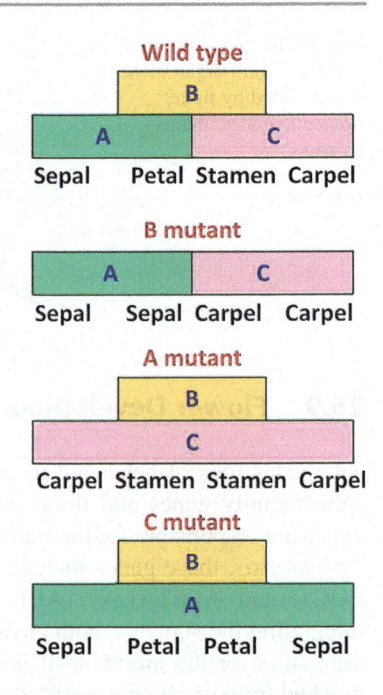

**Wild type**

| | B | |
|---|---|---|
| A | | C |

Sepal   Petal   Stamen   Carpel

**B mutant**

| A | C |
|---|---|

Sepal   Sepal   Carpel   Carpel

**A mutant**

| B |
|---|
| C |

Carpel   Stamen   Stamen   Carpel

**C mutant**

| B |
|---|
| A |

Sepal   Petal   Petal   Sepal

type B activity (encoded by AP3 and PI) results in the formation of sepals instead of petals in the second whorl and carpels instead of stamens in the third whorl since the genes belonging to this category control organ determination in the second and third whorl. Type C gene (AG) controls events in the third and fourth whorl. Loss of type C gene activity results in the formation of petals instead of stamens in the third whorl and replacement of fourth whorl by a new flower such that this whorl is occupied by sepals (Figs. 25.12 and 25.13).

These homeotic genes encode transcription factors which are consistent with their function in specifying organ identity. It is thought that each combination of A, B, or C class of transcription factors regulates a set of target genes required for the development of the corresponding organs. All homeotic genes, except AP2, encode MADS-domain transcription factors, which are characterized by a highly conserved, N-terminal DNA-binding domain unique to plants. Furthermore, expression of A, B, and C genes alone is insufficient for converting a leaf into floral organ. But, when they are expressed together with SEPALATA genes (SEP1, SEP2,SEP3), vegetative leaves are converted into floral organs dependent upon combination of A,B, and C class of genes co-expressed with SEP 1/2/3 (Fig. 25.14). The SEP genes are redundant, and only the triple SEP 1/2/3 are required for the normal development of petals, stamens, and carpels and are referred to as E class of genes. SEP genes also encode MADS-domain transcription factors that interact with other MADS-domain proteins. Thus, **ABCE** model was formulated based on genetic experiments in *Arabidopsis thaliana* and *Antirrhinum majus*. According to the ABCE model, carpel formation requires the activities of the C and E class of genes. However, a third group of MADS-box genes are required for ovule formation. These ovule-specific

## 25.9 Flower Development

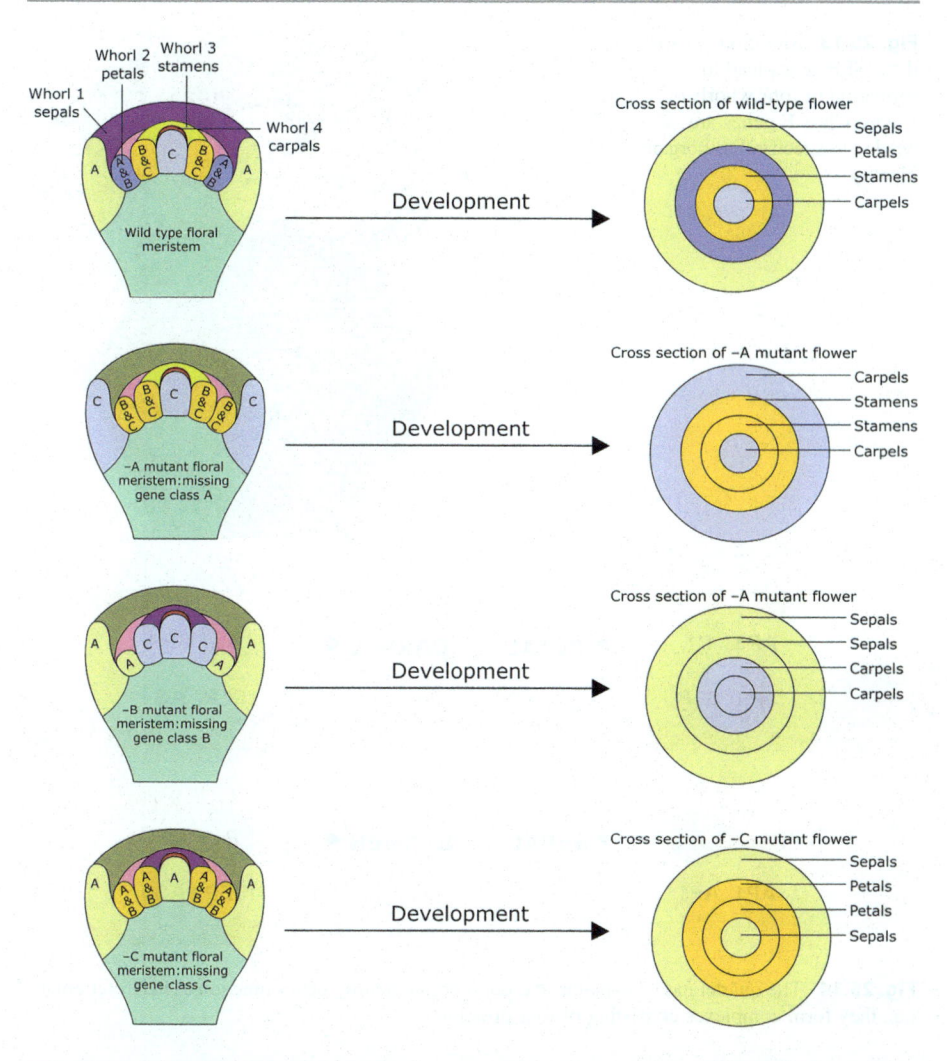

**Fig. 25.12** Letters within the whorls indicate active genes. In case of loss of function of A, the role of C expands to the first and second whorls; in case of loss of B gene activity, the outer two whorls will have function of A; loss of function of C, A expands into the inner two whorls

genes are called class D genes. Since ovule is a structure within the carpel, class D genes are not basically organ identity genes. They were first discovered in *Petunia*. Silencing of two MADS-box genes is known to be involved in floral development in *Petunia*, FLORAL BINDING PROTEIN 7/11 (FBP 7/11), results in the growth of style and stigma in a location normally occupied by ovule. When FBP11 is overexpressed in *Petunia*, ovule primordia are formed on the sepals and petals. Thus, class D genes are required for normal ovule development.

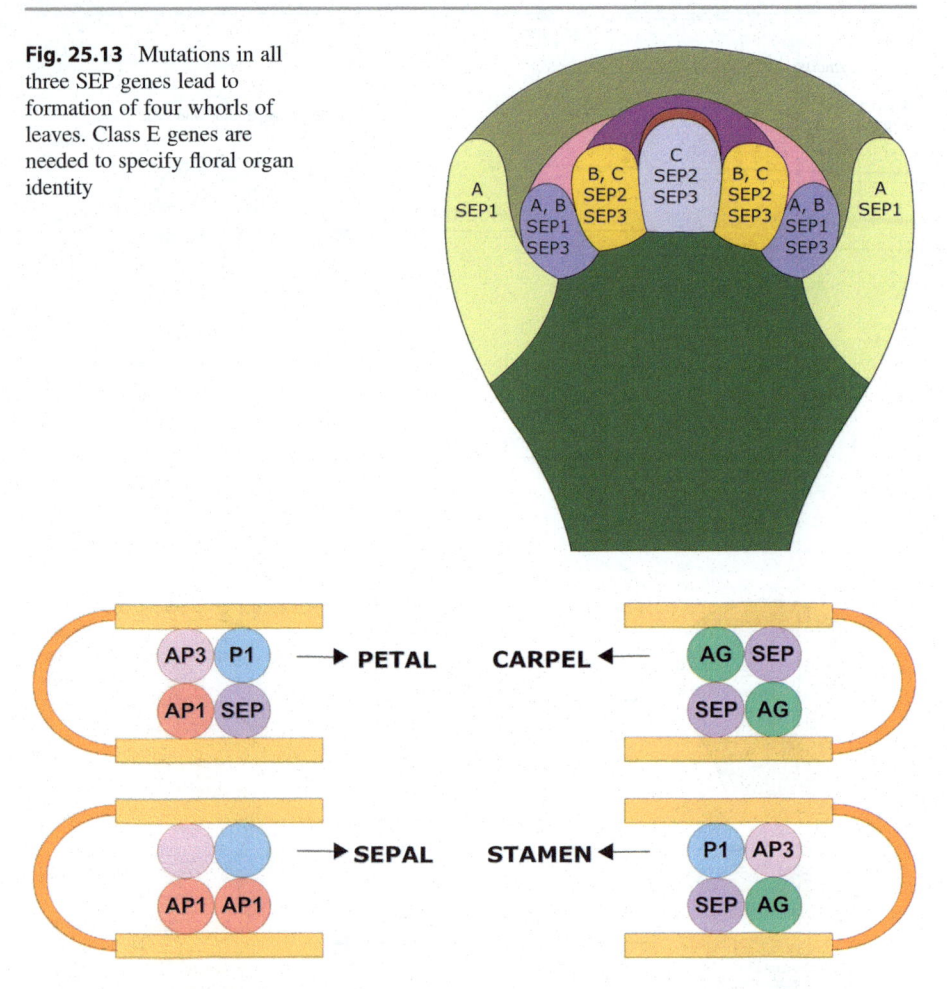

**Fig. 25.13** Mutations in all three SEP genes lead to formation of four whorls of leaves. Class E genes are needed to specify floral organ identity

**Fig. 25.14** The quartet model explains that floral organ identity genes function as heterotetramers, i.e., they form complexes consisting of four proteins

MADS-domain-containing transcription factors, encoded by floral organ identity, function as heterotetramers, i.e., they form complexes consisting of four proteins (**quartet model**). Each quartet consists of a set of MADS-domain proteins specifying a particular organ and controls a set of target genes required for the formation of specific organ. In other words, identity of different floral organs is determined by four combinations of floral homeotic proteins, known as MADS-box proteins. **MADS-box proteins** are transcription factors which operate by binding to the promoter region of target genes, which are activated or repressed for the development of floral organs. According to quartet model, two dimers of each tetramer recognize two different DNA sites on the same strand of DNA. These two sites are brought close by the bending of DNA. For example, the quartet

## 25.9 Flower Development

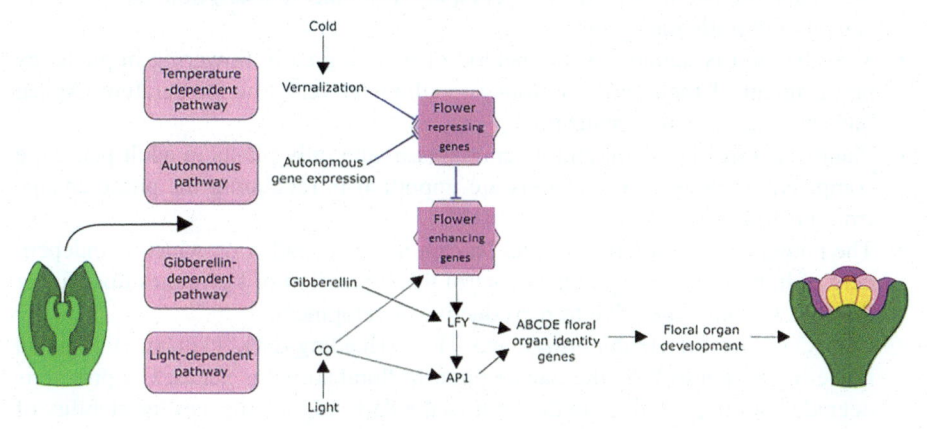

**Fig. 25.15** Temperature, light, and gibberellin-dependent pathways work through repression of floral inhibitors for flower formation as well as by activating floral meristem identity genes

directing petal development would contain A-class protein- AP1, B-class protein-PI and AP3, and a SEP protein.

To sum up, photoperiodism and vernalization facilitate plants to synchronize their life cycle with the time of the year. It is clear that the process of flower formation is an interplay of various transcriptional networks that regulate organ-specific gene expression. Such altered expression of floral homeotic genes also explains the floral diversity that we observe in our daily life. Flowering plants constitute an enormous range which needs to be explored with reference to gene networks that regulate floral development. Future challenge is to explore the variability found in nature which is due to gene network regulating the floral development (Fig. 25.15).

## Summary

- Flowering of plants depends on three basic requirements. Plants must be able to respond to an inductive signal. The signal is perceived in leaves and transmitted to the shoot apical meristem. The meristem responds by changing from vegetative phase to reproductive phase. Best understood factors that trigger flowering are duration and timing of light and dark periods (termed as photoperiodism) and temperature.
- Plants are classified into three classes according to their requirement of photoperiod, viz., long-day plants, short-day plants, and day-neutral plants.
- It is observed that leafless plants do not produce flowers. Leaf is the site of perception of photoperiodic signals. This means that some chemical agent (a flowering hormone) is synthesized in leaves and passed to the flowering apex. This hormone was named as florigen. This was later identified as FT (FLOWERING LOCUS T) which is a small globular protein which moves via phloem from leaves to the shoot apical meristem under inductive photoperiods. In

the shoot meristem, FT forms a complex with the transcription factor FD to activate floral identity genes.
- Vernalization is defined as the method of inducing early flowering in plants by pretreatment of their seeds or young seedlings at very low temperature. Apical buds are the sites of vernalization.
- Plants have distinct adult and juvenile phase, and only plants in adult phase are competent to flower. Gibberellins are important in regulating the phase change from juvenile to adult.
- The photoperiod, vernalization, and an autonomous pathway, which is independent of light, operate together to control the expression of key genes that trigger the switch from vegetative to reproductive development.
- CO (Constans in *Arabidopsis*) and Hd1 (Heading-date 1 in rice) regulate flowering by controlling the transcription of floral stimulus genes. CO protein is degraded at different rates in the light vs the dark. Light enhances the stability of CO, allowing it to accumulate during the day, and it is rapidly degraded in the dark.
- Flowers are made up of floral parts arranged in concentric whorls, with sepals and petals surrounding the inner reproductive parts. Formation of floral meristem requires active floral meristem identity genes, such as SOC1, AP1, and LFY, in *Arabidopsis*. Mutations in homeotic floral identity genes alter the types of organs produced in each of the whorls. The ABC model suggests that organ identity in each whorl is determined by the combined activity of set of three organ identity genes. A quartet model has been described to explain how transcription factors act together to specify floral organs. ABCE model with certain variations explains the diversity of angiosperm flower structure. MADS-box genes closely related to class C genes are required for ovule formation.

## Multiple-Choice Questions

1. Induction of flowering by low temperature treatment is termed as:
   (a) Vernalization
   (b) Photoperiodism
   (c) Cryopreservation
   (d) Defoliation
2. A gene that represses flowering in Arabidopsis:
   (a) LFY (Leafy)
   (b) FLC (Flowering locus C)
   (c) CO (Constans)
   (d) CRY2
3. The phenomenon of photoperiodism in plants was discovered by:
   (a) Chailakhyan
   (b) Borthwick and Hendricks
   (c) Skoog and Miller
   (d) Garner and Allard

4. Which of the following pigment plays a role in induction of flowering as identified by Chailakhyan:
   (a) Cytochrome
   (b) Vernalin
   (c) Florigen
   (d) Xanthophyll
5. According to ABC model in *Arabidopsis*, B and C genes are required for:
   (a) Petals
   (b) Stamens
   (c) Sepals
   (d) Carpels

**Answers**

1. a   2. b   3. d   4. c   5. b

## Suggested Further Readings

Davies B (2006) Floral meristem identity genes. In: Jordan BR (ed) The molecular biology and biotechnology of flower. CABI Publishers, Cambridge, MA, pp 81–99

van Dijk ADJ, Molenaar J (2017) Floral pathway integrator gene expression mediates gradual transmission of environmental and endogenous cues to flowering time. Peer J 5:e2724v1

# Pollination, Fertilization and Seed Development

## 26

Rashmi Shakya and Satish C Bhatla

Life cycle of plants is fundamentally different from that of animals. It is characterized by the presence of two distinct multicellular generations, referred as sporophytic (diploid) and gametophytic (haploid) generation which alternate with each other during the life cycle. Male and female reproductive organs in plants are stamens (androecium) and carpels (gynoecium), respectively. Both the reproductive structures produce haploid spores as a result of meiosis, namely, microspores (male) and megaspores (female). These spores undergo repeated mitotic divisions to produce male and female gametophytes, called as microgametophyte and megagametophyte, respectively. Development of male gametophyte takes place inside the anther, whereas female gametophyte develops inside the ovule. Upon maturity male and female gametophytes divide mitotically to produce male and female gametes, i.e., sperm and egg, which fuse to form zygote that develops to give rise to sporophytic plant (Fig. 26.1).

This chapter begins with description of male and female gametophyte. Later, transfer of pollen, followed by detailed account of events associated with pollen-pistil interaction, leading to double fertilization. Genetic basis of sexual incompatibility barrier has then been discussed describing three mechanisms worked out till date. At last, development of seed has been explained as a complex, coordinated process, including embryogenesis, endosperm development, and maturation of seed, conferring it to desiccation tolerance capability.

## 26.1 Development of Male Gametophyte

Pollen grain is the male gametophyte in flowering plants. Development of pollen grains occurs inside the anther which is the fertile portion of the stamen. Typically, an anther is composed of a well-defined anther wall which encloses a mass of sporogenous tissue inside the locules. The wall of anther is differentiated into four regions which are centripetally organized as epidermis (single layer), endothecium

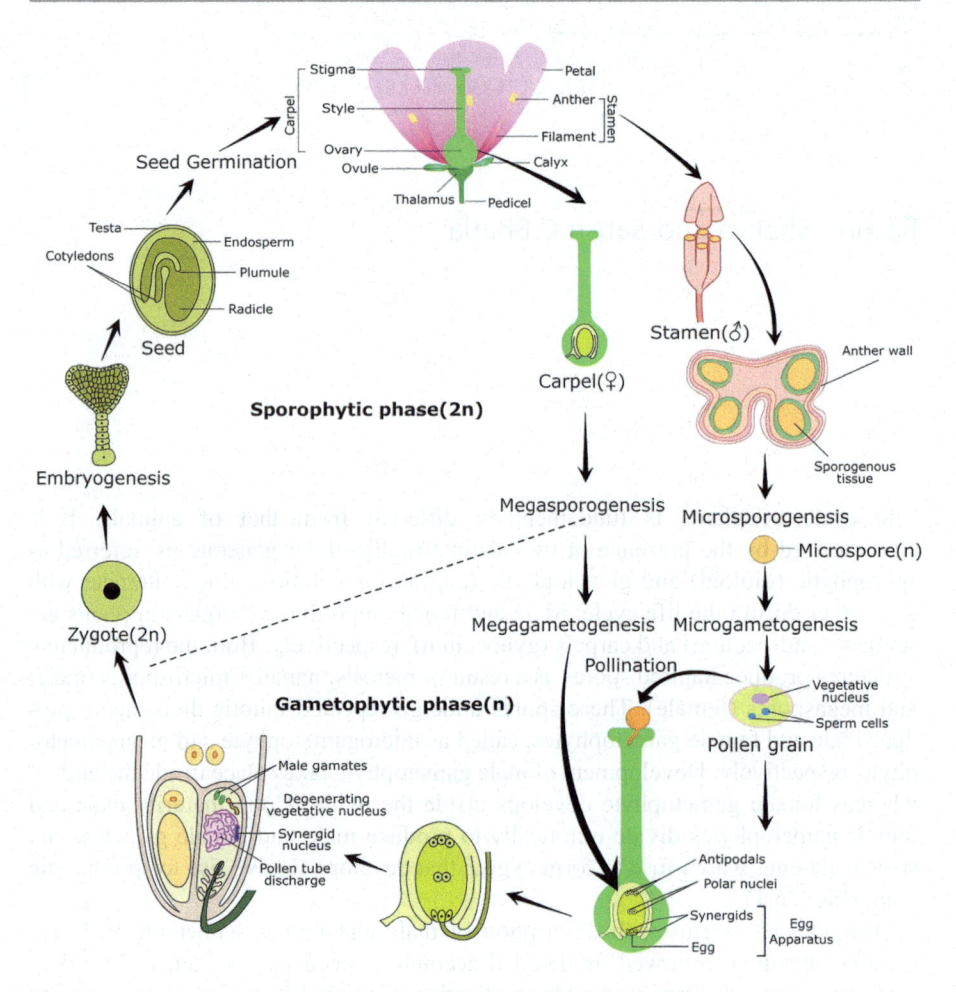

**Fig. 26.1** Alternation of sporophytic and gametophytic phases in the life cycle of flowering plants

(single layer), middle layers (2–3 layers), and tapetum (Fig. 26.2; Table 26.1). **Tapetum** consists of secretory cells which completely surround the inner sporogenous tissue. Sporogenous tissue inside the anther locule differentiates into microsporocytes, i.e., **pollen mother cells (PMCs)**, which undergo meiosis to form microspores. Microsporocytes undergo meiosis to give rise to haploid microspores which are joined to each other as tetrads through special cell wall (SCW), primarily made up of callose (Fig. 26.3). Secretory cells of the tapetal layer release callase and other cell wall-degrading enzymes which lead to the hydrolysis of SCW, thus separating the microspore tetrads into individual microspores. Each microspore develops into a **pollen grain**. During the development of pollen grains, the cytoplasm of microspores becomes highly vacuolated, and nucleus migrates to one side of the cell wall, conferring polarity to microspores. The

## 26.1 Development of Male Gametophyte

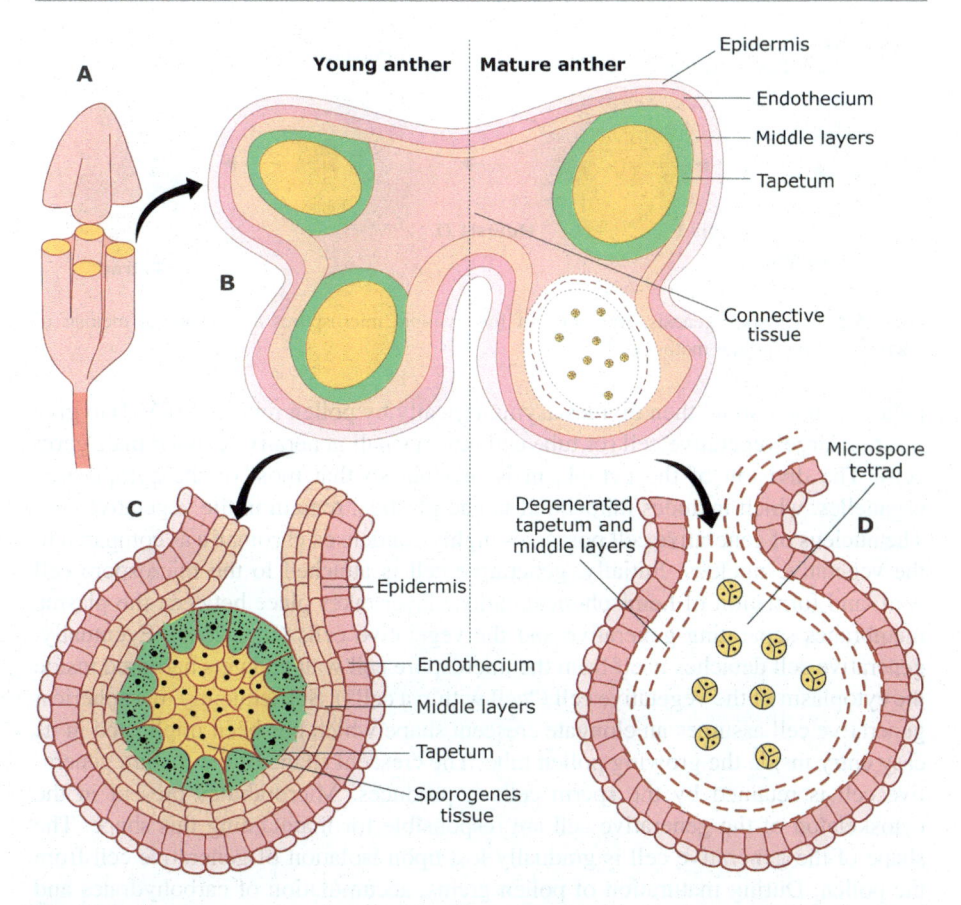

**Fig. 26.2** Transverse section of young (**a**) and mature anther (**b**) showing sporogenous tissue and microspore tetrads, respectively

**Table 26.1** Structure of stamen and carpel

| Stamen | Carpel |
|---|---|
| Filament | Stigma |
| Anther | Style |
|   Anther wall | Ovary |
|     Epidermis (single layered) |   Ovary wall |
|     Endothecium (single layered) |   Placenta |
|     Middle layers (2–3) |   Ovule(s) |
|     Tapetum (single layered) |     Funiculus |
|   Microspore mother cell |     Integuments and micropyle |
|   (microsporocyte) |     Nucellus |
|  |       Megaspore mother cell |
|  |       (megasporocyte) |

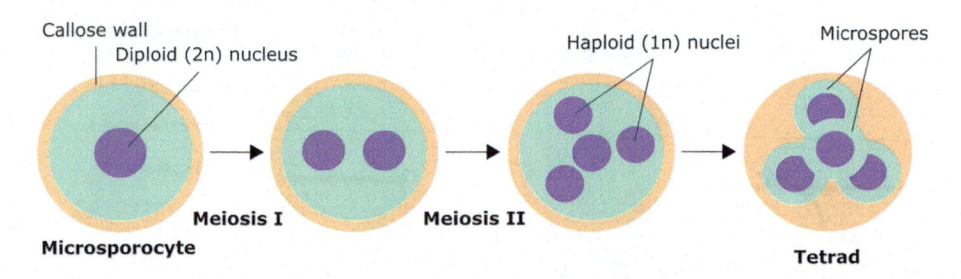

**Fig. 26.3** Microsporogenesis: formation of four haploid microspores as a result of meiosis of microsporocyte (pollen mother cell)

polarized microspore then divides asymmetrically by pollen mitosis I (PM I) to give rise to a large vegetative cell (or tube cell) and a small generative cell (or male germ cell). The division of the cytoplasm is unequal so that most of the cytoplasmic organelles, which includes mitochondria and plastids, remain in the vegetative cell. The nucleus of generative cell possesses highly condensed chromatin as compared to the vegetative nucleus. Initially, generative cell is attached to the microspore cell wall, and formation of hemispherical callose layer takes place between the plasma membranes separating generative and the vegetative cells (Fig. 26.4). Eventually, generative cell detaches itself from the microspore cell wall and gets harbored inside the cytoplasm of the vegetative cell ("cell within a cell"). Subsequently, the spherical generative cell assumes an elongate crescent shape which has been implicated in its easy entry inside the growing pollen tube. The crescent shape of the mature generative cell is retained by the sperm cells it produces. Microtubules present in the cytoskeleton of the generative cell are responsible for maintaining this shape. The shape of the generative cell is gradually lost upon isolation of generative cell from the pollen. During maturation of pollen grains, accumulation of carbohydrates and lipids takes place to support the upcoming pollen germination and pollen tube formation. Depending on the species, the tapetal layer may either remain intact at the periphery or may become amoeboid and migrate into the locule during pollen development. In both the cases, tapetum performs secretory function and provides nutrition to the developing pollen grains. Eventually, tapetal cells undergo programmed cell death (PCD), releasing their contents inside the anther locule. Tapetal cells perform significant role in the supply of nutrients, enzymes, and cell wall precursors to the developing pollen grains. For this reason, any defect in tapetum causes abnormal development of pollen grains, and consequent defects in their fertility. Mostly, anther dehiscence takes place at this stage when the pollen grains are two celled. Generative cell undergoes pollen mitosis II (PM II), and the timing of PM II varies from species to species. Mostly, generative cell undergoes PM II division to form two sperm cells when the pollen grains are still inside the anther locule. However, the generative cells may sometimes undergo PM II to form two male gametes inside each pollen grain after it has alighted onto stigma surface (e.g., *Holoptelea integrifolia*). PM II may take place after germination of pollen grains on stigma surface (e.g., *Zea mays* and *Nicotiana tabacum*). PM II in the generative cell

## 26.1 Development of Male Gametophyte

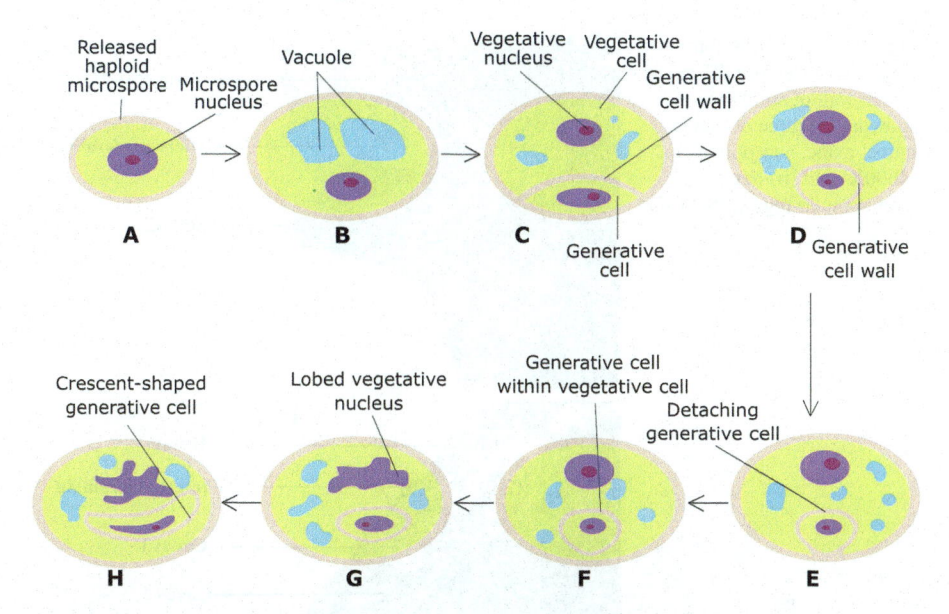

**Fig. 26.4** Microgametogenesis: formation of generative cell and vegetative cell due to mitotic division of the microspore

may take place in the pollen tube before it reaches the embryo sac (most common condition). Rarely, generative cell undergoes PM II after the pollen tube has reached the embryo sac (e.g., in *Euphorbia terracina*). The chromatin in the nucleus of sperm cells is highly condensed. Since generative cells are devoid of mitochondria and chloroplasts, the sperm cells also lack these organelles. Therefore, these organelles are inherited maternally in almost 90% of angiosperms.

Upon attaining maturity, the cell wall of pollen grains shows a great deal of variation in architecture which is ecologically important in the pollination process. Pollen wall is highly complex, and it mainly includes three domains which differ not only in their structure and chemical composition but also in biological and physiological significance. The three domains are exine, intine and pollen coat (tryphine) (Fig. 26.5). The process of cell wall formation in pollen is initiated immediately after meiosis. Deposition of ephemeral callose wall, i.e., SCW, is followed by the layers of sexine (ectexine), nexine (endexine), and finally the intine. Sporopollenin is a chemically inert biological polymer and has been suggested to consist of covalently linked phenolic and fatty acid-derived constituents. Initially sporopollenin precursors synthesized and secreted by the microspores contribute to the formation of exine layer. However, after dissolution of SCW, sporopollenin precursors are mostly provided by the secretory cells of the tapetal layer. The intine layer is primarily made up of cellulose and pectins. Mostly, pollen wall possesses elongated areas called as apertures or the germ pore region where exine is either missing or very thin. Through these germ pores, pollen tube emerges during pollen germination on the surface of stigma. In grasses, however, a pectin rich layer, known as Z-layer,

**Fig. 26.5** Transmission electron micrograph (TEM) of pollen wall of sunflower showing (**a**) three domains, intine, exine, and pollen coat (4600X) and (**b**) details of inter-spinular region (8400X)

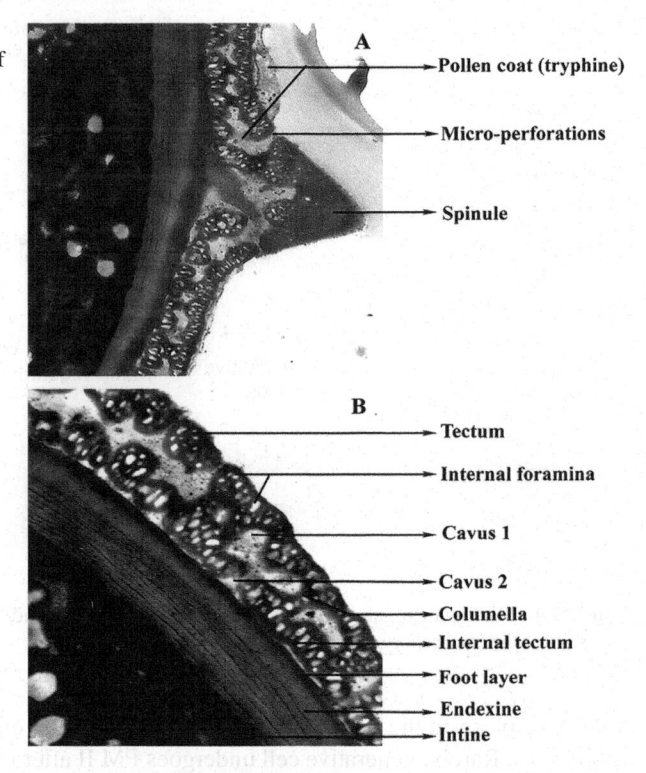

is present in between exine and intine. This layer is quite thick at germ pore areas and is referred to as "Zwischenkorper." Pollen wall diversity is often a characteristic feature of each species and is significant in assigning taxonomic position as well. In insect or self-pollinated species, like *Brassica*, *Arabidopsis* (Brassicaceae), and *Felicia* (Asteraceae), the pollen coat is thick. A large amount of pollen coat substances is present on the outer surface and interbacular cavities of ectexine. In wind- or cross-pollinated species, such as maize, the pollen coat is thin. Basically, two types of pollen coat materials are produced by tapetum layer of anther, namely **pollenkitt** and **tryphine**. Pollenkitt is the sticky substance present around most of the pollen grains pollinated by animals, whereas tryphine is exclusive to members of Brassicaceae which are entomophilous. The degeneration of tapetal cells at microspore stage of male gametophyte development leads to the formation of tryphine at the pollen surface. On the other hand, degeneration of tapetal layer at later stages causes pollenkitt deposition on the pollen wall. The nature of pollen coat substances is quite variable in different species. Pollen coat substances, mainly pollen coat proteins (PCPs) and lipids, have been implicated in several functions starting from holding the pollen before anthesis, at the time of pollination, and during initial stages of pollen-stigma interaction. PCP-B class of coat proteins have recently been reported to play a significant role in the hydration of pollen on the stigma papillae.

## 26.2 Development of Female Gametophyte

Carpel, the female reproductive structure in flowering plants, consists of stigma, style, and ovary. Ovules which contain the female gametophyte (embryo sac) are present inside the ovary. Ovule primordia arise along the placenta as projected rounded tips. During the early stage of ovule development, three regions can be identified. The basal proximal region gives rise to funiculus. The distal or micropylar region at the tip produces the nucellus. The central region, chalaza, gives rise to outer layers of the ovule called as **integuments**. The megaspore mother cell is characterized by its large size and large nucleus, and its dense cytoplasm gets differentiated in the nucellar tissue (Fig. 26.6a). The development of embryo sac is quite complex and diverse as compared to pollen. Approximately 15 types of embryo sac development patterns have been reported among angiosperm species. The most common pattern of embryo sac development observed in *Polygonum* and is known as ***Polygonum* type** of embryo sac development. The megaspore mother cell undergoes meiosis to form four megaspores out of which the three positioned toward the micropylar end degenerate (Fig. 26.6b–d). The functional megaspore divides three times by free nuclear division giving rise to a multinucleate cell called as **syncytium** (Fig. 26.6e–h). Of the eight nuclei present in the immature embryo sac, four move toward the chalazal end and the remaining four toward the micropylar end. Three nuclei, at the micropylar and chalazal ends, undergo cellularization process. One nucleus from each pole migrates toward the center. These are called as **polar nuclei**. Plasma membrane develops around both the polar nuclei and surrounding cytoplasm. Thus, mature "*Polygonum* type" of embryo sac contains

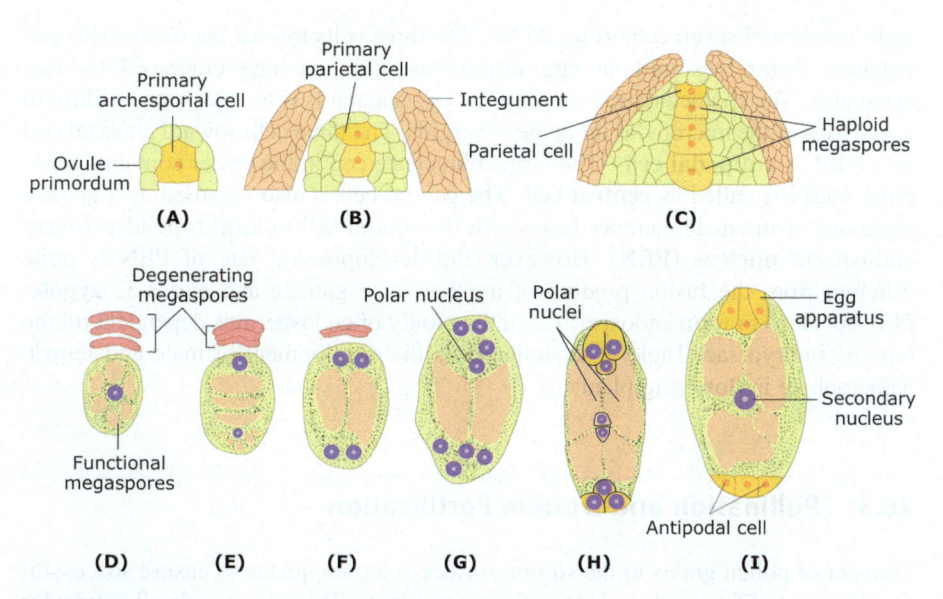

**Fig. 26.6** (**a–d**) Development of megaspore, i.e., megasporogenesis, and (**e–i**) development of megagametophyte, i.e., megagametogenesis to form embryo sac

**Table 26.2** Development of male and female gametophyte

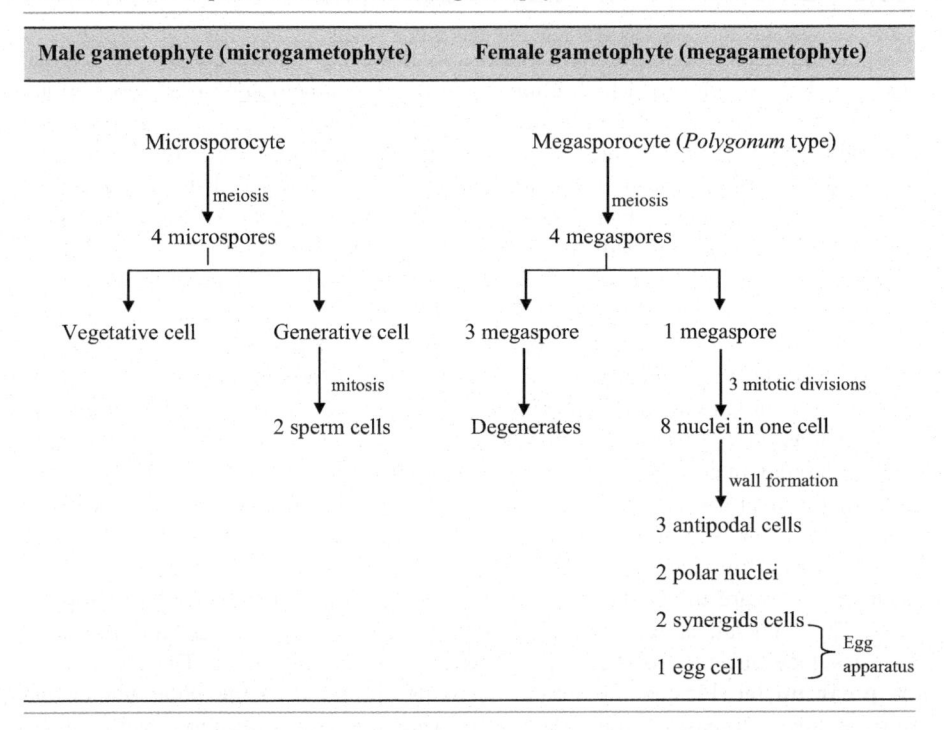

| Male gametophyte (microgametophyte) | Female gametophyte (megagametophyte) |
|---|---|
| Microsporocyte <br> ↓ meiosis <br> 4 microspores <br> ↓ ↓ <br> Vegetative cell    Generative cell <br>                ↓ mitosis <br>                2 sperm cells | Megasporocyte (*Polygonum* type) <br> ↓ meiosis <br> 4 megaspores <br> ↓        ↓ <br> 3 megaspore    1 megaspore <br> ↓          ↓ 3 mitotic divisions <br> Degenerates    8 nuclei in one cell <br>               ↓ wall formation <br>               3 antipodal cells <br>               2 polar nuclei <br>               2 synergids cells ⎫ Egg <br>               1 egg cell        ⎭ apparatus |

eight nuclei and seven cells (Fig. 26.6i). The three cells toward the micropylar end organize themselves into an **egg apparatus** with one **egg** embraced by two **synergids**. The characteristic feature of the egg apparatus is the presence of **filiform apparatus** at the micropylar tip in the synergids. The three cells toward chalazal end are called as **antipodal cells**. The large, binucleate cell in the center containing two polar nuclei is called as **central cell**. The central cell is also regarded as a gamete since one of the male gametes fuses with the central cell to form triploid **primary endosperm nucleus (PEN)**. However, the developmental fate of PEN is quite different from the fusion product of another male gamete and egg, i.e., **zygote**. PEN develops to form endosperm, and the ploidy of endosperm is dependent on the type of embryo sac. Table 26.2 summarizes the development of male and female gametophyte in flowering plants.

## 26.3 Pollination and Double Fertilization

Transfer of pollen grains to the stigma surface is a prerequisite to ensure successful development of fruit and seed set in flowering plants. This process of pollen transfer from anthers to the stigma surface is known as **pollination**. For the success of pollination, it is imperative that viability of pollen and period of stigma receptivity must coincide. Pollen may be transferred onto the stigma surface on the same flower (**self-pollination or autogamy**) or to the stigma of another flower present on the same plant (**geitonogamy**) or on a different plant (**xenogamy**). The latter two processes constitute **cross-pollination** whose success is dependent on the intervention of several external agencies and various factors, such as temperature. Pollen grains in certain plants, such as tomato, get damaged by heat, whereas some others can tolerate high temperature. After pollen has alighted onto the surface of stigma, a number of structural, biochemical, and physiological events occur as a part of cellular dialogue between pollen and pistil. Successful completion of these events leads to a complex phenomenon known as **double fertilization,** which is characterized by the fusion of one sperm nucleus with the egg cell to form zygote, and the second sperm nucleus fuses with polar nuclei or secondary nucleus to form PEN. Development of zygote results in the formation of embryo (embryogenesis), whereas PEN develops to form endosperm, the nutritive tissue that provides nourishment to the developing embryo.

### 26.3.1 Pollen Adhesion and Hydration

The nature of pollen coat and stigma surface determines the possibility of contact between these two surfaces. Stigma surface may be termed as dry or wet type depending on the presence of stigmatic exudates during the receptive period. In "wet-type" **stigmas**, the surface is characterized by exudates which are a complex mixture of polysaccharides, proteins, and lipids. In "dry-type" **stigmas**, papillae are covered by a proteinaceous pellicle, cuticle, and cell wall (Table 26.3). The surface in wet type of stigma, such as in *Nicotiana* and *Lilium*, is unable to discriminate between compatible and incompatible pollen grains. Thus, pollen from any species falling onto such stigma surface gets adhered to the exudates secreted during the receptive phase. However, in plants possessing dry-type stigma, such as *Arabidopsis*, pollen adhesion is a highly species-specific process. In *Arabidopsis*, self-pollen grains adhere more tightly to the surface of stigma as compared to the pollen from other unrelated plants, such as *Petunia hybrida* or other members of Brassicaceae, demonstrating a clear species preference during pollen adhesion. Upon landing of compatible pollen on the papillae of dry stigma, mobilization of pollen coat substances takes place on the papillae. This results in the formation of an interface (popularly referred to as "attachment foot") between pollen and stigmatic papillae where interaction between molecules from these two domains occurs.

**Table 26.3** Type of stigma surface and their chemical nature

| Type | Chemical constituents | Example |
| --- | --- | --- |
| Dry type | Proteinaceous pellicle and cuticle | *Arabidopsis*, sunflower, *Brassica*, *Senecio* |
| Wet type | Mixture of polysaccharides proteins and lipids | *Lilium*, *Petunia*, *Nicotiana*, *Papaver* |

Adhesion of pollen to stigmatic papillae is dependent on the biophysical and biochemical interactions between pollen coat substances and stigma surface proteins. Removal of pollen coat with organic solvents, such as cyclohexane, acetone, and diethyl ether, adversely affects pollen adhesion and its hydration. It has been reported that the strength of compatible pollination declines upon removal of the pollen coat and stigma waxes. Like pollen adhesion, hydration of pollen on wet stigma surface is not regulated, whereas it is highly regulated in dry stigma. In species possessing dry type of stigma, such as in *Arabidopsis* and sunflower (*Helianthus annuus*), exine components mediate the adhesion of pollen to the stigmatic papillae. This adhesion is highly species specific as the strength of adhesion is much stronger in intraspecific pollination as compared to interspecific pollination. It is thought that lipids and proteins present at the interface enable changes in the properties of the underlying papillae which lead to the transport of water and ions from the stigmatic papillae resulting in hydration of pollen. Two mechanisms have been implicated in pollen hydration. First, movement of water molecules through aquaporin channels located in the plasma membrane of papillae may take place. Second, targeted vesicular exocytosis to the papillae surface upon arrival of compatible pollen grain also facilitates pollen hydration among the members of Brassicaceae. Experiments have shown that defect in a gene required for normal exocytosis of Golgi vesicles leads to failure of pollen hydration on stigma surface. In species possessing wet type of stigma, such as *Lilium* and *Petunia*, pollen hydration is not a very specific requirement since water present in exudates supports hydration of pollen. In addition, stigmatic exudates contain lipids, proteins, sugars, and flavonoids that support pollen germination as well. It is believed that lipids present in the exudates substitute for the function of pollen coat and are involved in directional growth of pollen tube. Mutants lacking these exudates are female sterile and can be rescued by exogenous application of lipids on stigma surface.

### 26.3.2 $Ca^{2+}$ Triggered Polarization of Pollen Grain Before Emergence of Pollen Tube

Pollen grains land on the surface of stigma in a highly desiccated state and are inactive, i.e., they cannot germinate. Hydration of pollen on stigma surface renders it physiologically active. $Ca^{2+}$ plays a very crucial role in pollen germination and subsequent growth of pollen tube. Influx of $Ca^{2+}$ into the vegetative cell sets up reorganization of cytoskeleton which leads to polarization of pollen. Immediately following hydration, concentration of free cytosolic $Ca^{2+}$ ($Ca^{2+}_{cyt}$) increases below the germ pore region from where the emergence of pollen tube takes place. The concentration of $[Ca^{2+}_{cyt}]$ remains high until the pollen tube emergence. Accumulation of both actin microfilaments and secretory vesicles takes place below the germ pore area. Migration of the vegetative nucleus takes place in such a manner to facilitate the movement of sperm cells ahead of vegetative cell in the pollen tube during germination. Table 26.4 outlines the initial events prior to pollen tube emergence during pollen stigma interaction.

**Table 26.4** Initial events (prior to pollen tube emergence) during pollen-stigma interaction

| S. No. | Events during pollen-stigma interaction |
|---|---|
| I. | Adhesion |
|  | Formation of attachment foot |
|  | Molecular interactions of proteins and lipids |
|  | Changes in permeability patterns due to opening of aquaporin channels on stigmatic papillae |
| II. | Hydration of pollen |
|  | Caused by targeted vesicular exocytosis to the papillae |
| III. | $Ca^{2+}$ influx into the vegetative cell |
| IV. | Accumulation of actin and vesicles below the germ pore area |

**Table 26.5** Rate of growth of some tip-growing structures in plants

| Structure | Rate of growth | Examples |
|---|---|---|
| Pollen tube | 2.8 µm s$^{-1}$ | Maize |
|  | 0.2–0.3 µm s$^{-1}$ | Lily |
|  | 55 µm s$^{-1}$ | *Conospermum* species |
| Root hair | 0.01–0.04 µm s$^{-1}$ | *Arabidopsis* |
| Chloronema tip cell | 0.001625 µm s$^{-1}$ (5.85 µm h$^{-1}$) | *Physcomitrella patens* |
|  |  | *Funaria* |
| Fungal hyphae | 0.201 µm s$^{-1}$ | *Neurospora crassa* |
|  | 0.038 µm s$^{-1}$ | *Rhizoctonia solani* |

## 26.3.3 Apical Growth of Pollen Tube Tip and Its Regulation

Pollen tube elongates by tip growth following germination process. The growth rate of pollen tube is quite rapid (more than 5 µms$^{-1}$) as compared to that of root hair (10–40 nms$^{-1}$) which also grows by tip growth. In case of maize, pollen tubes attain length up to 40 cm. Table 26.5 summarizes the rate of growth in some tip-growing structures in plants. The growth of the pollen tube ceases at the surface of stigmatic papillae when rejection of pollen takes place in species exhibiting sporophytic self-incompatibility (SSI). However, in species showing gametophytic self-incompatibility (GSI), pollen tube penetrates through the cuticle layer of the papillar cell wall and penetrates stigma at the base of the papillae. Growing pollen tube also shows remarkable polarity which is generally evident in the form of four distinct regions (Fig. 26.7a). The apical region or tip of the growing pollen tube is free from any major organelles and is referred as "clear zone" (Fig. 26.7b). Within the clear zone, no cytoplasmic streaming can be observed, whereas in the subapical region, it does exist. The subapical region is highly vesicular but is devoid of cell organelles. Actin cytoskeleton supports the intracellular trafficking of secretory vesicles and other organelles along the axially oriented actin throughout the length of the elongating pollen tube. These organelles and secretory vesicles move toward the tip along the edge of the pollen tube, and after reaching the subapical region, they migrate backward toward the pollen through the center of the pollen tube. This type

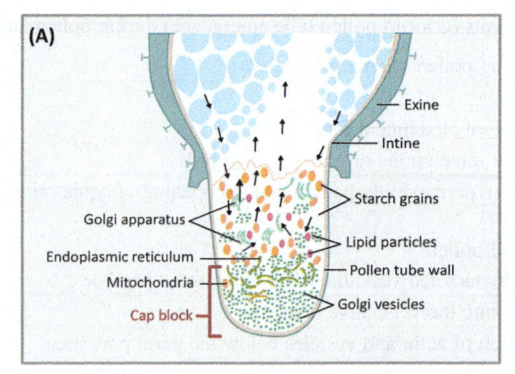

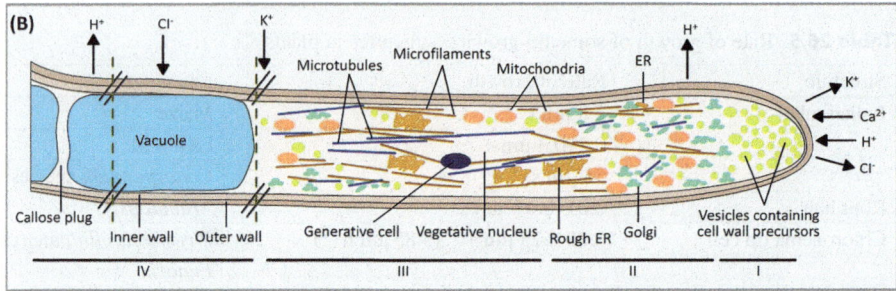

**Fig. 26.7** (a) Diagrammatic sketch of the tip of pollen tube showing "cap block" region in the just emerging germ tube and a pollen tube. (b) Four different zones (I–IV) in the pollen tube. I apical zone, II subapical zone, III nuclear zone, IV vacuolization zone

of movement gives rise to a reverse pattern of cytoplasmic streaming. Actin microfilaments extend to the subapical region but do not invade the clear zone (Fig. 26.8a). Clear zone contains short actin bundles. The base of the clear zone is characterized by the randomly oriented dense mesh of short actin filaments. This rapid actin remodeling at the subapical region is supposed to be important in directing reverse cytoplasmic streaming. Existence of clear zone seems to be related to the disruption and reorganization of actin microfilaments implicated in cytoplasmic streaming.

Actin dynamics is regulated by several actin-binding proteins. These include the G-actin-binding protein **profilin**; the G- and F-actin-binding proteins or the **actin-depolymerizing factors (ADFs)**, also known as **cofilins**; and others that affect different aspects of actin polymerization and higher order organization. In pollen tubes, increasing the level of profilin or ADF results in the disruption of the normal actin cytoskeleton organization and inhibition of pollen tube growth. The small secretory vesicles are involved in delivery of wall materials and membrane to the growing tip. Nuclear region containing two sperm nuclei and large cell organelles (such as mitochondria and endoplasmic reticulum) is present behind the subapical region. Lastly, vacuolar region contains a large vacuole to restrict the backward flow of cytoplasm and sperm nucleus inside the pollen. It is hypothesized that $Ca^{2+}$ and

## 26.3 Pollination and Double Fertilization

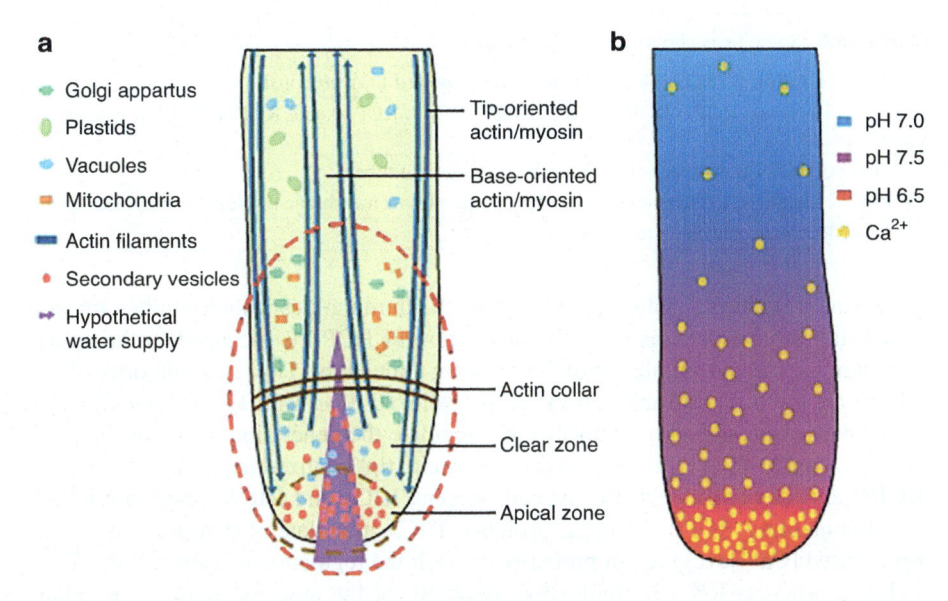

**Fig. 26.8** (a) Model to show the reverse pattern of cytoplasmic streaming. (b) Role of $Ca^{2+}$ and pH gradient in regulation of polarity inside growing pollen tube

pH gradients in the growing tip are involved in the regulation of polarity inside growing pollen tube (Fig. 26.8b). At the extreme tip of the pollen tube, the concentration of cytosolic $Ca^{2+}$ is high (3–10 µM) and drops drastically (0.2–0.3 µM) beyond 20 µm from the tip region. Additionally, the extreme tip region is characterized by slightly acidic cytosolic pH (pH 6.8) and an alkaline pH (7.5) at the interface of the apical and subapical regions. Periodic oscillation in $[Ca^{2+}]_{cyt}$ concentration and pH correlates with changes in the growth rate of pollen tube, pointing toward a possible link between them.

### 26.3.4 Signaling Events at the Tip of Growing Pollen Tube

A unique family of small guanine trinucleotide phosphatases (GTPases), known as Rho-like GTPase (ROP), is implicated in the regulation of pollen tube tip growth. Basically, GTPases act as molecular switches involving interconversion of GTP (bound active form) and GDP (bound inactive form). Guanine nucleotide exchange factors (GEFs) are responsible for the activation of the inactive GTPases by replacing GDP with GTP and triggering a downstream signaling cascade. GEFs themselves are activated by receptor-like kinases (RLKs). Inactivation of active GTPases is brought about by GTPase-activating proteins (GAPs). In *Arabidopsis*, seven (ROP1, ROP3, ROP5, ROP 8, ROP9, ROP10, and ROP11) out of total 11 different ROP genes are expressed in pollen grains. Levels of ROP1 transcripts are much higher as compared to ROP3 and ROP5, suggesting that ROP1 plays a

**Table 26.6** Some facts about ROP in *Arabidopsis*

| |
|---|
| 1. Total 11 isoforms (ROP1–ROP11) have been reported in *Arabidopsis* |
| 2. Seven isoforms (ROP1, ROP3, ROP5, ROP8, ROP9, ROP10, and ROP11) are expressed in pollen grains |
| 3. Level of ROP1 transcript is higher in the tip of pollen tube |
| 4. ROP1 is localized to the apical region close to the plasma membrane of the growing pollen tube. |
| 5. Other ROPs are mostly cytosolic |

dominant role during pollen tube tip growth. Immunolocalization studies on pea pollen tubes have demonstrated that a fraction of ROP1 is localized to the apical region of the plasma membrane of the growing pollen tube. However, majority of the ROPs are cytosolic (Table 26.6). It is believed that ROP1 activates at least two downstream signaling pathways that regulate the generation of the tip-focused $[Ca^{2+}]_{cyt}$ gradients and the assembly of dynamic, tip-localized actin microfilaments. ROP1 activates two direct downstream targets, RIC3 and RIC4, which are CRIB motif-containing ROP-interacting proteins. RIC3 modulates the formation of the tip-focused $[Ca^{2+}]cyt$ gradient probably through the regulation of extracellular $Ca^{2+}$ influxes, whereas RIC4 promotes the assembly of the apical F-actin. RLK, upon activation by an unidentified ligand, activates GEF which ultimately activates ROP1. Activated ROP1 stimulates NADPH oxidase (NOX) activity, resulting in the production of reactive oxygen species (ROS). In turn, ROS promotes influx of $Ca^{2+}$ from the extracellular space which enhances tip growth (Fig. 26.9). Actin dynamics is exhibited as periodic fluctuation of F-actin at the tip of the pollen tubes. Dynamics of the apical F-actin microfilaments is not only required for polarized growth of pollen tube but is probably also important for growth oscillations inside the pollen tube. In tobacco pollen tube, overexpression of RIC4 leads to stabilization of actin microfilaments at the tip region, resulting in depolarized growth along with loss of growth oscillations. Since, the activation of the RIC4-dependent actin pathway depends on interaction of RIC4 with active ROP1 at the plasma membrane of the tip region, it can be speculated that ROP1 activity at the tip controls actin dynamics. Cycling between GTP-bound active and GDP-bound inactive status of ROP1 is critical for normal tip growth of pollen tubes. Therefore, it can be proposed that periodic up- and downregulation of ROP1 activity might be required for the modulation of polarity and growth oscillations inside the growing pollen tube.

### 26.3.5 Directional Growth of Pollen Tube in the Pistil

Once the growing pollen tube penetrates the stigmatic tissue, the subsequent growth inside the pistil is required to be directional. This directional growth along the pistillar path is necessary for the pollen tube to enter the micropyle to effect fertilization. In fact, not one but several pollen tubes enter through the stigmatic tissue and compete for fertilization. Once fertilization takes place, all other pollen tubes stop growing further. The pollen tube growth guidance leading to fertilization

## 26.3 Pollination and Double Fertilization

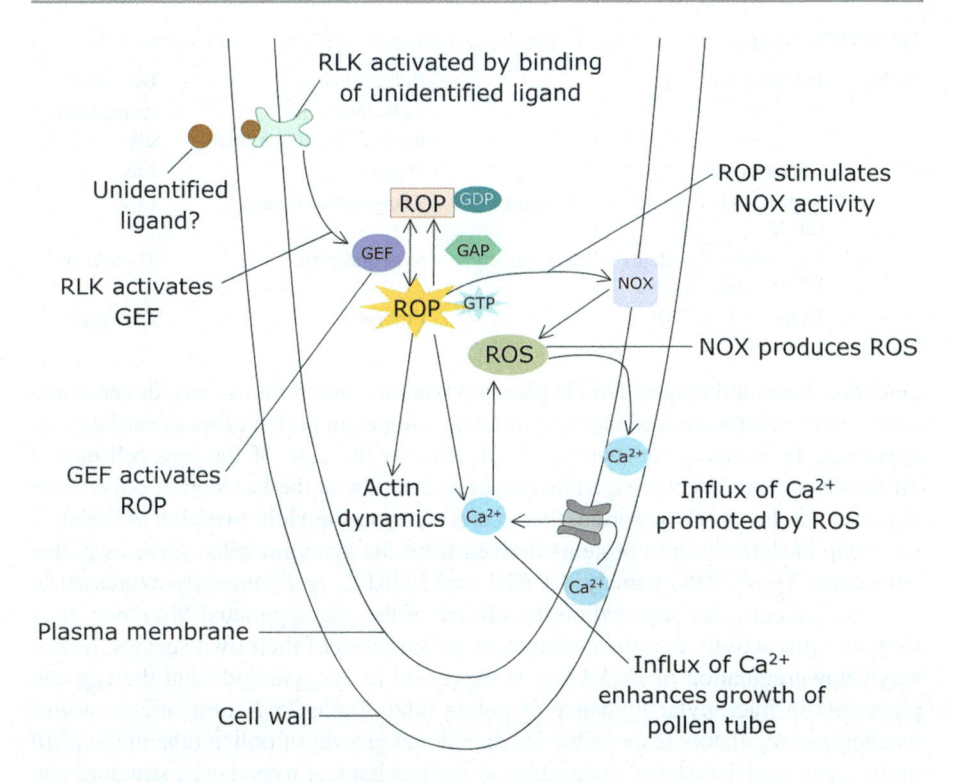

**Fig. 26.9** Signaling events at the tip of growing pollen tube

is dependent on the interaction of pollen tube with female tissues. A wide array of biomolecules has been implicated in this interaction process. The directional growth of pollen tube toward the female gametophyte has been explained with the help of two major models: **the chemotropic hypothesis** and the **mechanical hypothesis**. According to chemotropic model, a sequential arrangement of specific biomolecules directs pollen tube toward the ovule. In *Arabidopsis* and lily, plantacyanins (a type of phytocyanins, the blue copper proteins) have been reported to provide the initial **chemotactic guidance**. **Plantacyanins** are secreted by the stigmatic cells which line the transmitting tissue tract in the wet type of stigmas. Experiments involving in vitro pollen tube growth have shown that plantacyanins can reorient the growth of pollen tubes suggesting that they act as chemotactic cues. Overexpression of plantacyanins in the stigmatic papillar tissue makes pollen tube insensitive to the transmitting tissue. As a result, pollen tubes lose their orientation leading to loss of directional growth. Another member of phytocyanin family, **chemocyanin**, also acts as a directional cue. In lily, stigma/style **cysteine-rich adhesin (SCA)**, a lipid transfer protein, secreted by the transmitting tissue is also reported to guide growth of pollen tube. In *Arabidopsis*, embryo sac produces the chemotactic cues for pollen tube guidance. Mutants which lack female gametophyte show disrupted pollen tube

**Table 26.7** Biochemical regulators responsible for directional growth of pollen tube in the pistil

| S. No. | Biomolecules | Released from | Detected in |
|---|---|---|---|
| 1. | Plantacyanins | Cells lining the transmitting tissue tract | *Arabidopsis*, lily |
| 2. | Chemocyanin | Stigma | Lily |
| 3. | Stigma/style cysteine-rich Adhesin (SCA) | Transmitting tissue | Lily |
| 4. | Cysteine-rich proteins (CRPs), namely, LURE1 and LURE2 | Synergid cells | *Torenia* sp. |
| 5. | Defensin-like CRP | Embryo sac | *Zea mays* |

guidance. Laser ablation of female gametophyte has shown that synergids serve as a source of the chemotactic signals in attracting competent pollen tubes toward the egg apparatus. In *Torenia*, the two synergid cells on the side of the egg cell emit a diffusible, species-specific signal to attract pollen tube at the last step of pollen tube guidance. It has been reported that secreted, **cysteine-rich proteins (CRPs)**, a subgroup of defensin-like proteins derived from the synergid cells, serve as pollen attractants. Two CRPs, namely, LURE1 and LURE2, predominantly expressed in the synergid cells, are secreted on the surface of the egg apparatus. Moreover, they show in vitro activity to attract competent pollen tubes of their own species. In *Zea mays*, downregulation of *ZmEA1* gene expressed in the synergids and the egg cell plays role in micropylar guidance of pollen tube. Table 26.7 summarizes various biochemical regulators responsible for directional growth of pollen tube in the pistil in different model systems. According to the mechanical hypothesis, structure and organization of pistil tissue navigate the path of pollen tube using molecular factors. After entering the stigma surface, pollen tube interacts with **extracellular matrix (ECM)** of the transmitting tissue. ECM is a complex mixture of cell wall proteins, which include arabinogalactan proteins, hydroxyproline-rich proteins and proline-rich glycoproteins. According to mechanical guidance, these proteins serve as adhesive molecules to keep the pollen tube in place, and they also act as navigators for the growth of pollen tube toward the micropyle. ECM also provides nutrients for supporting metabolic activities inside the pollen tube.

Pollen tube reception in the micropyle and subsequent rupture inside the synergid require active communication between pollen tube and embryo sac. After entering the micropyle, pollen tube penetrates one of the synergids (designated as **receptive synergid**) through filiform apparatus and ruptures inside it to release two sperm cells (Fig. 26.10). The receptive synergid undergoes a programmed degeneration process, probably to set up cellular adjustments and reduces its turgor pressure to allow pollen tube discharge. The initiation of degeneration of receptive synergid takes place soon after pollination process, pointing toward a long distance signaling mechanism. In the beginning, the rupture of pollen tube inside the synergid was thought to be driven by the mechanical stimulus resulting from change in osmolarity, leading to pollen tube burst. **FER** gene in *Arabidopsis feronia* mutant encodes a **receptor-like kinase** which is expressed in the synergid cells and accumulates around the filiform apparatus. Probably, the ligand (as yet unknown) for FER is present on the surface

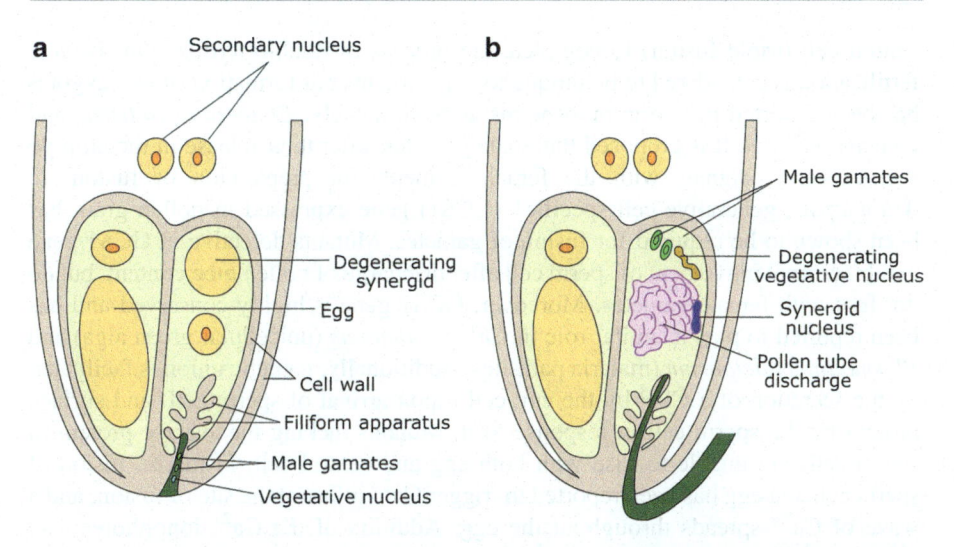

**Fig. 26.10** (a) Entry of the pollen tube inside the embryo sac through micropyle and its penetration into one of the synergids for delivery of sperm cells. (b) Delivery of male gametes and other pollen tube contents into the degenerating synergids

of pollen tube. Furthermore, effective fertilization also depends on the precise timing of pollen tube rupture. In *Arabidopsis*, two homologs of FER, the **ANXUR (ANX1/2) RLKs**, are exclusively expressed in pollen. In anx1/2 mutants, pollen tubes rupture before reaching the egg apparatus, suggesting that ANX1 and ANX2 are male factors controlling pollen tube behavior by directing rupture at proper timing. Furthermore, ANX1 and ANX2 have been reported to be the most closely related paralogs of FERONIA/SIRENE which control pollen tube behavior in the synergid cells. Rupture of pollen tube also depends on various ions, such as $Ca^+$ and $K^+$ which probably change the osmolarity of pollen tube leading to its lysis. In *Arabidopsis*, mutation in ACA9 gene affects the autoinhibition of $Ca^{2+}$-ATPase activity due to which pollen tube enters the micropyle but fails to burst and discharge its content, thus suggesting the involvement of $Ca^{2+}$ transport. In maize, ZmES4 (a defensin-like CRP) is exclusively expressed in the embryo sac. Experiments have shown that in vitro application of ZmES4 causes opening of the KZM1, an inward-rectifying $K^+$ channel. Downregulation of ZmES4 leads to failure of pollen tube discharge without affecting the guidance.

### 26.3.6 Double Fertilization

After bursting of the pollen tube inside the receptive synergid cell, the two sperm cells may either remain stationary at the boundary region between the egg and the central cell for some time or fuse with them immediately to initiate seed development. One sperm cell fuses with the egg (syngamy), and the other one fuses with the central cell (triple fusion) to complete the process of double fertilization. Double fertilization is considered to be unique to angiosperms but formation of two zygotes has been reported in two gymnosperms as well, namely, *Ephedra nevadensis* and *Gnetum gnemon*. It is proposed that male gametes, after their release in the embryo sac, exchange signals with the female gametes for preparation of fusion. In *Arabidopsis*, generative cell specific1 (GCS1) gene expressed in pollen grain has been shown to be required for fusion of gametes. Mutants defective in GCS1 gene exhibit normal movement of sperm cell after discharge of pollen tube content, but do not fuse with female gametes. Moreover, GCS1 gene is highly conserved and has been reported to play a similar role in *Chlamydomonas* (unicellular green alga) and *Plasmodium falciparum* (malaria parasite). Additionally, gametic union is facilitated by the secretion of a CRP by the egg cell (upon arrival of sperm cell) and surface protein by the sperm cell in response to it. Mutants lacking the surface protein in sperm cells are unable to fuse with both egg and central cell. In maize, fusion of sperm cell and egg has been reported to trigger $Ca^{2+}$ influx at the site of fusion, and a wave of $Ca^{2+}$ spreads throughout the egg. Addition of the $Ca^{2+}$ ionophores also causes $Ca^{2+}$ influx in addition to secretion of cell wall components, indicating that $Ca^{2+}$ is functionally important during the phase of fusion of gametes. The two sperms may or may not be equivalent when it comes to functionality. Which sperm cell will fuse with egg and which one with central cell is variable in different model systems. In *Plumbago zeylanica*, which exhibits sperm dimorphism (the two sperm cells differ morphologically in size and number of cell organelles), the smaller sperm cell always fertilizes the egg cell. In *Arabidopsis*, the isomorphic sperm cells are functionally equivalent.

## 26.4 Pre-zygotic Barriers to Self-Fertilization

Majority of the flowering plants (~85%) are bisexual. Thus, it is presumed that these must be self-fertilizing. However, floral morphology in most of the plants favors attracting pollinators which are instrumental in effecting cross-pollination and not self-pollination. Self-pollination is prevented in both bisexual and monoecious species through spatial and temporal features. Maturation of stamens and pistils at different times, known as **dichogamy**, prevents self-pollination. Dichogamy is of two types-**protandry** and **protogyny.** In protandry, stamens mature before pistils, whereas in protogyny maturation of pistils precedes that of stamens. Another spatial factor, **herkogamy**, i.e., spatial separation of pollen and stigma, reduces chances of self-pollination. An extreme case of spatial separation is exhibited by monoecious and dioecious species in which stamen and pistil are present in the same flower or in different flowers, respectively. In flowering plants, several steps prior to double fertilization are crucial for determining the reproductive success. Not only interspecific hybridization is prevented in many plants through various mechanisms but self-fertilization is also prevented by majority of species. Self-fertilization leads to inbreeding depression due to expression of deleterious traits as a result of high levels of homozygosity in a population. However, self-fertilization is beneficial at times as

## 26.4 Pre-zygotic Barriers to Self-Fertilization

it circumvents the necessity of a partner for effecting reproduction. Furthermore, many pioneer plants are self-fertilizing but phylogenetic studies have shown that most of the self-fertilizing species are evolutionarily young. The journey of pollen tube toward the female gametophyte depends on its intricate and highly regulated interaction with the pistillar tissue at various stages, beginning from adhesion and hydration of pollen grains on the stigma surface. This interaction proceeds with germination of pollen grain on stigma surface continuing with penetration of stigmatic surface and growth through the transmitting tissue and style and ultimately entry of pollen tube inside the embryo sac. Thus, pistil acts not only as a conduit for pollen tube growth but also acts as sieve for screening of incompatible pollen grains. This discrimination between compatible (nonself) and incompatible (self) pollen grains is exercised in highly prevalent self-incompatibility (SI) systems which ensure outcrossing by preventing selfing.

### 26.4.1 Genetic Basis of Self-Incompatibility (SI)

More than half of the angiosperm species (~125,000) have evolved self-incompatibility (SI) mechanisms. SI refers to the inability of the self-pollen to effect successful fertilization in the pistil. It encompasses a variety of diverse molecular and evolutionarily unrelated mechanisms that prevent self-fertilization. In 1925, E M East and A J Mangelsdorf, based on their studies on *Nicotiana*, reported that SI is determined by a single self-recognition locus, known as **sterility or S-locus**. The S-locus has several alleles ($S_1$, $S_2$, $S_3$,........$S_n$), and incompatibility results whenever pollen and pistil carry the same allele. S-locus is highly polymorphic such that every allele differs in sequence from the other. The genes implicated in SI recognition must be inherited together in a tight genetic linkage and by suppressing recombination between two components of S-locus. On the basis of genetic behavior, SI can be of two types: **gametophytic self-incompatibility (GSI)** and **sporophytic self-incompatibility (SSI)**. In GSI systems, SI is determined by the genotype of the haploid pollen grain itself, i.e., if the allele carried by the pollen matches with any one of the two S-alleles carried by the pistil, then it is rejected (Fig. 26.11a). In these systems, though pollen grain germinates and penetrates the stigmatic tissue, it gets rejected in the upper zone of the style. GSI systems have been reported in more than 60 families, including Solanaceae, Rosaceae, and Papaveraceae. In SSI systems, the incompatibility reaction is determined by the genotype of the diploid parent plant which produces pollen. In this case, either of the two alleles of the S-locus, which are present in the pollen parent, participates in the recognition reaction (Fig. 26.11b). Rejection reaction takes place if any of the two alleles present in the pollen parent match with the female plant. Majority of the SI systems are controlled by genes within a single S-locus. However, in *Secale cereale* and *Ranunculus acris* SI is dependent on two (loci S and Z) and four loci, respectively. Three main mechanisms have been discussed for the explanation of SI encoded by single S-locus. These mechanisms differ with respect to the manner in which self-pollen is rejected, i.e., whether this rejection involves programmed cell

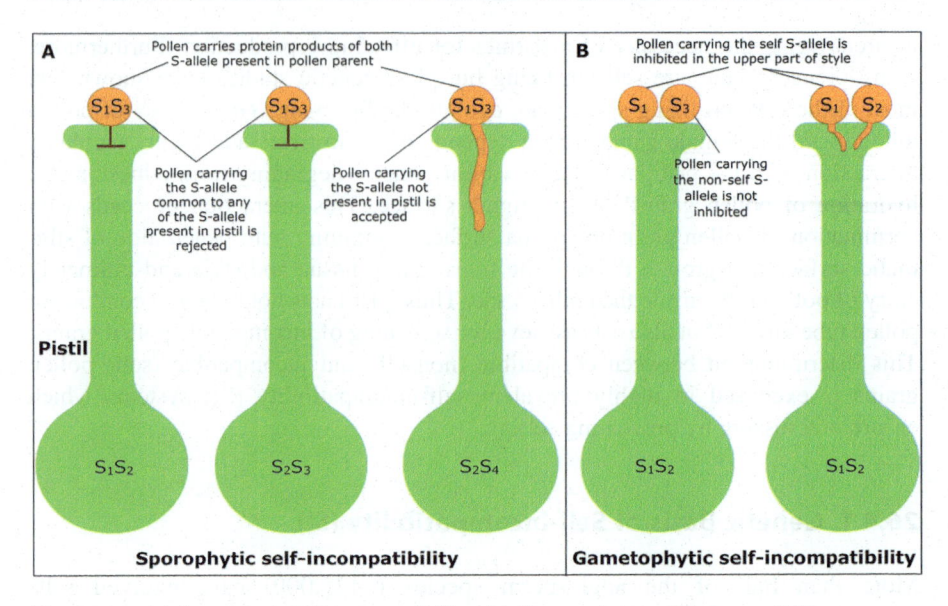

**Fig. 26.11** (a) Sporophytic self-incompatibility. (b) Gametophytic self-incompatibility

death (PCD). The phase and site of pollen rejection, i.e., whether it takes place during early or in the later phase of pollen tube growth through pistil, are determined by the attributes of the stigma surface. For instance, in Brassicaceae, self-pollen is rejected on the surface of dry stigma. Contrary to this, in members of families possessing wet type of stigma, such as in Solanaceae, self-pollen germinates and enters the stigmatic tissue, and inhibition of the pollen tube growth is accompanied by PCD of the pollen grains.

### 26.4.2 Receptor-Ligand Interactions Mediated Rejection of Self-Pollen at the Stigma Surface

In members of the family Brassicaceae, SI is operational at the level of interaction between pollen grain and stigma surface and is sporophytically determined, i.e., SSI system. The SI reaction is highly localized as it involves only the interface of pollen grain and stigmatic papilla and is observed within minutes of contact between pollen and stigma. This recognition reaction exhibits high degree of specificity such that a single papillar cell of stigma is capable of discriminating between a variety of genetically different pollen grains. As a result, germination of self-pollen grains is inhibited at the surface, whereas nonself pollen grains are allowed to proceed further to effect self-fertilization. Furthermore, incompatible pollen grains inhibited at the stigma surface are capable of forming pollen tubes upon transfer to the surface of compatible stigma. This shows that in members of Brassicaceae family, inhibition of self-pollen does not involve cell death of either pollen grain or stigmatic papilla

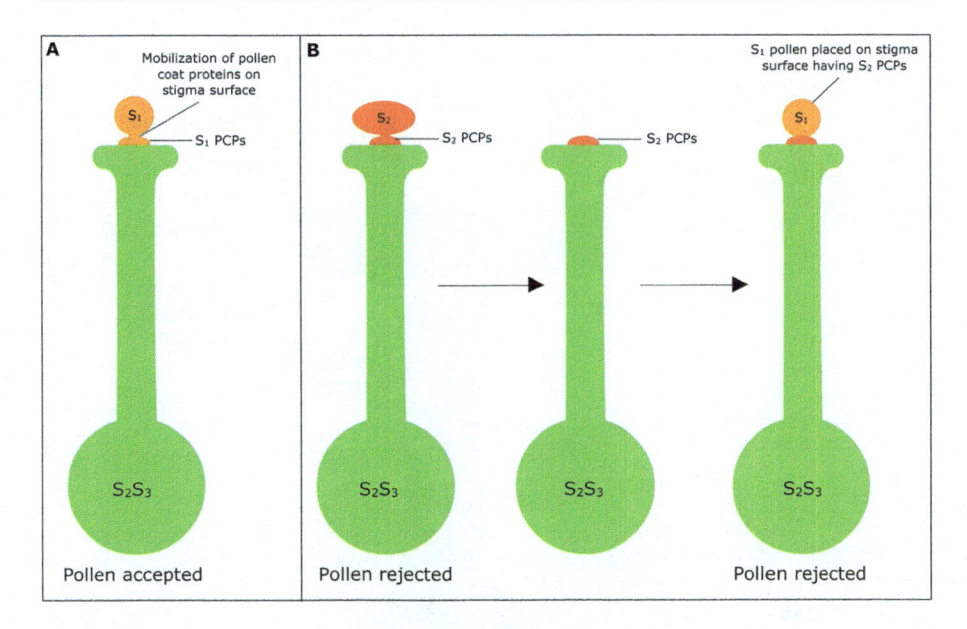

## 26.4 Pre-zygotic Barriers to Self-Fertilization

**Fig. 26.12** Role of pollen coat proteins (PCPs) in self- incompatibility response

(Fig. 26.12). Once a pollen grain lands on the stigma surface, a part of the pollen coat flows onto the papillar cell. The pollen coat contains the pollen factors that determine the specificity of the SI interaction. This can be demonstrated by placing a compatible pollen grain on to the same spot where an incompatible pollen grain was placed and then removed. The compatible pollen cannot germinate because the pollen coat material that was left behind, initiated an incompatible reaction. The presence of pollen factors in the pollen coat explains the sporophytic nature of SI system in Brassicaceae. The pollen coat contains molecules that were produced by the tapetum. Since the tapetum is diploid, it usually expresses both S-alleles of the pollen parent, such that two distinct pollen factors encoded by the two S-alleles are expected to be present in the pollen coat.

The mechanism of self-pollen recognition is based on the interaction between cell surface-localized receptors and ligands encoded by two S-locus genes. The S-locus receptor kinase (SRK) gene encoding a single-pass transmembrane serine/threonine kinase, localized in the plasma membrane of the stigmatic papilla, interacts with the S-locus cysteine-rich (SCR) gene (also known as S-locus protein 11, i.e., SP11), encoding a small peptide localized on the pollen coat (Fig. 26.13). SRK and SCR are highly polymorphic so that the amino acid sequence for different alleles can vary up to 35% and 70%, respectively. Initial contact between pollen and stigmatic papillae facilitates mobilization of SCR and other pollen coat substances to the papillae, resulting in the interaction of SRK and SCR, which act as ligand for binding with receptor kinase. The specific interaction between SRK and SCR variants encoded by the common S haplotype is responsible for species specificity of this SI system.

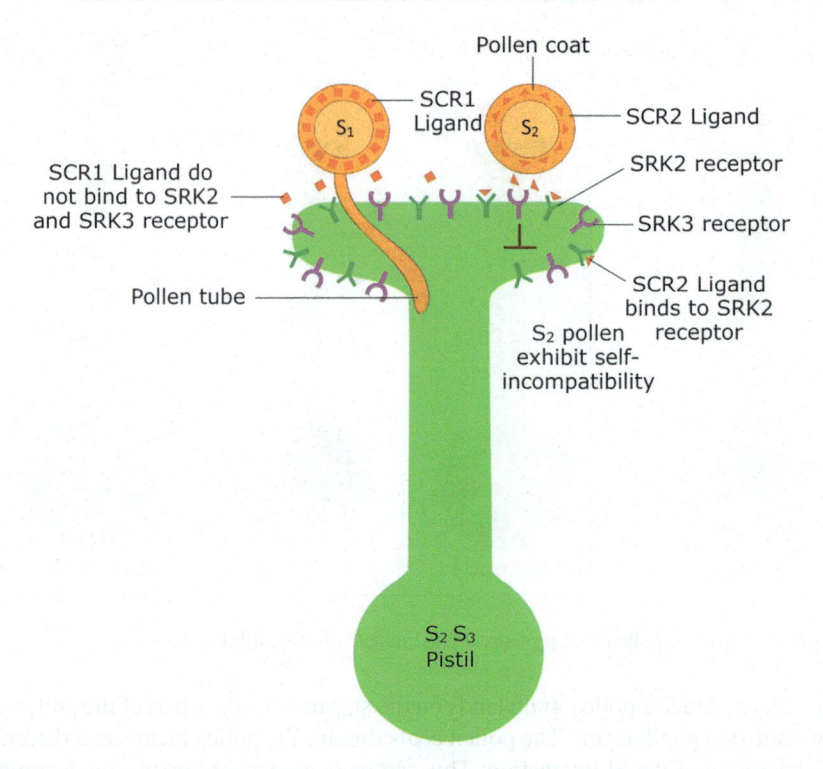

**Fig. 26.13** Interaction of receptor and ligand present on the surface of pollen and stigma, respectively, showing rejection of self-pollen ($S_2$) at stigma surface. SRK, S-locus receptor kinase; SCR, S-locus cysteine rich

### 26.4.3 Programmed Cell Death of the Pollen Tubes After Penetration into Stigma Surface

Contrary to SSI system in Brassicaceae in which SI response takes place early, *Papaver rhoeas*, which possess GSI system, shows SI response after pollen germination on stigma surface, and leads to the death of pollen tube after its entry into the stigma tissue. GSI system has been characterized in detail in *Papaver*; however, only the female determinant of the SI has been reported. The ligand reported to induce death of pollen tube is a 15 kDa glycoprotein secreted by stigma. This protein is designated as PrS (*Papaver rhoeas* stigma S) which exhibits polymorphism, causing inhibition of pollen tube growth in allele-specific manner. PrS does not show similarity to any protein with known function. However, self-PrS protein causes rapid influx of $Ca^{2+}$ just behind the tube tip which disturbs the cytosolic $Ca^{2+}$ gradient at the tip of pollen tube required for its growth. Initial inhibition of the pollen tube growth is due to actin depolymerization leading to disruption of actin cytoskeleton. This is followed by calcium-/calmodulin-dependent phosphorylation of a 26 kDa inorganic pyrophosphate (p26, IPP) and a 56 kDa mitogen-activated protein kinase (p56, MAPK) found in pollen. Subsequent activation of MAPK cascade, involving activation of caspases, leakage of cytochrome c from mitochondria into the cytosol and DNA fragmentation, leads to PCD of pollen tube. The putative receptor of PrS protein has been reported to be a novel transmembrane protein of 21 kDa and is designated as PrpS (Papaver rhoeas pollen S). Downregulation of PrpS reduces inhibition of pollen tube growth in an allele specific manner, pointing toward the crucial role it plays in rejection of self-pollen. Another integral membrane protein, S-protein-binding protein (SBP), has been reported to bind to PrpS in a non-allele specific manner and facilitate the inhibition of pollen tube growth.

### 26.4.4 Inhibition by Cytotoxic Stylar RNAses and Degradation of Pollen Tube RNA

Members of several families such as Solanaceae, Rosaceae, and Scrophulariaceae exhibiting GSI system show a different mechanism for inhibition of self-pollen grain. Germination of self-pollen occurs normally on the stigma surface. However, after entry of pollen tube into stigma, SI response is manifested in the upper region of the style where self-pollen tubes show reduction in the rate of their elongation, cell wall thickening, destruction of cell organelles, and loss of plasma membrane integrity, ultimately leading to bursting of pollen tube. These responses are mediated by non-specific ribonucleases, the **S-locus ribonucleases (S-RNases,** present in the pistil. S-RNase is abundant and highly polymorphic pistil-specific glycoprotein which is encoded by S- locus and secreted into the extracellular matrix present along the length of style through which pollen tube navigates its way to ovule. S-RNases are non-specifically taken up by both self and nonself pollen tubes, and allele-specific inhibition of self-pollen tubes takes place later on. The specificity of this SI response subsequent to RNAse uptake by the pollen tube is dependent on the interaction between S-RNase and S-locus-F-box (SLF) protein expressed in pollen tube. SLF, also known as S-haplotype- specific F-box, is a cytoplasmic protein belonging to F-box protein family and functions as a part of E3 ubiquitin ligase complexes. F-box protein confers specificity to the E3 ubiquitin ligase complexes by specific substrate binding and targets them for proteolysis. SLF binds to subunits of the E3 ligase complex and also interacts with S-RNases. The interaction of SLF and S-RNase is non-allele specific as SLF binds to both self and nonself S-RNases. However, interaction of SLF to nonself S-RNase is stronger as compared to that with self-RNAse. Thus, nonself RNAses are ubiquinated and degraded via proteasome pathway. As a result, growth of the nonself pollen tube is not inhibited. Experiments have shown that deletion of SLF gene does not cause constitutive rejection of the self and nonself pollen tubes. Thus, it has been suggested that an unknown RNAse inhibitor binds to all the RNAses, but binding of S-RNase with its cognate SLF does not permit inhibitor binding, rendering protected RNAse to degrade RNA. This model for self-pollen tube inhibition does not highlight the role of three stylar proteins reported to be required for inhibition of self-pollen. These three proteins are HT-B protein, 120 K glycoprotein, and factor 4936. Therefore, another model has been suggested for S-RNase-based inhibition in GSI systems. According to this model, HT-B protein along with 120 glycoprotein and factor 4936 enters the pollen tube along with the S-RNase. S-RNases are not degraded rather they are sequestered into endomembrane vesicles along with HT-B protein. These endomembrane vesicles break down subsequently during pollen tube growth and release S-RNases in self-pollen tube but remain intact in case of nonself pollen tubes. Breakdown of the endomembrane leads to release of sequestered S-RNases en masses which is too much for effective binding by inhibitor. Consequently, S-RNases are left active for RNA degradation.

## 26.5 Seed Development

In flowering plants, the diploid zygote develops into the embryo, whereas the fertilized central cell gives rise to endosperm which nourishes the embryo during its development. The process of seed development is quite complex, and for simplicity it can be discussed under three main processes. (1) **Embryogenesis** results in the formation of embryo from unicellular zygote through cell division, cell specialization, pattern formation and organized growth. (2) This is followed by the **development of endosperm** that takes place in parallel with embryogenesis and results in the development of specialized tissue for storage of nutritional reserves and formation of protective layers around the embryo. (3) The last process is desiccation or maturation drying in which the embryo prepares itself to survive long periods of metabolic inactivity in desiccated form.

### 26.5.1 Embryogenesis

Stages in the embryogenesis are quite variable among different species. In fact, the final overall appearance of the embryo is also species-specific. The **zygote** undergoes divisions in various planes to grow into a small cluster of cells, out of which some part develops into embryo proper and the other part grows into a short stalk-like structure known as **suspensor**. Suspensor pushes the embryo deep into the endosperm and gets crushed during the later stages of embryo development as it is usually ephemeral in nature. The cells at one end of the suspensor continue to divide mitotically and develop into an embryo. Initially these cells are arranged in the form of a sphere giving rise to the globular stage. Eventually, initiation of two primordia at the end of the embryo farther from suspensor marks the onset of heart-stage. Later, embryo assumes the torpedo stage (an elongate cylinder-like appearance) characterized by a short axis consisting of radical (embryonic root), epicotyls (embryonic stem), and hypocotyls (the root-shoot junction). Ultimately, vascular tissue differentiates within the embryo. The epicotyl may bear a pair of small leaves and radical and it often contains several primordia for lateral roots in its pericycle. After maturation, the embryo becomes quiescent and dehydrates partially. The funiculus may break which leave a small scar called hilum. *The detailed* *description of embryo development and its regulation has been discussed in detail in Chapter 24 (Embryogenesis, vegetative growth and organogenesis).*

## 26.5.2 Endosperm Development

The fusion product of second sperm cell with the central cell, i.e., primary endosperm nucleus, develops into endosperm which is a nutritive tissue required for supporting embryogenesis and seed germination. In flowering plants endosperm may be either transient or persistent in nature. *Transient endosperm* proliferates during early stage of seed formation. Later, however, it gets consumed by the developing embryo. In such cases, as exemplified by tomato, tobacco, and *Arabidopsis*, endosperm does not perform any major role as a storehouse of nutrition. Rather, embryo itself stores the nutrients essential for seed germination. In these cases, endosperm is represented by a single layer called as **aleurone layer** by analogy to cereal grains. In *persistent endosperm* (in case of cereals), endosperm contributes as much as 80% of the total seed mass and supports seed germination until the establishment of seedling. Development of embryo and endosperm occurs in a coordinated manner. Their development is influenced by seed coat which encloses both of them. For normal development of seed, the three components of the seed, namely, embryo, endosperm, and integuments, exchanges signals and coordinate development. However, very little is known about the molecular basis of interactions that takes place between them.

Three different types of endosperm development patterns are known in angiosperms. These are cellular, nuclear and helobial types. **Nuclear endosperm** is the most common type. A seed with nuclear type of endosperm development has an initial coenocytic phase and later cellular phase. The PEN (primary endosperm nucleus) undergoes several mitotic divisions which are not followed by simultaneous wall formation, thus resulting in the formation of **syncytial endosperm (coenocyte)**. Early mitotic divisions are synchronized but at later stage, divisions take place at variable rates resulting in differentiation of three distinct zones, namely, micropylar endosperm, peripheral endosperm, and chalazal endosperm (Fig. 26.14a–c). Expansion of the embryo sac takes place after fertilization, and enlargement of the central vacuole pushes the cytoplasm of the endosperm syncytium toward the periphery. At the globular stage of embryo development, the cytoplasm of the micropylar endosperm surrounds the developing embryo. The syncytial cytoplasm of the peripheral endosperm possesses evenly distributed nuclei.

## 26.5.3 Cellularization of Endosperm

In **cellular endosperm** development, the division of the primary endosperm nucleus is followed by subsequent wall formation. Thus, cellularization of endosperm is in place right from the initial stage of endosperm development. However, in nuclear type of endosperm, cellularization of endosperm takes place during later stages of

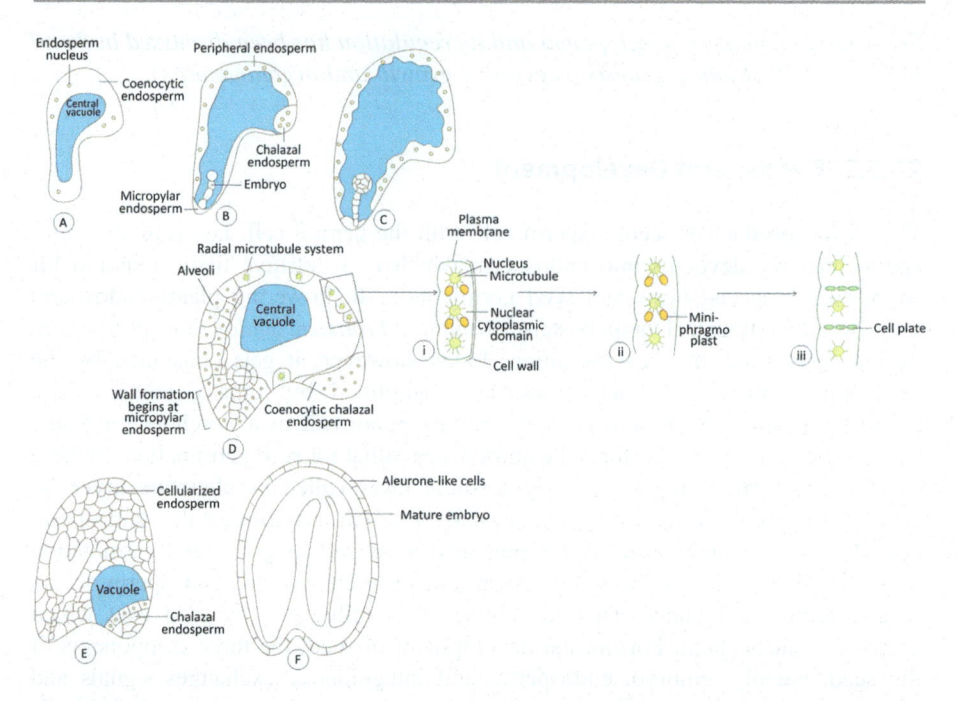

**Fig. 26.14** (**a–c**) Repeated divisions of primary endosperm nucleus without cell wall formation and the migration of free nuclei toward periphery of the coenocytic central cell due to enlargement of central vacuole. (**d**) *i-iii* Formation of cross wall in the peripheral endosperm. (**e–f**) Beginning of endosperm cellularization at the micropylar region and later progress toward chalazal region

development. This process of initiation of cell wall formation begins at the globular stage in *Arabidopsis*. Cellularization of endosperm begins in the micropylar embryo-surrounding region (ESR) and progresses toward the chalazal region. Cellularization occurs via formation of radial microtubule systems (RMS) and alveolation (Fig. 26.14d–f). Initiation of cellularization of the coenocytic endosperm begins with the formation of RMS on the surface of all nuclei. Region of the cytoplasm covered by these arrays around each nucleus is known as **nuclear cytoplasmic domain (NCD)**. Subsequently, microtubules from the adjacent nuclei meet to form interzones in which cell wall material, mainly callose, is deposited. Opposing microtubule arrays from the adjacent NCD are known as **cytoplasmic phragmoplasts**, which mediate the deposition of initial cell wall material. Initially, a tubelike structure, known as **alveolus**, is formed by the cell wall material deposited by cytoplasmic phragmoplasts around each nucleus. The open end of the alveolus faces toward the central vacuole. Later, RMS anchors the nuclei to the central cell wall and extends toward the central vacuole in a canopy of microtubules. The interzones between adjacent canopies of microtubules extend the alveoli toward the central vacuole. Ultimately, after first round of alveoli formation, the nuclei in each alveolus divide periclinally such that the orientation of new cell wall is parallel to central cell wall. These periclinal divisions segregate the alveoli into a peripheral cell and a new alveolus with its opening toward the central vacuole. Repetition of this process 4–5 times results in complete cellularization of the endosperm. At maturity, most of the endosperm is utilized by embryo for its nourishment, and single layer of endosperm is left. This single layer of persistent endosperm is referred to as aleurone layer by analogy to aleurone layer in cereal grains.

Unlike *Arabidopsis*, cellularization of coenocytic endosperm in the case of cereals occurs centripetally. Also, endosperm contributes to a major proportion of mature seed as it is not consumed during embryogenesis. During endosperm development, the triploid PEN undergoes repetitive mitotic divisions which are not accompanied by cell wall formation. This is followed by migration of nuclei toward the periphery of the central cell because of the enlargement of the central vacuole. Similar to *Arabidopsis*, cellularization in cereals proceeds through formation of radially arranged microtubules around each nucleus and alveoli. Formation of anticlinal wall takes place between adjacent nuclei giving rise to tubelike alveolar cells with open end toward the central vacuole. Nucleus of each alveolar cell undergoes 1–2 periclinal divisions accompanied by cytokinesis giving rise to daughter cells. This process takes place in such a manner that the innermost layer of daughter cells remains alveolar in nature. This layer undergoes same round of periclinal division until the cellularization process is complete. After completion of cellularization process, the innermost group of daughter cells undergoes further divisions so that initial cell file pattern is lost. This group of cells is the most important source of starchy endosperm. **Starchy endosperm** constitutes the bulk of endosperm in cereal grains. In addition to starch, starchy endosperm also contains storage proteins stored in protein storage vacuole. For proper development of starchy endosperm, endoreduplication of DNA play a very important role. In maize, endoreduplication leads to up to 96 times increase in DNA content. Often most of the part of cellular endosperm is consumed by the developing embryo. As a result, in a mature seed, embryo occupies the seed completely, and endosperm is represented by one or few layers (i.e., aleurone layer). In maize and wheat, aleurone layer is single layered, whereas barley has three-layered aleurone layer. In rice, the thickness of the aleurone layer varies from one to several layers. In addition to providing nourishment to developing embryo, aleurone layer also acts as major site of mineral storage and confers protection to the nutrient-rich endosperm by expression of stress- and pathogen-protective proteins such as **PR-4 (pathogen-related protein 4)**. During early stages of seedling growth, the aleurone layer mobilizes starch and storage protein from the starchy endosperm by the activity of α-amylase, proteases, and other hydrolases. During seed maturation, the cells of the starchy endosperm undergo programmed cell death. Contrary to this, cells of the aleurone layer survive and acquire desiccation tolerance due to action of abscisic acid. The production of these hydrolytic enzymes takes place in response to gibberellin production by the embryo.

## 26.5.4 Hormonal and Genetic Regulation of Aleurone Development in Cereal Grains

Early stages of aleurone differentiation are influenced by auxin and cytokinin. In maize, the transgene produced by inducing a mutation in the cytokinin biosynthetic enzyme gene-isopentenyl transferase (IPT) results in mosaic aleurone (interspersed patches of aleurone and starchy endosperm) on the crown regions of kernels. However, the role of endogenously produced cytokinin in controlling aleurone differentiation is still not clear. In contrast, auxin positively influences the aleurone layer. In maize, treatment with auxin transport inhibitor, N-1-naphthylphthalamic acid (NPA) leads to production of kernels with multiple-layered aleurone in contrast to single layer in untreated maize. NPA treatment leads to increased accumulation of auxin in the periphery of endosperm. ABA and GA act antagonistically to mediate the later stages of aleurone development. Aleurone maturation is promoted by ABA, whereas germination is promoted by GA.

Several genes implicated in the control of aleurone layer differentiation have been identified. DEFECTIVE KERNEL1 (DEK1) is essential for aleurone cell fate specification. It encodes for a large complex integral membrane protein localized on the plasma membrane. An extracellular loop present in its structure is suggestive of its potential to interact with extracellular molecules as well as signaling ligands. In maize, a loss-of-function mutation in DEK1 gene results in the production of seeds without aleurone layers. In wild-type maize plants, the cells of the aleurone layer are cuboidal and possess dense granular cytoplasm. In *dek1* mutant, the surface cells present in place of aleurone layer are similar to cells of starchy endosperm and also contain starch grains. Mutation in DEK1 gene results in similar effects in *Arabidopsis* and rice seeds. CRINKLY4 (CR4), a receptor-like kinase, also positively regulates the fate of aleurone layer. *Cr4* mutant, homozygous for recessive allele, shows sporadic patches that lack aleurone predominantly on the kernel. Phenotype of *cr4* mutants resembles those of *dek1-D*, a weak allele of *dek1*. Proteins encoded by these two genes are co-localized in the plasma membrane and in endocytic vesicles. Studies of genetic interactions *cr4* and *dek1* have suggested that these two genes function in overlapping biological processes. Mutation in the SUPERNUMERARY ALEURONE1 (SAL1) gene leads to the formation of multiple layer of aleurone in place of single layer indicating its role in negative regulation of aleurone layer. It exhibits resemblance to human CHMP1, a protein involved in vesicle trafficking. SAL1 protein is also co-localized in endocytic vesicles with DEK1 and CR4. Thus, it has been proposed that SAL1 acts as a negative regulator of DEK1 and CR4 by directing their retrograde cycling off the plasma membrane leading to inhibition of their signaling activity. *Des5* mutant in barley possess single layer of aleurone instead of normal three layers along with variations in characteristic features of aleurone cells, such as larger cell, less dense cytoplasmic contents, and thinner anticlinal walls. *Des5* mutants have drastically reduced cr4 transcript levels. However, reduction in dek1 transcripts was not found to be significant. This

**Table 26.8** Mutants defective in aleurone layer differentiation in cereals

| S. No. | Mutant | Plant | Phenotype defect |
|---|---|---|---|
| 1. | Dek1 | Maize, Arabidopsis, rice | Seeds without aleurone layers |
| 2. | Cr4 | Maize | Mosaic aleurone due to failure of endosperm to differentiate aleurone |
| 3. | Sal1 | Maize | Formation of multiple layers of aleurone |
| 4. | Des5 | Barley | Single layer of aleurone characterized by variations in aleurone cells, such as larger cell, less dense cytoplasmic content, and thinner anticlinal walls |

differential expression of CR4 and DEK1 genes in *des5* mutant indicates that these two genes might be independently regulated. Furthermore, expression of the SAL1 gene also decreased in *des5* mutant suggesting that regulation of aleurone cell layer number might be more complex than interplay between cr4 and sal1 functions. Interestingly, relationship between the aleurone layer and epidermis of the leaves has been observed. In maize mutation in CR4 gene not only disturbs the aleurone specification but also disrupts leaf epidermis variously, such as irregular cells with poorly developed cuticle and multiple-layered epidermis. Similarly, weak alleles of dek1 show pronounced effect on the leaf epidermis in maize, rice, and *Arabidopsis*. Table 26.8 summarizes the various mutants defective in aleurone layer differentiation, along with their respective phenotypic characteristics. Quantitative trait locus (QTL) mapping studies in barley have further indicated the possible involvement of some additional factors in determining the number of aleurone layers.

### 26.5.5 Development of Seed Coat

Following fertilization, differentiation of seed coat or testa takes place from the maternally derived ovule integuments over few weeks (~2–3 weeks in case of *Arabidopsis*). Seed coat forms the outer protective covering surrounding the embryo. At the time of anthesis, mature ovule in *Arabidopsis* consists of two layers of outer integument and three layered inner integument. An endothelium apparently does not form. Instead, the inner integument may be only one cell layer at the micropylar end grading to two and ultimately three cells thick at the chalazal end. Due to cell division and expansion, cells present in all the layers of both the integuments undergo dramatic phase of growth in the initial few days following fertilization. Subsequently, the five cell layers constituting both the integuments follow one of the four distinct pathways. The cells constituting the innermost layer of the integument synthesize proanthocyanidins. These compounds, also known as **condensed tannins**, accumulate in the central vacuole during first week following fertilization. Later, these compounds undergo oxidation and impart brown color to the seed coat.

Contrary to this, the remaining two layers of the inner integument do not differentiate and get compressed ultimately. During the initial phase of growth, cells constituting both layers of outer integument accumulate starch in the amyloplasts. Following this, the cells present in the two layers of integuments follow divergent paths. The subepidermal layer differentiates into palisade cells and produces thickened wall on the inner tangential side of the cells. The epidermal layer synthesizes and secretes abundant mucilage into the apoplast region at the junction of the outer tangential and radial walls. This mucilage is chiefly composed of pectin, and its major constituent is rhamnogalacturonan 1(RG1). This mucilage so deposited acts as specialized secondary cell wall. With the progression of mucilage deposition, contraction of vacuole takes place which leads to the formation of cytoplasmic column in the center of the cell, which is surrounded by donut-shaped apoplastic space filled with mucilage. Subsequently, the space occupied by the cytoplasmic column is replaced by the deposition of secondary cell wall material leading to the formation of columella. During later stages of seed development, the cells of all the layers of seed coat undergo programmed cell death. The structure of the outermost epidermal layer is preserved by mucilage and columella.

In flowering plants, development of seed occurs accompanying complex interactions between maternal tissues, embryo, and endosperm. Growth and differentiation of seed coat are initiated by fertilization, and its development is coordinated with the development of embryo and endosperm. It seems that one or more events during or following fertilization must be involved in transmission of signals to the seed coat that coordinates its development with embryo and endosperm. As already mentioned, in *fis* mutants the process of embryogenesis is blocked. but still development of endosperm and seed coat takes place almost normally. Thus, it appears as if a signal from coenocytic endosperm is sufficient for initiation of seed coat development in the integumentary cells. Furthermore, no significant growth of seed coat takes place in seeds where only egg cell is fertilized. Also, seeds in which endosperm is experimentally destroyed showed inhibition of seed coat development. Recent studies using various mutants have shown that endosperm influences both the growth and differentiation of the seed coat. The growth of seed coat is initiated soon after fertilization and involves both cell division and cell elongation. It has been shown that cell elongation, and not cell divisions, is significant for determining the size of seed. For instance, in *Arabidopsis*, mutation in the HAIKU gene is responsible for limited growth of coenocytic endosperm. This defect also affects the growth of the developing seed coat in such a manner that cell elongation in expanding seed coat is restricted. Thus, it has been suggested that the growing endosperm regulates the extent of cell elongation of the integument following initiation of seed coat development. Contrary to this, loss-of-function mutations in TRANSPARENT TESTA GLABRA (TTG2) gene in *Arabidopsis* restrict cell elongation of seed coat which in turn limits endosperm growth. This "crosstalk" between the developing endosperm and seed coat appears to coordinate growth between the endosperm and seed coat, ultimately establishing seed size.

## 26.5.6 Seed Maturation and Desiccation Tolerance

Maturation of seed is the final phase of seed development. Hallmarks of seed maturation include accumulation of storage compounds, acquisition of desiccation tolerance, growth arrest, and entry into dormancy period. In several species, **desiccation tolerance (DT)** is also acquired during maturation phase. DT involves drying of seed due to loss of water from seed through evaporation, and it allows the seed to remain dry for extended periods. Operationally, DT is defined as an organism's ability to dry to equilibrium with moderately dry air (50–70% and relative humidity at 20–30 °C) and resumption of normal function upon rehydration. It is basically tolerance to removal of almost total cellular water and its replacement by molecules that are capable of forming hydrogen bonds and substitute interactions in the absence of water molecules. DT is also correlated with seed longevity which determines the ability of the seed to retain its viability over long period of time in desiccated state. Those seeds which are able to tolerate desiccation and can be stored in a dry state for long periods (time period variable depending on species) are known as **orthodox seeds**. The orthodox seeds of *Phoenix dactylifera* were successfully germinated in 2005 after storage for 2000 years. Orthodox seeds acquire DT during seed development. It is commonly initiated along with accumulation of reserves and acquisition of dormancy and is usually fully established just before the drying phase at the end of seed maturation. Mostly, cultivated crops, such as rice, wheat, corn, and barley, produce desiccation-tolerant seeds. Contrary to this, seeds of some plants, such as mango, litchi, cocoa, avocado, and rubber tree, are shed with high water content and active metabolism. Such seeds, known as **recalcitrant seeds**, undergo deterioration upon drying and are not able to survive storage. Table 26.9 lists some examples of orthodox and recalcitrant seeds.

## 26.5.7 Molecular Basis of Desiccation Tolerance

Acquisition of desiccation tolerance in orthodox seeds is associated with several cellular processes. These include accumulation of disaccharides and oligosaccharides, synthesis of storage proteins, **late embryogenesis abundant (LEA) proteins**, synthesis of **small heat shock proteins (smHSPs)**, activation of antioxidative defense mechanisms, changes in the physical structure of the cell, and gradual increase in density. Variability in desiccation tolerance between different

**Table 26.9** Some examples of orthodox and recalcitrant seeds

| S. No. | Orthodox seeds | Recalcitrant seeds |
|---|---|---|
| 1. | *Citrus aurantifolia* | *Persea americana* |
| 2. | *Capsicum annum* | *Theobroma cacao* |
| 3. | *Hamelia patens* | *Cocos nucifera* |
| 4. | *Lantana camara* | *Artocarpus heterophyllus* |
| 5. | *Psidium guajava* | *Mangifera indica* |
| 6. | *Anacardium occidentale* | *Hevea brasiliensis* |

plant species is due to physical structure of the seed's internal matrix which involves interactions between sugar and protein complexes with salts, organic acids, and amino acids. Seeds do not possess stomata and have thick seed coat due to which uptake of oxygen and its availability for energy production become limited. Limitation of energy during the period of reserve accumulation becomes more problematic during seed desiccation due to increase in viscosity and molecular packing density of the cells. As a result of these changes, the cells of the embryo are transformed into a glassy matrix. Although seed desiccation refers to physical process of drying, the transition from period of reserve accumulation to seed desiccation is related to significant changes in gene expression as well. This indicates that seed desiccation is also quite an active phase in terms of gene expression. In *Arabidopsis*, expression of ~30% of total genome (6963 genes) changes significantly during desiccation period. Out of these, change in expression of ~21% genes (called as early expressed genes) already begins during the phase of reserve accumulation. Expression of ~43% early expressed genes is upregulated during desiccation, and these include HSPs and LEA genes that are associated with desiccation tolerance.

At molecular level, a strong correlation exists between protective mechanisms such as accumulation of LEA, HSPs, nonreducing sugars, and antioxidants that are activated during dehydration and play a central role in desiccation tolerance. In addition to structural and macromolecular protection, LEA proteins are also involved in the formation of intracellular glassy state and stabilization. Along with nonreducing soluble sugars, LEA proteins have been thought to play an important role in the control of viscosity and mobility properties of these biological glasses in dried state. **Biological glasses** can be defined as highly viscous liquid constituent of desiccated seeds with slow diffusion rate, thus limiting their participation in chemical reactions. Production of ROS is considered to be one of the main reasons for the damage caused due to dehydration. Thus, detoxification of ROS is a critical adaptive mechanism in desiccation tolerance. Many molecular antioxidants, such as ascorbate, glutathione, polyols, tocopherols, quinones, flavonoids, and phenolics, are believed to operate during drying and rehydration to alleviate oxidative stress imposed by desiccation. Several physiological and genetic studies have shown that synthesis of LEA proteins, storage proteins and lipids is promoted by ABA. ABA-deficient mutants fail to accumulate these proteins. ABA treatment induces synthesis of LEA proteins in vegetative tissue. ABA induces changes in cellular metabolism by activating a network of transcription factors either partially or indirectly. The expression of ABA-insensitive (ABI) gene, specifically ABI3, induces the synthesis of LEA and storage proteins through interaction with bZIP (basic leucine zipper) transcription factor such as ABI5. ABI5 gene has been reported to play key role by occupying the central position in the genetic regulatory network and is well connected with LEA and other genes implicated in DT. Thus, it has been proposed that ABI3 and ABI5 along with other genes are the core components of seed-specific ABA signaling pathway which regulate survival during desiccated state.

## 26.5 Seed Development

**Summary**

- Development of pollen grains takes place in anther locule as a result of two successive mitotic divisions of microsporocyte. During pollen mitosis I (microsporogenesis), formation of vegetative cell and generative cell takes place which is followed by mitotic division of generative cell to form two sperm cells (microgametogenesis). Structure of pollen wall is quite complex and consists of mainly exine (outer) and intine (inner) layer having a pollen coat or tryphine.
- Formation of egg takes place in the female gametophyte through megasporogenesis followed by megagametogenesis. Majority of angiosperms show *"Polygonum* type" of megagametophyte development in which diploid megaspore mother cell undergoes meiosis to produce four haploid megaspores out of which only one undergoes megagametogenesis.
- Transfer of pollen grains to the stigma surface takes place through various biotic and abiotic agencies. Post pollination, pollen grain germinates to form pollen tube which travels through the female gametophytic tissue to deliver sperm cells inside the embryo sac. Formation of pollen tube takes place only after recognition reaction between pollen and stigma. Pollen tubes grow by tip growth triggered by influx of $Ca^{2+}$. Activation of pollen-expressed receptor-linked kinases by a stigma-expressed unidentified ligand regulates a GTPase switch which enables polar cell expansion of the pollen tube. Directional growth of the pollen tube inside female gametophytic tissue is determined by physical and chemical cues from the pistil and embryo sac.
- After entry of pollen tube inside the embryo sac through micropylar end, the two sperm cells are delivered inside the embryo sac which fertilizes the egg cell and central cell. Double fertilization leads to the formation of diploid zygote (fusion product of sperm cell and egg cell) and primary endosperm nucleus (fusion product of sperm cell and central cell).
- Self-incompatibility (SI) prevents self-fertilization in flowering plants and is determined by S-locus having several alleles. Rejection of self-pollen at the stigma surface is mediated by interaction between S-locus receptor kinase (expressed in stigmatic papillae) and S-locus cysteine-rich protein (expressed in pollen coat). Gametophytic self-incompatibility (GSI), characterized by rejection of pollen tube in the upper part of style, is determined by the genotype of the pollen itself.
- Development of seed includes embryogenesis (development of embryo), development of endosperm, and desiccation tolerance. During embryogenesis diploid zygote undergoes divisions in various planes that eventually gives rise to a mature embryo with well differentiated parts, namely, cotyledon(s), plumule, and radical. Postfertilization, primary endosperm nucleus develops into endosperm which becomes multinucleate (coenocyte) and provides nourishment to the developing embryo. Cellularization of the coenocytic endosperm begins from the micropylar to the chalazal region whereas in cereals cellularization takes place centripetally.

Endosperm development is repressed until after fertilization by FIS proteins which are responsible for methylation and demethylation of DNA and histones in the endosperm. Differentiation of aleurone layer takes place from starchy endosperm cells, and few genes, such as DEK1, CR4, SAL1, and DES5, have been implicated. However, the exact mechanism is yet not understood. Seed coat develops from integuments, and its development is regulated by endosperm.
- Acquisition of desiccation tolerance is facilitated by expression of LEA proteins which are implicated in the formation of glassy state that confers stability to dessicated seed. Synthesis of LEA proteins is regulated by abscisic acid.

## Multiple-Choice Questions

1. The term "xenogamy" refers to:
   (a) Transfer of pollen on the stigma surface of the same flower
   (b) Transfer of pollen on the stigma surface of another flower present on the same plant
   (c) Transfer of pollen on the stigma surface of flower present on the different plant
   (d) Fusion of male gamete with the egg in the pistil present on the different plant
2. Which of the following type of stigma surface has not been characterized in flowering plants?
   (a) Dry type
   (b) Wet type
   (c) Semidry type
   (d) Semi-wet type
3. The most common type of embryo sac development observed in angiosperm species is:
   (a) *Oenothera* type
   (b) *Plumbago* type
   (c) *Plumbagella* type
   (d) *Polygonum* type
4. Which of the following statements is incorrect in context of pollen hydration and subsequent germination?
   (a) Influx of $Ca^{2+}$ into vegetative cells.
   (b) Concentration of free cytosolic $Ca^{2+}$ [$Ca^{2+}_{cyt}$] increases below germ pore region after pollen hydration.
   (c) Accumulation of actin microfilaments and secretory vesicles takes place below germ pore area prior to pollen tube emergence.
   (d) Following pollen germination, the elongation of pollen tube occurs by lateral growth.

5. The directional growth of pollen tube in pistil is not regulated by:
   (a) Plantacyanins secreted by the cells which line the transmitting tissue
   (b) Stigma/style cysteine-rich adhesin (SCA) secreted by the transmitting tissue
   (c) Cysteine-rich proteins (CRPs) derived from the transmitting tissue
   (d) Extracellular matrix (ECM) of the transmitting tissue
6. Which of the following is not a pre-zygotic barrier to self-fertilization?
   (a) Syngamy
   (b) Dichogamy
   (c) Herkogamy
   (d) Heterostyly
7. Which of the following mechanism is incorrect in context of SI mechanisms?
   (a) In sporophytic SI (SSI) systems, the rejection of self-pollen takes place on stigma surface due to interaction between S-receptor kinase (SRK) and S-locus cysteine rich (SCR) protein.
   (b) In some species exhibiting SSI, the programmed cell death of the pollen tubes takes place after penetration into stigma surface.
   (c) In some species exhibiting gametophytic SI (GSI), the programmed cell death of the pollen tubes takes place after penetration into stigma surface.
   (d) In some species exhibiting GSI, bursting of pollen tubes at the upper region of style due to responses mediated by stylar S-locus ribonucleases (S-RNases).
8. The most common type of endosperm development observed in angiosperms is:
   (a) Nuclear endosperm type
   (b) Cellular endosperm type
   (c) Helobial endosperm type
   (d) Ruminate endosperm type
9. Mutation in Dek1 gene results in:
   (a) Mosaic aleurone due to failure of endosperm to differentiate aleurone
   (b) Formation of multiple layers of aleurone
   (c) Seeds without aleurone layers
   (d) Single layer of aleurone
10. Which of the following species have recalcitrant seeds?
    (a) *Hamelia patens*
    (e) *Mangifera indica*
    (f) *Lantana camara*
    (g) *Capsicum annuum*

**Answers**

1. c   2. d   3. b   4. d   5. c   6. a   7. b
8. a   9. c   10. b

## Suggested Further Readings

Angelovici R, Galili G, Fernie AR, Fait A (2010) Seed desiccation: a bridge between maturation and germination. Trends Plant Sci 15(4):211–219

Hafidh S, Fíla J, Honys D (2016) Male gametophyte development and function in angiosperms: a general concept. Plant Reprod 29(1–2):31–51

Rea AC, Nasrallah JB (2004) Self-incompatibility systems: barriers to self-fertilization in flowering plants. Int J Dev Biol 52(5–6):627–636

# Fruit Development and Ripening

## 27

Rashmi Shakya and Manju A. Lal

Formation of fruits is a characteristic feature of angiosperms. Fruit types exhibit a great deal of diversity. They form an integral component of diet and provide not only vitamins and minerals but are also important source of antioxidants and fibers. Nutritional quality of the fruit is determined by various factors including those affecting the ripening stage. A fruit must have an optimum degree of ripeness so as to be consumed. Fruits are developed from less attractive immature stages to mature stages which attract seed-dispersing animals as well as human beings. Fruit ripening is a complex process, accomplished through several physiological, biochemical, and molecular mechanisms. These mechanisms also bring about changes in pigmentation due to loss of chlorophylls and a substantial increase in non-photosynthetic pigments, such as anthocyanins and carotenoids. Increase in the activity of cell wall hydrolases leads to fruit softening which is reflected in the texture of the mature fruit. Ripe fruits possess characteristic taste (due to elevated levels of sugars and depletion of organic acids) and aroma (due to biosynthesis of volatile compounds). Postharvest handling of the overripe fruits, however, is very difficult and adversely affects their storage and marketing.

This chapter begins with a brief account of different phases of fruit development. Detailed account of various physiological processes that take place during fruit ripening process, such as change in the color of fruit, change in cell wall architecture, intensification of flavors and fragrances, and synthesis of various bioactive compounds, has subsequently been presented. Both climacteric and non-climacteric patterns of fruit ripening have been dealt with to highlight the differences and common aspects among them. The role of ethylene and other phytohormones in regulating the ripening process has also been discussed. Genes implicated in the biosynthesis of ethylene and oxidative processes, especially those related to reactive oxygen species (ROS) have also been discussed. Fruit ripening is regulated by several genes that play a significant role in accumulation of antioxidants, sugars, and bioactive compounds, release of volatile compounds, and softening of cell wall. The role of transcription factors and genes functioning as

© Springer Nature Singapore Pte Ltd. 2018
S. C Bhatla, M. A. Lal, *Plant Physiology, Development and Metabolism*,
https://doi.org/10.1007/978-981-13-2023-1_27

master switches in fruit ripening has also been explained. Epigenetic studies on *colorless non-ripening (CNR)* mutant of the model system-tomato have been highlighted to discuss the importance of these modifications in affecting fruit phenotypes.

## 27.1 Stages of Fruit Development and Ripening Stages

Botanically, a fruit is defined as "a seed receptacle developed from an ovary." Many fruits arise from several ovaries, and sometimes associated floral parts are also involved in fruit development after fertilization. Development of fruits is an evolutionary adaptation which helps in seed dispersal through various agencies. Fruit development shows considerable variations and ranges from simple to aggregate and multiple fruits. Simple fruits develop from single ovary, whereas multiple fruits develop from fused ovaries of a single flower (Fig. 27.1). Development of aggregate fruits takes place from several single ovaries of a single flower, whereas multiple fruits develop from ovaries of several flowers in an inflorescence. Simple fruits can be either fleshy (e.g., tomato and apple) or dry, such as that of poppy. Aggregate and multiple fruits are mostly fleshy and sometimes, associated structures, such as a receptacle, become part of the fruit. Berries, such as tomato, possess numerous small seeds within the locule (Fig. 27.2). Drupes (e.g., peach) possess a single large seed which is surrounded by a fleshy mesocarp. In some varieties, such as those of banana, pineapple, and cucumber, the development of fruit takes place without prior fertilization, thereby resulting in seedless fruits (Box 27.1). A wide range of fruit types represent different morphologies which, to an extent, are associated with different metabolic pathways accompanying their maturation. However, many of the biochemical pathways involved are common to almost all fruits. Most of the physiological and biochemical studies on fruit development and ripening have been dealt with fleshy fruits. The pattern of growth curve of fruit development has been reported to be either sigmoidal or double sigmoidal in the majority of flowering plants. The double-sigmoidal curve is obtained sometimes due to the second burst of growth. On the basis of physiological and biochemical activities, fruit development process can be identified in four phases (Table 27.1). It is the physiological and biochemical activities occurring along the course of development which are responsible for the characteristic ripening of fruits. Ripening is the terminal stage of fruit development which is characterized by a change in size, color, and aroma of the fruit. Different ripening stages in model climacteric and non-climacteric fruits, i.e., tomato and strawberry, respectively, have been shown in Boxes 27.2 and 27.3. Table 27.1 highlights the events associated with fruit development and ripening. Phase I of fruit development is characterized by anther dehiscence and subsequent release of pollen grains. Phase I is known as fruit set which decides whether to abort or to go ahead with ovary development to form fruit. Phase II is marked by rapid divisions of cells. Since cessation of cell division takes place in phase III, fruit growth is mostly attributed to cell enlargement. Attainment of final fruit size and shape also occurs

## 27.1 Stages of Fruit Development and Ripening Stages

**A**      **B**      **C**

— leafy bracts

— individual fruits (berries)

Simple fruit    Aggregate fruit    Multiple fruit

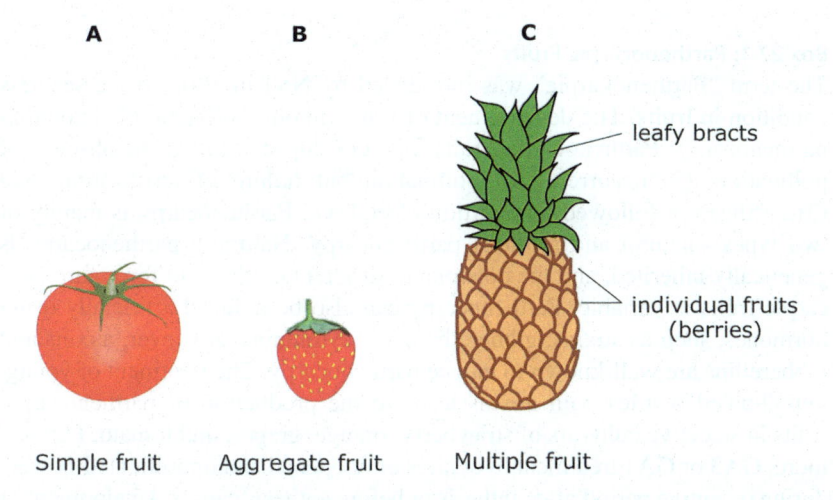

**Fig. 27.1** Different types of fruits. (**a**) A simple fruit develops from a single ovary (tomato). (**b**) An aggregate fruit develops from several individual ovaries of a single flower (strawberry). (**c**) A multiple fruit develops from fused ovaries of several flowers (pineapple)

**A**

— Locules
— Exocarp ⎤
— Mesocarp ⎬ Pericarp
— Endocarp ⎦
— Placenta
— Seed

**Berry (tomato)**

**B**

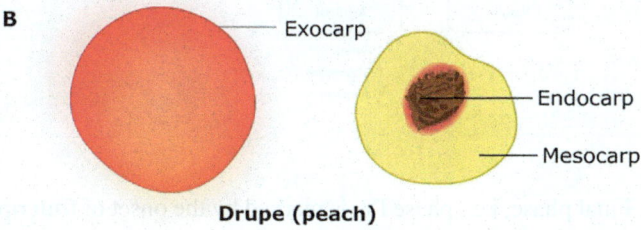

— Exocarp
— Endocarp
— Mesocarp

**Drupe (peach)**

**Fig. 27.2** (**a**) A berry has numerous small seeds within the locule (Tomato). (**b**) A drupe possesses a single large seed surrounded by fleshy mesocarp (Peach)

**Box 27.1: Parthenocarpic Fruits**
The term "Parthenokarpie" was introduced by Noll in 1902 to the seedless condition in fruits. The development of fruit without fertilization is known as parthenocarpy. Parthenocarpic fruits may develop due to (1) the absence of pollination, (2) occurrence of pollination but failure of fertilization, and (3) fertilization followed by abortion of embryo. Parthenocarpy is mainly of two types—natural and artificial parthenocarpy. Naturally parthenocarpy is genetically inherited, and the potential of different cultivars to form parthenocarpic fruits is variable. Parthenocarpy can also be induced artificially using hormones, such as auxins, gibberellins, and cytokinins. However, auxins and gibberellins are well-known to induce parthenocarpy. The treatment of young, unpollinated ovaries with auxins leads to the production of parthenocarpic fruits in selective cultivars of strawberry, orange, grapes, and tomato. Furthermore, GA3 or GA1 treatment to ovaries of unopened and unpollinated flowers during receptive period after anthesis or before anthesis causes development of parthenocarpic fruits in certain cultivars of tomatoes, blueberries, and *Arabidopsis*. The phenomenon of parthenocarpy is popular among horticulturists mainly for crops where seedless fruits are desirable, such as citrus, grapes, and bananas for the preparation of eatables like jam, jellies, sauces, etc. Also, seedless fruit has a longer shelf life than fruits with seeds because seeds produce hormones that trigger senescence. However, parthenocarpy has limited application in agriculture due to pleiotropic effects and occurrence of unfavorable characteristics, such as small fruits.

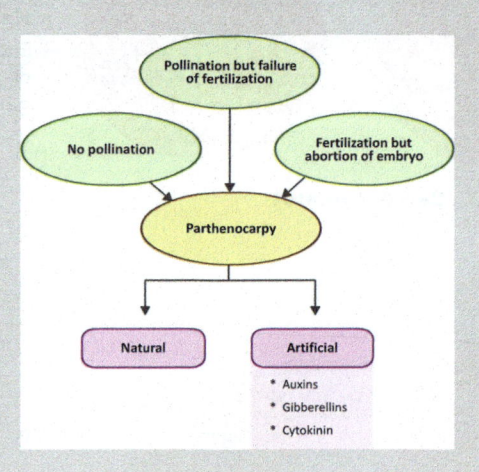

during this phase. Final phase, i.e., phase IV, is marked by the onset of fruit ripening. Sometimes, there could be a second burst of growth in some species which is responsible for getting double-sigmoidal growth curve.

## 27.1 Stages of Fruit Development and Ripening Stages

**Table 27.1** Events associated with different phases of fruit development and ripening

| Phase of fruit development | Events | Activity |
|---|---|---|
| Phase I | Fruit set | Ovary development, anthesis, and, following fertilization, decision whether to abort or to go ahead with fruit development |
| Phase II | Cell divisions | Embryo growth marked by rapid cell divisions |
| Phase III | Cell expansion | Cessation of cell division and cell enlargement, leading to attainment of final fruit shape and size |
| Phase IV | Onset of fruit ripening | Enhanced biosynthesis of carotenoid; changes in color, texture, and taste; second burst of growth in some species |

---

**Box 27.2: Tomato as a Model System for Understanding Fruit Development and Ripening**

Tomato (*Solanum lycopersicum*), belonging to Solanaceae family, has been considered a favorite model for understanding the physiology of fruit development and subsequent ripening. The pigmentation of tomato fruits is the most important external characteristic feature to assess its degree of ripeness and postharvest life. Red pigment accumulation is the result of chlorophyll degradation accompanied with the synthesis of lycopene and other carotenoids, as chloroplasts are converted into chromoplasts. On the basis of external color, six ripening stages of fruit ripening have been recognized in tomato:

I. Stage 1 or "Green stage" is characterized by 100% green pigmented fruit.
II. Stage 2 or "Breaker stage" is characterized by a noticeable break or change in color with less than 10% of nongreen color.
III. Stage 3 or "Turning stage" brings about 10–30% change in surface pigment to red.
IV. Stage 4 or "Pink stage." At this stage, about 30–60% fruit surface is red(ish).
V. Stage 5 or "Light red stage" is characterized by 60–90% red surface of the fruit.
VI. Stage 6 or "Red stage." During this stage, more than 90% of fruit surface appears red.

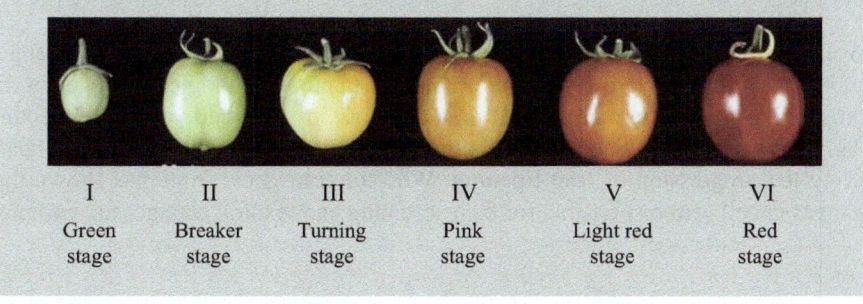

| I | II | III | IV | V | VI |
|---|---|---|---|---|---|
| Green stage | Breaker stage | Turning stage | Pink stage | Light red stage | Red stage |

> **Box 27.3: Strawberry: Prime Model for Non-climacteric Fruit Ripening**
> Studies in ripening of non-climacteric fruits are less in comparison to those done in climacteric fruits. Significant research has been done on some non-climacteric fruits, such as strawberry, citrus, and grapes. Strawberry is considered as the model of choice for understanding ripening in non-climacteric fruits. Some of the changes in strawberry ripening are common in climacteric fruits. This includes loss of chlorophyll and accumulation of anthocyanins, sugars, and volatiles. Biochemical analyses have revealed that ethylene concentration is relatively high at green stage, decreases in white fruits, and increases again at the red ripening stage. Increase in ethylene concentration is accompanied by enhancement in respiration rate similar to that in climacteric fruits before the onset of ripening. Exogenous application of ethylene causes downregulation of several cell wall-related genes for the activity of β-galactosidase, pectin methylesterase or β-xylosidase, and some more. Interestingly, expression of expansin gene FaEXP2 does not change. Transcriptomic and metabolomic studies in transgenic strawberries have shown the requirement of ethylene action during ripening.
>
> Relative differences in the concentration of endogenous ethylene during different developmental stages in strawberry. A. Green stage. B. White stage and C. Mature ripe red stage.

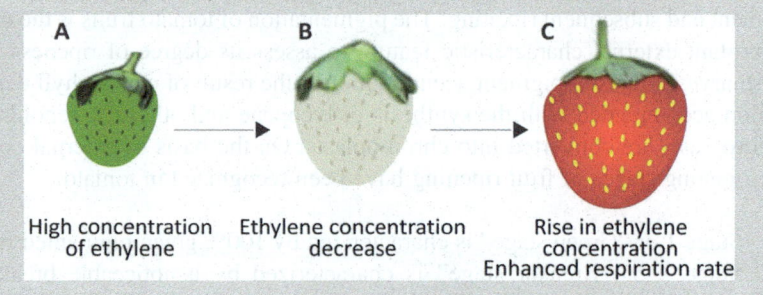

## 27.2 Physiological Changes During Fruit Ripening

### 27.2.1 Fruit Texture and Softening

Several metabolic events contribute to changes in the texture of fruits. These changes involve loss of turgor pressure, degradation of starch, physiological changes in the composition of membranes, and modifications in the cell wall architecture (Fig. 27.3). Changes in cell wall composition are considered highly significant during the progression of fruit ripening. Work on transgenic fruits has shown that changes in cell wall responsible for fruit softening and textural changes are complex

## 27.2 Physiological Changes During Fruit Ripening

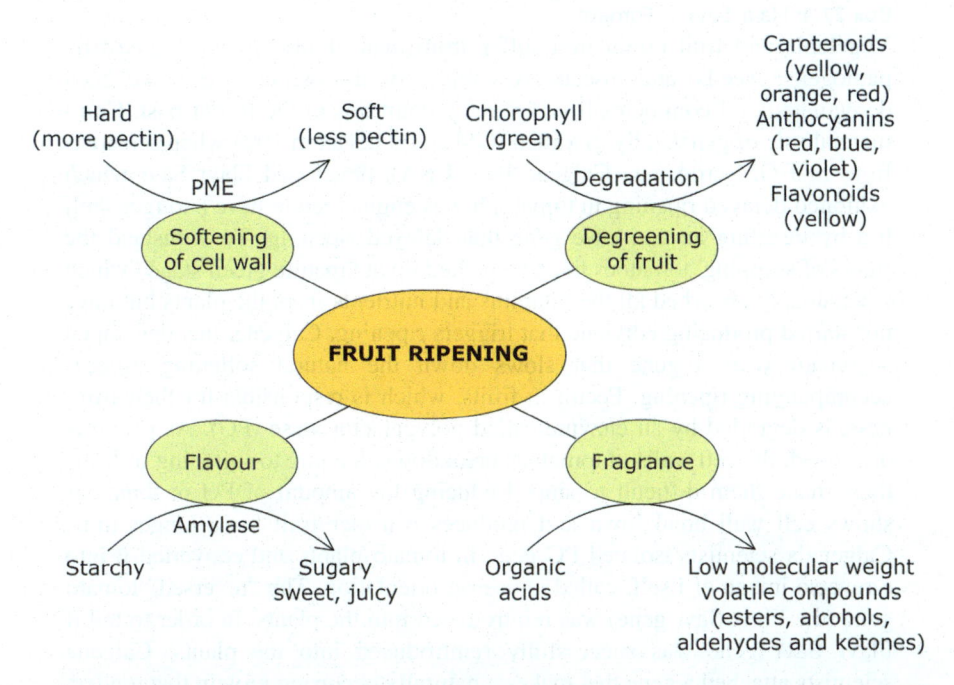

**Fig. 27.3** Different physiological changes that take place during fruit ripening process. PME, pectin methyl esterase

(Box 27.4). Coordination and synergistic action of a variety of cell wall-modifying enzymes are responsible for these changes. Structural changes in the cell wall are associated with the dissolution of middle lamella and disruption of the primary cell wall which leads to loss of firmness in mature fruits. These include depolymerization and solubilization of the polysaccharides and pectin components along with rearrangements of their associations. Modifications of polysaccharides may occur depending on the type of fruit. Table 27.2 summarizes the major modifications in the cell wall polysaccharides during ripening in some climacteric and non-climacteric fruits. In tomato, cell wall modifications include both depolymerization and solubilization of polyuronides and hemicelluloses. Cell wall-modifying enzymes are broadly categorized as pectolytic and non-pectolytic, depending on the specific polysaccharide used as substrate. Endo- and exo-polygalacturonases, pectin acetylesterases (PAE), pectate lyases (PL), pectin methylesterases (PME), β-galactosidases, and α-L-arabinofuranosidases are pectolytic enzymes. The activity of these enzymes leads to cleavage or modification of the polysaccharide backbone and removal of neutral sugars from the branched side chains. Removal of methyl ester groups from the fruit cell wall by PME facilitates access of polygalacturonase (PG) to its substrate. PME is expressed before ripening and is downregulated by ethylene during ripening. Repression of PME by antisense resulted in increased fruit viscosity. Non-pectolytic enzymes, such as endo-1, 4-β-glucanases (EGase), endo-1,4-β-xylanases, β-xylanases, xyloglucan endotransglucosylase/hydrolases (XTH),

**Box 27.4: Flavr Savr™ Tomato**

Vegetable and fruit production suffer from many biotic stresses caused by pathogens, weeds, and insects. A number of transgenic fruits have been deregulated and commercialized in many countries in the recent past. Commercial sale of genetically modified (GM) food began in 1994 when scientists from first GM crop, i.e., Calgene Inc. (USA), developed Flavr Savr which exhibited delayed ripening in tomato. It was engineered to have a longer shelf life by inserting an antisense gene that delayed ripening. To withstand the rigors of shipping, tomatoes must be picked up at "mature green stage" which have already absorbed all the vitamins and nutrients from the plants but have not started producing ethylene that triggers ripening. Calgene, Inc. developed a tomato with a gene that slows down the natural softening process accompanying ripening. Pectin in fruits, which is responsible for their firmness, is degraded by an enzyme called polygalacturonase (PG). As pectin is destroyed, the cell walls of tomatoes breakdown, leading to softening of fruits thus, make them difficult to ship. Reducing the amount of PG in tomatoes shows cell wall breakdown and produces a firmer fruit for a longer time. Calgene's scientists isolated PG gene in tomato plants and converted it into a reverse image of itself, called antisense orientation. The "reversed" tomato gene (the Flavr Savr gene) was reintroduced into the plants. In order to tell if Flavr Savr gene was successfully reintroduced into the plants, Calgene scientists attached a gene that makes a naturally occurring protein that renders plants resistant to the antibiotic kanamycin. By exposing the plants to the antibiotic, Calgene scientists could tell which plant had accepted that Flavr Savr gene. The ones unaffected by kanamycin grow to have designed traits of Flavr Savr. Once in a tomato plant, the Flavr Savr gene attaches itself to the PG gene.

Developing Flavr Savr Tomatoes

With the Flavr Savr gene adhering to it, the PG gene cannot give necessary signals to produce PG enzyme. Thus, these tomatoes retain their flavor. In 1994, US Food and Drug Administration (FDA) announced that Flavr Savr tomatoes are as safe as conventional tomatoes.

## 27.2 Physiological Changes During Fruit Ripening

**Table 27.2** Some climacteric and non-climacteric fruits which undergo polysaccharide depolymerization during ripening

| S. No. | Fruit | Climacteric/non-climacteric | Depolymerization of Pectin | Depolymerization of Hemicellulose |
|---|---|---|---|---|
| 1. | Tomato (*Solanum lycopersicum*) | Climacteric | Yes | Yes |
| 2. | Apple (*Pyrus malus*) | Climacteric | No | No |
| 3. | Strawberry (*Fragaria vesca*) | Non-climacteric | No | Yes |
| 4. | Grapes (*Vitis vinifera*) | Non-climacteric | Yes | Yes |
| 5. | Kiwifruit (*Actinidia deliciosa*) | Climacteric | Yes | Yes |

and expansins, are responsible for hemicellulose modifications. In tomato, softening and ripening are accompanied by about 2000-fold increase in PG mRNA. During softening in apples, cell wall polysaccharides do not undergo extensive depolymerization.

### 27.2.2 Changes in Pigmentation

Along with texture, fruit ripening is often accompanied by changes in flavor and aroma. During ripening, fruits exhibit transition from green to a variety of pigmentations, including red, yellow, orange, purple, and blue. The pigments involved in this process not only increase the visual appeal of the fruit but are also crucial for the taste and aroma. Generally, fruits contain a variety of pigments which includes chlorophyll (green color pigment), carotenoids (yellow, orange, and red colors), anthocyanins (red, blue, and violet colors), and flavonoids (yellow pigment). Degradation of chloroplasts takes place at the onset of ripening which is responsible for degreening of developing fruit. Chloroplasts are converted into chromoplasts because of degradation of chlorophyll and act as the sites for the accumulation of carotenoids. Chromoplasts are nongreen plastids that are responsible for the yellow, orange, and red colors of many fruits. They evolve during fruit ripening by the differentiation of other forms of plastids. In many fruits, such as tomato and bell pepper (*Capsicum annuum*), chromoplasts are derived from chloroplasts with the disintegration of the thylakoid membranes and the formation of new carotenoid-bearing structures. The colors of different kinds of bell pepper are a consequence of selection arrest of chloroplast to chromoplast transition at different stages (Box 27.5). Protein tyrosine nitration is also enhanced during fruit ripening in capsicum (Box 27.6). In other fruits, such as papaya, chromoplasts evolve from leucoplasts or proplastids, as no intermediary amyloplast or chloroplast transitions are encountered. The precursors for carotenoid biosynthesis are synthesized in plastids. The formation of the colorless molecule-phytoene, catalyzed by the enzyme phytoene synthase, is

**Box 27.5: Antioxidant Activity of Different Colored Bell Pepper Fruits**

Bell peppers (*Capsicum annuum* L.) are commonly available in green, yellow, orange, and red color. Seed source of all the bell peppers is the same, i.e., *Capsicum annuum*, but they have different characteristics depending upon the special variety of the plant. The two major factors responsible for the difference in a bell pepper's color are its time of harvesting and degree of ripening. A green bell pepper usually matures into a yellow/orange bell pepper and then grows on to gain red color. However, yellow, orange, and red bell peppers are always more ripe than green ones. They are costlier than green bell peppers because they require more time in the ground before they can be harvested. The green color of bell pepper is due to chlorophyll and the carotenoids typical of the chlorophyll. The yellow and orange colors of pepper are generally due to α- and β-carotene, zeaxanthin, lutein and β-cryptoxanthin. The red color is due to the carotenoids, such as capsanthin, capsorubin and capsanthin 5,6-epoxide. Differences exist not only in their colors but also in their nutrient compositions. Bell peppers are good sources of vitamins C and E, provitamin A and carotenoids. Additionally, they contain various antioxidants, such as phenolics and flavonoids.

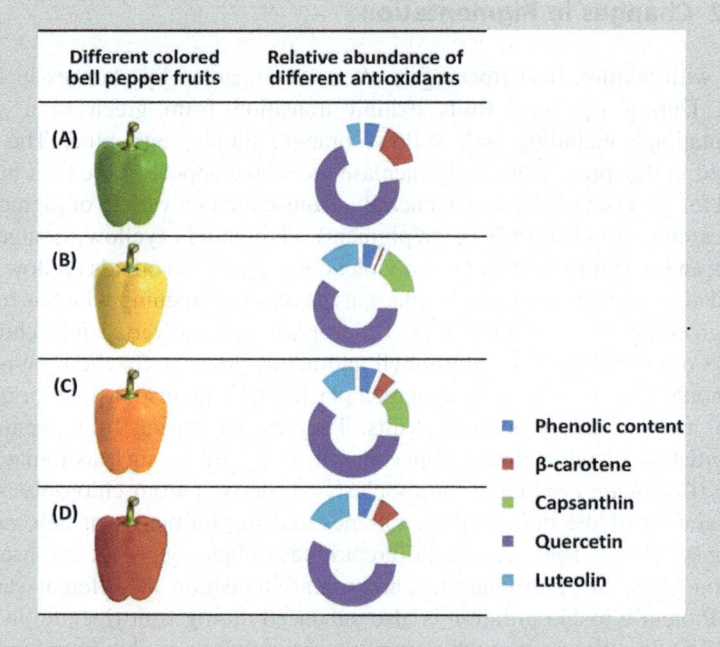

The figure depicts the relative differences in the abundance of various antioxidants in green, yellow, orange and red colored bell pepper fruits.

## Box 27.6: Enhancement of Protein Tyrosine Nitration Characterizes Fruit Ripening In Capsicum

Fruits of *Capsicum annuum* usually ripe within 3–8 days depending on the variety. After the attainment of the breaker point stage (BP stage), the process of ripening proceeds irreversibly in one direction. Exogenous application of NO gas initially delays fruit ripening without promoting protein nitration. However, once the early effect of NO is over as it is a highly reactive radical, it promotes nitration of tyrosine residues of certain proteins. The process of fruit ripening proceeds simultaneously with the conversion of protein tyrosine into 3-nitrotyrosine which involves peroxynitrite through the reaction of NO and superoxide radicals.

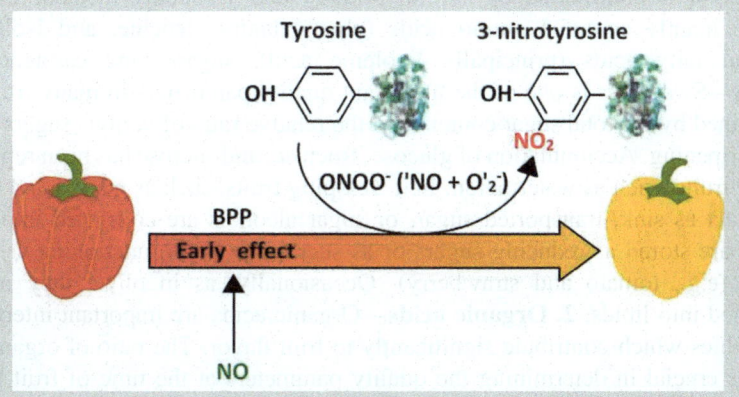

Model depicting proposed relationship between reactive nitrogen species (RNS) and the ripening of *Capsicum annuum* fruit.

the first committed step in this process. Phytoene synthase catalyzes a two-step reaction which results in the formation of phytoene (Eqs. 27.1 and 27.2). Through a series of reactions, phytoene is converted into red carotenoid-lycopene (Eqs. 27.3 and 27.4). During ripening in tomato, carotenoid concentration increases about 10–14 times due to the accumulation of lycopene in tomato. In fact, blueberries are known to accumulate more than a dozen different anthocyanins during ripening. These anthocyanins include malvidin-, delphinidin-, petunidin-, cyanidin-, and peonidin-glycosides. The synthesis of anthocyanins takes place through the phenylpropanoid pathway.

$$2\text{Geranylgeranyl diphosphate} \xrightarrow{\text{phytoene synthase}} PP_i + \text{prephytoene diphosphate} \tag{27.1}$$

$$\text{Prephytoene diphosphate} \xrightarrow{\text{phytoene synthase}} \text{phytoene} + PP_i \qquad (27.2)$$

$$\text{Phytoene} \xrightarrow{\text{phytoene desaturase}} \zeta - \text{carotene} \qquad (27.3)$$

$$\zeta - \text{carotene} + 2PQH_2 + 2O_2 \xleftrightarrow{\zeta-\text{carotene desaturase}} \text{Lycopene} + 4H_2O + 2PQ \qquad (27.4)$$

### 27.2.3 Flavor and Fragrance

Fruit flavor is contributed by several primary and secondary metabolites which include mainly essential amino acids (phenylalanine, leucine, and isoleucine), essential fatty acids (principally linolenic acid), sugars, and carotenoids. **1. Sugars**—Sweetness is one of the important quality parameters in many fruits. It is determined by the total sugar content and the relative ratio of various sugars present during ripening. Accumulation of glucose, fructose, and sucrose has been reported in several fruits, such as watermelon. In developing fruits, such as mango and banana, which act as sink, transported sugar, or sugar alcohols are converted into starch. Sugars are stored as reducing sugars or as sucrose in fruits functioning as storage organs (e.g., tomato and strawberry). Occasionally, as in olive, they may get converted into lipids. **2. Organic acids**—Organic acids are important intermediate metabolites which contribute significantly to fruit flavor. The ratio of organic acid: sugar is crucial in determining the quality parameters at the time of fruit harvest. Primary metabolic pathways, such as glycolysis and the tricarboxylic acid (TCA) cycle, determine the content of organic acids. Citrate and malate are the main organic acids, whereas oxalate, succinate, isocitrate, fumarate, and aconitate are minor ones. **3. Volatile compounds**—Several volatile compounds are produced during ripening process in fruits. Differences in their relative abundance exist. For instance, hexanal is usually present at about few micrograms level per gram fresh weight, whereas β-ionone is present in nanograms per gram. In ripe tomato, approximately 400 volatile compounds have been reported. Some of the volatiles are salicylaldehyde, guaiacol, eugenol, toluene, phenol, styrene, etc. Mostly, important volatile compounds related to flavor are derivatives of essential nutrients, such as phenylalanine, leucine, isoleucine, or linolenic acid.

### 27.2.4 Antioxidants and Bioactive Compounds

Fruits are rich sources of phytochemical compounds (also called as bioactive compounds) and are considered to be beneficial sources of antioxidants. Numerous bioactive compounds have been shown to play roles in preventing or ameliorating various chronic human diseases. Phytochemicals present in fruits have variable chemical structures and functions, and they are categorized into phenolic

## 27.3 Climacteric and Non-climacteric Fruit Ripening

**Phenolic compounds**
- Phenolic acids
- Flavonoids
- Lignans, tannins, coumarins

**Vitamin E**
- Tocopherols
- Tocotrienols

**Carotenoids**
- Carotenes
- Oxygenated derivatives

**Bioactive compounds**

**Vitamin C**
- Ascorbic acid
- Dehydroascorbic acid

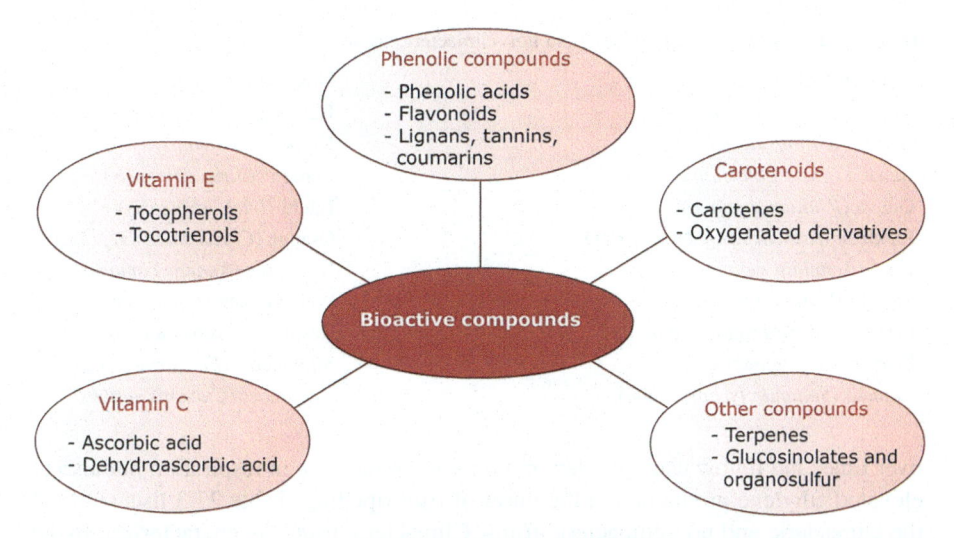

**Other compounds**
- Terpenes
- Glucosinolates and organosulfur

**Fig. 27.4** Different categories of bioactive compounds reported during fruit ripening

compounds, carotenoids, and vitamins (C and E) (Fig. 27.4). Major antioxidant and bioactive compounds found in the mature fruits include tocopherols (vitamin E), phenolic acid, flavonoids, tannins, coumarins, carotenoids, ascorbic acid, and terpenes. Noticeable differences in the antioxidant properties of tomato have been reported in different developmental stages. Phenolic compounds encompass a wide diversity of compounds. Mostly, they are derived from phenylalanine and tyrosine and exhibit variable degrees of hydroxylation, methoxylation, and glycosylation. Various factors influence the content of phenolic compounds in fruits. The most common phenolic acids in fruits are caffeic and gallic acid, and thus the total phenol content is usually determined in terms of caffeic or gallic acid. Flavonoids include flavonols, flavones, isoflavones, flavanols, flavanones, proanthocyanidins and anthocyanidins. Rutin, luteolin, and apigenin are the most common flavones. Quercetin, kaempferol, catechin, and epicatechin are typical flavonols present in fruits. Endogenous antioxidant activity is improved by phenolic compounds. They also protect the developing fruit against oxidative damage caused by free radicals. For this reason, fruits containing phenolics are generally associated with health-promoting properties and prevention of several degenerative diseases.

## 27.3 Climacteric and Non-climacteric Fruit Ripening

Two different types of fruit ripening patterns have been identified on the basis of the presence or absence of a characteristic rise in the rate of respiration associated with ripening. These are referred as climacteric and non-climacteric types, respectively. The term "climacteric" was used to describe the rapid increase in respiration associated with fruit maturation. However, the definition of climacteric has changed

**Table 27.3** Some of the climacteric and non-climacteric fruits

| Climacteric fruit | Non-climacteric fruit |
| --- | --- |
| Apple (*Malus domestica*) | Bell pepper (*Capsicum annuum*) |
| Apricot (*Prunus armeniaca*) | Grapes (*Vitis vinifera*) |
| Banana (*Musa balbisiana*) | Lemon (*Citrus chinensis*) |
| Guava (*Psidium guajava*) | Litchi (*Litchi chinensis*) |
| Jackfruit (*Artocarpus heterophyllus*) | Orange (*Citrus sinensis*) |
| Kiwi (*Actinidia deliciosa*) | Pineapple (*Ananas comosus*) |
| Papaya (*Carica papaya*) | Pomegranate (*Punica granatum*) |
| Passion fruit (*Passiflora edulis*) | Raspberry (*Rubus idaeus*) |
| Pear (*Pyrus communis*) | Strawberry (*Fragaria vesca*) |
| Tomato (*Solanum lycopersicum*) | Cucumber (*Cucumis sativus*) |

over time, and it now usually refers not only to an increase in respiration but also to elevated ethylene production at the onset of fruit ripening. Table 27.3 lists some of the climacteric and non-climacteric fruits. Climacteric fruits are characterized by an increase in respiration with a concomitant and rapid production of ethylene with the initiation of ripening. The considerable increase in ethylene production may take place before or just after the respiratory peak. In non-climacteric fruits, no rise in respiration rate is observed during ripening, and the process of ripening proceeds relatively slowly. Furthermore, climacteric fruits ripen even if harvested at an immature stage in contrast to non-climacteric fruits which do not ripe when plucked unripe. Figure 27.5 illustrates the comparative account of climacteric and non-climacteric patterns of fruit ripening processes in terms of ethylene production, $CO_2$ evolution, and the physical parameters, such as color and firmness of fruits. Tomato is one of the most widely studied model systems for fruit ripening in climacteric fruits (Box 27.2). In climacteric fruits, ethylene plays an important role in the fruit ripening process. In such fruits, initiation of ripening is coupled with a burst of ethylene production which further triggers its autocatalytic production. Subsequently, fruit ripening is accelerated, and dramatic changes in color, texture, and aroma of fruits become evident. Depending on the stage of fruit development, plants producing climacteric fruits exhibit two systems of ethylene production. System 1 operates in immature climacteric fruits and vegetative tissues in which ethylene inhibits its own biosynthesis by negative feedback. Mature climacteric fruits and senescing petals in some species exhibit system 2 for ethylene production in which ethylene stimulates its own biosynthesis. In system 2, the positive feedback for autocatalytic biosynthesis of ethylene ensures ripening of the entire fruit even after the commencement of ripening. Exposure to exogenous ethylene induces a rapid increase in autocatalytic ethylene production in mature climacteric fruits, thereby hastening the onset of the climacteric rise in ethylene and other changes associated with ripening. However, ethylene treatment of fruits at pre-climacteric stage results in gradual increase in the respiration rate due to rise in ethylene concentration, but it does not trigger further autocatalytic endogenous ethylene production or ripening. Treatment of fruits with inhibitors of the ethylene action,

**Fig. 27.5** Typical pattern of fruit ripening. (**a**) Climacteric fruits exhibit increased respiration rate concomitant with rapid increase in ethylene production during maturation phase. (**b**) In non-climacteric fruits, increase is neither observed in respiration rate nor in ethylene production. Increase in coloration and loss of firmness are relatively slower as compared to that in climacteric fruits. *DAA* Days after anthesis

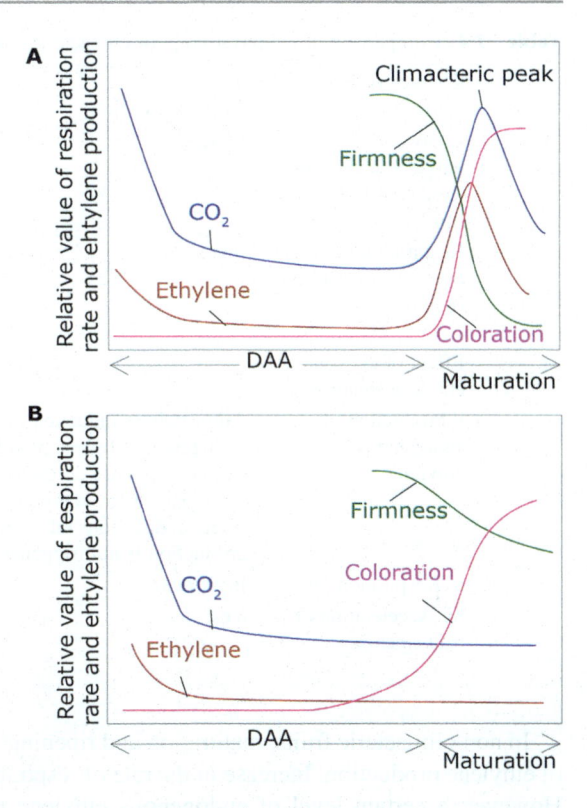

such as 1-methylcyclopropane (1-MCP), suppresses or delays fruit maturation. Transgenic tomatoes, in which ethylene production is inhibited by the antisense-induced repression of the 1-aminocyclopropane-1-carboxylic acid (ACC) synthase (ACS) and ACC oxidase (ACO) genes (which encode ethylene biosynthesis enzymes), exhibit delayed fruit maturation. These observations indicate the pivotal role played by ethylene in regulation of ripening process of fruits. Although continuous ethylene action is necessary for adequate ripening, the process in climacteric fruits is not necessarily regulated in an ethylene-dependent manner. In melon fruit, flesh softening and membrane degradation are regulated by both ethylene-dependent and -independent processes. However, some of the processes such as detachment of the abscission zone and color change of the rind, which are associated with chlorophyll degradation and an increase in yellow carotenoids, are regulated by ethylene. In papaya, the degreening of the skin and flesh softening are only partially dependent on ethylene, while production of volatile aromatic compounds is strictly dependent on ethylene. In case of pear, fruit softening is completely suppressed in the absence of ethylene, whereas in melon and papaya, the softening process shows only a partial dependence on ethylene. Thus, ripening of climacteric fruits is under both developmental and ethylene-dependent regulation. The degree of dependence on ethylene varies with respect to individual ripening parameters and the diversity of the species and cultivars.

**Table 27.4** Comparison of the physiology of climacteric and non-climacteric fruit ripening

| S. No. | Characteristic feature | Climacteric fruit | Non-climacteric fruit |
|---|---|---|---|
| 1. | Endogenous ethylene levels prior to start of fruit ripening | Increases | Does not increase |
| 2. | Respiration rate prior to onset of fruit ripening | Increases | Either remains unchanged or there is steady decline until senescence |
| 3. | Effect of exogenous ethylene treatment | | |
| | i. Production of endogenous ethylene | Autocatalytic induction of endogenous ethylene production if treated at mature climacteric fruit stage. However, at pre-climacteric stage, endogenous ethylene production is not triggered | Not triggered |
| | ii. Respiration rate | Increases | Increases |
| | iii. Acceleration of fruit ripening | Yes | No. In some fruits, such as citrus, degreening is seen |

In non-climacteric fruits, maturation and ripening processes occur without a burst of ethylene production. Increase in the rate of respiration has also not been observed. However, a certain level of endogenous ethylene is produced in non-climacteric fruits. Exogenous ethylene treatment increases respiration rate but does not trigger the production of endogenous ethylene, and fruit ripening is not accelerated in non-climacteric fruits. A comparative account of important events taking place in climacteric and non-climacteric fruits accompanying ripening process is summarized in Table 27.4. Recently, the involvement of ethylene in strawberry fruit ripening has been reported (Box 27.3). Respiration rate in strawberry fruit is stimulated by ethylene in a dose-dependent manner and results in slight acceleration of the cell wall softening process. There is an increase in the expression of ethylene receptor which increases the response to elevated ethylene levels in strawberry, indicating that a low level of ethylene is sufficient to trigger ripening process. Exogenous application of ethylene in citrus accelerates respiration rate and stimulates chlorophyll degradation and carotenoid biosynthesis. Treatment with ethylene antagonists, such as 2, 5-norbornadiene and silver nitrate, prevents degreening process.

## 27.4 Fruit Ripening: An Oxidative Phenomenon

Fruit ripening is a complex process and involves several oxidative processes, such as peroxidative damage, loss of membrane integrity, increase in ion leakage, and accumulation of lipid peroxides. Accumulation of reactive oxygen species (ROS),

## 27.4 Fruit Ripening: An Oxidative Phenomenon

**Table 27.5** Variations in the activity of various ROS-scavenging enzymes in ripe fruits

| S. No. | Name of the fruit | Antioxidant enzyme ||||  |
|---|---|---|---|---|---|---|
|  |  | SOD | CAT | APX | GR | POX |
| 1. | Tomato (*Solanum lycopersicum*) | ↓ | ↑ | ↑ | ---- | ↓ |
| 2. | Guava (*Psidium guajava*) | ↓ | ↑ | ↑ | ↓ | ↓ |
| 3. | Papaya (*Carica papaya*) | ↔ | ↑ | ↑ | ↑ | ------ |
| 4. | Cucumber (*Cucumis sativus*) | ↑ | ↓ | ↓ | ↑ | ↑ |
| 5. | Orange (*Citrus sinensis*) | ↓ | ↓ | ↓ | ↓ | ↓ |
| 6. | Ber (*Ziziphus sp.*) | ↔ | ↑ | ↑ | ↓ | ↓ |
| 7. | Mango (*Mangifera indica*) | ↓ | ↓ | ------- | ------ | ------ |

Abbreviations: *SOD* superoxide dismutase, *CAT* Catalase, *APX* Ascorbate peroxidase, *GR* Glutathione reductase, *POX* Peroxidase (↓ Decrease, ↑ Increase, ↔ Unaffected, ---- Not reported)

such as superoxide radicals ($O_2^{\bullet-}$), hydrogen peroxide ($H_2O_2$), and hydroxyl radicals ($OH^-$), brings about oxidative damage to the tissue. Accumulation of ROS leads to activation of the antioxidant system. Fruit tissue possesses a natural antioxidant system that includes enzymes, such as catalase (CAT), peroxidase (POX), superoxide dismutase (SOD), ascorbate peroxidase, and glutathione reductase (GR). Components of the antioxidant system act synergistically to scavenge $O_2^{\bullet-}$, $H_2O_2$ and $OH^-$ radicals. Table 27.5 summarizes the variations in the activity of various ROS-scavenging enzymes in ripe fruits. Variations in the pattern of antioxidant enzyme activities could be attributed to the differential activity of various isoenzymes. During early stages of fruit ripening, the antioxidant system protects fruits from the destructive effects of gradual oxidative stress. However, oxidative damage attributed to overproduction and accumulation of ROS takes place at later stages. Inability of the antioxidant system to eliminate ROS is also responsible for oxidative damages. These free radicals initiate the destructive changes during ripening. Fruit ripening is characterized by a gradual increase in oxidative processes, and it is considered a stressful process. Several fruits, such as tomato, mango, peach, and papaya, undergo oxidative stress during ripening. Free radicals induce the process of lipid peroxidation which is responsible for the initiation of ripening-associated deteriorative changes. Increase in lipid peroxidation is often associated with decreased CAT, GR, and SOD activity. Reduced activity of important antioxidant enzymes, such as CAT and SOD, leads to oxidative stress during ripening. However, it has been proposed that this increase in oxidative stress is necessary to facilitate the metabolic changes associated with fruit ripening.

### 27.4.1 Role of Alternative Oxidase (AOX)

Alternative oxidase (AOX) plays a role in ripening by mediating ethylene signaling as well as scavenging of ROS. AOX levels increase dramatically in ripe tomato fruits. Likewise, increased AOX levels have been reported in mango and apple fruits in which climacteric bursts are associated with an enhanced cyanide-insensitive respiration. The multigene family for the AOX expresses differentially during fruit ripening. Abundance of AOX protein peaks at the ripe stage. Pattern of protein accumulation indicates that an increase in cytochrome chain components plays an important role in facilitating the climacteric burst of respiration and fruit ripening. Transgenic tomato plants with reduced AOX levels, i.e., LeAOX1, show retarded ripening and reduction in carotenoid content, respiration, and ethylene production. Downregulation of ripening-associated genes has also been reported. Treatment of such mutants with 1-methylcyclopropane fails to induce ripening. However, inhibition of the AOX knockout mutant, i.e., 35S-AOX1a, is less effective. Mutant that overexpresses AOX levels, i.e., LeAOX1a, shows increased lycopene content and similar pattern of fruit ripening as reported in wild tomato. Therefore, it has been suggested that both AOX and ethylene are essential for fruit ripening in tomato, and inhibition of both is required to halt ripening process completely.

## 27.5 Role of Phytohormones

Abscisic acid (ABA) regulates the expression of ripening-related genes in strawberry. It has recently been implicated as a ripening factor both in climacteric (e.g., tomato, peach, and grapes) and non-climacteric fruits (e.g., apple, strawberry). In tomato, accumulation of endogenous ABA prior to ethylene burst has been reported, suggesting a possible role of ABA in the modulation of ethylene accumulation. Exogenous application of ABA promotes both ethylene biosynthesis and fruit ripening. However, application of nordihydroguaiaretic acid (NDGA), an inhibitor of ABA biosynthesis, suppresses fruit coloration and softening process. Four different phenomena are brought about by ABA during citrus fruit ripening:

1. Increase in endogenous ABA levels leads to ethylene production. Ethylene treatment, however, does not cause any increase in ABA levels.
2. Suppression of increase in ABA content accompanying fruit ripening by NDGA treatment whereas ethylene production remains unaffected.
3. Inhibition of ethylene production by ABA+1-MCP without any negative effect on ABA content.
4. Inhibition of fruit ripening and softening by treatment with NDGA and 1-MCP.

In tomato, both ABA and ethylene control the expression of genes encoding the cell wall- degrading enzymes. It has been suggested that the presence and perception of both ABA and ethylene are important for normal fruit ripening process. Auxin functions as a negative regulator of fruit ripening in a number of climacteric and non-climacteric fruits. Exogenous auxin delays the onset of ripening in avocado, pear, grape, and strawberry. In strawberry, the expression levels of genes encoding polygalacturonase and expansin (which are related to cell wall degradation) are repressed by auxin. High levels of cytokinins are often found in immature fruits as compared to mature fruits (Fig. 27.6). Exogenous application of cytokinin has been reported to delay fruit ripening in both climacteric and non-climacteric fruits.

**Fig. 27.6** Depiction of effect of exogenous application of phytohormones, such as auxin, gibberellins (GAs), brassinosteroids (BRs), and polyamines (PAs) on the process of fruit ripening

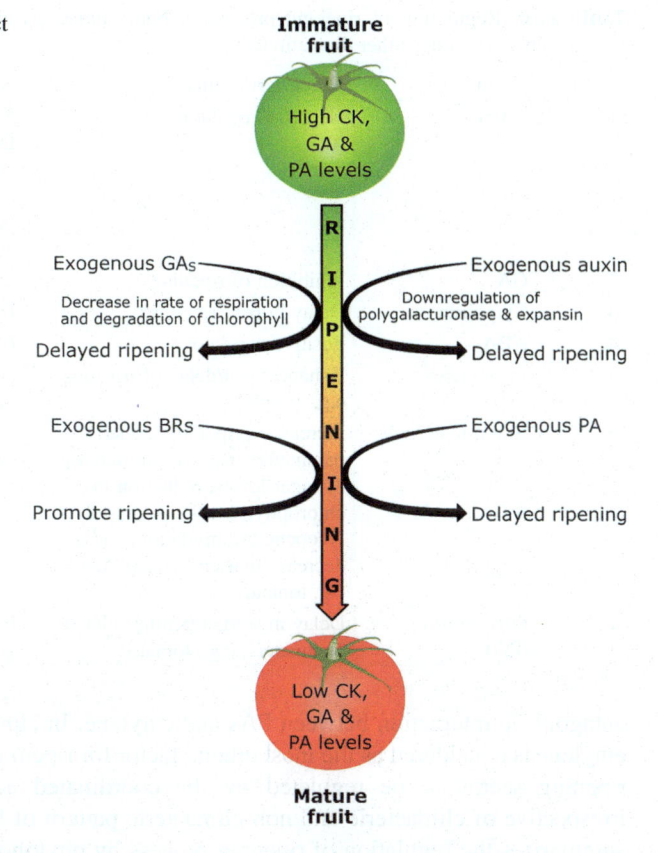

Increased concentration of GAs is generally evident during early stages of fruit development. Ripening process is characterized by decline in GA concentration. It seems that high levels of GA might be essential in promoting early fruit development and inhibiting ripening. Exogenous GA application leads to the delay in ripening process. GA application slows down respiratory activity, anthocyanin synthesis, and chlorophyll degradation and delays softening process in strawberry. Negative regulation of ripening process is often associated with the role of cytokinin and GA in increasing fruit size. Concentrations of jasmonates, such as jasmonic acid and methyl jasmonate, attain a peak just prior to ethylene burst in climacteric fruits. In non-climacteric fruits, however, higher levels of jasmonates are observed during early stages of fruit development which decrease during the ripening (later) phase. In grapes, levels of endogenous brassinosteroids (BR) increase with ripening. Exogenous application of BRs promotes ripening process in grapes and tomatoes. However, further work is required to ascertain the role of BRs in fruit ripening process. Polyamine (PA) levels have been reported to be high during early fruit development but decline following commencement of ripening. Putrescine and spermidine are usually the most and least abundant polyamines, respectively, in the developing fruits. Exogenous application of PAs delays ripening process, possibly due to the

**Table 27.6** Regulation of ripening process in both climacteric and non-climacteric fruits by various phytohormones other than ethylene

| S. No. | Hormone | Climacteric fruits | Non-climacteric fruits |
|---|---|---|---|
| 1. | Auxin | Negative regulator | Negative regulator. Downregulation of genes related to cell wall degradation, such as polygalacturonase and expansins |
| 2. | GA | Inhibition of ripening | Delay of fruit softening |
| 3. | Cytokinin | Delay in fruit ripening | Delay in fruit ripening |
| 4. | ABA | Positive regulator | Positive regulator |
| 5. | Jasmonates | Enhancement/delay of ripening, e.g., peach | Enhancement of ripening, e.g., blackberries |
| 6. | Brassinosteroids | Increase in ripening-related parameters, such as an increase in sugar levels, reduction in chlorophyll levels, increased lycopene accumulation, and a decrease in ascorbic acid levels, e.g. tomato | Enhancement of ripening, e.g., grape |
| 7. | Polyamines (PA) | Delay in some ripening-related parameters, e.g., tomato | Expression of PA biosynthetic gene increases, e.g., grape |

antagonistic interaction between PAs and ethylene. In climacteric fruits, the role of ethylene is considered as the most crucial factor for ripening process; however, fruit ripening seems to be regulated by the coordinated action of phytohormones irrespective of climacteric and non-climacteric pattern of fruit ripening. Table 27.6 summarizes the regulation of ripening process by phytohormones other than ethylene in both climacteric and non-climacteric fruits.

## 27.6 Nitric Oxide and Ethylene CrossTalk

In many fruits, especially tropical ones, the increase in ethylene production during ripening leads to induction of senescence which adversely affects their post-climacteric storage. As a result, the quality of these fruits in terms of nutrients, color, texture, and flavor is drastically reduced. There are many signals that regulate ethylene production and its perception in different parts of plants. Nitric oxide (NO) directly interferes with ethylene production during fruit ripening either by suppression of the activity of various enzymes implicated in ethylene biosynthesis or through direct stoichiometric inhibition. In ethylene biosynthesis pathway, the autocatalytic activity of ethylene results in the formation of S-adenosyl methionine (SAM) from methionine (MT) which is catalyzed by SAM synthetase. 1-aminocyclopropane-1-carboxylic acid synthase (ACC synthase) converts SAM into ACC which subsequently get oxidized to ethylene by the action of ACC

## 27.6 Nitric Oxide and Ethylene CrossTalk

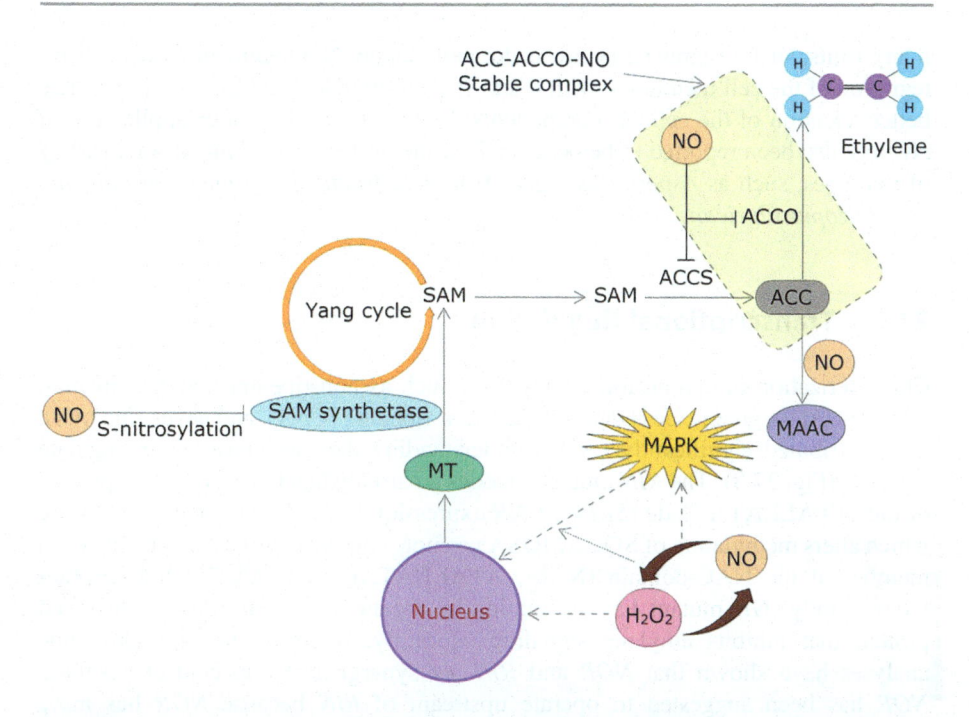

**Fig. 27.7** Schematic model depicting effect of NO on ethylene biosynthesis during fruit ripening. *NO* Nitric oxide, *MT* Methionine, *SAM* S-adenosyl methionine, *ACCS* 1-aminocyclopropane-1-carboxylic acid synthase, *ACC* 1-aminocyclopropane-1-carboxylic acid, *MAAC* 1-melonyl aminocyclopropane 1-carboxylic acid, *ACCO* 1-aminocyclopropane-1-carboxylic acid oxidase, $H_2O_2$ Hydrogen peroxide. $H_2O_2$ and NO reciprocally stimulate production of each other

oxidase. NO has been known to adversely affect ethylene biosynthesis due to S-nitrosylation of methionine adenosyltransferase (MAT), resulting in inhibition of SAM production. NO binds to ACC oxidase thereby inhibiting the autocatalytic biosynthesis of ethylene. This binding causes formation of ACCO-NO complex. Eventually, ACCO-NO complex forms a stable ternary complex, ACC-ACCO-NO thereby inhibiting ACO activity and consequent ethylene production (Fig. 27.7). NO also regulates ethylene production through interaction with hydrogen peroxide ($H_2O_2$). $H_2O_2$ effectively induce the biosynthesis of ethylene via gene transcription. Reciprocal interaction of NO and $H_2O_2$ also affects mitogen-activated protein kinase (MAPK)-mediated downstream components of ethylene biosynthesis. Being highly reactive, NO has the potential to directly trigger redox processes related to ROS production. Through post-translational modifications, such as S-nitrosylation and tyrosine nitration, NO targets transition metals, such as Cu, Zn, and Fe, which are components of various signaling molecules, receptors, enzymes, transcription factors, and proteins. NO reduces the production of ethylene leading to the delay in the expression of enzymes which are responsible for cell wall degradation, lignification, and pigmentation of fruits, thereby significantly extending the shelf life of fruits. Postharvest application of exogenous NO causes delayed ripening in many fruits, such as strawberry, avocado, and mango. NO treatment reduces disintegration of the cell membrane with reduced electrolyte leakage leading to facilitate better retention of the cellular components. Furthermore, exogenous application of NO has also been reported to be beneficial for the protection of fruits from a variety of microbes, such as *Aspergillus niger*, *Monilinia fructicola*, *Penicillium italicum*, and *Rhizopus nigricans*.

## 27.7 Transcriptional Regulation

Characterization of certain tomato mutants, such as non-ripening (*NOR*), ripening inhibitor (*RIN*), green-ripe (*GR*), colorless non-ripening *(CNR)*, and never-ripe (*NR*), has contributed significantly toward understanding the mechanisms that regulate ripening (Fig. 27.8). The *RIN* mutant encodes a partially deleted MADS-box protein of the SEPALLATA clade (*SEP4*). *CNR* expression is due to an epigenetic change which alters methylation of SQUAMOSA promoter-binding (SPB) protein. *NOR* is a member of the NAC-domain (NAM, ATAF1/ATAF2, and CUC2) transcription factor family. *GR* mutant has a mutation in a gene which encodes a conserved protein that inhibits ethylene signaling. Recently, transcriptome and proteome analyses have shown that *NOR* and *RIN* act synergistically to control ripening. *NOR* has been suggested to operate upstream of *RIN* because *NOR* has more profound effect on ripening-related gene expression. Chromatin immunoprecipitation (ChIP) and transcriptome analyses have revealed the interaction of *RIN* locus

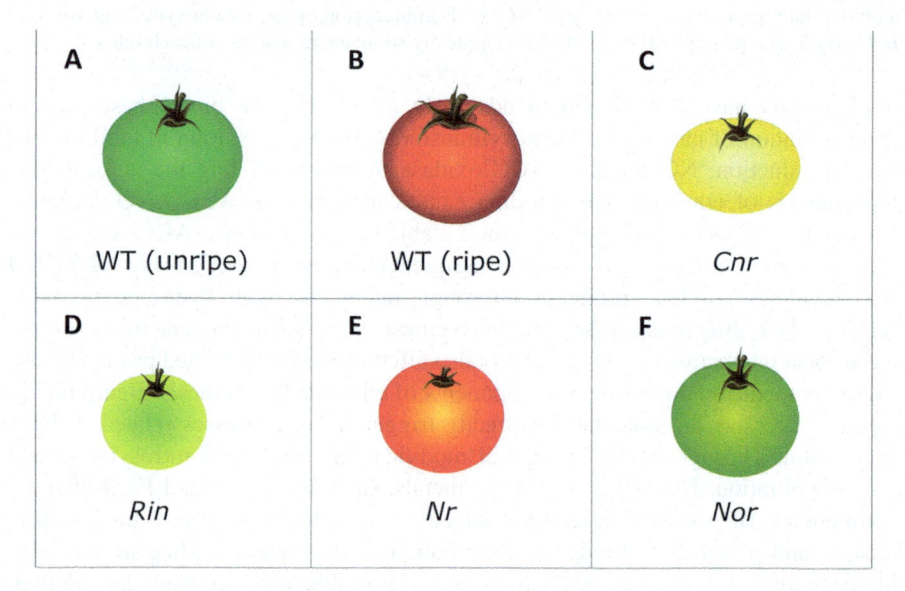

**Fig. 27.8** Wild type and different types of tomato mutants. (**a**) Wild-type unripe tomato. (**b**) Wild-type ripe tomato. (**c**) *Colorless non- ripening (Cnr)* mutant. (**d**) *Ripening inhibitor (Rin)* mutant. (**e**) *Never ripe (Nr)* mutant. (**f**) *Non- ripening (Nor)* mutant

## 27.7 Transcriptional Regulation

**Table 27.7** Different mutants of tomato and their respective modification exploited in fruit ripening studies

| S. No. | Name of the mutant | Modification |
|---|---|---|
| 1. | Non-ripening (NOR) | Member of the NAC-domain (NAM, ATAF1/ATAF2, and CUC2) transcription factor family |
| 2. | Ripening inhibitor (RIN) | Encodes a partially deleted MADS-box protein of the SEPALLATA clade (*SEP4*) |
| 3. | Green ripe (GR) | Mutation in a gene which encodes a conserved protein that inhibits ethylene signaling |
| 4. | Colorless non-ripening (CNR) | Epigenetic change which alters the promoter methylation of SQUAMOSA promoter-binding (SPB) protein |

with over 200 genes. This interaction modulates the expression of *RIN* target genes by either activation or repression. *CNR* and *NOR* are the *RIN* target genes, which are the major regulators of ripening process. *RIN* target genes are also implicated in pathways active during the transition from green unripe fruits to ripe fruits, such as carotenoid accumulation, chlorophyll breakdown, and ethylene biosynthesis. Table 27.7 shows the modification that has given rise to different tomato mutants.

MADS-box genes are involved in the regulation of fruit ripening in a variety of fruits, such as banana, tomato, and strawberry. Numerous fruit development and ripening-related genes have been isolated and characterized using differential gene expression patterns and biochemical functions, including the regulatory genes involved in ethylene biosynthesis. *RIN* and *CNR* encode certain transcription factors, which play important roles in the regulation of the ethylene biosynthesis pathway. A MADS-box transcription factor *MADS-RIN*, encoded by *RIN* locus, is induced at the onset of fruit ripening. In transgenic tomatoes, the suppression of *MADS-RIN* gene expression results in the formation of non-ripening. *RIN* mutants were rescued after complementation with MADS-RIN gene, thus demonstrating that *MADS-RIN* gene is essential for normal fruit ripening. *MADS-RIN* gene interacts with the promoter for *ACC synthase* gene suggesting its role in the regulation of ethylene biosynthesis. The expression of several ripening-related genes, such as *polygalacturonase (PG), expansins (EXP), lipoxygenase (LOXC)*, and *phytoene synthase (PSY1)*, is also controlled by *MADS-RIN* gene. These genes encode proteins involved in the metabolism of the cell wall and carotenoid biosynthesis, such as *PSY1*. Expression of some genes, such as *LOXC* and alcohol dehydrogenase, is also regulated by *MADS-RIN*. Strawberry, a non-climacteric fruit, has a fruit-specific *LeMADS-RIN* orthologue, which indicates the presence of ethylene-independent regulatory pathway involving MADS-box genes both in climacteric and non-climacteric fruit ripening. Furthermore, it has been suggested that *MADS-RIN* and ethylene signaling act synergistically for normal fruit ripening. Figure 27.9 summarizes the regulation of climacteric and non-climacteric patterns of fruit ripening.

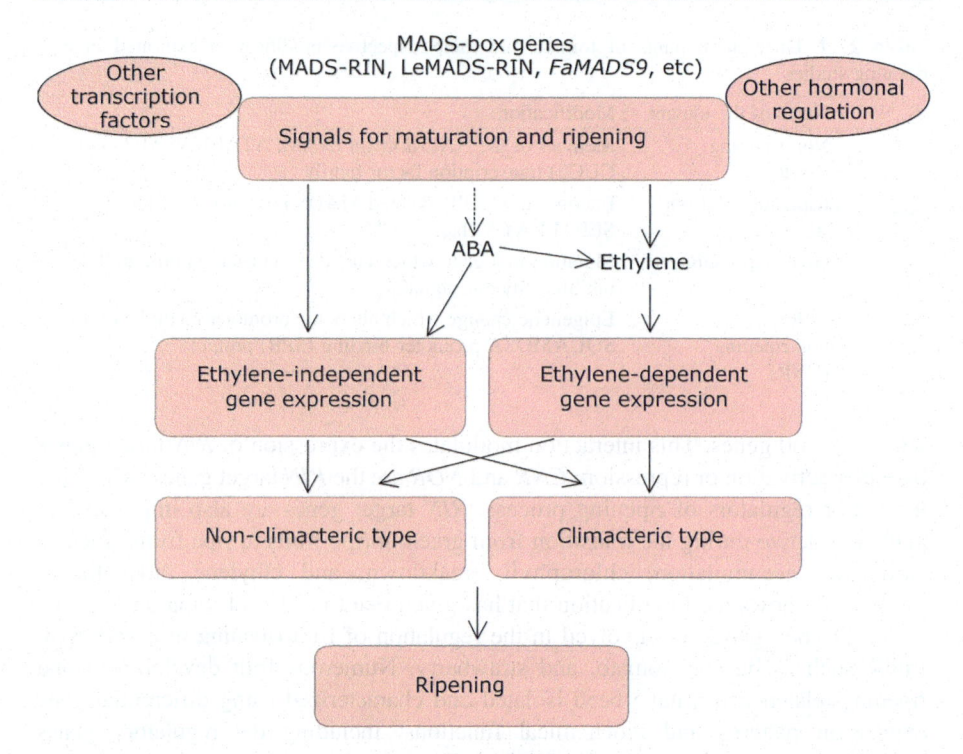

**Fig. 27.9** Schematic diagram illustrating the involvement of various genes, transcription factors, and hormones in regulation of climacteric and non-climacteric type of fruit ripening via ethylene-dependent and ethylene-independent pathways. *MADS-RIN* MADS-box transcription factor, *LeMADS-RIN* MADS-RIN in tomato, *FaMADS9* SEP1-/SEP2-like gene in strawberry, and *ABA* Abscisic acid

## 27.8 Epigenetic Regulation of Gene Expression

Epigenetic variations generally affect gene expression at the level of chromatin organization, mainly through DNA methylation, cytosine methylation, and histone modifications without affecting primary DNA sequence. Molecular nature of *CNR* mutant phenotype has effectively explained the impact of cytosine methylation on fruit ripening in tomato. *NR* phenotype is due to the heritable cytosine hypermethylation, including that in the *CNR* gene promoter region. During fruit development, in wild tomato, demethylation of *CNR* promoter takes place in specific region just before the onset of ripening. Thus, it has been hypothesized that DNA methylation seems to be crucial in the regulation of fruit ripening. Epigenomic studies have highlighted mainly three important aspects in the ripening process. 1. The degree of methylation of regions upstream the transcription start sites (TSS) decreases gradually along with fruit development. The decline in the degree of methylation has not been observed for *CNR* and *RIN*, whose CG methylation levels were constantly higher at TSS. 2. Promoters of typical ripening-related genes are gradually demethylated during fruit development. Analysis of methylation status of TSS has demonstrated their progressive demethylation during the transition from green unripe to red ripe fruits. 3. Analysis of epigenetic variation in *Arabidopsis* revealed that at least one-third of expressed genes are methylated in their coding region and only 5% of genes are methylated within promoter regions. However, the promoter-methylated genes have a higher degree of tissue-specific expression. It has been suggested that progressive demethylation of ripening-related gene promoters may be necessary for the binding of transcriptional regulators triggering the accumulation of ripening-related genes.

**Summary**

- Fruit development involves four different phases characterized by fruit set, rapid division of cells, cell enlargement, and onset of ripening. Different ripening stages have distinct physicochemical markers in different model systems.
- Fruits are generally categorized into climacteric and non-climacteric types on the basis of respiratory activity and are associated with ethylene biosynthesis during ripening process. Climacteric fruits are characterized by an increase in respiration rate and ethylene production at the onset of ripening, whereas non-climacteric fruits are characterized by the absence of ethylene production burst and respiratory rise. However, they have the capacity to produce ethylene endogenously. Climacteric fruits possess two systems for ethylene production depending on development stage of fruit. Initial stages of ripening have been characterized by the auto inhibition of ethylene biosynthesis by negative feedback (System 1), whereas at later stages, ethylene autocatalyzes its biosynthesis by positive feedback loop (System 2). Involvement of other phytohormones such as auxins, ABA, GA, jasmonates, and BRs in the regulation of the fruit ripening has also been reported.
- Cell wall architecture is progressively modified during fruit ripening, due to metabolic changes leading to softening (textural change) of the fruit due to the activity of various cell wall-degrading enzymes. Transition from green to a variety of colors due to development of different pigments, such as carotenoids, anthocyanins is responsible for the taste and aroma of the fruits. Sugars, organic acids, and volatiles are essential for intensification of flavors and fragrances.
- Oxidative processes during ripening involve an increase in ion leakage due to loss of membrane integrity and decrease in membrane fluidity. Antioxidant system becomes activated due to rise in the level of reactive oxygen species. NO inhibits ethylene biosynthesis via binding to ACC oxidase (ACCO). ACCS and ACCO genes are downregulated by NO leading to inhibition of ethylene.
- MADS-box genes, such as *MADS-RIN*, have been implicated in the regulation of fruit ripening process. Synergistic and coordinated actions of transcription factors and ethylene along with other phytohormones regulate the normal process of fruit ripening.

## Multiple-Choice Questions

1. A fruit is best defined as:
   (a) Ripened ovule enclosing one or more seeds
   (b) Ripened ovary enclosing one or more seeds
   (c) Ripened endosperm enclosing one or more seeds
   (d) Ripened nucellus enclosing one or more seeds
2. Which of the following statements is incorrect in the context of fruit development?
   (a) Phase I is characterized by anthesis and successful fertilization.
   (b) Phase II is characterized by rapid cell divisions during embryo development.
   (c) Phase III is characterized by cessation of cell divisions and cell enlargement.
   (d) Phase IV is characterized by enhanced carotenoid biosynthesis.
3. Which of the following is a non-climacteric fruit?
   (a) *Solanum lycopersicum*
   (b) *Musa balbisiana*
   (c) *Fragaria vesca*
   (d) *Artocarpus heterophyllus*
4. In climacteric fruits, exogenous application of ethylene results in:
   (a) Autocatalytic induction of endogenous ethylene production if treated at pre-climacteric stage
   (b) Autocatalytic induction of endogenous ethylene production if treated at mature climacteric fruit stage
   (c) Increase in respiration rate
   (d) Acceleration of fruit ripening
5. Which of the following statement is correct?
   (a) Exogenous application of gibberellic acid (GA) leads to decrease in rate of respiration and degradation of chlorophyll.
   (b) Exogenous application of auxin causes upregulation of polygalacturonase and expansin.
   (c) Exogenous application of cytokinin enhances fruit ripening.
   (d) Abscisic acid (ABA) is a positive regulator of fruit ripening.
6. Development of fruit without fertilization is called:
   (a) Apospory
   (b) Parthenogenesis
   (c) Polygamy
   (d) Parthenocarpy

## Answers

1.b  2.a  3.c  4.a  5.b  6.d

## Suggested Further Readings

Gapper NE, Giovannoni JJ, Watkins CB (2014) Understanding development and ripening of fruit crops in an 'omics' era. Horticl Res 23:14034–14044

Goulao LF, Oliveira CM (2008) Cell wall modifications during fruit ripening: when a fruit is not the fruit. Trends Food Sci Technol 19:4–25

Osorio S, Scossa F, Fernie A (2013) Molecular regulation of fruit ripening. Front Plant Sci 14:198–206

# Seed Dormancy and Germination

## 28

Renu Kathpalia and Satish C Bhatla

Plant life begins with seed formation and is renewed when seed germinates. Seeds are highly dehydrated, quiescent or resting structures, and carriers of the next generation in the life cycle of plants. Seeds have different shapes and sizes, ranging from the smallest orchid seed ($10^{-6}$ g) to the huge seed of the double coconut palm (30 kg). Some seeds are short-lived, e.g., willow seeds are viable for less than 1 week. Mimosa seeds can live up to ~200 years. Seeds of *Canna compacta* have been reported to remain viable up to 600 years, and seeds of lotus can survive up to 1000 years. To accomplish the remarkable feat of next generation, seed contains an embryo with reserve food. During seed germination, the embryo divides and differentiates into shoot-root axis. The tissue containing reserve food material (depending on the species) may persist and get reabsorbed later. Depending on the species, the reserve food material is stored either in the embryo or in the endosperm, and in some cases, it is present in both, embryo and endosperm. The degree of arrest is variable in different species. It may be true sleeping (**dormancy**) or **quiescent stage**, which only requires water for resuming the growth. A viable and nondormant seed is capable of germination after all the necessary environmental conditions are met, but dormant seeds do not germinate even when provided with favorable conditions. Different types of mechanisms impose seed dormancy in diverse climates and habitats. Abscisic acid (ABA) present in seeds induces dormancy, but the intensity of dormancy induction also depends on the genetic makeup of the plant and the environment in which it grows. Temperature, relative humidity, and day length also interact to modulate dormancy, thereby making it a complex phenomenon. Germination in different seeds is not synchronous, and stimuli required to promote germination vary widely. Prior to germination, seeds need to undergo imbibition, i.e., uptake of water by dry seed, followed by reactivation of metabolic activity and redifferentiation of embryonic tissue to mobilize the reserve food material stored in the seed and initiate meristematic activity. The transition from dry seed to seedling is highly sensitive to different environmental conditions, especially light, temperature, and availability of water. This response to

© Springer Nature Singapore Pte Ltd. 2018
S. C Bhatla, M. A. Lal, *Plant Physiology, Development and Metabolism*,
https://doi.org/10.1007/978-981-13-2023-1_28

environmental signals is mediated by one or more hormones whose signaling events lead to activation or de novo synthesis of hydrolytic enzymes. The emergence of radicle is the first visible step and indicates that seed is viable. If germination occurs in the dark, then root growth is slow. On the other hand, shoot growth accelerates. This behavior increases chances of seedling to obtain light so that it can turn green and start photosynthesizing. Once the seedling comes out of the soil, it turns green and starts producing new leaves.

## 28.1 Seed Morphology and Structure

All seeds are surrounded by a protective layer of dead cells (**testa**) forming seed coat. Sometimes testa is fused with fruit wall or **pericarp** (derived from ovary wall) (Fig. 28.1). In latter case, the "seed" is actually a fruit, e.g., maple, cereal grain, lettuce, and sunflower. The embryo is embedded in the tissue laden with reserve food materials. Most of the eudicot seeds have an embryo with an axis consisting of an embryonic root (the radicle) and an embryonic shoot (the plumule) with a pair of

**Fig. 28.1** (**a**) Bean seed with two massive cotyledons and an embryonal axis with radicle and plumule. (**b**) Castor seed with thin paper-like cotyledons and embryonal axis embedded in the massive endosperm. (**c**) A cereal grain with fruit wall fused with seed coat, single cotyledon along with embryonal axis having plumule covered by coleoptile and radicle covered by coleorhiza

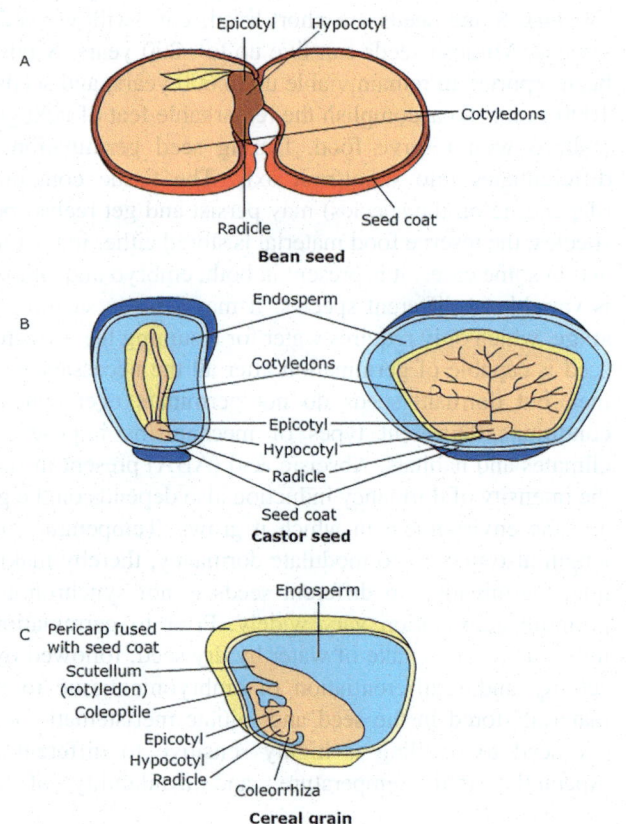

## 28.1 Seed Morphology and Structure

seedling leaves (the **cotyledons**). Monocot embryos have more variation in their structure with one cotyledon. In grass embryos, the single cotyledon is further modified to form an absorptive organ that lies adjacent to the endosperm. This is known as **scutellum**. In addition, plumule is surrounded by a sheath called as **coleoptile**, and radicle is surrounded by another sheath called as **coleorhiza**. The scutellum is an embryonic tissue and is covered by an epithelial layer, which is similar to aleurone tissue in wall characteristics and nature of reserved food material (Fig. 28.1). Cotyledon is an important part of the embryo within the seed and is formed prior to germination during the process of embryogenesis along with roots and shoots. Angiosperms are divided into two major groups on the basis of number of cotyledons. Plants with only one cotyledon are known as monocots and with two cotyledons are known as dicots. Cotyledons often contain food reserves which are used during the early stages of germination. In most plants, the cotyledons are brought out of the testa and above the ground where they become green and make food by photosynthesis. The cotyledons eventually fall off, usually after the first foliage leaves have been formed. Endosperm is a unique tissue present in angiosperm seeds, and such seeds are called **endospermous seeds**. Stored food in endosperm serves as fuel for the developing embryo. Many seeds, such as those of orchids, are an exception as they lack stored food reserves and their germination requires the participation of symbiotic endomycorrhizal fungi. Endosperm varies from a single layer in *Arabidopsis* to large, oil-storing endosperm of castor bean and starch-storing endosperm in cereals. Some of the eudicots, e.g., beans, which do not have endosperm are called as **non-endospermous seeds**. In non-endospermous seeds, food is stored in the fleshy cotyledons of the embryo. Endosperm is mainly differentiated into two structurally and functionally distinct tissues: the starchy endosperm and the **aleurone** layer (Fig. 28.2). Starchy endosperm contains enzymes but its cells are dead at maturity. The aleurone layer in mature seeds consists of living cells. It is about three to four cell layers thick in rice and barley and single-layered thick in wheat, maize, and oat. The cell wall of aleurone layer is thick and rich in **arabinans** and **arabinoxylans.** The cells of aleurone layer are interconnected with numerous plasmodesmata. Each cell is packed with protein bodies or protein storage vacuoles commonly referred as **aleurone grains**. In addition, aleurone cells also have storage bodies known as **globoids** which are rich in $Ca^{2+}$ and $Mg^{2+}$ salts of

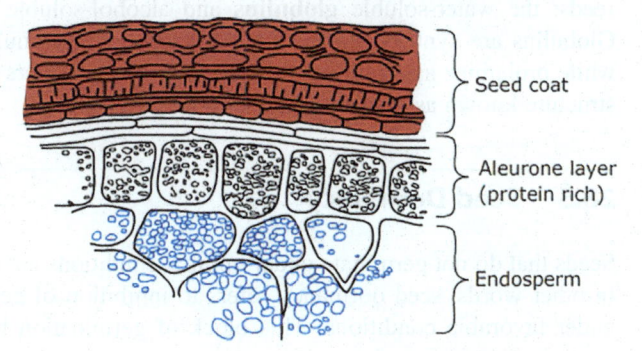

**Fig. 28.2** Differentiation of aleurone layer in a seed. The outer most layer of endosperm is rich in proteinaceous aleurone grains and is known as aleurone layer. Inner layers of endosperm are filled with starch

inositol hexaphosphoric acid (phytic acid). When seed germinates, hydrolytic enzymes are activated or synthesized in the presence of gibberellins, and protein bodies lose their reserved proteins. Both aleurone layer and scutellar epithelial cells are secretory tissues. The mature seed consists of protective seed coat layers of dead cells that protect the delicate embryo within. Seed dormancy limits the entry of water and oxygen in the seed. Many legumes and aquatic species have thick seed coat which provides physical barrier and prevents the growth of embryo or imbibition of water. The seed coat often contains phenolic compounds that impart red to brown color to many seeds. These phenolic compounds protect seeds from herbivore and pathogen attacks. In addition, phenolic compounds also protect seeds from harmful effects of short wavelength light and are responsible for imposing dormancy by making seed coat impervious to water and oxygen. Tannins and polyphenols in seed coats are responsible for astringent taste of many fruits and protection from herbivores. The presence of large number of double bonds in polyphenols (such as **proanthocyanidins**) makes these compounds effective in protecting the seeds from the mutagenic effect of short wavelengths of light. Seed coat is also beneficial for human health as a source of dietary fiber. Seed coat contains lignins and polysaccharides, which are largely indigestible and thus allow the passage of food through gut. It is also a rich source of antioxidants, mainly proanthocyanidins.

## 28.2 Food Reserves in Seeds

Seeds are an important source of food for human and livestock. More than 50% of human caloric intake and 47% of proteins come from consumption of seeds. Ninety percent of world's food is derived from less than 17 plant species. In addition to oils, polysaccharides, and proteins, seeds also store mineral elements. Mineral elements are stored as chelated complexes with inositol hexaphosphate in a compound called **phytate**. Starch grains and cell wall carbohydrates make up the bulk of storage polysaccharides as major food reserves in seeds. However, some of these storage compounds act as anti-nutrients for humans and animals. These include phytate and some storage proteins like lectins, enzyme inhibitors, and thionins. Seeds accumulate these proteins in defense against bacterial and fungal pathogens which also protect them from herbivores and insects. Cereals have two types of proteins in seeds: the water-soluble **globulins** and alcohol-soluble hydrophobic **prolamins**. Globulins are synthesized in rough ER and get accumulated in storage vacuoles, while prolamins assemble into disulfide-linked polymers in ER and form spherical structure known as **protein bodies**.

## 28.3 Seed Dormancy

Seeds that do not germinate under optimal conditions are considered to be dormant. In other words, seed dormancy refers to inhibition of germination of viable seeds under favorable conditions. This block of germination has evolved differently in different species through adaptation to the environment in which it is growing. Seed dormancy can be due to structural limitations or imbalance of growth substances. On the basis of time of appearance, seed dormancy can be referred as primary or secondary. Freshly harvested mature seeds with capacity of imbibition but unable to germinate are considered to have undergone **primary dormancy**. This type of dormancy is induced during seed development, and such seeds are released from the plant in a dormant state. It is induced due to accumulation of ABA to prevent germination while seeds are still attached to the mother plant. Primary dormancy can be overcome by keeping the seeds in dry state so that embryo gets afterripening period. Alternatively, primary dormancy can be removed by treatment of seeds in the imbibed state. This includes chilling, warm stratification, light, gibberellins, and some other chemical applications. If the conditions are unfavorable for germination at the time of their dispersal, seeds develop physiological dormancy related to seasonal cycles. This type of dormancy due to seasonal changes is referred as **secondary dormancy**. This is induced in mature seeds due to various stress factors mainly temperature after they are released from plants. For example, seeds of *Avena sativa* (oat) become dormant if the temperature at the time of their dispersal is higher than the optimum required for germination, whereas seeds of *Phacelia dubia* (small-flower scorpionweed) become dormant if temperature is below the optimal for germination. Endogenous dormancy prevents seed germination and is mainly due to the physiological status of the seed, while exogenous dormancy is due to seed coat and fruit wall, including endosperm or perisperm, and can be easily overcome. Coat-imposed dormancy is imposed by seed coat and other enclosing tissues, such as endosperm, pericarp, or extrafloral organs. It is associated with long-term viability, e.g., Indian lotus and date palm exhibit high longevity due to highly impermeable seed coat which prevents germination. Embryos of such seeds germinate once the seed coat is removed or damaged artificially or naturally. There are weak points (such as **strophiole** in leguminous seeds) from where damage results in water uptake (Fig. 28.3). Coat-imposed dormancy can be caused by one or more of the following factors: (i) *Prevention of water uptake*: Seed coat is a common cause of seed dormancy in plants found in arid and semiarid regions, especially in legumes. In the seeds of clover (*Trifolium* sp.) and alfalfa (*Medicago* sp.), waxy cuticles, suberized layers, and lignified sclereids combine to restrict water penetration into the seeds. Water impermeability cannot be overcome by **scarification** or passage

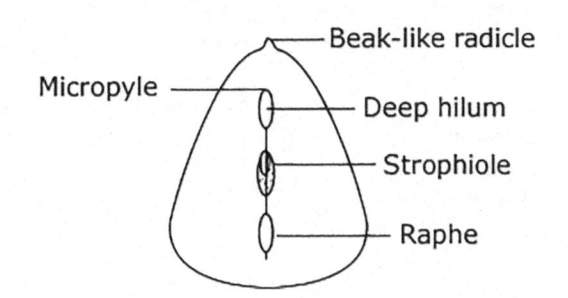

**Fig. 28.3** A gram seed showing the presence of strophiole, a small opening from where water enters into the seed during germination

through digestive tract of animals. (ii) *Mechanical constraint*: Seed coat may be too rigid for emergence of radicle, e.g., in nuts which have hard, lignified seed coat. In case of *Arabidopsis*, tomato, and tobacco, endosperm is very hard, and in such seeds the endosperm cell wall needs to be weakened for radicle emergence. (iii) *Interference with gaseous exchange*: Some seed coats are less permeable to gases. As a result, oxygen supply to embryo is reduced, e.g., making a hole in the seed coats of *Xanthium* seeds overcomes dormancy. In some seeds, more oxygen is consumed by the seed coat due to the presence of phenolics in the cell wall, resulting in reduction in oxygen availability to the embryos. (iv) *Retention of inhibitors*: Seed coat may prevent escape of inhibitors from the seeds, or inhibitors may diffuse into the embryo. The inhibitors in seed coat include unsaturated lactones (coumarin), phenolic compounds (ferulic acid), various amino acids, and cyanogenic compounds (i.e., compounds that release cyanide). In case of apple and other members of Rosaceae, seed coat releases cyanogenic compounds (Table 28.1). The *transparent testa (tt)* mutant of *Arabidopsis*, which contains reduced amounts of proanthocyanidins in its seed coat, shows less dormancy and reduced longevity. (iv) *Growth inhibitor production*: In some instances, seed coat and pericarp may contain high concentration of growth inhibitors that can suppress embryo germination. Growth inhibitor production plays important roles in preventing germination in seeds in many fleshy fruits. High concentration of solutes in fruit juices also inhibits seed germination. Besides, ABA has an essential function in preventing in situ germination. Seeds of ABA-deficient tomato mutant show in situ germination. Embryo dormancy is inherent in the embryo and is not due to any influence of the seed coat or other surrounding tissues. In some cases, embryo dormancy can be relieved by amputation of cotyledons, e.g., seeds of peach (*Pyrus* sp.) exhibit slow growth with cotyledons and normal growth without cotyledons. Seeds may also fail to germinate when embryo has not attained maturity, e.g., seeds of celery (*Apium graveolens*) and carrot (*Daucus carota*).

**Table 28.1** Inhibitors involved in seed dormancy and their localization in seeds

| Inhibitors | Location | Example |
|---|---|---|
| Coumaric acid | Seed coat | Legumes |
|  | Glume and hull | *Aegilops* |
| Ferulic acid | Fruit juice | Tomato |
|  | Embryo | Xanthium |
|  | Endosperm | Iris |
|  | Seed coat | *Cucurbita* |
| Aflatoxin | Seed | Barley |
| Amygdalin | Embryo | Rosaceae |
| Cinnamic acid | Chaff | *Parthenium argentatum (guayule)* |

## 28.4 Hormonal Regulation of Seed Dormancy

Abscisic acid (ABA) plays significant role/s in regulating the onset of seed dormancy and maintaining seeds in dormant state. ABA accumulation is low in developing seeds when storage reserves are being synthesized and its (ABA) level declines as seeds undergo maturation drying. ABA is synthesized in both dormant and nondormant seeds, but generally ABA catabolism is favored in nondormant seeds, while ABA synthesis is favored in dormant seeds, i.e., there is a shift in favor of ABA synthesis accompanying dormancy induction. The absence of ABA during seed development results in the production of viviparous or precocious germination of seeds. In mutants deficient in ABA, precocious germination is not prevented, and seeds exhibit reduced dormancy. ABA-insensitive mutants are programmed to germinate even in the presence of ABA and produce seeds that are intolerant to desiccation. Expression of many genes involved in dormancy can also be enhanced by ABA. Some of the transcripts present in the dormant and ABA-treated embryos encode LEA (late embryogenesis abundant)-type proteins which are synthesized during the later stage of seed development. These proteins are associated with survival during dehydration stress. ABA may also be responsible for the inhibition of radicle elongation during the later stage of seed germination. GA-deficient mutants of tomato and *Arabidopsis* require an exogenous supply of GA to complete germination. According to hormone balance theory, the ratio of ABA and GA is the main determinant of seed dormancy and germination. The tissue concentration of these two hormones is controlled by the rate of their synthesis, degradation and deactivation. Level of ABA decreases at the time of seed germination due to decrease in its biosynthesis and enhanced degradation. The rate-limiting step of ABA biosynthetic pathway involves the activity of 9-*cis*-epoxycarotenoid dioxygenase (NCED). The inactivation of ABA is controlled by ABA-8-′-hydroxylase (CYP707A2 gene). Red light breaks seed dormancy and stimulates seed germination by upregulation of CYP707A2 and downregulation of NCED gene both of which reduce ABA level. For GA, the rate-limiting step in its biosynthetic pathway is cataylzed by GA3-oxidase (GA3ox), and deactivating enzyme is GA2-oxidase (GA2ox), both of which negatively regulate seed germination. At gene level, the two biosynthetic and deactivation pathways are regulated by transcription factors. Promotion of seed germination by GA requires destruction of DELLA proteins. However, if ABA level increases, it promotes ABI—class protein phosphatase—regulated transcription factor and DELLA protein availability. DELLA protein, in turn, suppresses GA biosynthesis, thereby creating a positive feedback loop. The balance between the activities of ABA and gibberellin is controlled by the stage of seed development. During early stages of seed development, gibberellin sensitivity is low and ABA sensitivity is high. However, this gets reversed during seed germination. In addition to environmental conditions, other hormones like ethylene and brassinosteroids reduce the ability of ABA to inhibit seed germination.

**Table 28.2** Factors affecting seed dormancy and methods to overcome it

| Causes of dormancy | Examples | Methods to overcome dormancy |
|---|---|---|
| Rudimentary embryo | *Ginkgo biloba*, barley, wheat, and oat | Afterripening; stratification |
| Rudimentary endosperm | Tomato, tobacco, pepper, and *Datura* | |
| Impermeability of seed coat to water | *Chenopodium* and legumes | Scarification with acid |
| Impermeability of seed coat to gases | Apple, *Xanthium* (cocklebur) Graminaceae | Scarification |
| Inhibitors | Tomato, *Cucurbita* | Removal of pericarp or testa leach in running water |
| Hard seed coat | *Lepidium, Capsella, Acacia* sp., mustard | Scarification, treatment with enzymes and acids |

## 28.5 Mechanisms to Overcome Seed Dormancy

In nature, seeds overcome dormancy by different mechanisms as follows (Table 28.2):

- Microorganisms present in the soil weaken and decompose the hard seed coat.
- Digestive juices present in the alimentary canal of the fruit-eating birds make the seed coat soft.
- Mechanical abrasions weaken the tough and impermeable seed coat.
- Washing away of inhibitors by rain or irrigation water.
- Inactivation of growth inhibitors by high or low temperature.
- Synthesis of growth hormones.
- Maturation of embryo.

## 28.6 Release from Seed Dormancy

There are mainly two events taking place at the time of germination. *Primary events* are imbibition of water, leading to swelling and hydration. During hydration, dormancy-breaking signals (physical and environmental) are perceived and transduced. This is followed by a series of *secondary events* which overcome dormancy by lowering the level of ABA. The next event is induction of germination which results in the mobilization of reserve food materials in seed. During this phase, level of GA increases, and catabolic reactions increase more than anabolic reactions. There is depletion of tissue surrounding the embryo which is associated with

## 28.6 Release from Seed Dormancy

loosening of the cell wall. As a result, radicle shows extension growth. Seed dormancy is under the control of specific genes which get activated or inactivated due to changes in the levels of ABA and GA.

### 28.6.1 Afterripening

Freshly harvested seeds of many plants do not germinate or show poor germination even under optimal conditions. They require high temperature and dry storage for variable period of time. This phenomenon is known as **afterripening** and is observed in grasses and many dicots. Seeds that require afterripening may show coat-related dormancy, and in such cases, the removal of seed coat induces germination, e.g., in wheat, barley, oat, *Arabidopsis*, and tomato. In embryo-related dormancy, removal of seed coat does not induce germination, and such seeds may require afterripening. This method of breaking dormancy is usually performed in a drying oven. If seeds become too dry (5% or less water content), the effectiveness of afterripening is diminished. The biochemical changes that take place during afterripening are many, and in most cases GAs and ABA both are involved. In the seeds of wild oat, which show embryo-related dormancy, application of GA overcomes dormancy. Similar effect of GA is also observed in tomato and *Arabidopsis*. GA treatment can substitute for afterripening period. A correlation between GA treatment and low temperature is known for many physiological processes. The dormant seeds of many species require afterripening, whereas freshly harvested ones germinate better at low temperature.

### 28.6.2 Impaction

During seed development, when the seed water content is high, the hilum remains open so that water can evaporate. Once the water content decreases in the seed, then the hilum swells and blocks the entry of water from outside into the seed. In some other seeds, hilum has a plug called **strophiolar plug**, which must be mechanically removed before water and oxygen can enter. This plug also controls seed moisture content and maintains seeds in their dormant state. Impaction involves vigorous shaking of seeds to remove strophiolar plug in leguminous seeds, thereby facilitating their release from dormancy, e.g., sweet clover (*Melilotus alba*).

### 28.6.3 Scarification

The dormancy induced due to hard seed coat can be removed by mechanically disrupting or removing seed coat. This process is called **scarification**. In nature, scarification can be accomplished by microbial action, fire, leaching of inhibitors, rainfall, or passage of seeds through animal digestive track. Artificially it can be done by rubbing seeds with sandpaper, removing seed coat with knives, or treating them with acid, e.g., in case of some tropical leguminous seeds, cotton, and morning glory, seeds germinate by soaking in concentrated sulfuric acid for up to an hour.

## 28.6.4 Temperature

Many seeds require diurnal fluctuations in temperature for a season before they can germinate, e.g., *Rumex obtusifolius*. The temperature range over which germination occurs is an indication of the degree of dormancy of the seed. If the temperature range is narrow, then the seed is highly dormant. In many species, secondary dormancy is introduced by exposure to high temperature (20 °C) which causes large seasonal changes in the degree of dormancy. These seasonal changes restrict germination only in spring, which is the beginning of the most suitable season for growth in temperate climates. On the basis of temperature requirement, plants have been divided into summer and winter annuals. Summer annuals produce seeds in autumn and germinate in spring. In such cases, imbibed seeds are exposed to 1–4 months of low temperature (approximately 4 °C) to break dormancy. This process is termed **stratification**. In winter annuals, exposure to relatively high temperatures relieves seeds from dormancy without imbibition. In other words, in winter annuals, low temperature induces dormancy. This causes the seeds to germinate in autumn, which is the favorable condition for many Mediterranean climates.

## 28.6.5 Light

Seed dormancy is greatly affected by light. Light requirement prevents germination of seeds that are buried deep in soil. Seeds which specifically require light for germination are referred as **photoblastic** seeds. Such seeds are common in herbaceous plants which produce large number of seeds with very little reserve food material. Many of the photoblastic seeds require a brief exposure to red light. The photoreceptor involved is phytochrome, and breaking of seed dormancy exhibits red/far-red reversibility. For example, lettuce seeds germinate in response to red light exposure. In lettuce and *Arabidopsis* seeds, red light induces the transcription of gene encoding 3β-hydroxylases which catalyze the conversion of inactive GA into biologically active form. Since far-red light inhibits germination of light-requiring seeds, light-sensitive seeds under a canopy germinate only upon removal of stress by fire or destruction. Under a canopy, the seeds of many herbaceous plants receive only far-red light, as blue, green, and red lights are absorbed by the leaves of the canopy. This light filtering effect of canopies keeps the seeds in dormant state. All light-sensitive seeds exhibit seed coat dormancy, and light enables the radicle to rupture the seed coat by activating some enzymes involved in weakening of the enclosed tissues of the seed coat. Seeds must be in a hydrated state for phytochrome action. This has, for example, been observed in the seeds of lettuce (*Lactuca sativa*) and pepper grass (*Lepidium* sp.).

## 28.6.6 Light-Temperature Interaction

Seeds of lettuce (*Lactuca sativa*) and pepper grass (*Lepidium virginianum*) remain dormant at high temperature (35 °C) in the presence of light, but they germinate at low temperature (10–15 °C) in light. When temperature is increased, conversion of Pr to Pfr is not affected, but Pfr-triggered reactions are adversely affected, and dark conversion of Pfr to Pr speeds up. However, hydration of seeds is essential for light effect on Pr to Pfr conversion.

## 28.6.7 Leaching of Inhibitors by Rainfall

Seed coat of many plants has large amount of phenolics (trans-cinnamic acid, coumaric acid, and coumarin) as a defense against predation. Most of these phenolics are water soluble and are washed off during rainfall, thereby facilitating release from dormancy.

## 28.6.8 Leaching of Chemicals

Different species exhibit variable responses to a variety of chemicals, such as respiratory inhibitors, oxidants, sulfhydryl, and nitrogenous compounds, in facilitating the breaking of dormancy. Nitrate in combination with light is an obligate requirement for release of dormancy of *Sisymbrium officinale* (hedge mustard). Karrikinolide (a compound present in smoke) stimulates germination in many seeds. Nitric oxide (NO), a signaling molecule, can also break dormancy in some plants. The absence of NO reduces germination in *Arabidopsis* mutants, and the effect can be reversed by treating seeds with NO.

## 28.7 Sequence of Events Breaking Seed Dormancy

I. *Primary events*: The first step in breaking seed dormancy is imbibition leading to hydration of tissues. Seeds perceive dormancy-breaking signals from the environment, and these signals are then passed through different messengers to genes within the nucleus (Fig. 28.4).

II. *Secondary events*: Once the genes sense dormancy-breaking signals, they either trigger the degradation of ABA or enhance synthesis of GA. On the contrary, if the signals repress dormancy-breaking metabolism, then the level of ABA is maintained, and the level of GA is decreased (Fig. 28.5).

III. *Structural changes*: These changes involve weakening of structures surrounding the embryo and facilitate radicle emergence (Fig. 28.6).

IV. *Termination of dormancy*.

**Step : I** Primary event

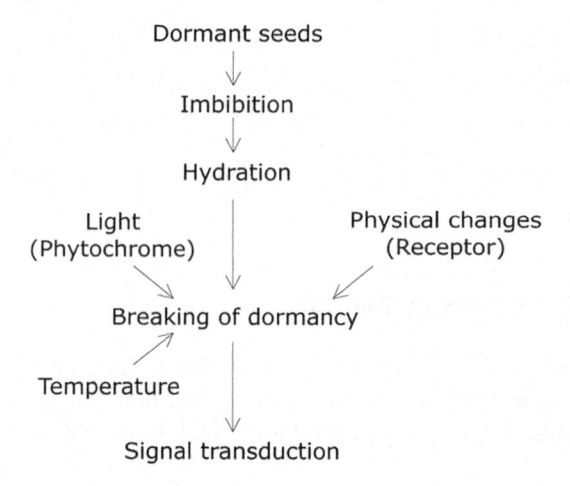

Dormant seeds
↓
Imbibition
↓
Hydration

Light (Phytochrome)     Physical changes (Receptor)
↘ ↓ ↙
Breaking of dormancy
↗
Temperature
↓
Signal transduction

**Step : II** Secondary event

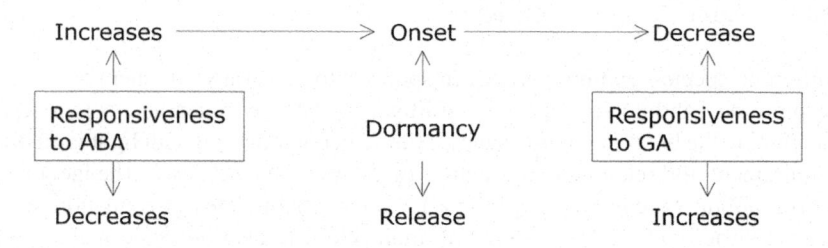

Increases ⟶ Onset ⟶ Decrease
↑          ↑          ↑

| Responsiveness to ABA | Dormancy | Responsiveness to GA |

↓          ↓          ↓
Decreases    Release    Increases

**Step : III** Structural changes

**Step : IV** Termination of Dormancy

**Fig. 28.4** A schematic representation of events involved in breaking seed dormancy

## 28.8 Significance of Seed Dormancy

### 28.8.1 Seasonal Synchrony

Seed dormancy ensures survival of a species. It is a kind of defense mechanism against unfavorable growth conditions. It benefits seed survival by synchronizing growth in favorable seasons. Thus, for example, a change in environmental temperature shifts the seeds of some plants in or out of the phase of dormancy, as in *Arabidopsis* and *Ambrosia* sp. (ragweed). Seed dormancy prevents germination in desert plants, which germinate only in rainy season. It helps seeds in sensing

## 28.8 Significance of Seed Dormancy

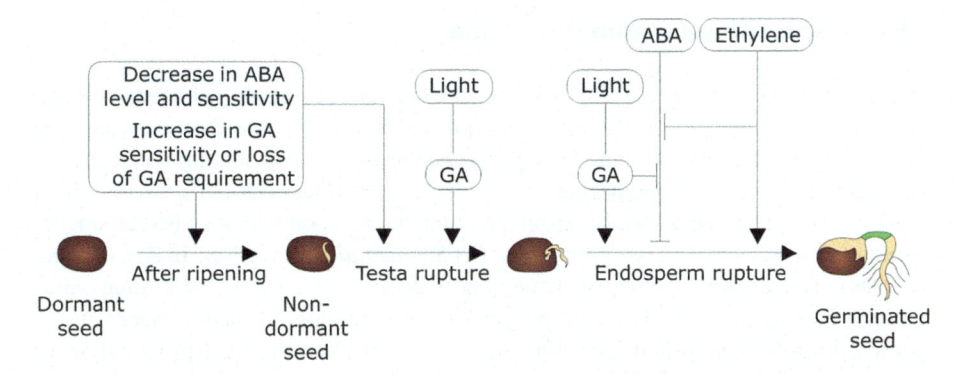

**Fig. 28.5** Secondary events in the process of release of seed dormancy. Following imbibition, there is lowering of ABA level and sensitivity. Endosperm and other tissues around the embryo undergo structural changes leading to emergence of radicle

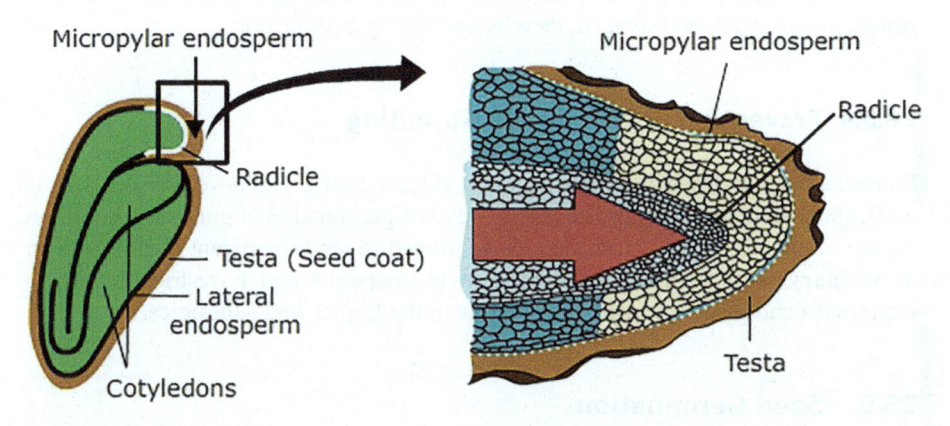

**Fig. 28.6** Cellular details showing weakening of endosperm and elongation of radicle

conditions conducive for germination. For example, cypress seeds germinate only in standing water. Seed dormancy also helps in adapting species to the seasonal characteristics of the environment. Thus, peach buds require cold period, so they are distributed only in temperate zone. Photoperiod affects dormancy. Seeds of *Begonia* germinate at photoperiods longer than 12 h. This would synchronize germination in summer.

### 28.8.2 Geographical Distribution

Seeds of tomato and many other fruits do not germinate until they have been separated from parental fruits. It helps them to achieve distribution/dispersal for their requirement of drying before termination of dormancy.

## 28.8.3 Spreading Germination Time

Plants develop one of the following mechanisms to spread germination time: (i) *Dormancy dimorphism*: Some weeds produce dimorphic seeds, i.e., seeds are both dormant and nondormant. This phenomenon is common among the members of Asteraceae and Chenopodiaceae. In *Chenopodium album* (fat hen), two kinds of seeds are produced: larger seeds germinate immediately upon release, whereas smaller ones germinate later at varying intervals. (ii) *Sequential dimorphism*: In *Avena fatua* (a grass) and another Compositae member *Asteriscus pygmaeus*, individual seeds along a fruiting axis develop with progressively increasing dormancy. Seeds in the panicle lose dormancy first, and after they are shed, then the next proximal seeds become nondormant. (iii) *Utilizing erratic opportunities to germinate*: Arabidopsis seeds are light-sensitive, and they germinate upon exposure to low fluence (2.5 µmolm$^{-2}$) of light, which is equivalent to moon light. Seeds of *Eucalyptus* sp. and *Albizia* sp. have a corky plug in the micropylar region. Fire disrupts this plug leading to penetration of water and oxygen, thereby facilitating germination.

## 28.8.4 Prevention of Pre-harvest Sprouting

Some degree of dormancy is advantageous, at least during seed development. This is particularly so in cereal crops because it prevents germination of grains while still on the ear of the parent plant. Germination of immature seeds on parent plant is known as **vivipary**. It is a rare phenomenon in angiosperms and is restricted only to mangroves and plants growing in estuaries in the tropics and subtropics.

## 28.9 Seed Germination

A seed is said to have germinated when the embryonic root emerges from the seed coat or, in other words, embryo growth resumes depending on environmental factors (Fig. 28.7). Seeds of some plants, such as rubber tree, sugar maple, and the willow, are viable for only a week or so, whereas many other seeds appear to retain their viability for several hundred years. The latter category includes seeds of *Mimosa glomerata* which, when kept in dry storage in a herbarium for 220 years, have been reported to germinate after being soaked in water. Similarly, seeds of *Albizia julibrissin* have been reported to germinate after 147 years. Seed germination in parasitic plants is dependent on host-plant signals. Obligate parasites, such as witchweed (*Striga*), must attach to a host within a few days of germination for their survival. Witchweed seeds germinate in response to hydroquinones released from the roots of its host (maize or sorghum). For some seeds, light is a major regulator of germination. Some of the other factors which affect seed germination include water for tissue hydration, oxygen to support aerobic respiration, and suitable physiological temperature. A delicate balance among the levels of various hormones is not only evident during early, mid, and late stages of seed development. Seed germination also exhibits modulation by IAA and GA (Fig. 28.8). Process of seed germination can be considered under five steps:

## 28.9 Seed Germination

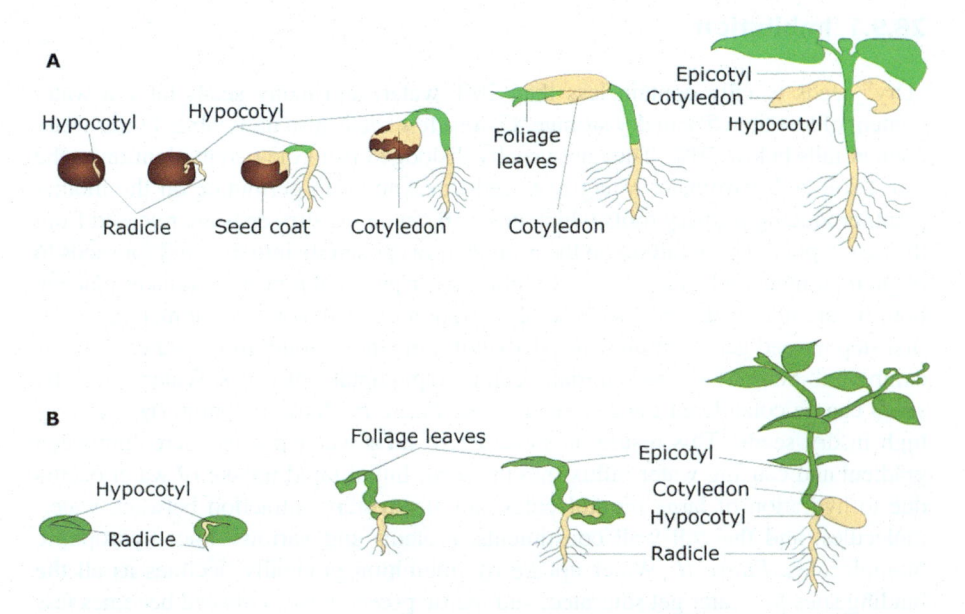

**Fig. 28.7** Stages of seed germination in dicots. (**a**) Epigeal type of germination in bean seed where hypocotyl elongates and pushes the cotyledonary leaves above the ground. (**b**) Hypogeal type of germination in pea seed where elongation of epicotyl leads to emergence of first foliage leaves above the soil

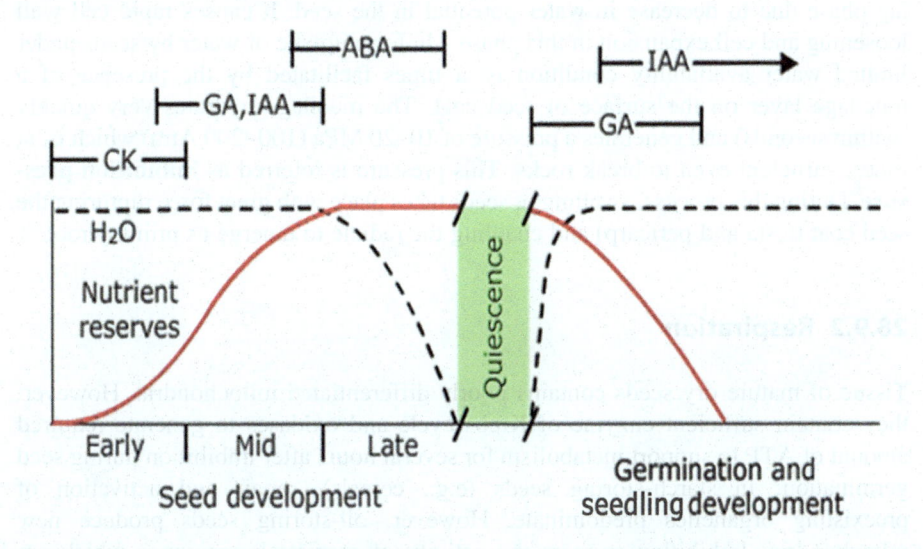

**Fig. 28.8** A graphical representation of the changes involving plant growth regulators, water, and nutrients, during seed development, germination, and seedling development. *ABA* abscisic acid, *CK* cytokinin, *GA* gibberellic acid, and *IAA* indole acetic acid

## 28.9.1 Imbibition

"Dry" seeds usually contain less than 20% water, and many seeds have a water content of less than 5% in dry storage. Quite a few seeds turn nonviable if their water content falls below 30%. They are usually dislodged from the parent plant on to the moist soil and germinate. Mangrove seeds are known to germinate on the mother plant, producing primary roots which grow 10 cm before the seeds are released from the parent plant. Upon falling on the ground, roots penetrate into the mud for seeds to begin their independent existence. Imbibition is a physical process and takes place in both living and dead seeds because of hydration of seed storage polymers. It is the first step in seed germination. Irrespective of dormant or nondormant state, all seeds exhibit imbibition. Dry seeds initially exhibit rapid uptake of water. Water uptake by seeds can be considered under three phases: *Phase I*: Matric potential ($\psi_m$) is very high in dry seeds. This results in lowering of their water potential creating steep gradient and causing water influx into the seed. Initial rapid uptake of water occurs due to hydration of large biomolecules and electrostatic attraction between water, molecules, and the cell wall constituents, proteins and various other hydrophilic biomolecules. *Phase II:* Water uptake by imbibition gradually declines as all the binding sites for water get saturated, and matric potential ($\psi_m$) of seed becomes less negative. This phase is considered as "**lag phase**" during which the solute potential ($\psi_s$) of the embryo becomes more negative due to breakdown of reserve food materials into osmotically active solutes (OAS), and metabolic processes, including transcription and translation, are initiated. *Phase III*: Uptake of water resumes after lag phase due to decrease in water potential in the seed. It causes rapid cell wall loosening and cell expansion in this phase. Uniform uptake of water by seeds under limited water availability condition is at times facilitated by the presence of a mucilage layer on the surface of seed coat. The mucilage hydrates very quickly (within seconds) and generates a pressure of 10–20 MPa (100–200 Atm) which is, at times, sufficient even to break rocks. This pressure is referred as **imbibition pressure**. During this process, swelling of seeds takes place with great force rupturing the seed coat (testa and pericarp) and enabling the radicle to emerge as primary root.

## 28.9.2 Respiration

Tissue of mature dry seeds contains poorly differentiated mitochondria. However, they contain sufficient enzyme of Krebs cycle and oxidases to generate required amount of ATP to support metabolism for several hours after imbibition during seed germination. In starch-storing seeds (e.g., cereals), repair and activation of preexisting organelles predominate. However, oil-storing seeds produce new mitochondria. Imbibition triggers the activity of respiratory enzymes which are conserved in the dehydrated state of the seeds. After an initial steep rise, a decline in respiration rate is observed until radicle penetrates the soil. Using preexisting mRNA transcript and ribosomes, protein synthesis also begins. This is followed by the release of hydrolytic enzymes and renewed cell division and cell elongation in the embryonic axis. The hydrolytic enzymes digest and mobilize stored food reserves. Glycolytic and citric acid pathways operate, and ATP is produced along with carbon skeleton required for the growth of developing embryo. Although initially seeds may respire anaerobically, but as soon as oxygen is available, they start respiring aerobically. Plowing aerates the soil and facilitates seed germination. Seeds implanted deep in soil may not germinate due to insufficient oxygen. Another reason may be the absence of light in case of photoblastic seeds.

### 28.9.3 Light Requirement

Light sensitivity of seeds for the release of dormancy and germination has been discussed in Sect. 28.6.5.

### 28.9.4 Mobilization of Reserves and Growth Regulators

Depending upon the type of seeds, the food reserves may either be stored in the endosperm or in the cotyledons. Regardless of site of storage, the food reserves of the seeds consist of starch, triglycerides, or storage proteins, all of them being insoluble in water. These molecules must be broken down into respective smaller subunits that are soluble in water and can be easily absorbed and transported to growing embryo. In cereal grains, the starch reserves are mobilized from the endosperm to embryo via a shield-like cotyledon (scutellum). The outermost layer of endosperm has specialized cells forming an aleurone layer. The cells of aleurone layer secrete hydrolyzing enzymes, such as $\alpha$-amylases and proteases, which facilitate the breakdown of stored starch and proteins in the endosperm. Sugars and amino acids so formed are transported to newly formed epicotyl, hypocotyl, radicle, and plumule of the embryo through scutellum. The initiation of the biosynthesis of hydrolytic enzymes is promoted by gibberellic acid released from embryo (Fig. 28.9). The proteases released during seed germination are categorized into four categories depending on their site of action on the polypeptide chain of enzyme:
1. Endopeptidases: These enzymes cleave peptide bonds to yield smaller peptides.
2. Aminopeptidases: They cleave terminal amino acid end of the polypeptide chain.
3. Carboxypeptidases: Amino acids are cleaved by these proteases from the carboxyl end of the chain. 4. Peptidases: They degrade small peptide fragments into amino acids.

Thioredoxin, a regulatory disulfide protein, regulates mobilization of both protein and carbohydrate reserves in developing seedlings. In cereal endosperm, NADPH and a flavin enzyme NADP-thioredoxin reductase reduce thioredoxin, which then reduces disulfide (S-S) groups of many stored enzyme proteins, thereby increasing their solubility which leads to their rapid proteolysis. Thioredoxin also regulates the activity of selected enzymes in cereal grains and legume seeds. Transgenic seeds with altered thioredoxin activity clearly manifest its role in seed germination. Cereal grains with higher activity of thioredoxin have been found to germinate faster, and

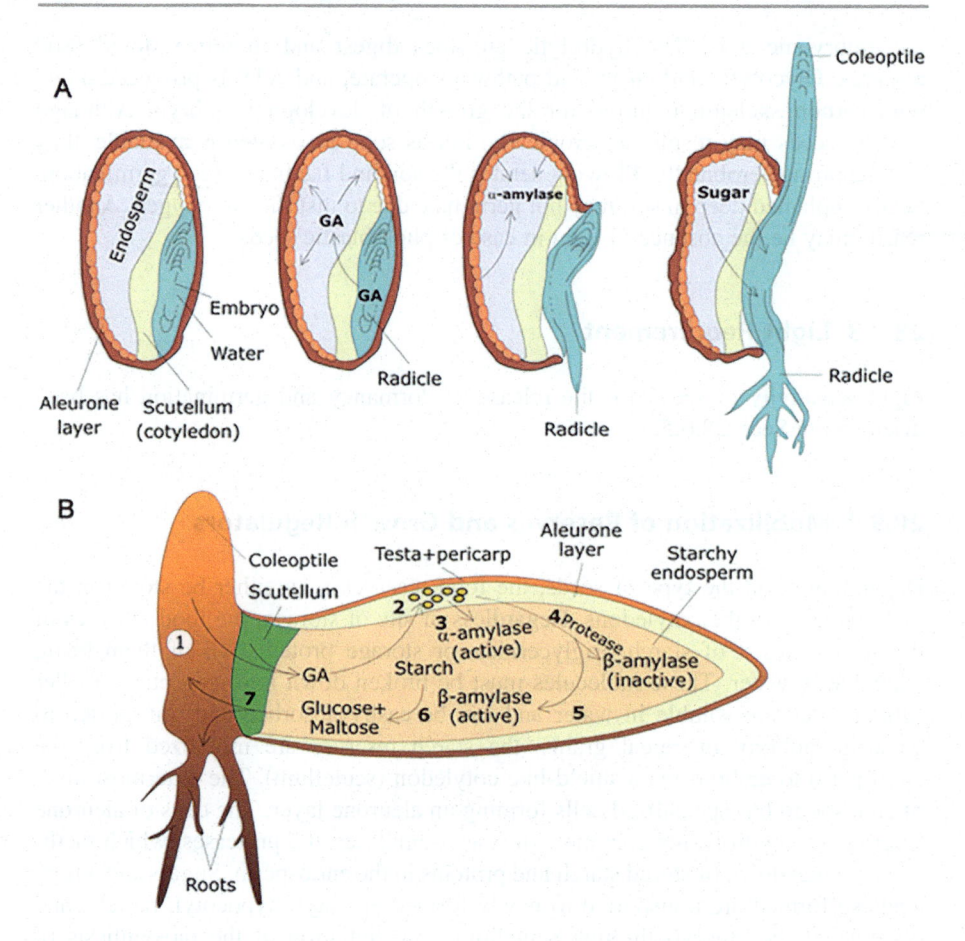

**Fig. 28.9** (a) Mobilization of reserves and release of growth regulators (GA) during seed germination in cereals. (b) The biochemical changes during seed germination in cereal seeds where GA induces release of amylase and other proteases from aleurone layer

downregulation of its activity slows down germination. Some of the proteases are synthesized de novo, while others may be present in the dehydrated seeds in inactive form. In cereals, acid hydrolases (hydrolytic enzymes active at pH between 4 and 5) are synthesized in the aleurone layer. Starch is a major carbohydrate reserve in most plants, consisting of amylose and amylopectin. It is cleaved by two types of enzymes: phosphorolytic enzymes that use Pi to break glycosidic bonds and hydrolytic enzymes that use water to break these bonds. Their actions have been discussed in Chap. 9.

The stored lipids in seeds, triacylglycerols (TAGs), are degraded and converted into glucose as plants are not able to transport fats to other tissues. It involves three organelles: (i) the oleosomes (the oil bodies), where triacylglycerols (TAGs) stored

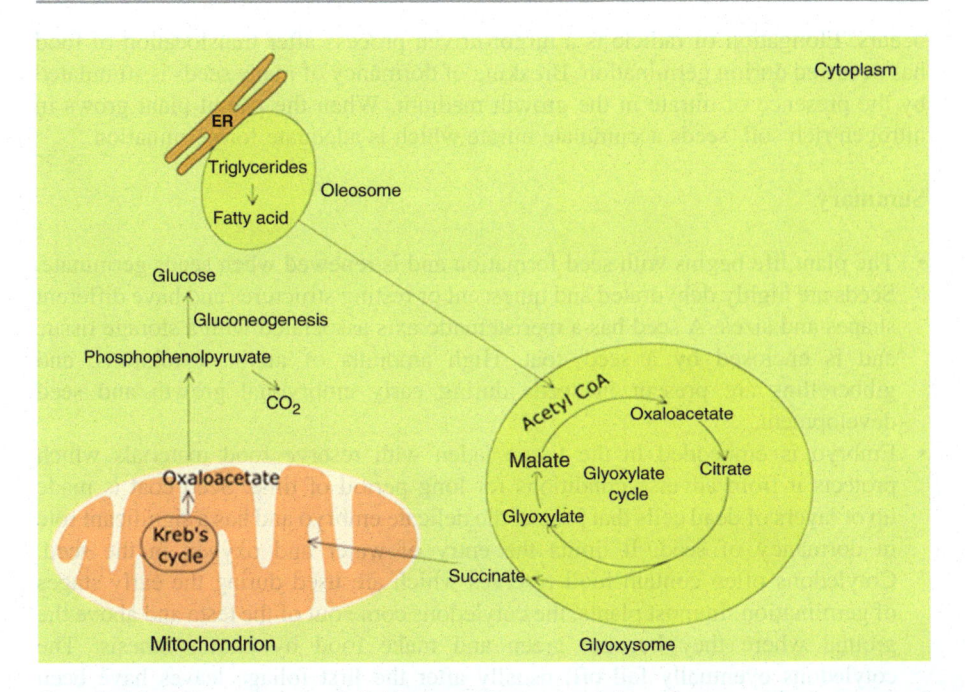

**Fig. 28.10** The conversion of lipids into glucose involving oleosome, glyoxysome, and mitochondria during germination in oil seeds

in oil bodies are converted into fatty acids and glycerol; (ii) glyoxysomes (specialized peroxisome), where fatty acids are converted into acetyl-coenzyme A (CoA) by β-oxidation and then to succinate by the glyoxylate cycle; and (iii) mitochondrion, where succinate is either respired or converted to malate which is transported to cytosol for further conversion to hexoses by gluconeogenesis (Fig. 28.10). Mineral elements stored in seeds as complex with phytic acid, namely, myo-inositol hexaphosphoric acid (IP$_6$), are hydrolyzed by enzyme phytase (a phosphatase that releases phosphate and chelated cations). During seed germination, phytase level increases, making phosphate and other minerals available to help embryo growth and development. The presence of phytates makes the seedling growth independent of external supply of essential elements for few days after germination.

## 28.9.5 Development of Embryonic Axis into Seedling

Radicle extension is the last event in seed germination. Two distinct phases of DNA synthesis occur in the radicle cells after imbibition. Immediately after imbibition, repair of damaged DNA (formed during dehydration) and synthesis of mitochondrial DNA take place. In the second phase, DNA synthesis associated with cell division occurs. Elongation of radicle is a turgor-driven process after translocation of food has occurred during germination. Breaking of dormancy of many seeds is stimulated by the presence of nitrate in the growth medium. When the parent plant grows in nitrogen-rich soil, seeds accumulate nitrate which is adequate for germination.

**Summary**

- The plant life begins with seed formation and is renewed when seeds germinate. Seeds are highly dehydrated and quiescent or resting structures and have different shapes and sizes. A seed has a meristematic axis associated with a storage tissue and is enclosed by a seed coat. High amounts of auxin, cytokinins, and gibberellins are present in seeds during early embryonal growth and seed development.
- Embryo is embedded in the tissue laden with reserve food materials which protects it from adverse conditions for long period of time. Seed coat is made up of layers of dead cells that protect the delicate embryo and has a significant role in dormancy of seed. It limits the entry of water and oxygen in the seed. Cotyledons often contain food reserves which are used during the early stages of germination. In most plants, the cotyledons come out of the testa and above the ground where they become green and make food by photosynthesis. The cotyledons eventually fall off, usually after the first foliage leaves have been formed. Almost all seeds contain stored food in endosperm. When seed germinates, hydrolytic enzymes present in the endosperm are activated or synthesized by GA action, resulting in breakdown of reserve food.
- Seed dormancy can be due to structural limitations or imbalance of growth substances. Primary dormancy is induced during seed development, and such seeds are released from the plant in a dormant state. On the other hand, secondary dormancy is induced in mature seeds after release from plants due to various stress factors. Seed dormancy can be coat or embryo-induced. Seed dormancy is affected by light, temperature, internal hormone level, and physiological status of the seed. The first step in breaking the dormancy is imbibition and hydration of tissues. The seed perceives dormancy-breaking signals from environment, and the signals are then passed through different messengers to genes in the nucleus, and seeds overcome dormancy.
- A seed is said to have germinated when the embryonic root emerges from the seed coat or, in other words, resumption of embryo growth is evident in response to different environmental factors. Imbibition is the first step of seed germination. It is a physical process and takes place in both living and dead seeds due to hydration of seed storage polymers. Subsequently, insoluble starch and proteins must be broken down chemically into smaller subunits that are soluble in water and can be easily absorbed and transported to the growing embryo. Radicle emergence is the last event in seed germination.

## Multiple-Choice Questions

1. The aleurone layer is specialized outer layer of endosperm, and it is rich in:
   (a) Arabinans and arabinoxylans
   (b) Cellulose and hemicellulose
   (c) Cellulose and pectin
   (d) Arabinans and cellulose
2. Phytates are major storage compound and mainly consist of:
   (a) Sulfur
   (b) Phosphorus
   (c) Iron
   (d) Copper
3. Absence of ABA in seeds leads to:
   (a) Seed dormancy
   (b) Nonviable seeds
   (c) Vivipary
   (d) Delayed germination
4. Photoblastic seeds are usually_____ in size:
   (a) Very small
   (b) Large
   (c) Medium
   (d) Variable
5. Strophiole, a special structure present in legumes, plays an important role in:
   (a) Uptake of water during imbibition
   (b) Gas exchange
   (c) Emergence of radicle
   (d) Blocking the entry of water so that seed does germinate precociously
6. Which of the following statement is true for seed?
   (a) Seed has high matric potential and low water potential.
   (b) There is no difference in water and matric potential in a seed.
   (c) Seed coat has low water potential and high matric potential.
   (d) Dry seed has high matric potential and low water potential.
7. The rate of respiration is _____ in germinating seeds:
   (a) Low
   (b) Average
   (c) High
   (d) Not affected
8. Which of the following seeds are non-endospermic?
   (a) Custard apple
   (b) Orchids
   (c) Wheat
   (d) Mango

9. Stratification of seeds is done to:
    (a) Stimulate natural winter conditions
    (b) Prolong shelf life of seed
    (c) Induce dormancy
    (d) Induce imbibition
10. Which of the following is not correct?
    (a) Scarification is the process of removing hard seed coat.
    (b) In nature, scarification can be accomplished by microbial action, fire, and leaching of inhibitors.
    (c) Scarification can also be done by treating the seeds at low temperature.
    (d) Scarification treatment is given to the seeds having seed coat-imposed dormancy.

## Answers

1. a  2. b  3. c  4. a  5. a  6. d  7. c
8. b  9. a  10. c

## Suggested Further Readings

Baskin JM, Baskin CC (2004) A classification system for seed dormancy. Seed Sci Res 14:1–16
Miransari M, Smith DL (2014) Plant hormones and seed germination. Environ Exp Bot 99:110–114
Roberts J, Downs S, Parker P (2002) Plant growth and development. In: Ridge I (ed) Plants. Oxford University Press, New York, pp 221–274

# Plant Movements

## 29

Rama Sisodia and Satish C Bhatla

Terrestrial plants are sessile and are incapable of changing their location. Yet, plants can sense their environment and can respond to sensory information through visible movements to optimize their survival, growth, and reproduction. The earliest findings published in this field were from Charles Darwin in 1880 in his book entitled *The Power of Movement in Plants* that explored the phenomenon of phototropism. Plant movements can be defined as the changes in the spatial orientation or conformation of an organ or its parts. Plant movements include movements toward light, opening and closing of flowers, growth of developing roots in search of water and nutrients, etc. In fact, inflorescence, fruits, and shoots are all capable of moving by sensing changes in environment, thereby adjusting their position, function, and behavior accordingly. Plant movements are extremely slow and often undetectable. Classical examples of visible plant movements include the response of sensitive plants like "touch-me-not" (*Mimosa pudica*) and the carnivorous Venus flytrap (*Dionaea muscipula*). In addition to whole organ, the movement of single cells is also possible which includes, for example, the movement of guard cells (which regulates the stomatal pore size), as well as growth of the pollen tube through the style. Such movements are caused by motor cells, driven by changes in pressure from within the cells exerted against the cell walls. Pressure causes changes in the dimensions of individual cells or tissues which results in distinctive movements. Plant movements are classified as:

1. **Tropic movements**: These movements occur in response to environmental signals. The direction of these movements is determined by vectorial signals, such as gravity and light.
2. **Nastic movements**: These predetermined movements occur independent of the direction of the stimulus. The location and structure of the motors driving these movements determine the direction of these movements.

## 29.1 Tropic Movements

Tropic movements are initiated by external directional signals that are perceived by the sensory receptors. These tropic responses are described as positive and negative tropisms based on whether the movement is toward or away from the signal, respectively. **Gravitropic** movements are directed by the perception of gravity by specialized sensory cells and **anisotropic** response of motor cells. **Phototropism** is guided by the direction of light. These include the movements of leaves and flowers that track the position of the sun. Such movements are also called as **heliotropic** movements, and the phenomenon is referred to as heliotropism. Climber plants have tendrils that wrap around their support in response to friction when they encounter mechanical support. Such touch (actually friction)-generated responses are called as **thigmotropic** responses. External signals can also be chemical signals that direct plant movements. Such movements are termed as **chemotropic** movements. They include movement of pollen tube toward the egg cell in response to chemical signals released from synergids. **Hydrotropism** refers to the growth movements in response to unilateral stimulus of water.

### 29.1.1 Phototropism

Several plant movements are driven by the necessity of plants to maximize capture of sunlight for driving photosynthesis. Light-directed movements, termed as phototropism, have been a subject of investigation since Charles Darwin published his book *The Power of Movement in Plants* in 1880. Among all environmental signals, light is one of the most versatile signals and varies along four parameters, namely, quality (wavelength), quantity (fluence, i.e., photons m$^{-2}$), direction, and duration (photoperiod). Phototropic responses can be termed as positive or negative depending upon whether the plant responds by growing toward or away from the light source. Aerial plant parts, including coleoptiles, hypocotyls, etc., are positively phototropic, while the tendrils of climbing plants are negatively phototropic. Interestingly leaves are **plagiotropic** organs orienting themselves at angles intermediate to the direction of light. Growth of roots is mostly non-phototropic, i.e., not directed by light. The stem of *Hedera helix* is negatively phototropic in the juvenile stage and prefers shade, while the mature plant is positively phototropic. The light-driven movements include subcellular movements as well as those of organs and whole plant. The opening and closing of stomata in the presence or absence of light are examples of light-driven movements at cellular level. Light is also known to direct the growth of seedlings toward photosynthetically active radiation (PAR) as seedlings transit from the use of stored reserve food (heterotrophic mode of nutrition) to becoming autotrophic. This is exemplified by the growth pattern of etiolated seedlings grown in dark and light (Fig. 29.1). Stem of an etiolated seedling is long and spindle-like as it grows out in search of light. Upon exposure to light, de-etiolation triggers reduction in shoot length and increase in radial growth. Shade growing plants are also known to accelerate their shoot growth towards

**Fig. 29.1** (a) An etiolated sunflower seedling with elongated stem and persistent apical hook. (b) A de-etiolated seedling with reduced stem growth and straightening of apical hook

ambient light. Nascent leaves are also affected by the intensity of light and reorient themselves to maximize the interception of light. Thus, most terrestrial plants use sunlight as signal to direct their growing organs for optimal utilization of light source. Phototropism is a growth response to a light gradient. Under laboratory conditions the growing plant organ bends toward unilateral light (Fig. 29.2). However, under normal light, as plant receives light from all sides, the bending is still seen as the fluence rate varies from one side to the other, creating a gradient. Differences in fluence rate influence phototropic response of bending of plant organ. The light gradient across a plant varies according to the optical properties of the tissue as well as the variation in incident light. The former properties relate to complex interactions between light and the pigments which cause scattering, reflection, or diffraction of light as it passes through the cells. The pulvinus located at the leaf base plays an important role in perception of the light signal for directing movement of lamina to face the light. Relocation of chloroplasts within the mesophyll cells in relation to light is also mediated by light. In high light intensities, plants prevent potential damage of the chloroplasts by aligning along the anticlinal walls, i.e., parallel to the incident light. In low light, on the other hand, chloroplasts are aligned along the periclinal walls, i.e., perpendicular to the incident light, in order to maximize absorption (Fig. 29.3). The relationship between fluence and response is best understood on analyzing fluence-response curves obtained by analyzing plant response to varying amount (fluence) of total light (Fig. 29.4). A classical phototropic response to unidirectional light shows an initial positive response (fPIPP—first pulse-induced positive phototropism) represented by a peak called the first peak or first positive curvature, and it occurs in response to a pulse of low fluence of light.

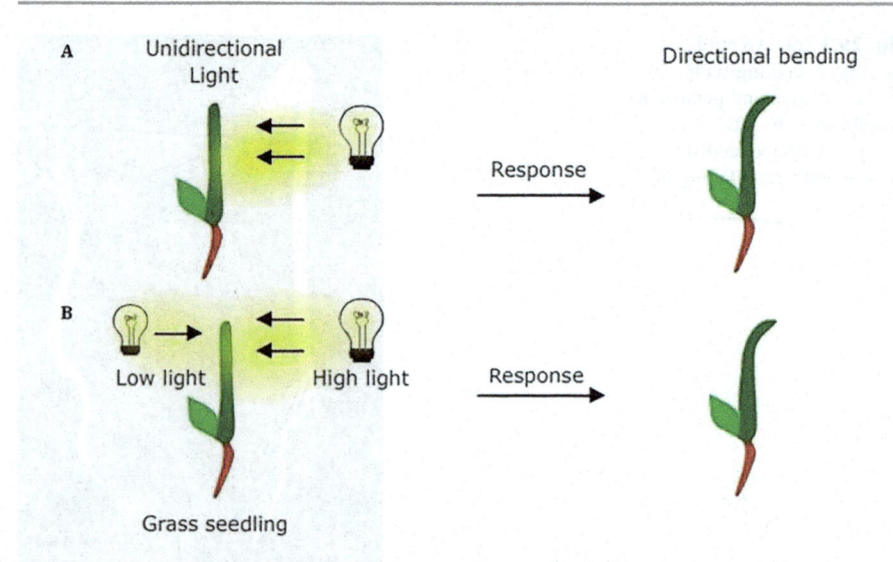

**Fig. 29.2** (a) Illumination by unilateral light causes the grass coleoptile to bend toward it; (b) illumination provided from all sides, but higher intensity of light from right also results in bending toward greater illumination

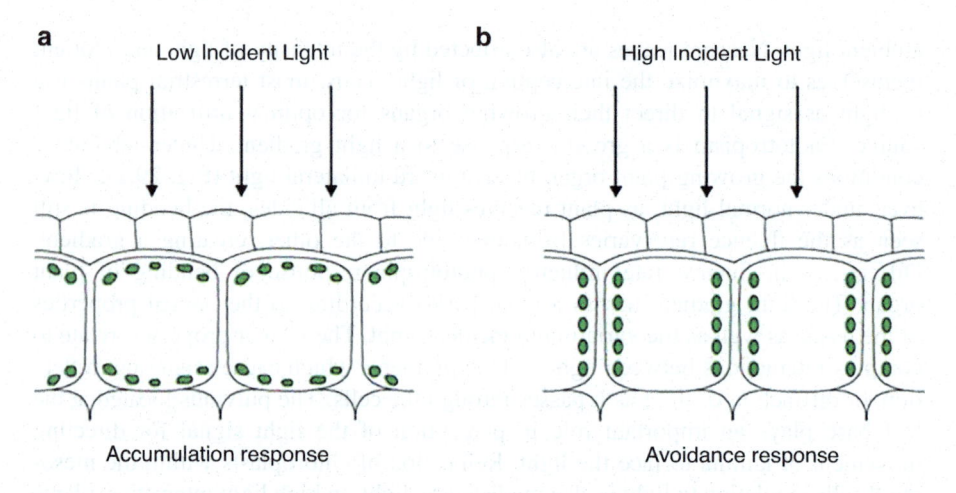

**Fig. 29.3** (a) In low light conditions, chloroplasts orient along the periclinal walls perpendicular to the incident light. (b) Under high light, the avoidance response involves chloroplast alignment along the anticlinal walls, parallel to the incident light, thereby minimizing exposure to light

There is again a rise in the response curve that forms a second positive response (sPIPP—second pulse-induced positive phototropism) on exposure to light of higher fluence. The third phase of phototropic bending response (time-dependent phototropism; TDP) occurs when the growing shoot is exposed to moderate to high

**Fig. 29.4** A typical phototropic fluence-response plot. fPIPP first pulse-induced positive phototropism, sPIPP second pulse-induced positive phototropism, TDP time-dependent phototropism

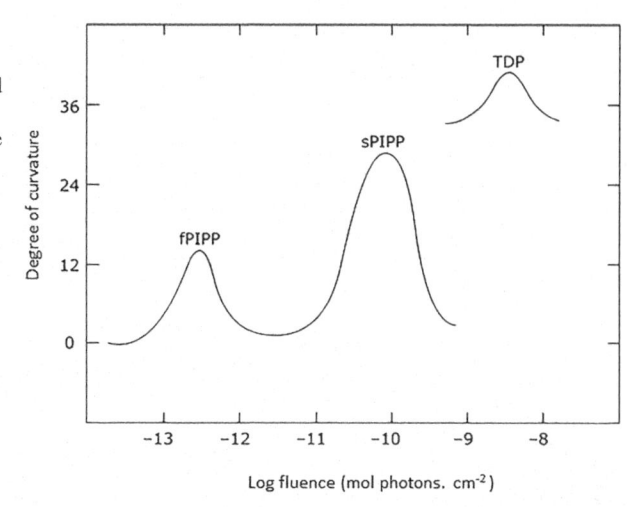

illumination (unidirectional) over an extended period. It is important to note that the photoreceptor responsible for transducing this light signal is located in the cytoplasm and the chloroplast movement is mediated by cytoskeleton. Analysis of action spectra for light mediated movement of coleoptile of plants such as *Avena sativa*, *Zea mays*, etc. has revealed that they share a common photoreceptor—a flavoprotein called **phototropin**. The absorption spectrum of phototropins shows a peak in the blue and UV-A regions of the spectrum. Phototropins mediate a number of blue light-induced responses such as phototropic movements, stomatal opening-closing, and redistribution of chloroplast in response to intensity of light. Phototropins have been studied in details in several plants including *Arabidopsis thaliana*, *Avena sativa*, and *Pisum sativum*. Initially, phototropin was characterized in an *Arabidopsis* mutant *non-phototropic hypocotyl* 1 or *nph 1*, which is unable to respond to the phototropic stimulus and lacks the 120 kDa membrane protein. It was later renamed as phototropin 1 (phot1) because of its role in phototropism. *Arabidopsis* contains two phototropins—phot1 and phot2—that have partially overlapping roles in regulating phototropism. Phototropins have two segments (Fig. 29.5): the N-terminus has the photosensory domain that contains two LOV domains that bind with flavin mononucleotide (FMN) making the phototropin sensitive to light. LOV domains share sequence homology to a number of eukaryotic and prokaryotic protein motifs involved in sensing light, oxygen, and voltage (hence the name LOV). The LOV domains consist of five antiparallel β-sheets and two α-helices that bind to FMN. The C-terminus has a serine-threonine kinase activity. In dark phototropin is unphosphorylated and inactive. The light signal initiates a photochemical reaction which is perceived by phototropin. This sets up a signal transduction chain beginning with the autophosphorylation of phototropin and initiates a phosphorylation cascade. In addition to phototropin, another photoreceptor called **neochrome** has been isolated from the fern *Adiantum capillus-veneris* that has the properties of both

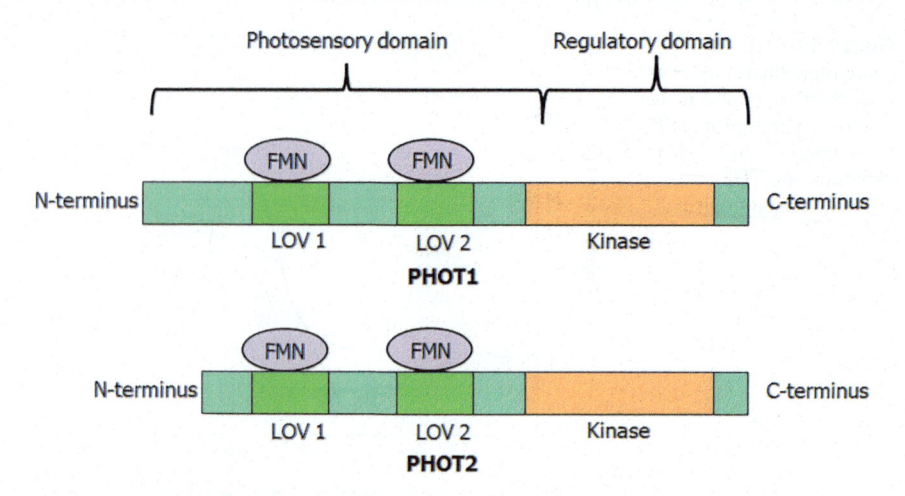

**Fig. 29.5** PHOT1 and PHOT2 of *Arabidopsis* contain two LOV (light, oxygen, voltage) domains. These bind to the cofactor flavin mononucleotide (FMN). C-terminal contains a kinase domain

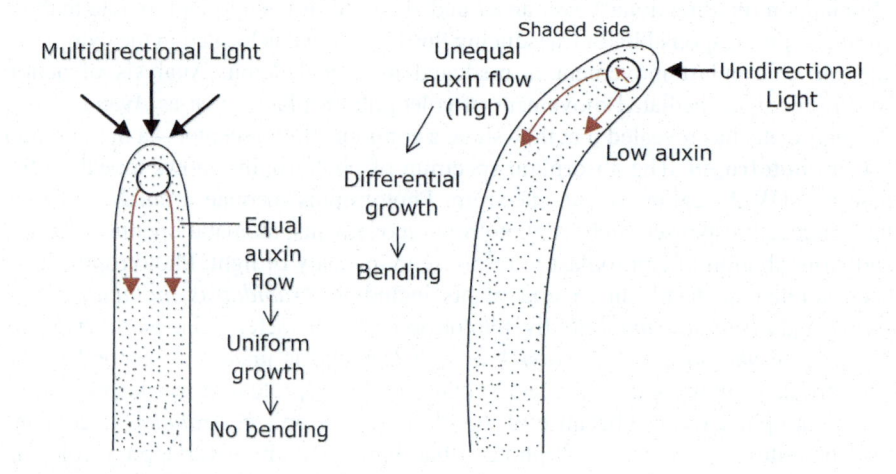

**Fig. 29.6** Differential auxin distribution as a response to unidirectional light leading to bending response. Increased auxin accumulation on the shaded sides enhances growth and consequent bending of the coleoptile toward light

phytochrome and phototropin and is regulated by both red and blue light. Both neochromes and phototropins are known to operate in ferns, mosses, and algae.

The phototropic response involves differential growth on one side of organs that responds to the light signal. The differential growth is mediated by lateral distribution of auxin as it moves from the apex basipetally to other areas. Upon unilateral illumination, an unequal distribution of auxin ensues and causes auxin accumulation and subsequent enhanced growth at the shaded side (Fig. 29.6). In *Zea mays*, the immediate tip area of the shoot apex (1–2 mm) is involved in auxin production, and the phototropic redistribution of the auxin occurs in the subapical region. The asymmetric distribution of auxin by phototropin is mediated by disrupting the distribution of auxin efflux facilitator protein called PIN. The latter is usually located at the basal end of cells associated with xylem cells and facilitates the vertical transport of auxin. Following phototropic stimulus, it has been experimentally shown that the basal location of PIN is disrupted in the cortical cells on the shaded side leading to accumulation of auxin and subsequent unequal growth and phototropic bending. As much as 20% higher concentration of auxin is known in *Brassica oleracea* hypocotyl on the shaded side with accompanying increase in auxin-regulated genes including those that code for α-expansins that regulate cell expansion.

## 29.1.2 Gravitropism

Gravity represents an omnipresent, constant force that acts on all organisms including plants, directing their growth and movements (Fig. 29.7). These movements are guided by gravity vector and are differentiated on the basis of vertical growth as **orthogravitropic**, as exhibited by primary root and shoot, or by the angle with which the organ grows (**plagiotropic**). The subterranean rhizomes exhibit a horizontal two-dimensional movement and are termed **diagravitropic**. The orthogravitropic movement in a single dimension facilitates the movement of the primary root vertically downward in search of water and minerals. Similarly, the young shoots grow vertically upward for optimal absorption of light and carbon dioxide for photosynthesis, and they are termed negatively geotropic in contrast to roots (positively geotropic). As the plant matures, the growth pattern shifts to plagiogravitropic mode allowing the plant to expand three-dimensionally, increasing

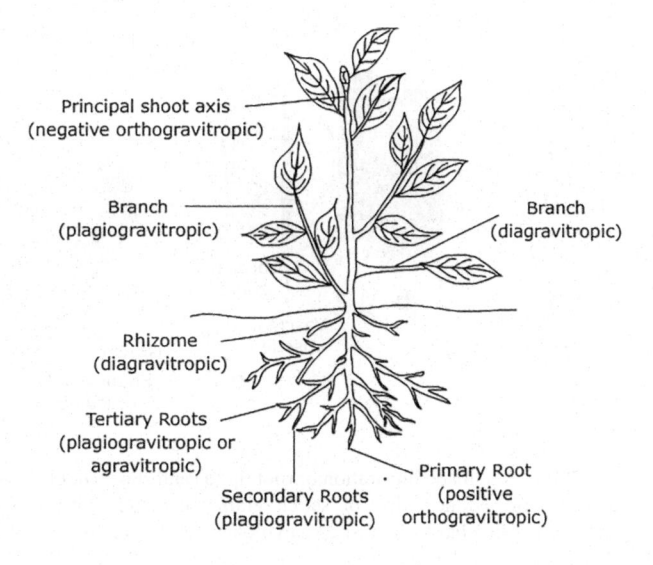

**Fig. 29.7** A variety of gravitropic responses direct growth and movements of plants

the supply of resources exponentially. The diagravitropic portions of the plants allow movements in two dimensions with the outgrowth of buds and leaves in case of shoots and adventitious roots growing out and thus expanding in the plant territory. Gravitropic movements are also modified by light with both the presence or absence of light and the direction of light modifying the response. Thus, young roots of *Zea mays* seedlings exhibit horizontal or diagravitropic response in dark. However, in the presence of red light, its roots grow vertically down changing their response to positive geotropic. In other cases, red light may increase or decrease the gravitropic response. Direction of light also modifies the gravitropic response. Unidirectional illumination from a direction opposite to the gravity signal results in an intermediate response curvature that depends on the magnitude and direction of the two vectors—light and gravity. In roots, the gravitropic response is also modified by the water potential gradient (hydrotropism) and also by mechanical stimulation (thigmotropism). The ability to perceive gravity signal resides in specialized cells called **statocytes**, which contain sedimentable starch grains called **statoliths**. Tissues that contain statocytes are called as statenchyma. These are located in the columella cells inside the root cap or calyptra. In stem and other aerial plant parts, the statocytes are located in the starch sheath. Statoliths consist of a group of starch grains enclosed within amyloplasts (Fig. 29.8). Usually amyloplasts are not mobile within the cell. However, they can move in the direction of gravity in columella cells of the root cap as well as in the starch sheath or endodermal cells. Amyloplasts' mobility is evident in the cells of the inner cortex of stems and the pulvini as well as in the motor tissue located at the nodes of grass stem. Starch grains are heavier than the cytoplasm and therefore tend to cluster under the influence of gravity near the lower surface (inner) of the statocyte (force sensor hypothesis). The weight and size of starch grains disturb the subcellular organization of the cytoskeletal elements and organelles such as endoplasmic reticulum, creating a physiological asymmetry in the cell or

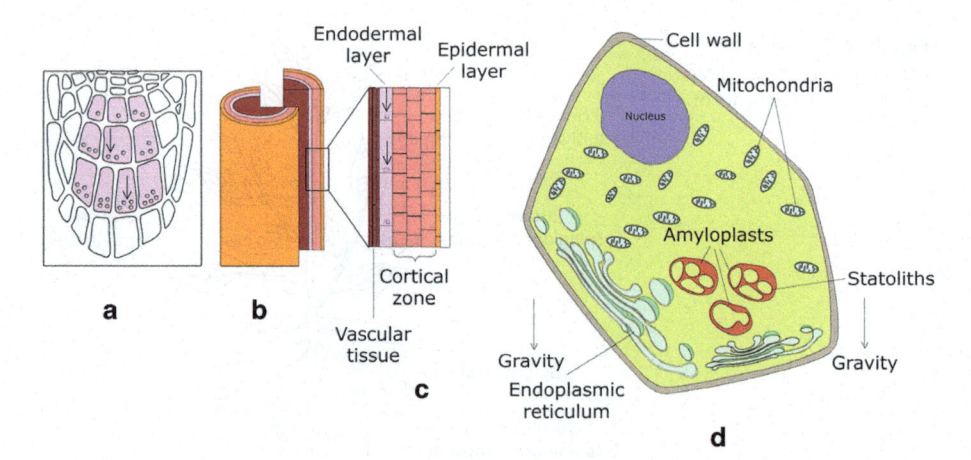

**Fig. 29.8** Schematic illustration of root tip (**a**) and shoot (**b**, **c**) in longitudinal section showing the direction of sedimentation of starch grains in response to gravity. (**d**) Statocytes containing amyloplasts that have established a cytoplasmic asymmetry in the cell in response to gravity

tissue. It has been shown that position (or inclination) of statoliths (and not their mass) is responsible for gravity perception in plant cells. Statoliths respond to gravity like liquid, and not like a granular material. They respond to the slightest change in angle as a liquid would do to attain a horizontal position. Thus, the hypothesis of statocytes as "force sensors" is now replaced by "position sensor" hypothesis. The gravity signal is perceived, transduced, and transmitted to generate a response in the zone of elongation of growing root and shoot (Fig. 29.9). The transmission and response of the gravitropic signal involve lateral redistribution of auxin in the elongation zone. The horizontal orientation of the plant shoot or root causes redistribution of auxin toward its lower side. This causes relatively more growth toward the lower side in comparison to the other side causing the negatively geotropic organ, such as the shoot or the coleoptile, to turn upward. In contrast, in positively geotropic organs, such as roots, the higher concentration of auxin in roots

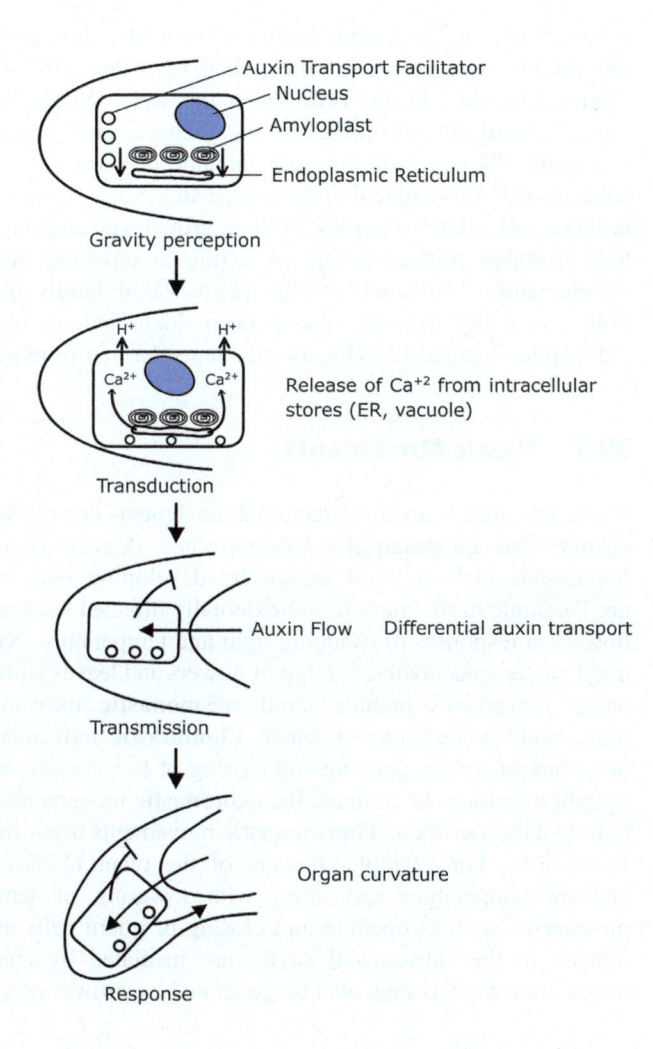

**Fig. 29.9** Phases of generation of gravitropic response: perception, transduction, transmission, and physiological response

inhibits elongation on the lower side in comparison to the upper side causing the root to grow downward. The path of auxin transport in root is often explained with the "auxin fountain model." Auxin synthesized in the shoot is transported basipetally down through the vascular tissue into the root and then acropetally toward the root tip. In the columella, the direction of the auxin flow is reversed, and it moves basipetally in a uniform manner into the cells of the root cortex in the zone of elongation. However, when roots are displaced horizontally, the auxin flow in the columella is redistributed laterally toward the lower side of the root. The higher concentration of auxin in the lower side inhibits elongation relative to the upper side, thus causing roots to turn down.

### 29.1.3 Chemotropism

External signals, including certain chemicals, are also known to drive plant movements. The growth of the pollen tube through the style is guided by the chemical secreted by the cells lining the style. Within the ovary the pollen tube moves toward the egg steered by the chemical cues provided by the degenerating synergids. These chemical signals include proteins like stigma/style cysteine-rich adhesion (SCA)—a lipid transfer protein that is secreted by the epidermal cells lining the style and is known to play a role in growth and adhesion of the tube in the stylar tract. Another protein that plays a role in directing the growth of the tube is **chemocyanin** (a member of the phytocyanin family of blue copper proteins). Pollen-attracting peptides have been identified in *Torenia* as cysteine-rich polypeptides, called **LUREs**, as attractants derived from synergid cells.

## 29.2 Nastic Movements

Nastic movements are nondirectional movements that are driven by growth or turgor changes and are genetically predetermined. Autonomic internally directed nastic movements include those shown by developing buds as they swell and open up. Paratonic nastic motions are externally directed such as those of the leaves and flowers in responses to changing light and temperature. **Nyctinastic** movements or night movements involve folding of flowers and leaves with the onset of night. Other nastic movements include rapid **seismonastic** movements and **thigmonastic** movements in response to touch. **Photonastic** movements such as folding and unfolding of leaves, opening and closing of flowers, etc. are synchronized by dark to light transition. In contrast, the **skotonastic** movements are synchronized by the light to dark transition. **Thermonastic** movements occur in response to temperature fluctuations. For example, flowers of the plant *Crocus* (Iridaceae) open under ambient temperature and close with lowering of temperature. **Chemonastic** movements such as opening and closing of guard cells in response to the carbon dioxide in the substomatal cavity are mediated by changes in the guard cell osmoticum. Signals may also be generated by growth processes, like movement of

## 29.2 Nastic Movements

root hairs through the soil, seed germination, initiation of leaf primordia, and shoot development.

### 29.2.1 Epinasty and Hyponasty

Nastic movements are also displayed by developing organs, such as leaves, petals, and sepals, which exhibit structural and physiological differences between their adaxial (upper) and abaxial (lower) sides. As these organs develop, they may exhibit unequal growth expansion on the two sides, resulting in upward (**hyponastic**) or downward (**epinastic**) growth (Fig. 29.10). Such bending movements are mediated by unequal growth and are part of normal development of the plant. For example, leaves of tomato plant growing in soil flooded with water exhibit epinastic movement of leaves. These are under endogenous control, and though the growth is irreversible, the bending of the organ is reversible. It has now been proved that hyponastic and epinastic movements are controlled by the differential distribution of auxin and ethylene. Similar epinastic and hyponastic movements are also exhibited by developing vegetative and floral buds. The dome-shaped vegetative apex initially produces leaf primordia by more rapid growth on the lower side, which becomes convex, while the upper side becomes concave. Such unequal growth results in the developing leaf primordia forming a sheath around the young shoot primordium, resulting in a compact bud formation (Fig. 29.11). As the growth proceeds, the lowermost leaves change from hyponastic to epinastic growth. Expansion of the upper side changes the leaf architecture from convex/concave to a typical flat leaf lamina. Elongation of the internode results in the separation of the newly flattened leaf lamina, which now becomes a part of the shoot. The sequential change of the leaf growth from hyponasty to epinasty is under endogenous control and enables leaves to expand and photosynthesize. Similarly, the transition from vegetative to reproductive growth results in the vegetative shoot apex transforming into floral bud

**Fig. 29.10** Epinastic movement in tomato leading to downward bending of leaves in response to stimulus (flooding)

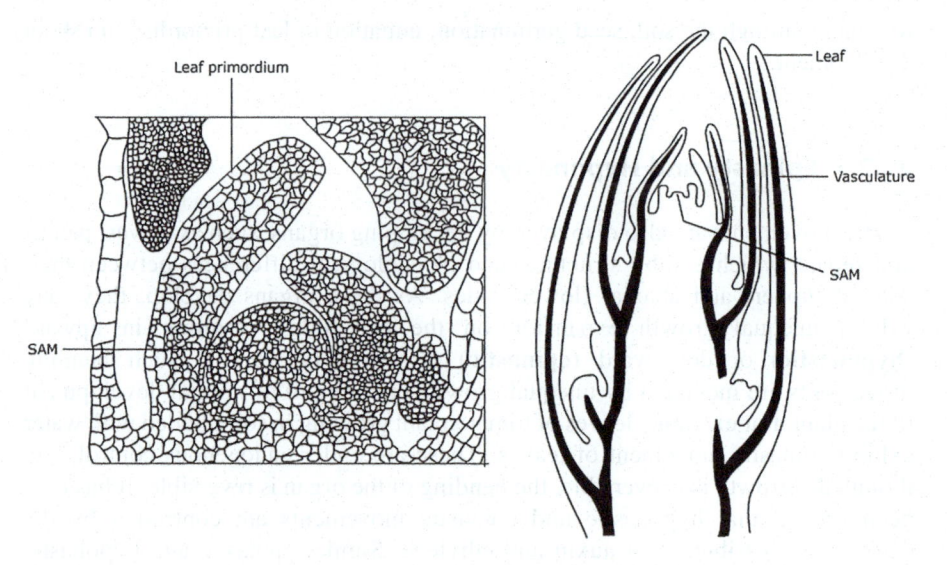

**Fig. 29.11** Developing shoot bud with ensheathing leaf primordia that transit from epinastic to hyponastic growth. *SAM* shoot apical meristem

**Fig. 29.12** Geotropic movement of flower stalks during development in *Papaver rhoeas* L.

primordium, and instead of leaves it produces flower parts: sepals, petals, stamens, and carpels. Since the function of flowers is to cause pollination, exposure of the stamens and ovary, which is a prerequisite to the process, occurs with the epinastic growth leading to unfolding of the calyx and corolla. In addition to these floral parts, other structures also exhibit epinastic movements. These include the epinastic movement of flower stalk of poppy plant as it curves down by differential growth forming a hook (Fig. 29.12). As the flower opens, growth is reversed resulting in hyponastic movement and straightening of the flower. Interestingly, the poppy flower responds to rain by downward bending of peduncle and loss of petal turgor which, therefore, causes closure of the flower, thereby preventing pollen loss.

**Fig. 29.13** (**a**) Nyctinastic movement of leaves in *Nicotiana tabacum*. Expanded/opened leaves during daytime. (**b**) Clasping of upper leaves at night

### 29.2.2 Nyctinasty

Nyctinastic movements (Gr. *nyctos*, night + *nastos* = closure) refer to the rhythmic sleep movements driven by reversible turgor pressure changes. During the day, plant leaves are typically in the horizontal or open position. The position changes during the night with the leaves acquiring a closed position. Leaves of plants like tobacco, clover, mint, etc. exhibit such movements (Fig. 29.13).

### 29.2.3 Thermonasty

Plant movements controlled by temperature fluctuations are called as thermonastic movements. Flowers of *Tulipa* sp. (Liliaceae) and *Crocus* sp. (Iridaceae) open on increasing the temperature and close on lowering the temperature by as less as 1–3 °C around the ambient temperature (Fig. 29.14). Closure of perianth is mediated by changes in the extension growth of the abaxial and adaxial surfaces of the perianth itself in response to temperature fluctuations.

### 29.2.4 Thigmonasty

"Touch-me-not" (*Mimosa pudica*) plant with its pinnately compound leaves is a popular example of a plant that exhibits thigmonastic movements, i.e., movements in response to touch (Fig. 29.15a, d). The touch stimulus is perceived by the **pulvinus** located at the base of each leaflet and is transmitted through sieve tubes. The

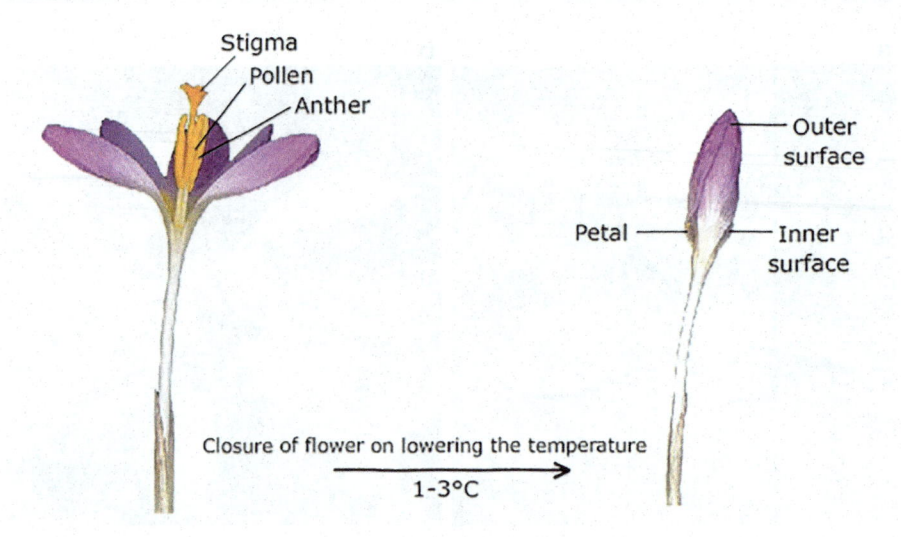

**Fig. 29.14** Thermonastic movement of *Crocus sativus* flower in response to temperature fluctuation. Flower petals close upon lowering the ambient temperature by 1–3 °C

stimulus includes ABA (abscisic acid), and ABA controlled ion transport leading to their diffusion into the upper region of the pulvinus causing efflux of $K^+$ ions and water into the intercellular space. The resulting loss of turgidity of the cells in the upper region of pulvinus causes the downward (closure) movement of the leaflets. The opening of the leaves takes place by the gain of turgor with active pumping of $K^+$ back into the cells (Fig. 29.15b, c). The inability of dark-grown plants to exhibit seismonastic responses suggests that these movements are ATP-dependent. $K^+$ efflux is regulated by $Ca^{+2}$ ions stored in vacuoles. An action potential is propagated from cell to cell causing neighboring leaves to respond in a similar manner. In *Arabidopsis*, four genes have been identified that are upregulated when the plant is touched (TCH1–4). Three of these encode proteins linked to $Ca^{+2}$ signaling, but TCH4 encodes xyloglucan endotransglucosylase, which modifies the structure of the xyloglucan components of the cell wall. It has been shown that treatment of tendrils with specific jasmonates causes them to coil in an identical manner to the touched tendrils. This effect was shown to be independent of ethylene and auxin.

## 29.3 Autonomous Movements

Interestingly there are few plant movements that are guided by both external and internal signals. These include movements that are under the control of circadian clock that are defined by the 24-hour diurnal rhythms of light and dark periods. Examples of such movements include the unfolding (and refolding) of flowers during the day, which synchronizes the opening of flowers for optimal periods of pollination. The circadian movements are synchronized and rephrased by environmental signals such as light or temperature.

## 29.3 Autonomous Movements

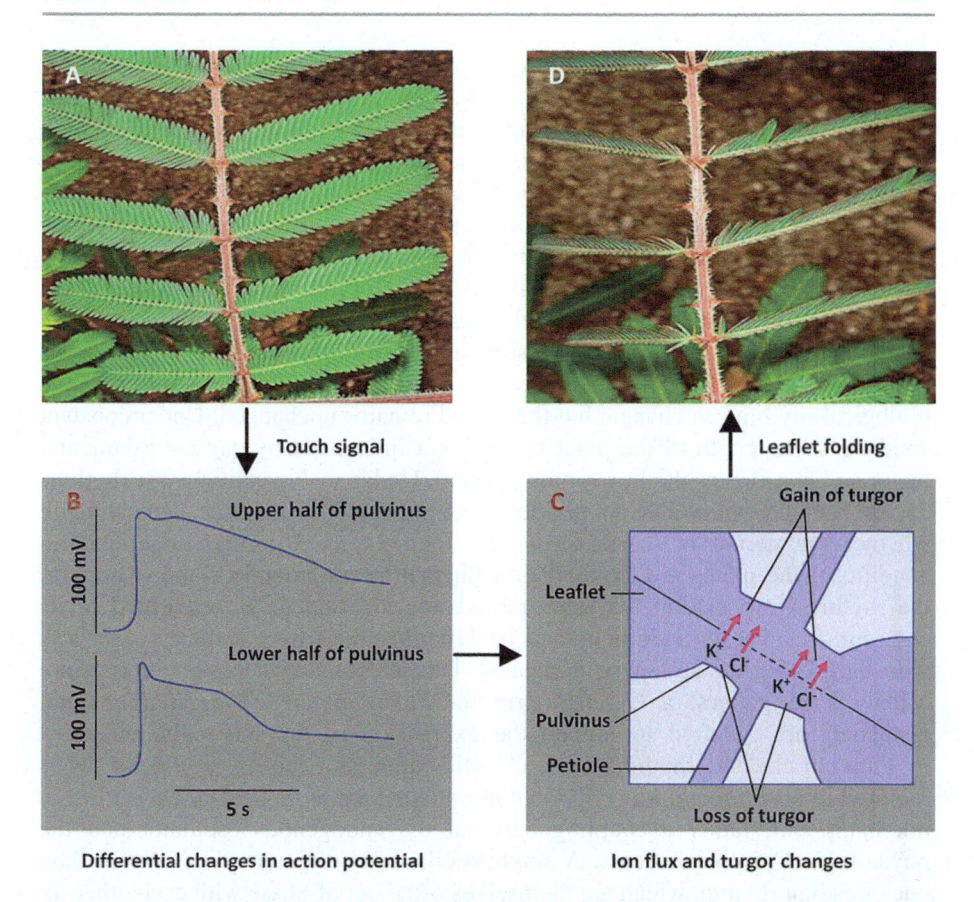

**Fig. 29.15** (a, d) Thigmonastic movements of leaflets of "touch-me-not" (*Mimosa pudica*) plant. (b, c) Mechanical stimulation activates ion channels that cause differential changes in action potential and closure of the leaflets

### 29.3.1 Diurnal Movements and Circadian Rhythms

Plants are routinely exposed to diurnal cycles of light and dark and exhibit rhythmic behavior in association with these changes. These include rhythmic movements of leaves and petals and opening and closing of stomata that influence the metabolic processes of photosynthesis and respiration. Flowers in the members of Onagraceae, Cactaceae, Convolvulaceae, and Oxalidaceae exhibit spectacular diurnal movements. Such flowers close by inward curvature of the perianth leaves and open by reversing the curvature. These movements are mediated by growth and take place as the flower/inflorescence develops. The rhythms alternate with a regular periodicity of 24 h and are called as **circadian rhythms** (derived from Latin, circa *diem* meaning approximately a day). Circadian rhythms are expressed continuously even in the absence of external controlling factors, e.g., upon transfer of plants from daily light and dark cycles to uniform conditions of either continuous dark or light. The rhythms are governed endogenously by an oscillator or a biological clock that is coupled to a variety of physiological processes. The biological clock is unaffected by temperature and can function normally under a variety of climatic conditions. It is, therefore, known to exhibit temperature compensation. In nature, the period of the circadian rhythms tends to be closer to 24 h (Fig. 29.16). Characteristic features define circadian rhythms. These include the period which refers to the time between two points in the repeating cycle, e.g., between consecutive peak (or maxima) and trough (or minima). Phase refers to any point on a cycle recognizable by its relationship to the rest of the cycle. Amplitude is the degree to which the observed response varies from the mean, and the range is the difference between the maximum and minimum values. The amplitude of a biological rhythm can change, but the period remains unchanged. Under constant conditions the length of the plant cycle is not influenced by any environmental factor and is called as the free-running period. Under such conditions the rhythms depart from the typical 24-hour period, either gaining or losing time, depending on whether the period is shorter or longer than 24 h. However, under natural conditions, the plants are exposed to a fluctuating environment, and, therefore, the rhythms also exhibit the same period and are said to be **entrained** to its environment. The **entrainment** can be brought about by a number of factors with light being the strongest modulator. The environmental signals or cues are called as **zeitgebers**, a German term for "time givers." The environmental zeitgebers are required to initiate the expression of the internally generated rhythms. In case the rhythms damp out with time, an exposure to a zeitgeber is required to restart them. It is important to note that the biological clock itself does not damp out, rather a coupling between the endogenous oscillator and the physiological response occurs. A single oscillator is often coupled to more than one circadian rhythm, which are themselves often out of phase with each other. In circadian rhythms, the operation of endogenous oscillator initiates response at a particular time of the day. A single oscillator can be coupled to multiple circadian rhythms, which may even be out of phase with each other. The entrainment of the molecular clock is mediated by specific photoreceptors. Both blue and red lights are effective in entrainment, indicating the role of both phytochromes and

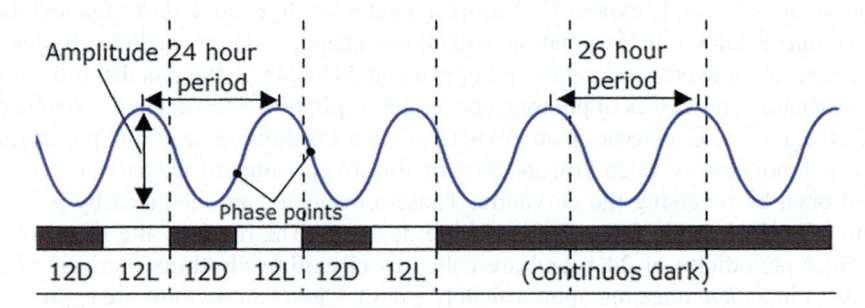

**Fig. 29.16** A typical circadian rhythm entrained to a 24-h light-dark cycle

cryptochromes in this process. In *Arabidopsis*, phytochrome C is involved in clock entrainment, whereas CRY1 and CRY2 proteins function as intermediates during the signaling of phytochrome-mediated clock entrainment.

### 29.3.2 Photoperiodism

The endogenous circadian clock of the plant enables it to monitor the time of the day and determines the specific time of the day at which a specific physiological or molecular event has to occur. This ability of a plant to detect day length and thus allow a seasonal response to occur is called as **photoperiodism**. Circadian rhythms and photoperiodism are modulated by cycles of light and dark periods. Geographic locations determine the length of the day and night. At equator, the length of the day and night is equal, and as one moves toward the poles, the days become longer in summer and shorter in winter (Fig. 29.17). Plants detect these seasonal changes and influence their photoperiodic responses. The understanding of plant photoperiodic responses was initiated with the work of Wightman Garner and Henry Allard in the 1920s as they conducted a series of experiments at the US Department of Agriculture, Beltsville, Maryland, on a mutant variety of tobacco Maryland Mammoth. The plants grew up to a height of about 5 m but failed to flower in the summer. However, the plants produced flowers in winter or on artificially providing short-day conditions during summer. It was, therefore, concluded that the length of the day was critical in determining the flowering response and later this could also be confirmed for a number of other plant species as well. Extensive experimentation

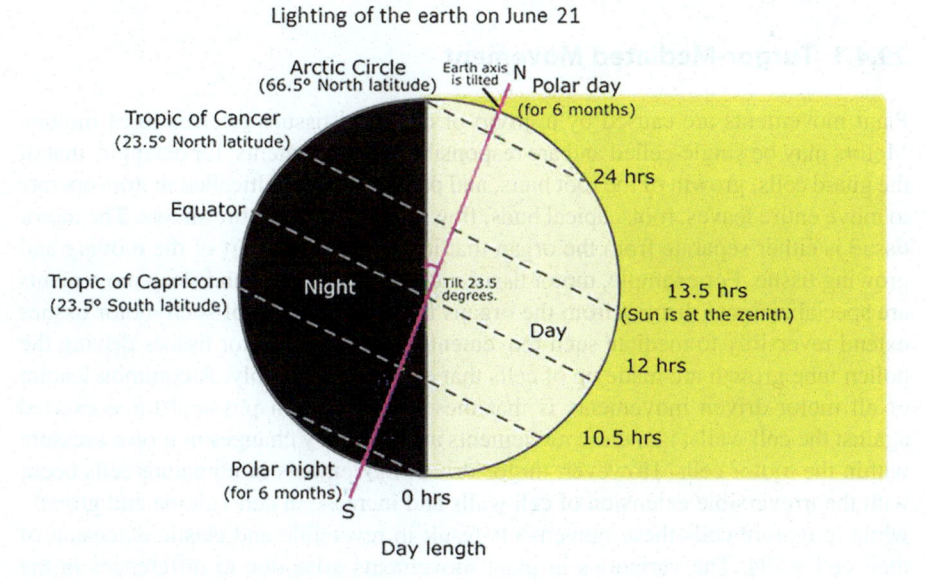

**Fig. 29.17** The relative length of day changes from equator to poles

lead to the establishment of the idea that flowering in plants is modulated by photoperiod and plants can be classified under different categories. 1. Short-day plants or SDPs: Plants flowering only under short days are called as qualitative SDPs. Incase flowering is accelerated by short days these plants are called quantitative SDPs. 2. Long-day plants or LDPs: Plants which flower only under long-day conditions are called as qualitative LDPs. Incase flowering is accelerated by long days the plants are called quantitative LDPs. Flowering in LDPs is promoted only when the length of the day exceeds a certain duration called as critical day length. The critical day length determines the flowering response in both LDPs and SDPs. Long-day plants flower when the day length exceeds or the night length is less than a critical duration in a 24-hour cycle. Short-day plants flower when the day length is less than or the night length exceeds a critical duration in a 24-hour cycle. LDPs, for example, many varieties of wheat (*Triticum aestivum*), monitor the lengthening of the day as spring progresses toward summer, and as the critical day length is reached, the plant flowers. SDPs, for example, *Chrysanthemum morifolium*, on the other hand, flower in fall when the day length is less than a critical day length. The absolute value of the critical day length, however, varies among plant species. A detailed account of photoperiodism has already been provided in Chap. 25.

## 29.4 Mechanisms of Movement

Plant movements can be caused by a variety of mechanisms mediated by turgor pressure changes, changing growth dynamics, movement by contraction, and change in conformation.

### 29.4.1 Turgor-Mediated Movement

Plant movements are caused by a group of cells and tissues referred to as motors. Motors may be single-celled and are responsible for movements, for example, that of the guard cells, growth of the root hairs, and pollen tubes. Multicelled motors operate to move entire leaves, roots, apical buds, flowers, fruits, and entire shoots. The motor tissue is either separate from the organ that it moves or it is part of the moving and growing tissue. For example, motor tissues driving heliotropic and sleep movements are specialized and distinct from the organs they move. Cells of such motor tissues extend reversibly to mediate such movements. In contrast motor tissues driving the pollen tube growth are made up of cells that expand irreversibly. A common feature of all motor-driven movements is that these are based on physical force exerted against the cell walls, i.e., these movements are driven by changes in turgor pressure within the motor cells. However, turgor-driven movements in immature cells occur with the irreversible extension of cell walls and increase in cell volume and growth, while in mature cells these movements result in reversible and elastic extension of their cell walls. The variations in plant movements arise due to differences in the signals that initiate them as well as due to the location of the motor cells and the mechanisms by which they respond. These movements rely on the turgor-mediated changes in a specialized motor organ called pulvinus that occurs at the base of the compound leaves in the families Leguminosae and Oxalidaceae. The motor cells of pulvinus are made up of large thin-walled parenchymatous cells that use turgor-mediated reversible changes in size and shape to alter the position of the leaf. The opposite sides of the pulvinus are called as flexor (upper) and extensor (lower) regions. At night, the extensor cells lose their turgor, while the flexor cells gain turgor resulting in the petiole turning toward the stem leading to closure of the leaf. During the day, the extensor cells regain their turgor, while the flexor cells lose their turgor resulting in the petiole turning away from the stem axis and leaf opening up. The relative swelling and shrinkage of the flexor and the extensor regions of the pulvinus thus regulate the movement of the leaves or leaflets. The relative position of the flexor and the extensor regions in the pulvinus is dependent on whether the closure movement is acropetal or basipetal. Nyctinastic movements are defined by blue light, phytochrome, as well as endogenous rhythms. The changes in the turgor pressure within the pulvinus are mediated by massive redistribution of potassium ion ($K^+$) between the symplast and apoplast regions of both the extensor and the flexor regions of the pulvinus. Depolarization of the membrane results in the $K^+$ channels to open and cause an influx of $K^+$ ions into the extensor cells, thereby increasing their turgor. A model integrating the role of phytochrome and secondary messengers, like inositol triphosphate, has been proposed to explain leaf movements of nyctinastic plants. Light signal activates the phytochrome and enhances the level of secondary messengers, inositol 1,4,5-triphosphate ($IP_3$) and diacylglycerol (DAG). The latter stimulates the release of $Ca^{+2}$ into the cytosol and causes phosphorylation of several cytoplasmic proteins that further causes the extrusion of protons from the cell. A proton motive force is generated thereof and causes the diffusion of $K^+$ ions into the cell.

### 29.4.2 Growth-Mediated Movement

Plants, being immobile, maximize their growth and survival by directing their growing active parts toward available resources. The growing root and shoot tips move forward through the soil and aerial environment, respectively, by means of motor tissues located in the elongation zone just behind the apex, with newer cells continually being added by the dividing cells of the apical meristems. Change in direction of the actively growing plant organs can also be achieved by growth-mediated curvature of the elongating shoot apex. Differences in growth rates in the opposite sides of the motor tissue as well as acceleration and inhibition of cell elongation in one region of the organ vs. the other side result in curvature and thereby bring about change of spatial orientation of the organ. Forward movement of root hairs, as it directs its growth around the soil particles, is affected by a shift in the pattern of growth of root tip toward the side opposite to the one that encounters the soil particles, thus growing around the obstacle and moving forward.

### 29.4.3 Movement by Change in Conformation

Change in the architecture and conformation also leads to movements such as those exhibited by the stomatal guard cells. The anisotropic structure and differentially thickened cell walls of the guard cells cause changes in their volume, thereby leading to opening and closing of stomata. The guard cells thus represent single-celled motors that operate by conformational changes. Multicellular motors, such as those that operate in the leaves of grasses, cause extensive changes in leaf conformation. These include motors located on the sides of the midribs along the major veins and which consist of a single row of large bulliform cells. As the grass leaves dry up, these bulliform cells dry up and cause the leaf lamina to roll along its length. When water becomes available, the cells inflate and thus flatten the leaves.

### 29.4.4 Movement by Contraction

In addition to movement effected by change in conformation, some movements are caused by contraction, for example, the movement of bulky storage organs, such as bulbs and corms, through the soil caused by specialized contractile roots. The contractile roots are characterized by the presence of an upper unbranched non-absorbing, thick fleshy part and a lower branched distal part. The radial expansion of the fleshy part of the roots caused by the growth of the parenchyma cells of the cortex creates a space in the soil facilitating the movement of the storage organ (Fig. 29.18).

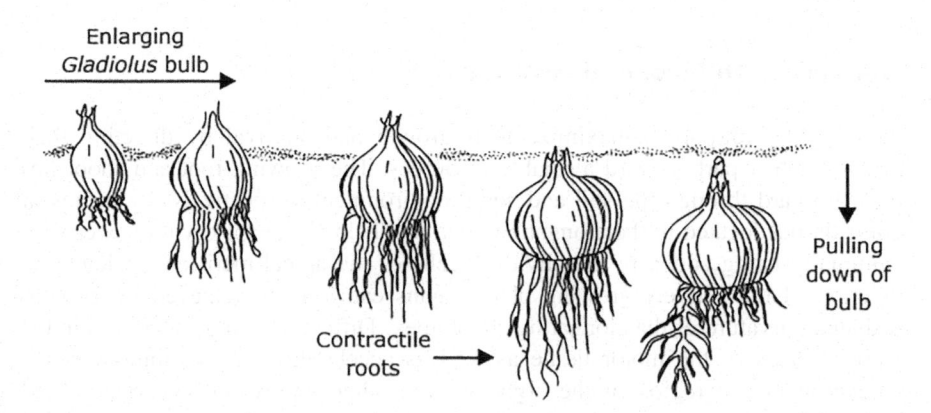

**Fig. 29.18** Enlargement-associated movement of the *Gladiolus* bulb through the soil. The contractile roots contract to pull down the attached bulb through the soil

## 29.4.5 Twining Plants

Other interesting examples of motors are those that operate in twining plants in which tendrils locate and attach to suitable support by forming extensive circumnutations. The tendrils may be modified branches as in *Bryonia* (Cucurbitaceae) and *Passiflora* (Passifloraceae) or even leaves as in *Lathyrus aphaca* or *Pisum sativum* (Fabaceae). The leaf lamina is highly reduced in such cases and is, in fact, limited to the midvein that functions as a tendril. The photosynthetic function of the leaves is performed by stipules. The tendrils resemble the torsion spring of mechanical watches, and differential contraction and expansion along the two sides of the tendril establish a dorsiventral asymmetry between the two flanks of the tendril resulting in progressive coiling of the tendril. Contraction of the tendril below the support results in drawing the plant toward the support. The thigmonastic response of the tendrils is rapid and highly selective. The epidermal protuberances found in Cucurbitaceae members, such as *Luffa cylindrica*, represent the receptors that elicit thigmonastic responses. The epidermal cells of the tendril are interconnected by plasmodesmatal connections that integrate the protoplasts into a symplast, thus allowing rapid transmission of the thigmonastic response.

## 29.5 Prey-Driven Movements

Several plants that lack chloroplasts are unable to follow an autotropic life and survive as parasites on autotrophic plants or are carnivorous that derive nutrition from their entrapped insect preys. These can be classified as:

- Holoparasites that are dependent on host plants for their organic nutrition, e.g., dodder (*Cuscuta*)
- Hemiparasites that have the ability to produce their own organic assimilates but are dependent on the host for absorbing water and solutes from the soil, e.g., mistletoe (*Phoradendron*)

Carnivorous plants are generally found in nitrogen-deficient habitats and obtain their nutrition by capturing and killing insects and absorbing the digested proteins of those insects.

### 29.5.1 Parasitic Plants

Parasitic plants locate their prey using a variety of mechanisms. For example, plants like the Santalaceae members—the mistletoes (*Viscum album* and *V. cruciatum*) and *Loranthus acacia* and *L. europaeus* from Loranthaceae—produce brightly colored and tasty fruits attracting birds. The birds eat the fruit, but the indigestible seed sticks onto their beaks and is generally wiped off against a tree branch where it gets deposited. On germination, the seedling invades the host tissue and uses a

**Fig. 29.19** *Cuscuta* or dodder plant coils around its host plant, drawing nutrition using a haustorium

haustorium to absorb nutrition. The parasitic dodder plant (*Cuscuta*, Convolvulaceae) is an obligate holoparasite that is dependent on its host plant to complete its life cycle. It produces innumerable seeds that germinate, and the seedling grows initially using the limited supply of nutrients available in the seed. The seedling twines around a suitable host attracted by the smell of a volatile compound or chemoattractants emitted by the host itself (Fig. 29.19). These chemical cues are volatile terpenoids such as α-pinene, β-myrcene, and β-phellandrene. As a contact is established, the parasitic epidermal and parenchymatous cells differentiate to form a secondary meristem and prehaustoria called as adhesive disk. Signals, like mechanical pressure, osmotic potentials, and phytohormones, such as auxins and cytokinins, initiate and control prehaustoria formation. Secretion of adhesive substances like pectins and other polysaccharides by the prehaustoria enhances adhesion which is further reinforced by the secretion of sticky substances such as arabinogalactan proteins by the host plant. Host cells also respond by an increase in the cytosolic calcium concentration initiating a signal transduction cascade. The prehaustoria develops into parasitic haustoria that penetrate the host stem by exerting mechanical pressure supported by degradation of host cell walls by the hydrolytic enzymes such as pectinases and cellulases. The haustoria invade the host tissue, establishing functional connections to the xylem and phloem of the host. *Cuscuta* forms a continuous symplastic connection with its host. The haustoria form an interface where nutrients, solutes, and carbohydrates as well as macromolecules, such as mRNA, proteins, and viruses, are exchanged between the host and parasite.

## 29.5.2 Carnivorous Plants

Carnivorous plants have developed strategy to entrap their mobile prey. The preys are attracted to the traps by means of sight or smell. Traps are made up of modified leaves and can be passive traps or active traps driven by mechanical triggers. Plants with mobile-active traps are the snap traps of Venus flytrap (*Dionaea muscipula*) and

## 29.5 Prey-Driven Movements

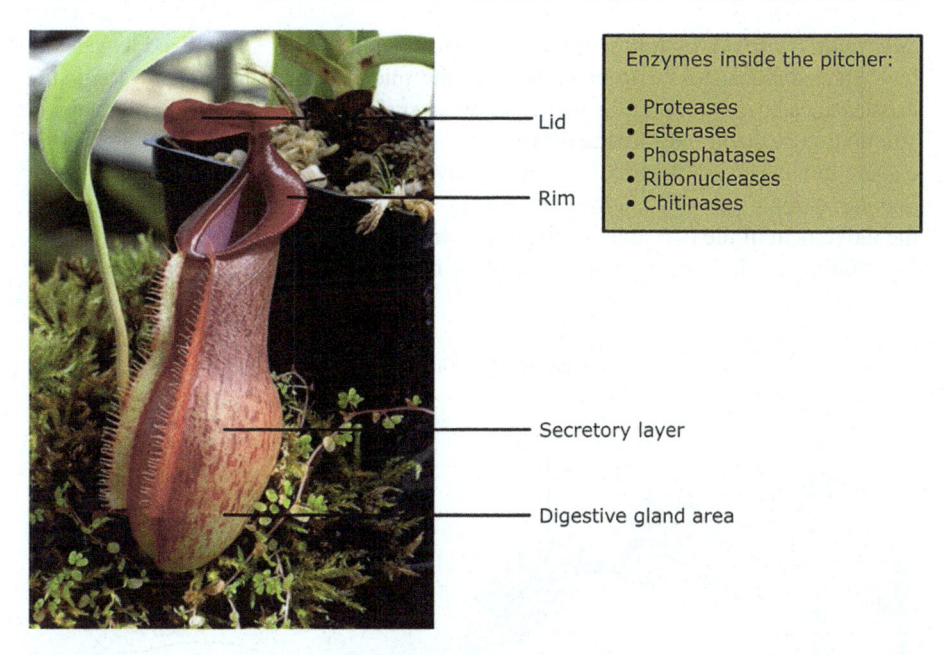

**Fig. 29.20** Pitcher plants (*Nepenthes* sp.) entrap the prey in the pitcher-like traps filled with digestive enzymes

Enzymes inside the pitcher:
- Proteases
- Esterases
- Phosphatases
- Ribonucleases
- Chitinases

Labels: Lid, Rim, Secretory layer, Digestive gland area

the suction traps of bladderworts (genus *Utricularia*). These are triggered by the prey and are activated by electrophysiological signaling processes. Passive traps do not move and use slippery surfaces, sticky secretions, etc. to trap their prey. Passive traps include those used by pitcher plants *Nepenthes* sp. (family Nepenthaceae), which has a pitcher-like container, formed by leaf lamina, and is filled with digestive enzymes secreted by specialized glands (Fig. 29.20). The prey attracted by the pitcher plant slips into the slippery neck of the trap and with its exit blocked by the stiff epidermal hairs pointing in one direction drowns into the enzyme soup and is digested away. Such traps are stationery and do not involve movements. However, carnivorous plants like *Pinguicula grandiflora* (family Lentibulariaceae) involve the use of movements of the leaves to entrap the prey in a cagelike structure formed by the folding of the leaf lamina as its cells loose turgor forming a bowl-like structure. Similarly, the sundew plant (*Drosera* sp. Family Droseraceae) uses traps capable of large-scale movements. The upper surface of the leaves in these plants bears long flexible tentacles each of which has a mucilage and enzyme-secreting gland at the top. The tentacles are arranged in concentric circles in the center of the lamina and are inclined at an angle to the surface. Those that are in the center are the ones that function to produce digestive enzymes, while those away from the center form the trap. Mechanical stimulation generated by the trapped prey in the tentacles is transmitted within seconds enabling rapid curvature of the tentacles toward the center where the prey drowns in the digestive enzymes secreted by the central tentacles forming mucilage-filled bowl. The trap of the Venus flytrap (*Dionaea muscipula*) is driven by turgor-mediated changes in lamina of specialized leaves that have two identical lobes on either side of the midrib. These lobes have long stiff bristles located perpendicular to the laminar surface (Fig.29.21a, b). Mechanical stimulation of the bristles by the prey activates mechano-sensitive ion channels. This causes induction of electrical signaling involving action potential, the transient depolarization of the plasma membrane (Fig. 29.21c). The action potential mediates the movement of the two lobes of the trap to move toward each other with the midrib functioning as a hinge and the bristles on the opposite lobes precisely interlocking to

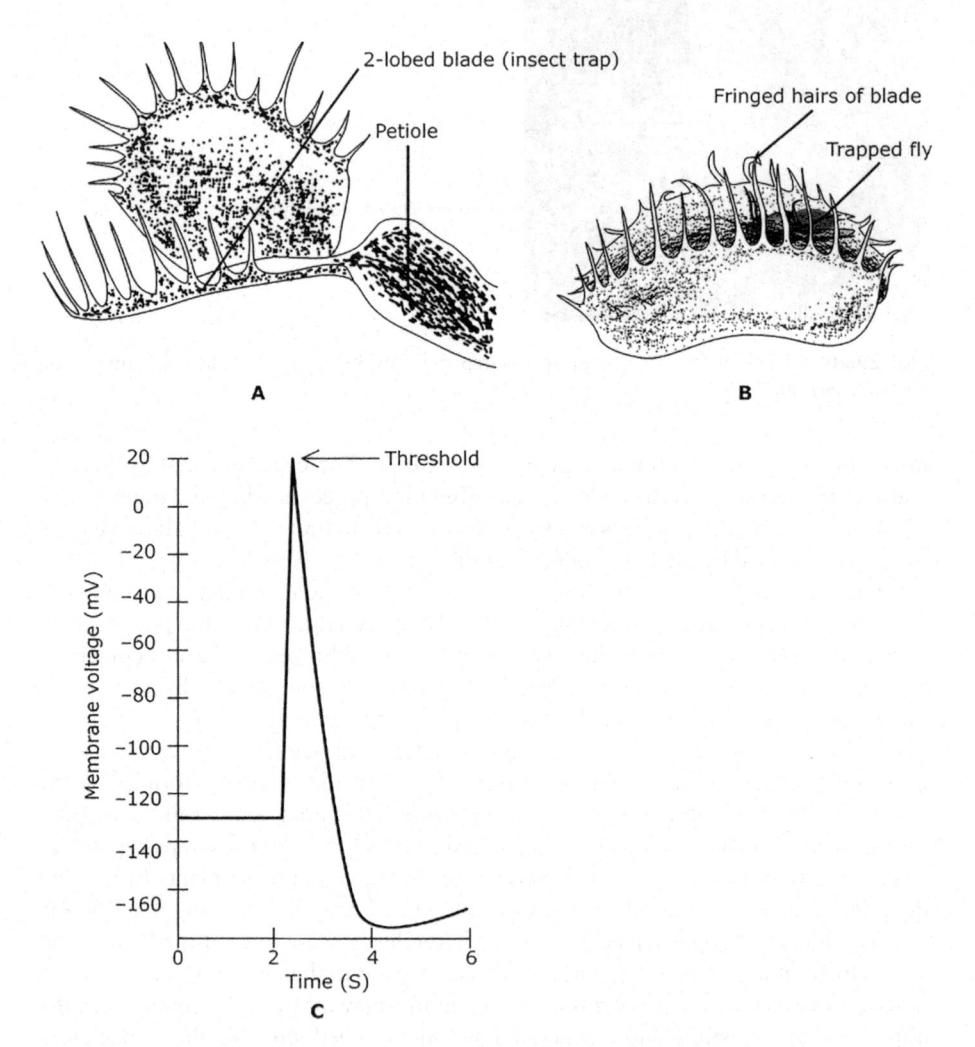

**Fig. 29.21** (**a**, **b**) Venus flytrap (*Dionaea muscipula*) has leaves modified to act as traps with bristles on edges that interlock upon excitation to entrap the prey. (**c**) Mechanical stimulation of bristles causes generation of action potential that mediates the movement of the two lobes toward each other trapping the prey

form a trap for the prey. For activation, the hairs must be touched within 20 s of each other. Each movement generates an action potential and is recorded by the plant. Induction of action potential causes the two lobes of the leaf to close and secrete digestive enzymes. Aquatic plant—*Utricularia* (Lentibulariaceae)—is known to possess traps with hinged trap doors. These traps are formed by leaves modified to form flattened pear-shaped bladders attached to the submerged branches of the plant by tiny stalks. At the entrance to the trap opening is a flap-like tissue that forms an inward opening door. Several bristle-like hairs surround the submerged trough, which entraps the prey preventing their exit.

## 29.6 Movements for Dispersal

Ability to undergo sexual reproduction and production of offsprings has been the primary reason why angiosperms have been able to dominate earth. Plants have devised their developmental programs to produce offsprings under optimal conditions to ensure their survival. Also, these offsprings, comprising of either seeds or spores, are packaged and dispersed using remarkably diverse mechanisms. Life cycle of all terrestrial plants comprises of a sexual generation that begins with unicellular spores produced within sporangia borne on specialized leaves called sporophylls. The sporangia of all seed plants, i.e., angiosperms and gymnosperms, are heterosporous, bearing numerous small male spores or microspores, also called as pollen grains borne within microsporangia (pollen sacs), and usually a single female spore, called megaspore located in the megasporangium. The megaspore develops into the female gametophyte and is enclosed within the protective megasporangium tissue and integument/s that forms the ovule. Upon fertilization, the fertilized egg develops into the diploid embryo, and the ovule becomes the seed enclosed within the ripened ovary now referred to as fruit. The fruit performs the dual function of protecting the seed and also aids in the dispersal of seed. A remarkable diversity of fruits is seen in angiosperms, and these are classified on the basis of characteristics such as whether it is dehiscent or indehiscent, dry or fleshy, etc. These features enable the fruits to facilitate seed dispersal.

### 29.6.1 Cohesion-Mediated Seed Propulsion

Passive dispersal of seeds is observed in plants that have dry dehiscent fruits such as capsules of *Lilium* (Liliaceae) and *Iris* (Iridaceae). The seeds in these plants are liberated by gravity along with the help of wind, rain, or animals. In many plants belonging to the families Acanthaceae, Balsaminaceae, and Euphorbiaceae, ballistic explosive movements of fruit coat at the time of dehiscence are responsible for dispersal of seeds. Other examples include *Alstroemeria* (Alstroemeriaceae), *Lotus* (Papilionaceae), and *Bauhinia* and *Cytisus scoparius* (Fabaceae). In *Ricinus communis* the capsules dehisce explosively as the fleshy exocarp dries and the endocarp dehisces along the three one-seeded valves with transverse fibers.

Progressive dehydration causes the capsule to recurve and split open catapulting the seeds to about 3–3.5 m away. Explosive dehiscence is also seen in the family Acanthaceae as the two-valved capsule that is enclosed tightly within two lignified sepals explode as they dry up. After rainfall, these sepals open up exposing the capsule. The capsule valves are under increasing strain as they dehydrate after maturation because of the structural differences between the outer and inner layers. On imbibing the moisture, the valves split open explosively, and the seeds are catapulted out of their capsules.

## 29.6.2 Turgor-Mediated Seed Propulsion

Plants, such as *Ecballium* (Cucurbitaceae), *Impatiens parviflora* (Balsaminaceae), *Cardamine impatiens* (Brassicaceae), and *Oxalis acetosella* (Oxalidaceae) have fruits that remain fleshy and turgid until they dehisce explosively. In such plants, the changes in the turgor mediate the explosive dehiscence of the fruits to propel the seeds out. Using this mechanism, the parasitic plant *Lathraea clandestina* (Orobanchaceae) can catapult its seeds to a considerable distance of 12–20 m. The fruits of the plant *Impatiens* are elongated capsules that have a fleshy pericarp with several valves joined margin to margin. The subepidermal cells of the capsule are large parenchymatous cells that are elongated in anticlinal orientation. During maturation, turgor pressure builds up in these cells exerting pressure along the fault lines (where the walls are weak) in the sutures. Eventually the strain splits the sutures, and the separation is immediate resulting in explosive inward curvature of the valves, thereby throwing the seeds to long distances.

**Summary**

- Movements in plants that occur in response to external stimuli are called as tropic movements. Based on the nature of the stimuli, these movements can be classified as phototropism, in response to light, and geotropism—in response to gravity. Other tropic movements include hydrotropism and chemotropism in response to water and specific chemicals, respectively.
- Nastic movements are independent of the direction of the stimuli and can be internally (autonomous) or externally directed (paratonic). These include nyctinasty or sleep movements; thermonasty, i.e., movements induced by temperature variations; and seismonastic movements that require a touch stimulus to occur.
- Central to the concept of movements in plants is the presence of cellulosic plant wall and the positive hydrostatic pressure that builds up in the cell. Movements can be attributed to change in growth or changes in turgor. Growth-mediated movements are caused by changes in the length of the opposite sectors. Relocation of auxin to a sector results in variation in growth on one side relative

to the other directing specific movement of the organ. Turgor-mediated movements on the other hand take place by contraction of one sector and the expansion of its opposite sector.
- In phototropism, the exposure to unilateral light affects the activity of auxin transporters activating in preference those farthest away from the light, and the resulting lateral transport of auxin causes the acceleration of the growth on the shaded side.
- Several plant movements are guided on a circadian cycle monitored by the length of the day. The universal biological oscillator functions as a "clock" and can be easily monitored in plants grown under constant environmental conditions of light, temperature, and humidity. Such plants exhibit autonomous movements with a remarkable constant rhythm. Examples of circadian movements include the sleep movements of leaves, the opening and closing of stomata, and the diurnal opening and closing of flowers.
- Parasitic plants exhibit many types of movements for trapping prey. The active traps of such plants have mobile elements that entrap the prey upon stimulation. The touch-induced closure of the traps involves induction of electrical signaling in the form of action potentials as voltage-gated ion channels cause transient depolarization of plasma membrane. Several plants also have passive traps that use slippery surfaces and sticky secretions to entrap their prey.
- Propulsion of seeds for dispersal is caused by passive cohesion as well as active turgor-mediated mechanisms.

> **Box 29.1 Transcription Regulation of Circadian Clock in Plants**
> The circadian clock system of plants operates with three essential processes—perception of the external stimulus, setting of the central clock or oscillator, and induction of specific physiological response. The perception of the external signal is mediated by photoreceptors such as phytochrome and cryptochrome. These modulate the phase of the oscillator in response to the external cues such as the day length. The final response is generated by the plant mediated by clock-controlled genes. Several clock-controlled genes have been investigated including *CAB* genes encoding chlorophyll a/b binding proteins of the light-harvesting complex, those for small subunit of Rubisco, etc. Transcription of *CAB* genes shows a cyclic increase at the beginning of the day and peaks few hours later, and the expression later goes down as the day proceeds. The *CAB* transcript level has been used to monitor circadian rhythms at the molecular level. The use of a *CAB2::luc* (luciferase) reporter gene cassette has been used to create a rhythmic light emission system facilitating genetic analysis of the clock. Several clock-associated genes have been isolated. These include the *TOC1* (*timing of CAB*), *CCA1* (*CIRCADIAN CLOCK ASSOCIATED 1*) and *LHY* (*LATE ELONGATED HYPOCOTYL*), *CAT3* (a gene encoding enzyme catalase), and *ELF3* (*early flowering 3*).
>
> (continued)

**Box 29.1** (continued)

*CCA1* is an important transcriptional regulator that binds to *CAB* promoter and links phytochrome with the expression of *CAB*. The biological clock is based on negative feedback system that relies on the mutual reciprocal feedback inhibition of *CCA1* and *TOC1*. The relative concentrations of the two rise and fall alternately during the day. These oscillations are kept in phase by the external light cue. The final physiological response is generated by the switching on and off *TOC1* and *CCA1* regulated genes. One of the important genes regulated by *TOC1* includes the *CO* (*Constans*) gene that regulates the flowering response.

## Multiple-Choice Questions

1. Pea tendril coiling around a support is an example of:
   (a) Thermotaxis
   (b) Thigmonasty
   (c) Thigmotaxis
   (d) Thigmotropic
2. Curving of plant toward light is called:
   (a) Thigmotropism
   (b) Gravitropism
   (c) Phototropism
   (d) Thigmonasty
3. Sleep movements are mediated by turgor changes in specialized motor cells located in:
   (a) Pulvinus
   (b) Stem
   (c) Leaf blade
   (d) Root
4. Roots perceive gravity by a mechanism that involves sedimentation of:
   (a) Chloroplasts
   (b) Statoliths
   (c) Ribosomes
   (d) Mitochondria
5. Which of the following is *not* true for nastic movements?
   (a) Depend on the direction of the stimulus
   (b) Are independent of the direction of the stimulus
   (c) Genetically predetermined
   (d) Driven by growth or turgor changes

6. Circadian rhythm is a biological rhythm with a cycle of:
   (a) 12 hours
   (b) 24 hours
   (c) 36 hours
   (d) 48 hours.
7. Curvature of the stem due to phototropic movement is due to uneven distribution of:
   (a) Cytokinin
   (b) Auxin
   (c) Phytochrome
   (d) Gibberellin
8. Folding of leaflets of *Mimosa pudica* is associated with the efflux of:
   (a) Calcium ions
   (b) Magnesium ions
   (c) Potassium ions
   (d) Chloride ions
9. _____ cells function as multicellular motors in the leaves of grasses cause extensive changes in leaf conformation.
   (a) Guard cells
   (b) Bulliform cells
   (c) Subsidiary cells
   (d) Lenticels
10. Carnivorous plants entrap their prey in traps driven by:
    (a) Turgor-mediated changes
    (b) Growth
    (c) Contraction
    (d) Locomotion

### Answers

1.b   2.c   3.a   4.b   5.a   6.b   7.b
8.c   9.b   10.a

## Suggested Further Readings

Darwin C (1898) The power of movements in plants. D. Appleton, New York
Koller D (2011) The restless plant. Harvard University Press, Cambridge, MA/London
Trewavas A (2014) Plant behaviour and intelligence. Oxford University Press, Oxford

# Senescence and Programmed Cell Death

## 30

Geetika Kalra and Satish C Bhatla

Both plants and animals go through onset and progress of certain processes leading to "aging" which ultimately causes death. **Aging** is defined as a degenerative biological change occurring over a period of time. Plants exhibit wide range of variations in life span, ranging from a week to few to many years. It is a common sight in temperate regions that the color of the leaves changes from green to yellow to orange or red before its final fall from the deciduous trees (Fig. 30.1). Such changes happen during the terminal phase of the life cycle of plants and are referred as senescence. **Senescence** is a self-digesting (autocatalytic) process controlled by environment and the genetic makeup of an organism. Changes taking place during this process are catabolic and thus irreversibly degenerative. Senescence is not just a passive decay of structural and biochemical machinery of cells; rather it is a precisely regulated series of events in which organelles, membranes, and macromolecules are broken down. Nutrients, like amino acids, sugars, and minerals, are reclaimed for export out of the senescing organ to other plant parts for later use. Nature is thus conservative as far as its precious resources are concerned. Another general term which is used for mechanisms underlying terminal events in the lives of a plant is **programmed cell death (PCD)**. PCD is also a genetically determined developmental event which leads to elimination of a cell or cells. Such eliminations determine the final shape and habit of a plant. PCD occurs in a wide range of developmental processes. For example, development of unisexual flowers where cells destined to form male or female parts are selectively deleted (Fig. 30.2). Unlike PCD, senescence is a phase of aging process where metabolic processes are catabolic and eventually terminate in death.

With reference to how plants meet their respective ends, they are categorized into **annuals** (which die off at the end of each season), **biennials** (plants living for two seasons), and **perennials** (which produce fruits year after year). Each kind has its

**Fig. 30.1** Forest showing green, yellow, orange, and red coloration (**a**) before its final fall from deciduous trees (**b**)

own benefit since annuals result in imparting greater genetic diversity. These plants have better survival strategy because they get new recombinant set of genes every year and they can afford to allocate their energies into producing seeds rather than saving metabolites for the plant to overwinter. On the other hand, perennials can become more robust with each passing year as they add roots, stem, and leaves to already existing structure. Such plants become dominant species in their habitat, and they tend to compete with other plants for light, water, nutrients, and space. Senescence-related changes in annuals occur during leaf maturation phase, and symptoms like abscission appear first in leaves. In evergreen perennials, leaves senesce and abscise after 2–3 years, but not at the same time. There are different factors and mechanisms which govern the senescence of annuals and perennials to be discussed later in the chapter. Senescence is also observed under adverse environmental conditions, like drought, nutrient deficiency, low and high temperatures, but such changes can be reversed when the conditions become congenial unlike natural senescence which is a part of the developmental program.

## 30.1 Patterns of Senescence

Senescence in plants shows wide range of patterns. Some plants are subjected to death of all aboveground parts at the end of growing season (**top senescence**) or loss of entire leafy cover leaving the stem and roots bare for stipulated time period (**deciduous senescence**); and still another form is the die back of oldest leaves in the normal development of annual plants (**progressive senescence**).

Senescence can be discussed under the following levels in plants.

30.1 Patterns of Senescence

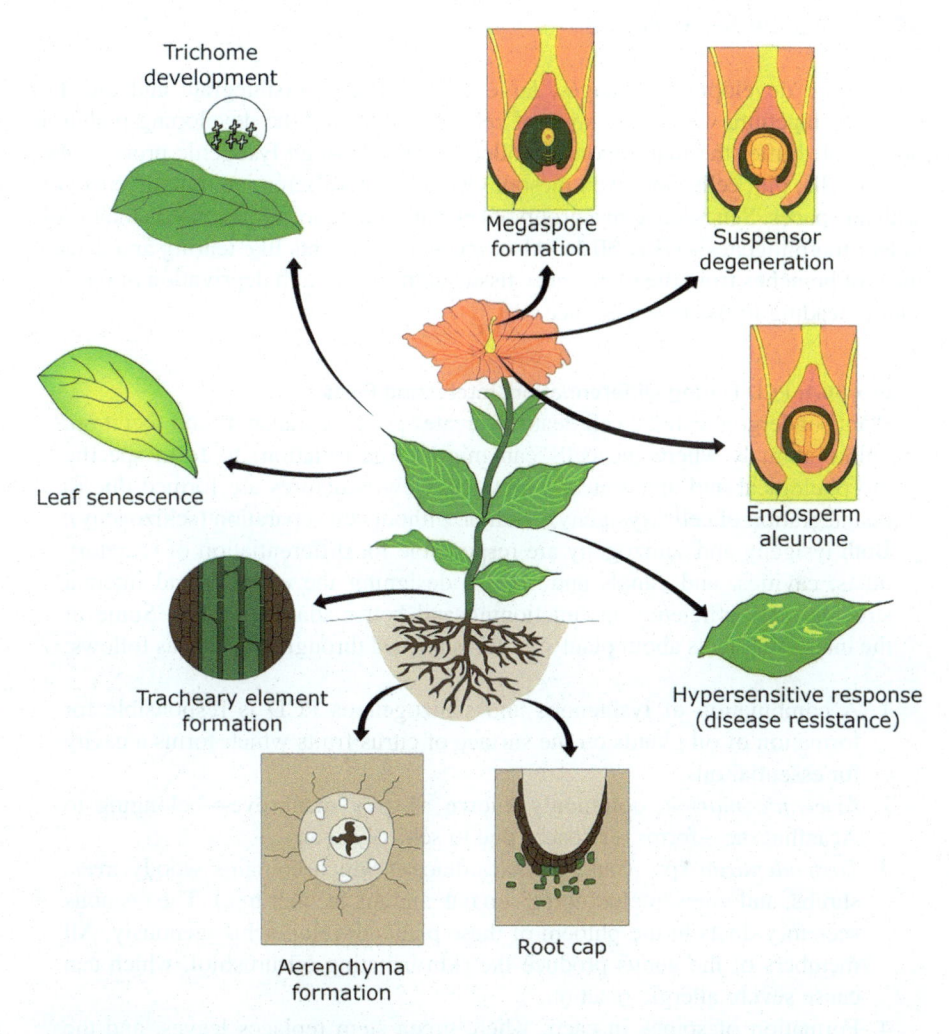

**Fig. 30.2** Range of plant processes occurring through programmed cell death (PCD)

## 30.1.1 Cellular Senescence

Individual cells or small number of cells of an organism undergo senescence. For example, during gametogenesis and embryo development, orderly sequence of events eliminates certain cells. During female gametophyte formation, three out of four cells produced by meiosis are degenerated. The synergids and the suspensor cells also senesce during embryo growth. Such cell elimination is also referred as PCD or programmed cell death.

## 30.1.2 Tissue Senescence

At specific developmental stages, large group of cells disintegrate and die. For example, tapetum, which is a layer of cells that surround the developing pollen, is degenerated after the maturation of pollen grains. Through lysogenic process, root cortex cells, pith cells, and even mesophyll cells may differentiate into aerenchyma with air spaces. Senescence in a group of cells at specific locations plays a major role in leaf morphogenesis (Box 30.1). Also, mechanical events like tearing and detachment of branches from the plant cause tissue damage through deprivation of nutrient supply leading to tissue senescence.

---

**Box 30.1: PCD During Differentiation: Interesting Facts**

Plants undergo localized cell death to create space for transport and secretion, unlike animals where the cells can migrate (gastrulation) to form specific morphological and anatomical structures. New structures are formed due to disintegration of cells (lysigeny), with or without cell separation (schizogeny). Both lysigeny and schizogeny are responsible for differentiation of secretory ducts, cavities, and canals and also in designing the external and internal structures for efficiency in functioning and better adaptive value. Some of the interesting facts about plant structures formed through PCD are as follows:

1. A combination of lysigenous and schizogenous PCD is responsible for formation of oil glands on the surface of citrus fruits which forms a cavity for essential oils.
2. *Avicennia marina*, commonly known as grey mangrove—belonging to Acanthaceae—forms air spaces due to schizolysigeny.
3. *Toxicodendron* spp. (family Anacardiaceae which contains woody trees, shrubs, and vines, including poison oak and the lacquer tree): The resinous secretory ducts in the phloem of these plants develop schizogenously. All members of the genus produce the skin-irritating oil urushiol, which can cause severe allergic reaction.
4. Formation of spines in cacti, where green stem replaces leaves, and the leaves are reduced to spines.

(continued)

**Box 30.1** (continued)

5. *Aponogeton madagascariensis* (commonly known as Madagascar laceleaf or lace plant) undergoes a unique form of leaf sculpting that creates a delicate lattice work. During early development, the lace plant leaf forms a pattern of equidistantly positioned perforations across the surface of the leaf, giving it a lattice-like appearance. Formation of these perforations is a result of PCD, where tonoplast membranes of lace plant ruptures and releases the vacuolar contents. The nuclei retain their identity and DNA becomes fragmented. The walls of the dying cells are also degraded by the action of cellulases and pectinases. After the formation of perforations, the walls of living cells surrounding the perforations get laden by suberin deposition which prevents further spread of PCD and protects the cells from invasion by microbes. Perforations, therefore, are formed at a set time in development and are localized between the interactions of longitudinal and transverse veins across the surface of the leaf. Cell death in this plant occurs at a predictable stage of the leaf development and at a precise location in relation to the completely formed leaf vein system. The lace plant is an ideal system for the study of developmentally regulated PCD in plants.

### 30.1.3 Organ Senescence

Leaves undergo senescence either on a seasonal basis or in response to environmental stress. Leaf senescence and eventually its fall represents an adaptation either to reduce the water evaporation surface or to allow better light perception by the plant.

## 30.2 Types of Cell Death

All plants and animals have evolved organized mechanisms for cell destruction which is a part of normal growth and development. PCD can be initiated by developmental signals or by external factors like pathogen attack. In animals, PCD is referred to as **apoptosis**, which is a highly regulated energy-dependent process. This process is characterized by chromatin condensation, nuclear and plasma membrane blebbing, and formation of apoptotic bodies which are eventually engulfed by neighboring phagocytes (Fig. 30.3). Chromosomes also undergo fragmentation as a result of **endonuclease** digestion of DNA between specific nucleosomes. This results in characteristic "laddered" fragments which are multiples of 180 bp. Another set of enzymes called **caspases** (cysteine-dependent aspartate-specific proteases) bring about breakdown of specific proteins which leads to controlled cell death. **Necrosis** is another way of bringing about cell death which generally results from trauma caused by external agents like herbicides. It occurs when the cells are damaged, which results in rupture of membranes, release of cellular contents, and tissue inflammation.

Plant cells differ from animal cells, as they possess cell wall. Some of the normal processes in plants where PCD is observed include degeneration of suspensor during embryo development in angiosperms, survival of one functional megaspore and degeneration of the rest of the three in *Selaginella*, production of unisexual flowers,

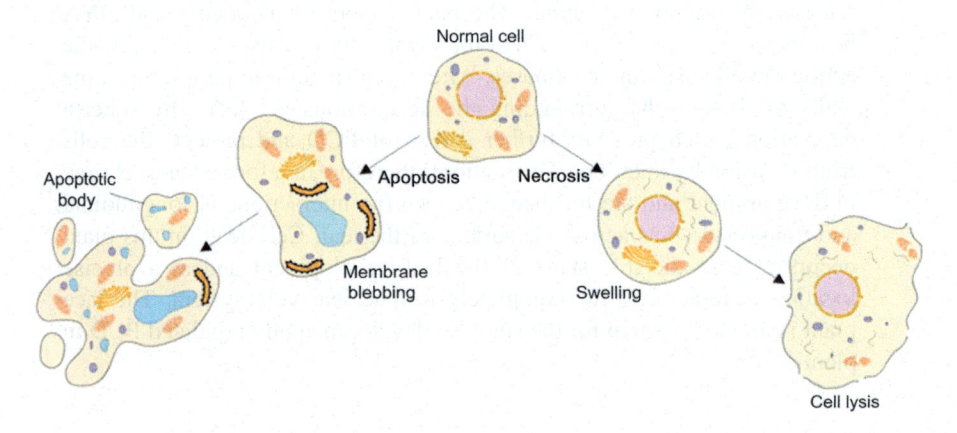

**Fig. 30.3** Cell death through apoptosis and necrosis

and differentiation of tracheids and vessel elements in xylem (Fig. 30.2). Two types of PCD pathways have been characterized in plants.

## 30.2.1 Vacuolar-Type PCD

This process is instrumental in processes like differentiation of xylem elements, leaf senescence, and megasporogenesis. During this process, the vacuole swells and ruptures thereby releasing hydrolases into the cytosol. This release causes breakdown of plasma membrane and complete or partial degradation of cell wall.

## 30.2.2 Hypersensitive Response-Type PCD

This gets activated in response to pathogen attack wherein the cells at the infection site isolate themselves and bring about self-degradation. This deprives the pathogen of much required nutrients and also checks its further spread. This process is characterized by vacuolar water loss and cell shrinkage, followed by degradation of nuclear DNA (Fig. 30.4). In plants, the general term used for dissolution of cytoplasm within the cell wall through the action of cell's own catabolic machinery is called **autolysis**.

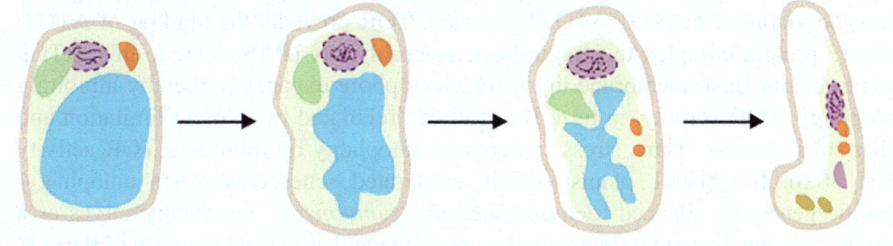

**Fig. 30.4** Two types of PCD (**a** and **b**) observed in plants

## 30.3  Autophagy

Cells have a defined life span, beyond which they are face organism's catabolic activity. Autophagy (self-eating) is one such mechanism whereby cellular components targeted for destruction are collected in lysosomes and degraded within it. Autophagy protects the cell from harmful effects of damaged proteins and organelles (Box 30.2). Moreover, in case of starvation, autophagic breakdown of cell constituents ensures recycling of cellular components which maintain energy levels. During this process vesicles are produced which engulf portions of the cell to be degraded. These vesicles are called **autophagosomes**. The formation of autophagosomes involves well-defined steps:

1. Vesicle induction: a double membrane, cup-shaped structure is pinched off from the endoplasmic reticulum. This is called **phagophore**.
2. Vesicle expansion: phagophore expands and engulfs the components targeted for destruction including the misfolded proteins and aging organelles.
3. Tonoplast docking and fusion: the phagophore becomes spherical after fusion of inner and outer phospholipid bilayer to form complete autophagosome. This fuses with the vacuolar membrane or tonoplast.
4. Digestion: after the fusion, there is generation of single-membrane vesicle called the **autophagic body** which is finally degraded in the vacuole (Fig. 30.5).

Hydrolytic breakdown of autophagic body generates monomers like amino acids and sugars which act as building blocks or energy source for cellular structures. The formation of autophagosomes is regulated by genes which are called **autophagy-related genes** or **ATG genes**. These genes are best described in yeast (*Saccharomyces cerevisiae*) after nutrient starvation and also in *Arabidopsis*. Specific proteins are localized on **phagophore assembly site** of the ER which play a major role in the initiation and growth of autophagosomes. A protein designated as ATG9 moves between phagophore assembly site and *trans*-Golgi network and supplies membrane components to the expanding phagophore. The movement of ATG9 is facilitated by ATG1/ATG13 complex and phosphatidylinositol 3-OH kinase complex. Autophagy regulation has been further understood with the identification of **TOR (target of rapamycin)**. TOR is a master switch which controls ATG genes, and it constitutes a serine/threonine protein kinase. It acts as a negative regulator of autophagy by phosphorylating the ATG1/ATG13 complex. TOR prevents the binding of ATG1/ATG13 protein complex to **phagophore assembly site (PAS)**. Due to this, ATG9 fails to obtain fresh membrane lipids for phagophore expansion, thereby inhibiting autophagy. TOR activity, in turn, is negatively regulated by nutrient limitation and other stress factors. Thus, stress can control autophagy by inhibiting TOR activity (Fig. 30.6). In general, plants exhibit accelerated senescence when autophagy-specific genes are silenced (demonstrated in *Arabidopsis*). Autophagy serves as a homeostatic mechanism to maintain the metabolic and structural integrity of the cell.

**Box 30.2: Deciphering the mechanism of autophagy bags Nobel Prize**
The 2016 Nobel Prize in Physiology or Medicine has been awarded to the Japanese cell biologist Yoshinori Ohsumi, a renowned name in the field of autophagy, for his commendable discoveries leading to understanding the intricacies of the process of cellular "self-eating" or "autophagy." **Autophagy** (derived from the Greek word "autóphagos" meaning "self-eating") refers to a highly regulated process of digestion of obsolete, dysfunctional, or damaged cellular components and recycling of macromolecules for renewal of cellular contents. This process essentially involves sequestration of portions of cytoplasm into specialized saclike membranous organelles called **autophagosomes**. The internalized contents are then delivered to the lysosomes or vacuoles through fusion where their degradation is carried out. It serves as a quality control mechanism that degrades nonessential cellular contents and detoxifies harmful materials. It plays vital roles during different stages of differentiation in a variety of tissues and provides a means to adapt to starvation or stressful conditions. First inkling toward uncovering this pathway of degradation of cellular constituents came in the 1950s with the discovery of specialized compartments in the cell which supposedly harbor enzymes that digest proteins, carbohydrates, and lipids. These distinct membranous structures were given the name **lysosomes** by Christian de Duve in 1955, and for this discovery, he was awarded a Nobel Prize in Physiology or Medicine in 1974. Soon thereafter electron microscopic studies carried out in the 1960s revealed the sequestration of portions of the cytoplasm, including organelles, into small vesicles, which were in various degrees of disintegration. It was then realized that these vesicles had the capacity to digest the internalized contents, and the term "autophagy" was coined by de Duve in 1963 to describe the process of degradation of the internalized contents in these sacs, and the sequestering organelle was named **autophagosome**. de Duve suggested a strong possibility of involvement of this mechanism in the targeted degradation of aberrant and redundant cellular contents. Thereafter, during the period from the 1970s to 1980s, constant efforts were made in explaining the working of the autophagic system. Around the same time, extensive work was being done in discerning another pathway that mediates intracellular protein degradation which eventually leads to characterization of the ATP-dependent **ubiquitin-mediated protein degradation** system via **proteasome** which is now recognized as a major pathway for efficient breakdown of most short-lived proteins. For their outstanding discovery of the ubiquitin-proteasome pathway, Aaron Ciechanover, Avram Hershko, and Irwin Rose were awarded the 2004 Nobel Prize in Chemistry. It was only in the 1990s when Yoshinori Ohsumi's group documented the identification of the gene *ATG1* (**autophagy-related gene 1**) in yeast, which the beginning of the molecular era in the field of autophagy was marked, an area which gained significant prominence in the forthcoming years.

(continued)

**Box 30.2** (continued)

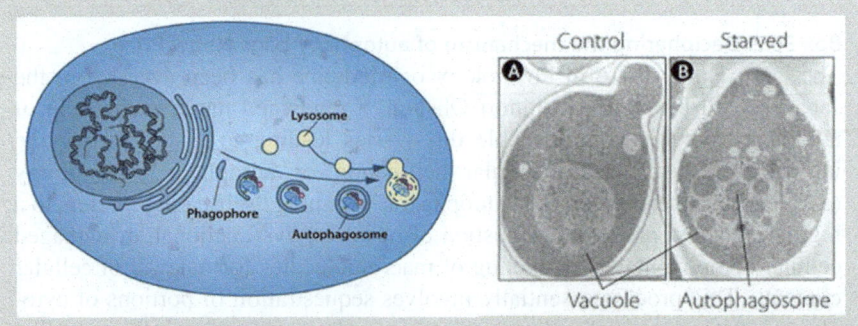

**Figure 1:** The cells harbor a machinery to dispose off redundant macromolecules and organelles via lysosomal degradation. In plant and yeast cells, vacuole corresponds to lysosomes in animal cells. Parts of cytoplasm are engulfed within the vesicles, known as autophagosomes, which then fuse with lysosome/vacuole, where the contents are degraded into smaller constituents (left panel). Prof. Ohsumi's experiments focused on generating mutant yeast cells lacking vacuolar degradation enzymes and demonstrated the accumulation of autophagosomes in the vacuole in starved yeast cells, thereby indicating the existence of autophagy (right panel)

Prof. Ohsumi's work centered around deciphering the progression of autophagy through different phases by simulating starvation in yeast cells, a condition already known to trigger autophagy. He also elucidated a similarity between the autophagy morphology in yeasts and mammals. By managing to disrupt the degradation process in the vacuole of the mutant yeast cells lacking vacuolar degradation enzymes, his group was able to demonstrate the accumulation of autophagosomes within the vacuoles, while the process of autophagy was active. A series of cloning experiments in yeast and mammalian cells carried out by Ohsumi's group led to the characterization of 15 key genes involved in this process and elucidation of the function of each of the encoded products in promoting distinct stages of autophagosome formation. These momentous investigations undertaken in yeast triggered subsequent analysis of autophagy in higher eukaryotes and have helped in the recognition and establishment of autophagy as a fundamental process in cell physiology. It is now established that the process of autophagy is an evolutionarily conserved process of recycling cellular constituents occurring ubiquitously in eukaryotic cells. Prof. Ohsumi's pioneering work has paved way for development of rational approaches to use autophagy in treating a number of neurodegenerative diseases and cancer effectively.

## 30.3 Autophagy

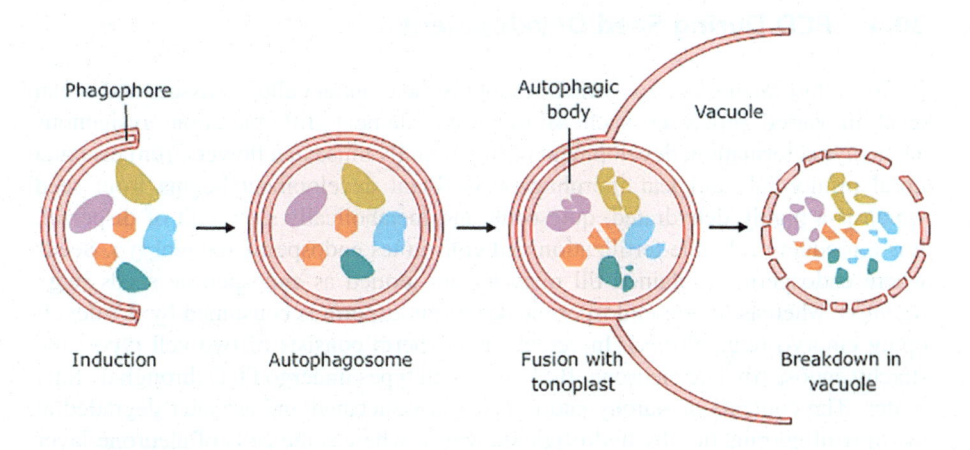

**Fig. 30.5** Steps involved in autophagy

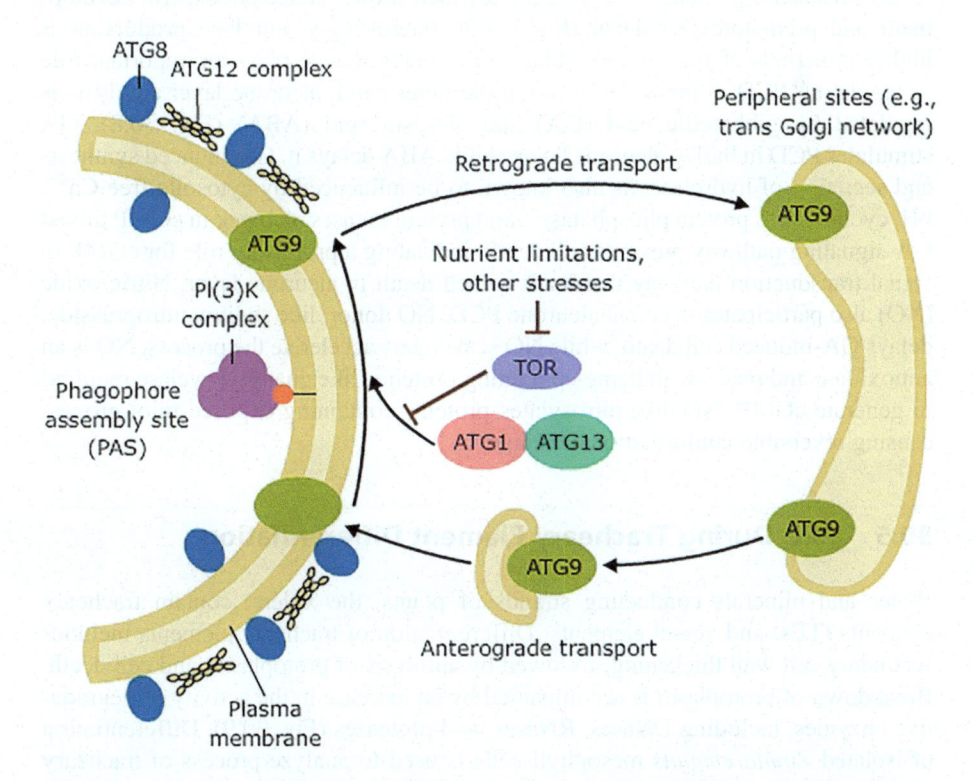

**Fig. 30.6** Scheme for autophagy. ATG9 drives the growth of phagophore by shuttling between phagophore assembly site and peripheral membrane sites. ATG9 is facilitated by ATG1/ATG13 and phosphatidylinositol 3-OH kinase complex. TOR promotes the negative regulation of autophagy by phosphorylating ATG1/ATG13 complex, thereby preventing their binding to phagophore assembly site (PAS). ATG, autophagy-related gene; TOR, target of rapamycin; PAS, phagophore assembly site

## 30.4 PCD During Seed Development

PCD is a fascinating process which remodels the plant at cellular, tissue, and organ level in varied processes such as tracheary element differentiation, lysigenous aerenchyma formation, development of functionally unisexual flowers from bisexual floral primordial, and leaf morphogenesis. Plant development begins from seed which is a highly dehydrated, quiescent, and metabolically slow unit of dispersal. It is a product of double fertilization and constitutes endosperm and embryo. Seeds where endosperm is retained till maturity are termed as endospermic seeds (e.g., *Ricinus*), whereas in non-endospermic seeds, endosperm is consumed by the developing embryo (e.g., *Pisum*). In cereals, endosperm consists of two cell types: the starchy endosperm and aleurone. Both these cell types undergo PCD through distinct routes. The contents of starchy endosperm get desiccated and are later degraded at the time of germination by hydrolytic enzymes, whereas the cells of aleurone layer remain alive until germination and endosperm mobilization. In maize, a mutant called *shrunken2* (*sh2*) has been identified which shows altered endosperm development and premature cell death (Fig. 30.7). Interestingly, ethylene production is higher in kernels of *sh2* mutants. This shows that ethylene plays an important role in regulating PCD in maize kernels. On the other hand, aleurone layer autolysis is regulated by gibberellic acid (GA) and abscisic acid (ABA) (Fig. 30.8). GA stimulates PCD in barley aleurone layer, while ABA delays it. GA-induced synthesis and secretion of hydrolases is also known to be influenced by cytosolic free $Ca^{2+}$, pH, cyclic GMP, protein phosphatases, and protein kinases. A block in cGMP linked GA-signaling pathway prevents cell death, indicating a promising role for cGMP in signal transduction pathway that leads to cell death in aleurone layer. Nitric oxide (NO) also participates in cereal aleurone PCD. NO donor, like sodium nitroprusside, delays GA-induced cell death, while NO scavengers accelerate the process. NO is an antioxidant and reacts with heme-containing proteins, like guanylyl cyclase required to generate cGMP. NO also nitrosylates proteins containing exposed thiol groups, causing reversible conformational change.

## 30.5 PCD During Tracheary Element Differentiation

Water and minerals conducting strands of plants, the xylem, contain tracheary elements (TEs) and vessel elements. Differentiation of tracheary elements includes secondary cell wall thickening, followed by autolysis of protoplasm, and cell death. Breakdown of protoplasm is accompanied by an increase in the activity of degradative enzymes, including DNases, RNases, and proteases (Fig. 30.9). Differentiation of isolated *Zinnia elegans* mesophyll cells is used to analyze process of tracheary element differentiation. In *Zinnia* cell cultures, individual mesophyll cells have been shown to transdifferentiate directly into tracheary elements without cell division, in the presence of phytohormones. Cell wall undergoes lignification, and all cell organelles degenerate to provide a clear passage for water movement. Calcium and calmodulin (along with auxins and cytokinins) also play a role in tracheary element

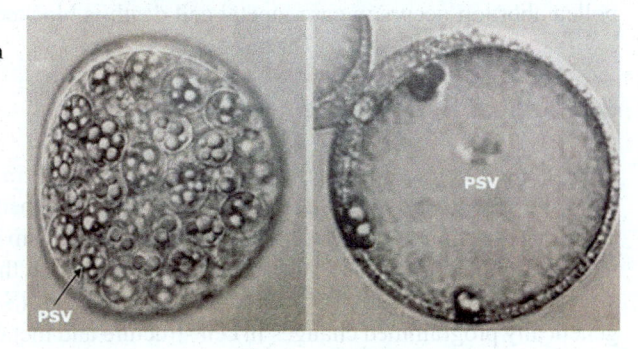

30.5 PCD During Tracheary Element Differentiation

**Fig. 30.7** PCD during development of maize aleurone layer. Mutant*, *shrunken 2 (sh2)* shows altered endosperm development and premature cell death as compared to the wild type (WT). En, endosperm; Sc, scutellum

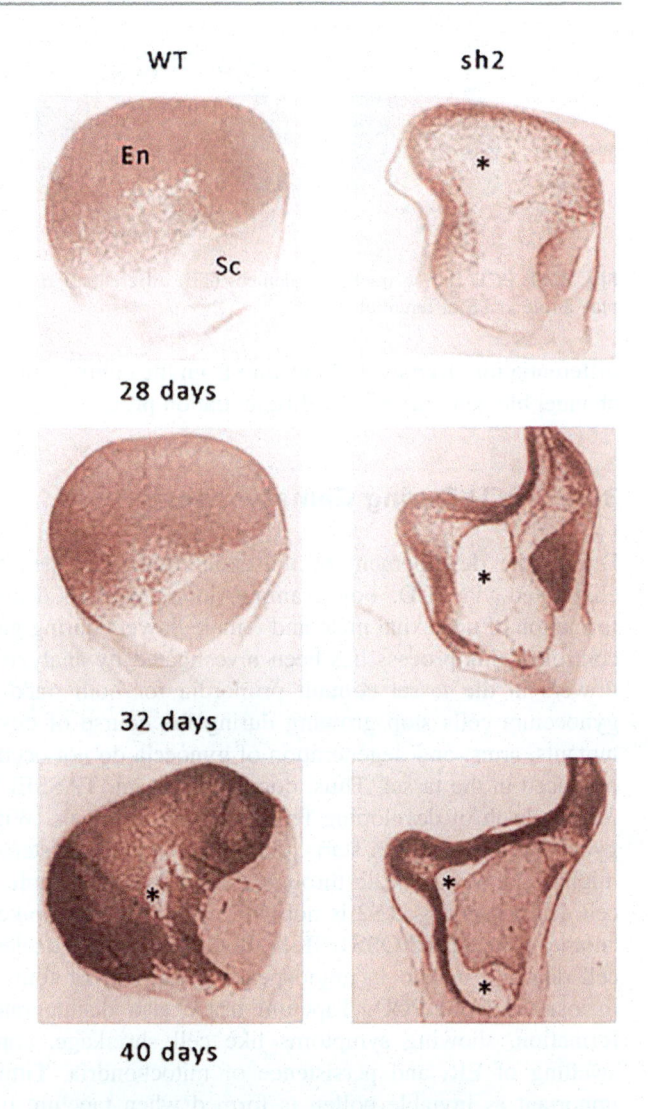

**Fig. 30.8** PCD in cereal aleurone showing vacuolation of cytoplasm without rupture of membranes. Membrane integrity is maintained until cell death. PSV, protein storage vacuole

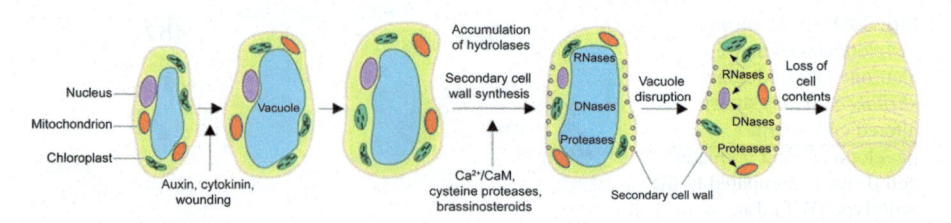

**Fig. 30.9** PCD during tracheary element (TE) differentiation. Auxins, cytokinins, and calcium play a role in TE differentiation

differentiation. Removal of calcium from the cultures and administration of calcium channel blockers inhibit the differentiation process.

## 30.6 PCD During Gametogenesis

During the development of unisexual flowers, either male or female parts are eliminated via PCD. For example, in maize, cell death takes place during the formation of unisexual male and female flowers during gametogenesis. The genetic control of this process has been investigated by analyzing mutants. Young maize flowers in the tassel contain primordia for both male and female flowers, but gynoecium cells stop growing during the course of development. In *tasselseed2* mutants, arrest and degeneration of gynoecia do not occur, and female flowers are produced in the tassel. Thus, dominant mutant, TASSELSEED2 (TS2) is required for the death of developing female organs in tassels. With the expression of TS2, gynoecial degeneration starts. TS2 encodes hydroxysteroid dehydrogenase, which might regulate cell death through a steroid-like molecule which acts as a switch in cell death pathway. TS2 is not expressed in female spikelet as it is suppressed by another gene SILKLOSS1(SK1). Both TS2 and SK1 selectively promote or inhibit cell death in maize. In angiosperms, three out of four megaspores formed after meiosis undergo PCD. Tapetum tissue also disintegrates at the time of pollen formation, showing symptoms like cell shrinkage, condensation of chromatin, swelling of ER, and persistence of mitochondria. Timing of cell death is also important as inviable pollen is formed when tapetum undergoes early PCD and pollen abortion occurs in case tapetal cell death is blocked.

## 30.7 Leaf Senescence

Leaf senescence is a final stage of development which involves nutrient relocation from the leaf to growing vegetative or reproductive parts of a plant. It involves coordinated action at cellular, tissue, organ, and organism levels. Our understanding regarding molecular basis of leaf senescence has come through characterization of senescence mutants and **senescence-associated genes (SAGs)**. Leaf cells undergo genetically programmed changes in cell structure and metabolism. It is thus regarded as a highly regulated ordered series of events where organelles, membranes, and macromolecules are broken down and nutrients like amino acids, sugars, and minerals are reclaimed from the leaf to be reused in other parts of the plants. Senescence in grains, like wheat or barley, is quite characteristic, i.e., the entire resources of the plant are mobilized and reclaimed to support grain or seed development. First symptom of senescence is a decline of photosynthesis and increased rate of respiration. Other changes are breakdown of chloroplast membranes, loss of chlorophyll, and protein and lipid metabolism. It is important for senescence that certain cell organelles and tissues remain intact and functional until after mobilization is completed. Hence, they are degraded in certain order. Chloroplasts are rich reservoirs of proteins (like Rubisco, chlorophyll a-/b-binding proteins) and membrane lipids. They are degraded first, while mitochondria, peroxisomes, and nuclei remain functional and transcriptionally active till late stages of senescence. Guard cells in the epidermis and phloem tissue also remain functional for effective gaseous exchange and transport until the metabolic breakdown of chloroplasts and export of metabolites is completed. Also, for intercellular traffic and movement of materials in and out of vacuoles, it is important that the plasma membrane and vacuolar membrane remain functional and selectively permeable till the end. Further, senescence is not altogether a degradative process. Many proteins are synthesized during the course of this terminal plant process, like those associated with synthesis of anthocyanins and carotenoids, which are formed after the loss of chlorophyll pigment. Leaf senescence is governed by the age of the plant under normal growth conditions, which takes place from the most mature to the youngest leaves (sequential leaf senescence). It may also take place in response to climate change, as in deciduous trees (seasonal leaf senescence). Both these types of senescence are part of developmental leaf senescence which consists of three distinct phases: **(1) Initiation phase**: at this stage, the decline in rate of photosynthesis takes place. **Blebbing**, a process that occurs in healthy cells, stops in senescing cells. In this process, membrane turnover takes place by the removal of lipid metabolites by forming lipid-protein particles which are shed by blebbing. This process stops during senescence. As a result lipid-protein particles accumulate between lipid bilayers, causing the membrane to become leaky. **(2) Degenerative phase**: this phase is marked by autolysis of cellular organelles and macromolecules. Many new genes are expressed, while those involved in photosynthetic activity are "turned off." **(3) Terminal phase**: during this phase, autolysis is completed, and cell separation takes place at the abscission layer (Table 30.1). All these phases merge to bring about leaf senescence by massive reprogramming of gene expression.

During chloroplast degradation, its primary degenerative products are extremely photoreactive and prove lethal to the cell. Thus, it is important to safely remove and dispose these toxic compounds. Thus, chloroplasts are transformed into **gerontoplasts** (a term introduced to define the unique features of plastids formed during leaf senescence). The formation of gerontoplasts from chloroplasts during senescence involves extensive structural modifications of thylakoid membrane with simultaneous formation of a large number of plastoglobules with lipophilic materials. The structural dismantling of grana is accompanied by a decline in

**Table 30.1** Phases of leaf senescence and accompanying events/factors

| Phase | Factor/event |
|---|---|
| Initiation phase | **Internal factors**: sugars, phase change, hormones (auxin, cytokinins, salicylic acid, jasmonic acid, ethylene, abscisic acid) |
| | **External factors**: shading, heat or cold, pathogen attack or wounding, UV or ozone, drought, nutrient limitation |
| Degenerative phase | **Cell degeneration**: salvage and translocation of nutrients (e.g., nitrogen and lipids), detoxification and defense (e.g., antioxidant production and activation of defense-related genes), chlorophyll loss, macromolecule degradation |
| Terminal phase | **Cell death**: disruption of nucleus and mitochondria, DNA laddering, breakdown of plasma and vacuolar membranes |

primary photochemical reactions. Gerontoplasts retain their ability to divide, and their development is reversible up to a certain threshold. This ability is lost as the cells enter the terminal phase of senescence. Chlorophyll breakdown proceeds by the loss of phytol tail, which is cleaved by a chlorophyllase action to yield chlorophyllide. Magnesium is removed from the porphyrin ring by an enzyme (magnesium dechelatase) that dechelates it. The resulting pheophorbide is degraded further in a two-step process to give rise to a colorless, straight-chain tetrapyrrole, and the chlorophyll-binding proteins are released for degradation. The remaining chlorophyll catabolites are exported from the chloroplast, modified slightly in the cytosol, and then imported into the vacuole, where further degradation occurs. Studies have demonstrated that the autophagy pathway is required for whole chloroplast breakdown during dark-induced leaf senescence.

A number of cDNAs have been cloned from plants like *Arabidopsis*, asparagus, maize, tomato, etc. during senescence. These are the genes whose expression is upregulated during senescence and are called **SAGs (senescence-associated genes)**. These genes are divided into two classes: (1) those which are expressed at a low basal level throughout most of the leaf development, but their expression is upregulated with the onset of senescence, and (2) those which are expressed only with the onset of senescence. Some of these genes are associated with biotic and abiotic stress conditions, response to reactive oxygen species (ROS), metal-ion binding, pectin esterase, and genes involved in lipid mobilization. Nitrogen is also translocated in phloem stream mainly in the form of amides, glutamine, and asparagine. Glutamine synthases (GS) are the enzymes that convert ammonia to glutamine. Two types of GS occur in plants. GS1 is located in cytosol, while GS2 is located in plastids. During senescence, the activity of GS2 decreases while that of GS1 increases. Several genes encoding GS1 are upregulated which suggests that ammonia released from the catabolism of amino acids may be reconverted to glutamine in the cytoplasm for transport. Further, gene coding for phosphoenolpyruvate carboxykinase (PEPCK), which converts oxaloacetate to PEP (a step in gluconeogenesis), is also upregulated. Genes encoding phospholipase D and a β-galactosidase, enzymes that are involved in the hydrolysis of membrane phospholipids and galactolipids, are also upregulated. Genes whose expression is suppressed by senescence are called **SDGs (senescence-downregulated genes)**.

## 30.8 Hormonal Regulation of Senescence

**Table 30.2** Metabolic pathways upregulated/downregulated during senescence in *Arabidopsis*

| Upregulated genes | Downregulated genes |
|---|---|
| Autophagy transport | Amino acid metabolism |
| Response to ROS | Chlorophyll biosynthesis |
| ABA signaling | Carotenoid biosynthesis |
| Metal-ion binding | Cytokinin-mediated signaling |
| DNA binding | Glycine metabolism |
| Protein binding | Photosynthesis |
| Carotene metabolism | Glutamine synthase 2 |
| Caspase activity | |
| Pectinesterase activity | |
| Ethylene signaling | |
| Lipid catabolism | |
| Glutamine synthase 1 | |

Table 30.2 depicts the list of metabolic pathways that are either upregulated or downregulated during senescence in *Arabidopsis*.

## 30.8 Hormonal Regulation of Senescence

Senescence is regulated by the developmental program of the plant, but it can be modulated by several hormones. It ensures efficient remobilization of nutrients to vegetative or reproductive sinks. Gibberellins, cytokinins, and brassinosteroids retard senescence, while ethylene, ABA, and jasmonates enhance senescence-related changes. However, it is observed that the same hormone can act as positive or negative regulator of senescence depending on the age of the leaf. At the same time, leaves should also develop competence to senesce, before they can respond to positive senescence regulators. There is a crosstalk between hormones which regulates leaf senescence, depicting a web of control mechanisms.

### 30.8.1 Cytokinins

They are known to delay senescence but do not prevent it indefinitely. It is observed that local application of cytokinin to mature green leaves delays senescence in the area where cytokinin is applied, while the rest of the leaf shows senescence. If *Arabidopsis* or tobacco plants are transformed with *ipt* gene from *Agrobacterium*, which encodes an isopentenyl transferase (this gene leads to an overproduction of cytokinin), the transformed plants show delayed senescence (Fig. 30.10). Gene transcripts involved in cytokinin biosynthesis decline, while transcripts of genes involved in degradation of cytokinins (like cytokinin oxidase) increase during senescence. These results suggest that cytokinins are natural regulators of senescence. The AHK3 is the receptor that regulates leaf senescence in *Arabidopsis*. Elevated levels of this receptor result in delay of senescence, while its disruption leads to premature leaf senescence. When *ipt* gene

**Fig. 30.10** Progressive loss of chlorophyll during tobacco leaf senescence. Note that the last areas to lose the green color are close to the veins, reflecting the fact that cells close to the veins need to remain active during nutrient export (*Contributed by: Neha Singh*)

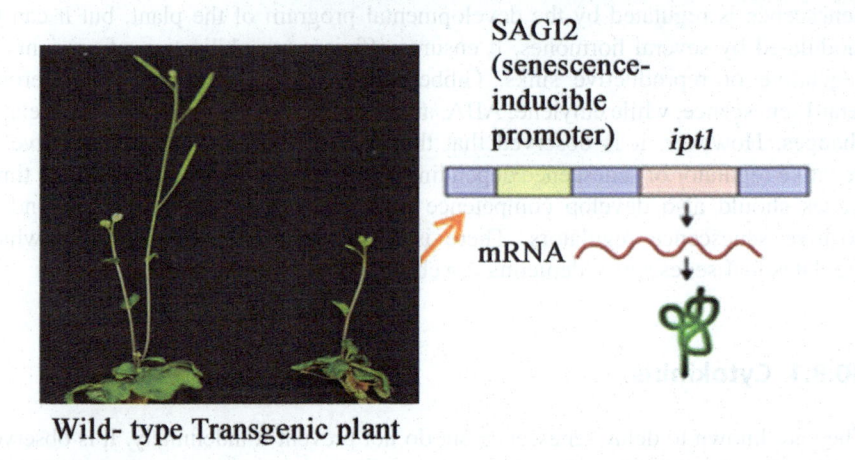

**Wild- type Transgenic plant**

**Fig. 30.11** Plants transformed with *ipt* gene leads to delayed senescence due to overproduction of isopentenyl transferase

is fused to the promoter of SAG12 (a gene specifically expressed in senesing tissues), it results in expression of the transgene at the time of senescence (Fig. 30.11). Cytokinin production is autoregulated as its production inhibits senescence, and this results in decreased expression of the transgene. In these plants, older leaves retain the ability to photosynthesize much longer than the leaves of control plants lacking the transgene. Such plants show significant increase in seed production and improved drought tolerance. It is believed that cytokinin represses leaf senescence by regulating nutrient mobilization and through regulation of sucrose levels. It also imparts enhanced sink strength to the organ concerned. Also, cytokinins are transported from roots to leaves. It is suggested that after flowering, cytokinins are redirected from roots to developing seeds instead of leaves, and this is instrumental in triggering leaf senescence. Cytokinin may also repress expression of key SAGs. For example, genes encoding a cysteine protease and a peroxidase are upregulated in *Petunia hybrida* callus cultures when transferred to low cytokinin medium. It is evident that cytokinins act at two levels: at a distance by promoting differentiation and strong sink activity and locally in senescing cells by mediating the initiation of senescence program. A very interesting action of cytokinin is seen during regreening of tobacco plants. The oldest leaf of *Nicotiana rustica* which has lost all chlorophyll can recover full greenness by removal of stem at the node or by treatment with benzylaminopurine (a synthetic cytokinin). Such leaf can differentiate its gerontoplasts into functional chloroplasts. Proteins characteristic of chloroplasts assembly in very young cells, like chlorophyll synthesis enzyme-protochlorophyllide oxidoreductase, are re-expressed (Fig. 30.11). This points at the fact that cytokinin is a key factor in delaying senescence, and also senescence is a reversible phenomenon even at an advanced stage.

### 30.8.2 Auxin

High auxin concentrations enhance ethylene production, which in turn, promote senescence in mature leaves. The level of auxin increases during leaf senescence, and many genes involved in auxin biosynthesis, such as tryptophan synthase, are upregulated during senescence. However, application of exogenous auxin in *Arabidopsis* leads to a decrease in expression of SAGs. Overexpression of YUCCA6, the flavin-containing monooxygenase that catalyzes the rate limiting step in auxin biosynthesis, delays leaf senescence and decreases SAG expression.

### 30.8.3 Gibberellins

Gibberellins play an important role in senescence as their active forms are known to decline in leaves with progressing age. The addition of $GA_3$ to leaf discs of *Taraxacum officinale* retards their senescence and delays decline in the levels of chlorophyll, protein, and RNA. Incorporation of radioactive leucine and adenine into protein and RNA, respectively, is increased by GA. This enhancement of protein and RNA synthesis does not occur if the discs are supplied with actinomycin D before treatment with gibberellin. However, if actinomycin D is added after the gibberellin treatment, then the stimulatory effect of the hormone is maintained. These results suggest that the retarding action of gibberellins on leaf senescence could be mediated through regulation of RNA synthesis, which is DNA dependent. In another experiment, gibberellic acid (at concentration $10^{-5}$ and $10^{-4}$ M) delayed the senescence of carnation flower, when applied continuously to flowers between the closed brush and fully open stages of development. Treatment with paclobutrazol, an inhibitor of GA biosynthesis, prevented tight buds from opening fully, reduced the longevity of partially open flowers, but was ineffective when applied to fully open flowers. GA-treated flowers did not show simultaneous petal in-rolling, a known indicator of senescence, and the time to complete petal drying was extended. These results provide evidence for a positive role for gibberellins in development of a flower and delaying senescence.

### 30.8.4 Jasmonic Acid

Jasmonates (JA) are oxylipins derived from linolenic acid in chloroplasts membranes. These hormones have been shown to reduce the photochemical efficiency of photosystem II and the levels of chlorophyll content of detached leaves when applied exogenously. A gene which is involved with senescence and jasmonic acid receptor in *Arabidopsis* has been designated as **coronatine-insensitive 1 (COI1)**. JA treatment accelerated leaf senescence in wild-type *Arabidopsis* plants but not in *coi1* mutants. Exogenous application of JA stimulates leaf senescence and controls the expression of a series of senescence-related genes.

### 30.8.5 Abscisic Acid

It promotes senescence and abscission of leaves, and the levels of the hormone increase dramatically in leaves that are undergoing senescence. Some of the key ABA biosynthesis genes are upregulated at the time of senescence, like NECD and aldehyde oxidase genes. It has been put forward that ABA is able to increase $H_2O_2$ levels (either by production or accumulation) in leaves, and this accelerates senescence process. Role of ABA in the regulation of senescence has been investigated in detached tobacco leaves (*Nicotiana rustica*). Leaves which senesce in darkness show a sharp rise in ABA level in the early stage of aging, followed by a rapid decline. The same trend is found when leaves are aged in light, but the rise in ABA occurs 4 days later than in darkness. Senescence is slower in light than in darkness. On the other hand, leaves treated with kinetin which senesce in light and darkness do not show an increase in ABA. Application of kinetin leads to a transformation from free to bound ABA. These results indicate that ABA and cytokinins are involved in a trigger mechanism to regulate senescence. The stage at which this trigger is activated determines the rate of senescence. ABA levels are also observed to rise under water or salt stress conditions. Senescing leaves dehydrate more rapidly than normal leaf because ABA-induced stomatal closure stops functioning. In senescing leaves, stomata stay open due to SAG113 gene which is induced by ABA. This gene encodes protein phosphatase 2C which inhibits stomatal closure in senescing leaves. Loss of function of SAG113 delays senescence, and its overexpression accelerates it. Before the onset of senescence, ABA induces stomatal closure to reduce loss of water. Contrary to this, ABA signaling changes to induce genes like SAG113 which inhibit ABA-induced stomatal closure and accelerates senescence. This also demonstrates how one hormone can have different effects at different phases of development.

### 30.8.6 Ethylene

Well known for its involvement in fruit ripening, ethylene is also involved in the control of senescence in leaves and flowers. Exogenous application of ethylene to leaves induces senescence-related changes in leaves and in flower petals. Endogenous ethylene levels are known to increase as the leaves get older, indicating its role in senescence-related changes. The use of inhibitors of ethylene action, such as silver thiosulfate or 1-methyl cyclopropane (1-MCP), prevents senescence of cut flowers and detached leaves. Expression of genes encoding ACC synthase (ACS) or ACC oxidase (ACO) in an antisense orientation in transgenic plants results in delayed senescence, whereas reverse happens if ethylene is overproduced by overexpression of an ACS gene. These effects of ethylene are brought about when the leaves or flowers are at the correct developmental stage, i.e., young leaves or young flower buds do not show senescence-related changes, despite the presence of ethylene. Interestingly, senescence of leaves and ripening of climacteric fruits share some common features, viz., both involve chlorophyll degradation, an increased activity of hydrolytic enzymes, and ethylene production. However, both are different processes, involve different sets of enzymes, and lead to different end results. In many plants (like *Petunia hybrida*), pollination triggers floral senescence, and ethylene production is detected within 20 min of pollination. The recognition of pollen by stigma is the trigger which induces ethylene production. Further evidence for the critical role of ethylene in senescence comes from *Arabidopsis etr1* mutants and *never ripe* tomatoes which are insensitive to ethylene due to mutation in an ethylene receptor protein. Both these mutants show delayed senescence, but it is also observed at the same time that leaves of *Arabidopsis* mutants eventually senesce, but tomato mutants never fully ripen. This indicates that ethylene promotes senescence in leaves, whereas it is an absolute requirement for tomato ripening.

### 30.8.7 Salicylic Acid

This hormone is known to play a role in plant responses to pathogens. It also influences the setting up of kinetics of hypersensitive response and in inducing tolerance for stress. It also plays a role in age-dependent senescence, and the levels of the hormone increase four times in leaves undergoing senescence. In *Arabidopsis* SAG12 and PR1a have been shown to be undetectable when SA is not present in the leaves. Interestingly, transcriptome changes that occur through the presence of SA are similar to those that occur naturally in age-dependent senescence. *Arabidopsis* plants defective in the SA-signaling pathway (*npr1* and *pad4* mutants and *NahG* transgenic plants) have been used to investigate senescence-enhanced gene expression, and a number of genes show altered gene expression pattern. The presence of SA induces the expression of cysteine protease gene SAG12. Such a change in gene expression delays yellowing and reduces necrosis in the mutant plants defective in SA signaling. This suggests a role for SA in cell death that occurs in final stages of senescence. A crosstalk has also been found between SA and JA with reference to senescence. Senescence-specific WRKY53 transcription factor interacts with the JA-inducible protein EPITHIOSPECIFYING SENESCENCE REGULATOR (ESR/ESP). The expression of these genes is antagonistically regulated in response to JA and SA, respectively, and they influence each other negatively which is most likely governed by the JA and SA equilibrium.

## 30.9 Developmental Regulation of Senescence

A number of external and internal factors influence various signaling pathways which alter gene expression resulting in sequentially organized leaf senescence. The external factors include seasonal changes and various biotic and abiotic factors, while internal factors include mainly the developmental phase of the plant (age of the leaf). The disposal of leaf is considered as an indirect selection for nutrient salvage. In other words, senescence syndrome is characterized by recruitment of nutrients from leaf tissues which is a part of genome optimization program. Thus, leaf senescence is a consequence of natural selection for genome reproduction and is initiated and progresses in age-dependent manner even if the plant is growing in adequate nutritional conditions away from pathogen attack as well as from biotic and abiotic stresses. Some of the pathways which are activated during leaf senescence include ROS-based signaling, the ubiquitin-proteasome pathway, etc. Epigenetic mechanisms also alter gene expression through histone and DNA modification and chromatin remodeling. Small RNAs also modulate gene expression at posttranscriptional level. The NAC and WRKY genes are the two most abundant families of differentially regulated transcription factors during senescence. NAC transcription factor contains a highly conserved N-terminal DNA-binding domain and a variable regulatory C-terminal domain. NAC gene was first discovered in cereals as a gene regulating senescence. The presence of a functional allele of NAC gene, called NAM-B1, causes earlier leaf senescence and nutrient translocation to the developing grains in a wild variety of wheat. This allows the grains to obtain benefit of reclaimed nutrients from the leaves. In another domesticated variety of wheat, a frameshift mutation results in loss of function of NAM-B1 allele which delays senescence. Early leaf senescence improves nutritional quality of the grain which points at important role of nutrient remobilization during leaf senescence for normal grain development. Another group of transcription factors, designated as WRKY, also plays a regulatory role in promoting leaf senescence. Leaf senescence is delayed in knockout mutants of WRKY53 gene in *Arabidopsis*. Expression of this gene is suppressed by light and promoted by darkness or ROS.

## 30.10 Role of ROS in Leaf Senescence

The process of senescence and abiotic stress is associated with overproduction of reactive oxygen species (ROS). ROS constitute $H_2O_2$, superoxide, singlet oxygen, and hydroxyl radicals, which cause oxidative damage to DNA, proteins, and membrane lipids. ROS contribute to progression of leaf senescence as the antioxidant property of the leaf declines. They also act as signals that activate genetically programmed pathways of gene expression that lead to regulated cell death events. ROS generally initiate cell death through lipid peroxidation. In this process hydrophilic moieties are introduced into lipid bilayers, leading to their disruption and the leakage of cytoplasmic contents. Further, degradation of membrane lipids results in free fatty acids which initiate oxidative deterioration by providing a substrate for the enzyme lipoxygenase, causing membrane lipid peroxidation.

## 30.11 Role of Sugar Accumulation in Leaf Senescence

Sugars are the building blocks for macromolecules and serve as energy source during various metabolic activities. They are now also known to serve as signaling molecules in regulating metabolic events. Investigations have shown that high concentration of sugars lower the photosynthetic activity and may also trigger leaf senescence if they exceed a certain threshold limit. Such sugar-induced leaf senescence is observed under low-nitrogen conditions.

## 30.12 Role of Pigment Composition in Senescence

Senescence of leaves is also characterized by color change which is related to nutrient mobilization and their reabsorption from leaf cells (Fig. 30.12). Pigment metabolism has adaptive significance as color of flowers and ripe fruits aid in pollination and effective seed dispersal. Pathways of chlorophyll breakdown and anthocyanin biosynthesis are specifically upregulated during senescence. Chlorophyll degradation starts with the release of chlorophyll from its association with pigment-binding proteins in the thylakoid membranes. This process is aided by a gene called **stay green (SGR)**. The gene has a highly conserved structure and is found across all groups of plants like mosses, algae, and prokaryotic cyanobacteria. SGR functions in the disassembly of thylakoid photosystem complexes, making chlorophyll available for degradation. In the next step, the magnesium of chlorophyll a is removed, resulting in production of phaeophytin a (Fig. 30.13). Phaeophytin a is hydrolyzed by the enzyme phaeophytinase to yield pheophorbide a. The other product of phaeophytinase action is phytol, which accumulates in the **plastoglobules** (lipid droplets) of gerontoplasts, largely in the form of esters. Chlorophyll b is converted to chlorophyll a before it can be subjected to breakdown pathway. Chlorophyll b reductase is instrumental in this breakdown and is known to be activated during senescence. Both phaeophytin and phaeophorbide retain

**Fig. 30.12** Regreening of tobacco plants by treatment with benzylaminopurine. Proteins characteristic of chloroplast assembly are re-expressed indicating a role of cytokinin in delaying senescence

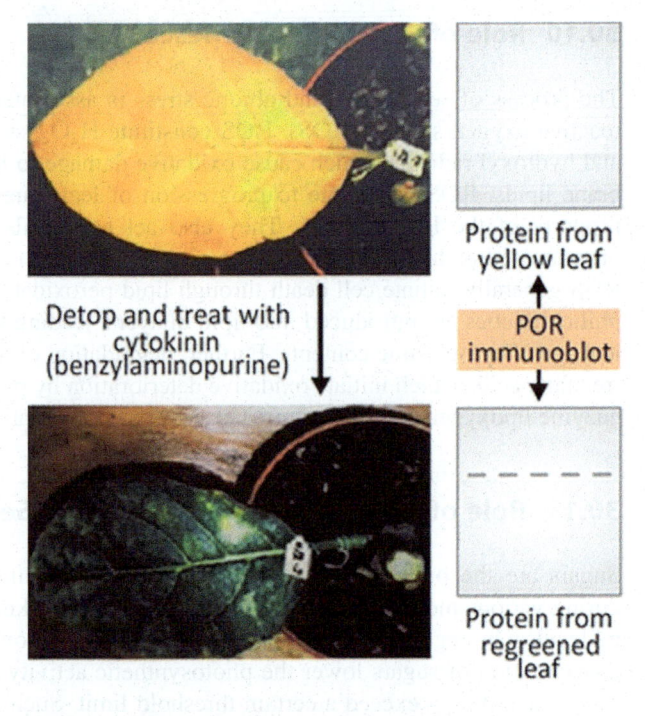

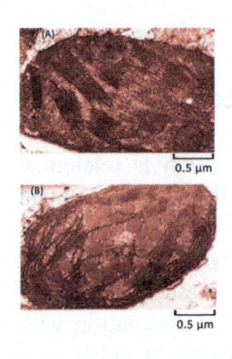

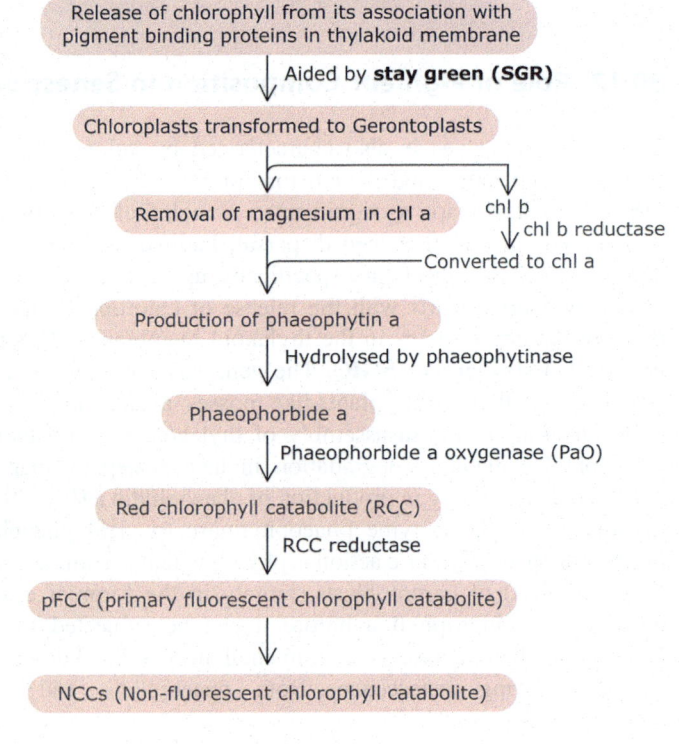

**Fig. 30.13** Steps in chlorophyll degradation during senescence

## 30.12 Role of Pigment Composition in Senescence

tetrapyrrole ring structure which is later opened to generate a colorless straight-chain tetrapyrrole. Enzyme *phaeophorbidea oxygenase* (PaO) aids this reaction which requires oxygen and Fe, operating in redox cycle driven by reduced ferredoxin. PaO uses pheophorbide a as a substrate which generates a red chlorophyll catabolite (RCC) that does not accumulate in plants but is metabolized further by RCC reductase. This enzyme catalyzes the ferredoxin-dependent reduction of a double bond in the pyrrole system of RCC to produce colorless tetrapyrrole with a strong blue fluorescence (primary fluorescent chlorophyll catabolite, pFCC). FCC is exported from the gerontoplasts to the cytosol by an ATP-dependent transporter located in the plastid envelop and enters the vacuole through ABC transporters in the tonoplasts. pFCC is then modified either by hydroxylation or by conjugation to form nonfluorescent chlorophyll catabolites (NCCs). In this way all carbon and nitrogen in the chlorophyll molecule end up in the vacuole, and chlorophyll molecule is abandoned after dismantling (Fig. 30.14).

**Fig. 30.14** Dismantling of chlorophyll molecule during senescence

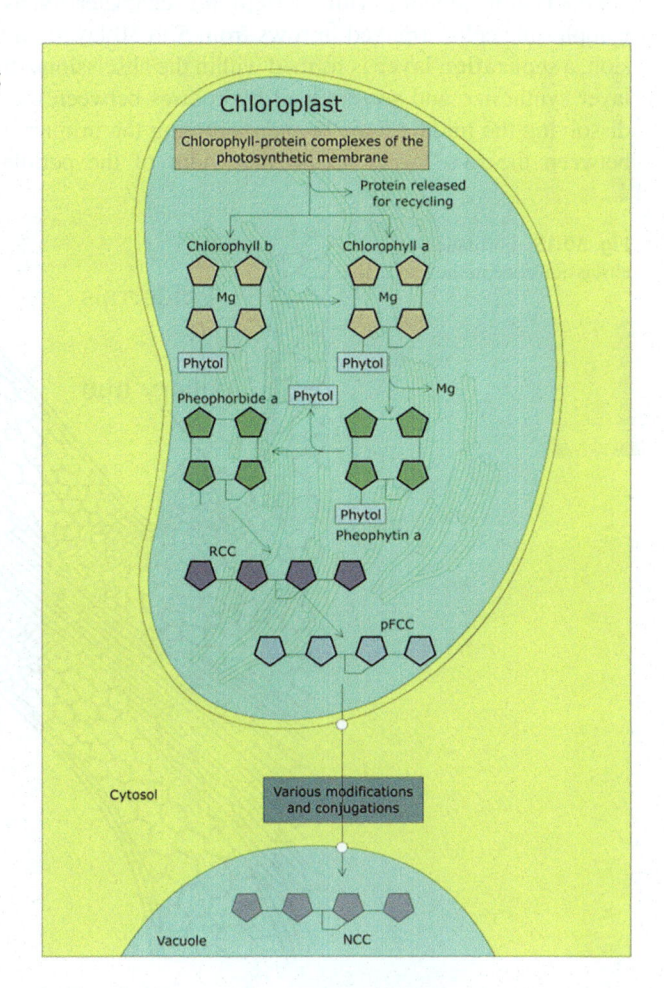

## 30.13 Leaf Abscission

Shedding of leaves in deciduous trees is an intriguing phenomenon, but for plants it is a process by which it gets rid of organs that are no longer required. Abscission usually follows senescence of the organs, but senescence is not essential for an organ to be abscised. The shed organs are generally leaves, flowers, floral parts, mature fruits, etc. It is a controlled process that is initiated in advance of the actual shedding of the organ. An **abscission zone** is formed in a region between the organ to be shed and the body of the plant (Fig. 30.15). This zone facilitates separation by the hydrolysis of wall materials between defined cell layers and to initiate the synthesis of materials that protect the body of the plant from water loss and from infection by microorganisms. Abscission zone can be morphologically identified as one or more layers of isodiametrically flattened cells. Location of abscission zone varies in different organs. For example, in leaves, it is usually formed at the base of the petiole near its junction with the stem. It is characterized as a band of small, densely cytoplasmic cells arranged in rows from 5 to 50 layers in thickness. Before abscission, a **separation layer** is formed within the abscission zone. Cells in the separation layer synthesize and secrete wall hydrolases between the two layers of cells, thus dissolving the middle lamella and disrupting the primary wall. The fracture occurs between the two layers across the width of the petiole. All cells of the layer

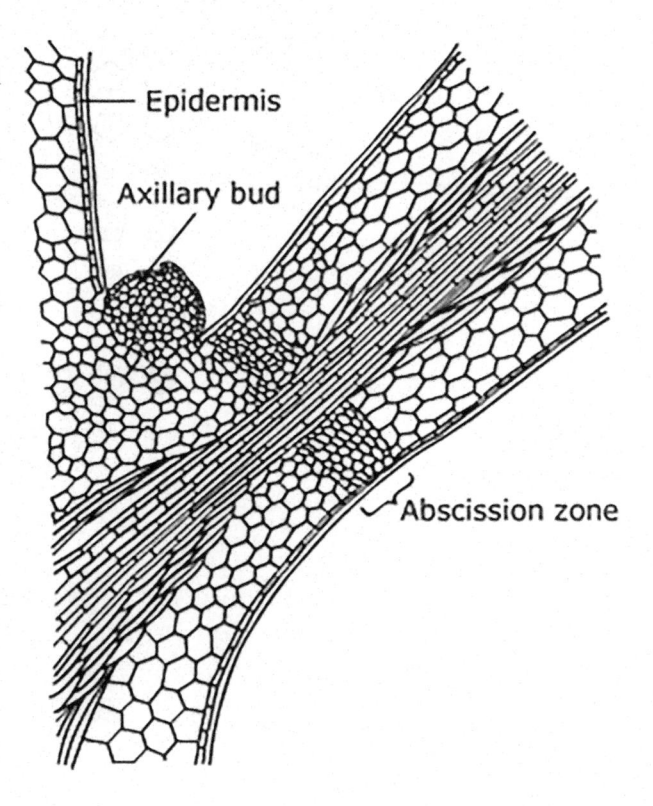

**Fig. 30.15** Formation of abscission zone in the leaf axil

participate in the separation except the dead tracheary cells and vessel elements which are broken mechanically. Later, broken vascular elements are plugged with tyloses or gums. A few layers of cells on the proximal side of the separation zone form the protective layer which synthesizes defense-related proteins to protect the exposed surface from pathogen infection.

## 30.14 Mechanism of Abscission

The young leaves abscise only when they become old. Ethylene and auxin are known to play a key role in regulation of abscission. It is observed that if the lamina portion of the leaf is removed, the leaf soon abscises. This suggests that some factor produced in leaves moves from the blade to the petiole to prevent abscission. Experimental evidence suggests that this factor is IAA. If auxin is applied to the cut end of debladed petiole, abscission is prevented. At the same time, application of auxin is effective if applied at the beginning of the lag phase. Delayed application of auxin may have little or no effect. This indicates that auxin controls leaf abscission. Further, endogenous concentration of auxin in the leaves falls at the time of abscission (Fig. 30.16). Ethylene also plays a key role regulation of abscission. Whole plants exposed to ethylene gas show enhanced rate of leaf or flower abscission. Ethylene production is found to be sufficiently high in abscising organs. Ethylene-insensitive mutants in *Arabidopsis* show delayed senescence and abscission. Investigations have also shown that the relative concentration of auxin on two sides of the abscission layer regulate the production of ethylene that

**Fig. 30.16** Effect of auxin application on abscission

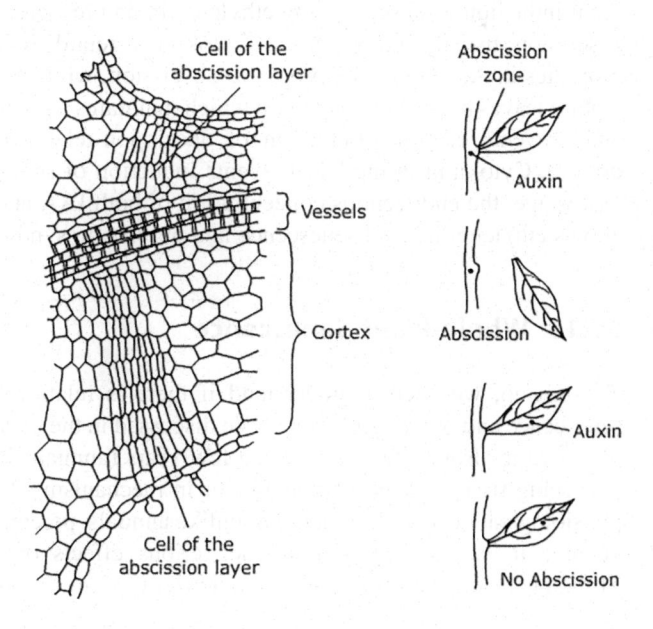

stimulates leaf abscission. The process of abscission is divided into three distinct phases during which there is increase in the sensitivity of the cells toward ethylene.

### 30.14.1 Leaf Maintenance Phase

At this stage there exists a gradient of auxin from the leaf blade to the stem, and the abscission zone remains in insensitive stage. There is no abscission.

### 30.14.2 Abscission Induction Phase

There is reduction in the auxin gradient in the leaf blade, and the abscission zone becomes sensitive to ethylene. Those treatments that enhance leaf senescence may do so by interfering with auxin gradient.

### 30.14.3 Abscission Phase

In this phase the cells of the abscission zone which are already sensitized respond to low concentration of endogenous ethylene. These cells synthesize and secrete wall-degrading enzymes resulting in cell separation and leaf abscission. This phase is marked by synthesis of new proteins. Inhibitors of transcription and translation (i.e., actinomycin D and cycloheximide, respectively) retard abscission if applied during early stage of development. They have little effect if applied at a later stage. Cell wall hydrolases (like endo-1,4-β-glucanases) and polygalacturonases are also synthesized upon induction of abscission by ethylene. In nature, senescence of an organ and its abscission usually go together. Cytokinins retard senescence, while ethylene promotes it. Also, ethylene promotes abscission, while auxin retards it. Young and mature leaves have abundant auxin which maintains a gradient in the abscission zone and prevents ethylene induction of abscission-related changes (abscission syndrome). Cytokinin in the leaf prevents induction of senescence in the leaf. As the leaves age, the endogenous concentration of both IAA and cytokinin drop, and this allows ethylene-induced senescence and abscission syndromes to be expressed.

## 30.15 Whole-Plant Senescence

To sum up, we need to understand if there is relationship between whole-plant senescence and organ senescence. We discussed in the beginning that the life span of plants varies from a few weeks to a few years (annuals, biennials, and perennials). Senescing tissues seem to be deficient in mechanisms that protect a plant against physiological decline. It is said that unlike annuals, perennials are better equipped to be able to survive through the deleterious effects of time (**mutational load**). Mutational load is the total genetic burden in a population resulting from accumulated deleterious mutations. It is defined as a balance between selection against a deleterious gene and its production by mutation. It occurs when cell replication mechanisms propagate errors over thousands of years. Mutation rate can also increase further due to ROS accumulation with increasing age of the plant. Another difference between annuals and perennials lies in the determinate nature of their apical meristems. In annuals, all indeterminate vegetative shoot apices become determinate floral apices and the entire plant senesces after seed dispersal. In contrast, perennials retain a population of indeterminate shoot apices as well as those apices that become reproductive and determinate. Interestingly, ability to delay senescence by removal of reproductive structures is a characteristic feature of annuals. It is observed that repeated de-podding of soybean plants is an inducing factor for the plant to remain vegetative. Earliest explanation for this is that vital nutrients are redistributed from vegetative sources to reproductive sinks. Resource redistribution along with alterations in source-sink relationship caused by flower development may induce a shift in hormonal and nutrient balance of the plant leading to senescence.

**Summary**

- Senescence involves irreversible and degenerative changes in an organism, leading to death. These changes are self-digesting processes controlled by environment and the genetic makeup of an organism.
- Programmed cell death occurs via vacuolar-type PCD (involves vacuolar swelling and cell rupturing) and via hypersensitive response-type PCD (involves vacuolar water loss and cell shrinkage). Vesicles formed during senescence that engulf portions of cells to be degraded are called **autophagosomes**. There formation is regulated by autophagy-related genes and specific proteins. PCD remodels the plant at cellular, tissue, and organ level in important processes like tracheary element differentiation, lysigenous aerenchyma formation, trichome development, functional megaspore formation, etc.
- Leaf senescence involves nutrient relocation from leaves to other parts of the plant thereby exhibiting coordinated chain of events at cell, tissue, and organ level. Senescence process is regulated by several hormones where gibberellins, cytokinins, and brassinosteroids retard the process, while ABA, ethylene, and jasmonates enhance it. Senescence is affected by both external and internal factors through alteration of gene expression (NAC and WRKY being the most abundant gene families regulating the process). The process of senescence has also been associated with overproduction of reactive oxygen species (ROS).
- Senescence is followed by abscission, which is shedding of leaves and other plant organs through formation of abscission zone.

## Multiple-Choice Questions

1. Catabolic process which is due to trauma caused by external agents and results in rupture of membranes and tissue inflammation:
   (a) Autophagy
   (b) Necrosis
   (c) Apoptosis
   (d) Degradation
2. A highly regulated energy-dependent process leading to programmed cell death in animals:
   (a) Necrosis
   (b) Autophagy
   (c) Apoptosis
   (d) Senescence
3. Which of the following cell organelle is degraded first during leaf senescence?
   (a) Peroxisomes
   (b) Mitochondria
   (c) Chloroplasts
   (d) Nuclei
4. The mechanism used by plants to prevent spread of infection by killing the cells around infection site to prevent its further spread is called:
   (a) Hypersensitive response
   (b) Vacuolar response
   (c) Autophagy
   (d) Camouflage

### Answers

1. b    2. c    3. c    4. a

## Suggested Further Readings

Taiz L, Zeiger E (2010) Plant physiology, 5th edn. Sinauer Associates Inc, Sunderland, pp 665–692

Thomas H, Ougham H, Mur L, Jansson S (2015) Senescence and cell death. In: Buchanan BB, Gruissem W, Jones RL (eds) Biochemistry and molecular biology of plants. Wiley-Blackwell, Chichester, pp 925–982

# Part IV

# Stress Physiology

Some examples of adaptations for abiotic stress tolerance in plants. More details are provided in Chap. 31, Sect. 31.1.3, Fig. 31.2

Chapter 31 Abiotic Stress
Chapter 32 Biotic Interactions

# Abiotic Stress

## 31

Satish C Bhatla

The conditions which allow plants to attain maximal growth and reproductive potential, as determined by their total biomass (comprising of plant height, weight, and seed number), can be considered as the ideal growth conditions for the particular plant. Since plants grow and reproduce in complex environmental conditions composed of a multitude of abiotic and biotic factors, they are frequently exposed to stress conditions. **Stress** can be described as environmental conditions which prevent plants from attaining their full genetic potential, and, hence, it adversely affects plant growth, development, and productivity. Environmental factors which result in stress may be divided into biotic and abiotic. **Biotic** stress is imposed by direct or indirect interactions with other organisms, whereas **abiotic** stress originates from excess or deficit in the physical, chemical, and energetic conditions to which plants are exposed. Environmental conditions which have deleterious effect on plants include drought, flood, excessive soil salinity, high or low temperature, excess or insufficient light, and inadequate mineral nutrients. Phytotoxic molecules, like ozone, can also cause damage to plant tissues. Stress affects numerous plant responses ranging from changes in cellular metabolism and gene expression to modulation of growth rate and crop yield. The response of a plant is influenced by the severity, duration, and rate at which stress is imposed. A single stress or a combination of varied stress conditions can result in different stress responses. Moreover, plant responses can be elicited directly by stress or indirectly through stress-induced injury. Various factors, such as genotype, developmental stage, species, and organ or type of tissue, influence the sensitivity or resistance of plants to stress condition(s).

## 31.1 Plant Responses to Abiotic Stress

### 31.1.1 Decline in Crop Yield

Losses due to biotic and abiotic stress conditions can result in a substantial decline in average crop productivity. Stress-tolerant crop plants, developed with the successful application of classical and biotechnological breeding techniques, can lead to an increase global food supplies, and, hence, they can be economically beneficial. Such crops can maintain yield under moderate environmental stress conditions or can enhance survival during intense or prolonged stress periods.

### 31.1.2 Physiological Adjustment

Stressed plants undergo flowering, fertilization, and seed set before attaining their full size, thereby resulting in smaller plants. The seeds produced may also be smaller and fewer in number due to limited leaves to provide photosynthates under stress. The specific developmental pathways adopted to maximize reproductive potential under stress conditions mainly depend on the life cycle of a plant. They are favorable for annual plants, which complete their life cycle in a single season, to adjust their developmental programs and metabolism in order to yield maximum number of viable seeds under prevailing environmental conditions. In contrast, perennial plants, which survive for over 2 years or more, tend to adjust their developmental programs and metabolism to secure requisite storage of food resources in order to survive the next season, even at the cost of seed production.

### 31.1.3 Resistance Mechanisms

Due to their sessile nature, plants have to adapt and develop unique mechanisms to combat varied stress conditions. Plants may deal with suboptimal environmental conditions using two general resistance mechanisms: tolerance or avoidance. **Stress tolerance** mechanisms enable the plants to withstand stress by development of resilient structures and physiological processes, whereas **stress avoidance** mechanisms prevent or minimize exposure to vulnerable stages of stress conditions by adopting life cycle strategies which limit growth to favorable periods in a varying environment (Fig. 31.1). The ability to survive varied degrees of stress may differ substantially depending upon plant species and genotype of a species. The **desert ephemerals** (short-lived weed plants growing in desert) germinate and complete their life cycles, while optimum water is available and, thus, avoids drought (Fig. 31.2). Another drought-avoidance mechanism is deep roots, which provides **phreatophytes** with improved access to groundwater and increased chances of survival during dry periods. Living organisms tend to maintain homeostasis by adjusting to stress and minimizing strain. Three modes of response to disturbance are observed in a homeostatic system. During *elastic response*, system bounces back

## 31.1 Plant Responses to Abiotic Stress

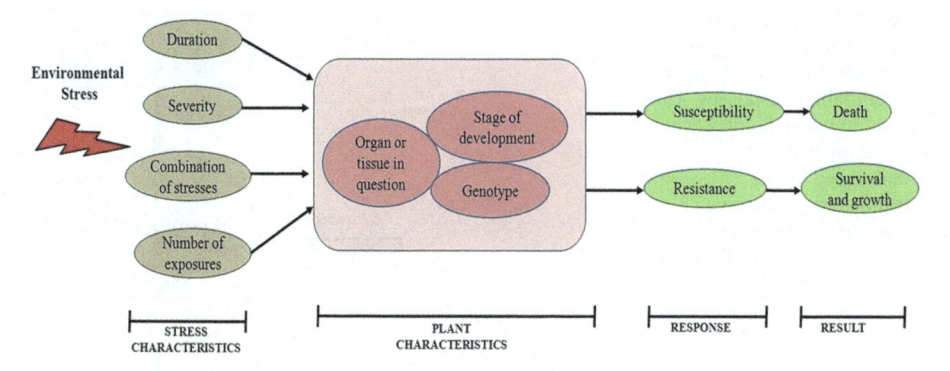

**Fig. 31.1** Factors affecting stress resistance in plants

**Fig. 31.2** Some examples of plant adaptations for stress tolerance. (**a**) *Cereus giganteus* (a drought-tolerant species showing succulent photosynthetic stem), (**b**) *Prosopis glandulosa* (a drought-avoiding species), (**c**) *Spinacia oleracea* (known to exhibit osmotic adjustment as acclimation mechanism), (**d**) *Monoptilon bellioides* (a desert plant which grows only in wet season), and (**e**) *Picea mariana* (a desert plant with needle-like leaves to tolerate freezing)

and regains its former state, and during *plastic response*, system is deformed and settles on a new stable configuration. A *catastrophic response* is observed when the limits of both elastic and plastic resilience are exceeded, during which the system enters into a stage of incoherence, entropy enhances, and the living system dies (Fig. 31.3). The homeostatic adjustment of an individual organism in response to fluctuating environmental factors is referred to as **acclimation**. Resistance may be conferred to an otherwise vulnerable plant by a period of acclimation before stress.

**Fig. 31.3** Biotic or abiotic stress factors invoke strain in the form of elastic, plastic, or system failure responses. The extent to which homeostasis can be maintained is conditioned by the acclimatory or adaptive capacities of the organ, individual, or population

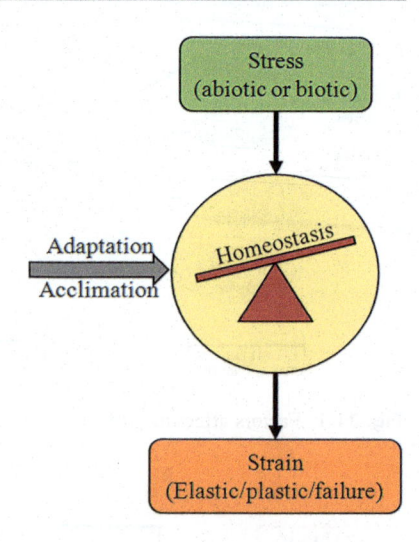

For example, in a process during gardening, *hardening off*, farmers grow plants indoors in pots under optimal conditions before moving them outdoors, so that they can be "toughened up" or "hardened" before exposure to nonlethal stress conditions. Another example of acclimatization is temperature stress, in which plants are unable to withstand freezing in summer, but some can acclimate in response to gradually declining temperatures in fall and may subsequently withstand even temperatures below −50 °C in winters. However, the physiological changes related to acclimation do not take place at genetic level and are generally reversible. Continuous or recurrent environmental stress conditions may exert selective pressure on the plants, resulting in evolution of fitness-enhancing traits. Certain morphological and physiological characteristics, such as crassulacean acid metabolism (CAM), water-storing succulent tissues, and spines, enable many **xerophytes** to tolerate stress. Irrespective of stress conditions, some constitutive, genotypically determined traits, such as light-reflective spines, deep roots, and sunken stomata, are expressed in some plants. Such evolutionary improvements, which enhance the fitness of the organism and occur over many generations and across entire populations, are termed **adaptations**. Stress resistance, based on either adaptation or acclimation, supports survival under fatal conditions, thereby maintaining crop productivity under adverse circumstances (Table 31.1).

### 31.1.4 Alteration in Gene Expression Patterns

Changes in the pattern of gene expression are often involved in stress-induced modifications in development and metabolism. The recognition of stress at the cellular level initiates a stress response in plants by activation of signal transduction pathways which transfer information within individual cells as well as intercellularly all over the plant. Eventually, the alterations in gene expression are amalgamated

## 31.1 Plant Responses to Abiotic Stress

**Table 31.1** Adaptation versus acclimation in response to stress

| | Adaptation | Acclimation |
|---|---|---|
| Level | Population | Individual |
| Caused by | Natural selection | Environmental conditions acting on physiological responsiveness |
| Reversibility | Irreversible | Reversible |
| Heritability | Genotypic | Generally non-heritable |
| Response of homeostasis to perturbation | Mostly plastic | Mostly elastic |
| Time scale | From generation time of the organism up to evolutionary | *Short-term* (minutes/hours) metabolic and physiological adjustments of existing components. No significant change in gene expression<br>*Long-term* (up to weeks or months) altered patterns of gene expression and reallocation of resources, morphological change |
| Deployment in the life cycle | Strategic | Tactical |

into a stress response by the entire plant, which modulates growth and development and can also influence its reproductive abilities. The timing and scale of response are governed by the severity and duration of the stress. Though bacterial and yeast proteins involved in initiation of signal transduction in response to stress (such as low osmotic potential) are known, similar information about stress recognition signaling molecules in plants is relatively new. Hormones, especially jasmonic acid (JA), ethylene, and abscisic acid (ABA), and second messengers, such as calcium ions, participate in the regulation of many abiotic stress responses in plants, but the intricate and intensive signaling pathways which are supposedly involved in changing the gene expression patterns are gradually being elucidated.

Enhanced or declined expression of certain genes and protein products may be exhibited in response to abiotic stress. Scientists are currently focusing on the functions of these proteins and the mechanisms involved in the regulation of their expression. Posttranscriptional regulatory mechanisms elevate levels of some protein-coding mRNAs or noncoding regulatory RNAs, increase translation, stabilize proteins, and modulate protein activity by different types of modifications. They can also influence the accumulation of gene products in addition to transcriptional activation of gene expression. Parallel or even overlapping approaches, such as activation or repression of gene expression, activation of transport activity, and production of proteins or compatible solutes, may also be adopted during plant's response to stresses. Plants are also frequently exposed to serial or concurrent abiotic stresses, such as flooding followed by drought or heat and drought. The synergistic or antagonistic interactions between varied kinds of stresses can influence the coordinated success of the activity of several genes, which enhance survival of crops in response to abiotic stress.

## 31.2 Oxidative Stress

### 31.2.1 Production of Reactive Oxygen Species (ROS)

Molecular oxygen is relatively unreactive in its ground state but can give rise to various toxic reactive forms, like free radicals, as a result of various environmental stresses. Reactive oxygen species (ROS) may result from the transfer of one, two, or three electrons to molecular oxygen ($O_2$) to form superoxide radical ($O_2^{\cdot-}$), hydrogen peroxide ($H_2O_2$), or hydroxyl radical ($^{\cdot}OH$), respectively, and from the excitation of $O_2$ to form singlet oxygen ($^1O_2$). The term ROS refers not only to free radicals but also to other toxic non-radicals as well. Reactive oxygen species (ROS) are constantly produced in all aerobic organisms as by-products of various metabolic pathways (e.g., during respiration, photorespiration, and photosynthesis) localized in various cellular compartments, predominantly chloroplasts, peroxisomes, and mitochondria (Table 31.2). One-electron reduction of $O_2$ results in the formation of superoxide anion ($O_2^{\cdot-}$), which further gets converted into $H_2O_2$ upon its dismutation by the enzyme superoxide dismutase (SOD). $H_2O_2$, a non-radical molecule, is neutralized by glutaredoxin enzyme system [glutathione reductase, glutathione (GSH), and glutathione peroxidase] and catalase into $O_2$ and $H_2O$ (Fig. 31.4). In plants, tripeptide glutathione (γ-glutamyl-L-cysteinylglycine, GSH) is one of the most common non-protein thiol metabolites. GSH serves as a major intracellular defense against ROS-induced oxidative damage. Under stress conditions, GSH concentration usually declines, and redox state becomes more oxidized, thereby causing damage to the system. So, the generation and maintenance

**Table 31.2** Mechanisms of generation of various types of ROS in a plant cells

| Molecule | Abbreviation | Biological source |
|---|---|---|
| Molecular oxygen (triplet ground state) | $O_2$. | Most common form of dioxygen gas |
| Singlet oxygen (first excited singlet state) | $^1O_2$. | Photoinhibition, PSII electron transfer |
| Superoxide anion | $O_2.^-$ | Mitochondrial electron transfer reactions, Mehler reaction (reduction of $O_2$ by iron-sulfur center of PSI), photorespiration in glyoxysomes, peroxisome reactions, plasma membrane, paraquat oxidation, nitrogen fixation pathogen, reaction of $O_3$ and $OH^-$ in apoplast, respiratory burst homolog (NADPH oxidase) |
| Hydrogen peroxide | $H_2O_2$ | Photorespiration, β-oxidation, proton-induced decomposition of $O_2.^-$, pathogen defense |
| Hydroxyl radical | $HO^-$ | Decomposition of $O_3$ in apoplast, pathogen defense, Fenton reaction |
| Perhydroxyl radical | $HO_2.^-$ | Reaction of $O_3$ and $HO^-$ in apoplast |
| Nitric oxide | NO | Nitrate reductase, nitrite reduction by the mitochondrial electron transport chain |

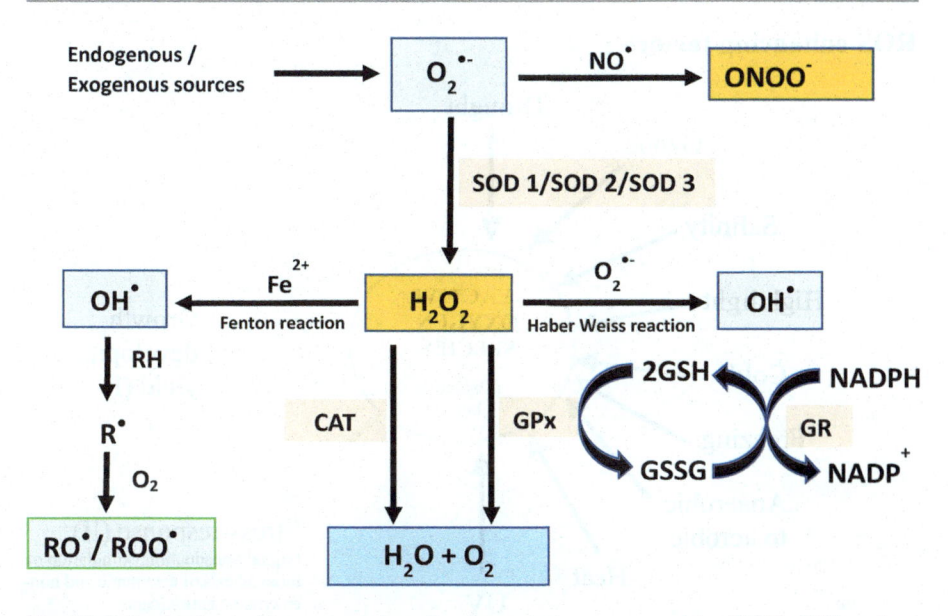

**Fig. 31.4** Production and interconversion of various reactive oxygen species (ROS). *CAT* catalase, *GR* glutathione reductase, *GPX* glutathione peroxidase, *SOD* superoxide dismutase, *GSH* reduced glutathione, *GSSG* oxidized glutathione

of reduced GSH pool are crucial for the cell. Glutathione peroxidase (GPx) reduces GSH into its oxidized form (GSSG), which is again regenerated by glutathione oxidase along with the oxidation of NADPH (ascorbate-glutathione cycle). Additionally, ˙OH, the most reactive chemical species in the biological world, can be produced in the presence of transition metals, such as iron and copper through the Fenton reaction or by the Haber-Weiss mechanism. $O_2^{\bullet-}$ can also form peroxynitrite [ONOO⁻, a kind of reactive nitrogen species (RNS)] upon reaction with another signaling free radical species, NO. Nitrogen-centered free radicals are termed as reactive nitrogen species (RNS). RNS include S-nitrosoglutathione (GSNO), peroxynitrite (ONOO⁻), dinitrogen trioxide ($N_2O_3$), and nitrogen dioxide ($NO_2$).

### 31.2.2 Dual Role of ROS

A rapid rise in ROS concentration results in oxidative stress in the cells, known as **oxidative burst**. Rate of ROS production and cellular ROS levels both increase significantly in plants subjected to abiotic or biotic stresses. Chemically, ROS are highly reactive and biologically toxic in nature. They exhibit much shorter half-life as compared to that of oxygen. It is established that ROS play dual roles in plant cells (Fig. 31.5). At normal concentrations under unstressed conditions, they play the role of signaling mediators for different cellular responses, such as stomatal closure, root gravitropism, seed germination, plant cell death, lignin biosynthesis, osmotic stress, and hypersensitive responses, which may be activated by certain external and

**ROS enhancing factors**

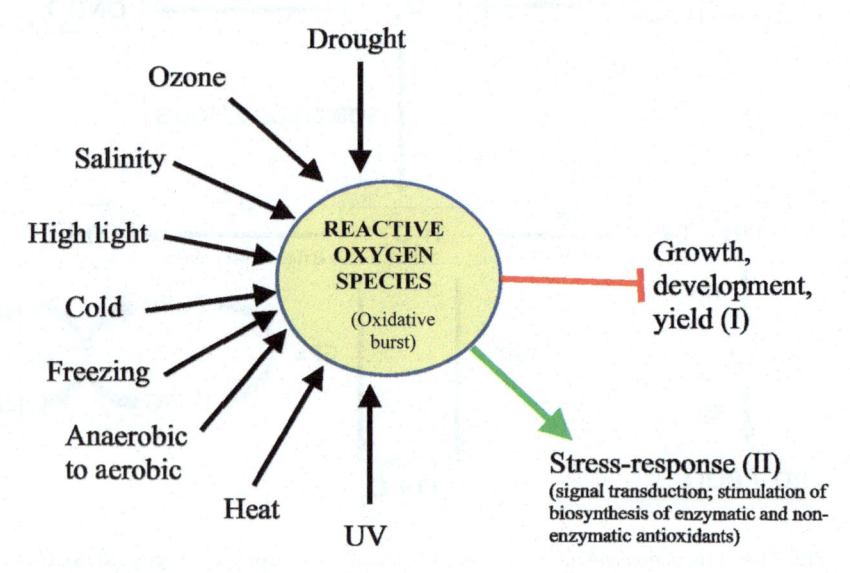

**Fig. 31.5** Dual role of ROS during abiotic stress

developmental stimuli. Excess of ROS produced in plant cells under stress conditions can lead to damage to biomolecules like proteins, lipids, carbohydrates, and DNA. These, in turn, can lead to alteration in the intrinsic membrane properties such as fluidity, loss of enzyme activity, ion transport, DNA damage, inhibition of protein synthesis, etc. Under physiological steady-state conditions, ROS concentration is maintained low by different antioxidative defense mechanisms which help in scavenging ROS. Various factors may disturb the equilibrium between production of ROS and their scavenging. As a result, intracellular ROS levels may rapidly rise.

### 31.2.3 Cellular Antioxidative Defense System

Under normal growth conditions, plants are well-adapted to deal with the deleterious effects of ROS. However, during stress, ROS production exceeds the quenching capacity of the antioxidant protective systems. In order to overcome oxidative damage caused due to high ROS concentration and for maintaining **redox homeostasis**, plants have a well-established defense system which includes both nonenzymatic and enzymatic antioxidants. The nonenzymatic antioxidant defense system consists of vitamins (A, C, E), glutathione, ascorbate, tocopherol, carotenoids, and phenols which directly react with ROS, leading to their scavenging. The enzymatic antioxidant defense system consists of enzymes, such as peroxidase (POD), ascorbate peroxidase (APX), glutathione reductase (GR), catalase (CAT), monodehydroascorbate reductase (MDHAR), dehydroascorbate reductase (DHR),

## 31.2 Oxidative Stress

**Table 31.3** Major antioxidant enzymes involved in ROS scavenging

| Enzyme | EC number | Reaction catalyzed |
|---|---|---|
| Superoxide dismutase (SOD) | 1.15.1.1 | $O_2^{\cdot-} + O_2^{\cdot-} + 2H^+ \longrightarrow 2H_2O_2 + O_2$ |
| Catalase (CAT) | 1.11.1.6 | $2H_2O_2 \longrightarrow O_2 + 2H_2O$ |
| Glutathione peroxidase (GPX) | 1.11.1.9 | $2GSH + R\text{-}OOH \longrightarrow GSSG + R\text{-}OH + H_2O$ |
| Ascorbate peroxidase (APX) | 1.11.1.11 | $AA + H_2O_2 \longrightarrow DHA + 2H_2O$ |
| Peroxidase (POD) | 1.11.1.7 | $Donor + H_2O_2 \longrightarrow oxidized\ donor + 2H_2O$ |
| Glutathione reductase (GR) | 1.6.4.2 | $NADPH + GSSG \longrightarrow NADP^+ + 2GSH$ |
| Phospholipid hydroperoxide glutathione peroxidase (PHGPX) | 1.11.1.12 | $2GSH + PUFA\text{-}OOH \longrightarrow GSSG + PUFA\text{-}OH + 2H_2O$ |

Abbreviations: GSSG-oxidized glutathione; GSH- reduced glutathione; AA- ascorbic acid; PUFA- polyunsaturated fatty acid; DHA- dehydroascorbate

and superoxide dismutase (SOD) which help in scavenging specific reactive oxygen species (Table 31.3) (Fig. 31.6). Various peroxidases, catalase, and the enzymes of ascorbate-glutathione cycle (APX, MDHAR, DHAR, and GR) quench peroxides through a series of coupled redox reactions (Fig. 31.7). These enzymes are nonuniformly distributed in various cellular compartments, and their isozymes are activated to varying extents upon exposure to stress. Oxidative stress can stimulate the biosynthesis of both nonenzymatic and enzymatic components of antioxidant machinery in many plants.

### 31.2.4 Ozone Exposure Leads to Oxidative Stress

Ozone, a normal component of stratosphere, protects Earth from hazardous UV radiation. However, since living organisms are increasingly exposed to harmful levels of ozone in the troposphere (lower atmosphere), ozone can cause damage through oxidative stress in a ROS-mediated manner. Ozone, as a pollutant, is generated in the lower atmosphere from reactions of molecular oxygen with anthropogenic hydrocarbons and oxides of sulfur ($SO_x$) and nitrogen ($NO_x$) under the effect of UV radiation. High concentration of ozone is one of the best characterized oxidative stress factors. Exposure of plants to damaging levels of ozone leads to impaired photosynthesis, reduction in growth of roots and shoots, and decrease in crop yield and leaf injury. Plants exhibit variable capacity to survive in high-ozone environments. They employ both avoidance and tolerance mechanisms to resist stress by closing stomata to exclude pollutants and by activating the antioxidant defense and varied repair mechanisms. Ozone-derived ROS, produced upon reaction

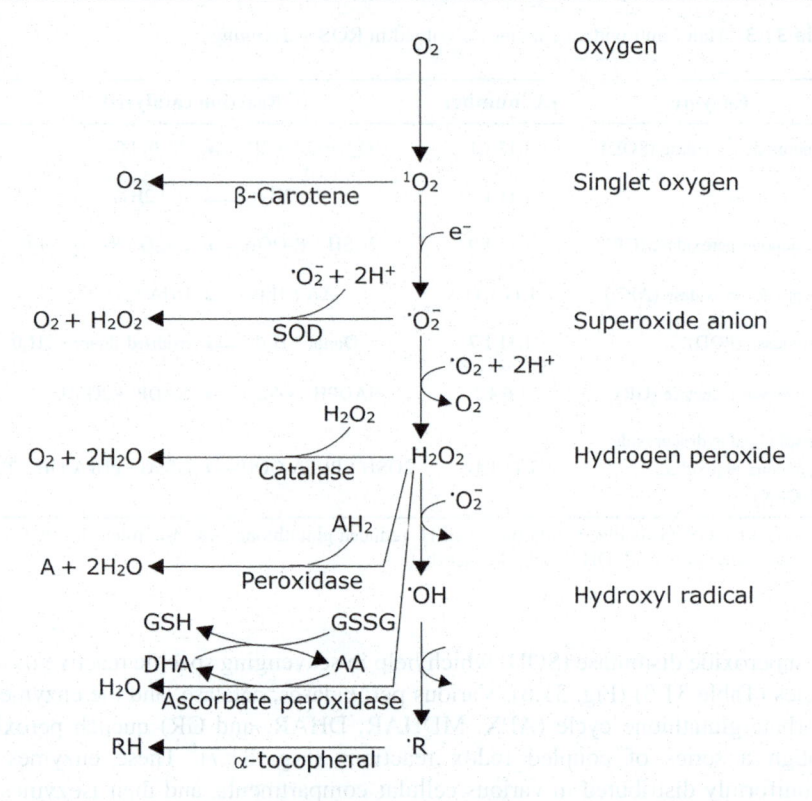

**Fig. 31.6** ROS formation and action of ROS scavenging enzymes. *AA* ascorbic acid, *DHA* dehydroascorbate, *GSH* glutathione, *GSSG* oxidized glutathione, *SOD* superoxide dismutase

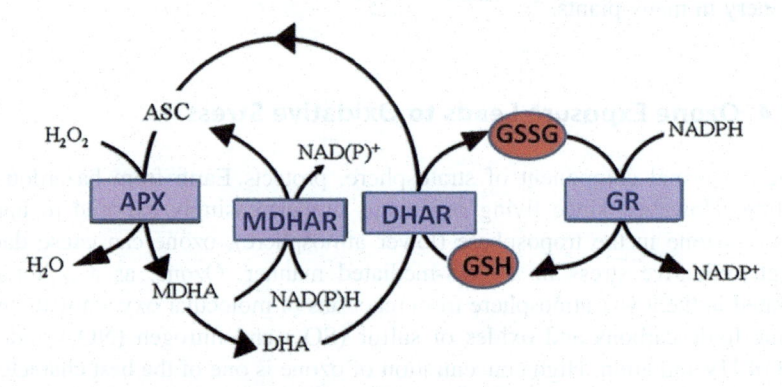

**Fig. 31.7** Ascorbate-glutathione cycle (Foyer-Halliwell-Asada pathway) for quenching peroxides. *APX* ascorbate peroxidase, *MDHAR* monodehydroascorbate reductase, *DHAR* dehydroascorbate reductase, *GR* glutathione reductase, *GSH* reduced glutathione, *GSSG* oxidized glutathione, *ASC* ascorbate, *MDHA* monodehydroascorbate, *DHA* dehydroascorbate

of ozone with ethylene and other alkenes in the apoplastic fluid, causes damage to membrane lipids, collapses membrane potential, increases $Ca^{2+}$ uptake, inhibits $H^+$-pump, and enhances membrane permeability. It can target proteins, carbohydrates, and nucleic acids as well. In the atmosphere, NO can react with excess ozone, thereby reducing its (ozone's) deleterious effects on plant growth.

ROS concentration can be significantly and rapidly enhanced due to disturbed equilibrium between ROS production and scavenging under varied abiotic stress conditions. ROS levels are stimulated, and membranes and proteins are rendered dysfunctional under temperature stress, such as chilling, excessive heat, or freezing. Chilling stress increases ROS production by inactivation of cytochrome c oxidase, while heat stress impairs the respiratory electron transport chain in mitochondria resulting in peroxidation of membrane lipids due to oxidative stress. In chloroplasts, ROS levels are enhanced due to damage to the photosynthetic machinery in response to freezing and chilling stresses. Varied stress conditions, such as drought, salinity, deoxygenation following flooding, and osmotic stress also result in increase in ROS and hence ROS-associated injury.

## 31.3 Salt Stress

One of the major widespread environmental stresses is caused by soil salinity. It can limit growth and development of plants due to excess chloride and sodium ions in soil, particularly in arid and semiarid regions. **Salinity** refers to an excessive accumulation of salts in soil solution. Soil salinity is a major threat to global food security. Up to 22% of the world's irrigated land, which produces one-third of the world's food, is salt affected. Global warming leads to more variations in atmospheric temperature, rise in temperature, and erratic rainfall leading to increase in salt concentration in top soil due to drying of soil, and a rise in sea levels results in salt water entering the rivers where water that is being used for irrigation is from. Terrestrial plants are rarely salt-tolerant due to their evolution from freshwater algae. The functions of $Na^+$ in marine ancestry are fulfilled by $K^+$ and $H^+$ in all terrestrial plants. When the concentrations of the sulfates or chlorides of Na, Mg, and Ca in the soil are enough to produce an electrical conductivity of at least 4 $dS.m^{-1}$ in the soil extract and a sodium absorption ratio [SAR; based on $Na/(Ca^{2+} + Mg^{2+}$] of less than 13, it is referred to as **saline soil**. Saline soils contain high levels of NaCl together with some fraction of $CaSO_4$, $MgSO_4$, and $Na_2SO_4$, while soils containing high $Na_2CO_3$ and high SAR due to low calcium and magnesium concentrations are called **sodic soils** (Table 31.4).

Table 31.4 Types and properties of salt-containing soils

| Soil type | Constituent salts | pH range |
|---|---|---|
| Saline soils | $CaSO_4$, $MgSO_4$, NaCl, $Na_2SO_4$ | pH 6–7 |
| Saline-sodic soils | NaCl, $CaSO_4$, $MgSO_4$ | pH < 8.5 |
| Sodic soils | High $NaCl/Na_2SO_4$ (most deleterious for plants) | pH 8.5–12 |

**Fig. 31.8** Some examples of halophytes. (**a**) *Atriplex nummularia*, (**b**) *Disphyma australe*, (**c**) *Suaeda maritima*

Plants develop mechanisms for salt exclusion or salt tolerance to deal with deleterious effects of salt stress. They are classified as halophytes or glycophytes, depending upon their capacity to survive on substrate containing high salt concentration. **Halophytes** (salt-tolerant plants) can grow at considerably high concentrations of salt, for example, *Atriplex nummularia*, *Disphyma australe*, and *Suaeda maritima* (Fig. 31.8). Depending upon the salt sensitivity, varied types of responses have been categorized in terrestrial plants: *Type I response*, plants which require some salt for adequate growth but are inhibited at seawater salinities, e.g., saltbush; *Type II response*, plants which exhibit significant decline in growth at EC of 4 $dS.m^{-1}$, such as rice and barley; and *Type III response*, salt-sensitive plant species which are unable to grow in saline soils, e.g., chickpea (Fig. 31.9). Salt-tolerant plants (halophytes) comprise of 0.25% of known angiosperms and include around 350 species mainly from Caryophyllales, Alismatales, Malpighiales, Poales, and Lamiales. Usually terrestrial plants are unable to tolerate salt stress and are referred as **glycophytes** (sweet plants). Most of the world's staple food crops are salt-sensitive, and some crops exhibit tolerance to mild salinity. Glycophytes and halophytes have strong differences in terms of their evolutionary and adaptational backgrounds. At low concentrations of around 1 mm $Na^+$ in the soil, some crops show $Na^+$-induced enhancement in growth, and also small amounts of sodium ions are essential for species with $C_4$ and CAM photosynthetic pathways. Thus, for these plants, sodium is a *beneficial element* (Table 31.5).

### 31.3.1 Disruption of Ionic Homeostasis due to Salt Stress

Salinity imposes stress both at cellular and whole plant levels. NaCl is an osmotically active compound and hence predominantly affects plant growth through *osmotic stress*. It reduces the water potential of soil, thereby restricting water uptake, and hence it leads to water deficiency. Osmotic stress so developed changes cell volume and results in loss of turgor pressure. Moreover, salinity causes *ionic stress* due to excessive uptake of $Na^+$ and $Cl^-$, leading to alterations in ion distribution

## 31.3 Salt Stress

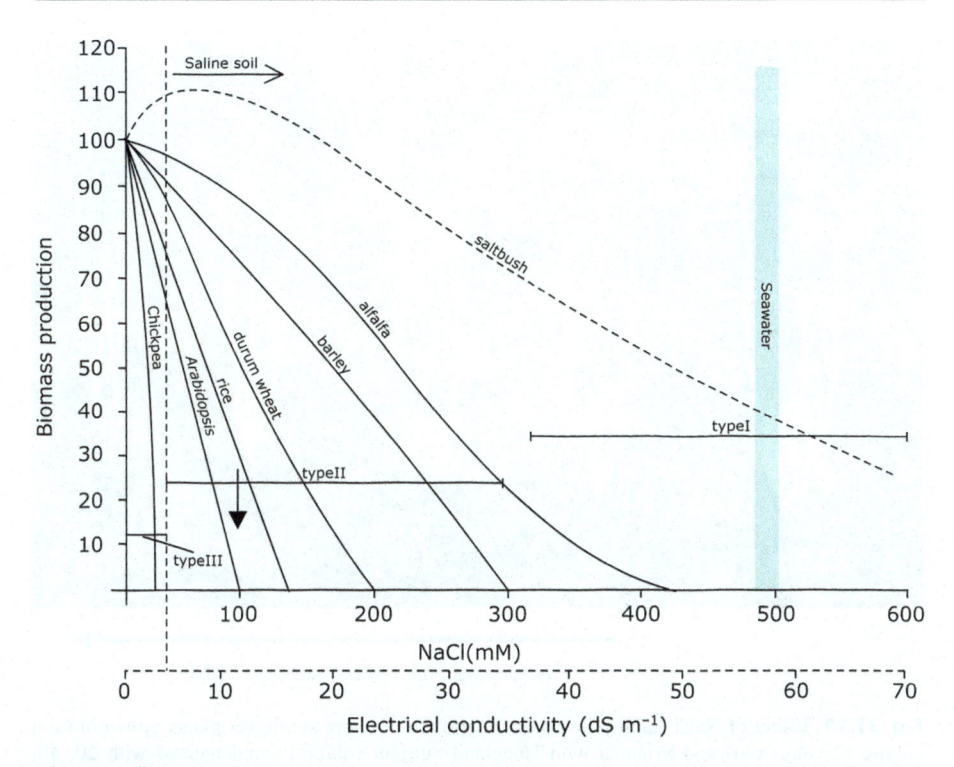

**Fig. 31.9** Salt-tolerance limits of selected plants in terms of biomass production with reference to electrical conductivity of the growth medium

**Table 31.5** Sodium as an essential element for plant development

| Plant | Effect |
|---|---|
| *Anabaena cylindrica* Lemm. | Growth enhancement |
| C4 plants | Aids in metabolism by regeneration of |
| *Cynodon dactylon* | Phosphenol pyruvate and its uptake in |
| *Amaranthus tricolor* | Chloroplasts |
| *Portulaca grandiflora* | |
| *Panicum miliaceum* | |

pattern (cytotoxicity) and aberrations in associated biochemical pathways. Potassium transporters take part in the maintenance of high concentration of $K^+$ in the plant cells and because of similarity of $Na^+$ to $K^+$ in terms of their ionic radii and ion hydration energies, these two ions compete for uptake via transporters under saline conditions. Plant growth is considerably reduced due to high $Na^+:K^+$ ratio, ultimately leading to toxicity (Fig. 31.10). High concentrations of sodium ($Na^+$) and chloride ($Cl^-$) ions in soil result in high salinity and, hence, cause both hyperionic and hyperosmotic conditions. A range of deleterious effects are caused by salinity stress. These include inhibition of photosynthesis rate, chlorophyll content, damage

**Fig. 31.10** Effect of NaCl salinity levels on the growth of entire sunflower plants grown in sand culture (25 days old) and irrigated with Hoagland nutrient solution supplemented with 20, 40, 80, and 160 mm of NaCl. C represents control plants (treated with only Hoagland nutrient medium)

to plasma membrane permeability, alteration in nutrient uptake, accumulation of toxic ions, osmotic and oxidative stress, and other metabolic disturbances in a number of plants.

## 31.3.2 Sodium Entry Through Symplastic and Apoplastic Pathways

Low cytosolic concentrations (1–5 mm) of sodium ions in plant cells lead to strong electrochemical driving force which facilitates inward entry of Na$^+$ through passive transport systems. Sodium influx into the root cells and its sequestration or long-distance transport are highly regulated. Entry of Na$^+$ into the epidermis of the root cells is followed by its apoplastic and symplastic movement to the cortical zone, endodermis, and xylem parenchyma cells (Fig. 31.11a). Development of Casparian strips is initiated in the differentiating zone of root endodermis, thereby restricting toxic solutes to enter the stelar cells. The suberized Casparian strip rarely acts as a complete barrier to apoplastic movement of Na$^+$. Such apoplastic bypass plays a crucial role in the entry of Na$^+$ in many species. Elongation zone of the roots is highly susceptible to Na$^+$ influx due to higher absorption capability in this region. In the plant roots, nonselective cation channels (NSCCs) are primarily responsible for Na$^+$ influx across the plasma membrane. Na$^+$/H$^+$ exchangers (NHXs) carried out the

## 31.3 Salt Stress

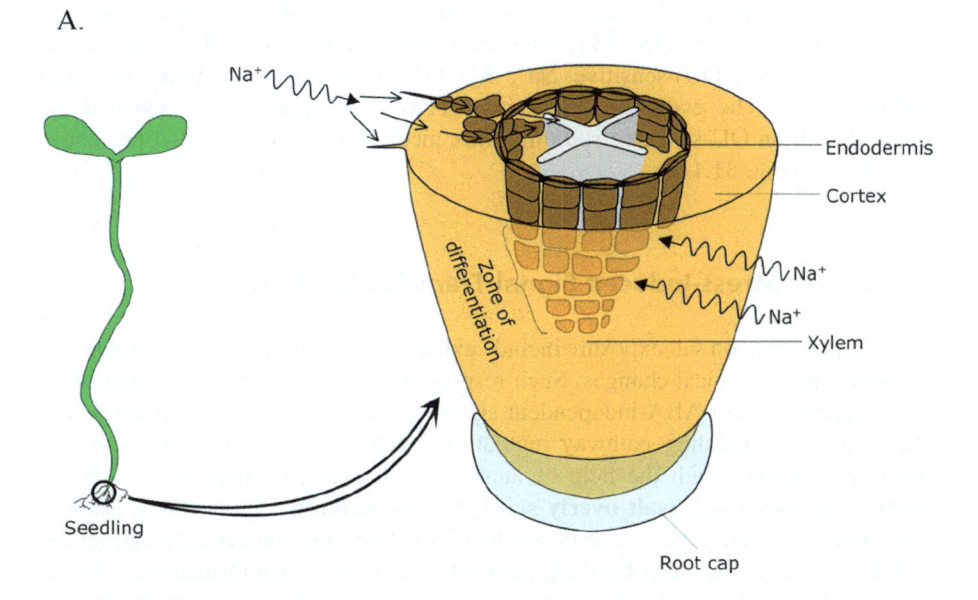

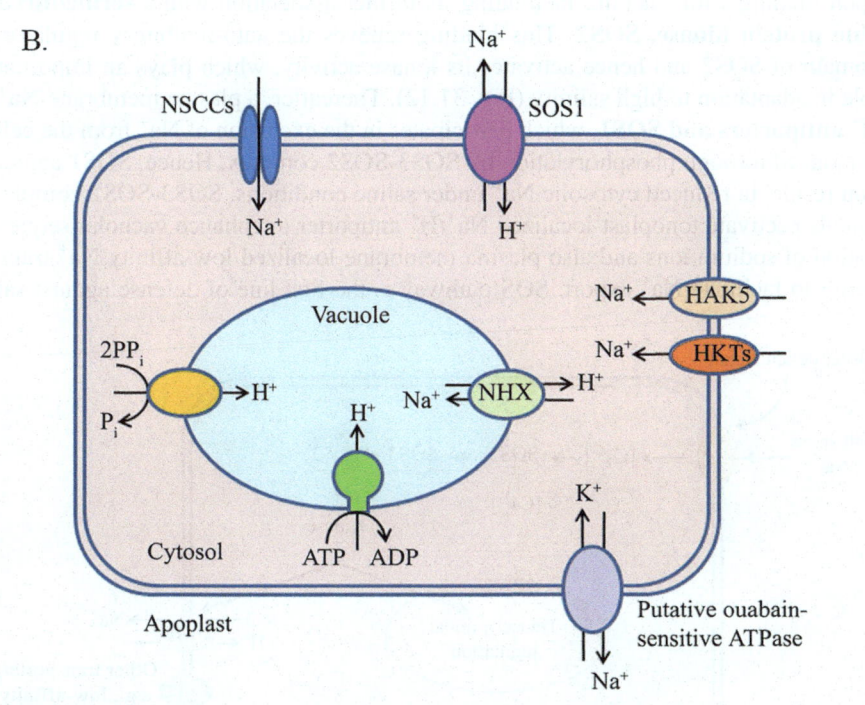

**Fig. 31.11** (a) Anatomical features of root apex and subapical region highlighting the zone for Na$^+$ uptake. (b) Mechanisms of Na$^+$ influx and efflux in plant cells

compartmentalization of Na⁺ into vacuoles. Na⁺ efflux from the roots is mediated by the salt overly sensitive 1 (SOS1) protein catalyzed activity of Na⁺/H⁺ antiporters. In animals, ouabain (OU)-sensitive Na⁺, K⁺-ATPase (a P-type ATPase) mediates sodium efflux. The evolution of P-type ATPases in higher plants revealed the possibility of an OU-sensitive sodium efflux mechanisms similar to that present in animal cells (Fig. 31.11b).

### 31.3.3 Salt Stress-Induced Signal Transduction Events

Plant responses upon salt exposure include extensive changes in gene expression and numerous physiological changes. Such responses to salt stress involve initiation of ABA-dependent and ABA-independent signaling pathways. The perception of salt stress involves signaling pathway regulating ion homeostasis. This pathway has been characterized with the help of analysis of *Arabidopsis* mutants, which are overly sensitive to salt (**salt overly sensitive**; *sos* **mutants**). Intracellular calcium concentration is enhanced in response to water deficiency induced by salt stress. **SOS3 protein (a calcium-binding protein)** undergoes a conformational change upon binding with calcium, facilitating its further association with a **serine/threonine protein kinase, SOS2**. This binding relieves the auto-inhibitory regulatory domain of SOS2 and hence activates its kinase activity, which plays an important role in adaptation to high salinity (Fig. 31.12). Thereafter, a plasma membrane **Na⁺/H⁺ antiporters and SOS1**, which participates in the exclusion of Na⁺ from the cell, is induced through phosphorylation by SOS3-SOS2 complex. Hence, SOS1 activation results in reduced cytosolic Na⁺ under saline conditions. SOS3-SOS2 complex can also activate tonoplast-localized Na⁺/H⁺ antiporter to enhance vacuolar sequestration of sodium ions and also plasma membrane-localized low-affinity Na⁺ transporter to facilitate Na⁺ export. SOS pathway is the first line of defense against salt

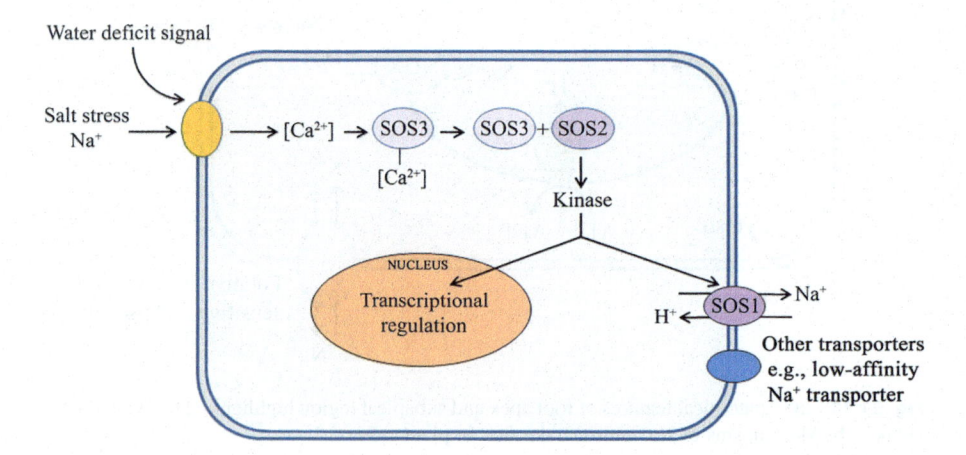

**Fig. 31.12** Signal transduction pathway showing response to salt stress in *Arabidopsis* roots

stress in terrestrial plants. However, since Na$^+$ efflux from the plant cells is not sufficient to survive under high sodium stress because of its limited capacity, varied tolerance and avoidance mechanisms have evolved as a prominent part of response to salt stress.

## 31.3.4 Salt Stress Tolerance Mechanisms

Salt stress signaling involves a series of physiological events which coordinate to maintain ion homoeostasis. Salt stress-induced signaling events can be classified as early and late responses. Early responsive events are associated with rapid fluctuations in the levels of cytosolic and organellar [Ca$^{2+}$] and its perception by calcium-dependent proteins. Late responsive events involve long-distance transport of Na$^+$ from root to the aerial organs and modulation of antioxidant machinery to prevent reactive oxygen species (ROS)-mediated oxidative damage to plant cells. Salt tolerance in plants depends upon the following factors: 1. accumulation and sequestration of Na$^+$ and Cl$^-$; 2. maintenance of K$^+$ and Ca$^{2+}$ pools as essential macronutrients in the presence of high Na$^+$ levels; 3. regulation of long-distance Na$^+$ transport and its accumulation in different aerial organs following transpiration; 4. accumulation of suitable osmolytes to prevent desiccation; and 5. modulation of efficient antioxidant machinery.

## 31.3.5 Accumulation and Sequestration of Na$^+$ and Cl$^-$

Under low external water potential during salt stress, water uptake is accomplished through a regulated uptake of Na$^+$ along with other counterbalancing ions, such as Cl$^-$, into the vacuoles. This ion uptake leads to water uptake in the cells. In the shoots of halophytes, cellular osmolarity is 2–3 times higher than the osmolarity of soil solution, but the cytoplasmic concentration of ions, including Na$^+$ and Cl$^-$, is maintained at nontoxic levels. This is attained by the action of tonoplast-localized Na$^+$ and Cl$^-$ importers, which sequester Na$^+$ and Cl$^-$ in the vacuoles (Fig. 31.13). The enhanced osmolarity in the vacuole is counterbalanced by accumulation of organic solutes in the cytoplasm. Apart from balancing of osmolarity, the organic solutes also function as osmoprotectant stabilizers of membranes, proteins, and free-radical scavengers. In halophytes, the uptake of Na$^+$ across the plasma membrane takes place by varied mechanisms. Usually Na$^+$ and Cl$^-$ uptake occurs through ion channels, but uptake by plasma membrane invaginations (pinocytosis) can also supplement ion uptake. Low-affinity K$^+$ transporters are another mode of Na$^+$ entry. The driving force for the active sequestration of Na$^+$ into the vacuoles by Na$^+$-H$^+$ antiporter is generated by tonoplast-localized H$^+$-ATPase and H$^+$-pyrophosphatases (Fig. 31.11b). This may be followed by passive Cl$^-$ entry through a specific tonoplastic uniport channel into the vacuoles. Na$^+$-H$^+$ antiporter may be expressed constitutively in some halophytes, and it can be induced rapidly in some salt-tolerant glycophytes under high external sodium stress. In glycophytes cells of

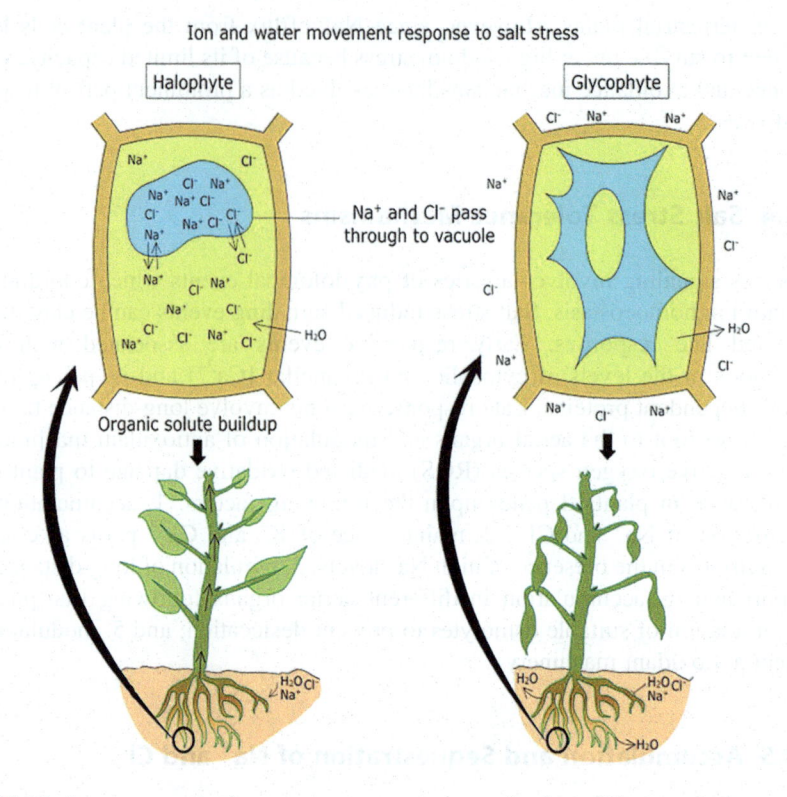

**Fig. 31.13** Ion movement and consequent water movement in plants. Diagrams depict responses to salt stress in a halophyte and a salt-sensitive glycophyte

roots respond to high Na$^+$ concentration in soil by loss of cytoplasmic water leading to plasmolysis.

## 31.3.6 Expulsion of Salt Ions

With the help of Na$^+$-H$^+$ antiporter and H$^+$ ATPase localized on the plasma membrane, halophytes are able to expel Na$^+$ and Cl$^-$ ions from the cytoplasm. When there are sudden salt shocks instead of long-term adjustment to high salt, expulsion of salt ions is used as a tolerance mechanism. This mechanism operates in an opposite manner to the mechanisms operating during salt sequestration in the vacuoles. This is so because cytoplasmic proteins and enzymes in halophytic plants are not tolerant to very high salt stress levels as the cytoplasm does not generally consists of high anion or cation concentrations.

### 31.3.7 Accumulation of Organic Solutes

In halophytes and salt-tolerant glycophytes, another distinctive feature is the accumulation of organic solutes in the cytoplasm in response to accumulation of salt ions in vacuoles. Accumulation of such nontoxic, "compatible" osmolytes provides osmotic balance and also protects the contents of cytoplasm. These osmoprotective compounds are synthesized under drought stress as well and are mainly categorized into amino acids, polyols/sugars, and quaternary ammonium ("onium") compounds. The most commonly produced quaternary ammonium compound is **glycine betaine**. Glycine betaine, derived from serine via a choline intermediate, is a ubiquitous compound in all plant species, whether present in trace amounts or accumulated under abiotic stress. Apart from its role as an osmoprotective compound, it also has cryoprotective effects as it leads to a decline in osmotic potential without causing disturbance to macromolecule-solvent interactions. Other betaines, such as β-alanine in *Limonium* sp. and β-dimethylsulfoniopropionate (DMSP) in *Wedelia biflora*, also accumulate in some plant species under salt stress. Accumulation of "onium" compounds is not a typical response to salt stress in all halophytic species. Some plant species, such as ice plant (*Mesembryanthemum crystallinum*), accumulate methylated inositols, namely, ononitol and pinitol (polyols), under low temperature and salt stress. Also, adjustment of gene expression to enhance encoding of water channels, aquaporins, is another mechanism to facilitate uptake of water in halophytes.

### 31.3.8 Morphological Adaptations to Salt Stress

Certain morphological features in some plants help their survival under high external salt stress. Since soil water potential in saline soils is as low as that of desert soils, stress responses of many halophytes are xeromorphic in nature, so as to help the plant to survive under osmotic and ionic challenges imposed by salinity. Numerous halophytes are succulents. Their thick and fleshy leaves facilitate storage of water. Under water-deficit conditions, water stored in tissue can maintain plant cells (especially in photosynthetic cells) in hydrated state to support carbon fixation. Majority of the leaf tissue contains large and thin-walled cells to store water in the central region in halophytic succulents, such as sea blite (*Suaeda maritima*) and prickly saltwort (*Salsola kali*). The photosynthetic mesophyll cells are localized around the periphery of this water-storing tissue. In certain plants like silver goosefoot (*Oblone portulacoides*), salt bladders (modification of leaf hairs) are formed from supplementation of the central water reserve in the leaf with swollen modified epidermal hairs (Fig. 31.14a). The underlying photosynthetic cells extract water from the bladder-like surface cells and the central aqueous tissue under high salinity conditions resulting in contraction of bladder cells. Salt bladder cells regenerate their water reservoir when the availability of water increases. In some plants, such as asparagus, succulence is developed only under salt stress. Extreme halophytes, such as *Salicornia* species, are stem succulents.

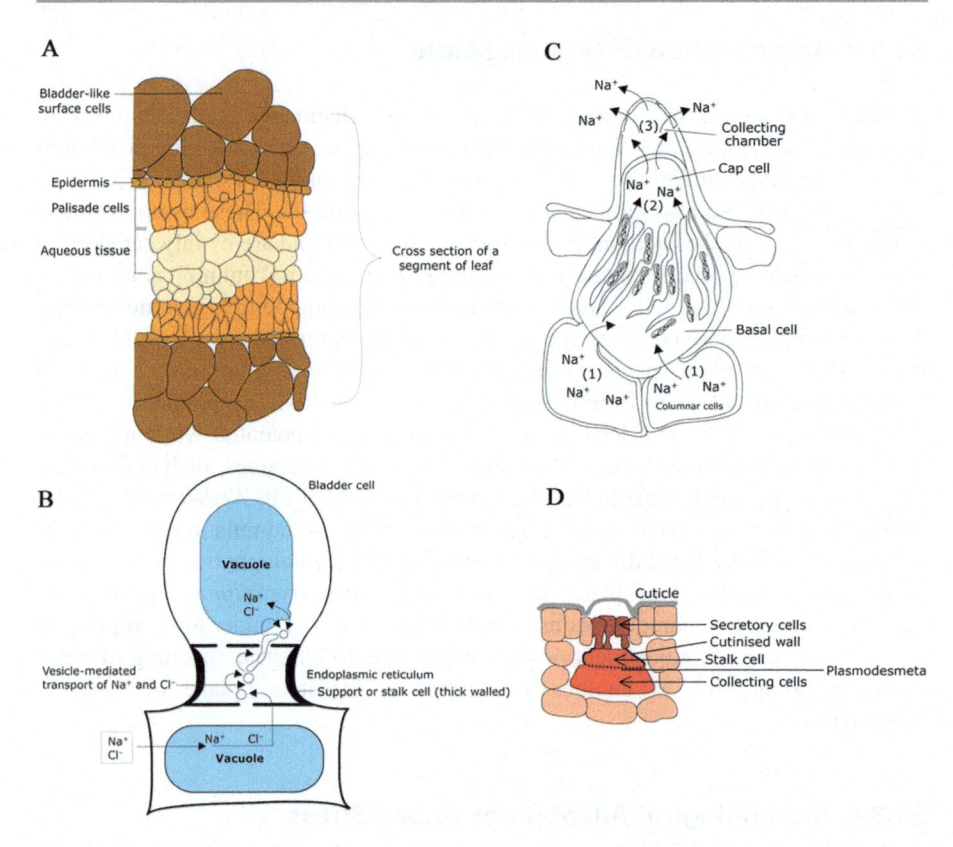

**Fig. 31.14** (a) Cross section of a part of the leaf of silver goosefoot (*Obione portulacoides*)—a halophyte, showing bladder-like cells on the leaf surface. (b) Salt gland of the salt brush (*Atriplex lentiformis*). Membrane transport followed by vesicular transport facilitates the influx of Na$^+$ and Cl$^-$ into the vacuole of the bladder cell. (c) The bicellular salt glands of the Poaceae. (1) NaCl is transported into the salt gland from cells with large vacuoles that are often found in columns beneath it and that probably help to conduct salt toward the gland. (2) NaCl solution is transported into the cap cell under pressure, probably via active processes in the mitochondria-rich basal cell. (3) Salt solution is squeezed from a collecting chamber through pores in the cuticle onto the leaf surface, where it can either crystallize or be washed away. Bicellular salt glands have been found in halophytic grasses in the genera *Aeluropus*, *Chloris*, *Distichlis*, *Cynodon*, *Spartina*, *Sporobolus*, and *Zoysia*. (d) Longitudinal section of the multicellular salt gland of mangrove (*Avicennia officinalis*)

Xeromorphic halophytes have extensive vacuolated parenchyma in leaves to assist in water storage and intracellular compartmentalization of NaCl. Even the stomatal aperture and density are low in halophytes relative to glycophytes. Stomata in halophytes are often localized in crypts below the leaf surface. They may also be surrounded by trichomes or papillae to prevent high transpiration rates. Thick cuticles and trichomes facilitate in the reflection of light from the surface. Under high saline environments, specialized secretory adaptations termed as salt glands, may assist in secretion of salt in halophytes. Salt secretion can take place via two mechanisms: secretion into vacuoles of the gland cells and direct external secretion through pores. Salt glands are not universally present in halophytes but are common in few mangrove plants, such as black mangrove (*Avicennia germinans*). There are various kinds of salt glands. Multicellular hairs consisting of 1–3 cells form salt glands in *Atriplex* species. Such salt glands comprise of a terminal cell, which acts as a bladder cell with a large vacuole for active salt transportation, and one to two thick-walled and cutinized support cells, which assist in maintenance of directional flow of salt and water. $Na^+$ ions accumulate in the vacuole of bladder cell via vesicles traversing from the vacuole of basal epidermal cells through the support cells (Fig. 31.14b). In halophytic NaCl-excreting grasses, simple bicellular salt glands are present. They consist of a large basal cell surmounted by a cap cell. A porous cuticular dome, called collecting chamber, is formed by detachment of the cuticle from the cell wall (Fig. 31.14c). The salt glands in black mangrove consist of 2–4 basal collecting cells with a stalk cell surrounded by radially arranged secretory cells. The gland is covered with cuticles containing numerous pores (Fig. 31.14d). The stalk cell is heavily cutinized on side walls and is linked to basal cells via numerous plasmodesmata. Salt is transported to the secretory cells via apoplastic pathway in the collecting cells and further symplastic pathways in the stalk cell. Cutinized walls of the stalk cell prevent the back flow and apoplastic flow of salt solutions. NaCl is actively secreted into the space beneath the cuticle from the secretory cells, and it is further squeezed out through the cuticular pores. The secretory cells show cytoskeletal and mitochondrial specialization that helps to actively excrete salt into a subcuticular cavity from which it seeps out onto the leaf. The collecting cells are not linked via plasmodesmata to the mesophyll cells but collect salt excreted into the apoplast by them. **Hydathodes** are formed during early leaf development and assist in removal of excess water which may be detrimental for the growth of young leaves in highly humid environment. They may be adapted to exclude excess salts as well in some halophytes. The rate of $Na^+$ secretion and density of salt glands is strongly related to salinity tolerance in halophytes containing salt glands.

### 31.3.9 Physiological Adaptations to Salt Stress

Stomatal responses under salt stress are regulated in numerous halophytic species, thereby influencing transpiration rate. $Na^+$ ions stimulate stomatal closure in halophytes and promote stomatal opening in glycophytes. Stomatal aperture movements are controlled by comparable sets of plasma membrane-localized ion channels, namely, $K^+$ inward- and outward-rectifying channels in glycophytes and halophytes. In guard cells of halophytes, continuous salt exposure leads to stomatal closure through inhibition of $K^+$ inward-rectifying channel. The opposite response with respect to stomatal aperture during salt stress in glycophytes and halophytes is probably due to signaling pathways involving cytosolic $Ca^{2+}$ which hinders $K^+$ inward-rectifying channel in halophytes.

## 31.4 Water Deficit

The deficiency of the basic necessity of life, water, is one of the most widespread repercussions of environmental stresses in plants. Salinity, drought, and low temperature result in shortage of water either through deficit of available water in environment or through decline in the external water potential, thereby preventing the entry of water into plant cells. The two most crucial physiological activities for crop yield, growth and photosynthesis, are adversely affected by water deficit. The global loss in possible agricultural yield due to water-deficit stress is considered to be more than the losses caused by other stresses together. The effects of these abiotic stresses are linked in terms of signaling mechanisms and metabolic and biochemical responses elicited. Deficiency of cellular water hinders plant growth and can cause membrane damage, alteration in solute concentration, protein denaturation, and loss of turgor. To prevent these detrimental changes and to restrict water loss via transpiration, plants exhibit certain physiological and cellular adaptations which are discussed below.

### 31.4.1 Avoidance of Water Stress

Some plants maintain a high water potential in the tissues to avoid water stress by reducing the loss of water to minimal level under conditions of acute water deficit. Many desert plants, like cacti, are included in the category of water stress avoiders. Desert ephemerals which emerge for a short time in the spring and die back to underground parts thereby live for most of their life span as dormant seeds. They are considered as supreme water stress avoiders. Usually these plants germinate following rainstorms and complete their life cycle very quickly while the water supply lasts. In climates with frequent dry season like Mediterranean, annuals synchronize their growth period annually with the water availability period, and perennials can store water in an underground organ during rainy season for successive growth. Water stress avoidance is crucial for plants especially in arid habitats. A xerophyte may develop water-conserving mechanism for defense against xeric habitat, but during extreme conditions of water deficit, plants need to develop tolerance. For example, resurrection plants usually avoid water stress, but desiccation-tolerant flowering plants are also observed.

### 31.4.2 Developmental Adaptations

Numerous flowering plants undergo a stage of low water content in the form of mature seeds (water content of 5–20%). Some seeds have been observed to survive drying conditions of water content below 1%. But all seeds are not highly resistant to water deficit. Thus, seeds of many species belonging to humid tropics are dehydration-sensitive. The resistance against desiccation of seeds is slowly lost during germination, and hence flowering plants are unable to endure water deficit for the rest of their life cycle. The **resurrection plants**, which can tolerate extreme desiccation and revive quickly under wet conditions in mature state, are an exception, for example, *Talbotia elegans*, *Borya nitida*, *Myrothamnus flabellifolia*, etc. Under severe drought conditions, the water potential values decline up to $-160$ MPa in the leaves of such plants, whereas the $\psi$ values vary between $-3$ and $-22$ MPa for leaves of most mesophytic species. A regular seasonal fluctuation in water stress tolerance is evident in the perennials of arctic and temperate climates, it being highest during winter months. Plants are exposed to water deficit in winter season due to freezing of soil. Consequently the water uptake capacity of root is reduced even if the soil contains aqueous water at a high $\psi$. The stomatal closure to maintain low transpiration rate results in reduced carbon fixation and hence limits plant growth. In numerous plants adapted to saline or water-deficit stress, a specialized metabolism, crassulacean acid metabolism (CAM), is found. CAM pathway is explained in detail in Photosynthesis chapter.

### 31.4.3 Specialized Xeromorphic Features

Xerophytes are the plants which grow in arid environments. Usually major xerophytic vegetation constitutes drought-tolerant perennials which have specialized morphology categorizing them into succulents or non-succulent perennials. **Succulents** store water to tolerate water stress conditions. Water storage can take place in the stem (e.g., in cacti and spurges), in leaves (e.g., *Aloe*, *Agave*), or in roots (e.g., *Fouquieria* spp.) (Fig. 31.15). Enlarged vacuoles in the parenchymatous cells are the sites of succulence. Moreover, succulent plants exhibit crassulacean acid metabolism, low transpiration rate, and thick cuticles to prevent water loss. Shallow root systems are found in certain cacti in order to assist them to tap the transitory surface moisture of the soil. This is unlike other xerophytes which have extensive root systems to tap deep water reserves. Some species are drought deciduous as well,

**Fig. 31.15** Succulent plants. (**a**) The ponytail palm (*Nolina recurvata*), (**b**) living stones (*Lithops gesincoe*), and (**c**) the candelabra tree (*Euphorbia candelabrum*)

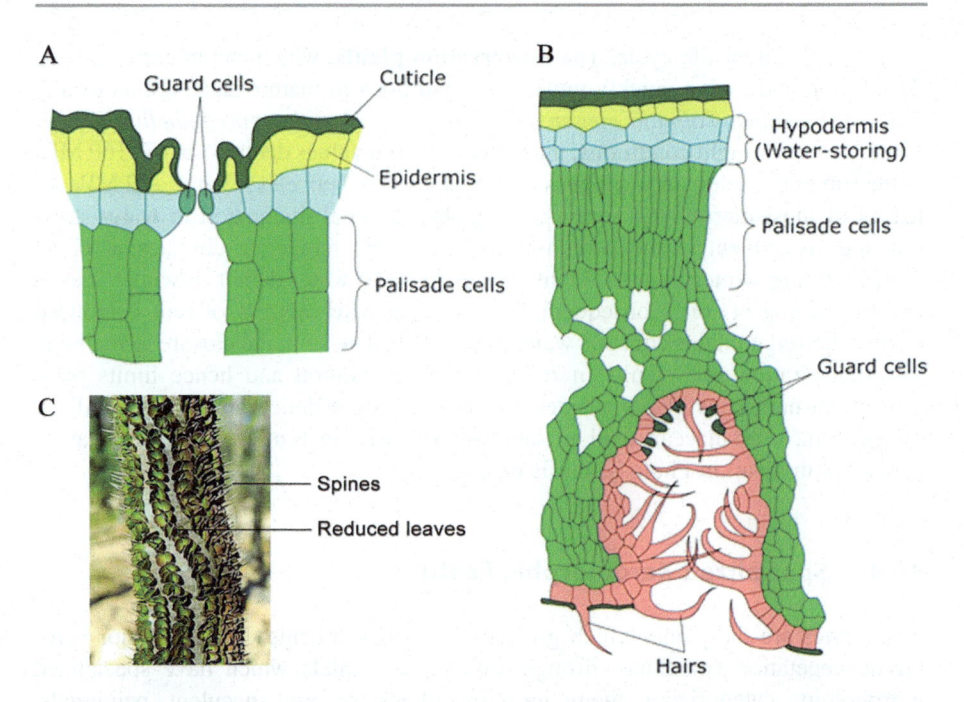

**Fig. 31.16** Morphological adaptations to drought. (**a**) Thick cuticle and sunken stomata observed in the section of leaf of Agave (*Agave americana*). (**b**) Thick cuticles and stomata confined to pits observed in the leaf of oleander (*Nerium oleander*). (**c**) Small leaves on a spiny tree (*Alluaudia procera*)

i.e., the rootlets abscise under water stress conditions and then regenerate quickly after rain. Numerous distinctive morphological adaptations, such as sunken stomata in grooves or pits, profuse production of trichomes (hairs), reduced leaf area, and thick cuticles, are shown by non-succulent perennials (Fig. 31.16). The presence of thick cuticle prevents water loss and also protects leaves from any damage or breakage due to wilting. Some cuticles have shiny surface to reduce the heat load of leaves by reflection of light. The copious production of trichomes assists in trapping a layer of air near to the leaf surface, and it also helps in reflection of light, so as to reduce leaf temperature. Reduced leaf area in non-succulent perennials assists in limitation of water loss from the leaf surface, such as asparagus. Some plants do not even possess leaves, and photosynthesis in such plants is restricted to petiole or stem. To prevent wilting, sclerenchymatous cells are reinforced in the tissues so as to prevent shrinkage. Another common xerophytic feature in plants is the presence of spines. Some drought deciduous plants can even lose their leaves under dry conditions. For example, *Salvia mellifera* loses 90% of leaves under water deficit. The retained leaves are generally smaller than those lost and rotate in such a manner that their white undersides are exposed so that they can reflect light to maintain low plant temperature. Certain other plants such as Kentucky bluegrass (*Poa pratensis*) retain their leaves under dry conditions, but the orientation or shape

**Fig. 31.17** (a) Folding of leaves of Kentucky bluegrass (*Poa pratensis*) in response to water deficit. The leaves fold when the cells in the grooves lose turgidity. (**b** and **c**) represent plants without roots, and these plants can obtain water from mist. (**b**) *Tillandsia purpurea*, positioned on sand or (**c**) epiphyte Spanish moss, *Tillandsia usneoides*. (**d**) *Welwitschia mirabilis*, growing in the desert

of leaves is modified. Under water deficit, large cells with enhanced water content (present along the longitudinal furrows of the leaves) lose turgidity leading to collapsing of furrows and folding of leaves (Fig. 31.17a). Some grasses undergo "aestivation" under dry conditions, i.e., their shallow root system completely dries out. The leaves die but do not undergo abscission so that the dead leaves protect the young buds at the soil surface. An epiphyte, *Tillandsia usneoides* (Spanish moss), uses dense, specialized nail-shaped trichomes to absorb water from the aerial sources (Fig. 31.17b, c).

Apart from perennials, an abundant flora of annual plants is present in dry regions with periods of rainfall. After several imbibitions during rain, seeds of annual plants germinate during favorable temperatures. Generally, the root system of annuals is shallow so as to absorb the surface moisture. Some plants, such as *Welwitschia mirabilis*, survive drought periods as subterranean perennating organs rather than as roots (Fig. 31.17d). Several annuals complete their life cycle by germination in the cool, wetter climates and organizing their vegetative leaves into a rosette. A warmer and humid microenvironment is created in a rosette as compared to 0–10 °C aboveground air temperature in winters, and as the temperature increases in spring, the stem undergoes elongation and produces cauline leaves. The cauline leaves and green stem perform photosynthesis, and the rosette leaves die.

### 31.4.4 Implications of Water-Deficit Stress

Water deficit affects almost all processes in plants due to the ubiquitous involvement of water in all cellular structures and processes. It is very difficult to differentiate between primary consequences of water deficit and secondary effects due to primary damage and to follow a series of events following primary perception to effects. The consequences of desiccation are time-dependent, i.e., damage increases with time. The extremity of water deficiency determines the physiological effects of desiccation stress. As soon as water content of the cells declines below the saturation level, growth begins to decrease due to extreme sensitivity of growth rate to water stress. This sensitivity can be due to decrease in cell turgor as the expansion growth is dependent on the yielding of cell wall under turgor pressure. The moderate level of water deficiency leads to rapid decline in the rate of photosynthesis in mesophytes. The decrease in photosynthetic rate coincides with the stomatal closure both in xerophytes and mesophytes. Under severe water stress, the reaction center of photosystem II degenerates, and the proportion of chloroplast proteins may decline as well. The decrease in photosynthetic activity may result in shortage of nutrients, thereby hindering plant growth. The excess of light energy absorbed, due to repressed fixation of $CO_2$, can lead to light-driven accumulation of reactive oxygen species (ROS). This may further impair photosynthetic potential of the plants due to ROS-mediated oxidative damage to the membranes. Depending upon concentration of the substrates, rates of metabolic reactions can be significantly affected as low cellular water content results in high solute concentrations. Extreme water deficit can result in an overall metabolic disturbance in the cell, thereby resulting in cell death. Moreover, drying out of cell affects structural organization at all levels from macromolecular structure to microscopically visible damage. With severe dehydration, the volume of protoplast contracts, resulting in tension and adherence to cell wall. Such physical stresses can result in tearing of the plasmalemma. The repetition of desiccation and rehydration causes cellular damage and even death at times. Protein denaturation enhances due to withdrawal of the hydration water during water stress. Also, hydrophilic heads of the lipid region of the membranes are held together by water of hydration, and water deficit can lead to semisolid state of lipids and hence loss of membrane integrity.

### 31.4.5 Water-Deficit Tolerance

Long-term exposure to conditions with low water can trigger the development of xeromorphic characters. Certain pressure sensor proteins on the plasma membrane, **osmosensors**, are involved in the primary perception of water deficit. In

## 31.4 Water Deficit

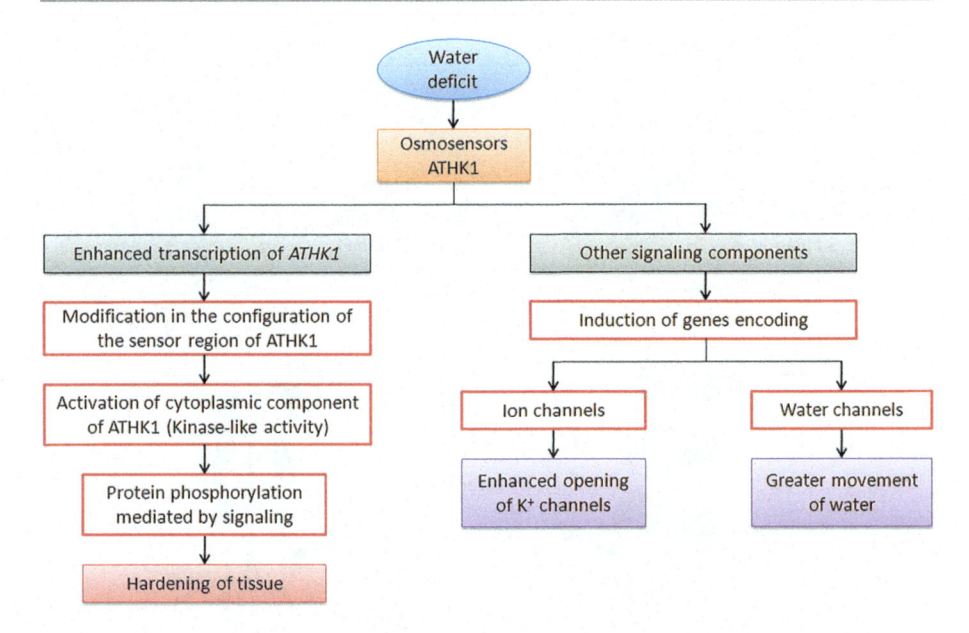

**Fig. 31.18** Mechanisms of water-deficit perception and signal transduction in plant cells

*Arabidopsis*, one such osmosensor, *ATHK1*, has been characterized. The transcription of *ATHK1* is enhanced under water stress, and presumably pressure changes modify the configuration of the sensor region leading to activation of cytoplasmic component (Fig. 31.18). The cytoplasmic component acts as a kinase to stimulate hardening via protein phosphorylation-mediated signal transduction chain. A significant change in the gene activity and protein synthesis is also evident under desiccation stress. Genes encoding ion channels and water channels (aquaporins) are also induced in response to water deficit. Plasma membrane-localized potassium channels permit enhanced Na$^+$ uptake to assist in osmotic adjustment upon induction by stress conditions. The expression of genes expressing aquaporins is elevated to facilitate the water movement into and out of cells.

### 31.4.5.1 Osmotic Adjustment and Accumulation of Solutes

Water movement across soil-plant-atmosphere continuum is possible only if water potential ($\psi$) decreases along that path. A biochemical mechanism which helps in plants acclimatization to dry or saline soil is osmotic adjustment. Several drought-tolerant plant cells accumulate solutes to regulate their solute potential and, hence, lower $\psi$ during transient or extended periods of water stress. This capacity is called **osmotic adjustment** (Fig. 31.19). This net increase in solute content is independent of the change in volume of the cell due to water loss. The decline in $\psi$ is usually restricted to about 1.2–1.8 MPa. Two processes are involved in osmotic adjustment, one takes place in the vacuole and the other in the cytosol. In order to increase the solute concentration in the root cells, ions can be taken up from the soil or can be transported from other plant organs to the root. Such a response is commonly

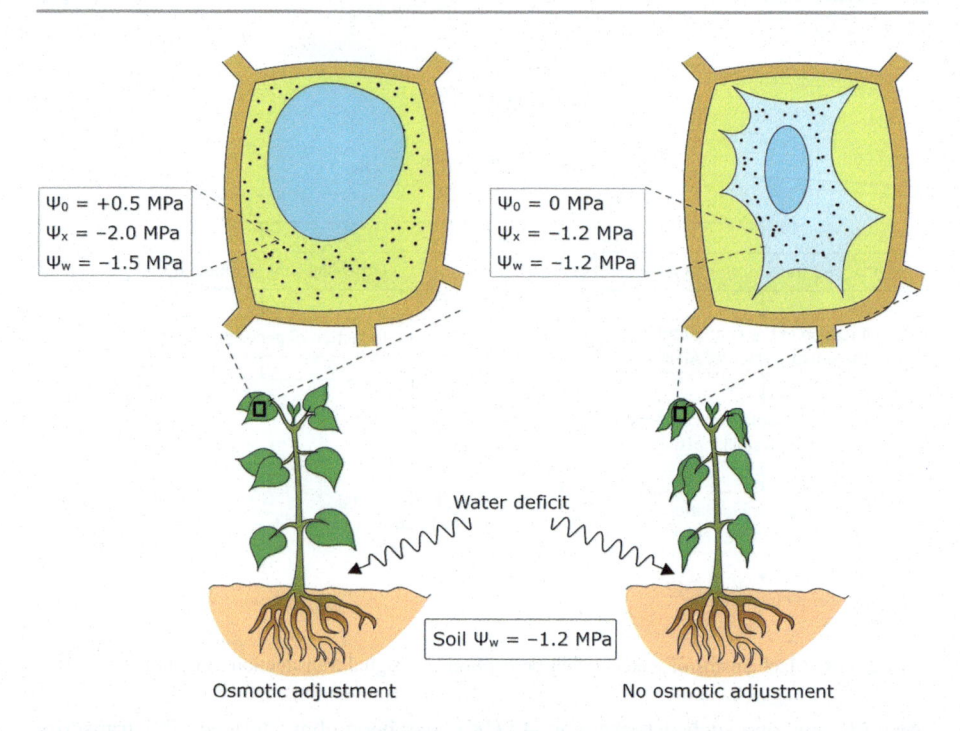

**Fig. 31.19** Positive turgor pressure in a cell caused by solute accumulation leads to osmotic adjustment. Ψ drops due to solute accumulation, thereby enhancing water flow into the cell. In cells that fail to adjust osmotically, solutes are concentrated passively, but turgor is lost

observed in plants growing in saline soils, which are rich in ions, such as potassium and calcium. These ions are readily available to the plant, but their uptake must be electrically balanced either by the synthesis and vacuolar accumulation of organic acids, such as malate or citrate, or by uptake of inorganic ions, such as $Cl^-$. Due to ion uptake, a potential problem is faced by plants, i.e., accumulation of high concentrations of elements like sodium and chloride, which can have a detrimental effect on cell metabolism. Hence, accumulation of ions is majorly limited to vacuoles so as to keep ions out of contact with cytosolic enzymes. Accompanying accumulation of high concentration of ions in the vacuole, certain solutes called **compatible solutes or osmolytes** accumulate in the cytosol in order to maintain the water potential equilibrium between the two compartments. Compatible solutes or osmolytes are osmotically active organic compounds in the cell which, even at high concentrations, do not interfere with the functions of enzymes or damage the membrane (and hence termed "compatible"). They consist of sugar alcohols, like sorbitol, quaternary ammonium compounds like glycine betaine, and amino acids like proline (Fig. 31.20). Some compatible solutes like proline may act as **osmoprotectant** as well. Osmoprotective effects include stabilization of protein and membrane structure, scavenging of free radicals and hence protecting against oxidative damage, and providing cellular source of carbon and nitrogen to the cell

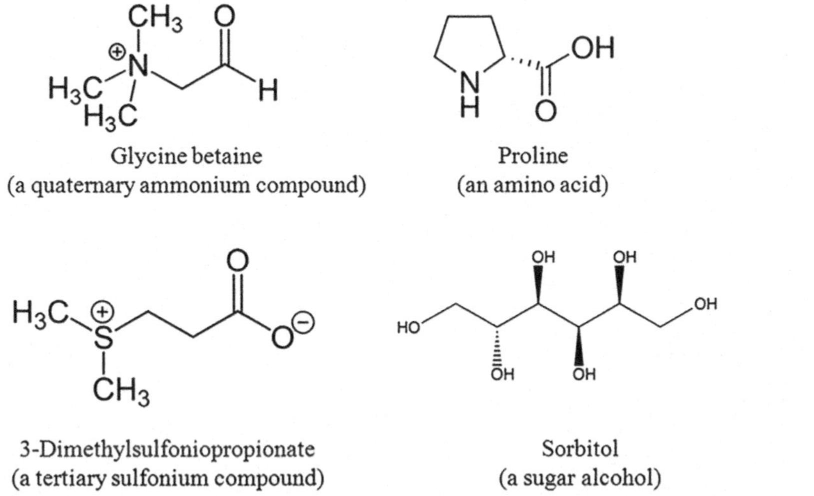

## 31.5 High Temperature Stress

**Fig. 31.20** Molecules often serving as compatible solutes in plant cells. These are amino acids, sugar alcohols, quaternary ammonium compounds, and tertiary sulfonium compounds. All these compounds are small and have no net charge

when conditions normalize. The synthesis of compatible solutes requires energy, and hence their production in plants reduces crop yield. Some plants upregulate the genes involved in encoding proline transporters to enhance the proline levels.

### 31.4.5.2 LEA Proteins

In addition to solutes, late embryogenesis abundant (LEA) proteins also accumulate in plant cells exposed to water stress. LEA proteins are hydrophilic globular proteins. They were initially characterized to accumulate in seeds during the process of maturation and desiccation. These proteins retain water and assist in stabilization and protection of proteins by acting as chaperones under water-deficit conditions. Chaperone action is crucial under water stress conditions to ensure correct folding of proteins, thereby preventing denaturation/aggregation (Fig. 31.21). In order to enhance tolerance against stress, salt or drought-sensitive plants can be transformed with genes encoding enzymes which are crucial for the production of osmoprotectants. Such plants can then be assayed for the increased accumulation of compatible osmolytes and ability to adjust osmotically under stress conditions.

## 31.5 High Temperature Stress

Temperatures over 45 °C are lethal for most plants, and temperatures over 30 °C are usually stressful. The high temperature tolerance range varies for different kinds of plant species, such as some Mediterranean species can survive at 48–55 °C, subtropical woody plants at 50–60 °C, and tropical trees at 45–55 °C (Fig. 31.22a, b).

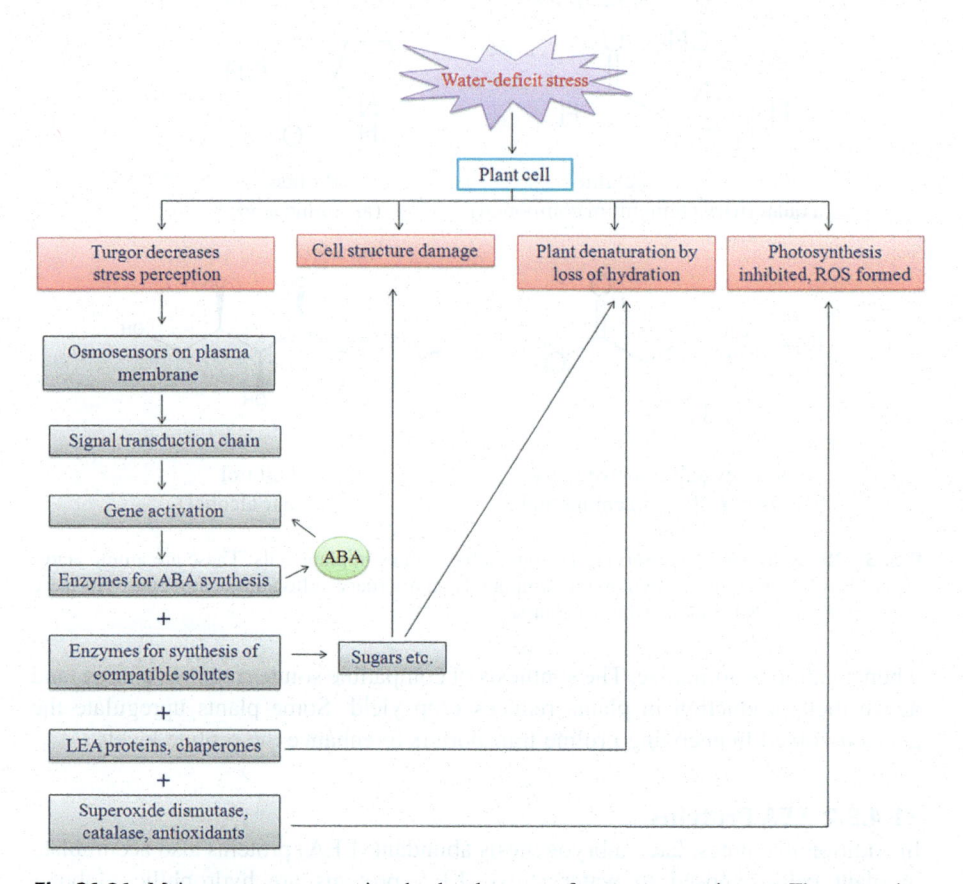

**Fig. 31.21** Main events accompanying the development of water stress resistance. The perception of stress initiates signal transduction chain leading to the synthesis of proteins. These include enzymes for synthesis of ABA, which can then activate other genes. The new proteins or their products counteract (upward arrows) the deleterious effects of water stress noted at the top of the diagram. *ROS* reactive oxygen species, *ABA* abscisic acid

Generally, an ambient temperature over 30 °C is stressful for plants even if they can survive in hot climates. If concurrent water shortage leads to stomatal closure, leaf temperature can rise up to 5–10 °C higher than the air temperature. Internal temperature of about 40–50 °C may be attained in fleshy leaves, while the ambient temperature is at 20–30 °C. Desiccated cells and tissues, such as pollen grains and seeds, have a much higher tolerance against high temperature as compared to higher plants. At times, soil temperature can rise quite high above the air temperature, leading to death of seedlings at soil level due to overheating.

**Thermal death point** of a plant is the heat-killing temperature depending upon the time of exposure. Higher temperatures can be endured if the exposure time is short, such as observed in the cells of *Tradescantia discolor* which are killed within 7 min upon exposure to 60 °C, in 4 h at 50 °C and in 22 h at 40 °C. Generally thermal

## 31.5 High Temperature Stress

**Fig. 31.22** Heat tolerant plants. (**a**) Purple coneflower (*Rudbeckia purpurea*), a heat-tolerant perennial. (**b**) *Cynoglossum magellense*, a member of the borage family. Its silvery-blue foliage reflects light and reduces the heat load of the leaves. (**c, d**) Leaves of wild olive (*Olea europaea*) are covered by hairs which give the foliage a silvery appearance. The hairs reflect light and reduce heat load

death point of hydrated organs of temperate plants lies between 45 and 55 °C upon exposure for some hours. Aquatic and shade plants have thermal death points of about 40 °C, and desert plants, like cacti, are able to survive over 60 °C. Usually high temperature stress is accompanied with periods of water shortage. Hence plants adapted to hot environments can survive both water and heat stress. Net carbon gain is sensitive to heat stress as the rate of photosynthetic carbon fixation elevates less quickly than the rate of photorespiration during high ambient temperatures. This results in reduced net carbon gain by plants. The compensation point is attained when carbon assimilation by photosynthesis is equivalent to the carbon loss via photorespiration. Plant growth is slowed down or ceased as the net carbon gain declines with elevated temperature toward or beyond the compensation point. Photosynthesis is also sensitive to heat stress. The photosynthetic protein complexes are dispersed between both the non-stacked and the stacked regions of thylakoid membranes. At high temperatures, organized structure of thylakoid membranes is damaged, thereby hindering effective electron transfer, detaching reaction centers from antennae pigments, and uncoupling photophosphorylation. Reactive oxygen species (ROS) production is also enhanced under high temperature stress.

### 31.5.1 Thermal Injury

A catastrophic collapse of cellular organization at high temperatures owing primarily to denaturation of proteins can result in death within a few seconds or minutes. At less extreme temperatures, a complex phenomenon of slower heat injury takes place when development of damage occurs over a period of few hours or some days. Protein denaturation occurs at a slower rate resulting in various disturbances in cell organization, including metabolic abnormalities and membrane organization. Thermal motion of molecules is increased causing enhanced membrane fluidity and extreme fluidity, which are detrimental for cell.

### 31.5.2 Developmental Adaptations to Heat Stress

#### 31.5.2.1 Heat Avoidance
Plants living in hot climates usually exhibit developmental adaptations which are advantageous in water-deficit conditions as well. These adaptations help plants to maintain temperature below the ambient temperature. High transpiration rate helps in dissipating heat as water evaporates and hence leads to cooling of leaves and protects against heat stress-induced injury. This mechanism is, however, observed in tropical plants which have ample water supply, and it is not suitable for plants in dry conditions undergoing high temperature stress. Orientation of leaves at steep angles to the incident light can help in reducing the temperature of leaves by 3–5 °C. This can prevent overheating of leaves. Such an adaptation is commonly found in plants adapted to hot and dry summers. The reflectance of incident light, by pale-colored leaf surface and also by profuse production of trichomes (epidermal hairs which reflect incident light), can lead to reduction in heat load of leaves (Fig. 31.22c, d). In hot, dry climates, some plants possess thick corky bark to reduce absorption of heat and also to provide insulation to the phloem and cambium from water loss.

#### 31.5.2.2 Heat Hardening and Heat Tolerance (Thermotolerance)
Tolerance to heat stress is majorly attributed to properties of cellular proteins and membrane lipids. A drastic change occurs in protein synthesis when a plant is exposed to a temperature above threshold. Although the biosynthesis of certain proteins declines in both prokaryotic and eukaryotic organisms in response to high temperature stress, a new set of proteins known as **heat shock proteins (HSPs)** are synthesized as a generalized physiological response (Fig. 31.23). The synthesis of HSPs is induced by higher than normal but sublethal temperatures. They are also produced upon gradual rise in temperature and result in hardening. HSPs provide

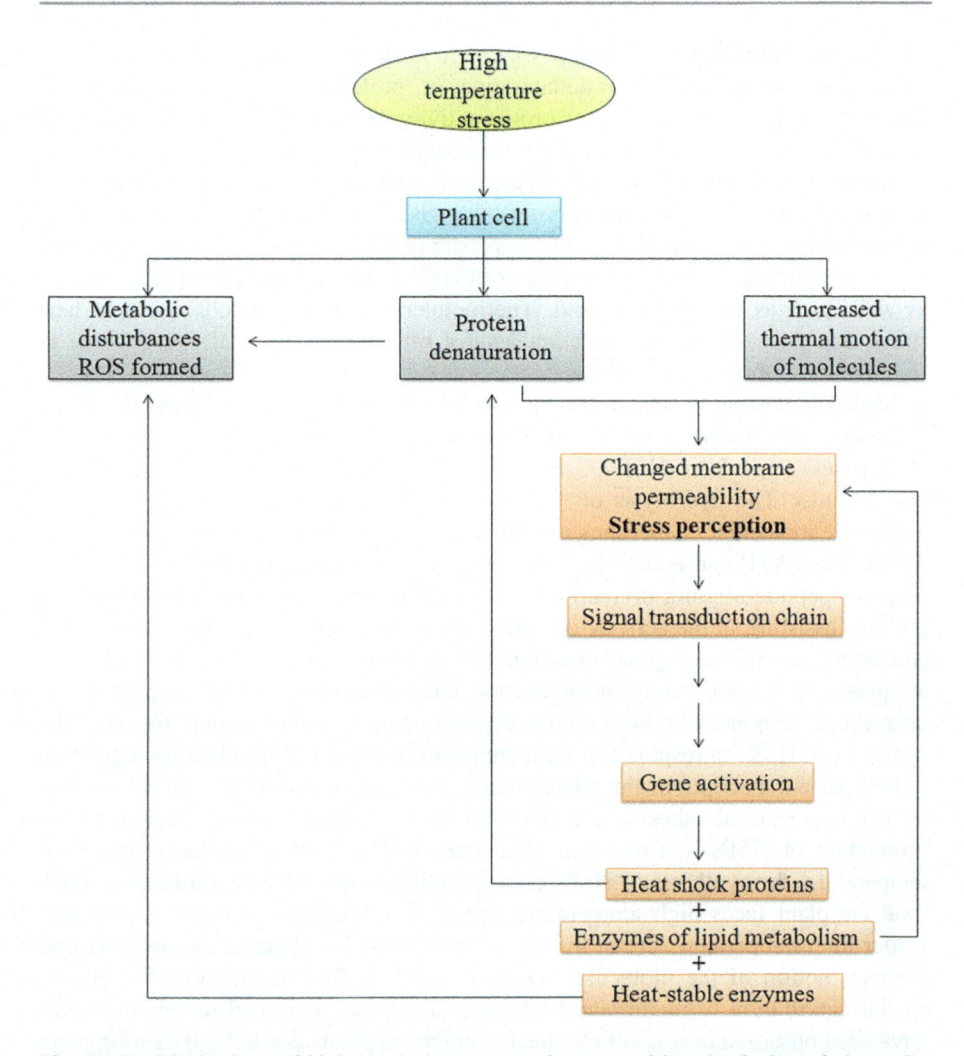

**Fig. 31.23** Mechanisms of high temperature stress tolerance and heat hardening of plant cells. Signal perception includes sensing of denatured proteins in the cytosol and effect on membranes. Upward arrows indicate the counteraction of deleterious effects of the high temperature by gene activation products (hardening process)

protection to organisms from acute damage, assist in acclimation by allowing metabolic and cellular activities to resume, and also increase the temperature tolerance range for survival. Depending upon the optimal growth temperatures for different plant species, heat shock response is invoked at specific temperature, for example, at 35 °C for temperate species (e.g., rye grass) and 45 °C for tropical cereals (e.g., millet and sorghum). Depending on molecular weight, the heat shock proteins from all type or organisms fall in five major groups: HSP100, HSP90, HSP70, HSP60, and small HSPs. All HSPs play the role of molecular chaperones by recognizing and binding to proteins which are in an unstable or inactive state. High temperature stress results in denaturation of proteins, i.e., loss of the three-dimensional structure essential for proper activity. HSPs prevent the manufacturing of inactive, misfolded proteins and assist in stabilizing and controlling the cycles of protein refolding. Many plants, which are previously exposed to smaller temperature increase, have been shown to survive otherwise lethally high temperatures. This thermotolerance is acquired in plants as a result of production of HSPs in response to pre-treatment. Interestingly, biosynthesis of HSPs is evoked by other abiotic stresses as well, such as drought and cold. Thermotolerance acquired subsequent to heat stress also provides cross protection against other environmental stresses due to stabilization of proteins by HSPs.

Molecular chaperons play a crucial role in both stressed and non-stressed cells by preventing misfolding of newly synthesized proteins. In all kinds of organisms, chaperones related to HSP90, HSP70 and HSP60 are functional at normal temperatures. The hydrolysis of ATP to ADP is essential for chaperone activity, and hence molecular chaperones usually consist of domains which bind to adenine nucleotides (ATP and ADP). In addition to ATPase domains, HSP70s contain an adjacent peptide-binding domain and also interact with co-chaperones, which are proteins assisting in the activity of HSPs to maintain target proteins in their native conformation. Although genes encoding the members of all five categories of HSPs are present in a wide variety of organisms, the role of specific HSP families in the heat shock responses or in acquired thermotolerance varies among species. The synthesis of HSPs in response to high temperature stress is controlled by regulation of HSP gene transcription, and the regulatory mechanism is similar in all eukaryotes.

The synthesis of other cellular proteins resumes after a certain period, as the production of HSPs is a transient phenomenon. Upon return to below threshold temperature, the synthesis of HSPs ceases leading to loss of thermotolerance. Each time the plant faces high temperature stress, new synthesis of HSPs is required. Under high temperature stress, fluidity of membrane is enhanced due to increased thermal motion of the molecules. Due to excessive fluid membrane, the precise spatial distribution of membrane proteins is disrupted. Saturated membrane lipids have been observed to confer more heat tolerance to plants due to high melting point which thereby contributes to greater rigidity of membranes.

## 31.6 Low Temperature Stress

### 31.6.1 Freezing Versus Chilling Stress

Exposure to suboptimal temperatures can lead to either chilling or freezing stress in plants depending upon the intermolecular arrangement of phospholipids in the cell membranes and physical state of water. Low temperature initially exhibits thermodynamic effects on rates of chemical reactions, and longer exposure results in more complex physiological changes. Chilling stress occurs in the temperature range of 0–15 °C, when water is in the liquid state, and freezing stress occurs when the ambient temperature leads to ice formation. Both these stresses have direct as well as indirect effects on plant growth, development, and metabolism. Cellular injury occurs due to both the stresses, but freezing stress has more severe indirect effects due to the formation of ice crystals in the intracellular spaces which further cause osmotic dehydration in the cell.

### 31.6.2 Low Temperature and Water Deficit

Apart from tropical plants, low temperature causes a major problem only when the freezing is induced. Due to lower solute content in the extracellular fluid in comparison to cells, freezing leads to ice formation in extracellular compartments. Ice formation results in reduced water potential outside the cell, thereby enhancing the movement of unfrozen water outside the cells, down the gradient. More than 90% of the osmotically active water moves into the intracellular spaces at $-10$ °C resulting in water deficit. Hence, many of the physiological adaptations in plants which make them survive low temperature stress are similar to adaptations during salt stress or drought.

### 31.6.3 Plant Species of Warm Climates

Temperate plants are usually able to survive chilling temperatures. However, plant species adapted to warm climates are damaged by low, non-freezing temperatures and hence are **chilling-sensitive**. The chilling-sensitive species include crops originated in tropical climates, such as tomato, soybean, cucumber, and maize. Low temperature limits productivity in these crops primarily by reducing reaction rates for carbon fixation.

### 31.6.4 Low Temperature Sensing Systems in Plants

Changes in temperature must be perceived by plants to undergo requisite adaptations to low temperature. *Synechocystis* PCC6803, a cyanobacterium, possesses a histidine kinase Hik33 which acts as a putative low temperature sensor. Changes in membrane fluidity lead to signal-mediated activation of Hik33, followed by autophosphorylation and transfer of phosphate group to Rre26, the response regulator. Hik33 is involved in the regulation of expression of numerous cold-regulated genes (Fig. 31.24). No Hik33 orthologs have been discovered in higher plants yet, but two-component systems consisting of a combination of His and Rre play a role in some phytohormone responses. The involvement of such two-component systems in sensing of low temperature in plants is yet to be deciphered. There are many possible candidates involved in sensing of low temperature in plants, but these have not been confirmed. Alteration in membrane fluidity is considered to participate in the sensing of temperature drop. In alfalfa cells, changes in membrane fluidity, due to benzyl

**Fig. 31.24** Signal transduction events associated with low temperature stress in a cyanobacterium (*Synechocystis* sp. PCC 6803). Similar events may be involved in higher plants as well

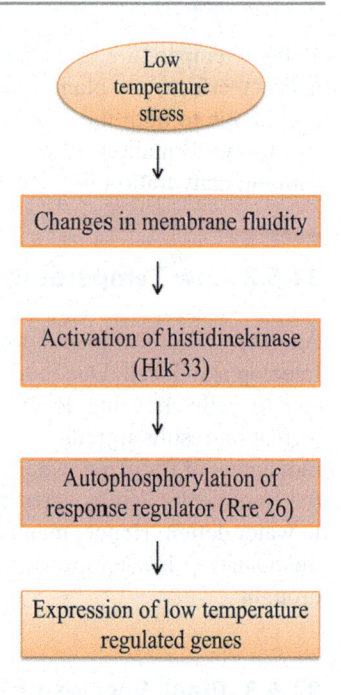

Low temperature stress
↓
Changes in membrane fluidity
↓
Activation of histidinekinase (Hik 33)
↓
Autophosphorylation of response regulator (Rre 26)
↓
Expression of low temperature regulated genes

alcohol treatment as a membrane stabilizer and dimethyl sulfoxide as a membrane fluidizer, have been shown to modulate both cold stress tolerance and cold-induced gene expression. Calcium influx and cytoskeleton organization have been observed to be affected as well under such conditions. Hence, alterations in membrane fluidity may play the role of cold stress sensor and trigger a chain of signal transduction events.

## 31.6.5 Membrane Destabilization as a Response to Chilling and Freezing

### 31.6.5.1 Chilling Stress

At optimal temperatures, membrane lipids exist in a liquid crystalline (fluid) state to ensure normal cellular functions. The lipids which have high melting temperatures undergo solidification to form a gel phase as the temperature declines and thus form a phase separation within the membrane (Fig. 31.25). The gel phase of membrane makes them leaky or even dysfunctional and hence intracellular solutes and water are lost. The reactions associated with membrane, such as enzyme-mediated processes, receptor functions, and carrier-mediated transport, are also disabled. One of the major effects of low temperature stress is impairing of electron transport chain, thereby limiting photosynthesis. Photoenergy reception is not altered by chilling stress, and, hence, exposure of chloroplasts to excess excitation energy leads to

**Fig. 31.25** (a) Diagrammatic cross-section of a phospholipid bilayer in a fluid (liquid crystal) state. The hydrophobic tails of the lipids point toward the middle of the bilayer, but the arrangement is not very regular and it changes from moment to moment owing to thermal motion of the molecules. (b) Bilayer in a gel state, as at low temperature. The tails are now regularly arranged and the structure is much more rigid, due to negligible thermal motion

A  Liquid crystal state of PM

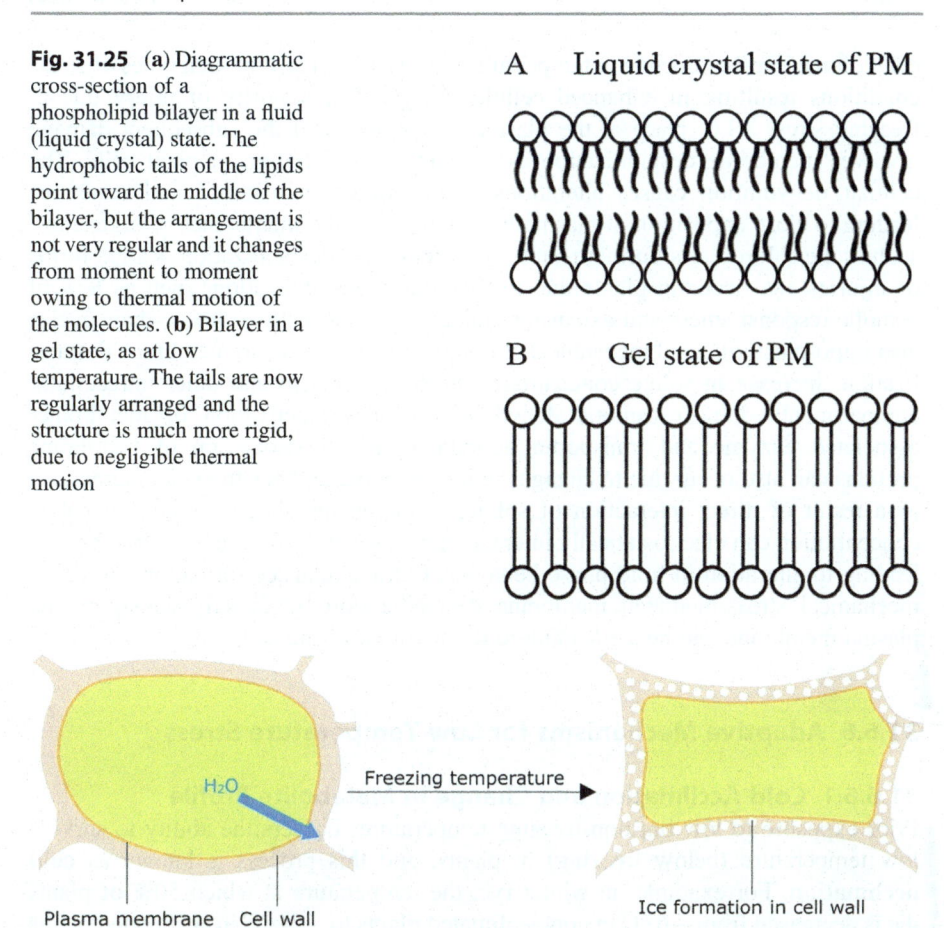

B  Gel state of PM

**Fig. 31.26** Effect of freezing temperatures in plants leading to cellular water deficit, as water travels down its potential gradient, crossing the plasma membrane into the cell wall and intercellular spaces. Slow rate of freezing prevents formation of ice crystals in the cytoplasm, cell dehydrates, and freezing occurs in the apoplast

photoreduction of oxygen molecules producing ROS. The oxidative stress further imposes damage on membrane proteins and lipids.

### 31.6.5.2 Freezing Stress

The water-deficit stress occurs due to the effect of freezing on cellular water. Ice has a relatively lower chemical potential than the unfrozen water. Also the extracellular ice has a lower vapor pressure than the cytoplasm- or vacuole-localized water. During ice formation in extracellular compartments, the cellular water moves across the plasma membrane toward the extracellular ice, down the water potential gradient (Fig. 31.26). In addition to distortion of cellular shape due to extracellular ice formation, this results in enhanced intracellular solute concentration and decreased cell volume. Further decline in temperature causes an increase in such types of stress conditions resulting in enhanced cellular injury. The majority of injury due to freezing stress takes place at the plasma membrane, and the subsequent damage involves membrane destabilization as a result of dehydration of plant cells. The cellular dehydration causes alterations in membrane interactions and structures leading to close apposition of plasma membrane with the membranes of organelles, such as the chloroplasts, further resulting in membrane destabilization. Varied forms of injuries can occur in plants due to membrane destabilization, such as loss of osmotic responsiveness and expansion-induced lysis. Repetitive freeze-thaw cycles also impose osmotic and mechanical stress which contributes to membrane destabilization. Increase in solute concentration in the cytoplasm and other intracellular compartments due to osmotic dehydration causes inactivation of membrane-associated enzyme and transporter activities. The disassociation of membrane proteins can also occur due to changes in hydrophobic and electrostatic interactions as a result of direct interaction of solutes with the membrane. Enhanced solute concentration can electrostatically interact with charged head groups of lipid bilayer leading to alteration in lipid phase behavior. Extracellular ice formation can cause mechanical stress-mediated membrane destabilization by directly damaging the plasma membrane and hence leading to deformation of the cell.

## 31.6.6 Adaptive Mechanisms for Low Temperature Stress

### 31.6.6.1 Cold Acclimation and Change in Metabolite Profile

Prior exposure to low but non-freezing temperatures confers the ability to survive low temperature (below freezing) in plants, and this process is known as **cold acclimation**. For example, in winter rye, the temperature at which 50% of plants die is decreased from $-6\,°C$ in non-acclimated plants to $-21\,°C$ in acclimated plants and in *Arabidopsis* from $-3\,°C$ to $-10\,°C$, when acclimated by 2 days exposure to $4\,°C$ before freezing. Cold tolerance is an accumulative process which can be reversed, restarted, or even stopped. For example, in cereals, the cold acclimation begins at the threshold temperature of $10\,°C$, with an optimum of $3\,°C$. The rate of cold acclimation is enhanced as the temperature falls below $10\,°C$. Cold acclimation may occur in different parts of a plant independently. Post-cold acclimation, the cold hardiness in plants, is retained as long as the temperatures remain below freezing. Also when temperature increases above $10\,°C$, cold hardiness is rapidly lost. So, even cold-tolerant species are damaged during summer frosts. In non-acclimated plants, primary damage due to exposure to low temperatures is caused by membrane damage. Membrane damage takes place due to loss of osmotic responsiveness, phase transition (an abrupt change in the physical properties of membrane, such as permeability and fluidity), and expansion-induced lysis (takes place during thawing). A series of physiological events take place during acclimation, which assist in protection of cellular membranes from freezing damage. During cold acclimation, the proportion of phospholipids (phosphatidylserine and phosphatidylcholine) is enhanced, while that of cerebrosides (lipids composed of ceramide and a sugar residue) is declined in the plasma membrane. This alteration in the proportion of lipids leads to a decline in tendency for membrane lipids to undergo fusion and rearrange to form pores and decrease in incidence of expansion-induced lysis. These changes have an additive effect, thereby diminishing the probability of freezing-induced membrane damage and cell death. During cold acclimation, expression of certain proteins is induced to stabilize the membranes by bringing down the incidence of phase transition. These proteins probably serve as nucleating sites for such changes. The levels of sugar-synthesizing enzymes are also raised to enhance membrane stability. To protect against water-deficit stress, novel hydrophilic protein and LEA proteins are also induced to reduce the probability of protein denaturation during cold acclimation. Protein denaturation resulting in functional loss is also reduced by the induction of chaperonins during cold acclimation.

In conjunction with freezing stress in many plants species, various compatible solutes, such as glycine betaine, proline, and simple sugars (glucose, raffinose, stachyose, fructose and sucrose), accumulate in the cells. The molecular functions of these compatible solutes during cold stress acclimation are similar to those in water-deficit stress, leading to osmotic and dehydration stress. Though sugars contribute toward enhanced freezing tolerance post-cold acclimation by protecting membranes, they alone are not sufficient for cold acclimation in plants.

### 31.6.6.2 Changes in Gene Expression

The alterations induced by cold acclimation are mediated by changes in the gene expression at low temperature. A group of genes, called **cold regulated genes** (COR), are induced during cold acclimation, and many of COR genes are induced during water-deficit stress as well. Each COR gene may contribute to freezing tolerance. COR genes involved in anthocyanin biosynthesis may also play an indirect role in cold acclimation. Though the functions of proteins encoded by COR genes are not yet deciphered, they are predicted to be highly hydrophilic in nature. The action of COR proteins on electrolyte leakage suggests their possible role in stabilizing plasma membrane during freezing stress, similar to LEA proteins. Overexpression of a spinach-derived novel hydrophilic COR proteins in transgenic tobacco conferred enhanced freezing tolerance by reducing cellular damage. Certain protein called as **antifreeze proteins (AFPs)** is also produced during cold acclimation. AFPs are secreted into the extracellular space/apoplast and prevent the nucleation of ice crystals or re-formation of ice crystals post a freeze-thaw cycle. Though accumulation of antifreeze proteins alone cannot determine the low temperature limit for plants, there is correlation between the accumulation of antifreeze proteins and cold tolerance. Interestingly, **molecular chaperones (chaperonins)** like HSP90 and HSP70.12 are also encoded by some COR genes to prevent protein denaturation during freezing. Proteins involved in low temperature signaling, like calmodulin-related proteins, certain transcription factors, and MAP kinase (MAPK) and MAP kinase kinase kinase (MAP3K), are also encoded by some COR genes. In *Arabidopsis*, three transcription factors have been deciphered which induce COR gene expression (Fig. 31.27).

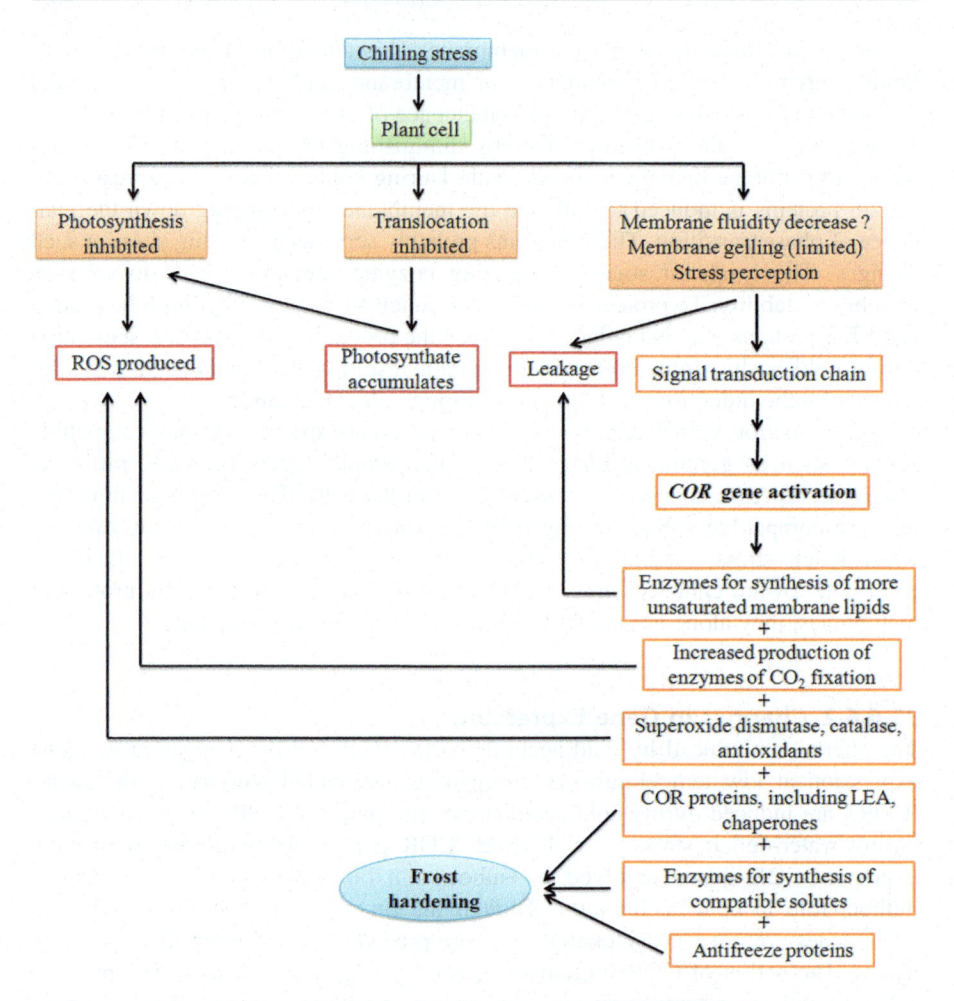

**Fig. 31.27** Mechanisms of chilling stress and frost hardening of plant cells. Stress perception leads to a signal transduction chain leading to gene activation. New proteins with upward arrows are active in counteracting deleterious effects of chilling. Others (listed at the bottom) are involved in frost hardening. *COR* gene products and compatible solutes can also act in chilling tolerance. *ROS* reactive oxygen species

### 31.6.6.3 Low Temperature Signaling

The series of signals involved in low temperature stress are not yet fully characterized. During low temperature, a rapid increase in the concentration of cytosolic free calcium is evident, primarily due to influx from extracellular sources. This abrupt rise in cytosolic calcium is essential for induction of COR genes and hence freezing tolerance. Protein phosphorylation is induced by increase in cytosolic calcium which further induces cold acclimation response. A MAPK cascade may also be involved in signaling pathway during low temperature acclimation as some MAP kinases are induced during cold temperature. Abscisic acid level also plays a role in enhancing response to low temperature in some species. Both ABA-dependent and ABA-independent signaling pathways take place during low temperature stress.

### 31.6.6.4 Vernalization and Cold Acclimation

Vernalization involves exposure of seeds to a period of low temperature to accelerate flowering and reproduction. To fulfill this requirement for vernalization, a few plant species are adapted for specific periods of growth at low temperature. Vernalization ensures that the temperate plants reach the stage of flowering at appropriate time as flowering is specifically sensitive to damage caused by low temperature. Vernalization is essential for flowering in winter wheat, whereas spring wheat does not have any such requirement and will flower irrespective of growth conditions. Vernalization and many cold acclimation traits, such as antifreeze protein accumulation, sugar accumulation, and flowering time, map to the same chromosomal region on chromosome 5A in wheat indicating that the two are functionally related.

## 31.7 Flooding (Anaerobic) Stress

Plants, like most eukaryotes, are **obligate aerobes**, i.e., they cannot live without oxygen. Excess of soil water (flooding) leads to environmental stress for land plants. Wetlands account for about 6% of the world's terrestrial habitats. Rice, one of the important crops of tropical and subtropical regions, is predominantly cultivated on flood, anaerobic soils. Waterlogging commonly occurs in winter in northern latitudes, and, in tundra, low oxygen levels may be created by ice crusts by prevention of diffusion of oxygen. Even in warmer climates, lack of oxygen in soil can occur due to higher rates of respiration by soil microflora. Bogs and swamps are known to exhibit permanent waterlogging. In plants growing in permanently flooded conditions, such as black mangroves (*Avicennia* sp.), negatively geotropic roots (pneumatophores) rise above the water level to allow passage of oxygen from the atmosphere to the roots down its concentration gradient via large air spaces. Oxygen supply to the root cells is affected by a number of factors such as soil temperature, water content, porosity, root density, and the presence of aerobic microorganisms and algae. Oxygen concentration in the root tissue may also be affected by the thickness and depth of root, volume of intercellular spaces, and metabolic activity. Plant or cellular oxygen status can be defined as normoxic (normal oxygen levels), hypoxic (reduced oxygen levels), or anoxic (no oxygen). Oxygen concentration in well-drained, porous soils is nearly equal to atmospheric concentration of 20.6%. Because of the available high concentration of oxygen in the atmosphere and its very rapid diffusion in the air, the atmosphere has a high oxygen-supplying capacity. Since oxygen diffuses slowly in water, so in saturated soil, the microbial demand for oxygen is generally much greater than the rate at which it can diffuse from the atmosphere. This results in depletion of oxygen in the soil for the plant roots. Submergence or flooding lowers $O_2$ supply all the more. Such a situation presents a significant challenge to the terrestrial plants. Soil aggregates with a

**Table 31.6** Some examples of wetland, flood-tolerant and flood-sensitive plants

| Wetland plants | Flood-tolerant plants | Flood-sensitive plants |
| --- | --- | --- |
| Barnyard grass (*Echinochloa phyllopogon*) | *Arabidopsis thaliana* | Pea (*Pisum sativum*) |
| Coral tree (*Erythrina caffra*) | Barnyard grass (*Echinochloa crus-pavonis*) | Soybean (*Glycine max*) |
| Common reed (*Phragmites australis*) | Barley (*Hordeum vulgare*) | Tomato (*Lycopersicon esculentum*) |
| Marsh dock (*Rumex palustris*) | Corn (*Zea mays*) | |
| Rice (*Oryza sativa*) | Potato (*Solanum tuberosum*) | |
| Rice grass (*Echinochloa crus-galli*) | | |
| Sweet flag (*Acorus calamus*) | | |

diameter greater than 30 mm can have waterlogged centers because microbial demand for $O_2$ is greater than the rate of penetration of oxygen into the aqueous phase around soil particles. Hypoxia (oxygen limitation) starts when gaseous oxygen level in soil falls below 50 mmol/m$^3$. Flooding/submergence lowers soil oxygen level by 60–95% within a day although the rate of depletion is influenced by soil type and temperature. The diffusion coefficient of oxygen in air is about 1000 times greater than in water. Oxygen is displaced from soil due to its low solubility and low rate of diffusion in water, and its loss is further accelerated due to oxygen consumption by soil microbes.

### 31.7.1 Flooding Sensitivity of Plants

Plants exhibit variations in their ability to tolerate flooding. Depending on their sensitivity to flooding, plants can be classified as wetland, flood tolerant, or flood sensitive (Table 31.6).

#### 31.7.1.1 Wetland Plants
As a consequence of various morphological, anatomical, and physiological features, wetland plants survive in waterlogged soil and in partial submergence. Such plants exhibit one or more of the following features (Fig. 31.28):

1. Elongation of stem or leaf petioles toward water surface.
2. Thinning of leaves to facilitate underwater photosynthesis.
3. Thick root hypodermis to reduce loss of oxygen to the anaerobic soil.
4. Aerenchyma formation to facilitate oxygen transport from aerial structures to submerged roots.
5. Adventitious roots formation, which develop aerenchyma to facilitate $O_2$ transport from aerial plant parts to the $O_2$-deprived tissue. Adventitious roots also develop in non-wetland species in the aerial parts under submerged conditions.
6. Pneumatophores.

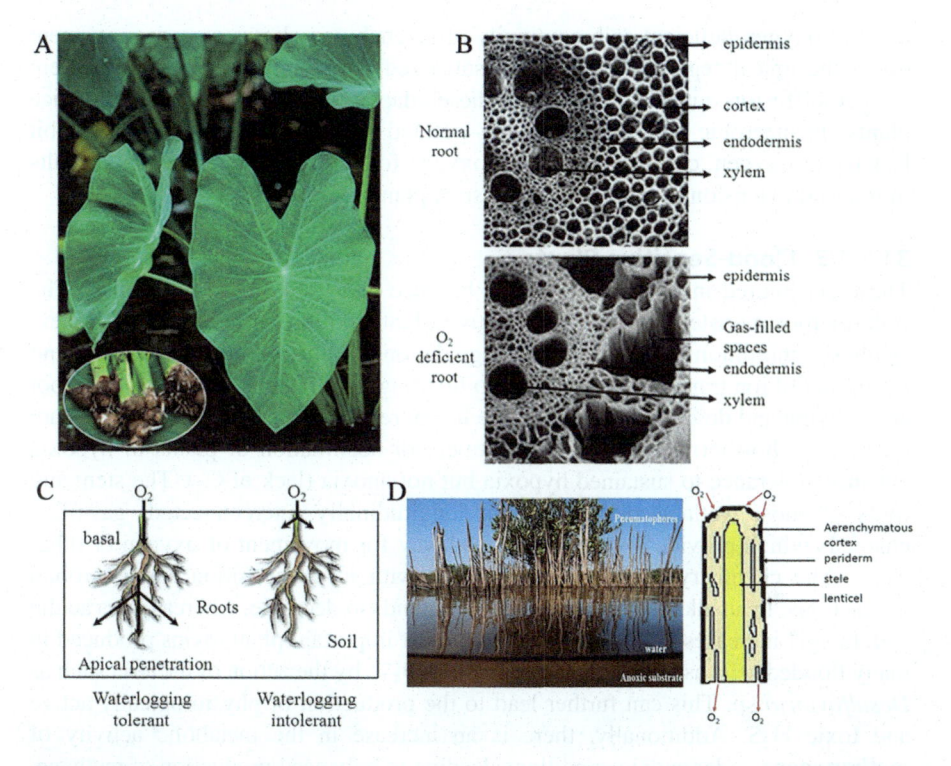

**Fig. 31.28** Adaptive features in plants in response to flooding. (**a**) *Colocasia esculenta* showing large cordate leaves (elephant ear) and swollen stem base known as corm. (**b**) Scanning electron micrographs of a normal and oxygen-deficient root. (**c**) Penetration of oxygen in the roots during waterlogging. (**d**) Pneumatophores with aerenchyma for gaseous exchange

7. Lenticels for gaseous exchange, e.g., in willow and alder trees.
8. Shallow roots.

### 31.7.1.2 Flood-Tolerant Plants

Flood-tolerant plants exhibit only temporary anoxic conditions under flooding. Some plants/tissues can tolerate anaerobic conditions in flooded soil for few weeks before developing aerenchyma. These include the embryo and coleoptile of rice and rhizomes of giant bulrush (*Schoenoplectus lacustris*) and *Typha angustifolia*. These rhizomes expand their leaves under anaerobic conditions for a long time. In the spring season, once leaves have expanded above the surface of water, oxygen diffuses down through aerenchyma into the rhizome. Metabolism then switches from anaerobic (fermentative) to aerobic mode. Similarly, in paddy fields, the coleoptile from germinating rice breaks through water surface, thereby creating a diffusion route for $O_2$ into the submerged roots. Although rice is a wetland species, its roots (as that of maize) are intolerant of anoxia. With the continued extension of roots in oxygen-deficient soil, continuity of aerenchyma allows oxygen movement up to the apical region. Additionally, suberized and lignified cell walls prevent oxygen diffusion outward to the soil. These adaptations allow root growth of rice plants in anaerobic soil. Roots of non-wetland species, such as maize, exhibit leakage of oxygen, causing insufficient oxygen for aerobic respiration. This results in reduced extension growth of roots of such plants as maize.

### 31.7.1.3 Flood-Sensitive Plants

These get injured in response to anoxia because of cytoplasmic acidification. The cells of root meristem of flooded sensitive plants exhibit reduced rate of protein synthesis, inhibition of cell division, degradation of mitochondria, cell death, and disruption of ion transport, in response to flooding. Such plants do not develop root aerenchyma and do not generally survive for more than 24 h. Most plant species are tolerant to short-term hypoxia. Pre-treatment or acclimation of plants to hypoxia enhances tolerance to sustained hypoxia but not anoxia (lack of $O_2$). The stem and roots of most wetland plants develop longitudinally interconnected, gas-filled channels which provide low-resistance pathway for movement of oxygen or other gases. The chemistry of flooded soil varies with its composition and microbial interactions. In an alkaline soil, the pH of soil tends to decrease, whereas in as acidic soil, the pH increases. Sulfide is one of the most important plant toxins produced in many flooded soils as a result of reduction of $SO_4^{-2}$ by the action of bacteria such as *Desulfovibrio* sp. This can further lead to the production of physiologically active and toxic $H_2S$. Additionally, there is an increase in the metabolic activity of **methanogens** under anoxic conditions, leading to enhanced production of methane.

## 31.7.2 Methane Emissions from Wetlands

Wetlands are often characterized by high organic matter and low redox potential. Such a region is very favorable for the multiplication of methanogenic microorganisms in soil. Almost one-third of methane released into the atmosphere globally comes from wetlands, and up to two-third of this is from rice fields alone. Landfill sites are another source of atmospheric methane. Methane produced in waterlogged soil exhibits a slow diffusion into the atmosphere because of high diffusive resistance. But plants growing in wetlands facilitate this process of methane release into the atmosphere because of rapid diffusion of methane as an uncharged molecule into and up through the plant to the atmosphere. Methanotrophic bacteria also use methane as a terminal electron acceptor, producing some toxic carbon compounds, both in anaerobic and aerobic conditions. The rate of methane release from wetlands into the atmosphere is, thus, determined by the combined metabolic activities of methanogenic and methanotrophic microbes in soil (Fig. 31.29). A decrease in the organic content of soil and control of the activity of aerobic methanotrophs are some of the management systems developed for rice fields to decrease methane emissions and manage the production of reduced organic compounds by methanotrophs. Methanotrophic bacteria from wetlands also

**Fig. 31.29** Methane in anoxic soils: Methane is produced from organic matter by anaerobic methanogens in waterlogged anoxic soils which further leads to the production of formaldehyde by methanotrophs at anoxic-oxic boundaries

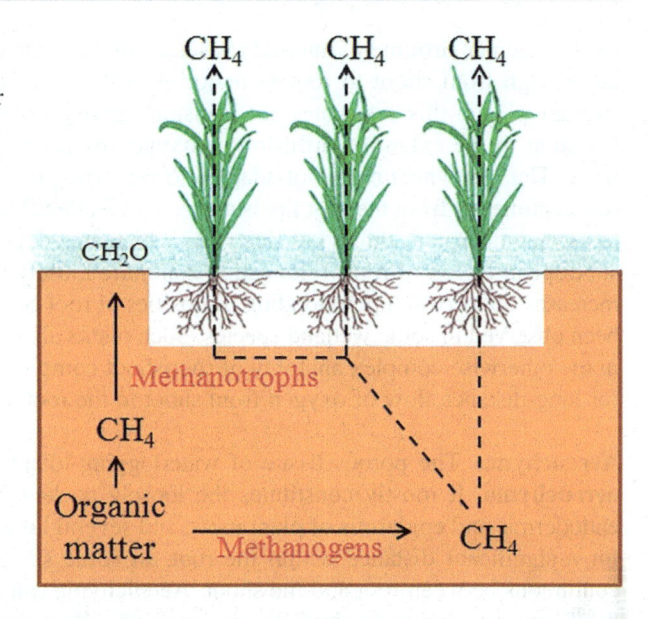

contribute to bioremediations due to their ability to oxidize a range of small carbon compounds and some organic pollutants.

### 31.7.3 Adaptations of Wetland Plants

Mangroves are known to inhabit environments which are flooded regularly. In order to provide support against tidal flow, mangrove trees develop secondary growth so as to develop prop roots. This inhibits oxygen influx to the cortex. Oxygen influx in mangroves is enhanced by the development of negatively geotropic pneumatophores which grow vertically upward above the ground. Pneumatophores have well-developed aerenchyma as long and continuous columns in the straightened cortex of the root system. Pneumatophores also provide structural support to the tree because of well-developed phelloderm that includes a layer of sclerified cells. The bark-like phellem provides barrier to the entry of oxygen. Extensive lenticels present on the surface of pneumatophores at the distal end facilitate entry of oxygen into the cortex from the atmosphere. Thus, pneumatophores both provide structural support and facilitate gaseous exchange in wetlands. Other roots (lateral roots, stilt roots, feeding roots) present in mangrove trees also have abundance of lenticels to allow influx of $O_2$ to aerenchyma. Pneumatophores are also known to develop colonies of cyanobacteria for nitrogen fixation.

#### 31.7.3.1 Changes in Root Anatomy

**Porosity** Plants exhibit an internal oxygen gradient if roots are growing in an anoxic environment in waterlogged soil in contrast with shoots exposed to oxygen-rich environment. Internal resistance to diffusion of oxygen down its gradient is high from shoot to root in majority of flooding intolerant plants. In flood-tolerant plants, this resistance to diffusion rapidly decreases. Plants exhibit this variation in the extent of diffusion of oxygen by altering the porosity of cells in roots. Thus, in wheat roots not adapted to wet soil, porosity between the cells just few centimeters from the root tip, is as low as 1%, thereby leading to high resistance to oxygen diffusion. In case of rice roots which are adapted to wet soil, the cells in subtropical region of roots (elongation zone) are loosely packed, and their porosity increases up to 10%, upon flooding. Porosities of root cells as high as 40–50% have been observed in some wetland species. Such plants also exhibit significant changes in the otherwise complex anatomy of root-shoot complex so as to increase porosity for long-distance flow of oxygen from shoot to the root under flooding.

**Aerenchyma** The porous tissue of waterlogging-tolerant plants is referred to as **aerenchyma**. It mostly constitutes the loosely packed cortical cells between the endodermis and epidermis of plant roots, and several **lacunae** extend longitudinally for a significant distance within the root. In some species, these lacunae can be continuous between root and the shoot. Aerenchyma can develop as a result of cell death and dissolution (lysigeny), separation of cells without collapse (schizogeny), or a combination of lysigeny and schizogeny (schizolysigeny). In most wetland plants, aerenchyma formation is a part of normal development and is initiated in young plants before flooding, and it gets further proliferated in response to flooding. Most of the aerenchyma, formed in the roots of plants in response to waterlogging, is the result of death and dissolution of targeted cortical cells by the process of lysigeny. Lysigeny is different from programmed cell death occurring in animal cells since it (lysigeny) is without phagocytic activity or removal of debris via a circulatory system. Instead, lysigeny involves digestion of cellulosic cell walls. Lysigenous aerenchyma normally occurs in the zone of cell elongation in roots and does not occur where lateral emerge. It is constitutive in the roots of rice and can be induced in response to waterlogging in flood-tolerant species (e.g., maize). During constitutive aerenchyma formation, lysis is evident in oxic conditions. Soil compaction (leading to decreased aeration) and nutrient deficiency are some other factors which can promote lysigenous aerenchyma formation in plant roots irrespective of waterlogging. Aerenchyma enhances the porosity of roots from the shoot to the roots.

**Hypodermis** A number of wetland plant roots have hypodermis—a layer of dense, hexagonally placed cells beneath the epidermis. In rice, hypodermis also contains sclerified fibers. Hypoxia-tolerant plant roots also exhibit deposition of suberin. Hypodermis helps to reduce the efflux of oxygen and influx of toxins as a result of suberization. It also provides structural support for root with aerenchyma in the cortex.

**Stele** Roots adapted to hypoxia also show much reduced stele. Thus, for example, in comparison with wheat roots in which stele occupies about 15% of cross-sectional area near the root tip. In the hypoxia-adapted rice roots, stele occupies nearly 5% of cross-sectional area to minimize oxygen demand at the center of the root. The root anatomy of waterlogging-tolerant plants is thus well-adapted to maximize oxygen supply.

**Lenticels and Lateral Roots** In the stem of some plants, gaseous flow is facilitated by lenticels. Some flood-tolerant woody plants are particularly rich in lenticels above the soil surface. They are used for oxygen supply to aerenchyma. In mangroves, the secondary growth in adventitious roots not only provides mechanical support; they are also rich in lenticels and aerenchyma to facilitate flow of oxygen. Lateral roots penetrate the hypodermis and provide a route for oxygen efflux. Laterals on adventitious roots can also significantly oxygenate the rhizosphere.

**Epinasty and Hyponasty** In some plants which are not adapted to flooding, waterlogging can lead to epinastic growth (Fig. 31.30). Nastic growth refers to growth response to a nondirectional stimulus, such as flooding. During **epinasty**, there is excess growth of the leaves on the upper (adaxial) surface than the lower (abaxial) surface, leading to downward curvature of leaves. Epinasty was first reported in tomato and is thought to be a characteristic of Solanaceae plants. Epinasty is generally not evident in flood-tolerant plants, indicating that their physiological adaptations help them avoid it. With increasing intensity or duration of flooding, many hypoxia-tolerant species exhibit upward growth of leaves (away from water), a phenomenon referred as **hyponasty**. This is caused by excess growth

**Fig. 31.30** Epinasty in plants under waterlogged soils

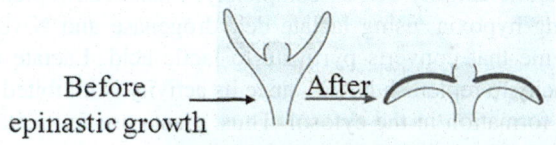

of leaves on the lower surface, directing leaves up to the water surface. The epinastic strategy may be adopted by plants during flooding by which downward directed leaves act as sails in wind. Such a bending of leaves facilitates faster removal of water vapors just below them, thereby facilitating faster aeration of top soil layer. Epinasty may also enhance transpiration rate. Both epinasty and hyponasty are triggered by enhanced ethylene available to the leaves. Flooded roots make ACC, the precursor to ethylene. ACC is transported from roots to the leaves along the xylem stream, where it gets converted into ethylene in the presence of oxygen. Depending on the environmental conditions, the leaf tissue can then respond differently to the available excess amount of ethylene. It is now believed that epinasty occurs through three different biochemical routes: (1) release of ethylene under root anoxic conditions; (2) high levels of salicylic acid during permanent flooding which keep stomata open, causing epinasty; and (3) by ABA during desiccation, when water circulation is still needed but transpiration is negligible.

### 31.7.3.2 Primary Metabolism

Under normal conditions, plants oxidize 1 mole of hexose sugar through glycolysis, citric acid cycle, and oxidative phosphorylation to produce 30–36 ATP. Under anoxic conditions, ATP is produced only by glycolysis (2–4 moles) and noncyclic mode of citric acid cycle (1 mole) per pyruvate metabolized. Under hypoxic conditions, the partial pressure of $O_2$ limits ATP production by oxidative phosphorylation. In waterlogging-sensitive plants, onset of hypoxia decreases flow of electrons to $O_2$, but glycolysis still occurs, using ADP and $NAD^+$ to produce some ATP and NADH. Without replenishment of $NAD^+$, glycolysis is inhibited, and metabolic activity ceases completely. Plants can replenish $NAD^+$ for a short time during hypoxia, using lactate dehydrogenase and NADH-dependent fermentation enzyme that converts pyruvate to lactic acid. Lactate dehydrogenase has limited capacity to replenish $NAD^+$ since its activity is inhibited by acidity created by lactic acid formation in the cytosol. Thus, plants use fermentation pathways to delay the effects of hypoxia.

### 31.7.3.3 Aquaporin Gating

Another early response to excess water is the modulation of gating of $PIP_2$ (plasma membrane intrinsic protein 2) aquaporin by cytosolic acidosis. This leads to a decrease in water uptake in response to flooding. In water-sufficient plants, $PIP_2$ are phosphorylated and open, while under hypoxia and in drought, dephosphorylation, protonation, and bound calcium close aquaporins.

### 31.7.3.4 Ion Toxicity

Many plants which are non-adapted to waterlogging exhibit ion toxicity caused by a reduction in their ability to select ions during uptake to the stele. When the water leading to flooding is saline, plants also lose their ability to discriminate $Na^+$ from $K^+$, hence leading to disturbance in K/Na ratio in the tissue.

## 31.7.4 Ethylene Action in Anaerobic Stress

The ethylene-mediated pathways are important for physiological responses as well as for morphological and anatomical adaptations to hypoxia. Ethylene is constitutively synthesized from methionine in plant cells. It triggers lysigenous aerenchyma production. Anaerobic stress in plant roots caused by flooding leads to enhanced ethylene levels in plants by one or more of the following reasons: (1) ethylene does not diffuse away from roots; (2) anoxic conditions stimulate expression of genes for ethylene biosynthesis; and (3) anoxic conditions promote enhanced ethylene synthesis. Oxygen is required for the activity of ACC oxidase which is one of the enzymes (the other being ACC synthase) required for ethylene biosynthesis. Under hypoxic conditions, however, ethylene levels increase due to ACC oxidase activity (Fig. 31.31). During flooding, temporal and spatial differences in ethylene concentration, ethylene receptors, and their activity are likely to help explain the different responses of tissues to ethylene, particularly in flood-tolerant plants. Figure 31.32 highlights the likely sequence of biochemical events triggered by ethylene in response to waterlogging, which leads to lysigenous formation of aerenchyma. Hypoxia brings about a rapid increase in cytoplasmic calcium concentration and altered expression of many genes for calcium signaling molecules, namely, calcineurin and CaM-like proteins, in hypoxic roots. An important target of increased $Ca^{2+}$ levels is an NADPH oxidase involved in cell lysis through production of $H_2O_2$. Genes encoding ROS scavengers are also downregulated in roots producing aerenchyma lysigenously. This indicates mitogen-activated protein kinase (MAPK) cascades, leading to cell death and dissolution, including chromatin

**Fig. 31.31** Response of aerial tissues to flood stress

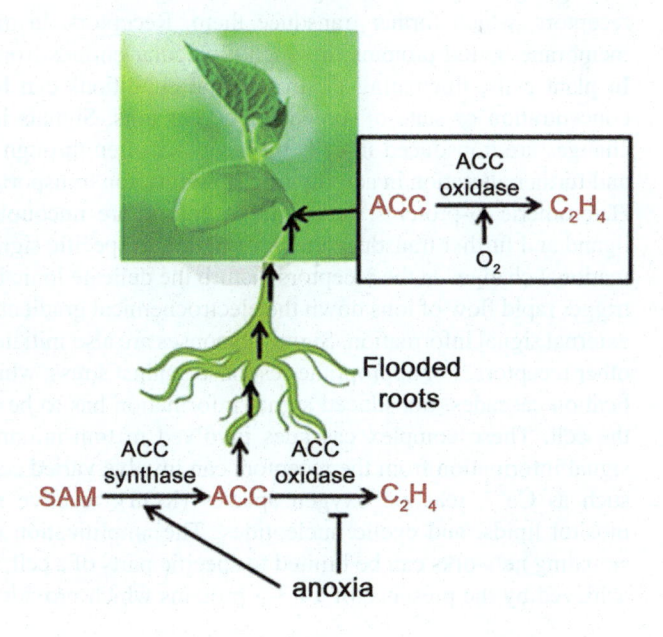

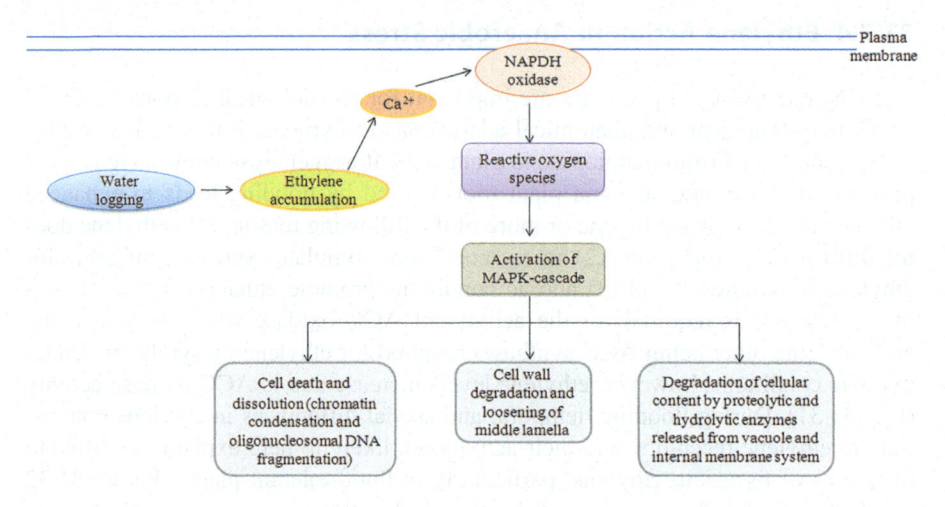

**Fig. 31.32** Process of lysigeny in plant cells

dissolution, hydrolysis and proteolysis of cell contents by enzymes released from the vacuole and ER, and loosening of middle lamella and degradation of cell wall.

## 31.8 Signaling Pathways in Response to Abiotic Stress Conditions

During signaling, diverse signals reflecting external conditions are perceived by receptors, which further transduce them. Receptors, in many cells, are generally membrane-bound proteins to which molecular entities from environment can bind. In plant cells, fluctuating environmental conditions can be detected by changing concentration or state of molecules and/or ions. Signals indicating environmental changes are transduced in cells by receptors often through change in conformation and further alteration in activity of G-proteins, ion transporters, enzymes, or kinases. Heteromeric G-proteins, bound to receptors, are uncoupled upon binding of the ligand and further transduce the information to specific signaling pathways. Conformational changes in the receptors disturb the delicate ionic balance in plant cells and trigger rapid flow of ions down the electrochemical gradient, thereby transducing the external signal information. Signal responses are also initiated by kinases linked with other receptors. For appropriate response against stress, which often involves amplification cascades, transduced signal information has to be relayed to other parts of the cell. These complex cascades involved in transmittance and amplification of signal information from the receptors can involve varied cellular entities/molecules, such as $Ca^{2+}$, reactive oxygen species (ROS), reactive nitrogen species (RNS), inositol lipids, and cyclic nucleotides. The amplification of signals through these signaling networks can be limited to specific parts of a cell. This confinement can be achieved by the presence of 14-3-3 proteins which provide scaffolding.

## 31.8.1 Rapid Signaling Stress Sensors

Various signaling molecules and transcription factors are known to play crucial roles in the maintenance of cellular homeostasis under stress conditions. Several possible sensors involved in early stress-sensing mechanisms have been identified (Fig. 31.33). A multitude of signal transduction pathways, involving calcium, reactive oxygen species (ROS), protein kinases, protein phosphatases, transcriptional regulators, and plant hormones, are transduced by rapid stress sensors as a downstream signal. These specific stress-responsive signals further lead to activation or suppression of different networks, which either allow continued growth and reproduction under stress or enable the survival of plants until the return of favorable conditions. Varied transcription factors can be phosphorylated or dephosphorylated by the activities of protein kinases and phosphatases, sensitized by the elevated levels of calcium and ROS during early stage of stress response. During abiotic stress, changes in the redox status of the cell can also be directly sensed by specific transcriptional regulators, thereby activating or inhibiting some transcription factors. Upon exposure to multiple stresses, an intensive crosstalk may take place between the protein kinases or phosphatases and hormones of individual stress response pathways in plant cells. For example, several abiotic stress responses are regulated by mitogen-activated protein kinases (MAPKs) in *Arabidopsis*. MAPK modules, constituting of MAP3K/MAP2K/MAPK cascades, regulate responses under varied abiotic stresses, such as salinity and osmotic stress, drought stress, oxidative stress, and temperature stress. Similar upstream signaling intermediates like ROS, phosphatidic acid, and calcium are involved in signaling response to the

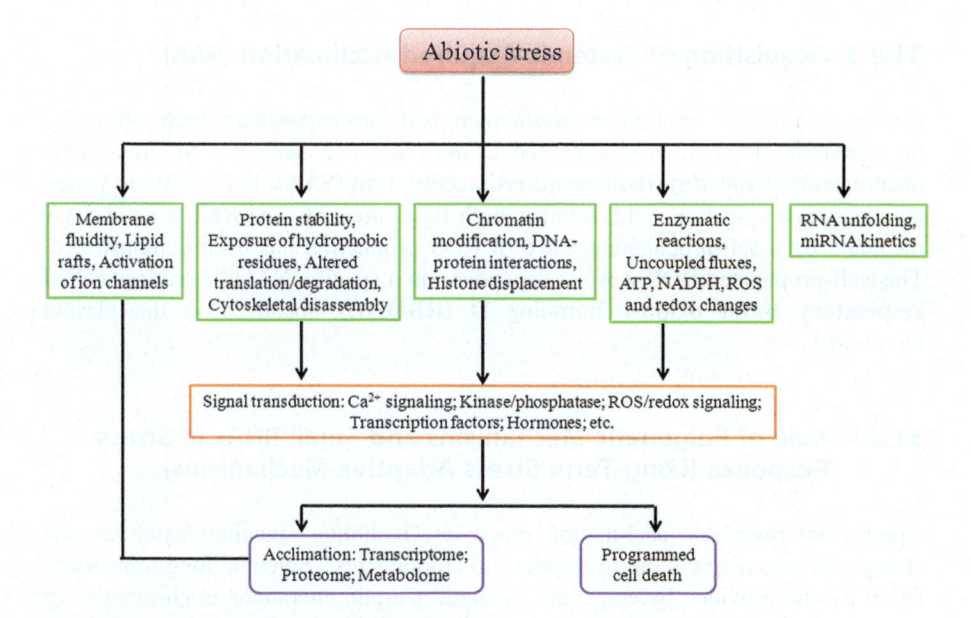

**Fig. 31.33** Early events associated with abiotic stress-sensing in plants

abovementioned four stresses, and hence production of these signaling intermediates during one of the stress response signaling mechanisms can also affect the responses of the other three stresses.

### 31.8.2 Involvement of Transcriptional Regulatory Networks (Regulons) During Stress Acclimation

Proteins which result in the activation or suppression of expression of various genes by binding to specific DNA sequences are termed as **transcriptional regulators or transcriptional factors**. A specific transcription factor can affect expression of hundreds of different genes simultaneously by binding to their promoters. A transcription factor also has the capacity to activate or suppress the expression of another transcription factor by binding to the promoter of the gene encoding the respective transcription factor. Hence, a cascade of transcriptional regulation of gene expression can take place. With the activation and suppression of some genes, a gene network can be created in response to a specific abiotic stress by combination of varied transcription factors. These transcriptional regulatory networks generated in response to abiotic stress are known as **stress-response regulons**. These regulons provide the advantage of simultaneous activation of stress-responsive pathways and suppression of other nonessential or even harmful pathways and hence are beneficial in regulating response of plants to a specific abiotic stress. For example, genes encoding photosynthetic antenna proteins might be suppressed, while other genes encoding antioxidant machinery might be activated under high light conditions.

### 31.8.3 Acquisition of Systemic Acquired Acclimation (SAA)

Acclimation can be acquired in specific plant parts not exposed to abiotic stress, by the transport of the signals generated in the plant part subjected to stress. This phenomenon is called **systemic acquired acclimation (SAA)**. In response to varied abiotic stresses, such as cold, salinity, high light intensity, and heat, a rapid SAA mediated by a self-propagating wave of ROS production has been demonstrated. This self-propagated ROS wave is dependent on a specific NADPH oxidase called **respiratory burst oxidase homolog D (RBOHD)**, localized in the plasma membrane.

### 31.8.4 Role of Epigenetic Mechanisms and Small RNAs in Stress-Response (Long-Term Stress Adaptive Mechanisms)

Apart from reversible acclimation processes, including signaling cascades and changes in gene expression, in response to abiotic stress, potential long-term adaptation can be provided by epigenetic changes. Certain alterations in chromatin are mitotically and meiotically heritable, and these epigenetic changes in response to stress might have evolutionary implications. The processes, such as stable or heritable DNA methylation and histone modifications, can be associated with certain abiotic stresses. Small RNAs, including microRNAs (miRNAs) and endogenous short-interfering RNAs (siRNAs), play important role in some plant responses against environmental stressors. MicroRNAs and siRNAs can result in posttranscriptional gene silencing through degradation of mRNA in the cytosol, mediated via RISC (RNA-induced silencing complex). siRNA can also result in suppression of gene expression by changes in chromatin properties in the nuclei through a ribonuclease complex, RNA-induced transcriptional silencing (RITS). Small RNAs have also been speculated to play a role in suppression of protein translation under stress conditions. They have been demonstrated to regulate gene expression during various abiotic stresses, such as nutrient deficiency, salinity, cold, dehydration, and oxidative stresses.

### 31.8.5 Regulation of Abiotic Stress Responses by Hormonal Interactions

A vast range of essential stress-responses involving adaptations are mediated by hormones in plants. One of the most rapid responses to abiotic stress in plants is the biosynthesis of abscisic acid (ABA). ABA biosynthesis or redistribution plays an effective role in closure of the stomata. Under water-deficit stress conditions, ABA accumulation in stressed leaves results in reduced water loss via transpiration. ABA synthesis has also been shown to be induced by cold stress, and exogenously applied ABA enhances cold tolerance in plants. Cytokinins also play crucial role in acclimation to different abiotic stresses. ABA and cytokinins are known to have opposite effects with respect to stomatal opening, transpiration and photosynthesis. ABA levels are enhanced, and cytokinin levels are declined during drought stress conditions. Though ABA is essential for closure of stomata to prevent excessive water loss, drought stress can lead to inhibition of photosynthesis and also premature leaf senescence. The effects of drought stress seem to be ameliorated by cytokinins by protecting the biochemical processes related to photosynthesis and delaying senescence. Gibberellic acid, salicylic acid, brassinosteroids, auxin, jasmonic acid, and ethylene are also known to play significant roles in stress responses in plants. Auxins play an essential role in acclimation to drought conditions in plants, and gibberellic acid and brassinosteroids can link growth regulation to responses against abiotic stresses. The comprehensive overlapping among various genes regulated by hormones elucidates the existence of a multiplex network involving crosstalk among varied hormonal pathways. The capacity of plants to acclimate to different abiotic stress conditions is greatly influenced by the coordination and mutual regulation of hormone biosynthetic pathways and also by synergistic or antagonistic nature of hormone action. Figure 31.34 depicts an interaction among ethylene, ABA, and GA in response to flooding, leading to rapid extension growth of rice seedlings.

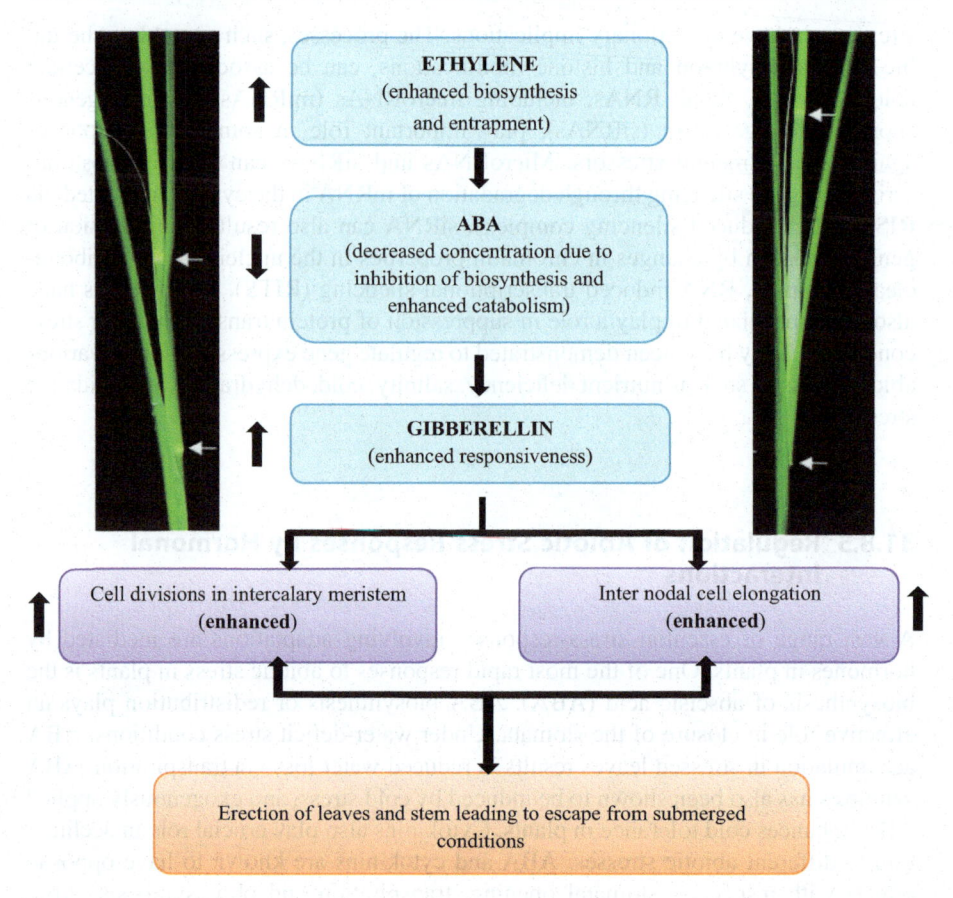

```
                    ETHYLENE
              (enhanced biosynthesis
                  and entrapment)
                         ↓
                       ABA
           (decreased concentration due to
             inhibition of biosynthesis and
               enhanced catabolism)
                         ↓
                   GIBBERELLIN
              (enhanced responsiveness)
                         ↓
        ┌────────────────┴────────────────┐
  Cell divisions in intercalary meristem    Inter nodal cell elongation
              (enhanced)                           (enhanced)
        └────────────────┬────────────────┘
                         ↓
     Erection of leaves and stem leading to escape from submerged
                         conditions
```

**Fig. 31.34** Hormonal interaction accompanying flooding-induced elongation of rice plants. Flooding induces extension growth up to as high as 25 cm in 24-h growth cycle in some varieties of rice

## 31.9 Summary

- Environmental factors which result in stress may be divided into biotic and abiotic. **Biotic** stress is imposed by direct or indirect interactions with other organisms, whereas **abiotic** stress originates from excess or deficit in the physical, chemical, and energetic conditions to which plants are exposed. Various factors, such as genotype, developmental stage, species, and organ or tissue type influence the sensitivity or resistance of a plant to stress condition(s).
- *Stress tolerance* mechanisms enable the plants to withstand stress by development of resilient structures and physiological processes, whereas **stress avoidance** mechanisms prevent or minimize exposure to vulnerable stages of stress

## 31.9 Summary

conditions by adopting life cycle strategies which limit growth to favorable periods in a varying environment.
- A rapid rise in ROS concentration results in oxidative stress in the cells, known as **oxidative burst**. Rate of ROS production and cellular ROS levels both increase significantly in plants subjected to abiotic or biotic stresses. In order to overcome oxidative damage caused due to high ROS concentration and for maintaining **redox homeostasis**, plants have a well-established defense system which includes both nonenzymatic and enzymatic antioxidants. High concentration of ozone is one of the best characterized oxidative stressors.
- *Salinity* refers to an excessive accumulation of salts in soil solution. When the concentrations of the sulfates or chlorides of Na, Mg, and Ca in the soil are enough to produce an electrical conductivity of at least 4 $dS.M^{-1}$ in the soil extract and a sodium absorption ratio [SAR; based on Na/(Ca$^+$Mg] of less than 13, it is referred to as **saline soil**. Plants develop mechanisms for salt exclusion or salt tolerance to deal with deleterious effects of salt stress. They are classified as halophytes or glycophytes, depending upon their capacity to survive on substrate containing high salt concentration. Elongation zone of the roots is highly susceptible to sodium influx due to higher absorption capability in this region. Salt tolerance in plants depends upon the following factors: 1. accumulation and sequestration of $Na^+$ and $Cl^-$, 2. maintenance of $K^+$ and $Ca^{2+}$ pools as essential macronutrients in the presence of high $Na^+$ levels, 3. regulation of long-distance $Na^+$ transport and its accumulation in different aerial organs following transpiration, 4. accumulation of suitable osmolytes to prevent desiccation, and 5. modulation of efficient antioxidant machinery.
- Some plants maintain a high water potential in the tissues to avoid water stress by reducing the loss of water to minimal level under conditions of acute water deficit. Many desert plants, like cacti, are included in the category of water stress avoiders. Water stress avoidance is crucial for plants especially in arid habitats. Succulents store water to tolerate water stress conditions. Moreover, succulent plants exhibit crassulacean acid metabolism, low transpiration rate, and thick cuticles to prevent water loss. Under severe water stress, the reaction center of photosystem II degenerates, and the proportion of chloroplast proteins may decline as well. The decrease in photosynthetic activity may result in shortage of nutrients, thereby hindering plant growth. Certain pressure sensor proteins on the plasma membrane, **osmosensors**, are involved in the primary perception of water deficit. Several drought-tolerant plant cells accumulate solutes to regulate their solute potential and, hence, lower $\psi$ during transient or extended periods of water stress. This capacity is called **osmotic adjustment**. Two processes are involved in osmotic adjustment: one takes place in the vacuole and the other in the cytosol. The accumulation of ions is majorly limited to vacuoles so as to keep ions out of contact with cytosolic enzymes. Accompanying accumulation of high concentration of ions in the vacuole, certain solutes called **compatible solutes or osmolytes** accumulate in the cytosol in order to maintain the water potential equilibrium between the two compartments. Some compatible solutes like proline may act as **osmoprotectant** as well. Osmoprotective effects include stabilization

of protein and membrane structure, scavenging of free radicals and hence protecting against oxidative damage, and providing cellular source of carbon and nitrogen to the cell when conditions normalize. In addition to solutes, late embryogenesis abundant (LEA) proteins also accumulate in plant cells exposed to water stress.
- Temperatures over 45 °C are lethal for most plants, and temperatures over 30 °C are usually stressful. **Thermal death point** of a plant is the heat-killing temperature depending upon the time of exposure. Thermal motion of molecules is increased causing enhanced membrane fluidity and extreme fluidity, which are detrimental for cell. High transpiration rate helps in dissipating heat as water evaporates and hence leads to cooling of leaves and protects against heat stress-induced injury. In hot, dry climates, some plants possess thick corky bark to reduce absorption of heat and also to provide insulation to the phloem and cambium from water loss. In response to high temperature stress, a new set of proteins known as heat shock proteins (HSPs) are synthesized as a generalized physiological response. HSPs provide protection to organisms from acute damage, assist in acclimation by allowing metabolic and cellular activities to resume, and also increase the temperature tolerance range for survival. The synthesis of HSPs in response to high temperature stress is controlled by regulation of HSP gene transcription, and the regulatory mechanism is similar in all eukaryotes.
- Temperate plants are usually able to survive chilling temperatures. However, plant species adapted to warm climates are damaged by low, non-freezing temperatures and hence are **chilling-sensitive**. During ice formation in extracellular compartments, the cellular water moves across the plasma membrane toward the extracellular ice, down the water potential gradient. The majority of injury due to freezing stress takes place at the plasma membrane, and the subsequent damage involves membrane destabilization as a result of dehydration of plant cells. Varied forms of injuries can occur in plants due to membrane destabilization, such as loss of osmotic responsiveness and expansion-induced lysis. Prior exposure to low but non-freezing temperatures confers the ability to survive low temperature (below freezing) in plants, and this process is known as **cold acclimation**. Post-cold acclimation, the cold hardiness in plants is retained as long as the temperatures remain below freezing. During cold acclimation, the proportion of phospholipids (phosphatidylserine and phosphatidylcholine) is enhanced, while that of cerebrosides (lipids composed of ceramide and a sugar residue) is declined in the plasma membrane. During cold acclimation, expression of certain proteins is induced to stabilize the membranes by bringing down the incidence of phase transition. A group of genes, called **cold regulated genes** (COR), are induced during cold acclimation, and many of COR genes are induced during water-deficit stress as well. Certain protein called as **antifreeze proteins (AFPs)** are also produced during cold acclimation. AFPs are secreted into the extracellular space/apoplast and prevent the nucleation of ice crystals or re-formation of ice crystals post a freeze-thaw cycle. **Molecular chaperones (chaperonins)** like HSP90 and HSP70.12 are also encoded by some COR genes to prevent protein denaturation during freezing. Proteins involved in low temperature signaling, like

calmodulin-related proteins, certain transcription factors, and MAP kinase (MAPK) and MAP kinase kinase kinase (MAP3K), are also encoded by some COR genes. During low temperature, a rapid increase in the concentration of cytosolic free calcium is evident, primarily due to influx from extracellular sources. This abrupt rise in cytosolic calcium is essential for induction of COR genes and hence freezing tolerance.

- *Flood-tolerant plants* exhibit only temporary anoxic conditions under flooding. With the continued extension of roots in oxygen-deficient soil, continuity of aerenchyma allows oxygen movement up to the apical region. Additionally, suberized and lignified cell walls prevent oxygen diffusion outward to the soil. **Flood-sensitive plants** get injured in response to anoxia because of cytoplasmic acidification. The cells of root meristem of flooded sensitive plants exhibit reduced rate of protein synthesis, inhibition of cell division, degradation of mitochondria, cell death, and disruption of ion transport, in response to flooding. Such plants do not develop root aerenchyma and do not generally survive for more than 24 h. The chemistry of flooded soil varies with its composition and microbial interactions. There is an increase in the metabolic activity of **methanogens** under anoxic conditions, leading to enhanced production of methane. The rate of methane release from wetlands into the atmosphere is, thus, determined by the combined metabolic activities of methanogenic and methanotrophic microbes in soil. Pneumatophores have well-developed aerenchyma as long and continuous columns in the straightened cortex of the root system. Extensive lenticels present on the surface of pneumatophores at the distal end facilitate entry of oxygen into the cortex from the atmosphere. Thus, pneumatophores both provide structural support and facilitate gaseous exchange in wetlands. Plants exhibit an internal oxygen gradient if roots are growing in an anoxic environment in waterlogged soil in contrast with shoots exposed to oxygen-rich environment. Porosities of root cells as high as 40–50% have been observed in some wetland species. The porous tissue of waterlogging-tolerant plants is referred to as **aerenchyma**. Aerenchyma can develop as a result of cell death and dissolution (lysigeny), separation of cells without collapse (schizogeny), or a combination of lysigeny and schizogeny (schizolysigeny). A number of wetland plant roots have hypodermis—a layer of dense, hexagonally placed cells beneath the epidermis. Roots adapted to hypoxia also show much reduced stele. In some plants which are not adapted to flooding, waterlogging can lead to epinastic growth. The epinastic strategy may be adopted by plants during flooding by which downward directed leaves act as sails in wind. Such a bending of leaves facilitates faster removal of water vapors just below them, thereby facilitating faster aeration of top soil layer. Both epinasty and hyponasty are triggered by enhanced ethylene available to the leaves. In waterlogging-sensitive plants, onset of hypoxia decreases flow of electrons to $O_2$, but glycolysis still occurs, using ADP and $NAD^+$ to produce some ATP and NADH. Another early response to excess water is the modulation of gating of $PIP_2$ (plasma membrane intrinsic protein 2) aquaporin by cytosolic acidosis. This leads to a decrease in water uptake in response to flooding. Many plants which are non-adapted to waterlogging exhibit ion toxicity caused by a reduction in their ability to select ions during uptake to the stele. Anaerobic stress in plant roots caused by flooding leads to enhanced ethylene levels in plants by one or more of the following reasons: (1) ethylene does not diffuse away from roots; (2) anoxic conditions stimulate expression of genes for ethylene biosynthesis; and (3) anoxic conditions promote enhanced ethylene synthesis. Oxygen is required for the activity of ACC oxidase which is one of the enzymes (the other being ACC synthase) required for ethylene biosynthesis. Under hypoxic conditions, however, ethylene levels increase due to ACC oxidase activity.

## Multiple-Choice Questions

1. Example of stress avoidance mechanism represented by desert plants:
   (a) Presence of deep root system
   (b) Crassulacean acid metabolism
   (c) Presence of spines
   (d) Presence of succulent tissues
2. A *catastrophic response* is observed when:
   (a) System bounces back and regains its former state.
   (b) System is deformed and settles on a new stable configuration.
   (c) System is deformed and leads to decrease in entropy and ultimately system dies.
   (d) System enters into a stage of incoherence, entropy enhances, and the living system dies.
3. Evolutionary improvements, which enhance the fitness of the organism and occur over many generations and across entire populations, are termed as:
   (a) Adaptations
   (b) Acclimation
   (c) Gene alterations
   (d) Both b and c
4. What happens to GSH concentration under stressed condition?
   (a) GSH concentration usually increases, and redox state becomes more reduced.
   (b) GSH concentration usually declines, and redox state becomes more oxidized.
   (c) GSH concentration usually declines; thus, redox state becomes more reduced.
   (d) GSH concentration remains the same though GSSG concentration increases.
5. Why $Na^+$ and $K^+$ ions compete for uptake via transporters under saline conditions?
   (a) Both have similar ionic radii and ion hydration energies.
   (b) Both have same transporters.

(c) Na⁺ ions have small ionic radii compared to K⁺ ions.
(d) Both b and c.
6. Compartmentalization of Na⁺ into vacuoles is carried out by?
   (a) Nonselective cation channels (NSCCs)
   (b) SOS3-SOS2 complex
   (c) Na+/H+ exchangers (NHXs)
   (d) P-type ATPases
7. Cellular osmolarity in the shoots of halophytes is 2–3 times higher than the osmolarity of soil solution however the cytoplasmic concentration of ions is maintained at nontoxic levels. How?
   (a) Action of tonoplast-localized Na+ and Cl− importers
   (b) Accumulation of organic solutes in the cytoplasm
   (c) Active sequestration of Na+ into the vacuoles by Na+-H+ antiporters
   (d) All of the above
8. Bladder cell is characteristic feature of:
   (a) Desert ephemerals
   (b) Saprophytes
   (c) Halophytes
   (d) None of the above
9. In which categories of plants minimum thermal death point has been observed?
   (a) Desert plants
   (b) Hydrated organs of temperate plants
   (c) Aquatic and shade plants
   (d) Both b and c
10. Biosynthesis of heat shock proteins (HSPs) is induced in response to:
    (a) High temperature stress
    (b) Cold stress
    (c) Drought stress
    (d) All of the above
11. In plants molecular chaperons are:
    (a) Histidine kinase Hik33
    (b) HSPs
    (c) LEA proteins
    (d) HSPs and LEA proteins
12. Role of antifreeze proteins (AFPs) under freezing stress:
    (a) Prevent protein denaturation
    (b) Act as a osmoprotectants
    (c) Prevent the nucleation of ice crystals
    (d) Prevent electrolyte leakage by stabilizing plasma membrane
13. miRNAs and siRNAs play important role in certain plant responses against environmental stressors via:
    (a) Post-translational gene silencing
    (b) RITS (RNA-induced transcriptional silencing) formation
    (c) Enhanced gene expression by ribonuclease complex formation
    (d) Both a and b

14. ABA and cytokinins exhibit opposite effects with respect to:
    (a) Stomatal closing and photosynthesis
    (b) Nutrient deficiency
    (c) Stomatal opening, transpiration, and photosynthesis
    (d) Photosynthesis
15. Flood-sensitive plants get injured in response to anoxia due to absence of:
    (a) Cell division
    (b) Aerenchyma
    (c) Ion transport
    (d) Protein synthesis
16. Both epinasty and hyponasty are triggered by enhanced production of:
    (a) ABA
    (b) GA
    (c) Ethylene
    (d) Jasmonic acid

**Answers**

1. a   2. d   3. a   4. b   5. a   6. c
7. d   8. c   9. c   10. d   11. d   12. c
13. b   14. c   15. b   16. c

## Suggested Further Readings

Shinozaki K, Uemura M, Bailey-Serres J, Bray EA, Weretilnyk E (2015) Responses to abiotic stress. In: Buchanan BB, Gruissem W, Jones RL (eds) Biochemistry and molecular biology of plants. Wiley-Blackwell, Chichester, pp 1051–1100

Smith AM, Coupland G, Liam D, Harberd N, Jones J, Martin C, Sablowski R, Amey A (2010) Plant biology, garland science. Taylor & Francis Group, New York, pp 437–498

Taiz L, Zeiger E (2010) Plant physiology, 5th edn. Sinauer Associates, Sunderland, pp 731–761

# Biotic Stress

**32**

Manju A. Lal, Renu Kathpalia, Rama Sisodia, and Rashmi Shakya

In nature plants seldom grow in isolation. Their growth and development are greatly influenced by abiotic and biotic factors to which they are continuously exposed. Unlike animals they are rooted in the soil, and consequently they cannot escape adverse environmental conditions in their vicinity (Fig. 32.1). Rather they have to develop the strategies to adapt themselves to the hostile conditions in order to survive and grow. Biotic interactions of plants are not always harmful but they can be beneficial too. Plant-pollinator interactions, rhizobia-legume interactions, or mycorrhizal interactions are examples of **mutualism** in which both the partners of the association are benefitted. In rhizobia-legume interactions, host plant provides **ecological niche** and nutrients which are required for the growth of prokaryotes, while the prokaryotes provide nitrogen to the host plant. Some plants are not able to excrete protons or organic acids in the soil required for uptake of nutrients like phosphates. They do so through symbiotic associations with **mycorrhizal fungi.** Associations, in which one of the partners is benefitted and the other one is not affected, are called **commensalism**. In some cases, the association may become beneficial as the benefitting partner may stimulate defense mechanism in the other organisms. There may also be harmful biotic interactions among plants and other organisms. These include interactions of plants with **pathogens**, with the **plant pests** and **parasitic associations** between the plants. Plant pathogens are the organisms that spend a part of their life cycle or complete their life cycle inside the plant. These include the microbial pathogen such as viruses, bacteria, or fungi. On the contrary, plant pests include the **herbivores**, such as insects, nematodes, or mammals which cause damage to plants by eating their vegetative tissue, fruits, or seeds (Fig. 32.2). It is interesting to note that plants growing in the wild rarely develop disease. However, domestication of economically important species leads to the development of **monocultures**, thereby resulting in genetic uniformity which makes them more susceptible to infections. Understanding the interactions of plants with their biotic environment greatly helps in reducing the use of agrochemicals which, in turn, would facilitate reduction in pollution and also in

© Springer Nature Singapore Pte Ltd. 2018
S. C Bhatla, M. A. Lal, *Plant Physiology, Development and Metabolism*,
https://doi.org/10.1007/978-981-13-2023-1_32

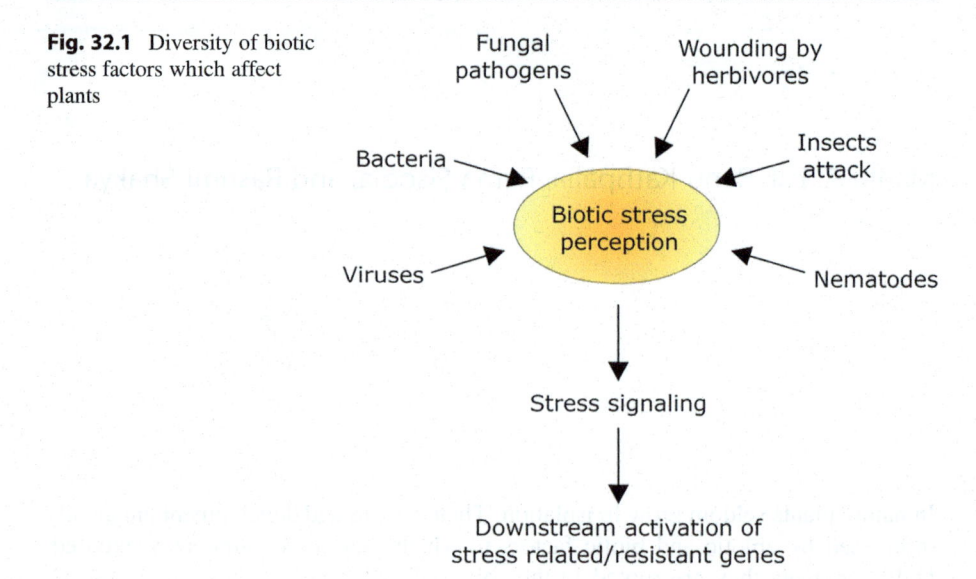

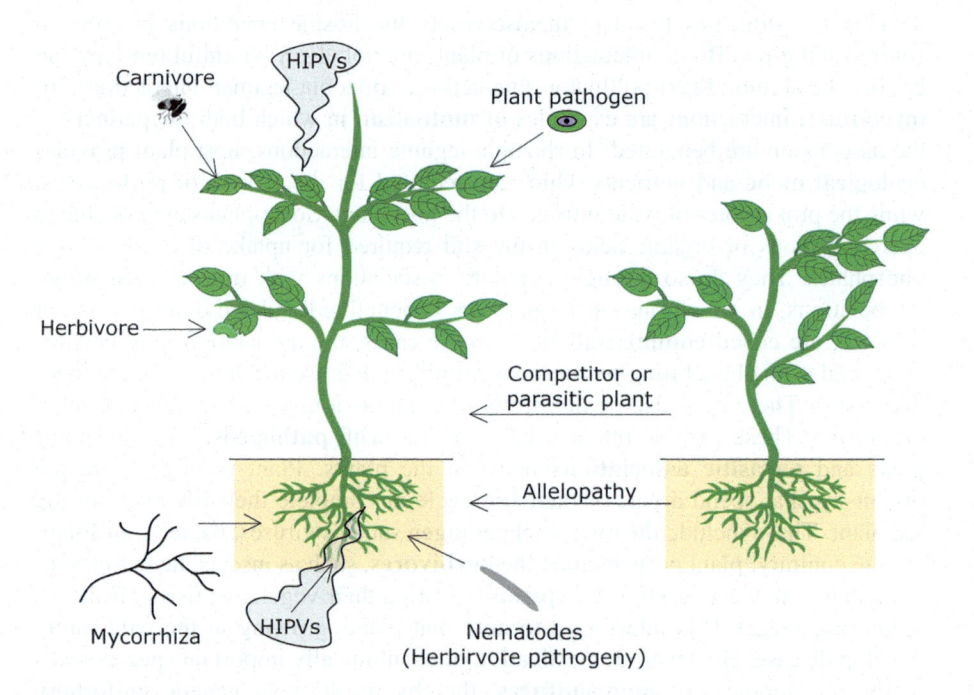

**Fig. 32.1** Diversity of biotic stress factors which affect plants

**Fig. 32.2** HIPVs (herbivore-associated plant volatiles) release associated with different organisms (signal receivers) around a damaged plant

the cost of energy required for their production. **Allelopathic** interactions occur between plants growing nearby. Growth of some plants is inhibited because of chemicals produced by the adjacent plants which results in a sort of a **chemical warfare**. In this chapter plant's interactions with biotic factors will be dealt with at physiological and molecular level.

## 32.1 Interactions with Pathogens

A plant disease can be defined as an abnormal growth and/or dysfunction of a plant, and disease-causing microbes are called **pathogens**. Those microorganisms which are unable to induce disease in a host are termed as nonpathogenic with respect to host. There is a huge genetic polymorphism among the phytopathogenic agents. The climatic factors further influence this polymorphism, causing evolution of aggressive strains or biotypes that alter host-pathogen interactions (Fig. 32.3). A particular microbe is considered as a pathogen only when the suspected pathogen has the following characteristic features: (i) it should be consistently associated with the same symptoms, (ii) it should be possible to culture the pathogen in pure form away from the host, and (iii) it should be possible to reinoculate the pathogen into a healthy host (test host), and the symptoms then developed should be identical to those from where it had been originally isolated. Pathogens are responsible for about 15% losses in global food production and are a major challenge in breeding resistant crops. Pathogens include bacteria, fungi, oomycetes, and viruses. More than 1600 species of bacteria have been found to be responsible for plant diseases. These include gram-negative bacteria like *Erwinia*, *Pseudomonas*, *Xanthomonas*, *Xylella*, and *Agrobacterium* and also gram-positive bacteria such as *Clavibacter* (*Corynebacterium*) and *Streptomyces* which commonly infect the plants. *Agrobacterium tumefaciens*, the cause of crown gall, is unique among plant pathogens as it transfers its DNA to the host. Mycoplasma-like organisms and spiroplasmas are cell wall-less prokaryotes known to cause about 200 diseases. More than 8000 species of fungi have also been found to infect plants. They have great diversity in their morphology and life cycle and develop complex relationship with the plants. Some of the microorganisms associated with plants are harmful though they may not have

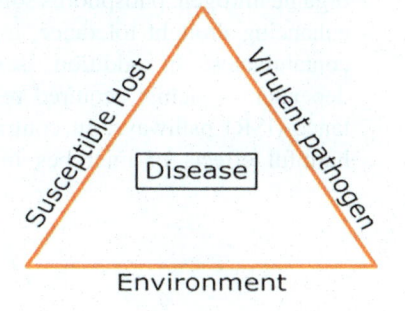

**Fig. 32.3** Disease triangle of plants. For the development of disease, the pathogen should be virulent, host should be susceptible, and environment should be conducive for pathogen multiplication

pathogenic relationship, e.g., bacterium *Pseudomonas syringae* is found on the surface of the plant, where it lives as a saprophyte (Box 32.1). Under normal conditions, its presence does not affect the plant or its growth in any detectable way. But as the temperature drops below the freezing point of water, the bacteria act as nuclei for the formation of ice crystals. This ice crystal formation leads to frost damage in the plant. When bacteria are not present on the leaf surface, no ice crystals are formed, and water remains as liquid not only at 0 °C but even at several degrees below freezing point of water as well. It was discovered that "frost injury" is due to an *ice* **nucleation active (INA)** protein formed by bacteria and the mutant strain for this protein is not so harmful for the plant.

---

**Box 32.1: Microbes in Rhizosphere and Phyllosphere**

Plants grow along with enormous microorganisms, termed as plant microbiota, which play a very significant role in plant growth and provide protection against pathogens. The two most important zones where microbiota proliferates are **rhizosphere** and **phyllosphere**. The region in the vicinity of roots which is rich in microbes is known as rhizosphere, and a similar region around leaves is known as phyllosphere. Rhizosphere is modified by nutrient-rich mucigel secreted by the roots known as **rhizosheath**. Many of the microbes found in this zone are non-specific and feed saprophytically on the root exudates. In absence of any host, microbes survive in soil as spores and other resting structures. Pathogenic microbes are also present as resting spores but are stimulated to grow toward the root surface by the chemicals in the root exudates. For example, *Sclerotinia cepivorum* causes white rot in onion and is attracted by the compounds secreted by roots of the members of the onion family. Ethanol, which is produced under waterlogged conditions during anaerobic respiration of roots, stimulates production of infection structures in the spores of *Phytophthora infestans*. Plant root exudates also contain **siderophores** (or **phytosiderophores**), which are low molecular mass, iron-binding agents produced under iron-limiting conditions. Phytosiderophores limit the growth of microbes in the vicinity of roots by reducing iron availability. The beneficial effects of rhizosphere are manifold, viz., decomposition of plant residue and organic matter, increasing plant nutrient availability, symbiotic mycorrhizal association, production of organic chelates, mineralization of organic nitrogen, phosphorus solubilization, protection against root pathogens, enhancing drought tolerance, and biodegradation of synthetic pesticides or contaminates. In addition, several rhizobacteria trigger salicylic acid-dependent systemic acquired resistance (SAR) and induced systemic resistance (ISR) pathways. In contrast to the beneficial effects, there are some harmful effects of microbes in the rhizosphere. Aerobic bacteria remove

(continued)

**Box 32.1** (continued)

oxygen or increase carbon dioxide, thereby reducing root elongation and development, as well as root hair formation. They also adversely affect the rate of nutrient and water uptake. In addition to nitrogen-fixing bacteria, other genera such as *Pseudomonas* and *Erwinia*, blue-green algae, fungi, and actinomycetes are commonly present on the leaf surface in the phyllosphere. These epiphytic microorganisms synthesize indole-3-acetic acid and help in nitrogen fixation, and their most important role is in inducing plant defense system to produce **phytoalexins**. Phytoalexins protect the plants from airborne microbes, e.g., infection of potato by *Phytophthora* sp. results in the production of many phytoalexins, such as caffeic acid, chlorogenic acid, scopolin, phytotuberin, etc.

A successful pathogen must enter the host plant to obtain nutrients after suppressing defenses of the host and finally reproduce to continue the life cycle (Box 32.2). Once pathogen enters the plant, it employs different strategies for survival (Fig. 32.4). When the plant cells are killed prior to infection so as to facilitate the pathogen to colonize it, it is referred as **necrotrophy**. Necrotrophs make use of products of enzymatic digestion of cell walls, simultaneously releasing nutrients to survive and colonize. Some of the necrotrophic pathogens produce host-selective toxins which specifically inactivate particular plant enzymes. For example, the toxin **fusicoccin** produced by *Fusicoccum amygdali* (a fungus) inactivates the plasma membrane localized $H^+$-ATPase. As a result, an irreversible opening of stomata followed by plant wilting occurs, resulting in cell death followed by fungal colonization. *Sclerotinia sclerotiorum*, a necrotrophic fungus that infects ~400 plant species, secretes the phytotoxin oxalic acid which helps in early phases of infection. Oxalic acid alters the redox status of invaded plant tissues and suppresses several early plant defenses, including the oxidative burst and callose deposition. Once the plant is infected, it induces plant to produce reactive oxygen species (ROS) which leads to death of the host tissues followed by fungal colonization and growth. Many necrotrophs attack on broad range of plant species, e.g., *Botrytis cinerea*, a fungus, can attack 1000 plant species, and *Erwinia*, a bacterium, causes rot in many fruits and vegetables. The second strategy is **biotrophy**, when plant cells remain alive throughout the infection, e.g., in mutualism and parasitism (Box 32.2). Colonization by biotrophic pathogens leads to alterations in metabolism and development of the plant. Alteration in the levels of phytohormones can induce senescence or abnormal growth of the infected tissue and will reduce yield. Biotrophic fungal conidia germinate to produce germ tube on the leaves of susceptible plant. On sensing the contact of the appropriate host, the germ tube stops growing and hooks itself. It forms a specialized infection structure called **appressorium**, which acquires water from the dewdrop because of its water potential lowered due to accumulation of

### Box 32.2: A Good Parasite Is a Poor Pathogen

When the invading microbes have some beneficial effects on the invaded partner, it is referred as symbiosis, but in case the balance shifts from mutualism to injurious effects on one partner, it is termed parasitism. Parasite is an organism living in intimate association with another living organism from which it derives some or all of its nutrients while giving no benefit in return. One of the most important differences between mutualism and parasitic association is the activity of $H^+$-ATPase on plasma membrane of the plant cells. In parasitic associations, $H^+$-ATPases are absent in the plasma membrane around the haustorium, while in mutualism the enzyme is present. This is mainly due to reciprocal movement of nutrients in mutualism, while there is unidirectional movement from the plants to the microbes in parasitism. When the parasitic interactions of host and microbes increase beyond nutrition, it results in disease. All parasites are potential pathogens, but not all pathogens are parasites. Any organism which is dependent upon another organism for its supply of nutrition might be expected to restrict its pathogenic effects to a minimum. Majority of parasites are nonpathogenic for their host but may be pathogenic to other organisms. However, there are parasites which become pathogenic to the same host due to changes in immediate environment. For example, most bacteria live as normal flora in the host, and with opportunity they become pathogens.

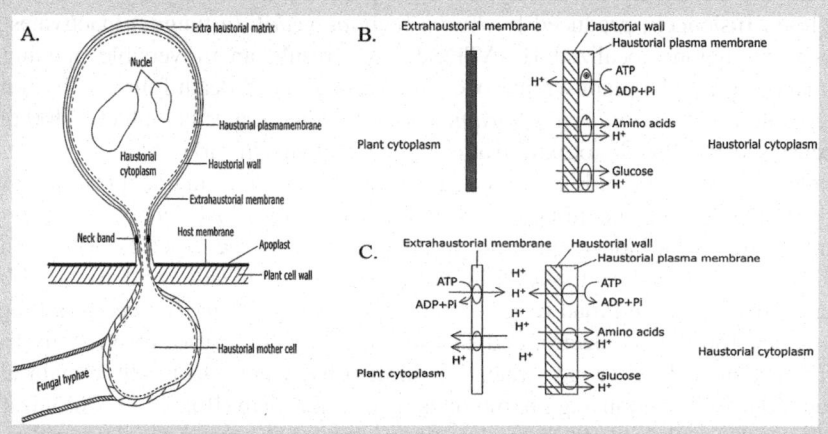

(**a**) A fungal hypha forming haustorium in the host cell. (**b**) Absence of $H^+$-ATPase pump in parasitic extrahaustorial membrane. (**c**) In mutualistic association $H^+$-ATPase pumps are present.

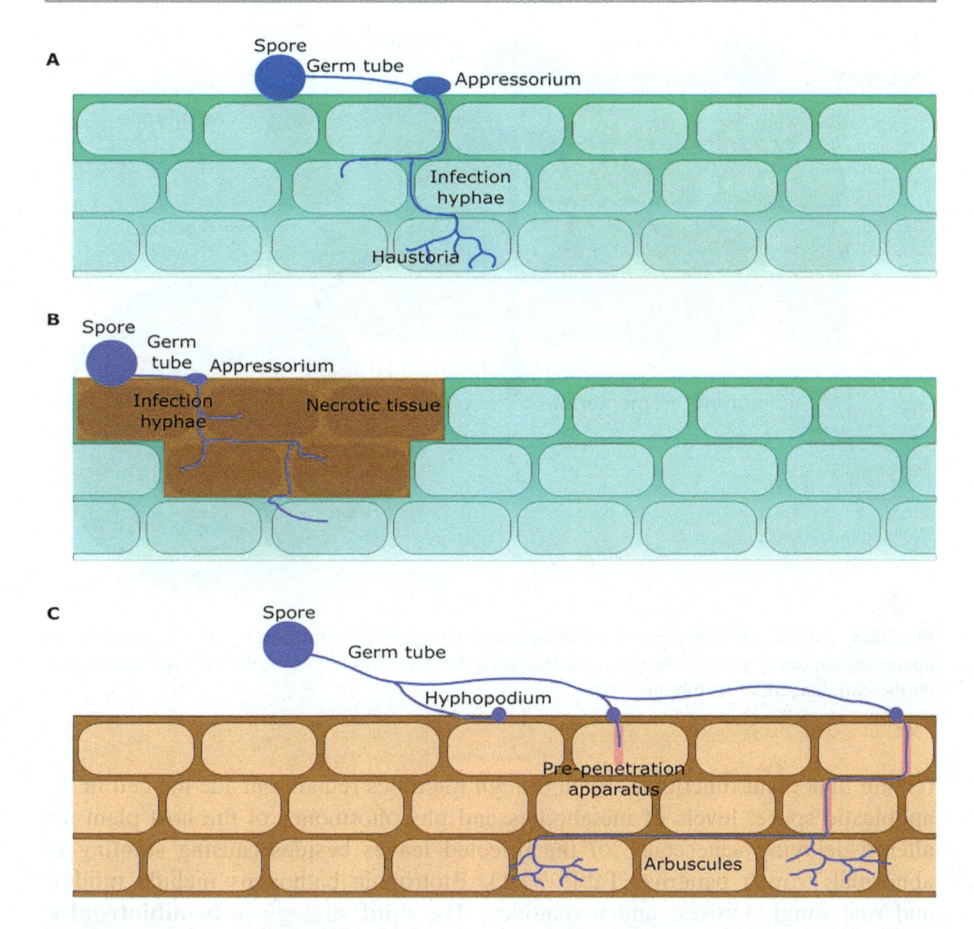

**Fig. 32.4** Infection and colonization pattern in (**a**) biotrophic, (**b**) necrotrophic, and (**c**) biotrophic mutualist pathogens

glycerol and other solutes. As a result, extremely high turgor pressure is generated. The turgor pressure of the penetration plug and of secondary germ tube produced by appressorium is strong enough to push even through inert non-biological material such as Teflon and is sufficient enough to pierce through the cuticle into the rigid plant cell wall (Fig. 32.5). Haustorium, which develops as an invagination in plasma membrane of the host plant, is covered by extrahaustorial membrane (EHM). EHM helps in the absorption of nutrients and water favoring pathogen growth. In biotrophic associations, whether the microbes enter through cell wall of the host plant or not, the plasma membranes of the two (microbe and the host plant)

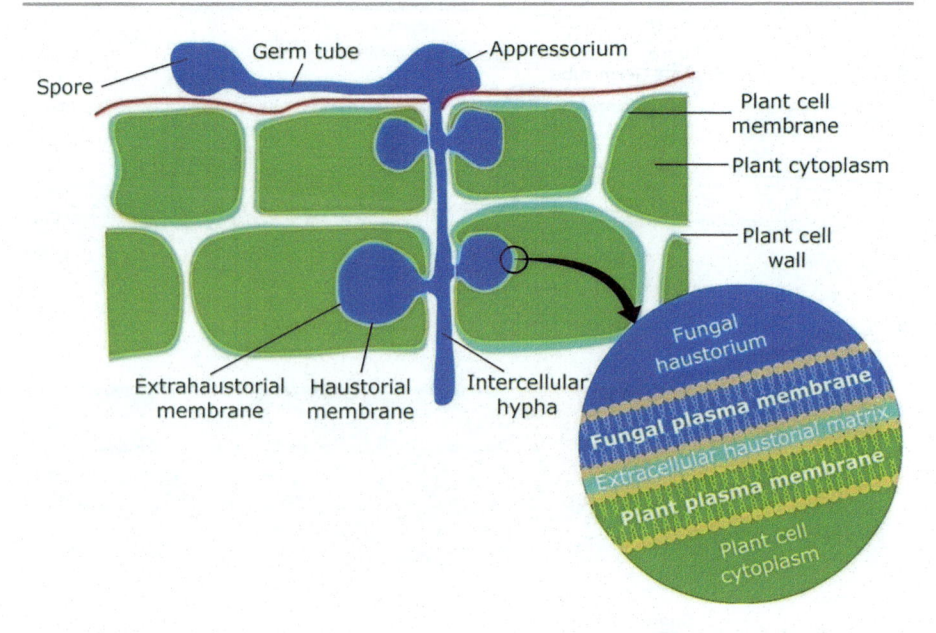

**Fig. 32.5** Infection by fungal spore, forming germ tube attached to the host by an adhesion pad. An appressorium develops over the penetration hypha; the haustorium invades the host cell surrounded by the extrahaustorial membrane

remain intact and functional. Even though microbes remain outside the cell in the apoplastic space, levels of metabolites and phytohormones of the host plant are altered delaying senescence of the infected leaves besides causing stunting or abnormal growth patterns (Table 32.1). Biotrophic pathogens include mildew and rust fungi, viruses, and nematodes. The third strategy is **hemibiotrophy**, where the plant remains alive initially but is followed by death of the tissue. Hemibiotrophic pathogens use both modes of nutrition, first as a biotrophic and then as necrotrophic. Switching to necrotrophic mode of nutrition by the pathogen is due to high nutritional demand as there is increase in pathogen biomass and occurrence of asexual reproduction. For example, in potato late blight caused by *Phytophthora infestans*, the potato leaf remains alive during initial phases of infection, but as the infection becomes severe, the tissue is killed. **Oomycetes** and fungi can be biotrophic, necrotrophic, or hemibiotrophic plant pathogens. *Puccinia* shows both biotrophic and hemibiotrophic plant-pathogen interactions. Many bacteria of the genus *Pseudomonas* also show both biotrophic and hemibiotrophic interactions, i.e., it colonizes initially in the living tissues, and later on the leaf tissues show dead patches (lesions).

**Table 32.1** Differences between necrotrophic and biotrophic pathogens

| Features | Necrotrophic | Biotrophic |
| --- | --- | --- |
| Biochemical | Host cells are not killed as few or no toxins/ enzymes are produced | Host cells are rapidly killed as more toxins/enzymes are produced |
| Morphological | Special parasitic structures, like haustoria, are formed<br>Pathogen penetrates directly or through natural opening | No special structures are formed<br>Pathogen penetrates through wounds or natural openings |
| Ecological | Host range is narrow<br>The pathogen cannot grow without host<br>Attack on healthy host | Host range is wide<br>The pathogen can grow saprophytically without the host<br>Attack at any stage of host |

## 32.2 Susceptibility and Resistance

The terms "susceptible" and "resistance" are used with reference to host. If a particular pathogen can cause disease on a particular plant, that plant is susceptible, and if disease is not caused, that plant is resistant to that particular pathogen. A plant, if susceptible to one pathogen, is not necessarily susceptible to another pathogen. Pathogens which can invade the host but are unable to cause disease are known as **avirulent**, and those which drastically affect the host and cause disease are known as **virulent**. There is variation in the degree of susceptibility and in the strength of resistance of a host. **Pathogenicity,** which refers to the ability of a pathogen to cause disease, is an all-or-none disease situation, while virulence describes the gradation of pathogenicity. Features of a pathogen, like rapid growth and releasing cell wall-degrading enzymes, may or may not be related to virulence. An interaction between host and pathogen, where symptoms are expressed clearly, is known as **compatible** disease reaction, and when symptoms are not expressed clearly and the reaction has minimum effect on the host, it is known as **incompatible** disease reaction.

**Specificity of a pathogen** describes the extent to which a pathogen is restricted to particular host plants. Most pathogens are restricted to one or a few host species, and most host plants are susceptible to only a few pathogens. For example, *Puccinia* is pathogenic to cereals and never to soybeans, and *Phytophthora megasperma* is pathogenic to soybean and never to wheat. Some pathogens, such as oomycetes genus *Pythium* and Ascomycetes genus *Sclerotinia*, attack hundreds of different plant species. Protection of a host against particular pathogen strain is known as **vertical resistance**. On the other hand, protection of a host against wide range of pathogens is known as **horizontal resistance**. The process of infection, colonization, and reproduction of pathogen in the host is known as **pathogenesis**.

### 32.2.1 Entry of Pathogen

There are three main routes of entry for pathogens in plants, viz., direct penetration through intact surface since most pathogens produce a wide variety of cell wall-degrading enzymes, entry through natural openings (stomata and lenticels), and entry through wound sites. Fungi can enter through all the three routes depending on species. Invasion of fungi involves landing of spores on the leaves followed by host recognition and germ tube emergence. Hyphae grow through cell wall and press against the cell membrane. Hyphae differentiate into haustoria and nutrient transfer begins (Fig. 32.6). Some rust fungi invade through stomatal opening. Bacteria rarely enter by direct penetration. They either enter through natural openings or depend on

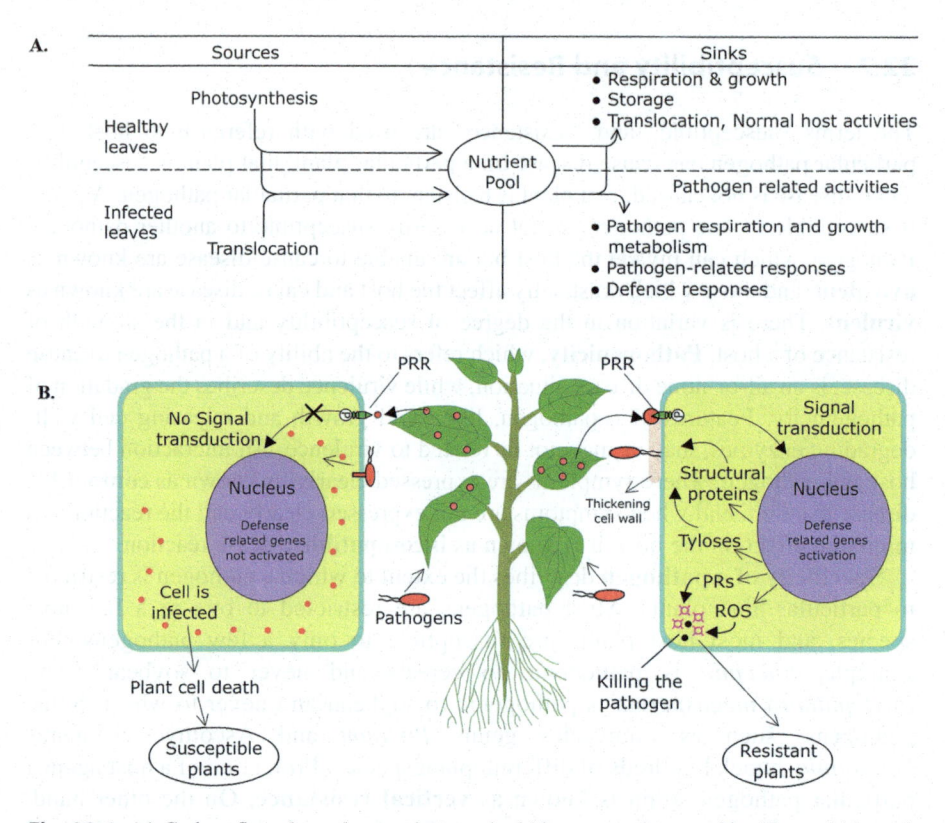

**Fig. 32.6** (a) Carbon flow through a nutrient pool within mature tissues of healthy and biotroph infected leaves. (b) Diagrammatic view of plant-pathogen interactions and molecular mechanisms involved in susceptible and resistance disease reactions by plant cell on pathogen attack. If plant cell PRRs are unable to identify pathogen, then no signals are transduced, and hence defense-related genes are not activated, and plant becomes susceptible. On the other hand, if plant cell PRRs recognize pathogen, they show hypersensitive responses (thickening of cell wall, tyloses, etc.), or defense-related genes are activated resulting in formation of PR proteins and ROS, which kill the pathogenic spores and plant becomes resistant. *PRR* pattern recognition receptors, *PRs* pathogenesis related, *ROS* reactive oxygen species

insects for entry into the plants. When the mouth part of an insect penetrates phloem for feeding (phloem feeders), bacteria are taken along the phloem sap which are transmitted to the vascular cells. Phytopathogenic bacteria colonize the apoplast region of host plants causing rots, spots, vascular wilts, cankers, and blights. Most of pathogenic bacteria are rod-shaped and belong to genera *Agrobacterium*, *Xanthomonas*, *Pseudomonas*, *Erwiniam*, and *Dickeya*. In plants, most bacteria remain in extracellular spaces and use a specialized **Type III secretion system** (T3SS) to deliver bacterial proteins into the plant cell (discussed later in the chapter). The sequential events leading to successful entry of pathogen can be summarized as follows: 1. The microorganism must adhere to the plant surface to initiate infection. Some of the pathogens synthesize adhesive chemicals and secrete them when they come in contact with the host. It involves relatively non-specific molecular interactions between some components of the cell surface of both host and the pathogen. 2. The pathogen must penetrate the epidermal layer of the plant cell if it cannot enter through the wound or stomata. Some fungi secrete cutinase enzyme that disintegrates waxy cuticle covering the epidermis. 3. Once inside, the host tissue invading fungi may grow intercellularly without further invading the cytoplasm of the living cells. They secrete enzymes, such as cellulases and hemicellulases, which enable them to digest the cellulose and other cell wall polysaccharides for their (fungi) growth and multiplication. Phytopathogenic bacteria colonize the apoplast to cause spots, vascular wilts, and blights. During the parasitic relationship, the bacteria reside in the intercellular spaces of various plant organs or in the xylem. The bacteria get surrounded by an extracellular polysaccharide material (EPS) and proliferate in close contact with the plant cell wall. Bacteria cause huge damage to the plant tissue by secreting toxins or cell wall-degrading enzymes. 4. Most of the pathogens secrete toxins in the host tissues. For example, *Erwinia amylovora* (the bacterium that causes fire blight of apple, pear, and other members of Rosaceae family) secretes a 44 kDa glycine-rich hydrophobic toxic protein that kills plant cells by disrupting the ion flux across the membrane and induces hypersensitive response.

### 32.2.2 Hypersensitive Response

Hypersensitive response is a localized response of plant cells to the microorganisms so as to prevent further invasion of the adjacent cells. In most cases microorganisms induce hypersensitive response in the plant without causing any disease. Most microbial pathogen attacks result in ROS formation, changes in cell wall composition and membrane permeability, callose deposition, and induction of phytoalexin or phenolic production leading to programmed cell death.

#### 32.2.2.1 Ion Fluxes
One of the earliest responses of a plant to any kind of invader is a rapid flux of ions. Penetration by pathogen leads to depolarization of the membrane in the cells of host plant followed by rapid movement of ions. Efflux of $H^+$ and an influx of extracellular $Ca^{2+}$ into the cytoplasm of the cell through activated calcium channels are also

evident. As a result, intracellular cytosolic concentration of calcium increases due to its transport from the apoplast or from the stored $Ca^{2+}$ in cell compartments. Increase in cytosolic $Ca^{2+}$ stimulates formation of wall thickenings by cross-linking and callose deposition leading to hypersensitive cell death besides stimulating phytoalexin synthesis.

### 32.2.2.2 Oxidative Burst

Oxidative or **respiratory burst** is the rapid release of ROS from different types of cells. The synthesis of ROS during oxidative burst occurs through NADPH oxidase and extracellular peroxidases. An increase in rate of respiration and activation of a membrane-bound NADPH oxidase is observed. This leads to the rapid generation of superoxide ($O_2^-$) anions which are converted into hydrogen peroxide ($H_2O_2$) either spontaneously or through the action of **superoxide dismutase**. ROS act as antimicrobial agents and strong oxidants which cause massive cell damage (Fig. 32.7). Hydrogen peroxide stimulates the phenylpropanoid pathway and ion fluxes. In addition, it activates genes responsible for synthesizing the enzymes that protect the host cells from these effects. These enzymes include catalase which removes $H_2O_2$ and glutathione S-transferase (GST), which removes dangerous radicals.

### 32.2.2.3 Changes in Cell Wall Composition

Some plants alter composition of the cell wall as soon as pathogen tries to enter. Deposition of lignin or suberin in the primary cell wall may block the entry of pathogen. Activity of enzyme phenylalanine ammonia lyase (PAL), which plays a key role in the synthesis of lignin precursors, is elevated during pathogenesis. Peroxidases play an important role in lignin synthesis as well as other cross-linking reactions in the cell wall of the infected cell. Cross-linking gives strength to the cell wall making it resistant to microbial attack. Tannins and lignin also bind to proteins and deactivate the degrading enzymes released from the fungus. Callose deposition on the inner side of the cell wall helps to slow down the entry of the pathogen. It helps in gene activation and de novo synthesis of defensive agents. Minute papillae composed of callose and lignin are formed beneath the penetration site of the biotrophic fungi to block fungal penetration into the plant cell. There is blockage of plasmodesmatal connections by callose deposition, thereby blocking cell-to-cell movement of viruses. Plants may also modify cell walls by synthesizing and depositing **hydroxyproline-rich glycoproteins (HRGPs)**, such as extensins and arabinogalactan proteins, in walls of the cells in vicinity of invading pathogen. Extracellular HRGPs contribute to the thickening of cell wall in two ways. 1. HRGPs rapidly form covalently cross-linked network generating negatively charged cushions at the membrane-cell wall interface, thereby acting as barriers against the entry of pathogen. 2. De novo HRGP synthesis initiates additional lignin polymerization to further strengthen the cell wall. These thickenings make cell wall resistant to microbial penetration and enzymatic degradation. This may prevent penetration of haustorium into the cell. Another class of extracellular defense-related proteins secreted by plants are **polygalacturonase-inhibiting proteins (PGIPs)**, which have a leucine-rich repeat (LRR) motif. Fungal hyphae, mainly of

## 32.2 Susceptibility and Resistance

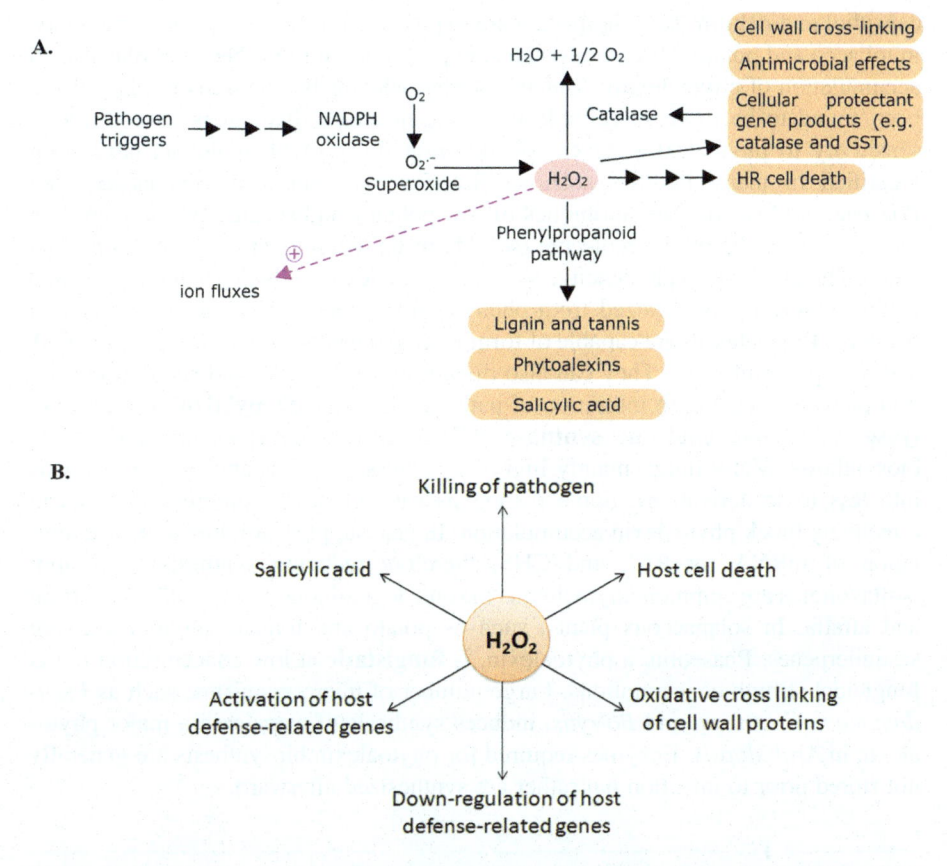

**Fig. 32.7** (a) Steps involved in oxidative burst and its role in inducing plant defense. Superoxide ions are converted into hydrogen peroxide ($H_2O_2$), which acts as a signal promoter in a number of defense responses, viz., cell wall cross-linking, phytoalexin synthesis, and hypersensitive responses (if signal is strong). $H_2O_2$ stimulates ion fluxes and removes itself through enhanced catalase. (+) sign indicates positive feedback. (b) Central role of $H_2O_2$ in defense response of plants to microbial pathogen infections. Response to $H_2O_2$ may be different in different plant species

necrotrophic pathogen, release a specific type of cell wall-degrading enzymes called polygalacturonases (PGs). These PGs cause release of oligogalacturonides with chain length of >8 units which act as elicitors. These are sometimes referred as **damage-associated molecular patters (DAMPs)** which interact with the membrane-localized receptors resulting in signal transduction and expression of plant defense proteins, including PGIPs.

### 32.2.2.4 Phytoalexins

In addition to chitinases and glucanases, plants synthesize phytoalexins in response to pathogen attack because of expression of genes. These include sesquiterpenoids, flavonoids, isoflavonoids, and phenylpropanoids. Genes for the synthesis of capsidiol and rishitin (sesquiterpenes) are expressed in response to pathogen attack in tobacco and potato. Disease resistance in plants has at times been correlated with accumulation of phytoalexins. Mutants of phytoalexins biosynthesis in plants have been found to be more susceptible to pathogen attack. Pathogens have different sensitivity to phytoalexins. Over 300 different types of phytoalexins have been identified in many families such as Brassicaceae, Fabaceae, Solanaceae, and Poaceae. Phytoalexins are antibiotics of low potency and specificity. They are low molecular mass lipophilic molecule (soluble in lipids) and hence can cross plasma membrane and exert their toxicity. Various phytoalexins, which are flavonoids and isoflavonoids, are synthesized from phenylalanine precursor via phenyl propanoid pathway. Phytoalexins are capable of forming free radicals, which can damage DNA and disrupt membranes. They can also disrupt mineral uptake and signal transduction pathways leading to inhibition of pathogen growth. **Phenylalanine ammonia lyase (PAL)** *and* **chalcone synthase (CHS)** are key enzymes in phytoalexins biosynthesis. Many fungi, mainly highly virulent strains, metabolize phytoalexins into less toxic derivatives. Some pathogens can synthesize suppressors that can effectively block phytoalexin accumulation. In pea, suppressor inhibits the accumulation of mRNA for PAL and CHS, therefore reducing synthesis of pisatin. Isoflavonoids are common phytoalexins in some leguminous plants, such as soybean and alfalfa. In solanaceous plants, such as potato and tomato, phytoalexins are sesquiterpenes. Phaseolin, a phytoalexin, is **fungistatic** at low concentration but is fungicidal at high concentrations. Large number of microorganisms, such as *Pseudomonas*, *Alternaria*, and *Botrytis*, induces synthesis of camalexin-a major phytoalexin in *Arabidopsis*. Enzymes required for phytoalexin biosynthesis are generally not stored prior to infection but rather are synthesized afterward.

## 32.3 Plant Defense Mechanisms

Within minutes of attack by pathogens or insects, plants exhibit a local response. An elaborate response is, however, activated within hours in plant tissues away from the initial site of infection or tissue damage. This induced systemic response is determined by the type of causal (attacking) organism. The induced systemic response is distinct in case of pathogens than the one induced by herbivores. In case of pathogens, this response is known as **systemic acquired resistance (SAR)**, which results in the synthesis of **pathogenesis-related proteins (PR proteins)** in plants. PR proteins include various enzymes, such as protease inhibitors (that inactivate the proteolytic enzymes secreted by the pathogens), or lytic enzymes such as β-1,3-glucanase or chitinase that degrade the invading pathogens, or enzymes which interfere with the growth of insects. Plants have innate (passive) immune system, unlike in animals where there are specialized cells which move to the infection site to kill the invading organisms, no such type of cells are reported in plants (Box 32.3). Rather, each cell needs to have its own system to fight the invading organism. Plant's immune responses include enabling other cells and the nearby plants for pathogenic infections in which salicylic acid plays an important role. Innate immunity is of greater efficiency and is the most common form of

## 32.3 Plant Defense Mechanisms

**Box 32.3: Immune System in Plants vs Animals**
Plants, unlike mammals, lack mobile defensive cells and a somatic adaptive immune system. Instead, they rely on the innate immunity of each cell and on systemic signals emanating from the infection sites. Innate immunity system is always present in the cell and is readily mobilized to fight microbes at the site of infection. In other words, innate immunity refers to the immunity that occurs naturally because of genetic factors or physiology and is not caused by infection or vaccination. **Adaptive immune system** (AIS) involves action against pathogens that are able to evade or overcome *innate* **immune system (IIS).** However, when activated, the components "adapt" to the presence of infection agents by activating, proliferating, and creating potent mechanisms for neutralizing or eliminating the microbes. Although plants neither have such a complicated AIS nor antibodies, they have the capacity to differentiate between self and nonself molecules. Plants do have cell surface receptors to identify certain patterns characteristic of pathogens. Such receptors, when activated, trigger the production of chemical substances, such as methyl jasmonates, that may induce both local as well as systemic defense responses. Humans and other vertebrates have a complex immune system, i.e., adaptive immune system. The main components of this system are the organism's ability to biochemically distinguish between its own and foreign cells and remember the specific features of the foreign cell. AIS uses foreign macromolecules as antigens, and using the characteristic features of these molecules, it generates specific antibodies. These antibodies tag the foreign molecules for destruction with the specialized blood cells, called lymphocytes. These specific antibodies then rapidly detect subsequent infection by a particular pathogen and show quick defense responses. Plants have unique and sensitive *chemoperception system* for substances derived from pathogens. It resembles olfactory (smell) perception in animals. Plants as well as animals are able to recognize common patterns on the surface of pathogens using **pattern recognition receptors (PRRs).** Similar to vertebrates, phytohormones play significant roles as signaling molecules as well as immunomodulators (altering the sensitivity of immune system) in the regulation of plant immune response.

resistance to microbes in plants which depends on the ability of plant to distinguish between self and nonself molecules. Plants remain disease-free or suffer from a disease depending on the balance between the pathogen's ability to cause disease and the host's ability to defend itself against it. Plant possesses a wide array of defense mechanisms to counter the undesirable invasion of pathogenic microbes. Some of the defense mechanisms are **constitutive**, which include the structural barriers (cell walls, epidermis layer, trichomes, thorns, etc.) and accumulation of chemical compounds (such as phenolics, nitrogen compounds, saponins, terpenoids, steroids, and glucosinolates) and proteins and enzymes with inherent antimicrobial properties. These compounds confer tolerance or resistance to biotic stresses by not only protecting the plants from invasion, but also by providing them strength and rigidity. The constitutive defense mechanisms are always present whether the plant is attacked or not. Plants may exhibit **qualitative defense** by producing low amount of highly toxic substances, or **quantitative defense**, which refers to production of less toxic compounds in much greater amount. Though qualitative defense requires less investment by plants, it is a more vulnerable strategy since chance for the pathogen to cope up is more in comparison to quantitative defense. On the contrary inducible defense mechanisms are apparent only when infection is underway. It includes the ability of the plant to detect the signal, transduction of the signal followed by the biochemical changes in the plants, which may respond by developing mechanical barriers, such as by triggering callose deposition or by producing metabolites which are toxic to invading pathogens. For example, a pathogen may be prevented to attack a nonhost plant part not only due to constitutive defense mechanisms but also because new cell wall material is synthesized and deposited at the site of infection as the induced response of plants to pathogenic infection. A fungal pathogen, *Gaeumannomyces graminis* var. *tritici*, is not able to attack oat because of synthesis of a toxic triterpene, avenacin, by the roots of oat plant, but it can infect wheat. However, a different form of fungus can infect oats because it has a mechanism to detoxify avenacin produced by the plant. A range of antibiotic compounds produced by plants inhibit or deter pathogens.

Inducible defense responses in the plants include **basal defense mechanisms** and **R gene-mediated defense mechanisms** which are also known as **microbial-associated molecular-pattern-triggered immunity** (MTI) or **pattern-triggered immunity (PTI)** and effector-induced resistance, often referred as **effector-triggered immunity (ETI)**, respectively. PTI is first-level immunity response triggered by plant **pattern recognition receptors (PRRs)** that sense molecular patterns of the pathogen. PTI is in response to the molecular patterns of the pathogen which are common among them, i.e., it is triggered in response to most of the pathogenic bacteria. ETI is the immune response of plant to specific pathogens and occurs in response to specific molecules secreted by the pathogenic bacteria which are known as effectors. ETI is the most successful means of controlling pathogens that are able to evade pattern-triggered immunity (PTI) (Table 32.2). ETI is much faster and quantitatively stronger than PTI. ETI is often associated with a localized cell death or hypersensitive response (HR) to restrict further spread of microbial attack. Cell death occurs through apoptosis and necrosis. The important feature of ETI is the plant's ability to sense microbe-mediated modifications of defense mechanism in the host, whereas PTI is able to sense infectious-self and nonself by guarding against pathogen. In other words, ETI is an efficient defense system for more progressed infections, whereas PTI is important for nonhost resistance and for basal immunity in susceptible host plant cultivars (Fig. 32.8a).

Inducible defense mechanisms include the following phases:

1. Most microorganisms have molecules at their surfaces that are recognized by receptors present on plant cell membrane, and once pathogen is identified, the activation of defense mechanism in the host plant is triggered. These are also

## 32.3 Plant Defense Mechanisms

**Table 32.2** Differences between pattern-triggered immunity (PTI) and effector-triggered immunity (ETI)

| Pattern-triggered immunity (PTI) | Effector-triggered immunity (ETI) |
|---|---|
| General type of defense response common among all pathogens | Highly specific type of defense response and is different among different pathogens |
| Plant can detect molecular signatures of all pathogens | It is gene-for-gene defense response. Plant can detect molecular signature of specific pathogen |
| The molecular signature of pathogen is recognized by pattern recognition receptors (PRRs) | It is initiated by the secretion of effector molecule called Avr (avirulence) protein and is detected by plant resistance gene R |
| Pattern recognition leads to mitogen-activated protein cascade, a signal transduction pathway common to all organism | The R gene encodes Ser/Thr protein kinases which interacts with nucleotide-binding leucine rich proteins (NB-LRRs) |
| The signal cascade results in the activation of several transcription factors which control defense genes | The signaling complex interacts with kinase cascade and results in the activation of hypersensitive response and rapid cell death around the site of infection |

referred as PAMP-triggered immunity (PAMP-pathogen-associated molecular patterns) or pattern-triggered immunity PTI. These defenses are also referred as basal defense mechanisms.
2. The basal defense mechanism puts a selective pressure on potential pathogens to produce effector molecules which interfere with basal defense mechanisms, further enabling pathogen to attack the plant. This is referred as **effector-triggered susceptibility (ETS)**.
3. Successful colonization of pathogen following ETS triggers further defense mechanisms in the host plant, which is known as R gene-mediated defense or effector-triggered immunity (ETI). This defense mechanism of the host plant is stronger than PTI. This defense mechanism is known as R gene-mediated defense, or effector-triggered immunity (ETI). This defense mechanism of the host plant is stronger than PTI. Receptors for these effector molecules are coded by resistance (R) genes of the host plant.
4. R proteins provide selective pressure on the pathogens to further produce effector molecules that are not recognized by the host plant. If a pathogen is able to produce such effector molecules not recognized by R gene-mediated defense mechanism, it will be able to attack and exploit the plant for its use, causing disease (Fig. 32.8b).

### 32.3.1 Pathogen- or Microbial-Associated Molecular-Pattern (PAMP/MAMP)-Triggered Immunity (PTI)

PTI (formerly called basal or horizontal disease resistance) can be considered as the primary driving force of plant-microbe interactions. The basal defense mechanisms are elicited by the molecules that are common in most of pathogenic species. This

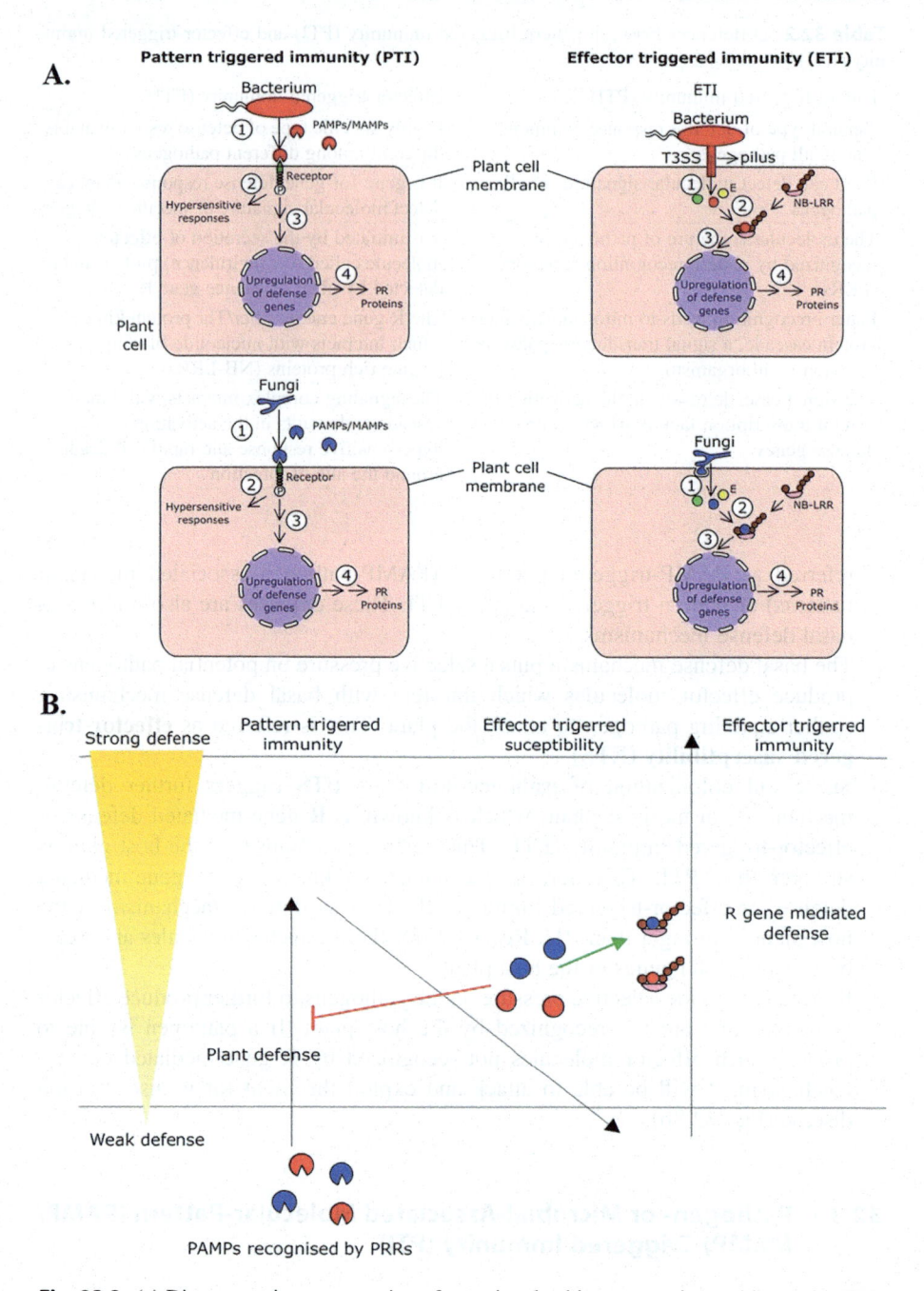

**Fig. 32.8** (a) Diagrammatic representation of steps involved in pattern-triggered immunity (PTI) and effector-triggered immunity (ETI), by bacterial and fungal pathogens, respectively. (b) Phases of inducible defense mechanism

includes 1. **microbial- or pathogen-associated molecular patterns (MAMPs/ PAMPs)**; these molecules are highly conserved across larger groups of pathogens and are indispensable for their survival. These molecules do not exist in the host, such as fungal chitin, xylanase or bacterial flagellin, lipopolysaccharides, and peptidoglycans, and 2. the other correspond to a compromised "self," also called **damage-associated molecular patterns (DAMPs)**. Both PAMPs and DAMPs stimulate the plant defense responses and are known as **elicitors**. An elicitor is a molecule that triggers defense responses in the host cells. Many molecules which are released either by the plants or pathogens act as elicitors. The elicitors released by plants are known as **endogenous elicitors**, while those released by pathogen are **exogenous elicitors**. Fungal wall fragments which are formed when plant enzyme (chitinase) degrades the fungal cell wall act as exogenous elicitors. Plant cell wall fragments which are formed by the enzymes (cutinase) secreted by pathogen act as endogenous elicitors. A large proportion of both kinds of elicitors are common to many plant-pathogen interactions. Elicitors can also be categorized in two different classes on the basis of their specificity, viz., **general or non-specific elicitors**, which provide general resistance and are not different in their effect on different plant species. Others are **specific elicitors**, which are formed by specialized pathogens and function only in plants carrying disease resistance genes. It is only when the plants show defense responses to such elicitors plant-pathogen interactions begin (Fig. 32.9). Elicitors are active at very low concentrations (in nanomolar range); they bind to receptors localized on plant cell membrane and specificity of the host-pathogen interaction lies in the chemical nature of the elicitor. Elicitor-receptor binding triggers the signal cascade which is responsible for expression of plant defense genes responsible for basal defense.

Basal defense response in plants is the first defense against pathogens triggered by conserved elicitor molecules (PAMPs and DAMPs). PAMPs include molecular patterns of the bacterial cells such as bacterial **flagellin** (flg22) and **EF-Tu** (elf18), **fungal chitin** (CEBiP), and oomycetes heptaglucan (Fig. 32.10). These microbial elicitors are recognized by receptor proteins, located on the plasma membrane of the plant cell, which are referred as pattern recognition receptors (PPRs). PRRs belong to either the receptor-like kinase (RLK) or receptor-like protein (RLP) families, which are membrane-bound extracellular receptors. RLPs resemble the extracellular domains of RLKs but lack the cytosolic signaling domain, whereas RLKs have both extracellular and intracellular kinase domains. PPRs, such as **CERK1** (chitin elicitor receptor kinase 1), FLS2, and EFR protein, have been identified in plants (rice, wheat, and *Arabidopsis*) for fungal chitin, flagellin (flg22), and elongation factor (EF-Tu), respectively (Fig. 32.10).

**Chitin**, a homopolymer of β-(1,4)-linked N-acetylglucosamine (GlcNAc) units, is a major constituent of fungal cell walls and is a classical PAMP. Chitin is an ideal point of attack during plant defense responses since glucosamine polymers are not found in plants. Upon pathogen contact, chitinases (hydrolytic enzymes) are released by the host plant which breaks down the fungal cell wall chitin polymers. These fungal chitin fragments are recognized by CEBiP (chitin elicitor-binding protein) and CERK1 (chitin elicitor receptor kinase 1) where both are located on the plasma

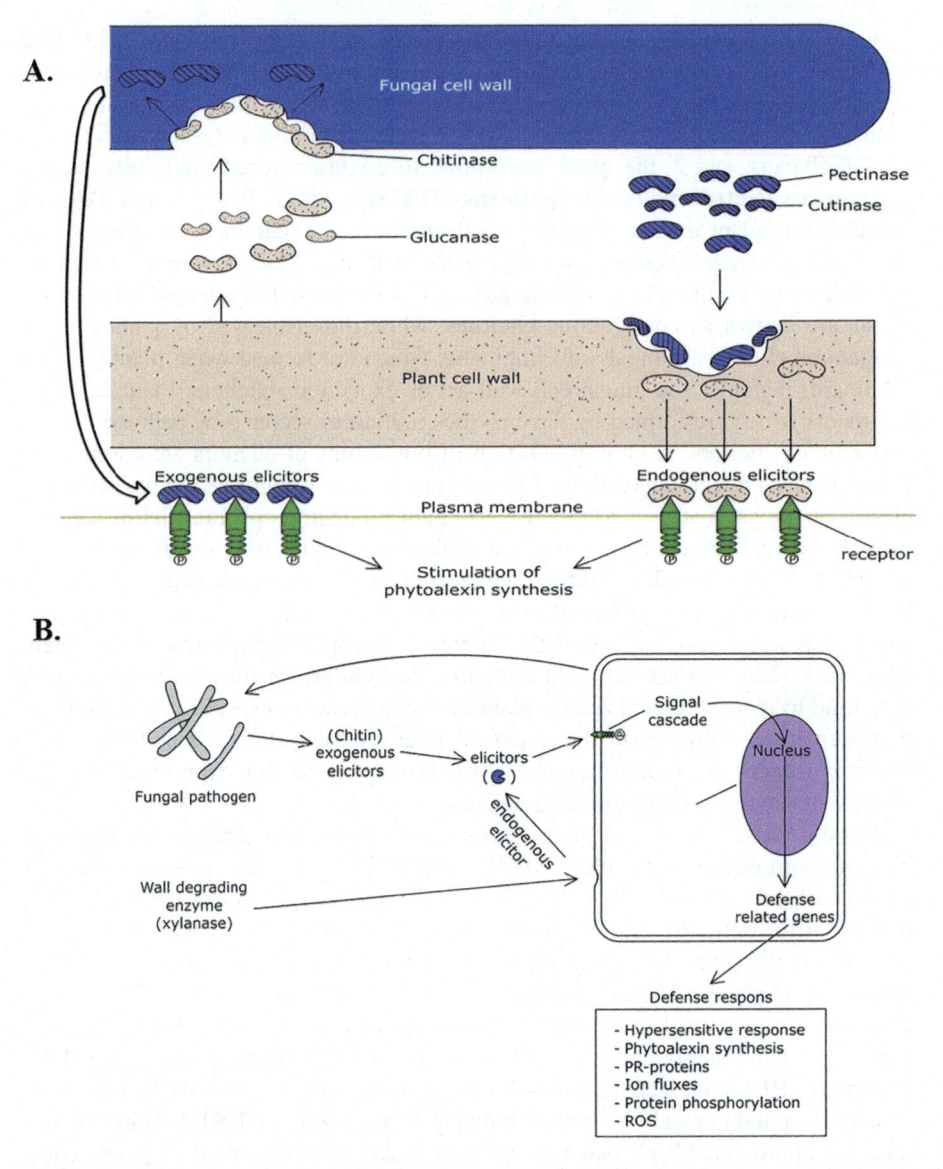

**Fig. 32.9** (a) Changes in the metabolism during plant-fungal interaction leading to production of elicitors and stimulation of phytoalexin synthesis. Plant cell releases glucanase and chitinase which lead to degradation of fungal cell wall. The fungal wall fragments act as elicitors (exogenous elicitors) and trigger phytoalexin synthesis (hypersensitive response). At the same time, fungi also release cutinase and pectinase which degrade plant cell wall resulting in the formation of elicitors (endogenous elicitors) and triggering hypersensitive response. (b) Diagram depicting elicitor-receptor binding which triggers signal transduction cascade and expression of plant defense genes responsible for basal defense

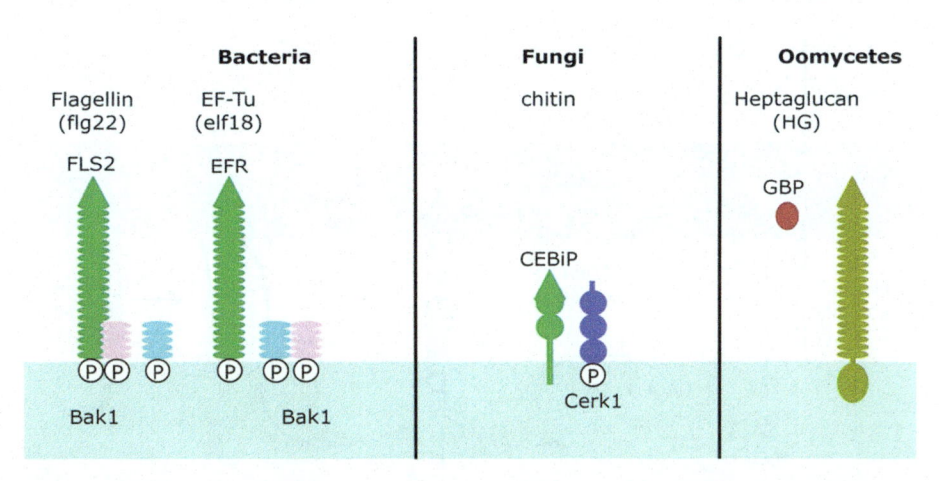

**Fig. 32.10** Pathogen-associated molecular patterns (PAMPs) of the bacterial cells such as bacterial flagellin (flg22) and EF-Tu (elf18), fungal chitin and oomycetes heptaglucan. These microbial elicitors are recognized by pattern recognition receptors (PRRs) such as FLS2, EFR protein, and CERK1

membrane of the plant and possess an extracellular domain that contains LysM motifs which bind chitin fragments. The first chitin-binding PRR was identified in rice as the *lysine motif (LysM)-RLP* and was named *chitin* elicitor-*binding* protein (*CEBiP*). *CEBiP* is a glycoprotein localized on the plasma membrane. Intracellular protein kinase domain of CERK1 is involved in downstream signaling. CERK1 of *Arabidopsis* is also involved in the recognition of bacterial peptidoglycan, a polysaccharide that has a repeating N-acetylglucosamine unit.

**Flagellin** is a bacterial flagellar protein with highly conserved N-terminal domain, flg22, which is a 22-amino acid peptide derived from N-terminal fragment of flagellin. In *Arabidopsis*, the flagellin-sensing FLS2 protein present in plasma membrane detects flg22. Another receptor in *Arabidopsis* is EF-Tu receptor (EFR). The EFR protein perceives acetylated N-terminal 18 amino acids (elf18) of elongation factor-Tu.

PAMPs induce early responses, within minutes to hours, including rapid ion fluxes across the plasma membrane, oxidative burst, activation of mitogen-activated protein kinases (MAPKs), and calcium-dependent protein kinases (CDPKs), inducing defense-related genes involving pathogen cell wall/cell membrane-lysing enzymes, e.g., chitinases, glucanases, and defensin. Other responses may include formation of antimicrobial phytoalexins, plant cell wall modifications, e.g., formation of papillae rich in callose, lignin biosynthesis or change in cell wall proteins, and pectic polysaccharide structure. Once recognition of elicitors by PRRs takes place, it triggers an influx of calcium ions into the cell and production of ROS such as superoxide ($O_2^-$) and hydrogen peroxide. The plasma membrane-spanning NADPH oxidase transfers electrons from NADPH within the cell across the plasma membrane and uses them to convert extracellular oxygen to the superoxide anions

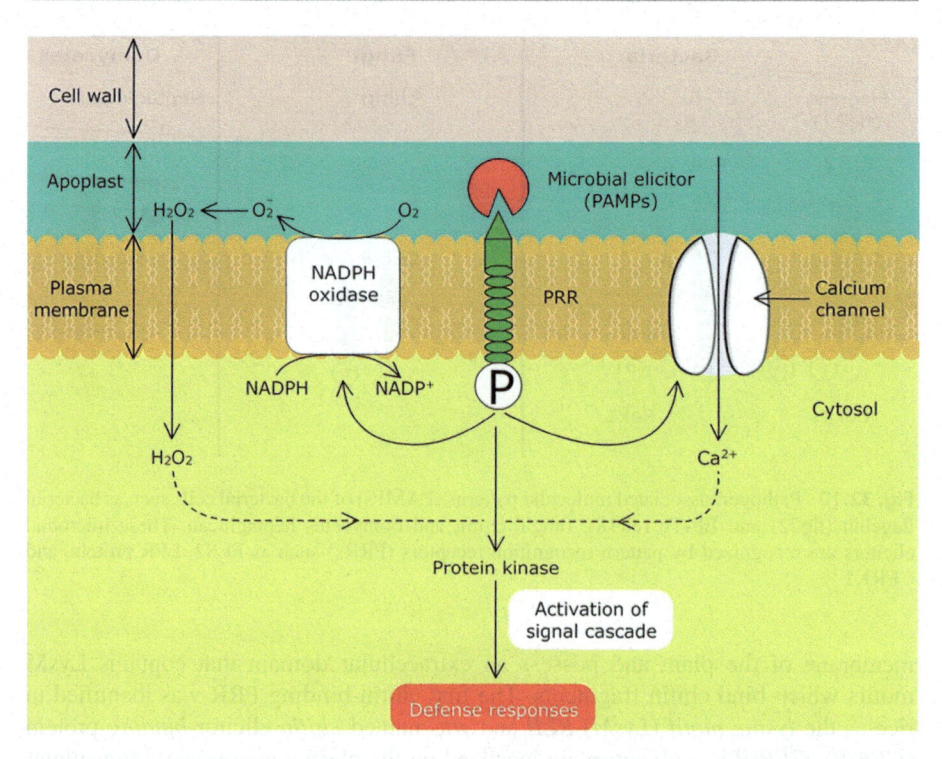

**Fig. 32.11** Events in recognition of elicitor or PAMP by a plant cell. The elicitor or PAMP is recognized by pattern recognition receptor (PRR)

($O_2^-$), which is rapidly converted into hydrogen peroxide ($H_2O_2$), either spontaneously or through the action of **superoxide dismutase**. Activation of NADPH oxidases involves the combined action of CDPKs and other protein kinases. The ROS generated by the extracellular peroxidases serve as antimicrobial agents and in cross-linking specific polymers to cell wall to improve rigidity. Thus, PAMPs trigger inducible defenses which include production of toxic chemicals and pathogen-degrading enzymes, and deliberate cell suicide is used conservatively by the plants (Fig. 32.11). Tomato plant produces tomatine, and only certain pathogens which produce a glucosidase that can toxify tomatine can infect tomatoes. For developing crops with higher resistance or inducible resistance, identification of several potential microbial molecules that act as PAMPs would increase chances of identifying more potential host plant PRRs.

## 32.3.2 Effector-Triggered Responses

Plant pathogens produce a wide range of molecules that enhance their successful colonization in the host plant. Any molecule made by the pathogen that enhances its

## 32.3 Plant Defense Mechanisms

**Table 32.3** Plant-pathogen effector molecules produced by pathogens and their role

| Organism | Effector molecule | Role of effector molecules in the regulation at cellular/molecular level |
|---|---|---|
| Tobacco mosaic virus | P50 | Viral replicase |
| Potato virus X | CP | Coat proteins |
| Phytophthora infestans | Avr3 | Prevention of cell death |
| Pseudomonas syringae | Avrpto3 | Protein kinase function to suppress defense response and enhance the bacterial growth and plant necrosis |

ability to overcome basal defense of the plant and make it to be able to colonize, grow, and reproduce in host plant is called an **effector molecule**. These molecules manipulate host's cell structure and function, facilitating growth and reproduction of pathogens besides stopping the basal defense response in the plant. These can be either toxic or act as elicitors. In case interaction between plants and pathogens results in disease, these are referred as "compatibility factors." On the contrary, if the interactions do not lead to disease, these are described as incompatible. Many effector molecules either alone or in combination with other molecules suppress host pathogen-triggered immunity (PTI) or effector-triggered immunity (ETI). These molecules are produced by pathogen usually only in the presence of host. Plant-pathogen interaction involves effector molecules (Table 32.3) which target distinct sites in plants, either apoplastic or cytoplasmic, and are known accordingly. Apoplastic effectors are released in extracellular space and target the membrane surface receptors, while cytoplasmic receptors are released in the cytoplasm of the host cell through specialized structures. These belong to diverse classes, such as enzymes, toxins, and growth regulators.

There are 10–30 types of effector molecules required by the plant to detect and respond to PAMPs, which facilitate infection by interfering signaling pathways of the host plants. Several pathogenic bacteria produce enzymes that degrade components of plant cell walls. These include hydrolytic enzymes such as cutinases, cellulases, xylanases, pectinases, proteases, etc. Some may produce enzymes that regulate protein activity in the cell through phosphorylation and dephosphorylation. One common activity of the pathogens is their ability to act and protect against hydrolytic enzymes produced by the plants. One of the examples is an effector produced by *Cladosporium fulvum*, an extracellular pathogen of tomato which does not produce any haustoria or haustoria-like structures. *C. fulvum* enters either through stomata or subsists on leaking nutrients. The fungus is protected by the effector molecules which are small cysteine-rich proteins thought to function exclusively in the apoplast. The effector inhibits cysteine proteinases that are important components of plant defense in tomato. Some of the effectors act by binding to chitin fragments, so that their detection is avoided by the PTI of the plant. Some pathogens produce cell wall-degrading enzymes only when bacterial populations on the host

**Box 32.4: Quorum Sensing**
Bacteria use a survival strategy by sending signals to other bacteria to act in group for their successful colonization in the host. This survival strategy is known as "**quorum sensing**." The capacity to respond collectively as a group has obvious advantages such as to be able to migrate to a more suitable environment for better nutrient supply and antibiotic production and to adopt new modes of growth, such as sporulation or biofilm formation. Quorum sensing bacteria release chemical signal molecules called **autoinducers** that increase in concentration as a function of cell density. Gram-negative bacteria (e.g., *Pseudomonas syringae* and *Erwinia carotovora*) use acylated homoserine lactones (AHL), and gram-positive bacteria (e.g., *Xanthomonas oryzae*) use processed oligopeptides (Ax21) as autoinducers to communicate. The pathogenicity of *E. carotovora* depends on the production of various plant tissue-degrading enzymes, including pectate lyases, polygalacturonase, cellulase, and proteases necessary for colonization of bacteria in the host. Therefore, *E. carotovora* uses quorum sensing to ensure that exoenzyme production does not occur until sufficient bacteria have accumulated for successful tissue destruction and evasion of plant defenses. Quorum sensing is a novel target for antimicrobial therapy, e.g., biofilm production by bacteria makes them resistant, if signal involved in quorum sensing of biofilm production is blocked.

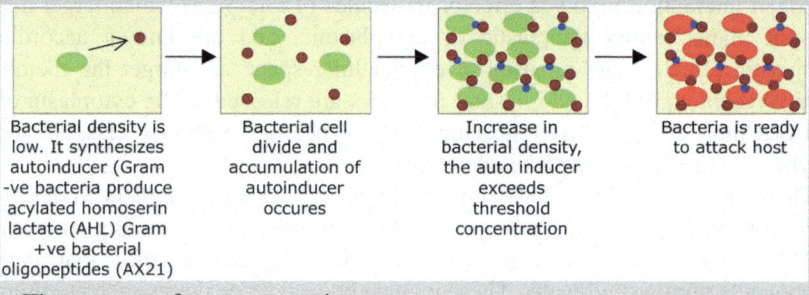

| Bacterial density is low. It synthesizes autoinducer (Gram -ve bacteria produce acylated homoserin lactate (AHL) Gram +ve bacterial oligopeptides (AX21) | Bacterial cell divide and accumulation of autoinducer occures | Increase in bacterial density, the auto inducer exceeds threshold concentration | Bacteria is ready to attack host |
|---|---|---|---|

The process of quorum sensing

reach a certain density. The characteristic feature of bacteria to monitor its own as well as density of other bacteria is known as **quorum sensing** (Box 32.4). Cell wall-degrading enzymes may act both as elicitors and as enzyme. Xylanase released by fungi causes necrosis, ethylene synthesis, and electrolyte leakage in tobacco. Fungal **endopolygalacturonase** cause both necrosis and induction of phytoalexins biosynthesis. It degrades plant cell wall and induce production of oligogalacturonides as well which are responsible for inducing plant defense. *Erwinia*, a necrotrophic bacterium, uses pectic enzymes (pectate and pectin lyase) and causes cell death

## 32.3 Plant Defense Mechanisms

and tissue disintegration by the hydrolysis of polygalacturonases. Pathogens produce phytotoxins that may be toxic to all plants or only to a particular plant species. Most pathogens produce at least one compound that is toxic to plants, while there are pathogens that produce many toxins. Majority of phytotoxins produced by phytopathogenic bacteria and fungi are non-specific. The site of action of non-specific toxins is known. Phytotoxins are mainly secondary metabolites though some are oligopeptides or proteins. Many elicitors of diverse nature are often phytotoxic. Some toxins are required for pathogenicity, whereas others elicit disease symptoms. For example, many of the bacteria damage plant tissues by secreting extracellular polysaccharides (EPSs) or cell wall-degrading enzymes. The extracellular polysaccharides (EPSs) secreted surround the growing bacterial colony aid bacterial virulence but are not required for pathogenesis. It may help in saturating intercellular spaces with water or blocking xylem to produce wilt symptoms.

Some pathogens such as ascomycetes fungal genera *Cochliobolus* (synonym *Helminthosporium*) and *Alternaria* produce host-specific toxins. *Cochliobolus carbonum* cause Northern leaf blight disease of maize. It makes a host-selective toxin called **HC toxin** (Fig. 32.12), which inhibits histone acetylase affecting expression of the genes involved in defense against the fungus. Mutant bacteria with impaired gene for the biosynthesis of HC toxin are not pathogenic. The TOX2 locus of *Cochliobolus carbonum* race 1 which controls production of HC toxin

**Fig. 32.12** Production of host-specific toxins (such as HC toxin) by nectrotrophic fungi. Toxin produced by *Cochliobolus carbonum* (a fungal pathogen on maize). **1.** HC toxin enters the responding plant cell and **2.** inhibits histone deacetylase activity which, in turn, inhibits transcription of defense genes. **3.** *Hm1*-resistant maize plants produce an HC toxin reductase which detoxifies HC toxin by reducing carbonyl group of the side chain of L-Aeo (2-amino-9,10-epoxy-8-oxo-decanoic acid)

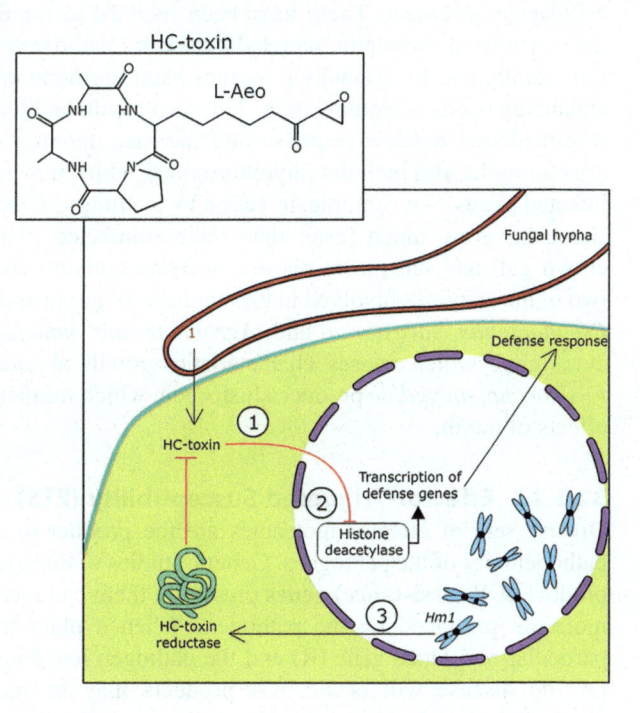

includes a gene encoding HC toxin synthetase. The complementary resistance gene in maize, called Hm, encodes an enzyme called HC toxin reductase that provides resistance to maize by reducing production of HC toxin. Mutant strains of *C. carbonum*, which do not produce toxin, are nonpathogenic. Maize genotype with Hm1 gene, which encodes an enzyme that can detoxify HC toxin, is resistant *to C. carbonum*. On the contrary, there are non-specific toxins, such as **fusicoccin**, which are produced by *Fusicoccum amygdale*. Fusicoccin inactivates membrane-bound $H^+$-ATPase proton pump. As a result, rate of proton pumping across plasma membrane of guard cells cannot be modulated. Because of this, stomata will not respond to environment stimuli, and irreversible opening of stomata leads to wilting of plants. Another compound tabtoxin is produced by pathovar strain of *Pseudomonas syringae* var. *tabaci*, which is a dipeptide and consists of two amino acids, i.e., threonine and a non-protein amino acid tabtoximine. A peptidase from the plant hydrolyzes the tabtoxin resulting in the release of tabtoximine from the dipeptide which is a potent inhibitor of glutamine synthetase, an important enzyme required for nitrogen metabolism. This enzyme is important for re-assimilation of ammonia released during photorespiration in green leaves. In some instances, toxins released by the pathogen create pores in the membrane resulting in the leakage of nutrients. Toxins produced by biotrophs are under different selection pressures. These have to avoid detection by the host plant while maintaining the defense suppressive function. Although effector molecules are produced by pathogen genes, they function in cellular environment. These have been selected to mimic plant molecules. One of the examples is coronatin, secreted by several pathogens of *Pseudomonas syringae*. Structurally and functionally it mimics plant hormone jasmonyl isoleucine, thereby enhancing bacterial colonization. This also results in blocking induction of salicylic acid-mediated defense response and increase opening of stomata. Third class of effector molecules includes phytohormones, which may result in abnormal growth in infected plants. For example, infection by the fungus *Gibberella fujikuroi* causes rice shoots to grow much faster than their uninfected plants. Agrobacterium causes crown gall and hairy root disease in many eudicots by inserting genes encoding two of the enzymes involved in the synthesis of auxin and cytokinins. Pathogens like *Pseudomonas savastanoi* and *Agrobacterium tumefaciens* produce auxin and cytokinins, which causes characteristic growth abnormalities in plants. Fungus *Fusicoccum amygdale* produces fusicoccin which mimics many of the physiological effects of auxin.

### 32.3.2.1 Effector-Triggered Susceptibility (ETS)

Diverse sets of effector molecules are the product of **avr genes** that determines pathogenicity of the pathogens. Genetic studies of diseased plants have revealed that product of **R (resistance) genes** present in them can recognize one specific effector molecule produced by the pathogen. When a plant has a dominant allele of a particular resistance gene (R) and the pathogen has dominant avirulence *avr* gene, i.e., no disease will occur. Avr products may be recognized intracellularly or extracellularly, depending on the pathogens. On the other hand, if there is no

## 32.3 Plant Defense Mechanisms

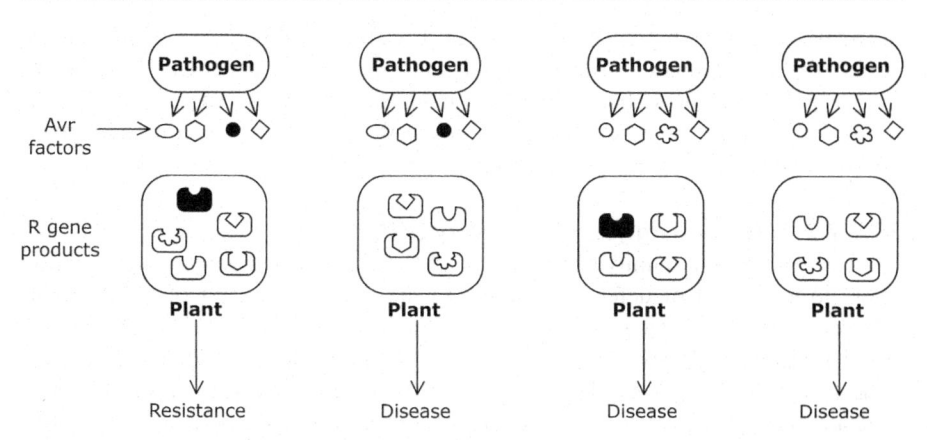

**Fig. 32.13** Hypothetical representation of gene-for-gene model. Strains of pathogens produce diverse types of effector molecules. The first strain produces Avr effector (product of *avr* gene shown by solid circle) compatible with R gene product of the plant. As a result, plant shows resistance. Though the second strain of bacteria may produce effector molecule similar to first strain but compatible reaction with the host does not occur in the absence of compatible R gene product of the host. The remaining two figures also indicate incompatible reactions of the bacteria and the host either because of the absence of compatible effector produced by the bacteria or the compatible R gene produced by the plant. As a result, plant shows susceptibility to the pathogen as depicted in the last three figures

recognition of product of two genes (i.e., Avr of bacteria and R gene product of the host), then the plant is susceptible to pathogen. Thus, the plants that lack R gene will not be able to recognize the effector molecule produced by bacteria and hence will not initiate defense responses and thus become diseased. Presence of *avr* gene, complementary to R gene product of host plant, converts a pathogenic bacterium into an avirulent. H. H. Flor described this as **gene-for-gene model**, which explains the genetic basis for vertical plant resistance to pathogen (Fig. 32.13). He stated that genes in the pathogen (*avr*) that determine the pathogenicity have a one-to-one corresponding match to specific gene in the host that determines resistance (R gene). He further found out that host resistance is usually dominant to virulence.

The bacterial Avr effectors delivered in plants have been identified to be enzymes like phosphatases, proteases, phosphoserine lyases, or E3 ubiquitin ligases which induce signaling of defense in the host. Another gene or gene clusters that are required for pathogenicity on compatible host are referred to as hrp (hypersensitive reaction and pathogenicity). Bacteria which need to develop a secretion system (Type III) so as to inject the effector molecule inside the host cell are known as T3SS (Type III secretions system). **Hypersensitive response and pathogenicity cluster** (hrp) genes produce proteins in T3SS bacteria which form Hrp secretion apparatus. The Hrp secretion apparatus encircles the outer and inner bacterial membranes, and a pilus is produced which penetrates into the plant cell. Once Avr effector is injected into the plant cell, Hrp proteins are not needed. The transcription of both *avr* and *hrp* genes is nutritionally regulated. The hrpN gene of *Erwinia amylovora* encodes a cell surface-associated protein called harpin that causes necrosis and other hypersensitive responses. The effectors produced by fungi may be secreted in the intercellular spaces and then taken up by the host cells. Some of the fungal pathogens have been found to induce synthesis of specific membrane transporters which have been found to aid in the secretion of non-protein toxins into the intercellular spaces of the plants.

### 32.3.2.2 Effector-Triggered Immunity (ETI)

Effector-triggered immunity (formerly called *R* gene-mediated or vertical resistance) refers to triggering of defense signaling in the plant as a result of incompatible interaction of Avr products of the pathogen and the R gene products in the host plant. It is based on highly specific, direct, or indirect interaction of pathogen and plant gene products (Box 32.5). The majority of R proteins (product of R gene) that mediate race-specific resistance and trigger ETI have diagnostic structural motifs. Different R proteins have two basic roles to play. First, they directly or indirectly recognize the effector molecule, and secondly, they activate downstream signaling, leading to induction of various defense responses. Most R genes are constitutively expressed at low level during normal growth. Once the pathogen attacks expression of R gene is upregulated. R proteins are involved in the detection of diverse pathogens, including bacteria, viruses, fungi, nematodes, insects, and oomycetes. ETI mostly relies on the endogenous NB-LRR protein products encoded by the R genes (resistance genes). The interaction is complicated as it involves two genomes, both of the plant and of pathogen. NB-LRR is named due to the presence of nucleotide-binding (NB) and leucine-rich repeat (LRR) domains. LRR regions are present in over 14,000 proteins in viruses, bacteria, archaea, and eukaryotes. Most

---

**Box 32.5: Host Strategies to Detect Pathogen Effector**

During infection, activity of the bacterial effectors on their target proteins goes unchecked in a susceptible host that lacks resistance to a particular pathogen. This renders the host susceptible to infection by that specific pathogen. On the contrary, in a resistant host, pathogen effectors can be sensed directly or indirectly by the products of **resistance (R) proteins** produced by the plant, which results in a protective immune response. However, many effector molecules are recognized indirectly by R proteins. According to **guard hypothesis**, the R proteins complex with the host target proteins. R proteins act as "guard" and the target molecules as guardee. The effector molecule released by the pathogen liberates the R protein (guard) from the complex, allowing the R protein to engage signaling pathways and induce a defense response. This hypothesis predicts that multiple pathogen effector proteins may interact with a common host target that can be guarded by more than one R protein. In *Arabidopsis*, RIN4 protein is targeted by three effector molecules delivered into cell by Type III secretion system. RIN4 is a 211-amino acid, plasma membrane-associated protein. It is guarded by at least two different R proteins.

LRR proteins are involved in protein-ligand and in protein-protein interactions. These LRR proteins include plant and mammalian immune response. Several classes contain LRR-containing receptor-like kinases (LRR-RLKs), LRR-containing receptor-like proteins (LRR-RLPs), nucleotide-binding site LRR (NBS-LRR) proteins, and PGIPs. They provide an early warning system for the presence of potential pathogens and activate protective immune signaling. In addition, they act as a signal amplifier in case of tissue damage. Most of R genes in plants encode nucleotide-binding site leucine-rich repeat (NBS-LRR) proteins, the largest proteins known in plants, ranging from about 860 to about 1900 amino acids. These proteins have at least four distinct domains joined by linker regions. These are N-terminal effector domain, central nucleotide-binding site (NBS), and C-terminal leucine-rich repeat (LRR) domains as well as variable amino- and carboxy-terminal domains, which largely vary in plants. Two major subgroups that have distinct N-terminal domains are generally recognized: (1) one group with a toll-interleukin 1 receptor (*TIR*) domain are called TNLs, and (2) those with a coiled-coil (CC) domain are called CNLs. The crystal structures of more than 20 LRR proteins have revealed that LRR domains characteristically contain a series of β-sheets that form the concave face, shaped like a horseshoe or banana. The LRR motif of cytoplasmic and extracytoplasmic plant R proteins contains several leucine or other hydrophobic amino acids at regular intervals over 23 or 24 amino acid lengths, respectively. Within the part of these repeats, leucine residues occur at every second or third position, i.e., X-Leu-X-X-Leu-X-Leu-X, where X is another amino acid. This part of the repeats forms a structure within protein called a parallel β-sheet, in which leucine is in the interior of the protein, while X residues are exposed to the exterior and interact with other proteins. Single amino acid variation on the outer surface can affect the ability of the LRR domain to bind target proteins. Plant NBS-LRR proteins act through a network of signaling pathways and induce a series of plant defense responses, such as activation of an oxidative burst, calcium ion fluxes, mitogen-associated protein kinase cascade, induction of pathogenesis-related genes, and the hypersensitive response.

### 32.3.3 Signal Transduction

Once a signal is perceived either through recognition of elicitors or effectors, it is transduced in the form of protein activation. Protein phosphorylation plays an important role during pathogen attack. At least two classes of protein kinases are activated: the **mitogen-activated protein kinase (MAPKs)** and **calcium-dependent protein kinases (CDPKs)**. The CDPK class of protein kinases is present in plants only, whereas the MAP kinases are signaling molecule in all eukaryotes. Signaling cascades of MAPKs and CDPKs activate the expression of first set of genes that encode regulatory and signaling proteins. These proteins, in turn, activate expression of second set of genes which encode defense proteins that prevent pathogen attack. The *Arabidopsis* genome contains 110 genes for MAPK pathway components. *Arabidopsis* contains over 70 WRKY proteins which are responsive to

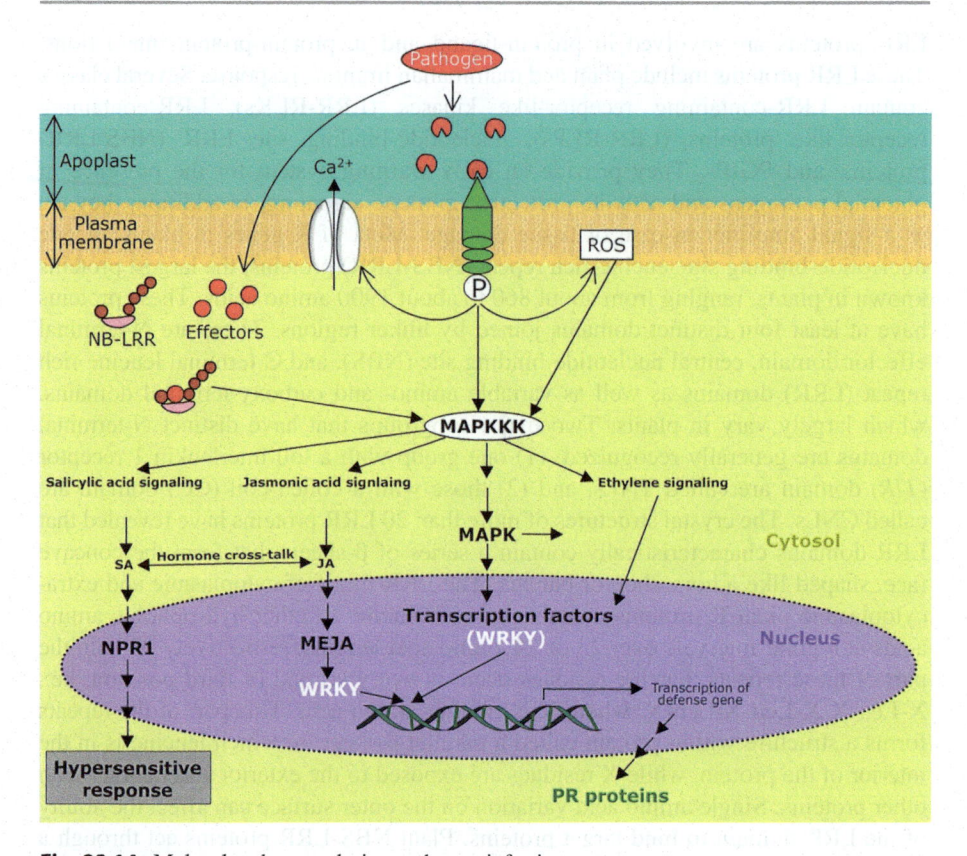

**Fig. 32.14** Molecular changes during pathogen infection

pathogen infection or salicylic acid treatment. **WRKY proteins** are a large class of WRKY domain-containing sequence-specific, DNA-binding transcription factors found in plants. WRKY proteins regulate growth, hormone signaling, secondary metabolism, response to various stresses, seed germination, leaf senescence, and the synthesis of phytoalexins and other defense mechanisms against pathogens in *Arabidopsis*. All WRKY proteins recognize and bind to the core W box sequence TTGACC/T in the promoter of the responsive genes (Fig. 32.14).

### 32.3.4 Systemic Acquired Resistance (SAR)

Sessile nature of plants requires an efficient signaling system capable of detecting, transporting, and interpreting signals produced at the plant-pathogen interface, and **systemic acquired resistance (SAR)** and **induced systemic resistance (ISR)** provide a practical means to confer a fitness advantage to plants in conditions of high disease pressure, since plants which are primed more quickly and effectively activate their defenses ahead of pathogen attack. In SAR, defense proteins accumulate not only at the site of infection but also systemically in uninfected tissues and/or uninfected plants. SAR provides long-term defense against a broad spectrum of pathogens. The SAR is analogous to the development of acquired immunity in animal systems. Another form of induced resistance which is similar to SAR in many aspects is ISR. SAR and ISR are two types of resistance which prepare plants against subsequent pathogen infection. Selected strains of plant growth-promoting rhizobacteria suppress infection by disease-causing pathogens by providing systemic resistance by antagonism, which is known as ISR. SAR and ISR resemble in the fact that in both the uninfected parts of the plant becomes resistant toward a broad spectrum of pathogens. SAR makes plant respond more quickly if it is attacked again. However, unlike human or mammalian immune response (antibodies persist in the body), SAR is neither very specific nor long lasting. In addition to SA production accompanying SAR, the accumulation of specific set of pathogenesis-related proteins (PRP) occurs both at the site of initial infection and in the uninfected tissue locally and at a distance from the site of infection. Induction of ISR may be due to induction of SA-triggered signaling pathway by the rhizobacteria. In ISR neither PR proteins nor SA seem to be involved. It is dependent on JA and ET (ethylene) signaling. MYC2 may be playing a potential role in signaling JA gene expression during rhizobacteria-mediated ISR against *Pseudomonas syringae* var. tomato DC3000. NPR1 gene is also needed besides MYC2 transcription factor. Combination of ISR and SAR can be commercially exploited to provide resistance against disease-causing agents.

### 32.3.5 Phytohormones in Plant Defense Response to Pathogens

Plant defense against pathogen attack involves many signal transduction pathways that are mediated by a network of phytohormones. Three most significant phytohormones for plant defense response against pathogens include salicylic acid (SA), jasmonic acid (JA), and ethylene (ET). Salicylic acid, a benzoic acid derivative, is an important phytohormone in regulation of the plant defense. It is derived from phenylpropanoid pathway and is similar in structure to aspirin (acetylsalicylate). It plays a role both in defense responses around the site of infection and in the establishment of SAR. In transgenic plants that constitutively express a bacterial *nahG* gene, which encodes salicylate hydroxylase responsible for converting SA into catechol and water, reduced level of SA is observed. Reduced level of SA results in weak R gene-mediated resistance. In *Arabidopsis*, activation of the SA pathway has been shown to be important in both basal and R gene-mediated biotrophic and hemibiotrophic pathogen defense. SA contributes HR-associated resistance. In tobacco, SA increases resistance in plants infected with TMV. Following treatment of *Nicotiana* plant with JA or SA, systemic resistance to TMV is enhanced. ETI can trigger SAR through both local and systemic synthesis of SA which, in turn, regulates transcription of genes encoding PR proteins. A change in amino acid homeostasis is responsible for SAR mediated by ETI. Amino acids are precursors of a large number of plant secondary metabolites involved in defense, including signaling molecules, cell wall components, and anthocyanins. Plants use amino acids and their derivatives for SA synthesis in order to survive after pathogen invasion. SA is also stored in vacuoles as nontoxic glucose esters, i.e., salicylate glucose ester (SGE) which can be hydrolyzed to release free SA during pathogen attack. Methylation of SA generates methyl SA (MeSA) which is the mobile form that can travel from the infected leaf to the systemic tissues of the plant where it activates resistance after being converted back to SA. MeSA is believed to be directly emitted into the atmosphere, and only a small amount is retained in the leaves. MeSA is converted to SA by the enzyme methyltransferase. Under stress conditions accumulation of SA alters plant metabolism so as to reduce HR response to the biotic stressors and induce SAR. Metabolic plasticity refers to the capacity of the plants to alter their metabolism when subjected to any kind of stress and is characteristic for their survival. Mitochondrial respiration is adversely affected under stress conditions, and accumulated SA triggers alternative oxidase (AOX) which results in lowering of ATP production and reduction in ROS production, thereby lowering HR response. Continuous flow of electrons through SA-triggered AOX will result in continued operation of pentose phosphate pathway, resulting in generation of erythrose 4-phosphate and NADPH, which are required for the synthesis of phytoalexins. Additional requirement of NADPH is met through the action of NADP-specific malate dehydrogenase.

Non-expressor of PR genes 1 (NPR1) is the key player of all the SA-mediated defense responses (Fig. 32.15). NPR1 protein is not the receptor of SA. Rather two proteins encoded by paralogues of the gene (i.e., related genes derived from gene duplication)—NPR3 and NPR4—are SA receptors which differ in their affinities for SA and function on the basis of SA concentrations. NPR1 promotes activation of

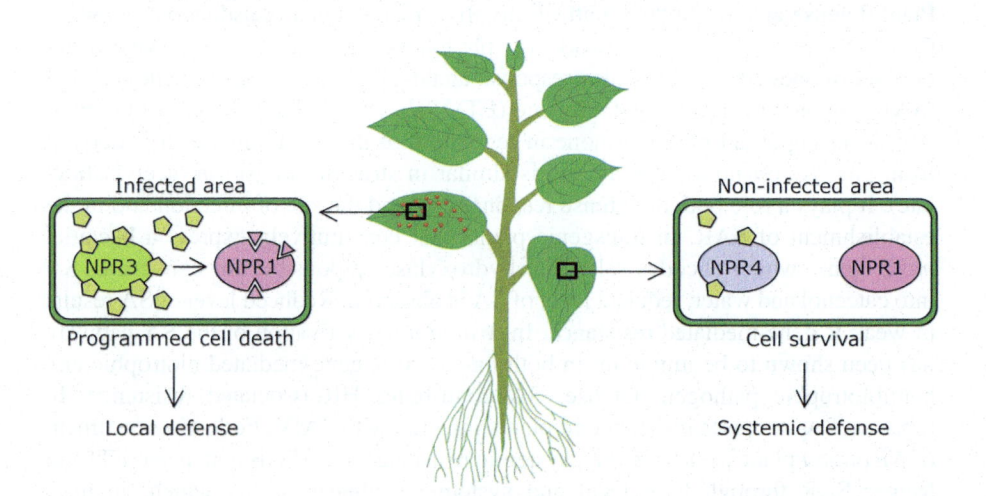

**Fig. 32.15** Non-expressor of PR genes 1 (*NPR1*) is the key player of all salicylic acid-mediated defense responses. *NPR1* promotes activation of salicylic acid responsive genes involved in defense by promoting degradation of repressor proteins

## 32.3 Plant Defense Mechanisms

salicylic acid responsive genes involved in defense by promoting degradation of repressor proteins. NPR1 exists in oligomeric or monomeric forms depending upon oxidizing and reducing conditions of cytoplasm, respectively. In the absence of infection, there will not be any SA accumulation and NPR1associated with NPR4 which is degraded via 26 proteasome pathway. As a result, no defense is triggered. In the infected tissue, SA alters the redox status of the cytoplasm. As a result, NPR1 is translocated to nucleus in monomeric form. In the infected cell, high concentration of SA in the nucleus promotes association of NPR1 with NPR3 promoting degradation of NPR1 through ubiquitin-proteasome pathway, triggering HR and ETI since at the distal site, SA concentration is low, it is unable to bind with NPR3, and cell death is blocked. Rather SA binds with high-affinity receptor NPR4, blocking degradation of NPR1. NPR1 upregulates expression of WRKY TFs which regulate many immunity-associated genes favoring cell survival and expression of genes associated with SAR (Fig. 32.16). SA also directly inhibits catalases resulting increased level of hydrogen peroxide induce peroxidase-catalyzed cross-linking of plant cell walls (Fig. 32.17). SA also interacts with other phytohormones, either synergistically or antagonistically. There is a connection between JA and SA signaling pathways and other defense pathways as part of the PR response. Plants with defective JA pathway fail to accumulate SA in the leaves or phloem and become highly susceptible to TMV. Conversely, SA pathway does not affect JA level but susceptibility is increased. SA-mediated signaling pathway is activated following infection by

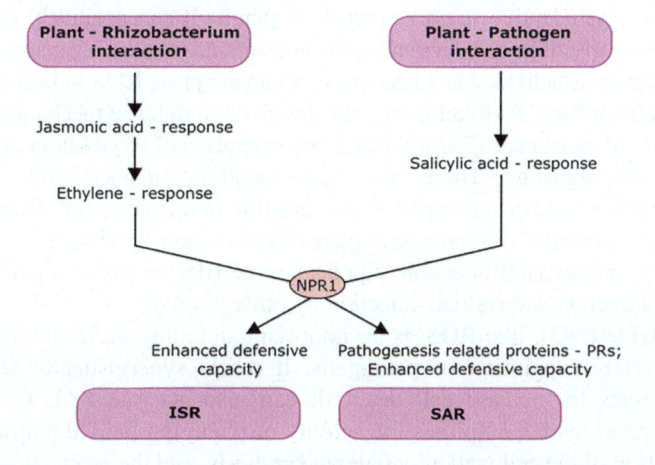

**Fig. 32.16** Plant-pathogen interactions and molecular mechanisms involved in susceptible and resistance disease reactions by plant cell upon pathogen attack. If plant cells PRRs are unable to identify pathogen, then no signals are transduced, and hence defense-related genes are not activated, and plant becomes susceptible. On the other hand, if plant cell PRRs recognize pathogen, it shows hypersensitive responses (thickening of cell wall, tyloses, etc.), or defense-related responses are activated resulting in formation of PR proteins and ROS, which kill pathogen spores, and plant becomes resistant. *PRR* pattern recognition receptors, *PRs* pathogenesis related, *ROS* reactive oxygen species

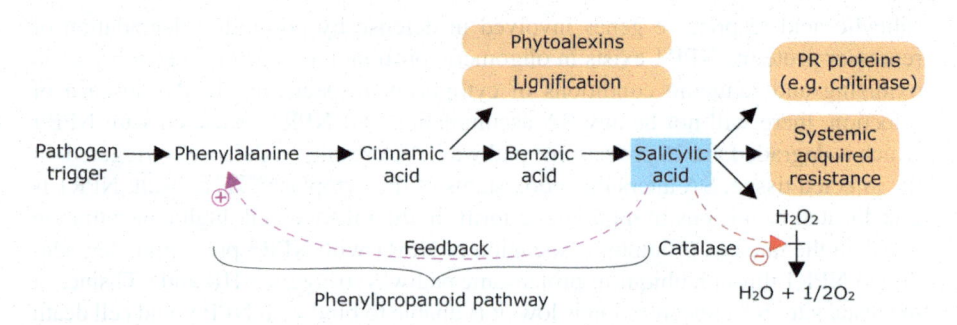

**Fig. 32.17** Role of salicylic acid (SA) in induced plant defense responses. SA is a product of the phenylpropanoid pathway and activates genes for PR proteins and the development of systemic acquired resistance. It stimulates the phenylpropanoid pathway directly but inhibits catalase

biotrophic pathogens, while attack by necrotrophic pathogens induces distinct pathway regulated by JA and ethylene. SA- and JA-/ET-mediated defense pathways undergo crosstalk, and their interactions are generally antagonistic (Fig. 32.16). Biosynthesis of jasmonic acid (JA) and ethylene is initiated as PAMPs/MAMPs are detected by the plant. Ethylene is produced within 10 min in response to flg22, while JA level rises slowly. Ethylene signaling activates resistance against necrotrophic pathogens. In addition, abscisic acid, gibberellic acids, auxins, brassinosteroids, and cytokinins also have roles to play in defense mechanisms. ABA regulates numerous development and adaptive stress responses in plants. It can positively regulate plant defense at the early stage of infection by closing stomata or inducing callose deposition. If ABA pathway is activated at a later stage, it can suppress ROS induction and SA or JA signal transduction. Cytokinins are also involved in defense mechanism, including the induction of resistance against viruses, but suppress HR. Cytokinins act synergistically with SA signaling. These have been found to enhance production of two antimicrobial phytoalexins, scopoletin and capsidol, in tobacco plant. Brassinosteroids (BRs) are also considered important in plant defense against pathogens. In potato BRs are effective against viral infection. Application of BRs on tobacco plants decreases TMV viral infection and restricts infection by other biotrophs.

Nitric oxide (NO), like ROS, is an important signaling molecule that is rapidly generated after recognition of pathogens. It reacts synergistically with reactive oxygen species to increase cell death. It can also activate PAL (phenylalanine ammonia lyase), which amplifies the activity of phenylpropanoid pathway, leading to lignification of the cell wall, phytoalexin synthesis, and the production of salicylic acid which, in turn, amplify defense responses such as systemic acquired resistance and production of pathogenesis-related proteins, such as chitinase and glucanases. Plants produce NO from the amino acid arginine by the activity of a putative NO synthase enzyme which is stimulated by ion fluxes (Fig. 32.18).

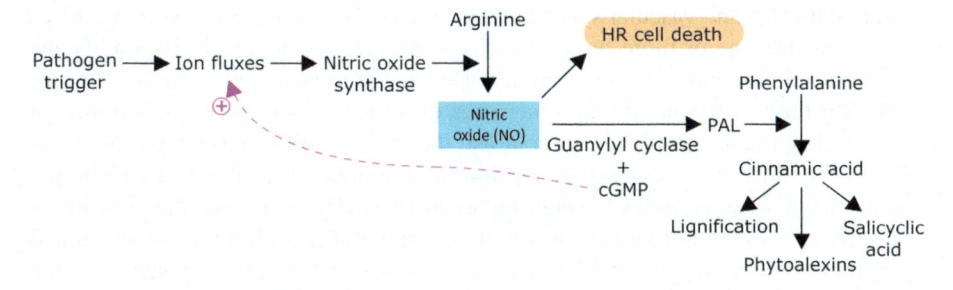

**Fig. 32.18** The production of NO is stimulated by ion fluxes. NO is involved in inducing cell death (hypersensitive response) and activating PAL (phenylalanine ammonia lyase) which enhances the activity of the phenylpropanoid pathway, leading to wall thickening, phytoalexins synthesis, and production of salicylic acid. Production of NO is enhanced by guanylyl cyclase and cGMP

### 32.3.6 Pathogenesis-Related Proteins

Proteins which are induced in the host plant upon pathogen attack are generally described as pathogenesis-related (PR) proteins. These are low molecular weight proteins (6–43 kDa). SA induces a set of pathogenesis-related proteins by regulating the transcriptional activation of many PR genes. Seventeen families of PR proteins have been classified in most plants. These include hydrolytic enzymes, such as chitinase and glucanase, which degrade fungal cell wall structural polysaccharides inhibiting fungal growth and spore germination. Other classes of PR proteins include cell wall-modifying enzymes and antifungal compounds and components of signal pathway. PR protein, such as **lipoxygenase,** may contribute by generating secondary signal molecules such as jasmonic acid (JA) and lipid peroxidase besides triggering synthesis of a number of volatile or nonvolatile compounds having antimicrobial properties. The PR proteins have typical properties that enable them to resist acidic pH and proteolytic cleavage and thus survive in harsh conditions. These proteins occur in vacuoles, cell wall, or intercellular spaces.

## 32.4 Viruses as Plant Pathogens

Another class of pathogens includes viruses and viroids. Both of them require living plant cells for their replication. They rarely kill their host but weaken the plant by reducing growth and seed yield. Viruses also exhibit three levels of interactions with their host. First is their activity within an infected cell, second is spread to neighboring cells, and last pertains to transmission of the virus from one host plant to another. Viruses and viroids enter through wounds and are transmitted from one host to another by insects. Plant viruses are obligate biotrophs and have small genomes encoding only three to ten proteins. These proteins help in replication, cell-to-cell movement, and symptom development. All viruses (except Gemini viruses and nano viruses) encode replicases (RNA or DNA polymerases), which help in replication of virus in host cell. Plant viruses face challenges during initial replication, movement to adjacent cells, and vascular system and in suppressing host defense systems. Most viruses encode one or more proteins called movement proteins, which modify the structure and function of plasmodesmata. During the movement of viruses, the size exclusion limit of plasmodesmata increases up to tenfolds to permit movement of large nucleoprotein complexes. In contrast, the dsDNA of cauliflower mosaic virus (CaMV) modifies the plasmodesmata into large tubular structures to facilitate the encapsulated virus particles through enlarged plasmodesmata. Another feature of cell-to-cell movement of virus is the involvement of the endoplasmic movement and membrane trafficking. In contrast to animal viruses, plant viruses never cross the plasma membrane of the infected cell. Many viruses also encode suppressors of the RNA-silencing mechanisms of the host. All these virus-encoded proteins can act as effector molecules similar to those of microbial pathogens. Once the virus enters the phloem cells, it can move at a speed of up to 1 cm/hour. Protein coat of virus and host-coated pectin methyl esterase are involved in the exit of the virus or viron from phloem into the distal leaves. Presence of certain protein motifs is a signal to plant cell about virus infection. NBS-LRR proteins are a major class of receptor proteins for recognition of virus infection. This signal reception leads to hypersensitive response. A 183 kDa protein from TMV interacts with a toll-like receptor in the host cell. This interaction of elicitor with the plant receptor results in the activation of a cascade of reactions leading to cell death. Plant mitochondria play an important role in detecting pathogen infection. Although viral patterns inducing PTI are well known from animal systems, there is no similar pattern reported for plants. Typical PTI cellular responses in plant-virus interactions include ion fluxes, ROS production, ethylene, salicylic acid (SA), MAPK signaling, and callose deposition. The virus-derived molecules (e.g., dsRNAs) act as PAMPs, which trigger PTI and RNA interference (RNAi). However, PTI is typically a form of innate immunity, whereas RNAi induces a form of adaptive immunity.

## 32.5 Plant Responses to Herbivory

Herbivores consume large quantities of plant leaves, seeds, fruits, etc. to acquire energy and nutrients. Some plant tissues serve as a better food source than others in terms of quantity and quality. For example, seeds and pollen have relatively high protein content. Similarly, meristematic tissues are made up of young, rapidly dividing cells that are energy and nitrogen-rich and are preferred by herbivores. Herbivore feeding causes mechanical damage to the plant tissue, and the degree of damage depends on how the food is taken up by the herbivore. Several classes of herbivores are known based on their feeding behaviors. Aphids and whiteflies, known as phloem feeders, insert their long narrow mouthparts into the conducting tissues of stem and leaves causing minimal damage to the epidermis and eliciting plant responses similar to pathogen attack. Phloem feeders also often act as vectors carrying viruses to other susceptible plants. In contrast, insects belonging to the orders Lepidoptera (moths and butterflies) and Orthoptera (e.g., caterpillars, grasshoppers and beetles) chew using their toothed mandibles to crush, cut and macerate, and cause extensive physical damage to plants. Piercing and sucking insects, such as mites, thrips, and nematodes have short stylets and feed on epidermal and mesophyll cells. Vertebrates such as mammals use plants/plant parts in their diet and are classified according to their dietary specialization. Nectarivores feed on nectar and pollen; gumivores feed on exudates from trees; browsers feed on stems, twigs, buds, and leaves; and grazers feed on grasses.

Plants perceive herbivory also on the basis of specific pattern of damage caused by the feeding herbivore. Chemicals released from herbivores in their saliva as well as those released from the damaged plant tissues, **oviposition** fluids, etc., elicit plant defense responses against herbivores. In response to herbivore attack, plants secrete various chemicals that may act either as deterrents for herbivores or attractants for predators of herbivores. Resistance to insect pests can also be classified as constitutive or inducible. Preformed constitutive defense barriers include defensive proteins and metabolites that are expressed irrespective of the level of herbivore threat. In contrast, induced defenses are invoked in herbivore-challenged plants at the site of attack as well as systemically in undamaged tissues. It is important to note that induced defense traits evolve, as they require lower resource allocation costs than constitutive defense traits. Moreover, herbivores could evolve strategies to overcome constitute defense mechanisms of plant. Plants are also known to activate sequestration of sugars in underground parts of the plant, allowing them to tolerate herbivory better. Several defense mechanisms are induced in the plant in response to an insect attack that leads to different defense responses (Fig. 32.19).

Plants may adopt direct or indirect defense strategies. Direct defense strategies include both preformed (constitutive) as well as inducible response, while indirect response includes release of volatile organic compounds (VOC) by the plant in response to herbivory (Box 32.6). These VOCs attract the predators of the herbivores. This interaction among the plant, herbivores, and their natural enemies is known as **tritrophic interactions**. Defense response to herbivores in plants, such

**Fig. 32.19** Phloem-mobile long-distance signal induces the expression of defense genes in distal parts of the plant

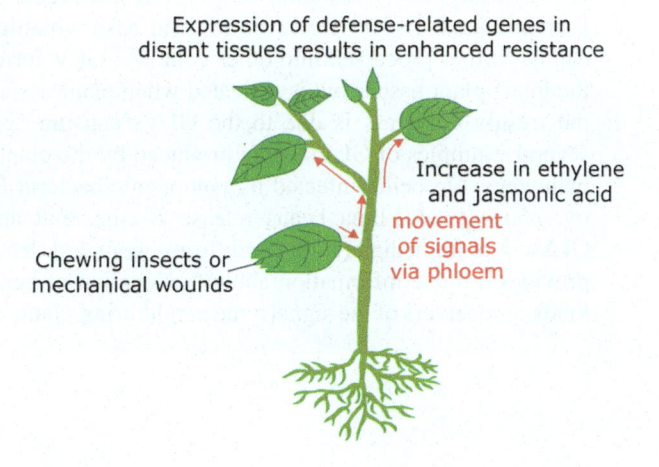

**Box 32.6: Herbivore-Induced Plant Volatiles: "The Watch Dogs"**
Volatile organic compounds (VOCs) are natural products emitted into the atmosphere from marine sources and terrestrial environment. Many of the VOCs are of biogenic origin (BVOCs). Since plants are constantly exposed to various threat (including insects, pathogens, and parasitic plants), they arm themselves with molecular weapons and adopt different mechanisms to protect themselves. They also alert the neighboring plants against the potential damage. Primary function of plant produced VOCs is to defend plants against herbivory, pathogens, besides attracting pollinators and seed dispersers. VOCs also serve as signals in plant-plant communications. Plant VOCs consist of GLVs (green leaf volatiles), terpenoids, MeJA, MeSA, methanol, ethylene, and other substances. Herbivore-induced plant volatiles (HIPVs) are released by the plants on being attacked by the herbivores. These are released from the exposed parts of the plants, such as leaves and flowers, and provide airborne info-signals that either attract predators and parasitoids (indirect defenses) or deter oviposition (direct defenses). VOCs emitted by herbivore-challenged plants of *Brassica oleracea* are known to reduce oviposition preference of adult female moths of *Pteris rapae* and larval performance. These VOCs also attract larval parasitoid *Cotesia glomerata*. VOCs can eavesdrop to prime the neighboring plants so that they are able to protect themselves better against the attacking herbivore. Both intraspecific and interspecific signaling by VOCs are known. Green leaf volatiles (GLVs) are secondary metabolites produced by most of the green plants. These are low molecular weight hydrocarbons formed from the bio-catalyzed conversions of omega-6 fatty acid linoleic acid. GLVs are volatile C6 aldehydes, alcohol, and their esters that are derived from the octadecanoid pathway. The pathway starts with the formation of linolenic acid which is the hydrolytic product of membrane phospholipids catalyzed by lipases. Linolenic acid is oxygenated to form C13-hydroxyperoxy linolenic acid (13-HP), a precursor for both GLV, and jasmonic acid. 13-HP is cleaved to form basic volatile C6 aldehydes, which can be further processed into other volatiles. GLV formation is suppressed in the intact plant tissues but is activated when plant tissue is damaged. Smell of the freshly cut grass is due to the GLVs causing "green odor." There are several examples of GLVs being produced by the plants on being infected by pathogens. On being infected by pathogenic bacteria *Pseudomonas syringae* pv. *phaseolicola*, lima beans release *E*-2-hexanal and *Z*-3-hexanal. Since GLVs are immediately released from wounded leaves, release of GLVs provides reliable information about the location of herbivores. There are two kinds of receivers of the signals: the neighboring plants and the distal organs of

(continued)

**Box 32.6** (continued)
the affected plant which trigger systemic defense. Within the same plant, the two types of signals (airborne signals as well as the one transmitted through vascular system) are responsible for triggering the **systemic defense**. Plasma membrane is the target of interactions with GLVs. On perceiving the signal, membrane is depolarized ($V_m$) due to activation of $Ca^{+2}$ channels. Major product in the wounded leaves of *Arabidopsis* is (Z)-3-hexanal, while (Z)-3-hexanol and (z)-3-hexanyl acetate are formed in partially wounded leaves. GLVs also have antimicrobial properties. These restrict herbivory by attracting predators of the herbivores. GLVs make certain caterpillars attractive to the "big-eyed bugs" as a result of reaction with their saliva. GLVs may be the precursors for ozone formation. Thus, they may contribute to photochemical smog in urban area.

| Plant volatiles | Chemical nature |
|---|---|
| Terpenes | 3E-4,8-dimethyl-1,3,7-nonatriene (DMNT) |
|  | 3E,7E-4,8,12-trimethyl-1,3,7,11-trideca-tetraene (TTMT) |
| Alkaloid indole or phenylpropanes | Methyl salicylate, methyl anthranilate |
| Jasmonates | *cis*-jasmone and methyl jasmonate |
| Green leaf volatiles (GLV) | Cleavage products of fatty acids, e.g., (*E*)-2-hexenal, (Z)-3-hexenal, (Z)-3-hexenol, and (Z)-3-hexenyl acetate |

as production of **proteinase inhibitors (PI)** and secondary metabolites by the plant, is induced either within a very brief period, known as short response, or the induction of response may occur after some time. Short-term responses occur within minutes to hours of herbivore attack and include reactions involving precursors already present in the leaf. Chewing of *Populus tremuloides* (quaking aspen) leaves causes enzymatic hydrolysis of two phenolic glycosides (salicocortin to salicin and tremulacin to tremuloidin) with the release of 6-HCH (6-hydroxycyclohex-2-ene-1-one) which is further converted to phenol or catechol (a potent toxin) in the gut of insects. As a result, insect cannot feed continuously on leaves; rather they must move constantly making them more vulnerable to predators. Variation in reallocation of resources in the plant also plays an important role. More carbon allocation is required by the plant for the production of the protective chemicals which possibly comes at the expense of investment of carbon in growth. Some plants cope up herbivory by attaining a high capacity to regrow. Plants that grow rapidly possibly invest less carbon in defense. Long-lived leaves of slow-growing plants should be well protected against herbivores to minimize tissue loss. According to optimum defense hypothesis, defense compounds should be concentrated more in those parts of plants where they are needed more. The secondary metabolites having a role in defense are concentrated more in younger leaves than in the older ones. The older leaves undergo senescence, and these are not of much value to plants because of deterioration in their productive value. Younger leaves are more valuable to plant, and they tend to be more responsive to herbivory. Tropical trees that grow on the infertile soil have higher concentration of secondary metabolites than do trees that grow on more fertile soils. Fast-growing species utilize more carbon for their growth and invest less carbon in secondary metabolites rather than slow growing ones, making them more susceptible because of less allocation of carbon for defense chemicals. So, plants growing more rapidly invest less carbon in defense. Factors that inhibit growth more than inhibiting photosynthesis enhance secondary metabolite production. Physical features, such as leaf toughness and trichomes, form a physical barrier against herbivores. Defense strategies, both direct and indirect, provide dual protection to the plant against insect herbivores in natural ecosystems.

### 32.5.1 Defense Mechanisms in Plants Against Herbivory

Plants have evolved a number of traits to protect or tolerate herbivory. These include constitutive as well as induced defense mechanisms. Constitutive defense mechanisms include development of specialized morphological features that physically deter feeding by the herbivores as well as production of secondary metabolites, proteins, and enzymes in the plants that have toxics and/or anti-nutritional effects on the herbivores. First line of defense against herbivory **(constitutive defense)** is played by *plant structure* that forms a physical barrier preventing the pests through formation of a waxy cuticle and/or formation of spines, thorns, or trichomes. These morphological and anatomical traits offer an advantage to the plant by directly inhibiting the herbivores from feeding. These structural traits include cell wall reinforcement by deposition of cuticle resulting in toughened and hardened leaves (sclerophylly), waxes, etc.; presence of spines, thorns, prickles (spinescence), and trichomes or hairs (pubescence); and granular mineral inclusions in plant tissues. Nonglandular trichomes prevent small insects from accessing the leaf surface and, therefore, physically limit their movement. Wounding due to herbivores leads to *enhanced synthesis of chemicals* such as lignins, cellulose, suberin, callose, phenolics, and silica particles. These enhance toughness of leaves, thereby forming an important physical barrier to prevent the mouthparts of piercing-sucking insects from penetrating the plant tissues and causing mandibular wear and tear in biting-chewing herbivores. An increase in density of the trichomes is seen in response to insects feeding on plants. This may also be classified as an **inducible response**. Trichomes, in addition to serving as barriers, also act as sensors that send electrical and/or chemical signals to induce defense responses. Presence of mineral crystals such as silica crystals, called as **phytoliths**, present among the members of grass family, adds toughness to the cell walls, making it difficult for insect herbivores to chew. Calcium oxalate crystals form needle-like structures called **raphides** (in specialized cells called as **idioblasts**) with sharp tips that can damage the soft tissue of the digestive tract of an herbivore. In addition to causing mechanical damage, raphides also facilitate the entry of toxic compounds produced by the plant through the damaged site. Latex-filled **laticifers** and **resin ducts** that store resins form a network of canals which constitute an anatomical defense mechanism in plants. Upon herbivore attack, the contents are exuded that entraps or poisons (such as toxic cardenolides) the herbivore. For example, the milkweeds (genus *Asclepias*) are known to exude latex that coagulates upon exposure to air and immobilizes the feeding insect larvae. Resin-based anatomical defense is seen in conifers. Resin is a mixture of terpenes that accumulate in resin duct network and are secreted out upon damage by herbivore attack. The highly volatile monoterpenes evaporate, trapping the insect in solidifying resins that also seals the wounded area. Sealing is important to prevent water loss and also to prevent bacterial and fungal infections at the site of infection.

A wide variety of **secondary metabolites**, including phenolics, terpenoids, alkaloids, cyanogenic glucosides, and glucosinolates, are secreted in response to herbivory that contributes toward plant defense against the pests. These metabolites inhibit herbivore activity by either reducing the nutritional value of plant food or by acting as deterrent to the feeding by herbivore or act as a toxin. A variety of **antinutritional proteins** are also induced in response to herbivory, which include serine proteinase inhibitors, several classes of proteases, oxidative enzymes, amino acid metabolizing enzymes, and lectins. Nitrogen-containing compounds produced by plants also act as defense molecules against pests. These include alkaloids, cyanogenic glycosides, glucosinolates, and some non-protein amino acids. Eating alkaloid-containing plants such as ragworts (*Senecio* spp. containing alkaloid senecionine) or lupin (*Lupinus* spp. containing alkaloid lupinine) can lead to poisoning among grazing animals.

Several alkaloids are known to be feeding deterrents against herbivores. Alkaloids derived from ornithine or arginine occur naturally in nontoxic forms in plants. These are reduced to toxic uncharged and hydrophobic tertiary alkaloids that can pass through membranes as they reach the alkaline digestive tracts of insect herbivores. Though stored in nontoxic forms, some compounds are converted to toxic forms when plant tissues are damaged. These include cyanogenic glycosides and glucosinolates. Cyanogenic glycosides and its hydrolytic enzymes are stored separately within the plant cells. Cyanogenic glycosides present in *Manihot esculenta* tubers enhance resistance against cassava burrower bug (*Cyrtomenus bergi*). *Sorghum* spp. which make dhurrin (a cyanogenic glycoside) is stored in the vacuoles of leaf epidermis, while the hydrolytic enzyme is stored in the mesophyll cells of the leaf. As the damage is caused to leaves, the two get mixed up resulting in release of hydrogen cyanide. Members of family Brassicaceae have got a characteristic smell which is due to glucosinolates. Similar to cyanogenic glycosides, these are also sulfur-containing nitrogenous compounds and are stored in vacuoles as conjugated with sugars, but sugar is attached to central carbon via sulfur atom. The enzyme myrosinase is stored separately, and both are released during mechanical damage resulting in production of toxic isothiocyanates and nitriles. The glucosides of another nitrogen-containing compound, such as benzoxazinoid compounds (e.g., **DIOBA** and **DIMBOA**), are also stored in vacuoles which themselves are nontoxic, but a specific glucosidase is present in chloroplasts, which is released when mechanical damage is done to the plant and can cleave the molecule to release a toxic compound benzoxazinoid. Colored flavonoids are responsible for the coloration in plants; colorless flavonoids such as rutin and isoquercitin are toxic to a number of insect species and act as feeding deterrents to insect herbivores. Accumulation of high levels of certain non-protein amino acids provides protection against herbivores. Non-protein amino acids have defense functions through their misincorporation during protein synthesis. **5-Hydroxynorvaline** is one such non-protein amino acid identified from the leaves of maize (*Zea mays*). It accumulates in leaves during insect herbivory as well as in response to treatments such as with jasmonates, salicylic acid, etc. t-RNA of herbivores which consume plants containing non-protein amino acids become charged with these amino acids leading to disruption of protein synthesis or formation of unstable proteins. Canavanine, another non-protein amino acid that resembles arginine, is synthesized in *Canavalia* sp. (Jack beans). Canavanine cannot be distinguished by arginyl-tRNA synthetase since it closely resembles arginine when ingested by the herbivores. As a result, protein synthesis gets disrupted in the herbivores. However, arginyl-tRNA synthetase present in the seeds of *Canavalia* can distinguish between arginine and canavanine. They thus remain unaffected. In jack beans, t-RNA can discriminate canavanine from arginine. However, when ingested by the herbivore, canavanine proves to be lethal.

The intake of non-protein amino acids by the herbivores interferes with their metabolism in the following ways: 1. substitution in place of normal amino acid resulting in formation of defective proteins; 2. inhibition of the activity of aminoacyl-tRNA synthetases and other steps of protein biosynthesis, e.g., azetidine-2-carboxylaic acid and 3,4-dehydroproline; 3. inhibition of amino acid biosynthesis either by competitive inhibition or end-product feedback inhibition of key enzymes of a metabolic pathway, e.g., azaserine, albizzine, and S-amino ethyl cysteine; 4. targeting DNA and RNA-related processes, e.g., canavanine and mimosine. Mimosine inhibits collagen biosynthesis, and L-hypoglycine inhibits β-oxidation of lipids. Non-protein amino acids are also metabolized to form antimetabolites that prove toxic to herbivores. For example, in *Allium* sp. certain non-protein amino acids derived from cysteine are not incorporated during protein synthesis but are metabolized to form allicin *syn*-propanethial *S*-oxide which has antimicrobial effect. It is important to note that contrary to herbivores, plants are able to distinguish between protein amino acids and their analogs; therefore, they are able to tolerate antimetabolites. However, non-adapted herbivores are not able to discriminate and are therefore susceptible to toxic effect. Two early events in plants associated with insect herbivory are mechanical damage inflicted to plant tissues and the elicitors by which plants are able to identify "self" from "nonself." These include **damage-associated molecular patterns (DAMPs)** in host plants which include structurally diverse groups of molecules. For example, oligosaccharide, such as oligogalacturonides, released from the cell wall acts as an elicitor. These appear in the apoplast and are involved in eliciting defense responses in plants. Another example of DAMP includes **systemin** which is a 20-amino acid peptide derived from prosystemin (200 amino acid long polypeptide). In the damaged leaf tissues of tomato, prosystemin is hydrolyzed to form systemin in phloem parenchyma. Systemin triggers a typical defense response in tomato plant. A 160 kDa plasma membrane-bound, systemin-binding LRR receptor kinase SR160/BRI1 has been identified on the surface of companion cells of phloem tissue in *Solanum peruvianum* which triggers the signal transduction pathway for jasmonate biosynthesis in CC-SE complexes. The second class of elicitors includes those which are generated in the oral secretions (OS) of insects. The resulting insect oral secretions (OS) recognized by plants are termed as herbivore-associated elicitors (**HAEs,** also called as **HAMPs**; herbivore-associated molecular patterns) that trigger and elicit defense responses in plants. One such elicitor is **violicitin**, a fatty acid-amino acid conjugate (FAC). Violicitin was initially discovered in oral secretions of *Spodoptera exigua* (beet armyworm), and it is composed of linoleic acid (hydroxylated at position 17) conjugated to glutamine. While the fatty acids are derived from the plant, an enzyme present in the gut of the insect facilitate conjugation of the fatty acids with insect-derived amino acids such a glutamine or glutamate after the insect ingests plant material containing these fatty acids.

### 32.5.2 Signal Transduction Triggering Plant Responses

As the insect feeds on the plant, a signal transduction cascade is initiated in the plant. It occurs both at the damaged site (local response) as well as a systemic response (at a site away from the damaged site). Signal transduction cascade is triggered after HAMPs/DAMPs are recognized. It involves intracellular changes such as $[Ca^{+2}]_{cyt}$ and changes in transmembrane potential leading to electrical signaling resulting in synthesis of hormones such as jasmonic acid (JA) and other defense responses of the plant.

**Electrical Signaling** First recognition of insect attack and initiation of signal transduction cascade occurs at the plasma membrane, which is in direct contact with the exterior. Evidences indicate involvement of electric signaling that is initiated in response to herbivore attack. Leaf injury caused by mechanical damage or herbivory induces the activity of two **glutamate-receptor-like (GLRs)** ion channels, such as calcium-ion channels, which results in generation and transmission of electrical signal, triggering defense responses at sites further away from injury. Double mutants from *Arabidopsis* for genes encoding two such channels (*glr3.3* and *glr3.6*) showed disruption of wound-induced electrical signaling, confirming the role of GLRs in wound-induced signal generation and transmission.

**Role of $Ca^{+2}$ as Secondary Messengers** Several signaling pathways are known to involve fluctuations in $Ca^{+2}$ concentrations following herbivore attack. Cytosolic concentration of $Ca^{+2}$ is much lower than that in the apoplastic fluid and in the cellular organelles, creating a driving force for the $Ca^{+2}$ to enter through the membrane channels. The calcium ions are later pumped back through $Ca^{+2}$-ATPase pumps. An increase in cytosolic $Ca^{+2}$ concentrations has been observed in response to oral secretions of *Spodoptera littoralis* (Egyptian cotton worm) feeding on *Phaseolus lunatus* (lima bean) and *Gingko biloba*. An increase in cytosolic $Ca^{2+}$ precedes membrane depolymerization through voltage-gated channels located in the plasma membrane and other intracellular membranes. Calcium signal activates calcium-sensing proteins, including calmodulin, calcineurin, and $Ca^{+2}$-binding protein kinases (CDPKs). This alters protein phosphorylation and gene expression patterns. CDPK signaling pathways further involve those that interact with mitogen-activated protein kinases (MAPKs) and form jasmonates (JAs).

**Intracellular Wound Signal** Mechanical damage of the leaf tissue induces the formation of systemin from prosystemin in the phloem parenchyma cell that functions as DAMP triggering response in the adjacent companion cell. Systemin enters into the apoplast and binds to a membrane-bound receptor of the companion cell and initiates an intracellular signaling cascade that activates a MAPK and, finally, leads to the biosynthesis of JA in the companion cells (Fig. 32.20). JA is then transported as methyl jasmonate to long distances via the phloem and activates target gene expression in distal undamaged leaves. Salicylic acid also plays a role in signaling in plants in response to aphids feeding through phloem. It acts as a negative modulator as it represses JA.

**JA-Mediated Induction of Defense Gene Expression as a Systemic Response** As mentioned earlier, defense responses are not only triggered at the damaged site but also in undamaged parts of the plant. While earlier systemin was proposed as the mobile signal, it is now known that systemin is involved only in initiating wound signaling and not in long-distance signaling. Long-distance signaling is mediated by JA (Fig. 32.21). Mutants deficient in JA are easily killed by insect pests and exogenous application of JA in such mutants has been shown to restore resistance. This indicates the vital role played by JA in herbivore-induced plant defense responses. JA is derived from its precursor α-linolenic acid via **octadecanoid pathway**. JA is the key regulator that modulates the expression of defense-related genes, such as those that code for proteins that interfere with the herbivore digestive mechanism. These include α-amylase inhibitors, lectins, proteases, and **proteinase inhibitors (PI)** (Fig. 32.22). Legumes produce α-amylase inhibitors. In some plants, defensive proteins such as lectins are produced that bind to oligoproteins. Upon ingestion by the herbivore, lectins bind to epithelial cells of the digestive tract and disrupt the absorption of nutrients. The gut epithelial membrane of the digestive tract is also directly affected by the ingestion of proteases that are produced by the plant in response to herbivory. Other important proteins that interfere with the digestion of the insect herbivore are the proteinase inhibitors,

## 32.5 Plant Responses to Herbivory

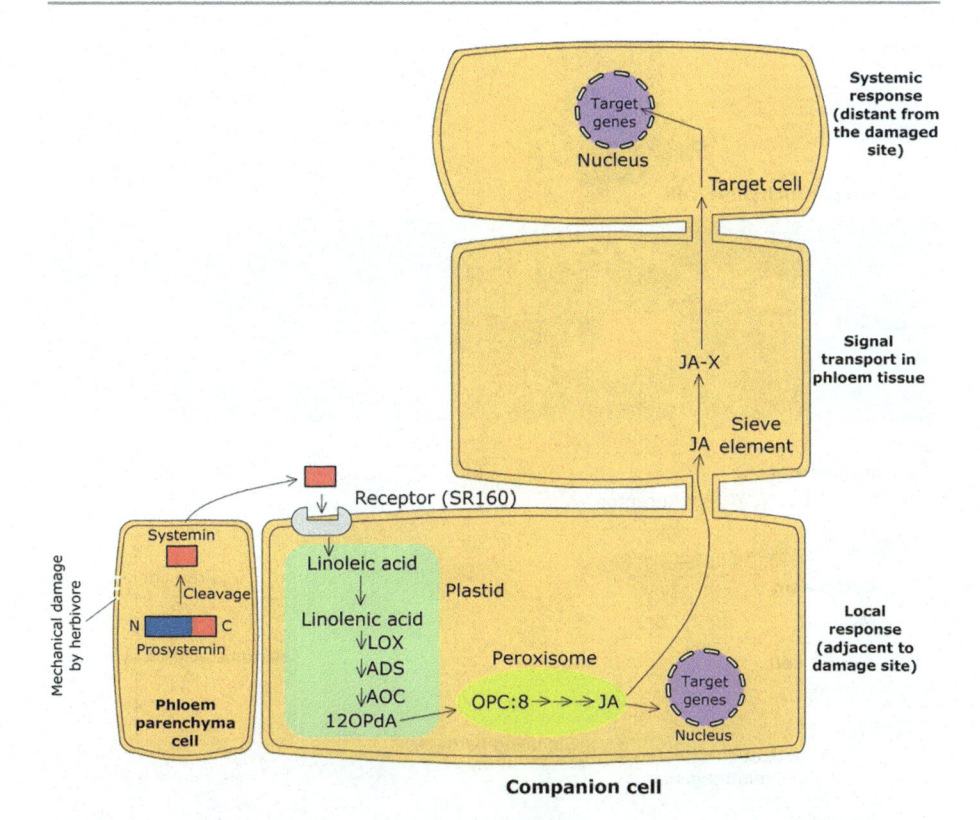

**Fig. 32.20** Schematic model showing the role of JA in systemic wound signaling. Chloroplastic and peroxisomal JA biosynthetic enzymes are located in vascular bundles of the leaf. Binding of systemin to its receptor (SR160) activates JA accumulation. JA produced in the companion cell-sieve element complex is transported in the phloem via plasmodesmata connections between cells. JA, or a covalently modified form of JA (JA-X, e.g., JA-Ile), activates target gene expression in distal undamaged leaves

which bind to and inhibit protein hydrolyzing enzymes such as trypsin and chymotrypsin in insect gut. This results in deficiencies causing delayed development, reduced fecundity, and mortality. In addition to inducing the synthesis of PI, it also activates the genes responsible for the synthesis of compounds such as terpenes, alkaloids, phenylpropanes, and glucosinolates which act as deterrent for herbivores. JA not only induces defense-related genes but also suppresses growth. Growth inhibition allows reallocation of resources for metabolic pathways involved in defense.

**Role of Ethylene** Ethylene controls the production of constitutive defense compounds after herbivore damage and stimulates the production of JA and volatile compounds [**Herbivore-induced plant volatiles (HIPV)**]. For example, it has been shown in *Medicago truncatula* that ethylene modulates herbivory-induced early

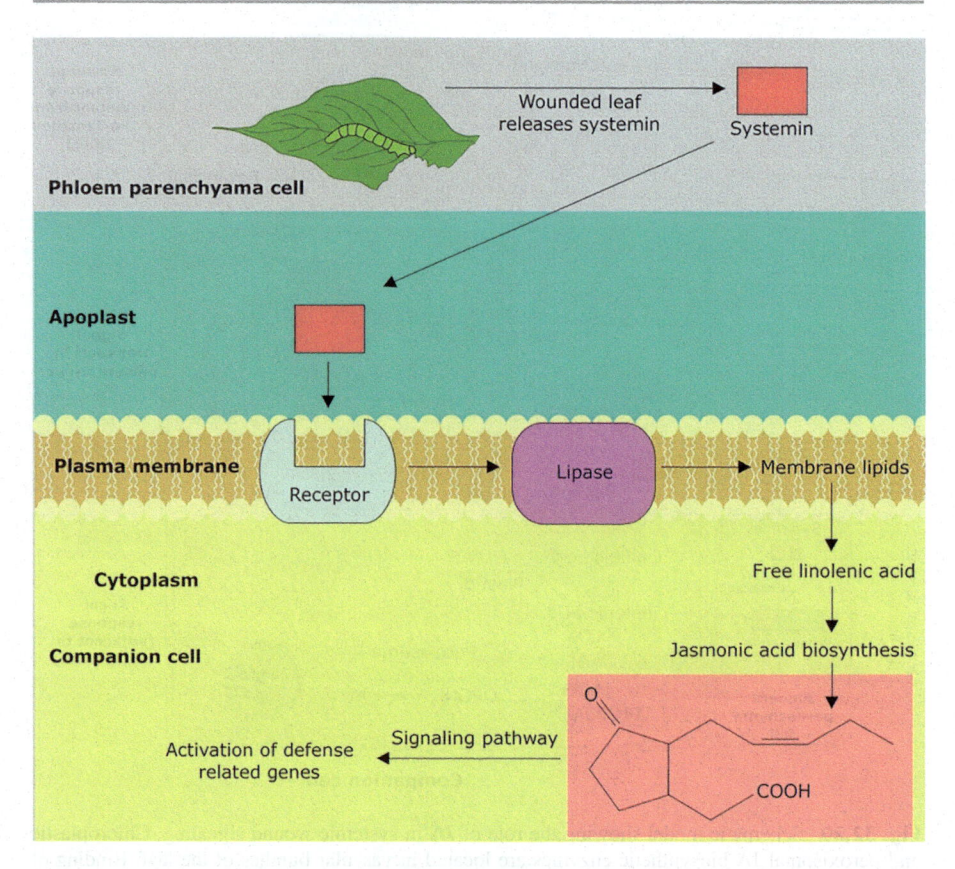

**Fig. 32.21** Wounded leaf signals the proteolytic cleavage of prosystemin to release systemin. Systemin signaling pathway is initiated upon binding of the peptide to a 160-kDa plasma membrane-bound receptor (SR160). Systemin binding activates lipase activity that releases linolenic acid, a JA precursor, from lipids in the plasma membrane. Following JA synthesis, defense-related genes are activated

signaling events such as $Ca^{+2}$ influx and downstream, JA-dependent biosynthesis of terpenoids. The crosstalk between JA and ethylene controls local cell expansion and growth following herbivore attack. This allows resource allocation toward induced defense responses against herbivores. Both JA and ethylene were found to suppress growth in attached leaves of tobacco. However, expression of trypsin proteinaceous inhibitors requires only JA signaling that is independent of ethylene. Ethylene modulates responses such as emission of specific VOCs, accumulation of phenolic compounds, and PI activity (Fig. 32.23). Ethylene also elicits several defensive proteins such as the enzymes polyphenol oxidase and peroxidase which form quinines. Quinines react with nucleophilic side chains of amino acids, thus causing formation of protein cross-linkages. Quinines interfere with digestion in insect gut.

## 32.5 Plant Responses to Herbivory

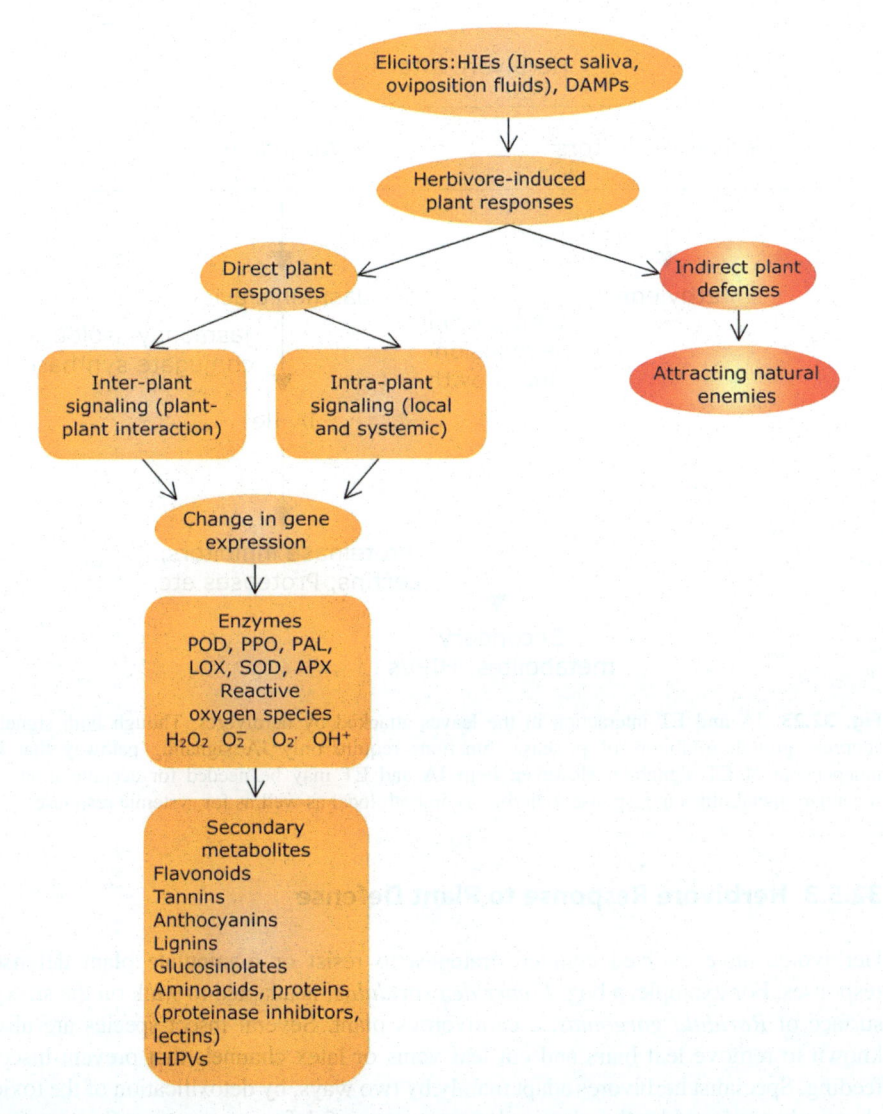

**Fig. 32.22** Classification of herbivore-associated responses in host plants and the resistance factors

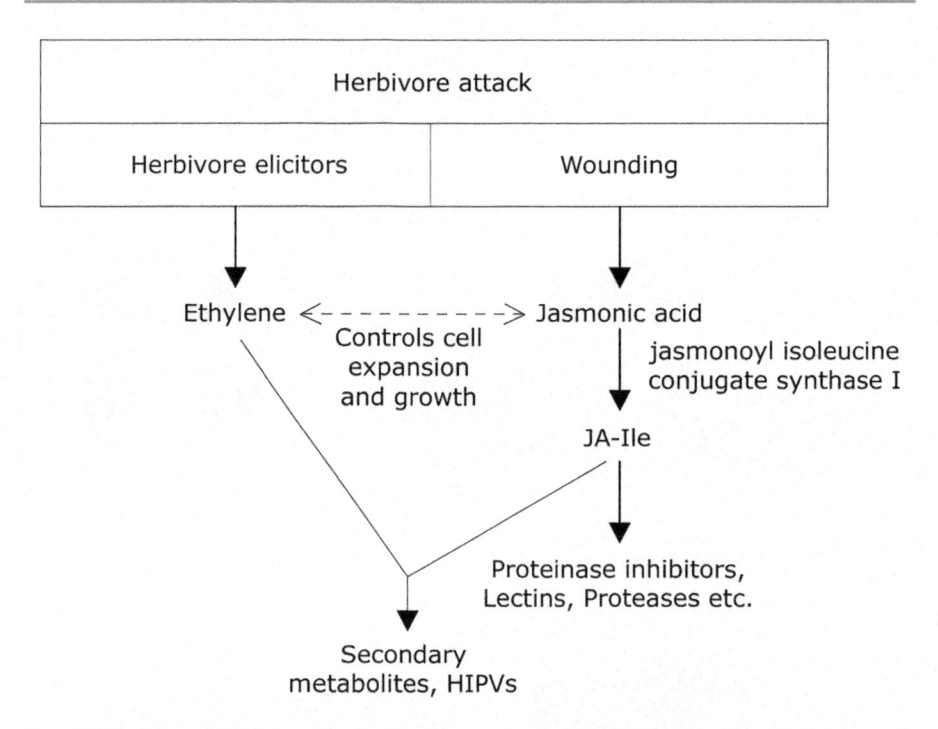

```
                    Herbivore attack
        ┌─────────────────────────┬─────────────────────┐
        │   Herbivore elicitors   │      Wounding       │
        └───────────┬─────────────┴──────────┬──────────┘
                    ▼                        ▼
        Ethylene <----------> Jasmonic acid
                  Controls cell      │ jasmonoyl isoleucine
                  expansion          │ conjugate synthase I
                  and growth         ▼
                                   JA-Ile
                                     │
                                     ▼
                        Proteinase inhibitors,
                        Lectins, Proteases etc.
                    ▼
                Secondary
              metabolites, HIPVs
```

**Fig. 32.23** JA and ET interaction in the leaves attacked by herbivores. Though both signals suppress growth inhibitors of protease, functions require only JA signaling pathway that is independent of ET signaling. However, both JA and ET may be needed for accumulation of secondary metabolites in response to herbivory in both local as well as for systemic response

### 32.5.3 Herbivore Response to Plant Defense

Herbivores have evolved counter strategies to resist or manipulate plant defense responses. For example, a bug, *Pameridea roridulae*, is adapted to walk on the sticky surface of *Roridula gorgonias*, a carnivorous plant. Several insect species are also known to remove leaf hairs and cut leaf veins or latex channels that prevent insect feeding. Specialist herbivores adapt mainly by two ways, by detoxification of the toxic chemicals produced by the plant or by prevention of defense activation. Suppression of host defenses often involves secretion of molecules called as effectors that modulate host defenses. The salivary secretions of insect *Helicoverpa zea* secrete the enzyme glucose oxidase (GOX) which acts as an effector molecule since it protects the herbivore against pathogen and also suppresses plant defense. Glucose oxidase oxidizes D-glucose from the host plant to D-gluconic acid and $H_2O_2$.

$$D - glucose + O_2 \rightarrow H_2O_2 + D - gluconic\ acid$$

GOX is a potent $O_2$ scavenger. GOX provides the initial oxidative burst of $H_2O_2$ which suppresses the induced defenses of the host plant. D-gluconic acid is known to scavenge free radicals and inhibit polyphenol oxidase. $H_2O_2$ is responsible for elicitation of salicylic acid induced response and associated negative crosstalk that attenuates JA and ET signaling. Herbivore-associated pathogens have also been associated with their roles as effectors. Since pathogens elicit SA-regulated defense, which often negatively crosstalk with jasmonate signaling, plants are unable to fully activate the jasmonate-mediated resistance against herbivores. Not only the chemicals of OS act as elicitors, eggs of some herbivores may cause local suppression of JA pathway via induction of SA pathway.

### 32.5.4 Nematodes

Nematodes are known to feed on plants using their stylets for obtaining food. Plant parasitic nematodes (PPN) can be classified as ectoparasites, migratory endoparasites, and sedentary parasites. All PPN possess a hollow protrusible stylet that punctures wall and is used for injecting secretions and ingesting nutrients from the plant cell. Ectoparasites remain outside the body of the host and use their stylets to feed on the content of the cell. This leads to wounding, necrosis and often gall formation in the host tissues. Migratory endoparasites have robust stylet that allows them to penetrate and continuously move through the root. Migration is made possible by the secretion of cell wall-degrading enzymes through the stylet. Infection by such nematodes results in extensive necrosis and formation of gall tissue. Sedentary endoparasites, such as the cyst nematodes and root-knot nematodes, induce the formation of permanent feeding cells inside the vascular tissue of the host plant after entering and migration. The young ones hatch close to the root tip and then migrate inter- and intracellularly toward the vascular cylinder. As they reach the vascular cylinder, they continuously feed on the cellular content while vigorously injecting the stylet secretions into the cell. This causes extensive changes in gene expression and morphology. The root tissue becomes hypertrophic and shows expansive proliferation of organelles (Fig. 32.24). Root nematodes result in the formation of six to seven giant multinucleated cells, while the cyst nematodes induce the formation of a **syncytium** after the initial feeding and fusion of adjacent cells. Unlike the migratory and ectoparasite nematodes, the endoparasitic nematodes maintain their feeding cells in a healthy and metabolically active state. Initially, as the PPN invade and remain inside the host for several weeks to months, they expose themselves to plant defense systems. However, to evade detection they adopt strategies such as repeated molting, creating a new challenge to the host immune system every time. The carbohydrate-rich outer surface coat is shed each time, and thereby the changes in composition create a variable target for plant immune system. In addition, PPN cover themselves with plant-derived carbohydrates to escape recognition as a nonself entity. In spite of these strategies, presence of PPN can be detected by the host plant which responds by the production of reactive oxygen species, cell wall modification enzymes, callose deposition, cell wall thickening, etc.

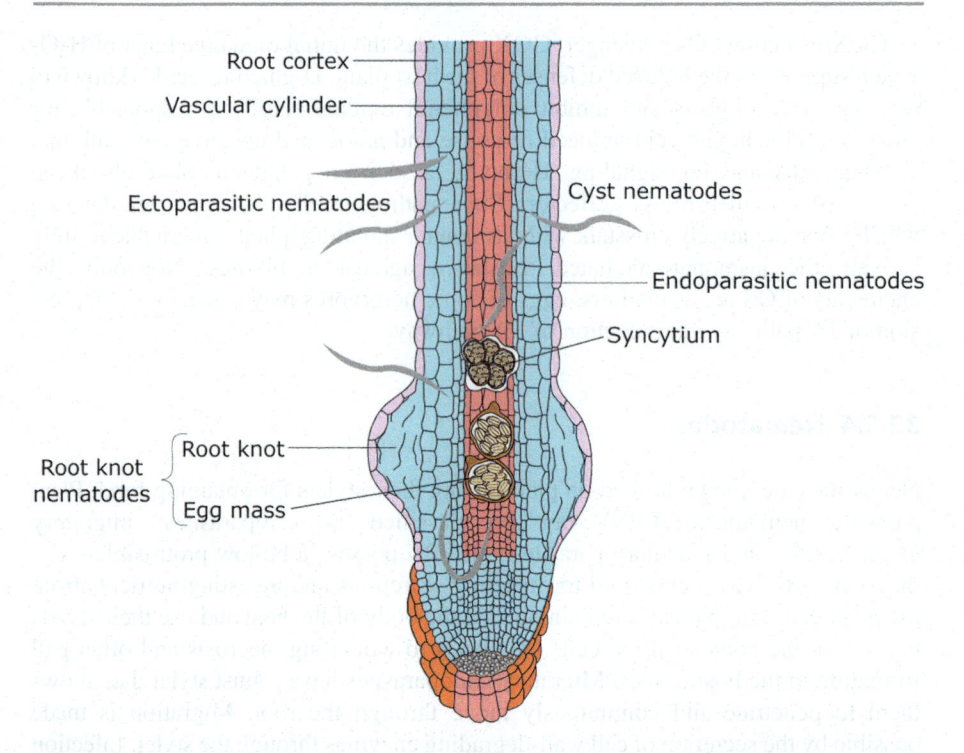

**Fig. 32.24** Plant parasitic nematodes are classified as ectoparasites and endoparasites. Ectoparasitic nematodes do not enter the plants but use their stylets to feed on plant tissue. They cause least damage to the root tissue. In contrast, the endoparasitic nematodes completely enter the host and may either move through various tissues (migratory endoparasites) or reside as sedentary endoparasites in the host tissue. The latter includes the root-knot nematodes and cyst nematodes in which the infective stage (juveniles) moves into the vascular cylinder and induces the formation of feeding structures for drawing the nutrients. The cyst nematodes induce the formation of syncytium of hundreds of fused cells. The root-knot nematodes on the other hand induce the formation of several coenocytic giant cells

It is yet to be deduced whether these responses resemble JA or PAMP-induced responses. A number of effector proteins are secreted by the nematodes, which includes cell wall-degrading enzymes such as cellulases, pectate lyases, polygalacturonases, xylanases, and expansions (Fig. 32.25). In addition to these, enzymes such as glutathione peroxides and peroxiredoxin secreted by the nematodes protect them from ROS formed in response to infestation. Effectors suppress SA or JA production or interfere with plants ubiquitin-proteasome pathway to suppress plant immune system.

## 32.6 Plant-Plant Interactions

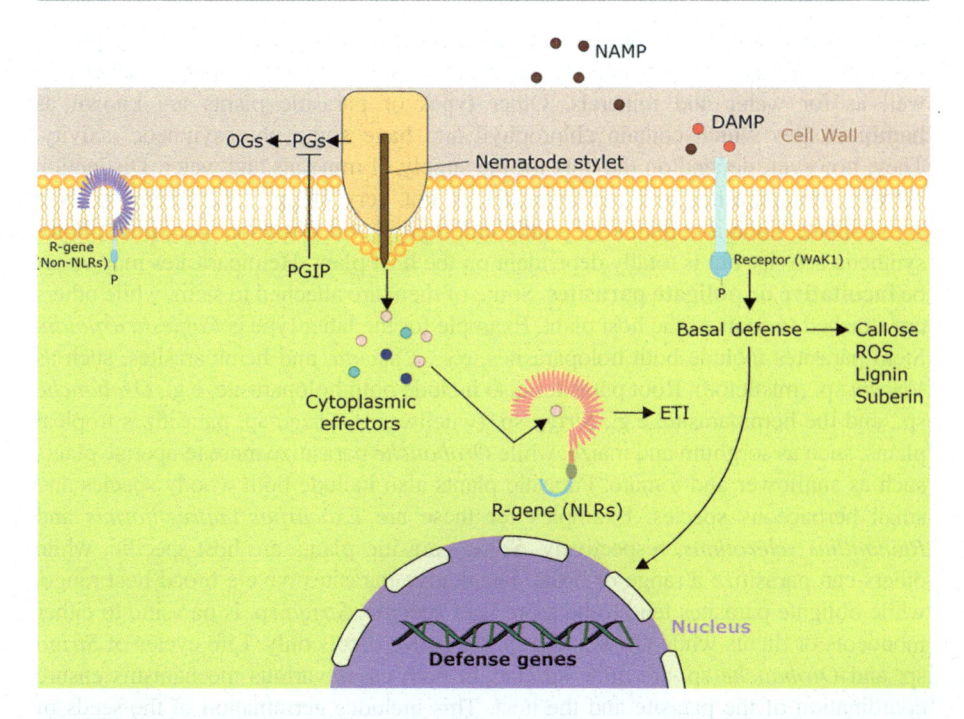

**Fig. 32.25** Molecular events during plant-nematode interaction. Invasion of the nematode caused cell wall damage producing damage-associated molecular patterns (DAMPs) that initiate plant defense responses by binding to DAMP receptors, such as WAK1. Nematodes also secrete cell wall-degrading enzymes such as polygalacturonases (PG) that hydrolyze cell wall. In response plant produces PG inhibiting proteins (PGIPs). Nematode-associated molecular patterns (NAMPs) released from the nematodes are recognized by plasma membrane receptors and trigger pattern-triggered immunity (PTI) that includes callose, reactive oxygen species (ROS), lignin, and suberin. Nematodes, in response to plant defenses, secrete effectors to counter the NAMP-/DAMP-based immunity. Plants, in turn, have R genes that recognize effectors and initiate effector-triggered immunity (ETI)

### 32.6 Plant-Plant Interactions

### 32.6.1 Parasitic Plants

Approximately 4000 species of plants mostly from distantly related families of dicotyledons are parasitic on other plants. These are biotrophic pathogens requiring living plants for their survival since these are either partially or totally dependent on host plants for their requirements including water, mineral nutrients, or photosynthates. Some of them have completely lost their capacity to photosynthesize as well as to assimilate inorganic nitrogen and are dependent on host plants for carbon and nitrogen nutrients. These are obligate parasites also called **holoparasites**. These types of plants have lost their total capacities to complete their life cycle without the host. They do not contain appreciable amount of chlorophylls and are unable to photosynthesize efficiently. Their $CO_2$ compensation point may be as high as 200 Pa. They also do not possess roots. One of the common examples is *Cuscuta* sp. which neither can photosynthesize nor has got true roots. It is dependent on host plants for photosynthates as well as for water and minerals. Other types of parasitic plants are known as **hemiparasites** which contain chlorophyll and have some photosynthetic activity. They, however, depend on the host for the supply of minerals and water. Distinction between holoparasites and hemiparasites is not very sharp, e.g., *Striga* sp. is a hemiparasite since it contains very little chlorophyll and possesses very little photosynthetic activity but is totally dependent on the host plant. Hemiparasites may either be **facultative or obligate parasites**. Some of them are attached to stem, while others are attached to roots of the host plant. Example for the latter type is *Rafflesia arnoldii*. Stem parasites include both holoparasites, e.g., *Cuscuta*, and hemiparasites, such as *Viscum* sp. (mistletoe). Root parasites also include both holoparasite, e.g., *Orobanche* sp., and the hemiparasite, e.g., *Striga* sp. (witchweed). *Striga* sp. parasitizes tropical plants, such as sorghum and maize, while *Orobanche* parasitize more temperate plants such as sunflower and tomato. Parasitic plants also include both woody species and small herbaceous species. Examples for these are *Exocarpos cupressiformis* and *Rhinanthus sclerotinus*, respectively. Some parasitic plants are host-specific, while others can parasitize a range of hosts. Facultative parasites have a broad host range, while obligate parasites tend to be more host-specific. *Striga* sp. is parasitic to either monocots or dicots while *Orobanche* sp. parasitize dicots only. Life cycles of *Striga* sp. and *Orobanche* sp. are quite similar. In both cases various mechanisms ensure coordination of the parasite and the host. This includes germination of the seeds of parasite plant, formation of haustoria to provide attachment, establishment of successful connection with vascular tissues of the host, and compatible interactions thereafter.

### 32.6.2 Establishment of Contact Between Parasitic and Host Plants

**Seed Germination** Seeds of *Striga* sp. and *Orobanche* sp. have very little reserves. They run the risk of dying if germinated without the availability of suitable host since their reserves would be exhausted before the plant starts getting nutrients. Interaction of parasite begins with release of a secondary metabolite by the roots of host plant. Seeds of *Striga* sp. germinate in response to a signal released from the roots of cotton (*Gossypium hirsutum*) plants even though it is not the host. The stimulant has been identified a sesquiterpene called strigol (Fig. 32.26). Strigol is active in concentrations as low as $10^{-12}$ M in soil solution. Many analogues of strigol have been synthesized and tested for their capacity to stimulate seed germination. Besides strigol, other stimulants have also been isolated and identified. Several stimulants for germination seeds of *Striga* sp. have been identified which collectively are called **strigolactones**. Strigol might be regulating synthesis or action of ethylene since there is absolute requirement of this for germination of seeds of *Striga* sp. Another member of the family, i.e., sorgolactone, has also been identified from sorghum. A gradient of germination stimulant may be responsible for the growth of radicle toward the host. Though the seeds of *Cuscuta* sp. which is an obligate stem parasite germinate even in absence of host, the seedlings are unable to survive for a longer period of time because of presence of limited reserves in the seeds. It is observed *Cuscuta* sp. perceives

## 32.6 Plant-Plant Interactions

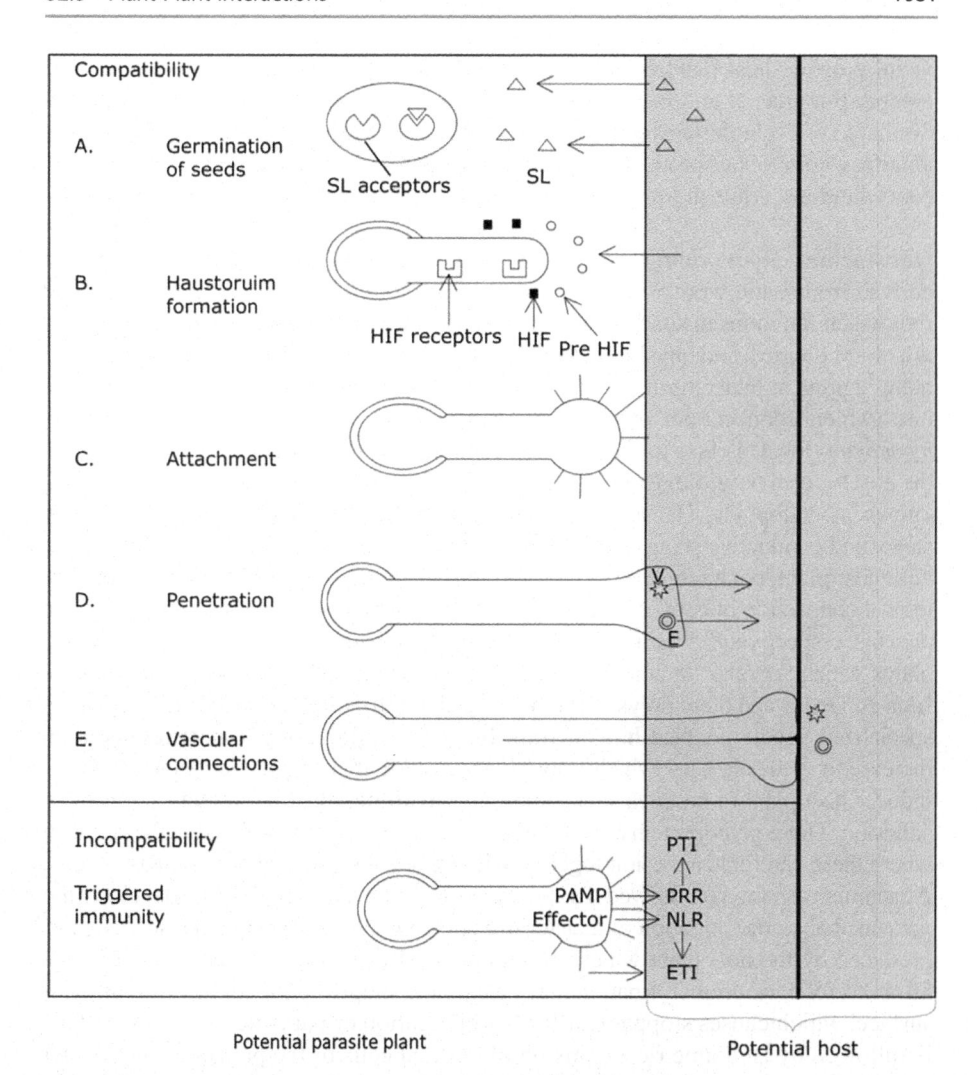

**Fig. 32.26** (a) Interaction between a parasitic plant and the potential host: *SL* strigolactone, *HIF* haustorium inducing factors, *V* virulence factor, *E* effector, *PTI* pattern-triggered immunity, *ETI* effector-triggered immunity; upper box represents development stages of parasitic plant (**a**–**e**) during a compatible interaction. Lower box represents host-invading stage during which molecules (PAMP and effector) trigger immunity in host plants. (**a**) Seeds of parasitic plants start germinating on receiving stimulus (SL) from the potential host plant; (**b**) oxidoreduction processes enable production of haustorium inducing factors (HIF) from precursors of HIFs (pre-HIF). Perception of active HIFs by HIF receptors activates the formation of haustorium; (**c**) haustorial hairs and certain chemicals enable attachment of parasitic plant to the host; (**d**) further growth of haustorium and probable secretion of unknown virulence factors (V, small compounds) and/or effector proteins (E); (**e**) vascular connection is established with the host roots for nutrient uptake. Lower box: demonstrating incompatibility, parasitic plant-derived molecules such as PAMPs and effectors may activate PTI and ETI by nucleotide-binding domain leucine-rich repeat domains (NLR) and by pattern recognition receptors (PRRs), respectively

vicinity of the host plant by sensing the volatile terpenoids and chemicals such as pinene, β-myrcene, and β-phellandrene. These compounds act as chemo-attractant. Similar to insect herbivores some parasites may locate preferred host plant based on the defensive compounds produced by them. These compounds serve as attractants at high concentrations, while at low concentrations they serve as defensive compounds.

**Haustorium** Haustorium is a defining feature of all parasitic plants. The word is derived from Latin word, "haurire" which means to drink. Haustorium is a modified root which functions in attachment, penetration as well as transfer water and solutes. It is a morphological and physiological bridge between host and the parasitic plant. Three main regions of haustorium are recognizable. These include a swollen-rounded structure (which connect root or stem of the host plant to the main haustorial tissue, *hymenium*) located close to the host tissue and the parasitic tissue that penetrates into the host to form physiological connections with the vascular tissues of the host. This is known as endophyte. Haustorial development requires an adhesive stage, intrusive stage, and conductive stage which includes development and adhesion of haustorium with host plant, penetration of haustorium into the host, and establishment of physiological connection in between the vascular tissues of host and conducting tissue of the parasite, respectively. After finding suitable host, prehaustoria develop in the parasitic plants which serve as an adhesive disk. Prehaustorium secretes adhesive substances, such as pectins and other polysaccharides, which reinforce adhesion and develop into a specialized structure called **haustorium.** In *Cuscuta*, proximity to host results in an increase in cytosolic $Ca^{2+}$ which lasts for about 48 h after the initial contact. *Cuscuta* induces host plant to produce sticky substance **arabinogalactan** proteins, to promote adhesion. These glycoproteins, produced by the host plant, are localized in the cell wall where these can force adhesion together with other sticky substances, such as pectins. Attachment phase is followed by penetration phase as prehaustoria develop into parasitic haustoria. Initiation of haustorium development occurs in response to signals produced by the host plant which are known as HIF (haustorium induction factors). Stimulus for haustorium formation may be initiated simply by the contact with the host surface, which causes stoppage of further elongation of parasitic root system. This is followed by isodiametric expansion of cortical cells of the parasitic roots. Inner cells divide first followed by cell division of peripheral cell resulting in formation of noticeable bump at or near the root tip meristem in parasites within 24 h. Epidermal cells elongate to form long, densely positioned haustorial hairs which are responsible for firm attachment of parasite with the host plant. After the parasite is firmly attached to host plant, a penetration peg invades host epidermis and cortex by a combination of physical and chemical processes until it reaches the vascular tissue of the host plant. In *Striga hermonthica*, haustorium formation is stimulated by phenolic acid, such as sinapic, vanillic, and syringic acids produced by the host plant. Enzymes (peroxidases and hydrolases) produced by the parasitic root system convert the inactive parahydroxyacids (by oxidative carboxylation) into active quinones and a functional haustorium is established. *Striga asiatica* is a hemiparasite which grows on the roots of plants such as *Vigna unguiculata* or *Sorghum bicolor*. In *Striga asiatica*, haustorium formation is induced in response to a signal, i.e., 2,6-dimethoxy-p-benzoquinone (DMBQ) which is produced by the host roots in response to an enzyme produced by the parasite. It is not produced by *Sorghum* sp. roots rather is a product of lignin decomposition by the enzyme laccase produced by root tip of *Striga* sp. The compound is probably not released in the root environment rather is tightly bound with the cell wall. Benzoquinones are produced in plants in shikimate acid pathway by the oxidative carboxylation of phenolics or by enzymatic degradation of cell wall phenols by peroxidases and laccase. It was demonstrated that $H_2O_2$ generated at *Striga* sp. radicle tip activates host plant peroxidases which convert cell wall phenols into benzoquinone. Proximity of host is ensured before haustorium development in *Striga* by this mechanism. The haustorium penetrates through the host tissue effected by mechanical pressure and is supported by the biochemical degradation of the host cell caused by the secreted hydrolytic enzymes, such as methylesterase, or lytic enzymes, such as pectinase or cellulose.

Walls of parasitic cells are thickened with polysaccharides rather than with lignin. Enzymes from the parasitic cells soften the surface tissues of the host, and haustorium penetrates the epidermis, bark, and parenchyma to reach vascular tissue to absorb nutrients. Vascular cells of parasite contact vascular tissues of the host. As a result, contents of xylem are diverted into the parasite. Cell walls of some of the endophytic cells rupture, producing holes through which constituents of the host xylem can pass through forming a conducting vessel between the host and the parasite. Water and solutes are transported to haustorium through xylem pits. Most of the hemiparasites are xylem tappers. Contrary to this holoparasites are phloem tappers. For phloem connections, conducting tissues of haustorium terminate some distance from the interface. As a result of this, solutes have to travel some distance through parenchymatous cells. After contact with sieve cells, hyphae grow around the cell-like fingers resulting in increase in the parasitic surface around 20 times. A cytoplasmic **syncytium** is built up between *Cuscuta* sp. and the host plant. These cells, which are adjacent to haustorium, are similar to transfer cells. Since no symplasmic connections have been observed between the two partners, transfer of solutes via haustorium may occur via apoplast. Presence of mitochondria, dictyosome, ribosome, and well-developed endoplasmic reticulum in the parenchymatous cells suggests a role of active transport in solute transfer from the host plant to the haustorium of the parasitic plant. Mannitol is present in xylem sap of *Striga hermonthica*, while it is not present in host plant *Sorghum bicolor*. Asparagine is the predominant nitrogenous compound in *Sorghum*, while in *Striga* sp. major nitrogenous compound is citrullin. Malate and citrate are the organic acids present in xylem sap of *Sorghum*, while shikimic acid is a predominant organic acid in xylem sap of *Striga*. Even carbohydrate concentrations have been found to be five times higher in xylem sap of parasitic plant than that of host plant.

## 32.6.3 Physiological Interactions Between the Parasite and Host

**Water and Mineral Relations** In order to maintain a continuous influx of water into the parasite, an osmotic gradient is required. Potassium seems to be a preferred osmoticum in some of the parasitic relationships. Movement of solutes and metabolites is mainly unidirectional. This is evident in case of phosphates which may be retained by the parasites despite of severe deficiency experienced by the host. Most of the xylem tapper parasitic plants do not have a mechanism to selectively import specific ions that arrive via xylem or to export the excess ions imported; these might be stored in parasitic plants. Parasitic plants have lower **water use efficiency (WUF)** than the host plants since these have higher transpiration rates in comparison to their counterparts. High rate of transpiration accounts for lowering of temperature around their leaves by as much as 7 °C. Water potential gradient in between the leaves and root is steep which accounts for their high transpiration rates. Sensitivity of parasitic species to ABA is much less in comparison to that of the host plant. Maize plants infected with the parasite *Striga hermonthica* have higher ABA content than the uninfected plants. High transpiration rates of parasitic plants allow rapid import of solutes through xylem. Reduced stomatal conductance in host plant after infection by the parasite may be attributed to change in their (host's) ABA content. Parasites generally maintain a water potential lower than their host plants by accumulating amino acids (e.g., proline, arginine), organic acids, carbohydrates, and mannitol besides accumulating immobile cations in the xylem.

**Carbon Relations** Carbon transfer from host to parasite is inversely proportional to photosynthetic efficiency of the parasite. Higher is the photosynthetic efficiency of parasite lower is the amount of carbon transferred from host to parasite. Obligate holoparasites obtain their entire carbon supply from the host since these do not have any photosynthetic efficiency. They predominantly import compounds from the sieve tubes of the host plants and have distinctly lower ratio of calcium to potassium in comparison to hemiparasites since calcium is present in lower concentrations in phloem sap. However, the value of carbon supply seems to be variable in case of facultative hemiparasites. In case of *Striga*-sorghum relationship, almost 30% of the parasitic carbon seems to be derived from host at low nitrogen supply, while in high nitrogen availability and at higher photosynthetic rates, the value declines to around 6%.

### 32.6.4 Competition for Resources

One of the ways parasites influence growth and reproduction in host is by reducing the amount of resources available for growth. Parasites rarely kill their hosts, but in most of the cases, host suffers reduced growth and reproductive performance. In many cases root parasites have the capacity to reduce crop production to zero. Parasite may provide a sink for photosynthates produced by the host plant, and in turn it increases the rate of photosynthesis in the host. The compensatory mechanisms in the host plant may include increase in leaf area, increased Rubisco content, and delayed leaf senescence or general reallocation of the carbon. Parasites prove to be a long drain on the nitrogen sources of the host plant and cause deficiency in the plant. A high demand for nitrogen by the parasites is explained by the observation that their growth is profuse when they are attached to legume crops. Accumulation of $NH_4^+$, $NO_3^-$, and amino acids in the host may occur due to demand by parasites. Upon attachment to host, cytokinin concentrations may increase in the parasitic plants which may diffuse into the host after infection.

Since crop plants are damaged by the parasites, they trigger immune responses in the host plant. Some plants are resistant to parasitic infection. Resistance to parasitic plants may occur before penetration of the host plant, during haustorium development, or during or after the establishment of vascular connections. Various defense responses, observed in the host plants leading to resistance to parasitic plants, include induction of immunity-related genes, ROS production, deposition of callose and other phenolic compounds, vessel blockage, as well as hypersensitive response which leads to arrest of invasion followed by necrosis of the parasitic structures. There may be involvement of multilayered surveillance system responsible for plant immune system similar to that employed against microorganisms, i.e., cell surface pattern recognition receptors binding to PAMPs, while nuclear and/or cytoplasmic resistance (R) receptors monitor the effectors. Identification and cloning of resistance protein suggest that the host plant may be following similar strategies against parasitic plants as those used against pathogens. Parasitic plants may be secreting effector proteins during infection. SA signaling pathway may also be playing an important role in resistance against parasitic plants besides upregulation of genes encoding PR proteins. Many resistant plants act by surrounding the invading endophyte and blocking its access to the vascular system of the host.

## 32.7 Allelopathy

Allelopathy is described as both beneficial and harmful biochemical interaction among plants and/or plants and microorganisms through the production of chemical compounds. Such chemical compounds, known as **allelochemicals,** escape into the environment and subsequently influence growth and development of the neighboring plants. Allelochemicals can affect several physiological processes, such as respiration, photosynthesis, and ion uptake. Allelopathy has been observed for over 200 years, and a document published in 300 BC mentioned about many crop plants inhibiting the growth of other plants. In 1937, the word allelopathy was coined in by Austrian plant physiologist Hans Molisch. The term originates from the Greek roots, *allelon*, meaning "mutual" or "among each other" and *pathos*, meaning "suffering" or "feeling." Mostly, allelopathy is used to refer to the harmful effect of one plant upon another. It was reinforced that the effects of one plant to another plant may be either stimulatory or inhibitory depending on the concentration of the released compounds. When a plant produces allelochemicals that are detrimental to the establishment of new seedlings of its own, the phenomenon is called **autotoxicity**. Autotoxicity is a specialized intraspecific form of allelopathy. Although allelopathic symptoms among plants have been observed for a long time but very few specific allelochemicals have been identified (Table 32.4). Allelochemicals are a suitable substitute for synthetic herbicides because they

**Table 32.4** Some allelopathic plants, their secondary metabolite secretions and effects.

| Plant | Allelochemicals produced | Effects |
|---|---|---|
| *Juglans nigra* L. | Juglone | Inhibit mitochondrial functions and electron transfer in photosystem II |
| *Sorghum bicolor* | Sorgoleone | |
| *Callistemon citrinus* | Leptospermone | Herbicide (low efficacy) |
| *Centaurea maculosa* (weed) | Catechins | Phytotoxic |
| *Actinidia chinensis* | Emodin | Feeding deterrent to a large spectrum of organisms |
| *Festuca rubra* | m-tyrosine | Phytotoxic |
| *Salvia leucophylla* | Mixture of monoterpenes (BVOCs[a]) | Toxic for seed germination and root growth |
| *Chrysanthemoides monilifera* | Mixture of sesquiterpenes (C15) and diterpenes (C20) [BVOCs] | |

[a]*BVOCs* Biogenic volatile organic compounds

(allelochemicals) do not produce residual or toxic effects, although efficacy and specificity of many allelochemicals may be limited.

### 32.7.1 Molecular Mechanisms of Allelopathy

#### 32.7.1.1 Allelopathic Action Through Root Exudates

Allelochemicals are released from the roots in the form of exudates into the rhizosphere and help in nutrient acquisition. They also serve as signals to a number of microorganisms. Some plants secrete about 5–20% of carbon fixed in photosynthesis into the rhizosphere indicating their importance for plant's survival. Allelopathic interactions are one of the important factors which contribute to the distribution of species and their abundance within plant communities. These interactions are also important in success of the invasive plants, such as *Eichhornia crassipes* (water hyacinth), *Centaurea maculosa* (spotted knapweed) and *Alliaria petiolata* (garlic mustard). *Centaurea maculosa* belongs to Asteraceae family and is native to Europe where it is neither dominant nor problematic. However, in North America it has become an invasive exotic weed and releases phytotoxic secondary metabolites into the soil. In Montana state of Northwestern United States, it has infested over 1.8 million ha (~4.4 million acres). It releases a racemic mixture of (±) catechin, a phytotoxic secondary metabolite. In susceptible species, it triggers accumulation of ROS in the root meristem which initiates a $Ca^{2+}$ signaling cascade. This leads to changes in genome-wide gene expression, causing death of root meristematic cells. In *Arabidopsis*, expression of approximately 1000 genes has been observed to double after 1 h of (±) catechin treatment.

Black walnut (*Juglans nigra*) produces juglone, a toxic naphthaquinone present in all parts of this species, but is released by living roots in high concentrations. *Sorghum bicolor* produces sorgoleone, a benzoquinone, which is specifically produced in and secreted by root hairs. Both juglone and sorgoleone are implicated in the inhibition of mitochondrial functions and electron transport in photosystem II. Thus, they affect respiratory and photosynthetic pathways. Consequently, secretion of juglone and sorgoleone as root exudates suppresses growth and establishment of other species in substantial area around these plants, thereby reducing competition.

In *Festuca rubra*, a grass species, a different mode of action of allelopathic compound secreted by roots has been reported. Roots of *Festuca rubra* exude *m*-tyrosine, an isomer of *p*-tyrosine and a non-protein amino acid, which is phytotoxic to root growth in different weed species. It has been proposed that *m*-tyrosine interferes with amino acid metabolism via its mis-incorporation instead of phenylalanine, leading to disruptions of protein regulation. Several compounds, such as flavonoids, quinines, cytokinin, and p-hydroxy acids secreted by host roots, have been reported to induce formation of haustorium. These molecules are also implicated in host-parasite recognition and mycorrhizal associations. In S*triga*, the secondary metabolites secreted by host roots act as signal to initiate the development of haustoria. The elongating *Striga* radical perceives haustorial initiation factors (HIFs) secreted by host roots and forms a functional attachment organ.

### 32.7.1.2 Allelopathic Action Via Biogenic Volatile Organic Compounds

Allelopathic actions are not only mediated through secretion of soluble root exudates. Several **biogenic volatile organic compounds (BVOCs)** have also been implicated in impeding growth of potential competing neighboring plants. *Salvia leucophylla* produces a mixture of volatile monoterpenes which suppress growth of neighboring species. Monoterpenes, such as camphor, camphene, 1,8-cineole, α-pinene, and β-pinene, suppress root growth in *Brassica campestris*. At high concentrations, monoterpenes inhibit seed germination in *Brassica campestris*. Production of α-pinene causes inhibition of root growth in many other plants as well. In *Cassia occidentalis*, exposure to α-pinene leads to enhanced lipid peroxidation and elevation of $H_2O_2$ levels, leading to severe membrane damage. It is quite possible that similar toxicity mechanisms are responsible for allelopathic mechanisms of other species as well to become invasive. For instance, *Chrysanthemoides monilifera*, invasive in Australia, emits a blend of sesquiterpenes (C15) and diterpenes (C20) which inhibit root growth and germination of the native sedge, *Isolepis nodosa*. BVOCs are particularly toxic to germination and root growth, whereas shoot growth seems more resistant. Roots are also known to emit a variety of BVOCs.

**Role of Microorganisms in the Rhizosphere** Various microorganisms in the rhizosphere of the plant may modify the action of allelochemicals either by enhancing or minimizing phototoxicity of allelopathic donor toward allelopathic receiver by degradation or transformation of toxic allelochemicals or by modulation of the defense mechanism operative in the allelochemical receivers. These may induce expression of stress responsive genes. Microbes may also act by degradation of nontoxic glycoside allelochemicals to toxic allelochemicals in the rhizosphere.

## 32.7.2 Plant Physiological and Biochemical Processes Affected by Allelochemicals

Allelochemicals adversely affect the shape and structure of plant cells. Treatment with hordenine and gramine (allelochemicals from barley roots) causes disorganization of organelles, increase in size and number of vacuoles, and cell autophagy. Citral, a volatile essential oil from *Cymbopogon citrates*, causes microtubule disruption in the roots of wheat and *Arabidopsis*. Monoterpenoids, such as camphor, 1,8-cineole, β-pinene, and α-pinene, affect cell proliferation and DNA synthesis in plant meristems. Sorgoleone results in polyploid nuclei. Rye allelochemicals, benzoxazolinone (BOA) and 2, 4-dihydroxy-1,4(2H)-benzoxazin-2-one (DIBOA), significantly inhibit the regeneration of cucumber root cap cells, thus inhibiting root growth. Allelochemicals significantly inhibit the activity of antioxidant enzymes and increase free radical levels causing greater membrane lipid peroxidation and membrane potential alteration. Citral has been reported to damage the membrane system of *Echinochloa crus-galli* L. (barnyard grass), causing lipid peroxidation and electrolyte leakage.

**Imbalance in the Antioxidant System** Balance of the redox state in the cell is instrumental in allelopathic effects. Upon exposure to allelochemicals, the activity of antioxidant enzymes such as superoxide dismutase (SOD), peroxidase (POD), and ascorbic acid peroxidase (APX) is altered to resist oxidative stress in the contact area of the recipient plants. Caffeic acid induces changes in the activities of proteases, PODs, and polyphenol oxidases (PPOs) during root development.

**Growth Regulator Homeostasis** Allelochemicals can cause alterations in the contents of plant growth regulators leading to their imbalances. This results in inhibition of growth and development in plants with respect to seed germination and seedling growth. Phenolic allelochemicals stimulate IAA oxidase activity and inhibit the reaction of POD with IAA and bound GA or IAA to influence endogenous hormonal levels. Salicylic acid has been reported to inhibit the synthesis of ethylene in cell suspension cultures of *Pyrus communis*.

**Enzyme Activity** Synthesis, functions, contents, and activities of various enzymes are affected by allelochemicals. Activity of enzyme **γ-phosphorylase,** involved in seed germination, might be inhibited by chlorogenic acid, caffeic acid, and catechol. Tannic acid suppresses the activity of POD (peroxidase), CAT (catalase), and cellulase. Phenolics can increase the activity of phenylalanine ammonia lyase (PAL) and β-glucosidase and reduce the activity of phenol-β-glucose transferase, thereby inhibiting root growth. The activity of protease, invertase, and succinic dehydrogenase (SDH) is also suppressed by allelochemicals.

**Respiration** Different stages of respiration, such as transport of electrons in mitochondria, oxidative phosphorylation, $CO_2$ generation, and ATP synthase activity, are affected by various allelochemicals. These can reduce oxygen intake which prevents NADH oxidation, thereby inhibiting the activity of ATP synthase, and reduces ATP formation, disturbs plant oxidative phosphorylation, and ultimately inhibits respiration. Sorgoleone blocks electron transport at the b-$c_1$ complex. Juglone has been implicated in the disruption of root oxygen uptake. α-pinene acts under at least two mechanisms: uncoupling of oxidative phosphorylation and inhibition of electron transfer. It results in strongly inhibiting mitochondrial ATP production, decreasing mitochondrial transmembrane potential and impairing mitochondrial energy metabolism.

**Photosynthesis** Impact of allelochemicals on plant photosynthesis mainly involves inhibition of or damage to the machinery involving acceleration of the decomposition of photosynthetic pigments. Consequently, photosynthetic pigment contents are decreased, which lowers energy and electron transfer, reduces activity of ATP synthase, and inhibits synthesis of ATP. Stomatal conductance is also affected inhibiting the overall photosynthetic process. Allelochemicals affect photosynthesis mainly by influencing the function of PS II. For example, sorgoleone inhibits the decay of variable fluorescence, blocks oxidation of the PSII and primary electron acceptor, $Q_A^-$ by PSII secondary electron acceptor $Q_B^-$ by displacing $Q_B^-$ from the $D_1$ protein of PSII, thus inhibiting photochemical reaction. Essential oil from lemongrass is found to decrease ratio of chlorophyll a/b and carotenoid content adversely affecting photosynthetic metabolism.

**Water and Nutrient Uptake** Several allelochemicals affect absorption of nutrients and induce water stress through long-term inhibition of water utilization. Allelochemicals can inhibit the activities of $Na^+/K^+$-ATPase involved in absorption and transport of ions at the plasma membrane. Subsequently, it leads to suppression of cellular absorption of $K^+$, $Na^+$, and other ions. The effect of allelochemicals on ion uptake depends upon their concentration. Sorgoleone and juglone have been reported to inhibit $H^+$-ATPase activity and $H^+$-pump across root cell plasma membrane in *Pisum sativum* which affects the solute and water uptake.

**Synthesis and Metabolism of Proteins and Nucleic Acids** Many alkaloids exhibit allelopathic effects. They inhibit DNA polymerase I preventing transcription of DNA and translation process. Integrity of DNA and RNA is affected by all phenolics. Allelochemicals have been shown to inhibit transport and absorption of amino acid, thereby interfering protein biosynthesis adversely affecting the cell growth. Protein biosynthesis has been reported to be affected by ferulic acid and cinnamic acid as well as many phenols and alkaloids. The observed allelopathic phenomenon may be partly a result of interaction of the allelochemicals with target molecules, such as DNA, RNA, and amino acids.

**Summary**

- Plants need to adapt themselves to environmental stress including the biotic environment. Not all biotic interactions are harmful. These can be beneficial also, as in commensalism, mutualism, or symbiotic associations. Plants develop defense mechanisms to counteract the biotic stresses which include pathogens (disease-causing microorganisms), viruses, tissue-damaging herbivores, nematodes, and parasitic plants and against allele chemicals produced by other plants.
- Disease-causing microbes are known as pathogens. Once pathogen enters the plant, it employs three different strategies for survival, viz., biotrophic, necrotrophic, and hemibiotrophic. Pathogens which can invade the host but are unable to cause disease are known as avirulent, and those which drastically affect the host and cause disease are known as virulent. The process of infection, colonization, and reproduction of pathogen is known as pathogenesis. Pathogens penetrate the epidermal layer of the plant cell or enter through the wound or stomata. Penetration involves molecular interaction between cell surfaces of both host and pathogen. Pathogen secretes enzymes such as cellulases and hemicellulases that enable them to digest cellulose and other cell wall polysaccharides for their growth and multiplication. Most of the pathogens also secrete toxins in the host tissues. In most cases microorganisms induce hypersensitive response in the plant without causing any disease. Hypersensitive response is a localized response of plant cells to the microorganisms so as to prevent further invasion in adjacent cells. Different hypersensitive responses are ROS formation, changes in cell wall composition and membrane permeability, callose deposition, and production of phytoalexins or phenolics leading to programmed cell death.
- Plants have innate immunity which depends on the ability of plant to distinguish between self and nonself molecules. Plants possess a wide array of defense mechanism which is broadly classified into constitutive and induced. The constitutive defense mechanisms are always present whether the plant is attacked or not. These include the structural barriers (cell wall, epidermis layer, trichomes, thorns, etc.) and accumulation of chemical compounds (such as phenolics, nitrogen compounds, saponins, terpenoids, steroids, and glucosinolates) and proteins and enzymes with inherent antimicrobial properties. Inducible defense mechanisms are apparent only when infection is underway. These include basal defense mechanisms or/and R gene-mediated defense mechanisms which are also known as microbial-associated molecular-pattern-triggered immunity (MTI) or pattern-triggered immunity (PTI) and effector-induced resistance often referred as effector-triggered immunity (ETI). PTI is a first-level immunity response triggered by plant pattern recognition receptors (PRRs) that recognize the molecular pattern of the pathogen. PTI develops in response to the molecular patterns of the pathogen which are common among them, i.e., it is triggered in response to most of the pathogenic bacteria. The basal defense mechanism puts selective pressure on potential pathogens to produce effector molecules which interfere with basal

defense mechanisms, therefore further enabling the pathogen to attack the plant. This is referred as effector-triggered susceptibility (ETS). ETI acts at the second level and is the most successful means of controlling pathogens that are able to evade basal defense. ETI is the immune response of plant to specific pathogens and occurs in response to specific molecules secreted by the pathogenic bacteria which are known as effectors, product of avr gene of bacteria. In plants, complementary to avr genes is the R (resistance) gene which produces R proteins. If pathogen is able to produce effector molecules which are not recognized by R gene-mediated defense mechanism, it will be able to attack and exploit the plant for its use causing disease.
- Once a signal is perceived either through recognition of elicitors or effectors, it is transduced in the form of protein activation. Protein phosphorylation plays an important role during pathogen attack. At least two classes of protein kinases are activated: the mitogen-activated protein kinase (MAPKs) and calcium-dependent protein kinases (CDPKs). This leads to release of WRKY33 transcription factor, which binds to the W box (region in promoters of various target genes) and induces pathogenesis-related (PR) genes. SAR (systemic acquired resistance) and ISR (induced systemic resistance) are two types of resistance which prepare plants against subsequent pathogen infection. In SAR, defense proteins accumulate not only at the site of infection but also systemically in uninfected tissues and/or uninfected plants. SAR provides long-term defense against a broad spectrum of pathogens. Selected strains of plant growth-promoting rhizobacteria suppress the infection by disease-causing pathogens by providing systemic resistance by antagonism which is known as ISR.
- Plant defense against pathogen attack involves many signal transduction pathways that are mediated by a network of phytohormones. Three most important phytohormones in plant defense response against pathogens include salicylic acid (SA), jasmonic acid (JA), and ethylene (ET). Proteins, which are induced in the host plant after pathogens have attacked, are generally described as pathogenesis-related (PR) proteins. SA induces PR proteins by regulating the transcriptional activation of many PR genes. Many plant viruses encode proteins that act as suppressors of RNA silencing and hence allow the virus to replicate in the plant.
- The host plant perceives herbivore attack by recognizing DAMPs (mechanical damage) and HAEs (oral secretions). Oral secretions of insect herbivores contain fatty acid-amino acid conjugate (FACs) such as volicitin, inceptin, etc. that are capable of eliciting defense responses in plants. Plants employ direct as well as indirect strategies against herbivory. Direct defense responses affect the interaction of the herbivore and the host plant directly. Direct defense strategies include resistance factors such as morphological and structural features (thorns, spines and trichomes, tissue toughness, secretory structures such as laticifers and resins) as well as biochemical (e.g., toxic and anti-nutritive metabolites and proteins). Defense responses can be constitutive or inducible depending on their mode of expression, i.e., the defense traits may be preformed resistance factors or

deployed after attack by herbivores. Indirect defense responses of plants rely on attracting natural enemies of the herbivores. Plants produce odors that attract predators of herbivores guiding them to their prey. Biochemical defense responses include production of secondary metabolites that act in multiple ways, as toxins, as feeding deterrents, as digestibility reducers, and also as volatiles in indirect defense. Such secondary metabolites include phenolics, terpenoids, alkaloids, cyanogenic glucosides, and glucosinolates. Anti-nutritional proteins are also synthesized such as proteases, proteinase inhibitors, lectins, etc. Herbivore induced VOCs function as external signals within and between plants. The defense responses are not only restricted to the site of attack but also extend to distal sites. Signaling molecules such as cell wall-derived oligogalacturonides, systemin, jasmonic acid, as well as electrical signaling have been implicated in the systemic signaling process. Sytemin is an 18-amino acid long peptide derived from proteolytic cleavage of a larger precursor prosystemin and acts as a primary wound signal generated in response to wounding triggers the production of JA via octadecanoid pathway. The long-distance signaling is mediated by JA-Ile conjugate that serves as a phloem mobile signal and induces the expression of defense-related genes in distal parts of the plant. In response to plant defenses, herbivores counterattack by secretion of molecules called effectors that modulate host cell defenses.
- Parasitic plants are biotrophic pathogens which depend on host plants for the nutrients, water, and minerals. Some parasitic plants lose the capacity to photosynthesize and are dependent on the host plant completely (holoparasites), while others retain some chlorophyll and are able to photosynthesize but are dependent on the host plants for water and minerals (hemiparasites). Some of the parasites are obligate, while others are facultative. Some are stem parasites; others are root parasites. *Striga* is an example of hemiparasite, while *Cuscuta* is holoparasite. Stimulant for germination of seeds of *Striga* has been identified to be a strigolactone which is for released from the host plant which is indication for the nearby presence of the host. On reaching the host prehaustoria develops, followed by the adhesion phase because of secretion of sticky substances such as arabinogalactones by the host plant. This is followed by development of haustoria from prehaustoria in response to haustoria induction factor (HIF) which is formed due to conversion of a compound from the host by the enzyme released by the parasitic plant. A quinone, BMBQ, has been identified which is responsible for development of haustorium inside the host tissues until a connection is established with the vascular tissues of the host plant.
- Allelopathic interactions include both beneficial and harmful interaction among plants through release of allelochemicals into the environment which affect various physiological and biochemical processes in plants. Mode of allelochemical action takes place either through release of soluble secretions in the form of root exudates in to the rhizosphere or via biogenic organic compounds.

## Multiple-Choice Questions

1. Interaction among plants, herbivores, and their natural enemies is known as:
   (a) Biotrophic interactions
   (b) Pathogenic interaction
   (c) Tritrophic interactions
   (d) Allelopathic interactions
2. A non-protein amino acid which has a role in defense is:
   (a) Homoserine
   (b) Canavanine
   (c) Citrulline
   (d) Ornithine
3. Systemin is an example of:
   (a) Pattern-associated molecular pattern
   (b) Damage-associated molecular pattern
   (c) An effector molecule
   (d) A molecule which is translocated away from the damaged site of the plant
4. Elicitor refers to the molecules:
   (a) Produced by the plant in response to pathogenic infection
   (b) Produced by the pathogen at the time of infection
   (c) Present on the surface of plant cells by which pathogens are recognized
   (d) By which plant is able to identify "self" from "nonself"
5. HAMPs are:
   (a) Components from the plant feed which are modified in the guts of the insects
   (b) Molecules produced in plants in response to the damage caused by herbivores
   (c) Molecular pattern on the plant surface recognized by the insects
   (d) Compounds in herbivores identified by the plant
6. An example of HAMP is:
   (a) Linoleic acid
   (b) Volicitin
   (c) Glutamine
   (d) Caeliferins
7. Holoparasites are the plants which:
   (a) Contain chlorophyll and have little photosynthetic capacity
   (b) Have lost their capacity to complete their life cycle without the host
   (c) Get complete nourishment from the host plant
   (d) Are dependent on the host plant for the supply of water and minerals
8. Strigol is a (tick mark which is not true):
   (a) Compound which inhibits germination of the seeds of *Striga* sp
   (b) Stimulant for the germination of cotton seeds
   (c) Secondary metabolite released from the roots of host plant
   (d) Sesquiterpene released from the roots of cotton plant

9. Sticky substance produced by the host plant when induced by *Cuscuta* include (tick mark which is not correct):
    (a) Arabinogalactan proteins
    (b) Polysaccharides
    (c) Parahydroxyacids
    (d) Pectins
10. Parasitic plants have:
    (a) Lower water use efficiency than the host plants
    (b) Higher water use efficiency than the host plants
    (c) Lower transpiration rates than the host plants
    (d) Maintain high temperature around their leaves
11. Catechin, a toxic secondary metabolite, is produced by:
    (a) *Eichhornia crassipes*
    (b) *Alliaria petiolata*
    (c) *Centaurea maculosa*
    (d) *Juglans nigra*
12. An interaction when both partners of the association are benefitted is known as:
    (a) Commensalism
    (b) Mutualism
    (c) Amensalism
    (d) Parasitism
13. Necrotrophy is the pathogenic infection in plants when the cells:
    (a) Are killed prior to infection
    (b) Remain alive throughout infection
    (c) Are immediately killed after the infection
    (d) Are killed much later on after the infection
14. An effector molecule is:
    (a) Produced by pathogens which triggers basal defense in plants
    (b) Present on the surface of pathogen by which it is identified
    (c) A molecule made by pathogen to overcome basal defense of the plant
    (d) Produced by the plant in response to pathogenic infection
15. Effector-triggered immunity refers to:
    (a) Defense strategies by host plant in response to the molecular pattern at the pathogen's surface
    (b) Suppression of host-triggered immunity by the effector molecules produced by pathogens
    (c) Triggering of defense response in plant as a result of incompatible interaction of factors produced by pathogens and R gene products
    (d) Immunity in host plants in response to elicitors
16. Fusicoccin is a non-specific toxin which:
    (a) Inhibits histone acetylase thus affecting gene expression involved in defense against pathogen
    (b) Mimics action of jasmonic acid isoleucine and helps in bacterial infection by activating JA signaling pathway
    (c) Acts by damaging plant cell walls
    (d) Inactivates membrane-bound $H^+$-ATPase proton pump

17. Toxicity of tabtoxin (a dipeptide) is due to:
    (a) Release of its hydrolysis product an amino acid threonine
    (b) Release of tabtoximine as a result of its hydrolysis, a potent inhibitor of glutamine synthetase
    (c) Due to pores created in the membrane resulting in leakage of nutrients
    (d) Irreversible opening of stomata which leads to wilting of plants
18. Effector molecules are:
    (a) Proteins required to develop Type III system in bacteria
    (b) Products of R genes in host plants
    (c) Molecules produced by pathogens that trigger infection of the host plant
    (d) Products of avr genes of pathogens
19. Systemic acquired resistance is:
    (a) Due to accumulation of defense proteins not only at the site of infection but also systemically in uninfected tissues and/or uninfected plants
    (b) Due to suppression of infection by disease-causing pathogens by providing systemic resistance by antagonism
    (c) Resistance in the host plant at the infected tissues by pathogens
    (d) Very specific and long lasting

## Answers

1. c   2. b   3. b   4. d   5. a   6. b   7. b
8. d   9. c   10. a   11. c   12. b   13. a   14. c
15. c   16. d   17. b   18. d   19. a

## Suggested Further Readings

Alba JM, Allmann S, Glas JJ, Schimmel CJ, Spyropoulou EA, Stoops MV, Kant MR (2012) Induction and suppression of herbivore-induced indirect defenses. In: Witzany G, Baluska F (eds) Biocommunication of plants. Springer, New York, pp 197–212

Engelberth J (2015) Plant resistance to insect herbivory. In: Witzany G, Baluska F (eds) Biocommunication of plants. Springer, New York, pp 303–326

Hammond KE, Jones JDG (2015) Responses to plant pathogens. In: Buchanan BB, Gruissem W, Jones RL (eds) Biochemistry and molecular biology of plants. Wiley Blackwell, UK, pp 984–1050

Smith AM, Coupland G, Dolan L, Harberd N, Martin C, Sablowski R, Amey A (2010) Interactions with other organisms. In: Plant biology. Garland Science, London, pp 499–572

Taiz L, Zeiger E, Moller IM, Murphy A (2015) Biotic interactions. In: Plant physiology and development, 6th edn. Sinauer Associates, Sunderland, pp 693–730

# Part V
# Applied Plant Physiology

α-Pinene

β-Pinene

Myrcene

Secondary metabolites in the resins from Pinus. More details are provided in Chap. 33, Sect. 33.1.4.2, Fig. 33.4

Chapter 33 Secondary Metabolites
Chapter 34 Plant Physiology in Agriculture and Biotechnology

# Secondary Metabolites

## 33

Satish C Bhatla

Plants synthesize an enormous array of organic compounds/metabolites, a large number of which are involved in facilitating the basic vital processes, such as growth, cell division, respiration, photosynthesis, reproduction, and storage. These metabolites are known as **primary metabolites**, while the rest of the compounds do not have any role in primary metabolism. These compounds perform varied functions and are collectively referred as **secondary metabolites**. At biosynthetic level, both categories of compounds are synthesized from same basic metabolic pathways and share many intermediates (Fig. 33.1). Secondary metabolites are also referred as **secondary products**, **specialized metabolites**, or **natural products**. Innumerable secondary metabolites have been identified till now. Primary metabolites are generally produced in large concentrations as they are required in vital processes. However, secondary metabolites are mostly produced in small quantities, are structurally diverse, and may have a restricted distribution in specific families, genera, or species. Due to their confined, specific distribution, secondary metabolites can be used as a diagnostic tool in chemotaxonomic studies. For many years, the significance of secondary metabolites was not well understood, and they were believed to be mere metabolic wastes or functionless intermediates or end products of primary metabolism. Later, investigations initiated in the nineteenth century, pioneered by organic chemists, revealed the importance of secondary metabolites in medicines (e.g., discovery of aspirin (acetylsalicylate), as defensive compounds (phytoalexins), role in restricting germination and growth of other plants growing in the vicinity (allelopathy), UV protection, etc. Therefore, these metabolites help in adaptation of plants according to their environment. Some of their major roles are:

- Protection of plants against the action of herbivores and microbial pathogens, thus playing a role in *defense mechanisms*.

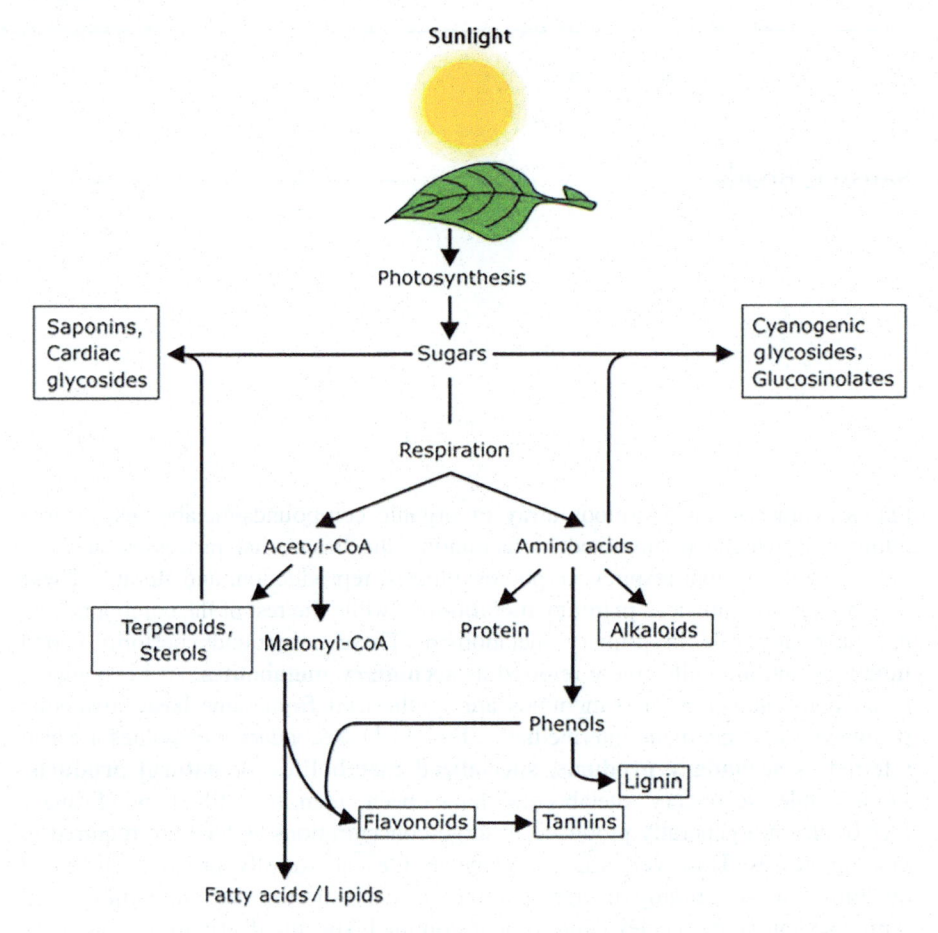

**Fig. 33.1** A simplified depiction of the interrelationship between the major primary and secondary metabolic pathways. Secondary metabolites are shown in boxes

- They may provide color, smell, or taste to the organ where they are present, thereby serving as *attractants for pollinators*, thus facilitating fruit and seed dispersal.
- Some of them may also have roles in *plant-plant competition* (allelopathy) and plant-microbe symbiosis in root nodules (as signaling molecules).
- A number of them are known to exhibit *medicinal properties*, thus making them immensely important pharmacologically active compounds (e.g., aspirin) for mankind.

Secondary metabolites can be categorized in many different ways such as on the basis of their chemical structure (presence of ring structure, sugar components, etc.), composition (nitrogen, sugar, etc.), solubility in various solvents, or according to the pathway by which they are synthesized (phenylpropanoid pathway, flavonoid pathway, etc.). The common categories of secondary metabolites are:

**Terpenes**—hormones, pigments, essential oils, steroids, and rubber
**Phenolics**—coumarins, flavonoids, lignin, and tannin
**Nitrogen-containing compounds**—alkaloids, glycosides (saponins, cardiac glycosides, and cyanogenic glycosides), glucosinolates, and non-protein amino acids

Diverse structural and chemical properties of secondary metabolites make them suitable substrates for biofuels, biomaterials, and pharmaceutical industries. For many years, various secondary metabolites have been used in the production of modern medicines (many of which are derived from phenols). Secondary metabolites are suitable for this purpose because they have the ability to cross the cell membranes and perform their biological activity. However in crude form, they are often toxic, non-nutritional, or allergenic in nature. High consumption of these compounds may sometimes have severe consequences. Heavy dosage of alkaloids leads to hallucination, blocking of ion channels and neurotransmission of signals, coordination problem, etc., while phenol consumption affects digestion, growth, and cell division and blocks enzyme activity and can be detrimental to humans and animals.

## 33.1 Terpenes

Terpenes constitute the largest and most diverse class of secondary metabolites having more than 25,000 compounds. Initially these compounds were called "terpene," because the first few compounds of this group were isolated from "turpentine" (pine) tree and were having the basic formula of $C_{10}H_{18}$. They are naturally produced among conifers, higher plants, and animals and are often characterized by their strong odor and volatile nature. Strong odor discourages herbivores and attracts their natural predators to protect plants from them, while their volatile nature provides the characteristic fragrance to plants and flowers. Chemically, terpenes are lipophilic products of 5-carbon (5-C) isoprene units (i.e., 2-methyl-1,3-butadiene) and are, therefore, also known as **isoprenoids**. Isoprene unit is a 5-C moiety having the isopentane carbon skeleton structure, obtained from the pyrolysis (dry distillation) of natural rubber (latex derived from *Hevea brasiliensis*) which, upon decomposition at high temperature, releases isoprene units.

$$\begin{array}{c} H_3C \\ \phantom{H_3C}\diagdown \\ \phantom{H_3C}\phantom{\diagdown}CH-CH_2-CH_3 \\ \phantom{H_3C}\diagup \\ H_3C \end{array} \qquad \begin{array}{c} H_3C \\ \phantom{H_3C}\diagdown \\ \phantom{H_3C}\phantom{\diagdown}CH-CH=CH_2 \\ \phantom{H_3C}\diagup \\ H_2C \end{array}$$

Isopentane          Isoprene unit

**Table 33.1** Terpenes in plant functions (some examples)

| Functions | Terpene constituents |
|---|---|
| Light harvesting (photosynthesis) | Phytol tail of chlorophyll |
| Electron transfer | Plastoquinone |
| Photoprotection | Carotenoids |
| Photosynthesis and respiration | Ubiquinone |
| Maintenance of cell membrane | Sterols |
| Pigmentation; antioxidant | Vitamin A |
| Various growth and development processes | Phytohormones; gibberellic acid (GA) and abscisic acid (ABA) |

Any modification in terpenes, like oxidation or rearrangement of carbon atoms, results in the formation of compounds called **terpenoids**. In other words, terpenes are hydrocarbon molecules, while terpenoids/isoprenoids are the terpenes that contain an additional functional group (hydroxyl, carbonyl, ketone, aldehyde, etc.) in their structure. However, in literature these terms are often used interchangeably. Terpenes have many biological functions, including their roles in light harvesting (as a constituent of phytol tail of the chlorophyll molecule), electron transfer (plastoquinone), and photoprotection (carotenoids) during photosynthesis and respiration (ubiquinone) processes, in maintenance of cell membrane structure and permeability (sterols), and as phytohormones (Table 33.1). Terpenoids also contribute to the flavors of clove (*Syzygium aromaticum*), cinnamon (*Cinnamomum verum*), and ginger (*Zingiber officinale*) and pigmentation in sunflower (*Helianthus annuus*) and tomato (*Solanum lycopersicum*). Gymnosperms, like pine and conifers, inhibit pathogen and herbivore attacks by secreting mixture of terpenoids which act as repellant and ultimately consolidate and function as a mechanical barrier and protect wounds. Both terpenes and terpenoids are basic constituents of essential oils (eucalyptus oil) and resins in gymnosperms and are generally insoluble in water due to their lipophilic nature (Table 33.2). Vitamin A is also a terpenoid. Synthetic derivatives of terpenes and terpenoids have expanded the range of aromas in perfume industry and flavors of food additives. Terpenoids are produced by a variety of plants, mainly conifers, citrus, and eucalyptus. Their strong smell and hydrophobic nature helps in protecting plants against parasites. Common examples are myrene in *Pinus*, menthol in *Mentha* leaves, cannabinoids in *Cannabis*, azadirachtin in neem, and ginkgolide and bilobalide of *Ginkgo biloba*. Rubber (cis-1,4-polyisoprene) obtained from *Hevea brasiliensis* is the longest terpene (Fig. 33.2).

**Table 33.2** Some important secondary metabolites and their functions

| Class | Examples | Sources | Effects/functions |
|---|---|---|---|
| *Terpenoids* | | | |
| Monoterpenes | Menthol, linalool | *Mentha* sp. (mint) and relatives | Block ion transport; interfere with neurotransmission; anesthetic |
| Sesquiterpenes | Parthenolide | *Parthenium* and relatives | Cause contact dermatitis |
| Diterpenes | Gossypol | *Gossypium* sp. (cotton) | Block phosphorylation; toxic |
| Triterpenes (e.g., cardiac glycosides) | Digitoxin, digoxin, gitoxin | *Digitalis* sp. | Stimulate heart muscle contraction; alter ion transport |
| Tetraterpenoids | Carotene | Almost all higher plants | Provide orange coloration; part of accessory pigments; antioxidant |
| Terpene polymers | Rubber | *Hevea* (rubber) trees | Manufacturing tires, gaskets, footwear, etc. |
| Sterols | Spinasterol | *Spinacia oleracea* (spinach) | Interfere with the action of sterol-based animal hormones |
| *Phenolics* | | | |
| Phenolic acids | Caffeic acid | Chlorogenical [chlorogenic acid (an ester of caffeic acid) containing] plants | Cause oxidative damage, responsible for browning in fruits and wine |
| Coumarins | Umbelliferone | Carrots (*Daucus carota*) and other Apiaceae members | Cross-link DNA; block cell division |
| Lignans | Podophyllin | Mayapple (*Podophyllum* sp.) | Cathartic, induce vomiting, cause allergic dermatitis |
| | Urushiol | Poison ivy (*Hedera helix*) | |
| Flavonoids | Anthocyanin | Almost all plants, especially in flowers | Provide color to leaves and flower petals; inhibit enzyme action; anti- and pro-oxidants; estrogenic |
| | Catechin | | |
| Tannins | Gallotannin, condensed tannin | Oak (*Quercus* sp.), hemlock (*Conium* sp.), bird's-foot trefoil, legumes; tea (*Camellia sinensis*) | Bind to proteins and enzymes; block digestion; antioxidants |
| Lignins | Lignin | All land plants | Provide mechanical support and structure; toughness; fibers |
| *Nitrogen-containing* | | | |
| Alkaloids | Cocaine | *Erythroxylon coca* (coca) | Block ion channels; inhibit enzyme action; interfere with neurotransmission |
| | Morphine | *Papaver somniferum* (opium poppy) | |
| | Nicotine | *Nicotiana* sp. (tobacco) | |
| | Theobromine | *Theobroma cacao* (chocolate) | |
| *Nitrogen- and sulfur-containing* | | | |
| Glucosinolates | Sinigrin | Members of Brassicaceae | Potential anticancerous properties |

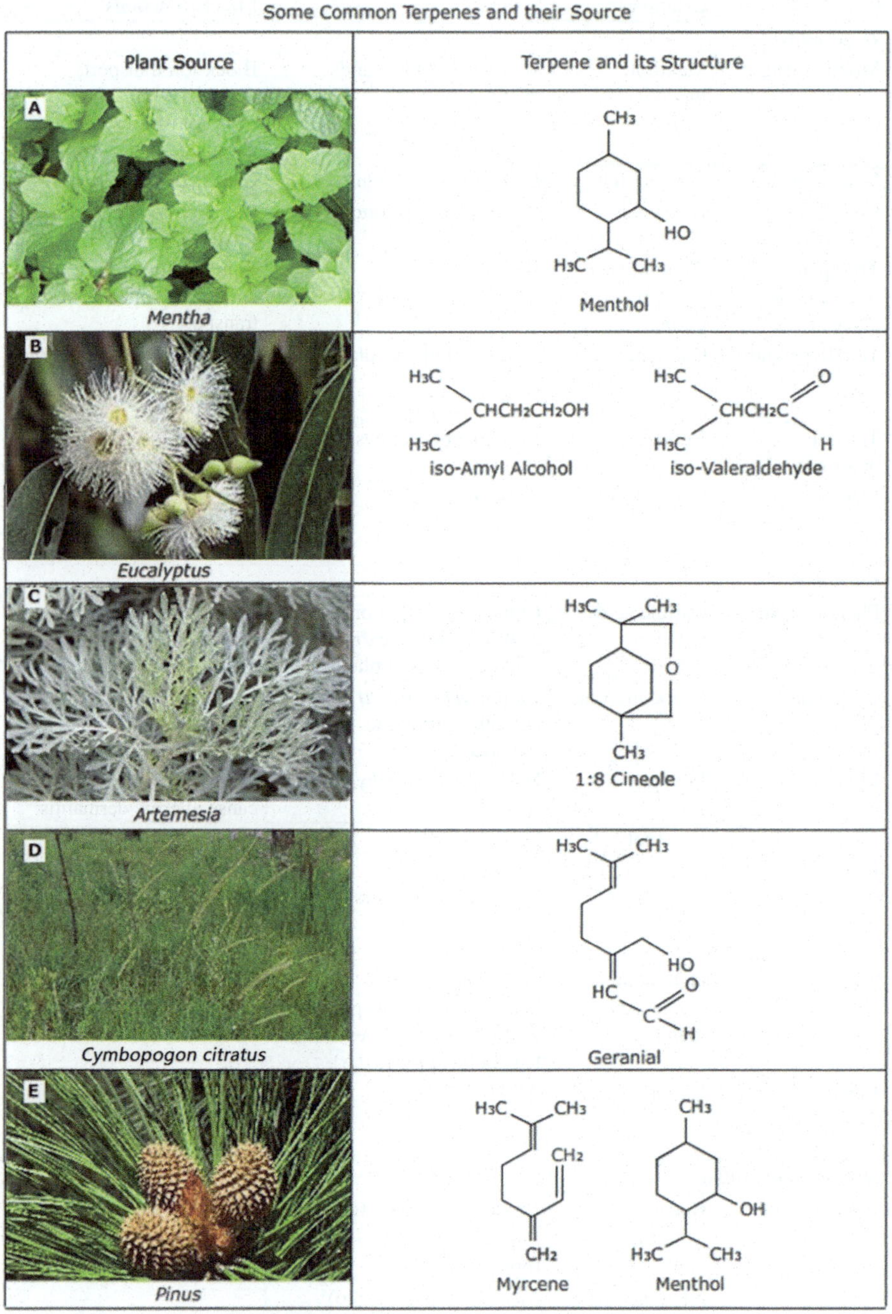

## Some Common Terpenes and their Source

| Plant Source | Terpene and its Structure |
|---|---|
| **A** Mentha | Menthol |
| **B** Eucalyptus | iso-Amyl Alcohol; iso-Valeraldehyde |
| **C** Artemesia | 1:8 Cineole |
| **D** Cymbopogon citratus | Geranial |
| **E** Pinus | Myrcene; Menthol |

**Fig. 33.2** Some common terpenes and their plant sources

33.1 Terpenes

| | |
|---|---|
| *Azadirachta indica* | Azadirachtin, a limonoid |
| *Polypodium vulgare* | α-Ecdysone, an insect molting hormone |
| Citrus (Lemon) | Limonene |
| *Taxus baccata* | Taxol-diterpene ($C_{20}$) |

**Fig. 33.2** (continued)

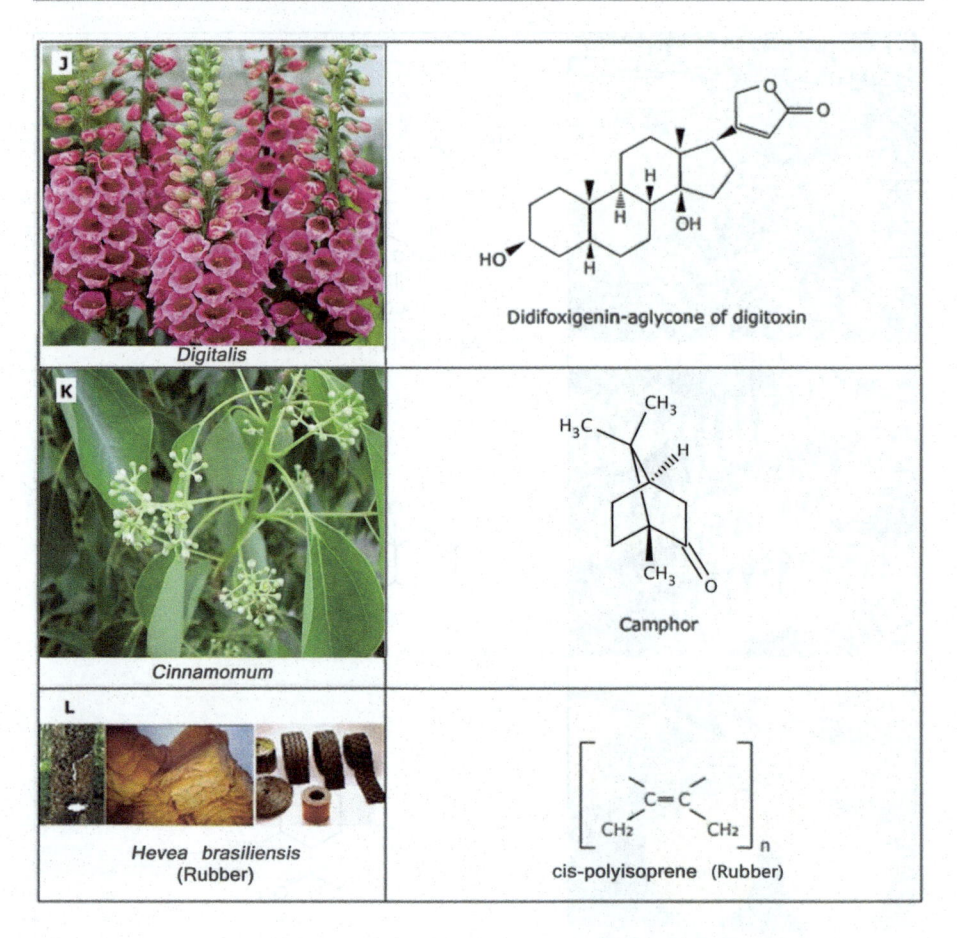

| | |
|---|---|
| Digitalis | Didifoxigenin-aglycone of digitoxin |
| Cinnamomum | Camphor |
| Hevea brasiliensis (Rubber) | cis-polyisoprene (Rubber) |

**Fig. 33.2** (continued)

### 33.1.1 Structure

Condensation of isoprene units [with the basic formula of $CH_2=C(CH_3)CH=CH_2(C_5H_8)$] leads to the formation of hydrocarbon compounds known as terpenes. Different structures of terpenes are multiples of the basic isoprene formula $(C_5H_8)_n$, n depicting the number of isoprene units linked together. This formulation is known as "isoprene rule." According to this rule, all terpenes and terpenoids are formed from one or more isoprene units, by "head-to-tail" linkage where the branched end of isoprene unit is known as the "tail" and the other end as "head." Exceptions to this rule are carotenoids, which are joined tail to tail, and some terpenoids, in which c-atoms are not multiples of five.

## 33.1 Terpenes

Myrcene
(a monoterpene)

### 33.1.2 Classification

Terpenes are classified on the basis of the number of $C_5$ isoprene units present in them (Table 33.3). Initially, 10-carbon terpenes were considered to be the smallest units and were thus named **monoterpenes**. Later, however, smaller 5-carbon terpenes were discovered and were then called **hemiterpenes** (i.e., half terpenes, e.g., tiglic acid) so as not to disturb earlier terminology. Although sometimes extensive structural modifications make classification of terpenes difficult, the major classification is based on the number of carbon atoms present in the structure and is tabulated in Table 33.3. These classes can be further divided into subclasses on the basis of number of rings present in their structure.

### 33.1.3 Biosynthesis

Terpene biosynthesis and storage are confined to distinctive organs, such as trichomes (glandular), cavities, resin ducts, and blisters. These organs depend on the neighboring tissues for carbon and energy supply required for the production of terpenes, as these organs are non-photosynthetic in nature. Terpenes are biosynthesized mainly by two pathways, viz., **mevalonic acid (MVA) pathway** and **methylerythritol phosphate (MEP) pathway** (Fig. 33.3). These two pathways give rise to DMAPP (dimethylallyl pyrophosphate) and IPP (isopentenyl pyrophosphate), the two major precursors for terpene biosynthesis. Mevalonic acid pathway has a eukaryotic origin, while methylerythritol phosphate pathway has prokaryotic origin. In the mevalonic acid pathway, two acetyl-CoA molecules combine to form acetoacetyl-CoA, followed by condensation of third acetyl-CoA to form 3-hydroxy-3-methyl-glutaryl-CoA (HMG-CoA) in a stepwise manner. HMG-CoA, upon reduction, provides a molecule of mevalonic acid (a 6-carbon intermediate) which, on further modification via pyrophosphorylation, forms mevalonic acid pyrophosphate (MVA-PP) by utilizing two molecules of ATP. In the last step, MVA-PP is decarboxylated, and subsequently water is removed, which results in the formation of **isopentenyl diphosphate (IPP)**, a 5-carbon unit. Sometimes, IPP is isomerized to

**Table 33.3** Classification of terpenes according to the number of 5-carbon isoprene units contained in them

| Number of carbon units | Class | Representative terpenes and their structures |
|---|---|---|
| 5 | Hemiterpenoids | Tiglic acid: CH₃–CH=CH–CH₂–COOH |
| 10 | Monoterpenoids | Geraniol: (CH₃)₂C=CH–(CH₂)₂–C(CH₃)=CH–CH₂OH |
| 10 | Cyclic monoterpenoids | Menthol (peppermint oil) |
| 15 | Sesquiterpenoids | Farnesol (widespread), HOCH₂– |
| 20 | Diterpenoids | Phytol (chlorophyll), –CH₂OH |
| 30 | Triterpenoids | Squalene (a steroid precursor) |

dimethylallyl pyrophosphate (DMAPP), another 5-carbon unit, by the action of an enzyme called isopentenyl diphosphate isomerase (IDI), and both IPP and DMAPP act as primary substrates in terpene production. In the MVA pathway, the step catalyzed by the enzyme 3-hydroxy-3-methylglutaryl-CoA reductase (HMGR) plays a decisive role in rate of biosynthesis of terpenes. Its gene is ER-targeted gene, and its differential expression in various tissues and stress results in production of various types of terpenes. IPP can also be synthesized via glyceraldehyde 3-phosphate (GAP) and pyruvate, intermediary products of glycolysis, through a unique chemical pathway called the methylerythritol phosphate (MEP) pathway. MEP pathway operates in plastids, and it begins with the condensation of pyruvate and glyceraldehyde 3-phosphate, which results in the formation of an intermediate, 1-deoxy-D-xylulose-5-phosphate (DXP). DXP then leads to the formation of MEP by reductive isomerization; MEP, on further phosphorylation, cyclization, and reductive isomerization in a stepwise manner, forms 4-hydroxy-3-methyl-butenyl 1 phosphate (HMBP). HMBP is finally converted to either of the two terpene precursors, IPP or DMAPP. Downstream pathway of terpene biosynthesis is similar in MVA and MEP pathways after the production of the building block, isopentenyl diphosphate (IPP), and its isomer, dimethylallyl pyrophosphate (DMAPP). Enzyme

## 33.1 Terpenes

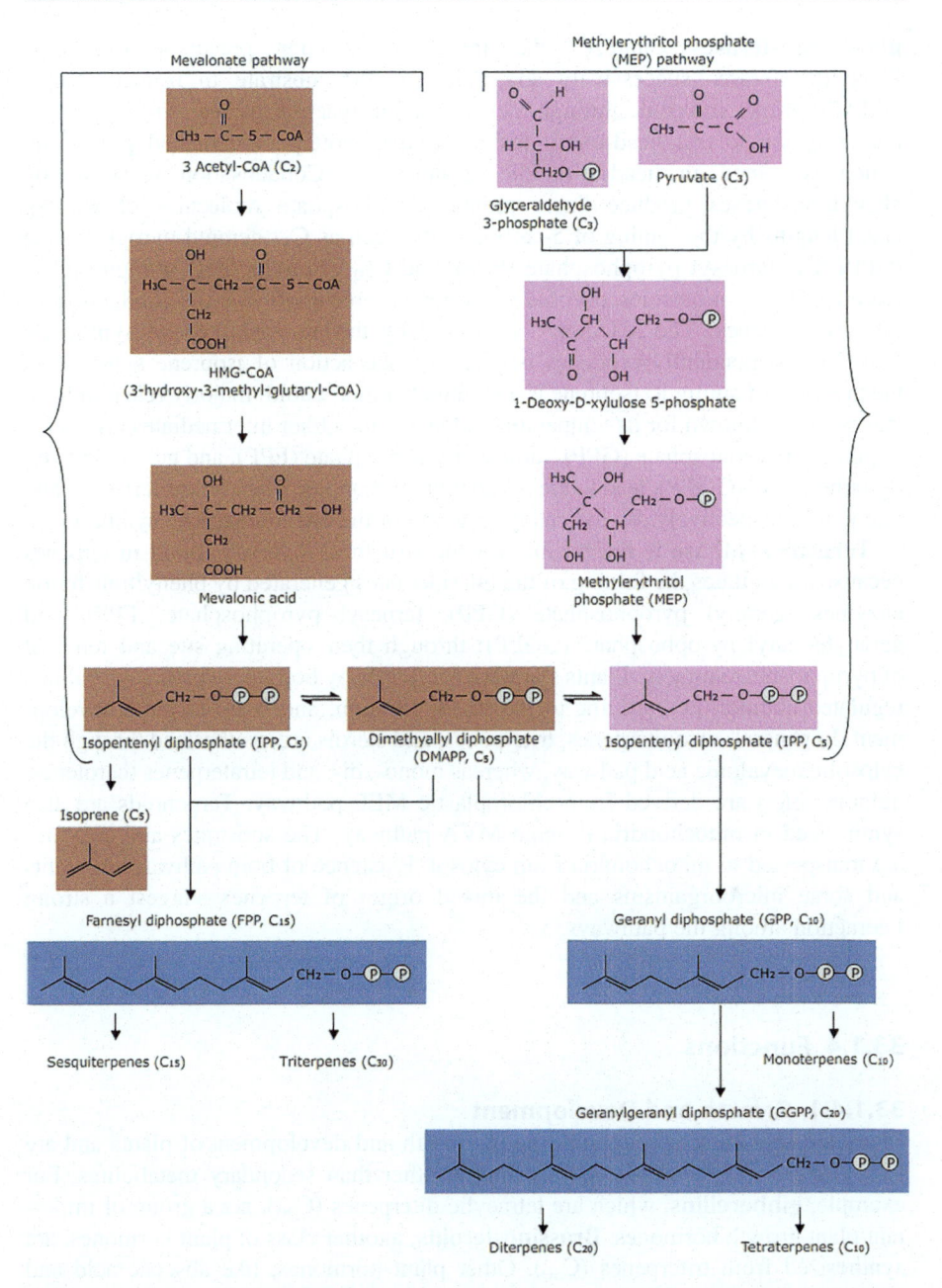

**Fig. 33.3** Terpene biosynthesis. Mevalonic acid pathway and MEP pathway are involved in synthesis of 5-C isopentyl pyrophosphate (IPP) and its isomer, dimethylallyl diphosphate (DMAPP). IPP and DMAPP combine to form geranyl diphosphate (GPP), which is the precursor for almost all monoterpenes, diterpenes, and triterpenes. GPP combines with another molecule of IPP to form farnesyl diphosphate (FPP) which serves as precursor to most sesquiterpenes and triterpenes

**phenyltransferases** catalyzes the initial steps after precursor formation. Phenyltransferases successively add units of IPP substrate to primary single DMAPP primer molecule through a condensation reaction in head-to-tail fashion. An exception to this head-to-tail fashion is a monoterpenoid named **pyrethrin**, which is joined in head-to-middle arrangement. Condensation reactions by phenyltransferases produce linear prenyl pyrophosphate molecules of varying chain lengths by the joining of 5-carbon units, such as $C_{10}$ geranyl pyrophosphate (GPP), $C_{15}$ farnesyl pyrophosphate (FPP), and $C_{20}$ geranylgeranyl pyrophosphate (GGPP). These molecules become the primary precursors for the production of different terpenes. The reaction is catalyzed by the enzyme **terpene synthases**. DMAPP independently produces isoprene by the action of isoprene synthase in the presence of sunlight. Isoprene is a volatile hemiterpenoid, majorly released from leaves, and is known for its temperature adjustments. Other intermediates also, such as geranyl pyrophosphate (GPP), farnesyl pyrophosphate (FPP), and geranylgeranyl pyrophosphate (GGPP), lead to the production of monoterpenes, sesquiterpenes, and diterpenes, respectively, by the catalytic action of the enzyme terpene synthase.

**Terpene synthase** is responsible for the structural diversity found in terpenes because it facilitates the folding of the intermediates generated by phenyltransferase enzymes (geranyl pyrophosphate (GPP), farnesyl pyrophosphate (FPP), and geranylgeranyl pyrophosphate (GGPP)) through their operating site and removal of pyrophosphate moiety. Plants synthesize terpenes by both these pathways but in a regulated manner, i.e., specific to the stress, location, and stage of plant development. In general, sesquiterpenes, triterpenes, and sterols are synthesized through the cytosolic mevalonic acid pathway, whereas mono-, di-, and tetraterpenes (carotenes, quinone, etc.) are derived from chloroplastic MEP pathway. Terpenoids are also synthesized in mitochondria through MVA pathway. The substrates and enzymes are transported to mitochondria from cytosol. Existence of both pathways in plants and some microorganisms and the mixed origin of terpenes suggest a strong interaction among the pathways.

### 33.1.4 Functions

#### 33.1.4.1 Growth and Development

Many terpenes play an essential role in growth and development of plants and are thus eligible to be considered as primary rather than secondary metabolites. For example, **gibberellins**, which are tetracylic diterpenes ($C_{20}$), are a group of important plant growth hormones. **Brassinosteroids**, another class of plant hormones, are synthesized from triterpenes ($C_{30}$). Other plant hormones, like abscisic acid and natural cytokinin, are hemiterpene ($C_{15}$) and sesquiterpene ($C_5$) by nature. **Sterols**, which are essential components of cell membranes, stabilize the membrane structure through their interaction with the phospholipid component of the membrane, which are also terpenes. They are made up of six chains of isoprene units (triterpenes, $C_{30}$). **Carotenoids** are a major class of terpenoid with tetraterpene ($C_{40}$) structure and are vital accessory pigments in energy transfer process during photosynthesis. They protect plants from the harmful effects of photooxidation and are responsible for various colors (red, yellow, orange) of flowers and fruits in some plants, which help to attract pollinators and seed dispersal animals. **Dolicols** (long-chain polyterpene alcohols) act as sugar carriers during the synthesis of glycoproteins. **Phytol** is also a diterpene ($C_{20}$) and forms the tail component of chlorophyll molecule and helps in the transportation of chlorophyll within the photosystem complex. It is present embedded in thylakoid membranes of chloroplasts. Terpenes are lipophilic by nature and hence are capable of mediating the interaction and adhesion of various molecules and compounds. Electron carriers, such as ubiquinones, plastoquinones, and phylloquinones, also adhere to their respective membranes with the help of 5–10 $C_5$ units.

### 33.1.4.2 Plant Defense

Terpenes have the ability to deter herbivores, insects, and animals as they are gluey, smelly, irritating, and toxic in nature. **Pyrethroids** are monoterpene esters, found in the leaves and flowers of *Chrysanthemum* species. They have strong insecticidal property due to their neurotoxicity toward insects which feed on them (Box 33.1). They are a popular ingredient in some commercial insecticides as well. In gymnosperms, such as pines and conifers, monoterpenes like **α-pinene**, **β-pinene**, and **limonene** accumulate in the resin ducts of needles, twigs, and trunks (Fig. 33.4). These monoterpenes have stinking smell and are toxic in nature. They thus play a major role in the defense against various insects, including bark bottles. These blends of many terpenes from plants are released immediately after the attack of herbivore or insects and repel herbivores and attract their natural predators, like wasps and arthropods. Some of the terpenes are nonvolatile in nature. **Limonoids** are a group of triterpenes, which are responsible for the bitterness of citrus fruits. Azadirachtin is one of the most powerful naturally occurring insect deterrents. It is a complex limonoid found in the seeds of neem tree (*Azadirachta indica*). Dry leaves of neem are commonly kept in wardrobes and are believed to be sufficient to keep insects away. It is considered to be a very powerful deterrent to insects, and its preparations are marketed for such purposes for application in agricultural fields and households. **Phytoecdysones** are plant steroids which were first reported from the fern *Polypodium vulgare*. They play an important role in defense against parasitic nematodes. Also, they disrupt molting and other developmental processes in insects with lethal consequences.

**Box 33.1: Secondary Metabolites as an Organic Tool to Control Pest**

Pyrethrin I (an insecticide)

Pests are a major cause for losses in agricultural productivity, and their effective management is important to overcome financial losses and to improve crop production. Pest management becomes more challenging for organic farmers who intend to avoid or minimize the use of insecticides or other chemicals. Among the different organic strategies adopted for effective pest control, one method is *trap cropping*. It involves planting of some trap plants (host), mostly peripheral to the main cash crop. *Chrysanthemum* is commonly used as *trap crop* for controlling *leaf miner* in bean crop (main cash crop). *Chrysanthemum* flowers (esp. ovaries) produce a monoterpene ester, pyrethrin, which acts as insect neurotoxin that interferes with the sodium channels of insect nerve cells. Also, the unopened flowers of *C. cinerariifolium* are dried, powdered, and used as organic insecticides in Japan and Eastern Africa. They are popular because of low mammalian toxicity and their perishable nature as they are inactivated upon oxidation.

[*A leaf miner is the larva of an insect that lives in and eats the leaf tissue of plants. The vast majority of leaf-mining insects are moths (Lepidoptera), sawflies (Symphyta, a type of wasp), and flies (Diptera), though some beetles also exhibit this behavior.*]

### 33.1.4.3 Essential Oils

Essential oils are naturally occurring volatile oils composed of aromatic compounds. These are responsible for emission of characteristic odors from plant's foliage or other parts and are mainly a volatile blend of monoterpenes and sesquiterpenes. The characteristic fragrance of lemon, mint, parsley, and peppermint is due to the presence of essential oils only. Limonene and menthol are the chief monoterpene constituents of lemon oil and peppermint oil, respectively. These essential oils endow plant with insect-resistant properties and protect the plant from insect attack as the glandular hairs or secretary cavities storing them are present on the epidermal surface and release its constituents on crushing.

**Fig. 33.4** Secondary metabolites present in *Pinus* sp. (**a**) Pines are rich source of resins. (**b, c**) The resin is collected from the trunk of pine tree using the tapping technique. Resins have high amounts of α- and β-pinene and myrcene

### 33.1.4.4 Sterols and Steroids

Sterols are steroids with an alcoholic group. They contain less than 30 carbon atoms and are synthesized from acyclic triterpene squalene. Plants contain more than 150 sterols, of which stigmasterol and sitosterol are the most abundant sterols in higher plants (Fig. 33.5). They comprise more than 70% of total sterols. Cholesterol, another sterol widespread in animal sources, is found in traces in plants as well. This is the reason why oils extracted from plants are preferred for dietary intake over animal fats. Sterols are constituents of cell membranes, in addition to phospholipids and proteins, which remain the major constituents. Since they are planar in structure, their molecules pack more tightly in the membranes than the phospholipids and thus enhance stability of membranes. Another group of triterpene glycosides are **saponins**. They are steroids and have soap-like detergent properties as they contain both lipophilic and hydrophilic chemical groups in the same molecule. Saponins

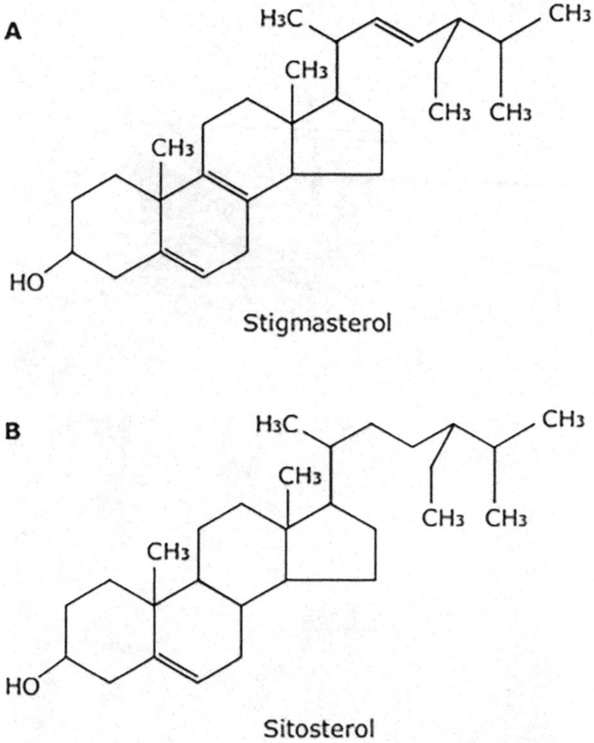

**Fig. 33.5** Structures of two sterols commonly found in plants

A

Stigmasterol

B

Sitosterol

form complexes with sterol and hinder the proper uptake of sterols in the digestive tract. They may also cause disruption of cell membranes when absorbed in the blood. These properties of saponins render them their toxic nature.

### 33.1.4.5 Rubber

Chemically, natural rubber is a polyterpene consisting of up to 15,000 units of isoprene units (Fig. 33.6). The polymer may exist in linear chains or cross-linked to form complex structures. These long polyterpene chains are coiled up like spring which gives rubber its elastic nature. Rubber occurs as microscopic particles suspended in sticky, milky-white aqueous suspension called **latex**. It accounts for 30–40% of latex. The rest of the latex is made up of terpenes, resins, proteins, sugars, and water. Latex is contained in interconnected vessels called as **laticifers**. Members of over 80 families, mainly Euphorbiaceae, Asteraceae, Asclepiadaceae, Apocynaceae, and Moraceae, comprising over 2000 species, are reported to produce latex. *Hevea brasiliensis* of Euphorbiaceae yields about 97% of the total of natural rubber produced worldwide. A single rubber tree may continue to produce rubber for up to 30 years of age. Rubber is extracted in the form of latex drawn out from the trunks by making incisions into the bark, using a technique called "tapping." The latex that oozes out is collected in vessels and then refined into rubber. **Gutta-percha**, or simply **gutta**, a substitute of rubber, is another polyterpene, and it differs

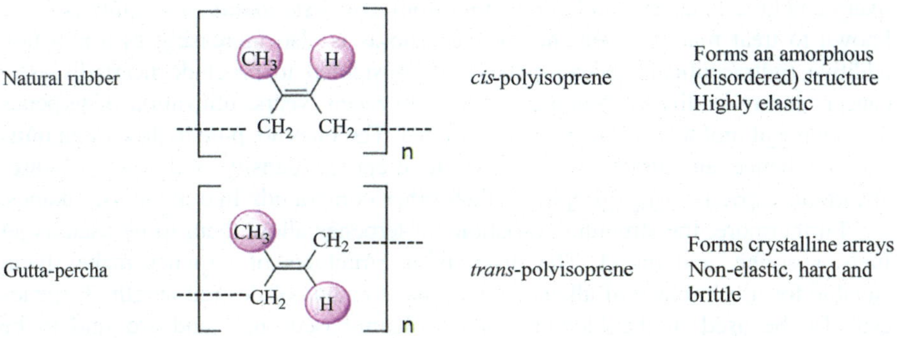

**Fig. 33.6** Basic moieties of rubber and gutta-percha

from rubber in the configuration of double bonds. Rubber is composed of all *cis*-linked isoprene units, forming an amorphous structure which makes it highly elastic. On the other hand, gutta-percha is a polymer of all *trans*-linked isoprene units, forming linear strands capable of arranging into crystalline arrays (Fig. 33.6). This difference in the configuration of the monomeric units leads to formation of a more rigid, inelastic structure, conferring plastic-like properties to gutta, rather than elastic properties of the *cis*-polyisoprene, i.e., rubber. Gutta is obtained from the sap of *Palaquium gutta* trees. The term "gutta" has been derived from the plant's name in Malay, i.e., "getah percha," which can be translated as "percha sap." Mexican guayule (*Parthenium argentatum*) is the best known source of gutta. In *P. argentatum*, latex is stored in vacuoles of cells in stem and root, instead of laticifers. Gutta-percha is impervious to water and is easily molded by heat and finds applications in the manufacture of doorknobs, basins, buckets, and products that have long-standing use underwater. It is also used in bubblegum industry. Rubber, on the other hand, being elastic and waterproof, has extensive applications in commercial production of numerous products having high resilience.

### 33.1.5 Industrial Applications

Structural and functional diversity of terpenes in plants has led to their commercialization for human benefits. Terpenes are used for various purposes, such as flavors, fragrances, beverages, pigments, soaps, toothpaste, and drug preparations. Therefore, terpenes are major components of food, agrochemicals, pharmaceuticals, and nutrition industries. For example, menthol, a monoterpene, is used as a flavoring agent and preservative in food industries and as an antibacterial component in pharmaceutical industry. Azadirachtin and pyrethrins are also utilized in production of insecticides because of their low toxicity to humans and less accumulation in the surrounding atmosphere (Box 33.1). Terpenes also have nutritional values, as many vitamins like vitamins A, D, E, and K are made up from several chains of C5 units. Taxol, a diterpene drug obtained from yew (*Taxus brevifolia*), is suspected to cure ovary and breast cancers, and artemisinin from *Artemisia annua*, a sesquiterpene, is known to treat malaria. Demand for carotenoids is also increasing as a nutrition additive (nutraceuticals) because they are considered to preclude heart diseases, cancer, and other life-threatening diseases. In recent years, utilization of terpenes as a source of biofuels is also being considered. Terpenes are pure hydrocarbon units and are hence an attractive source of high-energy density and low moisture-absorbing capacity (hygroscopicity) than other compounds like alcohols, alkanes, etc. Furthermore, the structural variations of terpenes allow them to be used as jet fuels, gasoline, and diesel. The diversity in structures of terpenes makes them suitable for use in place of alkane and aromatic chains of desired length. Terpenes can also be used in the blending of petroleum-based fuels and are said to be appropriate for use with existing transport fuel. Terpenes can also be employed in biodegradable plastics and polymers as replacement of petroleum-based plastics.

## 33.1.6 Metabolic Engineering in Terpenes

In recent years, popularity of terpenes has increased rapidly because of their increasing demand in food (fragrance, flavor), pharmaceutical (drugs), and agrochemical (insecticide) industries. But the major limitation is the trace amounts of terpenes produced by plants. For example, Taxol constitutes only 0.01–0.02% of bark dry weight of Pacific yew. Therefore, extraction of terpenes is not feasible as it requires sophisticated techniques and large natural resources to extract them in sufficient amount. Chemical synthesis of these compounds is also not economically feasible because of their complex structural diversity and high-energy requirements. However, terpene units (isoprene) are produced in large amounts by plants, and they account for about one-third global emission of volatile organic compounds (combining both natural and anthropogenic sources). One strategy to increase terpene production in plants is to either overexpress one or more genes or increase the rate of biosynthesis. Overexpression of genes often leads to production of novel type of terpene. Best known example of terpene bioengineering in plants is the development of **golden rice**. Golden rice is a variety of rice (*Oryza sativa*) having increased amount of β-carotene in its kernels. It was generated to treat the vitamin A deficiency among children and pregnant women. The series of reactions that produces β-carotene in plants initiates from the terpene precursor—isopentenyl diphosphate (IPP)—and the common intermediate of various IPP pathways, geranylgeranyl phosphate (GGPP), which is present in rice endosperm. But the conversion of geranylgeranyl phosphate (GGPP) to β-carotene requires a four-stage process assisted by four enzymes. This limits the production of vitamin A in plants. Overexpression of these four enzymes diverts GGPP to β-carotene through phytoene, resulting in 20-fold increase in carotene production than the normal rice, giving it a golden color. Another example of engineering of terpene metabolism is the production of artemisinic acid, the precursor of antimalarial sesquiterpene "artemisinin." In this case, the entire MAP pathway was incorporated in *E. coli* to enhance the production. In recent times, marine photosynthetic microbes are used to engineer terpene production as they can be cultured in salt water and require only $CO_2$ and sunlight for photosynthesis and metabolite production. This saves the energy and carbohydrate feedstock required to cultivate heterotrophic microorganisms and provides a greener route to generate metabolites.

## 33.2 Phenolic Compounds

Phenols are the omnipresent and varied group of secondary metabolites found in plants. Structurally, they are composed of one or more phenolic groups, i.e., six carbon (6-C) containing aromatic ring (benzene ring) with one or more hydroxyl functional groups as their basic chemical unit. Thus, they are collectively known as **phenolic compounds** or **polyphenols**. Phenolic hydroxyl group is slightly more acidic than other hydroxyl groups because it is located on a more stable structure of the benzene ring. Arenes can be easily stabilized by deprotonated oxygen substituent. (*Arenes refer to aromatic organic compounds that consist only of a conjugated planar ring system. Typical simple aromatic compounds are benzene and indole.*) These properties of phenols make them reactive molecules and suitable building blocks of long chain polymers, such as lignins and suberins and other protective compounds in plants. Phenolic compounds consist of heterogeneous chemicals, comprising of more than 10,000 compounds, and are widely distributed throughout the plant kingdom. Diversity of polyphenols ranges from simple molecules (e.g., phenolic acid, phenylpropanoids, flavanoids) to highly polymerized molecules (such as lignin, melanin, and tannin). Broadly, phenols in plants are divided into flavanoid, isoflavone, chalcone, stilbene, coumarin, furanocoumarin, lignin, tannin, and lignan subclasses, with flavanoids being the most common and abundant phenolic compounds in plants. Some phenols are ubiquitously distributed within the plant kingdom (such as chlorogenic acid), while the presence of others is restricted to particular classes/families/genera of plants, making them a worthy tool for taxonomic studies. Lignin, suberin, and pollen coat sporopollenin are some other examples of phenol polymers extensively distributed in plants. Bacteria, fungi, and algae lack phenolic compounds, but bryophytes consistently produce certain phenolic compounds such as flavanoids. At times phenolic units unite to form a dimer or polymer, generating a new group of polyphenols. For example, dimerization of gallic acid produces ellagic acid, which forms a class of ellagitannins. Ellagitannins are a diverse class of hydrolyzable tannins, a type of polyphenol formed primarily from the oxidative linkage of galloyl groups in 1,2,3,4,6-pentagalloyl glucose. Punicalagin isomers are ellagitannins found in the fruit, peel, and bark of pomegranates (*Punica granatum*). Sometimes, phenols form association with other chemical compounds like proteins, alkaloids, and terpenoids, and these groups of organic compounds are known as **phenylpropanoids**. They differ from each other in terms of structure and chemical properties, such as solubility in different solvents and their functional role in plants (Table 33.4). Their wide occurrence in plant products (vegetables, fruits, and beverages) makes them an integral part of human diet, and our daily phenylpropanoid intake ranges from 20 mg to 1 g. They are

**Table 33.4** Major groups of phenolic compounds, their functions, and examples of their major sources in plants

| Phenolic group | Functions | Specific example(s) and plant source |
| --- | --- | --- |
| Isoflavone | Fruit pigments | Osajin in *Maclura pomifera* |
| Anthocyanins |  | Petunidin glucoside in *Atropa belladonna* |
| Chalcones |  | Ocacin in *Kyllinga brevifolia* |
| Anthocyanins | Flower pigments | Cyanidin-3,5-diglucoside in *Rosa* |
| Chalcones |  | Coreopsin in *Coreopsis tinctoria* |
| Aurones |  | Aureusin in *Antirrhinum majus* |
| Flavones |  | Apigenin-7-glucoside in *Bellis perennis* |
| Yellow flavanoids |  | Gossypetin-7-glucoside in *Gossypium* |
| Phenylanthrenes | Phytoalexins | Orcinol in *Orchis militaris* |
| Stilbenes |  | Resveratrol in *Arachis hypogaea* |
| Isoflavanes |  | Vestitiol in *Lotus corniculatus* |
| Pterocarpanes |  | Pisatin in *Pisum sativum* |
| Phenylpropanoids |  | Coniferyl alcohol in *Linum usitatissimum* |
| Furocoumarins |  | Psoralen in *Petroselinum crispum* |
| Quinones | Protection against pests | Juglon in *Carya ovata* |
| Tannins |  | Gallotannin in *Quercus robur* |
| Flavonols |  | Quercetin glycosides in *Gossypium* |
| Isoflavones | Fungicide | Luteon in *Lupinus* |
| Phenolcarboxylic acids |  | Protocatechuic acid in *Allium* |
| Dihydrochalcones |  | Phloridine in *Malus pumila* |
| Quinones | Allelopathic substances | Juglon in *Juglans regia* |
| Phenols |  | Hydroquinone in *Arctostaphylos* |
| Phenolcarboxylic acids |  | Sialic acid in *Quercus falcata* |
| Hydrocinnamic acid |  | Ferulic acid in *Adenostoma* |

considered beneficial in human diet because of their antioxidant properties. Flavanoids and tannins are considered essential dietary phenolic compounds. Browning of apple after slicing or after exposure to air is due to the oxidation of phenolic compounds present in them (Box 33.2). Most of the phenolic compounds are soluble in water as they mostly occur in association with sugars, i.e., as glycosides, and also form conjugates with esters and exist as methyl ester derivatives. Due to their aromatic nature, they possess the ability to absorb most of the UV irradiation. Therefore the bulk of these compounds exists as constituents of the cell wall. Phenols, such as lignin, are one of the major constituents of cell wall and provide it with color, taste, stiffness, and strength. They also provide stability and vigor to the plant to enable it to stand in upright. Involvement of lignin and other phenolic compounds in providing structural stability forms the basis of shift of plant's habitat from aquatic to land. It is estimated that about 40% of biosphere's organic carbon is utilized in the production of phenols.

## Box 33.2: Why Do Cut Apples Turn Brown?

When an apple is cut or bruised, oxygen is introduced in the exposed and injured cells. The polyphenol oxidase (PPO, a mixture of monophenol oxidase and catechol oxidase) in these cells oxidizes phenolic compounds in the cells in the presence of oxygen, and they are converted to o-quinones, which are colorless precursors of brown-colored secondary products. o-Quinones react with amino acids or proteins, or they self-assemble into polymers resulting in brown-colored secondary products.

Coating freshly cut apples with sugar solution or simply dipping it in water can slow down its browning due to reduced oxygen diffusion. Lemon or pineapple juice can also prevent browning of cut apples since both these juices contain antioxidants, and also due to their acidic nature, they reduce PPO activity. Similar brown color of tea or coffee or cocoa is also developed by PPO-triggered enzymatic action during processing.

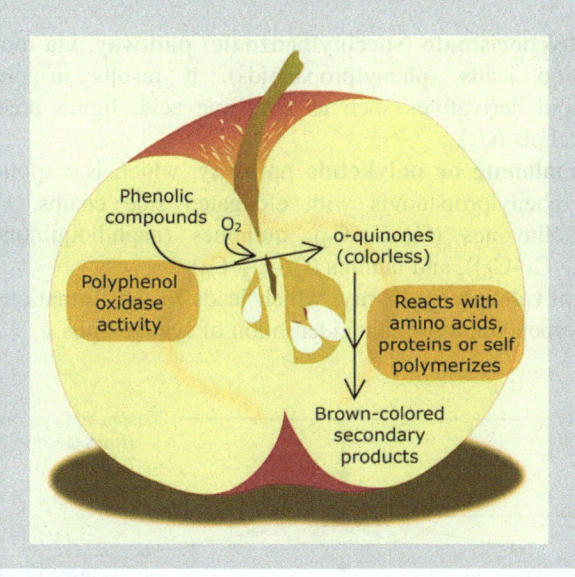

## 33.2.1 Functions

Biosynthesis of phenolic compounds in plants is attributed to various plant defense responses against biotic and abiotic stresses. They are often produced in response to wounding, pathogen/insect attack, high temperature, etc. For example, in high light conditions, biosynthesis and accumulation of phenols increase rapidly in chloroplasts and vacuoles, respectively. Nutrient (nitrogen, phosphorous, potassium, etc.) deficiency in plants also triggers production of phenylpropanoids (flavanoids) in some plants. They accumulate in subepidermal layers of plant tissue under such conditions. Some of the diverse roles assigned to these compounds are:

1. Protection of plants against herbivore and pathogen attack. They play dual role of repelling pathogens and herbivores and attracting their natural predators.
2. Coumarins and their derivatives act as anticoagulants.
3. Lignin provides mechanical strength to the plants.
4. Attract pollinators and facilitate dispersal of fruits and seeds.
5. Possess allelopathic roles, i.e., inhibit the growth of neighboring plants.
6. Protect plants against UV radiations.

### 33.2.2 Biosynthesis

Different metabolic routes are involved in the synthesis of a diverse array of phenolic compounds. Phenolics are synthesized naturally by three different pathways (Fig. 33.7):

(1) The **shikimate/chorismate (succinylbenzoate) pathway**, via modification of aromatic amino acids (phenylpropanoids). It results in production of phenylpropanoid derivatives such as cinnamic acid, lignin precursors ($C_6$–$C_3$)$_n$, and catechols ($C_6$)$_n$.
(2) The **acetate-malonate or polyketide pathway**, which is responsible for the synthesis of phenylpropanoids with elongated side chains, for example, flavanoids, isoflavones ($C_6$–$C_3$–$C_6$), quinones (naphthoquinones ($C_6$–$C_4$), xanthones ($C_6$–$C_1$–$C_6$)), and tannins ($C_6$–$C_3$–$C_6$)$_n$.
(3) The **acetate/mevalonate pathway**, which leads to the formation of aromatic phenols and terpenoids through condensation of acetate units.

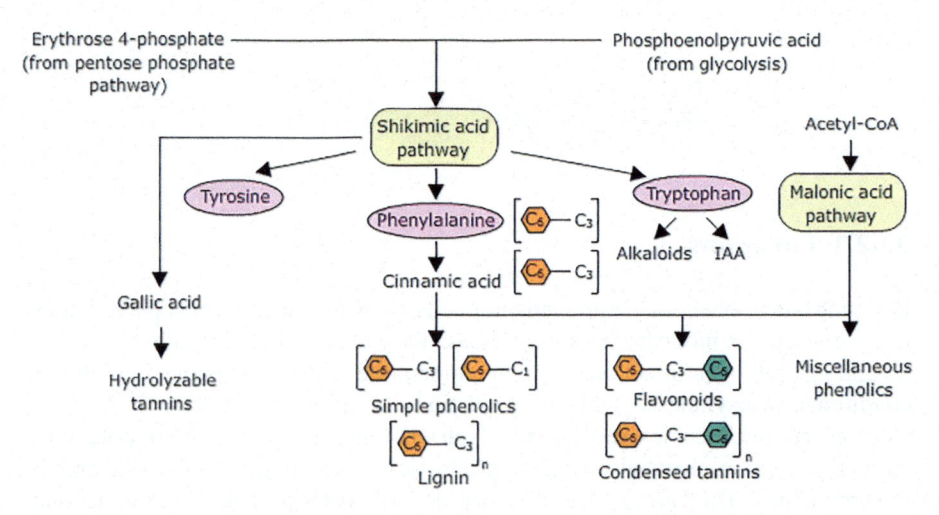

**Fig. 33.7** An overview of plant phenolic biosynthesis. Most phenolics are derivatives of phenylalanine formed in shikimic acid pathway. $C_6$: benzene ring; $C_3$: 3-carbon chain

## 33.2 Phenolic Compounds

Shikimic acid pathway starts with the condensation of erythose-4-phosphate and phosphoenolpyruvate, derived from the pentose phosphate pathway and glycolysis, respectively (Fig. 33.8). The resulting 7-carbon sugar is then cyclized and reduced to form shikimate or **shikimic acid** which is then converted to chorismate. Chorismate acts as a critical branch point, and with further modifications it leads to the formation of aromatic amino acids, i.e., phenylalanine, tyrosine, or tryptophan. Majority of simple phenolics are derived from phenylalanine. The simplest phenolic compounds are deaminated versions of the corresponding amino acids. In the first step toward the synthesis of most phenolic products, phenylalanine is deaminated to form cinnamic acid, and the reaction is catalyzed by the enzyme **phenylalanine ammonia lyase (PAL)** (Fig. 33.9). Subsequent series of reactions leads to incorporation of hydroxyl and other substituent groups in the structure and results in huge diverse group of phenolic compounds obtained in nature.

### 33.2.3 Classification

The major subgroups of phenolic compounds in plants include (Table 33.5):

Simple phenylpropanoids
Coumarins
Benzoic acid derivatives
Lignin precursors
Flavonoids and stilbenes
- Anthocyanins
- Flavones
- Flavonols
- Isoflavonols

Tannins

#### 33.2.3.1 Simple Phenylpropanoids
Free phenols are rare in plants. Some free phenols, such as hydroquinone, catechol, orcinol, and pyrogallol, are reported from some plants. Phenylpropanoids represent the largest pool of secondary metabolites (approx. 20% of carbon of terrestrial biosphere) and have a simple structure having 3-carbon side chain (linear side chain) attached to the aromatic ring (6-carbon phenyl ring) (Fig. 33.10). Approximately, more than 7000 phenylpropanoid compounds have been reported till now. They are biosynthesized from the aromatic amino acids, such as phenylalanine. They are formed by deamination of phenylalanine to cinnamic acid. This reaction is catalyzed by phenylalanine ammonia lyase (PAL). Examples are trans-cinnamic acid and p-coumaric acid, along with their derivatives, such as caffeic acid and ferulic acid. Simple phenylpropanoids act as precursors for the synthesis of complex phenylpropanoids, such as coumarins, flavonoids, lignin, and tannins.

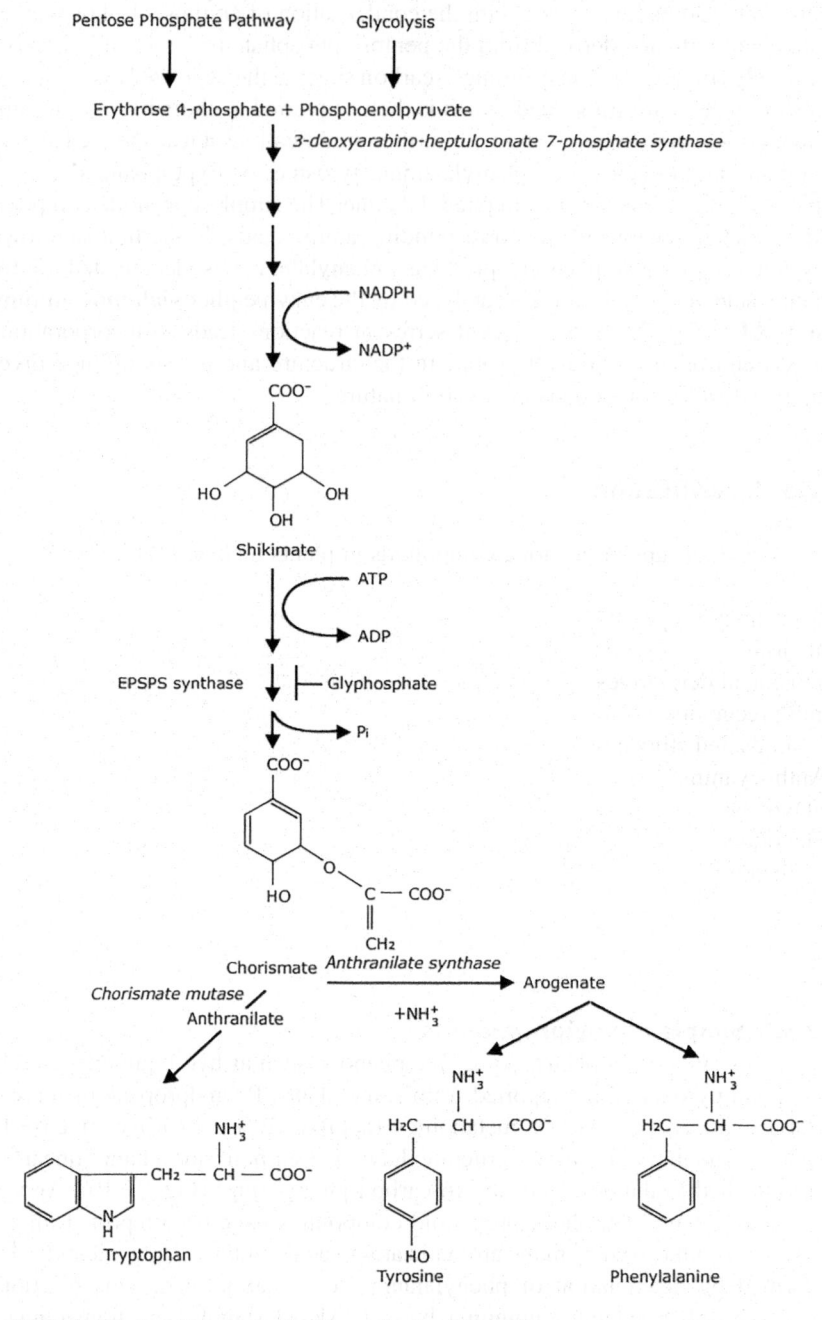

**Fig. 33.8** Shikimic acid pathway for biosynthesis of phenolics

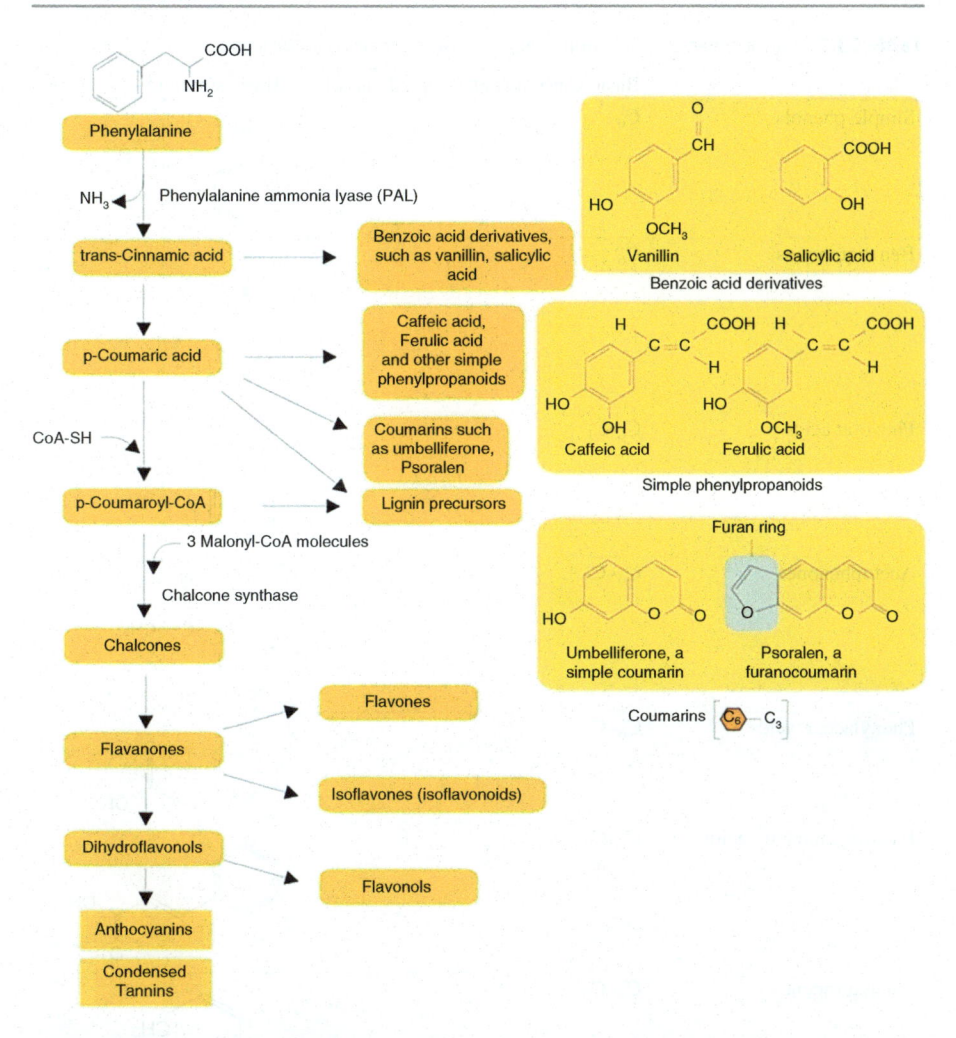

**Fig. 33.9** Various phenolic derivatives obtained from phenylalanine (a derivative of shikimic acid pathway of biosynthesis of phenolics)

### 33.2.3.2 Coumarins and Their Derivatives

They are cyclic esters, i.e., phenylpropanoid lactones (also called as benzopyranones), and constitute more than 1500 compounds having phenylpropanoid $C_6$–$C_3$ skeleton. Simplest examples are coumaric acid, umbelliferone (7-hydroxycoumarin), psoralen (furanocoumarin), scopoletin (6-methoxy-7-hydroxycoumarin), dicoumarol, and aflatoxin $B_1$ (Fig. 33.11). They are characteristic features of Umbelliferae family. These secondary metabolites function as antimicrobial agents, germination inhibitors, and feeding deterrents. Coumarin, a constituent of bergamot oil, is used to flavor food products such as

**Table 33.5** Various classes of phenolics on the basis of their basic structure

| Class | Basic Carbon skeleton made up of | Basic structure |
|---|---|---|
| Simple phenols | $C_6$ | Phenol (benzene with OH) |
| Benzoquinones | $C_6$ | 1,4-Benzoquinone |
| Phenolic acids | $C_6-C_1$ | Benzoate (benzene-COO⁻) |
| Acetophenones | $C_6-C_2$ | Benzene–C(=O)–CH₃ |
| Phenylacetic acids | $C_6-C_2$ | Benzene–CH₂–COOH |
| Hydroxycinnamic acids | $C_6-C_3$ | Benzene–CH=CH–COOH |
| Phenylpropenes | $C_6-C_3$ | Benzene–CH=CH–CH₃ |
| Coumarins, isocoumarins | $C_6-C_3$ | Coumarin structure |
| Chromones | $C_6-C_3$ | Chromone structure |

(continued)

## 33.2 Phenolic Compounds

**Table 33.5** (continued)

| Class | Basic Carbon skeleton made up of | Basic structure |
|---|---|---|
| Naphthoquinones | $C_6$–$C_4$ | 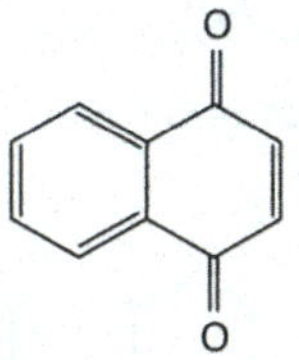 |
| Xanthones | $C_6$–$C_1$–$C_6$ | 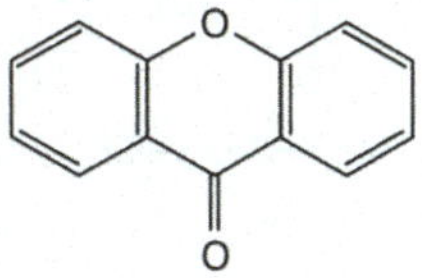 |
| Stilbenes | $C_6$–$C_2$–$C_6$ | 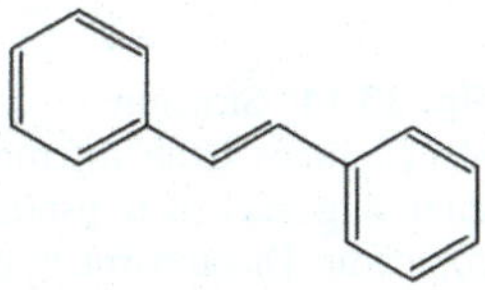 |
| Anthraquinones | $C_6$–$C_2$–$C_6$ | 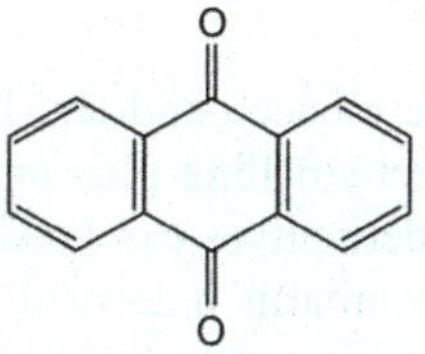 |
| Flavonoids | $C_6$–$C_3$–$C_6$ | 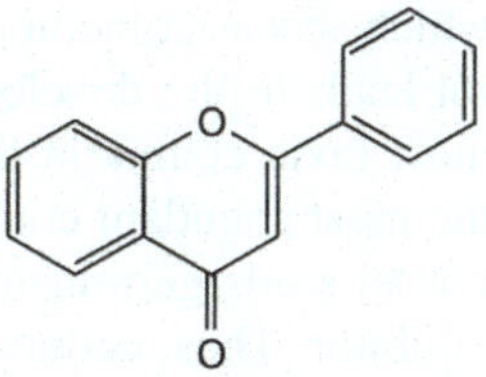 |

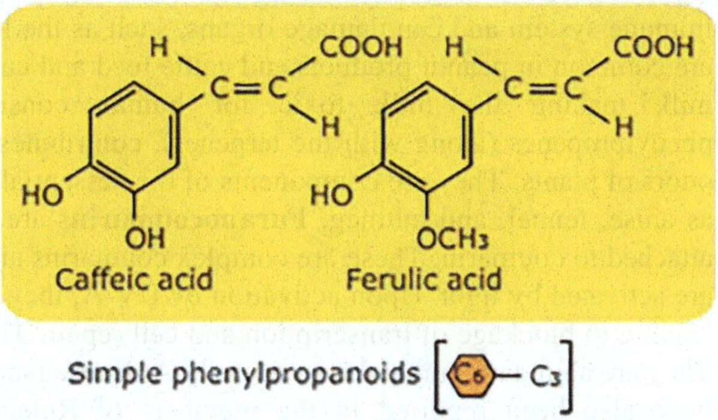

**Fig. 33.10** Basic structure of phenylpropanoids. Simple phenylpropanoids, such as caffeic acid and ferulic acid, are responsible for allelopathy

Caffeic acid

Ferulic acid

Simple phenylpropanoids $[C_6 - C_3]$

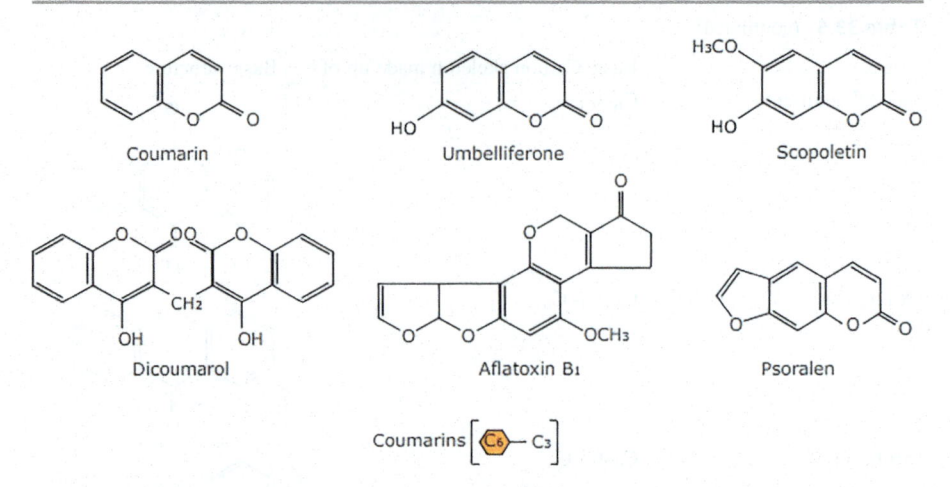

**Fig. 33.11** Structure of coumarins. More than 300 coumarins have been reported from different plant families. Umbelliferone is present in members of family Apiaceae. Coumarins may also have a furan ring, such as in psoralen which is toxic to insect herbivores. Scopoletin is the most common coumarin. Dicoumarol is a strong anticoagulant. Aflatoxin, a mycotoxin produced by fungi, is highly carcinogenic and toxic

pipe tobacco and tea. It is the most widespread coumarin. It is responsible for the sweet-smelling pleasant odor of new-mown hay. Though coumarin is mildly toxic, its derivatives can be very toxic. They also possess appetite-suppressing properties. Dicoumarin, a derivative of coumarin, present in moldy hay can cause fatal hemorrhage in cattle by inhibiting blood clotting. It inhibits the action of vitamin K, which acts as cofactor during the clotting of the blood. The discovery of dicoumarol leads to the development of warfarin, which is used as a rodent poison. The most toxic coumarin derivatives are mycotoxins produced by fungi. Scopoletin, the most abundant coumarin derivative, is present in some seed coats, and it does not let seeds germinate properly until it has leached out. It acts as a germination inhibitor. Thus, certain seeds germinate well when planted after washing. Fungus *Aspergillus* sp. produces the most potent natural toxins called **aflatoxins**, which may lead to death of organisms even when consumed in very minor quantities. Aflatoxins are mutagenic and are also known to be carcinogenic and impair immune system and can damage organs, such as the liver and kidney. Aflatoxins are common in peanut products and cattle feed and can even be secreted in cattle milk making the milk toxic for human consumption. Another group, phenylpropenes (along with the terpenes), contributes to the volatile flavors and odors of plants. They are components of the "essential oil" fraction of plants such as anise, fennel, and nutmeg. **Furanocoumarins** are coumarins with furan ring attached to coumarin. These are complex coumarins and are toxic only when they are activated by light. Upon activation by UV-A, they act as mutagens, ultimately leading to blockage of transcription and cell repair. This might lead to cell death. They are abundant in members of family Apiaceae (such as in celery, parsley) and have also been reported in the members of Rutaceae. The concentration of

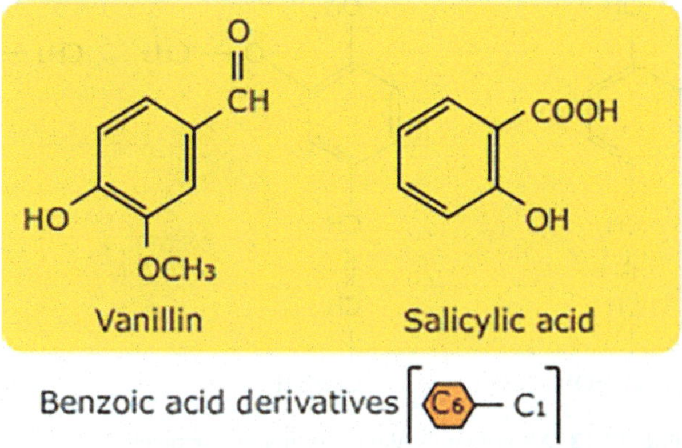

## 33.2 Phenolic Compounds

**Fig. 33.12** Benzoic acid derivatives include vanillin and salicylic acid which are involved in providing systemic resistance to plants against pathogens

Benzoic acid derivatives [C₆—C₁]

furanocoumarins increases up to hundredfold when plants are growing under stress. They may even lead to skin rashes when such stressed or diseased plants are handled.

### 33.2.3.3 Benzoic Acid Derivatives

They are formed by the cleavage of two carbon side chains from phenylpropanoid skeleton and consist of $C_6$–$C_1$ unit. Examples are vanillin, salicylic acid, caffeic acid, and ferulic acid (Fig. 33.12). Benzoic acid derivatives are reported to have allelopathic properties, i.e., the release of such secondary metabolites from one plant produces a negative (or sometimes positive) effect on another plant growing nearby. This improves plant growth by increasing its access to soil nutrients, light, and water. Benzoic acid derivatives, such as caffeic acid and ferulic acid, have negative effect on seed germination and growth of many plant species. Presence of benzene or phenol ring in these compounds facilitates their easy passage across biological membranes and exerts their biological activities.

### 33.2.3.4 Lignin

Lignin is a complex and branched polymer of simple phenolic alcohols, such as monolignols (p-coumaryl, coniferyl, and sinapyl alcohols). After cellulose, it is the most abundant organic substance in plants. Three phenylpropanoic alcohols (Fig. 33.13), i.e., coniferyl alcohol, sinapyl alcohol, and p-coumaryl alcohol, are believed to be the building blocks of lignin. These three building units are not linked in a simple repeating way but have a highly branched and complex arrangement to form lignin. These monolignols differ only in the number and positioning of methoxy ($CH_3O$-) groups attached to the phenyl ring. Upon incorporation into lignin polymer, the monolignols form their respective phenylpropanoid subunits (p-coumaryl: p-hydroxyphenyl (H), coniferyl: guaiacyl (G) and sinapyl alcohols: syringyl (S)). The proportion of these three monolignols in lignin varies depending on the species, plant organ, and even in different layers of the cell wall of the same cell. In gymnosperms, coniferyl alcohol is more abundant, whereas angiosperm lignin is a mixture of coniferyl and sinapyl monolignols. The three monomers link

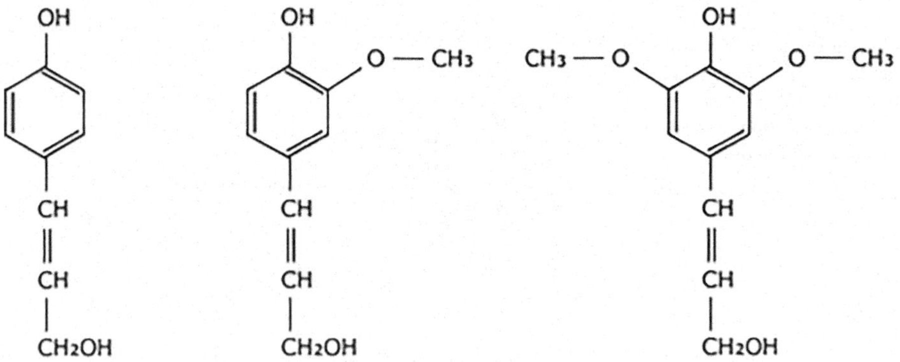

**Fig. 33.13** Monolignols, the major lignin monomers

together in various ways to form highly branched three-dimensional structures. The structure of lignin is obscure owing to the complexity involved in its extraction from various plant sources. Lignin sometimes also contains *p*- coumarate, ferulates, and *p*-hydroxybenzoates. Lignin provides strength and rigidity to woody plants by covalently binding with cellulose and other polymers in the cell wall. Lignin is chiefly deposited in the secondary cell wall of supporting and conducting tissues, mainly the tracheary elements. Sometimes, it is also found in the primary cell wall and middle lamella along with the cellulose and other polysaccharides. Mechanical strength provided by lignin to stem and vascular tissues leads to their vertical growth and allows conduction of water and mineral upward through xylem without collapsing of the tissue through the xylem. Lignin also provides protection to plants from herbivory as rigidity of lignified tissues makes it relatively indigestible and deters herbivores. Lignification and deposition of some phenolics also generally occur at the sites of pathogen attack and do not allow the further growth of pathogens.

*Lignans* are di- and oligomeric phenylpropanoids, structurally and biosynthetically similar to lignin. They are widely distributed in plant kingdom and function against the biotic and abiotic stress. They are commonly located in flowers, seeds, stem, bark, leaves, and roots.

### 33.2.3.5 Flavonoids

They are polyphenolic compounds having 15-carbon chain which are arranged in two aromatic rings by a 3-C bridge. The hydroxyl groups are located at positions 4, 5, and 7 of the ring. They are found throughout the plant kingdom with high concentration occurring in the epidermis of leaves and the skin of fruits. They are involved in UV protection, pigmentation, stimulation of nitrogen-fixing nodules, and disease resistance. The attractive colors of flower petals, bracts, fruits, and sometimes leaves as well, which appeal humans and plant pollinators, are mostly due to the presence of pigments known as anthocyanins, which belong to a bigger class of compounds, known as flavonoids. They have been used as a source of dyes since ages because of their bright colors. Flavonoids comprise a large class of plant

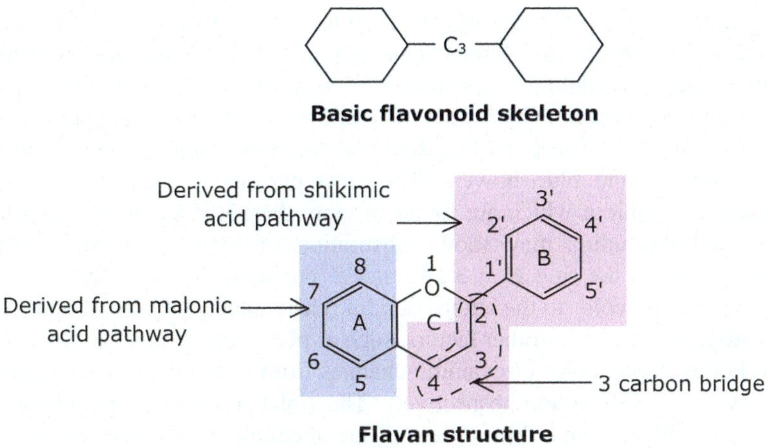

## 33.2 Phenolic Compounds

**Basic flavonoid skeleton**

**Flavan structure**

**Fig. 33.14** The basic flavonoid skeleton ($C_6$–$C_3$–$C_6$)

phenolics consisting of more than 4500 compounds (Fig. 33.14). The flavonoids are all structural derivatives of flavan, in which the C-3 link forms a heterocyclicpyrone ring. The heterocyclic ring is fully reduced in flavan, and different groups of flavonoids (at least 12) differ from each other depending on the oxidation state of this heterocyclic ring. Ring B and 3-carbon bridge making ring C are contributed via shikimic acid pathway (as phenylalanine and p-coumaric acid). Ring A is derived from malonic acid. The principal enzyme involved in flavonoid synthesis is chalcone synthase (CHS). Flavonoids include many groups among which four are major ones, viz., **anthocyanins**, **flavones**, **isoflavones**, **flavonols**, etc., depending upon the degree to which the $C_3$ bridge (pyrone ring) is oxidized. Various sugars, methyl ethers, and modified isopentyl units lead to huge diversity of compounds. Most flavonoids exist in nature as glycosides (complex with sugar moiety) and are water-soluble. The type of substituents attached to flavonoid determines its solubility. Hydroxyl groups and sugars make the flavonoids more hydrophilic, whereas attached methyl ethers and isopentyl units make them hydrophobic (lipophilic). Almost all flavonoids, particularly flavones and flavonols, absorb strongly in UV-B range. Most flavonoids are colored. They accumulate in vacuoles or sometimes in chromoplasts and chloroplasts. They perform a variety of functions and mainly provide coloration (pigmentation) and have defense roles.

**Anthocyanins** are the most strongly colored flavonoid pigments which commonly provide color to red, pink, purple, and blue flowers. They may also occur in fruits, stem, roots, bracts, and leaves. They are water-soluble and mostly stored in vacuoles. They are widely distributed within the plant kingdom. They also conjugate with organic acids, like malic acid, acetic acid, and hydroxycinnamates. Conjugation mostly occurs at the position $C_3$. Plants also have other pigments known as carotenoids, which are yellow, orange, and red terpenoid compounds, such as those present in tomato (lycopene) and yellow flowers. Anthocyanins help in attracting pollinators and in seed dispersal by making the flowers and fruits more visually appealing. Unlike carotenoids, anthocyanins do not serve as accessory pigments in photosynthesis. Anthocyanins are glycosides of **anthocyanidins**, having various sugars commonly at position 3 or elsewhere (Table 33.6). The aglycones rarely occur free in nature. Glycosides of cyanidin, delphinidin, and pelargonidin are most commonly found in plant kingdom and are responsible for pink, orange-red, scarlet, crimson, and blue flowers. The most prominent ones among these are glycosides of cyanidin which are found in more than 50% of flowers and 80% of leaves. Anthocyanidins may show substitution of hydroxyl group (-OH) and methoxy group in the ring B at 3', 4', and 5' positions. These substituent groups give characteristic color to the anthocyanidin (basic moiety). Color of anthocyanins is also influenced by many other factors, such as pH of the vacuole in which they are stored. For example, color of cyanidin changes from red to violet to blue in acidic, neutral and alkaline solution, respectively. The violet color of extract obtained from boiled red cabbage turns blue-green if the medium is alkaline. Color is also influenced by the occurrence of more than two different anthocyanins in the same

**Table 33.6** Structures of anthocyanidin and different anthocyanins

| (A) Anthocyanidin | | (B) Anthocyanin | |
|---|---|---|---|
| Anthocyanidin | + Glycoside | Anthocyanin | Some plant sources |
| Cyanidin (Red-purple) | 3, 5-diglucoside | Cyanin | Red rose petals |
| Delphinidin (Blue-purple) | 3-rhamnoglucoside | violanin | Violet (viola) flowers |
| Malvidin (Deep-purple) | 3-glucoside | oenin | Blue grapes |
| Pelargonidin (Orange-red) | 3, 5-diglucoside | pelargonin | Geranium (pelargonium) petals |

flower or plant organ and copigmentation, i.e., presence of anthocyanins along with other flavonoids, such as flavonols and flavones, and their existence as supramolecular complexes, i.e., association with chelated metal ions. Generally hydroxylation increases blueness, whereas presence of methoxy groups makes the color red. At least 12 different anthocyanidins are known. Of these, the most common are shown in Table 33.6. Anthocyanins provide useful information to plant taxonomists and help in classifying and determining the line of evolution. During evolution, flower color may be dictated by the need to attract different pollinators to maximize chances to survival. The abundance of anthocyanins (in vacuoles of epidermal cells) leads to masking of green color (due to chlorophyll) in *Coleus* and maple (*Acer*) leaves, resulting in leaves having characteristic colors. Matured red wine contains high amount of anthocyanins.

**Flavones and flavonols** are widely distributed in plants as copigments along with anthocyanins in flowers and leaves. This group of flavonoids generally absorbs light strongly in UV- B region. They are not perceived by the human eye. They are usually yellowish or ivory-colored pigments which are widespread in flower petals and contribute to flower color. They also occur in leaves and differ from anthocyanins with respect to the central ring of their molecules (Fig. 33.15).They are more perceived by insects, such as bees, which have the ability to see in the UV range of the spectrum, and act as attractants to such insects. These pigments (flavones and flavonols) are not visible to the human eye. They form symmetrical patterns, circles, spots, or stripes and indicate the position of pollen and nectar to insects, thus serving as nectar guides for the insects and facilitating pollination. Plants lacking flavonoids are more prone to photooxidative damages. Flavonols like kaempferol and quercetin have been reported to be involved in light-regulated stem growth of pea.

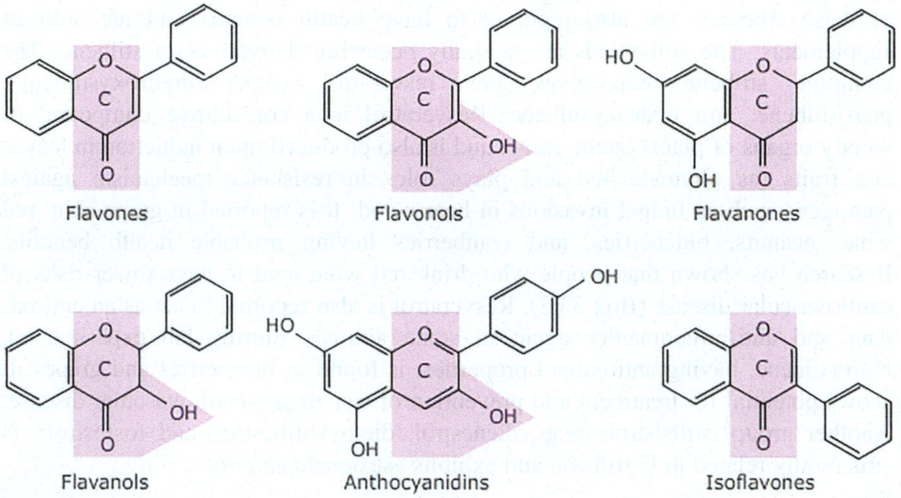

**Fig. 33.15** Main classes of flavonoids and their basic structures

Another group of flavonoids, known as isoflavones, differ from other flavonoids in structure as one aromatic ring "B" is shifted and is attached to third carbon of central ring instead to carbon at second position. They are found mostly in leguminous plants, with the highest amount found in soybean (*Glycine max*). They perform several biological functions, including insecticidal properties (such as rotenone), pesticides, and piscicides (fish poisons). Some flavonoids cause infertility in animals, such as sheep, due to their antiestrogenic properties, and hence are also known as **phytoestrogens**. Isoflavones, like genistein and daidzein from lucerne and clover (*Trifolium* sp.), mimic the structure of steroidal hormone "estradiol" which blocks the ovulation process. Dietary consumption of these compounds is known to reduce the risk of prostate and breast cancer in humans. They are also known for their antimicrobial activity. These are produced by plants in response to bacterial or fungal pathogens and limit their further spread. For example, phytoalexins are known for their antimicrobial features and are commonly involved in defense mechanisms. They are usually absent in uninfected plants or are present in low concentrations. They are rapidly synthesized upon bacterial or fungal invasion. A variety of biomolecules secreted by the pathogens act as elicitors for synthesis of phytoalexins. Isoflavonoids are predominant phytoalexins in Leguminosae, whereas in some other families such as Solanaceae, terpenes appear to play the role. Isoflavonoids are also held responsible for anticancerous benefits of soybeans.

**Stilbenes** are one of the non-flavanoid phenylpropanoid derivatives. They are a subgroup of polyphenols ($C_6$–$C_2$–$C_6$) which are not very common in plants in comparison with other phenolics. Stilbenes exist as two isomers, trans-1,2-diphenylethylene also known as trans-stilbene and cis-1,2-diphenylethylene (cis-stilbene). Stilbene biosynthesis involves condensation of malonyl-CoA subunits either with cinnamoyl-CoA or *p*-coumaroyl-CoA. This reaction is catalyzed by stilbene synthase. Stilbene itself is of not much value as such, but it serves as a precursor to the derivatives which can be used as dyes and optical brighteners. Some of these stilbenes are also reported to have health benefits and are sold as supplements. The stilbenoids are naturally occurring derivatives of stilbene. The common stilbene derivatives are resveratrol (3,5,4′-trihydroxystilbene), pterostilbene, and hydroxystilbene. Resveratrol is a constitutive compound of woody organs of plants (stem, roots) and is also produced upon induction in leaves and fruits (as phytoalexins) and plays roles in resistance mechanism against pathogens such as fungal invasions in heartwood. It is reported in grape skin, red wine, peanuts, blueberries, and cranberries having probable health benefits. Research has shown that people who drink red wine tend to have lower risks of cardiovascular disease (Box 33.3). Resveratrol is also reported to act as an antioxidant and anti-inflammatory agent in some animals (during lab experiments). Pterostilbene, having antioxidant properties, is found in blueberries and grapes. It shows potential for treatment and prevention of cancer and cardiovascular disease. Another group, stilbestrols (e.g., dienestrol, diethylstilbestrol, and fosfestrol), is structurally related to E-stilbene and exhibits estrogenic activity.

## Box 33.3: Resveratrol: Medicinally Important/Active Ingredient of Red Wine

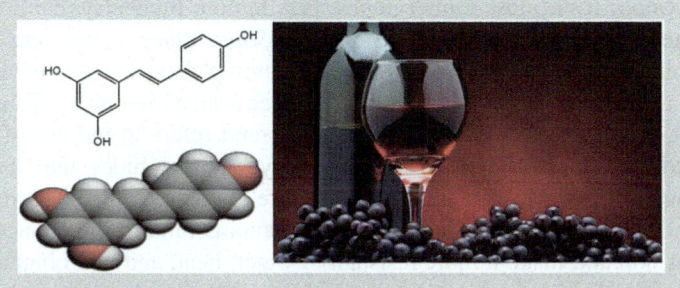

**Resveratrol**—a stilbene derivative, red grapes from *Vitis vinifera* (family—Vitaceae).

Resveratrol (3,5,4′-trihydroxy-*trans*-stilbene) is a stilbenoid, a derivative of *stilbene* (a type of natural phenol), and a *phytoalexin* produced by some plants as a response to environmental stress or pathogenic attack (*bacteria* or *fungi*). Resveratrol has been detected in at least 72 plant species. Some sources of resveratrol in food include the peel of *grapes*, *raspberries*, *blueberries*, and *mulberries*. Relatively high quantities of this compound are found in the grapes, probably due to the response of *Vitis vinifera* to fungal attack. Peel of fresh grapes contains approx. 50–100 μg of resveratrol per gram fresh weight, whereas the concentration of this compound in red wine is in the range of 1.5–3 mg/l. White and rosé wines have also been known to contain appreciable amounts of resveratrol. It is known to play a major role in the prevention of heart disease. It prevents damage to blood vessels and reduces LDL cholesterol. It has also been reported to inhibit aggregation of platelets and coagulation and alter eicosanoid synthesis and modulation of lipoprotein metabolism.

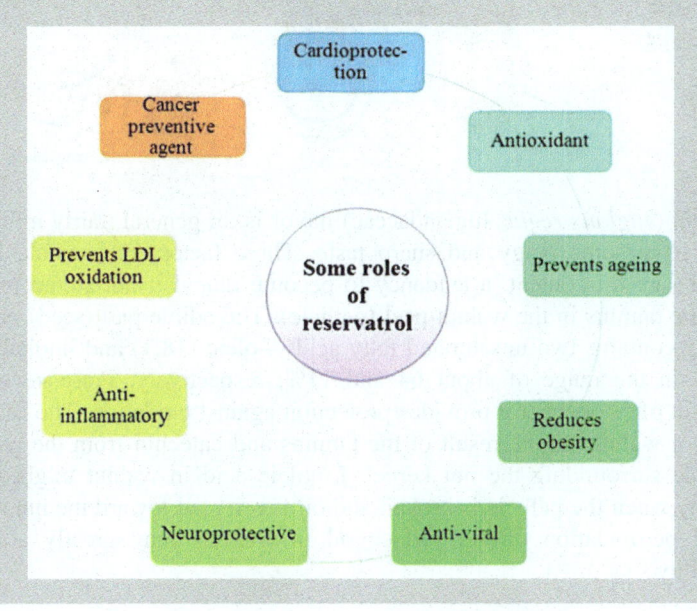

Some roles of reservatrol: Cardioprotection, Antioxidant, Prevents ageing, Reduces obesity, Anti-viral, Neuroprotective, Anti-inflammatory, Prevents LDL oxidation, Cancer preventive agent.

## 33.2.3.6 Tannins

Tannins are naturally occurring plant polyphenolic compounds, commonly found in bark, wood, fruit pods, fruits (grapes, persimmon, blueberry, etc.), tea, chocolate, legume forages, trees (*Acacia, Sesbania, Salix, Quercus* sp., etc.), and grasses (sorghum, corn). Tannin (tannic acid)-containing plant extracts have been used to "tan" animal hides to convert them into leather from earlier time. "Tannin" is derived from a German word "tanna" meaning oak and refers to the use of wood tannins derived from oak trees and is used to dye animal hides into leather (tanning technique). Chemically, tannins are a mixture of complex phenol derivatives having the ability to bind and denature proteins. Tannins bind to the collagen (protein) of animal skin and make it more resistant to water, heat, and microbial attack. Precipitation of gelatin (a protein) has been used to test the presence of tannin in any extract since long. Horvath in 1981 defined tannins as "Any phenolic compound of sufficiently high molecular weight containing adequate hydroxyl (-OH) and other suitable groups (i.e. carboxyls) to form effectively strong complexes with proteins and other macromolecules under particular environmental conditions." Tannins in bark protect plants from bacterial and fungal attack. They coagulate the cell wall-degrading enzymes and other enzymes secreted by bacteria and fungi, hence preventing their infection and growth. Bud scales on woody plants are the storage vesicles of tannins and protect leaf tissue being eaten by grazing animals. Tannins are also responsible for the astringent taste experienced by humans, on tasting wine or unripe fruits and enthralling colors of leaves and flowers in autumn (Box 33.4). The characteristic brown (tea) color of streams of watershed, dominated by conifers,

**Box 33.4: Walnut Flavor Factors**

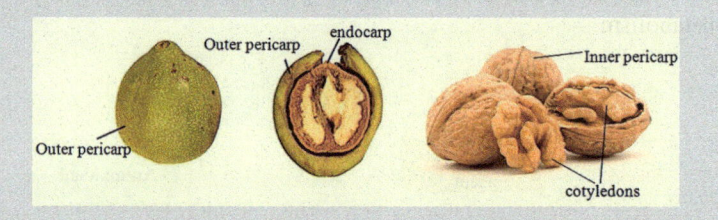

Walnut (*Juglans regia*; Juglandaceae) flavor is, in general, fairly mild, but it does have some tangy and sharp taste. Three factors responsible for this feature are oil content, a tendency to become rancid if not stored properly, and the tannins in the walnut peel (pellicle). The edible part (seed) predominantly contains two unsaturated fatty acids—oleic (18:1) and linoleic (18:2) acid—in the range of about 64 and 11%, respectively. Presence of large amount of γ-tocopherol provides protection against oxidation. The astringent taste of walnuts is the result of the tannins and catechin from the paperlike pellicle surrounding the nut kernel. Linoleic acid in walnut might oxidize readily when the pellicle is broken and oil is oriented toward the nut surface. Lipid peroxidation of the surface oil in walnuts may greatly accelerate rancidity.

is due to the leaching of tannins from the needles of these conifers. Lawsone or hennotannic acid present in henna leaves produces a strong stain on the skin by binding with keratin in the skin (Box 33.5). Tannins are categorized into two main groups, condensed tannins (proanthocyanidins) and hydrolyzable tannins, on the basis of their varied chemical structure and stability.

**Condensed tannins** are formed by the polymerization of flavonoid units, i.e., polymers of 15-carbon phenolic compounds with C6-C3-C6 carbon skeleton with strong C-C bonds (Fig. 33.16). They are not easily hydrolyzed. Anthocyanidins are released when condensed tannins are hydrolyzed by strong acid treatment. **Hydrolyzable tannins** are more easily hydrolyzed as compared to condensed tannins. They are heterogenous polymers relatively smaller in size than condensed tannins. Their basic structural unit consists of simple sugar glucose esterified to gallic acid (a phenolic acid).

Gallic acid residues are cross-linked extensively to form a polymer (Fig. 33.16). Gallic acid phenolic residues are known as gallotannins. Upon hydrolysis with strong acids, they often form anthocyanidins. So, they are sometimes also referred to as **proanthocyanidins**.

---

**Box 33.5: The Chemistry of Henna Dye Binding on the Skin**

Henna ("mehendi"), commonly used to create decorative designs on a person's body, is derived using an aqueous paste prepared from the leaves of henna (*Lawsonia inermis* L.) of Lythraceae family. The plant is a glabrous and multibranched flowering shrub or small tree. Natural henna leaves contain about 0.5–4% lawsone amounts, while the commercially grown henna contains 2–3%. The red-orange dye, lawsone or hennotannic acid (2-hydroxy-1,4-napthoquinone), present in henna leaves produces a strong stain on the skin or hair as a result of chemical interaction between lawsone from the leaf paste and keratin (protein) in the skin or hair through a process called Michael reaction. Lawsone is about the size of an amino acid (2.54 Angstroms) and hence easily breaches the skin cells. An extended period of drought followed by a brief flush of rain leads to new growth of henna leaves containing highest lawsone content.

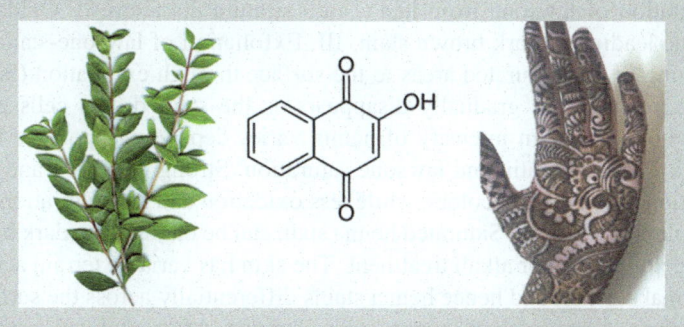

Leaves of *Lawsonia inermis*, structure of hennotannic acid (lawsone), and staining of the skin due to lawsone.

(continued)

**Box 33.5** (continued)

Lawsone is a weak organic acid, and its concentration in henna leaves typically varies from 1.3 to 1.5%. It has two carbonyl groups (C=O) and one acid group (OH) and hence has a very unstable negatively charged structure. In order to enhance the stability of lawsone, it is usually mixed with citric acid during processing of the "ready-to-use" henna powder. The excess protons in citric acid prevent ionization of the acid group or carbonyl groups. Thus, decomposition of lawsone molecules is prevented. The Michael reaction leading to pigmentation of the skin involves addition of a carbanion or another nucleophile to an unsaturated carbonyl compound. The abundant nucleophiles in proteins (keratin from cells of the skin) attack one of the beta carbons on lawsone. Both the carbons in lawsone are at alpha position with respect to one carbonyl group and at beta position with respect to the other carbonyl group. So, either carbon could be attacked. Usually, the lower beta carbon is attacked because of the steric hindrance of the top one. Thereafter, the electrons are "pushed" such that the top carbonyl is converted to an OH group and the double bond moves up. Thus, negatively charged carbonyl group of lawsone molecule reacts with positive amide group of keratin's peptide bond.

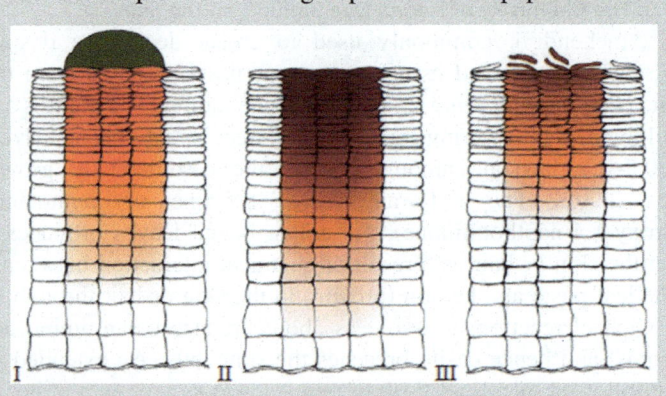

Process of henna staining on the skin and subsequent exfoliation: I. Migration of lawsone from henna into stratum corneum. II. Oxidation of lawsone leading to dark brown stain. III. Exfoliation of lawsone-stained skin and rising of less-saturated areas to the surface through exfoliation (fading).

The henna tattoo gradually disappears as the stained skin cells undergo exfoliation. The stain intensity of henna varies depending upon the level of oxidation in the keratin and lawsone saturation. Strong oxidation and intense saturation produce dark colors, while less oxidation and low saturation lead to pale colors of lawsone. Saturated henna stain can be darkened to dark brown to black color by heat or alkali treatment. The skin has variable terrain across the epidermal surface, and hence henna stains differentially across the surface. As the surface is replenished, henna stain changes. After application of henna paste to the skin, lawsone dye from henna migrates into the upper epidermal skin layer, breaching and saturating the cells. In 48 hours, henna stain darkens as it undergoes oxidation by contact with air or alkaline.

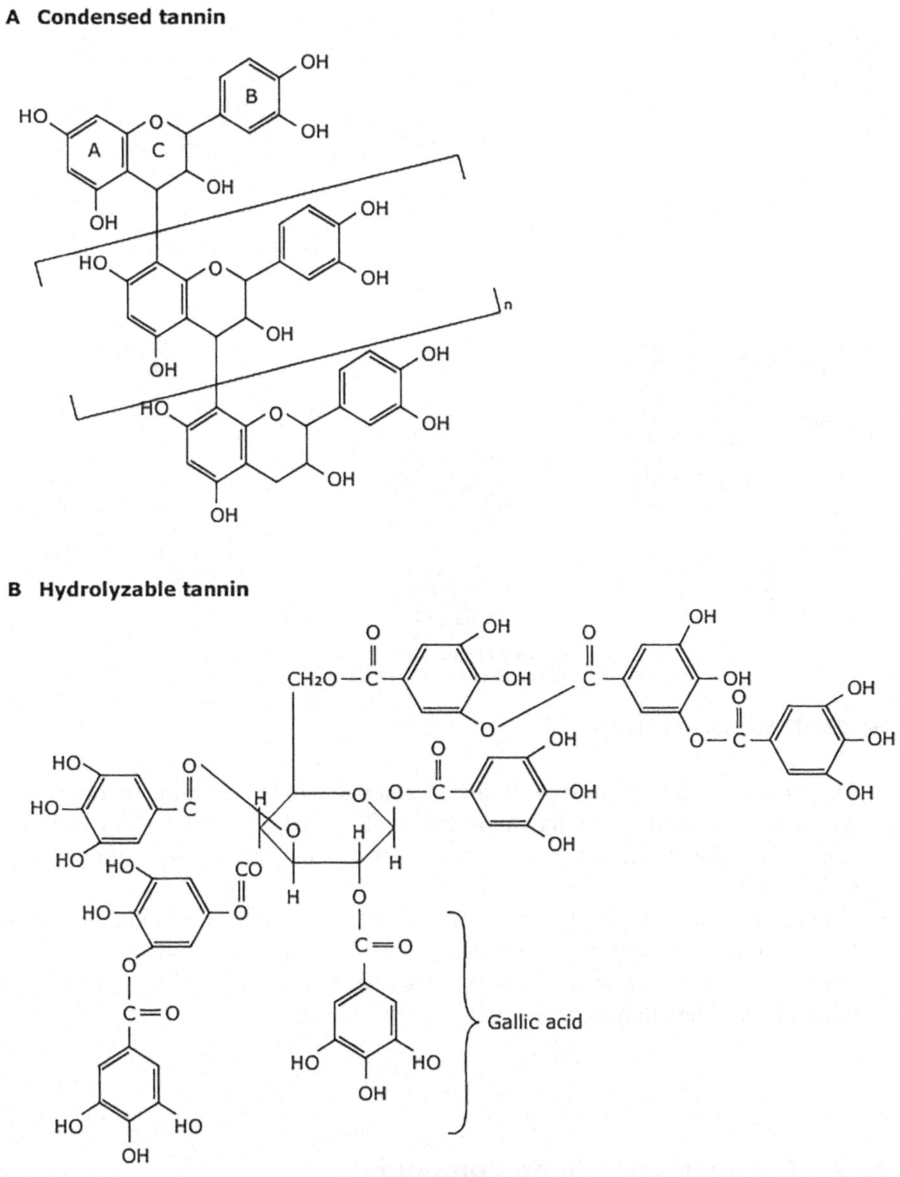

## 33.2 Phenolic Compounds

**A Condensed tannin**

**B Hydrolyzable tannin**

**Fig. 33.16** Structure of condensed and hydrolyzable tannins

Tannins have significant biological roles (Fig. 33.17):

1. They *deter feeding by herbivores*. Tannins bind with the gut proteins of herbivores non-specifically and diminish the digestibility of proteins consumed in food. They tend to reduce utilization of feed nutrients, growth rate, and survivor rate of herbivores.

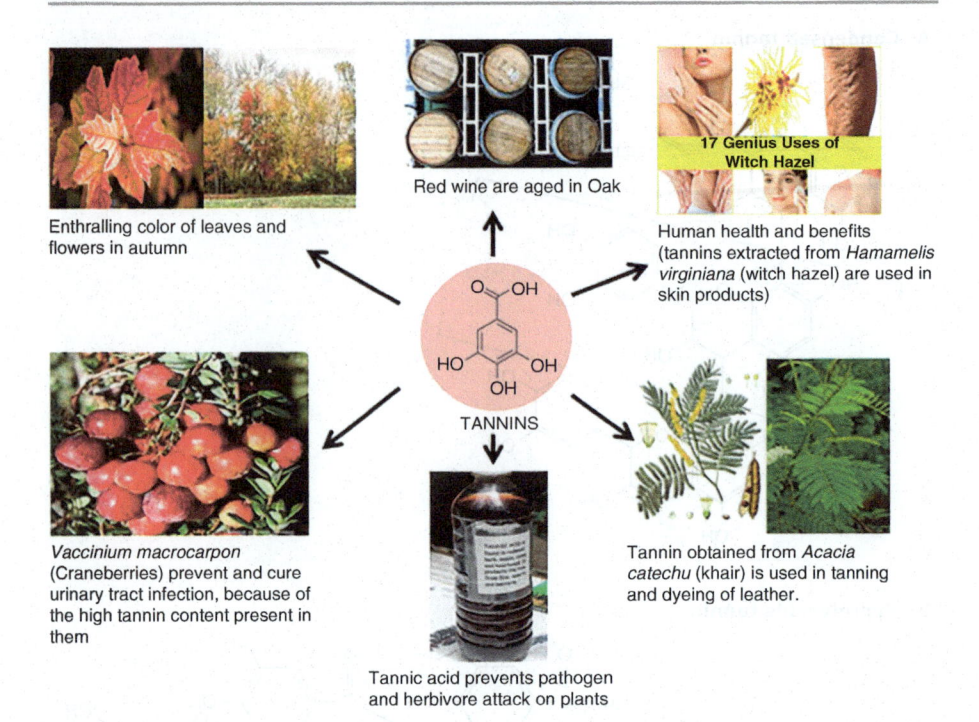

**Fig. 33.17** Some roles of tannins

2. They *act as astringents* as they produce a sharp and bitter sensation in the mouth. This property produces a flavor in tea, coffee, and red wine. Thus they are responsible for astringency of various foods such as apples, grapes, and blackberries.
3. Tannins *help in vasodilation* by blocking the formation of endothelin-1. The benefits associated with red wine consumption are due to this reason.
4. Some gallotannins *function as allelopathic compounds* and inhibit the growth of other plants. They may be toxic in other ways as well.

## 33.3 Nitrogen-Containing Compounds

Numerous secondary metabolites of plant origin have nitrogen in their structure. This category of natural products includes alkaloids, cyanogenic glycosides, glucosinolates, and some non-protein amino acids. Most of the N-containing secondary metabolites are principally synthesized from common amino acids, such as lysine, tyrosine, and tryptophan. A number of these metabolites are of significant importance because of their medicinal properties. Some of them can be toxic as well and hence have a functional role as anti-herbivores.

## 33.3.1 Alkaloids

Alkaloids constitute a large group of diverse biomolecules comprising more than 15,000 compounds. Use of alkaloids by human beings dates back to 1200 BC, and they find their origin in the name of the Arabic plant known as "al-qali," which was used to isolate soda at that time (Box 33.6). Literal meaning of alkaloids is "alkali-like." As their name suggests, most alkaloids are alkaline in nature, but some neutral alkaloids have also been reported. They are usually colorless and bitter in taste. They all have at least one nitrogen in their structure, usually as a part of a heterocyclic ring (e.g., codeine, nicotine, caffeine, morphine) as depicted in Table 33.7 and Fig. 33.18. The nitrogen atom of alkaloids is present in protonated form at the pH commonly found in the cytosol (pH 7.2) or in the vacuoles (pH 5-6). Alkaloids are primarily present in plants, and the alkaloid-producing plant species are dispersed throughout the plant kingdom, predominantly in angiosperms. They are found in around 20% of angiosperms, most of which are herbaceous dicots. Some of the important alkaloid-producing plants belonging to families are Magnoliaceae, Solanaceae, Papaveraceae, Fabaceae, Rubiaceae, Apocynaceae, and Ranunculaceae. The first alkaloid to be isolated and characterized from plants was **morphine** from opium poppy. Analysis of the alkaloid components of opium carried out by Friedrich Wilhelm Sertürner led to identification of morphine in 1803, the name being taken

---

**Box 33.6: Alkaloids in History**

Alkaloids have been used by humans for more than 3000 years. Use of *Papaver somniferum* (opium poppy) can be dated back to 1200 BC, in the eastern Mediterranean region. In India, the dried roots of *Rauvolfia serpentina* (sarpagandha, a medicinal plant) have been used to cure epilepsy, cardiac diseases, high blood pressure, intestinal disorders, and snake bite and as anthelmintic. One of the most potent alkaloid constituents of the plant is reserpine. Others include ajmaline, ajmalicine, serpentine, serpentinine, etc. Plant extracts containing large quantities of alkaloids have been used as constituents of poisons since long. During the execution of the great philosopher Socrates, he drank an extract of *Conium maculatum* plant (hemlock) which contains alkaloid coniine. Famous queen Cleopatra used the extracts of *Hyoscyamus* sp. (henbane), which contains atropine to appear more attractive, as it dilates pupils. Derivatives of tropane alkaloids (atropine) are used to dilate pupils during ophthalmological treatments. Theriac, a potion prepared from opium, wine, and snake meat, was devised in Greek-Roman culture and is one of the most long-standing medications still sometimes in use as antidote against snake poisoning and spider and scorpion bites. In rare cases, it is still administered for pain relief.

(continued)

**Box 33.6** (continued)

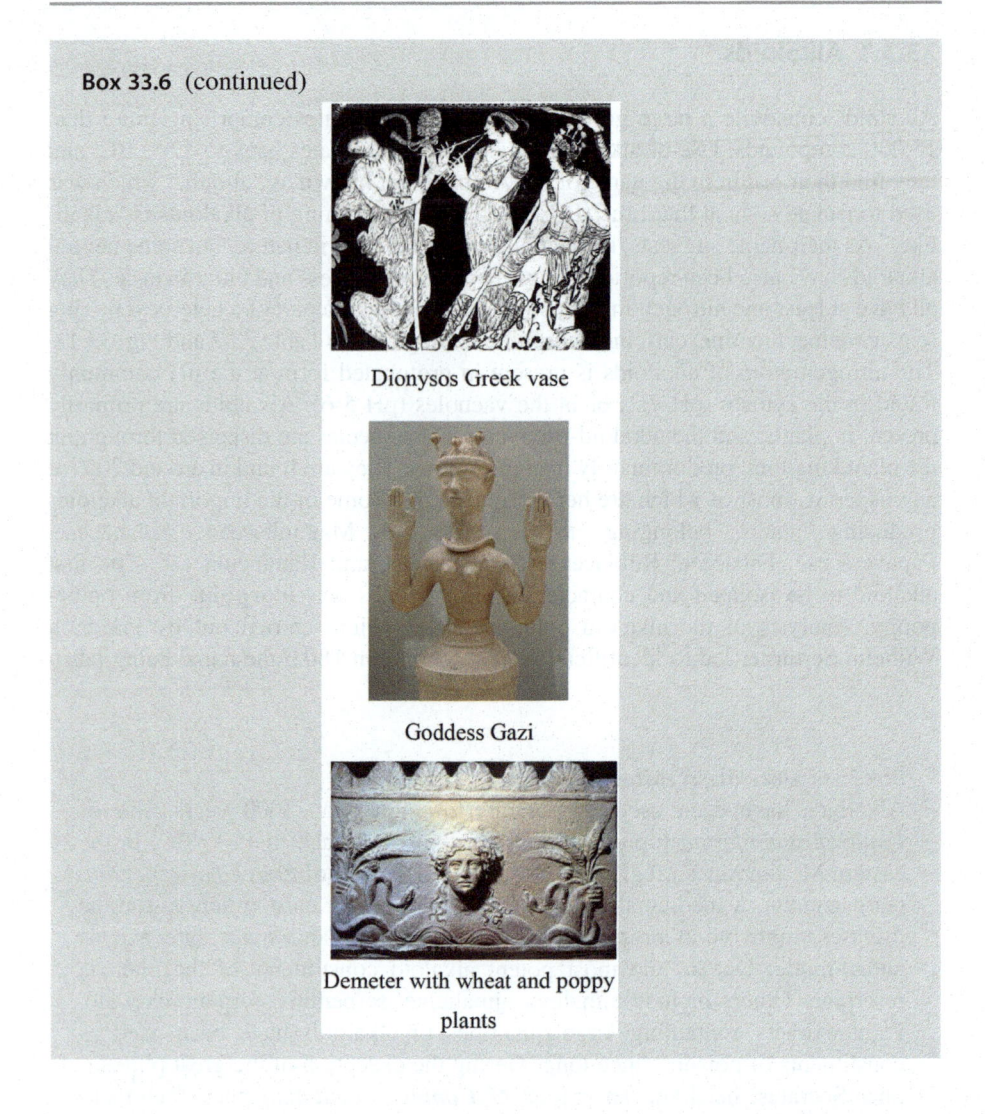

Dionysos Greek vase

Goddess Gazi

Demeter with wheat and poppy plants

after the Greek God *Morpheus* (the God of dreams because of its sleep- and dream-inducing properties). Isolation of morphine triggered intensive analysis of alkaloids in plants. Some of the well-known and highly documented alkaloids include opium from opium poppy (*Papaver somniferum*), cannabidiol from hemp (*Cannabis sativa*), nicotine from tobacco (*Nicotiana tabacum*), cocaine from coca (*Erythroxylon* sp.), coniine from hemlock (*Conium maculatum*), atropine from nightshade (*Atropa belladonna*), caffeine from coffee (*Coffea arabica*), quinine from cinchona (*Cinchona officinalis*), colchicine from meadow saffron (*Colchicum autumnale*), strychnine and brucine from strychnine tress (*Strychnos nux-vomica*), and vinblastine and vincristine from periwinkle (*Catharanthus roseus*). The

## 33.3 Nitrogen-Containing Compounds

**Table 33.7** Major classes of alkaloids, their basic structures, biosynthetic precursors, and representative examples

| Alkaloid class | Structure | Biosynthetic precursor | Examples |
|---|---|---|---|
| Pyrrolidine | | Ornithine (aspartate) | Nicotine |
| Tropane | | Ornithine or Argionine | Atropine |
| | | | Scopolamine |
| | | | Cocaine |
| | | | Hyoscyumine |
| Piperidine | | Lysine (or acetate) | Coniine |
| Pyrrolizidine | | Ornithine | Retrorsine |
| | | | Senecionine |
| Quinolizidine | | Lysine | Lupinine |
| | | | Cytisine |
| Isoquinoline | | Tyrosine | Papaverine, Codeine, Morphine, Barberine |
| Indole | | Tryptophan | Vindoline, Psilocybin, Vinblastine, Reserpine |

alkaloidal species may contain a single or a mixture of different alkaloids. Opium poppy is a source of more than 20 alkaloids including morphine, codeine, papaverine, and thebaine. These four are economically and medicinally important opium alkaloids, while others are of relatively little or no significance. *Rauvolfia serpentina* is another species that is a source of multiple alkaloids, producing more than 50 of them including ajmaline, ajmalidine and reserpine, and periwinkle (*Catharanthus roseus*) contains more than 100 different monoterpenoid indole alkaloids. The ability to produce alkaloids is not only confined to plants. They have also been isolated from animals like sponges, arthropods, amphibians, and mammals. However, these are believed to be of plant origin and acquired by animals through feeding on the

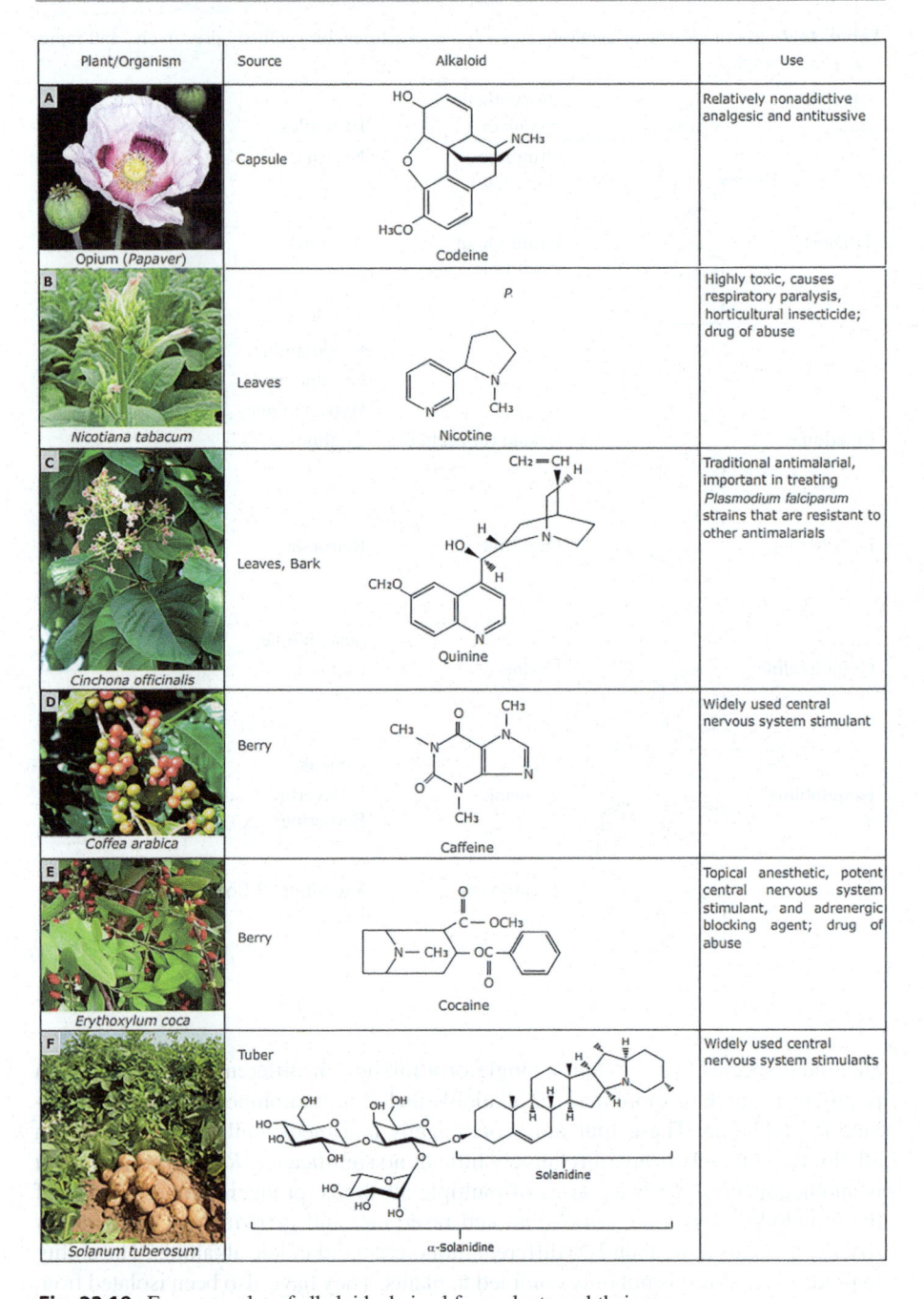

| Plant/Organism | Source | Alkaloid | Use |
|---|---|---|---|
| **A** Opium (*Papaver*) | Capsule | Codeine | Relatively nonaddictive analgesic and antitussive |
| **B** *Nicotiana tabacum* | Leaves | Nicotine | Highly toxic, causes respiratory paralysis, horticultural insecticide; drug of abuse |
| **C** *Cinchona officinalis* | Leaves, Bark | Quinine | Traditional antimalarial, important in treating *Plasmodium falciparum* strains that are resistant to other antimalarials |
| **D** *Coffea arabica* | Berry | Caffeine | Widely used central nervous system stimulant |
| **E** *Erythoxylum coca* | Berry | Cocaine | Topical anesthetic, potent central nervous system stimulant, and adrenergic blocking agent; drug of abuse |
| **F** *Solanum tuberosum* | Tuber | α-Solanidine (Solanidine) | Widely used central nervous system stimulants |

**Fig. 33.18** Few examples of alkaloids derived from plants and their uses

alkaloidal host plants. For example, morphine is derived mostly from the capsules of *Papaver somniferum*. However, it is also isolated from mammals such as mouse. Amphibians and insects particularly display remarkable diversity in accumulated alkaloids. Amphibians like frogs store one of the most noxious and toxic alkaloids in their skin and skin exudates, bufotalin present in *Bufo* (common toad) being one example (Box 33.7). In insects, these alkaloids may act as pheromones (attractants) and defensive agents. They also accumulate in several species of bacteria and fungi. Some of the alkaloids of fungal origin include chanoclavine-I, ergotamine, and ergonaline from *Claviceps purpurea*, oxaline from *Penicillium oxalicum*, and psilocybin from *Psilocybe* sp., while those produced by bacteria include pyocyanine from *Pseudomonas aeruginosa*. Interestingly, all alkaloids belonging to a plant are not always produced by the plant itself. A variety of different alkaloids are also synthesized by the endogenous fungal symbionts (endophytes) that grow in the apoplast of many plants especially grasses. Such grasses with fungal symbionts grow faster and have better defense against herbivores than those without symbionts. Sometimes, when the alkaloid content of these grasses (such as tall fescue) is too high, they may become toxic to livestock.

### 33.3.1.1 Biosynthesis

Most of the alkaloids are synthesized from common amino acids particularly lysine, tyrosine, or tryptophan. However, the carbon skeleton of some alkaloids is derived from mevalonic acid pathway (of terpene biosynthesis). Depending upon their biosynthesis, they can be classified as proto-, true, and pseudo-alkaloids. Proto- and true alkaloids are derived directly from amino acids, while pseudo-alkaloids are the ones which are present in conjugation with terpenoids. Alkaloids can also be classified on the basis of amino acid precursor from which they are derived. Table 33.7 lists the major classes of alkaloids, their biosynthetic precursors, and common examples. The alkaloids produced by a particular plant species are usually restricted in their accumulation to particular organs, such as leaves, roots, bark, etc. The zones that usually accumulate alkaloids are young-growing parts, epidermal and hypodermal cells, and latex vessels. In many instances, the site of production of the alkaloids is different from the site of accumulation. For example, in tobacco (*Nicotiana* sp.), nicotine is synthesized by the roots which is then translocated to leaves which is the storage site for the alkaloid.

### 33.3.1.2 Physiological Roles

Alkaloids form one of the most potent groups of plant products. They encompass intoxicating neuroactive substances, life-threatening poisons, as well as lifesaving medicines and anticancerous compounds. Earlier, they were presumed to be either nitrogenous biological wastes from plant systems (like urea and uric acid in animals) or nitrogen storage compounds. In spite of the widespread distribution of alkaloids in

**Box 33.7: Alkaloid Accumulation in Non-plant Species**
Alkaloids mediate a number of antagonistic ecological relationships between plants and herbivores. Alkaloid-producing plants have been known to deploy these secondary metabolites as toxins effectively serving as antibiotics, as herbivore deterrents, and also as measures against other biotic stresses. However, the occurrence of alkaloids has not essentially been restricted strictly to the plants. Alkaloids have also been found in organisms belonging to other classes, ranging from bacteria and fungi to higher animals like arthropods (ants, beetles, and butterflies), frogs, and mammals. There are numerous examples among arthropods which accumulate high concentrations of alkaloids. They exhibit an equally remarkable diversity in the kind of alkaloids they accumulate. Although, in many cases, the alkaloids accumulated by these animals have been found to be of plant origin and they acquire these compounds at different

stages of their life cycle from their host plants they feed on. For example, a number of aposematic butterfly and moth species acquire toxic alkaloids from their host plants and sequester them inside their bodies. Just like in plants, alkaloids serve as toxins and provide a means of self-defense against predators, deterring them from feeding on their toxin-accumulating prey. The warning coloration that they provide is specially efficient in driving the predators away, which gradually learn to avoid the prey with such conspicuous display of toxic compounds. Amphibians, specially frogs, exhibit production of an array of noxious or toxic alkaloids in the skin and secretory glands, which confer bright colors to their skin as visible signs of toxicity. Various species of poison dart frogs, for instance, are brightly colored, displaying aposematic patterns to warn potential predators. Their bright coloration is associated with their toxicity and levels of alkaloids. Insects, displaying accumulation of a wide variety of alkaloids, are also capable of employing the alkaloids acquired from host plants as a source of other useful substances. The alkaloids may either be stored without any change. Such accumulation makes them distasteful to predators. In other cases, they may be metabolized for the synthesis of pheromones necessary for gaining attention of opposite sex during courtship and compounds with defensive value. For example, larvae of cinnabar moth (*Tyria jacobaea*) gather pyrrolizidine alkaloids from their host plant *Senecio jacobaea*. These alkaloids are retained throughout metamorphosis, and their accumulation makes them develop warning coloration which keeps the predators away. The alkaloidal precursors obtained from their hosts are metabolized by the larvae of *Danaus* sp. (monarch butterfly) and are converted into structurally diverse compounds, such as danaidone, dannaidal,

(continued)

**Box 33.7** (continued)

and hydroxydanaidal, that serve as pheromones and aphrodisiac, playing important roles in self-advertisement and gaining special preference from female mates. Similar conversion is also seen in the Arctiid moth, *Utetheisa ornatrix*, larvae of which obtain their alkaloids from host plant *Crotalaria* sp. and retain them through metamorphosis. At the time of courtship, the male moths present their alkaloidal load to the females by disseminating the male courtship pheromone, hydroxydanaidal, as chemical signal from a pair of scent brushes (coremata). Females are able to differentiate between males that contain unequal quantities of hydroxydanaidal signal, and males having higher levels of the pheromone are apparently preferred over the others. Another group of insect, the adult Ithomiine butterflies, feeds on solanaceous host plants and ingests pyrrolizidine alkaloids and sequesters them in the form of N-oxides and monoesters inside their bodies, making them unpalatable to their predator, the giant orb spider. The spiders have learned to sense the unpalatability of the pyrrolizidine-feeding butterflies and release them from their webs while devouring the freshly emerged ones that have not yet had the opportunity to feed on their alkaloidal host. The adult males of this group also transfer a major load of the acquired alkaloids to the females during mating. Thus, the alkaloids also strongly influence the mating behavior among insects.

Larvae (**a, c**) of Cinnabar moth (*Tyria jacobaea*) and Arctiid moth (*Utetheisa ornatrix*) (**b, d**)

plants, their physiological roles in plants continue to be deciphered. The main functions of alkaloids are as follows:

1. Alkaloids can act as defense compounds and provide protection against pathogens and pests due to their toxicity and bitterness.
2. They may act as nitrogen reserves, serving to release nitrogen in case of nitrogen deficiency.
3. They may act as growth regulators or stimulants, especially as germination inhibitors.
4. They may help to maintain ionic balance due to their chelating power.
5. They may function as detoxicating agents by sequestering harmful or toxic compounds via methylation, condensation, and cyclization, rendering them harmless.

It is also established that they have significant medicinal value. Since early history, alkaloids have been frequently involved as active ingredients in many medicine systems. Generally, they exhibit dose-dependent cytotoxic effects. They can be lethal at high concentrations, but in small doses, they can serve as effective drugs for treatment of a range of medical conditions. Many synthetic drugs used in modern medicine are also designed on the basis of the structure of some natural alkaloids. For example, chloroquinone, an antimalarial drug, is an indole-derived alkaloid related to the natural product quinine.

Alkaloids, such as nicotine, caffeine, and cocaine, are used as stimulants and sedatives. Morphine is an analgesic, sedative, and pain reliever. It is a constituent of cough medicines, helps allay diarrhea and vomiting, and reduces blood pressure. Heroin, another alkaloid, is more powerful than morphine. New drugs with morphine-like activity, such as pethidine, dionin, metopon, etc., have now been commercially synthesized. Codeine is an analog of morphine. It is an antitussive and analgesic, which is slightly milder and is used in whooping cough medicines. Many alkaloids can block ion channels, inhibit various enzyme activities, and interfere with neurotransmission, producing hallucinations and causing loss of coordination, convulsions, vomiting, and even death in severe poisonings. For example, morphine and cocaine have high affinity for receptors of neurotransmitters and, thus, are potent hallucinogens. Vinblastine and vincristine have been recognized as anticancerous, being antineoplastic. Some of the important alkaloids and their medicinal uses have been listed in Table 33.8. Though alkaloids have huge medicinal properties, when taken in large quantities, almost all alkaloids are toxic to human beings.

Many alkaloids are believed to act as chemical defense against herbivores and insects. Plants may deploy them as antibiotics, as herbivore deterrents, and as measures against biotic stresses. By ingestion of alkaloid-rich plants, large numbers of livestock deaths are caused every year. Plants containing bitter-tasting alkaloids, such as *Lupinus* (lupines), which produces quinolizidine alkaloid lupanine; *Senecio jacobaea* (ragwort), which produces pyrrolizidine alkaloid senecionine; and *Delphinium* (larkspur), cause poisoning in grazing livestock animals when consumed in large quantities. Such alkaloids, such as senecionine, taste bitter and are thus avoided

## 33.3 Nitrogen-Containing Compounds

**Table 33.8** Some important alkaloids and their medicinal uses

| Alkaloid | Source | Effects | Uses |
|---|---|---|---|
| Ajmaline | *Rauvolfia serpentina* | Antiarrhythmic; inhibits glucose uptake by mitochondria in heart tissue | For identification of patients with Wolff-Parkinson-white syndrome |
| Atropine | *Atropine belladonna* | Anticholinergic | Antidote to nerve gas poisoning |
| Caffeine | *Coffea arabica* | Interferes with neurotransmission, stimulates central nervous system | Stimulant |
| Cocaine | *Erythroxylon coca* | Interferes with neurotransmission; stimulates central nervous system; adrenergic blocking agent | Topical anesthetic, stimulant, drug of abuse |
| Codeine | *Papaver somniferum* | Interferes with neurotransmission | Relatively nonaddictive analgesic and antitussive |
| Coniine | *Conium maculatum* | Causes paralysis of the motor nerve endings | Poison, used in small doses to treat ailments in homeopathy |
| Morphine | *Papaver somniferum* | Powerful narcotic; has strong affinity for neurotransmitter receptors | Analgesic, addictive drug of abuse |
| Nicotine | *Nicotiana tabacum* | Causes respiratory paralysis, interferes with neurotransmission | Toxin, insecticide, drug of abuse |
| Quinine | *Cinchona officinalis* | Anti-plasmodial | Antimalarial drug |
| Sanguinarine | *Eschscholzia californica* | Antiplaque activity | Antibacterial agent, used in toothpastes and oral washes |
| Strychnine | *Strychnos nux-vomica* | Poisonous /toxic | Used in pesticides, for killing small vertebrates like rodents |
| Vinblastine | *Catharanthus roseus* | Antineoplastic/anticancerous | Used to treat Hodgkin's disease and others |

by horses thereby protecting the plants against herbivores. It is believed that domestic animals (unlike wild animals) have not been subjected to natural selection to avoid such alkaloid-containing potentially toxic plants. Alkaloids pose potential health hazards for humans, too, as they are converted to highly toxic forms upon transformation by cytochrome P450 monooxygenase in the liver. They are also toxic to insects and serve as feeding deterrents. They may be constitutively produced by the plants or induced when the plants face a mechanical injury or threat. For example, production of nicotine, a potent insecticide produced in tobacco (*Nicotiana* sp.) plants, is rapidly stimulated, and its endogenous titers are boosted severalfold upon herbivory. Similarly, caffeine is another effective toxin against insects, which is found in leaves and seeds of coffee (*Coffea arabica*), tea (*Camellia sinensis*), and cocoa (*Theobroma cacao*). It is effective against insects at concentrations found in fresh coffee beans and tea leaves. However, several species of insects have become well-adapted to feed on the alkaloidal plants by having developed mechanisms to tolerate alkaloids acquired by the plants during feeding. They can feed on their alkaloid-containing host plants and remain rather unharmed. The acquired alkaloids can either be eliminated or transformed into harmless derivatives with the help of specific enzymes which are part of the detoxification machinery of the insects. Some insect species use alkaloids acquired from their hosts for their own benefit. They harbor specialized enzymes which catalyze the conversion of these alkaloids into pheromones that are essential for attracting mates or as defense compounds. Therefore, alkaloids serve important ecological functions besides their huge pharmacological applications.

## 33.3.2 Glycosides

They represent another class of widely distributed nitrogen-containing secondary metabolites. The name glycoside is derived from "glycosidic bond" which is formed when sugar molecules condense with the hydroxyl group of another molecule. Glycosidic bonds may be formed (a) between two sugars, as in polysaccharides, or (b) between a sugar and non-carbohydrate molecule such as amino acids or steroids. Glycosides comprise of two chemically and functionally independent portions: the aglycone (genin) and the glycone (saccharide) moieties, linked by a glycosidic linkage. They are derived from other secondary metabolites by the addition of sugar moieties catalyzed by enzymes called glycosyltransferases. The most common glycone or sugar constituent of glycosides is the monosaccharide, glucose, though other sugars, including rare ones and those of higher order (di-, tri-, or tetrasaccharide), are also often present. In addition to this, frequent modifications of the sugar moieties, such as acylation and oxidation, further add to the variation existing within glycosides. Most members of this class of secondary metabolites are known for their characteristic detergent-like properties. They are believed to play important roles in growth and development and signaling and claim a special reference in plant's defense responses. They are produced by plants in many instances of pathogen attack and herbivory as well as in response to fluctuation in environmental conditions. Glycosides can further be classified into different subgroups based on different classificatory systems. The most commonly used are those based on the type on glycosidic linkages, the aglycone structures, and the nature of the glycone moiety attached. Based on the kind of glycosidic linkages, they are divided as O-, N-, S-, and C-glycosides (Table 33.9). More accepted classification categorizes glycosides based on the nature of its aglycone part. The main representatives of such aglycone groups include phenolics, terpenes, sterols, and products of the phenylpropanoid pathway (Table 33.10). The most widely studied among these are saponins, glucosinolates, cardenolides (cardiac glycosides), and cyanogenic glycosides. However, nucleosides and antibiotic-like streptomycin are also included under glycosides.

## 33.3 Nitrogen-Containing Compounds

**Table 33.9** The classification of glycosides based on the kind of linkage

| Type | Representation | Linkage | Example | Structure |
|---|---|---|---|---|
| O-glycosides | -O-C$_6$H$_{11}$O$_5$ | Via oxygen | Rhein-8-glycoside | |
| S-glycosides | -S-C$_6$H$_{11}$O$_5$ | Via sulfur | Glucosinolates like sinigrin | |
| N-glycosides | =N-C$_6$H$_{11}$O$_5$ | Via nitrogen | Nucleosides | |
| C-glycosides | -C-C$_6$H$_{11}$O$_5$ | Via carbon | Aloin | |

**Table 33.10** Types of glycosides

| Type | Sources | Examples | Structure |
|---|---|---|---|
| *Phenol glycosides* | *Salix fragilis* | Salicin | |
| | *Populus* sp. | Populin | Salicin |
| *Coumarins* and *furanocoumarin glycosides* | *Ferula asafoetida* (asafoetida) | Skimmin (umbelliferone glycoside) | |
| | *Atropa belladonna* | Scopolin (scopoletin glycoside) | Skimmin |
| *Flavonoid glycosides* | | | |
| a. Flavone glycosides | *Apium petroselinum* | Apiin | |
| | | | Rutin |
| b. Flavanol glycosides | *Eucalyptus macrorhyncha* | Rutin | |
| c. Flavanone glycosides | *Citrus* sp. | Hesperidine | |
| d. Chalcone glycosides | *Tephrosia vogelii* | Tephrosin | |
| e. Isoflavonoid glycosides | *Prunus avium* | Prunetrin | |

(continued)

**Table 33.10** (continued)

| Type | Sources | Examples | Structure |
|---|---|---|---|
| f. Anthocyanidin glycosides | *Chrysanthemum* sp. | Cynidine-3-glycoside (chrysanthemin) | |
| *Anthracene (anthraquinone) glycosides* | *Cascara* sp. | Cascarosides A, B, C, and D | Cascaroside |
| *Cardenolides* [steroid (cardiac) glycosides] | | | |
| | *Digitalis purpurea* | Purpurea glycosides | |
| | *Digitalis lanata* | Digilanids | |
| *Aldehyde glycosides* | *Vanilla* sp. | Glucovanillin | |
| | *Cinnamomum verum* (*cinnamon*) | Cinnamic aldehyde | Glucovanillin |
| *Cyanogenic glycosides* | *Amygdala Amara* | Amygdalin | |
| | *Trifolium repens* | | |
| | *Manihot esculenta* | Linamarin | Amygdalin |
| *Glucosinolates* or *thioglycosides* | *Brassica juncea* | Sinigrin | |
| | *Sinapis alba* | Sinalbin | |
| | *Brassica napus* (rape) | Gluconapin | Sinigrin |
| *Saponin glycosides* | | | |
| a. Tetracyclic triterpenoid saponins | *Dioscorea* | Diosgenin | |
| | *Solanum khasianum* | Salosanine | |
| b. Pentacyclic triterpenoid saponins | *Panax ginseng* (ginseng) | Ginsenosides | Diosgenin |
| | *Bacopa monnieri* | Bacosides A and B | |
| *Bitter aldehydes* | *Gentiana lutea* (gentian) | Gentiopicrin | |
| | *Swertia chirata* (chirata) | Amarogentin and chiratin | Gentiopicrin |
| *Miscellaneous* | | | |
| a. Steroidal alkaloidal glycosides | Liliaceae and Solanaceae members | Solanine, rubijervine, and solanidine | |
| b. Antibiotics | *Streptomyces griseus* | | |
| c. Glycosidal resins | *Exogonium purga* | Streptomycin | |
| | *Convolvulus scammonia* | | |

### 33.3.2.1 Saponins

Saponins are glycosides of steroids, glycosides of steroid alkaloid, or triterpene glycosides exhibiting widespread occurrence in plants. Saponins may exist in either aglycone or glycosylated form in plant tissues. The aglycone forms of saponins are also known as sapogenins, which essentially are terpenes without any sugar group. Saponins get their name from a soapwort—*Saponaria*—which was earlier used as a substitute for soap. The name *Saponaria* has its origin from the Latin word "sapo" meaning "soap," indicating the surfactant properties possessed by the plant extract as well as saponins. Saponins are characterized by formation of stable soapy foam when agitated in aqueous medium. The soap-like property of saponins is due to their amphipathic nature, with the presence of both a hydrophilic sugar group and a hydrophobic part. *Saponaria officinalis* (soapwort) is among the richest sources of saponins. Among others are *Hedera helix* (ivy), *Medicago sativa* (alfalfa), *Asparagus officinalis* (asparagus), and *Ruscus aculeatus* (butcher's broom) (Fig. 33.19).

Saponins have been extensively studied for their wide array of properties like that as fungicide, pesticide, and insecticide. In plants, the main role of these metabolites is to protect plants against fungal pathogens. Many plant species store saponins in their roots where they serve as antimicrobial agents, protecting the plants from infections. They are believed to do so by forming complexes with sterol-containing unsubstituted 3-β-hydroxyl group. They react with membrane sterols of the invading fungal hyphae leading to membrane disruption and thus protecting the plant from

**Fig. 33.19** Some examples of saponin-producing plants

infection. For example, *Avena sativa* (oat) produces two kinds of saponins constitutively, namely, triterpene avenacins and steroidal avenacosides, in roots and leaves, respectively. The triterpenoid saponin, avenacin A-1, localized in the root hairs of *Avena* (oat) like any other saponins, may be present in either aglycone or glycosylated forms, the latter being the active form. Glycosylated avenacin A-1 gets accumulated in the epidermal cells of the root tips where it imparts potent fungicidal activity, thereby conferring resistance to a wide range of fungal pathogens and, in turn, contributing to disease tolerance. Additionally, its release in the rhizosphere in the form of secretions at varying concentrations further accounts for its role as phytoprotectant by providing resistance against other soilborne pathogens. In oat plants, action of this saponin is particularly evident in providing protection against fungal infection by *Gaeumannomyces graminis* var. *tritici*. However, many fungi have developed mechanisms to avoid such defense strategies of plants by synthesizing detoxifying enzymes for saponins. For example, *Gaeumannomyces graminis* var. *avenae* produces avenacinase, an enzyme which detoxifies avenacin A-1 (saponin) and makes the infection of oat, wheat, and rye by *G. graminis* var. *avenae* possible. The second kind of saponins produced by oat plants, i.e., steroidal avenocosides, is converted into their active forms upon damage of the plant tissue by invading pathogenic fungi. Disruption of membrane allows for hydrolysis and subsequent release of the glucose unit from the avenacosides by plant β-glucosidase resulting in their transformation into the active and toxic forms. In their active form, avenacosides also disrupt the fungal membrane by forming complex with ergosterols, the major sterol in fungal membranes, creating pores into it, eventually causing fungal cell death. Saponins also act as protecting agents against herbivores acting either as deterrents or toxins, majorly because of their hemolytic ability, i.e., capacity to lyse hematocytes and other cell types, via disruption of membranes due to detergent properties. Thus, they hemolyze RBCs if injected into the bloodstream. They also have a bitter and unpleasant taste and cause gastric irritations when consumed by animals. For example, alfalfa saponins are known to cause digestive problems and bloating in cattle. Their detergent and hemolytic properties find saponins having important roles in many industries. They have long been used as fish poisons. Commercially important saponins from *Quillaja saponaria* bark are added as emulsifiers in shampoos, detergents, toothpastes, and beverages. *Glycyrrhiza glabra* yields a saponin glycyrrhizin which has medicinal value and thus is pharmaceutically important. Drugs containing steroid saponins have expectorant effect and are employed for the treatment of purulent wounds and different dermatological diseases. They also serve as raw materials for the pharmaceutical industry, some being used as starting materials for the synthesis of sex hormones, cortisone, and vitamin D.

### 33.3.2.2 Cardenolides

Cardenolides are a subtype of **cardioactive glycosides** which are secondary metabolites that act on the contractile force of heart muscles, the other subtype being bufadienolides, which are structurally slightly different (Fig. 33.20). These should not be confused with **cardenolides**, which are a broader category of steroids of which cardenolides are a type. Structurally, they are similar to steroid saponins

## 33.3 Nitrogen-Containing Compounds

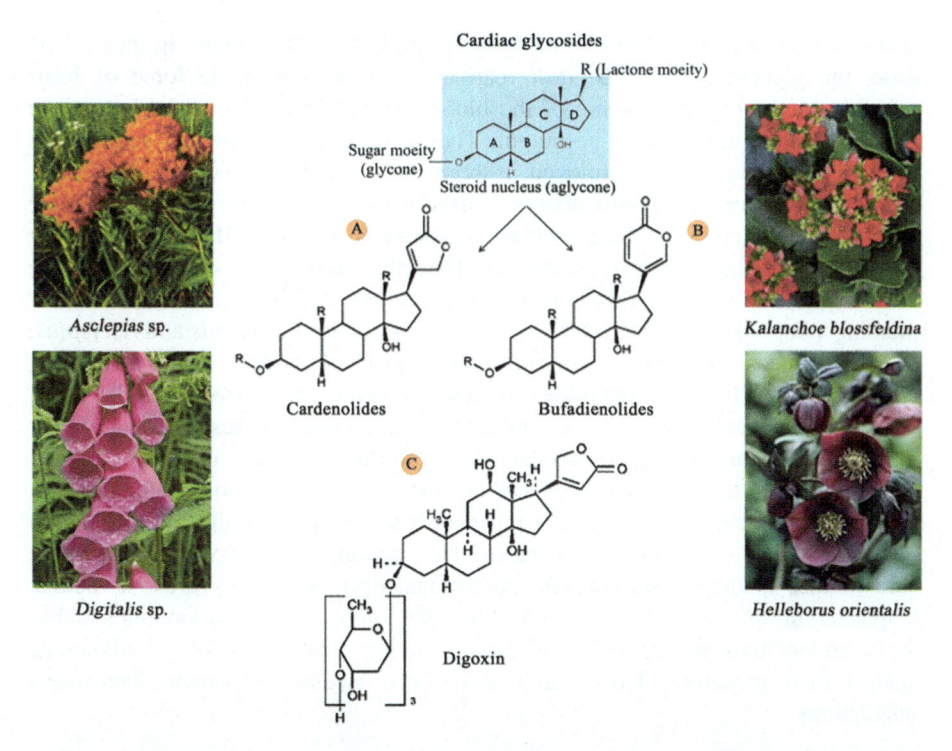

**Fig. 33.20** Plant sources and structures of cardiac glycosides. The basic structure consists of an aglycone moiety which is composed of a steroidal skeleton and a lactone ring attached at C-17. (**a**, **b**) The subtypes of cardiac glycosides, i.e., cardenolides and bufadienolides, differ at the lactone moiety. (**c**) The aglycones are glycosylated by addition of sugar moieties. Digoxin, a cardenolide present in *Digitalis* sp., has three sugar moieties attached to the steroid skeleton

and also possess soap-like properties. However, they can be distinguished by the presence of a lactone ring attached at carbon 17 and presence of rare sugar forming the glycoside part. A huge variety of cardenolides exist in plants with a widespread distribution in more than 200 species belonging to at least 55 genera, specially dominated by Apocynaceae family members. They are also reported in some insects which feed on the plants which produce cardenolides. In nature, they may occur as glycoside or in an aglycone form. The glycosylated and aglycone forms are interconvertible and are also known as primary and secondary cardenolides, respectively. The sugar components of cardenolides are majorly deoxy sugars and glucose. Cardiac glycosides are often toxic and have been reported to cause poisoning in human being, specifically affecting the vitality of the heart. They inhibit the $Na^+/K^+$-ATPase pumps of heart muscles in animals. The most well-known plants for its cardiac glycosides are *Digitalis purpurea* and *D. lanata*, which produce a mixture of glycosides such as digitoxin, digitonin, digitoxose, gitoxin, and gitalin. Since early times, *Digitalis* has been used for treatment of epilepsy, atherosclerosis, and other ailments. The first scientific information about *Digitalis* was published by William Withering in 1785. He reported curing of dropsy (condition in which unwanted accumulation of fluids occurs in the body) using leaves of the plant. In therapeutic dose, the glycosides from *Digitalis* can increase the contractile force of heart muscles, increasing the volume of the blood pumped into the circulatory system and thus improving the cardiac output. They can decrease heart rate and venous pressure. They have been employed in treatment of medical conditions caused by inefficiency of the heart due to atherosclerosis (hardening and narrowing of arteries) or dysfunctional cardiac valves, such as congestive heart failure. However, in higher doses these are highly toxic to vertebrates. The lethal properties of cardenolides have been exploited by African hunters to smear arrows with cardenolides and other piercing weapons of warfare. Other common sources are *Calotropis* and *Asclepias*, i.e., milkweeds, known so because they produce milky latex containing cardenolides. Although cardenolides are believed to have been evolved as defensive compounds in plants, they are also utilized by certain insect species in deterring their predators. The insects acquire cardenolides from the host plants they feed on, and accumulation of these cardenolides in their bodies causes them to become unpalatable to the predators. For example, milkweeds are hosts for ovipositing monarch butterflies (Fig. 33.21). Since the larvae feed on milkweed leaves, they acquire cardenolides in their bodies which are retained even when they grow as adults. Unpalatability of larvae and adult butterflies due to presence of cardenolides makes the predatory bird avoid feeding on them, offering them an ecological advantage against their predators. Thus, cardenolides also mediate important plant-insect interactions.

### 33.3.2.3 Cyanogenic Glycosides

Cyanogenic glycosides are another group of nitrogen-containing glycosidic compounds present in plants and have protective roles. Cyanogenic glycosides are called so because of their characteristic tendency to give rise to hydrogen cyanide (HCN), a poisonous gas, under certain circumstances, in a process known as cyanogenesis. HCN can cause inactivation of the enzyme cytochrome oxidase, inhibiting cellular respiration, which can eventually result in blocking of the central nervous system (CNS). Therefore, cyanogenic glycosides are considered potent

**Fig. 33.21** The larva (**a**) of monarch butterfly (**b**) feeds on cardenolide-rich milkweed plants and retains the acquired cardenolides throughout metamorphosis. The sequestration of these secondary metabolites provides warning coloration to both the larvae and adult butterflies which help to ward off predators

toxins. One of the earliest cyanogenic glycosides to be isolated was manihotoxin from *Manihot utilissima* (cassava), which was reported in 1830. In the same year, several others were also isolated from other cyanogenic plants, such as amygdalin from bitter almond (Box 33.8), linamarin from linseed, and phaseolunatin from the bean, *Phaseolus lunatin*. The estimated number of plant species which are cyanogenic is over 2600 belonging to about 110 families. Cyanogenic glycosides are widely distributed in family Fabaceae, Poaceae, Rosaceae, and Linaceae. Some cyanogenic plants include crops like sorghum (*Sorghum bicolor*), cassava (*Manihot esculenta*), and barley (*Hordeum vulgare*).

**Box 33.8: Why Do Stored Almonds Turn Rancid?**
Almond kernels (obtained from *Prunus dulcis*; Rosaceae) are considered to be a relatively low-moisture, high-oil-containing nuts with a long shelf life (up to 2 years) when properly handled. Lipid content in almonds is reported to range between 49 and 60 g/100 g. The major fatty acids in almonds are oleic acid (C 18:1) 60–70% of the total, linoleic acid (C 18:2) 14–26%, and α-linolenic acid less than 1% (10–12). At the end of their shelf life, almonds become unacceptable due to major changes in flavor and texture. The reactions causing these undesirable changes can be very complex. The factors that impact quality and shelf life of almonds are as follows:

Lipid oxidation (autoxidation) and rancidity: In oil-containing foods like almonds, the oxidation reactions lead to a loss of quality as the nuts turn "rancid" and odorous. Rancidity refers to unacceptable off-flavors and off-odors which develop with the breakdown of fats through various reactions. This is caused by lipid oxidation whereby oxygen reacts spontaneously with the fatty acids to form primary breakdown products (e.g., peroxides, conjugated dienes). As oxidation progresses, some other secondary products (e.g., volatile aldehydes, ketones) are also formed which are responsible for bad odor/flavor. When fats react with oxygen (lipid oxidation or autoxidation), it results in *oxidative* rancidity. *Hydrolytic* rancidity occurs when free fatty acids are released from fats by enzymatic lipid hydrolysis.

Roasting and microstructure changes: Roasted almonds have a shorter shelf life than raw almonds. Roasting causes changes to the cell microstructure that accelerates lipid oxidation and rancidity. The cell microstructure and the antioxidants present in raw almonds protect the oil from oxygen in the environment. Within each cell are many oil bodies or oleosomes, which contain oil. A membrane network separates the oleosomes from each other. During roasting, oleosomes burst as the membrane is damaged, and extracellular space increases. Because of this, oil within the cells is exposed to oxygen

(continued)

**Box 33.8** (continued)

all the more. Since lipid oxidation reactions can start immediately after roasting, almonds should be cooled immediately after roasting and kept protected from oxygen with suitable barrier packaging (e.g., in nitrogen gas).

Bitterness: An HCN-liberating compound (a cyanogenic glucoside) was discovered in bitter almonds by Robiquet and Boutron-Chalard. Because the compound was isolated from *Prunus amygdalus* (synonym *Prunus dulcis*), it was named amygdalin and is responsible for bitterness of almonds. Amygdalin has subsequently been found widespread in seeds of other members of the *Rosaceae*, like in apples (*Malus* spp.), peaches (*Prunus persica*), apricots (*Prunus armeniaca*), black cherries (*Prunus serotina*), and plums (*Prunus* spp.). Cyanogenic glucosides are now known to be present in more than 2500 plant species, including many important crop plants. In addition to their possible defense function, accumulation of cyanogenic glucosides in certain angiosperm seeds may provide a storage deposit of reduced nitrogen and sugar for the developing seedlings. Bitterness in almond is determined by the content of the cyanogenic diglucoside—amygdalin. Prunasin (the precursor of amygdalin) level decreases concomitant with the initiation of amygdalin accumulation in the cotyledons of the bitter almonds. In the sweet almonds, the inner epidermis in the tegument facing nucellus is rich in cytoplasmic, and vacuolar localized $\beta$-glucosidase activity is low in this layer.

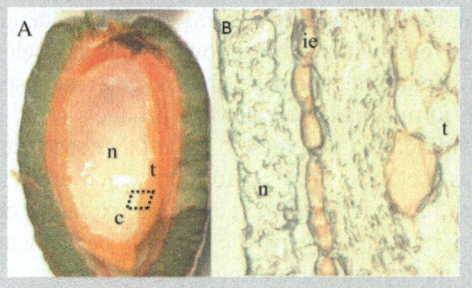

Anatomy of almond, i.e., inner epidermis; t, integument; n, nucellus; c, cotyledons

In the bitter genotype, prunasin synthesized in the tegument is transported into the cotyledon via the transfer cells and converted into amygdalin in the developing almond seeds, whereas in the bitter cultivar, the $\beta$-glucosidase-rich cell layer is in the inner epidermis of the tegument. Sweet kernel in almond has been shown to be a monogenic trait and the bitter kernel trait to be recessive. Prunasin is transformed into amygdalin during fruit ripening. Prunasin is

(continued)

**Box 33.8** (continued)
subsequently hydrolyzed by another β-glucosidase, named prunasin hydrolase (EC), to form mandelonitrile and glucose. Mandelonitrile is finally converted into benzaldehyde and HCN by the action of mandelonitrile lyase. HCN is an inhibitor of cell respiration and is detoxified by the action of the enzyme β-cyano-Ala synthase, which converts HCN and Cys into β-cyano-alanine.

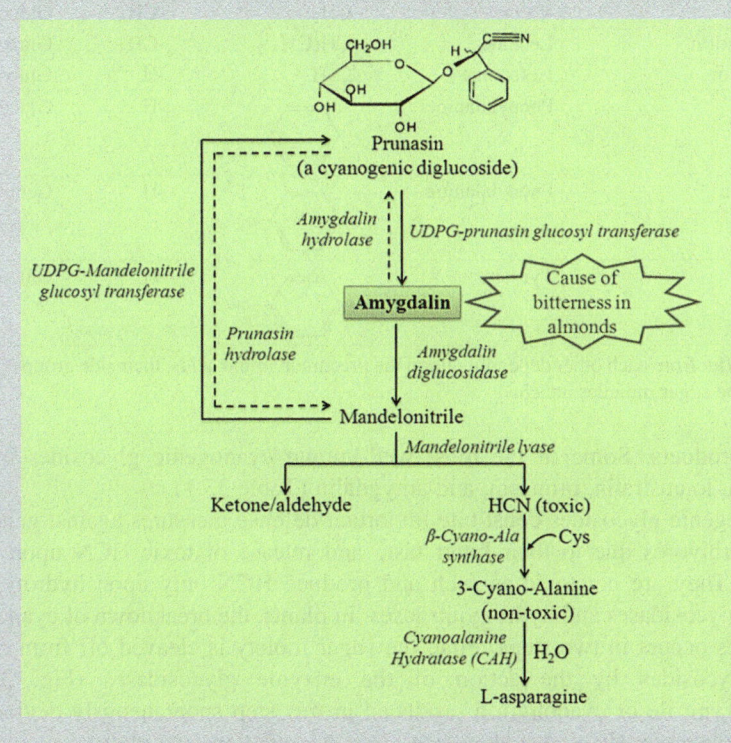

The cyanogenic glucoside metabolic pathway determining bitterness in almonds.

Chemically these are β-glycosides of α-hydroxynitriles (cyanohydrins). They are derived from amino acid precursors, from both proteinogenic and non-protein ones. Five precursors belong to the former category which are valine, isoleucine, leucine, phenylalanine, and tyrosine, while cyclopentenyl glycine belongs to the latter one. The biosynthesis of cyanogenic glycosides involves stepwise conversion of amino acids through a number of intermediates. Amino acids are first hydroxylated, and the resultant hydroxyl amino acids are then converted to aldoxime which in turn are converted to nitriles. Nitriles are subsequently hydroxylated to form α-hydroxynitriles which are further glycosylated to produce cyanogenic glycosides

**Table 33.11** Structures of some cyanogenic glycosides

$R_1$, $R_2$ attached to C, with O—β—Sugar and C≡N

| Basic moiety | | | | |
|---|---|---|---|---|
| Cyanogenic glycoside | Precursor | R1 | R2 | Sugar |
| Linamarin | Valine | -CH$_3$ | -CH$_3$ | Glucose |
| Heterodendrin | Leucine | -CH(CH$_3$)$_2$ | -CH$_3$ | Glucose |
| Lotaustralin | Isoleucine | -C$_2$H$_5$ | -H | Glucose |
| Prunasin | Phenylalanine | -C$_6$H$_5$ | -H | Glucose |
| Amygdalin | Phenylalanine | -C$_6$H$_5$ | -H | Gentiobiose |
| Dhurrin | Tyrosine | -C$_6$H$_4$-OH | -H | Glucose |

They all differ from each other depending upon the precursor amino acids, their side groups (R1 and R2), and the sugar moieties attached

as end products. Some of the most well-known cyanogenic glycosides include linamarin, lotaustralin, prunasin, and amygdalin (Table 33.11).

Cyanogenic glycosides constitute important defense measures against generalist insect herbivores due to their bitter taste and release of toxic HCN upon tissue damage. They are not toxic as such and produce HCN only upon hydrolysis by specific glycosidases and hydroxynitrilases. In plants, the breakdown of cyanogenic glycosides occurs in two steps. First, the sugar moiety is cleaved off from cyanogenic glycosides by the action of the enzyme glycosidases (Fig. 33.22). α-Hydroxynitrile or cyanohydrin produced in this step spontaneously degrades to release poisonous HCN at a slow rate. This degradation is further accelerated by enzyme, hydroxynitrilase. In intact plant tissues, the cyanogenic glycosides and the hydrolytic enzymes are spatially compartmentalized. The tissue-level compartmentalization prevents large-scale hydrolysis in intact plant tissue, so that plant itself is saved from the detrimental effects of overproduction of HCN (autotoxicity). However, liberation of HCN is brought about when the glycosides and enzymes are brought into contact upon infliction of damage on cyanogenic plant tissues, like in cases when bitten, chewed, or ingested by animals or insects or when the plant is crushed. For example, the cyanogenic glycoside dhurrin (*Sorghum* sp.) is found in the vacuoles of epidermal cells, whereas the hydrolytic and lytic enzymes are located in mesophyll cells. Therefore, in intact plants HCN is not formed, but when the plant is damaged during herbivore chewing, the cell contents of different tissues mix up and HCN is formed. Thus, cyanogenic glycosides in plants provide protective function by repelling insects and other herbivores. However, some specialist

## 33.3 Nitrogen-Containing Compounds

**Fig. 33.22** Metabolic degradation of cyanogenic glycosides

**Fig. 33.23** (**a, b**) *Zygaena filipendulae* and (**c, d**) *Histia flabellicornis* in larva and adult stage. The larvae release cyanogenic defense droplets from specialized glands present across their surface upon sensing a threat

herbivores have developed tolerance to high levels of HCN with the evolution of HCN detoxification and elimination methods and are adapted to feed on cyanogenic plants. Some insect species are also known to preferentially feed on cyanogenic plant hosts and store substantial quantities of cyanogenic glycosides which they acquire from their cyanogenic hosts and are capable of releasing HCN as a defense chemical. For instance, larvae of *Histia flabellicornis* secrete cyanogenic fluids in the form of droplets from the glands scattered across its body upon detecting any threat (Fig. 33.23). Similar cyanogenic droplets also known as defense droplets are also produced by larvae of *Zygaena filipendulae* which have the ability to both sequester cyanogenic glucosides (linamarin and lotaustralin) from feeding on their host *Lotus corniculatus* and synthesize these de novo. These are used as defense compounds during different stages of its life cycle.

## 33.3.3 Glucosinolates

Glucosinolates are a class of naturally occurring sulfur-containing nitrogen compounds in plants with pungent taste and flavors, like those characteristic of the members of the family Brassicaceae. These are also known as thioglycosides or mustard oil glycosides and break down to release isothiocyanates that act as defensive substances. They represent substituted esters of thio-amino acids, containing a side chain (R) and a sulfur-linked β-D-glucopyranose moiety. Different glucosinolates have different side groups. Mostly, glucosinolates are found in members of family Brassicaceae and some other plant families. The non-cruciferous glucosinolate-containing families include Capparaceae, Caricaceae, and Resedaceae. These compounds are responsible for the taste and smell of cruciferous vegetables, such as cabbage, radish, cauliflower, and broccoli. These vegetables contain different glucosinolates, each of which yields a different isothiocyanate upon hydrolysis (Table 33.12). Around 132 different glucosinolates are known to occur in plants. Methionine and cysteine serve as natural thiol donors during the biosynthesis of glucosinolates, and the original amino acid structure undergoes modifications resulting in the formation of different R side chains. Structurally they are very diverse, being derived from different amino acid precursors other than methionine, which can be any one among alanine, valine, isoleucine, leucine, tyrosine, phenylalanine, tryptophan, homomethionine (chain-elongated form of methionine), or homophenylalanine. The diversity is further expanded by the presence of double bonds, hydroxyl, carboxyl, or sulfur linkages in various oxidation states. Glucosinolates are strongly acidic and occur mainly as potassium salts under physiological pH. As with cyanogenic glycosides, glucosinolates also co-occur in plants with hydrolytic enzymes (myrosinase or β-thioglucosidases) but in separate compartments of intact plant tissue and are brought into contact only when the plant is attacked and physically severed. The disruption of plant integrity during herbivory (chewing or biting), hence, sets off the breakdown process of the substrate. The breakdown starts with action of hydrolytic enzyme called myrosinase or thioglucosidase (Fig. 33.24). It catalyzes the cleavage of glucose from glucosinolate to form an aglycone (non-sugar portion) entity. The aglycone intermediate is unstable and loses sulfate and undergoes spontaneous rearrangement to form chemically different reactive pungent products, such as isothiocyanates, nitriles, thiocyanates, epithionitriles, thiones, or indolyl compounds, depending upon the reaction conditions under which hydrolysis occurs (like pH), the presence of cofactors, the isozyme of myrosinase involved, and the type of glucosinolate substrate. However, isothiocyanates are the standard products of the hydrolysis, while the others only occur in certain conditions. Nitriles and isothiocyanates are recognized as biologically active products that act as feeding repellents and herbivore toxins, particularly against generalist herbivores. Mustard oil glycoside products have also been reported to have a strong antimicrobial activity, the most potent being isothiocyanates, followed by thiocyanates and nitriles.

## 33.3 Nitrogen-Containing Compounds

**Table 33.12** Glucosinolates known to cause unique taste and smell of well-known vegetables

| Glucosinolate | Precursor | Isothiocyanate | Major source | Structure of isothiocyanate |
|---|---|---|---|---|
| *Sinigrin* | Methionine | Allyl isothiocyanate | Broccoli, cabbage, radish | |
| *Glucoraphanin* | Methionine | Sulforaphane | Broccoli | |
| *Gluconasturtiin* | Phenyl alanine | Phenyl isothiocyanate | Watercress | |
| *Glucotropaeolin* | Phenyl alanine and cysteine | Benzyl isothiocyanate | Garden cress, cabbage | |
| *Gluconapin* | Methionine | 3-Butenyl isothiocyanate | Rape mustard | |

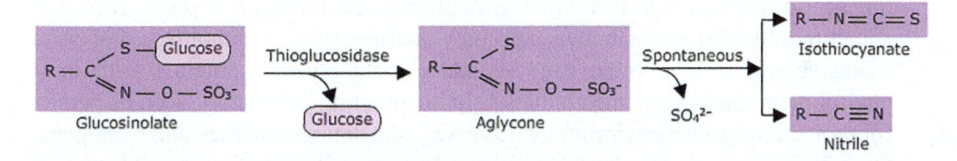

**Fig. 33.24** Metabolic degradation of glucosinolates

### 33.3.4 Non-protein Amino Acids

Many plants are known to contain some unusual amino acids that are usually not incorporated into proteins. In plants, there is a huge variety of at least 300 different non-protein amino acids which occur free in plants. Such amino acids, such as canavanine and azetidine-2-carboxylic acid, are present in free form in the cell and act as protective substances against herbivores. Structurally, these non-protein amino acids are usually similar to other amino acids. For example, canavanine is an analog of arginine, and azetidine2-carboxylic acid is similar to proline (Fig. 33.25). These non-protein amino acids may be toxic in various ways. Some may act to stop the synthesis or hinder the uptake of regular amino acids for protein synthesis, and others, such as canavanine, can be incorporated into proteins, by error. Once ingested, the enzymatic machinery of herbivores recognizes and binds canavanine to the arginine transfer RNA (tRNA). This results in a nonfunctional protein because either the tertiary structure or its catalytic site is disrupted. Plants that synthesize non-protein amino acids are not susceptible to the toxicity of these compounds because their protein synthesis enzymes can differentiate between arginine and canavanine. Some of the insects feeding on plant with non-protein amino acids might also have similar adaptations to distinguish between proteins.

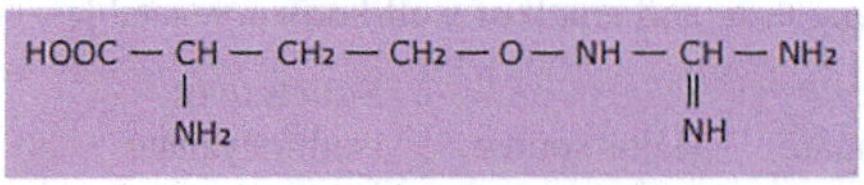

HOOC — CH — CH₂ — CH₂ — O — NH — CH — NH₂
       |                            ‖
       NH₂                          NH

Canavanine

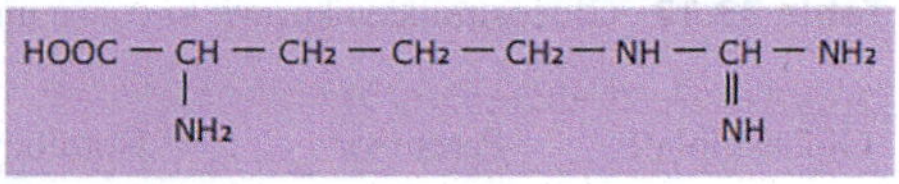

HOOC — CH — CH₂ — CH₂ — CH₂ — NH — CH — NH₂
       |                              ‖
       NH₂                            NH

Arginine

```
       CH₂
      /   \
   CH₂    CH — COOH
      \   /
       NH
```

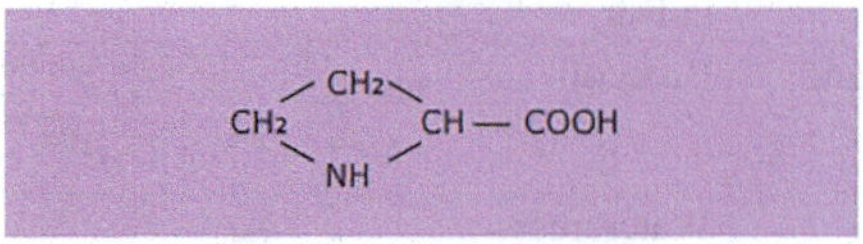

Axetidine-2-carboxylic acid

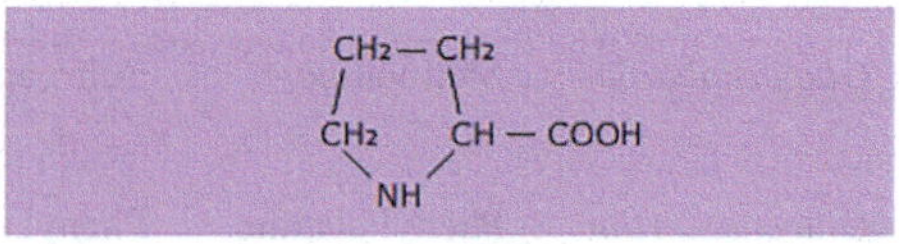

```
   CH₂ — CH₂
   |      \
   CH₂    CH — COOH
      \   /
       NH
```

Proline

**Fig. 33.25** Some non-protein amino acids

## Summary

- Secondary metabolites of plants are chemicals which are found in specific organs or tissues of plants or produced at particular developmental stages of plants, having no obvious role in normal growth and development of plant. However sometimes the distinction between primary and secondary metabolites is not easy. Plants, being sessile, often use secondary metabolites to interact with their environment. Secondary metabolites help in attracting pollinators, seed dispersal, for protection against environmental stresses, and deter herbivores and pathogens, thus playing protective roles for the plant. They contribute to the overall fitness of plant subjected to various biotic and abiotic stresses.
- Plant secondary metabolites are studied under three main groups, viz., terpenes, phenolics, and nitrogen-containing compounds. Terpenes are a diverse group formed from isoprenoid units, and some act as essential photosynthetic functions, electron transport chain components, plant growth regulators, wound-healing resins, and antibacterial. Some act as volatile signals, poisons, and feeding deterrent, thus playing a role in plant-animal interactions. Terpenoids such as Taxol have anticancer properties, terpenes such as carotenoid serve as precursor for biosynthesis of vitamin A, and yamogenin and gossypol have contraceptive properties which are of huge pharmacological importance to humans.
- Phenolics comprise a heterogenous group, synthesized via shikimic acid pathway. They have at least one phenol ring in their structure. These include lignin, flavonoids, anthocyanin, flavones, flavonols, isoflavonoids, and tannins. Their biological functions are equally diverse which include providing mechanical support to tissue, attracting agents for pollinators, deterring herbivores, and defending from pathogens and have role in symbiotic associations in legumes.
- Nitrogen-containing compounds are synthesized from amino acids and include alkaloids, cyanogenic glycosides, and glucosinolates. They serve a range of functions as poisons, feeding deterrents, antibiotics, and allelopathic agents. Most are toxic, but many are of commercial value in cancer treatment and also as food and drink additives such as caffeine.

- Genetic engineering has led to production of new products, such as decaffeinated plants and opium poppy having potential chemical constituents for treatment of baldness. A thorough knowledge and better understanding of the treasure of plant chemical diversity and its regulation by environmental factors at gene level hold immense potential to provide answers for many challenges faced by humankind.

## Multiple-Choice Questions

1. Phenolic compounds are produced by:
   (a) Plants and bacteria
   (b) Fungi
   (c) Algae and plants
   (d) Plants only
2. Association of phenols with various compounds such as proteins leads to the formation of new group of phenolic compounds, known as:
   (a) Polyphenols
   (b) Phenylpropanoids
   (c) Flavanoids
   (d) Stilbenes
3. Discoloration (browning) of apple after slicing or extended exposure to air occurs due to which of the following compounds?
   (a) Terpenoids
   (b) Alkaloids
   (c) Phenols
   (d) None of the above
4. Phenolic compounds are synthesized in plants via which of the following mechanism?
   (a) Shikimate/chorismate pathway
   (b) Acetate/malonate pathway
   (c) Acetate/mevalonate pathway
   (d) All of the above
5. The most abundant group of organic compounds in plants, after cellulose, is:
   a) Lignin
   b) Terpenes
   c) Alkaloids
   d) Flavanoids
6. Among the following, which describes the properties of flavanoids correctly?
   (a) Antiestrogenic, anticancerous, and antimicrobial
   (b) Anti-microbial and appetite suppressor
   (c) Mutagens and pigmentation in petals and bracts
   (d) Anticancerous, mutagens, and appetite suppressor

7. "Phytoalexins" produced by plants as a result of pathogen attack are:
   (a) Glycoproteins
   (b) Lipids
   (c) Phenolic compounds
   (d) Proteins
8. Which of the secondary metabolites listed below is a terpene?
   (a) Beta-carotene
   (b) Coumarins
   (c) Tannin
   (d) Glycosides
9. The first few compounds belonging to the category of terpenoids were isolated from?
   (a) Sunflower
   (b) Clove
   (c) Pine tree
   (d) Eucalyptus
10. Terpenes are _____ in nature, composed of ___ carbon units.
    (a) Lipophobic, five
    (b) Volatile, four
    (c) Lipophilic, five
    (d) Lipophobic, four
11. Condensation of isoprene units of 5-carbon leads to the formation of the terpenoids. In which of the terpenoids listed below are the carbon atoms not multiples of five?
    (a) Myrene
    (b) Carotenes
    (c) Tiglic acid
    (d) Steroids
12. The key enzyme in the mevalonic acid pathway (MVA) for the biosynthesis of terpenes is:
    (a) Isopentenyl diphosphatase
    (b) 3-hydroxyl-3-methylglutaryl CoA (HMGR)
    (c) Phenyltransferase
    (d) 1-deoxy-D-xylulose-5-phosphatase
13. Which of the following alkaloid is not a plant-derived alkaloid?
    (a) Caffeine
    (b) Colchicine
    (c) Bufotalin
    (d) Ajmaline
14. Which of the following groups of secondary metabolites is also known as proanthocyanidins?
    (a) Condensed tannins
    (b) Hydrolyzable tannins
    (c) Glycosides
    (d) None of the above

15. Which of the following is a synthetic antimalarial drug?
    (a) Quinine
    (b) Chloroquinone
    (c) Codeine
    (d) Heroin
16. First alkaloid to be isolated and characterized from plants is:
    (a) Caffeine
    (b) Morphine
    (c) Quinine
    (d) Cocaine
17. Which of the following is a characteristic of glycosides?

    (i) They comprise an aglycone and glycone moiety.
    (ii) Their synthesis involves the activity of glycosyltransferases.
    (iii) They are characterized by detergent-like properties.
        (a) Only i and ii
        (b) Only ii and iii
        (c) Only i and iii
        (d) All of the above

18. Anti-fungal activity of saponins is attributed to:
    (a) Hemolysis
    (b) Disruption of membrane
    (c) Coagulation of proteins
    (d) None of the above
19. Alkaloid with antineoplastic properties includes:
    (a) Vincristine
    (b) Morphine
    (c) Codeine
    (d) Nicotine
20. Which is not a cyanogenic glycoside?
    (a) Amygdalin
    (b) Lotaustralin
    (c) Gitalin
    (d) Linamarin
21. The cleavage of glucose from glucosinolate to form an aglycone entity is catalyzed by:
    (a) Glycosidase
    (b) Myrosinase
    (c) Hydroxynitrilase
    (d) Glucosidase
22. Amygdalin, a well-known cyanogenic glycoside, is isolated from:
    (a) Linseed
    (b) Bean
    (c) Cassava
    (d) Bitter almond

23. Which secondary metabolite is responsible for the pungent taste and smell of cruciferous vegetables?
    (a) Cardiac glycosides
    (b) Glucosinolates
    (c) Cyanogenic glycosides
    (d) Phenolics
24. Cyanogenic glycosides are so-called because under certain circumstances they give rise to:
    (a) Potassium cyanide
    (b) Sodium cyanide
    (c) Hydrogen cyanide
    (d) Both (a) and (b)

## Answers

1.a   2.b   3.c   4.d   5.a   6.a   7.c   8.a   9.c
10.c  11.b  12.b  13.c  14.a  15.b  16.b  17.d  18.b
19.a  20. c  21. b  22. d  23. b  24. c

## Suggested Further Reading

Kutchan TM, Gershenzon J, Moller BL, Gang DR (2015) Natural products. In: Buchanan BB, Gruissem W, Jones RL (eds) Biochemistry and molecular biology of plants. Wiley Blackwell, London, pp 1132–1206

# Plant Physiology in Agriculture and Biotechnology

## 34

Satish C Bhatla

Plant physiology has significantly contributed to the feeding of human beings. Applications of our knowledge of key concepts and processes in plant physiology are continuously evolving with time. Plants face a number of constraints in their optimal growth right from seed germination. They have to face challenges of weeds which compete with crop plants for limited availability of nutrients. Through our current understanding of the process of water and nutrient absorption and photosynthesis, farmers in the current era are able to optimize nutrient uptake and eradicate weeds in the crop fields. Early observations on the significance of ethylene in senescence and fruit maturation got extended into our present-day extensive knowledge of plant growth regulators and their derivatives in controlling fruit maturation, lodging, grapevine production, vegetable production, and fruit maturation. Of late, biotechnological approaches have also encouraged scientists to explore the possibility of producing transgenic fruits, reduce caffeine content in tea and coffee, and produce key metabolites in bulk for use by mankind. This chapter focuses on the key applications of the principles of plant physiology which have led to enhanced crop productivity.

## 34.1 Optimizing Nutrient and Water Uptake

An integration of the current knowledge about the water and nutrient uptake by plants and concepts of soil science has led to significant optimization of the techniques for water and nutrient mobilization by plants. Some of the major achievements in this aspect are presented below.

© Springer Nature Singapore Pte Ltd. 2018
S. C Bhatla, M. A. Lal, *Plant Physiology, Development and Metabolism*,
https://doi.org/10.1007/978-981-13-2023-1_34

## 34.1.1 Plant and Soil Elemental Analysis Facilitates Correct Application of Fertilizers

Fertilizers play a major role in increasing nutrient availability in soil for crop growth. Plant analysis, with improved diagnostic interpretation, has also played a key role in crop revolution. Assessment of nutrient status is based on the relationship between nutrient concentration in soil and yield and growth of a plant. The relationship between yield and nutrient concentration in plant tissues generally follows a bell-shaped curve. **Critical concentration (of an element)** is defined in plant analysis as the concentration of element that results in 90% of maximum yield or growth. Essential elements show two critical levels—"lower critical level," which is indicative of element deficiency, and "upper critical level," indicative of toxicity due to an element. The **critical toxicity level** of an element is the concentration of the element in tissue which results in 10% reduction in dry matter. NPK (nitrogen, phosphorus, potassium) fertilizers mixed with other micronutrients are commonly used. Foliar sprays are widely used to apply micronutrients in many crops. Major advantages of foliar sprays are (1) lower cost of compounds, (2) less utilization of nutrients, (3) uniform application, and (4) the nutrient deficiencies which can be corrected in the same growing season as the response to nutrient application is fast. However, there are certain disadvantages of the foliar application of nutrients as well. For example, high salt concentration may lead to leaf burn. In case of small leaf size, insufficient foliar adsorption can result in deficiency symptoms.

## 34.1.2 Managing Soil Acidity Through Liming for Better Nutrient Absorption by Plants

Mobility of nutrients within the soil is related to chemical properties of the soil, such as CEC (cation exchange capacity) and AEC (anion exchange capacity), as well as the soil conditions, such as moisture, pH, etc. Soil pH is a useful indicator of the relative acidity or alkalinity of the soil. Soil pH controls the nutrient availability of essential nutrients. When soil is flooded and becomes anaerobic, soil pH rises toward neutrality even when the pH was originally acidic. If the soil is subsequently drained and becomes more oxygen rich, then the pH will return to acidic state. However, care must be taken if the soil contains manganese oxide minerals, as flooding condition may lead to manganese toxicity. Acidic soils have low levels of basic ions like $Ca^{2+}$ and $Mg^{2+}$ but high levels of acidic ions especially $Al^{3+}$, $Mn^{2+}$, and $Fe^{2+}$. Some of these are not susceptible to leaching and occur at particularly high concentrations in acid soils, for example, aluminum. Under acidic conditions, phosphate is precipitated as $AlPO_4$ that increases $Al^{3+}$. Also, K and S availability decrease in acidic conditions. $Cu^{2+}$ and $Zn^{2+}$ are much more available in acid soils than in basic soils. Thus, in both acidic and basic soils, there is low availability of certain macronutrients, but the additional problem of high $Al^{3+}$ is found only in acidic soils. Attempts have been made to produce transgenic tobacco and papaya plants by expressing citrate synthase gene from the bacterium *Pseudomonas aeruginosa* since overproduction of citrate has been shown to enhance aluminum tolerance. In addition, acidic pH also limits oxidation of ammonium to nitrate ions. Soil pH also affects the growth of microorganisms and the type of plants which can grow in the soil. The relative amount of aluminum and hydrogen ions can change soil pH. Soil exhibits a buffering capacity. For example, removal of aluminum and hydrogen ions from the soil solution leads to replenishment of acid cations in soil solution. Minerals containing aluminum and hydrogen ions dissolve and release these cations as they are removed from the exchangeable pool. Finely textured clay soil tends to offer greater buffering capacity than coarse-textured soil. This has a great implication on **nutrient management** since buffering capacity determines the amount of resources, such as lime, that must be added to correct soil acidity. Soil with high buffering capacity requires large amounts of liming resources to raise the pH to a target value than soil with low buffering capacity. Heavy rainfall significantly lowers the soil pH by leaching out basic chemicals like calcium, magnesium, potassium, and sodium. The remaining elements like hydrogen, aluminum, and manganese contribute to an acidic soil profile. Soil erosion also contributes to elevation of soil acidity. Various liming materials may be added to the soil to neutralize or counteract soil acidity. **Liming materials** are bases that react with hydrogen ions in the soil solution to form water. Examples of common liming materials are limestone (calcium carbonate), dolomite (calcium/magnesium carbonate), hydrated lime (calcium hydroxide), and quicklime (calcium oxide). Calcium and magnesium silicates are also used as liming agents. There are four guidelines that help in determining the lime requirement of soil: (1) desired change in pH, (2) buffering capacity of soil, (3) type of lime material, and (4) the fineness or texture of the liming material. The optimal pH range for the growth of most plants is between 6.0 and 6.5. To avoid aluminum and manganese toxicity, a soil should be limed if the pH is less than 5.4. Since finer-textured soils have greater buffering capacity than coarse-textured soils, more lime must be added to the finer-textured soil to achieve the same effect and attain the target pH. Arbuscular mycorrhizal (AM) association with plant roots are also known to enhance nutrient uptake by plants in moderately acidic soils.

Addition of organic matter is another viable option to manage problems associated with the soil acidity. Organic matter increases the cation exchange capacity of the soil. Organic matter forms strong bonds, known as "chelates," with aluminum. Chelation reduces the solubility of aluminum and soil acidity. If soil is prone to manganese toxicity, it is advised not to add organic matter. Like organic matter, wood ash increases base saturation and forms chelates with aluminum.

### 34.1.3 Green Manure as an Alternative to Chemical Fertilizers

Green manure refers to the manure generated from crop plants by leaving them to wither and get incorporated into the soil, either uprooted or still rooted in the soil. Green manure, when turned into the soil, is broken down by soil-dwelling microorganisms to release nutrients and form humus. Humus is the dark organic matter which is composed of decaying remains of the plants, and animal tissue formed as a result of microbial action in the soil. It serves as a source of essential nutrients, such as nitrogen and phosphate. Green manure also generates humic acid and acetic acid and thus decreases the alkalinity of soils. Green manure has several advantages over the chemical fertilizers that have been long used extensively all over the world. Deep root systems of green manure crops keep the soil particles bound together and help improve the soil structure and prevent soil erosion and leaching of nutrients. In contrast, expensive inorganic fertilizers supply only nutrients and do not aid in improving the soil structure. The non-leguminous plants that are employed as green manure crops include wheat (*Triticum* sp.), mustard (*Brassica* sp.), carrot (*Daucus carota*), radish (*Raphanus sativus*), Jowar (*Sorghum vulgare*), and sunflower (*Helianthus annuus*), while the leguminous crops include river hemp (*Sesbania rostrata*), sann hemp (*Crotalaria juncea*), djainach (*Sesbania aculeata*), mung (*Phaseolus aureus*), cowpea (*Vigna catjang*), fenugreek (*Trigonella foenum-graecum*), sweet clover (*Melilotus officinalis*), lentil (*Lens esculenta*), senji (*Melilotus alba*), crimson clover (*Trifolium incarnatum*), and vetch (*Vicia sativa*). Fleshy green manures which are especially rich in nitrogen, such as leguminous plants, decompose quickly and release nitrogen and other nutrients fast allowing them to be available only after few weeks of incorporation into the soil, whereas other plants, like grasses, break down more slowly and thus are much slower to release nutrients.

The characteristics of plants used for green manure include fast growth to provide large yield within a short period, more leafy growth for a rapid decomposition, and deep root system to allow efficient nutrient acquisition from deeper parts of the soil. Deep root systems of plants also allow to bring up nutrients from deep layers of soil and make them available to next crop and also help to improve physical structure and properties of soil. Green manure plants grow quickly and are more efficient at acquiring nutrients from the soil, thus suppressing growth of weeds. Some green manures including many clovers and grazing rye, upon decomposition, release allelopathic chemicals in the soil and prevent the germination of seeds of weeds in the soil. Long-term green manures are generally grown for more than one season in order to prevent weed growth. They are grown as a part of crop rotation or intercropped with crop plants. During crop rotation, the plants are grown on the land when there is no crop. This helps to prevent weeds from growing on the land and nutrients from leaching out of the soil. It is particularly useful in building soil fertility before growing the crops which require nutrients in large quantities. During intercropping, the green manure plants are grown among the crop plants either at the same time as crop or sometimes slightly later to reduce competition between the green manure plants and the crop plants. Under sowing is one of the methods employed during intercropping in which green manure plants are sown under the crop plants such that the green manure is ready by the time crop is harvested and no extra time is required for preparing the soil and the next crop can be sown. Green manures have also been used in agroforestry where trees and shrubs are grown together along with crop plants and animals. The trees and shrubs provide large amounts of green leaves to be incorporated into the soil as green manure and provide nutrition during the crop growing period. They also provide food, fodder, and fuel wood along with other benefits. Green manure plants have been shown to reduce pest and controlling diseases in crops by providing habitats for natural predators of insect pests so as to reduce attack on the whole crop. Some green manure crops, when grown up to flowering stage, provide forage for pollinating insects. They also provide habitat for beneficial insects that act as natural predators for insect pests which allows for a reduction in the application of insecticides. Some green manure crops also provide fodder for livestock and can also be used for grazing.

The many advantages that green manures hold make them a sustainable and a rather inexpensive alternative to chemical fertilizers. Incorporation of green manures into agriculture thus can drastically reduce, if not eliminate, the need for supplementation of fertilizers, herbicides, and pesticides. They can also help to reclaim alkaline soil and to make the land usable for farming by improving its physical properties.

### 34.1.4 Antitranspirants for Enhancing Water Stress Tolerance in Ornamental Plants

Plants are often exposed to adverse environmental conditions, such as inadequate irrigation and high temperature, leading to acceleration of surface drying and plant wilting. Ornamental plants often lose water in large quantity during transportation, thereby causing considerable loss to floriculture industry. Water stress results in ABA production in the roots and its translocation to the leaves through the transpiration stream. In the leaves, ABA binds to ABA receptors causing activity of ion efflux and reduction of turgor pressure in the guard cells. Loss of turgidity in the guard cells leads to stomatal closure. As a result, transpiration is inhibited, and plants can withstand water stress by decreasing water loss. Floriculturists often use antitranspirants to reduce water loss by transpiration during shipping of ornamentals. Antitranspirants are a group of compounds which enhance water stress tolerance by preventing/reducing transpiration in plants. Antitranspirants can be either physical or physiological in nature. Physical antitranspirants include the use of polymers, resins, latex, and waxes to coat the leaf surface and minimize water loss by blocking the stomatal aperture. Physical antitranspirants have been successfully used to minimize loss of crop due to water stress in pepper (*Capsicum annuum*), peach (*Prunus persica*), and some herbaceous plants. Physiological antitranspirants reduce transpiration by inducing plants to close stomata. They include ABA or some chemicals which increase ABA concentration in the plants. Physiological antitranspirants have been successfully used in various horticultural crops, although plant responses to antitranspirants vary depending on the developmental stage, the species, and the concentration of the antitranspirant applied. Some sugar alcohol-based compounds (SACs) and biologically active ABA (*S*-ABA) have recently been used to enhance the shelf life of begonia (*Begonia semperflorens-cultorum*), petunia (*Petunia* x *hybrida*), marigold (*Tagetes erecta*), and impatiens (*Impatiens hawkeri*). *S*-ABA has been found to delay wilting symptoms in these plants as is also evident by the use of β-pinene polymer (βP) – a physical antitranspirant. To sum up, use of antitranspirants offers great potential in their exploitation by floriculturists to enhance the longevity of ornamental plants and maximize their aesthetic quality.

## 34.2 Eradicating Weeds

### 34.2.1 Parasitic Weeds in Crop Fields

Parasitic plants cause serious damage to the host plants resulting in substantial loss of yield. It is difficult to eradicate parasitic weeds because of enormous number of seeds produced by them, and these seeds generally have long life. *Orobanche* (broomrape), a genus from Orobanchaceae, consists of 10–60 cm tall, herbaceous parasitic plants with over 200 species which are known to parasitize tomato, eggplant, potato, bell pepper, cabbage, and beans by attaching haustoria to the host plants to draw nutrients. Another genus of Orobanchaceae, *Striga* (witchweed), is known to parasitize cereal crops. They are obligate hemiparasites of roots of the host plants. One of the crop practices employed is to stimulate germination of the parasitic seeds in the absence of the host, resulting in loss of seed bank of parasitic plants. If germinated in the absence of host, these seedlings will die because of lack of nutrients, thereby reducing population of parasitic plants. Germination in the absence of host is known as *suicide germination*. Both natural and artificial (man-made) compounds have been investigated as stimulants of suicide germination of seeds of parasitic plants. Such compounds should remain stable in soil for a long period of time. Strigol is the natural stimulant for germination of seeds of *Striga* sp. Low concentrations of synthetic analogs of strigol, such as GR24 and Nijmegen 1, stimulate germination of seeds of parasitic plants. However, high cost of production of these compounds at times limits their commercial use. Ethylene released into the soil has also been used as a measure to promote germination of weeds, triggering their germination in the absence of host, leading to their death. Inoculation of legumes with nitrogen-fixing bacteria coupled with ethylene-producing bacteria also has a great potential for stimulating "suicide germination" of weed seeds. Methyl jasmonate and some fungal toxins have also been found to stimulate germination of these seeds. One of the effective strategies seems to be planting nonhost crops to induce suicide germination.

### 34.2.2 Weed Control by Introducing Glyphosate-Resistant Crops

The enzyme 5-enolpyruvylshikimate-3-phosphate synthase (EPSPS) catalyzes the conversion of shikimate to chorismate in plants. Broad-spectrum systemic herbicide glyphosate (N-phosphonomethyl glycine) inhibits EPSPS activity, thereby blocking the conversion of shikimate-3-phosphate to 5-enolpyruvylshikimate-3-phosphate. This leads to accumulation of shikimate in plant tissues. The synthesis of aromatic amino acids, such as phenylalanine and tyrosine, is blocked in plants treated with glyphosate. Growth stops within hours and after few days leaves starts turning yellow. **Glyphosate** was discovered as an herbicide by Monsanto chemist JE Franz in 1970 and marketed by the name "Roundup" in 1974. Monsanto introduced glyphosate-resistant crops which enabled farmers to kill weeds selectively without killing the main crop. Weeds die in response to glyphosate spray as blocking of shikimic acid pathway leads to inhibition of biosynthesis of aromatic amino acids and their protein derivatives.

### 34.2.3 Blockers of Photosynthetic Electron Transport as Potential Herbicides

Herbicides have been used in agriculture since long to kill unwanted plants. Different herbicides act by disrupting the metabolism of the plants, such as blocking the synthesis of amino acids, carotenoids, or lipids. Photosynthesis inhibitors constitute a significant portion of the commercially important herbicides which are used to control annual and perennial broad-leaved weeds as well as grasses. These act by disrupting the photosynthetic ability of the susceptible plants via binding to specific sites within the chloroplast and interfering with or blocking electron transport. On the basis of site of their actions, herbicides can be classified into three categories. First category includes atrazine (2-chloro-4-ethylamino-6-isopropylamino-1,3,5-triazine) and DCMU (3-(3-4-dichlorophenyl)-1,1-dimethylurea) which are inhibitors of PSII. Second category includes the inhibitors which bind with the reducing site of PSI. As a result, they inhibit the reduction of ferredoxin, e.g., paraquat. Paraquat acts as an electron acceptor and intercepts electrons between ferredoxin and NADP. The third category of potent inhibitors of photosynthetic electron transport includes DBMIB (dibromothymoquinone) which blocks electron flow to cytochrome $b_6f$ at the $Q_P$ site of the complex. The photosynthetic blockers are used mainly for pre- and/or postemergence control of broad-leaved weeds. Grass crops, such as wheat and corn, benefit the most from photosynthetic inhibitors since these inhibitors primarily act on dicot weeds. DCMU is used as inhibitor of photosynthesis generally in laboratory and not commercially, since dosage required is high, and its degradation is slow. Atrazine works slowly and is a selective herbicide. Paraquat is a nonselective and broad-range herbicide and causes injury to the plants on contact. The visual injury symptoms to herbicides in weeds include chlorosis, desiccation, or browning of plant tissue. Susceptible plants exposed to photosynthesis-inhibiting herbicides show interveinal and veinal chlorosis. Necrosis of leaves begins around the edges of the oldest leaves, followed by similar damage to younger leaves. Sometimes, yellow spots may also appear on affected leaves. Chlorosis of leaf tissue results from photo-destruction of chlorophyll and other pigments. Excessive use of these herbicides has also been reported to induce oxidative damage in crop plants as well. Treatment of herbicides has been shown to result in accumulation of $O^{2-}$ and $H_2O_2$ in leaves, significant decrease in chlorophyll content, and inhibition of shoot and root growth and reduce the fresh weight. Apart from these damages to crop plants, incessant use of these herbicides has also been reported to cause low to high toxicity in humans.

## 34.3 Making Plants More Energy Efficient

Some microbial products help in protection of their host plants against a variety of stresses, such as drought, flooding, high salinity, heavy metal accumulation, and also against pathogens. One of the mechanisms to evade drought stress in plants is to produce more **trehalose** which helps in stabilizing the membranes and enzymes. Trehalose, a disaccharide formed by 1,1-glucoside bond between two α-glucose units, is synthesized and utilized by some bacteria, fungi, insects, and plants, as a source of energy. Its cleavage by the enzyme trehalase releases two molecules of glucose. This is double the efficiency of glucose released from starch. Instead of using biotechnological techniques for plants to produce more trehalose, it would be more effective to use bacteria which will provide surplus trehalose in association with plants.

## 34.4 Plant Growth Regulators (PGRs) in Agriculture and Horticulture

A wide variety of PGRs are currently in use in horticulture, viticulture, and agriculture in order to obtain varied advantages such as changes in plant constituents, ease of harvesting, enhancement of yield, and greater tolerance toward abiotic and biotic stress conditions (Table 34.1). Generally, PGRs are applied to crop plants using foliar sprays with water as a carrier. While doing so, due care has to be taken to ensure penetration of the active ingredient through epidermis and membranes by making an appropriate choice of the solvents and surfactants. Furthermore, application of the PGR treatment has to be given at the right stage of plant development and in "ideal" weather conditions.

**Auxins** are extensively used to stimulate root proliferation in the cuttings of a wide range of ornamental plants, vines, shrubs, and trees. 2,4-D is used to improve preharvest fruit retention in apples and pears. Auxins are also used to improved fruit set in tomatoes. 2,4-D and some other synthetic auxins also serve as herbicides. Auxin-type herbicides earned a bad name during the Vietnam War when they were used by the enemy to destroy crops. This included 2,4,5-trichlorophenoxy acetic acid (2,4,5-T) mixed with an extremely toxic 2,3,7,8-tetrachlorodibenzo-$p$-dioxin (TCDP). A combination of cyclanilide (an inhibitor of auxin transport) with ethephon is used to facilitate harvest of cotton bolls. Cyclanilide treatment promotes lateral branching in nursery trees.

Almost 140 gibberellins (*GA*) are known to occur in GA-producing fungi and in higher plants. Most of them are precursors and catabolites of active GAs, and very few of them possess biological activity. Commercial production of $GA_3$ and a

## 34.4 Plant Growth Regulators (PGRs) in Agriculture and Horticulture

**Table 34.1** Active ingredients of some PGRs and their major applications

| Plant growth regulators | Commercial products in use | Major applications |
|---|---|---|
| Auxins and related compounds | 2, 4-D (synthetic auxin) | Fruit setting and preharvest fruit retention |
| | Cyclanilide (transport inhibitor) | Promotes lateral branching |
| Cytokinins and related compounds | Thidiazuron (Diphenylurea type of cytokinin) | Defoliation in cotton |
| | Forchlorfenuron (synthetic cytokinin) | Increased size of fruits like kiwi and table grapes |
| | 6-Benzyladenine | Regulation of flowering and fruit set in grapevines |
| Ethylene and related compounds | Ethephon (commercial source of ethylene) | Fruit ripening (promoter), floral induction in pineapple |
| | Aviglycine (inhibitor of ethylene biosynthesis) | Delays harvest in pome fruits |
| | Silver nitrate (antagonist of ethylene action) | Prevention of abscission of flowers, leaves, and fruits |
| | Silver thiosulfate (ethylene receptor blocker) | Improved vase life of carnations |
| | 1-Methylcyclopropene (ethylene receptor blocker) | Delays preharvest ripening of pear and kiwi fruits and enhances fruit set in tomato and pepper |
| Gibberellic acid and related compounds | $GA_3$ | Seedless grapes |
| | Tebuconazole (inhibitor of GA biosynthesis) | Improves winter hardiness |
| | Metconazole (inhibitor of GA biosynthesis) | Anti-lodging agent |
| | Chlormequat chloride (inhibitor of GA biosynthesis) | Anti-lodging in cereals |
| | Mepiquat chloride (inhibitor of cyclases involved in GA biosynthesis) | Anti-lodging in barley and other cereals |
| | Paclobutrazol (inhibitor of GA synthesis) | Growth control in fruit trees and ornamentals, anti-lodging in rice |
| | Cycocel (inhibitor of GA biosynthesis) | Growth retardant in wheat, rye, and oats |
| Abscisic acid | S- ABA | Pigment (anthocyanin) enhancement in red grapes |
| Jasmonic acid | Prohydrojasmon | Anthocyanin production in apple and degreening of mango fruits |
| Others ("atypical" PGRs) | Maleic hydrazide, chlorpropham | Regulation of flowering and fruit set in grapevines |
| | Hydrogen cyanamide | Dormancy breaker in apples |

mixture of $GA_4$ and $GA_7$ (active GAs) is mostly undertaken by fermentation of the fungus *Gibberella fujikurai*. Commercially, $GA_3$ is most widely used followed by a mixture of $GA_4$ and $GA_7$. It is generally difficult to separate $GA_4$ and $GA_7$ during extraction by fermentation because of the structural similarity. Chemical synthesis of GAs is, however, a complex and expensive process for commercial use. Most of the GAs production worldwide originates from China. Applications of GA in agriculture and horticulture are mainly based on its unique ability to promote longitudinal growth in long-day plants, induce the activity of hydrolytic enzymes in germinating seeds, and promote the setting and development of fruits. Long-lasting effects of GAs are observed with the application of $GA_3$ or $GA_7$-$GA_4$mix, the latter being less persistent. More than 40 plant species have been reported to be exploited commercially through GA application. $GA_3$ is useful in the production of seedless grapes for increasing the fruit size in pears, for accelerating seed germination, and for improving the quality of citrus fruit. A mixture of $GA_4$ and $GA_7$ is, on the other hand, useful for reducing fruit **russetting** in apples (*russetting among apples is a disorder of the fruit skin due to microscopic cracks in the cuticle, leading to periderm formation. This phenomenon is also observed in pears*). GA inhibitors reduce longitudinal shoot growth and are commercially very useful in reducing the risk of lodging among cereals and oil seed rape and in rice cultivation. They are also used to control excessive vegetative growth in cotton plants, hedges, and ornamental trees. Some of the commercially useful GAs biosynthesis inhibitors are (1) chlormequat chloride and (2) mepiquat chloride (inhibit cyclases involved in early stages of GA metabolism). In Germany-based company, BASF introduced Cycocel (2-Chloroethyl-trimethylammonium chloride) in 1965 as a growth retardant, and it is still the most widely used PGR in cereal production, particularly in oats, rye, triticale, and wheat. In view of its extensive use among various cultivated plants, Cycocel remains one of the most extensively used PGR for commercial use on global scale. Mepiquat chloride (a quaternary ammonium compound) has also found extensive applications as a growth retardant particularly in cotton. Mepiquat has also been detected in processed roasted coffee beans, barley seeds, and crust of bread. Paclobutrazol, another established PGR, is used to control vegetative growth of fruit trees, such as mango, litchi, avocado, and also for controlling lodging in rice.

**Cytokinins** have been exploited commercially in agriculture and horticulture because of their unique ability to induce meristematic activity in plant cells, delay leaf senescence, promote photoassimilate transport, and antagonize auxin-induced apical dominance. Thidiazuron, a diphenylurea type of cytokinin, is commercially used for defoliation in cotton plants as a consequence of cytokinin-induced ethylene formation, leading to leaf abscission. 6-Benzyladenine is commercially used in grapevines for the regulation of flowering and subsequent fruit set. A synthetic cytokinin commercially available as for chlorofenuron [N1-(2-chloro-4-pyridyl)-$N_3$-phenylurea, CPPU] is known to influence fruit set among several fruit crop species such as water melon, kiwi, grapes, and apples. CPPU application has been observed to increase fresh weight of kiwi fruit through cell expansion and modulation of carbohydrate metabolism and water accumulation. In other words, CPPU application increases fruit sink strength (i.e., the capacity to attract carbohydrate).

**Abscisic acid** (ABA) plays significant role in inhibiting precocious seed germination and in protecting plants against abiotic stress. It is involved in stomatal closure both in normal growth conditions and under stress. Naturally occurring ABA is a (+)-*cis,trans*-isomer (=*S*-ABA). S-ABA is nowadays commercially produced from a phytopathogenic fungus *Botrytis cinerea*.

**Ethylene** is known to accelerate ripening of climacteric fruits. Gaseous application of ethylene is a difficult and an inefficient process. It is ideally applied as aqueous solution of acetylene or ethephon which decompose into ethylene upon getting absorbed by the plant. Ethephon is commercially used in agriculture for reducing stem elongation in cereal crops, particularly in barley, so as to reduce the risk of lodging. It is also useful as a defoliant and boll opener in cotton and to intensify the flow of latex from rubber trees. Ethephon is also commercially available as ethrel and florel. Aviglycine is a well-known inhibitor of ethylene biosynthesis which is produced on commercial scale from the fermentation of soil microorganism—*Streptomyces* sp. It is mainly used in horticulture for delaying the harvesting of pome fruits. Silver nitrate is an antagonist of ethylene action and is used to prevent abscission of flowers, leaves, and fruits although, in commercial applications, it has been found to exhibit relatively low mobility in plant tissues and is also phytotoxic. Silver thiosulfate does not show these limitations and is, therefore, used to enhance the shelf life of cut flowers. 1-methylcyclopropene (1-MCP) is a gaseous antagonist of ethylene with high affinity for ethylene receptors in plant tissues. 1-MCP complexed with α-cyclodextrin dissolved in water is commercially used for delaying preharvest ripening in pear and kiwi fruits, to improve fruit set in tomato and pepper and to enhance heat and drought resistance in maize, soybean, cotton, sunflower, rice, and wheat. A synthetic derivative of jasmonic acid called prohydrojasmon is relatively more stable than jasmonic acid in solution and is used to promote anthocyanin production in apple. Degreening of mango fruits is also enhanced by methyl jasmonates treatment. Jasmonate treatment can be used to enhance plant resistance to pests. This has been particularly observed in tobacco plants. JA can also be used to repel herbivores through volatile emissions in tomato. *Brassica rapa* treated with JA is less prone to attack by some insects.

Several compounds which do not directly interfere with plant hormone system are referred to as "atypical" PGRs because of their commercial applications. More than 40 "atypical" PGRs are currently in use either as a single component or in various combinations. Thus, maleic hydrazide (1,2-dihydro-3,6-pyridazinedione) and mefluidide [*N*-(2,4-dimethyl-5-{[(trifluoromethyl)sulfonyl]amino}phenyl) acetamide] are known to act by inhibiting cell division in the meristematic tissues. Chlorpropham [isopropyl (3-chlorophenyl)carbamate] and maleic hydrazide are commonly used to suppress shoot formation from the resting buds following the removal of flowers. These two compounds are also used to suppress sprouting of onions and tomatoes in storage. Among apples and grapevines, hydrogen cyanamide is commonly used as a dormancy breaker because of its ability to transiently inhibit respiration which leads to induction of α-amylase activity causing dormancy release.

## 34.4.1 Some Specific Applications of PGRs in Agriculture and Horticulture

**Lodging** Several crop plants with tall shoots, panicles, or other fruiting structures exhibit a tendency to fall over in response to wind and rain. This is commonly observed in rice, oilseed rape, linseed, and small grain cereals such as rye, oats, barley, and wheat. Use of stem-shortening PGRs on these crops reduces the risk of this kind of lodging. Stem lodging occurs due to break at the stem base in response to strong winds, intense rainfall, and thunder storms. Root lodging is generally caused after prolonged rainfall as the plant roots are unable to hold the weight of heavy aerial parts in water-soaked soils. Lodging can lead to loss of grain yield up to as high as 40%. Chlormequat chloride was the first PGR to be used as an anti-lodging agent. This compound increases the number of fertile tillers and also reduces stem length when applied to winter wheat crop at early stage of tillering. Barley crop is less responsive to chlormequat chloride than wheat, oats, and rye. Earlier varieties of rice were also prone to lodging although modern semidwarf cultivars are lodging-resistant. Loss of crop yield due to lodging can also be reduced by applying stem-shortening agents (triazole tebuconazole and metconazole) to oilseed rape (*Brassica napas*).

**Cotton Cultivation** Cotton is a perennial plant with indeterminate fruiting habit. It simultaneously produces vegetative and fruiting structures. The developing seeds inside the capsule (boll) develop cotton fibers on their surface. A mature boll ruptures to expose seeds with fibers for harvesting. Mepiquat chloride and chlormequat chloride are commonly used to control vegetative growth of cotton plants. Mature cotton bolls can be prompted to open by ethephon treatment, thereby increasing the harvest of cotton fibers. Prior to harvesting, cotton plants are generally subjected to defoliation by the application of some herbicides such as carfentrazone-ethyl and fluthiacet-methyl or PGR-type products containing ethephon or thidiazuron. None of these compounds kills cotton plants.

**Grapevine Production** Shoot growth in grapevines has to be controlled in order to have optimal production of berries. Hydrogen cyanamide, chlormequat chloride, and mepiquat chloride are commonly used to control vegetative growth in grapevine. $GA_3$ is used globally in the production of seedless grapes. Its application at the right time of plant growth is used to create space (by the extension growth of rachis), thereby allowing larger berries formation. In the case of red grapes, ethephon, or S-ABA, sprays are commonly used to intensify pigmentation.

**Pineapple Production** The production of pineapple fruits requires an average of 18–22 months from planting to harvesting. Pineapple plants are induced to flower with ethephon treatment. Aqueous solution of ethylene gas containing activated charcoal for delayed release of gas is also applied over the whole plant for floral induction in a gradual manner.

**Vegetable Production** GA$_3$ plays a major role in vegetable development. It effectively enhances seed germination, combats the problem of seed dormancy, induces flower formation, and elevates total fruit yield. Mepiquat chloride (N, N-dimethylpiperidinium chloride, well known as PIX) and chlormequat chloride (2-chloroethyltrimethylammonium chloride) are commonly used to enhance productivity and quality of bulbs in onion and garlic. While mepiquat chloride is an inhibitor of GA biosynthesis and regulates plant height and initiated defoliation, chlormequat enhances GA activity and is involved in enhancing the property of cell elongation. Use of synthetic gibberellin, auxins, and ethephon is known to enhance the yield of tomatoes. Pretreatment with GA$_3$, 2, 4-D, and ethephon enhances seed germination in tomato. Fruit ripening is enhanced by the use of ethephon-releasing compounds. Poor fruit set and yield is countered by soaking in a solution of GA$_3$, 2, 4-D, or thiourea or by foliar application of GA$_3$ or planofix (α-napthylactetic acid). This plant growth regulator is used for inducing flowering and preventing premature shedding of flower buds and unripe fruits. It helps in enlarging fruit size and increasing and improving the quality and yield of fruits.

**Parthenocarpy** Absence of need for pollination and fertilization for fruit development provides a great advantage in horticulture when the fruit set rate is slow because of unfavorable environmental conditions since pollen maturation and fertilization are affected by factors like light, humidity, temperature, etc. Use of parthenocarpy to promote fruit set under unfavorable environmental conditions could also improve the quality and quantity of pollinator-dependent fruit production by reducing the number of poorly formed fruits caused by insufficient pollination. The method of production of fruit bypassing pollination and fertilization is referred as "**parthenocarpy**" (such fruits are called virgin fruits). Parthenocarpy occasionally occurs as a mutation in nature (usually considered as a defect), as the plant can no longer sexually reproduce but may propagate by asexual means. It is different from "**stenospermocarpy**" in which seedless fruits are formed because of early abortion of seeds after fertilization. Parthenocarpy is often confused with **parthenogenesis** in animals. Parthenogenesis is a method of asexual reproduction, wherein embryo is formed without fertilization. However, parthenocarpy involves fruit formation without seed (embryo). In plants, the term equivalent to parthenogenesis is "**apomixis**." "Seedlessness" is a desirable condition for the production of some fruit plants, such as fruits with hard seeds, for example, pineapple, banana, oranges, and grapes. Furthermore, some seeds produce chemical substances that speed up the deterioration of fruits. For example, in eggplant fruit, the presence of seeds results in browning and texture reduction of the pulp. In dioecious plants, parthenocarpy is of great help as it increases fruit production at lesser cost because it eliminates the need for the plantation of staminate trees. Some plants also use parthenocarpy as a method of defense. Wild parsnip plants (*Pastinaca sativa*) produce both seeded and parthenocarpic fruits during their reproductive phase. Parthenocarpic fruits produced by parsnip plants are preferred by herbivores over their seeded fruits and thus act as a decoy (as in "decoy defense mechanism") in the scheme to protect the desired product (seeded fruits). Many agricultural practices employ phytohormones for the production of parthenocarpic fruit. Auxins, gibberellins, and cytokinins or mixtures of these phytohormones are effective in inducing fruit development in the absence of fertilization and increase productivity in various horticultural crops. Scientists are also increasingly finding ways to exploit genetic parthenocarpy. Approaches to genetic parthenocarpy have largely focused on selective breeding programs for seedlessness. For example, selective breeding of parthenocarpic sweet pepper, papaya, and summer squash varieties has all been shown to increase productivity. More recently, scientists have focused on genetic engineering approaches for parthenocarpic fruit set through modification of auxin synthesis, auxin sensitivity, auxin content, and auxin and gibberellin signal transduction.

## 34.5 Biotechnological Approaches

### 34.5.1 Transgenic Fruits

Vegetable and fruit production suffers from many biotic stresses caused by pathogens, weeds, and insects. Commercial sale of genetically modified (GM) food began in 1994 when scientists from Calgene Inc. (USA) developed Flavr Savr which exhibited delayed ripening in tomato. It was engineered to have a longer shelf life by inserting an antisense gene that delayed ripening. To withstand the rigors of shipping, tomatoes must be picked up at "mature green stage" which have already absorbed all the vitamins and nutrients from the plants but have not started producing ethylene that triggers ripening. Calgene, Inc. developed a tomato with a gene that slows down the natural softening process accompanying ripening. Pectin in fruits, which is responsible for their firmness, is degraded by an enzyme called polygalacturonase (PG). As pectin is destroyed, the cell walls of tomatoes break down, leading to softening of fruits, making them difficult to ship. Reducing the amount of PG in tomatoes slows cell wall breakdown and produces a firmer fruit for a longer time. Calgene scientists isolated PG gene in tomato plants and converted it into a reverse image of itself, called antisense orientation. The "reversed" tomato gene (the Flavr Savr gene) was reintroduced into the plants. In order to tell if Flavr Savr gene was successfully reintroduced into the plants, Calgene scientists attached a gene that makes a naturally occurring protein that renders plants resistant to the antibiotic kanamycin. By exposing the plants to the antibiotic, Calgene scientists could tell which plant had accepted that Flavr Savr gene. The ones unaffected by kanamycin grow to have designed traits of Flavr Savr. Once in a tomato plant, the Flavr Savr gene attaches itself to the PG gene. With the Flavr Savr gene adhering to it, the PG gene cannot give necessary signals to produce PG enzyme that destroys pectin. Thus, these tomatoes retain their peak flavor. In 1994, US Food and Drug Administration (FDA) announced that Flavr Savr tomatoes are as safe as conventional tomatoes.

A number of other transgenic fruit crops are being developed to enhance their host plant resistance to insects and plant pathogens for herbicide tolerance and to improve features such as slow ripening to enhance their shelf life. Transgenic papaya carrying coat-protein gene provides effective protection against papaya ringspot virus. The transgenic plum cultivar (honey sweet) provides germplasm for plum pox virus control. Transgenic bananas with enhanced resistance to *Xanthomonas campestris* have also been developed. In 2013, the USDA approved import of a GM pineapple that is pink in color and "overexpresses" gene derived from tangerine, increasing the production of lycopene. Subsequently, Arctic apples were approved in which gene silencing was used to reduce polyphenol oxidase activity, thus preventing the fruit from browning when cut. The European Union also undertook field tests for transgenic lemon, which exhibited fungal and kanamycin resistance.

## 34.5.2 Genetic Engineering and Conventional Breeding Approaches to Reduce Caffeine Content in Coffee and Tea

Flavor of coffee is due to caffeol, an essential oil. Likewise, tea is also an important beverage having distinctive character due to three components, i.e., essential oils, alkaloids, and polyphenols (tannins). The ethereal oil, theol is responsible for the aroma and flavor of tea, alkaloid theine gives refreshing and stimulating property, and tannins give the bitterness and astringency to tea. Worldwide the demand for decaffeinated coffee is increasing, and currently it accounts for about 10% of the sale. A similar market for tea is also emerging. Generally, to obtain decaffeinated coffee chemical methods using organic solvents (such as benzene) are employed. However, chemical methods lead to loss of flavor and antioxidants. So researchers have searched for plants which produce caffeine-free berries, so that the flavor is retained. For the plants to be economically viable, the berries must ripen synchronously and have a shape and size that can be harvested easily. In 2000, researcher Paulo Mazzafera and coffee breeder Silvarolla, in search of naturally low caffeine varieties of coffee, started working with plants collected from Ethiopia, and promising strains were discovered in 2003. These naturally decaffeinated plants contain only 0.08% caffeine in contrast with routinely used coffee which contain 1.2% caffeine in *Coffea arabica* beans and 2–3% in *C. canephora* beans. These plants have caffeine-free leaves as well as berries and are defective in final step of metabolic pathway which converts theobromine to caffeine. Subsequently, Japanese scientists were also able to produce coffee having 70% less caffeine than traditional varieties. Breeding experiments with these plants have been conducted to produce high-quality coffee beans which are decaffeinated naturally. These mutant varieties have mutation in the gene coding for caffeine synthase. Similar research is ongoing for tea as well. Another approach adopted recently is to produce genetically modified plants. Cloning of genes encoding enzymes for the biosynthesis of caffeine allowed generation of *C. canephora* seedlings having 70% less caffeine content. The pathway involves conversion of xanthosine to caffeine. Genes responsible for these steps have been isolated, and transgenic plants with reduced expression of theobromine synthase have been produced. The results were achieved using RNA interference technique. Both the coffee species have been successfully transformed. However, since theobromine and caffeine are involved in defense against herbivores, the ultimate aim is to generate plants having these metabolites in all tissues except beans. Otherwise such plants will be more susceptible to attacks from insects and herbivores, leading to loss in final production.

## 34.5.3 Bulk Production of Secondary Metabolites

In the quest for alternatives for higher production of medicinal compounds from plants, various biotechnological approaches have proved beneficial. One of the avenues explored is plant tissue culture, which holds immense potential for controlled and selective production of many useful secondary metabolites. Investigations on the production of secondary metabolites by callus and cell suspension cultures started as early as in 1956. Plant tissue culture offers an opportunity to exploit cells, tissues, organs, or the entire organisms by growing them in vitro and to genetically manipulate them to get desired compounds. These methods offer alternative to traditional ways for industrial production of plant secondary metabolites. Secondary metabolites can be produced by using different tissue culture approaches, such as callus, cell suspension, and organ cultures. These approaches have made the production of a variety of pharmaceuticals like alkaloids, terpenoids, steroids, saponins, phenolics, flavonoids, and amino acids possible in laboratories. Cell suspension culture systems are used for large-scale culture of plant cells from which secondary metabolites could be extracted. Cell cultures not only yield defined standard phytochemicals in large volumes but also eliminate the presence of interfering compounds that occur in the field-grown plants. This method can provide a continuous, reliable source of natural products. The major advantages of the cell cultures include synthesis of bioactive secondary metabolites, in controlled environment, independent of climate and soil conditions. A number of different types of **bioreactors** have been used for mass cultivation of plant cells. The first commercial application of large-scale cultivation of plant cells was carried out in stirred-tank reactors to produce shikonin by cell culture of *Lithospermum erythrorhizon*. Cells of *Catharanthus roseus*, *Dioscorea deltoidea*, *Digitalis lanata*, *Hypericum perforatum*, *Maackia amurensis*, *Panax ginseng*, *Taxus wallichiana*, and *Sophora flavescens* have been cultured in various bioreactors for the production of secondary plant products. Production of secondary metabolites in cell suspension cultures overcomes the problem of variable product quantity and quality from whole plants by avoiding the effects of different environmental factors, such as climate, diseases, and pests. *Maclura pomifera* cell suspension culture showed a greater level of metabolite accumulation (0.91%) than stems (0.26%), leaves (0.32%), and fruits (0.08%) of the parent plant. Non-embryogenic callus cultures, containing more or less homogenous clumps of dedifferentiated cells, are also used for production of flavonoids. The isoflavones and pterocarpans are produced by *Maackia amurensis* cultures. Elicitors markedly increase the production of flavonoids in comparison to the control cultures. The production of flavanones is stimulated up to five times by the addition of 2 mg/mL yeast extract. Moreover, the production of prenylated flavanones also can be

## 34.5 Biotechnological Approaches

increased, two to five times by addition of cork pieces. Organ cultures are relatively more stable for the production of secondary metabolites than cultures of undifferentiated cells, such as callus or suspension culture. Generally, two types of organ cultures are considered, i.e., root cultures and shoot cultures. Many of the secondary compounds, for example, the tropane alkaloids, hyoscyamine, and scopolamine, are produced quite well in root cultures. Root systems of higher plants, however, generally exhibit slower growth than the cultures of undifferentiated plant cells, and they are difficult to harvest. Therefore, alternative methods for the production of compounds synthesized in plant roots are being investigated. The most promising one of them is the use of plant hairy root cultures. Different types of bioreactors have been used for culture of plants roots and shoots. Till date, the only example of the commercial use of plant organ cultures for secondary metabolite production is the cultivation of ginseng (*Panax ginseng*, a medicinal herb) roots to obtain ginsenosides (a class of steroid glycosides).

Transformed roots have been widely used for in vitro production of secondary metabolites in many plant species. Hairy root cultures produce secondary metabolites over successive generations without losing genetic or biosynthetic stability. Also, production of two different secondary metabolites is possible simultaneously from adventitious root co-cultures. This vast potential of hairy root cultures as a stable source of biologically active chemicals has provided the exploitation of in vitro system through scaling up in bioreactors. Hairy root cultures of *Lithospermum erythrorhizon*, *Harpagophytum procumbens*, and adventitious roots of *Panax ginseng* and *Scopolia parviflora* have been analyzed in bioreactors to obtain shikonin, harpagide, ginsenosides, and alkaloids, respectively. Ginsenoside has also been produced in 5 L stirred-tank bioreactor using adventitious root culture. Hairy root culture of *Stizolobium hassjo* to yield 3,4-dihydroxyphenylalanine has been reported using 9 L mist bioreactor. Present scale-up technology dictates the use of stainless steel tanks for growth of plant cells on an industrial scale. Hairy root cultures continue to attract interest as a potential resource for large-scale production of commercially valuable compounds. *Catharanthus roseus* hairy root cultures are induced by pectinase and jasmonic acid and lead to enhanced accumulation of ajmalicine, serpentinine, tabersonine, and other secondary metabolites. Agrobacterium-transformed root cultures produce a wide variety of secondary metabolites and other bioactive compounds. Co-culture of hairy roots and microorganisms has been tested to produce novel secondary metabolites. For example, co-culture of *Hordeum vulgare* roots with *Glomus intraradices* (VAM fungi) induces synthesis and accumulation of terpenoid glycoside. The most famous plant-derived pharmaceuticals include digitalin from *Digitalis purpurea* which is prescribed for heart disorders, codeine obtained from *Papaver somniferum* (sedative), vinblastine and vincristine from *Catharanthus roseus* (for leukemia cancer treatment), artemisinin from *Artemisia annua* (for malaria treatment), quinine from *Cinchona officinalis* (for malaria), and paclitaxel (taxol from *Taxus brevifolia* and other species). Plant tissue culture methods have been developed for profitable production of these biochemicals. Such useful compounds have been reported to be successfully produced in callus as well as in suspension culture. Callus culture of *Catharanthus roseus* produced ajmaline (antihypertensive), and *Stizolobium hassjo* produced 1,2-dihydroxybenzene (also known as L-dopa, an antiparkinsonian drug). Suspension cultures of *Hyoscyamus niger* produced hyoscyamine derivative (anticholinergic). As sought, some of the secondary metabolites produced are in much higher concentrations in cultures as compared to intact plants from the same species. Such examples include *Panax ginseng* (ginsenoside) which has been reported to produce 27% of cell dry weight in culture, whereas only 4.5% is produced in whole plants, anthraquinones from *Morinda citrifolia* (18% in culture, 2.5% in plant) and shikonin from *Lithospermum erythrorhizon* (12% in culture, 1.5% in plants).

Promising results have been observed not only in research laboratory but large-scale suspension cultures have also been successfully executed. Mitsui Petrochemical Industries of Japan has developed plant cell cultures producing large quantities of berberine, ginseng, and shikonin. Reactors having a capacity of 75,000 liters are used in Germany for production of paclitaxel (produced by *Taxus* sp. in nature). Some of these compounds are now commercially available, such as shikonin and paclitaxel (Taxol). Developments in scale-up approaches and immobilization methods have enhanced the production of such compounds, and plant tissue culture represents the most promising area for further progress in pharmaceutical industry.

**Perspective**
Some of the aims of plant physiologists in human welfare are to produce varieties of crop plants with modified shoot architecture leading to ideal leaf shape, number and position for enhanced light interception, and varieties with improved panicle/ear. The "green revolution" of the last century leads to production of high-yielding crops such as dwarf and semidwarf varieties of cereals through crop genetic approaches. These cereals pool up most of their photoassimilates into grains and less in vegetative growth. They also provide plants which can resist damage due to wind. The dwarfing genes responsible for these traits include semidwarf genes (*sd*) in rice and reduced height (*Rht*) in wheat. Gibberellic acid (GA) is often referred as "green revolution hormone" since the *sd-1* gene in rice encodes for GA 20-oxidase which is responsible for oxidative elimination of carbon at position 20, leading to biosynthesis of 19 carbon bioactive gibberellins- $GA_1$ and $GA_4$. Future improvements in crop production will also include emphasis on better water use efficiency and tolerance to abiotic stress conditions through coordinated activities of various plant hormonal and nonhormonal molecules. Advances in our understanding of rhizobial and mycorrhizal symbiotic associations and ways to modify root architecture shall facilitate better nutrient use efficiency. Understanding defense hormonal signaling mechanisms will help in controlling tolerance to pests and pathogens. Keeping in view the ever-diminishing availability of agriculture land, future methods of enhancing crop productivity will benefit tremendously from our deeper understanding of all facets of plant development at physiological and molecular level.

## Summary

- Essential elements show two critical levels—"lower critical level," which is indicative of element deficiency, and "upper critical level," indicative of toxicity due to an element. The **critical toxicity level** of an element is the concentration of the element in tissue which results in 10% reduction in dry matter. Mobility of nutrients within the soil is related to chemical properties of the soil, such as CEC (cation exchange capacity) and AEC (anion exchange capacity), as well as the soil conditions, such as moisture, pH, etc. Under acidic conditions, phosphate is precipitated as $AlPO_4$ that increases $Al^{3+}$. Minerals containing aluminum and hydrogen ions dissolve and release these cations as they are removed from the exchangeable pool. Finely-textured clay soil tends to offer greater buffering capacity than coarse-textured soil.
- *Liming materials* are bases that react with hydrogen ions in the soil solution to form water. Calcium and magnesium silicates are also used as liming agents. To avoid aluminum and manganese toxicity, a soil should be limed if the pH is less than 5.4. Since finer-textured soils have greater buffering capacity than coarse-textured soils, more lime must be added to the finer-textured soil to achieve the same effect and attain the target pH. Addition of organic matter is another viable option to manage problems associated with the soil acidity. Organic matter forms strong bonds, known as "chelates," with aluminum.
- Green manure refers to the manure generated from crop plants by leaving them to wither and get incorporated into the soil, either uprooted or still rooted in the soil. Green manure, when turned into the soil, is broken down by soil-dwelling microorganisms to release nutrients and form humus. Humus is the dark organic matter which is composed of decaying remains of the plants and animal tissue formed as a result of microbial action in the soil. The characteristics of plants used for green manure include fast growth to provide large yield within a short period, more leafy growth for a rapid decomposition, and deep root system to allow efficient nutrient acquisition from deeper parts of the soil. Green manure plants have been shown to reduce pest and controlling diseases in crops by providing habitats for natural predators of insect pests so as to reduce attack on the whole crop. They can also help to reclaim alkaline soil and to make land usable for farming by improving its physical properties.
- Antitranspirants are a group of compounds which enhance water stress tolerance by preventing/reducing transpiration in plants. Antitranspirants can be either physical or physiological in nature. Physical antitranspirants include polymers, resins, latex, and waxes to coat the leaf surface and minimize water loss by blocking the stomatal aperture. Physical antitranspirants have been successfully used to minimize loss of crop due to water stress in pepper (*Capsicum annuum*), peach (*Prunus persica*), and some herbaceous plants. Physiological antitranspirants reduce transpiration by inducing plants to close stomata. They include ABA or some chemicals which increase ABA concentration in the plants. Physiological antitranspirants have been successfully used in various horticultural

crops, although plant responses to antitranspirants vary depending on the developmental stage, the species, and the concentration of the antitranspirant applied.
- Parasitic plants cause serious damage to the host plants resulting in substantial loss of yield. Both natural and artificial (man-made) compounds have been investigated as stimulants of suicide germination of seeds of parasitic plants. Methyl jasmonate and some fungal toxins have also been found to stimulate germination of these seeds. One of the effective strategies seems to be planting nonhost crops to induce suicide germination. Monsanto introduced glyphosate-resistant crops which enabled farmers to kill weeds selectively without killing the main crop. Weeds die in response to glyphosate spray as blocking of shikimic acid pathway leads to inhibition of biosynthesis of aromatic amino acids and their protein derivatives.
- A wide variety of PGRs are currently in use in horticulture, viticulture, and agriculture in order to obtain varied advantages such as changes in plant constituents, ease of harvesting, enhancement of yield, and greater tolerance toward abiotic and biotic stress conditions. Generally, PGRs are applied to crop plants using foliar sprays with water as a carrier. While doing so, due care has to be taken to ensure penetration of the active ingredient through epidermis and membranes by making an appropriate choice of the solvents and surfactants. Furthermore, application of the PGR treatment has to be given at the right stage of plant development and in "ideal" weather conditions.
- A number of other transgenic fruit crops are being developed to enhance their host plant resistance to insects and plant pathogens for herbicide tolerance and to improve features such as slow ripening to enhance their shelf life. Transgenic papaya carrying coat-protein gene provides effective protection against papaya ringspot virus. The transgenic plum cultivar (honey sweet) provides germplasm for plum pox virus control. Transgenic bananas with enhanced resistance to *Xanthomonas campestris* have also been developed.
- Plant tissue culture offers an opportunity to exploit cells, tissues, organs, or the entire organisms by growing them in vitro and to genetically manipulate them to get desired compounds. A number of different types of **bioreactors** have been used for mass cultivation of plant cells. Transformed roots have been widely used for in vitro production of secondary metabolites in many plant species. Hairy root cultures produce secondary metabolites over successive generations without losing genetic or biosynthetic stability. Also, production of two different secondary metabolites is possible simultaneously from adventitious root co-cultures.

## Multiple-Choice Questions

1. Parasitic weeds are difficult to be removed from crop fields because:
    (a) Enormous numbers of seeds are produced by them.
    (b) Seeds have long life.
    (c) Seeds are covered with water impervious material.
    (d) Both a and b.

2. Trehalose is a disaccharide formed by:
   (a) Two alpha glucose units
   (b) Glucose and fructose
   (c) Two fructose units
   (d) Glucose and galactose
3. Which of the following is used for reducing fruit russetting in apples?
   (a) Auxins
   (b) Mixture of GA4 and GA7
   (c) GA3
   (d) Cytokinins
4. Which of the following agents was used to destroy crops during the Vietnam War?
   (a) GA3
   (b) Zeatin
   (c) A mixture of 2,4-D and 2,4,5-T
   (d) Ethephon
5. Loss of crop yield due to lodging can be reduced by applying:
   (a) Chlormequat chloride
   (b) Metconazole
   (c) Triazole tebuconazole
   (d) All of the above
6. The production of fruit bypassing the need for pollination and fertilization is called as:
   (a) Parthenogenesis
   (b) Stenospermocarpy
   (c) Parthenocarpy
   (d) Vegetative propagation
7. The enzyme responsible for degradation of pectin in fruits is:
   (a) Trehalase
   (b) Phosphatase
   (c) Polyphenol oxidase
   (d) Polygalacturonase
8. The genetically modified Flavr Savr tomato was engineered to:
   (a) Increase the production of secondary metabolites
   (b) Prevent lodging
   (c) Have a longer shelf life
   (d) Produce pulpier fruits
9. The distinctive character of tea is due to:
   (a) Theol, theine, tannins
   (b) Tannins, caffeol, theine
   (c) Tannins, theobromine, caffeine
   (d) Theol, theine, caffeine

10. Glyphosate-resistant crops were first introduced by:
    (a) Monsanto
    (b) BASF
    (c) Calgene Inc.
    (d) Mitsui petrochemical industries

**Answers**

1. d   2. a   3. b   4. c   5. d   6. c
7. d   8. c   9. a   10. a

## Suggested Further Readings

Park S, Moon Y, Mills SA, Waterland NL (2016) Evaluation of antitranspirants for enhancing temporary water stress tolerance in bedding plants. Hortic Technol 26(4):444–452

Rademacher W (2015) Plant growth regulators: background and uses in plant. J Plant Growth Regul 34:845–872

Rohwer CL, Erwin JE (2008) Horticultural applications of jasmonates: a review. J Hortic Sci Biotechnol 83(3):283–304

# Glossary

**ABA receptors** Proteins that have been proposed to regulate responses to ABA upon binding it. Three classes of ABA receptors have been identified till date. These include (i) the plasma membrane-localized G-proteins GTG1 and GTG2, (ii) a plastid localized enzyme that coordinates nucleus to plastid signaling, and (iii) cytosolic ligand-binding proteins of START (Steroidogenic Acute Regulatory Protein-Related Lipid Transfer) domain superfamily.

**ABC model** A model that explains the development of floral organs in response to the pattern of gene expression of a set of floral meristem-identity genes. According to this model, three homeotic genes, namely, A, B, and C, are responsible for organ identity in each whorl of a flower. A unique combination of these three genes determines the organ formed in each whorl.

**ABC transporters** Members of a transport system superfamily which consist of multiple subunits, one or two of which are transmembrane ATPases.

**ABCB transporters** Transmembrane proteins that are phosphoglycoproteins in nature and belong to ABC transporter superfamily. They facilitate IAA efflux or influx across the plasma membrane (and also tonoplast), but unlike PIN proteins, they do not show preferential basal localization in the cells.

**Abiotic stress** Stress which originates from excess or deficit in the physical, chemical, and energetic conditions to which plants are exposed, for example, salinity stress, flooding, and drought.

**ABP1 (auxin-binding protein 1)** A glycoprotein auxin receptor localized in the lumen of the ER and plasma membrane. It is proposed to form a complex with a transmembrane docking protein which provides lipid solubility to anchor ABP1 to the membrane.

**Abscission** The process of natural detachment of plant parts, such as senesced leaves, flowers, and ripened fruits.

**Absorption spectrum** A graph depicting measurement of light absorption by a pigment (e.g., chlorophyll) as a function of wavelength.

**ACC (1-aminocyclopropane-1-carboxylic acid)** A disubstituted cyclic alpha-amino acid in which a three-membered cyclopropane ring is fused to the C-atom of the amino acid. It is an intermediary compound in ethylene biosynthesis.

© Springer Nature Singapore Pte Ltd. 2018
S. C Bhatla, M. A. Lal, *Plant Physiology, Development and Metabolism*,
https://doi.org/10.1007/978-981-13-2023-1

**Acid growth hypothesis** A hypothesis which explains plant cell enlargement on the basis of cell wall acidification resulting from proton extrusion across the plasma membrane into the apoplast by the action of plasma membrane-associated $H^+$-ATPases.

**Action spectrum** A graph depicting the magnitude of response of a biological system as a function of varying wavelengths of light, for example, rate of photosynthesis as a function of wavelength.

**Activation energy** The minimum amount of energy which the reacting species must possess to undergo specific chemical reaction or the difference in free energy of ground state and transition state of the substrate molecule.

**Active site** The catalytic region of an enzyme molecule at which it binds with the substrate.

**Active transport** An uphill transport of molecules across a membrane against their electrochemical gradient with the help of pumps that utilize ATP hydrolysis as a source of energy.

**Acyl carrier protein** A highly conserved protein which serves as the carrier of acyl intermediates during fatty acid and polyketide biosynthesis. It transports growing fatty acid chain between different enzymatic domains of fatty acid synthase.

**Adaptive immune system** Refers to actions against pathogens that are able to evade or overcome innate immune system. This system is activated in the presence of infectious agents and is proliferated to other cells for eliminating the microbes, for example, antibodies produced by lymphocytes. Plants lack this type of immunity.

**Adenine nucleotide translocase (ANT)** A mitochondrial protein that facilitates the exchange of ADP and ATP across inner mitochondrial membrane.

**Adenylate kinase** A phosphotransferase enzyme that catalyzes the conversion of two ADP molecules to ATP and AMP and vice versa.

**Adhesion** The force of attraction between two dissimilar substances. This force can be mechanical (sticking together) or electrostatic (attraction due to opposite charges). For example, wetting of any surface is due to the effect of adhesive forces between liquid and the surface.

**Aerenchyma** A spongy tissue with large air spaces which forms channels in the leaves, stem, or roots of some plants. It imparts buoyancy and facilitates exchange of gases in aquatic plants.

**Aeroponics** A technique of growing plants in which roots are suspended in growth chamber in air and roots are periodically wetted with a fine mist of nutrients.

**After-ripening** Prolonged dry storage of seeds at room temperature to overcome dormancy.

**Aging** Aging-Gradual loss of physiological and metabolic activities in an organism over a period of time, ultimately leading to death.

**Air-seeding hypothesis** The breakage of water column due to trapping of air in the xylem elements leading to embolism (blockage).

**Aleurone layer** The outermost layer of cells rich in protein bodies distinctly surrounding the starchy endosperm in cereal grains. It is mainly involved in the synthesis and release of hydrolytic enzymes to mobilize stored starch to support seed germination.

**Alkaloids** A large group of secondary metabolites which contain nitrogen in their structure mostly as heterocyclic ring, for example, nicotine, codeine, morphine, and caffeine. They are involved in plant defense and are of great medicinal importance to humans.

**Allelochemicals** Chemicals produced by the plant which are toxic in nature and detrimentally affect the growth and survival of neighboring plants when released into the environment.

**Allelopathy** A biological phenomenon of release of various chemicals (allelochemicals) into the environment by certain plants and inhibition of growth of neighboring plants.

**Allocation (ref. photoassimilate translocation)** The regulation of diversion of photoassimilates into various metabolic pathways in the source tissue.

**Allosteric inhibition** The inhibition of an enzyme caused by binding of an inhibitor molecule at the allosteric site. The inhibitor is capable of inducing changes in the conformation of the enzyme rendering it inactive.

**Allosteric site** A site present on the enzyme, other than the active site, where an effector (either negative or positive) molecule can bind to regulate the activity of the enzyme.

**α-Amylase** An enzyme which catalyzes the conversion of starch into low molecular weight sugars, i.e., glucose and maltose, via hydrolysis of the internal α-1,4-glycosidic linkages.

**Alternative oxidase (AOX)** The non-energy-conserving terminal enzyme which forms a part of mitochondrial electron transport chain in different organisms and catalyzes the cyanide-resistant oxygen reduction.

**Amensalism** Any relationship between organisms of different species in which one organism is negatively affected or destroyed while the other organism remains unaffected.

**Aminotransferases** A group of enzymes which catalyze the transfer of an amino group from one α-amino acid to an α-keto acid so that the donor amino acid is converted to its respective α-keto acid while the amino group acceptor α-keto acid is converted to its respective α-amino acid. They are also called transaminases.

**Ammonification** A process carried out by a variety of microorganisms present in the soil during which organic nitrogen of the soil (present in the soil due to decay of dead animals, plants, and excreta of animals) is broken down to ammonia which becomes available to plants.

**Amphibolic pathway** A metabolic pathway involving both anabolism and catabolism. For example, TCA cycle provides precursors for anabolic pathways while plays catabolic role in degradation of complex organic molecules to the simpler ones.

**Amphipathic molecules** Molecules containing both polar (hydrophilic) and nonpolar (hydrophobic) regions, for example, phospholipids which have a polar phosphate head and a nonpolar fatty acid tail.

**Amyloplasts** Starch-storing plastids found in storage tissues of shoots and roots which also serve as gravity sensors.

**Anion exchange capacity (AEC)** The ability of the soil colloids to hold and exchange anions. It increases as the soil pH decreases.

**Anisotropic** Having different properties along different axis, for example, optical anisotropy which causes polarization of light.

**Anoxia** A condition of complete depletion of oxygen in solutions or a condition characterized by the absence of oxygen to an organ or a tissue.

**Anoxygenic photosynthesis** A form of photosynthesis in which light-driven ATP and NADPH synthesis occurs with the utilization of sources other than water, such as hydrogen sulfide, as the electron source and the process are coupled with accumulation of free sulfur. This type of photosynthesis occurs in purple bacteria and green-sulfur bacteria.

**Antenna pigments** Pigments which are associated with subunit proteins to form light-harvesting complex. These serve as sites for absorption of light energy which is then supplied to the reaction center.

**Antiauxins** Compounds which inhibit auxin action by competing for the same receptors. They closely resemble auxin in structure but lack auxin-like activity, for example, α-p-chlorophenoxyl isobutyric acid (PCIB).

**Antiflorigen** A hypothetical hormone proposed to inhibit the formation of flowers in certain long-day plants under noninductive photoperiod.

**Antifreeze proteins** A group of proteins which are produced during cold acclimation and prevent the nucleation of ice crystals or re-formation of ice crystals post a freeze-thaw cycle.

**Anti-nutritional proteins** Chemicals present in food that reduce nutrient utilization or food intake, thereby contributing to impaired gastrointestinal and metabolic performance.

**Antioxidants** Molecules which inhibit the oxidation of other molecules, such as free radicals, that may damage cells through chain reactions, for example, vitamins A, C, and E.

**Antiport** A type of secondary active transport of solutes which couples the movement of two solutes but in opposite directions across a membrane. The movement of one species is down its electrochemical gradient which drives the active transport of the other against its gradient.

**Antiporters** Integral membrane proteins or cotransporters which facilitate movement of two solutes in opposite directions across the membrane (i.e., antiport), for example, $Na^+/Ca^{2+}$ antiporter that exchanges one $Ca^{2+}$ ion for extrusion of three $Na^+$ ions out of the cell.

**Antitranspirants** Chemicals which decrease water loss from surface of plant leaves due to transpiration. They can be either physical or physiological in nature. Physical antitranspirants include the use of polymers, resins, latex, and waxes to coat the leaf surface and minimize water loss by blocking the stomatal aperture. On the other hand, physiological antitranspirants reduce transpiration by inducing plants to close stomata, for example, ABA and phenyl mercuric acid (PMA).

**Apical dominance** The phenomenon of inhibition of axillary bud growth by the continued meristematic activity in the shoot apex. It is best observed in herbaceous plants and also in trees during first year of growth.

**Apoenzyme** The proteinaceous part of a holoenzyme without a cofactor.

**Apomixis** Asexual formation of seed from the maternal tissues of the ovule without the process of meiosis and fertilization.

**Apoplast** The continuous system of spaces that lies outside plasma membrane consisting of cell wall in plant cells and the intercellular spaces. It allows free diffusion of materials from one cell to another cell.

**Apoptosis** A type of programmed cell death in multicellular organisms, involving all biochemical events leading to characteristic cell changes and death.

**Appressorium** A flattened and thickened hyphal branch that facilitates penetration of fungal hyphae inside the host plant.

**Aquaporins** Integral membrane proteins in the phospholipid bilayer membranes, allowing cell-to-cell rapid transport of water molecules. Some plant aquaporins have now been reported to additionally transport other small, uncharged solutes, stress response factors, or signaling molecules. They are present on plasma membrane and tonoplast.

***Arabidopsis* histidine phosphotransfer (AHP) protein** A protein similar to bacterial and yeast histidine phosphoransferases (HPTs) which is a positive regulator of cytokinin signaling. It is responsible for cytokinin signal propagation from the plasma membrane receptor to the nucleus in a multistep signaling pathway.

***Arabidopsis* response regulators (ARRs)** Proteins identified in *Arabidopsis* that are similar to bacterial two component signaling proteins called response regulators. There are two classes of these proteins: Type-A ARRs, whose transcription is upregulated by cytokinin, and Type-B ARRs, whose expression is not affected by cytokinin.

**Arabinan** A polysaccharide, mostly a polymer of arabinose. It consists of a main chain of 1,5-α-linked L-arabinofuranosyl to which other L-arabinofuranosyl residues are linked by 1,3-α and 1,2-α linkage in either a comb-like or a ramified manner.

**Arabinogalactans (of extensin family)** A class of long, densely branched high-molecular polysaccharides consisting of monosaccharides—arabinose and galactose. These are often found attached to proteins forming arabinogalactan proteins (AGPs). Localized movement of AGP from plasma membrane to the cell wall is a component of mechanism of tip growth in plant cells. AGPs also serve as an intercellular signaling molecule and play a major role in cell-to-cell interaction during development and in sealing of wounds.

**Arabinoxylan** A hemicellulose consisting a linear β-1,4-xylose residues with arabinose substitution. It is found in both primary and secondary cell wall of plants.

**Arbuscular mycorrhizal (AM) fungi** A type of mycorrhizal association between a fungus (belonging to phylum *Glomeromycota*) and the roots of higher vascular plants in which the fungus penetrates and forms arbuscules (branched structures that are sites of nutrient exchange between the fungus and plant host) in the cortical cells of the roots.

**Ascent of sap** The upward movement of water and minerals through the xylem tissues from the roots to the aerial parts of the plant.

**Asparagine synthetase (AS)** An enzyme which catalyzes the formation of asparagine from aspartate.

**Aspartate aminotransferase** An enzyme that catalyzes reversible transfer of α-amino group between aspartate and glutamate and uses pyridoxal 5′-phosphate as the cofactor. It is also known as aspartate transaminase.

**ATP synthase** The multi-subunit protein complex that synthesizes ATP from ADP and $P_i$ in response to an electrochemical gradient created by proton movement in response to proton motive force during oxidative phosphorylation and photophosphorylation in mitochondria and chloroplasts, respectively.

**ATP synthasome** A complex of the ATP synthase with adenine nucleotide translocase (ANT) and phosphate translocase.

**Autocatalytic cycle** A metabolic cycle in which one of the products is also a reactant, for example, glycolysis, where two molecules of ATP are consumed and four are produced.

**Autogamy** The process of transfer of pollen onto the surface of stigma present in the same flower.

**Autolysis** Destruction of a cell through the action of its own hydrolytic enzymes.

**Autophagic body** The membrane-bound organelle derived from autophagosomes which enters the vacuole and releases its contents for degradation by hydrolytic enzymes.

**Autophagosomes** Double membrane-bound organelle that transfers cellular components to the vacuole for degradation.

**Autophagy-related genes (ATG genes)** A group of genes that code for proteins required for autophagy.

**Autophagy** A physiological process that is instrumental in maintaining cell homeostasis and involves transport of cellular macromolecules to lytic vacuoles for degradation.

**Autophosphorylation** The process of addition of phosphate group to serine, threonine, or tyrosine residues in the structure of a kinase protein by itself so as to regulate its own enzymatic activity.

**Autotoxicity** A form of allelopathy in which a species inhibits the growth of members of the same species through production of chemicals which are released into the environment.

**Autotroph** A living organism which is capable of manufacturing its own organic food from simple inorganic materials using external energy sources, for example, green plants, algae, and some bacterial species.

**AUX1** An auxin uptake carrier protein that functions in leaf vascular tissue and root apices and belongs to a family of proteins similar to permeases in prokaryotes.

**Avirulent** Strain of pathogen lacking virulence (the ability of an organism to cause disease).

**Avr gene** Refers to avirulence gene which is present in all pathogens. It codes for effector protein.

**Bacteriorhodopsin** A protein from halobacteria (a class of archaebacteria) which acts like a pump. It uses light energy to pump $H^+$ out of the cell.

**Bacteroid** The nitrogen-fixing form of *Rhizobium* which develops in the root nodules upon a signal from the host plant.

**Basal defense mechanism** The first level of preformed, inducible defense that protects plants against several types of pathogens.

**Beneficial elements** Mineral elements which are not essential to all plants but benefit certain plant species under defined conditions. Elements including silicon, aluminum, selenium, and cobalt are considered beneficial by various means such as acting as agents for stimulating the resistance mechanism, uptake of other nutrients, and detoxification of toxins and other metals.

**Bioactive compound** A compound that has an effect on a living organism, tissue, or cell.

**Bioassay** Quantitative estimation of a biologically active substance by measuring its effect on a living organism under standard conditions.

**Bioenergetics** The biological science which deals with the study of energy transduction and transformation within a living system and exchange of energy between living system and the environment.

**Biogenic volatile organic compounds (BVOCs)** Organic atmospheric trace gases (other than carbon dioxide and monoxide) produced by plants. They include isoprenoids (isoprene and monoterpenes), alkanes, alkenes, carbonyls, alcohols, esters, ethers, and acids.

**Biosensor** An analytical device used for detection of an analyte using a biological component and a physicochemical detector. In other words, it is a device that utilizes biological components, for example, enzymes, to detect a biomaterial, for example, surface plasmon resonance (SPR).

**Biotic stress** Stress which is imposed by direct or indirect interactions with other organisms, for example, infection by bacterial, fungal, and viral pathogens.

**Biotrophic pathogens** Plant pathogens that grow in or on the living cells and derive energy from the host.

**Bolting** A process observed in plants, such as spinach, lettuce, and cabbage, in which flowering stems are produced prematurely before harvesting. Exogenous application of gibberellin is known to promote bolting in a wide range of species. This leads to lengthening of stem, decrease in stem thickness and leaf size, and pale green coloration of leaves.

**Boom and bust cycle** The periodic reoccurrence of epidemics in crops due to control in disease (boom year) and failure of disease control (bust year).

**Boundary layer resistance** The resistance posed by the thin layer of air immediately around the leaf surface (known as boundary layer) to heat and vapor transfer from the leaf surface to the atmosphere during transpiration.

**Brassinosteroids (BRs)** A group of plant polyhydroxy steroidal hormones ($C_{27}$ to $C_{29}$) derived from castasterone. They are known to modulate de-etiolation and rolling of leaves among grasses.

**BRI1 (Brassinosteroid insensitive 1)** Brassinosteroid receptor located on plasma membrane belonging to the family of leucine-rich repeat (LRR) receptor and endosomes as a dimer. Its cytosolic loop exhibits LRRs for binding with BR molecule leading to signaling.

**Bulk flow** The concerted, long-distance movement of water and solutes in the tracheids and sieve-tube elements down a pressure gradient from the soil up through the plant. It is also applied to the movement of larger solutes in phloem.

**Bulliform cells** Enlarged epidermal cells present in leaves of grasses having a role in the rolling and unrolling of leaves.

**Calcium sensor proteins** Calcium-binding proteins which monitor temporal and spatial changes in $Ca^{2+}$ concentrations and undergo conformational changes upon calcium binding. They further trigger downstream signaling responses.

**Calcium/calmodulin-activated kinase (CCaMK)** A protein kinase which is a central regulator in plant root endosymbiosis and decodes $Ca^{2+}$ oscillations in the nucleus occurring in response to symbiosis. It adopts different modes of action upon binding with either free form of calcium or calcium bound with calmodulins.

**Calcium-dependent protein kinases (CDPKs)** A large family of serine/threonine protein kinases. They may function in perceiving intracellular changes in $Ca^{2+}$ concentration and translate them into enhanced protein kinase activity and initiate further downstream signaling processes.

**Callose** A $\beta$-1,3-glucan synthesized in the plasma membrane and deposited between the plasma membrane and cell wall. It plays important roles in plant development (e.g., development of pollen) and responses to biotic and abiotic stresses (e.g., by forming plugs at plasmodesmata).

**Callus** A mass of undifferentiated, growing, and dividing tissue produced during wound healing (to cover wound) or in tissue culture.

**Calmodulin** Calcium-binding conserved protein found in all eukaryotes which regulates many $Ca^{2+}$-driven metabolic reactions.

**Capillary action** The process by which water rises in the xylem elements of roots and stem of a plant body as a result of cohesive and adhesive forces between water molecules.

**Carbon dioxide compensation point** The concentration of carbon dioxide at which the rate of photosynthesis exactly matches the rate of respiration. As a result, there is no net change in concentrations of carbon dioxide and oxygen.

**Carotenoids** Naturally occurring pigments synthesized by plants, algae, and photosynthetic bacteria. They are responsible for yellow, orange, and red colors in leaves and fruits and are involved in quenching of excessive light energy to prevent photooxidative damage to chlorophyll and reaction center.

**Casparian bands** Bands of cell wall material particularly rich in suberin (waxlike, hydrophobic material) deposited on the radial and transverse wall of the endodermis. The Casparian strips prevent water from traveling through the apoplast once it reaches the endodermis.

**Catalase (CAT)** A heme-containing enzyme that acts directly on hydrogen peroxide and catalyzes its decomposition into water and oxygen, thus preventing cells from oxidative damage.

**Cation adsorption sites** Negatively charged sites on root surface due to carboxyl groups of polygalacturonic acid in the pectic substance of middle lamella which attract cations from soil solution, thereby causing their migration in the cell wall.

**Cation exchange capacity (CEC)** The ability of soil colloids to hold and exchange cations.

**Cavitation** The process of breakage of water column in xylem due to air entrapment.

**Cell autonomous responses** Signaling events in plants where signal perception and response occur in the same cell, for example, opening of stomatal aperture which involves activation of membrane-associated ion transporters in response to blue light, ultimately leading to swelling of guard cells via phototropin activation.

**Ceramide** A sphingolipid consisting of sphingosine and a fatty acid which forms component of sphingomyelin, a major lipid found in lipid bilayer. It plays an important role in coordinating cellular response to extracellular stimuli and stress.

**Chaperonins** Proteins which assist in the assembly and folding of other proteins in cells.

**Chemiosmotic mechanism** Process by which electrochemical gradient of protons is established across a membrane due to electron transport which drives ATP synthesis.

**Chemiosmotic model** A model proposed by Peter Mitchell which explains the establishment of an electrochemical gradient of protons across membrane in response to electron transport which drives ATP synthesis.

**Chemoautotroph** Any living organism which is able to synthesize carbon compounds for its growth, utilizing the energy released during the oxidation of the chemical compounds carried out by them.

**Chemocyanin** A small, secreted protein (chemotropic substance) in the style that acts as a directional cue for pollen tube growth toward ovule.

**Chemonastic response** A type of plant growth response or movement in response to an external chemical stimulus, either toward (positive chemotropic) or away from it (negative chemotropic), or a change in concentration of a chemical inside the plant, for example, growth of pollen tube through the style toward ovule in response to chemicals released by ovary.

**Chitin elicitor-binding protein (CEBiP)** It is a pattern recognition receptor (PRR) present on a host membrane. It has extracellular domain (with LysM motifs) which binds with chitin elicitor. It does not have an intracellular kinase domain and is always found in association with CERK1 (chitin elicitor receptor kinase 1) for signaling in PAMP-triggered immunity (PTI).

**Chitin** A polymer of N-acetyl-D-glucosamine and is the main component of fungal cell wall.

**Chlorine** A class of large heterocyclic tetrapyrroles derived from porphyrins consisting of three pyrroles at the core and one pyrroline coupled through four CH linkages, for example, chlorophyll.

**Chlorosis** Loss of chlorophyll leading to yellowing in leaves. It is caused by the deficiency of elements like potassium, magnesium, nitrogen, sulfur, iron, manganese, zinc, and molybdenum or due to destruction of chlorophyll by certain toxins, high alkalinity of soil, poor drainage, and infection by pathogens.

**Cholesterol** A sterol which is an essential structural component of all animal (not plant and bacterial) cell membranes. It is required both for structural integrity and fluidity of membranes.

**Chromophore** The prosthetic group of an organic molecule which is light-absorbing and non-proteinaceous, for example, the phycobilins serving as chromophores for phycobiliproteins. It changes conformation when it absorbs light.

**Circadian rhythms** Biological processes that display endogenous, entrainable oscillation of 24 hours.

**Climacteric** Stage of fruit ripening associated with increased ethylene production and a rise in cellular respiration. Apples, bananas, melons, and tomatoes are climacteric fruits which release large amounts of ethylene.

**Climacteric fruit ripening** The process of fruit ripening associated with increased ethylene production and rate of respiration.

**Coenzyme** A small organic non-protein molecule that binds with the protein molecule (apoenzyme) to form the active enzyme (haloenzyme). Coenzymes cannot function on their own. For example, $NAD^+$ is a coenzyme non-covalently bound to the protein part of many dehydrogenases.

**Coevolution** The evolution of complementary traits in two different species due to environmental interaction between them.

**Cofactor** The non-proteinaceous inorganic or organic component of a conjugated enzyme that assists during the catalysis of a reaction. For example, heme is a cofactor of hemoproteins.

**Coincidence model** A model for flowering in photoperiodic plants in which the circadian oscillator controls the timing of light-sensitive and light-insensitive phases during the 24-hour cycle.

**Commensalism** An association between two organisms in which one benefits and the other neither benefits nor is harmed, for example, bromeliads and orchids growing on trees.

**Companion cells** Metabolically active cells located in phloem which are closely associated with sieve elements through branched plasmodesmata and help in the transport of photoassimilates from the mesophyll cells into sieve elements and vice versa, thus regulating phloem loading and unloading.

**Compatible disease reaction** An interaction (host and pathogen) where symptoms are expressed clearly in the host.

**CONSTANS (CO)** A key component of the pathway that regulates the transcription of other genes and promotes flowering in *Arabidopsis* in long days.

**Constitutive defense** A type of defense which is always present irrespective of the attack by the microbes.

**Continuous-flow solution culture** A hydroponic technique in which plants are grown by constantly flowing nutrient solution over their root systems.

**COP1 (constitutive photomorphogenesis 1)** An E3 ligase that is a constitutive repressor of photomorphogenesis in dark and acts by promoting degradation of photomorphogenesis promoting factors such as HY5 via 26S proteasomal pathway.

**COP1-SPA (COP1 suppressor of PHY) complex** A protein complex that represses photomorphogenesis in dark by associating with and stabilizing PIF3 (phytochrome-interacting factor 3).

**Corpus** The core of an apical meristem consisting of cells that divide in various planes and contribute to growth in volume.

**Coumarins** Chemical compounds found in many plants which belong to the class of phenylpropanoid lactones, also known as benzopyrones, for example, scopoletin, aflatoxin, and umbelliferone.

**Crescograph** An instrument used to measure growth movements of plants in response to various stimuli. It was invented by Sir J.C. Bose.

**Critical concentration** The minimum concentration of a nutrient element that results in 90% of maximum yield or growth below which there is significant reduction in the yield. It is usually used in plant nutrient analysis.

**Critical day length** The minimum length of the day required for induction of flowering in a long-day plant.

**Critical elements** The elements in which soils are generally deficient, for example, nitrogen, phosphorus, and potassium. These are supplied to plants in the form of chemical fertilizers, called complete fertilizers.

**Critical toxicity level** The minimum concentration of a potentially toxic micronutrient (e.g., zinc, lead, and copper) in tissue which results in 10% reduction in dry matter.

**Cryptochromes** A class of flavoproteins that serve as sensitive photoreceptors to blue light and are implicated in many blue light responses, namely, promotion of cotyledon expansion, suppression of hypocotyl elongation, membrane depolarization, anthocyanin production, and circadian clock function. These are photolyase-like receptors having an N-terminal PHR (photolyase-homologous region) domain that binds chromophore FAD and the CCE (CRY C-dinucleotide) domain that is crucial for the regulation of their function.

**Cyanogenic derivatives** Non-alkaloid, nitrogen-containing compounds which produce hydrogen cyanide upon breakdown. They are involved in plant defense, for example, amygdalin in almonds.

**Cyclic electron flow** Cycling of electrons around photosystem I where electrons flow from electron acceptors through cytochrome $b_6f$ complex and return to photosystem I. In higher plants this results in the formation of a proton gradient due to proton pumping in the lumen. As a result, ATP is synthesized without any photolysis of water and reduction of $NADP^+$.

**Cyclic nucleotide-gated ion channels (CNGCs)** Nonselective cation channels which function in response to cyclic nucleotide binding and altered membrane potential. These channels are also modulated by calcium/calmodulin or phosphorylation events and thus triggering signaling cascades.

**Cyclin-dependent kinases (CDKs)** A class of protein kinases that regulate the transitions from $G_1$ to S and from $G_2$ to M (mitosis) phases during cell cycle.

**Cysteine synthase** An enzyme responsible for the formation of cysteine from O-acetyl serine and hydrogen sulfide with concomitant release of acetic acid.

**Cytochrome b5 (cyt b5)** A membrane-bound heme-containing protein which acts as a carrier of electrons for several membrane-bound oxygenases.

**Cytochrome oxidase** An oxidizing enzyme containing iron and porphyrin which is part of the electron transport chain. It is able to catalyze oxidation of cyt $c$ by transferring electrons from reduced cyt $c$ to molecular oxygen.

**Cytochrome P450 (cyt P450)** A family of heme-containing proteins belonging to the superfamily of monooxygenases. It is involved in the detoxification of potentially toxic compounds, for example, drugs, xenobiotics, and other toxins, processing and transport of proteins, and production of energy. 450 term is derived from spectrophotometric peak at 450 nm when it is in reduced state and is complexed with carbon monoxide.

**Cytochromes $b_6f$ complex** A large multi-subunit complex, consisting of two $b$-type hemes, one $c$-type heme (cyt $f$) and a Rieske Fe-S protein, distributed equally between grana and stroma lamellae.

**Cytokinin oxidase** Enzyme that catalyzes the catabolism of specific cytokinins to inactive products that lack the N 6-unsaturated side chains.

**Damage-associated molecular patterns (DAMPs)** Self-derived molecules produced in the cells and released in response to stress. They initiate immunity in the cell.

**Day-neutral plants** Plants in which flowering is not affected by day length, for example, sunflower, tomato, and maize.

**DELLA proteins** A family of proteins that are putative transcriptional regulators and act downstream of the GA receptor (GID1) and negatively modulate gibberellin signaling in plants and inhibit growth processes. Binding of GA with GID1 induces binding of GID1-GA complex to DELLA which, in turn, leads to tagging of DELLA by ubiquitination followed by its degradation via 26S proteasomal pathway.

**Desaturases (soluble)** A group of enzymes found in the stroma of plastids that catalyze the insertion of double bond into saturated fatty acids bound to ACP producing monounsaturated fatty acids.

**Determinate growth** Growth for limited duration which is characteristic of floral meristems and leaves.

**Development** A change in the pattern of growth of an organism or organ. For example, transition from leaf primordium to fully mature leaf or from vegetative to flowering condition.

**Dextrins** A group of low molecular weight carbohydrates composed of glucose molecules linked by α-1,4 and α-1,6 glycosidic bonds. These can be produced by hydrolysis of starch or glycogen through the action of amylases.

**Diacylglycerol** A glyceride containing two molecules of fatty acids covalently bonded to a glycerol molecule via ester linkages.

**Diagravitropic** Gravitropic response perpendicular to the stimulus as seen in rhizomes and stolons.

**Diaheliotropic leaves** Leaves which maximize light capture by heliotropism by continuous adjustment of their orientation with respect to the direction of sunlight.

**Dialysis** The process of separation of particles in a liquid through a membrane on the basis of differences in their size and concentration gradient.

**Dichogamy** The production of stamens and pistils at different times in hermaphroditic flowers in order to prevent self-pollination. It is of two types: protandry (anthers mature first) as seen in Apiaceae and Asteraceae and protogyny (pistil matures first) as observed in Brassicaceae and Rosaceae.

**Dieback** A symptom of copper deficiency in woody plants, particularly citrus, which brings about progressive death of shoots, branches, and roots beginning from the tip and spreading to rest of plant body. It is also caused by many fungal and bacterial pathogens.

**Differentiation** Processes by which distinct cell types arise from precursor cells, for example, differentiation of xylem and phloem from cambium.

**Diffusion** A spontaneous physical process of movement of any substance or molecule from the region of its higher concentration to the region of its lower concentration without the use of energy due to kinetic energy of the molecule. The rate of diffusion is also affected by the size of the molecule, its concentration gradient, viscosity of the medium, and temperature, for example, diffusion of carbon dioxide and oxygen via the stomata.

**Diffusion coefficient or diffusivity (Ds)** A proportionality factor that measures the ease of movement of a particular substance through a particular medium. Higher the diffusion coefficient of the substance with respect to the medium, faster will be its diffusion through the medium. The diffusion coefficient is a characteristic of a substance and depends on the medium.

**Diffusion potential** A potential that develops as a result of diffusion.

**Dimethylsulfide** An organosulfur compound produced by the breakdown of dimethysulfoniopropionate (DMSP). It is a volatile compound released during cooking of certain vegetables, such as cabbage and beetroot.

**Dimethysulfoniopropionate (DMSP)** An organosulfur compound found in marine microbes, phytoplanktons, and seaweeds which is broken down to produce two major volatile sulfur products, i.e., methanethiol and dimethylsulfide.

**Dormancy** A period of biological rest in an organism's life cycle when growth, development, and metabolic activity are temporarily stopped.

**Dry matter** A measure of the mass of a plant or animal tissue after complete drying. In plants, dry matter mainly consists of cell wall (polysaccharides and lignin) and protoplasmic components (proteins, lipids, amino acids, organic acids, and certain ions).

**Effector** Any molecule produced by a pathogen that enhances its (pathogen's) ability to overcome basal defense of the plant and enables it to colonize, grow, and reproduce within the host plant.

**Effector-triggered immunity (ETI)** Activation of defense response triggered by the effectors produced by a pathogen.

**Effector-triggered susceptibility (ETS)** Interference or suppression of defense response due to effector molecules resulting in susceptibility in the host.

**EFR protein** A leucine-rich receptor kinase which is a pattern recognition receptor (PRR) for EF-Tu.

**Einstein** A quantity representing one mole of photon.

**Electrochemical gradient** A type of potential energy developed across membrane because of development of gradient of ion and electrical charge across membrane.

**Elicitor** A molecule produced by pathogens which is extrinsic to the metabolism of host and triggers defense response in the host cells upon perception.

**Elongation factor Tu (EF-Tu)** The most abundant highly conserved bacterial protein which acts as a PAMP in *Arabidopsis* and other members of Brassicaceae.

**Embolism** The process of obstruction of the xylem elements due to trapping of air and breakage of water column in the vessels or tracheids.

**Embryogenesis** The development of a fully mature embryo from a zygote after effective fertilization (zygotic embryogenesis) or from a somatic cell (somatic embryogenesis).

**Emerson enhancement effect** The synergistic effect of deep red (680 nm) and far-red light (700 nm) on the rate of photosynthesis as compared with the sum of rates when the two wavelengths are delivered separately.

**Endoamylases** A glucanohydrolase that hydrolyzes internal glycosidic bonds in high molecular weight oligosaccharides. For example, $\alpha$-amylases hydrolyze internal $\alpha$-1,4 glycosidic linkages in starch, dextrins, and glycogen.

**Endomembrane system** The system composed of different membranes suspended in the cell cytoplasm which facilitates exchange of molecules by diffusion or transport vesicles. It consists of nuclear envelope, ER, Golgi, plasma membrane, vacuole, and various secretory vesicles.

**Endophyte** An organism which lives within a plant for a significant part of its life cycle asymptomatically without causing any disease, for example, arbuscular mycorrhizal fungi which live as fungal endophytes within plants.

**Endopolygalacturonase** Enzyme that catalyzes the fragmentation and solubilization of homogalacturonan (major component of primary cell wall). It is secreted by plants and fungal and bacterial pathogens to facilitate invasion into plant tissues.

**Endosomes** Intracellular membrane-bound vesicles created by endocytosis which function in the delivery of internalized material (through endocytosis) to the inside of the cell and transport of materials from Golgi apparatus to vacuoles and lysosomes. They increase the membrane area available for signaling interaction. Migration of endosomes by cytoplasmic streaming also enables their proximity to the nucleus to evoke a signaling response.

**Endospermous seeds (or albuminous seeds)** Seeds in which endosperm persists even after they mature and serves as storage tissue, for example, cereal grains.

**Energy fluence** A measure of the total amount of energy incident on unit area (unit = $J.m^{-2}$).

**ENOD (early nodulins)** Nodule-specific proteins synthesized by the plant significantly earlier during the process of nodulation (much before the expression of *nif* and *fix* genes).

**Entrainment** The phenomenon of synchronization or alignment of internal circadian rhythm to external cue, for example, entrainment of circadian rhythms to light-dark cycle.

**Epigenetics** Stable and heritable covalent modification of DNA, RNA, or protein molecules which result in the changes of function without altering their primary sequences.

**Epinasty** The phenomenon of downward curvature of plant parts, such as leaves, that occurs when the upper side of the petiole grows faster than the lower side. It is also exhibited by petals of a flower.

**Epiphytes** Plants which grow harmlessly on or attached to other plants for the physical support and derive the moisture and nutrients usually from the air, the rain, or the debris that collect on the supportive plant, for example, orchids.

**Epistatic gene** A gene that masks the expression of other genes.

**Epithem** The tissue surrounding the hydathodes. The cells of epithem are isodiametric in shape, are loosely arranged, have few or no chloroplasts, and enclose lot of intercellular spaces filled with water.

**Equilibrium constant ($K_{eq}$)** The constant expressing the ratio of products to reactants at equilibrium during a reaction.

**Equilibrium** A state in which there is no net change in the concentration of reactants and products over time.

**Ergosterol** A sterol found in cell membranes of fungi but absent in animals. Its function in membrane fluidity in fungi is similar to that of cholesterol in animal cells. It was first isolated from ergot which is the common name for members of the fungal genus *Claviceps*.

**Essential elements** The nutrients or elements necessary for the growth and development of a plant, for example, nitrogen, phosphorus, potassium, calcium, and magnesium. These are integral part of structure and metabolism of plants and their deficiency results in abnormal growth characterized by a set of deficiency symptoms.

**Exoamylase** A glucanohydrolase that hydrolyzes glycosidic bond near the nonreducing end of high molecular weight oligosaccharides. For example, β-amylase hydrolyzes terminal α-1,4 glycosidic linkages in starch, dextrins, and glycogen, releasing successive maltose units from the ends.

**Expansins** A group of nonenzymatic proteins occurring in plant cell walls which cause cell wall loosening in response to acidic pH and facilitate cell wall extension.

**Export (ref. photoassimilate translocation)** The movement of photoassimilates in the sieve elements away from the source tissues (e.g., from leaf to stem and root).

**Extensins** A group of hydroxyproline-rich glycoproteins present in the cell wall which are involved in the cell wall growth and expansion of cell.

**Extracellular matrix (ECM)** A collection of extracellular molecules secreted by cells that provides structural and biochemical support to the surrounding cells.

**Facilitated diffusion** A passive movement of specific solutes with the help of membrane transporter proteins whereby they diffuse across the membrane according to their concentration gradient without the use of energy. It involves the activities of uniporters and ion channels.

**Facultative parasites** Parasites which adopt parasitism but are able to complete their life cycle and survive on their own in the absence of host plant.

**Feed-forward regulation** Control of a metabolic pathway by a metabolite that acts in the same direction as the metabolic flux.

**Feedback regulation** A regulatory mechanism in which output of a process is used as an input to control the behavior of the process either in a positive or negative way.

**Ferredoxin reductase/oxidase** Enzyme responsible for reduction of $Fe^{3+}$ to $Fe^{2+}$ on the surface of the plasma membrane of the root cells.

**Ferredoxin** A small, water-soluble iron-sulfur (Fe-S) protein involved in electron transport in photosystem I (PS I).

**Ferritin** An iron-storage protein localized in plastids which is involved in the buffering of iron in chloroplasts.

**Fe-S proteins** Proteins having iron-sulfur clusters, which are involved in electron transport without themselves undergoing any significant structural changes.

**Fick's law of diffusion** A law stating that the rate of diffusion is directly proportional to the concentration gradient. It depends on the medium in which diffusion takes place.

**Filiform apparatus** A convoluted, thickened cell wall forming fingerlike projections that increases the surface area of the plasma membrane at the extreme micropylar end of a synergid cell.

**Flavocytochromes** Protein that contains both flavin and heme groups as cofactors. The flavin can be flavin adenine dinucleotide (FAD) or flavin mononucleotide (FMN), and the heme is usually heme-*b* or heme-*c*.

**Flavonoids** A class of plant secondary metabolites whose structure is similar to flavones. They consist of two phenyl rings and a heterocyclic ring that can be abbreviated as C6-C3-C6. They serve various functions in plants such as imparting color and fragrance to flowers and fruits, photoprotection, detoxification of harmful compounds, and antioxidants and also act as allelopathic and antimicrobial compounds.

**Flg22** A 22-amino acid flagellin peptide which has a core domain necessary for binding. It determines the specificity of the flagellum in eliciting the immune response.

**Floral meristem-identity genes** Genes encoding products which are responsible for development of individual whorls in a flower based on the pattern in which they are expressed.

**Floral pathway integrator genes** Genes whose expression transmits the environmental and endogenous signals in response to flowering time to enable adaptation to changing environmental factors such as light and temperature.

**Florigen** A hypothetical hormone which is synthesized by the leaf and translocated to the shoot apex to induce flowering under inductive conditions.

*FLOWERING LOCUS C (FLC)* A gene whose product inhibits flowering in *Arabidopsis* by repressing the activity of flowering promoting genes, namely, *FLOWERING LOCUS T (FT), FLOWERING LOCUS D (FD)*, and *SUPPRESSOR OF OVEREXPRESSION OF CONSTANS* 1 *(SOC1)*.

**FLOWERING LOCUS T (FT)** A key floral signal that promotes the transition from vegetative growth to flowering in *Arabidopsis*.

**Fluence** A measure of the quantity of radiant energy in terms of number of photons falling on a small sphere, divided by the cross section of the sphere. It is expressed either as photons or quanta (in moles, mol) or as amount of energy (in Joules, J).

**Fluorescence** The emission of light photons produced by certain compounds during a relatively slow process of return of electron from first excited singlet state to ground state upon absorption of light of a shorter wavelength. The emitted photon has lower energy than its excited form, and thus emission is in longer wavelength range. For example, chlorophyll, a molecule in solution, when excited by blue light absorption (430 nm), emits red fluorescence at 663 nm.

**Frataxin (FH)** A mitochondrial protein involved in making Fe-S clusters and biosynthesis of heme. It is a highly conserved protein which is required for efficient regulation of cellular iron homeostasis.

**Free space** Cell wall fraction of roots outer to Casparian strips in the endodermis which allows unhindered movement of water and solutes from soil by diffusion.

**Fungal chitin** Fragments of the fungal cell wall formed by the degradation of chitin (main component of fungal cell wall) via the action of chitinase released from plant cells during fungal infection. It acts as pathogen-associated molecular pattern (PAMP), and its recognition results in the activation of defense signaling pathway in the host cell.

**Fusicoccin** A diterpenoid glycoside produced by the fungus *Fusicoccum amygdali*. It activates the $H^+$-ATPase (proton pump) on the plasma membrane.

**Futile cycles** Two metabolic pathways running in opposite directions simultaneously with no net gain and dissipation of energy.

**G-protein-coupled receptors (GPCRs)** A family of membrane-associated proteins with an extracellular ligand-binding domain, a transmembrane domain, and an intracellular domain that interacts with inactive G-protein trimer. They bind extracellular ligands and transmit the signal via activation of G-proteins.

**G-proteins** A family of heterotrimeric GTP-binding proteins involved in signal transduction, whose activity is regulated by G-protein-coupled receptors (GPCRs).

G-proteins consist of three subunits—$G_\alpha$, $G_\beta$, and $G_\gamma$. They regulate responses to a variety of stimuli.

**GABA (gamma-aminobutyric acid)** A non-protein amino acid which gets accumulated in plant cells in response to variety of abiotic and biotic stress factors.

**Gametophytic self-incompatibility (GSI)** A type of self-incompatibility in which the incompatibility phenotype of the pollen is determined by genotype of pollen grain itself. A pollen will only grow in a pistil that does not contain the same allele. It occurs in members of Solanaceae, Rosaceae, and Liliaceae.

**Gating** The conformational shifts between permeable (or open) and non-permeable (or closed) states of ion channels. Gating is controlled either by voltage or ligands (chemicals that bind to channel proteins), such as hormones, $Ca^{2+}$ and G-proteins, and pH.

**Geitonogamy** The process of transfer of pollen onto the surface of stigma of a flower present on the same plant.

**Gene for gene model** A model for plant-pathogen interactions which explains the relationship between genes which encode effector molecules in pathogens and the genes that encode resistance proteins in hosts.

**General elicitor** An elicitor which is able to induce general resistance in both host and nonhost plants and is not different in its effect on different plant species.

**Gerontoplasts** Plastids found in senescing tissue exhibiting structural modification of thylakoid membrane with simultaneous formation of plastoglobuli with lipophilic materials.

**Gibberellins** A group of plant hormones which are tetracyclic diterpenoid acids and involved in the promotion of growth. They are known to stimulate cell elongation and influence various developmental processes like stem elongation, seed germination, dormancy, flowering, sex expression, enzyme induction, and leaf and fruit senescence.

**Girdling** Removal of a ring of bark from woody stem of a tree or shrub that severs the vascular system resulting in restriction of transport of nutrients between roots and aerial parts such that the aerial parts above the girdle become dead.

**Globulins** A major class of storage proteins in seeds which are rich in lysine. It is a source of nutrition for young embryo during seed germination.

**Gluconeogenesis** A metabolic pathway that results in the generation of glucose from noncarbohydrate carbon sources such as pyruvate.

**Glucosinolates** A subclass of secondary metabolites which are sulfur- and nitrogen-containing glycosides. They are particularly abundant in members of Brassicaceae. Breakdown of glucosinolates releases compounds such as isothiocyanate and nitrile which are responsible for plant defense against herbivory.

**Glutamate-like receptors** A highly conserved family of ligand-gated ion channels present in mammals and plants. They play roles in plant defense mechanisms.

**Glutathione** A tripeptide which is a non-specific reducing agent in the cell and is an important antioxidant in plants, animals, fungi, and certain bacteria.

**Glycophytes** Terrestrial plants which are unable to tolerate high concentration of salt.

**Glyoxylate** A two-carbon acid aldehyde which is an intermediate in the glyoxylate cycle.

**Glyoxylate cycle** A pathway of conversion of acetyl-CoA to succinate that occurs in glyoxysomes.

**Glyoxysomes** Specialized peroxisomes which harbor enzymes of glyoxylate cycle. These are found in fat storage tissues of germinating seed where they serve as sites for hydrolysis of fatty acids into acetyl-CoA by β-oxidation.

**GOGAT** Refers to glutamine oxoglutarate aminotransferase which, in association with glutamine synthetase, is crucial for assimilation of ammonia. It catalyzes the synthesis of two molecules of glutamate from glutamine and 2-oxoglutarate. It is also known as glutamate synthase.

**Grana lamellae (sing. granum)** The unstacked thylakoid membranes that connect grana.

**Granum (pl. grana)** Stacked thylakoid membranes in the chloroplast.

**Gravitational potential ($\psi_g$)** A measure of the influence of gravity on water potential of a system. It is measured in terms of the energy a substance possesses under the influence of gravity.

**Gravitropism** A differential growth response to gravity, either in the direction of gravity (positive geotropism) as shown by roots or in the opposite direction (negative geotropism) as shown by shoots.

**Green leaf volatiles** Volatile organic compounds released when plants suffer from tissue damage.

**Green manure** Refers to the manure generated from crop plants by leaving them to wither and get incorporated into the soil, either uprooted or still rooted in the soil. The crop plants that are employed as green manure crops can be either non-leguminous (e.g., wheat, mustard, and carrot) or leguminous (e.g., riverhemp, sannhemp, fenugreek, and crimson clover).

**Green window** Visible light spectrum not absorbed by green algae growing on top of the ocean such that the light reaching the deeper part of ocean mainly consists of green light.

**Growth** Increase in a measured parameter or attribute of a cell, tissue or organism as a function of time.

**GTPase-activating proteins (GAPs)** A family of regulatory proteins that inactivate GTPases by promoting GTP hydrolysis to GDP. For example, GAPs of heterotrimeric G-proteins inactivate the GTPase of the activated G-protein by promoting GTP hydrolysis, thereby switching off the signaling event.

**Guanine nucleotide exchange factors (GEFs)** A family of regulatory proteins that activate inactive GTPases by replacing GDP with GTP, for example, GEFs of heterotrimeric G-proteins which activate the GTPase of the inactivated G-protein by replacing GDP with GTP, thereby switching on the signaling event.

**Guttation** The process of extrusion of liquid droplets comprising xylem sap from tips or the margins of the leaves through special structures called hydathodes. It is common in grasses and small herbaceous plants like strawberry and tomato.

**Guttation burn** Leaf burn at the margin of leaves caused due to salts and minerals present in the droplet of water that remain after the evaporation of the liquid.

**Halophytes** Salt-tolerant plants which can grow at considerably high concentrations of salt, for example, *Spartina alterniflora* and *Salicornia*.

**Harvest index** Ratio of dry weight of harvestable part (economically important) of the plant to the total dry weight of the plant.

**Haustorium** Specialized sucking roots of a parasitic plant or a specialized branch of fungal hyphae formed inside a living cell of the host plant to obtain nutrients.

**Haustorium initiation factors (HIFs)** Host root-derived haustorium-inducing chemicals which trigger initiation of haustoria in the parasitic plants.

**HC toxin** A fungal toxin produced by *Cochliobolus carbonum* which is involved in inducing pathogenesis. It is a cyclic peptide which inhibits the activity of enzyme histone deacetylase in maize plants.

*Heading-date1* (*Hd1*) A gene found in rice which is homologous to *CONSTANS* and inhibits flowering.

*Heading-date3a* (*Hd3a*) A gene which stimulates flowering in rice. It is translocated via sieve tubes to the apical meristem.

**Heartrot** A symptom of boron deficiency occurring in plants, such as sugar beet, carrots, and turnips, whereby the young leaves turn black and die, resulting in a rosette of small dead leaves. The root becomes dark in the center due to death of tissue.

**Heat shock proteins (HSPs)** A set of proteins that are synthesized as a generalized physiological response in both prokaryotic and eukaryotic organisms to high temperature stress. These act as molecular chaperones to assist synthesis and folding of proteins.

**Heliotropism (or sun tracking)** Continuous adjustment of the orientation of leaves or other parts of a plant in the direction of sunlight such that their surfaces are perpendicular to sun rays and receive maximum light. For example, leaves of cotton, alfalfa, soybean, and lupine as well as young inflorescence of sunflower exhibit heliotropism.

**Heme** The iron-porphyrin prosthetic group of many heme proteins.

**Hemibiotrophic parasite** A pathogen that is initially parasitic (biotrophic) but over time leads to the death of the host tissue (necrotrophic), for example, *Phytophthora infestans* that causes late blight of potato.

**Hemiparasites** Parasites which rely partially on their host plants for resource acquisition but are able to photosynthesize, for example, mistletoe.

**Herbicides** Chemical substances used to kill the unwanted plants while leaving the desired crop rather unharmed, for example, 2,4-D and glyphosate.

**Herbivore-associated molecular patterns (HAMPs)** Molecules derived from herbivores which function as signal for triggering plant defense responses, for example, fatty acid-amino acid conjugates (FACs).

**Herbivore-induced plant volatiles** Volatile organic compounds produced by plants in response to herbivory.

**Herbivory** Feeding by animals on plant material. It includes different types of feeding strategies, such as sap-sucking, cell-content feeding, wood boring, leaf-mining, feeding on fruits, spores, etc.

**Heterostyly** The condition of having two or three different floral morphs characterized by different length of stamens and pistils. For example, it occurs in *Primula* and *Linum*.

**Heterotropic allosteric modulation** Allosteric regulation in which molecules other than substrate act as modulator of the activity of the enzyme. For example, ATP and CTP are heterotropic effectors of aspartate transcarbamoylase, acting to activate and inactivate the enzyme, respectively.

**Hexose monophosphate pool** The intracellular pool consisting of hexose monophosphates, i.e., glucose 1-phosphate, glucose 6-phosphate, and fructose 6-phosphate, which are interconvertible and can be removed or added to the pool according to the need in the cell.

**Histidine kinase receptors** Receptor proteins for signaling molecules which are multistep derivatives of bacterial two-component system. They may be membrane associated or soluble and exist as dimers. They play key roles in the two-component signal transduction and function via autophosphorylation.

**Histogenesis** A series of organized, integrated processes by which cells of the primary germ layers of an embryo differentiate and assume the characteristics of the tissues into which they will develop.

**Hoagland's solution** A nutrient solution used for plants that consists of all the essential elements which are necessary for growth of nearly all plants in a defined proportion. It was developed by D.R. Hoagland in 1933.

**Holoenzyme** A functionally active enzyme protein conjugated with cofactor (if any).

**Holoparasites** Parasites which are completely dependent on their host plants for photosynthates, water, and minerals.

**Homeostasis** Regulation of endogenous levels of biomolecules such as plant growth regulators, ions, and other molecules through modulation of their biosynthesis, catabolism, sequestration, conjugation, efflux/influx, and release from internal stores, such as ER and vacuoles.

**Homeotic genes** A set of genes responsible for development of specific organs and regulation of anatomical structures in various organisms. They encode transcription factors that regulate the expression of target genes and lead to establishment of pattern. Mutations in these genes result in displaced body parts.

**Homotropic allosteric modulation** Allosteric regulation in which substrate acts as modulator of the activity of the enzyme. For example, aspartate is a homotropic effector of aspartate transcarbamoylase.

**Horizontal resistance** Resistance that acts against a wide range of pathogens. It is also called polygenic resistance as it is governed by multiple genes.

**Hormone receptor** A hormone-binding protein which brings about a physiological effect after interacting with the hormone.

**Hormone sensitivity** The extent of response to a hormone. Sensitivity to a hormone depends on the concentration of hormone at the site of action and the abundance of receptor and hormone—receptor affinity.

**Humus** A dark, colloidal carbonaceous (organic) residue formed in the soil by the microbial decomposition of plant and animal tissues. It acts as a nutrient reservoir and increases soil fertility.

**Hydathodes** The specialized pores in the epidermal layer normally found at the tip and margins of the leaves, which are surrounded by a special parenchymatous tissue, called epithem. They are involved in extrusion of xylem sap in the form of droplets.

**Hydraulic conductance** The property which describes the ease with which a fluid (usually water) can move through pore spaces of a material. For example, the leaf hydraulic conductance represents the capacity of the water transport system in leaves.

**Hydraulic lift** The lifting of water from deep moist layers to shallow drier layers of soil by large woody trees. It influences the water content of the rhizosphere and delays the drying of soil.

**Hydraulic resistance** The property which describes the resistance faced by a fluid (usually water) while moving through porous spaces, for example, hydraulic resistance to bulk flow of water moving through the xylem elements.

**Hydroponics** A technique of growing plants in absence of soil in which roots of the plants are exposed to a nutrient solution consisting of all essential elements.

**Hydrotropism** Plant growth or movement determined by gradient of water.

**Hydroxyproline-rich glycoproteins (HRGPs)** A group of glycosylated hydroxyproline-rich proteins which are major constituents of plant cell walls rich in hydroxyproline. They contain arabinose and galactose in the attached oligosaccharides chains.

**Hygrometer** An instrument to measure moisture content in the atmosphere.

**Hyperaccumulators** Plants which accumulate high concentrations of metal elements from soil and store them in their aerial tissues. These are capable of growing in soils with high concentration of metals and can be used for phytoremediation, for example, *Agrostis castellana*, which is a hyperaccumulator of arsenic, manganese, lead, and zinc.

**Hypersensitive response** A localized response of plant cells to microorganisms characterized by rapid death of the infected cells and the cells surrounding the site of infection so as to prevent further invasion in adjacent cells. The invasion of pathogen is restricted due to deprivation of nutrients.

**Hypertonic solution** The solution with a higher concentration of solutes relative to that of another solution separated by semipermeable membrane. There is a movement of water into the hypertonic solution from the hypotonic solution across the membrane.

**Hyponasty** A nastic movement which involves upward bending of organs, such as leaves, petals, and petioles, caused by increased growth on their lower surface as compared to the upper surface. For example, young flowers remain tightly closed as a result of hyponasty.

**Hypotonic solution** The solution with a lower concentration of solutes relative to that of another solution separated by semipermeable membrane. There is a movement of water from the hypotonic solution into the hypertonic solution across the membrane.

**Hypoxia** A condition in which cells or tissues are deprived of oxygen.

**Ice nucleation active (INA) proteins** A family of proteins that promote nucleation of ice at relatively warm temperature by promoting alignment of water molecules (above $-5\,°C$). These proteins are localized on the outer surface of the membrane and can cause frost damage to many plants.

**Imbibition** The process by which the molecules of a liquid or gas diffuse into a solid substance causing it to increase in volume, for example, swelling of raisins and dry seeds in water.

**Immobile elements** Elements which upon absorption from soil cannot be relocated from the older to newly formed leaves, for example, calcium, iron, and copper. Plants deficient in immobile elements exhibit symptoms in the growing region, while the old leaves remain unaffected.

**Impaction** A process in which seeds with hard seed coat are subjected to vigorous shaking in order to remove the corky outgrowth blocking the opening for gaseous exchange.

**Import (ref. photoassimilate translocation)** The movement of photoassimilates through sieve elements into sink tissue.

**Incipient plasmolysis** The point at which the protoplast begins to pull away from the cell wall marking initiation of plasmolysis. However, for the estimation of osmotic potential of the cell, incipient plasmolysis is considered to have occurred when 50% of the cells have become plasmolyzed.

**Incompatible disease reaction** An interaction between host and pathogen with minimum effect on the host where symptoms are not expressed clearly.

**Indeterminate growth** Unrestricted or unlimited growth, as with a vegetative apical meristem that produces an unrestricted number of lateral organs indefinitely.

**Induced systemic resistance (ISR)** Ability of selected strains of plant growth-promoting microbes, like *Pseudomonas*, *Rhizobacteria*, *Bacillus*, and mycorrhizal species, to suppress the infection of disease-causing pathogens by providing systemic resistance.

**Inducible defense** The activation of defense mechanism after interaction with pathogen.

**Innate immune system** Refers to the immunity in the cell that occurs naturally because of genetic factors or physiology and is not caused by infection or vaccination. It is readily mobilized to fight microbes at the site of infection.

**Integument** The natural covering which forms a tough protective layer around the ovule. After the egg cell is fertilized, integument forms the seed coat.

**Intermediary cell** A type of companion cell with several plasmodesmata connecting the surrounding cells, particularly the adjacent bundle sheath cells.

**Intermediary metabolism** Combined activities of all metabolic pathways involved in interconversion of precursors, metabolites, and products of a pathway resulting in the formation of a series of intermediate compounds.

**Interveinal chlorosis** The yellowing of green leaf tissue between the veins.

**Ion channels** Integral membrane proteins which form a passageway for passive flow of ions according to a concentration gradient. They are driven solely by electrical potential difference across the membrane.

**Ionophores** Compounds capable of binding with metal ions and carrying them across the membrane. For example, valinomycin is known to facilitate transport of potassium ions across membranes.

**Iron-regulated transporters (IRT1)** Transporters localized in plasma membrane which are mainly responsible for iron uptake from roots.

**Irradiance** The flux of energy per unit area on a flat surface (units = $W.m^{-2}$).

**Isoamylase** A debranching enzyme which acts on α-1,6 glycosidic linkages in high molecular weight branched oligosaccharides, such as glycogen and amylopectin. It removes branches that are very close to other branches. It accelerates crystallization of nascent amylopectin molecule during starch synthesis.

**Isoprene** An unsaturated pentahydrocarbon (2-methyl-1,3-butadiene) derived from phenylpropanoid pathway which serves as the basic structural unit in terpene structure.

**Isotonic solution** Two solutions separated by a semipermeable membrane having same concentration of a solute in them. There is no net movement of water.

**JA receptors** F-box proteins which target downstream JAZ (JASMONATE ZIM DOMAIN) family of negative transcriptional repressors for degradation, thereby activating the expression of JA-sensitive genes.

**Jasmonic acid (JA)** A lipid-derived signaling molecule derived from α-linoleic. It participates in the regulation of a number of plant processes, including growth, reproductive development, photosynthesis, and responses to biotic and abiotic stress factors.

**Karrikins** A class of butanolide compounds derived from burnt plant material which can stimulate seed germination in a number of plants.

**Kranz anatomy** Wreath-like arrangement of mesophyll cells around the vascular bundles, forming a bundle sheath. This anatomical feature is typical of C4 plants.

**Late embryogenesis abundant (LEA) proteins** A family of hydrophilic and thermostable proteins which perform protective function against desiccation and oxidative stress. They have the ability to form viscous liquid with very slow diffusion and limited chemical reactions by forming hydrogen bonds with sucrose. They prevent other proteins from aggregation.

**Lateral heterogeneity** Unequal distribution of the components of photosynthetic electron transport systems in thylakoid membrane.

**Law of minimum** A law (proposed by Liebig) which states that yield is proportional to the amount of the most limiting nutrient and the growth is impaired if it is deficient even when all the other nutrients are available. If the deficient nutrient is supplied, yield is improved, but then some other nutrient may become limiting, and now the law of minima will be applicable to that nutrient.

**Leaf stomatal resistance** The resistance offered by guard cells on the exchange of carbon dioxide and water vapor through stomata.

**Lectins** Carbohydrate-binding proteins.

**Leghemoglobin** An oxygen carrier protein having chemical and structural similarity to hemoglobin present in humans. It is produced by the leguminous host plants in response to the symbiotic association with rhizobia. It facilitates the transport of oxygen for the respiration of symbiotic bacteria.

**Ligand** A molecule that binds to another (usually larger) molecule.

**Light-harvesting complex (LHC)** Antenna pigment-protein complex associated with photosystems which absorbs and funnels light energy to the reaction center.

**Lignin** A complex branched polymer of phenolic alcohols associated with cellulose in cell walls, especially deposited in the secondary walls and thickenings in the xylem elements. It is responsible for providing mechanical strength to plants. It is a type of phenylpropanoid.

**Liming** Treating soil with calcium and magnesium rich materials to reduce acidity and improve fertility or oxygen levels.

**Limit dextrinase or R-enzyme** An enzyme that hydrolyzes $\alpha$-1,6 glycosidic bonds in amylopectin and branched dextrins.

**Lipase** An enzyme that hydrolyzes triglycerides into glycerol and fatty acids. It is generally found in germinating oil seeds.

**Lipid transfer proteins** A group of highly conserved proteins of about 9 kDa found in tissues of higher plants which are responsible for shuttling of phospholipids and other fatty acid groups between cell membranes. They possess a tunnellike hydrophobic cavity which facilitates the binding of lipids.

**Lipidome** Complete lipid profile within a cell.

**Lipidomics** Large-scale study involving structures, pathways, and networks of cellular lipids in biological systems. It is a subset of metabolome.

**Lipochitooligosaccharides (LCOs)** Signaling molecules made up of acylated oligosaccharides that are released by rhizobia. They play a key role in the initiation of symbiosis between the legume and the symbiotic bacteria.

**Liposomes** Small spherical vesicles consisting of at least one phospholipid bilayer which are formed spontaneously when phospholipids are suspended in an aqueous buffer.

**Lipoxygenase** A family of non-heme iron enzymes that catalyze the oxidation of polyunsaturated fatty acids and lipids containing a cis,cis-1,4-pentadiene structure.

**Lodging** Displacement of stem or roots of grain crops from their vertical and proper placement which makes them difficult to harvest and significantly reduces yield.

**Long-day plant** A plant which flowers only in long days when the day length exceeds the critical day length, for example, spinach, lettuce, sugar beet, cabbage, and henbane.

**LUREs** Cysteine-rich polypeptides secreted from synergid cells that attract pollen tubes.

**Lycopene** A type of carotenoid pigment responsible for red color of fruit and vegetables.

**Lysimeter** An instrument to measure the amount of evapotranspiration in plants.

**Lysophosphatidate** A phospholipid derivative that can act as a signaling molecule in different signaling cascades.

**Lysophospholipids** Lipid molecules derived from the action of phospholipase A on fatty acids. They consist of a single carbon chain and a polar head group and are more hydrophilic than their corresponding phospholipids.

**Macroelements or macronutrients** Essential elements, for example, hydrogen, carbon, oxygen, nitrogen, potassium, calcium, magnesium, phosphorus, and sulfur, which are required in relatively large amounts for a plant to grow and remain healthy. They are present in plant tissue in quantities ranging from 0.2% to 4.0% of its dry weight.

**MAP kinases (MAPKs or mitogen-activated protein kinases)** Highly conserved family of serine/threonine protein kinases which are major components of signaling cascades in response to a variety of stimuli regulating developmental processes like embryogenesis, proliferation, and death.

**Matric potential ($\psi_m$)** A component of water potential expressed as the adsorption affinity of water to colloidal substances and surfaces in the plant cell, such as plasma membrane, and soil particles.

**1-MCP (1-methylcyclopropane)** A cyclopropane derivative used as a synthetic plant growth regulator which inhibits ethylene biosynthesis.

**Melatonin (or N-acetyl-5-methoxytryptamine)** A tryptophan-derived indoleamine which is pleiotropic in function and affects various plant physiological processes and also acts as a potent ROS scavenger.

**Metabolic channeling** Direct transfer of biosynthetic intermediates from one enzyme to another in a pathway minimizing their diffusion.

**Metabolic flux** Rate of turnover of metabolites in a particular metabolic pathway in a biological system or the movement of metabolites in a pathway.

**Metabolic plasticity** The capacity of plants to alter their metabolism when subjected to any kind of stress essential for their survival.

**Metabolic redundancy** A feature of metabolism of an organism whereby different pathways serving a similar function are operating and, therefore, can be replaced by each other without apparent loss in function. It is a common feature in plant metabolism.

**Metabolism** All the chemical reactions involved in maintaining the living state of a cell and an organism as a whole leading to production of energy, synthesis of new materials, and breakdown of others.

**Metabolites** Any substance produced in one of the various metabolic pathways or chemical intermediates taking part in the enzyme-catalyzed reactions taking place during metabolism.

**Metabolons** A temporary structural-functional complex formed as a result of close association among sequential enzymes of a metabolic pathway held together by non-covalent interactions and with the help of other proteins and cytoskeleton. This assembly permits the direct channeling of metabolites between the enzymes.

**Metallochaperone** A family of proteins involved in the movement of metal ions to target sites in a cell.

**Metallothioneins** A family of cysteine-rich proteins that bind metals using thiol groups. They are localized in the membrane of Golgi apparatus.

**Methanogens** Microorganisms which produce methane as a byproduct of metabolism under anoxic conditions, for example, *Methanobacterium*, *Methanospirillum*, and *Methanococcus*.

**Methionine sulfoximine (MSO)** An irreversible inhibitor of glutamine synthetase. It binds to the glutamate binding site and undergoes phosphorylation resulting in an irreversible non-covalent inhibition of the enzyme.

**Methylerythritol phosphate (MEP) pathway** The pathway involved in terpene biosynthesis. It leads to formation of isopentyl diphosphate (IPP) in plastids from glyceraldehyde 3-phosphate and pyruvate.

**Mevalonic acid pathway** The pathway involved in terpene biosynthesis. It leads to a stepwise condensation of three acetyl-CoA molecules to form mevalonic acid.

**MFS (major facilitator superfamily) transporters** A subfamily of membrane transport proteins that facilitate movement of small solutes across cell membranes in response to chemiosmotic gradients.

**Mg-protoporphyrin** Any plant porphyrin having magnesium in the center of the ring.

**Michaelis-Menten constant ($K_m$)** A constant value representing the substrate concentration at which the velocity of enzyme-catalyzed reaction is half of the maximum velocity. It gives a measure of the binding affinity of a given enzyme with its substrate (unit = mM).

**Microbial-associated molecular patterns (MAMPs)** Highly conserved molecules produced across large groups of pathogens as well as nonpathogenic microbes which are indispensable for their survival, for example, flagellin, elongation factor Tu (EF-Tu), chitin, lipopolysaccharides, and β-glucans.

**Microelements or micronutrient** Essential elements, for example, iron, manganese, boron, molybdenum, copper, zinc, and chlorine, which are required in relatively small amounts or traces for a plant to grow and remain healthy. They are present in plant tissue in quantities ranging from 5 to 200 ppm or less than 0.02% of its dry weight.

**Mitochondrial iron transporter (MIT)** A mitochondrial iron transporter essential for growth and function in rice plants.

**Mitogen-activated protein (MAP) kinase cascade** A series of kinases (proteins which are capable of phosphorylating other target proteins) which, upon binding of a ligand, phosphorylate each other in a defined sequential manner, leading to their activation, for example, MAPK, MAPKK, and MAPKKK. MAPK serves as the anchor of the relay sequence from cell surface to the nucleus.

**Mobile element** Element which upon absorption by the plant from soil can move from old leaves to younger plant parts, for example, nitrogen, phosphorus, and potassium. Plants deficient in mobile elements exhibit symptoms in the older leaves first as these are transported to growing parts.

**Monoculture** The cultivation of a single variety of crop on large scale.

**Monoglycerides** A glyceride in which each glycerol molecule is linked by an ester bond with one fatty acid molecule. It is also known as monoacylglycerol.

**Morphogenesis** The process of origin and development of the physical form and external structure of plants as a result of differentiation of tissues and organs.

**Morphogen** A biomolecule that plays a key role in providing positional cues in certain types of position-dependent development. They determine cell pattern formation in a concentration-dependent manner during morphogenesis. At different concentrations, they lead to different outputs. Typically, morphogens are produced from a source tissue and diffuse in neighboring tissue in an embryo (e.g., among mammals) to affect development.

**Motor cell** A cell which is capable of expanding or contracting, thereby facilitating movement of a plant part. It functions by adjusting the internal $K^+$ concentration to alter the turgidity.

**Movement protein** A protein encoded by virus genome in plant cell cytoplasm to facilitate the migration of virus from one cell to another through plasmodesmata.

**Multifunctional protein (MFP)** A protein having different catalytic functions residing in separate domains of the same polypeptide chain. For example, peroxisomal multifunctional protein possesses up to four enzymatic activities, each responsible for catalyzing a different step in the process of β-oxidation in peroxisome.

**Mutualism** The beneficial interaction between two organisms of different species in which each individual benefits from the activity of the other, for example, mycorrhiza which is a symbiotic relationship between some fungi and plant roots.

**Mycorrhiza** A symbiotic relationship of a fungus growing in association with roots of vascular plants. The fungal partner receives essential nutrients for its growth from the host and, in turn, increases the ability of the plant roots to absorb nutrients by increasing the surface area for absorption.

**Necrosis** Localized death of tissue that occurs in response to various agents including physical injury, exposure to toxins, deficiency of oxygen, and infection by pathogens.

**Necrotroph** Parasitic organism that invades and kills host organism and obtains nutrients from its dead tissues, for example, *Botrytis cinerea* on *Blumeria graminis*.

**Neochrome** A photoreceptor with LOV (light, oxygen, voltage) domain, found in ferns and algae, representing a fusion between phytochrome and phototropin. It has a phototropin-like protein sequence fused with a phytochrome chromophore-binding domain. Neochrome acts as a dual red/blue light photoreceptor. It is involved in regulation of chloroplast movement in *Mougeotia*, a filamentous green alga.

**Nitrite oxidoreductase** A membrane-bound enzyme involved in nitrification, containing iron-sulfur and molybdenum as the cofactors. It is present in the membranes of nitrifying bacteria, like *Nitrobacter*, and is a part of electron transport chain that channels electrons from nitrite to molecular oxygen.

**Nodulin** An organ-specific plant protein induced during symbiotic nitrogen fixation. It plays both metabolic and structural roles in infected and uninfected nodule cells.

**Non-cell autonomous responses** Signaling events where signal perception and response occur in different cells, for example, transcription and translation of SHORT-ROOT (SHR), a transcription regulator which occurs in cells of stele. The SHR protein moves to endodermis via plasmodesmata where it activates the expression of a cell plate regulator, SCARECROW (SCR).

**Non-climacteric ripening** Fruit ripening without increase in ethylene production and respiratory burst. Fruits, like citrus, grapes, and strawberries, are non-climacteric.

**Noncyclic photophosphorylation** The process of ATP synthesis which occurs as a result of proton gradient created across thylakoid membrane due to noncyclic electron transport involving both the photosystems. It is coupled with photolysis of water, $NADP^+$ reduction, and ATP synthesis.

**Non-endospermous seeds** Seeds in which endosperm is consumed during embryo development and does not persist till maturity, for example, pea, bean, and gram.

**Non-photochemical quenching** Harmless dissipation of excess light energy absorbed by plants as heat. It is a protective mechanism against excess light intensity.

**Non-protein amino acids** Unusual amino acids, such as canavanine and azetidine-2-carboxylic acid, which are normally not incorporated into proteins but have a role in plant defense.

**NRAMPs (natural resistance-associated macrophage proteins)** A family of metal ion transporters which play a major role in metal ion homeostasis. It is a novel family of functionally related proteins defined by a conserved hydrophobic core of ten transmembrane domains.

**Nuclear bodies** A cluster of small particles or speckles found in the nucleus which accumulate phy A and phy B. The number and size of these speckles is correlated with light responsiveness.

**Nucleophilic group** An electron-rich chemical species which donates electron to a positively charged species (electrophile) to form a chemical bond.

**Nutrient film technique (NFT)** A hydroponic cultivation technique in which dissolved nutrients required for normal growth of plants are continuously pumped over the roots through a channel or a tube.

**Nyctinasty** A nastic movement of periodic folding of leaves and closing of petals in flowers in response to onset of darkness. It is exhibited by many leguminous plants.

**Obligate parasite** Parasite which fully relies on its host to complete its life cycle and fails to reproduce in the absence of a suitable host.

**Octadecanoid pathway** Biosynthetic pathway for production of jasmonic acid from α-linolenic acid.
**Oleosin** A structural protein found in the phospholipid monolayer of oil bodies in oil seeds, for example, sunflower and soybean.
**Oleosomes** Specialized structures surrounded by phospholipid monolayer (half unit membrane) that serve as oil storage bodies for triacylglycerols in oil-bearing seeds. They are also known as oil bodies or sphaerosome.
**Oligomycin** An antibiotic that inhibits ATP synthesis by blocking channel ($F_o$) of ATP synthase.
**Organogenesis** Development of organized structures such as shoots, roots, and flower buds from cultured cells or tissues.
**Orthodox seeds** Seeds that can tolerate desiccation and remain viable when stored in a dry state, for example, pea, corn, and tomato.
**Ortho-gravitropism** The phenomenon whereby plant organs adopt vertical orientation with respect to the axis of the plant body in response to gravity.
**Osmolytes** Soluble compounds which affect osmosis in a cell and maintain cell volume and fluid balance, for example, betaine aldehyde and proline.
**Osmometer** An instrument to measure the osmotic strength of a solution.
**Osmoprotectant** A small polar molecule with neutral charge and low toxicity which helps survive extreme osmotic stress by maintaining osmotic balance.
**Osmoregulation** The maintenance of osmotic pressure in a cell or an organism by regulating internal water and solute concentration.
**Osmosensor** A pressure sensor protein on the plasma membrane which is involved in the primary perception of water deficit.
**Osmosis** A biological process in which solvent moves from the region of higher concentration (lower solute concentration) to the region of lower concentration (higher solute concentration) through a semipermeable membrane.
**Osmotic potential or solute potential ($\psi_s$)** A component of water potential expressed as the influence of dissolved solutes on the water potential of the cell. It is the potential of solute to bring about osmosis in a cell.
**Outcrossing** Cross-pollination of two plants belonging to different genotypes.
**Oxidative burst** Rapid release of reactive oxygen species (ROS), such as superoxide radical, hydrogen peroxide, and hydroxyl radical, within the cell.
**Oxidative deamination** A reaction of amino acid catabolism in which amino ($-NH_2$) group of α-amino acid is released as ammonia, and the amino acid is converted to its respective keto acid. Released ammonia is utilized in urea cycle, for example, conversion of glutamate to α-ketoglutarate catalyzed by glutamate dehydrogenase.
**Oxidative phosphorylation** Synthesis of ATP from ADP and $P_i$ which is coupled with electron transfer from a compound to molecular oxygen.
**Oxidative stress** A stressful condition created as a result of imbalance between the production of free radicals in the body and their detoxification.
**Oxygen Evolving Complex (OEC)** A complex of proteins and manganese ions associated with the reaction center of photosystem II. It is also known as water

splitting complex as it is the site for photooxidation of water during the light reaction of photosynthesis.

**Oxygenic photosynthesis** A type of photosynthesis in which light-driven ATP and NADPH synthesis occurs in organisms utilizing water as the reductant, and the process is coupled with release of free oxygen.

**PAMP-triggered immunity (PTI)** The defense mechanism in plants involving recognition of pathogen-associated molecular patterns (PAMPs) resent on the cell wall of microbes by the pattern recognition receptors (PRRs) present on the surface of the plant cells.

**Paraheliotropic leaves** Leaves which utilize heliotropic movement to avoid or reduce light capture in order to reduce the injuries sustained due to intense illumination.

**Parasite** An organism living in or on another organism (host) to obtain its nutrient supply.

**Parthenocarpy** Development of ovary into fruit without fertilization thus leading to the development of seedless fruits. It occurs naturally in banana and pineapple.

**Passage cells** Cells in the endodermis which are devoid of suberin (Casparian band) and allow free passage of water.

**Passive transport** The spontaneous transport of ions or molecules across a membrane down their electrochemical gradient without the utilization of energy.

**Patch clamp** An electrophysiological technique used to resolve the ionic activity (in the form of measurement of ion currents) of single or multiple protein molecules (channels) in individual intact cells or patches of membranes as they catalyze ion translocation across membrane.

**Paternally expressed genes (PEGs)** Genes for which only the paternal alleles are expressed.

**Pathogen** Disease-causing agent.

**Pathogen-associated molecular patterns (PAMPs)** Highly conserved molecules produced across large groups of pathogens which are indispensable for their survival. They are recognized by respective PRRs of the host plant.

**Pathogenesis** The process of infection, colonization, and reproduction of pathogen in the host.

**Pathogenesis-related proteins (PR proteins)** A group of small proteins encoded by host plants when attacked by pathogens. They function either as antimicrobial agents or initiate systemic defense responses.

**Pathogenicity** The ability of a pathogen to cause disease.

**Pathogenicity gene cluster (*hrp*)** Genes that control the ability of type III S bacteria to cause infection and elicit hypersensitive response in the host plant.

**Pattern recognition receptors (PRRs)** Receptors present in the plasma membrane of organisms which are important constituents of innate immunity system and recognize extracellular pathogens, such as bacteria or fungi via recognition of PAMPs (pathogen-associated molecular patterns), like lipooligosaccharides.

**Perception of a signal** Sensitization of plant cells to signals by employing sensor proteins (receptors).

**Permeability coefficient** A measure of resistance encountered by the transport of a material through a membrane per unit driving force per unit membrane. It is a constant that depends on pressure difference across the membrane and the size and solubility of molecule to be diffused.

**Peroxidases (POXs)** Heme-containing enzymes that utilize a wide variety of organic and inorganic substrates as electron donors and hydrogen peroxide as the electron acceptor to catalyze a number of oxidative reactions. They prevent cells from oxidative damage by reducing hydrogen peroxide to water.

**Phagophore** A double-membrane structure that encloses portions of cytoplasmic components during macrophagy. They develop via expansion of autophagosome through lipid acquisition.

**Phaseic acid (PA)** An inactive catabolite of ABA formed as a result of spontaneous cyclization of 8′-hydroxy ABA. PA does not exhibit any ABA-like activity. Very recently, however, it has been observed to activate a subset of ABA receptors.

**Phenolics** A subgroup of secondary metabolites which contain at least one phenol group, for example, caffeic acid, lignin, flavonoids, and tannin. They protect plants against herbivores; provide color; thus act as attractants, as UV protectants, and as fungicides; and are allelopathic.

**Phenylpropanoids** Diverse group of compounds derived from the carbon skeleton of phenylalanine.

**Phloem filament protein (PP1)** A phloem-specific protein synthesized in companion cells and transported into the sieve-tube elements. It is present in all developmental stages and forms filaments around the inner wall of sieve-tube elements.

**Phloem lectin protein (PP2)** A phloem-specific protein which is a dimeric lectin and forms the major constituent of phloem sap. It is translocated to sieve-tube elements upon complete maturation. It specifically binds to poly(β-1,4-N-acetylglucosamine) and is involved in formation of filaments along with phloem filament protein (PP1) that plug the damaged sieve elements.

**Phloem loading** The movement of photoassimilates from the mesophyll cells to the sieve elements of leaves via companion cells involving loading of sieve-tube elements.

**Phloem unloading** The movement of photoassimilates from sieve elements into the sink cells which store or metabolize them.

**PHO1** A phosphate transporter which uploads inorganic phosphate into xylem which is then transported from roots to the aerial parts of the plant.

**Phosphatidic acid** Simplest type of diacylglycerophospholipids which upon hydrolysis releases two molecules of fatty acid and one molecule of glycerol. It is a major constituent of membranes.

**Phospholipase A (PLA)** An enzyme responsible for cleaving one of the acyl ester bonds on the phospholipids, thereby releasing a fatty acid from the second carbon of glycerol and a lysophospholipid.

**Phospholipase C (PLC)** An enzyme which hydrolyzes the glycerophosphate bond of a phospholipid molecule to yield diacylglycerol and a phosphorylated head group, such as inositol triphosphate ($IP_3$).

**Phospholipase D (PLD)** An enzyme of phospholipase superfamily which hydrolyzes phosphatidyl choline to produce signal molecules of phosphatidic acid (PA) and soluble choline. Different phospholipase D isoforms are expressed in different cellular locations in plant tissues at various stages of development.

**Phosphorylation potential** A measure of the energy state of a cell in terms of actual free energy of ATP hydrolysis under intracellular conditions. It is the ratio of concentration of ATP to that of ADP and $P_i$ in the cytosol of a cell.

**Photoblastic seeds** Seeds in which germination is sensitive to light. They are of two types: positively photoblastic seeds (stimulated to germinate in the presence of light), for example, lettuce, and negatively photoblastic seeds (unable to germinate in presence of light), for example, *Phlox*.

**Photochemical reaction center** A part of the photosystem where the first charge separation event occurs and electron of the donor excited molecule is accepted by the primary electron acceptor.

**Photoinhibition** Damage to the photosynthetic capacity of leaves when exposed to excessive light.

**Photomorphogenesis** The developmental response of an organism to the information in light, such as quantity, quality (i.e., the wavelength), and direction, and the relative length of day and night (photoperiod).

**Photon fluence** Total number of photons incident on unit surface area (units = mol. $m^{-2}$).

**Photonastic response** A type of movement induced by light, such as opening and closing of flowers during day and night, respectively.

**Photonasty** A nastic movement representing the response of a plant organ upon providing light stimulus or changing light intensity.

**Photoperiodism** A biological phenomenon of flowering at specific times of the year in response to changing day length.

**Photophosphorylation** A reaction resulting in synthesis of ATP from ADP and $P_i$ in chloroplasts of photosynthetic cells using the light energy.

**Photoreceptors** Chromophore-containing biomolecules which absorb photons of a given wavelength and use this energy as a signal to initiate a photoresponse.

**Photosynthate or photoassimilate** The transportable forms of sugars or sugar derivatives that are formed during the process of photosynthesis.

**Photosynthetic induction period** Time interval required for the photosynthetic process to establish at constant velocity after the plant has been exposed to light.

**Photosynthetic photon flux density (PPFD)** The flux of light ($\mu mol.m^{-2}.s^{-1}$) of the photosynthetically active radiation (PAR) range, i.e., 400–700 nm incident on unit surface area for a specified amount of time.

**Photosystem I (PS I)** A specialized pigment-protein system that contains chlorophyll 700 in the reaction center and is involved in transfer of electrons from plastocyanin and reduction of ferredoxin. It absorbs maximally in the far-red region of light spectrum. It is also known as plastocyanin-ferredoxin oxidoreductase.

**Photosystem II (PS II)** A specialized pigment-protein system that contains chlorophyll 680 in the reaction center and is involved in capturing of photons and transfer of electrons from water to plastoquinone, with the release of oxygen. It absorbs poorly in the far-red region of light spectrum. It is also known as water-plastoquinone oxidoreductase.

**Phototaxis** Movement of unicellular organisms, like *Chlamydomonas reinhardtii*, toward or away from the direction of light. It is controlled by two rhodopsin-like molecules known as channel rhodopsins (ChR1 and ChR2) located in the eye spot.

**Phototropins** A class of blue light-sensitive flavoprotein photoreceptors containing two LOV (light, oxygen, and voltage) domains in the N-terminus that bind FAD and a serine/threonine kinase domain in the C-terminus. They are primarily associated with blue light responses, like phototropic bending response of shoots. They also modulate stomatal opening, ion transport, chloroplast movement, cotyledon, and hypocotyl growth.

**Phototropism** The process of directional movement of a light-responsive organ or an organism in response to unilateral light, either toward the source of light (positive phototropism) or away from it (negative phototropism), for example, bending of shoots in the direction of light.

**PHT 1** A family of high-affinity $H^+$-coupled phosphate symporters.

**Phycobilisomes** Protein complexes containing light-absorbing pigments called phycobilins. They are found in photosynthetic apparatus of cyanobacteria, red algae, and glaucophytes.

**Phyllosphere** The area surrounding plants where microorganisms proliferate.

**Physiology** A study of the activities and functions of cells, tissues, organs, or the whole organism.

**Phytic acid or phytate** A saturated cyclic acid, also known as inositol polyphosphate or hexakisphosphate, which may be found in conjugation with a salt. It is the main storage form of phosphorus in many plant tissues, especially in bran and seeds.

**Phytoalexins** A group of low molecular mass lipophilic compounds produced by higher plants in response to different types of stress factors that act antimicrobial agents, for example, capsidiol.

**Phytochelatins** Oligomers of glutathione found in plants, fungi, and nematodes, which act as chelators. They play an essential role in heavy metal detoxification.

**Phytoferritins** A superfamily of iron-storage and detoxification proteins present in plastids which play important roles in controlling cellular iron homeostasis.

**Phytoliths** Minute silica particles found in the epidermis of many plants, such as *Ficus* and *Artocarpus*.

**Phytosiderophores** Small iron chelators released by members of Poaceae family under iron and zinc deficiency. They chelate iron found in the rhizosphere and make it available to the roots for absorption.

**Phytosulfokines** A group of disulfated pentapeptide growth regulators that are involved in promotion of proliferation of cells.

**Phytotropins** Noncompetitive inhibitors of polar auxin transport, for example, TIBA (2,3,5-triiodobenzoic acid), morphactin (9-hydroxyfluorine-9-carboxylic acid), PBA (pyrenoyl benzoic acid), and NPA (N-1-naphthylphthalamic acid).

**Pigment** A molecule capable of interacting with photons from the visible range of the electromagnetic spectrum and absorbing light of certain wavelength, for example, chlorophyll which gives plants their green color.

**PIN proteins** Transmembrane proteins principally responsible for efflux of IAA across plasma membrane. They are mainly located in the basal region of IAA-transporting cells, mainly xylem parenchyma in both root and shoot, and, thus, cause directional and localized auxin transport.

**Plagiotropy** The orientation of long axis of roots or branches such that they are inclined away from the vertical line.

**Plant growth regulators (PGRs)** Naturally occurring or synthetic compounds which affect metabolic and developmental processes in higher plants, mostly at low concentrations. PGRs include plant hormones, their synthetic analogs, inhibitors of hormone biosynthesis, and blockers of hormone receptors.

**Plant hormones** A group of endogenous organic substances which influence various physiological and developmental processes in plants at low concentrations, for example, auxins, cytokinins, abscisic acid, gibberellins, etc.

**Plasmodesmata** Membrane-lined channels that connect adjacent cells through the cell wall and form a continuity of cytoplasm and a central rod (desmotubule) derived from ER. They allow movement of molecules from cell to cell through symplasm and exhibit size exclusion limit depending on the physiological state.

**Plasmolysis** The process of shrinkage of protoplast of a cell as it pulls away from the cell wall as a result of loss of water.

**Polyamines** A group of low molecular weight nitrogenous compounds which exist in positively charged state at physiological pH and bind to biological molecules bearing negative charge. They are involved in modulating various growth and developmental processes including regulation of organogenesis, embryogenesis, development of flowers and fruits, and senescence.

**Polycomb group proteins** A family of proteins discovered in *Drosophila* that mediate chromatin remodeling which leads to epigenetic gene silencing.

**Polygalacturonase** A hydrolase enzyme that catalyzes the hydrolysis of $\alpha$-1,4 glycosidic bonds between galacturonic acid residues.

**Polygalacturonase-inhibiting proteins (PGIPs)** Extracellular, defense-related proteins having a leucine-rich repeat (LRR) motif. They are secreted by plants and have the ability to inhibit the pectin-hydrolyzing activity of the pathogen secreted polygalacturonase enzyme.

**Polymer-trapping model** A model that explains the formation and symplastic accumulation of tri-, tetra-, and pentasaccharides in the sieve elements which is commonly observed in the members of Cucurbitaceae and Rosaceae.

**Porphyrin** A complex heterocyclic nitrogenous compound containing four substituted pyrroles covalently joined into a ring, often complexed with a central metal atom. It forms important constituent of many enzymes and molecules, for example, hemoglobin in human blood.

**POT (proton-dependent oligopeptide transporter)** A family of transporter proteins which are mainly involved in energy-dependent uptake of small peptides with the concomitant uptake of a proton. It is also known as peptide transporter (PTR) family.

**Potometer** A device to measure the rate of transpiration by a plant in terms of the rate at which water is drawn by the shoot of a plant.

**P-proteins** A set of proteins abundant in sieve elements of all dicots and many monocots. Their form (tubular, spheroidal, fibrillar, and granular) is dependent upon the species and maturity of the sieve-tube elements. They are involved in short-term sealing of damaged sieve elements via formation of filaments that plug the pores of the damaged sieve element and block translocation.

**Pressure potential ($\psi_p$)** A component of water potential expressed as the hydrostatic pressure of the solution.

**Pressure flow model** A model explaining translocation of solutes in the sieve elements driven by pressure gradient between source and sink. The pressure gradient is osmotically generated and results from photoassimilate loading at the source and unloading at the sink.

**Primary dormancy** A type of dormancy in which seeds are viable but are unable to germinate immediately after they are released from the parent plant.

**Primary response genes** A set of genes whose expression is rapidly upregulated independent of de novo protein synthesis as a part of early response to a stimulus.

**Programmed cell death (PCD)** The process of activating senescence program in cells, which brings about morphological and biochemical changes leading to cell death.

**Prolamins** A group of storage proteins having high proline and glutamine content typically found in cereals. In wheat, it is commonly known as gluten which is a complex mixture of 71–78 proteins.

**Prosthetic group** Inorganic or organic cofactor which is covalently bound to the enzyme protein.

**Prosystemin** The precursor molecule of 200 residues from which systemin is synthesized at the wounding site by cleavage of the C-terminus.

**Protease** An enzyme which hydrolyzes protein into smaller chains (peptides) or amino acids. It is also known as peptidase or proteinase.

**Protein bodies (or aleurone grains)** Highly specialized structures representing dry vacuoles which store proteins and hydrolytic enzymes in the storage tissues of seeds. The storage proteins are synthesized in ER and are co-translationally transported in ER lumen from where they are released into protein bodies after modifications.

**Proteinase inhibitors** Synthetic drugs which inhibit the activity of various protease enzymes. Proteinase inhibitors produced in some legumes and tomato block the activity of herbivore proteolytic enzymes such as trypsin and chemotrypsin.

**Protoderm** A single-layered meristematic tissue region that gives rise to the epidermis.

**Protonophores or proton translocators** Compounds which facilitate transport of protons across the lipid bilayer of membrane, for example, 2,4-dinitrophenol.

**PUFA (polyunsaturated fatty acid)** A type of fatty acid which has more than one double bond in its hydrocarbon backbone. Different types of PUFA are identified by the position of the last double bond in their structure.

**Pullulanase** A debranching enzyme which hydrolyzes the α-1,6 glycosidic linkages in pullulan (a polymer of maltotriose units) and amylopectin (a polymer of α-glucose units).

**Pulvinus** An enlargement at the base of the petiole of a leaf or petiolule of a leaflet. It has a role in the movement of leaf or leaflet.

**Pumps** Integral membrane proteins which actively transport solutes against their electrochemical gradient using ATP or pyrophosphate hydrolysis as a source of energy. The rate of solute transport by pumps is much faster than that by transporters.

**Pyrethroids** Monoterpene esters having high toxicity against insects. They form components of many insecticides.

**Pyridoxal 5′-phosphate** The active form of vitamin $B_6$ which serves as a coenzyme for several enzymes primarily involved in amino acid metabolism.

**Quanta (sing. quantum)** Discrete packets of energy contained in photons.

**Quantum flux or photon flux density (PED)** The number of photons of light striking the leaf per unit area per second. It is expressed in $mol.m^{-2}.s^{-1}$, where moles refer to the number of photons (1 mole of light = $6.02 \times 10^{23}$ photons; Avogadro's number).

**Quiescent center** A region in the apical meristem that has reached a state of relative metabolic inactivity and is capable of resuming meristematic activity upon damage of surrounding cells. It is common in roots.

**Quiescent stage** Suspended growth state of embryo or resting state of seed. Once favorable conditions arrive, seeds are able to germinate.

**Quorum sensing** A mechanism which is a characteristic feature of bacteria involving cell-to-cell communication by release of certain chemical signals which allows them to monitor density of their own strain as well as that of the other bacterial strains in the host tissue.

**Radial micellation** The radial arrangement of microfibrils in guard cells of stomata that facilitates opening and closing of stomatal pore.

**Radicle** The embryonic root which forms the basal continuation of the hypocotyl in an embryo and gives rise to root.

**Raffinose Family Oligosaccharides (RFOs)** A class of water-soluble polysaccharides which are alpha-galactosyl derivatives of sucrose, for example, raffinose (a trisaccharide), stachyose (tetrasaccharide), and verbascose (pentasaccharide). These are formed as a result of sequential addition of galactose units by the action of a set of galactosyltransferases.

**Raphides** Tiny needlelike crystals of calcium oxalate that are found in clusters in specialized cells called idioblasts within the tissues of plants mainly belonging to family Araceae and Commelinaceae.

**Reactive nitrogen species (RNS)** Various nitric oxide-derived products produced during metabolic events in cells. Peroxynitrite ($ONOO^-$) and peroxynitrous acid are two major reactive nitrogen species.

**Reactive oxygen species (ROS)** Oxygen containing chemically reactive molecules with free electrons, produced as a byproduct of normal metabolism. Incomplete reduction of molecular oxygen forms superoxide anion ($O_2^{\cdot-}$), which acts as a major precursor for all other ROS, such as peroxides (ROOR), hydroxyl radical (HO$^{\cdot}$), and singlet oxygen. ROS can act as signaling molecules at lower concentrations but cause damage to cellular components upon excessive accumulation.

**Recalcitrant seeds** Seeds with relatively high water content and active metabolism at the time of maturity. They lose viability rather rapidly and fail to survive storage, for example, avocado, cacao, coconut, mango, and lychee.

**Receptors** Specialized sensor proteins localized either on the membrane or in cytoplasm which perceive signals, leading to amplification of the cellular response. They play important parts in plant growth, development, and immunity.

**Receptor kinases** A family of protein receptors found in animals which are part of signaling pathways in response to various ligands (e.g., a hormone) and participate in eliciting cellular response by phosphorylating themselves or another target protein majorly at tyrosine residues.

**Receptor-like kinases (RLKs)** A family of putative transmembrane protein receptors in plants which have N-terminal extracellular domain and intracellular C-terminal with kinase activity. They are involved in signaling cascades leading to cellular responses to a variety of ligands in plant cells via phosphorylation of target proteins at serine and threonine residues.

**Rectifying channels** Ion channels which allow flow of ionic current only in one direction, either inward or outward, for example, potassium inward rectifiers and potassium outward rectifiers present in guard cells and a wide variety of other plant cells.

**Red drop phenomenon** A phenomenon characterized by decline in the quantum yield of photosynthesis when light of wavelength beyond 700 nm is absorbed by chlorophyll molecules.

**Reducing sugars** Sugars capable of reducing other compounds due to availability of an aldehyde or ketone groups for oxidation, for example, glucose and fructose.

**Reductive amination** The process by which ammonia is condensed with aldehydes or ketones to form imines, which are subsequently reduced to amines.

**Resistance genes (R-genes)** Genes responsible for regulation of gene expression, thereby providing resistance to a specific group of pathogens in plants. They encode proteins that play key roles in recognizing pathogen effectors (Avirulent proteins).

**Respirasomes** A supramolecular structural-functional complex formed by assembly of two or more different electron-transferring enzyme complexes of respiratory chain.

**Respiratory burst** The rapid and transient release of reactive oxygen species (superoxide radical and hydrogen peroxide) from different types of cells as one of the earliest feature of plant defense strategy against pathogens.

**Response** Biochemical, physiological, or developmental impact of one or more signals perceived and transduced by the sensing cells or tissues.

**Resurrection plants** Plants which can tolerate extreme desiccation and revive quickly under wet conditions in mature state, for example, *Selaginella lepidophylla*.

**Retrotranslocation** The reverse process of translocation.

**Reverse osmosis (RO)** A process by which a solvent moves into the direction opposite to that of natural osmosis through a membrane under the influence of pressure applied. It is used to demineralize or deionize water for making it potable. Pressure is applied on the water containing high amount of salts which forces it to move against the gradient across the semipermeable RO membrane, leaving behind dissolved salts. The water collected by this process has less salt and is safe for drinking.

**Revertants** Seeds produced by mutagenized plants that regain the previously lost ability to germinate after further mutation.

**Rhizosheath** A structure composed of mucilage secreted from plants and adherent soil particles which forms a cylindrical sheath around the root.

**Rhizosphere** The region of the soil around the roots which is influenced by the secretions of the plant roots and by the microorganisms associated with roots.

**Rhizotron** A specially designed camera to monitor the growth of root and its interaction with soil particles in vivo.

**Rieske Fe-S center** A type of protein having two Fe-S clusters in which one of the Fe ion is coordinated by two conserved cysteine residues while the other Fe ion is coordinated by two conserved histidine residues. It acts in many electron transfer reactions including photophosphorylation and oxidative phosphorylation. It is one of the components of cyt $bc_1$ and cyt $b_6f$.

**RNA interference** A posttranscriptional process triggered by the introduction of dsRNA in the cell cytoplasm. It is controlled by RNA-induced silencing complex (RISC) leading to gene silencing in a sequence-specific manner.

**Root pressure** The positive hydrostatic pressure (1–2 bars) developed within the roots due to the difference in solute potential between the soil solution and xylem sap. It drives the xylem sap upward into the xylem elements through the plant body.

**ROP (Rho-like GTPases of plants)** A group of plant-specific Rho GTPases that participate in a variety of processes, such as control of the cytoskeleton, vesicle trafficking, and cell growth.

**ROP-interactive CRIB motif-containing proteins (RICs)** Proteins that interact with ROP1 to regulate pollen tube growth and polarity.

**Rubisco (ribulose bisphosphate carboxylase/oxygenase)** An enzyme which, in the presence of carbon dioxide, utilizes ribulose 1,5-bisphosphate to produce two molecules of 3-phosphoglycerate and, in the presence of oxygen, produces one molecule each of 3-phosphoglycerate and phosphoglycolate.

**Russetting** A disorder of the fruit skin characterized by microscopic cracks in the cuticle, leading to periderm formation. It is observed in apples and pears.

**S-adenosylmethionine (SAM)** A common co-substrate involved in reactions including methyl group transfers, transsulfuration and aminopropylation. It is an intermediate in ethylene biosynthesis and plays an essential role in production of higher forms of polyamines from diamines like putrescine.

**Scarification** The process of mechanically removing or damaging the hard seed coat around the seeds of plants, such as morning glories, lupine, and sweet pea, to facilitate seed germination.

**Second messengers** A class of intracellular, diffusible small molecules and ions which are rapidly synthesized or released transiently as a concentrated pulse following signal perception by the receptors and modify the activity of target signaling proteins, for example, ROS and $Ca^{2+}$.

**Secondary active transporters (cotransporters)** Transmembrane proteins which utilize ion gradients established by primary active transport (facilitated by ATP hydrolysis) to move another solute against its electrochemical gradient. Thus, secondary active transporters facilitate the movement of two solutes simultaneously. One solute moves down its electrochemical gradient, whereas the second one moves up its electrochemical gradient. They include symporters and antiporters.

**Secondary dormancy** A type of dormancy which develops once seed is detached from the plant due to change in environmental conditions.

**Secondary metabolism** Metabolism, not directly related with maintaining the life. Products formed during secondary metabolism are called secondary metabolites.

**Secondary metabolites** Compounds which are generally not essential for basic growth or development of plants but are required for their survival in the environment by functioning as defenses against herbivores and microbial infection by microbial pathogens, attractants for pollinators, and seed-dispersing animals and as agents of plant-plant competition.

**Secondary response genes** Genes which bring about delayed response to a stimulus through de novo protein synthesis.

**Seismonastic response (or thigmonastic response)** A nastic response to a stimulus, especially mechanical stimulus, like touch and vibration irrespective of the direction of the stimulus, for example, drooping of leaflets in *Mimosa pudica* in response to touch stimulus and the shutting of Venus flytrap when an insect lands on it.

**Self-incompatibility (SI)** The inability of the pollen grain to effect fertilization in the pistil of the same flower. It is often induced in plant species to promote outcrossing.

**Senescence** A sequential process during which cells enter a state of growth arrest along with autocatalysis controlled by environment and genetic constitution of an organism. It is characterized by metabolic changes, such as hydrolysis of macromolecules and pigments.

**Senescence-associated genes (SAGs)** Genes which are upregulated with onset of senescence.

**Senescence-downregulated genes (SDGs)** Genes that exhibit decreased expression during senescence.

**Serotonin** A multifunctional tryptophan-derived indoleamine known to modulate various plant physiological processes including shoot morphogenesis, plant defense responses, and gene expression associated with auxin response pathways.

**Short-day plants** Plants which flower only when the day length does not exceed the critical day length, for example, chrysanthemum, poinsettias, rice, soybean, tobacco, and *Xanthium*.

**Short-distance transport** The transport of biomolecules over a distance of two or three cells, for example, the movement of sugars from mesophyll cells to the cells in the vicinity of the veins of the source leaf during the process of phloem loading.

**Sieve cells** Sieve elements of primitive type which are relatively unspecialized and characterized by narrow pores and absence of sieve plates.

**Sieve elements** The cells of phloem which transport sugars and other organic compounds throughout the plant body. They refer to both sieve-tube elements in angiosperms and sieve cells in gymnosperms.

**Sieve plates** Areas generally found in end walls of sieve-tube elements of angiosperms that have larger pores than other sieve areas.

**Sieve pores** The pores present on sieve elements.

**Sieve tube** Tubular strands formed by the joining together of individual sieve-tube elements at their end walls.

**Sieve-tube elements** Sieve elements of advanced type present in angiosperms which are highly differentiated and are characterized by presence of large pores and sieve plates at their end walls.

**Signal** Any environmental or intracellular input which initiates one or more responses in the cell/plant.

**Signal transduction pathway** A sequence of biochemical events that involves binding of a ligand (e.g., light or hormone) to a receptor protein, triggering changes in the downstream signaling molecules in the cell, thereby leading to a cellular response.

**Singlet state** The excited state of a molecule attained on absorption of a photon wherein all the electrons with opposite spins are paired. It has a natural life time of $10^{-9}$ seconds. Electron spin is zero because both the electrons in an orbital are spinning antiparallel to each other.

**Sink** Any tissue/organ which imports photoassimilates, for example, tubers, bulbs, and roots.

**Sink activity** The rate of uptake of photoassimilates per unit weight of the sink tissue.

**Sink size** The total weight of the sink tissue.

**Sink strength** The competitive ability of a sink to mobilize sugars toward it from other parts of the plant body.

**Siroheme** A heme-like prosthetic group found in many sulfite and nitrite reductases (enzymes that catalyze electron transfer in the process of reduction of nitrite and sulfite to ammonia and sulfide, respectively).

**Size exclusion limit** The molecular mass of the smallest solute that is excluded from movement through symplast across the plasmodesmic channels.

**Skotomorphogenesis** The development of a plant in dark.

**Skotonastic response** A type of movement that is synchronized by light to dark transition, for example, folding of leaves in dark.

**S-locus cysteine-rich protein (SCR)** A cysteine-rich protein located in the pollen coat that represents the male S-determinant in the members of Brassicaceae.

**S-locus receptor kinase (SRK)** A serine/threonine receptor kinase located in the plasma membrane of stigmatic cells that represents the female S-determinant in the members of Brassicaceae.

**Source** The plant organs which are capable of producing photosynthates and transporting them to other parts of the plant, for example, leaves. Storage areas such as roots and tubers also become sources under certain conditions.

**Sphingosine** An 18-carbon amino alcohol which is a primary component of sphingolipids (a class of phospholipids containing sphingomyelin).

**Sporophytic self-incompatibility (SSI)** A type of self-incompatibility where the incompatibility response or rejection of pollen is determined by the diploid genotype of the plant (sporophyte) that produces pollen. The pollen does not germinate on the stigma of flowers that contain either of the two alleles in the male sporophyte parent.

**Sporopollenin** Chemically inert biological polymer which consists of covalently linked phenolic acid and fatty acid-derived constituents. It forms the tough outer wall of pollen grains and provides protection from a variety of external agents.

**Standard free energy** The free energy associated with a reaction that occurs at physiological pH (7.0) at 25 °C and 100 kPa, when both reactants and products are at unit concentrations (1 M).

**Static solution culture** A hydroponic technique to culture plants in containers filled with nutrient solution which is aerated using a pump.

**Statocytes** A group of cells found in the root cap that contain statoliths. They are responsible for the gravitropic response exhibited by roots.

**Statoliths** Starch-containing plastids located in specialized cells known as statocytes found in the roots. They sediment to the bottom of the cells under the influence of gravity and, thus, regulate the gravitropic response in roots.

**Sterols** A group of lipids consisting of a steroid (comprising four hydrocarbon rings) with an attached hydroxyl group. They may exist as free sterols or with many modifications (acetylation, alkylation, and glycosylation), for example, campesterol and stigmasterol produced by plants.

**Stigma/style cysteine-rich adhesion (SCA) protein** A protein secreted by the transmitting tract of style in the members of Liliaceae that is involved in the growth and adhesion of pollen tubes.

**Stigmasterol** An unsaturated plant sterol which is similar to animal cholesterol. It is characterized by the presence of –OH group at C-3 of steroid skeleton and unsaturated bond at 5–6 position of ring B. It is found in fats and oils of soybean and many other legumes and vegetables.

**Stomatal frequency** The number of stomata per unit area of leaf. It plays an important role in determining the rate of exchange of gases and water vapor through a leaf surface.

**Stratification** The process of subjecting seeds to low temperature prior to sowing in order to simulate natural winter conditions.

**Stress** A condition which prevents plants from attaining their full genetic potential, thereby affecting plant growth, development, and productivity. It may be induced by abiotic or biotic factors.

**Stress response regulon** The transcriptional regulatory gene network generated in response to specific abiotic stresses by the action of different transcription factors that can cause activation of some genes and suppression of others.

**Strigolactones (SLs)** A class of plant growth regulators that are derived from carotenoids. They are involved in modulating plant development processes such branching and promoting symbiotic interactions between plants and microbes.

**Stroma** The fluidlike content of chloroplasts that surrounds photosynthetic membranes.

**Strophiole** An outgrowth of hilum region which restricts the movement of water in and out of the seeds with hard seed coat, for example, leguminous seeds.

**Substrate channeling** The movement of chemical intermediates in a series of enzyme-catalyzed reactions from the active site of an enzyme to that of the next enzyme in a pathway without dissociating from the surface of the protein complex.

**Substrate-level phosphorylation** A reaction resulting in the synthesis of ATP and GTP by direct transfer of phosphoryl group to ADP and GDP, respectively, from another phosphorylated compound. It is independent of electron transport.

**Succinoglycan** An extracellular acidic polysaccharide produced by microbes, which is composed of repeated units of an octasaccharide. The octasaccharide consists of galactose and glucose units in a ratio of 1:7, respectively, which are modified with acetyl, succinyl, and pyruvyl moieties.

**Suicide inactivators** Compounds which, instead of being converted to products in an enzyme-catalyzed reaction, are converted to highly reactive molecules that combine irreversibly with enzyme and inhibit its activity. For example, hydrogen peroxide is a suicide inactivator of peroxidases.

**Sunflecks** Patches of sunlight passing through small gaps in the canopy of trees in dense forests.

**Superoxide dismutase (SOD)** An enzyme that catalyzes the dismutation of the superoxide ($O_2^-$) radical into either molecular oxygen ($O_2$) or hydrogen peroxide ($H_2O_2$).

**Surface tension** The tension created on the surface of a liquid as a result of high cohesive forces among the liquid molecules. It pulls the surface layer of the liquid such that it appears stretched and tends to minimize the surface area.

**SUT1 (sucrose transporter 1)** A sucrose-$H^+$ symporter found in the plasma membrane of sieve elements that transports sucrose into the cell coupled with an uptake of proton.

**Symbiosis** A close and long-term association between two biological species for mutual benefit, for example, mycorrhiza which represents a relationship between certain fungi and roots of plants.

**Symplast** The continuous system formed by interconnection of the cytoplasm of one cell with that of the neighboring cells through plasmodesmatal connections. It plays an important role in transport of water, minerals, and other low molecular weight solutes.

**Symplastic pathway** The pathway by which water and other molecules travel from one cell to another through plasmodesmata.

**Symport** A type of secondary active transport of solutes involving the movement of two solutes in the same direction across a membrane. One species moves down its electrochemical gradient and drives the transport of the other against its gradient.

**Symporters** Integral membrane proteins which facilitate cotransport of two solutes across the membrane in the same direction. One molecule moves down the electrochemical gradient, while the other moves against the gradient, for example, sucrose-$H^+$ symporter.

**SymRK (symbiosis receptor-like kinase)** A membrane-bound kinase protein which consists of an extracellular leucine-rich repeat (LRR) domain, a transmembrane segment, and a functional intracellular kinase domain. It is required at the early stage of symbiosis between the host plant and the microsymbiont.

**Syncytium** A multinucleate cell resulting from the fusion of multiple uninucleate cells.

**Systemic acquired resistance (SAR)** A type of resistance occurring at whole plant level following pathogen attack due to induction of pathogenesis related genes by salicylic acid. It confers a long-term protection to a broad range of pathogens and pests.

**Systemic defense** A network of signal transduction and amplification that result in activation of defense genes and establish systemic resistance in the entire plant.

**Systemin** A small peptide signaling molecule biosynthesized concomitant to JA accumulation in some plants and plays important roles in short range intercellular communications. It is capable of inducing the synthesis of proteinase inhibitors and defense-related genes in response to insect attack.

**Tannins** Complex phenolic compounds involved in plant defense that have the ability to bind and denature proteins. They are exploited for tanning of hides in leather industry.

**Tricarboxylic acid (TCA) cycle** TCA cycle is also known as citric acid cycle or Krebs cycle during which acetyl-CoA is completely oxidized with simultaneous reduction of cofactors such as $NAD^+$ and FAD to NADH and $FADH_2$, respectively.

**Thermal death point** The temperature at which an organism is killed depending upon the time of exposure.

**Thermogenesis** The metabolic process in which heat is produced as a result of expenditure of energy. During thermogenesis, most of the electron flow in mitochondria gets diverted from cytochrome respiration pathway to cyanide-insensitive non-phosphorylating electron transport pathway which is unique to plant mitochondria. The energy released by electron flow through this alternative respiratory pathway is not conserved as chemical energy but is released as heat.

**Thermogenic plants** The plants which are capable of producing heat in order to raise their internal temperature above that of the environment, for example, lotus and arum. The heat is produced via the action of alternative oxidase in an alternate respiratory pathway.

**Thermonasty** A type of plant movement such as opening and closure of petals, due to differential growth in response to fluctuating temperature.

**Thigmotropism** A response of plant in the form of directional growth or movement toward the contact stimulus. For example, tendrils of a climbing plant twine around any support they touch.

**Thioredoxin** A small ubiquitous thiol-active protein that has a pair of active site cys-thiols which upon oxidation form a disulfide bond.

**Thylakoids** Flattened saclike membranes present in the stroma of plastids where photosynthetic pigments are localized. These may be present in stacked or unstacked form.

**Tonoplast intrinsic protein (TIP)** Intrinsic proteins of the tonoplast which are responsible for high permeability to water and various other molecules, including ammonium and urea.

**Totipotency** The potential of a single plant cell to divide and differentiate into various cell types and give rise to a whole plant.

**Toxicity** Accumulation of high levels of micronutrients in plants leading to deleterious effects.

**Trans fat (trans-unsaturated fatty acid)** A type of unsaturated fatty acid that contains one or more double bonds in trans geometric configuration. These are not commonly found in nature but produced from vegetable oil due to partial hydrogenation.

**Transaldolase** An enzyme of pentose phosphate pathway, which catalyzes reversible transfer of a three-carbon fragment from sedoheptulose 7-phosphate to glyceraldehyde 3-phosphate resulting in the formation of fructose 6-phosphate and erythrose 4-phosphate. The reaction provides a link between pentose phosphate pathway and glycolysis.

**Transcellular** Transport of water and solutes through a cell whereby they enter the cell from one side and exit from the other. The molecules to be transported by this mode have to cross the membrane twice.

**Transduction (of signals)** Conversion of a signal perceived by the receptor into another form so as to amplify the signal.

**Transition state** An intermediate, short-lived unstable state of a substrate molecule before being converted to the product in a chemical reaction.

**Transitory starch** Starch stored in chloroplasts during daytime and mobilized out of chloroplasts during night.

**Transketolase** An enzyme of both pentose phosphate pathway and Calvin cycle, which catalyzes transfer of a two-carbon fragment from a five-carbon ketose donor to an aldose acceptor.

**Translocation** The transport of photosynthates from source to sink tissue in the phloem.

**Transpiration** The evaporation of water in the form of water vapor from the aboveground parts of the plant, mainly through stomata.

**Transpiration pull** The pressure exerted on the xylem sap by transpirational force in the xylem tissue.

**Transpiration ratio** It is the ratio between the amount of water transpired to the mass of dry matter produced by a plant. It determines the efficiency of a plant to fix carbon dioxide at the expense of the amount of loss water by transpiration.

**Transpiration stream** The upward translocation of water in the form of continuous stream in xylem from root to leaves under the pressure of transpiration.

**Transporters** Specific transmembrane proteins present in the plasma membrane which facilitate transport of ions or molecules across the membrane.

**Triose phosphate pool** The intracellular pool of triose phosphates, i.e., 3-phosphoglyceraldehyde and dihydroxyacetone phosphate, from which either of the triose phosphates can be removed according to the need in cell metabolism.

**Triple response** Concentration-dependent morphological changes shown by dark-grown seedlings exposed to ethylene, characterized by the inhibition of hypocotyl and root elongation, a pronounced radial swelling of the hypocotyl and exaggerated curvature of the plumular hook.

**Triplet state** The excited state of a molecule attained when the molecule in singlet state loses some energy to environment as heat which is accompanied with the reversal of spinning of electron. The resulting electron spin resonance value is 1. This is the lowest energy state of the molecule having longer natural life time, i.e., $10^{-2}$–$10^{-4}$ s.

**Tritrophic interactions** Interactions that describe plant defense system against herbivores having impact across three trophic levels, i.e., the plant, the herbivore, and its natural enemies (parasitoids). In such an interaction, a plant upon being damaged by a particular herbivore releases volatile chemicals which, in turn, attract the parasitoids of the herbivore.

**Tryptophan decarboxylase** The key regulatory enzyme for mobilizing tryptophan in the biosynthesis of indole-3-acetic acid or serotonin/melatonin.

**Tunica** The peripheral layer in the shoot apical meristem consisting of cells that divide in anticlinal plane and contribute to the growth of the surface.

**Turgor pressure** The pressure exerted by water inside the cell per unit area of the plant cell wall. It pushes the plasma membrane outward and helps in maintaining the shape of the cell and provides force for cell expansion.

**Turnover number ($K_{cat}$)** The maximum number of chemical conversions of substrate molecules per second carried out by a single catalytic site of an enzyme at a given concentration.

**Type III secretion system (T3SS)** A type of secretory system present in gram-negative bacteria to deliver effector molecules into the host cell via the pilus (a hairlike appendage).

**Ubiquitination** The addition of ubiquitin to a protein targeted for degradation via the 26S proteasome pathway.

**Uncouplers** Compounds that uncouple ATP synthesis from electron transport, for example, 2,4-dinitrophenol.

**Uniporters** Integral membrane proteins facilitating transmembrane migration of specific solutes according to concentration gradient by binding one molecule at a time and transporting it in either direction across the membrane depending upon its concentration gradient.

**Ureides** Nitrogenous organic compounds which are acyl derivatives of urea. They are produced as a result of purine catabolism and contribute in translocation of nitrogen within the plant. Ureides include uric acid, allantoin, and allantoic acid.

**UVR8** Photoreceptor for sensing UV-B (280–315 nm) radiation in plants. It is a seven-bladed β-propeller protein without a prosthetic group.

**Vacuolar iron transporter (VIT)** A protein responsible for the transfer of iron from cytosol to vacuole for intracellular iron storage.

**Valinomycin** An antibiotic which functions as $K^+$-specific translocator and facilitates movement of potassium through lipid bilayer down the electrochemical gradient.

**Vapor density** Concentration of water molecules in vapor phase relative to that of hydrogen at the same conditions of temperature and pressure and is expressed as vapor mass per unit volume (unit = $g.m^{-3}$).

**Vapor pressure** The pressure exerted by vapors on the walls of the stomatal chamber.

**Variation potential (or slow-wave potential)** A temporary change in the membrane consisting of propagating electrical signals generated exclusively in plants cells in response to a change in hydraulic pressure wave or ligand transmitted in xylem. It can be generated by wounding, excision, or flame. In contrast with action potential, variation potential exhibits longer phase of repolarization, following depolarization.

**Vernalin** A hypothetical plant hormone produced in the shoot apex as a result of cold temperature treatment and is responsible for induction of flowering.

**Vernalization** A treatment of cold temperature required by the plant for the competence to photoperiodic stimulus for flowering.

**Vertical resistance** Specific resistance acquired by the host against a particular race of pathogen which is heritable and is passed down to the next generation. It is also called single-gene resistance as it is governed by a single gene.

**Virulent** Pathogens that have the ability to cause disease in host and induce drastic effects.

**Vivipary** A condition in which the seed germinates and embryo grows and develops while still attached to the parent plant. This phenomenon is observed in mangroves such as *Rhizophora* sp.

**Wall pressure** The pressure exerted by cell wall on the contents of a cell. It is equal and opposite to the force exerted by turgor pressure.

**Water potential ($\psi$)** A measure of the free energy of water per unit volume expressed as the difference between the chemical potential of water at any point in a system and that of pure water under standard conditions. It is a function of solute potential ($\psi_s$), pressure potential ($\psi_p$), matric potential ($\psi_m$), and gravitational potential ($\psi_g$).

**Wax synthases** A group of long-chain-alcohol o-fatty-acyltransferases that catalyzes the formation of wax esters.

**Whiptail** A deficiency symptom of molybdenum commonly seen in cauliflower which is characterized by poorly developed leaf blades with green and yellow margins.

**Wilting** A condition of a plant that arises when the plant body loses more water by evaporation than it is able to absorb from the soil, causing the cells to lose their turgidity and leading to the loss of rigidity of non-woody parts of the plants.

**WRKY proteins** A large class of WRKY domain-containing sequence-specific, DNA-binding transcription factors found in plants. They regulate many responses like biotic and abiotic stress, senescence, seed germination, seed dormancy, and some developmental processes.

**Xanthoxin** A natural growth inhibitor that has physiological properties similar to ABA. It leads to the formation of ABA via oxidative steps involving the intermediates ABA-aldehyde or xanthoxic acid in the cytoplasm.

**Xenobiotic** A chemical compound, generally synthetic, which is extrinsic to normal metabolism of an organism or to an ecological system. These include synthetic drugs or drug metabolites in the body or pollutants in environment such as synthetic pesticides, herbicides, industrial pollutants, etc.

**Xenogamy** The process of transfer of pollen onto the surface of stigma of a flower present on a different plant.

**Xylogenesis** The process of differentiation of xylem elements.

**Yang's cycle** The cycle responsible for prevention of methionine depletion and its regeneration in plants during ethylene production.

**Yellow stripe-like 1 (YSL1)** A member of oligopeptide transporter family predicted to be integral membrane proteins regulated in response to iron status in plants. It is involved in transport of iron ($Fe^{3+}$) complexed with phytosiderophores.

**Zeitgeber** An environmental cue that synchronizes the biological clock of an organism with the 24-hour diurnal and annual cycle.

**Zeitlupe (ZTL/ADO)** Photoreceptors involved in the targeted proteolysis of signaling components which control flowering time and circadian clock.